Geologic Time Scale

	Eon	Era	Period	Epoch	mya[1]
	Phanerozoic	Cenozoic	Quaternary	Recent	0.01
				Pleistocene	1.8
			Tertiary	Pliocene	5
				Miocene	23
				Oligocene	34
				Eocene	56
				Paleocene	65
		Mesozoic	Cretaceous		140
			Jurassic		200
			Triassic		245
		Paleozoic	Permian		290
			Carboniferous		350
			Devonian		410
			Silurian		440
			Ordovician		500
			Cambrian		570
Pre-cambrian	Proterozoic				2500
	Archean				3800
	Hadean				4600

[1]Millions of years ago (from beginning of interval)

INVERTEBRATE ZOOLOGY

INVERTEBRATE ZOOLOGY

A FUNCTIONAL EVOLUTIONARY APPROACH

seventh edition

Edward E. Ruppert
Clemson University, South Carolina

Richard S. Fox
Lander University, South Carolina

Robert D. Barnes
Late of Gettysburg College

Australia • Brazil • Japan • Korea • Mexico • Singapore • Spain • United Kingdom • United States

Invertebrate Zoology: A Functional Evolutionary Approach, Seventh Edition
Edward E. Ruppert, Richard S. Fox, Robert D. Barnes

Executive Editor: Nedah Rose
Development Editor: Shelley Parlante
Assistant Editor: Christopher Delgado
Editorial Assistant: Jennifer Keever
Technology Project Manager: Travis Metz
Marketing Manager: Ann Caven
Marketing Assistant: Sandra Perin
Advertising Project Manager: Linda Yip
Project Manager, Editorial Production: Belinda Krohmer
Print/Media Buyer: Kris Waller
Permissions Editor: Joohee Lee
Production Service: NOVA Graphic Services, Inc.
Text Designer: Kim Menning
Photo Researcher: Authors, NOVA Graphic Services, Inc.
Copy Editor: Nancy Elgin
Illustrator: John Norton
Cover Designer: Bill Stanton
Cover Image: Sea Anemone, by David Nardini, Getty Images
Compositor: Progressive Information Technologies

Chapter Opening Credits: Chapter 2, Dr. Donald Fawcett/Visuals Unlimited; Chapter 4, Dr. Yuuji Tsukii, Protist Information Server, http://protist.i.hosei.ac.jp/Protist_menuE.html; Chapter 8, © James McCullagh/Visuals Unlimited.

Library of Congress Control Number: 2003107287

ISBN-13: 978-0-03-025982-1
ISBN-10: 0-03-025982-7

Brooks/Cole
10 Davis Drive
Belmont, CA 94002-3098
USA

Cengage Learning is a leading provider of customized learning solutions with office locations around the globe, including Singapore, the United Kingdom, Australia, Mexico, Brazil, and Japan. Locate your local office at: **international.cengage.com/region**

Cengage Learning products are represented in Canada by Nelson Education, Ltd.

For your course and learning solutions, visit **academic.cengage.com**

Purchase any of our products at your local college store or at our preferred online store **www.ichapters.com**

Printed in Canada
4 5 6 7 8 11 10 09 08

For the next generation of
invertebrate zoologists

Contents

Preface

The seventh edition of *Invertebrate Zoology* is the first thoroughly revised and rewritten edition of the original classic by the late Robert Barnes. Since the first edition was published in 1963, invertebrate zoology has been advanced by a legion of knowledgeable and creative biologists. Many first studied invertebrate zoology from Robert Barnes's textbook or with the author himself at the Duke University Marine Laboratory, at the Bermuda Biological Station for Research, or at Gettysburg College. Their research and that of many others, much of it using new techniques and disciplinary approaches, are revolutionizing our knowledge of invertebrates. This new edition includes important results from all of these modern investigations.

Invertebrate zoology is for those who love animated life in all its myriad forms, for it includes most animals and even the animal-like protists. It is an opportunity to revel in the sheer joy of an astounding diversity, which ranges from the familiar to the bizarre, from the infinitesimal to the gigantic, and from the sluggish to the swift. Invertebrate zoology not only delights the eye with variety, it also challenges the mind to seek unity in diversity. Unifying principles from all disciplines in biology—indeed all sciences—are applied to the study of invertebrates, but invertebrate zoology also has its own unique core concept.

SPECIAL FEATURES OF THE SEVENTH EDITION

- **Five new chapters** introduce the functional-evolutionary approach, in which evolutionary innovations in animal form and function provide a unifying conceptual framework for invertebrate diversity.
- **New evolutionary trees (cladograms),** clearly labeled and illustrated, clarify evolutionary relationships.
- **Nearly 200 new figures** and numerous figure revisions increase both currency and clarity.
- **Extensive revision** of the text and reorganization of the chapters reflects current research, including recent changes in the classification of major taxa.
- **New taxa covered** include Cycliophora, Micrognathozoa, Tantulocarida, and Myxozoa.
- **Standardized uniform terminology** used for the same structure in different taxa minimizes confusion.
- **New annotated Web URLs** link the book to a wealth of resources, including photos, animations, current research, and additional detail on virtually all taxa.
- **A new Web site** with over 300 original color photos of living invertebrates and 300 key figures from the text provides additional teaching and learning resources.

New Chapters Introduce Functional-Evolutionary Approach

The central concept of invertebrate zoology is the evolution of animal form and function. It includes not only the reconstruction of evolutionary relationships, but also the history of successful innovations in form and function. These innovations constitute the ground plans of the animal taxa, and this edition explicitly identifies them, discusses their function, and presents current hypotheses for their evolutionary origin in what we call a *functional-evolutionary approach.* Although all chapters include functional-evolutionary content, we emphasize key innovations—the evolution of eukaryotic cells, multicellularity, epithelia, bilateral symmetry, and an impermeable, supportive exoskeleton—in separate, stand-alone, Introduction to . . . chapters.

For example, along with eukaryotes came large cell size, enhanced motility, and the ability to feed on their prokaryotic progenitors (Chapter 2, Introduction to Protozoa). Multicellularity resulted in another quantum leap in body size, enabling animals to feed on protozoa (as well as prokaryotes), but also conferring on them metabolic and other advantages over the unicellular protists (Chapter 4, Introduction to Metazoa). The evolution of epithelia allowed eumetazoans to physiologically regulate internal, extracellular compartments, such as the gut and coelom. This regulation not only improved the performance of digestive and other functions, but also liberated eumetazoans to some extent from the vagaries of environmental variations, thus enabling them to sustain relatively high levels of activity (Chapter 6, Introduction to Eumetazoa). The adoption of bilateral symmetry by eumetazoans resulted in a directionally polarized body equipped with sensory organs used for pursuing food and mates or escaping from enemies (Chapter 9, Introduction to Bilateria). The arthropods, especially the 20 million-plus species of insects, with their mobile appendages and protective, supportive exoskeletons, found the keys to the colonization of land and air (Chapter 16, Introduction to Arthropoda).

Emphasis on Predictive Science

The evolution of animal form and function is the conceptual backbone of this book. It not only provides the framework for invertebrate zoology, but also shifts it from a chiefly descriptive discipline to a dynamic predictive science. For example, if students have knowledge of an organism's body size and level of organization, they should be able to predict the occurrence and nature of its circulatory system, excretory system, gas-exchange system, locomotory tissues, and aspects of its reproduction. The few, but inevitable exceptions provide opportunities to test the predictions, discover new functional interactions, and propose new predictive models. It is our experience that this approach not only simplifies diversity by ordering it around predictive functional-evolutionary principles, but also encourages active hypothesis testing.

New Phylogenetic Trees

The functional-evolutionary approach relies on phylogeny to identify the key traits associated with each major evolutionary lineage. The seventh edition embraces cladistics as the method of choice for reconstructing phylogenetic trees (cladograms), and a simplified introduction to cladistics, with a set of illustrated exercises, is provided in Chapter 1 (Introduction to Invertebrates). Phylogenetic trees are drawn for most major taxa and we have designed the trees to be easily readable, thus encouraging students to study and test them. The terminal taxa are named, as always, at the branch tips, but sister taxa are also named at parallel levels within the tree. This convention readily identifies sister taxa and reinforces the chief goal of cladistics: to determine sister-taxon relationships by discovering synapomorphies.

The phylogenetic trees in this book are based on traditional and newly identified morphological characters as well as molecular data. The accompanying figure legends include not only the apomorphic characters associated, by number, with each branch, but also identify all taxa by name, thus providing a summary—a study guide—for the key traits of each taxon. When significant competing phylogenies have been published, we present trees illustrating the alternatives. Chapters also include a Phylogeny of . . . section that offers critical discussion of past and present phylogenetic hypotheses. All phylogenies are hypotheses, parts of the ongoing process of discovery. As new facts are discovered, students and teachers are encouraged to criticize and modify the trees in this book.

The widespread adoption of cladistics has resulted in "natural" classifications that reflect evolutionary relationships better than most other phylogenetic methods. Phylogenetic trees supported by synapomorphies are composed of sister taxa grouped pairwise, often at many hierarchic levels, or ranks. Such trees are highly informative, but the ranking of taxa often conflicts with the Linnean categories (kingdom, phylum, class, order, and so on) traditionally assigned to these same taxa. In addition, the sheer number of hierarchic ranks frequently exceeds the number of Linnean categories for them. Partly because of the large number of ranks in phylogenies based on cladistics, many modern systematists avoid assigning Linnean categories to them.

In accordance with modern practice, this edition of *Invertebrate Zoology* de-emphasizes the assignment of taxa to Linnean categories. Our own teaching experience indicates that students and professors welcome this change. To keep track of the relative ranks of taxa, we provide text figures of phylogenetic trees and a corresponding tabular hierarchy in a chapter section called Phylogenetic Hierarchy of . . . . For those who find the change to be unsuited to their background or style, we have retained the Linnean category names as superscript abbreviations on the names of taxa where those names appear in chapter outlines, chapter headings, and the Diversity of . . . sections of the chapters.

Standardization of Anatomical Terms

The functional-evolutionary approach is necessarily comparative, and we strive to encourage comparative as well as analytical thinking. To facilitate comparisons, many anatomical terms have been standardized in this edition. For example, despite being homologs, the filtration excretory organs of arthropods are traditionally assigned different names in different taxa. We note these specialized names, but describe the homologous organs under the single name *saccate nephridium.* Similarly, acknowledging the homology of hemichordate and echinoderm coelomic cavities, we use the hemichordate terms *protocoel, mesocoel,* and *metacoel* in reference to the embryonic and larval coelomic cavities of echinoderms. Such terminological changes, as well as some replacement terms coined for clarity, were adopted to help students learn invertebrate zoology and we hope that specialists will be patient with the changes. It is our experience that students who become invertebrate zoologists readily learn and adopt the classical terms applied to their organisms.

New Web Site with More than 600 Images

Invertebrates are not only curious and engaging animals, most are so colorful and beautiful, especially when seen alive. For many students, it is this first visual contact with a living snail, sea star, or squid that galvanizes their curiosity into a life-long devotion to invertebrate zoology and, often, biology in general. This edition gives the instructor access to an online bank of approximately 300 original color images of living, primarily marine invertebrates and to 300 of the text figures. Both the photos and the text images can be shared with students in a number of ways. For example, the instructor can provide access for viewing online or can download and customize the images into study sheets or lecture presentations.

Encouragement for Firsthand Discovery

Invertebrate zoology is a frontier for discovery open to students as well as experienced scientists. The overall goal of this new edition of *Invertebrate Zoology* is to encourage exploration of this frontier and provide some guideposts, but the best source of discovery is the animals themselves. To the extent that is possible for them, we encourage instructors to use living invertebrates in their teaching laboratories and motivated students to enroll in a summer invertebrate zoology course taught at a marine laboratory. The interaction between a curious mind and a living organism is a fertile medium for discovery. We have found that the mix of motivated students and living invertebrates invariably results in new observations and ideas.

Edward E. Ruppert
Richard S. Fox
June 2003

Acknowledgments

We thank our friend and collaborator, artist-biologist John Norton, who provided the new artwork for this and the previous edition of *Invertebrate Zoology.* John combines his impressive artistic talent and biological training with an instinct for clear and accurate presentation of concepts as well as organisms and their parts. Using traditional methods of illustration, he devoted countless hours to rendering our ideas, seeking criticism, and perfecting his drawings. We much appreciate his art and his good nature.

We are grateful to the following colleagues who reviewed one or more manuscript chapters and provided thoughtful criticism and advice: Elizabeth J. Balser (Illinois Wesleyan University), Brian Bingham (Western Washington University), Steven K. Burian (Southern Connecticut State University), Christopher Cameron (University of Victoria), Tamara Cook (Sam Houston State University), Ruth Ann Dewel (Appalachian State University), Ronald V. Dimock (Wake Forest University), Gonzalo Giribet (Harvard University), Jeremiah N. Jarrett (Central Connecticut State University), Donald A. Kangas (Truman State University), Robert E. Knowlton (George Washington University), Roger M. Lloyd (Florida Community College), Louise R. Page (University of Victoria), A. Richard Palmer (University of Alberta), John F. Pilger (Agnes Scott College), Pamela Roe (California State University, Stanislaus), William A. Shear (Hampden-Sydney College), George L. Shinn (Truman State University), Stephen M. Shuster (Northern Arizona University), Erik V. Thuesen (Evergreen State College), Seth Tyler (University of Maine), Elizabeth Waldorf (Mississippi Gulf Coast Community College), Mary K. Wicksten (Texas A & M University), and Patrick T. K. Woo (University of Guelph).

We also thank colleagues who provided us with expertise, bibliographic resources, results of their ongoing research, access to images, or discussion of ideas. These include: Geoffrey A. Boxshall (The Natural History Museum, London), Dale R. Calder (Royal Ontario Museum), Clayton Cook (Harbor Branch Oceanographic Institution), Roger D. Farley (University of California, Riverside), Jennifer E. Frick (Brevard College), Peter Funch (University of Copenhagen), Henrick Glenner (University of Copenhagen), Jens T. Høeg (University of Copenhagen), Dan R. Lee (Lander University), John Lee (City College of New York), Iain J. McGaw (University of Nevada, Las Vegas), John E. Miller, Gayle P. Noblet (Clemson University), Beth Okamura (University of Reading), Patrick Reynolds (Hamilton College), Scott Santagat (Smithsonian Marine Station at Ft. Pierce), Thomas Stach (Smithsonian Marine Station at Ft. Pierce), Joseph Staton (University of South Carolina), Sidney L. Tamm (Boston College), Lesly A. Temesvari (Clemson University), James M. Turbeville (Virginia Commonwealth University), A. P. Wheeler (Clemson University), Ward Wheeler (American Museum of Natural History), Betty H. Williams (Lander University), and John P. Wourms (Clemson University).

During the revision of this book, our longtime publisher, Saunders College Publishing, was acquired by Cengage Learning and a new publication team was assigned to the project. In the uncertainty associated with the transition, the project was kept alive by Lee Marcott, an avid "birder" and former developmental editor at Saunders. We are most grateful to Lee for her confidence and support.

The book manuscript evolved substantially during its preparation and we are thankful for the flexibility, guidance, and patience of our talented publication team at Cengage Learning. We much appreciate our creative and gracious developmental editor, Shelley Parlante, who helped us to clarify the book's major themes, kept us on track, and provided encouragement and ideas. She also adeptly managed the many format changes that occurred during the book's preparation. We thank executive editor Nedah Rose, who suggested a wide range of improvements and adapted the budget to improve the quality of the book. We also appreciate the contributions of project manager Belinda Krohmer and technology project manager Travis Metz.

The production team at NOVA Graphic Services was a dynamo of efficiency and talent. Editorial director Robin Bonner skillfully managed the demands of authors, publisher, and deadlines while maintaining a keen eye for accuracy. She also made several format improvements to the book. Copy editor extraordinaire Nancy Elgin substantially improved the text and heightened our awareness with her insightful observations. It has been a privilege to work with Robin and Nancy.

During our careers, we were inspired and supported by the following colleagues, to whom we express our enduring gratitude. They are: Peter Ax (University of Göttingen), the late Robert D. Barnes (Gettysburg College), Edward L. Bousfield (National Museum of Natural Sciences, Ottawa), Fu-Shiang Chia (University of Alberta), Richard A. Cloney (University of Washington), Kevin J. Eckelbarger (University of Maine), the late Robert L. Fernald (University of Washington), Richard Gude (Hartwick College), Frederick W. Harrison (Western Carolina University), the late Charles E. Jenner (University of North Carolina, Chapel Hill), Alan R. Kohn (University of Washington), Eugene N. Kozloff (University of Washington), H. Eugene Lehman (University of North Carolina, Chapel Hill), Fordyce G. Lux (Lander University), George O. Mackie (University of Victoria), Frank G. Nordlie (University of Florida), Mary E. Rice (Smithsonian Marine Station at Ft. Pierce), Rupert M. Riedl (University of Vienna), Reinhard M. Rieger (University of Innsbruck), Stephen E. Stancyk (University of South Carolina), Austin Williams (National Museum of Natural History), Robert M. Woollacott (Harvard University), and the late John Z. Young (University College, London).

1 Introduction to Invertebrates

The study of invertebrates is a gateway to the vast diversity of animal life. The astronomical numbers and myriad forms of invertebrates delight the eye, challenge the mind, and provide limitless opportunities for scientific discovery. Invertebrates account for more than 99% of all species of animals and, although fewer than 1 million living species have been named so far, the total number of animal species on Earth may exceed 30 million. Clearly, most invertebrate species remain to be discovered and described, while among those already identified, few have been studied in depth. Thus, any curious individual equipped with the simplest observational or experimental tools can make original and enduring contributions to science.

At first glance, the diversity of invertebrates may seem incomprehensible, like contemplating infinity, but the countless individuals and species are variations on a relatively few readily identifiable themes. One of these themes is the **ground plan** (basic design, or *Bauplan*) of each taxonomic group. For example, although there are millions of species of arthropods, all have a segmented body, an exoskeleton that is molted with growth, and jointed appendages. Knowing those few traits enables anyone to identify an arthropod as such and to appreciate the insect, crustacean, centipede, and spider variations on the arthropod theme. The chapters in this book have been written to highlight the ground plan of each major group and thus to provide a foundation on which to understand the thematic variations.

Another way to simplify the diversity of invertebrates is to establish the evolutionary relationships between the groups of animals, to draw an evolutionary tree. This is accomplished by identifying similarities in the themes of two or more groups and uniting them on the basis of their unique similarities. So, for example, crabs, lobsters, and shrimps (and a few minor groups) are closely related to each other because only they, among all crustaceans, share a design with five pairs of locomotory appendages. The evolutionary history of a group is called its **phylogeny** and the actual depiction of evolutionary relationships is known as a **phylogenetic tree** (or cladogram). Although a phylogenetic tree, as a scientific hypothesis, is subject to testing and change, it nevertheless provides a framework on which to organize and compare ground plans and thus summarize much of the factual content of each chapter. For those unfamiliar with the method for constructing or interpreting phylogenetic trees, an exercise in phylogenetic reconstruction is provided later in this chapter.

Yet another approach to invertebrate diversity is to view animal design in a framework of **functional principles.** Here we seek to understand how the fundamental principles of physics and chemistry impose limits—indeed, control—design. For example, flight requires an airfoil (wing) to produce lift. Although the composition of wings can differ (exoskeleton, skin, feathers, aluminum) and the animals having them may be unrelated, all necessarily have the shape of a wing. Thus, we can predict that any animal that evolves flight will have something that looks like a wing. At a slightly more derived level are the principles of physiology, development, and ecology. These concepts together—physical, chemical, biological—allow one to view living invertebrates as complex and fascinating expressions of a few principles. The subtlety and elegance of this approach breathes life into invertebrate zoology.

This book integrates invertebrate structure and function in an evolutionary context. The sequence of chapters, especially in the first part of the book, is progressive, and several chapters (those whose titles include "Introduction to") describe major turning points in animal evolution, such as the origin and significance of the multicellular body and, later, bilateral symmetry. At each of these "steps," the animal body evolved a radically new design, which created novel ecological opportunities and was inherited by all descendants. How those new designs might have evolved, how they work in the context of general functional principles, and how they enabled their possessors to exploit new adaptive zones are the subjects of the "Introduction to" chapters. This book, then, is a blend of factual descriptions and provisional syntheses that are subject to scientific testing and revision. Throughout the book, we emphasize that opportunities abound for discovery in this dynamic field.

To provide comprehensive coverage of invertebrates is a goal of this book, but a few taxa are described in less detail than others. Because most schools have specialized courses in protozoology, parasitology, and entomology, we provide only abbreviated coverage of unicellular organisms (protozoa), parasitic invertebrates, and insects (Hexapoda).

RECONSTRUCTION OF INVERTEBRATE PHYLOGENY

CLADISTIC METHOD

One of the noteworthy achievements of biological research has been the establishment of a **natural system,** the recognition of species and the arrangement of those species in a hierarchic pattern of relationships. This pattern of evolutionary or kinship relationships is called a **phylogeny,** or a "tree of life." The science that concerns itself with the discovery and depiction of phylogeny is known as phylogenetic systematics or **cladistics.** The goal of cladistics is to discover and portray the kinship relationships among species, ultimately in a **phylogenetic tree,** or cladogram.

In practice, one establishes a kinship relationship between two named groups, or **taxa** (the singular is **taxon**), by detecting a trait, or **character,** that is expressed in these taxa alone. Such a uniquely shared character is called a **synapomorphy** (= shared derived character), and taxa united by one or more synapomorphies are called **sister taxa.** Because a synapomorphy is shared by the sister taxa and no other taxon, it must have evolved in the immediate ancestor of the sister taxa. Thus the synapomorphy observed in the descendants actually originated in the ancestor as an evolutionary novelty, now called an **autapomorphy** (= self-derived character). As the sole descendants of that one immediate ancestor, the sister taxa constitute a **monophyletic** (= one origin) **taxon.** The ultimate goal of cladistics is to reconstruct a comprehensive tree of life based solely on monophyletic taxa.

An example of these concepts can be drawn from the arthropods (Fig. 1-1). For taxa, consider the crustaceans (such as shrimps, crabs, and lobsters) and tracheates (insects, milli-

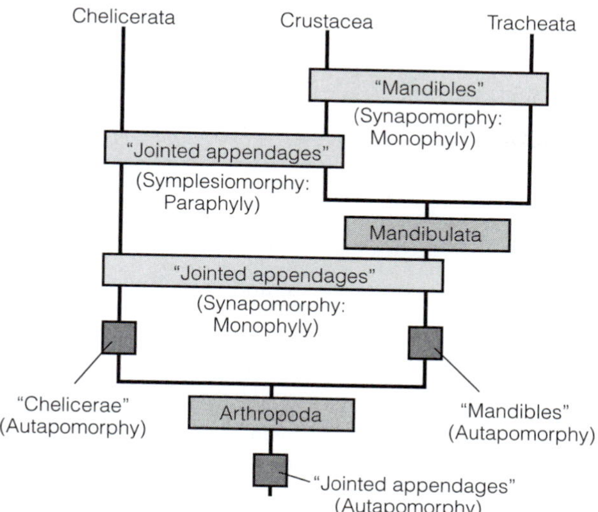

FIGURE 1-1 Cladistics. Example of the use of cladistics in phylogenetic reconstruction (see text). The observation of a uniquely shared character (synapomorphy), "mandibles," in Crustacea and Tracheata unites them as sister taxa in a monophyletic taxon, Mandibulata. Thus the stem species, or ancestor, of Mandibulata must have first expressed "mandibles" as an evolutionary novelty (autapomorphy). Similarly, the observation that "jointed appendages" occur in only chelicerates and mandibulates and nowhere else indicates that the character "jointed appendages" is a synapomorphy of the two sister taxa, which form the monophyletic taxon Arthropoda. The ancestral arthropod, therefore, first acquired "jointed appendages" as an evolutionary innovation, or autapomorphy. If an observer, for lack of complete knowledge, unites chelicerates and crustaceans into what he believes is a monophyletic taxon because both share "jointed appendages," the union is false because "jointed appendages" are not unique to chelicerates and crustaceans alone, but to all arthropods. Thus the character "jointed appendages" did not evolve in the immediate ancestor to chelicerates and crustaceans alone, but in a deeper ancestry that also gave rise to the tracheates. The chelicerates and crustaceans express this deeply ancestral trait as a shared primitive character, or symplesiomorphy. The false union based on a symplesiomorphy is called a paraphyletic taxon. The autapomorphy "chelicerae" is a pair of pincerlike appendages on the second segment of the head.

pedes, and centipedes, for example). Representatives of these two taxa show that all share the character "mandibles" (jaws derived from appendages on the third head segment), which occurs nowhere else and thus is a synapomorphy of crustaceans and tracheates. The synapomorphy indicates that the sister taxa (Crustacea, Tracheata) constitute a monophyletic taxon, the ancestor of which first evolved mandibles as an evolutionary novelty, or autapomorphy. Once the monophyletic taxon has been identified, it is given a formal name—in this case, Mandibulata.

Now suppose that, instead of crustaceans and tracheates, the study began with crustaceans and chelicerates (including horseshoe crabs, scorpions, and spiders) and found that they shared the character "jointed appendages," which was then hypothesized to be a synapomorphy of the two taxa (Fig. 1-1). The character "jointed appendages," however, is not shared by crustaceans and chelicerates alone, but also by the tracheates. This means that the "jointed appendages" of chelicerates and crustaceans is not a shared derived character but rather a shared ancestral trait, or **symplesiomorphy,** that evolved in the stem species (ancestor) of all arthropods. The erroneous union of crustaceans and chelicerates based on a symplesiomorphy is called a **paraphyletic taxon** (Fig. 1-1). A paraphyletic taxon contains some, but not all, of the descendants of the stem species. A major challenge of modern phylogenetic research is to identify and weed out paraphyletic taxa. In this example, "jointed appendages" is actually an autapomorphy of the monophyletic taxon Arthropoda, which includes the sister taxa Chelicerata and Mandibulata (Crustacea + Tracheata; Fig. 1-1).

A paraphyletic taxon fails to include all descendants of one ancestor, but a **polyphyletic taxon** includes the descendants of more than one ancestor. This mistake occurs when a shared similarity results from evolutionary **convergence** rather than common ancestry. Similarity attributable to common genetic inheritance is called **homology,** whereas the superficial similarity that arises from convergence is known as **homoplasy** (or analogy). Only homologous structures are useful in phylogenetic reconstruction based on monophyletic groups. An example of a polyphyletic taxon would be one that united birds, bats, and insects together because all share "wings." Bird wings, bat wings, and insect wings, however, are unrelated homoplasous structures, and the members of this polyphyletic taxon descended from three separate ancestors, each of which independently evolved its own unique wing.

The application of the cladistic method to species, groups of species, and groups of groups of species leads to the dichotomously branching hierarchic structure that typifies a phylogenetic tree. **Hierarchic structure** means that species are nested in larger, more inclusive taxa, which are contained in still more inclusive groups. So a species of shrimp is a kind of crustacean, which is a kind of mandibulate, which is a kind of arthropod, which is a kind of animal. The hierarchy shown in Figure 1-1 can also be summarized in tabular form in a **phylogenetic hierarchy:**

Arthropoda
    Chelicerata
    Mandibulata
        Crustacea
        Tracheata

A properly constructed phylogenetic tree is a deeply informative representation. The kinship relationships of species and higher taxa are shown graphically and the synapomorphies are indicated for each pair of sister taxa. The autapomorphies associated with stem species, moreover, offer insight into the form of the ancestors, which then may be compared with fossil evidence, if such exists.

Although we try to choose simple, factually sound examples to explain the method of phylogenetic reconstruction, in practice most reconstructions are more complex. It often happens, for example, that apparent synapomorphies are contradictory: Synapomorphy 1 unites taxa A and B and synapomorphy 2 unites taxa B and C. This results in two different and competing trees. In this situation, systematists choose the tree with the fewest branches, the tree that assumes the fewest number of character changes—that is, the most parsimonious tree. This principle of **parsimony** is one of the important underlying assumptions of phylogenetic systematics (and of science in general). It is an especially important tool in computer-aided cladistics. In such analyses, the software sifts through large amounts of data in the form of numerically coded morphological and molecular characters and, using different routines for joining or weighting them, generates more than one tree. From among these alternatives, a most parsimonious tree is typically calculated and drawn.

In this book, all trees of phylogenetic relationship are based on synapomorphies and, unless otherwise noted, are plotted by hand, as in Figure 1-1. This explicit and direct approach was adopted to allow you, the student, to evaluate the choice and validity of characters used in the reconstructions. Trees are sensitive to new facts and to reevaluation of assumptions about the nature of the characters themselves. As you acquire new knowledge applicable to phylogenetics, create your own trees and test them against those presented in the text.

DOWNPLAYING THE LINNEAN CATEGORIES

Modern cladistics is the method of choice for establishing kinship relationships among taxa. Once monophyletic taxa are identified and named on a phylogenetic tree or in a phylogenetic hierarchy, the hierarchic relationships among the taxa are obvious and clear (see Fig. 1-1 and the phylogenetic hierarchy set out in the last section). The question then is: "What is gained by assigning a Linnean category—phylum, class, order, and so on—to the taxon names already present?" Many systematists today have abandoned the Linnean categories because they are unnecessary, because they were established by Linnaeus for what he considered to be the immutable levels of Creation, and because new cladistic ranks of many taxa do not coincide well with the old Linnean categories. For some, liberation from the Linnean categories will shift emphasis away from pigeonholing ("To what class does this animal belong?") to science ("What are the postulated synapomorphies of Crustacea and Tracheata and how can they be tested?"). For others, however, the Linnean hierarchy of categories will continue to provide a familiar and constant frame of reference.

This book embraces the new cladistic approach to systematics and largely dispenses with Linnean categories. The relative ranks of taxa can be obtained from the phylogenetic trees, the text sections that include "Phylogenetic Hierarchy of" in their title, and the chapter headings. To soften the transition from the classical to modern approach, however, Linnean categories are assigned to taxa as superscript abbreviations when a taxon name is used as a text heading. The standard abbreviations are: ^{P} for phylum, ^{C} for Class, ^{O} for Order, and ^{F} for Family. An uppercase ^{S} prefixed to a category abbreviation identifies a supertaxon, a lowercase ^{s} identifies a subtaxon, and a lowercase ^{i} indicates an infrataxon. Super-, sub-, and infrataxa are ranked as follows:

Supertaxon
- Taxon (for example, phylum, class, order)
  - Subtaxon
    - Infrataxon

In the phylogenetic hierarchy given earlier, the superscripts would be: Arthropoda^{P}, Chelicerata^{sP}, Mandibulata^{sP}, Crustacea^{iP}, and Tracheata^{iP}.

AN EXERCISE IN CLADISTICS

Figures 1-2, 1-3, and 1-4 as well as the following text provide an exercise in cladistics for those who wish to practice building a phylogenetic tree. The exercise is divided into three steps, each of which is highlighted and explained here and in the figure legends.

- **Step 1: Study characters of a group of taxa in question and search for apomorphies, first within species (Fig. 1-2)**

Figure 1-2 shows one individual from each of six fictitious species. The first task is to establish (or not) the uniqueness (validity) of each species. To accomplish this, search for a character that is expressed uniquely in each species alone. Once identified, that character is assumed to have evolved as an evolutionary novelty in the ancestor of that species and is thus a species-level autapomorphy. An autapomorphy for each of the six species is shown in Figure 1-2. In this example, these species are then named appropriately according to their autapomorphies. So, the species with the autapomorphy "fan tail" is named *Fan tail,* the species with "long antennae" is named *Long antennae,* and so on as shown in Figure 1-2. The italicized species binomial always consists of a capitalized genus name (for example, *Fan*) and a lowercase species name (*tail*). In actual practice, both names would be latinized.

- **Step 2: Find synapomorphies that link species and groups of species pairwise into a most parsimonious tree (Fig. 1-3)**

Figure 1-3 shows the most parsimonious phylogenetic tree for the six species. The tree results from the recognition of a character or characters (synapomorphies) that are shared uniquely by two species or two groups of species. Notice that at least one synapomorphy supports the pairwise union of species (and higher groups) into sister taxa. For example, species *Shoulder leg* and *Fan tail* are united in a monophyletic taxon by the uniquely shared character (synapomorphy) "stalked eyes." This new monophyletic taxon has yet to be named and thus is temporarily designated **Nomen nominandum** (= new name), abbreviated **N. N.**

Proceeding deeper into the tree, notice that the monophyletic taxon that includes *Shoulder leg* and *Fan tail* is united with the species *Scissors tail* in another monophyletic taxon by the synapomorphies "three segments" and "pliers claw." This new, higher-level taxon is also temporarily designated N. N.

Figure 1-3 shows the progressive pairing of taxa into monophyletic groups based on shared derived characters. But a dilemma is encountered with *Paddle foot* at the base of the

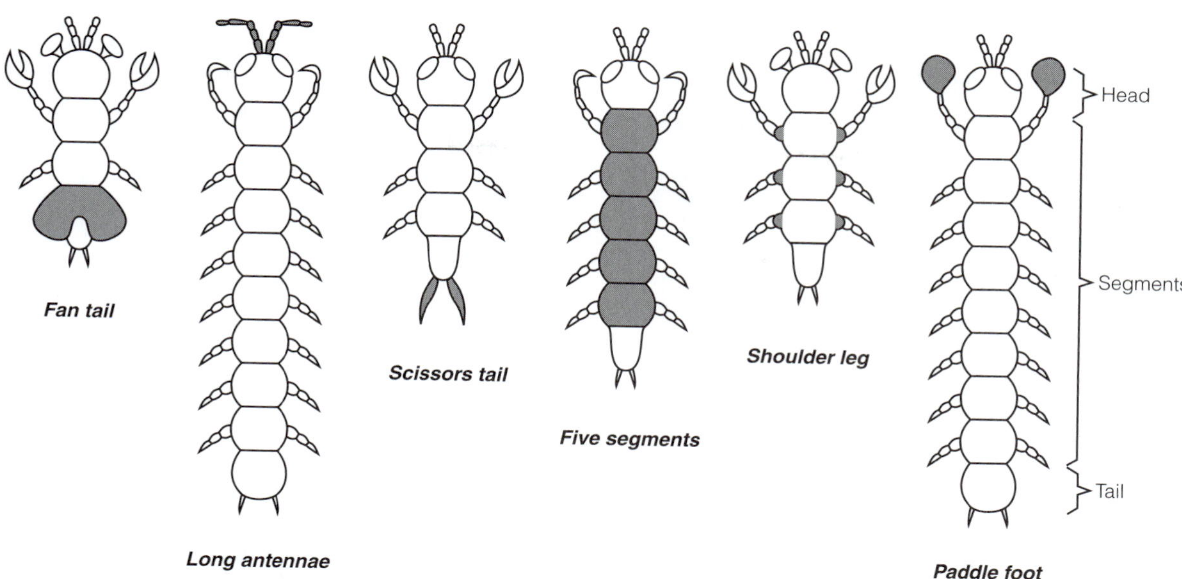

FIGURE 1-2 Exercise in cladistics, step 1 (see also text). Each of the six illustrated animals represents one fictitious species. The first step in the analysis is to identify one or more characters that is unique to each species (autapomorphy). The autapomorphies of each species in this example are shaded and reflected in the species' name.

tree. The dilemma is whether the characters associated with *Paddle foot* are plesiomorphies or apomorphies. Is the character "paddle foot" an autapomorphy of *Paddle foot* that is found in that species and nowhere else, or is it a symplesiomorphy occurring in *Paddle foot* and in other, as yet unstudied species? Similarly, is the character "eight segments" an autapomorphy of the original six species or is it a symplesiomorphy occurring in those six plus additional species? The questions boil down to this: How do we establish the apomorphy vs. plesiomorphy **polarity** of the characters? The monophyly or paraphyly of the group of six species hinges on the determination of character polarity. The practical solution to this dilemma is to discover whether or not "eight segments" and "paddle foot" occur outside the group of six species, and this requires the examination of additional species. The extension of the cladistic analysis beyond the taxa of immediate interest is called an **outgroup comparison.** The goal is to determine character polarity to resolve the plesiomorphy vs. apomorphy dilemma.

- **Step 3: Perform an outgroup analysis and name the monophyletic taxa (Fig. 1-4)**

Figure 1-4 resolves the uncertain polarity of characters "paddle foot" and "eight segments" by performing an outgroup analysis. The outgroup in this example is represented by a single unnamed species and it is compared with the ingroup consisting of the six original species. The outgroup species lacks the characters "paddle foot" (it has "normal" feet) and "eight segments" (it has nine segments). Thus, the outgroup comparison enables us to conclude that "paddle foot" is an autapomorphy of the species *Paddle foot* and "eight segments" is a synapomorphy of *Paddle foot* and the taxon (Cultrichela) that includes the remaining five ingroup species. Not all species of Cultrichela, however, have eight segments. Some have evolved a body with fewer segments, in this example, five and three segments. Or, alternatively, "eight segments" is an autapomorphy of a monophyletic taxon that includes the original six species. The character "eight segments," as both synapomorphy and autapomorphy, is shown in Figure 1-4.

In reference to the original analysis, the ingroup of six species, apomorphies now support the monophyly of all sister taxa. At this point, it is appropriate to replace the provisional N. N. designation with a formal Latin name. These formal names, which are derived from taxon autapomorphies, are shown on Figure 1-4. For example, the monophyletic taxon supported by the apomorphy "stalked eyes" is named Exophthalmia (= prominent eyes), the taxon supported by "three segments" is named Triannelida (= three rings), and the monophyletic taxon that includes all six species, based on the apomorphy "eight segments," is designated Octoannelida (= eight rings). The phylogenetic hierarchy of Octoannelida is:

Octoannelida
- *Paddle foot*
- Cultrichela
  - *Long antenna*
  - Urocopa
    - *Five segments*
    - Triannelida
      - *Scissors tail*
      - Exophthalmia
        - *Shoulder leg*
        - *Fan tail*

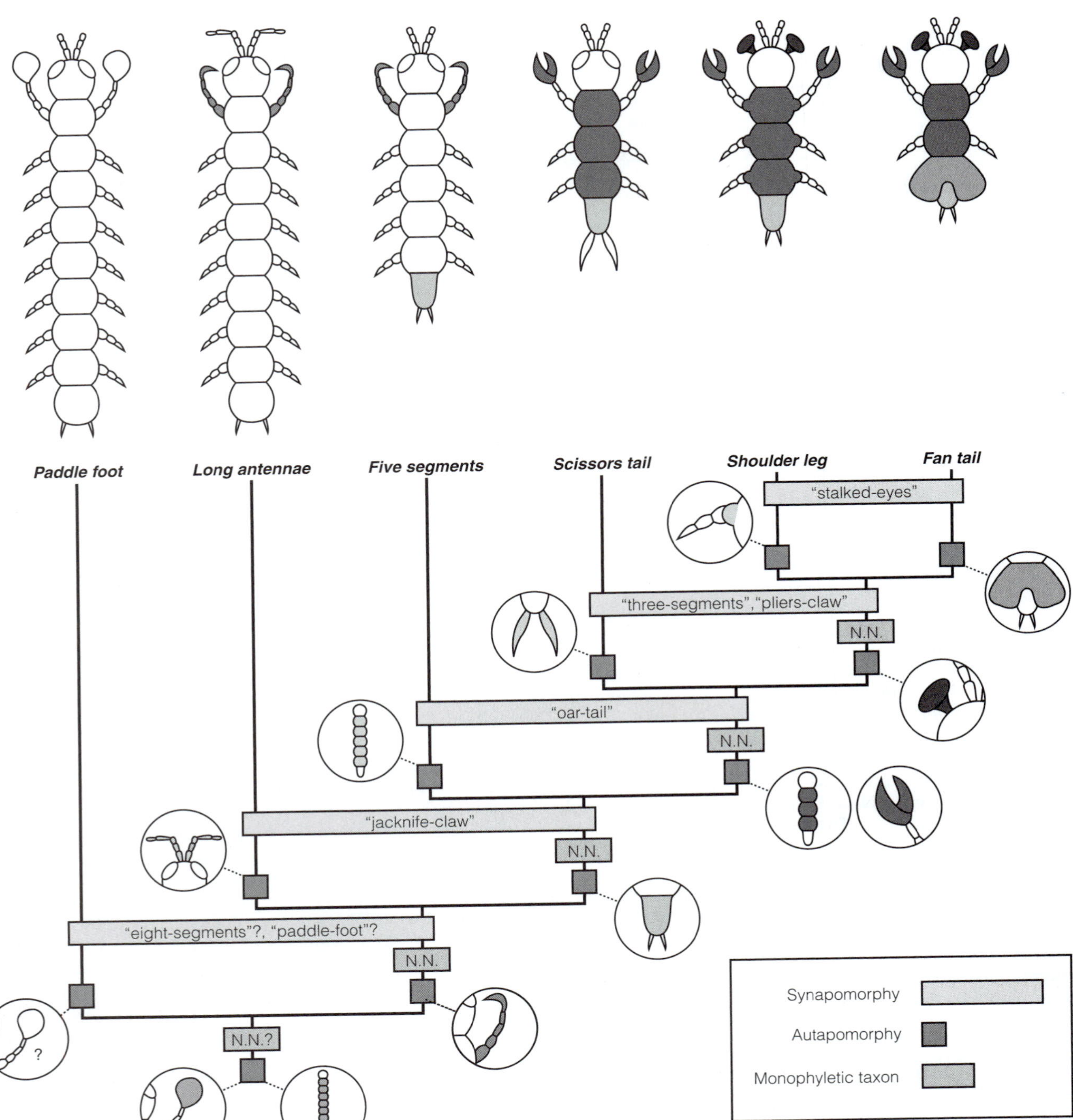

FIGURE 1-3 Exercise in cladistics, step 2 (see also text). The six fictitious species are united pairwise into monophyletic taxa on the basis of uniquely shared characters (synapomorphies). The observed synapomorphy of each pair of sister taxa (together a monophyletic taxon) evolved as an evolutionary novelty in the ancestor of the monophyletic taxon and is indicated on the tree as an autapomorphy. Each of the monophyletic taxa is given a general temporary designation, abbreviated N. N. Notice the uncertainty associated with the characters "eight segments" and "paddle foot." Both characters could be either apomorphies or plesiomorphies. If they are apomorphies, then the N. N. taxon that includes all six species is monophyletic; if they are plesiomorphies, then the taxon is paraphyletic. The plesiomorphic vs. apomorphic interpretation of these characters can be resolved only by comparison with a taxon or taxa outside this group of six species. Such an outgroup comparison is shown in Figure 1-4 and is described in the text.

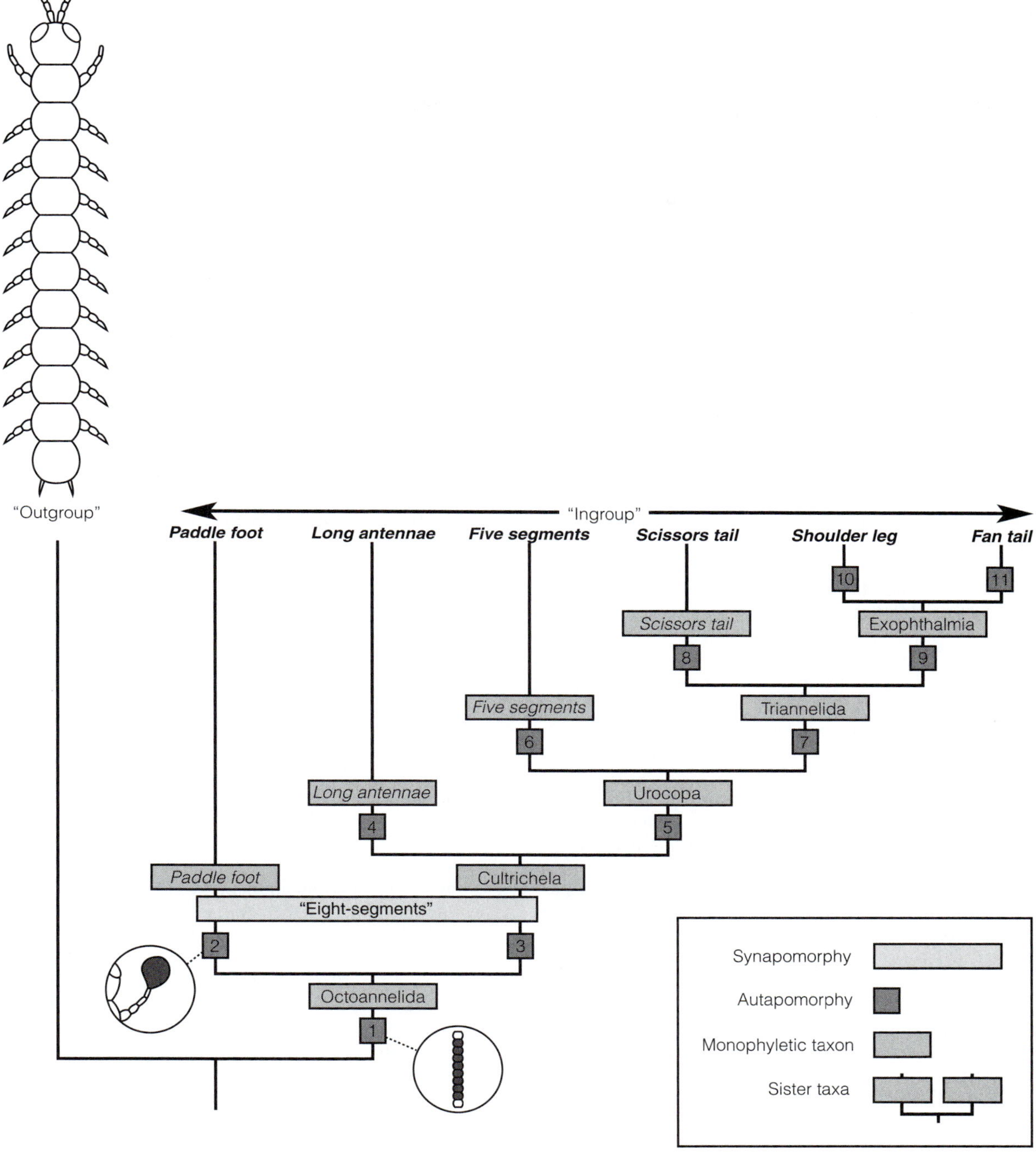

FIGURE 1-4 Exercise in cladistics, step 3 (see also text). An outgroup analysis shows that the character "paddle foot" is an autapomorphy of the species *Paddle foot.* It also indicates that the character "eight segments" is a synapomorphy of the taxon (species) *Paddle foot* and its sister taxon, Cultrichela, which includes all other species. Thus the synapomorphy "eight segments" establishes the monophyletic taxon Octoannelida for the six species in this study. The autapomorphies of the monophyletic taxa are: **1, Octoannelida:** "eight segments"; **2, *Paddle foot:*** "paddle foot"; **3, Cultrichela:** "jacknife claw"; **4, *Long antenna:*** "long antenna"; **5, Urocopa:** "oar tail"; **6, *Five segments:*** "five segments" (reduced from eight), "pliers claw" (modification of "jacknife claw"); **7, Triannelida:** "three segments" (reduced from eight); **8, *Scissors tail:*** "scissors tail"; **9, Exophthalmia:** "stalked eyes"; **10, *Shoulder leg:*** "shoulder leg"; **11, *Fan tail:*** "fan tail."

REFERENCES

BOOKS AND JOURNALS

Comprehensive

Beklemishev, V. N. 1969. Principles of Comparative Anatomy of Invertebrates. University of Chicago Press, Chicago. (Two volumes.)

Bronn, H. G. (Ed.): 1866–. Klassen und Ordnungen des Tierreichs. C. F. Winter, Leipzig and Heidelberg. (Many volumes; the series is incomplete.)

Fretter, V., and Graham, A. 1976. A Functional Anatomy of Invertebrates. Academic Press, London. 600 pp.

Grassé, P. (Ed.): 1948–1996. Traité de Zoologie. Masson, Paris. (Seventeen volumes.)

Harrison, F. W. (Ed.): 1991–1999. Microscopic Anatomy of Invertebrates. Wiley-Liss, New York. (Fifteen volumes.)

Hyman, L. H. 1940. The Invertebrates. Vol. 1. Protozoa through Ctenophora. McGraw-Hill, New York. 726 pp.

Hyman, L. H. 1951. The Invertebrates. Vol. 2. Platyhelminthes and Rhynchocoela: The Acoelomate Bilateria. McGraw-Hill, New York. 550 pp.

Hyman, L. H. 1951. The Invertebrates. Vol. 3. Acanthocephala, Aschelminthes and Entoprocta. McGraw-Hill, New York. 572 pp.

Hyman, L. H. 1955. The Invertebrates. Vol. 4. The Echinodermata. McGraw-Hill, New York. 763 pp.

Hyman, L. H. 1959. The Invertebrates. Vol. 5. Smaller Coelomate Groups. McGraw-Hill, New York. 783 pp.

Hyman, L. H. 1967. The Invertebrates. Vol. 6. Mollusca I. McGraw-Hill, New York. 792 pp.

Kozloff, E. N. 1990. Invertebrates. Saunders, Philadelphia. 866 pp.

Parker, T. J., and Haswell, W. A. 1951. A Text-Book of Zoology. Vol. 1. MacMillan, London. 770 pp.

Parker, S. P. (Ed.): 1982. Synopsis and Classification of Living Organisms. Vol. 1. McGraw-Hill, New York. 1166 pp.

Parker, S. P. (Ed.): 1982. Synopsis and Classification of Living Organisms. Vol. 2. McGraw-Hill, New York. 1236 pp.

Pearse, V., Pearse, J., Buchsbaum, M., and Buchsbaum, R. 1987. Living Invertebrates. Blackwell Scientific, Palo Alto, CA. 848 pp.

Russell-Hunter, W. D. 1979. A Life of Invertebrates. Macmillan, New York. 650 pp.

Westheide, W. and Rieger, R. M. (Eds.): 1996. Spezielle Zoologie. Erster Teil: Einzeller und Wirbellose Tiere. Gustav Fischer Verlag, Stuttgart. 909 pp.

Lab Manuals

Brown, F. A., Jr. 1950. Selected Invertebrate Types. John Wiley and Sons, New York. 597 pp.

Dales, R. P. 1981. Practical Invertebrate Zoology. 2nd Edition. John Wiley and Sons, New York. 356 pp.

Freeman, W. H., and Bracegirdle, B. 1971. An Atlas of Invertebrate Structure. Heinemann Educational Books, London. 129 pp.

Pierce, S. K., and Maugel, T. K. 1989. Illustrated Invertebrate Anatomy. Oxford University Press, Oxford. 320 pp.

Sherman, I. W., and Sherman, V. G. 1976. The Invertebrates: Function and Form. 2nd Edition. Macmillan, New York. 334 pp.

Wallace, R. L., Taylor, W. K., and Litton, J. R. 1988. Invertebrate Zoology. 4th Edition. Macmillan, New York. 475 pp.

Morphology

Abbott, D. P. 1987. Observing Marine Invertebrates. Edited by G. H. Hilgard. Stanford University Press, Stanford, CA. 380 pp.

Bereiter-Hahn, J., Matoltsy, A. G., and Richards, K. S. (Eds.): 1984. Biology of the Integument. Vol. 1: Invertebrates. Springer-Verlag, Berlin. 841 pp.

Kennedy, G. Y. 1979. Pigments of marine invertebrates. Adv. Mar. Biol. 16:309–381.

Welsch, U., and Storch, V. 1976. Comparative Animal Cytology and Histology. University of Washington Press, Seattle. 343 pp.

Functional Morphology and Physiology

Alexander, R. M. 1982. Locomotion of Animals. Chapman and Hall, London. 192 pp.

Atema, J., Fay, R. R., Popper, A. N., and Tavolga, W. N. (Eds.): 1987. Sensory Biology of Aquatic Animals. Springer-Verlag, New York. 936 pp.

Bullock, T. H., and Horridge, G. A. 1965. Structure and Function of the Nervous System of Invertebrates. W. H. Freeman, San Francisco. (Two volumes.)

Clark, R. B. 1964. Dynamics in Metazoan Evolution: The Origin of the Coelom and Segments. Clarendon Press, Oxford. 313 pp.

Elder, H. Y., and Trueman, E. R. (Eds.): 1980. Aspects of Animal Movement. Cambridge University Press, New York. 250 pp.

Highnam, K. C., and Hill, L. 1977. The Comparative Endocrinology of the Invertebrates. 2nd Edition. University Park Press, Baltimore. 357 pp.

Hughes, R. N. 1989. A Functional Biology of Clonal Animals. Chapman and Hall, London. 331 pp.

Johnsen, S. 2000. Transparent animals. Sci. Am. Feb.:83–89.

Laufer, H., and Downer, R. G. H. (Eds.): 1988. Invertebrate Endocrinology. Vol. 2. Endocrinology of Selected Invertebrate Types. Alan R. Liss, New York. 500 pp.

Rankin, J. C., and Davenport, J. A. 1981. Animal Osmoregulation. John Wiley and Sons, New York. 202 pp.

Russell, F. E. 1984. Marine toxins and venomous and poisonous marine plants and animals. Adv. Mar. Biol. 21:60–233.

Schmidt-Nielsen, K. 1990. Animal Physiology: Adaptation and Environment. 4th Edition. Cambridge University Press, Cambridge. 602 pp.

Trueman, E. R. 1975. The Locomotion of Soft-Bodied Animals. American Elsevier, New York. 200 pp.

Vernberg, F. J., and Vernberg, W. B. (Eds.): 1981. Functional Adaptations of Marine Organisms. Academic Press, New York. 347 pp.

Vogel, S. 1988. Life's Devices: The Physical World of Animals and Plants. Princeton University Press, Princeton, NJ. 384 pp.

Vogel, S. 1996. Life in Moving Fluids: The Physical Biology of Flow. Princeton University Press, Princeton, NJ. 484 pp.

Wainwright, S. A. 1988. Axis and Circumference: The Cylindrical Shape of Plants and Animals. Harvard University Press, Cambridge, MA. 176 pp.

Ecology

Carefoot, T. 1977. Pacific Seashores: A Guide to Intertidal Ecology. University of Washington Press, Seattle. 208 pp.

Crawford, C. S. 1981. Biology of Desert Invertebrates. Springer-Verlag, New York. 314 pp.; Clarendon Press, Oxford. 313 pp.

Daiber, F. C. 1982. Animals of the Tidal Marsh. Van Nostrand Reinhold, New York. 432 pp.

Eltringham, S. K. 1971. Life in Mud and Sand. Crane, Russak, New York. 218 pp.

Gage, J. D., and Tyler, P. A. 1991. Deep-Sea Biology: A Natural History of Organisms at the Deep-Sea Floor. Cambridge University Press, Cambridge. 504 pp.

Halstead, B. W. 1988. Poisonous and Venomous Marine Animals of the World. 3rd Edition. Darwin Press, Princeton, NJ. 1168 pp.

Hardy, A. C. 1956. The Open Sea. Houghton Mifflin, Boston. (Two volumes.)

Harris, V. A. 1990. Sessile Animals of the Seashore. Chapman and Hall, London. 379 pp.

Kerfoot, W. C. (Ed.): 1980. Evolution and Ecology of Zooplankton Communities. University Press of New England, Hanover, NH. 794 pp.

MacGinitie, G. E., and MacGinitie, N. 1968. Natural History of Marine Animals. 2nd Edition. McGraw-Hill, New York. 523 pp.

Marshall, N. B. 1979. Deep Sea Biology. Garland STPM Press, New York. 566 pp.

Newell, R. C. 1979. Biology of Intertidal Animals. 3rd Edition. Marine Ecological Surveys, Faversham, Kent, U.K. 560 pp.

Nicol, J. A. C. 1967. The Biology of Marine Animals. 2nd Edition. Wiley-Interscience, New York. 699 pp.

Schaller, F. 1968. Soil Animals. University of Michigan Press, Ann Arbor. 114 pp.

Stephenson, T. A., and Stephenson, A. 1972. Life between Tidemarks on Rocky Shores. W. H. Freeman, San Francisco. 425 pp.

Thorp, J. H., and Covich, A. P. (Eds.): 1991. Ecology and Classification of North American Freshwater Invertebrates. Academic Press, New York. 911 pp.

Thorp, J. H., and Covich, A. P. (Eds.): 2001. Ecology and Classification of North American Freshwater Invertebrates. 2nd Edition. Academic Press, New York. 1056 pp.

Yonge, C. M. 1949. The Seashore. Collins, London. 311 pp.

Reproduction, Development, Larvae, and Metamorphosis

Chia, F., and Rice, M. E. (Eds.): 1978. Settlement and Metamorphosis of Marine Invertebrate Larvae. Elsevier North Holland, New York. 290 pp.

Giese, A. C., and Pearse, J. S. 1974–1991. Reproduction of Marine Invertebrates. Academic Press, New York. (Six volumes.)

Harrison, F. W., and Cowden, R. R. (Eds.): 1982. Developmental Biology of Freshwater Invertebrates. Alan R. Liss, New York. 588 pp.

Gilbert, S. F., and Raunio, A. M. 1997. Embryology. Constructing the organism. Sinauer, Sunderland, MA. 537 pp.

Kume, M., and Dan, K. 1968. Invertebrate Embryology. Clearinghouse for Federal Scientific and Technical Information, Springfield, VA. 605 pp.

Young, C. M., Sewell, M. A., and Rice, M. E. (Eds.): 2001. Atlas of Marine Invertebrate Larvae. Academic Press, New York. 630 pp.

Cladistics and Evolution

Ax, P. 1996. Multicellular Animals. Vol. I. A New Approach to the Phylogenetic Order in Nature. Springer, Berlin. 225 pp.

Ax, P. 2000. Multicellular Animals. Vol. II. The Phylogenetic System of the Metazoa. Springer, Berlin. 395 pp.

Ax, P. 2001. Der System der Metazoa III. Ein Lehrbuch der Phylogenetischen Systematik. Spektrum Akademischer Verlag, Heidelberg. 283 pp.

Conway Morris, S., George, J. D., Gibson, R., et al (Eds.): 1985. The Origins and Relationships of Lower Invertebrates. Systematics Association Spec. Vol. 28. Clarendon Press, Oxford. 394 pp.

House, M. R. (Ed.): 1979. The Origin of Major Invertebrate Groups. Systematics Association Spec. Vol. 12. Academic Press, London. 515 pp.

Kitching, I. J. (Ed.): 1998. Cladistics: The Theory and Practice of Parsimony Analysis. 2nd Edition. Oxford University Press, Oxford. 240 pp.

Nielsen, C. 2001. Animal Evolution: Interrelationships of the Living Phyla. Oxford University Press, Oxford. 563 pp.

Wiens, J. J. (Ed.): 2000. Phylogenetic Analysis of Morphological Data. Smithsonian Institution Press, Washington, DC. 272 pp.

Paleontology

Boardman, R. S., Cheetham, A. H., and Rowell, A. J. 1986. Fossil Invertebrates. Blackwell Scientific, Boston. 713 pp.

Dyer, J. C., and Schram, F. R. 1983. A Manual of Invertebrate Paleontology. Stipes, Champaign, IL. 165 pp.

Moore, R. C. (Ed.): 1953–1966. Treatise on Invertebrate Paleontology. Geological Society of America, University of Kansas Press, Lawrence, KS. (Eighteen volumes.)

Field Guides

Campbell, A. C. 1976. The Hamlyn Guide to the Seashore and Shallow Seas of Britain and Europe. Hamlyn, London. 320 pp.

Colin, P. L. 1978. Caribbean Reef Invertebrates and Plants. T. F. H., Neptune City, NJ. 478 pp.

Fielding, A. 1998. Hawaiian Reefs and Tidepools. Booklines Hawaii, Mililani, HI. 103 pp.

Fotheringham, N., and Brunenmeister, S. L. 1975. Common Marine Invertebrates of the Northwestern Gulf Coast. Gulf, Houston. 175 pp.

Gosner, K. L. 1979. Peterson Field Guide Series: A Field Guide to the Atlantic Seashore. Houghton Mifflin, Boston. 329 pp.

Gunson, D. 1983. Collins Guide to the New Zealand Seashore. Collins, Auckland. 240 pp.

Hayward, P. J., and Ryland, J. S. (Eds.): 1991. The Marine Fauna of the British Isles and North-West Europe. Vol. 1: Introduction and Protozoans to Arthropods. Oxford University Press, Oxford. 688 pp.

Hayward, P. J., and Ryland, J. S. (Eds.): 1991. The Marine Fauna of the British Isles and North-West Europe. Vol. 2: Molluscs to Chordates. Oxford University Press, Oxford. 386 pp.

Hoover, J. P. 1999. Hawaii's Sea Creatures: A Guide to Hawaii's Marine Invertebrates. Mutual Publishers, Honolulu. 366 pp.

Hurlbert, S. H., and Villalobos-Figueroa, A. (Eds.): 1982. Aquatic Biota of Mexico, Central America and the West Indies. Aquatic Biota-SDSU Foundation, San Diego State University, San Diego. 529 pp.

Kaplan, E. H. 1982. Peterson Field Guide Series. A Field Guide to Coral Reefs of the Caribbean and Florida. Houghton Mifflin, Boston. 289 pp.

Kerstitch, A. 1989. Sea of Cortez Marine Invertebrates. Sea Challengers, Monterey, CA. 120 pp.

Kozloff, E. N. 1983. Seashore Life of the Northern Pacific Coast: An Illustrated Guide to Northern California, Oregon, Washington and British Columbia. Revised Edition. University of Washington Press, Seattle. 370 pp.

Kozloff, E. N. 1988. Marine Invertebrates of the Pacific Northwest. University of Washington Press, Seattle. 511 pp.

Meinkoth, N. A. 1981. The Audubon Society Field Guide to North American Seashore Creatures. Alfred Knopf, New York. 799 pp.

Morris, R. H., Abbott, D. P., and Haderlie, E. C. 1980. Intertidal Invertebrates of California. Stanford University Press, Palo Alto, CA. 690 pp.

Morton, B., and Morton, J. 1983. The Sea Shore Ecology of Hong Kong. Hong Kong University Press, Hong Kong. 350 pp.

Morton, J., and Miller, M. 1973. The New Zealand Sea Shore. 2nd Edition. Collins, London. 653 pp.

Newell, G. E., and Newell, R. C. 1973. Marine Plankton: A Practical Guide. Hutchinson Educational, London. 244 pp.

Peckarsky, B. L., Fraissinet, P. R., Penton, M. A., et al. 1990. Freshwater Macroinvertebrates of Northeastern North America. Comstock, Ithaca, NY. 442 pp.

Pennak, R. W. 1978. Fresh-Water Invertebrates of North America. 2nd Edition. John Wiley and Sons, New York. 803 pp.

Pennak, R. W. 1989. Freshwater Invertebrates of the United States. Protozoa to Molluscs. 3rd Edition. John Wiley and Sons, New York. 768 pp.

Ruppert, E., and Fox, R. 1988. Seashore Animals of the Southeast: A Guide to Common Shallow-Water Invertebrates of the Southeastern Atlantic Coast. University of South Carolina Press, Columbia. 429 pp.

Smith, D. G. 2001. Pennak's Freshwater Invertebrates of the United States: Porifera to Crustacea. 4th Edition. John Wiley and Sons, New York. 648 pp.

Smith, D. L., and Johnson, K. B. 1996. A Guide to Marine Coastal Plankton and Marine Invertebrate Larvae. 2nd Edition. Kendall/Hunt, Dubuque, IA. 221 pp.

Smith, R. I. (Ed.): 1964. Keys to Marine Invertebrates of the Woods Hole Region. Contribution No. 11. Systematics-Ecology Program, Marine Biology Laboratory, Woods Hole, MA. 208 pp.

Sterrer, W. E. (Ed.): 1986. Marine Fauna and Flora of Bermuda. Wiley-Interscience, New York. 742 pp.

Wickstead, J. H. 1965. An Introduction to the Study of Tropical Plankton. Hutchinson, London. 160 pp.

Wrobel, D., and Mills, C. 1998. Pacific Coast Pelagic Invertebrates: A Guide to the Common Gelatinous Animals. Sea Challengers, Monterey, CA. 108 pp.

INTERNET SITES

General

http://life.bio.sunysb.edu/marinebio/mbweb.html (State University of New York. Invertebrate images, references, career opportunities, and news regarding marine biology.)

www-marine.stanford.edu/HMSweb/careers.html (Marine biology links, careers, and more.)

www.ucmp.berkeley.edu (University of California at Berkeley Museum of Paleontology. Explore their invertebrate collections and follow links to impressive images of living invertebrates.)

www.ucmp.berkeley.edu/exhibit/phylogeny.html (Take the Web Lift to Taxa to discover the relationships that connect all organisms.)

www.nmnh.si.edu/departments/invert.html (Smithsonian National Museum of Natural History. Good coverage of certain groups of invertebrates, such as squids and their relatives.)

http://mbayaq.org (Monterey Bay Aquarium. The next best thing to being there. Explore their beach, rocky intertidal pool, open water, and kelp forest exhibits, several of which are three-dimensional and several that are interactive.)

www.bioimages.org.uk/index.htm (BioImages. A U.K. virtual field guide.)

www.biosis.org/zrdocs/zoolinfo/gp_index.htm (BIOSIS Internet Resource Guide for Zoology.)

Lab Manual

www.lander.edu/rsfox/310labindex.html (Lander University OnLine Invertebrate Lab. Illustrated anatomical descriptions of approximately 100 species in support of invertebrate teaching and research.)

Professional Societies

www.invertebrates.org (American Microscopical Society. Publishes the international journal *Invertebrate Biology*. The journal's Web site links to other sites on invertebrates and opportunities in the field for research, jobs, and education.)

www.museum.unl.edu/asp (American Society of Parasitologists. Much useful information and many images at their Web site.)

Cladistics

tolweb.org/tree/phylogeny.html (Tree of Life. Provides a phylogenetic classification of all taxa of living organisms based on molecular and morphological traits. Coverage is uneven and often outdated, but new information is being added constantly. Good source of contemporary references to invertebrate systematics.)

www.nhm.ukans.edu/downloads/CompleatCladist.pdf (An online version of Wiley, E. O., Siegel-Causey, D., Brooks, D. R., and Funk, V. A. 1991. *The Compleat Cladist: A Primer of Phylogenetic Procedures*. Special Publication No. 19. University of Kansas Museum of Natural History, Lawrence, KS.)

www.gwu.edu/~clade/faculty/lipscomb/Cladistics.pdf (Diana Lipscomb's 1998 workbook, *Basics of Cladistic Analysis* [George Washington University, Washington, DC.])

erms.biol.soton.ac.uk (European Register of Marine Species.)

www.nhm.ac.uk/hosted_sites/uksf (U.K. Systematics Forum. A database of taxonomists and the Web of Life.)

animaldiversity.ummz.umich.edu/index.html (University of Michigan Museum of Zoology Animal Diversity Web.)

2 Introduction to Protozoa

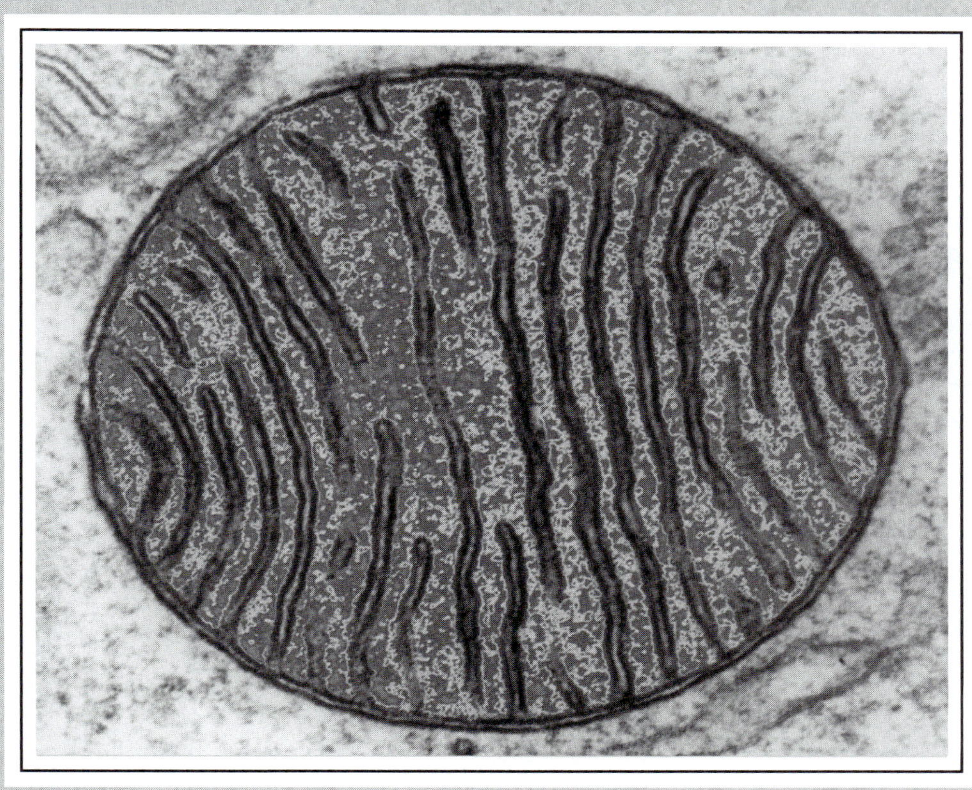

The unicellular **eukaryotes**—cells with a membrane-enclosed nucleus—are the atoms of the invertebrate world. Each cell is a complete organism adapted to meet the challenges of life, but some form colonies of cells, and from these evolved the world's multicellular organisms (fungi, algae, plants, animals). The entire assemblage of unicellular eukaryotes is known as Protista, and a large subgroup of mostly motile forms is called protozoa. The protozoa and its taxa will be discussed in Chapter 3. The purposes of this chapter are to describe the structure, function, and evolution of the eukaryotic cell with an emphasis on animal cells. As eukaryotes, protozoans have the same cellular components found in the cells of animals, plants, and fungi, but as cell-organisms, protozoans have specialized these parts into the functional equivalents of tissues and organs. These uniquely protozoan organelles and other structures will be described in Chapter 3. This chapter discusses the basic tool kit of eukaryotic cells and how it evolved.

EUKARYOTIC CELL STRUCTURE

Eukaryotic cells contain **organelles,** functionally specialized, regulated compartments surrounded by one or more membranes (Fig. 2-1). One organelle, the **nucleus,** contains the genome and is surrounded by a double membrane, thus segregating the genomic compartment from the metabolic machinery of the cell cytoplasm. Other organelles include mitochondria and chloroplasts, both of which are enclosed in two membranes. **Mitochondria** contain DNA and the enzymes for aerobic respiration. **Chloroplasts** also have DNA and are the sites for photosynthesis.

Apart from the cell membrane itself, the cytoplasm of eukaryotic cells has an **internal membrane system** that includes the endoplasmic reticulum, Golgi bodies, and lysosomes (Fig. 2-1). Arising from the outer nuclear membrane, the **endoplasmic reticulum** is a mazelike, tubular or lamellar network that functions in the synthesis of carbohydrates, lipids, and, when ribosomes are present, proteins. A **Golgi body** is a stack of flattened vesicles that receives products of

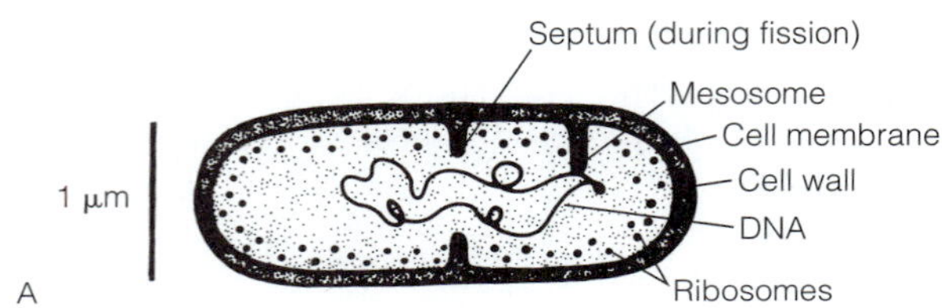

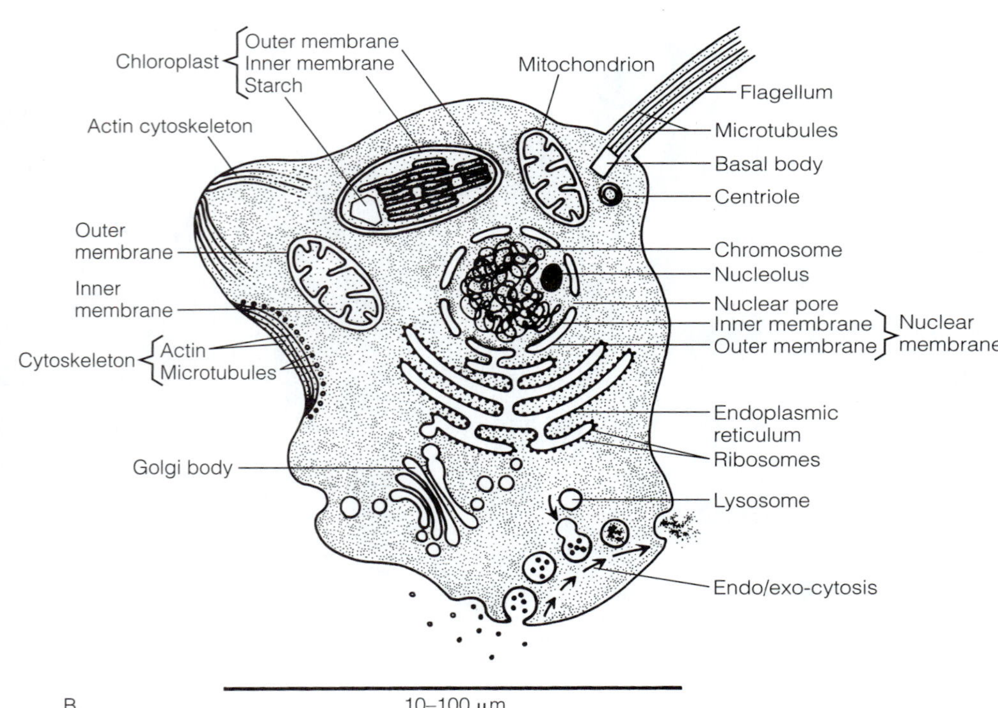

FIGURE 2-1 Diagrammatic comparison of prokaryotic and eukaryotic cell structure. **A,** prokaryotic; **B,** eukaryotic.

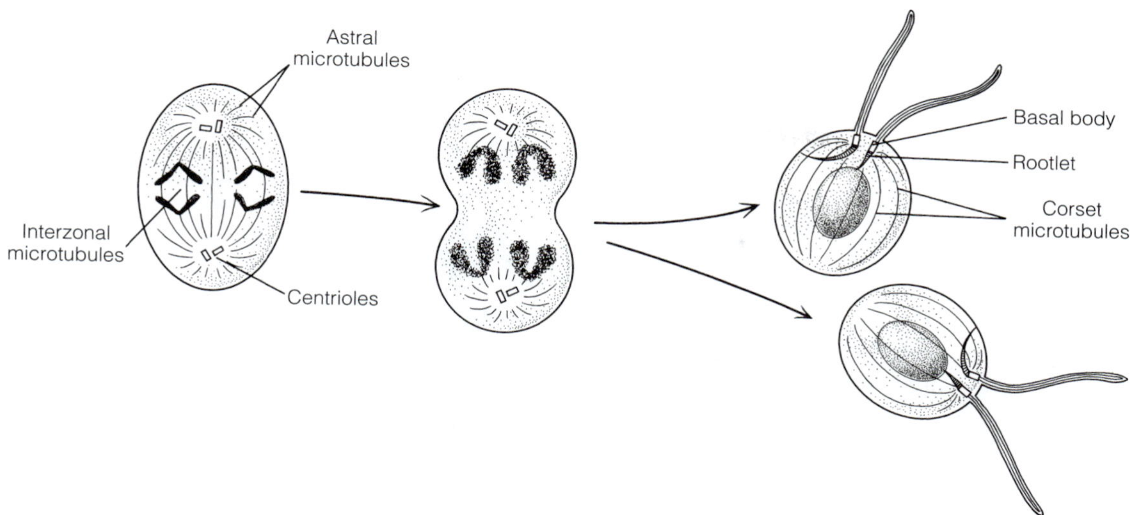

FIGURE 2-2 Eukaryotic cell structure. Relationship of centrioles and mitotic spindle fibers (astral and interzonal microtubules) to flagellar basal bodies and cytoskeleton (corset microtubules).

the endoplasmic reticulum, then modifies and releases them in vesicles for transport elsewhere, often to the surface of the cell. **Lysosomes** are Golgi-derived, membrane-bound vesicles that contain enzymes for intracellular digestion.

Unlike prokaryotes (bacteria), which support themselves with a cell wall, the eukaryotic cell has a **cytoskeleton** of protein filaments of different types and diameters. The most common of these are **actin** filaments (6 nanometers [nm] in diameter, also called **microfilaments**) and **microtubules** (15 nm in diameter; Fig. 2-2). Microfilaments are polymers of monomeric actin and microtubules are cylindrical polymers of the protein tubulin. The cytoskeleton typically has the form of a three-dimensional network and is responsible for the maintenance of cell shape. Often, it is well developed just below the cell membrane, where it strengthens the cell surface. The cytoskeleton, however, is not always a static or permanent fixture, but rather can be dynamic and temporary. Because polymerization of the cytoskeleton is reversible, the filaments or tubules can be locally assembled or disassembled to provide scaffolding for special structures, such as the transient spindle apparatus associated with mitosis (Fig. 2-2) or the outgrowth of semipermanent cilia and flagella from the cell surface (Fig. 2-1, 2-5).

The eukaryotic cytoskeleton is also essential for **cell motility.** As is true of other skeletons (see Chapter 4), the cytoskeleton can transmit force from one part of the cell to another, resulting in cell movement, or its filaments can serve as tracks along which vesicles and other structures are transported. In either case, the force for movement is generated by so-called **motor molecules,** such as myosin and dynein, which change shape in the presence of ATP. Typically, a motor molecule that is attached securely to one structure attaches temporarily to the cytoskeleton and flexes, moving the structure with respect to the skeleton (Fig. 2-5). The motor molecule then withdraws from its original attachment site, forms another attachment at a new position, and flexes once again. Repetition of this cycle is reminiscent of walking on a treadmill, and is referred to as **treadmilling. Dynein,** the motor molecule associated with microtubules, is important for the movement of cilia and flagella as well as for shuttling vesicles inside of the cell. **Myosin** binds to actin as well as to other structures and is responsible for ameboid movement (discussed later), streaming, and cyclosis (cytoplasmic circulation, also discussed later), cell division (cytokinesis), and muscle contraction in metazoans (Chapter 6).

The organelles and cytoskeleton of the eukaryotic cell are surrounded by a fluid **cytoplasm.** Cytoplasm, in turn, is enclosed by the **cell membrane,** a phospholipid bilayer that separates the internal environment of the cell from the exterior (Fig. 2-3). In doing so, it regulates the biochemical conditions of the cell's interior for the processes of life. The cell membrane controls what may enter and leave the cell, the responsiveness of the cell to external stimuli, the selectiveness with which the cell binds to other cells or to a substratum, and the maintenance of cell shape. The bilayered structure of the cell membrane results from the opposing phospholipids that compose it (Fig. 2-3). Proteins are also important membrane constituents and may span it or be attached to the inner or outer surfaces. The exposed outer surfaces of membrane proteins and lipids may have attached to them carbohydrates that radiate into the surrounding medium like tails. Together, these tails and especially their extracellular peripheral proteins form a surface coat, or **glycocalyx,** outside the cell. The glycocalyx is an important physiological barrier; it forms a template on which the exoskeleton is secreted and regulates binding to signal molecules and to surfaces, such as other cells. Membrane proteins may receive and transmit signals to the interior of the cell and serve as points of anchorage for cytoskeletal fibers. The cell membrane itself can also play a skeletal role. If the membrane lipids are largely unsaturated, like some vegetable oils used in cooking, the membrane is relatively fluid and flexible. If, on the other hand, the lipids

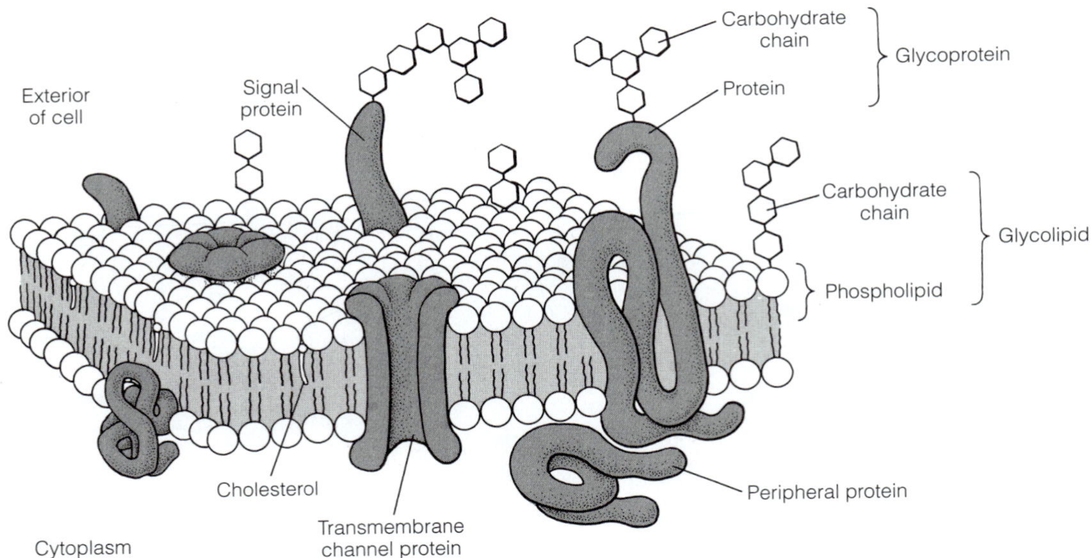

FIGURE 2-3 Eukaryotic cell structure. Diagrammatic ultrastructure of the cell membrane. The proteins and carbohydrate chains projecting from the exterior surface constitute the glycocalyx described in the text.

are mostly saturated, like lard or peanut butter, the membrane is less fluid and stiffer. Cholesterol, which is a common component of cell membranes, also stiffens the membrane.

In contrast to prokaryotes, eukaryotic cells ingest food in a wide range of sizes, including cells as large as themselves, by a process known as endocytosis. Eukaryotic cells are also able to detect, orient, and move toward food or prey and are capable of intercellular signaling. These attributes are discussed further in the following sections.

CELL MOTILITY

Ameboid Movement

Ameboid locomotion, a kind of cell crawling, not only typifies amebas and their relatives (see Chapter 3), but also occurs among certain cells of all animals (and fungi). These cells include embryonic mesenchyme cells, cells of the immune system, migrating cancer cells, and a variety of others.

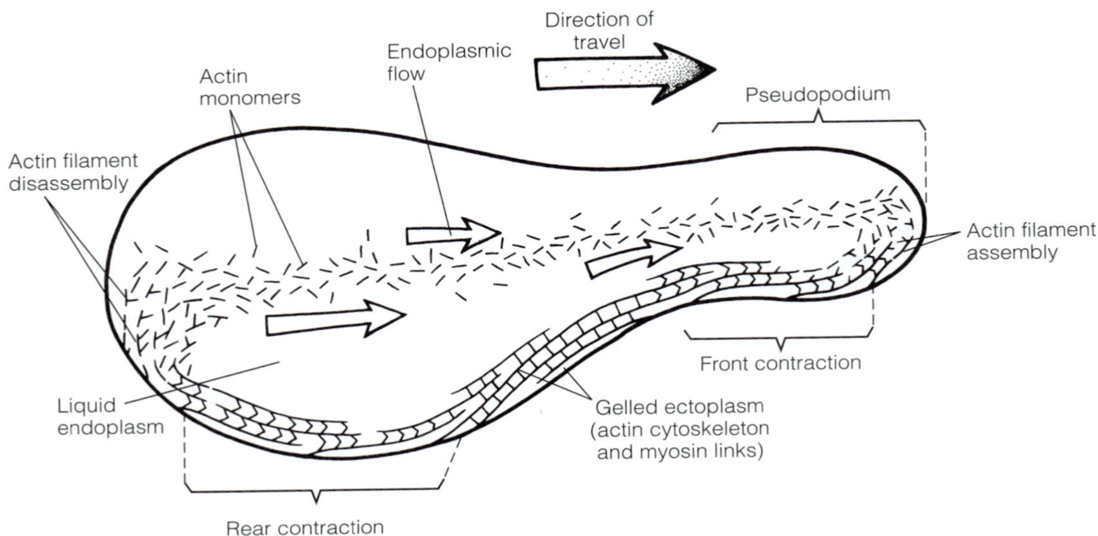

FIGURE 2-4 Cell motility: ameboid movement. The diagram embraces both rear-contraction and front-contraction hypotheses for cell movement. Myosin monomers are shown as cross-links between actin cytoskeletal filaments.

An ameba moves and captures food with flowing extensions of the cell, called **pseudopodia,** which protrude outward in the direction of travel (Fig. 2-4). Cytoplasm and cell organelles in the trailing part of the cell then flow forward into the pseudopodia, advancing the cell forward, and the cycle is repeated. The outflow of pseudopodia from the cell body involves a change of phase from solid to liquid and resembles melting ice. In fact, the outer cytoplasmic rind **(ectoplasm)** of an ameba is a stiff gel while the inner cytoplasm **(endoplasm)** is a fluid sol. A pseudopodium is formed at an ectoplasmic site where the gel liquefies and allows the outflow of fluid endoplasm. As this pseudopodium advances, its surface cytoplasm changes from liquid to gel (except at the pseudopodial tip) and forms a semirigid tube through which the trailing cytoplasm and organelles flow.

The liquid-gel (also known as sol-gel) phase transition of the cytoplasm is the result of assembly and disassembly of the actin cytoskeleton. Cytoskeleton disassembly, which converts gel into liquid cytoplasm, occurs at the trailing end of the advancing cell. The actin monomers at the rear then flow forward with the endoplasm and are used to assemble the new ectoplasmic sheath of the pseudopodium.

The current model of ameboid movement is analogous to muscle contraction. It involves a dynamic interaction between cytoskeletal actin and the motor molecule myosin. The myosin of ameboid cells is monomeric and is not present as the thick polymeric filaments that typify animal muscles. But monomeric myosin still forms cross-links with actin and, in the presence of Ca^{2+} and ATP, causes contraction. Thus, it is nearly certain that the ectoplasm of an ameba is contractile and this contraction provides the force to move the cytoplasm forward during ameboid movement. It is still unclear, however, whether the contraction occurs primarily in the rear of the cell (rear-contraction hypothesis) or in the wall of the advancing pseudopodium (front-contraction hypothesis). Rear contraction would force the endoplasm forward by compressing it. Front contraction would pull the cytoplasm forward by exerting tension, via the cytoskeleton, on the trailing part of the cell.

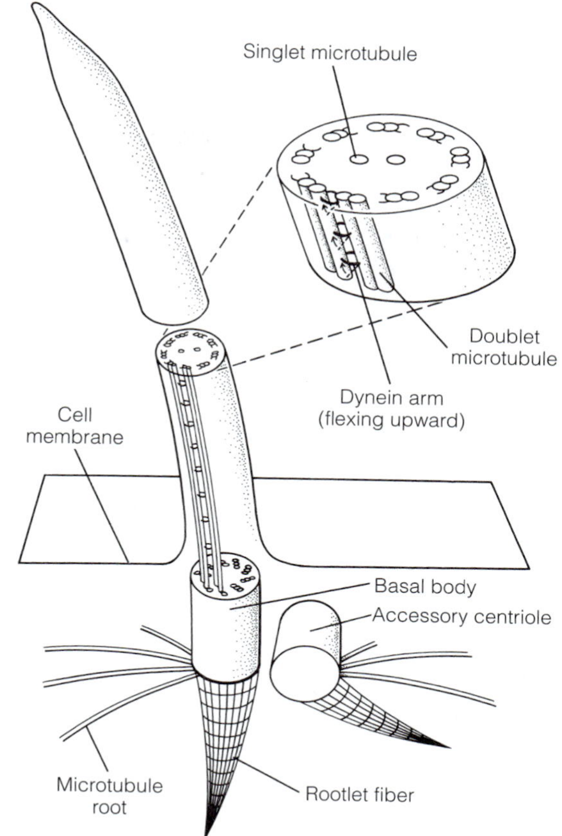

FIGURE 2-5 Cell motility: flagella and cilia. Ultrastructure of a flagellum or cilium as it occurs in a protozoan (Choanoflagellate), showing cytoskeletal anchoring microtubules and rootlet fibers. The exploded view shows the flagellar axoneme. Treadmilling is the cycle of attachment, flexion, and detachment of the dynein arms, causing one doublet to slide past its adjacent doublet.

Flagella and Cilia

Flagella and cilia are characteristic of many protozoan and metazoan cells. In general, **flagella** are typically long and their motion is a whiplike undulation. **Cilia,** on the other hand, are short and their motion is stiff and oarlike. The distinctions are not sharp, however, and many variations are common, especially in beat patterns.

Flagella and cilia are structurally identical: Each shaft, whether long or short, consists of a microtubular core **(axoneme)** enclosed by the cell membrane (Fig. 2-5). The axoneme consists of nine outer doublet microtubules that encircle two central singlet microtubules, forming the typical **9 × 2 + 2** pattern seen in cross sections. One microtubule of each doublet bears two rows of projections, or arms (containing the motor molecule dynein), directed toward the adjacent doublet.

Bending of the flagellum is caused by active sliding of adjacent doublets past each other. The dynein arms on the doublets provide the sliding force. In the presence of ATP, the dynein arms on one doublet attach to an adjacent doublet and flex, causing the doublets to move past each other by a one flex increment. Successive attachments and flexes cause the doublets to treadmill smoothly past one another over a distance sufficient to bend the flagellum.

Each flagellar or ciliary axoneme arises from and is anchored to a **basal body** that lies immediately below the cell membrane (Fig. 2-1B, 2-2, 2-5). Basal bodies resemble an axoneme except that the outer nine microtubules are triplets and the central singlets are absent **(9 × 3 + 0).** Two microtubules of each triplet are continuous with an axonemal doublet. Dynein arms are absent on the basal-body triplets. A basal body (and its cilium or flagellum) is usually anchored in the cell, often to the nucleus and cell membrane, by one or more cytoskeletal root structures (Fig. 2-5). These may be sprays of microtubules, tapered striated rootlets, or both. Some proteinaceous rootlet fibers are contractile and can, on contraction, pull the flagellum into a shallow pocket or alter its orientation. When basal bodies are distributed to daughter cells during mitosis, they typically arrange themselves at each pole of the mitotic spindle and are then designated **centrioles**

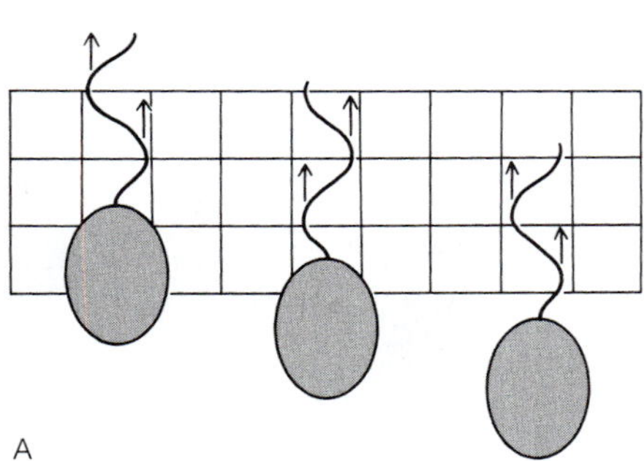

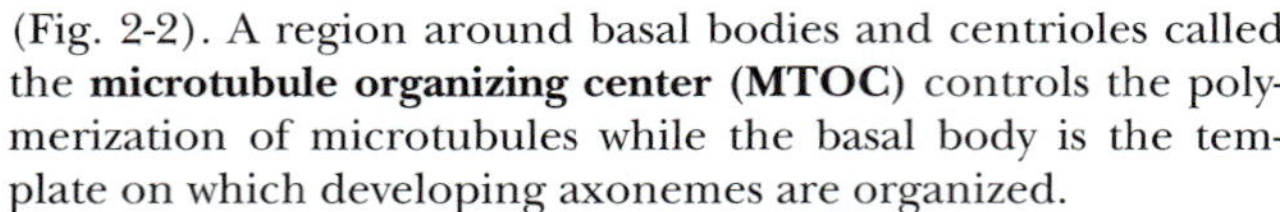

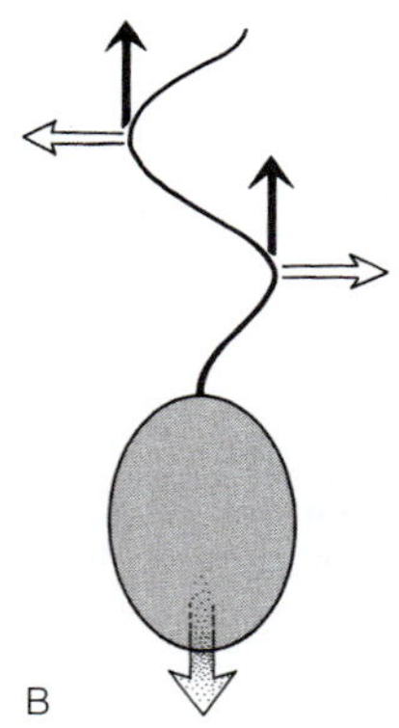

FIGURE 2-6 Cell motility: flagellar propulsion in a protozoan. **A,** base-to-tip wave propagation; **B,** forces generated by base-to-tip wave propagation. Lateral forces (outlined arrows) cancel each other. Longitudinal forces (solid arrows) combine to produce forward thrust.

(Fig. 2-2). A region around basal bodies and centrioles called the **microtubule organizing center (MTOC)** controls the polymerization of microtubules while the basal body is the template on which developing axonemes are organized.

In most protozoan and metazoan cells, the flagellum propagates an undulatory wave from the cell to the flagellar tip that pushes the cell in the direction opposite the flagellum or drives water away from the flagellar end of a stationary cell (Fig. 2-6). (We will encounter some exceptions later, in Chapter 3.) As an undulatory wave moves along the flagellum, the advancing wavefront, like a wave approaching a beach, generates a *longitudinal* pushing force (Fig. 2-6B). In the meantime, the sideways undulations of a flagellum generate *lateral* forces. Because the lateral undulations are usually symmetrical, the left-directed forces cancel the right-directed forces, and only the longitudinal force remains to move the cell.

Cilia are short, commonly numerous, densely arranged, and especially well represented in the ciliate protozoans such as *Paramecium* and related genera (Fig. 2-7A). During the effective stroke, the cilium is outstretched stiffly and moves in an oarlike fashion, perpendicular to the cell surface (Fig. 2-7B). In the recovery stroke, the cilium flexes and snakes forward parallel to the cell surface. As the organism moves through the medium, the ciliary beat is coordinated over the surface of the cell. The cilia in any cross row are all in the same stage of the beat cycle, while those in front are in an ear-

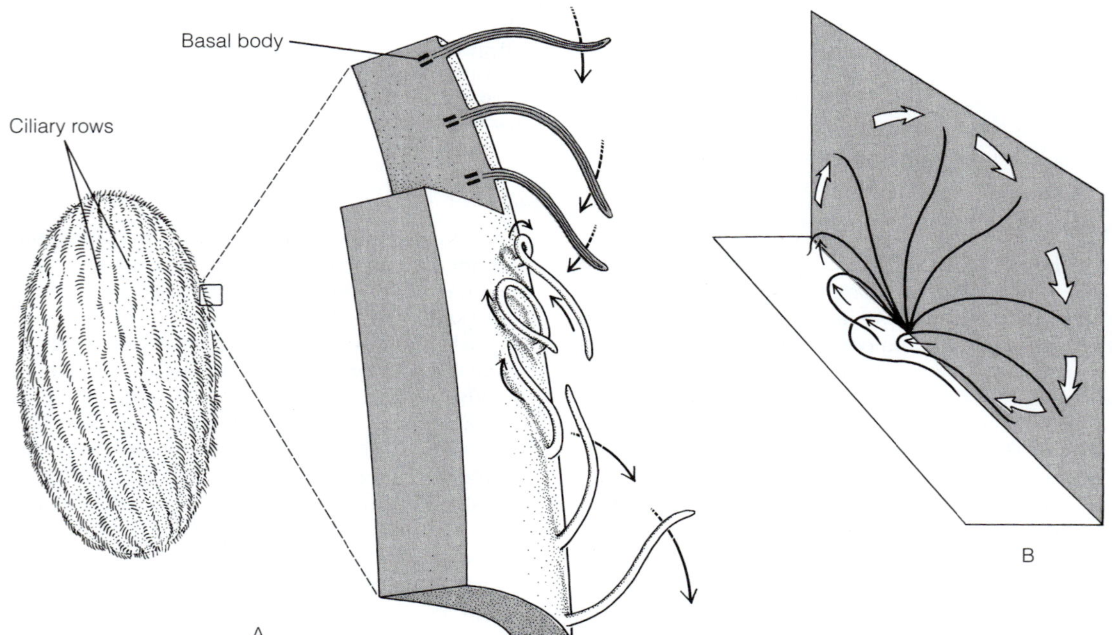

FIGURE 2-7 Cell motility: ciliary propulsion. **A,** Metachronal waves of cilia beating in a ciliated protozoan related to a *Paramecium* (left). Along the length of each row, adjacent cilia are in different phases of the beat cycle (right). **B,** The effective (outlined arrows) and recovery (solid arrows) strokes in the beat cycle of a single cilium.

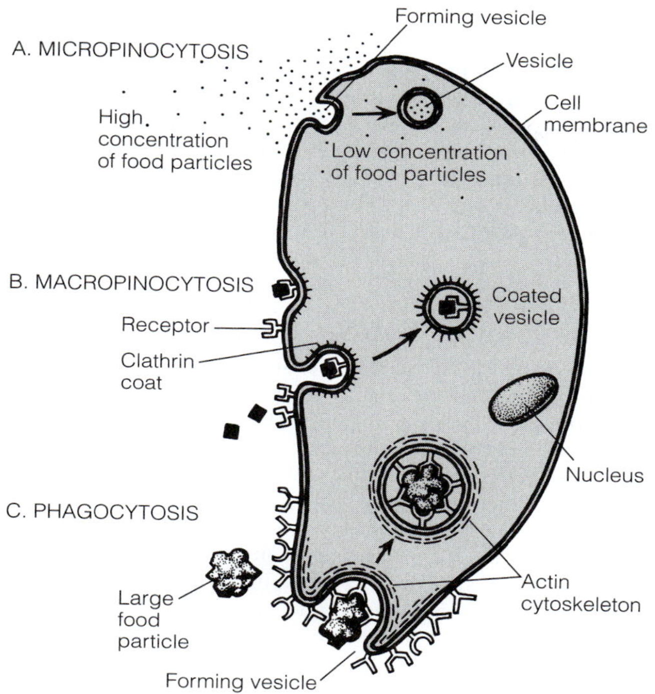

FIGURE 2-8 Endocytosis. **A,** micropinocytosis; **B,** macropinocytosis; **C,** phagocytosis.

lier stage and those behind are in a later stage (Fig. 2-7A inset). This phase shift is seen as waves, called **metachronal waves,** that pass over the surface of the cell like wind passes in waves over a wheat field.

UPTAKE BY CELLS

Substances enter the cells of protozoans and other eukaryotes in a variety of ways. The protein channels of the cell membrane provide for the passive diffusion of water, ions, and small molecules, such as sugars and amino acids. Some function as energy-requiring pumps, actively transporting certain molecules or moving ions in or out against their concentration gradient.

Some extracellular materials enter a cell in minute pits on the cell's membrane that later pinch off internally—a process called **endocytosis** (Fig. 2-8). **Micropinocytosis** is a nonspecific form of endocytosis in which the rate of uptake is in simple proportion to the external concentration of the material being absorbed (Fig. 2-8A). Water, ions, and small molecules may be taken in by micropinocytosis. **Macropinocytosis** brings in proteins and other *macro*molecules at a rate greater than predicted by the concentration gradient. These substances may or may not bind to, and be concentrated on, specific membrane receptors before they are internalized in vesicles, which are coated with a protein called clathrin (Fig. 2-8B). Larger particles, such as bacteria and protozoans, are taken up in large vesicles **(food vacuoles)** by **phagocytosis** (Fig. 2-8C). Phagocytosis requires binding of a particle to membrane receptors and dynamic alteration of the cell membrane involving the actin cytoskeleton.

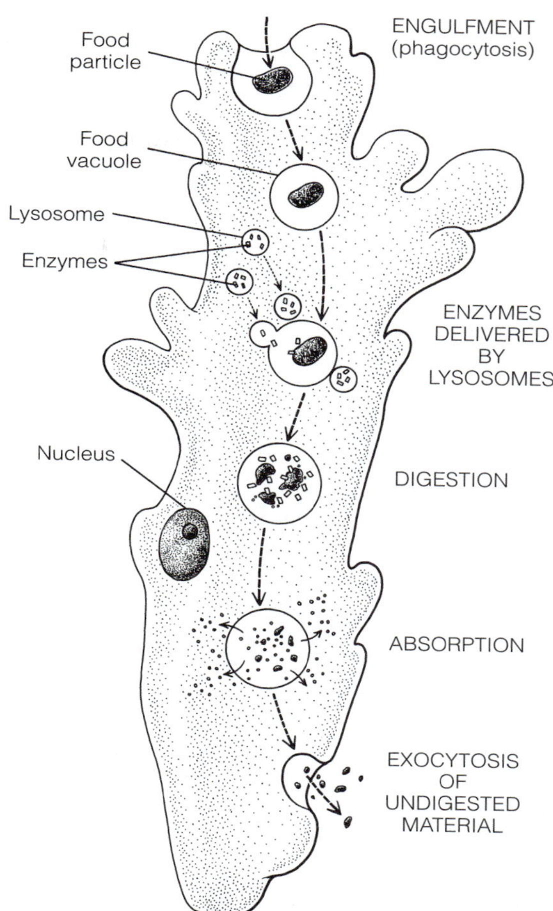

FIGURE 2-9 Intracellular digestion in an ameba-like protozoan.

INTRACELLULAR DIGESTION

Once food enters the cell, lysosomes fuse with the endocytic vesicles or food vacuoles. Lysosomes are membrane-enclosed organelles that originate from Golgi bodies and contain acids and hydrolytic enzymes (Fig. 2-9). Release of those biomolecules into the food vacuole initiates digestion. Eventually, the products of intracellular digestion diffuse across the vacuole membrane into the cytoplasm of the cell, where they may be used in metabolism or stored, after undergoing synthesis, in forms such as glycogen and lipids. Indigestible material is released from the cell to the exterior by fusion of the residual vacuole with the cell membrane in a process called **exocytosis** (Fig. 2-9).

CIRCULATION IN CELLS

Some protozoans have a definite cytoplasmic circulation. In general, circulatory systems are required when the supply of a substance by simple diffusion cannot keep pace with the metabolic demand for it. This limit is often reached as an organism becomes large, regardless of whether that results

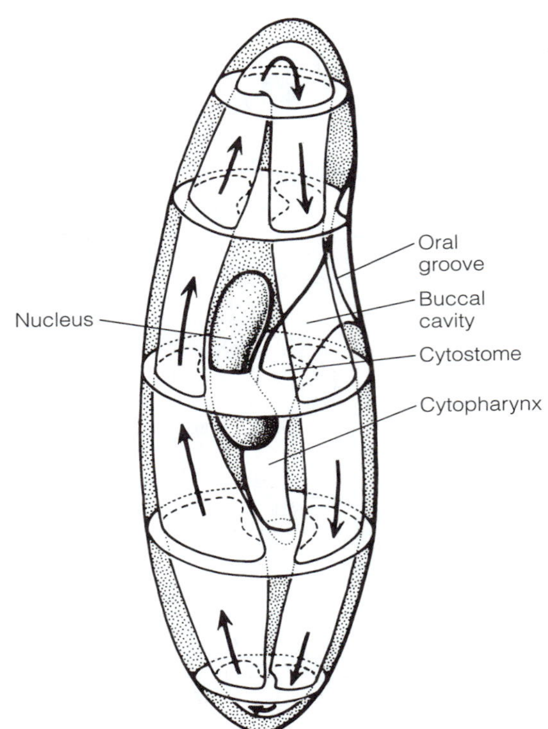

FIGURE 2-10 Circulation in cells. Cyclosis in *Paramecium.* Food vacuoles and vesicles move at 2–3 μm/s following the path indicated by the arrows. Stippled cytoplasm is noncirculating. Bulkheads, which do not exist, are drawn to provide three-dimensional perspective. *(Modified and redrawn from Sikora, J. Cytoplasmic steaming in* Paramecium. *Protoplasma 109:57; Haüsmann, K., and Hülsmann, N. 1996. Protozoolgy. Georg. Thieme, New York. 338 pp.)*

from developmental growth or evolutionary increase in body size. Protozoans with a cytoplasmic circulation are often large cells, such as some ciliates, or cells with long extensions (pseudopodia), such as forams. Directional transport within a pseudopodium or cell is referred to as **streaming,** or **shuttling** if in reference to the movement of vesicles along a cytoskeletal track. If flow is in a circuit, as in *Paramecium,* it is called **cyclosis** (Fig. 2-10). For additional discussion of circulatory systems, see Chapters 4 and 9.

CELL SECRETIONS

Many cells synthesize macromolecules, especially proteins and glycoproteins, that are exported to the outer surface of the cell. Synthesis typically involves the endoplasmic reticulum and Golgi bodies, the latter forming the vesicles that transport the secretory product to the cell membrane. Here, during exocytosis, the membrane of the vesicle fuses with the cell membrane and the product is expelled to the exterior (Fig. 2-9, 2-11). Many cell secretions, such as extracellular enzymes and pheromones, are exported away from the cell that produces them. One of the most widespread animal secretions, **mucus,** is a mucopolysaccharide with a large carbohydrate and a smaller protein component. Animals utilize mucus in a variety of ways: as an adhesive, a protective cover, and a lubricant.

Some cell secretions remain associated with the external surface of the cell membrane to form extracellular skeletal materials (Fig. 2-11), of which **chitin** is a good example. Chitin is a cellulose-like polysaccharide that is laid down as an exoskeleton around the bodies of some protozoans (for example, the protective retreat of the marine ciliate *Folliculina,* cysts of amebas) as well as metazoans (such as an

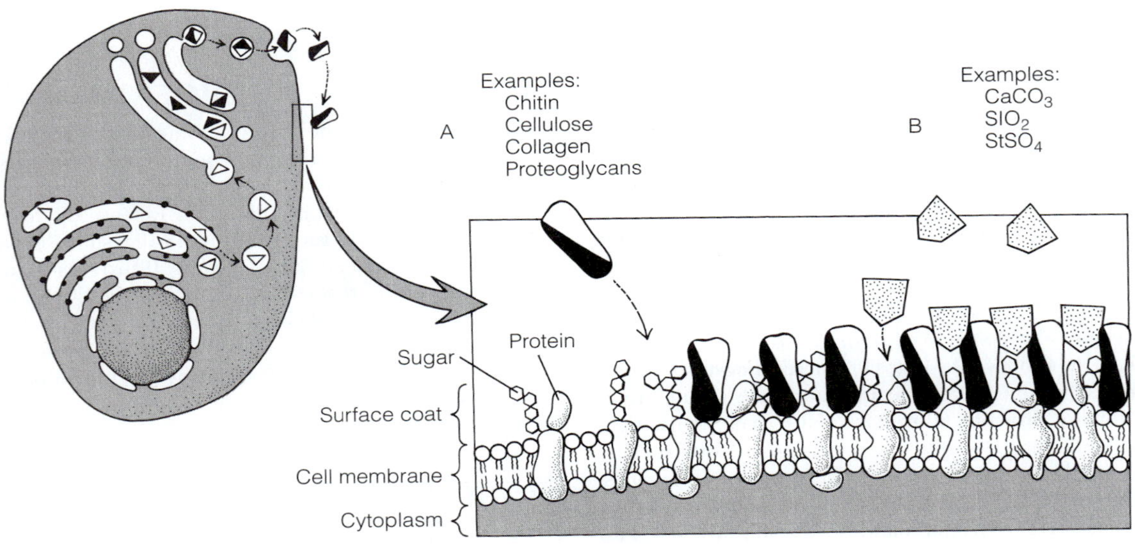

FIGURE 2-11 Cell secretions. Secretion of extracellular skeletal materials involves the nucleus (coding and transcription), ribosomes (black dots on endoplasmic reticulum; translation, protein synthesis), endoplasmic reticulum (addition of components), Golgi body (addition of components, packaging in vesicles for release at surface), and, following exocytosis, self-assembly of the exoskeleton. **A,** organic secretions; **B,** mineral secretions.

insect exoskeleton). Chitin and its associated proteins are exocytosed at the surface of the cell, where they assemble into the exoskeleton.

CELL COMMUNICATION

Protozoans respond to chemical and physical cues in ways that enable them to avoid adverse conditions, locate food, and find mates. In this sense, each protozoan cell must be both receptor and effector. In their receptiveness to environmental stimuli, protozoans resemble the sensory nerve cells of animals.

Like animal sensory receptor cells, protozoans often receive external stimuli as signal substances that bind to specific membrane molecules. Binding can cause a specific ion channel to open, allowing ions (often Na^+ and K^+) to flow down their concentration gradients (Na^+ in, K^+ out). Because the resting cell membrane is polarized with respect to the distribution of these ions, the opening of the ion channels depolarizes the membrane. (Depolarization can be measured as a change in electrical potential, or voltage, using electrodes and a voltmeter.) When the membrane is depolarized, Ca^{2+} channels open and calcium ions enter the cell. The entering calcium triggers other changes, such as a reversal in the ciliary beat, which causes the cell to withdraw from the disturbance. *Paramecium,* for example, has at least nine different ion channels, some of which are localized at the front and others at the rear of the cell. Such localized receptor fields differentiate "head" from "tail" and are thus analogous to the localization of receptor cells and organs in many metazoans. Intercellular chemical signaling (pheromones) in protozoans, in fact, often involves signal molecules, such as serotonin, β-endorphin, acetylcholine, and cyclic-AMP, which in animals function as neurotransmitters and internal messengers.

SYMBIOSIS BETWEEN CELLS

Animal-like eukaryotic cells (heterotrophs) often establish an endosymbiotic relationship with photosynthetic cells (autotrophs) to the benefit of both partners. The photosynthetic partner may be either a eukaryote or a prokaryotic cyanobacterium. When the photosynthetic symbionts are green unicellular algae or diatoms, they are referred to as **zoochlorellae** (both *os* in *zoo-* are pronounced), but the most commonly occurring symbionts are yellow or brown and are known as **zooxanthellae** (Fig. 7-11). These zooxanthellae are a nonmotile stage of flagellated protozoans called dinoflagellates, which will be described in Chapter 3. The photosynthetic member of the partnership is generally located intracellularly within a vesicle in the host cytoplasm, although in a few metazoans, it is found between cells.

This symbiosis has its evolutionary origin in the phagocytosis of photosynthetic cells by heterotrophic cells. Delayed digestion by the larger cell may have resulted in the captured cells continuing to live and photosynthesize. Use of any excess photosynthate by the larger partner would have created a positive selective pressure for it to maintain the autotroph alive within its cytoplasmic vesicle. This symbiosis evolved numerous times, considering the different sorts of autotrophs and their symbiotic partners.

The benefits of this symbiosis are probably similar wherever it occurs. The autotroph provides excess organic carbon from photosynthesis to the larger partner, which in return provides certain nutrients, such as CO_2, nitrogen, and phosphorus, as well as protection, to the autotroph. Rarely does the protozoan or metazoan rely entirely on its autotrophs for nutrition; typically, the benefits of symbiosis supplement heterotrophic nutrition to varying degrees.

EVOLUTIONARY ORIGIN OF EUKARYOTIC CELLS (INCLUDING PROTOZOA)

Our best hypothesis is that life began on an anoxic Earth some 3 billion years ago with the evolution of prokaryotic cells. Each of these tiny bacterium-like cells was enclosed in a membrane, but lacked internal membranes (organelles). In the absence of organelles, compartmentalization resulted from functional aggregations of biomolecules. Their food (energy for maintenance and reproduction) consisted of simple organic molecules that entered the cell and were distributed throughout by simple diffusion. In the absence of O_2, their central metabolic pathway was anaerobic (glycolysis), which resulted in limited ATP production and release of the energy-rich waste products, such as ethanol and lactic acid. As the competition for a limited supply of organic molecules intensified, some taxa evolved photosynthesis and were able to use sunlight energy to synthesize food from atmospheric CO_2 and N_2. The appearance of photosynthesis provided a new and renewable supply of organic molecules. The first photosynthetic microbes probably obtained the electrons to reduce CO_2 to carbohydrate from H_2S, the waste byproduct being elemental sulfur (S). Later, in taxa such as Cyanobacteria (blue-green bacteria), electrons were obtained from H_2O, resulting in the release and accumulation of O_2 in the atmosphere (Fig. 2-12). This newly available atmospheric O_2 set the stage for the evolutionary adoption of aerobic respiration, which enabled the complete breakdown of food for maximal ATP production and the release of the waste products CO_2 and H_2O.

Eukaryotic cells evolved about 1.5 billion years ago, nearly 2 billion years after the first prokaryotes and 1 billion years before the first animals. How did eukaryotic cells evolve from an ancestral prokaryote? As already noted, the small cells of prokaryotes lack internal membranes, except for the photosynthetic membranes of cyanobacteria and a fingerlike invagination of the cell membrane called a **mesosome,** to which the DNA is attached (Fig. 2-1A). In general, the cells of eukaryotes are 10 times larger than those of prokaryotes and may have required another level of compartmentalization to operate effectively. That new level, beyond the organized cytoplasm, was the organelle. What was the evolutionary origin of these organelles? Some seem to have evolved by modification of preexisting prokaryotic structures and others from entire prokaryotic cells that were engulfed by another cell and became permanent residents. The establishment of one cell inside of another, for their mutual benefit, is called an **endosymbiosis** (Fig. 2-13). A possible scenario for the endosymbiotic origin of eukaryotic organelles follows.

As oxygen was liberated by photosynthesis on the early Earth, the anaerobic prokaryotes presumably were faced with

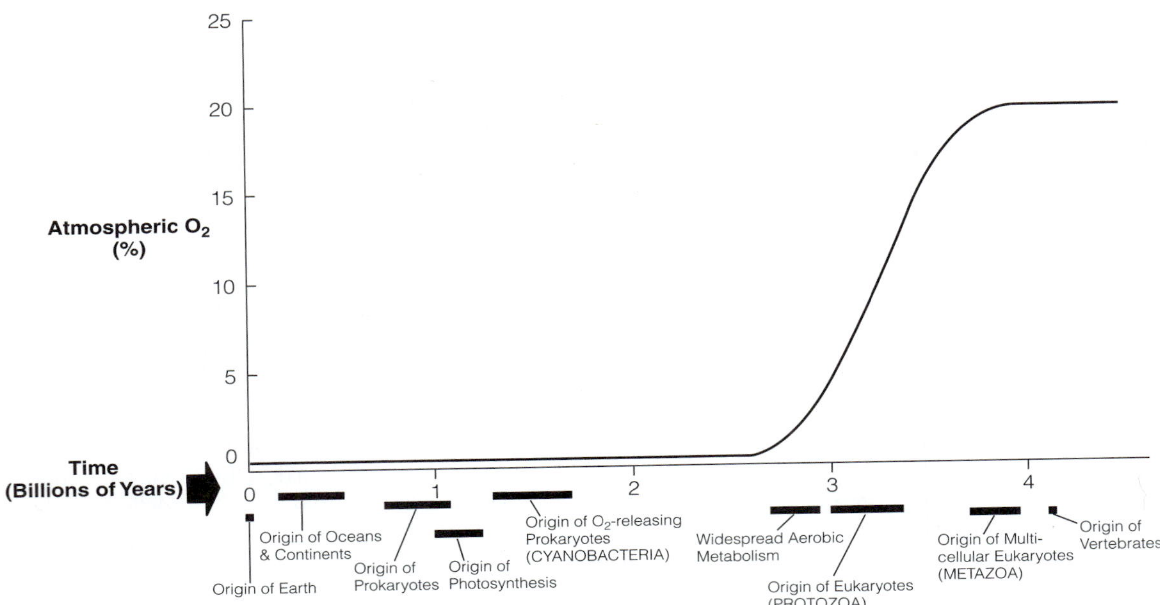

FIGURE 2-12 Evolution of life in relation to Earth history and oxygen availability. Note the billion-year lag between the first oxygen-producing photosynthetic cyanobacteria and the rise of atmospheric oxygen. Geological deposits of massive amounts of iron oxide suggest that the first free molecules of oxygen combined with ferrous ions in the sea until these ions were depleted, presumably requiring a billion years. *(Modified and redrawn from Alberts, B., Bray, D., Lewis. J., Raff, M., and Watson, J. D. 2002. Molecular Biology of the Cell. Garland Publishing, New York. 1616 pp.)*

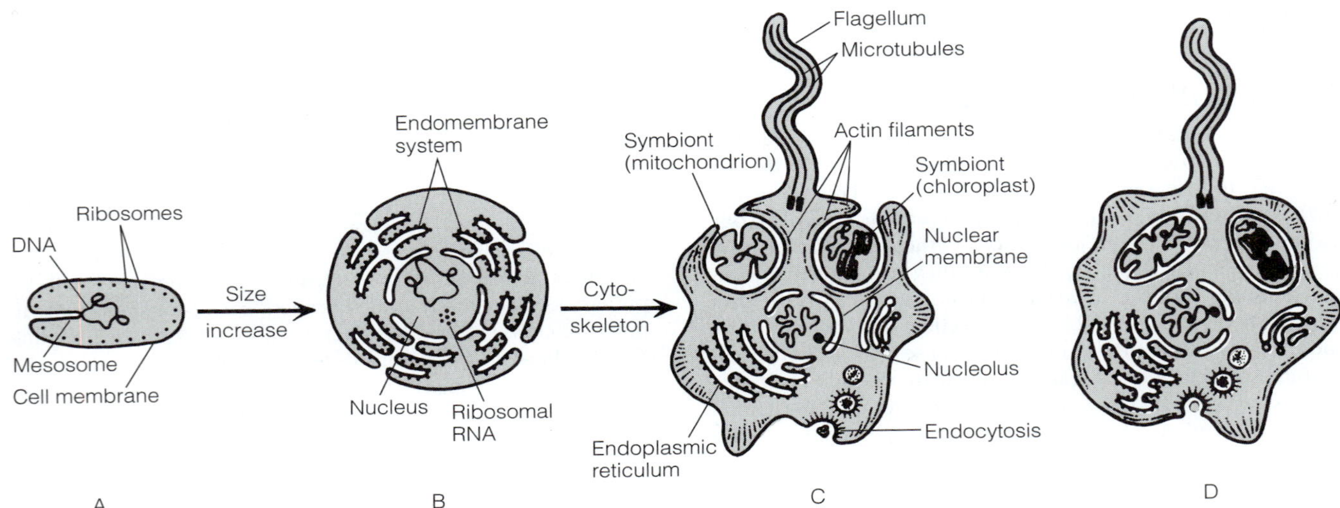

FIGURE 2-13 Evolution of eukaryotic cells. A scenario for the evolution of a eukaryotic cell from a prokaryotic cell. **A,** A hypothetical ancestral prokaryote. **B,** Increase in cell size and the origin of internal membranes. The nuclear membrane and endomembrane system may have evolved from a series of mesosome-like invaginations of the cell membrane. The endomembrane system of eukaryotes increases the surface area on which proteins are synthesized (ribosomes). **C,** Origin of cytoskeleton (actin, microtubules) and motor molecules, allowing flagellar (ciliary) and ameboid motion, as well as endocytosis. Acquisition of mitochondrion by phagocytosis of a prokaryotic aerobe and chloroplast by phagocytosis of a photosynthetic prokaryote. **D,** A eukaryotic cell.

a challenge: Adapt to the presence of O_2 or face extinction. Undoubtedly, some found anoxic refuges, perhaps deep within waterlogged sediments, while others, through variation and natural selection, evolved aerobic respiration and took advantage of the newly available O_2. At this point, the competition between aerobes and anaerobes may have been intense, with the advantage going to the aerobes as O_2 levels rose. Perhaps during this time, some *anaerobic* cell with the capacity for phagocytosis engulfed an *aerobic* prokaryote that was not digested, but rather was permanently sequestered as an endosymbiont. The host cell retained its cytoplasmic anaerobic pathway (glycolysis), providing the end products (lactate, pyruvate) to the symbiont as food. Using aerobic respiration, the symbiont then converted that food-energy into ATP, which was shared with the host, eventually releasing CO_2 and H_2O as waste products. The aerobic endosymbiont, of course, eventually became the mitochondrion (Fig. 2-13). Phagocytosis of a *photosynthetic* prokaryote, followed by the evolution of a mutualism, probably established the chloroplast.

The evidence for these hypotheses stems from several sources. One is that both mitochondria and chloroplasts are enclosed by two membranes. If the endosymbiotic hypothesis is correct, then the outer membrane should represent the membrane of the original phagocytic vesicle while the inner membrane should correspond to the original cell membrane of the prokaryote. In support of the hypothesis, the biochemistry of the outer membrane of mitochondria and chloroplasts resembles that of a eukaryotic cell membrane whereas the inner is similar to a prokaryotic cell membrane. Pharmaceutical evidence provides further support, as both mitochondria and chloroplasts are susceptible to antibacterial antibiotics. Mitochondria and chloroplasts also have DNA and ribosomes that are similar to those of prokaryotes. Further support is provided by *Pelomyxa palustris,* a large ameba that lacks mitochondria but has aerobic endosymbiotic bacteria that carry out oxidative metabolism. Although *Pelomyxa* is not the actual intermediate between the ancestral, mitochondria-free eukaryote and its descendants with typical mitochondria—it's an example of an evolutionary parallelism—it nevertheless indicates the plausibility of the endosymbiotic scenario.

Similar to mitochondria, the eukaryotic cell nucleus also is surrounded by a double membrane, but this does not seem to indicate an endosymbiotic origin for the nucleus. Instead, both nuclear membranes resemble a eukaryotic cell membrane. Perhaps the evolutionary origin of the nuclear membranes was by modification of one or more mesosome-like infoldings of the ancestral cell's surface (Fig. 2-13). If so, the blind ends of these infoldings may have expanded around the centrally located DNA, forming the nuclear envelope, while the infoldings themselves became a rudimentary endomembrane system from which the endoplasmic reticulum, Golgi bodies, and other structures eventually differentiated.

At present, a model for the evolution of eukaryotic cells is incomplete. Few plausible hypotheses, for example, have been formulated for the origin of the cytoskeleton and related structures. It is generally assumed, however, that the evolution of the microtubular mitotic spindle, including its centrioles, is closely linked with the origin of cilia and flagella, which use centrioles as basal bodies.

REFERENCES

GENERAL

Alberts, B., Bray, D., Lewis, J., Raff, M., Roberts, K., and Watson, J. D. 2002. Molecular Biology of the Cell. Garland Publishing, New York. 1616 pp.

Allen, R. D., and Naitoh, Y. 2002. Osmoregulation and contractile vacuoles of Protozoa. Int. Rev. Cytol. 215:351–394.

Corliss, J. O. 1989. Protistan diversity and origins of multicellular/multitissued organisms. Boll. Zool. 56:227–234.

Fenchel, T., and Finlay, B. J. 1994. The evolution of life without oxygen. Am. Sci. 82:22.

Fukui, Y. 1993. Toward a new concept of cell motility: Cytoskeletal dynamics in amoeboid movement and cell division. Int. Rev. Cytol. 144:85–127.

Grain, J. 1986. The cytoskeleton in protists: Nature, structure and functions. Int. Rev. Cytol. 104:153.

Grebecki, A. 1994. Membrane and cytoskeleton flow in motile cells with emphasis on the contribution of free-living amoebae. Int. Rev. Cytol. 148:37–80.

Heppert, M., and Mayer, F. 1999. Prokaryotes. Am. Sci. 87: 518–525.

Karsenti, E. 1999. Centrioles reveal their secrets. Nature Cell Biol. 1:E62–E64.

Lindemann, C. B., and Kanous, K. S. 1997. A model for flagellar motility. Int. Rev. Cytol. 173:1–72.

Lodish, H., Berk, A., Zipursky, S. L., Matsudaira, P., Baltimore, D., and Darnell, J. 2000. Molecular Cell Biology. 4th Edition. W. H. Freeman, New York. 1084 pp.

Margulis, L. 1993. Symbiosis in Cell Evolution. 2nd Edition. W. H. Freeman, New York. 452 pp.

Margulis, L. 2000. Symbiotic Planet. Basic Books, New York. 160 pp.

Margulis, L., Sagan, D., Thomas, L. 1997. Microcosmos: Four Billion Years of Evolution from Our Microbial Ancestors. University of California Press, Berkeley. 304 pp.

Patterson, D. J., and Sogin, M. L. 1993. Eukaryote origins and protistan diversity. In Hartman, H., and Matsuno, K. (Eds.): The Origin and Evolution of the Cell. World Scientific Publishing, Singapore. pp. 13–46.

Sikora, J. 1981. Cytoplasmic streaming in *Paramecium.* Protoplasma 109:57.

Sogin, M. L. 1991. Early evolution and the origin of eukaryotes. Curr. Opin. Genet. Devel. 1:457–463.

Weatherbee, R., Anderson, R. A., and Pickett-Heaps, J. D. (Eds.): 1994. The Protistan Cell Surface. Springer Verlag, Vienna. 290 pp.

Whatley, J. M. 1993. The endosymbiotic origin of chloroplasts. Int. Rev. Cytol. 144:259–299.

INTERNET SITES

www.cco.caltech.edu/~brokawc/Demo1/BeadExpt.html (Image of microtubule sliding in an active flagellum of a sea urchin sperm.)

http://cellbio.utmb.edu/cellbio/cilia.htm (Transmission electron micrographs and diagrams of cilia/centriole structure.)

http://cas.bellarmine.edu/tietjen/images/origin_of_mitochondria_in_eukary.htm (Concise description of ideas and facts regarding the evolution of mitochondria.)

3 Protozoa

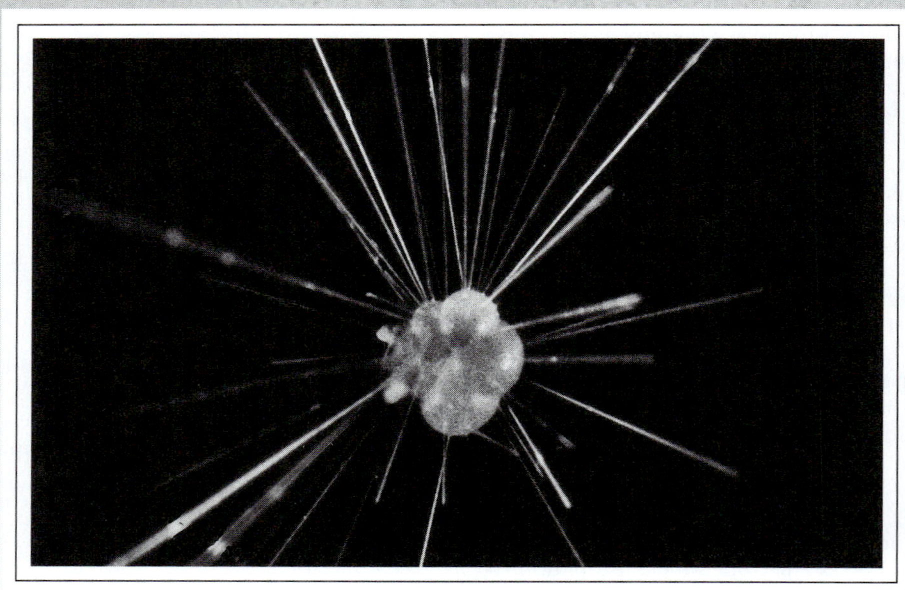

Formerly known as infusoria or animalcules, protozoans are motile, eukaryotic, unicellular organisms. United by the common possession of motility, an attribute that undoubtedly evolved independently in many groups, protozoa is a polyphyletic taxon with an unclear boundary. Historically, protozoa included nearly every group of what we now consider Protista—funguslike, animal-like, plantlike, and other unicellular eukaryotes. The name *protozoa* means "first animals," and it was natural for early biologists to seek the ancestor of Metazoa (animals) from among the free-moving protists. But it has now been established that only one protozoan taxon, the collared flagellates (Choanoflagellata), is the sister taxon of Metazoa and truly qualifies as the "first animal." The remaining protozoan taxa are either unique groups with no significant multicellular descendants or they are related closely to plants or fungi. Remarkably, one protozoan taxon, Myxozoa (formerly Myxosporidia), is actually a group of animals related to cnidarians (anemones and jellyfishes). Thus *protozoa* is the name for a grade within a scheme of organization, a loose confederation of eukaryotic taxa, rather than for a monophyletic taxon.

Protists exhibit astounding diversity and play a significant ecological role. The nearly 215,000 described species equal in number the vascular plants and are 10 times more plentiful than the number of bacterial and viral species combined. Of the total number of protist species, slightly less than half (around 92,000) are protozoans, one-quarter of which live as symbionts of other organisms. Protozoan parasites, for example, have an enormous impact on humans: Millions of people die yearly from malaria and other parasitic protozoans, and protozoans that sicken and kill livestock, poultry, fish, and wildlife cost economies several hundred million dollars annually. But the other face of protozoan ecology is beneficial. The mutualism between photosynthetic protozoans and corals underlies the coral-reef ecosystem, one of the most diverse on earth. Myriads occupy aquatic environments and soils and play essential roles in food chains, including the control of bacterial populations and the recycling of nutrients. The protists as a whole, including the photosynthetic protozoa, account for 40% of global primary productivity.

The great diversity of protozoa necessarily restricts coverage in this chapter to the most common and significant freshwater, marine, and parasitic taxa. Omitted are many "algal" taxa that other biologists consider to be protozoans. This chapter's goals are to provide an overview of protozoan diversity, to examine the functional adaptations of cells as organisms, and to identify living examples of how cell-organisms might have evolved into those multicellular creatures we call animals. One of the best models to illustrate that evolutionary transition is *Volvox* and its relatives. Although the multicellular *Volvox* is clearly a green alga related to land plants and thus provides only a parallel example for the evolution of multicellularity, it is included in this chapter because of its easy availability for study.

FORM AND FUNCTION

The body of most protozoans consists of a single cell, although many species form colonies. Cell size ranges from approximately 10 μm, as in choanoflagellates, to several centimeters in some dinoflagellates, forams, and amebas.

The protozoan body is usually enclosed only by the cell membrane. The rigidity or flexibility of the body and its shape are largely determined by the cytoskeleton, which typically is located just below the cell membrane. The cytoskeleton and cell membrane together form the **pellicle,** a sort of protozoan "body wall." The **cytoskeleton** often is composed of protein filaments (actin, for example), microtubules, vesicles (such as alveoli), or combinations of all three. The protein filaments may form a dense mesh in the outermost cytoplasm (Fig. 3-1A) as in, for example, *Euglena.* More conspicuous cytoskeletal structures are pellicular microtubules that occur in flagellates, apicomplexans, and ciliates. They can be arranged as a microtubular corset (Fig. 3-1B) or, as in some flagellates, the microtubules can originate on the flagellar basal bodies and radiate rearward to the opposite extremity of the cell as a sort of axial skeleton (axostyle; Fig. 3-1C). Such microtubules resemble the

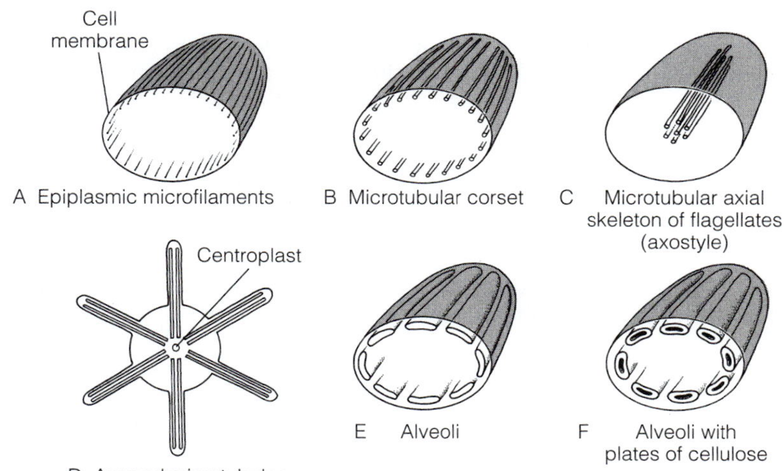

FIGURE 3-1 Protozoa: Cytoskeletons of actin microfilaments, microtubules, and alveoli. Examples: **A,** amebas, euglenoids; **B,** euglenoids; **C,** axostylates; **D,** heliozoans; **E,** ciliates, apicomplexans; **F,** dinoflagellates.

microtubules of a mitotic spindle, which radiate from centrioles and form the mitotic apparatus (Fig. 2-2). In other protozoa, such as the spherical radiolarians and heliozoans, bundles of microtubules radiate from a centroplast (an MTOC) at the cell's center and then extend into and support a raylike projection of the cell's surface (axopod; Fig. 3-1D). The centroplast and its microtubules resemble the starlike asters that form around centrioles at the poles of the mitotic spindle.

Vesicles, known as **alveoli,** occur immediately below the cell membrane in many protozoans, such as dinoflagellates, apicomplexans, and ciliates (together forming the Alveolata). "Empty" alveoli, like those that occur in ciliates, may be turgid and help to support the cell, but they also store Ca^{2+}, which can be released to trigger cellular responses (Fig. 3-1E). In some dinoflagellates, plates of cellulose secreted into the alveolar vesicles form a rigid endoskeleton (Fig. 3-1F).

Protozoan skeletons, like those of metazoans, can also be endo- or exoskeletons. A skeleton that forms a more or less complete covering, whether internal or external, is called a **test** (or a lorica, theca, or shell).

The protozoan locomotor organelles may be flagella, cilia, or flowing extensions of the cell known as pseudopodia (described in Chapter 2). The undulatory waves of flagella pass from base to tip and drive the organism in the opposite direction (Fig. 2-6). The flagella of many protozoans bear fine lateral "hairs" called **mastigonemes** (Fig. 3-2). The mastigonemes cause the flagellum to pull rather than push as the flagellar waves pass from base to tip. Flagellar, ciliary, and pseudopodial specializations characterize many of the protozoan taxa.

All types of nutrition occur in protozoa. Some protozoa rely on photosynthesis, others absorb dissolved organic material from the environment, and many digest food particles or prey intracellularly in food vacuoles. Food enters the vacuole by phagocytosis, often at a definite cell mouth, or **cytostome.** The vacuole then may be shuttled to the interior along a specialized microtubular tract called a **cytopharynx.** Macromolecules enter by micro- and macropinocytosis, which may occur over the entire surface of the cell. Intracellular digestion has been most studied in amebas and ciliates, and, for the most part, it follows the general pattern described in Chapter 2. Digestive specializations of ciliates will be described later in this chapter.

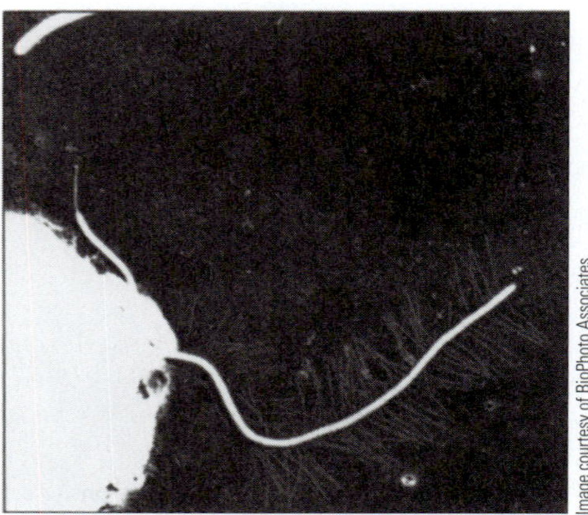

FIGURE 3-2 Protozoa: Flagellar mastigonemes. Phytoflagellate with one short smooth flagellum and one long flagellum bearing mastigonemes.

Diffusion is important for internal transport in all protozoans and may be the sole mechanism in small cells. Some large protozoans and those with long pseudopodia have active mechanisms of internal transport. The inner, fluid cytoplasm of *Paramecium* circulates, via cyclosis, in a closed loop (Fig. 2-10). In forams and actinopods, **bidirectional shuttling** of vesicles occurs on tracks of microtubules in the axis of each slender pseudopodium.

Most protozoans are aerobes that rely on diffusion for the uptake of oxygen and release of CO_2. A few protozoans, however, are obligate anaerobes, especially those that live as symbionts in the digestive tract of animals. Aquatic species associated with decomposing organic matter may be facultative anaerobes, using oxygen when it is present but also capable of anaerobic respiration. In general, the changing availability of food and oxygen associated with decomposition results in a successional sequence of protozoan species. Because of their short generation time, protozoan community structure changes rapidly with environmental change and can be used to monitor aquatic systems for pollution.

Many freshwater protozoa osmoregulate to remove excess water (volume regulation) and to adjust the concentration and proportions of their internal ions (ionic regulation). Excess water enters by osmosis when the internal osmotic concentration exceeds that of the surrounding water. Additional water may enter with food in vacuoles and pinocytotic vesicles. For example, an ameba fed on a protein solution imbibes, by macropinocytosis, a quantity of water equivalent to one-third of its body volume.

Osmoregulation is accomplished by active ion transport at the cell membrane and by a system of water- and ion-pumping organelles called the **contractile vacuole complex** (Fig. 3-3). The complex is composed of a large spherical vesicle—the **contractile vacuole** proper—and, surrounding it, an array of cytoplasmic vesicles or tubules termed the **spongiome.** The spongiome collects fluid from the cytoplasm and conducts it to the contractile vacuole. The contractile vacuole then contracts and discharges the fluid to the outside of the organism through a temporary or permanent pore. The rate of discharge depends on the osmotic concentration of the external medium. *Paramecium caudatum,* which lives in fresh water, can complete a cycle of vacuole filling and discharge as rapidly as every 6 s and expel a volume equivalent to its entire body every 15 min. The basis for contraction may differ among groups of protozoans. In dinoflagellates, a flagellar rootlet branches to form a contractile sheath around the vacuole. In *Paramecium,* microtubules provide a scaffold around the vacuole (Fig. 3-3), but actin and myosin or elastic energy stored in the stretched vacuolar membrane may be responsible for contraction.

Freshwater protozoans excrete a hypotonic urine. Although the mechanism of contractile vacuole function is not fully understood, it is likely that ions are pumped from the cytoplasm into the spongiome tubules, establishing an osmotic gradient. Cytoplasmic water enters the tubules down the

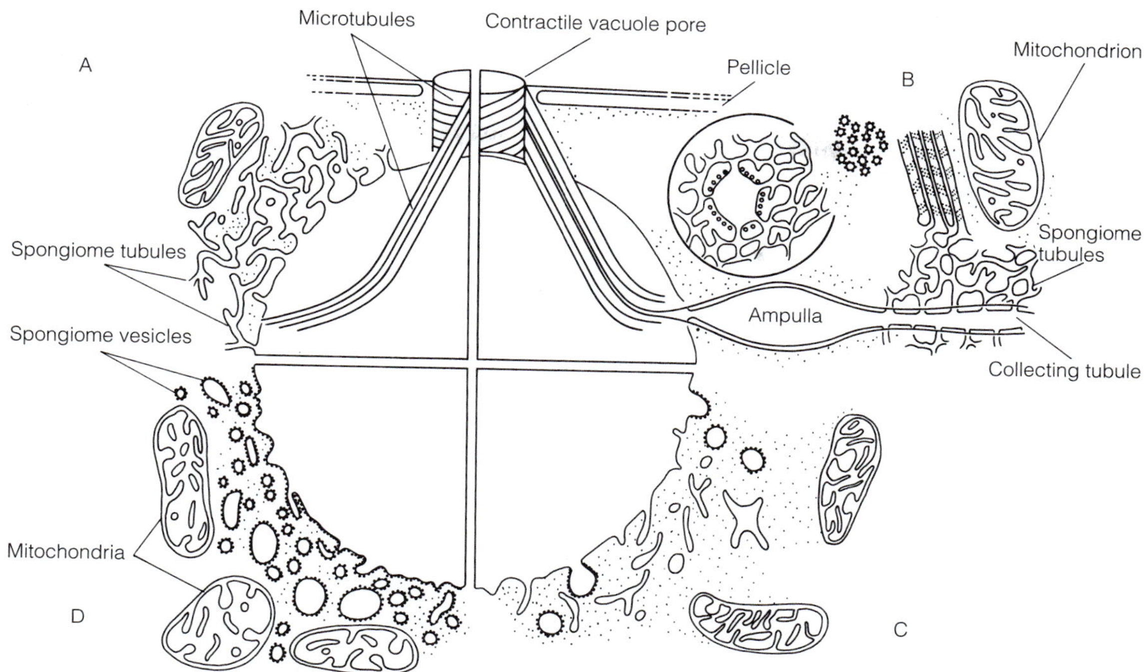

FIGURE 3-3 Protozoa: Diagram of four types of contractile vacuoles. Types **A** and **B** are from ciliates, in which the spongiome is composed of irregular, fluid-filled tubules. Actin filaments (not shown) wind around the pore and extend over the vacuole surface. **A,** The network of spongiome tubules empties directly into the vacuole. **B,** The network of irregular tubules first empties into ampullae, which dilate and then contract, discharging fluid into the vacuole, as occurs in *Paramecium.* **C,** Typical of flagellates and small amebas, the spongiome contains small vesicles and tubules. **D,** Arrangement found in large amebas. *(After Patterson, D. J. 1980. Contractile vacuoles and associated structures; their organization and function. Biol. Rev. 55:1–46. © Copyright Cambridge University Press, reprinted by permission.)*

osmotic gradient. As water and ions flow along the tubules, ions and perhaps other substances are selectively reabsorbed before the urine is discharged to the exterior. The contractile vacuole system is of no particular significance in removing metabolic wastes, such as ammonia and CO_2, as these simply diffuse to the outside of the organism.

REPRODUCTION AND LIFE CYCLES

Clonal (asexual) reproduction by mitosis occurs in most protozoa and is the only known mode of reproduction in some species. Division of the parent into two or more daughter cells is called **fission.** When this process results in two similar progeny cells, it is termed **binary fission;** when one progeny cell is much smaller than the other, the process is called **budding.** Division of the parent into more than two daughter cells is known as **multiple fission. Schizogony** is a specialized form of multiple fission in which repeated divisions of the nucleus precede the cell divisions. With few exceptions, clonal reproduction involves some replication of organelles before or after fission.

The mitotic division of the protozoan cell nucleus differs, in most cases, from that of an animal cell. In animal cells undergoing mitosis, the nuclear membrane disintegrates during mitosis as the chromosomes condense and attach to the mitotic spindle, located in the cytoplasm of the cell. Because the nuclear membrane breaks down, this form of mitosis is said to have an **open spindle.** Later in mitosis, after the chromosomes have separated, a new nuclear membrane is assembled around each nucleus. Among most of the protozoans described in this chapter, however, the nuclear membrane does not break down during mitosis and the spindle forms within the nucleus itself. As the chromosomes separate, the intact nucleus stretches and then constricts, pinching off two new nuclei. Protozoans with this arrangement have a **closed spindle.** The closed spindle is regarded as the primitive form of mitosis in eukaryotic cells. Intermediates between closed and open spindles occur in chlorophytes *(Chlamydomonas, Volvox)* and apicomplexans. In these taxa, the nuclear membrane remains largely intact, but breaks occur that allow cytoplasmic spindle microtubules to enter the nucleus and attach to the chromosomes.

Sexual reproduction is widespread but not universal in protozoans, and life cycles are diverse. Many well-studied protozoans lack sexual reproduction entirely. In some species this absence may be primitive, whereas in others it may be a secondary loss. The primitive protozoan life cycle may have been sex free: a haploid (N) individual reproduced solely by fission, as in the living kinetoplastids (Fig. 3-4A).

The three general forms of sexual life cycles in protozoans are haploid dominance, diploid dominance, and haploid-diploid codominance. A **haploid-dominant** life cycle

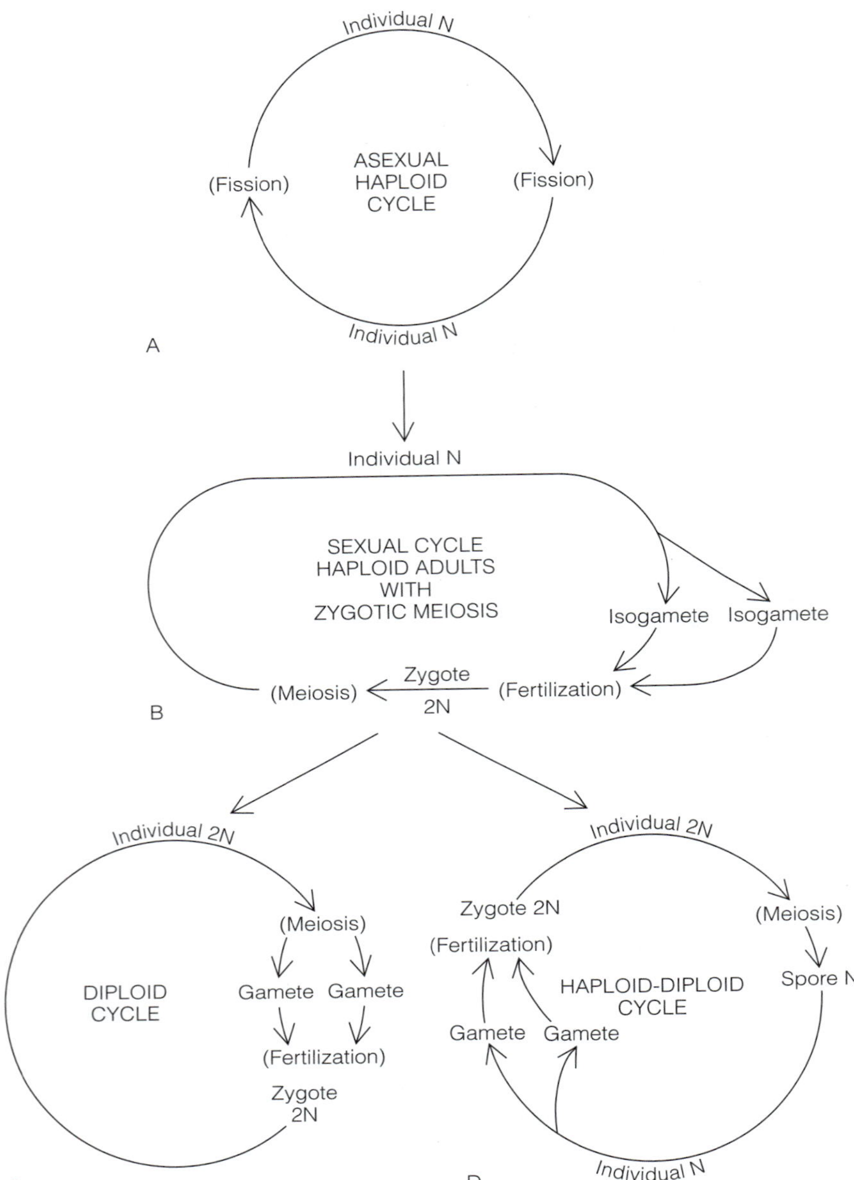

FIGURE 3-4 Protozoa: Life cycles. N = haploid, 2N = diploid. **A,** Haploid asexual life cycle: New individuals arise directly by fission (mitosis), as illustrated by the kinetoplastids. **B,** Haploid-dominant life cycle: Two N individuals mitotically produce isogametes, which fuse to form a diploid zygote. The zygote then undergoes meiosis to form haploid individuals. Examples include the Volvocida, many dinoflagellates, axostylates, and apicomplexans (sporozoans). **C,** Diploid-dominant life cycle: 2N individuals meiotically produce N gametes, which fuse to restore a 2N individual, as happens in some axostylates, heliozoans, many green algae, diatoms, and ciliates (and multicellular animals). Ciliates, however, do not form gametes, but exchange haploid nuclei, which fuse. **D,** Haploid-diploid codominant life cycle: 2N individuals meiotically produce N spores that develop into N individuals that mitotically form N gametes that fuse to restore the 2N individuals; includes many forams, and many algae (and multicellular green plants).

includes haploid individuals that either transform into gametes or produce them by mitosis. Fusion of the haploid gametes results in a diploid zygote that soon undergoes meiosis to form four new haploid individuals (Fig. 3-4B). The haploid-dominant life cycle typifies apicomplexans. In a **diploid-dominant** life cycle, the 2N individuals undergo meiosis to produce N gametes (or gamete nuclei), which fuse into a 2N zygote individual (Fig. 3-4C). This type of life cycle occurs, for example, in ciliates (and animals). In the **haploid-diploid codominant** life cycle, an asexual generation (N or 2N) alternates with a sexual generation (2N or N; Fig. 3-4D). This pattern is characteristic of forams (and plants).

Encystment is characteristic of the life cycle of many protozoa, including the majority of freshwater species. In forming a cyst, the protozoan secretes a thickened envelope about itself and becomes inactive. Depending on the species, the protective cyst is resistant to desiccation or low temperatures and encystment enables the cell to pass through unfavorable environmental conditions. The simplest life cycle includes only two phases: an active phase and a protective, encysted phase. However, the more complex life cycles are often characterized by encysted zygotes or by formation of special reproductive cysts in which fission, gametogenesis, or other reproductive processes take place.

Protozoa may be dispersed over long distances in either the active or encysted stages. Water currents, wind, and mud and debris on the bodies of waterbirds and other animals are common means of dispersal.

DIVERSITY OF PROTOZOA

EUGLENOZOA[P]

Euglenoidea[C]

The euglenoid species of *Peranema* and *Euglena* are among the most familiar of all flagellates. The body is elongate with an invagination, the **reservoir,** at the anterior end (Fig. 3-5). The cytostome lies at the base of the reservoir and joins a cytopharynx. A contractile vacuole discharges into the reservoir in freshwater species, and two flagella, each bearing a row of mastigonemes, arise from the reservoir wall.

In *Euglena,* one flagellum is very short and terminates at the base of the long flagellum (Fig. 3-5A). A pigmented **eyespot,** or stigma, shades a photosensitive bulge, the **paraflagellar body,** at the base of the long flagellum. In the colorless heterotroph *Peranema,* both flagella are long, but one trails backward and can be used to catch food or temporarily attach to something (Fig. 3-5B). The long locomotory flagellum is thickened, up to five times the normal flagellar diameter, and stiffened along most of its length by a **paraxial rod** located to one side of the axoneme (Fig. 3-5B,C). Only the mobile terminal end of the flagellum lacks the rod.

Seen in cross section, the euglenoid pellicle is thrown into rounded ridges alternating with narrow grooves, which together wind helically around the cell. The ectoplasm has a dense skeletal mesh of fibrous proteins. Pellicular microtubules situated below this mesh may be responsible for the peristaltic movements of the cell known as **euglenoid movement** (or metaboly).

Approximately two-thirds of the 1000 species of the marine and freshwater Euglenoidea are colorless heterotrophs and

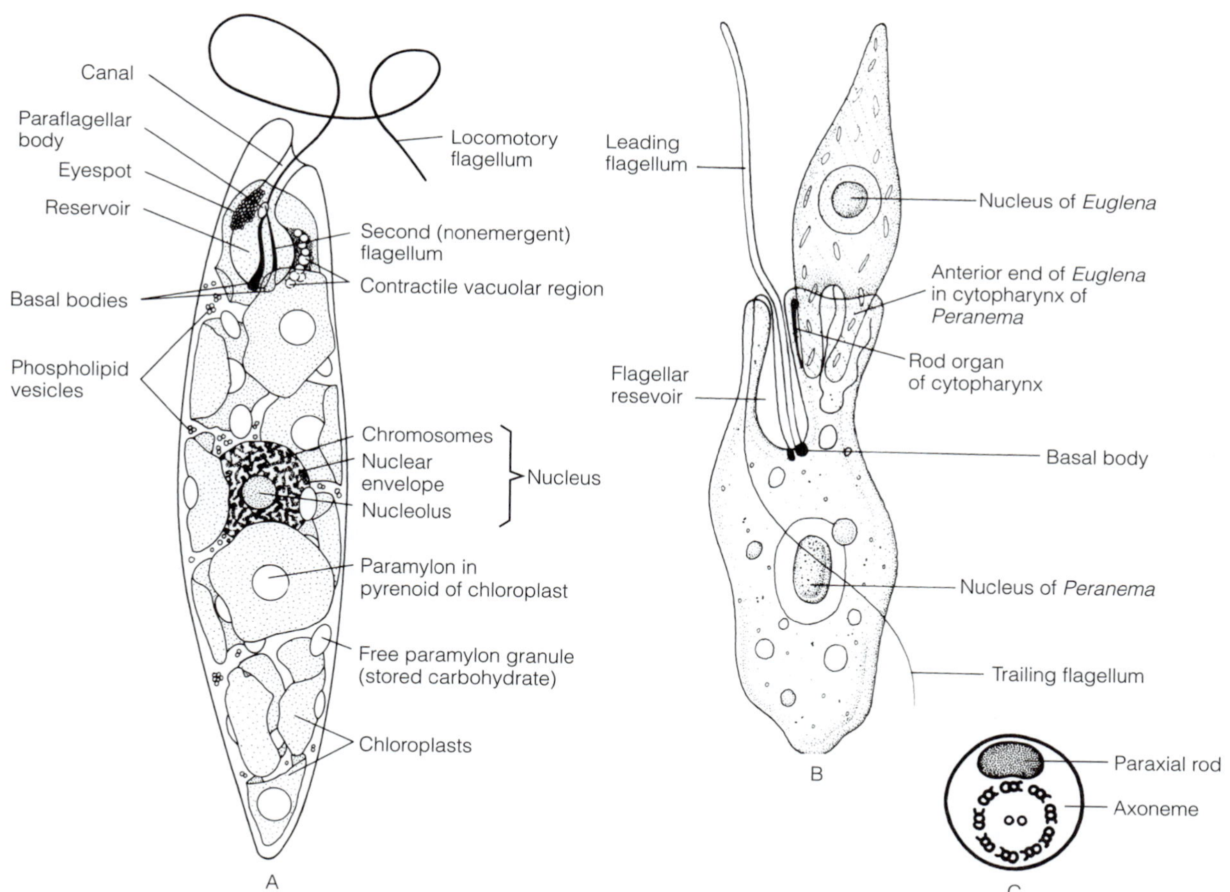

FIGURE 3-5 Euglenozoa: Euglenoidea. **A,** Structure of the photosynthetic *Euglena gracilis.* **B,** The colorless heterotroph *Peranema* swallowing a *Euglena.* **C,** Cross section of leading flagellum of *Peranema* showing paraxial rod. *(A, From Leedale, G. F., 1967, Euglenoid Flagellates. Prentice-Hall, Inc. Englewood Cliffs, N.J.; B, Modified after Chen)*

one-third are green photoautotrophs, such as the species *Euglena.* The chloroplasts of photosynthetic species contain chlorophylls a and b. Photosynthetic euglenoids rotate around their longitudinal axis as they swim toward light. As long as they maintain this orientation, the photosensitive paraflagellar body receives constant illumination. But if they deviate from their head-on approach to a light source, the rotating eyespot periodically shades the paraflagellar body and elicits a course correction. The heterotrophic mode of nutrition is primitive in euglenoids. Chloroplasts were acquired secondarily within the taxon and independently of other photosynthetic flagellates.

The green, photosynthetic species such as *Euglena* store food energy as a unique starchlike carbohydrate called **paramylon.** Paramylon is synthesized in a specialized region, the pyrenoid, of the chloroplast, but stored as free granules in the cytoplasm (Fig. 3-5A). The large paramylon granules may also have a skeletal function, as in *Cyclidiopsis acus,* whose longitudinally aligned granules form an intracellular "backbone." The chloroplasts of euglenoids are surrounded by *three* membranes, not two as in green algae and plants. For this reason, euglenoids are believed to have acquired their chloroplasts by phagocytosis of an entire eukaryotic algal cell, probably a chlorophyte (discussed later), which then became an endosymbiont. If so, the outermost membrane of the euglenoid chloroplast may correspond to the cell membrane of the chlorophyte cell, or the cell membrane fused with the membrane of the phagocytic vesicle.

Food for the colorless heterotrophs consists of organic compounds absorbed from the surrounding water, bacteria, and other protozoan cells. *Peranema* seizes prey with a unique rod organ associated with its cytopharynx and cytostome. The **rod organ** (Fig. 3-5B) consists of two stiff, parallel rods (microtubule bundles) and other intracellular structures called "vanes." (*Euglena* has a rudimentary rod organ, an indication of its heterotrophic ancestry.) *Peranema* feeds on a wide variety of living organisms, including *Euglena,* and the cytostome can be greatly distended to permit phagocytosis of large prey. While feeding, the rod organ is protruded, attaches to the prey, and then retracts, pulling the prey into the cytostome and cytopharynx. (Fig. 3-5B). The prey is swallowed (phagocytosed) whole and digested in a food vacuole.

Sexual reproduction has not been observed in euglenoids, but clonal reproduction occurs by longitudinal binary fission (Fig. 3-6). The two flagella and their basal bodies, as well as the nucleus, replicate before the cell itself divides.

Kinetoplastida[C]

Kinetoplastid flagellates are colorless heterotrophs. A few of the 600 species are free living, but most are important parasites. All share the flagellar paraxial rod with their euglenoid relatives, but uniquely have a conspicuous mass of DNA, called a **kinetoplast,** located within a single, large mitochondrion (Fig. 3-7D). Most of the kinetoplast DNA sequences code for the morphogenesis of mitochondria. The large kinetoplast genome probably is related to the cyclical differentiation and regression of mitochondria as parasitic kinetoplastids adapt their energy metabolism to alternating aerobic and anaerobic host environments. The one or two flagella arise from a reservoir-like pit, which bears a cytostome that leads into a cytopharynx. Their basal bodies are located on or near the kinetoplast mitochondrion.

Species of the free-living biflagellate *Bodo* (Fig. 3-7D) commonly are found in brackish and fresh water and in soil, where they feed on bacteria. The trypanosome kinetoplastids are

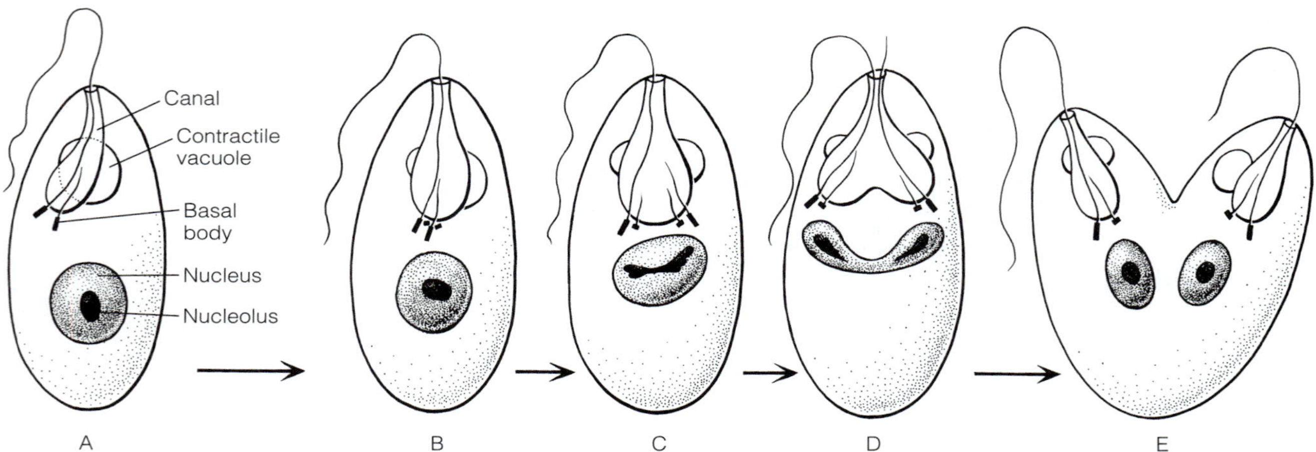

FIGURE 3-6 Euglenozoa: Euglenoidea. Clonal reproduction by longitudinal fission (symmetrogenic division) in *Euglena.* **A,** An interphase cell with two flagella and two basal bodies. **B,** A premitotic cell: Two new basal bodies have formed adjacent to the parental basal bodies. **C,** Early mitosis: The parental basal bodies separate, each becoming associated with one new basal body. All four basal bodies bear a flagellum. Elongation of the nucleolus indicates the onset of nuclear division. Unlike mitosis of animal cells, the euglenoid nuclear membrane remains intact (closed spindle) during the entire division cycle and the flagella do not regress. **D,** Late mitosis: Each separate pair of flagella consists of a parental and a daughter basal body. The nucleus is dividing by constriction, the contractile vacuole has divided, and the reservoir (gullet) is undergoing division. **E,** The anterior end is dividing following duplication of organelles. *(Modified and redrawn from Ratcliffe, 1927, and Triemer, www.lifesci.rutgers.edu/~triemer/flagellar_appt/flagellarapparatus.html)*

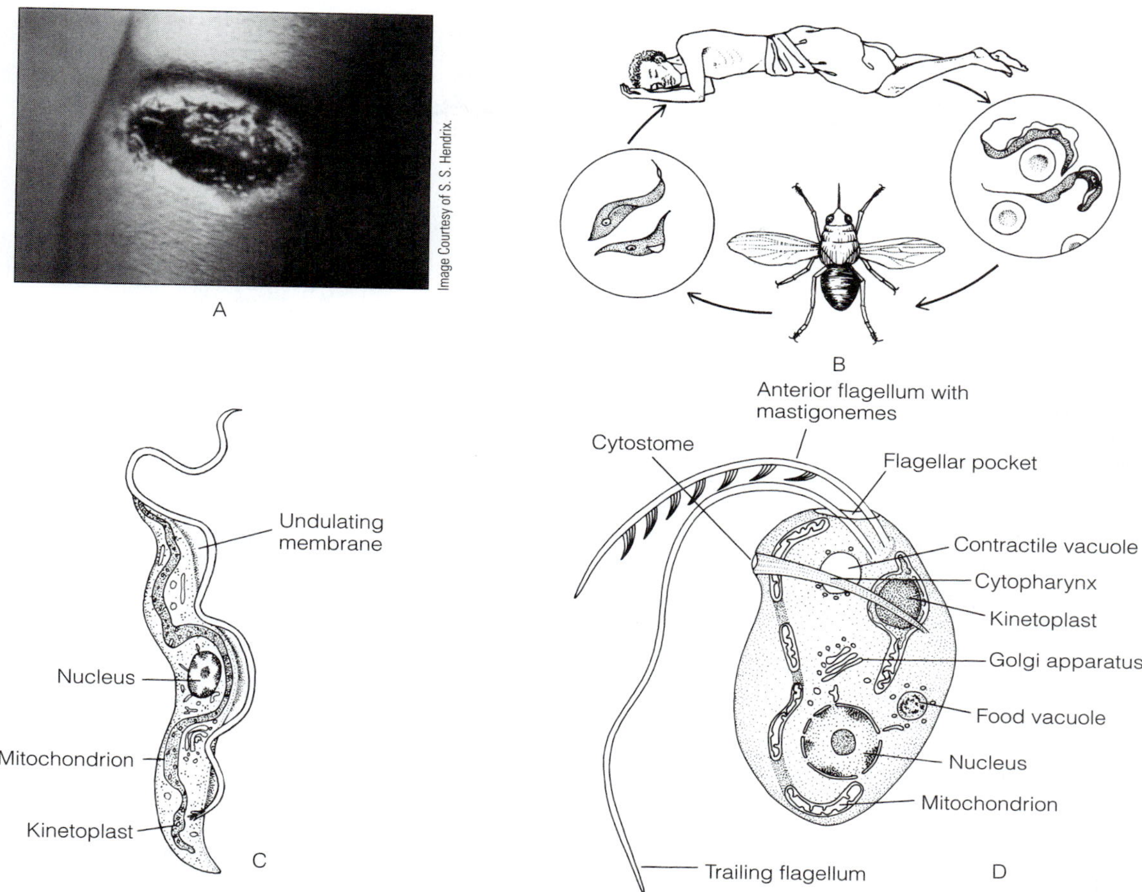

FIGURE 3-7 Euglenozoa: Kinetoplastida. **A,** Skin lesion on a boy's wrist, one of the symptoms of kala-azar disease caused by *Leishmania.* **B,** Life cycle of *Trypanosoma brucei:* The bite of a tsetse fly introduces infective-stage cells that migrate into the human's cerebrospinal fluid to cause the lethargy associated with "sleeping sickness." Life cycle is completed when fly takes another blood meal and ingests the parasite. **C,** Structure of *Trypanosoma brucei.* **D,** *Bodo saltans,* a free-living member of the Kinetoplastida. Only sectioned parts of the long, single looping mitochondrion are shown. *(B, From Sleigh, M. A., 1973, The Biology of Protozoa. Edward Arnold, London. p.141; C, Modified after Brooker from Farmer, J. N. 1980. The Protozoa: Introduction to Protozoology. C. V. Mosby Co., St. Louis, p. 214.)*

gut parasites of insects and blood parasites of vertebrates. Usually only the anterior flagellum is present (Fig. 3-7C), the second flagellum being represented only by a basal body. Commonly, the flagellum trails and is joined to the side of the body by an **undulating membrane.** The pellicle has a thick glycocalyx, whose protein composition is controlled by up to 1000 genes, roughly 40% of the cell's large nuclear genome (120 chromosomes in *Trypanosoma brucei*). Differential gene expression (and protein synthesis) during the various infective stages changes the antigen signature of the glycocalyx, enabling the parasite to elude the host's immune system.

Species of the trypanosome genera *Leishmania* and *Trypanosoma* are agents of numerous diseases of humans and domesticated animals in subtropical and tropical regions of the world. Part of the life cycle is passed within or attached to gut cells of blood-sucking insects, mostly various kinds of flies, and another part of the cycle is spent in the blood plasma or in white blood cells and lymphoid cells of the vertebrate host, although other tissues may be invaded. Intracellular stages are aflagellate, but during the life cycle, motile, extracellular, flagellated stages occur in the vertebrate bloodstream or in the invertebrate host (Fig. 3-7B,C).

Leishmania is the agent of the widespread kala-azar and related diseases of Eurasia, Africa, and the Americas. They cause skin lesions (Fig. 3-7A) and interfere with immune responses, among other effects. Tiny biting flies known as sand flies or no-see-ums (Ceratopogonidae) are the blood-sucking insect host of this protozoan.

Chagas's disease of tropical America, which probably accounted for Darwin's chronic ill health following the voyage of the *Beagle,* is caused by *Trypanosoma cruzi* and is transmitted by blood-sucking bugs. Extensive damage to the human host occurs when the parasite leaves the circulatory system and invades the liver, spleen, and heart muscles.

Trypanosoma brucei rhodesiense and *T. b. gambiense* cause African sleeping sickness and are transmitted by the tsetse fly

(Fig. 3-7B,C). The parasite invades the cerebrospinal fluid and brain, producing the lethargy, drowsiness, and mental deterioration that mark the terminal phase of the disease. Various trypanosome diseases of horses, cattle, and sheep are of considerable economic importance.

Kinetoplastids undergo binary fission. Like euglenoids, sexual reproduction has not been observed, but its occurrence is suspected.

CHLOROPHYTA[P]

Volvocida[O]

Volvocida is a taxon of green algae (Chlorophyta), in which the large, cup-shaped chloroplasts contain chlorophylls a and b and a pyrenoid that synthesizes starch (amylopectin) as a food storage product. Many chlorophytes are nonmotile marine and freshwater algae, such as the filamentous *Spirogyra* of fresh water. The cells of Volvocida, however, are permanently flagellated: Each cell bears two, four, or occasionally eight flagella lacking mastigonemes. An eyespot and two contractile vacuoles also may be present. The cells are enclosed in a gel matrix composed of glycoproteins and glycoaminoglycans and are interjoined by cytoplasmic bridges.

Among the flagellated species, some are solitary, such as *Chlamydomonas* (Fig. 3-8A), and others are colonial. The colonies of *Gonium* (Fig. 3-8B) are flat plates of 32 to 40 cells, but other genera form hollow spheres: *Pandorina* (16 to 32 cells), *Eudorina* (32 cells), *Pleodorina* (64 to 128 cells), *Volvox* (2000 to 6000 cells).

Chlamydomonas reproduces clonally by longitudinal binary fission. In *Volvox,* only specialized, large, aflagellate cells **(gonidia)** are capable of asexual and sexual reproduction. During clonal reproduction, a gonidium undergoes multiple fission and forms a hollow sphere within the parent colony (Fig. 3-8C). The cell polarity of this sphere, however, is opposite that of the parent—the future flagella-bearing ends of the cells face the interior of the young colony. To correct its reversed polarity, the daughter colony inverts and reforms a sphere, now with flagella on the outer surface. The daughter colonies usually escape by rupturing the wall of the parent colony.

The volvocids have a haploid-dominant life cycle with postzygotic meiosis (Fig. 3-4B). In most species of *Chlamydomonas,* the two structurally identical cells act as gametes (isogametes), fuse, and form a zygote. Other species show the beginnings of sex differentiation by having gametes that differ slightly in size (anisogametes). In *Pleodorina,* the size distinction is pronounced, but the large macrogametes still retain flagella and are free swimming. Finally, in *Volvox,* true eggs and sperm develop from gonidia at the posterior of the colony. The egg is stationary and is fertilized within the parent colony by a sperm packet released from another colony. Colonies may be either hermaphrodites or one or the other sex.

Although closely related to plants and not to animals, *Volvox* nevertheless illustrates how multicellularity might have evolved in the first animals. Beginning as a single cell, subsequent mitoses result in a symmetrical colony composed of hundreds of cells. These cells then specialize functionally into somatic cells and reproductive cells (gonidia).

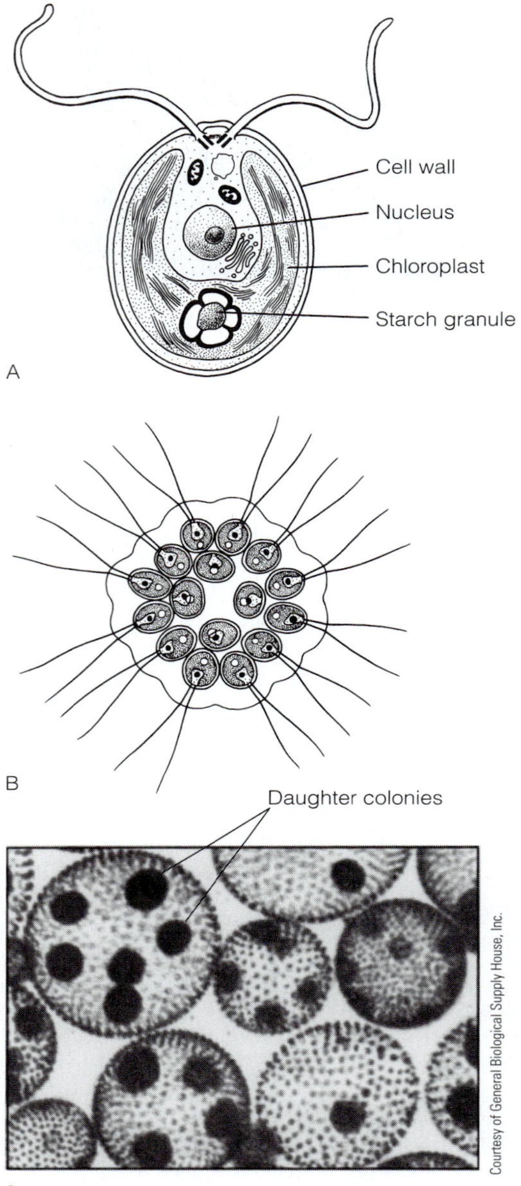

FIGURE 3-8 Chlorophyta: Volvocida. **A,** *Chlamydomonas reinhardtii,* a noncolonial solitary species. **B,** *Gonium pectorale. Gonium* species form colonies in the form of a flat, square plate in which all cells are embedded in a common gelatinous envelope. **C,** *Volvox* colonies are hollow spheres. Note daughter colonies within parent colonies. *(A, From Sleigh, M. 1989. Protozoa and Other Protists. Edward Arnold, London, p.140; B, Courtesy of General Biological Supply House, Inc.)*

CHOANOFLAGELLATA[P]

Surprising as it may seem, the marine and freshwater choanoflagellates are the sister taxon of animals (Metazoa). Both choanoflagellate and primitive monociliated animal cells bear a single flagellum, which bears a bilateral **vane** of

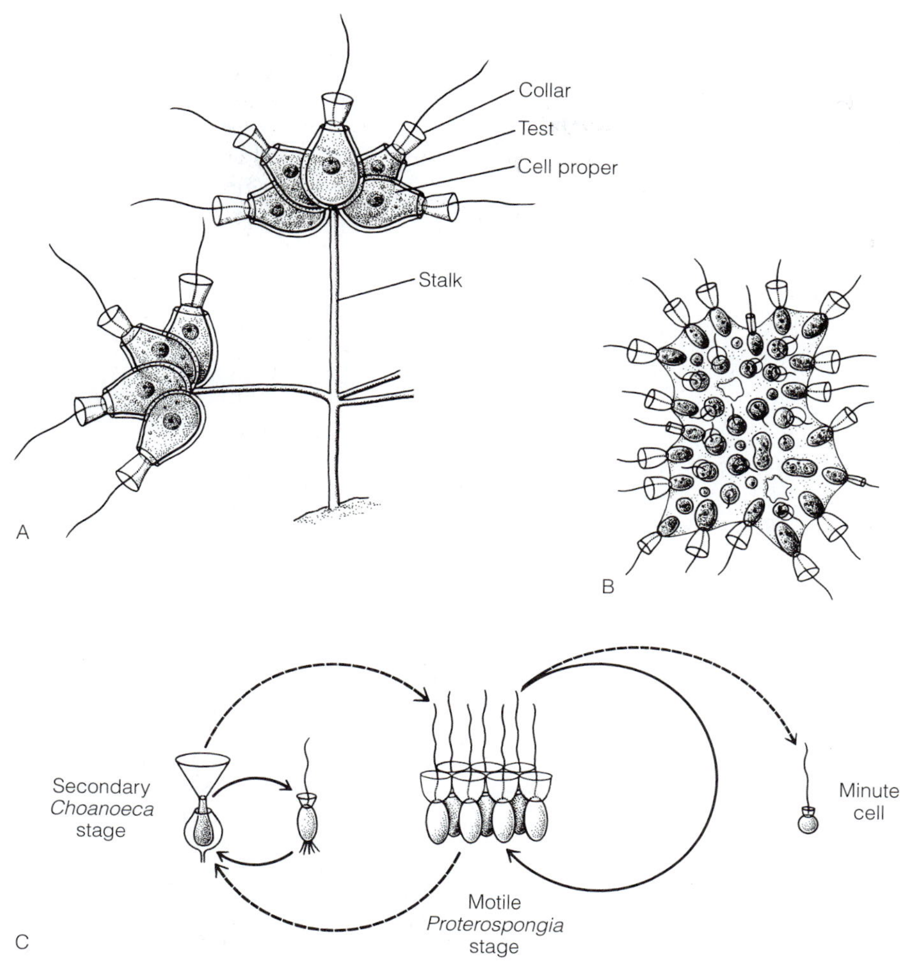

FIGURE 3-9 Choanoflagellata. Choanoflagellates have one flagellum surrounded by a collar of microvilli. **A,** A stalked colonial species. The stalk is an extension of the vaselike test that surrounds each cell. **B,** *Proterospongia,* a colonial species with cells united in a gelatinous matrix. **C,** *Proterospongia choanojuncta* has both a sessile and planktonic stage. *(A, From Farmer, J. N. 1980. The Protozoa: Introduction to Protozoology. C. V. Mosby Co., St. Louis.; B and C, From Leadbeater, B. S. C. 1983. Life-history and ultrastructure of a new marine species of Proterospongia. Jour. Mar. Biol. Assoc. U.K. 63:135–160.)*

mastigoneme-like filaments and is surrounded by a cylindrical **collar** of microvilli (Fig. 4-2A). This synapomorphy, along with support provided by rDNA sequences, unites the choanoflagellates and metazoans as sister taxa in a monophyletic taxon (see Chapter 1 for cladistic terms and method).

The 600 species of choanoflagellates are mostly tiny and inconspicuous, usually not in excess of 10 μm in diameter (Fig. 3-9, 4-12A,B, 4-13A). While feeding, the flagellum creates a water current from which the collar filters bacteria and organic particulates. Bacteria trapped on the collar are ingested by phagocytosis.

Choanoflagellates may be solitary or colonial, attached or free swimming. Some sessile species are attached by a stalk, part of a vaselike test (Fig. 3-9A). The test is composed of interconnected, extracellular, siliceous rods. The individuals of colonial planktonic forms, such as species of *Proterospongia,* are united by a jellylike extracellular matrix or by their collars (Fig. 3-9B,C, 4-12A,B). In the latter case, the colony may resemble a plate, with all of the collars and flagella located on the same side, or a sphere on which the flagellated collars radiate from the surface (Fig. 4-12A). The marine *Proterospongia choanojuncta* was found to include both a colonial planktonic stage and a solitary, aflagellate attached stage (Fig. 3-9C).

RETORTAMONADA[P] AND AXOSTYLATA[P]

These two taxa of heterotrophic flagellates have from four to thousands of flagella organized in functional groups. A few of the 700 species are free living *(Hexamita)* in anoxic habitats, but most live anaerobically in the guts of vertebrates and insects, especially wood roaches and termites. Because they live in oxygen-free environments, mitochondria are either absent or atypical, the cells being specialized for glycolysis rather than aerobic respiration. Even when mitochondria are absent, as in *Giardia,* certain mitochondrial genes and proteins do occur, suggesting that the lack of mitochondria is secondary rather than primary.

Retortamonads, such as *Giardia lamblia,* have four flagella, one of which trails behind the leading three and the cell body, and lack Golgi bodies as well as mitochondria. *Giardia lamblia,* which can cause a bloody diarrhea, is a common intestinal parasite in the United States. It frequently occurs in toddlers and child-care workers, but also can be acquired by drinking from seemingly pristine mountain streams. The axostylate *Trichomonas vaginalis* is a small parasite with four anterior flagella (Fig. 3-10A) that inhabits the urogenital tract of

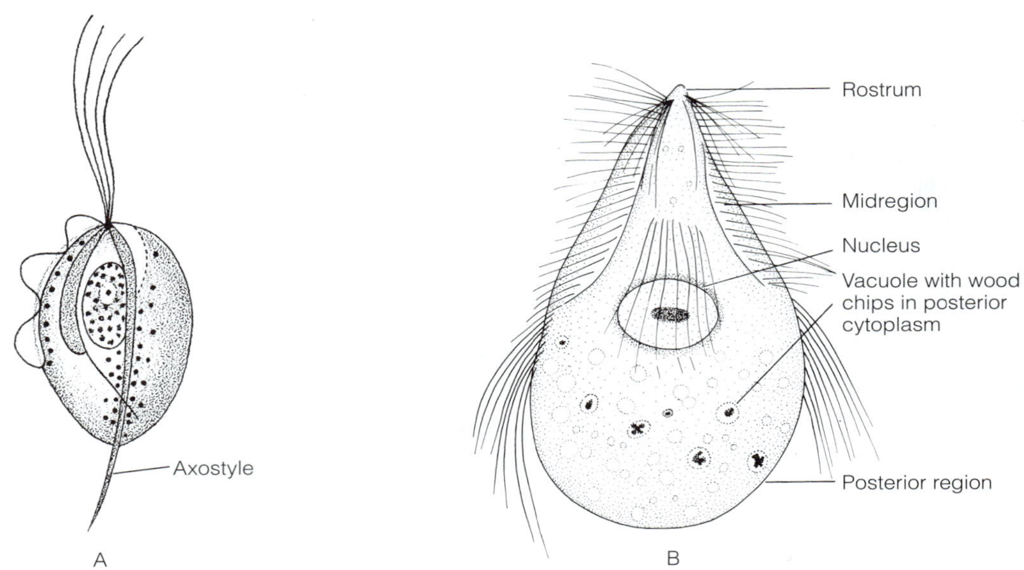

FIGURE 3-10 Axostylata. **A,** *Trichomonas vaginalis,* a trichomonad parasite in the human vagina and male reproductive tract. In addition to the four anterior flagella, a trailing flagellum borders a looping undulating membrane. An axial skeleton, the axostyle, originates at the flagellar basal bodies, passes through the body of the cell, and protrudes posteriorly. **B,** The hypermastigid *Trichonympha campanula* lives in the gut of termites. *(A, After Wenrich; B, from Farmer, J. N. 1980. The Protozoa: Introduction to Protozoology. C. V. Mosby Co., St. Louis. p. 266.)*

humans and causes a widespread sexually transmitted disease. Living tissues can be invaded and the vaginas of seriously infected women produces a greenish yellow discharge.

The axostylates have a bundle of microtubules called an **axostyle** that extends the length of the cell. It most species, it is skeletal in function, like an intracellular backbone, but in some primitive species, it undulates and imparts a snaky motion to the cell. The derived axostylates, such as the hypermastigid mutualists in the gut of termites and wood roaches, have hundreds or thousands of flagella and astounding internal complexity; *Trichonympha* (Fig. 3-10B) is a good example. Most have a saclike or elongated body usually bearing an anterior rostrum. Axostylates lack mitochondria, but have Golgi bodies.

Many termites and wood-eating cockroaches are dependent on their hypermastigids for the digestion of wood. The flagellates, however, rely on intra- and extracellular bacteria and spirochetes for the actual breakdown of cellulose. The nutrients released from the wood are used by bacterium, flagellate, and insect. The termite host loses its gut mutualists with each molt of its exoskeleton, but by licking other individuals, by rectal feeding, or by eating cysts passed in feces (in the case of roaches), a new innoculation is obtained. In wood-eating cockroaches, the life cycles of the flagellates are closely tied to the production of molting hormones by the late nymphal insect.

Diversity of Retortamonada

Retortamonadea[C]: Two or four anterior flagella, one of which is associated with the cytostome, which is elongate longitudinally as a body furrow; mitochondria are absent. *Chilomastix* sp. (plural, spp.) cause diarrhea in humans, poultry; *Retortamonas.*

Diplomonadea[C]: Cell with eight flagella, two cytostomes, two nuclei (twinned, diplozoic cell); mitochondria are absent. Free-living *Hexamita* and parasitic *Giardia,* with their attachment disc and long flagellar axonemes.

Diversity of Axostylata

Oxymonadea[C]: Four posterior flagella; no cytostome, mitochondria, or Golgi bodies. Undulatory axostyle, with intracellular bacteria and surface-attached spirochetes. Anaerobes in the gut of termites and wood roaches. *Oxymonas, Pyrsonympha.*

Parabasalea[C]: Cells have from a few to thousands of flagella and aggregates of large Golgi bodies (parabasal bodies). Axostyle is skeletal, single, replicated, or lost; mitochondria are absent; gut symbionts. Trichomonadida[O] has two to six flagella; a recurrent flagellum forms an undulating membrane; axostyle projects posteriorly to form the attachment site. *Trichomonas* and *Mixotricha paradoxa* in the termite gut have surface-attached spirochetes whose motion propels the flagellate. Hypermastigida[O] has many flagella at the anterior end of the cell (sometimes also elsewhere). Most have surface-attached symbiotic bacteria. In the gut of termites and wood roaches. *Barbulanympha, Lophomonas, Trichonympha.*

ALVEOLATA[P]

Three taxa, Dinoflagellata, Ciliophora, and Apicomplexa (Sporozoa) constitute the Alveolata. Alveolates are united on the basis of having similar ribosomal DNA sequences and pellicular alveoli.

Dinoflagellata[sP]

Approximately one-half of the 4000 marine and freshwater species of dinoflagellates have chloroplasts and are important primary producers, especially in the sea. The xanthophyll pigment **peridinin** colors them red-brown or golden brown. Their chloroplasts are surrounded by three membranes and have chlorophylls a and c, but lack chlorophyll b. Dinoflagellate chloroplasts are diverse, having originated as endosymbionts from at least three different taxa of photosynthetic cells. Heterotrophic dinoflagellates lack plastids and are colorless. Like euglenoids, dinoflagellates originated as colorless heterotrophs that independently acquired chloroplasts by endosymbiosis, probably more than once. A few dinoflagellates are endoparasites of other protozoans, crustaceans, and fishes. The cell nucleus contains permanently condensed (thickened) chromosomes having relatively small amounts of protein, and each chromosome is permanently attached to the nuclear membrane.

Typical dinoflagellates have two flagella. One is attached a short distance behind the middle of the body, is directed posteriorly, and lies in a longitudinal groove **(sulcus)** (Fig. 3-11B). Its surface is smooth or it may have two rows of mastigonemes. The other flagellum is transverse and located in a groove **(cingulum)** that either rings the body once or forms a spiral of several turns. The transverse flagellum, which bears a unilateral row of mastigonemes, causes both rotation and forward movement. The longitudinal flagellum drives water posteriorly and contributes to forward motion. The dinoflagellate contractile vacuole, called a **pusule,** opens to the exterior near the bases of the flagella. The pusule is surrounded by contractile myonemes.

Dinoflagellates have a complex skeleton, or **theca,** which often contains deposits of skeletal cellulose in alveoli. Where the theca is thin and flexible, as in the common freshwater and marine genus *Gymnodinium,* the dinoflagellate is said to be unarmored, or naked (Fig. 3-11A). Armored dinoflagellates have a thick theca composed of a few to several plates (Fig. 3-11B) formed by cellulose-filled alveoli. Frequently the armor is sculptured, and often long projections or winglike extensions protrude from the body, creating bizarre shapes (Fig. 3-11C). The large, colorless, and aberrant *Noctiluca* (Fig. 3-11D) and many smaller species are the principal contributors to planktonic bioluminescence. At night on a quiet sea, their greenish light sparkles in the wake of a boat or as startled fish streak away like shooting stars.

Dinoflagellates are either pigmented photoautotrophs or colorless heterotrophs, but some pigmented species exhibit both modes of nutrition. The prey is usually captured with pseudopodia and ingested through an oral opening associated with the longitudinal flagellar groove. *Noctiluca* is a predator that uses a single contractile tentacle, containing myonemes, to catch prey and convey it to its cell mouth (Fig. 3-11D). Among the symbiotic dinoflagellates, the mutualistic zooxanthellae of corals, without which the coral-reef ecosystem probably would not exist, are primarily one dinoflagellate species, *Symbiodinium microadriaticum.*

Myriad dinoflagellates occur in marine plankton as important contributors to oceanic primary production, especially in the tropics. Marine species of the genera *Gymnodinium, Gonyaulax,* and others are responsible for outbreaks of the so-called red tides (Fig. 3-11E). Under ideal environmental conditions and perhaps with the presence of a growth-promoting substance, populations of certain species increase astronomically. Red tides, however, are not always red. The water may be yellow, green, or brown, depending on the predominant pigments of the blooming organisms. Concentrations of toxic alkaloids produced by the dinoflagellates can reach such high levels that other marine life may be killed. The 1972 red tides off the coasts of New England and Florida killed thousands of birds, fish, and other animals and wreaked havoc on the shellfish industry by infecting clams and oysters that fed on the dinoflagellates.

Pfiesteria piscicida, the cell from Hell, is the dinoflagellate responsible for fish kills in estuaries along the middle Atlantic and southeastern coasts of the United States. Under conditions of organic enrichment, either from human pollution or the feces of schooling fish, the normally nontoxic cells release a waterborne toxin that causes skin lesions in fish. The dinoflagellates then attack the sores and consume the fish. *Pfiesteria* is a colorless heterotroph that feeds by phagocytosis on a variety of organisms. When feeding on unicellular algae, it can digest the prey-cell but retain its chloroplasts intact and then use them to provide itself with photosynthate. The *Pfiesteria* life cycle includes several stages besides the typical biflagellated planktonic cell. These include a benthic ameba and encysted stages, as well as a planktonic form that superficially resembles a heliozoan (see Heliozoa later in this chapter).

Ciguatera food poisoning in humans is caused by a marine dinoflagellate that lives attached to multicellular algae. Ciguatoxin is acquired by grazing herbivorous fish that concentrate the toxin in their tissues and pass it up the food chain. The toxin can reach such high levels in the tissues of carnivorous fish that, when eaten by humans, it produces serious poisoning and even death. In addition to gastrointestinal symptoms such as diarrhea and nausea, there may be respiratory problems, muscle weakness, and long-lasting, strange skin sensations.

Dinoflagellates undergo longitudinal binary fission. Cysts are formed in many flagellate groups, including dinoflagellates. In addition to the ameboid form of *Pfiesteria* already noted, some dinoflagellates can adopt the form of a naked, nonflagellated ball called a **palmella.** Fission often transforms the unicellular palmella into a cluster of cells. The dinoflagellates that inhabit corals as zooxanthellae do so in the palmella stage.

Ciliophora[sP]

Ciliophora is a monophyletic taxon of animated and engaging cell-organisms. Most seem like diminutive animals because of their sophisticated cellular organelles and the complexity of their behavior. Many animal tissues and organs, such as muscle and gut, have analogs in the cellular anatomy of ciliates. The 8000+ described species are widely distributed in fresh water, the sea, and in the water film around soil particles. All ciliates are heterotrophs, but about one-third of them are ecto- or endocommensals or parasites.

FORM AND FUNCTION

Diverse body forms occur among the ciliates and, despite their motility and fixed anterior-posterior polarity, most are asymmetric. A few, however, are radially symmetric with an anterior

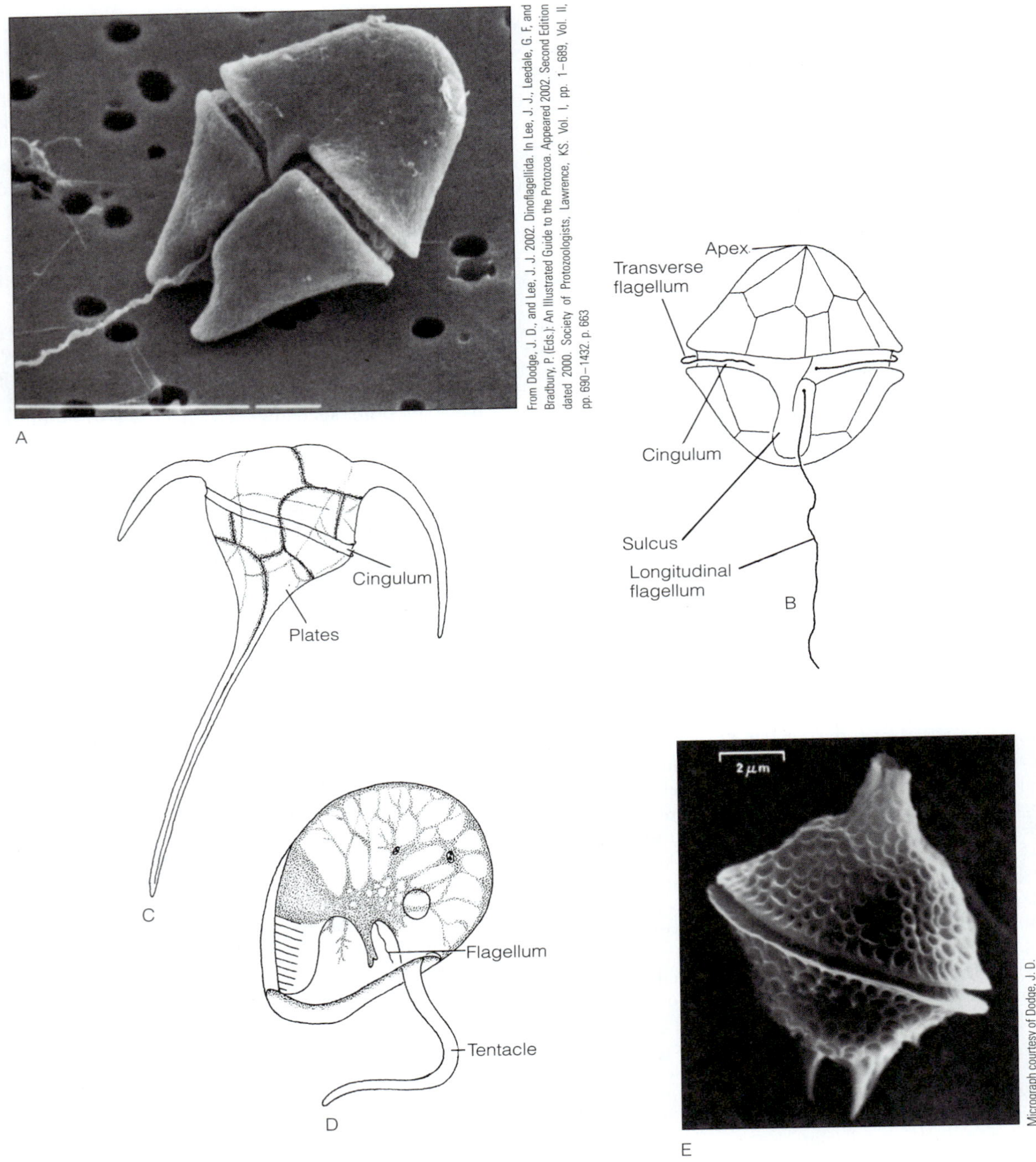

FIGURE 3-11 Alveolata: Dinoflagellata. **A,** The naked *Gymnodinium.* **B,** A freshwater armored species, *Glenodinium cinctum.* **C,** The armored, *Ceratium.* **D,** *Noctiluca,* a bioluminescent carnivore with a prehensile tentacle. Only one small flagellum occurs in an "oral" depression. **E,** *Gonyaulax digitale,* a marine species that causes red tides. *(B, After Pennak, R. W. 1978. Freshwater Invertebrates of the United States. 2nd Edition. John Wiley and Sons, New York; C, After Jorgenson; D, After Robin)*

mouth (Fig. 3-12). Most ciliates are solitary and motile, but some species form colonies and are sedentary. Most ciliates are "naked," but tintinnids, some heterotrichs, peritrichs, and suctorians are housed in a test of secreted organic material or of cemented foreign matter (Fig. 3-13). Ciliate cell size ranges from 10 μm to 4.5 mm.

The surface cilia are specialized into a **somatic ciliature** on the general body surface and an **oral ciliature** associated with

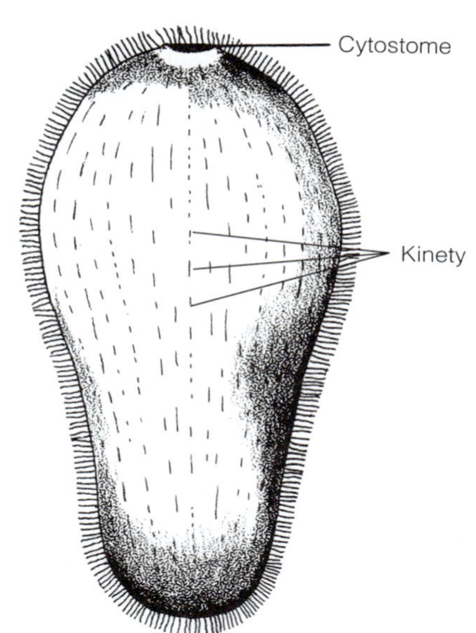

FIGURE 3-12 Alveolata: Ciliophora. *Prorodon,* a radially symmetrical ciliate. *(After Fauré-Fremiet from Corliss, 1979)*

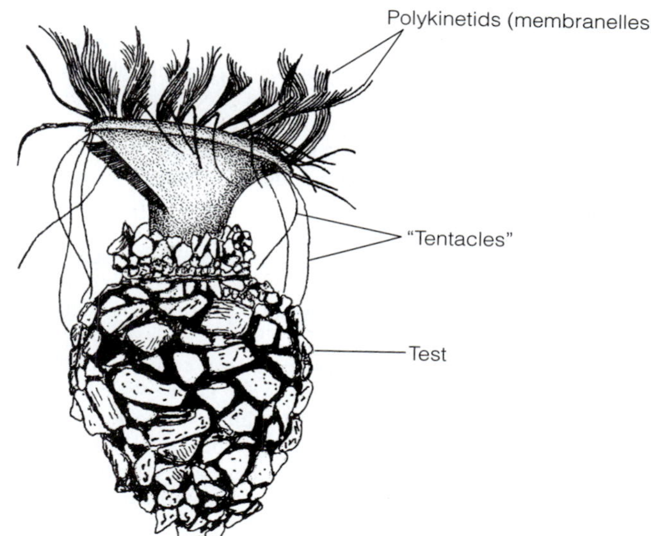

FIGURE 3-13 Alveolata: Ciliophora. *Tintinnopsis,* a marine ciliate (tintinnid) with a test composed of foreign particles. Note conspicuous polykinetids (membranelles) and tentacle-like organelles interspersed between them. *(After Fauré-Fremiet from Corliss, 1979)*

the mouth region. Distribution of body cilia varies between species. In some, cilia cover the entire cell and are arranged in longitudinal rows, each called a **kinety** (Fig. 3-12), but in more specialized taxa the cilia are restricted to regions of the body (Fig. 3-13, 3-18).

A kinety is a row of repeating **kinetids,** each comprising a cilium, basal body, and associated fibers (Fig. 3-14). One of the fibers attached to the basal body is a striated rootlet, which is oriented anteriorly. The rootlet fibers from all basal bodies in a row may combine, like wires in a cable, to form a single **kinetodesma,** which runs the length of the row (Fig. 3-14). The other fibers associated with each basal body are ribbons of microtubules. A postciliary microtubular ribbon extends posteriorly from each basal body. A transverse

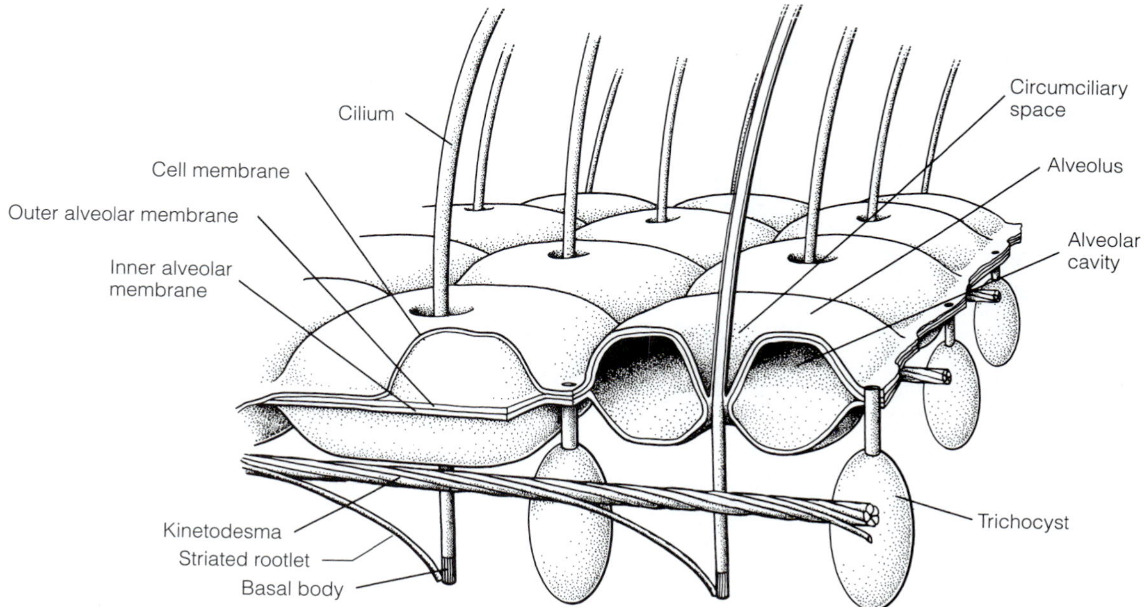

FIGURE 3-14 Alveolata: Ciliophora. The pellicle of *Paramecium.* *(After Ehret and Powers from Corliss, 1979)*

microtubular ribbon extends from the left side of each basal body. All kinetid fibers are thought to be skeletal in function, either for ciliary anchorage or maintenance of cell shape.

The unitary kinetid described in the preceding paragraph is called a **monokinetid** (Fig. 3-14). In some ciliates, the monokinetid is doubled into a **dikinetid** and the cilia occur in pairs along the kinety. In **polykinetids,** multiple cilia function together in a compound unit. If that unit is a tuft, it is called a **cirrus** (plural, **cirri**) (Fig. 3-18B), and if it is a short row, then the paddlelike unit is known as a **membranelle** (Fig. 3-18A,B,F). Kinetids typify all ciliates, even groups such as the Suctoria, which lack cilia as adults but retain the intracellular components of the kinetids.

The ciliate body is typically covered by a complex pellicle. Below the outer cell membrane is a single layer of small membranous sacs, the alveoli, each of which is moderately to greatly flattened (Fig. 3-1E, 3-14). Cilia emerge from between adjacent alveoli, as do trichocysts and other extrusomes discussed later (Fig. 3-14). Alveoli have a skeletal function and also store Ca^{2+} ions. Following an appropriate stimulation of the cell, these ions are released into the cytoplasm, where they can initiate changes in ciliary beat or discharge of extrusomes.

Extrusomes are secretory bodies specialized for rapid release at the surface of the cell. In *Paramecium* and other ciliates, bottle-shaped extrusomes, **trichocysts,** alternate with the alveoli (Fig. 3-14). In the undischarged state, a trichocyst is perpendicular to the body surface. At discharge, the trichocyst rapidly ejects a long, striated, threadlike shaft surmounted by a barb (Fig. 3-15). The shaft is not evident in the undischarged state and probably polymerizes during discharge. Trichocysts appear to function in defense against predators. **Toxicysts** are extrusomes found in the pellicle of *Dileptus* and *Didinium.* A toxicyst discharges a long thread with a bulbous base containing a toxin. Toxicysts are used for defense and for capturing prey. They are commonly restricted to the parts of the ciliate body that contact prey, such as around the cytostome in *Didinium* or the anterior body region of *Dileptus* (Fig. 3-16). **Mucocysts** are arranged in rows like trichocysts and discharge a spray or network of mucoid filaments. These may function in the formation of protective cysts or provide a sticky surface for prey capture. They occur in many ciliates, including *Didinium.*

LOCOMOTION

The ciliates are the fastest protozoans, achieving velocities in the range of 0.4 to 2 mm s^{-1} (or approximately eight body-lengths s^{-1} for a typical *Paramecium*). The fastest flagellates, on

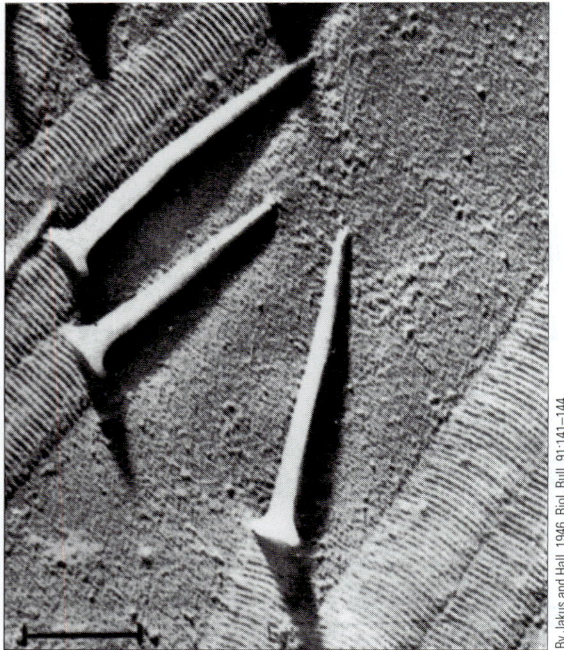

FIGURE 3-15 Alveolata: Ciliophora. Discharged trichocysts of *Paramecium* (electron micrograph). Note golf-tee-shaped barb and part of long striated shaft.

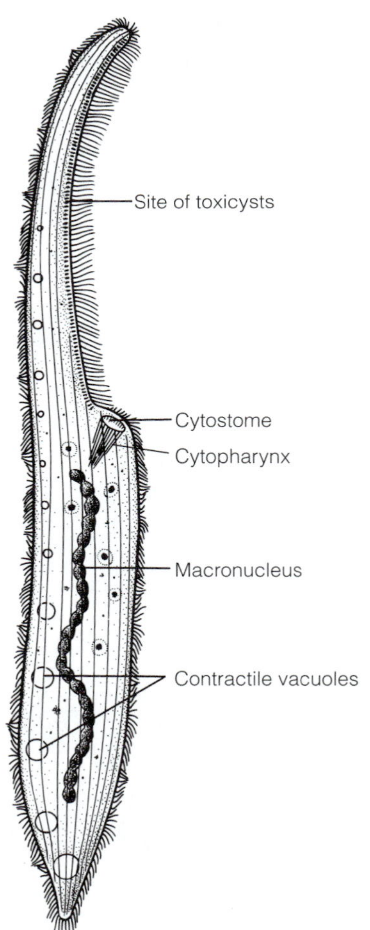

FIGURE 3-16 Alveolata: Ciliophora. *Dileptus anser,* a carnivorous ciliate with a long row of toxicysts in front of the cytostome. *(After Sleigh, M. 1989. Protozoa and Other Protists. Edward Arnold, London. p. 198)*

the other hand, reach only 0.2 mm s^{-1}. On average, ciliates move faster than flagellates because of the numerous cilia on their surfaces.

Metachronal waves (Chapter 2 and Fig. 2-7A) pass over the surface of active ciliates, approximately 10 waves at any moment on the body of a *Paramecium.* The metachronal coordination of cilia is thought to be controlled by water motion. The water movement created by one cilium initiates movement in the next cilium, like a sequence of falling dominoes. The kinetodesmal fibers are not regarded as a conducting system in ciliary-beat coordination.

In genera such as *Paramecium,* the direction of the ciliary effective stroke is oblique to the long axis of the body (Fig. 3-17A). This causes the ciliate to swim in a spiral course and simultaneously to rotate around its longitudinal axis. To change direction, *Paramecium* instantaneously reverses the direction of ciliary beat, retreats, stops, turns, and then proceeds forward in a new direction (Fig. 3-17B). This turning sequence is known as an **avoidance reaction.** Mechanical stimuli may be detected by long, stiff, nonmotile (sensory) cilia. The direction and intensity of the beat are controlled by changing levels of Ca^{2+} and K^{+} ions released from alveolar stores in the pellicle.

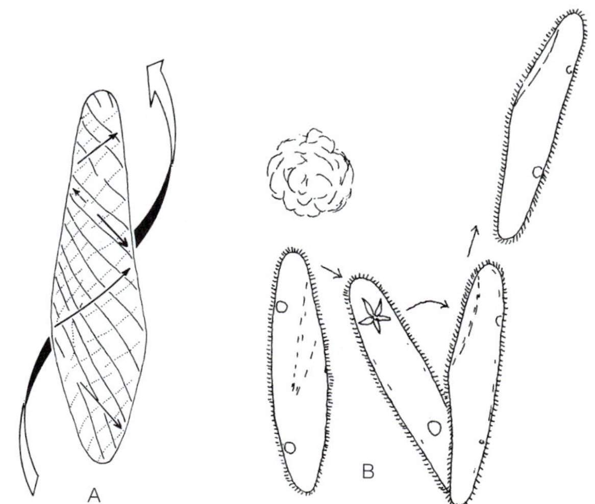

FIGURE 3-17 Alveolata: Ciliophora. Locomotion in *Paramecium.* **A,** Metachronal waves in *Paramecium* during forward swimming. Wave crests are shown by diagonal lines (dotted on ventral surface), and their direction is shown by the small solid arrows. Rotational forward movement of the ciliate indicated by large arrow. **B,** The avoidance reaction of *Paramecium.* When *Paramecium* contacts an object, the cell membrane is depolarized, allowing an influx of Ca^{2+} into the cytoplasm, which causes reversal of the ciliary beat. As Ca^{2+} pumps are reactivated and cytoplasmic Ca^{2+} levels begin to drop, the ciliary beat becomes uncoordinated and the cell turns as a result. When cytoplasmic Ca^{2+} levels reach their normal level, forward motion resumes. The alveoli are the sites of Ca^{2+} uptake, storage, and release. *(A, From Machemer, H. 1974. Ciliary activity and metachronism in Protozoa. In Sleigh, M. A. (Ed.): Cilia and Flagella. Academic Press, London. p. 224. B, After Hyman, L. H. 1940. The Invertebrates, Vol. 1. McGraw-Hill Book Co., New York)*

The highly specialized stichotrichs and hypotrichs, such as *Urostyla, Stylonychia,* and *Euplotes* (Fig. 3-18A,B), have bodies differentiated into distinct dorsal and ventral sides. Cilia have largely disappeared except on localized ventral areas that bear cirri. The cilia of each cirrus are synchronized and the cirrus beats functionally as a single large, forceful unit.

Some ciliates, such as the elongate karyorelictids that live between sand grains on marine beaches or common sessile species of *Vorticella* or *Stentor,* are highly contractile and withdraw rapidly from potential predators. Contraction results from the shortening of striated protein fibers called **myonemes.** *Stentor* shortens its entire body with pellicular myonemes, but in *Vorticella* and the colonial *Carchesium,* the myonemes extend into the stalk as a single large, spiral fiber, the **spasmoneme** (Fig. 3-18C,D). This spasmoneme contracts rapidly, in a few milliseconds, presumably as an escape response. Re-extension of the spasmoneme is slow and may result from the elastic recoiling of the extracellular sheath around the stalk and the beating of the oral cilia. Myonemes are not composed of actin and myosin, as in animal muscle, but rather of another protein called **spasmin** that requires Ca^{2+}, but apparently not ATP, for contraction.

NUTRITION

Free-living ciliates may be detritivores, bacteriovores, herbivores, or predators. Predators may be raptorial, actively pursuing their prey, or ambush predators that lie in wait for their quarry. The predators feed on other protozoans, including other ciliates, and even small animals such as rotifers. Many small ciliates move in search of food—bacteria, diatoms, detritus—and ingest it after making contact. Others, usually larger-bodied species, may use their body cilia to suspension feed on similar foods. The preoral cilia of suspension feeders is usually complex, whereas ciliates that feed by direct interception have less complex oral regions.

Most ciliates have a cytostome, a dedicated endocytic area of the cell membrane that is free of cilia, infraciliature, and alveoli. In some groups the cytostome is anterior (Fig. 3-12), but in most ciliates it has been displaced more or less posteriorly (Fig. 3-16, 3-19). In its least complex form, the cytostome lies directly over a cytopharynx, a cylinder of microtubules located in the cytoplasm (Fig. 3-16). Food is ingested at the cytostome by phagocytosis and the cytopharynx conveys the food vacuole inward.

The oral structures may consist solely of the cytostome and cytopharynx (Fig. 3-16, 3-18F), but in most ciliates the cytostome is preceded by a preoral chamber that aids in food capture and manipulation. The preoral chamber, called a **vestibule,** may be lined only with simple cilia derived from somatic cilia. In other, more complex ciliates, the preoral chamber differs from a vestibule by containing compound ciliary organelles (polykinetids) instead of simple cilia and is then designated a **buccal cavity** (or peristome; Fig. 3-18A,D). In *Paramecium,* the preoral chamber is divided into an outer vestibule and an inner buccal cavity (Fig. 3-19). The polykinetids of its buccal cavity create a current that transports bacteria or small protozoans into the cavity.

Among predators, species of *Didinium* have been carefully studied. These barrel-shaped ciliates feed on other ciliates,

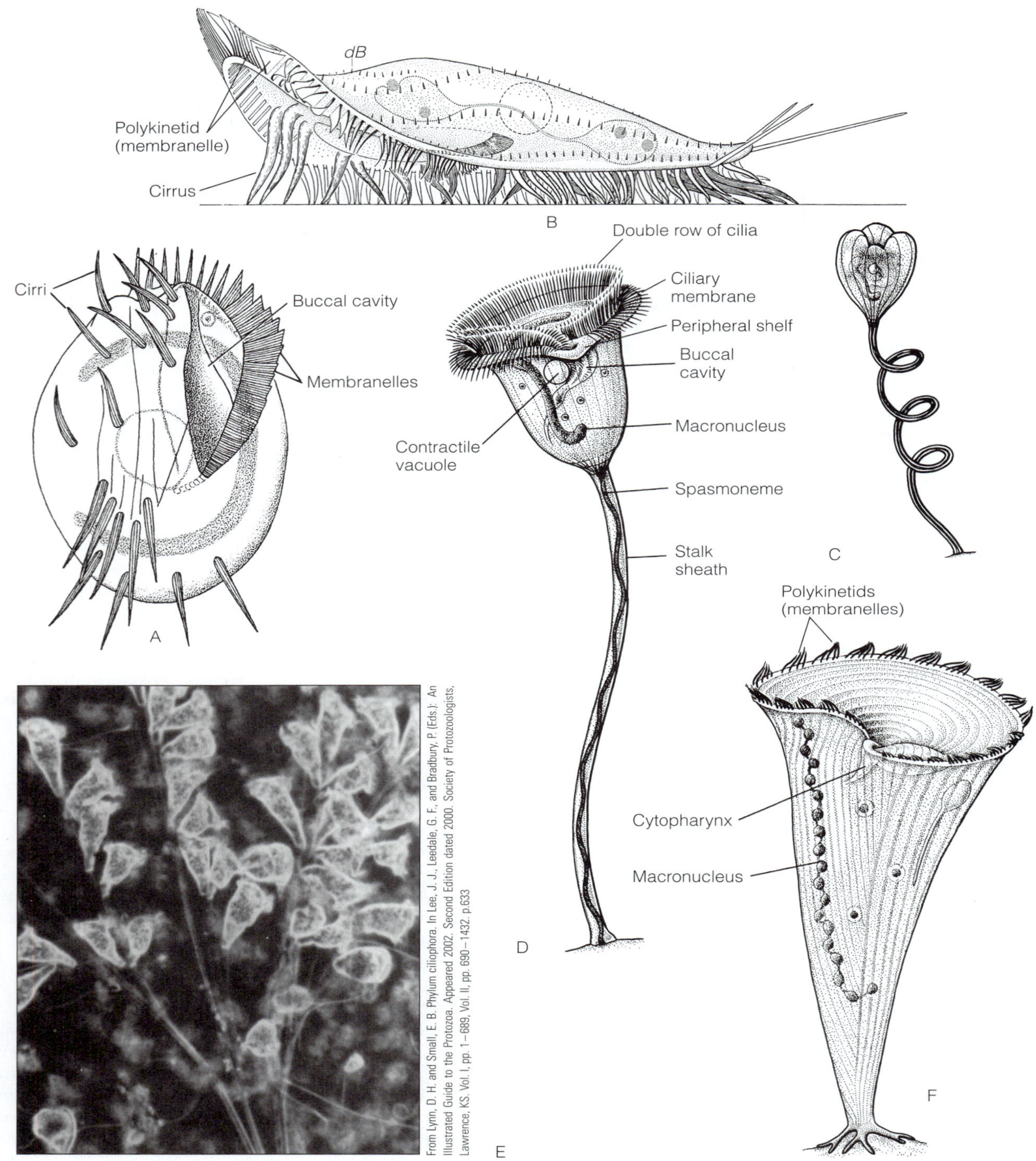

FIGURE 3-18 Alveolata: Ciliophora. **A,** Ventral view of the hypotrich *Euplotes.* **B,** Lateral view of the stichotrich *Stylonychia mytilus.* The arrangement of organelles on the ventral side is similar to that of *Euplotes.* **C** and **D,** *Vorticella convallaria* (Peritrichia) in contracted state (**C**) and extended state (**D**). **E,** *Carchesium polypinum,* a colonial peritrich similar to *Vorticella.* **F,** *Stentor coeruleus* (Heterotrichia). Note the large macronucleus in *Vorticella* and *Stentor*, both of which are large cells (up to 2 mm). *(A, After Pierson from Kudo. B, By Machemer, H. In Grell, K. G. 1973. Protozoology, Springer-Verlag, New York, p. 304. C, D, and F, from Sleigh, M. 1989. Protozoa and Other Protists. Edward Arnold, London. pp. 211 and 213.)*

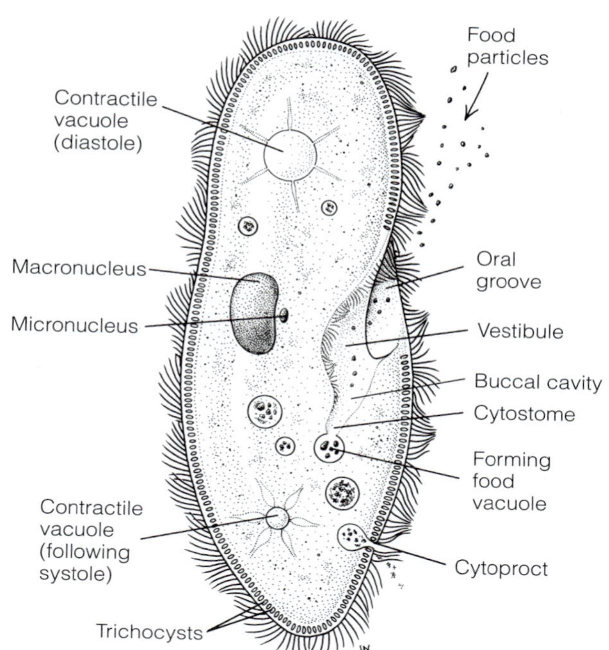

FIGURE 3-19 Alveolata: Ciliophora. Structure of *Paramecium*. *(After Mast from Dogiel; B, After Clakins from Hyman; C and D, from Sleigh, M. A. 1973. The Biology of Protozoa. Edward Arnold Publishers, London. p. 64. Based on micrographs of Rudzinska, Bardele, and Grell.)*

particularly *Paramecium* (Fig. 3-20A). When *Didinium* attacks a *Paramecium,* it discharges toxicysts into the *Paramecium* and the proboscis-like anterior end attaches to the prey through the terminal cytostome, which can open almost as wide as the diameter of the body. Once seized, the *Paramecium* is ingested by phagocytosis.

The free-living members of the Suctoria are ambush predators that resemble tiny, carnivorous sundew plants (Fig. 3-20B). Unlike other ciliates, suctorians lack cilia, except in immature stages. Suctorians are sessile and most are attached by a stalk to the surface of marine and freshwater invertebrates. Stiff tentacles radiate outward from the body and may be knobbed at their tips or shaped like long, pointed spines (Fig. 3-20B). Each tentacle is supported internally by a cylinder of microtubules and bears special attachment extrusomes called **haptocysts** at the tentacle tips (Fig. 3-20C,D). When prey organisms, including other ciliates, strike the tentacles, the haptocysts are discharged into the prey, anchoring it to the tentacles. The contents of the prey are then "sucked" into the tentacle, entering a long food vacuole that eventually extends into the body of the suctorian. "Suction" is actually a rapid phagocytosis, accelerated by the microtubular cylinder, which functions as a cytopharynx in the axis of each tentacle.

Suspension feeders typically have a buccal cavity. Food is brought to the body and into the buccal cavity by the compound ciliary organelles. From the buccal cavity the food particles are driven through the cytostome and into the cytopharynx. When the particles reach the cytopharynx, they are collected in a food vacuole.

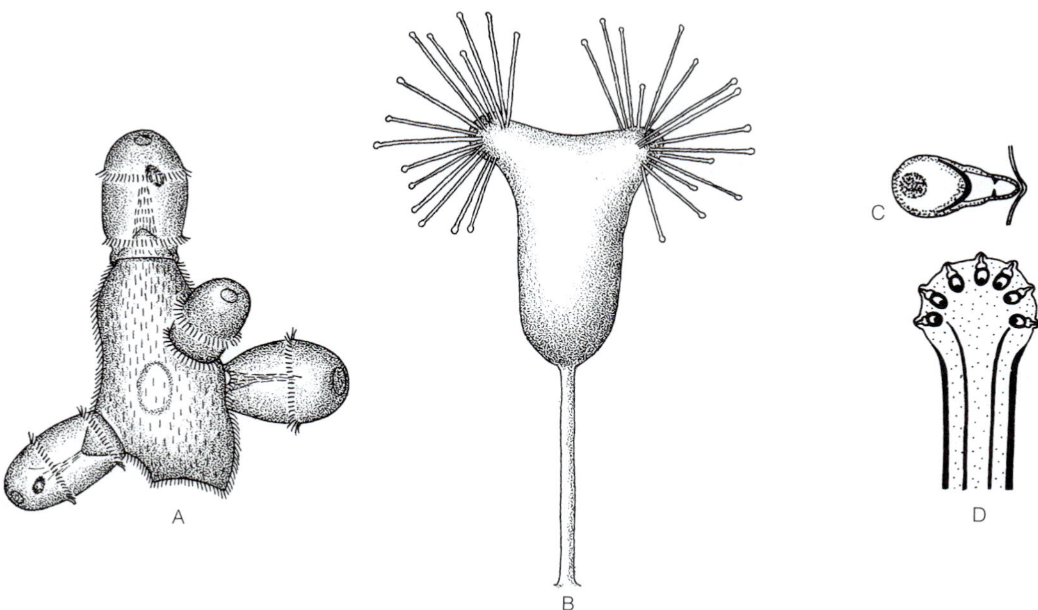

FIGURE 3-20 Alveolata: Ciliophora. Predatory ciliates. **A,** Four *Didinium* attacking one *Paramecium.* **B,** *Acineta,* a suctorian. **C,** A single undischarged haptocyst below the surface of a tentacle cell membrane. **D,** Several haptocysts in a tentacle tip; the two lines below the haptocysts are a section through the microtubular cylinder in the tentacle. *(A, After Mast from Dogiel; B, After Clakins from Hyman, 1940; C and D, From Sleigh, 1973)*

In the filter-feeding Peritrichia, whose members possess little or no body cilia, the buccal ciliary structures are highly developed and are part of a disklike area at the oral end of the body. In *Vorticella,* a peripheral shelf (Fig. 3-18D) closes over the disk during retraction (Fig. 3-18C). The buccal cilia are in a groove between the edge of the disk and the peripheral shelf. These cilia form an outer membrane of fused cilia and an inner double row of unfused cilia. Both membrane and ciliary rows wind counterclockwise around the margin of the disk and then turn downward into the funnel-shaped buccal cavity (Fig. 3-18D). The inner ciliary rows generate the water current, and the outer membrane acts as the filter. The food, mostly bacteria, is transported between the membrane and ciliary rows into the buccal cavity.

Food is ingested by phagocytosis at the cytostome and the food vacuole is transported inward by the cytopharynx. When the food vacuole reaches a certain size, it breaks free from the cytopharynx and a new vacuole forms at the cytostome. Detached vacuoles then begin a more or less circulatory movement through the endoplasm.

Digestion follows the general pattern described in the Introduction to Protozoa, but is peculiar in that it develops a very low initial pH. In *Paramecium,* following the formation of the food vacuole (Fig. 3-21), acidic vesicles (acidosomes) fuse with the vacuole and some cell membrane is removed. As a result, the vacuole becomes smaller and the pH drops to 3. Lysosomes now join the vacuole, but the contents are too acid for effective enzymatic action. For reasons still unknown, the pH rises, and at pH 4.5 to 5, digestion occurs. This is the same pH characteristic of intracellular digestion in other organisms that have been studied. Following digestion, the waste-laden food vacuole moves to a fixed exocytosis site, the **cytoproct,** fuses with the cell membrane and expels its contents. Residual vacuolar membrane breaks up into small vesicles that move to the region of the cytostome and then reconstitute a food vacuole.

About 15% of ciliate species are parasites, and many are ecto- and endocommensals. Many suctorians are commensals and a few are parasites. Hosts include fishes, mammals, various invertebrates, and other ciliates. *Endosphaera,* for example, is parasitic within the body of the peritrich *Telotrochidium.* The commensal hypotrich *Kerona* and peritrich *Trichodina* both occur on the surface of *Hydra. Balantidium* species are endocommensals or endoparasites in the gut of insects and many vertebrates. *Balantidium coli* (Trichostomatia) occupies the intestine of pigs and is passed by means of cysts in the feces. This ciliate has occasionally been found in humans, where, in conjunction with bacteria, it erodes pits in the intestinal mucosa, causing pathogenic symptoms. Other trichostomes are mutualists in the digestive tract of ruminants. Like the flagellate symbionts of termites and roaches, some of them ingest and break down the cellulose of the vegetation eaten by their hosts. The products of digestion are utilized by the host.

Some ciliates harbor symbiotic algae. The most familiar of these is *Paramecium bursaria,* in which the endoplasm har-

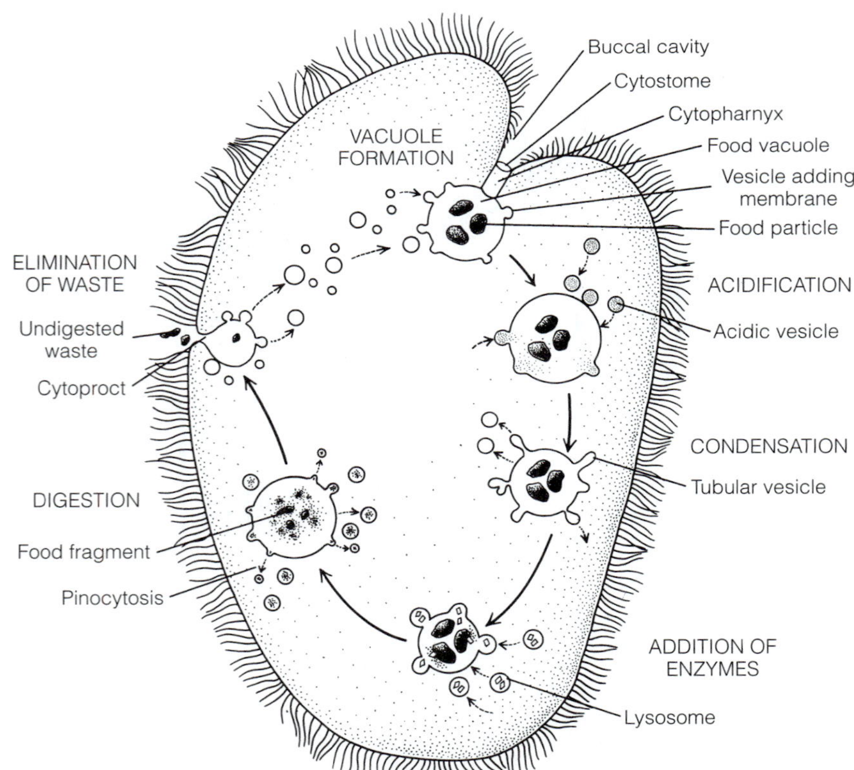

FIGURE 3-21 Alveolata: Ciliophora. Intracellular digestion in ciliates. A forming food vacuole enlarges by fusion with small vesicles that add membrane to the vacuole wall. Fusion of acidic vesicles (acidosomes) causes a drop in pH. Fusion of lysosomes adds digestive enzymes. Eventually, the vacuole shrinks as it blebs off small vesicles to be added to a new food vacuole.

bors green zoochlorellae. A mouthless marine species of *Mesodinium* has algal symbionts.

EXCRETION

Excretion in ciliates is largely a matter of volume regulation. Contractile vacuoles are found in both marine and freshwater species, but in the latter they discharge more frequently. In some species a single vacuole is located near the posterior, but many species have more than one (Fig. 3-16). In *Paramecium,* a vacuole is located at both the posterior and anterior ends of the body (Fig. 3-19). The vacuoles are always associated with the innermost region of the ectoplasm and empty through one or two permanent pores that penetrate the pellicle. The spongiome contains a network of irregular tubules that may empty into the vacuole directly or by way of collecting tubules (Fig. 3-3).

NUCLEAR DIMORPHISM

In contrast to most other protozoan classes, ciliates have two types of nuclei (heterokaryosis): a **micronucleus** that is inactive except during cell division and houses the master copy of the genome, and a **macronucleus** whose genes are actively transcribed for the daily synthetic activities of the cell. Each cell typically has 1 to 20 diploid micronuclei and 1 to many polyploid macronuclei; the numbers vary by species. The macronucleus is sometimes called the vegetative nucleus because it is not essential in sexual reproduction. Instead it is necessary for normal metabolism and the control of cell differentiation. The macronucleus contains hundreds to thousands of times more DNA than does the micronucleus because of duplications following the micronuclear origin of the macronucleus. But many of the DNA sequences of the micronucleus (up to 98% in *Stylonychia*) are eliminated during macronucleus formation. Furthermore, macronuclear DNA is organized not in chromosomes, but rather in small subchromosomal or gene-sized units, some of which are amplified up to 1 million times. The macronuclei contain many nucleoli in which ribosomal RNA is synthesized. The amplification of genes in the macronucleus and the multiple nucleoli probably increase the rate of synthesis of proteins to be used in the assembly of the complex and numerous ciliate organelles.

Macronuclei may assume a variety of shapes (Fig. 3-16, 3-18D,F). The large macronucleus of *Paramecium* is oval or bean-shaped and located just anterior to the middle of the body (Fig. 3-19). In *Stentor* and *Spirostomum,* the macronuclei are long and arranged like a string of beads (Fig. 3-18F). Not infrequently, the macronucleus is in the form of a long rod bent in different configurations, such as a **C** in *Euplotes* or a horseshoe in *Vorticella* (3-18D). The unusual shape of many macronuclei may be an adaptation to reduce the diffusion distance between the nucleus and the cytoplasm of these large cells.

CLONAL REPRODUCTION

Clonal reproduction is by binary transverse fission, with the division plane cutting across the kineties (Fig. 3-22A) in contrast to the longitudinal fission of flagellates (Fig. 3-6). Many sessile ciliates, for example, *Vorticella,* reproduce asexually by budding (Fig. 3-22B).

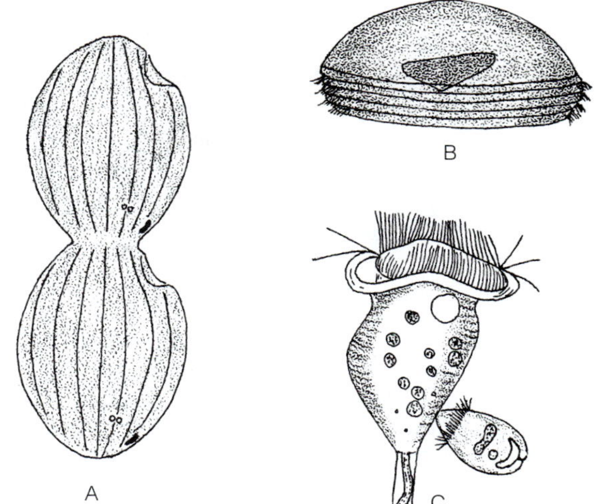

FIGURE 3-22 Alveolata: Ciliophora. **A,** Transverse fission, in which the plane of division cuts across the kineties. **B,** Detached bud of *Dendrocometes.* **C,** Conjugation in *Vorticella.* Note the small, motile microconjugant. *(A, After Corliss, 1979; B, After Pestel from Hyman, 1940; C, After Kent from Hyman, 1940)*

The micronucleus divides by mitosis with a closed spindle. Division of the macronuclei is amitotic and is usually accomplished by constriction. When several macronuclei are present, they may first combine as a single body before dividing.

SEXUAL REPRODUCTION

Sexual reproduction in ciliates is a direct exchange of genes without first packaging them in either egg or sperm cells. To accomplish this, two sexually compatible ciliates fuse along a shared surface, the membrane between them disappears, and a mutual exchange of genes occurs (Fig. 3-23A-F). This process is known as **conjugation** and the two fused ciliates are called **conjugants.** Conjugants may be blissfully fused for several hours. Only the micronuclei function in conjugation; the macronucleus disintegrates during the sexual process.

The steps leading to the exchange of genes between the two conjugants are fairly constant in all species. After two meiotic divisions of the micronuclei, all but one degenerate. This one then divides, producing two haploid gametic micronuclei that are genetically identical. One is stationary while the other migrates into the opposite conjugant. Once the migratory nucleus arrives, it fuses with the partner's stationary nucleus to form a 2N zygote nucleus, or **synkaryon.** Shortly after nuclear fusion the two ciliates separate, and each is then called an **exconjugant.** Each exconjugant undergoes mitotic nuclear divisions to restore the species-specific number of cell nuclei. This event usually, but not always, involves cell divisions. For example, in species normally with a single macronucleus and a single micronucleus, the synkaryon divides once. One of the nuclei forms a micronucleus; the other becomes the macronucleus. In this case, the normal nucleus number is restored without any cell divisions.

But in *Paramecium caudatum,* which also has a single nucleus of each type, the synkaryon divides three times, producing

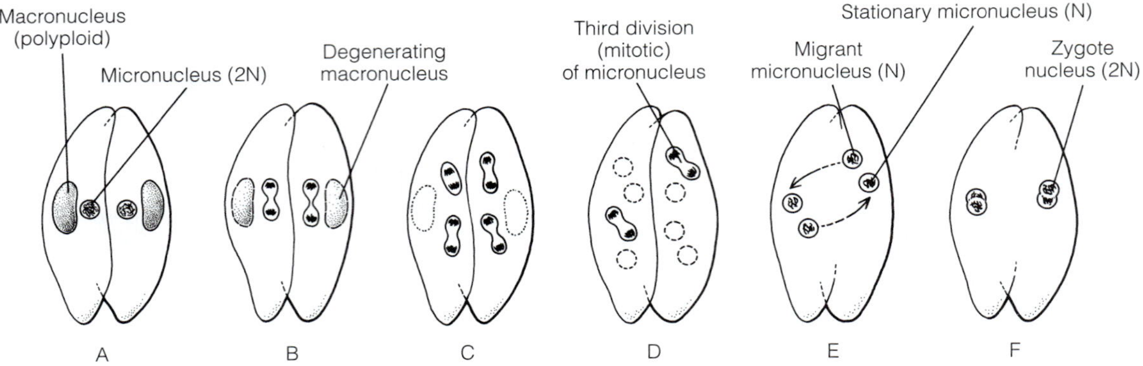

FIGURE 3-23 Alveolata: Ciliophora. Conjugation (sexual reproduction) in *Paramecium caudatum,* a species with one macronucleus and one micronucleus. **A,** Two individuals are united in conjugation. **B–D,** The micronucleus of each conjugant undergoes three divisions, the first two of which (**B** and **C**) are meiotic. **E,** Migrant micronuclei are exchanged between conjugants. **F,** The migratory nucleus fuses with the stationary micronucleus of the opposite conjugant to form a synkaryon, or "zygote nucleus." Note that the micronuclear membrane does not break down during meiosis (or mitosis) in *Paramecium* (or other ciliates).

eight nuclei. Four become micronuclei and four become macronuclei. Three of the micronuclei degenerate. The remaining micronucleus divides during each of the two subsequent cell divisions and each of the four resulting offspring cells receives one macronucleus and one micronucleus. In those species that have numerous nuclei of both types, there is no cell division; the synkaryon merely divides a sufficient number of times to produce the appropriate number of macronuclei and micronuclei.

In some of the more specialized ciliates, the conjugants are a little smaller than nonconjugating individuals, or the two members of a conjugating pair are of strikingly different sizes. Such *gonochoric* macro- and microconjugants occur in *Vorticella* (Fig. 3-22C) and are an adaptation for conjugation in sessile species. The macroconjugant remains attached while the small bell of the microconjugant breaks free from its stalk and swims about. On contact with an attached macroconjugant the two bells adhere. A synkaryon forms only in the macroconjugant from one gametic N nucleus contributed by each conjugant. The conjugal bond is permanent and fatal to the microconjugant, which degenerates after contributing its gamete nucleus. In the sessile attached Suctoria, conjugation takes place between two adjacent individuals that lean together like lovers on a park bench.

The frequency of conjugation varies from once every few days to not at all (or not yet observed). In some species a period of "immaturity," in which only fission occurs, precedes a period during which individuals are capable of conjugation. Numerous factors, such as temperature, light, and food supply, are known to induce or influence conjugation.

In some ciliates, sex is rejuvenating and necessary for additional bouts of clonal fission. For example, some species of *Paramecium* are limited to only 350 clonal generations and die out in the absence of conjugation. Sex restores asexual capacity.

Most ciliates are capable of forming resistant cysts in response to unfavorable conditions, such as lack of food or desiccation. Encystment enables the species to survive cold or dry periods and provides a form for dispersal by wind or attachment to animals.

DIVERSITY OF CILIOPHORA

Karyorelictea^{C}: Freshwater *Loxodes* and marine interstitial *Geleia, Remanella,* and *Trachelоraphis,* all highly contractile. Macronuclei and micronuclei both diploid; somatic dikinetids.

Spirotrichea^{C}: Ciliates with oral membranelles (polykinetids) that wind clockwise to the cytostome; somatic dikinetids or polykinetids. Includes Heterotrichia^{sC}, the contractile *Blepharisma, Folliculina, Spirostomum, Stentor;* Oligotrichia^{sC}, the tintinnids, *Halteria* with somatic cirri; Stichotrichia^{sC}, with ventral cirri, such as the dorsoventrally flattened *Stylonychia* (Fig. 3-18B); Hypotrichia^{sC}, which are flattened with cirri on the ventral surface and have postciliary microtubule (MT) ribbons, such as bacterivorous *Aspidisca* and *Euplotes* (Fig. 3-18B).

Litostomatea^{C}: Somatic monokinetids; MTs from circumcytostomal dikinetids form basketlike cytopharynx and have a transverse ribbon of MTs from ciliary basal bodies and laterally directed kinetodesmal fibers. Includes Haptoria^{sC}, mostly predators with lateral, ventral, or posterior cytostome and toxicysts, *Didinium, Dileptus, Mesodinium* (with endosymbiotic dinoflagellates); and Trichostomatia^{sC}, mutualists in the gut of ruminants that assist in breakdown of cellulose, *Balantidium* and *Entodinium.*

Prostomatea^{C}: Oral region similar to that of litostomates, but some polykinetids are also present; have somatic monokinetids with radially arranged MT ribbons and a cytostome at the anterior end of the cell; toxicysts are common. Marine and freshwater *Coleps, Prorodon.*

Phyllopharyngea^{C}: Leaflike ribbons of MTs surrounded by longitudinal bundles of MTs that form a basket-shaped cytopharynx (cyrtos); somatic monokinetids. Phyllopharyngia^{sC}, of which *Chilodonella* is flattened, ciliated ventrally, and found in sewage; Chonotrichia^{sC}, which are sessile, nonciliated filter feeders with a spiral oral end that attach to crustaceans; Suctoria^{sC}, which are sessile, cilia-free predators with prey-catching tentacles, resemble miniature sundews and include *Allantosoma* (in horse colon), *Ephelota, Heliophrya, Tokophrya.* Marine and fresh water.

Nassophorea^{C}: Transverse MT ribbons tangential to the basal bodies; well-developed kinetodesma; MT bundles form a complex, basket-shaped cytopharynx (nasse); somatic mono- or dikinetids. Peniculida^{O} has an oral apparatus that is an elastic slit and three oral membranelles (peniculus) on its left side and an undulating membrane on the right; a nasse is absent; includes the slipper ciliate, *Paramecium.*

Oligohymenophorea^{C}: A few oral polykinetids, usually three, on left side of the cytostome; somatic monokinetids with MT ribbons that radiate from the basal bodies. Hymenostomatia^{sC}, oral apparatus like that of Nassophorea. The best-known ciliate is the free-living *Tetrahymena; Ichthyophthirius,* the cause of "ich" disease of freshwater fishes; *Pleuronema, Uronema.* Peritrichia^{sC}, a ciliary ring on its oral rim that winds helically counterclockwise to the cytostome and then splits into three membranelles; somatic cilia are reduced; often have contractile stalks (or bodies) and are mostly sessile and attached, but some can detach and swim: *Carchesium, Epistylis, Trichodina, Urceolaria, Vorticella.*

Colpodea^{C}: Kidney-shaped cells with spiral kineties and somatic dikinetids: *Bursaria, Colpoda.*

Apicomplexa^{sP} (Sporozoa)

The some 5000 species of apicomplexans are widespread and common parasites of such animals as worms, echinoderms, insects, and vertebrates. Depending on the species, they may be extra- or intracellular parasites or both at different stages of the life cycle. Apicomplexans also are responsible for malaria, the number-one parasitic disease of humankind, as well as similar debilitating diseases of livestock.

Apicomplexans are so named because motile infective stages (sporozoites, merozoites) bear an anterior **apical complex** that attaches to or penetrates into host cells. A fully developed apical complex consists of an anterior conoid, one or two polar rings, 2 to 20 flask-shaped glandular structures (rhoptries), and numerous membranous Golgi-derived tubules (micronemes) (Fig. 3-24). The conoid is open at both ends and encircled by the polar rings, which link to subpellicular microtubules. The micronemes contain enzymes presumably used for host-cell penetration, but the functions of the other components are unclear. Apicomplexans lack cilia, but flagella occur on their microgametes. Pseudopodia also are absent. Infective stages move by gliding, which may result from microscopic undulations of the pellicle. One or more feeding pores, called **micropores,** are located on the side of the body (Fig. 3-24). The apicomplexan pellicle consists of the outer cell membrane and two additional membranes below it. The two inner membranes are actually the outer and inner walls of a flattened alveolus, which completely encloses the subpellicular cytoplasm except for breaks anteriorly (apical complex), laterally (micropores), and posteriorly (site of exocytosis).

The extraordinary life cycles of apicomplexans achieve mind-challenging complexity in species that infect more than one host. The basic life cycle, however, is reasonably straightforward. Its sexual and clonal stages are haploid, except for the zygote (haploid-dominant cycle; Fig. 3-4D). The motile infective stage is called a **sporozoite.** The haploid sporozoite enters the body of the host, takes up host nutrients, grows, and differentiates into a **gamont,** or gamete-producing cell. Generally, male and female gamonts pair, become enclosed in a common envelope (cyst), and each produces many gametes via multiple fission within the cyst. Once full grown, these gametes fuse to form diploid zygotes, each of which secretes a protective extracellular capsule and is then called a **spore.** Within the spore, the zygote nucleus undergoes meiosis to restore the haploid chromosome number and then mitosis to produce eight cells, which differentiate into sporozoites. The encapsulated sporozoites are liberated from the spore after it is ingested by a host. In this life cycle, **gamogony,** the production of gametes, refers to the period from the pairing of the gamonts to the fusion of gametes. **Sporogony,** the production of spores, refers to the period beginning with meiosis of the zygote to the differentiation of sporozoites within the spore.

The basic life cycle is illustrated by the gregarine (Gregarinea) *Monocystis lumbrici,* which parasitizes seminal vesicles of the earthworm, *Lumbricus terrestris* (Fig. 3-25). Worms become infected when they ingest soil containing spores. Within the earthworm's gizzard, the spores hatch and release

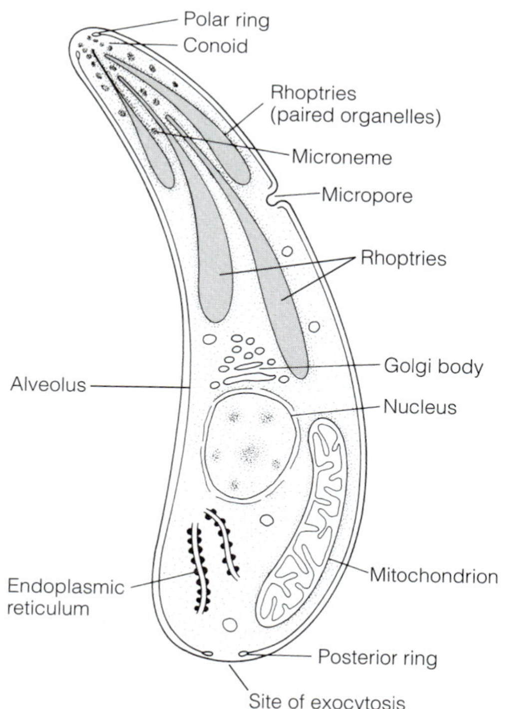

FIGURE 3-24 Alveolata: Apicomplexa. Lateral view of a generalized sporozoan. The polar ring, conoid, micronemes, and rhoptries are parts of the apical complex. *(From Farmer, J. N. 1980. The Protozoa: Introduction to Protozoology. C. V. Mosby Co., St. Louis. p. 360)*

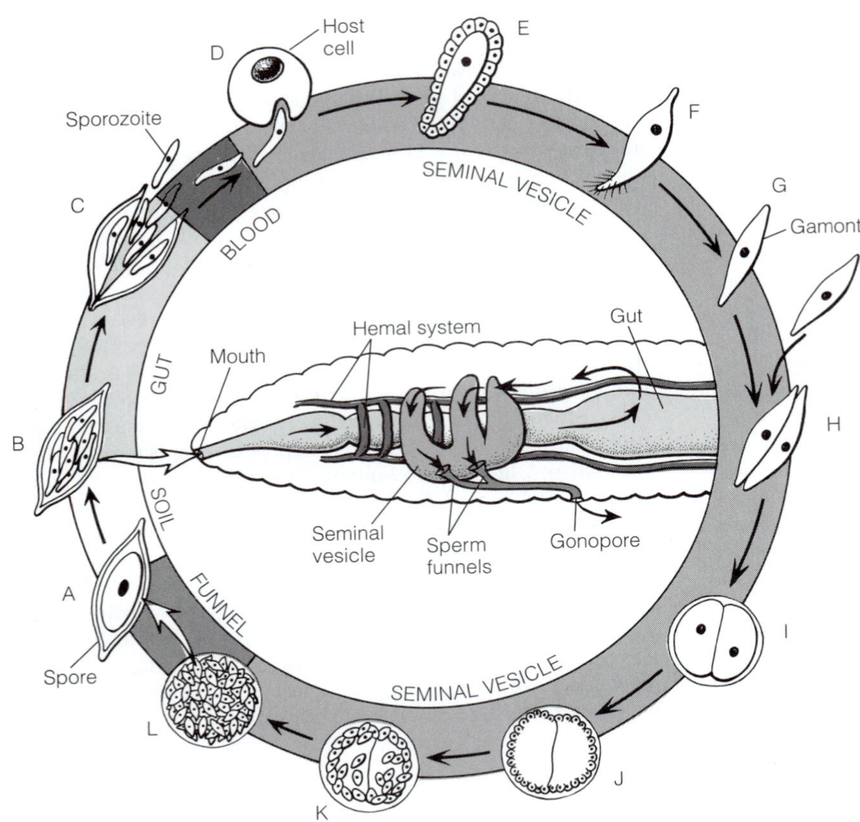

FIGURE 3-25 Alveolata: Apicomplexa. Life cycle of the gregarine *Monocystis lumbrici*, a parasite of earthworm seminal vesicles. **A,** Spore containing a 2N zygote, which undergoes meiosis and then mitosis to generate N sporozoites. **B,** Sporozoites in the spore. **C,** Sporozoites emerge from the spore in the gizzard. **D,** Sporozoite enters sperm-forming cell in the wall of the seminal vesicle. **E,** Sporozoite grows at the expense of the developing spermatocytes (small cells). **F,** Sporozoite enters the cavity of the seminal vesicle bearing remnants (tails) of aborted host sperm and transforms into a gamont. **G,** Gamonts pair. **H,** Paired gamonts. **I–K,** Encysted gamonts mitotically produce micro- and macrogametes. **L,** Gamete fusion produces zygotes, each one enclosed in a spore. *(Modified and redrawn from Janovy, J., and Roberts, L. S. 2000. Foundations of Parasitology. 6th Ed. McGraw-Hill Co., NY, 688 pp.)*

sporozoites that penetrate into the circulatory system, eventually entering the seminal vesicles. Here they penetrate and enter sperm-forming cells in the vesicle, wall, parasitizing them at the expense of the developing spermatocytes. The enlarged sporozoites then emerge from the host cells, enter the cavity of the vesicle, and transform into gamonts (trophozoites) approximately 200 μm in length. Male and female gamonts attach to the funnels of the worm's sperm ducts, pair, and encyst. Within the cyst, multiple gametes of each sex are produced. Each gamete-pair fuses to form a zygote that becomes encapsulated as a spore. Eventually, eight sporozoites are generated in each spore. Either the cyst or liberated spores exit the host's sperm ducts and are deposited in the soil where they await a feeding worm, the next host.

Other gregarines are extracellular or intracellular parasites of the gut and other organs of invertebrates, especially annelids and insects. Some reach 10 mm in length. The body of a feeding-stage gregarine (trophozoite) is elongate and the anterior part sometimes bears hooks, one or more suckers, or a simple filament or knob for anchoring the parasite into the host's cells.

Compared with gregarines, the malaria-causing *Plasmodium* and relatives (Hematozoea and Coccidia) are small cells, and sexual reproduction typically occurs *within* a host cell. For a given species, there may be only one host, as in gregarines, but many require two hosts to complete the life cycle.

These parasites add one or more rounds of multiple fission (schizogony) to the basic life cycle described above (Fig. 3-26).

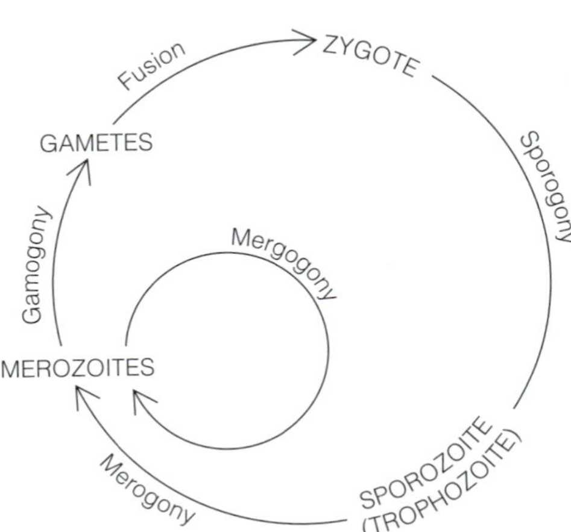

FIGURE 3-26 Alveolata: Apicomplexa. Life cycle of coccidian and haematozoen sporozoans. All stages are haploid except the zygote, which undergoes meiosis in the formation of spores (sporogony). The ability of merozoites to produce more merozoites (merogony) constitutes a clonal cycle within the sexual life cycle. *(From Levine, N. D. 1985. Phylum Apicomplexa. In Lee, J. J., et al. (Eds.): Illustrated Guide to the Protozoa. Society for Protozoology, Lawerence, KS. p. 325)*

Each of these additional rounds, called **merogony,** results in the production of motile, reinfective **merozoites.** The typical life cycle includes a sporozoite that infects a host cell, grows, and transforms into an ameboid trophozoite. The trophozoite undergoes merogony to form merozoites, each of which resembles a sporozoite. The merozoites infect other host cells in which they undergo another round of merogony or transform into gamonts, initiating gamogony. Each female gamont transforms into one macrogamete, but a male gamont, via multiple fission, produces many biflagellated microgametes. After fertilization, the zygote undergoes sporogony to produce sporozoites, which are encapsulated as an oocyst. This encysted zygote undergoes meiosis, then mitosis, to form several encapsulated spores. Later, sporozoites differentiate within each of the spores.

The most notorious hematozoeans are four species of *Plasmodium* that cause malaria, one of the worst scourges of humankind. Originally restricted to the Old World tropics, malaria was introduced into the New World by European colonists. Currently, about 300 million people (1 in 50) worldwide are believed to be infected each year, and the annual death rate is about 1% of those infected. Left untreated, the disease can be long-lasting, debilitating, and fatal.

Malaria has played a major but often unrecognized role in human history. The name means literally "bad air," because originally the disease was thought to be caused by the fetid air of swamps and marshes. Although malaria had been recognized since ancient times, the causative agent was not discovered until 1880, when Louis Laveran, a physician with the French army in North Africa, identified the parasite *Plasmodium* in the blood cells of a malarial patient. In 1887, Ronald Ross, a physician in the British army in India, determined that a mosquito was the vector.

The malarial parasite is introduced into a human host by the bite of *Anopheles* mosquitoes, which inject saliva and sporozoites into the capillaries of the skin (Fig. 3-27). The sporozoite is carried by the bloodstream to the liver, where it invades a liver cell and becomes a feeding trophozoite. After further development, the trophozoites reproduce clonally by merogony to form thousands of merozoites. These merozoites reinvade host liver cells and undergo another round of merogony. After a week or so, merozoites leave the liver cells and invade red blood cells. Within the red blood cell the merozoites transform into trophozoites, which increase in size and again undergo merogony to form yet more merozoites that reinvade other red cells. After a few days, merozoite release occurs in discrete pulses as their developmental cycles become synchronized. The periodic release of the merozoites, along with cell fragments and metabolic byproducts, causes chills and fever—the typical symptoms of malaria. Serious damage results from the blocking of capillaries by infected and less pliable red blood cells. While in the host's red cells, the trophozoites phagocytose protein (hemoglobin) at their micropores.

Eventually, some of the merozoites transform into gamonts (gametocytes) within the red blood cell, but these do not unite in pairs. Instead, each separately produces gametes only after being ingested by the mosquito. Once the mosquito imbibes infected blood, the gamonts are released from the red blood cells and produce gametes in the gut lumen. After fertilization, the zygote transforms directly into a motile cell (with apical

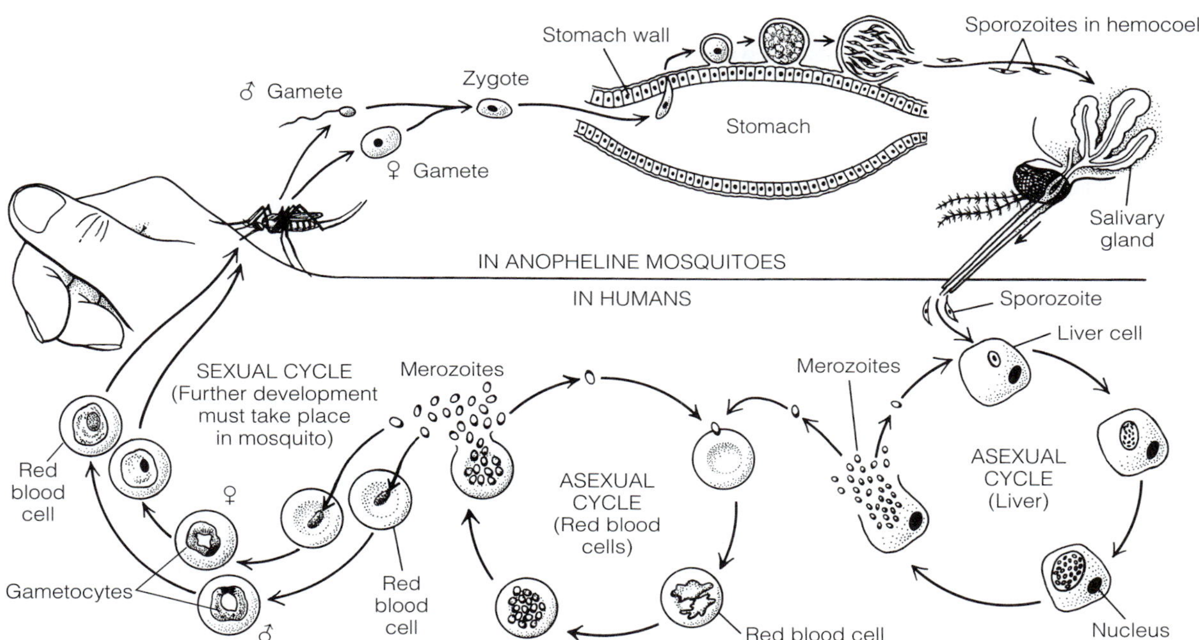

FIGURE 3-27 Alveolata: Apicomplexa. Malaria: the life cycle of *Plasmodium* in mosquito and human. Reinvasion of liver cells in humans, as shown in this figure, does not occur in *Plasmodium falciparum*. *(Redrawn and modified from Blacklock and Southwell)*

complex) that penetrates and encysts in the gut wall. Sporogony within the cyst eventually results in the release of thousands of sporozoites into the mosquito's hemocoel. The sporozoites migrate into the insect's salivary glands, from which they will be injected into the next victim of the mosquito's bite.

Related parasites (Coccidia) cause diseases in domesticated animals. Species of the genus *Eimeria,* for example, affect chickens, turkeys, pigs, sheep, and cattle.

Two other taxa of spore-forming parasites, the Microsporidia and the Myxosporidia, were formerly considered to be close relatives of the apicomplexans. Now, the microsporidians are classified either with the fungi or placed near the base of the eukaryotes because they lack, and presumably never had, flagella, mitochondria, and Golgi bodies. The myxosporidians, currently called myxozoans, are multicellular organisms with cnidae (stinging capsules) that are now classified with the metazoan taxon Cnidaria (corals, anemones, and jellyfishes) and are described in Chapter 7.

DIVERSITY OF APICOMPLEXA (SPOROZOA)

Gregarinea^{C}: Life cycle with one host; multiple fission by both male and female gamonts; constriction separates gamont body into anterior protomerite and posterior deutomerite, with epimerite (such as hooks) at tip of protomerite; gamont cells unite (syzygy) before encystment and move by gliding; most stages are extracellular parasites of echinoderms, molluscs, annelids, and especially arthropods. Species identification based on epimerite structure. *Gregarina, Monocystis.*

Coccidia^{C}: Each macrogamont forms but one macrogamete; gamonts encyst; most species are intracellular parasites of invertebrates and vertebrates in one or two hosts. *Cryptosporidium, Eimeria* (in poultry), *Haemogregarina, Toxoplasma* (causes toxoplasmosis in cats and sometimes in humans).

Hematozoea^{C}: Blood parasites that alternate between vertebrate (intermediate) and arthropod (final) hosts; sporozoites infect vertebrates and motile micro- and macrogametes are transferred to arthropods; gamonts do not pair or encyst. *Haemoproteus, Leucocytozoon* (causes turkey malaria), *Plasmodium* (causes malaria).

AMEBOID PROTOZOA

The ameboid protozoans, traditionally placed in the taxon Sarcodina (subphylum), have flowing extensions of the body called pseudopodia. Included here are the classroom amebas as well as many other marine, freshwater, and terrestrial taxa with pseudopodia. Pseudopodia are used for capturing prey and, in benthic taxa, also for locomotion. Ameboid movement may be a primitive character of eukaryotic cells. If so, it is a symplesiomorphy of the amoeboid protozoans that cannot unite them in a monophyletic taxon. Within this paraphyletic or, more likely, polyphyletic assemblage, some subgroups are monophyletic and others are not. Taxon names with a superscripted notation are probably monophyletic, while stand-alone names are not.

Ameboid protozoa are mostly asymmetric, but some with skeletons exhibit radial symmetry. In general, small-bodied species have one nucleus whereas large species have many and, in one taxon (forams), heterokaryotic nuclei occur, as in ciliates. Ameboid protozoa have relatively few specialized organelles and in this respect are among the simplest protozoa. The skeletal structures that occur in the majority of species, however, reach a complexity and beauty that is surpassed by few other organisms. The three principal groups of ameboid protozoa are the amebas (Caryoblasta, Heterolobosa, and Amoebozoa), the forams (Foraminiferea), and the actinopods (Radiolaria, Acantharea, and Heliozoa).

Amebas

Amebas (or amebae) may be naked or enclosed in a test. The naked amebas, which include *Amoeba,* live in the sea, in fresh water, and in the water film around soil particles (Fig. 3-28A). The shape, although constantly changing, is characteristic in different species. Some giants, such as *Pelomyxa palustris* or *Chaos carolinense,* can be 5 mm in length and are multinucleated cells. The cytoplasm in amebas is divided into a stiff, clear, external ectoplasm and a more fluid internal endoplasm (Fig. 3-28A). The pseudopodia adopt one of two general forms. **Lobopodia,** which are typical of many amebas, are wide and rounded with blunt tips (Fig. 3-28A,B). They are commonly tubular and composed of both ectoplasm and endoplasm. **Filopodia,** which occur in many small amebas, are slender, clear, and sometimes branched, but the branches do not interjoin extensively to form nets (Fig. 3-28C).

In shelled (testate) amebas, which are largely inhabitants of fresh water, damp soil, and mosses, either a radial or bilateral extracellular test is secreted by the cytoplasm. The test is an organic matrix to which secreted siliceous elements or foreign materials are attached. The ameba is attached by cytoplasmic strands to the inner wall of the test. Pseudopodia, which may be either lobopods or filopods, protrude through an opening in the test. In *Arcella* (Fig. 3-29A,B), one of the most common freshwater amebas, the brown or straw-colored protein test has the shape of a squat dome with the aperture in the underside center. In *Euglypha* the organic test bears overlapping siliceous scales (Fig. 3-29C). *Difflugia* has a test composed of mineral particles that are first endocytosed, then exocytosed and embedded in the organic matrix (Fig. 3-29D).

Marine species commonly lack contractile vacuoles; freshwater species have one to several, and, at least in larger naked species, contractile vacuoles form and fill anteriorly and discharge at the trailing end of the cell.

DIVERSITY OF AMEBAS

Caryoblasta^{P}: Cell has one to several nonmotile flagella, each with a basal body from which MTs radiate and surround a nucleus; mitochondria and Golgi bodies are absent, respiratory organelles are endosymbiotic bacteria. *Pelomyxa palustris* is 1 to 5 mm in diameter with hundreds of nuclei and several flagella; *Mastigamoeba* and *Mastigella* have one flagellum and one nucleus. In micro-oxic freshwater sediments. Some systematists regard caryoblasts as premitochondrial eukaryotes, and thus among the most primitive of living protists.

Heterolobosa^{P}: Cells have an inducible flagellated stage with two to four functional flagella; lobopodia seem to erupt during ameboid movement. Encystment occurs under adverse

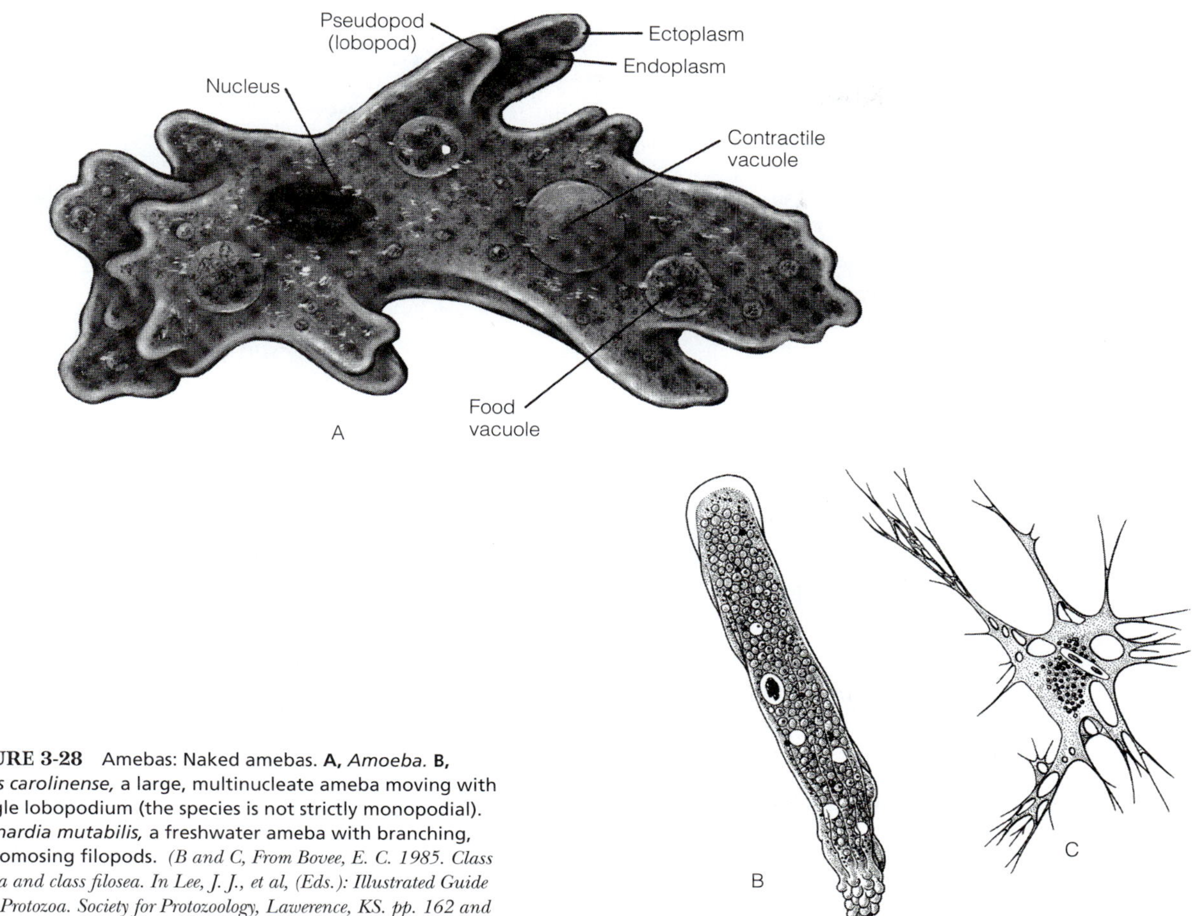

FIGURE 3-28 Amebas: Naked amebas. **A,** *Amoeba.* **B,** *Chaos carolinense,* a large, multinucleate ameba moving with a single lobopodium (the species is not strictly monopodial). **C,** *Penardia mutabilis,* a freshwater ameba with branching, anastomosing filopods. *(B and C, From Bovee, E. C. 1985. Class lobosea and class filosea. In Lee, J. J., et al, (Eds.): Illustrated Guide to the Protozoa. Society for Protozoology, Lawerence, KS. pp. 162 and 230)*

conditions. Freshwater and marine sediments. *Naegleria* (two flagella; cause of primary amebic meningoencephalitis), *Tetramitus* (four flagella). Contemporary systematists include this taxon with the flagellates.

Amoebozoa: Polyphyletic taxon: cells with pseudopodia that lack MTs. MTs are associated only with the mitotic spindle; flagella and centrioles are absent; freshwater, marine, terrestrial, and symbiotic habitats. "Lobosea," with lobopodia, includes the naked (atestate) amebas—*Acanthamoeba, Amoeba, Chaos, Entamoeba, Vannella*—and the testate amebas—*Arcella, Difflugia.* "Filosea," with rapidly forming filopodia, includes testate and atestate species: *Euglypha, Gromia, Vampyrella.*

Foraminiferea[P]

The large taxon Foraminiferea (forams or foraminifers) is primarily marine. The countless filiform pseudopodia, called **reticulopodia,** actively branch and interconnect (anastomose) to form a complex threadlike mesh, usually known as a **reticulopodial network** (Fig. 3-30B). Each reticulopodium has an axis of microtubules that shuttles vesicles bidirectionally to and from the cell body. The abundant vesicles confer a granular appearance on the reticulopods. Locomotion in creeping forams, such as *Allogromia,* results from extension of the reticulopodial network, anchorage on the substratum, and retraction of the net, which pulls the cell body forward. Movement of the reticulopodial net involves lengthening and shortening of the axial microtubules.

Forams construct an extracellular test of organic material, cemented foreign mineral particles, or calcium carbonate secreted onto the organic matrix. Calcareous tests are common and well preserved in the fossil record; 40,000 of the 45,000 described species of forams are fossil species. The largest forams, members of the deep-sea Xenophyophorea, are several centimeters in diameter (the size of a clenched fist).

A few foram species occupy a test of one chamber, but most have multichambered calcified tests. Multichambered forams begin life in a single chamber, but as the organism increases in size, reticulopods extend from the aperture of the original chamber, arrange themselves in the appropriate shape, and secrete the new chamber. This process continues throughout

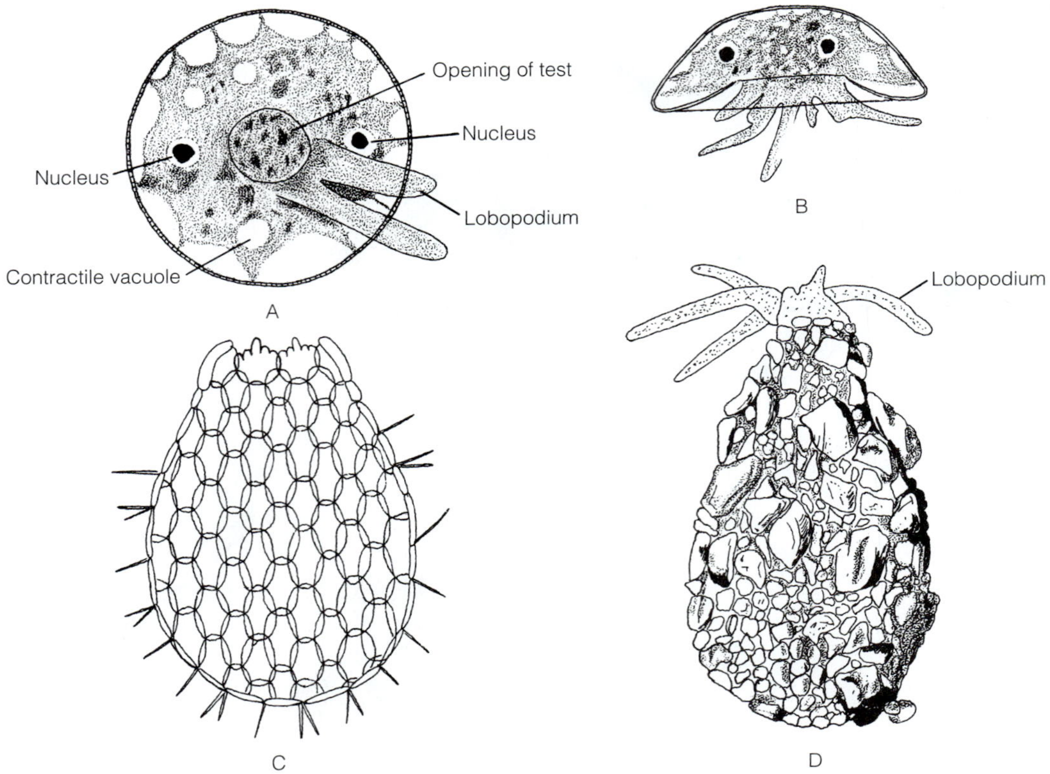

FIGURE 3-29 Amebas: Testate amebas. **A,** *Arcella vulgaris,* apical view. **B,** Side view. **C,** Test of *Euglypha strigosa,* composed of siliceous scales and spines. **D,** *Difflugia oblonga* with test of gathered mineral particles. *(A, B, and D, After Deflandre, G. 1953. In Grassé, P. Traité de Zoologie, Masson and Co., Paris. Vol. I, pt II. C, After Wailes)*

life and results in a series of many chambers, each of which may be larger than the preceding one. Because the addition of new chambers follows a symmetrical pattern, the tests have a distinctive shape and arrangement of chambers (Fig. 3-30).

The entire test is filled by one cell that extends from one chamber to the next. An extension of the cell from the aperture also creates a thin layer outside the test. Reticulopodia may be restricted to the aperture or they may arise from the test layer (Fig. 3-30B). In some species they emerge through test pores, but others have blind pores that do not penetrate the test.

Forams cast their extensive reticulopodial nets widely over surfaces, into the water, or between grains of sand in search of food. The net is dynamic, with its shape and extent changing constantly as reticulopods shorten, lengthen, fuse, and arise or regress spontaneously anywhere in the net. No crevice is too small to be probed by the myriad tentacles of the net. Once a diatom, bacterium, or other small prey is contacted, it adheres to a reticulopod and is transported along it, as if on an escalator, to the cell body waiting like an orb spider at the net's hub. On reaching the cell body, food is ingested by phagocytosis.

Most forams are benthic, but species of *Globigerina* and related genera are common planktonic forms. The chambers of these species are spherical, but spirally arranged (Fig. 3-30B,D). Planktonic forams have more delicate tests than do benthic species and the tests commonly bear spines, which slow the rate of sinking. The spines are so long in some species that the foram is visible to the naked eye and can be scooped with a jar by a scuba diver. A few forams are sessile. *Homotrema rubrum* forms large, red, calcareous tubercles about the size of a wart on the underside of coral heads. The pink sands of the beaches of Bermuda result from the accumulation of *Homotrema* tests.

Several forams harbor an unusual diversity of endosymbiotic photosynthetic protists—chlorophytes, diatoms, dinoflagellates, or unicellular red algae, depending on the foram. One taxon harboring zooxanthellae, the Soritidae (which includes mermaid's pennies), averages about 1 cm in diameter and is common on coral reefs.

Forams first appeared in the Cambrian period and have fossilized throughout geological history. Extensive accumulations of tests occurred during the Mesozoic and early Cenozoic eras and contributed to the formation of great limestone and chalk deposits in different parts of the world. The White Cliffs of Dover in England and the quarries that provided stone for the Egyptian pyramids are composed predominantly of foram tests.

Their widespread fossil occurrence and their long geologic history make forams useful as index fossils. Because sedimentary rock containing the same taxa of forams was deposited at the same time, geologists use these index species to identify oil-containing strata. In some species of *Globigerina,* the coiling direction of the test is influenced by water temperature: Left-hand (sinistral) coiling is associated with low temperatures

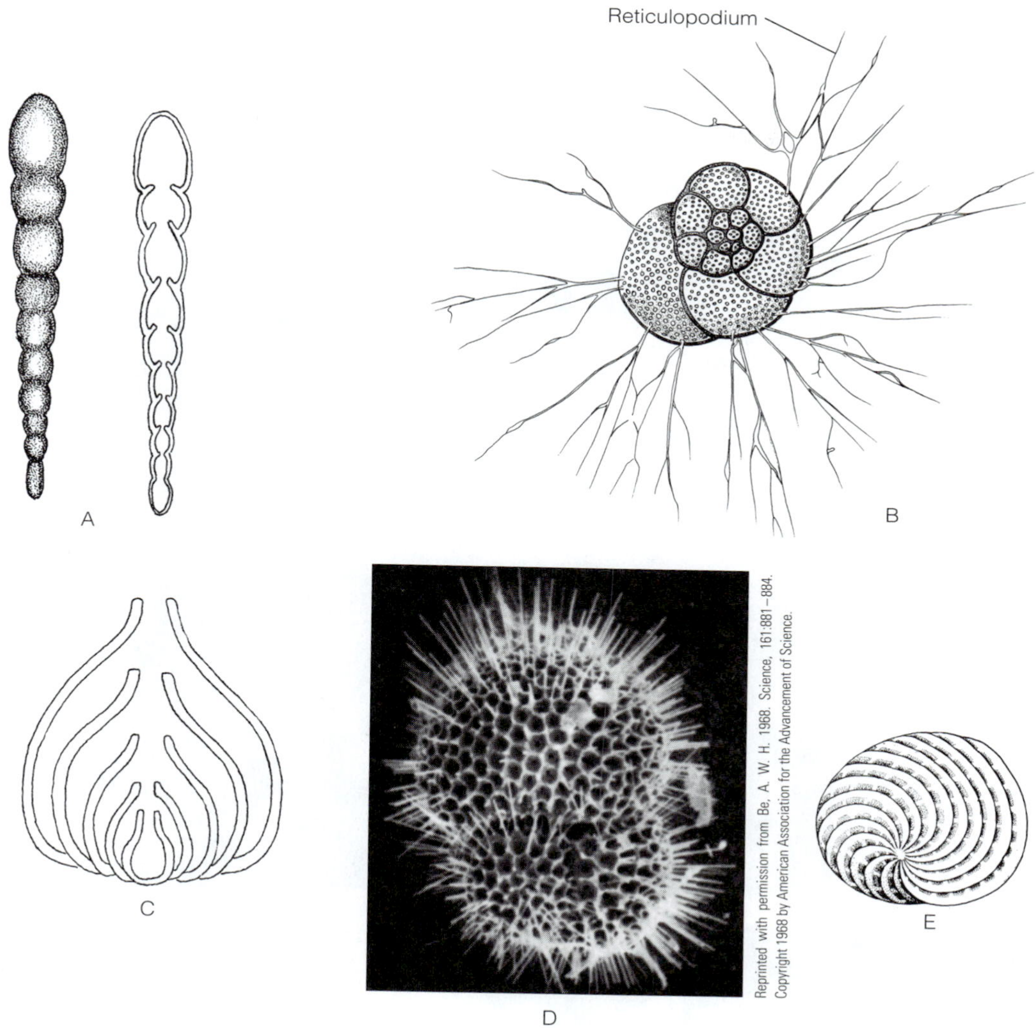

FIGURE 3-30 Foraminiferea. **A,** Test of the foram *Rheophax nodulosa,* entire and in section. **B,** Living *Globigerina bulloides.* **C,** Test of an ellipsoidinid foram, in section. **D,** Cleaned test of *Globigerinoides sacculifer,* a tropical planktonic foram with spines. **E,** *Archaias* sp., a common benthic foram of shallow tropical seas. *(A, After Brady. B, drawn form a photograph in Grell, K. G. 1973. Protozoology. Springer-Verlag, Berlin, p. 285)*

whereas right-hand (dextral) coiling is associated with high temperatures. Thus, the coiling direction of certain fossils provides a record of past cold and warm periods. The varying ratios of oxygen isotopes in foram tests from deep-sea sediments also provide clues about global temperature change and glacial ice accumulation.

DIVERSITY OF FORAMINIFEREA

Until recently, Foraminiferea was included in Granuloreticulosa, a phylum-level taxon of three major subgroups—Athalamida, Monothalamida, and Foraminiferida—all sharing the character reticulopodia. In contrast to forams, however, the other two taxa lack an alternation of generations in their life cycle. Athalamids lack a test and occur in fresh water; monothalamids have an organic or calcareous test of one chamber and occur primarily in fresh water, although some species are marine. Recent molecular studies suggest that athalamids are forams modified for life in fresh water. Here we consider only taxa traditionally considered to be forams.

Allogromiina[C]: Organic test is flexible and sometimes has attached foreign matter. *Iridia, Myxotheca, Nemogullmia.*

Textulariina[C]: Organic test made rigid by adding foreign particles. *Allogromia, Ammodiscus, Astrorhiza, Clavulina, Textularia.*

Miliolina[C]: Calcareous test resembles porcelain. *Amphisorus* (mermaid's penny), *Pyrgo* (ooze former), *Quinqueloculina, Sorites.*

Rotaliina[C]: Calcareous test is glassy (hyaline) and has pores. *Bulimina, Discorbis, Globigerinoides* (planktonic), *Homotrema, Lagena, Marginulina, Rotaliella.*

Actinopoda[P]

The actinopods are primarily spherical, planktonic cells with long, stiff, needlelike pseudopodia called **axopodia** that radiate outward like spikes on a mace (Fig. 3-31, 3-32, 3-33). The axis of each axopodium contains a supportive axial rod that is a bundle of microtubules originating in the cell body. Bidirectional shuttling of vesicles occurs in axopodia, as in reticulopodia, and they can shorten rapidly after contacting

Neg./Trans. No. 318863. Courtesy Department of Library Services, American Museum of Natural History

From Cachon, J. and Cachon, M. 1985. Class Polycystinea. In Lee, J. J., et al. (Eds.): Illustrated Guide to the Protozoa. Society for Protozoology, Lawerence, KS, pp. 288 and 296

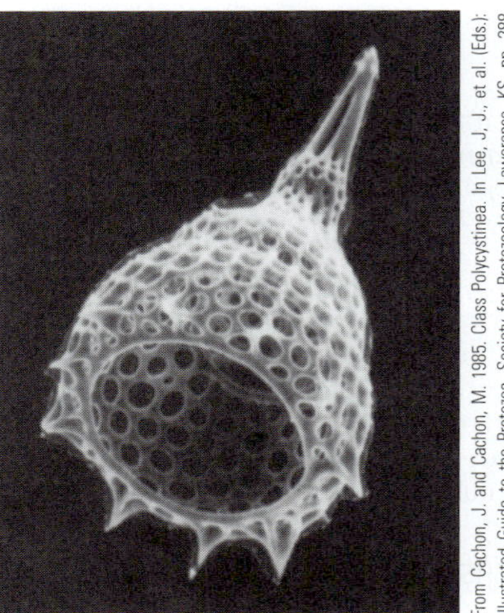

From Cachon, J. and Cachon, M. 1985. Class Polycystinea. In Lee, J. J., et al. (Eds.): Illustrated Guide to the Protozoa. Society for Protozoology, Lawerence, KS, pp. 288 and 296

FIGURE 3-31 Radiolaria. **A,** Glass model of a colonial radiolarian, *Trypanosphaera transformata.* Note the radiating axopodia, thick vacuolated cortex, and medulla overlaid by a skeletal grid. **B,** Spherical siliceous test of *Hexacontium.* **C,** Conical test of *Lamprocyclas.*

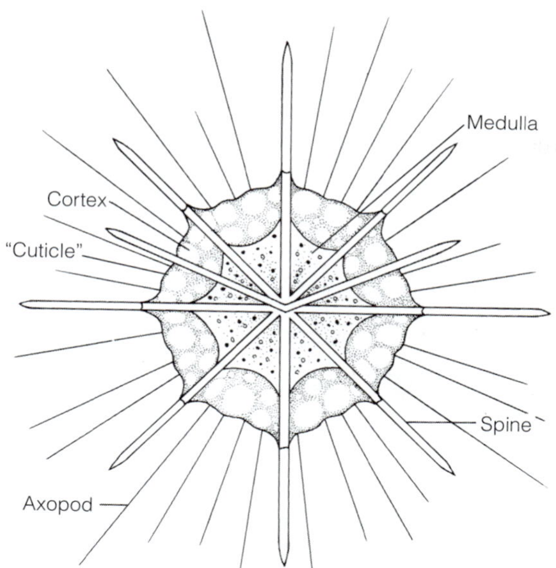

FIGURE 3-32 Acantharea. *Acanthometra,* with mineralized test of radiating strontium sulfate spines. Note extracellular cuticle enclosing cell and attached to spines. *(From Farmer, J. N. 1980. The Protozoa: Introduction to Protozoology. C. V. Mosby Co., St. Louis, p. 353)*

and adhering to prey. The axopods are used for prey capture, flotation, locomotion, and attachment to surfaces.

Most actinopods are enclosed in a perforated, organic, extracellular test (central capsule). The actinopodia and other types of pseudopodia emerge through the test pores. Unlike the pseudopodia of forams and testate amebas, however, those of actinopods permanently extend over and beyond the test and do not withdraw into it. The non-actinopodia pseudopods, which are often reticulopodia, filopodia, or vacuolated pseudopodia (for flotation), form a thick shroud over the test. This pseudopodial shroud is called the **cortex** (calymma, or ectoplasm) and the cell body is the **medulla** (central capsule, or endoplasm; Fig. 3-32, 3-33A, 3-34, 3-35). The perforated test (capsular membrane), when present, separates the medulla from the cortex. If the cortex is experimentally removed, a new one is regenerated from the medulla.

Medulla and cortex compartmentalize the actinopod cell. The cortex encounters the external environment, captures and digests prey, conveys nutrients to the medulla, provides flotation, and often bears photosynthetic endosymbionts. The medulla houses the nucleus and the synthetic machinery of the cell, as well as nutrients held in storage.

In addition to the organic test, most actinopods also have a mineral skeleton, usually made of silica, that may be intracellular, extracellular, or both, according to the taxon. In heliozoans, the siliceous skeleton is apparently restricted to the cortex, whereas in acanthareans and some radiolarians, it occurs in the medulla and cortex (Fig. 3-35).

RADIOLARIA[C]

With a siliceous test that looks like a crystal starburst, radiolarians are among the most elegant protozoans (Fig. 3-31). Entirely marine and primarily planktonic, radiolarians are relatively large protozoa: A few solitary species are millimeters in diameter, and some colonial species attain a length of up to 20 cm *(Collozoum)*. The radiolarian cell is usually spherical and divided distinctly into medulla and cortex by a perforated organic test (Fig. 3-31A, 3-35C). Highly specialized

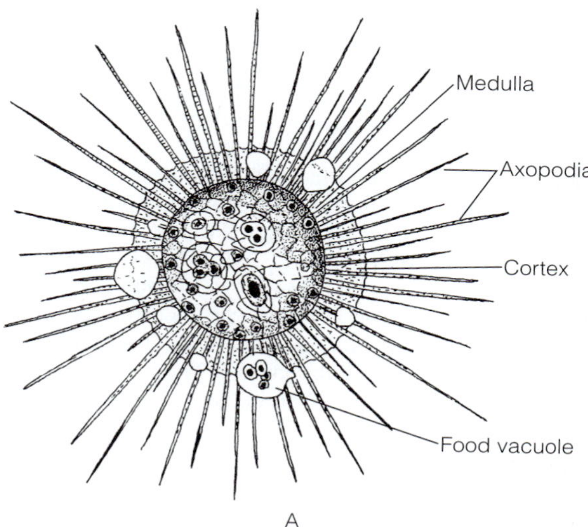

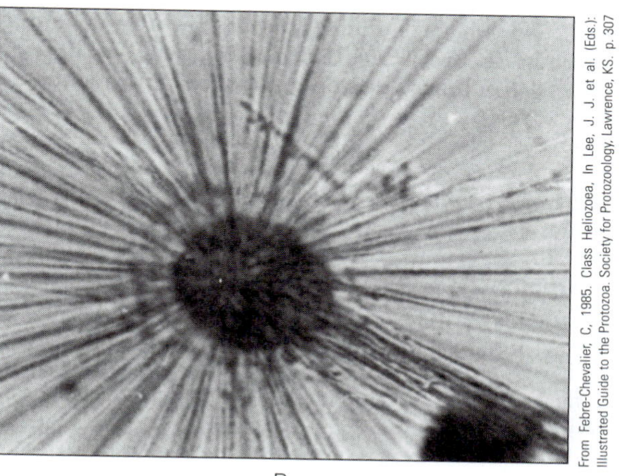

From Febre-Chevalier, C. 1985. Class Heliozoea. In Lee, J. J. et al. (Eds.): Illustrated Guide to the Protozoa. Society for Protozoology, Lawrence, KS. p. 307

FIGURE 3-33 Heliozoa: Atestate heliozoans. **A,** A multinucleate heliozoan, *Actinosphaerium eichorni.* **B,** A living, sessile, stalked heliozoan. Stalk extends toward the lower right corner. Medulla, cortex, and axopodia are visible. *(A, After Doflein)*

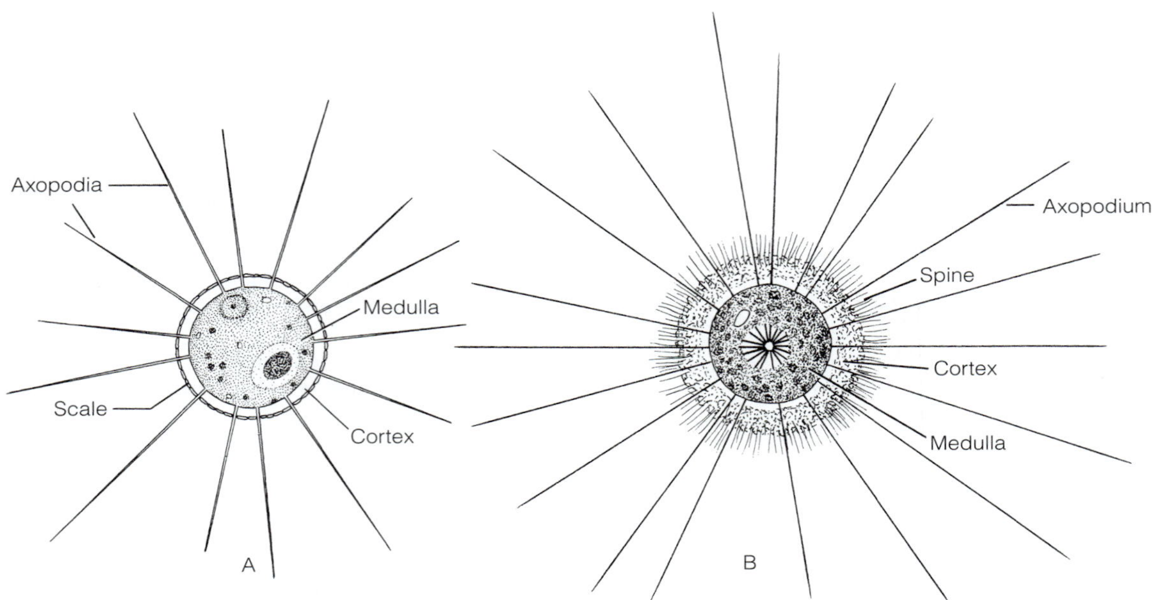

FIGURE 3-34 Heliozoa: Testate heliozoans. **A,** *Pinaciophora fluviatilis* with a test of scales. **B,** *Heterophrys myriopoda* with cuticle-bearing spines. *(A and B, After Penard from Hall)*

perforations (fusules) in the test allow for the passage of the axopodia as well as vacuolated filopodia and reticulopodia that form the cortex. The cortex, sometimes highly vacoluated (Fig. 3-31A), functions in flotation, prey capture, and intracellular digestion as well as often bearing symbiotic dinoflagellates (zooxanthellae) or other photosynthetic protists. The medulla, which is often vacuolated peripherally, contains one to many nuclei and nutritional reserves, such as lipid drops (also used for buoyancy).

In addition to the organic test, radiolarians also have an intracellular mineral test of silica (SiO_2) synthesized in the reticulopodial network of the cortex and sometimes also in the medulla. The siliceous test itself is a network of slender, interconnected rods that resembles a geodesic dome, often with radiating spines (Fig. 3-31). In its more complex forms, it can consist of two or three interconnected concentric spheres of striking symmetry and beauty. In some radiolarians the siliceous skeleton is rudimentary or even absent, but if absent, the organic test is still present.

The planktonic radiolarians display a distinct vertical stratification from the ocean surface down to 5000 m depths. A testimony to the enormous population densities of planktonic radiolarians is provided by the thick accumulation of their tests, after death, on many parts of the ocean floor. In some of these areas, where tests account for 30% or more of sediment composition, the sediment is called **radiolarian ooze.** Similarly, **foraminiferan ooze,** from accumulated foram tests, characterizes other parts of the ocean floor. At depths below 4000 m, however, the great pressure tends to dissolve the calcareous foram tests.

Diversity of Radiolaria

Polycystinea[O]: Taxon contains the majority of familiar radiolarians. All have perforated siliceous skeletons and are solitary and colonial species, 30 μm to 2 mm in size. *Collozoum, Eucoronis, Thallasicola.*

Phaeodarea[O]: Taxon of deep-sea radiolarians having a siliceous test with hollow spines and incorporated organic matter. Central capsule has three openings, one oral and two for axopods. Yellow-brown pigment mass (phaeodium) near the oral opening. *Astracantha, Coelodendrum, Phaeodina.*

ACANTHAREA[C]

Acanthareans are planktonic marine actinopods that superficially resemble radiolarians (Fig. 3-32). The cell is a radially symmetric sphere from which project slender axopodia and 10 to 20 stout skeletal spines. The intracellular spines unite in the center of the medulla, radiate outward, and project beyond the periphery of the cell. Myonemes attach to the spine bases and, on contraction, expand the cortex to the tips of the spines. Acantharean spines are composed of strontium sulfate and not the silica of radiolarians. Strontium sulfate is soluble in seawater and, after cell death, the skeleton soon dissolves and leaves no trace either as a bottom ooze or in the fossil record. The cell is divided by a perforated organic test into a medulla and vacuolated cortex. The cortex is overlaid with a flexible extracellular cuticle. The medulla contains several nuclei and, often, zooxanthellae.

HELIOZOA[C]

The spherical actinopods called heliozoans (sun animalcules) occur in the sea, in still bodies of fresh water, and in mosses. In aquatic habitats, they may be floating or, more commonly, located in bottom debris. Some benthic species are stalked (Fig. 3-33B). Numerous slender axopodia radiate stiffly from the

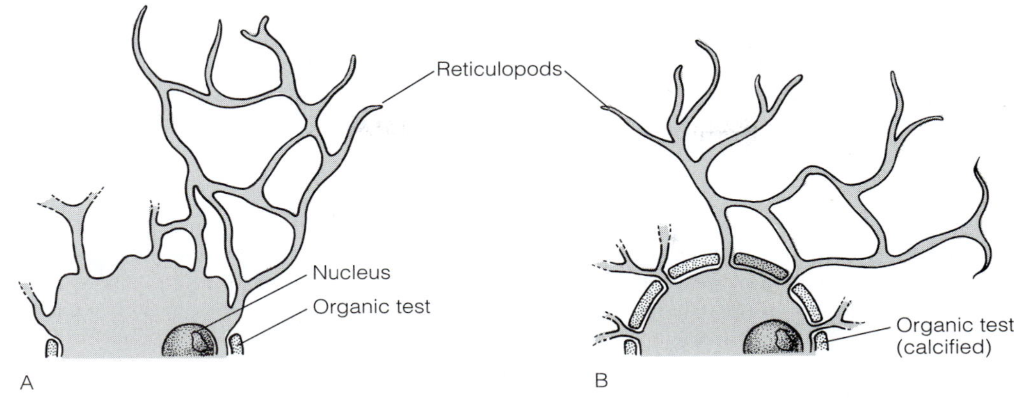

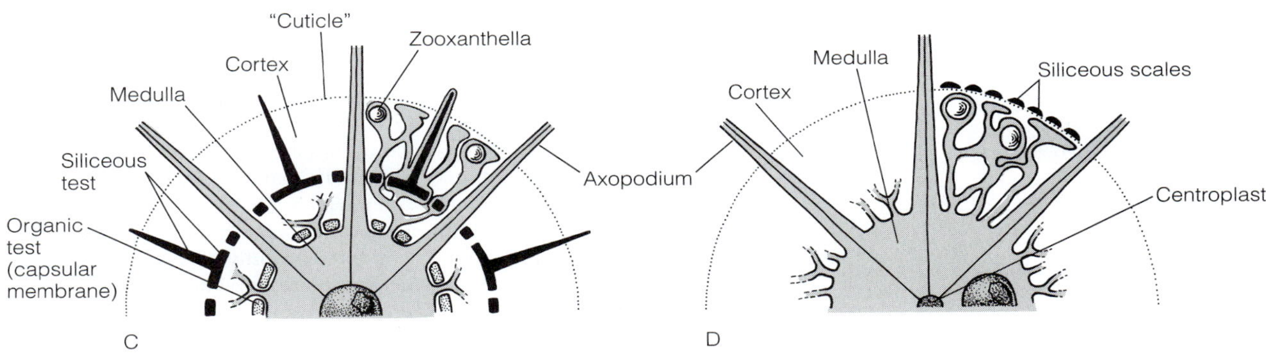

FIGURE 3-35 Anatomy of forams, radiolarians, and heliozoans. **A,** A single-chambered foram; **B,** A multichambered foram with test pores; **C,** A radiolarian; **D,** A heliozoan. **B–D,** Partly hypothetical.

surface of the cell (Fig. 3-33A,B), but can shorten or even "melt" as the axial microtubules depolymerize. Some heliozoan species have long, delicate filopodia in addition to the axopodia.

A heliozoan cell, like that of other actinopods, is divided into a cortex of vacuolated filopods and a medulla, but an organic test is absent between the two regions (Fig. 3-33A). Some heliozoans, however, secrete an extracellular cuticle (gel) over the surface of the cortex, as do acanthareans and perhaps some radiolarians. Discrete skeletal structures are attached to this cuticle in some species. These may be organic or siliceous spicules or incorporated foreign materials. The diverse siliceous spicules may be scales (Fig. 3-34A), spines that radiate from the cell surface (Fig. 3-34B), or structures of other shapes. Contractile vacuoles occur in the cortex of freshwater species. The medulla contains one to many nuclei, the bases of the axial rods, and sometimes symbiotic green algae (zoochlorellae).

Diversity of Heliozoa

Actinophryida[O]: Axopodial MTs originate on the nuclear membrane and form two intertwined spirals; uni- and multinucleate species are capable of encystment; marine, freshwater, terrestrial (peat). *Actinophrys, Actinosphaerium, Camptonema.*

Desmothoracida[O]: Sessile, mostly stalked species; irregularly arranged axopod MTs; filopodia are present. *Clathrulina, Hedriocystis, Orbulinella.*

Ciliophryida[O]: Similar in form to actinophryids, but adult bears a single flagellum with pinnate mastigonemes; axopods with few MTs. *Actinomonas, Ciliophrys, Pteridomonas.*

Taxopodida[O]: Bilaterally symmetric with stout siliceous spines in rosettes; axopodal MTs in a hexagonal pattern; marine. *Sticholonche.*

Centrohelida[O]: Numerous slender and long axopods arise from a central point (centroplast); axopods bear extrusomes used in prey capture; often have surface covering of extracellular siliceous scales or spinelets; axopod MTs in hexagonal or triangular arrays. *Acanthocystis, Gymnosphaera, Hedraiophrys, Heterophrys.*

Nutrition

Ameboid protozoa are heterotrophs. Their food consists of small organisms such as bacteria, algae, diatoms, protozoans, and even small multicellular animals, including rotifers and roundworms (nematodes). The prey is captured by pseudopodia and ingested by phagocytosis.

In amebas, pseudopodia extend around the prey, eventually enveloping it completely, or the body surface invaginates to form a food vacuole (Fig. 2-8, 2-9). In forams, heliozoans, and radiolarians, the numerous radiating pseudopodia are traps for prey. Any organism that comes in contact with them adheres and is quickly paralyzed, presumably by toxins. Enzymes released from the pseudopods predigest prey, such as small crustaceans, before they are phagocytosed by the cell. The axial rods of some heliozoans contract, drawing the prey into the cortex, or the axopods may coalesce and surround the food, forming a vacuole at the site of capture. The vacuole then moves inward. Digestion occurs in the cortex of heliozoans and radiolarians. Egestion (exocytosis) can take place at any point on the body surface, but in crawling amebas, wastes are usually released at the trailing end of the cell.

Some naked amebas are parasites. The majority of these are endoparasites in the digestive tract of annelids, insects, and vertebrates. Several species occur in the human digestive tract, but of these only *Entamoeba histolytica,* which is responsible for amebic dysentery, is ordinarily pathogenic. The life cycle of these intestinal amebas is direct, and the parasites are usually transmitted from the digestive tract of one host to that of another by means of cysts that are passed in feces.

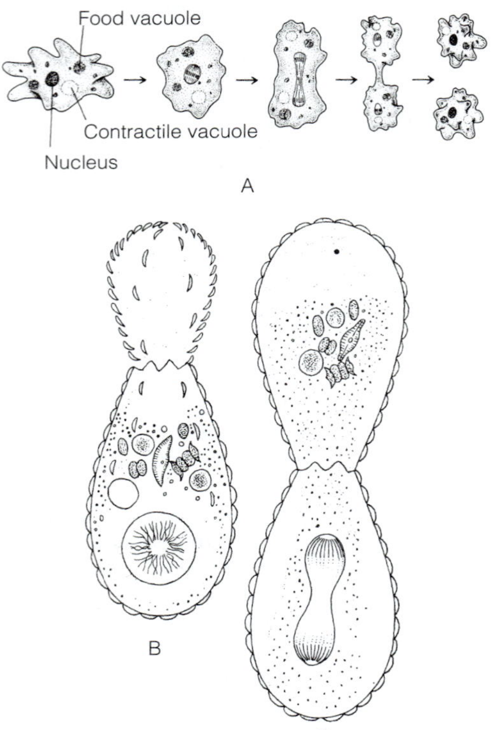

FIGURE 3-36 Amebas: Clonal reproduction. **A,** Fission in a naked ameba. **B** and **C,** Two stages in the division of *Euglypha,* a testate ameba. B, Formation of test plates on a cellular mass protruding from the aperture. **C,** Division of the nucleus. One of the daughter nuclei will move into the new cell. Note that the nuclear membrane remains intact during mitosis. *(B and C, After Sevajakov from Dogiel)*

Reproduction

Clonal reproduction in most amebas, heliozoans, and radiolarians is by binary fission (Fig. 3-36A). In amebas with a thin organic test, the test divides into two parts and each daughter cell forms a new half. When the test is dense and continuous, as in *Arcella* and *Euglypha,* an outgrowth of the cell emerges from the test aperture prior to division and secretes a new test (Fig. 3-36B). The double-test cell then divides (Fig. 3-36C). Multiple fission is common in multinucleated amebas, forams, and heliozoans. The fission products of some, such as heliozoans, may be biflagellated swarmers that disperse and transform first into amebas and then into the form of the parent. Division in the radiolarians is similar to that in the testate amebas: Either the test itself divides and each progeny cell forms the lacking half, or one offspring retains the test and the other secretes a new one.

Sexual reproduction has been observed infrequently in amebas. Among the heliozoans, sexual reproduction is known in some genera, such as *Actinosphaerium* and *Actinophrys* (diploid-dominant life cycle, Fig. 3-4C). Sexual reproduction in forams commonly involves an alternation of clonal and sexual generations (haploid-diploid codominance, Fig. 3-4D).

PHYLOGENY OF PROTOZOA

The evolutionary origin of protozoa is discussed in Chapter 2, Introduction to Protozoa. The phylogenetic relationships among the protozoan taxa have been notoriously difficult to unravel, but progress is ongoing thanks to the application of microanatomical and molecular techniques (Fig. 3-37). In general, the ameboid taxa (and some flagellates) are thought to represent primitive taxa, although their interrelationships are presently unclear. At least five monophyletic taxa have been proposed. One of these is the Euglenozoa, which includes the euglenoids and kinetoplastids. Another is the Chlorophyta, which includes the Volvocida, green algae, and multicellular green plants. Ciliates, dinoflagellates, and apicomplexans form another monophyletic taxon (Alveolata) based on their common possession of alveoli and mitochondria with tubular cristae. From the perspective of animal evolution, the most interesting monophyletic taxon is the Opisthokonta, characterized by a posterior flagellum on motile cells as well as gene-sequence similarities. Included in this taxon are the sister taxa Choanoflagellata and Metazoa (multicellular animals) as well as Fungi and perhaps Microsporidia. The only other opisthokont taxon is the former protozoan group Myxosporidia, which is now included among the metazoans as Myxozoa.

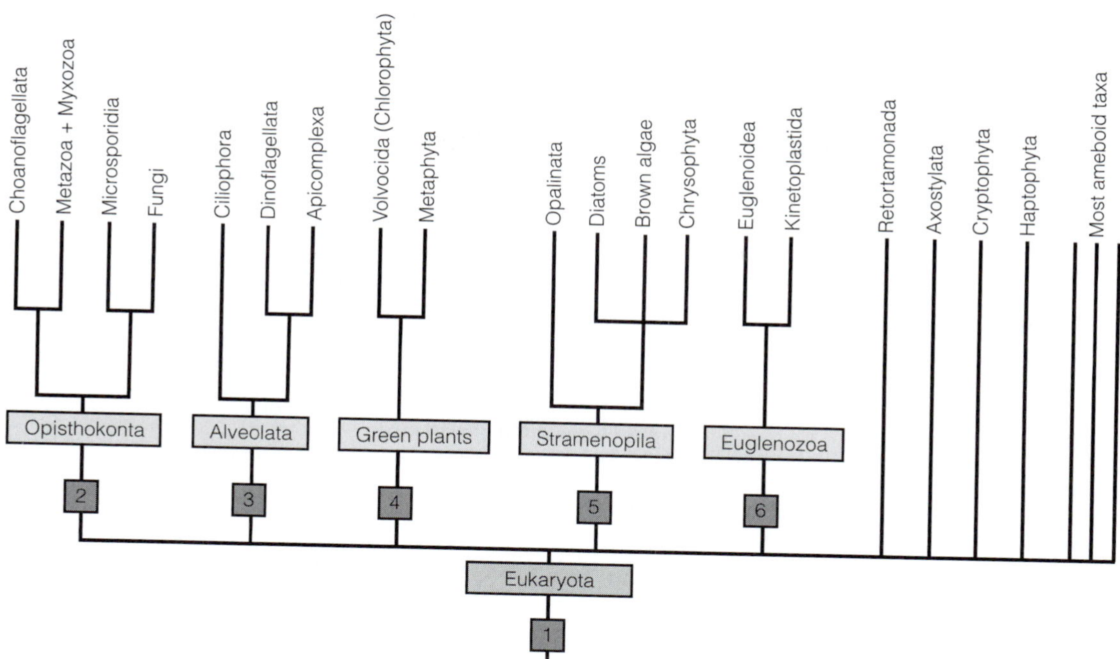

FIGURE 3-37 Phylogeny of Protozoa based on gene-sequence data and morphology. Morphological characters: **1, Eukaryota:** Nucleus is enclosed in a nuclear membrane. **2, Opisthokonta:** Locomotory flagellum is posterior on motile cells. **3, Alveolata:** Alveoli are below the cell membrane, mitochondrial cristae are tubular. **4, Green plants:** Chloroplasts are enclosed in two membranes and include chlorophylls a and b, starch is produced as a storage product, and a cellulose cell wall is present. **5, Stramenopila:** Flagella have three-part, tubular mastigonemes. **6, Euglenozoa:** One or two anterior flagella emerge from a pit, a paraflagellar rod is present, and mitochondrial cristae are discoid.

REFERENCES

GENERAL

Additional references on protozoa can be found in the References section of Chapters 1 and 2.

Anderson, O. R. 1983. Radiolaria. Springer-Verlag, Berlin. 355 pp.

Anderson, O. R. 1987. Comparative Protozoology: Ecology, Physiology, Life History. Springer-Verlag, Berlin. 482 pp.

Be, A. W. H. 1968. Shell Porosity of Recent Planktonic Foraminifera as a Climatic Index. Science 161:881–884.

Be, A. W. H. 1982. Biology of planktonic Foraminifera. Stud. Geol. 6:51–92.

Borror, A. C. 1973. Protozoa: Ciliophora. Marine Flora and Fauna of the Northeastern United States. NOAA Tech. Report NMFS Circular 378. U.S. Government Printing Office. 62 pp.

Bouck, G. B., and Ngo, H. 1996. Cortical structure and function in euglenoids with reference to trypanosomes, ciliates, and dinoflagellates. Int. Rev. Cytol. 169:267–318.

Boynton, J. E., and Small, E. B. 1984. Ciliates by the slice. Sci. Teach. Feb.:35–38.

Capriulo, G. M. (Ed.): 1990. Ecology of Marine Protozoa. Oxford University Press, 366 pp.

Coombs, G. H., Vickerman, K., Sleigh, M. A., and Warren, A. (Eds.): 1998. Evolutionary Relationships among Protozoa. Kluwer Academic, Dordrecht, The Netherlands. 486 pp.

Corliss, J. O. 1979. The Ciliated Protozoa: Characterization, Classification, and Guide to the Literature. 2nd Edition. Pergamon Press, New York. 455 pp.

Corliss, J. O. 1994. An interim utilitarian ("user-friendly") hierarchical classification and characterization of the protists. Acta Protozool. 33:1–51.

Elliott, A. M. (Ed.): 1973. Biology of *Tetrahymena*. Dowden, Hutchinson, and Ross, Stroudsburg, PA. 508 pp.

Farmer, J. N. 1980. The Protozoa: Introduction to Protozoology. C. V. Mosby, St. Louis. 732 pp.

Fenchel, T. 1987. Ecology of Protozoa: Biology of Free-Living Phagotrophic Protists. Science Tech, Madison, WI. 197 pp.

Galloway, J. 1987. A cause for reflection? Nature 330:204–205. (Discusses water temperature and coiling direction in forams.)

Grell, K. G. 1973. Protozoology. Springer-Verlag, Berlin. 554 pp.

Hammond, D. M., and Long, P. L. (Eds.): 1973. The Coccidia. University Park Press, Baltimore, MD. 482 pp.

Harrison, R. W., and Corliss, J. O. (Eds.): 1991. Microscopic Anatomy of Invertebrates. Vol. 1. Protozoa. Wiley-Liss, New York. 508 pp.

Harumoto, T., and Miyake, A. 1991. Defensive function of trichocysts in *Paramecium*. J. Exp. Zool. 260:84.

Hausmann, K., and Hülsmann, N. 1996. Protozoology. Georg Thieme, New York. 338 pp.

Haynes, J. R. 1981. Foraminifera. John Wiley and Sons, New York. 434 pp.

Kreier, J. P. 1991–1995. Parasitic protozoa. 2nd Edition. Academic Press, New York. (Ten volumes.)

Laybourn-Parry, J. 1984. A Functional Biology of Free-Living Protozoa. University of California Press, Berkeley, CA. 218 pp.

Laybourn-Parry, J. 1992. Protozoan Plankton Ecology. Chapman and Hall, New York. 231 pp.

Leadbeater, B. S. 1983. Life-history and ultrastructure of a new marine species of *Proterospongia*. J. Mar. Biol. Ass. U.K. 63:135–160.

Lee, J. J., and Anderson, O. R. 1991. Biology of Foraminifera. Academic Press, London. 384 pp.

Lee, J. J., Leedale, G. F., and Bradbury, P. (Eds.): 2000. An Illustrated Guide to the Protozoa. Vol. 1. 2nd Edition. Society of Protozoologists, Lawrence, KS. 683 pp.

Lee, J. J., Leedale, G. F., and Bradbury, P. (Eds.): 2000. An Illustrated Guide to the Protozoa. Vol. 2. 2nd Edition. Society of Protozoologists, Lawrence, KS. 1432 pp.

Leedale, G. F. 1967. Euglenoid Flagellates. Prentice-Hall, Englewood Cliffs, NJ. 242 pp.

Lynn, D. H. 1981. The organization and evolution of microtubular organelles in ciliated protozoa. Biol. Rev. 56:243–292.

Margulis, L., Corliss, J. O., Melkonian, M., et al. (Eds.): 1990. Handbook of Protoctista. Jones and Bartlett, Boston. 1024 pp.

Noble, E. R., and Noble, G. A. 1982. Parasitology. 5th Edition. Lea & Febiger, Philadelphia. 522 pp.

Ogden, C. G., and Hedley, R. H. 1980. An Atlas of Freshwater Testate Amoebae. British Museum, Oxford University Press, Oxford. 222 pp.

Patterson, D. J. 1980. Contractile vacuoles and associated structures: Their organization and function. Biol. Rev. 55:1–46.

Patterson, D. J., and Larsen, J. 1991. The Biology of Free-Living Heterotrophic Flagellates. Oxford University Press, New York. 505 pp.

Prescott, D. M. 1994. The DNA of ciliated protozoa. Microbiol. Rev. 58:233–267.

Sarjeant, W. A. S. 1974. Fossil and Living Dinoflagellates. Academic Press, London. 1002 pp.

Sleigh, M. A. 1989. Protozoa and Other Protists. Edward Arnold, London. 342 pp.

Sleigh, M. A. 1991. Mechanisms of flagellar propulsion. A biologist's view of the relation between structure, motion and fluid mechanics. Protoplasma 164:45–53.

Spoon, D. M., Chapman, G. B., Cheng, R. S., et al. 1976. Observations on the behavior and feeding mechanisms of the suctorian *Heliophyra erhardi* (Reider) Matthes preying on *Paramecium*. Trans. Am. Micros. Soc. 95:443–462.

Taylor, F. J. R. (Ed.): 1987. The Biology of Dinoflagellates. Blackwell Scientific, Oxford. 785 pp.

Wichterman, R. 1986. The Biology of *Paramecium*. 2nd Edition. Plenum, New York. 599 pp.

INTERNET SITES

General

www.mdsg.umd.edu/pfiesteria/ (Maryland Sea Grant page on *Pfiesteria.*)

www.med.cmu.ac.th/dept/parasite/framepro.htm (Light microscope images of parasitic protozoa.)

http://megasun.bch.umontreal.ca/protists/gallery.html (Images of free-living protozoa.)

www.microscopy-uk.org.uk/mag/wimsmall/flagdr.html (Color images of living and preserved protozoa.)

www.durr.demon.co.uk/ (Steve Durr's page of color images of living protozoa.)

www.uga.edu/~protozoa/ (Society of Protozoolgists home page.)

Euglenozoa

http://bio.rutgers.edu/euglena/ (The Euglenoid Project page.)

http://taxa.soken.ac.jp/WWW/PDB/Images/Mastigophora/Peranema/ (Color images of living *Peranema.*)

www.microscopyu.com/moviegallery/pondscum/protozoa/peranema/ (Movies of living *Peranema.*)

http://tryps.rockefeller.edu/crosslab_intro.html (Introduction to trypanosomes and disease.)

Retortamonada and Axostylata

www.geocities.com/CollegePark/Lab/4551/ (*Giardia* page.)

http://vm.cfsan.fda.gov/~mow/chap22.html (U.S. Food and Drug Administration page on *Giardia.*)

www.utoronto.ca/forest/termite/flagella.html (List of flagellate species found in the gut of termites.)

http://comenius.susqu.edu/bi/202/ProtistPix/parabasalotista/trichonympha.htm (Images of *Trichonympha* from a termite gut.)

Volvocida

http://megasun.bch.umontreal.ca/protists/chlamy/appearance.html (Image of living *Chlamydomonas* and other information.)

http://taxa.soken.ac.jp/WWW/PDB/Images/Chlorophyta/Volvox/ (Color images of living *Volvox.*)

Choanoflagellata

http://thalassa.gso.uri.edu/rines/ecology/choanofl.htm (Image of living choanoflagellates.)

http://protist.i.hosei.ac.jp/taxonomy/Zoomastigophora/Choanoflagellida.html (Color images of living choanoflagellates.)

Dinoflagellata

www.geo.ucalgary.ca/~macrae/palynology/dinoflagellates/dinoflagellates.html (Dinoflagellate images and information.)

Ciliophora

www.uoguelph.ca/~ciliates/ (The Ciliate Resource Archive.)

www.micrographia.com/specbiol/protis/cili/peri0100.htm (Color images of peritrichs.)

http://members.magnet.at/p.eigner/ (Information on and images of hypotrichs.)

http://taxa.soken.ac.jp/WWW/PDB/images/Protista/Ciliophora.html (Color images of living ciliates.)

Apicomplexa

www.saxonet.de/coccidia/ (Life cycle of *Eimeria*.)

www.biosci.ohio-state.edu/~parasite/plasmodium.html (Malaria page.)

www-micro.msb.le.ac.uk/224/Bradley/Biology.html (Another malaria page with animation and images.)

Amebas

http://micro.magnet.fsu.edu/moviegallery/pondscum/protozoa/amoeba/ (Movies of ameboid motion.)

www.microscopy-uk.org.uk/intro/illu/dark.html (Images of living *Amoeba proteus*.)

http://taxa.soken.ac.jp/WWW/PDB/Images/Sarcodina/ap/intactcell2.html (Images of living *Amoeba*.)

http://taxa.soken.ac.jp/WWW/PDB/Galleries/Uruguay1999/Arcella/ (Color images of living *Arcella*.)

http://taxa.soken.ac.jp/WWW/PDB/Images/Sarcodina/Difflugia/ (Images of *Difflugia*.)

Foraminiferea

http://cushforams.niu.edu/Forams.htm (Images of living forams and skeletons.)

Radiolaria and Acantharea

www.radiolaria.org/ (Images and information on fossil radiolarians.)

http://caliban.mpiz-koeln.mpg.de/~stueber/haeckel/radiolarien/index.html (Ernst Haeckel's 1862 color illustrations of Radiolaria and Acantharea.)

www.cladocera.de/protozoa/rhizopoda/imgal_radiolaria.html (Color image of a living colonial radiolarian.)

Heliozoa

www.biol.kobe-u.ac.jp/labs/suzaki/heliozoa/heliozoa-E.html (Research page on contraction mechanism of axopodia.)

www.cladocera.de/protozoa/rhizopoda/imgal_heliozoa.html (Color images of living heliozoans.)

www.microscopyu.com/moviegallery/pondscum/protozoa/actinophrys/ (Movies of *Actinophrys* moving and feeding.)

Phylogeny

http://tolweb.org/tree?group=Eukaryotes&contgroup=Life (Patterson and Sogin's Tree of Life page.)

4 Introduction to Metazoa

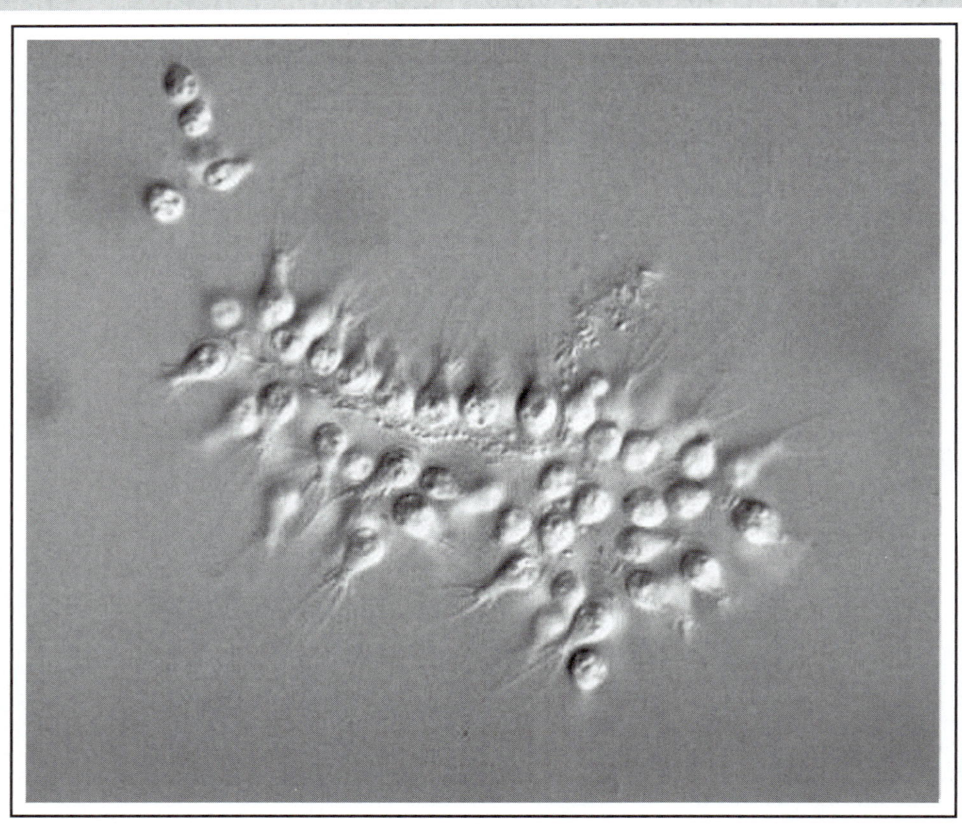

Metazoans are the multicellular organisms we commonly identify as animals. In the past, protozoans and metazoans were both considered to be animals, and the name Metazoa (= later animals) reflects their evolution from a protozoan (= first animal) ancestor. Now, however, the word *animal* is applied solely to the multicellular Metazoa, also known as Animalia. As noted in Chapter 3, protozoans are unicellular eukaryotic organisms that often form multicellular colonies. In doing so, protozoans knocked at the evolutionary door of multicellular plants and animals. Once the door opened and metazoans evolved, there followed such a prolific evolutionary radiation—an estimated 1 million to 30 million species in 29 major taxa (phyla)—that we are led to inquire into the causes of this astounding success. What were the keys that unlocked that evolutionary door and how were those keys forged? What innovations evolved in metazoans, what were their functional implications, and from what preexisting structures did they evolve? These are the topics of this chapter.

GROUND PLAN

Metazoans are multicellular, heterotrophic, motile eukaryotes. The body is polarized along an anterior-posterior locomotory axis. In contrast to protozoans, metazoans generally are large organisms.

CELLS, TISSUES, AND SKELETONS

The metazoan body is composed of functionally specialized cells, each type of which is dedicated to one or a few functions. In protozoans, on the other hand, all the functions of life are housed in a single cell. Metazoan cells are not distributed randomly in the body, but are bound together in specialized layers called **tissues.** Minimally, the metazoan body consists of two tissues, at least one more than any colonial protozoan. The somatic cells of the protozoan colony *Volvox,* for example, constitute the sole tissue layer of the body. Some species of colonial choanoflagellates may have two tissue layers, but Choanoflagellata is the sister taxon of Metazoa (Fig. 3-37).

The two basic types of metazoan tissue are epithelial and connective. The simplest metazoans, and developmental stages of many primitive invertebrates, consist solely of these two layers. Thus, epithelial and connective tissues may be the primary (original) tissues of metazoans, and both are important in the functional organization of animals.

An **epithelium** consists of cells that adjoin one another to form a sheet that covers the body or lines an internal cavity. The epithelium that covers the body (the epidermis) typically has on its external surface a secreted extracellular matrix (ECM), or **cuticle** (Fig. 4-1). In its simplest form, the cuticle consists of little more than a thin surface coat of glycoproteins. The epithelial cells secrete and rest on a **basal lamina,** a thin, dense, fibrous ECM. Epithelial cells are bound to each other and to the basal lamina by **cell-adhesion molecules.** Although there are many of these proteins, they fall into distinct classes, two of which are integrins and cadherins. Integrins bind cells to the basal lamina and cadherins bind cells to each other. Epithelial cells are commonly ciliated. There may be one or many cilia per cell; the monociliated state is believed to be primitive because it occurs in all primitive metazoan taxa, some apomorphic groups, and, notably, in Choanoflagellata, the sister taxon of Metazoa (Fig. 4-13A,B). Chapter 6 thoroughly describes the structure and function of epithelial tissue.

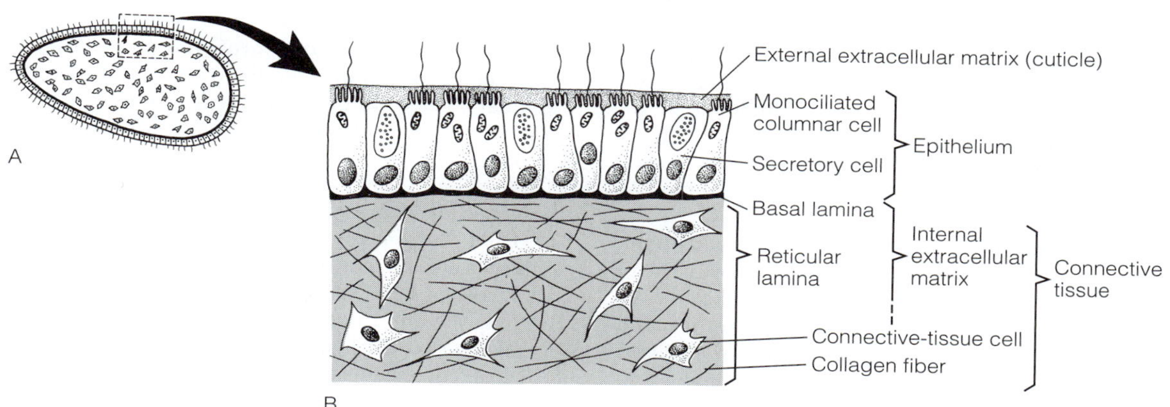

FIGURE 4-1 Cells, tissues, and skeletons. **A,** A hypothetical metazoan with the two primary metazoan tissues: a monociliated, columnar epithelium overlying connective tissue. **B,** Enlargement of boxed area in **A.** The epithelium is an unbroken layer of cells that provides a physiological barrier between the interior of the organism and the exterior environment. It is anchored between a basal lamina below and a secreted, protective cuticle above. Secretory cells are interspersed among the monociliated cells. The connective-tissue layer (or compartment) consists of a proteoglycan gel containing collagen fibers and individual connective-tissue cells. Both the cuticle and internal ECM function as a skeleton.

Connective tissue consists of widely separated, nonadjoining cells in an ECM (Fig. 4-1). The ECM component, which may be more or less voluminous, consists largely of a proteoglycan gel with embedded protein fibers. The most common of these proteins is **collagen,** probably the most important structural protein in animals (Fig. 4-1). In humans, for example, collagen accounts for one-quarter of all protein in the body. Collagen fibers associated with connective tissue typically have a skeletal function (described below). The ECM also provides attachment sites for cells, pathways over which cells move, and the appropriate environment for cell differentiation. Because of these attributes, the ECM plays an important role in organizing and maintaining tissues of the metazoan body.

Both internal and external ECMs can be modified to form a skeleton. A **skeleton** is any structure that supports the body and transmits the force of muscular contraction. Skeletons also provide protection from predation, injury, infection, or environmental challenge. Metazoan skeletons usually are novel tissue-level structures, unlike those of plants and fungi, which are compounded from preexisting, individual *cell* walls. An **exoskeleton** is a thickened cuticle that becomes rigid by chemically cross-linking its proteins (as in insects) or by adding mineral to the organic matrix (as in crabs). Similarly, all or part of the internal ECM, or connective tissue, can be stiffened by chemical cross-linking (as in cartilage) or by mineral secretion (as in bone) to produce an **endoskeleton.** (Other types of endoskeletons and derivatives of connective tissue are described in Chapter 6.) Most metazoans have an endoskeleton, exoskeleton, or both.

REPRODUCTION AND DEVELOPMENT

Metazoans employ both sexual and clonal (asexual) reproduction. Clonal reproduction is accomplished by fragmentation, fission, budding, or parthenogenesis. **Fragmentation** occurs when the body breaks up irregularly into several pieces. **Fission** is a more orderly division of the body along the longitudinal or transverse axis. Following fragmentation and fission, the daughter animals regenerate missing parts. **Budding** is the differentiation of a daughter individual before it detaches from the parent body. **Parthenogenesis** (= virgin birth) is the development of an individual from an unfertilized egg or other totipotent cell.

Metazoans also reproduce sexually. They are diploid organisms in which meiosis is restricted to the formation of haploid gametes, male **sperm** and female **egg,** both specialized, polarized cells. Sexual union of a sperm and egg at **fertilization** restores diploidy and produces a polarized **zygote,** or fertilized egg. The polar axis of the zygote runs between **animal** and **vegetal poles** (Fig. 4-2A). In most cases, the animal pole corresponds to the egg's apical end, the former site of a flagellum and microvillar collar that regressed during differentiation of the egg (Fig. 4-13B-D).

The zygote divides by mitosis to produce a multicellular **embryo,** the developmental stage of an organism before it adopts the recognizable form of a juvenile or larva. The early divisions of the zygote are called **cleavages** and the resulting cells are known as **blastomeres.** The first two division planes of the zygote are typically parallel to the animal-vegetal axis (meridional) and the entire zygote is cleaved into two and then four blastomeres. Such complete cleavage is termed **holoblastic** cleavage (Fig. 4-2A,B).

Eggs contain various amounts of yolk distributed in different ways in different taxa. Because yolk increases the size of the egg and tends to interfere with cleavage, both the amount and distribution of yolk profoundly affect the cleavage pattern and the shape of the embryo. **Microlecithal eggs** are small cells with little yolk distributed evenly throughout the cytoplasm. Such eggs typically undergo holoblastic cleavage and produce blastomeres of equal size (equal cleavage; Fig. 4-2A). **Mesolecithal eggs** are medium-size cells that contain a moderate amount of yolk restricted to the vegetal hemisphere. They undergo holoblastic, but unequal, cleavage (Fig. 4-2B). In these embryos, the blastomeres of the animal hemisphere are smaller **(micromeres)** than those of the yolky vegetal hemisphere **(macromeres). Macrolecithal eggs** are large and very yolky, have unfavorable (low) area-to-volume ratios, and do not cleave holoblastically. Instead, incomplete cleavage, called **meroblastic cleavage,** occurs in a specific region of the egg's surface. Initially, the cleavage-furrow membranes are incomplete and do not fully isolate the cytoplasm of the blastomeres from the underlying yolk mass. Later, the completed membranes separate a thin cap of cells, the **blastodisc,** from the yolk (Fig. 4-2C). Because it is thin and flat, the blastodisc is provisioned by the diffusion of yolk nutrients from below and gases from above (Fig. 4-2C). Macrolecithal eggs and meroblastic cleavage occur in squids, octopuses, certain fishes, birds, and reptiles. Another type of incomplete cleavage, called superficial cleavage, typifies the macrolecithal eggs of insects and many other arthropods. Superficial cleavage will be described in Chapter 16.

Early embryos typically form a one-cell-layer-thick ball of cells called a **blastula,** which is either hollow (coeloblastula; Fig. 4-2A,B) or solid (stereoblastula). After continued cell division, some blastomeres move into the interior of the blastula by a process called **gastrulation** to form a two-layered **gastrula.** Gastrulation is the first event in **morphogenesis,** the gradual conversion of a more or less uniform embryo into a complex, multilayered adult (Fig. 4-15). In most animals (but perhaps not sponges), the primary germ layers (outer **ectoderm** and inner **endoderm**) are established by gastrulation. Between ectoderm and endoderm is a gelatinous layer of ECM known as the **blastocoel** (Fig. 4-2). Ectoderm, endoderm, and blastocoel are the embryonic precursors of adult tissues. Later in development, ectoderm and endoderm become the outer (epidermis) and inner (gut) layers, respectively, whereas the blastocoel is the developmental precursor of connective tissue (Fig. 4-14). After the gastrula stage, the developing embryo increases in complexity (forms organs, for example) and gradually adopts the anatomy typical of its species.

The events that unfold during the developmental period from zygote to adult of any individual constitute its **ontogeny** (= origin of being). Because ontogeny has a time course and because adaptations in development affect the reproductive

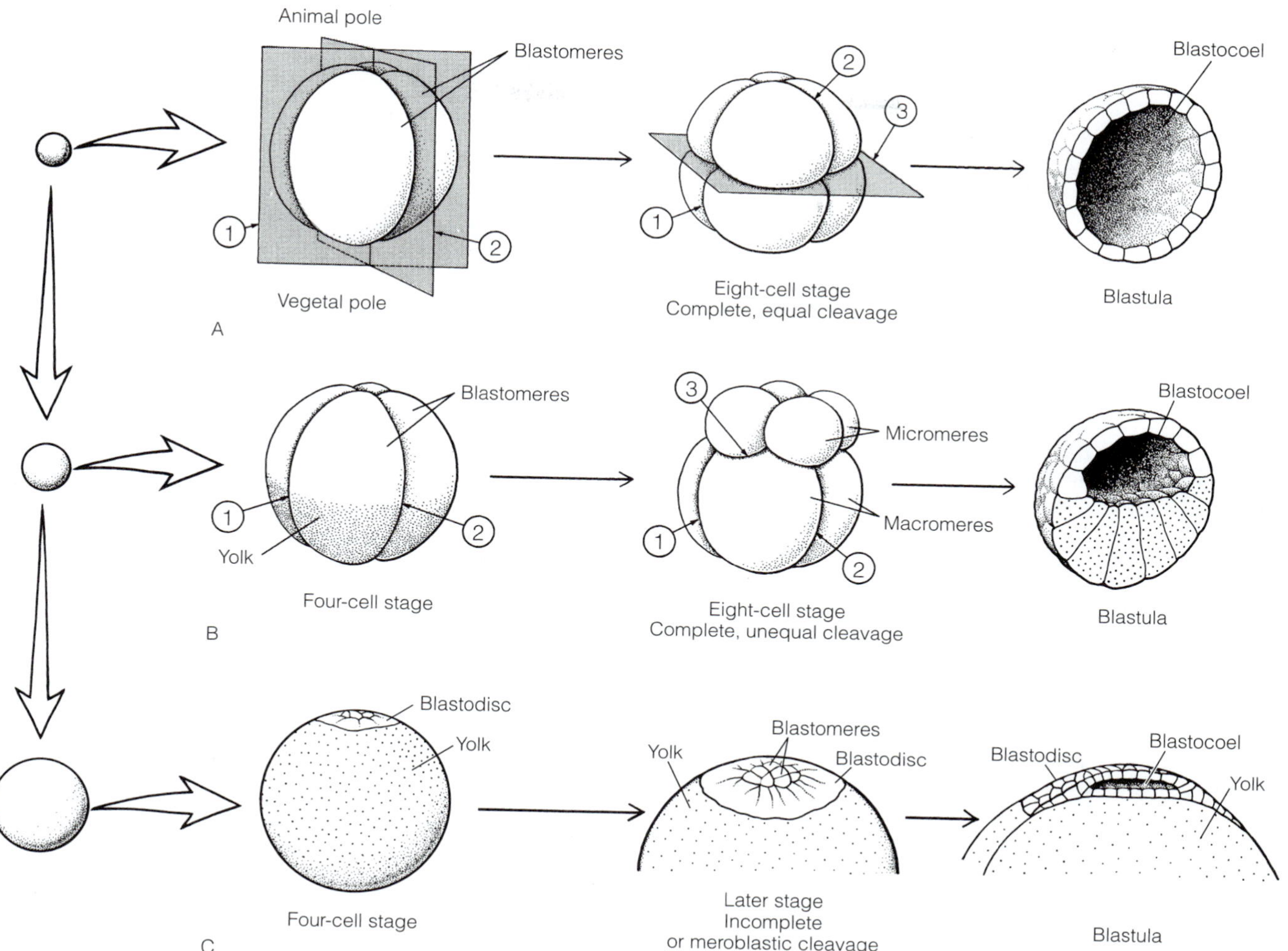

FIGURE 4-2 Reproduction and development: metazoan eggs, early cleavage stages in relation to yolk content, and blastulae. **A,** A small microlecithal egg with a uniform distribution of yolk. **B,** A medium-size mesolecithal egg with yolk concentrated in the vegetal hemisphere. The first two cleavage planes—numbers 1 and 2 in **A** and **B**—are meridional. The third cleavage plane (number 3) is equatorial. **C,** A large macrolecithal egg with only a small apical blastodisc free of yolk. Cleavage occurs only in the blastodisc. Expanding vertical arrows indicate increasing egg size and yolk content.

success of the adults, natural selection has modified ontogenetic as well as adult stages of metazoans. One common ontogenetic modification is a larva.

A **larva** is a developmental stage of distinctive form that persists for a period of time, lives independently of the adult, differs in form from the adult, and occupies a separate ecological niche. Animals with a life cycle that includes a larval stage have **indirect development.** Many aquatic invertebrates, especially those in the sea, have a bottom-dwelling **(benthic)** adult and a floating **(planktonic)** or swimming larva, linked together in a **biphasic life cycle.** The larva is anatomically different from the juvenile and adult, reflecting its adaptation to the larval, not the adult, ecological niche. Thus, at the close of the larval period, the larva must locate and **settle** in a habitat suitable for the adult and then change its form to that of the **juvenile,** an immature form of the adult. This transformation is called a **metamorphosis** (= change of form). Metamorphosis ranges in extent from the gradual addition of new structures and slight rearrangement of tissues to a complete remaking of the juvenile body from undifferentiated **embryonic reserve cells** (or set-aside cells) housed in the larval body. The process by which a juvenile differentiates from larval embryonic reserve cells resembles budding. The extent of metamorphosis, whether gradual, cataclysmic, or intermediate, depends on how far the larval and adult body designs have diverged from each other.

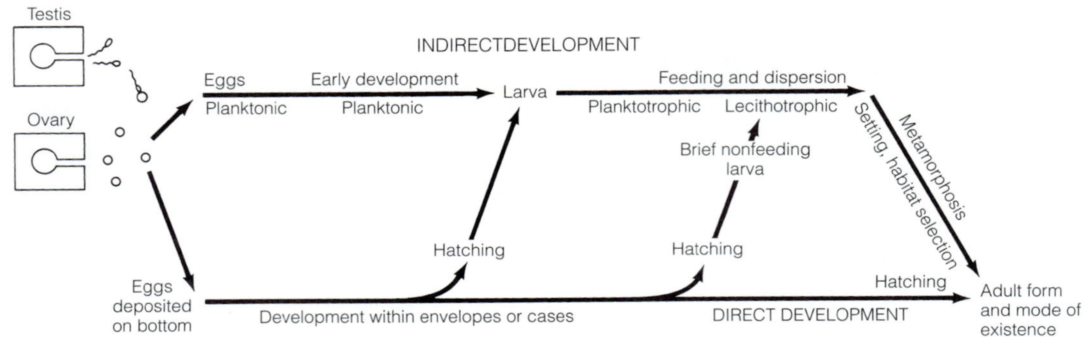

FIGURE 4-3 Reproduction and development: life history patterns of metazoans.

The larvae of aquatic invertebrates may be planktonic or, less commonly, benthic (Fig. 4-3). The extent of the larval period may be short (a few hours) or long (lasting from months to over a year). Short-lived larvae that commonly feed on the maternal supply of yolk that carries over from the egg into the embryonic endoderm are called **lecithotrophs** (= yolk feeders). Most long-lived larvae initially are lecithotrophic, but soon a functional gut differentiates and the larva then sustains itself by feeding on plankton. Such larvae are said to be **planktotrophs.**

Indirect developers whose life cycle includes a planktotrophic larval stage typically produce many microlecithal eggs. Because the larvae feed themselves in a different habitat (the water column), they do not compete with the benthic adults for food and, if they are long-lived, can disperse widely. A long planktonic period increases the risk of predation and death, but the large number of eggs produced offsets these losses. Indirect developers whose life cycle includes a planktonic lecithotrophic larval stage usually produce fewer eggs than do species with planktotrophic larvae, but each egg is large and has a substantial amount of yolk. Because the larva that arises from such an egg has a fixed amount of food, its life span is relatively short, but it is well provisioned. The short planktonic existence, moreover, reduces the risk of predation, although it allows for only limited dispersal. Some other species that have **direct development** bypass a larval stage altogether and their embryos develop directly into juveniles (Fig. 4-3). An indirect life cycle with external fertilization and planktonic development is believed to be the primitive pattern in animals.

All animals have evolved reproductive adaptations that enhance the likelihood of fertilization and the survival of the embryo. Adaptations that enhance the likelihood of fertilization focus largely on increasing synchrony (male and female gametes are produced and released at the same time) and proximity (gametes are released near each other). Synchronous production and release of gametes are triggered largely by environmental signals such as temperature, light, and tides. Proximity has been achieved in several ways. **Hermaphroditism,** the presence of both male and female gonads in the same individual, is a common adaptation where population densities are low or where the adults live permanently attached to surfaces. In these circumstances, hermaphroditism assures that any nearby individual is a potential mate. Most hermaphroditic animals cross-fertilize rather than self-fertilize because "selfing" limits genetic variation and enhances expression of deleterious recessive alleles. **Gonochorism** (dioecy), the occurrence of separate male and female individuals, is the opposite of hermaphroditism.

Many adaptations contribute to the survival of offspring, but maternal provisioning of nutrients and physical protection of eggs and embryos are among the most important. **Oviparous** species that spawn eggs before or immediately after fertilization provide the shortest period of parental protection. **Viviparous** species have internal fertilization, retain **(gestate)** embryos in the maternal body, and release differentiated offspring, usually larvae or juveniles. The embryos of viviparous species receive nutrients either **matrotrophically** (directly from the mother via a placenta, for example) or **lecithotrophically** from yolk stored in the egg. **Brooding** occurs when eggs are released from the mother, but then are retained on or taken back into her body.

FUNCTIONAL CONSEQUENCES OF BODY SIZE

The evolution of multicellular animals from unicellular protozoans resulted in many new traits, most of which, including a tissue-level skeleton, are related to the larger body size of metazoans. Living organisms occupy a size range extending from less than 1 μm to many meters, a span of roughly eight orders of magnitude. Within this size range, prokaryotic organisms occupy the region from <1 μm to 2 μm, unicellular protozoans from 2 μm to 0.5 mm (although some exceptional species are larger), and multicellular animals from 0.5 mm to >1 m. These numbers suggest that fundamental changes in body *design* might have occurred at approximately 2 μm and 0.5 mm. The transition at 2 μm was from prokaryotic to eukaryotic cells, while another redesign occurred at the 0.5 mm threshold, marking the transition from protozoa to Metazoa. The following sections describe the structural changes and functional consequences of these two transitions. A discussion of body size and locomotion is postponed until Chapter 6, the Introduction to Eumetazoa, which were the first animals to evolve muscle and nerve. For an intuitive grasp of the different body sizes of prokaryotes, protozoans, and metazoans, consider a 1 μm bacterium (such as *Pneumococcus*), a 100 μm protozoan (dinoflagellate), and a 10 cm metazoan (sea urchin), all within a modest range of six orders of magnitude. If we scale up these organisms to the size of familiar objects while retaining their sizes relative to each other, they would correspond approximately to a mustard seed *(Pneumococcus)*, a grapefruit (dinoflagellate), and the Astrodome (sea urchin). Starting with the same basic set of building materials, how would you design a functional organism at each of these three scales?

SIZE AND COMPARTMENTALIZATION

Metazoan cell specialization segregates functions (divides labor) into separate compartments, similar to the division of labor among people in a family, corporation, or society. Each individual becomes proficient at a specific essential task instead of being a "jack-of-all-trades and master of none." Mastery improves **efficiency,** the ratio of work output to work input. A skilled (specialized) fruit picker, for example, picks more fruit per unit time and with greater economy of motion than a less skilled (more generalized) novice. But the cost of occupational specialization is a loss of self-sufficiency: Specialists in different occupations depend on each other for the delivery of essential goods and services they themselves cannot provide. The efficiency of cell specialization in the metazoan body is inevitably linked with the interdependence of those specialized, functionally compartmentalized cells. Thus, it seems that evolution of large, active metazoans relied on simultaneous functional compartmentalization at a new cellular level and the functional integration of those cells.

Functional compartmentalization and integration, however, are not limited to metazoans. Among protozoans, the compartments are intracellular organelles, such as the nucleus, Golgi body, or mitochondrion. Presumably, the division of labor at the level of organelles and the integration of their specialized functions by intracellular chemical messengers supported the greater size and activity of eukaryotic cells as compared with prokaryotic cells.

The still-larger metazoans inherited the organelles of the eukaryotic cell, but then evolved a new level of compartmentalization—the specialization of cells within the multicellular body. At the next higher level, functions such as skeletal support became associated with aggregates of similarly specialized cells known as tissues. Eventually, in the largest, most active animals, the hierarchy of functional compartments included not only organelles, cells, and tissues, but also organs and organ systems. At whatever level, the chemical environment of the compartment is regulated by its borders. Marching hand-in-hand with each new level of function are systems of control and integration, such as the nervous, endocrine, and circulatory systems. When muscle tissue first evolved, for example, nervous tissue accompanied it to control its contraction.

Functional compartmentalization and integration, in general, are important and recurring themes in the evolution of metazoans and the remainder of this book. As the subsequent chapters unfold, try to correlate the levels and degrees of functional compartmentalization and integration with the organism's body size and activity level.

SIZE, SURFACE AREA, AND VOLUME

For any organism of constant shape, a change in body size disproportionately changes the body's surface area in relation to its volume. This area-to-volume ratio *(S:V)* is important because any exchange of substances (gases, nutrients, and wastes) between an organism and its external environment occurs across a surface, and the metabolic demand of a body is related to its volume. As a solid three-dimensional body becomes larger, its surface area increases in proportion to the radius squared (r^2), but its volume increases more rapidly, in proportion to the radius cubed (r^3; Fig. 4-4). Under these conditions, the enlarging body will eventually reach a size at which its surface area (supply) is inadequate to serve its volume (demand).

What options are available to ensure that supply meets demand as body size increases? Consider the different sizes of a prokaryotic bacterium, a eukaryotic protozoan, and a metazoan, all of cubical shape (Fig. 4-4). For a typical 1 μm bacterium, the *S:V* is 6:1. For a protozoan of 10 μm, the ratio drops to 0.6:1, a 10-fold decrease in area for each unit of volume. The ratio for a 1 cm metazoan is 0.006:1, a further 100-fold drop.

The calculated decrease in *S:V* for the eukaryotic cell as compared with a prokaryotic cell is more fiction than fact. This is because eukaryotic cells rarely retain smooth surfaces or perfect geometric shapes. Instead, their surfaces are folded or bear outgrowths such as microvilli and pseudopodia that significantly increase area. Because of these specializations, the *S:V* of a mammalian liver cell is 200,000:1, but it is no longer an unadorned cube! Perhaps such measurements indicate that eukaryotic animal cells, with their enhanced surfaces, can supply their volume metabolic demand at a greater rate than can prokaryotic cells. In this same sense, perhaps some organelles of eukaryotic cells, such as endoplasmic reticulum and Golgi bodies, are surface ingrowths (Fig. 2-13B) that increase area for the metabolic machinery, thus amplifying metabolic capacity (Fig. 4-4B).

The earliest metazoans inherited surface-enhanced eukaryotic cells, but their larger bodies required yet another level of improvement in the area-to-volume ratio. The most primitive metazoans, similar to choanoflagellate colonies, may have achieved this by arranging their cells in a thin two-dimensional sheet over a metabolically inert core of ECM (Fig. 4-4C inset, 4-12D). Many others, such as sponges or earthworms, arrange the cell sheets in the form of hollow tubes, or tubes in tubes. Still others, for example flatworms, roundworms, and slime molds (Fig. 4-6), are flat and thin or long and slender as a means of improving the area-to-volume ratio (Fig. 4-5).

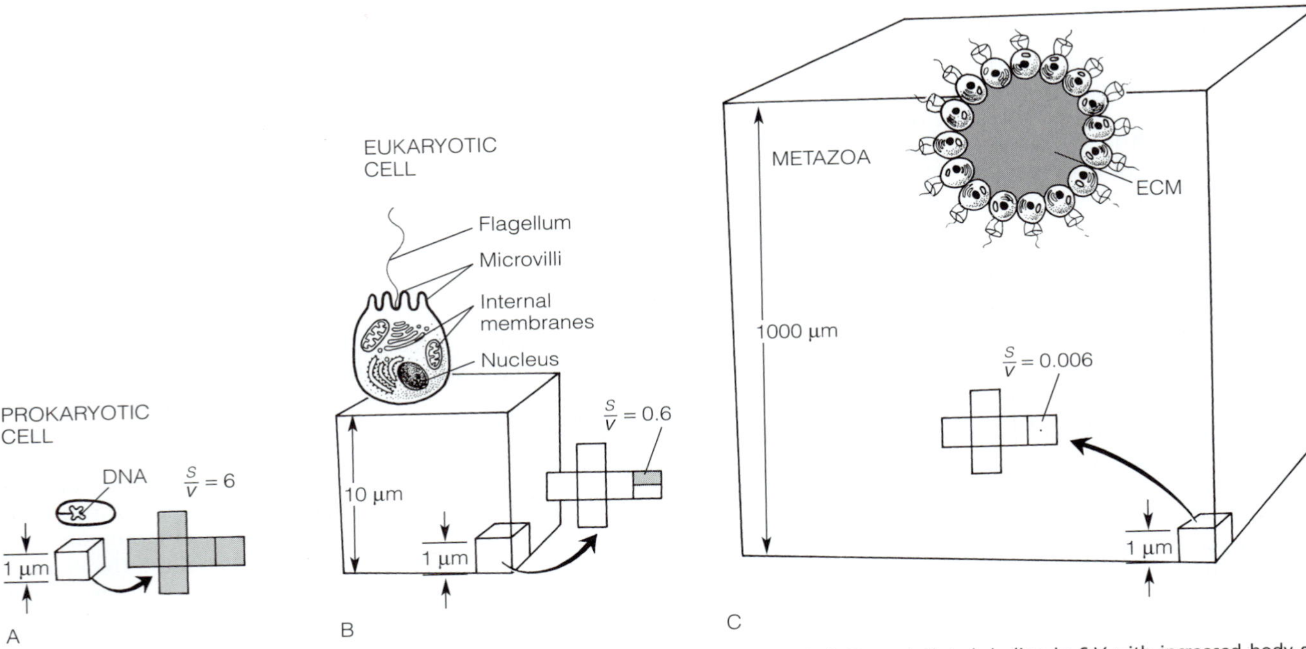

FIGURE 4-4 Size, surface area, and volume for organisms of three different body sizes. **A,** prokaryotic cell; **B,** eukaryotic cell; **C,** metazoan. For ease of calculation, each organism is represented geometrically by a cube and assumed to be metabolically active from surface to center. The surface area available to supply the 1 µm^3 body volume of prokaryote cube **A** is 6 µm^2 (shaded part of unfolded cube); in the larger eukaryotic cell cube, **B,** the surface area to supply each 1 µm^3 of body volume is 0.6 µm^2 (shaded part of unfolded cube); in metazoan cube **C,** each 1 µm^3 of volume has only 0.006 µm^2 of supply area (dot in unfolded cube). The calculated decline in *S:V* with increased body size, however, is not realized in actuality. As body size increases, organisms redesign their bodies to improve the *S:V* ratio. Eukaryotic cells **(B)** enhance surface area with microvilli, internal membranes, and other structures. Metazoans are composed of surface-enhanced eukaryotic cells and typically arrange tissues in two-dimensional sheets (epithelia), as in this example **(C)**, in which the sheet covers a core of metabolically inert ECM.

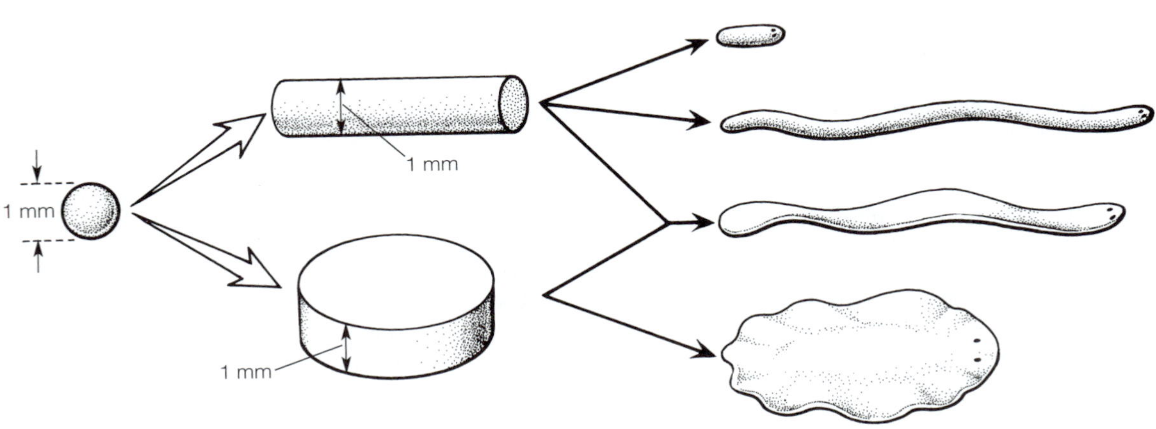

FIGURE 4-5 Size, surface area, and volume: body forms that increase the area-to-volume ratio. These two shapes—long and slender, thin and flat (center)—greatly improve the *S:V* ratio over that of a sphere (left), but restrict growth to one or two dimensions (right) because of the limits imposed by diffusion.

SIZE AND TRANSPORT

Diffusion is a process by which molecules in a liquid or gas enter and leave cells and tissues. Diffusion results from the random movement of heated (above absolute zero) molecules of a liquid or gas. Although the motion is random, if a particular substance, say O_2, is in higher concentration outside of a cell than inside, then on average, more O_2 will be entering the cell than leaving it at any moment. The concentration difference results in a net movement of O_2 into the cell.

Diffusion transports substances rapidly at the size scale of most cells and certain simple tissues, but is slow over greater distances. This is because the time required for diffusion to achieve a given concentration *(C)* is proportional to the square of the distance from the source. So, for example, the time needed to achieve concentration *C* at 10 times the distance from the source is 10^2, or 100 times as long. Because of this relationship, diffusion is effective only over short distances. For aquatic animals, the effective diffusion distance is roughly 0.5 mm (Fig. 4-7A). This means that a solid, metabolically active, cellular body will be diffusion limited at a diameter of approximately 1 mm. Notice how close this value is to the size at which metazoans departed from protozoans!

Some metazoans and many protozoans rely on diffusion alone for internal transport and for transport between themselves and the external environment (Fig. 4-5). Among the protozoans, many of these diffusion-limited organisms are tiny—well below 1 mm in diameter. But larger protozoans and some larger metazoans have altered the shapes of their bodies to remain within the limits imposed by diffusion, becoming either slender and threadlike or flat and thin (Fig. 4-5, 4-6). In both cases, the body radius is small, within the scope of diffusional transport, and the surface area of the body is large in relation to its volume. Among protozoans, for example, are large forams with numerous radiating threadlike reticulopods (Fig. 3-30B). Giant multinucleated amebas are flat and thin. In fact, the largest known cell (a multinucleated ameba) is the slime mold *Physarum polycephalum,* one specimen of which had an area of 5.54 m^2 but was only 1 mm in thickness ($S{:}V$ = 1:1, Fig. 4-6). Metazoan flatworms, such as the common planaria, can be as flat and thin as tissue paper. Some of the other metazoan worms whose cylindrical bodies are cellular from surface to core (roundworms) are as thin or thinner than sewing thread.

For a body that exceeds 1 mm in diameter, diffusion alone probably cannot keep pace with metabolic demand. As a result, the cells farthest from the surface may die unless there is a more effective means of transport (Fig. 4-7B). A **circulatory system** is an internal transport system that provides a mass flow of fluid (called convection) and frees metazoans from the body-size and -shape limitations imposed by simple diffusion (Fig. 4-7C). Among animals, cilia, muscles, or both circulate the fluid. Because the fluid circulates, it can rapidly transport metabolites between supplies and demands (Fig. 4-7C), thus allowing body sizes far in excess of 1 mm, up to several or even tens of meters.

The metazoan circulatory system may have evolved from internalized surface tissue that formed circulatory vessels. This initial circulatory system could have been similar to the water-filled channels (aquiferous system) that course through some sponges (Fig. 4-8A)or the simple invagination of one side of the body, as in *Hydra* and its relatives (the coelenteron; Fig. 4-8B), and during gastrulation in the development of most metazoans (the archenteron; Fig. 4-14E). Later, bilaterally symmetrical animals (Chapter 9) evolved functionally specialized circulatory systems called hemal and coelomic systems. Whatever its original form, a circulatory system was a precondition for the evolution of large body size.

Slime mold

After Achenbach, from Hausmann, K., and Hülsmann, N. 1996. Protozoology. Georg Thieme, New York. 434 pp.

FIGURE 4-6 Size, surface area, and volume: a giant specimen of the multinucleate plasmodium (single-celled ameba) of the slime mold *Physarum polycephalum,* shown as an extreme example of a diffusion-limited body. This specimen has an enormous surface area, in excess of 5.5 m^2, but is only 1 mm in thickness ($S{:}V$ = 1). The slime mold in the photograph has a wavy outline and is growing on nutrient medium in the form of a **W.**

SIZE AND METABOLISM

In considering the changes that occurred as protozoans evolved into large metazoans, we might ask how the power-consumption (metabolic) rate of metazoans compares with that of protozoans. Is larger body size accompanied by a discount or a tax on power consumption? What, if any, relationship exists between body size and metabolic rate?

Generally, metabolic rate increases with body size—bigger animals consume more power per unit time than smaller animals (Fig. 4-9 on p. 67). But as body size increases, each additional unit of mass requires less than one additional unit of power. For example, a large metazoan composed of 10,000 cells consumes more energy per unit time than a single-celled protozoan does, but not 10,000 times more energy. If it did, the slope of the lines in Figure 4-9 would be 1.0 instead of 0.75. Stated another way, the metabolic rate of that single protozoan cell exceeds that of any 1 of the 10,000 metazoan cells. So, although a huge elephant consumes power at a greater rate than a tiny shrew, 1 gram (g) of shrew tissue consumes more power than 1 g of elephant tissue does. On average, small animals have a greater mass-specific

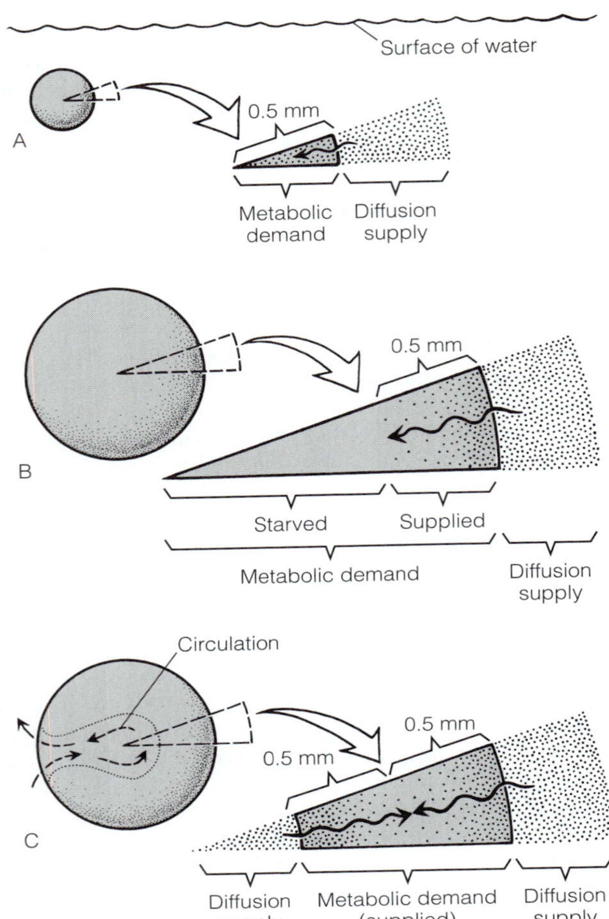

FIGURE 4-7 Size and transport: diffusion and circulation in aquatic organisms. Limit to diffusion in aquatic organisms. The spheres in **A** and **B** represent an aquatic metazoan composed of metabolically active tissue from surface to center. Wedges in **A–C**: The supply concentration of an essential substance is indicated by the density of the dots outside of the organism. Wedge in **A**: As that substance diffuses at a fixed rate toward the center of the body, the tissues along the diffusion path consume the substance and reduce its concentration. Wedge in **B**: Because the tissues consume the substance along the entire path, at some distance from the surface the concentration drops to zero. Beyond that point, diffusional supply cannot keep pace with the demand of the deeper tissues. Although the actual distance over which diffusion can effectively transport metabolites depends on the initial concentrations of those metabolites, the metabolic rate of the tissue, and other variables, the rule of thumb is that the effective diffusion distance for aquatic organisms is approximately 0.5 mm (wedge in **A**). **C** shows how a circulatory system offsets the size limit imposed by diffusion. Circulation of fluid transports substances to within the diffusion range (0.5 mm) of tissues.

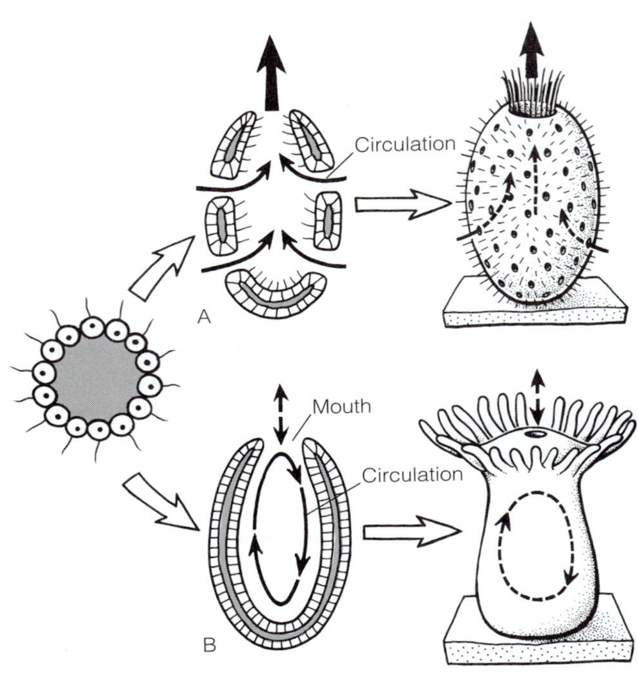

FIGURE 4-8 Size and transport: two types of circulatory systems in primitive metazoans. **A**, A unidirectional flow-through system in a sponge. **B**, Bidirectional, or circuitous, circulation in a polyp, such as a *Hydra* or sea anemone; exchange at the mouth is infrequent and intermittent.

power consumption than large animals. So large animals might be said to receive a mass-specific discount in power consumption.

Further examination of Figure 4-9, however, suggests another important relationship. Although the lines representing protozoa, cold-blooded Metazoa (poikilotherms), and warm-blooded Metazoa (homeotherms, the birds and mammals) share the same slope (0.75), they are offset from each other because each has a different y-intercept. The offset between the protozoans and metazoan poikilotherms represents the 8-times-greater power consumption of a metazoan as compared with a protozoan of equal mass. Similarly, the offset between the poikilotherm and homeotherm lines shows the 29-times-greater power consumption of a homeotherm vs. a poikilotherm of equal weight. What accounts for these stepwise increases in power consumption?

The 29-fold increase at the transition from poikilotherms to homeotherms represents the cost of maintaining a constant body temperature. Homeotherms pay more in fuel costs than poikilotherms, but by maintaining a constant body temperture they can remain active over a wide range of environmental temperatures. Why does power consumption jump 8-fold with the evolution of metazoans from protozoans? Why does it require 8 times more energy to power a metazoan than a protozoan of equivalent size? To arrive at a hypothesis, we should search for a uniquely metazoan feature that creates a new and constant power demand. Of all the metazoan traits reviewed in this chapter, the one that qualifies best is the extracellular circulatory system. Metazoans may pay a surcharge for the power needed to circulate fluid using cilia, muscles, or both, and ultimately for the evolution of large body size.

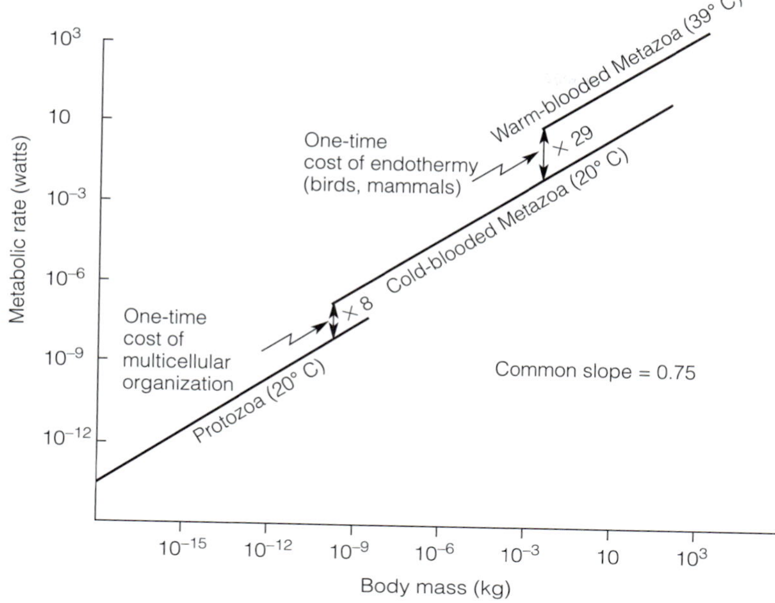

FIGURE 4-9 Size and metabolism: relationship between metabolic (power-consumption) rate and body weight (mass) in protozoans, cold-blooded metazoans, and warm-blooded metazoans. Power consumption increases with body size, but at a rate of less than 1 unit of power for each additional unit of weight; thus the slope is less than 1.0 (0.75). The 8-times increase in metabolic rate at the transition from protozoans to metazoans and the 29-times increase from cold-blooded to warm-blooded metazoans probably reflect the costs of a circulatory system and warm-bloodedness (homeothermy), respectively. See text for further explanation. *(Modified and redrawn from Peters, R. H., 1983. The Ecological Implications of Body Size. Cambridge University Press, Cambridge. 329 pp.)*

Because that surcharge is not insignificant, we might now ask, What benefits result from the evolution of large body size?

ADVANTAGES OF LARGE BODY SIZE

The large body size of metazoans confers a mass-specific discount on metabolic rate and presumably other advantages over small-bodied protozoans. Large body size may have lowered metazoan vulnerability to protozoan predation. Although protozoans might still have grazed on metazoans, the likelihood is slim that a protozoan would engulf an entire animal. (Some very small metazoans, such as rotifers and gastrotrichs, are eaten by protozoans, however.) It is more likely that metazoans, because of their larger bodies and tissue specializations, became efficient at eating protozoans, thereby gaining an important advantage. Metazoan sponges and protozoan choanoflagellates, for example, both filter feed with similar collar cells, but sponges can consume particulates, including eukaryotic cells, in the submicrometer to 50 μm (or larger) size range, whereas choanoflagellates are restricted to the smallest particulates. Sponges can exploit this wider range of food sizes because their food filter is not restricted to the collar cell alone, but also includes a tissue-level (multicellular) component not available to the protozoan. Large metazoans can also move at greater speeds than small organisms can (see Movement and Body Size, Chapter 6).

Another advantage, shared with colonial protozoans, is that multicellularity buffers the individual against environmental damage and loss. Multicellularity and regenerative ability enable the body to tolerate the loss of some cells to, for example, novel toxins ingested with food or grazing predators, which damage or destroy some cells, but do not kill the individual.

ONTOGENY AND PHYLOGENY

Unlike protozoans, metazoans have an ontogeny, a developmental time course over which growth and differentiation convert the zygote to a multicellular adult body. The fact that metazoan ontogeny consists of developmental stages, the variations of which are subject to natural selection, creates a near-limitless potential for the evolution of new forms. For example, consider the life cycle of a segmented worm that includes a benthic adult and a planktonic larva. Random variation resulted in sexually mature larvae or juveniles that reproduced, but never grew to the adult form. Such a variant was favored by natural selection, perhaps because it exploited a novel environment, and the new species resembled the larva or juvenile, not the adult, of the ancestral species (Fig. 4-10). Or, consider a curious example from the toothed whales (dolphins, orcas, sperm whales). During tooth morphogenesis, there occurred variants in which one incisor tooth grew either at a faster rate or for an extended period of time, resulting in an enormous unicorn-like tusk. That tusk proved adaptive and was selected for, perhaps because it was a potent sexual signal, and a new whale species evolved (the narwhal) with one tooth that far exceeded the size of the equivalent tooth in the ancestor. In both worm and whale examples, the variants arose from a change in timing of developmental events. In the worm, sexual maturity

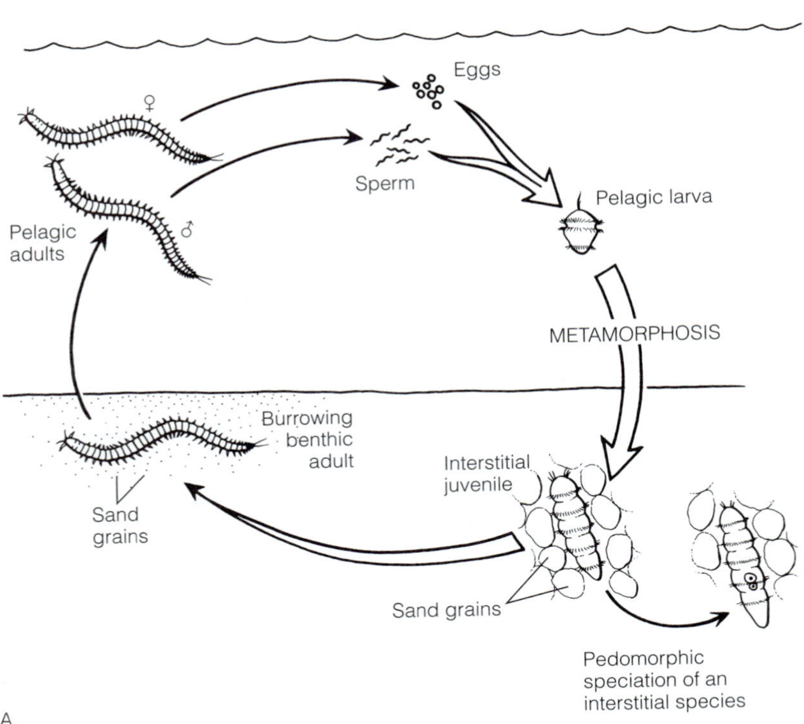

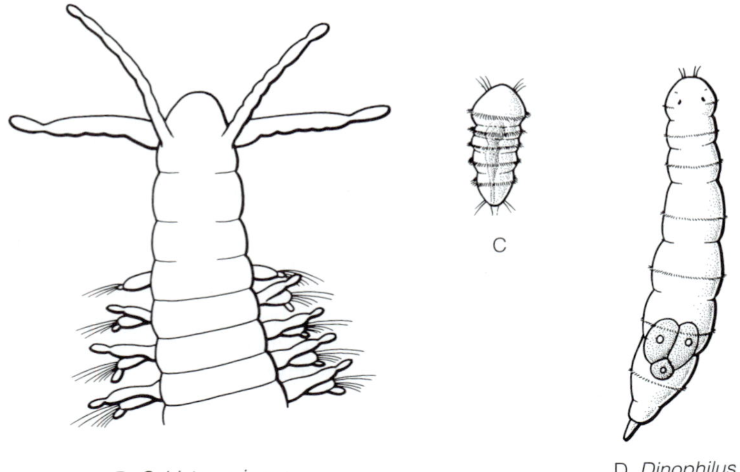

FIGURE 4-10 Ontogeny and phylogeny: heterochrony. **A,** Pedomorphosis in the evolution of a small-bodied species associated with colonization of the pore spaces between sand grains (an interstitial habitat). **B,** Pedomorphosis in the evolution of a small-bodied descendant that retains larval ancestral traits (lack of appendages, ciliated whorls) in a family of segmented worms (Dorvelliidae, Polychaeta). Cladistic analysis (not shown here) indicates that the ancestor had a large body and appendages similar to those of the animal *(Schistomeringos)* shown in **B.** The pedomorphic descendant species *(Dinophilus),* shown in **D,** has a small body lacking appendages, but bearing whorls of cilia similar to the ciliary whorl of a larval stage **(C).** *(A, B, and D from Westheide, W. 1987. Progenesis as a principal in meiofauna evolution. J. Nat. Hist. 21:843–854. © 1987 Taylor & Francis, Ltd.; C from Dawydoff, C. 1928. Traité d'Embryologie Comparée des Invertébrés. Masson, Paris. 930 pp.)*

appeared sooner rather than later in the life cycle; in the whale, there was an increase in the rate or period of development of an organ or structure.

Such changes in the timing of developmental events are described by a phenomenon called **heterochrony** (= changed time). Because heterochrony can occur at any stage in the developmental period and affect any component of the body, including body size, it permits radical (as well as moderate) evolutionary change. Heterochrony is divided into two general categories based on the form of descendants in relation to their ancestors. If the trait of the descendant species resembles an ancestral developmental trait and thus is larval or juvenile in form, the type of heterochrony is called **pedomorphosis** (= restrained shaping). The segmented worm described above and in Figure 4-10 is an example of pedomorphosis (or paedomorphosis). On the other hand, if the descendant's trait develops to an extent beyond that found in the ancestor, it is

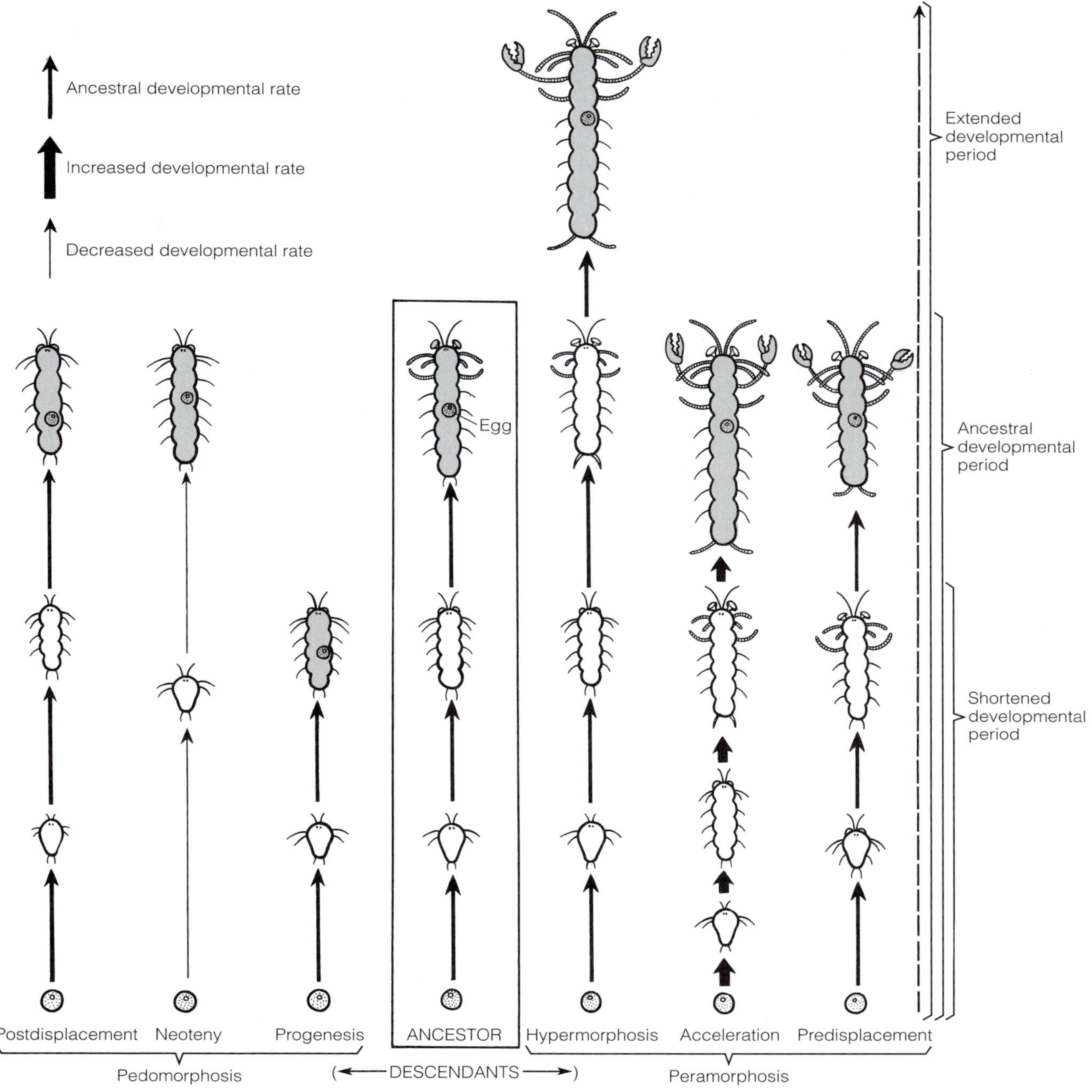

FIGURE 4-11 Ontogeny and phylogeny: heterochrony. The forms of pedo- and peramorphosis, as illustrated by the descendants of a fictitious ancestral metazoan (boxed, center). All six forms of heterochrony arise as timing variations in the ancestral ontogeny. Specifically, variations occur in the rate of development, the period of development, or the onset (start time) of one or more traits. In this example, four traits are affected: body segments, eyes, appendages, and gonad differentiation (maturity). All changes are in reference to ancestral ontogeny. **Progenesis:** developmental rate is unchanged, but an early onset of gonad differentiation arrests further body (somatic) development, thus shortening the developmental period. **Neoteny:** the developmental period and onset are unchanged and the rate of gonad differentiation also is unchanged, but the rate of somatic differentiation is slowed. **Postdisplacement:** the period and rate of development are unchanged, and the onset of gonad differentiation also is unchanged, and the onset of somatic differentiation is delayed. **Hypermorphosis:** developmental rate and somatic onset are unchanged, but the developmental period is extended and the onset of gonad differentiation is delayed. **Acceleration:** developmental period and onset are unchanged, but rate of somatic differentiation is increased. **Predisplacement:** developmental period and rate are unchanged, but onset of differentiation is early.

called **peramorphosis** (= extended shaping). The narwhal's enormous tooth is an example of peramorphosis.

Pedomorphosis often results in descendants that are smaller and anatomically simpler than their ancestors. Small body size and short generation time (because the developmental program has been truncated) typify many species that have colonized unpredictable or changeable environments such as temporary pools, and perhaps the plankton community. Pedomorphosis is also common among symbiotic species, such as certain parasitic barnacles (Fig. 19-83). In these, the nonpedomorphic female is modified to absorb nutrients from the host and to produce the myriad eggs and offspring necessary to infect new hosts. Compared with the female, the pedomorphic male is a dwarf, little more than a sexually mature larva, parasitic on the female, whose only role is to provide sperm to fertilize the female's eggs. Being tiny, the male does not compete substantially with the female for the host's nutritional resources, allowing her to direct more food-energy to egg production. Finally, pedomorphic descendants of large-bodied ancestors have colonized entirely new habitats. As shown in Figure 4-10, heterochrony in a family of segmented worms (polychaetes) enabled the small-bodied pedomorphic descendant to colonize the minute water-filled spaces between sand grains (an interstitial community).

Peramorphosis typically results in descendants that are larger and anatomically more complex than their ancestors. Much of the story of metazoan evolution, as we have indicated throughout this chapter, is the achievement of large body size. Large body size and long generation times should be favored in constant or predictable environments such as the deep sea, upwelling areas of the sea, and coral reefs. Other advantages of large body size were discussed earlier in this chapter.

Throughout this book, we will use the terms pedomorphosis and peramorphosis to describe significant and appropriate examples of heterochrony. You should be aware, however, that each of these categories is divided into three subcategories based on the exact form of heterochronic change. In the worm example, the morphogenesis of a functional gonad in the larval stage of the ancestral life cycle was a case of pedomorphosis. But the occurrence of a gonad at that early developmental stage could have resulted from progenesis, a normal (ancestral) rate of development with an early onset of sexual maturity, thus arresting further development of the nonreproductive (somatic) parts of the body; neoteny, a normal (ancestral) duration of development leading to sexual maturity with a decreased developmental rate of somatic structures, meaning that differentiation of the body lags permanently behind reproductive development; or postdisplacement, a normal (ancestral) rate and period of development with a delayed onset (postponed start of differentiation) of somatic structures, resulting again in a sexually mature juvenile body. These three forms of pedomorphosis and the similar categories of peramorphosis are summarized in Figure 4-11. Although they are easy to define in theory, these categories are difficult to distinguish in practice and, for that reason, we use only the general terms pedo- and peramorphosis in this book.

EVOLUTIONARY ORIGINS

ORIGIN OF METAZOA

Most zoologists agree that metazoans share a common ancestry with some unicellular organism. The classical **colonial theory,** in which Metazoa is derived from a colony of flagellated protozoa, is the most widely accepted hypothesis among contemporary zoologists (Fig. 4-12). An alternative hypothesis, the **syncytial theory,** proposes that metazoans evolved from a multinucleate but unicellular plasmodium similar to a slime mold or perhaps a ciliate protozoan. Later, membranes evolved to produce a cell boundary around each of the nuclei. The syncytial theory receives some support from the development of organisms such as slime molds or insects *(Drosophila),* in which an early multinucleated stage is followed by cellularization to form a multicellular body. However, phylogenetic analysis based on morphology and gene sequences as well as the developmental patterns of most animals contradict the syncytial theory and favor the colonial theory. For these reasons, we consider only the colonial theory here.

A modern version of the colonial theory states that the premetazoan (a protozoan) consisted of a small spherical colony bearing a surface layer of flagellated cells that was used for locomotion and feeding (Fig. 4-12C). The colony originated from a cell that divided repeatedly by mitosis, but the daughter cells did not separate after cell division. Those daughter cells were surrounded and held together by a proteinaceous ECM in which they were deeply embedded. The gelatinous ECM also occupied much of the interior of the sphere. Similar to extant choanoflagellates, the cells each bore a single collared flagellum. A few nonflagellated cells, capable of giving rise to flagellated cells and gametes, were scattered in the subsurface ECM (Fig. 4-12C).

Several facts support such a flagellate ancestry. Flagellated, or monociliated, collar cells—cells with a single flagellum and collar of microvilli—are widespread in metazoans, and the ciliated cells of lower metazoans, particularly sponges, jellyfishes, sea anemones, and corals, are exclusively monociliated (Fig. 4-13B). Molecular and morphological data both indicate that choanoflagellates and metazoans are sister taxa. An ECM to which cells are attached and through which they move is nearly universal in metazoans. Cellular specializations, such as eggs and sperm, have evolved in some colonial flagellates, such as the spherical, colonial *Volvox.*

Although it is unrelated to metazoans, *Volvox* is an analog for an ancestral metazoan because it demonstrates how a multicellular organism evolved from a unicellular ancestor, in this case a *Chlamydomonas*-like cell. *Volvox* is not the ancestor of Metazoa, but rather is an autotrophic organism with plantlike cells. rRNA sequence data indicate that multicellularity in *Volvox* evolved 50 to 75 million years ago, far too late for it to have been a progenitor of metazoans, which had their origin at least 600 million years in the past. Thus, the volvocids evolved multicellularity *in parallel* with the metazoans (and with at least four other groups: fungi, brown algae, red algae, and green plants).

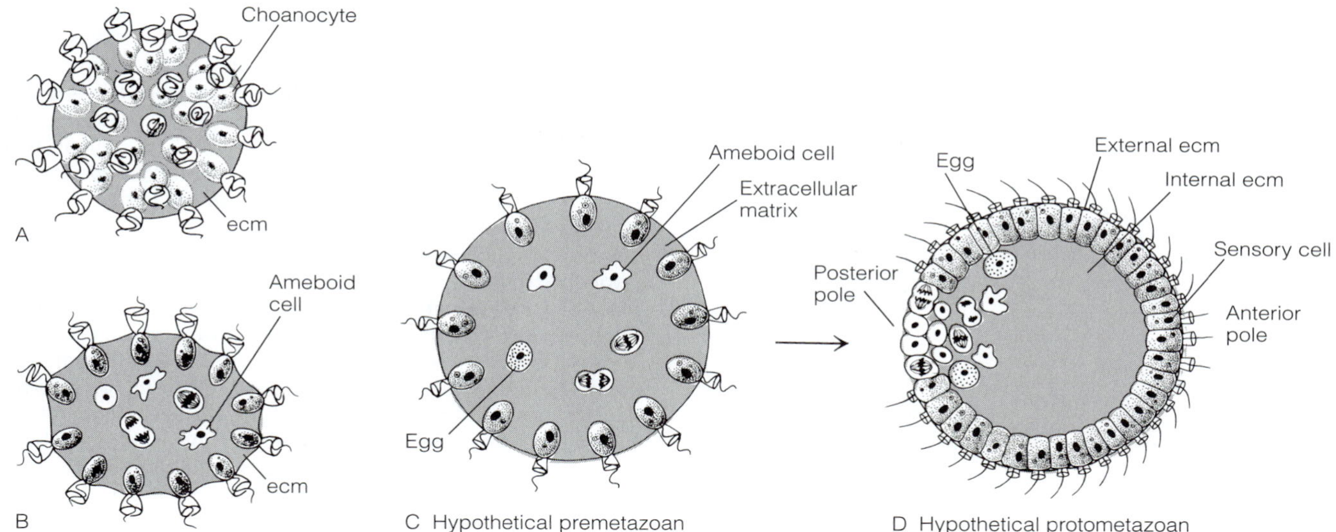

FIGURE 4-12 Origin of Metazoa: choanoflagellate colonies and metazoan evolution. **A** and **B,** different forms of the choanoflagellate colony *Proterospongia haeckeli.* **C,** A hypothetical premetazoan based on the choanoflagellates in **A** and **B. D,** A hypothetical protometazoan showing cellular specialization along an anterior-posterior axis. Surface cells are bound together and in mutual contact, thus allowing physiological regulation of the interior of the body.

The actual sister taxon of Metazoa is most likely Choanoflagellata. Among the extant choanoflagellates, colonies of *Proterospongia haeckeli* (Fig. 4-12A,B) closely resemble the hypothetical premetazoan (Fig. 4-12C). The colony consists of flagellated collar cells embedded in the surface of a gelatinous ECM. Cells undergoing division and lacking flagella occur deeper in the ECM (Fig. 4-12B). A surface sheetlike layer of cells and an ECM containing individual free cells foreshadow the two primary metazoan tissues, epithelial and connective, described earlier in this chapter. At the cellular level, choanoflagellate collar cells are virtually identical to collar cells (choanocytes) found in the metazoan sponges. A choanoflagellate cell and sponge choanocyte both have a single flagellum surrounded by a collar of microvilli and a flagellar shaft with a bilateral finlike vane (Fig. 4-13). In both, the flagellum is anchored in the cell by microtubules, which radiate from the flagellar basal body.

The hypothetical first metazoan, the protometazoan (Fig. 4-12D), may have differed from the premetazoan (Fig. 4-12C) in several ways. First, the surface cells probably closely adjoined or were in contact with each other, thus facilitating intercellular communication and providing a regulatory barrier between the external environment and the ECM in the interior of the body. Second, the close association of adjacent cells largely excluded the ECM from between the cells, thus separating the ECM into external and internal layers, each of which could then adopt independent functions (Fig. 4-12D). Third, the body was polarized along the anterior-posterior axis. Fourth, the separation of layers and body polarity promoted cell specialization (Fig. 4-12D).

ORIGIN OF POLARITY AND CELL SPECIALIZATION

Most motile protozoans are polarized cells that have leading (anterior) and trailing (posterior) ends or, if sessile and attached, they have *oral* free ends and *aboral* attached ends. Metazoans are similarly polarized, depending on whether they are motile or sessile, but how did the polarity of the multicellular metazoan body evolve from the unicellular polarity of protozoans? A clue to the answer is found in the eggs of several groups of metazoans. During oogenesis in these groups, the eggs express a rudimentary flagellum and a collar of microvilli at a site on the cell surface that corresponds to the animal pole of the zygote (Fig. 4-13B–D). In certain invertebrate taxa, the animal pole corresponds to the anterior end of the larva (although it is the posterior end in others). Additional research is needed, but the current evidence suggests a line of descent from the polarity of a choanocyte to the primary anterior-posterior polarity of the metazoan body (Fig. 4-13).

The protometazoan probably was polarized along an anterior-posterior axis, but what environmental conditions might have selected for the evolution of such polarity? For an aquatic metazoan, the environment presents itself in gradients of light, temperature, oxygen, and food availability. If, for

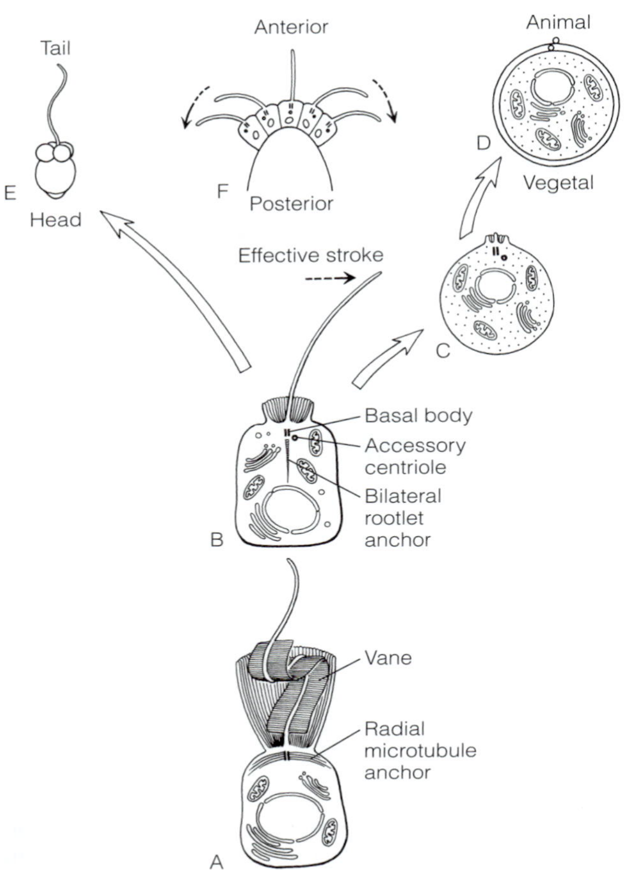

FIGURE 4-13 Origin of metazoan polarity. **A,** Primitive collar cell with a flagellar vane, as occurs in protozoan choanoflagellates and metazoan sponges. **B,** Choanocyte-like monociliated cell from the surface epithelium (epidermis) of an anemone larva (Cnidaria), showing its power stroke in the direction of the accessory centriole. **C,** Early oocyte of a sea cucumber (Echinodermata) showing a low collar and rudimentary cilium; **D,** Sea cucumber zygote in the same orientation as the oocyte in **C.** Note that the animal and vegetal poles correspond to the apex and base, respectively, of the epithelial cell in **B. E,** A typical metazoan sperm showing its polarity in relation to an epithelial spermatogonium from which it arose. **F,** Anterior end of a simple larva (Cnidaria). Note the position of the accessory centrioles, which indicate posteriorly directed ciliary effective strokes and thus forward movement in an anterior direction. *(B, Modified and redrawn from Rieger, R. M. 1976. Monociliated epidermal cells in Gastrotricha: Significance for concepts of early metazoan evolution. Z. zool. Syst. Evolut.-forsch. 14:198–226; C–E, Modified from Frick, J. E., and Ruppert, E. E. 1996. Primordial germ cells of* Synaptula hydriformis *(Holothuroidea: Echinodermata) are epethelial flagellated-collar cells: Their apical-basal polarity becomes primary egg polarity. Biol. Bull. 191:168–177, and Frick, J. E., and Ruppert, E. E. 1997. Primordial germ cells and oocytes of* Branchiostoma virginiae *(Cephalochordata, Acrania) are flagellated epithelial cells: Relationship between epithelial and primary egg polarity. Zygote 5:139–151.)*

example, the availability of food is related to light (through photosynthesis), natural selection would favor any variant individual capable of detecting and following a light gradient. This is best accomplished with a direction-sensitive sensory capability coupled with a directional (polarized) locomotory system. Thus, selection may have favored any organism capable of tracking a resource concentration gradient. Alternatively, developmental biologist Lewis Wolpert suggests that body polarity evolved from attachment to a substratum. Attachment to a rock in water, for example, places an organism at an interface, a very steep gradient. Once attached, variants would be favored that adhere well at the attached end and perform other tasks, such as feeding, at the opposite end. This again leads to a polarized body.

Once polarity was established, movement would create an environmental gradient along the locomotory axis that would favor differential expression of traits (Fig. 4-12D). For example, enhanced membrane sensitivity to environmental stimuli might be favored in cells at the anterior end of the body because they are the first to encounter changes in environmental conditions. Similarly, enhanced flagellar growth, density, or activity might be favored in cells at the equator, or widest part of the body, since those locations best contribute to locomotion. Cells with a capacity for division, leading to growth, might be favored at the posterior end, because in that position they contribute to and interfere least with locomotion. Thus, motility along a polar axis may itself promote cellular specialization because the cells occupy different fixed positions in an environmental gradient (Fig. 4-12D).

According to a hypothesis by developmental biologist Leo Buss, the origin of metazoan cell specialization may be related to a conflict between the demands for growth and locomotion. *Volvox, Proterospongia,* and planktonic blastula stages of metazoans require flagellated surface cells for locomotion, but most flagellated cells cannot divide by mitosis because the centrioles needed to form the mitotic spindles are already in use as the flagellar basal bodies. A metazoan flagellated cell can divide only *after* the flagellum regresses and its basal bodies are freed to form the mitotic apparatus. Thus, the options for growth in a premetazoan composed solely of flagellated cells (for example, the species of *Proterospongia* in Fig. 4-12A) are: (1) enlargement of existing cells; (2) disassembly of flagella, cell division, and then flagellar reassembly; (3) division of a few cells scattered throughout the embryo while others retain flagella; or (4) division of a few localized cells set aside for growth (Fig. 4-12B–D). Option 1, because of the restricted number of cells, limits ultimate body size, although *Volvox* daughter colonies and a few postembryonic micrometazoans grow by cell enlargement only. Option 2, by requiring the regression of flagella, compromises locomotion. Options 3 and 4 both permit growth and locomotion, but most invertebrate metazoans have adopted option 4. The setting aside of mitotically active cells enables metazoan growth without compromising locomotion (Fig. 4-12B–D). (A related evolutionary event was the setting aside of germ cells capable of meiosis.)

Surface flagella are a functional necessity for locomotion in the protometazoan, but what is the optimal position for the set-aside growth cells? Buss suggests that placement at the surface might have resulted in overgrowth of locomotory cells or growth of a tumorlike appendage, either of which would negatively affect motility. Internalizing the growth cells, however, would neither distort the body surface nor interfere with the locomotory cells. Perhaps for these reasons the protometazoan did internalize its growth cells and the process was preserved, as gastrulation, in its descendants (Fig. 4-14;

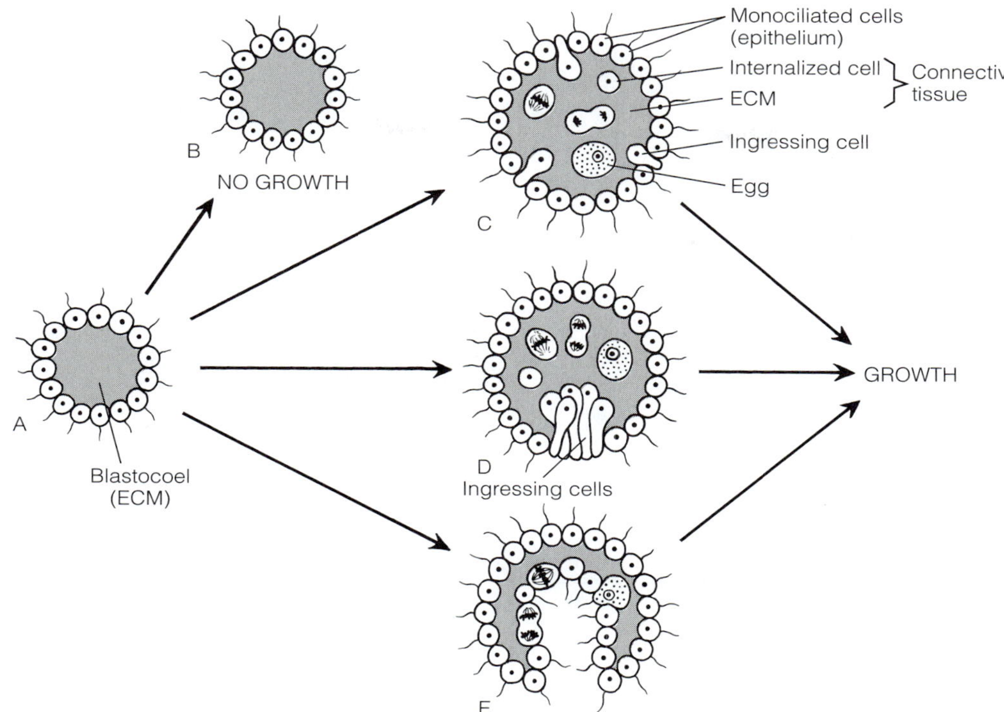

FIGURE 4-14 Origin of cell and tissue specialization: segregation patterns of growth cells in hypothetical protometazoans. **A,** Blastula-like developmental stage of a hypothetical organism; **B,** Hypothetical blastula-like organism that does not grow because its cells retain locomotory cilia and thus cannot divide by mitosis. **C–E,** Growth by internalizing nonflagellated mitotically active cells in three simple hypothetical metazoans. The patterns in **C–E** also represent three basic forms of gastrulation in actual animals. **C,** multipolar ingression; **D,** unipolar ingression; **E,** invagination (or emboly). Note how the embryonic blastocoel (ECM) is equivalent to adult connective tissue. *(Based on ideas in Buss, L. W. 1987. The Evolution of Individuality. Princeton University Press, Princeton, NJ. 201 pp.)*

also see Reproduction and Development in this chapter). An internalization of cells capable of mitosis also evolved in *Volvox* and *Proterospongia haeckeli* (Fig. 4-12B). A colony of *P. haeckeli* grows by the addition of new surface choanocytes that originate by mitosis of cells in the subsurface gel. Similarly, colonies of *Volvox* consist of a thousand or more flagellated surface cells and invaginated subsurface pockets of nonflagellated dividing cells (gonidia). The gonidia can give rise to either new colonies or germ cells.

The simultaneous requirement for locomotion and growth in early metazoans favored a body with two layers of differentiated cells (Fig. 4-12D). The layering of cells established tissues (as described earlier in Cells, Tissues, and Skeletons) and an exterior-interior gradient. The surface cells had direct access to gases and other raw materials, but were constrained by the requirements for locomotion and for interfacing with the external environment. The interior cells, liberated from direct interaction with the environment, became specialized for reproduction, nutrient storage, and, with the evolution of a gut, the digestion of food.

ORIGIN OF COMPLEXITY

Evidence supports the hypothesis that metazoans evolved from protozoan colonies in which initially similar cells became specialized for different functions. If so, the evolution of Metazoa can be described as a replication of similar units (cells) followed by unit specialization and integration into an organism at a new, higher level of complexity. This **replication-specialization-integration of units** sequence is a general pattern in the evolution of large body size and complexity. An example, as we have seen already, is the replication, specialization, and integration of cilia on the body of many ciliates, which are among the largest, most diverse, most active, and most complex protozoan cells. Among metazoans, one example is a body composed of a series of similar segments, as in earthworms or crustaceans. Later, these segments become specialized and integrated into regions, such as the head, thorax, and abdomen. Examples at all levels of biological complexity are illustrated and described in Figure 4-15.

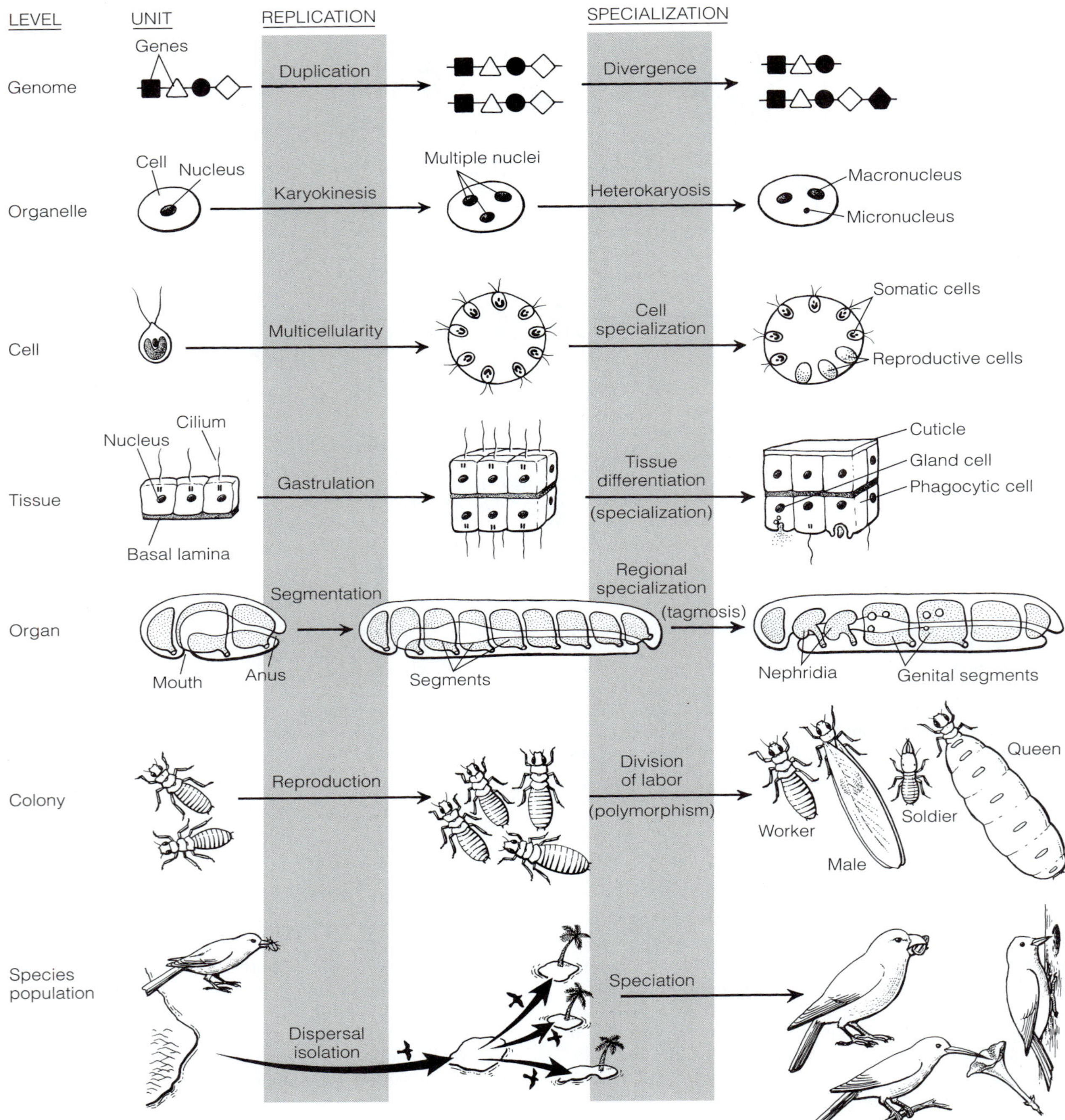

FIGURE 4-15 Origin of complexity at different levels of organization by replication and specialization of standard units.

REFERENCES

GENERAL

Bonner, J. T. 1988. The Evolution of Complexity. Princeton University Press, Princeton, NJ. 260 pp.

Bonner, J. T. 1993. Life Cycles. Princeton University Press, Princeton, NJ. 209 pp.

Buss, L. W. 1987. The Evolution of Individuality. Princeton University Press, Princeton, NJ. 201 pp.

Dawydoff, C. 1928. Traité d'Embryologie Comparée des Invertébrés. Masson, Paris. 930 pp.

Fawcett, D. W. 1981. The Cell. 2nd Edition. W. B. Saunders, Philadelphia. 862 pp.

Frick, J. E., and Ruppert, E. E. 1996. Primordial germ cells of *Synaptula hydriformis* (Holothuroidea; Echinodermata) are epithelial flagellated-collar cells: Their apical-basal polarity becomes primary egg polarity. Biol. Bull. 191:168–177.

Frick, J. E., and Ruppert, E. E. 1997. Primordial germ cells and oocytes of *Branchiostoma virginiae* (Cephalochordata, Acrania) are flagellated epithelial cells: Relationship between epithelial and primary egg polarity. Zygote 5:139–151.

Gerhart, J., and Kirschner, M. 1997. Cells, Embryos, and Evolution. Blackwell Science, Malden, MA. 642 pp.

Gilbert, S. F., and Raunio, A. M. 1997. Embryology: Constructing the Organism. Sinauer, Sunderland, MA. 537 pp.

Gould, S. J. 1977. Ontogeny and Phylogeny. Harvard University Press, Cambridge, MA. 501 pp.

Hausmann, K., and Hülsmann, N. 1996. Protozoology. 2nd Edition. Georg Thieme, Stuttgart. 338 pp.

Hemmingsen, A. M. 1960. Energy metabolism as related to body size and respiratory surfaces and its evolution. Rep. Steno Mem. Hosp. (Copenhagen) 9:1–58.

Hibberd, D. J. 1975. Observations on the ultrastructure of the choanoflagellate *Codosiga botrytis* (Ehr.) Saville-Kent with special reference to the flagellar apparatus. J. Cell Sci. 17:191–219.

Krogh, A. 1941. The Comparative Physiology of Respiratory Mechanisms. University of Pennsylvania Press, Philadelphia. 172 pp.

McKinney, M. L., and McNamara, K. J. 1991. Heterochrony: The Evolution of Ontogeny. Plenum Press, New York. 437 pp.

McMahon, T. A., and Bonner, J. T. 1983. On Size and Life. Scientific American, New York. 255 pp.

McNamara, K. J. 1997. Shapes of Time. Johns Hopkins University Press, Baltimore, MD. 342 pp.

Peters, R. H. 1983. The Ecological Implications of Body Size. Cambridge University Press, Cambridge. 329 pp.

Remane, A. 1963. The evolution of the Metazoa from colonial flagellates vs. plasmodial ciliates. In: Dougherty, E. C. (Ed.): The Lower Metazoa. Berkeley University Press, Berkeley, CA. pp. 23–32.

Nielsen, C. 2001. Animal Evolution: Interrelationships of the Living Phyla. 2nd Edition. Oxford University Press, New York. 563 pp.

Raff, R. A. 1996. The Shape of Life: Genes, Development, and the Evolution of Animal Form. University of Chicago Press, Chicago. 520 pp.

Rice, S. H. 2001. The role of heterochrony in primate brain evolution. In Minugh-Purvis, N., and McNamara, K. J. (Eds.): Human Evolution through Developmental Change. Johns Hopkins University Press, Baltimore. 508 pp.

Rieger, G. E., and Rieger, R. M. 1977. Comparative fine structure of the gastrotrich cuticle and aspects of cuticle evolution within the Aschelminthes. Z. zool. Syst. Evolut.-forsch. 15:81–124.

Rieger, R. M. 1976. Monociliated epidermal in Gastrotricha: Significance for concepts of early metazoan evolution. Z. zool. Syst. Evolut.-forsch. 14:198–226.

Ruppert, E. E. 1991. Introduction to the aschelminth phyla: A consideration of mesoderm, body cavities, and cuticle. In Harrison, F. W., and Ruppert, E. E. (Eds.): Microscopic Anatomy of Invertebrates. Vol. 4. Aschelminthes. Wiley-Liss, New York. pp. 1–17.

Salvini-Plawen, L. V. 1978. On the origin and evolution of the lower metazoa. Z. zool. Syst. Evolut.-forsch. 16:40–88.

Schmidt-Nielsen, K. 1984. Scaling: Why Is Animal Size So Important? Cambridge University Press, Cambridge. 241 pp.

Vogel, S. 1992. Vital Circuits: On Pumps, Pipes, and the Workings of Circulatory Systems. Oxford University Press, Oxford. 315 pp.

Vogel, S. 1996. Life in Moving Fluids: The Physical Biology of Flow. 2nd Edition. Princeton University Press, Princeton, NJ. 484 pp.

Wainwright, S. A. 1988. Axis and Circumference: The Cylindrical Shape of Plants and Animals. Harvard University Press, Cambridge, MA. 132 pp.

Westheide, W. 1987. Progenesis as a principle in meiofauna evolution. J. Nat. His. 21:843–854.

Willmer, P. 1990. Invertebrate Relationships: Patterns in Animal Evolution. Cambridge University Press, Cambridge. 400 pp.

Wolpert, L. 1994. The evolutionary origin of development: Cycles, patterning, privilege and continuity. Development (Suppl.):79–84.

Zeuthen, E. 1953. Oxygen uptake as related to body size in organisms. Q. Rev. Biol. 28:1–12.

INTERNET SITES

http://zygote.swarthmore.edu/germ8.html (Concise, illustrated discussion of egg-cell polarity.)

http://pantheon.yale.edu/~sean/hetero.html (Advanced, illustrated discussion of heterochrony.)

www.neoteny.org (Provides access to a database of quotations and sources of information on heterochrony.)

Porifera^{P} and Placozoa^{P}

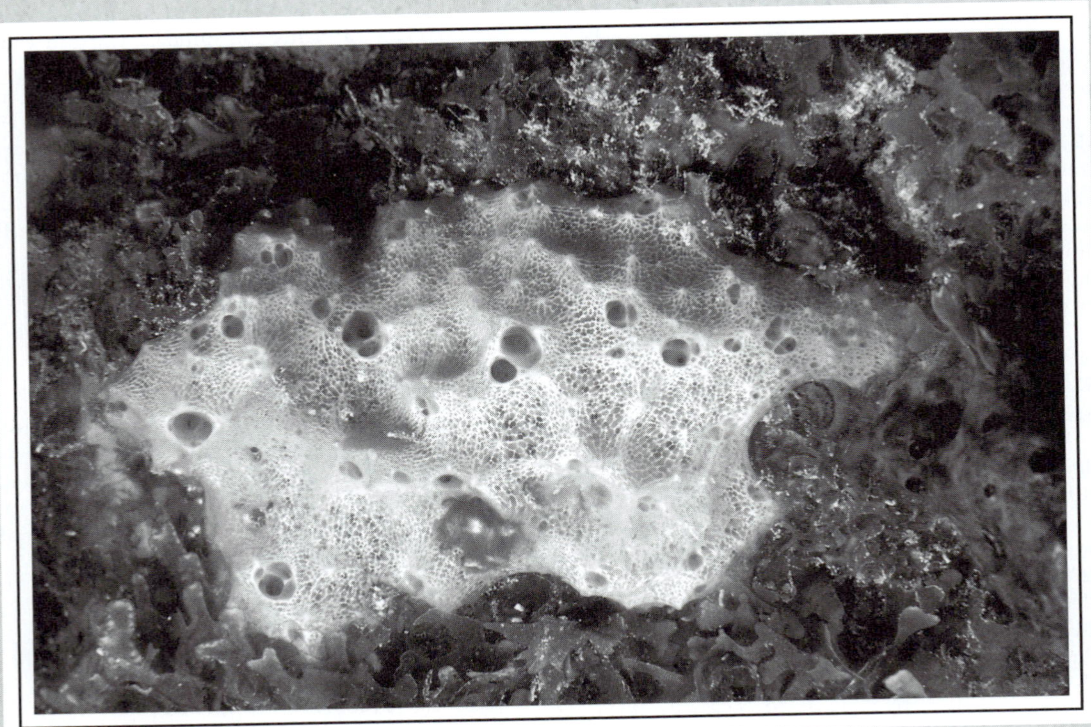

PORIFERA[P]

Sponges are a conspicuous and colorful component of many seascapes. The attached, often upright sponges are to coral reefs, sea grottoes, and floats what stalagmites and stalactites are to terrestrial limestone caves, except that sponge colors are as vivid and varied as those of van Gogh's flowers. When we look at sponges underwater in tropical seas, it seems a stretch to admit these motionless organisms with their irregular, often branched bodies to the pantheon of animals. Yet despite their superficial similarity to plants, they are indeed animals, but, like plants, they capture and concentrate dilute resources using their large surface area. Instead of relying on leaves and roots to trap light, CO_2, and water for photosynthesis, sponges have expanded their surfaces to catch the organic food particles suspended in seawater. Other, higher metazoans also evolved the ability to suspension feed, but sponges were surely the first to do so, and they continue to enjoy undiminished success. This chapter explores the functional design and diversity of these strange but engaging animals.

Sponges evolved a multicellular body uniquely specialized for **filter feeding,** the separation of suspended food particles from water by passing them through a mesh that strains out the food. The body is unique because it continuously remolds itself to fine-tune its filter-feeding system. This constant rearrangement of tissues is brought about by the ameboid movements of cells that wander throughout the sponge, adopt new positions, and change from one differentiated form to another. Such **dynamic tissues** and **totipotent cells** suggest that sponges are an intermediate form between protozoan colonies and other metazoans in which tissue and cell specializations tend to be more permanent. Because of its intermediate evolutionary status, Porifera generally is considered to be the sister taxon of the remaining Metazoa (Eumetazoa).

As the name Porifera (= pore bearers) suggests, the sponge body is exceptionally porous. Water enters through the pores and flows throughout the body in a system of flagellated canals. Food and other metabolites are removed from the water flow for use by the sponge. Adult sponges are sessile and attached organisms, although some are capable of limited movement of the body or its parts. The connective tissue is well developed and typically forms a complex and often elegant skeleton. Sponges range in size from a few millimeters to more than one meter in diameter and height (as in, for example, loggerhead sponges). The body symmetry may be radial (sphere, cone, cylinder), but asymmetry predominates. **Indeterminate growth,** enlargement without a fixed upper size limit, is common. The growth forms may be massive (thick), erect, branching, or encrusting, depending on the species and environmental conditions (Fig. 5-1). Many species are brightly colored red,

FIGURE 5-1 Porifera: growth forms of sponges and their relationship to microhabitat. **A,** Two massive sponges on top of a calcareous rock require an exposed surface, but their elevated form enables them to utilize water well above the substratum. The attachment area is a relatively small part of their total body surface. **B,** The encrusting sponges below the rock use much of their surface for attachment, but their low encrusting form enables them to exploit the limited space of crevices. **C,** The sponge on the cutaway vertical rock surface at the left actually utilizes space *in* the substratum. Arrows indicate water flow through the sponges.

purple, green, yellow, or orange, but some are brown or gray. The color results from cell pigments or endosymbionts.

Approximately 8000 species of sponge have been described. Most are marine, and they abound in seas wherever rocks, shells, submerged timbers, or corals provide firm sites of attachment. Some species, however, anchor in sand or mud. Most prefer relatively shallow water, but some taxa, including most glass sponges, live in deep water. Some 150 species have colonized fresh water.

FORM

The filter-feeding body of a sponge is built around one of three anatomical designs: asconoid, syconoid, or leuconoid. The simplest of these, the **asconoid** design, is a hollow cylinder attached by its base to the substratum (Fig. 5-2A, 5-3A). The body surface is covered by a monolayer of flat cells called the **pinacoderm** (= platter skin). The hollow interior, the **atrium** or spongocoel, is lined with a monolayer of

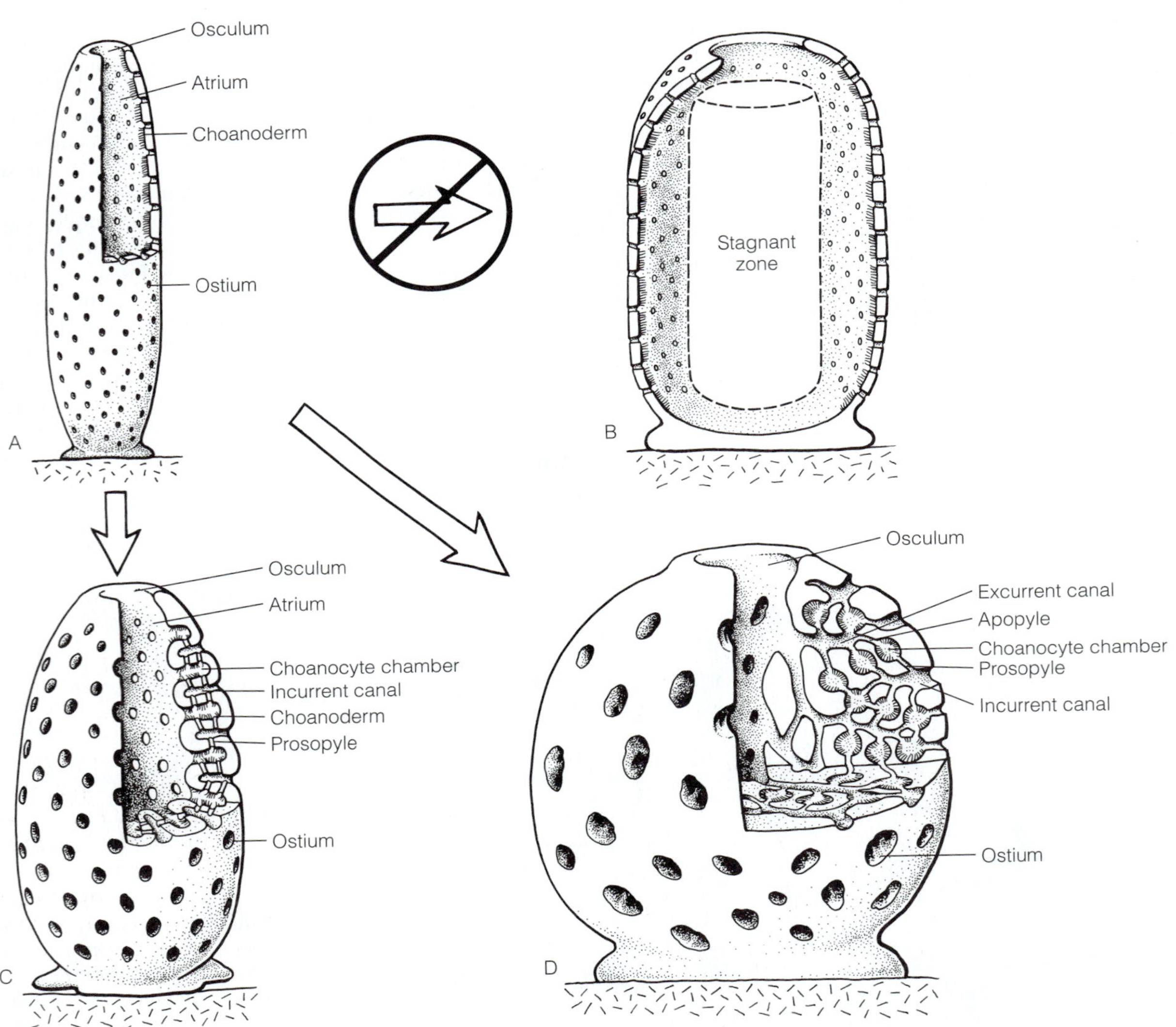

FIGURE 5-2 Porifera: relationship between body design and body size. **A,** Asconoid design, a simple cylinder approximately 1 mm in diameter, limits sponges to small body sizes. Water flow is created by the flagellated choanoderm. Because these flagella move water effectively only near the choanoderm surface, an increase in body diameter, as shown in **B,** would create a volume too large to be pumped by the flagella unless wall thickness increased or the body was redesigned. Large body size resulted from a change in *design.* **C,** The syconoid design, which results in sponges in the centimeter size range, improves the area-to-volume ratio by alternating surface inpockets with interior outpockets. **D,** In the leuconoid design, the flagellar pumps occur in thousands of spherical chambers and the water volume is restricted to minute capillary-like vessels that enter and leave the chambers. Leuconoid sponges can reach large sizes, in excess of 1 m.

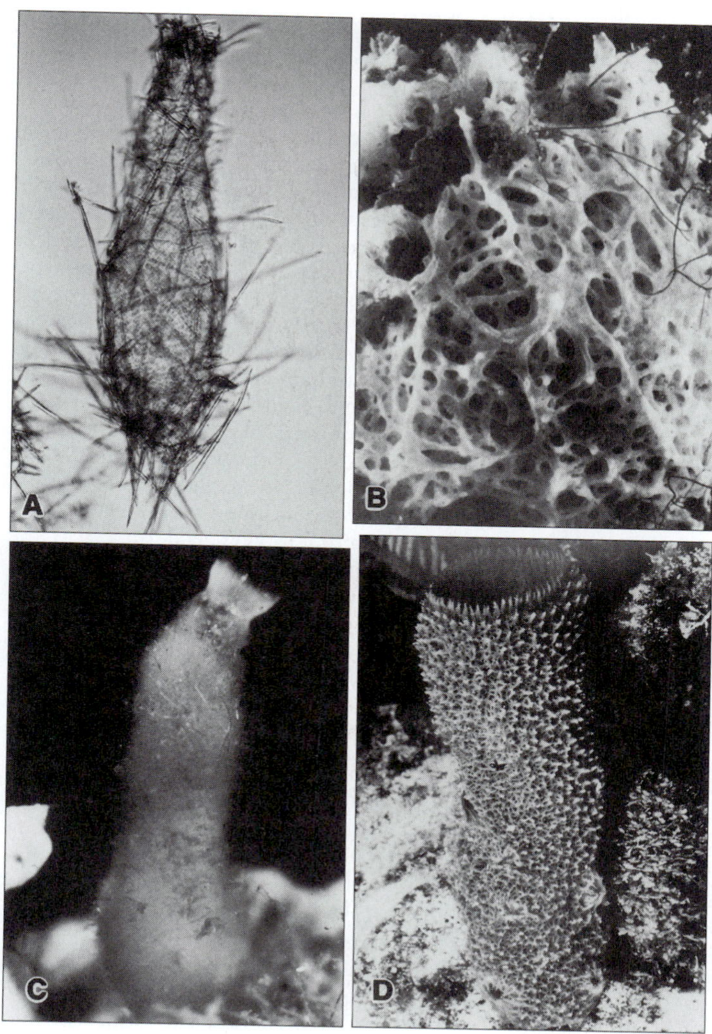

FIGURE 5-3 Porifera: living sponges of asconoid, syconoid, and leuconoid design. **A,** The asconoid calcareous sponge, *Leucosolenia* sp. (length ~1 cm); **B,** the asconoid calcareous sponge *Clathrina coriacea* (each tube in network ~1 mm in diameter); **C,** the syconoid calcareous sponge *Sycon barbadensis* (~3 cm in height); **D,** the leuconoid demosponge *Niphates digitalis* (~30 cm in height).

flagellated collar cells called **choanoderm** (= collar skin). Many small pores, known as **ostia** (sing., ostium) perforate the cylinder wall. A larger opening, the **osculum** (pl., oscula), is situated at the upper, free end of the body. The flagellated choanoderm creates a unidirectional water flow that enters the ostia, passes over the choanoderm en route to the atrium, and exits through the osculum. This circulatory system of choanoderm, pores, and chambers is called an **aquiferous system.**

All asconoid sponges are small and have cylindrical, or tubular, bodies, which typically do not exceed a diameter of 1 mm. Species of *Leucosolenia* may be a single tube or a cluster of tubes joined at their bases, whereas *Clathrina* species form a network of tubes (Fig. 5-3B). The asconoid design limits body size because growth in diameter produces an unfavorable area-to-volume ratio. The amount of water flow through the asconoid atrium (a volume) depends on the flagellated choanoderm (a surface). Because volume increases faster than area with growth, the body volume soon exceeds the pumping capacity of the choanoderm (Fig. 5-2B; see Chapter 4 for a general discussion of area and volume). Thus the attainment of larger body sizes in sponges required a change of design.

One such innovation, the **syconoid** design (Fig. 5-2C, 5-3C), increased surface area and reduced atrial volume by forming alternating inpockets and outpockets of the body wall. The arrangement can be visualized by interdigitating the extended fingers of your two hands and imagining that your fingers are hollow. The outpockets of the choanoderm are called **choanocyte chambers** (or radial canals). The inpockets of the pinacoderm are **incurrent canals.** Incurrent canals discharge into the choanocyte chambers via numerous small openings known as **prosopyles** (= front gates). At the outer surface of the sponge, water may enter the incurrent canals directly or first pass through a narrow ostium formed by a secondary growth of tissue. Thus, water flow in the syconoid aquiferous system generally follows the route: ostia → incurrent canals → prosopyles → choanocyte chambers → atrium → osculum. This new design decreases the volume of the atrium and increases the area of flagellated choanoderm. As a result, syconoid sponges generally are larger than asconoid sponges. Their diameters are typically in the range of one to a few centimeters. Syconoid sponges include species in familiar genera such as *Grantia* and *Sycon* (previously called *Scypha;* Fig. 5-3B).

Sponges of **leuconoid** design achieve the largest body sizes, ranging from a few centimeters to more than one meter (Fig. 5-3D, 5-4B–E). In leuconoid sponges, the aquiferous system is a complex network of water vessels that permeate a solid, spongy body. It consists of spherical choanocyte chambers that lie at the intersection of incurrent and excurrent canals (Fig. 5-3D, 5-6). Small-diameter **excurrent canals** and often multiple oscula replace the relatively voluminous atrium and single osculum of asconoid and syconoid sponges. Water enters a leuconoid sponge via surface ostia before flowing into incurrent canals (Fig. 5-2D). From the incurrent canals, water passes through prosopyles into choanocyte chambers. Water exits each choanocyte chamber through an **apopyle** (= back gate) and then flows through the excurrent canals, which become progressively larger in diameter as they join with other excurrent canals. Large excurrent canals discharge exhaust water to the exterior via one or more oscula.

The choanocyte chambers of leuconoid sponges occur in high densities. In *Microciona prolifera,* for example, there are approximately 10,000 chambers/mm^3, each 20 to 39 μm in diameter and containing approximately 57 flagellated cells (choanocytes). The leuconoid design vastly increases the area of flagellated choanoderm while minimizing the water volume that must be moved.

Most shallow-water marine sponges and all freshwater sponges are built on the leuconoid design. Leuconoid sponges achieve large body size because water pumping is decentralized: Any growth increment produces a sufficient number of new choanocyte chambers to ventilate the new increment. Common leuconoid growth forms include crusts, mounds, branches, fingers, plates, tubes, and vases (Fig. 5-3D, 5-4B–E).

BODY WALL

The body wall is thin in asconoid sponges, thick in most leuconoid sponges, and intermediate in thickness in syconoid species, but a clear-cut distinction occurs between the glass sponges (Hexactinellida) and all others (Demospongiae and Calcarea). In demosponges and calcareous sponges (among which are most sponges), the body wall is cellular, but the glass-sponge body wall is a syncytium. A **syncytium** is a large or extensive multinucleated cytoplasm enclosed by an external membrane but not divided into cells by internal membranes.

Cellular

The bodies of demosponges and calcareous sponges (which together make up the Cellularia) are composed of cells organized into two types of tissue, epithelioid and connective (Figures 5-5, 5-6). **Epithelioid** tissue resembles epithelium (see Chapter 4), but it lacks epithelium's intercellular junctions and hemidesmosomes and is not underlaid by a basal lamina (see Chapter 6). Sponge epithelioid tissues are the pinacoderm, which covers the outer surface of the body **(exopinacoderm)** and lines the incurrent and excurrent canals **(endopinacoderm),** and the flagellated choanoderm, which forms the atrial lining (asconoid design) and the lining of the choanocyte chambers (syconoid and leuconoid designs). The connective-tissue layer between the pinacoderm and the choanoderm is called the **mesohyl** (= middle wood) because it forms a bushy, fibrous network that is especially obvious in bath sponges.

The pinacoderm consists chiefly of two types of differentiated cells. By far the most common of these is the **pinacocyte** (= platter cell). Pinacocytes are flattened (squamous) cells that abut each other edge-to-edge to form a skinlike cellular pavement over the body surface and line the incurrent and excurrent canals (Fig. 5-5, 5-6). Pinacocytes generally lack flagella, except for species of *Plakina* and *Oscarella* that have a flagellated endopinacoderm lining their canals. A less common but nevertheless important pinacoderm cell is the **porocyte** (Fig. 5-5, 5-6). Porocytes form the ostia of all asconoid as well as many syconoid and leuconoid sponges. They also constitute the prosopyles and apopyles of many syconoid and leuconoid sponges, although in others the ostia are gaps between adjacent pinacocytes and the prosopyles can be simple gaps in the choanoderm. Each porocyte surrounds a pore, the diameter of which is regulated by contraction of cytoplasmic filaments. Thus, porocytes are miniature sphincter valves.

The mesohyl is the only layer of the sponge body wall that typically is not bathed with environmental water. In this sense, the mesohyl is the sole internal compartment of the body. As a connective tissue, the mesohyl is composed of a proteinaceous, gel-like matrix that contains differentiated and undifferentiated cells as well as skeletal elements (Fig. 5-5, 5-6). Among the many cells present in the mesohyl are macrophage-like **archeocytes** (= progenitor cells), which are large ameboid cells bearing a conspicuous nucleus and numerous large lysosomes. Archeocytes are totipotent and can differentiate into any other type of sponge cell. They are also phagocytic and play a role in digestion and internal transport. **Lophocytes** (= crest cells) are archeocyte-like ameboid cells that secrete collagen fibers from their trailing end as they move through the mesohyl. They produce and maintain the fine collagen fibers of the mesohyl. **Spongocytes,** which occur only in the taxon Demospongiae, resemble archeocytes, but secrete collagen that polymerizes into thick skeletal fibers known as **spongin** (see Skeleton, the next section). **Sclerocytes** (= hard cells) secrete the mineralized skeletal spicules of many sponges (see Skeleton). **Myocytes** (= muscle cells) are musclelike cells containing actin and myosin that aggregate around the oscula of some demosponges. They regulate the size of the oscular aperture and thus help to control water flow through the sponge. Finally, **oocytes** and **spermatocytes,** which will be described later in more detail, are reproductive cells that undergo gametogenesis in the mesohyl to form sperm and eggs.

The choanoderm consists of flagellated collar cells, or **choanocytes** (= collar cells), that generate the water flow through the sponge. Choanocytes have an apical collar of long microvilli around a single flagellum (Fig. 5-5). The collar is in

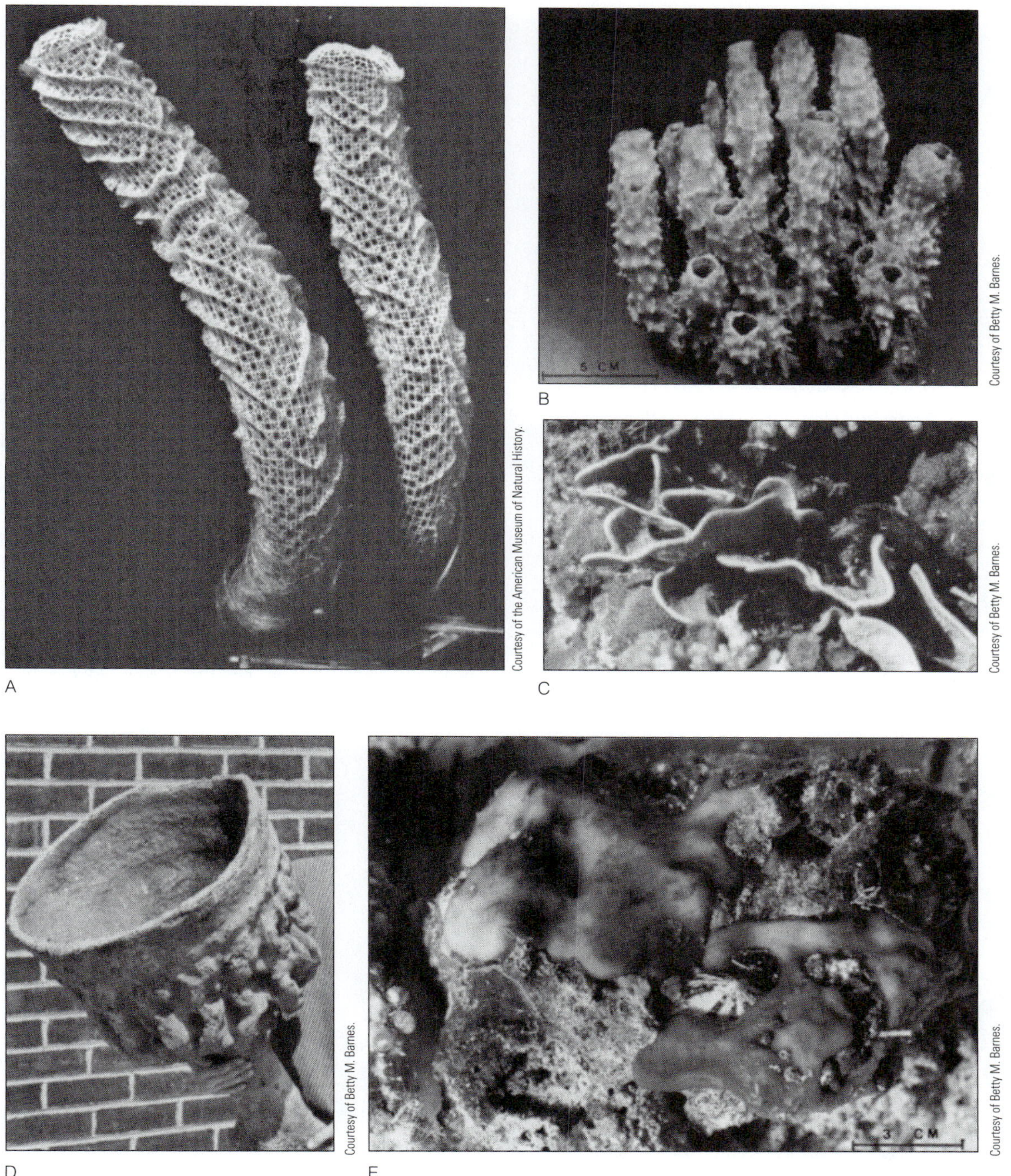

FIGURE 5-4 Porifera: living **(B, C, and E)** and preserved sponges **(A and D)**. **A,** *Euplectella aspergillum,* Venus's flower basket, a glass sponge (Hexactinellida) in which the siliceous spicules are fused to form a lattice. **B,** *Callyspongia vaginalis,* a tropical leuconoid sponge (Demospongiae) with a tubular body form. **C,** *Phyllospongia,* a leaflike sponge (Demospongiae) on a reef flat in Fiji. **D,** *Poterion,* a large goblet-shaped leuconoid sponge (Demospongiae). **E,** *Chondrilla nucula,* a common West Indian sponge (Demospongiae) with a tough, cartilaginous spongin skeleton.

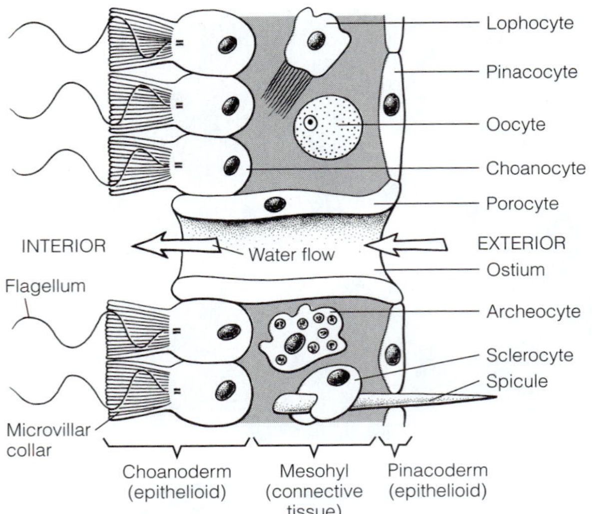

FIGURE 5-5 Porifera body wall: section through an asconoid sponge. *(Modified and redrawn from Rigby, J. K. 1987. Phylum Porifera. In Boardman, R. S., Cheetham, A. H.,and Rowell, A. J. (Eds.): Fossil Invertebrates, Blackwell Science, Cambridge, MA. pp. 116–139.)*

the form of a cylinder or an inverted cone. The basal part of the choanocyte flagellum of many species (such as *Microciona* sp. and *Grantia compressa*), if not all, bears a bilateral **vane,** as in the choanoflagellates (Fig. 4-13A).

Syncytial

The glass sponge's body wall lacks the sheetlike pinacoderm pavement that covers the body and lines the aquiferous system of cellularian sponges. Instead, the living tissue in hexactinellids is arranged in three-dimensional, cobweb-like strands called a **trabecular syncytium** or network (Fig. 5-7). The membranes that normally separate cells are absent and the cytoplasm is continuous and uninterrupted throughout the syncytium.

A cellular choanoderm also is absent; in its place is another syncytium, the **choanosyncytium.** Individual **collar bodies,** each with a collar and flagellum but lacking a nucleus, arise from the surface of the choanosyncytium. Each group of collar bodies occupies a syconoid-like pocket that is supported by the trabecular network. The many collar bodies of each pocket arise developmentally as outgrowths of a single nucleated stem cell, the choanoblast.

Each strand in the trabecular syncytium surrounds and encloses an axis of mesohyl. The mesohyl contains bundles of collagen fibers, spicules, and *cells*—sclerocytes, archeocytes, and presumably germ cells. In *Rhabdocalyptus dawsoni,* the mesohyl cells are joined to the trabecular syncytium by slender cellular extensions and are themselves partially syncytial (Fig. 5-7B), but in *Dactylocalyx pumiceus,* these cells are reported to be independent of the syncytium. Thus, the glass sponges may be constituted of a unique combination of cellular and syncytial tissues.

The gross anatomy of hexactinellids is syconoid, but the aquiferous system, because it has both incurrent and excurrent canals, resembles the leuconoid design. Water in

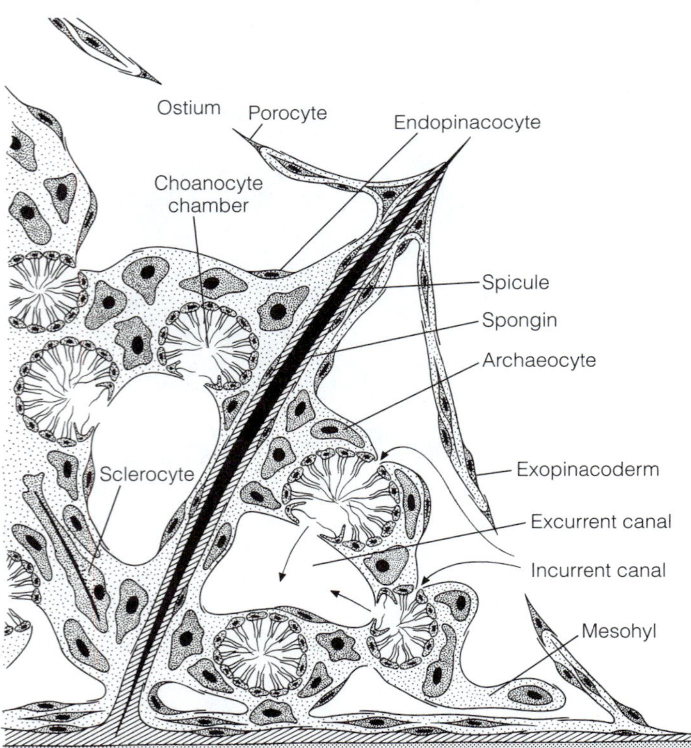

FIGURE 5-6 Porifera body wall: section through the body wall of a freshwater leuconoid sponge. *(From Ax, P. 1996. Multicellular Animals: A New Approach to the Phylogenetic Order in Nature. Springer-Verlag, Berlin, Heidelberg, NY. p. 71.)*

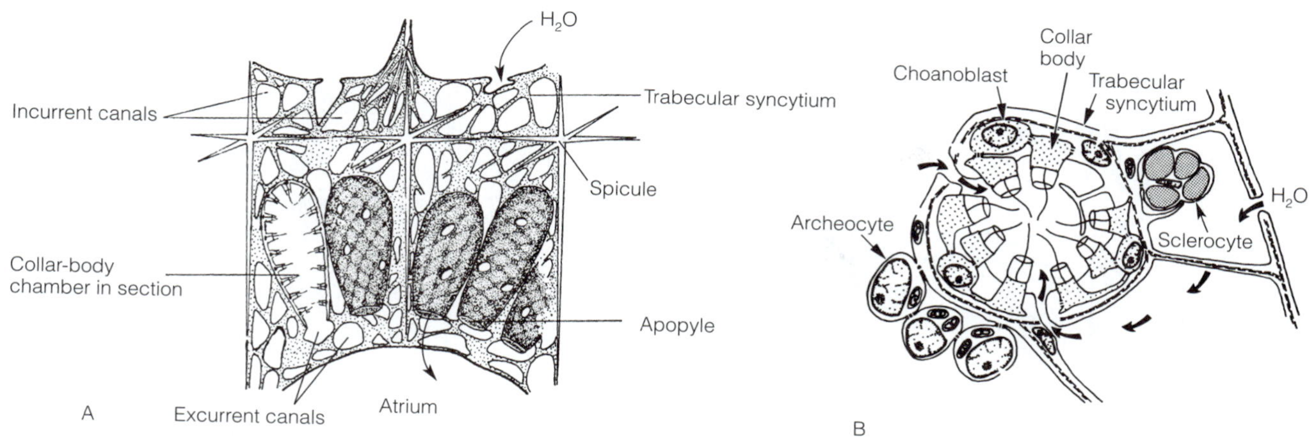

FIGURE 5-7 Porifera body wall: glass (hexactinellid) sponges. **A,** *Euplectella aspergillum* (Venus's flower basket). **B,** *Rhabdocalyptus dawsoni.* *(A, After Schulze, in Bergquist, P. R. 1978. Sponges. Hutchinson, London. pp. 59 and 26; B, From Leys, S. P. 1995. Cytoskeletal architecture and organelle transport in giant syncytia formed by fusion of Hexactinellid sponge tissues. Bio. Bull. 188:241–254)*

hexactinellids flows from: surface openings in the trabecular network → incurrent canals → collar-body chambers → excurrent canals → atrium → osculum.

WATER PUMPING

The volume of water pumped by a sponge is impressive. In general, a sponge pumps a volume of water equal to its body volume once every 5 seconds. Because water is incompressible, the volume entering must be equal to the volume exiting the sponge at any moment. The flow velocity is fastest through the osculum and slowest in the choanocyte chambers, because these two regions have, respectively, the smallest and the largest total cross-sectional areas in the aquiferous system (Fig. 5-8). Many sponges, however, can slow the overall flow rate or even stop it entirely to avoid the intake of silt. They do this by regulating the size of the osculum (using contractile myocytes), closing the ostia (sometimes using tubular porocytes), or adjusting the choanocyte flagellar beat (as in glass sponges, which lack myocytes and porocytes).

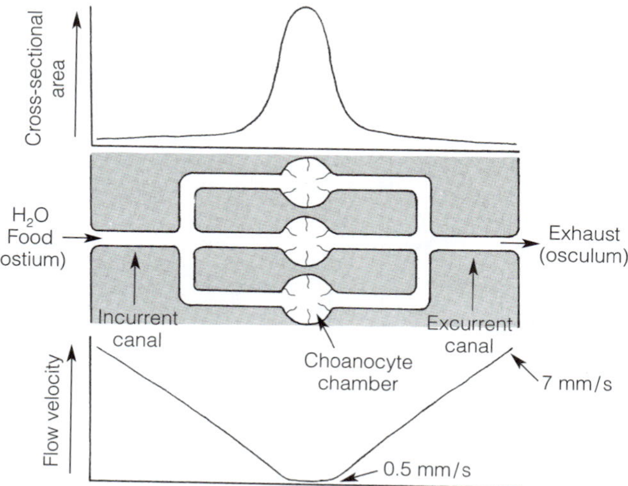

FIGURE 5-8 Porifera: water pumping and physiological exchange. The 30-times greater cross-sectional area of choanocyte chambers as compared with ostia and oscula causes the water flow velocity to slow in the chambers. The slow flow improves the effectiveness of food capture by the choanocytes. *(Based on data in Reiswig, 1975b)*

The water current is produced by the activity of the choanocyte flagella (Fig. 5-5). The undulatory beat of each flagellum is in a single plane. The flagellar vane, which is restricted to the collar region of the flagellum, may help to "pump" water from the collar. In at least one species *(Trochospongilla pennsylvanicus),* the beating plane of the choanocyte flagellum shifts every few seconds, eventually rotating around 360°. The choanocyte flagella and collars are oriented away from the ostia (in the asconoid design) or prosopyles (in the syconoid and leuconoid designs), and each flagellum beats from base to tip, driving water toward the excurrent canals and osculum.

The oscula of many sponges are situated on chimneys well above the main body and ostia (Fig. 5-1A). That elevated position exposes the oscula to environmental water currents faster than those that occur near the base of the sponge. The higher-velocity flow over the chimneys lowers the pressure at the oscula in relation to the ostia and induces a flow from the high-pressure to low-pressure (ostium-to-osculum) end of the system. Because most sponges are exposed to significant ambient water currents, such induced flows undoubtedly supplement flagellar pumping and conserve metabolic energy.

SKELETON

Whatever their growth form, most sponges live in moving water and support themselves with a well-developed skeleton. The skeleton is chiefly a mesohylar endoskeleton, but an

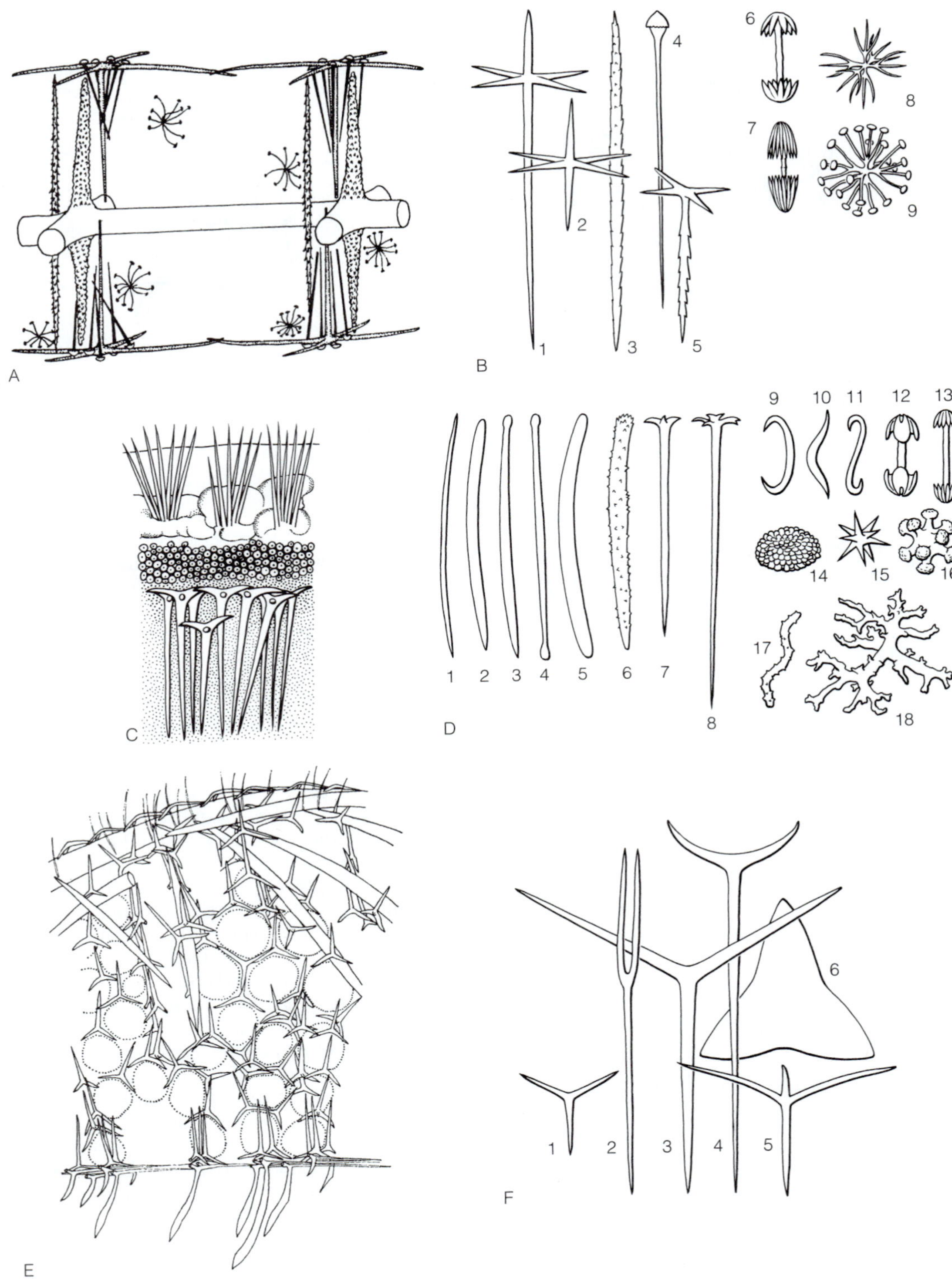

FIGURE 5-9 Porifera: spicular skeletons and spicules. **A,** Body-wall skeleton of *Farrea sollasii* (Hexactinellida). **B,** Hexactinellid megascleres and microscleres. Megascleres: 1–5, microscleres: 6–9. **C,** Body-wall skeleton of *Geodia* (Demospongiae). **D,** Demosponge megascleres and microscleres. Megascleres: 1–8, microscleres: 9–18. **E,** Body-wall skeleton of the leuconoid sponge *Afroceras ensata* (Calcarea). **F,** Selection of calcarean spicules. *(A and E, From Bergquist, P. R. 1978. Sponges. Hutchinson, London. 268 pp.; C, Modified and redrawn from Bergquist, 1978)*

exoskeleton also may occur regionally or over the entire body. The stiffness or rigidity of the skeleton varies widely among species and growth forms. Among soft encrusting forms, such as *Oscarella* and *Halisarca,* the sole skeleton is the gelatinous mesohyl supported only by fine collagen fibers. More commonly, the mesohylar matrix is supplemented with mineral **spicules,** spongin, or both. Although spicules occur principally in the mesohyl, they can project freely through the surface pinacoderm, thus affording the sponge some protection (Fig. 5-3A,C). Such projecting spicules commonly guard the oscula and sometimes the ostia.

Spicules stiffen the mesohyl to varying degrees depending on their density, arrangement, and the extent to which they fuse or interlock. In the extreme, a spicular skeleton can be a rigid, brittle, three-dimensional lattice or framework, as in the glass sponge *Euplectella aspergillum* (Fig. 5-4A, 5-9A) or the calcareous sponge *Minchinella* sp. The relict sphinctozoan sponge, *Vaceletia crypta,* has a chambered calcareous *exo*skeleton. The calcifying demosponges ("sclerosponges") secrete a massive basal exoskeleton of $CaCO_3$ on which the body rests. These sclerosponges also secrete siliceous spicules in the mesohyl (Fig. 5-10).

Some sponges lack spicules, but secrete organic spongin (Fig. 5-11A,B). Such sponges, for example the bath sponges *(Spongia, Hippospongia),* are often compressible, elastic, and "spongy." A high density of spongin produces a skeleton that is firm, tough, and rubbery, as in the tropical chicken-liver sponge, *Chondrilla nucula* (Fig. 5-4E).

Spongin and spicules occur together in most species of sponges. In some species of *Haliclona,* spongin welds together the tips of spicules to form a skeletal network (Fig. 5-11C). In other parts of the skeleton, spicules are embedded into the spongin fibers themselves (Fig. 5-11D). Sometimes, foreign material, such as sand grains, is incorporated into the skeleton as a substitute for the spicules, as happens in the tropical ethereal-blue sponge *Dysidea etheria* and other species (Fig. 5-11E). In *Dysidea janiae,* the sponge produces no spicular skeleton of its own, but instead uses the calcareous skeleton of its symbiotic red alga *(Jania).* The result of these and other combinations of spicules and spongin is a wide variety of skeletal properties, from soft and spongy to hard and brittle.

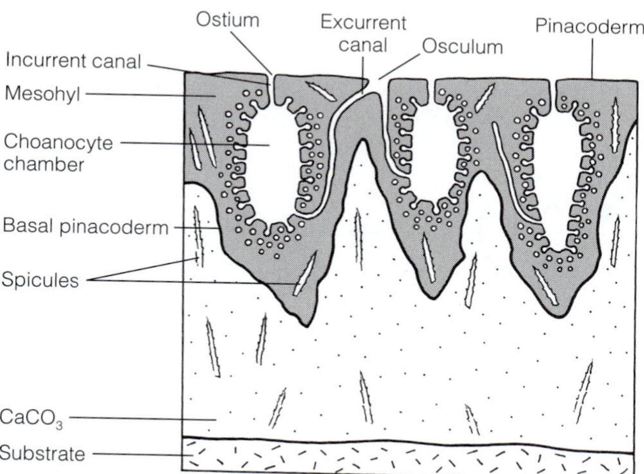

FIGURE 5-10 Porifera: body wall and skeleton of a calcifying sclerosponge (Demospongiae: Ceratoporellidae). The calcareous exoskeleton is secreted by the basal exopinacoderm and contains embedded siliceous spicules. Siliceous spicules also occur in the mesohyl of the living tissue. *(Modified and redrawn after Hutchinson from Bergquist, P. R. 1978. Sponges. Hutchinson, London. 268 pp.)*

Spicules are siliceous (SiO_2) or calcareous ($CaCO_3$) elements whose composition, size, and shape are used at all levels in the classification of sponges (Fig. 5-9). As a result, an extensive nomenclature describes the forms and sizes of spicules. At the most general level, spicules are separated into two size classes, large **megascleres** and small **microscleres.** Megascleres typically form the principal skeletal framework, whereas the considerably smaller microscleres may support the pinacodermal lining of the canal system or, in high density, toughen the body wall (Fig. 5-9A,C,E). Megasclere names are based on the spicule's number of axes or number of rays or points. The suffix *-axon* refers to the number of axes; *-actine* indicates the number of points. A monaxon spicule, for example, has one axis and is shaped like a needle or rod, although it may be straight or curved, with pointed, knobbed, or hooked ends (Fig. 5-9D1-6). Triaxons have either three rays (triactines; Fig. 5-9F) or six (hexactines; Fig. 5-9B).

Spicules are secreted extracellularly by sclerocytes in calcareous sponges, intracellularly in sclerocytes in demosponges, and intrasyncytially in glass sponges. From one to several sclerocytes typically secrete a single spicule in the calcareous sponges. A three-rayed spicule, for example, originates between three sclerocytes derived from a single stem cell (scleroblast; Fig. 5-11G). Each member of the trio then divides and one ray of the spicule is secreted between each pair of daughter cells. The three rays fuse at their bases. Each of the three pairs of sclerocytes now moves outward along a ray, one cell lengthening its end while the other cell thickens its base (Fig. 5-11G). The secretion of a siliceous monaxon spicule is initiated around an organic filament in an intracellular vesicle (Fig. 5-11F). As the spicule crystallizes and grows, the cell first elongates and then divides into two cells, each of which adds additional silica to a growing tip of the spicule.

LOCOMOTION AND DYNAMIC TISSUES

Although sponges are basically sessile animals, some species have a limited capacity for locomotion. Both freshwater *(Ephydatia)* and marine *(Chondrilla, Hymeniacidon, Tethya)* species can move over a substratum at rates of 1 to 4 mm/day (Fig. 5-12A). The movement apparently results from the collective ameboid movements of pinacodermal and other cells. Other sponge movements include whole-body contraction *(Clathrina coriacea)* and, in many species, constriction of oscula by myocytes. These movements probably arrest or limit flow through the aquiferous system in response to an

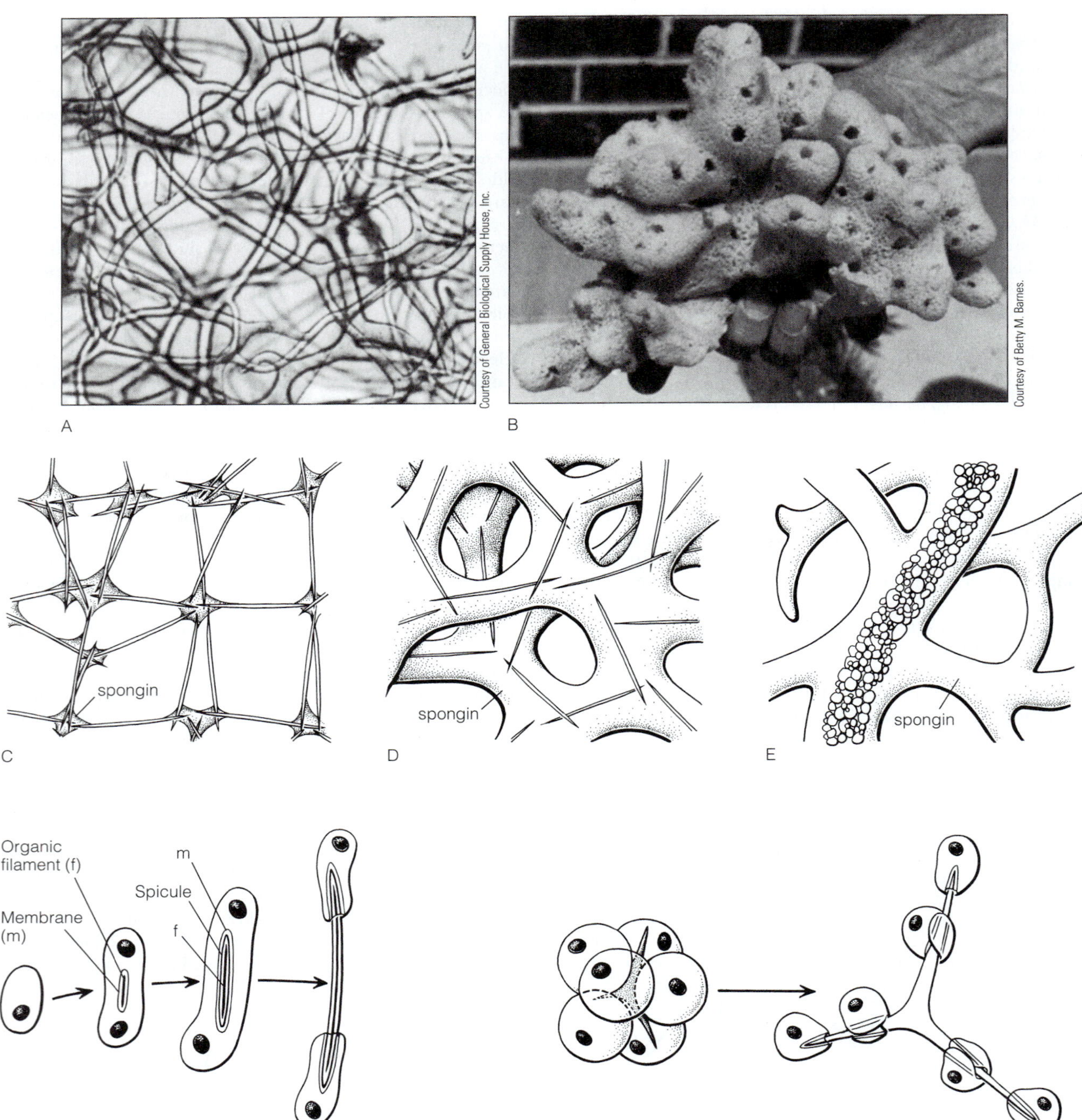

FIGURE 5-11 Porifera: spongin skeleton, spongin-composite skeletons, and spicule secretion. **A,** Photomicrograph of spongin fibers (they appear translucent). **B,** Spongin skeleton of a commercial bath sponge, *Spongia officianalis,* from the Mediterranean. The large openings are oscula. **C,** Siliceous spicules (oxeas) glued together at their tips with spongin to form a network in *Haliclona rosea.* **D,** Spongin fibers with embedded siliceous spicules (oxeas) form a reinforced network in *Endectyon.* **E,** Spongin network of *Hippospongia communis* stiffened by the incorporation of foreign material, especially sand grains, into the spongin fibers. **F,** Secretion by sclerocytes of a siliceous monaxon (oxea). **G,** Secretion by sclerocytes of a calcareous triaxon. *(C, Modified and redrawn from Hartman, W. D. 1958. Natural history of the marine sponges of southern New England. Bull. Peabody Mus. Nat. Hist. Yale Univ. 12:1–155. D and E, Redrawn from Kaestner, A.1980. Lehrbuch der Speziellen Zoologie 1(1): Wirbellosen Tiere. Gustav Fischer Verlag, Stuttgart. 318 pp.)*

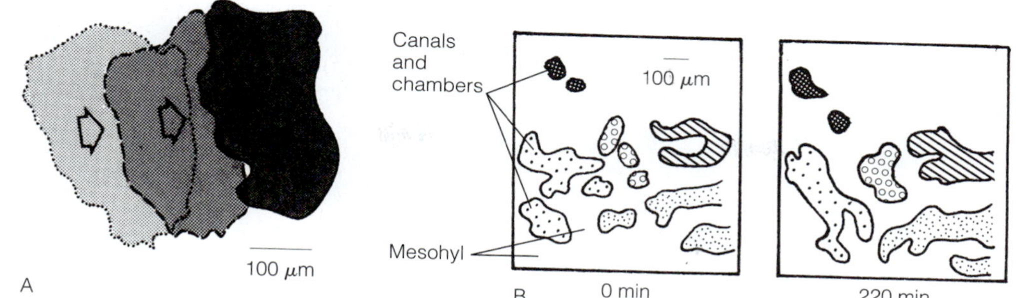

FIGURE 5-12 Porifera: locomotion and tissue dynamics of the freshwater sponge *Ephydatia fluviatilis*. **A,** Locomotion of an isolated sheet of exopinacocytes. **B,** Dynamic remodeling of the aquiferous system. Each of the shaded areas outlines canals and choanocyte chambers of a living sponge, as seen under a microscope. Between time 0 and approximately 4 hours later, canal fusion and reshaping are evident. *(Modified and redrawn from Bond, C. 1992. Continuous cell movements rearrange anatomical structures in intact sponges. J. Exp. Zool. 263:284–302.)*

environmental challenge, such as a sudden increase in waterborne silt.

A hallmark of sponges is the dynamism of their tissues. Mesohyl cells, all of which are ameboid, are more or less in constant motion. Similarly, endopinacocytes and choanocytes can move about to remodel the aquiferous system (Fig. 5-12B). This remodeling, which involves the addition or fusion of flagellated chambers and the merging and branching of canals, may "fine-tune" the system to optimize water flow as the sponge grows or as it encounters changes in environmental water currents. The absence of intercellular junctions, basal lamina, and hemidesmosomes in sponge tissues allows these independent and frequent cell movements.

PHYSIOLOGICAL COMPARTMENTALIZATION

The physiological importance of water flow through the aquiferous system of a sponge cannot be overstated. This single system accomplishes the tasks of gas exchange, food acquisition, waste disposal, and the release of sperm and larvae. The functions associated with, for example, the mammalian trachea and lungs, alimentary canal, circulatory vessels, kidneys, and gonoducts are, in sponges, combined in this one multifunctional system. Such a low level of physiological compartmentalization in sponges has two implications. First, because there is little segregation of function, integrating systems, such as the nervous or endocrine systems, are not well developed or necessary. In fact, sponges lack nervous tissue (discussed later in this chapter). Second, because of functional overlapping in tissues and cells, the efficiency of any individual function is likely to be low in comparison to the efficiency in animals such as ourselves, which have a specialized compartment for each function. This minimal efficiency manifests itself as a low level of activity. Few animals are less mobile or more plantlike than sponges.

We should not think that, because of their low level of compartmentalization and integration, sponges are somehow at a disadvantage. On the contrary, their low level of organization allows them to adaptively remold their bodies, to regenerate readily after damage, and to clone themselves.

NUTRITION

Sponges filter food particles from water flowing through their bodies. Generally, the filtered particles range in size from 50 μm to 1 μm or less. This range includes unicellular plankton, such as dinoflagellates and bacteria, viruses, small organic debris, and perhaps even dissolved organic material. In tropical seas, where sponges are abundant, the smallest fractions of food are approximately seven times more available than the larger size classes. All sponge cells can ingest particles by phagocytosis.

The food-trapping filters are the incurrent canals, which progressively decrease in diameter as they penetrate inward, and choanocytes (Fig. 5-13). Food and other particles are filtered as they lodge in different parts of the system, depending on their diameter. The largest particles, exceeding about 50 μm in diameter, are too large to enter an ostium and can be phagocytosed by cells of the exopinacoderm. Particles in the size range of 5 to 50 μm lodge in an incurrent canal and may be phagocytosed by endopinacocytes or by archeocytes that have entered the canal (either between pinacocytes or through porocytes in the canal lining). The smallest, bacteria-size particles enter the choanocyte chambers and are removed by phagocytosis or pinocytosis on the choanocyte surface. The choanocyte collar of microvilli, and its extracellular matrix, may be the mesh that traps the finest material.

Both choanocytes and archeocytes engulf and digest particles in vesicles, but the choanocyte often transfers particles to the archeocyte for digestion. The archeocytes probably also store nutrients such as glycogen or lipids.

Carnivorous species occur in the demosponge family Cladorhizidae. These sponges trap crustaceans and other small animals on sticky cellular threads that extend out from the surface of the sponge. Once an animal is trapped, the threads shorten, drawing the prey onto the body surface,

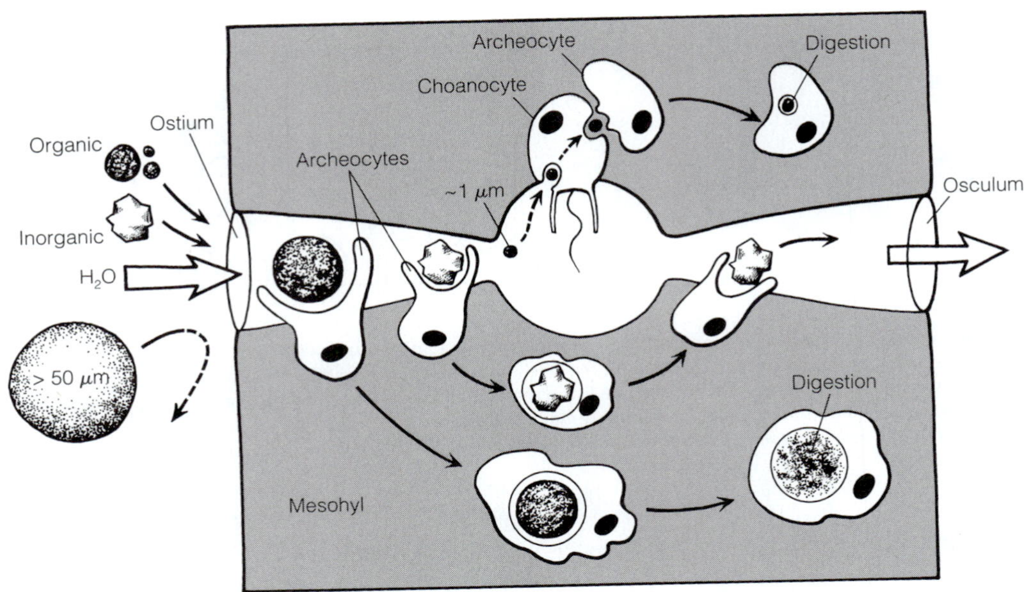

FIGURE 5-13 Porifera: diagrammatic summary of filter feeding, digestion, and egestion. See text for explanation.

which slowly overgrows the prey and consumes it, presumably with archeocytes. These strange sponges lack choanocytes and an aquiferous system.

The two sources of particulate wastes in sponges are indigestible products of intracellular digestion and inorganic mineral particles that enter the sponge in the water stream. Mineral particles must be removed from the incurrent canals, which they would otherwise block and inactivate. An inorganic particle that lodges in an incurrent canal is phagocytosed by an archeocyte, which transports it to the downstream side of the canal system and then exocytoses it into an excurrent canal (Fig. 5-13). In those species that incorporate foreign material into their skeletons, the archeocyte may transport the intercepted particle to a site of skeleton secretion.

Many sponges, both marine and freshwater, harbor photosynthetic endosymbionts in their tissues and derive a nutritional benefit from the photosynthate. Freshwater sponges typically harbor green algae (zoochlorellae) in archeocytes and other cells. Marine sponges—both calcareous and demosponges—may host dinoflagellates (zooxanthellae) or, more commonly, Cyanobacteria. One species, *Mycale laxissima* in Belize, incorporates both green and red algae in the spongin fibers of its skeleton. The cyanobacterial symbionts of some sponges, including *Verongia*, may constitute up to one-third of the sponge's biomass. Such sponges live in shallow, well-lit habitats and may have symbiotic bacteria restricted to the outer layers of the body. Excess photosynthate in the form of glycerol and a phosphorylated compound are utilized by the sponge. Some sponges studied on Australia's Great Barrier Reef obtain from 48 to 80% of their energy requirements from their Cyanobacteria. Sponges frequently contain intra- and extracellular bacteria in addition to the Cyanobacteria and other symbionts. The significance of such bacteria, however, is unknown.

INTERNAL TRANSPORT, GAS EXCHANGE, AND EXCRETION

Because the aquiferous system ventilates the entire body to within 1 mm of all cells, simple diffusion accounts for the transport of gases and metabolic wastes (largely ammonia) between the body and environmental water in the aquiferous system. Nutrients, too, probably diffuse from the widespread sites of intracellular digestion, although archeocytes, by ameboid movement, deliver nutrients to developing gametes and tissues throughout the body. At least one species of *Aplysina* (also known as *Verongia*) has specialized internal fibers that serve as tracks for the movement of nutrient-laden archeocytes.

Internal transport of food in glass sponges is intrasyncytial. Once ingested by the collar bodies, food-containing vesicles are transported by dynein motor molecules on bundles of microtubules that extend throughout the syncytium of the sponge. This mode of transport is identical to vesicular transport in radiolarian and foram pseudopodia (Chapter 3), as well as in the nerve-cell axons of higher animals.

The near absence of intercellular junctions in pinacoderm and choanoderm suggests that these layers constitute a poor regulatory barrier between the mesohyl and water of the external environment. Physiologically speaking, sponges are said to be "leaky" animals. Accordingly, the composition of their interstitial fluid (the fluid between cells) is likely to be similar to that of the environmental water, even among freshwater species. Most cells of freshwater sponges contain contractile vacuoles, but those vacuoles are osmoregulating for individual cells and not for the sponge body as a whole.

INTEGRATION

Sponges lack nerve cells and nervous tissue, though some are capable of limited impulse conduction. In most cases, this conduction is a slow "epithelial" spread of electrical activity over a few millimeters that results in a local myocyte contraction in response to a local stimulus. Such impulse conduction is slow because specialized intercellular junctions (gap junctions), which promote epithelial conduction, are absent. Thus, the membranes between cells tend to isolate rather than conduct the wave of depolarization. An exception to this generalization, however, occurs in the syncytial tissues of glass sponges. In *Rhabdocalyptus dawsoni,* electrical impulses (action potentials) are propagated rapidly along the syncytial strands from a point of stimulation to all parts of the sponge. This activity arrests flagellar beating and shuts down water pumping by the sponge.

BIOACTIVE METABOLITES AND BIOLOGICAL ASSOCIATIONS

Many sponges produce metabolites that may prevent settlement of other organisms on their surfaces or deter grazing predators. Nine out of 16 Antarctic sponges and 27 of 36 Caribbean species were found to be toxic to fish. The fish toxins, however, did not necessarily discourage nonfish grazers, and some fish, such as angelfishes, filefishes, and the moorish idol, are specialized spongivores. Turtles, especially the Hawksbill turtle, commonly feed on sponges and up to 95% of their feces may consist of siliceous spicules. A taxon of sea slugs (dorid nudibranchs) specializes on sponge species in a manner similar to certain caterpillars on their host plants. Some sponges use metabolites to compete for space with other organisms. For example, the Caribbean chicken-liver sponge *(Chondrilla nucula)* releases compounds that kill adjacent stony corals, allowing the sponge to overgrow their skeletons. Some species have distinctive odors, such as the garlic sponge, *Lissodendoryx isodictyalis.* A few, such as the Caribbean fire sponge, *Tedania ignis,* cause a severe rash when handled. Various sponge biochemicals are being investigated to determine their potential medical and commercial benefits.

Many sponges harbor endosymbionts that occupy space in the aquiferous system and take advantage of the water flow and protection afforded by their host. Some large leuconoid sponges are veritable apartment houses for shrimps, amphipods, and brittle stars. One investigator collected over 16,000 snapping shrimps from within the water canals of one large loggerhead sponge. Certain worms (spionid polychaetes) infest, eat, and thereby adopt the color of their sponge host. Decorator crabs attach sponges, algae, and other sessile organisms to their backs to form a microcommunity. The community grows on this mobile substratum, providing the crab with an effective camouflage. Certain other crabs (Dromiidae) cut out a cap of sponge and affix it to their back, or attach a sponge fragment that then grows, covers, and camouflages the crab.

BIOEROSION

Species of the demosponge family Clionidae play an important role in the breakdown of calcareous shell and coralline rock in the sea (Fig. 5-14). *Cliona celata,* for example, bores

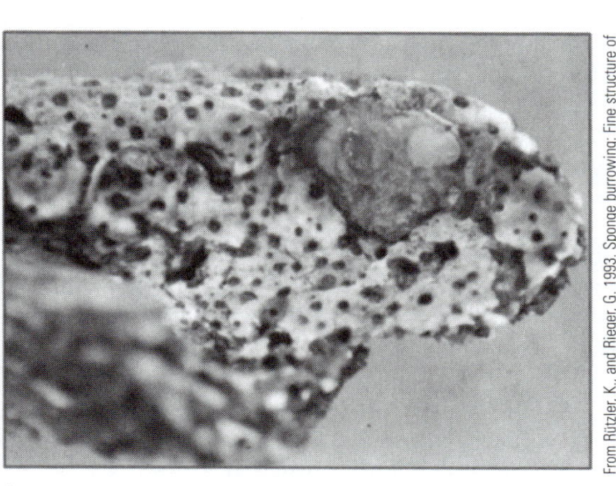

A

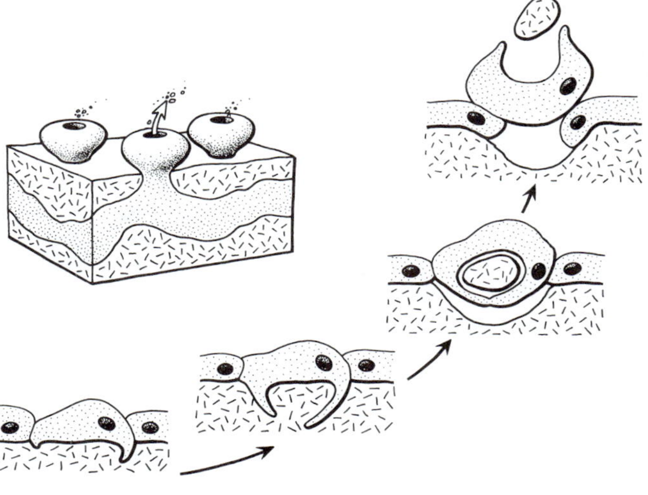

B

FIGURE 5-14 Porifera: bioerosion by boring demosponges (Clionidae). **A,** Remains of a clamshell that has been riddled by *Cliona lampa.* **B,** The shell-boring sequence: etching cell penetrates shell by chemical dissolution; shell chip is cut away from shell; chip is released into the aquiferous system; chip is discarded in exhalant water flow. *(B, Modified and redrawn after Hatch from Pomponi, S. A. 1980. Cytological mechanisms of calcium carbonate excavation by boring sponges. Int. Rev. Cytol. 65:301–319.)*

into mollusc shells and creates a network of subsurface tunnels. At regular intervals, a papilla of sponge tissue erupts through a hole in the surface of the shell (Fig. 5-14B). Some papillae bear ostia, others oscula. Shells washed ashore at this stage appear to be peppered with birdshot (Fig. 5-14A). Eventually, the sponge overgrows the entire shell and finally destroys it completely. Specialized archeocytes called **etching cells** are responsible for eroding the shell. Each etching cell chemically cuts away a small chip of shell. The freed chip is released into the aquiferous system and eventually discharged at the surface through an osculum. Divers regularly see these sponges "spit" particles. Because it affords the sponge physical protection from grazing predation, boring into shell or rock may enhance the survival of juvenile and adult sponges.

REPRODUCTION

Clonal Reproduction

Sponges reproduce clonally (asexually) by fragmentation, budding, and by the formation of overwintering propagules called **gemmules.** Fragmentation primarily results from current or wave damage, and perhaps from damage done by grazing carnivores. The dislodged fragments rely on their remodeling capacity for regeneration. The fragment soon attaches to the substratum and reorganizes itself into a functional sponge. An extreme form of fragmentation—dissociation of a sponge into individual cells or clumps of cells—can be accomplished in the laboratory by squeezing a piece of sponge through finely woven cloth. This experiment was first conducted by the zoologist H. V. Wilson early in the 20th century. Since that time, it has been repeated frequently to demonstrate species recognition at the cellular level, to model morphogenesis, and to study the mechanisms by which cells recognize and adhere to each other.

Budding is uncommon but does occur in a few sponges. In *Clathrina*, for example, the free ends of the asconoid tubes are said to swell into buds, break free, and then attach and grow into another sponge. Some species of *Tethya* produce stalked buds. Species of *Oscarella* and *Aplysilla* are reported to produce papillae that self-amputate and grow into new sponges.

Many freshwater sponges and a few marine species produce hundreds to thousands of sporelike gemmules, typically in the fall of the year (Fig. 5-15A). The autumn gemmules of freshwater species may enter **diapause,** a state of near metabolic arrest, and then require a period of very cold temperature before they are activated, germinate, and differentiate into a new sponge, usually in the spring. While the gemmule is in diapause, it is resistant to environmental extremes of temperature, salinity, and desiccation. A standard practice of sponge biologists, in fact, is to keep a humidified jar of gemmules in a refrigerator. When sponges are needed for observation or experimentation, the gemmules are germinated by "seeding" some into a container of pond water.

Gemmules are produced in the mesohyl of a dying sponge around a cluster of nutrient-laden archeocytes. Spongocytes secrete a spongin shell around the cellular mass. The shell may also contain spicules secreted by sclerocytes. The shell completely encloses the cell mass except at one pole where an opening, the **micropyle,** remains. The completed gemmule consists of a shell and its enclosed archeocytes, each of which soon becomes spherical, resembles an embryonic cell, and then is called a **thesocyte** (Fig. 5-15A).

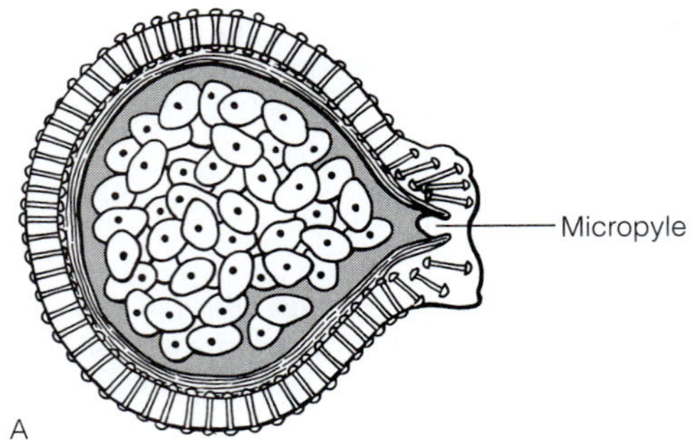

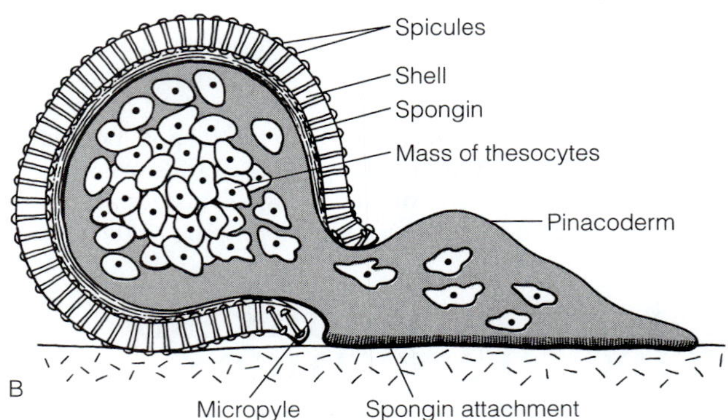

FIGURE 5-15 Porifera clonal reproduction: gemmules. **A,** A vertical section through a full-grown gemmule. **B,** Hatching of a gemmule. *(A, After Evans from Hyman, L. H. 1940. The Invertebrates. Vol. 1. Protozoa through Ctenophora. McGraw-Hill Book Co., NewYork. 726 pp.; B, Modified and redrawn from Fell, P. E. 1997. Poriferans, the sponges. In Gilbert, S. F., and Raunio, A. M. (Eds.): Embryology: Constructing the Organism. Sinauer, Sunderland, MA. pp. 39–54.)*

During the spring gemmule "hatch," the peripheral thesocytes differentiate into a pinacoderm that balloons out, like a bubblegum bubble, through the micropyle (Fig. 5-15B). This pinacoderm bubble makes contact with and attaches to the substratum. Next, the deeper thesocytes issue from the micropyle into the bubble and establish, after differentiation, the interior of the juvenile sponge. An interesting variation on this theme, which challenges the notion of individuality, is that thesocytes from gemmules of the same or different parentage (but of the same species) can intermingle during germination to form one "individual" sponge.

Sexual Reproduction and Development

Sponges, with few exceptions, are hermaphrodites. At the appropriate time, sperm are spawned from one sponge and transported by water currents to another, in which fertilization occurs internally. A few species (such as *Cliona*) are oviparous and release zygotes into the water, where they complete their development. Most sponges are viviparous, retaining the zygotes in the parent's body and releasing larvae (sometimes called larviparity). Embryos and larvae are lecithotrophic.

Sponges are said to lack genital organs (gonads), and germ cells occur in either simple clusters (sperm) or individually

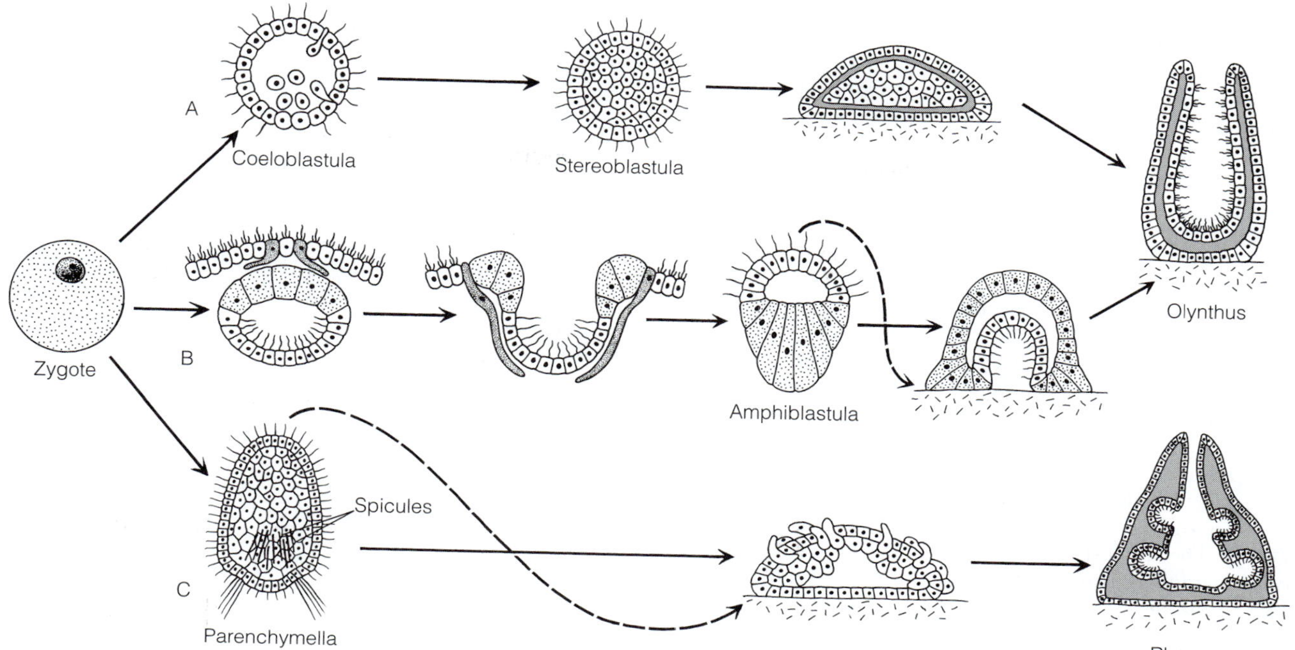

FIGURE 5-16 Porifera sexual reproduction: larval development and metamorphosis. In **A,** *Clathrina* (Calcarea: Calcinia), a hollow blastula develops in the mesohyl and is released into the water. After release, cellular ingression converts the coeloblastula into a stereoblastula, the settling stage. In **B,** *Sycon* (Calcarea: Calcaronea), the egg is derived from a choanocyte that loses its flagellum and sinks into the mesohyl below the choanoderm. After fertilization, the egg (zygote) divides to form an internally flagellated sphere that resembles a choanocyte chamber. Upon release from the sponge, the sphere inverts to position its flagella on the surface of the larva. During metamorphosis, the external flagellated cells, after losing their flagella, return to the interior of the body and differentiate into choanocytes. The metamorphosed juvenile of a calcareous sponge is called an olynthus. In **C,** *Haliclona* (Demospongiae), a differentiated parenchymella is released from the sponge into the plankton. After settlement, it undergoes a complex metamorphosis to form a juvenile, or rhagon (see text).

(eggs) throughout the mesohyl. Sperm arise from choanocytes or entire choanocyte chambers that sink into the mesohyl and become enclosed in a thin cellular wall to form a **spermatic cyst.** Eggs arise from archeocytes or dedifferentiated choanocytes (in some calcareous sponges). Each egg generally accumulates its yolk by phagocytosis of adjacent nurse cells. The egg and nurse cells together may be enclosed in a **follicle** of ensheathing cells.* Because the aquiferous system supplies all parts of the body equally, the germ cells also are widely distributed throughout the mesohyl of the body, but always within diffusion distance of a canal or chamber.

During spawning, sperm rupture the wall of the spermatic cyst, enter the excurrent canals (or atrium), and are released from the oscula. Certain tropical species, known to scuba divers as smoking sponges, suddenly spew sperm in milky clouds from their oscula. Such sudden sperm release may be typical for most sponges.

When the spawned sperm drift into contact with another sponge, they are swept into its aquiferous system by the incurrent water flow. Once in the aquiferous system, sperm are transported to the choanoderm or choanocyte chamber and are phagocytosed, but not digested, by a choanocyte. The choanocyte then loses its flagellum and collar, becomes ameboid, and transports the sperm head (nucleus) to the egg. The transformed ameboid choanocyte is called a **carrier cell.** After the carrier cell reaches an egg in the nearby mesohyl, it either transfers the sperm nucleus to the egg or the carrier cell and sperm nucleus together are phagocytosed by the egg. In either case, fertilization occurs internally in the "ovary" of the sponge.

The sperm of most sponges lack an acrosome, the structure responsible for penetrating the egg-cell membrane during fertilization in most other animals. An acrosome probably is unnecessary because the sperm nucleus enters the egg by phagocytosis. Acrosomal sperm do occur in *Oscarella lobularis,* suggesting that it has a conventional means of egg fertilization, but the reproductive details of this species are unknown.

The zygote cleaves holoblastically into equal-size blastomeres. The pattern of cells that results from cleavage, however, varies among species of sponges. The larvae that develop from embryos are also diverse and are described under the names coeloblastula, amphiblastula, parenchymella, and trichimella larvae.

A **coeloblastula larva** is produced by calcareous sponges, such as species of *Clathrina* (Calcinea; Fig. 5-16A). This larva is

* By definition, an organ is composed of two or more tissues. If these cyst or follicle cells are shown to have a different tissue origin than the germ cells, then the spermatic cyst and egg follicle would be organs (gonads).

a hollow sphere composed of a single layer of flagellated cells. While in the plankton, some of the surface cells lose their flagella, become ameboid, and enter the blastocoel, eventually obliterating it. This converts a hollow coeloblastula into a solid stereoblastula.

An **amphiblastula larva** occurs in other calcareous sponges, for example, *Grantia, Sycon,* and *Leucosolenia* (Calcaronea, Fig. 5-16B). An amphiblastula larva develops as a hollow ball composed of two types of cells, anterior flagellated cells and posterior nonflagellated granular cells. Initially, within the mesohyl of the parent, the flagella are directed into the blastocoel, but a break soon develops in the granule-cell surface of the larva and it turns itself inside-out (inverts) through that opening. After inversion, the flagella project outward from the surface of the larva, enabling it to swim. It is released from the parent at that stage. Inversion is correlated with eggs that arise from choanocytes: After fertilization, the cells divide as if to form new choanocyte chambers, with the flagella directed toward the interior of the chamber. The demosponge genera *Oscarella* and *Plakina* also produce amphiblastula larvae, but these form secondarily, after passing through a parenchymella stage.

A **parenchymella larva** is characteristic of most demosponges (Fig. 5-16C). In this case, the embryo develops directly into a solid mass of cells, forming a stereoblastula. The outer layer is composed of widespread flagellated cells interspersed with occasional vesicle-containing cells that lack flagella. The larval interior houses many types of differentiated cells—sclerocytes, collencytes, pinacocytes, even choanocyte chambers—and archeocytes. To a certain degree, then, parenchymella larvae are prefabricated juveniles specialized for swimming.

Trichimella larvae typify the glass sponges. These are stereoblastulae that bear a band of flagellated cells around the equator of the larval body. The interior is occupied by yolk-bearing cells, sclerocytes (spicules), other cells, and choanocyte chambers.

All sponge larvae are lecithotrophic and therefore relatively short-lived. Typically, they are released at dawn in response to a light cue. After a period of a few hours to a few days, the larvae settle and creep over the bottom in search of a suitable site for attachment. Once a site is found, the larva metamorphoses into a juvenile sponge, which differs somewhat for each of the larval types (Fig. 5-16). Because the metamorphosis involves a rearrangement of cells into more or less definite layers, it is frequently compared with gastrulation in other metazoans, but the ingression of cells that results in the so-called stereoblastulae of many sponges might also be regarded as a form of gastrulation (see Chapter 4).

Immediately prior to metamorphosis, the cells of the coeloblastula, now a stereoblastula, dedifferentiate into a mass of totipotent cells (Fig. 5-16A). Once attached, this mass spreads over the substratum, the surface cells become pinacoderm, and the deeper cells differentiate into other typical sponge cells. Gaps that form between the interior cells merge together to form the atrium as the interior cells undergo rearrangement.

The amphiblastula larva settles and attaches on its flagellated end (Fig. 5-17B). Those flagellated cells, now attached to the substratum, lose their flagella, migrate internally, and form the sponge interior. The granular cells become the pinacoderm. When the juvenile sponge becomes functional, begins to feed, and is a miniature asconoid in design, it is called an **olynthus** (Fig. 5-16B).

Metamorphosis of parenchymella larvae differs among species. In general, following larval attachment, the interior cells differentiate and rearrange themselves to build most, if not all, of the sponge body. The question is, What, if any, contribution to the juvenile body is made by the larval flagellated cells? In one species, *Mycale contarenii,* the flagellated cells contribute to the formation of choanocytes, as might be expected (Fig. 5-16C). But in other species (such as some freshwater sponges and *Microciona prolifera*), the flagellated cells are phagocytosed by archeocytes and do not contribute directly to the juvenile body. In any case, the metamorphosed juvenile sponge often initially has an asconoid or syconoid design, but with thick walls, before transforming into a leuconoid sponge. This early juvenile stage is called a **rhagon** (Fig. 5-16C).

Sponges in temperate zones may live for from 1 to a few years, but some tropical species and perhaps many in the deep sea can be long-lived, up to 200 or more years. Some sponges do not reproduce sexually until they are several years old, whereas others begin to reproduce when they are only 2 or 3 weeks old. Some of the calcified demosponges grow very slowly, at a rate of only 0.2 mm/year. If that growth rate is constant, these reef sponges, which can reach 1 m in size, may be 5000 years old.

DIVERSITY OF PORIFERA

Symplasma^sP (Hexactinellida)

Glass sponges; have syncytial tissues; spicules are siliceous triaxonal hexactines that form intracellularly (sclerocytes are cellular, not syncytial). Many species have elongate bundle ("root") of monaxons that anchor the sponge in mud bottoms; trichimella larva resembles modified parenchymella. Marine; approximately 400 extant species. *Euplectella, Dactylocalyx, Hyalonema, Monoraphis, Rhabdocalyptus.*

Cellularia^sP

Porifera with cellular tissues.

DEMOSPONGIAE^C

Cellularia of leuconoid design; 80 to 90% of all described species. Skeleton of siliceous spicules, spongin, spicules and spongin, or mesohyl only; fused calcareous basal exoskeleton in some relict species. Megascleres: monaxons, triaxons, tetraxons; all spicules secreted intracellularly; mesohyl well developed; choanocytes typically smaller than pinacocytes and archeocytes. Marine and fresh water. (Currently, there is lack of consensus over the classification of subtaxa of Demospongiae.)

Homoscleromorpha^sC: Demospongiae lacking distinction between mega- and microscleres. Spicule types not localized in body; siliceous spicules are di-, tri-, and tetractines; spongin mostly absent. Larviparous with coeloblastula larva. *Octavella* and *Oscarella* have only mesohylar skeleton, lack spongin and spicules. *Plakina* (syconoid).

Tetractinomorpha^sC: Demospongiae with tetraxons, asterose microscleres, and mostly without spongin. Oviparous.

Acanthochaetes, Ceratoporella, Merlia, all "sclerosponges" with a basal calcareous exoskeleton as well as siliceous spicules; *Chondrilla nucula* (chicken-liver sponge of West Indies); *Cliona* spp. (boring sponges); *Geodia; Suberites ficus* (fig sponge); *Tethya actinia* (tangerine sponge); *Tetilla.*

Ceractinomorpha[sC]: Demospongiae with distinct mega- and microscleres, if microscleres are present; spongin is often well developed and several taxa ("keratosa") have spongin only; spicule types localized to specific tissues or regions. Larviparous with parenchymella larva. *Aplysilla longispina* (sulfur sponge); *Asbestopluma* (Cladorhizidae), carnivorous sponges; *Callyspongia vaginalis* (tube sponge); *Dysidea etherea* (ethereal blue sponge); *Ephydatia fluviatilis, Spongilla lacustris, Trochospongilla pennsylvanicus,* freshwater sponges; *Halichondria bowerbanki* (bread sponge); *Haliclona; Halisarca* lacks spicular and spongin skeleton; *Hymeniacidon heliophila* (sun sponge); *Hippospongia, Spongia* (bath sponges); *Niphates digitalis* (vase sponge); *Lissodendoryx isodictyalis* (garlic sponge); *Microciona prolifera* (red-beard sponge); *Mycale; Ophlitaspongia; Neofibularia nolitangere* (touch-me-not sponge); *Spheciospongia vesparia* (loggerhead sponge); *Tedania ignis* (fire sponge); *Vaceletia crypta,* a relict sphinctozoan with a chambered calcareous exoskeleton; and *Verongia* (also called *Aplysina*).

CALCAREA[C]

Cellularia including species of asconoid, syconoid, and leuconoid design; spicules are calcite, mostly unfused triaxons, tetraxons and monaxons; each spicule formed extracellularly by more than one sclerocyte. Mesohyl is thin; choanocytes relatively large, same size as pinacocytes and archeocytes. Larva a hollow blastula. Marine; 500 extant species.

Calcinea[sC]: Calcarea with a choanocytic flagellum not in close association with the nucleus; nucleus in base of cell; triaxons have equiangular rays of equal length. Coeloblastula larva. *Clathrina* (asconoid) forms tubular network; *Murrayona* has a reticulate rigid skeleton of fused calcareous bodies (sclerodermites).

Calcaronea[sC]: Calcarea with a choanocyte flagellum close to the nucleus; nucleus is apical in cell; triaxons are inequiangular, one ray longer than other two. Amphiblastula larva. *Grantia; Leucandra* (leuconoid); *Leucosolenia* (asconoid), single tubes or tube cluster from a stolon, among algae; *Minchinella,* rigid skeleton of fused spicules and cement; *Petrobiona,* skeleton is a solid calcareous mass; *Sycon* or *Scypha* (syconoid), cylindrical to nearly spherical body, common under rocks.

PALEONTOLOGY AND PHYLOGENY OF PORIFERA

The fossil record of the three major taxa of extant sponges—Hexactinellida, Demospongiae, and Calcarea—is rooted in the Cambrian or Ordovician periods. Two extinct taxa of

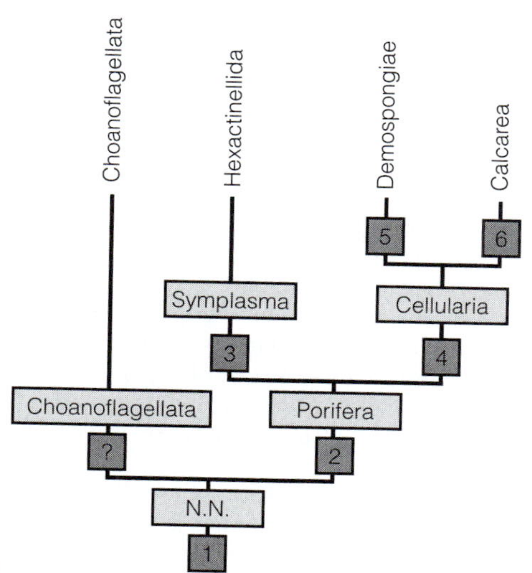

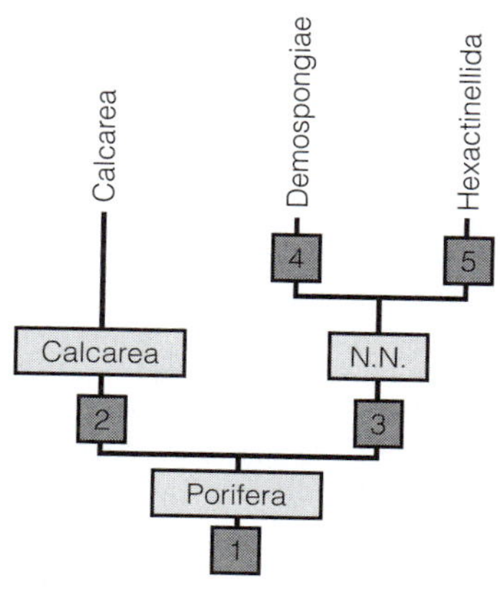

FIGURE 5-17 Porifera: competing phylogenies. **A,** Phylogeny adopted in this book. **1, N.N.:** Small choanocytes with flagellar vane; microtubular flagellar root; intracellular secretion of siliceous spicules; archeocytes. **2, Porifera:** Sessile adult; pinacoderm, mesohyl, and internal aquiferous system; dynamic tissue remodeling; archeocytes; sclerocytes; siliceous spicules secreted intracellularly around organic axial filament; stereoblastula larva. **3, Symplasma (Hexactinellida):** Syncytial trabecular network, choanosyncytium; siliceous hexactines; secondary silicification. **4, Cellularia:** Cellular tissues (possible plesiomorphy); porocytes; extracellular calcification. **5, Demospongiae:** Siliceous tetraxons; spongocytes and spongin. **6, Calcarea:** Large choanocytes; calcareous spicules; loss of siliceous spicules; coeloblastula larva. **B,** A traditional phylogeny. **1, Porifera:** As in **A,** above, plus porocytes in pinacoderm; contractile cells; blastula larva. **2, Calcarea:** Calcareous spicules secreted extracellularly by sclerocytes. **3, N.N.:** Siliceous spicules secreted by sclerocytes intracellularly around organic axial filament; parenchymella larva. **4, Demospongiae:** Spongocytes; tetraxon spicules. **5, Hexactinellida:** Syncytial pinacoderm, choanoderm; loss of porocytes. *(B, Redrawn after Böger from Ax, P. 1996. Multicellular Animals. Vol. 1. A New Approach to the Phylogenetic Order in Nature. Springer, Berlin. 225 pp.)*

PHYLOGENETIC HIERARCHY OF PORIFERA

Porifera
- Symplasma (Hexactinellida)
- Cellularia
  - Demospongiae
  - Calcarea

organisms often considered to be sponges are Archaeocyatha (Cambrian) and Stromatoporata (Ordovician to Cretaceous). The archeocyathan body consisted of a double-walled, porous, calcareous skeleton in the form of an inverted hollow cone with radially arranged septa. Stromatoporates resembled present-day calcifying demosponges (sclerosponges) in that they had a massive basal calcareous skeleton with internal tubes, but unlike sclerosponges, they lacked siliceous spicules. Another taxon, Sphinctozoa (primarily Ordovician to Triassic), had a porous calcareous skeleton that was annulated like a string of pearls. The extant relict sphinctozoan *Vaceletia crypta* indicates that the skeleton was external to the soft tissue. Similarities in its soft tissues with those of demosponges suggest that *Vaceletia,* and perhaps some other sphinctozoans, should be classified in Demospongiae.

Archaeocyathans, sphinctozoans, stromatoporates, and demosponges were important reef builders in Cambrian and Mesozoic seas. Some sponge biologists, accordingly, have suggested that an evolutionary trend in sponges has been a reduction of the massive reef-building skeleton (Fig. 5-10) in favor of a spicular skeleton. The slow growth rate of extant sponges with massive skeletons, as compared with stony corals, may be a reason for the decline of reef-building sponges. On the other hand, the choanoflagellates, which share a common ancestor with sponges (Fig. 5-17), produce a siliceous spicular skeleton.

Sponge classification is controversial, even at the highest levels. Most of the recent phylogenetic discussions have included the sclerosponges (formerly Sclerospongiae) in Demospongiae, and that change is adopted here. Another recent suggestion is the establishment of two subtaxa, Symplasma (Hexactinellida) and Cellularia, which formally recognize the distinction between hexactinellid syncytial organization and the cellular bodies of Calcarea and Demospongiae. Current systematic discussions center on the phylogenetic relationships of the three extant taxa (Fig. 5-17), the position of the sclerosponge species, and the systematic relationships among Archaeocyatha, Stromatoporata, and Sphinctozoa.

PLACOZOA[P]

In 1883, a minute metazoan superficially resembling a large ameba (Fig. 5-18) was discovered in an Austrian seawater aquarium and named *Trichoplax adhaerens.* It has since been collected in the sea in various parts of the world and cultured numerous times.

The flattened body, which reaches 2 to 3 mm in diameter but only 25 μm in thickness, is enclosed in a layer of cells, one cell thick, that resembles an epithelium, particularly because typical intercellular junctions join the adjacent cells (Fig. 5-18B). The epithelioid layer, however, lacks a basal lamina, which is a typical epithelial characteristic (see Chapter 6). The cells on the upper surface of the body differ from those on the lower surface. The upper cells are flat and monociliated, and each usually contains a large, spherical lipid droplet. The lower surface is a creepsole composed of gland and monociliated cells with microvilli. Because these cells are tall and slender, the individual cilia are close together, producing a densely ciliated surface for locomotion. Between the upper and lower cell layers is a connective tissue of watery extracellular matrix and a syncytial network, the **fiber syncytium,** the fourth type of "cell" in the placozoan body. The multiple nuclei of the fiber syncytium are separated from each other by intracellular septa, which are not membranes. Such septa are common in the syncytial network of hexactinellid sponges and in fungi. The fiber syncytium, which is thought to be contractile, contains actin (and presumably myosin) and microtubules.

Trichoplax resembles a large macroscopic ameba in form and locomotion (Fig. 5-18A). The animals change shape more or less constantly as they glide slowly over the substratum. Apart from having differentiated upper and lower surfaces, *Trichoplax* is not polarized. As a result, it can move in any direction without turning. Sometimes it moves in two directions simultaneously and may pull itself apart in the process.

Trichoplax feeds on algae and other material on the substratum. It digests its food extracellularly and extracorporeally (outside of its body) between its ventral surface and the substratum or it can arch upward to produce a pocket in which food is digested. The lower cell layer absorbs the products of digestion.

The predominant mode of reproduction is clonal by fragmentation, as mentioned earlier, and by budding. The buds, which are more or less spherical bodies, appear to emerge from the upper surface of the body, but they contain cells from the upper and lower surfaces as well as connective tissue. The flagellated buds are released from the surface and swim away. Sexual reproduction has not been observed with certainty. Eggs have been described in laboratory individuals whose bodies were swollen, spherical, and detached from the substratum. Apparently, the eggs arise from cells of the ventral surface that dedifferentiate and ingress into the connective-tissue space. Definite sperm have not been observed. If eggs are confirmed and sperm discovered, then the number of specialized cells in placozoans rises to six. The DNA content of *Trichoplax* is smaller than that determined for any other animal.

The taxon Placozoa was created for *Trichoplax adhaerens,* which, like sponges, is probably an early evolutionary line among Metazoa (although the fiber syncytium and extracorporeal digestion are reminiscent of fungi). Placozoans, being composed of only four types of cells, are indeed simple metazoans. Their small, flat bodies enable them to rely on simple diffusion for transport, thus avoiding the complexity associated with a circulatory system. In some respects, placo-

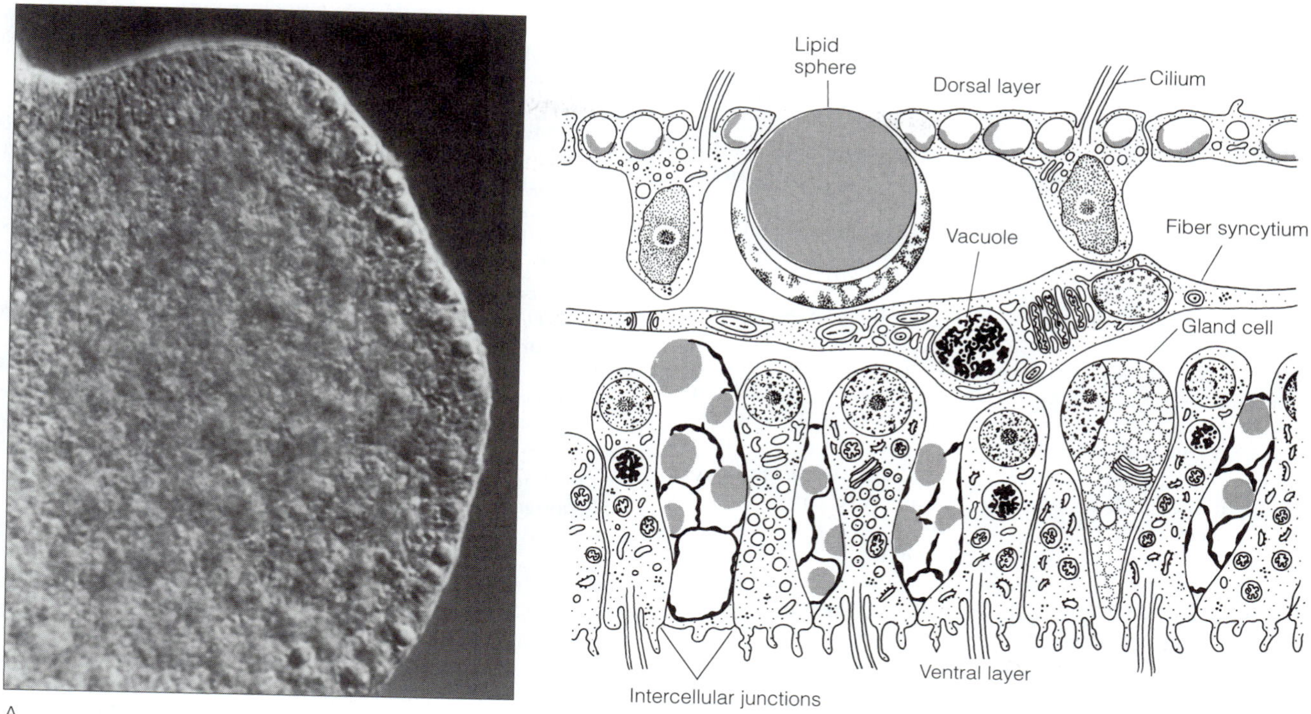

FIGURE 5-18 Placozoa. **A,** Part of a living *Trichoplax adhaerens* in dorsal view. **B,** Diagrammatic vertical section through *Trichoplax* showing part of the upper (dorsal) and lower (ventral) cell layers and the middle fiber syncytium. *(B, After Grell, K. G. 1981. Trichoplax adhaerens and the origins of Metazoa. International congress on the origin of the large phyla of metazoans. Accademia Nazionale dei Lincei. Atti dei Convegni Lincei. 49:113)*

zoans are intermediate between sponges and the remaining metazoans. They resemble the hypothetical protometazoan (Fig. 4-12D) that has adopted a benthic crawling existence and has differentiated its upper and lower surfaces accordingly. The monociliated cells resemble collar cells in which the collars have been reduced to low microvilli, perhaps correlated with locomotion over a substratum and the abandonment of filter feeding. The outer epithelioid layer is one step closer than that of sponges to a true epithelium, which appears fully formed in Cnidaria (Chapter 7). The lower cell layer is reminiscent of the digestive epithelium of the guts of other animals.

There remains the problem of a lack of anterior-posterior polarity in placozoans. If such polarity occurred in the protometazoan, then placozoans, at least the adults, must have abandoned it. Perhaps there is some merit to this idea, because anterior-posterior polarity is expressed only in the larval stages of sponges; larvae have not yet been observed in placozoans. A phylogenetic tree depicting the possible relationships of sponges, placozoans, and the remaining metazoans is shown in Figure 5-19. Finally, 18S RNA gene-sequence data also suggest that placozoans are intermediate between sponges and cnidarians.

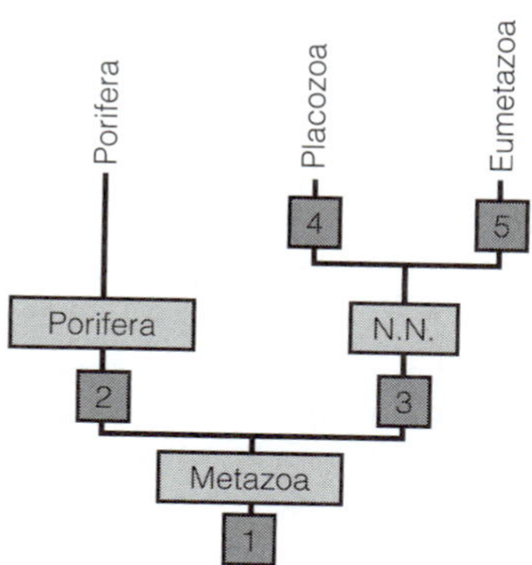

FIGURE 5-19 Phylogeny of lower Metazoa. **1, Metazoa:** Anterior-posterior body polarity, monociliated collar cells, connective tissue, epithelioid tissue in which adjacent cells are in mutual contact. **2, Porifera:** Filter feeders with an aquiferous system. **3, N.N.:** Monociliated cells with short microvilli or with microvilli absent, epithelioid tissue in which adjacent polarized cells are joined by an intercellular junction (belt desmosome), extracorporeal digestion. **4, Placozoa:** Highly flattened body, fiber-syncytium, questionable loss of anterior-posterior polarity. **5, Eumetazoa:** True epithelium composed of a sheet of polarized cells, interjoined by junctional complexes, that rests on a collagenous basal lamina; internal digestive epithelium.

REFERENCES

PORIFERA

General

Afzelius, B. A. 1961. Flimmer flagellum of the sponge. Nature 191:1318–1319.

Ax, P. 1996. Multicellular Animals. Springer-Verlag, Berlin. pp. 68–76.

Ayling, A. L. 1983. Growth and regeneration rates in thinly encrusting Demospongiae from temperate waters. Biol. Bull. 165:343–352.

Bavestrello, G., Burlando, B., and Sara, M. 1988. The architecture of the canal systems of *Petrosia ficiformis* and *Chondrosia reniformis* studied by corrosion casts (Porifera, Demospongiae). Zoomorphology 108:161–166.

Benavides, L. M., and Druffel, E. R. M. 1986. Sclerosponge growth rates as determined by ^{210}Pb and Δ^{14}C chronologies. Coral Reefs 4:221–224.

Bergquist, P. R. 1978. Sponges. Hutchinson, London. 268 pp.

Bergquist, P. R. 1985. Poriferan relationships. In Conway Morris, S., George, J. D., Gibson, R., et al. (Eds.): The Origins and Relationships of Lower Invertebrates. Systematics Association Spec. Vol. 28. Clarendon Press, Oxford. 344 pp.

Böger, H. 1988. Versuch über das phylogenetische System der Porifera. Meyniana 40:143–154.

Bond, C. 1992. Continuous cell movements rearrange anatomical structures in intact sponges. J. Exp. Zool. 263: 284–302.

Bond, C., and Harris, A. K. 1988. Locomotion of sponges and its physical mechanism. J. Exp. Zool. 246:271–284.

Brauer, E. B. 1975. Osmoregulation in the freshwater sponge, *Spongilla lacustris*. J. Exp. Zool. 192:181–192.

Brien, P., Levi, C., Sara, M., et al. 1973. Spongiaires: Traité de Zoologie. Vol. 3. Pt. 1. Masson, Paris. 716 pp.

Cobb, W. R. 1969. Penetration of calcium carbonate substrates by the boring sponge, *Cliona*. Am. Zool. 9:783–790.

De Vos, L. 1991. Atlas of Sponge Morphology. Smithsonian Institution Press, Washington, DC. 117 pp.

Fell, P. E. 1997. Poriferans, the sponges. In Gilbert, S. F., and Raunio, A. M. (Eds.): Embryology: Constructing the Organism. Sinauer, Sunderland, MA. pp. 39–54.

Frost, T. M., Nagy, G. S., and Gilbert, J. J. 1982. Population dynamics and standing biomass of the freshwater sponge *Spongilla lacustris*. Ecology 63:1203–1210.

Frost, T. M., and Williamson, C. E. 1980. *In situ* determination of the effect of symbiotic algae on the growth of the freshwater sponge *Spongilla lacustris*. Ecology 61:1361–1370.

Fry, W. G. (Ed.): 1970. The Biology of Porifera. Academic Press, New York. 512 pp.

Green, G. 1977. Ecology of toxicity in marine sponges. Mar. Biol. 40:207–215.

Harrison, F. W., and Cowden, R. R. 1976. Aspects of Sponge Biology. Academic Press, New York. 354 pp.

Harrison, F. W., and De Vos, L. 1990. Porifera. In Harrison, F. W., and Westfall, J. A. (Eds.): Microscopic Anatomy of Invertebrates. Vol. 2. Wiley-Liss, New York. pp. 29–89.

Hartman, W. D. 1958. Natural history of the marine sponges of southern New England. Bull. Peabody Mus. Nat. Hist. Yale Univ. 12:1–155

Hooper, J. N. A., and Van Soest, R. W. M. (Eds.): 2002. Systema Porifera: A Guide to the Classification of Sponges. Sea Challengers, Danville, CA. 1700 pp.

Jaeckle, W. B. 1995. Transport and metabolism of alanine and palmitic acid by field-collected larvae of *Tedania ignis* (Porifera, Demospongiae): Estimated consequences of limited label translocation. Biol. Bull. 189:159–167.

Kaestner, A. 1980. Lehrbuch der speziellen Zoologie 1(1): Wirbellose Tiere. Gustav Fischer Verlag, Stuttgart. 318 pp.

Koehl, M. A. R. 1982. Mechanical design of spicule-reinforced connective tissue: Stiffness. J. Exp. Biol. 98:239–267.

Lawn, I. D., Mackie, G. O., and Silver, G. 1981. Conduction system in a sponge. Science 211:1169–1171.

Ledger, P. W., and Jones, W. C. 1977. Spicule formation in the calcareous sponge *Sycon ciliatum*. Cell Tissue Res. 181:553–567.

Leys, S. P. 1995. Cytoskeletal architecture and organelle transport in giant syncytia formed by fusion of hexactinellid sponge tissues. Biol. Bull. 188:241–254.

Leys, S. P. 1999. The choanosome of hexactinellid sponges. Invert. Biol. 118:221–235.

Leys, S. P., and Lauzon, N. R. J. 1998. Hexactinellid sponge ecology: Growth rates and seasonality in deep water sponges. J. Exp. Mar. Biol. Ecol. 230:111–129.

Leys, S. P., and Reiswig, H. M. 1998. Transport pathways in the neotropical sponge *Aplysina*. Biol. Bull. 195:30–42.

Leys, S. P., Mackie, G. O., and Meech, R. W. 1999. Impulse conduction in a sponge. J. Exp. Biol. 202:1139–1150.

McClintock, J. B. 1987. Investigation of the relationship between invertebrate predation and biochemical composition, energy content, spicule armament and toxicity of benthic sponges at McMurdo Sound, Antarctica. Mar. Biol. 94:479–487.

Mehl, D., and Reiswig, H. M. 1991. The presence of flagellar vanes in choanomers of Porifera and their possible phylogenetic implications. Z. zool. Syst. Evolut.-forsch. 29:312–319.

Moore, R. C. (Ed.): 1955. Treatise on Invertebrate Paleontology. Vol. E. Archaeocyatha, Porifera. Geological Society of America, University of Kansas Press, Lawrence, KS. 122 pp.

Palumbi, S. R. 1986. How body plans limit acclimation: Responses of a demosponge to wave force. Ecology 67: 208–214.

Pavans de Ceccatty, M. 1974. Coordination in sponges. The foundations of integration. Am. Zool. 14:895–903.

Pomponi, S. A. 1980. Cytological mechanisms of calcium carbonate excavation by boring sponges. Int. Rev. Cytol. 65:301–319.

Porter, J. W., and Targett, N. M. 1988. Allelochemical interactions between sponges and corals. Biol. Bull. 175:230–239.

Reitner, J., and Keupp, H. (Eds.): 1991. Fossil and Recent Sponges. Springer-Verlag, New York. 595 pp.

Reiswig, H. M. 1971a. *In situ* pumping activities of tropical Demospongiae. Mar. Biol. 9:38–50.

Reiswig, H. M. 1971b. Particle feeding in natural populations of three marine demosponges. Biol. Bull. 141:568–591.

Reiswig, H. M. 1975a. Bacteria as food for temperate-water marine sponges. Can. J. Zool. 53:582–589.

Reiswig, H. M. 1975b. The aquiferous systems of three marine Demospongiae. J. Morphol. 145:493–502.

Reiswig, H. M., and Mackie, G. O. 1983. Studies on the hexactinellid sponges. III. The taxonomic status of Hexactinellida within the Porifera. Phil. Trans. R. Soc. Lond. B. 301:419–428.

Rigby, J. K. 1987. Phylum Porifera. In Boardman, R. S., Cheetham, A. H., and Rowell, A. J. (Eds.): Fossil Invertebrates. Blackwell Science, Cambridge. MA. pp. 116–139.

Rützler, K. 1990. New Perspectives in Sponge Biology. Smithsonian Institution Press, Washington, DC. 533 pp.

Rützler, K., and Rieger, G. 1973. Sponge burrowing: Fine structure of *Cliona lampa* penetrating calcareous substrata. Mar. Biol. 21:144–162.

Saller, U. 1989. Microscopical aspects on symbiosis of *Spongilla lacustris* and green algae. Zoomorphology 108:291–296.

Schultz, B. A., and Bakus, G. J. 1992. Predation deterrence in marine sponges: Laboratory versus field studies. Bull. Mar. Sci. 50:205–211.

Simpson, T. L. 1984. The Cell Biology of Sponges. Springer-Verlag, New York. 662 pp.

Stearn, C. W. 1975. The stromatoporoid animal. Lethaia 8:89–100.

Teragawa, C. K. 1986. Particle transport and incorporation during skeleton formation in a keratose sponge: *Dysidea etheria.* Biol. Bull. 170:321–334.

Vacelet, J., and Boury-Esnault, N. 1995. Carnivorous sponges. Nature 373:333–335.

Vogel, S. 1974. Current induced flow through the sponge, *Halichondria.* Biol. Bull. 147:443–456.

Wiedenmayer, F. 1977. Shallow-Water Sponges of the Western Bahamas. Birkhauser Verlag, Basel, Switzerland. 287 pp.

Wielsputz, C., and Saller, U. 1990. The metamorphosis of the parenchymula larva of *Ephydatia fluviatilis.* Zoomorphology 109:173–177.

Willenz, P. 1980. Kinetic and morphological aspects of particle ingestion by the freshwater sponge *Ephydatia fluviatilis.* In Smith, D. C., and Tiffon, Y. (Eds.): Nutrition in the Lower Metazoa. Pergamon Press, Oxford. pp. 163–178.

Willenz, P., and Hartman, W. D. 1989. Micromorphology and ultrastructure of Caribbean sclerosponges. I. *Ceratoporella* and *Nicholsoni* and *Stromatospongia norae.* Mar. Biol. 103:387–401.

Woollacott, R. M. 1990. Structure and swimming behavior of the larva of *Halichondria melandocia* (Porifera: Demospongiae). J. Morphol. 205:135–145.

Internet Sites

www.biology.ualberta.ca/facilities/multimedia/index.php?Page=252 (Animations of basic sponge design; flows in asconoid, syconoid, and leuconoid sponges; and water flow and particle capture by choanocytes.)

www.ucmp.berkeley.edu/porifera/porifera.html (General information, some photographs.)

PLACOZOA

General

Grell, K. G. 1981. *Trichoplax adhaerens* and the origin of the Metazoa. In Origine dei Grandi Phyla dei Metazoi. Accad. Naz. Lincei Covegni Lincei 49:107–121.

Grell, K. G., and Ruthmann, A. 1991. Placozoa. In Harrison, F. W., and Westfall, J. A. (Eds.): Microscopic Anatomy of Invertebrates. Vol. 2. Placozoa, Porifera, Cnidaria, and Ctenophora. Wiley-Liss, New York. pp. 13–27.

Pearse, V. B. 1989. Growth and behavior of *Trichoplax adhaerens:* First record of the phylum Placozoa in Hawaii. Pac. Sci. 43:117–121.

Ruthmann, A. 1977. Cell differentiation, DNA-content and chromosomes of *Trichoplax adhaerens* F. E. Schulze. Cytobiologie 15:58–64.

Wenderoth, H. 1986. Transepithelial cytophagy by *Trichoplax adhaerens* F. E. Schulze (Placozoa) feeding on yeast. Z. Naturforsch. 41:343–347.

Internet Sites

www.microscopy-uk.org.uk/mag/artoct98/tricho.html (Richard Howey's observations of living *Trichoplax* found in detritus from an aquarium shop.)

www.ucmp.berkeley.edu/phyla/placozoa/placozoa.html (Images and description of *Trichoplax.*)

6 Introduction to Eumetazoa

For most people, the concept of an animal includes irritability (the ability to detect and respond to stimuli), locomotion, a sense of direction, and the capture and ingestion of food. The two metazoan taxa discussed so far, Porifera and Placozoa, fall short of these qualifications and it may be a stretch to admit them to the fellowship of animals. The sponges, although sometimes large and colorful, are firmly attached to the seabed. Their motionless bodies are elaborate skeletal frameworks through which water is inconspicuously pumped and filtered. Only their microscopic motile larvae seem to hint at future possibilities. Placozoans move, but are little more than flattened blastulae that lack a head, so any point on the body circumference can become the leading end, like a computer mouse moving over its pad. Furthermore, as saprobes, placozoans feed more like a fungus than an animal. To a significant extent, the marginal animal status of sponges and placozoans can be attributed to the absence of a gut, muscles, and nerves. Although they probably are animals, these absences distinguish sponges and placozoans from all other metazoans. For this reason, sponges and placozoans are sometimes classified as Parazoa (= near animals), the sister taxon of Eumetazoa (= true animals), which includes all other animals.

In contrast to sponges, eumetazoan tissues and organs, such as gut, nervous system, musculature, and gonads, are arranged and localized along one or more definite body axes. The functional parts of a sponge, on the other hand, are uniformly distributed throughout the body: A tissue sample taken from anywhere in the body will always yield the same parts, such as the aquiferous system and mesohyl cells.

Compared with sponges, the eumetazoan's greater degree of differentiation applies not only to the regions of the body, but also to the tissues and cells. Eumetazoans inherited connective tissue but evolved epithelial tissue, which enabled them to isolate and physiologically regulate internal compartments of the body. Some epithelial tissues, moreover, evolved into muscle and nervous tissues, which together allowed eumetazoans to evolve large, mobile, and complex bodies (see also Chapter 4). These and other eumetazoan innovations are the subjects of this chapter.

EPITHELIAL TISSUE

Epithelial tissue was a key innovation that facilitated the extraordinary diversification of animals. An epithelium is a cell layer that covers surfaces and lines internal cavities and spaces.

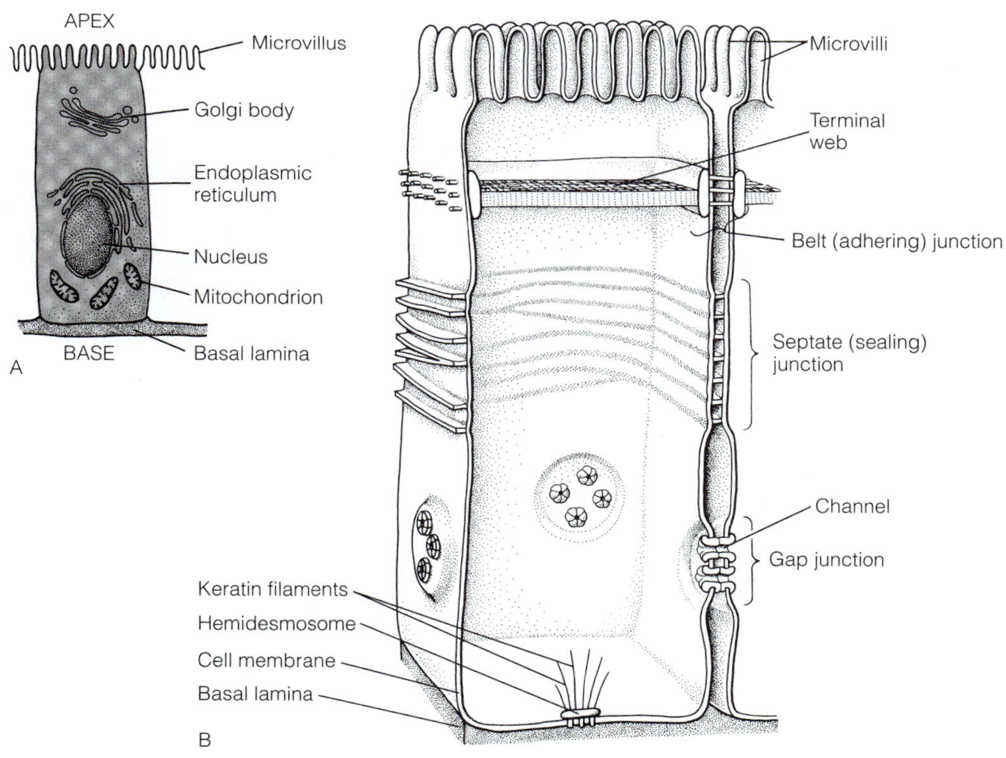

FIGURE 6-1 Epithelial tissue. **A,** A typical invertebrate epithelial cell with microvilli on its exposed surface. **B,** Types of junctions between epithelial cells. At sealing septate junctions, apposing cell membranes are bound together by surface proteins occluding the intercellular space. At belt junctions, the cells are united by actin filaments that emerge from a beltlike thickening on the cytoplasmic side of each cell. At gap junctions, the apposing cell membranes are close together and the narrowed intercellular space is bridged by membrane channels. Hemidesmosomes anchor epithelial cells to the basal lamina.

It separates the external environment from the internal environment of the body or separates internal body compartments of different chemical composition. The skin, lining of the gut, and linings of many other body cavities are all epithelia.

Epithelia are essential for the physiological regulation of internal extracellular compartments, such as a gut cavity. Eumetazoan bodies are collections of isolated compartments, each chemically regulated for its specific function. The stomach requires a different chemical environment than the intestine, for example, and the pericardial cavity must be different from the lumen of the heart. Once regulated independently, each body compartment can then adopt one or more specialized functions, allowing for the evolution of ever-more complex bodies (see Chapter 4 and Fig. 4-15). Such division of labor promoted homeostasis, which eventually allowed animals to leave the sea and colonize the much more physiologically challenging environments of fresh water and land (see Chapter 1).

Four characteristics of an epithelium are responsible for its function: Its component cells are arranged in a continuous, unbroken layer, have apical-basal polarity, rest on a basal lamina, and are joined by intercellular junctions. For an epithelium to maintain a chemical difference between adjacent compartments, it must be in the form of an unbroken layer.

Apical-basal polarity means that the cell apex, which faces one compartment, differs functionally from the base, which faces another compartment. For example, the apical surface may absorb macromolecules and the basal surface pump ions. The **basal lamina** is a thin, dense ECM that contains collagen (type IV) secreted by the epithelium (Fig. 6-1). Epithelial cells anchor to the basal lamina via specialized junctions called **hemidesmosomes,** which are composed partly of cell-adhesion molecules (integrins). The basal lamina also may limit transport across the epithelium to ions and small molecules. Intercellular junctions link the cell membranes of adjacent epithelial cells and, in combination with the basal lamina, add strength and structural integrity to the epithelium. Some, called **adhering junctions** (spot desmosomes and belt junctions), are mechanical in function and bind together the cells of the epithelial layer. **Sealing junctions** (septate junctions and tight junctions) seal the spaces between cells, thus preventing the unregulated passage of ions and molecules across the epithelium. **Gap junctions** create small transmembrane channels ("gaps") that join the cytoplasms of adjacent cells and allow the free, unregulated passage of ions and molecules between them. Epithelial (and other) cells linked by gap junctions behave like a single cell, especially with respect to nervous impulse conduction. Impulse conduction is very rapid along such an epithelium because there is no delay at intercellular junctions, as is the case when conduction requires the release and diffusion of neurotransmitters at a synapse.

EPIDERMIS, GASTRODERMIS, AND GUT

The epithelium that covers the surface of the eumetazoan body is the **epidermis** (= outer skin). Its primary function is to present a barrier to the external environment and so to allow internal regulation and homeostasis.

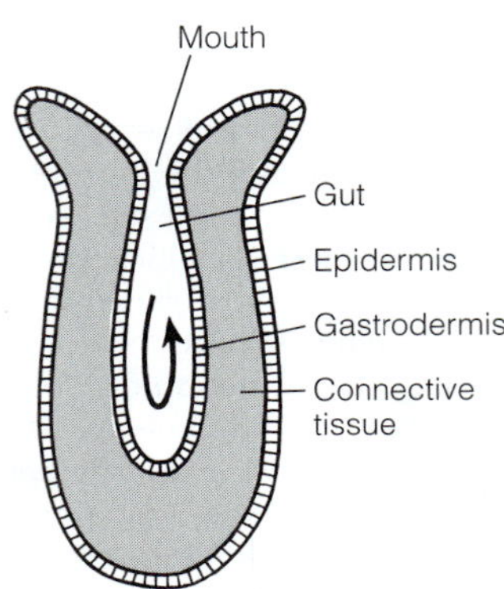

FIGURE 6-2 Gastrodermis and gut. The gut is an internal body compartment specialized for the extracellular digestion of large foods, circulation of food molecules (arrow), and absorption of nutrients.

Another eumetazoan innovation was a **gut,** an internal, epithelium-lined cavity for the digestion and absorption of food. Except for sponges and placozoans, almost all animals have an internal gut cavity that opens to the outside through an opening called the **mouth.** The gut cavity is lined by an epithelium, called the **gastrodermis** (= stomach skin), that joins the epidermis at the mouth (Fig. 6-2).

The **gut cavity,** or lumen, isolated from the external environment, makes possible the use of large foods. **Extracellular digestion** occurs in most eumetazoan guts and is a function of that cavity's volume. Thus, animals that digest food in an extracellular gut compartment require less gut surface area than do intracellular digesters to break down (hydrolyze) an equivalent amount of food. During the process of extracellular digestion, secretory gastrodermal cells release enzymes into the gut lumen and within this confined space the enzymes hydrolyze large foods into small particles and molecules that are then absorbed by other cells of the gastrodermis.

Absorption of nutrients, however, is a function of the area of the absorptive surface. Thus, the absorptive part of the eumetazoan gut typically has a large area, which often results from branching or folding of the gut wall.

In several taxa of eumetazoans, the gut is also a circulatory system for transport of nutrients and other substances (see Chapter 4 and Fig. 4-8, 6-2). Cilia or muscles associated with the gastrodermis mix the gut contents during digestion and distribute nutrients, sometimes widely, for absorption and use by tissues and organs of the body.

CONNECTIVE TISSUE

The **connective-tissue compartment,** which may be thin or thick, is situated between the epidermis and gastrodermis (Fig. 6-2). Sometimes it consists only of epidermal basal lamina and gastrodermal basal lamina, but sometimes it is the two laminae plus looser ECM called **reticular lamina** (Fig. 4-1). The reticular lamina usually contains fibroblast cells, which secrete protein components, as well as other types of cells, such as those of the sponge mesohyl. By definition, a **connective tissue** consists of both ECM and cells (see Chapter 4); if cells are absent, then the layer is simply ECM. This middle layer, however, is probably homologous in all animals and perhaps should be designated as the connective-tissue compartment, whether or not it meets the technical definition of connective tissue in all instances. The connective-tissue compartment of sponges is the mesohyl; of placozoans, the unnamed cavity containing the fiber syncytium; and of metazoan embryos in general, the blastocoel. The connective-tissue compartment of *Hydra,* sea anemones (Cnidaria), and comb jellies (Ctenophora) is called mesoglea, whereas the equivalent layer in higher animals is known simply as connective tissue.

Motile ameboid cells that arise from epithelia and enter the embryonic blastocoel are called **mesenchyme,** or mesenchymal cells, but there is no collective term for cells in the connective-tissue compartment of adults apart from "connective-tissue cells" or even "mesodermal cells." Few zoologists, however, would favor the term *mesoderm* (= middle skin) for cells of the sponge mesohyl, for example. Instead, *mesoderm* generally is reserved for middle-layer cells that originate from a specific, identifiable germ layer, namely the endoderm (see Development later in this chapter and Chapter 9).

SKELETONS

A tissue-level skeleton typifies metazoans. A skeleton is any structure that maintains body shape, supports or protects a body, and transmits contractile forces (see Chapter 4). The skeleton may be an exoskeleton, such as the cuticle of an insect, that encloses the body, or an endoskeleton, such as the sponges' spicular skeleton that is secreted by connective-tissue cells, or the bones of vertebrates. Although eumetazoans exploit the potential of both skeletal types, they also introduce to the endoskeletal theme a new variation, a fluid-filled body cavity used as a skeleton.

FLUID SKELETON

Curiously, fluids, such as water, can be effective skeletons. A water-filled skeleton is called a **hydrostatic skeleton,** or hydrostat. An inflated sea anemone is a good example of an animal supported by a hydrostatic skeleton, although many other animals also use them. The bodies of these animals are supported by the slightly pressurized water within them, much as a balloon maintains its shape when inflated with water (or air) (Fig. 6-3A,B). Because water is virtually incompressible at the pressure generated by muscles, hydrostats are constant in volume and any localized increase in pressure as a result of muscular contraction will be transmitted equally throughout the hydrostat. Not only can the body be supported by such a mechanism, a force generated by the displacement of water in one region can be used to do work in another. (Using your hand or your imagination, constrict the middle of a water-filled, cylindrical balloon and observe the effect on the two free ends.) Such fluid movements are used to inflate and extend body parts, such as sea anemone tentacles and spider legs, to anchor in crevices or sediments by swelling the embedded part of the body, and to produce coordinated waves along the surface of the body (peristalsis) for swimming, burrowing, or ventilating a burrow.

Many animals with hydrostatic skeletons, such as sea anemones and worms, have more or less cylindrical bodies, and often the body wall is reinforced with a mesh of inelastic fibers similar to the windings in the wall of a garden hose or the wire belts of a radial tire. The fibrous mesh toughens the body wall but also prevents uncontrollable bulges, or aneurisms, from developing as the hydrostat is pressurized (Fig. 6-3C). To prevent aneurisms, the meshwork fibers may be oriented in either an **orthogonal pattern,** in which one set is parallel and the other perpendicular to the long axis of the body (Fig. 6-3D), or in a **crossed-helical pattern,** in which the fibers wind around the body in two helixes, one left-handed and one right-handed (Fig. 6-3E).

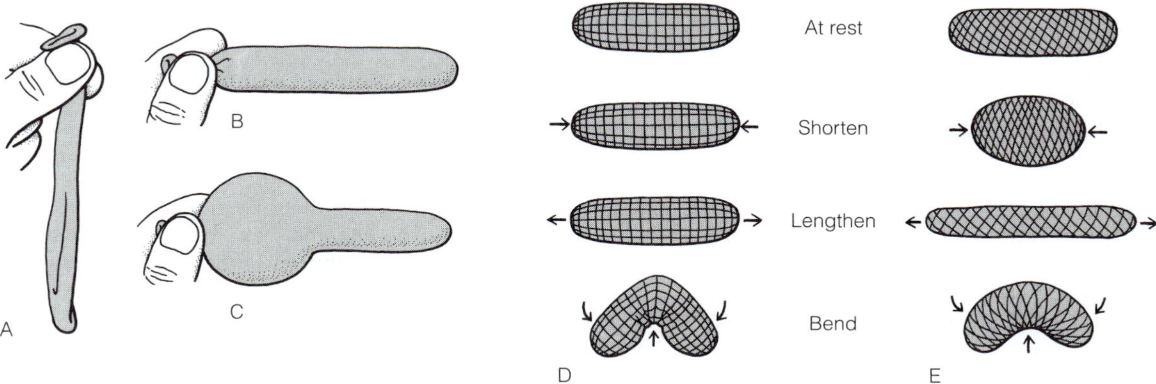

FIGURE 6-3 Hydrostatic skeleton: fiber reinforcement. A balloon **(A),** when filled **(B),** is supportive and approximates a hydrostatic skeleton. **C,** Addition of more air or water causes an aneurism as the balloon begins to inflate. Although both orthogonal **(D)** and crossed-helical **(E)** meshes protect against aneurisms, only the crossed-helical pattern prevents kinks and allows the body to elongate and shorten. *(Redrawn from Wainwright, S. A. 1988. Axis and Circumference. Harvard University Press, Cambridge, MA. 132 pp.)*

Although both orthogonal and crossed-helical patterns prevent aneurisms, only the crossed-helical pattern of fiber reinforcement is suitable for animal body walls. The orthogonal pattern (Fig. 6-3D) fails because the orientation of the *in*elastic fibers does not allow the length or diameter of the body to change. The orthogonal pattern also fails because it does not resist kinks as the body bends. Kinks, like aneurisms, are maladaptive for animals. The crossed-helical pattern, in contrast, resists kinks and permits changes in body length and diameter because the fibers are always at an angle to the primary axes of the body (Fig. 6-3E). The behavior of helical fibers in the body wall of an anemone is roughly similar to that of a coil spring; the spring is composed of inelastic steel wire, yet it can be freely extended and shortened and does not kink when bent.

Helical reinforcing fibers are often proteins, such as collagen, embedded in invertebrate cuticles, as in earthworms (annelids), or in the connective tissue (reticular lamina) beneath the epidermis of the body, although they can also occur around such specialized body structures as tentacles, blood vessels, and notochords. In humans, for example, failure of the fibers to contain internal pressures may result in arterial aneurisms, varicose veins, ruptured ("slipped") vertebral discs, and hernias.

SOLID SKELETON

Eumetazoans also can have endoskeletons and exoskeletons composed of solid materials. These solid skeletons can be further classified as either pliant or rigid skeletons. A **pliant skeleton** consists of materials that are rubbery and elastic, such as cartilage. A **rigid skeleton** is made of materials such as bone or shell that resist any change of shape.

Pliant skeletons, like hydrostatic skeletons, deform when stressed, but unlike them, they are **elastic** and spring back to their original shapes when the stress is removed. Among invertebrates, pliant skeletal material is used in the hinge of clams and in the supportive structures of horny corals, such as the sea fans. Pliant skeletons may be either endoskeletons or exoskeletons. Although there are a few specialized rubbery proteins, most pliant skeletons are composites formed of proteins, polysaccharides, and water. They range in consistency from watery gels, as in the mesohyl of sponges and the connective tissue (mesoglea) of a few comb jellies, to the stiff, springy jelly (mesoglea) of many jellyfishes. One obvious advantage of having a pliant skeleton is that muscle power is required only for the initial change of shape, not to restore the body to its starting position. Energy is stored by the pliant skeleton and used later to restore the original shape. Pliant skeletons can be made stiffer by the addition of rigid pieces, such as mineral spicules, sand grains, or organic fibers that include collagen (spongin), chitin, or cellulose.

Rigid skeletons may form a supportive platform as in stony corals, a lattice arrangement as in glass sponges, or a framework of beams and levers as in the appendages of insects and vertebrates. Rigid skeletons may be endoskeletons or exoskeletons, but they always are composed of composite materials. Although there are exceptions, rigid skeletons tend to be more common in terrestrial animals, which must support themselves in air; in animals that move rapidly in water, such as crustaceans and fishes; and in exposed sessile and slow-moving benthic animals like barnacles and snails, for whom the skeleton is not only a supportive structure but also a protective retreat. Rigid skeletal frameworks, in conjunction with specialized muscles and ligaments, provide leverage for the precise application of large forces by certain appendages and the rapid, but less forceful movements of others.

MOVEMENT AND BODY SIZE

Movement is a distinguishing characteristic of animals. Although some animals, like sea anemones, oysters, and barnacles, live attached to rocks and other objects, they can still move parts of their bodies. Even adult sponges are capable of limited movement. Animals use two fundamentally different structures for movement: cilia and muscle.

Cilia and flagella are surface structures that function in an aquatic medium. Because they occur on surfaces, their use in locomotion is effective in animals that have a large surface area in relation to their volume. These are organisms of small body size, such as ciliated protozoa or metazoan larvae, or very flat body form. In both cases, the ciliated area of the body is large in relation to its volume (or mass), and the sinking speeds are slow. Such small animals typically move at speeds of no more than a few millimeters per second, but this can mean several body lengths per second. Ciliary motion also is useful for moving fluid and small particles over surfaces such as the linings of small tubes, as when water is moved over gills, particulate food is moved through the gut, or dust is moved from a human trachea.

A striking feature of locomotion in tiny animals is that their forward motion stops immediately when the ciliary beat is arrested; they do not glide or coast. For them the water is not very liquid, but rather as viscous and sticky as molasses is to us. The reasons for this are again related to surface area and volume. Surface friction, or viscous drag, between the water and the body is large because the body's surface area is so large. Forward momentum and the tendency to coast are infinitesimal forces because the body's volume-mass is tiny and swimming speed slow. Thus surface friction overwhelms momentum: Movement ceases the instant ciliary beat stops.

The aquatic locomotion of lilliputian animals is dominated by surface friction. Using their cilia (and sometimes other appendages, too), they move through water like a raft being poled over a mucky shoal. The cilium (or pole) works in propulsion because of the friction between it and the medium. Forward progress depends entirely on the difference in friction between the "power" stroke and the "recovery" stroke. For this reason, a cilium is extended stiffly during its power stroke, which maximizes friction, and flexed close to the body during the recovery stroke to minimize friction. Under these conditions, the organism (or raft) does not develop momentum and thus stops immediately at the conclusion of the ciliary (or pole's) power stroke.

Among bigger animals, the volume-mass of the body is large in relation to its area, the opposite of the relationship seen in tiny animals. For this reason, the surface-area friction tends to be small and the volume-mass momentum large.

Thus momentum overwhelms friction and the animals glide or coast when their locomotory appendages stop moving. Because of their relatively small surface area and large volume-mass, large animals have abandoned locomotory cilia. Instead, they are propelled by the contraction of a musculature, which occupies a *volume* in the body. The momentum developed by a large massive body in motion dominates the retarding effect of the viscous forces, but substantial pressure drag can develop if the flow of water around the body becomes turbulent instead of smooth. Large fast-moving animals, like squid and fish, minimize this turbulence by becoming streamlined. The fastest swimmers are typically the largest, because as body size increases, the body's volume, which produces the musculature, increases faster than its surface area, which creates the surface friction.

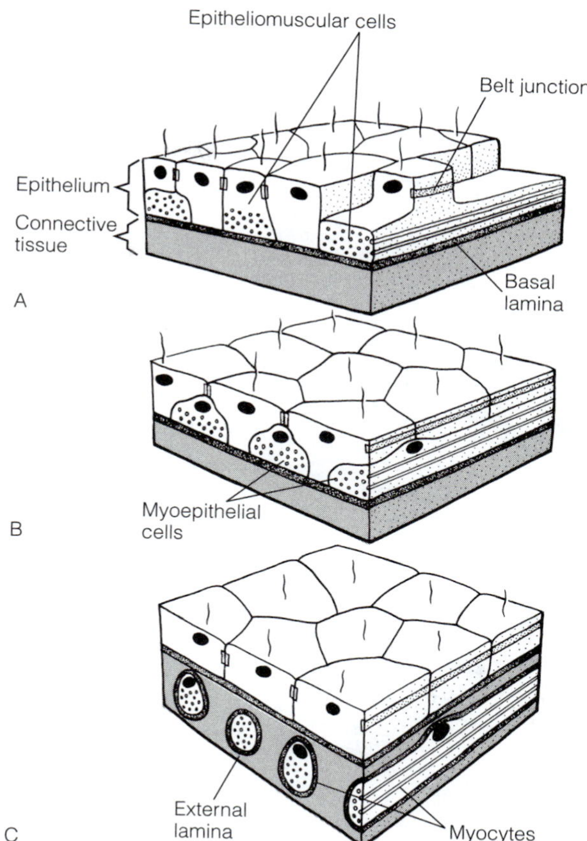

FIGURE 6-4 Muscle tissue: three evolutionary grades of organization. Epitheliomuscular cells **(A)** and myoepithelial cells **(B)** are both part of a multifunctional epithelium. In **C,** myocytes, sometimes called true muscle cells, are below the epithelium in the connective-tissue compartment and constitute a distinct muscle tissue separate from the overlying epithelium.

MUSCLE CELLS AND TISSUE (MUSCULATURE)

All animals move using muscular contractions even if the chief locomotory force is ciliary. Elongating and shortening the body or a body part is a simple form of movement that usually involves changes in diameter. Restriction of elongation or shortening to one side of the body or appendage causes it to bend. Such contractions enable sessile, attached animals such as the sea anemone to extend upward, bend toward food, or retract from predators. Small motile eumetazoans, for example the larvae of sea anemones, swim with cilia, but can bend and thus turn using muscular contraction. As surprising as it may seem, the ability to turn purposefully while moving forward is a eumetazoan innovation. For a cumbersome alternative, recall how *Paramecium,* in the absence of turning muscles, changes course while swimming. Its "avoidance reaction" requires it to stop, back up, turn (using cilia), and again move forward.

Bodywide contractile movements became possible with the evolution of muscle fibrils (actin and myosin filaments in fixed arrangement) located within the basal part of epithelial cells that have other functions besides contraction. Such cells, called **epitheliomuscular cells** (Fig. 6-4A) are characteristic of *Hydra,* sea anemones, and their relatives and also appear scattered elsewhere in Eumetazoa. **Myoepithelial cells** are similar, but the apical ends of the cells are reduced and are not exposed at the surface of the epithelium (Fig. 6-4B). For example, human sweat glands, mammary glands, and the iris of the eye all contain myoepithelial cells. True muscle cells, or **myocytes,** have lost their epithelial characteristics, abandoned the epithelium for the connective-tissue compartment, and are specialized solely for contractile function (Fig. 6-4C).

Three basic types of muscles—smooth, cross-striated, and obliquely striated—occur in eumetazoans (Fig. 6-5). **Smooth muscle** (Fig. 6-5A,D) generally contracts slowly, but develops tension over a wide range of stretch lengths (excursions). Its presence is often correlated with an extensible body or appendage, for example, a highly extensible and contractile tentacle. **Cross-striated muscle** (Fig. 6-5B) contracts rapidly, but develops tension over a limited excursion. Cross-striated muscle is often associated with a body or appendage that moves quickly over a fixed distance, such as a swimming undulation or the snapping of jaws or claws. **Obliquely striated muscle** (Fig. 6-5C) is intermediate in performance between smooth and cross-striated muscle. It contracts relatively quickly and operates over a wider excursion than cross-striated muscle. Not surprisingly, it is often associated with soft-bodied and thus extensible animals, such as earthworms, that also can make rapid movements.

Muscles are often organized as an **antagonistic set** that acts on a skeleton. One muscle of the set moves the body or appendage in one way, whereas its antagonist moves it in the opposite direction to restore it to the original position. Because muscles can only contract, antagonism is essential to restore contracted muscles to their extended condition so that they can contract again. The biceps and triceps muscles of mammals, for example, constitute an antagonistic pair. The biceps flexes the forearm and the triceps extends it. Invertebrates with a hydrostatic skeleton, such as sea anemones, usually have antagonistic circular and longitudinal muscles that extend and shorten the body, respectively.

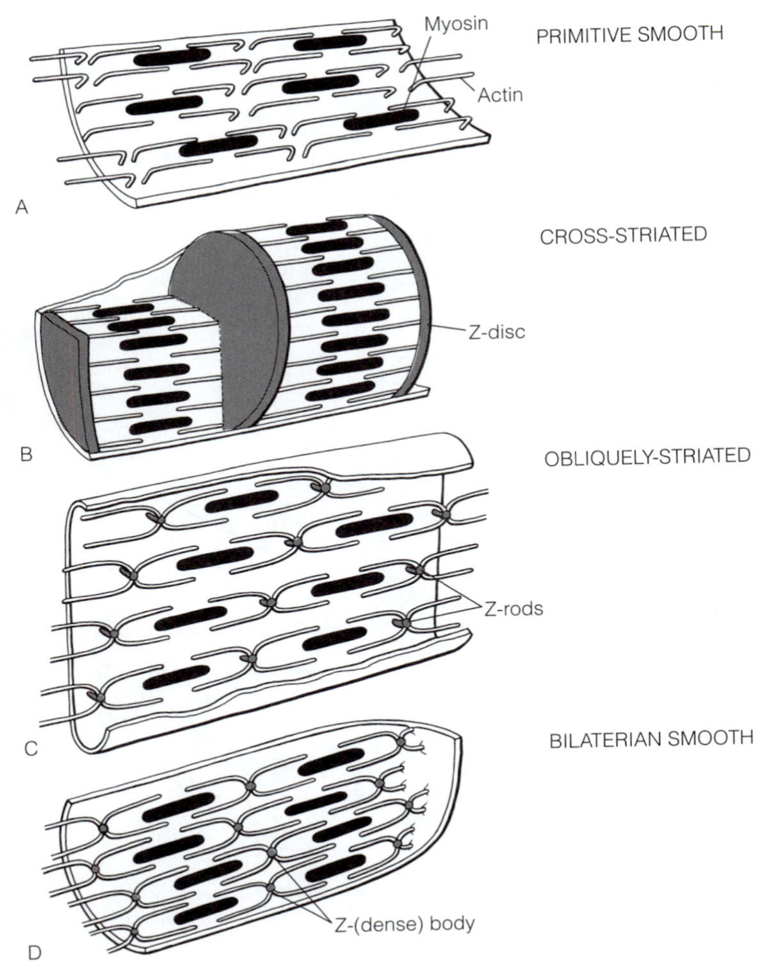

FIGURE 6-5 Muscle cells: functional designs. **A,** Smooth muscle found in cnidarians and ctenophores. Actin (thin) filaments anchor directly to the cell membrane. These smooth muscle cells operate over a wide excursion and sometimes coil as they shorten. **B–D,** Actin filaments anchor to the cell membrane and cytoplasmic proteins (Z material). **B,** Cross-striated muscle shortens rapidly over a fixed distance. It is associated with the twitchlike swimming pulsations of cnidarian jellyfish, movement of appendages in arthropods and vertebrates, retraction of tentacles in bryozoans, and similar movements in other taxa. **C,** Obliquely striated muscle shortens more slowly than cross-striated does, but it contracts forcefully over a wider excursion. It tends to occur in soft animals, such as ribbon worms and annelids, whose bodies may be stretched substantially. **D,** Bilaterian smooth muscle is common in flatworms, molluscs, and the gut musculature of many animals. Depending on details of its design, such as filament length and thickness, it may shorten moderately quickly to slowly and develop moderate to high levels of tension. As in **A** and **C,** this design can operate over a wide excursion.

NERVE CELLS (NEURONS) AND TISSUE (NERVOUS SYSTEM)

For animals to move rapidly in response to external stimuli, they must coordinate their body's movements through muscle contraction. **Neurons,** or nerve cells, evolved in conjunction with muscle and other effectors as a means of meeting those needs (Fig. 6-6). Neurons respond to stimuli and transmit information in the form of a rapid wave of depolarization (a nerve impulse) along the neuron membrane (axon) to the target cells and tissues (effectors). The evolution of neurons exploited the capacity of all cells to regulate the ion concentrations on the inner and outer sides of the cell membrane, thereby creating a difference, or polarity, in the electrical charge on the two sides. This charge difference creates an electrical potential, which is measured and expressed in millivolts. The nerve impulse (action potential) is simply a propagating loss of this charge difference (depolarization) along the length of the membrane.

When an action potential reaches the end of a neuron, it can be transmitted to the next neuron or to an effector such as a muscle in one of two ways. If the two cells are joined by a gap junction, the action potential spreads unimpeded and without a time lag to the second cell. Such cells are said to be electrically coupled. If the two cells lack a gap junction, the action potential must be conducted by the release and diffusion of a chemical neurotransmitter from the first cell to the second. After a short diffusional time lag, the neurotransmitter initiates a new action potential in the second cell. This specialized, chemically coupled junction is called a **synapse.**

The capacity of cells to secrete neurotransmitters such as acetylcholine or serotonin (Chapter 2) and to conduct an action potential preceded the evolution of neurons. The available evidence suggests that neurons, like myocytes, originated from epithelial cells but then migrated to the connective-tissue compartment (Fig. 6-6). Some of these neurons, the superficial sensory cells, remained in the epithelium and retained epithelial characteristics. Others adopted a subsurface position, developed long processes, and linked the effectors with the sensory cells. Unique to neurons is their wirelike shape, making possible the conduction of signals over a long distance, and their arrangement into networks and circuits, which allowed centers of information processing and distribution to evolve.

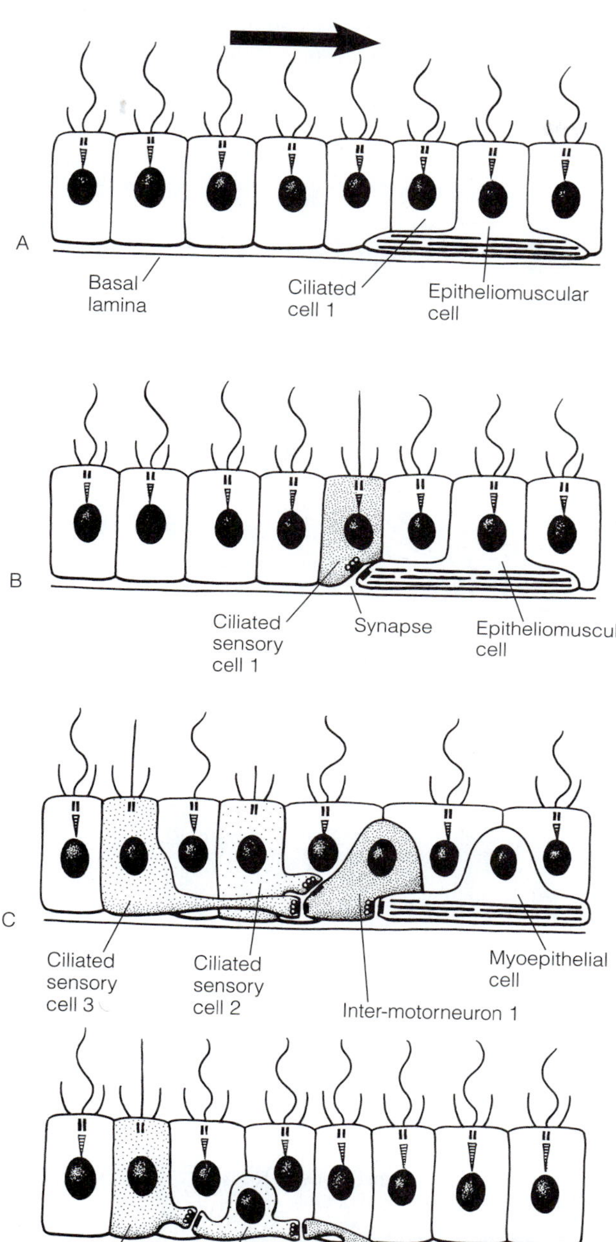

FIGURE 6-6 Evolution of the neuromuscular system (hypothetical). **A,** Epithelial conduction (arrow) of stimulus to an epitheliomuscular cell (or other effector). **B,** Ciliated sensory receptor cell 1 evolved from a ciliated epithelial cell and forms a synapse directly with the effector. **C,** Sensory cell 1 has evolved into a combination inter-motorneuron that receives and integrates stimuli from receptor cells 2 and 3 and then stimulates the myoepithelial cell. **D,** Neuromuscular system consisting of sensory cell 3, an interneuron (modified sensory cell 2), and a motorneuron (modified sensory cell 1), which innervates the myocyte.

SENSE CELLS AND ORGANS

FORM AND FUNCTION

Animals, like other living organisms, are more likely to be successful if they respond appropriately to environmental variations. These variations include the availability of food, water, mates, and other resources, as well as the presence of predators. Short-term adaptive responses, such as movement toward prey or away from predators, are mediated through the nervous system, whereas long-term responses, such as a seasonal onset of reproduction, are mediated through the endocrine system by the release of hormones.

Animals obtain a dynamic picture of their environment by using sensory structures, each of which is specialized to detect a particular type of stimulus. Three classes of stimuli can be identified: electromagnetic energy (light, including infrared), mechanical energy (sound, vibrations, touch, pressure, gravity), and chemical stimuli (taste, smell).

In the simplest cases, a sensory neuron may be linked directly to an effector, such as a chemoreceptive sensory cell that makes a synapse with a stinging cell (cnidocyte) of a *Hydra.* This circuit results in a local action in response to a local stimulus. At the opposite extreme, the entire body can respond to a complex pattern of stimuli. For example, photoreceptive cells can be grouped together with other cells to form an eye. The eye of some animals contains accessory structures, such as a lens, which can focus an image on the sensory neurons. The sensory neurons encode and transmit the image data as impulses to optic neurons, which convey the information to the central nervous system (CNS). The CNS, in turn, decodes the information and projects it as a mental image, a faithful representation of the external world, on which an adaptive response can be based.

Direction and distance to the source of stimuli are two of the most important sensory determinations because they define the spatial position of the organism in relation to objects and events in the environment. Eumetazoan animals have receptors for all three classes of stimuli that provide directional and distance information. Generally, all three types of receptors occur in each organism. But in many cases, one sensory modality dominates the others, as with vision in primates, hearing in bats, and smell in sharks.

Invertebrate receptor neurons evolved from ciliated epithelial cells, and their sensory surfaces are typically modifications of cilia or microvilli once on the apex of the ancestral cell (Fig. 6-6). The simplest receptors are individual receptor neurons situated in the epidermis (or gastrodermis) of an organism. Later, groups of photoreceptor or chemoreceptor neurons became locally concentrated in patches and cuplike depressions in the epithelium. The sensory cups typically have only a single window through which to admit stimuli such as light or chemical molecules, and that organization imparts directional sensitivity. Mechanoreceptors also may be clustered in patches, or even in rows representing x-, y-, and z-axes, but rarely among invertebrates do they form subsurface organs as in the lateral-line system of fishes or the mammalian inner ear. One exception is a gravity receptor called a statocyst, which is of common but sporadic occurrence.

The gravity receptors and eyes described next are **organs:** functionally specialized structures composed of two or more different tissues. These two organs typically consist of nervous and nonnervous (somatic) tissues. Along with a gonad, which consists of reproductive and nonreproductive tissues, gravity receptors and eyes are the first definite organs to appear in eumetazoans. It is tempting to suggest that organs evolved first in eumetazoans, but remember that the spermatic cysts and

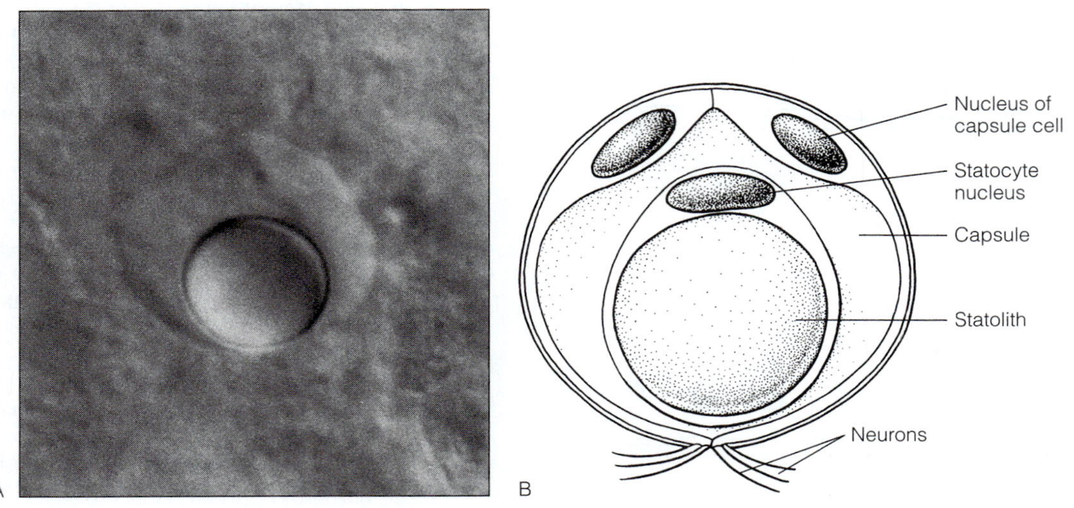

FIGURE 6-7 Sense organs: statocyst of an acoel flatworm. **A,** Microscopic view of a statocyst of a living acoel. **B,** Micro-anatomy of the acoel statocyst. *(B, Modified and redrawn from Ehlers, 1991)*

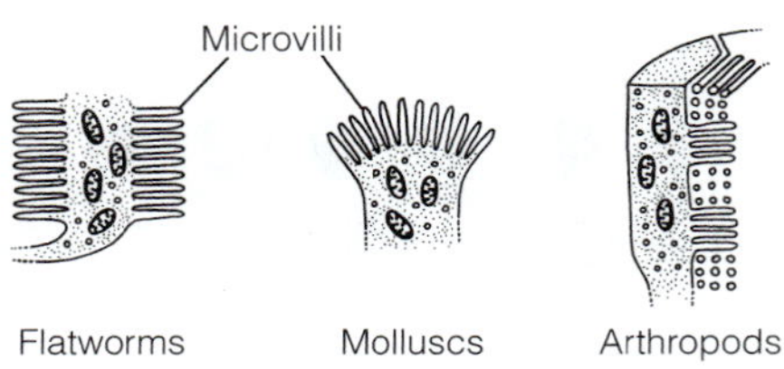

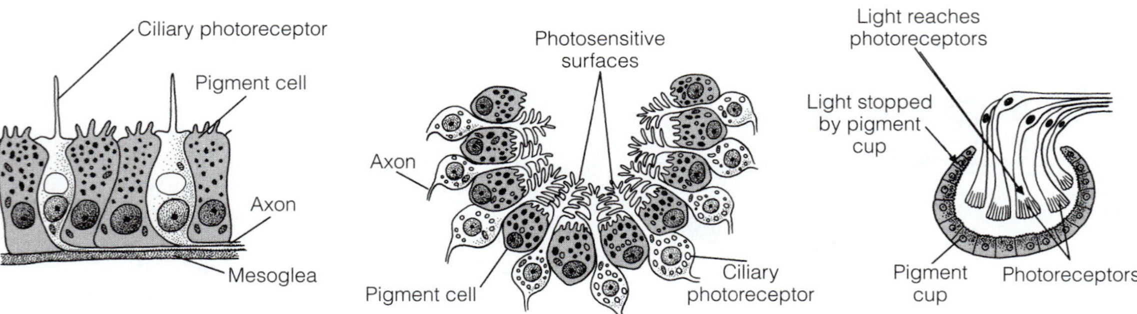

FIGURE 6-8 Sense organs: photoreceptors and ocelli. **A,** Ciliary photoreceptors. **B,** Rhabdomeric photoreceptors. **C,** A cnidarian pigment spot ocellus. **D,** An everted pigment cup ocellus. **E,** An inverted pigment cup ocellus. *(A and B, Modified from Eakin, R. M. 1968. Evolution of photoreceptors. In Dobzhansky, T. et al. (Eds.): Evolutionary Biology. Vol. 2. Appleton-Century-Crofts, New York. p. 206; C and D, adapted from Singla, C. L. 1974. Ocelli of hydromedusae. Cell Tiss. Res. 49:413–429.)*

egg follicles of sponges also meet the criteria of a gonad organ (see Sexual Reproduction and Development, Chapter 5).

GRAVITY RECEPTORS

A **statocyst** is a hollow capsule lined with mechanoreceptive cells that has a dense particle, the **statolith,** at its center (Fig. 6-7). As the orientation of the animal changes, the direction of gravitational pull on the statolith changes and stimulates different areas of the statocyst lining, thus conveying to the animal a sense of its orientation in relation to up and down.

PHOTORECEPTORS AND EYES

Of all the sensory organs expressed by invertebrates, the most conspicuous and best known are eyes. The photoreceptor cells of most animals have evaginations on some part of their apical surface, providing greater surface area for photochemical reactions. The two principal types of photoreceptors, ciliary and rhabdomeric, are both derived from monociliated cells. In **ciliary photoreceptors** the photosensitive surface is derived from the membrane of a cilium (Fig. 6-8A), and in **rhabdomeric photoreceptors** it is derived from the microvilli of the cell surface (Fig. 6-8B). Both types certainly evolved numerous times in the animal kingdom, and epidermal cells probably were their precursors; epidermal cells usually bear cilia and microvilli, and epidermal cells are the first cells through which light would pass on entering an animal's body.

The photoreceptors of some animals, such as earthworms, are dispersed as individual cells over the integument. Thus, earthworms can respond to light in the absence of eyes. Most animals, however, have photoreceptors concentrated in an **eye.** An eye enables an animal to utilize other information provided by light besides general light intensity.

The simplest eye, called an **ocellus,** is organized as a pigment spot or pigment cup and is usually associated with the integument. A **pigment spot ocellus** is a patch of photoreceptors interspersed with pigment cells (Fig. 6-8C). In a **pigment cup ocellus** the pigment cells form a cup into which the photoreceptor elements project (Fig. 6-8E). When the photoreceptors project between the pigment cells into the lumen of the cup and face the light, the ocellus is said to be **everted** (direct) (Fig. 6-8D); this arrangement is typical of integumental ocelli. In flatworms, the pigment cup lies below the epidermis and the photoreceptors project into the cup opening, not between the pigment cells (Fig. 6-8E), and face away from the light. An ocellus of this type is said to be **inverted** (indirect). In all eyes, the pigment cells partially shade the photoreceptors, enabling the animal, depending on the number of photoreceptors shaded, to determine the direction of the light source.

DEVELOPMENT

As in all metazoans, eumetazoan development begins with a fertilized egg, or zygote. Cleavage of the zygote leads to a blastula, which consists of a single layer of cells forming a solid mass (stereoblastula) or layer of cells over a "hollow" blastocoel (coeloblastula) (Fig. 6-9).

In eumetazoans, the singled-layered blastula is converted into a two-layered gastrula during gastrulation. A **gastrula** is the embryonic stage composed of the two primary germ layers, the ectoderm and endoderm (Fig. 6-9). Later in development,

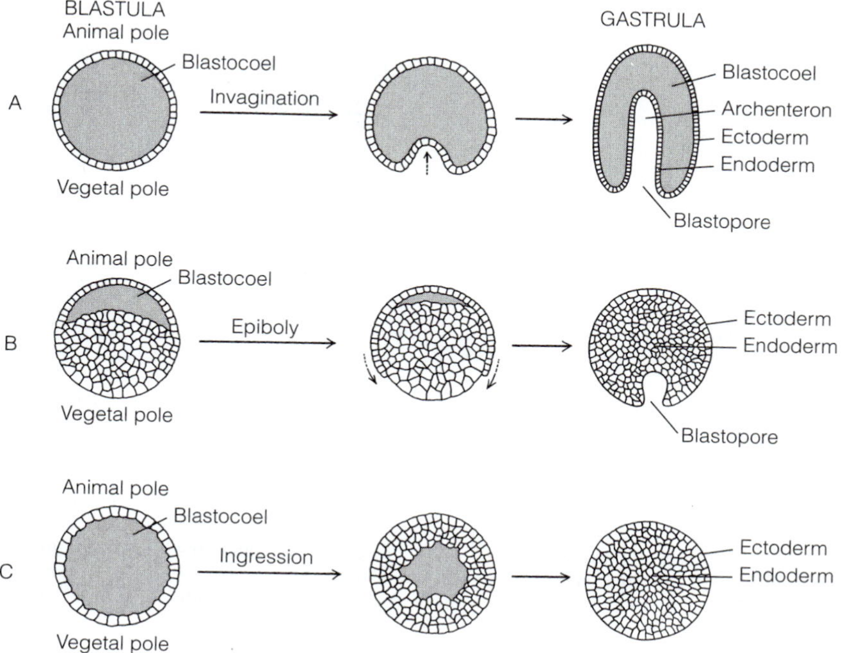

FIGURE 6-9 Gastrulation: invagination, epiboly, ingression. **A,** Invagination: vegetal-pole cells fold inward to form an archenteron. The opening to the exterior is the blastopore. **B,** Epiboly: animal-pole cells grow down and over the yolk-laden cells of the vegetal hemisphere. **C,** Ingression: cells cleaved inward from the blastula wall fill the interior of the blastocoel. The interior mass of the cells later hollows out to form the cavity of the archenteron.

the ectoderm will give rise to the epidermis and the nervous system and the endoderm to the gastrodermis and gut derivatives. The extracellular matrix between the ectoderm and endoderm is the blastocoel, which is the embryonic precursor of connective tissue (or connective-tissue compartment). The cavity formed by the endoderm is the **archenteron** (= primitive gut) that will become the adult gut cavity. Often, the archenteron opens at a surface **blastopore.** The blastopore can become the mouth, anus, or both, depending on the taxon.

Gastrulation, a critical step in eumetazoan development, is a period of cell movements in which the primary germ layers are established. Three common forms of gastrulation are invagination, epiboly, and ingression (Fig. 4-14, 6-9). During **invagination** (or emboly), cells of the vegetal hemisphere fold inward into the interior (Fig. 6-9A) to form the endoderm and archenteron, leaving the ectoderm on the outside. In gastrulation by **epiboly,** the cells of the animal half of the embryo overgrow those of the vegetal half (Fig. 6-9B). During **ingression,** cells are proliferated from the blastular wall into the interior (Fig. 4-14, 6-9C). In general, very small eggs with little yolk gastrulate by ingression, those of intermediate size by invagination, and the largest with abundant yolk by epiboly or other means.

The adult body of a few primitive eumetazoans, such as *Hydra,* consists only of the two epithelial layers, the epidermis and gastrodermis. These arise directly from the embryonic germ layers of the ectoderm and endoderm, respectively, and for this reason such animals are said to be **diploblastic** (= two germs). The embryos of other primitive eumetazoans, as well as all apomorphic taxa, have a third germ layer, the **mesoderm,** between the ectoderm and endoderm, and are therefore **triploblastic.** Mesoderm may be either epithelial or mesenchymal and may originate developmentally from either of the two primary germ layers. The origin and derivatives of mesoderm are discussed further in Chapter 9.

GROWTH

Eumetazoans are differentiated regionally into functionally specialized parts, such as feeding appendages, mouth, gut, gonads and other structures. Together, these parts constitute an integrated functional unit that we generally think of as the body of an animal. This body is established at the close of the developmental period, but once development ends and the juvenile form is attained, two different patterns of growth can occur (Fig. 6-10) to bring about further changes.

If the juvenile grows by a general increase in body size, then the result is a large **solitary adult** (Fig. 6-10A,B). On the other hand, the juvenile may enlarge only slightly and then replicate itself by clonal budding. If the differentiated buds remain attached to one another, then growth is by the addition of new functional units, or modules, each of which resembles the juvenile and is called a **zooid** (= tiny animal, pronounced ZOE-oid). Growth by the addition of zooids, called **modular growth,** results in a **colonial adult** (Fig. 6-10A,C).

Although both solitary and modular growth can result in a large body, growth in solitary organisms is accompanied by a disproportionate, or **allometric** (= different measure), increase in area of one or more surfaces of the body. These amplified surfaces may be gills, feeding appendages, the gut lining, or other structures. In the absence of such enhanced surface growth, the area for the supplying of food, gases, and other substances might not keep pace with the demand of the body's greatly increased volume (Fig. 4-6, 6-10A,B).

The modular growth of colonial organisms, on the other hand, is additive, or **isometric** (= similar measure), and thus preserves the favorable area-to-volume ratio ($S{:}V$) of the small-bodied juvenile. This relatively large surface area is retained because growth is by the addition of small-bodied zooids, each of which has a large $S{:}V$. The large surface area, the broad distribution of multiple mouths and feeding appendages over

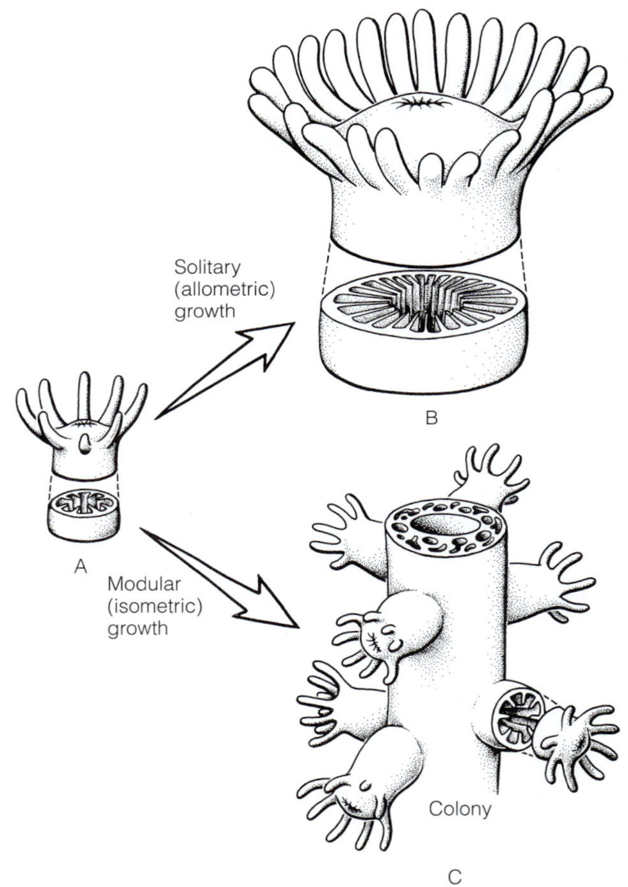

FIGURE 6-10 Growth: solitary and modular. In **A** and **B,** the juvenile body **(A)** undergoes a general enlargement resulting in a solitary adult **(B).** This pattern of growth is accompanied by a change in body proportions (allometry) to ensure appropriately increased surface area. In this example, growth is accompanied by an increased number of tentacles for feeding and gas exchange and by gastrodermal folds (septa) for digestion and absorption. In **A** and **C,** the juvenile **(A)** grows by budding to form a colonial adult. In this modular pattern of growth, the juvenile body does not increase in size, but rather in number, and its proportions change little if at all (isometry). As a result, such colonial adults retain a favorable area-to-volume ratio.

the surface of the body, and the small size of the zooids all predispose colonial animals to suspension feeding (or light capture for photosynthesis). In contrast, gut enlargement with growth in solitary animals adapts them to feed on large prey. Suspension feeding remains an option for solitary animals, but only if the area of the food-collecting surface is amplified.

Solitary growth is correlated with locomotion using a whole-body musculature, although some solitary animals, such as sea anemones, are nearly sessile. A large body and muscular locomotion, in turn, require a complex nervous system for sensory input, integration, and motor control. In contrast, colonial animals, with their small zooids and multiple body axes, are usually sessile and attached organisms, but a few are motile. Motile colonies, such as the Portuguese man-o'-war, are usually planktonic and at the mercy of wind and currents. Such colonies never match the locomotory performance of similarly sized solitary animals.

Among solitary animals, growth and increasing complexity typically occur along one primary body axis, which generally coincides with the axis of the gut. Colonies often grow along multiple axes and only rarely does one of these correspond to the gut axis of the unbudded juvenile. Instead, modular growth can produce either a flat mat from which zooids arise or a stolon or a network of stolons bearing zooids, or the interconnected zooids themselves may give rise to a variety of forms, including branching networks, tufts, and crusts. Because modular growth resembles growth in many plants, colonial organisms are often plantlike in appearance and are often confused with seaweeds or mosses.

Although modular growth can result in a large animal with little more structural complexity than a single zooid, some colonial animals have achieved moderate levels of differentiation and functional integration. In those few cases, such as the Portuguese man-o'-war, the differentiation occurs along a single body axis and thus the colony individual comes to resemble a solitary individual. Differentiation during modular growth is achieved by specialization of zooids, which results in zooidal **polymorphism,** or zooids of different forms and functions. These polymorphic zooids are comparable to the specialized organs of a solitary organism. Polymorphism within a colonial body is another example of the replication-specialization-integration pattern for the evolution of complexity discussed in Chapter 4 (Fig. 4-15).

Animals with either a solitary or colonial body are capable of self-repair after damage, but a solitary body can be disabled until regeneration is complete if one of its crucial organs is damaged. A colony, being composed of multiple functional units, is less likely to be disabled after being damaged, because the undamaged intact zooids can continue to function normally. One consequence is that colonial animals typically survive grazing, whereas solitary animals often are killed.

REFERENCES

EUMETAZOA

Ax, P. 1996. Multicellular Animals: A New Approach to the Phylogenetic Order in Nature. Vol. 1. Springer, New York. pp. 80–82.

Nielsen, C. 2001. Animal Evolution: Interrelationships of the Living Phyla. 2nd Edition. Oxford University Press, New York. pp. 51–58.

EPITHELIA

Caplan, M. C., and Rodriguez-Boulan, E. 1997. Epithelial cell polarity: Challenges and methodologies. In Hoffman, J. F., and Jamieson, J. D. (Eds.): Handbook of Physiology. Section 14: Cell Physiology. Oxford University Press, New York. pp. 663–688.

Lodish, H., Berk, A., Zipursky, S. L., and Matsudaira, P. 2000. Molecular Cell Biology. W. H. Freeman, New York. 1084 pp.

Lane, N. J., and Chandler, H. J. 1980. Definitive evidence for the existence of tight junctions in invertebrates. J. Cell Biol. 86:765–774.

Larsen, W. J. 1983. Biological implications of gap junction structure, distribution, and composition. Tiss. Cell 15:645–671.

Lord, B. A., and diBona, D. R. 1976. Role of the septate junction in the regulation of paracellular transepithelial flow. J. Cell Biol. 71:967–972.

Mackie, G. O. 1984. Introduction to the diploblastic level. In Bereiter-Hahn, J., Matoltsky, A. G., and Richards, K. S. (Eds.): Biology of the Integument. Vol. 1. Springer-Verlag, Berlin. pp. 43–46.

Reuss, L. 1997. Epithelial transport. In Hoffman, J. F., and Jamieson, J. D. (Eds.): Handbook of Physiology. Section 14. Cell Physiology. Oxford University Press, New York. pp. 309–388.

Welsch, U., and Storch, V. 1976. Comparative Animal Cytology and Histology. University of Washington Press, Seattle. 243 pp.

SKELETONS

Clark, R. B. 1964. Dynamics in Metazoan Evolution. Oxford University Press, London. 313 pp.

Koehl, M. A. R. 1982. Mechanical design of spicule-reinforced connective tissue: Stiffness. J. Exp. Biol. 98:239–267.

Wainwright, S. A. 1988. Axis and Circumference. Harvard University Press, Cambridge, MA. 132 pp.

Vogel, S. 1988. Life's Devices: The Physical World of Animals and Plants. Princeton University Press, Princeton, NJ. 367 pp.

MUSCULATURE

Hernandez-Nicaise, M. L., and Amsellem, J. 1980. Ultrastructure of the giant smooth muscle fiber of the ctenophore *Beroë ovata.* J. Ultrastruct. Res. 72:151–158.

Hernandez-Nicaise, M. L., Nicaise, G., and Malaval, L. 1984. Giant smooth muscle cells of the ctenophore *Mnemiopsis leidyi:* Ultrastructural study of *in situ* and isolated cells. Biol. Bull. 167:210–228.

Lanzavecchia, G. 1981. Morphofunctional and phylogenetic relations in helical [obliquely striated] muscles. Boll. Zool. 48:29–40.

Lanzavecchia, G., and Arcidiacono, G. 1981. Contraction mechanism of helical [obliquely striated] muscles: Experimental and theoretical analysis. J. Submicros. Cytol. 13:253–266.

Vogel, S. 2002. Prime Mover: A Natural History of Muscle. W. W. Norton, New York. 384 pp.

NERVOUS SYSTEM

Csaba, G. 1994. Phylogeny and ontogeny of chemical signaling: Origin and development of hormone receptors. Int. Rev. Cytol. 155:1–47.

Eakin, R. M. 1968. Evolution of photoreceptors. In Dobzhansky, T., et al. (Eds.): Evolutionary Biology. Vol. 2. Appleton-Century-Crofts, New York. pp. 194–242.

Ehlers, U. 1991. Comparative morphology of statocysts in the Plathelminthes and Xenoturbellida. Hydrobiologia 227:263–271.

Horridge, G. A. 1968. The origins of the nervous system. In Bourne, G. H. (Ed.): The Structure and Function of Nervous Tissue. Vol. 1. Structure 1. Academic Press, New York. pp. 1–31.

Lentz, T. L. 1968. Primitive Nervous Systems. Yale University Press, New Haven, CT. 148 pp.

Mackie, G. O. 1964. Conduction in the nerve-free epithelia of siphonophores. Am. Zool. 5:439–453.

Mackie, G. O. 1990. The elementary nervous system revisited. Am. Zool. 30:907–920.

Parker, G. H. 1919. The Elementary Nervous System. Lippincott, Philadelphia. 229 pp.

Singla, C. L. 1974. Ocelli of hydromedusae. Cell Tiss. Res. 149:413–429.

Singla, C. L. 1975. Statocysts of hydromedusae. Cell Tiss. Res. 158:391–407.

Westfall, J. A. 1973. Ultrastructural evidence for a granule-containing sensory-motor-interneuron in *Hydra littoralis*. J. Ultrastruct. Res. 42:268–282.

GASTRULATION AND GERM LAYERS

Lankester, E. R. 1900. A Treatise on Zoology. Part 2. The Porifera and Coelentera. Adam & Charles Black, London. pp. 1–37.

Korschelt, E., and Heider, K. 1895. Text-Book of the Embryology of Invertebrates. Swan Sonnenschein, London; Macmillan, New York. pp. 1–9.

INTERNET SITES

http://academic.reed.edu/biology/courses/BIO351/movie.html (Movie of epibolic gastrulation in a mesolecithal amphibian embryo from confocal images.)

www.uoguelph.ca/zoology/devobio/210labs/gastrulation2.html (Labeled color micrographs of epibolic gastrulation in a mesolecithal amphibian embryo; good accompaniment to preceding Web page.)

http://uoguelph.ca/zoology/devobio/210labs/gastrulation1.html (Color micrographs of invagination gastrulation in a microlecithal echinoderm embryo.)

SOLITARY AND MODULAR GROWTH

Harvell, C. D. 1994. The evolution of polymorphism in colonial invertebrates and social insects. Q. Rev. Biol. 69:155–185.

Jackson, J. B. C. 1979. Morphological strategies of sessile animals. In Larwood, G., and Rosen, B. R. (Eds.): Biology and Systematics of Colonial Animals. Academic Press, New York. pp. 499–555.

Jackson, J. B. C. 1985. Distribution and ecology of clonal and aclonal benthic invertebrates. In Jackson, J. B. C., Buss, L. W., and Cook, R. E. (Eds.): Population Biology and Evolution of Clonal Organisms. Yale University Press, New Haven, CT. pp. 297–355.

Schmidt-Nielsen, K. 1984. Scaling: Why Is Animal Size So Important? Cambridge University Press, Cambridge. 241 pp.

Thompson, D. W. 1961. On Growth and Form. Abridged edition edited by J. T. Bonner. Cambridge University Press, London. 345 pp.

7 Cnidaria[P]

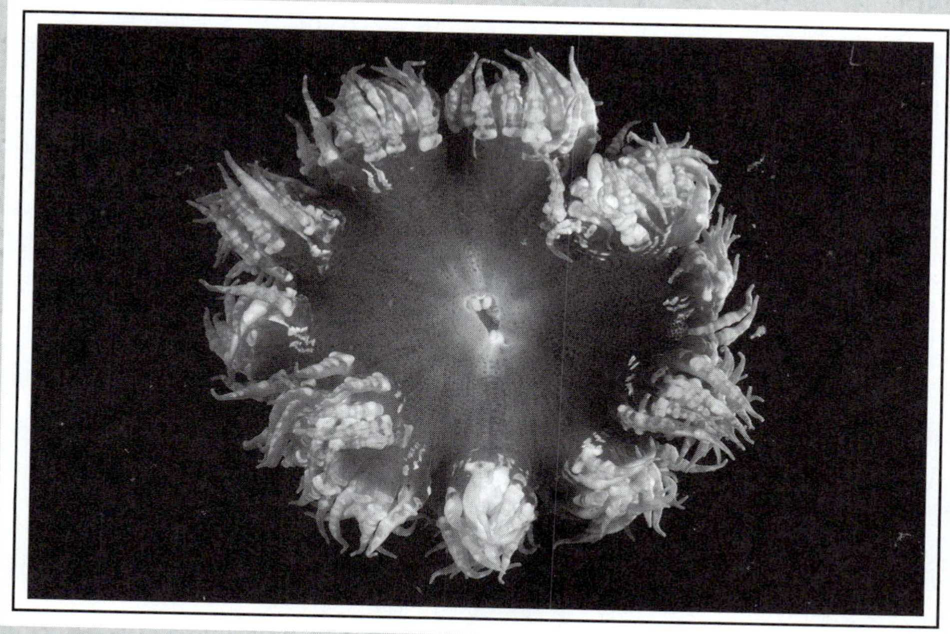

GENERAL BIOLOGY

The colorful, radially symmetric cnidarians are the flowers of the sea. A few, like the familiar *Hydra,* have colonized fresh water, but none live on land. Cnidaria includes fleshy polyps—graceful freshwater hydras, colorful sea anemones, sculptured stony corals, plantlike soft corals, and weedy hydroids—as well as medusae, or jellyfishes, whose pulsating bells resemble disembodied hearts.

Many cnidarians, corals for example, are colonies that undergo modular growth and may reach several meters in size. Solitary individuals also can be large, like the jellyfish *Cyanea capillata* (lion's mane jelly), which is reported to reach 2 m in diameter, or the 1.25-m-wide anemone *Stichodactyla mertensii* from Australia's Great Barrier Reef. The stony corals, with their massive skeletons and photosynthetic zooxanthellae, are the foundation of the coral-reef community, one of the most spectacular ecosystems on earth. Coral reefs are rivaled in diversity only by tropical rainforests, but the reef outstrips the rainforest in color and activity, as well as in the taxonomic range of animals present. Cnidarian diversity reaches it zenith on coral reefs, but cnidarians are widespread and common worldwide in shallow seas, where they attach to hard objects, burrow in sediments, or drift in the plankton. Of the approximately 10,000 known species, only 20 occur in fresh water; the remainder are marine. Historically, cnidarians were classified alone, or with ctenophores, as Coelenterata, but most zoologists no longer accept this group. Fossil cnidarians are known from the Precambrian period, and a rich fossil record dates from the Cambrian.

The delicate beauty of cnidarians conceals a hard fact: Most are carnivores that catch and disable prey or discourage aggressors with stinging or sticky tubules fired from uniquely specialized cells called cnidocytes. To humans, venomous cnidarian stings feel like a surprise injection of hot needles, similar to the burning sensation produced by contact with the stinging nettle plant, after which Cnidaria is named (cnid- = nettle). Cnidarian stings are typically fatal to their prey. Stings may even be fatal to humans if the cnidarian is a box jelly or other equally venomous species, but most encounters result only in skin irritation.

FORM AND SYMMETRY OF SOLITARY INDIVIDUALS

General

In basic form, the saccate cnidarian body resembles a gastrula, which consists of a gutlike cavity enclosed by a solid body wall (Fig. 7-1). The cavity, called the **coelenteron,** or gastrovascular cavity, opens to the exterior via a mouth surrounded by one or more whorls of **tentacles.** Sensory organs, when present, are distributed at or near the bases of the tentacles.

The cnidarian body exhibits **radial symmetry** around one axis, the **oral-aboral axis,** that extends from mouth to base. The oral end of the body, which bears the mouth, is at one end of the axis, and opposite it is the aboral end. A body is radially symmetric when a vertical section along an axis or through any radius produces mirror-image halves. This means that differentiated parts, such as the petals of a daisy or the tentacles of a sea anemone, are duplicated in each radius and distributed 360° around the periphery. Thus, radially symmetric animals simultaneously face all compass directions with their sensory organs and tentacles, which, like sentries around the perimeter of a fort, are poised for action from any quarter. Radial symmetry may be useful when an abundant but diffuse resource—plankton, light (or insect pollinators for flowers)—or danger has an equal probability of arriving from any or all directions

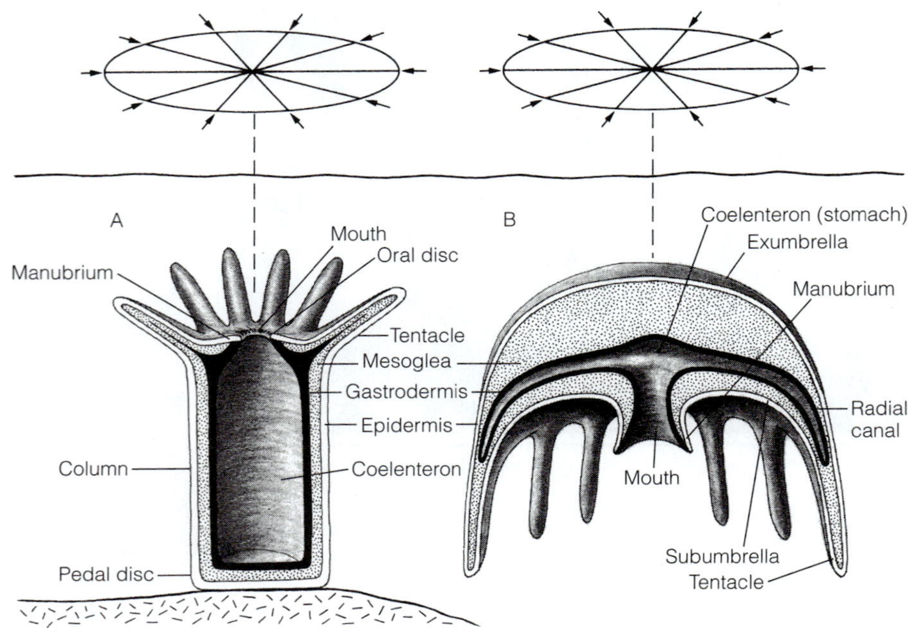

FIGURE 7-1 Cnidaria: solitary body forms. **A,** Radially symmetric polyp. **B,** Radially symmetric medusa. The diagrams above the polyp and medusa indicate their radial symmetry, which is adapted to receive inputs (arrows) from all compass directions.

(Fig. 7-1). This applies to most cnidarians because they do not stalk their food, but instead feed opportunistically on prey as it inadvertently swims or drifts into contact with them.

Polyps and Medusae

Two different body forms, polyp and medusa, occur in cnidarians, often in the life cycle of one individual. A **polyp** resembles a flower and its stalk (Fig. 7-1A). The stalk is a cylindrical elongate **column** arising from an aboral **pedal disc.** At the opposite end of the column, the **manubrium** (or hypostome)—an elevation with the mouth at its summit—is situated in the center of an **oral disc.** The margin of the oral disc bears a whorl of tentacles. The pedal disc and column of the polyp, but not the oral end, may secrete a chitinous exoskeleton, the **periderm,** which provides for protection and attachment to the substratum. Polyps are chiefly attached, sessile, benthic animals with a mouth-up orientation.

A **medusa** has the shape of an umbrella or a bell (Fig. 7-1B). The oral surface (disc) is called the **subumbrella** and the opposite, aboral side is the **exumbrella.** The mouth is at the tip of a mobile appendage, the manubrium, that resembles an elephant's trunk. Tentacles arise from the margin of the bell and surround the mouth. Generally, medusae swim by bell pulsations and orient themselves in a mouth-down position. Medusae are commonly known as jellyfish because their connective tissue (mesoglea) is thick, gelatinous, and buoyant; polyp mesoglea is thin, often composed of little more than the combined epidermal and gastrodermal basal laminas (Fig. 7-1, 7-2).

Tissues and Body Compartments

The cnidarian body wall is composed of three tissue layers (Fig. 7-2): an outer epithelium, the **epidermis;** an inner epithelium, the **gastrodermis,** which lines the gutlike coelenteron and joins the epidermis at the mouth; and, between these two epithelia, a gelatinous extracellular matrix called **mesoglea.** Mesoglea, which is comparable to the mesohyl of sponges, often contains ameboid cells and is thus a connective tissue (Fig. 4-1). Most cells of the cnidarian body, however, are confined to the epithelia, the general characteristics of which are described in Chapter 6. Both epidermis and gastrodermis contain

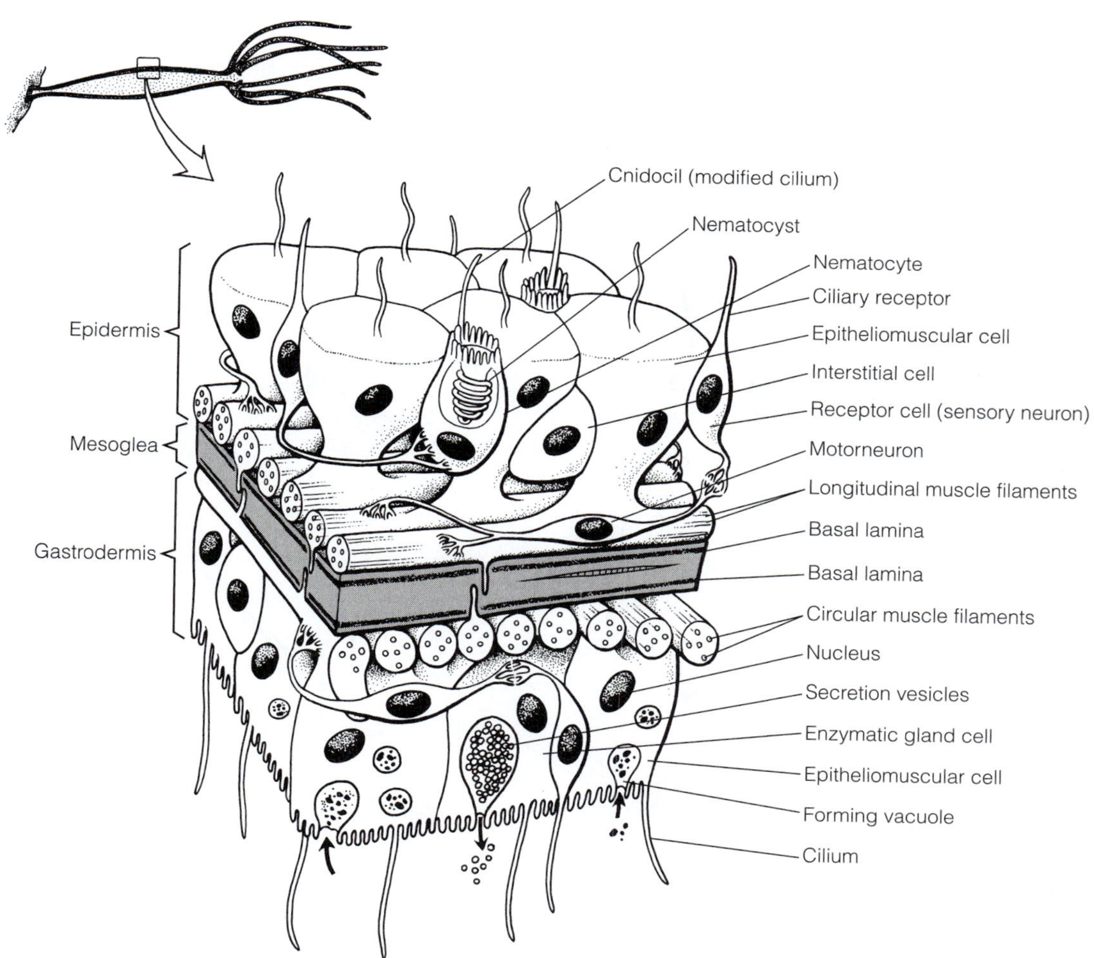

FIGURE 7-2 Cnidaria: body wall of a polyp *(Hydra).*

cnidocytes, muscle, nerve, glandular, interstitial, and ciliated cells, although their specific functions may differ in the two epithelia. Sensory cells occur in the epidermis, germ cells in the gastrodermis. These cells are more fully described in later sections of this introduction.

Cnidarians were the first animals to exploit a body design based primarily on two-dimensional sheets of epithelial tissue (Chapter 6; Fig. 6-1). Although they have a connective-tissue mesoglea, its function is chiefly structural. Cells concerned with communication, movement, digestion, internal transport, and reproduction are part of the epithelia, not the connective tissue. As largely epithelial animals, cnidarians contrast with the sponges, in which the body is organized around a functionally diverse connective tissue. Nevertheless, cnidarians begin an evolutionary trend for certain cells to abandon the epithelia for the underlying connective tissue. Included among these epithelial expatriots are muscle, nerve, and germ cells, the first two having their evolutionary origin in the cnidarians. So within Cnidaria we find living examples of how muscle and nervous tissue evolved from epithelial tissue (Fig. 6-4, 6-6, 7-2).

Because both the mesoglea and the coelenteron are enclosed by epithelia, both can be extracellular compartments physiologically tuned for distinct functions. All life's functions, however, must be accommodated in one or the other of these two compartments, because there are no others. Mesoglea and coelenteron, therefore, are multifunctional compartments. This is seen clearly in the coelenteron, which always has digestive, circulatory, and absorptive functions and sometimes also is a hydrostatic skeleton or a brood chamber for developing embryos, and may collect and eliminate some excretory products. The mesoglea has a skeletal function, may be important in buoyancy (in jellyfish), and may provide a stable or nutritive milieu for the proper operation of muscles and nerves, as well as the growth of germ cells. Physiological regulation of extracellular body compartments may be rudimentary in cnidarians, but they are nevertheless among the first animals to tinker with its possibilities. The most obvious of these is extracellular digestion, which allowed for the use of large food and discontinuous feeding, and a neuromuscular system, which enabled the body to make directed movements in response to environmental events.

FORM OF COLONIAL INDIVIDUALS

Many cnidarian species adopt a colonial growth form, with the colonies consisting of polyp zooids, medusa zooids, or both. Colonial species dominate several major taxa, such as corals, hydroids, and siphonophores. Colonial growth forms are almost

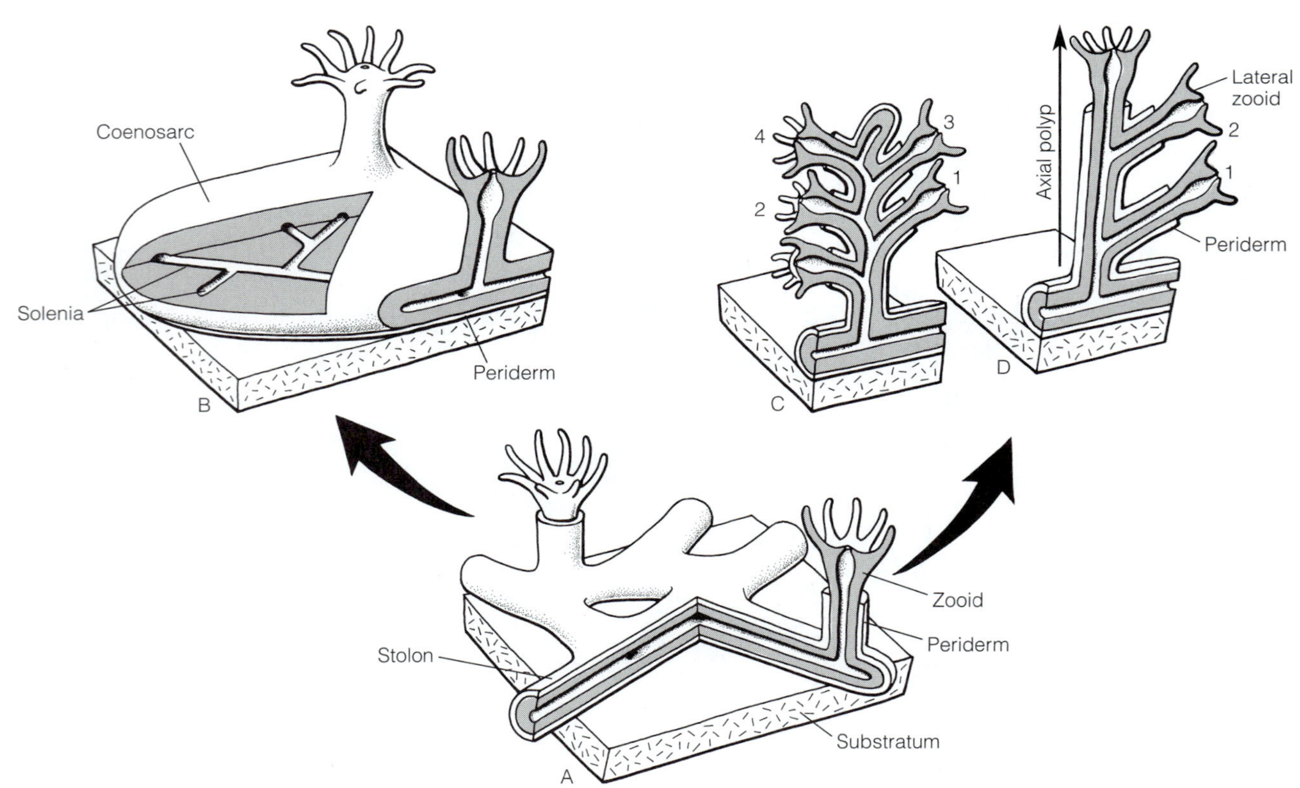

FIGURE 7-3 Cnidaria: colonial body forms. **A,** Stolonate colony. **B,** Coenosarcal colony (stolonal mat). **C–D,** Fruticose (shrubby) colonies. Bases of colonies, attached to substratum, can be either stolons (shown) or coenosarc (not shown), depending on taxon. **C,** Fruticose colony resulting from fixed-length budding. **D,** Fruticose colony grows by indeterminate growth of an axial polyp, which buds a succession of lateral zooids. Numbers indicate sequence of bud formation.

infinitely varied, but most can be reduced to one of three themes: stolonate colonies, coenosarcal colonies, and fruticose colonies.

A **stolonate colony** (Fig. 7-3A), like a strawberry runner, consists of a prostrate stolon, single or branched in a two-dimensional network, that attaches to the surface of the substratum and bears upright, unbranched zooids. The **stolon** is a tubular body wall outgrowth that includes the coelenteron, from which zooids bud at intervals along its length. Typically, the stolon is enclosed by periderm. Most species that form stolonate colonies are small and inconspicuous.

Colonies of several taxa consist of zooids that arise from the surface of a continuous sheet of tissue called a stolonal mat, or **coenosarc** (= common flesh; pronounced SIN-oh-sark) that unites the zooids and attaches the colony to the substratum (Fig. 7-3B). Periderm is restricted to the undersurface of the coenosarc, in contact with the substratum, or it may be replaced by a secreted mineral skeleton, as in stony corals. The coenosarc ranges from a thin, membranelike sheet, as in *Hydractinia* (Fig. 7-61) and most stony corals (Fig. 7-23A,B), to a thick, fleshy mass in which the zooids are embedded except for their oral ends, as in soft corals (Fig. 7-38). Between the upper and lower epidermal surfaces of the coenosarc is a network of hollow gastrodermal tubes called **solenia,** which course through the mesoglea and interconnect the zooidal coelenterons (Fig. 7-3B). The solenia resemble stolons embedded between the upper and lower epidermal layers of the coenosarc sandwich, and new zooids can bud from solenia near the surface of the colony. Unlike stolons, however, solenia lack epidermis and periderm and consist solely of gastrodermis around a coelenteric channel.

The colonies described so far are more or less simple, consisting of zooids borne either directly on stolons (Fig. 7-3A) like bulbs on a string of Christmas lights or individually on a coenosarc (Fig. 7-3B). More complex colonies arise not only by the budding of zooids from stolons or coenosarc, but also from other zooids. With few exceptions, zooidal budding produces an upright, plantlike, **fruticose colony** that may be grassy, bushy, shrubby, or feathery in appearance. In one form of zooidal budding, **fixed-length budding,** the first zooid grows to a fixed length and then buds another zooid from its column below the whorl of tentacles (Fig. 7-3C). The new bud differentiates into Zooid 2 and grows to the same length as Zooid 1, but extends above it. Zooid 2 then buds Zooid 3, which extends above Zooid 2. In this stepwise fashion, like adding sections to a stovepipe, the colony grows upright above the substratum. The stems of these colonies consist of a series of sections or segments, each one being the "column"

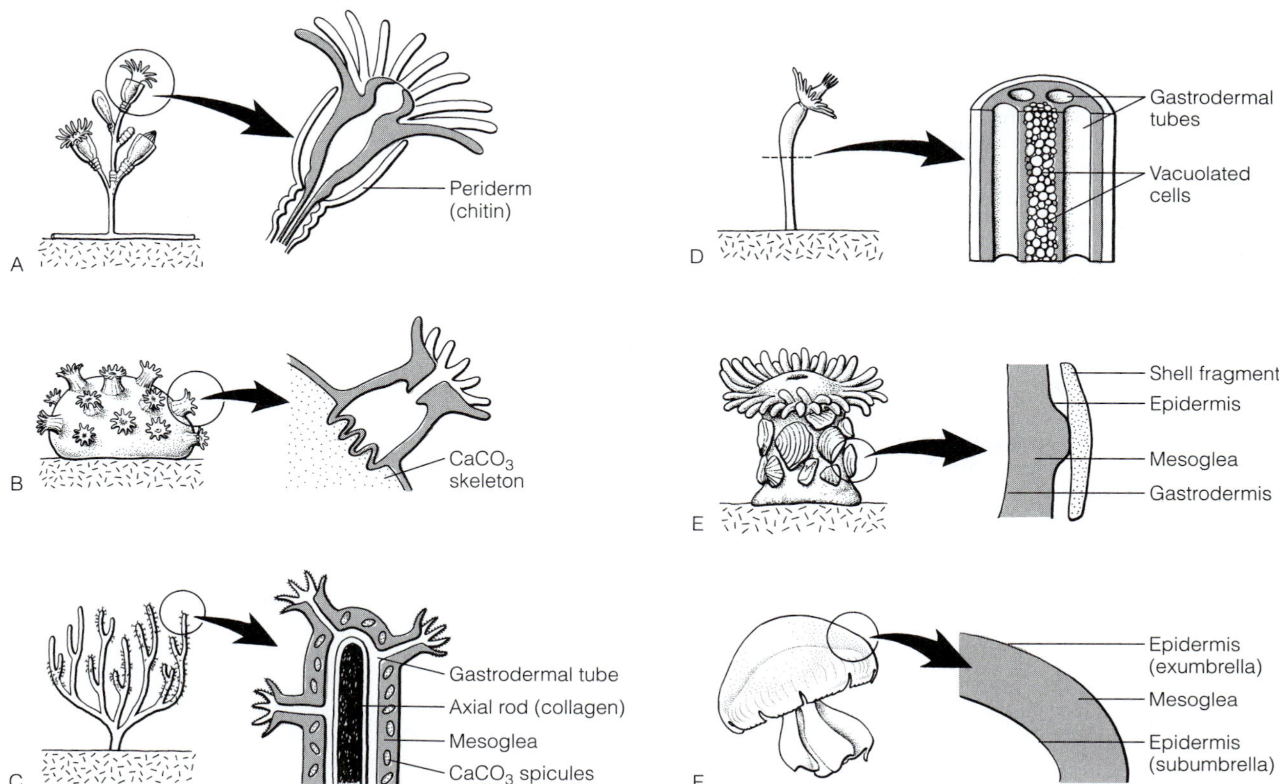

FIGURE 7-4 Cnidaria: skeletons. **A,** Periderm exoskeleton (chitin) in a hydrozoan. **B,** Calcareous exoskeleton (aragonite) in a stony coral. **C,** Organic endoskeletion (collagen) in a sea plume. **D,** Turgid (vacuolated) cellular endoskeleton in a hydrozoan. **E,** Skeleton fabricated of foreign material (shell fragments) in a sea anemone. **F,** An elastic endoskeleton (mesoglea collagen) in a medusa.

of a single zooid. This pattern of budding typifies some hydrozoan colonies, such as *Obelia* (Fig. 7-64A), which attach to the substratum by stolons.

In contrast to fixed-length budding, the main stem of some fruticose colonies arises from the column of a single **axial polyp,** which grows in length more or less indefinitely (Fig. 7-3D). The mouth of the elongate axial polyp opens at the growing tip, often a considerable distance from the pedal end of the polyp. As the axial polyp grows, it buds lateral zooids from its elongating column. The lateral buds are proliferated sequentially from a zone below the tentacular whorl and thus usually the youngest zooids are at the oral end of the axial polyp, the oldest at its pedal end. Sometimes, one of the lateral zooids transforms into an axial polyp, elongates, creates a branch, and buds its own lateral zooids. Lateral zooids, which are often numerous and dense, cover the sides of the main stem and branch stems. Such **axial-polyp budding** typifies colony growth in many taxa, including staghorn corals (Fig. 7-25D), sea pens (Fig. 7-40B), sea pansies (Fig. 7-40C,D), many hydroids (Fig. 7-62), and siphonophores (Fig. 7-69). Fruticose colonies with axial-polyp budding attach to the substratum via stolons or coenosarc.

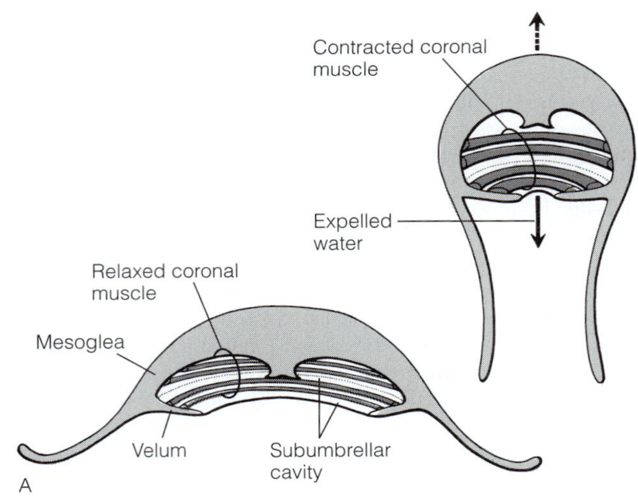

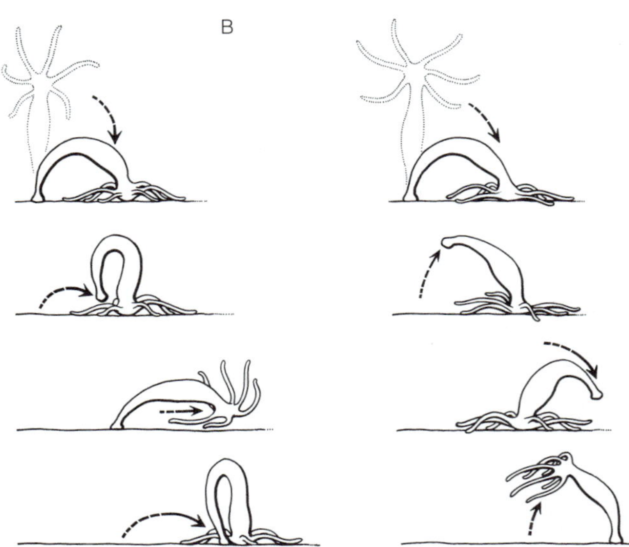

FIGURE 7-5 Cnidaria: locomotion. **A,** Swimming (jet propulsion) in a medusa. Left: coronal muscle relaxed, subumbrellar cavity expanded and filled with water; right: coronal muscle contracted, subumbrella constricted, forcing a locomotory water jet from the subumbrellar cavity. When the coronal muscle relaxes, the elastic mesoglea restores the body to its original form (left). **B,** *Hydra:* inchworm-like locomotion (left column); somersaulting (right column).

SKELETON

The startling variety of cnidarian skeletons, especially evident in polyps, exceeds even that of sponges. Exoskeletons may be thin cuticles of chitinous periderm, as found in many small-bodied solitary polyps and colonies (Fig. 7-4A), but not *Hydra* or most anemones, which are naked. The polyps of stony corals secrete a hard calcareous exoskeleton that in some reef species can, over many generations, reach a diameter of 1 m or more and weigh several tons (Fig. 7-4B). Still other polyps, such as sea plumes and sea fans, rely on endoskeletons for support. Like sponges, they have calcareous spicules and horny organic fibers in their mesoglea (Fig. 7-4C). Yet other polyps support their bodies with columns of cells containing turgid vacuoles, as in the large, tall, and slender hydroid *Tubularia* (Fig. 7-4D). For sea anemones and some other polyps, including *Hydra,* the principal means of support is a hydrostatic skeleton. A few polyps even collect foreign matter—sand grains and shell fragments—to use for protection or to stiffen their bodies (Fig. 7-4E). Among these are colonial anemones (zoanthids), some of which embed sediment particles in their mesoglea, and certain species of sea anemones that attach shell fragments to their surface for protection and possibly for support.

In contrast to the skeletal diversity found in polyps, the mesoglea alone is the sole skeleton of medusae. As such, the form of the mesoglea determines the overall body shape of the medusa (Fig. 7-4F). But the mesoglea is also an elastic gel that springs back to its original shape after being deformed by the swimming musculature.

MUSCULATURE AND MOVEMENT

One general form of musculature is associated with polyps and another with medusae. In polyps, the muscles are distributed in antagonistic sheets of longitudinal and circular fibers. Typically, the epidermal musculature is longitudinal and the gastrodermal is circular (Fig. 7-2, 7-7). The muscles of both layers are smooth (Fig. 6-5). The principal muscle of medusae is the halo-shaped, circular **coronal muscle** on the subumbrellar surface (Fig. 7-5A). The cross-striated coronal muscle is antagonized by the elastic mesoglea.

Cnidarian muscles are primarily epidermal and gastrodermal epitheliomuscular cells (Fig. 6-4A, 7-2, 7-7), but in anthozoans and scyphozoans, some of these cells have abandoned the epithelia, entered the mesoglea, and transformed into myocytes, or "true" muscle cells (Fig. 6-4C).

Cnidarians perform a wide variety of movements. Polyps can extensively shorten and extend or bend their bodies, whereas medusae jet-swim by rapidly constricting their bells (Fig. 7-5A). Hydras and some sea anemones can detach and shift locations by creeping on their bases, by inchworm-like crawling, by somersaulting (Fig. 7-5B), or by floating. Both polyps and medusae use muscles to move tentacles and other appendages, especially while feeding. Some polyps, such as sea anemones and corals, and most medusae also have radial muscles that extend from near the body axis toward the margin, like spokes on a wheel. In medusae, contraction of these muscles causes the subumbrella to roll inward, crumpling the smooth surface of the bell. Sometimes this action is accompanied by withdrawal of the manubrium, retraction of the tentacles, and a flash of bioluminescence. It is thought to be a means of avoiding contact with and potential damage from plankton.

NERVOUS SYSTEM

The cnidarian nervous system, like that of all eumetazoans, consists of superficial sensory neurons that monitor the environment, motorneurons that activate effectors such as muscles or cnidocytes (Fig. 7-2), and interneurons that join the sensory

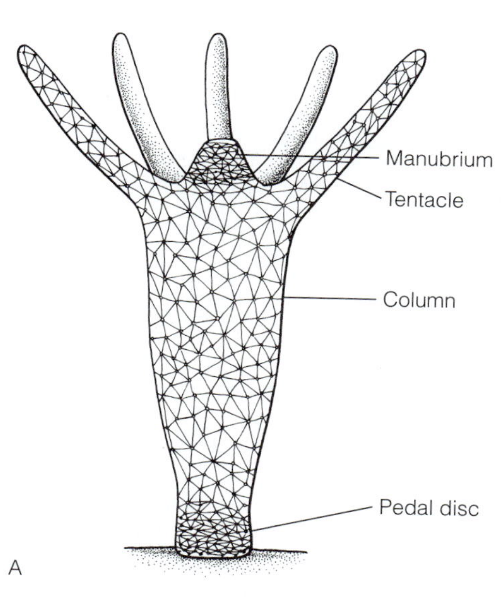

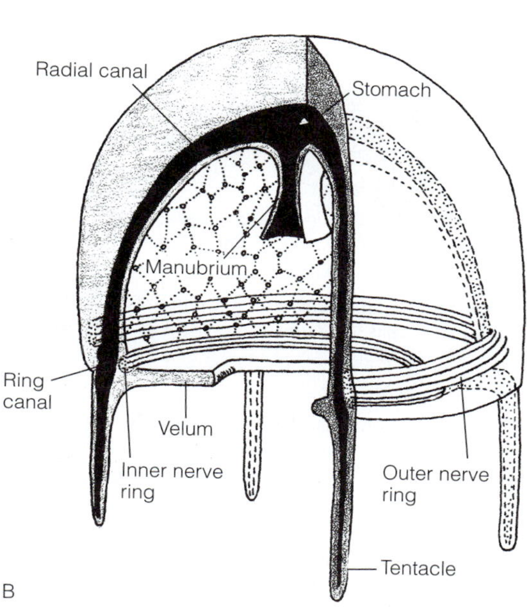

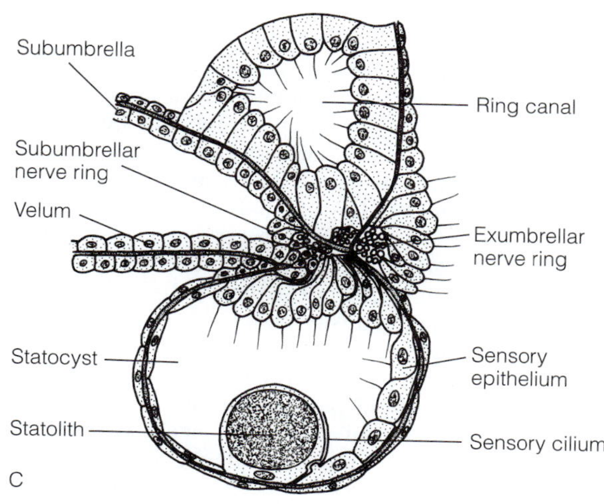

FIGURE 7-6 Cnidaria: nervous system and sense organ. **A,** Nerve net of a polyp. Note oral and pedal concentrations. **B,** Nervous system of medusa. **C,** Statocyst on the bell margin of hydromedusa (vertical section). *(B, After Bütschli from Kaestner, A. 1984. Lehrbuch der Speziellen Zoologie. 2. Teil. Gustav Fischer Verlag, Stuttgart, 621 pp.; C, From Singla, C. L. 1975. Statocysts of hydromedusae. Cell Tissue Res. 158:391–407.)*

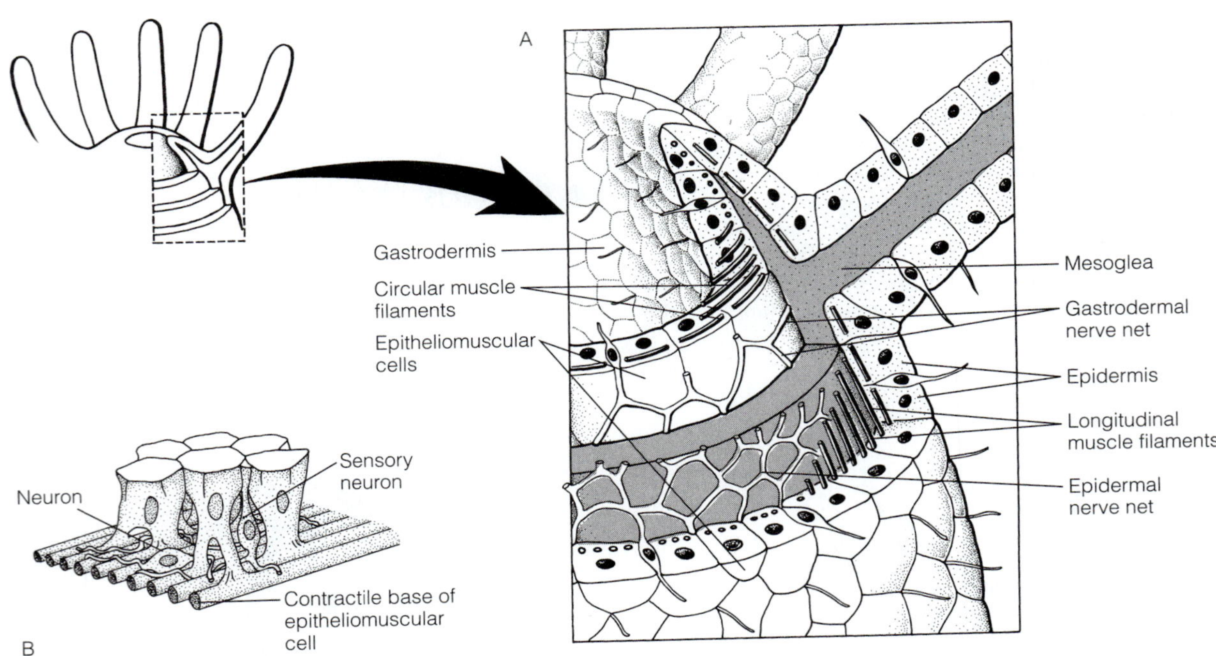

FIGURE 7-7 Cnidaria: neuromuscular anatomy and histology. **A,** Cutaway view of polyp body wall. **B,** Isolated patch of epitheliomuscular cells. *(B, From Mackie, G. O., and Passano, L. M. 1968. Epithelial conduction in hydromedusae. J. Gen. Physiol. 52:600.)*

receptors to the motorneurons. Interneurons not only link receptors to effectors, they form networks and ganglia and so control the intensity and quality of signal conduction. Ultimately this control manifests itself as reflex and complex behavior. In cnidarians, sensory neurons are usually bipolar, having a receptive dendrite at one end and a transmitting axon at the other. Interneurons and motorneurons may be bipolar or multipolar. If multipolar, each cell typically has several dendrites and one axon.

In Cnidaria, interconnected neurons form a pair of complex, two-dimensional **nerve nets** (Fig. 7-6A,B; 7-7). One net lies in the base of the epidermis, another in the base of the gastrodermis, and the two nets are joined by neurons that bridge the mesoglea. The junctions in each nerve net are synapses between neurons. Thus each nerve net resembles a map of city streets intersecting each other. Like vehicles traveling along roads, nervous impulses can travel in any direction through the net. This is possible because the synapses of some neurons conduct in both directions, like two-way streets, or, if they conduct in one direction only, they are paralleled by opposing one-way streets. Impulses arising from a point stimulus typically radiate through the net like ripples from a pebble tossed in a pond. Such **diffuse conduction** is characteristic of nerve nets. For animals with radial sensory structures and a sheetlike musculature, diffuse or multidirectional conduction is the logical form of impulse conduction. Two-dimensional nerve nets, muscular sheaths, and diffuse conduction also occur in medusae (Fig. 7-6B), as well as in the gut wall and other epithelial (sheetlike) organs of higher animals, but in cnidarian polyps the *entire* nervous system is in the form of a net.

Although medusae have nerve nets, their nervous system also includes concentrated nerve rings (Fig. 7-6B) and ganglia around the margin of the bell. These are associated with marginal sensory *organs*—statocysts, ocelli (Fig. 6-7, 6-8)—as well as patches of mechano- and chemoreceptive cells (Fig. 7-6C) and the swimming musculature. Each of the organs communicates with the whole-body nerve nets, but also is associated with a ganglion and the nerve rings that encircle the subumbrellar surface and innervate the swimming muscle.

A **ganglion** is a concentration of neurons that serves as a brainlike integration center. As such, it receives sensory input, integrates that information with other inputs, and generates motor output. The ganglion associated with each marginal sensory organ, for example, receives information from that organ (and from elsewhere via the net) and generates motor output that affects the contraction rate (pace) of the swimming muscle. The marginal ganglion with the highest frequency of motor stimuli overrides the other ganglia and sets the contraction pace of the bell.

Motor impulses are conducted via the nerve ring to the medusa's coronal (swimming) muscle (Fig. 7-5A). Because the entire coronal muscle must contract simultaneously to produce an effective thrust, the motor impulse must simultaneously reach all parts of the muscle. One way that jellyfish (hydrozoans) achieve a simultaneous contraction of the coronal muscle is to dispense with chemical synapses, which cause time delays, between neurons in the nerve ring and instead couple them electrically with gap junctions (Chapter 6). Such conduction is very rapid and results in nearly instantaneous transmission of the nerve impulse around the ring and to all parts of the muscle.

CNIDOCYTES AND CNIDAE

A unique and defining feature of cnidarians is the **cnidocyte:** a combined sensory-effector cell (Fig. 7-8, 7-9, 7-10) that plays a central role in prey capture and defense. Each cnidocyte houses a **cnida,** a fluid-filled membranous capsule containing a long tubular invagination of the capsule wall (Fig. 7-8). When the cnidocyte receives appropriate stimulation, the tubule everts explosively to the exterior (Fig. 7-9). Depending on the type of cnida, the everted tubule may sting or paralyze the prey by penetrating the integument and releasing toxins, adhere by sticking to the surface of an animal or substrate, or perform other tasks. Cnidocytes often bear a single sensory cilium and thus appear to be independent sensory-effector cells (Fig. 7-8; but see below). Cnidocytes typically occur throughout the epidermis, but are in especially high density on the tentacles and other prey-catching or defensive structures on or near the mouth. Many cnidarians also have cnidocytes in localized parts of the gastrodermis to quell swallowed prey.

Three general types of cnidae—nematocysts, spirocysts, and ptychocysts—occur in cnidarians and are housed in three types of cnidocytes—nematocytes, spirocytes, and ptychocytes, respectively. Thick-walled **nematocysts,** which occur in all groups of cnidarians, often have spines or barbs on the surface of the everted tubule (Fig. 7-8, 7-9, 7-10A–C). Like a jack-in-the-box, the firing site may be covered by a hinged lid called an **operculum** (Fig. 7-9A, 7-10B; in scyphozoans and hydrozoans) or by three apical flaps (in anthozoans). The sensory cilium of the nematocyte (a cnidocyte that houses a nematocyst) is mechanoreceptive, may be nonmotile and stiff, and is called a **cnidocil** (Fig. 7-8B; in scyphozoans and hydrozoans), or it may be motile and called a **ciliary cone** (Fig. 7-8A; in anthozoans). All stinging and toxic cnidae are nematocysts, but some types of nematocysts have other, nonvenomous functions; at least 30 different types of nematocysts have been identified. A **spirocyst** is a thin-walled cnida in which the undischarged tubule is coiled like a spring (Fig. 7-10D–E, 7-11). The discharged spirocyst tubule lacks barbs or spines, but the minute, sticky threads radiating from the wall of the tubule like bristles on a bottle brush are adhesive. Uniquely, a spirocyte lacks a sensory cilium. Spirocytes occur only in anthozoans. Similar to a spirocyst, the **ptychocyst** releases an adhesive tubule, but it lacks the spirocyst's radiating sticky threads and, when undischarged, is stored differently within the capsule (Fig. 7-10F). Instead of being coiled within the capsule, the ptychocyst tubule is folded in a compact zigzag, like a collapsed fire hose in its storage cabinet. The ptychocyte bears a motile cilium. Ptychocysts occur uniquely in the epidermis of anthozoan tube anemones (Ceriantharia). Their everted tubules are interwoven to form a tough, feltlike protective tube in which the polyp lives.

The mechanism of cnida discharge has been determined largely from research on hydrozoan nematocysts (Fig. 7-8B,

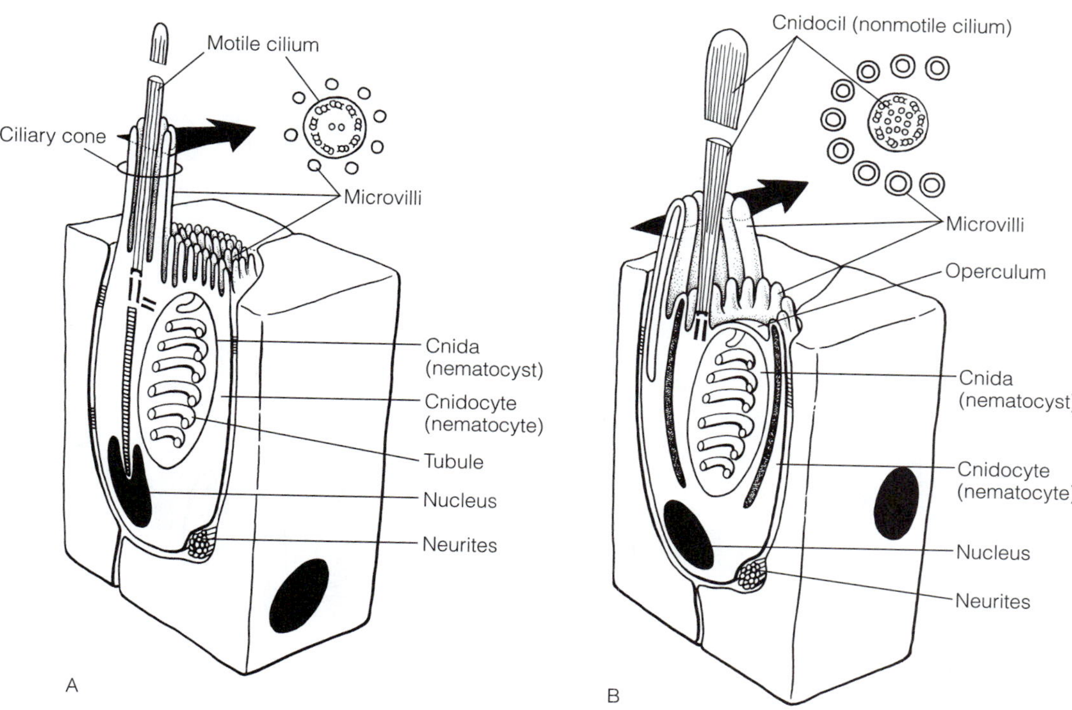

FIGURE 7-8 Cnidaria: cnidocytes and cnidae. **A,** Anthozoan nematocyte. Note that the ciliary cone includes a motile cilium with a normal $9 \times 2 + 2$ axoneme (cross section). **B,** Hydrozoan (or scyphozoan) nematocyte. Note that cnidocil is a nonmotile (sensory) cilium whose axoneme departs from the normal $9 \times 2 + 2$ pattern (cross section). Neurites are unidentified nerve-cell processes, either axons or dendrites. *(Modified and redrawn from Ax, P. 1996. Multicellular Animals. Vol. I. A New Approach to the Phylogenetic Order in Nature. Springer, Berlin. 225 pp.)*

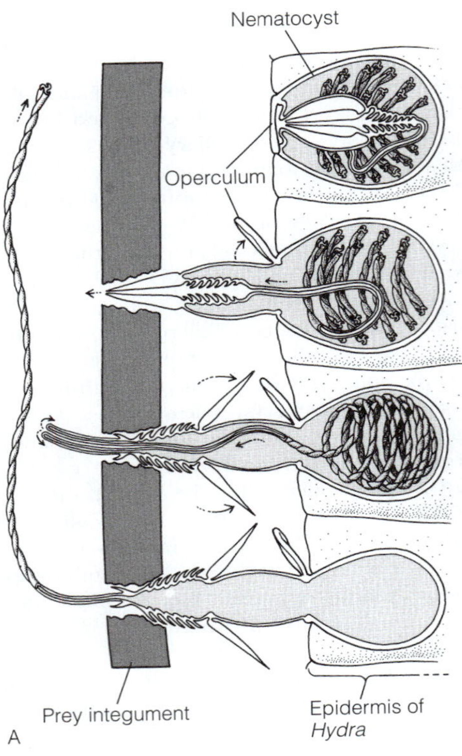

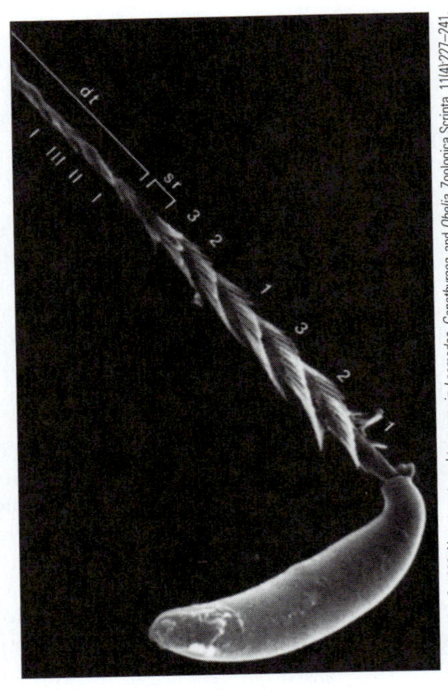

FIGURE 7-9 Cnidaria: nematocyst dynamics. **A,** Discharge and prey body-wall penetration by a *Hydra* nematocyst (stenotele or "penetrant"). **B,** Scanning electron micrograph of a discharged nematocyst tubule of the hydrozoan *Laomedea.* *(A, Modified and redrawn from Tardent, P., and Holstein, T. 1982. Morphology and morphodynamics of the stenotele nematocyst of Hydra attenuata Pall. Cell Tissue Res. 224:269–290.)*

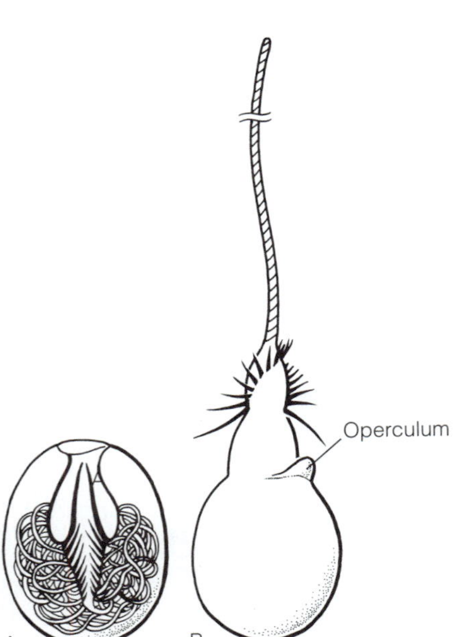

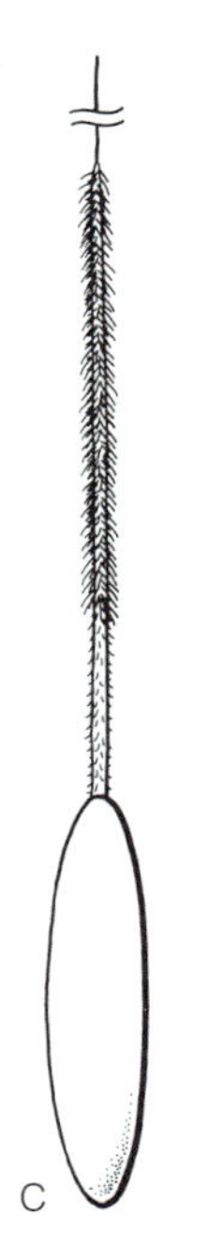

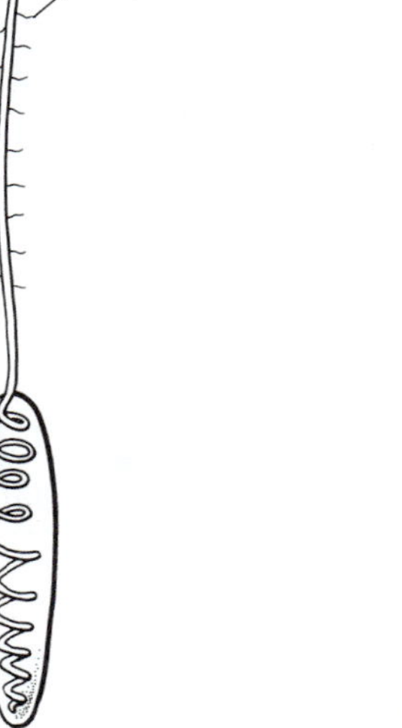

FIGURE 7-10 Cnidaria: major types of cnidae. **A** and **B,** Nematocyst (stenotele or "penetrant") of a hydrozoan, undischarged **(A)**, discharged **(B)**. **C,** Discharged nematocyst (macrobasic mastigophore) of an anthozoan. **D** and **E,** Spirocyst of an anthozoan, undischarged **(D)**, partly discharged **(E)**. **F,** Partly discharged ptychocyst of a tube anemone (Anthozoa: Ceriantharia). *(F, Redrawn from denHartog, J. C. 1977. Descriptions of two new marine Ceriantharia from the Caribbean region. Biol. Meded. 51:211–242.)*

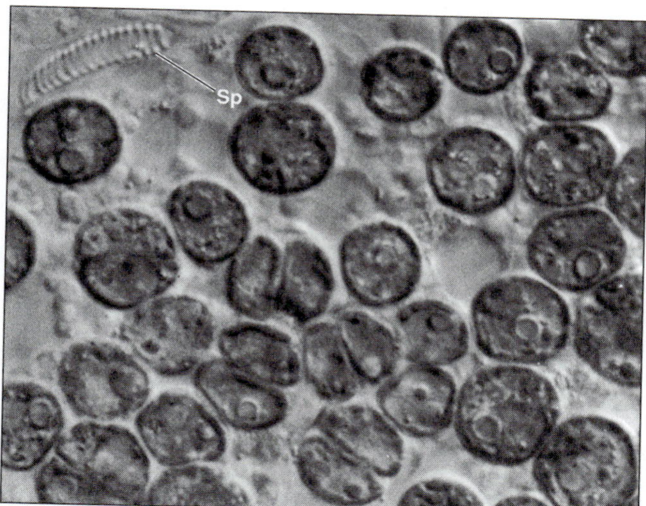

FIGURE 7-11 Cnidaria: zooxanthellae and a spirocyst. Photomicrograph of several cells of the zooxanthella *Symbiodinium microadriaticum* and a single spirocyst (sp) in the tentacle gastrodermis of a sea anemone *(Aiptasia pallida).*

7-9A). The nematocyst wall is thickened and stiffened by a layer of collagen. Prior to discharge, the hollow inverted tubule lies coiled and folded within the capsule and is surrounded by fluid that contains a high concentration of ions, amino acids, and proteins. If the nematocyst is venomous, some of these proteins are toxins, including some that interfere with sodium pumps and the integrity of prey-cell membranes. Many of the ions, especially Ca^{2+}, are complexed with venoms and other proteins to form large aggregates that create little osmotic pressure. But when the nematocyte is appropriately stimulated, these ions dissociate from their macromolecules to create a high intracapsular osmotic concentration. This causes water from the cytoplasm to rush into the capsule, pressurizing it and everting the tubule forcibly and rapidly. Presumably, the thick wall of the nematocyst prevents it from swelling appreciably during the water inrush. Intracapsular osmotic pressures of up to 140 atmospheres ($=2058$ lb/in^2 $= 10^5$ mmHg), nearly the pressure of a filled scuba air tank, have been recorded. The tubule may be fully everted in less than 2 msec, but may require as long as a few seconds, depending on the type of nematocyst. Once the nematocyst is discharged, the nematocyte degenerates and is replaced by a newly differentiated cell. About 25% of the nematocysts of *Hydra littoralis* are lost from the tentacles in the process of eating a brine shrimp. The discharged nematocysts are replaced within 48 h.

At a cost of one cell for each tubule fired, it is not surprising that cnidarians have evolved ways to reduce the number of accidental or inconsequential discharges. Ideally, nematocyst discharges should be reserved for catching and immobilizing prey, defense, and other adaptive tasks, and should not occur when the animal brushes against the substratum or is struck by an inert or harmless object. Most cnidarians seem to exert this control over their nematocytes by requiring two modes of stimulation, mechanical and chemical, before the capsule will fire, although contact alone often results in the discharge of a *few* nematocysts. Further control is provided by heightened sensitivity to particular types of mechanical and chemical stimulation. For example, the mechanoreceptors (including the cnidocil) are especially receptive to vibrations at particular frequencies, perhaps the frequencies at which favored prey move their appendages. Chemical stimuli, such as certain amino acids and sugars associated with mucus, also enhance sensitivity and are normally associated with prey or predators. A particular chemical stimulus, moreover, can change the sensitivity of the mechanoreceptors from one vibration frequency to another, perhaps tuning the receptor-nematocyst complex to a specific class of prey. Finally, the physiological state of the animal, as communicated through the nervous system, can raise or lower the threshold of nematocyst discharge.

INTERSTITIAL CELLS

Interstitial cells (I-cells) are multipotent stem cells that originate from embryonic endoderm, but migrate into all three tissue layers of the body (Fig. 7-2). I-cells are known to occur only in hydrozoans and have been especially well studied in *Hydra,* but probably occur in the other cnidarian classes. In hydrozoans, I-cells differentiate into neurons, gland cells, gametes, and cnidocytes. Thus the I-cells form a pool from which spent cnidocytes are replaced following the developmental path: I-cell → cnidoblast → cnidocyte. Among anthozoans with skeletal spicules in the mesoglea, I-cells differentiate into sclerocytes, which secrete the spicules. I-cells' replacement of spent cnidocytes in Anthozoa and Scyphozoa needs confirmation. It is uncertain whether gametes in these two taxa arise from ciliated gastrodermal cells or I-cells.

COELENTERON: NUTRITION AND INTERNAL TRANSPORT

The **coelenteron** is a blind, saclike cavity lined by gastrodermis and opening to the exterior via the mouth (Fig. 7-1). In large polyps, the cavity may be more or less partitioned by septa, which increase the surface area of the gastrodermis. Among medusae, it is often regionally specialized into a central **stomach** from which **radial canals** (Fig. 7-1B) extend to join a marginal **ring canal.** In most polyps and medusae, a branch of the coelenteron extends into each tentacle. The coelenteron is a multifunctional compartment with roles in extracellular digestion, circulation, excretion, reproduction, and hydrostatic skeletal support.

Feeding in carnivorous cnidarians begins when prey contacts tentacles or other prey-capturing appendages. As the prey is seized by discharged cnidae, the cnidae may wound, paralyze, and even initiate digestion by injecting proteolytic enzymes into the prey. Fluid loss from the prey elicits an ingestive response from the predator. The cnidarian responds to amino acids, especially glutathione, by opening its mouth and stuffing its tentacles, with prey attached, inward (Fig. 7-12C). If the manubrium is long and mobile, it may bend toward and grasp the food like an elephant's trunk. Once the prey is in the coelenteron, **enzymatic gland cells** in the gastrodermis

(Fig. 7-2) release enzymes, primarily proteases, that digest it extracellularly to a slurry of juice and fragments. The slurry is circulated throughout the coelenteron to be absorbed by epitheliomuscular cells, germ cells, and other gastrodermal cells (Fig. 7-2, 7-20E). Large food fragments are phagocytosed and digested intracellularly. This process breaks down lipids and carbohydrates as well as completing the digestion of proteins.

The extracellular phase of digestion is often completed in a few hours and is followed by absorption of the slurry by the gastrodermis in 8 to 12 h. Intracellular digestion usually requires a few days to complete. Indigestible material, often consolidated with mucus into a fecal mass, is egested through the mouth.

In many cnidarians, the gastrodermal cells contain mutualistic algae. Some species of *Hydra* and anemones such as *Anthopleura* harbor green **zoochlorellae,** as do certain freshwater sponges. The mutualistic algae of most marine cnidarians, however, are yellow-brown **zooxanthellae** (Fig. 7-11). Both zoochlorellae and zooxanthellae provide their cnidarian hosts with photosynthate, sometimes accounting for up to 90% of their nutrition, and in return receive nutrients, CO_2, and a substratum deployed in the sun.

The coelenteron is the internal transport system of cnidarians (Fig. 7-12). Circulation of fluid is accomplished by the ciliated gastrodermis, muscular contraction, or both. In general, flows follow definite and sometimes complex patterns that differ among taxa, polyps and medusae, and solitary and colonial growth forms. In some taxa, localized ciliary (siphonoglyphs and asulcal septa; Fig. 7-16, 7-33) or muscular pumps (stomach and tentacle bulbs in the hydromedusa *Sarsia princeps*) function as heart analogs.

GAS EXCHANGE AND EXCRETION

The tentacles and general body wall are gill surfaces, and circulation of water over the body surface by ciliated epidermal cells facilitates gas exchange (Fig. 7-12C).

The excretory product of cnidarians is ammonia, which readily dissolves in water, diffuses across the body wall, and is dispersed by currents. The freshwater *Hydra* concentrates K^+ and removes Na^+ from its cells. Some of the Na^+ is released into the coelenteric fluid, elevating its osmotic concentration above that of the environmental water. The resulting osmotic influx of water into the coelenteron may help to pressurize the hydrostat, or the excess water (and Na^+) may periodically be discharged through the mouth.

REPRODUCTION AND DEVELOPMENT

Regeneration and Clonal Reproduction

Cnidarians have an impressive ability to heal and regenerate missing parts after injury. In *Hydra,* for example, loss of the oral end is followed by regrowth and differentiation of a new mouth and tentacles. Living specimens of the common anemone *Aiptasia pallida* can be dissected, pinned open for observation, and then returned to an aquarium, where they will heal after a few days. Medusae also heal and regenerate missing parts after damage, as do planula larvae (described later).

Clonal (asexual) reproduction is widespread, if not universal, among cnidarians (Fig. 7-13). Both polyps and medusae undergo clonal reproduction, but it is more common among polyps. The forms of clonal reproduction include transverse (uncommon) and longitudinal (common) fission (Fig. 7-13B,

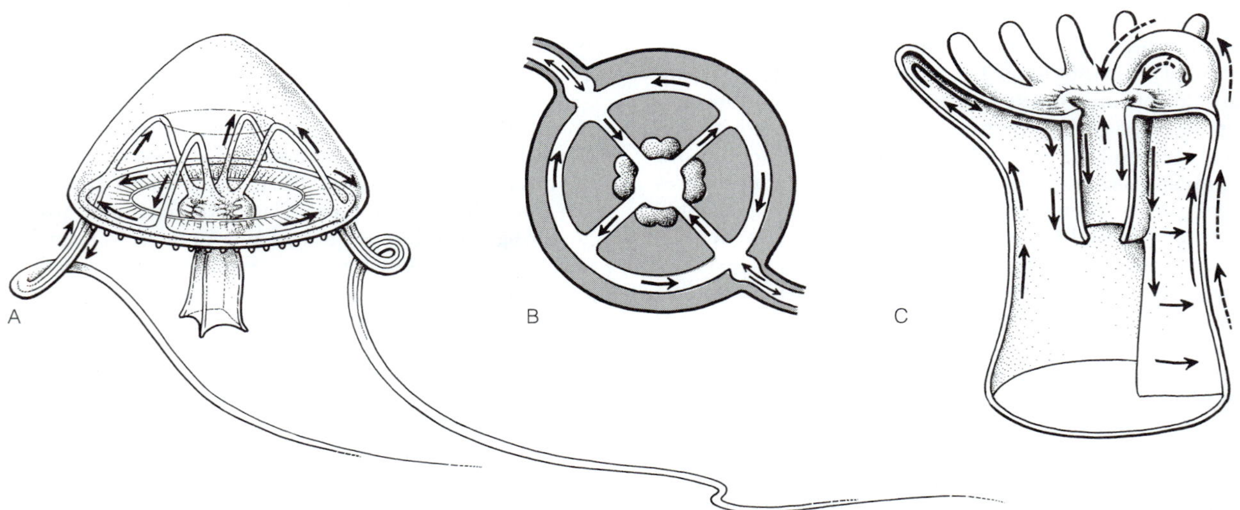

FIGURE 7-12 Cnidaria: circulatory system. Patterns of ciliary circulation in the radial canals, ring canal, and tentacles of a medusa *(Stomotoca atra)* in side view **(A)** and from above **(B).** Note the four-square manubrium and gonads attached to it. **C,** Pattern of circulation over (broken arrows) and in (solid arrows) a sea anemone, including flows in the pharynx. Right side shows flow pattern on a septum; left side between septa. Flow from base to tip of tentacle allows inedible material to be rejected or food to be passed into the mouth as the tentacle bends oralward.

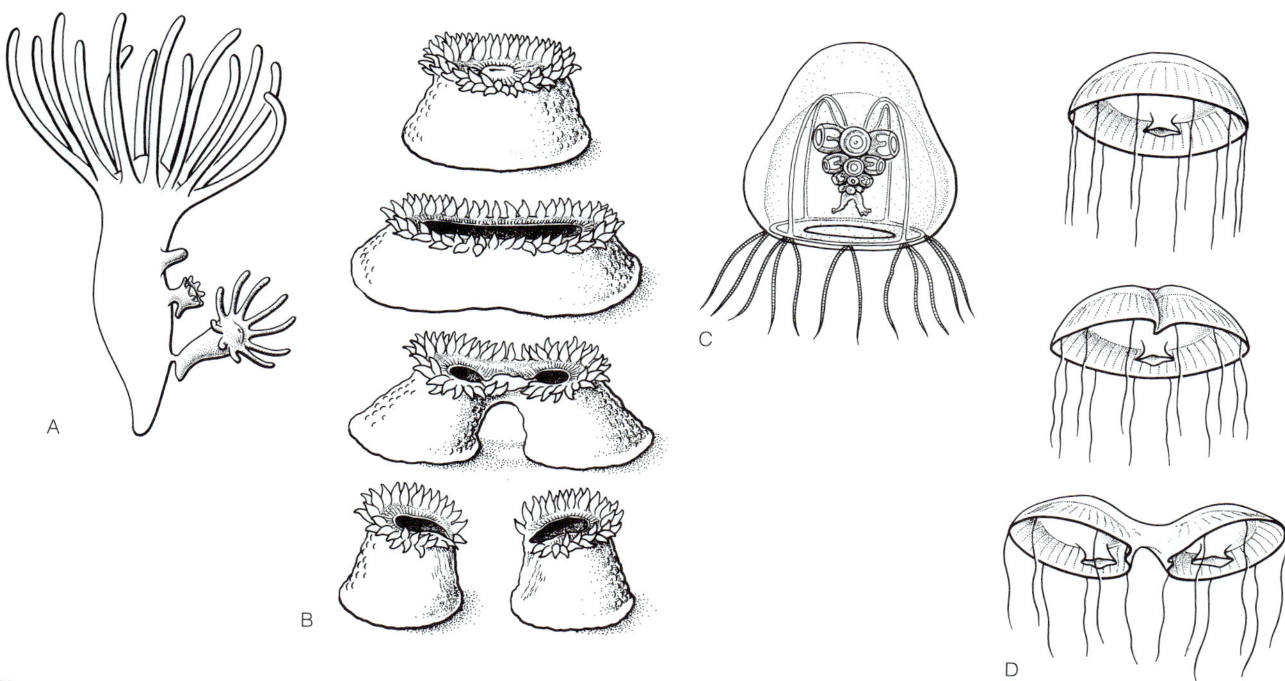

FIGURE 7-13 Cnidaria: clonal (asexual) reproduction. **A,** Budding (polyp). **B,** Longitudinal fission (polyp). **C,** Budding (medusa). **D,** Longitudinal fission (medusa). *(B, Redrawn from Sebens, K. P. 1983. Morphological variability during longitudinal fission of the intertidal sea anemone,* Anthopleura elegantissima *(Brandt). Pac. sci. 37:121–132; C and D, Drawn from a photograph in Stretch J. J., and King, J. M. 1980. Direct fission: An undescribed reproductive method in Hydromedusae. Bull. Mar. Sci. 30:522–525.)*

D), budding (Fig. 7-13A,C), and fragmentation. Anthozoan polyps exhibit all forms of asexual reproduction, scyphozoan polyps undergo budding and transverse fission, and hydrozoan polyps only bud. Some hydrozoan medusae undergo fission (Fig. 7-13C,D).

Sexual Reproduction and Life Cycles

Cnidarian adults are generally gonochoric, but certain taxa, such as stony corals, have many hermaphroditic species. Germ cells originate in the endoderm and generally grow and differentiate in the gastrodermis, except in some hydrozoans, in which they migrate into the epidermis. Ripe individuals spawn their gametes and, primitively, fertilization is external in the water. The zygote cleaves holoblastically to form a blastula (Fig. 7-14A–D). Later, the embryo gastrulates to form ectoderm and endoderm (Fig. 7-14E,F), which become epidermis and gastrodermis, respectively. The blastocoel becomes the mesoglea while the blastopore, when present, gives rise to the mouth-anus (Fig. 7-14G). Gastrulation in cnidarians, whether by invagination or ingression (Fig. 6-9), occurs at the animal-pole end of the zygote (Fig. 7-14E), in contrast to the vegetal-pole position of gastrulation in higher (bilateral) animals (Fig. 6-9A).

Further development leads to a planktonic larva known as a **planula** (Fig 7-14G). Its epidermis is monociliated and its gastrodermis is often yolky. The fusiform planula swims with its aboral (vegetal) end leading. After a brief swimming existence, the planula settles to the bottom, attaches at its aboral end, and metamorphoses into a juvenile (Fig. 7-14H).

The primitive life cycle of ancestral cnidarians, as inferred from anthozoans, consists of a sessile, benthic adult polyp and a planktonic planula larva that metamorphoses into a juvenile polyp (Fig. 7-14H). The anthozoan planula may be planktotrophic or lecithotrophic (often very yolky) and some anthozoan larvae harbor zooxanthellae. These nutritional adaptations enable many anthozoan planulae to sustain larval life long enough to disperse widely. All larvae of Scyphozoa and Hydrozoa, on the other hand, are lecithotrophs, and most lack zooxanthellae. As a result, their larval life typically is short and their dispersal range limited. Perhaps for this reason, a new planktotrophic stage, the medusa, which arises by budding from the polyp, has evolved in these taxa (Fig. 7-15)

Inserted in the life cycle between the polyp and the planula, the medusa not only provides a stage capable of wide dispersal, but also appropriates the role of sexual reproduction from the polyp. This relieves the polyp of the energetic cost, mostly from egg production, of sexual reproduction, because the medusa can feed itself and grow. A polyp's reproductive energy can then be channeled into clonal reproduction or modular growth, both of which enable rapid colonization of attachment surfaces. Thus a life cycle with a planktonic, sexual medusa and a benthic, asexual polyp partitions the competing energy demands of sexual and asexual reproduction to life stages in different ecological niches. Lacking a medusa, the vast majority of adult anthozoans may directly confront this energetic challenge by supplementing their normal nutritional input with photosynthate from zooxanthellae and zoochlorellae.

Blastocoel
Blastopore
A B C D E F
Endoderm
Ectoderm

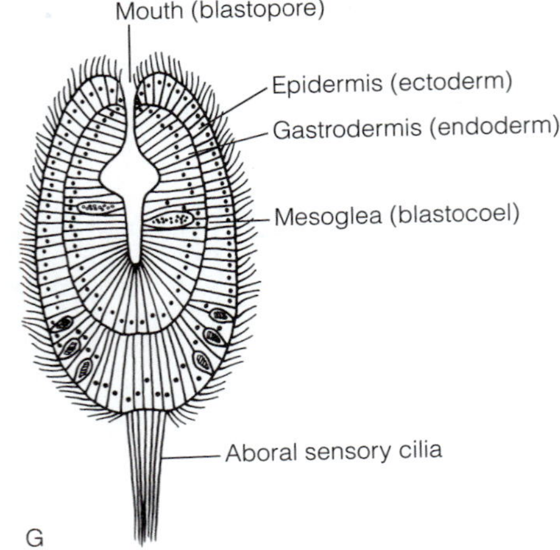

FIGURE 7-14 Cnidaria: development, larva, and life cycle. **A–F,** Embryonic development. **A,** Cleavage initiates at the animal pole (indicated by polar body), the future oral end of the body. Following the blastula stages (**C** and **D**), gastrulation (**E** and **F**) establishes the two germ layers, ectoderm and endoderm, with the blastocoel becoming the mesoglea. Further differentiation results in a planula larva (**G,** longitudinal section of an anthozoan planula). **H,** The presumed primitive life cycle, based on Anthozoa and beginning with the adult, is: polyp → planula → polyp. Note that the planula swims with the aboral end (future pedal end) leading. *(G, Redrawn after Widersten from Kaestner, A. 1984. Lehrbuch der Speziellen Zoologie. 2. Teil. Gustav Fischer Verlag, Stuttgart. 621 pp.)*

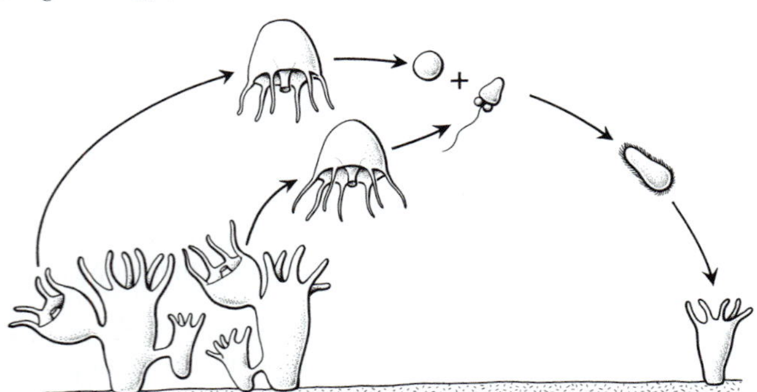

FIGURE 7-15 Cnidaria: presumed apomorphic (medusozoan) cnidarian life cycle: polyp → medusa (adult) → planula → polyp.

ANTHOZOA[C]

Anthozoa, or "flower animals," includes the exclusively marine sea anemones, corals, sea fans, sea pens, and sea pansies. In their stunningly wide range of forms and colors, anthozoans create and accent coral-reef seascapes and tidepools with splashes of emerald, ruby, and amethyst. Anthozoa is the largest cnidarian taxon, containing over 6000 solitary and colonial species. Anthozoans are polyps only, and a medusa stage is absent. Thus the polyp is responsible for both sexual and, when present, asexual reproduction. As compared with the polyps of other cnidarian classes, anthozoans are large polyps that range in diameter from approximately 0.5 cm to 1 m. Anthozoan body volume is variable because most have inflatable and retractile polyps. Anthozoan mesoglea contains amebocytes and is thus a true connective tissue. Anthozoa is the only cnidarian taxon that has all three types of cnidae—nematocysts, spirocysts, and ptychocysts.

POLYP FORM

The body consists of a tubular column surmounted by a wide plateau, the oral disc (Fig. 7-16). The oval or slitlike mouth is in the center of the oral disc and a whorl of tentacles occurs at its margin. In solitary individuals, the base of the column is often expanded into a **pedal disc** that adheres to the substratum.

The mouth leads into a tubular but laterally compressed **pharynx** that descends below the oral disc and opens into the coelenteron via an internal opening (Fig. 7-16). Because of

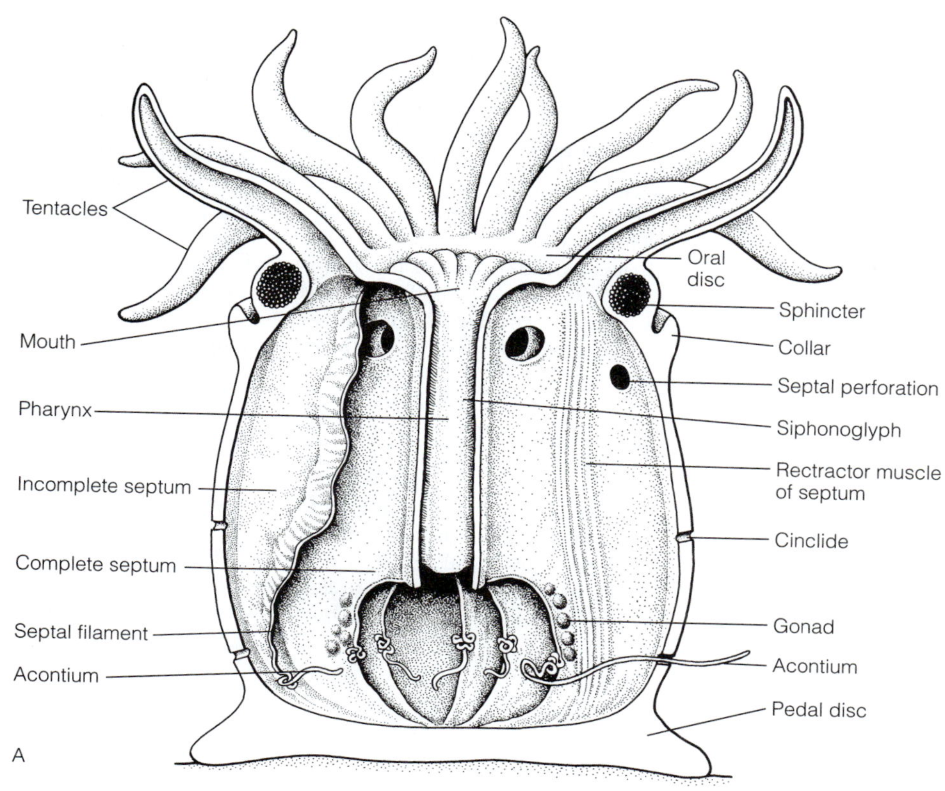

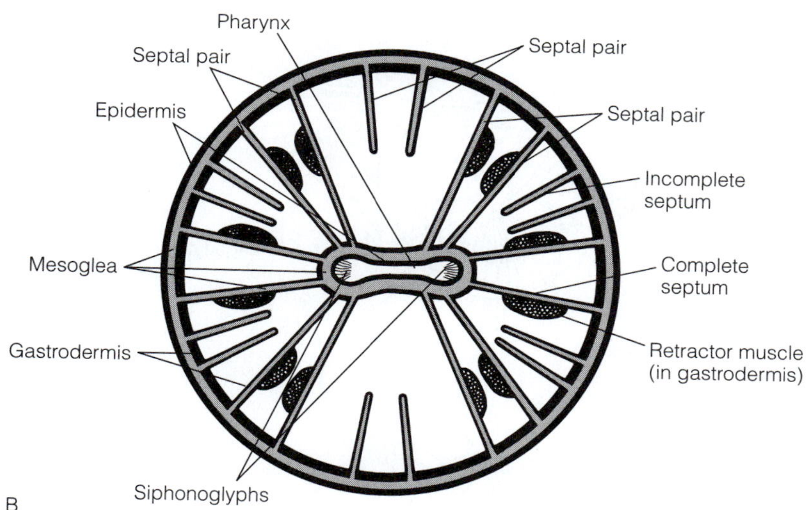

FIGURE 7-16 Anthozoa: anatomy. **A,** Cutaway view of a sea anemone (Actiniaria). **B,** Cross section of body at level of pharynx.

pharyngeal compression, the mouth and internal pharynx opening are elongated slits. The two opposite corners of the mouth extend inward along the length of the pharynx as ciliated grooves called **siphonoglyphs** (Fig. 7-16). The compressed pharynx and siphonoglyphs confer either a biradial (two siphonoglyphs; Fig. 7-16B) or bilateral (one siphonoglyph; Fig. 7-18B) symmetry on the anthozoan body. Developmentally, the pharynx originates as an ectodermal ingrowth, or stomodeum, at the site of the blastopore. Thus the larval mouth is carried inward to become the pharyngeal opening into the coelenteron. In some anemones, the body wall of the column is perforated by pores, the **cinclides,** through which coelenteric fluid can be released to the exterior (Fig. 7-16A).

The coelenteron is divided by vertical partitions called **septa** (= mesenteries) into radial compartments. Each septum is an outfold of gastrodermis and mesoglea, forming a sandwich of mesoglea between two layers of gastrodermis (Fig. 7-16, 7-17, 7-18). Septa that span the coelenteron and join the pharynx are **complete septa;** those that do not are **incomplete septa** (Fig. 7-16B). In addition to forming junctions with the pharynx, septa also insert on the pedal and oral discs. Each tentacle originates as a hollow outgrowth of body wall between a pair of septal insertions on the oral disc. The number of septa generally varies with body size (Fig. 7-76A–C): Large polyps have many and small polyps have few, but no adult anthozoan has fewer than 6 septa; the largest polyps have 192 or more. The number and pattern of septa are important characteristics used in anthozoan classification.

The inner free margin of each septum is convoluted and has a swollen edge called a **septal filament,** which resembles piping on a pillow seam (Fig. 7-16A, 7-17B–D, 7-18A). Microscopically, each septal filament is composed of either one or three lobes. When trilobed, the middle lobe, called the **cnidoglandular band,** bears cnidocytes (mostly nematocytes) and enzymatic gland cells (Fig. 7-17B,C). The two lateral lobes, the **flagellar bands,** are densely flagellated (Fig. 7-17B,C). The effective strokes of the flagella beat toward the central axis of the body. The middle section of a unilobed septal filament bears the cnidoglandular cells whereas the lateral parts are flagellated.

In some sea anemones, one region of the cnidoglandular band of each septal filament known as an **acontium** extends away from its attachment to the septum as a long thread, usually heavily armed with cnidae (Fig. 7-16A). Acontia sprout from the lower end of the septal filaments and their free ends lie variously coiled in the coelenteron. The outrush of water that accompanies a sudden deflation of the body causes acontia to spew from the mouth, tentacle tips, and, if present, cinclides. Many anthozoans that lack acontia, such as stony corals, nevertheless extrude the free margins of their septa (with *attached* septal filaments) through the mouth or other openings in the body wall. Extruded acontia and septal filaments are used for defense, prey capture, and extracorporeal digestion of prey.

MUSCULATURE AND NERVOUS SYSTEM

Most anthozoan muscles are arranged in sheets of epidermal or gastrodermal epitheliomuscular cells (Fig. 7-18). However, the mesogleal sphincter muscle (Fig. 7-16A), which closes over the withdrawn oral disc and tentacles of some anemones and zoanthids, can be composed of myocytes. The

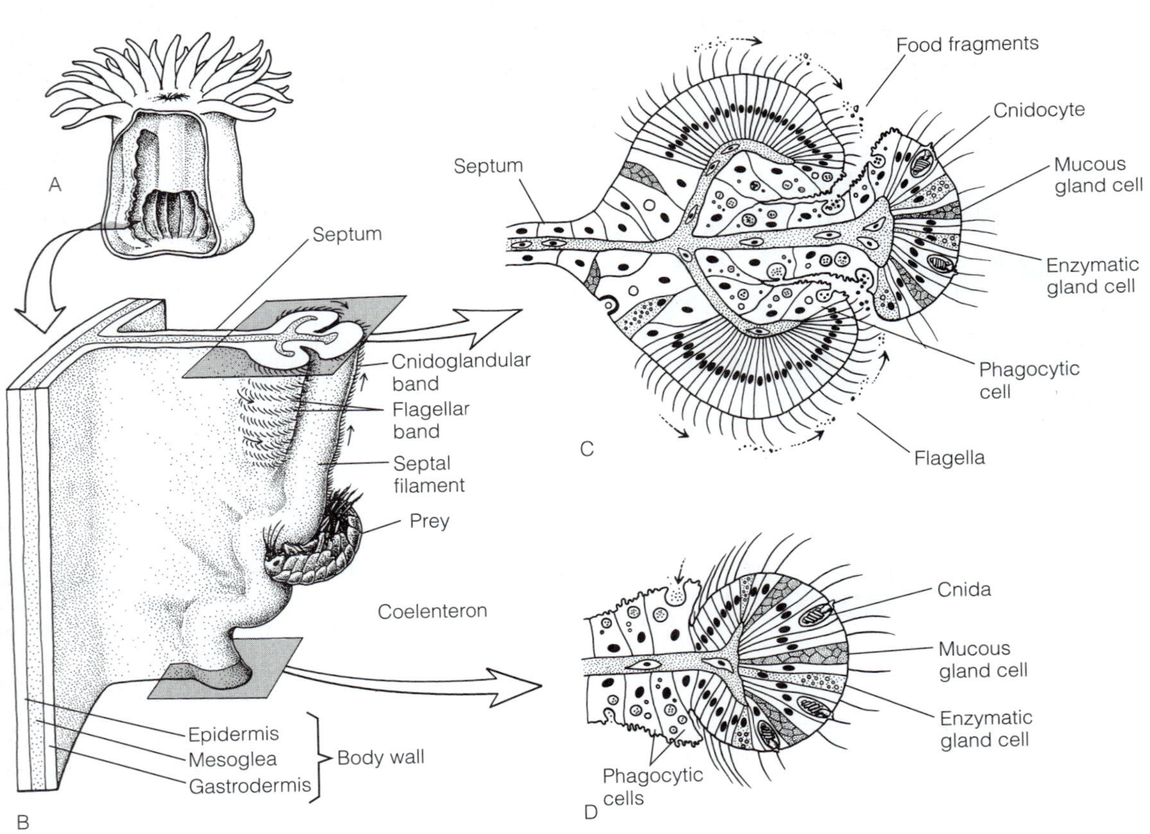

FIGURE 7-17 Anthozoa: septal filament. **A,** Cutaway view of column of sea anemone (Actiniaria), showing septa. **B,** Exploded view of designated area in **A,** showing details of body wall, septum, and septal filament. **C,** Cross section enlargement of upper, trilobed part of septal filament. **D,** Cross section enlargement of lower, unilobed part of septal filament. Arrows indicate direction of ciliary flows of fluid and food. *(Modified and redrawn from Van-Praet, M. 1985. Nutrition of sea anemones. Adv. Mar. Biol. 22:65–99.)*

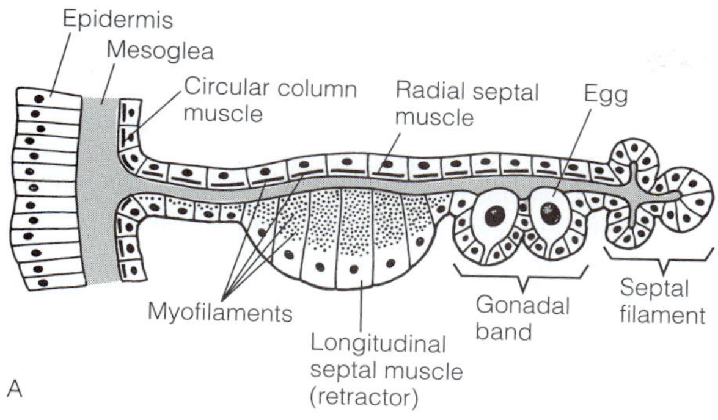

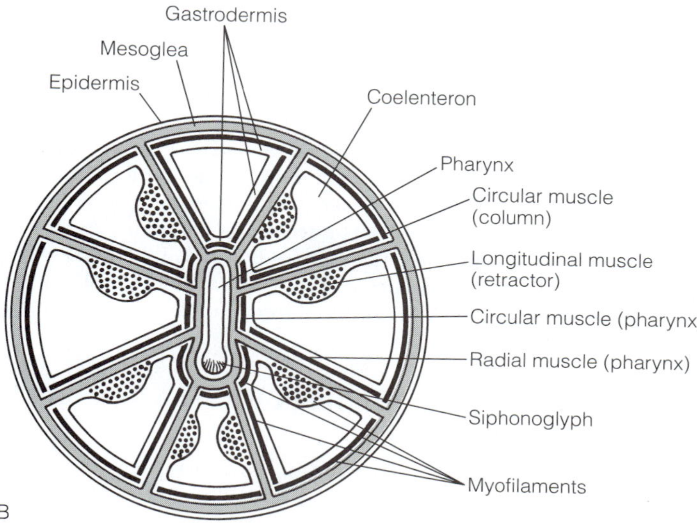

FIGURE 7-18 Anthozoa: musculature. **A,** Cross section of anemone body wall and septum showing arrangement of muscles and gonad. **B,** Cross section through column and pharynx of a soft coral (Alcyonaria) showing arrangement of body wall and septal muscles.

epidermal musculature is largely confined to the tentacles (longitudinal) and oral disc (radial), except in tube anemones (Ceriantharia) and black corals (Antipatharia), which have longitudinal musculature in the column epidermis. The remaining muscles are gastrodermal (Fig. 7-18). These consist of a sheet of circular muscle in the column wall, longitudinal and radial septal muscles, and a circular musculature around the pharynx. Of these, the longitudinal septal muscles, called **retractors** to indicate their polyp-retraction function, are strongly developed and conspicuous, especially in sea anemones.

A single retractor muscle occurs on one of the two faces of each septum and runs its length from pedal to oral disc (Fig. 7-16A). A retractor is a local concentration of enlarged epitheliomuscular cells that each contain a high density of myofilaments, so the muscle bulges from the surface of the septum as a conspicuous band (Fig. 7-18A). Retractor muscles on adjacent septa may face each other (apex to apex), face away from each other (base to base), or may orient apex to base (Fig. 7-18B). These differing muscle patterns, as well as the number and morphogenetic pattern of the septa themselves, are autapomorphies of the major anthozoan taxa.

The nervous system consists of at least two intraepithelial nerve nets—one epidermal and one gastrodermal—that join together at the pharynx, at septal insertions on the pedal and oral discs, and across the mesoglea. The epidermis contains the highest density of sensory cells, but never sensory tissues or organs. Sensory cells such as nematocytes (ciliary cones) and chemoreceptors bear a single cilium surrounded by a collar of microvilli.

RETRACTION AND EXTENSION

Most anthozoan polyps can strongly retract their tentacles, oral disc, and column toward their pedal attachment (Fig. 7-19). During retraction, many anthozoans simultaneously invaginate the oral disc and tentacles into the column and deflate the tentacles and column by expelling coelenteric fluid from the mouth and, in some anemones, cinclides (Fig. 7-16A).

The septal retractor muscles are responsible for retraction and invagination of the oral disc and tentacles. These powerful longitudinal muscles originate on the pedal disc, which is anchored to the substratum, and insert on the underside of the oral disc. Thus contraction of these muscles not only shortens the column, it also bows the oral disc inward and then pulls it (and the tentacles) into the column. Once the disc and tentacles are invaginated, the purse-string-like sphincter muscle constricts the oral-disc margin.

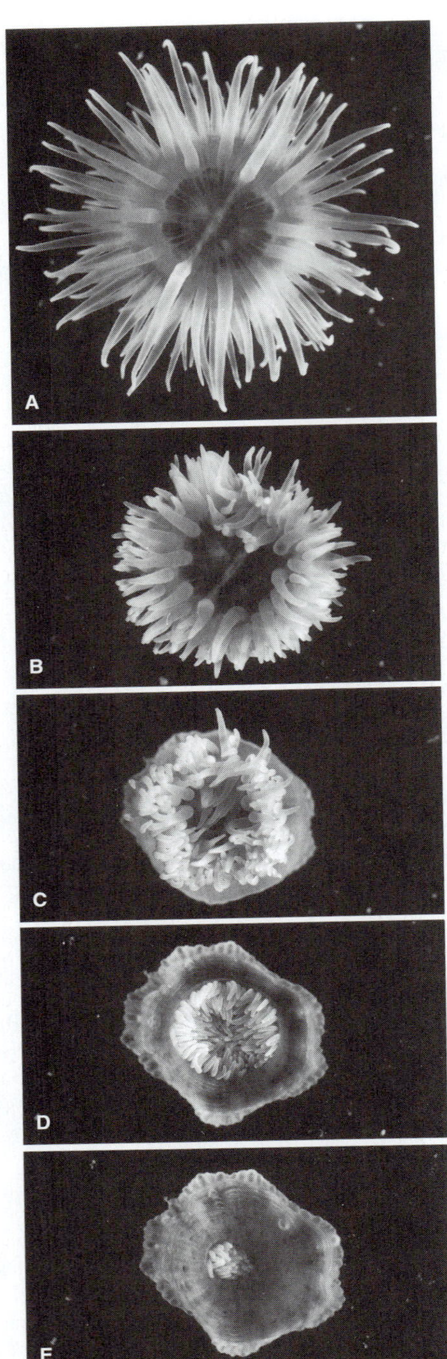

FIGURE 7-19 Anthozoa: retraction-deflation sequence. **A–E,** Retraction-deflation of the sea anemone *Aiptasia pallida.*

Loss of fluid during deflation of the body requires the active contraction of two sets of muscles: radial and longitudinal. Contraction of radial muscles in all of the septa and the oral disc opens the pharynx and mouth. With the pharynx and mouth held open, contraction of longitudinal muscles retracts the polyp, evacuating the coelenteric fluid through the mouth. This fluid loss also causes the tentacles to collapse like deflated balloons, permitting them to be more easily accommodated as they are withdrawn into the column.

Polyp deflation results from a partial to complete loss of fluid. This near-total loss of body volume requires a fluid-recovery mechanism for reinflation that involves uniquely anthozoan structures. The pharyngeal siphonoglyphs are the water pumps responsible for polyp inflation. Siphonogylph cilia always beat with their effective strokes directed inward, so as soon as the body-wall muscles of the retracted polyp relax, the siphonoglyph incurrent slowly refills the coelenteron, expanding the body to its extended posture and appropriate volume. With the body fluid restored, coelenteric fluid pressure tends to push together the opposite, flat sides of the pharyngeal wall. This seals the pharynx, preventing backflow of water from the coelenteron to the exterior. Thus, the compressed form of the anthozoan pharynx is important to its function as a nonreturn valve.

Many anthozoan polyps extend to feed and retract for protection from disturbances, including predators. Inflation can also have other purposes, such as anchoring the polyp in a burrow, tube, or crevice. The sea-onion anemone, *Paranthus rapiformis,* uses inflation as an aid to dispersal. When dislodged from its burrow, it inflates its body into a sphere and is then transported, tumbleweed-fashion, by currents.

NUTRITION AND INTERNAL TRANSPORT

Most anthozoans are opportunistic carnivores that catch prey as it drifts or swims into contact with their tentacles or oral disc. Once caught and immobilized by a combination of mucus, spirocysts, and nematocysts, the tentacles bend toward the oral disc and push large prey into the mouth. Sometimes the border of the mouth is inflated into lips that pucker upward to help seize the prey. Smaller prey, in the size range of plankton, are transported by cilia to the tentacle tips, which are then inserted into the mouth (Fig. 7-12C). Upon entering the mouth, which can open widely to accommodate large foods, the prey is mixed with mucus and swallowed slowly by the pharynx. Swallowing results from ciliary action and peristaltic movements of the pharyngeal wall. Once released into the coelenteron, the food mass contacts the cnidoglandular bands of the septa. Extracellular digestion is initiated by nematocyst discharge and enzyme release, chiefly proteases and lipases, from the enzymatic gland cells. The nematocysts may paralyze active prey by injecting enzymes or venom into the prey's body. This first phase of digestion reduces the food to particles and molecules, which are circulated throughout the coelenteron for phagocytic uptake by gastrodermal cells, including the gametes (Fig. 7-17B–D, 7-20E). The flagellar bands and flagella elsewhere on the gastrodermis circulate the coelenteric fluid (Fig. 17-17B).

Most anthozoans supplement their nutrition with photosynthate provided by zooxanthellae, occasionally zoochlorellae, or both. These intracellular, mutualistic algae occur especially in the gastrodermal cells of the tentacles (Fig. 7-11) and oral disc, but sometimes also in the column and septa.

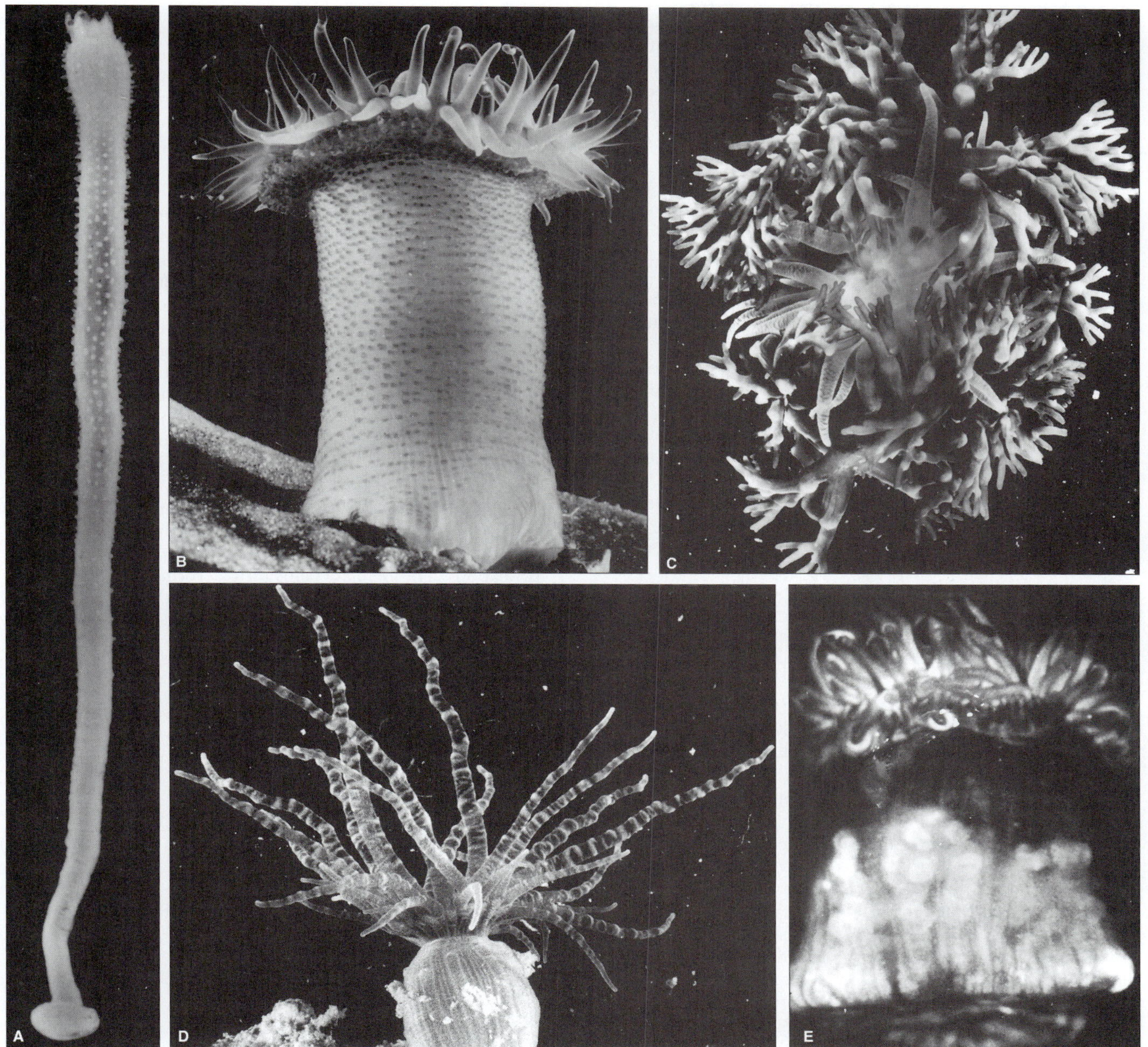

FIGURE 7-20 Zoantharian diversity: Actiniaria (sea anemones). **A–D,** Living animals. **A,** Freshly excavated burrowing anemone, *Haloclava producta*. Note anchoring physa at pedal end and solid papillae on column. **B,** *Bunodosoma cavernata* attaches to hard substrata with a pedal disc. Note verrucae (adhesive papillae) on column. **C,** Oral view of the anemone *Lebrunia danae,* which has branched pseudotentacles bearing zooxanthellae and unbranched tentacles lacking them. **D,** With its graceful, ringed tentacles, *Bartholomea annulata* is a common zooxanthellate species of the West Indian region. **E,** A gravid female of *Aiptasia pallida* that was incubated for 4 h in fluorescent dextran (a carbohydrate), fixed, and then photographed with a fluorescence microscope. The bright fluorescence associated with septa, gonads, and tentacles indicates gastrodermal sites of intense nutrient uptake.

GAS EXCHANGE AND EXCRETION

Exchange of gases and release of ammonia occur by diffusion across the tentacles and remainder of the body. This exchange is facilitated by the ciliary flow of fluid over the gastrodermis and a similar ciliary flow over the epidermis (Fig. 7-12C).

REPRODUCTION AND GROWTH

Clonal reproduction is widespread and diverse in anthozoans. Fission (Fig. 7-13B), fragmentation, and budding of polyps are the primary modes, but planula larvae of some corals can also arise asexually.

Both gonochorism and hermaphroditism are common in anthozoans. Gonads, which are mere aggregations of germ cells, occur in the septa behind the septal filaments (Fig. 7-16A, 7-18A). Mature gametes, which arise from endoderm, are shed into the coelenteron and spawned through the mouth in most species. Fertilization is external in seawater except among viviparous species, in which it occurs in the coelenteron. The zygote develops into a planula larva that often bears a well-developed tuft of sensory cilia at its leading (aboral) end (Fig. 7-14G,H). The planula larvae of most species are planktonic lecithotrophs, but some are planktotrophs that filter feed by trailing sticky mucous threads from their mouths. Metamorphosis generally begins with the early morphogenesis of tentacles, septa, and pharynx before larval settlement on the aboral end (Fig. 7-14H).

DIVERSITY OF ANTHOZOA

Zoantharia[sC] (Hexacorallia)

Zoantharia includes approximately 4000 species of sea anemones, stony corals, coral anemones, mat anemones, and black or thorny corals. Zoantharian symmetry usually is hexamerous, with the septa occurring in multiples of six. Septa, which may be complete or incomplete, typically are not single, but instead are closely set doubles called **pairs** (Fig. 7-16B). Only Antipatharia (black corals) and Ceriantharia (tube anemones) have single septa. A juvenile polyp usually has six pairs of septa and 12 tentacles, 1 tentacle extending from each interseptal space. As the polyp increases in size, a second round of six paired septa may be added along with an additional 12 tentacles. Two types of cnidae are present, nematocysts (six types) and spirocysts, the latter being unique to Zoantharia. The gonads form a longitudinal band behind the septal filament of each fertile septum (Fig. 7-16A, 7-18A). Zoantharians are represented by colonial and solitary species.

ACTINIARIA[O]

Sea anemones are generally large, solitary polyps. Most range from 1.5 to 10 cm in length and from 1 to 5 cm in diameter, but *Tealia columbiana* on the North Pacific coast of the United States and *Stichodactyla mertensii* and *Heteractis magnifica* on the Great Barrier Reef of Australia may exceed a diameter of 1 m at the oral end. The height record of 1 m is held by *Metridium farcimen* of the Pacific Northwest coast of the United States. Sea anemones are often brightly colored white, green, blue, orange, red, or an entire palette of colors. Sea anemones inhabit deep or shallow coastal waters worldwide, but are particularly diverse in tropical oceans. Approximately 1350 species are estimated to exist worldwide. They commonly attach to rocks, shells, and submerged timbers or burrow into mud or sand. A few establish symbioses with other animals, such as hermit crabs, fishes, and shrimps.

Most anemones are highly inflatable and thus of variable shape and size, depending on circumstances (Fig. 7-19). Species on hard surfaces attach with a pedal disc (Fig. 7-16A, 7-20B,E). Burrowers usually lack a pedal disc, but the basal end expands to form a bulbous or mushroomlike swelling (physa) for anchorage in sediment (Fig. 7-20A, 7-21). In the remarkable tropical genus *Minyas,* some species of which are open-ocean blue in color, the expanded pedal disc encloses an air-filled chitinous sac, and the anemones float inverted at the surface of the sea. This adaptation to neustonic life parallels that of the hydrozoans *Velella* and *Porpita* (Fig. 7-67).

Anemones have either six or more than eight simple, tapered tentacles (Fig. 7-20B,D,E), which often bear a terminal pore. In some anemones, however, the tentacles are branched (Fig. 7-20C), knobbed, or reduced to numerous low nubbins scattered uniformly over the entire oral disc, as in "sun" anemones (*Stoichactis* spp.). Some anemones, such as species of *Actinia* and *Anthopleura,* ward off competitors with specialized tentacle-like projections called **acrorhagi** that lay below and near the bases of the true tentacles and are inflated and used as weapons. The nematocyst-bearing acrorhagi are called into action when contact occurs between different species or even genetically different clones of the same species. Following such

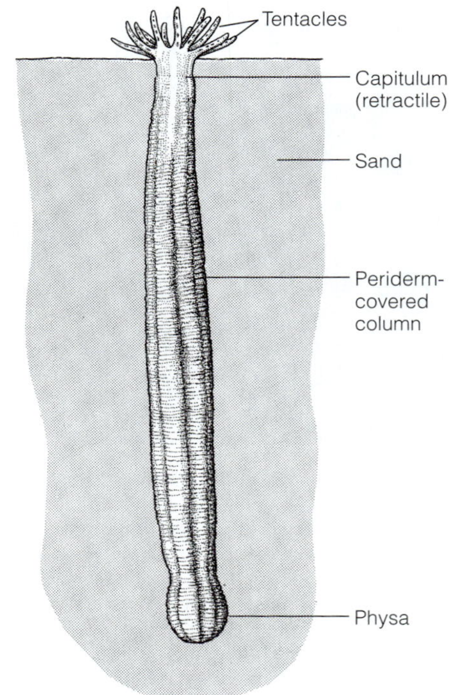

FIGURE 7-21 Zoantharian diversity: Actiniaria. A sea anemone with a well-developed periderm. The primitive burrowing species *Edwardsia elegans.* *(From Ruppert, E. E, and Fox, R. S. 1988. Seashore animals of the Southeast. University of South Carolina Press, Columbia, SC. 429 pp.)*

an interaction, one or both combatants may withdraw, with either or both suffering tissue damage.

The appearance of the column generally is uniform along its entire length. In certain species, however, the upper part of the column, immediately below the oral disc and tentacles, is a thin-walled, necklike introvert, or **capitulum** (Fig. 7-21). The column wall below the introvert is usually thicker and the transition between the two regions often bears a collarlike fold (parapet; Fig. 7-16A), as in species of *Actinia, Metridium,* and *Urticina.* During polyp retraction, as the oral disc, tentacles, and capitulum retract into the column, the transitional region constricts and the parapet covers and protects the opening. The constriction is caused by the epidermal or mesogleal sphincter muscle.

The column wall may be more or less smooth and undifferentiated or it may bear specialized structures. **Solid papillae** cover the columns of *Haloclava producta* (7-20A) and *Bunodosoma cavernata* (Fig. 7-20B). Rows of **adhesive papillae** (verrucae) occur on the columns of other anemones, such as *Anthopleura, Urticina, Bunodosoma* (Fig. 7-20B), and *Bunodactis.* These anemones use the papillae for attachment of protective sand and shell fragments (Fig. 7-4E). Some anemones have cinclides in the column wall through which water and acontia, when present, exit during polyp retraction (Fig. 7-16A). A few, such as species of *Bunodeopsis,* have individual or clusters of thin-walled, zooxanthellae-bearing **vesicles** protruding from the column.

Some anemones have one siphonoglyph, but most have two (Fig. 7-16). Both complete and incomplete septal pairs are usually present and never number fewer than 12, and often many more. Acontia may be present or absent. Anemones that bear acontia, such as *Aiptasia, Bartholomea,* and *Metridium,* are said to be acontiate. The longitudinal muscle bands in the septa are especially well developed and anchored to the oral disc above and the pedal disc below. They are primarily responsible for the infolding of the oral disc and tentacles, as well as the retraction of the column during polyp deflation.

The chief anemone skeleton is the hydrostatic coelenteron, but a few species secrete a chitinous periderm, probably more protective than supportive, and usually restricted to the pedal disc or the column below the introvert (Fig. 7-21). Chitin secretion is best developed in the pelagic *Minyas* (mentioned earlier) and in a remarkable group of deep-water anemones called cloak anemones, which are described later in this section.

Anemones attached to surfaces can move about by a slow, muscular gliding on their pedal discs. Sediment-dwelling species burrow using peristalsis with the pedal-end leading. A few can walk on their tentacles and the hydra-size *Gonactinia prolifera* swims by thrashing its tentacles. The large anemone *Stomphia,* though normally attached to hard surfaces, will release and swim by swinging the lower half of its column when threatened by the predatory sea star.

Sea anemones generally feed on various invertebrates, but large species feed on crabs, wave-dislodged clams, and even fish. The prey is caught with the tentacles, paralyzed or secured by cnidae, and carried to the mouth. The lips can be inflated and may aid in prey capture. Anemones with numerous tentacles, such as *Metridium, Radianthus,* and *Stichodactyla,* are suspension feeders, but the sun anemone, *Stichodactyla helianthus,* is reported to catch sea urchins by enveloping them in its muscular oral disc. Suspension feeders trap plankton in mucus on the surface of the column and tentacles (Fig. 7-12C). Cilia on the surface of the column beat toward the oral disc, and cilia on the tentacles beat toward the tentacle tips. The tentacles then bend to deposit the food in the mouth.

Many anemones harbor gastrodermal zooxanthellae, zoochlorellae, or both, especially in the tentacles and oral disc (Fig. 7-11). The individual color variations of *Anthopleura elegantissima* are determined chiefly by the relative predominance of zoochlorellae and zooxanthellae. The tropical anemone *Lebrunia danae* has two sets of tentacles, one branched whorl of zooxanthellae-bearing "pseudotentacles" and a whorl of simple tentacles for catching prey (Fig. 7-20C). The photosynthetic pseudotentacles are extended by day, the raptorial tentacles at night.

Clonal (asexual) reproduction, either fission or fragmentation, is common in anemones. A specialized form of fragmentation called **pedal laceration** occurs in clonal species such as *Aiptasia pallida, Haliplanella luciae,* and *Metridium senile* (Fig. 7-22). Small pieces from the margin of the pedal disc detach and remain behind as the animal moves, or the pieces may move away from the stationary anemone to form a "fairy ring" around the base of the parental column. In either case, the detached fragments soon develop into small sea anemones. Many sea anemones reproduce asexually by longitudinal fission, and a few species, such as *Gonactinia prolifera* and *Nematostella vectensis,* do so by transverse fission.

Species of sea anemones may be either gonochoric or hermaphroditic. Each septal gonad is a longitudinal, cushionlike **gonadal band** between the septal filament and retractor muscle (Fig. 7-16A, 7-18A). The eggs may be fertilized and develop in the coelenteron or fertilization may occur externally in the sea. The planula larva, which may be planktotrophic or lecithotrophic, has a variable life span, depending on its species. Larval metamorphosis is as it was described for the Anthozoa.

FIGURE 7-22 Zoantharian diversity: Actiniaria. Pedal laceration in the anemone *Aiptasia pallida.* The small individuals differentiating into anemones arose from pedal-disc fragments of the large polyp.

Anemones and hermit crabs are a common mutualism in the sea. Typically, one or more anemones are associated with each crab. Benefits to the anemone are thought to include a surface (shell) for attachment, transport to food sources, including shreds of food from the feeding crab, and protection from predators. Encounters between hermit crabs may provide reproductive opportunities for the anemones as well as the crabs. The crab, on the other hand, receives both passive (camouflage) and active (nematocyst) protection, especially from such hermit-crab predators as octopuses and box crabs, which are repelled by the anemones. When the crab outgrows its shell and switches to a new, larger home, it helps the anemone move to the new shell. To accomplish this, the crab strokes the anemone's column to induce the anemone to release its pedal-disc attachment, and the crab then transfers the anemone to its new shell. In a few species, the anemone itself effects the transfer by somersaulting from the old to the new shell.

With growth, most hermit crabs seek out and transfer into ever-larger snail shells. Colonization of new shells poses risks for crabs, such as exposure to predation and combat with other hermits for the limited supply of shells. The anemones *Stylobates* spp. have an expanded pedal disc that secretes a chitinous, surrogate "snail" shell for its hermit crab, which probably has few shells available to it in the deep sea. The crab also avoids shell-transfer risks because the anemone both produces and enlarges the shell. This may also benefit the cloak anemone, which avoids being abandoned during shell exchange. The hermit crab also may discourage the anemone's enemies and inadvertently share food with it.

In the Indo-Pacific, little fishes mostly of the genus *Amphiprion* (clownfish or anemonefish) live as mutualists among the tentacles of large sea anemones. Juvenile fish are recruited to their sea anemone by species-specific attractant substances released by the anemone. The mucous coating the surface of the fish lacks nematocyte-triggering compounds, making it possible for the fish to live in an otherwise lethal habitat. The sea anemone provides protection and some food scraps; the fish may attract food (other fish) to the sea anemone, protect the sea anemone from some predators (butterflyfish), remove necrotic tissue, and, by its swimming and ventilating movements, reduce fouling of the anemone by sediment. Other symbionts of sea anemones include amphipods, cleaning shrimps, snapping shrimps, arrow crabs, and brittle stars.

SCLERACTINIA^O

Closely related to sea anemones (Actiniaria) are the stony, or scleractinian, corals (also called madreporarian corals), which constitute the largest taxon of anthozoans with approximately 3600 species. In contrast to anemones, stony corals produce a calcium carbonate exoskeleton. Some corals, such as the Indo-Pacific reef-inhabiting *Fungia* and some deep-sea species, are solitary and have polyps as large as 50 cm in diameter (Fig. 7-24D), but the majority are colonial, with small polyps averaging 1 to 3 mm in diameter (Fig. 7-23A, 7-25). A colony, however, may grow to several meters in height, weigh tons, and be composed of more than 100,000 polyps. Coral polyps are very similar in structure to sea anemones, including their possession of paired septa. Most, however, lack obvious

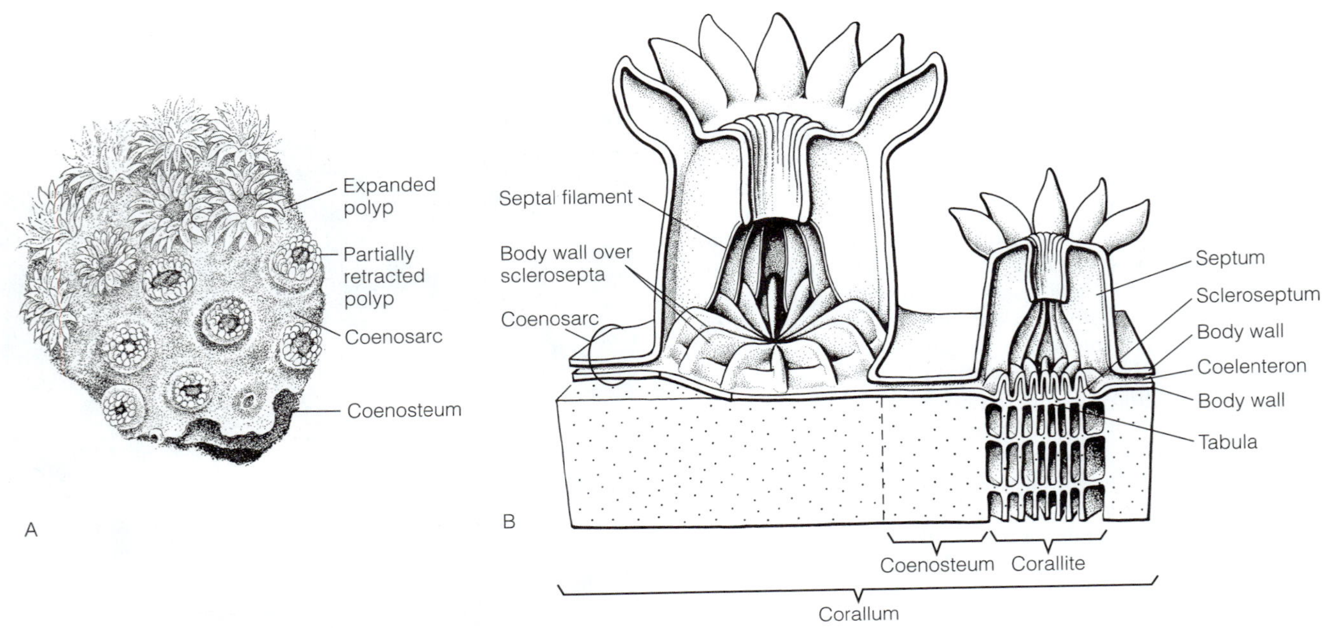

FIGURE 7-23 Zoantharian diversity: Scleractinia. Anatomy of stony corals. **A,** Surface of a colonial coral showing expanded and contracted polyps, coenosarc, and underlying calcareous exoskeleton (corallum). **B,** Cutaway view of a colonial coral.

FIGURE 7-24 Zoantharian diversity: Scleractinia. **A,** Corallites of ivory coral *(Oculina arbuscula);* **B,** staghorn coral *(Acropora cervicornis);* **C,** starlet coral *(Siderastrea radians);* **D,** the solitary mushroom coral (*Fungia* sp.). **A–C** are West Indian species, **D** is Indo-Pacific. In **B,** note the corallite of the axial polyp at the branch tip; the other corallites belong to lateral polyps.

siphonoglyphs (but see Fig. 7-25A) and their retractor muscles are sheetlike, not localized and bulging as in anemones. Despite these apparent handicaps, coral polyps inflate and deflate like anemones. Presumably, the pharynx has inward-beating cilia that are not organized into specialized siphonoglyphs.

The skeleton is composed of calcium carbonate (aragonite) secreted by the epidermis of the lower half of the column and the pedal disc in solitary species. This process produces a skeletal cup, or **corallite,** which the polyp inhabits and retracts into for protection. The wall of the corallite is the **theca.** The floor of the corallite bears radiating, calcareous vanes—the **sclerosepta** (= hard septa; Fig. 7-23B, 7-24A,C,D). The thin edge of each scleroseptum projects upward into the polyp's pedal disc, infolding it between each pair of septa. In some colonial species, such as the lettuce corals *(Agaricia),* sclerosepta spill over the rim of the coral cups and extend onto the surface of the intervening skeleton as **costae.** In

FIGURE 7-25 Zoantharian diversity: Scleractinia. Living zooxanthellate corals. **A,** Ivory coral *(Oculina diffusa).* Note siphonoglyph-like opening at each corner of closed mouths. **B,** Starlet coral *(Siderastrea radians);* **C,** Finger coral *(Porites porites);* **D,** Staghorn coral (*Acropora cervicornis;* note axial polyp at branch tips).

many species, the floor of the corallite bears a central projection, the **columella.**

The polyps of colonial corals arise from a coenosarc, which is continuous with the wall of the column (Fig. 7-23). The coenosarc contains a broad extension of the coelenteron or coelenteric tubes (solenia) sandwiched between the upper and lower layers of epidermis. The undersurface of the coenosarc secretes the skeleton that lies between the corallites. This is known as **coenosteum** (= common bone; Fig. 7-23). The corallites and coenosteum together comprise the coral skeleton, or **corallum** (Fig. 7-23B). The living coral tissue, polyps and coenosarc, lies entirely above the corallum and completely covers it (Fig. 7-23B).

Sclerosepta, a hallmark of stony corals, probably have several functions, but only one, anchorage, is established with certainty. Specialized anchoring cells (desmocytes) attach the polyp to its skeleton. Sclerosepta may also help the coral to resist predation. As coral polyps retract tightly into their corallites, some of the living tissue withdraws into the protective base of the corallite and the valleys between the sclerosepta, but the remainder is stretched over the scleroseptal ridges. The low proportion of tissue-to-mineral on the projecting sclerosepta may discourage all but the most specialized grazers, such as pufferfish with their ratlike incisors. Finally, because many scleractinians with zooxanthellae are retracted during daylight hours, the draping of the coral's photosynthetic tissue over the sclerosepta, rather than bunching it together in the base of a sclerosepta-free cup, may help expose it to light. The surface of the white skeleton also may help to reflect sunlight onto the zooxanthellae.

As long as a coral colony is alive, new calcium carbonate is deposited beneath the living tissues, polyps and coenosarc alike. This deposition increases the diameter and thickness of the corallum as well as the height of the sclerosepta, thus deepening the corallites below the polyps. Each polyp, however, has a fixed adult size, so periodically it reduces the depth of the corallite by lifting its base and secreting a new floor **(tabula),** sealing off the old floor and a small space above it (Fig. 7-23B). Over time, repetition of this process creates a tenement of abandoned floors below the living polyps, which occupy the top floor only.

The diversity and beauty of coral skeletons results from the growth form of the colony and the arrangement of polyps in the colony (Fig. 7-24, 7-26A). Some species form flat or rounded skeletal masses whereas others, especially species of *Acropora,* have a fruticose growth form (Fig. 7-24B). Some are large and heavy, others small and delicate. When the polyps and their corallites are well separated, the coral skeleton has a pitted appearance, as in species of the ivory coral *Oculina* (Fig. 7-24A), the star coral *Astrangia,* and the reef coral *Montastrea.* The coalesced corallites of brain corals are arranged in curving rows (Fig. 7-26A). The rows are well separated, but the polyps in each row are fused together and their corallites are confluent. As a result, the skeletal surface of meandering valleys and ridges resembles the appearance of a human brain.

All colonial corals grow by the budding of new polyps, which adds new zooids and increases the area of the colonial body. Depending on the species, the buds may arise from the oral disc, column, or base of parent polyps, or they may originate directly from the surface of the coenosarc (Fig. 7-27). Among coensarcal species, **intratentacular budding** refers to buds that arise on the oral disc within the whorl of tentacles (Fig. 7-27A); in **extratentacular budding,** buds arise outside the tentacular whorl (Fig. 7-27B). Growth by intratentacular budding can produce colonies of individually separated polyps (Fig. 7-24A) or a row of incompletely separated polyps whose merged corallites form the meandering channels typical of brain corals (Fig. 7-26). Each row of polyps in a living brain coral, such as species of *Diploria* or *Manicina,* is a continuous, elongate oral disc bearing a series of mouths (Fig. 7-26C). Tentacles occur on the margins of the oral disc on each side of the line of mouths, but do not encircle each mouth. Extratentacular budding always results in a colony of individually separated polyps. Species of *Acropora* grow into fruticose colonies by lateral budding from axial polyps (Fig. 7-24B, 7-25D), which form the main stem and branch stems.

Calcium carbonate deposition and growth vary diurnally and seasonally with temperature and light. Thus, many coral skeletons exhibit seasonal growth bands like tree rings, which can show up on x-ray images and be used to determine the age and growth rate of the coral. Many dome and plate corals grow only 0.3 to 2 cm per year. Some species of *Acropora,* however, increase the length of their branches by as much as 10 cm per year.

Although coral-reef species are variously adapted to live on particular parts of a reef, many compete for space with other species. When a coral is contacted by another species, it may extrude septal filaments from its mouth and damage the intruder's tissues or, as in some sea anemones, it may use specialized, extensible (sweeper) tentacles. Aggressiveness varies among species.

A mutualism between zooxanthellae and most corals, and the reciprocal metabolic exchanges between them, have enabled both coral and alga to flourish in an environment marginal to either alone. The environment in which a coral reef develops is the sun-drenched, nutrient-poor, shallow waters of the tropics. Corals that house zooxanthellae in their gastrodermal cells are said to be **zooxanthellate;** those that lack them, **azooxanthellate** (aposymbiotic). The algae can reach a concentration of up to 5×10^6 cells per square centimeter of coral tissue. All reef-building (hermatypic) corals are zooxanthellate, but deepwater and some cold-water species lack zooxanthellae. The zooxanthellae use the coral body as a protective habitat, for exposure to light, and as

FIGURE 7-26 Zoantharian diversity: Scleractinia. The brain-coral relative, *Manicina areolata.* **A,** Corallum; **B,** alive retracted; **C,** alive expanded; note one (arrow) of the several mouths surrounded by the common margin of tentacles.

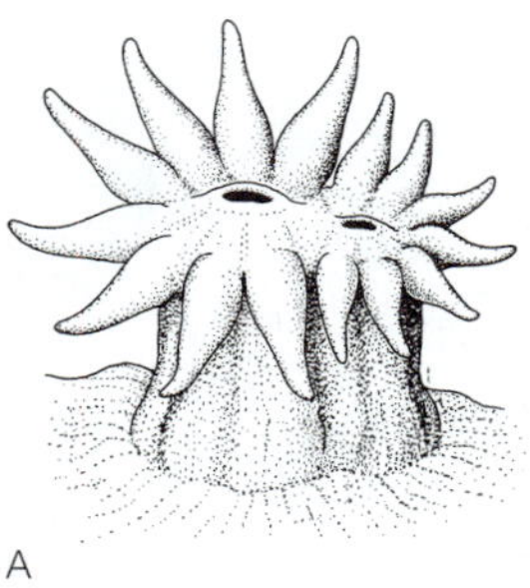

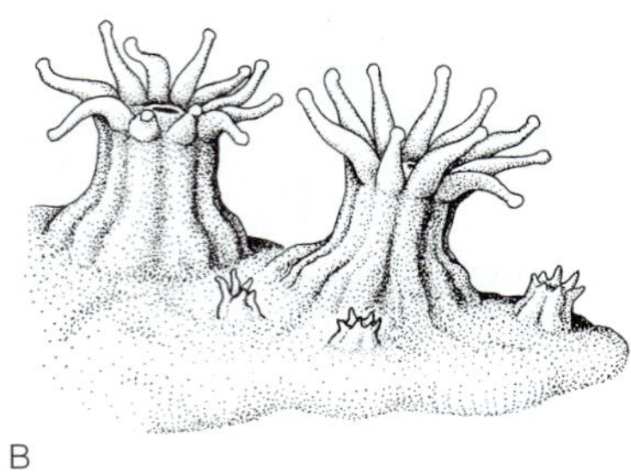

FIGURE 7-27 Zoantharian diversity: Scleractinia. **A,** Intratentacular budding. **B,** Extratentacular budding.

a source of nutrients, especially the PO_4, NH_3, and CO_2 by-products of coral metabolism. The coral, in turn, receives from algal photosynthate some of its N, C, and energy needs (up to 50%), as well as O_2. The mutualism also increases the rate of coral-skeleton growth because of the photosynthetic nutritional supplementation and removal of CO_2 by the algae.

Like sea anemones, both zooxanthellate and azooxanthellate corals feed on animals, and depending on the size of the polyps, the prey ranges from zooplankton to fish. When expanded, the outstretched tentacles of adjacent polyps present a broad, continuous mesh difficult for the prey to avoid (Fig. 7-25A,C,D). In addition to zooplankton, many corals also trap fine particles on mucous sheets or strings, which are then driven by cilia to the mouth. Some, such as the foliaceous agaricids, which have reduced or no tentacles, are entirely mucous suspension feeders. Corals produce large quantities of mucus. *Acropora acuminata* has been estimated to release in mucus as much as 40% of the daily fixed carbon received from its zooxanthellae.

Although there are many exceptions, Caribbean corals feed primarily at night and are retracted during the day, and some species display a persistent endogenous rhythm of expansion and retraction, even when kept in constant darkness or light. Many Indo-Pacific corals feed during both day and night.

Under adverse conditions, zooxanthellate corals expel some or all of their algae. With the loss of the pigmented algae, the white calcareous skeleton is visible through the now transparent tissue and the coral is said to have undergone **bleaching.** The stresses causing bleaching may be under- or overillumination, excess UV exposure, salinity fluctuation, and low or high temperatures. Because many zooxanthellate corals are thought to live at water temperatures near the lethal upper limit, a temperature rise of only 1°C may be sufficient to induce bleaching. Episodes of extensive coral bleaching recorded in the past 20 years may be linked to global warming. The loss of zooxanthellae from a coral, however, is not always complete and does not always result in coral mortality, especially if the period of environmental stress is not prolonged.

Many variants—perhaps many species—of zooxanthellae appear to exist. These variants are distributed among species of corals, or even in a single coral individual. When environmental stress reaches a threshold for a particular coral and its zooxanthella variant, the coral bleaches, but may then acquire another algal variant, which confers a higher degree of stress tolerance to the coral-algal mutualism. Thus *small*-scale bleaching may be an adaptive mechanism by which a coral selects the algal variant best suited to function under the changed environmental conditions.

Clonal reproduction is highly developed in the zooxanthellate mushroom corals (Fungiidae), which are primarily large (up to 50 cm in diameter), solitary, and semimotile polyps common in the coral-reef lagoons of the Indo-Pacific. The expanded polyps, which conceal the dome-shaped skeleton (Fig. 7-24D), are easily confused with sea anemones. Using their tentacles, fungiids can move over the bottom, right themselves, and even regain the surface after burial in sand. Juveniles are attached to a hard substratum by a stalk. Periodically, the differentiated corallite "cap" detaches from the stalk and moves away. The stalk, which is a permanent feature, then regenerates another cap and the cycle is repeated.

Sexual reproduction is similar to that of sea anemones, and both gonochoric (one-third of the total) and hermaphroditic (the other two-thirds) species occur. Between 70 and 85% of the species spawn their gametes; the remainder brood their eggs in the coelenteron and release short-lived planula larvae. Recent evidence suggests that adults release zooxanthellae as they spawn eggs or release larvae, inoculating them with the maternal zooxanthellae. After larval settlement and metamorphosis, the juvenile polyp begins to deposit a skeleton and, in colonial species, buds additional polyps for colony growth.

Corals are subject to injury or death from storms, extremely low tides, predation, and disease. White-band and black-band diseases are caused by microorganisms and can produce tissue death and permanent scarring. Disease is exacerbated when pollution raises nutrient levels in the surrounding water.

Zoanthidea[O]: 200 largely tropical and common reef-inhabiting species; do not secrete calcium carbonate, rely instead on periderm, hydrostat, mesoglea, sometimes embedded foreign matter for skeletal support, protection. Most 1–2 cm diameter, colonial, joined by stolonal network or coenosarc (Fig. 7-28A, 7-29); some solitary and large; thick, fleshy polyps tall (Fig. 7-28A, 7-29) or short; fringe of short marginal tentacles on broad oral disc; column and coenosarc covered by thick periderm (Fig. 7-28B), many species (such as *Palythoa*) have embedded sand, other debris in periderm and mesoglea; thick mesoglea with solenia that open into coelenteron and perhaps column surface (Fig. 7-28B); solenia circulatory channels for thick body wall; one siphonoglyph; coelenteron partitioned by complete and incomplete, unpaired septa (Fig. 7-28B). *Zoanthus:* zooxanthellate reef species in low patches; *Palythoa:* thickly encrust rocks, toxic (polytoxine), native Hawaiians poisoned arrowheads by rubbing with *Palythoa toxica; Epizoanthus* and

A

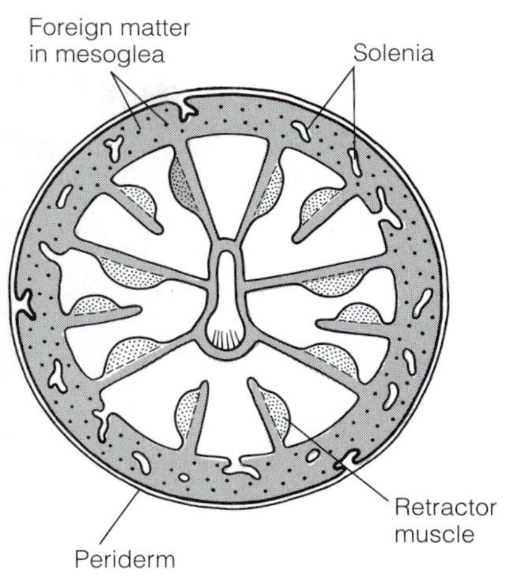

B

FIGURE 7-28 Zoantharian diversity: Zoanthidea. **A,** Photograph of a living mat anemone, *Palythoa texaensis.* Note expanded and partly retracted zooids. **B,** Cross section of column and pharynx. Note complete and incomplete septa, thick body wall with solenia, foreign matter, and periderm. *(B, Modified and redrawn after Hyman, L. H. 1940. The Invertebrates. vol 1. Protozoa through Ctenophora. McGraw-Hill Book Co., New York. 726 pp.)*

Parazoanthus: common commensals of sponges, hermit crabs, other animals.

Corallimorpharia[O]: Coral anemones; 50 species, all solitary; lack calcareous exoskeleton; most are broad, squat polyps, often larger than dinner plates (up to 1 m diameter); on coral reefs; expanded oral disc, with many short tentacles (Fig. 7-29) or no tentacles; tentacles usually house zooxanthellae, but polyps also feed on large prey such as fish, sea urchins; prey enveloped by muscular oral disc, similar to prey capture by Venus's flytrap. Everted nematocyst and spirocyst tubules of azooxanthellate species longest known in Anthozoa; lack, or perhaps have two weakly developed, siphonoglyphs; coelenteron partitioned by paired complete and incomplete septa. *Corynactis, Discosoma.*

Antipatharia[O]: 150 spp.; black, or thorny, corals; exclusively colonial; slender fruticose colonies to 5 m height; on deep, vertical, seaward walls of coral reefs; also elsewhere in deep water (Fig. 7-30A); stem and branches with dark, thorny, flexible **axial skeleton** inside thin mantle of polyps and coenosarc (Fig. 7-30B); skeleton unique, noncollagenous scleroprotein (antipathin); polyps bear six slender tentacles, two opposite longer than others (Fig. 7-30B); 6, 10, or 12 complete, unpaired septa (Fig. 7-30D), each having longitudinal muscle fibers on one face (radial on other), but poorly developed, do not bulge from septal surface; weakly developed longitudinal musculature in column epidermis; polyps slightly or nonretractile; pharynx may have two siphonoglyphs, one **hyposulculus** (below one siphonoglyph, the two specialized, inward-beating ciliary bands on free margin of the two [directive] septa). Feed by casting plankton-trapping mucous strings, reeled in by cilia. Collected commercially for black skeleton, pieces polished and worn as jewelry. *Antipathes, Dendrobrachia, Schizopathes.*

Ceriantharia[O]: Tube anemones; 75 species; large, solitary, mostly burrowing polyps permanently occupy secreted feltlike tubes; tubes can exceed 1 m in length (7-31A,B); tubes typically buried in sediment except for top few centimeters (Fig. 7-31A,B); tube fabric tough, feltlike from discharged, woven, ptychocyst tubules; also have nematocysts and spirocysts, thus all three types of cnidae. Suspension feed with two whorls of tentacles; when disturbed, tentacles bundle together and withdraw, with column, into tube; swollen pedal end (physa) provides anchorage for polyp retraction; epidermal longitudinal musculature for rapid withdrawal into tube; physa has terminal, "anal" pore; all septa complete, unpaired, lack retractor muscles (Fig. 7-31C), thus cerianthids cannot invaginate oral disc and tentacles into the column for protection; single siphonoglyph and corresponding hyposulculus. *Cerianthus, Ceriantheopsis, Pachycerianthus.*

Alcyonaria[sC] (Octocorallia)

The exclusively marine and mostly colonial Alcyonaria (or Octocorallia) includes the colorful sea plumes, fans, pens and pansies, as well as organ-pipe coral, precious red coral, blue coral, and soft corals. Of the 2000+ species, most are associated with the coral-reef community (Fig. 7-32A), but alcyonarians also are commonly encountered in shallow-water areas of the oceans worldwide and even in the deep sea. In general, reef-dwelling alcyonarians tend to be more tolerant of environmental extremes than stony corals and will colonize areas marginal for scleractinians. Most alcyonarians harbor zooxanthellae and many are thought to rely on their algae to supply much of their nutrition. In contrast to the stony corals, which defend themselves with nematocysts and by retracting into their sharp-edged corallites, many alcyonarians use chemical defenses, primarily terpenoids, to poison or discourage would-be predators.

As a group, the colonial alcyonarians have widely diverse growth forms (Fig. 7-32A), but uniformly similar octamerous

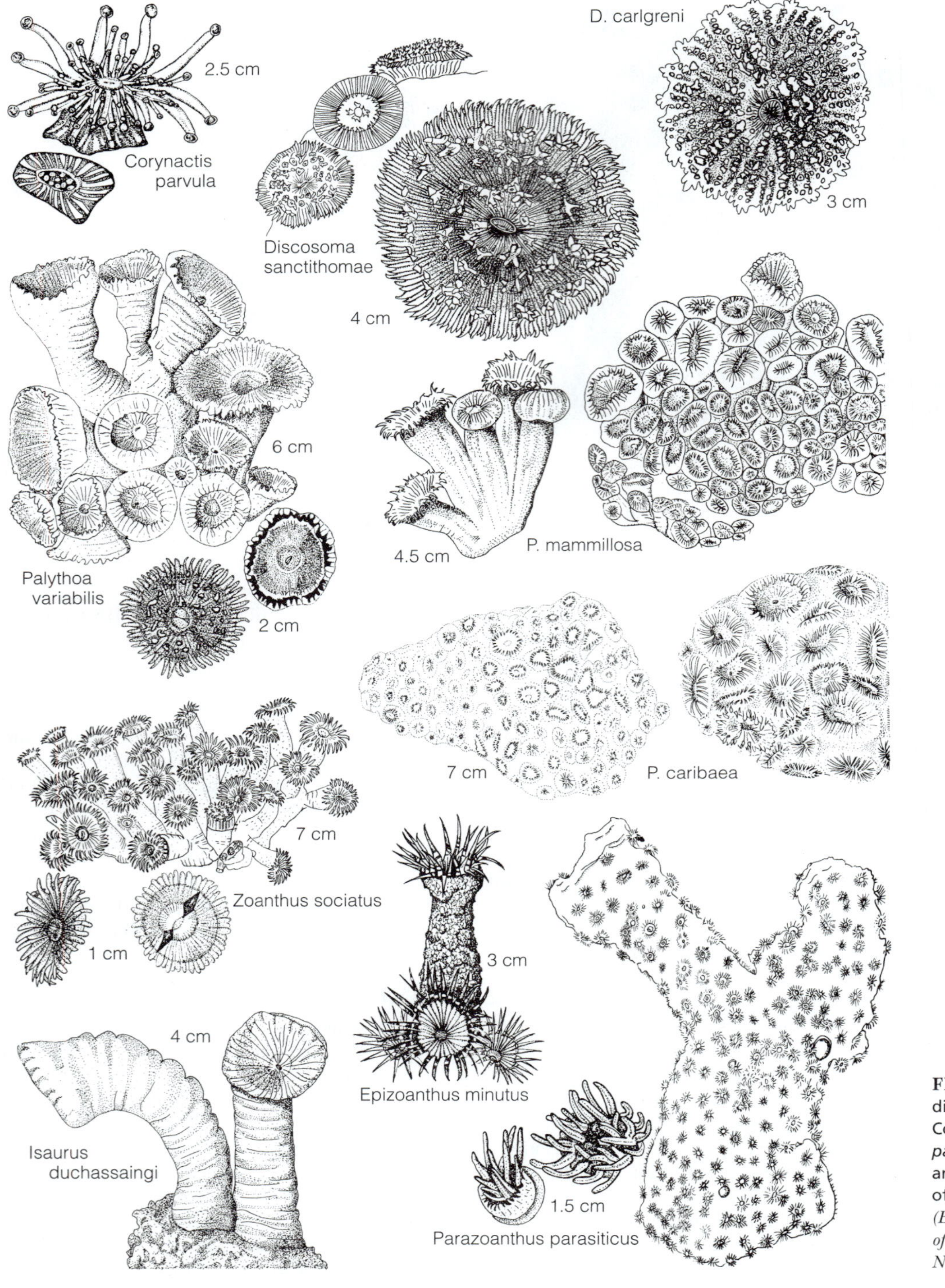

FIGURE 7-29 Zoantharian diversity: Three species of Corallimorpharia *(Corynactis parvula, Discosoma sanctithomae,* and *D. calgreni)* and seven species of Zoanthidea. *(From Sterrer, W. (Ed.): 1986. Marine Fauna and Flora of Bermuda. John Wiley and Sons, New York. 742 pp.)*

polyps (Fig. 7-32B, 7-33A). The small retractile polyp, 0.5 mm to 2 cm in diameter, has eight pinnate tentacles, each bearing side branches called **pinnules.** The pharynx has one siphonoglyph (Fig. 7-33B). Internally, the eight complete unpaired septa bear unilobed septal filaments and retractors that lie on the siphonoglyph (sulcal) side of each septum (Fig. 7-33B). The two septa opposite the siphonoglyph, called **asulcal septa,** bear long, densely ciliated, septal filaments that create an orally directed water flow (Fig. 7-33A,C). This flow, in conjunction with the aboral flow produced by the siphonoglyph, creates a circulation in the polyp and contributes to colonywide circulation of nutrients and gases.

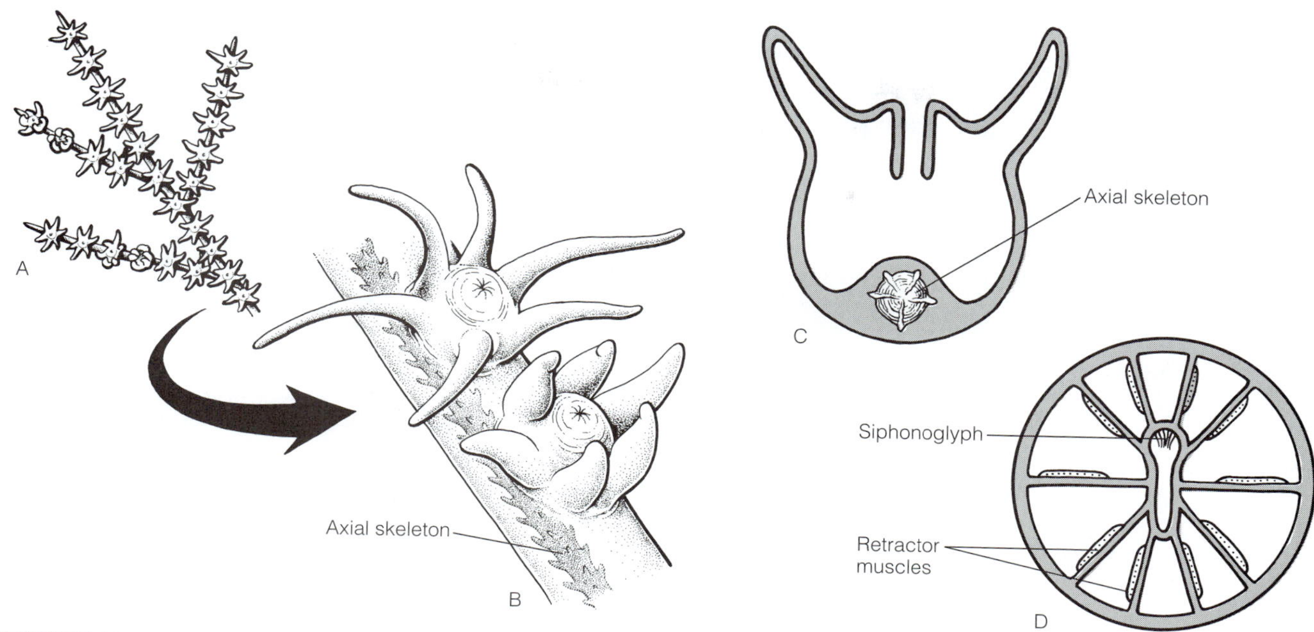

FIGURE 7-30 Zoantharian diversity: Antipatharia. **A,** Piece of living black coral. **B,** Enlargement of branch showing two zooids and thorny skeletal axis. Two opposite tentacles, perpendicular to the branch, are longer than the other four. **C,** Vertical section through a polyp (septa not shown) and cross section through branch and axial skeleton. The developmental origin of the axial skeleton is unknown; it may be an invaginated periderm. **D,** Cross section through column and pharynx; neither siphonoglyphs nor septal muscles are well developed. *(Modified and redrawn from Delage Y., and Hérouard, E. 1901. Les Coelentérés. Schleicher Frères, Paris. 848 pp.)*

A few alcyonarian colonies are stolonate (Fig. 7-34), but most are united by a coenosarc (also called coenenchyme, pronounced SIN-in-kime; Fig. 7-35, 7-36) and at least some of the fruticose forms develop from axial polyps (Fig. 7-37). A coenosarc may be a more or less two-dimensional mat that unites the polyps (Fig. 7-35, 7-36), similar to the arrangement in most stony corals, or it may be much thicker. If the coenosarc is thick, the polyps may be deeply embedded with only their oral-most ends **(anthocodia)** extending above the surface (Fig. 7-32B, 7-33A). The pharynx and septa, except for the asulcal septa, often are restricted to the anthocodia (Fig. 7-33A). The coelenteron of each polyp may end just below the coenosarcal surface (Fig. 7-33, 7-39) or, free of septa, it may extend deeply into the coenosarc as a simple **gastrodermal tube** (Fig. 7-38). In most coenosarcal species, a network of ciliated solenia course through the coenosarcal mesoglea and interconnect the coelenteron and gastrodermal tubes, thus creating a colonial circulatory system (Fig. 7-33A, 7-36, 7-39).

Alcyonarian skeletons are diverse and often complex. A periderm is associated with the prostrate stolons (Fig. 7-34), coenosarc, and upright stems of many primitive species, but is generally absent in the others. Blue coral, like stony corals, has a calcareous exoskeleton (Fig. 7-36), whereas the calcified skeleton of organ-pipe coral is internal (Fig. 7-35). The mesoglea that surrounds the solenia and lower ends of the polyps also has a skeletal function in alcyonarians. It contains calcareous spicules of various shapes and colors (Fig. 7-32B, 7-33A, 7-35B, 7-39) and organic fibers that toughen and harden the body wall. Many fruticose species, such as sea plumes and precious red coral, also have either an organic or calcified axial skeleton to support their stems and branches (Fig. 7-41D–F).

Colonies with a thick coenosarc often have dimorphic polyps called autozooids and siphonozooids. **Autozooids** are typical feeding polyps, but **siphonozooids** are highly modified for ciliary water pumping, colony inflation, circulation, and, in some cases, sexual reproduction (perhaps because of the importance of a respiratory water flow over the gonads) (Fig. 7-40C,D). Siphonozooids lack tentacles and do not extend above the surface of the coenosarc. They have an enormously developed siphonoglyph and little else, except for gonads in certain species. Siphonozooids receive their nutrition, via solenia, from autozooids.

Clonal reproduction is by fragmentation or fission. Most alcyonarians are gonochorists that release their gametes in mass spawnings similar to those used by many stony corals, but a few viviparous species release planula larvae. The gonads occur on all septa except the asulcal septa, which are sterile. Planula larvae, which settle and metamorphose into polyps, are often zooxanthellate.

Stolonifera^O: Stolonate or, despite the taxon name, coenosarcal. *Cornularia cornucopae:* stolonate (Fig. 7-34, 7-41A); mesoglea thin, lacks spicules; stolon and polyp bases enclosed in periderm. *Clavularia* (some species): prostrate stolons fused into coenosarc; upright tubular polyps interconnected at various levels by coenosarcal bridges with solenia (Fig. 7-41B); mesoglea has spicules. *Tubipora musica:* organ-pipe coral (Indo-Pacific); parallel upright, tubular polyps ("organ pipes") from basal coenosarc; horizontal coenosarcal platforms containing

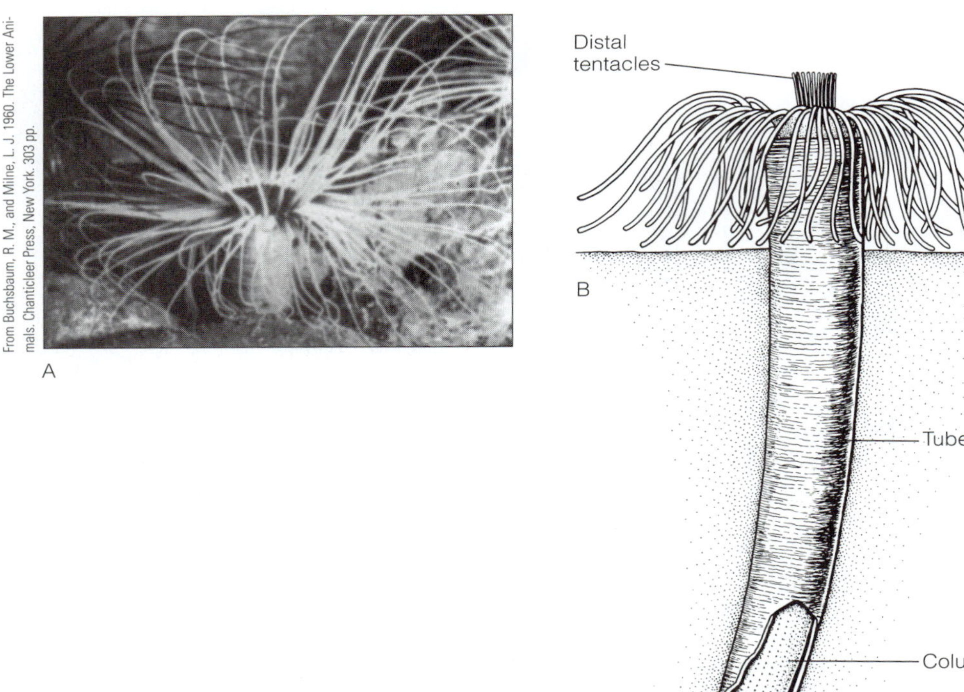

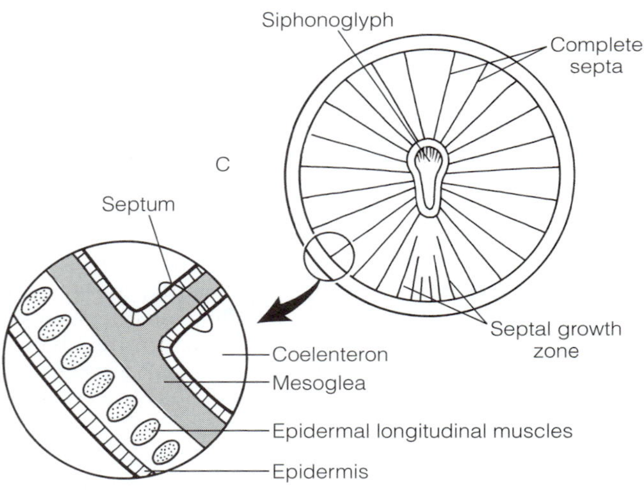

FIGURE 7-31 Zoantharian diversity: Ceriantharia. **A,** Photograph of living tube anemone *Cerianthus.* **B,** Cutaway view of a cerianthid and its tube. **C,** Cross section through the column and pharynx. Enlargement shows well-developed longitudinal musculature in epidermis. *(B, Modified and redrawn from Kaestner, A. 1984. Lehrbuch der Speziellen Zoologie. 2. Teil. Gustav Fischer Verlag, Stuttgart. 621 pp.; C, Modified and redrawn from Delage, Y., and Hérouard, E. 1901. Les Coelentérés. Schleicher Frères, Paris. 848 pp.)*

FIGURE 7-32 Alcyonaria. **A,** Sea plumes (Gorgonacea: Holaxonia) photographed underwater on a Florida coral reef. **B,** Close-up of a living branch of the sea plume *Lophogorgia hebes* with expanded and contracted anthocodia (autozooids). Note pinnules on tentacles and spicular coenosarc between zooids.

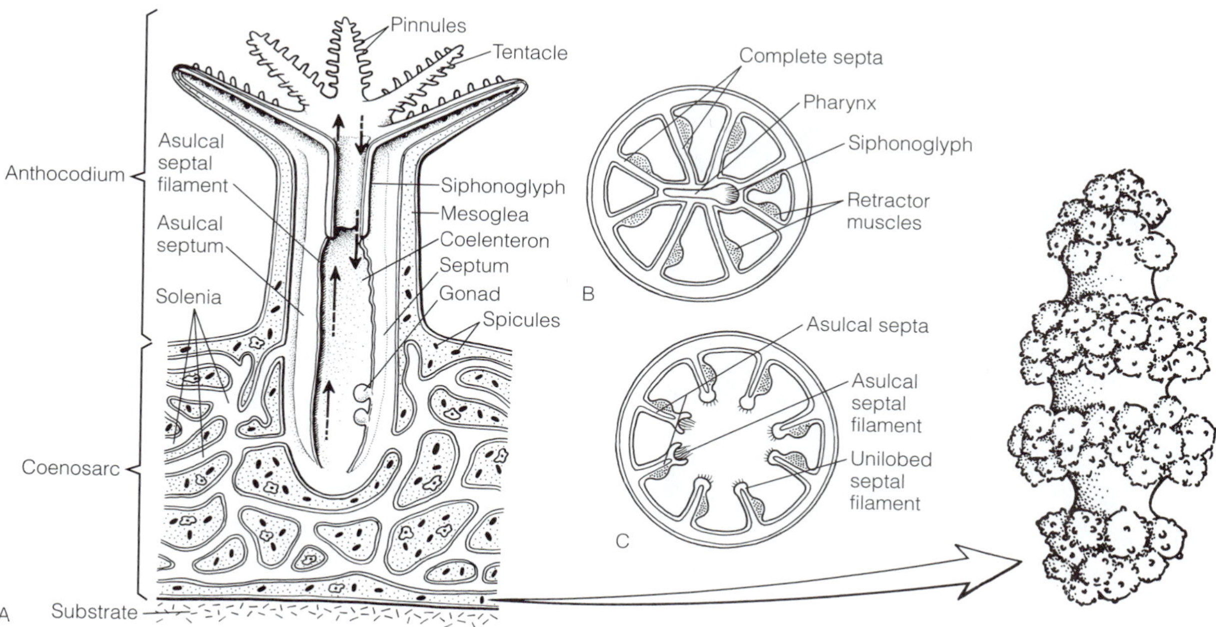

FIGURE 7-33 Alcyonaria: polyp anatomy. **A,** Cutaway view of polyp and coenosarc. Expanding arrow: typical calcareous spicule from mesoglea. **B,** Cross section through polyp at level of pharynx. **C,** Cross section of polyp below the pharynx. *(A, Modified and redrawn from Kaestner, A. 1984. Lehrbuch der Speziellen Zoologie. 2. Teil. Gustav Fischer Verlag, Stuttgart. 621 pp; B and C, Modified and redrawn from Hyman, L. H. 1940. The Invertebrates. vol 1. Protozoa through Ctenophora. McGraw-Hill Book Co., New York. 726 pp.)*

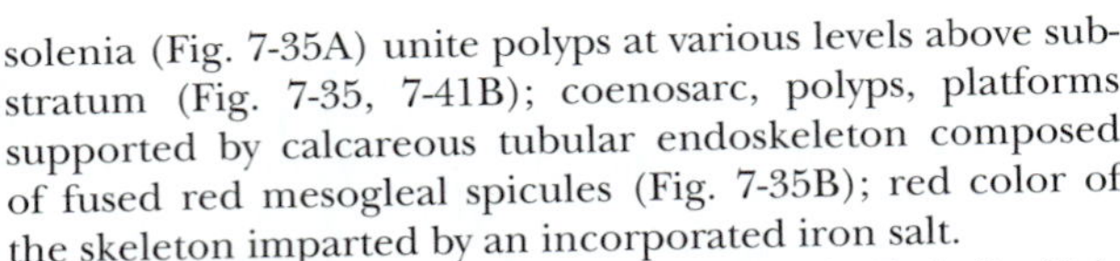
solenia (Fig. 7-35A) unite polyps at various levels above substratum (Fig. 7-35, 7-41B); coenosarc, polyps, platforms supported by calcareous tubular endoskeleton composed of fused red mesogleal spicules (Fig. 7-35B); red color of the skeleton imparted by an incorporated iron salt.

Helioporacea^O: *Heliopora coerulea:* blue coral (Indo-Pacific), the sole reef-building alcyonarian; zooxanthellate. Secretes massive calcareous exoskeleton (aragonite) from underside of matlike coenosarc, 2–3 mm thick (Fig. 7-36); corallum blue because of iron salts in skeleton; tiny polyps, 1 mm diameter, extend singly from circular corallites that lack sclerosepta (Fig. 7-36B); polyp coelenterons united by coenosarcal solenia. Coenosarcal undersurface bears numerous blind vertical tubules (diverticula), which extend into pits between corallites (Fig. 7-36A); with growth, bases of polyps and diverticula both produce tabulae, as in stony corals (Fig. 7-36A); diverticula increase surface area, secrete $CaCO_3$.

Telestacea^O: Mostly species of *Telesto:* small upright branching colonies of polyps that arise from a coenosarc; main stem and branches are elongate axial polyps with short, laterally budded zooids (Fig. 7-37, 7-41G); thick stem and branches encased in spicular coenosarc; coenosarcal solenia interjoin polyps' coelenterons.

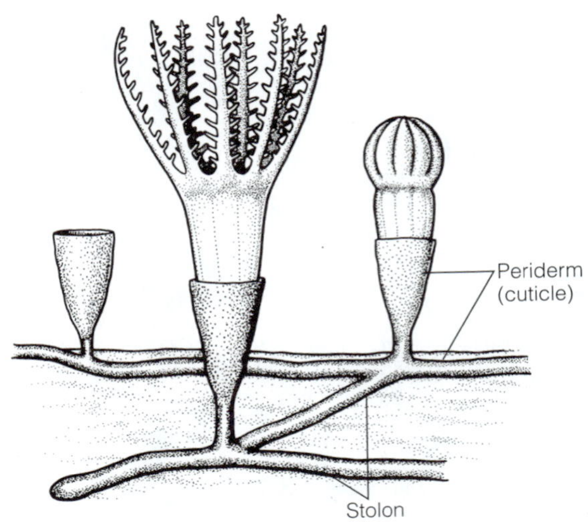

FIGURE 7-34 Alcyonarian diversity: Stolonifera. The stolonate *Cornularia cornucopae* lacks spicules and has a well-developed periderm. *(Modified and redrawn from Kaestner, A. 1984. Lehrbuch der Speziellen Zoologie. 2. Teil. Gustav Fischer Verlag, Stuttgart. 621 pp.)*

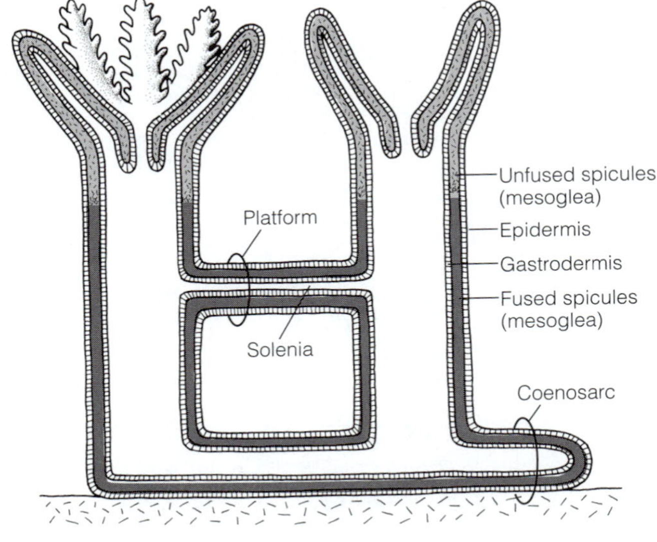

FIGURE 7-35 Alcyonarian diversity: Stolonifera. The organ-pipe coral, *Tubipora musica*, a zooxanthellate species from reefs of the Indo-Pacific. Colony base is a coenosarc. **A,** Endoskeleton of *Tubipora* is composed of fused mesogleal spicules. Note solenial canals in fractured platforms (horizontal stolons). **B,** Anatomy of *Tubipora* in vertical section. *(B, Modified and redrawn from Kaestner, A. 1984. Lehrbuch der Speziellen Zoologie. 2. Teil. Gustav Fischer Verlag, Stuttgart. 621 pp.)*

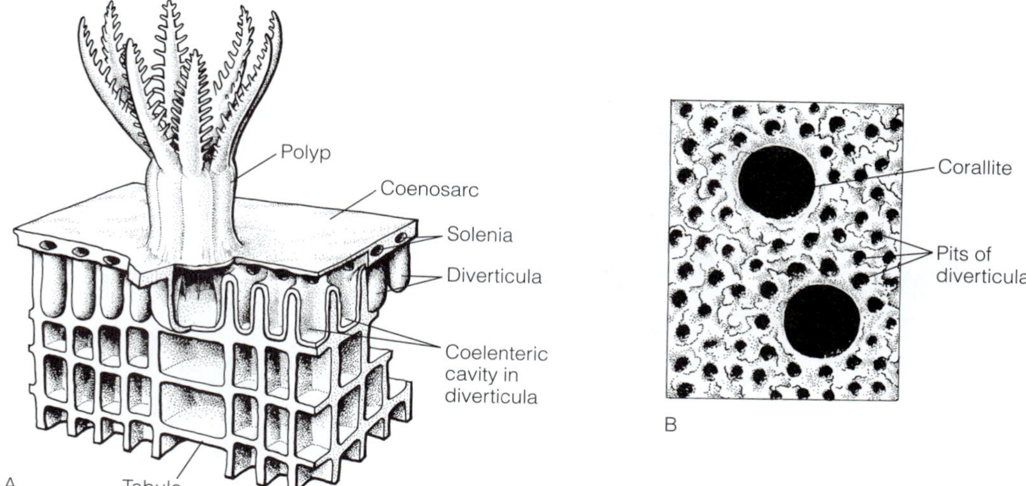

FIGURE 7-36 Alcyonarian diversity: Helioporacea. Blue coral, *Heliopora coerulea,* a zooxanthellate, reef-building species from the Indo-Pacific. The nonspicular exoskeleton, like that of stony corals, is secreted from the undersurface of the coenosarc. **A,** Cutaway view of polyp, coenosarc, and skeleton. **B,** Surface view of skeleton. *(A and B, Modified and redrawn after Bourne from Delage, Y., and Hérouard, E. 1901. Traité Zoologie Concrète. Vol. 2. Les Coelentérés. Schleicher Frères, Paris. 848 pp.)*

Alcyonacea[O]: Soft, leather, or mushroom corals; high diversity on Indo-Pacific and Red Sea coral reefs, but some (for example, *Alcyonium, Gersemia*) in temperate waters; with or without zooxanthellae. Elongate, upright, parallel, tubular polyps completely embedded in thick fleshy coenosarc except for oral-most ends (anthocodia; Fig. 7-38, 7-41C); polyp gastrodermal tubes (coelenterons), interjoined by solenia, typically reach to attached colony base; polyp growth indeterminate (like axial polyp), keeps pace with growth of coenosarc, which may be branched or lobed; anthocodia often restricted to lobes or branch ends; spicular coenosarc, sometimes also with organic fibers. Very thick coenosarc requires ventilation for gas exchange: siphonozooids in *Heteroxenia, Sarcophyton; Xenia* and others lack siphonozooids, but contract autozooids rhythmically, like hearts, to pump water.

FIGURE 7-37 Alcyonarian diversity: Telestacea. Part of a fruticose colony of *Telesto riisei* from Florida. Greatly elongate axial polyps, which bud short polyps laterally, grow upright to form the main stems and branches. An axial polyp is situated at the branch tip on the left side of the photograph. Its growth and budding zone lies just below the crown of tentacles.

Gorgonacea[O], Scleraxonia[sO]: Includes precious red coral, *Corallium rubrum* (Mediterranean Sea, Sea of Japan), and other fruticose species with lateral anthocodia on coenosarcal stems and branches, each supported by a skeletal axis; axis not enclosed in or secreted by its own specialized epithelium; skeleton an axial concentration of coenosarcal spicules and organic fibers (as in *Briareum*) or spicules alone (as in *Corallium;* Fig. 7-41D); coenosarcal solenia in skeletal axis (except in *Corallium*). *Corallium:* calcified skeletal axis surrounded by monolayer of gastrodermal tubes (possibly specialized solenia or axial-polyp coelenterons) parallel to axis and perpendicular to anthocodia; tubes join anthocodia directly or via solenia; auto- and siphonozooids present; siphonozooids alone bear gonads; calcified skeletal axis collected for centuries for jewelry and other decorations.

Gorgonacea[O], Holaxonia[sO]: Large, plantlike, often colorful sea plumes, sea fans, sea whips that sway gracefully in currents on coral reefs and elsewhere (Fig. 7-32A); resemble candelabra, cattails, or ostrich plumes; few are single upright stems, most are branched and, in sea fans, the branches are cross-connected to form a mesh; most tropical species zooxanthellate, those in colder, more productive waters lack algae. Numerous anthocodia borne laterally

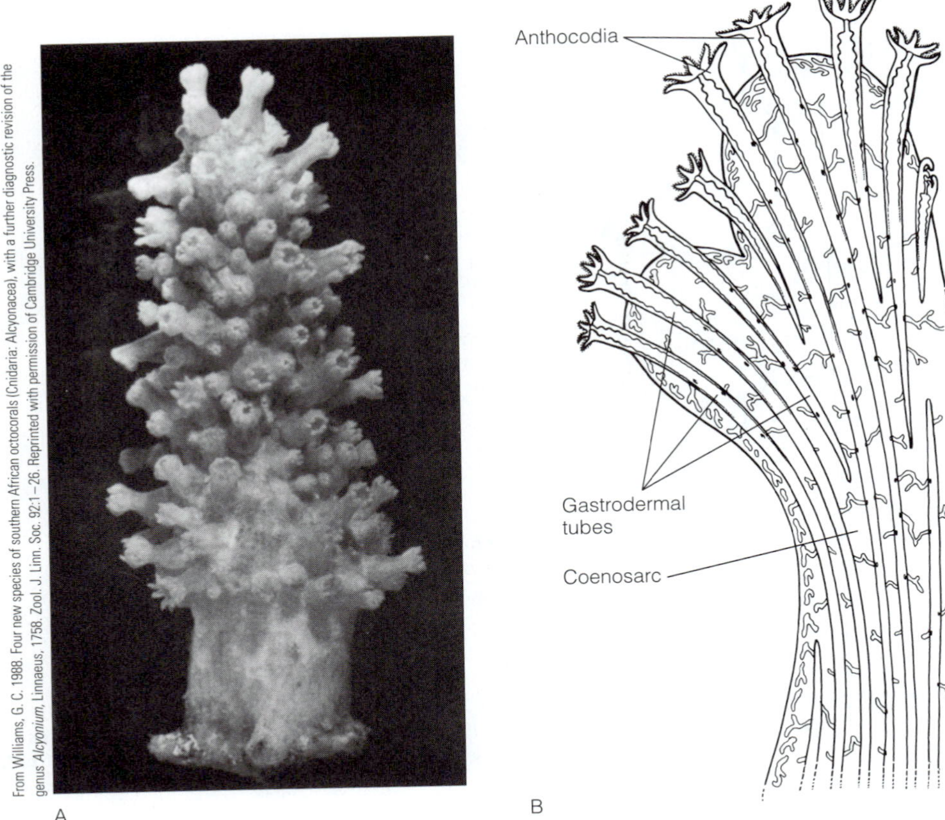

FIGURE 7-38 Alcyonarian diversity: Alcyonacea. **A,** Living soft, or mushroom, coral *(Alcyonium khoisanianum)* from South Africa. **B,** Cutaway view of the soft coral *Alcyonium distinctum,* also from South Africa. Note the thick, fleshy, spicular coenosarc and the gastrodermal tubes, one for each anthocodium, that reach the base of the coenosarc. *(B, Modified and redrawn from reference A)*

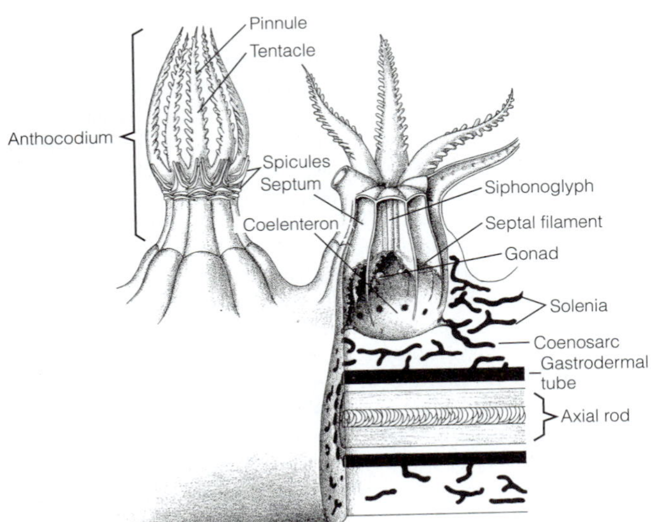

FIGURE 7-39 Alcyonarian diversity: Gorgonacea: Holaxonia. Anatomy of a sea plume (or sea whip). *(Modified and redrawn from Bayer, F. M. 1956. Octocorallia. In Moore, R. C. (Ed.): Treatise on Invertebrate Paleontology. Part F. Coelenterata. Geological Society of America and University of Kansas Press, Lawrence. pp.166–230.)*

around coenosarcal stems and branches; siphonozooids absent (Fig. 7-32B); expanded anthocodia (autozooids) densely distributed, stems and branches often "furry"; when disturbed, anthocodia retract into spicule-toughened, relatively thin coenosarc; on upright stems and branches, siphonoglyph side of each polyp faces base of colony; primary skeletal support is an organic **axial rod,** enclosed by a specialized secretory epithelium **(axial epithelium),** in stems and branches (Fig. 7-39); axial rod springy, virtually unbreakable, firmly attached to substratum; rod composed of highly cross-linked collagen (and other proteins) called **gorgonin** and more or less calcified; monolayer of parallel gastrodermal tubes (possibly specialized solenia or axial-polyp coelenterons) covers axial rod and runs along its length (Fig. 7-39, 7-41E), as in scleraxonian *Corallium;* lateral anthocodia join these tubes, either directly or via solenia.

Pennatulacea[O]: Sea feathers, sea pens, and sea pansies; polymorphic colonies, often bilaterally symmetric, anchor in soft bottoms of sand or mud (Fig. 7-40A,B); many resemble a feather stuck upright in sediment, its broad surface perpendicular to the current (Fig. 7-40B), but

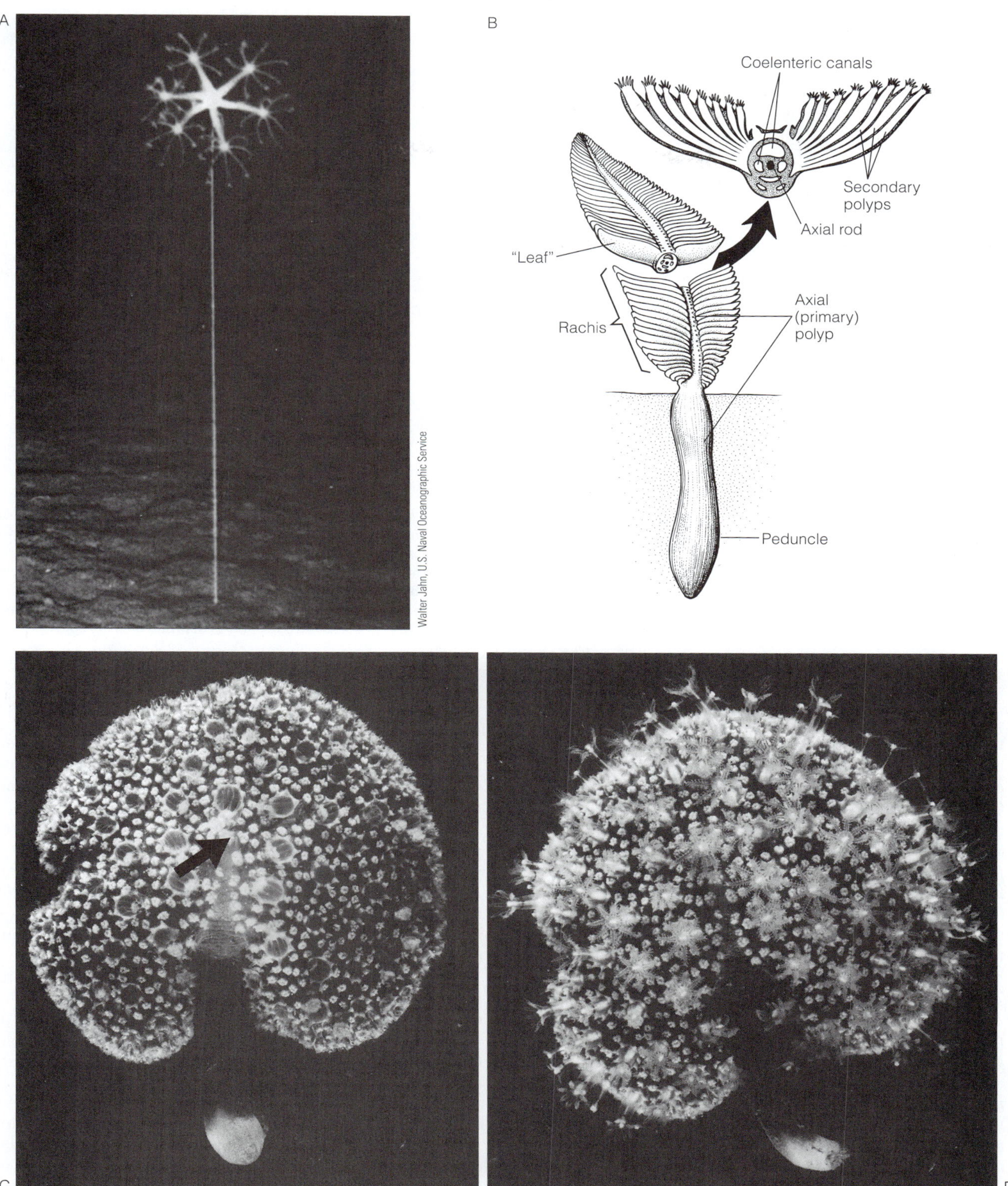

FIGURE 7-40 Alcyonarian diversity: Pennatulacea. **A,** Deep-sea (5000 m), south Atlantic species of *Umbellula* with 1 m-long peduncle. **B,** Anatomy of a sea pen similar to *Pennatula* or *Ptilosarcus.* **C,** Deflated sea pansy *(Renilla reniformis)* from South Carolina. The dark circles rimmed in white are the retracted autozooids; the small white spots are inhalant siphonozooids; the arrow points to the *single* exhalant siphonozooid. **D,** Inflated sea pansy. *(B, Modified and redrawn from Kaestner, A. 1984. Lehrbuch der Speziellen Zoologie. 2. Teil. Gustav Fischer Verlag, Stuttgart. 621 pp.)*

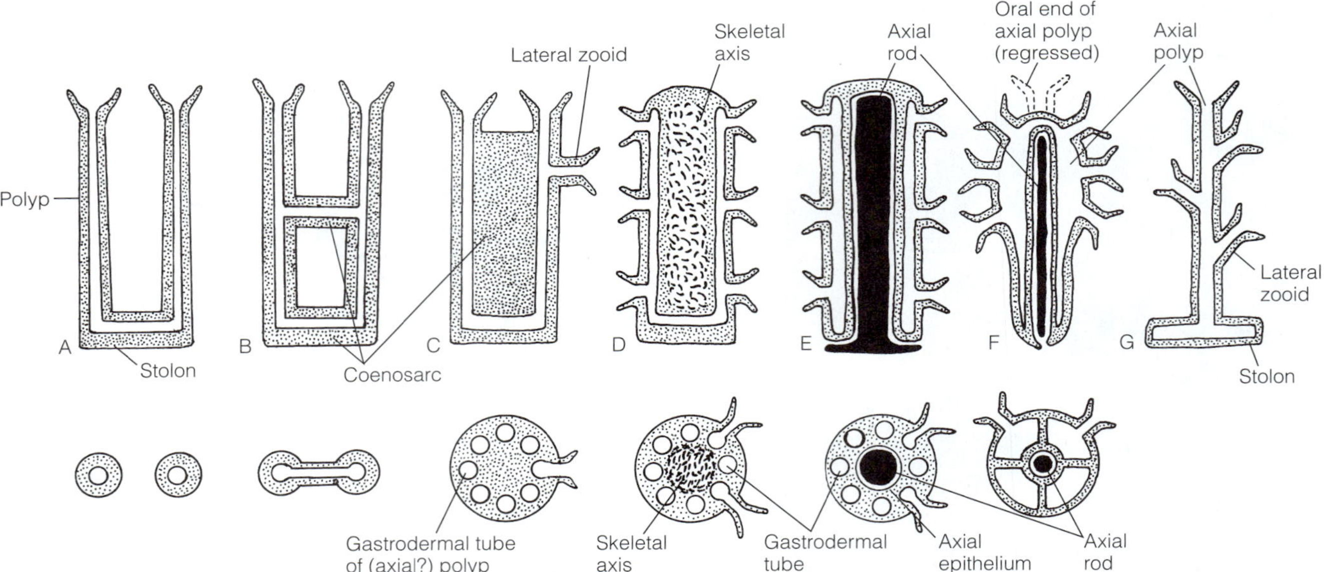

FIGURE 7-41 Alcyonarian diversity: body designs. Diagrammatic vertical and cross-sectional views. Solenia not shown. **A,** Stolonifera: simple stolonate species such as *Cornularia.* **B,** Stolonifera: species such as *Tubipora* with elevated horizontal stolons (platforms). **C,** Alcyonacea: species with thick coenosarc in which elongate zooids (anthocodia and gastrodermal tubes) are embedded. **D,** Gorgonacea, Scleraxonia: fruticose species such as precious red coral *(Corallium rubrum)* with an axial skeleton of concentrated spicules, organic fibers, or both. **E,** Gorgonacea, Holaxonia: fruticose sea plumes and sea fans with an axial rod enclosed in an epithelium and composed of gorgonin with or without nonspicular calcium carbonate. **F,** Pennatulacea: sea pens and pansies; body consists of a single large axial polyp bearing lateral zooids and an axial rod. **G,** Telestacea: fruticose colonies of *Telesto;* main and branch stems are axial polyps that bud lateral zooids.

sea pansies *(Renilla),* though rooted in sand, flop over to lie flat on sediment surface (Fig. 7-40C,D) and can move slowly over it; zooxanthellae absent, feed entirely on plankton, perhaps other suspended organic material. Colonies composed of three (occasionally four) types of polyps in a thin spicular coenosarc; main axis and bulk of body is one enlarged, elongate axial polyp (= primary polyp; Fig. 7-40B), more than 3 m long in *Chunella* (similarly elongate *Umbellula* in Figure 7-40A); with growth, axial-polyp tentacles, mouth, and pharynx regress and disappear, but column persists and functionally is divided into two parts; swollen lower part, the **peduncle,** for burrowing and anchorage, is devoid of zooids (Fig. 7-40B); **rachis,** or upper part of column, is budding zone and bears auto- and siphonozooids, both oriented perpendicular to axis of primary polyp (Fig. 7-40B); auto- and siphonozooid siphonoglyphs generally face attached end of colony; polyp coelenterons interconnect directly, but solenial connections also present; with their integrated polymorphic polyps, bilateral symmetry, and motility *(Renilla),* pennatulacean body is at organ-level of organization similar to that of bilaterally symmetric animals. *Funiculina, Ptilosarcus:* primary polyp encloses a horny axial rod (Fig. 7-41F) surrounded by axial epithelium, similar or identical to gorgonin, but designated "pennatulin"; *Stylatula, Virgularia:* axial rod mineralized with nonspicular calcium carbonate; *Renilla:* lacks axial rod.

PHYLOGENY OF ANTHOZOA

Anthozoan phylogeny is not yet well understood, and several competing models have been proposed. The phylogenetic trees shown here for Anthozoa (Fig. 7-42) and Alcyonaria (Fig. 7-43) are both based solely on morphological characters. At present, there is little consensus among trees derived from molecular analyses.

Here we recognize two major taxa of Anthozoa: Alcyonaria and Zoantharia (Fig. 7-42). Within Zoantharia, united by their hexamerous symmetry and especially by the spirocyst synapomorphy, the sea anemones, stony corals, and coral anemones may constitute a monophyletic taxon, but the relationships among them are unclear. Scleractinians and corallimorphs, however, may be more closely related to each other than either is to the sea anemones. The systematic positions of the zoanthids and especially the antipatharians are also uncertain. Our analysis suggests that the zoanthids may be the sister taxon of the Actiniaria–Scleractinia–Corallimorpharia and Antipatharia is the sister taxon of the Ceriantharia.

For the Alcyonaria in particular, the phylogenetic tree (Fig. 7-43) suggests that primitive members were stolonate or stolonal-mat colonies that lacked spicules. The sister taxon, representing most other alcyonarians, evolved a complex coenosarc bearing spicules as well as zooidal dimorphism (autozooids and siphonozooids). The apomorphic taxa form fruticose colonies that arise from the growth and budding of axial polyps and are supported by an axial rod. The homology of some characters used in this phylogeny, however, are

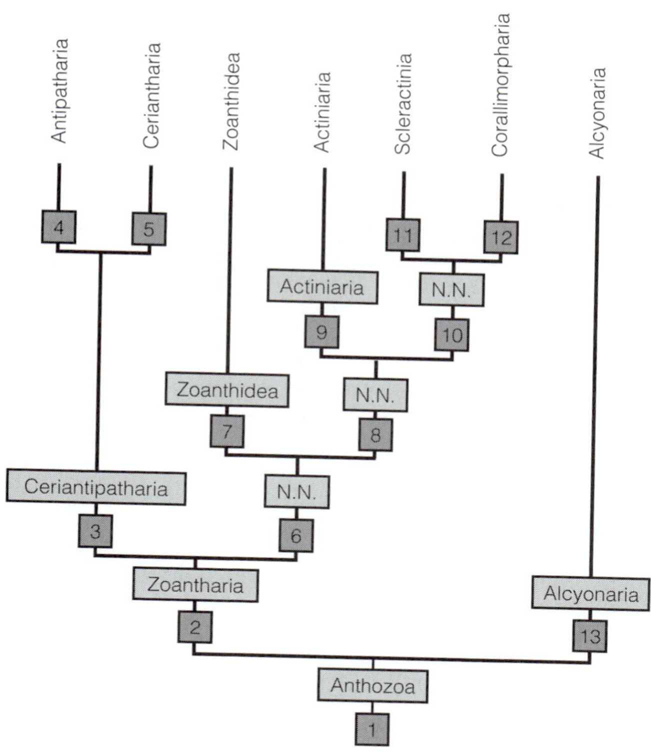

FIGURE 7-42 Phylogeny of Anthozoa and Zoantharia. **1, Anthozoa:** polyp with pharynx, one siphonoglyph, unpaired complete septa, septal filaments, each septum with sheetlike radial and longitudinal musculature on opposite faces. **2, Zoantharia:** hexamerous symmetry, new septa added as bilateral singles from localized growth zone, spirocysts. **3, Ceriantipatharia:** hyposulculus (aboral continuation of "dorsal" siphonoglyph [sulculus] onto directive septa), nonretractile tentacles. **4, Antipatharia:** spiny axial skeleton (noncollagen "antipathin"). **5, Ceriantharia:** tube composed of ptychocysts, well-developed longitudinal column musculature (epidermal), two whorls of tentacles. **6, N.N.:** complete and incomplete septa, tentacles and oral disc retractile into column. **7, Zoanthidea:** column with periderm, mesoglea with solenia. **8, N.N.:** paired septa, new septa added as bilateral doubles (pairs) from generalized growth zone; mesogleal sphincter muscle. **9, Actiniaria:** well-developed septal retractor muscles, solitary, two siphonoglyphs. **10, N.N.:** reduced siphonoglyphs (possibly even absent). **11, Scleractinia:** corallum with corallites, sclerosepta. **12, Corallimorpharia:** multiple tentacles in each interseptal space, muscular oral disc. **13, Alcyonaria:** see Fig. 7-43.

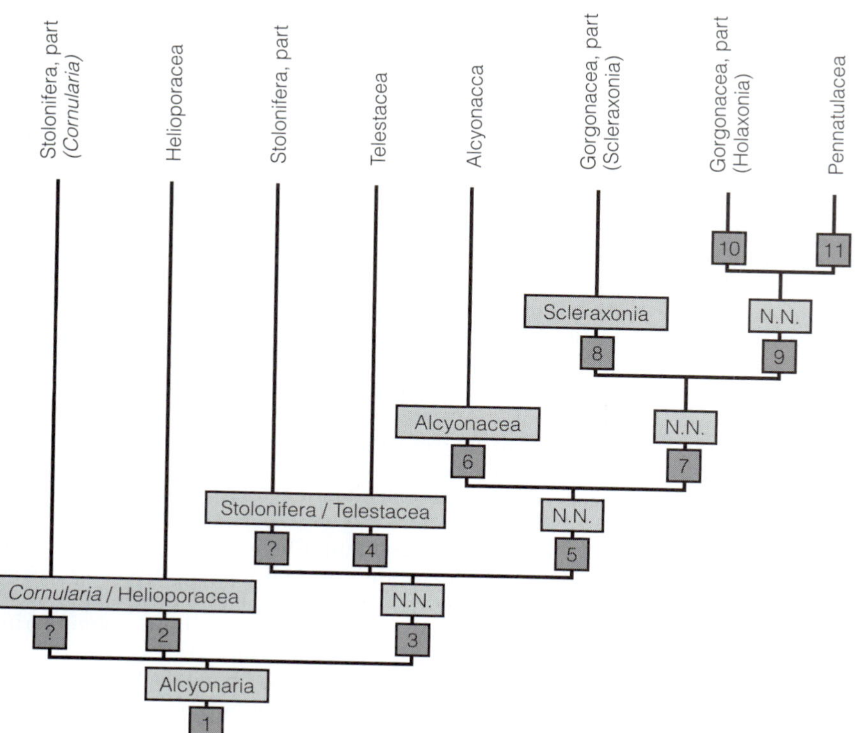

FIGURE 7-43 Phylogeny of Alcyonaria. **1, Alcyonaria:** eight pinnate tentacles, eight unpaired complete septa, asulcal septa specialized for water pumping. **2, Helioporacea:** massive nonspicular calcareous exoskeleton (aragonite). **3, N.N.:** calcareous spicules in mesoglea. **4, Telestacea:** growth by budding from an axial polyp. **5, N.N.:** siphonozooids, organic fibers (probably collagen) in mesoglea. **6, Alcyonacea:** very thick coenosarc, deeply embedded elongate polyps extend to colony base. **7, N.N.:** fruticose colony, axial skeleton, lateral budding of short anthocodia. **8, Gorgonacea (Scleraxonia):** skeletal axis of spicules, organic fibers, or both; axis lacks enclosing epithelium. **9, N.N.:** axial rod (of collagen and other scleroproteins) with or without nonspicular calcium carbonate, rod bound by an axial epithelium. **10, Gorgonacea (Holaxonia):** autozooids arranged radially around stems and branches, siphonozooids absent. **11, Pennatulacea:** auto- and siphonozooids arise from surface of single enlarged axial polyp.

PHYLOGENETIC HIERARCHY OF ANTHOZOA

Anthozoa
- Alcyonaria
  - Stolonifera (part)-Helioporacea
  - N.N.
    - Stolonifera (part)-Telestacea
    - N.N.
      - Alcyonacea
      - N.N.
        - Scleraxonia
        - N.N.
          - Holaxonia
          - Pennatulacea
- Zoantharia
  - Ceriantipatharia
    - Antipatharia
    - Ceriantharia
  - N.N.
    - Zoanthidea
    - N.N.
      - Actiniaria
      - N.N.
        - Scleractinia
        - Corallimorpharia

uncertain. For example, the axial rod of gorgonians may not be homologous with that of pennatulaceans and the gastrodermal tubes in the stem and branch axes of gorgonians may not be axial polyps, but rather upright stolons or highly organized solenia. Such uncertainties highlight the need for renewed basic research in comparative and developmental morphology.

MEDUSOZOA

Cnidarians with a medusa stage in their life cycle constitute the taxon of Medusozoa. The basic life cycle, beginning with the adult, is: medusa → planula → polyp → medusa (Fig. 7-15). Medusozoans have tetramerous symmetry, based on four or multiples of four, and a linear mitochondrial DNA molecule, rather than the circular form found in Anthozoa and all other animals. Cnidae are present as nematocysts only; spirocysts and ptychocysts are absent. Medusozoa includes two major taxa, Scyphozoa (large jellies) and Hydrozoa (small jellies, *Hydra,* hydroids, and relatives). We begin with the large, conspicuous scyphozoans.

SCYPHOZOA[C]

The exclusively marine Scyphozoa includes the box jellies, stalked jellies, flag-mouth jellies, and root-mouth jellies. The taxon, which consists of only 200 species, is best known for its large medusae, which are seen washed ashore on beaches, adrift in nearshore currents, or pulsating rhythmically near the water's surface. Because the medusae are conspicuous and common, references to "jellyfish" most often apply to these animals.

Form and Function

Scyphozoans have small, funnel-shaped polyps, **scyphistomae,** that are solitary or colonial, depending on species (Fig. 7-44, 7-45, 7-76D–F). The wide oral end of each polyp tapers sharply to about the middle of the column. Below this level, the column is a narrow stalk that extends to the small pedal disc. Typical polyps reach a few centimeters in height, but rarely more than 2 mm in diameter. The polyps attach to the substratum with a chitinous periderm that can embrace part or all of the column (Fig. 7-44, 7-45). The tetramerous symmetry of scyphopolyps is indicated by four equidistant, gastrodermal septa partitioning the coelenteron (Fig. 7-45). These septa are widest at their insertion on the oral disc, then narrow aborally and disappear before reaching the pedal end of the polyp. Anthozoan-like septal filaments are absent, but each septum has a **septal funnel** (subumbrellar funnel) that originates as an invagination of the oral disc and extends deeply into the septum (Fig. 7-45). The four hollow funnels end blindly within the septa, but open on the surface of the oral disc. Epitheliomuscular cells in the wall of each funnel form a longitudinal retractor muscle. In contrast to anthozoan polyps, scyphopolyps lack a pharynx and oral sphincter.

The large scyphomedusae (Fig. 7-46A), which occur in all oceans of the world, have a typical range of bell diameters from 2 to 40 cm, but some are much larger, up to 2 m in *Cyanea capillata.* Their colors are often striking, especially when the gonads and other internal structures (deep orange, pink, or other colors) are visible through the transparent or delicately tinted bell. Tetramerous symmetry is often apparent on the specialized manubrium, which may be square or elongate and divided into four **oral arms** that stream below the bell like masthead pennants on old sailing ships (Fig. 7-46A). Sometimes, the marginal tentacles are arranged in four or eight groups, but often they are uniformly distributed around the perimeter of the bell (Fig. 7-47A).

Four equidistant septa divide the coelenteron into a central stomach and four **gastric pockets** (Fig. 7-46B). The thick septa, which persist from the polyp stage, each contain a septal funnel that opens onto the oral surface (Fig. 7-46B, 7-47A). These funnels are in close proximity to the eight gonads (four pairs), which, as in anthozoans, are borne on the sides of the septa. A water flow, probably ciliary in origin, enters and leaves the funnels, perhaps supplying oxygen to the gonads. The free margins of the septa bear a cluster of slender, threadlike **gastric filaments,** with the unattached end of each filament extending freely into the stomach (Fig. 7-46B, 7-47A). Reminiscent of the cnidoglandular band and acontia of anthozoans, the gastric filaments bear nematocytes and enzymatic gland cells and are important in subduing prey and digestion. In some scyphozoans, the stomach region gives rise to radial canals, often repeatedly branched, that create a coelenteric network in the thick mesoglea and often join a marginal ring canal (Fig. 7-46C, 7-47A). Gastrodermal cilia circulate the coelenteric fluid in a definite circuit. If the stomach is considered to be the "heart" of this circulatory system, then some radial canals are "arteries," carrying fluid away from the stomach,

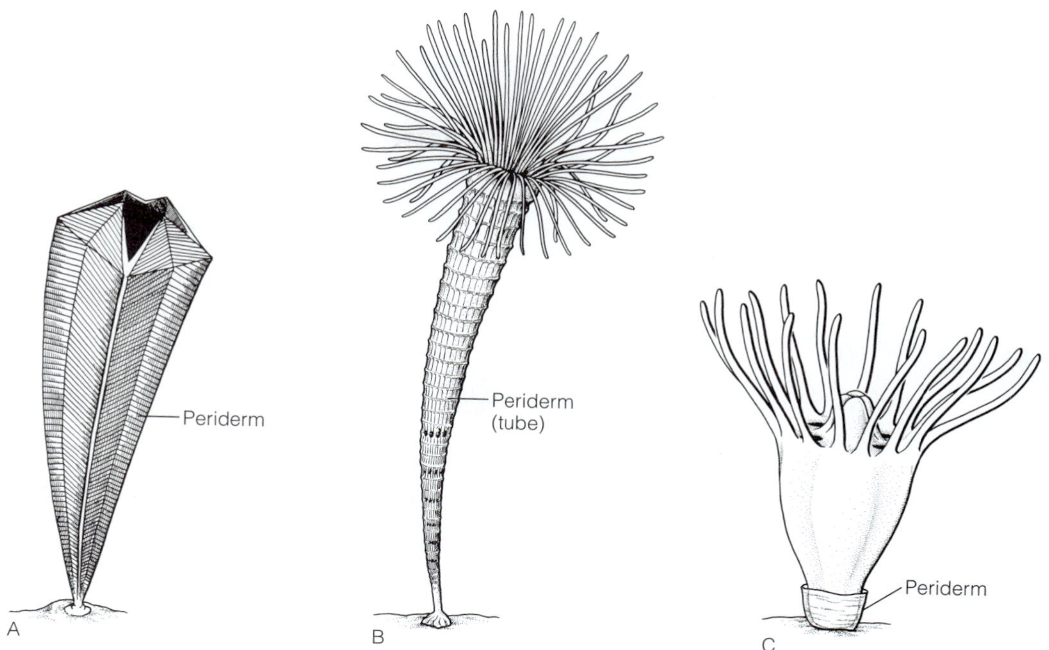

FIGURE 7-44 Scyphozoa: polyps. **A,** Conulata: fossil periderm tube of an extinct Paleozoic specimen. **B,** Coronatae: living polyp in its annulated periderm tube. **C,** Semaeostomeae: living polyp (scyphistoma) with periderm at pedal end only.

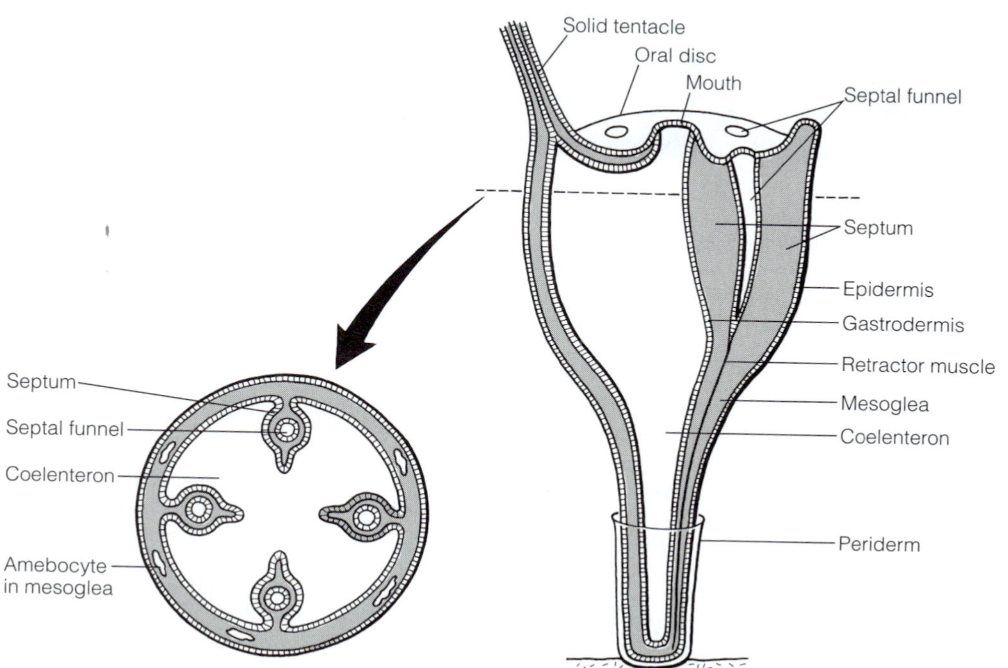

FIGURE 7-45 Scyphozoa: polyp anatomy (based on Semaeostomeae and Rhizostomeae).

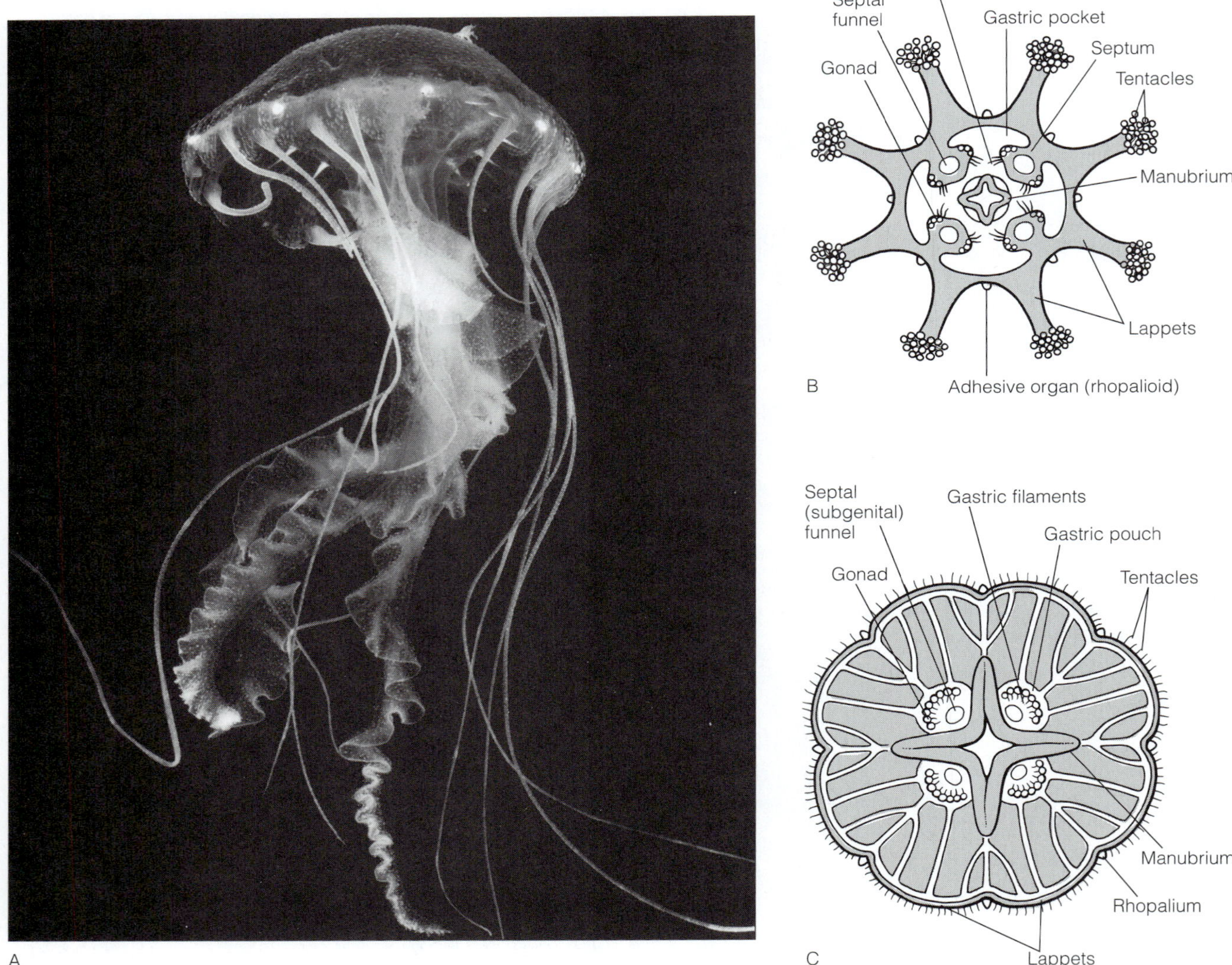

FIGURE 7-46 Scyphozoa: medusae. **A,** The sea nettle *Chrysaora quinquecirrha* (Semaeostomeae). Note oral arms, tentacles, and rhopalia (white spots on bell margin). **B,** Anatomy of a stalked jelly such as *Haliclystus* (Stauromedusae). **C,** Anatomy of a flag-mouth jelly such as *Aurelia* (Semaeostomeae). Note that the gastric pouches in **C** correspond in position to the septa in **B.**

whereas others are "veins" (Fig. 7-47B). The thick mesoglea contains amebocytes that perhaps are functional fibroblasts.

The umbrellar margin is often scalloped into rounded lobes called **lappets,** which provide flexion points during bell contraction (Fig. 7-46B,C, 7-47A). At equal intervals around the margin of the bell, in the niches between the lappets, are knoblike sensory lobes known as **rhopalia** (Fig. 7-46A,C, 7-47). The rhopalia are sensory *organs* that have gastrodermal, mesogleal, and epidermal components (Fig. 7-48C). The sensory structures in each rhopalium are a gastrodermal statocyst, a mechanoreceptor, probably a chemoreceptor, and sometimes a photoreceptor. Although rudimentary in comparison with human sensory capabilities, each rhopalium provides information related to gravity-equilibrium, waterborne vibrations (hearing), as well as odor and light detection. A ganglion associated with the rhopalial base receives and integrates the sensory information with a nerve net and nerve ring. Rhopalial ganglia have pacemaker neurons whose burst rate is affected by input from the rhopalia, thus controlling swimming rate and direction. The subumbrellar locomotory muscles are a cross-striated coronal muscle and radial muscles.

Adult scyphozoans feed on small animals, especially crustaceans, but many also eat other jellies. Some scyphozoans feed on fish; however, larval fish of a number of species swim with certain species of scyphozoans for protection. As the medusa gently swims or slowly sinks, prey is captured on con-

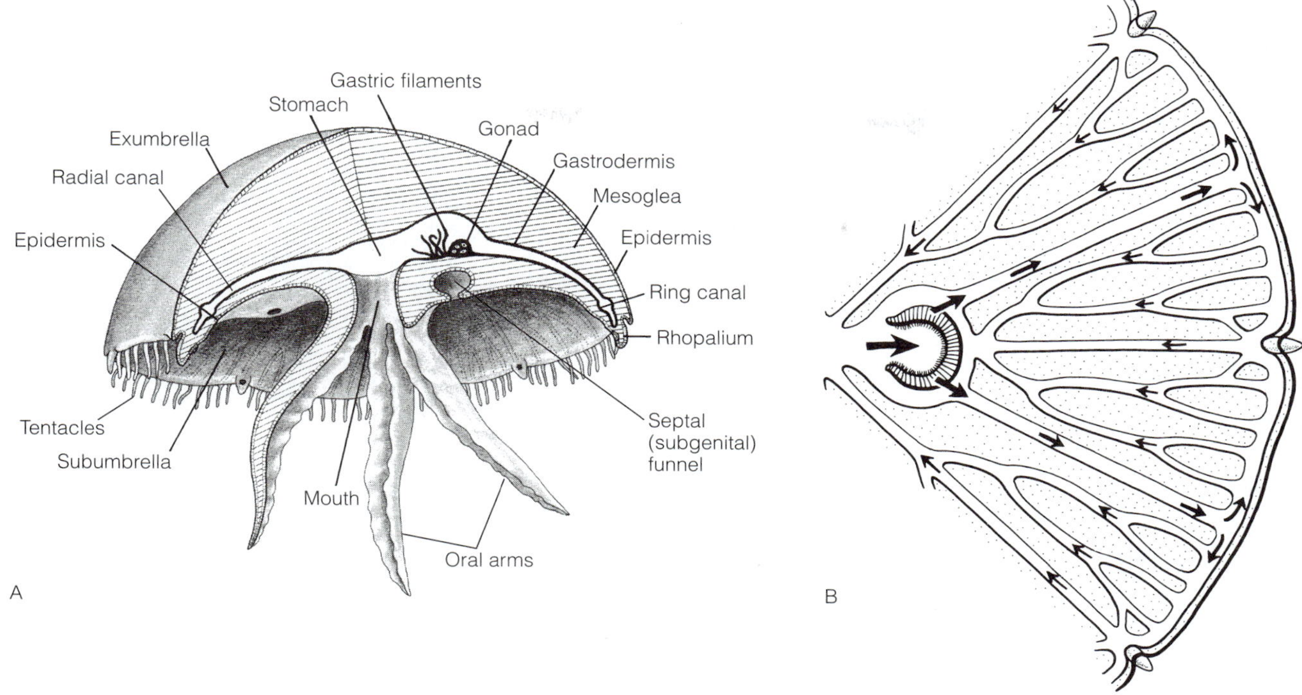

FIGURE 7-47 Scyphozoa: anatomy and internal transport (*Aurelia:* Semaeostomeae). **A,** Cutaway view of adult. **B,** Ciliary flows in the coelenteric network. *(B, Modified and redrawn after Southward, A. J. 1955. Observations on the ciliary currents of the jellyfish* Aurelia aurita. *JMBA (UK) 34:201–216.)*

tact with the tentacles or the manubrium. The tentacles may bend or contract to bring the prey to the vicinity of the manubrium. Scyphozoan nematocysts occur in the epidermis of polyps and medusae, and also in the gastrodermis (gastric filaments) of medusae.

In general, scyphozoan polyps (scyphistomae) are asexual and medusae sexual. In scyphistomae, budding is responsible for growth in colonial species and clonal reproduction in solitary forms. The buds arise either on the column wall or from a stolon (as in *Aurelia*). At the appropriate time of year, under the influence of hormonal and environmental factors, scyphistomae asexually produce young medusae in a process known as **strobilation** (Fig. 7-49). A disclike medusa differentiates from the oral end of the strobilating scyphistoma, now called a **strobila,** and then separates from it by transverse fission. The newly released, free-swimming, juvenile medusa is an **ephyra.** The minute ephyra has a deeply incised bell margin, which flaps vigorously as the jelly swims (Fig. 7-49). Depending on species, a strobila produces either one medusa (monodisc strobilation) or a succession of many (polydisc strobilation). In polydisc-strobilating forms such as *Aurelia* and other common scyphopolyps, the differentiating medusae resemble a stack of saucers, the oldest and first released being oral-most in the stack (Fig. 7-49). After strobilation, the scyphistoma may resume its polypoid existence until the following year, when formation of ephyrae is repeated. A scyphistoma may live for one or several years.

Some ephyrae take two years to grow into sexually reproducing adult medusae, whereas others are relatively short lived. The ephyrae of *Aurelia aurita* on the west coast of the United States are produced in March and reach sexual maturity by June.

Most adult scyphomedusae are gonochoric with eight gastrodermal gonads, one on each surface of the four septa (Fig. 7-46B). Gametes are spawned through the mouth, but some species brood eggs on the surface of the body. Zygotes develop into planulae that, after a brief free-swimming existence, settle to the bottom, become attached by their anterior end, and metamorphose into polyps.

Diversity of Scyphozoa

Scyphozoa is divided into five taxa. Some traditional classifications, however, consider one of these taxa, Cubomedusae (as Cubozoa), to be the sister taxon of Scyphozoa. Our cladistic analysis, however, does not support this interpretation. Instead, the placement of Cubomedusae within Scyphozoa seems to be supported by several synapomorphies. These include septa, gastric filaments, eight gonads, gastric pockets, rhopalia, and medusa differentiation from the oral end of the polyp.

Semaeostomeae[O]: The flag-mouth jellies are large medusae 10–30 cm in diameter (but can be up to 2 m wide), in which the manubrium is elongate and divided into four oral

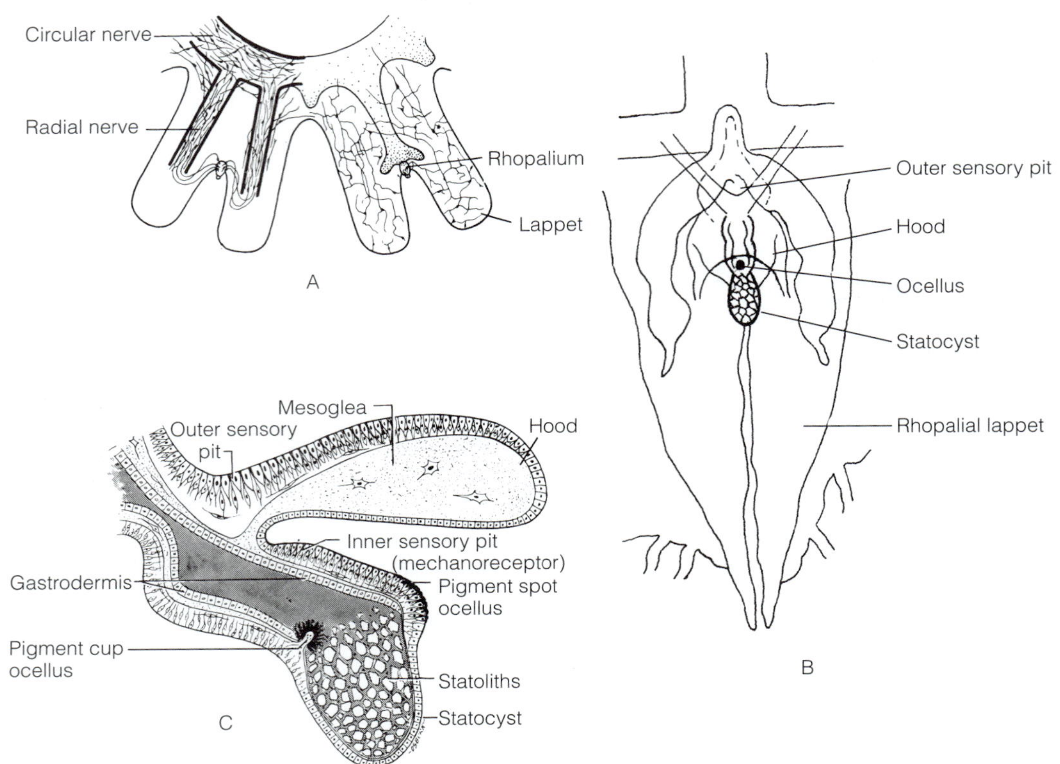

FIGURE 7-48 Scyphozoa: nervous system and sense organs (*Aurelia:* Semaeostomeae). **A,** Bell margin (ephyra) showing nerve net, lappets, and rhopalia. **B,** Exumbrellar view of adult rhopalium. **C,** Vertical section of rhopalium. *(B and C, Modified and redrawn from Hyman, L. H. 1940. The Invertebrates, Vol. 1. McGraw-Hill Book Co., New York.)*

arms. Most of the 50 species are coastal in distribution, including *Aurelia aurita,* the moon jelly; species of *Chrysaora,* often called sea nettles (Fig. 7-46A); *Cyanea capillata,* the winter or lion's mane jelly; and the fried-egg jelly, *Phacellophora camtschatica. Pelagia noctiluca,* however, lives offshore in the open ocean, and *Stygiomedusa fabulosa* is a deep-water form.

The bell of semaeostomes (sem-ME-oh-stomes) bears hollow marginal tentacles, lappets, and usually eight rhopalia. In *Aurelia,* the bell has eight lappets, eight rhopalia, and a marginal fringe of short tentacles (Fig. 7-46C, 7-47A). The mouth leads into a large four-lobed stomach, the lobes being called **gastric pouches** (Fig. 7-46C). Unlike the gastric pockets, which primitively occur in the four interseptal spaces (in cubo-, stauro-, and coronate medusae), the gastric pouches of semaeostomes (and rhizostomes) occupy the positions of the septa themselves (compare B and C of Fig. 7-46). Septa are absent in semaeostome medusae, but in their locations are the new gastric pouches, within each of which are all the typical septal structures except the actual septum. These septal structures include gonads, gastric filaments, and septal funnels (Fig. 7-46C, 7-47). Each septal funnel primitively projected into a septum (Fig. 7-45, 7-46B), but now projects into the center of a gastric pouch (Fig. 7-46C), which has replaced the septum. The gonads, still faithful to the original septal positions, now lie in the walls of the pouches, but with no septa to separate the members of each pair, fuse from eight into four gonads. Each gonad is closely associated with the blind inner end of a septal funnel, now called a **subgenital funnel** (Fig. 7-46C, 7-47A). As in other scyphomedusae, the bidirectional water flow in the funnels probably ventilates the gonads for gas exchange. From the central stomach, the coelenteron breaks up into a complex **coelenteric network** in the large, thick mesoglea of the bell (Fig. 7-46C, 7-47B). This network, with discrete arterial and venous flows, is the circulatory system for the peripheral parts of the body (Fig. 7-47B).

Aurelia is a suspension feeder that traps plankton in mucus on the ciliated subumbrellar surface. Cilia then sweep the food to the bell margin, where it is scraped off by the oral arms. Ciliated grooves on the oral arms carry the food to the mouth and stomach. Other semaeostomes also feed on plankton, but supplement their diet with larger prey, including fishes, worms, and other jellies.

In *Aurelia,* four horseshoe-shaped gonads lie in the oral wall of the stomach, one in each gastric pouch (Fig. 7-46C, 7-47A). When mature, the gametes usually rupture into the pouch cavity, where fertilization occurs, and zygotes exit through the mouth. The zygotes become lodged in brood pits on the oral arms. Eventually, planula larvae are released from these pits

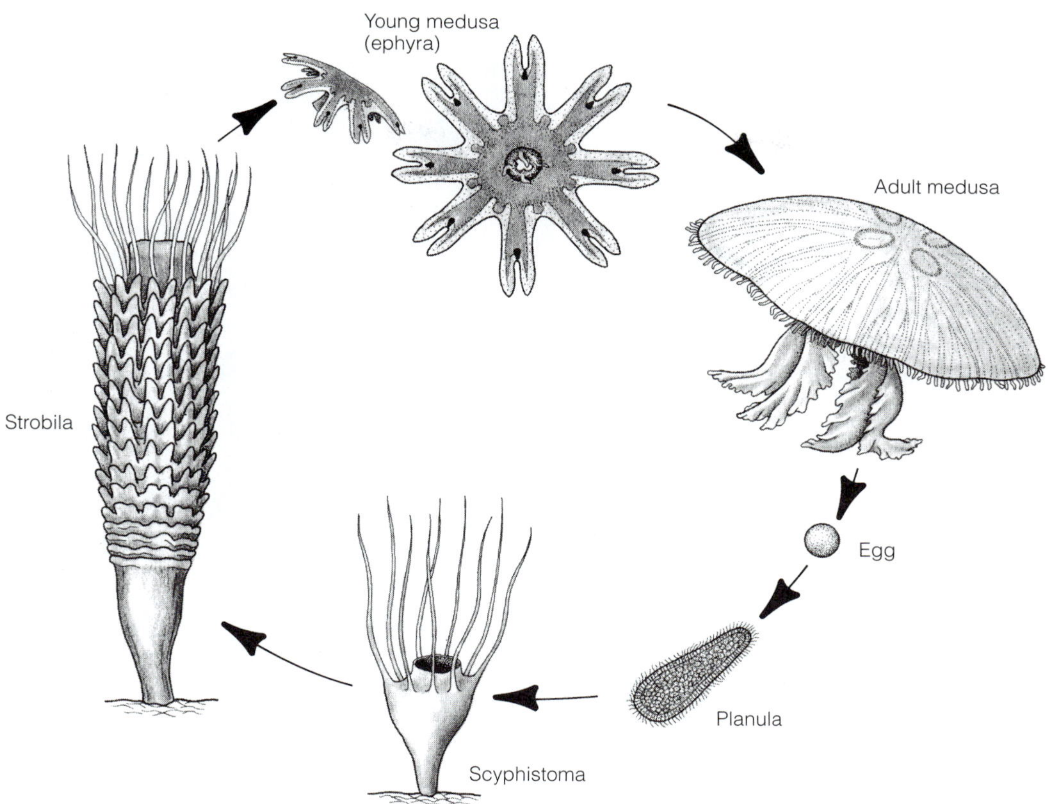

FIGURE 7-49 Scyphozoa: life cycle (*Aurelia:* Semaeostomeae).

(Fig. 7-49). In *Cyanea,* the eggs are fertilized in the gonad and released at the gastrula stage. The planula larva settles and attaches after a brief planktonic existence, then metamorphoses into a funnel-shaped scyphistoma complete with septa and septal funnels. These scyphistomae live for several years, bud other scyphistomae from the column or stolons, and may go through several seasonal rounds of strobilation. Semaeostomes produce medusae by strobilation, as described above for Scyphozoa (Fig. 7-49). Strobilation is polydisc in species of *Aurelia, Chrysaora,* and *Phacellophora.* The open-ocean species *Pelagia noctiluca* lacks a polyp stage; its planula metamorphoses directly into a medusa.

Rhizostomeae[O]: 80 species, the most diverse scyphozoan taxon, includes *Cassiopeia xamachana* (upside-down jelly), *Stomolophus meleagris* (cannonball jelly, cabbage-head, jellyball), *Rhizostoma octopus* (cauliflower jelly). Collectively known as "root-mouth jellies" because rootlike manubrium develops from branched or fused oral arms (Fig. 7-50); numerous additional manubrial outgrowths greatly increase surface area; complex coelenteric canal network in manubrium and its surface projections, as well as in mesoglea surrounding stomach; in many, original mouth closes and terminal ends of manubrial coelenteric canals break through as secondary mouths; like a colony of feeding zooids, the multiple manubrial mouths and huge surface area loaded with nematocytes and mucus-secreting cells allows rhizostomes to trap plankton from their own water jet, thus coupling feeding with locomotion; *Cassiopeia* (Fig. 7-50B) and *Mastigias* with manubrial zooxanthellae occur in shallow, protected, sun-drenched tropical lagoons; *Cassiopeia* normally lies upside down on bottom, pulsating slowly while exposing its eight branched but unfused oral arms to sunlight; a few rhizostomes with fused oral arms feed on fish, apparently by reducing them to a slurry on the manubrial surface before ingesting the fragments with secondary mouths. Like semaeostomes, septa absent, but gastric filaments and subgenital funnels associated with four gonads in stomach wall; marginal tentacles absent; sexual reproduction, larval development, settlement, and metamorphosis all similar to those of semaeostomes; scyphistomae, as strobilas, undergo polydisc or monodisc (*Cassiopeia*) strobilation.

Cubomedusae[O]: Box jellies: distinctive cubic bell, colorless transparent body, and highly venomous nematocysts; admired for their beauty and feared for their stings (Fig. 7-51); 15 species inhabit tropical and warm-temperate seas; strong swimmers, feed primarily on fish; *Chironex fleckeri:* notorious sea wasp (Australia's Great Barrier Reef) with bell diameter up to 30 cm and 2 m-long tentacles, one of world's most venomous animals; on average, two bathers die per year, usually in minutes, from sea-wasp stings; nematocyst venom causes excruciating pain, toxic to nerve, muscle, and

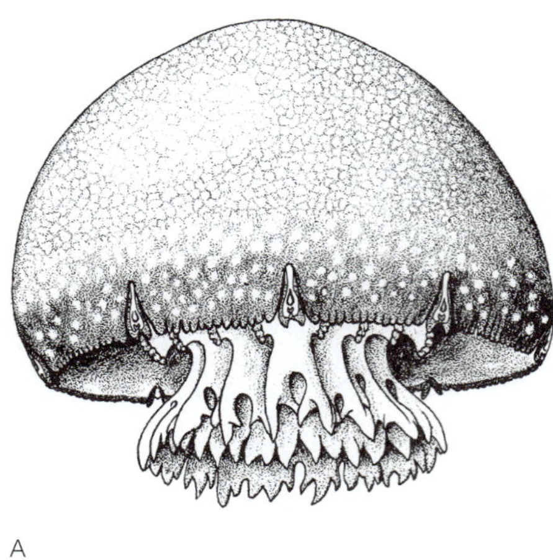

A

B

FIGURE 7-50 Scyphozoan diversity: Rhizostomeae. **A,** The cannonball jelly—also known as a cabbage-head jelly or jellyball—*Stomolophus meleagris,* from South Carolina. This species trails sticky mucous strings from its short, stiff manubrium. Plankton in the water jet are trapped on the mucus and ingested by the jelly. **B,** Oral (subumbrellar) view of the upside-down jelly, *Cassiopeia xamachana,* from Florida. The highly branched oral appendages house zooxanthellae. *(A, Redrawn after Mayer, 1910.)*

heart; *Chiropsalmus quadrumanus:* common box jelly of southeastern United States; few centimeters in size (Fig. 7-51A), responsible for at least one known human fatality. Medusa has long, slender, hollow tentacles on outgrowths (pedalia) of four bell corners (Fig. 7-51A, 7-52A); pendant, prehensile, four-square manubrium (Fig. 7-52A); shelflike outfold of subumbrellar margin, the **velarium,** restricts bell aperture, creates powerful water jet with each bell pulsation (Fig. 7-52A); each flat outer side of bell bears single rhopalium set in a pocket; often are two highly developed rhopalial ocelli that resemble vertebrate image-forming eyes, one eye directed upward, other inward toward manubrium (Fig. 7-52B); coelenteron partitioned by four low, broad septa, which contain septal funnels, bear gastric filaments aborally, and produce the eight gonads, two per septum. Gonochoric medusae spawn gametes; fertilization external; zygotes become planula larvae that settle and metamorphose into tiny solitary polyps, usually 1 mm long and 0.5 mm in diameter, that have knobbed (capitate) tentacles; minute polyps lack septa and septal funnels; in one species, base of polyp enclosed in chitinous periderm (Fig. 7-53); each polyp can bud secondary polyps that creep over substratum, eventually attach; single medusa differentiates from the oral end of a polyp, process resembles monodisc strobilation; juvenile medusa detaches from the stalklike pedal end of polyp column, which persists as a vestige, later degenerates (Fig. 7-53).

Coronatae^O: 35 species; mostly deepwater, few in shallow tropical seas; *Linuche unguiculata* (thimble jelly): common, in shallow water, zooxanthellate; *Periphylla periphylla* (Fig. 7-54): widely distributed in deep water. Most bell diameters less than 5 cm, but largest reach 20 cm; distinctive feature is circumferential constriction, the **coronal groove,** around middle of exumbrella; below coronal groove, bell is radially creased and lobed, each lobe (pedalium) bearing a solid, nonretractile tentacle; coronal groove and creases provide flexion zones on an otherwise stiff bell; below pedalia and tentacles, on bell margin are four or eight rhopalia, depending on species, in niches between lappets (Fig. 7-54); large mouth on short, four-sided manubrium; four well-developed septa bearing funnels, gastric filaments, and eight gonads divide coelenteron into four gastric pockets. Many coronates feed with outstretched tentacles directed aborally; when prey is contacted, tentacles flex stiffly oralward, pushing food into mouth. Adult medusae free-spawn gametes; planula larvae settle and metamorphose into slender polyps resembling heraldic trumpets; solitary or colonial polyp secretes annulated chitinous tube enclosing all but oral disc and tentacles (Fig. 7-44B); polyps have four septa, lack septal funnels, undergo polydisc strobilation.

Stauromedusae^O: (STORE-oh-meh-doo-see); 25 species; sessile, attached, resemble sundew plants; often called "stalked jellies" because medusoid body attaches to substratum by an aboral stalk; anatomy and development indicate adult is an overgrown polyp whose oral end is partially differentiated into a medusa; a stauromedusan can be viewed as a large, sexually mature, monodisc strobila with its ephyra still attached (Fig. 7-55); "exumbrellar stalk" actually a column and "basal adhesive plate" is a

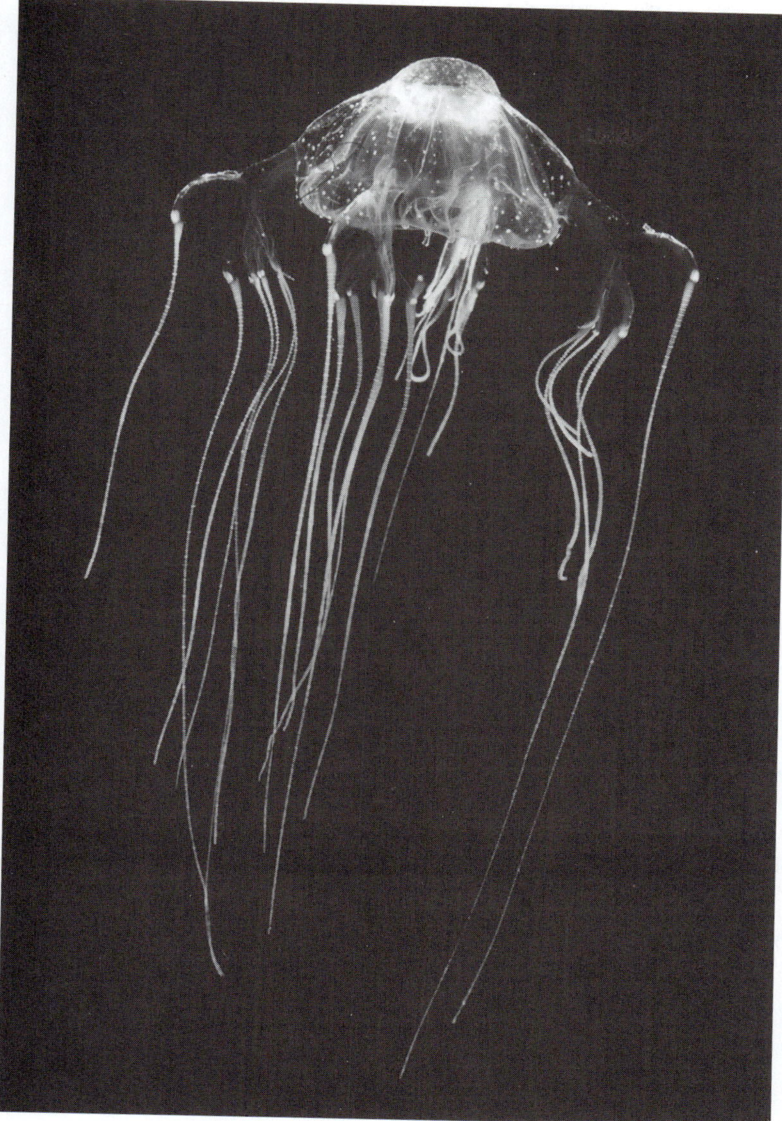

A

B

From Dorit, R. L., Walker, W. F., Jr., and Barnes R. D. 1991. Zoology. Saunders College Publishing, Philadelphia.

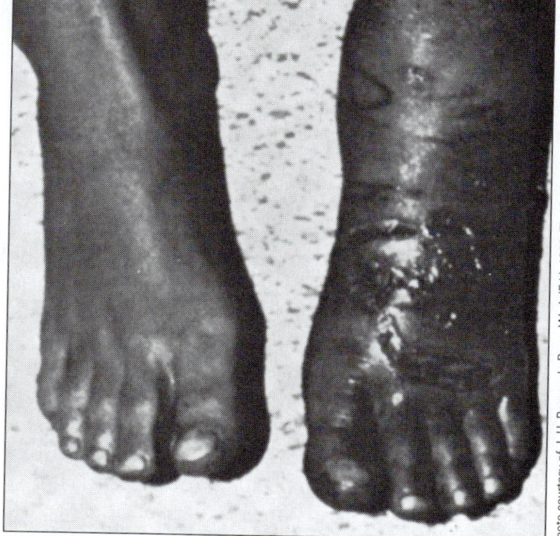

C

Photo courtesy of J. H. Barnes. In Rees, W. J. (Ed.): 1966. The Cnidaria and Their Evolution. Zoological Society of London.

FIGURE 7-51 Scyphozoan diversity: Cubomedusae. **A,** Photograph of a living box jelly, *Chiropsalmus quadrumanus.* Note long tentacles borne on fleshy pedalia. **B,** Warning to swimmers on a beach in Queensland, Australia. "Stingers" are the highly venomous sea wasp, *Chironex fleckeri.* **C,** The result of a minor sting from the sea wasp, *Chironex fleckeri.* The skin lesions are five days old.

pedal disc; expanded oral end broad, up to 8.5 cm diameter, with weakly developed coronal muscle, a medusoid trait, but which probably functions to flex tentacle-bearing lobes (lappets) toward mouth during feeding and does not produce swimming pulsations; attaches by pedal disc to sea grasses, algae, or rocks in temperate coastal areas; species associated with algae or plants typically greenish brown; *Haliclystus, Lucernaria, Thaumatoscyphus* are common taxa. A four-sided mouth and manubrium in center of oral disc, margin of which is drawn out into eight armlike lappets, each with a cluster of short, stiffly radiating, knobbed, capitate tentacles (Fig. 7-55B); terminal knob of each slender tentacle bears nematocyst battery; using lappets and tentacles, *Haliclystus* feeds on amphipod crustaceans; between the lappet bases on oral-disc margin are eight suckerlike **adhesive organs** (Fig. 7-55B) used for temporary attachment and somersaulting locomotion; adhesive organs (rhopalioids) may be homologous with rhopalia as both are modified tentacle bases (bulbs); internally, four septa extend between oral and

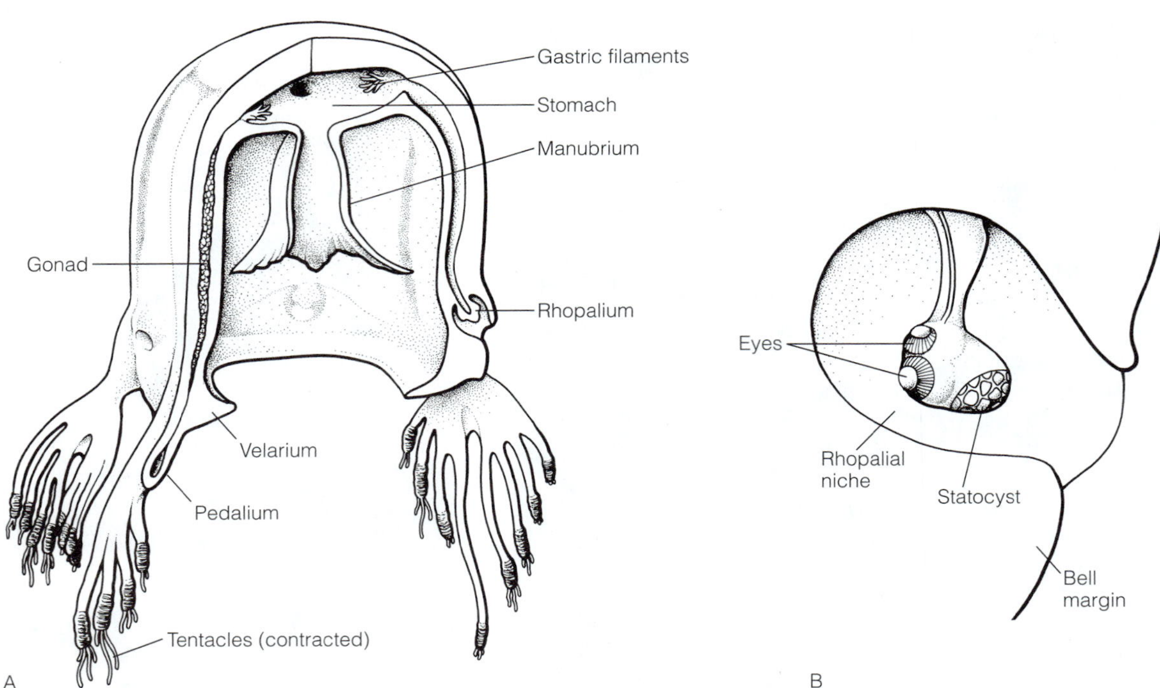

FIGURE 7-52 Scyphozoan diversity: Cubomedusae. Anatomy of a box jelly. **A,** Cutaway view of adult medusa. **B,** Close-up of a rhopalium in its niche. One eye looks inward, the other upward. *(A, Modified and redrawn after Conant from Hyman, L. H. 1940. The Invertebrates. Vol 1. Protozoa through Ctenophora. McGraw-Hill Book Co., New York, 726 pp.)*

pedal discs and partition coelenteron into four gastric pockets (Fig. 7-46B); septum bears septal funnel, longitudinal muscle, gastric filaments, and two gonadal bands, one on each side of septum; the eight gonadal bands restricted to oral (medusa) end of body; gonochoric adults spawn gametes; each zygote develops into a lecithotrophic, nonswimming planula that creeps on substratum; some planulae bud additional larvae; planula attaches to surface and metamorphoses into tiny polyp with four knobbed tentacles and slightly expanded oral disc; further growth and differentiation establishes adult form.

Phylogeny of Scyphozoa

The scyphozoan phylogeny shown in Figure 7-56 is consistent with most traditional and modern phylogenetic trees, except for the inclusion of Cubomedusae in Scyphozoa. Most contemporary accounts consider Cubomedusae (as Cubozoa) to be the sister taxon of Scyphozoa (or Hydrozoa) and cite the cubic body form and mode of medusa formation as unique to them. Cubomedusae are reported to arise by a metamorphosis of the entire polyp and not by strobilation. But most, if not all, cubomedusan species form medusae by transformation of the oral end of the polyp only and a remnant of the pedal end remains. This mode of medusa formation seems to be a variation on monodisc strobilation, which is an autapomorphy of Scyphozoa. Aspects of cubomedusan anatomy, however, remain open to various interpretations in the absence of new research facts.

PHYLOGENETIC HIERARCHY OF SCYPHOZOA

- Scyphozoa
  - Cubomedusae
  - N.N.
    - Coronatae
    - N.N.
      - Stauromedusae
      - N.N.
        - Semaeostomeae
        - Rhizostomeae

HYDROZOA[C]

With some 3000 species, the diverse Hydrozoa is made up of chiefly colonial medusozoans in which the life cycle may include polyps, medusae, or both. Hydrozoans are the only cnidarians to form colonies combining both polyp-zooids and medusa-zooids. Colonial species include the seaweedlike hydroids, the pelagic Portuguese man-of-war, and reef-building fire and rose corals. Among the exclusively solitary species are a few marine jellies and the freshwater *Hydra*. Hydrozoa is the only taxon of Cnidaria with freshwater species, as Anthozoa and Scyphozoa are entirely marine.

In contrast to those in other cnidarians, hydrozoan nematocysts are restricted to epidermal structures. The functional types of nematocysts are, however, diverse: Of the approximately 30 known forms of nematocysts, hydrozoans have 23. Some of these, called glutinants, evert sticky threads and are

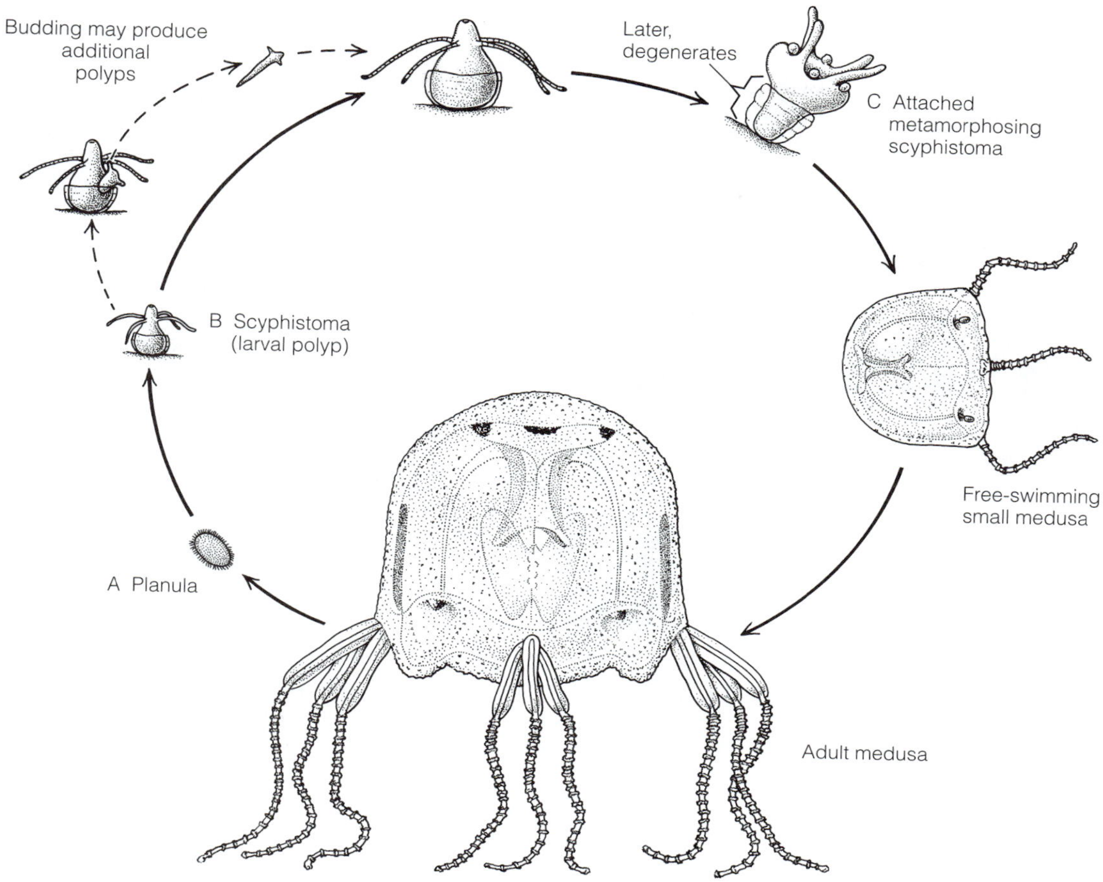

FIGURE 7-53 Scyphozoan diversity: Cubomedusae. Life cycle of the box jelly *Tripedalia cystophora*. The planula **(A)** metamorphoses into a solitary polyp **(B)**. Later, the oral end of the polyp transforms into a medusa **(C)** and the pedal remnant of the polyp and its periderm degenerate (not shown). The scyphistoma also can bud additional polyps. *(From Werner, B. 1973. New investigations on systematics and evolution of the class Scyphozoa and the phylum Cnidaria Proc. 2nd Internat. Symp. Cnidaria, Publ. Seto Mar. Biol. Lab. 20:35–61.)*

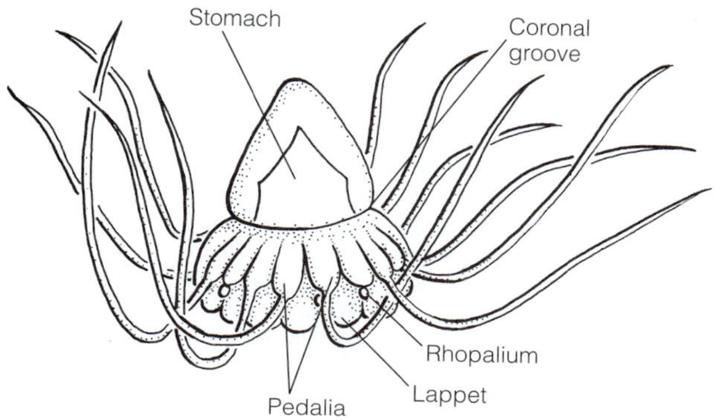

FIGURE 7-54 Scyphozoan diversity: Coronatae. The coronate medusa, *Periphylla periphylla*. *(Modified and redrawn after Hyman, L. H.1940. The Invertebrates, vol.1 Protozoa through Ctenophora. McGraw-Hill Book Co., New York. 726 pp.; and Wrobel D., and Mills, C. E. 1998. Pacific Coast Pelagic Invertebrates: A Guide to the Common Gelatinous Animals. Sea Challengers, Monterey, CA. 108 pp.)*

functionally similar to anthozoan spirocysts, but most, known as penetrants, are venomous to a greater or lesser extent.

Although hydrozoans in general have both polyps and medusae, one stage or the other, usually the medusa, is reduced or absent in many species. Obviously, when the life cycle of a species includes either a polyp only or a medusa only, then that form becomes the sexual stage. When a polyp and a free-swimming medusa occur in the life cycle of one individual, the medusa is sexual and the polyp asexual. In many species, however, the polyp appears to be the sexual stage, because the medusa, which originates as a bud on the polyp, is not always liberated, but remains attached to the polyp. While attached, this medusa, or a specialized variant of it, effectively becomes the polyp's gonad. Sometimes these attached medusae resemble free-swimming jellies, but in other species the medusa traits are detectable but reduced, and *Hydra* lacks any trace of a medusa, except its gonads. The

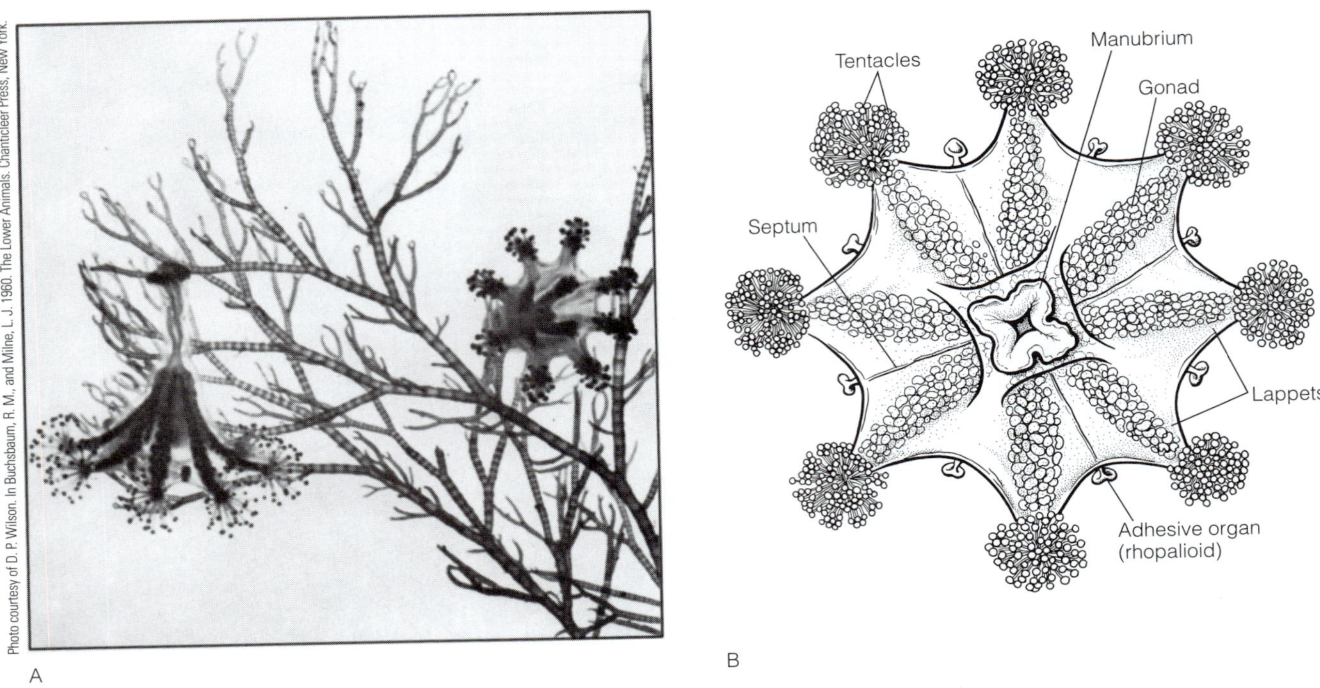

FIGURE 7-55 Scyphozoan diversity: Stauromedusae. **A,** Left: the stalked jelly *Haliclystus auricula* attached to an alga by its pedal disc; right: a second species, *Craterolophus convolulus.* Each animal is approximately 2 cm in diameter. **B,** Oral view of a stauromedusan. *(B, After Wilson from Buchsbaum and Milne, 1960.)*

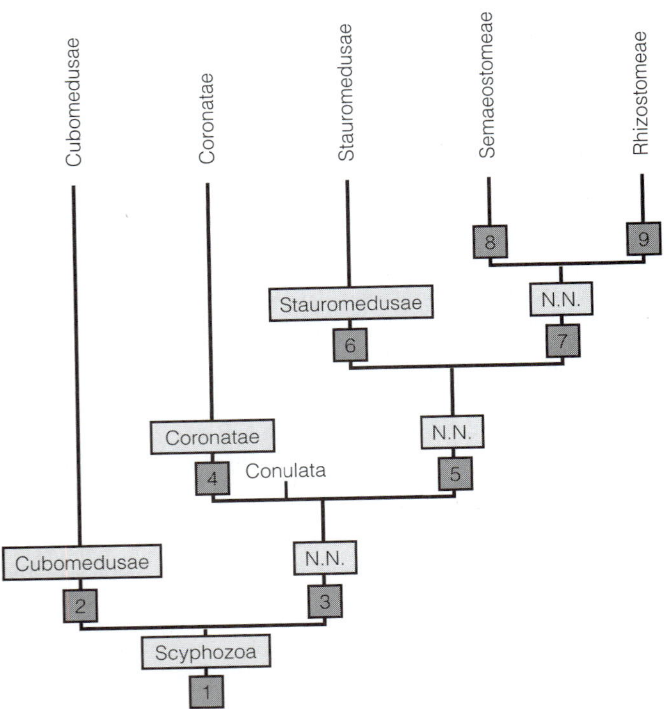

FIGURE 7-56 Phylogeny of Scyphozoa. **1, Scyphozoa:** funnel-shaped polyp, oral end of polyp transforms into medusa (monodisc strobilation), slender basal end (stalk) with periderm, medusa with four septa and septal funnels, gastric (septal) filaments, eight gonadal bands on septa, four gastric pockets, rhopalia. **2, Cubomedusae:** small polyps lacking septa, cubic medusa, tentacles on pedalia at four corners of bell, four rhopalia, complex rhopalial ocelli, velarium, highly venomous nematocysts. **3, N.N.:** polyps with four septa, medusa with octoradial lappets and rhopalia. **4, Coronatae:** polyps trumpet-shaped without septal funnels, all but oral disc and tentacles enclosed by periderm, medusa with coronal groove (some 200 species of Conulata, extinct since the Triassic, had a peridermal tube similar to coronate polyps and may have been scyphozoans, Fig. 7-44A).
5, N.N.: polyps with septa and septal funnels.
6, Stauromedusae: rhopalia are specialized marginal adhesive organs, adult polyp and medusa (sexually mature strobila).
7, N.N.: septa of polyps do not reach pedal disc, septa absent in medusae and stomach pouches (not gastric pockets) occupy original site of septa, four gonads in walls of pouches, septal funnels transformed into subgenital funnels, coelenteric network in mesoglea, branched manubrium, tentacles arise directly from margin of bell.
8, Semaeostomeae: four oral arms. **9, Rhizostomeae:** branched oral arms more or less fused with secondary mouths, marginal tentacles absent.

nearly limitless variations on this theme among hydrozoans are evident in diverse life cycles that are challenging to master for beginners and experts alike.

Hydrozoan polyps, medusae, and colonies seem to fall into one of two contrasting designs, perhaps reflecting a deep evolutionary division within the taxon. Thus, medusae are either tall and bullet-shaped or squat and saucerlike; polyps are covered either partly or completely by periderm. Although these contrasting designs occur throughout Hydrozoa, they are best illustrated in reference to two hydrozoan taxa called Anthoathecatae and Leptothecatae. In the following overview of hydrozoan anatomy, these differing designs will be designated as A-form and L-form and summarized in Table 7-1. Later, in the diversity and phylogeny sections, this fundamental division will be described in an evolutionary context.

Polyps

As a general rule, hydrozoans have tiny polyp zooids, often not more than 1 mm in height and a fraction of a millimeter in diameter. Because of their small size, hydropolyps have a large surface area and a small volume, especially when compared with anthozoan polyps. Perhaps for this reason, hydropolyps lack gastrodermal septa and the nematocyte-bearing septal (or gastric) filaments that accompany them in anthozoans and scyphozoans (Fig. 7-76G,H).

The solitary *Hydra* is a conventional polyp with a mouth, manubrium, tentacles, column, and pedal disc (Fig. 7-57). In contrast, most other hydropolyps are members of a colony and have a modified form that resembles a flower and its stem. The slender column is usually called a **pedicel** (or stem or stalk), while the "flower" is known as the **hydranth** (= water flower) (Fig. 7-60). Each hydranth has a well-developed manubrium (or hypostome), which may be fingerlike, conical, bulbous, or low and rounded (Fig. 7-58). The swollen manubrium of many species, or a short region

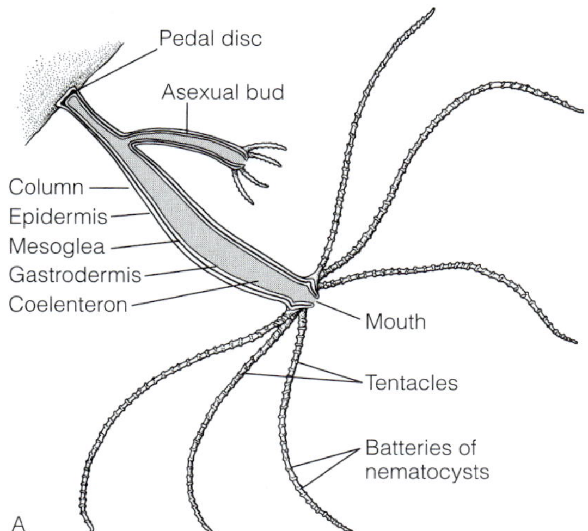

FIGURE 7-57 Hydrozoa: polyp anatomy *(Hydra)*. *(Redrawn from Hyman, L. H. 1940. The Invertebrates: Protozoa through Ctenophora. Vol. 1. McGraw-Hill Book Co., New York.)*

immediately below it, houses the stomach (Fig. 7-58C-E). Surrounding the base of the manubrium is a whorl of solid tentacles (hollow in *Hydra*) (Fig. 7-58D,E); sometimes a second (distal) whorl encircles the mouth rim at the tip of the manubrium (Fig. 7-58A), and in other species, such as *Halocordyle* (also known as *Pennaria*), tentacles may be scattered over the entire surface of the manubrium (Fig. 7-58C). Tentacles may be either **capitate,** usually short and ending in a knob (Fig. 7-58B,C), or **filiform,** long and threadlike (Fig. 7-58D). The mouth, manubrium, tentacles, and stomach together constitute the hydranth.

Medusae

Hydrozoan medusae arise as lateral buds from the hydranth or from other parts of the colony. When on hydranths, they never originate by transformation and transverse fission of the oral end of the polyp (except perhaps in the Trachylina), as they do in strobilating scyphopolyps. Even as full-grown adults, most hydromedusae are small, in the size range of 1 mm to 2 or 3 cm, although a few species, such as *Aequorea aequorea,* achieve a bell diameter of 20 cm and *Rhacostoma* up to 50 cm.

A-form bells are deeply arched and tulip-shaped (Antho- = flower) whereas the bodies of L-form medusae are shallow and saucerlike (Lepto- = thin). Below the bell, on the subumbrellar margin, is an iris diaphragm called the **velum** that creates a powerful water jet as the jelly swims (Fig. 7-59). It is functionally identical to the cubomedusan velarium, but differs from it anatomically. Both are shelflike outfolds of epidermis and mesoglea, but the gastrodermal canals in the cubomedusan velarium are absent in the hydrozoan velum. A cross-striated ring muscle occurs in the velum and on the subumbrellar surface (coronal muscle). The nerve-ring neurons that innervate these muscles transmit action potentials across gap junctions rather than chemical synapses. Gap junctions avoid the time delays associated with signal transmission across synapses and result in a simultaneous contraction of fibers in the coronal and velar muscles (see Chapter 6).

A tubular or four-square manubrium, long to short, hangs below the center of the bell (Fig. 7-59). From the mouth, the manubrial canal dilates slightly to form a stomach from which four equidistant radial canals extend to the margin of the bell (Fig. 7-59). In some species, more than four radial canals are present, and the large-bodied *Aequorea* has many. The radial canals join a marginal ring canal (Fig. 7-59C), which generally supplies each of the hollow tentacles. The circulatory flows in the canals may be ciliary or muscular in origin (Fig. 7-12A,B).

Hydromedusae usually have four tentacles, one at the end of each radial canal, but the tentacle number ranges from zero to many, depending on species. Statocysts or ocelli, but usually not both in one species, occur on the margin of the bell. Statocysts are usually found between the bases of the tentacles of L-form medusae (Fig. 7-6C); ocelli occur in the tentacle bases of A-form bells. In contrast to Anthozoa and Scyphozoa, the mesoglea of hydromedusae (and hydropolyps) is acellular and free of ameboid and other cells.

Hydrozoan germ cells originate from endodermal I-cells. The germ cells may then remain in the gastrodermis or migrate into the epidermis and aggregate to form gonads. Thus hydrozoans have either gastrodermal or epidermal gonads, depending on

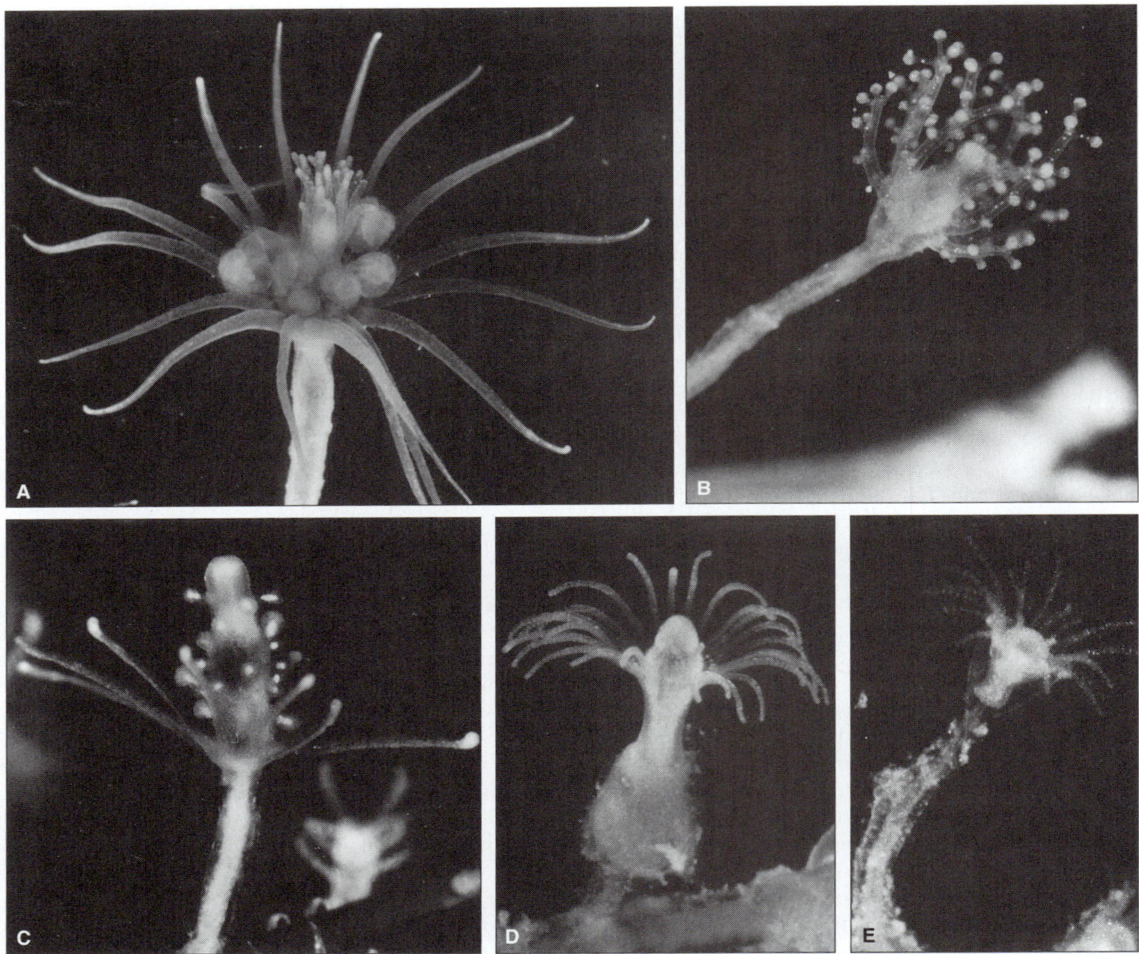

FIGURE 7-58 Hydrozoa: living polyps (hydroids). **A–C,** Athecate (A-form) hydranths. **A,** *Tubularia crocea.* **B,** Unidentified athecate hydranth with capitate tentacles. **C,** *Halocordyle (Pennaria) disticha.* **D–E,** Thecate (L-form) hydranths. **D,** *Thyroscyphus ramosus* (the theca is the dark, fuzzy cup). **E,** *Obelia dichotoma* (transparent wineglass-shaped theca).

species. As a general rule, the gonads of A-form medusae are found on the manubrium while those of L-forms occur on the underside of the radial canals (Fig. 7-59A,C). Regardless of gonadal position, gametes are spawned directly to the exterior. In species with gastrodermal gonads, the developing gametes bulge into the epidermis, eventually rupturing it.

Colonies

Most species of hydrozoans with a polyp stage in the life cycle are colonial, and these colonies often include both polyp-zooids and medusa-zooids. Sessile benthic colonies are called **hydroids,** and most pelagic colonies are siphonophores. Hydroids resemble seaweeds and grow rapidly and profusely on rocks, pilings, seawalls, docks, shells, and boat hulls, thus constituting an important component of the so-called fouling community.

Hydrozoan colonies may be stolonate, coenosarcal, or fruticose (Fig. 7-3). In stolonate colonies, each zooid arises directly from a stolon (hydrorhiza), which is often part of a stolonal network (Fig. 7-60A). Some stolonate taxa, such as fire corals (*Millepora* spp.), secrete a massive, calcareous exoskeleton (corallum) around their stolonal networks and retracted zooids (Fig. 7-66) and contribute to the growth of coral reefs. The zooids of coenosarcal colonies, such as *Hydractinia* (Fig. 7-61), are borne singly on the surface of the coenosarc and joined inside by solenia. In *Hydractinia,* the coenosarc originates developmentally from coalesced stolons of a stolonal network (Fig. 7-61C). Fruticose colonies are composed of single or branched **stems** (hydrocauli), each bearing multiple zooids (Fig. 7-62, 7-63). In the benthic hydroids, the stems are borne on prostrate stolons attached to the substratum whereas in the pelagic siphonophores, the stem constitutes the major (locomotory) body axis. Stem composition, however, differs in A-form and

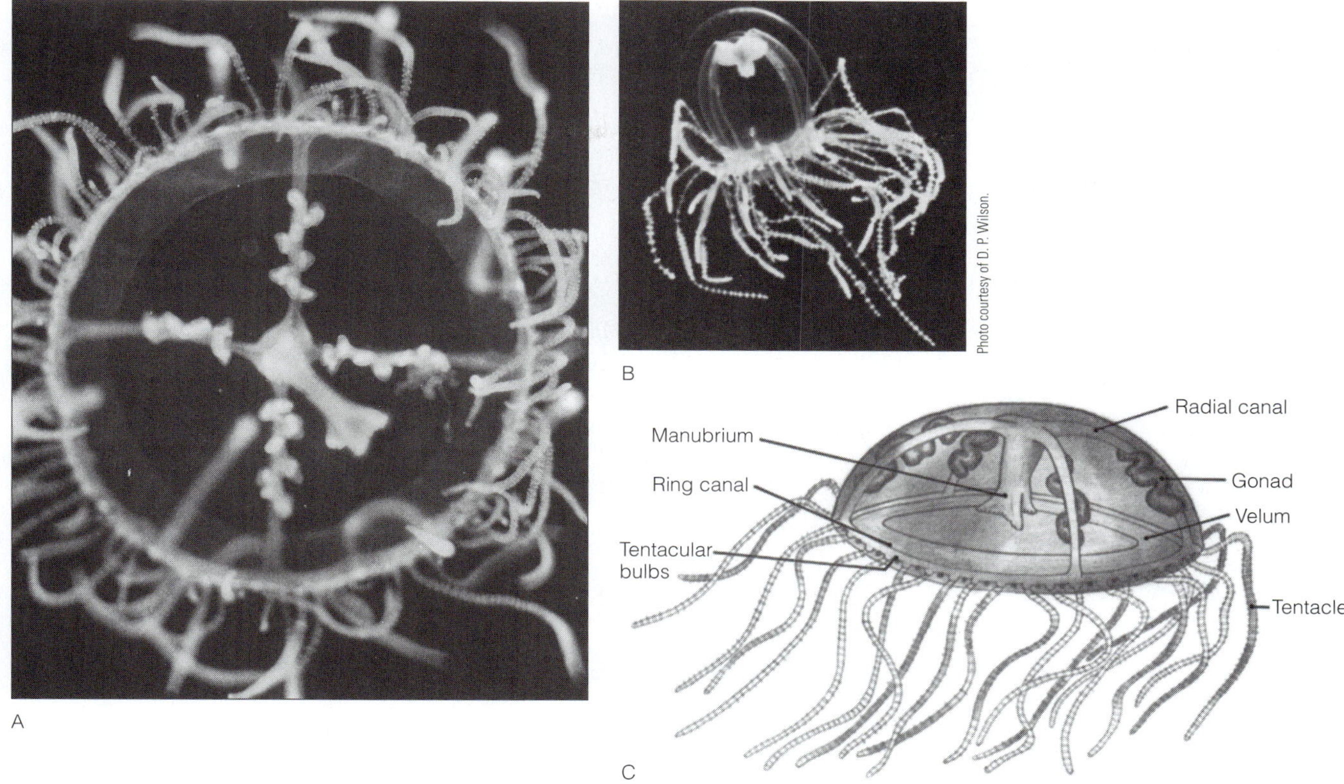

FIGURE 7-59 Hydrozoa: medusa anatomy. **A,** Oral view of *Vallentinia gabriella* (Limnomedusae) showing four-square manubrium, numerous marginal tentacles, four gonads, one on each radial canal. **B,** Photograph of *Gonionemus vertens* (Limnomedusae). **C,** Anatomy of *Gonionemus.* *(C, Redrawn from Meyer.)*

L-form hydrozoans. The stem of an A-form colony, such as the hydroids *Halocordyle* or *Eudendrium* or siphonophores, is an axial polyp that elongates more or less indefinitely and, with growth, buds a succession of new zooids laterally (Fig. 7-3D, 7-62, 7-69). The stem of an L-form colony, which grows by fixed-length budding, is composed of a series of segmentlike sections, each a component zooid (Fig. 7-3C, 7-60C, 7-63).

Periderm may either embrace the hydranth up to the tentacle bases to form the wineglass-shaped **theca** characteristic of L-forms (hydrotheca; Fig. 7-58E, 7-60, 7-63) or end at the point of attachment of the hydranth to its pedicel, as is typical of A-forms (Fig. 7-58A–D, 7-62). Hydroids with a theca are said to be **thecate** (L-forms); those without, **athecate** (A-forms). Thecate hydroids can withdraw their hydranths into the theca for protection (Fig. 7-63). This withdrawal involves a longitudinal shortening of the hydranth body and withdrawal of tentacles over the oral end of the polyp. In some thecate hydroids, such as species of *Sertularia,* the theca has a hinged lid (operculum) that seals the thecal opening as the hydranth retracts within.

Hydrozoan colonies may be monomorphic or polymorphic. A **monomorphic colony,** such as *Halocordyle,* consists only of **gastrozooids,** the typical feeding hydranths described earlier (Fig. 7-58C). Monomorphic colonies release free-swimming medusae that originate from **medusa buds** (Fig. 7-58C, 7-62). **Polymorphic colonies** have gastrozooids and at least one other form of permanently attached zooid, either a modified hydranth, a modified medusa, or both. A widespread medusa zooid is the **gonophore** (Fig. 7-58A, 7-62). Gonophores are incompletely differentiated medusae that produce gonads, but lack some or most other medusan traits. Although medusa buds and gonophores both originate as buds, sometimes on gastrozooids or other parts of the colony, they are often borne only on specialized gastrozooids called **gonozooids.** Gonozooids differ from typical gastrozooids in having more or less reduced tentacles, mouth, and manubrium (Fig. 7-61B, 7-62, 7-63). When all of these are absent, the fingerlike gonozooid with its lateral medusa buds is called a **blastostyle** (Fig. 7-62, 7-63). In L-forms *(Obelia, Orthopyxis),* the blastostyle is enclosed in a theca **(gonotheca)** and the entire structure is known as a **gonangium** (Fig. 7-60, 7-63, 7-64A). Highly polymorphic colonies not only have gastrozooids and gonophore-bearing gonozooids, but also protective or food-catching polyps called **dactylozooids** that lack mouth and tentacles, but have batteries of nematocytes and may be strongly retractile. If used in feeding, dactylozooids seize prey with discharged nematocysts, then transfer it to the mouth of a gastrozooid. The dactylozooids of A-forms such as *Hydractinia* are long and slender (Fig. 7-61B) whereas those of L-forms are shorter and are called **nematophores** (Fig. 7-60C). Other specialized zooids, especially those of siphonophores, will be discussed under Diversity of Hydrozoa.

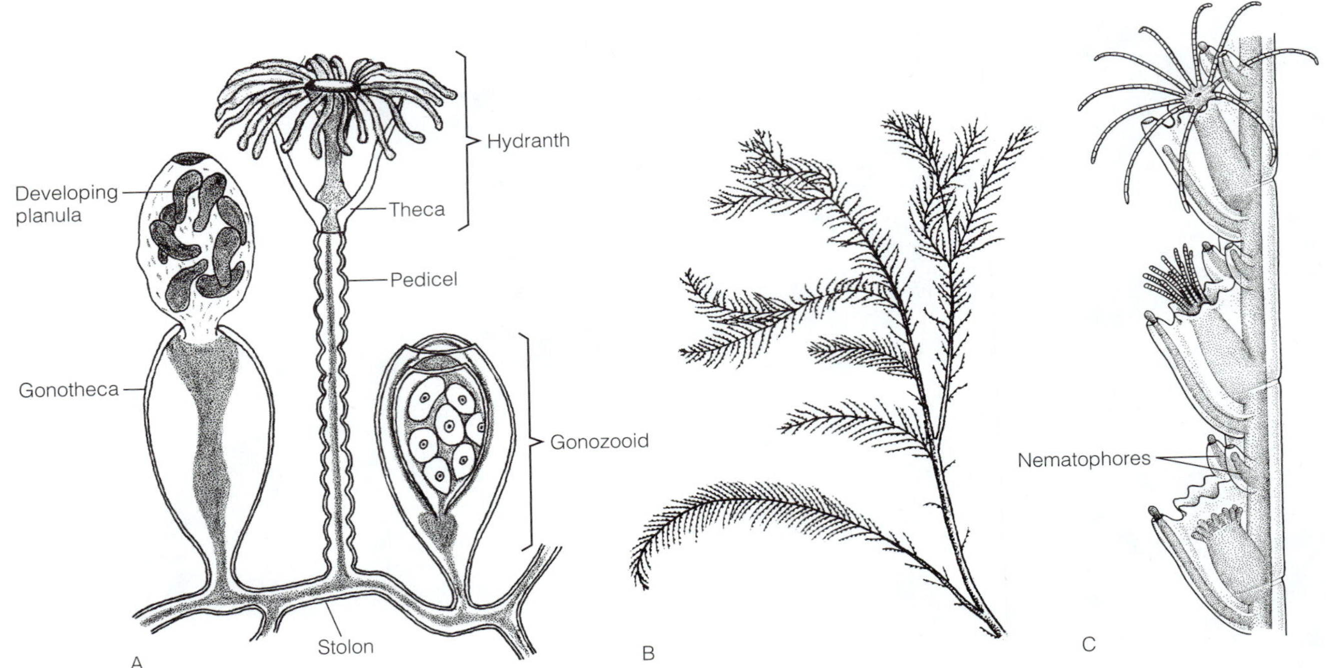

FIGURE 7-60 Hydrozoa: L-form colonies of hydroids. **A,** Stolonate species of *Orthopyxis* (Leptothecatae). **B,** Fruticose colony of a species of *Aglaophenia,* which grows by fixed-length budding (Leptothecatae). **C,** *Aglaophenia*: close-up of three gastrozooids each associated with a pair of nematophores. *(A, Redrawn from Hyman, L. H. 1940. The Invertebrates. Protozoa through Ctenophora. Vol. 1. McGraw-Hill Book Co., New York.)*

Obelia is an example of a hydroid that produces free medusae from medusa buds (Fig. 7-64A). Once released, the medusae grow, attain sexual maturity, and spawn their gametes. The planktonic zygotes develop into planula larvae, settle, and metamorphose into primary polyps that bud to form the colony. Species of hydroids with gonophores (sessile medusae) may free-spawn gametes, but often the eggs are retained and fertilized internally. After a period of embryonic development, a planula is released (as in *Eudendrium*) that, after a brief planktonic existence, settles and metamorphoses into a polyp. In other cases, for example in *Tubularia,* even the planula is sessile and develops into an **actinula** before escaping from the gonophore (Fig. 7-64B). The actinula, which may be creeping or planktonic, is short along the oral-aboral axis and has a whorl of four or more slender tentacles as well as a manubrium and a mouth. The squat body is a medusa trait, but in other respects the actinula is similar to a polyp. In fact, the dualistic actinula can differentiate directly into either polyp or medusa, depending on species. In the hydroid *Tubularia,* the benthic actinula transforms into a polyp (Fig. 7-64B), but in *Aglaura hemistoma* and *Liriope tetraphylla* (Trachylina), species that lack a polyp in their life cycles, the planktonic actinula transforms directly into a juvenile medusa (Fig. 7-65C).

Clonal (asexual) reproduction, by budding, is common among solitary A-form polyps such as *Hydra* (Fig. 7-57, 7-65B) and several free-swimming A-form medusae (Fig. 7-13C). The L-form jelly *Aequorea* can reproduce asexually by fission.

Diversity of Hydrozoa

We divide Hydrozoa into five major taxa, but several other schemes are commonly used in traditional classifications. The A-form taxa (Table 7-1) are represented by the Anthoathecatae and Siphonophora, and the L-form taxa by Leptothecatae, Limnomedusae, and Trachylina.

The general discussion of Hydrozoa in the preceding sections was based largely on characteristics of two taxa of hydroids, Anthoathecatae and Leptothecatae. Their diversity of approximately 2600 species approaches that of the stony corals of tropical seas. The 150 species of colonial siphonophores (Siphonophora) are important members of the pelagic realm, along with solitary hydromedusae (with about 825 species of Anthoathecatae, Leptothecatae, and Trachylina). A few hydrozoans, such as fire and rose corals, contribute to coral reefs, but that environment is dominated by colonial anthozoans.

A-FORMS

Athecate polyps, budding from axial polyps, tall medusae with ocelli and manubrial gonads.

Anthoathecatae[O] (Anthomedusae, Athecata): Mostly colonial A-form hydroids (Table 7-1; Fig. 7-62); divided into two groups: Capitata and Filifera, based on structure of tentacles.

Capitata[sO]: Polyp tentacles **capitate,** or terminate in knoblike swellings, with high nematocyte concentration;

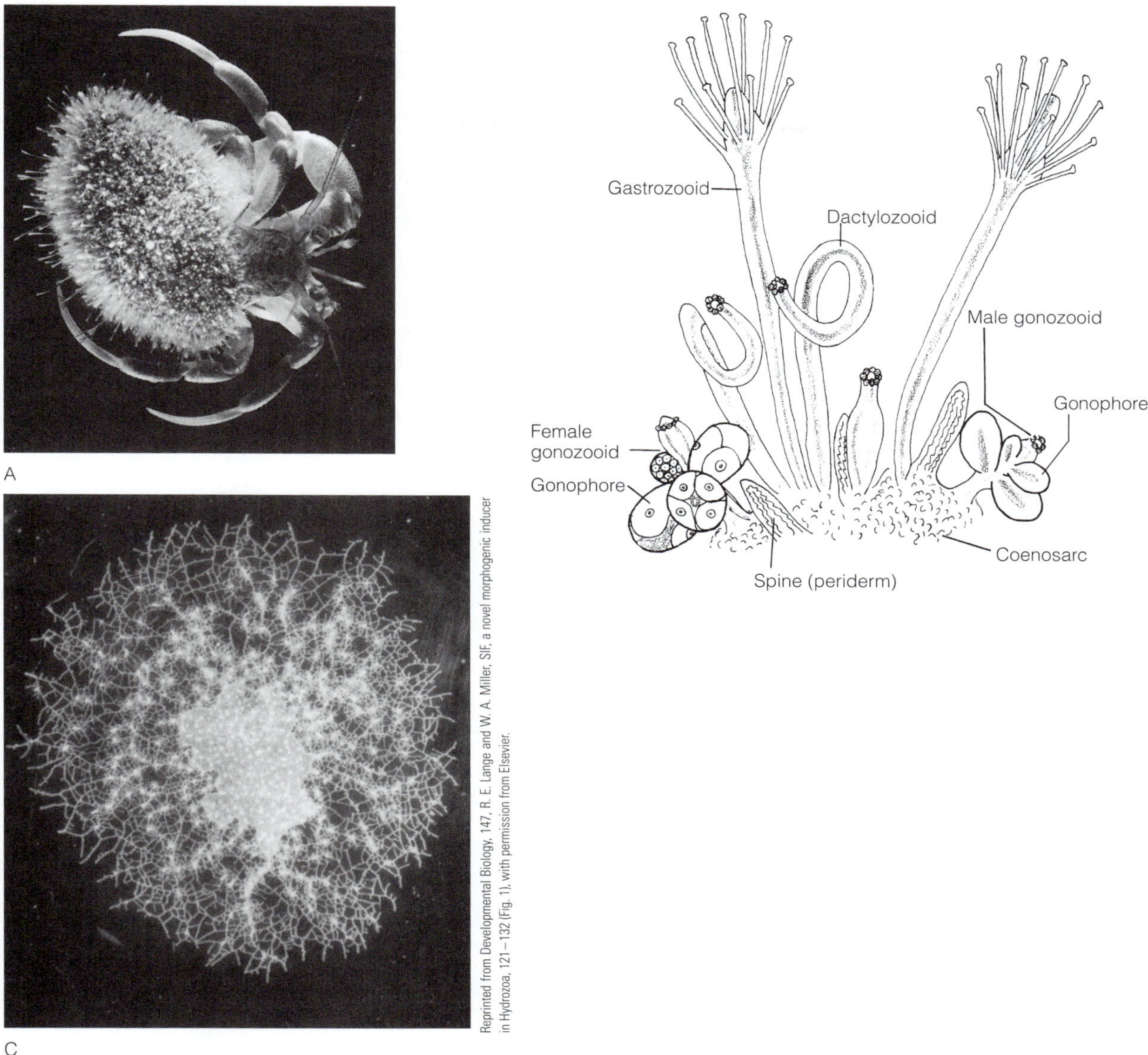

FIGURE 7-61 Hydrozoa: colonies of *Hydractinia* (A-form). Zooids arise singly from a coenosarc. **A,** Living colony of *H. symbiolongicarpus* on the shell of the hermit crab, *Pagurus longicarpus*. **B,** Polymorphic zooids of *Hydractinia*. **C,** Overview of a young colony of *H. echinata*. In the central region, the stolonal network is fusing into a coenosarc. As the coenosarc expands, the gastrodermal tubes of the stolons are incorporated into the coenosarc as solenia. *(B, From Hyman, L. H. 1940. The Invertebrates Protozoa through Ctenophora. Vol. 1. McGraw-Hill Book Co., New York.)*

some species with capitate tentacles only in juveniles, later replaced by filiform tentacles. *Halocordyle* (Fig. 7-58C): feather-like; *Tubularia* (Fig. 7-58A), *Ectopleura*, and *Corymorpha:* large axial polyps borne on long pedicels, pedicels of *Tubularia* and *Ectopleura* may attain 20 cm length and attach to hard substrates, but *Corymorpha* and similar deep-sea *Branchiocerianthus* anchor in soft sediments with rootlike outgrowths; pedicels of *Branchiocerianthus imperator* (5000 m depth off Japanese coast) reach 2.5 m length, 5 cm or more diameter; thick pedicels of such large polyps have either gastrodermal septa (*Tubularia;* Fig. 7-76I) or solenia-like gastrodermal tubes (*Corymorpha, Branchiocerianthus;* Fig. 7-4D), both size-related increases in area for uptake and distribution of gases and nutrients; *Tubularia* polyps have gonophore-bearing blastostyles on manubrium that release creeping actinula larvae (Fig. 7-58A, 7-64B); similarly positioned medusa buds of *Ectopleura* liberate medusae.

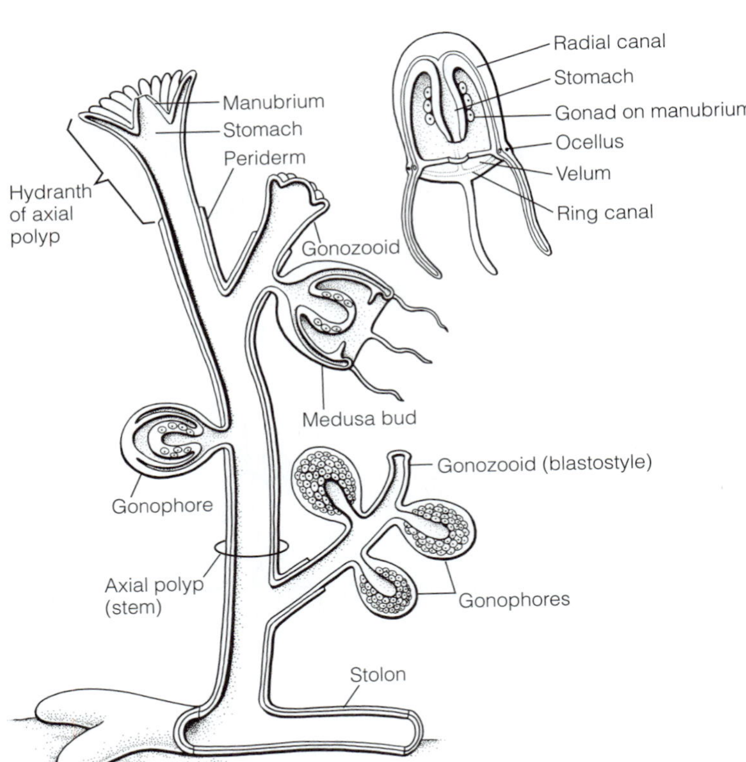

FIGURE 7-62 Hydrozoa: anatomy of athecate hydroids and their medusae (Anthoathecatae and other A-form hydrozoans). *(Modified and redrawn from Delage, Y., and Hérouard, E. 1901. Traité de Zoologie Concrète. Vol. 2. Les Coelentérés. Schleicher Frères, Paris. 848 pp.*

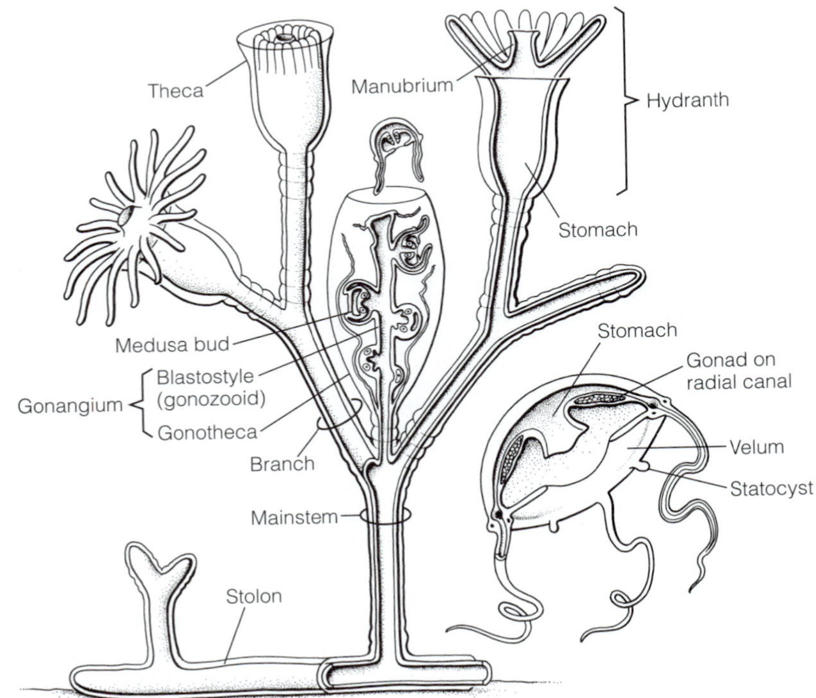

FIGURE 7-63 Hydrozoa: anatomy of thecate hydroids and their medusae (Leptothecatae and other L-form hydrozoans). *(Modified and redrawn from Delage, Y., and Hérouard, E. 1901. Traité de Zoologie Concrète. Vol. 2. Les Coelentérés. Schleicher Frères, Paris. 848 pp.*

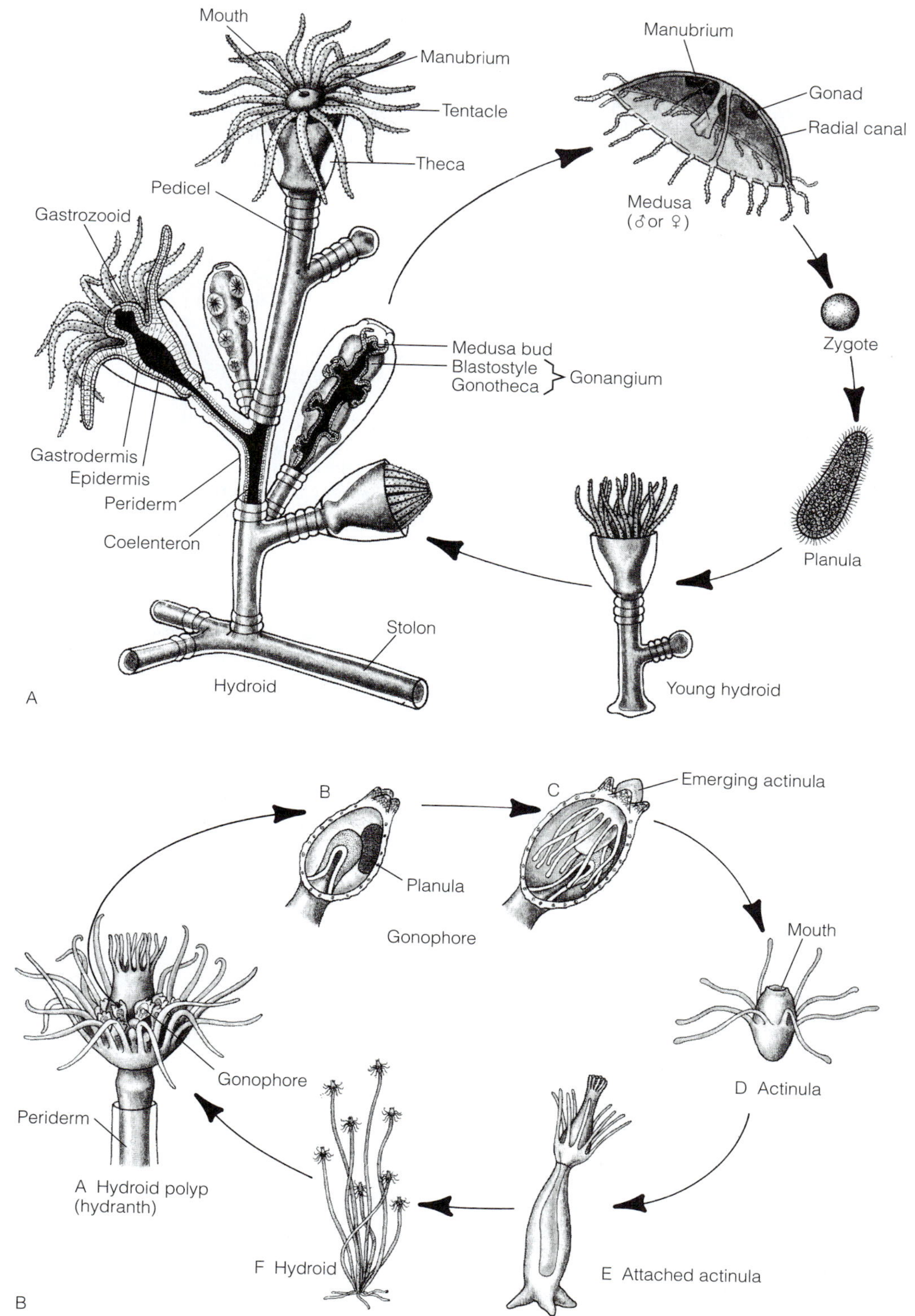

FIGURE 7-64 Hydrozoa: life cycles. **A,** *Obelia* (Leptothecatae): life cycle includes polyp, free-swimming medusa, and planula. **B,** *Tubularia* (Anthoathecatae): zygote through planula stages are retained in gonophore; actinula released, which metamorphoses into a polyp. *(A and B, After Allman from Bayer, F., and Owre, H. B. 1968. The Free-Living Lower Invertebrates. Macmillan, New York. 229 pp.)*

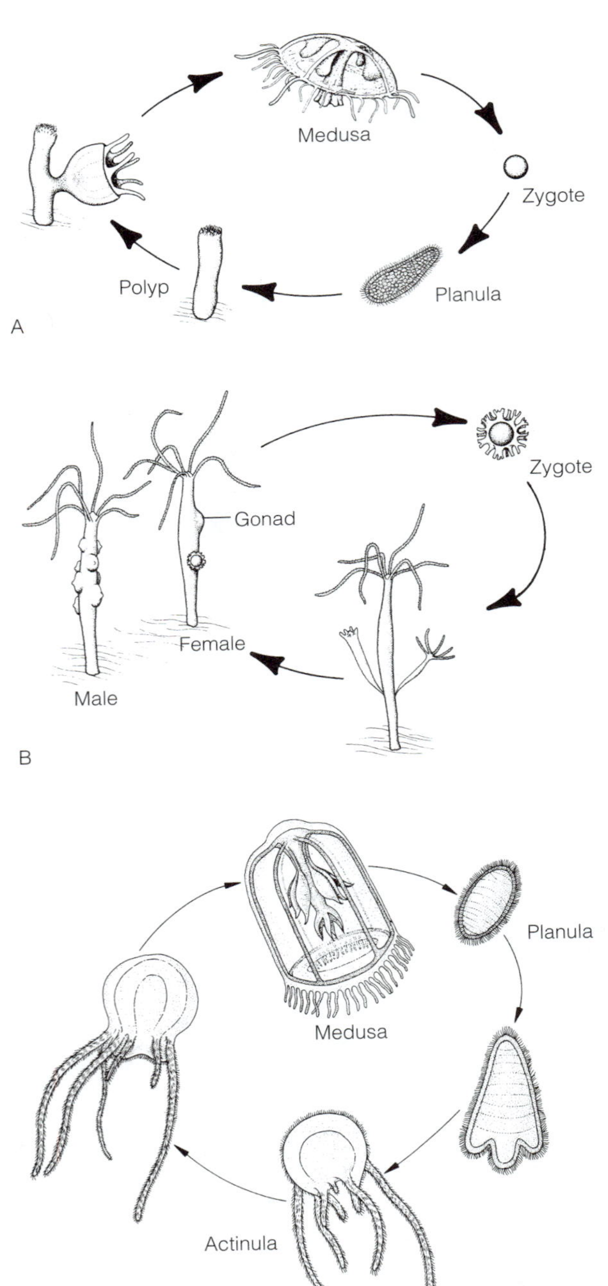

FIGURE 7-65 Hydrozoa: life cycles. **A,** The freshwater *Craspedacusta* (Limnomedusae; life cycle of the marine *Gonionemus* is similar): polyp solitary, lacks periderm, and has short tentacles. **B,** Freshwater species of *Hydra* (Anthoathecatae): medusa and planula stages absent. **C,** The open-ocean *Aglaura* (Trachymedusae): polyp absent; medusa develops directly from actinula. *(From Bayer, F., and Owre, H. B. 1968. The Free-Living Lower Invertebrates. Macmillan, New York. 229 pp.)*

Hydra (freshwater), *Protohydra, Psammohydra* (marine): slender tentacles (Fig. 7-57); lack periderm, medusa, and planula; zygotes develop directly into polyps (Fig. 7-65B); polyps solitary (except while budding), move by creeping or somersaulting (Fig. 7-5B); *Hydra* can secrete and hold gas bubble on pedal disc, float to surface, and drift; some *Hydra* species, such as *H.* (formerly *Chlorohydra*) *viridissima,* harbor zoochlorellae; oddly, no permanent mouth in hydras: in nonfeeding animals, opposing lips are sealed together with septate junctions until the lips are torn apart by muscle contraction; sealing the mouth with impermeable junctions may isolate coelenteron from freshwater environment and allow its more effective use in digestion and, especially, osmoregulation; most *Hydra* species gonochoric, sexual reproduction chiefly in autumn; germ cells aggregate in column epidermis as ovaries or testes (Fig. 7-65B); one I-cell becomes single egg in each ovary, which other I-cells surround and provide nutrition for; as egg enlarges, a rupture in overlying epidermis exposes but does not release egg; testis a conical swelling with nipplelike pore through which sperm escape; spawned sperm penetrate exposed egg surface, fertilizing egg *in situ;* zygote undergoes cleavage, simultaneously is covered by chitinous shell; when shell formation complete, shelled embryo drops off parent and diapauses through winter; with advent of spring, shell softens and young polyp emerges; because each individual may bear several ovaries, many eggs can be produced each season.

Fire corals, *Millepora alcicornis* (tropical Atlantic and eight Indo-Pacific species): zooxanthellate; reef-builders; secrete massive, brittle, calcareous skeleton around stolonal network (apparently not solenia; Fig. 7-66); colony growth form depends chiefly on shape of colonized substratum; in Caribbean, *Millepora* colonizes bare axial rods of dead gorgonians, often overgrowing adjacent living tissue, and adopts gorgonian's fruticose form; trimorphic zooids arise from stolonal network in the skeleton and emerge from tiny cups called pores on surface of corallum (Fig. 7-66A,B,D); large pores (gastropores) occupied by gastrozooids, each a short polyp bearing four capitate tentacles; more numerous small pores (dactylopores) house long, threadlike dactylozooids with scattered short, capitate tentacles; extended dactylozooids form dense surface fringe (Fig. 7-66C); skin contact with dactylozooids results in a severe sting that burns like fire; third type of pore, the ampulla, contains a medusa bud that releases tiny, less than 1 mm, free-swimming medusae; after medusa release, ampullae are filled in by skeleton deposition.

Velella velella (by-the-wind sailor; Fig. 7-67A) and *Porpita porpita* (blue button; Fig. 7-67B): zooxanthellate, intensely blue, raftlike, float at surface of warm seas worldwide; highly modified, upside-down, *Tubularia-* or *Corymorpha*-like gastrozooid (axial polyp) bearing gonozooids (thus a colony); body consists of upper float, middle central mass (coenosarc), and undersurface with zooids suspended mouth down; discoid float of concentric, chitin-lined, air-filled chambers for buoyancy; develops as pedal inpocket of axial-polyp; float opens to air at small pores on upper surface; blind-ending tracheae extend from air chambers into central mass; undersurface with large central tentacle-free gastrozooid, marginal tentacles, and gonozooids bearing medusa buds between the gastrozooid and tentacles; zooids

TABLE 7-1 Characteristics of A-Form and L-Form Hydrozoans

	Polyp	Medusa			Colony	
A-Forms	**Athecate**	**Tall**	**Ocelli**	**Manubrial Gonad**	**Axial-Polyp Budding**	**Dactylozooid**
Anthoathecatae	X	X	X	X	X	X
Siphonophora	X	X	—	X	X	X
	Polyp	**Medusa**			**Colony**	
L-Forms	**Thecate**	**Short, Flat**	**Statocysts**	**Radial-Canal Gonad**	**Fixed-Length Budding**	**Nematophore**
Leptothecatae	X	X	X	X	X	X
Limnomedusae	Reduced polyp	X	X	X	—	—
Trachylina	—	X	X	X (or stomach wall if radial canals absent)	—	—

interjoined by solenia in central mass and periphery of float; medusa buds release thimble-shaped, free-swimming, zooxanthellate medusae with manubrial gonads, two capitate tentacles. *Velella* (Fig. 7-67A): parallelogram body shape, triangular sail at angle to long axis of float; in large oceanic population, sails of one-half of individuals angled right, those of other half angled left; in sustained wind, contrasting sail orientations disperse animals in opposite directions, assuring that, if close to shore, only half the population will be stranded; *Porpita* (Fig. 7-67B): body a perfect circle, lacks sail. The taxonomic separation of these species from Siphonophora is problematic and open to future research.

A few of the capitate hydroids are better known for their (solitary) medusae than their polyps. Among these are *Sarsia (Coryne):* four long tentacles and equally long prehensile manubrium; *Polyorchis*: with many tentacles and long, gonad-bearing, manubrium-like peduncle (see Trachymedusae); *Cladonema:* algae-dwelling jelly with branched capitate tentacles, some modified for adhesion.

Filifera[sO]: Filiform tentacles lacking terminal knobs: *Bougainvillea, Clava, Eudendrium* (Fig. 7-58D), *Turritopsis.* Snail fur, *Hydractinia* spp. (*H. symbiolongicarpus* in United States; Fig. 7-61): common commensals on hermit-crab shells, but grow elsewhere on hard substrates; polymorphic polyps grow singly on coenosarc that covers shell or other surface; solenia interconnect zooids; chitinous periderm, secreted by lower epidermis, provides adhesion to substratum; below coenosarc, extra periderm secretion produces spines, which poke up between polyps (Fig. 7-61); *Hydractinia* feeds with gastrozooids, produces gametes on male and female gonozooids; dactylozooids, called spiral zooids, restricted to colony margin above shell aperture from which hermit crab emerges; spiral zooids reported to snatch hermit-crab eggs from crab's brood sponge and transfer them to gastrozooids.

Proboscidactyla (sometimes classified with Limnomedusae): best known for colonial polyp stage only on circular tube rims of feather-duster worms; simple tubular gastrozooids, each with two outstretched tentacles, resemble worshipping elves; colonies also include dactylozooids and gonozooids that release medusae.

Rose corals, 150 species, occur at all latitudes and depths, most common in tropical Indo-Pacific where zooxanthellate species contribute to coral reefs. Like fire corals (Fig. 7-66B), rose corals secrete limestone exoskeleton around a stolonal network; colonies with gastrozooids, dactylozooids, gonophores; gonophores release planula larvae. Most are species of rose-red *Stylaster:* each skeletal pore resembles a stony-coral corallite with radial sclerosepta around cup margin (Fig. 7-68); "sclerosepta," however, are partitions between adjacent dactylozooids, which encircle central gastrozooid to form a cyclosystem; gonophores and ampullae arise as subsurface buds that bulge from corallum surface but apparently lack openings; sperm and planula larvae apparently released through cyclosystem pores. *Distichopora:* blue or purple colonies.

Siphonophora[O]

Depending on their form, the pelagic siphonophores, with their trailing clusters of transparent zooids, resemble strings of diamonds or crystal chandeliers cruising through inner space. All siphonophores, including the familiar *Muggiaea* and *Physalia* (Portuguese man-of-war), are colonial, polymorphic, A-form hydrozoans. Siphonophores have six kinds of nematocysts and some of these can be large, numerous, and virulent. Of the 150 marine species, the largest exceed 3 m in length, and tentacles can be much longer.

Each siphonophore originates as an inverted axial polyp (Fig. 7-69). In all but one taxon (Calycophorida), the pedal (anterior) end of the axial polyp is inpocketed to form a periderm-lined float, or **pneumatophore,** which is gas-filled and provides buoyancy. The opposite, oral end of the axial polyp bears a mouth and sometimes a single tentacle. The colonial zooids arise as lateral buds from the elongated column (stem or **siphonosome**) of the axial polyp. Buds arise in sets, called **cormidia,** each cormidium consisting usually of a gastrozooid, gonozooid, dactylozooid, and bract (described below), but other combinations can occur. As the colony grows,

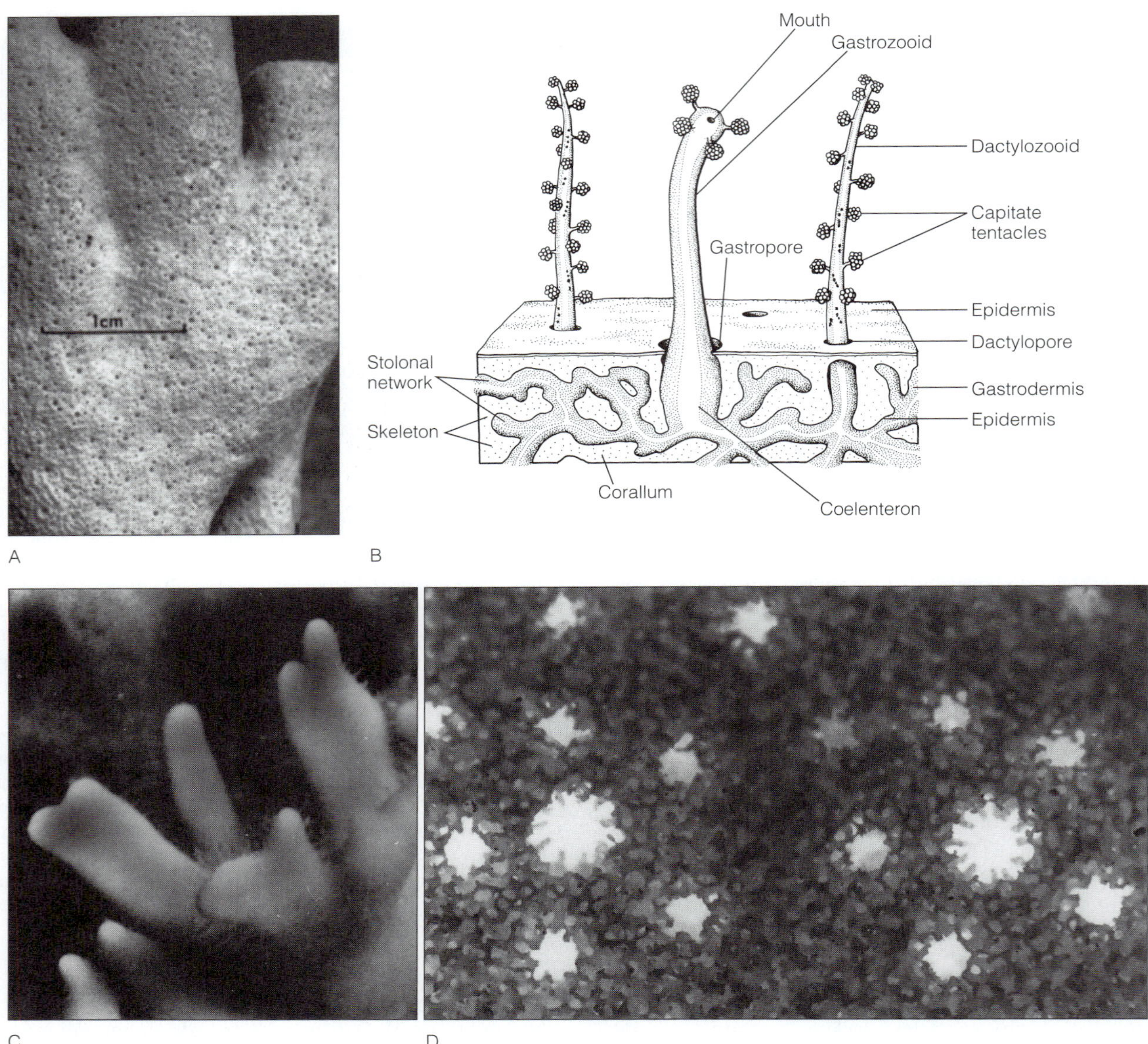

FIGURE 7-66 Hydrozoan diversity: Anthoathecatae, Capitata. The fire coral *Millepora alcicornis.* **A,** Calcareous skeleton of *Millepora* showing large gastropores and smaller dactylopores. **B,** Cutaway view of living *Millepora.* **C,** Photograph of living *Millepora* with extended dactylozooids. **D,** Microscopic view of skeleton showing two gastropore-dactylopore systems. *(B, Modified and redrawn from de Kruijf, H. A. M. 1975. General morphology and behavior of gastrozooids and dactylozooids in two species of Millepora. Mar. Behav. Physio. 3:181–192. © Gordon and Breach Science Publishers, Inc.)*

cormidia are added sequentially, the oldest being situated nearest the oral (posterior) end of the axial polyp (Fig. 7-69).

The polyp-zooids may be gastrozooids, dactylozooids, or gonozooids (Fig. 7-69). The long tubular gastrozooids (siphons) either lack tentacles or bear a single long, branched tentacle packed with nematocysts. Dactylozooids (palpons or tasters) are similar to gastrozooids but lack a mouth, and their single tentacle is unbranched. Gonozooids, often branched, bear either medusa buds or gonophores, depending on taxon. In some species, however, the thimble-shaped gonophores are fully differentiated bells that bear gonads on the manubrium, as in Anthoathecatae.

The medusa-zooids are the swimming bells (nectophores) and gonophores, and sometimes also structures called bracts

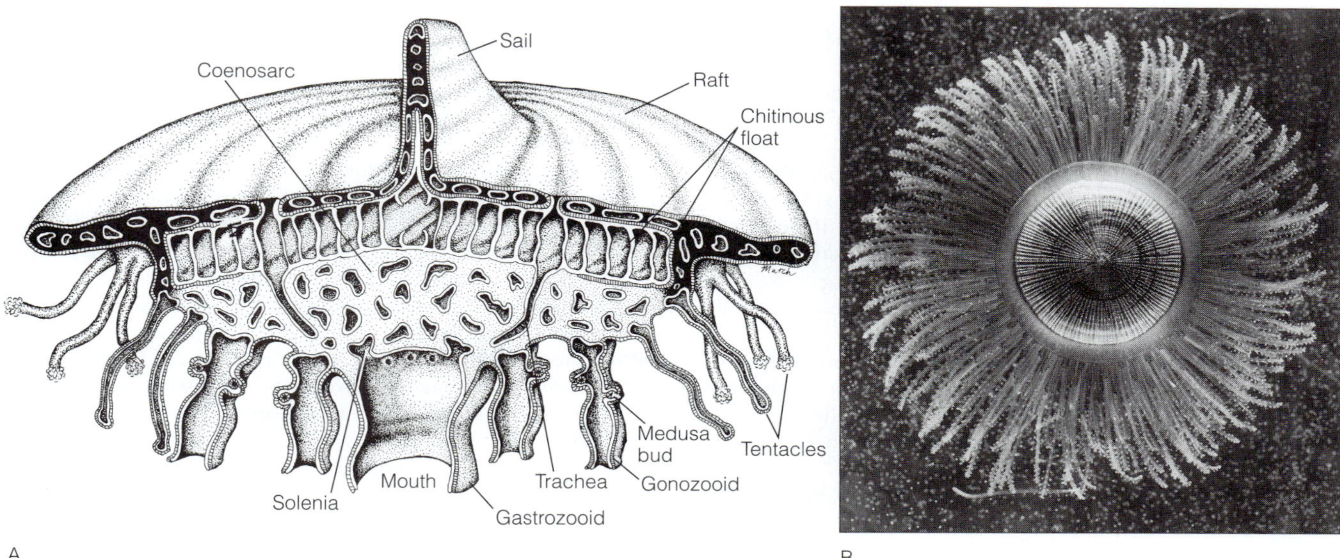

FIGURE 7-67 Hydrozoa diversity: Anthoathecatae, Capitata. *Velella* and *Porpita.* **A,** Cutaway view of the anatomy of the by-the-wind sailor, *Velella velella.* **B,** Aboral view of a living blue button, *Porpita porpita.* The white dots in the water around the animal are newly released medusae. *(A, Modifed and redrawn from Delage, Y., and Hérouard, E. 1901. Traité de Zoologie Concrète. Vol. 2. Les Coelentérés. Schleicher Frères, Paris. 848 pp.)*

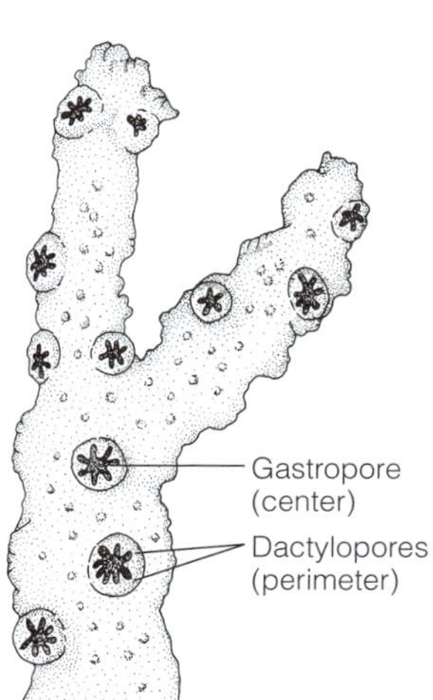

FIGURE 7-68 Hydrozoan diversity: Anthoathecatae, Filifera. The rose coral *Allopora. (After Moseley from Bayer, F. M., and Owre, H. B. 1968. The Free-Living Lower Invertebrates. Macmillan Co., New York.)*

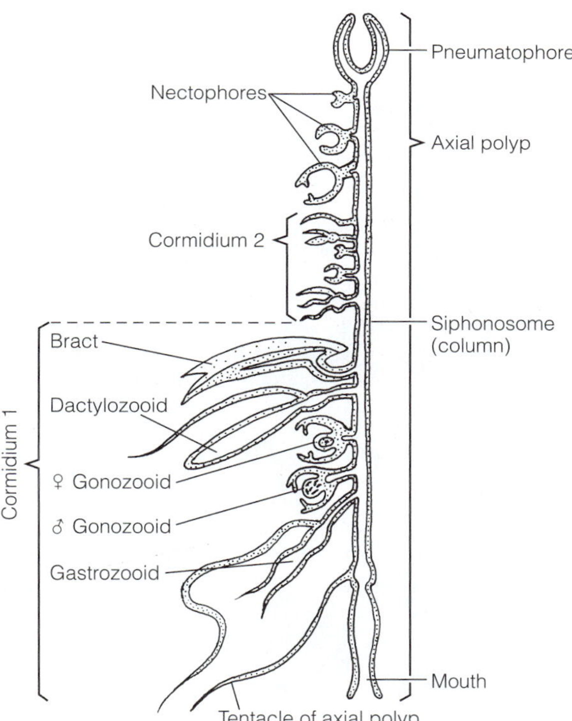

FIGURE 7-69 Hydrozoan diversity: Siphonophora. Diagrammatic anatomy of a siphonophore (Physonectida). Note that the siphonophore is an inverted axial polyp that produces lateral buds. The pneumatophore is the invaginated, gas-containing, pedal end of the axial polyp. *(Modified and redrawn from Delage, Y., and Hérouard, E. 1901. Traité de Zoologie Concrète. Vol. 2. Les Coelentérés. Schleicher Frères, Paris. 848 pp.*

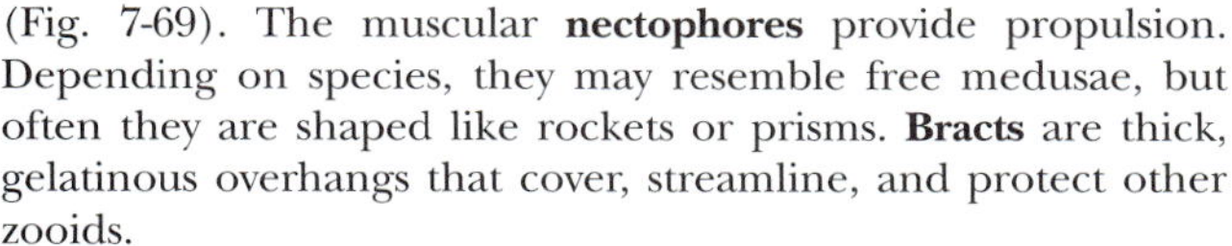
(Fig. 7-69). The muscular **nectophores** provide propulsion. Depending on species, they may resemble free medusae, but often they are shaped like rockets or prisms. **Bracts** are thick, gelatinous overhangs that cover, streamline, and protect other zooids.

The degree of functional specialization and integration of zooids along the siphonophore body axis is often so impressive that colonies appear and behave as coordinated individuals. Despite their pelagic lifestyle and integrated bodies, however, siphonophores lack the sensory organs found in other pelagic hydrozoans.

Siphonophores feed on animal plankton, especially crustaceans, and even small fish. Prey capture is on the highly retractile tentacles of gastro- and dactylozooids; if on dactylozooids, then prey is transferred to a gastrozooid for ingestion. Prey too large to be ingested by individual gastrozooids is wrapped with nematocyst-bearing tentacles and partially digested to smaller fragments.

Siphonophores are hermaphrodites, the gonozooids of each colony bearing both male and female gonophores. Ripe gametes, both eggs and sperm, are spawned into the water and fertilization is external. The zygote develops into an oval or circular planula.

Siphonophora is divided into three taxa: Cystonectida, Physonectida, and Calycophorida.

Cystonectida[sO]: Medusa zooids absent, except for gonophores and perhaps "jelly polyps" of *Physalia*, which may be modified nectophores. Notorious purple-blue *Physalia physalis* (Portuguese man-of-war) cruises surface of warm seas with its sail-like float (Fig. 7-70A); most 10 cm length, but up to 30 cm and tentacles over 50 m; bloated float (pneumatophore) filled with normal atomospheric gases plus CO, which can account for up to 13% of total; curiously, boatlike siphonosome of *Physalia* is oriented horizontally, oral end forward and pedal end aft; cormidium arise from underside of siphonosome hull and hang down like a keel; each cormidium includes tentacle-free gastrozooids, dactylozooids with a long or very long tentacle on each one, highly branched gonozooids bearing gonophores, and other zooids (Fig. 7-70B); subsurface species of cystonects, which have a long *vertical* siphonosome attached to a pneumatophore, drift vertically in currents.

Physonectida[sO]: Physonects, such as *Nanomia bijuga* and *Physophora hydrostatica*, often have long stems (up to 3 m in *Apolemia uvaria*), a pneumatophore, and nectophores (Fig. 7-71A); nectophores arise from budding zone immediately below float while cormidia bud from stem below zone of nectophores; each cormidium includes a bract, gastrozooid, dactylozooid, and a branched gonozooid bearing gonophores; each female gonophore produces but one egg.

Calycophorida[sO]: Calycophores, such as *Muggiaea* (Fig. 7-71B), and *Diphyes*, lack a pneumatophore and the leading end of the colony bears one (in *Muggiaea*), two (in

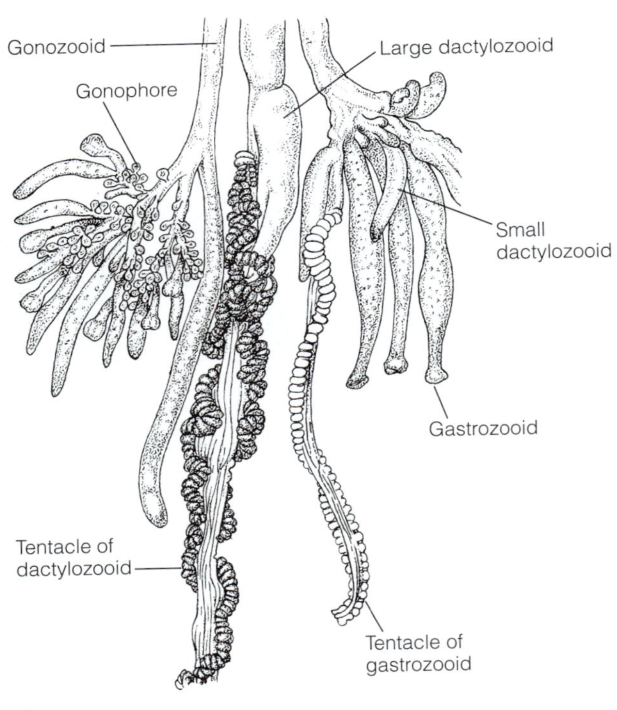

FIGURE 7-70 Hydrozoan diversity: Siphonophora. **A,** The Portuguese man-of-war, *Physalia physalis* (Cystonectida). The axial polyp is horizontal, corresponding with the locomotory body axis, and gives rise to lateral buds that hang below the pneumatophore. **B,** Cormidium of *Physalia*.

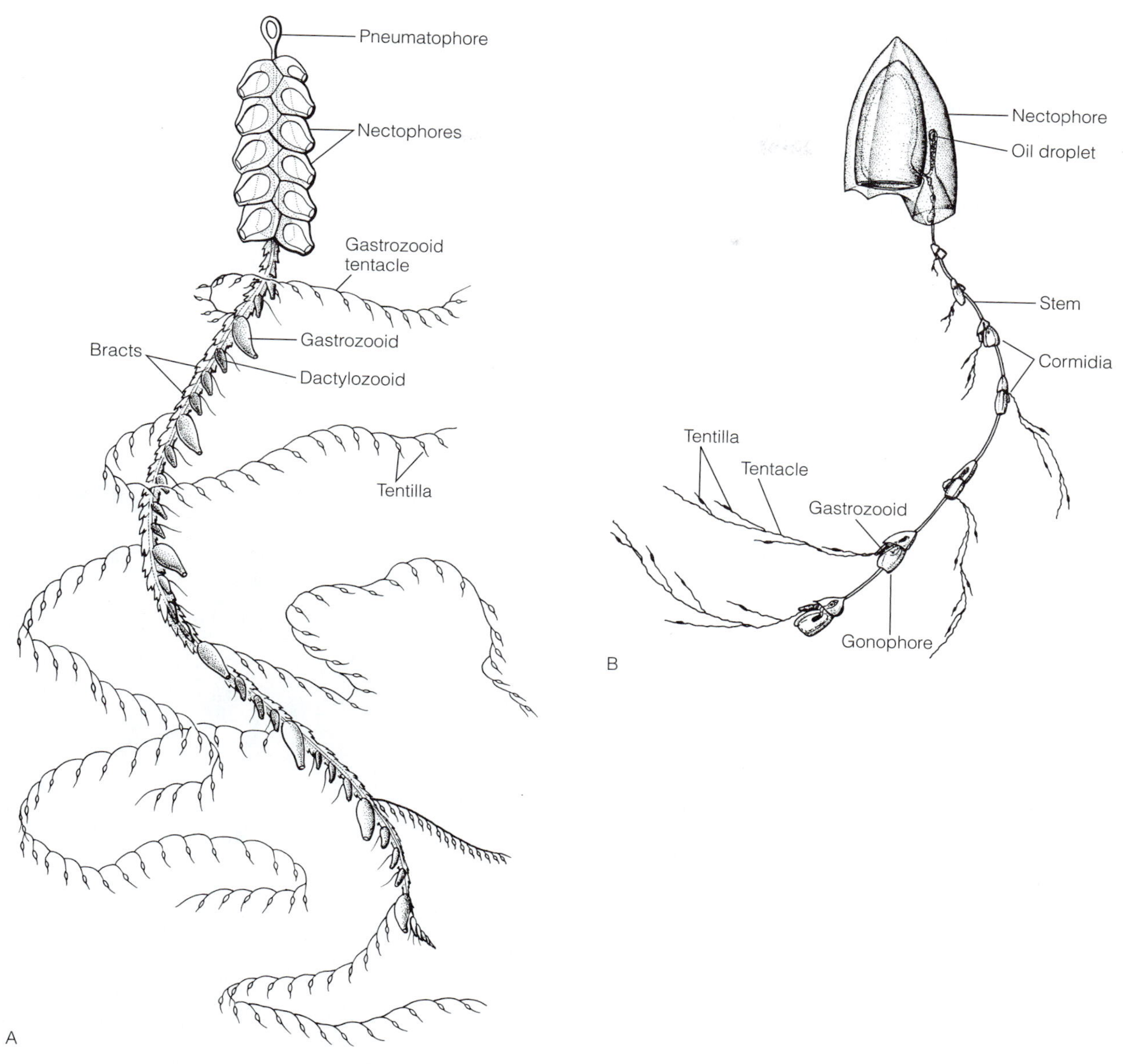

FIGURE 7-71 Hydrozoan diversity: Siphonophora. **A,** *Nanomia cara* (Physonectida). Compare with Fig. 7-69. **B,** *Muggiaea kochii* (Calycophorida). *(A, Modified and redrawn from Mackie, G. O. 1964. Analysis of locomotion in a siphonophore colony. Proc. Roy Soc. Lond. (B) 159:366–391.)*

Diphyes), or more rocket-shaped nectophores; with such thrusters, *Muggiaea* and others are strong, acrobatic swimmers; long stem originates in a buoyant, oil-filled pocket in the anterior nectophore, then extends beyond it and buds a series of cormidia; each cormidium has a helmet-like bract that covers a gastrozooid bearing bell-shaped gonophores; specialized gonozooids and dactylozooids are absent; oldest cormidia, at tail end of stalk, break free, adopt an independent existence, and are then called **eudoxids;** only eudoxid gonozooids become sexually mature.

L-FORMS

Thecate or reduced polyps, colony growth by fixed-length budding, medusae short and flat with statocysts; gonads on radial canals or stomach wall.

Leptothecatae[O] **(Leptomedusae, Thecata):** L-form hydroids (Table 7-1, Fig. 7-63); represented by *Obelia* (Fig. 7-58E, 7-64A), *Orthopyxis* (Fig. 7-60A), both with smooth-rimmed, wineglass-shaped theca; *Clytia* with tooth-rimmed theca; *Aglaophenia, Plumularia, Sertularia* form feather-shaped colonies, theca usually with operculum. Thecate polyps often tiny, but medusae include largest among Hydrozoa; *Mitrocoma:* few centimeters diameter; *Aequorea:* up to 20 cm; open-ocean *Rhacostoma atlanticum:* one-half meter.

Limnomedusae^{O}: Life cycle with both polyp and medusa; shallow, cup-shaped medusa is dominant stage; inconspicuous polyp athecate, usually solitary, lacks periderm and sometimes tentacles (Fig. 7-65A); polyps bud medusae and planula-like **frustules** that creep over the substratum and differentiate into polyps. A typical representative is the medusa of *Gonionemus vertens:* medusa semisessile, coastal, marine, associated with eelgrass and algae (Fig. 7-59C); numerous tentacles bearing rings of nematocyte batteries, four gonad-bearing radial canals, numerous marginal statocysts between tentacle bases; swollen tentacle bases house nematocyte battery; adhesive organ situated at kink, or elbow, on each tentacle. *Craspedacusta:* common, but sporadic, in freshwater lakes and ponds; medusae appear intermittently, often in large numbers, in late summer; 2-cm jellies resemble *Gonionemus* medusae, but lack adhesive organs and have three sets of tentacles instead of one; polyps lack tentacles, may form colony of a few zooids. *Vallentinia gabriella* (Fig. 7-59A): fresh water, similar to *Craspedacusta,* but has short adhesive tentacles interspersed with long feeding tentacles.

Trachylina^{O}: 115 species; life cycle devoid of polyp stage, medusae only (Fig. 7-65C); when present, statocysts borne on stalks, statoliths originate from endoderm; most species open-water marine, occur at various depths from surface to deep sea; planula metamorphoses into actinula, which differentiates into medusa; most range from 1–2 cm, but *Solmissus* to 10-cm diameter; three trachyline taxa: Trachymedusae, Narcomedusae, and Actinulida.

Trachymedusae^{sO}: Medusae tall or hemispherical with well-developed velum, usually many solid marginal tentacles, club-shaped statocysts, gonads on radial canals; many species, including *Aglantha digitale* (Fig. 7-72A), have subumbrella drawn out into manubrium-like peduncle that bears terminal, four-lipped mouth; peduncle is body-wall outgrowth that includes proximal (stomach) ends of radial canals; strong swimmers with well-developed coronal and velar muscles. *Aglantha digitale:* inhabits cold northern waters; usually eight radial canals each with pendant, saclike gonad; taxon also includes *Aglaura* and *Liriope.*

Narcomedusae^{sO}: Medusae relatively flat with thin, lobed margin; central part of the bell thick and lenslike, whorl of thick, solid tentacles usually on exumbrellar surface; tentacles held stiffly, often above bell rim; club-shaped statocysts on bell margin; manubrium and radial canals reduced or absent, but stomach large; gonads in stomach wall (radial canals absent); some species viviparous, supply nutrition to young (actinulas) in maternal stomach *(Cunina prolifera);* others *(C. octonaria, C. proboscidea)* release parasitic larvae, attack manubrium or stomach of other medusae, including the anthoathecate *Turritopsis nutricola. Aegina, Cunina,* and *Solmissus* are well-known genera (Fig. 7-72B); possibly related to this group are *Polypodium hydriforme,* an intracellular parasite of sturgeon eggs, and myxozoans (see below).

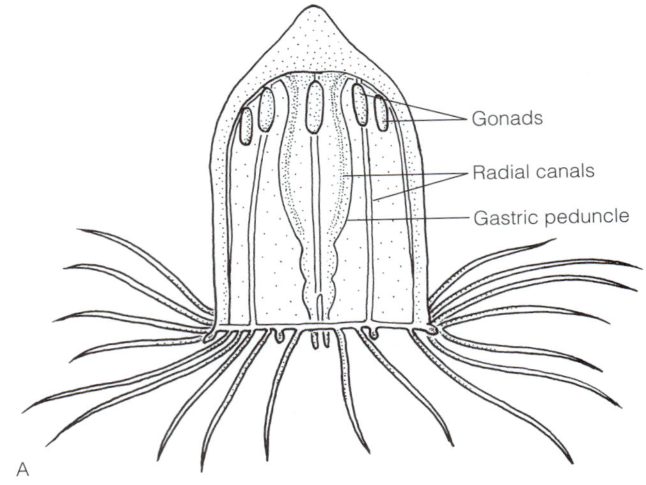

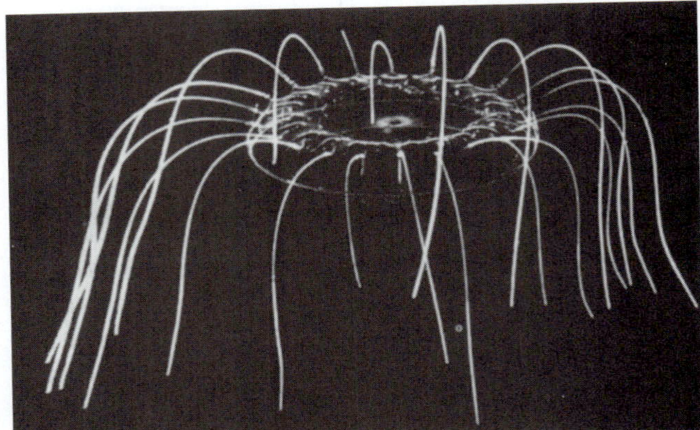

From Larson, R. J., Mills, C. E., and Harbison, G. R. 1989. *In situ* foraging and feeding behavior of Narcomedusae (Cnidaria: Hydrozoa). J. Mar. Biol. Assoc. U. K. 69:785–794 (Fig. 1D). Reprinted with permission of Cambridge Univer-

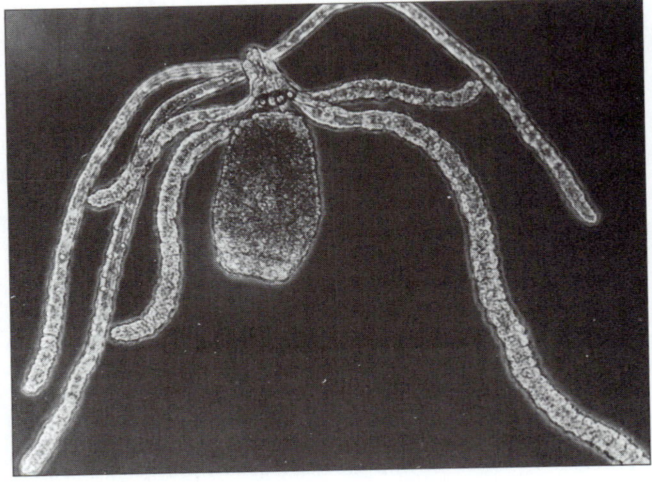

FIGURE 7-72 Hydrozoan diversity: Trachylina. **A,** *Aglantha digitale* (Trachymedusae) from Washington. **B,** *Solmissus incisa* (Narcomedusae). Note stiff exumbrellar tentacles and absence of radial canals. **C,** *Halammohydra* sp. (Actinulida) from North Carolina. Note conspicuous spherical statocysts. *(A, Drawn from a photograph provided by C. E. Mills)*

Actinulida[sO]: Adults tiny medusoid animals adapted for life in seawater-filled, interstitial spaces between sand grains (Fig. 7-72C); resemble actinula larvae and may have evolved, via pedomorphosis, from actinula stage of a trachyline ancestor; body slender and ciliated, as are the long, solid tentacles; cilia, rather than muscles, used to cruise through passageways between sand grains; four marginal, club-shaped statocysts; bulbous part of body (Fig. 7-72C), called gastric tube, bears the mouth at one end; gastric tube may be manubrium or subumbrellar peduncle (radial canals are absent); gonads in gastric-tube mesoglea, fertilization internal; either an actinula or planula released; if planula, it transforms directly into actinula. *Halammohydra:* aboral adhesive organ; *Otohydra.*

Myxozoa (taxon of uncertain rank; may not even be cnidarians): 1200 species; cell and tissue parasites of connective tissue, muscle, and internal organs of invertebrates, fishes (Fig. 7-73A); within host, parasite is motile, multinucleated syncytium; periodically, certain nuclei

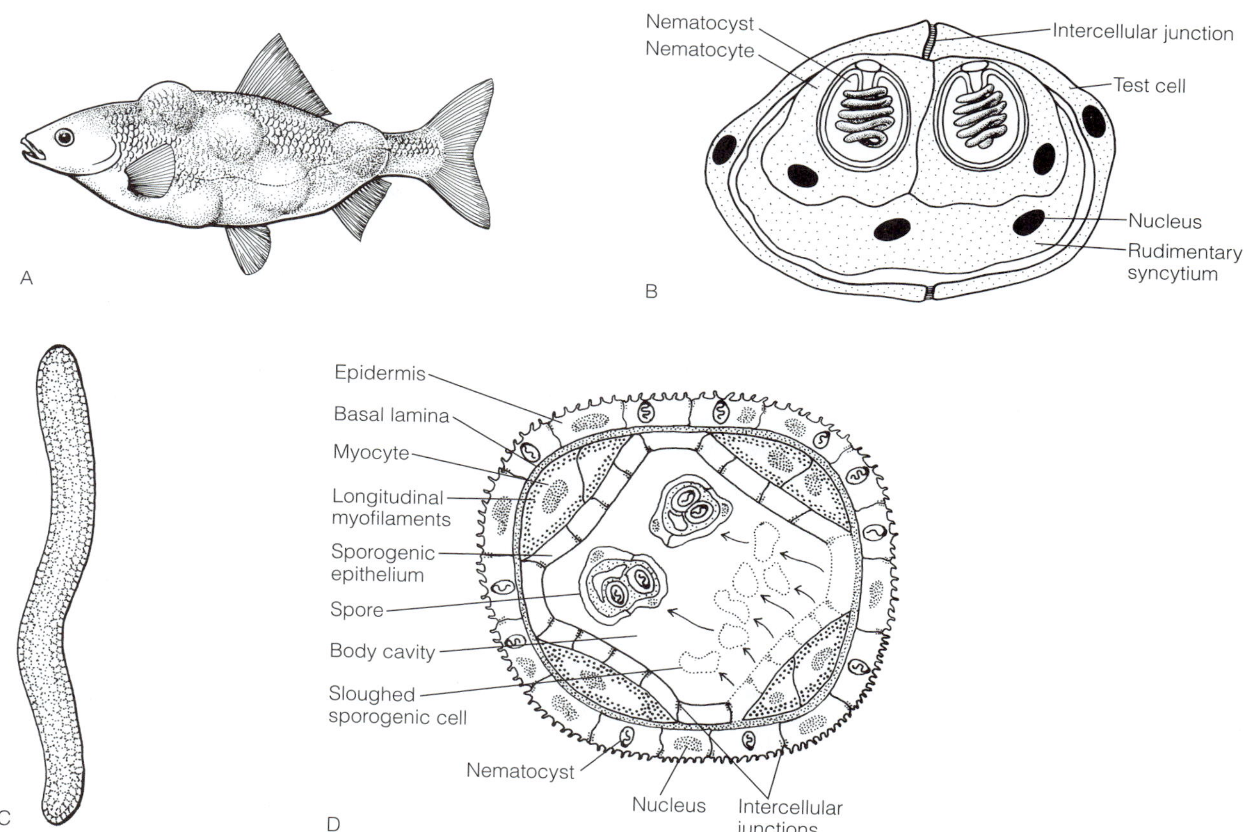

FIGURE 7-73 Hydrozoan diversity: Myxozoa (placement in Hydrozoa is uncertain). **A,** A fish with tumorlike swellings caused by infection with *Myxobolus.* **B,** Multicellular spore of *Myxobolus.* Note intercellular junctions (metazoan character) and cnidae (cnidarian character). **C,** External view of a living *Buddenbrockia plumatellae,* a worm-shaped myxozoan that parasitizes the coelom of freshwater bryozoans (Phylactolaemata). The worms writhe slowly inside of the host's coelom, but once released into the water, they contract into a corkscrew shape. **D,** Cross section of *B. plumatellae.* The inner sporogenic epithelium (possibly gastrodermis) and the bundles of myocytes are tetraradially symmetric, as in Medusozoa. As the worm "matures," all cells of the sporogenic epithelium separate from each other and individually enter the body cavity. These sloughed sporogenic cells divide and differentiate into multicellular spores, each with four nematocysts, similar to the spore shown in **B.** (The transition from epithelium to spores is shown, using arrows, for only part of the sporogenic epithelium.) The body cavity of the "mature" worm is filled with spores that are liberated through breaks in the parasite's degenerating epidermis. The cnidarian affinity of *Buddenbrockia* (and other myxozoans) is uncertain. The scientists whose publications provided the data for **C** and **D** regard *Buddenbrockia* as highly derived bilaterians based on the arrangement of muscles in four quadrants, as in the bilaterian roundworms (Nematoda). In the absence of any trace of bilateral symmetry, however, the case for *Buddenbrockia* as a bilaterian, although provocative, is not yet firmly established. *(A, After Grell from Hausmann, K., and Hülsmann, N., 1996. Protozoology. Georg Thieme, New York. 434 pp.; B, From Hausmann and Hülsmann, 1996; C and D, Drawn from photo- and electronmicrographs and descriptions in Okamura, B., Curry, A., Wood, T. S., and Canning, E. U. 2002. Ultrastructure of* Buddenbrockia *identifies it as a myxozoan and verifies the bilateral origin of Myxozoa. Parasitology. 124: 215–223, and Canning, E. U., Tops, S., Curry, A., Wood, T. S., and Okamura, B. 2002. Ecology, development and pathogenicity of* Buddenbrockia plumatellae *Schröeder 1910 (myxozoa, malacosporea; syn.* Tetracapsula bryozoides*) and establishment of* Tetracapsuloides *n. gen. for* Tetracapsula bryosalmonae. *J. Eukaryot. Microbiol. 49:280–295.)*

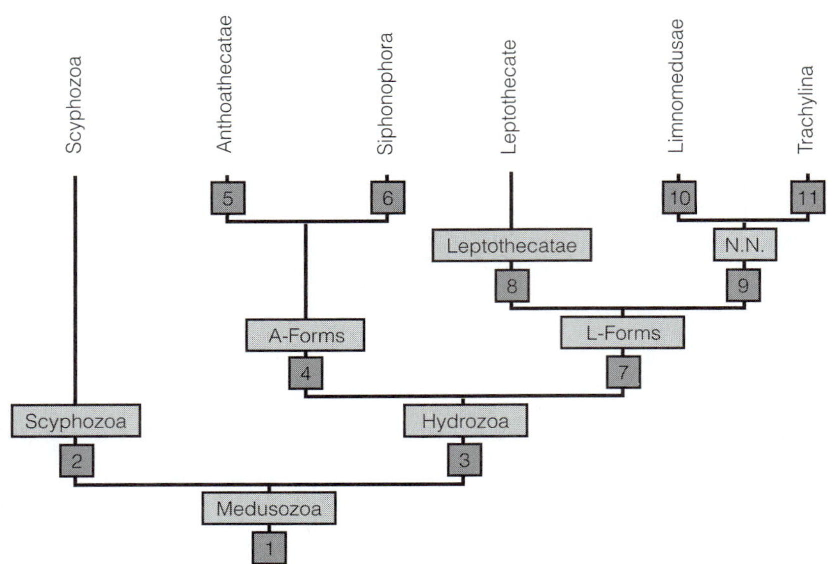

FIGURE 7-74 Phylogeny of Medusozoa and Hydrozoa. **1, Medusozoa:** tetramerous symmetry, cnidocytes with cnidocil, linear mtDNA, sense organs (ocelli, statocysts). **2, Scyphozoa:** rhopalia, gastric filaments, solid tentacles, oral end of funnel-shaped polyp transforms into medusa, medusa has septa. **3, Hydrozoa:** medusa with velum, hollow swellings (bulbs) at bases of tentacles, nematocytes restricted to epidermis, germ cells shed directly from gonad to exterior. **4, A-Forms:** periderm does not enclose hydranth (athecate), fruticose colonies arise by budding from axial polyp, manubrial gonads. **5, Anthoathecatae:** medusae with ocelli in tentacular bases (bulbs), desmoneme nematocysts. **6, Siphonophora:** pelagic, pneumatophore, cormidia. **7, L-Forms:** periderm encloses polyp in theca (thecate), fruticose colonies arise from fixed-length budding, medusae have statocysts, gonads on radial canals. **8, Leptothecatae:** ectodermal statoliths. **9, N.N.:** endodermal statoliths, polyp stage reduced, tentacle bulbs infilled with endoderm. **10, Limnomedusae:** polyp tentacles reduced, the numerous medusa tentacles are ringed (annulated). **11, Trachylina:** polyp stage absent, manubrium reduced.

(germinal nuclei) in syncytium divide, become cellular, and form infective spores, which are ingested with food by next host; formerly considered a protozoan taxon related to Apicomplexa, myxozoan spores attach to host tissues and cells with one or more eversible "polar filaments" now known to be nematocysts (atrichous isorhizas), each in its own nematocyte; spores are multicellular, include two or four nematocytes, one or two cells that will become a new parasitic syncytium, and two test (valvogenic) cells forming spore wall (Fig. 7-73B); test cells joined together by septate and adherens-type junctions typical of metazoans, unknown in protozoans; extracellular matrix containing collagen, also typical of metazoans, between some cells; placement of myxozoans here based on tetramerous symmetry of *Buddenbrockia* (Fig. 7-73C,D) and similar occurrence of cell and tissue parasitism in some Narcomedusae. *Buddenbrockia* and *Tetracapsuloides* in freshwater (Phylactolaemate) bryozoans; *Myxobolus:* in fishes; *Triactinomyxon:* in *Tubifex* annelid.

Phylogeny of Hydrozoa

It is an ongoing challenge for systematists to draw a comprehensive phylogeny of Hydrozoa. To make matters worse, specialists who primarily study either hydroids or jellies have proposed separate classifications based on either polyps or medusae. To date, these competing classifications have not been satisfactorily tested or resolved into a single diagram of phylogenetic relationship. Our phylogeny is tentative and offered solely as a framework on which to initiate a better understanding of hydrozoan evolution (Fig. 7-74). The basic division separates the five Linnean orders into two major taxa based on A-form and L-form attributes (Table 7-1). The A-form taxa are the Anthoathecatae and Siphonophora; L-form taxa are Leptothecatae, Limnomedusae, and Trachylina (Trachymedusae, Narcomedusae, and Actinulida).

PHYLOGENETIC HIERARCHY OF HYDROZOA

Hydrozoa
- A-Forms
  - Anthoathecatae
  - Siphonophora
- L-Forms
  - Leptothecatae
  - N.N.
    - Limnomedusae
    - Trachylina

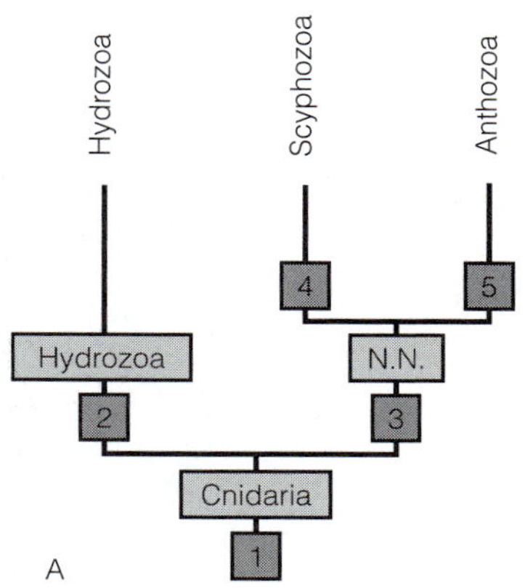

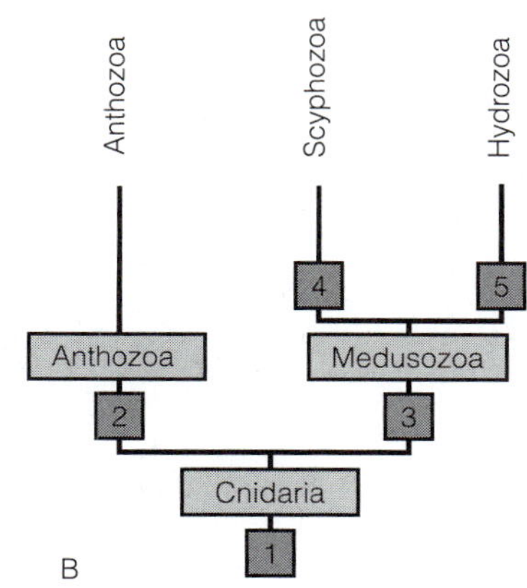

FIGURE 7-75 Cnidaria: competing phylogenies. **A,** Traditional phylogeny based on morphology. **1, Cnidaria:** tetramerous medusa (adult), polyp (juvenile), planula (larva); ectodermal nematocytes; ecto- or endodermal gonads, gametes spawned directly from gonad to exterior; ecto- and endodermal nerve nets; longitudinal musculature in ectoderm, circular musculature in endoderm; periderm; budding. **2, Hydrozoa:** medusa with velum. **3, N.N.:** endodermal gonads and nematocytes; gametes released into coelenteron, then spawned through mouth; transverse and longitudinal fission. **4, Scyphozoa:** medusa with rhopalia, gastric filaments; strobilation. **5, Anthozoa:** medusa absent; polyp with pharynx, septa, septal filaments. **B,** Competing current phylogeny based on morphology and molecules. **1, Cnidaria:** polyp (adult), planula (larva); ecto- and endodermal nematocytes; nematocytes with motile cilium; ecto- and endodermal nerve nets; longitudinal musculature in ectoderm, circular musculature in endoderm; endodermal gonads; periderm; budding, transverse, and longitudinal fission; mtDNA circular. **2, Anthozoa:** polyps with pharynx, siphonoglyph, septa, septal filaments. **3, Medusozoa:** medusa; mtDNA linear; tetramerous symmetry. **4, Scyphozoa:** medusa with rhopalia, gastric filaments; strobilation. **5, Hydrozoa:** medusa with velum; endodermal nematocytes absent; gametes spawned directly from gonad to exterior; nerve rings with gap junctions. *(Based on several sources)*

PHYLOGENY OF CNIDARIA

Cnidarian species fall into three major taxa: Anthozoa, Scyphozoa, and Hydrozoa. Representative anthozoans are corals and anemones; scyphozoans are large jellyfishes, and hydrozoans are hydroids, *Hydra,* siphonophores, fire corals, and small jellyfishes.

The traditional view of the evolutionary relationships among cnidarian taxa (Fig. 7-75A), from primitive to derived, is: Hydrozoa → Scyphozoa → Anthozoa. This suggests that the hydrozoan life cycle of medusa → planula → polyp → medusa is primitive for Cnidaria. Within this life cycle, the sexual medusa is interpreted as the ancestral adult stage, the planula its larva, and the asexual polyp a persistent juvenile. The absence of a medusa in anthozoans and the internal complexity of their polyps, as compared with the hydrozoan polyp, are both viewed as autapomorphies of anthozoans.

Modern molecular and morphological analyses of cnidarian relationships (Fig. 7-75B) reverse the evolutionary sequence to: Anthozoa → Scyphozoa → Hydrozoa. This arrangement suggests that a sexual polyp was the ancestral adult, the planula its larva, and a medusa was absent from the original life cycle (polyp → planula → polyp). At first glance, this phylogeny appears to have a glaring weakness: Although polyps occur in all four classes, those of the primitive anthozoans are the most complex, those of the derived hydrozoans the least complex. Because we generally equate evolution with increasing levels of complexity, an explanation is required for the apparent evolution of simplicity in cnidarian polyps. Among several explanations for this paradox, perhaps the best is the relationship between body size and complexity (Fig. 7-76).

The largest and most complex polyps occur in Anthozoa (Fig. 7-76A,C) whereas the smallest and least complex polyps are found in Hydrozoa (Fig. 7-76G,I). Furthermore, the complexity of anthozoan polyps results chiefly from gastrodermal outfolds (septa), which radiate from the body wall into the coelenteron. These septa increase the surface area of the gastrodermis and reduce the volume of the coelenteron, exactly as predicted from the discussion of area-volume relationships in Chapter 4. Thus, by virtue of the fact that anthozoan polyps are large, they have to be complex, and by the same token, the evolution of small polyps and a more favorable area-to-volume ratio, as in hydrozoans, relieves them of the necessity to fold their body wall, resulting in a simplified body.

PHYLOGENETIC HIERARCHY OF CNIDARIA

Cnidaria
- Anthozoa
- Medusozoa
    - Scyphozoa
    - Hydrozoa

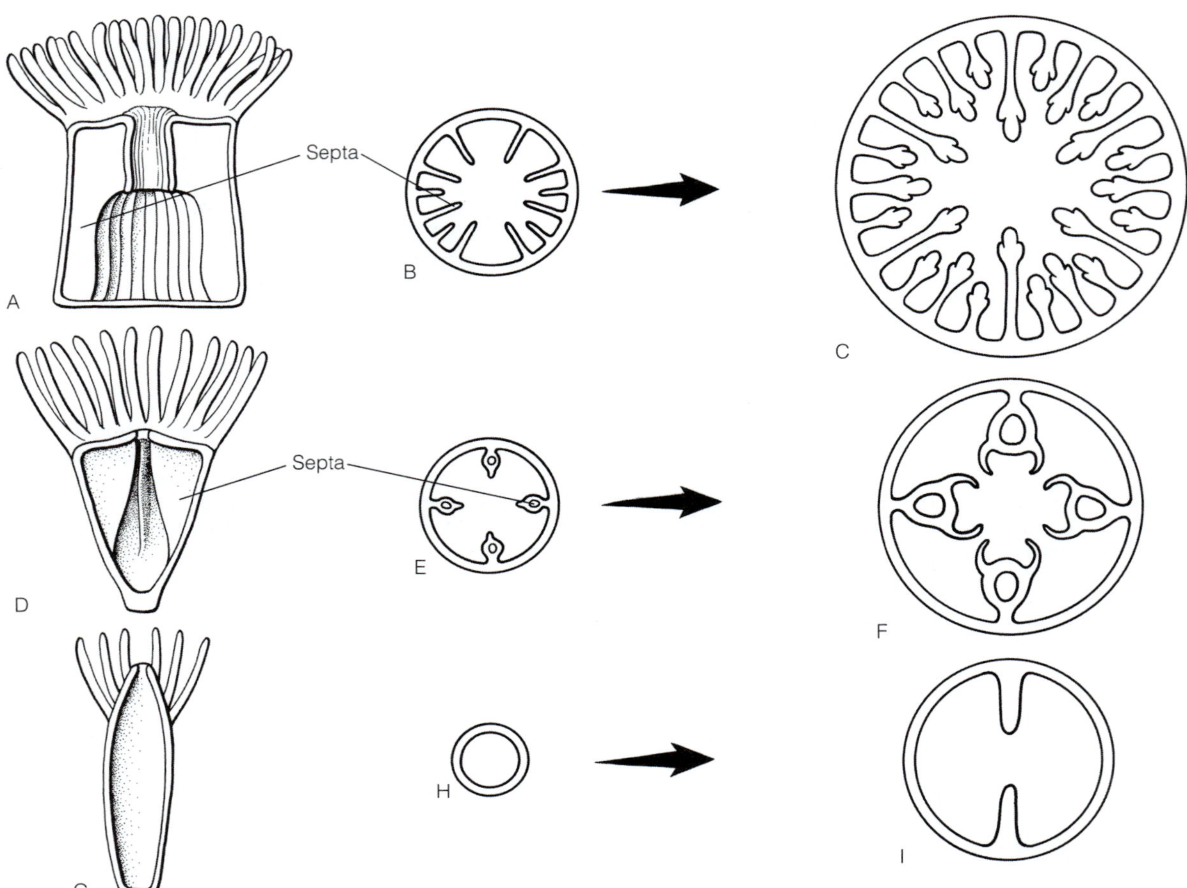

FIGURE 7-76 Cnidaria: body size and design in polyps. In general, the body is largest in anthozoan polyps **(A)**, smallest in hydrozoan polyps **(G)**, and intermediate in scyphozoan polyps **(D)**. In keeping with surface-area-to-volume considerations, the anthozoan polyps have many tentacles **(A)** and the area of the gastrodermis is enlarged by septa **(B)**. The number of septa and tentacles increases with growth **(C)**. Intermediate-size scyphopolyps have fewer tentacles **(D)** than large anthozoan polyps **(A)** and never more than four gastrodermal septa **(E and F)**. Small hydropolyps have the fewest tentacles **(G)** and gastrodermal septa are typically absent **(H)**. The largest hydropolyps, however, may have two septa **(I)** or other adaptations to increase the gastrodermal area.

REFERENCES

CNIDARIA

Bayer, F., and Owre, H. B. 1968. The Free-Living Lower Invertebrates. MacMillan, New York. 229 pp.

Boardman, R. S., Cheetham, A. H., and Oliver, W. A. (Eds.): 1973. Animal Colonies. Dowden, Hutchinson and Ross, Stroudsburg, PA. 603 pp.

Bridge, D., Cunningham, C. W., DeSalle, R., and Buss, L. W. 1995. Class-level relationships in the phylum Cnidaria: Molecular and morphological evidence. Mol. Biol. Evol. 12:679–689.

Buchsbaum, R. M., and Milne, L. J. 1960. The Lower Animals. Chanticleer Press, New York. 303 pp.

Campbell, R. D. 1974. Cnidaria. In Giese, A. C., and Pearse, J. S. (Eds.): Reproduction of Marine Invertebrates I. Academic Press, New York. pp. 133–200.

Cook, C. B. 1983. Metabolic interchange in algae-invertebrate symbiosis. Int. Rev. Cytol. 14(Suppl.):177–209.

Dorit, R. L., Walker, W. F., and Barnes, R. D. 1991. Zoology. Saunders, Philadelphia. 1009 pp.

Fautin, D. G. 1992. Cnidaria. In Adiyodi, K. G., and Adiyodi, R. G. (Eds.): Reproductive Biology of Invertebrates. Vol. 5. Sexual Differentiation and Behavior. Oxford and IBH, New Delhi. pp. 31–52.

Finnerty, J. R., and Martindale, M. Q. 1999. Ancient origins of axial patterning genes: Hox genes and ParaHox genes in the Cnidaria. Evol. Devel. 1:16–23.

Gladfelter, W. G. 1973. A comparative analysis of the locomotory system of medusoid Cnidaria. Helgol. Wiss. Meeresunters. 25:228–272.

Grimmellikhuijzen, C. J. P., Leviev, I., and Carstensen, K. 1996. Peptides in the nervous system of cnidarians: Structure, function, and biosynthesis. Int. Rev. Cytol. 167:37–89.

Hand, C. 1959. On the origin and phylogeny of coelenterates. Syst. Zool. 8:191–202.

Hargitt, G. T. 1919. Germ cells of coelenterates. VI. General considerations, discussion, conclusions. J. Morphol. 33:1–59 (This paper debunks Weismann's idea that the germ cells of hydrozoans always originate in ectoderm. As of 1919, research on 44 species indicated 23 with ectodermal gonads and 31 with endodermal gonads.)

Harrison, F. W. (Ed.): 1991. Microscopic Anatomy of Invertebrates, Vol. 2. Placozoa, Porifera, Cnidaria, and Ctenophora. Alan Liss, New York. 436 pp. (This volume contains chapters on each of the classes of cnidarians.)

Hessinger, D. A., and Lenhoff, H. M. (Eds.): 1988. The Biology of Nematocysts. Academic Press, San Diego. 600 pp.

Holstein, T., and Tardent, P. 1984. An ultrahigh-speed analysis of exocytosis: Nematocyst discharge. Science 223:830–833.

Johnsen, S. 2000. Transparent animals. Sci. Am. Feb.:80–89.

Kaestner, A. 1984. Lehrbuch der Speziellen Zoologie. 2 Teil. Gustav Fischer Verlag, Stuttgart. 621 pp.

Mackie, G. O. 1986. From aggregates to integrates: Physiological aspects of modularity in colonial animals. Phil. Trans. Roy. Soc. Lond. (B) 313:175–196.

Mackie, G. O. 1990. The elementary nervous system revisited. Amer. Zool. 30:907–920.

Mackie, G. O. 1999a. Coelenterate organs. Mar. Freshwat. Behav. Physiol. 32:113–127.

Mackie, G. O. 1999b. Nerve nets. In Adelman, G., and Smith, B. H. (Eds.): Elsevier's Encyclopedia of Neuroscience.. Elsevier Science B.V., Amsterdam. pp. 1299–1302.

Mackie, G. O. (Ed.): 1976. Coelenterate Ecology and Behavior. Plenum Press, New York. 744 pp.

Martin, V. 1997. Cnidarians. In Gilbert, S. F., and Raunio, A. M. (Eds.): Embryology: Constructing the Organism. Sinauer, Sunderland, MA. pp. 57–86.

Muscatine, L., and Lenhoff, H. M. (Eds.): 1974. Coelenterate Biology. Academic Press, New York. 501 pp.

Rees, W. J. (Ed.): 1966. The Cnidaria and Their Evolution. Academic Press, New York. 449 pp.

Spencer, A. N. 1982. The physiology of a coelenterate neuromuscular synapse. J. Comp. Physiol. 148:353–363.

Sterrer, W. (Ed.): 1986. Marine Fauna and Flora of Bermuda. John Wiley and Sons, New York. 742 pp.

Tardent, P., and Holstein, T. 1982. Morphology and morphodynamics of the stenotele nematocyst of *Hydra attenuata* Pall. Cell Tiss. Res. 224:269–290.

Taylor, D. L. 1973. The cellular interactions of algal-invertebrate symbiosis. Adv. Mar. Biol. 11:1–56.

Watson, G. M., and Mire-Thibodeaux, P. 1994. The cell biology of nematocysts. Int. Rev. Cytol. 156:275–300.

Werner, B. 1973. New investigations on systematics and evolution of the class Scyphozoa and the phylum Cnidaria. Proceedings of the Second International Symposium on Cnidaria. Publ. Seto Mar. Biol. Lab. 20:35–61.

ANTHOZOA

General

Bayer, F. M. 1956. Octocorallia. In Moore, R. C. (Ed.): Treatise on Invertebrate Paleontology. Part F. Coelenterata. Geological Society of America and University of Kansas Press, Lawrence, KS. pp. 166–230.

Bayer, F. M. 1961. The Shallow-Water Octocorallia of the West Indian Region. Martinus Nijhoff, The Hague. 373 pp.

Bigger, C. H. 1982. The cellular basis of the aggressive acrorhagial response of sea anemones. J. Morph. 173:259–278.

Buddemeier, R. W., and Fautin, D. G. 1993. Coral bleaching as an adaptive mechanism. BioScience 43:320–326.

Cairns, S. 1976. Guide to the Commoner Shallow-Water Gorgonians (sea whips, sea feathers, and sea fans) of Florida, the Gulf of Mexico, and the Caribbean Region. Sea Grant Field Guide Series, no. 6. University of Miami, Miami, FL. 74 pp.

Chen, C. A., Odorico, D. M., ten Lohuis, M., Veron, J. E. N., and Miller, D. J. 1995. Systematic relationships within the Anthozoa (Cnidaria: Anthozoa) using the 5'-end of the 28S rDNA. Mol. Phylog. Evol. 4:175–183.

Clayton, W. S., Jr., and Lasker, H. R. 1982. Effects of light and dark treatment on feeding by the reef coral *Pocillopora damicornis*. J. Exp. Mar. Biol. Ecol. 63(3):269–280.

Cook, C. B., D'Elia, C. F., and Muller-Parker, G. 1988. Host feeding and nutrient sufficiency for zooxanthellae in the sea anemone *Aptasia pallida*. Mar. Biol. 98:253–262.

Cook, C. B., Logan, A., Ward, J., et al. 1990. Elevated temperatures and bleaching on a high latitude coral reef: The 1988 Bermuda event. Coral Reefs 9:45–49.

denHartog, J. C. 1977. Descriptions of two new Ceriantharia from the Caribbean region. Biol. Meded. 51:211–242.

Dunn, D. F. 1981. The clownfish sea anemones: Stichodactylidae (Coelenterata: Actiniaria) and other sea anemones symbiotic with pomacentrid fishes. Trans. Amer. Phil. Soc. 71:1–115.

Dunn, D. F. 1982. Cnidaria. In Parker, S. P. (Ed.): Synopsis and Classification of Living Organisms, Vol. 1. McGraw-Hill, New York. pp. 669–706.

Dunn, D. F., Devaney, D. M., and Roth, B. 1980. *Stylobates:* A shell-forming sea anemone (Coelenterata, Anthozoa, Actiniidae). Pac. Sci. 34:379–388.

Fadlallah, Y. H. 1983. Sexual reproduction, development and larval biology in scleractinian corals. A review. Coral Reefs 2:129–150.

Fautin, D. G. 1991. Developmental pathways of anthozoans. Hydrobiologia 216/217:143–149.

Fautin, D. G. 1992. A shell with a new twist. Nat. Hist. (4):50–57. (Sea anemone–hermit crab symbiosis.)

Fautin, D. G., and Lowenstein, J. M. 1992. Phylogenetic relationships among scleractinians, actinians, and corallimorpharians (Coelenterata: Anthozoa). Proc. 7th Int. Coral Reef Symp. 2:665–670.

Fautin, D. G., and Mariscal, R. N. 1991. Cnidaria: Anthozoa. In Harrison, F. W., and Westfall, J. A. (Eds.): Microscopic Anatomy of Invertebrates. Vol. 2: Placozoa, Porifera, Cnidaria, and Ctenophora. Wiley-Liss, New York. pp. 267–358.

Fautin, D. G., Spaulding, J. G., and Chia, F.-S. 1989. Cnidaria. In Adiyodi, K. G., and Adiyodi, R. G. (Eds.): Reproductive Biology of Invertebrates. Vol. 4, Pt. A: Fertilization, Development, and Parental Care. Oxford and IBH, New Delhi. pp. 43–62.

France, S. C., Rose, P. E., Agenbroad, J. E., Mullineaux, L. S., and Kocher, T. D. 1996. DNA sequence variation of mitochondrial large-subunit rRNA provides support for a two subclass organization of the Anthozoa (Cnidaria). Mol. Mar. Biol. Biotech. 5:15–28.

Francis, L. 1973. Intraspecific aggression and its effect on the distribution of *Anthopleura elegantissima* and some related sea anemones. Biol. Bull. 144:73–92.

Gattuso, J.-P., Allemand, D., and Frankignoulle, M. 1999. Photosynthesis and calcification at cellular, organismal and community levels in coral reefs: A review on interactions and control by carbonate chemistry. Am. Zool. 39:160–183.

Gladfelter, E. H., Monohan, R. K., and Gladfelter, W. G. 1978. Growth rates of five species of reef-building corals in the northeastern Caribbean. Bull. Mar. Sci. 28(4):728–734.

Godknecht, A., and Tardent, P. 1988. Discharge and mode of action of the tentacular nematocysts of *Anemonia sulcata.* Mar. Biol. 100:83–92.

Goldberg, W. M. 1976. Comparative study of the chemistry and structure of gorgonian and antipatharian coral skeletons. Mar. Biol. 35:253–267.

Isdale, P. 1977. Variation in growth rate of hermatypic corals in a uniform environment. Proc. 3rd Int. Coral Reef Symp. 2:403–408.

Kastendiek, J. 1976. Behavior of the sea pansy *Renilla kollikeri* Pfeffer and its influence on the distribution and biological interactions of the species. Biol. Bull. 151:518–537.

Knutson, D. W., Buddemeier, R. W., and Smith, S. V. 1972. Coral chronometers: Seasonal growth bands in reef corals. Science 177:270–272.

Lewis, D. H., and Smith, D. C. 1971. The autotrophic nutrition of symbiotic marine coelenterates with special reference to hermatypic corals. Proc. Roy. Soc. Lond. Biol. 178:111–129.

Lewis, J. B., and Price, W. S. 1975. Feeding mechanisms and feeding strategies of Atlantic reef corals. J. Zool. (Lond.) 176:527–544.

Lewis, J. B., and Price, W. S. 1976. Patterns of ciliary currents in Atlantic reef corals and their functional significance. J. Zool. (Lond.) 178:77–89.

Manuel, R. L. 1981. British Anthozoa. Synopses of the British Fauna, no. 18. Academic Press, London. 250 pp.

Mariscal, R. N., Conklin, E. J., and Bigger, C. H. 1977. The ptychocyst, a major new category of cnida used in tube construction by a cerianthid anemone. Biol. Bull. 152:392–405.

Mariscal, R. N., McLean, R. B., and Hand, C. 1977. The form and function of cnidarian spirocysts. 3. Ultrastructure of the thread and the function of spirocysts. Cell Tiss. Res. 178:427–433.

Opresko, D. M. 1972. Redescriptions and reevaluations of the antipatharians described by L. F. Pourtales. Bull. Mar. Sci. 22:950–1017.

Ottaway, J. R. 1980. Population ecology of the intertidal anemone, *Actinia tenebrosa:* 4. Growth rates and longevities. Aust. J. Mar. Freshwat. Res. 31(3):385–396.

Richardson, C. A., Dustan, P., and Lang, J. C. 1979. Maintenance of living space by sweeper tentacles of *Montastrea cavernosa,* a Caribbean reef coral. Mar. Biol. 55:181–186.

Sebens, K. P. 1983. Morphological variability during longitudinal fission of the intertidal sea anemone, *Anthopleura elegantissima* (Brandt). Pac. Sci. 37:121–132.

Shick, J. M. 1991. A Functional Biology of Sea Anemones. Chapman and Hall, London. 395 pp.

Schuchert, P. 1993. Phylogenetic analysis of the Cnidaria. Z. zool. Syst. Evolut.-forsch. 31:161–173.

Smith, F. G. W. 1971. Atlantic Reef Corals. University of Miami Press, Coral Gables, FL. 164 pp.

Song, J.-I., Kim, W., Kim, W. K., and Kim, J. 1994. Molecular phylogeny of anthozoans (phylum Cnidaria) based on the nucleotide sequences of 18S rRNA gene. Kor. J. Zool. 37:343–351.

Song, J.-I., and Won, J. H. 1997. Systematic relationship of the anthozoan orders based on the partial nuclear 18S rDNA sequences. Kor. J. Biol.. Sci. 1:43–52.

Stanley, G. D., Jr., and Fautin, D. G. 2001. The origins of modern corals. Science 291:1913–1914.

Steele, R. D., and Goreau, N. I. 1977. The breakdown of symbiotic zooxanthellae in the sea anemone *Phyllactis osculifera.* J. Zool. (Lond.) 181:421–437.

Stricker, S. A. 1985. An ultrastructural study of larval settlement in the sea anemone *Urticina crassicornis* (Cnidaria, Actiniaria). J. Morphol. 186:237–253.

Thorington, G. U., and Hessinger, D. A. 1990. Control of cnida discharge: III. Spirocysts are regulated by three classes of chemoreceptors. Biol. Bull. 178:74–83.

Van-Praet, M. 1985. Nutrition of sea anemones. Adv. Mar. Biol. 22:65–99.

Veron, J. E. N., Pichon, M., and Wijsman-Best, M. 1976–1984: Scleractinia of Eastern Australia. Pts. 1–4. Australian Institute of Marine Science Monograph Series. Australian Institute of Marine Science, Canberra.

Watson, G. M., and Hessinger, D. A. 1989. Cnidocyte mechanoreceptors are tuned to the movements of swimming prey by chemoreceptors. Science 243:1589–1591.

Wellington, G. M., and Glynn, P. W. 1983. Environmental influences on skeleton banding in eastern Pacific corals. Coral Reefs 1:215–222.

Williams, G. C. 1988. Four new species of southern African octocorals (Cnidaria: Alcyonacea), with a further diagnostic revision of the genus *Alcyonium,* Linnaeus, 1758. Zool. J. Linn. Soc. 92:1–26.

Williams, R. B. 1978. Some recent observations on the acrorhagi of sea anemones. J. Mar. Biol. Assoc. U.K. 58:787–788.

Wineberg, S., and Wineberg, F. 1979. The life cycle of a gorgonian: *Eunicella singularis.* Bijdr. Dierjunde. 48:127–140.

Won, J. H., Rho, B. J., and Song, J. I. 2001. A phylogenetic study of the Anthozoa (phylum Cnidaria) based on morphological and molecular characters. Coral Reefs 20:39–50.

Wood, E. M. 1983. Corals of the World. T. F. H., Neptune City, NJ. 256 pp.

Internet Sites

http://tolweb.org/tree?group=Anthozoa&contgroup=Cnidaria (D. G. Fautin and S. Romano's illustrated Web page for Anthozoa and Zoantharia.)

http://hercules.kgs.ku.edu/hexacoral/anemone2/index.cfm (D. G. Fautin's "Hexacorallians of the World" site is an up-to-date source of systematic information.)

http://mars.reefkeepers.net/movie.html (Online movie showing growth of axial polyps and budding of lateral polyps in *Acropora.*)

www.state.gov/www/global/global_issues/coral_reefs/990305_coralreef_rpt.html (U.S. government report on coral bleaching.)

http://porites.geology.uiowa.edu/florlist.htm (Illustrated key to common shallow-water Caribbean reef corals.)

SCYPHOZOA

General

Arai, M. N. 1997. A Functional Biology of Scyphozoa. Chapman and Hall, London. 316 pp.

Berrill, N. J. 1949. Developmental analysis of Scyphomedusae. Biol. Rev. 24:393–410.

Calder, D. R., and Peters, E. C. 1975. Nematocysts of *Chiropsalmus quadrumanus* with comments on the systematic status of the Cubomedusae. Helgol. Wiss. Meeresunters. 27:364–369.

Chapman, D. M. 1978. Microanatomy of the cubopolyp, *Tripedalia cystophora* (Class Cubozoa). Helgol. Wiss. Meeresunters. 31:128–168.

Costello, J. H., and Colin, S. P. 1994. Morphology, fluid motion and predation by the scyphomedusae *Aurelia aurita*. Mar. Biol. 121:327–334.

Costello, J. H., and Colin, S. P. 1995. Flow and feeding by swimming scyphomedusae. Mar. Biol. 124:399–406.

Eckelbarger, K. J., and Larson, R. J. 1993. Ultrastructural study of the sessile scyphozoan, *Haliclystus octoradiatus* (Cnidaria: Stauromedusae). J. Morphol. 218:225–236.

Fancett, M. S. 1988. Diet and prey selectivity of scyphomedusae from Port Phillip Bay, Australia. Mar. Biol. 98:503–509.

Hartwick, R. F. 1991. Observations on the anatomy, behaviour, reproduction and life cycle of the cubozoan *Carybdea sivickisi*. Hydrobiologia 216/217:171–179.

Jarms, G., Båmstedt, U., Tiemann, H., Martinussen, M. B., and Fosså, J. H. 1999. The holopelagic life cycle of the deep-sea medusa *Periphylla periphylla* (Scyphozoa, Coronatae). Sarsia 84:55–65.

Kirkpatrick, P. A., and Pugh, P. R. 1984. Siphonophores and Velellids. E. J. Brill/Dr. W. Backhuys, London. 154 pp.

Kramp, P. L. 1961. Synopsis of the medusae of the world. JMBA (UK) 40:1–469.

Larson, R. J. 1976a. Cnidaria: Scyphozoa. Marine Flora and Fauna of the Northeastern United States. NOAA Tech. Report NMFS Circular 397. U.S. Government Printing Office, Washington, DC. 18 pp.

Larson, R. J. 1976b. Cubomedusae: Feeding-functional morphology, behavior and phylogenetic position. In Mackie, G. O. (Ed.): Coelenterate Ecology and Behavior. Plenum Press, New York. pp. 237–245.

Lesh-Laurie, G. E. and P. E. Suchy. 1991. Cnidaria: Scyphozoa and Cubozoa. In Harrison, F. W., and Westfall, J. A. (Eds.): Microscopic Anatomy of Invertebrates. Vol. 2. Wiley-Liss, New York. pp. 185–266.

Mayer, A. G. 1910. Medusae of the World. III. Scyphomedusae. Carnegie Institution, Washington, DC. 236 pp.

Olesen, N. J., Purcell, J. E., and Stoecker, D. K. 1996. Feeding and growth by ephyrae of scyphomedusae *Chrysaora quinquecirrha*. Mar. Ecol. Progr. Ser. 137:149–159.

Sandrini, L. R., and Avian, M. 1989. Feeding mechanism of *Pelagia noctiluca,* laboratory and open sea observations. Mar. Biol. 102:49–55.

Shih, C. T. 1977. A Guide to the Jellyfish of Canadian Atlantic Waters. Natural History Museum of Canada. University of Chicago Press, Chicago. 90 pp.

Southward, A. J. 1955. Observations on the ciliary currents of the jelly-fish *Aurelia aurita*. JMBA (UK) 34:201–216.

Spangenberg, D. B. 1968. Recent studies of strobilation in jellyfish. Oceanogr. Mar. Biol. Ann. Rev. 6:231–247.

Thiel, H. 1966. The evolution of Scyphozoa. A review. In Rees, W. J. (Ed.): The Cnidaria and Their Evolution. Academic Press, London. pp. 77–117.

Werner, B. 1973. New investigations on systematics and evolution of the class Scyphozoa and the phylum Cnidaria. Proceedings of the 2nd International Symposium on Cnidaria. Publ. Seto Mar. Biol. Lab. 20:35–61.

Werner, B. 1975. Structure and life history of the polyp of *Tripedalia cystophora* (Cubozoa, class. nov., Carybdeidae) and its importance for the evolution of the Cnidaria. Helgol. Wiss. Meeresunters. 27:461–504.

Werner, B. 1979. Coloniality in the Scyphozoa: Cnidaria. In Larwood, G., and Rosen, B. R. (Eds.): Biology and Systematics of Colonial Organisms. Academic Press, New York. pp. 81–103.

Internet Sites

www.biology.ualberta.ca/facilities/multimedia/index.php?Page=252 (Animation of *Aurelia* life cycle by A. R. Palmer and H. Kroening.)

http://faculty.washington.edu/cemills/Stauromedusae.html (C. E. Mills's photographs of living stauromedusans and a list of the world's species.)

www.ucmp.berkeley.edu/cnidaria/cubozoa.html (Introduction to Cubomedusae with excellent photographs.)

www.pharmacology.unimelb.edu.au/PHARMWWW/avruweb/jellyfi.htm (Introduction to box-jelly pharmacology with natural history information about several species.)

www.pbrc.hawaii.edu/bekesy/angel/video/ (Video of swimming cubomedusan.)

http://thechesapeakebay.com/jellyfish_facts.shtml (Information about sea nettles in Chesapeake Bay.)

HYDROZOA

General

Benos, D. J., and Prusch, R. D. 1972. Osmoregulation in freshwater hydra. Comp. Biochem. Physiol. 43A:165–171.

Berrill, N. J., and Liu, C. K. 1948. Germplasm, Weissmann, and Hydrozoa. Quart. Rev. Biol. 23:124–132.

Biggs, D. C. 1977. Field studies of fishing, feeding and digestion in siphonophores. Mar. Behav. Physiol. 4:261–274.

Calder, D. R. 1991. Shallow-water hydroids of Bermuda. The thecatae, exclusive of Plumularioidea. Life Sci. Contri. Roy. Ontario Mus. 154:1–140.

Canning, E. U., Tops, S., Curry, A., Wood, T. S., and Okamura, B. 2002. Ecology, development and pathogenicity of *Buddenbrockia plumatellae* Schröder 1910 (Myxozoa, Malacosporea) (syn. *Tetracapsula bryozoides*) and establishment of *Tetracapsuloides* n. gen. for *Tetracapsula brysalmonae*. J. Eukaryot. Microbiol. 49:280–295.

Christensen, H. E. 1967. Ecology of *Hydractinia echinata.* I. Feeding biology. Ophelia 4:245–275.

de Kruijf, H. A. M. 1975. General morphology and behavior of gastrozooids and dactylozooids in two species of *Millepora.* Mar. Behav. Physiol. 3:181–192.

Fraser, C. 1954. Hydroids of the Atlantic Coast of North America. University of Toronto Press, Toronto. 451 pp.

Gierer, A. 1974. Hydra as a model for the development of biological form. Sci. Am. 231:44–54.

Lange, R. G., and Müller, W. A. 1991. SIF, a novel morphogenetic inducer in Hydrozoa. Dev. Biol. 147:121–132.

Larson, R. J., Mills, C. E., and Harbison, G. R. 1989. *In situ* foraging and feeding behavior of Narcomedusae (Cnidaria: Hydrozoa). JMBA (UK) 69:785–794.

Mackie, G. O., and Passano, L. M. 1968. Epithelial conduction in hydromedusae. J. Gen. Physiol. 52:600.

Mackie, G. O., Pugh, P. P., and Purcell, J. E. 1987. Siphonophore biology. Adv. Mar. Biol. 24:97–262.

Mills, C. E. 1981. Diversity of swimming behaviors in hydromedusae as related to feeding and utilization of space. Mar. Biol. 64:185–189.

Okamura, B., Curry, A., Wood, T. S., and Canning, E. U. 2002. Ultrastructure of *Buddenbrockia* identifies it as a myxozoan and verifies the bilaterian origin of the Myxozoa. Parasitology 124:215–223.

Ostman, C. 1982. Nematocysts and taxonomy in *Laomedea, Gonothyraea* and *Obelia.* Zool. Scripta 11:227–241.

Pardy, R. L., and White, B. N. 1977. Metabolic relationships between green hydra and its symbiotic algae. Biol. Bull. 153:228–236.

Petersen, K. W. 1990. Evolution and taxonomy in capitate hydroids and medusae (Cnidaria: Hydrozoa). Zool. J. Linn. Soc. 100:101–231.

Purcell, J. E. 1981. Feeding ecology of *Rhizophysa eysenhardti,* a siphonophore predator of fish larvae. Limnol. Oceanogr. 26:421–432.

Purcell, J. E. 1984. The functions of nematocysts in prey capture by epipelagic siphonophores. Biol. Bull. 166:310–327.

Satterlie, R. A., and Spencer, A. D. 1983. Neuronal control of locomotion in hydrozoan medusae: A comparative study. J. Comp. Physiol. 150:195–206.

Siddall, M. E., Martin, D. S., Bridge, D., Cone, D. K., and Desser, S. S. 1995. The demise of a phylum of protists: Myxozoa and other parasitic Cnidaria. J. Parasitol. 81:964–967.

Singla, C. L. 1975. Statocysts of hydromedusae. Cell Tiss. Res. 158:391–407.

Soong, K., and Cho, L. C. 1998. Synchronized release of medusae from three species of hydrozoan fire corals. Coral Reefs 17:145–154.

Spadinger, R., and Maier, G. 1999. Prey selection and diel feeding of the freshwater jellyfish, *Craspedacusta sowerbyi.* Freshwat. Biol. 41:567–573.

Stretch, J. J., and King, J. M. 1980. Direct fission: An undescribed reproductive method in hydromedusae. Bull. Mar. Sci. 30:522–525.

Swedmark, B., and Teissier, G. 1966. The Actinulida and their evolutionary significance. In Rees, W. J. (Ed.): The Cnidaria and Their Evolution. Symp. Zool. Soc. Lond. 16:119–133.

Thomas, M. B., and Edwards, N. C. 1991. Cnidaria: Hydrozoa. In Harrison, F. W., and Westfall, J. A. (Eds.): Microscopic Anatomy of Invertebrates. Vol. 2. Wiley-Liss, New York. pp. 91–193.

Totton, A. K. 1965. A Synopsis of the Siphonophora. Trustees of the British Museum (Natural History), London. 230 pp.

Internet Sites

www.biology.ualberta.ca/facilities/multimedia/index.php?Page=252 (Animation of life cycle of *Obelia* by A. R. Palmer and H. Kroening.)

http://faculty.washington.edu/cemills/Hydromedusae.html (C. E. Mills's Web page of photographs and information about hydromedusae.)

www.mbari.org/~kraskoff/medusae2.htm (K. Raskoff's stunning photographs of midwater hydromedusae from Monterey Bay.)

8 Ctenophora^P

The mostly planktonic ctenophores are gelatinous marine animals known as comb jellies. Ctenophores are voracious predators of zooplankton, such as crustaceans, jellyfish, fish eggs, and often each other. Of the approximately 80 species, several are seasonally abundant in coastal waters and thus play an important role in the planktonic food web. The transparent, fragile body is generally spherical or egg-shaped and superficially resembles that of some cnidarian jellyfish (Fig. 8-1). Similar to jellyfish, the main axis of the body is oral-aboral and the mesoglea is thick and buoyant. But ctenophores differ from jellyfish in two striking ways: They use cilia instead of muscles for locomotion and they lack cnidae. Because of these differences and some unique attributes of their own, ctenophores are classified separately from cnidarians. Historically, however, ctenophores and cnidarians were combined in the now defunct taxon Coelenterata, although the informal term "coelenterate" continues to be used in reference to both taxa.

Ranging in size from a few millimeters to 30 cm or more (the ribbonlike *Cestum* reaches 1.5 m), ctenophores are the largest solitary animals to use cilia for locomotion. The smooth ciliary locomotion of ctenophores may cloak them from vibration-sensitive predators and prey, thus allowing them to slip stealthily through the water.

Transparency and lack of color typify most planktonic ctenophores and account for their near invisibility in water.

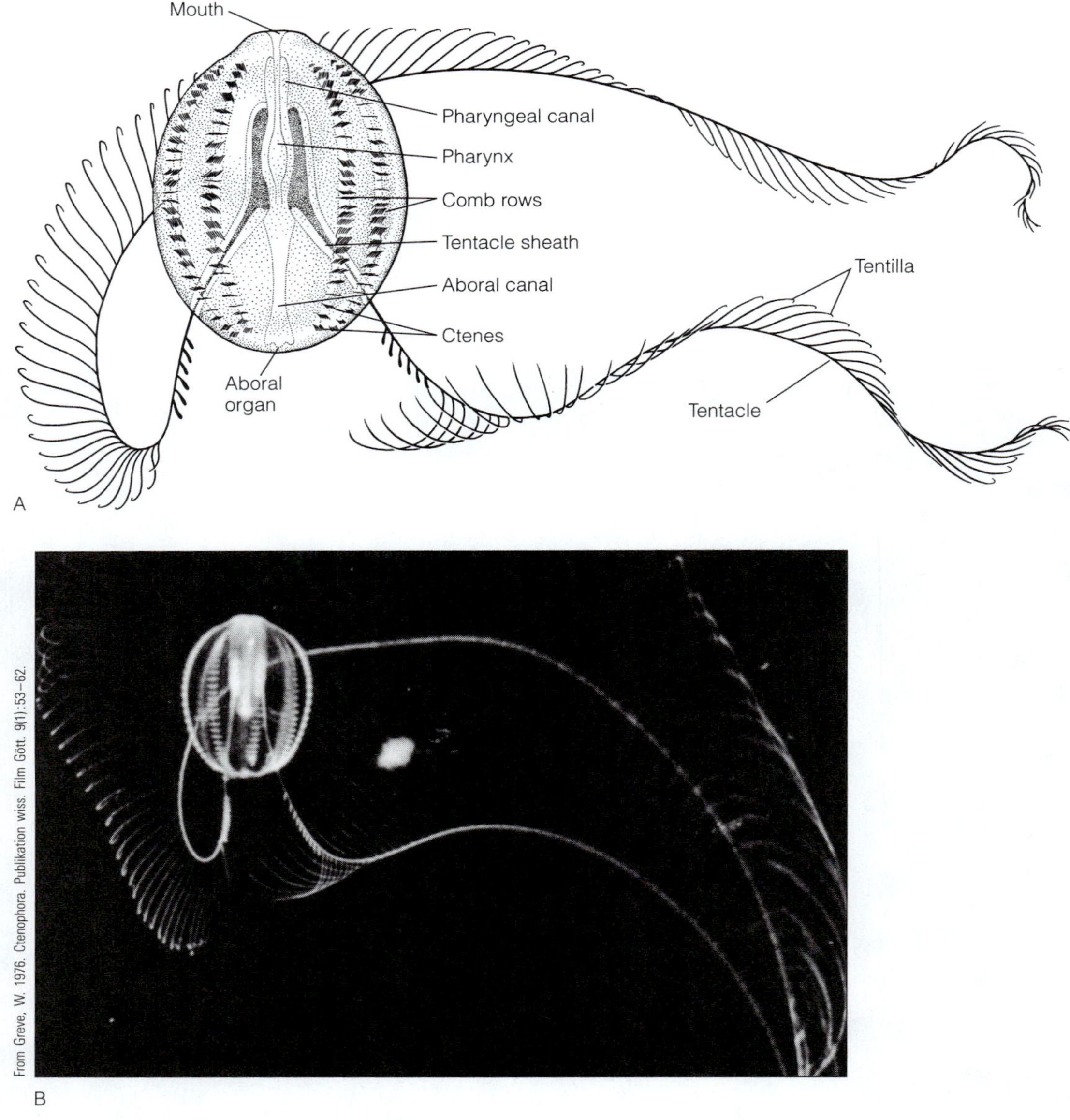

FIGURE 8-1 Ctenophora: external form and feeding orientation. **A,** *Pleurobrachia,* tentacular-plane view. **B,** Living *Pleurobrachia pileus* "fishing" with expanded tentacles. *(A, Modified and redrawn from Hyman, L. H. 1940. The Invertebrates, Vol. 1. McGraw-Hill Book Co., New York.)*

A few species, however, commonly those of *Beroë* (ba-ROE-ee), are pale pink to nearly red, but more deeply pigmented species generally occur in deep water in the absence of light. Pigmentation also occurs in unusual benthic species (platyctenids), which attach to substrates such as mangrove roots and the bodies of sessile animals and use color patterns to camouflage themselves. At night, ctenophores (except for species of *Pleurobrachia* and a few others) are noted for their striking bioluminescence, which emanates from cells below the ciliated bands (comb rows). When disturbed, the tropical *Eurhamphaea vexilligera* releases a sparkling, reddish, bioluminescent ink that may distract predators.

GENERAL BIOLOGY

The ctenophore ground plan is expressed by adults of a few species, especially those of *Pleurobrachia,* and by juveniles. Like cnidarians, the ctenophore body is arranged radially around an oral-aboral axis. The oral end of the body bears the mouth and a brainlike organ occurs at the aboral pole.

Between the two poles are eight rows of extraordinary cilia that can move the large body at speeds of up to 50 mm/s. The cilia in each row are exceptionally long (up to 2 mm) and grouped together in paddles known as **ctenes,** or combs (Fig. 8-1, 8-2); each ctene is composed of

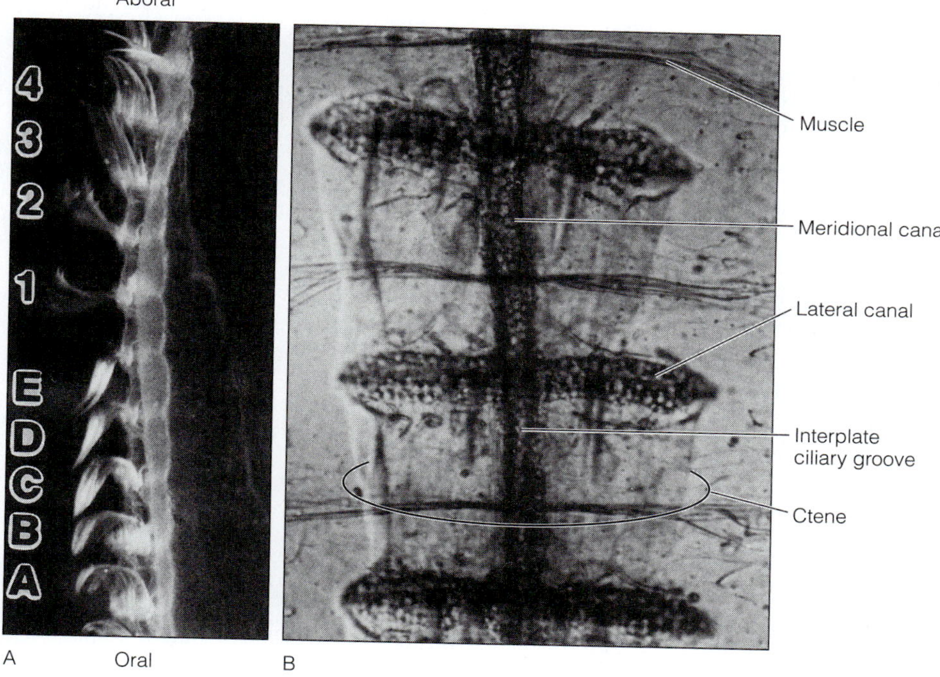

FIGURE 8-2 Ctenophora: functional anatomy of comb rows. Images of living *Mnemiopsis mccradyi* (Lobata). **A,** Lateral view of beating ctenes. Numbers 1–4 indicate ctenes in successive stages of power stroke; letters A–E, stages of recovery stroke. **B,** Surface view of two ctenes and associated structures.

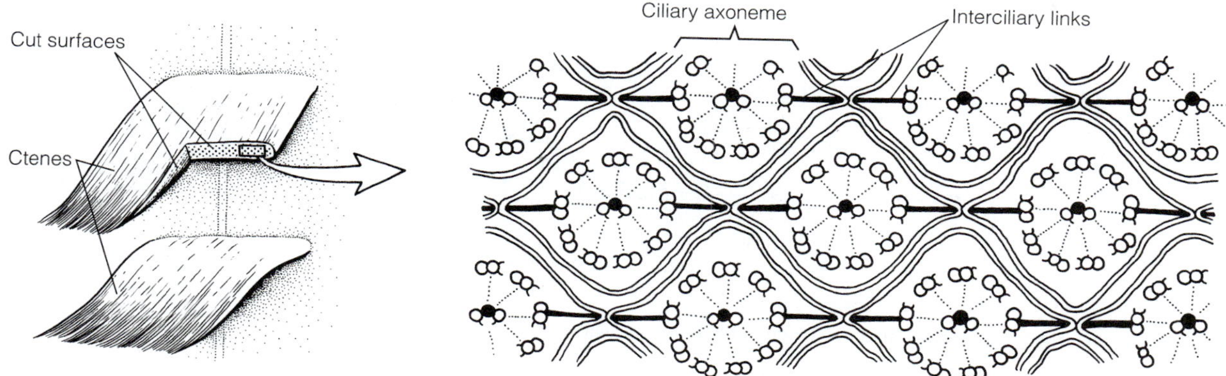

FIGURE 8-3 Ctenophora: ctene microanatomy. Parallel axonemes of ctene cilia are linked together in the beat plane of the comb. *(Redrawn and modified from Tamm, S. L. 1980. Cilia and ctenophores. Oceanus 23:50–59.)*

several thousand cilia (Fig. 8-3). The ctenes are aligned single-file in each of the eight **comb rows** (Figs. 8-1A, 8-4A), which originate at the aboral pole but end before reaching the mouth at the opposite end (Fig. 8-1). The paddlelike ctenes beat with effective strokes directed aborally (Fig. 8-2A). So ctenophores, unlike jellyfish, swim with the mouth forward. Characteristically, light diffracted by the moving ctenes produces a shimmering spectrum of color. In addition to their ctenes, ctenophores use other, highly modified cilia as specialized devices, such as pacemakers (balancers) to set the beat rate of the ctenes and even as teeth (macrocilia in *Beroë*).

Two long **tentacles,** one on each side of the body, originate on the aboral hemisphere and bear a lateral row of threadlike filaments called **tentilla** (sing., tentillum) (Fig. 8-1, 8-4A, 8-5A). Each highly extensible and contractile tentacle emerges from the bottom of a deep, ciliated pouch, the **tentacle sheath** (Fig. 8-1, 8-4A).

Much of the body volume is occupied by the skeletal **mesoglea,** an elastic gel that contains collagen and cells of various types. Internal to the mouth, the coelenteron consists of a large, compressed pharynx that opens into a short canal from which additional canals arise to supply the comb rows, tentacle sheaths, and other structures (Fig. 8-4A).

Ctenophores have a curious **biradial symmetry** about two imaginary planes, each of which divides the body into mirror-image halves. The **tentacular plane** passes along the oral-aboral axis and through both tentacle sheaths. The **pharyngeal plane** also passes along the oral-aboral axis, but at a right angle to the tentacular plane (Fig. 8-4A). Thus, in contrast to most cnidarians, which have many radial planes of symmetry, ctenophores have only two.

BODY WALL AND COLLOCYTES

The ctenophore body wall, including the pharynx and tentacle sheaths, includes an epidermis, which is coated externally by a protective layer of mucus, and a well-developed mesoglea (Fig. 8-5C). Muscle and nerve cells occur both in the epidermis and in the mesoglea.

The ctenophore epidermis appears to be a bilayered epithelium. The *outer* layer consists of multiciliated cells, inter-

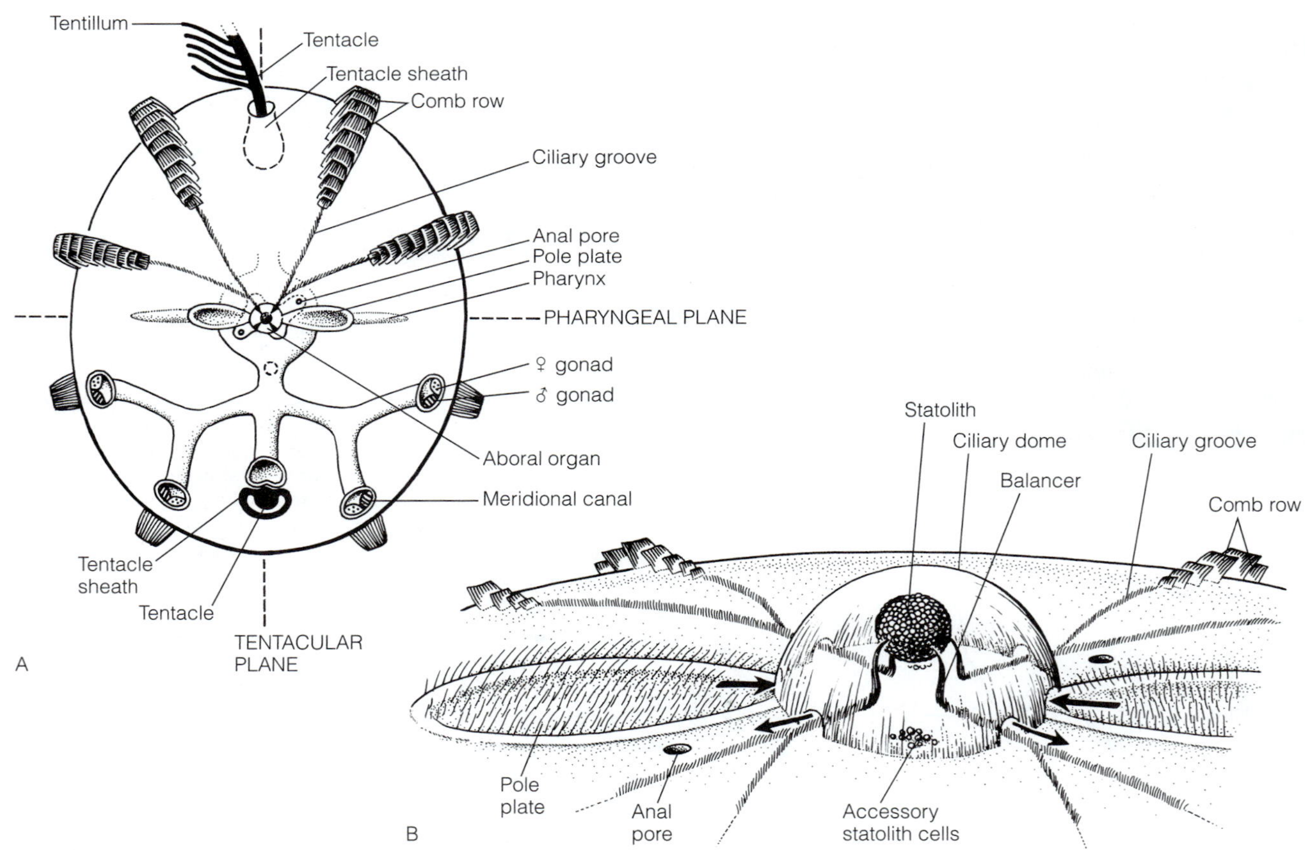

FIGURE 8-4 Ctenophora: symmetry and anatomy (Cydippida). **A,** Aboral view (top half of figure), sectional view (bottom half of figure) of *Hormiphora* (similar to *Pleurobrachia*). **B,** Perspective view of aboral organ of **A.** Arrows indicate ciliary-beat direction and water flow. *(A, Modified and redrawn from Delage Y., and Hérouard, E. 1901. Traité de Zoologie Concrète. Vol. 2. Les Coelentérés. Schleicher Frères, Paris. 848 pp.; B, Modified and redrawn from Kaestner, A. 1984. Lehrbuch der Speziellen Zoologie. 2. Teil. Gustav Fischer Verlag, Stuttgart. 621 pp.)*

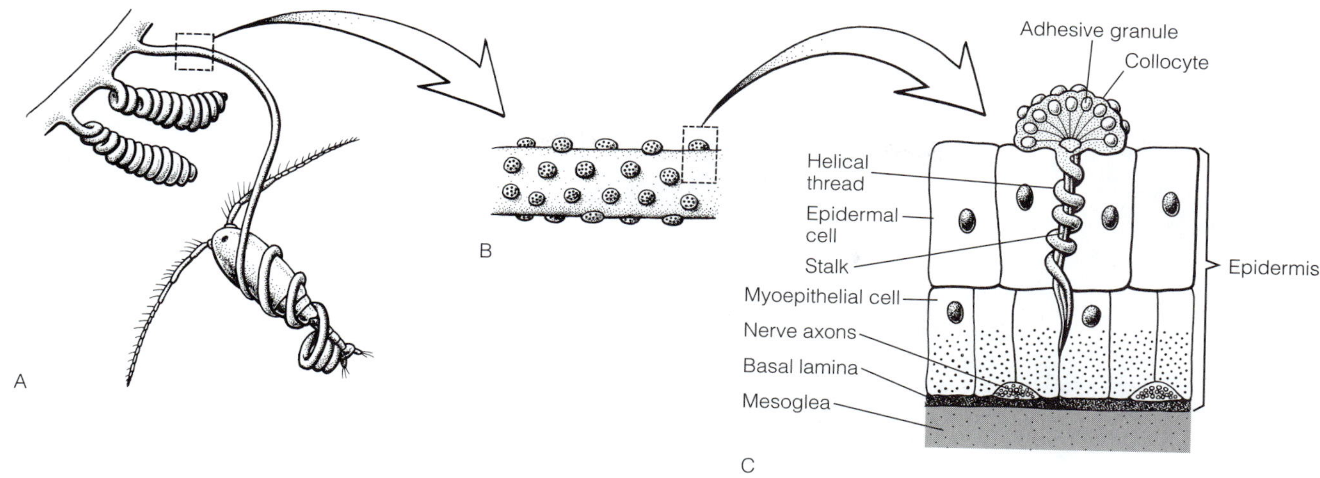

FIGURE 8-5 Ctenophora: prey capture using tentilla and collocytes. **A,** Prehensile tentillum of *Euplokamis dunlapae* uncoils rapidly to seize prey (copepod). **B,** High magnification of tentillum showing protruding collocytes. **C,** Section through collocyte and wall of tentillum. The muscle-cell nuclei degenerate in the fully formed tentillum. *(A and C, Modified and redrawn from Mackie, G. O., Mills, C. E., and Singla, C. L. 1988. Structure and function of the prehensile tentilla of* Euplokamis *(Ctenophora, Cydippida). Zoomorphology 107:319–337; B, Modified and redrawn from Hyman, L. H. The Invertebrates, Vol. 1. Mc Graw-Hill Book Co., New York.)*

stitial cells, mucus-secreting cells, collocytes (see below), and various sensory cells. Of these cells, the mucus-secreting cells, which produce the protective mucous covering of the body, and the sensory cells are widespread; the other cells are localized to specific structures, such as the ctenes, pharynx, and tentacle sheaths. The *inner* layer of the epidermis contains myoepithelial cells and a nerve net. Beneath the bilayered epidermis, the mesoglea houses myocytes as well as nerve, pigment (*Beroë*), and ameboid cells, the latter of which secrete and maintain the extracellular matrix of the mesoglea.

The epidermis of each tentacle and tentillum has numerous specialized cells, called **collocytes,** that are unique to ctenophores (Fig. 8-5C). Collocytes release a sticky substance and adhere to prey. Like a cnidarian cnidocyte, each collocyte differentiates from an interstitial cell in the epidermis. The differentiated collocyte, which resembles a mushroom, consists of a swollen apical cap borne on a stalk. The cap bears numerous vesicles that contain the adhesive substance (Fig. 8-5C). The stalk anchors the collocyte in the epidermis and in the deeper muscle layer or mesoglea. A helical thread resembling a coiled spring surrounds the stalk and is attached to the root tip at one end and the cap at the other (Fig. 8-5C). Although the function of this peculiar structure is uncertain, it may absorb stress when an attached prey tugs on the collocyte, thus preventing the collocyte from being torn away. Apparently, each collocyte is used only once and then replaced.

MUSCULATURE

Muscles occur throughout the epidermis and mesoglea. The myoepithelial epidermal musculature may include longitudinal muscles, circular muscles, or both, depending on species. When both orientations are well developed, as in the prehensile oral lobes of lobate ctenophores, the fibers form a crisscrossed mesh (Fig. 8-6). The mesogleal musculature, composed of myocytes, may be oriented circularly, longitudinally, or radially. Circular and radial fibers of mesogleal muscle, for example, act together with *epidermal* longitudinal fibers to open and close the mouth and pharynx of *Beroë*. In *Mnemiopsis*, two well-developed bands of longitudinal muscle (retractor muscles) insert on the underside of the aboral organ, then extend through the mesoglea to their origin at the oral end of the body (Fig. 8-7A). These muscles draw in the aboral organ to protect it. Similarly, radial muscles attached to the underside of the comb rows (Fig. 8-2B) contract to create a furrow into which the fragile combs are withdrawn for protection.

FIGURE 8-6 Ctenophora: musculature. Surface view of the crisscrossed muscle fibers in the oral lobe of a living lobate ctenophore, *Mnemiopsis mccradyi.* This muscle grid enables *Mnemiopsis* to cup its oral lobes over its mouth.

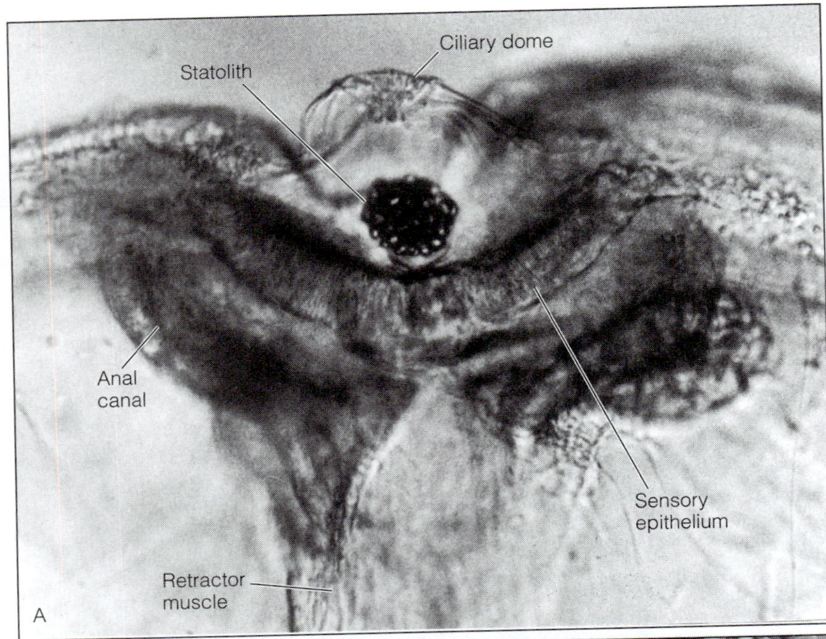

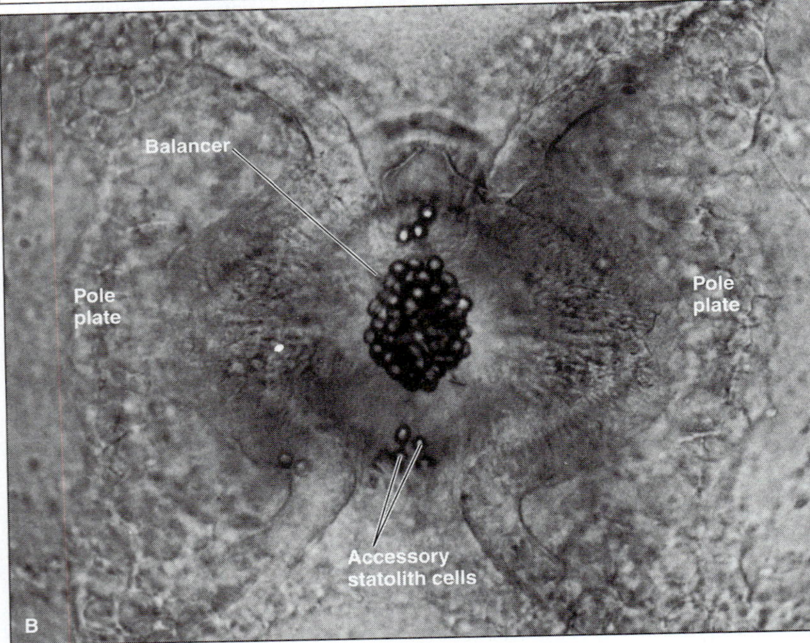

FIGURE 8-7 Ctenophora: aboral organ of a living *Mnemiopsis mccradyi* (Lobata) in lateral **(A)** and surface **(B)** views.

With few exceptions, all ctenophore muscles are smooth; cross-striated muscle is described only in *Euplokamis dunlapae,* where it causes a rapid uncoiling of the tentilla during prey capture (Fig. 8-5A). Ctenophores apparently lack the epitheliomuscular cells characteristic of most cnidarians. Instead, their epidermal fibers are considered to be myoepithelial cells and their mesogleal fibers, myocytes (Fig. 6-4C). The mesogleal myocytes are surrounded with basal lamina (called external lamina), which betrays their probable epithelial origin. The smooth muscle fibers of both cnidarians and ctenophores, however, differ from smooth fibers of "higher" (bilateral) animals in one surprising respect: Their actin (thin) filaments anchor only to the cell membrane. Among bilateral animals, such as flatworms and molluscs, the actin filaments also anchor to proteins (such as α-actinin) within the cell cytoplasm (Fig. 6-5A).

NERVOUS SYSTEM

The ctenophore nervous system forms an epidermal nerve net that is concentrated in association with complex structures, such as the comb rows, pharynx, tentacles, and aboral organ. Neurons also occur in the mesoglea and innervate the mesogleal musculature. Although the net forms a ring around the mouth (associated with sensory cells on the lips), the principal nervous center is the aboral organ.

The **aboral** (or apical) **organ,** a rudimentary brain, is a combination receptor-effector that controls the locomotory comb rows (Fig. 8-4, 8-7). Its most conspicuous component is a glassy spherical **statocyst** perched atop the flared end of the aboral coelenteric canal like a diminutive crystal ball. Remarkably, the transparent dome of the statocyst is formed from long, nonmotile cilia, which arch over and enclose a **statolith** composed of aggregated cells bearing reflective granules. Basally, the statolith is supported on four equidistant tufts of motile cilia, the so-called **balancers,** that arise from cells of the statocyst floor. Each balancer controls ciliary activity in one **quadrant,** or two adjacent comb rows, of the body. From each balancer, a fine **ciliary groove** radiates orally, then bifurcates to join two adjacent comb rows (Fig. 8-4B). The beat rate of the ctenes in each quadrant is set by the beat rate of its balancer, which behaves as a pacemaker. The floor of the statocyst is a depression or dimple in a sensory epithelium, a specialized patch of epidermis. This sensory epithelium extends orally beyond the statocyst as two opposite, narrow, ciliated bands, the **pole plates,** in the pharyngeal plane of the body (Fig. 8-4). The pole-plate cilia create a ventilatory flow toward and over the statocyst. Other pole-plate cells may be sensory in function. Among some ctenophores, such as *Beroë,* branched, fingerlike outgrowths called **papillae** project from the pole-plate perimeter (Fig. 8-14B). In addition to gravity reception (by the statocyst), the aboral organ may function in photo-, pressure-, and chemoreception.

LOCOMOTION

Garden-variety ctenophores such as *Mnemiopsis, Pleurobrachia, Bolinopsis,* and *Beroë* rely primarily on ctenes for locomotion. Some less common species (cestids, platyctenids, some lobates), however, use muscular undulations of the body to swim or to crawl like a flatworm (Fig. 8-12A,C).

During ciliary locomotion, the power stroke of the paddlelike ctenes is from oral to aboral. Activation of the ctenes, however, is the opposite, beginning aborally and proceeding to the oral end of each comb row. Curiously, the beat stimulus is initiated by the balancer and transmitted to the comb rows by cilia in the ciliary grooves rather than by nervous conduction. Each beat of a balancer provides a *mechanical* stimulus, which propagates along the comb rows aborally to orally, in a manner similar to a cascade of falling dominoes. In *Pleurobrachia* and other cydippids, the mechanical stimulus is thought to be the fluid disturbance caused by the beating ctene. In many other ctenophores, however, the ciliary grooves do not end at the first ctene in each comb row, but continue along the row as **interplate ciliary grooves** (Fig. 8-2B), which link the separate ctenes and aid in conducting the beat stimulus.

The beat frequency of the balancer is controlled by the weight of the statolith bearing on it. When the ctenophore is oriented vertically, the weight is equal on all four balancers and their beat rates are the same. If the ctenophore tilts away from vertical, the weight is greater on the lower balancers than the upper ones (Fig. 8-8A). This load imbalance causes the upper and lower balancers to beat at different rates, which, in turn, causes beat-rate changes in the comb rows. Greater statolith weight on the lower balancers, however, may increase *or* decrease the beat rate, depending on the ctenophore's "**mood.**" If its mood is positive (+ geotaxis), increased weight on the lower balancers slows their beat rate and the beat rate of the lower comb rows. In the meantime, the beat rate of the upper balancers and their comb rows increases. Because the upper ctenes beat frequently and the lower ctenes hardly at all, the ctenophore turns its leading mouth end downward and swims away from the surface of the water. When a comb jelly's mood is negative (− geotaxis), the response of its balancers to gravitational load is reversed and it swims toward the surface of the water (Fig. 8-8A).

A ctenophore's "mood" is controlled by sensory information received and processed by the nervous system. The nervous system also controls the direction of effective strokes of the ctenes. Although normally the effective stroke is toward the aboral pole, it can be reversed in two or more pairs of comb rows.

COELENTERON

The coelenteron consists of distinct compartments and canals (Fig. 8-9) that supply the organs of the body. In *Pleurobranchia,* the slit-shaped mouth leads into a laterally compressed pharynx that joins a short **aboral canal,** the first part of which is sometimes called a **stomach** (Fig. 8-9). Immediately below the aboral organ, the aboral canal branches into four short **anal canals** that embrace and supply the organ. Two of the four anal canals, 180° apart, open to the exterior at tiny **anal pores.** The remaining bodily organs and structures are supplied by canals that originate from the small stomach. A **pharyngeal canal** extends over each of the two flattened surfaces of the pharynx; a **tentacular canal** lies along each of the two tentacular sheaths; a **meridional canal** is located beneath each of the eight comb rows; and a **lateral canal** underlies each ctene (Fig. 8-2B). The meridional canals end blindly in *Pleurobrachia,* but in many species, including *Beroë,* they join a circumoral ring canal that unites them. The arrangement of coelenteric canals varies among the species of ctenophores, but most can be interpreted as modifications to the layout of *Pleurobrachia* (Fig. 8-9).

Ciliated cells occur throughout the gastrodermis, but are especially well developed on the innermost part of the pharynx near its junction with the aboral canal. These pharyngeal cilia occur in four longitudinal ridges *(Mnemiopsis)* resembling low septa and in a cluster at the junction with the stomach. Secretory gland cells also are prominent on the pharyngeal ridges and elsewhere. Only the mouth and pharynx have a well-developed musculature in addition to cilia.

With the exception of the aboral canal, all coelenteric canals are located near the body or gut surface, immediately below the organs being supplied, and the gastrodermis on the organ side of the canal differs from the gastrodermis on the opposite side. The organ-side gastrodermis is composed of tall, vacuolated **nutritive cells** (Fig. 8-10A, 8-14C) interspersed with germ cells and **photocytes,** the cells responsible for bioluminescence. The gastrodermis on the opposite side of the canal consists of a low, ciliated epithelium periodically interrupted by pores. Each pore is surrounded by two ciliary whorls, one atop the other, of six cells each. The cilia of the inner whorl project into the coelenteron, while those of

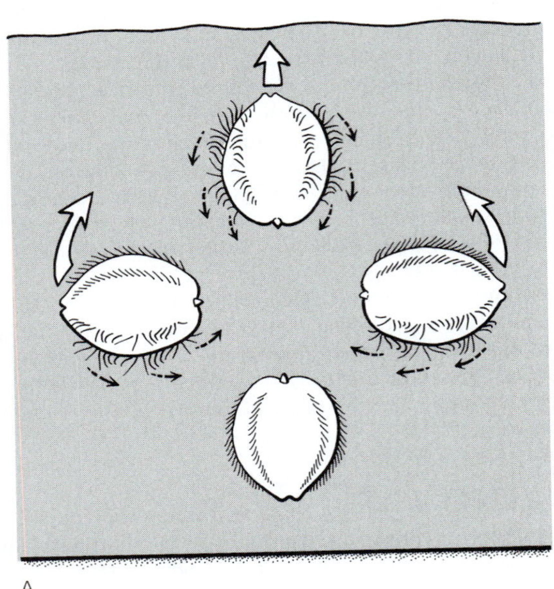

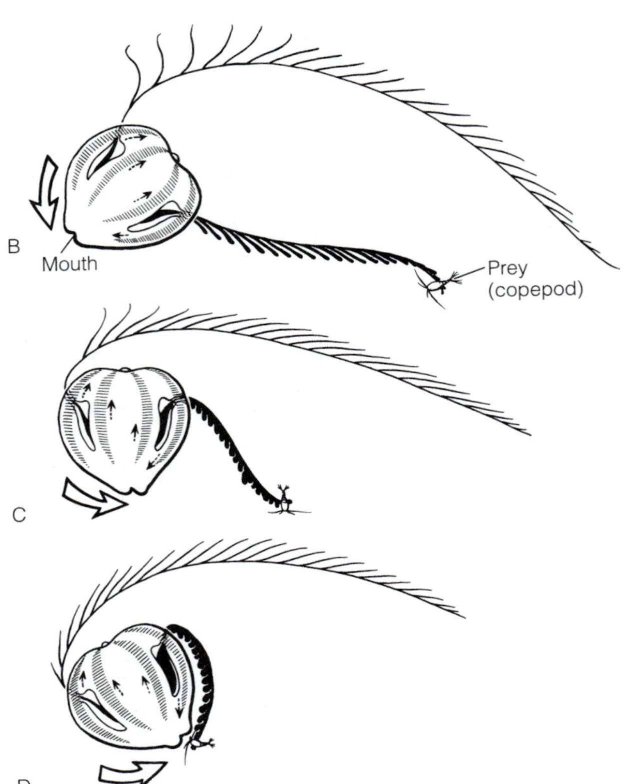

FIGURE 8-8 Ctenophora: orientation behavior in *Pleurobrachia* (Cydippida). **A,** *Pleurobrachia* in negative "mood" (– geotaxis), responding to departures from preferred orientation (top). The active course corrections (left and right) are controlled by the aboral organ. **B–D,** Prey capture and ingestion ("spin capture"). As prey contacts a tentacle, one pair of comb rows reverses the direction of its power strokes to rotate the ctenophore toward the capture tentacle **(B)**. The body continues to rotate **(C)** as the tentacle shortens and brings the prey into contact with the mouth **(D)**, which engulfs it. *(Modified and redrawn from Tamm, S. L. Cilia and Ctenophores. Oceanus 23:50–59.)*

the outer whorl extend into the mesoglea. The entire pore is called a **ciliary rosette** (Fig. 8-10, 8-14C). Its function is discussed under Excretion and Buoyancy Regulation.

PREY CAPTURE, DIGESTION, INTERNAL TRANSPORT

Ctenophores are carnivores that feed chiefly on planktonic crustaceans and other jellies. Crustacean-ingesting ctenophores such as *Pleurobrachia* use tentacles and tentilla to capture prey, whereas jelly consumers use either their simple tentacles, which lack tentilla, or direct seizure with the mouth. An example of a ctenophore that uses the latter method is *Beroë,* which feeds on other ctenophores, lacks tentacles altogether, and actively pursues its prey.

Swimming mouth forward, *Pleurobrachia* feeds by trailing its outstretched tentacles and tentilla through the water like a drift net. If a small crustacean contacts one of the tentacles, it is seized by one or more adhesive tentilla, which collocytes have made as sticky as spider's silk. Then the tentacle shortens, pulling the prey forward and toward the mouth. Reversing the beat of the appropriate pairs of comb rows, the ctenophore rotates its body toward the capture tentacle to bring its mouth into contact with the prey (Fig. 8-8B–D). This form of feeding is known as **spin capture.**

Once the prey is swallowed, muscular churning and enzyme release by the pharynx begin digestion within minutes. The pharynx reduces the prey to a slurry of fine particles that stream rapidly from the ciliated ridges into the stomach. (The large, indigestible parts of the prey body, such as the exoskeleton, are egested through the mouth and do not enter the canal system.) Once in the stomach, the food is distributed by ciliary flow to all the canals of the coelenteron. Particles flow bidirectionally in the canals until they are endocytosed by the nutritive cells, in which they presumably undergo intracellular digestion. Storage of nutrients probably occurs in the nutritive cells. The location of nutritive cells immediately below the ctenes, comb rows, aboral organ, and other structures undoubtedly facilitates transfer of nutrients to those tissues and organs. The pores of ciliary rosettes open into the myocyte-containing mesoglea and may be routes of nutri-

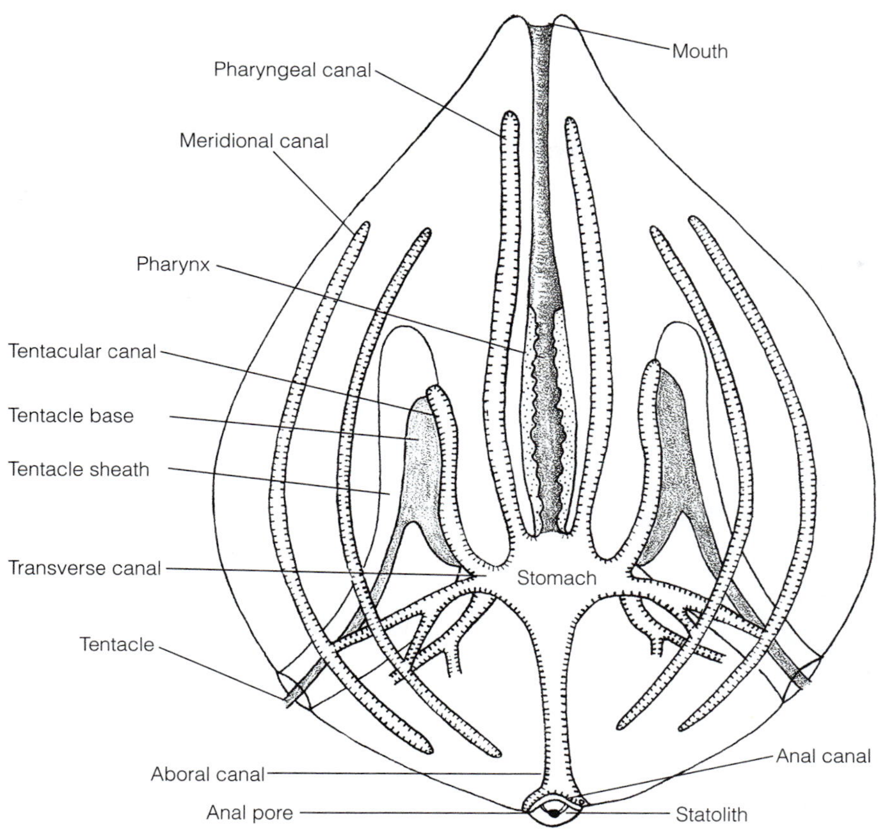

FIGURE 8-9 Ctenophora: digestive system of a cydippid ctenophore similar to *Pleurobrachia*. *(Modified from Hyman, L. H. 1940. The Invertebrates, Vol. 1. McGraw-Hill Book Co., New York)*

ent transport to those muscles. Although some rejected particles leave the body via the anal pores, most are eliminated through the mouth.

EXCRETION AND BUOYANCY REGULATION

Little is known about ctenophore excretion. The by-products of cellular metabolism, such as ammonia from proteins, probably diffuse across the general body surface, although some may leave with coelenteric fluid from the mouth and anal pores. The excretory role, if any, of the ciliary rosettes is unknown.

Maintenance of neutral buoyancy by planktonic ctenophores is probably the result of passive osmotic adjustment to water of different densities. The ciliary rosettes, however, may assist in this accommodation process by pumping water into or from the mesoglea, the ctenophore's flotation tissue. Because ctenophore body fluids are isosmotic to seawater, which is denser than fresh water, ctenophores tend to sink as they enter the brackish water of coastal estuaries. In this situation, the mesogleal ciliary tuft of the rosettes may actively pump the less-dense brackish water into the mesoglea to increase its volume and decrease its density, thus approaching neutral buoyancy (Fig. 8-10, 8-14C). Alternatively, when a brackish-water-adapted ctenophore moves into full-strength seawater, water may be pumped by the coelenteric ciliary tuft from the mesoglea into the coelenteron and eventually to the outside through the mouth. This loss of water may increase the density of the body to that of full-strength seawater. Salt (ionic) regulation has not been studied.

REPRODUCTION AND DEVELOPMENT

Clonal (asexual) reproduction, uncommon in ctenophores, occurs only in the benthic platyctenids, which can excise fragments from the margin of their flattened bodies. Like the pedal lacerates of some sea anemones (Fig. 7-22), these liberated fragments differentiate and grow into new individuals.

With the exception of two gonochoric species of *Ocyropsis,* all species of ctenophores are hermaphrodites. In response to light cues, gametes are spawned and the eggs are fertilized in seawater. Cross-fertilization probably is typical of most ctenophores, but self-fertilization occurs in species of *Mnemiopsis.* An interesting aspect of the reproductive biology of lobate and cydippid ctenophores is that their juveniles (cydippids) can precociously attain sexual maturity and successfully reproduce. This phenomenon, an example of pedomorphosis (Chapter 4), is known as **dissogeny.**

Generally, a pair of gonads, one ovary and one testis, is situated in the body-surface-side wall of each meridional canal (Fig. 8-4A). In this position, the gonads are in contact with the nutritive cells of the gastrodermis. Within the ovary, each egg is partially fused by cytoplasmic bridges to many nurse cells, which supply nutrients for very rapid growth. At

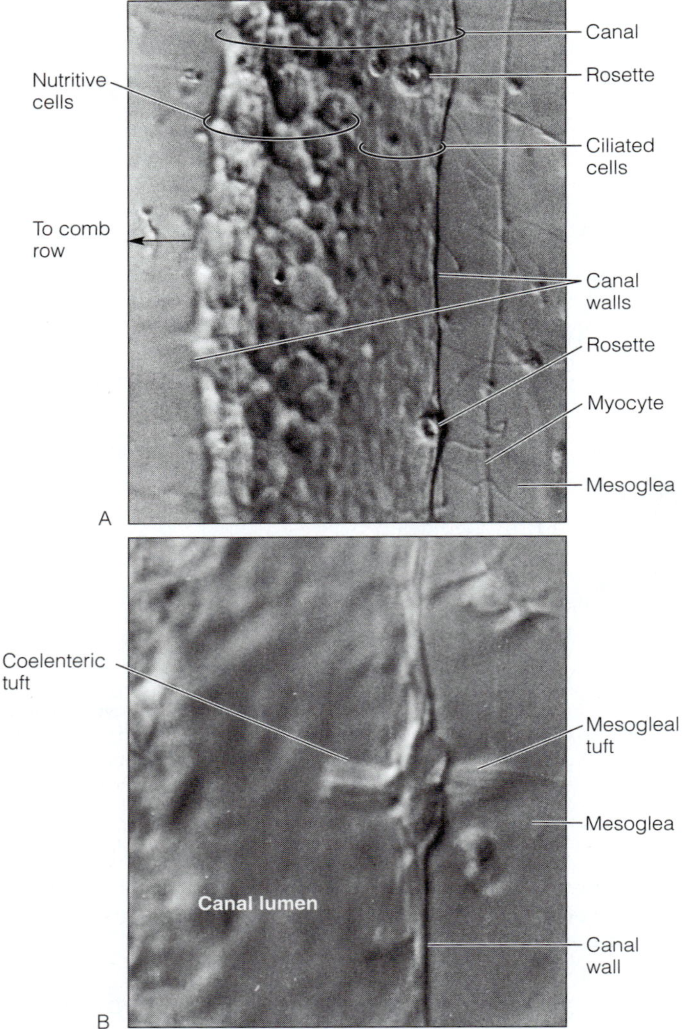

FIGURE 8-10 Ctenophora: ciliary rosettes. Microscopic images of a coelenteric canal and rosettes in a living *Mnemiopsis mccradyi* (Lobata). **A,** Canal showing two ciliary rosettes, the upper in a face-on view. **B,** High magnification of a rosette in optical section. As this photograph was taken, the planar ciliary movement of the rodlike coelenteric tuft blurred the image (the fan-shaped mesogleal tuft was stationary).

spawning, the gametes are shed from the gonads into short canals that lead to a series of gonopores, each located between two adjacent ctenes.

The glass-clear yolky zygote undergoes determinate (mosaic) development, in which the fate of each blastomere is determined early in development (Fig. 8-11). The first cleavage furrow begins on one side of the zygote and progresses around its circumference, eventually dividing it into two equal cells. The point at which the first cleavage is initiated corresponds to the oral end of the embryo and adult. The second cleavage is at 90° to the first and results in an embryo composed of four equal cells. Seen from the oral pole, the crossed planes of the first and second cleavages correspond to the future pharyngeal and tentacular planes, respectively (Fig. 8-11C). Additional cleavages result in an embryo composed of a few large oral macromeres and many small aboral micromeres (Fig. 8-11E). During gastrulation, the aboral micromeres spread orally (by epiboly), and some are internalized (by invagination) at the oral pole. The aboral micromeres form the ectoderm and will differentiate into epidermis, pharynx, tentacle sheaths, and gonads, which are thus of ectodermal origin. Before the macromeres give rise to the endodermal coelenteron, they produce another set of micromeres at the *oral* pole. These oral micromeres are mesoderm and give rise to the mesogleal muscles and other cells.

Embryonic development is direct and leads to a planktonic juvenile, called a **cydippid,** that is of similar form in nearly all groups of ctenophores. It resembles a miniature of *Pleurobrachia* with eight comb rows and a pair of tentacles (Fig. 8-11I). The juvenile of *Beroë* is similar, but lacks tentacles and tentacle sheaths, as does the adult. As the cydippid grows, it gradually adopts the specific form of its parent. Among certain groups of ctenophores, such as the benthic platyctenids, the cydippid is a true larva that occupies a different ecological niche than the parent and undergoes a metamorphosis as it adopts the unusual body form of the adult.

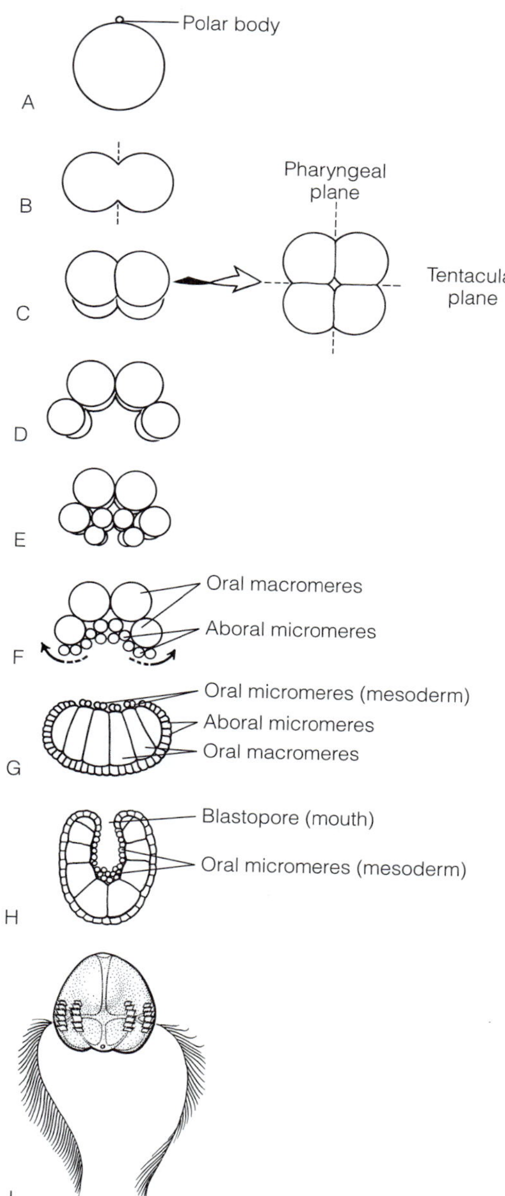

FIGURE 8-11 Ctenophora. Development **(A-I)** is characterized by an early determination of the fate of blastomeres (mosaicism) and biradial symmetry, which is established at the four-cell stage **(C)**. Epiboly of the aboral micromeres occurs in **F** (arrows), marking the onset of gastrulation; **G,** early gastrula; **H,** gastrula; **I,** cydippid. *(Redrawn and modified from several sources, primarily Martindale, M. Q., and Henry, J. 1997. Ctenophorans, the comb jellies. In Gilbert, S. F., and Raunio, A. M. (Eds.): Embryology, Constructing The Organism. Sinauer, Sunderland, MA. pp.87–111.)*

DIVERSITY OF CTENOPHORA

For a taxon of fewer than 100 species, the body forms and lifestyles of ctenophores are surprisingly diverse. Those most commonly seen and studied have spherical or oval bodies and occur in the shallow-water plankton. Many other ctenophores have bodies that are flattened in the tentacular plane, pharyngeal plane, or along the oral-aboral axis, depending on the taxon. A few have adopted a medusoid body form and resemble cnidarian jellyfish. Some ctenophores live in the open ocean, others inhabit deep water, and still others are benthic creeping species.

Cydippida and Platyctenida have either nearly spherical or laterally compressed bodies. If compressed, the pharyngeal axis is short (compressed in the tentacular plane). Tentacles are the primary means of prey capture. The pharyngeal and meridional canals either end blindly (cydippids) or they branch to form a complex network (platyctenids). Lobata, Ganeshida, Cestida, Thalassocalycida, and Beroida have compressed bodies with a short tentacular axis (compressed in the pharyngeal plane) and a somewhat long to very long pharyngeal axis. None use tentacles primarily, if at all, for prey capture. The meridional and pharyngeal canals unite with a circumoral ring canal.

Cydippida^{O}: Planktonic spherical or oval ctenophores; few species compressed; well-developed tentacles and tentacle sheaths; interplate ciliary grooves absent. *Callianira:* aboral body wall bears two crests; *Pleurobrachia:* two long tentacles with tentilla (Fig. 8-1); *Euplokamis:* coiled tentilla uncoil rapidly during prey capture; *Haeckelia:* tentacles bear nematocysts obtained from and used on its cnidarian prey; *Hormiphora* (similar to *Pleurobrachia*): two types of tentilla; *Lampea:* juveniles (possibly parasitic) attach to salps with the everted and distended lining of their pharynx; *Mertensia:* compressed body (short pharyngeal axis).

Platyctenida^{O}: The odd ducks among ctenophores. Unlike other comb jellies, the flattened, benthic platyctenids resemble some flatworms and sea slugs (Fig. 8-12A,B). Comb rows absent (one exception); two well-developed, tentilla-bearing tentacles capture prey. A few species have four instead of two pole plates, and their larvae, coincidentally, have only six comb rows. Ovaries normal, but testes in discrete sacs on coelenteric canals; embryos brooded in aboral, occasionally oral, sacs; normal planktonic cydippids released. Body length 5–25 mm. As the cydippid settles and metamorphoses, comb rows are lost, body flattens profoundly in oral-aboral axis, proportion of mesoglea to other tissues decreases, and coelenteric canals branch to form a complex whole-body network. Creeping sole arises from everted and distended pharynx lining, seen also in the cydippid *Lampea; Ctenoplana:* planktonic and facultatively benthic, retains comb rows throughout life; *Coeloplana* (Fig. 8-12A,B): found on alcyonarian corals; similar to *Ctenoplana* but lacks comb rows (replaced by simple papillae containing a blind meridional-canal outgrowth [Fig. 8-12A]); *Vallicula:* on submerged mangrove roots, folds body (oral) margins around bases of its tentacles to form "chimneys"; flue of each chimney is a passageway from tentacle to mouth; *Lyrocteis* and *Tjalfiella:* on pennatulaceans, marginal folds at the bases of tentacles fuse to form permanent "chimneys."

Lobata^{O}: Common planktonic comb jellies; two expanded muscular **oral lobes** flank mouth and tentacles, creating a "subumbrellar" cavity (Fig. 8-13); from base of each reduced tentacle, two **auricular grooves,** bearing tentilla, extend over subumbrellar surface of each lobe; associated

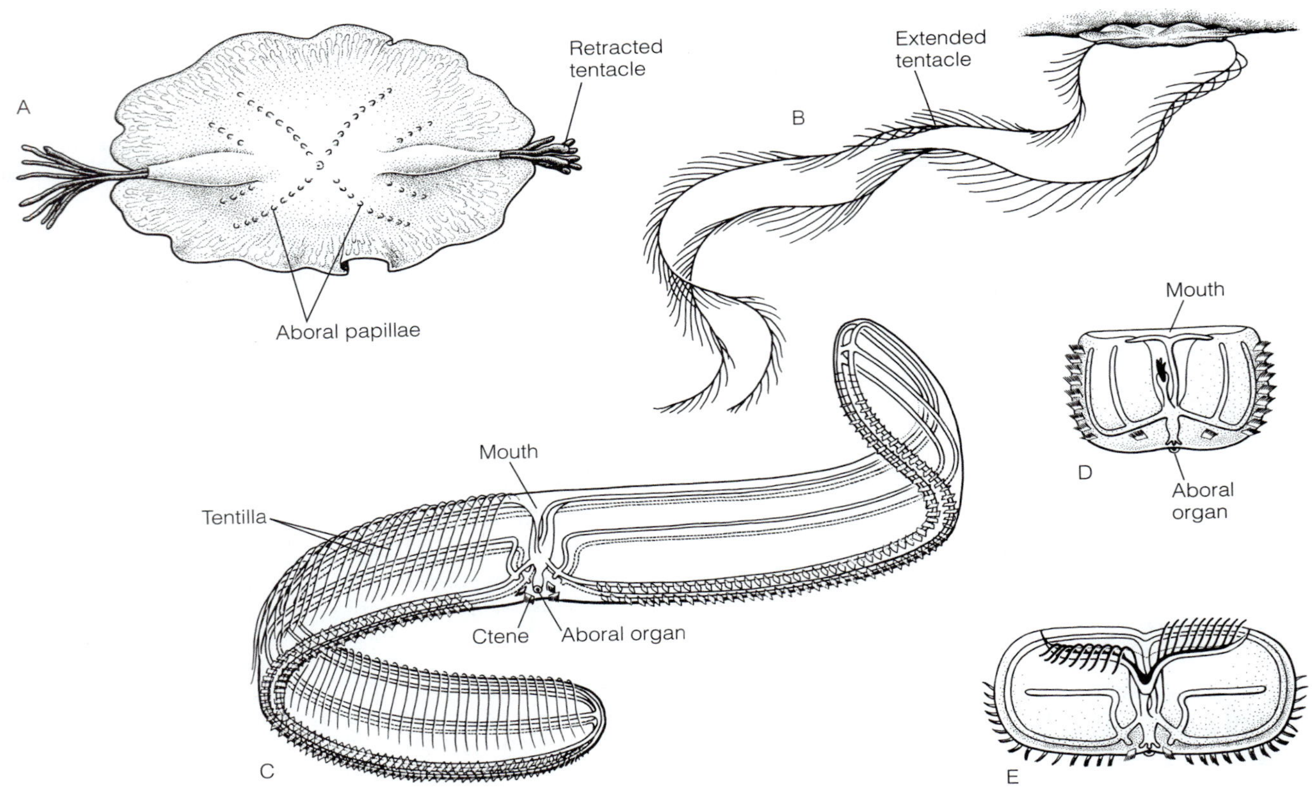

FIGURE 8-12 Ctenophora: diversity. **A** and **B**, The platyctenid *Coeloplana agniae:* **A**, aboral view with tentacles retracted; **B**, lateral view, attached to underside of substrate with tentacles extended. **C**, The cestid *Cestum veneris* in surface view with tentilla extended (left side of figure) and internal view (right side of figure). **D–E**, juvenile development of *Cestum* showing beginning of elongation in the pharyngeal axis and arrested development of comb rows flanking the tentacles. *(A and B, Modified and redrawn from Dawydoff, C. 1928. Traité d'Embryologie Comparée des Invertébrés. Masson, Paris. 930 pp; C, Modified and redrawn after Mayer, A. G. 1912. Ctenophores of the Atlantic Coast of North America. Carnegie Institution of Washington, Publication 162. 58 pp; D and E, Modified and redrawn from Kaestner, A. 1984. Lehrbuch der Speziellen Zoologie. 2. Teil. Gustav Fischer Verlag, Stuttgart. 621 pp.)*

with auricular grooves are four **auricles,** each the oral extension of a comb row that bears long whiskerlike cilia instead of paddle-shaped ctenes. The mouth-forward locomotion, aided by cilia, draws water and prey inward to the auricles, the cilia of which create a swirling current over the outstretched tentilla; prey adhere to tentilla and are transferred to mouth. Lobe system an alternative to tentacles, allows for uninterrupted feeding on suspended prey. Interplate ciliary grooves are present. *Leucothea:* long tentacles, body-wall papillae, large muscular oral lobes used for locomotion; *Mnemiopsis:* rudimentary tentacles and long, deep auricular grooves (Fig. 8-13); *Bolinopsis:* rudimentary tentacles and short auricular grooves; *Ocyropsis:* large prehensile oral lobes for prey capture and swimming, tentacles absent, gonochoric.

Ganeshida[O]: Two known species of *Ganescha:* found in tropical seas, superficially resemble lobate juveniles. Mouth flanked by two small oral lobes, a pair of tentacles present, but body is more or less circular in cross section and not appreciably compressed; coelenteric canals as in lobates; oral end of pharynx, however, is continuous with subumbrellar surface of oral lobes and thus is similar to that of platyctenids.

Cestida[O]: Remarkable, tropical-subtropical, planktonic ctenophores with body a transparent colorless ribbon or cummerbund (Fig. 8-12C–E). Tentacular axis short and pharyngeal axis extremely long. Four subtentacular comb rows rudimentary; remaining four along aboral edge of ribbon. Interplate ciliary grooves present. Bases of reduced tentacles give rise to four tentilla-bearing grooves, two on each side of mouth; grooves extend length of oral edge of body. During feeding, extended tentilla stream aborally, like curtains, over flat surfaces of body (Fig. 8-12C). Ribbon-shaped body another means by which suspension-feeding ctenophores increase feeding surface area in the near-absence of tentacles. Besides comb-row hovering or gliding with oral edge leading, cestids swim by muscular body undulations. Two species, *Cestum veneris* (Venus's girdle): length to 1.5 m; *Velamen parallelum:* to 90 cm.

Thalassocalycida[O]: Monotypic medusoid *Thalassocalyce inconstans* superficially resembles a hydromedusa, but larger, up to 15 cm. Tentacular body axis short; short aboral comb rows; interplate ciliary grooves present. Prey caught in subumbrel-

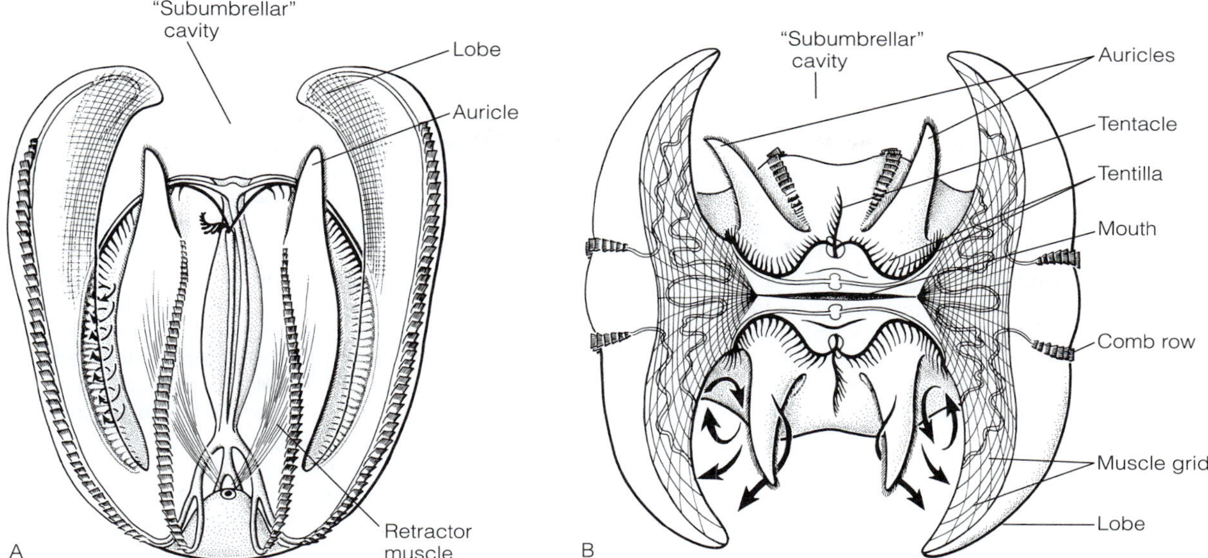

FIGURE 8-13 Ctenophora: diversity. The lobate *Mnemiopsis leidyi*. **A,** Pharyngeal plane view. **B,** Oral view. *(Modified and redrawn after Mayer, A. G. 1912. Ctenophores of the Atlantic Coast of North America. Carnegie Institution of Washington, Publication 162. 58 pp.)*

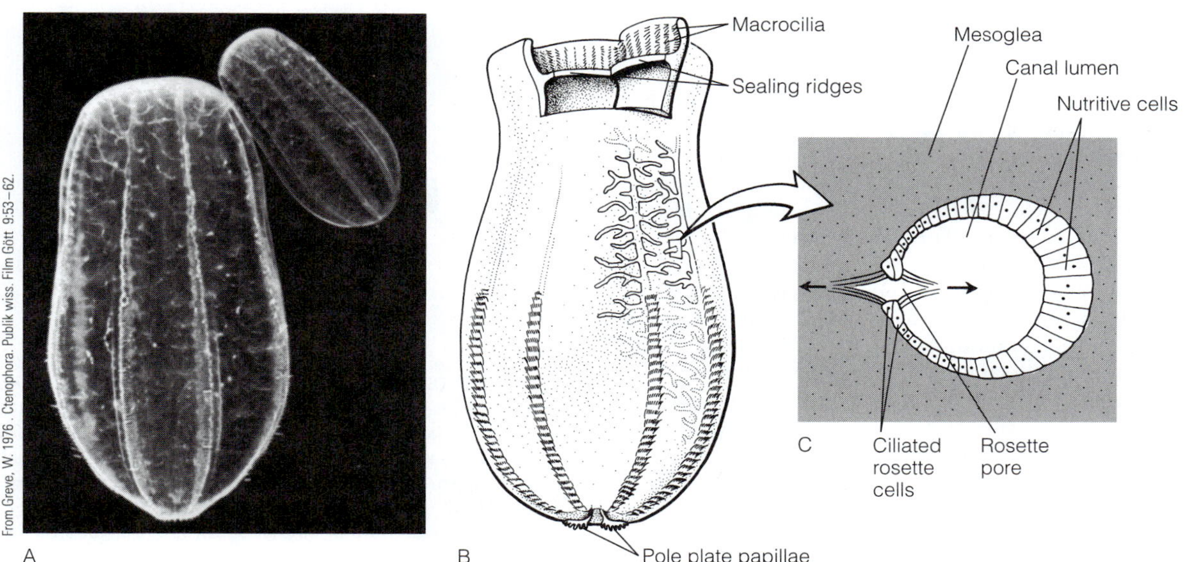

FIGURE 8-14 Ctenophora: diversity. **A,** Pharyngeal plane view of living *Beroë cucumis;* **B,** Anatomy of *Beroë* showing partially dissected mouth and pharynx. The sealing ridges, on opposite sides of the pharynx, make contact and adhere tightly when the mouth is closed. The zipperlike adhesion results from temporary intercellular junctions between cells of the opposing ridges (similar to the *Hydra* mouth). The tightly sealed mouth and lips create a streamlined leading edge as the ctenophore swims in pursuit of prey. **C,** Cross section of a coelenteric branch showing a ciliary rosette. Arrows show the presumed direction of water flow associated with the activity of the two ciliary tufts. *(B, Modified and redrawn from Tamm, S. L., and Tamm, S. 1993. Dynamic control of cell-cell adhesion and membrane-associated actin during food-induced mouth opening in* Beroë. *J. Cell. Sci. 106:355–364.)*

lar cavity using movements of bell and perhaps two short tentacles bearing tentilla. Mouth small, suspended on manubrium-like cone in subumbrellar cavity.

Beroida[O]: Melon or miter jellies, up to 30 cm in length, all species of *Beroë* and *Neis*. Lacking tentacles, often slightly compressed, saclike body resembles a ski cap (Fig. 8-14). Pole-plate margins have papillae. Interplate ciliary grooves absent. Mouth and pharynx compressed, large; meridional canals with branched outgrowths in interradial parts of body; some outgrowths join pharynx, others end blindly. Specialized conical bundles of cilia **(macrocilia)** on pharynx lining immediately inside of mouth (Fig. 8-14B); macrocilium composed of several thousand cilia bound together in functional unit, similar to ctene. Large prey engulfed whole by mouth and pharynx; tissue chunks bitten from oversize prey using macrocilia as teeth. Feed almost exclusively on ctenophores.

PHYLOGENY OF CTENOPHORA

In Linnean classifications, Ctenophora is divided into the class Tentaculata, which includes *Pleurobrachia* and all other ctenophores except Beroida (*Beroë* and *Neis*), which lacks tentacles and comprises the class Nuda. If the absence of tentacles is an autapomorphy of Beroida (Nuda), then the character "tentacles" is likely to be a symplesiomorphy of the remaining ctenophore taxa. This means that Tentaculata is a paraphyletic taxon and, for this reason, we do not recognize it (or Nuda, which is equivalent to Beroida) in the preceding section. As an alternative, some systematists suggest that the atentaculate beroids are primitive and the acquisition of tentacles is an autapomorphy of the tentaculate ctenophores. At present, the distribution of known morphological characters does not allow the construction of a phylogenetic tree that would reject either of these alternatives, or any other for that matter. In the absence of a tree supported by apomorphies, other arguments have been advanced to choose between tentaculate vs. atentaculate ground plans. One of the most compelling of these is that beroids feed on tentaculate ctenophores, and thus the tentaculates likely evolved before their atentaculate predators.

Traditionally, ctenophores are considered to be evolutionary intermediates between the radial cnidarians and the bilateral flatworms (Platyhelminthes). The cnidarian alliance was based largely on the gelatinous ("medusoid") body and the occurrence of cnidae in the tentacles of certain ctenophores. A transparent, gelatinous body, however, is widespread among planktonic animals and is probably an adaptation to buoyancy and invisibility in water. Behavioral studies of cnidae-bearing ctenophores, such as *Haeckelia rubra*, have shown that these animals do not produce the cnidae themselves, but appropriate them from their cnidarian prey.

Still, cnidarians and ctenophores share an embryonic axis that becomes the oral-aboral axis, radial body symmetry, and smooth-muscle actin filaments that anchor only to the cell membrane (all of which are probably plesiomorphies). The correspondence of the animal pole of a ctenophore zygote with the site of gastrulation is a trait shared only with cnidarians (based on current knowledge). In all higher animals (bilaterians), the gastrulation begins 180° away from the animal pole at or near the vegetal pole of the egg. In most bilaterians, the sensory organs are associated with the mouth (and head), not 180° away from the mouth at the aboral pole, as in ctenophores. This aboral position of the nerve center also is shared with cnidarians, especially with the planula larvae of some anthozoans. In these larvae, a concentration of sensory cells occurs at the aboral pole and forms a conspicuous sensory tuft. Perhaps the ctenophores diverged, by pedomorphosis (Chapter 4), from such a planula-like ancestor?

A ctenophore relationship with flatworms (or bilaterians in general) has been suggested because both have a mesodermal musculature (myocytes in the connective-tissue compartment = mesoglea; Fig. 8-10A) and because the platyctenid ctenophores superficially resemble certain flatworms. Although myocytes, as we have seen, also occur in cnidarians (such as cubopolyps), those of ctenophores and most higher animals originate from endomesoderm. The body similarity of platyctenids and platyhelminths is probably superficial. The creepsole of platyctenids arises from the distended lining of the pharynx, whereas the sole of platyhelminths receives no contributions from the pharyngeal lining.

At present, molecular systematic studies of ctenophores are inconclusive. Some suggest that ctenophores are the sister taxon of sponges while others, in agreement with comparative morphology, place ctenophores between cnidarians and bilateral animals (Fig. 8-15).

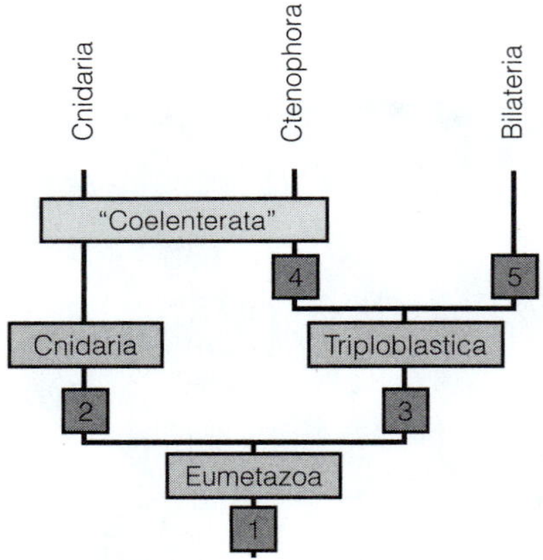

FIGURE 8-15 Phylogeny of Eumetazoa. **1, Eumetazoa:** two embryonic germ layers (ectoderm, endoderm)—diploblastic; two adult epithelia (epidermis, gastrodermis); embryonic axis corresponds to adult oral-aboral axis; multifunctional coelenteron (digestion, absorption, circulation, volume regulation, skeletal function [hydrostasis]) with single opening to exterior (mouth/anus/gonopore); nerve net; musculature of epitheliomuscular or myoepithelial cells; endodermal gonads. **2, Cnidaria:** cnidae; polypoid body form; planula larva. **3, Triploblastica:** endomesodermal myocytes; three embryonic germ layers (ectoderm, mesoderm, endoderm)—triploblastic. **4, Ctenophora:** biradial symmetry; collocytes on pair of retractile tentacles; aboral organ; ctenes in comb rows; ciliary rosettes; photocytes. **5, Bilateria:** bilateral symmetry; adult anterior-posterior axis typically departs from embryonic axis; dorsal and ventral surfaces structurally and functionally distinct; nervous concentration associated with anterior end; filtration excretory organs (protonephridia and metanephridia, see Chapter 9).

REFERENCES

GENERAL

Anctil, M. 1985. Ultrastructure of the luminescent system of the ctenophore *Mnemiopsis leidyi.* Cell Tiss. Res. 242:333–340.

Costello, J. H., and Case, J. F. 1998. Planktonic feeding and evolutionary significance of the lobate body plan within the Ctenophora. Biol. Bull. 195:247–248.

Dawydoff, C. 1928. Traité d'Embryologie Comparée des Invertébrés. Masson, Paris. 930 pp.

Delage, Y., and Hérouard, E. 1901. Traité de Zoologie Concrète. Vol. 2. Les Coelentérés. Schleicher Freres, Paris. 848 pp.

Fricke, H. W., and Plante, R. 1971. Contribution à l'étude de cténophores platycténides de Madagascar: *Ctenoplana* (*Diploctena* n.s. gen.) *neritica* n. sp. et *Coeloplana* (*Benthoplana* n.s. gen.) *meteoris* (Thiel 1968). Cah. Biol. Mar. 12:57–75.

Greve, W. 1976. Ctenophora. Publik. wiss. Film Gött. 9:53–62.

Harbison, G. R. 1985. On the classification and evolution of the Ctenophora. In Conway Morris, S., et al. (Eds.): The origins and relationships of lower invertebrates. Systematics Association Spec. Vol. 28. Clarendon Press, Oxford. pp. 78–100.

Harbison, G. R., and Miller, R. L. 1986. Not all ctenophores are hermaphrodites: Studies on the systematics, distribution, sexuality and development of two species of *Ocyropsis.* Mar. Biol. 90:413–424.

Hernandez-Nicaise, M.-L. 1991. Ctenophora. In Harrison, F. W., and Westfall, J. A. (Eds.): Microscopic Anatomy of Invertebrates. Vol. 2. Placozoa, Porifera, Cnidaria, and Ctenophora. Wiley-Liss, New York. pp. 359–418.

Kaestner, A. 1984. Lehrbuch der Speziellen Zoologie. Wirbellosen Tiere. Vol. 2. Gustav Fischer Verlag, Stuttgart. 621 pp.

Larson, R. J. 1988. Feeding and functional morphology of the lobate ctenophore *Mnemiopsis mccradyi.* Est. Coast. Shelf Sci. 27:495–502.

Mackie, G. O., Mills, C. E., and Singla, C. L. 1988. Structure and function of the prehensile tentilla of *Euplokamis* (Ctenophora, Cydippida). Zoomorphology 107:319–337.

Mackie, G. O., Mills, C. E., and Singla, C. L. 1992. Giant axons and escape swimming in *Euplokamis dunlapae* (Ctenophora: Cydippida). Biol. Bull. 182:248–256.

Martindale, M. Q., and Henry, J. 1997. Ctenophorans, the comb jellies. In Gilbert, S. F., and Raunio, A. M. (Eds.): Embryology, Constructing the Organism. Sinauer, Sunderland, MA. pp. 87–111.

Martindale, M. Q., and Henry, J. Q. 1999. Intracellular fate mapping in a basal metazoan, the ctenophore *Mnemiopsis leidyi,* reveals the origin of mesoderm and the existence of indeterminate cell lineages. Dev. Biol. 214:243–257.

Matsumoto, G. I., and Hamner, W. M. 1988. Modes of water manipulation by the lobate ctenophore *Leucothea* sp. Mar. Biol. 97:551–558.

Matsumoto, G. I., and Harbison, G. R. 1993. *In situ* observations of foraging, feeding, and escape behavior in three orders of oceanic ctenophores: Lobata, Cestida, and Beroida. Mar. Biol. 117:279–287.

Mayer, A. G. 1912. Ctenophores of the Atlantic Coast of North America. Carnegie Institution of Washington, Publication 162. 58 pp.

Mills, C. E., and Miller, R. L. 1984. Ingestion of a medusa *(Aegina citrea)* by the nematocyst-containing ctenophore *Haeckelia rubra* (formerly *Euchlora rubra*): Phylogenetic implications. Mar. Biol. 78:215–221.

Pianka, H. D. 1974. Ctenophora. In Giese, A. C., and Pearse, J. S. (Eds.): Reproduction of Marine Invertebrates. Vol. 1. Acoelomate and Pseudocoelomate Metazoans. Academic Press, New York. pp. 201–265.

Reeve, M. R., and Walter, M. A. 1978. Nutritional ecology of ctenophores: A review of recent research. Adv. Mar. Biol. 15:249–287.

Stretch, J. J. 1982. Observations on the abundance and feeding behavior of the cestid ctenophore, *Velamen parallelum.* Bull. Mar. Sci. 32:796–799.

Swanberg, N. 1974. The feeding behavior of *Beroë ovata.* Mar. Biol. 24:69–76.

Tamm, S., and Tamm, S. L. 1995. A giant nerve net with multi-effector synapses underlying epithelial adhesive strips in the mouth of *Beroë.* J. Neurocytol. 24:711–723.

Tamm, S. L. 1980. Cilia and ctenophores. Oceanus 23:50–59.

Tamm, S. L., and Moss, A. G. 1985. Unilateral ciliary reversal and motor responses during prey capture by the ctenophore *Pleurobrachia.* J. Exp. Biol. 114:443–461.

Tamm, S. L., and Tamm, S. 1993. Dynamic control of cell-cell adhesion and membrane-associated actin during food-induced mouth opening in *Beroë.* J. Cell Sci. 106:355–364.

Winnepenninckx, B. M. H., Van der Peer, Y., and Backeliau, T. 1998. Metazoan relationships on the basis of 18S rRNA sequences: A few years later . . . . Am. Zool. 38:888–906.

Wrobel, D., and Mills, C. 1998. Pacific Coast Pelagic Invertebrates: A Guide to the Common Gelatinous Animals. Sea Challengers and Monterey Bay Aquarium, Monterey, CA. 108 pp.

INTERNET SITES

http://faculty.washington.edu/cemills/Ctenophores.html (List of all known ctenophore species with good color photos of several, including the platyctenid *Lyrocteis.*)

http://scilib.ucsd.edu/sio/nsf/fguide/ctenophora.html (Good images of living Antarctic comb jellies, including a platyctenid.)

http://seaslugforum.net/ctenopho.htm (Image of a spectacular but unidentified living platyctenid from northern Australia.)

9 Introduction to Bilateria

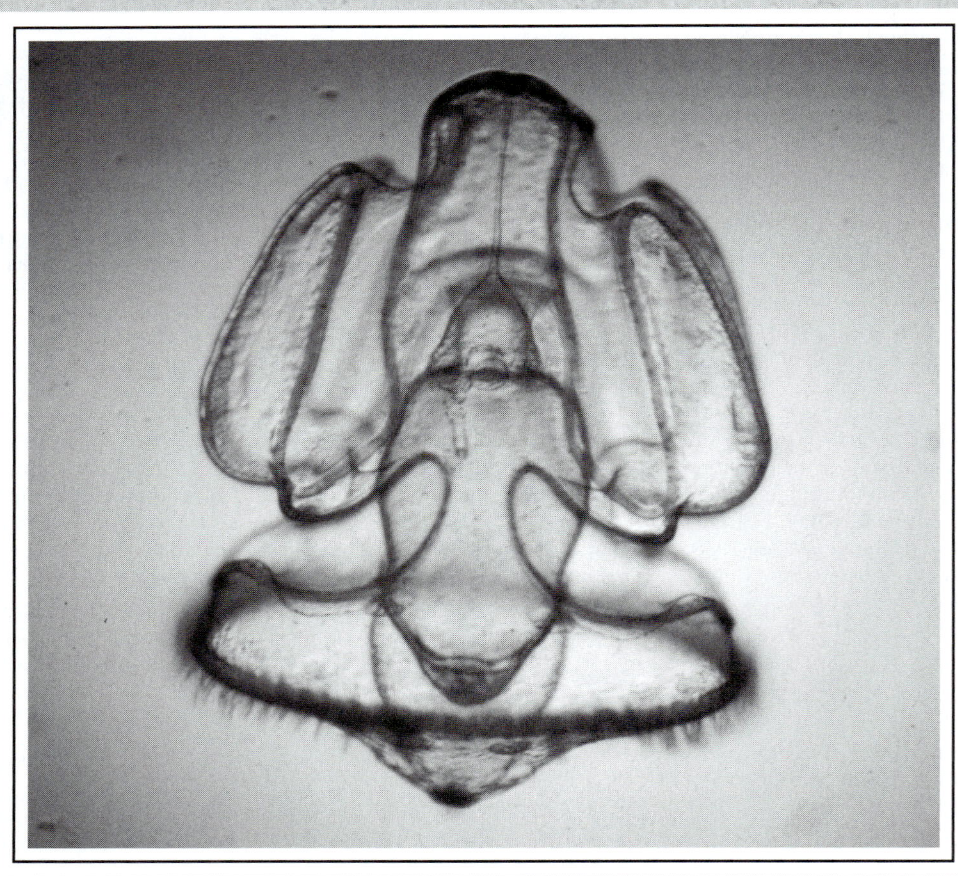

The vast majority of animals on Earth, and certainly the most familiar, are those that exhibit bilateral symmetry. In this huge taxon, known as Bilateria, are found the flatworms, molluscs, crustaceans, insects, echinoderms, and chordates, to name only a few. As compared with sponges, cnidarians, and ctenophores, bilaterians enjoy an unparalleled diversity and adaptive success. They account for over 99% of all animal species and have radiated and flourished in the sea, in fresh water, on land, in air, and recently one species, our own, has entered outer space. Their sweep of adult body sizes extends from less than 100 μm, the size range of larger protozoans, to 15 m (giant squid) and beyond 30 m (blue whale). In general, the microscopic species are of simple design, some not exceeding the tissue level of organization, while many large-bodied forms have multiple organs and complex organ systems. The astonishing range of body size and complexity, moreover, includes both solitary and colonial growth forms, the latter achieving complexity through zooid polymorphism. Unique to bilaterians is the evolution of social behavior, which emerged independently in insects and vertebrates, and contributes to their success. The potential of social organization was further realized some 10,000 years ago when humans established a mutualism—agriculture—with plants and animals and developed civilization.

The goal of this chapter is to lay a foundation of facts, trends, and correlations on which to build a framework of principles pertaining to the bilaterian radiation. The descriptions of anatomy and physiology are reasonably well established, but the broad correlations of structure and function and ideas about evolution are first approaches, hypotheses to be examined by students and specialists alike. Test these hypotheses using the facts available in the remaining chapters of this book and elsewhere, but remember that the best sources of reliable information are the animals themselves.

This chapter focuses on the key motifs of the bilaterian blitzkrieg. These are the nature and functional significance of bilateral symmetry, the enhancement of the neuromuscular system and locomotion, the relationship between body size and functional design, the developmental origin of the bilaterian ground plan, and the phylogeny of bilaterians.

BILATERAL SYMMETRY—ENCOUNTERING RESOURCES

An animal's body exhibits **bilateral symmetry** if only one plane of bisection, the **midsagittal plane** or midplane, produces left and right mirror-image halves (Fig. 9-1). Another characteristic of bilaterians is that the body is polarized along two perpendicular axes, the **anterior-posterior** (head-tail) **axis** and the **dorsal-ventral** (back-belly) **axis**. A third polarity axis, the **proximal-distal axis,** extends away from the midsagittal axis. The equivalence of left and right halves of the body means, developmentally, that bilateral symmetry is the default state and any departures from pure bilateral symmetry, such as the enlargement of either the right or left claw of a fiddler crab, requires a new developmental command and additional genetic information.

To arrive at a hypothesis for the adaptive significance of bilateral symmetry, consider a contrast with the radially symmetric cnidarians. Whether a benthic polyp or pelagic medusa, most cnidarians are oriented vertically along the surface-to-bottom, substrate-to-water, or solar-gravitational environmental gradient (Fig. 9-2A,B) and their bodies are polarized along this gradient: For polyps, the feeding structures are above, the attachment surface below; for medusae, the buoyant bell is above, the feeding structures below. On the other hand, the environment is more or less uniform horizontally around the circumference of the body. For this reason, body parts are distributed regularly around the body circumference, and this is the hallmark of radial symmetry. Among cnidarians, departures from radial symmetry occur most clearly in animals such as sea pens, which suspension feed in a unidirectional current (Fig. 7-40B). These animals are confronted with two *perpendicular* gradients, the vertical substratum-water gradient and the horizontal upstream-downstream current gradient. As a result, the body is polarized along both gradients to cope with the consistently different challenges and opportunities provided by each. Such polarization of structure and function along two perpendicular axes is correlated with the bilateral symmetry of sea pens and bilaterians.

Bilaterian bilateral symmetry may have evolved when radial animals began to move along a surface, probably the interface between water and substratum. Because an interface is actually a very steep gradient, a body moving along such an interface has a consistent orientation with respect to two perpendicular gradients: (1) the horizontal locomotory gradient along the interface and (2) the vertical (water-substratum) gradient (Fig. 9-2C). Polarization of the body along the locomotory gradient resulted in the anterior-posterior body axis. The anterior end was adapted to sense the oncoming environment and to transmit information posteriorly. Polarization along the vertical water-substratum gradient produced the dorsal-ventral axis. The dorsal side was adapted to conditions in the overlying water whereas the ventral side was specialized to interact with the substratum. Perhaps the dorsal surface was protective and the ventral surface locomotory, adhesive, or both (Fig. 9-2C). Once these two axes—anterior-posterior and dorsal-ventral—evolved and their new genetic determinants were established, the mirror-image proximal-distal axes of bilateral symmetry were the immediate and inevitable result of the new body geometry.

The hypothetical transition from radial to bilateral symmetry is foreshadowed in several bilateral taxa of Cnidaria and Ctenophora. The lobate and beroid ctenophores swim across, rather than along, the vertical gradient in search of prey. The platyctenid ctenophores and sea pansies (*Renilla;* Fig. 7-40C,D) move across the substratum-water gradient on benthic surfaces. The cnidarian by-the-wind sailor, *Velella,* and the Portuguese man-of-war, *Physalia* (Fig. 7-67A, 7-70A) sail across the vertical gradient on the air-water interface at the surface of the sea.

The ecological significance of movement across rather than along a vertical environmental gradient may be an increased

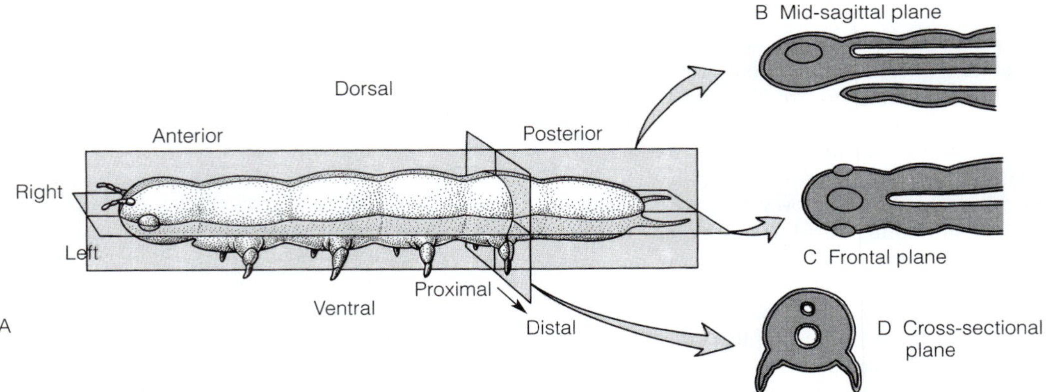

FIGURE 9-1 Bilateria: symmetry planes and body axes. Animals are bilaterally symmetric when only one plane (the vertical plane, **A**) divides the body into mirror-image halves and when the body is polarized along two perpendicular axes: the anterior-posterior axis and the dorsal-ventral axis. Anterior refers to the head end, posterior to the tail, dorsal to the upper surface, and ventral to the belly side. Any plane that divides the body into left and right sides is called a sagittal plane. When a sagittal plane passes through the midline of the body, it is called a midsagittal plane, or midplane **(B)**. A proximal-distal axis occurs on each side of the midplane. The midplane is the only plane that produces mirror-image body halves. The term lateral is used in reference to the left or right side of the body. A plane that divides across the long axis of the body, as you would slice a loaf of bread, is the cross-sectional plane **(D)**. If the body is divided into dorsal and ventral halves, as you would slice a submarine sandwich, the surfaces are cut in the frontal plane **(C)**.

probability of encountering food. Organisms in general, and food organisms in particular, are not distributed uniformly but in isolated patches, especially on substrates and other interfaces. Horizontal movement probably improved the chance of locating these patches.

If bilateral animals first evolved on the sea bottom, they quickly radiated into all environments and habitats. Among these environments, bilateral symmetry is most striking in taxa that move horizontally across a strong vertical gradient, such as fish swimming in water, worms crawling over the sea

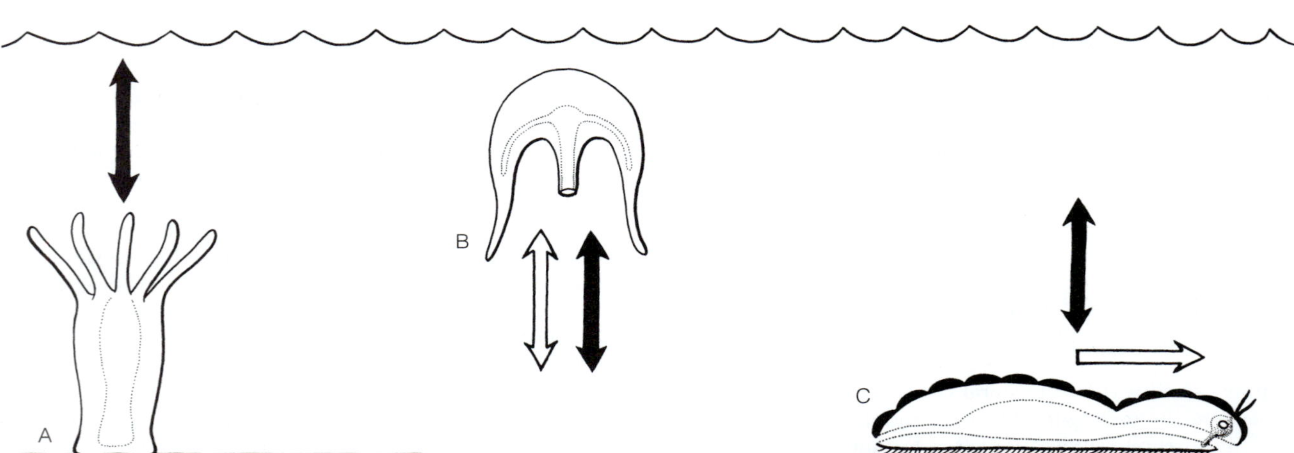

FIGURE 9-2 Bilateria: functional origin of bilateral symmetry. Bilateral symmetry may have evolved in relation to environmental (solid arrows) and locomotory gradients (open arrows). **A,** The body of a sessile cnidarian polyp is polarized along an oral-aboral axis that parallels the environmental gradient. **B,** In a motile medusa, the body also is polarized along only one axis because the environmental and locomotory gradients coincide. **C,** Motile bilaterians, especially those on surfaces, move perpendicular to the environmental gradient. Thus their bodies are polarized along the locomotory (anterior-posterior) gradient and the perpendicular environmental (dorsal-ventral) gradient. The hallmark of anterior-posterior polarity is cephalization, the concentration of sensory organs and nervous tissue at the leading end of the body. Movement across the environmental gradient may have improved the chance of encountering food or mates. The dorsal surface is often modified for protection while the ventral surface may be adapted for locomotion and adhesion.

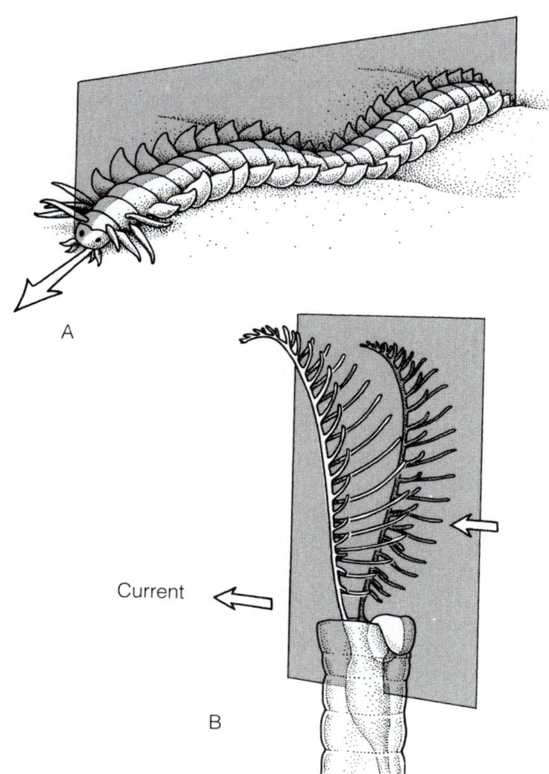

FIGURE 9-3 Bilateria: bilateral symmetry and locomotion. Bilateral symmetry is correlated with unidirectional locomotion, as shown in **A** for a segmented worm, or with a unidirectional current delivery of food to a sessile suspension-feeding animal, such as the pterobranch hemichordate shown in **B.** In motile bilaterians **(A),** cephalization is apparent, whereas sessile bilaterally symmetrical animals **(B)** usually are not cephalized.

bottom (Fig. 9-3A), animals walking on land, or birds and insects flying through air. In each of these cases, the dorsal and ventral surfaces are distinctively specialized to cope with the different challenges and opportunities from above and below. Bilateral symmetry is less obvious and sometimes masked in bilaterians that are secondarily either sessile suspension feeders or burrowers. Sessile suspension feeders, for example moss animals (bryozoans), feather-duster worms (annelids), or sea lilies (echinoderms), have all superimposed radial symmetry on bilateral symmetry (Fig 9-4). This is because radial symmetry works best for sessile suspension feeders unless the ambient water currents are unidirectional (and therefore predictable), and then a sea-pen-like bilateral symmetry is common (Fig. 9-3B). Similarly, bilaterians that burrow actively in sediments escape the steep gradient at the surface of the sediment, find themselves in a more uniform environment, and adopt various degrees of wormlike radial symmetry. Burrowers such as roundworms (nematodes), earthworms (annelids), and legless amphibians and lizards are good examples of bilaterians with a superficial, functional, external radial symmetry.

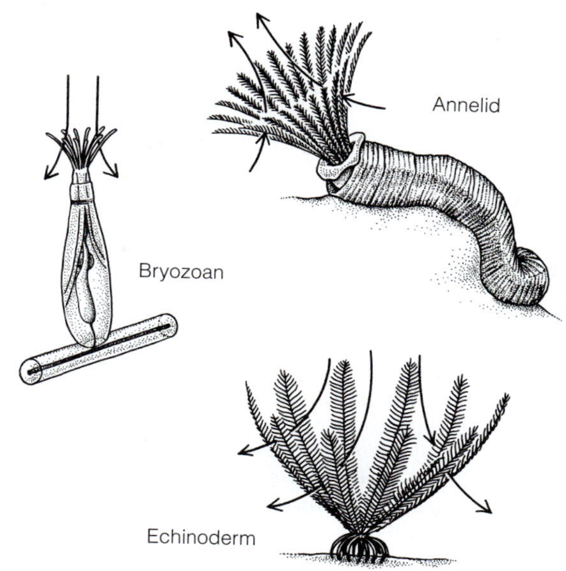

FIGURE 9-4 Bilateria: radial symmetry and functional design. Sessile bilaterians that have an equal chance of obtaining food from all sides have evolved a secondary radial symmetry to meet the environment equally in all directions. In these animals, suspended food particles are distributed more or less uniformly in all directions and a radial collection system, which shows no directional preference, is an effective adaptation for gathering food. Although bilaterians mask their bilateral symmetry, it is nevertheless evident in many larval and some adult structures. Some motile burrowers, such as earthworms, that live in more or less uniform soils also have evolved a superficial radial symmetry.

CEPHALIZATION—TARGETING RESOURCES

MOTILE BILATERIA

In contrast to most motile cnidarians and ctenophores, which rely on chance encounters with food and each other, bilaterians actively detect and pursue food, mates, and shelter with a battery of sensory organs and ganglia on the anterior end of the body. This anterior concentration of nervous tissue and sensory organs is called **cephalization** (= head development). Cephalization is associated with a **central nervous system** consisting of anterior brain and longitudinal nerve cords. The **brain,** the central processing unit, is a bilateral concentration of neurons for the integration of sensory signals and their translation into motor commands. These commands, when mediated through the musculature, cause the body to turn toward food or mates and away from predators or unfavorable environmental conditions. By coupling cephalization and directional locomotion, bilaterians can detect targets and then move toward or away from them.

The sensory organs, which usually are paired and thus adapted for targeting by triangulation, represent the typical sensory modalities of photoreception, mechanoreception, and chemoreception. The mouth typically is located at or near the anterior end, where it can ingest food that has been detected by the sensory organs. Bilateral **longitudinal nerve cords,** usually paired, transmit information along the length of the body. In large or long bilaterians, these cords typically include one or more large-diameter axons, called **giant axons,** that are specialized for rapid transmission. Their large diameter and low resistance allows giant axons to conduct impulses rapidly, usually to facilitate an escape response.

SESSILE BILATERIA

Sessile, suspension-feeding bilaterians are not cephalized, and their feeding apparatus, usually a crown of tentacles, is more or less radially symmetric, like that of cnidarian polyps. With few exceptions, however, bilaterian suspension feeders, unlike cnidarians, *actively* create a water flow, usually with cilia, over their tentacles or other suspension-feeding surface. Cnidarian hydroids and bilaterian bryozoans (moss animals) are both suspension feeders of similar size, form, and habitat. Beginners often confuse them, but they are readily distinguishable by their activity. Using specialized cilia, bryozoans actively pump water and particles over their tentacles whereas hydroids passively wait for food to bump into them.

MUSCULATURE—PURSUING RESOURCES

The detection and active pursuit of food and fortune requires, especially in large animals, a musculature dedicated to motion along the anterior-posterior axis. In nearly all cases, the locomotory muscles are obliquely or cross-striated and contract rapidly, unlike the smooth muscles of cnidarian polyps, which usually are slow. The body-wall musculature of soft-bodied bilaterians (worms) is arranged in circular and longitudinal muscle layers (Fig. 9-5). When the body is viewed in cross section, the **outer circular musculature** almost always encloses the **inner longitudinal musculature** (Fig. 9-5). Longitudinal muscles allow animals to bend and turn toward or away from their targets. Many worms swim or burrow through loose sediments by using their longitudinal muscles to generate undulatory waves along their bodies. In a few instances, the undulations are fishlike, but often they are wild writhing or lashing movements. A longitudinal musculature is used by some sedentary tube and crevice dwellers (Fig. 9-4) to retract into their refuges for safety and by others to produce swimming undulations (Fig. 9-6C).

The alternate contraction of longitudinal and circular muscles around an enclosed fluid space produces locomotion. Contraction of the circular musculature causes elongation, whereas contraction of the longitudinal musculature results in shortening (Fig. 9-5A–C). These alternating movements typify soft-bodied animals that burrow using peristalsis, such as earthworms, in which both the circular and longitudinal muscles are more or less equally developed.

Specialized muscles that result in precise movements are widespread in bilaterians, especially those with rigid exo- and endoskeletons. Most of these muscles will be described later, as they are encountered in particular taxa, but a few widespread muscles are noted here. A vertical **dorsoventral musculature** flattens the body and lowers its profile in many animals that creep over surfaces (Fig. 9-6A). In others, it flattens the body to improve thrust for swimming (Fig. 9-6B). Dorsoventral muscles typically are arranged in a series along the length of the body (Fig. 9-6A). Some soft-bodied bilaterians twist using a **helical musculature,** which winds around the body in two helices, one left-handed and one right-handed (Fig. 9-6D).

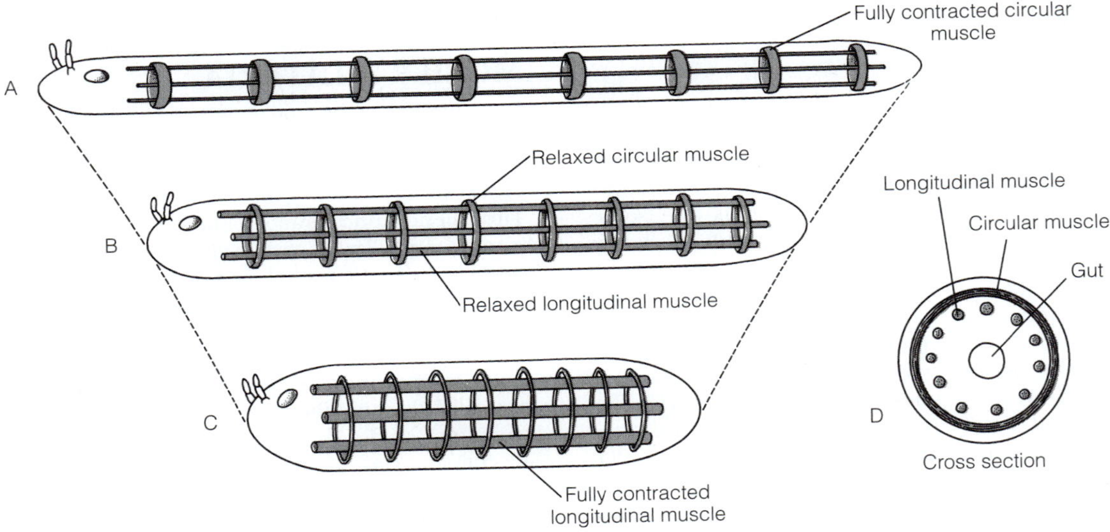

FIGURE 9-5 Bilateria: body-wall musculature. The basic arrangement of body-wall muscles in soft-bodied bilateral animals, as shown in **B** and the cross section **(D),** is an outer circular and an inner longitudinal musculature. These two layers have antagonistic actions: Contraction of the circular musculature causes elongation of the body **(A),** whereas contraction of the longitudinal musculature causes shortening **(C).** Longitudinal muscles alone allow the animal to bend and turn. The circular body wall muscles typically are positioned outside of the longitudinal muscles because the effectiveness of their action (elongation or peristalsis) depends on compression of the bodily tissues, including the longitudinal musculature.

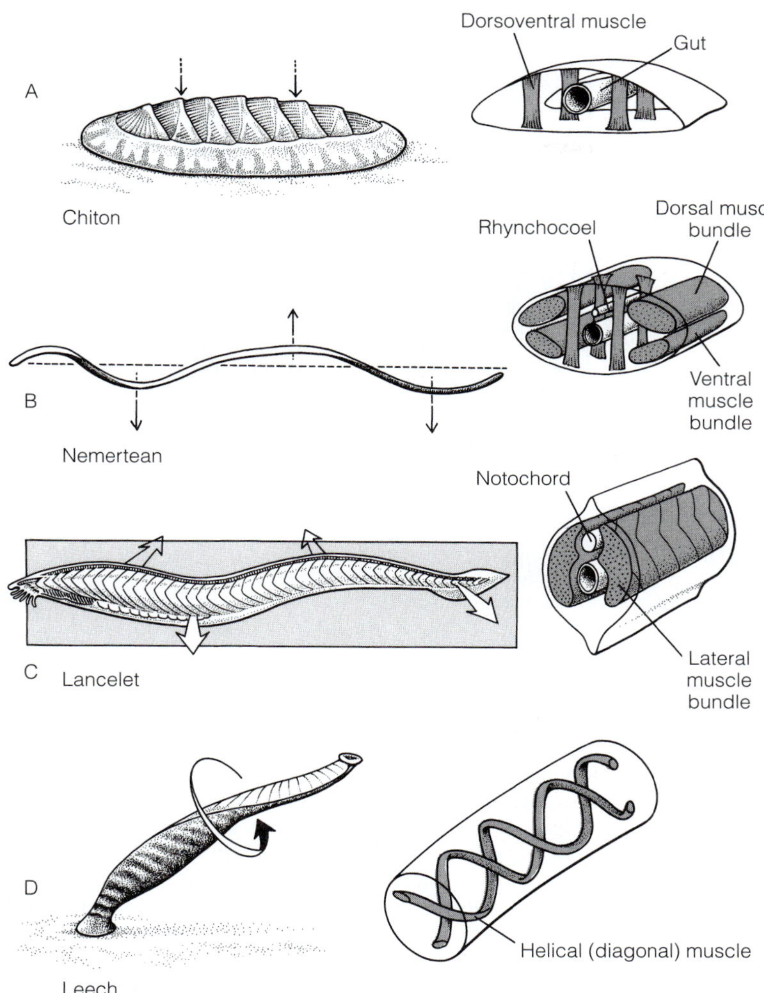

FIGURE 9-6 Bilateria: specialized muscles. **A** and **B,** Dorsoventral muscles flatten the body in animals such as chitons **(A)** and nemerteans **(B).** Flattening expands the surface in contact with the substratum in creeping animals and thus promotes adhesion. In addition, some nemerteans and leeches flatten while swimming to expand their thrusting surfaces **(B).** A longitudinal musculature can produce swimming undulations if bundles on opposite body sides contract alternately and if an axial skeleton is present to prevent shortening of the body. Under these conditions, alternate contractions of dorsal and ventral muscle bundles produce vertical undulations **(B);** alternate contractions of lateral bundles results in horizontal undulations, as in lancelets and fishes **(C).** The incompressible axial skeleton of lancelets is the notochord whereas in nemerteans, the rhynchocoel may serve the same function. In leeches **(D)** and other animals, twisting movements of the body are accomplished using helical muscles.

BURROWING MECHANICS

Burrowing in moist or submerged sediments is a common behavior of many invertebrates such as sea anemones and earthworms. Any animal that burrows must anchor itself posteriorly in order to push the sediment aside without backslipping. This kind of anchor is called a **penetration anchor** because it holds one part of the body in place as another part penetrates and advances into the sediment (Fig. 9-7A). Once the leading end of the body has pushed forward into the sediment, it too must anchor itself so that the trailing end can be drawn forward without the leading end slipping. The anchor at the leading end is called a **terminal anchor** (Fig. 9-7B). Continuous burrowing thus requires sustained and coordinated alternating use of penetration and terminal anchors.

Peristalsis is a simple way to coordinate alternation of penetration and terminal anchors (Fig. 9-8) and is common in soft-bodied animals. As a peristaltic wave passes along the length of a burrowing worm, longitudinal and circular muscles contract alternately. At points where the longitudinal

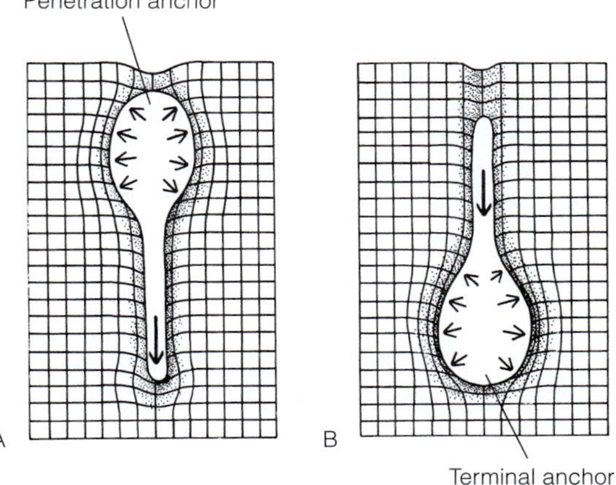

FIGURE 9-7 Bilateria: burrowing mechanics. Burrowers must establish a penetration anchor **(A)** to prevent backslipping while pushing the body into the sediment. Once the body enters the sediment, a terminal anchor **(B)** allows the trailing part of the body to be pulled forward into the burrow.

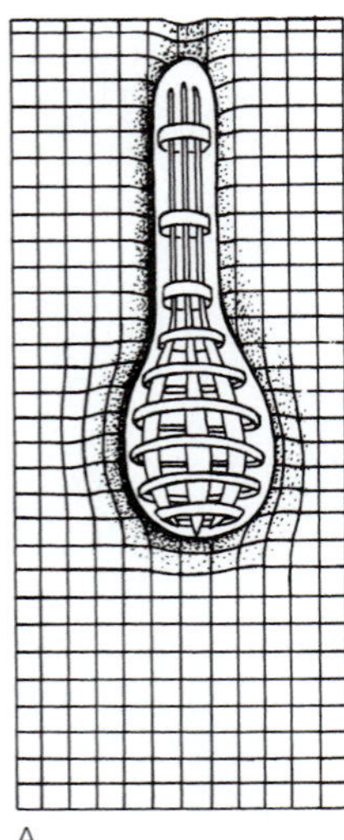

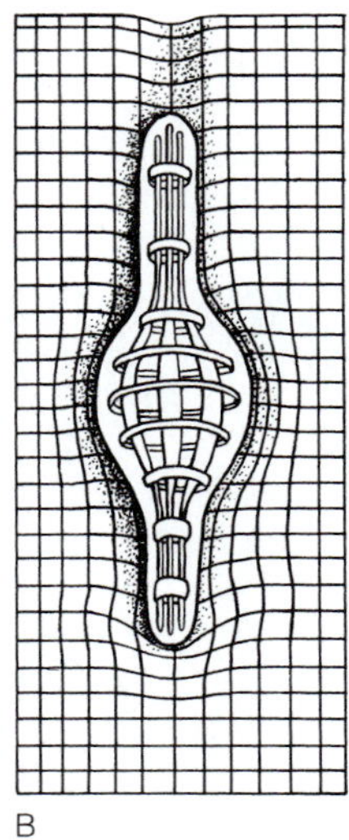

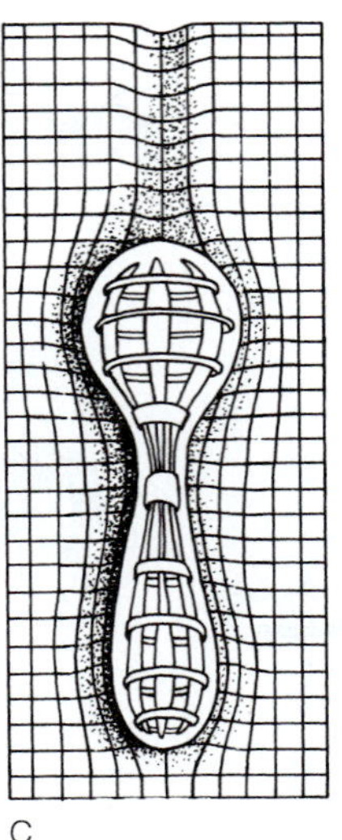

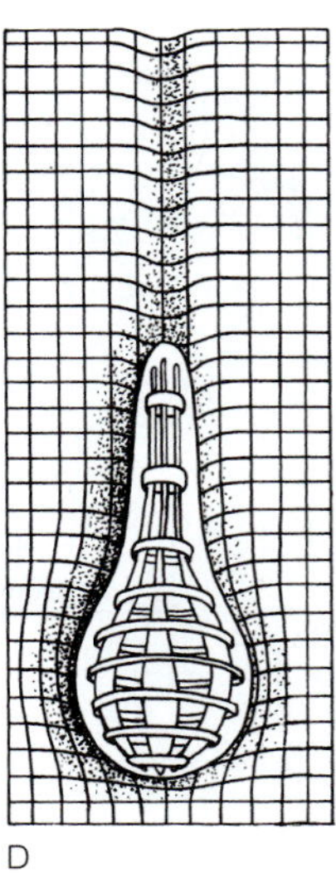

FIGURE 9-8 Bilateria: peristaltic burrowing. Peristalsis is a common means of coordinating penetration and terminal anchors while burrowing. During peristaltic burrowing, alternating waves of longitudinal and circular muscle contraction pass rearward along the body. This figure shows one complete wave of longitudinal and circular muscle contraction.

musculature is fully contracted, the body shortens and widens to form both the penetration and terminal anchors. At points where the longitudinal musculature is only partly contracted, the body shortens without anchoring, thus pulling a section of the body toward a terminal anchor. In body regions where the circular musculature is contracted, the body elongates, pushing forward into the sediment.

Not all burrowers use whole-body peristalsis. Instead, some use **extension-retraction** cycles of an eversible (able to be turned inside out) or protrusible (able to be made to bulge outward) appendage, such as a proboscis, pharynx, foot, or introvert (Fig. 9-10, 11-4, 12-5, 12-113, 13-10, 22-31). An **introvert** is a retractile head, or the anterior end of the body, that can be either eversible or protrusible (Fig. 9-10D,E). Animals that burrow with an extensible appendage usually form the penetration anchor while the appendage thrusts forward and enters the sediment. The flow of body fluid into the hollow appendage or the unfurling of spines causes the advancing tip to expand into a terminal anchor. As the appendage is retracted, usually by specialized retractor muscles, the trunk is pulled forward toward the terminal anchor.

COMPARTMENTALIZATION—PHYSIOLOGICAL REGULATION AND SPECIALIZATION

CNIDARIA

An early milestone in the evolution of metazoans was the organization of cells into epithelia. As a general rule, an epithelium separates and physiologically regulates extracellular body compartments for specialized functions. Before epithelia evolved in metazoans, such physiological control was more or less limited to regulation within cells, as probably occurs in sponges. Regulation of extracellular compartments, such as fluid-filled cavities, the gut, and the connective tissue, was probably a precondition for the evolution of specialized muscular, nervous, digestive, and other systems.

Epithelial pioneers, the cnidarians invented epidermis and gastrodermis and used them to regulate the first two extracellular compartments, the coelenteron and the mesoglea. Gastrodermis encloses and regulates the coelenteron chiefly for digestion. Epidermis and gastrodermis together sandwich the mesogleal compartment and regulate the passage of nutrients, gases, ions, and other substances (Fig. 9-9A,B). This control establishes a protected extracellular milieu in the mesoglea that is more attuned to muscle contraction, nerve conduction, and growth of gametes than is the case in either epithelium itself. Probably because of this, germ, nerve, and muscle cells of cnidarians and ctenophores tend to lose their association with the surface epithelia and invade the regulated mesoglea. Similarly, the regulatory gastrodermis controlled the type and concentration of substances, such as digestive enzymes, in the coelenteron and permitted the extracellular digestion of large food items.

Although cnidarians (or ctenophores) first evolved a cavity for extracellular digestion, the coelenteron remains a slave to multiple competing functions and is not free to specialize in digestion alone. In addition to digestion and absorption, coelenteric functions include internal transport (circulation), hydrostatic support, excretion, and reproduction. Thus the coelenteron could not evolve into a specialized gut, as we generally think of it, until it divested itself of circulatory, hydrostatic, excretory, and reproductive functions. Where, then, would these nondigestive functions be housed?

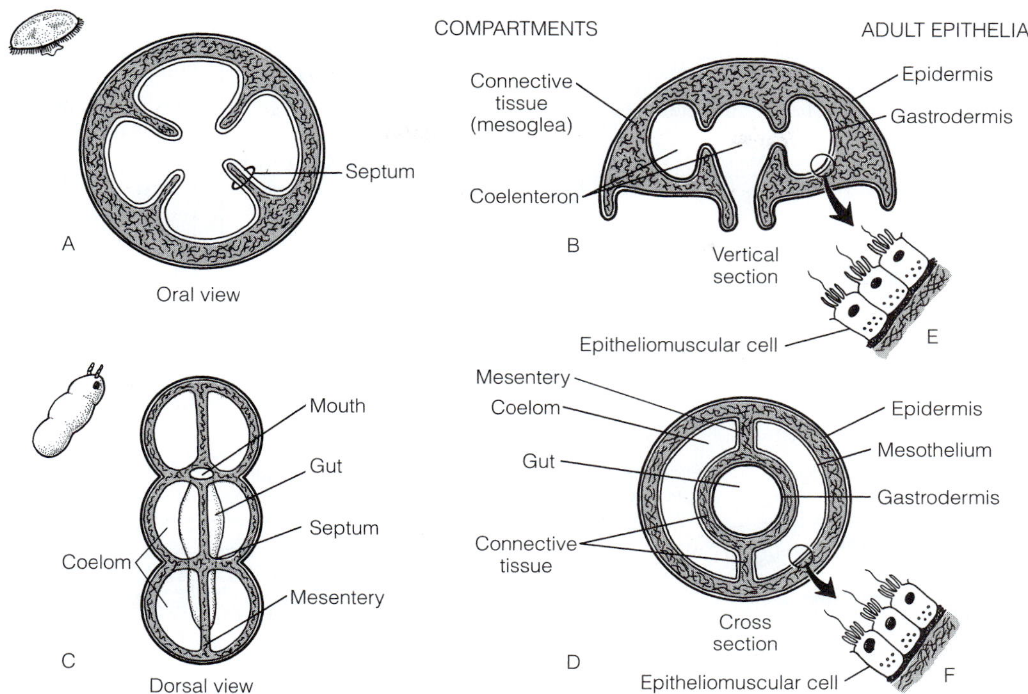

FIGURE 9-9 Bilateria: body compartmentalization. The body of radial animals **(A and B)** consists of two epithelia, the outer epidermis and inner gastrodermis that enclose the coelenteron and mesoglea. This arrangement of epithelia allows for physiological regulation and specialization of these two extracellular body compartments. In contrast to cnidarians and ctenophores, coelomate bilaterians specialize and regulate three extracellular compartments, the connective tissue, the gut lumen, and the coelom. The additional compartment is made possible by a body design that consists of three, rather than two, epithelia. These three are the epidermis, gastrodermis, and mesothelium **(C and D)**. The fact that the epithelial musculature of cnidarians is identical to the coelomic epithelial musculature of bilaterians **(E and F)** suggests that bilaterian muscles in general evolved from a coelomic epithelium.

BILATERIA

Large-bodied bilaterians evolved a new internal compartment separate from the gut that is called a coelom. It functions as a hydrostatic skeleton and has roles in circulation, reproduction, and excretion (Fig. 9-9C,D). Because the coelom adopted these functions, the gut was free to specialize in digestion and absorption. In the meantime, part of the connective-tissue compartment became the hemal system (blood-vascular system), with roles in circulation and excretion, sometimes in hydrostatic support (Fig. 9-12, 9-13), and in locomotion (in spiders). Each of these compartments is defined and discussed in the following sections and in Table 9-1. The body compartments of small bilaterians vary and are described separately.

Gut

LARGE BILATERIA

The bilaterian **gut** is typically a complete tube that opens to the exterior at both ends. It consists of mouth, foregut, midgut, hindgut, and anus (Fig. 9-10A). Because the gut is open at both ends, food can move unidirectionally through it, permitting regions to specialize, like workers on a disassembly line, in different steps in the digestive process. Each region of the gut is itself a different functional regulated compartment. The mouth and **foregut** are specialized for ingestion of food but may also grind the food and secrete digestive enzymes and mucus. The **midgut** is the site of digestion (hydrolysis) and absorption and the **hindgut** is usually specialized for the formation, storage, and egestion of feces through the anus. In terrestrial animals, the hindgut may also function in water reclamation and ion regulation.

The foregut and hindgut develop from embryonic ectoderm (the **stomodeum** and **proctodeum,** respectively; Figure 9-10A) that unites with the endodermal midgut to form the complete digestive tube. In animals such as arthropods and roundworms, which have an exoskeleton secreted by ectoderm, the foregut and hindgut also are lined by exoskeleton. These cuticular skeletal linings may be specialized as grinding, piercing, or filtering teeth in the foregut and as valves and other structures in the hindgut. The midgut develops from embryonic endoderm, which becomes the adult gastrodermis. The gastrodermis lacks a cuticle, permitting adaptation of the midgut for secretion of digestive enzymes and absorption of nutrients.

TABLE 9-1 Characteristics of Bilaterian Circulatory Systems

System	Compartment	Developmental origin	Lining	Anatomy	Pump	Fluid
Coelomic	Coelom	Mesoderm-coelom	Mesothelium	Small or large cavities, vessel-like canals	Cilia, body-wall musculature	Coelomic
Hemal	Connective tissue	Blastocoel	Basal lamina	Tubular vessels, hemocoelic sinuses, or both	Contractile vessels, heart, body-wall musculature	Blood, hemolymph
Gastro-vascular	Gut	Endoderm-archenteron	Gastrodermis	Gut tube and ceca	Cilia, gut musculature	Chyme

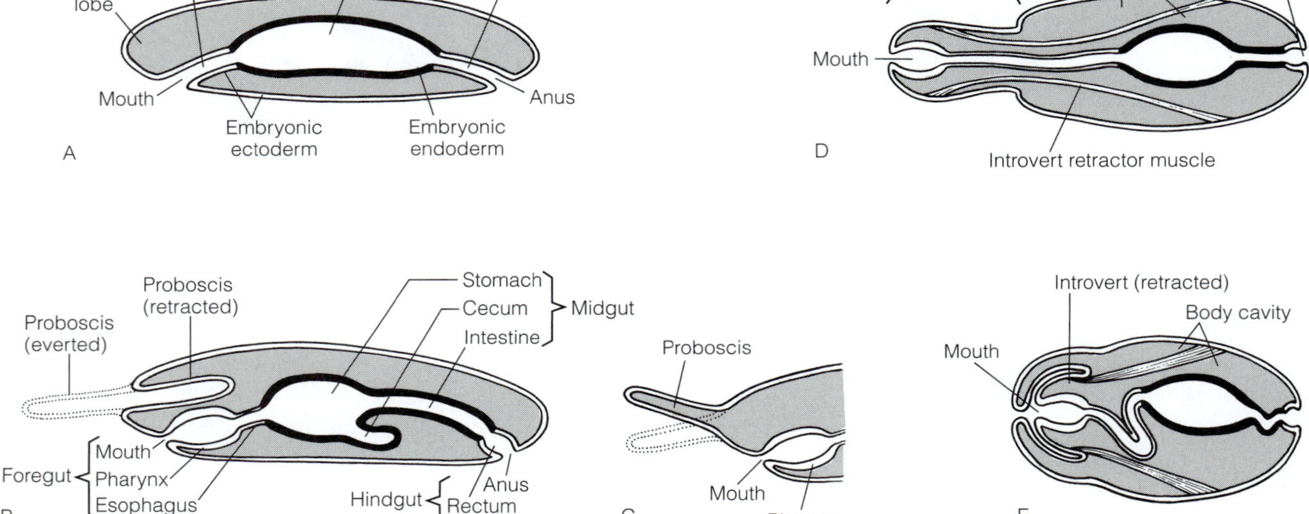

FIGURE 9-10 Bilateria: gut, proboscis, and introvert. **A,** The bilaterian gut typically is divided into three general regions. Foregut and hindgut develop from invaginated ectoderm, whereas the midgut originates from endoderm. **B,** Gut specializations and proboscis. A proboscis **(B** and **C)** is a modified, mobile, preoral lobe that is eversible (turned inside out as it extends; **B**) or permanently extended **(C)**. **D** and **E,** An introvert is an anterior end, including the mouth, that can be retracted into the trunk. It is either protrusible (shown) or eversible depending on taxon.

The foregut is commonly specialized into two or more regions. The mouth sometimes opens into a **buccal cavity,** an anterior chamber, small or large, that receives the food and may bear teeth. This is followed by a **pharynx,** a long or short, often muscular tube (Fig. 9-10B). In many taxa, the pharynx is a feeding or digging organ that can be everted (turned inside-out) or protruded (bulged outward) through the mouth and then retracted. Like the buccal cavity, it sometimes bears cuticular teeth. A **proboscis** is sometimes confused with a pharynx, but it is not part of the gut. Rather, it is a *pre*oral, prehensile appendage and usually the anteriormost part of the body (Fig. 9-10B,C). An elephant's trunk is a familiar example of a proboscis. A proboscis is always a mobile appendage and, in some instances, also is retractable into the body (Fig. 9-10B). If retractable, then it either is everted or protruded for use. The proboscis functions in food collection, locomotion, and defense. Food grasped by or collected on the proboscis is then transferred to the mouth. In some taxa, however, the proboscis secondarily establishes a connection to the gut. The last section of the foregut is the **esophagus,** which links foregut and midgut and is modestly muscular and often ciliated (Fig. 9-10B).

Midgut regions commonly include an enlarged **stomach** for digestion and a tubular **intestine** for the formation of feces and junction with the hindgut. A digestive **cecum** (pl., ceca; = digestive diverticulum, digestive gland) is a simple or branched outpocket of the midgut, either the stomach or intestine (Fig. 9-10B). The stomach receives food from the foregut and is typically the site of extracellular digestion. When present, the digestive ceca are sites of intracellular digestion, absorption, and nutrient storage.

An enlargement of the hindgut immediately before its junction with the anus is called a **rectum** (Fig. 9-10B). When the rectum receives not only indigestible wastes from the gut, but also gonoducts or excretory ducts, it is known as a **cloaca** (= sewer). Use of the names of gut regions is not standardized and may vary with taxon.

SMALL BILATERIA

The gut of most small bilaterians has a well-developed mouth, pharynx, and midgut, but the hindgut and anus sometimes are either weakly differentiated or absent. When an anus is absent, feces are egested through the mouth.

Coelom

LARGE BILATERIA

A **coelom** is a fluid-filled cavity or canal lined by a mesoderm-derived epithelium known as **mesothelium** (= middle epithelium; Fig. 9-9D). The coelom is unique to Bilateria and provides a new, third internal compartment for physiological regulation and specialization. It also functions as a hydrostatic skeleton, especially in soft-bodied worms, and in internal transport, excretion, and reproduction. The coelom occupies some of the connective-tissue space between epidermis and gastrodermis (Fig. 9-9D), but during its development, it displaces the connective tissue peripherally. Its cavity is filled with **coelomic fluid,** which is circulated by mesothelial cilia or contraction of body-wall muscles of mesothelial origin.

Animals with a coelom are known as **coelomates.** The simplest and probably most primitive coelomic cavities are lined by a monolayered mesothelium composed chiefly of monociliated, contractile epitheliomuscular cells (Fig. 6-4A). This single epithelium is simultaneously the body musculature and the coelomic lining (Fig. 9-9D,F, 9-11A,B). A more complex, bilayered coelomic lining occurs in some invertebrate bilaterians and in the vertebrates. These coelomic cavities are lined by a thin, noncontractile epithelium called **peritoneum** (Fig. 9-11E), which separates the coelomic fluid from the retroperitoneal (= behind the peritoneum) muscle and connective tissue. Developmental and comparative evidence indicate that this bilayered coelomic lining that includes the peritoneum evolved by stratification of the monolayered mesothelium (Fig. 9-11B–E). The epitheliomuscular mesothelium separates into an epithelial peritoneum and a muscular myocyte layer. The presence of peritoneum is usually part of textbook definitions of a coelom, but few invertebrate coelomates actually have a vertebrate-like peritoneum and instead have only a mesothelium.

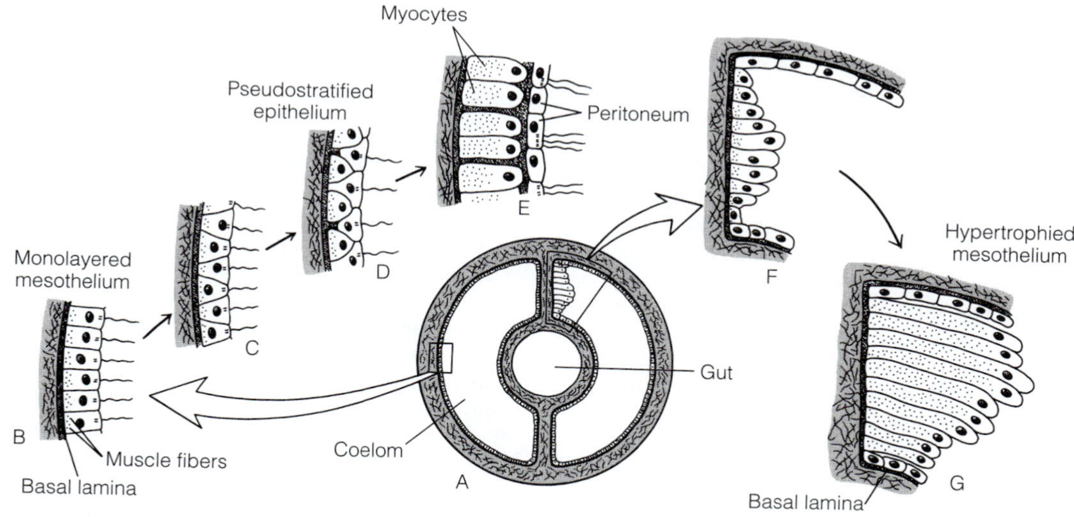

FIGURE 9-11 Coelomate Bilateria: evolution of coelomic lining and body-wall musculature. **B–E, pattern 1,** Mesothelium initially is a monolayer of epitheliomuscular cells **(A** and **B).** Later **(C** and **D),** noncontractile epithelial cells arch up over the epitheliomuscular cells and separate them from direct contact with coelomic fluid. This arrangement is termed a pseudostratified epithelium **(D)** because, although the cells are in two layers, all cells remain attached to a common basal lamina. Still later **(E),** the noncontractile cells have separated completely from the contractile cells, now myocytes, and formed a new epithelium called peritoneum. Peritoneum separates the body-wall musculature, which often is well developed, from the coelomic fluid. **A** and **F–G, pattern 2,** A powerful musculature also can evolve directly from a monolayered mesothelium **(A).** In this case, the epitheliomuscular cells enlarge (hypertrophy) and eventually fill and obliterate the coelomic cavity **(G).** As a longitudinal musculature, such hypertrophied muscles may partly occlude the coelomic cavity, as in many annelids and acorn worms (Hemichordata), or completely fill it, as in lancelets (Cephalochordata). *(A–E, Modified and redrawn from Rieger, R. M., and Lombardi, J. 1987. Ultrastructure of coelomic lining in echinoderm podia: Significance for concepts in the evolution of muscle and peritoneal cells. Zoomorphology 107:91–208.)*

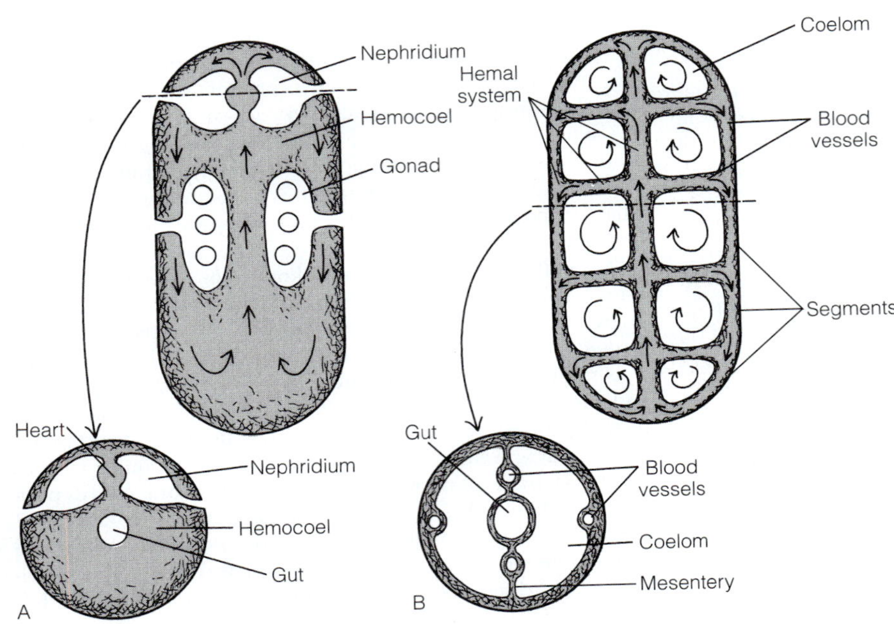

FIGURE 9-12 Coelomate Bilateria: internal fluid-transport systems. **A,** When a hemocoel (hemal system) is present, it constitutes the main body cavity and the coelomic cavities are small and specialized as gonadal or nephridal sacs. **B,** When coelomic cavities are well developed, they constitute the body cavity and the hemal system is confined to vessels and small sinuses.

Bilateral pairs of coelomic cavities contribute to the formation of **segments,** or somites, a series of similar modules along the length of the body (Fig. 9-12B). Typically, there is one pair of coelomic cavities per segment, as in each of the many segments of an earthworm. Left and right cavities surround and approach each other above and below the gut. At the point of contact, the abutting walls of the two cavities form a longitudinal partition in the midsagittal plane called a **mesentery** (Fig. 9-9C,D). A dorsal mesentery is situated above the gut, a ventral mesentery below. A **septum** (Fig. 9-9C) is a transverse partition, resembling a bulkhead in a ship or an airplane, between successive pairs of cavities. Septa and mesenteries divide the segmented coelom into separate fluid-filled compartments (Fig. 9-12B), each of which may be regulated, by virtue of its lining epithelium, for a particular specialized function. Specialized coelomic cavities, for example, often enclose gonads as genital coeloms (gonocoels) or enclose a heart as a pericardial cavity. The pleural, pericardial, and abdominal cavities of mammals are also specialized coelomic cavities.

Coelomic fluid has not been as well analyzed as blood, but, in general, it contains water, ions, low-molecular-weight solutes, and cells. Overall, it differs from blood in having a low, rather than a high, protein concentration. In some coelomates, germ cells are released from the gonad into the coelomic fluid. There they take up nutrients, grow, and complete their differentiation into gametes. The mesothelium is responsible for regulating the composition of the coelomic fluid for the growth and maintenance of the gametes. Other free cells of various functions, collectively called **coelomocytes,** originate from the mesothelium and circulate in the coelomic fluid. Phagocytic coelomocytes police the coelomic fluid for bacteria and other foreign or unwanted material. Respiratory proteins such as hemoglobin often are found in the coelom, but in contrast with their occurrence in invertebrate blood, they are invariably intracellular in coelomocytes.

SMALL BILATERIA

Small-bodied bilaterians generally lack a coelom altogether or the coelom is represented only as the gonadal cavity (gonocoel) or cavity of the excretory organs (nephrocoel). Because a large, fluid-filled cavity is absent and the body is more or less solid, these animals are described as **compact,** or acoelomate.

Hemal System

LARGE BILATERIA

A **hemal system** is a specialized part of the connective-tissue compartment that consists of interjoined vessels and cavities through which a fluid, the **blood,** circulates (Fig. 9-12B, 9-13A). Because the hemal system (and the connective tissue of which it is a part) is sandwiched between the bases of opposing epithelia, the vessels and cavities are lined solely by the basal lamina of those epithelia (Fig. 9-13A). In cnidarians and ctenophores, the connective tissue (mesoglea) is always gelatinous, not fluid, and is skeletal in function (Fig. 9-9A,B). Gas and nutrient transport through mesoglea is by diffusion, not convection. Only bilaterians have a hemal system in which there is convective transport (flow) of liquid connective-tissue fluid (blood). Connective tissue in general, and the bilaterian hemal system in particular, arise developmentally from the embryonic blastocoel.

Hemal vessels and cavities range widely in size. Small tubular channels are called **vessels** (or blood vessels, including arteries, veins, capillaries), saclike cavities are **sinuses** (Fig. 9-13A), and a large sinus that becomes the principal body cavity is known as a **hemocoel** (= pseudocoel; Fig. 9-12A). In large bilaterians the presence of a hemocoel is correlated with an exoskeleton, probably because the exoskeleton appropriates skeletal function from the coelom, reducing its degree of development. Coelomic reduction is offset by enlargement of

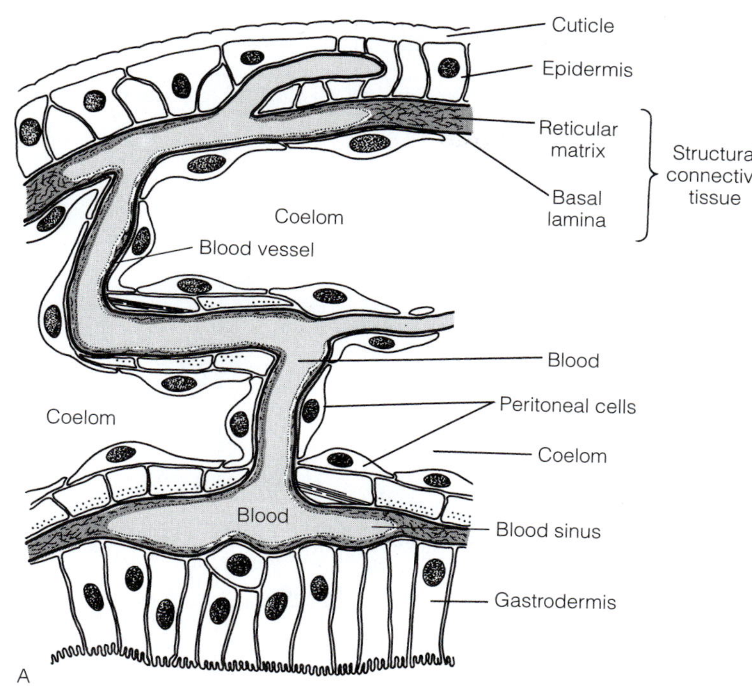

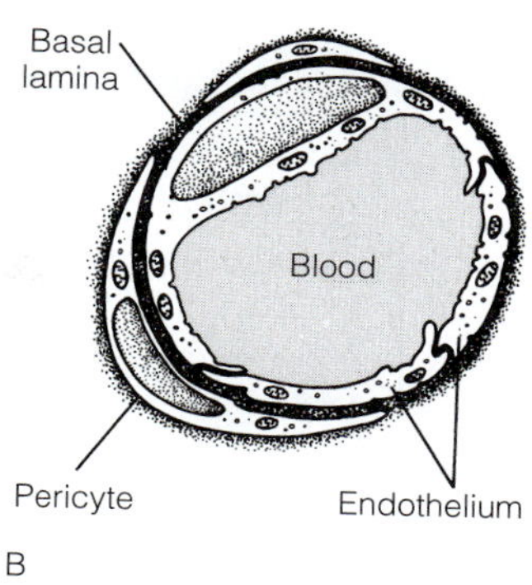

FIGURE 9-13 Bilateria: hemal system and blood vessel histology. Invertebrate and vertebrate blood vessels are both fluid-filled channels in the connective-tissue compartment. **A,** The blood of most invertebrates is in direct contact with structural connective tissue, especially the basal lamina of the overlying epithelial cells. **B,** The blood of vertebrates, cephalopod molluscs, and a few other taxa is separated from the basal lamina by a new endothelium that lines the vessels, controls transport across the vessel wall, and facilitates flow within the vessels. *(A, Modified from Nakao, T. 1974. An electron microscopic study of the circulatory system in* Nereis japonica. *J. Morphol. 144:217–236; B, Modified from Welsch, U., and Storch, V. 1976. Comparative Animal Cytology and Histology. University of Washington Press, Seattle. 343 pp.)*

the connective-tissue compartment, which often becomes an elaborate hemocoel. Structurally, vessels, sinuses, and hemocoels differ only in size. All of them are lined only by basal lamina and are filled with blood (Fig. 9-13A), sometimes called hemolymph in bilaterians with a hemocoel. Vertebrates and a very few invertebrates such as squids have evolved a secondary epithelium, the **endothelium** (Fig. 9-13B), that lines their blood vessels.

Where blood vessels pass through mesenteries and septa, they are overlain by mesothelium, from which derives a circular musculature (Fig. 9-17). Peristaltic contraction of this musculature pushes the blood along in waves. Often, the musculature is localized and thickened into one or more specialized **hearts** (Fig. 9-15). As a general rule, muscular contractions circulate blood, whereas cilia circulate coelomic fluid.

Blood consists of noncellular **plasma,** which contains water, ions, proteins, and other substances. When present, cellular **hemocytes** (corpuscles) circulate with the plasma. The blood may be colorless or pigmented if respiratory pigments are present. Respiratory pigments, such as hemoglobin, are sometimes small intracellular molecules contained in hemocytes, but usually are large extracellular molecules dissolved in the plasma. The protein concentration of blood is higher than that of coelomic fluid chiefly because of the plasma proteins and extracellular respiratory pigments.

SMALL BILATERIA

A hemal system is either absent in small bilaterians or present as a hemocoel, especially in taxa with an introvert. In the latter, the hemocoel not only provides a cavity for internal transport, but also a space into which the introvert can be retracted and a hydraulic mechanism for introvert extension (Fig. 9-10D,E).

INTERNAL TRANSPORT

LARGE BILATERIA

The fluid-transport systems of large bilateral animals are the coelomic and hemal systems. They often co-occur in animals with bodies composed of two or more segments (or zooids in colonial organisms; Table 9-2, Fig. 9-14C,D).

Circulation of coelomic fluid, by cilia or muscle, often is important for internal transport. In rare instances when the coelom is large, undivided, and extends throughout the body, it constitutes the sole circulatory system (Fig. 9-14A,B).

TABLE 9-2 Bilaterians: Correlations of Size and Design

Size and Form	Body Cavity			
	Compact	Hemocoel	Coelom	Coelom + Hemal System
Mostly small, no introvert	Gnathostomulida Platyhelminthes Mesozoa Gastrotricha Nematoda[1] Nematomorpha Annelida (e.g., Dinophilidae)			
Mostly small, introvert		Priapulida[2] Loricifera Kinorhyncha Rotifera Acanthocephala Tardigrada[3] Cycliophora Kamptozoa[4]		
Mostly large, introvert, no segments			Nemertea Sipuncula Annelida (Glyceridae)	
Mostly large, with or without introvert, two or more segments				Mollusca[5] Annelida (most spp.) Echiura Panarthropoda Phoronida Brachiopoda[6] Bryozoa[7] Hemichordata Echinodermata Chordata Chaetognatha[8]

[1]Only large-bodied nematodes, such as *Ascaris,* have a hemocoel. [2]Depending on species, priapulids may be microscopic or large. [3]Tardigrades lack an introvert, but have four pairs of retractile, telescoping legs. [4]Kamptozoans (entoprocts) lack an introvert, but have hollow, retractile tentacles. [5]Segmentation in molluscs is a matter of ongoing debate. [6]Brachiopods may not be segmented. [7]Bryozoan zooids are tiny, but the colonial bodies can be large. [8]Chaetognaths are small animals.

In most coelomates, however, the coelom is divided into two or more segmental coelomic cavities by septa and mesenteries. As a result of these divisions, coelomic fluid is restricted to each compartment and cannot flow between them (Fig. 9-12B). Substances, such as nutrients and gases, dissolved in the coelomic fluid cross between cavities only by diffusion or active transport across their epithelial walls (Fig. 9-14C,D). Thus, when multiple coelomic cavities occur in one organism, coelomic circulation is effective for local transport only. Another system, one adapted for whole-body transport, is required.

The hemal system is the through-transport system of the body. Its blood-filled vessels and sinuses are in the connective-tissue compartment, which is continuous around and between all tissue layers of the body (Fig. 9-12B, 9-13A, 9-14C). Each blood vessel or sinus passes *between* epithelia—in the body wall between the epidermis and mesothelium, in the gut wall between the gastrodermis and mesothelium, or in mesenteries and septa between two sheets of mesothelium (Fig. 9-12B, 9-13A, 9-14C). The blood itself, although part of the connective tissue, lacks the fibers and molecules typical of the structural connective tissue. As a result, the blood is fluid and it flows (Fig. 9-12B).

The bilaterian hemal system adopts many forms, but there are two general types of circuits: parallel and series. In a **parallel circuit,** the major vessels form an elongate loop—dorsal vessel above, ventral vessel below—in the midsagittal plane (midplane) of the body (Fig. 9-15A). Between the dorsal and ventral vessels is a sequence of parallel vertical vessels and capillaries supplying the various organs of the body. As blood is pumped along the loop vessels, some is shunted into each of the parallel capillary beds. A parallel

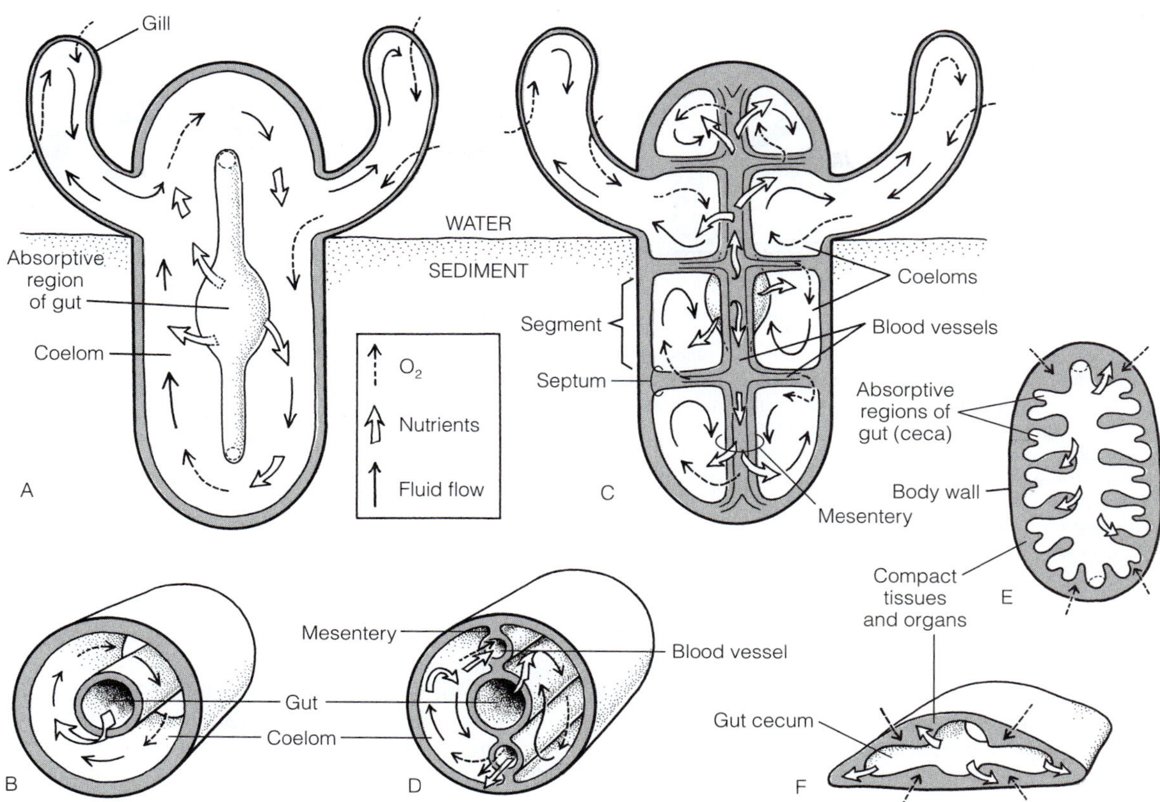

FIGURE 9-14 Bilateria: internal transport by convection and diffusion. **A–D, large bilaterians:** Functional significance of two convective circulatory systems, coelomic and hemal, in one organism. **A** and **C** are two large coelomate bilaterians partly buried in oxygen-free (anoxic) sediment. The two appendages are gills extending into the oxygen-rich water overhead. Oxygen (broken arrows) enters the body by diffusion across the gills and nutrients (open arrows) enter from the localized absorptive region of the gut. Once inside, simple diffusion is not rapid enough to transport the oxygen and nutrients throughout the body to the consuming tissues. The animals instead require a convective circulatory system (solid arrows) for internal transport. **A** and **B** (cross section of **A**), A few bilaterians have a continuous, unpartitioned coelom throughout the body. In these the coelomic fluid circulates globally, reaching within diffusion distance of all body tissues. Here the coelom is the sole circulatory system and a hemal system is absent. In most coelomate bilaterians, however, the coelom is partitioned by septa and mesenteries into segments, as shown in **C** (cross section in **D**), and these restrict the coelomic circulation to local segmental circuits. For whole-body (global) transport, these animals have evolved a hemal system, which consists of blood-filled vessels ands sinuses in the connective tissue (shaded) of septa, mesenteries, and body wall. **E** and **F, small, compact bilaterians:** Small size and a flat, two-dimensional shape permit a diffusional supply of oxygen across the body (gill) surface and both coelomic and hemal transport systems are absent. Lacking a circulatory system other than ciliary flow in the gut, the absorptive regions of the gut branch throughout the body within diffusion distance of tissues and organs.

circuit occurs chiefly in hemal systems composed of vessels and small sinuses. A **series circuit** is associated with bilaterians, such as molluscs and arthropods, having a hemocoel (Fig. 9-15B). The hemocoel constitutes the body cavity and surrounds the gut and other organs, which essentially extend through or bulge into it. If a heart is present, blood flows chiefly in an elongate loop in the midplane of the body. As the blood circulates, it encounters each of the organs one after the other, in series, before returning to its starting point, usually a heart. Along the way, nutrients acquired by blood from the intestine, for example, are delivered first to organ 1, then organ 2, then organ 3, and so on. This means that organ 1 is first in line for the highest concentration of nutrients, organ 2 receives less, organ 3 still less. In contrast to a parallel circuit, which delivers equal concentrations of substances to each of its organs, the delivery is unequal in a series circuit. As a result, some bilaterians with a series circuit undergo periodic **heartbeat reversal,** which reverses the direction of blood flow and places the organ that was previously last in the series first. Heartbeat reversal may be an adaptation to average the distribution of substances to organs arranged in series.

SMALL BILATERIA

A fluid-transport system may be absent in small-bodied bilaterians or present as a gastrovascular system or a hemocoel. Many small, compact, cylindrical bilaterians lack both coelomic and

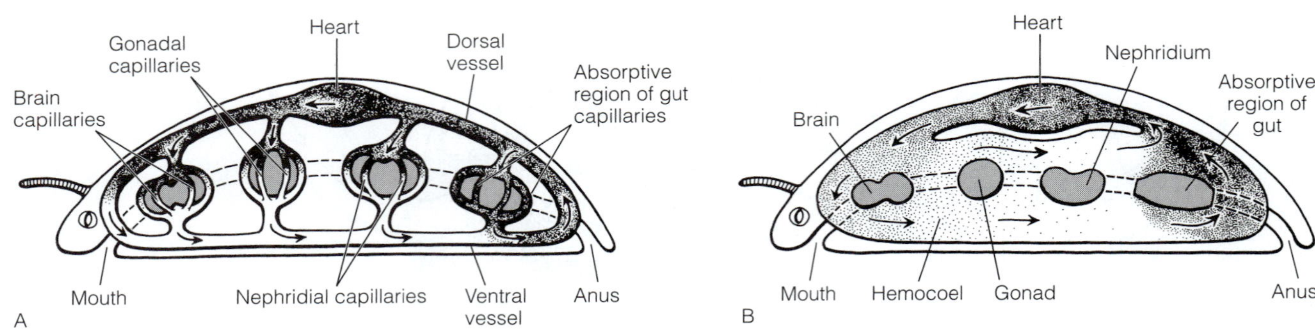

FIGURE 9-15 Bilaterian hemal system: parallel and series blood circuits. **A,** In a parallel circuit, the main loop consists of a dorsal and ventral vessel joined together by a sequence of parallel shunts, each of which supplies a major organ or tissue. In this example, the blood acquires nutrients from the gut and delivers them in equal concentrations (shading) to other organs in the circuit. **B,** In a series circuit, the hemocoel (a large hemal sinus) is the body cavity and organs bulge into it (brain, gonad, nephridium) or extend through it (gut). As blood flows (as shown by the arrows), it encounters the organs, in series, one after the other. Nutrients acquired from the gut are distributed in declining concentrations (shading) first to the heart, then the brain, and so on. Some bilaterians with a hemocoel correct this unequal distribution of nutrients by periodically reversing the direction of heartbeat and blood flow.

hemal systems, and diffusion delivers gases and nutrients to all parts of the body. The fluid-filled extracellular spaces, collectively called **interstitium,** are minute crevices in which there is no convective circulation.

The tissue-thin body of compact, flat bilaterians is suited for gas transport by diffusion, but for nutrients, the distance from the central gut to the body margin exceeds the limit of simple diffusion. As a result, the gut of flat, compact bilaterians typically develops outpockets called **ceca** (sing., cecum) that deliver nutrients to the periphery of the body (Fig. 9-14E,F). The gut itself becomes a circulatory system, called a **gastrovascular system,** that transports fluid and nutrients, together known as **chyme** (pronounced kime). Gastrovascular circulation results from the action of gastrodermal cilia or contraction of the gut musculature. Developmentally, a conventional gut or gastrovascular system arises from the embryonic archenteron. The coelenteric circulatory system of cnidarians and ctenophores is an example of a gastrovascular system that we have already discussed in earlier chapters.

Tiny bilaterians with an introvert usually have a hemocoel (Table 9-2). In these, a heart is typically absent and the blood is circulated chiefly by muscles of the gut and body wall.

GAS EXCHANGE AND RESPIRATORY PIGMENTS

LARGE BILATERIA

Large bilaterians have enlarged, specialized surfaces for gas exchange. Among aquatic bilaterians, these are usually evaginations of the body surface, or gills, although there are exceptions. Terrestrial animals lack gills, but usually have lungs or other systems of ventilated air-filled tubes (tracheae).

Oxygen that diffuses across a surface may be transported in physical solution in plasma or coelomic fluid or it may bind to a respiratory pigment in those fluids. A respiratory pigment increases the oxygen-carrying capacity of the blood or coelomic fluid. Three different respiratory pigments—hemoglobin, hemerythrin, and hemocyanin—appear sporadically in Bilateria. Hemoglobin is widespread among bilateral animals and even occurs in plants, hemocyanin is limited to molluscs and arthropods, and hemerythrin is rare, being found only in a few taxa (inarticulate brachiopods, magelonid polychaetes, sipunculans, and priapulids). All respiratory pigments have the ability to react reversibly with oxygen, binding it when oxygen tension is high and releasing it when tension is low.

A **hemoglobin** molecule consists of two parts: **heme,** a porphyrin ring with an iron atom at its center, and a protein called **globin** (Fig. 9-16A). The number of globins and hemes varies widely in different hemoglobins, but in all of them a molecule of oxygen binds with each heme. An important function of globin is to prevent oxygen from binding too tightly to the heme so that the reaction is reversible and oxygen can be released to the tissues. Cellular hemoglobin commonly occurs in coelomic cavities and extracellular hemoglobin is in blood plasma, but there are a few exceptions. Extracellular hemoglobins typically are large, high-molecular-weight molecules whereas intracellular hemoglobins are small molecules. The extracellular occurrence of a few large, osmotically active molecules rather than many small ones may be necessary to keep the osmotic concentration of the blood within reasonable limits, but perhaps also to prevent loss of the molecules during ultrafiltration (see below). Most hemoglobin is red when oxygenated, but one variant, chlorocruorin, is green.

A molecule of **hemerythrin,** like hemoglobin, uses iron atoms to bind oxygen, but unlike hemoglobin, there is no heme and the two iron atoms are bound directly to the

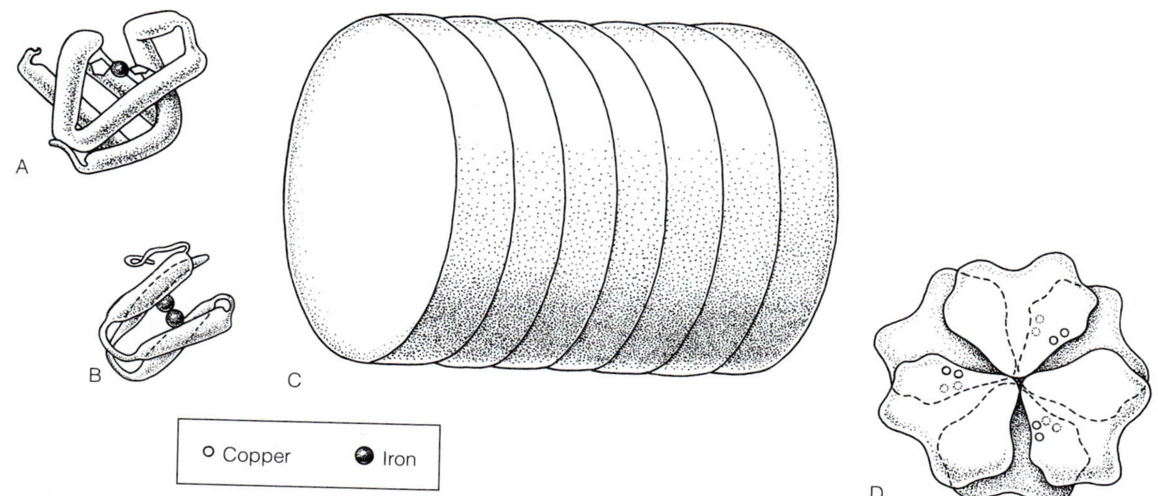

FIGURE 9-16 Bilateria: respiratory pigments. Respiratory pigments are all complexes of metal (iron or copper) and protein. Although oxygen actually binds to the metal and not the protein, the protein regulates the tightness of the bond and thus the delivery of oxygen to tissues. Hemoglobin **(A)** and hemerythrin **(B)** both contain iron and are similar in size and shape (myoglobin and myohemerythrin are shown). In hemoglobin, one iron atom is bound to heme (pentagons), whereas in hemerythrin two iron atoms are bound directly to the protein. The amino acid sequences of the two molecules are also different. Both molluscs and arthropods have large, copper-based hemocyanins, but the pigments are not homologous in the two taxa. Molluscan hemocyanin **(C)** is a large cylindrical molecule, whereas arthropod hemocyanin **(D)** is a smaller (but still large) hexamer (shown) or dodecamer. *(A, Redrawn from Dickerson, R. E., and Geis, I. 1983. Hemoglobin. The Benjamin Cummings Publ. Co., Inc., Menlo Park, NJ. 176 pp.; B, Modified and redrawn from Klippenstein, G. L. 1980. Structural aspects of hemerythrin and myohemerythrin. Am. Zool. 20:39–51; C, Drawn from micrograph in Magnum, C. 1985. Oxygen transport in invertebrates. Am. J. Physiol. 248:R505–R514, D, Redrawn from Linzen, B. et al. 1985. The structure of arthropod hemocyanins. Science. 229:519–524.)*

protein (Fig. 9-16B). Hemerythrin always occurs in cells and is never free in plasma or coelomic fluid. It is pink or violet when oxygenated and colorless when deoxygenated.

Hemocyanin binds a molecule of oxygen between a pair of copper atoms. The many pairs of copper atoms are bound directly to the protein part of the molecule and a heme is absent. Hemocyanin only occurs as large *extra*cellular molecules and is never found in cells (Fig. 9-16C,D). It is ethereal blue when oxygenated and colorless when deoxygenated.

Because they chemically bind oxygen, respiratory pigments increase the capacity of blood, coelomic fluid, or tissue to transport or store oxygen. Such enhanced oxygen transport is important when metabolic oxygen demand is not met by the supply of oxygen dissolved in blood plasma, coelomic fluid, or cellular tissues. Thus respiratory pigments tend to be associated with animals that have a high oxygen demand (active animals), occupy oxygen-poor habitats (burrowers), or have metabolically active tissues, such as muscle **(myoglobin)** or nerve **(neuroglobin).**

Functionally different forms, with different oxygen affinities, of the same respiratory pigment often occur in different body compartments of the same animal. For example, there may be one hemoglobin in the blood, another in the coelom, and yet another in a tissue such as muscle or nerve, each with a different protein component (globin) that determines the affinity of the hemoglobin for oxygen. Being part of the connective tissue below the skin, the blood is closest to the environmental source of oxygen, the coelom is next closest, and the target tissue is most distant (Fig. 9-14C). For oxygen to be transported from blood hemoglobin to coelomic hemoglobin to muscle hemoglobin, for example, the different hemoglobins must be arranged in order of increasing oxygen affinity. Muscle hemoglobin has the highest affinity for oxygen, blood the lowest, and coelomic is intermediate. Thus blood binds environmental oxygen, coelomic hemoglobin removes it from blood, and muscle strips it away from the coelomic pigment. This stepwise arrangement of affinities, from low to high as one moves from outside to inside the animal, creates a cascade down which oxygen is transferred from blood to coelom to tissues. The electron carriers of muscle mitochondria have even higher affinities for oxygen. Hydrogen itself, at the end of the electron-transport chain, has the highest affinity of all.

SMALL BILATERIA

Small bilaterians lack specialized gills and use the entire body surface for gas exchange. Respiratory pigments occur rarely, only as hemoglobin, and usually in fixed, noncirculating cells. In the few studied cases, the hemoglobin functions in oxygen storage for use when the animal enters an anoxic environment.

EXCRETION

GENERAL

Excretion maintains the body's internal constancy (homeostasis) by the elimination of excess substances (or their metabolic by-products) that enter the body. Such substances may be water, salts, and by-products of cellular metabolism, especially nitrogenous wastes. In most aquatic invertebrates, nitrogen from protein metabolism is excreted as ammonia. Although highly toxic, it is soluble in water, diffuses across the body surface (often across gills), and is diluted and carried away by currents.

Water tends to enter or leave the body of aquatic organisms by osmosis. The solutes in coelomic fluid and blood, including plasma and respiratory proteins, are osmotically active. Together, these solutes may raise the osmotic concentration of extracellular body fluids above that of the surrounding water, even if it is seawater. So water enters by osmosis and the body must work to offset its water load. Bilaterians accomplish this by pumping the water back out with ciliated tubules called **nephridia** (= little kidneys). This outflow is also used to discard metabolic wastes, excess ions, spent hormones, and perhaps toxins ingested with food. The result is a waste product called **urine.**

LARGE BILATERIA

Several types of nephridia—filtration, secretion, and storage—occur in bilaterians. Bilaterians in which urine formation includes an ultrafiltration step are said to have **filtration nephridia**. In these, urine is formed in three steps: ultrafiltration, modification, and release. **Ultrafiltration** produces a protein- and cell-free filtrate of blood called **primary urine** by forcing blood, under pressure generated by heart or vessel contraction, through a macromolecular filter. The primary urine then enters the coelom, which is drained by paired ciliated tubules, typically one pair per segment. These tubules, known as **metanephridia,** open from the coelom at a ciliated funnel, the **nephrostome,** and to the exterior at a **nephridiopore.** The coelomic mesothelium and the cells lining the metanephridium modify the primary urine by selective reabsorption and secretion to secondary, or **final, urine,** which is released at the nephridiopore.

Ultrafiltration occurs at a specialized hemal site or sites, which, if well developed, are called **glomeruli** (sing., glomerulus). At such sites, the mesothelium includes modified cells called **podocytes** (= foot cells; Fig. 9-17) that create gaps in the otherwise unbroken mesothelium. Pressurized blood would flow freely across these gaps into the coelom were it not for the mesothelial basal lamina, which is the sole barrier to

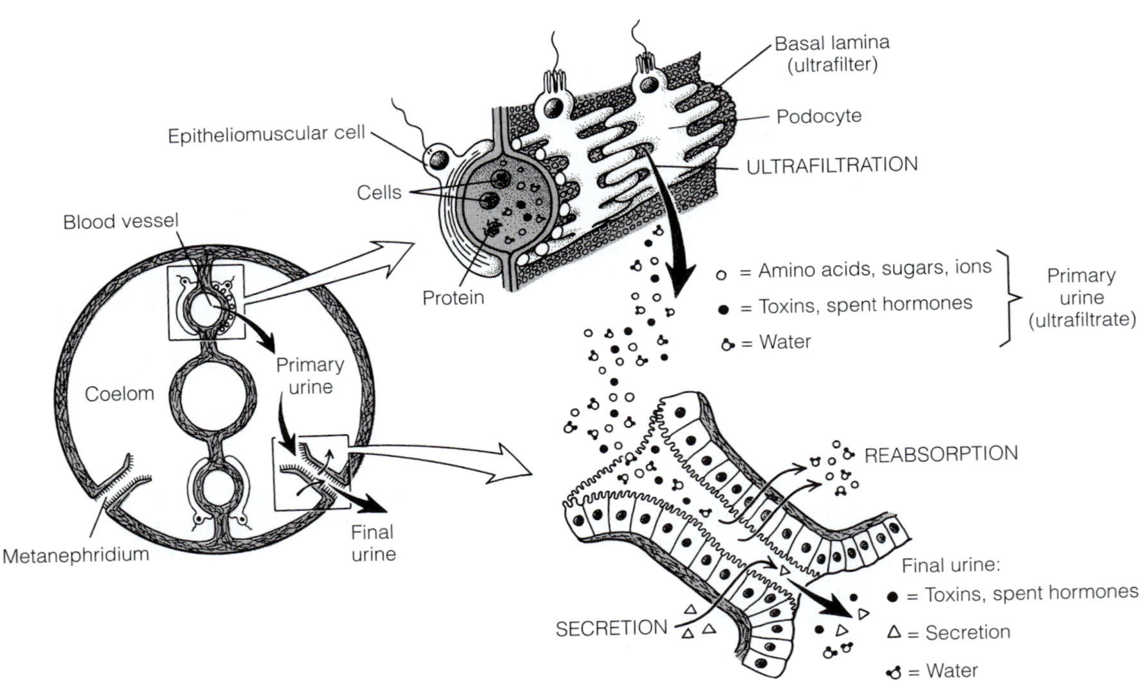

FIGURE 9-17 Bilaterian excretion: metanephridial system. A metanephridial excretory system consists of an ultrafiltration site on a blood vessel (indicated by podocytes), a coelom, and a tubule (metanephridium) that leads to the exterior. Muscular contraction of the blood vessel pressurizes the blood, forcing a cell- and protein-free ultrafiltrate (primary urine) across the podocyte layer and basal lamina (ultrafilter) into the coelom. Coelomic cells, including gametes, probably reabsorb nutrients from the primary urine, but the ciliated metanephridia are the chief sites for selective reabsorption and secretion, where the primary urine is converted to final urine. Urine is released from the nephridiopore.

this flow. This mesh of collagen fibers constitutes the ultrafilter. It retains cells and proteins in the blood, but allows primary urine—water, ions, and small molecules, including nutrients, hormones, and toxins (in other words, plasma minus proteins)—to enter the coelom and ultimately the metanephridia.

The primary urine is modified to final urine by reabsorption and secretion. These processes occur chiefly in the metanephridium, but reabsorption also may occur in the coelom. **Reabsorption** is the selective recovery, by pinocytosis and macropinocytosis, of useful metabolites from the primary urine. Other metabolites, including ingested toxins, are ignored and thus exit the body without the cost of active transport. Primary urine also is modified by the active **secretion** of specific wastes into it (Fig. 9-17), for example uric acid (a breakdown product of nucleic acids). The metanephridium typically is a secretion site for urates, organic acids, and other waste compounds. Because secretion, unlike ultrafiltration, moves a specific metabolite and very little water from blood to urine, it is often employed as the predominant or *sole* mode of urine formation, replacing ultrafiltration, in semiterrestrial and terrestrial animals. Such secretion kidneys will be discussed later.

A filtration nephridium that includes a hemal filtration site with podocytes, a coelom, and a metanephridium is called a **metanephridial system.** Metanephridial systems typify most large bilaterians, including the terrestrial molluscs and vertebrates (Fig. 9-18), but not insects, which have evolved a unique secretory kidney, the Malpighian tubule.

SMALL BILATERIA

The excretory organs of small bilaterians, although they are filtration nephridia, differ in design from a metanephridial system and are called protonephridia (Fig. 9-19). A **protonephridium** is a ciliated excretory tubule that opens to the exterior at

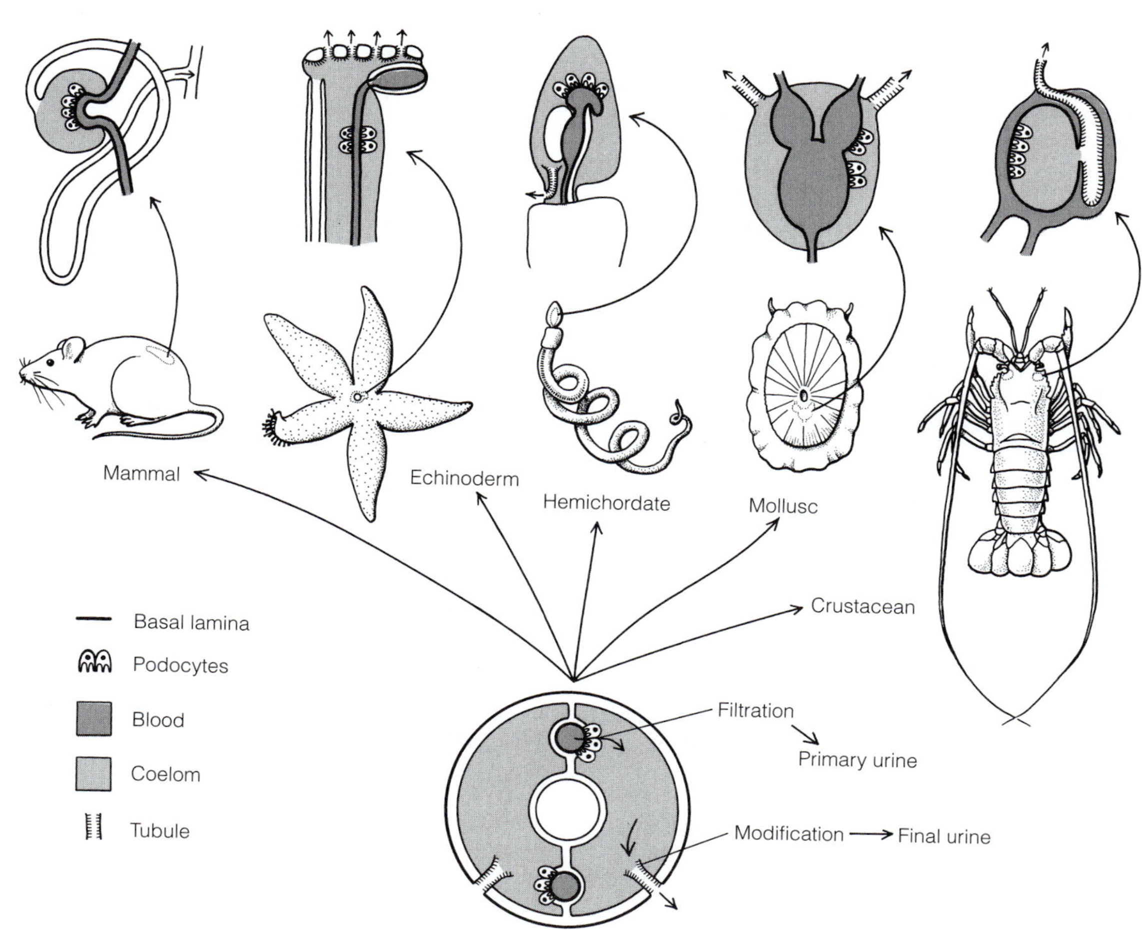

FIGURE 9-18 Bilaterian excretion: body design and the occurrence of metanephridial systems. Metanephridial systems usually occur in large bilaterians with both coelomic and hemal compartments. *(From Ruppert, E. E., and Smith, P. R. 1988. The functional organization of filtration nephridia. Biol. Rev. 63:231–258. Reprinted with the permission of Cambridge University Press.)*

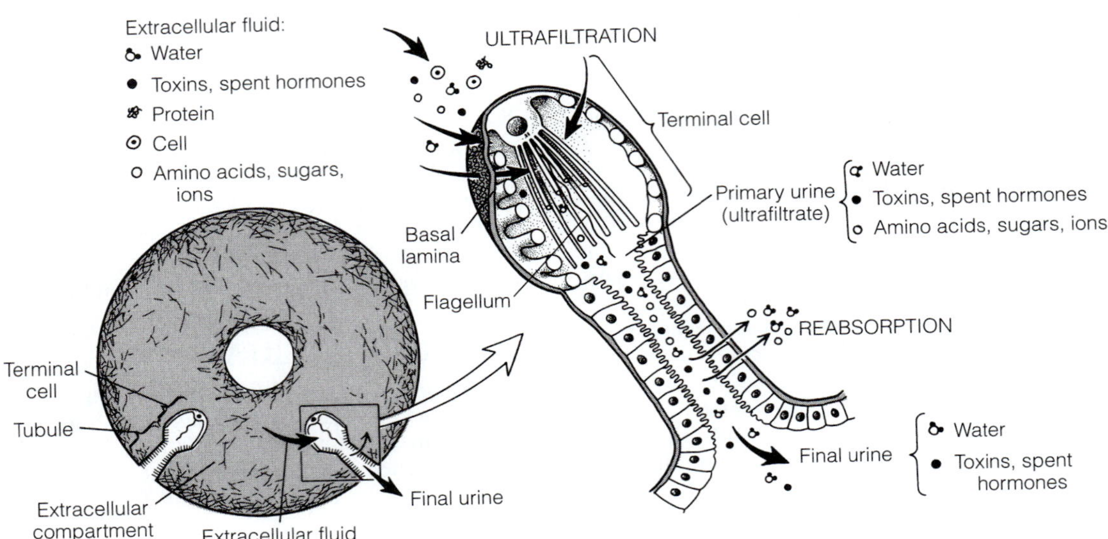

FIGURE 9-19 Bilaterian excretion: protonephridia. As in metanephridial systems, protonephridial excretion is a two-step process involving ultrafiltration and modification (reabsorption and secretion) to form final urine. In protonephridia, ultrafiltration occurs across the wall of a terminal cell, which resembles a single podocyte attached to the inner end of the nephridial tubule. The flagellum (or flagella) on the terminal cell is believed to create the pressure difference for ultrafiltration. As ultrafiltrate (primary urine) enters the tubule, it is modified by reabsorption to final urine, which is transported to the exterior by cilia on the tubule lining.

a nephridiopore but is capped at its internal end by a specialized, flagellated **terminal cell.** The terminal cell resembles a podocyte grafted onto the inner end of a nephridial tubule. The modified surface of the terminal cell creates gaps, bridged by basal lamina, between the extracellular body compartment and the lumen of the protonephridial tubule. The terminal-cell flagellum (or flagella) and cilia in the tubule move fluid from the body, across the basal lamina ultrafilter, and into the tubule. Once in the tubule, the primary urine is modified to final urine and released at the nephridiopore.

Protonephridia occur in animals that lack blood vessels, a coelom, or both. These are typically bilaterians with only one unpartitioned, extracellular body cavity. Most are small-bodied animals with an interstitium or hemocoel as the sole body cavity, but a few are large animals in which one body cavity (usually the coelom) is continuous and unpartitioned (Fig. 9-19). Because ultrafiltration requires the juxtaposition of *two* fluid-filled compartments separated by an ultrafilter, those with only one body cavity position the ultrafilter on the inner end of the nephridial tubule between the body and nephridial lumen (Fig. 9-19).

A protonephridium may be capped by one or many terminal cells, and each terminal cell can bear one or many flagella. The flagella beat in undulatory, planar waves propagated from base to tip. When multiple flagella are present, their synchronized beat resembles a minute, flickering candle flame. A terminal cell with one flagellum is called a **solenocyte,** those with many are either a flame cell or flame bulb. A **flame bulb** has its nucleus offset to one side of the flame and is typical of rotifers (Fig. 9-20). A **flame cell** bears its nucleus at the base of the flame and can be found in many flatworms (Fig. 9-20).

Filtration nephridia are unique to Bilateria. Their origin was probably necessitated by the commitment of tissues and organs—muscles, gonads, coeloms, nerves—to the middle layer of the body, between epidermis and gastrodermis. Sandwiched between these two epithelia, which form a diffusion barrier between the middle layer and the exterior, nephridia evolved to provide a flow from the isolated interior to the outside. Again, this is foreshadowed by the ctenophores, which have a mesogleal musculature and ciliary rosettes that link the mesoglea to the exterior.

REPRODUCTION AND DEVELOPMENT

SEXUAL REPRODUCTION

Bilaterian germ cells and gonads originate from mesoderm. The gonads adopt three general forms depending on body design and size (Fig. 9-21). (1) If the gonad occurs in a body region with a coelom and metanephridium, it resides in the mesothelium, releases gametes into the coelom, and spawns them through the metanephridium (Fig. 9-21A). (2) If the gonadal region is a hemocoel, compact cellular tissue, or a coelom without a metanephridium, the gonad is enclosed in its own sac. The saccate gonad is lined by a mesodermal epithelium including germ cells that opens to the exterior via its own gonoduct (Fig. 9-21B,C). (3) Gonads of some very small, compact-bodied taxa are mere aggregations of germ cells that lack an enclosing epithelium. In these, the gametes

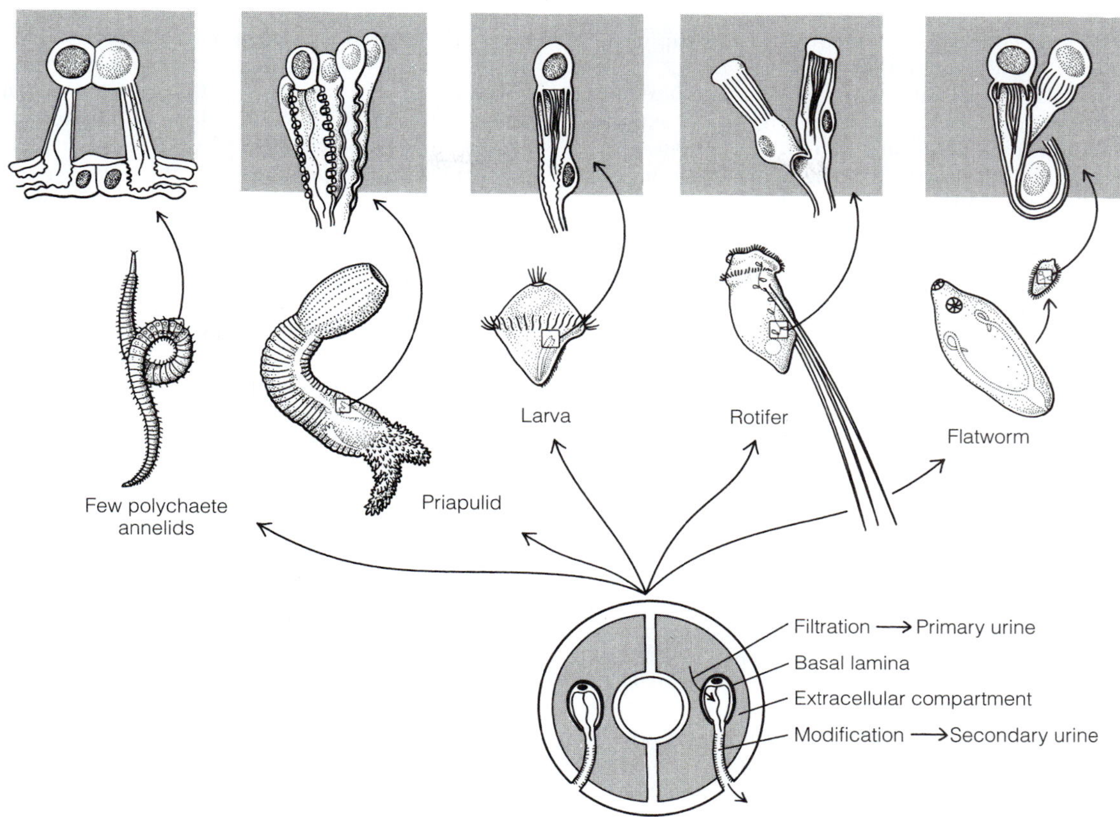

FIGURE 9-20 Bilaterian excretion: body design and the occurrence of protonephridia. Bilaterians that lack a hemal system, a coelomic system, or both have protonephridia instead of a metanephridial system. Typically these are small animals that rely on simple diffusion for internal transport. Protonephridia also occur in a few large animals such as priapulids that have *one* fluid-filled body cavity, usually an unpartitioned coelom or hemocoel. *(From Ruppert, E. E., and Smith, P. R. 1988. The functional organization of filtration nephridia. Biol. Rev. 63:231–258. Reprinted with the permission of Cambridge University Press.)*

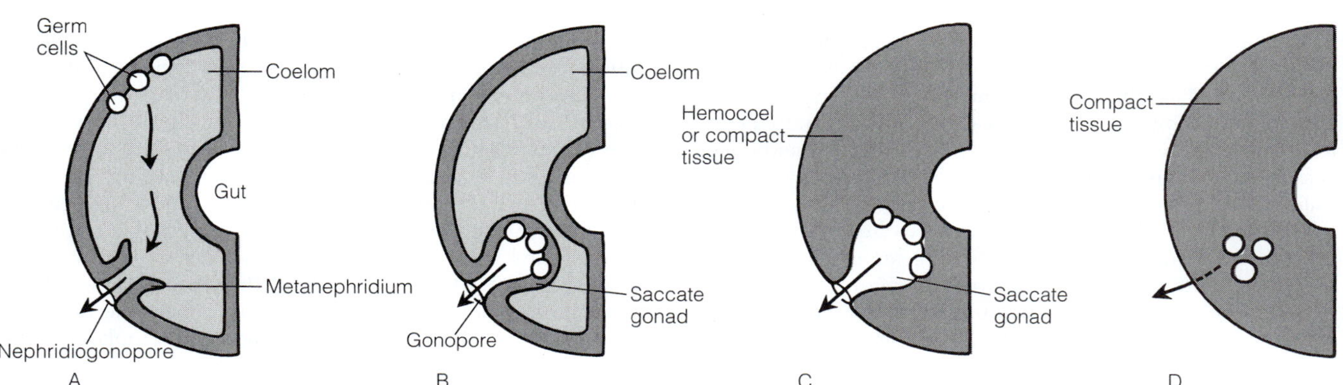

FIGURE 9-21 Bilateria: gonadal anatomy and oviposition. **A–D,** Half cross sections. **A,** Gonad in body region with coelom and metanephridium: gonads in mesothelium, gametes released into coelom, discharged through metanephridium (Lophophorata, Trochozoa). **B** and **C,** A saccate gonad with its own gonoduct occurs in coelomates that lack a metanephridium (**B,** deuterostomes) and in taxa with either a hemocoel or a compact (acoelomate) body **(C)**. **D,** Gonad associated with compact tissue, mostly in small-bodied bilaterians: gonad absent, replaced by simple aggregations of germ cells, ripe gametes exit through gonoducts (not shown) or rupture through body wall.

are released either through a gonoduct or by rupturing through the body wall (Fig. 9-21D).

Among animals in general, adaptations have evolved to release sperm and eggs synchronously and in close proximity, thus improving the chance of fertilization. In invertebrates with external fertilization, day length, moonlight, and pheromonal cues can draw mates together and synchronize gamete maturation and spawning. **Internal fertilization** results from deposition of sperm in the female reproductive tract. It not only increases the proximity of eggs and sperm for fertilization, but also allows for sperm storage for the fertilization of eggs over an extended period of time. Internal fertilization occurs in all small and many large-bodied bilaterians. Tiny bilaterians produce few germ cells and cannot tolerate the inevitable mortality of gametes and zygotes that is associated with external fertilization.

Reproductive systems may be simple in species that free-spawn their gametes into the water, but often they are complex in species with internal fertilization, especially hermaphrodites. The gonoduct of internal fertilizers typically is modified into accessory reproductive organs that store, nourish, transport, and encapsulate the gametes. Males usually have secretory glands that nourish and assist movement of sperm. A **spermatophore gland** secretes a protective capsule called a **spermatophore** around a cluster of sperm in some taxa. The spermatophore may be attached to the substratum and later picked up by the female or it may be attached directly to her. Other male accessory organs include a **seminal vesicle,** which stores sperm prior to transmission, and a **penis** or specialized copulatory appendage that aids in the transmission of sperm to the female. Common female accessory organs are a copulatory **bursa** and a **seminal receptacle** (spermatheca), pouches or chambers that receive and store sperm for short and long periods, respectively. When fertilization is internal in large bilaterians, the zygotes can be released into the water, **gestated** internally, **brooded** externally, or deposited in a protective **capsule,** or case. If brooded or deposited in a capsule, the eggs are typically yolky. Tiny bilaterians rarely release zygotes into the water, occasionally are viviparous, and frequently attach zygotes individually or in capsules to the substratum.

The amount of yolk in each egg not only affects the cleavage pattern in developing embryos (Chapter 4, Fig. 4-2), but also the pattern of larval development in species with indirect development. Embryos developing from small, relatively nonyolky eggs often become long-lived, planktotrophic larvae. Planktotrophic larvae are common in Bilateria and occur elsewhere in Anthozoa only rarely. These bilaterian larvae typically have specialized appendages and ciliary bands with which to actively suspension feed. Their metamorphosis is often complex and involves loss of many larval structures and extensive remodeling of others. Embryos developing from large, yolky eggs may become lecithotrophic larvae that rely on yolk reserves, or they may develop directly into juveniles. Very small bilaterians are invariably direct developers with short generation times.

CLEAVAGE PATTERNS AND DEVELOPMENTAL DETERMINISM

In all bilaterians the late embryo (postgastrula) is bilaterally symmetric, but this stage is usually preceded by radial symmetry of the blastula and often an early gastrula, which may reflect the original metazoan pattern of symmetry. This initial radial symmetry results from two widespread cleavage patterns: radial and spiral (Fig. 9-22A,B). In **radial cleavage,** which occurs in some cnidarians, lophophorates, and deuterostomes, the cleavage planes are either parallel or perpendicular to the polar axis of the egg (Fig. 9-22A). The resulting tiers of blastomeres arrange themselves radially around the polar axis and corresponding blastomeres of different tiers are located directly above and below each other. In **spiral cleavage,** on the other hand, the cleavage planes are oblique to the polar axis of the egg. Successive tiers of blastomeres arrange themselves radially about the polar axis, but corresponding blastomeres of different tiers are offset with respect to those above and below. A blastomere in an upper tier rests in the furrow between two blastomeres in the next lower tier (Fig. 9-22B). Spiral cleavage typifies the large bilaterian taxon Spiralia, which includes flatworms, molluscs, and segmented worms, among others.

Sooner or later after the onset of gastrulation, the radial symmetry of the blastula is replaced by bilateral symmetry. Some invertebrate embryos, such as those of roundworms and sea squirts, are bilaterally symmetrical from the beginning of cleavage and never show radially symmetric or spiral patterns. Such embryos have a **bilateral cleavage** pattern (Fig. 9-22C). In cases in which the egg accumulates large amounts of yolk (various species of almost all bilaterian taxa), the original cleavage pattern becomes modified and obscured. For example, arthropods have yolky eggs, like those of birds, that cleave superficially (meroblastic cleavage).

Because all animals develop from embryos composed of undifferentiated cells into adults composed of differentiated cells in specific locations, at some point during development the fate of embryonic blastomeres must be established. The specification of cell fate during development results from two general processes that are roughly correlated with the cleavage patterns previously described. One of these processes is called **determinate,** or mosaic, **development** and is associated with spiral and bilateral cleavage patterns. In such embryos, cell fate is determined early in development and results primarily from the action of specific factors that are unevenly distributed, like pieces of a mosaic, in the cytoplasm of the uncleaved egg. As the egg divides, some blastomeres incorporate the determining factors whereas others do not. A clear example of determinate development occurs in Spiralia. In the early embryos of these animals, an unidentified cytoplasmic factor is segregated into one blastomere, the **mesentoblast** (Fig. 9-24C), whose presence there causes the cell and its progeny to form mesoderm. (The mesentoblast is sometimes called the "4d" cell in the formal system of labeling each blastomere.) Thus the fate of the mesentoblast is fixed early in development by inheritance of a maternal mesodermalizing factor that is already present in the cytoplasm of the uncleaved egg.

Another form of embryonic fate determination called **regulative development** predominates in animals that have radial cleavage. Cell fates in these animals (for example, sea urchins) are determined by a network of cellular communication in the embryo. The fate of a particular cell results from its position in the network and not from a cytoplasmic determinant that specifies its fate regardless of embryonic location. Because cell fate determination in regulative embryos requires the presence

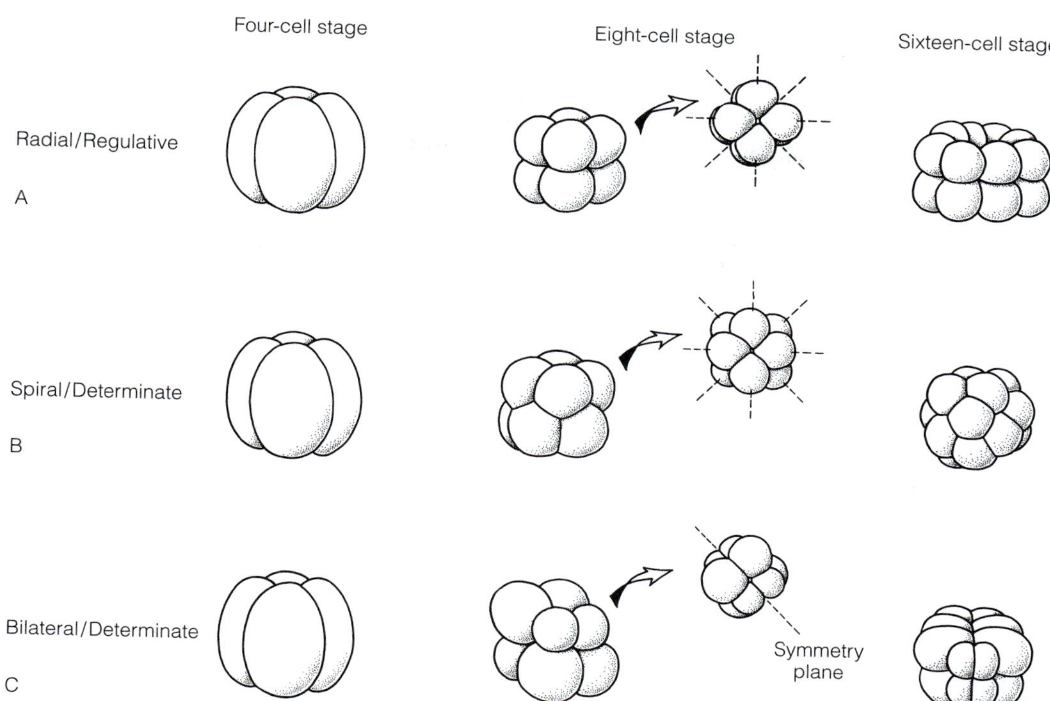

FIGURE 9-22 Bilaterian development: three basic cleavage patterns shown in 4-, 8-, and 16-cell stages (lateral views). Arrows point to eight-cell stages in animal-pole view. Dashed lines indicate mirror-image symmetry planes that establish radial symmetry in early radial and spiral embryos and bilateral symmetry in bilaterally cleaving embryos. Later in development, the embryos of all bilaterians adopt bilateral symmetry. *(Modified and redrawn from Siewing, R. 1969. Lehrbuch der vergleichenden Entwicklungsgeschichte der Tiere. Verlag Paul Parey, Hamburg, Berlin. pp. 46, 60, and Conklin, E. G. 1905. The organization and cell-lineage of the ascidian egg. J. Am. Nat. Soc. 13:1–119.)*

of several interacting cells, it tends to occur later in development (after more cleavages) than in determinate embryos. It must be noted, however, that although determinate development predominates in spirally and bilaterally cleaving embryos and regulative development predominates in radially cleaving embryos, both forms of determination occur simultaneously, but in different proportions, in the development of *all* animals.

GASTRULATION

All bilaterian zygotes gastrulate to form embryos with three germ layers; these are, from outside to inside, **ectoderm, mesoderm,** and **endoderm** (Fig. 9-23A,H). Such a three-layered **(triploblastic)** embryo is not unique to bilaterians—ctenophores also have three layers—but bilaterian animals, in general, diversify and specialize the layers to a greater extent than do their radial relatives. In most bilaterians, each of the embryonic germ layers differentiates into an adult epithelium: Ectoderm becomes **epidermis,** mesoderm becomes **meso-thelium,** and endoderm becomes **gastrodermis** (Table 9-1). Epidermis encloses and covers the body, secretes the cuticle (exoskeleton), and typically contains sensory, glandular, ciliated, and other cells, including absorptive cells. Much of the nervous system arises from, and often is restricted to, the epidermis. Mesothelium lines coeloms; forms the musculature, septa, and mesenteries; and, together with the gametes, constitutes the gonads. The mesothelium often includes muscle, ciliated, secretory, absorptive, and storage cells. It also gives rise, through an epithelium-to-mesenchyme transition, to nonepithelial blood corpuscles, coelomocytes, and connective-tissue cells. The gastrodermis lines the midgut and contains ciliated, secretory, absorptive, and storage cells. The **blastocoel,** which is the embryonic connective-tissue compartment, differentiates into blood vessels, sinuses, hemocoel, blood (Fig. 9-23E–I), and structural connective tissue.

MESODERM SEGREGATION

Although ectoderm and endoderm usually are epithelia (as are their adult derivatives), mesoderm may be epithelium or mesenchyme. Epithelial mesoderm usually originates as outfolds of the wall of the archenteron, as in most deuterostomes. Each of these outfolds (the number varies) is thus an epithelium from the outset. Eventually each outpocket pinches free of the archenteron and becomes an independent coelomic compartment. This process of coelom and mesoderm formation is known as **enterocoely** (Fig. 9-23G–I, 9-24A).

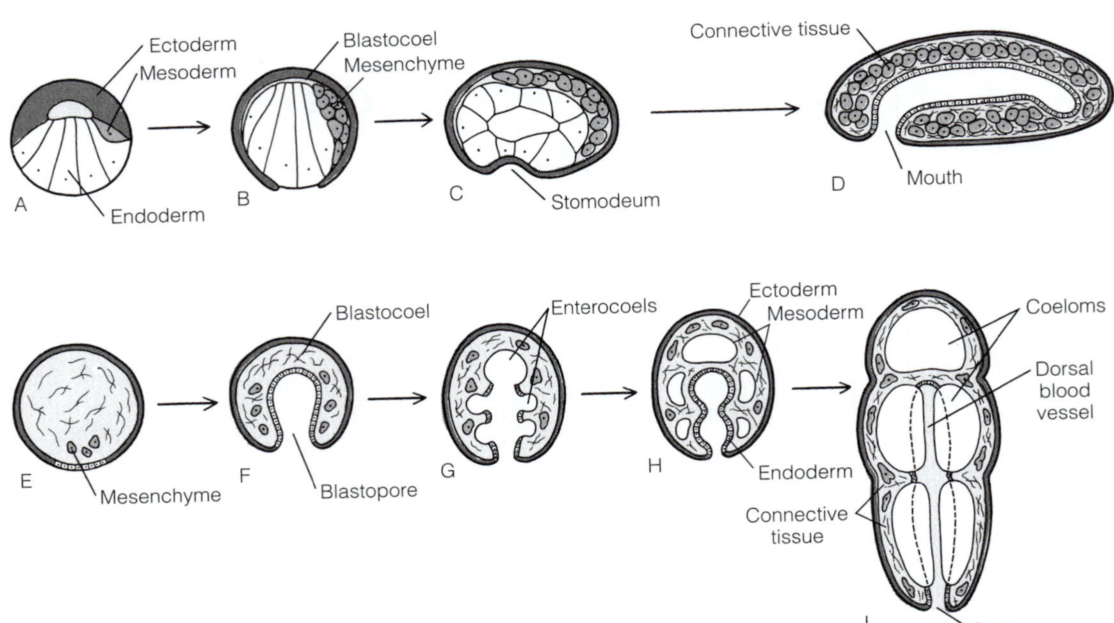

FIGURE 9-23 Bilaterian development: gastrulation, mesoderm segregation, and morphogenesis of connective tissue. **A–D,** A small, compact (acoelomate) flatworm. Gastrulation occurs by epiboly **(A and B)** and results in a stereogastrula **(B).** The endoderm hollows to form the gut and joins the stomodeum to form the mouth **(C and D).** Mesoderm is segregated as mesenchyme into the blastocoel **(B and C)** and differentiates into gonads, musculature, and connective tissue **(D). E–I,** A large, coelomate acorn worm (deuterostome). Gastrulation occurs by invagination. Enterocoelous outpockets from the archenteron pinch off to form the mesothelium-lined coelomic cavities while mesenchyme invades the blastocoel **(F–I).** Musculature and gonads arise from mesothelium; hemal system and structural connective tissue arise from mesenchyme and blastocoel. Note that the blastopore becomes the anus.

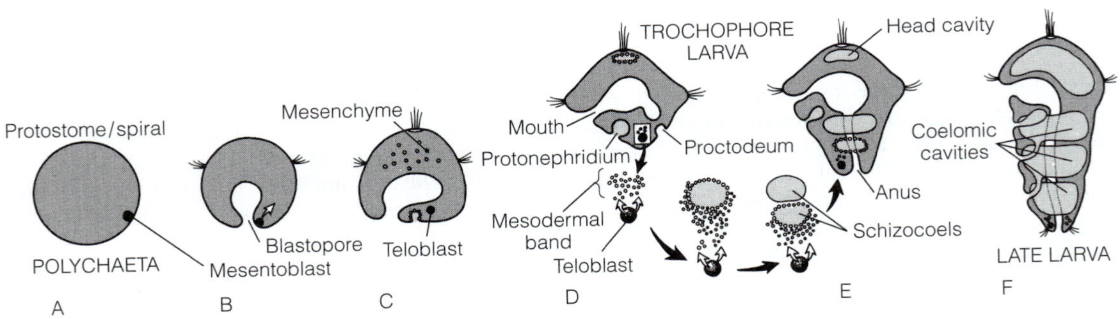

FIGURE 9-24 Bilaterian development: larval development, mesoderm segregation, and coelom formation in Trochozoa (for example, Annelida and Polychaeta). Some larval mesoderm arises from mesenchyme **(C),** but most mesoderm and the segmented coelomic cavities originate from a single cell, the mesentoblast **(A;** = 4d cell), whose fate is determined early in development. The mesentoblast divides into left and right teloblast cells **(C)** that become located near the posterior end of the body. The two teloblasts divide repeatedly and form two longitudinal, mesodermal bands **(D and E).** These compact, epithelial bands cavitate, in a process called schizocoely, to form the segmental coelomic cavities.

In spiralians, such as flatworms, the mesentoblast produces mesenchyme (Fig. 9-23B). Mesenchymal cells are individual, mobile cells located in the blastocoel (Fig. 9-23B,C), or when a blastocoel is absent, they lie between the ectoderm and endoderm. Later in development, they differentiate into germ, muscle, skeletal, and connective-tissue cells (Fig. 9-23D).

In coelomate spiralians, such as segmented worms (annelids), mesoderm destined to form a coelomic lining originates from daughters of the mesentoblast cell. The mesentoblast first divides into two bilateral, right and left, **teloblast cells** (Fig. 9-24C). Each teloblast then divides repeatedly to produce an elongate, solid **mesodermal band** of cells, one on each side of the archenteron. The cylindrical mesodermal band is a monolayered epithelium around a central longitudinal axis. Later in development, fluid accumulates in this centerline, expanding it into a coelomic cavity (Fig. 9-24D). The process is called **schizocoely** (schiz = split) because the initial cavitation is a split in the mesodermal band.

FATES OF THE BLASTOPORE

The blastopore of some radially cleaving embryos (deuterostomes) becomes the adult anus and the mouth forms as a new structure. In most protostomes, the blastopore either becomes the mouth (and the anus forms anew; Fig. 9-24D,E) or gives rise to both mouth *and* anus (as in some molluscs, polychaetes, and onychophorans). When the latter occurs, the blastopore first forms a long furrow. Later, the furrow margins converge and fuse between the two extremes, one of which remains open as the mouth and the other as the anus. The blastopore of several taxa closes and disappears during early development and a new mouth and anus form later. In such cases, the assignment of a taxon to either Protostomia or Deuterostomia, based on this character, depends on where the new mouth or anus opens in relation to the old blastopore. If the mouth opens near the site of the closed blastopore, then a protostome alliance is assumed; if the anus coincides with the old blastopore, then a deuterostome relationship may be likely.

Morphologists have long speculated that the developmental origin of both mouth and anus from the blastopore was primitive for Bilateria. This hypothesis is now supported by expression of the T-box gene, *Brachyury*. In cnidarians, the gene is expressed in the region of a sea anemone blastopore and around the mouth of *Hydra*. In bilaterians, its expression occurs in the embryonic mouth *and* anus of a polychaete (protostome) and a hemichordate (deuterostome).

PHYLOGENY OF BILATERIA

CONSENSUS AND CONFLICTS

Morphological and molecular systematics agree on two broad divisions of Bilateria. These two major taxa are **Protostomia** and **Deuterostomia** (Fig. 9-25). Deuterostomes include hemichordates, echinoderms, chordates, and sometimes chaetognaths, whereas protostomes include the remaining bilaterians, such as molluscs, arthropods, annelids, and flatworms. Protostome cleavage is mostly spiral, the embryonic blastopore usually persists to become the adult mouth (or mouth and anus), and the mesothelium, when present, is typically schizocoelous in origin. In deuterostomes, cleavage

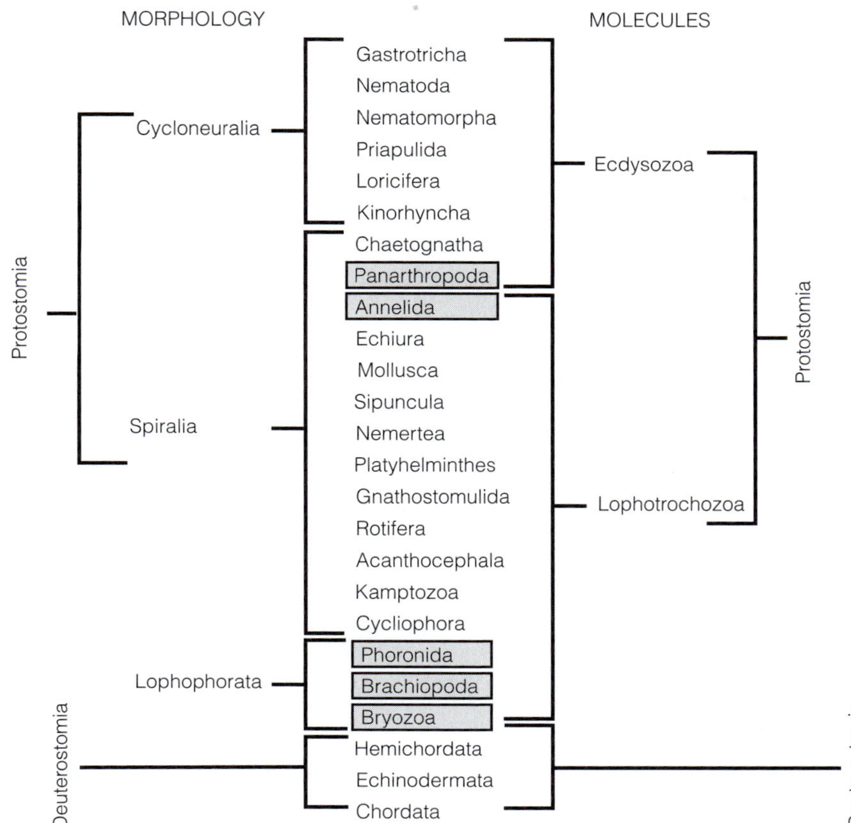

FIGURE 9-25 Bilateria: competing classifications based on morphology and molecules. The taxon names in shaded boxes identify the most important conflicts between the two classifications. *(Modified and redrawn after Nielsen, C. 2001. Animal Evolution. Interrelationships of the Living Phyla. 2nd Edition. Oxford University Press, New York. 563 pp.)*

is usually radial, the blastopore typically becomes the anus, the mouth forms anew from the surface ectoderm, and the mesothelium arises by enterocoely.

Traditional morphology and contemporary molecular systematics are in agreement regarding deuterostomes, but disagree over the membership and major subdivisions of protostomes (Fig. 9-25, 9-26), and these disagreements are part of a lively debate and ongoing research. Morphology usually recognizes two protostome taxa, Spiralia and Cycloneuralia (also called Aschelminthes, Pseudocoelomata, or Nemathelminthes; Fig. 9-25, 9-26A). Spiralians are united by the synapomorphy "spiral cleavage" and include the flatworms, molluscs, annelids, and arthropods, among others. Annelida, Onychophora, and Arthropoda together constitute the Articulata, so-called because the segmented worms, velvet worms, and arthropods are all articulated, meaning that they share a segmented body and similar pattern of growth. Cycloneuralians are mostly small-bodied animals, such as gastrotrichs and nematodes, whose cleavage pattern differs from spiral and whose adult traits tend to isolate them from the remaining protostomes. A third assemblage, Lophophorata (brachiopods, phoronids, bryozoans), is considered to be

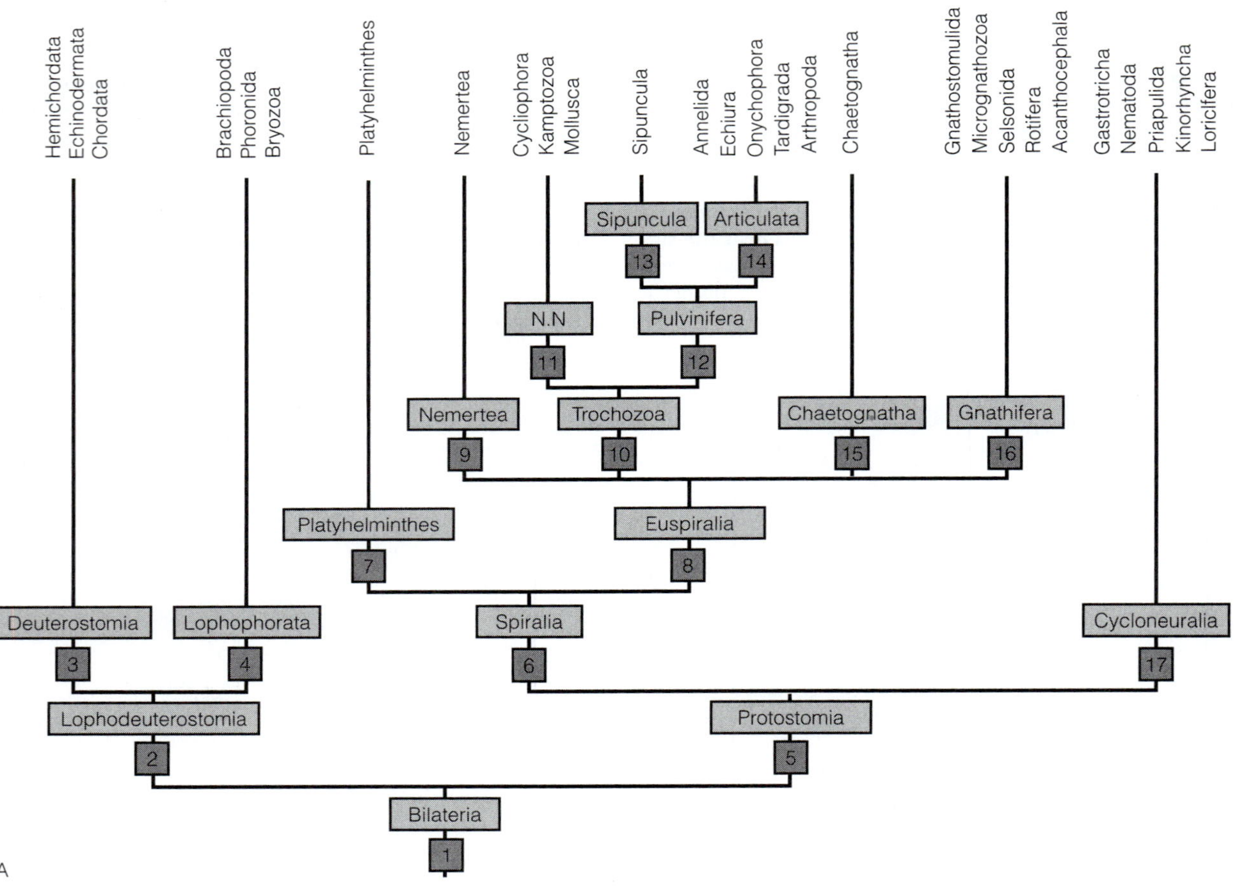

FIGURE 9-26 Bilateria: competing phylogenies. The few, often uncertain homologies that support alternatives **A** and **B** reflect the tentativeness of these trees. **A,** A more or less traditional phylogeny that divides bilaterians into protostomes, lophophorates, and deuterostomes and recognizes a sister-taxon relationship between lophophorates and deuterostomes. **1, Bilateria:** bilateral symmetry, filtration nephridia, blastopore becomes mouth and anus. **2, Lophodeuterostomia:** lophophore (possible plesiomorphy), U-shaped gut; upstream suspension-feeding system. **3, Deuterostomia:** trimeric, protocoelic metanephridial system, blastopore becomes anus. **4, Lophophorata:** possibly dimeric, mesosome secretes body covering (tube, shell, cuticle). **5, Protostomia:** determinate development. **6, Spiralia:** spiral cleavage, mesentoblast (4d cell). **7, Platyhelminthes:** possibility of epidermal replacement cells, hermaphroditism, internal fertilization. **8, Euspiralia:** complete gut with anus. **9, Nemertea:** rhynchocoel plus proboscis. **10, Trochozoa:** trochophore larva, trochoblast cells, downstream suspension-feeding system. **11, N.N.:** No known autapomorphy. **12, Pulvinifera:** crisscrossed collagen fibers in cuticle (replaced by chitin in panarthropods), egg envelope (shell) becomes incorporated into juvenile cuticle (replaced by hatching in panarthropods). **13, Sipuncula:** introvert, J-shaped gut, dorsal anus. **14, Articulata:** segmentation, appendages, teloblastic growth. **15, Chaetognatha:** grasping spines, postanal testes. **16, Gnathifera:** jaws. **17, Cycloneuralia:** terminal mouth, ringlike circumpharyngeal brain. **B,** A contemporary phylogeny, inspired by molecular data, recognizes Protostomia and Deuterostomia, but includes lophophorates among the protostomes. It also recognizes Ecdysozoa, a taxon of molting animals, and separates

intermediate between protostomes and deuterostomes, but to have closer ties to the deuterostomes (Fig. 9-25, 9-26A).

Molecules also divide Protostomia into two groups, but taxon membership in them differs from the traditional scheme just described (Fig. 9-25, 9-26B). One taxon, called Ecdysozoa, unites all animals that periodically molt an exoskeleton. It includes arthropods and their allies (together Panarthropoda) and cycloneuralians, but not annelids, which do not molt. The second taxon, Lophotrochozoa, encompasses all other protostomes, including the lophophorates. The Ecdysozoa grouping supports the monophyly of the panarthropods and provides a home for the cycloneuralians, but divorces the Annelida, casting it into the Lophotrochozoa, thus rejecting the longstanding annelid-arthropod alliance (Articulata). Because annelids and arthropods share a segmental body composed of many segments, their placement in separate evolutionary lines means that each evolved segmentation independently of the other, a conclusion vigorously opposed by most morphologists. Molecular systematics also views lophophorates as protostomes with no indication of a deuterostome alliance (Fig. 9-25, 9-26B).

Current evidence is insufficient to choose between these two alternative phylogenies, and we await the discovery of new characters. In the interim, we have arranged the sequence of this book's chapters to be compatible with either evolutionary viewpoint. The sequence is more or less traditional—beginning with platyhelminths and ending with deuterostomes—but the taxa of molting animals (Ecdysozoa) are grouped together in the middle of the book.

THE ANCESTOR: SMALL OR LARGE?

Figure 9-26 shows two widely contrasting views of bilaterian evolution. A more or less traditional phylogenetic tree (Fig. 9-26A), rooted in morphology, considers the small-bodied

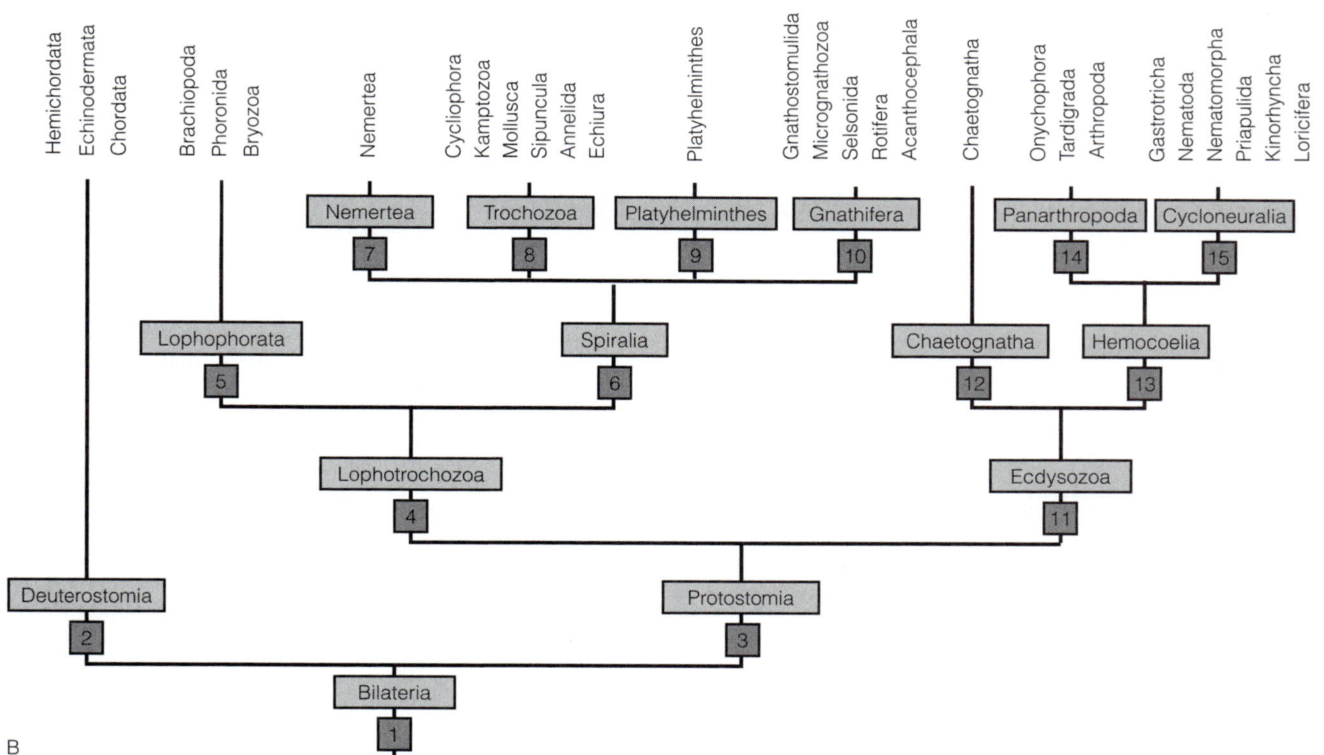

the segmented animals that molt (panarthropods) from those that do not (Annelida). If this phylogeny is correct, then segmentation probably evolved independently in these two taxa. **1, Bilateria:** bilateral symmetry, filtration nephridia, blastopore becomes mouth and anus. **2, Deuterostomia:** trimeric (possibly a plesiomorphy), protocoelic metanephridial system, blastopore becomes anus. **3, Protostomia:** no known autapomorphy. **4, Lophotrochozoa:** larval protonephridia, chaetae, β-chitin. **5, Lophophorata:** U-shaped gut, lophophore, metasomal secretion of body covering (cuticle, shell, tube). **6, Spiralia:** spiral cleavage, mesentoblast (4d cell). **7, Nemertea:** rhynchocoel and proboscis. **8, Trochozoa:** trochophore larva, trochoblast cells, downstream suspension-feeding system. **9, Platyhelminthes:** epidermal replacement cells (uncertain autapomorphy), hermaphroditism, internal fertilization. **10, Gnathifera:** jaws. **11, Ecdysozoa:** molted exoskeleton (ecdysis; Gastrotricha do not molt but their embryos hatch (molt?) from their eggshells), setae (uncertain autapomorphy), α-chitin. **12, Chaetognatha:** grasping spines, postanal testes. **13, Hemocoelia:** body cavity a hemocoel (except in compact-bodied Gastrotricha). **14, Panarthropoda:** segmentation. **15, Cycloneuralia:** terminal mouth, ringlike circumpharyngeal brain, sucking pharynx. *(After several sources)*

Platyhelminthes (flatworms) to be the primitive bilaterian taxon. If so, the stem species of Bilateria probably was a small ciliated animal lacking coelom, hemal system, and anus and having protonephridia as excretory organs. Most likely, it had internal fertilization and direct development (Table 9-3). From this small-bodied ancestor evolved large-bodied descendants: peramorphosis was the heterochronic trend (Chapter 4, Fig. 4-11). Most biologists agree with this perspective. In fact, a small-bodied ancestor is inferred for every major taxon in which there is a wide range of body sizes among species. Vertebrates, for example, are believed to have evolved from an amphioxus-like ancestor only a few centimeters in length, molluscs from a small, flatwormlike ancestor, and insects from a tiny, wingless predecessor.

Challenging this near-consensus is a persistent minority opinion that the ancestor to Bilateria was a large organism. As such, it relied on muscles for locomotion and probably had coelom, hemal, and metanephridial systems, as well as a specialized gill surface for gas exchange. It may have spawned, had external fertilization, and produced a free-swimming, perhaps planktotrophic, larva. From this large ancestor evolved small-bodied descendants (Table 9-3). The heterochronic trend here was pedomorphosis (Chapter 4, Fig. 4-10, 4-11). This idea has been revived recently and independently in the publications of a morphologist who also considered molecular and paleontological data (R. A. Dewel), a molecular systematist (G. Balavoine), and two systematists who combine morphology and molecules in their analyses (K. J. Peterson and D. J. Eernisse). Dewel, moreover, regards the ancestor as a colonial rather than a solitary species.

Perhaps significantly, molecular systematics occasionally recognizes taxa of small, anatomically simple animals that may have evolved from large, complex ancestors. For example, the application of molecular techniques to cnidarians is largely responsible for the current consensus that the large, complex anthozoans are primitive while the mostly small hydrozoans are derived. Similarly, the very simple and tiny mesozoans (Orthonectida and Dicyemida) are usually placed at or near the origin of Metazoa by morphologists, but molecular systematists assign them to the Bilateria (near flatworms).

Whether or not pedomorphosis is thought to play a role in the evolution of major taxa may depend on methods employed by systematists. Traditional cladistics, as used in this book, depends on morphological novelty—the evolutionary elaboration or compounding of preexisting traits—as a criterion for the recognition of synapomorphies and the establishment of monophyletic taxa. A synapomorphy is a shared *derived* character, one almost always more complex than the character-state from which it evolved. If apomorphies are complex and plesiomorphies simple, then apomorphic taxa are more likely to be large-bodied animals than plesiomorphic taxa (Table 9-3). Thus traditional cladistics may be biased toward a phylogeny from small (simple) to large (complex). Molecular systematics, on the other hand, because of its equal accounting of both the presence and absence of sequence information, may be better able to detect pedomorphic trends in evolution.

TABLE 9-3 Bilateria: Correlations with Body Size

Structure/Function	Body Size	
	Small	Large
Surface:volume ratio	Large	Small
Locomotion	Cilia[1]	Muscle
Gas exchange surface	Body wall	Body wall + gills (tentacles, other outgrowths, gill slits)
Gut	Unbranched	Branched (including digestive ceca)
Nervous system	No giant axons	Giant axons
Internal transport	Diffusion[2]	Convection + diffusion
Circulatory system	None or hemal (without heart)	Coelom, hemal, or both (often with heart)
Excretory system	Protonephridia	Metanephridial system
Egg output	Few	Many
Fertilization	Internal	External or internal
Copulatory organs	Present	Absent or present
Larva	Absent	Present or absent
Life span	Short	Long or short
Heterochronic trend	Pedomorphosis	Peramorphosis

[1]Small bilaterians with rigid exoskeletons—nematodes, kinorhynchs, microcrustaceans—lack surface cilia and use muscle for locomotion. [2]Small bilaterians with retractile body parts, such as an introvert or telescoping legs, have a hemocoel.

PHYLOGENETIC HIERARCHY OF BILATERIA

(Molecules)
Bilateria
  Deuterostomia
  Protostomia
    Lophotrochozoa
      Lophophorata
      Spiralia
        Nemertea
        Trochozoa
        Platyhelminthes
        Gnathifera
    Ecdysozoa
      Chaetognatha
      Hemocoelia
        Panarthropoda
        Cycloneuralia

(Morphology)
Bilateria
  Lophodeuterostomia
    Deuterostomia
    Lophophorata
  Protostomia
    Spiralia
      Platyhelminthes
      Euspiralia
        Nemertea
        Trochozoa
          N.N.
          Pulvinifera
            Sipuncula
            Articulata
        Chaetognatha
        Gnathifera
    Cycloneuralia

REFERENCES

EVOLUTIONARY AND FUNCTIONAL MORPHOLOGY

General

Bartolomaeus, T. 1994. On the ultrastructure of the coelomic lining in the Annelida, Sipuncula, and Echiura. Microfauna Marina 9:171–220.

Bartolomaeus, T., and Ax, P. 1992. Protonephridia and metanephridia—their relation within the bilateria. Z. zool. Syst. Evolut.-forsch. 30:21–45.

Clark, R. B. 1964. Dynamics in Metazoan Evolution. The Origin of the Coelom and Segments. Clarendon Press, Oxford. 313 pp.

Fransen, M. E. 1988. Coelomic and vascular systems. In Westheide, W., and Hermans, C. O. (Eds.): Ultrastructure of the Polychaeta. Microfauna Marina 4:199–213.

Kirschner, L. B. 1967. Comparative physiology: Invertebrate excretory organs. Ann. Rev. Physiol. 29:169–196.

Kümmel, G. 1975. The physiology of protonephridia. Fortschr. Zool. 23:18–32.

Lankester, E. R. 1900. The Enterocoela and the Coelomocoela. In Lankester, E. R. (Ed.): A Treatise on Zoology. Part 2. The Porifera and Coelentera. Adam and Charles Black, London. pp. 1–37.

Mackie, G. O. 1984. Introduction to the diploblastic level. In Bereiter-Hahn, J., Matoltsky, A. G., and Richards, K. S. (Eds.): Biology of the Integument. Vol. 1. Invertebrates. Springer-Verlag, Berlin. pp. 43–46.

Nakao, T. 1974. An electron microscopic study of the circulatory system in *Nereis japonica*. J. Morphol. 144:217–236.

Oglesby, L. C. 1981. Volume regulation in aquatic invertebrates. J. Exp. Zool. 215:289–301.

Pantin, C. F. A. 1959. Diploblastic animals. Proc. Linn. Soc. Lond. 171:1–14.

Riedl, R. 1970. Water movement. Animals. In Kinne, O. (Ed.): Marine Ecology: A Comprehensive, Integrated Treatise on Life in Oceans and Coastal Waters. Wiley-Interscience, London. pp. 1085–1150.

Riegel, J. A. 1972. Comparative Physiology of Renal Excretion. Hafner, New York. 204 pp.

Rieger, R. M., and Lombardi, J. 1987. Ultrastructure of coelomic lining in echinoderm podia: Significance for concepts in the evolution of muscle and peritoneal cells. Zoomorphology 107:191–208.

Rieger, R. M., Haszprunar, G., and Schuchert, P. 1989. On the origin of the Bilateria: Traditional views and recent alternative concepts. In Simonetta, A., and Conway Morris, S. (Eds.): The Early Evolution of Metazoa and the Significance of Problematic Taxa. Cambridge University Press, Cambridge. pp. 107–112.

Ruppert, E. E. 1991. Introduction to the aschelminth phyla: A consideration of mesoderm, body cavities, and cuticle. In Harrison, F. W., and Ruppert, E. E. (Eds.): Microscopic Anatomy of Invertebrates. Vol. 4. Wiley-Liss, New York. pp. 1–17.

Ruppert, E. E., and Carle, K. J. 1983. Morphology of metazoan circulatory systems. Zoomorphology. 103:193–208.

Ruppert, E. E., and Smith, P. R. 1988. The functional organization of filtration nephridia. Biol. Rev. 63:231–258.

Schmidt-Nielsen, K. 1983. Animal Physiology: Adaptation and Environment. Cambridge University Press, Cambridge. 619 pp.

Trueman, E. R., and Ansell, A. D. 1969. The mechanisms of burrowing into soft substrata by marine animals. Oceanogr. Mar. Biol. Ann. Rev. 7:315–366.

Vogel, S. 1988. Life's Devices. The Physical World of Animals and Plants. Princeton University Press, Princeton, NJ. 367 pp.

Welsch, U., and Storch, V. 1976. Comparative Animal Cytology and Histology. University of Washington Press, Seattle. 343 pp.

Willmer, P. 1990. Invertebrate Relationships. Patterns in Animal Evolution. Cambridge University Press, Cambridge. 400 pp.

Wilson, R. A., and Webster, L. A. 1974. Protonephridia. Biol. Rev. 49:127–160.

Internet Site

www.biology.ualberta.ca/facilities/multimedia/index.php?Page = 252 (Animations by A. R. Palmer and H. Kroening of ultrafiltration, transport, and resorption in a protonephridium and metanephridial system; also form and locomotion [burrowing] in an earthworm.)

DEVELOPMENT

General

Conklin, E. G. 1905. The organization and cell-lineage of the ascidian egg. J. Am. Nat. Soc. 13:1–119.

Costello, D. P., and Henley, C. 1976. Spiralian development: A perspective. Am. Zool. 16:277–291.

Gilbert, S. F., and Raunio, A. M. 1997. Embryology. Constructing the Organism. Sinauer, Sunderland, MA. 537 pp.

Korschelt, E., and Heider, K. 1895. Text-Book of the Embryology of Invertebrates. Swan Sonnenschein, London. (Four volumes.)

Kume, M., and Dan, K. (Eds.): 1968. Invertebrate Embryology. NOLIT, Belgrade. 605 pp.

MacBride, E. W. 1914. Text-Book of Embryology. Vol. 1. Invertebrata. MacMillan, London. 692 pp.

Siewing, R. 1969. Lehrbuch der vergleichenden Entwicklungsgeschichte der Tiere. Verlag Paul Parey, Hamburg. 531 pp.

Smith, P. R. 1986. Development of the blood vascular system in *Sabellaria cementarium* (Annelida, Polychaeta): An ultrastructural investigation. Zoomorphology 105:67–74.

Strathmann, R. R. 1989. Existence and functions of a gel filled primary body cavity in development of echinoderms and hemichordates. Biol. Bull. 176:25–31.

Internet Site

www.luc.edu/depts/biology/dev/radspirl.htm (Simple but effective animation contrasting spiral and radial cleavage.)

RESPIRATORY PIGMENTS

Dickerson, R. E., and Geis, I. 1983. Hemoglobin. Benjamin Cummings, Menlo Park, NJ. 176 pp.

Klippenstein, G. L. 1980. Structural aspects of hemerythrin and myohemerythrin. Am. Zool. 20:39–51.

Linzen, B., et al., 1985. The structure of arthropod hemocyanins. Science 229:519–524.

Mangum, C. P. 1985. Oxygen transport in invertebrates. Am. J. Physiol. 248:R505–R514.

Mangum, C. P. 1992a. Respiratory function of the red blood cell hemoglobins of six animal phyla. Adv. Comp. Environ. Physiol. 13:117–149.

Mangum, C. P. 1992b. Physiological function of the hemerythrins. Adv. Comp. Environ. Physiol. 13:173–192.

LIFE HISTORY, ECOLOGY, AND EVOLUTION

Jägersten, G. 1972. Evolution of the Metazoan Life Cycle. A Comprehensive Theory. Academic Press, New York. 282 pp.

Nielsen, C. 1998. Origin and evolution of animal life cycles. Biol. Rev. 73:125–155.

Pechenik, J. A. 1999. On the advantages and disadvantages of larval stages in benthic marine invertebrate life cycles. Mar. Ecol. Progr. Ser. 177:269–297.

Strathmann, R. R., and Strathmann, M. F. 1982. The relationship between adult size and brooding in marine invertebrates. Am. Nat. 119:91–101.

Stearns, S. C. 1976. Life history tactics: A review of the ideas. Q. Rev. Biol. 51:3–47.

PHYLOGENY

General

Aguinaldo, A. M. A., Turbeville, J. M., Linford, L. S., Rivera, M. C., Garey, J. R., Raff, R. A., and Lake, J. A. 1997. Evidence for a clade of nematodes, arthropods and other moulting animals. Nature 387:489–493.

Aguinaldo, A. M. A., and Lake, J. A. 1998. Evolution of the multicellular animals. Am. Zool. 38:878–887.

Ax, P. 1995–2001. Multicellular Animals: A New Approach to the Phylogenetic Order in Nature. Vol. 1–3. Springer-Verlag, Berlin.

Balavoine, G. 1997. The early emergence of platyhelminths is contradicted by the agreement between 18S rRNA and *Hox* genes data. Compt. Rend. Acad. Sci. Paris Sci. Vie 320:83–94.

Balavoine, G. 1998. Are platyhelminthes coelomates without a coelom? An argument based on the evolution of *Hox* genes. Am. Zool. 38:843–858.

Conway Morris, S. 2000. Evolution: Bringing molecules into the fold. Cell 100:1–11.

Dewel, R. A. 2000. Colonial origin for eumetazoa: Major morphological transitions and the origin of bilaterian complexity. J. Morphol. 243:35–74.

Eernisse, D. J., Albert, J. S., and Anderson, F. E. 1992. Annelida and Arthopoda are not sister taxa: A phylogenetic analysis of spiralian metazoan morphology. Syst. Biol. 41:305–330.

Jenner, R. A. 2000. Evolution of animal body plans: The role of metazoan phylogeny at the interface between pattern and process. Evol. Devel. 2:208–221.

Jenner, R. A., and Schram, F. R. 1999. The grand game of metazoan phylogeny: Rules and strategies. Biol. Rev. 74:121–142.

Nielsen, C. 2001. Animal Evolution. Interrelationships of the Living Phyla. 2nd Edition. Oxford University Press, New York. 563 pp.

Peterson, K. J., and Eernisse, D. J. 2001. Animal phylogeny and the ancestry of bilaterians: Inferences from morphology and 18S rRNA gene sequences. Evol. Devel. 3:170–205.

Schmidt-Rhaesa, A., Bartolomaeus, T., Lemburg, C., Ehlers, U., and Garey, J. R. 1998. The position of the Arthropoda in the phylogenetic system. J. Morphol. 238:263–285.

Siewing, R. 1981. Problems and results of research on the phylogenetic origin of coelomata. Atti Convegni Lincei 49:123–160.

Technau, U. 2001. *Brachyury*, the blastopore and the evolution of the mesoderm. BioEssays 23:788–794.

Internet Sites

http://tolweb.org/tree?group=Bilateria&contgroup=Animals (Tree of Life: A more or less traditional phylogeny, but extensive list of modern references to systematics of Metazoa and Bilateria.)

www.biology.ualberta.ca/courses.hp/biol606/OldLecs/Lecture2K.09.Penney.html (Two critiques of the Ecdysozoa.)

10 Platyhelminthes^P, Orthonectida^P, and Dicyemida^P

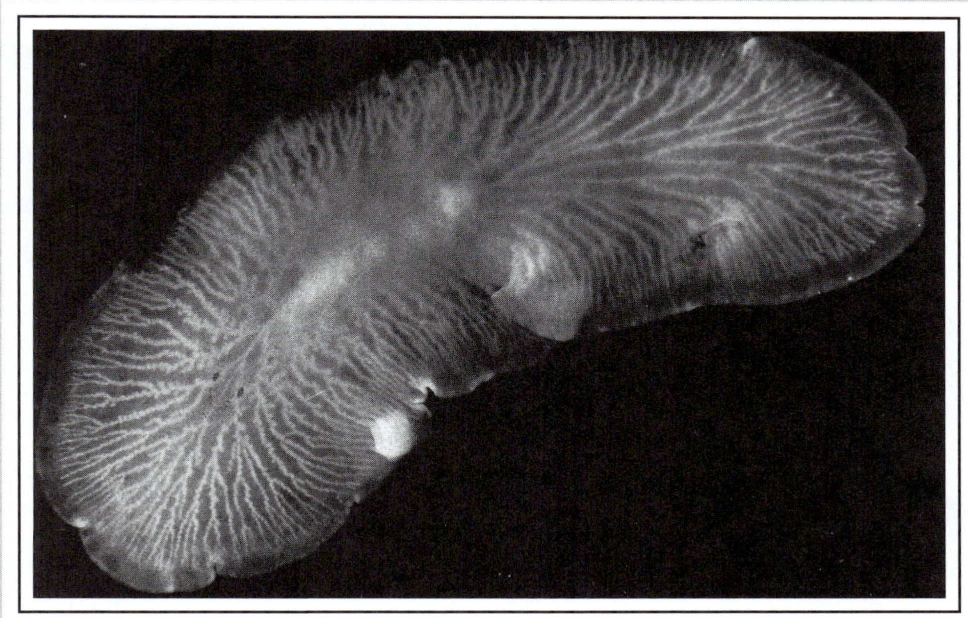

PLATYHELMINTHES[P]

Platyhelminths are mostly small, soft-bodied, aquatic worms that do not burrow, but instead move over and between rocks, sediment particles, detritus, algae, and in the tissues of their prey. They include the free-living flatworms (Turbellaria) and the parasitic flukes (Trematoda) and tapeworms (Cestoda). All lack a coelom, hemal system, cuticle, and, in the parasitic taxa, epidermis and sometimes gut. These absences result in an uncomplicated anatomy, except for a complex reproductive system. Great variety exists, however, in this relatively simple design, as represented by some 20,000 species, many higher taxa, and myriad lifestyles.

Flatworm diversity is one of the most curious and instructive chapters in functional and evolutionary biology. By degrees, the platyhelminths make an evolutionary transition from free-living flatworms to parasitic flukes and tapeworms, several of which are scourges of humankind. What are the attributes that enabled flatworms to cross the line between friend and foe? Certainly one factor was small body size (most platyhelminths are in the millimeter-size range), if only because endoparasites are necessarily smaller than their hosts. Another is that flatworms, being tiny or flat, are adapted to live in tight places, including the cavities and tissue-spaces of other organisms. Most free-living flatworms, for example, live in crevices between rocks, shells, and attached organisms or occupy the water-filled interstitial spaces between sand grains. These and other traits that predisposed flatworms to parasitism will be highlighted in this chapter.

Flatworms lack a circulatory system—a hemal system and coelom are absent—and thus are diffusion-limited animals. Under this limitation, distances between sources and sinks must be short in all transport systems, including those for gases, nutrients, and excretory wastes.

Knowing that platyhelminths rely on simple diffusion rather than circulation for internal transport, try to predict their diffusion-based anatomy using the principles discussed in Chapters 2, 4, and 6. What body shape would you expect to find in platyhelminths in the millimeter-size range? In the centimeter range? Would they have protonephridia or metanephridial systems? How, in the larger-bodied species, would nutrients be distributed in the absence of coelom and hemal system? Would you predict reproduction by free-spawning and external fertilization or internal fertilization after copulation?

Perhaps the deepest question about flatworms is whether the absence of hemal and coelomic systems is primary or secondary. Did flatworms evolve, via pedomorphosis, from large-bodied coelomate ancestors, or does their relatively simple anatomy indicate primitive status? Most morphologists easily accept the primitiveness of flatworms. But historically, a few have challenged this opinion, and recent results from molecular systematics have also questioned the traditional interpretation. At stake is our conception of the ancestral bilaterian: Was it big or small? Coelomate or acoelomate? Or did

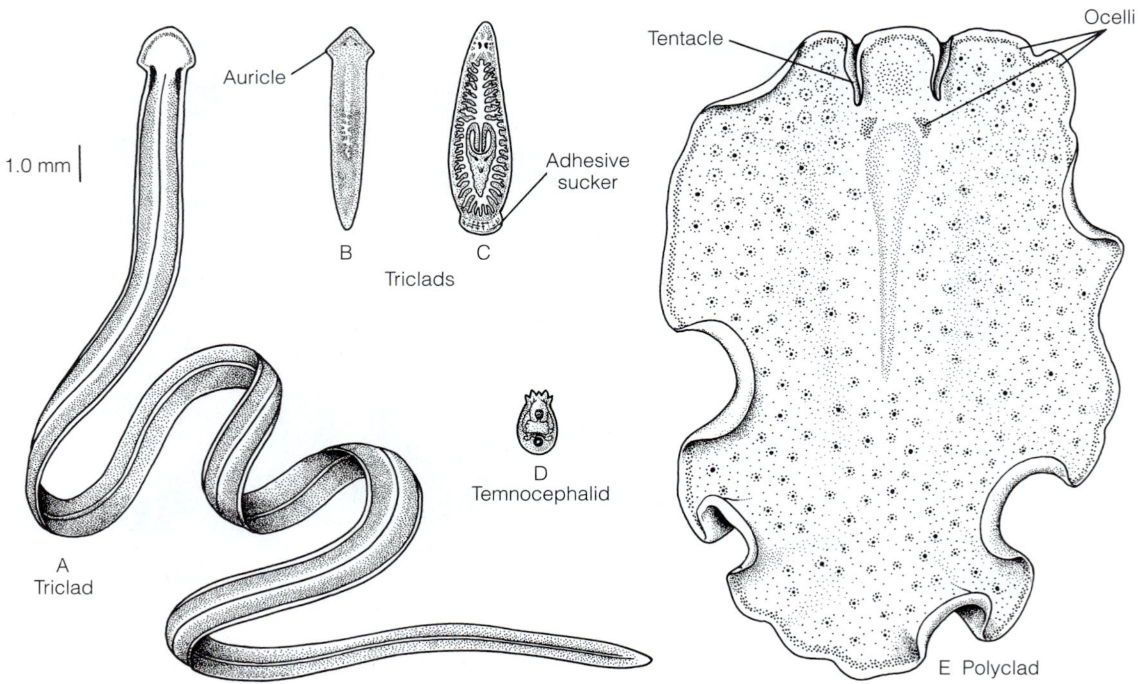

FIGURE 10-1 Turbellaria: diversity of macroturbellaria. **A,** The terrestrial triclad *Bipalium adventitium.* **B,** The freshwater triclad *Dugesia tigrina.* **C,** The horseshoe-crab commensal triclad *Bdelloura candida.* **D,** The temnocephalid from Figure 10-2H, for scale. **E,** The polyclad *Oligoclado floridanus.* *(A and C, From Hyman, 1951; D, Redrawn from Williams, J. B. 1980. Morphology of a species of* Temnocephala *(Platyhelminthes) ectocommensal on the isopod* Phreatoicopsis terricola. *J. Nat. Hist. 14:183–199.)*

its life cycle include a large coelomate adult and a small acoelomate larva? Front and center in this controversy is Platyhelminthes. Finding its sister taxon and true position among bilaterians are ongoing quests of contemporary phylogenetic research.

TURBELLARIA[C]

For many, the first contact with free-living flatworms is with planarians, those small, brownish, cross-eyed worms that live in freshwater streams under rocks. The lucky few who have explored a rocky seashore or a coral reef may have seen large, tissue-thin, stunningly colored marine flatworms that glide and shuffle over the substratum, conforming their bodies to every surface irregularity. Some fortunate others may have watched a sandy beach turn green as zoochlorellae-bearing turbellarians surfaced with the receding tide to expose themselves to the rays of the sun. These are the large, visible Turbellaria, or **macroturbellaria,** in the centimeter-size range (Fig. 10-1). The largest of all macroturbellarians is a planarian, *Rimacephalus arecepta,* from Lake Baikal in Russia, that reaches a length of 60 cm. All macroturbellarians are flat and thin (as you may have predicted) and, in general, the larger the worm, the more pronounced the flattening.

Although macroturbellarians are conspicuous and useful for class study, most of the 4500 species of Turbellaria are microscopic animals rarely exceeding a length of a few millimeters, with the smallest being approximately 0.5 mm in length. These **microturbellaria** (Fig. 10-2) may be flat, but commonly they are either cylindrical or arched above and flattened below to form a creepsole.

Turbellarians are chiefly aquatic and the great majority are marine. Most are bottom dwellers that live in sand or mud, under stones and shells, or on seaweed, but a few species are pelagic. Many microturbellaria are common members of the interstitial fauna that occupy and move like snakes in a rock pile through the water-filled spaces between sand grains. Freshwater turbellarians are benthic inhabitants of lakes,

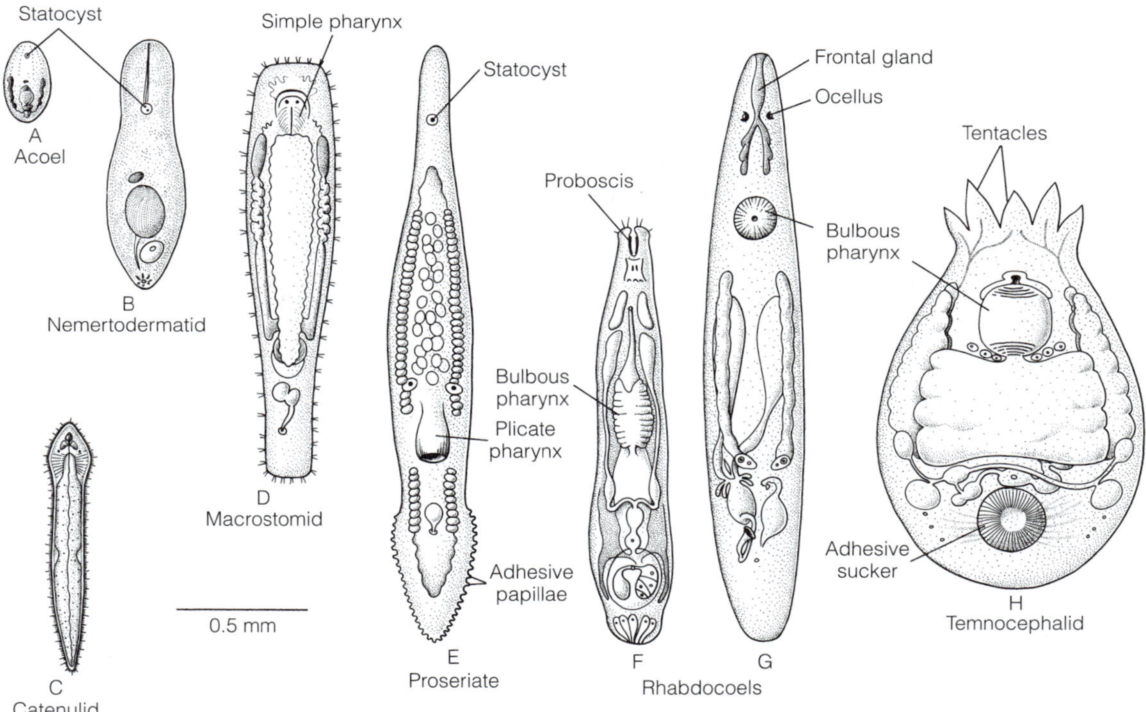

FIGURE 10-2 Turbellaria: diversity of microturbellaria. **A,** The acoel *Pseudactinoposthia parva.* **B,** The nemertodermatid *Flagellophora.* **C,** The catenulid *Stenostomum virginianum.* **D,** The macrostomid *Macrostomum appendiculatum.* **E,** The proseriate *Monocelis galapagoensis.* **F,** The kalyptorhynch rhabdocoel *Karkinorhynchus tetragnathus.* **G,** The typhloplanoid rhabdocoel *Ceratopera bifida.* **H,** The temnocephalid *Temnocephala geonoma.* *(All figures redrawn from original sources. A, From Ehlers, U., and Dörjes, J. 1979. Interstitielle Fauna von Galapagos. XXIII. Acoela (Turbellaria). Mikrofauna Meeresbodens 72:1–74; B, From Sterrer, W. 1966. New polylithophorous marine Turbellaria. Nature 210:436; C, From Nuttycombe, J. W. 1931. Two new species of Stenostomum from the southeastern United States. Zool. Anz. 97:80–85; D, From Ferguson, F. F. 1937. The morphology and taxonomy of* Macrostomum virginianum *n. sp. Zool. Anz. 119:25–32; E, From Ax, P., and Ax, R. 1977. Interstitielle Fauna von Galapagos XIX. Monocelididae (Turbellaria, Proseriata). Mikrofauna Meeresbodens 64:1–40; F, From Ax, P., and Schilke, K. 1971.* Karkinorhynchus tetragnathus *nov. spec. ein Schizorhynchier mit zweigeteilten Russelhaken (Turbellaria, Kalyptorhyncia). Mikrofauna Meeresbodens 5:1–10; G, From Ehlers, U., and Ax, P., 1974. Interstitielle Fauna von Galapagos. VIII. Trigonostominae (Turbellaria, Typhloplanoida). Mikrofauna Meeresbodens 30:1–33; H, From Williams, J. B. 1980. Morphology of a species of* Temnocephala *(Platyhelminthes) ectocommensal on the isopod* Phraetoicopsis terricola. *J. Nat. Hist. 14:183–199.)*

ponds, streams, and springs. A few turbellarians are terrestrial, but these are confined to humid areas and usually retreat below logs and leaf mold during the day, emerging only at night to feed. They are mostly large tropical species, but some, such as the introduced North American *Bipalium kewense, B. adventitium* (Fig. 10-1A), and related species, live in temperate regions.

Body Wall

The turbellarian body is covered by a monolayered, ciliated epidermis in which each ciliated cell bears many cilia (Fig. 10-3). The swirling motion of microscopic particles close to the ciliated epidermis is responsible for the name Turbellaria, which means "whirlpool." In some macroturbellarians and a few others, cilia are confined to or predominate on the ventral surface. Short microvilli cover the epidermal surface between the cilia, but a cuticle is absent. In a few turbellarians the epidermis is syncytial, and cellular boundaries are only partially present or absent (Fig. 10-3D). Beneath the epidermis is a basal lamina or basement membrane, except in Acoela, which has reduced extracellular matrix (Fig. 10-3A).

In the absence of a cuticle, turbellarians typically use the basal lamina and intracellular fibers to support the body wall. The most common intracellular skeleton is a weblike sheet of actin filaments (a terminal web) within the epidermis itself (Fig. 10-3). This sheet apparently helps the epidermis to bear stresses. Evidence for a skeletal role for basal lamina is found in unusual marine turbellarians with calcareous spicules in this layer (Fig. 10-4). In one taxon (Acoelomorpha), a network of interconnected ciliary rootlets, the fibrous structures anchoring each cilium in the cell, may also bear stress, perhaps compensating for their reduced extracellular matrix. In some species, the gut or the mesodermal parenchyma (discussed below) also has a skeletal function.

Turbellarians have a richly glandular epidermis (Fig. 10-5, 10-6, 10-13). Although the gland cells may be entirely within the epidermis, they commonly are submerged into or below the muscle layers, with only the neck of the gland penetrating the epidermis (or gastrodermis, as in Fig. 10-5). The glands may secrete adhesives, mucus, and other substances.

Typical of most turbellarians are numerous membrane-bounded, rod-shaped secretions known as rhabdoids that are released to the surface of the epidermis, where they expand to form mucus. The most common kind of rhabdoid is the **rhabdite,** which is characterized by a specific, layered ultrastructure (Fig. 10-5). Rhabdites are secreted by epidermal gland cells that usually are submerged, except for their slender necks, below the epidermis. Several functions have been attributed to rhabdites, such as slime production for locomotion, cocoon formation, and predator repellant, but none has been experimentally tested. Other, structurally different rhabdoids co-occur with rhabdites in individuals and are widespread among turbellarian taxa.

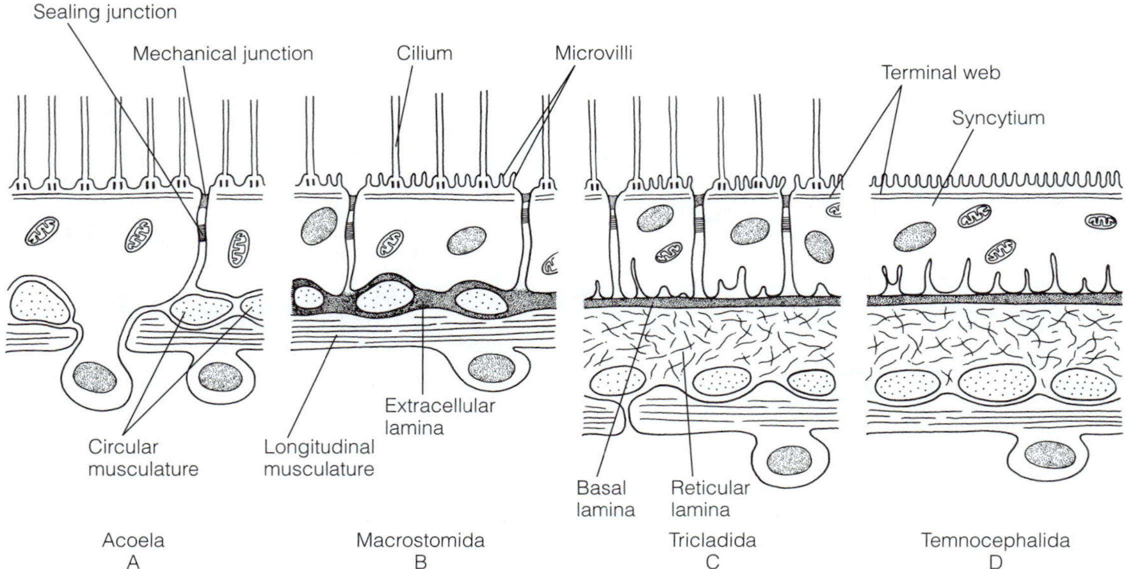

FIGURE 10-3 Turbellaria: body-wall diversity. **A,** Acoela; **B,** Macrostomida; **C,** Tricladida (planarians); **D,** Temnocephalida. The amount of extracellular matrix tends to increase with body size from **A** to **D.** A syncytial epidermis occurs in some rhabdocoels but is typical of the ectocommensal temnocephalids **(D).** A similar syncytial neodermis is characteristic of trematodes, monogeneans, and tapeworms. *(Adapted from Tyler, S. 1984. Turbellarian platyhelminths. In Bereiter-Hahn, J. Matoltsky, A. G., and Richards, K. S. (Eds.): Biology of the Integument. Vol. 1. Invertebrates. Springer-Verlag. Berlin. pp. 112–131. And Rieger, R. M. 1981. Morphology of the Turbellaria at the ultrastructural level. Hydrobiologia. 84:213–229.)*

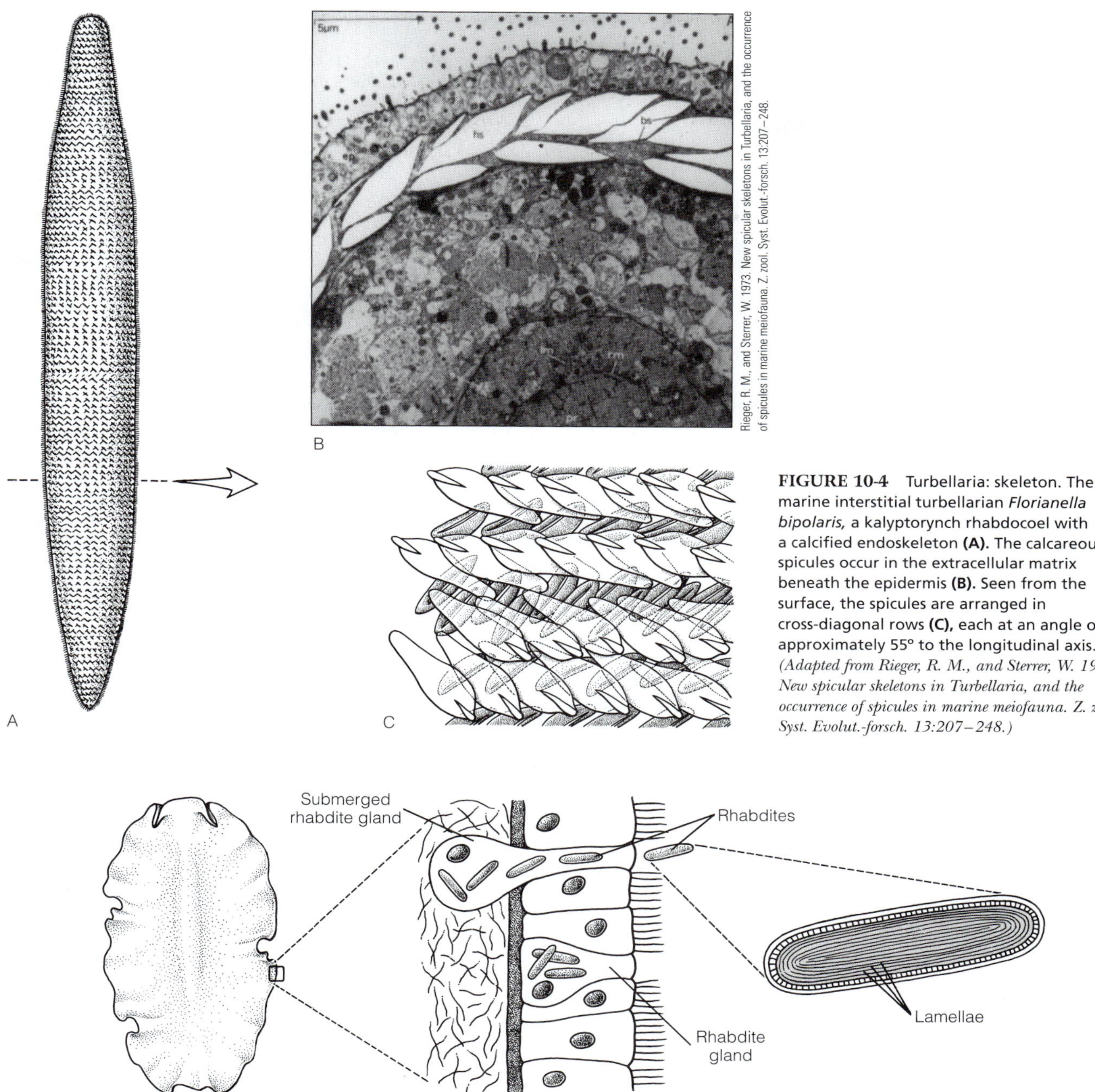

FIGURE 10-4 Turbellaria: skeleton. The marine interstitial turbellarian *Florianella bipolaris,* a kalyptorynch rhabdocoel with a calcified endoskeleton **(A).** The calcareous spicules occur in the extracellular matrix beneath the epidermis **(B).** Seen from the surface, the spicules are arranged in cross-diagonal rows **(C),** each at an angle of approximately 55° to the longitudinal axis. *(Adapted from Rieger, R. M., and Sterrer, W. 1973. New spicular skeletons in Turbellaria, and the occurrence of spicules in marine meiofauna. Z. zool. Syst. Evolut.-forsch. 13:207–248.)*

FIGURE 10-5 Turbellaria: rhabdites. One of the characteristic turbellarian (rhabditophoran) secretions is a rhabdite, a rod-shaped secretion composed of successive microscopic lamellae. *(Enlarged granule drawn from Smith, J. P. S. III, Tyler, S., Thomas, M. B. et al. 1982. The morphology of turbellarian rhabdites: Phylogenetic implications. Trans. Am. Micros. Soc. 101:209–228.)*

An anterior aggregation of secretory cells called a **frontal gland** is characteristic of most turbellarians (Fig. 10-2G, 10-6). Its function, however, is unknown, although roles in defense, slime production for locomotion, and adhesion (in larvae) have been suggested. Some turbellarians have aggregations of gland cells at the posterior end of the body (Fig. 10-2F). In *Bdelloura,* which lives as a commensal on the book gill of the Atlantic horseshoe crab, the glands form an adhesive plate (Fig. 10-1C).

Temporary adhesion to the substratum is made possible by adhesive glands, adhesive cilia, or muscular suckers

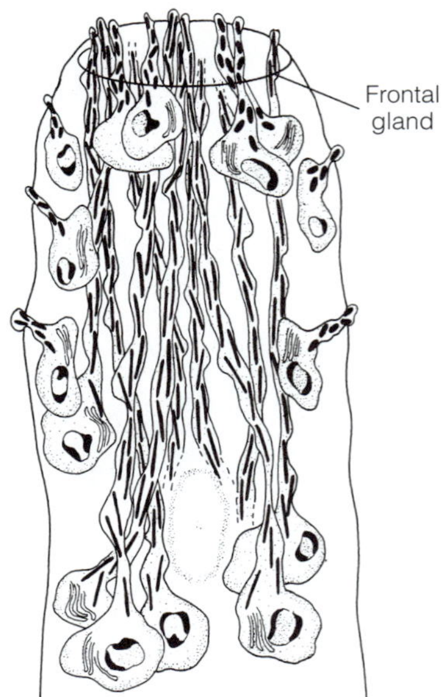

FIGURE 10-6 Turbellaria: frontal gland. The frontal gland of the macrostomid *Paramalostomum coronum.* The individual gland cells opening on the sides of the head are not part of the frontal gland. *(From Klauser, M. D., and Tyler, S. 1987. Frontal glands and frontal sensory structures in the Macrostomida (Turbellaria). Zool. Scripta. 16:95–110.)*

(Fig. 10-7). Many interstitial marine species adhere to the sand grains with glandular adhesive organs known as **duo-gland organs,** each of which may project from the body surface as a papilla (Fig. 10-2E, 10-7B, 10-9E). A duo-gland organ is composed of two different kinds of gland cells. One, designated a **viscid gland,** secretes the adhesive and cements the animal to the sand grain, and the other, called a **releasing gland,** secretes the de-adhesive, the substance that breaks the attachment (Fig. 10-7B). The gland necks of both viscid and releasing glands pass through a third cell, an epidermal **anchor cell,** which is specialized to bear the strain of attachment. Using groups of these duo-gland adhesive organs, interstitial turbellarians rapidly stick to and release from sand grains. Macroturbellarians (Polycladida) and ectocommensals on crustaceans (Temnocephalida) attach to surfaces using a well-developed ventral sucker (Fig. 10-1C, 10-2H, 10-7C). At least one tiny worm, *Paratomella rubra* (Acoela), is able to stick to surfaces using the flattened tips of specialized adhesive cilia (10-7A).

Musculature and Locomotion

As small animals without appendages, turbellarians rely primarily on their cilia to glide over surfaces. But, surprisingly, many turbellarians have a well-developed and complex musculature that enables them to make a wide variety of movements.

The musculature consists of a gridlike arrangement of outer circular and inner longitudinal fibers. Between these two layers, many species also have two sets of **diagonal muscles** that crisscross each other (Fig. 10-8). Some of the large, flat species also have dorsoventral muscles and additional layers of circular and longitudinal muscles. As might be expected in soft-bodied worms lacking a rigid skeleton, smooth or obliquely striated muscle fibers typify turbellarians.

Turbellarians have a wide variety of locomotory adaptations that range from ciliary gliding over surfaces and through water to muscular creeping and swimming. Other body move-

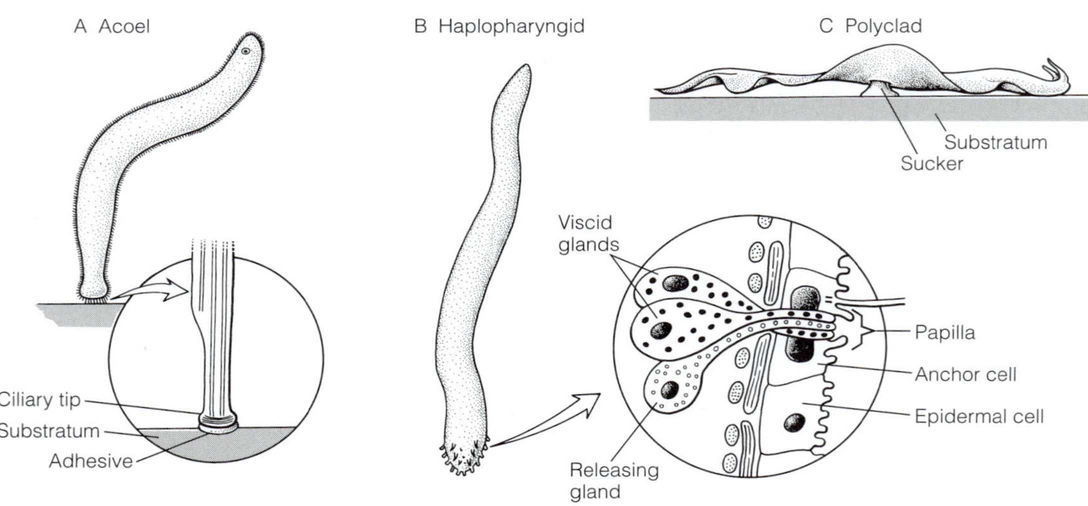

FIGURE 10-7 Turbellaria: temporary adhesion. **A,** Adhesive cilia in the acoel *Paratomella rubra.* **B,** A duo-gland adhesive papilla (inset) in *Haplopharynx* (Macrostomorpha) and many other turbellarians. **C,** An adhesive sucker in the polyclad *Oligoclado floridanus* and its cotylean relatives. *(A, Modified from Tyler, S. 1973. An adhesive function for modified cilia in an interstitial turbellarian. Acta Zool. 54:139–151; B, Modified and redrawn from Tyler, S. 1976. Comparative ultrastructure of adhesive systems in the Turbellaria. Zoolmorphologie 84:1–76.)*

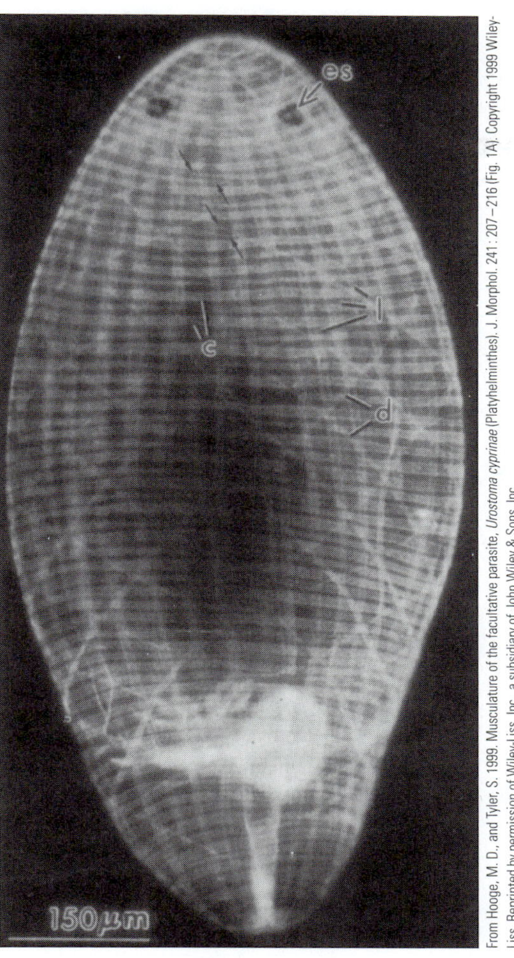

FIGURE 10-8 Turbellaria: musculature. In most turbellarians, large or small, the body-wall musculature consists of the usual outer circular, middle diagonal, and inner longitudinal muscles, as shown in this epifluorescence micrograph of *Urostoma cyprinae* (Prolecithophora).

ments include retraction and extension, peristalsis, twisting, turning, waving, and somersaulting (Fig. 10-9). In general, small aquatic turbellarians use cilia for locomotion (Fig. 10-9A), whereas large turbellarians use muscular movements (perhaps aided by cilia) of the entire body or along a specialized ventral sole (Fig. 10-9C,D). In some small species, body-wall peristalsis is common but not used in locomotion. Instead, these movements may be important in mixing the contents of the gut, which often has a weakly developed musculature (Fig. 10-9G). Some large marine species (Polycladida) swim using dorsoventral undulations of the lateral body margins (Fig. 10-10).

Nervous System and Sense Organs

The layout of the turbellarian nervous system is variable, especially with respect to the number and arrangement of nerve cords. In general, and perhaps primitively, it consists of a subepidermal ringlike brain from which one to several nerve cords, depending on taxon, extend posteriorly through the body. When several pairs of longitudinal nerves issue from the brain, usually they are equidistant from each other and impart radial symmetry on the nervous system (Fig. 10-11A). The longitudinal nerve cords join a nerve net located internally to the body-wall musculature. This submuscular net, in turn, joins two other, more peripheral nets—one between the epidermis and musculature and another within the epidermis. Although some nervous tissue is concentrated into the brain and longitudinal cords, the nervous system as a whole is diffuse, netlike, and reminiscent of that of cnidarians or hemichordates (Fig. 10-11B). A nerve net is a logical arrangement for innervating the sheetlike, two-dimensional body-wall musculature. A specialized nerve net also may be associated with the muscular pharynx and midgut of some turbellarians.

Many turbellarians have abandoned the ringlike brain and multiple nerve cords in favor of a concentrated bilateral brain and two ventrolateral, longitudinal nerve cords, as occurs in the common planarian *Dugesia* for example. Although *Dugesia* retains a peripheral nerve net (and associated muscle grid), the longitudinal cords are joined at regular intervals by transverse commissures, which together with the cords give the nervous system a segmented, ladderlike appearance (Fig. 10-11C). Such a highly organized nervous system suggests a level of hierarchic order and control not present in netlike systems. In all turbellarians, the nervous system is relatively primitive in its lack of ganglia, except in the brain, but typical sensory, motor, and interneurons are present.

Pigment-cup ocelli are common in most turbellarians (Fig. 10-12B). The usual number is two (Fig. 10-2G), but two or three pairs are not uncommon. In macroturbellarians, including the land planarians, many eyes may occur in clusters over the brain, in tentacles, or distributed uniformly around the body margin (Fig. 10-1E, 10-12B). The eyes function largely in orienting to light, and most turbellarians are negatively phototactic. Strong light directed onto a swimming polyclad, for example, arrests locomotion, causing it to sink to the bottom, where it crawls away in search of cover.

Other than eyes, statocysts are the most conspicuous of the turbellarian sensory organs (Fig. 10-12A), but they occur only in a few taxa (primarily catenulids, acoelomorphs, and seriates). Turbellarian statocysts are unpaired and located medially near the brain. Each statocyst consists of a capsule that encloses a fluid-filled cavity and a central concretion called a statolith. Sometimes more than one statolith is present; nemertodermatids, for example, have two. Because of their general similarity to the statocysts of some cnidarian medusae and ctenophores, turbellarian statocysts are presumed to be gravity receptors. Unlike the statocysts of diploblastic animals, however, turbellarian statocysts lack sensory cilia, and the statolith makes contact with the unspecialized wall of the capsule. The mechanism of sensory reception is unknown.

Single-celled ciliary receptors (Fig. 10-12C), many of which are probably mechanoreceptors, are distributed over the entire body, but are particularly concentrated on tentacles, auricles (Fig. 10-1B), and body margins. Specialized pits or grooves

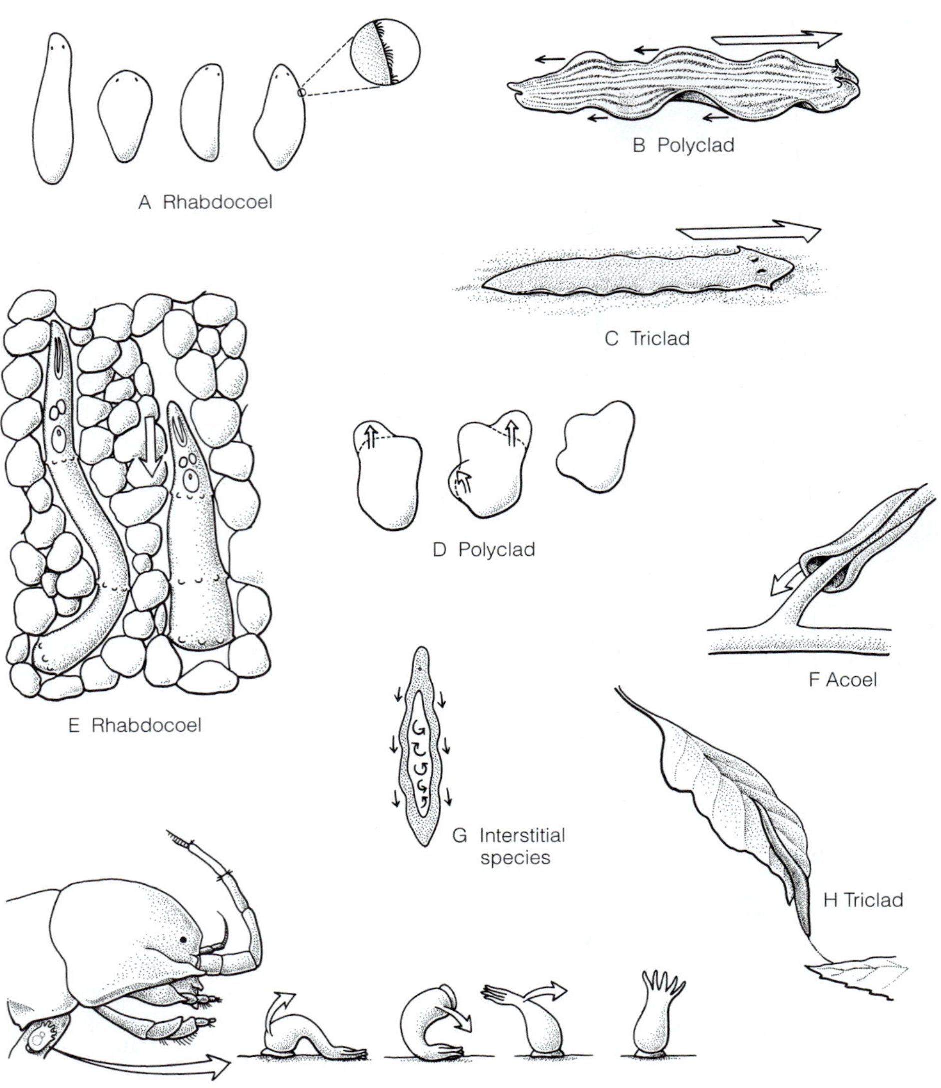

FIGURE 10-9 Turbellaria: locomotion. Ciliary locomotion is common in small turbellarians, such as the rhabdocoel *Kytorhynchella meixneri* **(A)** and the acoel *Convoluta convoluta* **(F)**. The large polyclad *Pseudoceros crozieri* **(B)** swims using undulations of the body margins. Planarians produce muscular waves along their ventral surface to move over substrata **(C)**, whereas polyclads such as *Corondena mutabilis* shuffle forward ditaxically **(D)** like some snails. Retraction movements are common among turbellarians, shown here for the interstitial rhabdocoel *Proschizorhynchus anophthalmus* **(E)**, which not only withdraws from a disturbance but also simultaneously anchors itself using girdles of duo-gland adhesive papillae. While stationary, some interstitial species produce peristaltic waves along the body surface, presumably to help mix the gut contents **(G)**. The terrestrial triclad *Rhynchodemus terrestris* casts a mucous thread and then uses it as a suspension bridge to cross between two leaves **(H)**. Some temnocephalids, which are ectocommensals on crustaceans, can somersault by alternately attaching the posterior sucker and anterior tentacles **(I)**. *(A, Modified from Rieger, R. M. 1974. A new group of Turbellaria-Typhloplanoida with a proboscis and its relationship to Kalyptorhynchia. In Riser, N. W., and Morse, M. P. (Eds.): Biology of the Turbellaria. McGraw-Hill Book Co., New York. pp. 23–62; E, Redrawn and modified from L'Hardy, J.-P. 1965. Turbellaries Schizorhynchidae des sables de Roscoff. II. Le genre* Proschizorhynchus. *Cah. Biol. Mar. 6:125–161; F, Redrawn from Apelt, G. 1969. Fortpflanzungsbiologie, Entwicklungszyklen und vergleichende Frühentwicklung aceoler Turbellarian. Mar. Biol. 4:267– 325; H, Redrawn and modified from Reisinger, E. 1923. Turbellaria. Strüdelwurmer. Biol. Tiere Deutsch. Lief. 6:1–64; I, Redrawn and modified from Williams, J. B. 1980. Morphology of a species of* Temnocephala *(Platyhelminthes) ectocommensal on the isopod* Phreatoicopsis terricola. *J. Nat. Hist. 14:183–199, and after Haswell from de Beauchamp, P. 1961. Classe des Turbellaries. In Grassé, P. -P. (Ed.): Traité de Zoologie. Vol. 4. Masson et Cie, Paris. pp. 216.)*

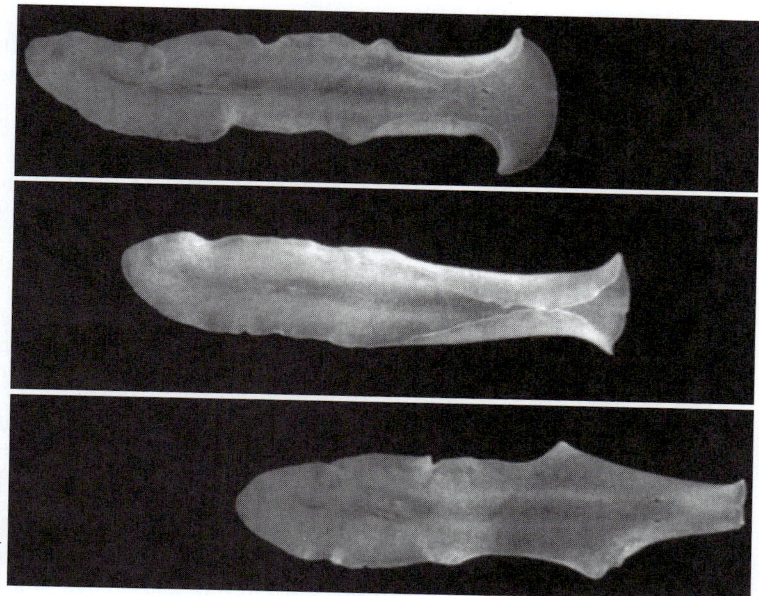

FIGURE 10-10 Turbellaria: locomotion. The polyclad *Stylochoplana floridana* is a 20 mm-long flatworm that swims rapidly (15 mm/s) by flapping its anterior body margins in the manner of a sea hare.

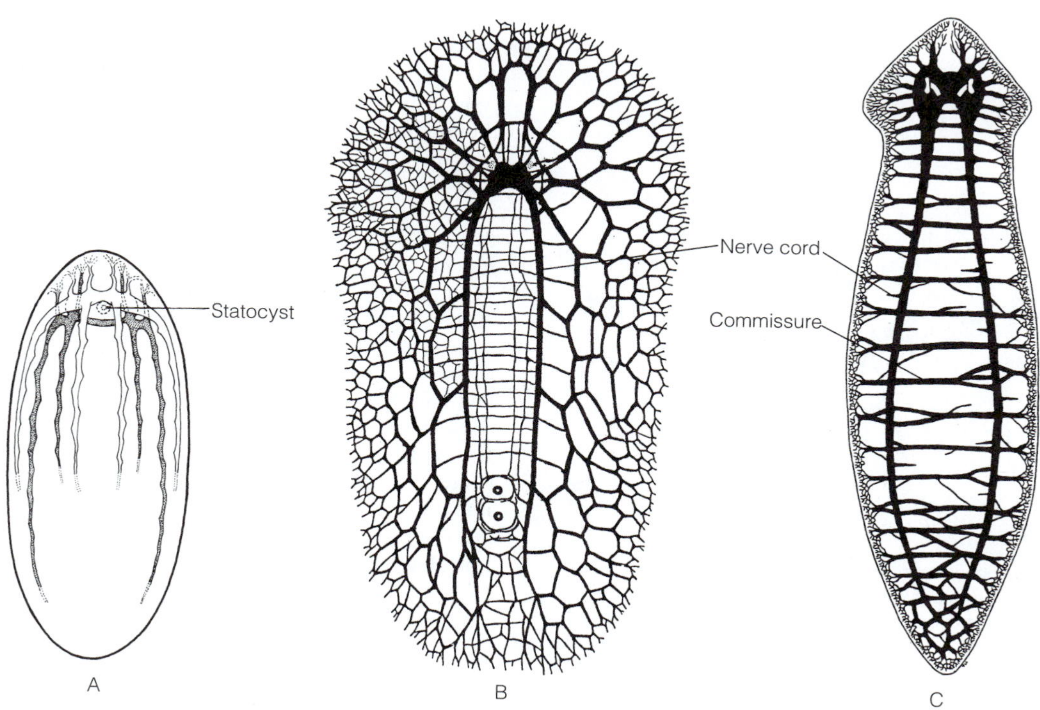

FIGURE 10-11 Turbellaria: nervous system. The central components include a brain and a variable number of nerve cords. In the acoels **(A),** the brain and nerve cords are more or less radially disposed (peripheral nerve net not shown). The netlike arrangement of peripheral nerves is clear in polyclads (**B,** ventral nervous system) and triclads **(C),** but in triclads a regular series of transverse commissures imparts a segmental pattern on the nervous system. *(A, After Westblad, B.; B, After Hadenfeldt from Stummer-Traunfels, R. V. 1933. Polycladida. In Bronn, H. G. (Ed.): Klassen und Ordnungen des Tierreichs; C, From Lentz, T. L. 1968. Primitive nervous systems. Yale University Press, New Haven. pp. 72 and 77. Reprinted with permission.)*

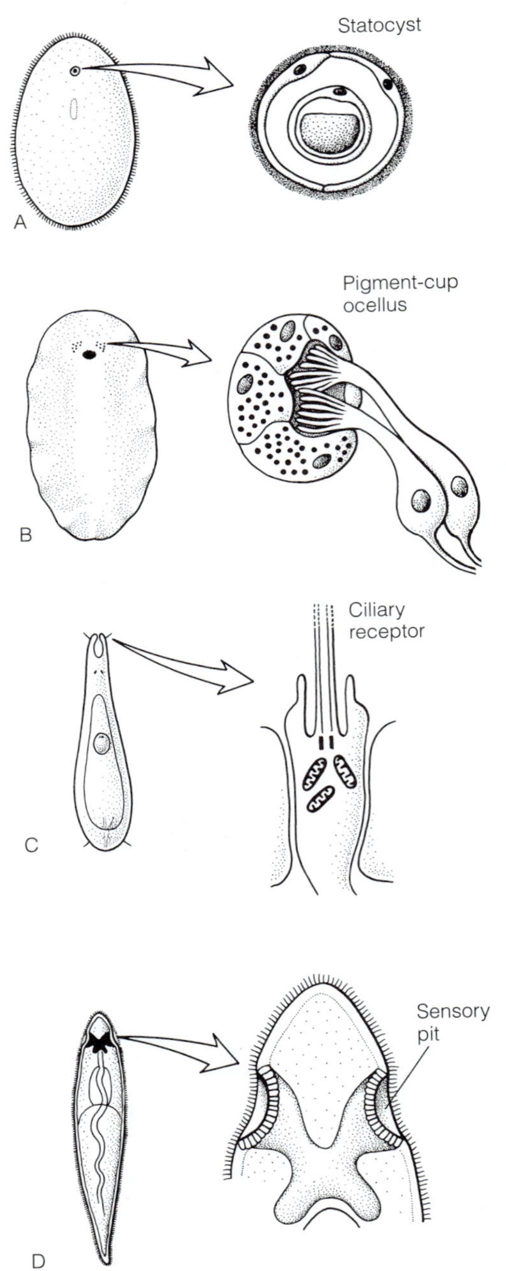

FIGURE 10-12 Turbellaria: sensory organs. **A,** Statocyst of an acoel; **B,** pigment-cup ocellus of a polyclad; **C,** simple ciliary receptor of a rhabdocoel, probably a mechanoreceptor; **D,** sensory pit of a catenulid, probably a chemoreceptor. *(A–C, Modified and redrawn from Rieger, R. M., Tyler, S. III, et al. 1991. Platyhelminthes: Turbellaria. In Harrison, F. W., and Bogitsh, B. (Eds.).: Microscopic Anatomy of Invertebrates. Vol. 3. Wiley-Liss, Inc. New York. pp. 7–140; D, Modified and redrawn from Nuttycombe, J. W., and Waters, A. J. 1938. The American species of the genus* Stenostomum. *Proc. Am. Philos. Soc. 79:213–300.)*

on the head contain sensory cells that are probably chemoreceptors (Fig. 10-12D), which may be used in locating food or mates.

Parenchyma

The connective-tissue compartment between the body-wall musculature and gut is called the **parenchyma.** Like the typical connective tissue described in Chapter 4 (Fig. 4-1), the parenchyma of most *macro*turbellarians is composed of cells in a fibrous extracellular matrix (Fig. 10-13B). Departures from this organization occur in two directions. First, the parenchyma of *micro*turbellarians contains little extracellular matrix, and in one taxon (Acoela), an extracellular matrix is nearly absent and the parenchyma is chiefly cellular (Fig. 10-13A). Second, among several freshwater catenulids, the extracellular matrix part of the parenchyma is well developed and fluid rather than fibrous and forms a hemocoel (Fig. 10-13C). The hemocoel may play a role in internal transport and serve as a hydrostatic skeleton.

Flatworm parenchymal cells are diverse, and only those with well-established functions are mentioned here. **Epidermal replacement cells** migrate from the parenchyma to the body surface and replace any damaged or destroyed epidermal cells. This unusual means of replacement is necessitated by the absence of mitosis in the adult epidermis. The epidermal replacement cells are situated immediately below the body wall and each contains a cluster of centrioles (which later become the ciliary basal bodies). Many turbellarians have a population of totipotent cells called **neoblasts,** which are important in wound healing and regeneration (Fig. 10-13B). They may also give rise to the epidermal replacement cells. Another common cell of the parenchyma is the **fixed parenchymal cell,** a large, branched cell that makes gap junctions with other parenchymal cells as well as epidermal and gastrodermal cells (Fig. 10-13B), thus linking together all tissue layers of the body. Gap junctions are intercellular channels for low-resistance transport of metabolites, and their presence indicates that the network of cells linked by them are physiologically coupled. In the absence of a circulatory system, perhaps this network constitutes a specialized intercellular transport system.

Some planarians have parenchymal **pigment cells** (Fig. 10-13B) and **chromatophores,** the latter of which enable the animal to lighten and darken when the pigment in the cell is concentrated or dispersed, respectively. Chromatophores are controlled by the brain, for the posterior half of a bisected dark worm does not lighten until it has regenerated a brain. In at least one interstitial acoel *(Paratomella rubra)* and some rhabdocoels, hemoglobin-containing parenchymal cells impart a red color to the body and probably function as an oxygen store for use when the animal wanders into oxygen-poor layers of the sand.

Digestive System and Nutrition

The digestive cavity, or gut, of turbellarians is typically a blind sac and the mouth is used for both ingestion and egestion (Fig. 10-14). An anus or multiple anuses occur only in some very long worms and in some turbellarians with highly branched guts. In these, the normal return of undigested wastes to the mouth is apparently complicated by the extreme length or complex branching of the gut. The wall of the gut is single-layered and composed of phagocytic and gland cells. Primitively, the

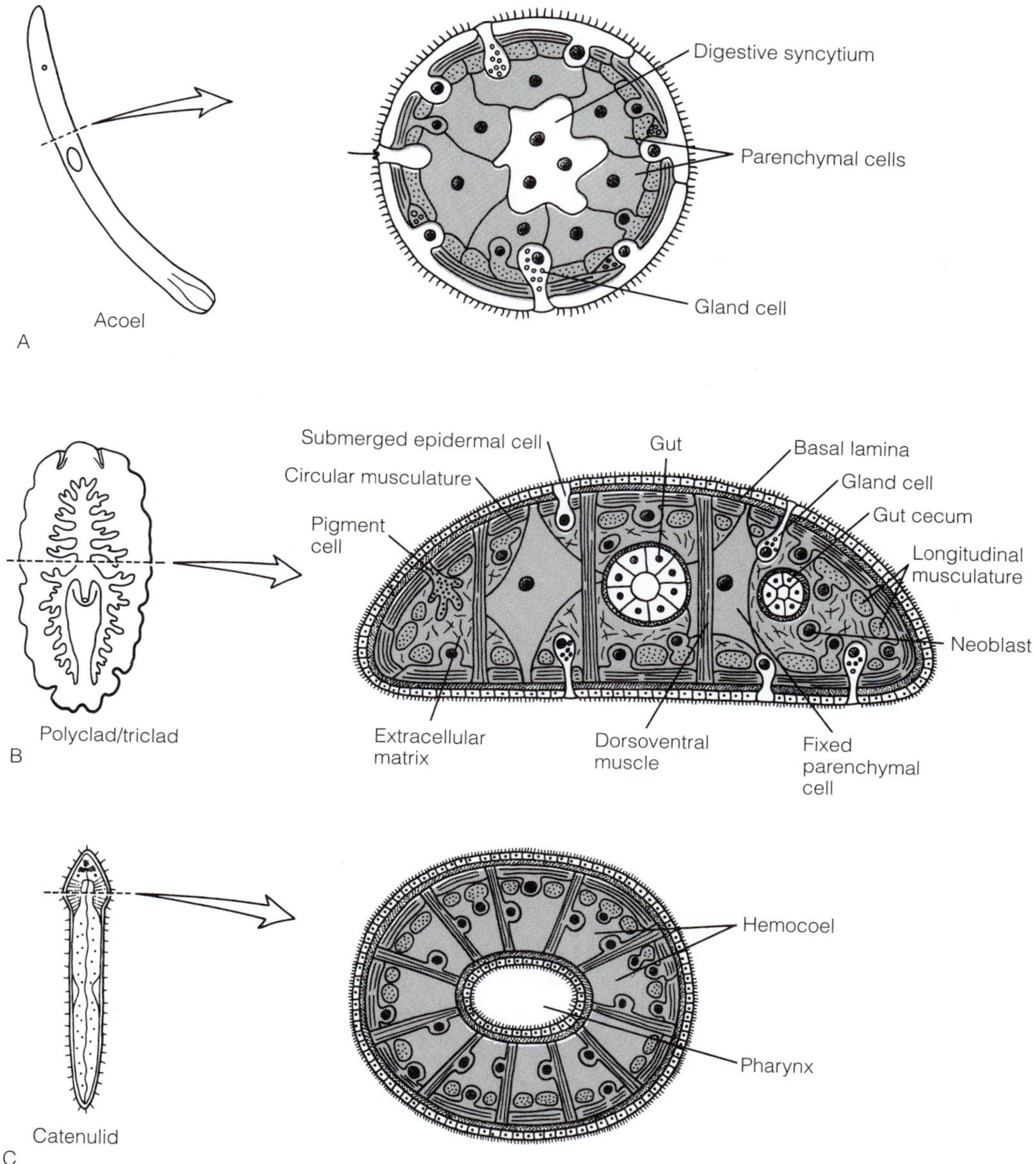

FIGURE 10-13 Turbellaria: anatomy and parenchyma (connective-tissue compartment). **A,** In acoels such as *Diopisthoporus longitubus,* the parenchyma is entirely cellular and lacks an extracellular matrix. **B,** In most other turbellarians, including the polyclads and triclads, the parenchyma is like typical connective tissue, containing both cells and extracellular matrix. **C,** The extracellular matrix of some catenulids is fluid and forms a hemocoel (pseudocoel). *(A and B, Modified and redrawn from Smith, J. P. S. III, and Tyler, S. 1985. The acoel turbellarians: Kingpins of metazoan evolution or a specialized offshoot? And Rieger, R. M. 1985. The phylogenetic status of the acoelomate organization within the Bilateria: A histological perspective. Both in Conway Morris, S., George, J. D., Gibson, R., et al. (Eds.): The Origins and Relationships of Lower Invertebrates. Clarendon Press, Oxford. pp. 123–142, 101–122.)*

gut is ciliated, as in some major taxa (Catenulida, Macrostomida, and some Polycladida), but in most others it lacks cilia. The gut of acoel turbellarians lacks a lumen altogether and is usually a syncytium enclosed by a common cell membrane (Fig. 10-13A, 10-14F). The taxon name, "Acoela", refers to this lack of a gut cavity.

The shape of the gut is in part related to the size of the worm (Fig. 10-13A, 10-14F). The microturbellarian gut typically is a simple, unbranched sac or blind-ended tube (Fig. 10-14A–C). The flat macroturbellarians usually have a gut with lateral branches (ceca) that extend to the margin of the body (Fig. 10-14D,E, 10-31A,B). These branches not only

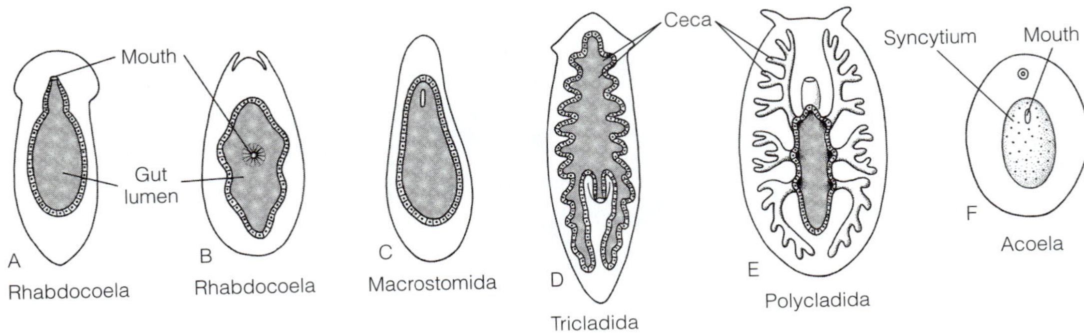

FIGURE 10-14 Turbellaria: gut. The midgut is a more or less simple sac in microturbellarians **(A–C)**, such as most rhabdocoels **(A and B)** and macrostomids **(C)**. In the acoels **(F)**, the gut is usually a lumenless syncytium. The macroturbellarians, such as triclads **(D)** and polyclads **(E)**, have vessel-like ceca that increase the area of the gut and distribute nutrients to all parts of the body. *(A–F, Modified and redrawn after Hyman from Rieger, R. M., Tyler, S., Smith, J. P. III, et al. 1991. Platyhelminthes: Turbellaria. In Harrison, F. W., and Bogitsh, B. (Eds.): Microscopic Anatomy of Invertebrates. Vol. 3. Wiley-Liss, New York. pp. 7–140.)*

provide a large area for digestion and absorption, they also transport nutrients to all parts of the body and thus function as a gastrovascular system.

The mouth commonly is located on the midventral surface, but may be situated anteriorly, posteriorly, or anywhere along the midventral line, depending on taxon (Fig. 10-15). Most turbellarians also have a specialized glandular pharynx, but its structure varies widely. Only one taxon (Acoela) lacks a pharynx **(zero pharynx),** and the mouth opens directly into the cytoplasm of the digestive syncytium (Fig. 10-14F). Among the others, the pharynx ranges from a simple ciliated tube called a **simple pharynx** (as in Macrostomida and Catenulida; Fig. 10-16A, 10-17) to complex protrusible organs in higher turbellarians (polyclads and all neoophorans). One such pharynx is the folded **plicate pharynx** of polyclads, triclads, and proseriates (Fig. 10-2E, 10-16B), which are chiefly macroturbellarians with branched intestines. The plicate pharynx is a long, muscular tube that, when retracted, has a fold or pleat in its wall and is enclosed in a sheathlike cavity. The extra tissue in the pleat allows the pharynx to be protruded outward through the mouth during feeding (Fig. 10-18). The plicate pharynx may project backward, as in the common freshwater planarians (Fig. 10-31A,C), or it may be attached to the cavity posteriorly and extend forward. The **bulbous pharynx** of Rhabdocoela is a muscular sucking bulb separated from the parenchyma by a septum (Fig. 10-2F–H, 10-16C). This separation isolates the pumping action of the pharynx from the damping effect of the inertial parenchyma. In many species, the bulbous pharynx can be protruded from the mouth.

Turbellarians with a zero or simple pharynx tend to feed on unicellular algae, and perhaps on bacteria and protozoans. The plicate pharynx is associated with predatory species, and the bulbous pharynx occurs in predators and the parasitic flukes. As such, the evolution of a sucking, bulbous pharynx in free-living flatworms is one trait that predisposed them to parasitism.

Turbellarians are largely carnivorous and prey on various invertebrates that are small enough to be captured (Fig. 10-17), as well as on the dead bodies of animals that sink to the bottom. Feeding behavior is elicited, at least in some species (such as planarians), by substances emitted from the

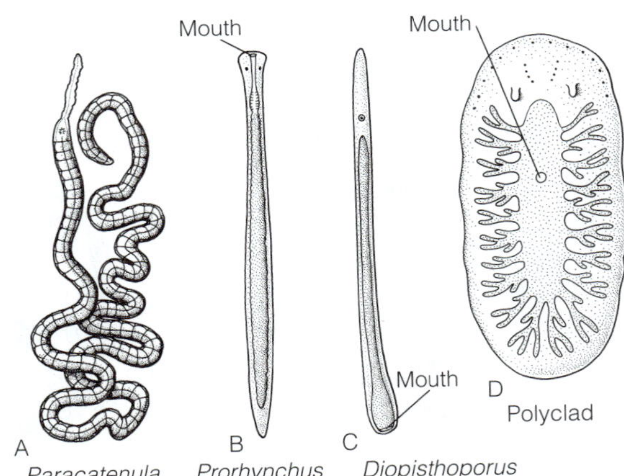

FIGURE 10-15 Turbellaria: mouth. The presence and position of the mouth is variable in turbellarians. In at least one rare marine interstitial species **(A)**, the catenulid *Paracatenula urania,* a mouth is absent and the animal derives its nutrition directly from prokaryotic symbionts found in its rudimentary gut. When present, the mouth position ranges from anterior in the prolecithophoran *Prorhynchus stagnalis* **(B)** to posterior in the acoel *Diopisthoporus gymnopharyngeus* **(C)** to midventral in many others, such as this polyclad **(D)**. *(A, Redrawn from Sterrer, W., and Rieger, R. M. 1974. Retronectidae—A new cosmopolitan marine family of Catenulida (Turbellaria). In Riser, N. W., and Morse, M. P. (Eds.): Biology of the Turbellaria. McGraw-Hill Book Co., New York, p. 152.; B, Redrawn from Hyman, L. H. 1951. The Invertebrates: Platyhelminthes and Rhynchocoela. The Acoelomate Bilateria. Vol. II. McGraw-Hill Book Co., New York, p. 152; C, Drawn from a photograph in Smith, J. P. S. III, and Tyler, S. 1985. Fine-Structure and Evolutionary Implications of the Frontal Organ in Turbellaria Acoela. 1.* Diopisthoporus gymnopharyngeus *sp. n. Zool. Scripta 14:91–102.)*

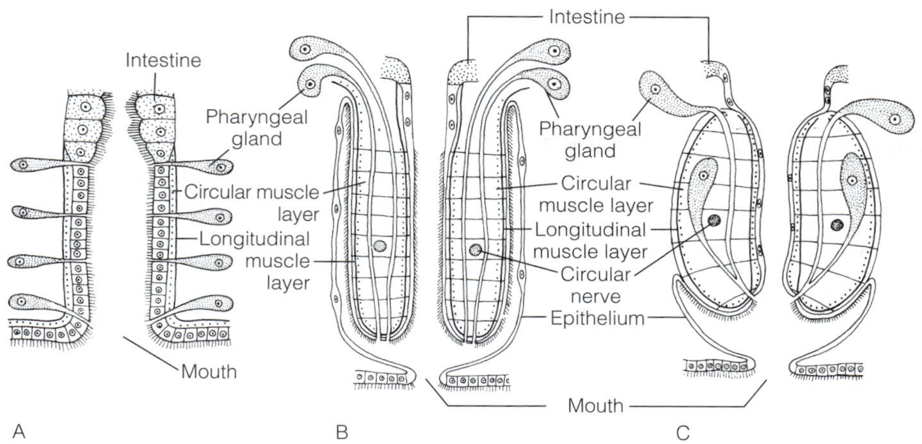

FIGURE 10-16 Turbellaria: pharynx. **A,** The simple pharynx of macrostomorphs, catenulids, and a few acoels. **B,** The plicate pharynx of polyclads, proseriates, and planarians. **C,** The bulbous pharynx of rhabdocoels. *(A–C, Modified from Ax, P. 1963. Relationships and phylogeny of the Turbellaria. In Dougherty, E. C. (Ed.): The Lower Metazoa. University of California Press, Berkeley. pp. 191–224.)*

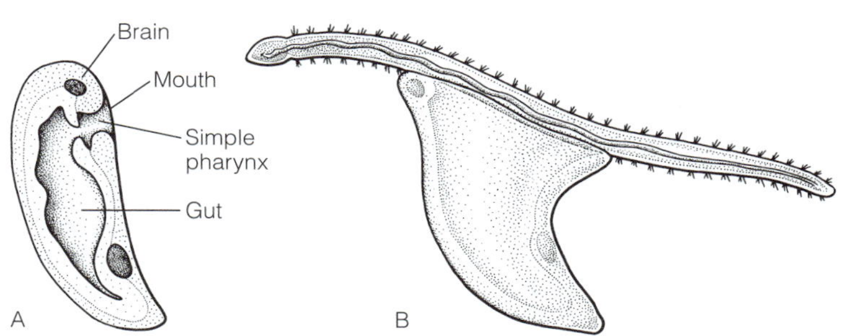

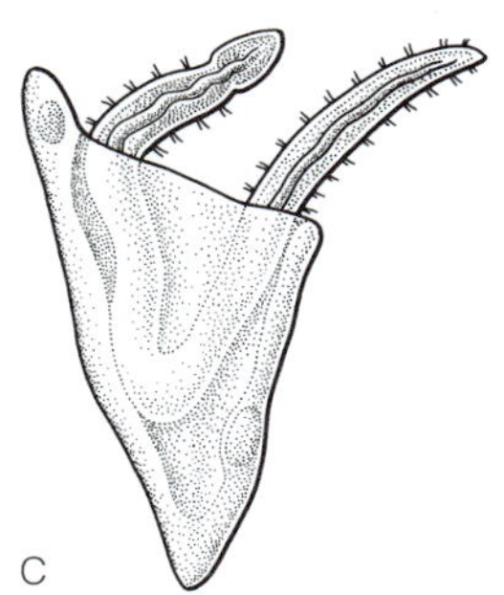

FIGURE 10-17 Turbellaria: feeding. The freshwater *Microstomum caudatum* (Macrostomida) uses its simple pharynx and mouth, which undergo a surprising dilation, to engulf an oligochaete worm. *(All lateral views. Modified and redrawn from Kepner, W. A., and Helvestine, F., Jr. 1920. Pharynx of Microstoma caudatum. J. Morphol. 33:309–316.)*

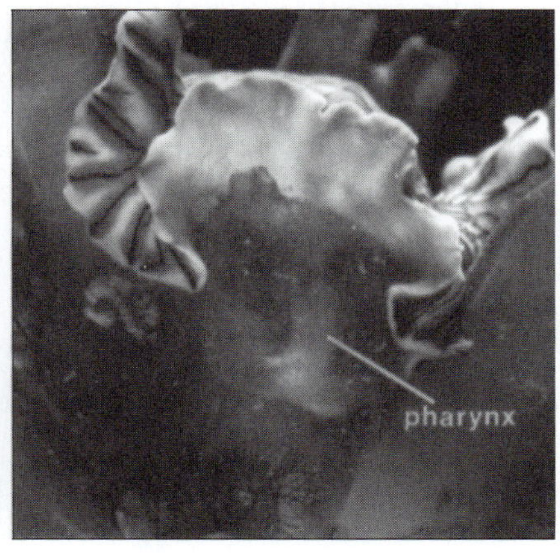

FIGURE 10-18 Turbellaria: feeding. The polyclad *Pseudoceros crozieri* crawls over (left photograph) and preys on (right photograph) one zooid of the colonial tunicate *Ecteinascidia turbinata*. While feeding, the worm thrusts its everted plicate pharynx into one of the sea squirt's siphons (right photograph).

food. Protozoa, rotifers, insect larvae, small crustaceans, snails, and small annelid worms are common prey, but some marine species feed on sessile animals, such as bryozoans and small tunicates. The polyclad *Stylochus frontalis* feeds on living oysters and is nicknamed the "oyster leech," *Stylochus triparitus* preys on barnacles, and the West Indian *Pseudoceros crozieri* feeds on the colonial tunicate *Ecteinascidia turbinata* (Fig. 10-18).

As already indicated, not all turbellarians are predators. Some acoel, macrostomid, and polyclad species feed on algae, especially diatoms, while other species that are predaceous as adults feed on diatoms as juveniles. Several acoel species harbor zoochlorellae, zooxanthellae, or diatoms in their parenchyma. At least one of these species *(Symsagittifera roscoffensis)* relies on its mutualists for nutrition and does not feed when algae are present in its tissues.

Many turbellarians capture prey by wrapping themselves around it, entangling it in slime, or pinning it to the substratum by means of their adhesive organs. Species of *Mesostoma* paralyze their prey with toxic mucus. A few species are known to stab prey with their penis, which terminates in a hardened stylet and projects from the mouth; the interstitial kalyptorhynch rhabdocoels have an anterior, raptorial proboscis that may have either a sticky tip or grasping hooks (Fig. 10-19). The proboscis of these species is independent of the mouth and bulbous pharynx.

Food is swallowed whole or in pieces. Turbellarians with a simple pharynx swallow food whole (Fig. 10-17). Predators and scavengers typically use the pharynx to ingest whole prey, but others swallow only fragments. In the triclads, for example, the extended pharyngeal tube is inserted into the body of the prey or the carrion and the contents are pumped in (Fig. 10-31D). The exoskeleton of crustaceans is penetrated at thin areas, such as the articular membranes between body segments. Penetration by the pharynx and ingestion of the body tissues of the prey are aided by proteolytic enzymes (endopeptidases) produced by pharyngeal glands that open onto the tip of the pharynx. The partially digested and liquified contents are then pumped into the gut by pharyngeal peristalsis.

Studies on the acoel *Convoluta convoluta* showed that small prey is captured and engulfed by the digestive syncytium, which is partially protruded through the mouth. Larger prey is pressed into the mouth and swallowed. In both of these cases, prey probably enters the syncytium by phagocytosis.

Digestion is first extracellular. Hydrolysis of the ingested food is initiated by pharyngeal enzymes, and additional enzymes (endopeptidases) are supplied by gland cells of the gut. The resulting food fragments are then engulfed by phagocytic gut cells and intracellular digestion is initiated by endopeptidases in vesicles at low pH. About 8 to 12 h following phagocytosis, the vesicles become alkaline, which marks the appearance of the exopeptidases, lipases, and carbohydrases necessary to complete digestion.

Freshwater planarians as well as many other turbellarians can withstand prolonged periods of experimental starvation. In extreme cases, they reabsorb and metabolize part of the gut and all of the parenchyma and reproductive system tissues.

Although parasitism in flatworms is usually associated with flukes and tapeworms, several turbellarians are either commensals or parasites. These are largely freshwater and marine rhabdocoels (Dalyellioida) as well as the freshwater taxon Temnocephalida (Fig. 10-2H). Commensals include species that live in the mantle cavity of molluscs and on the gills of crustaceans. For example, the triclad *Bdelloura* is an ectocommensal on the book gills of horseshoe crabs that shares food collected by the host. Parasitic species inhabit the guts and body cavities of molluscs, crustaceans, and echinoderms, as well as the skin of fishes. Species of the rhabdocoel family Fecampiidae are endoparasitic in the hemocoel of crustaceans. These turbellarians lack a digestive tract and absorb host nutrients across the naked body wall.

Internal Transport

Turbellarians are either tiny and more or less cylindrical or large and dorsoventrally flattened. In either case, the diffusion distance for gas transport is short and oxygen is absorbed across the general body wall. In small turbellarians, nutrients diffuse from the central gut to the nearby tissues, but in large, *flat* worms the distance from the central gut to the lateral body margin exceeds the range of simple diffusion. As a result, large, flat turbellarians have gut ceca that transport nutrients via ciliary currents to the peripheral parts of the body. In this functional sense, the gut of large turbellarians is a gastrovascular transport system similar to that of cnidarians and ctenophores. Nutrients may also be transported intracellularly

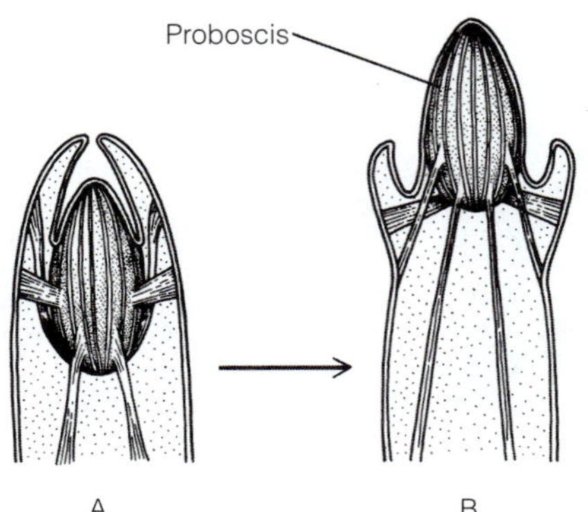

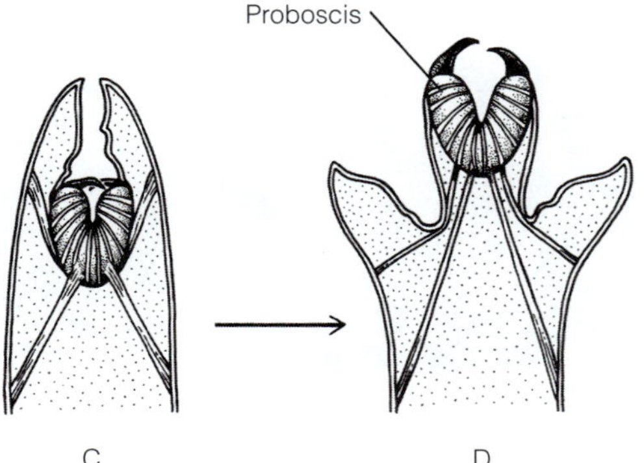

FIGURE 10-19 Turbellaria: proboscis. A protrusible proboscis occurs in several groups of turbellarians, but is particularly well developed in the kalyptorhynch rhabdocoels shown here. One type **(A and B)** is a muscular mass covered by a sticky surface to which prey adhere. Another form **(C and D)** is a grasping organ that in many cases bears hooks. *(Redrawn and modified from de Beauchamp, P. 1961. Classe des Turbellaries. In Grassé, P. -P. (Ed.): Traité de Zoologie. Vol. 4. Masson et Cie, Paris. p. 172.)*

from gastrodermal to fixed parenchymal cells via gap junctions. Convective circulation may occur in the fluid-filled hemocoel of freshwater catenulids.

Excretion

Turbellarians eliminate nitrogen from protein metabolism in the form of ammonia, which diffuses across the body surface, but they release excess water and probably other waste metabolites using protonephridia that bear multiciliated terminal cells (Fig. 10-20). Because there is no circulatory system to deliver excess water and other wastes from all parts of the body to a centralized kidney, turbellarian terminal cells typically are scattered widely throughout the body so that they are within diffusion distance of all tissues (Fig. 10-20A). (An exception occurs in the catenulids, which have a single anterior protonephridium suspended in a hemocoel with circulating blood.) The terminal cells then empty into anastomosing and sometimes ciliated ducts that eventually open to the exterior at one or more pores, depending on species. Unlike all other turbellarians, acoelomorphs, which are almost exclusively marine, lack nephridia.

Reproduction

REGENERATION

The phenomenon of regeneration has been studied intensively in the freshwater planarians and with good reason: The regenerative ability of planarians is extraordinary, matched only by cnidarians such as *Hydra*. As long ago as 1825, biologists realized that a planarian whose head had been bisected longitudinally would soon heal the cut surfaces and regenerate two complete heads. Similarly, if an animal was severed into two pieces, either transversely or longitudinally, each of the two halves would regenerate the missing parts and become a whole worm. In fact, a fragment as small as 1/300th of the body will regenerate an entire worm. Such experiments continue to provide classes with an introduction to planarian regeneration, but contemporary research centers on answering two questions: What controls the reestablishment of structural pattern and polarity? What is the source of the regenerating cells?

When a planarian is cut or wounded, the adjacent epidermis spreads over and seals the wound. A dome-shaped mass of neoblasts called a **blastema** then forms beneath the epidermis. Eventually, the missing parts of the body differentiate from blastema cells. What is the source of the blastema? The undifferentiated cells of the blastema may arise from differentiated cells, such as muscle cells, by a process of dedifferentiation, or reversion of the cell to its totipotent, embryonic, undifferentiated state, or from a permanent pool of totipotent, undifferentiated neoblast cells similar to sponge archeocytes or cnidarian interstitial cells. At present, there is uncertainty over which of these sources predominates in a regenerating turbellarian.

The control of regeneration is a complex and lengthy topic and only one classical example is given here. One of the earliest controls to be studied was regulation of head-tail polarity in regenerating transverse slices. If an intact worm is cut transversely into a series of slices of equal length, a new head always regenerates on the front end of each slice, but the new heads

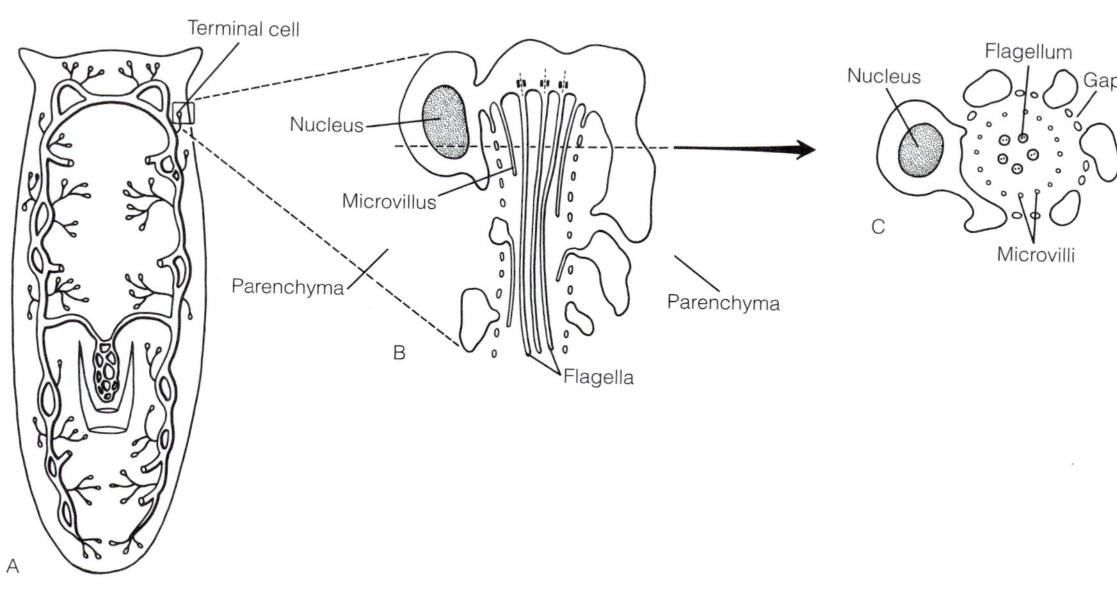

FIGURE 10-20 Turbellaria: protonephridia. **A,** The terminal cells of protonephridia in many turbellarians are scattered throughout the parenchyma to bring them within diffusion distance of all the tissues. **B,** Longitudinal and **C,** transverse sections of a typical terminal cell. *(A, Modified and redrawn after Ijima from Benham, W. B. 1901. The Platyhelmia, Mesozoa, and Nemertini. In Lankester, E. R. (Ed.): A Treatise on Zoology. Part 4. Adam and Charles Black, London. p. 20 ; B and C, Redrawn after McKanna from Rieger, R. M., Tyler, S., Smith, J. P. S. III, et al. 1991. Platyhelminthes: Turbellaria. In Harrison, F. W., and Bogitsh, B. (Eds.): Microscopic Anatomy of Invertebrates. Vol. 3. Wiley-Liss, New York. p. 82.)*

do not regenerate at the same rate on all the slices. Heads regenerate faster on slices from the anterior end of the worm than on those from the posterior end. Such experiments suggest that the factor (or factors) that controls the rate of head regeneration is distributed along the worm in an anterior-to-posterior concentration gradient (Fig. 10-21A), with more of the factor being present in the head and less in the tail. The concentration difference between any two points along the length of the worm may be the control of head-tail polarity. As long as a significant difference in concentration exists between the two ends of the slice, a head will regenerate on the end having the higher concentration. Thus there should be some minimum concentration difference below which head-tail polarity is lost. This minimum difference should correspond to a minimum distance that can be established experimentally by cutting progressively thinner transverse slices. Eventually, when a slice is too thin to contain a minimum concentration difference, polarity is lost and a head is regenerated at both ends (Fig. 10-21B). Such two-headed monsters are called Janus heads, after the Roman god of doors, who is depicted as a head with two opposite faces.

Modern research seeks the mechanisms that control regeneration, for example, the growth factors that might initiate regeneration, the Hox genes that control anterior-posterior polarity, and the region-specific markers (proteins) that may direct anterior-posterior and dorsal-ventral polarity. Once these control factors have been identified, it is hoped that some may have a therapeutic role in stimulating regeneration of, among other things, nervous tissue in humans with damaged spinal cords.

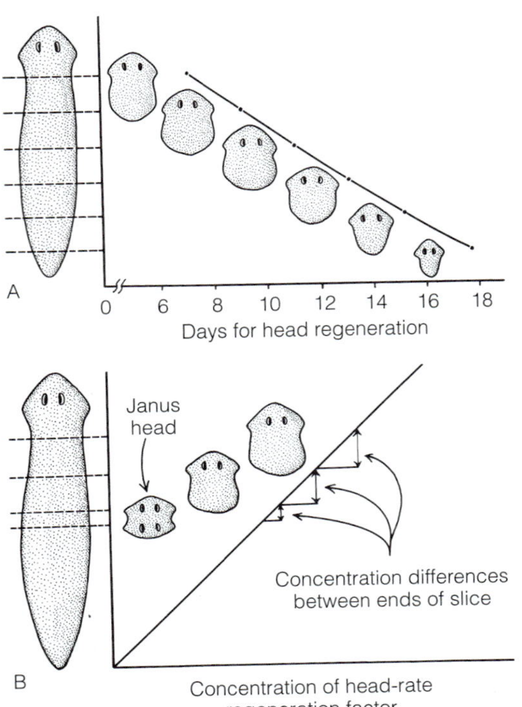

FIGURE 10-21 Turbellaria: regeneration. **A,** Regeneration rate of a planarian head at different body levels. **B,** The thickness of the slice affects maintenance of head-tail (anterior-posterior) polarity. Very thin slices lose polarity and form Janus heads. *(A, Adapted from Dubois, F. 1949. Contribution à l'étude de la régénération chez planaires dulcicoles. Bull. Biol. 83:213–283; B, Adapted from Goss, R. 1969. Principles of regeneration. Academic Press, New York.)*

CLONAL REPRODUCTION

Many turbellarians, especially freshwater species, reproduce clonally by means of transverse or, rarely, longitudinal fission and by budding (Fig. 10-22). The large freshwater planarians typically divide into two pieces and then regenerate missing parts after separation. The fission plane usually forms behind the pharynx and separation appears to depend on locomotion: The posterior end of the worm clings to the substratum while the anterior half continues to move forward until the two regions pull apart. Among species of *Catenula, Stenostomum, and Microstomum,* the parent's body differentiates into a chain of zooids before fission separates them into new individuals in a process called **paratomy** (par-AT-oh-me; Fig. 10-22A). This form of reproduction resembles strobilation in scyphomedusae.

If differentiation (or regeneration) occurs after fission, the process is known as **architomy** (are-KIT-oh-me). Architomy of the body into several fragments occurs in a few species of freshwater planarians, such as members of the genus *Phagocata,* and some land planarians (Fig. 10-22B). In *Phagocata,* each fragment forms a cyst in which regeneration takes place. Later, a small, complete worm emerges from the cyst.

Budding occurs primarily in acoels, such as *Convolutriloba,* which buds offspring from any lobe of its trilobed posterior end (Fig. 10-22C). The anterior-posterior polarity of the differentiating bud is 180° opposite that of the parent. The opposing ciliary movement of parent and bud causes the bud to break free.

Clonal reproduction may be controlled by day length and temperature. Freshwater planarians, for example, almost all of which inhabit temperate regions, reproduce clonally by fission during the summer and sexually during the fall under the stimulus of shorter day lengths and lower temperatures. Asexually reproducing laboratory cultures of *Catenula* have been maintained for as long as six years without the occurrence of sexual reproduction. Parthenogenesis is also an important reproductive strategy in most Catenulida.

The common planarian *Dugesia dorotocephala,* which is easily maintained in laboratory culture and has been studied extensively, undergoes fission only at night. During the day the brain produces a substance that inhibits fission, and the production of the inhibitor appears to be under photoperiodic control.

SEXUAL REPRODUCTION AND DEVELOPMENT

All turbellarians, except for a few specialized parasitic species, are hermaphrodites, reproducing by way of copulation and internal fertilization. Because turbellarians are small animals and because eggs have a lower natural-size limit of approximately 50 μm, turbellarians produce relatively few eggs, which are never recklessly spawned. Development is usually direct, but in many large-bodied marine species (Polycladida), successive

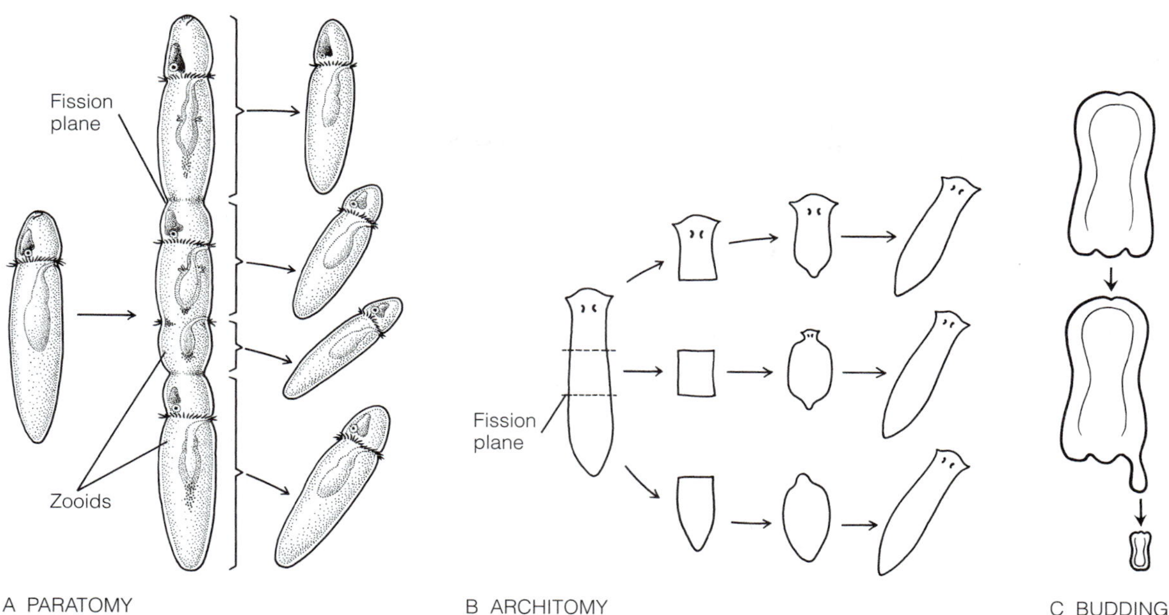

FIGURE 10-22 Turbellaria: clonal (asexual) reproduction. **A,** Paratomy in the catenulid *Catenula lemnae.* **B,** Architomy in a planarian. **C,** Budding in the acoel *Convolutriloba. (B, After Marcus from de Beauchamp, P. 1961. Classe des Turbellaries. In Grassé, P. -P. (Ed.): Traité de Zoologie. Vol. 4. Masson et Cie, Paris; C, From Tyler, S. 1999. Platyhelminthes. In Knobil, E., and Neill, J. D. (Eds.): Encyclopedia of Reproduction, Vol. 3. Academic Press, San Diego, pp. 901–908.)*

batches of numerous eggs develop indirectly into planktotrophic larvae.

Except in acoels and a few others (Fig. 10-23A), the gonads are saccate and enclosed in an epithelium and thereby separated from the surrounding parenchyma (Fig. 10-23B). The male and female systems are complex and variable, but can be generalized as follows. The male part of the system, transporting sperm *out,* consists of paired testes, each of which leads into a sperm duct, a seminal vesicle (a storage sac for its own sperm), and a penis (copulatory organ). The penis may be armed with a stylet and

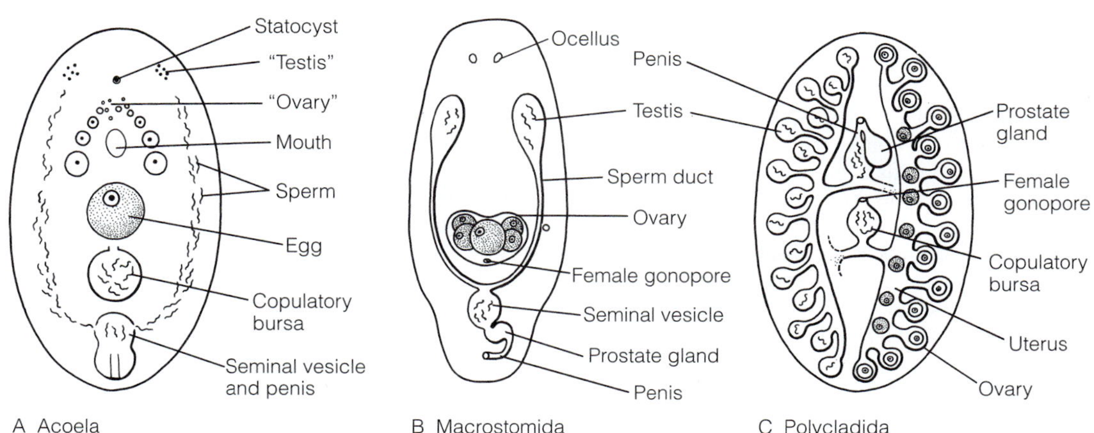

FIGURE 10-23 Turbellaria: reproductive system ("archoophora"). **A,** Acoela: discrete ovaries and testes are absent and germ cells mature in parenchyma. **B,** Macrostomida: small animals with ovaries, testes, and well-developed accessory reproductive organs. **C,** Polycladida: multiple ovaries and testes scattered between gut ceca (not shown); accessory organs are well developed. *(A, Modified and redrawn from Rieger, R. M., Tyler, S., Smith, J. P. S. III, et al. 1991. Platyhelminthes: Turbellaria. In Harrison, F. W., and Botish, B. (Eds.): Microscopic Anatomy of Invertebrates. Vol. 3. Wiley-Liss, New York. p. 98; B, Modified and redrawn from Schmidt, P., and Sopott-Ehlers, B. 1976. Interstitielle Fauna von Galapagos. XV. Macrostomum O. Schmidt, 1848 und Siccomacrostomum triviale nov. gen. nov. spec. (Turbellaria, Macrostomida). Mikrofauna Meeresbodens 57:1–44; C, Redrawn and modified after von Graff from Benham, W. B. 1901. The Platyhelmia, Mesozoa, and Nemertini. In Lankester, E. R. (Ed.): A Treatise on Zoology. Part 4. Adam and Charles Black, London. p. 37.)*

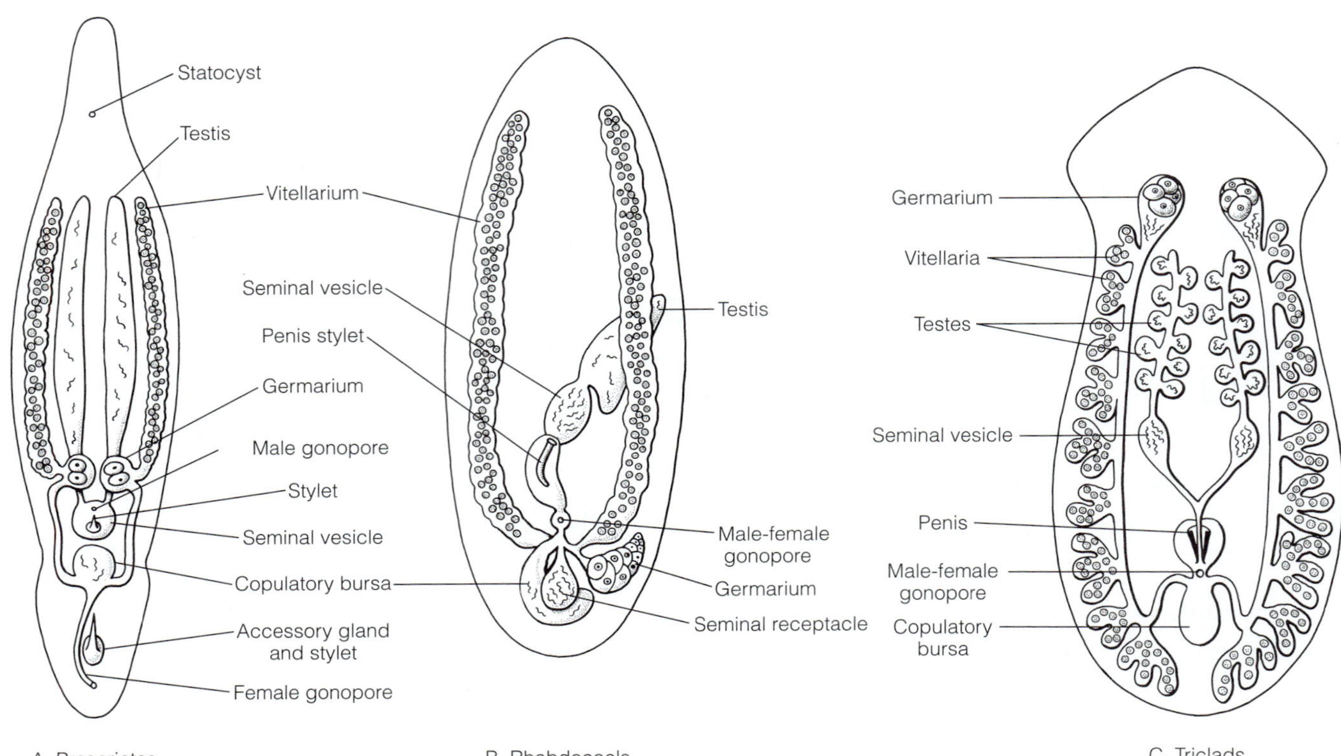

FIGURE 10-24 Turbellaria: reproductive system (Neoophora). Proseriate **(A)**, rhabdocoel **(B)**, and triclad **(C)** turbellarians with heterocellular ovaries. In all three examples, the vitellaria, which contain the yolk-forming vitellocytes, lie near the wall of the gut and its branches. Vitellaria, with their large volume and small vitellocytes, may be a specialization for rapid yolk synthesis. *(A, Modified and redrawn from Ax, P., and Ax, R. 1977. Interstitielle Fauna von Galapagos. XIX. Monocelidae (Turbellaria, Proseriata). Mikrofauna Meeresbodens 64:1–44; B, Modified and redrawn from Noldt, U., and Reise, K. 1987. Morphology and ecology of the kalyptorhynch* Typhlopolycystis rubra *(Platyhelminthes) an inmate of lugworm burrows in the Wadden Sea. Helgol. Wiss. Meeresunteers. 41:185–199; C, Modified and redrawn from de Beauchamp, P. 1961. Classe des Turbellaries. In Grassé, P. -P. (Ed.): Traité de Zoologie. Vol. 4. Masson et Cie, Paris. p. 83.)*

often receives secretions from a prostate gland. The female part of the system, through which sperm move *in* from the partner, is specialized as a gonopore (vagina), copulatory bursa, and seminal receptacle, the latter two for short- and long-term storage, respectively, of the partner's sperm (Fig. 10-23, 10-24). The female system also produces eggs and transports them *out* from paired ovaries via an oviduct to the gonopore. The eggs are oviposited singly or cemented down in masses (Fig. 10-27A,B). Some turbellarians have more than two testes or ovaries.

Another female accessory reproductive organ, the **uterus,** is used as a temporary storage sac for ripe eggs. The uterus may be a blind sac, as is the case in some rhabdocoels, or it may be merely a dilated part of the oviduct, as in polyclads (Fig. 10-23C). Most turbellarians, however, lack uteri because only a few eggs are laid at a time.

In acoels and catenulids, the female system is less well developed than in other turbellarians. Some have no female ducts at all, not even a gonopore (Fig. 10-23A). In others, there are no oviducts, but a short, blind vagina for receiving the penis leads from a female gonopore. Some zoologists believe that this condition in acoels is secondarily derived from a more complex system, whereas others consider it to be primitive. Separate male and female gonopores are characteristic of most macrostomids, many acoels, and the polyclads, and this is probably the primitive turbellarian condition. Many turbellarians, however, including the common planarians, have a single gonopore and genital atrium into which both male and female systems open.

Depending on taxon, turbellarians have either homocellular or heterocellular ovaries and archoophoran or neoophoran reproductive systems, respectively. The **homocellular ovary** of primitive turbellarians (= "Archoophora"; Fig. 10-23) produces eggs in which yolk is an integral part of each egg's cytoplasm **(entolecithal eggs),** similar to the eggs of most other animals (Fig. 10-25E). The **heterocellular ovary** of the derived Neoophora (= new egg bearers), however, is divided into two specialized regions, a **germarium** (or ovary proper) for the production of eggs and a **vitellarium** (or yolk gland) for the production of yolk-filled cells called **vitellocytes** (Fig. 10-24B,C; 10-25F). The germarium and vitellarium can be united into a **germovitellarium** or they may be separate organs linked together by a common duct (Fig. 10-24B). In either case, after

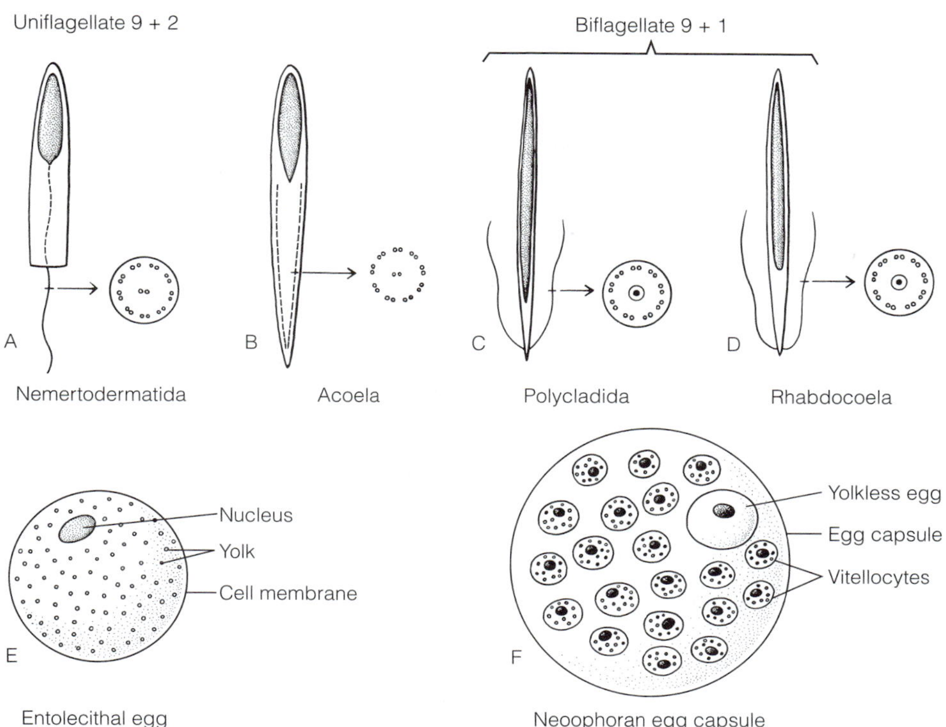

FIGURE 10-25 Turbellaria: sperm and eggs. **A,** Uniflagellate 9 + 2 sperm of the Nemertodermatida. **B,** Sperm of some Acoela without free flagella (two 9 + 2 axonemes incorporated into cell). **C** and **D,** Biflagellated sperm of Polycladida and Rhabdocoela, both with 9 + 1 axonemes. **E,** Entolecithal (yolk-containing) egg of archoophoran turbellarians. **F,** Encapsulated egg + vitellocytes (together an ectolecithal "egg") of neoophoran turbellarians (and Neodermata). In some examples, the capsule may contain several eggs and hundreds of vitellocytes. *(A-D, Redrawn from Hendelberg, J. 1986. The phylogenetic significance of sperm morphology in the Platyhelminthes. Hydrobiologia. 132:53–58; E and F, Combined and redrawn from several sources.)*

the yolkless egg is released from the germarium, it is surrounded by several vitellocytes and together egg and vitellocytes are enclosed in a **capsule** (Fig. 10-25F, 10-27C,D) secreted, in part, by a specialized region of the reproductive duct. The encapsulated egg and vitellocytes are sometimes called an **ectolecithal "egg."** In the remainder of this chapter, the term *egg* refers to the female reproductive cell (oocyte or ovum), whereas the term *"egg"* means the encapsulated ovum (or zygote or embryo) plus the vitellocytes.

The division of the heterocellular ovary into egg- and vitellocyte-producing regions may be an adaptation to increase the rate of yolk synthesis and thus the number of eggs provisioned or the amount of yolk allocated to each egg. During yolk synthesis in the vitellarium, the small yolk-precursor molecules enter the vitellocyte across its membrane. The rate at which the precursors enter is probably determined, in part, by the surface area of the cell. Because many small vitellocytes have a larger total surface area than one large, yolk-filled egg of equal volume, the vitellocytes should be able to synthesize yolk at a greater rate than a single large egg. Furthermore, by scattering the many vitellocytes throughout the body in vitellaria (Fig. 10-24C), the yolk-forming cells have access to nutrients provided by the entire surface of the gut and its ceca. At least two predictions follow from this untested hypothesis: Flatworms with a heterocellular ovary should produce eggs at a greater rate than worms with homocellular ovaries and entolecithal eggs, or they should provide more nutrition to each egg, resulting in larger-bodied juveniles. Either outcome—more offspring or larger, better-provisioned offspring—may enhance fitness. All parasitic platyhelminths have heterocellular ovaries and produce the large number of offspring necessary to locate their hosts.

With the exception of one taxon (Nemertodermatida), turbellarian sperm are highly modified for internal fertilization. Nemertodermatids have uniflagellate sperm with a 9 + 2 flagellar axoneme (Fig. 10-25A), as is typical of most other animals. The sperm of most turbellarians, however, are biflagellate. Acoel sperm are biflagellate, but the 9 + 2 flagellar axonemes are embedded in the body of the sperm and do not enter a tail (Fig. 10-25B). Each of the two sperm flagella of higher turbellarians (Trepaxonemata) has a 9 + 1 axoneme (Fig. 10-25C,D); the sperm of a few turbellarians lack flagella altogether.

Sperm transfer in turbellarians is by copulation and is usually reciprocal. In most, the penis is inserted into the female gonopore or common female-male gonopore of the partner. During copulation the worms orient themselves in a variety of ways with the ventral surfaces around the genital region pressed together and elevated (Fig. 10-26A,B). **Hypodermic impregnation** occurs in some acoels, macrostomids, rhabdocoels, and polyclads. The penis, which bears stylets, is pushed through the body wall of the copulating partner, depositing sperm into the parenchyma (Fig. 10-26B,C). The sperm then migrate to the ovaries. Self-fertilization probably occurs only in some exceptional species.

Except for the macroturbellarian polyclads, turbellarians with homocellular ovaries generally release only a few eggs at any one time (Fig. 10-27A). Polyclads lay their numerous eggs in gelatinous strings or masses (Fig. 10-27B), and one individual may lay several egg masses in succession. Although vitellaria are absent, the fecundity of polyclads probably results from large

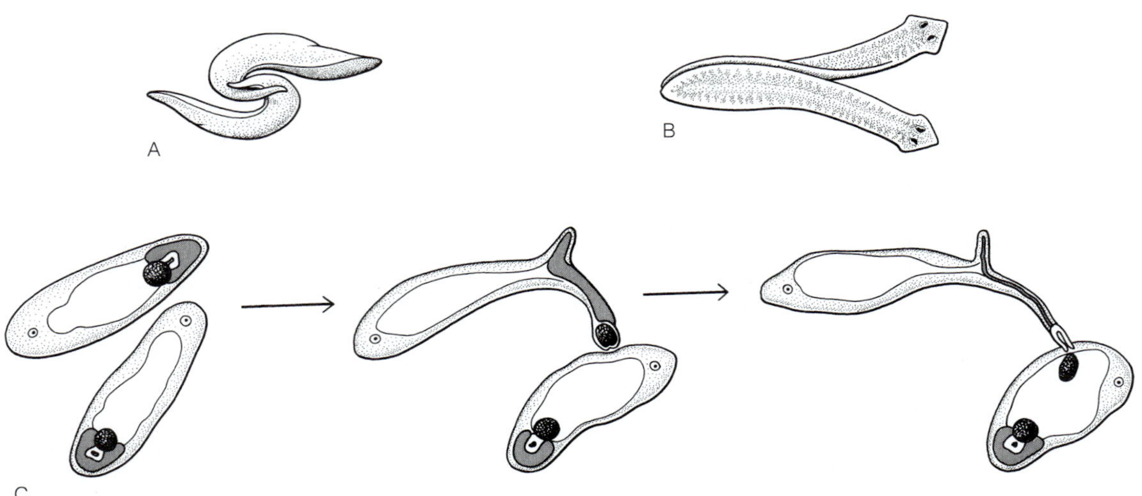

FIGURE 10-26 Turbellaria: copulation. **A,** The acoel *Polychoerus carmelensis;* **B,** The freshwater triclad *Dugesia;* **C,** The acoel *Pseudaphanostoma psammophilum.* *(A, Redrawn from Costello, H. M., and Costello, D. P. 1938. Copulation in the acoelous turbellarian* Polychoerus carmelensis. *Biol. Bull. 75:85–98; B, Redrawn from Hyman, 1951; C, Redrawn from Apelt, G. 1969. Fortpflanzungsbiologie, Entwicklungszyklen und vergleichende Frühentwicklung acoeler Turbellarien. Mar. Biol. 4:267–325.)*

body size and division of the ovary into numerous small follicles, each of which is supplied with nutrients from its own gut cecum (Fig. 10-23C). The eggs of many polyclads also develop into planktotrophic larvae, and thus the parent may not invest as much energy in each egg as would a direct-developing flatworm. Acoels lack oviducts and release their eggs through the mouth or by temporary rupture of the body wall (Fig. 10-27A).

Neoophoran turbellarians, including planarians, typically deposit their eggs and vitellocytes in capsules attached to the substratum. Each capsule may house one or more eggs. In many species, including *Dugesia,* each oval, brownish capsule is attached by a slender stalk and resembles a balloon (Fig. 10-27C,D). One worm can produce several capsules and many embryos may develop in each. The protective egg capsules of neoophoran turbellarians may have preadapted them to resist the digestive enzymes of grazing animals, thus setting the stage for the adoption of endoparasitism.

Some freshwater turbellarians produce two types of "eggs": **summer "eggs,"** which are enclosed in a thin egg capsule and hatch in a relatively short period, and resting "eggs," which have a thicker and more resistant capsule (Fig. 10-28). Resting "eggs" are usually produced in the fall, remain dormant during the winter, resist freezing and drying, and hatch in the spring with the rise in water temperature. In *Mesostoma ehrenbergii,* the generation time is between 16 and 75 days, depending on water temperature, and one individual produces about 15 summer "eggs" or 45 resting "eggs." The life span is between 65 and 140 days.

Early development of entolecithal eggs is by spiral cleavage like that occurring in other spiralian taxa (Fig. 10-29A,B,

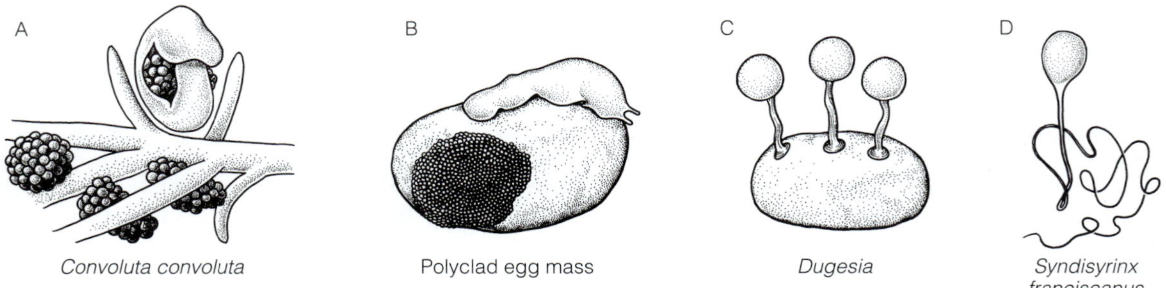

FIGURE 10-27 Turbellaria: oviposition. **A,** Egg masses attached to algae by the marine acoel *Convoluta convoluta.* **B,** Egg mass attached to a stone by a marine polyclad. **C,** Egg capsules ("eggs") attached to a stone by the freshwater triclad *Dugesia.* **D,** Egg capsule ("egg") released into the intestine of a sea urchin by the parasitic marine rhabdocoel *Syndisyrinx franciscanus.* *(A, Redrawn from Apelt, G. 1969. Fortpflanzungsbiologie, Entwicklungszyklen und vergleichende Frühentwicklung acoeler Turbellarien. Mar. Biol. 4:267–325; C, Redrawn from Pennak, 1978; D, Drawn from a photograph in Shinn, G. L., and Cloney, R. A. 1986. Egg capsules of a parasitic flatworm: Ultrastructure of hatching sutures. J. Morphol. 188:15–28.)*

FIGURE 10-28 Turbellaria: resting "eggs." Thick-walled resting "eggs" in the freshwater rhabdocoel *Mesostoma lingua*. *(From Mac-Fira, V. 1974. The turbellarian fauna of the Romanian littoral waters of the Black Sea and its annexes. In Riser, N. W., and Morse, M. P. (Eds.): Biology of the Turbellaria. McGraw-Hill Book Co., New York. p. 263.)*

9-22B). Gastrulation is by epiboly and produces a stereogastrula. The mouth and pharynx form from an ectodermal invagination near the site of the closed blastopore. This invagination joins the archenteron, which is a solid mass that later becomes hollow. Mesoderm originates, as mesenchyme, from both ectoderm and endoderm. It migrates into the region between the two primary germ layers and later gives rise to the musculature, parenchyma, and germ cells (Fig. 10-29A,B).

Turbellarians in general have direct development, but some polyclads produce uniformly ciliated, planktotrophic larvae: **Goette's larvae** with four arms and **Mueller's larvae** with eight (Fig. 10-30). Cilia on the arms are longer than those elsewhere on the body and are used in locomotion and feeding, like the prototroch of trochophore larvae (Chapter 12). Anteriorly, there is a ciliary tuft on an apical organ. The larvae swim about for several days, gradually absorb their arms, and then settle to the bottom as young worms.

The developmental pattern of neoophoran flatworms probably evolved from spiral cleavage, but in most species the presence of vitellocytes has altered the cleavage pattern of the egg, disguising its original form (Fig. 10-29C). Neoophoran turbellarians lack a free-swimming larva, development is direct, and the young worms emerge from the capsule in a few weeks.

Diversity of Turbellaria

In traditional classifications, Turbellaria (free-living flatworms) is a class of Platyhelminthes containing numerous orders, such as Acoela, Tricladida, and Polycladida, all of equal rank. Modern cladistic analysis preserves these groups, but not with equal ranks (Fig. 10-32). The "class" Turbellaria, moreover, is a paraphyletic taxon characterized solely by plesiomorphic traits such as a ciliated epidermis, compact body, or free-living (nonparasitic) lifestyle.

Homocellular ovaries and entolecithal eggs typify the primitive taxa of Turbellaria, such as Catenulida, Acoelomorpha, Macrostomorpha, and Polycladida, which previously were united in the paraphyletic "Archoophora" (= primitive egg bearers). The remaining flatworms, including the parasitic taxa, share the apomorphy of heterocellular gonads and ectolecithal eggs and are Neoophora (= new egg bearers). Most of the taxa listed below are the well-defined, longstanding Linnean orders, the working units of turbellarian systematists. Cladistic analyses of Turbellaria are recent, and many of the new cladistic ranks are supported by few or questionable apomorphies. Some of these higher taxa are indicated below without a Linnean superscript and ranked in the Phylogenetic Hierarchy of Platyhelminthes and in Figure 10-32.

Catenulida^O: Small body usually long and slender (Fig. 10-15A). Density of epidermal cilia low (two per cell) as compared with remaining turbellarians. Head commonly has a statocyst *(Catenula)* with one statolith (uniquely behind brain), two ciliated pits (*Stenostomum;* Fig. 10-12D), and often two colorless eyes. Internally, head of some *(Stenostomum, Rhynchoscolex)* houses a hemocoel (Fig. 10-13C) in which protonephridial terminal cells are suspended. Nephridioduct extends posteriorly and opens on tail. Ventral (or dorsal) mouth opens into simple pharynx that leads into ciliated, blind, tubular gut. Gonads are unpaired, anterior and dorsal, male gonopore opens dorsally above pharynx. Sperm lack flagella. Oviducts absent. Many reproduce clonally by paratomy, and individuals composed of chains of zooids are not uncommon (Fig. 10-22A). Marine, common in freshwater lakes and ponds. *Catenula, Stenostomum* (Fig. 10-2C), *Rhynchoscolex.*

Euplatyhelminthes: Sister taxon of Catenulida, including all other turbellarians. High density of epidermal cilia; primitively with frontal glands.

Acoelomorpha: Epidermal ciliary rootlets interjoined to form skeletal network. Epidermal cilia have step or shelf near their truncated tips (Fig. 10-7A). Extracellular matrix is nearly absent (Fig. 10-3A, 10-13A); protonephridia are absent. Includes Nemertodermatida and Acoela.

Nemertodermatida^O: Small; resemble acoels, but have uniflagellate sperm (Fig. 10-25A), gut with lumen lined by cellular, nonciliated gastrodermis, statocyst with two statoliths (Fig. 10-2B). Marine. *Nemertoderma.*

Acoela^O: Small, usually less than 2 mm in length. Statocyst with one statolith (Fig. 10-11A, 10-12A). Pharynx usually absent (zero pharynx); mouth opens into nonglandular central syncytium; cellular gut absent (Fig. 10-13A, 10-14F). Gonads lack epithelium; individual germ cells in parenchyma (Fig. 10-23A). Sperm biflagellate, flagellar shafts not free, instead embedded in the body of the sperm. Oviducts absent, eggs released from mouth or body-wall rupture. Spiral cleavage by duets (Fig. 10-29A) rather than quartets as in other spiralians. Common, marine; found under rocks, among algae, in mud, interstitially in sand, several species pelagic, few in fresh water (*Oligochoerus;* Fig. 10-2A). Some species

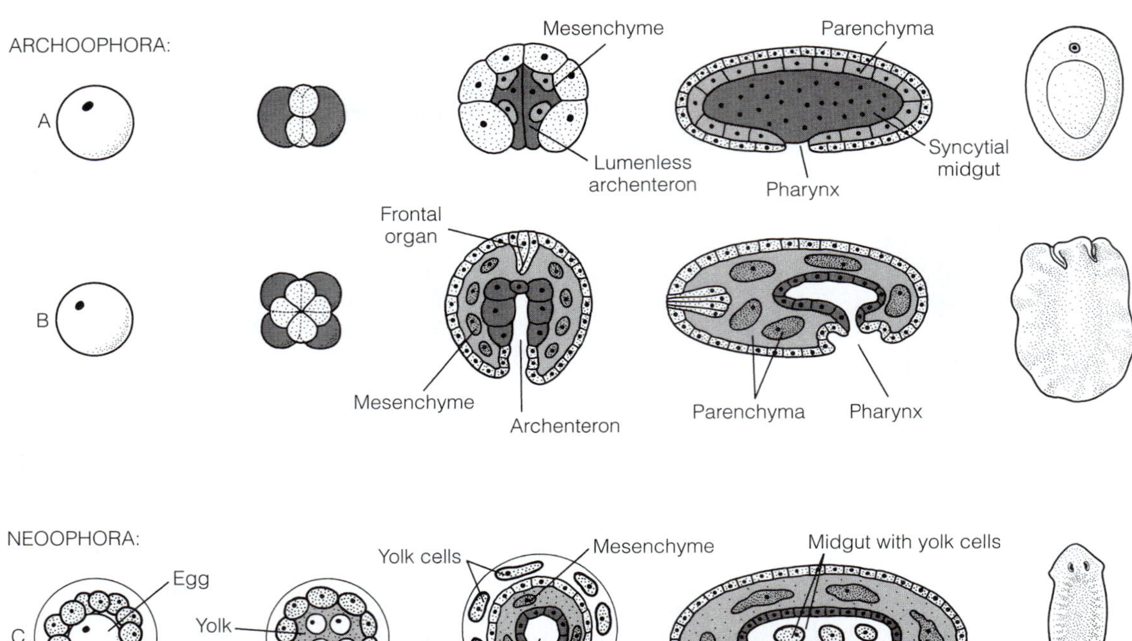

FIGURE 10-29 Turbellaria: development (cleavage, gastrulation, and mesoderm segregation). **A,** Spiral cleavage by duets and development in Acoela. **B,** Spiral cleavage by quartets (typical of other spiralians) and development in Polycladida. **C,** Modified cleavage and development in Neoophora—in this example in a triclad, *Dendrocoelum* **(C)**—in which some of the vitellocytes liberate their yolk into the embryonic mass (later the parenchyma) and others are ingested by the archenteron. Mesenchyme is the embryonic precursor of the parenchyma. *(A, Modified and redrawn after Bresslau from de Beauchamp, P. 1961. Classe des Turbellaries. In Grassé, P. -P. (Ed.): Traité de Zoologie. Vol. 4. Masson et Cie, Paris. p. 183; B, Modified and redrawn after Kato from Thomas, M. B. 1986. Embryology of the Turbellaria and its phylogenetic significance. Hydrobiologia 132:105–115; C, Modified and redrawn after Hallez from Korschelt, E., and Heider, K. 1895. Textbook of the Embryology of Invertebrates. Part 1. Sawn Sonnenschein & Co., New York. pp. 170–172.)*

commensal in intestine of various echinoderms. *Amphiscolops, Archaphanostoma, Convoluta, Diopisthoporus, Paratomella, Polychoerus, Symsagittifera.*

Rhabditophora: Sister taxon of Acoelomorpha. With rhabdites (Fig. 10-5) and duo-gland adhesive structures (Fig. 10-7B), primitive with simple pharynx surrounded by nerve ring near mouth. Includes all remaining turbellarians.

Macrostomorpha: Sister taxon of Trepaxonemata; includes Macrostomida and Haplopharyngida; paired lateral nerve cords joined behind mouth by a commissure.

Macrostomida^O: Small; paired ocelli; simple, saclike ciliated gut; simple pharynx (Fig. 10-2D, 10-14, 10-17, 10-23B). One pair of ventrolateral nerve cords with nerve commissure behind mouth. Sperm lack flagella. *Microstomum* reproduces asexually by paratomy, forms chains of zooids (similar to catenulid in Fig. 10-22A). Marine and fresh water. *Macrostomum, Microstomum.*

Haplopharyngida^O: Small, similar to macrostomids, but with proboscis and temporary anus (Fig. 10-7B). Marine. *Haplopharynx.*

Trepaxonemata: Sister taxon of Macrostomida plus Haplopharyngida (Macrostomorpha). With biflagellate sperm, the flagella of each with 9 + 1 axoneme (Fig. 10-25C,D). All have a muscular protrusible pharynx (Fig. 10-16B,C). Includes all remaining turbellarians.

Polycladida^O: Large (centimeter-size range; largest 30 cm), marine. More or less oval body highly flattened, tissue thin (Fig. 10-1E, 10-5). Primarily benthic (Fig. 10-9B), but many species swim by undulating body margins (Fig. 10-10). Often have one pair of anterior marginal, or dorsal, chemosensory tentacles. Eyes numerous over brain, sometimes also along body margin, in tentacles (Fig. 10-1E, 10-12B). Ventral attachment sucker in Cotylea^sO (Fig. 10-7C), absent in Acotylea^sO.

Long central gut with many ceca (Polycladida = many branched; Fig. 10-14E). Plicate pharynx either anteriorly directed tube or pendant from roof of pharyngeal cavity (Fig. 10-14E, 10-18). Parenchyma well developed (Fig. 10-13B). Testes and homocellular ovaries divided into small follicles scattered throughout body among gut branches (Fig. 10-23C). Entolecithal eggs with spiral cleavage by quartets (Fig. 10-29B). Planktotrophic Mueller's and Goette's larvae in many species (Fig. 10-30).

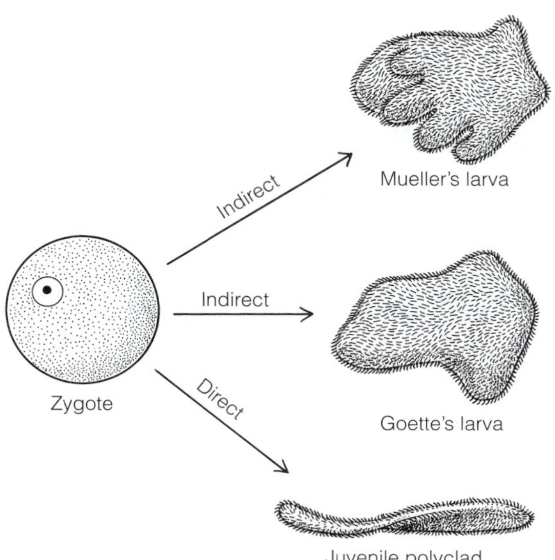

FIGURE 10-30 Turbellaria: direct development and larval development in flatworms with homocellular ovaries ("archoophora"). The indirect development of many Polycladida results in either planktotrophic Mueller's or Goette's larvae. During metamorphosis, these free-swimming larvae reabsorb their arms and soon resemble the juvenile at the bottom of the figure. *(Redrawn from Ruppert, E. E. 1978. A review of metamorphosis of turbellarian larvae. In Chia, F. -S., and Rice, M. E. (Eds.): Settlement and Metamorphosis of Marine Invertebrate Larvae. Elsevier/North-Holland, Inc., New York. pp. 65–82.)*

Brightly colored polyclads, which are stunningly beautiful, feed in plain view on sea squirts, other sessile invertebrates. Bright colors probably aposematic, warning would-be predators. Some species poisonous with tetrodotoxin or staurosporine; others mimic toxic nudibranch molluscs (sea slugs) and gain protection by association. *Stylochus frontalis* (oyster leech), southeastern United States, grazes on flesh of living oysters as they gape open to feed. *Gnesioceros, Leptoplana, Notoplana, Pseudoceros, Stylochus.*

Neoophora: Sister taxon of Polycladida. Flatworms (including parasitic platyhelminths) with heterocellular ovary (Fig. 10-24). Ovary divided into two regions, egg-producing germarium (ovary) and yolk-cell- (vitellocyte-) producing vitellarium (Fig. 10-24); ectolecithal "eggs" (Fig. 10-25F) with some egg-capsule components contributed by glands associated with egg-laying duct. Development greatly modified from ancestral spiral pattern (Fig. 10-29C).

Lecithoepitheliata^O: Each egg surrounded by an epithelium-like monolayer of yolk cells. Germarium and vitellarium combined in a mixed germovitellarium. Mouth and complex pharynx are at anterior end of body; the gut is simple, unbranched (Fig. 10-15B). Marine and freshwater species. In freshwater *Geocentrophora* and *Prorhynchus,* male copulatory apparatus, which bears a stylet, opens into oral cavity. Slender, white *Prorhynchus stagnalis* is one of the most common freshwater flatworms.

Prolecithophora^O: Small; plicate or bulbous pharynx; simple gut, sometimes with ruffled outline. Heterocellular ovary divided spatially into separate germarium and vitellarium. In many species, but not in *Plagiostomum* or *Hydrolimax,* the male copulatory organ, which usually lacks a stylet, opens into the oral cavity (*Urostoma;* Fig. 10-8); sperm lack flagella. Marine and fresh water.

Seriata: Plicate pharynx an elongate tube in longitudinal body axis (Fig. 10-2E, 10-31A). Testes and vitellaria divided into follicles (Fig. 10-24C). Includes Proseriata, Tricladida.

Proseriata^O: Small; statocyst with one statolith. Pharynx plicate and tubular, but gut unbranched. Mostly marine, many interstitial (Fig. 10-2E). *Otoplana* common, interstitial, fast-moving species in swash zone of high-energy beaches. *Monocelis, Nemertoplana.*

Tricladida^O: Large; tubular gut branches into three parts—one anterior, two posterior (Tricladida = three branched; Fig. 10-14D, 10-31A). Each of three branches has ceca. Three main gut branches unite in midbody, anterior to mouth and pharynx, to form common chamber where pharynx and mouth are located. Triclad embryos uniquely develop a transitory pharynx for feeding on yolk cells. Among marine species, *Bdelloura* (Fig. 10-1C) is commensal on book gills of horseshoe crabs. Freshwater planarians include *Dendrocoelum, Dugesia* (Fig. 10-1B), *Phagocata, Planaria, Polycelis, Procerodes, Procotyla,* and *Rimacephalus. Dugesia* spp., widely studied in classrooms and research labs, is probably the best-known turbellarian (Fig. 10-9C, 10-11C, 10-21, 10-24C, 10-27B, 10-31). Land planarians include *Bipalium, Geoplana, Orthodemus.*

Rhabdocoela^O: Small; bulbous pharynx (Fig. 10-2F,G, 10-14B, 10-16C), simple gut, one pair of nerve cords. Marine and fresh water. Includes remaining taxa.

Typhloplanoida^sO: Mouth and bulbous pharynx near middle of body (Fig. 10-2G); pharyngeal axis dorsoventral, pharynx (dorsal view) resembles a doughnut. Kalyptorhynchia: mostly marine, interstitial species with anterior terminal proboscis (Fig. 10-19). The protrusible proboscis often bears hooks or teeth and is used to catch prey; mouth and pharynx anterior to middle of body. Marine, fresh water. *Gnathorhynchus, Gyratrix, Mesostoma* (Fig. 10-28).

Dalyellioida^sO: Marine and fresh water, some free-living *(Dalyellia, Provortex),* others commensal or parasitic on and in snails, clams, sea urchins, and sea cucumbers (for example, *Anoplodiera, Kronborgia* (Fecampiidae), *Syndesmis*).

Temnocephalida^sO: Posterior ventral surface differentiated into adhesive disc, anterior margin bears fingerlike projections by which worm creeps leechlike on its host (Fig. 10-9I). Syncytial epidermis (Fig. 10-3D). Commensal and parasitic on crustaceans, molluscs, and turtles (Fig. 10-1D, 10-2H). *Temnocephala.*

Neodermata: Parasitic flukes and tapeworms; to be discussed later in this chapter.

Phylogeny of Turbellaria

Few evolutionary questions have prompted more speculation and controversy than the origin of Turbellaria. Although several theories propose to trace their ancestry, only two are actively discussed by current researchers. The first is the **planula theory,** which was devised by L. von Graff in the nineteenth century and championed in the United States by the late Libbie Hyman. It states that the turbellarians and cnidarians arose from a common ancestor, the planuloid, which resembled a cnidarian planula larva. The planuloid ancestor would have had an outer

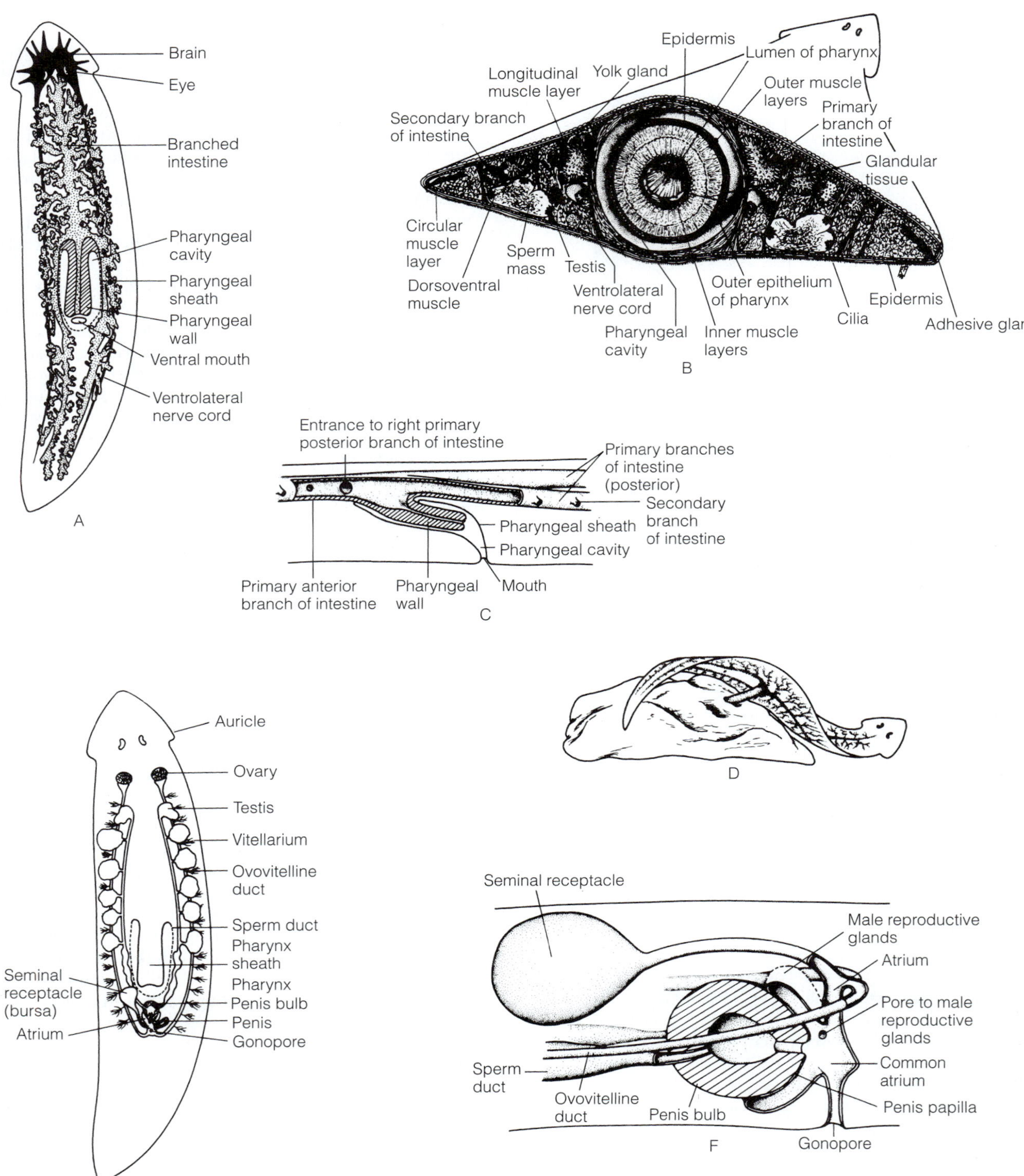

FIGURE 10-31 Turbellaria: anatomy. The common freshwater (and classroom) planarian *Dugesia.* **A,** Dorsal view of digestive and nervous systems. Commissures between cords are not shown. **B,** Cross section of the body at the level of the pharynx and gut. **C,** Vertical section through the pharynx at its junction with the gut. **D,** Lateral view of the pharynx protruded into food mass. **E,** Dorsal view of the reproductive system. **F,** Vertical cutaway view of the reproductive system. *(From several sources)*

epidermal epithelium, an inner gastrodermal epithelium, and connective tissue (mesoglea) in between. According to this theory, the flatworm gut and parenchyma correspond to the coelenterate coelenteron and mesoglea, respectively.

An alternative idea is that turbellarians are not primitive bilaterians, but instead have evolved from a coelomate ancestor by anatomical simplification. This hypothesis, the **coelomate theory** (part of a broader enterocoel theory), receives indirect support from the complexity of the turbellarian reproductive system and epidermis, neither of which has a counterpart in diploblastic animals. The ciliated epidermal cells of turbellarians, for example, always bear many cilia each, whereas those of diploblastic animals primitively and predominantly have only one cilium per cell.

Although it is relatively easy to imagine how an organism like a planula larva could evolve into a simple turbellarian by the addition of complexity, it is more difficult to see how a coelomate animal such as a segmented worm (annelid) or an acorn worm (hemichordate) could, by simplification, transform into an acoelomate animal. Modern proponents of this latter theory believe that reduction in body size could have been the crucial factor in such a transformation. It has been noted already that small body size is correlated with ciliary rather than muscular modes of locomotion, an absence of internal fluid transport systems (including a coelom), and the occurrence of protonephridia (Chapter 9). The actual disappearance of a coelom could occur by a filling in of the coelomic cavity with enlarged mesothelial cells to form a cellular parenchyma (Fig. 9-11F,G).

Another possibility is that the Turbellaria evolved from a coelomate ancestor not by reduction, but by pedomorphosis—that is, from an early developmental stage of such a coelomate ancestor before differentiation of the coelom. Larvae of coelomates are acoelomate (compact), similar to flatworms, and if they developed reproductive organs precociously, before the coelom differentiated, a flatwormlike animal would result (Fig. 9-23F).

On the other hand, advocates of the planula theory rationalize some of the structural discrepancies between turbellarians and diploblastic animals by suggesting that the acoelomate and monociliated Gnathostomulida (Chapter 23) forms an evolutionary bridge between the two taxa.

The uncertain systematic position of Turbellaria (and Platyhelminthes) has so far eluded molecular systematics and computer-based numerical taxonomy. Some studies place flatworms at or near the origin of Bilateria whereas others, based on the 18S ribosomal gene, position only the Acoela or Nemertodermatida in such a basal position as a separate taxon, leaving the remaining Platyhelminthes as a derived taxon of spiralians.

A phylogeny of platyhelminth taxa, based on morphology, is shown in Figure 10-32.

PHYLOGENETIC HIERARCHY OF PLATYHELMINTHES

- Platyhelminthes
  - Catenulida
  - Euplatyhelminthes
    - Acoelomorpha
      - Nemertodermatida
      - Acoela
    - Rhabditophora
      - Macrostomorpha
        - Macrostomida
        - Haplopharyngida
      - Trepaxonemata
        - Polycladida
        - Neoophora
          - "Lecithoepitheliata"*
          - "Prolecithophora"
          - Seriata
            - Proseriata
            - Tricladida
          - Rhabdocoela
            - "Typhloplanoida"
            - Doliopharyngiophora
              - "Dalyellioida"
              - Neodermata

NEODERMATA

Few experiences are more memorable than firsthand encounters with parasitic flatworms, such as pulsating tapeworm proglottids emerging from a cat's anus or a large fluke in the liver of a deer. The worms elicit feelings of revulsion but engage our morbid curiosity, and deserve admiration for their adaptive success. These parasites, which include flukes, tapeworms, and their relatives, are biologically remarkable animals. Most are endoparasites and many colonize a succession of different hosts in different environments before they complete their life cycle. Not only must they adjust physiologically to life in each of these abiotic and host environments, but also locate and infect each new host. Their fascinating biology is related to how they meet these challenges. From the standpoint of human ecology, these worms attack and debilitate economically important wildlife and domestic animals, and humans themselves. Schistosomiasis, a fluke affliction of 300 million humans, is the third most common parasitic disease of humans, behind malaria and hookworm infections.

The parasitic flatworms inherited the sucking, bulbous pharynx from free-living turbellarian ancestors and evolved a new trait that helped adapt them to endoparasitism. That new trait is the partial or complete replacement of the cellular epidermis by a new, nonciliated, syncytial layer called **neodermis,** or tegument (Fig. 10-33). During development, the embryonic epidermis is sloughed off and then replaced from below by necklike extensions of parenchymal neoblasts. The neoblast necks penetrate the epidermal basal lamina, spread over its outer surface, and fuse laterally with the spreading necks of other neoblasts to form the syncytial neodermis. The neoblast cell bodies, called **cytons,** each containing a nucleus, remain below the basal lamina in the parenchyma. Thus the neodermis encloses the body in a syncytium, which extends via its cytons deep into the parenchyma.

The functional attributes of a syncytium are related to its structure. A syncytial layer, because it is one multinucleate cell, lacks intercellular spaces. The absence of intercellular spaces means that substances cannot slip unregulated between cells to

*The names of paraphyletic taxa are enclosed in quotation marks.

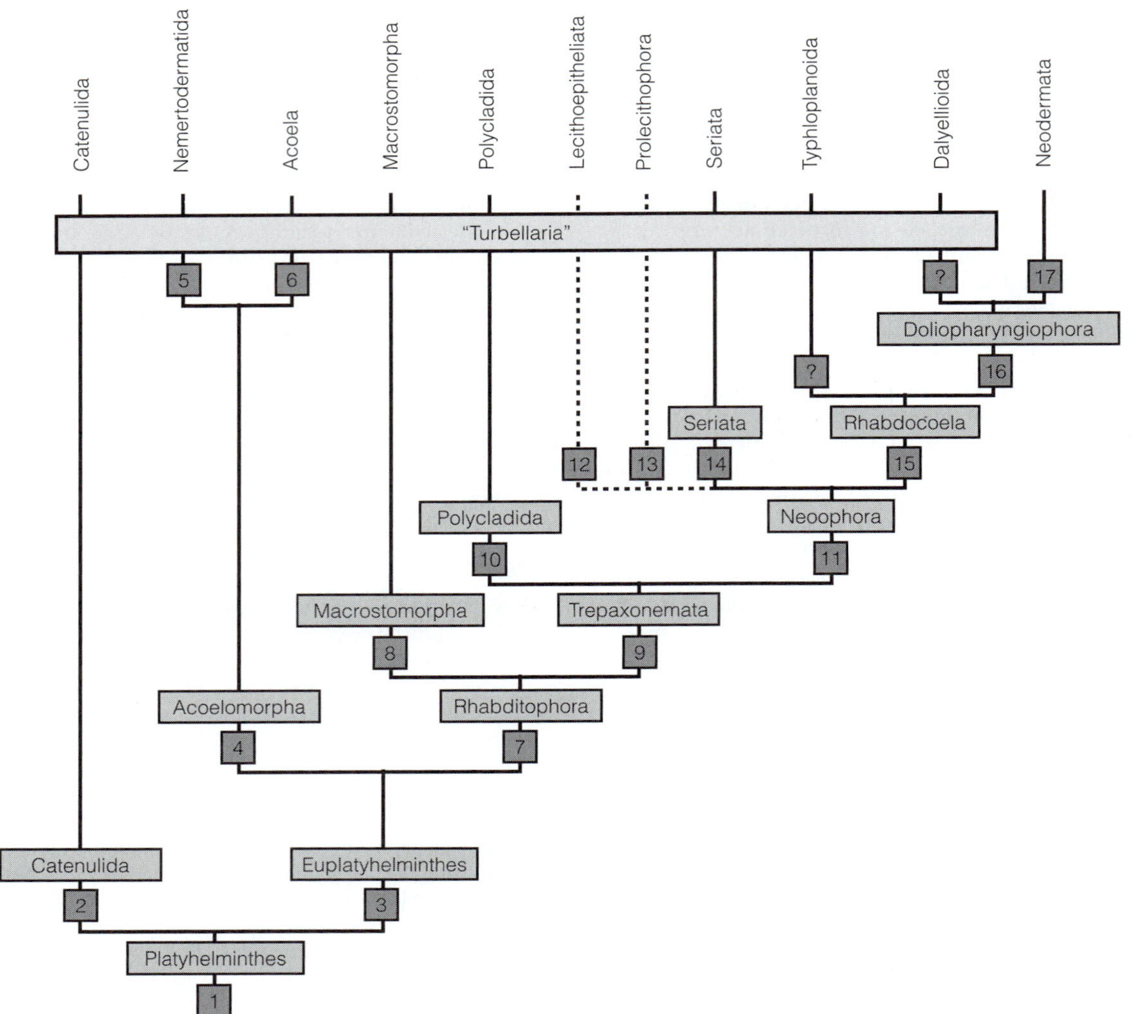

FIGURE 10-32 Phylogeny of Platyhelminthes with emphasis on Turbellaria. **1, Platyhelminthes:** multiciliated epidermis and gastrodermis; each protonephridial terminal cell has two cilia. **2, Catenulida:** mid-dorsal testes and anterior mid-dorsal male gonopore; single mid-dorsal protonephridium, protonephridial terminal cell anterior in head, nephridiopore posterior; head has well-developed interstitial spaces or hemocoel; nonmotile sperm lack flagella. **3, Euplatyhelminthes:** concentration of gland cells is a frontal "organ." **4, Acoelomorpha:** protonephridia absent; ciliary rootlets of epidermal cells interconnected to form a fibrous skeletal network; each epidermal cilium has a step or shelf near its tip. **5, Nemertodermatida:** statocyst bears a bilateral pair of statoliths, seven cells line the statocyst capsule; sperm has a single flagellum (a plesiomorphy). **6, Acoela:** statocyst has one lens-shaped statolith and two cells lining the capsule; gland cells absent from gut, gut partly or completely syncytial and lacking a lumen (food digested intrasyncytially); basal lamina basement membrane (extracellular matrix) nearly absent (correlated with presence of highly interconnected ciliary rootlets); spiral cleavage by duets. **7, Rhabditophora:** duo-gland adhesive system, rhabdites; terminal cell of protonephridia has four cilia. **8, Macrostomorpha:** paired ventrolateral nerve cords joined post-orally by a prominent commissure; sperm lack flagella. **9, Trepaxonemata:** sperm have two flagella each, the two central singlets of each sperm axoneme (9 + 2 pattern) are replaced by an unpaired axial rod (9 + 1 arrangement); muscular protrusible pharynx. **10, Polycladida:** large (centimeter range), dorsoventrally flattened body; extensively branched gut reaches all parts of body; testes and ovaries subdivided into many follicles and dispersed throughout the body (supplied by the gut ceca); well-developed parenchyma. **11, Neoophora:** heterocellular ovary (germovitellarium), ectolecithal "eggs." **12** and **13:** These two taxa have a germovitellarium in which the oocytes and vitellocytes are intermixed (a so-called "mixed gonad," presumably a plesiomorphy). **12, Lecithoepitheliata:** vitellocytes form an epithelial cortex over the oocyte. **13, Prolecithophora:** sperm lack flagella. **14, Seriata** (includes the Proseriata and Tricladida): original plicate pharynx modified into a muscular tube that when retracted lies in the long axis of the body with its opening directed posteriorly; testes and vitellaria subdivided into follicles and scattered in the body. **15, Rhabdocoela:** bulbous pharynx. **16, Doliopharyngiophora:** specialized bulbous pharynx located anteriorly with terminal opening. **17, Neodermata:** during development, replacement of ciliated epidermis by nonciliated syncytial neodermis (tegument) derived from neoblasts. *(Based on Ehlers, U. 1986. Comments on a phylogenetic system of the Plathelminthes. Hydrobiologia 132:1–12; Ax, P. 1996. Multicellular Animals: A New Approach to the Phylogenetic Order in Nature. Vol. 1. Springer, Berlin. 255 pp.; Tyler, S. 1999. Systematics of flatworms: Libbie Hyman's influence on current views of the Platyhelminthes. American Museum Novitates No. 3277. American Museum of Natural History, New York. pp. 52–66.)*

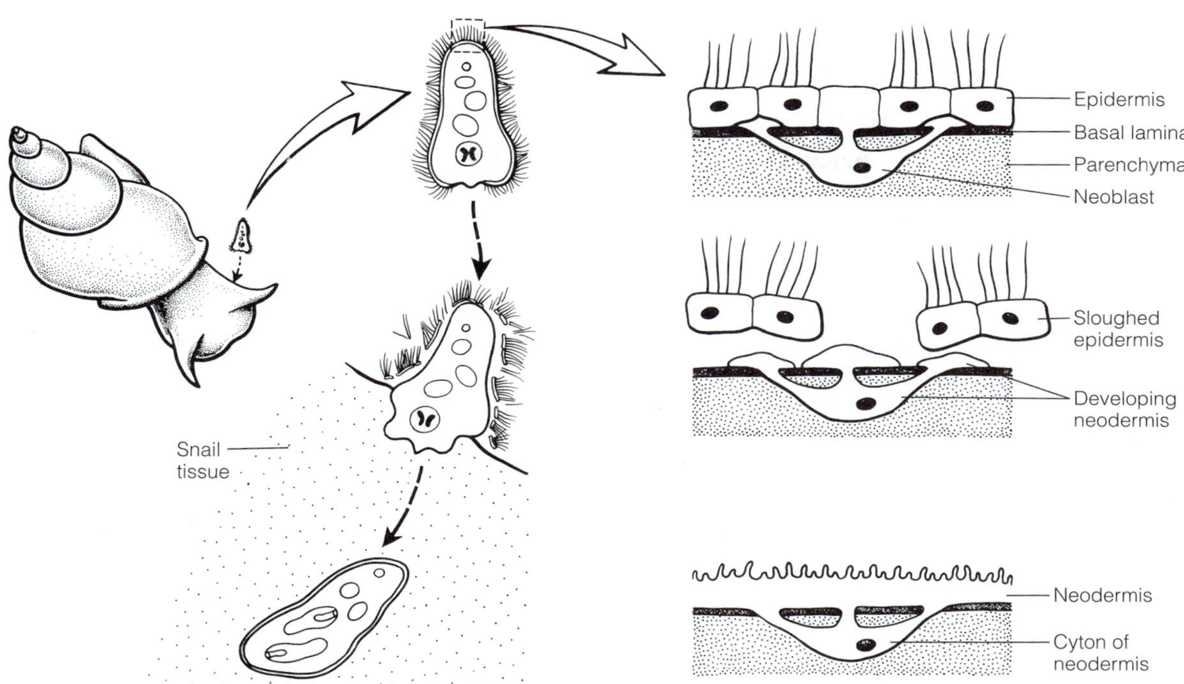

FIGURE 10-33 Neodermata: developmental origin of the neodermis (tegument). As the infective larval stage (miracidium) of the parasite penetrates the body wall of the snail, it sloughs off its embryonic epidermis and replaces it, from below, with a syncytial neodermis derived from parenchymal cells. *(Modified and redrawn after Xylander from Westheide, W., and Rieger, R. M. (Eds.): 1996. Spezielle Zoologie. Teil 1. Einzeller und Wirbellose Tiere. Gustav Fischer Verlag, Stuttgart. 909 pp.)*

enter or leave the body. Any substance crossing the body wall must pass intracellularly through the syncytium. Thus a syncytial epidermis may enable the organism to better regulate what enters and leaves its body, restricting some substances and accelerating transport of others. Uptake of host nutrients through the neodermis is highly regulated, for example, in flukes and tapeworms. The neodermis may also enable the parasite to cope with the different osmotic challenges encountered during its life cycle, which may include a freshwater aquatic phase and one or more hosts.

Trematoda[C]

Trematoda includes two subtaxa of closely related parasitic flatworms, Digenea, a large, economically and medically important taxon, and Aspidogastrea, a small taxon of no medical or economic importance.

DIGENEA[sC]

General Biology and Life Cycle

Digeneans, known as flukes, are common endoparasites of all major vertebrate taxa: fishes, amphibians, reptiles, birds, and mammals. Some flukes cause debilitating disease in livestock and humans. With some 11,000 species, more than all other flatworms combined, digeneans are second in diversity only to the roundworms (Nematoda) among metazoan parasites. The life cycle includes two or more hosts and at least two infective stages, which accounts for the name Digenea (= two generations). The **first intermediate host** (Fig. 10-34, 10-35A) is typically a gastropod mollusc (snail); if a **second intermediate host** occurs, it is usually an arthropod or fish; and a vertebrate is the final or **definitive host** (Fig. 10-34).

Although life-cycle variations occur among species, a generalized digenean life cycle (Fig. 10-34, 10-35) consists of the following stages: zygote → miracidium larva → sporocyst → redia → cercaria → metacercaria → adult (Table 10-1). The cycle begins when an adult fluke releases "eggs" (each an encapsulated zygote plus vitellocytes) that leave the vertebrate host in its feces, urine, or sputum. If deposited on land, these "eggs" may be ingested by a terrestrial snail, but if released into water, they hatch, releasing a ciliated, swimming **miracidium** larva that penetrates the body wall of an aquatic snail. The miracidium sloughs off its epidermis, which is replaced by the neodermis, during or after host penetration. Once inside the snail, the miracidium metamorphoses into a saclike, gutless **sporocyst** that contains several embryos (germ balls; Fig. 10-35B). Each embryo develops into another sporocyst or into a **redia,** which has a mouth, pharynx, and gut as well as its own embryos (Fig. 10-35C). Within the redia, each embryo develops into a **cercaria** (Fig. 10-35D,E). The cercaria has a digestive tract, suckers, and a tail. A cercaria leaves the snail host and swims in search of a second intermediate host or encysts as a **metacercaria** (Fig. 10-35F) on aquatic vegetation or an inanimate object in the

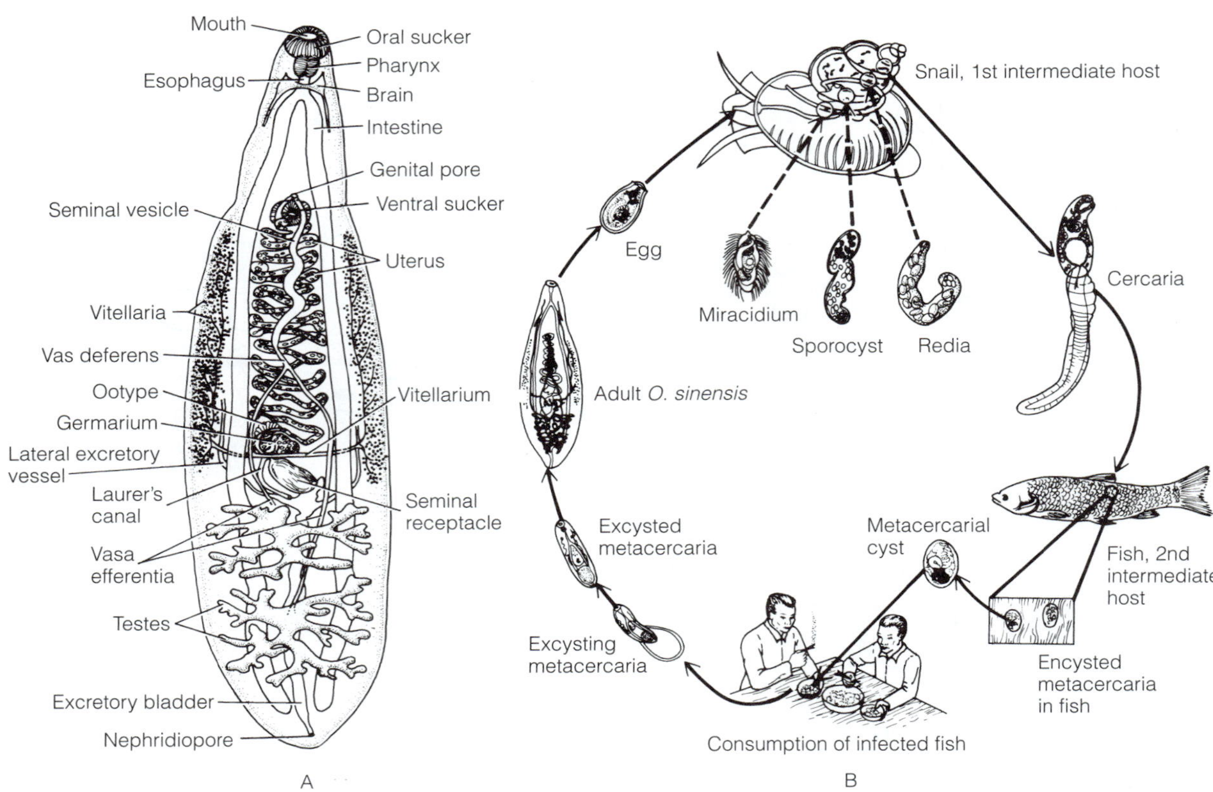

FIGURE 10-34 Digenea: life cycle. The Chinese liver fluke *Opisthorchis sinensis.* Millions of Asians are believed to be infected with liver flukes transmitted from uncooked fish. One worm can live as long as eight years. **A,** Dorsal view of an adult worm. **B,** Life cycle. *(A, After Brown from Noble, E. R. and Noble, G. A.: 1982: Parasitology. 5th Edition. Lea and Febiger, Philadelphia. B, After Yoshimura from Noble, E. R. and Noble, G. A. 1982. Parasitology. 5th Edition. Lea and Febiger, Philadelphia.)*

water, awaiting ingestion by the definitive host. If the life cycle includes a second intermediate host, commonly an arthropod or fish, the cercaria penetrates its skin, casts off its tail, and encysts as a metacercaria in the host's tissues. When the free metacercaria or intermediate host plus metacercaria is eaten by the definitive vertebrate host, the metacercaria escapes from its cyst (excysts), migrates to a characteristic location in the host body, and grows into a sexual adult. In the definitive host, sites of infection include the hemal system, gut, and other endodermal organs: intestine, liver, bile ducts, pancreatic ducts, and lungs.

Form and Function

Adult digeneans range in size from approximately 0.2 mm to 6.0 cm in length. They are typically dorsoventrally flattened, but some are thick and fleshy, others long and threadlike. An **oral sucker** surrounds the mouth, and many, but not all, have a midventral or posterior **ventral sucker** (acetabulum; Fig. 10-34A, 10-37). The suckers are important organs that prevent dislodgment and aid in feeding (oral sucker).

The digenean body is covered by a neodermis overlying consecutive layers of circular, longitudinal, and diagonal muscle (Fig. 10-36). The neodermis plays a vital role in the physiology of digeneans. It provides protection, especially against the host's enzymes in gut-inhabiting species. Nitrogenous wastes diffuse to the exterior through it (protonephridia are also used), and it is the site of gas exchange. The neodermis, as well as the gut, absorbs glucose and some amino acids. Protein synthesis for growth, yolk production, and clonal reproduction require a substantial supply of amino acids.

The mouth is surrounded by a powerful, muscular oral sucker used for attachment and to aid transport of food into the mouth. The bulbous pharynx ingests cells and cell fragments, mucus, tissue fluids, or blood of the host on which the parasite feeds. The pharynx joins a short esophagus leading into two blind intestinal ceca that extend posteriorly along the length of the body (Fig. 10-34A, 10-37). Digestion is primarily extracellular in the ceca. Nutrition in the preadult life cycle stages is summarized in Table 10-1.

Digeneans are facultative anaerobes and both blood and liver flukes derive most of their energy from glycolysis. Little is known about the energy metabolism of larval stages. Digeneans have protonephridia, especially well developed in the miracidium, cercarial, and adult stages. Typically, there is a pair of longitudinal collecting ducts that empties into a single posterior bladder that releases its contents through a nephridiopore (Fig. 10-34A). The nephridia excrete water and waste metabolites, such as unwanted iron from hemoglobin in blood-feeding flukes.

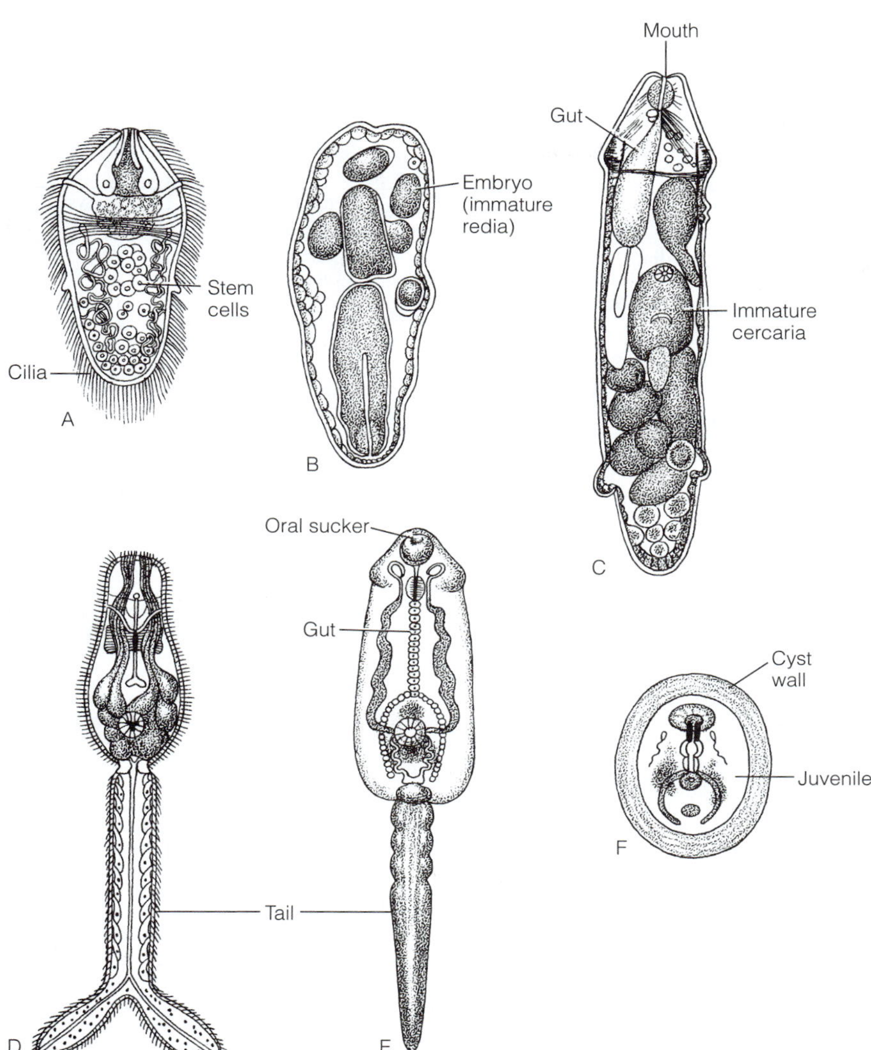

FIGURE 10-35 Digenea: life-cycle stages. **A,** Miracidium; **B,** Sporocyst; **C,** Redia; **D** and **E,** Cercariae; **F,** Metacercaria. *(From the U.S. Naval Medical School Laboratory Manual)*

The nervous system is essentially like that of turbellarians. The brain is a pair of anterior cerebral ganglia from which several longitudinal nerve cords extend posteriorly; the ventral pair is typically the best developed. The fluke body surface has a variety of sensory papillae. Ocelli occur in many miracidia and some cercariae.

Reproduction

Flukes have a highly organized neoophoran reproductive system with a heterocellular ovary (Fig. 10-37). Perhaps because of this and the more or less steady supply of host nutrients, "egg" production is estimated to be 10,000 to 100,000 times greater than that of any free-living flatworm.

The male system (Fig. 10-34A, 10-37) typically consists of two testes and accessory reproductive organs. Sperm ducts, one from each testis, unite anteriorly and may expand into an external seminal vesicle before entering into the **cirrus sac** (or pouch). This sac contains the internal seminal vesicle, prostate glands, and an eversible copulatory **cirrus,** which protrudes into a common **genital atrium** shared with the female reproductive system. Sperm, on leaving the testes, are stored in the seminal vesicle. The single gonopore opens from the common genital atrium and is usually located midventrally anterior to the ventral sucker. Many variations on this general plan occur.

The female reproductive system (Fig. 10-34A, 10-37) usually consists of a single ovary (germarium) and an oviduct that leads into a small sac, called the **ootype** (oogenotop), in which a capsule of tanned proteins is secreted around the fertilized egg and vitellocytes. Unicellular gland cells, collectively called **Mehlis's gland,** open into the ootype and may secrete egg-capsule components, but most are thought to be produced by the vitellocytes. En route to the ootype, the oviduct receives a duct from the seminal receptacle and a common duct from the right and left vitellaria. Downstream from the ootype is the **uterus,** which runs anteriorly to the genital atrium and gonopore. In most digeneans, a short, inconspicuous canal

TABLE 10-1 Characteristics of the Digenean Life Cycle

Stage	Habitat	Role	Anatomy	Nutrition	Reproduction	Progeny
"Egg" (encapsulated zygote)	Water or vegetation	Dispersal; direct infection	Encapsulated embryo + vitellocytes	Yolk	Cleavage	Miracidium
Miracidium (larva)	Water	Dispersal; infection (1st intermediate host)	Gutless; ciliated epidermis; protonephridia; sensory and penetration structures; stem cells/embryos	Yolk	Metamorphosis: neodermis replaces epidermis	Sporocyst
Sporocyst	Snail (1st intermediate host)	Replication	Gutless sac with stem cells/embryos	Endocytosis by neodermis	Clonal (parthenogenesis?)	Sporocysts or rediae
Redia	Snail (1st intermediate host)	Dispersal in host; replication	Functional gut and musculature; stem cells/embryos	Gut ingestion of host tissues; endocytosis by neodermis	Clonal (parthenogenesis?)	Rediae or cercariae
Cercaria (larva)	Water	Dispersal; direct infection (2nd intermediate or definitive host); encystment	Swimming tail; gut; nonciliated epidermis; sensory and penetration structures; protonephridia; germ cells	Nonfeeding	Metamorphosis: neodermis replaces epidermis; cyst wall secreted by neodermis	Metacercaria
Metacercaria (encysted juvenile of sexual adult)	Water; aquatic vegetation, inanimate objects; arthropod (2nd intermediate host); vertebrate (definitive host)	Dispersal; persistence; infection	Like cercaria, but encysted and without tail or epidermis	Nonfeeding or endocytosis by neodermis	Excystment (in definitive host); growth	Sexual adult
Adult	Vertebrate (definitive host)	Replication	Functional gut; neodermis; protonephridia; hermaphroditic gonad	Ingestion of host tissues; endocytosis by neodermis	Sexual reproduction	"Eggs"

(Laurer's canal), perhaps a vestigial vagina, extends from the duct of the seminal receptacle to the dorsal surface of the worm, where it may open at a minute pore (Fig. 10-37).

Most flukes are cross-fertilizing hermaphrodites, but self-fertilization occurs in rare cases. During copulation, the cirrus of one worm is inserted into the partner's gonopore and sperm are ejaculated into the uterus. The prostate gland provides nutrients for sperm survival. Sperm travel up the uterus and through the ootype to be stored in the seminal receptacle.

As eggs leave the ovary and enter the oviduct, the partner's stored sperm are released from the seminal receptacle and fertilize the eggs in either the oviduct or ootype. Each zygote and its vitellocytes are then combined in a capsule; the package typically is referred to as an "egg." Encapsulated zygotes ("eggs") enter the uterus, where development usually begins, and encapsulated embryos, also known as "eggs," are released from the gonopore.

Digenean development is not adequately described and its interpretations are controversial. The zygote divides into two distinguishable cells, a **somatic cell** and a **stem cell.** The somatic-cell line ultimately establishes the body (in all life cycle stages) whereas the stem cells eventually establish the **germ line,** the cells committed to sexual (adult) and asexual (sporocyst, redia) reproduction. Initially, each division of the stem cell results in a somatic cell, which contributes to the morphogenesis of the body, and another stem cell. Sooner or later, however, subsequent divisions of the stem cell yield only more stem cells. As the embryo develops into a miracidium, the mass of stem cells occupies the posterior half of the body. Each of the miracidium's stem cells, though unreduced and unfertilized, cleaves in the manner of a sexual zygote, forming an embryo of somatic and stem cells. As the miracidium undergoes metamorphosis, these asexually produced embryos, or

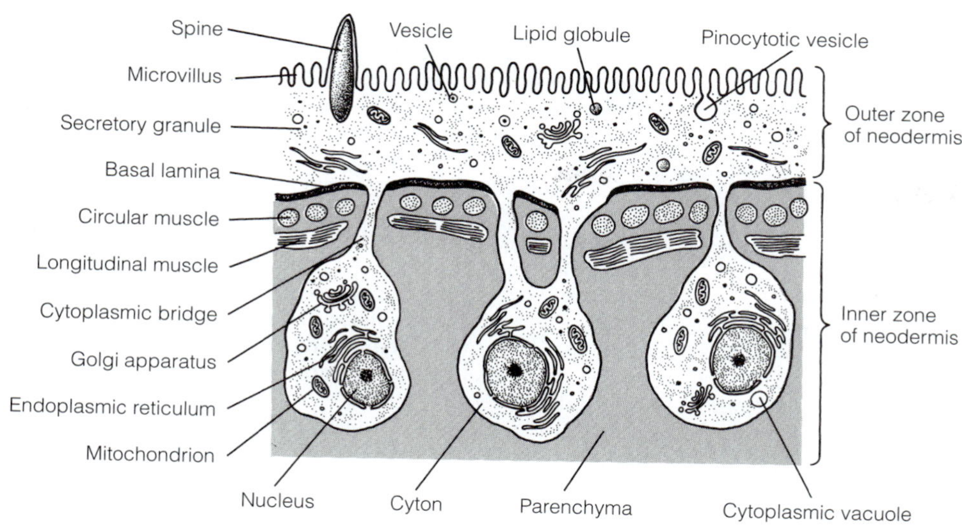

FIGURE 10-36 Digenea: histology. Section through the neodermis (tegument) and body wall of a generalized digenean. *(Modified and redrawn from Cheng, T. C. 1973. General Parasitology. Academic Press, New York. 965 pp.)*

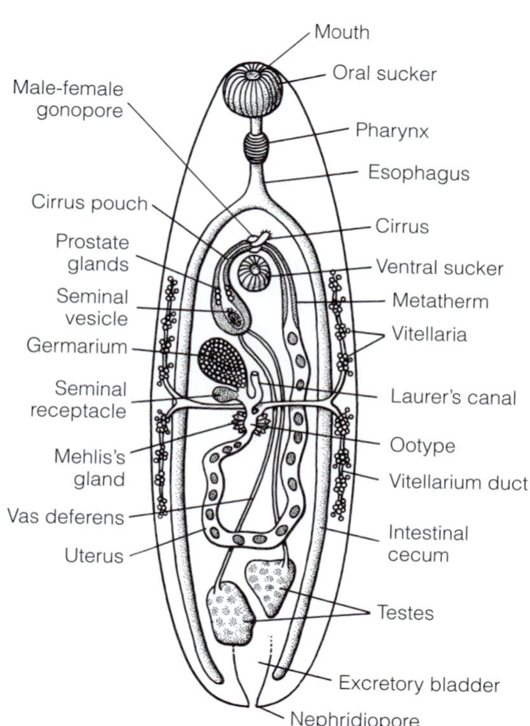

FIGURE 10-37 Digenea: generalized anatomy. *(After Cable, 1949, from Smyth, J. D. 1976. Introduction to Animal Parasitology. 2nd Edition. Hodder & Stoughton, London. 486 pp.)*

germ balls, are incorporated into the body of the sporocyst. The embryos complete development in the sporocyst, differentiating into either more sporocysts or rediae. These progeny, in turn, house their own stem cells, each of which develops into a cercaria whose stem cells eventually differentiate into the adult gonads. The controversy revolves around the interpretation of the reproductive stem cells in the clonal sporocyst and redial stages. Parasitologists consider the origin of these stem cells as similar to twinning, the division of the zygote into two or more blastomeres, each developing into a complete organism, and describe the phenomenon as **polyembryony** (= many embryos). Another interpretation is that these stem cells are eggs that develop by **parthenogenesis** (without fertilization) in the stem-cell masses (gonads) of the sporocyst and redial stages. If this interpretation is correct, then sporocyst and redia are parthenogenetic adults in the first intermediate host, alternating with the sexual adult in the definitive host. The miracidium would then be the larval stage of the sexual adult and the cercaria the larval stage of the parthenogenetic adult, and the life cycle would be: sexual adult (in definitive host) → zygote → miracidium larva (in water) $\xrightarrow{\text{(metamorphosis)}}$ parthenogenetic adult (sporocyst-redia in intermediate host) → cercaria larva (in water) $\xrightarrow{\text{(metamorphosis)}}$ encysted juvenile (metacercaria) $\xrightarrow{\text{(excystment)}}$ sexual adult (in definitive host).

Life Cycle Examples

The Chinese liver fluke, *Opisthorchis* (or *Clonorchis*) *sinensis,* infects the bile ducts of over 20 million humans and their domesticated animals, especially dogs, in eastern Asia (Fig. 10-34). Adult flukes reach 2.5 cm in length and may live as long as eight years, with each fluke producing as many as 4000 "eggs" per day for up to six months of the year.

If feces containing liver-fluke "eggs" wash or drop into water, the "eggs" are consumed by certain species of aquatic snails, the first intermediate host (Fig. 10-34B). Once ingested by the snail, the miracidium penetrates the wall of the intestine and metamorphoses into a sporocyst in the host's tissues. The sporocyst produces the redia generation, which gives rise to cercariae that escape into the water. Each cercaria swims to the surface using its muscular tail and then sinks slowly until it contacts an inert object or is disturbed by water turbulence. Either occurrence reactivates the cercaria, which repeats the swim-sink cycle to position it for a likely encounter with the second intermediate fish host. On contact with a fish, the cercaria attaches with its suckers, sheds its tail, and bores through the host's skin. Within the fish's subcutaneous tissue or musculature, the cercaria encysts to form a metacercaria. When humans or other mammals eat the uncooked fish, young worms emerge from their cysts in the small intestine and migrate into the bile ducts.

The parasite load of an infected a person is generally low, but even modest infections (20 to 200 flukes per liver) can cause jaundice, gallstones, general debilitation, and perhaps liver cancer. In one unfortunate person, over 21,000 flukes were found at autopsy. Although drugs are partially effective against the disease, prevention is the best approach, but it is often resisted by cultural habits or economic limitations. Transmission of the disease in rural areas results from building toilet facilities (privies) over fishponds. In urban areas, especially in Asia, uncooked fish, which are sometimes glistening with metacercarial cysts, are considered to be a delicacy, but in rural areas their consumption can be a necessity because fuel for cooking is either unavailable or prohibitively expensive.

Three digenean families inhabit the blood of their hosts, but certainly the best-known blood flukes belong to Schistosomatidae, which contains species producing the human disease schistosomiasis. *Schistosoma mansoni,* one of several species parasitic in humans, occurs in Africa and tropical areas of the New World (Fig. 10-38). The adult, like that of other species of schistosomes, inhabits the intestinal veins. In contrast to most other flatworms, schistosomatids are gonochoric, with a male and female permanently paired throughout life. The male is 6 to 10 mm in length and 0.5 mm in width. A ventral groove along the body of the male embraces the long, slender female. When laying "eggs," the female protrudes from the groove of the male worm and deposits the "eggs" in intestinal venules. Eventually, using spines and enzymes, the "eggs" work their way into the intestinal lumen and are released in the host's feces. If defecation occurs in the water, the thin-walled "eggs" hatch and miracidia escape. The miracidia are well supplied with sensory receptors and seek a particular species of freshwater snail. After penetration, they transform into sporocysts, which eventually give rise to a second generation of sporocysts. The sporocysts produce cercariae without an intermediate redial stage. The cercariae leave the molluscan host and, on contact with human skin, penetrate using enzymes and muscular boring movements. The now tailless "schistosomules" are carried by the bloodstream first to the lungs, then to the liver, and finally to the veins of the intestine or bladder. During this migration, the schistosomules gradually transform into adults.

Schistosomiasis is a seriously debilitating disease that can be lethal. "Egg" penetration through the intestinal wall and bladder, aberrant lodging of "eggs" in various organs, and the developmental stages in the lung and liver can result in inflammation, necrosis, or fibrosis, depending on the degree of infection. The pathogenic immunological response to the "eggs" is generally more serious than that to the schistosomules or adults. The percentage of the population infected in endemic areas is enormous, and globally some 300 million people are estimated to be infected by one of the three *Schistosoma* species. Schistosomiasis, malaria, and hookworm infections are the three great parasitic scourges of humankind.

Other members of Schistosomatidae infect various birds and mammals, including domestic species. "Swimmer's itch," which occurs commonly in North America, is an irritation produced by the incomplete penetration into human skin by cercariae of blood flukes of birds.

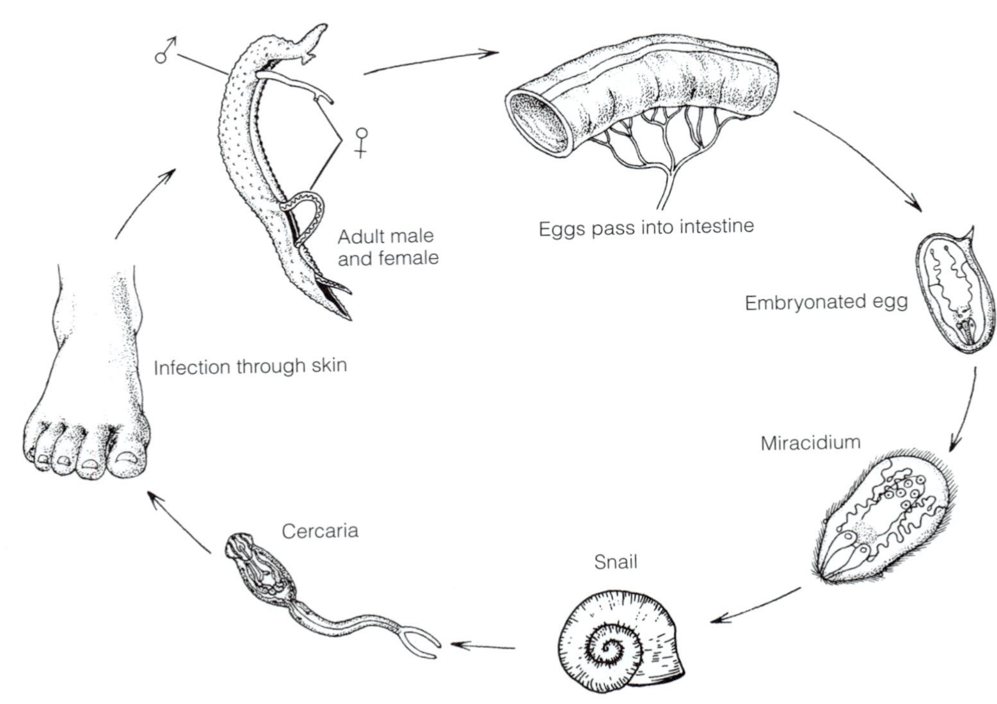

FIGURE 10-38 Digenea: life cycle of *Schistosoma mansoni.* The *eggs* in this figure are egg capsules ("eggs") and an *embryonated egg* is an encapsulated embryo.

ASPIDOGASTREA[sC]

Aspidogastrea (or Aspidobothrea) is a small trematode taxon with similarities to other parasitic flatworms, but more closely related to Digenea. The distinguishing trait is a huge **adhesive organ,** which is either a single, septate sucker covering the entire ventral surface or a longitudinal row of suckers (Fig. 10-39). The digestive tract contains a single intestinal cecum. The reproductive system is essentially like that of Digenea, but there is typically only one testis.

Aspidogastreans are chiefly endoparasites in the gut of fish and turtles and in the pericardial and renal cavities of bivalve molluscs. Different species have life cycles with a ciliated swimming larva and one or two hosts.

Cercomeromorpha

The sister taxon of Trematoda, these important parasitic flatworms have crescent-shaped hooks on a posterior appendage, the **cercomer,** of their ciliated oncosphere or oncomiracidium larvae.

MONOGENEA[C]

General Biology

The approximately 1100 species of monogeneans are chiefly host-specific ectoparasites, or occasionally endoparasites, of aquatic vertebrates, especially fishes, but amphibians and reptiles are also hosts. Because most attach to the skin of a fast-moving host, monogeneans are dorsoventrally flattened and have a large, posterior attachment organ, the **haptor,** that bears hooks and suckers, allowing the parasite to cling tenaciously to the host (Fig. 10-40). Most are 1 to 5 mm in length, but some reach 20 mm. The monogenean life cycle differs from that of Digenea in that there is no intermediate host, one "egg" gives rise to only one adult worm, and clonal reproduction is absent, hence the name Monogenea (= one generation; Fig. 10-43). The "egg" develops first into a ciliated, hooked **oncomiracidium** (= hooked miracidium) that has two pairs of pigment-cup ocelli (Fig. 10-41).

The adult body is composed of a head, trunk, and haptor (Fig. 10-40, 10-42). The head may have a muscular oral sucker around the mouth (Fig. 10-40) or a sucker can be absent and **adhesive glands** (also called head organs; Fig. 10-41, 10-42) are used for attachment. Alternate attaching and detaching of adhesive glands and haptor enables some monogeneans to creep like inchworms; others may be permanently attached to the host. The neodermis is covered with microvilli, which increase the area for uptake of nutrients across the body wall to supplement uptake through the gut.

The digestive system is similar to that of digeneans, but the pharynx of some monogeneans secretes a protease that digests the host's skin, allowing the parasite to ingest blood and cellular debris. Many monogeneans, however, graze exclusively on mucus and superficial cells of the host skin. Some gut-absorbed nutrients are transported, via fixed parenchymal cells and gap junctions, to the vitellaria for yolk synthesis.

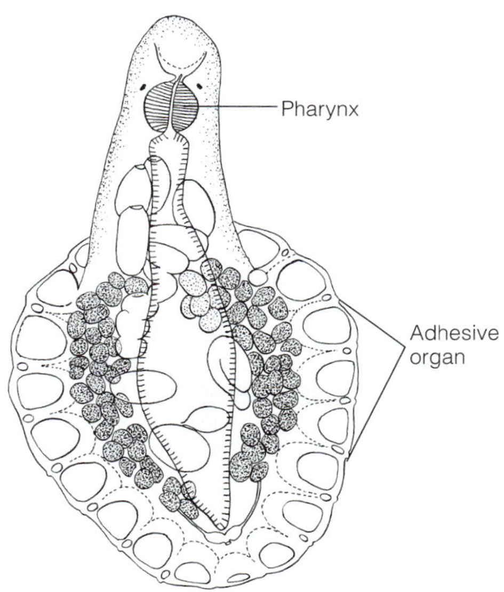

FIGURE 10-39 Aspidogastrea: dorsal view of *Cotylaspis insignis,* a fluke parasite of freshwater mussels. Note the large ventral adhesive organ (sucker) with subdivisions. *(After Hendrix and Short.)*

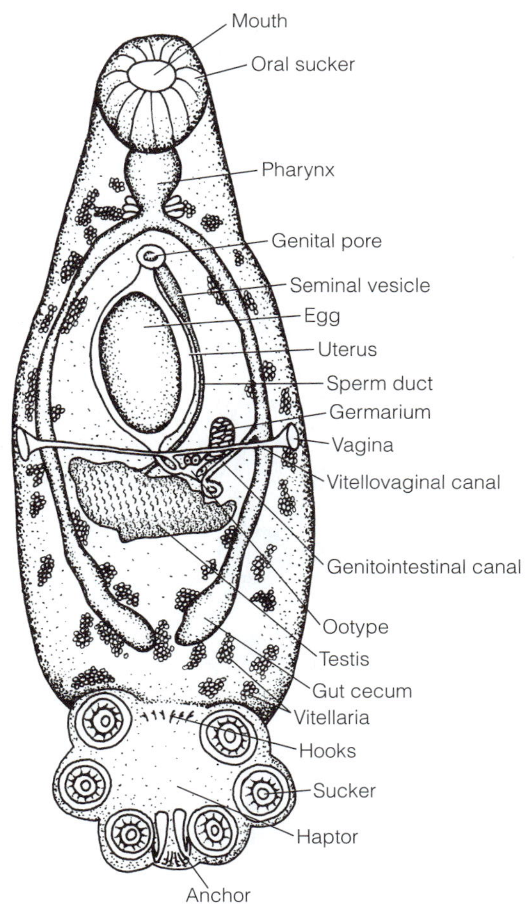

FIGURE 10-40 Monogenea: *Polystomoidella oblongum,* a parasite of the urinary bladder of turtles. *(After Cable, 1949, from Smyth, J. D. 1976. Introduction to Animal Parasitology. 2nd Edition. Hodder & Stoughton, London. 486 pp.)*

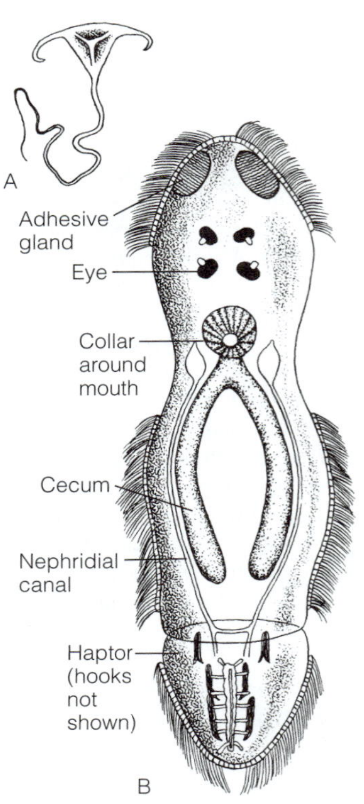

FIGURE 10-41 Monogenea: **A,** Egg capsule and **B,** oncomiracidium of *Benedenia melleni,* an ectoparasite of fish. *(After Jahn and Kuhn.)*

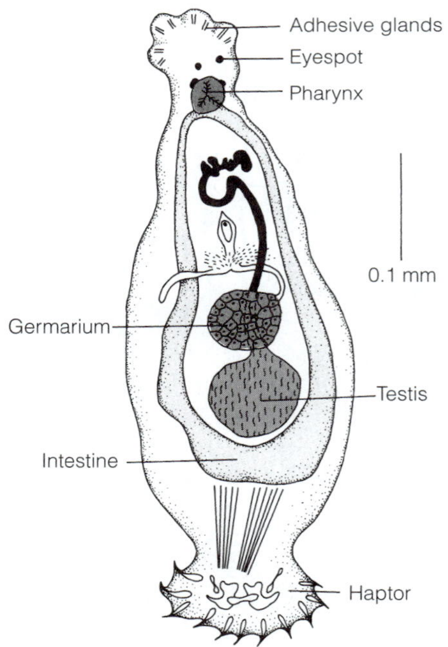

FIGURE 10-42 Monogenea: *Dactylogyrus vastator,* an ectoparasite on the gills of freshwater fishes. *(After Bychowsky from Schmidt, G. D., and Roberts, L. S. 1989. Foundations of Parasitology, 4th Edition. Times Mirror/Mosby College, St. Louis.)*

Monogeneans have inconspicuous protonephridia that consist of scattered terminal cells and their collecting tubules. The latter open to the exterior at two dorsolateral pores. In some species, a urinary bladder is associated with each pore.

In contrast to the endoparasitic digeneans, monogenean ectoparasites have aerobic metabolism. They have no medical significance.

Reproduction

The reproductive system of monogeneans resembles that of other neoophorans (Fig. 10-40). The male reproductive system usually has only one circular or oval testis, but a few species have two or more. The sperm duct usually leads into the base of the copulatory organ, which is a protrusible penislike structure that may be armed with hooks. During copulation a mutual exchange of sperm occurs, with seminal fluid being passed into the single or paired vagina of each partner. Sperm generally are stored in a seminal receptacle situated near the ovary and oviduct. There is a single heterocellular ovary with its paired but extensive vitellaria in close proximity to the gut ceca.

Examples of both cross- and self-fertilization are known among the hermaphroditic monogeneans. One species, however, has evolved the ultimate guarantee of cross-fertilization and spousal fidelity. During the development of *Diplozoon paradoxum,* a parasite of European freshwater fishes, larvae fuse in pairs. As their reproductive structures differentiate during metamorphosis, the sperm duct of each worm fuses with the vagina of the other, joining them permanently *in copula.*

Life Cycle Examples

The various species of *Dactylogyrus* are common ectoparasites on the gills of various freshwater fishes (Fig. 10-42). *Dactylogyrus* can be a serious problem in fish hatcheries, causing high mortality in young fish from secondary infection, smothering by excess mucus production, or loss of blood. The "eggs" are released and drop to the bottom. Eventually, they hatch to liberate the oncomiracidia. On contacting a suitable host fish, each larva transforms into the adult worm. "Egg" production increases with rising water temperature so that the population of monogeneans builds up over the summer. Some "eggs" can overwinter to begin new cycles of infestation in the spring, and the adult *Dactylogyrus* can persist on a host over the winter.

Polystoma integerrimum (Fig. 10-43) is found in the bladders of Old World frogs and toads and is an example of a remarkable synchronization of the parasite's life cycle with that of the amphibian host. (A similar species, *P. nearcticum,* occurs in tree frogs in the United States.) The "eggs" are produced and stored until the frog or toad returns to the water to breed, at which time they are released. The oncomiracidium attaches to the gills of the tadpoles. When the tadpoles metamorphose, the parasite leaves the gill chamber, crawls over the host's belly, and enters the bladder. When the tadpole is very young, some of the larvae may attain a precocious sexual maturity and produce "eggs." This ectoparasitic generation dies when metamorphosis occurs.

CESTODA[C]

Cestoda is the most evolutionarily derived of the flatworm taxa, and all 3400 species are endoparasites of the gut of vertebrates. A gut is absent and the neodermis is highly specialized

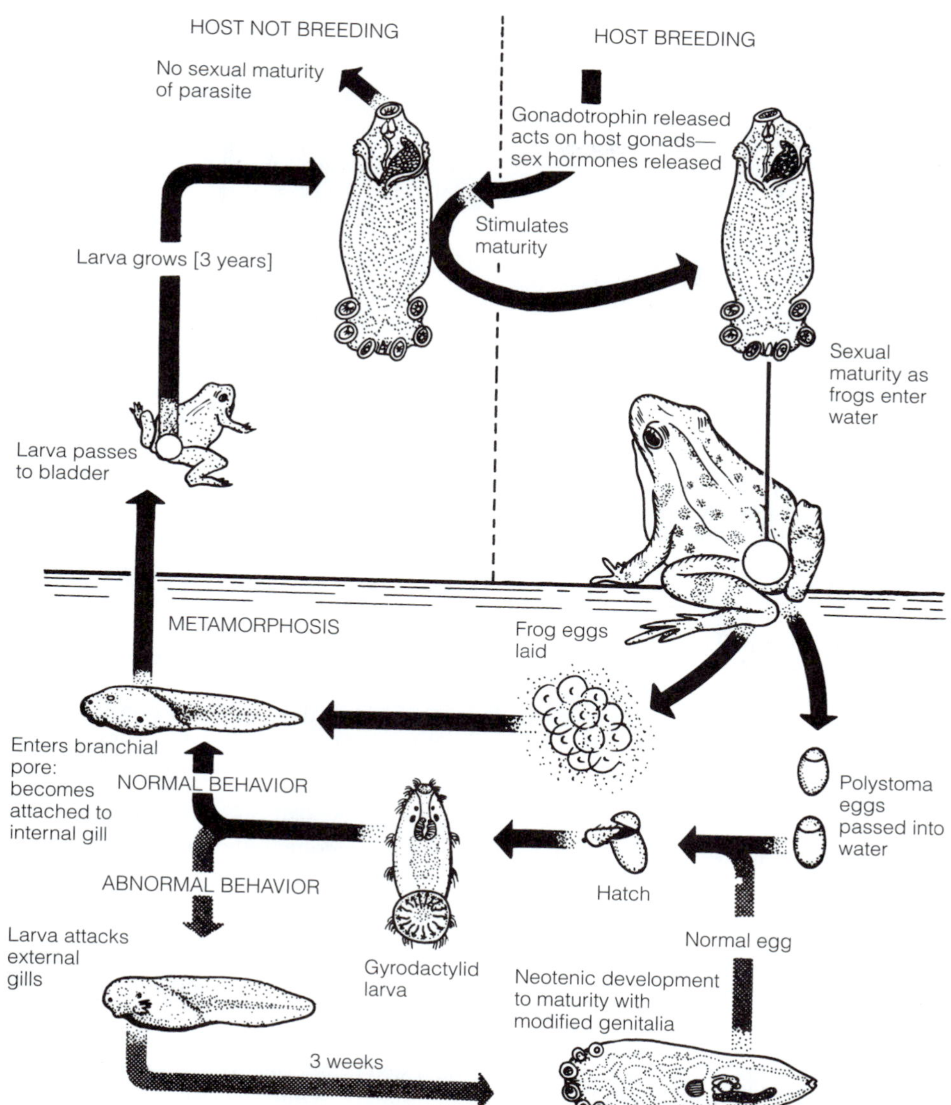

FIGURE 10-43 Monogenea: life cycle of *Polystoma integerrimum,* a parasite in the bladder of frogs. Diagram also shows a pedomorphic population parasitic on the gills of the tadpole. *(From Smyth, J. D. 1976. Introduction to Animal Parasitology. 2nd Edition. Hodden and Stoughton Educational, Kent. p. 139.)*

for nutrient uptake from the host. Most species of Cestoda occur in Eucestoda[sC], which is described here.

Form and Function

Cestodes are the tapeworms. The ribbonlike adult body is divided into a **scolex** (Fig. 10-44, 10-47), which is adapted for attachment to the host, a narrow **neck** (growth zone with stem cells), and the segmented trunk, or **strobila,** which constitutes most of the worm's body (Fig. 10-44B). Each of the trunk segments is called a **proglottid** (Fig. 10-47). Tapeworms are generally long, with some species reaching lengths of 25 m. The scolex, neck, and strobila are regarded as a single individual, not a colony.

Compared with the mature proglottids, the scolex is often minute. It generally is a four-sided knob having suckers or hooks for attachment to the host gut wall. Although there typically are four large suckers arranged around the sides of the scolex, as in commonly encountered tapeworms such as *Taenia* (Fig. 10-44A), the scolex is a more complex structure in other representatives. The main attachment organs may be leaflike or ruffled, and there may be terminal accessory suckers in place of or in addition to hooks (Fig. 10-44B,C).

The neck is a short region behind the scolex that produces each proglottid by mitotic growth followed by a transverse constriction (Fig. 10-44B). The youngest proglottids are thus at the anterior end immediately behind the scolex; they increase in size and maturity as they are displaced toward the posterior end of the strobila.

The tapeworm neodermis plays a vital role in the active uptake of host carbohydrates and amino acids. It is also important in evading the host's immune response. The outer plasma

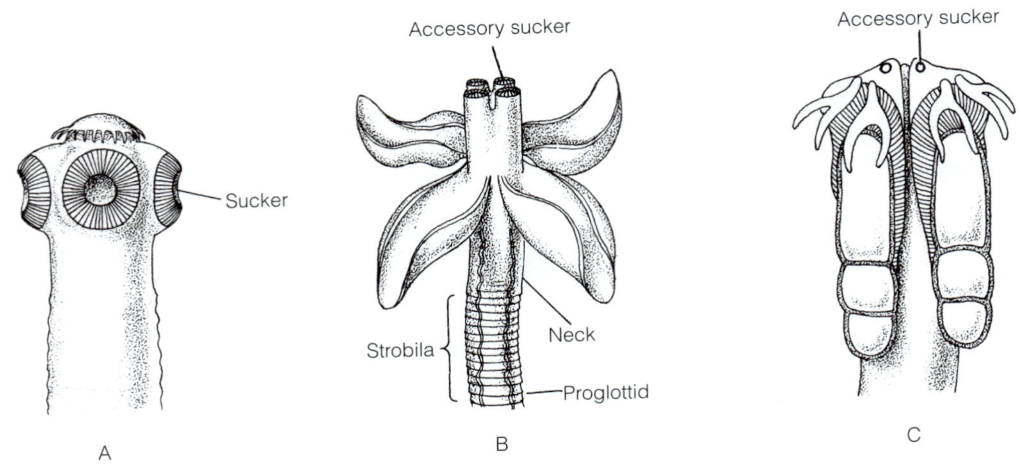

FIGURE 10-44 Cestoda: the scolex of three different tapeworms. **A,** *Taenia,* with four principal suckers. **B,** *Myzophyllobothrium,* with suckerlike adhesive flaps and small accessory suckers. **C,** *Acanthobothrium,* with accessory suckers and hooks. *(A and C, After Southwell; B, After Shipley and Hornell)*

membrane of the neodermis is thrown into specialized microvilli with dense, spiny tips, the **microtriches** (sing. **microthrix;** Fig. 10-45). Anaerobic metabolism apparently predominates in tapeworms, but is not their exclusive mode of metabolism.

Tapeworm musculature consists of the typical body-wall circular and longitudinal layers, but additional inner layers of longitudinal, transverse, and dorsoventral fibers enclose the interior parenchyma.

The nervous system, protonephridial system, and longitudinal musculature extend uninterrupted through the chain of proglottids. An anterior nerve mass lies in the scolex, and two lateral, longitudinal cords extend posteriorly through the strobila (Fig. 10-47). There also can be a dorsal and ventral pair of cords and accessory lateral cords. Ring commissures connect the longitudinal cords in each proglottid.

Protonephridial terminal cells and tubules in the parenchyma drain into four longitudinal collecting canals, two of which are dorsolateral and two ventrolateral (Fig. 10-47). The ventral canals are usually connected by a transverse canal in the posterior end of each proglottid. After the proglottids begin to be shed, the collecting ducts open to the exterior through the terminal proglottid.

Reproduction

The continuous, serial formation of proglottids from the growth zone in the neck is called strobilation. It enables one worm to produce an enormous number of "eggs" over a long period of time.

A complete neoophoran reproductive system occurs in each proglottid. The tapeworm reproductive anatomy (Fig. 10-47) is similar to that of Monogenea. The uterus, usually a blind sac extending from the ootype into the parenchyma of the proglottid, functions solely in storing developing encapsulated zygotes called "eggs."

Cross-fertilization is probably the rule where there is more than one worm in the host's gut, but self-fertilization can occur between two proglottids of one worm or even in one proglottid. At copulation the cirrus is everted into the gonopore of the proglottid of an adjacent worm. Sperm are stored in the partner's seminal receptacle, then released to fertilize eggs in the oviduct. The zygotes then enter the ootype, where they are encapsulated together with vitellocytes. These "eggs" enter the uterus, in which development begins and the "egg" capsule is hardened. In many cases, mature terminal proglottids packed with "eggs" break away from the strobila. The "eggs" are freed from the proglottid either in the host's feces or in the next host's intestine after it swallows a proglottid.

Life Cycle Examples

Tapeworms are gut endoparasites of all major vertebrate taxa. Their life cycles rarely include one host and usually two or more, typically arthropods and vertebrates. The definitive host is always a vertebrate. The basic life cycle includes zygote → oncosphere larva (ingested by intermediate or definitive host) $\xrightarrow{\text{(metamorphosis)}}$ extraintestinal juvenile (metacestode) → intestinal adult (in definitive host). Two life cycles are included here: In one, transmission is through an aquatic food chain; in the other, transmission is terrestrial.

Diphyllobothrium spp. are widely distributed in northern latitudes and are parasitic in the gut of many fish-eating carnivores, including humans *(Diphyllobothrium latum).* If the "eggs" in feces are deposited in water, each "egg" develops into a ciliated, free-swimming **oncosphere** (coracidium) bearing three pairs of small hooks, as is typical of all eucestodes (Fig. 10-46A). The oncosphere is ingested by a copepod crustacean. It penetrates the copepod's intestinal wall, sheds its ciliated epidermis (which is replaced by neodermis), enters the hemocoel, and develops into a **procercoid,** which retains the oncosphere's hooks but shifts them posteriorly to the newly formed cercomer (Fig. 10-46B). When the copepod is eaten by a freshwater fish, the procercoid penetrates the fish's gut and migrates to the striated muscles, where it transforms into a juvenile called a **metacestode,** known in this example as the **plerocercoid.** After the fish is eaten by a suitable

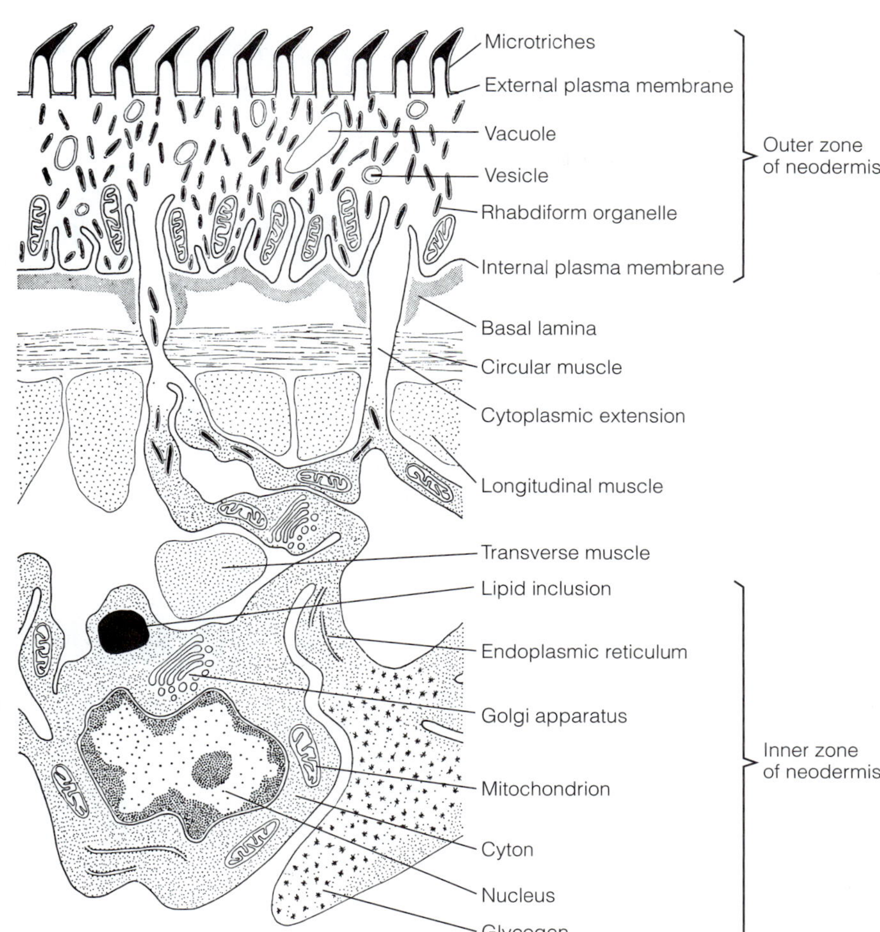

FIGURE 10-45 Cestoda: Section through the neodermis and body wall of the tapeworm *Caryophyllaeus.* Other than the complex microtriches on the surface membrane, note the similarity to the digenean body wall shown in Figure 10-36. *(After Beguin from Smyth, J. D. 1969. The Physiology of Cestodes. W. H. Freeman and Co., San Francisco.)*

warm-blooded definitive host, the plerocercoid, which looks like a small, unsegmented tapeworm complete with a miniature scolex, develops into an adult tapeworm in the host's gut.

Species of Taeniidae are among the best-known tapeworms because they infect humans and domesticated animals. *Taenia saginata,* the cosmopolitan beef tapeworm, is one of the most commonly occurring species in humans, where it lives in the intestine and may reach a length of over 20 m; typical lengths are 3 to 5 m (Fig. 10-47). A large specimen has thousands of proglottids, each containing "eggs" that are eliminated through the anus, usually with feces. If an infected person defecates on a pasture, the "eggs" may be eaten by grazing cattle. On hatching in the gut of the intermediate host, the oncosphere bores into the intestinal wall, where it is picked up by the circulatory system and transported to striated muscle. Here the oncosphere develops into the metacestode, in this case called a **cysticercus.** The cysticercus is an oval worm of about 10 mm in length that has its scolex retracted and inverted, like a nemertean proboscis, into the body. If raw or insufficiently cooked beef is ingested by humans, the cysticercus is freed, the scolex evaginates and attaches to the intestinal lining, and an adult worm develops.

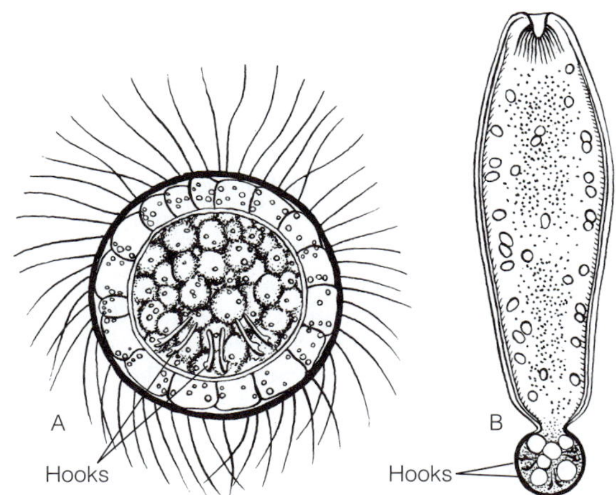

FIGURE 10-46 Cestoda: two stages in the life cycle of the fish tapeworm *Diphyllobothrium latum.* **A,** A ciliated oncosphere. **B,** A mature procercoid. *(A, After Vergeer; B, After Brumpt)*

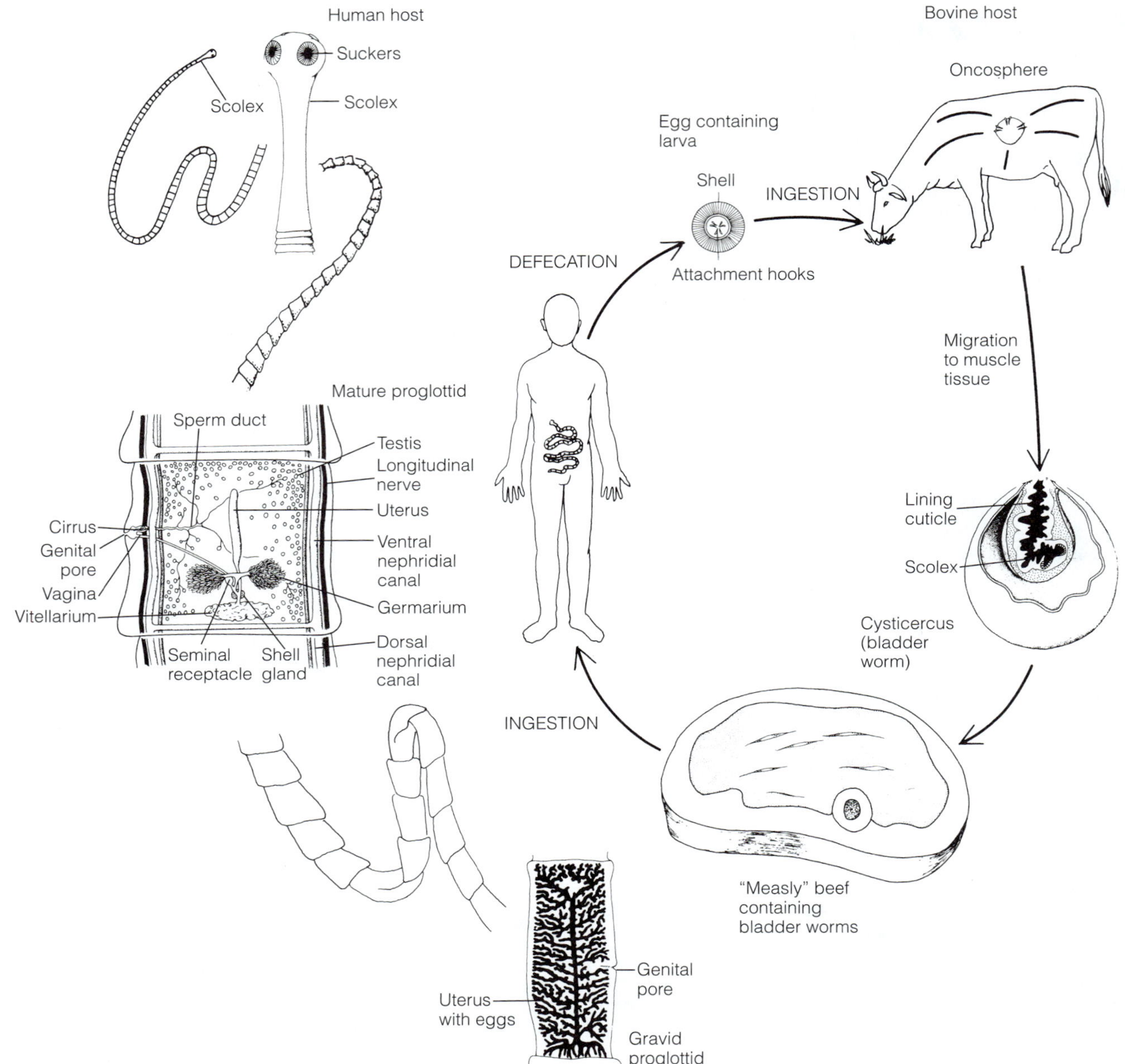

FIGURE 10-47 Cestoda: anatomy and life cycle of the beef tapeworm *Taenia saginata.*

Taenia solium, the pork tapeworm, is also a parasite of humans, but the intermediate host is a pig and the cysticercus is obtained from inadequately cooked pork. *Taenia pisiformis* occurs in cats and dogs, with rabbits as the intermediate hosts.

A severe infection of adult tapeworms may cause diarrhea, weight loss, and adverse reactions to the worm's toxic wastes. The worms can be easily eliminated with drugs. Much more serious are the "accidental" cysticercus infections of humans in the role of intermediate host. In the case of the pork tapeworm, *Taenia solium,* and the dog tapeworm, *Echinococcus granulosus,* serious disease may result in humans. The adult *Echinococcus* is minute, with only a few proglottids present at any one time. Many different mammals, especially grazing herbivores but occasionally humans, serve as intermediate hosts. In humans, the *Echinococcus* cysticercus, which buds internally to produce multiple scolices, develops mostly in host lung or liver, but can occur elsewhere. Pork tapeworm cysticerci develop in subcutaneous connective tissue as well as in the eye, brain, heart, and other organs. The cystercerci (but not the adults) of both species are life threatening when they grow in such places as the brain and can do much damage elsewhere. Such cysticerci, called hydatids in *Echinococcus,* swell with fluid, sometimes over 10 liters, reach enormous sizes, contain millions of scolices, cause serious pathology, and if they burst, instantaneous death. Hydatid cysts can be removed only by surgery.

Origin and Phylogeny of Neodermata

Several trends, culminating in the higher rhabdocoels, may have predisposed the turbellarians to a parasitic lifestyle. These attributes include a small (or flat) body size, an interstitial habitat; a permeable body wall; a heterocellular ovary with protected encapsulated eggs plus vitellocytes; a specialized sucking, bulbous pharynx; and clonal reproduction. Furthermore, turbellarian epidermal replacement cells, originating from parenchyma, foreshadow the neodermatan neodermis. From the worms' perspective, the steps to endoparasitism led to the benefits of a sheltered, if not necessarily benign, environment and a constant and abundant supply of food. The principal hurdles to be cleared before reaping these benefits were osmotic variation, low oxygen availability, host immune-system challenges, and the task of finding hosts and completing the life cycle. The improbability of colonizing new hosts is met, largely at the host's expense, by the production of myriad infective offspring. The osmotic swing between external and host environments is probably overcome by the well-developed protonephridia. A flexible metabolism with a capacity for both aerobic and anaerobic pathways allowed the rapid transition between oxic and anoxic environments. Less is known about how the worms cope with host antibodies, but one mechanism is the rapid replacement by the parasite of surface membrane damaged by the host.

PHYLOGENETIC HIERARCHY OF NEODERMATA

Neodermata
- Trematoda
  - Aspidogastrea
  - Digenea
- Cercomeromorpha
  - Monogenea
  - Cestoda

The parasitic platyhelminths (Neodermata) probably evolved from a dalyellioid-like rhabdocoel ancestor (Fig. 10-32). Within Neodermata, aspidogastreans and digeneans (together Trematoda) have long been perceived as allied to the dalyellioids on the basis of similarities in larvae, female reproductive system, and preference for a molluscan host (the intermediate host in digeneans). The sister taxon of the Trematoda, Cercomeromorpha, includes the monogeneans and cestodes. This alliance is based on the similarity of attachment hooks on the larval haptor (cercomer) of the two groups. A phylogeny of the Neodermata is shown in Figure 10-48.

"MESOZOA"

Mesozoans are small, parasitic marine worms of simple anatomy that inhabit the nephridia of octopuses and cuttlefishes (Dicyemida) as well as the tissues and body cavities of other invertebrates (Orthonectida). The cylindrical body is more or less radially symmetric, except during some bilateral

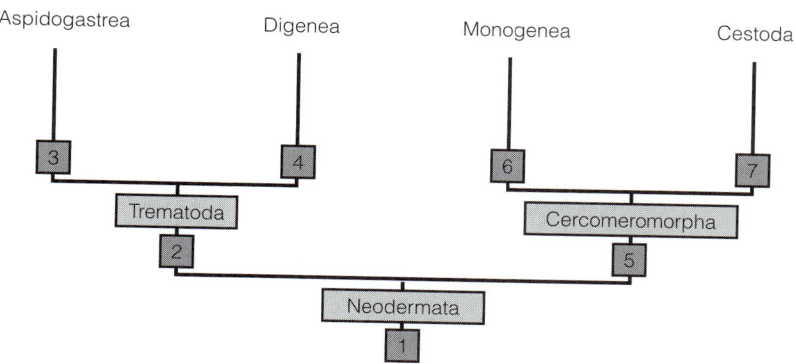

FIGURE 10-48 Phylogeny of Neodermata. **1, Neodermata:** neodermis. **2, Trematoda:** larval epidermis is a combination of ciliated cells and syncytial neodermis (tegument); Laurer's canal associated with female reproductive system (possibly a plesiomorphy); molluscs are primary hosts. **3, Aspidogastrea:** larva (cotylocidium) has scattered tufts of epidermal cilia and posterior ventral sucker; large, complex ventral adult sucker. **4, Digenea:** larva (miracidium) with transverse bands of ciliated cells alternating with neodermis; life cycle includes intermediate (mollusc) and definitive (vertebrate) hosts; hermaphroditic adults reproduce sexually in a vertebrate to produce miracidia; miracidium infects mollusc and reproduces clonally (as sporocyst) to produce rediae; redia in mollusc reproduces clonally to produce cercariae that infect vertebrate host. **5, Cercomeromorpha:** anchoring hooks on modified posterior end (cercomer) of larva. **6, Monogenea:** larva (onchomiracidium) has three widely separated groups of epidermal ciliated bands separated by a nonciliated epidermal syncytium (not a neodermis); onchomiracidium has a pair of pigment-cup ocelli. **7, Cestoda:** gut absent and food molecules absorbed through neodermis; ciliated larval epidermis syncytial (not a neodermis). *(Based on Ehlers, U. 1986. Comments on the phylogenetic system of the Plathelminthes. Hydrobiologia 132:1–12; Brooks, D. R. 1989. The phylogeny of the Cercomeria (Platyhelminthes: Rhabdocoela) and general evolutionary principles. J. Parasitol. 75:606–616; Ax, P. 1996. Multicellular Animals: a new approach to the phylogenetic order in nature. Vol. 1. Springer, Berlin. 225 pp; Tyler, S. 1999. Systematics of flatworms: Libbie Hyman's influence on current views of the Platyhelminthes. American Museum Novitates No. 3277. American Museum of Natural History, New York. pp. 52–66.)*

developmental stages. The compact body lacks gut, coelom, hemal system, nephridia, and neurons, and most species also lack muscle. The body wall consists solely of a monolayered, multiciliated epidermis around a core of one to several cells, including stem cells. The epidermis functions in locomotion and host nutrient uptake whereas the interior cells are reproductive. Mesozoan life cycles are complex, diverse, and, in most cases, not fully documented.

Because of their shared simplicity and parasitic habit, Orthonectida and Dicyemida were previously assigned to the now defunct "Mesozoa" (= middle animals), a taxon once regarded as a link between protozoa and Metazoa. Most zoologists consider their simple anatomy to be the result of convergent adaptations to small body size and endoparasitism, and the two taxa are now treated independently.

ORTHONECTIDA[P]

The 20-some species of orthonectids are endoparasites of turbellarians, nemerteans, polychaetes, gastropod and bivalve molluscs, brittle stars, and tunicates. The tiny adults, less than 1 mm in length (Fig. 10-49A,B), bear a series of epidermal **ciliary rings** that encircle the body. These rings alternate with rings of nonciliated cells in a species-specific fashion. The ciliary rootlets of adjacent cilia are interjoined in an intracellular network. A thin cuticle covers the body. Below the cuticle and epidermis are an extracellular matrix and myocytes containing very thick, longitudinally oriented filaments (Fig. 10-49D) composed of the protein **paramyosin.** Internal to the muscle sheath is a mass of gametes, either sperm or eggs.

Orthonectids are gonochoric, except for the hermaphroditic *Stoecharthrum,* and fertilization is internal after copulation. Females are larger than males and the interior of the body is occupied by a mass of individual eggs not enclosed in an ovarian epithelium (Fig. 10- 49A). In *Rhopalura ophiocomae,* a female gonopore is situated just posterior to the midbody. The male body is tripartite (Fig. 10-49B). The epidermis of the anterior region bears refractile granules of unknown composition, and the middle section houses an unpaired testis enclosed in a sheath of muscle cells. The third region bears the gonopore and encloses a notochord-like skeletal axis of four turgid muscle cells containing paramyosin (very thick) filaments (Fig. 10-49D). Nine longitudinal muscle cells surround the turgid cells and extend anteriorly into the first

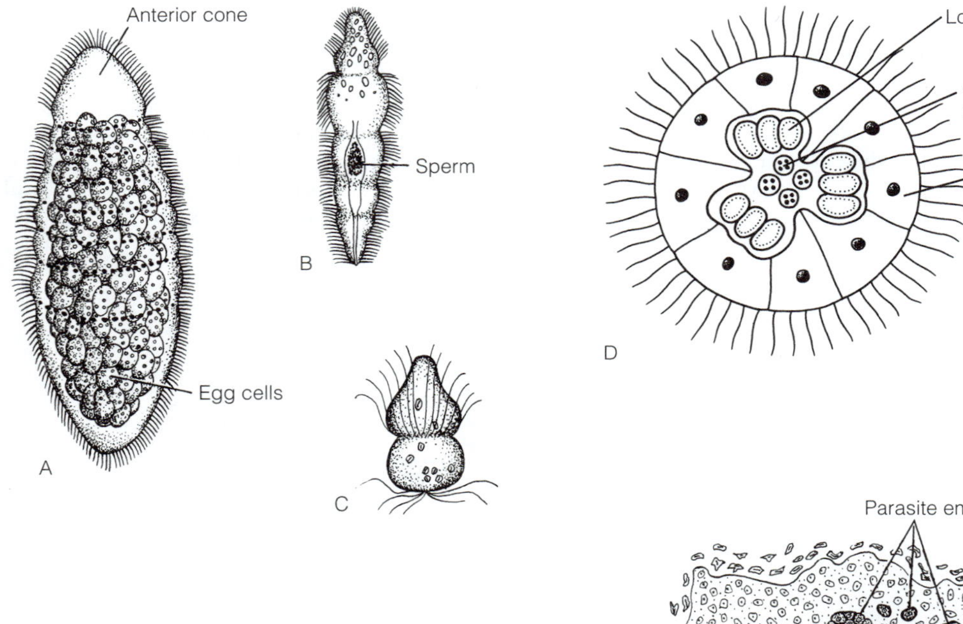

FIGURE 10-49 Orthonectida: anatomy and life cycle *(Rhopalura).* **A,** Mature female; **B,** male; **C,** ciliated larva of the clam parasite *R. granosa.* **D,** Simplified cross section of tail region of male of the brittle star parasite *R. ophiocomae.* **E,** Male plasmodium of another species of *Rhopalura.* *(A–C, After Atkins, 1933; D, Drawn from micrographs in Kozloff, E. N. 1969. Morphology of the orthonectid* Rhopalura ophiocomae. *J. Parasitol. 55:171–195; E, After Caullery and Mesnil, 1901)*

body region. Musculature and skeleton together enable the body to bend stiffly.

Rhopalura ophiocomae parasitizes the gestation sacs (genital bursae) of the viviparous brittle star, *Amphipholis squamata (Axiognathus squammata)*. The parasite renders the host's gonads nonfunctional and presumably absorbs nutrients intended for the brittle star's offspring. Full-grown parasites exit the host's genital bursae, copulate in the sea, and fertilize the eggs internally. Inside the viviparous female, zygotes develop into simple, ciliated larvae consisting of few cells (Fig. 10-49C). Presumably, the larvae are released from the female and somehow reenter a new host. Once inside the new host, the larvae attach to the bursal lining and develop into adults. The host tissue around the attached larvae becomes distinctively disorganized and is called a **plasmodium** (Fig. 10-49E). Other species of *Rhopalura* occur in clams and snails. Other orthonectid taxa are *Intoshia*, *Ciliocincta*, and *Stoecharthrum*.

DICYEMIDA[P]

The 75 species of dicyemids are delicate, slender worms that parasitize the kidneys of octopuses and cuttlefishes, but usually not squids. They occur, typically in high densities, in the renal sac and pericardial cavity, both of which are urine-containing compartments. One end of the parasite attaches to the nephridium's absorptive lining while the body endocytoses low-molecular-weight nutrients from the primary urine (Fig. 10-50A,D). It thus competes with the reabsorptive epithelium of the nephridium (and host).

The long, threadlike adult, called a **nematogen** (or vermiform), ranges in length from one to several millimeters (Fig. 10-50A,D). The multiciliated epidermis usually consists of 40 to 50 cells. At the anterior end, two rows of cells, called the **calotte,** bear short cilia that interdigitate with and adhere to the microvillar absorptive lining of the host's nephridium

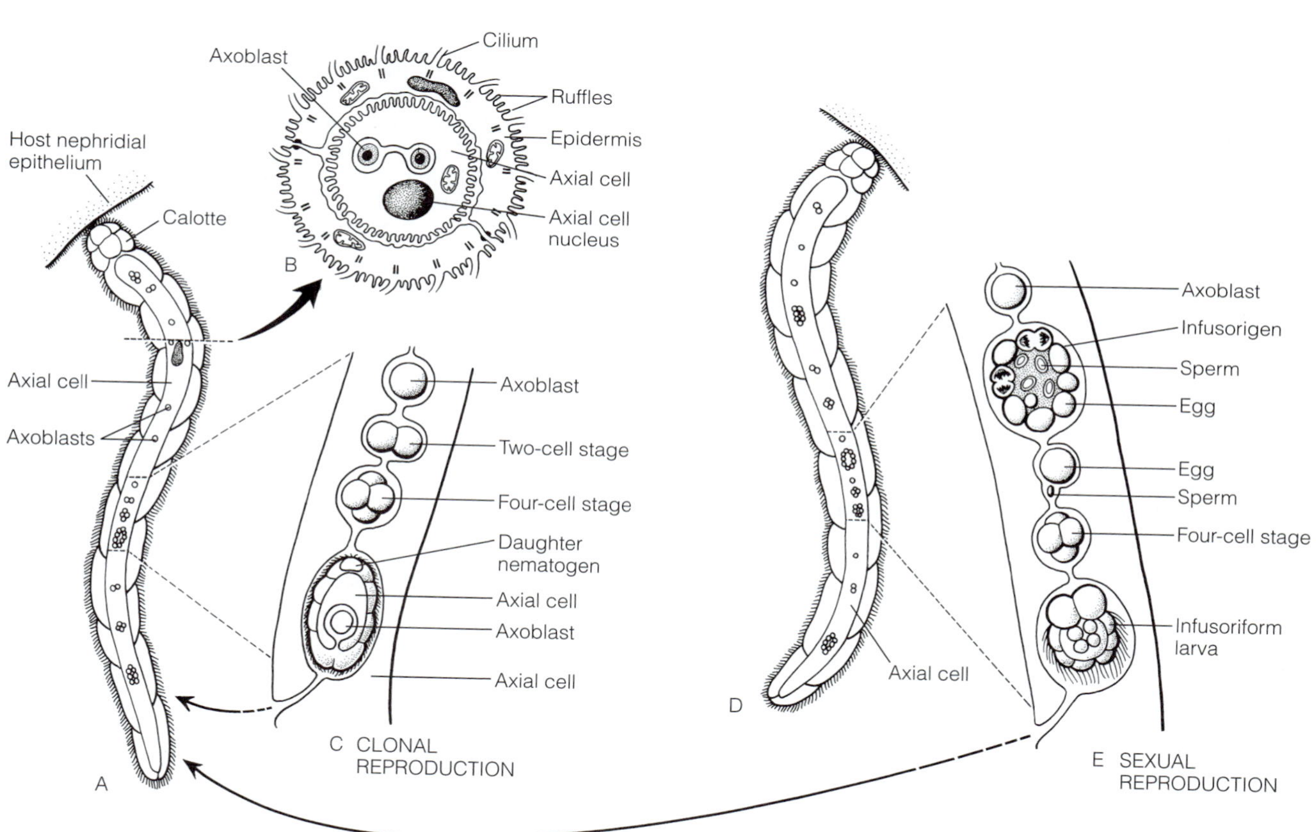

FIGURE 10-50 Dicyemida: anatomy and life cycle. **A,** Nematogen attached to host nephridial lining. **B,** Cross section of **A.** Epidermis and surface of axial cell both specialized for endocytosis of host nutrients. **C,** Clonal reproduction (possibly parthenogenesis) in axial cell of the nematogen shown in **A.** Daughter nematogen (juvenile) leaves body of parent and attaches to the nephridial wall of the host. **D,** Nematogen attached to host nephridial lining. **E,** Sexual reproduction in axial cell of the nematogen (now a rhombogen) shown in **D.** The infusorigen is a dwarf hermaphroditic individual. Its surface cells undergo meiosis to become (possibly haploid) eggs while the axoblasts in its axial cell become tailless sperm. The hermaphrodite self-fertilizes and the zygote undergoes spiral cleavage to become an infusoriform larva. The infusoriform larva leaves the nematogen and host via the latter's urine to somehow infect a new host. Once in the new host, the larva, or each of its interior cells, becomes a nematogen.

(Fig. 10-50A). Between the cilia, the epidermal surface is ruffled, increasing its area for uptake of host nutrients (Fig. 10-50B). The core of the body is occupied solely by a long **axial cell** with a single large nucleus (Fig. 10-50A,B); muscle cells are absent. Stem cells, each called an **axoblast,** occur inside the axial cell. Each axoblast or its cell-division product (a cluster of axoblasts) is enclosed in a vesicle and thus separated from the axial-cell cytoplasm (Fig. 10-50). The axial cell houses and transfers nutrients to the axoblasts.

Nematogens reproduce clonally and sexually. Once the parasite is established in the host's nephridium, clonal reproduction quickly increases the number of nematogens. When parasite densities reach a certain threshold, the nematogens reproduce sexually and release infective larvae. How infection spreads is unknown, but infective larvae discharged in the host's urine may infect the host's own offspring, which are ventilated by a jet of water (containing urine) from the parent's funnel. Other means of infecting new hosts are possible.

During clonal reproduction (Fig. 10-50A,C), an axoblast cell in the axial cell of the parent nematogen develops parthenogenetically into a daughter nematogen. Several daughter nematogens can be produced simultaneously in each parent. Once cilia develop on the daughter nematogen, it squeezes between the parental cells and emerges in the host's urine, where it attaches to the nephridial lining.

The onset of sexual reproduction triggers an axoblast (or several axoblasts) of the parent nematogen to divide and differentiate into a self-fertilizing, hermaphroditic individual called an **infusorigen** that remains permanently in the parent nematogen's axial cell (Fig. 10-50D,E). The infusorigen body differentiates into an outer sheath of eggs around an inner cluster of aflagellate sperm, the latter inside the infusorigen's axial cell. Meiosis probably occurs during spermatogenesis and oogenesis. Sometimes the infusorigen is interpreted as a hermaphroditic gonad. Following fertilization, cleavage is holoblastic, unequal, and spiral, the pattern reminiscent of spiral cleavage by duets as seen in acoel turbellarians. Subsequent development leads to a ciliated, bilaterally symmetric **infusoriform larva** (Fig. 10-50E). The larvae rupture from the parent nematogen, leave the host during urination, and infect new hosts.

PHYLOGENY OF ORTHONECTIDA AND DICYEMIDA

Although Orthonectida and Dicyemida share many characters—multiciliated absorptive epidermis and the absence of gut, coelom, hemal system, and nephridia—all of these traits, except the absence of nephridia, are common in small-bodied bilaterians and endoparasites. In the absence of a synapomorphy, it is appropriate to regard them as unrelated taxa.

Both taxa, however, are likely to be bilaterians. Orthonectid myocytes contain paramyosin, a protein that occurs sporadically in the muscles of many bilaterian taxa but is unknown elsewhere. Also similar to many bilaterians, orthonectids copulate and have permanent gonopores. Dicyemids also qualify as bilaterians. Their infusoriform larvae are bilaterally symmetric. The spiral cleavage pattern indicates not only a bilaterian alliance, but specifically a membership in Spiralia, which includes Platyhelminthes, Mollusca, and relatives. If they truly have duet spiral cleavage, Acoela may be the sister taxon of Dicyemida. Clonal reproduction in which parthenogenetic embryos develop viviparously in the body of the parent is another similarity between Dicyemida and Platyhelminthes (Digenea). Most molecular systematic studies place Dicyemida (Orthonectida has not been evaluated) in Bilateria, but do not identify a sister taxon.

REFERENCES

PLATYHELMINTHES

General

Apelt, G. 1969. Fortpflanzungsbiologie, Entwicklungszyklen und vergleichende Frühentwicklung acoeler Turbellarien. Mar. Biol. 4:267–325.

Arne, C., and Pappas, P. W. (Eds.): 1983. Biology of the Eucestoda. Academic Press, London. (Two volumes.)

Ax, P. 1963. Relationships and phylogeny of the Turbellaria. In Dougherty, E. C. (Ed.): The Lower Metazoa. University of California Press, Berkeley. pp. 191–224.

Ax, P. 1977. Life cycles of interstitial Turbellaria from the eulittoral of the North Sea. Acta Zool. Fennica 154:11–20.

Ax, P. 1996. Multicellular animals: A new approach to the phylogenetic order in nature. Vol. 1. Springer Berlin. 225 pp.

Ax, P., and Ax, R. 1977. Interstitielle Fauna von Galapagos. XIX. Monocelididae (Turbellaria, Proseriata). Mikrofauna Meeresbodens 64:1–40.

Ax, P., and Borkett, H. 1968. Organisation und Fortpflanzung von *Macrostomum romanicum* (Turbellaria, Macrostomida). Zool. Anz. Suppl. 32:344–347.

Ax, P., and Schilke, K. 1971. *Karkinorhynchus tetragnathus* nov. spec. ein Schizorhynchier mit zweigeteilten Russelhaken (Turbellaria, Kalyptorhynchia). Mikrofauna Meeresbodens 5:1–10.

Baguña, J. 1998. Planarians. In Ferretti, P., and Géraudie, J. (Eds.): Cellular and Molecular Basis of Regeneration: From Invertebrates to Humans. John Wiley and Sons, Chichester. pp. 135–165.

Balavoine, G. 1998. Are Platyhelminthes coelomates without a coelom? An argument based on the evolution of Hox genes. Am. Zool. 38:843–858.

Benham, W. D. 1901. The Platyhelmia, Mesozoa, and Nemertini. In Lankester, E. R. (Ed.): A Treatise on Zoology. Adam & Charles Black, London. pp. 204.

Bronsted, D. 1969. Planarian Regeneration. Pergamon Press, Oxford. 216 pp.

Brooks, D. R. 1989. The phylogeny of the Cercomeria (Platyhelminthes: Rhabdocoela) and general evolutionary principles. J. Parasitol. 75:606–616.

Bush, L. 1975. Biology of *Neochildia fusca* n. gen., n. sp. from the northeastern coast of the United States. Biol. Bull. 148:35–48.

Costello, H. M., and Costello, D. P. 1938. Copulation in the acoelous turbellarian *Polychoerus carmelensis*. Biol. Bull. 75:85–98.

Crezée, M. R. 1984. Turbellaria. In Parker, S. P. (Ed.): Synopsis and classification of living organisms. Vol. 1. McGraw-Hill, New York. pp. 718–740.

de Beauchamp, P. 1961. Classe des Turbellaries. In Grassé, P.-P. (Ed.): Traité de Zoologie. Vol. 4. Masson, Paris. pp. 35–213.

Doe, D. A. 1981. Comparative ultrastructure of the pharynx simplex in Turbellaria. Zoomorphologie 97:133–193.

Ehlers, U. 1985a. Das phylogenetisches System der Plathelminthes. Verlag Paul Parey, Stuttgart. 317 pp.

Ehlers, U. 1985b. Phylogenetic relationships within the Plathelminthes. In Conway Morris, S., George, J. D., Gibson, R., et al. (Eds.): The Origins and Relationships of Lower Invertebrates. Clarendon Press, Oxford. pp. 143–158.

Ehlers, U. 1986. Comments on a phylogenetic system of the Plathelminthes. Hydrobiologia 132:1–12.

Ehlers, U. 1995. The basic organization of the Plathelminthes. Hydrobiologia 305:21–26.

Ehlers, U., and Ax, P. 1974. Interstitielle Fauna von Galapagos. VIII. Trigonostominae (Turbellaria, Typhloplanoida). Mikrofauna Meeresbodens 30:1–33.

Ehlers, U., and Dörjes, J. 1979. Interstitielle Fauna von Galapagos. XXIII. Acoela (Turbellaria). Mikrofauna Meeresbodens 72:1–74.

Erasmus, D. A. 1972. The Biology of Trematodes. Crane, Russak, New York. 312 pp.

Ferguson, F. F. 1937. The morphology and taxonomy of *Macrostomum virginianum* n. sp. Zool. Anz. 119:25–32.

Fournier, A. 1984. Photoreceptors and photosensitivity in Platyhelminthes. In Ali, M. A. (Ed.): Photoreception and Vision in Invertebrates. Plenum Press, New York. pp. 217–240.

Freeman, R. 1973. Ontogeny of cestodes and its bearing on their phylogeny and systematics. Adv. Parasitol. 11:481–557.

Goss, R. 1969. Principles of Regeneration. Academic Press, New York. 287 pp.

Harrison, F. W., and Bogitsh, B. J. (Eds.): 1991. Microscopic Anatomy of Invertebrates. Vol. 1. Platyhelminthes and Nemertinea. Wiley-Liss, New York. pp. 1–347.

Heitkamp, U. 1977. The reproductive biology of *Mesostoma ehrenbergii.* Hydrobiologia 55:21–32.

Hendelberg, J. 1986. The phylogenetic significance of sperm morphology in the Platyhelminthes. Hydrobiologia 132:53–58.

Henley, C. 1974. Platyhelminthes (Turbellaria). In Giese, A. C., and Pearse, J. S. (Eds.): Reproduction of Marine Invertebrates. Vol. 1. Acoelomate and Pseudocoelomate Metazoans. Academic Press, New York. pp. 267–343.

Hermans, C. O. 1983. The duo-gland adhesive system. Oceanogr. Mar. Biol. Ann. Rev. 21:283–339.

Highnam, K. C., and Hill, L. 1977. The Comparative Endocrinology of the Invertebrates. 2nd Edition. University Park Press, Baltimore, MD. 357 pp.

Hooge, M. D. 2001. Evolution of body-wall musculature in the Platyhelminthes (Acoelomorpha, Catenulida, Rhaditophora). J. Morphol. 249:171–194.

Hooge, M. D., Haye, P. A., Tyler, S., Litvaitis, M. K., and Kornfield, I. 2002. Molecular systematics of the Acoela (Acoelomorpha, Platyhelminthes) and its concordance with morphology. Mol. Phylog. Evol. 24:333–342.

Hooge, M. D., and Tyler, S. 1999. Musculature of the facultative parasite *Urostoma cyprinae* (Platyhelminthes). J. Morphol. 241:207–216.

Jennings, J. B. 1957. Studies on feeding, digestion, and food storage in free-living flatworms. Biol. Bull. 112:63–80.

Jennings, J. B. 1968. Nutrition and digestion in Platyhelminthes. In Florkin, M., and Scheer, B. T. (Eds.): Chemical Zoology. Vol. 2. Academic Press, New York. pp. 305–327.

Jennings, J. B. 1974. Digestive physiology of the Turbellaria. In Riser, N. W., and Morse, M. P. (Eds.): Biology of the Turbellaria. McGraw-Hill, New York. pp. 173–197.

Joffe, B. I., and Cannon, L. R. G. 1998. The organization and evolution of the mosaic of the epidermal syncytia in the Temnocephalida (Platyhelminthes: Neoophora). Zool. Anz. 237:1–14.

Karling, T. G. 1974. On the anatomy and affinities of the turbellarian orders. In Riser, N. W., and Morse, M. P. (Eds.): Biology of the Turbellaria. McGraw-Hill, New York. pp. 1–16.

Kenk, R. 1972. Freshwater Planarians (Turbellaria) of North America. Biota of Freshwater Ecosystems. Identification Manual No. 1. Environmental Protection Agency, Washington, DC. 81 pp.

Kepner, W. A., and Helvestine, F. Jr. 1920. Pharynx of *Microstoma caudatum.* J. Morphol. 33:309–316.

Klauser, M. D., and Tyler, S. 1987. Frontal glands and frontal sensory structures in the Macrostomida (Turbellaria). Zool. Scripta 16:95–110.

Koopowitz, H. 1974. Some aspects of the physiology and organization of the nerve plexus in polyclad flatworms. In Riser, N. W., and Morse, M. P. (Eds.): Biology of the Turbellaria. McGraw-Hill, New York. pp. 198–212.

Korschelt, E., and Heider, K. 1895. Text-Book of the Embryology of Invertebrates. Pt. 1. Sawn Sonnenschein, New York. pp. 170–172.

Kozloff, E. N. 1972. Selection of food, feeding and physical aspects of digestion in the acoel turbellarian *Otocelis luteola.* Trans. Am. Microsc. Soc. 91:556–565.

Lauer, D. M., and Fried, B. 1977. Observations on nutrition of *Bdelloura candida,* an ectocommensal of *Limulus polyphemus.* Am. Midl. Nat. 97:240–247.

Lentz, T. L. 1968. Primitive Nervous Systems. Yale University Press, New Haven. 148 pp.

L'Hardy, J.-P. 1965. Turbellaries Schizorhynchidae des sables de Roscoff. II. Le genre *Proschizorhynchus.* Cah. Biol. Mar. 6:135–161.

Littlewood, D. T. J., Rohde, K., and Clough, K. A. 1999. The interrelationships of all major groups of Platyhelminthes: Phylogenetic evidence from morphology and molecules. Biol. J. Linnean Soc. 66:75–114.

Livaitis, M. K., and Rohde, K. 1999. A molecular test of platyhelminth phylogeny: Inferences from partial 28S rDNA sequences. Invert. Biol. 118:42–56.

Llewellyn, J. 1965. The evolution of parasitic Platyhelminthes. In Taylor, A. E. R. (Ed.): Evolution of Parasites. 3rd Symposium for the British Society for Parasitology. Blackwell, London. pp. 47–78.

Mac-Fira, V. 1974. The turbellarian fauna of the Romanian littoral waters of the Black Sea and its annexes. In Riser, N. W., and Morse, M. P. (Eds.): Biology of the Turbellaria. McGraw-Hill, New York. pp. 248–290.

Moraczewski, J. 1977. Asexual reproduction and regeneration of *Catenula.* Zoomorphologie 88:65–80.

Morita, M., and Best, J. B. 1984. Effects of photoperiods and melatonin on planarian asexual reproduction. J. Exp. Zool. 231:273–282.

Muscatine, L., Boyle, J. E., and Smith, D. C. 1974. Symbiosis of the acoel flatworm *Convoluta roscoffensis* with the alga *Platymonas convolutae*. Proc. Roy. Soc. Lond. 187:221–234.

Noldt, U., and Reise, K. 1987. Morphology and ecology of the kalyptorhynch *Typhlopolycystis rubra* (Platyhelminthes), an inmate of lugworm burrows in the Wadden Sea. Helgol. Wiss. Meeresunters. 41:185–199.

Nuttycombe, J. W. 1931. Two new species of *Stenostomum* from the southeastern United States. Zool. Anz. 97:80–85.

Nuttycombe, J. W., and Waters, A. J. 1938. The American species of the genus *Stenostomum*. Proc. Am. Philos. Soc. 79:213–300.

Palladini, G., Medolago-Albani, L., Margotta, V., et al. 1979. The pigmentary system of planaria: 2. Physiology and functional morphology. Cell Tissue Res. 199:203–211.

Prusch, R. D. 1976. Osmotic and ionic relationships in the freshwater flatworm *Dugesia dorotocephala*. Comp. Biochem. Physiol. 54A:287–290.

Reisinger, E. 1923. Turbellaria. Strüdelwurmer. Biol. Tiere Deutsch. Lief. 6:1–64.

Reuter, M., and Halton, D. W. 2001. Comparative neurobiology of Platyhelminthes. In Littlewood, D. T. J., and Bray, R. A. (Eds.): Interrelationships of the Platyhelminthes. Taylor & Francis, London. pp. 239–249.

Reuter, M., Raikova, O. I., Jondelius, U., Gustafsson, M. K. S., Maule, A. G., and Halton, D. W. 2001. Organization of the nervous system in the Acoela: An immunocytochemical study. Tissue Cell 33:119–128.

Reynoldson, T. B., and Sefton, A. D. 1976. The food of *Planaria torva*, a laboratory and field study. Freshwat. Biol. 6:383–393.

Rieger, R. M. 1974. A new group of Turbellaria-Typhloplanoida with a proboscis and its relationship to Kalyptorhynchia. In Riser, N. W., and Morse, M. P. (Eds.): Biology of the Turbellaria. McGraw-Hill, New York. pp. 23–62.

Rieger, R. M. 1981. Morphology of the Turbellaria at the ultrastructural level. Hydrobiologia 84:213–229.

Rieger, R. M. 1985. The phylogenetic status of the acoelomate organization with the Bilateria: A histological perspective. In Conway Morris, S., George, J. D., Gibson, R., et al. (Eds.): The Origins and Relationships of Lower Invertebrates. Clarendon Press, Oxford. pp. 101–122.

Rieger, R. M., and Sterrer, W. 1973. New spicular skeletons in Turbellaria, and the occurrence of spicules in marine meiofauna. Z. zool. Syst. Evolut.-forsch. 13:207–248.

Rieger, R. M., Tyler, S., Smith, J. P. III, et al. 1991. Turbellaria. In Harrison, F. W., and Bogitsh, B. J. (Eds.): Microscopic Anatomy of Invertebrates. Vol. 3. Platyhelminthes and Nemertinea. Wiley-Liss, New York. pp. 7–140.

Riser, N. W., and Morse, M. P. (Eds.): 1974. Biology of the Turbellaria. McGraw-Hill, New York. 530 pp.

Ruppert, E. E. 1978. A review of metamorphosis of turbellarian larvae. In Chia, F. S., and Rice, M. (Eds.): Settlement and Metamorphosis of Marine Invertebrate Larvae. Elsevier North Holland Biomedical Press, New York. pp. 65–81.

Schell, S. C. 1970. How to Know the Trematodes. W. C. Brown, Dubuque, IA. 355 pp.

Schmidt, P., and Sopott-Ehlers, B. 1976. Interstitielle Fauna von Galapagos. XV. *Macrostomum* O. Schmidt, 1848 und *Siccomacrostomum triviale* nov. gen. nov. spec. (Turbellaria, Macrostomida). Mikrofauna Meeresbodens 57:1–44.

Schockaert, E. R., and Ball, I. R. (Eds.): 1981. The Biology of the Turbellaria. Dr. W. Junk, The Hague. 300 pp.

Shinn, G. L. 1987. Phylum Platyhelminthes with emphasis on marine Turbellaria. In Strathmann, M. F. (Ed.): Reproduction and Development of Marine Invertebrates of the Northern Pacific Coast. University of Washington Press, Seattle. pp. 114–128.

Shinn, G. L., and Christensen, A. M. 1985. *Kronborgia pugettensis* sp. nov. (Neorhabdocoela: Fecampiidae), an endoparasitic turbellarian infesting the shrimp *Heptacarpus kincaidi* (Rathbun), with notes on its life history. Parasitology 91:431–447.

Shinn, G. L., and Cloney, R. A. 1986. Egg capsules of a parasitic flatworm: Ultrastructure of hatching sutures. J. Morphol. 188:15–28.

Slais, J. 1973. Functional morphology of cestode larvae. Adv. Parasitol. 11:395–480.

Smith, J. P. S. III, and Tyler, S. 1985a. The acoel turbellarians: Kingpins of metazoan evolution or a specialized offshoot? In Conway Morris, S., George, J. D., Gibson, R. et al. (Eds.): The Origins and Relationships of Lower Invertebrates. Clarendon Press, Oxford. pp. 123–142.

Smith, J. P. S. III., and Tyler, S. 1985b. Fine-structure and evolutionary implications of the frontal organ in Turbellaria Acoela. 1. *Diopisthoporus gymnopharyngeus* sp. n. Zool. Scripta 14:91–102.

Smith, J. P. S. III, Tyler, S., and Rieger, R. M. 1986. Is the Turbellaria polyphyletic? Hydrobiologia 132:13–21.

Smith, J. P. S. III, Tyler, S., Thomas, M. B., et al. 1982. The morphology of turbellarian rhabdites: Phylogenetic implications. Trans. Amer. Micros. Soc. 101:209–228.

Smyth, J. D., and Halton, D. W. 1983. The Physiology of Trematodes. 2nd Edition. Cambridge University Press, Cambridge. 446 pp.

Sterrer, W. 1966. New polylithophorous marine Turbellaria. Nature 210:436.

Sterrer, W., and Rieger, R. M. 1974. Retronectidae: A new cosmopolitan marine family of Catenulida (Turbellaria). In Riser, N. W., and Morse, M. P. (Eds.): Biology of the Turbellaria. McGraw-Hill, New York. pp. 63–82.

Stummer-Traunfels, R. V. 1933. Polycladida. In Bronns, H. G. (Ed.): Klassen und Ordnungen des Tierreichs. Vol. 4. Akademische Verlagsgesellschaft, Leipzig. pp. 3485–3596.

Tempel, D., and Westheide, W. 1980. Uptake and incorporation of dissolved amino acids by interstitial Turbellaria and Polychaeta and their dependence on temperature and salinity. Mar. Ecol. Prog. Ser. 3:41–50.

Thomas, M. B. 1986. Embryology of the Turbellaria and its phylogenetic significance. Hydrobiologia 132:105–115.

Tyler, S. 1973. An adhesive function for modified cilia in an interstitial turbellarian. Acta Zool. 54:139–151.

Tyler, S. 1976. Comparative ultrastructure of adhesive systems in the Turbellaria. Zoomorphologie 84:1–76.

Tyler, S. 1984. Turbellarian platyhelminths. In Bereiter-Hahn, J., Matoltsky, A. G., and Richards, K. S. (Eds.): Biology of the Integument. Vol. 1. Invertebrates. Springer-Verlag, Berlin. pp. 112–131.

Tyler, S. 1999. Platyhelminthes. In Knobil, E., and Neill, J. D. (Eds.): Encyclopedia of Reproduction. Vol. 3. Academic Press, San Diego. pp. 901–908.

Tyler, S. 1999. Systematics of the flatworms: Libbie Hyman's influence on current views of the Platyhelminthes. American Museum Novitates No. 3277. American Museum of Natural History, New York. pp. 52–66.

Tyler, S. 2001. The early worm: Origins and relationships of the lower flatworms. In Littlewood, D. T. J., and Bray, R. (Eds.): Interrelationships of the Platyhelminthes. London: Taylor & Francis. pp 3–12.

Tyler, S. (Ed.): 1986. Advances in the biology of turbellarians and related platyhelminths. Hydrobiologia 132:1–357.

Tyler, S. (Ed.): 1986. Advances in the Biology of Turbellarians and Related Platyhelminths. Dr. W. Junk, Dordrecht, The Netherlands. 357 pp.

Tyler, S. (Ed.): 1991. Turbellarian Biology. Dr. W. Junk, Dordrecht, Netherlands. 398 pp.

Tyler, S., and Rieger, R. M. 1975. Uniflagellate spermatozoa in *Nemertoderma* and their phylogenetic significance. Science 188:730–732.

Tyler, S., and Rieger, R. M. 1999. Functional morphology of musculature in the acoelomate worm *Convoluta pulchra* (Platyhelminthes). Zoomorphology 119:127–141.

Tyler, S., and Tyler, M. S. 1997. Origin of the epidermis in parasitic platyhelminths. Int. J. Parasitol. 27:715–738.

Westheide, W., and Rieger, R. M. (Eds.): 1996. Spezielle Zoologie. Teil 1. Einzeller und Wirbellose Tiere. Gustav Fischer Verlag, Stuttgart. 909 pp.

Williams, J. B. 1980. Morphology of a species of *Temnocephala* (Platyhelminthes) ectocommensal on the isopod *Phreatoicopsis terricola*. J. Nat. Hist. 14:183–199.

Wright, C. A. 1971. Flukes and Snails. Science of Biology Series 4. Allen and Unwin, London. 168 pp.

Parasitology Textbooks

Baer, J. G. 1971. Animal Parasites. McGraw-Hill, New York. 256 pp.

Chappell, L. H. 1980. Physiology of Parasites. John Wiley and Sons, New York. 230 pp.

Cheng, T. C. 1973. General Parasitology. Academic Press, New York. 965 pp.

Cox, F. E. G. 1982. Modern Parasitology. Blackwell Scientific, Oxford. 346 pp.

Noble, E. R., and Noble, G. A. 1982. Parasitology. 5th Edition. Lea & Febiger, Philadelphia. 522 pp.

Olsen, O. W. 1967. Animal Parasites: Their Biology and Life Cycles. Burgess, Minneapolis. 431 pp.

Read, C. P. 1970. Parasitism and Symbiology. Ronald Press, New York. 316 pp.

Roberts, L. S., and Janovy, J. Jr. 2000. Gerald D. Schmidt and Larry S. Roberts' Foundations of Parasitology. 6th Edition. McGraw-Hill, New York. 670 pp.

Smyth, J. D. 1976. Introduction to Animal Parasitology. 2nd Edition. John Wiley and Sons, New York. 466 pp.

Whitfield, P. J. 1979. The Biology of Parasitism: An Introduction to the Study of Associating Organisms. University Park Press, Baltimore, MD. 277 pp.

Internet Sites

http://devbio.umesci.maine.edu/styler/turbellaria/ (Seth Tyler's taxonomic database for Turbellaria.)

http://hooge.developmentalbiology.com/acoel/ (Matt Hooge's illustrated guide to the taxonomy of marine acoels.)

www.rzuser.uni-heidelberg.de/~bu6/flatintr.htm#introdu (A site devoted to the colorful polyclads.)

http://citd.scar.utoronto.ca/EESC04/SCMEDIA/INVPHYLO/Platyhelminthes/Table.html (Color images of a variety of platyhelminths.)

MESOZOA

Furuya, H., Hochberg, F. G., and Tsuneki, K. 2001. Developmental patterns and cell lineages of vermiform embryos in dicyemid mesozoans. Biol. Bull. 201:405–416.

Hochberg, F. G. 1982. The "kidneys" of cephalopods: A unique habitat for parasites. Malacologia 23:121–134.

Horvath, P. 1997. Dicyemid mesozoans. In Gilbert, S. F., and Raunio, A. M. (Eds.): Embryology: Constructing the Organism. Sinauer, Sunderland, MA. pp. 31–38.

Katayama, T., Wada, H., Furuya, H., Satoh, N., and Yamamoto, M. H, 1995. Phylogenetic position of the dicyemid Mesozoa inferred from 18S rDNA sequences. Biol. Bull. 189:81–90.

Kobayashi, M., Furuya, H., and Holland, P. W. H. 1999. Evolution: Dicyemids are higher animals. Nature 401:762.

Kozloff, E. N. 1969. Morphology of the orthonectid *Rhopalura ophiocomae*. J. Parasitol. 55:171–195.

Kozloff, E. N. 1971. Morphology of the orthonectid *Ciliocincta sabellariae*. J. Parasitol. 57:585–597.

Kozloff, E. N. 1993. The structure and origin of the plasmodium of *Rhopalura ophiocomae* (Phylum Orthonectida). Acta Zool. (Stockh.) 75:191–199.

Kozloff, E. N. 1997. Studies on the so-called plasmodium of *Ciliocincta sabellariae* (Phylum Orthonectida), with notes on an associated microsporan parasite. Cah. Biol. Mar. 38:151–159.

Lapan, E. A., and Morowitz, H. 1972. The Mesozoa. Sci. Am. 227:94–101.

Matsubara, J. A., and Dudley, P. L. 1976a. Fine structural studies of the dicyemid mesozoan *Dicyemmenea californica* McConnaughey. I. Adult stages. J. Parasitol. 62:377–389.

Matsubara, J. A., and Dudley, P. L. 1976b. Fine structural studies of the dicyemid mesozoan *Dicyemmenea californica* McConnaughey. II. The young vermiform stage and the infusoriform larva. J. Parasitol. 62:390–409.

Stunkard, H. W. 1954. The life history and systematic relations of the Mesozoa. Q. Rev. Biol. 29:230–244.

Nemertea[P]

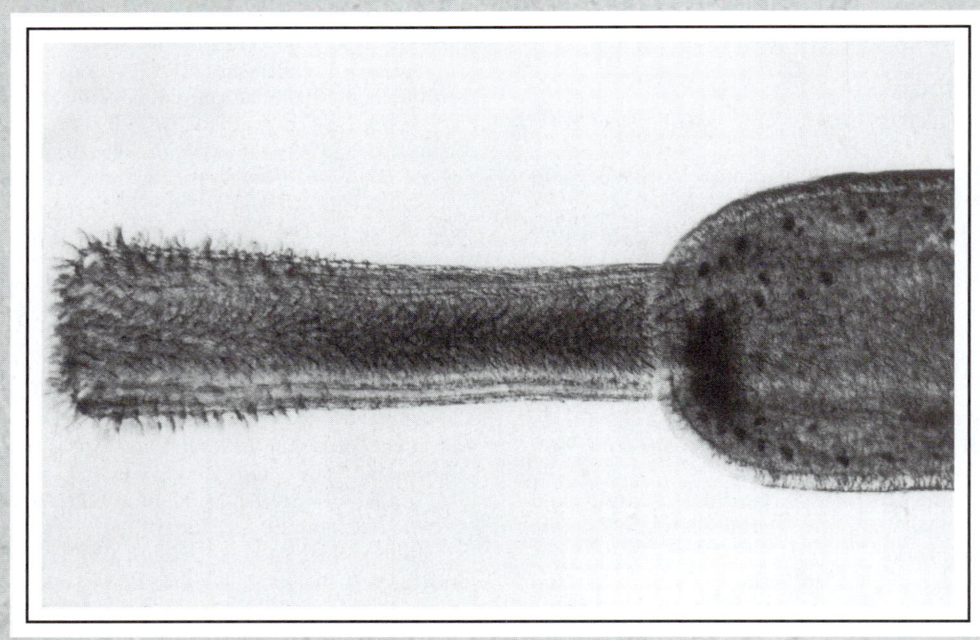

The 1150 species of ribbon worms are colorful ambush predators that harpoon or lasso their prey with a venomous, penetrating, or sticky proboscis. The proboscis, which is everted rapidly like a giant nematocyst tubule, is of prodigious length, often longer than the entire animal. Nemerteans are long, slender worms, with several species exceeding 1 m. The length record of 180 feet (54 m) is held by a boot-lace worm, *Lineus longissimus,* the longest animal on Earth, that washed ashore after a storm in St. Andrews, Scotland. Nemerteans commonly burrow in marine sediments, lurk (or lie in tangles) in crevices among shells, stones, and the rootlike holdfasts of algae and sessile animals, but some are deepwater pelagic species with gelatinous bodies. Others form semipermanent burrows lined with mucus or even produce unique cellophane-like tubes. A few species live as ectosymbionts on crabs, in the mantle cavity of bivalve molluscs, or in the atrium of tunicates. Approximately 12 species occur in fresh water and some 15 terrestrial species are confined primarily to the humid tropics and subtropics. The relatively few animals that eat nemerteans include bottom-feeding fishes, some shorebirds, and other invertebrates such as horseshoe crabs, as well as fellow nemerteans. Two large ribbon worms, the North American *Cerebratulus lacteus* and the South African *Polybrachiorhynchus dayi,* known locally as "tapeworms," are collected and sold as fish bait. They, of course, are unrelated to true tapeworms and are not parasites.

GENERAL BIOLOGY

FORM

Nemerteans resemble flatworms and annelids, but tend to be more robust and elongate than most flatworms and lack the segmentation of annelids (Fig. 11-1). The epidermis of a few species is annulated and internal organs, such as gonads and nephridia, are serially arranged, but this segmentation does not extend to muscles, nerves, or other internal structures (Fig. 11-1A,B,E). The anterior end is commonly pointed, rounded, or spatula-shaped. Most species are less than 20 cm in length and some are only a few millimeters long, but *Cerebratulus* and *Lineus* may exceed a meter in length. As you would predict from a consideration of body size, surface area, and volume, the smallest nemerteans are circular or only slightly flattened in

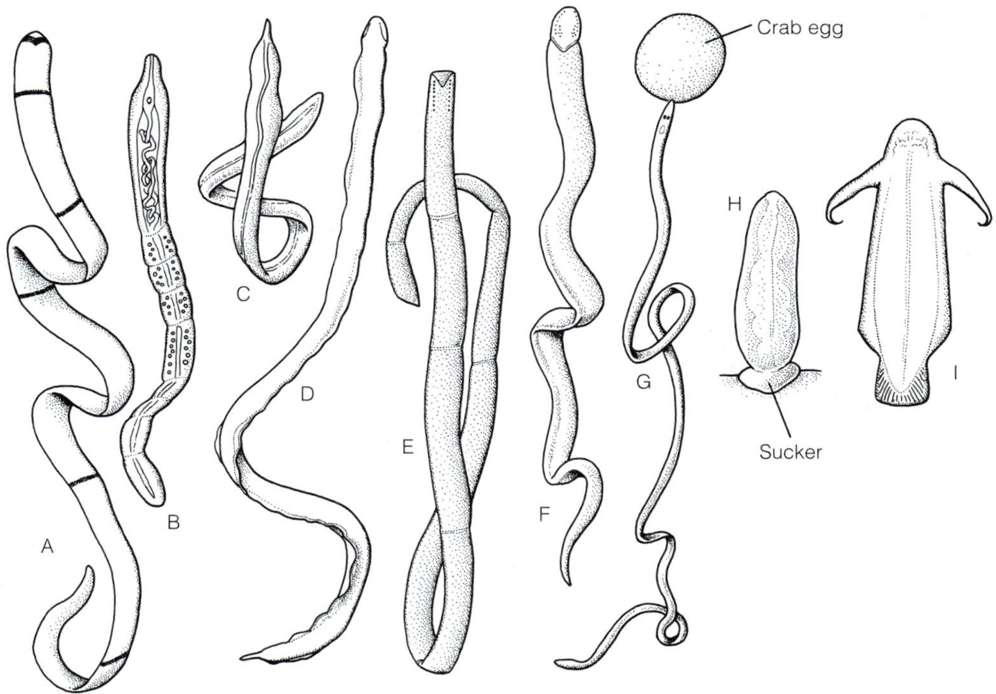

FIGURE 11-1 Nemertea: diversity **A,** *Tubulanus rhabdotus,* a tube-dwelling palaeonemertean with dark crossbands. **B,** An unidentified annulated, sand-dwelling palaeonemertean. **C,** *Carinoma tremaphoros,* a burrowing palaeonemertean. **D,** *Cerebratulus lacteus,* a large, burrowing heteronemertean. **E,** *Lineus socialis,* a crevice-dwelling heteronemertean. **F,** *Paranemertes peregrina,* a burrowing hoplonemertean. Note ocelli on heads of **E–G. G,** *Carcinonemertes carcinophila,* a hoplonemertean parasitic on crabs (shown feeding on egg). **H,** *Malacobdella grossa,* a bdellonemertean commensal of clams. **I,** *Nectonemertes mirabilis,* a pelagic hoplonemertean. *(F, Redrawn from Coe, W. R. 1901. The nemerteans. Wash. Acad. Sci. Proc. 3:1–110; H, Redrawn from Coe, W. R. 1951. The nemertean faunas of the Gulf of Mexico and southern Florida. Bull. Mar. Sci. 1:149–186.; I, Redrawn from Coe, W. R. 1926. The pelagic nemerteans. Mem. Mus. Comp. Zool. 49:1–244.)*

cross section, whereas larger-bodied species are dorsoventrally flattened (ribbon-shaped). Nemerteans often are boldly pigmented and patterned in yellow, orange, red, and green, but some are pale and nondescript. Many deepwater pelagic species, which live in darkness, are bright red, orange, or yellow.

BODY WALL, LOCOMOTION, AND EXTENSIBILITY

The body wall (Fig. 11-2C) is composed of a densely ciliated, glandular epidermis, a layer of connective tissue, and a thick musculature. A cuticle is absent and each ciliated cell bears many cilia and microvilli. Nemerteans lack a protective cuticle or exoskeleton, but the epidermis secretes a sticky, toxic mucus to discourage predators. The bright and contrasting colors of some species are believed to be aposematic, alerting would-be predators to the nemertean's distastefulness. Beneath the epidermis is the well-developed musculature, which usually consists of at least three layers. The outermost layer is circular, the innermost is usually longitudinal and often very thick (Fig. 11-2C), with crisscrossing helical muscles between the circulars and longitudinals. The helical muscles allow the worms to twist and coil. A series of dorsoventral muscles, which flatten the body, is also present. A connective-tissue layer, developed to a greater or lesser extent, lies between the musculature and the gut of some species. In pelagic nemerteans, this connective tissue is gelatinous and buoyant.

Most nemerteans use their epidermal cilia to glide over the substratum on a trail of slime, some of which is secreted by **cephalic glands** on the head (Fig. 11-2A,B). The larger species, such as *Cerebratulus,* use muscular waves to crawl over surfaces. Burrowing nemerteans such as *Cerebratulus, Carinoma,* and *Zygeupolia* do not use cilia for locomotion, but instead employ a muscular peristalsis (Fig. 11-3). The peristaltic waves are most powerful anteriorly, where the fluid-filled rhynchocoel of the proboscis apparatus functions as a hydrostat. Burrowing nemerteans often have extremely muscular body walls. Indeed, branches from the mesodermal muscles of some nemerteans enter the epidermis to form extra muscle layers. Still other ribbon worms, such as *Cerebratulus,* swim using dorsoventral undulations of the body. In these, the dorsoventral musculature, which flattens the body, is particularly well developed.

Most nemerteans are remarkably extensible worms, a trait they share only with a few annelids. Some, such as *Micrura, Lineus,* and *Tubulanus,* can extend to at least 10 times the

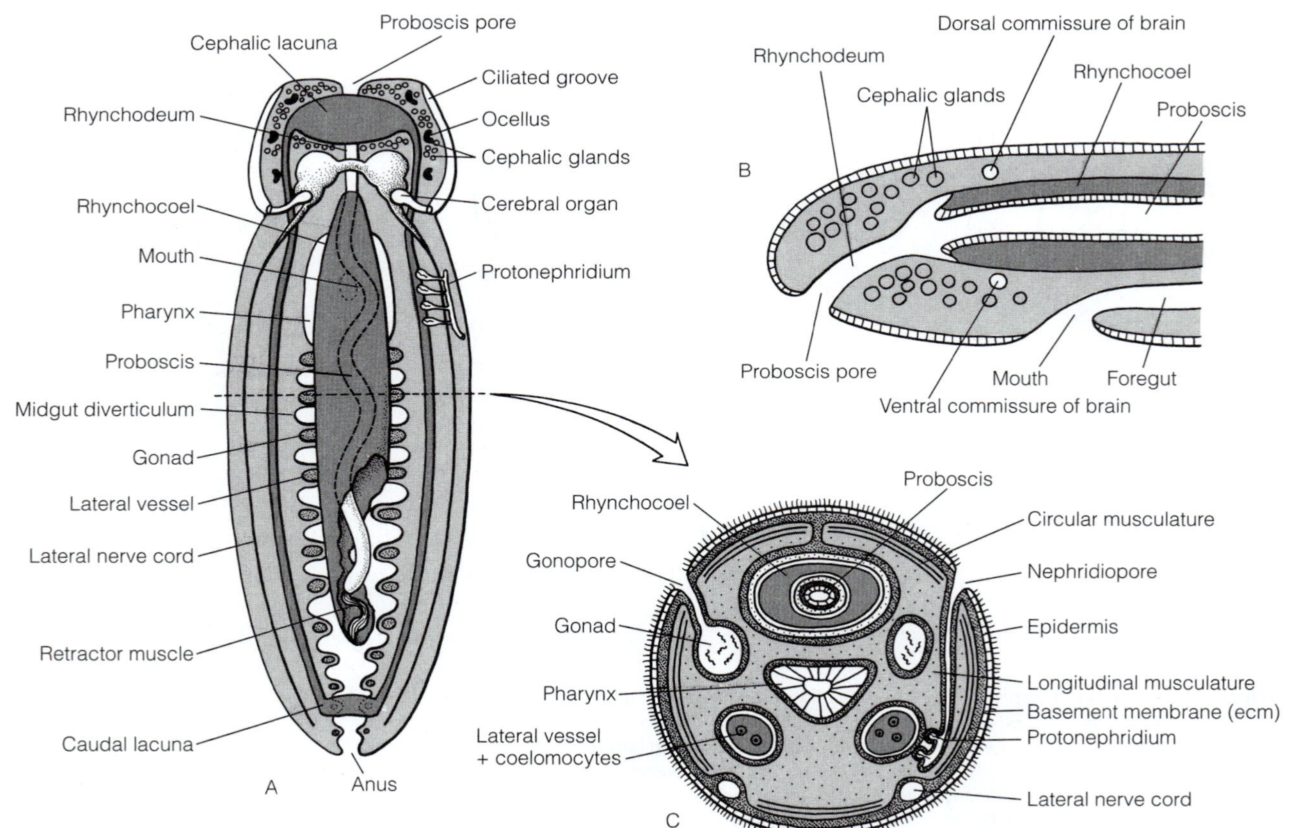

FIGURE 11-2 Anatomy of unarmed nemerteans. **A,** Dorsal view with rhynchocoel wall partly broken away to show the proboscis within. **B,** Sagittal section of head of a palaeo- or heteronemertean. **C,** Cross section of a palaeonemertean trunk.

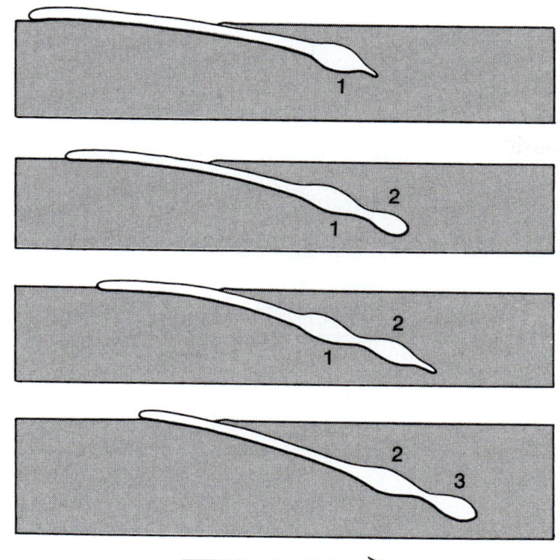

FIGURE 11-3 Nemertea: peristaltic burrowing in *Carinoma tremaphoros.* Peristaltic waves (numbered) originate anteriorly and progress rearward as the animal burrows. Peristalsis is restricted to the section of the body in which the rhynchocoel (fluid skeleton) is well developed. *(Modified and redrawn from Turbeville, J. M., and Ruppert, E. E. 1983. Epidermal muscles and peristaltic burrowing in* Carinoma tremaphoros: *Correlates of effective burrowing without segmentation. Zoomorphology 103:103–120.)*

resting body length. The muscles are smooth (in palaeonemerteans) or obliquely striated (in all others) mesodermal myocytes (Fig. 6-4C, 6-5C,D).

PROBOSCIS AND RHYNCHOCOEL

Nemerteans have a characteristic **proboscis apparatus** used to capture prey and sometimes to burrow (Fig. 11-2A,B, 11-4). The **proboscis** is a long, extensible, muscular tube lying in a fluid-filled coelomic cavity called the **rhynchocoel.** The inner end of the proboscis is blind and attached to the rear of the rhynchocoel by a greatly extensible **retractor muscle** (Fig. 11-2A) that can stretch up to 30 times its resting length. Anteriorly, the lumen of the proboscis joins a short canal known as the **rhynchodeum,** which begins at approximately the level of the brain and extends forward to a **proboscis pore** on the anterior tip of the worm (Fig. 11-2A,B). The proboscis everts (turns inside out) through the proboscis pore as muscles in the rhynchocoel wall compress the rhynchocoel (Fig. 11-4D); the retractor muscle pulls the proboscis back into the rhynchocoel. Both rhynchodeum and proboscis develop as body-wall invaginations, and their walls are similar to the body wall.

In some nemerteans (Anopla), the proboscis is a simple unbranched or branched tube (Fig. 11-4A,B), but in others, called armed nemerteans (Enopla), the proboscis usually bears one calcareous barb called a **stylet** (some species have many stylets), which is attached to the proboscis wall by a bulbous, secreted structure called the **basis** (Fig. 11-4C; 11-5A,C,D). The point of stylet attachment is not at the end of the proboscis,

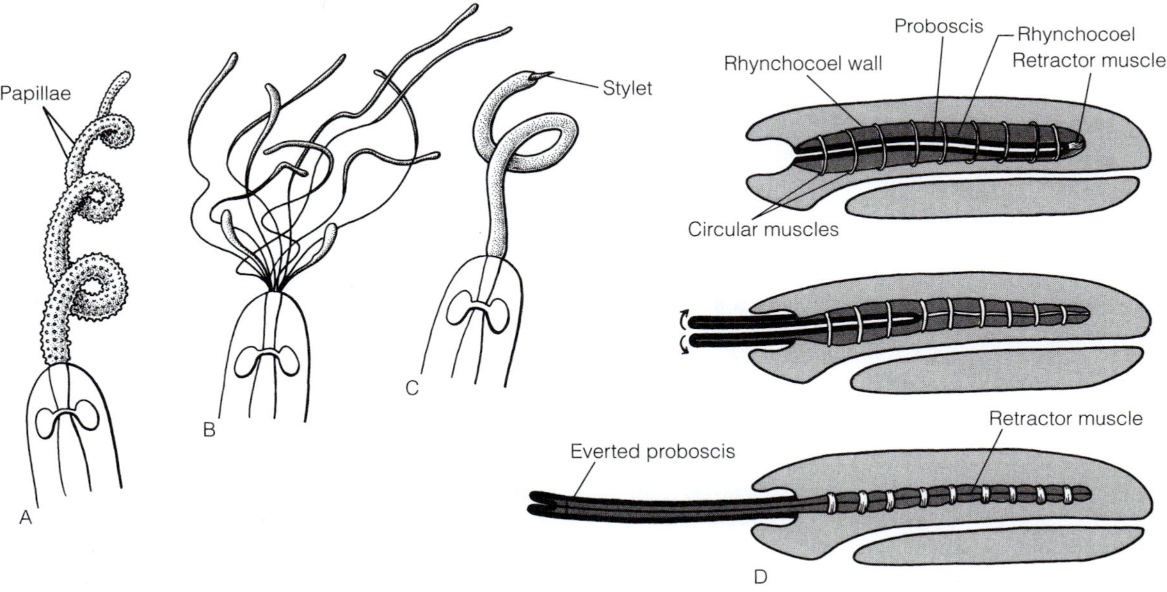

FIGURE 11-4 Nemertea: proboscis types and proboscis eversion. **A,** Palaeo- and heteronemerteans have a proboscis that lacks a calcified stylet and is typically covered with sticky papillae. **B,** A branched proboscis occurs in species of the heteronemerteans *Gorgonorhynchus* and *Polybrachiorhynchus.* In these, the proboscis erupts from its pore like a mass of sticky spaghetti and entangles the prey. **C,** An armed proboscis of a monostiliferan hoplonemertean. **D,** Contraction of circular muscles in the rhynchocoel wall raises rhynchocoelic fluid pressure, causing the proboscis to evert. Proboscis eversion stretches the proboscis retractor muscle. On contraction, the retractor muscle will pull the proboscis back into the rhynchocoel. *(B, Redrawn from Sterrer, W. 1986. Marine Fauna and Flora of Bermuda. John Wiley and Sons, Inc., New York. p. 210.)*

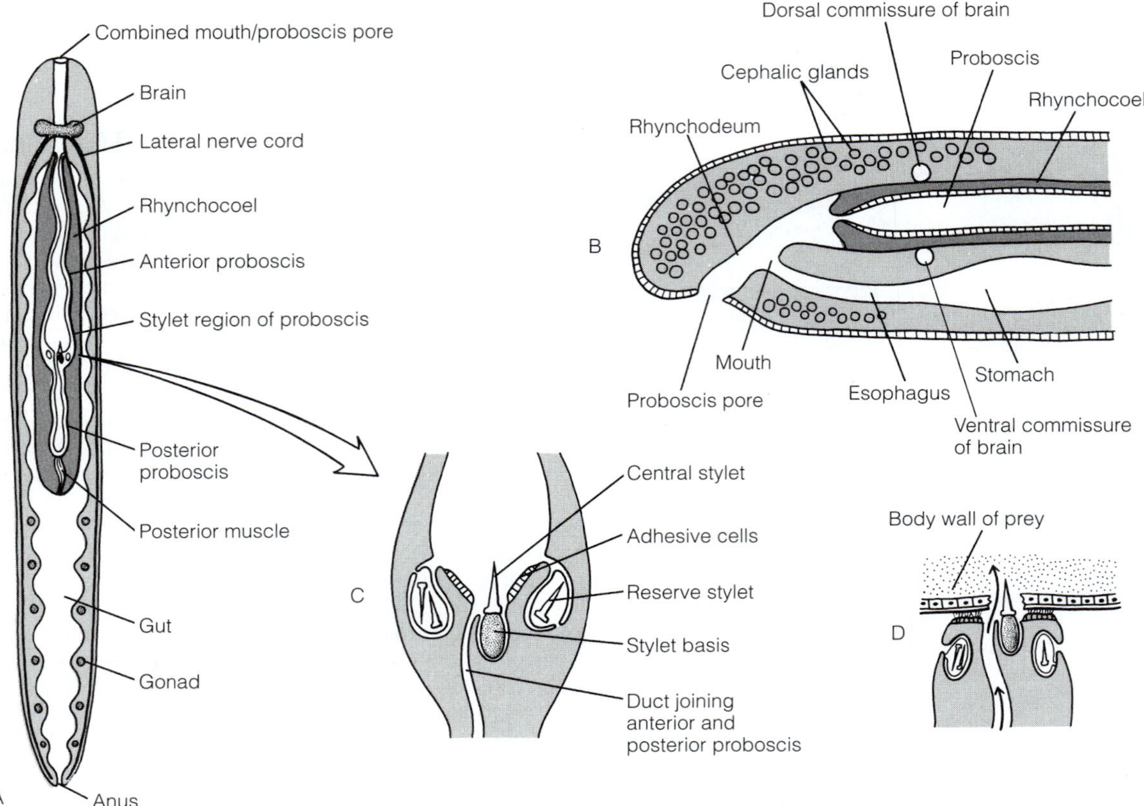

FIGURE 11-5 Anatomy of armed nemerteans (Monostilifera). **A,** Dorsal view (frontal section). **B,** Sagittal section of head showing union of foregut and rhynchodeum. **C** and **D,** Detail of the stylet apparatus before **(C)** and after **(D)** eversion. Much of the toxin that enters the prey originates in the anterior proboscis, but some also may enter the wound from the posterior proboscis (arrows in **D**). *(B, Modified and redrawn from Stricker, S. A. 1982. The morphology of* Paranemertes sanjuanensis *sp. n. (Nemertea, Monostilifera) from Washington, U.S.A. Zool. Scripta 11:107–115; C, Modified and redrawn from Stricker, S. A., and Cloney, R. A. 1981. The stylet apparatus of the nemertean* Paranemertes peregrina: *Its ultrastructure and role in prey capture. Zoomorphology. 97:205–223.)*

but rather approximately two-thirds of the distance from the anterior end of the body. **Reserve stylets** are held on each side of the active central stylet to provide replacements as the animal increases in size or when the main stylet is lost during feeding (Fig. 11-5A,C,D). In armed nemerteans, the proboscis everts only enough to expose the stylet. The everted part of the proboscis is called the **anterior proboscis** and the uneverted part behind the stylet is the **posterior proboscis.**

NUTRITION AND DIGESTIVE SYSTEM

Nemerteans are carnivores that feed chiefly on annelids and crustaceans, but a few may scavenge for dead animals. The discharged proboscis of unarmed nemerteans coils around the prey while sticky, toxic secretions aid in holding and immobilizing it (Fig. 11-4A). In the armed nemerteans, the proboscis is everted to expose the stylet at the proboscis tip (Fig. 11-5D). The stylet stabs the prey repeatedly, which allows the neurotoxic (tetrodotoxin and others) and cytolytic secretions to enter the prey's body. Some additional toxin may be pumped into the wound from the posterior proboscis chamber (Fig. 11-5A,C,D). The immobilized prey is either swallowed whole or, after partial digestion, its tissues are sucked directly into the mouth.

Nemerteans feed on a variety of prey, including worms, clams, and crustaceans. Some large, bivalve-feeding species can devastate clam beds. The large, unarmed *Cerebratulus lacteus* of the eastern coast of the United States enters the burrow of the razor clam, *Ensis directus,* from below and swallows the clam as it withdraws downward. *Paranemertes peregrina,* an armed, intertidal nemertean found on the Pacific coast of the United States, feeds on polychaete annelids. This nemertean leaves its burrow to feed and can follow the mucous trails of prey, but it must touch the prey to initiate the feeding response. Once contact occurs, the everted proboscis wraps around and stabs the prey repeatedly. The paralyzed prey is swallowed as the retracting proboscis drags it into the mouth. After feeding, the nemertean finds its home burrow by backtracking along its own mucous trail. Other armed nemerteans feed on small crustaceans such as amphipods. They kill the prey with a piercing stylet strike to the ventral exoskeleton and then wedge their head into the breech.

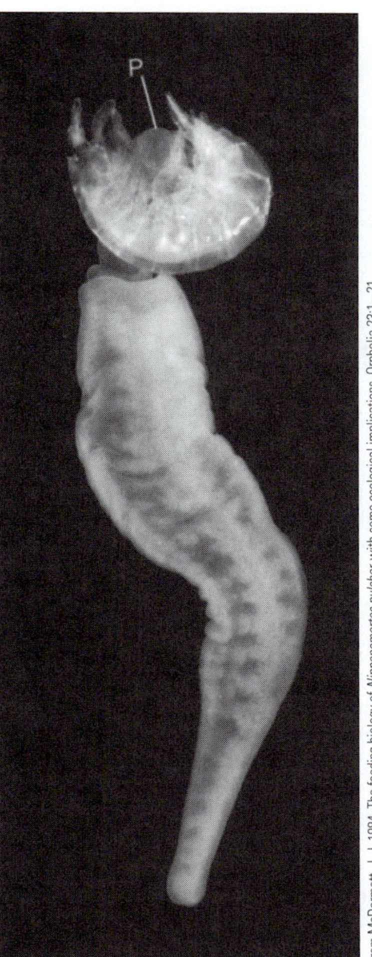

From McDermott, J. J. 1984. The feeding biology of *Nipponemertes pulcher*, with some ecological implications. Orphelia 23:1–21.

FIGURE 11-6 Nemertea: the armed *Nipponemertes pulcher* attacking an amphipod crustacean. The proboscis (P) is wrapped around the amphipod from behind and is striking the ventral surface.

The esophagus is everted and the contents of the prey are sucked out and digested (Fig. 11-6). Species of *Carcinonemertes* are economically important predators on the external egg "sponges" of brooding crabs. The short proboscis everts to just beyond the proboscis pore, where its stylet punctures the eggshell to allow the contents to be sucked out (Fig. 11-1G).

The digestive system consists of mouth, foregut, stomach, intestine, and anus. The ventral mouth is at the anterior end of the body near the level of the brain. It opens into a foregut, which is often subdivided into buccal cavity, pharynx, and glandular stomach. The foregut joins a long intestine with lateral **intestinal diverticula** (Fig. 11-2A). In some species, another intestinal diverticulum, the cecum, extends anteriorly past the foregut junction. The intestine opens at the **anus,** located at the tip of the tail (Fig. 11-2A, 11-5A).

In many armed nemerteans, the mouth has disappeared and the pharynx unites with the rhynchodeum so that pharynx and proboscis share the proboscis pore (Fig. 11-5B). A similar union occurs in the commensal bdellonemerteans (Fig. 11-1H), but a rhynchodeum is absent. In all other nemerteans, the digestive system is completely separate from the proboscis apparatus.

Digestion is initially extracellular in the intestinal lumen, including the diverticula. Partially digested particulates are phagocytosed by gastrodermal cells and digestion is completed intracellularly. Nutrient storage also occurs in these cells.

GAS EXCHANGE, INTERNAL TRANSPORT, AND EXCRETION

Nemerteans lack specialized gills, and gas exchange occurs across the surface of the long, sometimes flattened body. Nemerteans, like other large animals with thick body walls, use fluid circulation rather than simple diffusion to transport substances throughout their bodies. Nemerteans have a **coelomic circulatory system** consisting of two components: the central **rhynchocoel** and peripheral **vessels** (Fig. 11-2A,C, 11-7). The rhynchocoelic fluid transports substances to and from the proboscis, as well as performing a fluid-skeletal role in proboscis eversion and burrowing. Fluid in the vessels transports substances throughout the body, including to and from the rhynchocoel. The vessels provide whole-body circulation and the rhynchocoel, specialized local circulation.

In basic design, the vessels form a simple loop consisting of two lateral vessels, one on each side of the gut, joined together anteriorly and posteriorly (Fig. 11-7A). In many nemerteans, additional longitudinal and transverse vessels have been added to the circulatory loop. Commonly, a branch from each lateral vessel extends into the rhynchocoel wall. The vessels are lined by a mesothelium, and some cells are epitheliomuscular and bear a cilium (Fig. 11-7D,E).

The vessels are contractile, although fluid flow depends on contraction of both vessels and body-wall musculature. Circulation is intermittent in some species, and fluid ebbs and flows in the two lateral vessels. In other species, such as *Amphiporus cruentatus,* which has the basic loop plus a dorsal longitudinal vessel, fluid flows anteriorly in the dorsal vessel and posteriorly in the lateral vessels (Fig. 11-7B).

The vessel fluid is usually colorless, but in many species it contains cells that are yellow, orange, green, or occasionally red (hemoglobin). The hemoglobin-containing cells bind and transport oxygen, but the function of the other pigments is unknown. In addition to the pigmented corpuscles, the vessel and rhynchocoelic fluid also contain colorless amebocytes.

Communication between the vessel and rhynchocoelic coeloms occurs across specialized exchange sites called **vascular plugs.** A vascular plug occurs at the blind end of a vessel that distends the rhynchocoel wall and bulges into the rhynchocoel. In at least one species, the plug is overlaid with podocytes, across which exchange of fluid, ions, nutrients, and other substances may occur between the two compartments.

The excretory system consists of two or more protonephridia, each bearing many terminal cells (Fig. 11-2A). The protonephridia generally are restricted to the anterior foregut region of the body, not scattered throughout as in most flatworms. The terminal cells project into the wall of the lateral vessels (Fig. 11-8) and modify the fluid delivered to them by the circulatory system. In a few cases, the vessel lining is interrupted

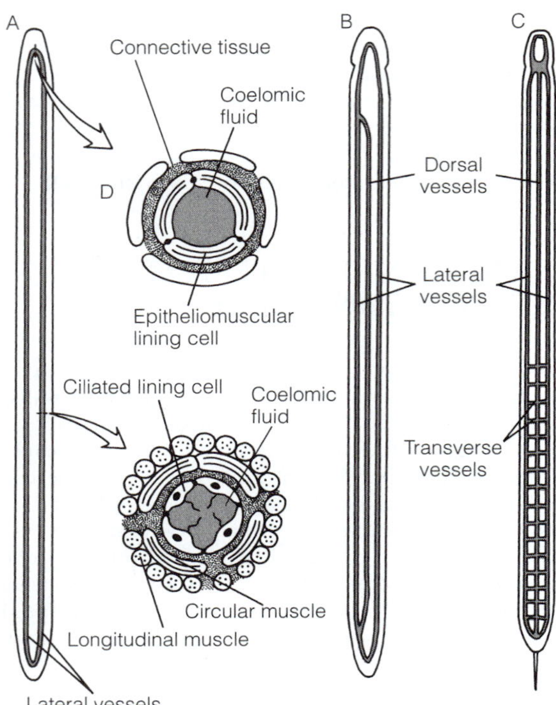

FIGURE 11-7 Nemertea: coelomic circulatory system. **A,** The palaeonemertean *Cephalothrix* sp., **B,** Hoplonemertean *Amphiporus cruentatus,* and **C,** Heteronemertean *Cerebratulus* spp. Fluid flows anteriorly in the median dorsal vessel of *Amphiporus* and posteriorly in the two lateral vessels. The cross sections of the vessels in *Cephalothrix* show epitheliomuscular **(D)** and ciliated **(E)** mesothelial lining cells. Such cells do not occur in hemal vessels, but are of common occurrence in coelomic linings. These facts indicate that the nemertean circulatory system is, like the rhynchocoel, a modified coelom. The coelomic fluid of *Amphiporus cruentatus* is red because it includes hemoglobin-containing coelomocytes. *(D and E, Modified and redrawn from Turbeville, J. M. 1991. Nemertinea. In Harrison, F. W., and Bogitsh, B. J. (Eds.): Microscopic Anatomy of Invertebrates. Vol. 3. Wiley-Liss, Inc., New York. pp. 285–328.)*

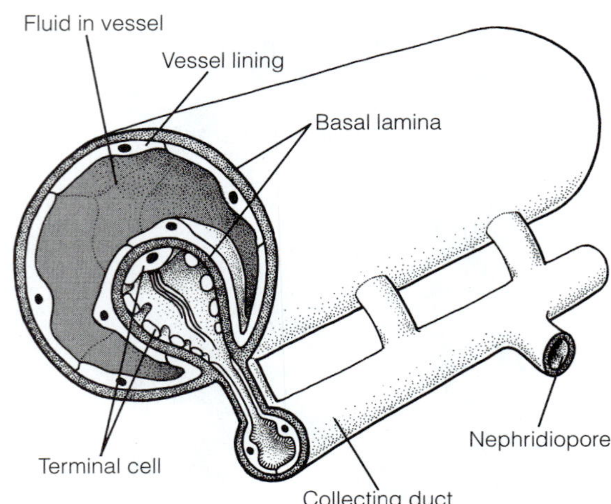

FIGURE 11-8 Nemertea: excretion. Protonephridia are intimately associated with the lateral coelomic vessels. This association allows the nephridia to be anatomically localized while functionally serving the entire body.

at the sites of contact so that the basal lamina around the terminal cells is bathed directly in fluid. The many nephridial tubules unite to form larger collecting ducts before opening to the exterior. A nephridiopore, or pores, is located on each side of the body at the level of the foregut. Protonephridia play a role in osmoregulation because semiterrestrial and freshwater species have many more terminal cells, sometimes thousands more, than do their marine counterparts.

NERVOUS SYSTEM AND SENSE ORGANS

The central nervous system consists of a brain and a pair of longitudinal nerve cords. The brain is a ring of four anterior ganglia around the rhynchodeum or anterior rhynchocoel. The two lateral nerve cords are large and nonganglionated (Fig. 11-2A,C, 11-5A). Additional longitudinal cords are often present, including a dorsal cord. The nerve cords and especially the brain are commonly pink or red because they contain a noncirculating hemoglobin **(neuroglobin).** This respiratory protein stores oxygen, thus extending the time period of peak neuromuscular functioning, or even normal functioning, under conditions of environmental anoxia (such as burrowing in oxygen-free sediments). Hemoglobin is absent in muscles and only rarely occurs in the circulatory system. Perhaps for these reasons, most nemerteans are limited to short bursts of vigorous movement.

Sense organs consist of sensory epidermal pits, pigment-cup ocelli, ciliated cephalic slits and grooves, cerebral organs, and eversible frontal organs (Fig. 11-9A–D) The last three of these are probably chemoreceptors. Cephalic slits and grooves are shallow, ciliated furrows underlaid by neurons. The **cerebral organs** are a pair of blind sacs lined by nerve and neuroendocrine cells and associated with the cerebral ganglia. A ciliated canal leads from each sac to the exterior (Fig. 11-9A,C). The external openings of the canals are in the cephalic slits or grooves, or in a pair of pits over the brain area. Bidirectional water currents created by cilia in the canals appear to be activated in the presence of food. The cerebral organs also may have a neuroendocrine role in osmoregulation. The cerebral organs of worms placed in low-salinity water release a substance into the blood that may increase the pumping rate of the nephridia.

REPRODUCTION AND DEVELOPMENT

Nemerteans regenerate readily and reproduce clonally as well as sexually. Fragmentation is common in large-bodied species, especially when irritated. Because of this tendency, collecting large, intact specimens is often difficult. Sometimes, too, the proboscis becomes detached when everted. The worm soon regenerates a lost proboscis, but the ability of body fragments to regenerate new worms varies greatly, depending on species. Some, including certain members of the genus *Lineus,* routinely reproduce by fragmentation, and even posterior sections

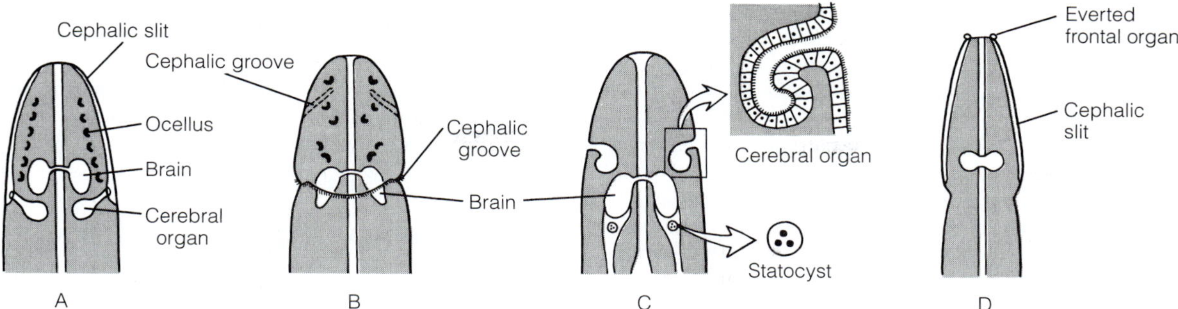

FIGURE 11-9 Nemertea: sense organs. **A** and **D,** Lateral cephalic slits (Heteronemertea). **B,** Transverse or oblique cephalic grooves (Hoplonemertea). **A** and **B,** Photoreceptive ocelli (all taxa) range in number from one to many pairs. **A, C,** and **top inset,** Cerebral organs—most likely chemoreceptors—are closely associated with the brain (in most nemerteans). **C** and **bottom inset,** Statocysts (in a few tiny interstitial nemerteans). **D,** Eversible frontal organs (in several taxa): **A,** *Lineus socialis;* **B,** *Amphiporus ochraceous;* **C,** *Ototyphlonemertes pallida;* **D,** *Cerebratulus lineolatus.* *(C, Modified and redrawn from Mock, H. 1978. Ototyphlonemertes pallida (Keferstein, 1862). Mikrofauna Meeresbodens. 67:559–570.)*

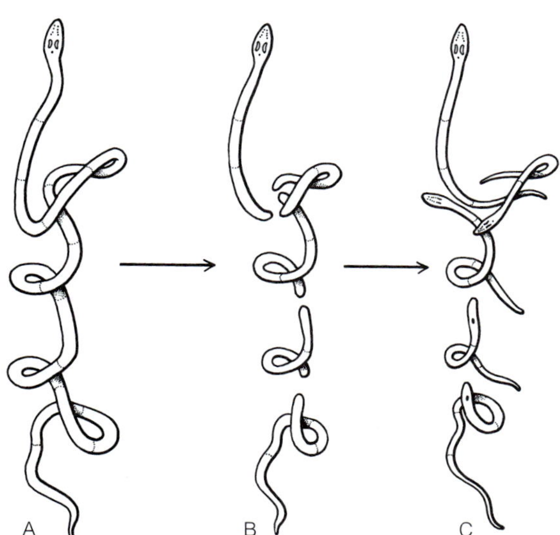

FIGURE 11-10 Nemertea: clonal reproduction. Fragmentation occurs in *Lineus,* as well as in other nemerteans. Annulations on the body surface **(A)** are preformed fragmentation zones. After fragmentation **(B),** each fragment regenerates missing parts and becomes a complete worm **(C).** *(Modified and redrawn from Coe, W. R. 1943. Biology of the nemerteans of the Atlantic coast of North America. Trans. Conn. Acad. Arts Sci. 35:129–328.)*

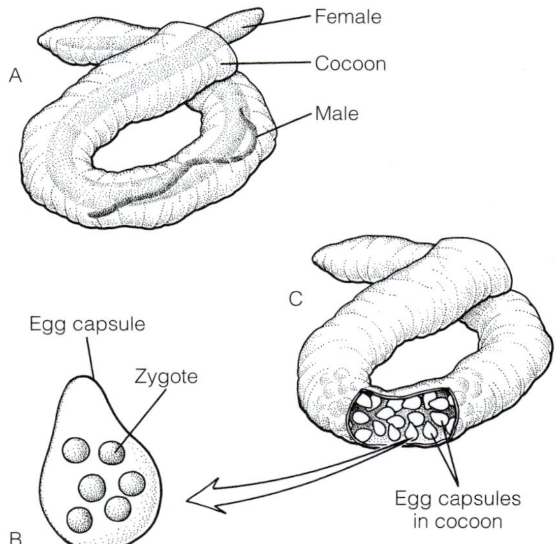

FIGURE 11-11 Nemertea: sexual pairing, fertilization, and egg capsules. Preceding fertilization in *Lineus viridis,* the female constructs a tubular cocoon (which resembles the permanent tube of the palaeonemertean *Tubulanus*) around herself and the smaller male **(A).** After an hour or two, the male releases sperm, which are believed to enter the female and fertilize her eggs internally. She then releases fertilized eggs in capsules **(B)** that are attached to the tube lining and surrounded by mucus **(C).** *(Modified and redrawn from Bartolomaeus, T. 1984. Zur Fortpflanzungbiologie von* Lineus viridis *(Nemertini). Helgol. wiss. Meeresunters. 38:185–188.)*

of the body are capable of total regeneration (Fig. 11-10), which takes place in a mucous cyst.

Most nemerteans are gonochoric and the reproductive system is simple. Gametes develop from stem cells that aggregate and become enclosed in an epithelium to form a gonad. The gonads alternate with the series of intestinal diverticula on each side of the body (Fig. 11-2A, 11-5A). Each gonad is tucked into the pocket formed by two diverticula and probably receives nutrients directly from them.

After maturation of the gametes, a short gonoduct grows from each gonad to the outside, allowing the gametes to escape. Each ovary produces 1 to 50 eggs, depending on the species. The shedding of eggs or sperm does not necessarily require contact between two worms, although some species aggregate at the time of spawning or a pair of worms may occupy a common burrow or secreted cocoon (Fig. 11-11).

Fertilization is external in most nemerteans, and the eggs are either shed and dispersed into seawater or deposited within the burrow, or tube, or in gelatinous strings (Fig. 11-11). A few species are viviparous and gestate their embryos internally.

Cleavage in nemerteans is spiral, development is determinate, and most mesoderm arises from a mesentoblast (4d) cell. The lateral blood vessels originate as a pair of mesodermal bands that undergo schizocoely. Similar to flatworm embryos, but different from annelids, molluscs, and relatives (Trochozoa), nemertean embryos lack trochoblast cells, the cells that give rise to the characteristic ciliary band (prototroch) of the trochophore larva of trochozoans (Chapter 12).

The embryos of most nemertean taxa develop directly or via a short-lived, lecithotrophic larval stage. Many heteronemerteans, however, pass through a free-swimming, planktotrophic larval stage called a **pilidium** (Fig. 11-12). The helmet-shaped pilidium has an apical tuft of cilia (partly sensory), a band of locomotory cilia, and a thick, gelatinous blastocoel for buoyancy. Mouth and gut are well developed, but an anus is absent. Among planktotrophic larvae in general, the pilidium is atypical for its lack of excretory organs. After a free-swimming existence, the pilidium undergoes a complex metamorphosis (Fig. 11-12E–H) in which the larval body is largely discarded and eaten by the juvenile, which differentiates from imaginal discs. *Paranemertes peregrina* has a life span of about one and a half years. Spawning occurs in spring and summer, and adults die in the winter. Juveniles resulting from the spring and summer spawn attain sexual maturity the following spring and summer.

FUNCTIONAL DESIGN OF NEMERTEA

The functional design of nemerteans is similar to that of flatworms, except for their greater length, proboscis apparatus, and the occurrence of a circulatory system. Like flatworms, they rely on diffusion for oxygen transport across the body wall and diffusion of nutrients from gut to gonads, which are intimately associated with the absorptive and storage region of the gut. In such long worms, the circulatory system is important for convective transport of nutrients to body regions distant from the intestine and for hormone and gas transport, at least in some species. The circulatory system, via vascular plugs, also supplies the rhynchocoel and proboscis. The contact and exchange between the circulatory system and protonephridia allow the nephridia to be concentrated regionally rather than being widely dispersed throughout the body, as in flatworms.

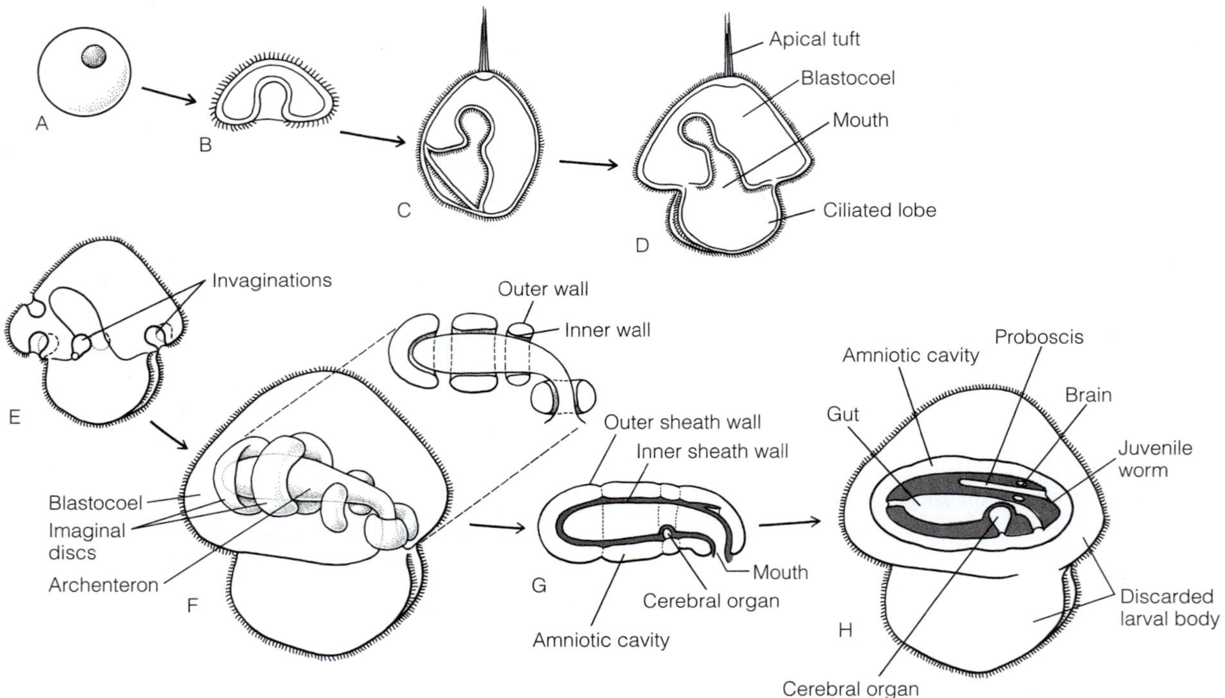

FIGURE 11-12 Nemertea: larval development and metamorphosis in heteronemerteans. **A–D,** Development of the pilidium larva **(D).** Metamorphosis is initiated when seven invaginations of larval ectoderm **(E)** pinch off into the blastocoel to form seven hollow vesicles (imaginal discs). These vesicles grow to surround the larval endoderm **(F)** and eventually fuse together **(G)** to form a continuous, double-walled sheath over the larval gut. The inner wall of the sheath becomes the epidermis of the developing worm. Inner and outer walls enclose a fluid-filled amniotic cavity **(G** and **H).** Eventually, the juvenile worm ruptures from the amnionic enclosure and the larval body to adopt an independent existence. *(A–D, Modified and redrawn from Iwata, F. 1985. Foregut formation of the nemerteans and its role in nemertean systematics. Am Zool. 25:23–36.)*

DIVERSITY OF NEMERTEA

Anopla[C] (= unarmed): Simple, unarmed proboscis without stylet, mouth posterior to brain (both plesiomorphies, thus Anopla is paraphyletic).

Palaeonemertea[O]: 100 marine species. Body-wall musculature arranged in outer circular, inner longitudinal layers (Fig. 11-2C). *Carinoma, Cephalothrix, Tubulanus* (Fig. 11-1A,C). *Carinoma tremaphoros:* in addition to two muscle layers below epidermis, it has an additional circular and longitudinal layer within epidermis; four-layered musculature apparently related to rapid burrowing using peristalsis (Fig. 11-3).

Heteronemertea[O]: Approximately 400 common and familiar species; majority are marine, but 3 species in fresh water. Four layers of body-wall muscles: an outer, weakly developed circular layer; a powerful longitudinal layer; another circular layer; and an innermost longitudinal layer. Nerve cords in the innermost circular muscle layer. Includes strongest nemertean swimmers, notably *Cerebratulus* (Fig. 11-1D). *Gorgonorhynchus, Polybrachiorhynchus:* proboscis tip branched and when everted functions as a sticky cast net (Fig. 11-4B). *Baseodiscus, Lineus* (Fig. 11-1E), *Micrura, Zygeupolia,* and many others.

Enopla[C] (= armed): Armed nemerteans (except Bdellonemertea); mouth anterior to brain. Nerve cords located inside body-wall musculature.

Bdellonemertea[O]: Seven species of *Malacobdella,* six commensals in mantle cavity of large marine clams, including *Macoma, Mercenaria, Mya,* and *Siliqua,* one in mantle cavity of freshwater snail. Steal food from filter-feeding host. Short, wide, flat, leechlike (Fig. 11-1H). Lack stylet, but have unique sucking pharynx and posterior sucker for attachment to host. Move in inchworm fashion, alternately attaching mouth and sucker. *Malacobdella grossa:* Europe and eastern North America; *M. siliquae* and *M. macomae:* Pacific northwest of North America.

Hoplonemertea[O]: Most diverse nemertean taxon, 650 benthic, pelagic, commensal, parasitic marine species as well as freshwater and terrestrial species. Proboscis armed with stylet.

Monostilifera[sO]: 500 benthic species. Proboscis with stylet bulb and single central stylet (Fig. 11-4C, 11-5A,C,D). *Emplectonema, Nemertopsis:* nearshore marine; individuals aggregate, intertwine threadlike bodies in a Gordian knot that only they can untie. Marine *Amphiporus, Paranemertes* (Fig.11-1F), *Prosorhochmus* (viviparous), and *Zygonemertes* also common inshore. Marine *Carcinonemertes* are parasites (egg predators) on crabs (Fig. 11-1G). *Ototyphlonemertes:* tiny, marine, interstitial: move among spaces between sand grains on beaches; the only nemerteans with statocysts (Fig. 11-9C). Freshwater *Prostoma:* edges of ponds, lakes, associated with aquatic algae, submerged plants. Terrestrial *Geonemertes, Pantinonemertes,* others: in warm, humid, coastal areas under leaves, tree bark, logs. Typically circular cross section to minimize surface area, water loss. Some use proboscis for locomotion as well as for prey capture. For locomotion, extended proboscis tip adheres to object and worm is then pulled forward as proboscis retracts. Most freshwater and terrestrial nemerteans are hermaphrodites.

Polystilifera[sO]: About 100 pelagic and 50 benthic marine species. Instead of stylet bulb and stylet, have padlike structure that bears many thorny stylets. *Nectonemertes* (Fig. 11-1I): pelagic, gelatinous, buoyant, at depths of 650–1700 m.

PHYLOGENY OF NEMERTEA

Traditionally, nemerteans were considered to be the sister taxon of Platyhelminthes based on the common occurrence of a flatwormlike shape, cuticle-free, ciliated epidermis, rhabdites, parenchyma, follicular gonads, protonephridia, and an eversible proboscis (present in some flatworms). Assuming that platyhelminths are the most primitive bilaterians, nemerteans were regarded as the first animals to evolve a coelom (albeit an unusual one in the form of a rhynchocoel), a hemal system (an earlier interpretation of the coelomic vessels), and a close association between circulatory and excretory systems (as in, for example, the vertebrate kidney).

Most of the characters shared by Platyhelminthes and Nemertea, however, are probably symplesiomorphies or convergences that do not support a sister-taxon relationship. Nemertean rhabdites are now known to be unlike and unrelated to those of flatworms; nemerteans probably lack flatwormlike parenchyma; the flatworm proboscis is dissimilar and probably unrelated to that of nemerteans; and both a ciliated epidermis and protonephridia are widespread in coelomate and acoelomate animals, not just in flatworms and nemerteans. Recent evidence also indicates that the nemertean circulatory system is not a hemal system, but rather a modified coelom similar to the rhynchocoel. This conclusion is based on several characters that nemertean vessels share with coeloms, including the lateral position of the vessels, a mesothelial lining (absent in invertebrate blood vessels), the occurrence of ciliated epitheliomuscular cells in the vessel linings, and their morphogenesis by schizocoely from the embryonic mesodermal bands (derivatives of the mesentoblast).

Nemerteans and platyhelminths lack any synapomorphies and only one, the presence of an anus, unites nemerteans with trochozoans. To this one character might be added the common occurrence of a coelom, and these two characters together, as synapomorphies, would establish a sister-taxon relationship between Nemertea and Trochozoa (Fig. 9-26A). The nemertean coelom (vessels and rhynchocoel), however, is highly specialized and its relation with other organs differs from that of trochozoans. In nemerteans, gametes are shed to the exterior directly from the gonads, whereas in trochozoans, they are released first into the coelom. Nemertean protonephridia bulge into the coelomic vessels but are not in direct contact with the coelomic fluid. Trochozoan protonephridia are suspended in and exposed to the coelomic fluid. These comparisons suggest that nemertean and trochozoan coeloms are, at best, homologous as mesothelium-lined cavities derived from mesoderm, but their specific adult forms probably are independently derived.

Vigorous phylogenetic research on Nemertea is in progress, but results are not yet available. The cladistic analysis of Nemertea shown in Figure 11-13 supports the monophyly of Enopla, but not Anopla, which may be a paraphyletic taxon.

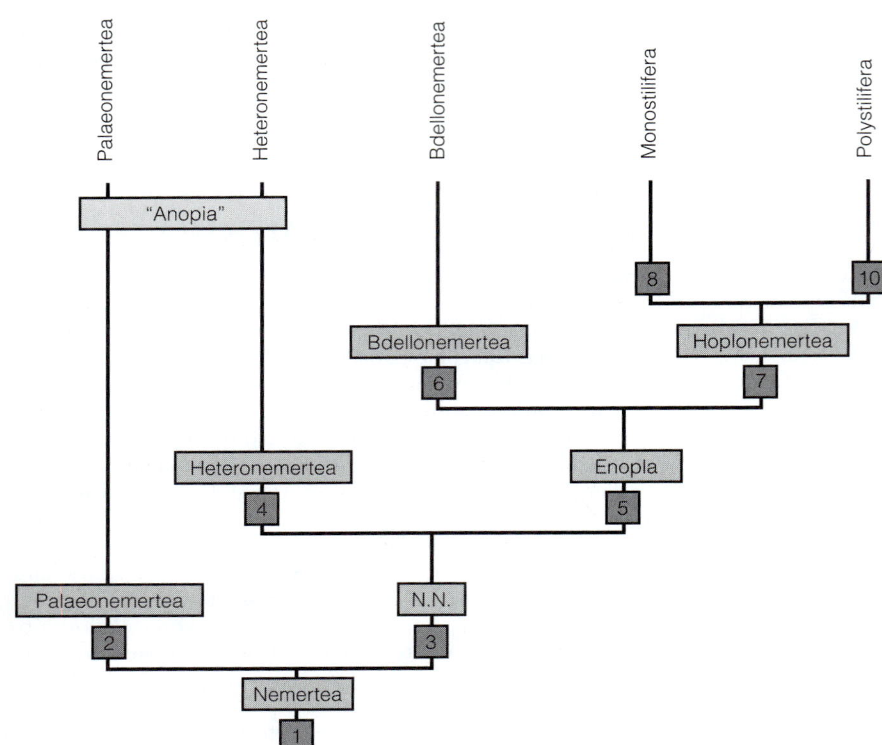

FIGURE 11-13 Nemertea: phylogeny. **1, Nemertea:** eversible proboscis + rhynchocoel, coelomic lateral vessels, vascular plugs, protonephridia concentrated and associated with lateral vessels, complete gut, cephalic organs, toxic epidermal secretions. **2, Palaeonemertea:** (perhaps a paraphyletic taxon). **3, N.N.:** obliquely striated muscle. **4, Heteronemertea:** outer longitudinal muscle layer of body wall well developed, pilidium larva. **5, Enopla:** nerve cords internal to body-wall musculature. **6, Bdellonemertea:** union of proboscis with pharynx, rhynchocoel absent, posterior sucker. **7, Hoplonemertea:** proboscis stylet, tripartite pharynx. **8, Monostilifera:** union of proboscis with pharynx, single stylet on bulbous base. **9, Polystilifera:** multiple stylets on pad.

PHYLOGENETIC HIERARCHY OF NEMERTEA

- Nemertea
  - Palaeonemertea
  - N.N.
    - Heteronemertea
    - Enopla
      - Bdellonemertea
      - Hoplonemertea
        - Monostylifera
        - Polystilifera

REFERENCES

GENERAL

Amerongen, H. M., and Chia, F.-S. 1982. Behavioral evidence for a chemoreceptive function of the cerebral organs in *Paranemertes peregrina* Coe (Hoplonemertea: Monostilifera). J. Exp. Mar. Biol. Ecol. 64:11–16.

Amerongen, H. M., and Chia, F.-S. 1987. Fine structure of the cerebral organs in hoplonemertines (Nemertini) with a discussion of their function. Zoomorphology 107:145–159.

Bartolomaeus, T. 1984. Zur Fortpflanzungsbiologie von *Lineus viridis* (Nemertini). Helgol. Wiss. Meeresunters. 38:185–188.

Bartolomaeus, T. 1985. Ultrastructure and development of the protonephridia of *Lineus viridis* (Nemertini). Microfauna Marina 2:61–83.

Cantell, C. E. 1969. Morphology, development, and biology of the pilidium larvae from the Swedish west coast. Zool. Bidrag fran Uppsala 38:61–111.

Coe, W. R. 1901. The nemerteans. Wash. Acad. Sci. Proc. 3:1–110.

Coe, W. R. 1926. The pelagic nemerteans. Mem. Mus. Comp. Zool. 49:1–244.

Coe, W. R. 1943. Biology of the nemerteans of the Atlantic coast of North America. Trans. Conn. Acad. Arts Sci. 35:129–328.

Coe, W. R. 1951. The nemertean faunas of the Gulf of Mexico and southern Florida. Bull. Mar. Sci. 1:149–186.

Feller, R. J., Roe, P., and Norenburg, J. L. 1998. Dietary immunoassay of pelagic nemerteans by use of cross-reacting polyclonal antibodies: Preliminary findings. Hydrobiologia 365:263–269.

Ferraris, J. D. 1985. Putative neuroendocrine devices in the Nemertina—An overview of structure and function. Am. Zool. 25:73–85.

Gibson, R. 1972. Nemerteans. Hutchinson University Library, London. 224 pp.

Gibson, R. 1974. Histochemical observations on the localization of some enzymes associated with digestion in four species of Brazilian nemerteans. Biol. Bull. 147:352–368.

Gibson, R., and Moore, J. 1976. Freshwater nemerteans. Zool. Linn. Soc. 58:117–218.

Gibson, R. 1995. Nemertean genera and species of the world: An annotated checklist of original names and description citations, synonyms, current taxonomic status, habits and recorded zoogeographic distribution. J. Nat. Hist. 29:271–562.

Henry, J., and Martindale, M. Q. 1997. Nemertines, the ribbon worms. In Gilbert, S. F., and Rauni, A. M. (Eds.): Embryology: Constructing the Organism. Sinauer, Sunderland, MA. pp. 151–166.

Henry, J. J., and Martindale, M. Q. 1998. Conservation of the spiralian developmental program: Cell lineage of the nemertean *Cerebratulus lacteus*. Devel. Biol. 201:253–269.

Iwata, F. 1985. Foregut formation of the nemerteans and its role in nemertean systematics. Am. Zool. 25:23–36.

Jespersen, A., and Lutzen, J. 1988. Ultrastructure and morphological interpretation of the circulatory system of nemerteans (Phylum Rhynchocoela). Vidensk. Meddr. Dansk Naturh. Foren. 147:47–66.

Kozloff, E. N. 1991. *Malacobdella siliquae* sp. nov. and *Malacobdella macomae* sp. nov., commensal nemerteans from bivalve molluscs on the Pacific coast of North America. Can. J. Zool. 69:1612–1618.

McDermott, J. J. 1976a. Predation of the razor clam *Ensis directus* by the nemertean worm *Cerebratulus lacteus*. Chesapeake Sci. 17:299–301.

McDermott, J. J. 1976b. Observations on the food and feeding behavior of estuarine nemertean worms belonging to the order Hoplonemertea. Biol. Bull. 150:57–68.

McDermott, J. J. 1984. The feeding biology of *Nipponemertes pulcher*, with some ecological implications. Ophelia 23:1–21.

McDermott, J. J. 2001. Status of Nemertea as prey in marine ecosystems. Hydrobiologia 456:7–20.

McDermott, J. J., and Roe, P. 1985. Food, feeding behavior and feeding ecology of nemerteans. Am. Zool. 25:113–126.

Miyazawa, K., Higashiyama, M., Ito, K., Moguchi, T., Arakawa, O., Shida, Y., and Hashimoto, K. 1988. Tetrodotoxin in two species of ribbon worm (Nemertini), *Lineus viridis* and *Tubulanus punctatus*. Toxicon 26:867–874.

Mock, H. 1978. *Ototyphlonemertes pallida* (Keferstein, 1862). Mikrofauna Meeresbodens 67:559–570.

Moore, J., and Gibson, R. 1985. The evolution and comparative physiology of terrestrial and freshwater nemerteans. Biol. Rev. 60:267–312.

Norenburg, J. L. 1985. Structure of the nemertine integument with consideration of its ecological and phylogenetic significance. Am. Zool. 25:37–51.

Norenburg, J. L. 1988. Nemertina. In Higgins, R. P., and Thiel, H. (Eds.): Introduction to the Study of Meiofauna. Smithsonian Institution Press, Washington, DC. pp. 287–292.

Norenburg, J. L., and Roe, P. 1998. Reproductive biology of several species of recently collected pelagic nemerteans. Hydrobiologia 365:73–91.

Norenburg, J. L., and Roe, P. 1998. Unusual features of the musculature of pelagic nemerteans. Hydrobiologia 365:109–120.

Norenburg, J. L., and Stricker, S. A. 2002. Phylum Nemertea. In Young, C. M., Rice, M., and Sewell, M. A. (Eds.): Atlas of Marine Invertebrate Larvae. Academic Press, New York. 630 pp.

Riser, N. W. 1974. Nemertinea. In Giese, A. C., and Pearse, J. S. (Eds.): Reproduction of Marine Invertebrates. Vol. 1. Academic Press, New York. pp. 359–389.

Roe, P. 1970. The nutrition of *Paranemertes peregrina*. I. Studies on food and feeding behavior. Biol. Bull. 139:80–91.

Roe, P. 1976. Life history and predator–prey interactions of the nemertean *Paranemertes peregrina*. Biol. Bull. 150:80–106.

Roe, P., and Norenburg, J. L. 1999. Observations on depth distribution, diversity and abundance of pelagic nemerteans from the Pacific coast off California and Hawaii. Deep-Sea Res. 46:1201–1220.

Rowell, T. W., and Woo, P. 1990. Predation by the nemertean worm, *Cerebratulus lacteus* Verrill, on the soft-shell clam *Mya arenaria* Linnaeus 1758, and its apparent role in the destruction of a clam flat. J. Shellfish Res. 9:291–297.

Sterrer, W. 1986. Marine Fauna and Flora of Bermuda. John Wiley and Sons, New York. 742 pp.

Stricker, S. A. 1982. The morphology of *Paranemertes sanjuanensis* sp. n. (Nemertea, Monostilifera) from Washington, U.S.A. Zool. Scripta 11:107–115.

Stricker, S. A. 1985. The stylet apparatus of monostiliferous hoplonemerteans. Am. Zool. 25:87–97.

Stricker, S. A. 1987. Phylum Nemertea. In Strathmann, M. F. (Ed.): Reproduction and Development of Marine Invertebrates of the Northern Pacific Coast. University of Washington Press, Seattle. pp. 129–137.

Stricker, S. A. 1986. An ultrastructural study of oogenesis, fertilization, and egg laying in a nemertean ectosymbiont of crabs, *Carcinonemertes epialti* (Nemertea, Hoplonemertea). Can. J. Zool. 64:1256–1269.

Stricker, S. A., and Cavey, M. J. 1986. An ultrastructural study of spermatogenesis and the morphology of the testis in the nemertean worm *Tetrastemma phyllospadicola* (Nemertea, Hoplonemertea). Can. J. Zool. 64:2187–2202.

Stricker, S. A., and Cloney, R. A., 1981. The stylet apparatus of the nemertean *Paranemertes peregrina:* Its ultrastructure and role in prey capture. Zoomorphology. 97: 205–223.

Stricker, S. A., and Cloney, R. A. 1982. Stylet formation in nemerteans. Biol. Bull. 162:387–403.

Stricker, S. A., Smythe, T. L., Miller, L., and Norenburg, J. L. 2001. Comparative biology of oogenesis in nemertean worms. Acta Zool. 82:213–230.

Stricker, S. A., and Reed, C. G. 1981. Larval morphology of the nemertean *Carcinonemertes epialti* (Nemertea, Hoplonemertea). J. Morphol. 169:61–70.

Sundberg, P., and Hylbom, R. 1994. Phylogeny of the nemertean subclass Palaeonemertea (Anopla, Nemertea). Cladistics 10:347–402.

Turbeville, J. M. 1986. An ultrastructural analysis of coelomogenesis in the hoplonemertine *Prosorhochmus americanus* and the polychete *Magelona* sp. J. Morphol. 187:51–60.

Turbeville, J. M. 1991. Nemertinea. In Harrison, F. W., and Bogitsh, B. J. (Eds.): Microscopic Anatomy of Invertebrates. Wiley-Liss, New York. pp. 285–328.

Turbeville, J. M. 1998. Nemertea. In Encyclopedia of Reproduction. Vol. 3. Academic Press, New York. pp. 21–29.

Turbeville, J. M., Field, K. G., and Raff, R. A. 1992. Phylogenetic position of phylum Nemertini inferred from 18S rRNA sequences: Molecular data as a test of morphological character homology. Mol. Biol. Evol. 9:235–249.

Turbeville, J. M., and Ruppert, E. E. 1983. Epidermal muscles and peristaltic burrowing in *Carinoma tremaphoros:* Correlates of effective burrowing without segmentation. Zoomorphology 103:103–120.

Turbeville, J. M., and Ruppert, E. E. 1985. Comparative ultrastructure and the evolution of nemertines. Am. Zool. 25:53–72.

INTERNET SITES

www.ucmp.berkeley.edu/nemertini/nemertini.html (Brief introduction to nemertines with a few good color images.)

http://nemertes.si.edu/ (J. Norenburg's page on nemertean research.)

www.mbl.edu/BiologicalBulletin/EGGCOMP/pages/25.html (Online version of Costello and Henley's instructions for the collection and maintenance of adult *Cerebratulus lacteus* and for rearing their embryos.)

12 Mollusca[P]

Mollusca is an enormous taxon second only to Arthropoda in number of living species. It includes many familiar animals such as clams, oysters, mussels, snails, slugs, octopods, and squids in seven living classes. In spite of their obvious dissimilarities, these animals are all molluscs and share many basic features.

There are probably about 100,000 described species of Recent molluscs. Estimates range from 50,000 to 150,000, with the uncertainty being due largely to the many species that have been described several times. There are also about 35,000 described extinct species. Mollusca has a long evolutionary history and, thanks to the easily preserved calcareous shell, a rich fossil record that dates back to the Cambrian. Molluscs are chiefly marine, and all seven classes are present and most successful in the sea. Some bivalves and gastropods are found in freshwater habitats, but only the gastropods are present on land.

Mollusca includes the taxa Aplacophora, Polyplacophora, Monoplacophora, Gastropoda, Cephalopoda, Bivalvia, and Scaphopoda (Fig. 12-125), which are considered to be classes in traditional Linnean classifications. Aplacophora consists of small, wormlike molluscs with numerous calcareous spicules rather than a shell. Polyplacophorans, or chitons, have eight shell plates. Monoplacophorans have a low, conical, limpetlike, shell and live in deep water. (The **limpet** shape appears often in molluscan evolution and includes an uncoiled, caplike, low-profile shell and a large creeping foot). Gastropods are the snails and slugs and have a one-piece, often coiled shell and are found in the sea, in fresh water, and on land. Cephalopods include the squids and octopods and mostly have reduced, absent, or internalized shells. The bivalves are the clams and their relatives with a shell divided into two pieces. Scaphopods are the tusk molluscs with a one-piece, tubular shell.

GENERALIZED MOLLUSC

The seven major taxa of living molluscs share a common general design, but each has modified one or more aspects of that design in a manner characteristic of the taxon. For that reason, no single taxon fully represents the common molluscan layout and none can be considered an exemplar of Mollusca. Ideally, a molluscan ground plan, derived from phylogenetic research, would be used to describe basic molluscan form and function; however, such a ground plan remains a matter of debate and no single model has been universally adopted. One such model will be discussed with the phylogeny of molluscs at the end of the chapter. For the time being, we construct a generalized mollusc whose form and function provide a convenient starting point on which to build an understanding of molluscan biology. The actual molluscan common ancestor could have resembled this generalized mollusc, but the latter was manufactured as a teaching tool and not in strict accordance with the rules of phylogenetic research.

The generalized mollusc is a marine benthic animal, bilaterally symmetric, dorsoventrally depressed, and ovoid in outline (Fig. 12-1A). The body is divided into a small, poorly defined anterior head, a large dorsal **visceral mass,** and a broad, flat ventral **foot.** This animal is adapted for life on hard, rocky substrates, where it uses its rasplike **radula** to graze on the biofilm of microscopic algae and other small sessile organisms. The animal attaches to the substratum by the muscular foot, which it uses to move slowly about in search for food. There is a dorsal shieldlike **shell,** which can be clamped against the substratum to protect the soft parts from predators.

MANTLE

Dorsally, the body wall over the visceral mass is elaborated to form the **mantle** (Fig. 12-1A,B), or pallium, which is characteristic of all molluscs. The mantle epidermis secretes protein, calcium salts, and mucus and is also sensory.

SHELL

The shell is a simple, low, conical cap covering the dorsum of the generalized mollusc (Fig. 12-1A,B). In most living molluscs, however, the shell has become a spacious retreat. The mantle epithelium secretes protein and calcareous material that in most molluscs take the form of a calcareous shell and, occasionally, needlelike spicules. The shell probably originated in the ancestral mollusc as a thick dorsal cuticle of chitin and protein, to which calcium salts were later added.

The eumolluscan shell consists of three basic layers, one organic and two calcareous. The outermost layer is the **periostracum,** composed of the protein **conchiolin.** Immediately under this layer is the calcareous **ostracum** and inside that is the **hypostracum.** These two inner layers are composed of calcium carbonate deposited over an organic matrix. In the gastropods, cephalopods, and bivalves, the hypostracum may be nacreous (pearly). All shell layers are secreted by the mantle epidermis and through its activity the shell increases in size as the animal grows.

MANTLE CAVITY

A dorsal peripheral outfold of the mantle, the **mantle skirt,** partially encloses a pocket, the **mantle cavity,** that is continuous with the surrounding seawater (Fig. 12-1A,B). This space is a central and characteristic feature of the molluscan body and one on which many important processes depend. The outer epidermis of the mantle skirt underlies the shell, whereas the inner epidermis lines the mantle cavity, secretes mucus, and is often ciliated. The epidermis forming the edge of the skirt is secretory and sensory and is responsible for producing the protein and calcium carbonate necessary for increases in shell length. The outer epidermis secretes the calcium carbonate required for increasing shell thickness. In the ancestral mollusc, the mantle cavity was probably posterior and extended anteriorly as a pair of lateral pallial grooves, one on each side of the body. In specific molluscs, it may be posterior, lateral, anterior, or even reduced or lost. The gills, nephridiopores, anus, gonopores, and sensory osphradia are located in the mantle cavity. It is sometimes involved in feeding and, when anterior, may provide space for the retracted head and foot.

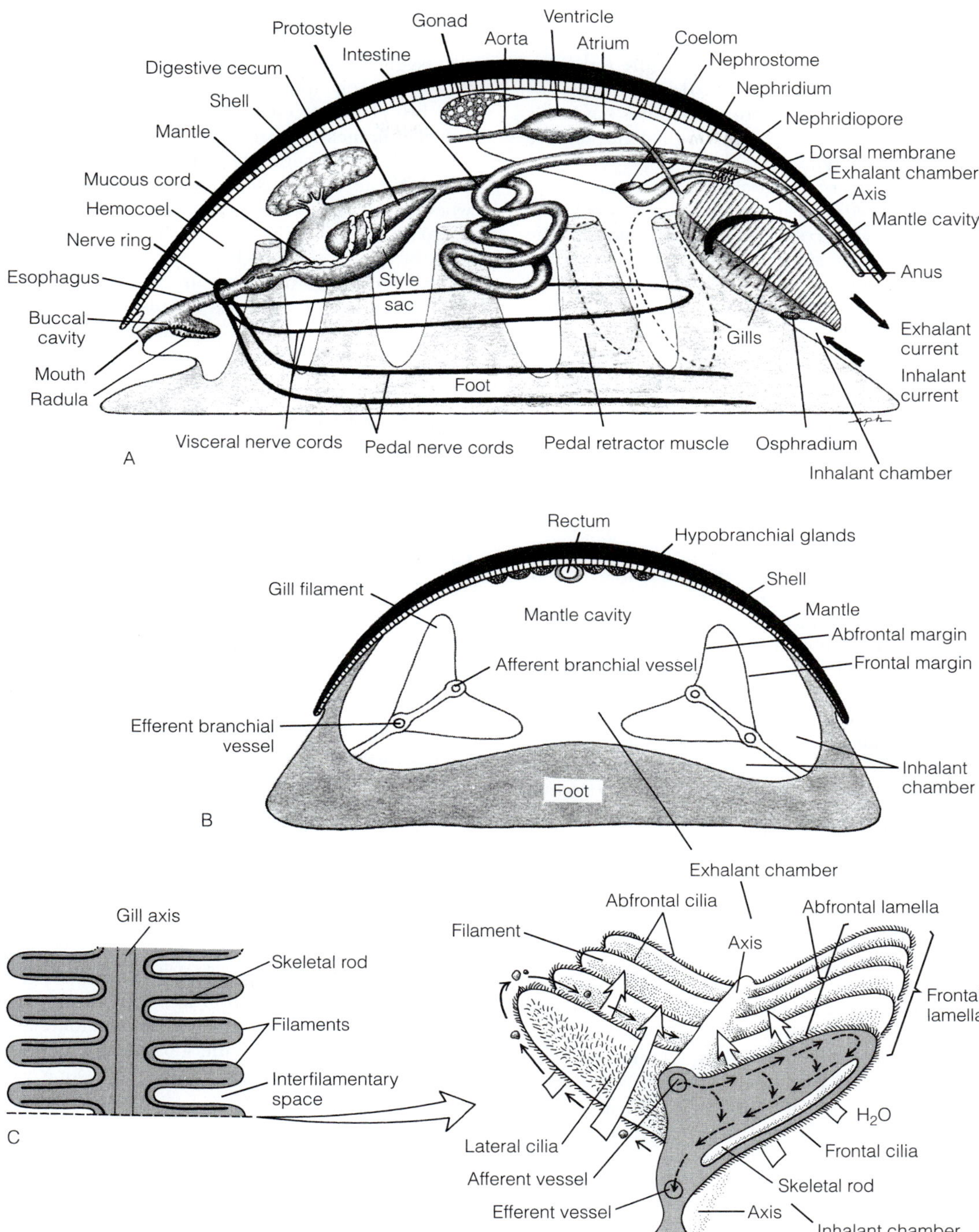

FIGURE 12-1 The generalized mollusc. **A,** Lateral view. Arrows indicate path of water current through mantle cavity. The ventral current is inhalant, the dorsal exhalant. **B,** Transverse section through the body of the generalized mollusc at the level of the mantle cavity. **C,** Frontal section through the primitive bipectinate gill, showing alternating filaments and supporting chitinous rods. **D,** Transverse section through the gill of the primitive gastropod *Haliotis.* Large outlined arrows indicate the direction of water current over the gill filaments; small solid arrows indicate the direction of the cleansing ciliary currents; small broken arrows indicate the direction of blood flow within the gill filaments. *(B–D, Modified and redrawn from Yonge, C. M. 1947. The pallial organs of the aspidobranchiates Mollusca, a functional interpretation of their structure and evolution. Phil. Trans. Roy. Soc. Lond. B. 443–518.)*

GILLS

The generalized mollusc has several pairs of **gills** (ctenidia; ctene = comb), one pair of which is located in the posterior mantle cavity (Fig. 12-1A,B). Most modern molluscs have only one pair or, sometimes, only one gill. The original gills may be lost or replaced by secondary gills.

A generalized gill consists of a **central axis,** which is attached to the mantle along one edge (Fig. 12-1D). The axis contains muscles, blood vessels, and nerves. From opposite sides of the axis arise two rows of leaflike **gill filaments.** Each filament has an upstream **frontal margin** on one side and a downstream **abfrontal margin** on the opposite side (Fig. 12-1B). The two rows of filaments are usually staggered with respect to each other (Fig. 12-1C) so that their bases alternate on opposite sides of the axis. Because there are two rows of filaments, these primitive gills are referred to as **bipectinate** (Fig. 12-1C,D). Many living molluscs, however, have **monopectinate** gills in which there are filaments on only one side of the axis, like teeth on a comb (pectin = comb).

The filaments lie close together but are separated from their neighbors by **interfilamentary water spaces** (Fig. 12-1C). The gills divide the mantle cavity into ventral **inhalant** (infrabranchial) and dorsal **exhalant** (suprabranchial) **chambers** (Fig. 12-1A,B,D). The exposed gill surfaces formed by the combined edges of the filaments are known as **lamellae.** That portion facing the inhalant chamber, the **frontal lamella,** is composed of the combined frontal margins of the gill filaments. Similarly, the **abfrontal lamella** faces the exhalant chamber and is composed of the combined abfrontal edges of the filaments. Cilia on the filaments generate a water current that enters the mantle cavity ventrally, flows between the filaments into the exhalant chamber, and then exits the mantle cavity dorsally (Fig. 12-1A,D).

The epidermis of the filaments bears three types of cilia, only one of which—the powerful **lateral cilia** (Fig. 12-1D)—generates the respiratory current through the mantle cavity (Fig. 12-1A,D). Gas is exchanged between the flat surfaces of the filaments, facing the interfilamentary spaces, and the current as the water passes through these spaces.

The two remaining types of cilia move mucus and particles rather than water, and keep the gills free of particulate material (sediment and detritus) that might interfere with the respiratory flow of water between the filaments. The upstream frontal margin of the filament bears frontal cilia and the downstream abfrontal margin bears abfrontal cilia (Fig. 12-1D). Secretory cells of the gill epithelium secrete mucus, which entangles particles from the respiratory current. The **frontal cilia** generate a current moving away from the gill axis toward the tip of the filament. This current transports particles trapped in mucus off of the frontal lamella and keeps the gill surface clean, preventing the particles from clogging the interfilamentary spaces. The cilia move the particles along the edge of the filament to the exhalant side, where they are released as **pseudofeces** into the exhalant water current, which sweeps them out of the mantle cavity. Pseudofeces are strings or pellets of mucus and sediment. The **abfrontal cilia** participate in transporting pseudofeces. On the mantle roof are two patches of mucus-secreting epithelium, called **hypobranchial glands** (Fig. 12-1B). They lie downstream of the gills and trap sediment in the exhalant current.

Oxygen is transported from the gills to the tissues by the blood (hemolymph). An **afferent branchial vessel** in the gill axis brings unoxygenated blood to the gill filaments and an **efferent branchial vessel** removes oxygenated blood from the gill and transports it to the heart (Fig. 12-1D). Blood circulates through the filaments in an exhalant-to-inhalant direction, whereas the respiratory water current moves in an inhalant-to-exhalant direction (Fig. 12-1D). The resulting countercurrent flow facilitates the efficient extraction of oxygen from the water. Use of countercurrent gas-exchange mechanisms is common in aquatic animals because the low solubility of oxygen in water necessitates the efficient uptake of oxygen by the gills.

OSPHRADIA

The presence of a pair of sensory **osphradia** in the inhalant water current is widespread among molluscs (Fig. 12-1A). Its receptor cells monitor the water entering the mantle cavity for chemicals and perhaps for sediment. Detection of undesirable chemicals or, presumably, a high sediment load, results in cessation of ciliary beating and water flow over the gills until conditions improve.

FOOT

The ventral surface of the generalized body is occupied by a large, muscular foot with a broad, flat, ciliated creepsole (Fig. 12-1A). The foot attaches to hard substrata and moves the animal over their surfaces. The foot is abundantly supplied with mucous gland cells that produce a lubricating mucus to facilitate locomotion. The foot contains a blood sinus, the **pedal hemocoel.**

Several pairs of dorsoventral **pedal retractor muscles** extend from the inner surface of the shell to the foot (Fig. 12-1A). Contraction of these muscles pulls the shell toward the foot, or vice versa. In the generalized mollusc, the foot is usually attached to the substratum, and the action of the pedal retractors is to clamp the shell over the visceral mass, head, and foot to protect them. In contrast, most living molluscs, when threatened, release the hold on the substratum and use the retractor muscles to withdraw (retract) the foot, along with the head, into the shell.

NUTRITION

The generalized mollusc is a **microphagous browser** that uses its radula to scrape microscopic algae, other organisms, and detritus from hard substrata. The gut is adapted for separating and processing the mixture of fine organic particles (food) and fine mineral particles (nonfood). Such a diet requires a complicated gut capable of sorting particulate matter into nonfood and food particles, discarding the former, and digesting and absorbing the latter. Unique among the molluscs is the use of a variety of external and internal ciliary sorting fields for this purpose. Many living molluscs, such as suspension feeders or microphagous browsers, also feed on mixtures

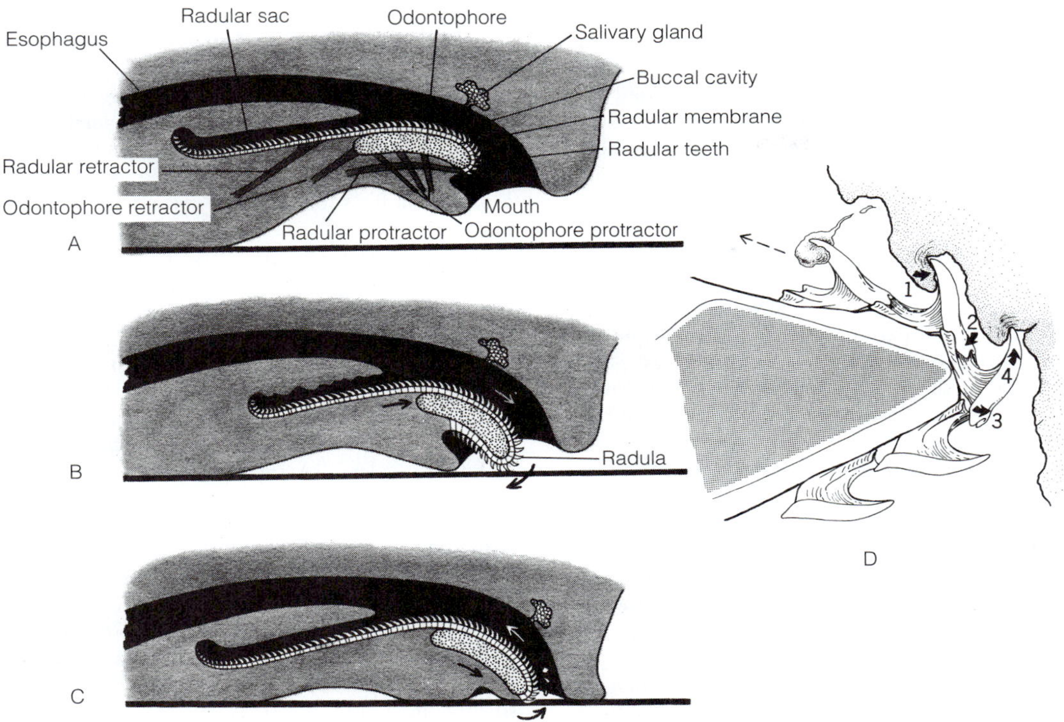

FIGURE 12-2 Molluscan radula. **A,** Mouth cavity, showing radula apparatus in lateral view. **B,** Protraction of the radula against the substratum. **C,** Retracting movement, during which the substratum is scraped by the radula teeth. **D,** The cutting action of the radula teeth when they are erected over the end of the odontophore during radula retraction. *(D, From Solem, A. 1974. The Shell Makers: Introducing Mollusks. Reprinted by permission of John Wiley and Sons, New York. pp. 135 and 150)*

of fine organic and mineral particles and may retain many of the features of the generalized gut. Many other living molluscs, however, are macrophagous herbivores or carnivores that feed on large pieces of organic material. These have radically different, simpler digestive systems in which sorting mineral and organic particles is not necessary.

The gut consists of the foregut, midgut, and hindgut. The foregut (mouth, buccal cavity, pharynx) and hindgut (rectum, anus), as ectodermal derivatives, are lined by cuticle, but the midgut (esophagus, stomach, digestive ceca, intestine) is endodermal and unlined.

The mouth is anterior and opens into a cuticularized buccal cavity. The radular apparatus lies in the floor of the buccal cavity and includes the radula itself; the connective tissue **odontophore,** which supports the radula; and the complex of muscles that operate the radula and odontophore (Fig. 12-2). The **radular sac** is a ventral evagination of the floor of the buccal cavity. It secretes and contains the **radula,** which is a flexible, longitudinal ribbon of transverse rows of tiny chitinous teeth (Fig. 12-2, 12-41). The radula is unique to molluscs and is present in all taxa except the bivalves, where it has been secondarily lost, presumably because it has no function in these filter-feeding animals. Its teeth are composed of α-chitin and tanned proteins that are sometimes hardened by iron or silicon compounds. The teeth are in transverse rows and exhibit a variety of shapes adapted for the food utilized by various taxa (Fig. 12-40, 12-41, 12-42). Although its original function was as a scraper for microphagous browsing, it has been adapted for use in a variety of feeding modes and its morphology varies accordingly.

The radula is supported by and sits on top of thick rods of firm connective tissue known as the odontophore (odonto = tooth, phore = to carry). The odontophores of most molluscs are said to be chondroid (cartilage-like), although they differ chemically and histologically from vertebrate cartilage.

The odontophore and radula are equipped with numerous **protractor** and **retractor muscles** whose function is to move the odontophore and radula anteriorly and posteriorly (Fig. 12-2A–C). During feeding, the odontophore protractor muscles pull the odontophore anteriorly, causing it to protrude slightly from the mouth in a position close to the substratum on which the animal wishes to feed (Fig. 12-2B). The radula, resting atop the odontophore, is thus moved into the feeding position. Next the radular protractor and retractor muscles contract alternately to move the radula forward (Fig. 12-2B), then backward (Fig. 12-2C), over the odontophore. This moves the radula against the substratum, like a woodworker's rasp. The recurved radular teeth cut into the substratum and scrape away algae on the return (backward) stroke (Fig. 12-2D). Algae and accompanying mineral particles are pulled into the buccal cavity by this stroke. The odontophore is returned to its resting position by contraction of the odontophore retractor muscles.

When at rest within the radular sac the lateral margins of the radula tend to roll up, but as the odontophore is projected out of the mouth over the substratum, the changing tension causes the radular belt to flatten as it bends around the odontophore tip. The flattening in turn brings about the erection of the teeth. You could imitate the action of a radula by pressing your lips against a surface (a Popsicle, perhaps) with your mouth slightly open and then licking the surface repeatedly with your tongue.

The radular teeth are secreted by epithelium at the posterior end of the sac. The radula grows anteriorly from this tissue, so the most-anterior teeth are the oldest and consequently the most worn from abrasion against rocks and mineral particles. The worn, distal teeth are shed and replaced with fresh teeth as the radula grows anteriorly. The radula grows at a rate of one to five rows of teeth per day.

One or more pairs of **salivary glands** secrete mucus into the buccal cavity (Fig. 12-2A). The mucus mixes with the food and mineral particles in the buccal cavity to form a mucous string that can be transported posteriorly by cilia. The esophagus exits the posterior buccal cavity and extends to the stomach. Cilia lining the esophagus and stomach move the mucous string, with its trapped food and mineral particles, posteriorly into the stomach.

The stomach of the generalized mollusc is a pear-shaped sac in the anterior visceral mass. The esophagus enters at its anterior, expanded end and the intestine exits at its narrow, tapered, posterior end (Fig. 12-3). A pair of extensive, highly branched **digestive ceca** (also called digestive glands, liver, digestive diverticula, or hepatopancreas) connects by ducts with the anterior region of the stomach. Digestive enzymes are secreted by the gastrodermis of the digestive ceca and transported to the stomach, where extracellular digestion occurs.

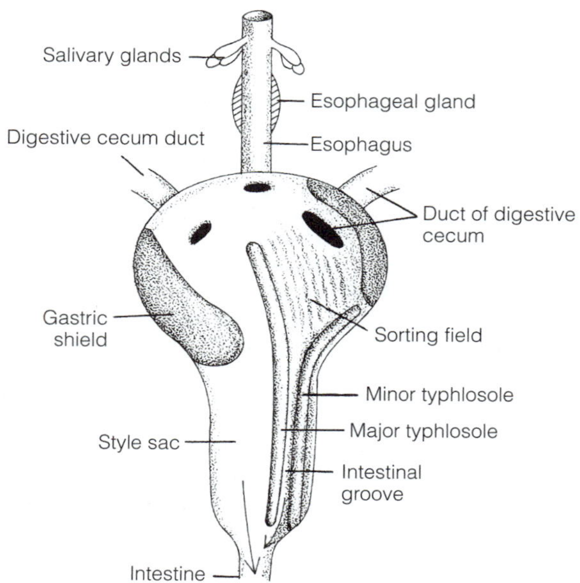

FIGURE 12-3 A primitive molluscan stomach. *(Modified from Owen.)*

Absorption occurs in the digestive ceca, as do phagocytosis, intracellular digestion, and storage. Digestive ceca are present in all eumolluscan taxa, but are absent in Aplacophora.

The epithelium of the anterior stomach is partly lined by a chitinous **gastric shield** and bears a **sorting field** of ciliated ridges and grooves (Fig. 12-3). The gastric shield protects the stomach gastrodermis from abrasion whereas the sorting field separates indigestible mineral particles from nutritious organic particles.

The posterior end of the stomach is the narrow, ciliated **style sac.** A long, deep **intestinal groove** extends from the sorting field posteriorly through the style sac to the intestine. It is flanked by two ciliated ridges, the **major** and **minor typhlosoles,** that extend from the sorting field to the opening of the intestine at the posterior end of the sac.

In the stomach and style sac, mucus and embedded fecal particles form a stiff mass known as the **protostyle,** or fecal rod (Fig. 12-1A). (A **crystalline style** composed of digestive enzymes, rather than mucus, is present in some conchiferans, most notably the bivalves and some gastropods.) The protostyle is rotated by the cilia of the style sac. Rotation of the protostyle helps pull the mucous string through the esophagus into the stomach, where it is wound into the anterior end of the protostyle. The acidity (pH = 5 to 6) of the stomach lowers the viscosity of the mucus and frees particles from its grip. Freed particles are swept away from the protostyle onto the sorting field, the ciliated ridges and grooves of which grade them by size. Large organic particles tend to be sent to the digestive ceca for digestion, and smaller mineral particles become incorporated in the protostyle, although the sorting process is far from perfect. Contractions of circular muscles periodically pinch **fecal pellets** from the posterior end of the protostyle into the intestine. The stomach of macrophagous molluscs lacks the sorting equipment, style sac, gastric shield, and protostyle, and is usually a simple sac.

The long, coiled intestine is not involved in digestion and materials entering it are wastes destined for defecation. Its function is the modification and storage of fecal pellets. The anus is located dorsally in the mantle cavity, in the exhalant current, where fecal pellets can be swept from the mantle cavity without fouling the gills (Fig. 12-1A).

COELOM

The reduced coelom of molluscs, although small, retains many of its ancestral functions and is still an important space. Its original function as a skeleton has been assumed by the hemocoel and shell, but it continues to be an integral component of the excretory and reproductive systems. The molluscan coelom is the **pericardial cavity** and **gonocoel,** which contain the heart and gonad, respectively (Fig. 12-1A). A short segment of the posterior intestine passes through the pericardial cavity.

INTERNAL TRANSPORT

The hemal system of the generalized mollusc consists of a heart, a dorsal aorta, blood vessels, blood (hemolymph), and a hemocoel (Fig. 12-1A). The blood contains amebocytes as well as the respiratory pigment hemocyanin. The hemal system, which lacks a cellular lining, lies in the connective-tissue compartment in the spaces between the basal laminae of the major epithelia, namely the epidermis, mesothelium, and gastrodermis.

The heart consists of one or more pairs of **atria** (auricles), which receive oxygenated blood from the gills via the efferent branchial vessels. The atria empty into the median unpaired, muscular **ventricle,** which is continuous with the **aorta.** The aorta branches into smaller arteries that deliver blood to the several hemocoelic sinuses or, rarely, capillary beds. The major sinuses are in the head, foot, and visceral mass. Blood bathes the tissues in and around the sinuses and then returns to the heart after passing over the nephridia and through the gills (Fig. 12-52). Blood is delivered to the gills by the afferent branchial vessels. Such a system is sometimes characterized as "open," meaning that arterial outflow and venous return are connected by large sinuses. In contrast, a "closed" hemal system, such as that of annelids, has small capillaries connecting arteries and veins.

EXCRETION

Excretion in molluscs is accomplished by the **heart-kidney complex,** which is derived from and functions like the hemal-coelom-metanephridial system of other coelomate animals. In the generalized mollusc there is a pair of metanephridia (kidneys, or nephridia) with their nephrostomes opening from the pericardial cavity (coelom) and their nephridiopores into the exhalant chamber of the mantle cavity (Fig. 12-1A). In specific molluscs the number varies, but usually there is either one pair or a half-pair.

In most molluscs, the kidney is a large sac with thick secretory and absorptive walls surrounded by blood in the visceral sinus of the hemocoel, with which it exchanges materials. The connection with the pericardial cavity is the **renopericardial canal** (Fig. 12-52), whose opening from the pericardial cavity is the nephrostome. Podocytes in the walls of the atria make possible the formation of an ultrafiltrate (primary urine) in the pericardial cavity. The primary urine enters the nephrostome and travels through the renopericardial canal to the nephridium, where it is modified by secretion and absorption to and from the surrounding blood in the visceral sinus. The nephridial epithelium reabsorbs valuable materials from the urine and returns them to the blood while simultaneously secreting toxins and wastes from the blood into the urine. The final, modified urine leaves the kidney via the nephridiopore and enters the exhalant current of the mantle cavity.

NERVOUS SYSTEM AND SENSE ORGANS

The molluscan central nervous system consists of several pairs of ganglia, some of which are associated with an esophageal nerve ring, and two pairs of longitudinal nerve cords (Fig. 12-17A). A **ganglion,** or neuromere, consists of a cortex of cell bodies and a medulla of axons known as the neuropil. The ganglion is enclosed in a connective-tissue capsule. Ganglia are connected to sensory receptors by afferent, or sensory, neurons and to muscles or endocrine organs by efferent, or motor, neurons.

The **cerebral ganglia** (brain) receive sensory nerves from the eyes, tentacles, and statocysts (Fig. 12-17A). They are linked by connectives with the pedal, buccal, and pleural ganglia. The **buccal ganglia** are located in the buccal mass and innervate the muscles of the radula and odontophore. Motor nerves from the **pedal ganglia** innervate the foot muscles. The **pleural ganglia** innervate the mantle.

The ganglia of any given pair are connected to each other by a **commissure,** whereas the ganglia of different pairs are joined by **connectives.** In general, commissures are transverse and connectives are longitudinal. Connectives and commissures are usually named for the ganglia they connect. For example, the two cerebral ganglia are connected by the cerebral commissure whereas the cerebral and pedal ganglia are linked by the cerebropedal connective. Most paired ganglia are joined by a commissure, but the two pleural ganglia are not. The pleural and pedal ganglia are joined by a pair of connectives. Another pair of connectives, the **visceral nerve cords,** exit the pleural ganglia and extend posteriorly to the **visceral ganglia,** which innervate the organs of the visceral mass. A pair of **esophageal** (the intestinal, or parietal) **ganglia** occur on the visceral cords. The esophageal ganglia supply the gills, osphradium, and mantle.

The cerebral and pedal ganglia along with their connectives and commissures form the circumesophageal **nerve ring** (Fig. 12-1A). A pair of **pedal nerve cords** extends from the pedal ganglia into the foot (Fig. 12-1A), and the two **visceral nerve cords** run from the nerve ring to the visceral ganglion. Because there are four nerve cords, the system is said to be **tetraneurous.** In primitive molluscs there are transverse connections between the nerve cords, and the system is ladderlike. Molluscs exhibit a strong tendency to cephalize the nervous system by shortening connectives and commissures to bring all ganglia close together in the vicinity of the nerve ring.

The typical molluscan sense organs include chemomechanoreceptive **cephalic tentacles** on the head, a pair of eyes on the head, a pair of **statocysts** in the foot, and a pair of osphradia in the inhalant chamber of the mantle cavity. In many molluscs there are sensory tentacles on the mantle margins.

REPRODUCTION

The generalized mollusc is gonochoric with external fertilization in the sea. A pair of gonads is closely associated with the pericardial cavity (coelom). Gametes are shed by the gonads into the coelom. They enter the nephrostomes of the nephridia, which serve as **gonoducts,** and exit via the nephridiopore into the exhalant chamber of the mantle cavity. They are carried into the sea by the exhalant current. In some molluscs there are separate gonoducts independent of the nephridia. Fertilization is internal in many molluscs, and some are hermaphroditic. Both hermaphroditism and internal fertilization require elaboration of the reproductive system.

DEVELOPMENT

Spiral, holoblastic cleavage is a feature of the generalized mollusc. The developing blastula bears the characteristic **molluscan cross** of cells at its animal pole (Fig. 14-15C). Gastrulation is by invagination in species with small, microlecithal eggs and by epiboly in species with large, yolky eggs. The blastopore becomes the mouth. Mesoderm arises from the 4d mesentoblast cell, which divides to produce right and left mesodermal bands, and these cavitate to produce the coelom by schizocoely. The gastrula develops into a trochophore larva, which in primitive molluscs is the hatching stage. In most species, however, the trochophore is suppressed and a later stage, often a **veliger,** hatches. Some molluscs have yolky eggs and direct development. Development is similar to that of other spiralians.

Trochophore Larva

The **trochophore** is the characteristic larva of molluscs, annelids, sipunculans, echiurans, and many other protostomes (Fig. 12-4). A hallmark of this top-shaped larva is a conspicuous preoral girdle of cilia, the **prototroch,** that encircles the body near its middle, but anterior to the mouth. The gut is a complete tube with the mouth opening ventrally just posterior to the prototroch and the anus at the posterior end. Additional rings of cilia are often present. In many molluscs, polychaetes, and sipunculans a second ciliary girdle (this one post-oral), the **metatroch,** develops just posterior to the mouth. In some taxa a third ring, the **telotroch,** is found at the posterior end. An **apical tuft** of cilia is located at the anterior end. Internally, the blastocoel persists as a large, gelatinous connective-tissue layer between the gut and the outer ectoderm (Fig. 12-4B). Larval muscle bands cross the blastocoel and a pair of protonephridia form, one on each side of the gut. The adult mesodermal structures develop from a pair

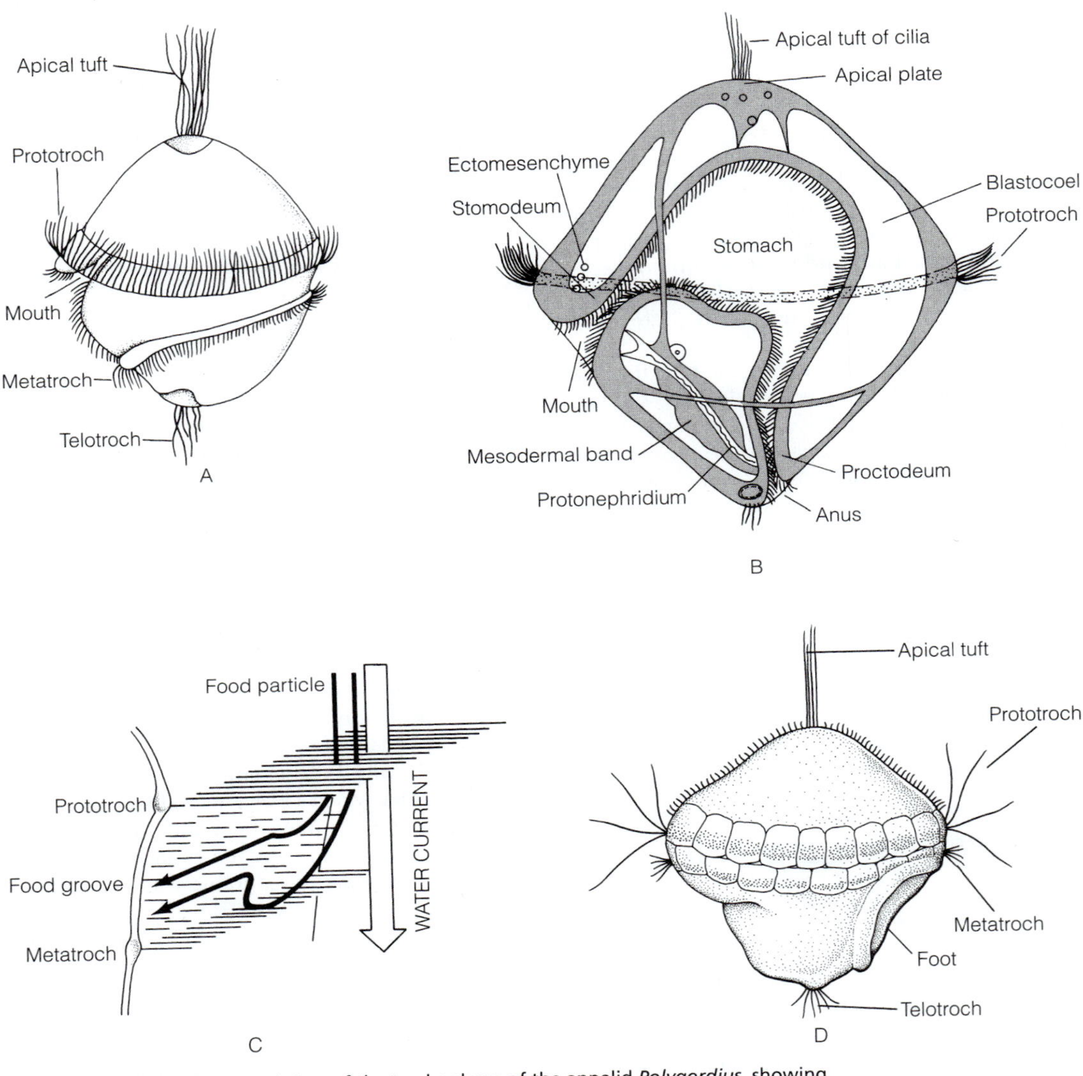

FIGURE 12-4 Trochophore larvae. **A,** External view of the trochophore of the annelid *Polygordius*, showing characteristic ciliary bands. **B,** Internal anatomy of an annelid trochophore. **C,** Two-band suspension feeding mechanism of a trochophore larva. **D,** Trochophore larva of the primitive patellogastropod *Notoacmaea scutum*. *(A, After Dawydoff; B, After Shearer from Hyman; C, From Strathmann, R. R., Jahn, T. L., and Fonesca, J. R. 1972. Suspension feeding by marine invertebrate larvae: Clearance of particles by ciliated bands of rotifer, pluteus and trochophore. Bio. Bull. 142:505–519; D, Redrawn from a sketch by G. L. Shinn.)*

of ventrolateral **mesodermal bands.** The rosette cells of the molluscan cross differentiate into an anterior ectodermal **apical plate,** from which arises the apical tuft. The cerebral ganglia develop from the apical plate. The cross cells, along with other cells, form the pretrochal epidermis.

A fully developed trochophore consists of three regions. The **pretrochal region** consists of the apical plate, prototroch, and the area surrounding the mouth. The **pygidium** is the region consisting of the telotroch and the area around the anus. The **growth zone** lies between the mouth and the telotroch. New tissue is added by proliferation of cells in the mesodermal bands so that the pretrochal region and pygidium are pushed farther apart as the animal grows.

Trochophores are usually planktonic and feed on tiny suspended particles, living or dead. Such larvae are **planktotrophic** and typically have long planktonic lives. Alternatively, some trochophores depend on yolk, and are **lecithotrophic** and usually have a short planktonic life. The prototroch is the swimming organ and, in planktotrophic species, is also responsible for collecting food particles. A food groove (Fig. 12-4C) lies between the prototroch, which is preoral, and the metatroch, which is post-oral. The longer cilia of the prototroch drive water and suspended particles posteriorly. The shorter cilia of the metatroch beat at right angles to the prototroch and particles are driven ventrally into the food groove between the two ciliary bands and transported to the mouth.

The trochophore (Fig. 12-4D) is followed in the development of many living molluscs by a more advanced stage, the veliger larva, as is found in gastropods (Fig. 12-60), bivalves (Fig. 12-121B), and scaphopods (Fig. 12-123C). The **hatching stage** may be the trochophore or the veliger if the trochophore is suppressed in the egg. The veliger has a foot, shell, and two lateral ciliated lobes together called the **velum,** with which it swims and feeds. The ciliated velum develops from the prototroch of the trochophore. At the end of larval life, it settles to the bottom and metamorphoses into a benthic adult. Some molluscs, such as the cephalopods and many gastropods, have macrolecithal eggs and **direct development,** without a larva (Fig. 12-87).

APLACOPHORA^C

Aplacophora (Fig. 12-5) consists of two taxa, Neomeniomorpha, and Chaetodermomorpha, of about 300 species of small, strange, wormlike marine molluscs. Aplacophorans are usually less than 5 mm in length, but some reach 30 cm. They differ from all other molluscs in the presence of calcareous spicules instead of a shell, in the absence or reduction of the foot, and in the cylindrical, wormlike shape of the body. The mantle cavity is posterior. A radula is present in most species. They exhibit many characteristics of the generalized mollusc (radula, nervous system, gonocoel-pericardial connection, mantle and mantle cavity), but in some respects are specialized (cylindrical body, reduction of foot). They are not ancestral to other molluscs and are probably an early offshoot from the stem molluscs (Fig. 12-125).

Aplacophorans are found in all oceans to depths of 7000 m and most are found between 200 and 3000 m. Most specimens have been collected by dredging in deep water, but some occur in shallow coastal or even intertidal waters. Their biology is still little known.

FORM

The dorsal body wall is the mantle**,** which extends laterally to cover the entire animal except for the foot, if present. The mantle epidermis secretes a glycoprotein cuticle that contains embedded calcareous (aragonite) **spicules** of various shapes. Unlike in other molluscan taxa, a shell is absent. A connective-tissue dermis lies below the epidermis. Successive layers of circular, oblique, and longitudinal muscles lie beneath the dermis. The body cavity, or hemocoel**,** is adjacent to the innermost (longitudinal) muscle layer.

The foot, when present, is a longitudinal fold (sometimes multiple folds) filled with the pedal hemocoel. Its epithelium is ciliated and it is not lined with cuticle. It is not muscular and locomotion is ciliary. The foot lies in the **pedal groove,** which is lined by cuticle.

The mantle cavity is posterior (Fig. 12-5A) and contains the anus and the openings of the gonoducts (coelomoducts). In chaetodermomorphs, it contains a pair of bipectinate gills. The gill epithelium, like that of the mantle cavity, is ciliated, and afferent and efferent vessels enter and leave the gills. Under normal circumstances the gills protrude from the mantle cavity (Fig. 12-5B), but they can be withdrawn by retractor muscles. In neomeniomorphs, the mantle epithelium is folded to form secondary gills (respiratory papillae) that are not true gills.

The gut consists of the mouth, cuticularized buccal cavity, esophagus, midgut, intestine, rectum, and anus. The buccal cavity usually contains a radula and the anus is in the mantle cavity. There may be a large digestive cecum. In about 20% of neomeniomorphs, the radula is absent and instead the buccal cavity functions as a sucking pump. Some derived chaetodermomorphs have a style sac, gastric shield, and rotating mucous protostyle.

The hemal system includes a posterior heart similar to that described for the generalized mollusc. The ventricle forms as an invagination of the dorsal pericardium and usually remains attached to it by a membrane. The hemocoel is divided into a dorsal perivisceral hemocoel and a ventral pedal hemocoel by a longitudinal, muscular, horizontal septum. The pedal hemocoel is present in all Aplacophora, even if there is no foot.

The excretory system is poorly known. There are no recognizable nephridia, but podocytes occur on the pericardial wall. Presumably the system operates as it does in other molluscs (and annelids). If so, an ultrafiltrate of the blood would be formed in the pericardial cavity and modified as it passed through the renopericardial ducts to the mantle cavity, where the final urine would be released. Some epidermal gland cells are suspected of having an excretory function.

The nervous system is ladderlike. There is a pair of cerebral ganglia dorsal to the anterior gut. These connect with two pairs of longitudinal nerve cords, the lateral visceral nerves and the ventral pedal nerves, which extend the length of the body and are connected with each other by

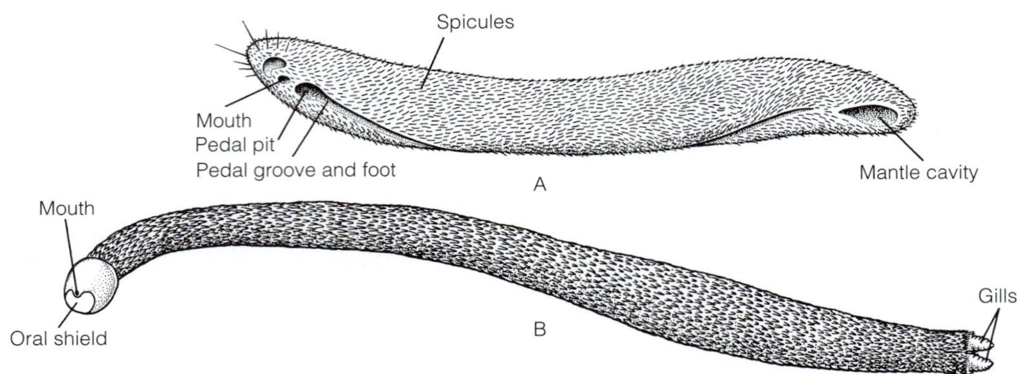

FIGURE 12-5 External anatomy of aplacophoran molluscs. **A,** A neomeniomorph. **B,** A chaetodermomorph. *(Redrawn and modified from Salvini-Plawen, L. V. 1972. Zur Morphologie und Phylogenie der Mollusken: Die Beziehungen der Caudofoveata und der Solenogastres als Aculivera, als Mollusca und als Spiralia. Z. wiss. Zool. 184:205–394)*

commissures. The cords of the right and left sides connect via a suprarectal commissure that arches *over* the posterior gut. A posterior chemoreceptive dorsoterminal sense organ is a characteristic feature of the aplacophoran sensory system. It has been suggested, but not demonstrated, that it is homologous to the osphradium of the other molluscan taxa. In the neomeniomorphs the mucus-secreting, ciliated **pedal pit** has a sensory function and there is also a dorsofrontal sensory pit above the mouth. In the chaetodermomorphs, an **oral shield** surrounds the mouth. This ring of mantle epidermis and thickened cuticle is penetrated by sensory cilia. The oral shield is innervated by the cerebral ganglia.

The gonads are paired and release their gametes directly into the pericardial cavity (coelom) via gonopericardial ducts. The pericardial cavity is connected with the mantle cavity by a pair of elaborate coelomoducts (gonoducts). This condition probably is primitive for the molluscs. Most aplacophorans are hermaphroditic, but chaetodermomorphs are gonochoric. Eggs may be retained, in which case development is direct, or spawned, whereupon a lecithotrophic larva latches from the egg.

DIVERSITY OF APLACOPHORA

Neomeniomorpha[sC] (Solenogastres; Fig. 12-5A): The larger taxon. Reduced foot consists of midventral ridge in a longitudinal groove. Pedal pit present at anterior end of foot. Gills lost. Creep over hydroids and alcyonarian corals or across the sea bottom by using their narrow foot. Carnivores feed on the cnidarians on which they live. Hermaphroditic. Seminal receptacles, seminal vesicles, copulatory spicules are present. Copulation with internal fertilization. 25 families. *Eleutheromenia, Lyratoherpia, Neomenia.*

Chaetodermomorpha[sC] (Caudofoveata; Fig. 12-5B): Small taxon of three families. Foot completely lost. A pair of gills in the posterior mantle cavity. Oral shield partly or completely encircles the mouth. Selective deposit feeders or carnivores burrowing in soft substrates. Gut with a digestive cecum. Gonochoric. Fertilization probably external. *Chaetoderma, Epimenia, Falcidens, Limifossor, Scutopus.*

POLYPLACOPHORA[C]

Members of Polyplacophora are chitons, or coat-of-mail molluscs (Fig. 12-6) found only in the sea. Although some features of their structure and development are primitive, chitons are highly adapted for a life adhering to rocks and shells. Many live in shallow coastal waters, where they cling tenaciously in powerful surf and surge. About half of modern species live in the intertidal or shallow subtidal zones. Other species are found in deeper, quieter water. The elongate, oval body is strongly depressed dorsoventrally (Fig. 12-7). Chitons are distinctive in having a shell composed of eight overlapping plates, which permits flexibility in conforming to the shape of the irregular, rocky substratum. Chitons lack cephalic eyes or tentacles and the head is poorly developed and indistinct. The mantle is thick and the foot is broad and flat to facilitate adhesion to hard substrates.

Approximately 800 extant species are known, but the fossil record is sparse. Some 350 fossil chitons are described, beginning at least by the Ordovician and perhaps as early as the Cambrian. Polyplacophora is the sister taxon of Conchifera and in Linnean classifications is ranked as a class.

Chitons range in size from 3 mm to 40 cm, the largest being the giant Pacific gumshoe chiton, *Cryptochiton stelleri* (Fig. 12-7A). Most species, however, are 3 to 12 cm in length. Chitons typically are drab shades of red, brown, yellow, or green.

MANTLE

The mantle covers the entire dorsal surface of a chiton and secretes a thin glycoprotein cuticle. The mantle is a thick, stiff, heavy dorsal layer extending to the lateral margins of the body and partially (or entirely) covering the shell valves. It extends well beyond the lateral margins of the valves to overhang the lateral mantle cavity (Fig. 12-8A). The lateral portion of the mantle, peripheral to the valves, is sometimes referred to as the **girdle.** The cuticle may be smooth or bear scales, bristles, or calcareous spicules similar to those of the Aplacophora (Fig. 12-6C).

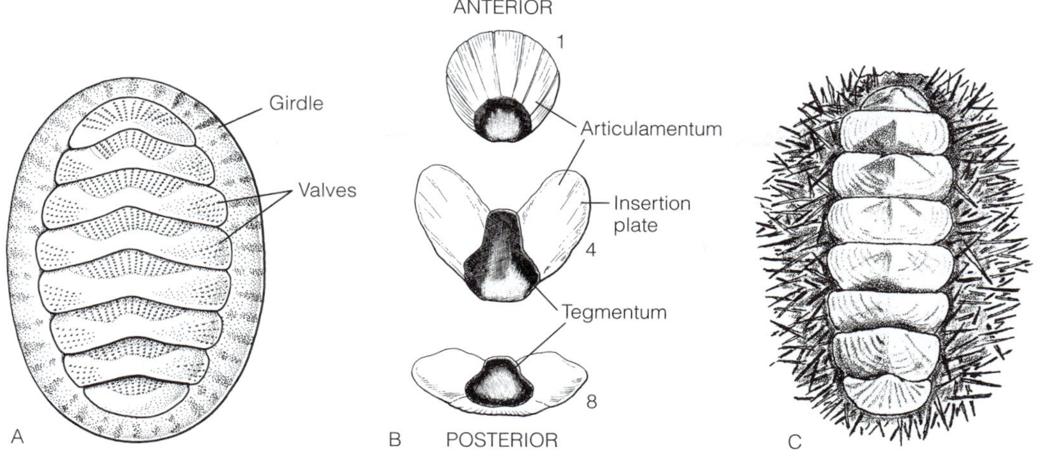

FIGURE 12-6 Polyplacophora. **A,** Common American Atlantic coast chiton *Chaetopleura apiculata,* which is only a few centimeters long. **B,** Representative valves of *Katharina,* numbered from anterior to posterior. The intermediate valves 2 to 7 are similar to 4. **C,** A chiton with spicules in the girdle. *(A, Redrawn after Pierce, M. E. 1950,* Chaetopleura apiculata. *In Brown, F. A. (Ed): Selected Invertebrate Types. John Wiley and Sons, New York. pp. 318–319; C, After Borradaile, L. A., and Potts, F. A. 1959. The Invertebrata: A Manual for the Use of Students. Cambridge, 795 pp.)*

FIGURE 12-7 Polyplacophora. **A,** Two species of chitons from the northwest Pacific coast. The valves of the larger species *(Cryptochiton)* are completely covered by the mantle; those of the smaller species *(Katharina)* are partially covered. **B,** The West Indian chiton *Chiton tuberculatus* exposed on rocks at low tide.

SHELL

By far the most distinctive feature of a chiton is its shell, which is divided into eight overlapping transverse **shell valves,** or shell plates (Fig. 12-6A). Polyplacophora means "bearer of many plates." Except for the overlapping posterior edge, the margin of each valve is embedded deeply in mantle tissue, but the extent of coverage varies among different species. Typically, most of the valve width is exposed (Fig. 12-6A). In *Katharina,* however, only the midsection of each valve is uncovered, and in the exceptional *Cryptochiton,* the shell is completely covered by the mantle (Fig. 12-7A). The posterior edge of each valve overlaps the anterior edge of the valve posterior to it. A large, winglike **insertion plate** (Fig. 12-6B) extends laterally and anteriorly from each valve into the surrounding mantle and underneath the preceding valve to anchor the shell securely. A pair of strong, double pedal retractor muscles extends from the foot to each of the valves. (Monoplacophorans also have eight pairs of pedal retractor muscles, but theirs are attached to an undivided shell.)

Each valve consists of four layers. The outermost, exposed layer is the thin organic periostracum. Under it lies the **tegmentum** (Fig. 12-6B), composed of the protein conchiolin plus calcium carbonate. The tegmentum contains sense organs known as esthetes and the neurons that serve them. It is probably a derived feature that is not homologous with any part of the conchiferan shell. The **articulamentum** is a purely calcareous

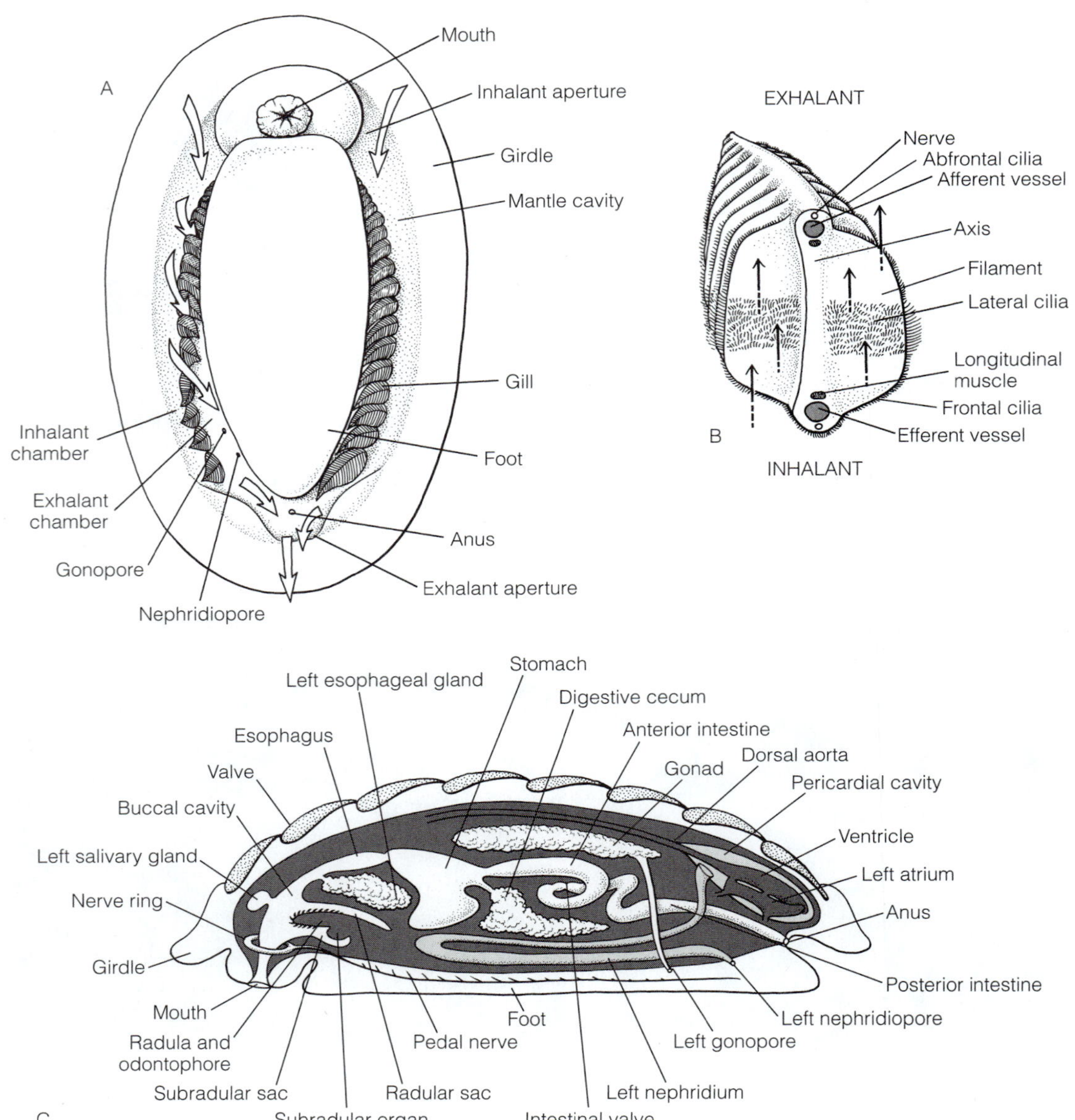

FIGURE 12-8 The chiton *Lepidochitona cinerea.* **A,** A ventral view showing the pallial groove and direction of the respiratory water currents. The arrows show water entering through the two anterolateral inhalant apertures and exiting through the single median, posterior exhalant aperture. **B,** Transverse section through the gill axis. Arrows indicate direction of water current over the filament surface. **C,** A lateral view showing the digestive tract, hemal system, excretory system, and nervous system. *(A and B, Redrawn after Yonge, C. M. 1939. On the mantle cavity and its contained organs in the Loricata. Quart. J. Micro. Sci. 81:367–390.)*

(aragonite) layer that lies under the tegmentum in the center of the valve but extends laterally free of the tegmentum to form the insertion plates (Fig. 12-6B). A second calcareous layer, the hypostracum**,** lies under the articulamentum. The tegmentum, covered by the periostracum, is the only part of the valve visible externally. The articulamentum is hidden either by the tegmentum or the mantle. The insertion plates consist of articulamentum and hypostracum, but no tegmentum.

FOOT AND LOCOMOTION

The broad, flat, muscular foot occupies most of the ventral surface and is used for locomotion and for adhering to the substratum (Fig. 12-8A), which is usually rock or shell. Chitons use waves of muscular contraction to creep very slowly over the substratum. The division of the shell into articulating valves enables chitons to conform to an irregularly curved surface of almost any topography. If accidentally dislodged, chitons can use a pair of longitudinal **enrollment muscles** to roll into a ball, and although this may be a defense mechanism, it also enables the animal to right itself and reattach the foot to the substratum.

The animal adheres to its substratum using both foot and mantle. Under normal circumstances only the foot is used, but when threatened or otherwise disturbed, the mantle is brought into play. A chiton forewarned is almost impossible to remove from a rock without damaging the animal, so tenaciously does it adhere when both foot and mantle are employed. When in use, the outermost margin of the mantle is brought into contact with the substratum, and the inner margin is elevated slightly. This creates a suction that enables the animal to tightly grip the substratum.

Chitons are common rocky intertidal inhabitants and, like limpets, most species are motionless at low tide. When the rock surface is submerged or splashed, they move about to feed. They are usually negatively phototactic and thus tend to position themselves under rocks and ledges. They are most active at night if they are submerged by the tide. Like limpets, some species exhibit homing behavior.

MANTLE CAVITY AND VENTILATION

The mantle cavity is lateral and consists of a ventral groove on each side of the body between the foot and the mantle (Fig. 12-8A). The two lateral cavities are continuous posteriorly. Each mantle cavity contains multiple gills, a gonopore, and a nephridiopore. The anus is situated medially in the area where the two lateral cavities join posterior to the foot. Each mantle cavity contains a longitudinal row of 6 to 88 ciliated, bipectinate gills. The number of gills is variable between and within species, depending, in part, on the size of the chiton.

The row of bipectinate gills forms a curtain that divides the mantle cavity into a lateral inhalant chamber and a medial exhalant chamber (Fig. 12-8A). On each side of the body an anterior area of girdle can be lifted from the substratum to form an **inhalant aperture.** The **respiratory current,** generated by the gill cilia, is pulled into the mantle cavity (Fig. 12-8A) through this opening. The water flows posteriorly in the inhalant chamber and then passes through the curtain of gills into the exhalant chamber. Gas exchange between water and blood occurs as the water flows between the gill filaments. Once in the exhalant chamber, the water continues to flow posteriorly, being pushed by the gill cilia. At the posterior end of the foot, the right and left exhalant currents coalesce on the midline and flow out of the mantle cavity via a single, median **exhalant aperture.** The structure, ciliation, ventilation, and blood flow of the gills are similar to those of the generalized mollusc (12-8B).

NUTRITION

Most chitons are microphagous browsers that feed on fine algae and other organisms that they scrape with the radula from the surface of rocks and shells. This is probably the ancestral molluscan feeding mode and the original function of the radula. The scraping of the radula indiscriminately removes algae and associated animals, along with an abundance of sediment particles that have accumulated in the interstices among the algae. For example, the gut contents of three chitons from the Maine coast included 14 algal and animal species, but about 75% of its volume was sediment.

Some chitons are **macroherbivores** that feed on larger algae (seaweeds), and one taxon, *Placiphorella,* on the western coast of the United States, is a carnivore that uses its raised and flaring anterior mantle to trap small crustaceans and other invertebrates. A few chitons are detritivores. The intertidal species *Nuttallina californica,* on the southern California coast, excavates and inhabits deep depressions in the rock substratum. The chiton never leaves its depression, which it uses to collect the seaweed detritus upon which it subsists. Successive generations inhabit the same depressions, which in some cases are estimated to be a thousand years old.

The mouth opens into the chitin-lined buccal cavity (Fig. 12-8C). A long radular sac that projects posteriorly from the back of the buccal cavity contains the radula. The radula (Fig. 12-8C), which is very long compared to that of other molluscs, bears 17 teeth in each transverse row. Some of these teeth are mineralized to harden them and increase their usable life span. Some are capped with **magnetite,** an iron-containing mineral. The great length of the radula and its hardened teeth are both adaptations to the wear imposed by almost continuous scraping against the surface of rocks. The radula is supported by an odontophore ventral to it, as was described for the generalized mollusc. The subradular sac is a smaller evagination of the buccal cavity ventral to the radula. It contains the tonguelike, chemosensory subradular organ. The radula, odontophore, and subradular organ are all equipped with protractor and retractor muscles and can be extended and retracted from the mouth at will.

During feeding, the subradular organ is protruded and pressed against the rock. If it detects food, the odontophore and the radula are extended from the mouth and the radula scrapes the surface of the rock, as described for the generalized mollusc. Periodically, scraping ceases as the subradular organ again tests the substratum.

The scrapings are pulled into the buccal cavity by the recurved teeth of the radula on each return stroke (Fig. 12-2C,D). Mucus is secreted into the buccal cavity from a pair of salivary glands (buccal glands) and mixed with the particles to form a food string that cilia continuously propel posteriorly into the esophagus. The string is moved through the esophagus to the stomach (Fig. 12-8C). During this passage, the food particles are mixed with amylase from a pair of large **esophageal glands** (sugar glands, or pharyngeal glands) whose ducts open into the anterior esophagus.

The esophagus empties into the anterior stomach, where the food is mixed with proteolytic enzymes from the digestive cecum. The style sac typical of conchiferans is absent. Digestion is almost entirely extracellular and takes place in the digestive cecum, stomach, and anterior intestine.

The **anterior intestine** makes a loop and then joins a large, coiled **posterior intestine,** where fecal pellets are formed. The two regions of the intestine are separated by a valve. The anus opens at the midline just beyond the posterior margin of the foot and the egested fecal pellets are swept away by the exhalant current (Fig. 12-8A).

INTERNAL TRANSPORT

The pericardial cavity (coelom) is posterior, located under shell valves 7 and 8, and large (Fig. 12-8C). It is enclosed by the **pericardium** (Fig. 12-9A). The coelom contains the heart, which consists of a median ventricle and one pair of lateral atria. A median dorsal aorta exits the anterior end of the ventricle and leads to the hemocoel of the head after giving off paired lateral channels along the way. The hemocoel is divided into sinuses connected by channels that may be well defined and lined with connective tissue (but lack an endothelium). Blood from the efferent vessels of the gills drains into the atria and then passes to the ventricle, which pumps it to the hemocoel. Unlike aplacophorans, chitons have no connection between the gonad and the pericardial coelom.

EXCRETION

The excretory organs are two very large, glandular, and secretory nephridia situated laterally in the hemocoel (Fig. 12-9A). Each connects with the pericardial cavity by a renopericardial canal. Each nephridium opens via a nephridiopore in the exhalant chamber of the posterior mantle cavity (Fig. 12-8A) to release the final urine. Urine formation is as it was described for the generalized mollusc.

NERVOUS SYSTEM AND SENSE ORGANS

The nervous system (Fig. 12-9B) is like that of aplacophorans except that ganglia are absent. A circumenteric nerve ring surrounds the anterior gut. The system is tetraneurous and ladderlike, with transverse runglike commissures connecting the four longitudinal cords. The nerve cords are a pair of ventral pedal nerve cords and a pair of lateral visceral nerve cords, probably homologous to the similar cords of the aplacophorans. A suprarectal commissure arches over the gut to connect the right and left lateral cords, as it does in aplacophorans.

The chief sense organs are the subradular organ, mantle receptors, and the esthetes. **Esthetes** (also spelled *aesthetes*), which are unique to chitons, consist of groups of epidermal cells specialized for sensory reception, secretion, and support (Fig. 12-10). Esthetes are both small (**micresthetes,** or microaesthetes) and large (**megalesthetes,** or macresthetes). They are housed in tiny epidermally lined vertical canals in the tegmentum of the valves. Each canal extends from the base of the tegmentum to terminate beneath a cap on the shell surface. The mantle epidermis extends into the shell to line the canals. Esthete densities of 1750 canals/mm^2 of exposed shell surface have been recorded for *Lepidochitona cinerea.* Although the structure of esthetes is well known and they have been well studied, their function remains disputed. They apparently are both sensory and secretory. In some chitons the esthetes include ocelli, which, because of their well-developed lenses and electrical response to photostimulation, are assumed to be photoreceptors. Several thousand ocelli may be concentrated on the anterior valves. Other sensory cells may be chemoreceptors or mechanoreceptors. Secretory functions may include production of the periostracum, discouraging fouling or predation, or retarding desiccation.

A variety of sensory receptors, including mechanoreceptors and photoreceptors, are associated with the mantle. A so-called osphradium is present in each mantle cavity, but in some species it is in the inhalant chamber and in others in the exhalant chamber.

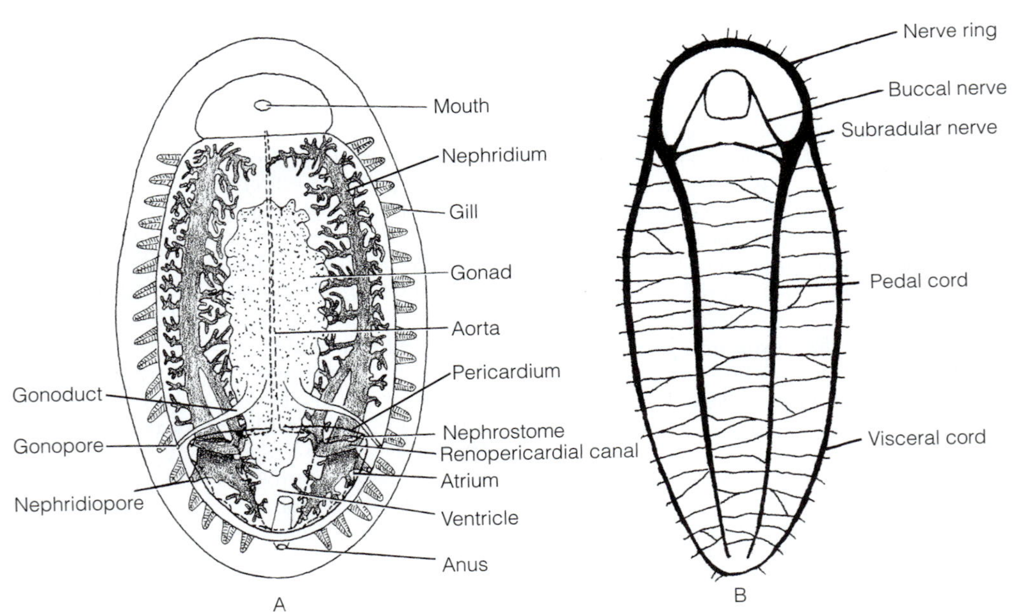

FIGURE 12-9 Polyplacophora. **A,** Internal structure of a chiton in dorsal view. **B,** Ladderlike nervous system of a chiton in dorsal view. *(A, After Lang and Haller; B, After Thiele from Parker and Haswell, 1951)*

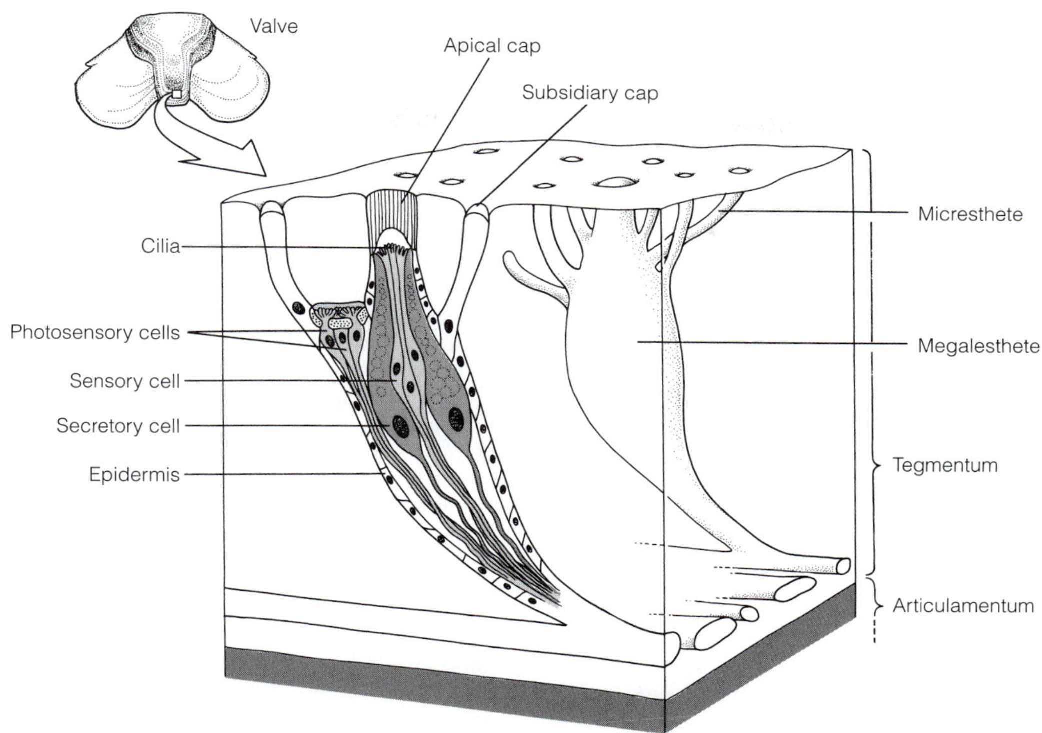

FIGURE 12-10 Chiton esthetes. Section of shell showing esthete canals and surface terminations. The larger canal houses a megalesthete; the smaller, a micresthete. *(Modified and redrawn from Boyle, P. R. 1974. The aesthetes of chitons. II. Fine Structure in* Lepidochiton cinereus. *Cell Tiss. Res., 153:383–398, and other sources.)*

REPRODUCTION AND DEVELOPMENT

Almost all chitons are gonochoric. A single median gonad develops from paired embryonic gonads in the dorsal hemocoel anterior to the pericardial coelom. This is the same spatial relationship between gonad and coelom as is seen in aplacophorans except that in chitons there is no connection between these two subdivisions of the coelom (Fig. 12-8C, 12-9A). The gametes are transported directly to the outside by two gonoducts (Fig. 12-8C, 12-9A) and do not pass through either the pericardial cavity or the nephridia. A gonopore occurs in the exhalant chamber of the mantle cavity just anterior to the nephridiopore (Fig. 12-8A).

Fertilization is external and there is no copulation. Instead, males release sperm into their exhalant respiratory currents. The eggs, which are enclosed in a spiny envelope, are usually shed into the sea either singly or in strings. Fertilization occurs in the sea or within the mantle cavity of the female. The gregariousness of chitons makes fertilization by this mechanism feasible since the gametes need not disperse over large distances. A lecithotrophic planktonic trochophore with ocelli, prototroch, mantle, shell gland, and foot emerges from the egg in most species (Fig. 12-11A). There is no veliger. In about 30 species, the eggs are brooded in the mantle cavity of the

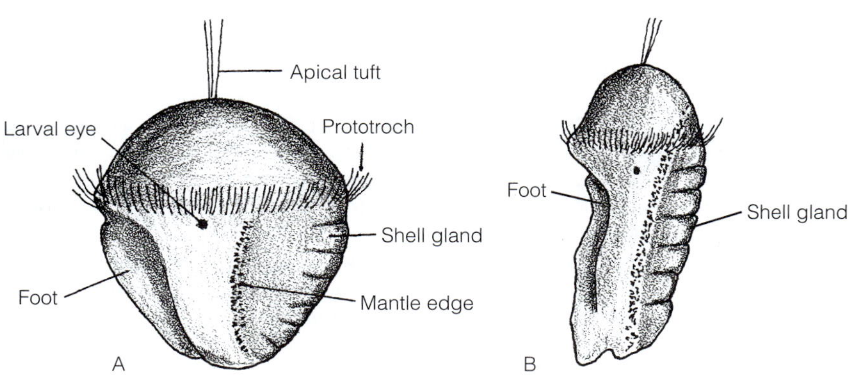

FIGURE 12-11 Chiton larvae in lateral views. **A,** Trochophore of *Ischnochiton.* **B,** Metamorphic stage of *Ischnochiton.* *(A and B, After Heath from Dawydoff.)*

female and development is direct. During development the post-prototrochal region elongates to become most of the body. A shell gland with seven regions develops on the dorsum and will secrete the valves (Fig. 12-11B). The seventh region divides to produce the seventh and eighth valves. The prototroch degenerates and the animal metamorphoses and settles to the bottom as a young chiton. The larval ocelli are retained for some time after metamorphosis.

DIVERSITY OF POLYPLACOPHORA

Paleoloricata^O: Extinct. Shell plates lack articulamentum.

Neoloricata^O: All living chitons. Shell plates with articulamentum.

Lepidopleurina^sO: Small, mostly deepwater chitons less than 50 mm in length. Articulamenta lack or have poorly developed insertion plates. Gills relatively few in number and confined to the posterior ends of the pallial grooves. *Hanleya, Lepidopleurus.*

Ischnochitonina^sO: Most chitons belong to this taxon. Much of the shell plate (tegmentum) is exposed. *Chaetopleura, Chiton, Katharina, Lepidochitona, Mopalia, Nuttallina, Placiphorella.*

Acanthochitonina^sO: Toothed insertion plates. Relatively small part, sometimes none, of tegmentum is exposed. Last (eighth) plate has large insertion plates. *Acanthochitona, Cryptochiton, Cryptoconchus.*

MONOPLACOPHORA^C

The remaining molluscs constitute the monophyletic taxon Conchifera, consisting of Monoplacophora, Gastropoda, Cephalopoda, Bivalvia, and Scaphopoda (Fig. 12-125). These five taxa probably arose from a common ancestor similar to a monoplacophoran.

Monoplacophorans resemble the generalized mollusc and are small, flattened, marine molluscs that inhabit deep water. Living representatives are 3 mm to little more than 3 cm long and are bilaterally symmetrical (Fig. 12-12). They were known from Cambrian to Devonian fossils but were thought to be extinct until 1952, when 10 living specimens of *Neopilina galatheae* were dredged from a deep ocean trench off the Pacific coast of Costa Rica. Since this discovery, specimens belonging to about 20 species and three genera have been collected from waters 1800 to 7000 m deep in the northern and southern Atlantic, Indian, and eastern Pacific Oceans. The only known shallow-water species was found off the California coast at about 200 m. The modern species are thought to be the remnants of a formerly large and abundant taxon of molluscs.

As the name implies, monoplacophorans have a single bilaterally symmetrical shell that varies in shape from a flattened, shieldlike plate to a short cone. It is secreted by the mantle and its apex is directed anteriorly (Fig. 12-12B). The shell is composed of an outer organic periostracum made of the protein conchiolin, a middle ostracum, and an inner hypostracum. That these layers are like those of other conchiferans and unlike those of chitons supports the hypothesis that Polyplacophora and Conchifera arose from a shell-less common ancestor and evolved their shells independently of each other.

The mantle cavity is a pair of lateral groovelike mantle cavities separating the edge of the wide, flat foot from the mantle on each side (Fig. 12-12A). The two grooves coalesce posteriorly. As in chitons, eight pairs of pedal retractor muscles extend from the foot to the inside surface of the shell. One of the most striking features of monoplacophoran anatomy is the serial repetition of organs in several systems, including the retractor muscles. The mantle cavity contains three, five, or six pairs of monopectinate gills (Fig. 12-12A,E). The respiratory current probably enters laterally and passes through the gills to the medial side of the mantle cavity and then posteriorly to exit, as in chitons.

Monoplacophorans are microphagous browsers that feed on microorganisms and detritus on hard substrates. Examination of stomach contents has revealed diatoms, foraminiferans, and sponge spicules. The mouth is located on the ventral midline anterior to the foot and the anus is in the mantle cavity at the posterior end of the body (Fig. 12-12A). The mouth opens into a cuticularized buccal cavity, which contains a radula and a subradular organ. The esophagus extends posteriorly from the buccal cavity to the stomach and contains a heavily ciliated food channel bordered by a pair of ridges. In most species a pair of enormous esophageal pouches opens from the anterior esophagus. The stomach may contain a style sac, protostyle, and gastric cecum. Two large digestive ceca (midgut glands) are broadly connected with the anterior end of the stomach. The long intestine exits the posterior stomach and lies coiled counterclockwise in the posterior hemocoel. The rectum connects the last loop of the intestine with the anus.

The coelom consists of the gonocoels surrounding the gonads and the pericardial cavity surrounding the heart. There are three, four, six, or seven pairs of nephridia located laterally (Fig. 12-12D,E). Multiple pairs of nephridiopores open into the mantle cavity. The existence of renopericardial ducts connecting the nephridia with the pericardial cavity is uncertain.

The hemal system consists of a hemocoel and a heart in the posterior pericardial cavity (Fig. 12-12E). The rectum passes through the pericardial cavity and ventricle and divides both into right and left portions, although the halves remain continuous. Two pairs of atria open into the ventricle. The two sides of the ventricle pump blood into a single dorsal aorta, which empties into the hemocoel.

The tetraneurous nervous system resembles that of aplacophorans and polyplacophorans (Fig. 12-12E). It consists of a pair of weakly developed cerebral ganglia, a circumenteric nerve ring from which emerges a pair of visceral nerve cords to the mantle lobe and a pair of pedal nerve cords to the foot. Transverse commissures connect the pedal and visceral nerves of each side and the visceral cords are connected posteriorly by a suprarectal commissure. The two pedal nerves are connected by anterior and posterior commissures to form a loop. The radular muscles are served by buccal ganglia. No eyes or osphradia are found in any monoplacophoran, but a large subradular organ is always present. A pair of statocysts lies beside the anterior pedal commissure.

Almost all monoplacophorans are gonochoric, and two pairs of gonads are located in the middle of the body (Fig. 12-12E).

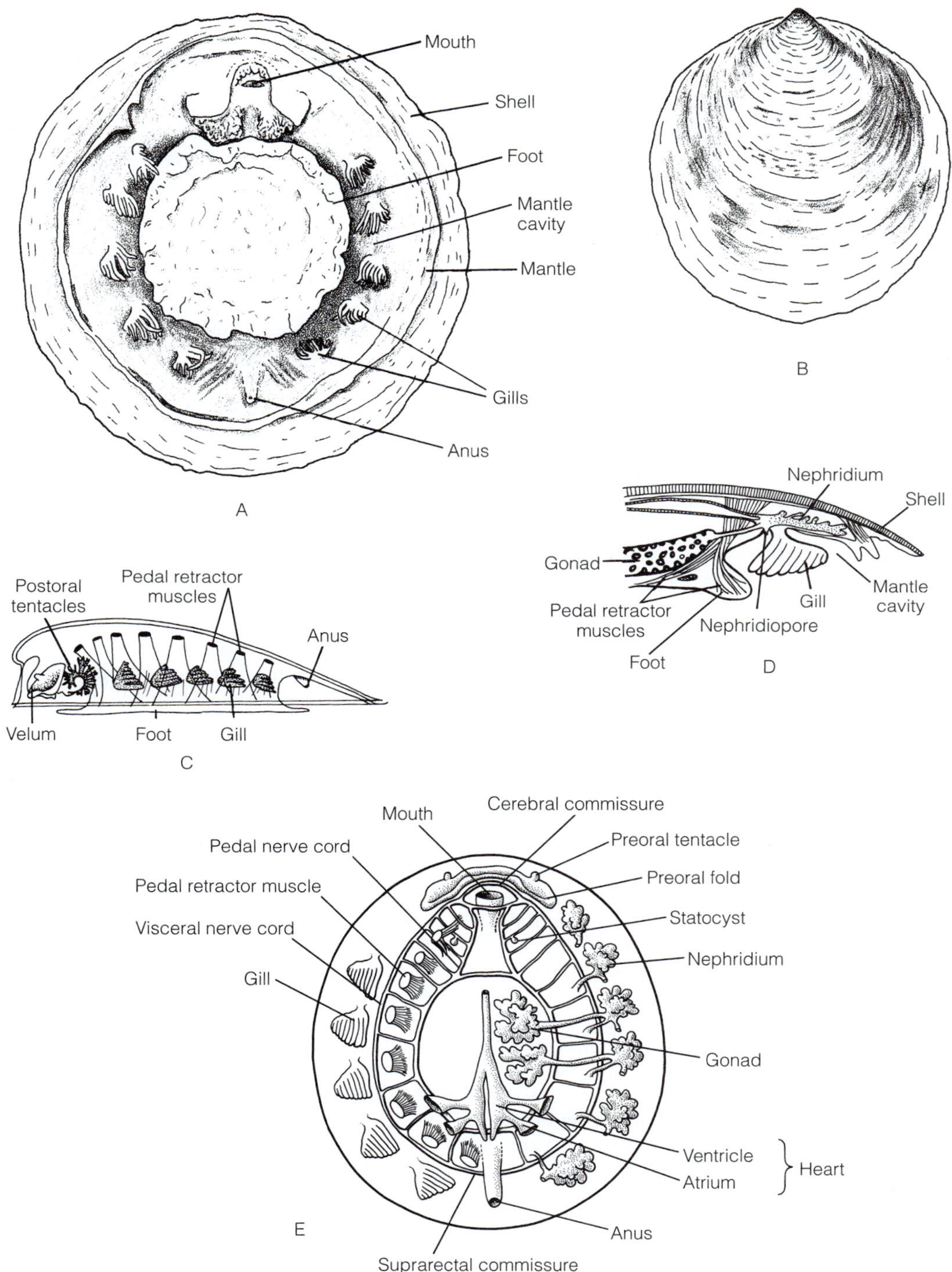

FIGURE 12-12 The monoplacophoran *Neopilina*. **A,** Ventral view. **B,** Dorsal view of shell. **C,** Side view. **D,** Transverse section through one half of the body showing the spatial relationship among the nephridia, coelom, gonad, and mantle cavity. **E,** Internal anatomy in ventral view. *(Adapted from Lemche, H., and Wingstrand, K. G. 1959. The anatomy of* Neopilina galatheae. *Galathea Rep. 3:9–71)*

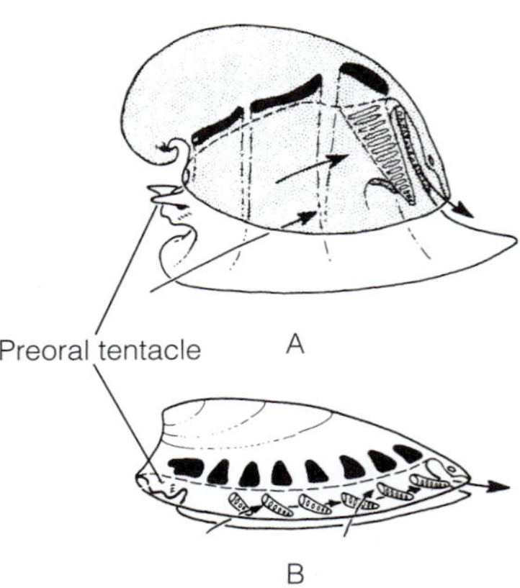

FIGURE 12-13 Monoplacophora. **A,** Reconstruction of an extinct cyclomyan monoplacophoran with an anteriorly coiled shell. Black areas indicate attachment scars of retractor muscles on shell. Arrows indicate possible ventilating current. **B,** Reconstruction of an extinct tergomyan monoplacophoran. *(From Stasek, C. R. 1972. In Florkin, M., and Scheer, B. T. (Eds.): Chemical Zoology. Vol. VII. Academic Press, New York. p. 21.)*

Each gonad is enclosed in a gonocoel and connects via a gonoduct to one of the middle nephridia (Fig. 12-12D,E). Fertilization is assumed to be external because the anatomy of the reproductive system is not consistent with internal fertilization, there being no intromittent organ or seminal receptacles. Nothing is known of the development of these elusive deepwater animals.

The survival of the few species of monoplacophorans is probably correlated with their living at great depths, where perhaps they escaped competition, predation, habitat change, or other forces that led to the extinction of other members of their taxon. Fossil monoplacophorans evolved along two lines. In the taxon Cyclomya, there was a lengthening of the dorsoventral axis of the body leading to a symmetrically coiled shell and a reduction in the number of gills and pedal retractor muscles (Fig. 12-13A). Although they disappeared from the fossil record in the Devonian period, cyclomyans may have been ancestral to the gastropods and cephalopods, if those taxa arose from coiled ancestors. The other taxon, Tergomya, retained a flattened, uncoiled, limpetlike shell with five to eight pairs of retractor muscles and five or six pairs of gills (Fig. 12-13B). Although this taxon was thought to have become extinct in the Devonian, when it disappeared from the fossil record, the recently discovered living species may belong to it. If the ancestors of gastropods were not coiled, tergomyans are likely candidates.

GASTROPODA[C]

Gastropoda is the largest and most diverse molluscan class. The gastropods are the snails and slugs, and the taxon includes an enormous diversity of these animals. The gastropod fossil record begins in the early Cambrian period and extends unbroken to the present. No other molluscan class has experienced the extraordinary adaptive radiation seen in the gastropods during this period. Gastropods originated in the sea, but have colonized fresh water and are the only molluscs found in terrestrial habitats. Among metazoan animals, only the vertebrates, crustaceans, and perhaps the annelids have enjoyed similar success in all three major environments. Estimates of the number of species vary widely from 40,000 to 100,000 species, but there are probably around 60,000 described, extant gastropods and another 15,000 known fossil species. Gastropods are primarily benthic and have become adapted to life on all types of bottoms, but some are adapted to a pelagic life. The pulmonate snails and slugs, as well as several other groups, have conquered land by eliminating the gills and converting the mantle cavity into a lung. Three major groups, the prosobranchs, pulmonates, and opisthobranchs, have tended to specialize along different lines. The prosobranchs are primarily benthic marine animals, whereas pulmonates have capitalized on the ecological possibilities of aerial respiration and opisthobranchs have experimented with the advantages of losing the shell.

Due to their great variety, it is difficult to describe a generalized gastropod. The only characteristic common to all gastropods is **torsion,** which is a 180° counterclockwise rotation of the visceral mass with respect to the foot. It will be discussed later. Many have a shell, which usually is asymmetrically coiled (Fig. 12-14A). The shell is univalve and external or, in some, internal or even lost. Gastropods retain the broad, flat, creeping foot of the generalized mollusc but are much more active and mobile than the chitons and monoplacophorans. Accordingly, the nervous system is cephalized and served by a well-developed sensory system. The head is distinct and bears one or two pairs of sensory tentacles and a pair of eyes. The visceral mass is large and is usually coiled into the shell. The mantle cavity is anterior, lateral on the right, posterior, or lost, and it usually contains one gill. The mantle cavity, shell, and visceral mass are asymmetric. There is usually one atrium and one nephridium and the gut is U-shaped. One gonad, the right, is retained. The tetraneurous nervous system consists of the several pairs of ganglia characteristic of the Conchifera, a nerve ring, commissures, and connectives, but it is twisted by torsion (streptoneurous) or secondarily untwisted (euthyneurous). The hatching stage is usually a veliger, and a larval operculum is present. Many gastropods have lost their shell, have lost their mantle cavity, have secondary gills or no gills, are not coiled, are secondarily symmetrical, or have partially reversed the effects of torsion.

PREVIEW OF GASTROPOD SYSTEMATICS

The phylogeny and classification of the gastropods, which were relatively stable for most of the 20th century, are currently being reconsidered in light of a wealth of new data from a variety of sources. The discovery of new gastropod taxa from deep-sea hydrothermal vents, new gross morphological information, ultrastructural data from electron microscopy, nucleotide sequences from molecular systematics, and the new techniques and philosophies of phylogenetic systematics have stimulated a vigorous reevaluation of the evolution of the gastropods. It is now clear that the old classification

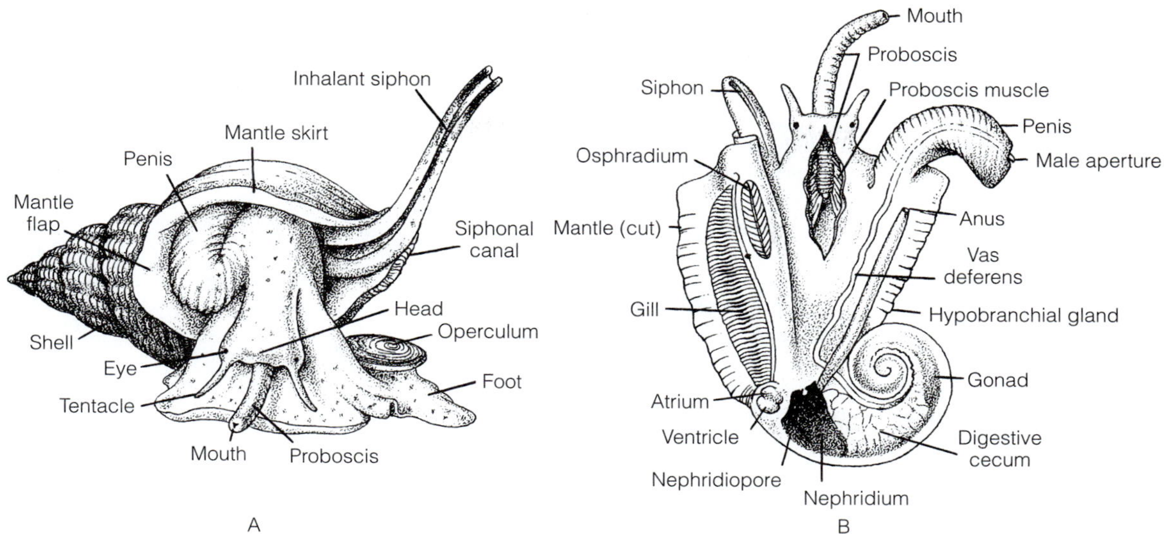

FIGURE 12-14 External anatomy of a marine gastropod, the prosobranch *Buccinum undatum.* **A,** The snail is crawling with both proboscis and siphon protruded. **B,** Dorsal view, with shell removed and wall of mantle cavity cut and reflected to the left. A small slit in the dorsal body wall provides a view of the proboscis retractor muscles. *(After Cox, L. R. 1960. Gastropoda. In Moore, R. C. (Ed.): Treatise on Invertebrate Paleontology. Vol. I, pt. 1. Geological Society of America and University of Kansas Press, Lawrence. pp. 189 and 191.)*

system did not accurately reflect the evolution and must be replaced. Laboratories worldwide are collaborating in the effort to construct a new gastropod phylogeny and classification, but the work is still in progress and consensus is yet to be reached. Pending the consolidation of these efforts, this text employs a provisional phylogeny and classification based on many sources. For the present, it seems best to arrange the gastropods in monophyletic taxa (orders) without grouping them into higher categories. We will, however, informally use the old taxonomic names of the higher categories for purposes of discussion. Informal names are not capitalized and do not have Latin endings.

For example, Prosobranchia, Opisthobranchia, and Pulmonata traditionally have been viewed as subclasses, but since no apomorphies separate and define these taxa and all three appear to be paraphyletic or polyphyletic, this classification is no longer acceptable. Nevertheless, since these old groups are useful pedagogically and there are as yet no accepted alternatives, they are employed here informally as prosobranchs, opisthobranchs, and pulmonates.

The primitive prosobranchs are marine, freshwater, and terrestrial, mostly gonochoric snails in which the mantle cavity and its organs are located at the anterior end of the body. They usually have a shell and operculum and are coiled. Prosobranchs were once divided into Archaeogastropoda, Mesogastropoda, and Neogastropoda, but the archaeogastropods and mesogastropods are not natural groups (although Neogastropoda is) and this division is not recognized in modern phylogenetic classifications. Nevertheless, the old groups remain convenient for purposes of discussion and are used here informally. Archaeogastropods are the most primitive prosobranchs and the ones most like the ancestral gastropod. The mesogastropods and neogastropods, known as the higher gastropods, are descended from the archaeogastropods.

Opisthobranchs (sea slugs, sea hares, sea butterflies, bubble snails, and others) and pulmonates (terrestrial and freshwater snails and slugs plus a few shallow-water marine species) exhibit a strong tendency to reduce, uncoil, or lose the shell and adopt a secondary bilateral symmetry. Gastropod classification is considered in more detail in the Diversity and Evolution of Gastropoda section of this chapter.

ORIGIN AND EVOLUTION OF THE GASTROPOD DESIGN

The gastropods arose when a monoplacophoran ancestor underwent torsion, a radical 180° twist of the visceral mass. All gastropods exhibit some degree of torsion and its consequences, and this is the single unifying characteristic of the taxon.

The familiar torted, spirally coiled, and asymmetric gastropod body evolved in a series of sequential processes, each contributing an essential feature to the overall design. These processes are: the elongation of the shell to make a portable retreat, flexure of the gut, planispiral coiling, torsion, and conispiral coiling and asymmetry. The order in which coiling and torsion occurred in the sequence is the subject of debate.

Torsion and coiling—both "twists" of the visceral mass—are sometimes confused with each other, but they are two quite different and independent events in gastropod evolution. Coiling is the twisting of the shell and visceral mass into the conspicuous and characteristic spiral associated with a typical snail (Fig. 12-14) and results from differential growth of the two sides of the shell. Torsion is the reversal of polarity of the visceral mass above the foot, and while it is a more profound change, it is not so conspicuous or noticeable externally. Both are typical of gastropods, but whereas torsion is characteristic of all gastropods, the more obvious coiling is not.

Coiling is usually considered to have occurred in the monoplacophoran ancestor and to have preceded torsion. Coiling would have produced a group of symmetrically and exogastrically (anteriorly) coiled monoplacophorans such as the cyclomyans (Fig. 12-13A), which later underwent torsion to become gastropods. In this scenario, the sequence of critical events would be, (1) Elongation of the visceral mass and lengthening of the shell to form a conical retreat associated with flexure of the gut; (2) planispiral coiling of the shell and visceral mass; (3) torsion; (4) conversion of the symmetrical planispiral shell to the asymmetrical conispiral shell. It has recently been proposed that torsion occurred before coiling and that the earliest gastropods were torted but limpetlike, nearly symmetrical, and uncoiled. There is evidence supporting both viewpoints. In the following account, it is assumed that coiling preceded torsion but, if desired, the order can be rearranged by reading the torsion section first, followed by those on elongation, coiling, and asymmetry.

Elongation of the Shell and Visceral Mass

The shell of the typical monoplacophoran (Fig. 12-12B) is a low, symmetrical, conical shield that protects the dorsal surface of the animal but does not provide a deep "retreat" into which the entire body can be withdrawn. When threatened, a monoplacophoran contracts the pedal retractor muscles and clamps the shell against the substratum to provide protection to its soft tissues. Because of the shell's low profile, there is no room for expansion of the visceral mass to accommodate an enlarged gonad or digestive system.

In the monoplacophorans that preceded the gastropods (progastropods), the height of the shell increased and the size of the aperture decreased to convert the low shield into a more spacious cone (Fig. 12-15A). These changes provided additional space within the confines of the shell that allowed for expansion of the soft tissues, especially the visceral mass, and its use as a retreat into which the animal could completely withdraw when threatened.

Associated with the increase in the height of the shell came the flexure of the gut. Prior to flexure, the gut extended more or less straight posteriorly from the anterior mouth to the posterior anus (Fig. 12-16A,C). Following flexure, the gut is U-shaped and arches up into the newly formed cone (Fig. 12-16B,D). The mouth and anus must remain outside the cone, of course, and are still ventral, but the stomach is dorsal. This flexure moves the gut and associated viscera away from the longitudinal head-foot axis and up into the shell. Dislocating the visceral mass away from the longitudinal axis is a prerequisite to taking advantage of the developing space in the elongate shell.

Planispiral Coiling

With increases in the height of the cone, the shell quickly became unwieldy, awkward to transport, and difficult to move into small spaces and crevices or through beds of macroalgae. This problem was alleviated by the coiling the shell over the head as it became higher (Fig. 12-18A). Eventually the shell

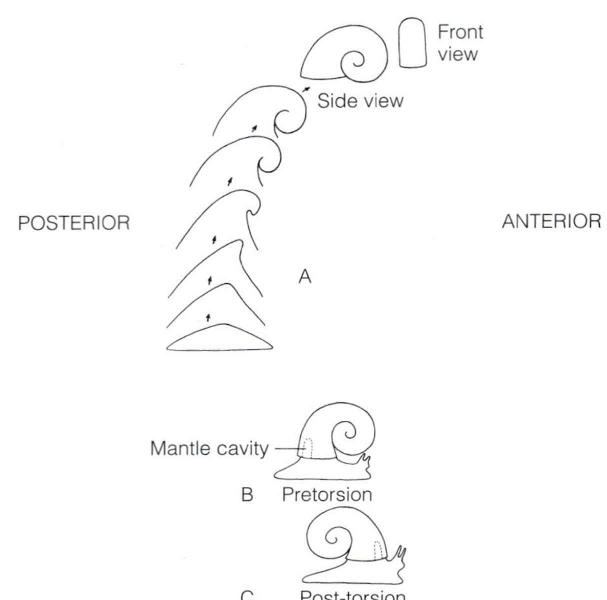

FIGURE 12-15 Gastropoda. **A,** Evolution of the planispiral shell beginning with an ancestral limpetlike monoplacophoran possessing a low, shieldlike shell. The height of the shell increases to provide a retreat in the face of danger and for the expansion of the visceral mass. To accommodate further increases in shell height, the apex of the cone is coiled anteriorly (exogastrically) and ventrally into a bilaterally symmetrical planispiral, as seen in cyclomyan monoplacophorans. The aperture is reduced and the snail can withdraw into its spiral shell, which is more compact and less awkward to carry than a straight conical shell would be. **B,** A progastropod monoplacophoran with a planispirally coiled shell before torsion. **C,** A gastropod showing the relationship between shell and head-foot after torsion. Anterior is to the right, posterior to the left in all.

consisted of several complete coils, or **whorls,** forming a spiral around a central node. In these first coiled shells the whorls all lay in a single plane so that the shell was **planispiral** and the animal and its shell were bilaterally symmetrical (Fig. 12-18A,B,C). In a planispiral shell (Fig. 12-15B,C), each coil is located completely outside of the one preceding it and in the same plane, like a hose coiled flat on the ground, or a rope coiled on a dock. The shell has an axis that passes through the center of the coil and is perpendicular to the plane of the coil (Fig. 12-19A). The resulting shell is **exogastrically coiled,** meaning that the direction of coiling is anterior, over the head (Fig. 12-15B). The opposite is **endogastric coiling,** in which coiling is posterior, over the foot (Fig. 12-15C). Although no modern gastropod has a true planispiral shell, many are functionally planispiral with flat, disclike shells but whorls that are not quite in a single plane. Nautiloid cephalopods have a true planispiral shell (Fig. 12-65).

Torsion

The progastropod monoplacophoran was untorted and we assume it had a planispiral shell that coiled exogastrically (anteriorly) so that its bulk and weight were over its head

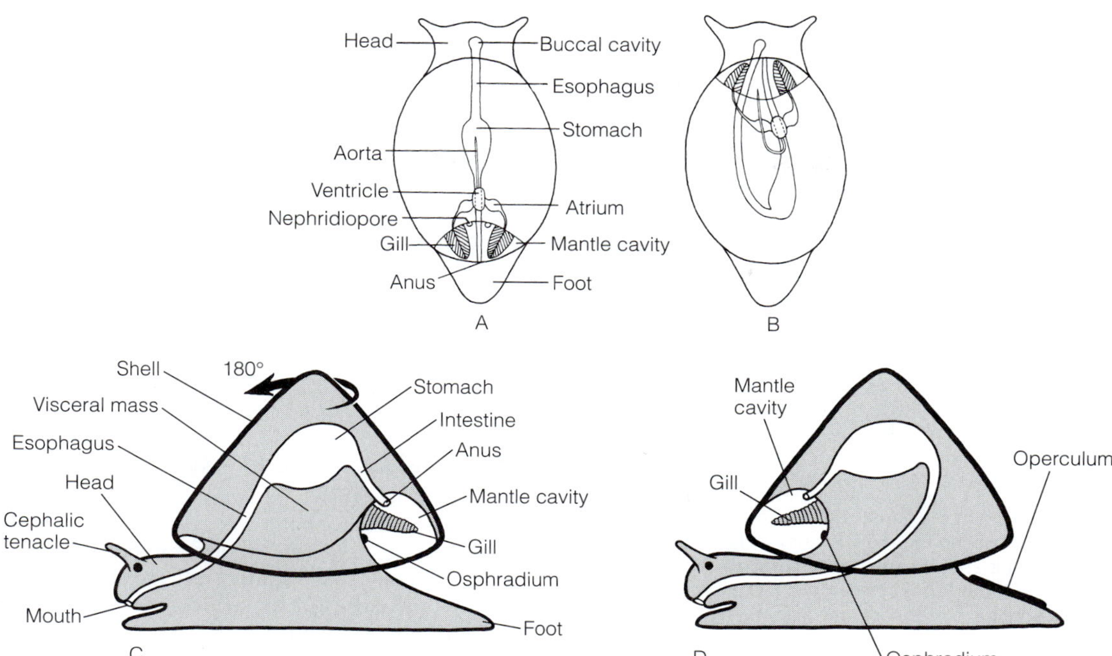

FIGURE 12-16 The effect of torsion on spatial relationships in gastropods. **A** and **B,** Dorsal views. **A,** The monoplacophoran ancestor prior to torsion. **B,** The early gastropod after torsion. **C** and **D,** Lateral views from the left. **C,** The monoplacophoran ancestor prior to torsion. **D,** The early gastropod after torsion. Coiling is omitted from the drawings for clarity. *(A and B, Modified from Graham. C and D, Modified from Fretter, V. and Graham, A. 1994. British Prosobranch Molluscs. Second Edition. Ray Society. Vol. 161. Ray Society, London. 820 pp.)*

(Fig. 12-15B). The mantle cavity was posterior. Because the progastropod was not torted, it was not a gastropod. It did not have an operculum.

Torsion is a 180° counterclockwise rotation, or twisting, of the visceral mass, shell, mantle, and mantle cavity with respect to the head and foot (Fig. 12-16). It is a developmental event that occurs in the larva, not the adult. The head and foot remain unaltered by this rotation and, except for the change in orientation, the visceral mass is largely unaltered. In contrast, structures such as nerves and gut that pass between the head-foot and the visceral mass are twisted. Torsion is a defining characteristic of the gastropods and one of the synapomorphies uniting the taxon. All gastropods are torted or had ancestors that were torted.

Prior to torsion, the mantle cavity and its organs (two gills, two nephridiopores, two hypobranchial glands, two osphradia, and an anus) are posterior. Following torsion, the mantle cavity and organs are anterior, immediately behind and above the head. Internally, the digestive tract is looped into another U superimposed on that resulting from flexure. The stomach is now posterior and dorsal, whereas the mouth and anus are anterior and ventral (Fig. 12-16B,D). The nervous system has been twisted into a figure eight, with the nerve ring at the top of the 8 and the visceral ganglion at the bottom (Fig. 12-17B). The head and foot retain the original untorted bilateral symmetry and the shell remains unchanged except that its anteroposterior polarity is now reversed.

Torsion heralded the origin of gastropods from monoplacophorans, but it is, and was, a developmental event that continues to occur in the life cycle of each individual gastropod. The bilaterally symmetrical, untorted trochophore larva develops into an untorted veliger larva. The early veliger has an uncalcified shell and, early in development, no mantle cavity. Its gut undergoes flexure to arch dorsally into the visceral mass so that both mouth and anus are ventral to the stomach. Torsion occurs in two phases of differential growth of the right and left sides of the body, resulting in rotation of the body first 90° and then 180° counterclockwise. Differential muscle contraction may contribute to torsion in some molluscs.

Many derived gastropods (Euthyneura; opisthobranchs and pulmonates) have undergone partial **detorsion** by about 90° to move the mantle cavity and/or its contents to the right side of the body. This is accomplished by the reversal of one phase of torsion.

There is no obvious, convincing reason for torsion to have occurred and it creates serious problems, so it must have a compensating adaptive advantage, either to the pelagic larva, benthic adult, or both. Several hypotheses have been suggested regarding the adaptive advantage of torsion, but none is compelling. Torsion places the mantle cavity anteriorly so that, when danger threatens, in either the larva or adult, the head can be retracted into the safety of the shell first, followed by the foot. The operculum on the posterior foot then forms a door to occlude the opening to the mantle cavity, and the animal presents an unbroken, hopefully impenetrable shell, closed by the operculum to the outside world. The operculum appears early in gastropod evolution, at about the same time

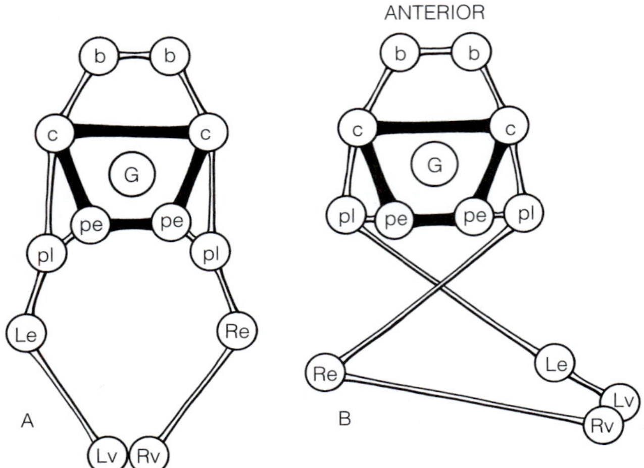

FIGURE 12-17 Evolution of the gastropod streptoneurous nervous system. Both views are dorsal, so the animal's right and left are those of the image. Both systems are hypoathroid, with the pleural and pedal ganglia adjacent to each other. The circumesophageal nerve ring is shaded. b = buccal ganglion, c = cerebral ganglion, G = gut, Le = left esophageal ganglion, pe = pedal ganglion, pl = pleural ganglion, Re = right esophageal ganglion, v = visceral ganglion. **A,** The pretorsional, symmetrical, ancestral nervous system, such as that of a monoplacophoran. This pattern is similar to the detorted, euthyneurous system of the pulmonates and opisthobranchs. **B,** A torted, unconcentrated, streptoneurous nervous system such as that of a prosobranch gastropod. *(Redrawn and adapted from Haszprunar, G. 1988. On the origin and evolution of major gastropod groups, with special reference to the Streptoneura. J. Moll. Stud. 54:367–441.)*

as torsion, suggesting a relationship between the two. Considering the near-microscopic size of the larva, however, it is difficult to see how this would be a benefit, because most larval predators consume the entire larva, shell and all. It could, however, be of value to the adult.

The anterior position of the mantle cavity might facilitate ventilation of the mantle cavity and gills. Forward motion of a torted snail would tend to rock the shell back and open the anterior mantle cavity for improved ventilation. A posterior mantle cavity, on the other hand, would tend to be closed by such backward rocking of the shell. It is also possible that an anterior position provides the mantle cavity with water free of sediments stirred up as the foot passes over the substratum.

Torsion places the sensory osphradium anteriorly, where it can sample the water the animal is entering rather than the water the animal has just left. In this position it might supplement the usual cephalic senses in testing the environment ahead of the animal. It is known that some carnivorous gastropods use the osphradium to locate prey. Torsion may also shift the center of gravity of the shell and visceral mass to a more advantageous location.

A final possibility is suggested by the fact that the gastropods are the only mollusc taxon with a coiled shell and rapid, creeping locomotion. Other conchiferans exhibit coiled or slightly curved shells, but the curvature is usually exogastric. The gastropods are endogastrically coiled (after torsion) and only the gastropods creep rapidly. It may be that gastropod-style creeping locomotion is inefficient with exogastric coiling. Such a possibility is suggested immediately by comparing the appearance of exogastrically (Fig. 12-15B) and endogastrically (Fig.12-15C) coiled animals. The center of gravity of an exogastric shell would cause the shell to fall forward, onto or in front of the head so that the animal would have to push the shell forward into sediment or vegetation. Torsion and the resulting endogastric coiling would correct this awkward and inefficient arrangement.

Whatever the advantage of torsion, some of the resulting changes in spatial relationships must have created serious sanitation problems for the early gastropods. Torsion places the anus, nephridiopore, and gonopore over the head and mouth, where their products could foul the mouth or gills. Planispiral coiling created other problems for the gastropods. The last major steps in gastropod evolution deal with solving these problems.

Conispiral Coiling and Asymmetry

Because each whorl of a planispiral shell lies entirely outside the circumference of the preceding whorl, spirals with large diameters are required to accommodate relatively modest increases in biomass. Such large, flat, disc-shaped shells are not compact and are difficult to drag through restricted spaces, vegetation, or sediment. Furthermore, the high center of gravity of a planispiral shell makes it difficult to hold the shell erect above the foot, so it tends to flop to the side.

At some point in gastropod evolution, this problem was solved by converting the symmetrical planispiral shell to an asymmetrical **conispiral** shell in which each whorl partially or completely overlaps and encloses the preceding whorl (Fig. 12-18D). First, the shell is extended to the right to form an apex, or spire, on the right of the body (Fig. 12-18D, 12-19B). The shell is now neither planispiral nor bilaterally symmetrical; instead, it is an asymmetric conispiral. This forms a more compact shell, but exacerbates weight distribution problems by moving the center of gravity to the right, away from the midline.

To bring the center of gravity of the shell and visceral mass over the head and foot, the shell axis shifts so its posterior end moves to the left and ventrally (Fig. 12-18E,F; 12-19C) and approaches the body axis. The anterior end of the shell axis shifts correspondingly, to the right and dorsally. The shell axis is now oblique to the body axis. These changes in symmetry and center of gravity occurred simultaneously and resulted in the compact, asymmetric, coiled shell with a low center of gravity over the body. Such a shell is more compact and stronger than a planispiral shell. With the exception of various limpetlike groups, the shells of modern gastropods are asymmetrical as described.

In almost all prosobranch gastropods, the mantle cavity is asymmetrical. This asymmetry is probably the result of a combination of reduction in the size of the aperture, coiling, the loss of symmetry, development of the visceral mass and digestive gland on the right, and the shift in the shell axis away from perpendicular. The result is a constriction of the mantle cavity on the right and its expansion on the left.

Sanitation

Moving the mantle cavity to the anterior end of the body placed it, the anus, gonopores, and nephridiopores in close proximity to the head and mouth and introduced potential

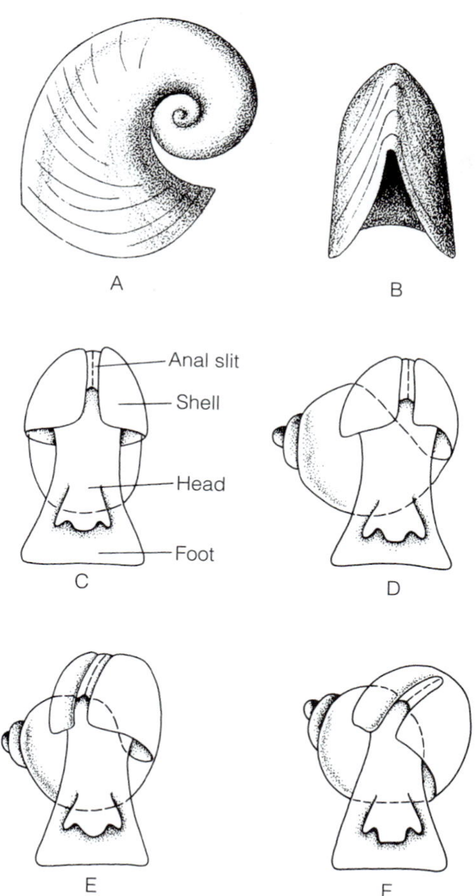

FIGURE 12-18 Side view **(A)** and front view **(B)** of the planispiral shell of *Strepsodiscus* of the fossil Bellerophontida, possibly the earliest known gastropod. **C–F,** Evolution of the asymmetrical gastropod shell. Cleft in shell for exhalant water current marks location of mantle cavity. **C–E,** Hypothetical stages. **C,** Ancestral post-torsion gastropod with planispiral shell. **D,** The apex of the spiral is drawn out, producing a more compact shell. **E,** The position of the shell over the body is shifted, providing a more equal distribution of weight. **F,** Final position of the shell over the body, typical of most living shelled gastropods. The axis of the shell is oblique to the long axis of body and the mantle cavity is on the left side. The right side is compressed by the shell. *(A and B, After Knight, J. B., Cox, L. R., Keen, A. M., Batten, R. L., Yochelson, E. L., and Robertson, R. 1960. Systematic descriptions. In Moore, R. C. (Ed): Treatise on Invertebrate Paleontology. Vol. 1, pt.1. Geological Society of America, University of Kansas Press, Lawrence, KS. pp. I169–I330; C–F, After Yonge)*

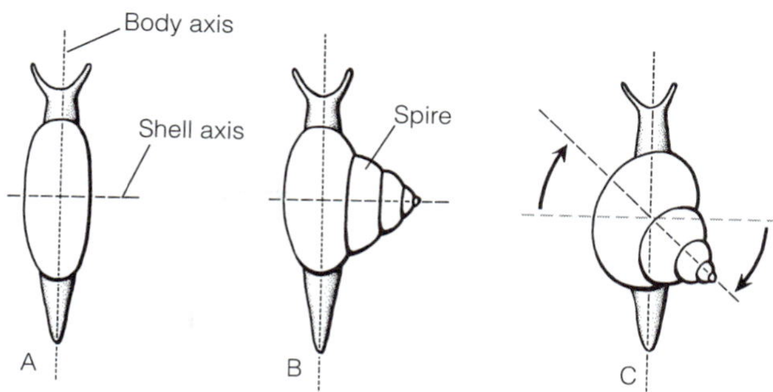

FIGURE 12-19 Evolution of the asymmetrical conispiral gastropod shell. Diagrammatic gastropods in dorsal view. **A,** A bilaterally symmetrical, top-heavy planispiral shell. **B,** An unbalanced shell with the spire extended to the right. **C,** A balanced asymmetrical gastropod shell with the spire shifted posteriorly.

conflicts between sanitation and ventilation of the gills. Two different flow patterns could have accomplished the simultaneous ventilation of the two gills of early gastropods. In one, the water enters bilaterally, on the right and left, and exits medially, above the head (Fig. 12-20A). In this pattern, clean water would flow first over the gills and then the nephridiopores and anus, where it would entrain urine and feces. It would then turn medially and anteriorly to exit the mantle cavity over the head, potentially fouling the mouth with waste. In the alternative pattern (Fig. 12-20B), water enters medially and flows first over the mouth, then the nephridiopores and anus, before turning laterally, both right and left, carrying wastes. It would then flow through the two gills, which would be fouled, before the water exited bilaterally. Understandably, neither alternative is acceptable and no known gastropod exhibits either pattern of water flow.

A third possibility entails significant modification of the shell. Although this modification is found in the primitive vetigastropods (archaeogastropods) and was probably employed by the ancestral gastropod, it is not characteristic of higher gastropods. Both primitive vetigastropods and the ancestral gastropod possessed right and left gills in a more or less symmetrical mantle cavity (Fig. 12-20C). Vetigastropoda includes the slit snails, keyhole limpets, and abalones, all of which have symmetrical respiratory currents that enter laterally, on both the right and left, and move medially. They also have one or more additional openings in the shell that permit the respiratory current to exit after it passes the anus (Fig. 12-20C). These **anal pores** (or perforations or slits) permit fouled water to exit dorsally or posteriorly and thus circumvent the mouth, even though the water flow is symmetrical. The rectum and anus are displaced posteriorly from the anterior edge of the mantle cavity and open beneath the dorsal anal pore. The respiratory current enters the mantle cavity from right and left at the anterior end of the body, passes between the gill filaments of the right and left gills, then over the anus, and finally dorsally and posteriorly to exit through the anal pore or slit.

The deepwater slit snails *(Scissurella, Pleurotomaria)* have typical spirally coiled shells, but the anterior margin of the shell and the underlying mantle are cleft by a deep slit (Fig. 12-21C). The abalones (*Haliotis;* Fig. 12-21A,B) and the keyhole limpets (*Diodora;* Fig. 12-22) are intertidal and shallow-water inhabitants of wave-swept rocks. The low, broad shells of both groups are designed to minimize water resistance and to serve as protective shields when the animal is clamped against rock. These vetigastropods have perforations rather than a slit for the exit of the respiratory current. Perforations do not weaken the shell like a long slit does and are better suited to animals living in rough water.

The low, shieldlike abalone shell is asymmetrically coiled, and most of it is a single large body whorl (Fig. 12-21A). The asymmetrical mantle cavity that is displaced to the left side

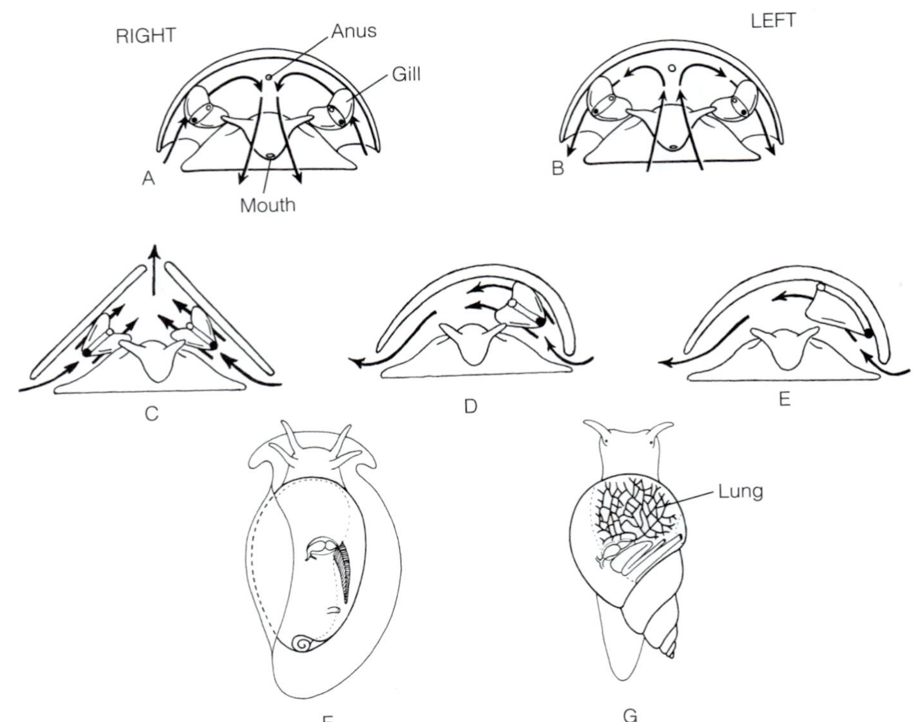

FIGURE 12-20 Gastropod mantle cavity evolution. **A** and **B,** Two possible patterns of symmetrical water flow through a mantle cavity with two gills. Conflicts between sanitation and ventilation are created by placing a symmetrical mantle cavity over the head and mouth. **A,** The ventilating current enters the mantle cavity laterally, bringing clean water to the gills. The water then flows over the anus and exits medially, where it may bring feces to the vicinity of the mouth. **B,** The ventilating current enters medially and flows over the anus, where it entrains feces that are then carried into the gills. Neither alternative is desirable. **C,** Primitive vetigastropod archaeogastropods with two bipectinate gills (keyhole limpet). Water enters on both the right and left and then flows over the right and left gills, respectively. It then passes over the anus and exits via the anal pore, pores, or slit. **D,** Derived vetigastropods and neritimorphs with one bipectinate gill and oblique water flow. Water enters on the left, flows first over the left gill, then over the anus, and then exits on the right. **E,** Mesogastropods and neogastropods with one monopectinate gill and oblique water flow, as in **D. F,** Opisthobranchs, with varying degrees of reduction of gill and mantle cavity. **G,** Pulmonates, which have replaced the gill with a lung.

of the body contains two gills. In *Haliotis tuberculata,* the shell above the mantle cavity is perforated by a row of holes. The mantle is cleft along the line of the anal pores (Fig. 12-21B) and the edges of the mantle project into the perforations to form a lining for each pore. The respiratory current enters the mantle cavity under the anterior edge of the shell (and also through the anteriormost pores) and passes beneath the shell to the left of the head. It exits through the posterior pores (Fig. 12-21A). The anus and nephridiopores lie beneath one of the posterior pores (Fig. 12-21B). New pores are added as the shell grows and old ones are plugged. Each pore begins as a short slit at the anterior margin of the shell and eventually becomes enclosed to form a circular perforation.

The keyhole limpets have a conical, uncoiled, symmetrical shell that has either a slit at the anterior margin *(Emarginula)* or, more commonly, an anal pore at the apex (*Diodora;* Fig. 12-22). The anal pore originates as a notch along the shell margin during the early stages of development. The notch then becomes enclosed and through differential growth it gradually assumes a position at the apex of the shell, reminiscent of the crater atop a volcano. Water enters the mantle cavity anteriorly, flows over the two gills, and issues as a powerful stream from the anal pore. The anus and urogenital openings are located just beneath the posterior margin of the pore.

In each of these three examples, an anal pore of some description is present in the shell. Regardless of the final condition of the aperture (slit, row of pores, or single apical pore), it begins in development as a slit in the anterior margin of the shell that is later modified to the adult form. Consequently, an anterior anal slit is thought to have been present in the ancestral gastropod.

The extinct Bellerophontida is symmetrical and planispirally coiled, and some have an anal slit on the dorsal margin of the aperture (Fig. 12-18A,B). Because they are fossil, nothing is known of their soft anatomy, and their position in the evolutionary sequence is conjectural. The confusion, however, may be because unrelated fossils have been unintentionally lumped together in this taxon. Those with an anal slit are probably torted, with mantle cavity and anal

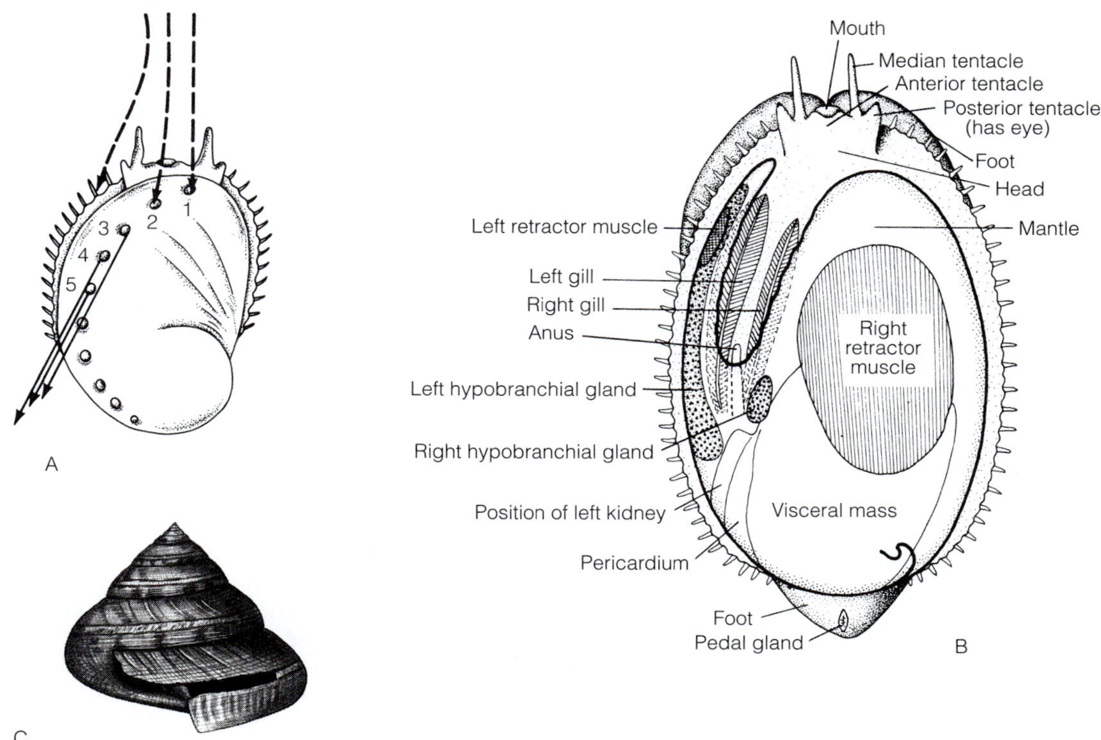

FIGURE 12-21 Vetigastropods: abalones and slit snails. **A,** Dorsal view of the abalone *Haliotis kamtschatkana* showing the path of the ventilating current. Part of the inhalant water stream enters the mantle cavity beneath the shell to the left of the head and part enters through the anterior shell perforations. The exhalant water stream leaves the mantle cavity through posterior perforations. The most posterior, sealed, nonfunctional perforations appear as a curved line of depressions on the shell. **B,** Dorsal view of *Haliotis* with the shell removed and the mantle cavity exposed. **C,** The slit snail *Pleurotomaria,* a gastropod with an anal slit for the exhalant current. *(A, Based on Voltzow, J. 1983. Flow through and around the abalone* Haliotis kamtschatkana. *Veliger. 26(1):18–21; B, Redrawn from Bullough, 1958; C, Drawn from a photograph by Abbott)*

slit anterior, and thus are early gastropods. Those without a slit could be untorted monoplacophorans and not bellerophonts (or gastropods) after all. They would have a posterior mantle cavity and no need for an anal slit. An anal slit is needed only if the anus is close to the mouth and thus would be expected only in animals that had undergone torsion (gastropods).

Obviously, there must be yet another solution to the sanitation problem if anal perforations are restricted to a relatively small number of gastropods. The solution characteristic of almost all prosobranchs (some archaeogastropods [Fig. 12-23], all mesogastropods, and all neogastropods) is a unidirectional flow of water through an asymmetrical mantle cavity. The right gill, along with the right atrium, right kidney, and right nephridiopore, are lost, presumably in conjunction with the constriction of the right side of the mantle cavity associated with asymmetry. The fouling problem is eliminated if there is only one gill to ventilate. Water enters on the left, flows over the left gill, anus, nephridiopore, and gonopore and then exits on the right (Fig. 12-20D,E). The mouth is out of the path and the solitary gill is upstream of everything that might foul it. A similar pattern is seen in some living patellogastropod limpets. In *Acmaea, Notoacmaea, Collisella,* and *Lottia,* only the left gill is present, but it extends to the right side of the body. As in all limpets, the mantle and shell overhang produces a distinct pallial groove on each side between the foot and the mantle edge (Fig. 12-24A), similar to the lateral mantle cavity of the monoplacophorans and polyplacophorans. The inhalant respiratory current enters the mantle cavity anteriorly on the left side and exits on the right. In some species the exiting current, or part of it, may flow posteriorly in the lateral mantle grooves. Some patellogastropod limpets, *Patella* for example, have lost the original gills. Instead, folds of the mantle form secondary gills, which project into the pallial groove along each side of the body (Fig. 12-24B).

Another modification associated with the respiratory current of many meso- and neogastropods is the development of a flexible, extensible inhalant siphon (Fig. 12-14). A portion of the mantle skirt on the left side, beside the head, elongates and rolls to form the **siphon** (Fig. 12-25B). The siphon gives the snail some selectivity in choosing the source of respiratory water and allows it to test the water ahead before crawling

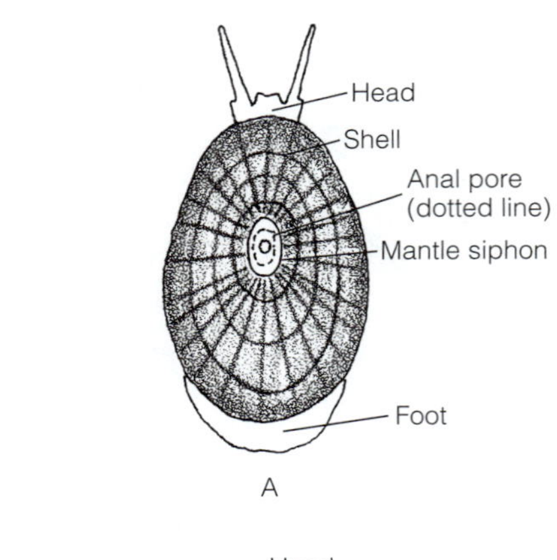

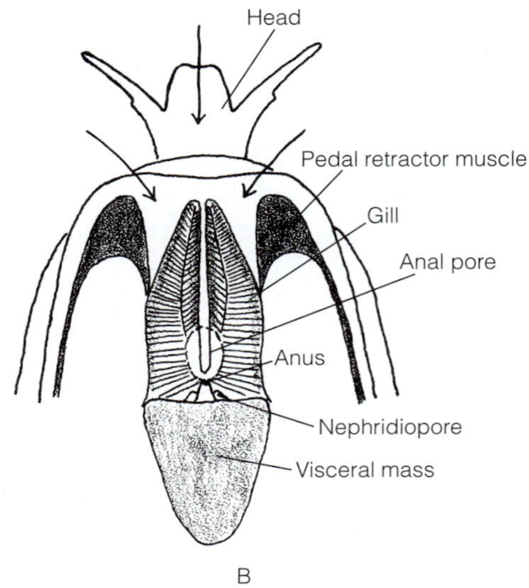

FIGURE 12-22 Vetigastropods: keyhole limpets. **A,** The keyhole limpet *Diodora* in dorsal view. **B,** Dorsal view of the exposed mantle cavity of *Diodora,* showing paired, bipectinate gills. Arrows indicate the path of the respiratory current. *(After Yonge)*

into it. It may also provide burrowing species with access to water, assist snails in following a chemical trail, or, if equipped with various receptors, function as a sense organ, especially in carnivores.

The Ancestral Gastropod

Torsion by some ancestral coiled (but perhaps uncoiled), cyclomyan monoplacophoran (Fig. 12-13A) produced the first gastropod. The resulting protogastropod gastropod may have resembled the living patellogastropod limpets or the primitive vetigastropods or it may have been a bellerophont. Its shell was probably coiled, although recently it has been suggested that it was an uncoiled, symmetrical limpet (Fig. 12-26). The shell had an anal slit on its anterior margin and water flow through the mantle cavity was probably bilateral, entering on the right and left and exiting medially through the anal slit (Fig. 12-20C). Two pedal retractor muscles were present; the loss of the left pedal retractor in higher gastropods is a derived condition.

The mantle cavity was anterior and contained a pair of bipectinate gills, a pair of mucus-secreting hypobranchial glands, two nephridiopores, a pair of osphradia, and the anus. The heart was anterior and enclosed in the pericardial cavity (pericardial coelom). It had a ventricle and two atria, one to receive blood from each gill. Such a **diotocardian** heart (with two atria) is found among living gastropods only in the primitive vetigastropods (keyhole limpets, abalones, slit snails, top snails, and turbans). The vetigastropods have retained this primitive character, but all other gastropods, including the otherwise primitive patellogastropods, have lost the right atrium to become **monotocardian.**

The ancestral gastropod was a microphagous browser. The gut was U-shaped, with the esophagus entering the stomach posteriorly and the intestine exiting anteriorly. The mouth was at the anterior end of the head, and the anus, also anterior, was dorsal to the head in the mantle cavity. A radular sac with radula and subradular organ was present in the buccal cavity.

The stomach was a pear-shaped sac with a chitinous gastric shield, ciliated sorting field, protostyle, typhlosole and intestinal groove extending from the esophagus to the intestine, and a pair of openings to the digestive ceca (Fig. 12-39A). The intestine looped several times through the visceral mass, passed through the pericardial cavity, and emptied via the anus into the right side of the mantle cavity. In derived gastropods, the character of the gut depends on feeding habits.

There were two nephridia surrounded by the hemocoel and connected with the pericardial cavity by renopericardial ducts. Distally, the nephridia opened into the mantle cavity via nephridiopores. In living gastropods with two nephridia, the right one is dedicated to nitrogen excretion and the left to ion regulation, as is assumed to also have been true of the archetype. Most gastropods have only one nephridium, the right one having been lost.

A single gonad**,** on the right, was located in a coelomic space, the gonocoel. The gonoduct exiting the gonad joined the right nephridial duct so that gametes exited the body through the right gonoduct, the right kidney, and the right nephridiopore (which was also the gonopore). All living gastropods have only a right gonad.

The nervous system was tetraneurous, **streptoneurous** (twisted; Fig. 12-17B), and **hypoathroid** (pleural and pedal ganglia touching, with pleural ganglia distant from the cerebral ganglia). It consisted of the cerebral and pedal ganglia in a ring around the esophagus that connected to the pleural, esophageal, and visceral ganglia by connectives. The two visceral cords formed a visceral loop, which was torted. The two pedal cords extended to the foot. The sensory system included a pair of cephalic tentacles innervated by the cerebral ganglia and a pair of simple eyes located laterally at the base of the cephalic tentacles. A pair of lateral statocysts that arose from ectodermal invaginations was innervated by the cerebral ganglia.

The early gastropod was probably gonochoric with external fertilization of macrolecithal eggs. The larva was planktonic and lecithotrophic. The veliger had an operculum.

A

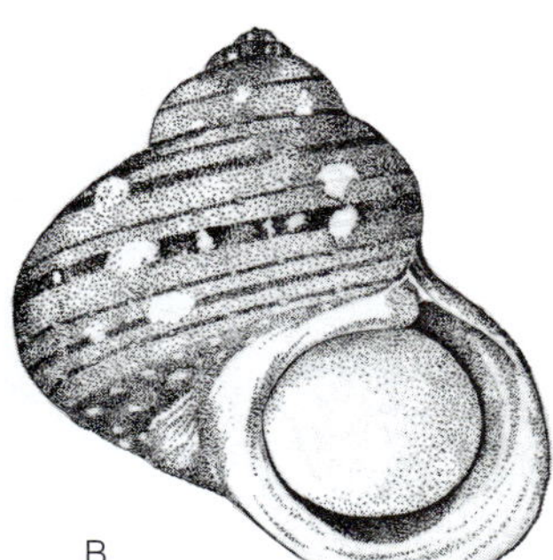

B

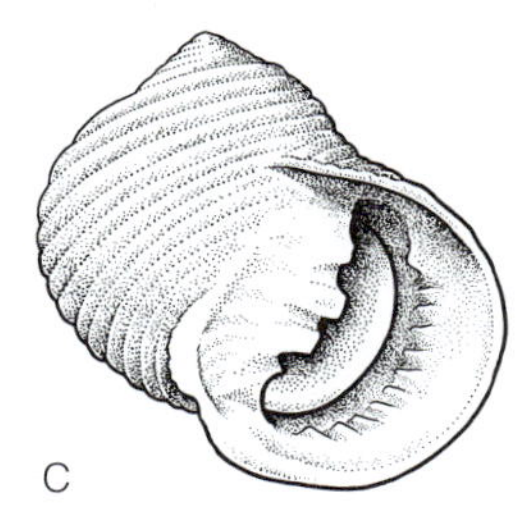

C

FIGURE 12-23 Archaeogastropods with a single gill (left) and an oblique ventilating current. **A,** Top snail shells (Trochoidea). The large specimen is *Trochus niloticus,* a common species of South Pacific islands. **B,** A turban snail (Trochoidea), *Turbo,* a common tropical Pacific and Indian Ocean taxon. The heavy, calcareous operculum of the South Pacific species, which often washes up on beaches in large numbers, is called a cat's eye. **C,** The nerite *Nerita tesselata* from south Florida.

SHELL

The gastropod shell is a hollow cone coiled around a central axis known as the **columella** (Fig. 12-27A). After coiling, the shell is still a hollow, tapering tube with a large base and a small apex, but it no longer looks like a cone.

The base of the cone is a large opening, the **aperture,** through which the head and foot can be extended or retracted (Fig. 12-27B). The visceral mass is coiled to fit into the cone. In all gastropods (except the vetigastropods), the aperture is the only opening. The outside edge of the aperture is the **outer lip** and the edge lying against the rest of the shell is the **inner lip.** In many species, the anterior margin of the shell aperture is indented to form a **siphonal notch** (Fig. 12-27B,C) or drawn out as a **siphonal canal** (Fig. 12-27A) to accommodate the siphon.

Each complete revolution, or coil, of the cone around the columella is a whorl. Successive whorls partially or entirely overlie their predecessors. The outermost whorl at the base of the cone is the **body whorl.** The largest of the whorls, it houses most of the visceral mass and the head and foot. The remaining whorls together form a **spire** that rises above the body whorl and includes the **apex.** The height of the spire varies widely with species. In some the spire is so low the shell appears to be planispiral (Fig. 12-28B), whereas in others it is so tall that it superimposes a conical shape onto the entire shell. In the cowries, each whorl completely encloses the preceding whorl and there is no spire (Fig. 12-29).

The original shell of the veliger is a spiraled **protoconch.** From this beginning the adult snail constructs the adult shell, or **teloconch,** by adding new shell material to the outer lip of the aperture to lengthen the whorl and increase its diameter. Sometimes the protoconch remains at the apex of the adult shell, but typically it is fragile and quickly wears away.

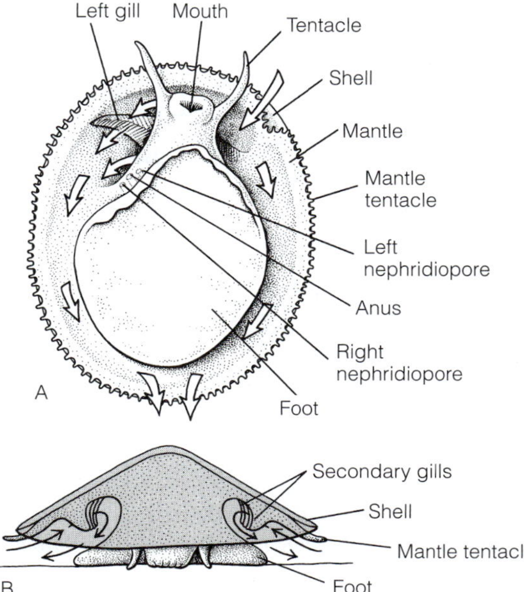

FIGURE 12-24 Patellogastropod limpets. **A,** *Acmaea* (ventral view). **B,** Cross section through a patellid limpet showing secondary gills. Arrows indicate the respiratory current.

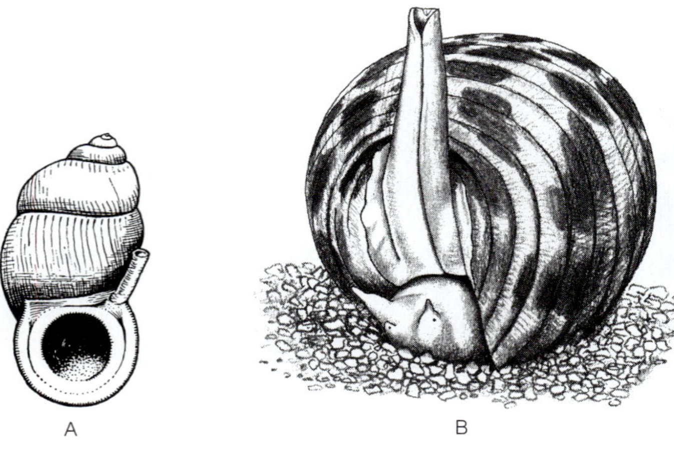

FIGURE 12-25 Miscellaneous prosobranchs. **A,** Shell of the cyclophorid land snail *Rhaphaulus chrysalis*, showing the breathing tube, a secondary opening into the mantle cavity that permits gas exchange when the animal is withdrawn and the aperture is closed by the operculum. Cyclophorids are architaenioglossans in which the gill has been lost. **B,** Anterior end of the neogastropod *Mitra,* showing the folded origin of the siphon. *(A, After Rees, 1964, from Purchon, R. D. 1977. The Biology of the Mollusca. 2nd Edition. Pergamon Press, Oxford; B, Based on a photograph by Paul Zahl.)*

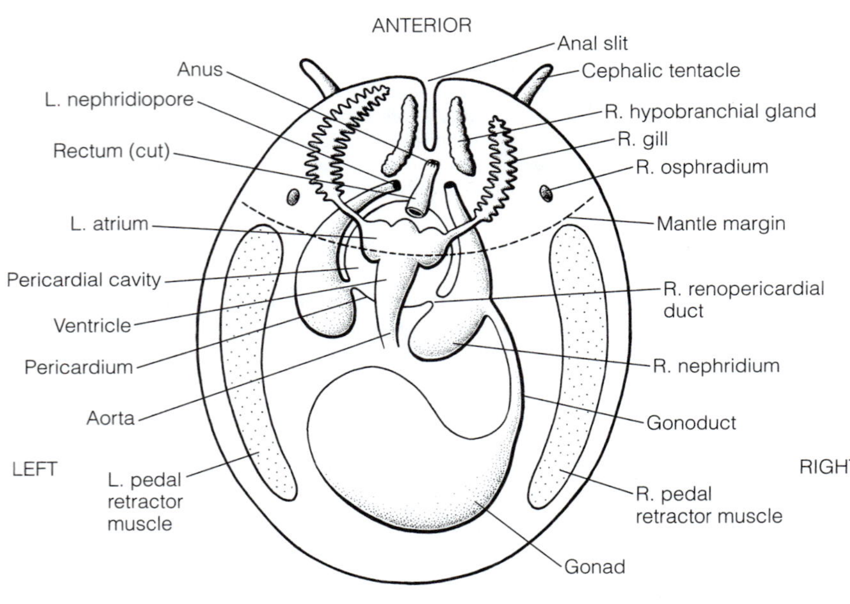

FIGURE 12-26 Dorsal view of a proposed ancestral gastropod based on a patellogastropod model. The animal is torted and the mantle cavity is anterior. The posterior margin of the mantle cavity is indicated by the broken line. *(Redrawn and modified from Haszprunar, G. 1988. On the origin and evolution of major gastropod groups, with special reference to the Streptoneura. J. Moll. Stud. 54:367–441)*

A gastropod shell may be spiraled clockwise (right-handed or **dextral**) or counterclockwise (left-handed or **sinistral**). A shell is right-handed if the aperture is on the right of the columella when the shell is held with the spire up and the aperture facing the observer, as in Figure 12-27B, and left-handed when it opens on the left. Most gastropods are right-handed, a few are left-handed, and some species have both right-handed and left-handed individuals.

The gastropod head and foot are withdrawn into the shell by the **columellar muscle** that originates on the columella and inserts on the operculum in the foot, if there is one. The columellar muscle is homologous to the right pedal retractor muscle. Primitively, two pedal retractor muscles, the right and left, are present (Fig. 12-26). This paired arrangement is found in a few living species, although the left muscle is usually very small. In most gastropods, however, the left pedal retractor muscle has disappeared and only the right one remains as the columellar muscle.

The foot of most prosobranchs (and the larvae of some opisthobranchs and pulmonates) bears a horny disc called the **operculum** on its posterior dorsal surface (Fig. 12-27C). The operculum is usually a flexible proteinaceous disc, but in some cases it is calcified and rigid (Fig. 12-23B). The doorlike operculum closes the aperture when the foot and head are retracted. It protects the snail from predation and desiccation.

Unlike monoplacophorans and chitons, which clamp themselves tightly onto the substratum for protection, most gastropods withdraw the large head and foot through the narrow aperture of the shell and break contact with the substratum when threatened. This is accomplished partly by

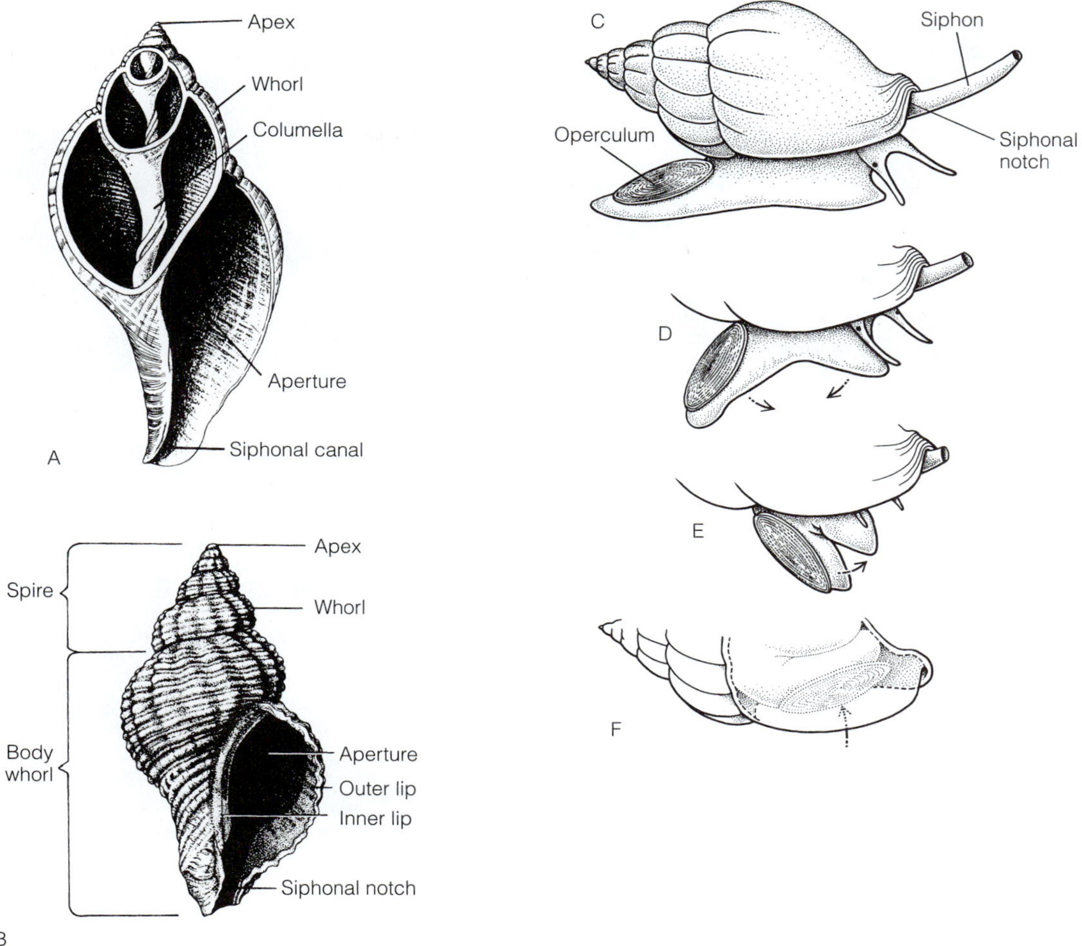

FIGURE 12-27 Gastropod shells. The shells are dextral. **A,** Longitudinal section through a typical shell showing the columella. **B,** Intact shell of the oyster drill *Urosalpinx cinerea.* **C,** Gastropod with an operculum. **D–F,** Withdrawal into the shell and closure of the aperture by the operculum. *(B, After Turner)*

compressing and partly by folding the soft tissues as they are withdrawn (Fig. 12-27E). The head is withdrawn first, followed by the foot. As the foot withdraws, the posterior half folds forward, like a hinged lid (Fig. 12-27E), against the anterior half so the soles of the two halves are in contact with each other. This leaves the operculum (if present) as the last part to be withdrawn, and it forms a door to block the aperture.

A gastropod shell consists of a variable number of organic and calcareous layers. The outer periostracum is composed of a quinone-tanned, horny protein material known as conchiolin, although its composition varies. The periostracum usually is thin but may be entirely absent, as in the cowries, or thick and bristly, as in some whelks.

The inner shell layers consist of calcium carbonate crystals, either aragonite or calcite, embedded in a protein matrix. The matrix holds the crystals together, retards crack proliferation, and guides the growth of the crystals into the proper size, shape, and orientation. The shape and arrangement of the crystals varies in different layers. Of the several types of crystal organization, prismatic, crossed-lamellar, and nacreous are the most common.

The outermost calcareous layer (ostracum) is usually a prismatic layer with its crystals deposited in columns perpendicular to the surface of the shell. The prismatic layer can be either calcite or aragonite.

The inner calcareous layers (hypostracum) may have aragonite or calcite crystals held in an organic matrix to form sheets, or lamellae, in a **crossed-lamellar** structure. The crystals are arranged with their axes oriented in different directions in different lamellae or even in different areas of the same lamella, similar to a piece of plywood. This is the most common type of shell material in the molluscs, and in some species the entire mineral portion of the shell is crossed-lamellar. The lamellae are usually oriented perpendicular to the surface, and there may be more than one layer.

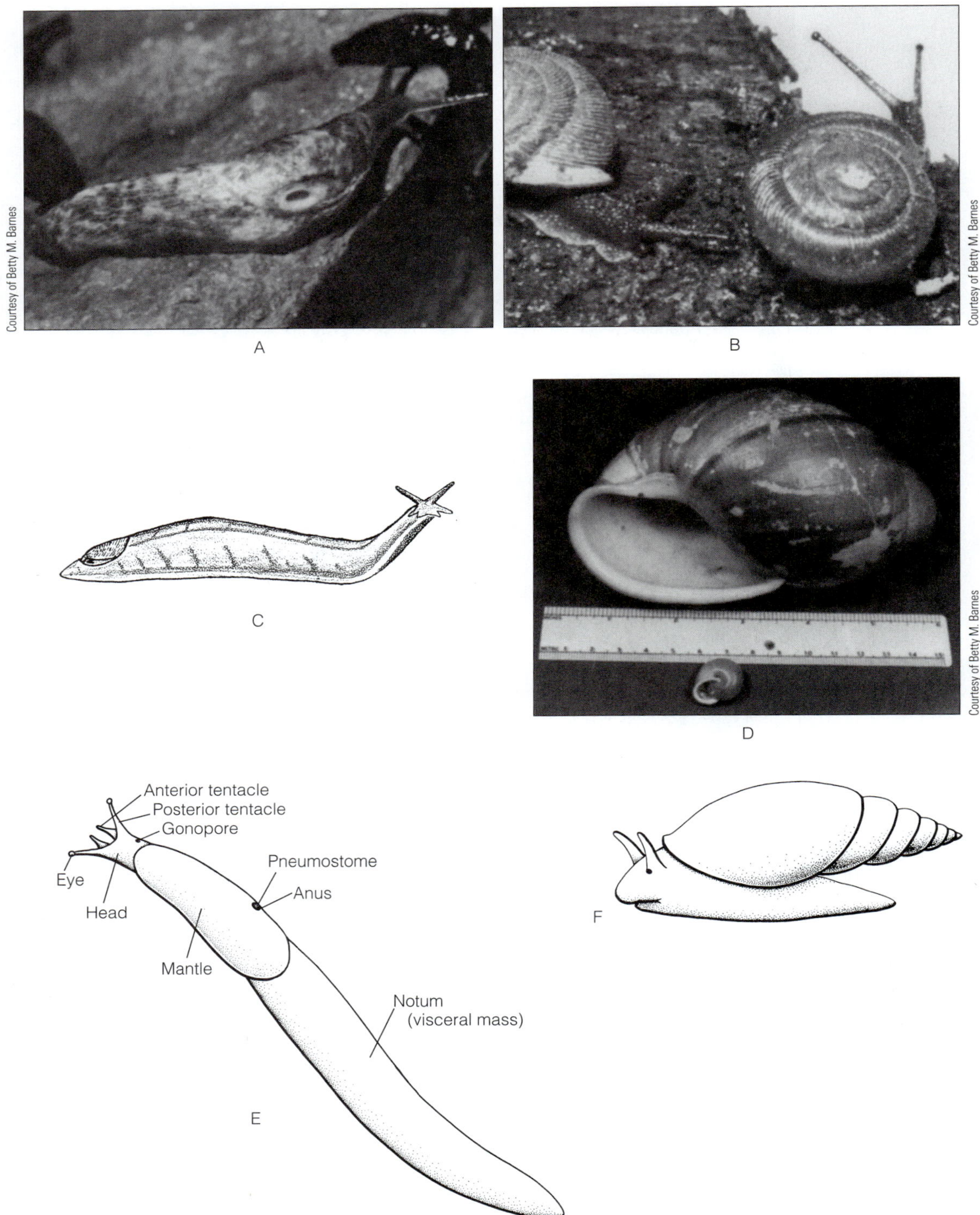

FIGURE 12-28 Representative pulmonates. **A–E,** Stylommatophorans. **A,** A terrestrial slug. The pneumostome is obvious at the lower right edge of the faintly visible, saddlelike mantle. **B,** A land pulmonate with a superficially planispiral shell and eyes at the tips of the posterior tentacles. **C,** *Testacella,* a land slug with a reduced posterior shell. **D,** Shell of the large South American pulmonate *Strophocheilus* compared with two common temperate North American land snails. *Polygyra,* the larger of these two, is below the centimeter rule, and the tiny *Retinella* is on the rule. **E,** Dorsal view of the great slug, *Limax maximus.* **F,** A generalized basommatophoran snail. *(C, After Baker)*

FIGURE 12-29 Cowrie shells. The shells of these tropical gastropods are superficially bilateral because the body whorl completely encloses the previous whorls. The specimen on the right is viewed from the aperture side. The specimens on the left and in the middle are viewed from the side opposite the aperture, and the one on the left has been cut away to show the older whorls inside.

In the archaeogastropods there may be a **nacreous layer** in which the crystals are disc-shaped tablets arranged in sheets parallel to the surface. This shell material is known as nacre or mother-of-pearl, and it is smooth and lustrous. Nacre is always made of aragonite and usually forms the innermost layer, although sometimes it is a middle layer. Nacre occurs in the monoplacophorans, some gastropods, cephalopods, and bivalves, but is not present in the higher prosobranchs. Nacre is the strongest shell material, whereas crossed-lamellar is the hardest. Nacre is energetically the most expensive to produce. Cracks propagate most readily through the prismatic layer and least through a nacreous layer.

The color of the shell results from pigments in the periostracum or pigments deposited in the calcareous layers as the shell is secreted. Pigments may originate in the diet or be synthesized by the snail. The incorporation of pigments from the diet may result in crypsis, with the snail having the same color as its prey.

The shell grows through the addition of new material, both organic and mineral, secreted by the outer edge of the mantle and applied to the lips of the aperture. The process is similar to that in bivalves, in which it has been more extensively studied. The difference in the rate of mineral deposition to the inner and outer lips results in the spiral growth of the shell as the outer lip grows faster than the inner. Growth usually is not continuous, and the intervals between growth periods can often be recognized by growth lines, as in bivalves, and by the sculpturing of the shell surface. In most gastropods, shell growth rate declines with age.

Gastropod shells display a great variety of colors, patterns, shapes, and sculpturing, but only two of the more widespread modifications in shell form will be mentioned. In many gastropods, the spiral nature of the shell is obvious only in the juvenile stages and obscured in the adult. The coiling ceases with age and the adult shell consists of a single, large, expanded body whorl that does not appear to be spiraled. In the abalone, *Haliotis* (Fig. 12-21A), and in the slipper snails, *Crepidula* (Fig. 12-57D), the shell uncoils but remains asymmetrical. In the keyhole limpets, the shell has secondarily lost all traces of coiling and become secondarily symmetrical. It looks like a Chinese straw hat or a volcano (Fig. 12-22A). The limpet shape has evolved independently several times in the gastropods.

A second important and often repeated event in gastropod shell design is the reduction or complete loss of the shell. A reduced shell often becomes embedded in the mantle and internalized. Other shell modifications will be described later in connection with ventilation, movement, and habitat.

FOOT, LOCOMOTION, AND HABITAT

Foot

Locomotion is the primary function of the gastropod foot, but it may also be responsible for activities such as prey capture, reproduction, and defense. In gastropods, the foot is a mass of muscle and connective tissue with a broad, flat, creeping sole. There is no large blood sinus in the gastropod foot and its hydrostatic skeleton is a combination of small blood spaces and the muscle tissue itself.

The columellar muscle is the dorsal and central part of the foot, whereas the tarsos muscle is ventral and peripheral. The columellar muscle, which consists of large muscle bundles wrapped in connective tissue, is responsible, of course, for the retraction of head and foot, and also for their extension, and for twisting the foot with respect to the shell. Although there is no fluid-filled hemocoel in the columellar muscles, the muscle bundles themselves function as hydrostatic skeletons and antagonize each other. Consequently the columellar muscle can both extend and retract the foot.

The **sole** of the foot is the **tarsos muscle,** which is composed of very fine muscle bundles. These are also invested in connective-tissue sheaths, but extend in many directions. The tarsos is responsible for locomotion, prey capture, and, in females, the molding of egg cases. The sole of higher gastropods is sometimes divided by transverse grooves into an anterior **propodium,** middle **mesopodium,** and posterior **metapodium.** The mesopodium is the locomotory region and the metapodium bears the operculum. The propodium is reduced or absent in most prosobranchs except the moon snails, where it is greatly expanded.

Locomotion

The foot of almost all gastropods is involved in locomotion. The surface of the sole of the typical gastropod foot is broad and flat and variously adapted for locomotion over a variety of substrates. In most gastropods, the sole is ciliated and provided with abundant secretory gland cells or, in the pulmonates, with a large pedal gland. The glands secrete mucus to create a mucous trail over which the animal glides. Some very small snails, as well as some species that live on sand and mud bottoms, move by ciliary propulsion, but most rely on muscles.

Most hard-bottom gastropods, terrestrial pulmonates, and even large, soft-bottom species move rapidly by propelling waves of muscular contraction that sweep along the sole of the foot. The sole is firmly anchored to the substratum by sticky, gelatinous mucus except in a wave, where the mucus is liquefied to permit the foot to slide forward. In species with direct waves (see the discussion below), the mucus changes very rapidly from firm gel to liquid sol as a consequence of the

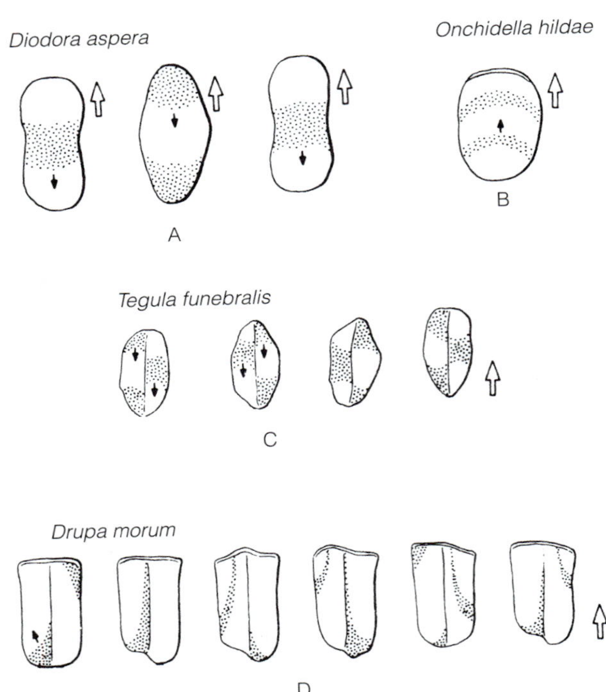

FIGURE 12-30 Patterns of pedal waves in gastropods. **A,** Retrograde monotaxic waves in the keyhole limpet *Diodora aspera.* **B,** Direct monotaxic waves in the intertidal pulmonate slug *Onchidella hildae.* **C,** Retrograde ditaxic waves in the archaeogastropod *Tegula funebralis.* **D,** Direct ditaxic waves in the neogastropod *Drupa morum.* Wave sequence is from left to right in each case. Large, outlined arrows indicate the direction of the animal's movement and small, black arrows the direction of the waves. *(From Miller, S. L. 1974. The classification, taxonomic distribution, and evolution of locomotor types among prosobranch gastropods. Proc. Malacol. Soc. London. 41:233–272.)*

shearing force produced by contractions of the fine muscle bundles of the tarsos. Each wave performs a small step in liquefied mucus.

There are many types of waves and wave-driven locomotion in gastropods. In some species **monotaxic** waves extend across the entire width of the foot (Fig. 12-30A,B). In others, **ditaxic** waves extend only half of the width of the foot (Fig. 12-30C,D). Wave motion on the right side alternates with that on the left. **Direct waves** propagate from posterior to anterior in the same direction in which the snail is moving. **Retrograde waves,** on the other hand, pass from anterior to posterior, in a direction opposite that of the animal's movement.

Direct waves involve contraction of longitudinal and dorsoventral musculature beginning at the posterior end of the foot. Successive sections of the foot are in effect pushed forward. Retrograde waves involve contraction of transverse muscles, which, along with blood pressure, extends the front of the foot forward. This backward-moving wave of elongation is followed by the contraction of longitudinal muscles, and successive areas of the foot are pulled forward. Direct and retrograde waves may be associated with either the monotaxic or the ditaxic condition so that locomotion may be direct monotaxic, direct ditaxic, retrograde monotaxic, or retrograde ditaxic. The most common pattern among prosobranchs is retrograde ditaxic. Most pulmonates use direct monotaxic waves. In general, pedal waves, whether direct or retrograde, occupy about one-third of the foot length and only one or two waves are present simultaneously. Retrograde ditaxic waves are often oblique (Fig. 12-32D).

Habitat

Gastropods evolved on rocky marine bottoms but have radiated into a wide variety of other habitats. Natural selection has resulted in close correlation among shell shape, locomotion, and habitat. In general, shells with low spires are more stable and better adapted for carriage upside down or on the vertical surfaces of rocks and vegetation. Shells with long spires are carried horizontally or even dragged over soft bottoms. Spines and other shell projections and sculpturing may contribute to shell strengthening, protection, stabilization in soft bottoms, burrowing, or even landing right-side-up if the animal is knocked off a rock.

Limpets, abalones, and slipper snails are admirably adapted for clinging to rocks and shells, even in surf or strong currents. All have low, broad shells that offer little resistance to flowing water. The large, mucus-covered foot (Fig. 12-31B) functions as an adhesive organ as well as a locomotory organ. The surrounding mantle skirt, which may bear tentacles, is an important sense organ.

Homing behavior has evolved in several unrelated intertidal limpets, including *Collisella, Patella, Fissurella,* and *Siphonaria.* The limpet inhabits a slight depression, or "home," in the rock surface, over which the edges of the shell fit snugly. When immersed by a flooding tide the animal wanders 10 to 150 cm away from its home as it browses on algae. It then returns to its home as the tide ebbs. Experimental studies indicate that homing ability depends primarily on chemical cues in the mucus laid down by the limpet as it makes its feeding excursions. Homing may reduce both intraspecific competition by establishing a grazing territory and desiccation and predation because of the tight fit into the home site. The mucus of some territorial homing species stimulates algal growth in the territory.

Other marine snails, such as *Ilyanassa obsoleta* (Fig. 12-48A) and certain species of *Nerita* (Fig. 12-23C), use mucous trails to follow conspecifics. Should they cross a trail, they follow it to reach aggregations of conspecifics. They are able to determine the directionality of the trail from its mucous components. Some homing limpets backtrack their trails to the starting (home) position. Many terrestrial snails and slugs return to shelters beneath logs and stones, especially when the shelter is occupied by other individuals of the same species. The homing cue appears to be a pheromone component of mucous trails or airborne pheromones from fecal pellets.

Some prosobranchs and cephalaspid opisthobranchs (Fig. 12-34A) that live on soft sand bottoms have become adapted for burrowing. In the moon snails *Natica* and *Polinices,* the anterior region of the gastropod foot, called the propodium, is hypertrophied and acts like a plow and anchor while a dorsal, flaplike fold of the foot covers the head as a protective shield (Fig. 12-32A). The conch *Strombus* uses its

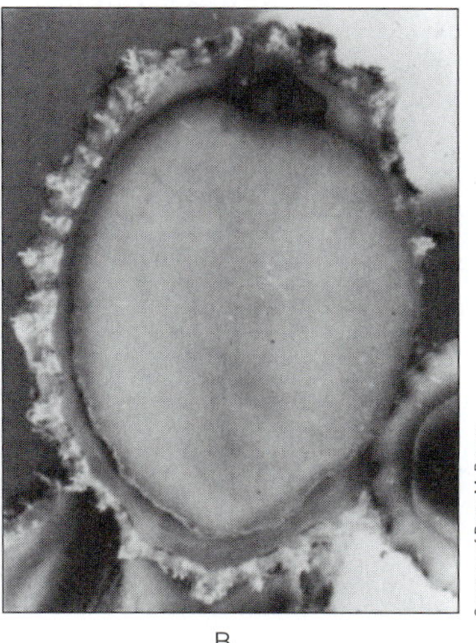

FIGURE 12-31 Limpets. **A,** Photograph of a small area of intertidal rock on the west coast of Scotland. Three specimens of the neogastropod *Thais,* which is not a limpet, are feeding on barnacles. Four limpets belonging to the genus *Patella* can be seen to the right and above the snails. Numerous barnacles have settled on the limpets' shells and on the rock. **B,** Ventral view of the keyhole limpet *Fissurella,* showing the broad adhesive foot and the sensory tentacles on the mantle margin.

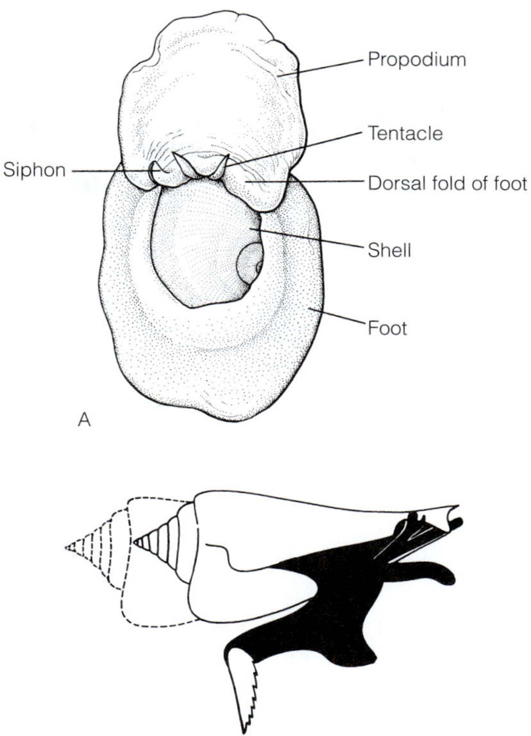

FIGURE 12-32 Benthic gastropods. **A,** *Polinices,* a burrowing gastropod (dorsal view). **B,** Lateral view of *Strombus.* The foot, indicated in black, is attached to the bladelike operculum. The operculum is inserted into the substratum and remains stationary so that extension of the foot propels the animal forward (to the right in the image). The dotted outline indicates the position of the snail before its leap. *(B, From Morton, J. E. 1964. In Wilbur, K. M., and Yonge, C. M. (Eds.): Physiology of Mollusca. Vol. 1. Academic Press, New York.)*

foot to move over sand, but the mechanism is very different than that used by other gastropods. The large, stiff, clawlike operculum digs into the sand and the animal "poles" forward by rapidly extending the foot (Fig. 12-32B).

A small number of gastropods are adapted for a sessile existence firmly attached to a substratum. The worm snails, members of three unrelated mesogastropod families (Vermetidae, Turritellidae, and Siliquariidae), have the customary tightly coiled larval and juvenile shells, but as the animal grows, the whorls become completely separated. If the coils are regular the shell looks like a corkscrew, but often the shell is irregular (Fig. 12-33A). Worm snails attach their shells to sponges, other shells, or rocks, with the separated whorls providing greater surface area for attachment. The foot is reduced, but an operculum is present (Fig. 12-33B). The turitellids are filter feeders that gather food with the gill and transfer it to the mouth in a mucous string. Some vermetids *(Serpulorbis)* use mucous strands up to 30 cm in length to trap plankton.

A few gastropod taxa have independently become adapted for a pelagic life. These include the heteropods (mesogastropods) and the thecosome (Fig. 12-34) and gymnosome opisthobranchs, known as sea butterflies. In most of these taxa, the foot has become a finlike swimming organ. Heteropods may have a large, gelatinous "body" and there may or may not be a shell (Fig. 12-35D). Ventrally there is a fin and small sucker, apparently derived from the foot. The mouth is at the anterior end and the animal swims upside down with the ventral fin uppermost. As raptors, heteropods have especially well-developed sense organs. The largest pelagic gastropod is *Carinaria cristata,* which reaches lengths of 50 cm.

Most of the pelagic opisthobranchs use a pair of lateral, winglike expansions of the foot known as **parapodia** for swimming. The pelagic pteropods, or sea butterflies, have anterior

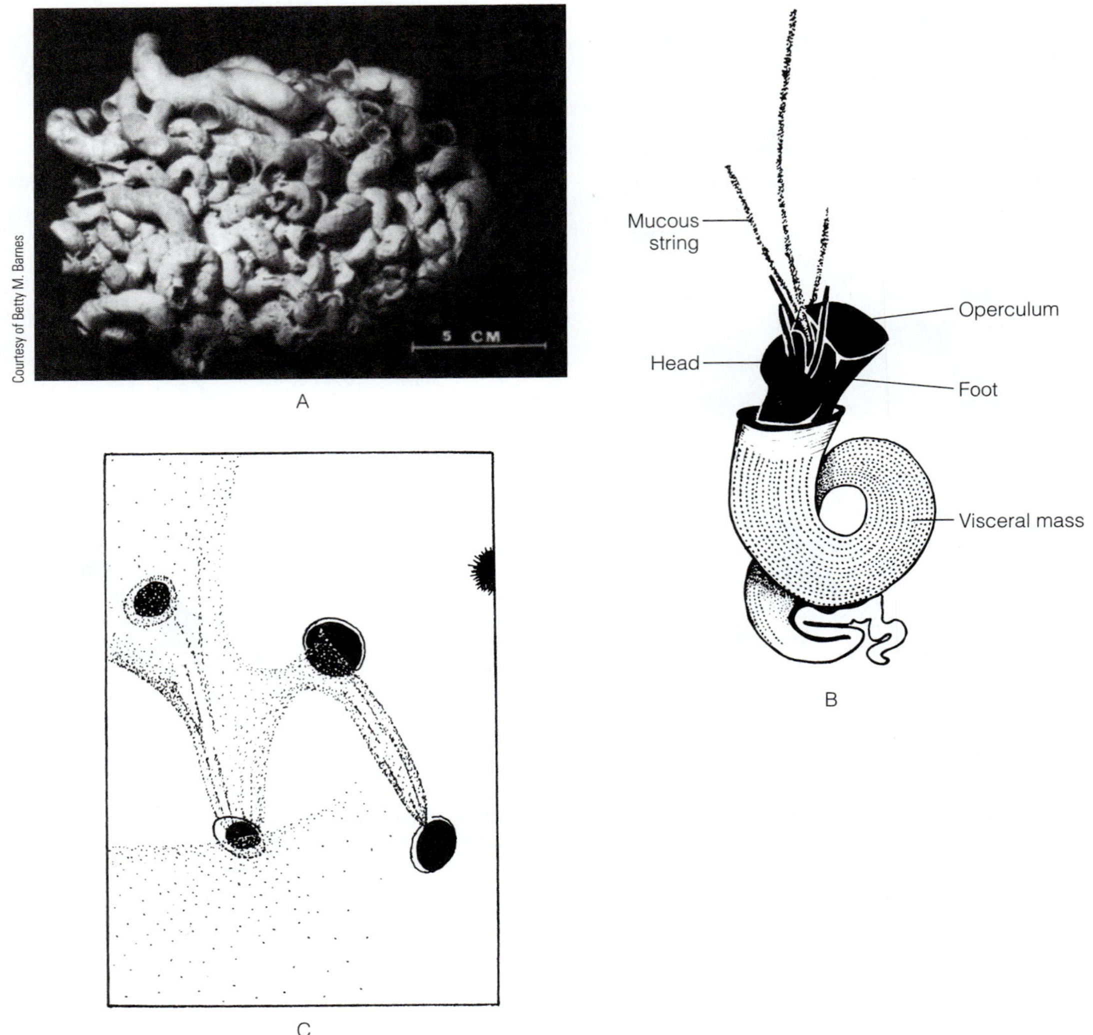

FIGURE 12-33 Worm snails. **A,** A mass of vermetid worm snails. **B,** The worm snail *Serpulorbis* with three mucous threads. The shell has been removed. **C,** Interconnecting mucous nets extending from the apertures of adjacent worm snails. Only the apertures of the snails are drawn. *(B, After Morton from Hyman, L. H. 1967. The Invertebrates. Vol. VI Mollusca. Pt. I. McGraw-Hill Book Co., New York; C, From Hughes, R. N., and Lewis, A. H. 1974. On the spatial distribution, feeding and reproduction of the vermetid gastropod* Dendropoma maxima. *J. Zool. London. 172(4):539.)*

parapodia that function as oars (Fig. 12-35A–C), and the animals swim upside down. In the benthic sea hares, which swim intermittently, the large parapodia arise from the middle of the body and are very broad (Fig. 12-36A, 12-37).

The common but minute species of the prosobranch *Caecum,* which live in sand, have short, tusklike shells. The pelagic violet snails, *Janthina,* float beneath a raft of bubbles secreted by the foot (Fig. 12-38A). The planktonic sea slugs, *Glaucus* and *Glaucilla,* hold a bubble of air in the stomach to maintain positive buoyancy. Coralliophilids, such as *Magilis,* which often have oddly shaped shells, bore into coral (Fig. 12-38B). The carrier snail, *Xenophora,* attaches foreign objects, including other gastropod and bivalve shells, to its own shell with a secretion from foot glands (Fig. 12-38C). There are also a few minute, naked, interstitial opisthobranchs (Fig. 12-28C).

NUTRITION AND DIGESTION

Gut

Virtually every possible feeding mode is found in gastropods, and the morphology and physiology of the digestive system vary widely. Gastropods can be microphagous browsers, herbivores, carnivores, omnivores, scavengers, deposit feeders, suspension feeders, cytoplasm suckers, and parasites. Although there is no typical gastropod in terms of feeding habits and gut morphology, it is possible to recognize several features that are common to most or all gastropods.

1. The gut consists of a mouth, buccal cavity, esophagus, stomach, intestine, rectum, and anus.
2. A radula is usually employed in feeding.

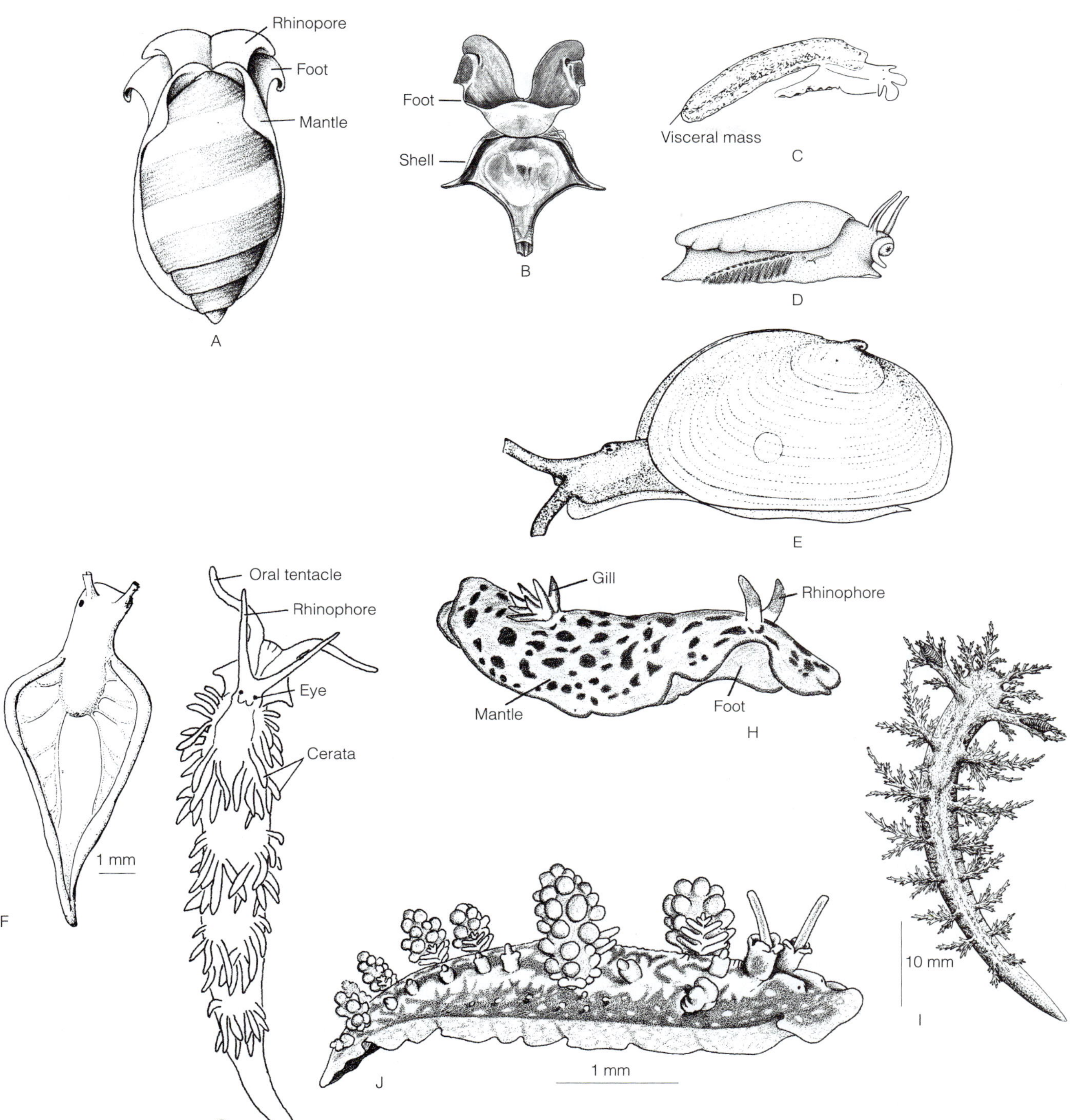

FIGURE 12-34 Representative opisthobranchs. **A,** The bubble snail *Hydatina,* a cephalaspidean. **B,** *Cavolinia,* a shelled sea butterfly (pteropod; Thecosomata). **C,** The interstitial *Microhedyle* (Acochlidiacea). **D,** The slug *Pleurobranchus* (Notaspidea). **E,** *Berthelinia,* a gastropod with a bivalved shell (Sacoglossa). **F,** *Elysia viridis,* a sluglike sacoglossan. **G,** *Aeolidia papillosa,* a nudibranch with cerata (dorsal view). **H,** *Glossodoris,* a nudibranch with anal gills and no cerata. **I,** *Dendronotus frondosus,* a nudibranch with branched, treelike cerata. **J,** *Doto chica,* a nudibranch with cerata that look like clusters of grapes. *(A, Modified from several authors; B, after a photograph by Abbott; C, After Odhner from Hyman; D, After Vayassière from Hyman, 1967; E, After Kawaguti from Hyman, 1967; F, After Gascoigne, T. 1975. Methods of mounting sacoglossan radulae. Microscopy 32:513; G, After Pierce; H, After a photograph by P. Zahl; I, From Thompson, T. E., and Brown, G. H. 1976. British Opisthobranch Molluscs. Academic Press, London. p. 67.)*

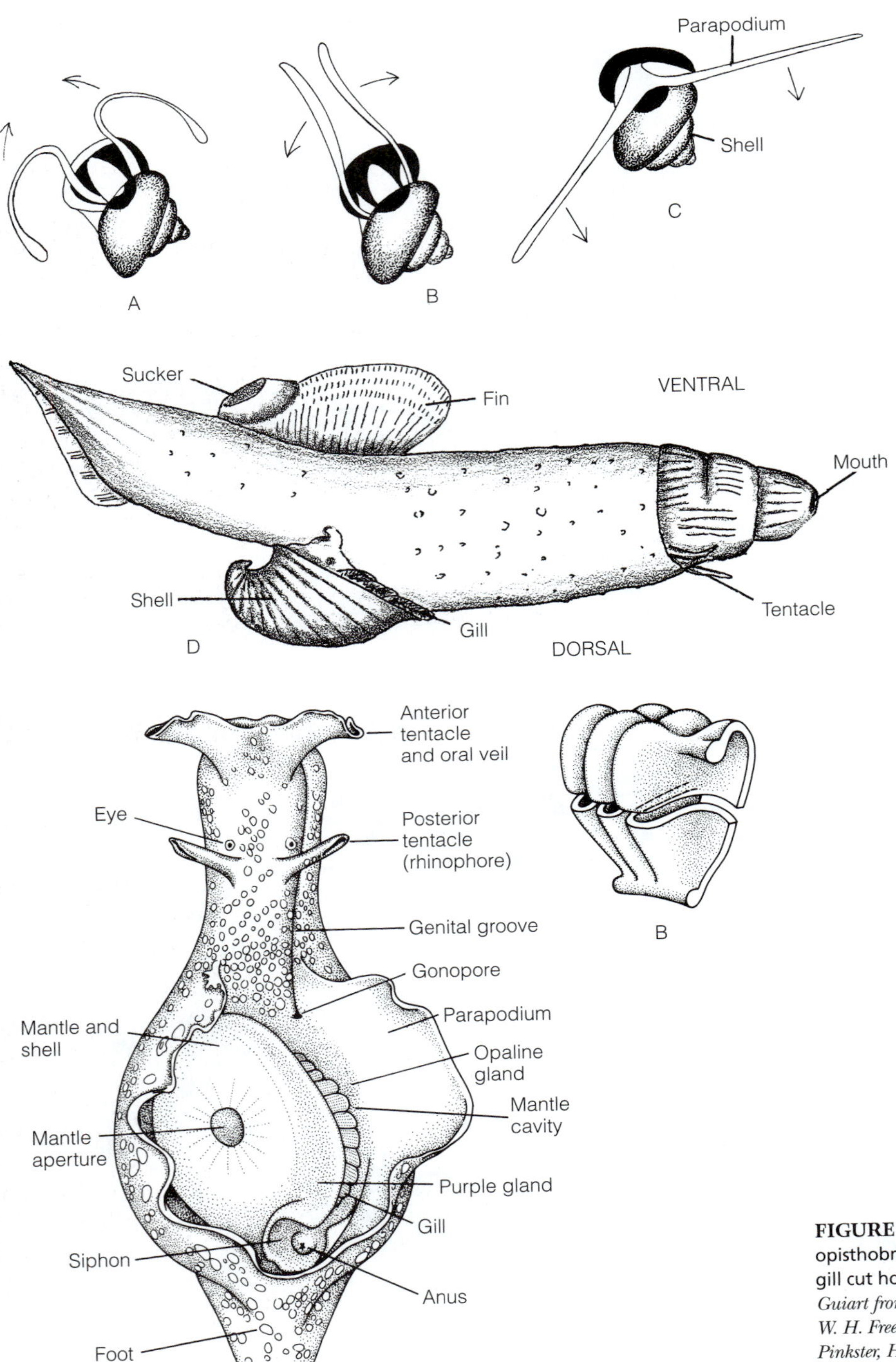

FIGURE 12-35 Pelagic gastropods. **A–C,** Swimming in the shelled (thecosome) sea butterfly *Limacina.* Arrows indicate the direction of movement of the parapodia. **A,** The recovery stroke; **B,** the beginning of the effective stroke; and **C,** the middle of the effective stroke. **D,** *Carinaria,* a pelagic heteropod in its upside-down swimming position. *(A–C, After Morton, J. E. 1967. Molluscs. 4th Edition. Hutchinson University Library, London; D, After Abbot, 1954.)*

FIGURE 12-36 The sea hare *Aplysia*, an anaspidean opisthobranch. **A,** Diagrammatic dorsal view. **B,** Part of the gill cut horizontally to show its folded condition. *(A, After Guiart from Kandel, E. R. 1979. Behavioral Biology of* Aplysia. *W. H. Freeman and Co., San Francisco; B, Modified from Carew, T., Pinkster, H. M., Rubinson, K. and Kandel, E. R. 1974. Physiological and biochemical properties of neuromuscular transmission between identified motor neurons and gill muscle in* Aplysia. *J. Neurophysiol. 37:1020–1040)*

3. Digestion is always at least partly extracellular.
4. With few exceptions, the enzymes for extracellular digestion are produced by the salivary glands, esophageal pouches, digestive ceca, or a combination of these structures.
5. The stomach is the site of extracellular digestion and the digestive ceca are the sites of absorption and of intracellular digestion, if such digestion takes place.
6. As a result of torsion, the stomach has been rotated 180°, so the esophagus enters it posteriorly and the intestine leaves

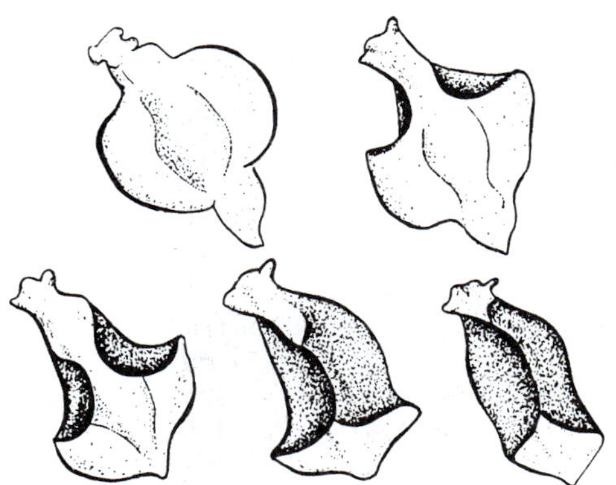

FIGURE 12-37 Swimming by the sea hare *Aplysia.* The lateral natatory parapodia are extensions of the foot. *(After Pruvot-Fol from Farmer, W. M. 1970. Swimming gastropods (Opisthobranchia and Prosobranchia). Veliger 13(1):73.)*

anteriorly (Fig. 12-16B,D, 12-39A). In the higher gastropods there has been a tendency for the esophageal opening to migrate forward toward a more customary anterior position.

The mouth on the anterior head opens into the buccal cavity (Fig. 12-39A) and is typically surrounded by fleshy lips. In carnivorous species it is often at the distal tip of an eversible, tubular proboscis (Fig. 12-14). **Jaws** composed of hardened, chitinous plates are sometimes present just inside the mouth, especially in pulmonates and opisthobranchs. The buccal cavity contains the radular sac and radula. If present, buccal glands secrete mucus into the buccal cavity. One or two pairs of salivary glands secrete mucus via ducts into the buccal cavity. Accessory salivary glands may be modified as venom glands, as in the cone snail *Conus.*

In most gastropods the radula is a highly developed feeding organ that acts as a grater, rasp, brush cutter, grasper, harpoon, or conveyor. The total number of teeth varies from 16 to thousands and are almost always arranged in a longitudinal ribbon of transverse rows (except in the cone snails). Usually, each row bears a **median tooth** (rachidian), and to each side of the median tooth, **lateral** and **marginal teeth** appear in succession (Fig. 12-40). The median, lateral, and marginal teeth usually differ from one another in shape and structure and have different functions. The shape and arrangement of the radular teeth are species-specific and often important in classification and identification.

Retracted, the typical radula is rolled along its edges to form a narrow band. When extended, the radula flattens out and acts as a broad rasp to scrape algae from rocks and mix it with mucus. When the radula is retracted, its lateral margins roll medially to transform the broad fan into a longitudinal

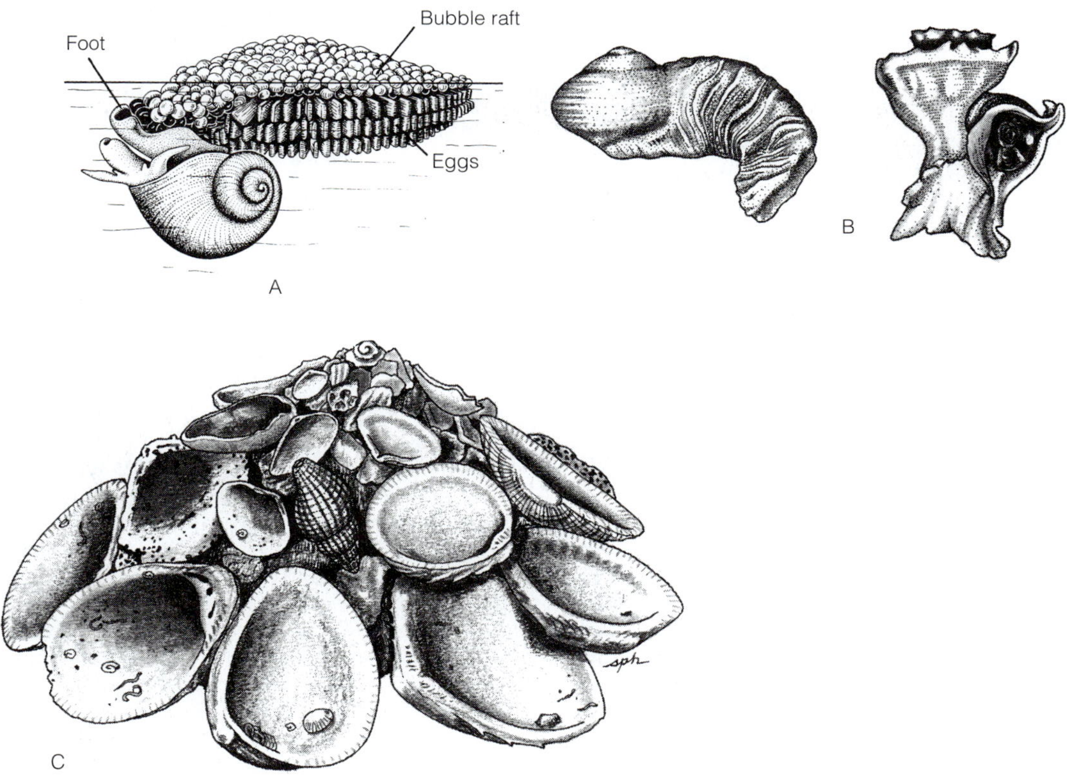

FIGURE 12-38 **A,** The prosobranch violet snail *Janthina,* which floats beneath a raft of bubbles secreted by the foot. **B,** *Magilus,* a member of the Indo-Pacific Coralliophilidae that live embedded in coral. **C,** The carrier snail *Xenophora,* which attaches shells and other foreign objects to its own shell. *(A, After Fraenkel; B and C, Based on photographs by T. Abbott.)*

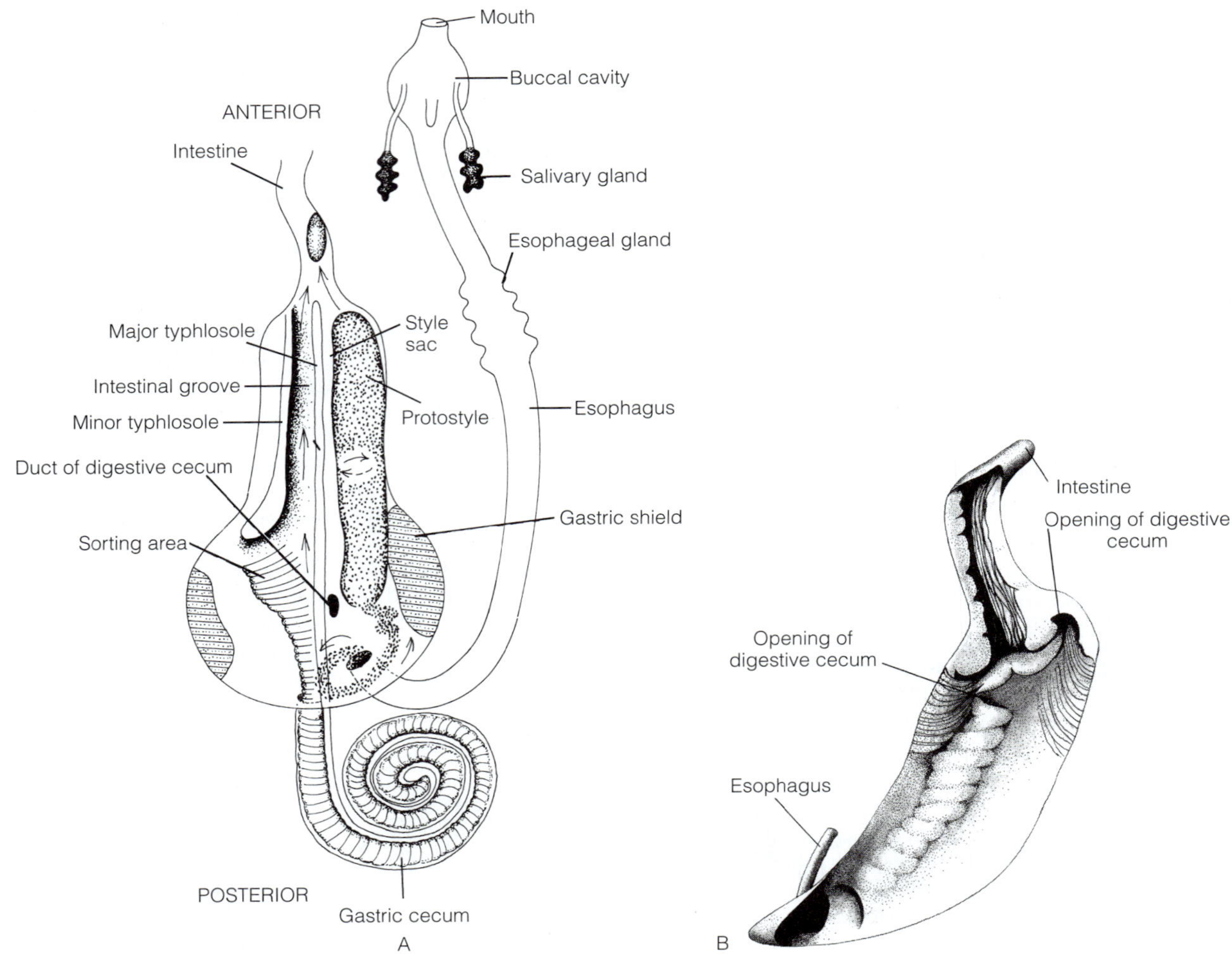

FIGURE 12-39 Prosobranch gut. **A,** Diagram of the digestive tract of a typical archaeogastropod (Trochidae). Arrows show ciliary currents and rotation of the mucous mass (protostyle) within the style sac. **B,** Saclike stomach of a carnivorous prosobranch *(Natica)*. The stomach has been opened dorsally. *(A, After Owen; B, From Fretter, V., and Graham, A., 1994. British Prosobranch Molluscs. Second Edition. Ray Society Vol. 161. Ray Society, London.)*

gutter into which the food, mixed with mucus, is channeled. A string of mucous and food is formed within the gutter and pulled through the esophagus and into the stomach by the rotating protostyle.

The esophagus exits the buccal cavity and connects it with the stomach. Its grooved walls often secrete mucus. It consists of three regions, the anterior, middle, and posterior. The mid-esophagus is twisted by torsion so that its formerly ventral surface becomes dorsal. Esophageal glands present in some species secrete digestive enzymes into the lumen of the esophagus. In neogastropods, a gland of Leiblein secretes digestive enzymes into the mid-esophagus, which is separated from the anterior esophagus by a valve.

The gastropod stomach resembles that of the generalized mollusc except that its polarity is reversed by torsion. Extracellular digestion occurs in the stomach, using enzymes from the esophageal pouches and the digestive cecum. In primitive species the stomach is complex and equipped with ciliated sorting fields and transport tracts (Fig. 12-39A). In carnivores and herbivores, it is simpler (Fig. 12-39B). A crystalline style may be present in species that feed on fine particulates, but is not found in macroherbivores, carnivores, or those that suck cytoplasm. Species with a crystalline style usually do not have esophageal glands. The style is a rod of digestive enzymes secreted by the epithelium of the style sac.

The digestive cecum consists of right and left lobes, each composed of masses of tubules ultimately connected by a duct with the stomach lumen. Its epithelium is secretory (digestive enzymes) and absorptive. It secretes enzymes to be used for extracellular digestion in the stomach, phagocytoses particulates for intracellular digestion, and absorbs digested molecules for transfer to the hemocoel for storage.

The intestine extends from the anterior end of the stomach through the visceral mass to open via the anus on the right side of the mantle cavity. Its terminal portion is the rectum. Its role is chiefly in feces formation, packaging, and storage.

Feeding and Systematics

ARCHAEOGASTROPODS

The most primitive feeding habit and digestive tract are found in the archaeogastropods. These are microphagous browsers, grazing on fine algae, sponges, or other organisms growing on rocks or kelps. Accordingly, the gut resembles that of the ancestral mollusc, which was also a microphagous browser (Fig. 12-39A). In the primitive patellogastropod limpets, the radula (Fig. 12-40A) is **docoglossate** (doco = beam or spear, glossa = tongue) and each row contains 13 or fewer teeth. The teeth are stout and impregnated with iron and silicon. In most other archaeogastropods, however, the radula is **rhipidioglossate** (rhipidio = fan) and each row contains an abundance of teeth (Fig. 12-40B). Grazing by large populations of intertidal limpets may greatly depress the growth of algae.

The stomach (Fig. 12-39A) is complex and resembles that described for the generalized mollusc except for the effects of torsion and the presence of a **gastric cecum** (not the digestive cecum). The esophagus enters the stomach on its right, posterior end near the ducts from the digestive cecum. The intestine exits at the anterior end. The surface area of the sorting field of most archaeogastropods is expanded by extending it into the long, spiraled gastric cecum. A longitudinal ridge, the major typhlosole, extends the length of the stomach from the gastric cecum to the intestine. It lies to the right of the longitudinal intestinal groove. The minor typhlosole is on the left of the intestinal groove. The typhlosoles and intestinal groove are ciliated. The cuticular gastric shield covers much of the wall of the posterior end of the stomach. Anteriorly, the stomach narrows to form a style sac from which the intestine exits. A protostyle composed of mucus and mineral particles is found in the style sac. Cilia of the epithelium of the style sac rotate the protostyle in the sac.

The mucous food string extends from the radula through the esophagus into the stomach, where it is continuous with the protostyle so that rotation of the protostyle helps pull the string into the stomach. Within the stomach, food and mineral particles are released from the food string and mixed with enzymes from the esophageal glands and digestive cecum. Ciliary activity and muscular contractions mix the food particles and enzymes. The cilia of the sorting field tend to move organic particles into the digestive cecum, whereas the indigestible particles move anteriorly along the typhlosoles to

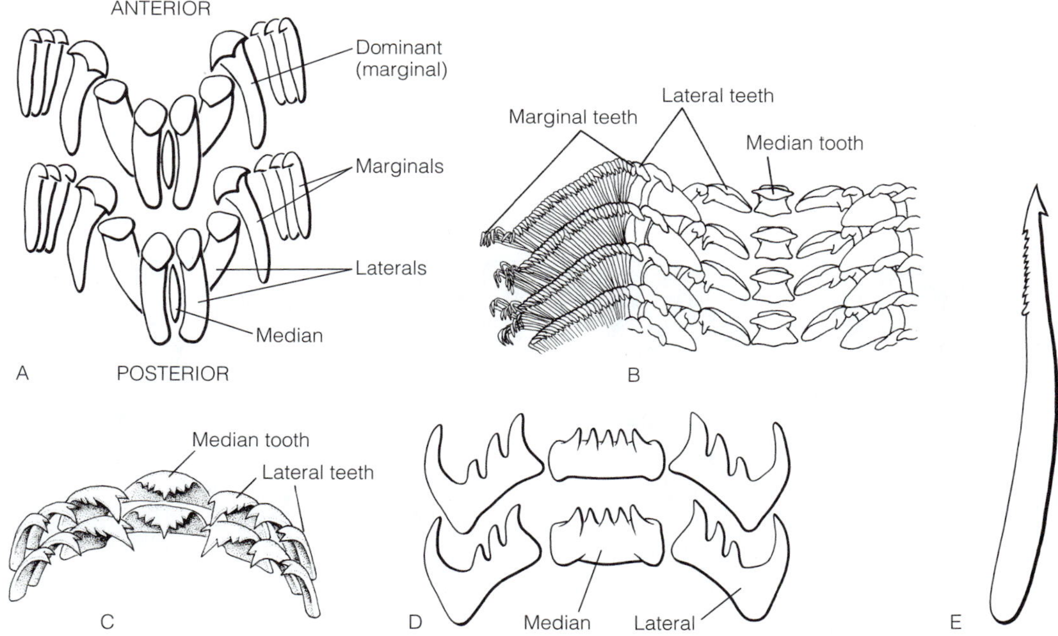

FIGURE 12-40 Major types of prosobranch radulae. **A,** Docoglossate radula. Dorsal view of two transverse tooth rows of the limpet *Patella* (Patellogastropoda). The median tooth is small or absent in this type of radula. All other teeth have a single cusp except the dominant marginal, which is pluricuspid (with multiple cusps). The laterals and dominant marginals are pigmented (black) apically. The median and outer marginals are not pigmented. **B,** Rhipidioglossate radula: four transverse rows of the radula of the abalone *Haliotis,* a zeugobranch archaeogastropod. The marginal teeth on one side have been omitted. The median tooth is well developed and there are five laterals on each side. The needlelike marginals can be exceedingly numerous. **C,** Taenioglossate radula: two transverse rows of the mesogastropod *Lanistes.* **D,** Rachiglossate radula: two transverse rows of the radula of the neogastropod *Buccinum.* There are only three teeth to a transverse row, but the teeth bear several cusps. There are no marginals. **E,** Toxoglossate radula: a highly modified tooth from the radula of a cone snail, *Conus.* *(A, B, D, and E, Redrawn from Fretter, V., and Graham, A. 1994. British Prosobranch Molluscs. Second Edition. Ray Society Vol. 161. Ray Society, London; C, Modified from Turner)*

the style sac and intestine. Extracellular digestion is completed in the digestive cecum and absorption and intracellular digestion occur here.

HIGHER GASTROPODS

In association with their adaptive radiation into soft-bottom and pelagic habitats, the diets and feeding habits of the higher gastropods became extremely diverse, especially among the mesogastropods and opisthobranchs. Many higher gastropods are macrophagous animals that feed on large food items. Digestion has become entirely extracellular, taking place in the stomach. Because macrophagous feeding does not require the elaborate sorting equipment of the primitive molluscan stomach, it has been lost. The stomach is a simple sac without a gastric shield, sorting field, style sac, gastric cecum, or protostyle (Fig. 12-39B). Enzymes are supplied by the digestive cecum or by glands associated with the esophagus or buccal region.

The highly adaptable **taenioglossate radula** (taenio = band) of mesogastropods has seven teeth in each transverse row (Fig. 12-40C). The neogastropods are mostly carnivores and have specialized radulae. Most neogastropods (Muricoidea) have **rachioglossate radulae** (rachio = spine) with only three teeth (sometimes just one) per transverse row, but the teeth are heavy and usually bear several cusps (Fig. 12-40D). The outer, hooked lateral teeth collect torn or detached particles and bring them into the center when the radula is retracted. The Conoidea (cones), which are also neogastropods, have a specialized **toxoglossate radula** (toxo = archer or bow) that delivers poison to the prey. It will be discussed in a following section on carnivory. Only marginal teeth are present, and these are highly modified to form hollow, venom-filled harpoons (Fig. 12-40E, 12-47).

The largely macroherbivorous pulmonates have radulae with the largest number of teeth of any gastropod. There may be up to 750 small teeth per transverse row (Fig. 12-41A). Opisthobranch radulae are highly variable. The efficiency of the radula results not only from the adaptive design of particular teeth, but also from the complex ways in which the teeth interact with each other.

Feeding Ecology

HERBIVORES

The many herbivorous gastropods include some marine prosobranchs, the freshwater prosobranchs, the operculate land snails, a variety of opisthobranchs, and a majority of the pulmonates. Most marine herbivores are either microphagous browsers that feed on fine algae or macrophages specializing in macroalgae, such as kelps and other seaweeds. Freshwater and terrestrial species consume the tender parts of aquatic and terrestrial vascular plants, decaying vegetation, or fungi. A few terrestrial snails and slugs are serious agricultural pests. The giant African snail *Achatina fulica,* introduced into Hawaii and the continental United States, can be very destructive, and considerable effort has been expended to prevent its spread.

Members of the mesogastropod family Littorinidae (periwinkles or winkles) are found on rocky shores, mangroves, and even marsh grasses throughout the world. These widespread and often abundant snails live in the intertidal zone, each species occupying a characteristic level between the low-tide mark and the supratidal splash zone. At low tide the animal withdraws into its shell behind the operculum, attaching the lip of the shell to the substratum with mucus. At high tide it emerges to graze on microalgae, including endolithic species that penetrate into the rock surface. The gut contents include large quantities of mineral material that must be separated from the organic particles.

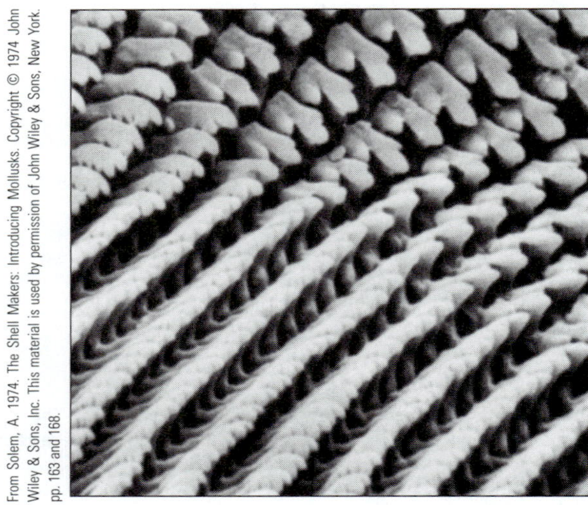

A

B

FIGURE 12-41 Scanning electron micrographs of pulmonate radulae. **A,** The herbivorous snail *Diastole conula.* **B,** The carnivorous *Euglandina rosea.* Each tooth, a long cone, is erect when the radula is in a functional position.

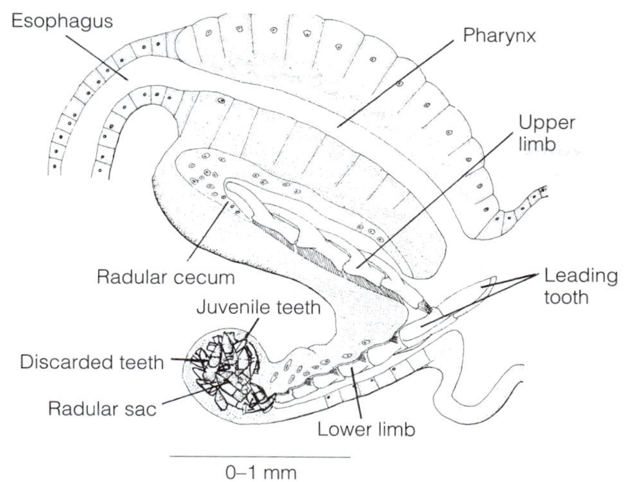

FIGURE 12-42 Lateral section of the radula and buccal mass of the shell-less sacoglossan *Limapontia.* The single row of radula teeth is adapted for puncturing algal cells. Used teeth are discarded and stored in a sac. *(From Gascoigne, T., and Sartory, P. K. 1974. The teeth of three bivalved gastropods and three other species of the order Sacoglossa. Proc. Malacol. Soc. London. 41:11.)*

The sacoglossan opisthobranchs (Fig. 12-34E,F) are specialized herbivores. In these tiny, sluglike gastropods the radula is reduced to a single, longitudinal row of teeth. The needlelike radular teeth used to penetrate algal cell walls are discarded into a sac, or ascus, after use (Fig. 12-42). The contents of the cells are then sucked out by a muscular pharynx. Sacoglossans tend to be specific in the species of algae they use for food. Members of the sacoglossan taxon Elysiidae retain and cultivate the undamaged chloroplasts from their food and harvest the photosynthetic product. Epithelial cells in the digestive cecum phagocytose the chloroplasts but do not digest them. Instead, the photosynthetic organelles are kept alive in the gastrodermal cytoplasm and are provided with the raw materials for photosynthesis. The resulting carbohydrates are shared with the host. Nudibranch sea slugs have long been known to sequester the nematocysts of their cnidarian prey. It has now been shown that some species that feed on octocorals also retain and nurture the zooxanthellae (photosynthetic endosymbiotic dinoflagellates) from their prey and utilize the photosynthetic product.

Limpets living near hydrothermal vents in the deep sea maintain populations of filamentous sulfide bacteria on their gills. These prokaryotes are symbiotic chemoautotrophs with the metabolic machinery to utilize the energy of reduced sulfur compounds to reduce carbon dioxide. They then synthesize organic compounds from the reduced carbon using the same enzyme (RuBisCO) that is used by plants in the Calvin cycle. The resulting food molecules are shared with the limpet.

In many macroherbivorous species, the esophagus or anterior part of the stomach is modified as a crop and gizzard. The gizzard may be lined with cuticle (as in the sea hares, which feed on large pieces of algae) or contain sand grains (convergent with many freshwater snails and also with birds). Amylases and cellulases are produced by the esophageal pouches or the digestive cecum. Terrestrial pulmonates, such as *Helix,* have a powerful array of digestive enzymes. Most digestion

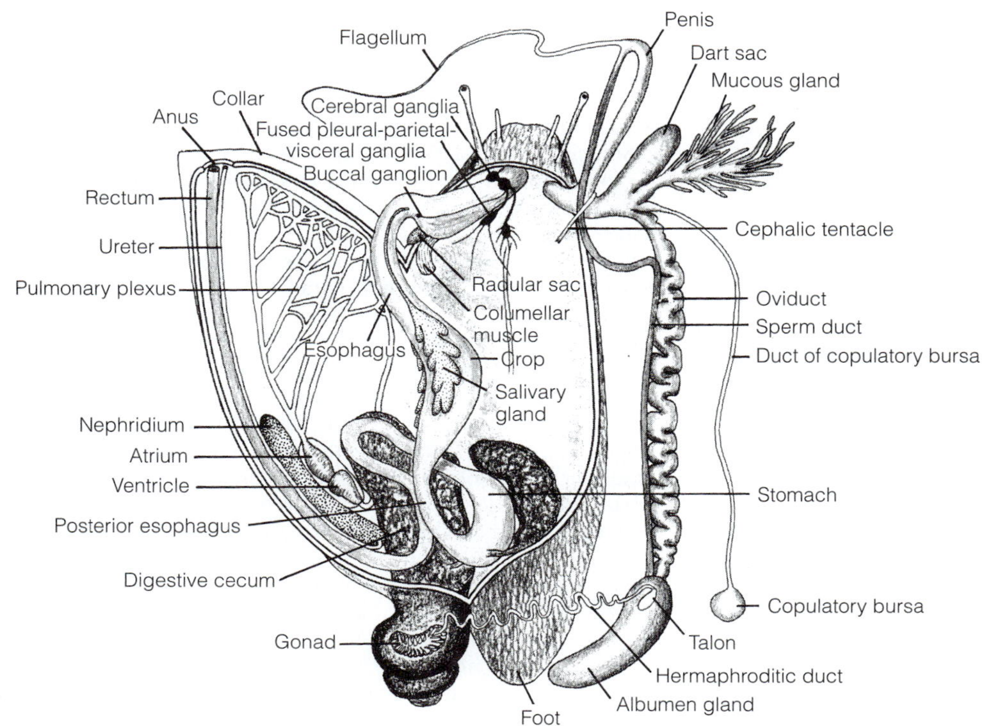

FIGURE 12-43 Dissection of the land snail *Helix.* The mantle cavity (lung) has been opened on the right and its roof reflected to the left. The visceral mass has been opened to reveal the viscera. The hermaphroditic reproductive system has been displaced to the right. *(Adapted from several sources.)*

occurs in the large crop, which to a great extent replaces the stomach (Fig. 12-43). There is no gizzard. Cellulase has been reported, but it may be synthesized by bacteria in the crop or intestine.

CARNIVORES

Prosobranchs

Most carnivorous gastropods are marine prosobranchs or opisthobranchs, although there are a few pulmonates that feed on earthworms or other pulmonates. The radula is variously modified for cutting, grasping, tearing, scraping, or transporting, and jaws are sometimes present. The most common adaptation of carnivorous prosobranchs is a highly extensible tubular **proboscis** with the mouth at its tip (Fig. 12-14), which enables the animal to reach and penetrate vulnerable areas of the prey. The proboscis is part of the digestive tract and contains the esophagus, buccal cavity, and radula housed in a **proboscis sheath,** or sac (Fig. 12-44, 12-45B). It should not be confused with the siphon, which is part of the mantle and is used as a snorkel to conduct water to the osphradium and gills (Fig. 12-14). In feeding, blood pressure extends the proboscis from the opening of the proboscis sheath. Specific proteins liberated by prey or carrion detected in the respiratory current aid in locating the food source and elicit protrusion and search with the proboscis. The whelk *Buccinum undatum* can identify a food source 30 m upstream, but it takes several days to reach it.

Bottom-dwelling carnivores, especially large species of neogastropods and some mesogastropods, commonly feed on bivalve molluscs, other gastropods, sea urchins, sea stars, polychaetes, crustaceans, and even fish. Many burrow into the sand to reach their prey. Bonnets *(Phalium)*, tuns *(Tonna)*, and tritons *(Cymatium)* incapacitate their prey with salivary secretions containing sulfuric acid or other toxins. Some helmets use a combination of sulfuric acid and the radula to cut a small window in the calcareous test of the sea urchins they feed on. They extend the proboscis into the window and remove the tissue with the radula. The pelagic *Janthina* (Fig. 12-38A) feeds on pelagic, floating siphonophores at the water surface. Some prosobranchs specialize in feeding on foraminiferans.

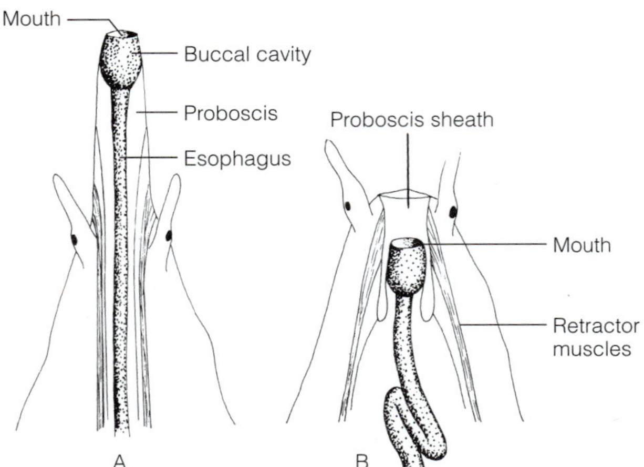

FIGURE 12-44 Dorsal view of the head of a carnivorous prosobranch, showing the proboscis. The gut is stippled. **A,** Proboscis protracted. The proboscis sheath disappears when the proboscis is protracted and the mouth is at the tip of the proboscis. **B,** Proboscis retracted. Note that the opening of the proboscis sheath is not the same as the mouth. *(After Fretter, V., and Graham, A. 1994. British Prosobranch Molluscs. Second Edition, Ray Society Vol. 161. Ray Society, London.)*

Some olives and volutes smother the prey with their large foot. Some whelks *(Buccinum, Busycon, Fasciolaria, Murex)* may grip a bivalve with the foot, pulling or wedging the two valves apart with the outer lip of its aperture or siphonal canal (Fig. 12-46A). To accomplish the wedging, the gastropod may first erode the valve margin with the lip of its own shell.

A number of prosobranchs are adapted for drilling holes in the shells of limpets, barnacles, and other molluscs, especially bivalves. The two families best known for this are the neogastropod Muricidae *(Urosalpinx, Murex, Thais, Nucella, Eupleura)* and the mesogastropod Naticidae (moon snails, *Natica, Polinices*). The mechanism has been most extensively studied in the Muricidae, and particularly in *Urosalpinx,* which causes great damage in oyster beds. Both the American oyster drill, *Urosalpinx,* and the Japanese drill, *Rapana,* have been introduced into other parts of the world with shipments of oysters.

Drilling is a combination of mechanical abrasion and chemical breakdown of the shell. The anterior sole of the foot contains an eversible accessory boring organ that produces an acidic secretion. The snail everts the gland and holds it against the shell for about 30 minutes. The acid reacts with the insoluble calcium carbonate of the shell and dissolves it. It also degrades and weakens the protein matrix of the shell. The snail then uses its radula for about a minute to remove the softened shell material, then repeats the cycle until the shell is penetrated. Penetration is primarily a result of the chemical reaction rather than the drilling by the radula. Approximately 8 h are required to penetrate a shell 2 mm thick, and penetration to a depth of 5 mm has been recorded. When drilling is completed, the proboscis is extended into the hole and the soft tissues of the prey are torn by the radula and ingested. Some muricids inject a muscle relaxant or toxin through the borehole. In the naticids the shell-softening gland is located at the proboscis tip rather than on the foot. Naticid boreholes have beveled sides, whereas those of muricids have parallel sides.

One of the most remarkable taxa of carnivores is the toxoglossan neogastropods, including the turrids, augers, and cone snails. Cone snails (Conidae) are tropical and subtropical and are found mainly in the western Atlantic and Indo-Pacific Oceans. They feed primarily on polychaete worms, other gastropods, and even fish, which they stab and poison with their radular teeth. The odontophore has been lost and the radula is greatly modified (Fig. 12-47A). The teeth are long, grooved, hollow, and barbed at the end (Fig. 12-40E, 12-47B). They are attached to the radular membrane by a fragile, slender cord of tissue. The teeth are loosely arranged in the radular sac and there is no conventional radular ribbon. A large, muscular bulb pumps poison to the buccal cavity and radular sac. The toxin is secreted not by the bulb, but by the epithelium of its duct. The teeth in the radular sac are immersed in the venom, which fills and coats them. The teeth are disposable, and each tooth is used only once and replaced by another tooth from

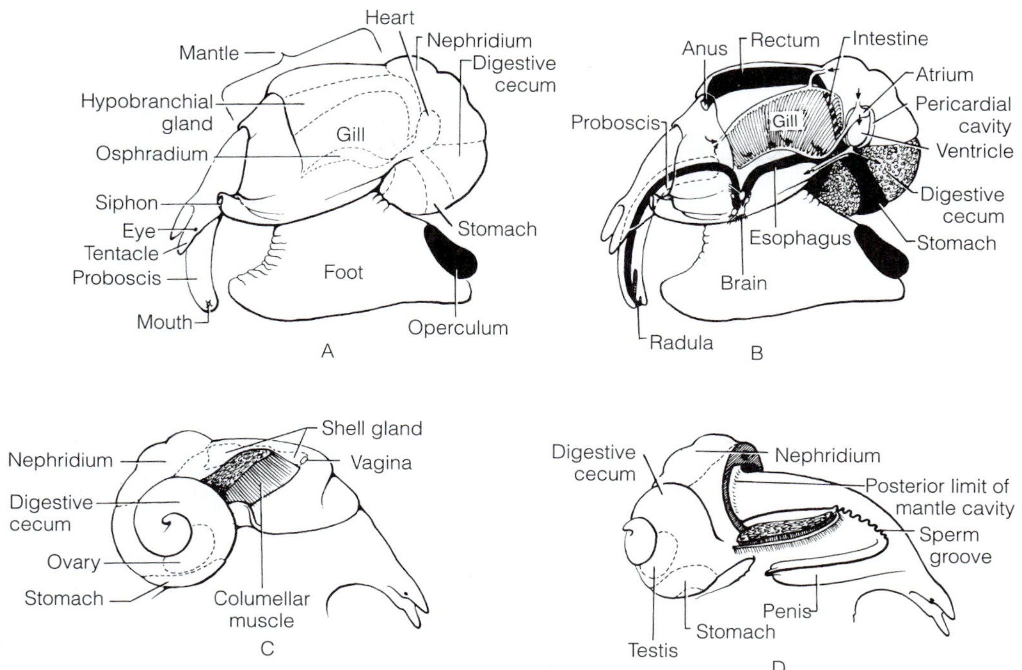

FIGURE 12-45 Gastropoda. Several views of the whelk *Busycon canaliculatum,* showing aspects of its soft anatomy. The shell has been removed. **A,** Left side, showing external organs and internal organs visible through the integument. **B,** Left side, with digestive, respiratory, circulatory, and nervous systems emphasized. **C,** Female, showing a portion of the right side with the mantle and columellar muscle intact. **D,** Male, showing a portion of the right side with the mantle and columellar muscle removed. In **A** and **B,** the proboscis is extended. In **C** and **D,** it is withdrawn.

the radular sac. Only one tooth is in use at any given time and the current active tooth is attached to the tip of the long, highly maneuverable proboscis (Fig. 12-47A).

The proboscis is protracted rapidly so that its tip strikes the prey with force, embedding the single active tooth in the prey. In most species the tooth falls free of the proboscis and remains embedded in the prey. In those species that feed on fish (Fig. 12-47C), the snail lies buried in the sand and does not strike until the fish pauses over the bottom. The harpoon is then thrust into the soft belly of the prey, and the cone

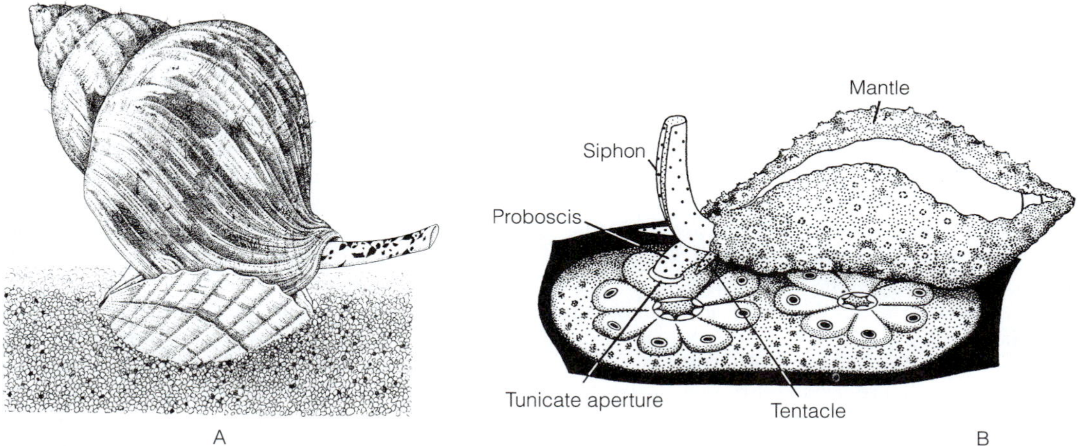

FIGURE 12-46 Prosobranch feeding. **A,** *Buccinum* using the edge of its shell to pry open a cockle. **B,** The triviid *Erato voluta* feeding on a colonial tunicate. The proboscis is thrust into the buccal opening of the tunicate. The shell is partially covered by the reflexed mantle. *(A, From Nielsen, C. 1975. Observations on* Buccinum undatum *attacking bivalves and on prey responses, with a short review on attack methods of other prosobranchs. Ophelia 13:87–108; B, From Fretter, V. 1951. Some observations on British cypraeids. Proc. Malacol. Soc. London. 29:15.)*

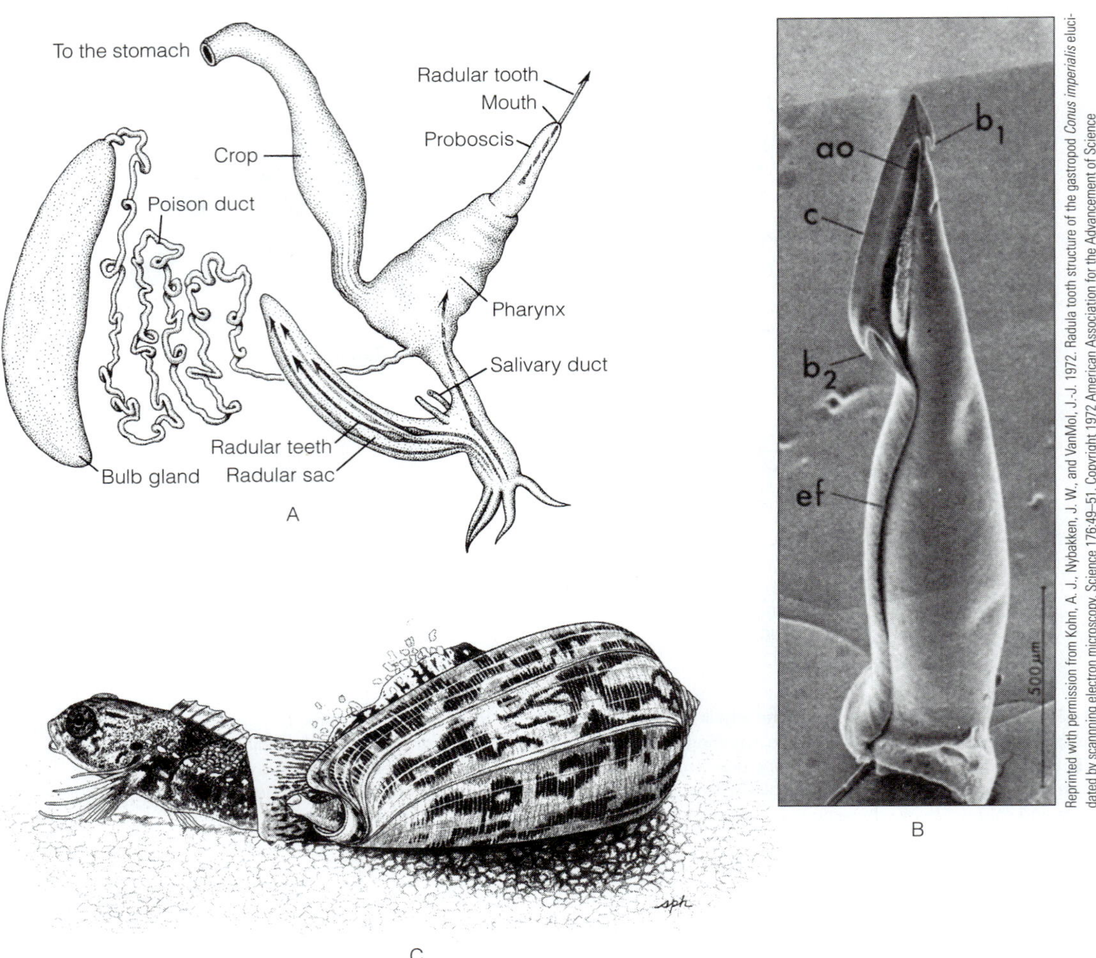

FIGURE 12-47 Feeding mechanism of cone snails *(Conus).* **A,** Buccal apparatus of *Conus striatus.* **B,** Scanning electron micrograph of the harpoonlike radula tooth of *Conus imperialis.* Note the folded structure and barbed end. **C,** A cone snail swallowing a fish. *(A, Modified after Clench; C, Based on a photograph by Robert F. Sisson and Paul Zahl.)*

retains a grip on the end of the tooth. The victim is quickly immobilized by the neurotoxic poison, which enters the wound through the hollow cavity of the tooth. The toxins differ with species, but are small oligopeptides effective in minuscule concentrations. The bite, or sting, of some South Pacific species is dangerously toxic to humans and a few deaths have been reported, in one case within 4 h.

The pelagic heteropods (Fig. 12-35D) feed on other small, pelagic invertebrates and are the prosobranch ecological equivalents of the opisthobranch sea butterflies (gymnosomes). The cowries, which are mesogastropods, resemble the opisthobranch nudibranchs in being grazers that feed on sessile invertebrates. The cowrie *Erato voluta* feeds on the small, individual zooids of colonial tunicates (Fig. 12-46B). The cowrie crawls over the surface of the colony, inserts its proboscis into the zooid, and removes the tissues with the radula.

Opisthobranchs

There are many carnivorous opisthobranchs, the principal raptorial taxa of which are the naked sea butterflies (gymnosomes) and many bubble snails (bullomorphs). The former prey on shelled sea butterflies (thecosomes) and the latter on bivalves and gastropods, which are seized with the hooked teeth of the radula and swallowed whole.

The nudibranchs are grazing carnivores that feed on sessile animals such as hydroids, sea anemones, soft corals, bryozoans, sponges, ascidians, barnacles, and fish eggs. Each family of nudibranchs is typically specialized for one type of prey. There usually is no proboscis, but jaws are commonly present. In the Aeolidiidae (Fig. 12-34G), most of which feed on hydroids and sea anemones, the two bladelike jaws are used to cut small pieces of tissue from the prey.

Nudibranchs that prey on cnidarians usually ingest, save, and utilize the prey's undischarged nematocysts. Ironically, the

A

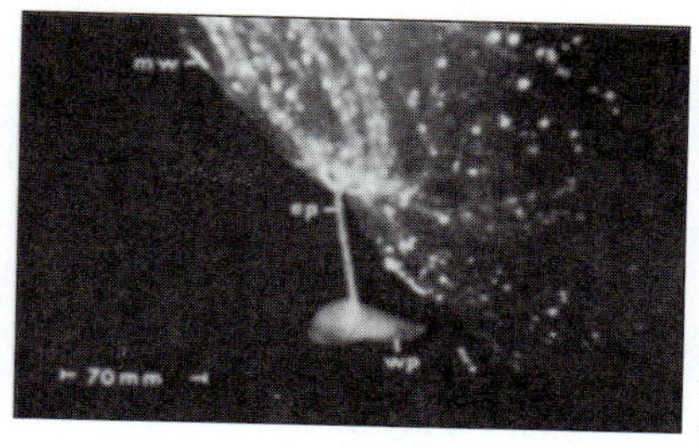

B

FIGURE 12-48 Gastropoda. **A,** An aggregation of the mud snail *Ilyanassa obsoleta* feeding at low tide. This intertidal prosobranch scavenger and deposit feeder occurs in enormous numbers on protected beaches on the east coast of the United States. **B,** The sea butterfly *Gleba cordata* feeding from its large, delicate, mucous web (mw). A winglike parapodium (wp) and the long proboscis (ep) are the only visible parts of the snail.

nematocysts, which failed to protect their rightful owner, are used by the predator for its own defense. The nudibranch eats the cnidarian and ingests the nematocysts without discharging them. Ciliary tracts in the stomach and digestive ceca carry undischarged, even immature, nematocysts to the cerata, where they are stored in **cnidosacs** at the distal tips of the cerata. The nudibranch discharges the stored nematocysts for defense by some unknown mechanism. Although cnidarians typically possess several types of nematocysts, the nudibranchs select and use only a subset of those available. The pelagic nudibranch *Glaucus,* which feeds on pelagic cnidarians including the Portuguese man-of-war, can inflict a sting as serious as that of the man-of-war itself. Nudibranchs with stinging capability are typically **aposematically** pigmented (a bright warning coloration) and are known for their flamboyant color patterns.

SCAVENGERS AND DEPOSIT FEEDERS

A scavenging habit has been adopted by numerous gastropods, of which *Nassarius* and *Ilyanassa* are notable examples. The feeding habits of these little neogastropods cover the range from carnivory to deposit feeding. On quiet, protected beaches along the Atlantic coast of the eastern United States, the mud snail *Ilyanassa obsoleta* may occur in enormous numbers at low tide, feeding on organic material deposited in the intertidal zone (Fig. 12-48A). The snails aggregate in herds and are not uniformly distributed over the flat. This species is also a facultative carrion feeder and consumes the flesh of fresh fish.

Along the coasts of Britain and Europe the tiny, deposit-feeding mesogastropod *Hydrobia* occurs in densities as high as $30{,}000/\text{m}^2$ on muddy, intertidal flats. Other deposit feeders include many common species of horn snails *(Cerithium, Cerithidea, Batillaria).* The conch *Strombus* and the related burrower *Aporrhais* are deposit feeders or grazers on fine algae, with a large, mobile siphon that sweeps across the bottom like a vacuum cleaner.

SUSPENSION FEEDERS

Although more typical of bivalves, suspension feeding by various mechanisms has evolved several times in the gastropods. Any suspension feeder must have some mechanism for concentrating suspended particles by separating them from a much larger volume of water. Filtering is one such mechanism, but there are others. Gills or mucous nets may be used to filter a water current, or particles may be allowed to settle passively onto a mucous sheet. Cilia are often, but not always, involved in suspension feeding.

The mesogastropod slipper snail, *Crepidula,* is a filter-feeding suspension feeder. The gill has been enlarged by lengthening its filaments to form a filter (Fig. 12-49). Water enters the mantle cavity on the left, as in other prosobranchs, and passes between the gill filaments to the right. Particulates, such as phytoplankton, are trapped in a sheet of mucus on the upstream side of the filaments. Cilia move the mucus and its trapped particles to the right edge of the gill, where they are transferred to a longitudinal, ciliated groove on the body. This groove transports the mucus string anteriorly toward the mouth to be ingested.

Many of the sessile worm snails (Turritellidae and Vermetidae) use their gills as a food-trapping surface. Others secrete a net or veil of mucous threads in the vicinity of the shell aperture and allow suspended particles to settle from quiet water onto the mucus, which is then ingested. The net is produced by the pedal gland, laid down by pedal tentacles, and spread by wave action (Fig. 12-33B,C), and may cover as much as 50 cm of area in *Serpulorbis squamigerous* of the American Pacific coast. In the Red Sea, *Dendropoma maxima* takes 2 min to haul in the net with its radula, which it does every 13 min.

The shell-bearing sea butterflies (thecosomes) are suspension feeders that trap food particles in the mucus covering the parapodia on their mantle cavities. Some sea butterflies, such as *Gleba* and *Corolla,* secrete enormous, floating mucous nets up to 2 m in diameter (Fig. 12-48B). The animal hangs beneath the net by its extended proboscis, the cilia of which pull in the food particles trapped in the net. There is no radula.

A crystalline style is found in many suspension feeders, deposit feeders, and grazers, including *Crepidula, Struthiolaria,* many worm snails, *Strombus, Nassarius, Cerithium,* and some sea butterflies. Like filter feeding, the style is more typical of bivalves than gastropods. The presence of a crystalline style in gastropods is associated with the more or less continuous feeding on small particles of phytoplankton or organic detritus.

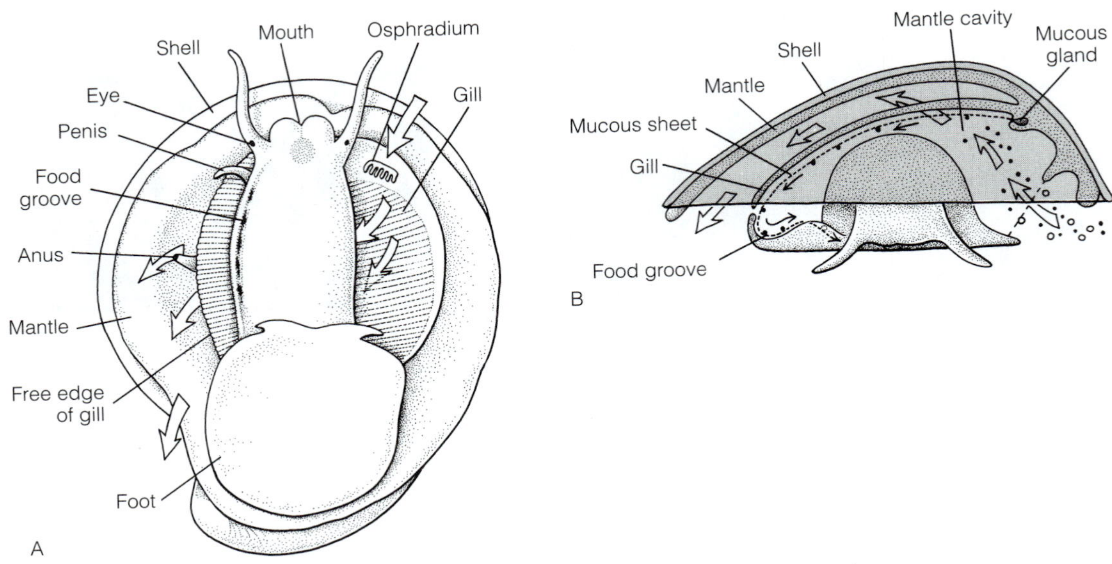

FIGURE 12-49 Suspension feeding. **A,** The slipper snail *Crepidula fornicata,* a ciliary feeder (ventral view). **B,** Anterior view of *Crepidula,* showing the direction of the water current (large, outlined arrows) through the mantle cavity. Ciliary currents (small arrows) carry food particles. Dashed lines indicate filtering sheet of mucus. Coarse particles are removed by the mucous web at the entrance to the mantle cavity; fine particles are trapped by the mucous sheet on the gill. The latter is driven by the frontal gill cilia to a food-collecting groove along the side of the body, where it is rolled up into a string and periodically pulled into the mouth by the radula. *(B, Redrawn and modified from Morton.)*

PARASITES

Parasitism has evolved independently in several gastropod taxa, and it includes several levels of specialization—epizoic, ectoparasitic, and endoparasitic relationships. The heterostroph Pyramidelloidea is the major parasitic taxon of gastropods. They are small (1 to 3 mm) ectoparasites on bivalves and polychaetes that have required relatively minor modification to suit them to parasitic life. Their adaptations are chiefly of the buccal region and digestive system. They lack a radula and gill but have a long proboscis, chitinous jaws, stylets, and a pumping pharynx for sucking blood from the host (Fig. 12-50A). The spiral shell and operculum are distinctly snail-like.

The other major taxon of gastropod parasites is Eulimoidea, whose members live on or in echinoderms. Members of Eulimidae, such as *Stilifer,* are snails with a heavy proboscis that is extended through the host's test deep into its body cavity (Fig. 12-50B). The presence of the snail induces the host to secrete a gall that encloses the snail. The eulimids are more highly modified for parasitism than are the pyramidellids, but they are still easily recognizable as snails.

The most modified parasitic snails are members of Entoconchidae, also in Eulimoidea, which live in the perivisceral coelom of sea cucumbers and are so highly modified for parasitism that the wormlike adults are unrecognizable as molluscs (Fig. 12-50C). *Entoconcha* has a rudimentary mouth and gut and attaches to a blood vessel, but *Enteroxenos* has no mouth or gut and absorbs nutriment through the body wall. Female *Enteroxenos* individuals may reach lengths of 15 cm and are themselves host to parasitic dwarf males, which fertilize their eggs. Culminating the evolutionary sequence is *Parenteroxenos,* which reaches lengths of 130 cm and is a hollow sac with no organ systems other than the reproductive. Development of eulimnodeans includes a veliger larva with a typical coiled gastropod protoconch. Young snails probably enter the perivisceral coelom of the host by boring through the body wall with the proboscis. They then transform into an adult parasite, which is unrecognizable as a mollusc.

EXCRETION

The marine gastropods are mostly ammonotelic, meaning that ammonia is the end product of their protein metabolism, and much nitrogen is lost by diffusion of ammonia across the body and gill surfaces. Freshwater and amphibious species may be ammonotelic or ureotelic; terrestrial species usually are uricotelic. Intertidal species may alternate between being uricotelic when the tide is out and ammonotelic when immersed by the flooding tide.

The ancestral gastropod had two nephridia (Fig. 12-26), as do the living vetigastropods and neritimorphs (Fig. 12-51). The right nephridium has been lost by almost all living gastropods, although a small section of it is retained as part of the gonoduct (Fig. 12-51). Only the patellogastropods retain the right kidney instead of the left.

As a result of torsion, the nephridium is located anteriorly in the visceral mass (Fig. 12-26). It is a blind sac surrounded by the hemocoel. Its walls are folded to increase the surface area for secretion and absorption to and from the surrounding blood. At the end of the sac, near the nephridiopore, the

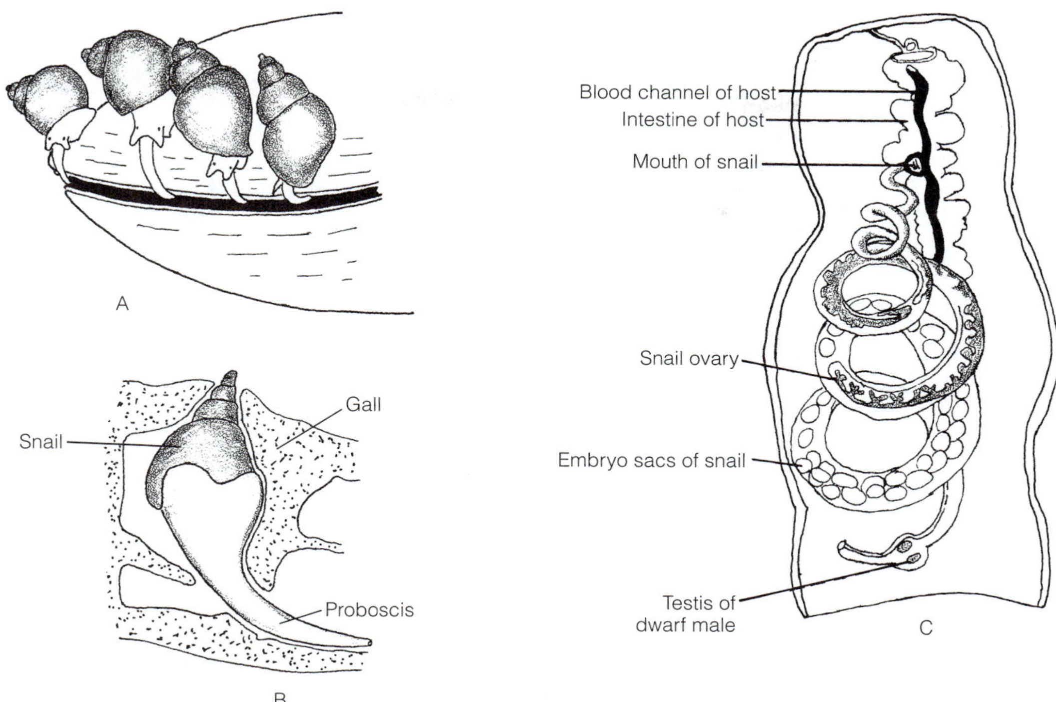

FIGURE 12-50 Parasitic gastropods. **A,** The pyramidellid *Brachystomia,* an ectoparasite, feeding on the body fluids of a clam. **B,** The parasite *Stilifer* embedded in the body wall of a sea star. **C,** The endoparasite *Entoconcha* in the body cavity of a sea cucumber. *(A and B, After Abbott, 1954; C, After Baur from Hyman, 1967)*

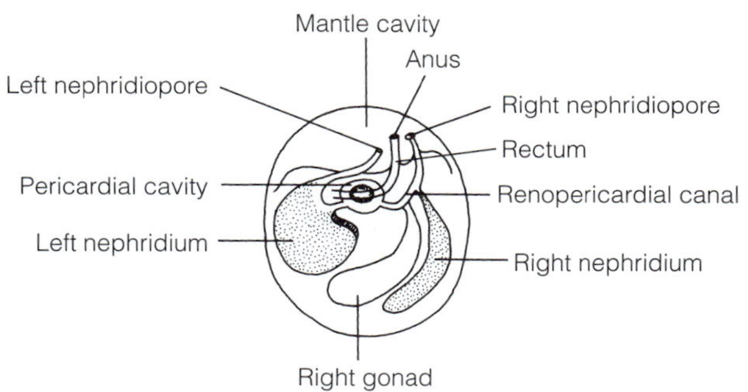

FIGURE 12-51 Diagrammatic dorsal view of an archaeogastropod (Trochacea) showing the relationship of the right gonad to the nephridia. The gonad opens into the renopericardial canal of the right nephridium. In the higher gastropods the right nephridium is lost, but its duct is retained by the gonad as the middle region of the genital duct. *(From Fretter, V., and Graham, A. 1962. British Prosobranch Molluscs. Second Edition. Ray Society Vol. 161. Ray Society, London. p. 283.)*

kidney connects with the pericardial cavity via a renopericardial canal (Fig. 12-52, 12-26, 12-51). The nephridiopore of the prosobranchs and the lower opisthobranchs opens at the rear of the mantle cavity, on the downstream side of the gills, and wastes are removed by the respiratory water current.

Urine formation is like that in other molluscs. Pericardial glands with podocytes in the walls of the atrium and sometimes the ventricle produce an ultrafiltrate in the pericardial cavity, which then flows through the renopericardial duct into the nephridium (Fig. 12-52). Selective absorption and secretion by the nephridial epithelium modifies the primary urine and converts it to the final urine, which is released from the nephridiopore in the mantle cavity.

In terrestrial pulmonates urine cannot be released into the posterior end of the mantle cavity because no water circulates to carry wastes away. As a result, the nephridial tubule has lengthened to form a **ureter** extending along the right wall of the mantle to open near the anus and the pneumostome (Fig. 12-43).

Freshwater gastropods maintain a low blood osmolarity and the nephridia reabsorb salts so that copious hyposmotic urine is released. Terrestrial pulmonates and operculate land snails conserve water by converting ammonia to relatively insoluble uric acid. Because terrestrial pulmonates have no pericardial glands, they do not form an ultrafiltrate. The renopericardial canal is small, and a concentrated urine results largely from nephridial secretion, with scant water. This is also true of some intertidal prosobranchs, such as the littorinids.

Terrestrial pulmonates cannot avoid losing water with the large amounts of slime they produce for locomotion and by evaporation from the body and lung surfaces. Many pulmonates, however, can survive extensive desiccation. The escargot, *Helix,* can tolerate a water loss equal to 50% of its body weight and the slug, *Limax,* can survive an 80% loss. Nevertheless, the majority of pulmonates require a humid environment. They tend to be nocturnal and inhabit humid

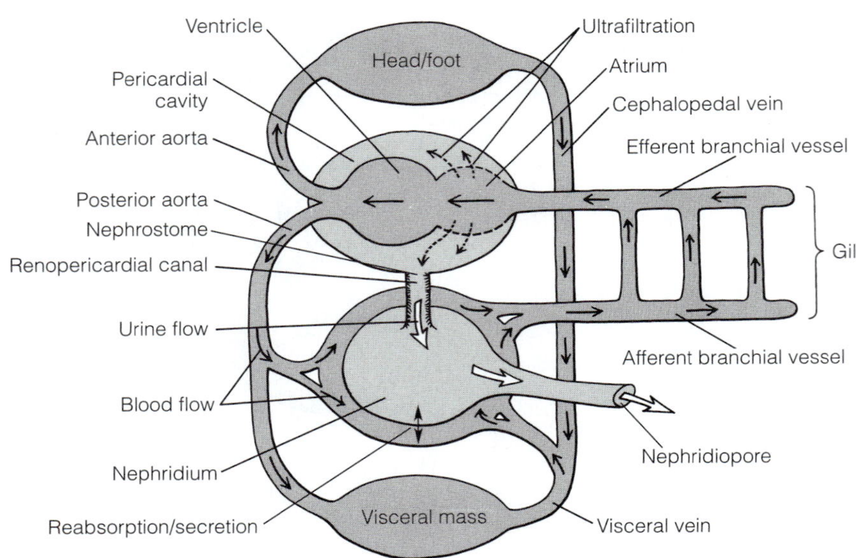

FIGURE 12-52 Blood flow in the periwinkle *Littorina littorea.* The hollow arrows indicate urine flow, the solid arrows blood flow, and the dashed arrows ultrafiltration. *(Simplified and redrawn from Andrews, E. B., and Taylor, P. M. 1988. Fine structure, mechanism of heart function and haemodynamics in the prosobranch gastropod mollusc* Littorina littorea *(L.). J. Comp. Physiol. B 158:247–262.)*

environments, such as areas beneath logs or in leaf litter on forest floors. Some pulmonates, however, inhabit dry, rocky areas; sand dunes; and even deserts, but are active only at night or following rain.

During periods of unfavorable weather, terrestrial gastropods may become torpid, either by hibernating in cold weather or estivating in hot, dry weather by becoming inactive and lowering the metabolic rate. Before entering torpor, they may first burrow into humus or soil or climb into vegetation and attach the aperture edge of the shell with dried mucus. Estivation in elevated positions is probably an adaptation to avoid ground-dwelling predators such as mice, reptiles, and beetles. Terrestrial prosobranchs close the aperture with the operculum, but pulmonates have no operculum and secrete a substitute known as an **epiphragm,** which is made of mucus or, occasionally, calcium carbonate. The edges of the mantle are drawn together in front of the aperture and a sheet of mucus is secreted over the opening. The sheet hardens when it dries. Sometimes a thin calcareous membrane is secreted to protect the aperture. Several epiphragma may be secreted in succession as the snail withdraws farther and farther into the shell.

Freshwater snails also estivate when temporary ponds dry up and hibernate when the water freezes. Estivation may last a number of months, and there are records of snails estivating for several years. Reactivation commonly results from temperature change, a rise in humidity, or mechanical disturbance, such as raindrops striking the shell. Handling can also cause the snails to break estivation.

INTERNAL TRANSPORT

The gastropod hemal system is similar to that of the generalized mollusc except that torsion has moved the pericardial cavity and heart from the original posterior location to a new position in the anterior visceral mass (Fig. 12-16B) and there is only one atrium in most species. In the early gastropods the heart was diotocardian (Fig. 12-26) and each atrium received oxygenated blood from one of the two gills. Among living gastropods, only vetigastropods and neritimorphs are diotocardian, and in all higher gastropods the right gill and right atrium are lost (Fig. 12-14B, 12-43). The ventricle and atrium are muscular, the ventricle more so than the atrium. The atrial wall contains podocytes and is the site of ultrafiltration. Arteries and veins consist of circular muscle and connective tissue. The hemocoel consists of two major spaces, the cephalopedal and visceral hemocoels. The hemocoelic spaces are referred to as being "open" because they are not lined by an endothelium, although they may be large sinuses or small, vessel-like, or even capillary-like channels. Veins drain blood from the hemocoelic spaces back to the heart.

In a typical gastropod, such as the mesogastropod *Littorina,* the ventricle receives oxygenated blood from the atrium and pumps it into a system of major vessels, the aortae (Fig. 12-52). The posterior aorta (visceral aorta) supplies the **visceral hemocoel** and the anterior aorta (cephalic aorta) feeds into the **cephalopedal hemocoel** of the head and foot. The aortae may deliver blood to small vessels or to large blood sinuses where exchange with the tissues occurs. Unoxygenated blood returning to the heart passes over the nephridium and then enters the gill through the afferent branchial vessel. Following oxygenation, the blood exits the gill through the efferent branchial vessel to the atrium. In the pulmonates, all blood returning to the atrium passes through the pulmonary plexus (capillary network) in the roof of the lung (Fig. 12-43).

The morphology of the blood spaces is partly related to the local function of the blood. Gastropod blood serves as a medium for internal transport, but it is also a hydrostatic skeleton. In regions where the major role of the blood is hydraulic, the hemocoel is likely to be a large sinus, as in the head/foot of the abalone, *Haliotis.* Where the blood is mostly involved in exchange of materials with the tissues, the hemocoel is likely to be capillary-like, as in the visceral hemocoel of *Haliotis.*

The respiratory pigment hemocyanin is dissolved in the plasma of prosobranchs and pulmonates. The freshwater

Planorbidae (pulmonates) have hemoglobin instead of hemocyanin in the plasma. The tiny mesogastropod commensal snail, *Cochiolepis parasitica,* which lives under the scales of a large polychaete, has hemoglobin and its tissues are bright red and easily visible through its transparent shell. Myoglobin is frequently present in muscles, especially those of the buccal mass, and gives them a red or orange color. Opisthobranch gas transport mechanisms are poorly known, but some species of sea hares (such as *Aplysia*) have hemocyanin whereas others lack it.

NERVOUS SYSTEM

The gastropod nervous system is similar to that of the generalized mollusc except that it is torted, at least in prosobranchs, and it tends to be cephalized. Prior to torsion, the nervous system is bilaterally symmetrical and consists of the cerebral, pedal, pleural, buccal, esophageal, and visceral ganglia and their commissures and connectives (Fig. 12-17A).

Although torsion twists the nervous system, it affects only the parts that pass from the head-foot to the visceral mass. Thus the anterior nervous system, consisting of cerebral, pleural, and pedal ganglia, is unaffected, but the system posterior to the pleural ganglia, consisting of the visceral and esophageal ganglia and the visceral cords, is altered. The visceral cords are twisted into a **visceral loop** that resembles a figure eight (Fig.12-17B). Further, torsion twists the gut on its long axis in the vicinity of the mid-esophagus so that the stomach and visceral ganglia, which were originally ventral to the esophagus, are now dorsal to it (Fig. 12-53). The right esophageal ganglion is moved dorsally and to the left of the esophagus to become the **supraesophageal ganglion** and the left esophageal ganglion moves ventrally and to the right to become the **subesophageal ganglion.**

This twisted, or streptoneurous, condition is a distinctive feature of the primitive gastropod nervous system that is common to all prosobranchs. The ganglia of the primitive nervous system are separated by long connectives and are distant from each other, so the system is not concentrated (condensed, cephalized). The length of the connectives delays communication between the cerebral and other ganglia. Such a nervous system is found, with some modifications, in many early prosobranchs.

In the higher gastropods, the primitive asymmetrical streptoneurous pattern is obscured by two strong evolutionary trends. First, there has been a tendency toward both cephalization and fusion of ganglia by shortening the connectives between ganglia. This creates a concentrated, cephalized central nervous system with all the major ganglia closely associated with the nerve ring and adjacent to the major sense organs in the head (Fig. 12-54A). Communication time between the cerebral ganglia and the other ganglia is minimized.

The second evolutionary trend is the tendency for the ganglia and cords to adopt a secondary bilateral symmetry (Fig. 12-54B) known as **euthyneury.** In most pulmonates and opisthobranchs the nervous system has detorted and the visceral loop has untwisted and become symmetrical once more, similar to the condition shown in Figure 12-17A. Because they share this pattern, the pulmonates and the opisthobranchs are grouped together in the taxon Euthyneura. Euthyneury (detorted) is the opposite of streptoneury (torted).

Prosobranchs are streptoneurous, but even so, there is a tendency in the derived taxa to exhibit an approximation of bilateral symmetry. This functional bilateral symmetry, called **zygoneury,** is accomplished by adding a new connective between the right esophageal ganglion and the left pleural ganglion and another between the left esophageal ganglion and the right pleural ganglion. If you pencil these new connectives onto Figure 12-17B, the resulting bilateral symmetry will be apparent, but be sure to notice that the result is not the same as true euthyneury, in which the system detorts and comes to resemble Figure 12-17A.

SENSE ORGANS

Gastropod sense organs include eyes, tentacles, osphradia, statocysts, and a magnetic sense. The simplest gastropod eyes, found only in the primitive patellogastropods such as *Patella,*

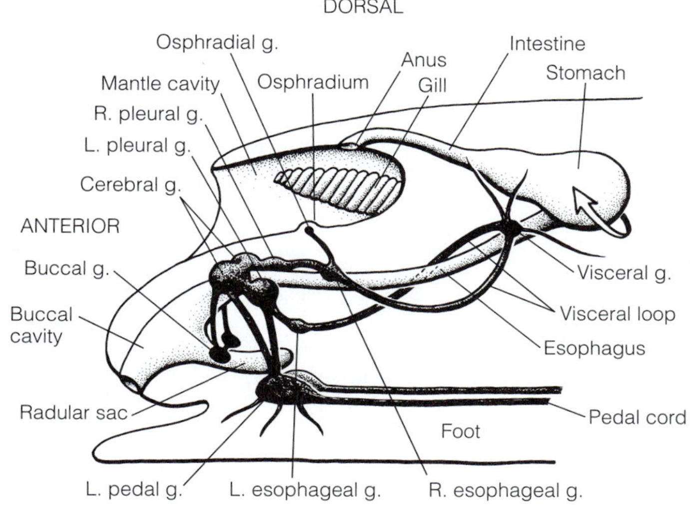

FIGURE 12-53 Lateral view of a generalized mesogastropod showing the effect of torsion on the nervous system. Anterior to the pleural ganglia, the system is unaffected by torsion and is bilaterally symmetrical. Only the visceral and esophageal ganglia and the visceral loop connecting them are altered and have become asymmetrical. The nervous system's pretorsional symmetry can be imagined by rotating it and the gut as indicated by the arrow. This would untwist the esophagus and place the visceral ganglion back in its original position ventral to the gut. The right cerebropedal and pleuropedal connectives have been omitted for clarity. *(Redrawn from Fretter, V., and Graham, A. 1994. British Prosobranch Molluscs. Second Edition. Ray Society Vol. 161. Ray Society, London.)*

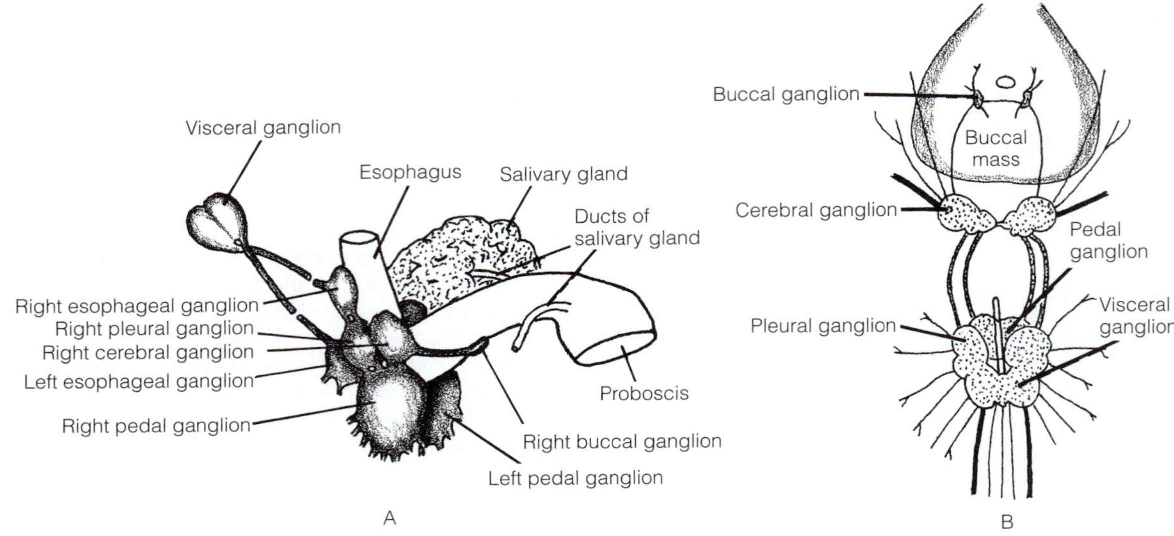

FIGURE 12-54 **A,** Concentrated and cephalized nervous system of the neogastropod *Busycon* viewed from the right side. The visceral ganglia are remote from the cephalized complex of ganglia in the head. **B,** Secondarily symmetrical (euthyneurous) nervous system of the pulmonate *Helix* viewed dorsally. The visceral cords are shortened and the system is highly concentrated. *(A, After Pierce, M. In Brown, F. A. (Ed.): 1950. Selected Invertebrate Types. John Wiley and Sons, New York. pp. 318–319; B, Redrawn from Bullough, 1958)*

are simple, open pigment-cups on the surface of the head. This eye has a retina of photoreceptors enclosed by a pigment-cup, but it is open to the environment and there is no lens (Fig. 12-55A). In the higher gastropods the cup is closed and is a spherical vesicle with a lens and cornea, in addition to the retina and pigment-cup (Fig. 12-55B). Scattered photoreceptors that provide a sensitivity to light and dark independent of the more highly organized eyes often occur in the epidermis, especially on the siphon.

The eyes may be situated at the base of the cephalic tentacles, but often they are on **eyestalks.** The stalks frequently fuse with the cephalic tentacles so the eye appears to be on the side of the tentacle (Fig. 12-14), or they may be independent of the cephalic tentacles and form a second set of posterior tentacles, as occurs in the stylommatophoran pulmonates (Fig. 12-28E).

The eyes of most gastropods appear to function only in the detection of light, but those of some, such as the conch *(Strombus)* and the pelagic heteropods, are probably capable of forming an image and are used to detect predators, or, in the case of the heteropods, prey.

A pair of closed vesicular statocysts is located in the foot near the pedal ganglia. They are innervated from the cerebral ganglia, however, not the pedal. Each statocyst is a fluid-filled capsule lined with a ciliated epithelium of sensory hair cells. It contains a calcareous statolith resting on the cilia of the hair cells. Statocysts are absent from sessile species, such as slipper and worm snails. They arise in the embryo as invaginations of the ectoderm (as does the inner ear of vertebrates).

The osphradium is a characteristic feature of most prosobranchs and is also present in some opisthobranchs and aquatic pulmonates, although it is usually absent in terrestrial species. It is thought to have a chemosensory function, but there is speculation, without evidence, that it also plays a role in detection of sediment in the water.

The evolution of the osphradium of gastropods closely parallels that of the gills. In the stem gastropod and the primitive vetigastropods, an osphradium is present for each of the two gills. In all other prosobranchs the right osphradium is lost along with the right gill. The single remaining osphradium is located in the mantle cavity anterior to the attachment of the

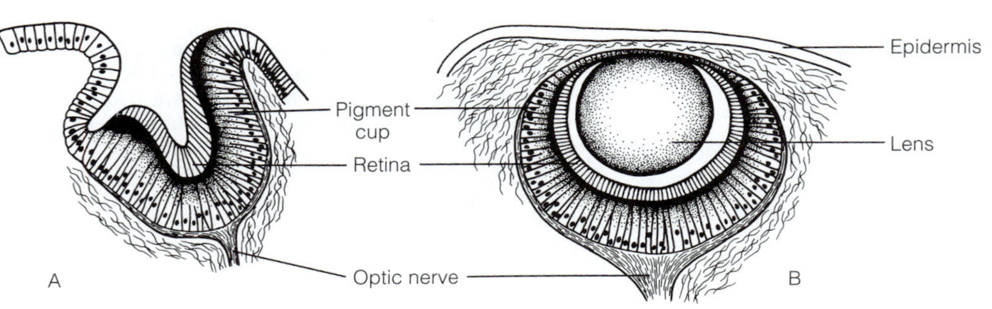

FIGURE 12-55 Eyes of two marine prosobranchs. **A,** The open pigment cup of *Patella,* a limpet. **B,** The closed pigment cup eye, with lens, of *Murex.* *(Both after Helger from Parker and Haswell, 1951)*

gill (Fig. 12-14B, 12-49A, 12-53). In the archaeogastropods and mesogastropods the osphradium is a simple ridge, but in the neogastropods it becomes a complex, bipectinate, gill-like structure with an expanded surface area. These prosobranch carnivores and scavengers can locate carrion, animal juices, or prey from large distances, as much as 2 m in the case of the scavenging whelk, *Cominella,* and 30 m in another whelk, *Buccinum undatum.* Higher gastropods continually wave their siphons about as they move over the bottom, selecting and monitoring water from various parts of the environment.

The various cephalic, ocular, epipodial, and pallial **tentacles** of gastropods are all chemoreceptors, sometimes exclusively and sometimes in addition to having such other roles as photoreception or mechanoreception. In opisthobranchs, the posterior pair of head tentacles is often elaborate, chemosensory **rhinophores** (Fig. 12-36A, 12-34A,G,H). Other areas of the epidermis, such as the foot margins and siphon, may also have chemoreceptors.

Magnetoreceptors sensitive to a magnetic field have been found in the neogastropod mud snail, *Ilyanassa obsoleta,* and may also be present in some opisthobranchs.

REPRODUCTION

Most prosobranchs are gonochoric whereas pulmonates and opisthobranchs are hermaphroditic. The left gonad has been lost and only the right gonad is present. Primitively, the gonad occupied a coelomic space, the gonocoel, and connected with the exterior by a coelomoduct or modified metanephridium, the gonoduct. In Recent gastropods the original gonoduct has become a more complicated structure known as the **genital duct,** only part of which is the original gonoduct. The genital duct ranges from very simple in gonochoric gastropods with external fertilization (primitive prosobranchs) to extremely complex in hermaphroditic species with internal fertilization (pulmonates and opisthobranchs). The genital duct evolved in conjunction with the right nephridium, and in all gastropods it remains associated with the right kidney duct.

Prosobranchs

In the prosobranchs the gonad, either ovary or testis, is located in the spirals of the visceral mass near the digestive cecum (Fig. 12-14B), and gametes are delivered to the mantle cavity by the genital duct. The gastropod genital duct, either oviduct or vas deferens, is derived from three sources, and its evolution can be traced in the phylogeny of the prosobranchs.

In the ancestral gastropod and most vetigastropods, both nephridia are present and functional, but they are asymmetrical (Fig. 12-51). The left kidney empties directly into the mantle cavity and is independent of the reproductive system. The gonoduct exits the right (and only) gonad and joins the right nephridium or its renopericardial duct (Fig. 12-26). Gametes are released from the gonoduct into the right nephridium and delivered by it to the mantle cavity. The right nephridiopore—also the gonopore—is located posteriorly in the mantle cavity. The genital duct of primitive prosobranchs thus has two origins, the right gonoduct and the right renal duct. In this type of reproductive system, the eggs are provided, at most, with gelatinous envelopes produced by the ovary or the nephridium. There is no copulation, and external fertilization takes place in the sea after the gametes are swept out of the mantle cavity by the respiratory current.

In all other gastropods, fertilization is internal with copulation, and the genital duct is correspondingly complex. Copulation requires, at a minimum, that the female genital duct have a space to receive the penis and a receptacle for the storage of sperm and that the male have an intromittent organ, or penis. If the system is hermaphroditic, it is even more complex, as equipment for both sexes must be present in each individual. If, as is usually the case, the eggs are enclosed in layers of nutritive material or protective envelopes, then still more specialization of the female duct is required.

The complex genital duct developed in the higher prosobranchs and was inherited by pulmonates and opisthobranchs. In higher prosobranchs, the right kidney is lost except for a short segment, the **renal duct,** that remains as part of the genital duct. Furthermore, the genital duct is considerably lengthened distally by a third addition, derived from the mantle epithelium, that extends the genital duct across the floor of the mantle cavity from the original posterior gonopore to a new gonopore at the anterior edge of the mantle cavity. This third section, the **pallial duct,** originated as a ciliated groove beginning at the posterior nephridiopore/gonopore, and it remains a groove in many gastropods, but in many others it has become a closed tube. The resulting complete genital duct consists of the proximal gonoduct, middle renal duct, and distal pallial duct or groove (Fig. 12-56). Although the gonoduct and renal duct remain relatively unspecialized, the pallial duct undergoes elaborate differentiation to produce a number of specialized regions that serve a variety of purposes, including sperm storage, egg capsule formation, and secretion of nutritive substances and mucus. The freeing of the right nephridium from excretory functions and the subsequent development of a complex reproductive system probably was an important contribution to the dramatic success of the mesogastropods and neogastropods.

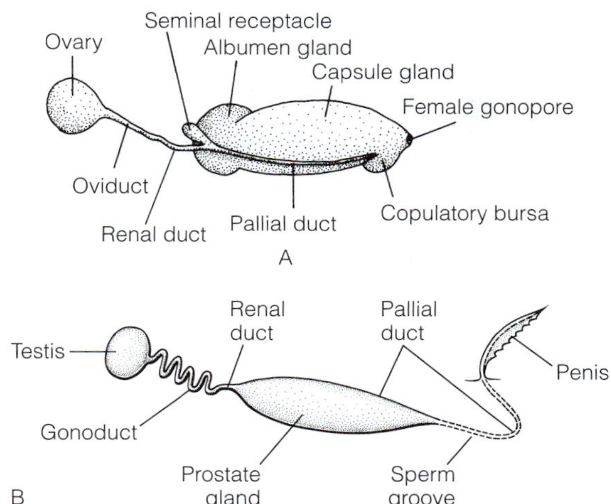

FIGURE 12-56 Reproductive systems of the mesogastropod *Littorina.* **A,** Female. **B,** Male. *(A, from Fretter and Graham, 1962; B, Redrawn from Fretter, V., and Graham, A. 1994. Second Edition. Ray Society Vol. 161. British Prosobranch Molluscs. Ray Society, London.)*

Gastropod reproductive anatomy is highly variable. In males the genital duct is a vas deferens (sperm duct). In a representative species, such as the mesogastropod periwinkle, *Littorina* (Fig. 12-56B), the gonoduct becomes a convoluted **proximal vas deferens** for sperm storage and transport. The renal portion of the vas deferens is a very short, unspecialized tube that connects the proximal vas deferens with the pallial region of the duct. The pallial duct is elaborated into a large **prostate gland,** which secretes prostatic fluid, and a ciliated sperm groove that extends across the floor of the mantle cavity. The margin of the mantle skirt is elaborated to form a tentacle-like intromittent organ, the **penis,** located on the right side near the cephalic tentacle (Fig. 12-14). The **sperm groove** extends to the tip of the penis. The pallial portion of the vas deferens is an open groove in some prosobranchs (neritids, littorinids, cerithiids) and a closed tube in others. Representatives of all three gastropod grades transfer sperm in spermatophores. For example, the sessile worm snails release spermatophores to drift in the water, where they are caught in the mucous nets of other individuals.

In the female the genital duct is the oviduct. The pallial section of the oviduct of *Littorina* is secretory and modified to form a variety of glands as well as chambers for the reception and storage of sperm (Fig. 12-56A). Among these are **albumen, jelly,** and **capsule glands.** These glands apply layers of nutritive and protective materials to the outside of the fertilized egg as it passes through this region of the oviduct. The albumen gland secretes a layer of nutritive albumen over the outside of the egg. Species such as *Cerithium* embed the eggs in jelly masses produced by the jelly gland, but in many of the higher prosobranchs, the eggs are enclosed in a protein capsule. There is a **copulatory bursa** and/or a seminal receptacle for the initial reception and long-term storage of sperm, respectively. Where both are present, the bursa receives the sperm from the penis. The sperm then move to the seminal receptacle for storage until fertilization at some later time. Fertilization occurs as the eggs leave the ovary and pass the seminal receptacle and before they enter the secretory, glandular regions where the various layers are applied.

Neogastropods often have a tough, proteinaceous **egg case** whose shape is characteristic of the species (Fig. 12-57A,B). The fluid, plastic protein is secreted over the eggs as they pass through the capsule gland. The still malleable capsule, with eggs inside, exits the gonopore and passes to a **pedal gland** on the foot where it is molded into the shape characteristic of the species, allowed to harden, and then attached to the substratum. The capsules typically contain a characteristic number of eggs. Development occurs in the capsule and young snails eventually emerge from a weak spot in the capsule wall. The oyster drill *Urosalpinx,* for example, deposits from 7 to 96 such cases containing 4 to 12 eggs each during a breeding season.

The mesogastropod moon snails *Natica* and *Polinices* produce a thin "collar" about 10 cm in diameter that is composed of sand grains held together by jelly. Within the sand are small, jelly-filled chambers containing eggs or embryos (Fig. 12-57C).

A few prosobranchs, such as many of the patellogastropod limpets and the mesogastropod slipper snails *Crepidula* are protandric hermaphrodites that begin life as males and change to female as they age and grow. The sessile slipper snails *Crepidula fornicata* (Fig. 12-49) live in stacks of as many as 12 individuals, in which they maintain a constant position, facing in the same direction with the right margins aligned above each other (Fig. 12-57D). Juxtaposition of the right margins facilitates the insertion of the penis of one individual into the female gonopore of an individual lower in the stack. In the immature ovotestis (the hermaphroditic gonad), the male cells mature first, so young individuals are functionally male. Thus, the young specimens on top of the stack begin life as males having the responsibility of fertilizing the older females below. As the upper individuals age, their female cells mature and they become functional females. They can then be fertilized by any younger males that have settled above them. The sex of each individual is influenced by the sex ratio of the stack, presumably controlled by

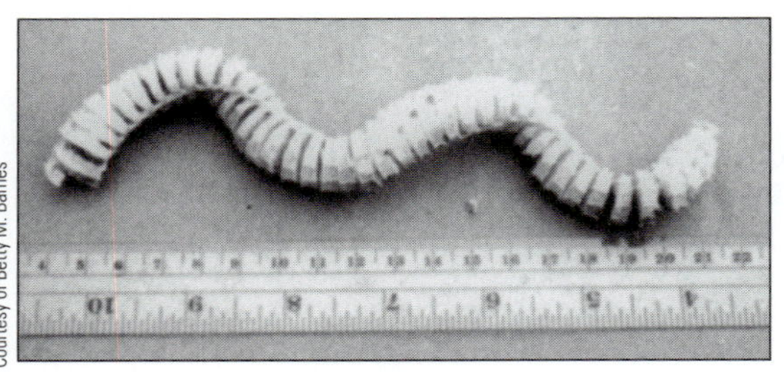
Courtesy of Betty M. Barnes

A

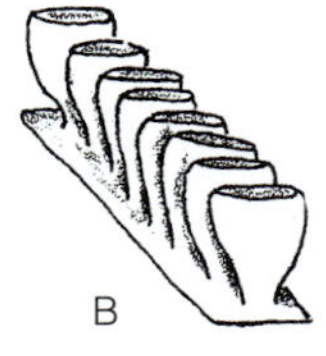

B

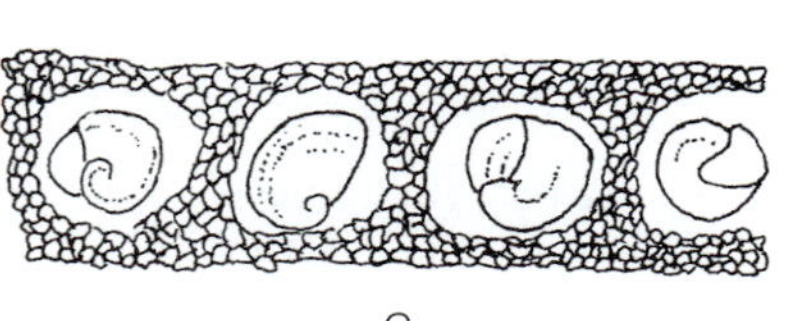

C

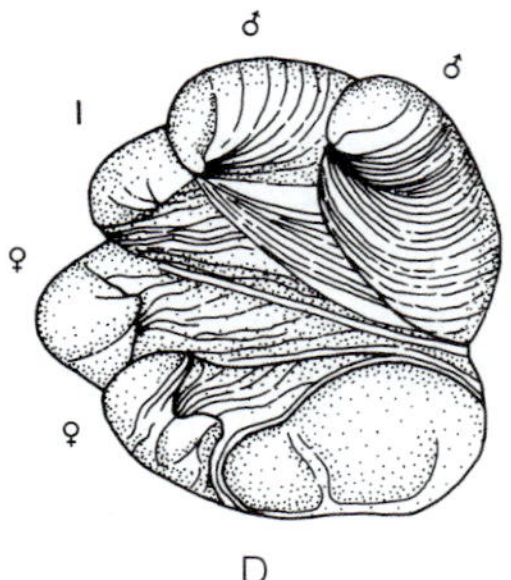

D

FIGURE 12-57 Prosobranch reproduction. **A,** Protein egg cases of the whelk *Busycon carica.* The many individual cases, each shaped like an oval pill box about 3 cm in diameter and up to 8 mm thick, are connected by a protein cord. There may be as many as 100 eggs per disk, but the average is about 40 and there are typically 80 to 130 disks on the cord. Development is direct and the little whelks emerge from a hole at the margin of each disk. Although the empty cords of cases often wash up on beaches, the cases are produced while the female whelk is buried in the sand and remain anchored in the sand until the young snails depart. **B,** Protein eggs cases of the cone snail, *Conus.* **C,** Section through the sand egg case of a moon snail showing the small, jelly-filled egg chambers in the sand matrix. **D,** A stack of slipper snails, *Crepidula fornicata,* in life position. *(B and C, After Abbott, R. T. 1954. American Sea Shells. Van Nostrand Co., Princeton, N. J.; D, From Hoagland, K. E. 1979. The behavior of three sympatric species of* Crepidula *from the Atlantic, with implications for evolutionary ecology. Nautilus 94(4):143–149.)*

pheromones. An older male will remain male longer if it is attached to a female. If such a male is removed or isolated, it will change into a female. A scarcity of females influences some of the males to change sex. The transformation of the reproductive system is controlled by the nervous and endocrine systems. Once an individual becomes female, it remains so.

Pulmonates and Opisthobranchs

Pulmonates and opisthobranchs (Euthyneura) are mostly simultaneous hermaphrodites, although the hermaphroditic gonad, the **ovotestis,** may not produce sperm and eggs at the same time. Copulation, with reciprocal sperm transfer, is typical in Euthyneura. The reproductive systems are very complicated and display endless variations (Fig. 12-43, 12-59B).

In land pulmonates the sperm are usually exchanged in spermatophores and copulation is commonly preceded by a courtship involving circling, oral and tentacular contact, and intertwining of the bodies. Bizarre sexual behavior occurs in some species, particularly slugs, but also in some snails. In the shelled Helicidae (*Helix,* the garden snail or escargot), the vagina contains an oval dart sac, which secretes a calcareous dart (Fig. 12-43). When two snails are intertwined, one snail drives its dart into the body wall of the other. Copulation follows this rather drastic form of courtship. The dart shooter is stimulated to copulate, whereas the behavior of the dart receiver appears relatively unmodified. The process may have evolved as a means of inducing one individual to act as a male and the other as a female.

Copulating limacid slugs *(Limax)* hang intertwined, like amorous bungee jumpers, from a mucous cord attached to a tree trunk or branch. The penes of the two partners are extended to lengths of 10 to 25 cm (sometimes as much as 85 cm) and are twisted together at the tips (Fig. 12-58A). Spermatophores are exchanged by the penial tips and carried back to the body by the retracting penes for storage prior to fertilization. Some pulmonates produce eggs with a calcareous shell (Fig. 12-58B).

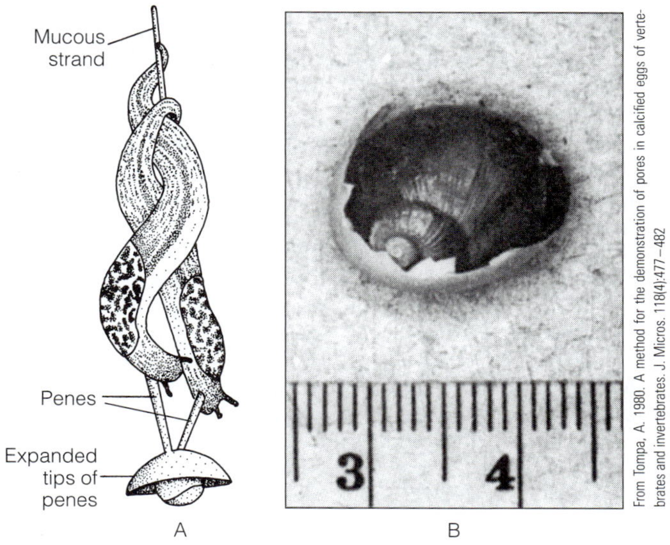

The reciprocal transfer of sperm has been carefully studied in the opisthobranch sea hare *Phyllaplysia taylori* (Fig. 12-59A,B). The two partners face each other head to head with their right sides juxtaposed. Each inserts its penis through the common gonopore and into the **hermaphroditic duct** (combined male and female genital duct) of its partner. **Autosperm** (sperm produced by an individual for transfer to its partner) has previously been produced by spermatogenesis in the ovotestis and stored in the seminal vesicle. During copulation an individual ejaculates autosperm from the vesicle into the hermaphroditic duct of its partner (Fig. 12-59B, heavy arrows). At about the same time, the individual receives an ejaculation of **allosperm** (sperm received by an individual from its partner) from the mate. The allosperm are stored in the seminal receptacle until fertilization and oviposition take place at a later time. The copulatory bursa (gametolytic gland) probably destroys superfluous sperm. The two partners separate and each later uses the stored allosperm from its own seminal receptacle to fertilize its eggs prior to oviposition.

During **oviposition** by the sea hare *Aplysia,* eggs leave the ovotestis and pass down the hermaphroditic duct. They receive a coating of albumen and are fertilized by allosperm ejected from the seminal receptacle. Thick layers of mucus and jelly are applied to the eggs and formed into spaghetti-like strings that are extruded from the common gonopore and attached to the substratum (Fig. 12-59C,D). Allosperm in the seminal receptacle are active, mobile, and capable of fertilizing eggs. Autosperm in the seminal vesicle are inactive, immobile, and incapable of fertilization. This insures that eggs en route from the ovary to the gonopore will not be self-fertilized by autosperm from the vesicle.

Mucous strand

Penes

Expanded tips of penes

A

B

From Tompa, A. 1980. A method for the demonstration of pores in calcified eggs of vertebrates and invertebrates. J. Micros. 118(4):477–482

FIGURE 12-58 Pulmonate reproduction. **A,** Copulating pair of *Limax* pulmonate slugs. **B,** The very large (about 16 mm) egg of the pulmonate *Strophocheilus.* The calcareous eggshell has been partially removed to reveal the young snail within. *(A, After Adams from Hyman, 1967)*

DEVELOPMENT

Archaeogastropods are broadcast spawners with external fertilization and planktonic trochophore larvae. Deposition of macrolecithal eggs in gelatinous strings, ribbons, or masses is characteristic of most mesogastropods and opisthobranchs, but neogastropods and some mesogastropods embed their eggs in an albumen mass surrounded by a protein capsule or case, which is usually attached to the substratum. The size and shape of the case, the nature of the wall (which may be leathery or gelatinous), and the number of cases attached together are extremely variable and characteristic of the species (Fig. 12-57A,B).

Most aquatic pulmonates deposit their eggs in gelatinous capsules. Terrestrial species produce a relatively small number of eggs, each enclosed with albumen within a separate capsule. The eggs are usually laid in a heap in soil. Among the largest eggs, up to 16 mm in diameter and enclosed in a calcareous shell, are laid by a 15 cm South American species of *Strophocheilus* (Fig. 12-58B).

In over half of the terrestrial pulmonate families and even in some prosobranch land snails, the capsule wall either contains calcite crystals embedded in mucus or consists of a calcareous shell. The shell not only supports and protects the egg contents,

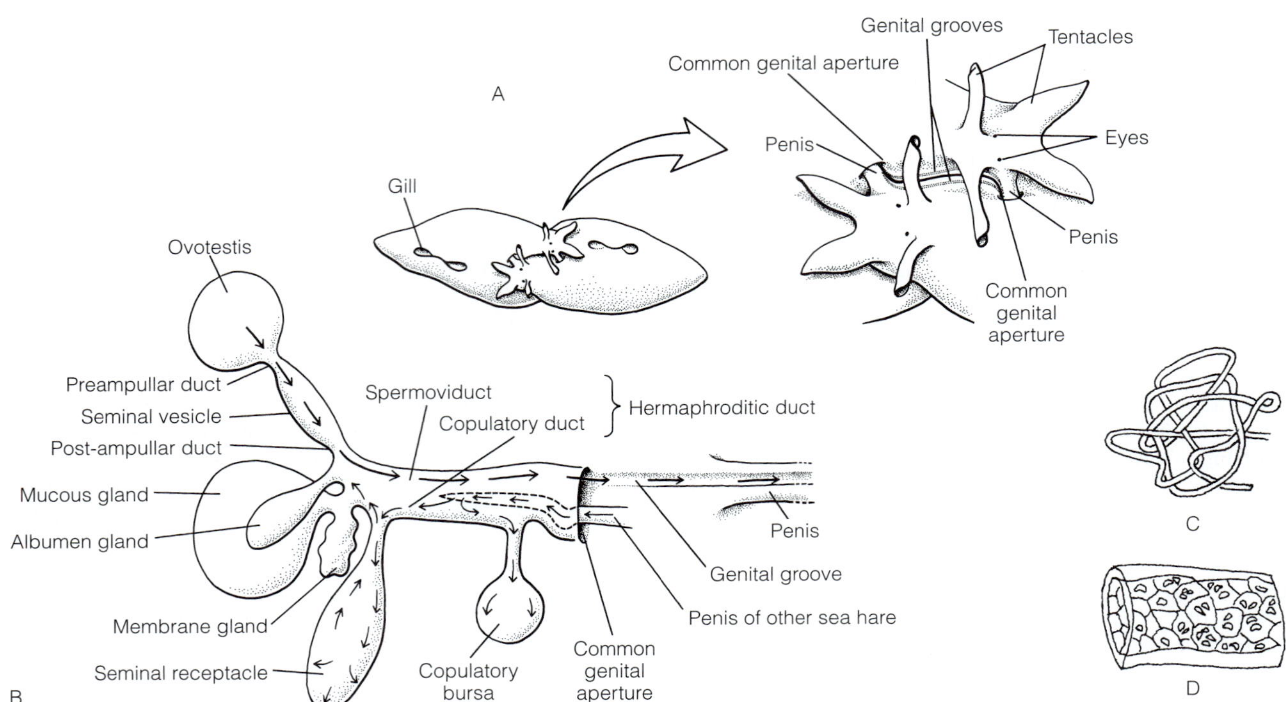

FIGURE 12-59 Opisthobranch reproduction. Reciprocal sperm transmission in the hermaphroditic sea hare *Phyllaplysia taylori*. **A,** Dorsal view of copulating pair. **B,** Sperm movement in one partner of a copulating pair. Long arrows indicate the path of the autosperm and short arrows the path of allosperm. **C,** Gelatinous egg string of a sea hare, *Aplysia*. **D,** Enlarged view of the egg string showing eggs in the hollow jelly tube. *(A and B, Redrawn from Beeman, R. D. 1970. An autoradiographic study of sperm exchange and storage in a sea hare,* Phyllaplysia taylori, *a hermaphroditic gastropod. J. Exp. Zool. 175(1):130; C and D, After Abbott, R. T. 1954. American Sea Shells. Van Nostrand Co., Princeton, N. J.)*

it also serves as a source of calcium for the shell of the developing snail within. A contractile, vesicle-like extension of the embryonic foot, called a **podocyst,** functions in absorption of albumen as well as in gas exchange and excretion. The podocyst is thus analogous to the allantois of reptile and bird embryos.

A free-swimming trochophore larva is found only in archaeogastropods that spawn their eggs directly into the sea. In all the other gastropods, the trochophore stage is suppressed and passed before hatching. Indirect development may then continue with a free-swimming veliger larva. Some marine prosobranchs, especially the neogastropods, nearly all freshwater prosobranchs, and almost all pulmonates undergo direct development and have no free-swimming larvae. At hatching, a tiny snail emerges from the protective shell or case (Fig. 12-58B).

The veliger larva is derived from a trochophore but is a later, more developed stage (Fig. 12-60) characteristic of gastropods, bivalves, and scaphopods. The veliger is named for its swimming organ, the velum, which consists of two large, semicircular, ciliated lobes. The **velum** forms as an outward extension of the prototroch of the trochophore. The foot, eyes, and tentacles differentiate from the body of the embryo. The veliger has a shell, the protoconch, which is spirally coiled in most gastropods but not in the patellogastropods or, presumably, the stem gastropod. It may remain at the apex of the adult shell for some time until it eventually erodes and disappears. In the shell-less sea slugs, a shell that is present in the veliger is later lost during metamorphosis.

Some gastropods have feeding, planktotrophic veligers with a larval life that may last as long as three months whereas others have short-lived, yolk-laden, nonfeeding, lecithotrophic veligers. The long cilia of the velum of planktotrophic veligers function not only in locomotion, but also in suspension feeding by bringing fine plankton in contact with the shorter cilia of the subvelar food groove. Within the food groove, particles become entangled in mucus and are conducted to the mouth.

During the veliger stage, torsion twists the shell and visceral mass 180° in relation to the head and foot. Torsion may be very rapid (about 3 min in the marine limpet *Acmaea*) or it may be a gradual process. As development proceeds, the swimming veliger reaches a point at which the foot is sufficiently formed to allow creeping, and at this time settling and metamorphosis occur. The velum is lost and the final features of the adult form are attained. Choosing the correct settling site is of critical importance for survival, and many species can delay metamorphosis until the required substratum is found.

The mode of development may vary greatly even within a group of closely related species. For example, among the common intertidal periwinkles *(Littorina),* some release planktonic egg capsules from which veliger larvae hatch. Others

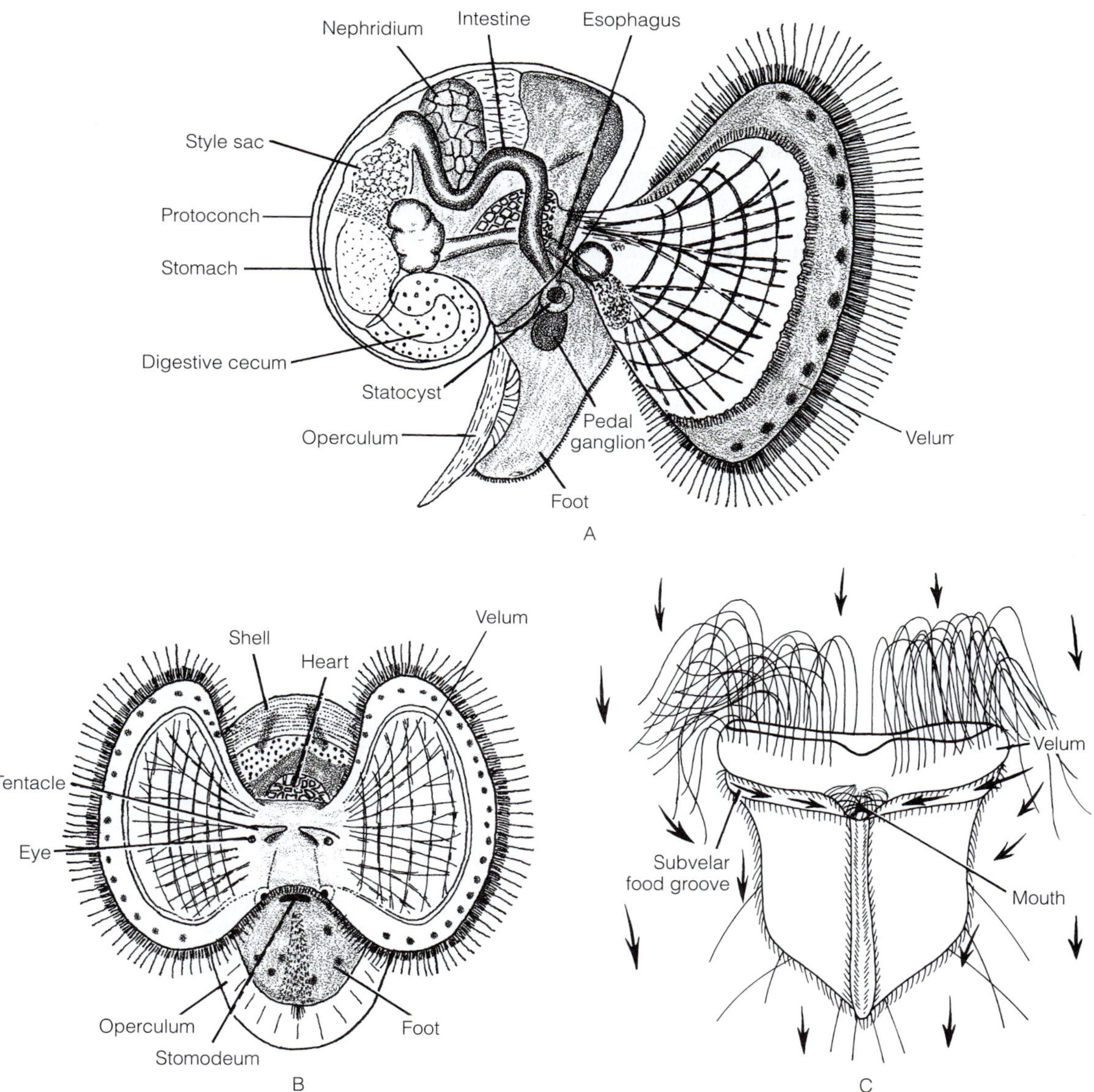

FIGURE 12-60 Veliger larva of the slipper snail *Crepidula.* **A,** Lateral view. **B,** Anterior view. **C,** Suspension feeding in the early veliger of the nudibranch *Archidoris.* Arrows indicate direction of movement of food particles. *(A and B, Redrawn from Hyman, 1967, after Werner; C, After Thompson.)*

attach gelatinous egg cases to the substratum, from which either veligers or juvenile snails emerge. Still others are viviparous and brood the eggs, releasing either veliger larvae or juvenile snails. The mode of each species is related to its ecological situation, especially in the intertidal zone where the timing of reproductive events may be tied to tidal cycles.

Many gastropods reach adult size and sexual maturity at 6 months to 2 years, but growth may continue at a slower rate, and larger species may not reach maximum size for many years. In general, growth is more rapid in tropical species than in temperate ones. The life span is highly variable: 5 to 16 years for the limpet *Patella vulgata;* 4 to 10 years for the periwinkle *Littorina littorea;* 1 to 2 years for many freshwater pulmonates; 5 to 6 years for the land snail *Helix aspersa;* and only 1 year for many nudibranchs. Mortality among larvae and juveniles is high. Few juvenile gastropods survive to sexual maturity. Of the 1000 eggs produced each year by the muricid prosobranch *Thais,* not more than 10 reach an age of one year.

Settlement and Metamorphosis

A planktonic larval stage is a common feature of the life cycle of many benthic marine animals. Indeed, most metazoan phyla have at least a few species with pelagic larvae. A benthic species benefits from having a life stage capable of dispersal. A planktonic larva is especially adaptive when the adults are slow moving (many polychaetes and molluscs) or sessile (barnacles). Some species with long-lived planktotrophic larvae may be carried great distances on ocean currents. The larvae of some marine snails, for example, are believed to be transported for thousands of kilometers. The lecithotrophic, nonfeeding larvae of many other marine species are planktonic for only hours or a few days. Even in this short time, they may be moved considerable distances and well away from the parents.

Settlement is a crucial stage in the life of the animal because most species are specialized for life in specific habitats. Developmental processes occurring in the larva may force it to settle at a specific time, whether or not it has located a suitable substratum. In the absence of the proper habitat, such larvae may settle before or shortly after metamorphosis. Many species have the ability to test the substratum and postpone settling for a time if testing proves it to be unsuitable. Such larvae typically respond to specific attributes of the surface and settle only if those features are present.

The larval response to favorable surface cues is settling and subsequent metamorphosis to the adult body form. Light intensity, surface texture, and substratum or water chemistry can be important signals. For example, many larvae require a bacterial biofilm over the surface to trigger settling and metamorphosis. Those invertebrates that live on kelps have larvae that respond to specific compounds produced by the kelp. The juveniles of the red abalone of California, *Haliotis rufescens,* as well as many other abalone species, graze on coralline red algae. The larvae of these molluscs stop swimming upon detecting a small protein characteristic of these algae. In the absence of this chemical signal, swimming continues and settling can be delayed for as long as a month. Premature settling is also undesirable because the larvae will not have completed its development, nor will it have dispersed as far as might be desired. Red abalone larvae must remain in the plankton for at least seven days because their receptors for the algal protein are not functional until then. The planula larvae of corals of the genus *Agaricia* also settle on coralline algae, but the different species of *Agaricia* are restricted to certain species of coralline algae and respond to surface oligosaccharides.

Larvae of invertebrates that live in aggregations with conspecifics, such as barnacles and reef-building tubeworms *(Phragmatopoma),* are stimulated to settle by surface proteins produced by adults of their own species. Sometimes settling is induced by a combination of factors. The sea slug *Rostanga pulchra* feeds on an encrusting sponge that lives in rocky crevices. The sponge provides a chemical cue to initiate settling, but the larva is also negatively phototaxic, which increases the chances that it will choose sponges in crevices.

DIVERSITY AND EVOLUTION OF GASTROPODA

The current level of understanding of Gastropoda does not support construction of a phylogenetic classification or tree. The provisional classification presented here is abbreviated and traditional orders are not arranged into higher taxa, although informal groups are indicated.

Prosobranchs (formerly Prosobranchia[sC])

The more than 20,000 species of prosobranch gastropods occur in marine, freshwater, and terrestrial habitats, where they may be abundant and diverse, but most are inhabitants of marine benthic habitats. Their remarkable diversity is the result of several episodes of adaptive radiation in their evolutionary history.

Prosobranchs are completely torted and most are snail-shaped with a spirally coiled, asymmetrical visceral mass and shell; broad creeping foot; and well-developed head. The mantle cavity, with one or more gills, a gonopore, nephridiopore, and anus, is located at the anterior of the body. The head is well developed and bears two cephalic tentacles, usually with a lateral eye at the base of each. The gill of most is monopectinate, but primitively it is bipectinate. Usually, there is an operculum. The gut is U-shaped, with both the mouth and anus anterior. The anterior gut contains a radula. The prosobranch heart consists of a ventricle and one (monotocardian) or two (diotocardian) atria. The great majority of them are monotocardian and have only the left atrium, left kidney, left gill, and right gonad. The heart is surrounded by a pericardial cavity that connects with the left kidney via a renopericardial duct. The right pedal retractor muscle has become the columellar muscle. The nervous system is streptoneurous (twisted as a result of torsion) and the well-developed sense organs include eyes, chemoreceptors, statocysts, and an osphradium. Most prosobranchs are gonochoric and have a veliger larva. The gonad occupies a coelomic gonocoel, which communicates with the mantle cavity through a genital duct derived from several sources, including the right nephridium or its remnants. Prosobranchs range from a few millimeters to over 70 cm in length.

The archaeogastropods in the broad sense *(sensu lato)* include the primitive gastropods, such as limpets, slit snails, abalones, nerites, top and turban snails, and others. Because of the structure of their bipectinate gill, archaeogastropods are largely restricted to the clean water associated with the surfaces of rock and seaweeds.

The limpet shape has evolved several times in the molluscs, and there are several types of prosobranch (and pulmonate) limpets. Limpets have a low, conical shell with a large basal opening, or aperture, from which the head and foot extend (Fig. 12-22, 12-31A,B), and they resemble monoplacophorans. Most limpets evolved from coiled ancestors and became secondarily symmetrical, but the uncoiled condition of patellogastropod limpets may be primary.

PATELLOGASTROPODA[O]

Also called Docoglossa. The largest assemblage of limpets. Found from shallow water to the deep sea; many are common in the rocky intertidal areas. Evolved independently of keyhole limpets. Shell a noncoiled, symmetrical cone; symmetry may be primary. Perhaps the most primitive living gastropods. Includes Patellidae (*Cellana, Helcion, Patella;* Fig. 12-24), Acmaeidae *(Pectura = Acmaea),* Lottidae *(Lottia, Notoacmaea),* and Nacellidae *(Nacella).*

VETIGASTROPODA^O

Archaeogastropods *sensu lato.* Primitive diotocardians, two nephridia and bipectinate gills. Right gill may be reduced or absent, osphradium is a simple ridge. Fissurellidae (keyhole limpets): secondarily symmetrical conical shell with a single anal aperture and two bipectinate gills; Fig. 12-22; *Diodora, Emarginula, Fissurella.* Pleurotomariidae (slit snails): shell with an anal slit, two gills; Fig. 12-21C; *Entemnotrochus, Pleorotrochus, Pleurotomaria.* Haliotidae (abalones): shell with a row of anal pores opening from the mantle cavity, two gills; Fig. 12-21A,B; *Haliotis.* Trochidae (top snails; Fig. 12-23A) and Turbinidae (turban snails; Fig. 12-23B): both families have coiled, conical shells, an operculum; Only a bipectinate left gill, two nephridia, and two atria; some trochids: *Calliostoma, Gibbula, Margarites, Monodonta, Tegula, Trochus;* some turbinids: *Astraea, Turbo.*

NERITIMORPHA^O

Includes many common, rocky intertidal species, such as the semitropical and tropical species of *Nerita.* Diotocardian, but have either a single monopectinate gill or no gill. Eye with lens, penis present. Some have invaded fresh water *(Theodoxus)* and, perhaps from some freshwater ancestor, there arose a family of tropical land snails, the Helicinidae. Neritidae (nerites; Fig. 12-23C): globose operculate snails with one nephridium and a single bipectinate gill, no gill, or secondary gills. Reproductive system complex. *Nerita, Neritina, Theodoxus.*

CAENOGASTROPODA^O

Includes the mesogastropods and neogastropods and thus most living prosobranchs. The great adaptive diversity of these two grades is in part due to the evolution of the monopectinate gill, which may have facilitated expansion from the ancestral rocky coasts into silty, especially soft-bottom, habitats. The dorsal and ventral membranes (Fig. 12-1A) that suspend the archaeogastropod bipectinate gill are easily fouled by sediment carried by the respiratory current. In the mesogastropods and neogastropods the membranous suspension has been lost and the gill axis is attached directly to the mantle wall (Fig. 12-20E). The filaments on the side of the attachment have disappeared, leaving those of the opposite side to project into the mantle cavity. The loss of filaments from one side of the archaeogastropod bipectinate gill results in the monopectinate gill characteristic of the meso- and neogastropods. A well-developed inhalant siphon is typical of many meso- and neo-gastropods.

Mesogastropods (formerly Mesogastropoda^sO)

Snails with a single monopectinate gill, one atrium (monotocardian), one nephridium. Osphradium simple (ridgelike). Reproductive system is complex, usually with a penis. Radula is taenioglossate, seven teeth in each transverse row (Fig. 12-40C). Chiefly marine, but with many freshwater and terrestrial taxa. Browsers, detritivores, suspension-feeders, parasites, and carnivores.

Cerithioidea^SF: Marine and freshwater snails with high spires; Turritellidae (turret snails; *Turritella*), Batillariidae *(Batillaria),* Cerithiidae (ceriths; *Bittium, Cerithium*), Siliquariidae *(Siliquaria),* Pleuroceridae *(Elimia, Pleurocera).*

Vermetoidea^SF: Worm snails, with long, loosely, irregularly coiled shells and no operculum; Fig. 12-33; Vermetidae *(Dendropoma, Macrophragma, Petaloconchus, Serpulorbis, Vermetus).*

Littorinoidea^SF: Littorinidae (intertidal periwinkles; *Lacuna, Littorina, Tectarius*).

Rissoidea^SF: Large assemblage of several families of small marine, freshwater, and terrestrial snails, including Caecidae *(Caecum),* Hydrobiidae *(Hydrobia),* Truncatellidae *(Truncatella),* Vitrinellidae *(Cochliolepis, Cyclostremiscus, Vitrinella).*

Stromboidea^SF: Mostly large gastropods having shells with siphonal canals and a flaring lip; Fig. 12-32B; Strombidae (conchs; *Lambis, Strombus, Tibia*), Aporrhaidae (pelican-foot snails; *Aporrhais*), Struthiolariidae (ostrich-foot snails, *Struthiolaria*).

Xenophoroidea^SF: Most attach stones or shells to the shell (Fig. 12-38C). Xenophoridae (carrier snails; *Xenophora*).

Calyptraeoidea^SF: Filter-feeding, protandric snails with mostly caplike or limpetlike shells; Calyptraeidae (slipper snails and shelf limpets; *Calyptraea, Crepidula;* Fig. 12-49, 12-57D), Capulidae (cap snails; *Capulus*).

Cypraeoidea^SF: Spire enclosed within the last whorl of the shell; Cypraeidae (cowries; *Cypraea;* Fig. 12-29), Ovulidae (egg snails; *Cyphoma, Simnia*).

Velutinoidea^SF: Velutinidae *(Velutina),* Triviidae, Fig. 12-46B (trivias; *Trivia, Erato*).

Naticoidea^SF: Burrowing moon snails with globose shells and a drilling radula; Fig. 12-32A (Naticidae; *Natica, Polinices, Sinum*).

Tonnoidea^SF: Marine snails with heavy, often large shells; Cassidae (helmet snails; *Cassidarius, Cassis,* bonnets; *Phalium*), Bursidae (frog snails; *Bofonaria*), Ranellidae (tritons; *Cymatium*), Tonnidae (tuns; *Tonna, Malea*), Ficidae (fig snails; *Ficus*).

Carinarioidea^SF (Heteropoda): Pelagic snails with finlike foot and reduced or absent shell; Atlantidae has well-developed shell *(Atlanta, Oxygyrus);* Carinariidae has small shell (*Carinaria,* Fig. 12-35D; *Pterosoma*); Pterotracheidae has no shell *(Pterotrachea).*

Epitonioidea^SF: Janthinidae, pelagic (violet snails, *Janthina,* Fig. 12-38A; *Recluzia*), Epitoniidae (wentletraps, *Epitonium, Opalia*).

Eulimoidea^SF: Predators, commensals, and echinoderm parasites; Eulimidae, Fig. 12-50B *(Eulima, Melanella, Stilifer);* Entoconchidae, Fig. 12-50C (sea cucumber endoparasites; *Enteroxenos, Entocolax, Entoconcha*).

Mesogastropods are well represented in fresh water as a result of several independent invasions. The majority of freshwater mesogastropods are tropical, but many taxa, such as *Elimia (Goniobasis), Pleurocera, Viviparus, Campeloma,* and *Valvata,* include temperate species. All are operculate, in contrast with the freshwater pulmonates. There are no freshwater or terrestrial neogastropods.

Cyclophoridae (Fig. 12-25A) and Pomatiasidae are two large families of mesogastropod land snails. Like the archaeogastropod Helicinidae, they are largely tropical and operculate and have no gill, and gas exchange occurs across a vascularized mantle wall within the mantle cavity (lung). A notch or a breathing tube in the shell aperture of some species permits air to enter when the operculum is closed.

Neogastropoda^sO

Snails that share many features with the mesogastropods but are entirely marine. Most are highly specialized carnivores. This is one of the largest and most successful molluscan taxa. Monotocardian with one nephridium, one monopectinate gill. Radula has no more than three teeth per row (Fig. 12-40D); specialized for predation, often used with an eversible proboscis (Fig. 12-14). Osphradium is well developed, bipectinate, gill-like. Many have poison glands. Stomach is a simple sac; gut includes gland of Leiblein, valve of Leiblein, anal gland. Reproductive system is complex. Some are large. Neogastropods have about twice as many chromosomes as meso- or archaeogastropods, suggesting the possibility of an origin by polypoidy. Neogastropods are the most derived prosobranchs. Three higher taxa:

Muricoidea^SF: Radula with three teeth. Heavy, conical, sculptured shells with long siphonal canal; Muricidae (drills; *Coralliophila* [Fig.12-38B], *Eupleura, Morula, Murex, Nucella, Purpura, Rapana, Thais* [Fig. 12-31A], *Urosalpinx* Fig. 12-27B), Buccinidae (*Buccinum* [Fig. 12-46A], *Cominella*), Melongenidae (whelks; *Busycon, Melongena*), Columbellidae (dove snails; *Anachis, Columbella, Mitrella*), Fasciolariidae (tulip snails; *Fasciolaria*), Marginellidae (marginellas; *Marginella*), Mitridae (miter snails; *Mitra, Vexillum*), Nassariidae (mud snails; *Ilyanassa* [Fig. 12-48A], *Nassarius*), Olividae (olive snails; *Oliva, Olivella*), Harpidae (harp snails; *Harpa*), Volutidae (volutes; *Cymbium, Voluta*), Turbinellidae *(Turbinella)*.

Cancellarioidea^SF: Radula has only the central tooth, which is used to suck fluid from prey; Cancellariidae (nutmeg snails; *Admete, Cancellaria, Sveltia*).

Conoidea^SF: Predatory snails with poison gland and highly modified toxoglossate radula having only one tooth per row (or radula absent). Conidae (cone snails; *Conus*, Fig. 12-47C), Turridae, the largest mollusc family, with several thousand species (turret snails; *Mangelia, Polystira, Turris*), Terebridae *(Terebra)*.

HETEROSTROPHA^O

Snails once were thought to be opisthobranchs, but now are believed to be closer to prosobranchs. Includes the important **Pyramidelloidea^SF**, a large taxon of small carnivorous ectoparasites of bivalve molluscs and polychaetes (Fig. 12-50A). Shell is well developed, coiled, and multiwhorled, with operculum. Proboscis is long, with a stylet instead of a radula. A true radula and gill are absent.

Opisthobranchs (formerly Opisthobranchia^sC)

The remaining two groups of gastropods, the opisthobranchs and pulmonates, are sometimes combined in Euthyneura, which is marked by the replacement of the twisted (streptoneurous) prosobranch nervous system with a secondarily bilaterally symmetrical (euthyneurous) system. In the euthyneuran nervous system, the effects of torsion are reversed by detorsion and the ganglia are highly concentrated and cephalized (moved close together in the head).

The two euthyneuran groups are thought to have arisen from ancestral mesogastropod prosobranchs and may share a common ancestor, which would have been torted and streptoneurous and monotocardian with a single monopectinate gill, epibenthic, and marine, and have had a spirally coiled shell. The two lines of descent took advantage of environments poorly or not at all exploited by prosobranchs. One, the pulmonates, developed a lung and radiated in terrestrial, freshwater, and intertidal marine habitats where the advantages of aerial respiration could be exploited. The other, the opisthobranchs, initially exploited burrowing marine habitats made habitable by reduction of the shell and detorsion, and they later expanded to other marine, including pelagic, habitats. In both groups there are tendencies to reduce the shell, reverse torsion, and become secondarily bilaterally symmetrical. Both groups are hermaphroditic and most lack an operculum. The sister taxon of Euthyneura is Streptoneura (with a twisted nervous system), which is roughly equivalent to the prosobranchs.

The original opisthobranchs diverged into several distinct lines, often exhibiting parallel or convergent evolution with each other. Opisthobranchs are gastropods that have specialized in the reduction and loss of the shell and radiated into niches suitable for shell-less gastropods. Opisthobranchs are polyphyletic, and it is very difficult to distinguish homology from homoplasy within the group. As a result, it has so far been impossible to construct a natural classification. Opisthobranchs explored several interesting evolutionary pathways made possible by loss or reduction of the shell (burrowing, swimming, chemical defenses). The 3000-some species are almost entirely marine, and the majority are benthic. They vary in size from tiny interstitial species to the largest sea hares reaching 60 cm in length. Some of the most colorful, graceful, and beautiful molluscs are opisthobranchs.

Opisthobranchs probably arose from a torted, coiled, asymmetrical, monotocardian, prosobranch ancestor with a single gill in an anterior mantle cavity. Primitive opisthobranchs resemble mesogastropod snails (Fig. 12-34A) and are difficult to distinguish from prosobranchs. They are asymmetrical and have a more or less typical spirally coiled, prosobranch-like shell. An operculum is present in primitive species and in many larvae, but is lacking in most adult opisthobranchs. Common trends, however, are a tendency toward shell reduction and loss, reduction of the mantle cavity and the accompanying loss of the original gill, detorsion, loss of coiling and development of secondary bilateral symmetry, appropriation and reuse of organelles such as nematocysts and chloroplasts from the prey, and the development of secondary defenses to compensate for the loss of the shell. Opisthobranch taxa are noted for their extensive parallel evolution, and each of these trends has occurred more than once. The shell may be well developed, with thick calcareous layers and a high coiled spire; thin and bubblelike, with a vestigial spire; a flat or gently curved internal or external plate; or absent. All stages of detorsion occur, including fully torted, partially torted, and fully detorted, but most are partially detorted. Accordingly, the mantle cavity may be anterior, on the right (Fig. 12-36A), or posterior. Many opisthobranchs have become sluglike.

Opisthobranchs, especially the sea slugs, have evolved a variety of defenses that do not require a heavy, inert shell, and each has evolved more than once. Escape swimming, often employing large, lateral expansions of the foot (**parapodia;** Fig. 12-36A), is common in the opisthobranchs. Many have skin glands that produce sulfuric acid or a nonacidic but noxious substance that repels potential predators, especially fishes. Some utilize nematocysts from cnidarian prey. Some have spicules embedded in the mantle. Many employ color patterns in their defense. Through **cryptic coloration,** many defenseless nudibranchs are colored and patterned to be nearly invisible when in the feeding position on the surface of their prey, often a sponge, alcyonarian coral, or bryozoan. Those with toxins or nematocysts are usually aposematically colored.

The opisthobranch gill, when present, is structurally unlike that of prosobranchs, and its surface area is provided by mantle folds, or plications (Fig. 12-36B), rather than by the lamellae characteristic of prosobranch gills. Nevertheless, prosobranch and most opisthobranch gills are probably homologous, with opisthobranchs presumably inheriting the gill from their prosobranch-like ancestors. Gills are lost in some opisthobranchs and in others they may be located on the body surface in the absence of a mantle cavity. The gills of some nudibranchs are secondary inventions derived independently of those of other opisthobranchs. Opisthobranchs are monotocardian with one atrium and one nephridium.

Primitive opisthobranchs have a prosobranch-like, streptoneurous nervous system, but in most there is a euthyneurous system with a strong tendency toward cephalization to the extent that the ganglia are often fused to form a brain. Most opisthobranchs have a second pair of sensory head tentacles called **rhinophores** that are located posterior to the first pair and sometimes surrounded at the base by a collarlike fold (Fig. 12-34A,G,H,J). A pair of statocysts is present. There is a tendency for the eyes to decrease in importance. Opisthobranchs are hermaphroditic and the larva is a veliger.

Cephalaspidea^{sO} (Bullomorpha, the bubble snails): Largest, most primitive taxon. The ancestors of most other opisthobranchs probably were cephalaspidean-like. In general, they are specialized for burrowing in soft sediments, and detorsion has moved the mantle cavity to the right or even the posterior end. The shell is usually reduced (as in *Bulla*), but primitive genera such as *Acteon* and *Hydatina* (Fig. 12-34A) more closely resemble their prosobranch ancestors and have not yet become fully adapted to burrowing. In *Acteon,* the most primitive known opisthobranch, the nervous system is still twisted (streptoneurous) and the shell is well developed, spiral, and closed by an operculum. More-derived bubble snails crawl on the surface or burrow into soft bottoms. Displacement of the mantle cavity to the right or the posterior end reduces the likelihood of its being fouled with sediment during burrowing. Some cephalaspidean families and genera are Ringiculidae *(Ringicula),* Acteonidae *(Acteon, Pupa, Rictaxis),* Hydatinidae (Hydatina; Fig. 12-34A), Bullidae *(Bulla),* Haminoeidae *(Haminoea, Atys),* Cylichnidae *(Acteocina, Cylichna, Scaphander,),* and Retusidae *(Volvulella, Cylichnina).*

Anaspidea^{sO} (sea hares): The largest (60 cm in length and 2 kg) opisthobranchs. Feed on seaweeds in shallow water. Shell reduced, buried in the mantle, or completely lost. Body secondarily and imperfectly bilaterally symmetrical (Fig. 12-36A). Mantle cavity and gill are present and the posterior edge of the mantle can be rolled to form an exhalant siphon. Some sea hares swim using rhythmic undulations of winglike lateral parapodia (Fig. 12-37). Some are capable of jet propulsion. When disturbed, many release a defensive purple ink derived from the pigments of the red algae on which they feed. Aplysiidae *(Aplysia, Bursatella, Dolabrifera),* Akeridae *(Akera).*

Thecosomata^{sO} (shelled pteropods): Pteropods (wing-foot) are two taxa of small, swimming opisthobranchs specialized for a pelagic life. Shelled pteropods have a shell (Fig. 12-34B), often have an operculum, and swim using large parapodia. Spiratellidae *(Limacina, Spiratella),* Cavoliniidae (*Cavolinia* [Fig. 12-34B], *Clio*), Cymbullidae (*Cymbulia, Gleba* [Fig. 12-48B]).

Gymnosomata^{sO} (naked pteropods): Shell-less carnivores that specialize in eating thecosomes. Gill is absent in most. Mantle cavity is absent. Gas exchange occurs across the general body surface. *Cliopsis, Pneumoderma.*

Nudibranchia^{sO} (sea slugs): Among the most spectacular, colorful, and beautiful molluscs. Shell, mantle cavity, and sometimes gills absent. Body secondarily bilaterally symmetrical and the anus posterior. Many bear numerous dorsal, fingerlike projections called **cerata** (Fig. 12-34G), which can be filamentous, club-shaped, branched (Fig. 12-34I), resemble a cluster of grapes (Fig. 12-34J), or exhibit variations on these themes. Each ceras contains a branch of the digestive cecum. Some nudibranchs lack cerata on the dorsum, but have a circle of gills surrounding a posterior anus (Fig. 12-34H). This, the largest and most heterogeneous opisthobranch taxon, includes several higher taxa.

Doridoidea^{SF}: Gills encircle anus on posterior dorsum; Fig. 12-34H; 26 families, including *Chromodoris, Dendrodoris, Doris, Glossodoris, Onchidoris.*

Dendronotoidea^{SF}: Simple or branched nematocyst-sequestering cerata, secondary gills; Fig. 12-34I,J; 10 families, including *Dendronotus, Doto, Hancockia, Scyllaea, Tethys, Tritonia.*

Arminoidea^{SF}: Platelike gills beneath mantle edge; nine families, including *Armina, Hero.*

Aeolidoidea^{SF}: Simple dorsal, nematocyst-sequestering cerata; Fig. 12-34G; 21 families, including *Aeolidia, Berghia, Coryphella, Cuthona, Eubranchus, Facelina, Fiona, Flabellina, Glaucus, Spurilla.*

Sacoglossa^{sO} (Ascoglossa): Herbivores that feed on cytoplasm sucked from macroalgae. Shell may or may not be present; a few (Juliidae: *Berthelinia, Julia*) have a bivalved, clamlike shell (Fig. 12-34E), but inside the two valves resides a typical snail. These are the only bivalved gastropods. Others are Elysiidae (Fig. 12-34F; *Elysia, Tridachia*), Stiligeridae *(Alderia, Hermaea, Stiliger).*

Acochlidiacea^{sO}: Small, shell-less, gill-less, interstitial opisthobranchs that live in coarse sediments, sometimes in low salinities or fresh water (Fig. 12-34C). Includes Acochlidiidae *(Acochlidium, Microhedyle).*

Notaspidea^{sO} (Pleurobrancheomorpha): Diverse group of epifaunal carnivores, many of which feed on sponges or sea squirts (Fig. 12-34D). Shell may be present or absent, internal or external. Bilateral symmetry is superficial, a bipectinate gill is present, and the rhinophores have a longitudinal slit. Includes Umbraculidae *(Umbraculum),* Tylodinidae *(Tylodina),* Pleurobranchidae (Fig. 12-34D; *Berthella, Berthellina, Pleurobranchus).*

Pulmonates (formerly Pulmonata[sC])

The pulmonates contain the highly successful land snails, as well as many freshwater forms and a few intertidal marine species. Most freshwater snails and limpets and most terrestrial snails and slugs are pulmonates. Estimates of the number of described species vary widely from 16,000 to 30,000, but there is no doubting the success of this taxon. Pulmonates are not as diverse or heterogeneous as the prosobranchs, but they have prospered in habitats where competition from prosobranchs is reduced. They are widely distributed in both tropical and temperate regions throughout the world. The most primitive pulmonates are aquatic, both marine and freshwater, but the derived pulmonates are terrestrial and the grade has specialized in ecological niches open to air-breathing gastropods. The aquatic species, being air breathers, are largely restricted to shallow water and the marine species are usually intertidal or supratidal.

Pulmonates have the same tendency toward partial detorsion as the opisthobranchs and in most, the mantle cavity has moved to the right side. The most distinctive pulmonate feature, and the one from which the group takes its name, is the loss of the gill and conversion of the mantle cavity into a lung. The mantle skirt grows to the animal's back (notum) to completely enclose the mantle cavity except for a small opening on the right side called the pneumostome (Fig. 12-28A,E). The roof of the mantle cavity is highly vascularized. Passive ventilation by diffusion of gases through the open pneumostome is an important respiratory mechanism, especially for small, inactive species. Active ventilation is accomplished by elevating and depressing the floor of the lung to create a tidal flow of air through the pneumostome.

Pulmonates are monotocardian, with one atrium and one nephridium. The typical pulmonate shell is spirally coiled and lacks an operculum. In some, the shell is lost entirely (as in slugs), in a few it is a small external plate, and in some it is conical and limpet-shaped. Many spiral shells secondarily approximate a planispiral condition (Fig. 12-28B). Although mucus is important in the biology of all animals, it is especially so in pulmonates. It is important in both ciliary and muscular locomotion, where it lubricates the passage of the foot over the substratum. It is a defense against bacterial infection and against predators, and it is used to avoid desiccation. Many terrestrial snails estivate during hot, dry weather, becoming quiescent and metabolically inactive. The snail withdraws into the shell and secretes a mucous plug that functions as an operculum to reduce water loss. Because most are freshwater and terrestrial animals, osmoregulation and water balance are important physiological concerns for pulmonates. Terrestrial species have sensory and neural equipment for detecting and responding appropriately to humidity gradients. Most pulmonates are herbivores, and some are agricultural pests. The nervous system is euthyneurous. Pulmonates, like opisthobranchs, are hermaphroditic with internal fertilization and thus have complicated reproductive systems. Development is usually direct.

The first pulmonate land snails appeared in the Carboniferous period, but their origin is obscure. They probably evolved from some group of operculate prosobranchs that had a single gill, and they may share a common ancestor with the opisthobranchs. The pulmonate line of euthyneurans could have evolved from ancestral forms that inhabited estuarine marshes and mud flats. The pulmonate condition may have evolved as a means of gas exchange when the animals were confined to small, stagnant puddles or to wet but exposed surfaces out of the water, conditions similar to those postulated for the origin of amphibians.

Basommatophora[sO] (lower pulmonates): Have one pair of tentacles with an eye situated laterally at the base of each (Fig. 12-28F). Aquatic. Includes all marine and freshwater pulmonates.

There are a few primitive marine pulmonates, all of which live at the edge of the sea on intertidal rocks or in estuarine habitats. Tropical marine limpets of the genus *Siphonaria* (Siphonariidae) and the temperate *Melampus* and *Detracia* (Melampidae) of salt marshes and drift rows are among the few pulmonates that have a veliger larva. The presence of a veliger in primitive pulmonates indicates that the marine habitat is the original pulmonate home, rather than having been adopted secondarily by terrestrial ancestors.

Amphibola (Amphibolidae), another marine pulmonate, has a typical shell but is unusual in having an operculum, indicating that the ancestral pulmonate was operculate, as was the ancestral opisthobranch. In all other pulmonates the operculum is lost during the course of development. All freshwater and terrestrial pulmonates lack an operculum, but prosobranchs living in these habitats are always operculate, making it easy to distinguish between them. Unfortunately, for identification purposes, many marine prosobranchs have lost the operculum, so the criterion is useful only with freshwater snails.

Many freshwater basommatophorans, such as the cosmopolitan *Lymnaea* (Lymnaeidae) and *Physa* (Physidae), come to the surface to obtain air for gas exchange. In *Lymnaea,* the edges of the mantle cavity can be extended as a long, tubular snorkel for this purpose. The snail can remain submerged for up to an hour with the pneumostome closed.

Some deep-lake lymnaeids that live at depths that prohibit periodic respiratory journeys to the surface have secondarily converted the lung back into a water-filled mantle cavity. Some freshwater pulmonates, such as planorbids (Planorbidae, including *Armiger, Helisoma, Planorbis*) and ancylids, develop secondary gills (pseudobranchs) from folds of the mantle near the pneumostome. Ancylids (such as *Ancylus*) are limpets adapted for life in fast-running streams.

Basommatophorans usually have a well-developed, calcareous shell. In most it is an asymmetric spiral, but in the ancylids (freshwater limpets) it is a symmetrical cone. The conical limpet shape has appeared many times in molluscan evolution, beginning with the monoplacophorans. Locomotion in aquatic pulmonates is usually via ciliary gliding using the cilia on the sole of the foot.

Some species of freshwater pulmonate snails are important hosts for human parasites. The African taxon *Bulinus,* for example, is the intermediate host for the trematodes that cause schistosomiasis.

Stylommatophora[sO] (higher pulmonates): Including almost all terrestrial species, this is a far larger taxon than the aquatic Basommatophora. Two pairs of tentacles (Fig.

12-28E), although the anterior, lower pair may be inconspicuous. The two eyes are located at the distal tips of the posterior, upper pair.

Most terrestrial pulmonates have a calcareous shell, although it is usually not as heavily calcified as those of marine snails. The degree of calcification varies with the availability of calcium in the soil and diet. In general, the distribution of terrestrial and freshwater molluscs tends to be related to the availability of calcium. Calcium is always available in seawater but is often scarce, and frequently limiting, in terrestrial and freshwater habitats. The periostracum protects the calcareous layers from dissolution by humic acid and may also function as a water repellent.

In many small species the shell aperture is partially occluded by large teeth or ridges, which exclude such predators as insects but allow the soft body of the snail to protrude. The largest shells, 23 cm in height, belong to members of the African species *Achatina fulica,* but the South American strophocheilids (Fig. 12-28D) are also large, reaching 15 cm in height. Many species of land snails have shells that measure less than 1 cm. Although such species are found throughout the world in leaf mold and beneath bark and stones, they are especially abundant on oceanic islands in the Pacific.

Shell reduction or loss has occurred independently a number of times within the higher pulmonates (as in Stylommatophora), and such naked, symmetrical species are called **slugs** *(Arion, Limax, Philomycus),* as are similarly shaped opisthobranchs (Fig. 12-28A,C,E). Like the limpet, the slug shape has appeared several times in molluscan evolution. Slugs are elongate and superficially bilaterally symmetrical. The shell is generally absent or reduced and buried within the mantle, but in *Testacella* a small, external shell is perched on the posterior notum (Fig. 12-28C).

The slug pneumostome usually is a conspicuous opening on the right side of the body (Fig. 12-28A). The evolution of the shell-less terrestrial slugs is perhaps an adaptive response to a low availability of calcium. The original centers of slug distribution are restricted to areas of high humidity and low soil calcium, but slugs have now been introduced into many parts of the world from which they were originally absent. Locomotion in terrestrial species is via rhythmic contractions of muscles in the broad sole of the foot. Not all terrestrial snails are pulmonates. Prosobranch land snails, although a smaller group (4000 species of terrestrial prosobranchs, compared with some 20,000 species of terrestrial pulmonates), are very common in the tropics, and their adaptations for life on land parallel those of pulmonates in many ways. Unlike pulmonates, they are operculate.

CEPHALOPODA[C]

Cephalopoda is an ancient, highly specialized taxon of about 700 extant and 10,000 extinct species, including the chambered nautilus, cuttlefish, squids, and octopods as well as the fossil nautiloids, ammonoids, and belemnoids. Cephalopods are entirely carnivorous and most are swift, active, agile predators (Fig. 12-61). They occupy ecological niches for fast pelagic carnivores similar to those occupied by fishes, with which they compete and share many convergent characteristics.

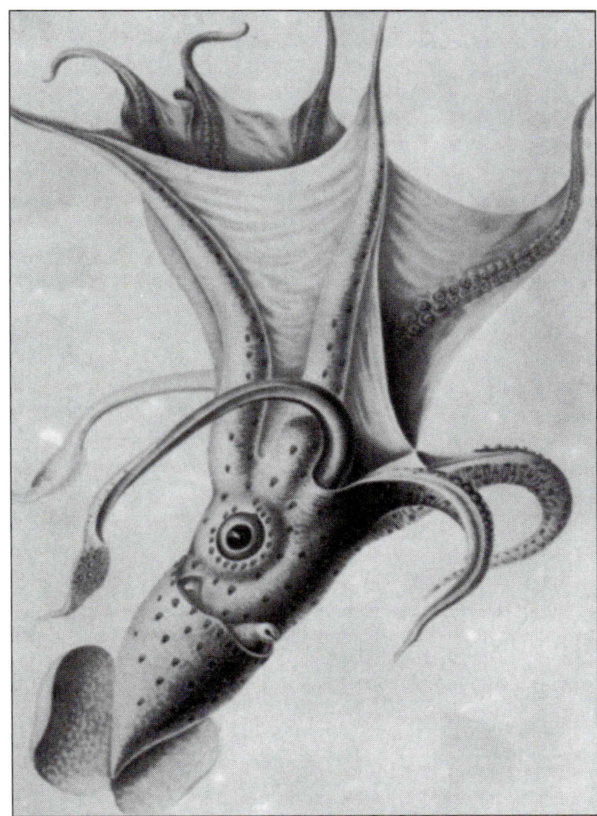

FIGURE 12-61 The bathypelagic squid *Histioteuthis bonelli,* a squid with webbed arms. The body is covered with photophores. *(After Chun from Lane, F. W. 1960. Kingdom of the Octopus. Sheridan House, New York.)*

All cephalopods are marine and most are pelagic. Although some cephalopods, such as *Octopus,* have secondarily assumed a bottom-dwelling habit, the class as a whole is adapted for swimming, suspended in the water column by one of several buoyancy compensation mechanisms. Cephalopods are easily the most active molluscs and have highly developed nervous, sensory, and locomotory systems.

FORM

Major Axes and Orientation

Cephalopods are a homogenous taxon whose members share a common, immediately recognizable body plan. Cephalopods arose from limpetlike monoplacophoran ancestors with a broad, creeping ventral foot and a high dorsal shell. Early in their evolution, cephalopods elongated the dorsoventral axis to make it the major body axis, replacing the traditional anterior-posterior axis (Fig. 12-62A). In the cephalopods, head and foot are at one end of an elongate body. That end, while

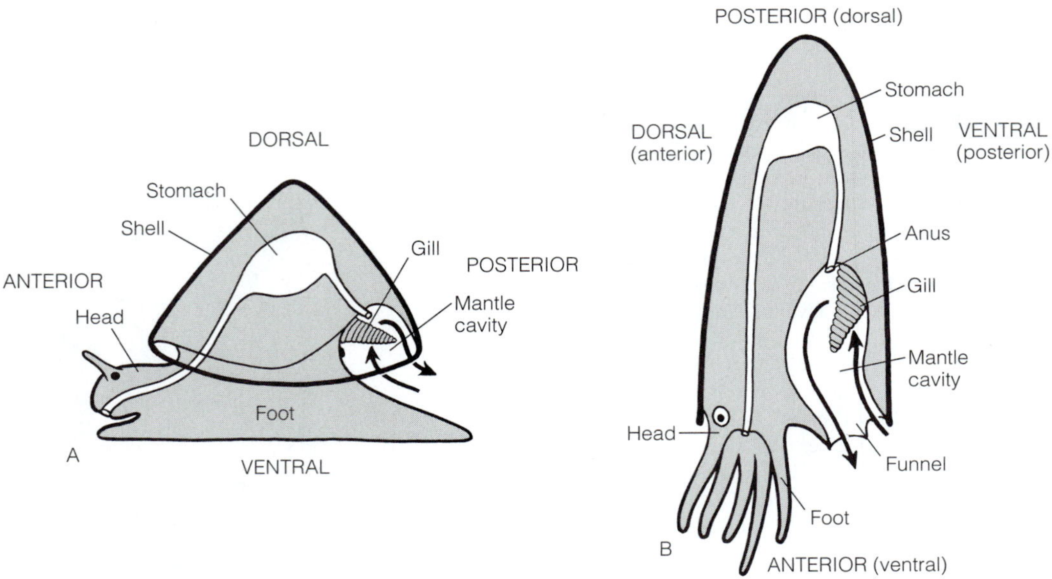

FIGURE 12-62 Reorganization of major body axes in cephalopods. **A,** A monoplacophoran. **B,** A generalized early cephalopod. Accompanying the lengthening of the dorsoventral axis is a shift in the primary functional axis of the cephalopod body. The dorsoventral axis of the ancestor lengthens and becomes the functional anterior-posterior axis of the cephalopod. The former anterior-posterior axis has shortened and becomes the functional dorsoventral axis. Right and left remain unchanged. In both drawings, morphological orientation is noted in lowercased words, functional orientation in uppercase. *(A, Redrawn from Fretter, V., and Graham, A., 1994. British Prosobranch Molluscs. Second Edition, Ray Society Vol. 161. Ray Society, London.)*

morphologically ventral, is functionally anterior and leads the animal into new environments. This is a potential source of confusion for those who, reasonably enough, expect an animal's head to be at its anterior (rather than ventral) end. To minimize confusion, it is customary to rename the major axes on the basis of function, rather than morphology, and that custom will be followed here. Ventral therefore becomes anterior, dorsal becomes posterior, anterior becomes dorsal, and posterior becomes ventral. Right and left are not affected (Fig. 12-62B).

Body Regions

Like that of other molluscs, the cephalopod body consists of a foot, head, and visceral mass. There is also a mantle, mantle cavity, radula, paired gills, and usually a reduced shell. The head, at the functional anterior end, includes the mouth, anterior gut, radula, and brain (Fig. 12-63). Dorsally, the foot has been transformed into a set of flexible, usually suckered prehensile **appendages** (arms or tentacles) that surround the mouth (Fig. 12-64). This close association of head and foot is responsible for the name cephalopod (= head-foot). A ventral region of the foot forms a tubular siphon, or **funnel,** exiting the mantle cavity (Fig. 12-63, 12-64). The visceral mass is elongated to form a conical posterior cone known as the **visceral hump** (Fig. 12-64). The mantle cavity is anteroventral and contains a pair of gills (Fig. 12-64). The mantle walls are heavily muscularized. The shell of all living cephalopods except *Nautilus* (Fig. 12-65) is internal and embedded in the mantle (Fig. 12-63, 12-64) or absent.

Cephalopods have attained the greatest size of any invertebrate and the largest of them rival the largest vertebrates. Although the majority range from 6 to 70 cm in length, including the arms and tentacles, some species reach giant proportions. The largest cephalopods are the giant squids, *Architeuthis* (Fig. 12-69B), which probably achieve a length of 20 m, including the tentacles. This is about the length of a sperm whale, but considerably shorter than a blue whale, at over 30 m. Giant octopods with arms 10 to 15 m long have been reported by divers in the Sea of Japan, but no specimens have been collected. The dorsal mantle length of the giant Pacific coast octopus *Enteroctopus dofleini*, one of the largest known octopods, does not usually exceed 36 cm, although its rather slender arms may be five times the body length. The record is a specimen with a 9.6 m arm span. The squid *Idiosepius pygmaeus*, at 1 cm, is the smallest cephalopod.

Cephalopoda first appeared in the Cambrian period as molluscs became adapted for life in the pelagic habitat. Twice, once during the Paleozoic era and once in the Mesozoic, they underwent episodes of adaptive radiation resulting in the formation of many species.

SHELL

Diversity and Evolution

The origin and evolution of Cephalopoda centers on the development of the shell as a buoyancy compensation device and its subsequent replacement by other buoyancy adjusting mecha-

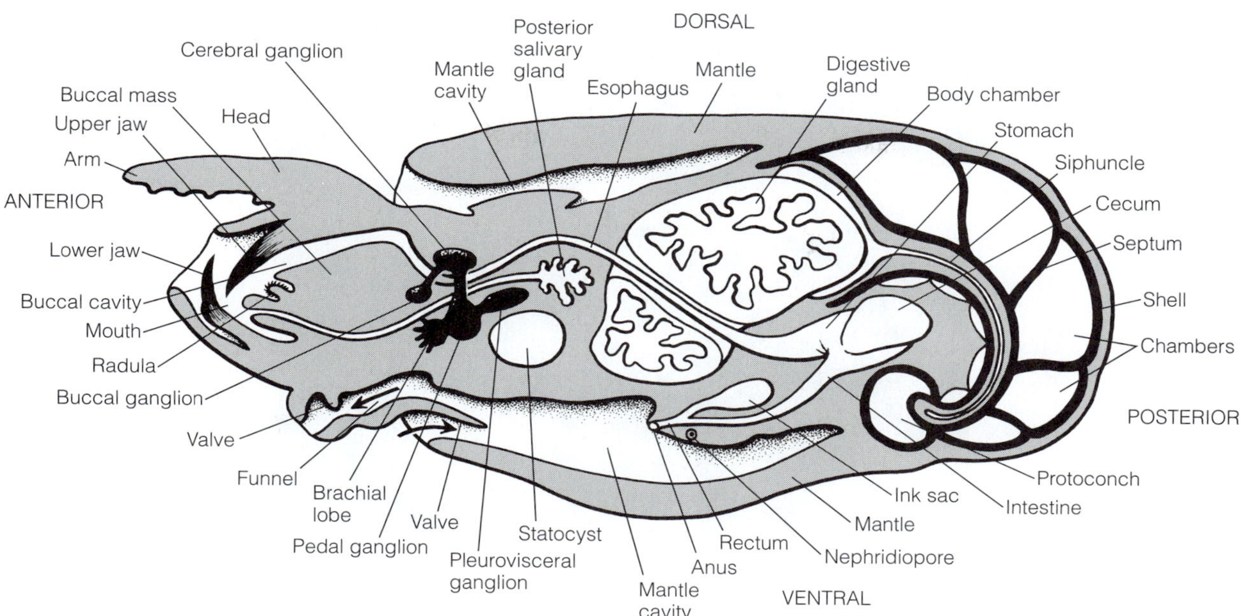

FIGURE 12-63 Sagittal section of *Spirula* showing the major anatomical features of the cephalopods. Compare the direction of shell coiling with that of *Nautilus.* The shell of *Spirula* is endogastric and internal. The body chamber of *Spirula* contains only a small portion of the visceral mass and the remainder of the animal is outside the shell. Scale bar = 5 mm. *(Redrawn and modified from Budelmann, B. U., Schipp, R. and von Boletzky, S. 1997. Cephalopoda. In Harrison, F. W., and Kohn, A. J. (Eds.): Microscopic Anatomy of Invertebrates. Vol. 6A: Mollusca II. Wiley-Liss, New York. pp. 119–414, after Chun.)*

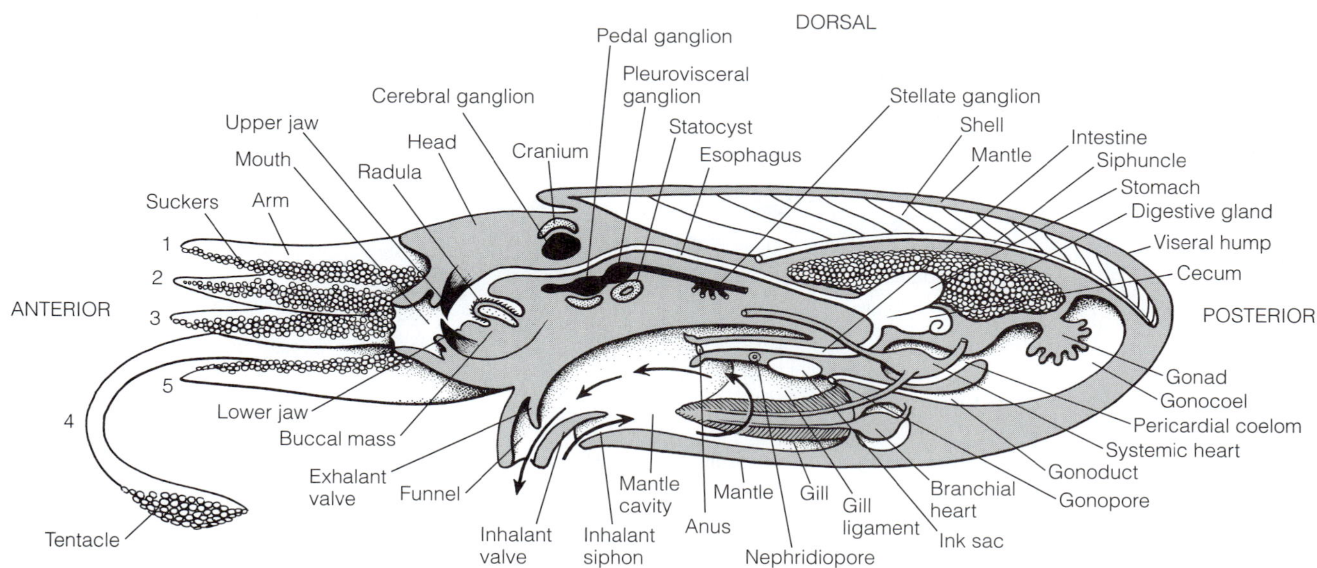

FIGURE 12-64 Sagittal section of a generalized coleoid cephalopod. Blood vessels have been truncated for clarity. *(Redrawn and modified from several authors including Budelmann, B. U., Schipp, R. and von Boletzky, S. 1997. Cephalopoda. In Harrison, F. W., and Kohn, A. J. (Eds.): Microscopic Anatomy of Invertebrates. Vol. 6A: Mollusca II. Wiley-Liss, New York. pp. 119–414, after Chun; and Pearse et al. 1987)*

nisms. The presence of a mechanism that permits an organism to be neutrally buoyant frees it from the need to expend energy to remain at any desired level in the water column. This ability is a prerequisite for any successful invasion of pelagic habitats. Based on shell characteristics, Cephalopoda is divided into two large groups. One, the **ectocochleate** cephalopods, is characterized by a well-developed, calcareous external shell. The ancestral cephalopods belonged to this group but almost all are now extinct. The other group, the **endocochleate** cephalopods, is characterized by a reduced

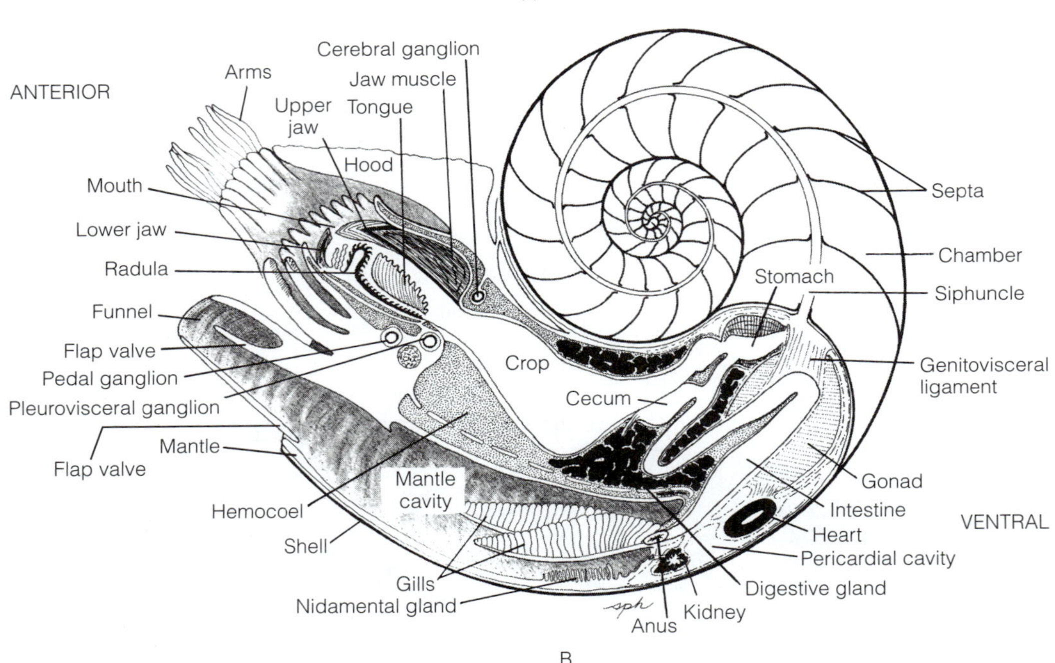

FIGURE 12-65 *Nautilus.* **A**, Side view in swimming position. **B**, Sagittal section. *Nautilus* is coiled exogastrically, over the head. *(B, After Stenzel.)*

internal or absent shell and comprises a single taxon, Coleoidea. At present, coleoids are the only successful taxon, representing almost all living cephalopods.

ECTOCOCHLEATE CEPHALOPODS

Two major taxa of ectocochleate cephalopods are known, **Nautiloidea** and **Ammonoidea.** Nautiloids first appeared in the Cambrian and would be extinct today were it not for four species of *Nautilus* living in the Indo-Pacific region. Ammonoidea appeared in the Silurian period and became extinct at the end of the Cretaceous. The earliest known cephalopod is the small (1 cm) Upper Cambrian *Plectronoceras,* a nautiloid with a slightly curved, conical, septate shell. The shell lengthened in its descendants to a long cone.

The early cephalopods are known only from their fossilized shells, and consequently what is known of their morphology and evolution, as well as their classification, is based on shell characteristics. We know little of the soft parts of fossil ammonoids and nautiloids, but assume they were similar to the living species of *Nautilus* (Fig. 12-65). Ammonoids and nautiloids both had well-developed external shells and left abundant fossil records.

The ancestral cephalopod shell is exemplified by that of the ammonoids and nautiloids. The shell of a primitive cephalopod, such as *Nautilus,* consists of three layers. The outermost is a thin organic periostracum. Under the periostracum are two calcareous layers of aragonite crystals deposited on an organic matrix. The outermost calcareous layer is the prismatic layer, which contains aragonite in prism-shaped crystals. Innermost is the nacreous layer composed of tabular aragonite crystals.

In both taxa the shell is a long, septate cone partitioned by transverse, calcareous **septa** into successively larger chambers (or cameras) (Fig. 12-65B, 12-66), with the oldest and smallest chambers being at the posterior end. The animal occupies the newest, largest, anteriormost region of the shell, known as the **body chamber.** The anterior end of the animal, bearing the head and tentacles, extends from the opening, or aperture, at the large (anterior) end of the shell.

Growth of the body necessitates an increase in the size of the body chamber that houses it. The shell increases in length as new shell material is secreted by the mantle margin around the shell aperture. As the body chamber lengthens, the animal slowly shifts anteriorly, leaving the posterior region of the chamber vacant. Fluid fills this space as the animal vacates it. The forward migration of the animal necessitates the gradual relocation of muscle insertions and that the continuity of the siphuncle be maintained. Eventually, the mantle secretes a new septum to wall off the fluid-filled posterior portion of the body chamber and create a new chamber.

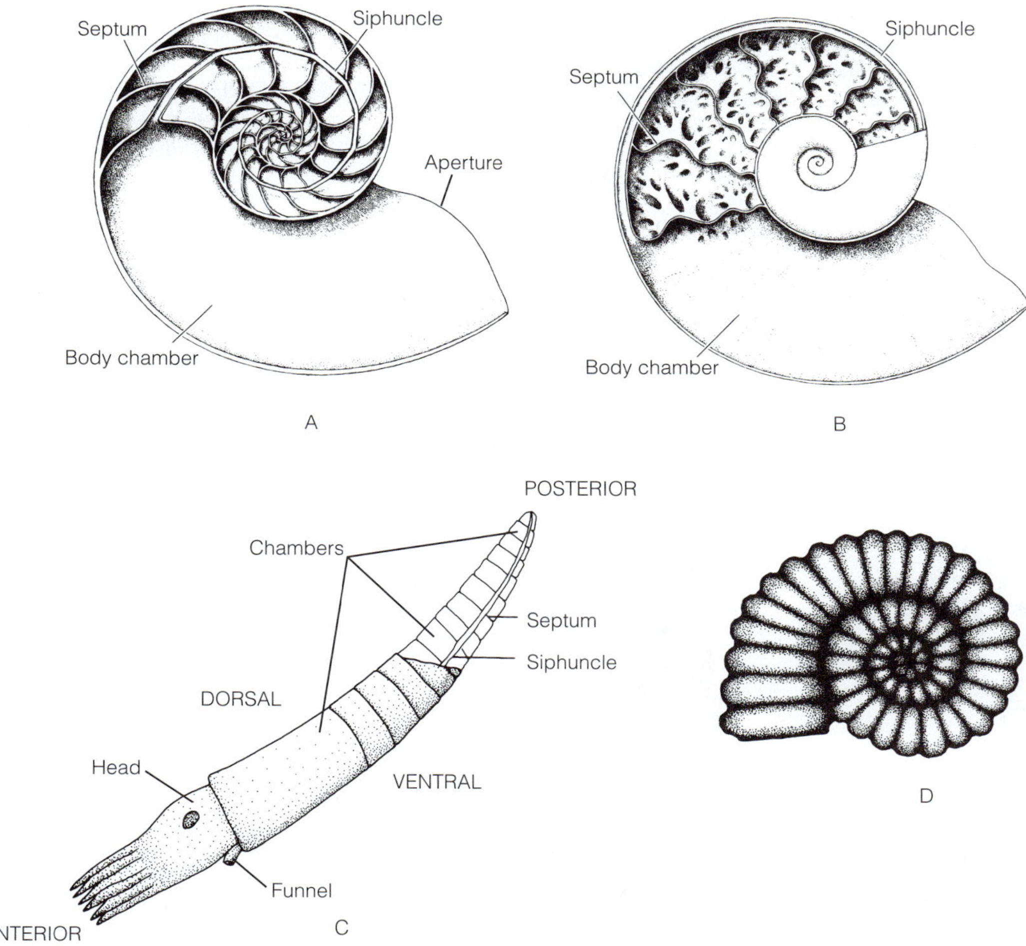

FIGURE 12-66 Fossil cephalopod shells. **A,** Inner view of a coiled nautiloid shell, showing the central position of the siphuncle within the simple curved septa. **B,** View of the interior of an ammonoid shell. Siphuncle passes along the outer wall and the septal junctions with the outer wall of the shell are complex. **C,** A straight-cone nautiloid with a slight exogastric curve. The shell has been cut away to show the septa and siphuncle. **D,** The planispiral ammonoid *Douvilleiceras. (A and B, From Ward, P. 1983. The extinction of the ammonites. Sci. Am. 249(4):136–147; D, From Mutch, A.)*

The septa are perforated near the middle or at the edge. A strand of tissue enclosed in a delicate calcareous tube, together forming the **siphuncle,** extends posteriorly from the visceral mass and passes through the septal perforations to reach the upper end of the shell (Fig. 12-65B). The siphuncle functions as an osmotic pump to remove liquid from the chambers and replace it with gas. The shell chambers and siphuncle form a buoyancy compensation device that enables the animal to maintain neutral buoyancy. The presence of air-filled chambers in the shell is a cephalopod autapomorphy that separates them from all other molluscs.

Cephalopods probably evolved from monoplacophorans with high, conical, septate shells (Fig. 12-62A). The septations were probably unrelated to buoyancy, however, and presumably were present to strengthen the shell. The addition of a siphuncle that could secrete gas into empty chambers was an innovation made by the earliest cephalopods, such as *Plectronoceras.* Initially, such gas-filled chambers may only have aided in keeping the shell upright as the monoplacophoran-like animal moved about over the bottom. Swimming and permanent invasion of the pelagic habitat were probably later developments.

The first cephalopod shells are believed to have been gently curved cones from which arose several lineages with both straight (Fig. 12-66C) and coiled (Fig. 12-66A,B,D) shells. The straight shells of some species from the Ordovician period were more than 5 m in length, with an aperture 36 cm in diameter. Most coiled shells were planispirals (Fig. 12-66A,B,D), but one fossil taxon had asymmetrically or irregularly coiled shells (Fig. 12-67). The largest fossil species with a coiled shell was *Pachydiscus seppenradensis* from the Cretaceous period, which had a shell diameter of 3 m, but some fossil cephalopods were small species with shells only 3 cm in diameter.

Nautilus is the only living cephalopod with an external shell, which is well developed and similar to that of its ancestors. It is planispiral, with the spire coiled anteriorly above the head (Fig. 12-65), a condition known as **exogastric** coiling. The opposite, **endogastric** coiling, in which the shell coils posteriorly, away from the head, is characteristic of gastropods.

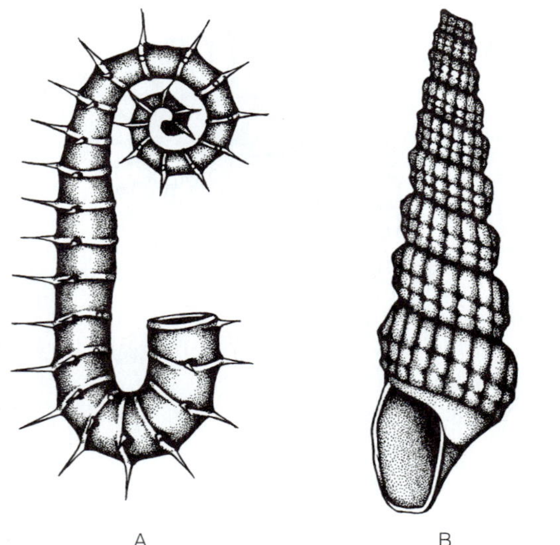

FIGURE 12-67 Two ammonoids with highly modified shells, believed to be planktonic. **A,** *Ancyloceras.* **B,** *Turrilites.* *(From Mutch, A.)*

Ammonoids and nautiloids are distinguished from each other on the basis of shell characteristics, since nothing is known of ammonoid soft anatomy. Nautiloid shells may be straight or coiled, whereas those of ammonoids are always coiled. The **sutures** formed by the junction of each transverse septum with the outer tubular shell wall are important in distinguishing ammonoids from nautiloids. The simplest suture lines were straight or slightly waved, and are characteristic of nautiloids such as *Nautilus* (Fig. 12-66A,C). The ammonoids, however, developed elaborate sutures that were zigzagged or minutely crinkled (Fig. 12-66B). Such sutures reflect a corresponding complexity in the nature of the septal junction and increased the strength of the shell to compensate for the somewhat thinner ammonoid shell. Furthermore, the nautiloid siphuncle perforates the septum near its center (Fig. 12-66A), whereas in ammonoids it is peripheral (Fig. 12-66B).

ENDOCOCHLEATE CEPHALOPODS

Except for the four species of *Nautilus,* all living cephalopods are endocochleate coleoids in which the shell is always internal and reduced or absent.

Coleoidea is a highly successful modern taxon whose members compete ecologically with the teleost fishes, which they resemble in size, habitat, ecology, anatomy, and behavior. A. Packard has observed that, functionally, cephalopods are fish. Both the teleosts and the coleoids began episodes of adaptive radiation in the mid-Mesozoic that continue today. The teleost radiation produced about 20,000 extant species, with many in fresh water, whereas the 600 Recent coleoids are all marine.

The ancestor of the coleoids and their close relatives the extinct belemnoids was probably nautiloid-like with a straight, external, septate shell that became completely enclosed by the mantle. Belemnoids and the ancestral coleoids had an internalized three-part shell. One part was the **phragmocone** (Fig. 12-68). This region was septate, equipped with a siphuncle, and used for buoyancy control. Another part was the heavy **rostrum,** which partially enclosed the phragmocone and supported it. Its weight probably adjusted the center of gravity to give the animal the desired horizontal orientation in the water. The third part of the shell, the **proostracum,** supported the mantle, served as a skeleton for muscle attachment, and protected the visceral mass.

From such ancestors arose four distinct evolutionary lineages, each leading to one or more Recent taxa (Fig. 12-68). In a line leading to **Spirulida^sO,** the rostrum and proostracum were lost. The phragmocone became coiled and was reduced, although still septate, and displaced to the posterior end of the visceral hump. The body itself is far larger than the body chamber and cannot be accommodated by it (Fig. 12-63). The siphuncle is peripheral (Fig. 12-63, 12-69B). Coiling is endogastric, opposite that of *Nautilus.*

In **Sepiida^sO** (cuttlefish), the second line, the septate phragmocone persists as the calcareous, dorsal **cuttlebone** (Fig. 12-69A). The other parts of the shell are absent or vestigial. The cuttlebone, which is underlaid by a heavily

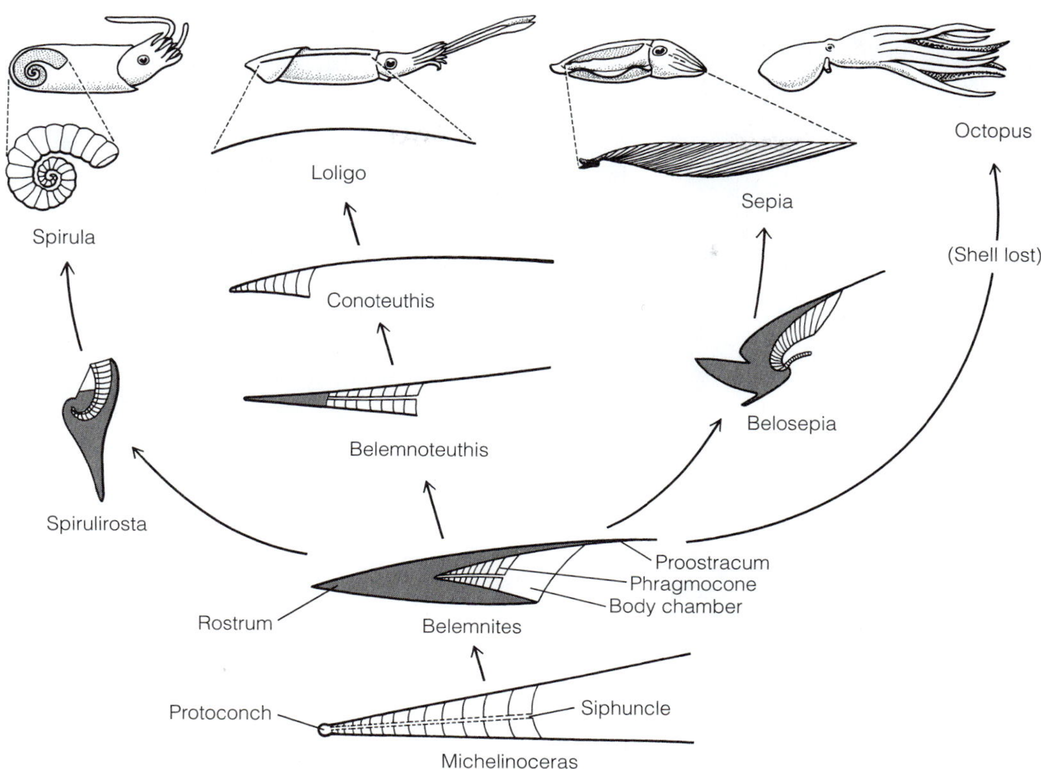

FIGURE 12-68 Evolution of Coleoidea and their shells. Coleoids arose from ancestors similar to belemnoids and radiated into four extant taxa. In Spirulida, represented by *Spirula,* the shell is coiled, septate, and posterior. In Teuthida, represented by *Loligo,* the shell is a straight, nonseptate, chitinous pen in the dorsal body. Sepiida, represented here by *Sepia,* have a straight, septate, calcareous, dorsal shell. Octopoda, represented by *Octopus,* have lost the shell. *(Redrawn after Shrock, R. R., and Twenhofel, W. H. 1953. Principles of Invertebrate Paleontology. McGraw-Hill, New York. 516 pp)*

vascularized epithelium that functions as a siphuncle, is used for buoyancy compensation and is a honeycomb of small, gas-filled spaces.

In the third line, **Teuthoidea[O]** (squids), the shell is reduced to a longitudinal chitinous **pen** (gladius) located dorsally along the midline, where it plays a role similar to the chordate notochord in stiffening the body and resisting longitudinal deformation. The pen is derived from the proostracum and is not septate, coiled, or calcareous, nor is there a siphuncle. Buoyancy compensation is chemical (discussed later in the chapter). Members of the taxa Vampyromorphida and Sepiolida also have pens.

In **Octopoda[O],** the fourth line, the shell is vestigial. Interestingly, females of one octopod genus (the paper nautilus, *Argonauta*) secrete a thin, calcareous, shell-like case that is used as a brood chamber and float. Entirely external, it is neither homologous to the shell nor even attached to the body, so that the female must hold it in place with her arms. The much smaller male often shares the female's case.

Buoyancy

Invasion of the pelagic realm required the development of structures and mechanisms analogous to the swim bladders of fishes to achieve neutral buoyancy by an otherwise negatively buoyant animal. In the absence of such equipment, would-be pelagic animals would be forced to expend prohibitive amounts of energy to counter their tendency to sink when they wished to remain at a given depth in the water. Cephalopods have developed a succession of buoyancy compensation mechanisms, beginning with the chambered, gas-filled external shell equipped with a siphuncle. Among living cephalopods, only *Nautilus* retains a heavy external shell, but such a shell was characteristic of the early cephalopods. Its weight is countered by its numerous gas-filled chambers. A siphuncle/gas chamber shell system is used today by nautiloids, spirulids, and sepiids for buoyancy compensation. Because of the very large density difference between gas and water, gas is the most effective counter to negative buoyancy; no other mechanism could provide the lift required to compensate for the heavy mineral shell of the early cephalopods.

In cephalopods with septate shells, gas is added to the most recently constructed chamber as soon as a septum is constructed to isolate it. The epithelium of the siphuncle maintains a high intracellular salt concentration and is hyperosmotic to the chamber fluid (cameral fluid). When the new septum is strong enough to withstand the substantial hydrostatic pressure generated on an air-filled space, the siphuncle actively absorbs ions from the cameral fluid, creating an

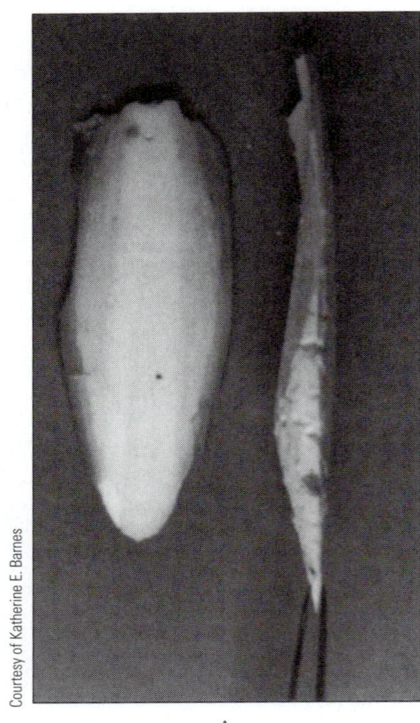
Courtesy of Katherine E. Barnes

A

Courtesy of Katherine E. Barnes

B

FIGURE 12-69 Internal shells of coleoids. **A,** The cuttlebone of the cuttlefish *Sepia,* viewed dorsally (left) and laterally (right). **B,** *Spirula,* viewed from the left side (left) and the face of a septum (right). Note the peripheral siphuncle at the bottom of the septum in the end view. The shell of *Sepia* is located on the dorsal side of the body, whereas that of *Spirula* is on the posterior end of the body.

osmotic gradient from the chamber into the siphuncle. Water follows this gradient and leaves the chamber to enter the blood in the siphuncle. As water is removed from the chamber, gas dissolved in the blood of the siphuncle diffuses into the chamber and replaces the water. Fluid exchange is restricted to the more anterior chambers, and fluid is gradually removed to compensate for the gradual growth of the shell. This mechanism does not allow for buoyancy regulation for short-term changes in depth.

In contrast, by regulating the relative amounts of fluid and gas in its cuttlebone, *Sepia* can adjust the degree of buoyancy over the short term. Light is an important factor controlling the regulating mechanism. During the day, *Sepia* lies buried in the sea bottom, but at night the animal becomes active, swimming and hunting. Accordingly, buoyancy decreases when the animal is exposed to light and increases in the dark.

The loss of the septate shell by the coleoids necessitated the development of alternative mechanisms for countering the tendency to sink. Fortunately, in the absence of the heavy shell, achieving neutral buoyancy becomes much easier. Many pelagic squids maintain, in the coelom or connective tissue, large reservoirs of fluid in which the heavy cations typical of seawater are replaced by low-molecular-weight ammonium ions (produced by ammonotelic nitrogen metabolism). The enormous, fluid-filled coelom of cranchiid squids, for example, which accounts for about two-thirds of the volume of the animal, is a chemical buoyancy chamber with a high concentration of ammonium ions. This substitution reduces the specific gravity of the squid and, along with the reduction in shell weight, confers neutral buoyancy on the animal. About half of the 28 families of squids, including the giant squid *Architeuthis* (Fig. 12-70B), the cranchiids (Fig. 12-71C), and the chiroteuthids (Fig. 12-71A,B), are "ammoniacal" squids that use this chemical mechanism to achieve neutral buoyancy.

Some pelagic squids, however, are negatively buoyant and must expend energy to maintain position in the water column. They remain in nearly constant motion and rely on the dynamic lift produced by the locomotory equipment (fins and funnel) to prevent sinking.

Cirrate octopods use the surface area of the extended webbing, parachute-like, to retard sinking (Fig. 12-72B). The secondarily pelagic octopod *Argonauta* uses its thin, lightweight, calcareous case as a gas-filled swim bladder. Benthic cephalopods, such as *Octopus,* have no buoyancy compensation mechanisms and spend no energy countering the tendency to sink (Fig. 12-73). In fact, they depend on negative buoyancy to remain in place on the sea floor.

Cartilage

Many semirigid structural materials in invertebrates are referred to as "cartilage" even though most lack the histological properties of vertebrate cartilage. The **cartilage** of cephalopods closely resembles that of vertebrates, however, and is yet another addition to the remarkable list of conver-

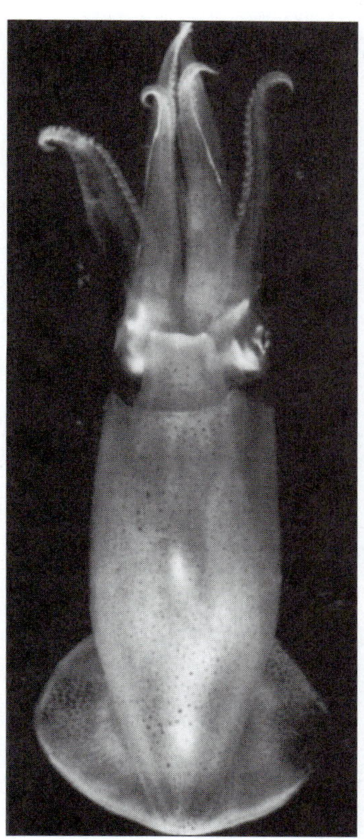

A

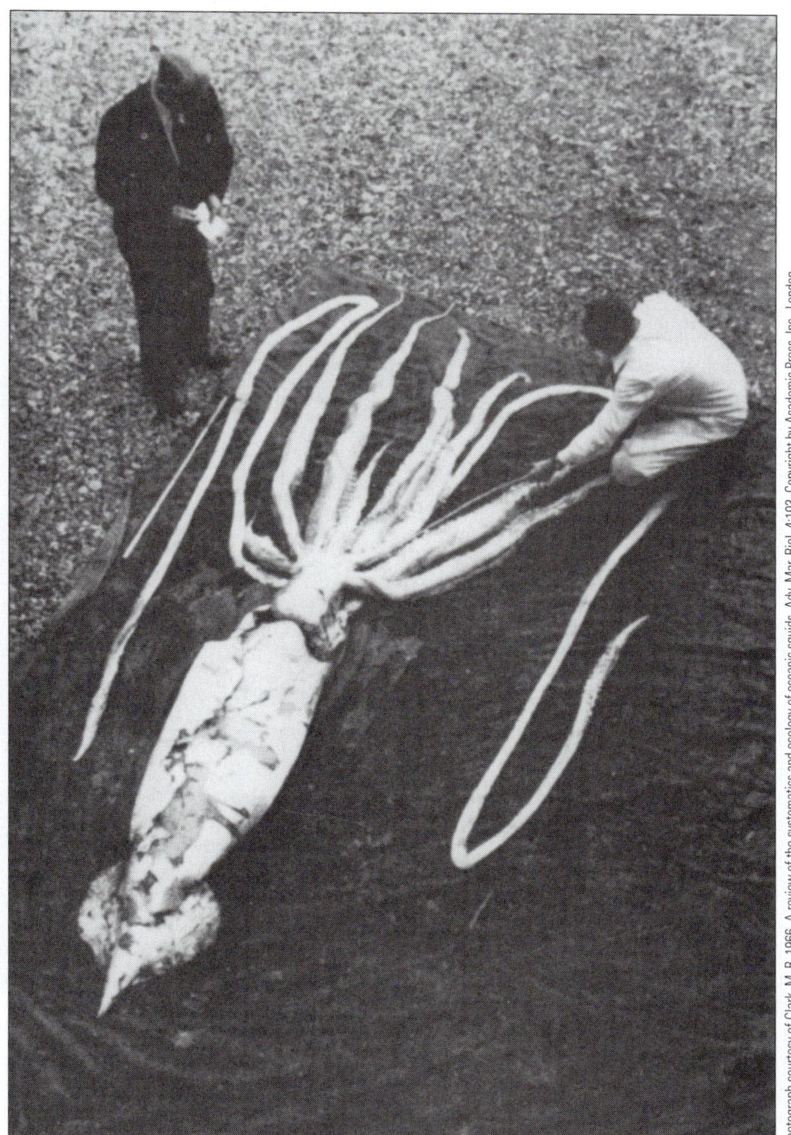

B

FIGURE 12-70 Coleoidea. **A,** The small myopsid squid *Lolliguncula brevis,* common in tidal creeks in the southeastern United States (ventral view). **B,** A giant squid of the oegopsid genus *Architeuthis* stranded at Rahneim, Norway, in 1954.

gences between these two taxa. Cephalopod cartilage, like that of vertebrates, has chondrocytes embedded in a hyaline ground substance of chondroitin sulfate and hyaluronic acid interlaced with a matrix of fibrous proteins, including collagen and elastic fibers.

LOCOMOTION

Cephalopods employ several different locomotory mechanisms, including walking, several types of swimming, and even flying.

Fin Swimming

The decapods and cirrate octopods have a pair of muscular, lateral fins whose slow, rhythmic undulations propel the animal forward through the water (Fig. 12-74).

Mantle Cavity and Jetting

The most common type of locomotion uses the muscles of the mantle wall to eject a jet of water from the funnel. The mantle consists of a thick layer of circular muscle penetrated by smaller radial muscles (Fig. 12-75). The muscles surround the mantle

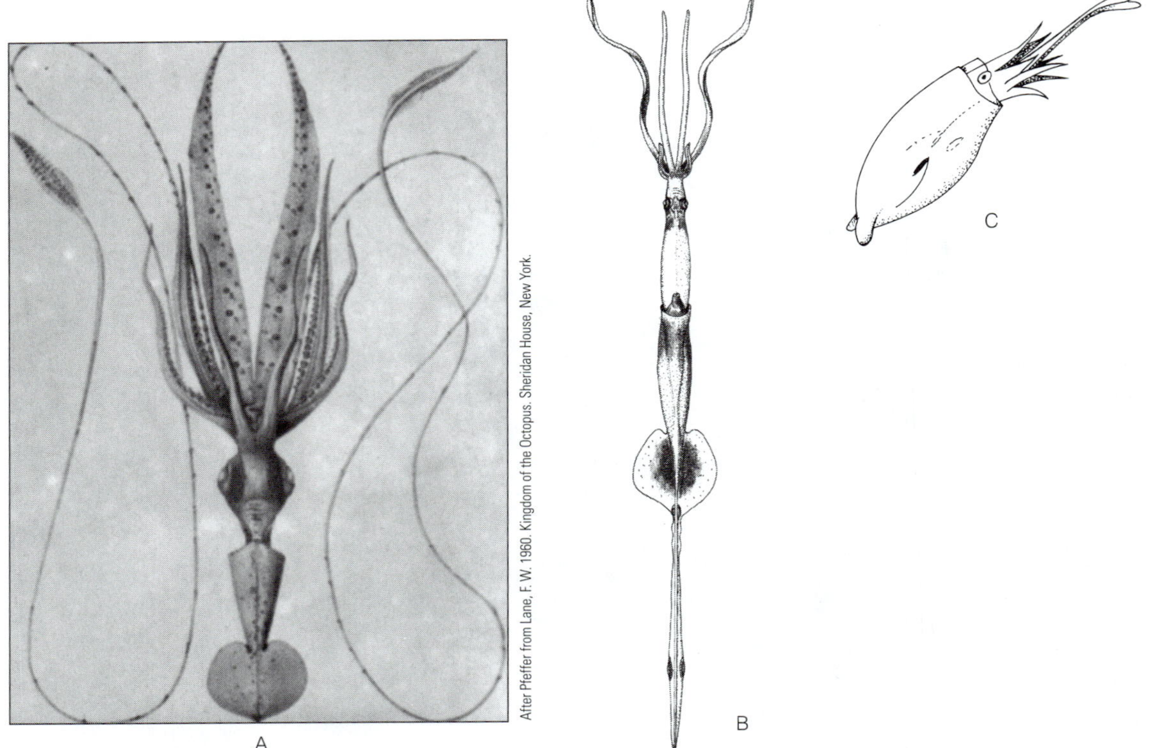

FIGURE 12-71 Bathypelagic oegopsid squids. **A** and **B,** The squid *Chiroteuthis veranyi.* Mature **(A)** and immature **(B)** individuals. **C,** The squid *Cranchia.* *(B, After Pfeffer from Lane, F. W. 1960. Kingdom of the Octopus. Sheridan House, New York; C, After Morton, J. E. 1967. Molluscs. Fourth Edition. Hutchinson University Library, London.)*

cavity and lie between inner and outer **collagen tunics.** The radial muscles extend from the inner tunic to the outer tunic. There is frequently a stiff shell remnant (pen or cuttlebone) to counter the tendency of the body to lengthen during muscle contraction.

The inhalant and exhalant mantle channels are equipped with one-way **flap valves** (Fig. 12-64). Because of these valves, water cannot exit the inhalant channel (siphons) or enter the exhalant siphon (funnel).

Contraction of the circular muscles constricts the mantle cavity and forces water through the funnel. The funnel can be pointed in almost any direction to control the direction of movement, which is opposite that of the water jet. The mantle muscles and valves can be used for two types of jetting movement, slow and fast.

SLOW JETTING

Most cephalopods use slow jetting for ordinary locomotion (and ventilation of the gills). Contraction of the circular muscles expels water from the mantle cavity through the funnel and also compresses the connective tissue in the mantle wall (Fig. 12-75A). The resulting elevation of water pressure opens the exhalant valve, closes the inhalant flap valves, and locks the edges of the mantle tightly around the head. The water jet from the funnel propels the animal in the opposite direction. The elastic recoil of connective tissue in the mantle antagonizes the circular muscles. When the circular muscles relax, the elastic fibers return to their original conformation, the circular muscles are stretched, and the mantle cavity inflates, creating a negative pressure that opens the inhalant valve, closes the exhalant valve, and causes water to enter the cavity. The radial muscles apparently do not participate in slow swimming.

During a contraction cycle, water flows into the mantle cavity; over the gills, gonopore, nephridiopores, and anus; then exits via the funnel (Fig. 12-64). The swimming current, which is also the respiratory current, is generated by muscles rather than cilia, in contrast with other molluscs, but like vertebrates.

FAST JETTING

Squids are also capable of a fast jetting, an escape response in which both circular and radial muscles participate. Contraction of the radial muscles hyperextends the mantle cavity so it contains more water than it does during slow jetting (Fig. 12-75B).

To initiate inhalation, the circular muscles relax and the radial muscles contract and hyperextend the mantle wall (Fig. 12-75B). Contraction of the radials pressurizes the circular muscle mass, which acts as an incompressible hydrostatic skeleton whose volume cannot change. Pressurization of the circular muscle layer results in its deformation, as it must accommodate the tendency of the radial muscles to pull the two tunics closer together and reduce its thickness. The obvious and intuitive result would be for the circular muscle

From Roper, C. E. F., and Brundage, W. L. 1972. Cirrate octopods with associated deep sea organisms: New biological data based on deep benthic photographs. Smithson. Contrib. Zool. 21:1–46.

A

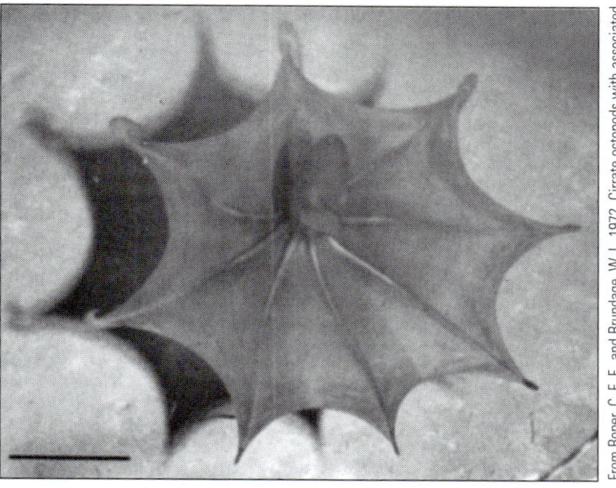

From Roper, C. E. F., and Brundage, W. L. 1972. Cirrate octopods with associated deep sea organisms: New biological data based on deep benthic photographs. Smithson. Contrib. Zool. 21:1–46.

B

FIGURE 12-72 Deep-sea photographs of a cirrate octopod, *Cirroteuthis,* which lives near the bottom. **A,** A view of the animal perhaps at the beginning of a downward stroke, with arm web closed and fins folded dorsally. **B,** A view onto the interbrachial web. Cirri on top of the arms (not visible in photo) may have a sensory function. Scale bar = 30 cm. Photographs taken with a deep-sea camera at 3000 m in the Virgin Islands Basin.

layer, along with the mantle and mantle cavity, to lengthen. This cannot happen, however, because the pen, if present, and collagen tunics prevent elongation of the body. The alternative, less obvious possibility is for the diameter of the mantle cavity to increase, and this is what happens. The mantle cavity hyperextends and can now hold more water than during normal slow jetting. The increase in the diameter of the mantle cavity creates a negative pressure that draws water in through the peripheral inhalant siphons. The influx of water closes the exhalant flap valve in the funnel and opens those in the inhalant siphons.

Exhalation is more intuitive. Relaxation of the radial muscles and contraction of the circular muscles increases the thickness of the circular muscle layer and decreases the diameter of the mantle cavity (Fig. 12-75A). This places the water under positive pressure, and the water seeks an outlet from the mantle cavity. The pressure closes the inhalant flap valve but opens the exhalant flap valve in the funnel, and water jets from the funnel with great force.

FIGURE 12-73 Lateral view of *Octopus,* an incirrate octopod lacking both cirri on top of the arms and webbing between them.

During fast jetting the funnel is pointed anteriorly so the animal moves rapidly in a posterior direction. The more powerful contractions of escape swimming are produced by one type of circular muscle fiber, whereas the rhythmic, less powerful contractions used for ventilation and slow swimming are produced by another.

Other

The benthic octopods crawl using their suckered arms. Cirrate octopods with webbed arms swim like jellyfish, using pulsation of the bell of webbing. During fast jetting, some squids escape the water surface and glide through the air for distances of up to 50 m to escape predators. The "flying squids" (Teuthoidea: Onycoteuthidae), which have long, tapered bodies and highly developed fin vanes and funnels, shoot out of the water during escape jetting and glide, like flying fishes, for considerable distances. There have been reports of squids accidentally flying onto the decks of ships 4 m above the water's surface.

ADAPTIVE DIVERSITY

Species of *Nautilus* are mobile epibenthic animals found from the surface to depths as great as 600 m. Because they require cool water temperatures, they are not found in the warmer waters above 100 m. Moreover, they ascend at night and descend during the day. Using radio telemetry, one specimen tagged in Palau, north of New Guinea, was found to ascend to 150 to 100 m at night and then move down to 250 to 350 m during the day. When resting, they attach to rubble or the walls of crevices with their appendages. Whether the animal is swimming or resting, the gas-filled chambers keep the shell upright.

Except when feeding, *Nautilus* swims backward, at about the same speed as a person doing a slow breaststroke. The locomotor mechanism is similar to that described for squids, except that ejection of water through the funnel is not caused by contraction of circular muscles. Instead, the animal partially retracts the body into the shell and partially retracts the funnel. Both actions compress the mantle cavity and eject water from the funnel. Contracting the mantle in imitation of

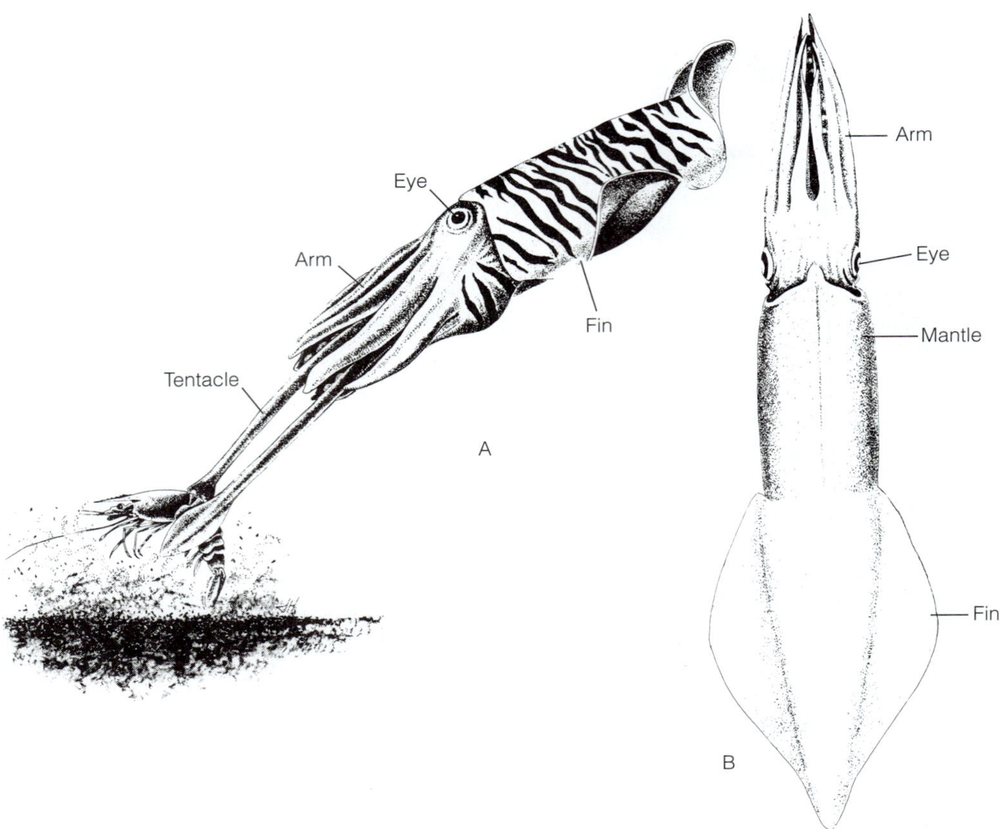

FIGURE 12-74 Coleoidea. **A,** The cuttlefish, *Sepia,* seizing a shrimp with its two tentacles. **B,** Dorsal view of the squid *Loligo* in swimming position. The tentacles and arms are held together, acting as a rudder.

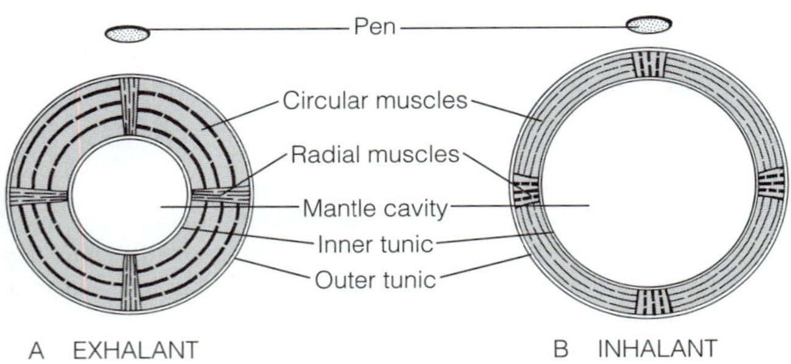

FIGURE 12-75 Cross sections of a generalized coleoid mantle. **A,** Contraction of circular mantle muscles has constricted the mantle cavity and forced water from the funnel, as in both slow and fast jetting. **B,** In fast jetting, contraction of radial muscles hyperextends the mantle wall to fill the mantle cavity with a larger than normal volume of water in anticipation of its powerful expulsion by the circular muscles.

the coleoids would probably be difficult for an animal enclosed in a rigid shell. The extinct ammonoids and nautiloids probably used the same mechanism that *Nautilus* does.

Unlike the funnel of coleoids but like the siphons of gastropods, the nautiloid funnel is a rolled sheet, not a complete tube. It can be extended and directed up or down and from side to side. Lift is obtained by directing the funnel downward.

Squids and cuttlefishes are agile swimmers that make full use of their water jets. They can hover, accomplish delicate adjustments in position, attitude, and speed; cruise slowly; or dart rapidly. Squids attain the greatest swimming speeds of any aquatic invertebrate, up to 40 km/h. The body of a typical squid is long, tapered posteriorly, and has a pair of posterior lateral fins that function as stabilizers and rudders and are used for swimming at slow speeds (Fig. 12-74B).

The giant architeuthid squids may inhabit depths of 300 to 600 m over the continental slopes (Fig. 12-70B) but they are not rapid swimmers. Two other interesting taxa of deepwater squids are the bathypelagic chiroteuthids and the cranchiids. The chiroteuthids have very long, slender bodies and long, whiplike tentacles (Fig. 12-71A,B). The cranchiids (Fig. 12-71C) are planktonic and many are strangely shaped.

The body of a cuttlefish, such as *Sepia,* tends to be short, broad, and flattened (Fig. 12-74A). Cuttlefish are versatile swimmers but they are not as fast as the more streamlined squids (Fig. 12-74B).

The deepwater *Spirula,* which lives as deep as 1000 m, swims with its anterior end and tentacles hanging downward. When the animal dies, the little, gas-filled shell floats to the surface and is commonly washed ashore (Fig. 12-69B). Their abundance on tropical and semitropical beaches attests to the large numbers of this cephalopod. The smallest cuttlefish are species of the sepiolid genus *Idiosepius,* which are about 15 mm long. They live in tide pools and possess a dorsal disc on the mantle for attaching to algae.

Living cephalopods such as *Nautilus, Spirula,* and *Sepia* that use gas-filled shells to maintain neutral buoyancy are restricted to depths at which the shell strength can withstand the pressure, for below that depth the shell collapses. Extinct cephalopods would have been limited in the same manner, and it is reasonable to assume they were all inhabitants of relatively shallow water. Replacement of the gas-filled septate shell with an ammoniacal buoyancy compensation mechanism released the squids from these depth restrictions.

During the late Paleozoic and Mesozoic, cephalopods were the dominant and most highly developed pelagic animals until fishes excluded them from many pelagic niches. The fast, agile coleoid design may have evolved as a result of competition with fish and may account for the many convergent features, such as complex eyes, shared by the two taxa. With further modification, the coleoid design permitted invasion of the deep and open ocean, from which the ammonoids and nautiloids were excluded.

Approximately 50% of cephalopod genera inhabit open ocean as opposed to coastal habitats and are found at all levels in the water column. Most, however, inhabit the uppermost 1000 meters, near the boundary between the mesopelagic (200–1000 m) and bathypelagic (1000–4000 m) zones, but there are species that live at deeper levels. Many epipelagic (0–200 m) and mesopelagic cephalopods undergo diel vertical migration (DVM), moving upward during the night and returning to greater depths during the day. Squid may have high population densities and an important role in pelagic food webs. The stomachs of slaughtered sperm whales often contain the beaks of 2500 to 4000 squids, and one whale feasted on 14,000.

Most incirrate octopods have returned to benthic habitats. The body is globular and baglike, and there are no fins (Fig. 12-73). The mantle edges are fused dorsally and laterally to the head, resulting in a much more restricted aperture into the mantle cavity. Although octopods are capable of jetting backward with arms trailing, they usually crawl about over the rocks around which they live. The arms have adhesive suction discs that are used to pull the animal along or anchor it to the substratum. Species of *Octopus* usually occupy a den or retreat, from which they make feeding excursions.

A very different mode of existence is exhibited by a number of families of deepwater cirrate octopods, some bathypelagic and some abyssobenthic. These animals, which have fins and webbed, umbrella-like arms, swim like jellyfish, by pulsations of the arms, but also use water jets from the funnel (Fig. 12-72).

NUTRITION

Appendages

Cephalopods are highly adapted for raptorial feeding and a strictly carnivorous diet. Prey is located with the efficient, image-forming eyes and captured with the unique, prehensile, circumoral appendages. Squids and cuttlefish have 10 of these appendages, arranged in five pairs around the head (Fig. 12-64). The eight **arms** are short and heavy, but the fourth pair down from the dorsal side are long, retractile **tentacles** (Fig. 12-74A).

The inner surface of each arm is flattened and covered with stalked, cup-shaped, muscular **suckers** (Fig. 12-64). The rim of the cup is applied to the substratum (the prey) to form a seal, whereupon contraction of radial muscles in the cup wall expands the cup volume to create a negative pressure and a strong suction. Contraction of circular muscles reduces the volume of the cup and releases the suction.

The rim of the sucker may be toothed, and the inner wall in some species has hooks (Fig. 12-76). Some squids have curved hooks instead of suckers. Although suckers are present for the entire length of the arms, they occur only on the flattened spatulate ends of the tentacles. The highly mobile tentacles are extended rapidly to seize prey. The arms aid in

A

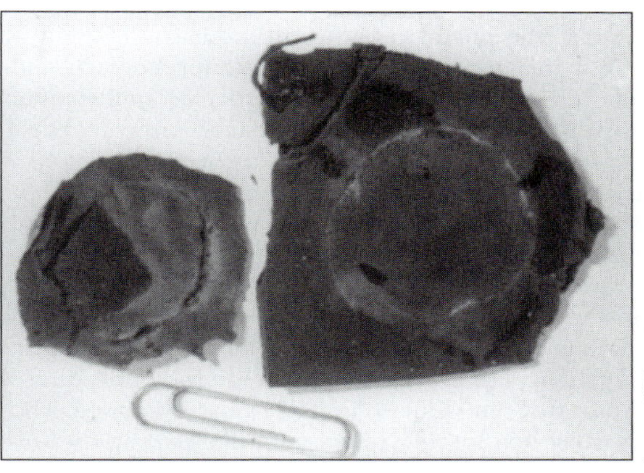

B

Courtesy of C. Roper

FIGURE 12-76 Suckers. **A,** Toothed suckers of the squid *Lolliguncula brevis.* **B,** Scars from the suckers of the giant squid *Architeuthis* on the skin of a sperm whale. Sperm whales are the principal predators of the giant squid.

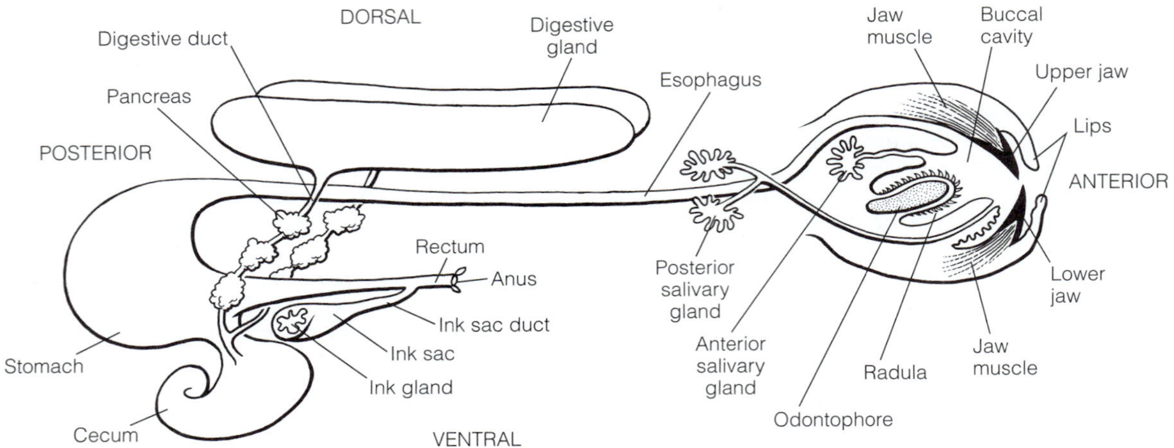

FIGURE 12-77 Diagrammatic lateral view of the digestive tract of the cuttlefish, *Sepia,* viewed from the right. The buccal mass is sectioned sagittally. *(Adapted from Budelmann, B. U., Schipp, R., and von Boletzky, S. 1997. Cephalopoda. In Harrison, F. W. and Kohn, A. J. (Eds): Microscopic Anatomy of Invertebrates. Vol. 6A, Molluscs II. Wiley-Liss, New York, pp. 119–414; after Mangold, K., and Bidder, P. M. 1989. L'appareil digestif et la digestion. In Mangold, K. (Ed.): Céphalopodes. Traité de Zoologie, no. 5. Paris, Masson; and Kozloff, E. N. 1990. Invertabrates. Saunders College Publishing, Philadelphia, PA. 866 pp; after Tompsett, D. H. 1939.* Sepia. *Liverpool Mar. Biol. Comm. Mem. 32:1–184)*

holding and manipulating the prey after capture by the tentacles (Fig. 12-74A).

Octopods have eight arms and no tentacles. The arms are similar to those of squids, except that the suckers are sessile (stalkless) and lack horny rings and hooks.

Nautilus has some 90 arms arranged around the head (Fig. 12-65). Many are chemosensory or tactile receptors, but some are prehensile and used to bring food to the mouth. *Nautilus* arms lack adhesive suckers or discs, but have transverse friction ridges to help hold the prey. The bases of the arms are united and form a cephalic sheath that encircles the buccal area. Dorsally, the cephalic sheath is continuous with a large, leathery, protective hood that acts like an operculum and covers the aperture when the animal withdraws.

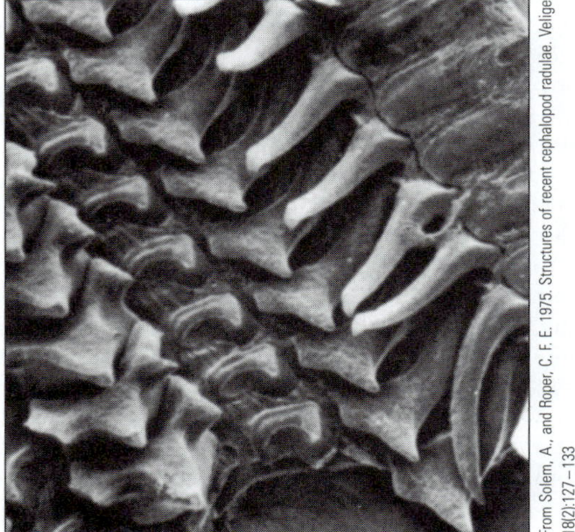

From Solem, A., and Roper, C. F. E. 1975. Structures of recent cephalopod radulae. Veliger 18(2):127–133

FIGURE 12-78 Radula of *Octopus briareus.*

Digestive System

Although there is variability in the cephalopod digestive system, most conform to a common plan (Fig. 12-77). The mouth opens into the buccal cavity and the long esophagus extends posteriorly from the buccal cavity to the stomach. The gut is U-shaped, and the intestine exits the posterior stomach and extends anteriorly to the anus. Several diverticula join the gut in the region of the esophagus-stomach-intestine junction. Their size, morphology, function, and position vary by taxon.

The mouth is surrounded by the ring of appendages. It opens into the buccal cavity and is equipped with a pair of large dorsal and ventral jaws forming a parrotlike **beak** (Fig. 12-63). The hard jaws, composed of tanned protein and chitin, are equipped with powerful muscles that form a large buccal mass occupying much of the interior of the head. The beak bites off large pieces of prey tissue, which are then pulled into the buccal cavity by the radula (Fig. 12-77) and swallowed. The buccal mass is located within a blood sinus that permits the animal to rotate the entire buccal apparatus at will and apply the beak with great dexterity to the prey.

The posterior wall of the buccal cavity bears a radular pouch and a radula (Fig. 12-77) that consists of transverse and longitudinal rows of teeth (Fig. 12-78). Each transverse row includes several types of teeth, and the number and types of teeth in each row vary with taxon. In *Octopus,* for example, 9 teeth are in each row, and in *Nautilus* there are 13. Each longitudinal row is composed of the same type of teeth.

Two pairs of salivary glands and a submandibular gland empty into the buccal cavity. The anterior salivary glands (Fig. 12-77) are either in or near the buccal mass and secrete mucus into the buccal cavity. An unpaired submandibular gland has a similar function. The posterior salivary glands (venom glands) are situated farther posteriorly and secrete poison, proteolytic enzymes, hyaluronidase, and other compounds. The poison, which varies with species, enters the hemal system of the prey through wounds inflicted by the beak. Vasodilators and heart-exciting neurotransmitters accompany the toxin and hasten its spread by the hemal system of the prey. The little blue-ringed octopus *Hapalochlaena maculosa,* which feeds on crustaceans in shallow water in the Indo-Pacific, is extremely venomous (its toxin is a tetrodotoxin) and, although reluctant to bite, it has been responsible for a few human fatalities.

The long esophagus conducts food posteriorly from the buccal cavity to the stomach (Fig. 12-77). Its walls contain circular and longitudinal muscle layers, and food transport is by muscular peristalsis rather than cilia. This differs radically from food transport in all other molluscs, but is the same as in vertebrates. In most cephalopods the esophagus is a narrow tube leading posteriorly to the muscular stomach (Fig. 12-63), but in some, such as *Nautilus* (Fig. 12-65B) and *Octopus,* it is expanded to form a crop. Digestion begins in the esophagus.

Important diverticula, including the digestive gland, pancreas, and cecum, attach to the gut in the vicinity of the stomach. The **cecum** is a large, coiled, thin-walled absorptive pouch and the **digestive gland** is a thick-walled tube. Part of the digestive gland is differentiated into a **pancreas** (Fig. 12-77) and both of these structures empty into the cecum via the **digestive ducts** (hepatopancreatic ducts). Like the vertebrate liver, the digestive gland is the largest organ in the body. It performs many of the same functions as the vertebrate liver.

Digestion in cephalopods is entirely extracellular and occurs primarily in the stomach and cecum. Food enters the stomach from the esophagus and is mixed with enzymes from the digestive gland and pancreas, which arrive via the digestive duct. A system of sphincter valves and grooves routes the chyme in a complicated pathway between the stomach, cecum, and intestine as digestion and absorption occur. Most absorption takes place across the walls of the cecum (as in *Loligo*) or the digestive gland (as in *Sepia* and *Octopus*).

The intestine exits the anterior wall of the stomach and passes anteriorly to empty via the anus in the mantle cavity. The distal portion of the intestine is the rectum. Resorption of valuable molecules probably occurs in the intestine. Large, indigestible items in the stomach are passed directly into the intestine without entering the cecum or digestive gland. Wastes leave the anus near the funnel and are carried away with the exhalant water jet. The digestive modifications of cephalopods, particularly squids, are probably adaptations for the rapid digestion required by an active pelagic life and a carnivorous diet.

Ink Sac Complex

An **ink sac complex** is present in most coleoids but absent in nautiloids. It consists of an ink gland, ink sac, and ink sac duct (Fig. 12-77) derived from a diverticulum of the rectal wall. The gland secretes a dark brown, black, or blue ink composed of a suspension of melanin granules. The ink is stored in the ink sac, which is connected to the lumen of the rectum by a duct. Contraction of the muscular walls of the sac ejects ink from the anus into the exhalant water current. Outside the mantle cavity, the ink rapidly disperses to form a dark cloud that may confuse a predator or hide the escape maneuvers of the cephalopod. The alkaloid nature of the ink may be objectionable to predators, particularly fish, whose chemoreceptors it may anesthetize.

Diet

The diet of cephalopods varies depending on their habitat. Pelagic squids, such as *Loligo* and *Alloteuthis,* feed on fish, crustaceans, and other squids. *Loligo* darts into a school of young mackerel, seizes a fish with its tentacles, and quickly bites out a chunk behind the head or bites off the head. The fish is devoured with small bites of the beak until only the gut and tail remain. These parts are then dropped to the bottom. Cuttlefish swim over the bottom and feed on epibenthic invertebrates, especially shrimps and crabs. *Sepia* may rest on the bottom, lying in wait for passing prey.

Octopods live in dens located in crevices and holes and make excursions in search of food or lie in wait for it near the entrance of their lair. Clams, snails, and crustaceans are common prey, and the diet of a single species may include as many as 55 different prey items, although a few items predominate. Some of the prey are eaten during the foraging excursion and others are taken back to the den to be eaten there. Octopods commonly have shell middens around their dens.

Enteroctopus dofleini, found along the western coast of the United States, is most active at night and has foraging areas of up to 250 m surrounding the den. The reef-inhabiting, Indo-Pacific *Octopus cyanea* is active during the day; it may make hour-long hunting excursions over distances of up to 100 m. Movement is by a combination of swimming and crawling, and the animal may perch momentarily on a rock during the course of an expedition. The octopus leaps on motile prey, such as a crab, enveloping it in the outstretched arm web. It may also leap on algal clumps and other objects and then feel beneath the web to see if it caught anything. It paralyzes its prey with venom from the salivary glands and may take several paralyzed prey home for later consumption.

In contrast to cuttlefish and squids, which tear prey with the beak, the feeding habit of octopods resembles that of spiders. The prey is injected with poison, with or without a bite from the beak, and then flooded with enzymes. The partially digested and liquefied tissues pass into the gut and the indigestible remains are eventually discarded. To remove gastropods from their shells, octopods drill a hole through the shell with the radula or a toothed salivary papilla and inject poison directly into the soft tissue. A chemical softening agent is secreted to assist in the drilling process. A few deep-sea octopods are suspension feeders convergent with jellyfish.

Nautilus is a scavenger and benthic predator specializing in decapod crustaceans, especially hermit crabs. When feeding, it swims forward, searching the bottom with its extended arms. Efforts to capture nautiloids with baited traps are successful only when placed on the bottom, never when suspended in the water column.

GAS EXCHANGE

Gas exchange in most cephalopods is accomplished by a pair of bipectinate, nonciliated gills (Fig. 12-64, 12-79), but the general body surface also contributes. Cephalopod gills are thought to be homologous to those of the ancestral mollusc. They are attached to the dorsal wall of the mantle cavity by a gill ligament and their filaments are twice folded to create a large surface area. The locomotory flow of water through the mantle cavity also ventilates the gills. Nautiloids have four gills (tetrabranchiate), whereas coleoids have only two (dibranchiate).

Muscular contractions of the mantle create a posterior flow of water into the ventral mantle cavity that turns dorsally to pass between the gill filaments and then flows anteriorly out the

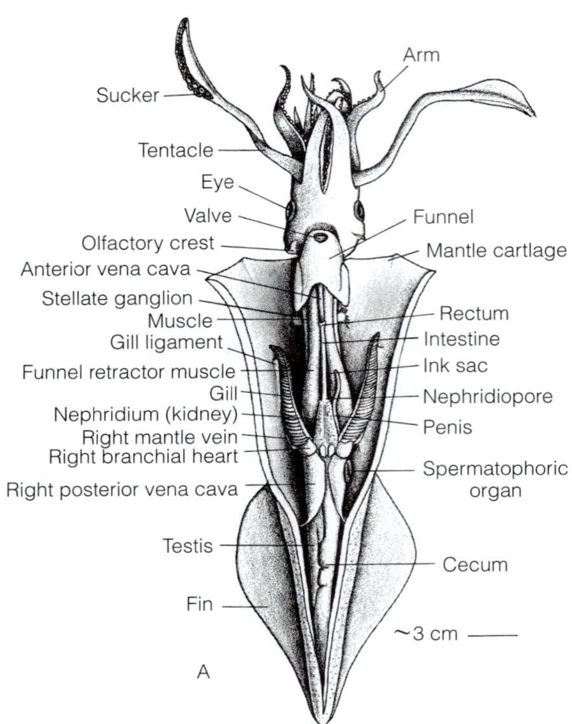

FIGURE 12-79 Anatomy of the squid *Loligo* (male). Ventral view, with the mantle opened along the ventral midline to reveal organs within the mantle cavity. *(By Mary Ann Nelson.)*

funnel (Fig. 12-64). Flow over the filaments is usually considered to be concurrent, but in some regions it may be countercurrent. The loss of countercurrent exchange efficiency is compensated for by the large folded gill surface area; supplementary use of the body surface; the powerful, swift, muscle-generated respiratory current; and an efficient, pressurized blood supply to the gills. Even so, most cephalopods are restricted to cold water, which has a higher oxygen solubility, and will perish quickly in warm water. Cephalopods that swim with webbed arms often have vestigial gills, and gas exchange takes place through the enhanced surface area of the webbing.

INTERNAL TRANSPORT

The coleoid hemal system is adapted to support an active, fast predator with a high metabolic rate. Unlike that of other molluscs, but like that of vertebrates, the system is "closed," with the blood contained in a system of vessels, including capillaries, rather than in hemocoelic sinuses. Arteries and veins are connected by capillary beds (and some sinuses), where exchange with the tissues occurs. Capillaries provide for a vastly greater area of contact between the hemal system and the tissues than is provided by large sinuses. Furthermore, the blood vessels are lined by a cellular **endothelium** similar to that of vertebrates and very different from other invertebrates. There is a **systemic heart** dedicated to pumping blood to the body and two auxiliary **branchial hearts** whose function is to provide the gills with pressurized blood (Fig. 12-80A). The systemic heart is a heavily muscularized organ consisting of a central ventricle and a pair of atria, one for each gill. Both atria and ventricle are muscular and contractile.

The walls of the hearts consist of an outer epithelial epicardium, a middle muscular myocardium, and an inner epithelial endocardium (endothelium), as do those of vertebrates. Also like the vertebrates, the major blood vessels are composed of three layers: an outer tunica adventitia consisting primarily of connective tissue with elastic fibers, a middle tunica media that is largely muscular, and a tunica intima consisting of the endothelium and its basal lamina. Most of the venous system, including veins and lacunae, is contractile to facilitate the return of blood to the hearts. Cephalopod blood contains hemocyanin to assist in the transport of oxygen.

The central systemic heart of coleoids pumps oxygenated blood through the aortae to the tissues (Fig. 12-80A,B). Unoxygenated blood from the tissues flows through the vena cavae to a pair of branchial hearts, one for each gill. Contractions of the branchial hearts force blood through the afferent branchial vessels and into the gills (Fig. 12-80B). Oxygenated blood exits the gills via the efferent branchial vessels to the systemic heart to complete the circuit. The possession of auxiliary hearts is not as strange as it may seem. They are present in mammals and birds, for example. The right heart of mammals is an auxiliary heart (viz. a "pulmonary" heart) dedicated to providing the lungs with pressurized unoxygenated blood. The left heart is the systemic heart that pumps oxygenated blood to the rest of the body. In mammals the two hearts happen to be stuck together but that is an accident of their evolution and ontogeny and is not a functional necessity.

Unlike that of coleoids, the hemal system of nautiloids is not completely closed, and it has more in common with that of other molluscs. The nautiloid systemic heart has four atria and there are no branchial hearts.

EXCRETION

Nitrogen metabolism is ammonotelic. The major excretory system of coleoids is similar to that of other molluscs in that an ultrafiltrate of the blood is produced into the pericardial cavity and then modified by secretion and absorption before being released as final urine (Fig 12-80A,C). The process is modified in the coleoids to improve its efficiency and compensate for the absence of an open hemocoel in which to bathe the nephridium. The nephridia are associated with the branchial hearts rather than the systemic heart. Four nephridia are present in nautiluses, two in coleoids.

The nephridium, known as the **renal sac,** and the renopericardial duct connect the pericardial cavity of the branchial heart with the mantle cavity (Fig. 12-80A,C). On its way to the branchial heart, the surface of the vena cava is expanded to form **renal appendages,** which are in contact with the thin, permeable walls of the renal sac. A region of the branchial heart wall (known as the branchial heart appendage; Fig. 12-80A) that is equipped with podocytes is the site of ultrafiltration. The primary urine is formed in the pericardial cavity and modified by selective adsorption and secretion between the renal appendages and renal sac to become the final urine. Thus, contractions of the branchial heart pressurize the blood

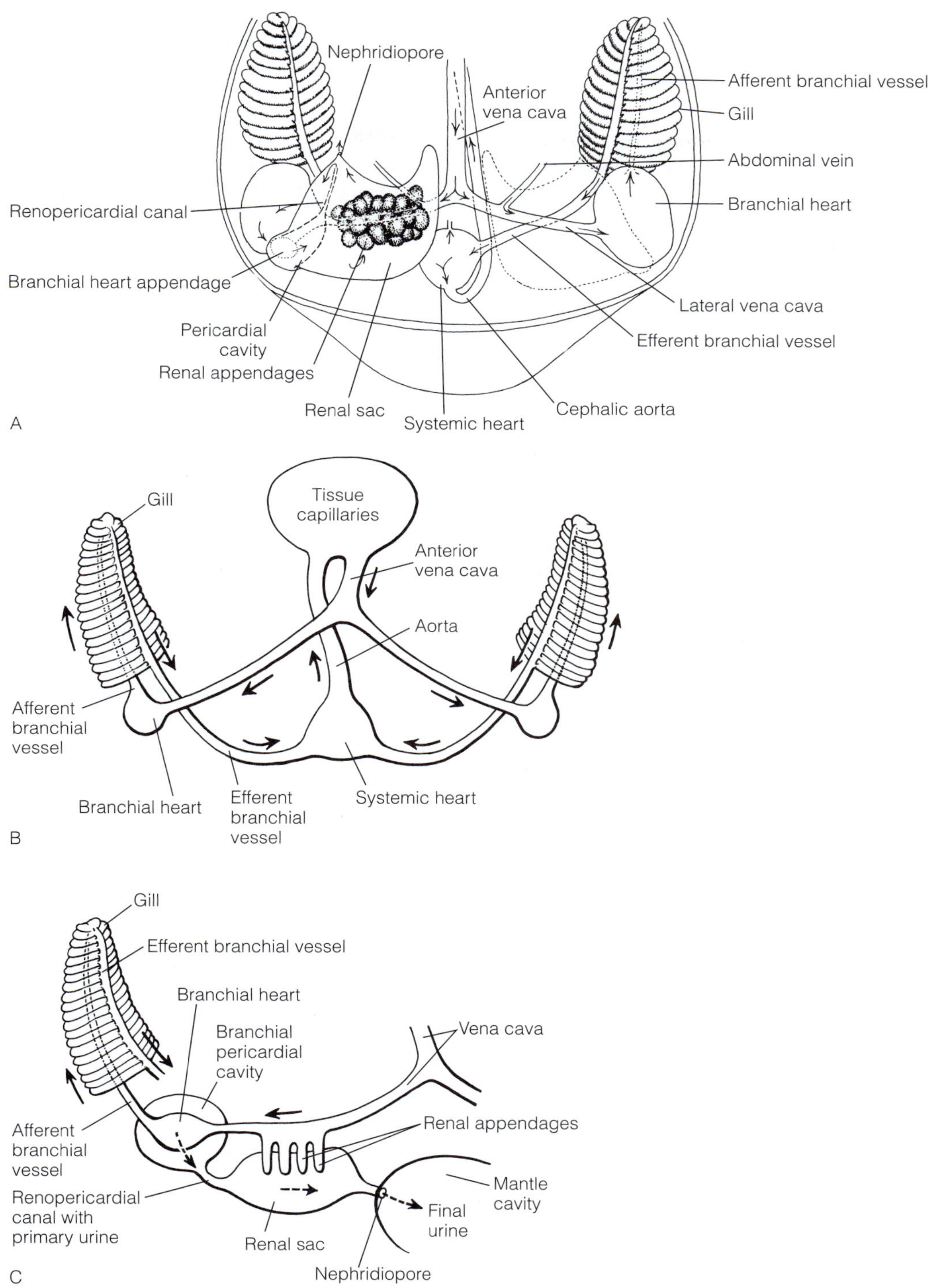

FIGURE 12-80 Cephalopod hemal and excretory systems. **A,** *Enteroctopus dofleini.* The arrows indicate the flow of blood and urine. **B,** Schematic diagram showing a simplified pattern of blood flow in a coleoid cephalopod. **C,** Schematic diagram depicting the spatial relationships between the components of the cephalopod excretory system. Solid arrows indicate blood flow and dashed arrows are urine. *(A, After Potts, W. T. W. 1965. Ammonia excretion in* Octopus dofleini. *Comp. Biochem. Physiol. 16:479–489.)*

and force ultrafiltrate across its wall into the pericardial cavity and renopericardial duct. This primary urine flows into the renal sac, where it bathes the surfaces of the renal appendages and is modified, chiefly by secretion of ammonia from the renal appendages, but also by selective absorption of components of the urine back into the renal appendages. The final urine is excreted via the nephridiopores into the mantle cavity. The gills and pancreas are also involved in excretion.

NERVOUS SYSTEM

The highly developed cephalopod nervous system is unequaled among invertebrates and comparable to that of vertebrates. It is correlated with the locomotor dexterity necessary for the raptorial feeding of these animals. The central nervous system is strongly concentrated, cephalized, and bilaterally symmetrical. It is larger, in reference to body size, than the brain of fishes. It consists of numerous ganglia, their connectives, commissures, and nerves. It is supplied with information by a highly developed sensory system. Homologs of the typical molluscan ganglia, plus many more, are present but coalesced into a compact and complex brain enclosed in a cartilaginous **cranium.** The brain is a nerve ring that encircles the esophagus (Fig. 12-63), as in other molluscs, but its connectives and commissures are unusually short so that the ganglia, even the visceral, are very close together (Fig. 12-81). The brain is divided into a supraesophageal region that is primarily sensory and integrative and a subesophageal region that is primarily motor.

The two cerebral ganglia (Fig. 12-81), located dorsal to the esophagus, process sensory input from the eyes, statocyst, and olfactory epithelium. The eyes and statocyst are enclosed in cartilaginous optic and otic capsules, respectively. The buccal ganglia control the muscles of the buccal mass and are connected to the cerebral ganglia by connectives. The two pedal ganglia ventral to the esophagus connect with the two cerebral ganglia by a pair of circumesophageal connectives, and the four ganglia and their connectives and commissures make up the nerve ring, as they do in all molluscs. As a derivative of the foot, the funnel is served by the pedal ganglia. Similarly, each appendage is served by a **brachial nerve** from a pedal ganglion.

The two pleural ganglia are combined with the visceral ganglia but are not discernible externally. The combined **pleurovisceral ganglia** (also known as the palliovisceral ganglia) are connected with the pedal ganglia by short connectives. A pair of large pallial nerves from the pleurovisceral ganglia extend to the **stellate ganglia** (Fig. 12-64), through which they supply the swimming muscles of the mantle. Ordinary swimming and respiratory contractions of the mantle musculature result from impulses conveyed through a system of many small motor neurons radiating from the stellate ganglia, one in each side of the mantle wall.

The rapid escape movements of fast-jetting cephalopods such as squids result from a highly organized system of **giant motor axons** that bring about powerful synchronous contractions of the circular muscles of the mantle. Since speed of transmission of nerve impulses is directly related to the diameter of the neuron, the giant axons conduct information rapidly and at a rate proportional to their diameter. The longest axons, extending to the posterior mantle cavity, have the greatest diameter, whereas the neurons with shorter distances to travel have proportionately smaller diameters. A command sent from the visceral ganglion to contract will thus travel fastest on the neurons that have the longest distance to traverse and slowest in the short neurons. This assures that the signals carried on all neurons will arrive at their target muscles simultaneously, as in a "time on target" artillery barrage. All mantle muscles receive the command at the same time and contract simultaneously to produce a powerful, efficient expulsion of water from the funnel.

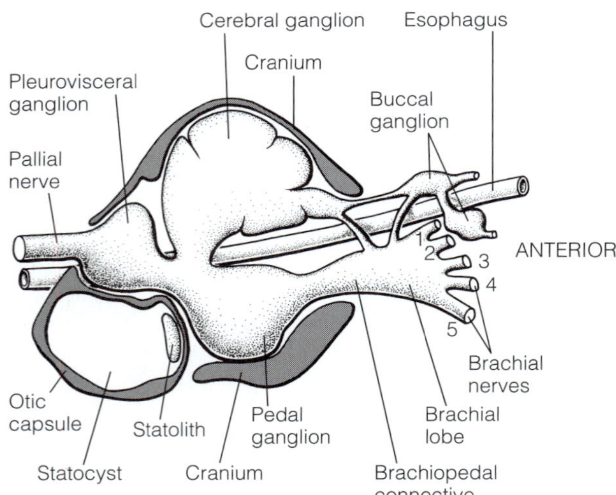

FIGURE 12-81 Sagittal section of the brain of the squid *Loligo*. *(Redrawn from Budelmann, B. U. Schipp, R. and von Boletzky, S. 1997. Cephalopoda. In Harrison, F. W., and Kohn, A. J. (Eds.): Microscopic Anatomy of Invertebrates. Vol. 6A: Mollusca II. Wiley-Liss, New York. pp. 119–414; after Young, J. Z. 1976. The nervous system of* Loligo. *II. Suboesophageal centres. Phil. Trans. Ray. Soc. Lond. B 274:101–167.)*

SENSE ORGANS

The well-developed cephalopod sensory system is an adaptation for raptorial feeding and rapid locomotion. The three primary cranial senses of vertebrates—olfactory, optic, and otic—are present, closely associated with the brain, and housed in sense capsules associated with the cranium, as they are in vertebrates. Other senses are also present.

Vision

The sophisticated camera-style eyes of coleoid cephalopods (Fig. 12-82) are strikingly similar in structure to those of fishes but are unusually large. Each eye is housed in a cartilaginous optic capsule that is fused with the cranium. The epidermis covering the eye laterally is a transparent **cornea.** The **iris,** with a slit-shaped **pupil** in its center, lies inside the cornea. The iris is a muscular diaphragm that adjusts the size of the pupil to regulate the entry of light. The lens is suspended by the ciliary muscle just inside the pupil. The **retina** covers the back of the eye and consists of closely packed rhabdomeric photoreceptor cells. Light passes through the cornea, pupil, and lens before striking the retina. A pigmented choroid and a tough connective tissue sclera lie under the retina, enclosing the eye and giving it shape. The axons of the photoreceptors extend to the large **optic ganglion** lying beside the cerebral ganglion.

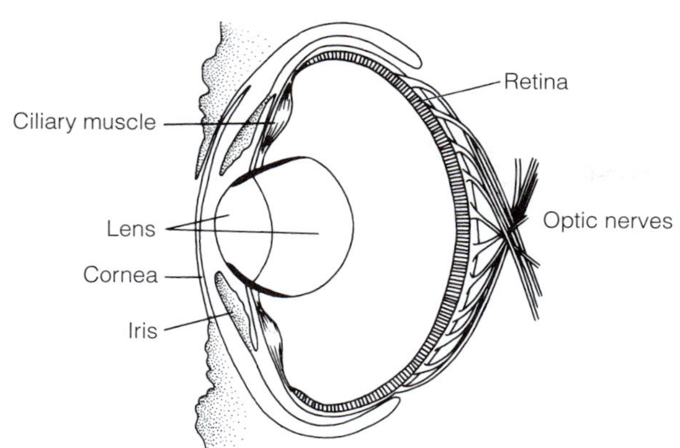

FIGURE 12-82 The eye of *Octopus* in cross section. *(After Wells, M. J. 1961. What the octopus makes of it; our world from another point of view. Adv. Sci. London. 20:461–471.)*

The retina consists of a single layer of photosensitive **retinula cells** whose parts are arranged into three layers, with the light-sensitive region closest to the light source. Each retinula cell has a rhabdomere with up to 700,000 rhodopsin-bearing microvilli. The rhabdomeres of four adjacent retinula cells form a rhabdome similar to that of the arthropod ommatidium. The cephalopod retina is part of a direct pigment cup with its receptors facing the source of light. Due to a happenstance of its evolutionary history, the vertebrate eye is an indirect pigment cup with its photoreceptors facing away from the light source. The cephalopod pigment cup forms during ontogeny as an invagination of the ectoderm of the head whereas that of the vertebrates forms as an evagination of the lateral wall of the brain, which itself formed as an invagination of the head ectoderm. The orientation of the cephalopod eye is controlled by from 4 to 14 **extrinsic eye muscles** similar to the six such muscles of the vertebrate eye.

The cephalopod eye undoubtedly forms an image, and the optical connections appear to be especially effective in analyzing vertical and horizontal movement of objects in the visual field. It is sensitive to polarized light. Experimental studies indicate that *Octopus* can discriminate objects as small as 0.5 cm from a distance of 1 m, which is comparable to the abilities of fishes. With few exceptions cephalopods have only one visual pigment and they cannot discriminate among colors. The lens of shallow-water cephalopods contains a yellow pigment that absorbs ultraviolet radiation and protects the retina from its harmful effects. Species from deeper water lack the yellow pigment.

The spherical lens has a depth of focus from about 3 cm to infinity and there is little need for accommodation, although the ciliary muscles can accomplish limited focusing by changing the position of the lens. Like that of fishes, the cephalopod cornea has a refractive index near that of water and is unable to refract light passing through it, and so is of little use in focusing.

The cephalopod eye adjusts to changes in light intensity by modifications in the pupil's size and by the migration of pigment in the retina. In bright light the pupil is constricted and screening pigment in the retinula cells moves distally to isolate adjacent cells much as it does in a light-adapted (apposition) arthropod ommatidium. In dim light the pupil dilates and the retinal screening pigment is withdrawn so that more light reaches the sensory cells.

Squids of the taxon Histioteuthidae, which live in the mesopelagic zone, have bilaterally dimorphic eyes. The large left eye is directed upward and responds to faint light from the surface, and the small right eye is directed forward and responds to bioluminescent light.

The large eyes of *Nautilus* are carried at the end of short stalks. Although the eye contains a large number of photoreceptors, it lacks a lens and its interior is open to the external seawater through a small aperture, the pupil. It presumably functions like a pinhole camera, but without a lens the resolving power and image-forming capability must be very limited. Its primary function is probably simple light detection.

Most cephalopods, but not *Nautilus,* have accessory photoreceptive vesicles that resemble simple eyes. These apparently assess the intensity of light passing down in the water column from above. Such information might be useful in adjusting countershading or for coordinating diel vertical migrations. For countershading, adjustments in chromatophore deployment match the brightness of the ventral surface with the intensity of down-welling light so that the animal is invisible from below.

Equilibrium

A system of mechanoreceptors mimics many of the functions of the lateral line and inner ear (acousticolateralis) system of vertebrates. In fishes this system consists of neuromasts (hair cells) associated with cupulae extending into water-filled endolymphatic spaces. They detect gravity, acceleration, turning motions, equilibrium, and vibrations in water.

The cephalopod statocyst is the analog of the fish inner ear. Although present in nautiloids, statocysts (Fig. 12-63) are best developed in coleoids, in which they are large and embedded in cartilaginous **otic capsules** on each side of the braincase (Fig. 12-81). They provide information about body position in relation to gravity and, like the semicircular canals of the vertebrate inner ear, detect angular acceleration. Without its statocysts, a cephalopod cannot keep the pupil slits of the eyes horizontal or discriminate between horizontal and vertical surfaces. The statocyst is a cavity lined with ciliated mechanoreceptors and a central statolith. Some ciliated cells respond to deformation by the statolith, and others, equipped with cupulae, are deformed by fluid movement resulting from acceleration.

Lateral Line Analog

The head and arms of decapods have **epidermal lines** of hair cells sensitive to weak water movement and pressure waves. These are analogous to and have the same function as the lateral line system of fishes. The organization of the hair cells in lines confers directionality on the signal. The squid can compare the time of arrival of the pressure wave at opposite ends of the line of receptors to infer the direction of origin of the wave. The epidermal lines are used to detect vibrations originating with prey and perhaps also from predators or conspecifics.

Chemoreception

Cephalopods have both contact (gustation, taste) and distance (olfaction, smell) chemoreceptors, but the chemical senses do not appear to be nearly as important as vision, at least not in coleoids, which are strongly visually oriented animals. *Nautilus,* with its poorer vision, apparently relies more heavily on chemoreception. Coleoids have patches of **olfactory epithelium** (the olfactory organs) on the sides of the head. The axons from these receptor cells form an olfactory nerve extending to a cerebral ganglion. The suckers and lips of coleoids and the tentacles of *Nautilus* are equipped with contact chemoreceptors. The arms, and especially the sucker epithelium, are liberally supplied with tactile cells and chemoreceptor cells, particularly in the benthic hunting octopods. Osphradia are absent in all cephalopods except *Nautilus,* in which they are located in the mantle cavity upstream of the gills.

INTEGUMENT AND CHROMATIC ORGANS

The cephalopod integument is a thin outer epidermis underlaid by a connective-tissue dermis. The epidermis is a simple columnar epithelium including mucous and sensory cells and having a basal lamina. The dermis is composed of collagen fibers, ground substance, and abundant **chromatic organs** of various types. The variable and complex external coloration of most cephalopods (other than *Nautilus*) is caused by the interplay of the dermal chromatic organs, which include chromatophores, leucophores, reflector cells, iridiocytes, and photophores.

Chromatophores are multicellular organs consisting of a central pigment cell and numerous small radial muscles under nervous control (Fig. 12-83A). The plasma membrane of the pigment cell is extensively folded to allow for expansion and most of the cell's interior is occupied by a large, flattened, elastic **pigment sac** filled with pigment granules. The radial muscles lie in the equatorial plane of the flat chromatophore, originate in the dermis, and insert on the elastic wall of the pigment sac. Contraction of the muscles dilates the pigment sac and converts it to a large, flat plate. This disperses the pigment over a larger surface, making it more visible. Pigment cells may increase their diameter by a factor of 20 when expanded. Relaxation of the radial muscles allows the elastic sac to return to its original smaller size and once again concentrates the pigment in a smaller area, rendering it less visible. Each pigment cell contains a single pigment, which may be red, orange, brown, black, yellow, or blue.

Chromatophores are controlled by the nervous system and each muscle fiber receives at least one motor axon. There may also be hormonal control. Changes in chromatophore dispersal typically result from visual stimulation. The extent of color change and the nature of the required stimulus vary considerably. The cuttlefish *Sepia officinalis* displays complex color changes and may imitate the hues and patterns of sand, rock, and other backgrounds. Most changes, however, are correlated with behavior. Many species exhibit color change when alarmed. For example, the littoral squid *Loligo vulgaris* is normally very pale, but darkens when disturbed. On the other hand, the Caribbean reef squid *Sepioteuthis sepioidea* lightens its coloration when alarmed. A frightened octopus may flatten its body and present an elaborate "defensive" color display, including waves of colors flowing over the body and large dark spots around the eyes. Color displays associated with courtship in many cephalopods are described in the next section.

Iridiocytes also contribute to the color of the animal. These are individual cells located in the dermis, iris, and ink sac. They are elliptical cells whose extensively folded plasma membrane is a diffraction grating that refracts light of specific wavelengths

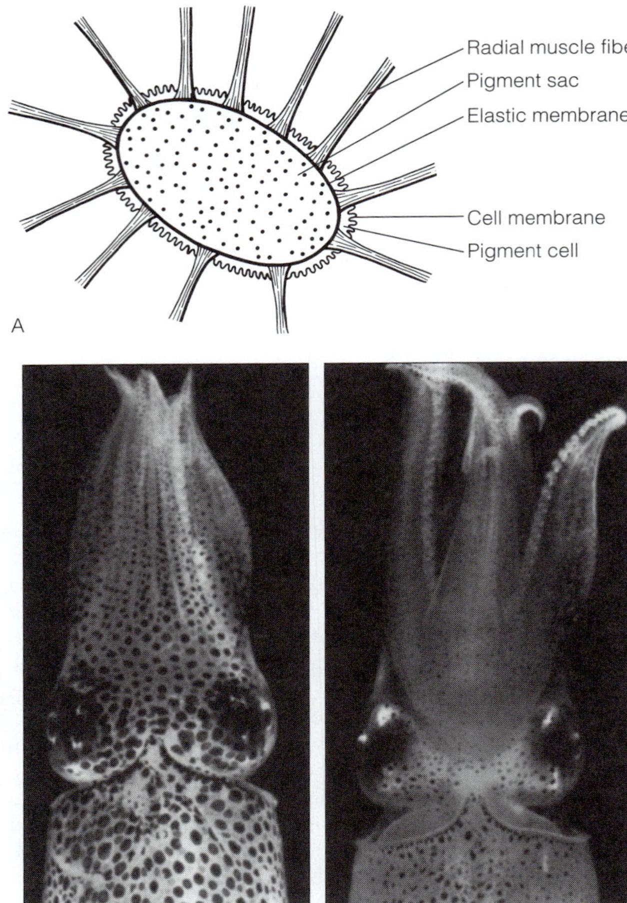

FIGURE 12-83 Chromatophores. **A,** Polar view of a partially expanded chromatophore. The radial muscles are confined to the equatorial plane. **B,** The squid *Lolliguncula brevis* with most of its chromatophores expanded. The central pigment cell of each expanded chromatophore is clearly visible as a large circular or oval plate of up to 1.5 mm in diameter. **C,** The squid *Lolliguncula brevis* with most of its chromatophores contracted. *(A, Drawn from a photograph in Budelmann, B. U., Schipp, R., and von Boletzky, S. 1997. Cephalopoda. In Harrison, F. W., and Kohn, A. J. (Eds.): Microscopic Anatomy of Invertebrates. Vol. 6A: Mollusca. Wiley-Liss, New York. pp. 119–414; after Florey, E. 1969. Ultrastructure and function of cephalopod chromatophores. Am. Zool. 9:429–442.)*

although no pigments are present. Iridiocytes can be silver, pink, green, blue, yellow, red, or gold. Their colors are typically iridescent and metallic. One type, the passive iridiocyte, produces a single color, whereas active iridiocytes are variable. The coloration of the skin at any particular time is due to the combined effects of the chromatophores and iridiocytes.

Bioluminescence is common in deep-sea teleost fishes, crustaceans, and cephalopods. About half of coleoid genera have bioluminescent organs, or **photophores,** that produce light (Fig. 12-84, 12-61). Photophores may be bacterial, in which the light is produced by symbiotic bacteria, or intrinsic, in which the light is produced by the mollusc itself. The structure of cephalopod photophores is variable, but they typically consist of a cup-shaped **reflector,** often made of iridiocytes, that

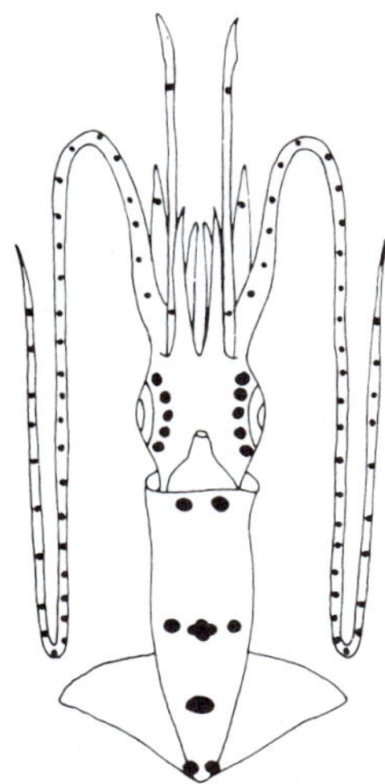

FIGURE 12-84 Light organs (indicated by spots) in the squid *Nematolampas.* Eye and body light organs serve for counterillumination; those on arm tips are of unknown function. *(From Herring, P. J. 1977. Luminescence in cephalopods and fish. In Nixon, M., and Messenger, J. B. (Eds.): The Biology of Cephalopods. Symposia of the Zoological Society of London. 38. Academic Press, London. pp. 127–159.)*

faces the body surface and contains the **photogenic cells,** which produce the light, within its cup. A lens may be present over the mouth of the cup. Superficially, photophores resemble eyes but emit light rather than receive it. They occur on the ink sac, head, appendages, and mantle and are largely restricted to decapods. The two primary roles of photophores are counterillumination of the ventral surface to match the intensity of down-welling light and intraspecific communication.

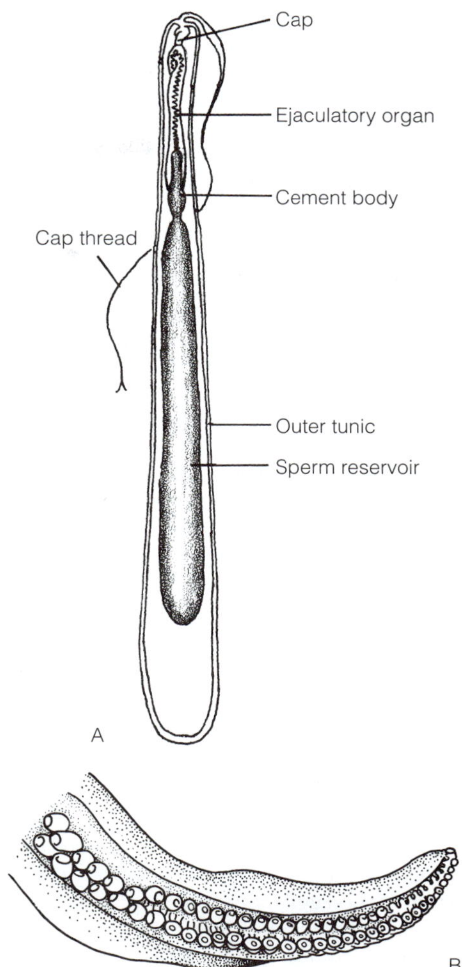

FIGURE 12-85 **A,** Spermatophore of *Loligo.* **B,** Hectocotylus arm of the squid *Loligo roperi.* *(B, From Nesis, K. N. 1987. Cephalopods of the World. T. F. H. Publications, Inc., Neptune City, N. J. 351 pp.)*

REPRODUCTION

Cephalopods are gonochoric and, like the gastropods, have a single posterior gonad associated with the coelom. Fertilization is usually external (usually in the mantle cavity) but sometimes internal; sperm transfer is indirect with copulation and involves spermatophores and an intromittent organ. In cephalopods the coelom consists of the pericardial cavity and a large gonocoel (Fig. 12-64). The gonad, either ovary or testis, bulges into the gonocoel and releases gametes into it. In each sex the gonocoel connects with the mantle cavity via a gonoduct derived proximally from a coelomoduct and distally from an ectodermal invagination of the mantle epithelium.

In the male the gonoduct is unpaired and consists of the convoluted vas deferens and a seminal vesicle with ciliated, grooved walls. In the seminal vesicle the sperm are rolled together to form large, elongate spermatophores (Fig. 12-85A), which are stored in a large **spermatophoric sac** (Needham's sac). The gonoduct extends as a tube, known as the penis, from the spermatophoric sac into the mantle cavity (Fig. 12-79). Although the male gonopore is at the distal end of the penis, the penis is not an intromittent organ. Sperm transfer is indirect.

In females the oviduct may be single or paired and extends from the gonocoel to the **oviductal gland** and then continues to the gonopore in the mantle cavity. A pair of large ectodermal **nidamental glands** opens into the mantle cavity. Accessory nidamental glands of unknown function but containing populations of symbiotic bacteria are also present. There may be a seminal receptacle for the storage of spermatophores or sperm. In octopods the seminal receptacle is part of the oviduct, but in most it is in the mantle cavity and independent of the duct.

Courtship and Fertilization

Fertilization may take place within the mantle cavity or in the sea outside, but in either case it always involves copulation and indirect sperm transfer. One or more of the arms of the male is modified to serve as an intromittent organ known as a **hectocotylus** (Fig. 12-85B). The degree of modification of the hectocotylus varies but typically involves adaptation of the tip of an arm, and its suckers, to pick up and transport spermatophores. In *Sepia* and *Loligo* several rows of suckers are smaller and form an adhesive area for handling spermatophores. In *Octopus* the tip of the arm carries a spoonlike depression and in *Argonauta* and others there is actually a cavity or chamber where the spermatophores are stored.

Before copulation a male cephalopod performs various displays that serve to identify it as a potential mate to the female. The male cuttlefish *Sepia* presents a striped color pattern and establishes a temporary bond with a female, swimming above her. The display is also directed toward intruding males and, if there is a confrontation, the weaker male departs. In pelagic cephalopods copulation occurs while the animals are swimming, the male grasping the female head-on (Fig. 12-86). During copulation the hectocotylus picks up spermatophores from the spermatophoric sac. The hectocotylus is then inserted into the female's mantle cavity to deposit the spermatophores in the appropriate location for the species. This may be on the mantle wall near the gonopores or, as in *Octopus,* into the genital duct itself. In *Loligo* and other taxa of the same family, the hectocotylus may be inserted into the female's horseshoe-shaped seminal receptacle, located in a fold beneath the mouth. The buccal membrane alone receives the spermatophores in *Sepia.*

The spermatophore is shaped like a club or baseball bat and consists of an elongate sperm mass (sperm rope), a cement body, a coiled, springlike, ejaculatory organ, and a cap with a thread attached (Fig. 12-85A). Upon transfer to the female, the cap falls off and the sperm reservoir is evaginated by an increase in pressure between the outer tunic and the reservoir, created by the swelling of the jelly that fills this space. The sperm reservoir attaches to the female mantle wall by the cement body. Sperm are released and used immediately to fertilize eggs or stored until oviposition occurs at a later time.

As the eggs pass through the oviductal gland, they are enclosed in a capsule composed of its gelatinous secretions. Further coatings may be added in the mantle cavity by the nidamental glands. In *Loligo,* secretions from the nidimental glands surround the eggs in a gelatinous mass. In the mantle cavity, the eggs are fertilized by sperm from the remains of the spermatophore or from the seminal receptacles. The eggs and their coatings leave the mantle cavity, often though a funnel formed by the arms. The female attaches the fertilized eggs to the substratum in clusters or strings (Fig. 12-87A). The gelatinous covering of each mass hardens on exposure to seawater and the individual egg capsules swell to several times the original diameter. Large numbers of *Loligo* come together to copulate and spawn at the same time, and a "community pile" of egg strings may be formed on the bottom. Adults usually die shortly after spawning.

Sepia deposits its eggs singly but attaches each by a stalk to seaweeds or other objects. *Octopus* egg clusters resemble a bunch of grapes and are attached in rocky recesses. Female benthic octopods remain to ventilate the eggs after they are deposited. The females die after brooding their eggs. The egg masses of deepwater and some pelagic squids, such as the oegopsids, may be free-floating rather than attached.

A remarkable adaptation for egg deposition occurs in the pelagic genus *Argonauta,* commonly known as the paper nautilus. The two dorsal arms of the female secrete a beautiful, calcareous case into which the eggs are deposited. The case is carried about and serves as a brood chamber. The posterior of the female usually remains in the case and when disturbed, she withdraws completely into the retreat. During copulation, the male hectocotylus breaks off in the female mantle cavity and remains active and wiggling. It was thought by early biologists to be a parasitic worm and was described under the name *Hectocotylus.* The name is now applied to the intromittent appendage of all cephalopods.

Reproduction is assumed to be under hormonal control, although endocrinological studies have been largely restricted to octopods. Hormones are produced by a pair of spherical optic glands associated with the optic tracts. The secretions not only regulate the production of eggs and sperm but also, following spawning, cause the female to cease feeding and to brood her eggs. Death follows the reproductive period in both sexes. If the optic glands of brooding females are removed, they stop brooding and resume feeding, and the life span is extended. In contrast to coleoids, *Nautilus* can breed annually for a number of years.

DEVELOPMENT

Cephalopod eggs are yolky (macrolecithal) and larger than most mollusc eggs. The eggs of some, such as *Sepia* and *Ozaena,* are especially large and may reach 15 mm in diameter. Cleavage is superficial and meroblastic, with only the cytoplasm at the animal pole undergoing cleavage. This results in the formation of a **blastodisc** of cells at the animal pole (Fig. 12-87B) that develops into the embryo (Fig. 12-87C). During gastrulation the margin of the disc grows down (by epiboly) and around the yolk mass to enclose it in a yolk sac that is eventually incorporated into the gut. Above the yolk mass, the disc differentiates upward to become the embryo with a dorsal shell gland, mantle, gills, and lateral eyes. The funnel and arms develop ventrally on the disc as part of the foot and then migrate to more dorsal

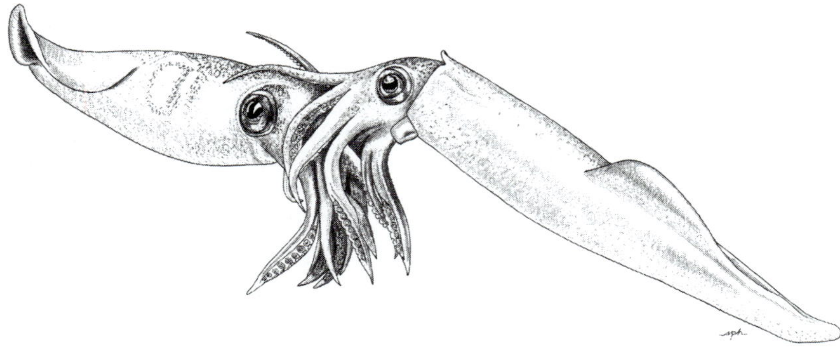

FIGURE 12-86 Copulation in the squid *Loligo.* *(Based on a photograph by Robert F. Sisson.)*

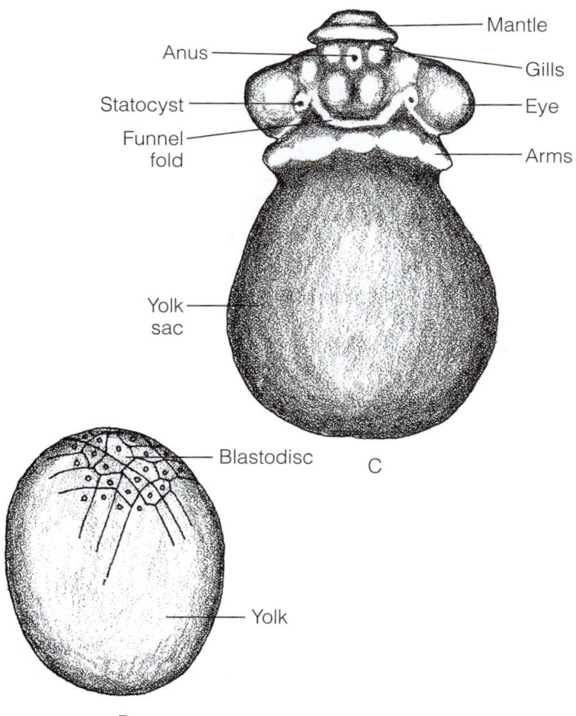

FIGURE 12-87 Development. **A,** Egg cluster of the Caribbean reef squid *Sepioteuthis sepioidea.* One of the young has just hatched, and others can be seen within the eggs. **B,** Discoidal meroblastic cleavage in *Loligo.* **C,** Embryo of *Loligo.* *(B, After Watase from Dawydoff; C, After Naef from Dawydoff.)*

positions, with the arms forming a circle around the mouth and the funnel remaining ventral to it. The yolk is gradually absorbed as it nourishes the developing embryo.

Although development is direct and there is no trochophore or veliger larva, the hatchlings may be planktonic for a time. This is true even of octopods that do not take up a benthic existence until they have reached a larger size. In some pelagic squids, the juveniles live at higher levels than do the adults. The shell of newly hatched *Nautilus* is about 25 mm in diameter and contains seven septa.

Many cephalopods have short lifespans. Squids (such as *Loligo*) live only one to three years, depending on the species, and usually die after a single spawning. *Octopus vulgaris* also dies after one brood, when the animal is two years old. *Nautilus,* however, may have a life span of 15 or more years, taking as long as 10 years to reach sexual maturity, after which there is little growth of the shell.

DIVERSITY OF CEPHALOPODA

Nautiloidea^{sC}: Small taxon with *Nautilus* being the only living genus (Fig. 12-65, 12-66A,C). Soft anatomy of fossil nautiloids is unknown. The four living species have two pairs of gills (tetrabranchiate), two pairs of nephridia, an external (ectocochleate) straight or planispiral chambered shell, abundant arms, pinhole camera eye with no lens, no ink sac, no chromatophores. Shell sutures simple and the many arms lack suckers. Shells of all Recent nautiloids are planispiral. *Nautilus, Plectronoceras.*

Ammonoidea^{sC}: Extinct and known solely from their shells, which are coiled and external (ectocochleate), with complex interlocking sutures (Fig. 12-66B,D, 12-67). Ammonoids appeared in the seas of the Silurian period and persisted to the Cretaceous. *Ancyloceras, Ceratites, Pachydiscus, Scaphites, Turrilites.*

Coleoidea^{sC} (Dibranchiata): Includes squids, cuttlefish, octopods, and vampyromorphs, all having 10 or fewer arms, two gills, two nephridia, an internal or vestigial shell, chromatophores, ink sac, and large brain. Coleoids first appear in the fossil record in the Jurassic period but may have been present earlier.

Belemnoidea^{SO}: Extinct. The first known cephalopods with an internal, three-part shell. Soft anatomy unknown. May have been early coleoids or coleoid ancestors.

Decabrachia^{SO}: Squids and cuttlefishes, which are most cephalopods. Eight arms and two longer tentacles, an internalized shell.

Sepioidea^{O}: Shell is internal and chambered in most. Arms lack hooks.

Spirulida^{sO}: Only one species, *Spirula spirula* (Fig. 12-63, 12-69B), with internal, calcified, chambered, planispiral shell. Sometimes included in Sepiida.

Sepiida^{sO}: Cuttlefish (Fig. 12-74A). Calcified, internal, chambered, straight, septate shell, or it may be reduced or lost. Body usually is short and broad or saclike. Benthic or pelagic. Sepiidae is the only family, with about 100 species, including *Metasepia, Sepia, Sepiella.*

Sepiolida^{sO}: Three families and about 40 species of benthic bobtailed squids with small internal, chitinous pens. *Idiosepius, Rossia, Sepiola.*

Teuthoidea[O]: Squids. By far the largest and most heterogeneous cephalopod taxon. Eight arms, two long tentacles. Shell is an internal dorsal, flattened pen. Body is usually elongate and tubular. Photophores present in many species.

Myopsida[sO]: Transparent corneal membrane over eye. Although there are relatively few species, most are coastal and as a consequence tend to be more familiar than the deep-sea oegopsids. Two families. *Alloteuthis, Loligo* (Fig. 12-74B), *Lolliguncula* (Fig. 12-70A), *Sepioteuthis.*

Oegopsida[sO]: Eyelids and circular pupil, but no corneal membrane. Most squids, and hence most cephalopods, are oegopsids. Largely pelagic in deep water. Twenty-three families. *Abralia, Abraliopsis, Architeuthis* (Fig. 12-70B), *Bathyteuthis, Chiroteuthis* (Fig. 12-71A,B), *Cranchia* (Fig. 12-71C), *Ctenopteryx, Gonatus, Histioteuthis* (Fig. 12-61), *Illex, Ommastrephes, Onychoteuthis.*

Octopodiformes[SO]: Octopods. Eight arms, usuallly without tentacles. Body is a globular sac.

Vampyromorpha[O]: Characteristics are intermediate between squids and octopods and taxon has been classified with both. Represented in fossil record by several species, but only one, the deep-sea *Vampyroteuthis infernalis,* is alive today. Small, deepwater octopod-like cephalopods with eight arms united by an interbrachial web and two reduced, retractable tentacles. Ink sac and chromatophores have been lost.

Octopoda[O]: Eight arms, no tentacles, globular, saclike body, vestigial shell.

Cirrata[sO]: Figure 12-72; small taxon of finned octopods in three families; pelagic, deepwater octopods with gelatinous fins and arms with cirri; arms joined by a broad interbrachial web; ink sac and chromatophores lost and radula reduced. *Cirrothauma, Cirroteuthis,* and *Opisthoteuthis.*

Incirrata[sO]: Figure 12-73; includes most octopods (eight families). Web and fins lost. Most are benthic. *Amphitretus, Argonauta, Eledonella, Enteroctopus, Hapalochlaena, Octopus, Ozaena, Vitrelodonella.*

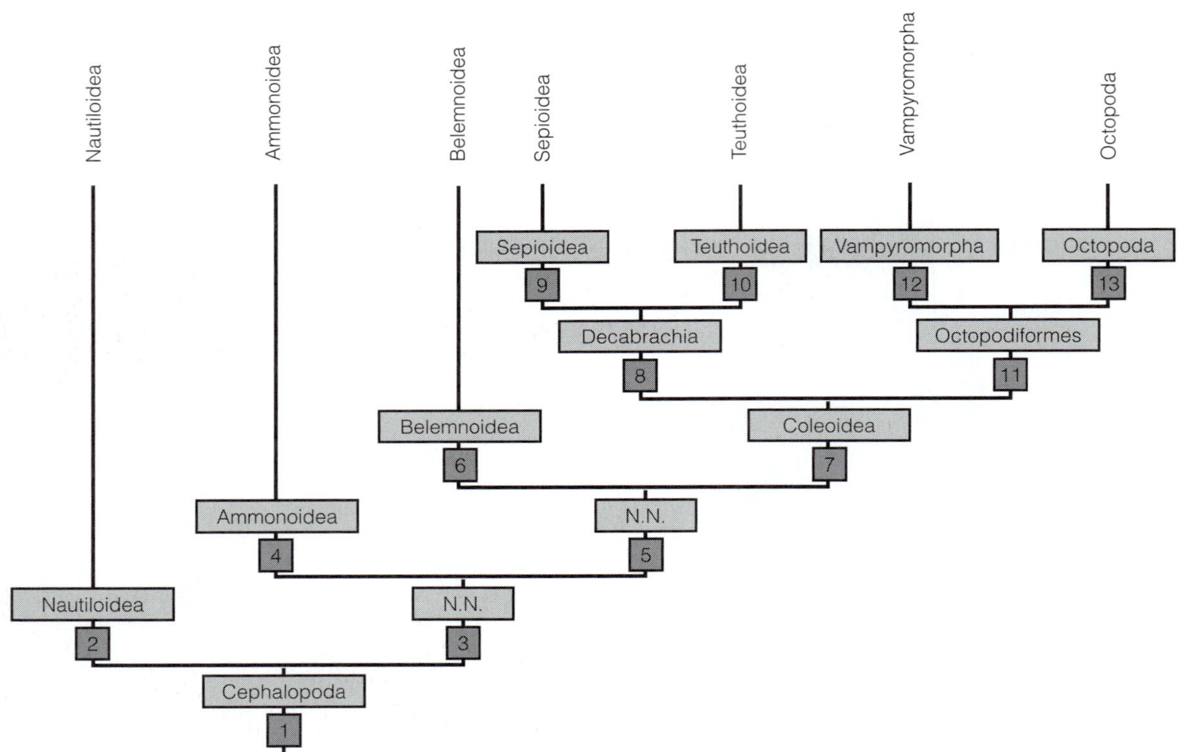

FIGURE 12-88 A phylogeny of Cephalopoda. **1, Cephalopoda:** pelagic habitat with hydrostatic organ and reorientation of major axis; large, beaklike jaws and raptorial feeding; circumoral arms; jet propulsion with funnel; cephalization and concentration of central nervous system; elaborate pinhole camera eyes; yolky eggs and direct development with trochophore lost. **2, Nautiloidea:** two pairs of gills and nephridia. **3, N.N.:** number of circumoral arms reduced. **4, Ammonoidea:** marginal siphuncle, elaborate interlocking sutures. **5, N.N.:** maximum of 10 arms; ink sac; internalization of shell, shell with rostrum; cartilaginous mantle-funnel interlocks; fins; eye with lens. **6, Belemnoidea:** soft anatomy unknown; shell like that of primitive coleoids. **7, Coleoidea:** arms have suckers; hectocotyli for spermatophore transfer. **8, Decabrachia:** fourth pair of arms specialized for prey capture; stalked, movable suckers have a horny rim. **9, Sepioidea:** chambered shell for buoyancy compensation. **10, Teuthoidea:** shell reduced to a pen (gladius). **11, Octopodiformes:** arms reduced to four pairs, connected by interbrachial webbing. **12, Vampyromorpha:** large interbrachial membrane; ink sac lost; chromatophores lost; hectocotylus lost; two arms reduced to filaments. **13, Octopoda:** sacciform body; shell vestigial. *(Adapted from Ax, 2000; Berthold, T., and Engeser, T. 1987. Phylogenetic analysis and systematization of the Cephalopoda (Mollusca). Verh naturwiss. Ver. Hamburg (NF) 29:187–220; Voss, G. L. 1977. Classification of Recent Cephalopods. In Nixon, M., and Messenger, J. B. (Eds): The Biology of Cephalopods. Symposia of the Zoological Society of London, No. 38. Academic Press, London. pp. 75–579; and Hanlon R., and Messenger, J. B. 1996. Cephalopod Behaviour. Cambridge University Press, New York. 232 pp.)*

PHYLOGENY OF CEPHALOPODA

Gastropoda and Cephalopoda are sister taxa (Fig. 12-125) whose common ancestor was probably benthic. The cephalopods have abandoned the bottom to exploit and radiate in open water pelagic habitats, although some have secondarily returned to the benthos. The earliest cephalopods were small molluscs less than 20 mm in length that appeared in the Upper Cambrian about 510 mya (million years ago) and from which arose the nautiloids, ammonoids, belemnoids, and coleoids. Invasion of the open-water habitat necessitated the extensive changes in locomotion, axis orientation, buoyancy, nutrition, sensory equipment, and nervous system that have been discussed in this chapter (Fig. 12-88).

Recent cephalopods consist exclusively of species in the sister taxa Nautiloidea and Coleoidea. The nautiloids have the apomorphies of two pairs of gills and nephridia, as opposed to the single pair of the common ancestor. The chambered external shell, multiple arms, radula, and pinhole-camera eye are plesiomorphies. In Coleoidea the shell is internal, the number of arms has been reduced to about 10, the arms have suckers, the camera eye has a lens, chromatophores and an ink sac are present, the funnel is a closed tube, a hectocotylus arm is present for spermatophore transfer, and auxiliary branchial hearts are present. Plesiomorphies include a single pair each of gills, atria, and nephridia.

Coleoidea includes the sister taxa Decabrachia and Octopodiformes. In the Decabrachia, the fourth pair of arms is lengthened and specialized for prey capture. The arms are equipped with stalked, movable suckers with a horny rim. Their 10 arms are plesiomorphic. Octopodiformes is united by the loss or reduction of the second pair of arms, resulting in a total of only eight functional arms. The arms are interconnected by an interbrachial web.

Octopodiformes consists of the sister taxa Vampyromorpha and Octopoda. Vampyromorpha retains the second pair of arms, albeit greatly reduced, and the well-developed interbrachial web of the ancestor. The loss of the ink sac and the addition of chromatophores and the hectocotylus are autapomorphies. Octopoda share the transformation of the body from an elongate cone to a globular sac. The shell is lost or vestigial, and the second pair of arms is absent.

The phylogenetic relationships of the coleoid orders are not well understood. Octopods may have evolved from a teuthoidlike ancestor with the vampyromorphids diverging early from the octopod line. Sepiids may also have arisen from teuthoidlike ancestors and sepiolids and spirulids from a sepioidlike form.

PHYLOGENETIC HIERARCHY OF EXTANT CEPHALOPODA

- Cephalopoda
  - Nautiloidea
  - Coleoidea
    - Decabrachia
      - Sepioidea
        - Sepiida
        - Sepiolida
        - Spirulida
      - Teuthoidea
        - Myopsida
        - Oegopsida
    - Octopodiformes
      - Vampyromorpha
      - Octopoda
        - Cirrata
        - Incirrata

BIVALVIA^C

Bivalvia (Pelecypoda) includes such common molluscs as clams, oysters, mussels, scallops, and shipworms (Fig. 12-89, 12-107, 12-115, 12-117). Bivalvia includes about 8000 described extant species, of which about 1300 live in fresh water and the remainder are marine. Bivalves range in size from the tiny freshwater fingernail clams, some of which are less than 2 mm in length, to the giant clam *Tridacna* of the South Pacific, which attains a length of over a meter and may weigh almost 300 kg. The common name for bivalve is "clam," and any bivalve is correctly called a clam, even if it is a scallop or oyster.

Bivalvia contains three major morphological groups, the protobranchs, lamellibranchs, and septibranchs, distinguished by differences in their gills (Fig. 12-93) and feeding modes. In the past these groups were considered to be subclasses, but in current classifications the lamellibranchs, which are the overwhelming majority of bivalves, do not receive formal taxonomic recognition. Conceptually, however, the three groupings continue to be useful as levels of organization, and we will make frequent reference to them in this context, using the anglicized, uncapitalized form of the name. Most bivalves are suspension-feeding lamellibranchs, but the deposit-feeding protobranchs are believed to be the most primitive existing bivalves. Septibranchs are specialized carnivores.

Bivalves are highly derived molluscs adapted for exploitation of the infaunal benthic habitat, where they are relatively safe from predators. Unique among the molluscs, bivalves have lost the radula and almost all rely on the gills for food capture. Lateral flattening of foot and shell facilitates burrowing in soft sediments. Filter feeding permits feeding while buried in the sediment, and siphons permit access to fresh food- and oxygen-bearing water without leaving the sediment.

Although modern bivalves have invaded other habitats, the original adaptations for burrowing in mud and sand have taken bivalves so far down the road of specialization that they have become largely chained to a sedentary existence. Within these constraints, bivalves have exploited and radiated extensively into several benthic habitats suitable for sedentary molluscs. Most occupy the ancestral infaunal habitat in soft sediments. Many, however, are adapted for epibenthic life and live on the surface. Others are adapted for exploitation of infaunal habitats accessed by boring into hard substrata such as calcareous rock, shells, clay, peat, or wood. Some are commensal with other invertebrates, such as polychaetes, echinoderms, or crustaceans. A few are parasites and some can swim.

FORM

The bilaterally symmetrical bivalve body is highly modified from that of the ancestral mollusc and is strongly compressed laterally (Fig. 12-90). The head is vestigial in these relatively immobile animals and the sensory equipment is

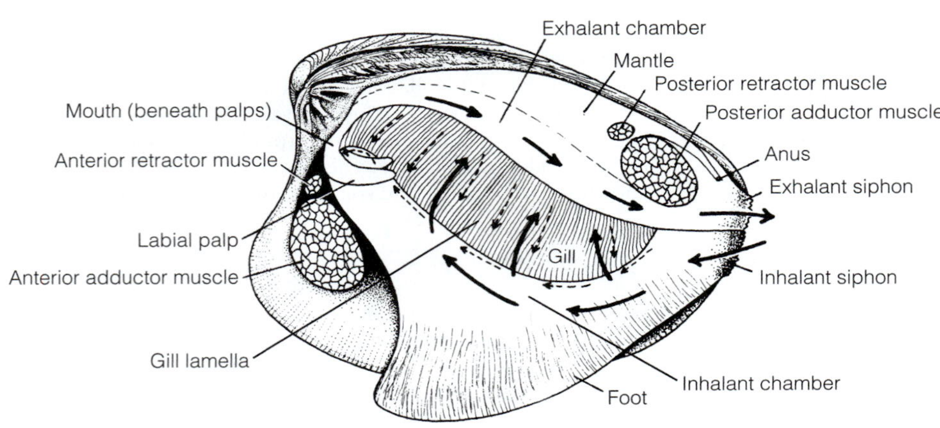

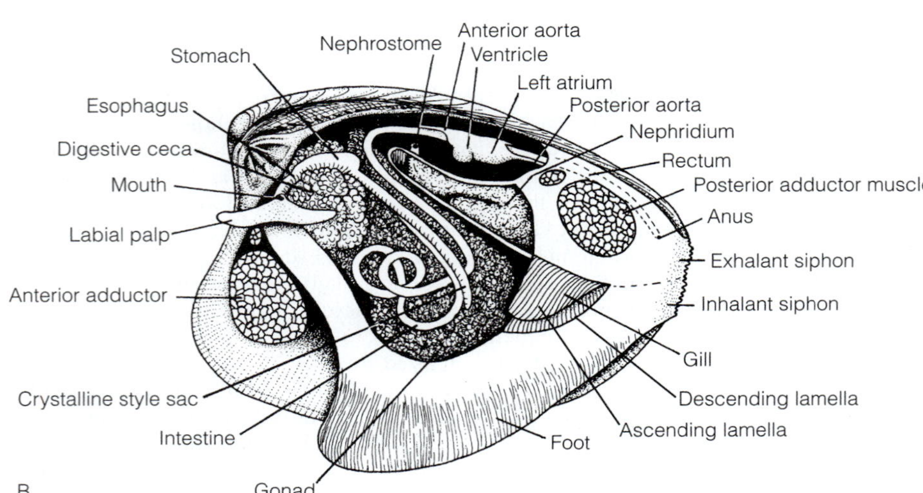

FIGURE 12-89 Anatomy of *Mercenaria mercenaria*. **A,** Interior of the left side with the left valve and left mantle skirt removed. **B,** Partial dissection, showing some of the internal organs.

located elsewhere, chiefly on the mantle margins. The large mantle encloses a pair of lateral mantle cavities, each of which contains a gill. The gills are usually very large, and in most species are responsible for filter-feeding in addition to gas exchange. The unique two-part shell is hinged dorsally and encloses the body. The visceral mass is dorsal under the hinge and a bladelike foot, usually adapted for digging, is attached to its ventral border.

MANTLE

The mantle covers the visceral mass dorsally, but its most conspicuous feature is the pair of enormous lateral lobes, the mantle skirts (Fig. 12-90). These evaginations, or folds, of the dorsal body wall extend laterally and ventrally to the ventral midline, where they may touch, and sometimes fuse together. The skirts enclose the body and a capacious water space, which is, of course, the mantle cavity. The lateral mantle cavity in the bivalves is the most spacious of the molluscs. The skirts are attached to the body dorsally but anteriorly, posteriorly and ventrally they are unattached and terminate in an edge, the mantle margin.

The mantle margin is divided into three longitudinal ridges, or folds, separated by two grooves (Fig. 12-91). The three folds have important but different functions. The **inner fold** is muscular and contains the **pallial** (mantle) **muscles.** These muscles originate in the mantle and insert on the inner surface of the shell. Most of them insert along a curved **pallial line** that extends from anterior to posterior parallel to the margin of the shell (Fig. 12-92B). The **middle fold** is sensory and may bear photo-, mechano-, or chemoreceptors (Fig. 12-115A). The **outer fold,** in conjunction with the entire outer mantle epithelium, secretes the shell. Posteriorly the two mantle skirts form a pair of siphons that channel water into and out of the mantle cavity (Fig. 12-89).

SHELL

A typical bivalve shell consists of two similar (equivalve), more or less oval, usually convex, calcareous **valves** that articulate with each other (Fig. 12-92B) along a dorsal **hinge.** An **equivalve** shell is bilaterally symmetrical, as is the soft anatomy, and the plane of symmetry passes through the hinge and between the right and left valves. Some derived bivalves—oysters for

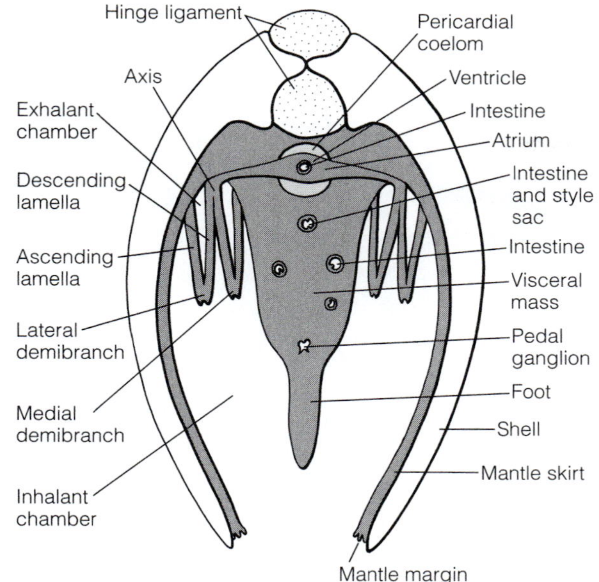

FIGURE 12-90 Cross section of a generalized lamellibranch bivalve.

example—have **inequivalve** shells, in which the right and left valves are dissimilar. The two valves are joined dorsally by a two-part, proteinaceous **hinge ligament** derived from the periostracum (Fig. 12-92). The flexible hinge ligament unites the two valves along the dorsal midline and is part of the shell consisting of the periostracum alone, without the rigid ostracum and hypostracum.

Two transverse **adductor muscles,** one anterior and one posterior, extend from valve to valve (Fig. 12-89A, 12-92A). The hinge ligaments and adductor muscles antagonize each other and work together to open (abduct) and close (adduct) the valves. There are no abductor muscles, however, and the ligament is responsible for opening the valves. Bivalve evolution has produced a predator-resistant alternative to an abductor muscle.

The ligament is composed of inner and outer masses of the elastic protein conchiolin, one on either side of the fulcrum of the hinge where the calcareous valves touch each other (Fig. 12-92A). The **outer ligament** (tensilium) is readily visible outside the shell, but the inner is hidden. The **inner ligament** (resilium) may be housed in a cup- or spoon-shaped depression in the hinge known as the chondrophore. When the valves are open, the ligaments are relaxed and under neither tension nor compression. When the adductor muscles contract, however, the outer ligament is stretched and under tension, whereas the inner ligament is compressed. Both ligaments are deformed, but they are elastic and, if given the opportunity, will return to their original resting shape. Their elasticity allows them to store some of the energy of the muscle contraction. When the adductor muscles relax, the ligaments are free to return to their original shapes, the outer ligament by contracting and the inner ligament by expanding. Both responses abduct the valves and, equally important, re-extend the adductor muscles so they are ready to contract again when needed.

In most bivalves the anterior and posterior adductor muscles are similar and about the same size. This is the **isomyarian** (iso = same, myo = muscle) condition. Sometimes, however, the anterior adductor is reduced, as in marine mussels and pen clams, and is much smaller than the posterior adductor. This is the **anisomyarian** condition. In other bivalves, such as scallops, the anterior adductor is completely lost and only the posterior adductor, which is very large, is present. This is the **monomyarian** condition. The posterior adductor muscle, by the way, is the edible part of the scallop.

The attachment positions of the muscles are visible on empty valves as **muscle scars** (Fig. 12-92B). The valves also bear the scars of other muscles, such as the pedal retractor and pallial muscles.

Adductor muscles must be capable of two quite different actions. On one hand, they must be capable of rapidly

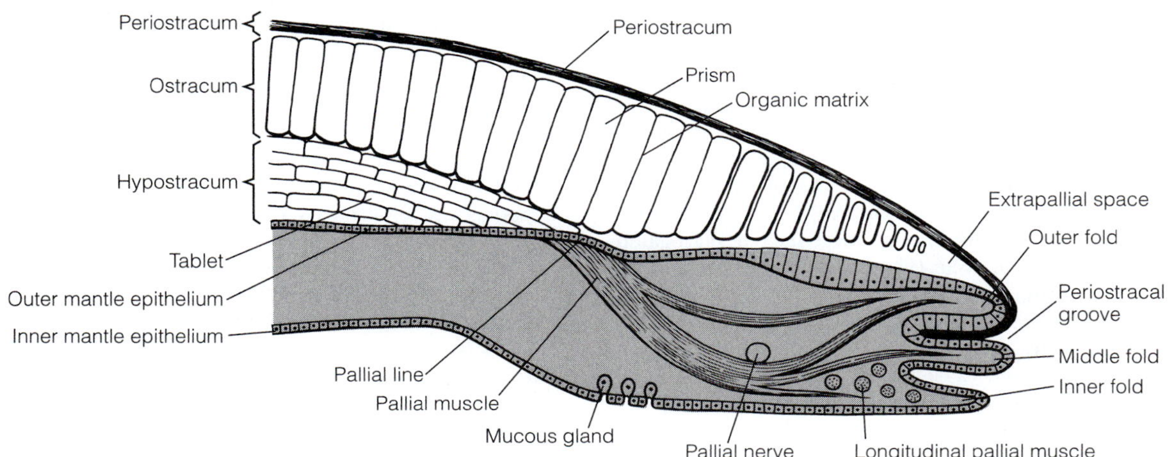

FIGURE 12-91 Diagrammatic transverse section of the edge of a bivalve valve and mantle margin. *(Redrawn and modified after several authors, including Wilbur K. H., and Saleuddin, A. S. M. 1983. Shell formation. In Saleuddin, A. S. M., Wilbur, K. H. (Eds): The Mollusca. Vol. 4. Pt.* I. *Physiology. Academic Press, New York. pp. 235–287; and Kennedy, W. J., et al., 1969. Environmental and biological controls on bivalve shell minerology. Biol. Rev. 44:499–530.)*

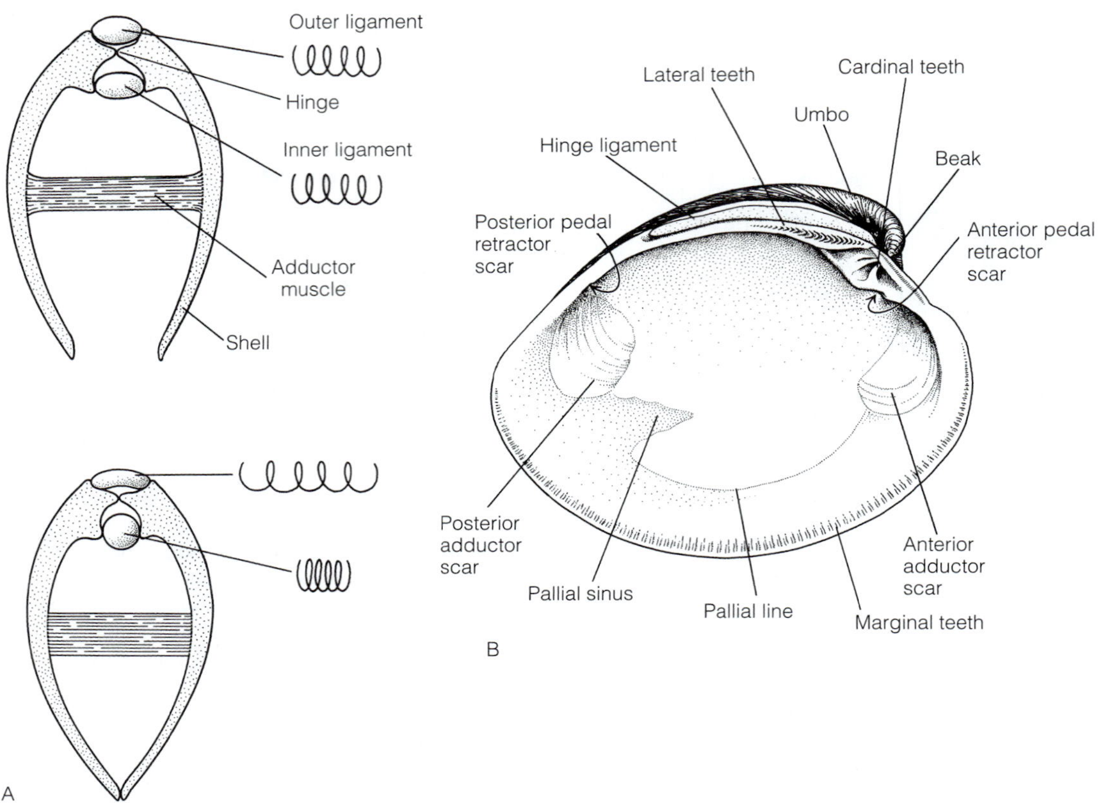

FIGURE 12-92 Bivalve shell. **A,** Transverse section showing antagonistic functions of hinge ligaments and adductor muscles. When the valves are closed by the adductor muscles, the outer hinge ligament is stretched and the inner ligament is compressed. **B,** Inner surface of the left valve of the marine clam *Mercenaria.*

contracting to close the valves when threatened or to expel pseudofeces. They must also be capable of maintaining the sustained contractions required to prevent desiccation when the tide is out or to foil a persistent predator. To accomplish these tasks, bivalve adductor muscles are composed of two fiber types. **Quick muscles** are cross-striated fibers capable of rapid contractions, but they fatigue quickly and cannot remain contracted. **Catch muscles,** on the other hand, are smooth fibers that respond slowly but are capable of remaining contracted for extended periods. These may be segregated in separate parts of the muscle (Fig. 12-115B) or they may be mixed together throughout the muscle.

The adductors typically are large, powerful muscles capable of contracting with great force. Were they not prevented from doing so, the force would cause the valves to shear anteriorly and posteriorly past each other. Such slippage is prevented by complementary **teeth** and sockets on the valves that interdigitate. **Hinge teeth,** such as the cardinals and laterals, are associated with the hinge, whereas **marginal teeth** occur along the entire periphery of the valves (Fig. 12-92B). In primitive bivalves the hinge teeth are **taxodont,** meaning numerous and similar, but in derived taxa the number is reduced and the teeth are specialized **(heterodont).** Heterodont teeth are usually divided into cardinal and lateral teeth. **Cardinal teeth** typically are tooth-shaped projections near the center of the hinge (Fig. 12-92B). **Lateral teeth** are anterior or posterior to the cardinal teeth and usually are longitudinal ridges that may not resemble teeth.

Beside the hinge, each valve bears a dorsal protuberance called the **umbo** (beak), which rises above the hinge. The umbo is the oldest part of the shell, and each valve grows outward in concentric rings from the umbo.

Bivalve shells exhibit great variety in size, shape, surface sculpturing, and color. When the addition of new shell material occurs equally around the umbo, the result is a valve that is symmetrical around a vertical (dorsoventral) axis through the umbo. Such a valve is said to be **equilateral** (Fig. 12-109A). If the growth rate differs around the periphery of the valve, then it is **inequilateral** and the anterior end is not a mirror image of the posterior (Fig. 12-108A).

As in gastropods, surface sculpturing may contribute to traction, protection, or shell strength. The familiar ribbing, or corrugations, of cockle and some scallop shells, for example, increases the shell strength (Fig. 12-115A).

Shell Structure

A typical bivalve shell includes a minimum of three layers, one organic and two calcareous. The outermost layer is the proteinaceous periostracum, consisting of quinone-tanned proteins generally known as conchiolin. The periostracum may be very thick, as in many large, freshwater clams, or very thin, as in the edible marine quahog, *Mercenaria mercenaria.* The periostracum probably plays a role in the secretion of the outer calcareous layer of the shell. It protects the underlying calcium carbonate from dissolution in acidic and soft water (low Ca^{2+}) and may also facilitate formation of a tight seal when the valves are adducted.

Under the periostracum are two to four layers of crystalline calcium carbonate. Two basic layers are always present, but there may be others. In these layers calcium carbonate is deposited in crystals. Crystals may be formed in a variety of shapes, including prisms, granules, tablets, lenses, or more complex forms. The basic crystal units are always deposited within an organic framework, which together with the periostracum may account for 12% to 72% of the dry weight of the shell. Although shell structure is not uniform throughout Bivalvia, it is constant and characteristic within different subtaxa. The calcareous layers may be entirely aragonite in primitive bivalves or a mixture of aragonite and calcite in derived taxa. Calcite has the advantage of being less soluble than aragonite.

Of the two calcareous layers, the outermost is the ostracum (prismatic layer) composed of calcium carbonate crystals in a prismatic structure in which the crystals are deposited in columnar prisms perpendicular to the mantle epithelium (Fig. 12-91). The crystals are secreted by columnar cells of the outside surface of the outer mantle fold distal to the pallial line.

The inner crystalline layer is the hypostracum (lamellar layer), secreted by cuboidal cells covering the outer surface of the mantle skirt, proximal to the pallial line (Fig. 12-91). The crystals are deposited in sheets (lamellae) parallel to the mantle epithelium.

In many bivalves (pteriomorphs and paleoheterodonts) the hypostracum is composed of layers of thick tabular crystals and is said to be nacreous. **Nacre** (NAKE-ur), or mother-of-pearl, is the smooth, pearly, lustrous inner lining of many shells. A nacreous hypostracum is composed of aragonite, and the crystal units are flat tablets arranged in parallel sheets. A nonnacreous hypostracum composed of calcite crystals is characteristic of the higher bivalves (heterodonts). Nacre is found in the monoplacophorans, gastropods, cephalopods, and bivalves.

Shell Secretion

Molluscan shell secretion is best understood in bivalves, where it has been most carefully studied. In all molluscs the mantle epithelium is responsible for the secretion of all shell layers. The inner surface of the outer fold (of the mantle margin) secretes the periostracum. The secretory cells responsible for this are deep in the **periostracal groove** (Fig. 12-91), between the outer and middle folds. The protein is secreted in soluble plastic form that is then sclerotized (hardened) and darkened by quinone tanning, a process in which fibrous proteins are cross-linked to form a strong, dark, insoluble layer.

The outer surface of the outer fold secretes the ostracum. The entire outer mantle epithelium, proximal to the pallial line, secretes the hypostracum. There is a fluid-filled **extrapallial space** between the mantle epithelium and periostracum. Calcium salts and the proteinaceous matrix are first secreted into this compartment and from there are deposited in the shell. The periostracum seals the extrapallial space and isolates it from the surrounding environment.

PEARLS

Despite the muscular and periostracal attachment of the mantle to the shell, occasionally some foreign object, such as a sand grain or a parasite, enters the extrapallial space. The object becomes a nucleus around which concentric layers of nacreous shell are deposited, resulting in the formation of a **pearl.** If the object moves frequently during secretion, the pearl may be spherical or ovoid. More commonly, however, the developing pearl adheres to or becomes completely embedded in the shell.

Pearls can be produced by most shell-bearing molluscs, but only those with a nacreous hypostracum produce pearls of commercial value. The finest natural pearls are produced by the pearl oysters *Pinctada margaritifera* and *P. mertensi,* which inhabit the tropical and subtropical Pacific.

There are two methods, seed and bead, of producing cultured pearls. Some cultured pearls are initiated with a microscopic "seed" ground from the shell of a freshwater (unionid) mussel and placed into the extrapallial space of a pearl oyster. The oyster coats the seed with a layer of nacre to produce a year-old seed pearl, which is then transplanted into another oyster. A pearl of marketable size is obtained three years after transplantation.

Bead pearls are produced more quickly by starting with a spherical shell bead, also cut from the shell of a freshwater mussel. These beads are much larger than the seeds used in seed pearls. They are, in fact, only slightly smaller than the finished pearl. The bead is enclosed in a wrapping of mantle tissue and transplanted into the mantle or soft tissue (often the gonad) of another pearl oyster. A nacreous veneer approximately 1 mm thick is then laid down around the bead.

About 12 species of native North American freshwater mussels, occurring from Wisconsin to Alabama, supply about 95% of the demand for bead stock from the bead pearl industry. In 1994, 2700 tons of native mussels were harvested for this purpose in Tennessee alone. Recently, new techniques have been developed for culturing high-quality pearls in freshwater mussels, but unfortunately, the new methods continue to rely on beads cut from other freshwater mussels. The extremely diverse native mussel fauna of the southeastern United States is threatened by a number of human activities, including harvesting, siltation, impoundments, pollution, and introduction of exotics.

FOOT

In the evolution of bivalves from the ancestral conchiferan, the broad, flat, creeping foot became laterally compressed, blade-like, and directed anteriorly as an adaptation for burrowing in soft sediments (Fig. 12-89, 12-90). It is muscular and contains the large pedal sinus of the hemocoel. A combination of blood pressure and the action of pedal retractor muscles effect foot movement, which will be described in detail later in connection with burrowing. These latter muscles, which are homologous to the pedal retractors of other molluscs, form the lateral sides of the foot and extend to the opposite valve, where they are usually attached to the shell at scars near those of the adductor muscles (Fig. 12-92B, 12-112A). Other specializations of the bivalve shell, mantle, and foot will be described later in connection with different adaptive groups.

GILLS AND THE EVOLUTION OF BIVALVE FEEDING

Most modern bivalves are suspension feeders that use the gills to filter particulate organic matter, usually phytoplankton, from the water. The ancestral bivalves, however, were probably selective deposit feeders that used the gills only for gas exchange. Bivalve evolution is essentially the story of the evolution of filtering gills from the purely respiratory gills of the ancestral bivalve.

As in other microphagous molluscs, utilization of fine particulate food requires ciliary sorting fields that separate organic food particles from indigestible mineral particles. In the bivalves, these fields are both internal, in the stomach, and external, on the gills and labial palps. The internal fields are inherited with little modification from the ancestral mollusc, but the adaption and modification of external ciliary fields is an important aspect of the evolution of bivalve feeding.

Protobranchs

The ancestral bivalve was a protobranch similar to living protobranchs in the taxa *Nucula, Nuculana, Yoldia, Malletia,* and *Solemya.* It probably lived partially buried in soft sediments. It had a single pair of posterolateral bipectinate gills, from which the name (protobranch = first gills) is derived (Fig. 12-93A, 12-94A). The sole of the protobranch foot is divided into right and left halves and is often fringed with papillae. From this ancestor evolved the lamellibranch (meaning "sheet gill") bivalves, with large, flat, filter-feeding gills and an undivided foot lacking papillae.

Most living protobranchs, and presumably the ancestral ones as well, are selective deposit feeders, which many zoologists believe was the feeding mode of the ancestral bivalves. The ciliated papillae of the foot, perhaps in conjunction with long, ciliated tentacles, were used to locate food particles in the sediment and transfer them to the mouth. The radula had no role to play in this process and was lost. The gills were small and used solely for respiration. The major respiratory current probably entered the mantle cavity anteriorly, passed across the gills, and exited posteriorly (Fig. 12-94A), similar to the pathway in monoplacophorans, polyplacophorans, and the generalized mollusc. Minor water flows could enter anywhere around the periphery of the valves. In its life orientation this ancestor probably laid on its side, only partially buried in the sediment.

Protobranch gills are located in the posterior mantle cavity and are small in comparison with those of lamellibranchs. They are similar to those of the stem mollusc and primitive living molluscs such as chitons in being bipectinate with a central axis, from which arise two rows of flat filaments (Fig. 12-94B). The apposing surfaces of adjacent filaments bear lateral cilia and the margins have frontal and abfrontal cilia (Fig. 12-1D). The lateral cilia create the respiratory current through the gills whereas the frontal and abfrontal cilia remove unwanted particles that accumulate on the gill surface.

In the ancestral molluscs, the mouth rested against the hard bottom over which the animal crawled and a radula was used to scrape particles from it. When bivalves abandoned this ancestral habitat and became adapted for burrowing in soft sand or mud, the mouth was elevated above the substratum. This occurred as the body became laterally compressed and the height of the dorsoventral axis increased. Modern protobranchs maintain feeding contact with the substratum with a pair of elongate **palpal tentacles** (proboscides) at the posterior end of the mantle cavity (Fig. 12-94A). Each tentacle is associated with a large, bilobed **labial palp,** located laterally, one on each side of the mouth and extending anteriorly to the mouth. The labial palp is a large sheet of ciliated tissue folded on itself like a menu or greeting card. When in its normal position, the "menu" is closed, so its two lobes face each other. The palp shown in Figure 12-94D, however, is open to reveal the details of its inner surface. The apposing surfaces of the two lobes are covered with ciliated ridges and grooves that sort mineral and organic particles, discarding the

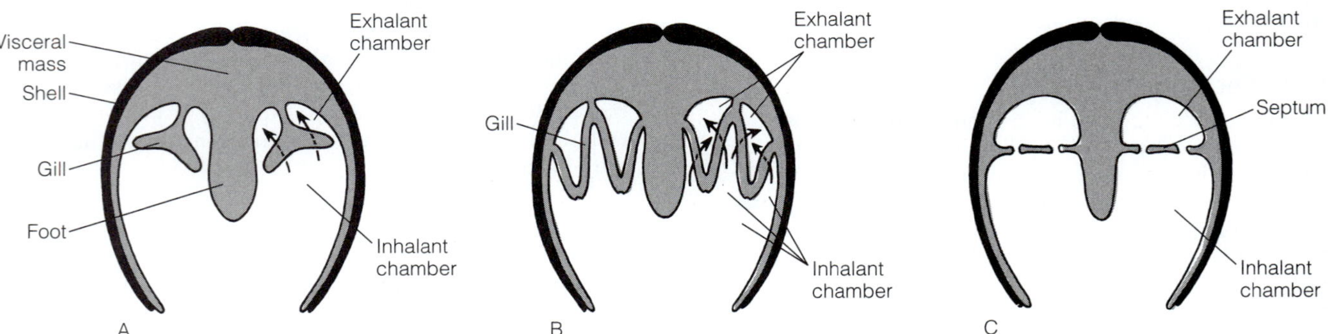

FIGURE 12-93 Diagrammatic cross sections of the three grades of bivalve molluscs emphasizing the gills and their relationship with the mantle cavity. Arrows show direction of water flow. **A,** Protobranch. **B,** Lamellibranch. **C,** Septibranch.

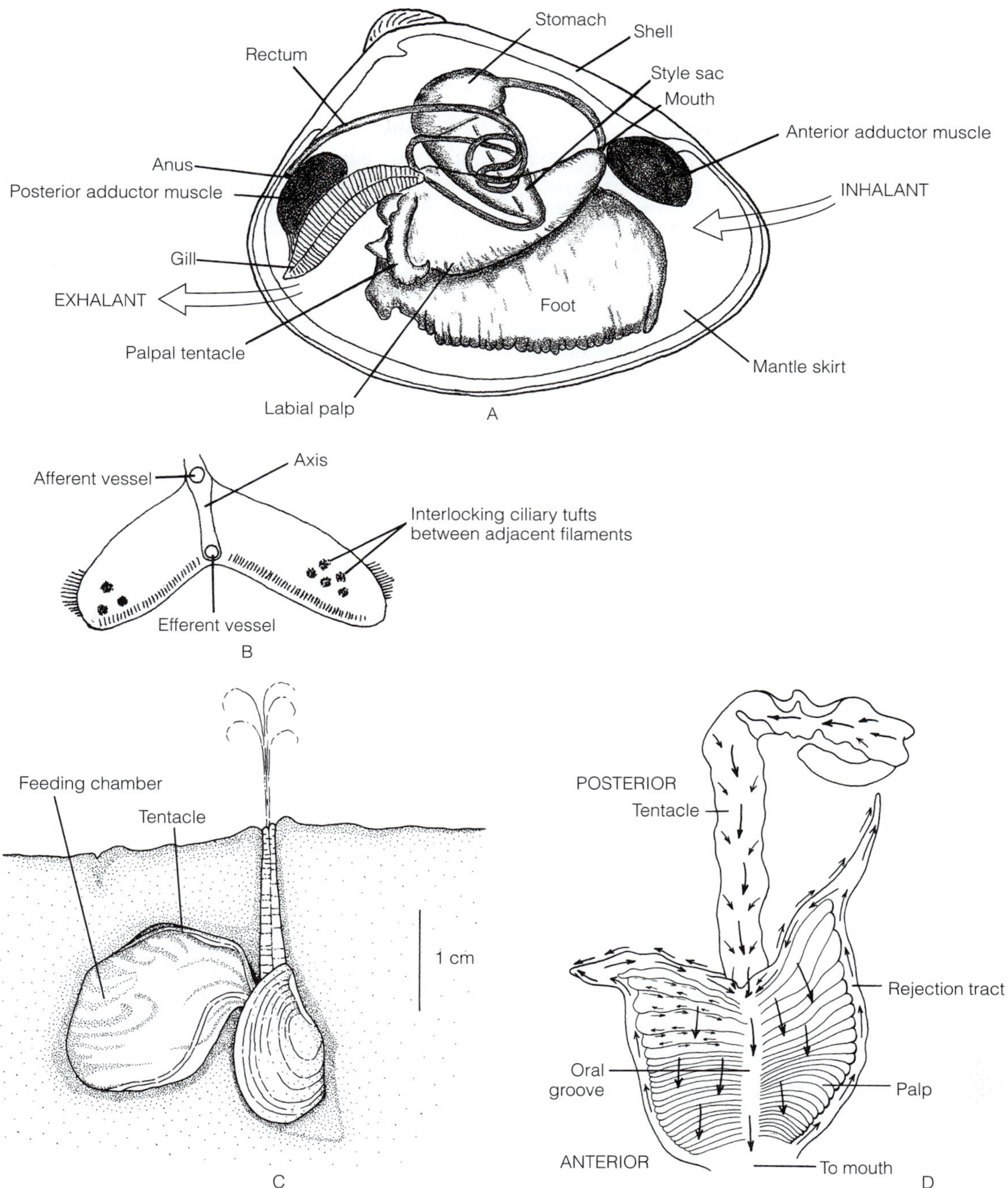

FIGURE 12-94 Feeding in protobranchs. **A,** Lateral view of *Nucula* with right valve and mantle removed. Anterior is to the right. **B,** Transverse section of the gill of *Nucula,* showing lateral filaments and central axis. **C,** The protobranch *Yoldia limatula* deposit-feeding within a chamber excavated below the surface. The palpal tentacles are visible in the chamber. Unlike *Nucula, Yoldia* has siphons that reach to the surface. This species may also extend its palpal tentacles to the surface and feed on particles it finds there. **D,** One labial palp and its palpal tentacle from *Nuculana minuta.* Apposing lobes of the palp have been pulled apart to reveal the ciliated sorting field. Arrows indicate the direction of ciliary currents. *(A and B, After Yonge, C. M. 1939. The protobranchiate Mollusca, a functional interpretation of their structure and evolution. Phil. Trans. Roy. Soc. London B 230:79–147; C, From Bender, K., and Davis, W. R. 1984 The effect of feeding by* Yoldia limatula *on bioturbation. Ophelia 23(1):91–100; D, After Atkins.)*

former and sending the latter to the mouth. The particles are gathered by the palpal tentacles and transferred to the palpal sorting field by cilia.

During feeding, the tentacles are extended into the bottom sediments. Particles, both mineral and organic, adhere to the mucus-covered surface of the tentacle and are transported by cilia from the posterior end of the palp (Fig. 12-94D) to the sorting field, where cilia located on the ridges, in the grooves, and on the slopes between the ridges and grooves sort incoming particles, more or less on the basis of size (Fig. 12-95). **Ridge cilia** generate a current perpendicular to the long axis of the ridges (across the ridges) and thus toward the mouth (Fig. 12-94D). **Groove cilia** generate a current parallel to the long axis of the groove and thus toward the periphery of the palp. **Slope cilia** beat downslope to move particles of the proper sizes into the grooves. Coarser particles cannot fit into the grooves and remain on the ridges, where they tend to be moved by ridge cilia toward the mouth. Fine particles small enough to be swept by slope cilia into the grooves are moved to the palp margins by the groove cilia. The sorting is by no means perfect and many mineral particles are sent to the mouth.

Cilia along the free margins of the palps beat posteriorly and form **rejection tracts** (Fig. 12-94D). Small particles, mostly mineral, from the grooves move to the rejection tracts and are transported, entangled in mucus, posteriorly to the tips of the palps. The clumps of mucus and mineral particles fall off the palps into the mantle cavity. This waste, known as pseudofeces because of its resemblance to egested material from the anus, is periodically ejected from the mantle cavity by forceful adduction of the valves. Note, however, that it has never been ingested and is not truly feces.

The distance between ridges can be adjusted by expanding or contracting the palps. Changing the distance between adjacent ridges changes the size criteria for acceptance and rejection. If the ridges are closer together, smaller particles are accepted, whereas if the ridges are moved farther apart, larger particles will be rejected.

In other sorting fields, such as those of the gills, the criteria for acceptance and rejection may differ from those of the palps. For example, coarse material could be rejected and fine material accepted. A mixture of coarse sand particles, intermediate food particles, and fine clay particles, for example might be sorted into coarse (sand) particles for rejection by one sorting field, with fine (food and clay) particles being sent for further sorting by a second field. Using the opposite criteria, the second field would accept coarse particles (food) and reject fine (clay).

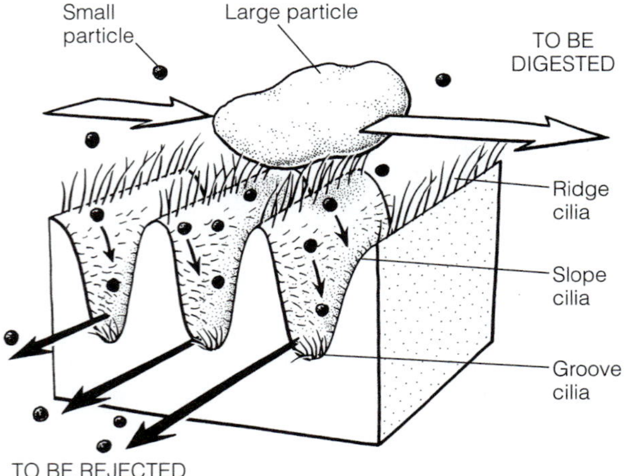

FIGURE 12-95 Diagrammatic cross section of the ciliary sorting field of the labial palps. In this model, large organic particles (food) cannot fit into the grooves and are moved by ridge cilia to the mouth. Small mineral particles (nonfood) slip into the grooves and will be discarded.

Lamellibranchs

EVOLUTION OF LAMELLIBRANCH GILLS

Selective deposit feeders, such as the protobranchs, and suspension feeders, such as the lamellibranchs, are both microphages specializing in tiny food particles. Much of the particle-handling equipment of the protobranchs could be adapted to process particles collected by suspension feeding. The early protobranchs were ideally suited morphologically and behaviorally to developing a suspension-feeding mechanism utilizing their gills. As gill-breathing, infaunal deposit feeders, they already had a foot capable of burrowing into soft sediments and a mechanism for creating a flow of water through the mantle cavity and across the gill. Furthermore, they had a gill that, although used exclusively for gas exchange, was constructed in such a way that it inadvertently trapped particles too large to pass between the filaments on its upstream (frontal) surface. What's more, there existed a ciliary mechanism for moving these particles off the frontal face of the gill. There was even a sorting device, the labial palps, capable of separating food particles from mineral particles and sending the food to the mouth and the waste elsewhere.

With all this equipment already in place, it was probably inevitable that some ancient protobranch would begin to take advantage of the organic particles unintentionally collected by its gills. Gradually some protobranchs came to rely more on suspension feeding and less on deposit feeding until eventually the lamellibranch bivalves, with their large, efficient, filter-feeding gills, evolved. The simple protobranch gill was modified to improve its efficiency as a filter while retaining its original function in gas exchange. When the lamellibranchs abandoned deposit feeding in favor of suspension feeding they also abandoned the palpal tentacles, but they retained the labial palps and continued to use their sorting fields. They also developed sorting fields on the gills themselves.

What might not have been so predictable was the great success this innovation would enjoy. The new feeding mechanism allowed the lamellibranchs to burrow into soft sediments, where they were relatively safe from predators, while exploiting the rich supply of phytoplankton and oxygen in the water above the sediments. Natural selection favored the lamellibranchs and they underwent an extensive adaptive radiation to exploit most of the possibilities of the infaunal biotope as well as subsequent expansion into many epifaunal benthic habitats suited for sedentary molluscs. Lamellibranchs now dominate the bivalve fauna, and most them are suspension feeders.

GILL MORPHOLOGY

The protobranch gill is not large enough to collect sufficient food to support the animal by suspension feeding alone. In the first modification of the gill, its surface area was increased so it could filter more water and trap more food particles. The gill surface area was enlarged in two ways. An enormous number of additional filaments were added to lengthen the gill so that it eventually extended over most of the length of the mantle cavity (Fig. 12-89A). Oysters, for example, have over 8000 filaments per gill. In addition, the filaments themselves were lengthened (Fig. 12-93B, 12-96) to provide even more surface area. The space between the visceral mass and the mantle skirt could not accommodate the longer filaments and they had to be folded to fit.

In the ancestral molluscan gill filament, the function of which is solely for gas exchange, the important surface is the lateral surface, the one that bears the lateral cilia. The lateral surfaces face the interfilamentary spaces between the filaments and are the major sites of gas exchange (Fig. 12-96B). In the filter-feeding lamellibranch gill, on the other hand, the important surface is the frontal surface with the frontal cilia, for it is here that food particles are trapped. Accordingly, a third modification contributing to the evolution of the lamellibranch gills was the reduction in the lateral surface and enhancement of the frontal surface. The filaments became slender and filamentous (Fig. 12-96C,D), unlike the flat lamellar filaments of the protobranchs. The lateral surface is reduced but, because the gill is now so much larger, there is still ample area for a gas-exchange space for the still-important lateral cilia. The combined frontal surfaces of the filaments form large sheets known as lamellae (hence the name lamellibranch), through which water passes. The filaments, which in protobranchs alternate along the central axis, became opposite (Fig. 12-96D).

The resulting lamellibranch gills are complex, and specialized terminology has developed to facilitate discussion of them. The gill filaments are attached, as in the ancestor, to the central axis, which is itself attached to the mantle in the angle between the visceral mass and the mantle skirt (Fig. 12-96D).

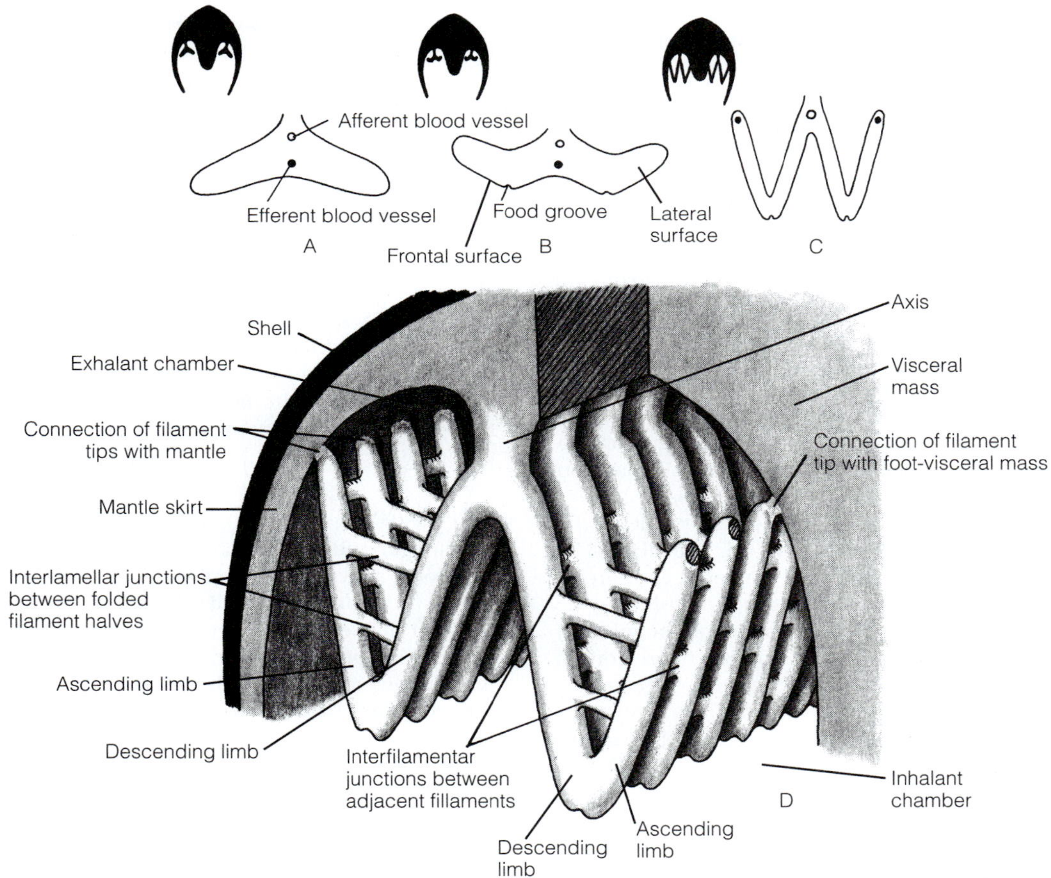

FIGURE 12-96 Evolution of lamellibranch gills. **A,** Primitive protobranch gill (position relative to the foot/visceral mass and mantle indicated in cross section). **B,** Development of the ventral food grooves in a hypothetical intermediate condition. **C,** Folding of filaments at the food groove to produce the lamellibranch condition. **D,** Tissue connections that provide support for the folded lamellibranch filaments. In **C** and **D,** the distance between the descending and ascending limbs is exaggerated for clarity.

Each filament is folded into a V and, since the gills are bipectinate, there are two Vs, one medial and one lateral to the central axis. Together the Vs form a W (Fig. 12-96C,D). Each filament drops ventrally from the central axis and then climbs dorsally. The part directed downward is the **descending limb,** whereas the upward part is the **ascending limb** (Fig. 12-96D, 12-90). Each filament has one of each. Both descending limbs are attached to the central axis. The ascending limbs extend to either the mantle skirt (lateral filaments) or the visceral mass (medial filaments). There they may attach by ciliary or tissue junctions or they may be unattached.

So far we have considered the filaments separately, but they cannot function as filters individually. Each filament is closely associated with the adjacent filaments of its limb, and these filaments together form the lamella that is the filtering device (Fig 12-89A). Adjacent filaments are held close together by junctions, which may be either ciliary or tissue. Water flows through the lamellae by passing between the filaments, which filter it. Each gill has four of these lamellae, each formed of joined filaments. The combined descending limbs of all the lateral filaments together form a **descending lamella** (Fig. 12-90) and the ascending filaments form an **ascending lamella.** Similarly, the medial filaments also form ascending and descending lamellae. Thus there are two ascending and two descending lamellae for each gill.

Remember that there are two gills, one on the left and one on the right, in the mantle cavity. Each gill is a **holobranch,** or whole gill, composed of two half gills, or **demibranchs.** A demibranch is composed of two lamellae, one ascending and one descending (Fig. 12-90). Each demibranch gives the impression of being a complete gill but is not, for each is formed of either all the lateral filaments or all the medial filaments, but not both. A holobranch consists of two demibranchs (lateral and medial) and a total of four lamellae (two ascending and two descending). Since there are two gills, one right and one left, a typical clam has two holobranchs, four demibranchs, and eight lamellae.

As in other molluscs, lamellibranch gills divide the mantle cavity into a ventral inhalant chamber (infrabranchial chamber) and a dorsal exhalant chamber (suprabranchial chamber; Fig. 12-90). In the lamellibranchs, however, the separation is so effective that water cannot pass from the inhalant side to the exhalant side without passing through the lamellae. The lamellae, being composed of side-by-side filaments, are porous, and water can flow through them. Particles, however, may be too large to make this passage and are retained on the inhalant side of the lamella (the frontal side) and are thus filtered out of the water current and entangled in mucus. An important part of gill evolution is an increase in the ability of adjacent filaments to function as a filter. This included mechanisms to ensure the proper spacing between filaments.

CILIA

In the lamellibranch gill, the frontal and lateral cilia function largely as they do in the protobranchs and ancestral molluscs and are located in similar positions on the filaments (Fig. 12-96B). The lateral cilia, located on the lateral surfaces of the filaments, create the respiratory and feeding current. This current enters the inhalant chamber and passes between the filaments to the exhalant chamber. The frontal cilia, located on the frontal surface of the filament, transport particles dorsally or ventrally over the surface of the lamellae to longitudinal, ciliated food grooves. Abfrontal cilia usually are not present.

A new set of cilia, the **laterofrontal cilia,** has been added to the ciliary complement of lamellibranch gills. These are large cilia fused together to form **laterofrontal cirri** located at the edges of the filaments, between the frontal and lateral cilia (Fig. 12-97A,B). Each cirrus is a bundle of adhering cilia, as can be clearly seen in Figure 12-97A. The laterofrontal cirri extend partly across the gap between filaments and overlap each other. Opposing cirri form a fine mesh that filters particles from the water entering the gill. Very small particles that otherwise would pass through the interfilamentary gaps are intercepted by the cirri and shifted to the frontal cilia, which may transport them to a food groove and then to the labial palps and mouth.

WATER FLOW

During their evolution, the protobranchs shifted the flow of the respiratory current from a more or less anterior to posterior one-way flow (Fig. 12-94A) to a bidirectional flow in which water enters and exits posteriorly. This new pathway facilitated the exploitation of infaunal habitats by the lamellibranchs. In it, the anteriorly directed inhalant feeding and respiratory current enters the ventral part of the mantle cavity (inhalant chamber) at the posterior end of the animal, flows between the filaments into the dorsal exhalant chamber, and finally flows out posteriorly.

In lamellibranchs the posterior margins of the right and left mantle skirts combine to form a pair of well-defined apertures, the ventral inhalant and dorsal exhalant siphons, to channel water into and out of the two chambers of the mantle cavity (Fig. 12-89, 12-107, 12-108). The inhalant siphon conducts water into the inhalant chamber ventral to the gills, whereas the exhalant siphon expels water from the exhalant chamber. Siphons are absent in most protobranchs (*Yoldia* is an exception), in which water enters the mantle cavity almost anywhere around its periphery.

JUNCTIONS

The longer and thinner lamellibranch gill filaments required additional structural support to maintain the appropriate spacing for filtering between filaments. Furthermore, the spacing between filaments had to remain constant to prevent the passage of food particles. Several types of structural supports arose in response to these needs. **Interlamellar junctions** spanned the exhalant chambers to connect the two lamellae of each demibranch (Fig. 12-96D) and **interfilamentar junctions** connected adjacent filaments and maintained the proper mesh size of the filter.

A third type of junction exists between the tips of the filaments (ascending lamellae) and the mantle or visceral mass. These junctions ensure that the gills form a complete barrier between the inhalant and exhalant chambers so that water is forced to pass between the filaments. The character of these connections varies in different lamellibranch taxa and often defines the several types of lamellibranch gills. There may be no junction, ciliary junctions, or tissue junctions.

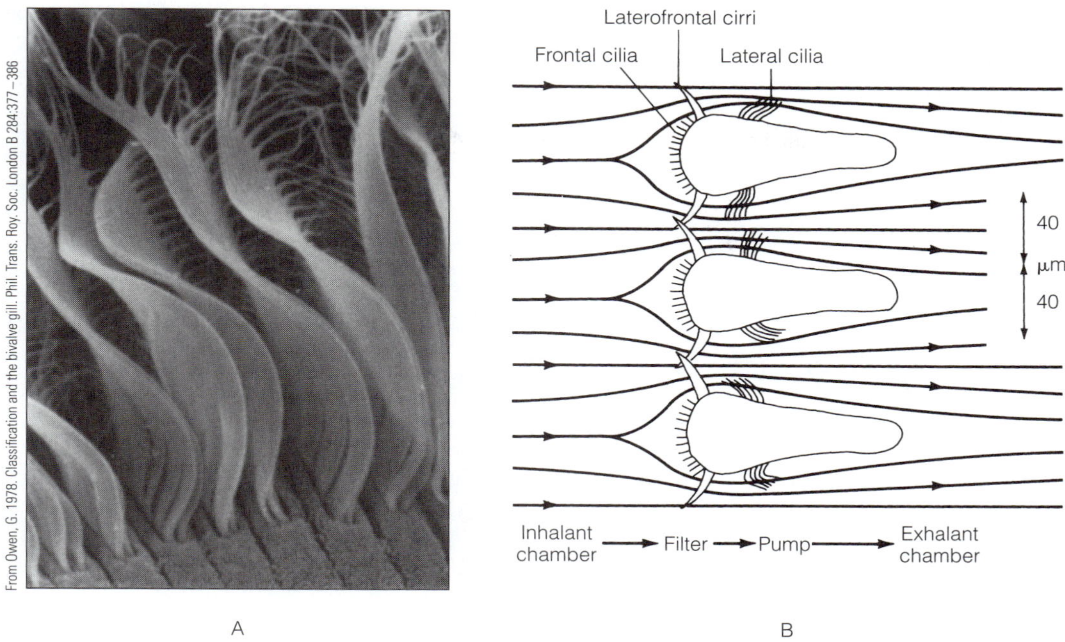

FIGURE 12-97 Gill cilia. **A,** Laterofrontal cirri of *Venus casina.* Each cirrus is composed of two rows of fused cilia on a single cell. The distal ends of the cilia are unfused. The two cilia rows are clearly evident at the base of each cirrus. **B,** Diagrammatic cross section through three gill filaments of *Mytilus.* The frontal cilia beat perpendicular to the plane of the page. The effective stroke of the laterofrontal cirri is to the left, and the effective stroke of the lateral cilia is to the right of the image. Arrows indicate the direction of the feeding/ventilating current produced by the lateral cilia. *(B, From Silvester, N. R., and Sleigh, M. A. 1984. Hydrodynamic aspects of particle capture by* Mytilus. *J. Mar. Biol. Assoc. 64:859–879. Copyrighted and reprinted by permission of Cambridge University Press.)*

GRADES OF LAMELLIBRANCH GILLS

The three grades in the evolution of lamellibranch gills differ in the structure and degree of development of the interfilamentar and interlamellar junctions. The most primitive lamellibranch gills are **filibranch gills,** in which the individual filaments remain more or less independent of each other. Adjacent filaments are held together not by permanent tissue connections, but by tufts of interlocking cilia (Fig. 12-98A,B). The opposite lamellae of a demibranch, however, are connected by tissue interlamellar junctions that hold the two limbs of the filaments permanently in a V shape. The tips of the filaments are either not connected or are only weakly connected with the mantle or visceral mass. Filibranch gills are found in pteriomorph bivalves such as arks *(Arca),* scallops *(Pecten),* and mussels *(Mytilus).*

Some other pteriomorph bivalves, such as the oysters *(Crassostrea, Ostrea)* and pen clams *(Pinna),* have **pseudolamellibranch gills.** In these, adjacent filaments are held together by combinations of both tissue and ciliary interfilamentar junctions. Ciliary junctions predominate, however.

Most bivalves have **eulamellibranch gills.** These are the most specialized, and they exhibit several important advances over filibranch and pseudolamellibranch gills. Both interfilamentar and interlamellar junctions are extensively developed permanent tissue connections (Fig. 12-98C,D). The interlamellar junctions are solid, vertical partitions of tissue that extend the entire length of the lamellae and divide the exhalant chamber into vertical **water tubes.** Interlamellar junctions are not present on all filaments, however, so each water tube spans several adjacent filaments.

The interfilamentar junctions are also permanent and more extensive. They form nearly solid sheets of tissue extending between adjacent filaments, but these are penetrated by occasional pores known as **ostia** (Fig. 12-98C,D). The cilia of the filament are not affected by this fusion, and lateral, laterofrontal, and frontal cilia are present.

The inhalant side of an eulamellibranch gill lamella is distinctly ridged, with each filament forming a ridge and each interfilamentar junction a groove between adjacent ridges. The lateral cilia generate a water current that flows along a groove until it encounters an ostium, enters it, and passes into a water tube. Oxygenation of the blood takes place over the inhalant surface of the gill and in the water tubes. From the water tubes, water flows posteriorly through the exhalant chamber and out the exhalant siphon.

The tips of the ascending limbs of the filaments have become permanently fused with the upper surface of the mantle on the outside and the visceral mass on the inside, thus separating the inhalant chamber from the exhalant chamber. In order for water to move from the inhalant to the exhalant chamber, it must pass through the ostia.

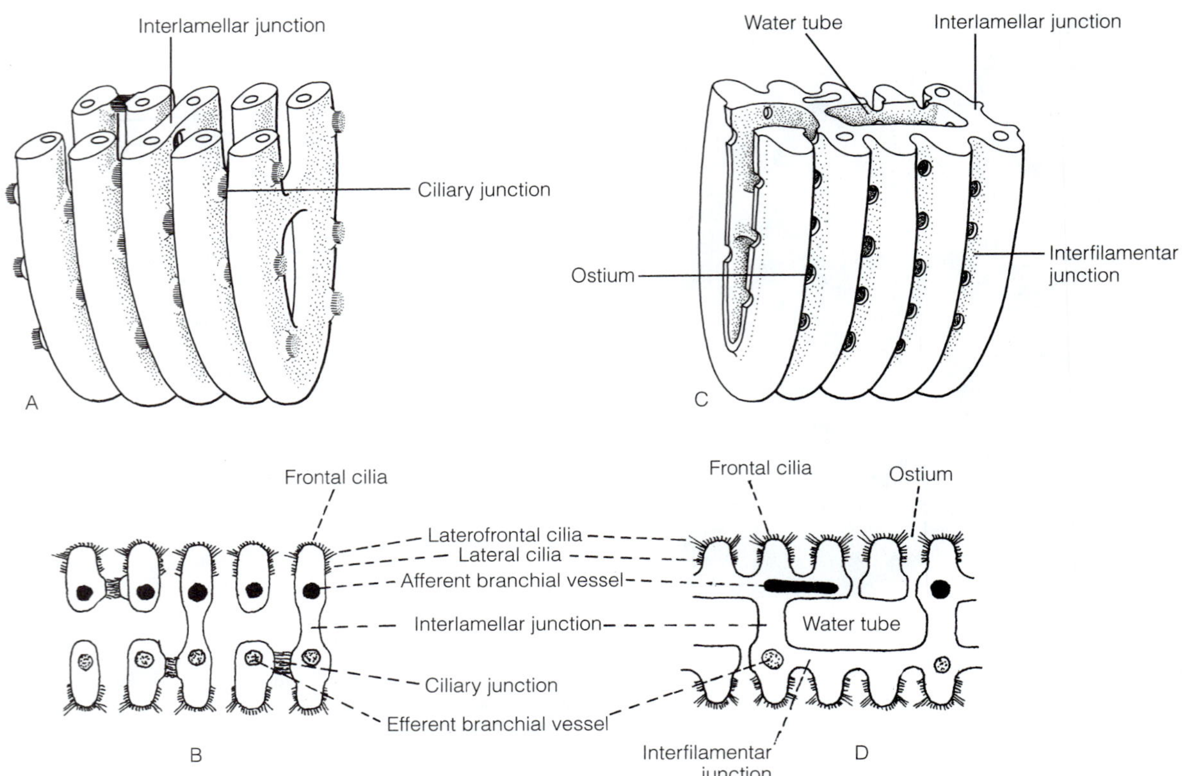

FIGURE 12-98 Filibranch and eulamellibranch gills. **A** and **B**, A filibranch gill. **A**, An oblique view of five adjacent filaments at the edge of a demibranch. **B**, Frontal section of a demibranch. **C** and **D**, A eulamellibranch gill. **C**, An oblique view of five adjacent filaments at the edge of a demibranch. **D**, Frontal section of a demibranch.

BLOOD VESSELS

In primitive molluscan gills, including those of protobranchs, the efferent blood vessel lies in the axis of the filament beneath the afferent vessel (Fig. 12-96A). With the elongation and folding of the filament in eulamellibranch gills, the old efferent vessel was lost from the gill axis and has been replaced by a pair of vessels at the junctions of the tips of the ascending limbs of each filament with the body (Fig. 12-96C). This results in a one-way flow of blood through the filament.

FOOD GROOVES AND PREPALPAL SORTING

Most lamellibranchs feed on fine plankton and suspended detritus. Food particles, in some cases as small as 1 μm, are removed from the water currents as they pass between filaments or enter the ostia. The particles are then moved onto the frontal cilia, where they are entangled in mucus and moved up or down the filament to a longitudinal groove. At some stage in the evolution of lamellibranch gills, longitudinal ciliated **food grooves** (acceptance tracts) formed along the dorsal and ventral edges of the lamella (Fig. 12-99). A maximum of five such grooves, two ventral and three dorsal, can be associated with each holobranch. Cilia in these grooves transport food and mucus anteriorly to the labial palps or, in some instances, rejected particles may be transported anteriorly or posteriorly for ejection as pseudofeces.

The primitive lamellibranch has five longitudinal grooves for transporting particles (Fig. 12-99). Three of the grooves are located at the top of the gills between and outside the demibranchs. The other two are located ventrally, one along the ventral margin of each demibranch.

The frontal cilia are arranged into separate tracts, one carrying particles upward and one downward. This system is capable of presorting food particles from mineral particles before the particles reach the labial palps, where additional sorting occurs. Such a two-way vertical tract system with five grooves is found in oysters and scallops (Fig. 12-99C). Food particles are moved up into three food grooves that transport food anteriorly to the labial palps and ultimately to the mouth. Other cilia move sediment particles down to two rejection tracts.

From such a primitive condition, several variations evolved, a few of which are illustrated in Figure 12-99. In mussels and pen clams (Fig. 12-99A), food particles and sediment particles are transported both up and down (mostly down) by frontal cilia to any of five anteriorly directed food grooves. There is little or no sorting on the gills. In arcs and jingles, on the other hand (Fig. 12-99B), sorting occurs as tracts of frontal cilia transport food particles up to three anteriorly directed food grooves

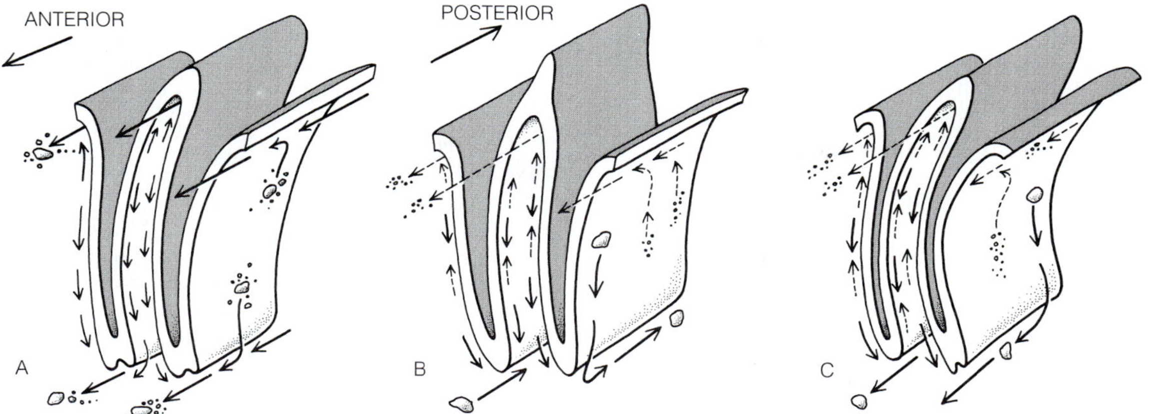

FIGURE 12-99 Transverse sections of representative lamellibranch gills, showing the direction of the frontal cilia beat and the positions of longitudinally moving food tracts. **A,** Mytilidae and Pinnidae. **B,** Arcidae and Anomiidae. **C,** Ostreidae and Pectinidae. In **B** and **C,** broken arrows indicate the fine frontal cilia carrying food particles upward; solid arrows indicate the coarse frontal cilia carrying sediment particles ventrally. The inner demibranch is on the right in all cases. *(Redrawn and modified from Atkins, D. 1937. On the ciliary mechanisms and interrelationships of lamellibranchs. II. Sorting devices on the gill. Quart. J. Microsc. Sci. 79:339–370.)*

while sediment particles move down to posteriorly directed rejection tracts. Sediment and mucus from the rejection tracts are ejected as pseudofeces. The dual-tract system and prepalpal sorting has disappeared in most advanced lamellibranchs.

LABIAL PALPS

The lamellibranch palpal lamellae supposedly have the same sorting and conveying function as in protobranchs (Fig. 12-94D). In this model, particles are sorted by size and food particles are carried across the palpal surface on the crests of the ciliated ridges. Sediment particles, which are rejected, are carried to the edge of the lamellae in the grooves between ridges and fall, as pseudofeces, into the mantle cavity. Studies on the scallop *Placopecten magellanicus* and the oyster *Crassostrea gigas,* however, raise questions about various aspects of this classical account. The palps do not receive free particles from the gills, but rather a cord of particles bound in mucus that travels in the oral groove at the junction between the two palpal lamellae (Fig. 12-100). In the normal position, in which the two lamellae are appressed, there does not appear to be any particle selection by the crests of the ridges.

PSEUDOFECES

The gills and the labial palps produce pseudofeces, which consists primarily of sediment particles and mucus. This material is moved posteriorly by a ventral ciliary tract on the inner surface of the mantle skirt to a position near the inhalant siphon. Occasional rapid contractions of the adductor muscles compress

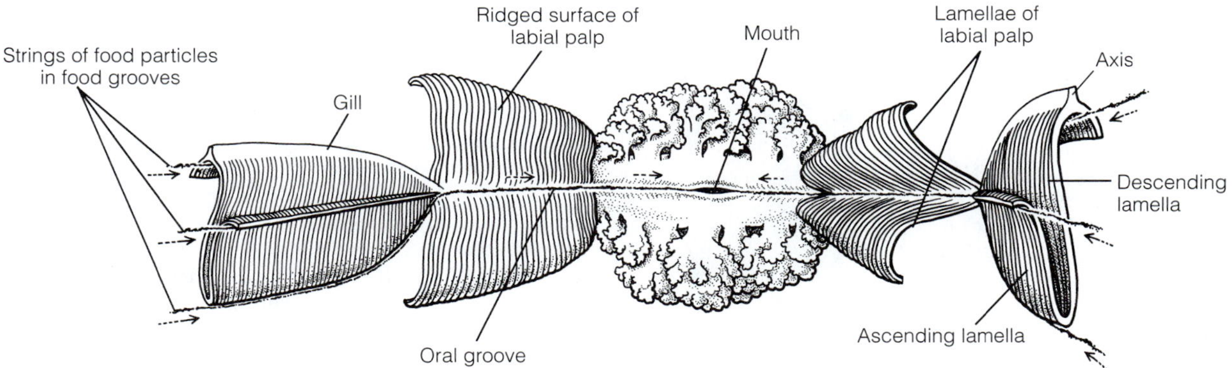

FIGURE 12-100 Relationship of gills, labial palps, and mouth. Anterior view of the deep-sea scallop *Placopecten magellanicus.* The mouth is surrounded by bushy folds. The palps and gills are drawn as if they are in the plane with the mouth when in actuality they would extend posteriorly, into the plane of the page. The palps are shown with their lamellae open. Only the lateral demibranchs of the gills are included. *(Redrawn and modified from Benninger P. G., Auffret, M., and Le Pennec, M. 1990. Peribuccal organs of* Placopecten magellanicus *and* Chlamys varia: *structure, ultrastructure and implications. Mar. Biol. 107:215–223.)*

the mantle cavity and force a jet of water and pseudofeces out the inhalant siphon. Note that this is opposite the usual direction of flow in the inhalant siphon.

FLOW RATE

The animal can regulate the rate of water flow through the filtering system by changing the size of the inhalant aperture and expanding or contracting the gill. Adjustment of the size of the gill regulates the amount of water that can flow between the filaments. The flow's direction can be temporarily reversed or it can be stopped completely.

Septibranchs

The third grade of bivalves, after the protobranchs and lamellibranchs, is the septibranchs. This small group of small, derived carnivores, scavengers, or detritivores is closely related to the Anomalodesmata and presumably arose from a lamellibranch-like ancestor. Most septibranchs are raptorial carnivores that capture individual prey using suction generated by the mantle cavity.

Septibranch gills have been highly modified to form a pair of perforated muscular septa, which separate the inhalant and exhalant chambers of the mantle cavity (Fig. 12-93C). The gill is greatly reduced and there is no surface area expansion for either gas exchange or filtering. Gas exchange occurs over the entire exposed mantle surface. The septate gills are used only to create suction used in feeding.

Rapid contraction of the septal muscles closes the septal perforations and elevates the septum. This draws water through the enlarged inhalant siphon into the inhalant chamber. Any organic particle, be it a small crustacean, a bit of detritus, or a small carcass, is likely to be caught in the suction and drawn into the chamber (Fig. 12-101). Prey items are grasped by a pair of muscular labial palps, transferred to the mouth, and ingested. During recovery, the septal muscles relax, the perforations open, and the septum returns slowly to its original position. The inhalant siphon of *Poromya* is large, hoodlike, and equipped with sensory tentacles for the detection of prey. The hood is rapidly clamped over the prey, which is then sucked into the mantle cavity. Another taxon, *Cuspidaria,* rapidly extends its inhalant siphon to capture prey, also by suction.

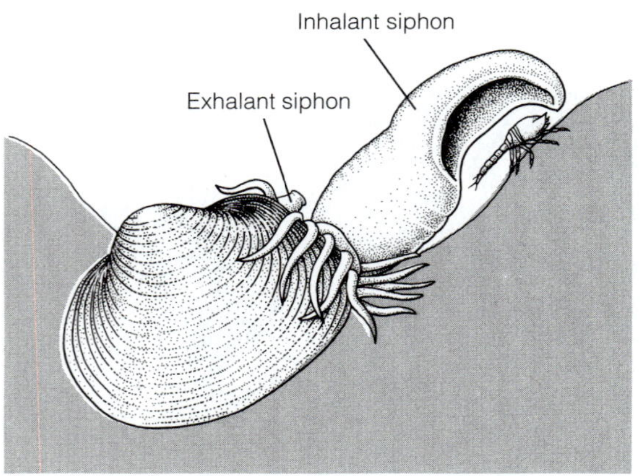

FIGURE 12-101 The septibranch *Poromya granulata* capturing a crustacean (a cumacean) with its hoodlike inhalant siphon. *(Modified and redrawn from Morton, B. 1981. Prey capture in the carnivorous septibranch* Poromya granulata. *Sarsia 66:241–256.)*

NUTRITION

The bivalve digestive system consists of a mouth, esophagus, stomach, intestine, rectum, and anus. The mouth is anterior on the visceral mass, between the labial palps. Branched digestive ceca extend into the hemocoel as diverticula of the stomach (Fig. 12-89B). The intestine typically makes several loops through the visceral mass before extending posteriorly through the pericardial cavity to the anus. The anus is posterior and empties into the exhalant chamber.

Feeding Diversity

PROTOBRANCHS

With the exception of the septibranchs, bivalves are microphages that consume small organic particles collected either by deposit or suspension feeding. The ancestral protobranch gut, like that of monoplacophorans, polyplacophorans, and primitive gastropods, is equipped to process small particles. The stomach, lined by a chitinous cuticle, has a ciliary sorting field. There is an amorphous mucous protostyle in a large style sac. Digestion in most protobranchs is extracellular in the stomach, and absorption occurs in the digestive ceca.

SEPTIBRANCHS

The stomach of the carnivorous septibranch is a simple, heavily muscled sac with a small sorting field and vestigial style. It is lined by an extensive chitinous cuticle, and the stomach functions as a gizzard for crushing the prey. Proteases from the digestive ceca initiate extracellular digestion in the stomach lumen. The products of this digestion are conveyed into the ducts of the digestive ceca, where digestion is completed intracellularly.

LAMELLIBRANCHS

As suspension feeders, lamellibranchs are also microphages, but they tend to utilize smaller food particles than do protobranchs. The digestive system has undergone modification and the stomach is more complex than that of protobranchs. The lamellibranch stomach contains an enzymatic crystalline style, has numerous connections with the digestive ceca, and has a reduced chitinous shield. The intestinal groove extends out of the intestine into the stomach. Digestion is both extracellular and intracellular, with the former occurring in the stomach and the latter in the digestive ceca. There are three different types of lamellibranch stomach. The following description is a generalization.

The mucous food string from the labial palps enters the mouth and is transported by cilia posteriorly through the esophagus to the stomach. The stomach is a bulbous pouch in the anterior dorsal visceral mass (Fig. 12-89B). The esophagus enters anteriorly and the intestine and style sac exit posteriorly. The left anterior stomach wall bears a small, chitinous gastric shield (Fig. 12-102). Several ducts exit the anterior stomach

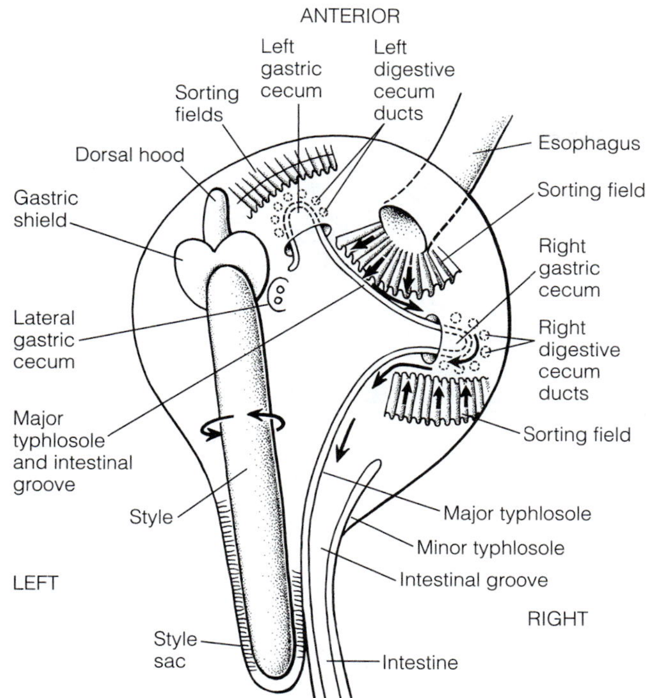

FIGURE 12-102 Dorsal view of the interior of a generalized eulamellibranch stomach. *(Redrawn and modified from Purchon, R. D. 1977. The Biology of Mollusca. Second Edition. Pergamon Press, New York. 596 pp.)*

and extend to the digestive ceca. These ducts arise from one to three short **gastric ceca** that evaginate from the stomach walls.

There is one or more ciliary sorting field, consisting of the usual ciliated ridges and grooves, close to the entrances to the digestive ceca ducts. The sorting fields function like those of the labial palps and move large particles over the crests perpendicular to the long axis of the ridges. Small mineral particles in the grooves are moved parallel to the long axis and ultimately to rejection tracts. In general, the sorting fields of the stomach move mineral particles to rejection tracts leading to the intestine, but food particles tend to be resuspended in the fluid in the stomach. The process thus constantly removes mineral particles and increases the proportion of food particles in the stomach fluid.

A large, ciliated intestinal groove extends from the stomach into the intestine (Fig. 12-102). The left side of the groove is bordered by a large ciliated ridge, the major typhlosole, which arches over the groove and partially isolates it from the stomach. Another ridge, the minor typhlosole, lies on the right side of the intestinal groove. The major typhlosole and intestinal groove arise in the left gastric cecum and extend across the anterior stomach into the right gastric cecum. They exit this cecum and extend into the intestine, accompanied by the minor typhlosole. The typhlosole and intestinal groove form the rejection tract that leads a continuous stream of mineral particles out of the stomach and into the intestine.

A large style sac with a ciliated secretory epithelium extends posteriorly from the stomach. Its epithelium secretes the crystalline style, which is probably homologous to the protostyle of the protobranchs. The style consists of a protein matrix onto which are secreted a variety of enzymes, including amylases, glycogenases, lipases, and cellulases. In some bivalves with cellulases, the style contains high population densities of spirochete bacteria, which perhaps are endosymbionts that provide cellulase. There are no proteases in the stomach (they would digest the proteinaceous enzymes of the style), and protein digestion is intracellular in the digestive ceca. The length of the style varies, but it is remarkably long in comparison with the size of the animal. A 12-cm *Tagelus,* for example, may have a 5-cm style and a 1-m *Tridacna* may have a 36-cm style.

The style sac lies beside the intestine, and the lumina of the two may be continuous (Fig. 12-89B, 12-103A) or separate. The style sac cilia cause the style to rotate and rub against the gastric shield (Fig. 12-102). This abrades the tip of the style and releases digestive enzymes into the stomach lumen. New enzymes are added to the posterior end of the style by secretion from the style sac epithelium. Rotation of the style stirs the contents of the stomach, and the style probably functions as a windlass to pull the mucous food string through the esophagus into the stomach. Extracellular digestion of carbohydrates and lipids begins in the stomach. Most protein digestion occurs intracellularly within the digestive ceca.

Acid released from the abrading style and from the digestive ceca reduces the pH of the stomach contents, and in this acidic environment the viscosity of the mucus is reduced. Food particles that were held by viscous mucus are released when the viscosity drops, and they move onto the sorting fields.

As in all molluscs, the stomach is connected with the two digestive ceca by ducts. In the lamellibranchs, several primary ducts exit the gastric ceca to branch into secondary ducts that terminate in small blind pouches, or **acini** (Fig. 12-103B), in the digestive ceca. The acinar epithelium is responsible for secretion, absorption, pinocytosis, and intracellular digestion. The digestive ceca are composed of ducts and, chiefly, acini. Food particles are moved from the stomach lumen into the gastric ceca, then to the ducts, and then into the acini. Furthermore, waste particles exocytosed from acinar epithelial cells are returned to the stomach lumen and sent to the intestine to be incorporated into feces. The ducts are divided into two channels. The exhalant channel is lined by cilia, but the inhalant channel lacks them, although it bears a brush border of microvilli and is absorptive. All ciliary currents associated with the digestive ceca are exhalant—that is, they move wastes out of the ceca but do not move food in. These exhalant currents connect directly with the typhlosoles and intestinal groove and remove wastes from the digestive ceca. The constant outward flow of fluid driven by the exhalant cilia creates a negative pressure in the digestive ceca that sucks stomach fluid into the acini. Since the sorting process has enriched the stomach fluid with food particles and has selectively removed mineral particles, most of the particles entering the digestive ceca are organic and can be digested extracellularly or phagocytosed and digested intracellularly. Soluble organic molecules, produced by extracellular digestion in the stomach, are absorbed by acinar cells and stored or transferred to the blood. Cellular wastes are exocytosed and voided as described.

In some intertidal bivalves, such as the American oyster, *Crassostrea virginica,* and the cockle *Cardium edule,* that feed

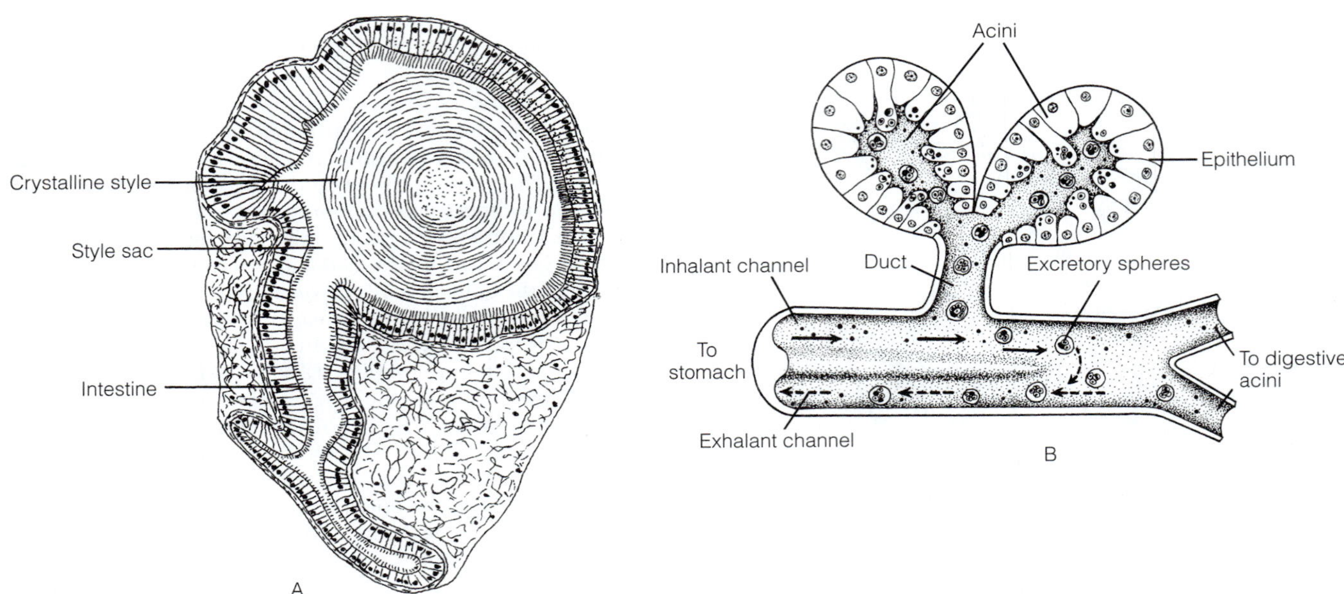

FIGURE 12-103 **A,** Cross section of the style sac and intestine of the freshwater clam *Lampsilis anodontoides.* **B,** Diagram of a section of digestive cecum, showing the absorption and intracellular digestion of material passed inward from the stomach (solid arrows) and the outward passage of waste (dashed arrows). *(After Owen, G. 1974. Feeding and digestion in the Bivalvia. Adv. Comp. Physiol. Biochem. 5:1–35.)*

only at high tide when they are covered with water, the digestive processes display a tidal rhythm. The crystalline style is dissolved at low tide when the animal is not feeding and is secreted anew when the tide returns.

Studies on energy expenditure in *Mytilus edulis* reveal that the cost of pumping water and transporting food by the gills is slight. At high food-intake levels, about half of the energy cost is in general body maintenance and the remainder is partitioned between the cost of digestion and absorption and growth. A significant fraction of ingested food may be expelled undigested, partly due to inefficient sorting.

The primary responsibility of the intestine is feces formation, although there are indications that some absorption may occur there. The intestine passes through the pericardial coelom and is completely surrounded by the ventricle (Fig. 12-89B, 12-90). It then passes dorsally over the posterior adductor muscle to terminate at the anus, which opens into the exhalant chamber at the base of the exhalant siphon.

ADAPTIVE RADIATION OF LAMELLIBRANCHS

The evolution of filter feeding and the bidirectional flow of water through the mantle cavity freed lamellibranchs from their dependence on deposited particles on or near the sediment surface and made possible the colonization of many habitats that were not available to their protobranch ancestors. The success of this adaptive radiation is reflected in the fact that of 8000 or so described species and 75 families of bivalves, almost all are lamellibranchs. The first lamellibranchs were adapted for life in soft sediments, and most continue to exploit this ancestral habitat. The bivalves have undergone an extensive adaptive radiation, however, that has led them repeatedly into other aquatic habitats suitable for sedentary filter feeders.

Lamellibranch diversity is best studied from the standpoint of ecological rather than taxonomic groups. The major ecological groups of lamellibranchs are soft-bottom infauna, borers in hard substrata, attached epifauna, unattached epifauna, symbiotic association with other metazoans, and fresh water. Colonization and adaptation to each type of habitat has been achieved independently by multiple families or superfamilies. Occupation of a similar habitat accomplished through similar adaptations is not by itself evidence of a close relationship. We will consider, in turn, each of the major ecological groups and the associated adaptations.

Soft-Bottom Infaunal Burrowers

The ancestral lamellibranchs invaded infaunal soft-bottom habitats to escape predation pressure on the surface while simultaneously taking advantage of the abundant phytoplankton and oxygen available in the overlying water. The laterally compressed digging foot of the protobranch ancestors proved to be a valuable preadaptation for the invasion of this new habitat. The addition of suspension feeding made it possible to utilize phytoplankton as food. Modification of the unidirectional anterior-to-posterior water flow of the protobranchs to a bidirectional flow through the posterior mantle cavity made it possible to burrow deeper into the sediment. The development of siphons allowed the bivalves to maintain efficient contact, for feeding, respiratory, and reproductive purposes, with the overlying water, even when deeply buried. Lamellibranchs radiated extensively to exploit this habitat, and this is where most modern species are found.

BURROWING

Burrowing into soft sediments is accomplished by the laterally compressed, muscular **foot** operating in conjunction with the shell, adductor and pedal retractor muscles, and hemocoel. Two pairs of pedal retractor muscles, anterior and posterior (Fig. 12-112A), extend from origins on the shell near the adductor muscles to insertions on the sides of the foot. In fact, the side of the foot consists primarily of the long posterior pedal retractor muscle (Fig. 12-104D) and a layer of connective tissue. The anterior pedal retractors, which are shorter and insert laterally on the outer surface of the posterior retractors, function primarily in pedal retraction, but they also play an indirect role in the initial protrusion of the foot and are sometimes referred to as pedal protractor muscles. Transverse pedal muscles extend from right to left across the lumen of the pedal hemocoel, and circular pedal muscles encircle the hemocoel. The pedal hemocoel, may function as a hydrostatic skeleton or, in some bivalves, the muscle itself is the hydrostat.

Burrowing is initiated with the adductor muscles and pedal retractor muscles relaxed and the shell gaping (Fig. 12-105A) with the siphons open and water flowing through the mantle cavity. Contraction of the proximal transverse muscles, circular muscles, and anterior retractor muscles compresses the pedal hemocoel and forces blood from the visceral hemocoel into the foot, bringing about its elongation. The foot probes and pushes into the surrounding sand. During this process, the weight and friction of the slightly gaped shell serves as a penetration anchor, preventing the clam from moving backward.

When the foot is protruded, the siphons close, the adductor muscles contract suddenly, and the valves are pulled together, forcing water out of the gape and putting pressure on the hemocoel. This forces additional blood into the distal foot, causing it to dilate and form a terminal anchor. The water expelled through the gape between the valves loosens the sediment in front of the shell and facilitates penetration (Fig. 12-105B).

Then the anterior pedal retractor muscles contract, causing the shell to rock dorsally, in the direction of the hinge, and to simultaneously move toward the terminal anchor (Fig. 12-105C). The loosened sediment facilitates movement of the shell.

Next, the powerful posterior pedal retractor muscles contract and cause the shell to rock ventrally, toward the gape (Fig. 12-105D). As the shell rocks, the combined pull of the anterior and posterior retractors draws the shell farther into the sediment toward the terminal anchor.

Alternating contractions of the anterior and posterior pedal retractors rock the shell and move it closer and closer to the terminal anchor until it can move no farther. At this time the entire process is repeated, beginning with the extension of the foot while the shell serves as a penetration anchor. The cycle is repeated until the clam is at the desired depth in the sediment.

Once at the desired depth, the foot is retracted, the adductor muscles relax, the valves gape, the siphons are extended to the surface, and feeding can begin.

Some bivalves can back out of the burrow by pushing against the anchored foot.

In primitive protobranchs such as *Nucula, Solemya,* and *Yoldia,* the foot has a flattened sole (Fig. 12-104B), and during the penetration phase of digging, the two sides of the sole are

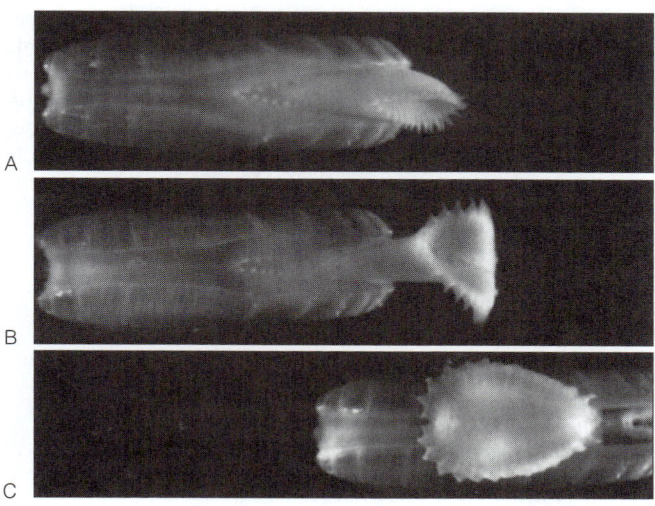

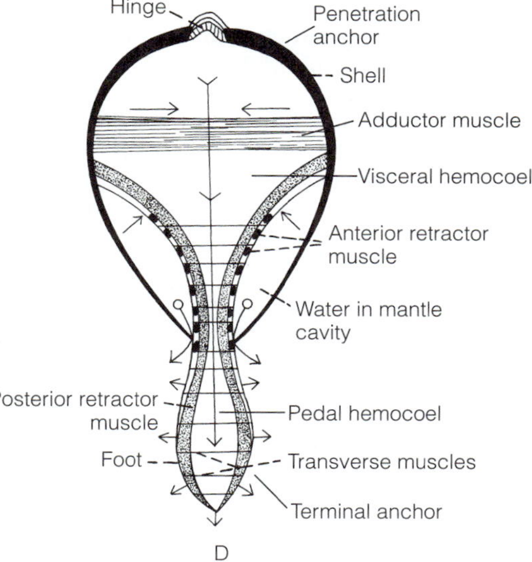

FIGURE 12-104 Operation of the foot of the protobranch *Solemya velum.* **A,** The end of the foot is folded and partly extended. **B,** The foot is fully extended, opened, and anchored. **C,** The body advances by contraction of the longitudinal pedal muscles of the foot. **D,** Diagrammatic cross section of a bivalve showing the hydrostatic forces that produce dilation of the foot. Central vertical arrow indicates the flow of blood from the visceral hemocoel into the foot. *(D, Modified after Trueman, 1966.)*

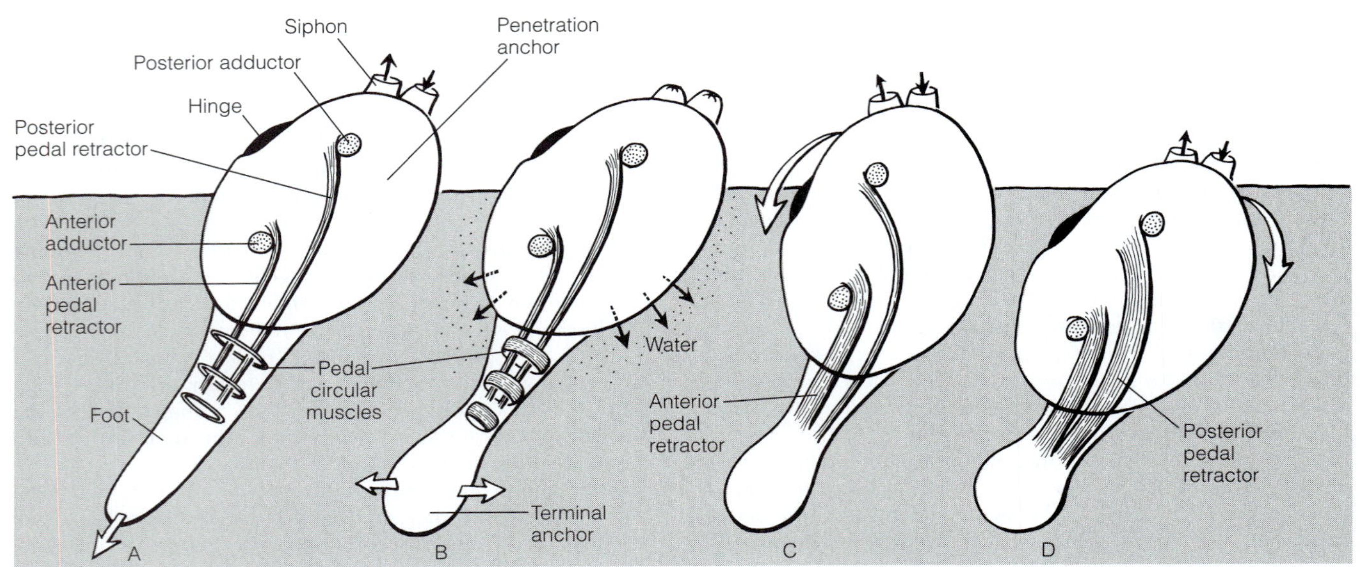

FIGURE 12-105 Burrowing by bivalves. **A,** The siphons are open and water flows through the mantle cavity. The adductor muscles are relaxed and the valves are spread, widening the shell to create a penetration anchor. The pedal retractor muscles are relaxed. Contraction of the pedal circular muscles extends the foot into the sediment. **B,** Further contraction of the proximal pedal circular muscles forces blood into the distal foot, which expands to form a terminal anchor. The siphons close and the adductor muscles contract partially, forcing water out of the gape to loosen the surrounding sediment. **C,** The anterior pedal retractor muscle contracts and rocks the shell dorsally (to the left in the image), pulling the anterodorsal corner of the shell into the loosened sediment, toward the terminal anchor. **D,** The posterior pedal retractor muscle contracts and rocks the shell ventrally, pulling the anteroventral edge of the shell into the sediment toward the terminal anchor. The process begins again with **A** and is repeated until the clam reaches the desired depth. *(Redrawn and modified from Russell-Hunter, W. D. 1979. A Life of Invertebrates. Macmillan Publishing Co. New York. 650 pp; and Trueman, E. R. 1966. Bivalve mollusks: Fluid dynamics of burrowing. Science 152:523–525.)*

folded together, producing a bladelike edge (Fig. 12-104A). The foot is then thrust into the mud or sand and the sole opens and flares outward to form an anchor (Fig. 12-104B).

MANTLE FUSION

The influx of sediment particles into the mantle cavity is a persistent problem for burrowing bivalves. The valves must gape slightly when the animal is feeding or burrowing, and the opening provides an entry for sediment. Sediment also may be introduced by the inhalant water flow.

Natural selection has favored a variety of solutions to the problem. Increased blood pressure in the mantle swells its margins so the right and left mantle margins touch and form a seal to exclude particles, even when the valves are slightly gaping. Furthermore, there has been a tendency for the right and left mantle margins to grow together over the midline to seal the mantle in places where openings are not necessary. The most frequent point of fusion is between the inhalant and exhalant openings (Fig. 12-106). This fusion forms a distinct dorsal exhalant aperture. A second point of fusion often occurs immediately below the ventral inhalant opening, forming a permanent inhalant aperture that is completely surrounded by fused mantle. Bivalves that lack permanent fusion have temporary openings that form whenever the left and

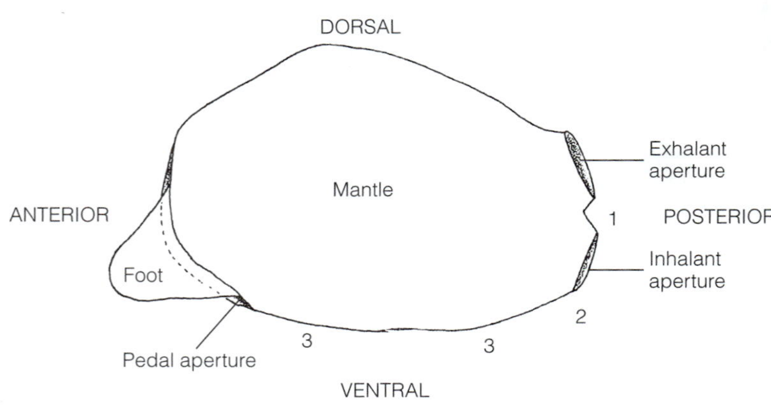

FIGURE 12-106 Areas of mantle fusion in various bivalves: **1,** Between inhalant and exhalant apertures or siphons, the most common point of fusion. **2,** Below inhalant aperture or siphon. **3,** Between inhalant aperture and foot, leaving only a pedal aperture for the extension of the foot.

right margins of the mantle are pressed together below the inhalant aperture.

Additional mantle fusion has occurred in some species, especially deep burrowers, so that most of the ventral margin between the inhalant aperture and the foot is sealed. A clam with maximal fusion at all three positions would have three openings in the mantle: the exhalant aperture, inhalant aperture, and the pedal aperture. Extensive mantle fusion not only reduces fouling of the mantle cavity, it also facilitates maintenance of the hydraulic pressure within the mantle cavity that is conducive to burrowing.

SIPHONS

Fusion of opposing mantle folds in the vicinity of the inhalant and exhalant apertures creates well-defined, permanent apertures known as the inhalant and exhalant siphons. The inhalant siphon is ventral and the exhalant siphon dorsal. The mantle margins surrounding the apertures may be elongate to form tubes of varying lengths (Fig. 12-107, 12-108, 12-109, 12-112A) or they may be simple, thickened areas of the mantle margin (Fig. 12-89). Infaunal bivalves with tubular siphons can be completely buried, with only the siphon tips projecting above the sediment and vulnerable to predation. Although the siphons can be withdrawn rapidly by **siphon retractor muscles,** they are subject to grazing by fishes, but can be regenerated when lost. The siphons are extended by blood pressure or by water pressure within the mantle cavity when the valves are closed. The two siphons may be fused or separate (Fig 12-108A,B).

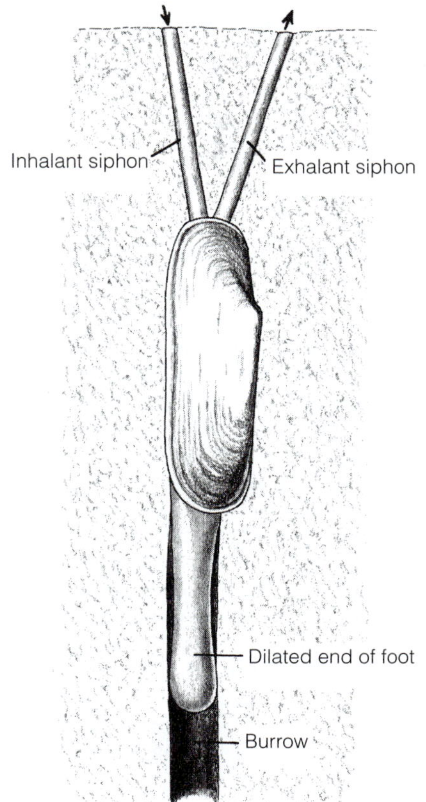

FIGURE 12-107 The razor clam *Tagelus plebius,* an infaunal bivalve from the southeastern coast of the United States, where it may occur in enormous numbers.

Most modern soft-bottom-dwelling filter feeders have siphons, and those that do not are shallow burrowers. The siphon length is positively correlated with the depth at which the species lives in the sediment. Some bivalves, such as the quahog (CO-hawg), *Mercenaria mercenaria,* live at the sediment surface and have unfused mantle margins (Fig. 12-89) which function as short siphons. Most, however, live deeper in the sediment and have well-developed siphons. The geoduck (GOOEY-duck), *Panopea generosa,* of the Pacific coast of North America is among the deepest burrowers, going down more than a meter. Its siphons are so large they cannot be fully retracted (Fig. 12-108B).

Some bivalves, especially the deeper burrowers, tend to have semipermanent or permanent burrows. The burrow walls are coated with mucus, which stabilizes the sediments and reduces contamination of the mantle cavity with particles from crumbling burrow walls.

Burrowing bivalves often have sensory receptors at the tips of the siphons—the only parts of the clam exposed to the overlying water, which is the source of danger and food. Photoreceptors, chemoreceptors, and mechanoreceptors may be found on the tips of the siphons. Siphons with photoreceptors may be rapidly withdrawn when covered by a shadow.

Large, well-developed siphons are usually associated with a deep posterior indentation, or **pallial sinus,** in the pallial line (Fig. 12-92B). This shell feature is the scar made by the attachment of the siphon retractor muscles to the shell.

SPECIALIZATIONS

The ancestral lamellibranch soft-bottom burrower is thought to have had an equivalve and equilateral shell with unfused mantle margins and more or less circular valves. Its adductor muscles were isomyarian. Specializations in different groups of infaunal bivalves have modified the ancestral plan in various ways, of which mantle fusion has already been discussed. In many groups, the valves, too, are modified. Species of *Donax* (the coquina clam), which are inhabitants of surf-swept, high-energy beaches, have shells that are pointed anteriorly and blunt posteriorly (Fig. 12-108A) to facilitate rapid burrowing. *Donax* backs out of the sand as a wave advances onto the shore and rapidly reburrows as the wave recedes. The opening of the inhalant siphon is protected by papillae, which exclude sand grains. Some permanent inhabitants of deep burrows, such as the geoduck with its fused siphons and sealed mantle margins, have reduced valves (Fig. 12-108B).

The razor clams *Ensis* and *Tagelus,* two unrelated but similar bivalves, have compressed, elongate valves and an elongate foot, which enable them to dig efficiently and move rapidly within their more or less permanent burrows. *Tagelus* (Fig. 12-107) has long, unfused siphons, each of which has a separate opening to the surface. *Ensis,* which has short siphons, comes to the surface to feed, temporarily leaving the deeper, more protected part of its burrow.

Some members of the large taxon Tellinoidea have reverted to deposit feeding. *Scrobicularia,* for example, extends

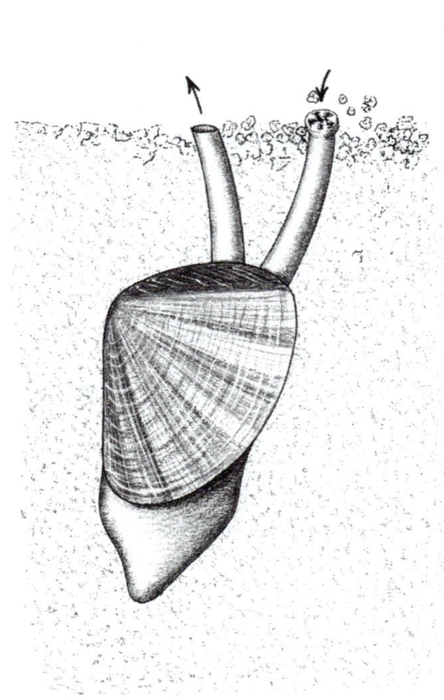

FIGURE 12-108 Infaunal bivalves. **A,** *Donax variabilis,* a common inhabitant of surf beaches. Rapid burrowing is facilitated by the thin, pointed foot. The opening of the inhalant siphon is frilled, preventing the entrance of sand grains. **B,** The geoduck, *Panopea generosa,* a giant Pacific North American bivalve, the body and siphon of which cannot be enclosed within valves. *(A, From Milne and Milne, 1959. Animal Life. Prentice-Hall, Englewood Cliffs, N.J.)*

its inhalant siphon above the surface and uses it like a vacuum cleaner to suck in deposit material (Fig. 12-109), which is then sorted on the gills. In Tellinoidea, deposit feeding is generally an addition to, rather than a substitute for, filter feeding.

Many Lucinoida *(Lucina, Divaricella, Codakia)* have lost the inhalant siphon entirely and use the foot to construct a mucus-lined canal extending to the surface. The feeding and respiratory current enters anteriorly by this passageway and leaves by the exhalant siphon.

Finally, almost all freshwater bivalves are infaunal in soft bottoms, have relatively short siphons, and may live with the posterior end of the shell exposed above the sediment.

Attached Epifauna

A number of evolutionary lines of lamellibranchs have abandoned the infaunal habitat and become adapted to life attached to the surfaces of hard substrata such as rock, shell, wood, coral, seawalls, jetties, hulls, and wharf pilings. These bivalves typically have a reduced or absent foot, because burrowing is no longer necessary. The anterior adductor muscle usually is reduced (anisomyarian) or absent (monomyarian). The mantle margins are not fused and the siphons are absent or reduced. Attachment is accomplished in one of two ways, either by a byssus or by cementing one valve to the substratum.

BYSSAL ATTACHMENT

More than any other innovation, it was the exploitation of the byssal complex that made possible the extensive radiation of bivalves in the epibenthic habitat. The **byssus** is a bundle of strong protein threads extending from the base of the foot to the substratum (Fig. 12-110B). Eighteen evolutionary lines of bivalves, including the marine mussels, arcs, pen clams, jingles, file clams, scallops, giant clams, and many others, have a byssus as adults, and many others have one as juveniles.

Construction of the byssus by the marine mussels (Mytilidae) has been carefully studied (Fig. 12-110B). The reduced, wormlike foot is pressed against the substratum and the **byssal glands** at the base of the foot secrete liquid conchiolin into a longitudinal **byssal groove** on the posterior edge of the foot. Here the protein is molded into a thread and its distal end stuck to the substratum, where it is sclerotized by tanning. After a few minutes the foot withdraws, leaving behind a byssal thread attached distally to the substratum by a tiny disc. The foot and byssal gland may then form additional threads, and collectively the threads constitute the byssus.

Proximally the byssal threads are attached to eight pairs of **byssal retractor muscles.** These homologs of the pedal retractor muscles extend from the byssus at the base of the foot to a row of attachments along the dorsal edge of each valve. Once the

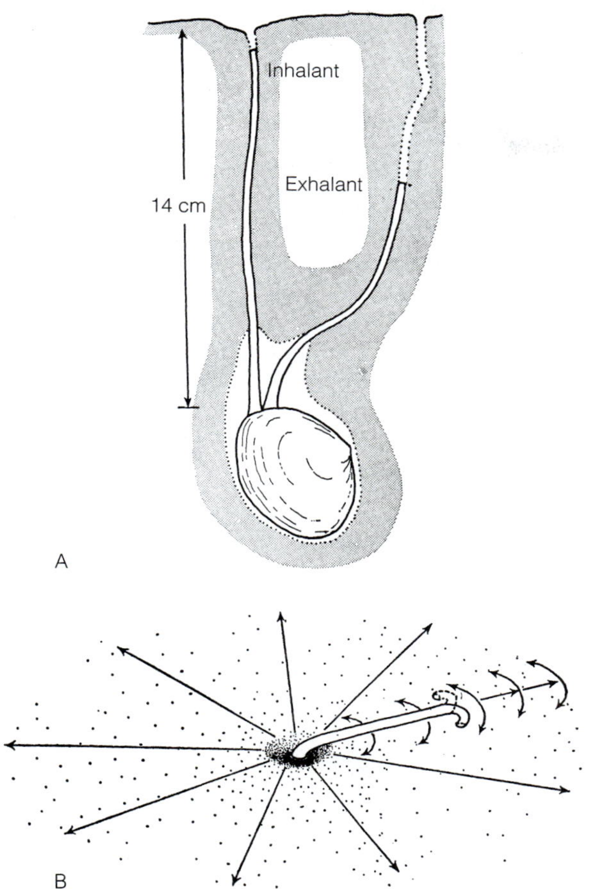

FIGURE 12-109 Deposit feeding in *Scrobicularia*. **A,** Animal in burrow with inhalant siphon withdrawn. **B,** Inhalant siphon at low tide, extended and feeding. *(From Hughes, R. N. 1969. J. Mar. Biol. Assoc. 49: 807. Copyrighted and reprinted by permission of Cambridge University Press)*

byssus is constructed, contraction of the retractor muscles pulls the animal to the substratum.

Among the many surface dwellers attached by byssal threads, the widely distributed and common marine mussels (Mytilidae) are perhaps the most familiar. They live attached to wharf pilings, seawalls, and rocks, or among oysters, often in dense beds. The hairlike threads often radiate outward like guy wires (Fig. 12-110A,B,E). Young individuals use byssal threads to climb vertically. Mytilids are widely used as food in many parts of the world, and commonly are farmed on ropes suspended from floats.

Other surface inhabitants attached by byssal secretions include many of the heavy-bodied arks (Arcidae; Fig. 12-110C), which are very common on tropical coralline substrates; mangrove oysters *(Isognomon),* which hang in clusters from mangrove roots; and winged oysters *(Pteria),* which live attached to sea fans and other gorgonian corals.

The mytilid taxa *Modiolus* and *Geukensia* live partially buried in peat or sediments (Fig. 12-110E) and the pen clams *Pinna* and *Atrina* (Fig. 12-110D) occupy a similar position in sand, attaching the byssal threads to small stones. Pen clams may reach impressive lengths of up to 60 cm and produce very long, fine byssal threads that were once woven by Mediterranean cultures into a beautiful, durable, golden cloth reserved for use by the nobility. A line in the film *The Ten Commandments* that refers to a "golden gown spun from the beards of shellfish" alludes to this practice. Small items such as gloves, scarves, and hats are still produced from this cloth in Italy, primarily for the tourist trade. The only freshwater bivalves with a byssus (as adults) are the dreissenids, which include the zebra mussel, *Dreissena polymorpha.* Attachment of *Dreissena* to the shells of native North American freshwater mussels has contributed to the local extinction of some native populations.

Tridacnidae, which includes the Indo-Pacific giant clams *(Tridacna),* includes infaunal and both types of epifaunal (attached and unattached) species. The smallest of the seven species is only 10 cm long but the largest, *Tridacna gigas,* reaches 1.37 m and weighs 270 kg (Fig. 12-111). All orient vertically, upside down with the hinge side down. Early in life the larvae are attached by byssal threads. As adults, some species retain the byssus, but some lose it and rest on the bottom, held in place by the weight of the massive shell. One species, *Tridacna maxima,* bores into coral or coralline rock so that the valve margins are flush with the substratum surface.

Tridacnids are highly modified as a consequence of the byssus, epibenthic habitat, and especially by their reliance on endosymbiotic zooxanthellae to supplement their nutritional requirements. They have undergone a major morphological reorganization to maximize the exposure of their photosynthetic endosymbionts to sunlight (Fig. 12-112B). The tridacnids maintain dense populations of photosynthetic zooxanthellae in their greatly expanded siphonal tissues. The gape of all tridacnids is directed upward with the hypertrophied siphon surface deployed across the gape, which is scalloped to increase the surface area exposed to light (Fig. 12-111B). Blood sinuses within the siphon tissue contain a garden of symbiotic zooxanthellae that are continuously harvested to provide the clam with an auxiliary source of nutrition. The hemal system transports harvested zooxanthellae to the digestive ceca, where they are digested and assimilated.

The mantle tissue contains brilliant green, blue, red, violet, or brown pigments that are probably sunscreens whose function is to reduce light intensity and protect the photosynthetic endosymbionts from photodamage by ultraviolet radiation. Large populations of tridacnids, such as the small boring species *Tridacna maxima,* can contribute dramatically to the beauty of an Indo-Pacific coral reef.

The utilization of byssal threads by adult bivalves is believed to represent retention of a larval adaptation by attached epifaunal species. The larvae of many unattached burrowing forms produce a byssus for temporary anchorage upon settling (Fig. 12-110F).

CEMENTED ATTACHMENT

Some attached epifaunal bivalves cement one valve permanently to the substratum so that the valve conforms to its shape and is intimately attached to it. Such bivalves lie on one side, fixed to the substratum by either the right or the left valve, depending on the taxon. There are at least eight evolutionary

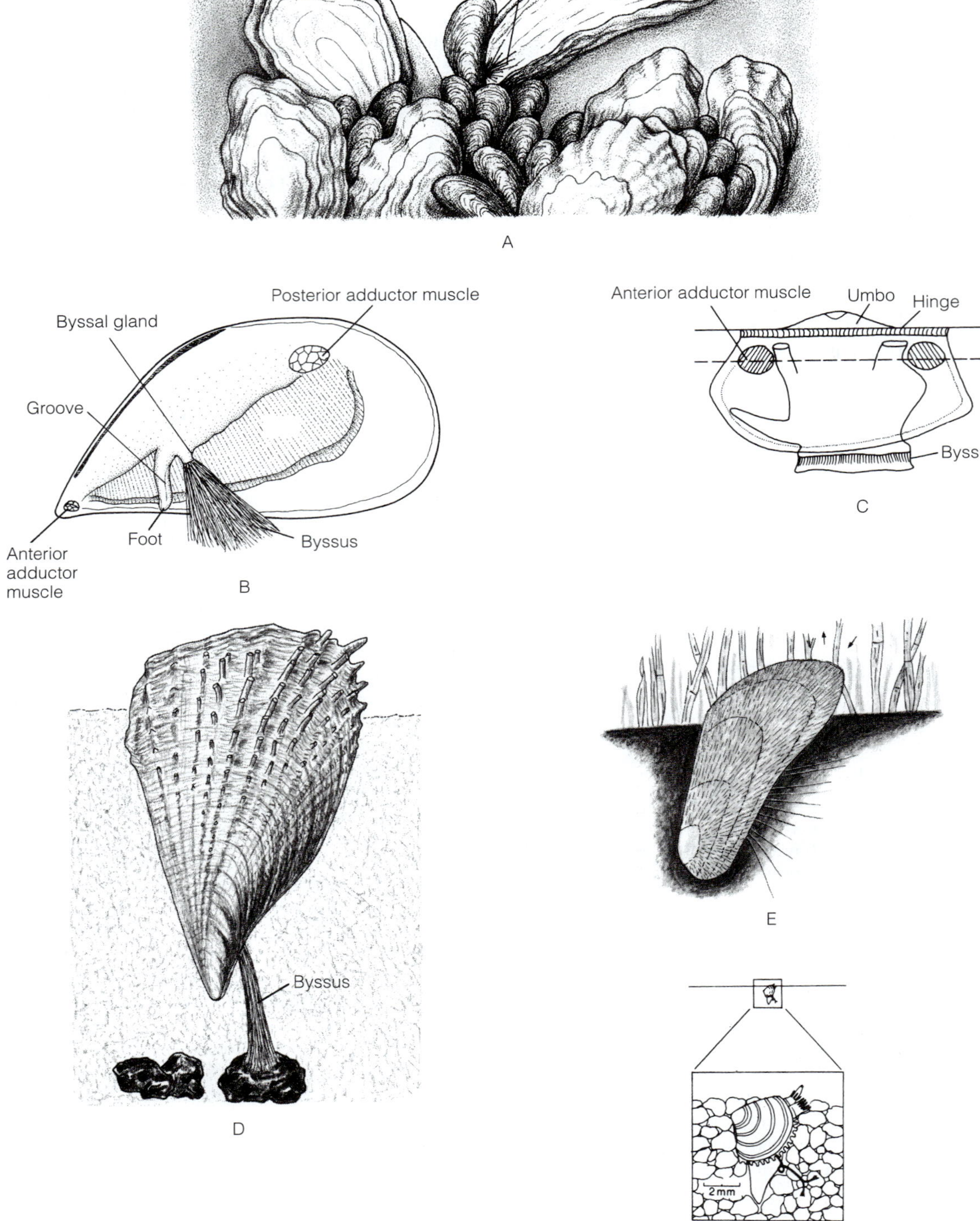

FIGURE 12-110 Sessile bivalves attached by byssal threads. **A,** Mussels *(Mytilus)* anchored by byssal threads to oysters *(Crassostrea).* **B,** Lateral view of a mussel with the left valve removed. **C,** Diagrammatic lateral view of the ark, *Arca.* **D,** A pen clam, *Atrina rigida.* **E,** The mussel *Geukensia* partially buried among intertidal *Spartina* marsh grass. **F,** A newly settled clam, *Mercenaria mercenaria,* anchored in sand by byssal threads. Slightly larger *Mercenaria* lose the byssus and live unattached in shallow burrows in soft sediments. *(A–C, Modified from Yonge, C. M. 1953. Phil. Trans. R. Soc. London B. 287:365; E, After Yonge from Stanley, S. M. 1972. J. Paleontol. 46(2):165–212; F, From Stanley, S. M. 1972. J. Paleontol. 46(2):165–212.)*

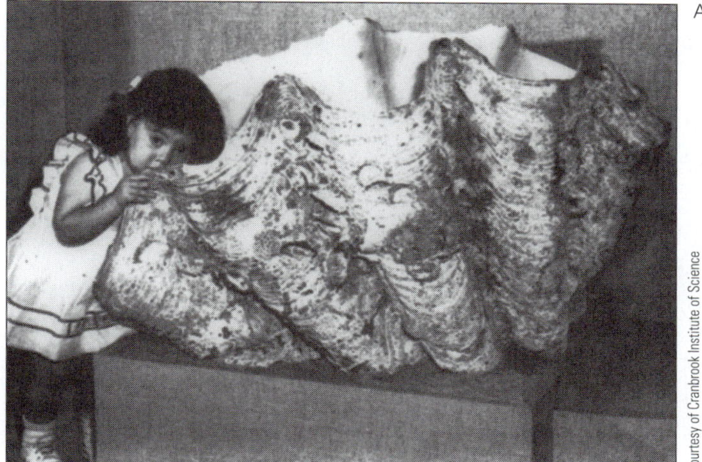

FIGURE 12-111 The giant clam, *Tridacna*. **A,** Shell. **B,** Looking down at an expanded specimen of *Tridacna derasa*. The mantle extends over the fluted shell margins. The white-lipped, oval aperture is the exhalant siphon.

lines of attached, cemented, epifaunal bivalves, of which oysters are the most familiar. The name *oyster*, however, is applied to a wide variety of unrelated taxa, many of which attach by byssal threads and some by cementation. Members of Ostreidae (true oysters), including the edible American east coast oyster *Crassostrea virginica* (Fig. 12-113) and the European *Ostrea edulis*, attach by cementation to the substratum. The metamorphosing veliger is initially anchored with an organic adhesive produced first by the foot and then by the mantle. The mantle margin then secretes the left valve against the substratum.

The attachment of one valve to the substratum has led to varying degrees of inequality in the size of the two valves. The shells of cemented epifaunal bivalves tend to be both inequivalve and inequilateral. In most cases the lower, attached valve is the larger and is concave so that it forms a cup to contain the body. The upper, unattached valve is usually relatively flat and forms a lid for the container. In the true oysters (Ostreidae), the large left valve is attached to the substratum and the flat right valve is the upper lid (Fig. 12-113, 12-110A). In the tropical jewel boxes *(Chama)* the left valve forms a deep box attached to the substratum (Fig. 12-114A). An extreme condition was reached in the extinct Mesozoic rudists, in which the lower valve was shaped like a deep tube or horn (Fig. 12-114B). The rudists often occurred in reeflike aggregations.

The common jingle clams, in the family Anomiidae, have features of both the byssally attached and the cemented bivalves. They lie on their side with the right valve tightly appressed against the substratum, but they are actually anchored by a large, calcified byssal thread (peduncle) that passes through a large opening in the right valve to reach the substratum, where it is attached (Fig. 12-114C). The left valve—the upper valve—lacks the byssal opening.

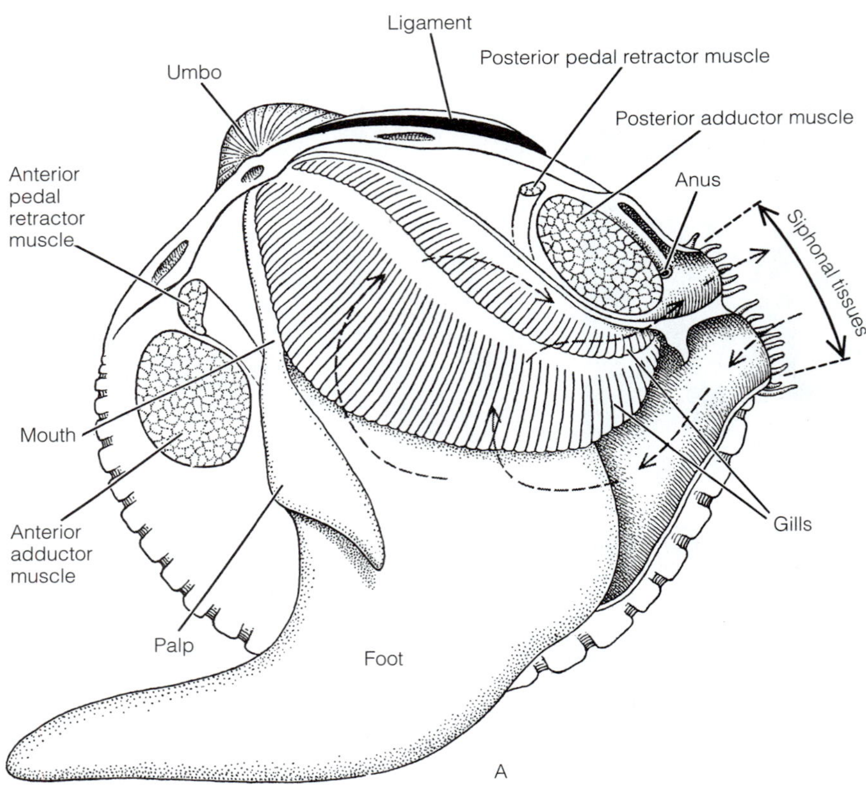

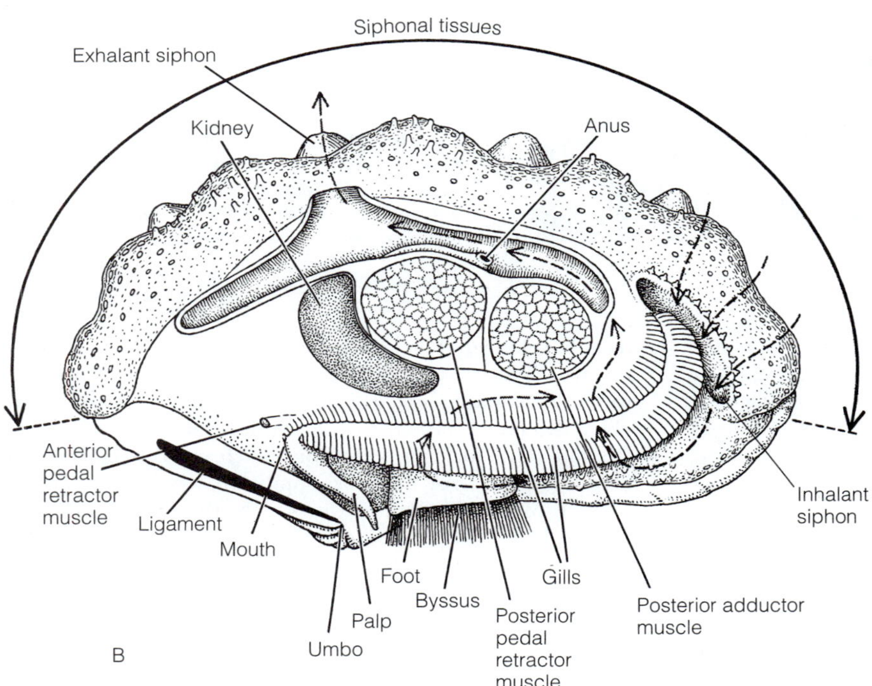

FIGURE 12-112 Comparison of the structures of **(A)** a cockle, a shallow burrower in soft bottoms with typical bivalve anatomy, with **(B)**, the related but morphologically aberrant and extensively reorganized giant clam *(Tridacna)*, which is byssally attached and oriented with the hinge side down. Both views are from the left and are oriented with the foot ventral, in its correct position. In the cockle, the remaining parts are in their accustomed locations with respect to the foot. In the giant clam there has been, in effect, a counterclockwise rotation (when viewed from the left) of structures around the foot, which remains in its original ventral position at 6 o'clock on a clock face. The formerly dorsal hinge, ligament, and umbo are now ventral, having moved from 12 o'clock to about 7 o'clock. The formerly posterior exhalant siphon has moved to near the midpoint of the gape, on top of the animal (from 3 to 11 o'clock). The posterior pedal retractor muscle has moved anterior to the middle of the clam and has hypertrophied enormously. The mouth, anterior pedal retractor muscle, gills, posterior adductor muscle, and inhalant siphon are in approximately their original positions. A result of this rotation is that the expanded siphonal tissue, with its population of zooxanthellae, now occupies most of the gape and is facing upward, toward the sun. *(From Yonge, C. M. 1975. Giant clams. Sci. Am. 232(4):96–105.)*

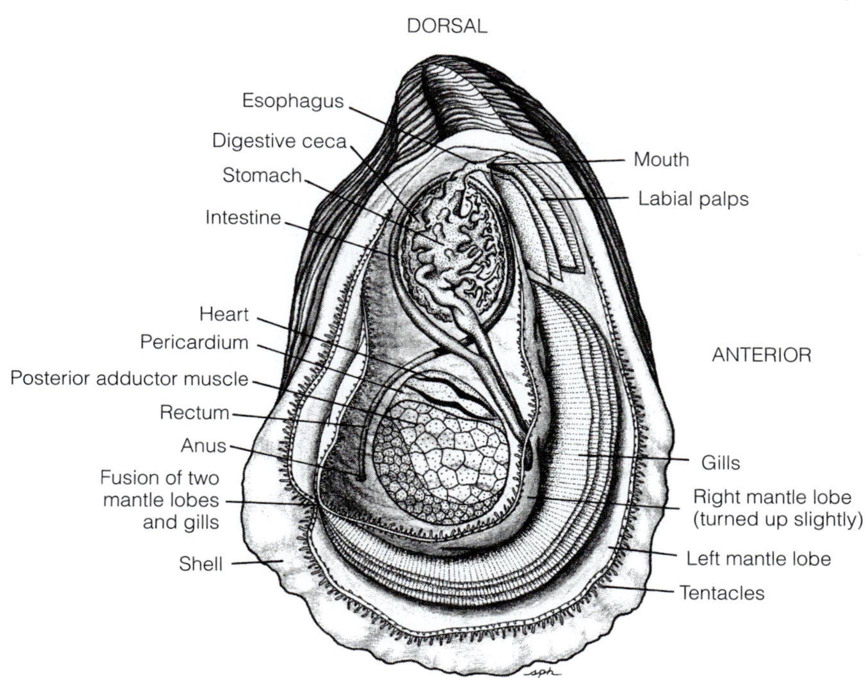

FIGURE 12-113 Anatomy of the American oyster *Crassostrea virginica.* The animal is shown in its attached left valve with the right valve removed. The visceral mass is drawn as if transparent to reveal features of the internal anatomy. *(Modified after Galtsoff, 1964)*

SPECIALIZATIONS

Epifaunal bivalves share a number of convergent features even though the taxa may be unrelated. Predictably, the foot is reduced by varying degrees, usually being present but small in byssally attached species and completely lost in cemented species. There is a tendency to reduce the anterior end, so that the muscles are anisomyarian or monomyarian and the valves are inequilateral. Anterior reduction in mussels is perhaps an adaptation to facilitate elevation of the posterior end above the substratum, thereby reducing the likelihood of obstruction that the dense aggregations of these bivalves tend to create.

Many epibenthic bivalves have a well-developed middle mantle lobe with tentacles and eyes. Obviously the relationship between an epibenthic bivalve and its environment differs from that of an infaunal bivalve and its environment. Epibenthic species encounter food and danger on all sides, not just at the tips of the siphons, and it is advantageous to have sense organs along the entire length of the mantle margin.

Mantle fusion and siphon formation have not occurred in epifaunal bivalves since they live above the surface and generally on hard substrata, where fouling by sediment particles is less likely. However, oysters and mussels, which occur in dense beds, are subject to self-fouling and depend on the cleansing action of tidal currents to prevent them from becoming completely buried in their own feces and pseudofeces. Young *Mytilus edulis* (mussels) clean the exterior of the shell with their precociously elongate foot.

Unattached Epifauna

Of the few bivalves that live on the surface without attaching to it, the best known are the scallops (Pectinidae; Fig. 12-115) and the file clams (Limidae). Both families contain species that attach using a byssus, but many species are unattached. These bivalves always rest with the right valve down, against the substratum. The foot is reduced, being used only to clean the mantle cavity, and the anterior adductor muscle is lost. The large, well-developed posterior adductor muscle occupies a more central position and is divided into catch (smooth) and fast (striated) portions (Fig. 12-115B). Rapid contractions of the striated fibers are used for swimming, whereas the sustained contractions of smooth fibers provide for prolonged closure of the valves.

The unattached species of file clams and scallops swim by clapping the valves together. This forces a jet of water from the mantle cavity that propels the animal away from a predator. The muscular inner lobe of the mantle margin, when appressed against the lobe on the opposite mantle surface, controls the direction of the water jet, permitting it (in scallops) to exit on either side of the hinge line or opposite the hinge line. The scallop can, to a limited extent, control the direction in which it swims.

The swimming ability of scallops and file clams is used primarily to escape predators. For example, if a predatory starfish, or even one tube foot of such a starfish, touches the mantle margins of a scallop, the swimming response is evoked, and the scallop will swim a meter or so away at a much faster rate than the starfish can follow. Some scallops use the water jets to blow out a depression in the sand surface into which they settle. File clams nest in crevices beneath stones and swim only when disturbed.

Boring

The ability to penetrate and live beneath the surface of firm, consolidated substrata, such as peat, clay, sandstone, shell, coral, calcareous rock, and wood, has evolved in five lamellibranch

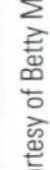

A

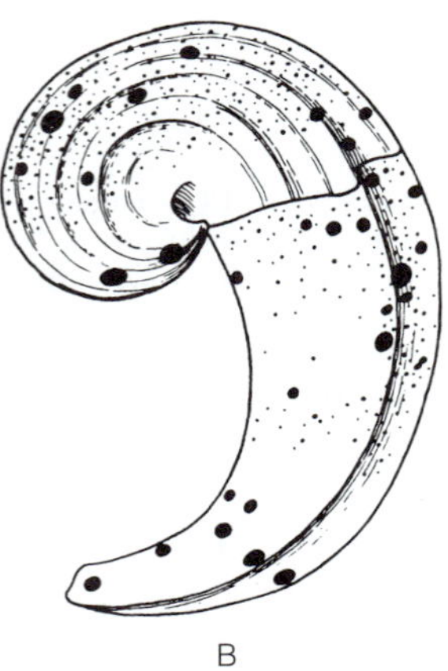

B

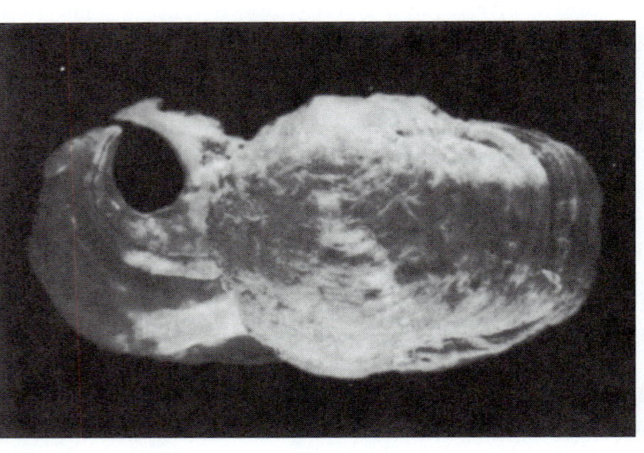

Courtesy of J. B. Gates, from Andrews, J. 1971. Sea Shells of the Texas Coast. University of Texas Press, Austin. p. 168

C

FIGURE 12-114 **A,** The jewel box, *Chama,* a common, tropical sessile bivalve whose upper (right) valve forms a lid over the attached lower (left) valve. **B,** A rudist, a Mesozoic bivalve in which a caplike upper valve covers a deep, hornlike lower valve. **C,** A jingle, or toenail, clam (Anomiidae). The lower (right) valve contains a large hole for the calcareous peduncle, homologous to the byssus of other bivalves. The right valve of the clam is on the left in the photograph. *(B, From Kauffman, E. G., and Sohl, N. F. 1973. Verh. Nat. Ges.)*

families. These belong to two distinct taxonomic and evolutionary groups but are collectively known as the boring bivalves. One, composed of some members of Myoida and Veneroida, arose from soft-bottom infaunal burrowers that gradually evolved the ability to bore into successively firmer substrata. Pholadidae (Fig. 12-116), containing the piddocks, martesias, and angel wings, are the most conspicuous and best-known examples of this group, but also included are the gastrochaenids, teredinids, and petricolids. Boring species with burrowing ancestors usually attach to the burrow wall by the foot, which has developed a suckerlike ventral surface for this purpose (Fig. 12-116B).

The ancestors of the second group, represented by the boring mussels (Mytilidae), were probably byssally attached epifaunal bivalves. Boring bivalves with epifaunal ancestors attach to the side of their burrow by byssal threads.

Each boring bivalve begins excavation soon after the larva settles on a suitable substratum and slowly enlarges and deepens the burrow as it grows. Once the burrow is constructed, the animal is forever enclosed in it and communicates with the outside world via the siphons extended from a small opening at the surface. A boring bivalve removed from its chamber cannot excavate a new burrow and is doomed to quick consumption by a predator.

In the great majority of species, boring is a mechanical process employing the valves as drills to abrade the substratum and produce a burrow. The anterior surfaces of the valves are frequently roughened or toothed to increase their abrasive ability (Fig. 12-116A). The drilling movements are adaptations of the locomotory (burrowing) movements of their nonboring ancestors.

Drilling rates vary. Over an 18-month period, the piddocks *Penitella penita* and *Chaceia ovoidea* excavated soft shale at rates of 2.6 mm and 11.4 mm per month, respectively. Some boring bivalves rotate on their longitudinal axis within the burrow, and as a result the burrow cross section is round (Fig. 12-116C). Others remain attached to one wall of the burrow and maintain a constant orientation with respect to their longitudinal axis, so the burrow tube reflects the cross-sectional shape of the shell. Some of these species, such as *Gastrochaena,* create separate openings for the inhalant and exhalant siphons at the surface of the rock by secreting calcareous material between the two siphons. Much of the particulate waste produced by drilling is transferred to the mantle cavity and then ejected with the pseudofeces through the inhalant siphon.

One of the date-pit mussels, *Lithophaga bisulcata, a* very common, cigar-shaped, byssally attached borer in shell and coral, excavates chemically. A mucoprotein-chelating agent, secreted by glands in the middle fold of the mantle margin, softens the calcareous substratum, which is then scraped away by the valves. Those species that bore into live coral have, in addition, glands whose secretions prevent the coral from depositing calcium carbonate over the bivalve's bore hole and still other glands that inhibit discharge of the coral's nematocysts.

Many boring bivalves inhabit wood. The wood-boring pholadids, such as *Martesia* and *Xylophaga,* are adapted in much the same way as the many rock- and shell-boring members of the same family. Wood panels planted 1830 m deep on the sea bottom were completely riddled by *Xylophaga* and a related genus when recovered 104 days later.

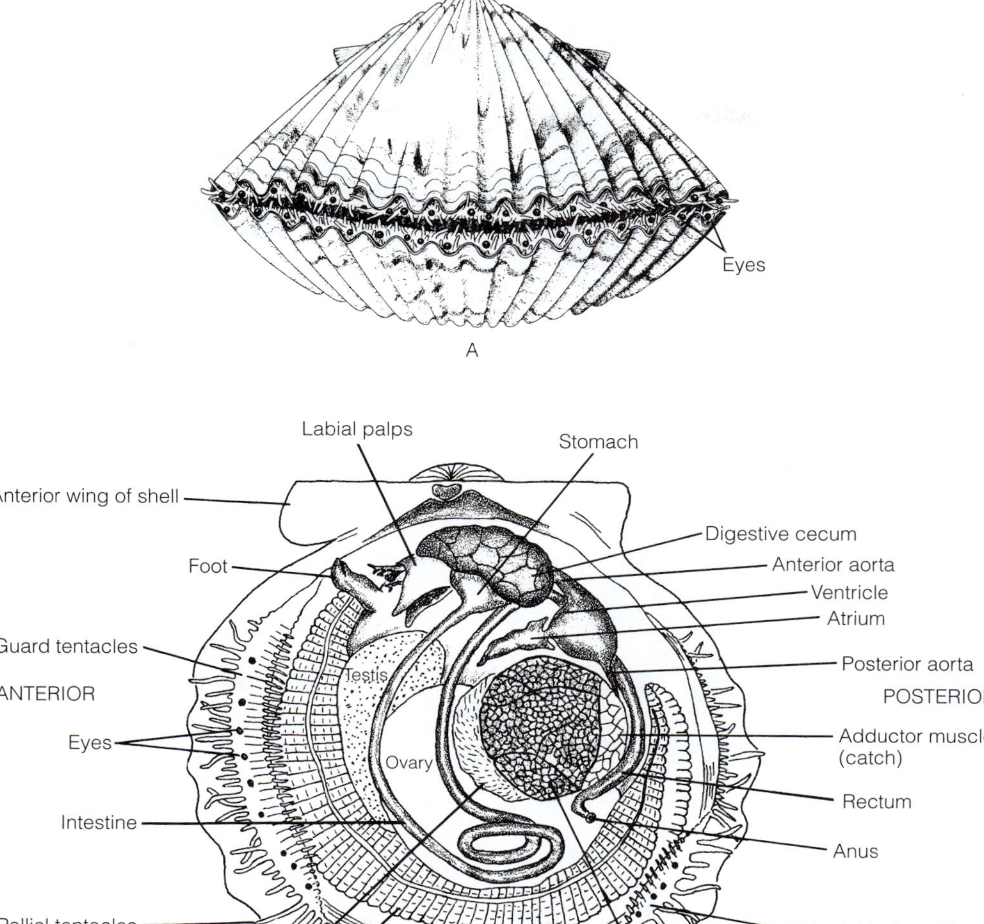

FIGURE 12-115 The scallop *Pecten.* **A,** Gape view, showing eyes and tentacles on the sensory fold of the mantle margin. **B,** Internal structure viewed from the left side, with the left valve removed. Anterior is on the viewer's left. *(After Pierce, M. 1950. In Brown, F. A. (Ed.): Selected Invertebrate Types. John Wiley and Sons, New York.)*

The most specialized wood borers are the shipworms, members of the taxon Teredinidae. The natural habitats of the 60 species of this widely distributed taxon are mangrove prop roots and timber swept into the sea by rivers, and they play an important ecological role in the degradation of wood in the sea. In an economic context, they are destructive animals that damage or destroy piers, pilings, hulls, and other wooden structures placed by humans in the sea, and much expense and research have been devoted to their control. Timbers can quickly become completely riddled with tunnels (Fig. 12-117B). They were an even more serious economic problem in the days of wooden sailing vessels.

Shipworms are the most morphologically derived of the boring bivalves, perhaps of all the bivalves. The body is greatly elongate, cylindrical, and wormlike (Fig. 12-117A). The shell is reduced to two small anterior valves, which are used to rasp into the wood. Cutting of the wood is accomplished by opening and rocking motions of the valves while the body is attached to the burrow by the small foot. The elongate mantle, enclosing most of the body behind the valves, secretes a calcareous lining for the burrow. The long, delicate siphons open at the surface of the wood and the burrow entrance is plugged by calcareous pallets when the siphons are retracted. Burrow size increases with the growth of the shipworm and

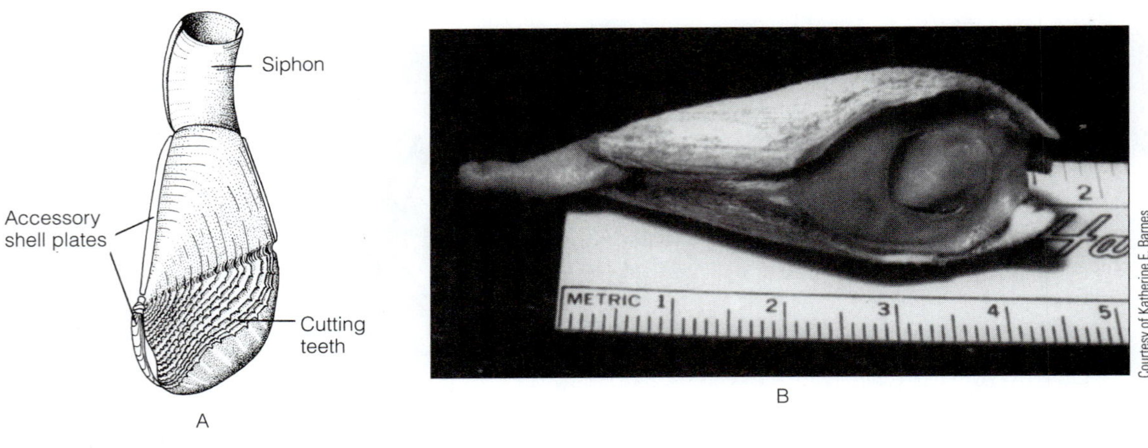

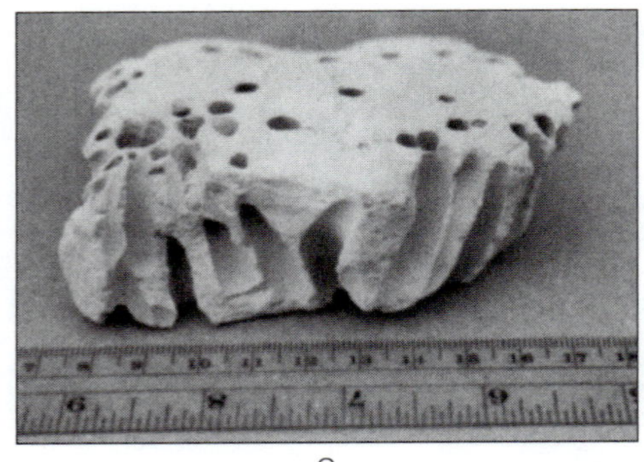

FIGURE 12-116 Boring bivalves. **A,** Structure of a pholadid. **B,** Ventral view of a pholadid, showing the suckerlike foot in the pedal aperture. Siphon is to the left. **C,** Surface openings and part of the burrows of pholadid bivalves in a piece of hard clay.

may reach a length of 18 cm to 2 m, depending on the species. The life span is one to several years, again depending on the species.

Unlike the piddocks, shipworms use the excavated sawdust for food, although many also filter feed. Most boring organisms use their substratum as a safe habitat but do not consume it for food. Shipworms and termites are important exceptions, however, and their great ecological and economic importance is a consequence of this unusual ability. Because they burrow for food, the extent of damage is far greater than it would be if the excavations were only a place to live. The shipworm stomach has a large cecum for sawdust storage. Symbiotic bacteria housed within a special organ that opens into the esophagus not only digest cellulose, but also, by fixing nitrogen, compensate for the low-protein diet.

Commensals and Parasites

A small number of bivalves belonging almost exclusively to the veneroid taxon Lasaeidae, have evolved commensal and parasitic relationships with other invertebrates. Most commensals are related to free-living epibenthic forms, such as *Kellia* and *Lasaea,* which nestle in crevices. Most attach by byssus threads, but some crawl on their foot like a snail. The hosts are usually burrowing echinoderms (such as heart urchins, brittle stars, and sea cucumbers), burrowing polychaetes, and burrowing, shrimplike crustaceans. *Entovalva mirabilis,* which lives in the gut of sea cucumbers, is the only known parasitic bivalve. *Paramya subovata,* a commensal with the echiuran *Thalassema hartmani,* and *Cryptomya californica,* a commensal with the echiuran *Urechis caupo,* are myoids, not veneroids.

Commensal bivalves tend to be highly adapted for life in the burrows of other organisms. They are always very small and often have a reduced and internalized shell, an expanded mantle, a byssus, and a one-way flow of water with anterior inhalant and posterior exhalant siphons. They exhibit a variety of reproductive adaptations, including hermaphroditism and dwarf, parasitic males; fertilization in the mantle cavity; and brooding in the mantle cavity, and one species even has seminal receptacles.

Freshwater Habitats

Except for occasional representatives from a variety of marine families such as the oysters, shipworms, piddocks, mussels, and others that can tolerate low salinities and have invaded brackish estuaries and marshes, the freshwater bivalves are restricted to nine families variously represented in different parts of the world.

Native North American freshwater bivalves belong to Unionidae, Margaritiferidae, and Sphaeriidae. All are

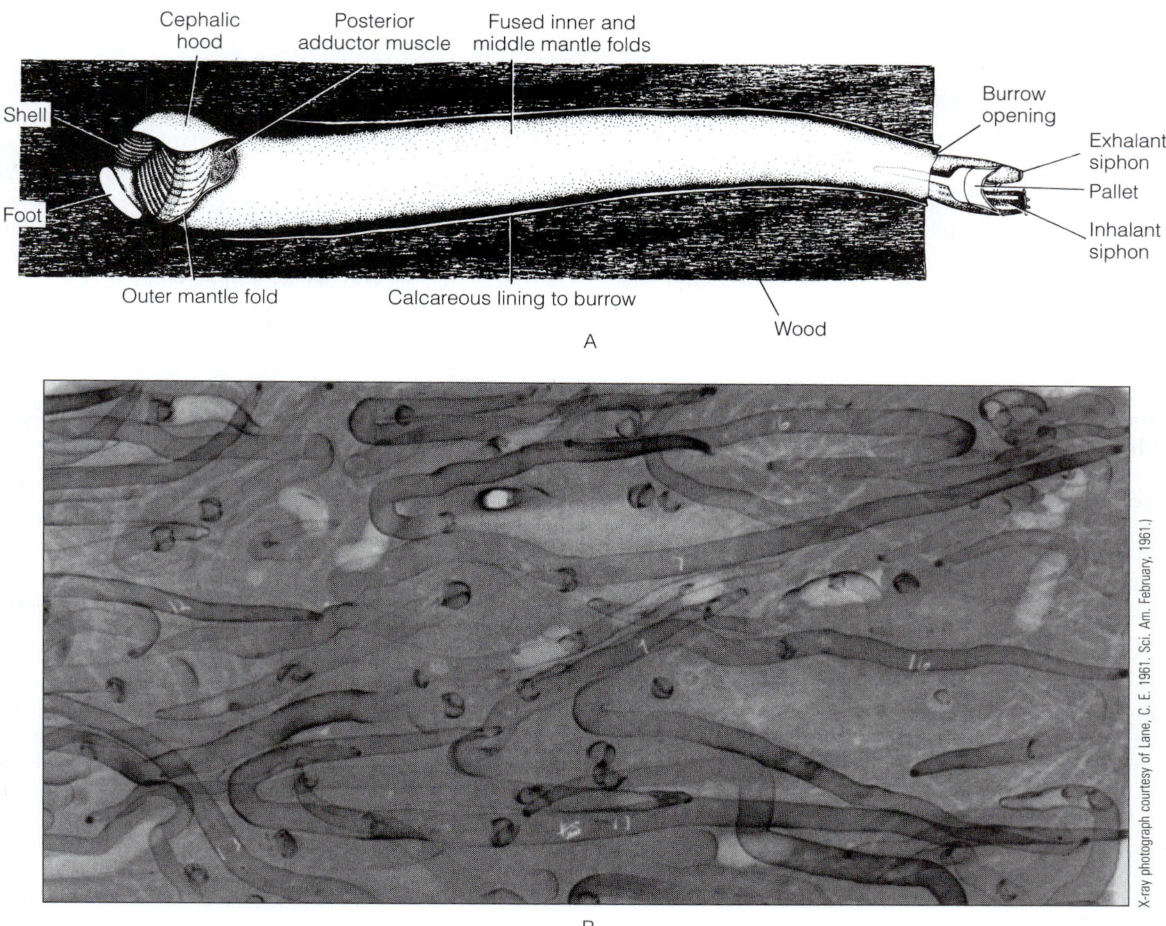

FIGURE 12-117 More boring bivalves. **A,** A shipworm, a wood-boring bivalve. **B,** X-ray photograph of a marine timber section showing shipworms. *(A, By Brian Morton.)*

ancient taxa, well adapted to life in freshwater habitats. As an important adaptation to freshwater life, all three have lost the planktonic veliger larva. Planktonic larvae are usually a disadvantage in freshwater animals, partly because currents sweep them downstream, away from optimal habitat for the species. Sphaeriids (fingernail clams) have direct development and brood their juveniles in the mantle cavity. Many fingernail clams are adapted for living in temporary bodies of fresh water. The unionids (freshwater mussels), with 110 extant genera, have been especially successful, most notably in eastern North America, where their diversity is very high. Unionids, and the less diverse margaritiferids, have a specialized larva that is parasitic on fishes (Fig. 12-121C). They brood the embryos in the gills before releasing the larvae. Development in freshwater mussels is discussed in more detail in a later section.

The Asian clam, *Corbicula fluminea,* was introduced to the west coast of North America in 1925 and has since spread across the continent so that it is now present in most fresh water south of about 40° N latitude, including streams, lakes, and reservoirs. Population densities of 10 to $3000/m^2$ are typical, but $130{,}000/m^2$ have been reported. *Corbicula* competes successfully with native mussels and excludes them from their habitats. It clogs irrigation ditches and canals and degrades the quality of river sand and gravel for construction purposes.

Another exotic species with the potential for serious ecological and economic harm is the zebra mussel, *Dreissena polymorpha,* which has been transported across the Atlantic in the ballast water of cargo ships filled and emptied in freshwater ports. The zebra mussel is believed to have entered the St. Lawrence Great Lakes in 1986 with the discharge of ballast water originating in a Baltic port and has since undergone an explosive population increase. This small, epifaunal mussel attaches by byssal threads and forms dense mats containing as many as 700,000 individuals per square meter. In addition to altering the ecosystem by removing a large amount of plankton and overgrowing and smothering native clams, the zebra mussel clogs the intake pipes of municipal water plants and the cooling water intakes of various industrial facilities, including nuclear reactors. *Dreissena* is unusual among freshwater clams in having a free-swimming veliger larva.

INTERNAL TRANSPORT

The hemal system is similar to that of other conchiferans. It consists of a heart in a pericardial cavity, aortae leading to hemocoelic blood sinuses, and numerous veins returning the blood to the heart (Fig. 12-52). The heart consists of a central ventricle that receives blood from a pair of lateral atria that drain oxygenated blood from the gills. There are some modifications of the typical molluscan circulatory path in bivalves, including a well-developed circuit through the mantle skirts, which are auxiliary respiratory organs. In most bivalves, the intestine passes through the pericardial cavity and ventricle (Fig. 12-89B).

The blood (hemolymph) contains hemocytes and sometimes respiratory pigments. In addition to its transport functions, bivalve blood is a component of the hydrostatic skeleton that plays a role in digging and about 50% of the soft-tissue volume of a bivalve is blood.

GAS EXCHANGE

Gas exchange occurs over the gills, the inner surface of the mantle skirts, the foot, and other exposed epithelia. Removal of oxygen from the respiratory current is inefficient compared with that of other molluscs. For example, scallops remove 2.5 to 6.8% of the oxygen whereas abalones (gastropods) remove 48 to 70%. This low rate of oxygen assimilation is made possible by the unusually large volume of water passing through the unusually large gills. Lamellibranch gills are disproportionately large because of their role in filter feeding, and there is much more surface area and more water flow than needed for gas exchange. Moreover, in some bivalves, such as the intertidal mussel *Geukensia demissa,* there is substantial uptake of oxygen across the mantle skirts. At low tide, when the mussel is exposed and the gills are collapsed, it utilizes atmospheric oxygen absorbed by the mantle rather than the gills.

Bivalve blood usually lacks respiratory pigments, but in about 20 species, including arks *(Noetia, Arca, Anadara)* and *Calyptogena* (of deep-sea hydrothermal vents), the blood contains intracellular or extracellular hemoglobin. In some species, such as *Tellina alternata,* neuroglobin is present in the ganglia, which consequently are yellow or orange. Hemoglobin and muscle myoglobin may give tissues a bright red color. In the ark *Noetia,* at least, blood hemoglobin functions both in oxygen transport and in oxygen storage.

Hemocyanin has been found in two taxa of protobranchs. This is probably additional evidence that hemocyanin is the primitive molluscan respiratory pigment that has been lost in most bivalves.

EXCRETION

The bivalve excretory apparatus is a heart-kidney complex similar to that of other molluscs (Fig. 12-89B, 12-118). The pericardial glands are evaginations of the pericardium into some part of the hemal system, typically the atria. The epithelium of these evaginations separates the pericardial cavity from the blood, is equipped with abundant podocytes, and is the site of ultrafiltration, which forms a primary urine in the pericardial cavity. Each of the two nephridia (kidneys) is a large, elaborated nephridium connected with the pericardial cavity by the ciliated renopericardial duct, which opens from the cavity by the nephrostome. The nephridium is surrounded by the renal vein and its walls are composed of secretory and absorptive cells. The nephridium, the site of modification of the primary urine, is a long, sinuous tube with an enhanced surface area in contact with the blood of the surrounding renal vein. Each nephridium empties into the exhalant chamber via a nephridiopore.

The operation of the heart-kidney complex should be familiar by now. Blood enters the pericardial glands (Fig. 12-118) from the atria and is ultrafiltered across the pericardium to form the primary urine in the pericardial cavity. The primary urine passes through the nephrostome into the renopericardial duct to the nephridium. The composition of the primary urine is altered by selective secretion and absorption by the nephridial epithelium vis-à-vis the blood of the surrounding renal vein. With modification complete, the final urine exits the nephridiopore into the exhalant chamber of the mantle cavity and is removed from the clam by the exhalant current.

Freshwater bivalves, like their gastropod counterparts, excrete large volumes of water through the nephridia. The urine is very hyposmotic to the blood, and salts are reabsorbed by the nephridium to maintain very low concentrations in the final urine. The mantle and gill epithelia absorb salts from the respiratory water current.

NERVOUS SYSTEM

The nervous system is similar to that of other conchiferans, but it is less cephalized due to the reduction of the head as a center for sensory reception in these sedentary animals. The head, such as it is, is inside the shell and is not well situated for sensory functions. Rather than being centralized, the sense organs are located in peripheral regions such as the foot, mantle margins, and siphons that come in contact with the environment. The usual molluscan ganglia are present, but they tend to be widely separated by long connectives and commissures.

The nervous system is bilaterally symmetrical with three major pairs of ganglia and three pairs of nerve cords (Fig. 12-119). The paired **cerebropleural ganglia** are composed of the cerebral, pleural, and buccal ganglia of the conchiferan ancestor. They are located dorsolaterally beside the esophagus and are connected across the midline, dorsal to the esophagus, by the cerebral commissure. The cerebropleural ganglia serve the anterior adductor muscle, anterior mantle, cerebral eyes, labial palps, anterior gut, and statocysts. Coordination of foot and adductor muscle activity is a function of the cerebral ganglia.

The two pedal ganglia are coalesced on the midline, well ventral to the esophagus, and near the junction between the visceral mass and the foot (Fig. 12-90). Motor nerves from the pedal ganglia control the anterior pedal retractor muscles, byssal retractor muscles, and the foot. The pedal and cerebropleural ganglia are included in the nerve ring and are connected by pedal nerves.

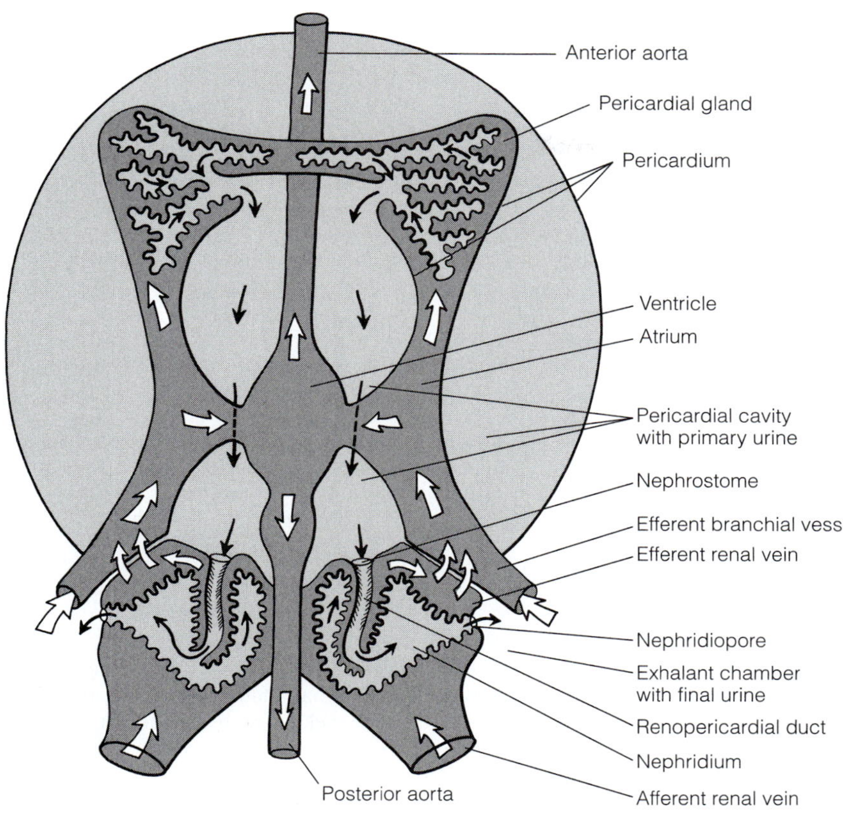

FIGURE 12-118 Diagrammatic dorsal view of the heart-kidney complex of the bivalve *Scrobicularia plana*. Solid arrows indicate urine flow, outlined arrows indicate blood flow. Ultrafiltration takes place in the pericardial glands and produces primary urine in the pericardial cavity. This ultrafiltrate flows through the renopericardial duct to the lumen of the nephridium, where it is modified by exchanging materials with the surrounding blood in the renal veins. Valuable materials are reabsorbed by epithelial cells and returned to the blood, whereas additional toxins and wastes are transferred from the blood to the urine. Upon reaching the nephridiopore, the proportion of wastes to valuable materials in the urine is high and the urine is in its final condition ready for elimination with the exhalant water stream. The rectum is omitted for clarity. *(Redrawn and modified from Morse, M. P., and Zardus, J. D. 1997. Bivalvia. In Harrison, F. W., and Kohr, A. J. (Eds): Microscopic Anatomy of Invertebrates. Vol. 6A, Mollusca II. Wiley-Liss, New York. pp. 7–118; after Andrews, E. B., and Jennings, K. H. 1993. The anatomical and ultrastructural basis of primary urine formation in bivalve molluscs. J. Moll. Stud. 59:223–257.)*

Two long visceral nerves, situated dorsally in the visceral mass, extend posteriorly from the cerebropleural ganglia to terminate at the visceral ganglia (composed of the fused visceral and esophageal ganglia of other molluscs) on the anteroventral surface of the posterior adductor muscle (Fig. 12-119). The visceral ganglia serve the mantle and siphons, posterior adductor muscle, posterior pedal retractors, gills, osphradia, and most of the viscera (heart, gonad, nephridia, and intestine).

Two **pallial nerves** make a large loop from the cerebropleural to the visceral ganglia ventrally in the visceral mass. The pallial nerves lie in the periphery of the skirts close to the mantle folds (Fig. 12-91). A pair of **siphonal ganglia,** located on the pallial nerves close to the visceral ganglia, receives the abundant sensory nerves of the siphons.

SENSE ORGANS

The margin of the mantle, particularly the middle fold and the siphons, is the location of most bivalve sense organs. In many species the mantle edge bears **pallial tentacles** (Fig. 12-115) with primary ciliary receptor cells that are mechanoreceptors and perhaps chemoreceptors.

A pair of statocysts is usually found in the foot near or within the pedal ganglia, although they are innervated by the cerebral, not the pedal, ganglia. Each statocyst consists of a sac of ciliated cells containing a calcareous statolith. The statocysts of some attached species, such as oysters, are reduced.

Pallial eyes (ocelli) are features of the mantle margin and may be on the middle mantle fold or on the siphons. They enable a surface-dwelling bivalve such as a scallop or cockle or an infaunal clam with a siphon to detect sudden changes in light intensity, such as the shadow cast by a predator. The exposed mantle tissue of giant clams contains several thousand ocelli. In most cases, bivalve ocelli are simple pigment spots or pigment cups. In arks, however, the ocelli are compound, and in scallops and thorny oysters, the ocelli are well-developed eyes.

The unusual eyes of scallops *(Pecten)* are located on the middle mantle fold (Fig. 12-115) and have an outer epithelial cornea that covers a large lens composed of cells resting on a basal lamina. Under the basal lamina are two distinct layers of photoreceptor cells, the distal and proximal retinas. The **distal retina** is a direct pigment cup, whereas the **proximal retina** is an indirect cup. The distal retina is composed of ciliary receptors that face the light source. The proximal retina is composed of rhabdomeric receptors facing away from the light source. Below the two retinas is a layer of reflective pigment **(tapetum)** followed by a layer of opaque pigment. Light that passes the two retinas is reflected back to the photoreceptors, where it may be detected. The advantage of this arrangement is unclear. Perhaps the tapetum reflects light back to the proximal retina, whose receptors face it. The innermost layer of opaque pigment is the pigment cup, whose function is to prevent the entry of light from any direction other than the lens.

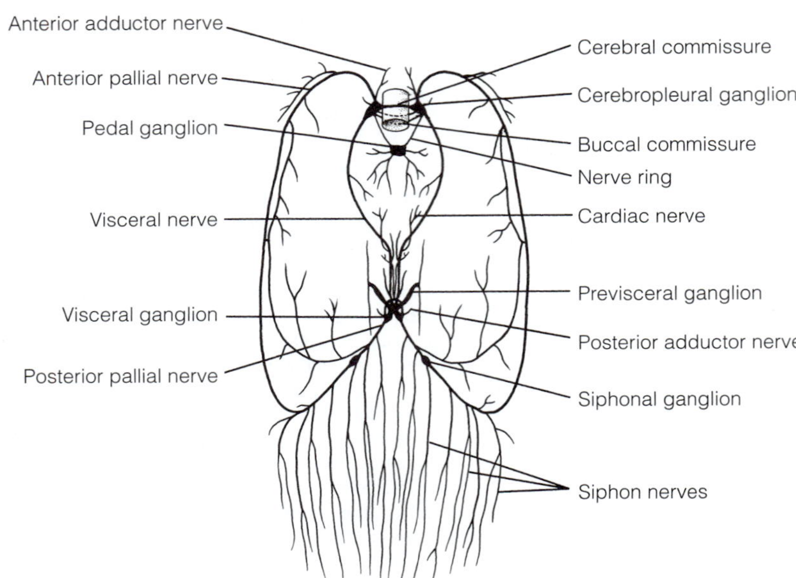

FIGURE 12-119 Dorsal view of the nervous system of *Pholas,* a siphonate eulamellibranch that bores into hard substrata. *(After Forster from Bullock, T. H., and Horridge, G. A. 1965. Structure and Function in the Nervous System of Invertebrates. Vol. II. W. H. Freeman, San Francisco. p. 1391.)*

Cerebral eyes (cephalic eyes) are direct pigment cups present in veliger larvae and some adult mussels (such as *Mytilus edulis*). They are paired and one is located at the anterior end of each gill axis. A small translucent area of the overlying shell permits light to reach the eye. Cerebral eyes are probably homologous to the eyes of gastropods.

Immediately beneath the posterior adductor muscle in the *exhalant* chamber is a patch of sensory epithelium usually called an osphradium. The bivalve osphradium is assumed to be a chemoreceptor for monitoring the water passing through the mantle cavity. Its position in the exhalant chamber, however, means it is unable to sample the incoming water current before it reaches the gills, which casts some doubt on its supposed role.

REPRODUCTION

Most bivalves are gonochoric and fertilization is almost always external, although it sometimes occurs in the mantle cavity. The gonads are paired and expand to surround the intestinal loops in the visceral mass (Fig. 12-89B). The two gonads are usually so close together they cannot be distinguished. In these gonochoric animals with no copulation, there is no reason for the gonoducts to be more than simple conveyances for gametes and the reproductive system is simple.

In the ancestral bivalves, as in the ancestral mollusc, the gonoduct opened into the pericardial cavity and the gametes were released from there through the nephridium to the mantle cavity (Fig. 12-120A). In intermediate bivalves, such as protobranchs and filibranchs, the gonoduct joins the nephridium somewhere along its length and the gametes enter the mantle cavity through the nephridiopore (Fig. 12-120B). In most eulamellibranchs the gonoduct and nephridium are entirely separate and have independent pores into the exhalant chamber (Fig. 12-120C).

Hermaphroditic bivalves include the shipworms (Teredinidae), the freshwater fingernail clams (Sphaeriidae), a few Unionidae, some species of cockles, oysters, scallops, and the Asian clam, *Corbicula fluminea.* In simultaneously hermaphroditic scallops, the gonad is divided into a ventral ovary and a dorsal testis, both of which lie on the anterior side of the adductor muscle (Fig. 12-115B). The European oyster, *Ostrea edulis,* like other species of *Ostrea,* is a protandric hermaphrodite. *Ostrea edulis* not only shifts from male to female, but also changes back from female to male. An individual may exhibit active male and female phases each year. Species of *Crassostrea,* including the American east coast oyster, *Crassostrea virginica,* are mostly gonochoric.

DEVELOPMENT

In most bivalves, gametes are shed into the environment with the exhalant current. Some bivalves, however, brood their eggs in the exhalant chamber, as do some shipworms, or within the demibranchs, as does *Ostrea edulis,* the freshwater Unionidae (mussels), and Sphaeriidae (fingernail clams). Brooded eggs are usually fertilized by sperm brought in with the feeding current or, in the case of some shipworms, the male uses the exhalant siphon as an intromittent organ to deliver sperm into the female's inhalant siphon.

Cleavage is spiral and a molluscan cross is formed in the blastula. In most marine bivalves a free-swimming trochophore develops from the gastrula and is succeeded by the typical bivalve veliger (Fig. 12-121A,B). The bilaterally symmetrical veliger is enclosed by two valves, has a ciliated velum, foot, gut, gill rudiment, and two adductor muscles, and resembles a miniature clam. Protobranchs have a trochophore, sometimes known as a pericalymma, but no veliger.

Like gastropods, some marine bivalves have long-lived, planktotrophic (feeding) veligers capable of dispersal over long

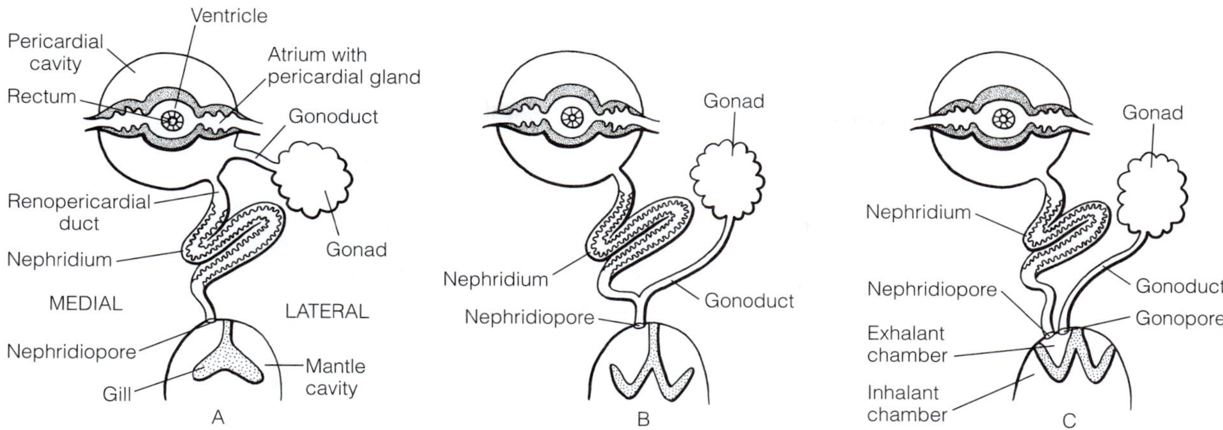

FIGURE 12-120 The evolution of independent bivalve gonoducts. Diagrammatic cross sections showing the right mantle cavity of representative bivalves. **A,** An ancestral bivalve with the gonoduct opening directly into the pericardial cavity. Gametes pass to the mantle cavity through the nephridium. **B,** An intermediate condition like that in the filibranch *(Mytilus)* with the gonoduct sharing part of the nephridial tube. **C,** The condition found in most eulamellibranchs, with the nephridium and gonad having separate ducts and openings in the exhalant chamber. In hermaphroditic bivalves the sequence and relationships are the same, but the gonads are divided into male and female regions. *(Redrawn and modified from Mackie, 1984)*

distances, but some have short-lived, lecithotrophic (nonfeeding) veligers sustained by yolk reserves. The planktotrophic veligers of oysters (Ostreidae) are capable of dispersion over distances of up to 1300 km. Some bivalves, such as *Ostrea* and some species of shipworms, are **larviparous** and brood the embryos until they become veligers, at which time they are released.

Metamorphosis is characterized by the sudden shedding of the velum. Settling may require extensive testing of the substratum and delayed metamorphosis until an acceptable habitat is found. *Ostrea edulis* swims upward and attaches to the shaded underside of objects. Shipworm veligers settle only on wood.

Freshwater Bivalves

With the exception of *Dreissena* and *Nausitoria,* which have free-swimming veliger larvae, freshwater bivalves invariably employ modified developmental patterns designed to circumvent the many problems encountered by planktonic larvae in freshwater environments. All North American freshwater bivalves (except *Dreissena*) brood their larvae in **marsupia** in the modified water tubes between the lamellae of the demibranchs. Sperm enter via the inhalant current and fertilize the eggs in the marsupium or exhalant chamber.

For several reasons, freshwater bivalves have eliminated or modified the ancestral planktonic larva. Most freshwater mussels have very specific habitat requirements, and a unidirectional river current would be certain to move larvae inexorably downstream from optimal to suboptimal habitat. Because there is no compensating mechanism for upstream recolonization, the population would be doomed. Further, because of the low specific gravity of fresh water, it is difficult for planktonic larvae to remain suspended in the water column, especially in summer, when water density is lowest.

Sphaeriidae, Unionida, Mutelidae, and Mycetopodidae have solved this problem by eliminating the typical planktonic marine veliger. Sphaeriids have direct development and brood their eggs in marsupial sacs between the gill lamellae to release tiny clams. Unionida, Mutelidae, and Mycetopodidae retain indirect development with larvae, but their highly specialized larvae are not planktonic. These bivalves brood the eggs in the exhalant water tubes between the gill lamellae, where they develop to the veliger stage (Fig. 12-121D). The veligers, however, are highly modified for a parasitic existence on freshwater fishes. As a parasite, the larva relies on its host fish for sustenance and, equally importantly, for dispersal within the stream reach preferred by both fish and mussel. By this mechanism, the mussel larva avoids being carried downstream by the current and can remain in optimal habitat.

The parasitic larva of Unionida (freshwater mussels) is the **glochidium** (Fig. 12-121C). This specialized veliger has two valves lined by the larval mantle with peripheral clusters of sensory bristles. Primordial adult structures such as the foot, gills, and gut are present, and some species have a long, adhesive **larval thread.** The valves are hinged and closed by a single larval adductor muscle. There is no mouth, anus, or digestive tract. Mature glochidia range in size from 50 to 500 μm, depending on the species.

Release of glochidia is related to the habits of the host fish. In some mussel species the glochidia are dispersed over the bottom in spidery mucous strands and inadvertently acquired by fishes with benthic nesting or feeding habits. Other mussels release their glochidia in the form of colored masses that look like worms, insects, or juvenile fish. The mass is ingested by the fish host and the glochidia attach to the gills. One species lures the host with a lifelike model fish fashioned from the mantle margin (Fig. 12-121E). When the host fish investigates the undulating lure, the mussel squirts a cloud of glochidia in

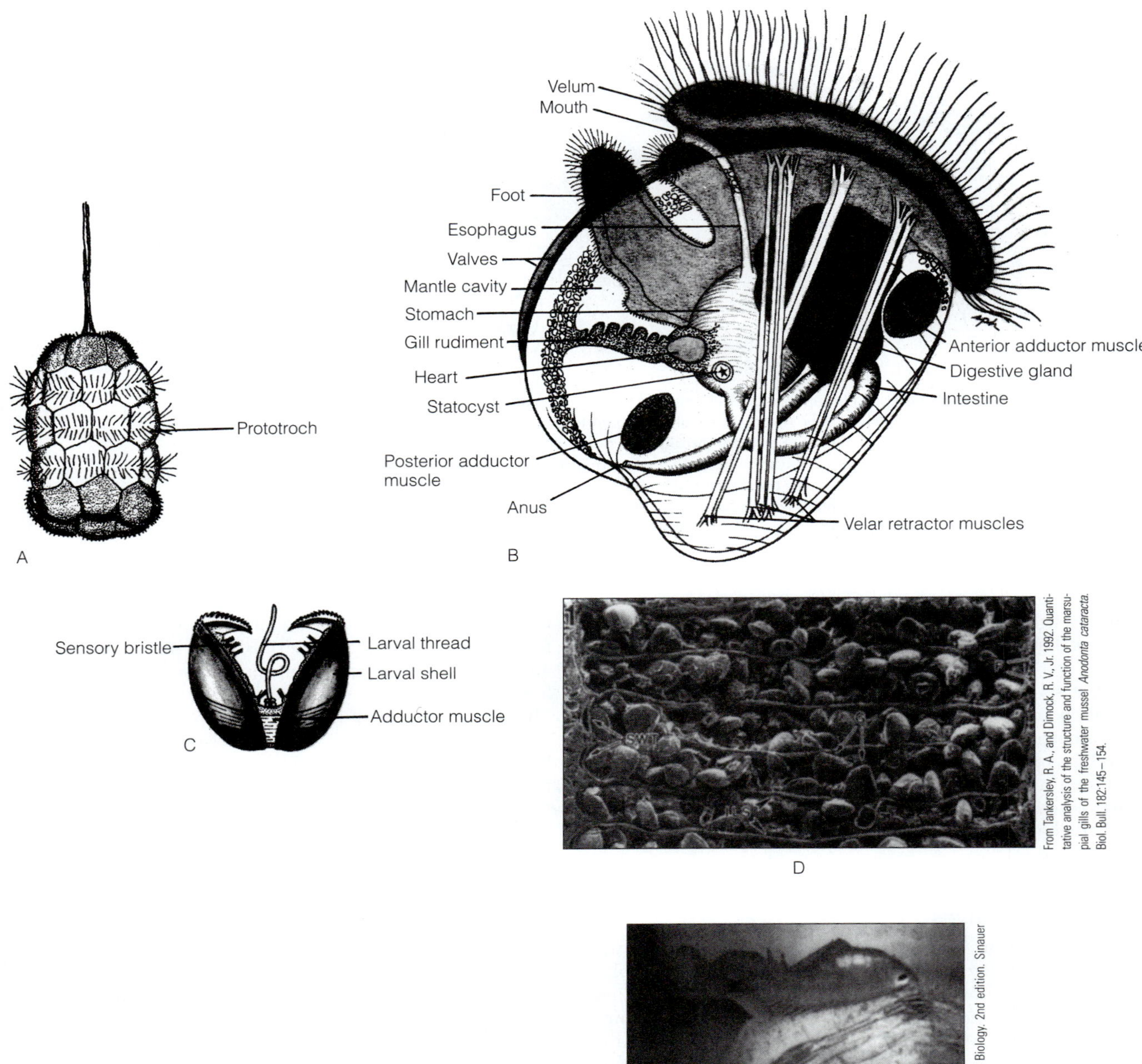

FIGURE 12-121 **A,** Trochophore of *Yoldia limatula.* **B,** A fully developed veliger larva of an oyster. **C,** Glochidium of the freshwater clam *Anodonta.* **D,** Scanning electron micrograph of the interior of the marsupial gill of *Anodonta cataracta* showing glochidia larvae (G). Interlamellar septa (ILS), secondary water tubes (SWT), and a gill filament (F) are also visible. **E,** A fishlike lure produced by an extension of the inner fold of the mantle of brooding females of *Lampsilis ventricosa.* Presumably, the lure attracts predaceous fish to the clam, which then releases its glochidia larvae for attachment to the fish's gills. *(A, After Drew from Dawydoff; B, After Galtsoff, 1964; C, Redrawn from Harms.)*

its face. Some mussel species are highly host selective, but most parasitize more than one species of fish and some fish species may be the host for a number of mussel species.

The hooked valves of *Anodonta* glochidia (Fig. 12-121C) immediately clamp onto the fins and other parts of the body surface of the fish. The long, sticky larval thread aids in initial contact and adhesion. Clasping by the valves is a response to specific compounds in the fish mucus. The hookless glochidia of other mussels are picked up by the respiratory or feeding currents of the fish and attach to the gills. In either case, fish cells migrate to the glochidium and encapsulate it in a cyst. Phagocytic cells in the larval mantle degrade and absorb the fish tissue to nourish the developing clam. During this parasitic period, which lasts from 10 to 30 days, or sometimes several months, many of the larval structures, such as the sensory bristles on the mantle, the larval thread, the larval adductor muscle, and the larval mantle, disappear and adult organs begin developing. Eventually, a juvenile mussel breaks out of the cyst, falls to the bottom, and burrows in the sediment. Here development is completed and the adult habit is gradually assumed.

Some of the larger freshwater mussels may produce as many as 17,000,000 glochidia annually, and a single fish has been reported to support up to 3000 glochidia. Adult fish apparently are not harmed by the parasitic glochidia, but fry may die from secondary infections. Obviously, most glochidia perish and do not become adult mussels.

The Asian clam, *Corbicula,* broods embryos in the medial demibranchs. Its eggs are microlecithal and the developing embryos may receive some nourishment from the gill tissues. Within the brood chambers a trochophore develops and becomes a veliger. The veliger loses its velum and becomes a juvenile, or **pediveliger,** and at this stage it is released from the brood chamber. The pediveliger is benthic, not planktonic, and uses its foot to crawl over the bottom and invade new habitat upstream or downstream.

Growth and Life Span

As in gastropods, the rates of growth and life span of bivalves vary greatly. The common mussel of the California coast *Mytilus californianus* may reach a length of 86 mm within one year. In general, most bivalves grow most rapidly during their early years. Ages of 20 to 30 years are now known to be common for bivalves, and for some species there are records of individuals reaching ages of 150 years.

The growth stages of commercially farmed species are well known. Oysters (Ostreidae), for example, reach marketable size in one to three years depending on the species, latitude, and various environmental conditions. Newly settled oysters, called **spat,** are collected on tiles, twigs, or other objects and allowed to grow to a few centimeters in length. These seed oysters are then distributed over a managed bed, where they grow until harvested. In a natural oyster reef, the average life span is uncertain. Certainly some live longer than 10 years. Small scallops of the genus *Argopecten* have a life span of only 1 to 2 years, but the deep-sea scallops *(Placopecten)* are about 10 years old when they reach their maximum size of 15 cm.

DIVERSITY OF BIVALVIA

Protobranchia[sC]: Chiefly isomyarian deposit feeders with large labial palps and palpal tentacles.

Nuculoida[O] (Paleotaxodonta): Nut clams. Equivalve shells with a row of short, similar teeth (taxodont) along the hinge. Nuculidae *(Nucula),* Nuculanidae *(Nuculana, Yoldia),* Malletiidae *(Malletia).*

Solemyoida[O] (Cryptodonta): Valves thin, equivalve, elongate. Periostracum extends well beyond the edges of the valves and is responsible for the common name, awning clam. No hinge teeth. Solemyidae (awning clams; *Solemya*).

Metabranchia[sC]: Gills adapted for filter feeding (unless secondarily modified).

Filibranchia[SO]: Suspension-feeding bivalves with filibranch gills.

Trigonioida[O]: Small taxon with five species in a single genus, *Trigonia* (formerly *Neotrigonia*), in Trigoniidae. Valves triangular. Hinge teeth strong, consisting of one long, grooved tooth anterior and one posterior to the beak.

Pteriomorpha[O]: Large taxon of epibenthic bivalves attached by byssus or cementation. Mantle margins not fused. Mytilidae (marine mussels; *Bathymodiolus, Brachidontes, Geukensia, Lithophaga, Modiolus, Mytilus*), Arcidae (arks; *Anadara, Arca, Barbatia, Glycymeris, Noetia*), Pectinidae (scallops; *Aequipecten, Argopecten, Chlamys, Pecten, Placopecten*), Ostreidae (oysters; *Crassostrea, Lopha, Ostrea*), Anomiidae (jingles; *Anomia*), Pinnidae (pen clams; *Pinna, Atrina*), Pteriidae (winged oysters; *Isognomon, Malleus, Pinctada, Pteria*), Limidae (file clams; *Lima*).

Eulamellibranchia[SO]: Most bivalves belong to this taxon. Derived, efficient eulamellibranch gills.

Unionoida[O] (Palaeoheterodonta): Important freshwater bivalves. Shells are equivalve, have few hinge teeth. A nacreous layer covers the interior of the shell and the siphons are usually poorly developed. Unionidae (pearly mussels; *Anodonta, Elliptio, Lampsilis, Unio*), Margaritiferidae (freshwater pearl mussels; *Cumberlandia, Margaritifera*), Mutelidae (African; *Mutela*), Etheriidae (freshwater oysters), and Hydriidae (Australasian; *Hydridella, Velesunio*).

Veneroida[O] (Heterodonta, in part): Shell usually is equivalve and lacks a nacreous layer. Most are burrowers with siphons. Lasaeidae (mostly commensals; *Entovalva, Kellia, Lasaea, Lepton, Montacuta*), Chamidae (jewel boxes; *Chama*), Cardiidae (cockles; *Cardium, Dinocardium, Laevicardium, Trachycardium*), Tridacnidae (giant clams; *Hippopus, Tridacna*), Mactridae (surf clams; *Mactra, Spisula*), Solenidae (razor clams; *Ensis, Siliqua, Solen*), Tellinidae (deposit-feeding tellins; *Macoma, Strigella, Tellina*), Semelidae *(Abra, Scrobicularia, Semele),* Vescomyidae *(Calyptogena),* Donacidae (coquinas; *Donax*), Solecurtidae (razor clams; *Solecurtus, Tagelus*), Sphaeriidae (Pisidiidae; freshwater fingernail clams; *Eupera, Pisidium, Sphaerium*), Dreissenidae (freshwater and estuarine clams, including the zebra mussel, *Dreissena*), Veneridae (Venus clams; *Callista, Chione, Dosinia, Gemma, Macrocallista, Mercenaria, Venus*), Petricolidae (rock borers; *Petricola*), and Corbiculidae (includes the Asian clam, *Corbicula*).

Lucinoida[O] (Heterodonta, in part): Specialized bivalves similar to veneroids. Inhalant siphon lost, foot used to construct a substitute siphon of mucus. Anterior adductor muscle scar has a distinctive fingerlike ventral extension lying beside pallial line. Hinge has two cardinal teeth. Pallial sinus absent. Valves circular or oval. Lucinidae (lucines; *Codakia, Divarecella, Linga, Lucina, Myrtea*).

Myoida[O] (Heterodonta, in part): Most are thin-shelled borers with well-developed siphons. Hinge with one or two cardinal teeth but no laterals. Mantle margins fused and shell has no nacreous layer. Myidae (soft-shelled clams; *Cryptomya, Mya, Paramya*), Corbulidae (*Corbula*), Hiatellidae (geoducks; *Hyatella, Panopea*), Gastrochaenidae (rock borers; *Gastrochaena*), Pholadidae (piddocks, angel wings, *Martesia*, borers in rock, clay, wood, mud; *Barnea, Chaceia, Cyrtopleura, Penitella, Pholas, Xylophaga, Zirphaea*), Teredinidae (shipworms; *Bankia, Teredo*).

Anomalodesmata[O]: Small group with no hinge teeth. Adductor muscles are isomyarian. Shell has a nacreous layer. Pandoridae (*Pandora*), Lyonsiidae (*Lyonsia*), Pholadomyidae (*Pholadomya*).

Septibranchia[O]: Carnivorous with septate gills and enlarged inhalant siphon. Poromyidae (*Cuspidaria, Poromya*).

PHYLOGENY OF BIVALVIA

Bivalves and scaphopods are thought to have arisen from rostroconch ancestors, which themselves were descended from laterally compressed monoplacophorans. Rostroconchs are an extinct class of molluscs that, like bivalves, were enclosed within two shell lobes, but there was no flexible dorsal hinge. The halves of the shell were continuous with each other across the dorsal surface. In many species the ventral shell margins touched except for an anterior gape for the foot, and commonly there was a tubular, rostrumlike posterior extension of the shell through which circulating water may have passed.

Bivalves presumably arose with the loss of the radula and the division of the univalve rostroconch shell into right and left valves. This was accomplished through the loss of the calcareous layers of the shell along the dorsal midline. The periostracum on the midline, however, was retained as a flexible hinge connecting right and left valves. Bivalves retained one pair of gills.

The most primitive known bivalve is *Pojetaia*, from the Lower Cambrian of Australia. This clam is similar to the presumed

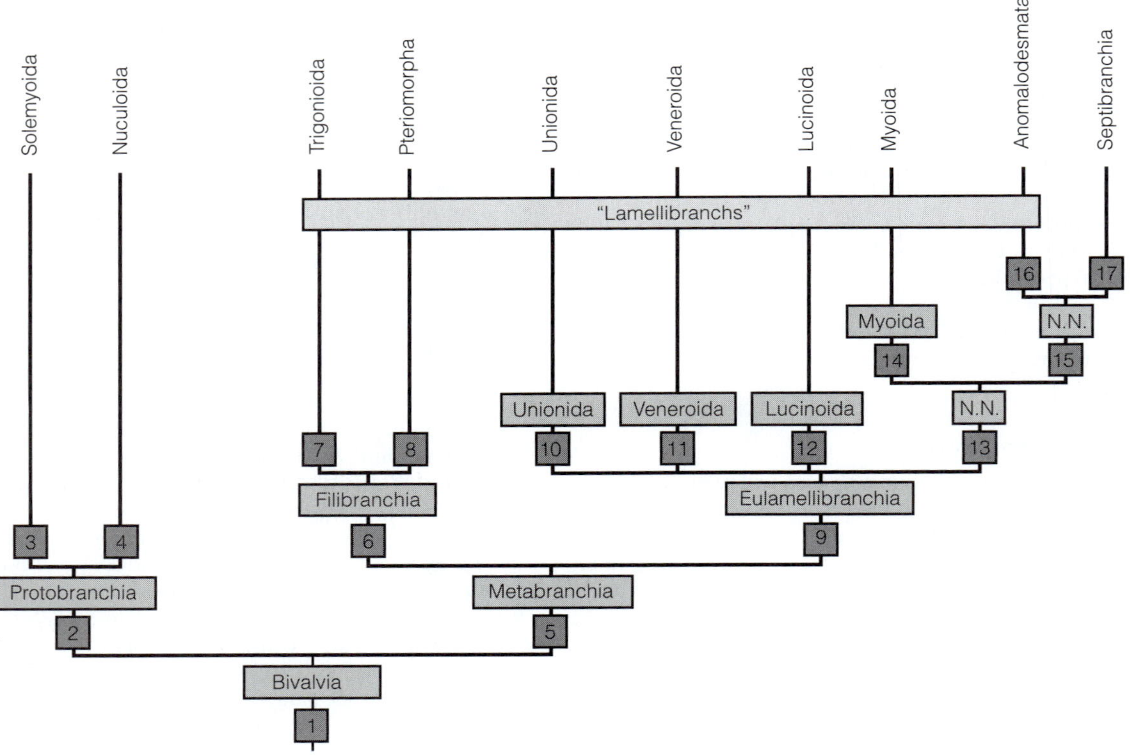

FIGURE 12-122 A phylogeny of Bivalvia based on 42 morphological characters. **1, Bivalvia:** shell is divided into two valves; radula is lost; one pair of gills. **2, Protobranchia:** gills are oriented diagonally; an adoral sense organ is present. **3, Solemyoida:** ligament is external and posterior to the umbos (opisthodetic); abfrontal cilia are absent; anterior adductor muscle is larger. **4, Nuculoida:** palps are large; palp appendages are present. **5, Metabranchia:** gills are filibranch and adapted for filter feeding; water intake to mantle cavity is posterior. **6, Filibranchia:** abdominal sense organ, byssus in larva and adult. **7, Trigonioida:** two pairs or less of dorsoventral muscles; ligament is opisthodetic and external. **8, Pteriomorpha:** posterior adductor muscles enlarged, anteriors reduced. **9, Eulamellibranchia:** gill filaments fused to form solid lamellae. **10, Unionida:** ligament is internal. **11, Veneroida:** ligament is both internal and external. **12, Lucinoida:** inhalant flow is anteroventral. **13, N.N.:** mantle fusion is greater than 50%. **14, Myoida:** chondrophore is present. **15, N.N.:** lithodesma is present; ligament is reduced; hermaphroditic. **16, Anomalodesmata:** posterior adductor muscle larger than the anteriors. **17, Septibranchia:** gills are replaced by a horizontal septum. *(Based on Salvini-Plawen and Steiner, 1996, and Ax, 2000)*

ancestral bivalve and is the oldest known protobranch (palaeotaxodont). Of the living bivalves, the protobranch *Nucula* is probably most like the stem bivalve. Another Lower Cambrian bivalve, *Fordilla,* is the oldest known pteriomorph (mytiloid).

A recent cladistic analysis (Fig. 12-122), based on 42 morphological characteristics, reveals Bivalvia to consist of two monophyletic sister taxa, Protobranchia and Metabranchia. Although Protobranchia is a morphologically recognizable taxon, there are not many synapomorphies to unite its members. The gills are oriented obliquely and an adoral sense organ is present. Plesiomorphies include the broad ventral sole, a respiratory current that enters anteriorly, gills with abfrontal cilia, and a protostyle. The Metabranchia, however, undergo extensive modification of the gills for filter feeding, resulting in a broad suite of associated apomorphies.

The two basic types of filter-feeding gills are present in the two sister taxa of Metabranchia, which are Filibranchia and Eulamellibranchia. In Filibranchia the gill filaments are independent of each other, whereas in Eulamellibranchia the filaments are joined to form solid, but ostiate, lamellae. Eulamellibranchia includes several taxa (Unionida, Veneroida, Lucinoida, Myoida, Anomalodesmata, and Septibranchia). Anomalodesmata and Septibranchia clearly are sister taxa, but the relationships between Unionida, Veneroida, Lucinoida, and Myoida currently are not understood and require further study.

The new cladistic classification coincides rather closely with an older, widely used Linnean classification in which the above taxa are arranged in a series of subclasses. These subclasses are Protobranchia (composition as above), Pteriomorpha (as above), Palaeoheterodonta (Unionida and Trigonioida), Heterodonta (Veneroida, Lucinoida, and Myoida), and Anomalodesmata *sensu lato* (Anomalodesmata *sensu stricto* and Septibranchia).

The new classification also corresponds reasonably well with the organization of bivalves into groups. Protobranchia is a natural unity coinciding with the protobranch grade. The lamellibranch grade, on the other hand, is not monophyletic, for it consists of most, but not all, of the Metabranchia. Septibranchia is a highly derived monophyletic taxon in Metabranchia that corresponds one to one with the septibranch grade.

PHYLOGENETIC HIERARCHY OF BIVALVIA

- Bivalvia
  - Protobranchia
    - Solemyoida
    - Nuculoida
  - Metabranchia
    - Filibranchia
      - Trigonioida
      - Pteriomorpha
    - Eulamellibranchia
      - Unionida
      - Veneroida
      - Lucinoida
      - N.N.
        - Myoida
        - N.N.
          - Anomalodesmata
          - Septibranchia

SCAPHOPODA^C

Scaphopoda contains about 500 described extant species of burrowing, exclusively marine molluscs popularly known as tusk, or tooth, molluscs because of their shape (Fig. 12-123). They are the most recent molluscan higher taxon to appear in the fossil record and are highly derived. They probably are most closely related to the bivalves, although they bear little superficial resemblance to them. Scaphopods inhabit a tubular shell with an opening at each end and have a radula but lack gills. Like bivalves, they are bilaterally symmetrical and have a foot adapted for digging. Scaphopods live buried in sand, head downward and with the body steeply inclined. The small, posterior aperture usually projects slightly above the surface, but sometimes the animal may be completely buried to up to 30 cm deep in the sediment. The majority of scaphopods are microcarnivores that burrow in sand at depths greater than 6 m and up to 7000 m. Although relatively common, they are not often encountered in shallow water, although at least one species occurs in the low intertidal. Empty shells often wash ashore on some coasts.

FORM

The scaphopod body is greatly elongate along the anterior-posterior axis (Fig. 12-124) and, like that of other molluscs, consists of a head, foot, and visceral mass. The head and foot are anterior and can be projected from the larger anterior aperture of the shell.

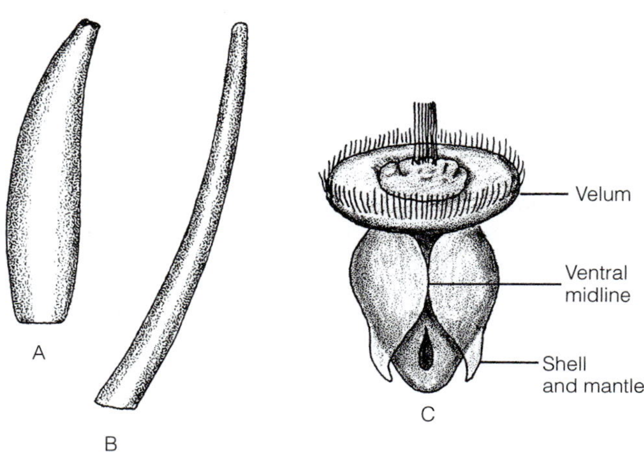

FIGURE 12-123 Scaphopoda. **A,** Shell of *Cadulus*, a gadilidan scaphopod. **B,** Shell of *Dentalium*, a dentaliidan scaphopod. **C,** Veliger larva. *(A and B, After Abbott, R. T. 1954. American Sea Shells. Van Nostrand Co., Princeton, N.J.; C, After Lacaze-Duthiers from Dawydoff.)*

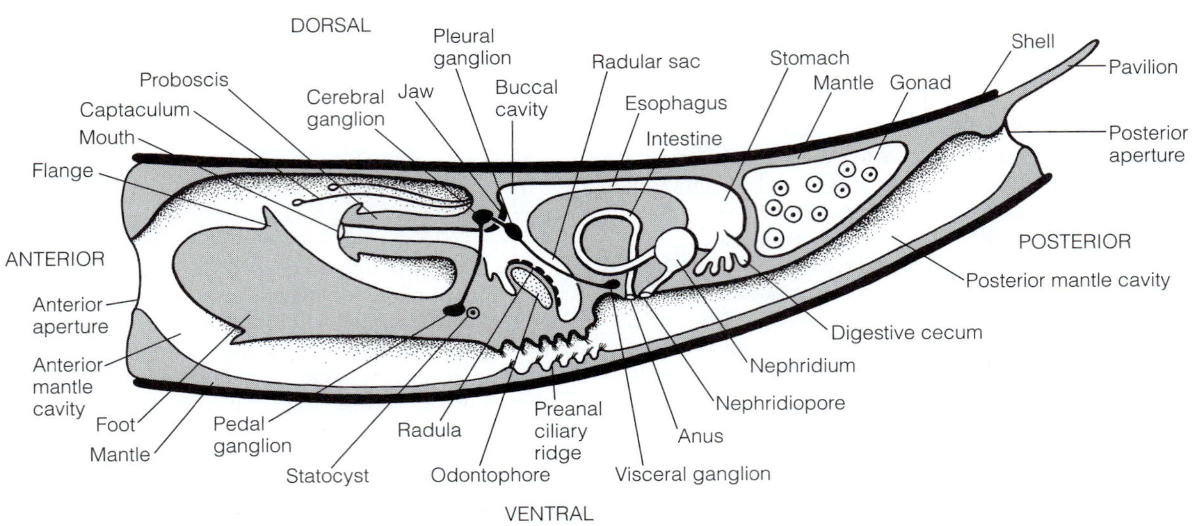

FIGURE 12-124 Sagittal section of a generalized scaphopod with its foot, head, and captacula retracted into the shell. The ganglia, captacula, nephridia, and digestive ceca are lateral to the midsagittal plane. *(Redrawn and modified from Shimek, R. L., and Steiner, G. 1997, Scaphopoda. In Harrison, F. W., and Kohn, A. J. (Eds). Microscopic Anatomy of Invertebrates. Vol. 68., Mollusca II. Wiley-Liss, New York. pp. 719–781; and Reynolds, P. O. 2002. The Scaphopoda. Adv. Mar. Biol. 42:137–236.)*

The head is reduced to a short, mobile proboscis, which bears the mouth at its anterior end (Fig. 12-124). The visceral mass occupies the middle and posterior regions of the shell.

The foot is ventral to the proboscis at the anterior end. In most scaphopods, it is conical and equipped with a central pedal hemocoel. Like the bivalve foot, it is adapted for burrowing. The pedal hemocoel is a hydrostatic skeleton used to extend the foot into the sand. Once penetration is achieved, lateral flanges expand at or near the tip of the foot to form a terminal anchor and one or two pairs of pedal retractor muscles pull the shell and body toward the anchored foot. The foot of the more derived gadilidan scaphopods is worm-shaped and has an expansible terminal disc.

MANTLE AND MANTLE CAVITY

In scaphopods the mantle has two large lateral folds (skirts) that extend the length of the body, as is also the case in bivalves. In scaphopods, however, the skirts reach to the ventral midline, where they fuse to form a tube that encloses the entire body, including a constricted, tubular mantle cavity extending the length of the body (Fig. 12-124). The mantle extends posteriorly as an awning-like dorsolateral pavilion (also spelled *pavillon*) equipped with sensory receptors. The mantle cavity is open to the sea at both the anterior and posterior ends, but not ventrally. There are no gills and no osphradia.

The respiratory current is generated by dense rows of cilia on a series of 4 to 15 **preanal ciliary ridges** that encircle the mantle cavity just anterior to the anus. The current enters through the posterior aperture. Some of this flow continues anteriorly and exits the anterior aperture, but much, perhaps most, of it reverses direction and is ejected from the posterior aperture by periodic retraction of the foot. This vigorous activity also ejects feces, gametes, and rejected food or mineral particles from the mantle cavity. The preanal ridges are heavily vascularized and may be the primary gas exchange surfaces. It has been suggested that they are homologous to gills but, appealing as it is, there is no support for this hypothesis. The general mantle surface is probably a gas exchange surface also.

SHELL

The common names of tusk mollusc and tooth mollusc are derived from the shape of the shell, which is an elongate cylindrical tube usually shaped like an elephant's tusk (Fig. 12-123B) or a canine tooth (Fig. 12-123A). Both ends of the tube are open via anterior and posterior apertures, with the anterior aperture being larger. Scaphopod shells average 3 to 6 cm in length, but *Cadulus mayori,* found off the Florida coast, is less than 4 mm in length and the Japanese species *Dentalium vernedei* reaches a length of 15 cm. A fossil species of *Dentalium,* however, has a shell 30 cm long with a maximum diameter of well over 3 cm. The shells of most scaphopods are white or yellowish, but one East Indian species of *Dentalium* is a brilliant jade green and others are pink or tan. Some scaphopod shells are slightly curved toward the anterior (slightly exogastric) so that the concave edge of the shell is dorsal (Fig. 12-123B).

The mantle epithelium secretes the shell. An increase in length is accomplished by the anterior mantle margin and growth occurs at the anterior end. Increasing thickness is due to secretion by the general mantle surface. There are two to four calcareous (aragonite) layers with an outer proteinaceous periostracum sometimes present. As the animal grows, the posterior aperture must be enlarged to accommodate the required increase in the volume of the respiratory current and the need to remove a greater volume of wastes and gametes. The progressive removal of the posterior apex of the shell as the shell grows in the opposite direction accomplishes this.

NUTRITION

Scaphopods are selective **microcarnivores** that feed on microscopic organisms, especially foraminiferans, but also juvenile bivalves, kinorhynchs, and other small animals, in the surrounding sand and water. Some are selective detritivores.

Food items are selected and transported to the mouth by the **captacula,** which are unique to scaphopods. These are threadlike, ciliated, muscular tentacles borne in large numbers on a pair of lobes above and beside the proboscis (Fig. 12-124). Captacula are expendable and replaced when lost.

Each captaculum consists of a slender stalk with a bulb at its tip. The bulb bears a secretory and presumably sensory pit on one surface. Stalk, bulb, and pit are ciliated, and the stalk is extended into the sediment or water by the activity of these cilia. The pit is brought into contact with potential food items and apparently determines the suitability of the object as food, although sensory receptors have not been found. The selected food particle adheres to the bulb with adhesive secretions from the pit. Contractions of muscles within the stalk bring the particle back to the mantle cavity. Alternatively, particles may be transported to the mantle cavity by ciliary tracts on the stalk. Some species of *Dentalium* are deposit feeders that collect fine particles by means of a ciliated tract on the captacular stalk. The particles accumulate in the mantle cavity, where further sorting occurs prior to ingestion. Probing and pushing actions of the foot excavate a feeding cavity in the sediment, and in some scaphopods particles may be transported to the mantle cavity by the foot, but the primary feeding responsibility lies with the captacula.

The gut extends from the anterior mouth to the anus located on the ventral midline at approximately the middle of the body (Fig. 12-124). The mouth is located at the tip of the tubular or conical proboscis (buccal tube), which bears lips and buccal pouches that assist in sorting potential food particles. The mouth opens into the buccal cavity, which contains a median, presumably chitinous, jaw on its dorsal wall, a well-developed radula with large, flattened median teeth, and a sensory subradular organ. Neither the radula nor the subradular organ is eversible. The radula is located in a radular sac and supported by an odontophore. The morphology of the sac is similar to that of gastropods but, relative to body size, the scaphopod radula is the largest of any mollusc. It is not used for scraping, and instead may grind against the jaw to crush food particles and/or move them into the esophagus.

Ciliary tracts in the esophagus move food to the stomach. One or two secretory and absorptive digestive ceca (midgut glands) open from the stomach and occupy much of the middle region of the body. Extracellular digestion begins in the stomach, using enzymes secreted by the ceca, and is completed in the ceca. Absorption and intracellular digestion also occur in the ceca. Waste materials are moved by cilia from the stomach through the intestine and to the anus. The intestine exits the stomach anteriorly, and consequently the gut is U-shaped (Fig. 12-124). The anus opens near the middle of the mantle cavity.

INTERNAL TRANSPORT

The hemal system, which is highly derived in scaphopods, is reduced to a series of blood sinuses. One of these, the perianal sinus, associated with a pericardium, is the chief pumping organ and may be homologous to the ventricle of other molluscs. The perianal sinus is located near the middle of the body in the vicinity of the anus. Since there are no gills, there are no atria. A pair of small slits of unknown function opens from the hemocoel into the mantle cavity.

EXCRETION

Podocytes are present in the pericardium (which is associated with the perianal sinus), and probably are responsible for producing the primary urine. A pair of nephridia presumably connects with the pericardial cavity via renopericardial ducts, but these have not been convincingly demonstrated. The primary urine is probably modified by selective secretion and absorption as it passes through the kidneys, and the final urine is released into the mantle cavity by the nephridiopores near the anus.

NERVOUS SYSTEM AND SENSE ORGANS

The nervous system exhibits the typical tetraneurous conchiferan plan, including a nerve ring around the anterior gut and paired pedal and visceral nerves. Paired cerebral ganglia are located at the base of the proboscis dorsal to the gut (Fig. 12-124). Pedal, visceral, pleural, and buccal ganglia are also present and linked by commissures and connectives in typical molluscan fashion.

There are no eyes, sensory tentacles, or osphradia. The subradular organ is chemosensory (gustatory) and evaluates food while it is being processed by the radula, before it is committed to the esophagus. A pair of statocysts beside the pedal ganglion detects spatial orientation. The mantle margins associated with the anterior and posterior apertures are equipped with sensory receptors. The captacula are well supplied with nerves and presumed to have a sensory function. Each captacular bulb is equipped with a ganglion, but sensory receptors have not been identified.

REPRODUCTION AND DEVELOPMENT

Scaphopods are gonochoric. The unpaired gonad fills most of the posterior part of the body (Fig. 12-124). Gametes are discharged into the right nephridium, through which they reach the mantle cavity. Spawning has been observed only rarely and never in nature. In the laboratory, eggs are extruded in gelatinous strings. Fertilization is external, presumably in the mantle cavity.

Scaphopod development is much like that of the marine bivalves. Cleavage is unequal and follows the typical molluscan spiral pattern to produce a coeloblastula that gastrulates by invagination. There is a free-swimming lecithotrophic trochophore larva, succeeded by a bilaterally symmetrical, shelled veliger (Fig. 12-123C). The larval mantle and shell begin as an unpaired dorsal shell gland that grows laterally. It becomes bilobed as a pair of lateral evaginations form and extend ventrally. The two mantle lobes, and then the shell lobes they secrete, fuse along their ventral margins to form the tubular mantle cavity and shell, which has anterior and posterior apertures.

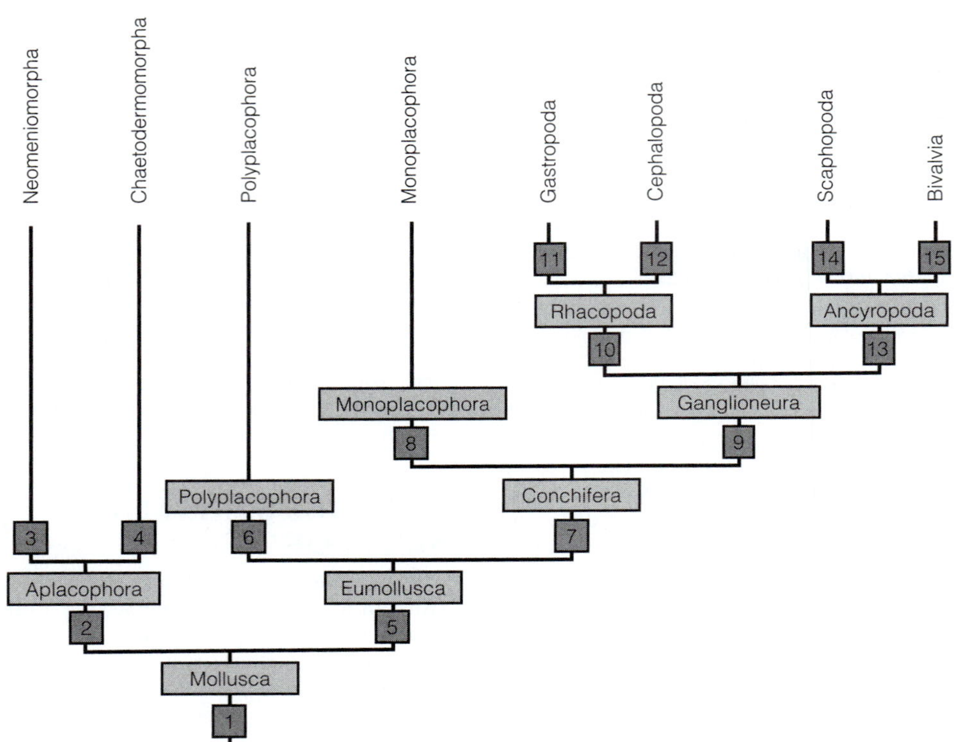

FIGURE 12-125 A phylogeny of Mollusca. **1, Mollusca:** dorsal mantle has secreted chitin-protein cuticle and calcareous spicules; radula; foot and paired pedal retractor muscles; bipectinate gills; tetraneural nervous system. **2, Aplacophora:** cylindrical vermiform body; foot reduced or lost; mantle with cuticle and calcareous spicules; dorsoterminal sense organ; gonads open into pericardial coelom. **3, Neomeniomorpha:** gills lost; reduced foot is present; hermaphroditic. **4, Chaetodermomorpha:** foot lost; oral shield is present; gills are present; gonochoric. **5, Eumollusca:** gut with digestive ceca; gut coiled; complex radular musculature; odontophore with chondroid supports. **6, Polyplacophora:** eight dorsal, overlapping valves; valves consist of tegmentum and articulamentum; tegmentum with esthetes. **7, Conchifera:** one-piece shell with periostracum and calcareous layers; mantle margins with three folds; crystalline style; mantle spicules lost; eight pairs of dorsoventral pedal retractor muscles; subrectal commissure. **8, Monoplacophora:** monopectinate gills; multiple pairs of gills, atria, nephridiopores, and gonoducts; connections between nephridia and pericardial cavity lost. **9, Ganglioneura:** neuron cell bodies concentrated in paired ganglia; a single pair of foot retractor muscles. **10, Rhacopoda:** posterior mantle cavity (anterior after torsion); extensible head; tall, conical shell. **11, Gastropoda:** torted; usually coiled, asymmetrical; mantle cavity is anterior; streptoneurous nervous system; operculum is present; left gonad lost. **12, Cephalopoda:** pelagic; chambered shell; hydrostatic organ; carnivorous with beaks and tentacles for prey capture; extreme concentration and cephalization of nervous system; camera-style eyes; mantle cavity is muscular for ventilation and swimming; crystalline style lost; macrolecithal eggs with discoid cleavage. **13, Ancyropoda:** mantle, gills, and mantle cavity developed laterally; bivalve shells open anteriorly and posteriorly and are not conical; valves are joined dorsally and free ventrally; foot is adapted for digging. **14, Scaphopoda:** bivalve shell with valves fused dorsally and ventrally to form a one-piece, tubular, conical shell open at both ends and without a hinge; proboscis and captacula; gills lost; water enters mantle cavity posteriorly. **15, Bivalvia:** bivalve shell with right and left valves joined dorsally at a movable hinge and free ventrally; adductor muscles; radula lost. *(Based on Nielsen, C. 2001. Animal Evolution. 2nd Ed. Oxford University, Press. 563 pp; Ax, 2000; and Götting, K. J. 1996. Mollusca. In Westheide, W., and Rieger, R. M. (Eds.): Spezielle Zoologie. Erster Tiel: Einzellen and Wirbellose Tiere. Gustav Fischer Verlag, Stuttgart. pp. 276–336.)*

DIVERSITY OF SCAPHOPODA

Dentaliida^O: Shell is tusk shaped (Fig. 12-123B), reaching its maximum diameter at the anterior aperture. Foot is conical. Two digestive ceca. *Anulidentalium, Dentalium, Episiphon, Fustiaria.*

Gadilida^O: Shell usually resembles a canine tooth, with its greatest diameter posterior to the anterior aperture (Fig. 12-123A), although some are tusklike. Foot is vermiform and capable of distal expansion to form a terminal disc. Only one digestive cecum. *Cadulus, Costentalina, Entalina, Gadila, Polyschides, Pulsellum, Solenoxiphus.*

PHYLOGENY OF SCAPHOPODA

Scaphopods are the most recent molluscan class to evolve, and their evolutionary relationships are still unresolved. They usually are considered to be the sister taxon of the bivalves (Fig. 12-125), although relationships with the gastropods and cephalopods have also been suggested. Scaphopods share many features with the bivalves, and like them may have evolved from rostroconchs. Many rostroconchs had shells in which the ventral margins were in contact. If they fused, the shell would have become tubular with a posterior aperture (rostral) and an anterior aperture (pedal). The reduction of the head, the burrowing habit with a digging foot, the symmetrical veliger, and the embryonic bilobed mantle and shell are strikingly similar to bivalves. The absence of gills is unlike bivalves, however, and the presence of a radula argues against a bivalvelike ancestor. It is most likely that scaphopods and bivalves arose from a common ancestor with a radula and gills and that the gills were secondarily lost in the scaphopods, whereas the radula was lost in the bivalves.

PHYLOGENY OF MOLLUSCA

MOLLUSCAN GROUND PLAN

The first molluscs with calcareous shells appear in the fossil record during the so-called Cambrian Explosion, when fossils of multicellular animals first appear. Presumably they were preceded by shell-less Upper Precambrian species that left no record. These ancestral molluscs would have shared a common ground plan from which the body plans of the seven extant molluscan higher taxa are derived. As always, the ground plan consists of plesiomorphic characters inherited from ancestors that were not molluscs and apomorphic characters that are unique to molluscs.

The ancestral mollusc was probably a soft-bodied, benthic marine animal with a planktonic larva. It and its descendants shared a group of apomorphies unique to Mollusca. These characteristics, except when secondarily lost, are present in all extant molluscan higher-level taxa. These include the mantle, which secretes a mucopolysaccharide cuticle and calcareous spicules. The mantle invaginates, or folds, peripherally to enclose a mantle cavity continuous with the sea. Gas exchange occurs on paired, ciliated, bipectinate gills in the mantle cavity that are ventilated by a ciliary current. A creeping, ciliated foot occupies the ventral surface and can be withdrawn by pedal retractor muscles attached to the mantle and cuticle. Ventrally, the anterior head is separated from the foot by a transverse groove. A radula is used for microphagous browsing or feeding on detritus. A pair of sensory osphradia monitor the water entering the mantle cavity. The nervous system consists of a pair of cerebral ganglia and is tetraneurous. There is some (hotly debated) evidence that the ancestor may have been metameric, with eight segments.

The stem mollusc also possesses a suite of plesiomorphies inherited from its non-molluscan ancestors. It is a protostome bilaterian with a reduced coelom, which retains its roles as pericardium, kidney, and gonad but does not function as a hydrostatic skeleton. The hemal system is open and consists of a heart, arteries, and hemocoelic sinuses. It is probably gonochoric, gametes were free-spawned into the sea, and fertilization was external. Eggs are microlecithal and cleavage is holoblastic and spiral. The hatching stage of the ancestor is probably a planktonic trochophore larva, or a later stage derived from a trochophore. Mesoderm arose from 4d mesentoblast cells and coelom formation was by schizocoely within a mass of cells derived from the mesentoblast.

ORIGIN OF MOLLUSCA

Molluscs share a number of features with annelids, sipunculans, and echiurans and are generally thought, by both morphologists and molecular systematists, to share a common ancestor with them in the taxon Trochozoa. Molluscs and other coelomate trochozoans have coelomic spaces associated with the heart. The gonads release gametes into the coelom, and the hemal system ultrafilters a primary urine into it. Podocytes are involved in the formation of the ultrafiltrate. Coelomoducts (metanephridia) transport the gametes and urine to the exterior. The trochophore larvae of the taxa are similar. Cleavage is spiral, and the 4d cells produce mesentoblasts from which arise lateral mesodermal bands. Schizocoely produces a coelomic cavity within these bands. Nucleotide sequences of 18S rRNA support a close relationship between molluscs and trochozoans.

The presence of a flat, ciliated foot, a ciliary gliding mode of locomotion, and similarities with Aplacophora have suggested to some zoologists that the molluscs arose from turbellarian-like ancestors. If this hypothesis is true, then the gonocoel-pericardial cavity-metanephridial system of molluscs is convergent, not homologous, with that of the eucoelomate annelids.

There are indications that Trochozoa, including molluscs, arose from a segmented ancestor. Chitons have eight shell valves with eight pairs of pedal retractor muscles. Monoplacophorans have multiple pairs of many organs and structures, including three to seven pairs of nephridiopores, one to three pairs of gonoducts, three to six pairs of gills, two pairs of atria, and eight pairs of pedal retractor muscles. Furthermore, some fossil bivalves have eight pairs of pedal retractor muscles, as do some modern mussels. The cephalopod *Nautilus* has two pairs of atria, two pairs of gills, and two pairs of nephridia. Seven or eight transverse dorsal rows of spicules occur in larval aplacophorans, larval polyplacophorans, and adult polyplacophorans. Some zoologists believe the molluscs arose from ancestors in which the body, but not the coelom, consisted of eight segments. In these epibenthic ancestors there was never a segmented coelom to serve as a hydrostatic skeleton for burrowing, as there was in the annelids. The coelom of all molluscs is small and is never segmented. Those molluscs that burrow use the hemocoel, not the coelom, as a hydrostatic skeleton. According to this hypothesis, the ancestral trochozoan was segmented, making segmentation an autapomorphy of coelomate Trochozoa. The ancestral segmentation was retained by annelids but lost by molluscs, echiurans, and sipunculans.

Other invertebrate zoologists believe that the lack of agreement in the number of pairs of these organs, the complete lack of segmentation of the coelom, and the lack of consistent segmentation in the taxon argue against a segmented ancestry for Trochozoa and Mollusca. In this hypothesis, the ancestral trochozoan was nonsegmented and molluscs retained the plesiomorphous nonsegmented condition. Annelids, echiurans, and sipunculans, having diverged after the evolution of segmentation, form a monophyletic taxon united by the synapomorphy of segmentation.

EVOLUTION WITHIN MOLLUSCA

Aplacophorans were probably an early offshoot of the ancestral molluscs (as exemplified by the molluscan ground plan), and they retained the shell-less condition (Fig. 12-125). The absence of a shell is probably primary. The aplacophorans kept the calcareous spicules of the ground plan but lost or reduced the foot and developed a vermiform body. Some lost the radula and most lost the gills. Aplacophora is the sister taxon of Eumollusca (Polyplacophora + Conchifera) and its closest living relatives are the Polyplacophora. Both taxa have a cuticle, spicules, and suprarectal commissures.

Within Eumollusca, Polyplacophora develop a shell divided into eight dorsal valves whereas Monoplacophora have an undivided shell. Both have eight pairs of dorsoventral pedal retractor muscles. There is no evidence that monoplacophorans evolved from polyplacophorans by losing seven valves or that polyplacophorans evolved from monoplacophorans by shell division. The alternative, then, is that both Polyplacophora and Monoplacophora arose from a shell-less common ancestor and each independently invented a solid shell plate secreted by the mantle epithelium. This epithelium was already equipped for secreting calcium carbonate spicules and it is not unreasonable to suppose that the ability to secrete a solid calcareous shell arose independently in these two lines. One line terminated with Polyplacophora, but the other line led to Conchifera and all the remaining molluscs.

Conchifera (including Monoplacophora, Gastropoda, Cephalopoda, Scaphopoda, and Bivalvia) is a monophyletic taxon that arose from monoplacophoran ancestors or from ancestors close to monoplacophorans. Conchifera is the sister taxon of Polyplacophora. The generalized mollusc described early in this chapter is similar to what we imagine the common ancestor of the Conchifera to be. All have a single shell, but in most the multiple retractor muscles are lost. The shell is composed of three layers. There is a crystalline style, and the mantle margins have three folds. The gastropods and cephalopods shared an ancestor with a high, conical shell like that still seen in most gastropods (although coiled) and in primitive cephalopods. The ancestral mantle cavity and gills were posterior. The gastropod ancestor underwent torsion and developed a coiled, asymmetric shell and anterior mantle cavity. The cephalopods developed into pelagic raptors with efficient and effective neuromuscular systems.

PHYLOGENETIC HIERARCHY OF MOLLUSCA

- Mollusca
  - Aplacophora
    - Neomeniomorpha
    - Chaetodermomorpha
  - Eumollusca
    - Polyplacophora
    - Conchifera
      - Monoplacophora
      - Ganglioneura
        - Rhacopoda
          - Gastropoda
          - Cephalopoda
        - Ancyropoda
          - Scaphopoda
          - Bivalvia

The common ancestor of bivalves and scaphopods had a single shell that was open along its anterior, ventral, and posterior margins but closed dorsally, even if only by its periostracum. The mantle cavity of the ancestor was lateral, as were its gills. In the bivalves the calcareous parts of the shell lost their dorsal continuity and the shell divided into two lateral valves held together dorsally by a strip of periostracum known as the hinge. Ventrally, the shell remained open. The radula was lost. In scaphopods the calcareous shell remained intact dorsally and its ventral margins fused together along the midline to form a conical tube open anteriorly and posteriorly. The gills were lost.

REFERENCES

GENERAL

Abbott, R. T. 1974. American Seashells. Second Edition. Van Nostrand Reinhold, New York. 663 pp.

Andrews, J. 1971. Sea Shells of the Texas Coast. University of Texas Press, Austin. 298 pp.

Barnes, R. D., and Harrison, F. W. 1994. Introduction to the Mollusca. In Harrison, F. W., and Kohn, A. J. (Eds.): Microscopic Anatomy of Invertebrates. Vol. 5: Mollusca I. Wiley-Liss, New York. pp. 1–12.

Barnes, R. D., and Harrison, F. W. 1997. Introduction to the Mollusca. In Harrison, F. W., and Kohn, A. J. (Eds.): Microscopic Anatomy of Invertebrates. Vol. 6A: Mollusca II. Wiley-Liss, New York. pp. 1–6.

Bourne, G. B., Redmond, J. R., and Jorgensen, D. D. 1990. Dynamics of the molluscan circulatory system: Open versus closed. Physiol. Zool. 63:140–166.

Ghiselin, M. T. 1988. The origin of molluscs in the light of molecular evidence. In Harvey, P. H., and Partridge, L. (Eds.): Oxford Surveys in Evolutionary Biology. Vol. 5. Oxford University Press, Oxford. pp. 66–95.

Götting, K.-J. 1996. Mollusca. In Westheide, W., and Rieger, R. M. (Eds.): Spezielle Zoologie Erster Tiel: Einzeller und Wirbellose Tiere. Fischer Verlag, Stuttgart. pp. 276–336.

Harrison, F. W., and Kohn, A. J. (Eds.): 1994. Microscopic Anatomy of Invertebrates. Vol. 5: Mollusca I. Wiley-Liss, New York. 390 pp.

Harrison, F. W., and Kohn, A. J. (Eds.): 1997. Microscopic Anatomy of Invertebrates. Vol. 6A: Mollusca II. Wiley-Liss, New York. 414 pp.

Harrison, F. W., and Kohn, A. J. (Eds.): 1997. Microscopic Anatomy of Invertebrates. Vol. 6B: Mollusca II. Wiley-Liss, New York. 413 pp.

Haszprunar, G. 1996. The Mollusca: Coelomate turbellarians or mesenchymate annelids. In Taylor, J. D. (Ed.): Origin and Evolutionary Radiation of the Mollusca. Malacological Society of London and Oxford University Press, Oxford. pp. 1–28.

Morse, A. N. C. 1991. How do planktonic larvae know where to settle? Am. Sci. 79:154–167.

Morse, M. P., and Reynolds, P. D. 1996. Ultrastructure of the heart-kidney complex in smaller classes supports symplesiomorphy of molluscan coelomic characters. In Taylor, J. D. (Ed.): Origin and Evolutionary Radiation of the Mollusca. Malacological Society of London and Oxford University Press, Oxford. pp. 89–97.

Morton, J. E. 1967. Molluscs. Fourth Edition. Hutchinson University Library, London. 244 pp.

Purchon, R. D. 1977. The Biology of the Mollusca. Second Edition. Pergamon Press, New York. 596 pp.

Rosenberg, G. 1992. The Encyclopedia of Seashells. Dorset Press, New York. 224 pp.

Salvini-Plawen, L. v., and Steiner, G. 1996. Synapomorphies and plesiomorphies in higher classification of Mollusca. In Taylor, J. D. (Ed.): Origin and Evolutionary Radiation of the Mollusca. Malacological Society of London and Oxford University Press, Oxford. pp. 29–51.

Solem, A. 1974. The Shell Makers: Introducing Mollusks. John Wiley and Sons, New York. 289 pp.

Strathmann, R. R., Jahn, T. L., and Fonesca, J. R. 1972. Suspension feeding by marine invertebrate larvae: Clearance of particles by ciliated bands of rotifer, pluteus, and trochophore. Biol. Bull. 142:505–519.

Taylor, J. D. (Ed.): 1996. Origin and Evolutionary Radiation of the Mollusca. Malacological Society of London and Oxford University Press, Oxford. 392 pp.

Trueman, E. R. 1983. Locomotion in molluscs. In Saleuddin, A. S. M., and Wilbur, K. M. (Eds.): The Mollusca. Vol. 4, Pt. 1. Physiology. Academic Press, New York. pp. 155–198.

Vermeij, G. J. 1993. A Natural History of Shells. Princeton University Press, Princeton, NJ. 207 pp.

Wilbur, K. M. (Ed.): 1983–1988. The Mollusca. Academic Press, New York. (Twelve volumes.)

Yonge, C. M. 1947. The pallial organs in the aspidobranch Gastropoda and their evolution throughout the Mollusca. Phil. Trans. Roy. Soc. Lond. B 232:443–518.

Internet Sites

http://coa.acnatsci.org/conchnet/classify.html (Conchology 101: A Classification of the Mollusca, by Gary Rosenberg, Academy of Natural Sciences, Philadelphia.)

www.conchology.uunethost.be (Conchology.be, by G. T. Poppe.)

www.sunderland.ac.uk/MalacSoc (Malacological Society of London.)

www.ucmp.berkeley.edu/mologis/mollia.html (Mollia: Information for Malacologists.)

www.molluscan.com/home.html (Southeast Asian Molluscan Pictures from C. Sow-Yan.)

APLACOPHORA

Reynolds, P. D., Morse, M. P., and Norenburg, J. 1993. Ultrastructure of the heart and pericardium of an aplacophoran mollusc (Neomeniomorpha): Evidence for ultrafiltration of blood. Proc. Roy. Soc. Lond. B 254:147–152.

Salvini-Plawen, L. v. 1972. Zur Morphologie und Phylogenie der Mollusken: Die Beziehungen der Caudofoveata und der Solenogastres als Aculifera, als Mollusca und als Spiralia. Z. wiss. Zool. 184:205–394.

Scheltema, A. H. 1978. Position of the class Aplacophora in the phylum Mollusca. Malacologia 17:99–109.

Scheltema, A., Tscherakassy, M., and Kuzirian, A. M. 1994. Aplacophora. In Harrison, F. W., and Kohn, A. J. (Eds.): Microscopic Anatomy of Invertebrates. Vol. 5: Mollusca I. Wiley-Liss, New York. pp. 13–54.

Internet Site

www.whoi.edu/science/B/aplacophora (Aplacophora Home Page, maintained by C. Schander, Woods Hole Oceanographic Institution.)

POLYPLACOPHORA

Boyle, P. R. 1974. The aesthetes of chitons. II. Fine structure in *Lepidochiton cinereus*. Cell Tiss. Res. 153:383–398.

Eernisse, D. J., and Reynolds, P. D. 1994. Polyplacophora. In Harrison, F. W., and Kohn, A. J. (Eds.): Microscopic Anatomy of Invertebrates. Vol. 5: Mollusca I. Wiley-Liss, New York. pp. 55–110.

Friedrich, S., Wanninger, A., Brückner, M., and Haszprunar, G. 2002. Neurogenesis in the mossy chiton, *Mopalia muscosa* (Gould) (Polyplacophora): Evidence against molluscan metamerism. J. Morphol. 253:109–117.

Pierce, M. E. 1950. *Chaetopleura apiculata*. In Brown, F. A. (Ed.): Selected Invertebrate Types. John Wiley and Sons, New York. pp. 318–319.

Yonge, C. M. 1939. On the mantle cavity and its contained organs in the Loricata. Quart. J. Micro. Sci. 81:367–390.

Internet Site

http://home.inreach.com/burghart (Chitons.com.)

MONOPLACOPHORA

Borradaile, L. A., and Potts, F. A. 1959. The Invertebrata: A Manual for the Use of Students. Cambridge University Press, Cambridge. 795 pp.

Haszprunar, G., and Schaefer, K. 1997. Monoplacophora. In Harrison, F. W., and Kohn, A. J. (Eds.): Microscopic Anatomy of Invertebrates. Vol. 6B: Mollusca II. Wiley-Liss, New York. pp. 415–457.

Lemche, H., and Wingstrand, K. G. 1959. The anatomy of *Neopilina galatheae* Lemche, 1957. Galathea Rept. 3:9–71.

Stasek, C. R. 1972. The molluscan framework. In Florkin, M., and Scheer, B. T. (Eds.): Chemical Zoology. Vol. 3. Academic Press, New York. pp. 1–44.

Wingstrand, K. G. 1985. On the anatomy and relationships of recent Monoplacophora. Galathea Rep. 16:7–94.

GASTROPODA

Andrews, E. B., and Taylor, P. M. 1988. Fine structure, mechanism of heart function and haemodynamics in the prosobranch gastropod mollusc *Littorina littorea* (L.). J. Comp. Physiol. B 158:247–262.

Beeman, R. D. 1970. An autoradiographic study of sperm exchange and storage in a sea hare, *Phyllaplysia taylori*, a hermaphroditic gastropod. J. Exp. Zool. 175:125–132.

Bevelander, G. 1988. Abalone: Gross and Fine Structure. Boxwood Press, Pacific Grove, CA. 80 pp.

Bieler, R. 1992. Gastropod phylogeny and systematics. Ann. Rev. Ecol. Syst. 23:318–338.

Branch, G. M. 1981. The biology of limpets: Physical factors, energy flow and ecological interactions. Oceanogr. Mar. Biol. Ann. Rev. 19:235–380.

Carew, T., Pinkster, H. M., Rubinson, K., and Kandel, E. R. 1974. Physiological and biochemical properties of neuromuscular transmission between identified motoneurons and gill muscle in *Aplysia*. J. Neurophysiol. 37:1020–1040.

Carriker, M. R. 1981. Shell penetration and feeding by natacean and muricacean predatory gastropods: A synthesis. Malacologia 20:403–422.

Cox, L. R. 1960. Gastropoda. In Moore, R. C. (Ed.): Treatise on Invertebrate Paleontology. Vol. I, Pt. 1. Geological Society of America, University of Kansas Press, Lawrence, KS. pp. *I*84–*I*169.

Farmer, W. M. 1970. Swimming gastropods (Opisthobranchia and Prosobranchia). Veliger 13:73–89.

Fretter, V. 1951. Some observations on British cypraeids. Proc. Malacol. Soc. Lond. 29:14–20.

Fretter, V., and Graham, A. 1994. British Prosobranch Molluscs. Second Edition. Ray Society Vol. 161. Ray Society, London. 820 pp.

Gascoigne, T. 1975. Methods of mounting sacoglossan radulae. Microscopy 32:513.

Gascoigne, T., and Sartory, P. K. 1974. The teeth of three bivalved gastropods and three other species of the order Sacoglossa. Proc. Malacol. Soc. Lond. 41:109–126.

Gilmer, R. W. 1972. Free-floating mucus webs: A novel feeding adaptation for the open ocean. Science 176:1239–1240.

Gosliner, T. M. 1994. Gastropoda: Opisthobranchian. In Harrison, F. W., and Kohn, A. J. (Eds.): Microscopic Anatomy of Invertebrates. Vol. 5: Mollusca I. Wiley-Liss, New York. pp. 253–355.

Haszprunar, G. 1988. On the origin and evolution of major gastropod groups, with special reference to the Streptoneura. J. Moll. Stud. 54:367–441.

Hoagland, K. E. 1979. The behavior of three sympatric species of *Crepidula* from the Atlantic, with implications for evolutionary ecology. Nautilus 94:143–149.

Hughes, R. N. 1986. A Functional Biology of Marine Gastropods. Johns Hopkins Press, Baltimore, MD. 245 pp.

Hughes, R. N., and Lewis, A. H. 1974. On the spatial distribution, feeding and reproduction of the vermetid gastropod *Dendropoma maximum*. J. Zool. (Lond.) 172:531–548.

Kandel, E. R. 1979. Behavioral Biology of *Aplysia*. W. H. Freeman, San Francisco. 463 pp.

Knight, J. B., Cox, L. R., Keen, A. M., Batten, R. L., Yochelson, E. L., and Robertson, R. 1960. Systematic descriptions. In Moore, R. C. (Ed.): Treatise on Invertebrate Paleontology. Vol. I, Pt. 1. Geological Society of America, University of Kansas Press, Lawrence, KS. pp. *I*169–*I*330.

Kohn, A. J., Nybakken, J. W., and VanMol, J.-J. 1972. Radula tooth structure of the gastropod *Conus imperialis* elucidated by scanning electron microscopy. Science 176:49–51.

Lalli, C. M., and Gilmer, R. W. 1989. Pelagic Snails: The Biology of Holoplanktonic Gastropod Mollusks. Stanford University Press, Stanford, CA. 289 pp.

Lindberg, D. R. 1988. The Patellogastropoda. Malacol. Rev. Suppl. 4:35–63.

Luchtel, D. L., Martin, A. W., Deyrup-Olsen, I., and Boer, H. H. 1997. Gastropoda: Pulmonata. In Harrison, F. W., and Kohn, A. J. (Eds.): Microscopic Anatomy of Invertebrates. Vol. 6B: Mollusca II. Wiley-Liss, New York. pp. 459–718.

Miller, S. L. 1974. The classification, taxonomic distribution, and evolution of locomotor types among prosobranch gastropods. Proc. Malacol. Soc. Lond. 41:233–272.

Morton, J. E. 1964. Locomotion. In Wilbur, K. M., and Yonge, C. M. (Eds.): Physiology of Mollusca. Vol. I. Academic Press, New York. pp. 383–423.

Nielsen, C. 1975. Observations on *Buccinum undatum* attacking bivalves and on prey responses, with a short review on attack methods of other prosobranchs. Ophelia 13:87–108.

Ponder, W. F., and Lindberg, D. R. 1996. Gastropod phylogeny: Challenges for the 90s. In Taylor, J. D. (Ed.): Origin and Evolutionary Radiation of the Mollusca. Malacological Society of London and Oxford University Press, Oxford. pp. 135–154.

Ponder, W. F, and Lindberg, D. R. 1997. Towards a phylogeny of gastropod molluscs. Zool. J. Linn. Soc. 119:83–265.

Rees, W. J. 1964. A review of the breathing devices of operculate land snails. Proc. Malacol. Soc. Lond. 36:55–67.

Rosenberg, G. 1992. Encyclopedia of Seashells. Dorsett Press, New York. 224 pp.

South, A. 1992. Terrestrial Slugs. Chapman & Hall, London. 428 pp.

Taylor, J. D., and Norris, N. J. 1988. Relationships of neogastropods. Malacol. Rev. Suppl. 4:167–179.

Thiriot-Quievreux, C. 1973. Heteropoda. Oceanogr. Mar. Biol. Ann. Rev. 11:237–261.

Thompson, T. E. 1976. Biology of Opisthobranch Mollusca. Vol. I. Ray Society Vol. 151. Ray Society, London. 206 pp.

Thompson, T. E., and Brown, G. H. 1976. British Opisthobranch Molluscs. Academic Press, London. 203 pp.

Thompson, T. E., and Brown, G. H. 1984. Biology of Opisthobranch Mollusca. Vol. II. Ray Society Vol. 156. Ray Society, London. 229 pp.

Tillier, S. 1989. Comparative morphology, phylogeny and classification of land snails and slugs. Malacologia 30:1–303.

Todd, C. D. 1981. The ecology of nudibranch molluscs. Oceanogr. Mar. Biol. Ann. Rev. 19:141–234.

Tompa, A. S. 1980. Studies on the reproductive biology of gastropods: Part III. Calcium provision and the evolution of terrestrial eggs among gastropods. J. Conch. 30: 145–154.

Underwood, A. J. 1979. The ecology of intertidal gastropods. Oceanogr. Mar. Biol. Ann. Rev. 16:111–210.

Voltzow, J. 1983. Flow through and around the abalone *Haliotis kamtschatkana*. Veliger 26(1):18–21.

Voltzow, J. 1994. Gastropoda: Prosobranchia. In Harrison, F. W., and Kohn, A. J. (Eds.): Microscopic Anatomy of Invertebrates. Vol. 5: Mollusca I. Wiley-Liss, New York. pp. 111–252.

Internet Sites

http://grimwade.biochem.unimelb.edu.au/cone/index1.html (Cone snails and conotoxins, from B. Livett, University of Melbourne.)

www.flmnh.ufl.edu/natsci/malacology/fl-snail/snails1.htm (Field Guide to the Freshwater Snails of Florida: F. G. Thompson, Florida Museum of Natural History.)

www.gastropods.com/ (Hardy's Internet Guide to Marine Gastropods.)

http://manandmollusc.net (Man and Mollusc resource site.)

www.applesnail.net (Apple Snails, by J. S. Ghesquiere.)

http://medslugs.de/E/Photographers/Wolfgang_Seifarth.htm#photo (Mediterranean nudibranch photos from W. Seifarth.)

http://divegallery.com (Jeff's Nudibranch Site and Coral Reef Gallery.)

http://rfbolland.com/okislugs/index.html (Okinawa Slug Site by R. Bolland.)

http://slugsite.tierranet.com (Slugsite, by S. Long.)

CEPHALOPODA

Berthold, T., and Engeser, T. 1987. Phylogenetic analysis and systematization of the Cephalopoda (Mollusca). Verh. naturwiss. Ver. Hamburg (NF) 29:187–220.

Budelmann, B. U., Schipp, R., and von Boletzky, S. 1997. Cephalopoda. In Harrison, F. W., and Kohn, A. J. (Eds.): Microscopic Anatomy of Invertebrates. Vol. 6A: Mollusca II. Wiley-Liss, New York. pp. 119–414.

Clark, M. R. 1966. A review of the systematics and ecology of oceanic squids. Adv. Mar. Biol. 4:91–300.

Ellis, R. 1998. The Search for the Giant Squid. Penguin, New York. 322 pp.

Florey, E. 1969. Ultrastructure and function of cephalopod chromatophores. Am. Zool. 9:429–442.

Gilbert, D. L., Adelman, W. J., and Arnold, J. M. (Eds.): 1990. Squid as Experimental Animals. Plenum Press, New York. 516 pp.

Gosline, J. M., and DeMont, M. E. 1985. Jet-propelled swimming in squids. Sci. Am. 252:96–103.

Hanlon, R., and Messenger, J. B. 1996. Cephalopod Behaviour. Cambridge University Press, New York. 232 pp.

Herring, P. J. 1977. Luminescence in cephalopods and fish. In Nixon, M., and Messenger, J. B. (Eds.): The Biology of Cephalopods. Symposia of the Zoological Society of London, no. 38. Academic Press, London. pp. 127–159.

Holme, N. A. 1974. The biology of *Loligo forbesi* Steenstrup in the Plymouth area. J. Mar. Biol. Assoc. U.K. 54:481–503.

Lane, F. W. 1960. Kingdom of the Octopus. Sheridan House, New York. 300 pp.

Lehmann, U. 1981. The Ammonites. Cambridge University Press, Cambridge. 246 pp.

Mangold, K., and Bidder, A. M. 1989. L'appareil digestif et la digestion. In Mangold, K. (Ed.): Céphalopodes. Traité de Zoologie, no. 5. Paris: Masson. pp. 321–373.

Moynihan, M., and Rodaniche, A. F. 1982. The Behavior and Natural History of the Caribbean Reef Squid *(Sepioteuthis sepioidea)*. Paul Parey, Hamburg. 150 pp.

Nesis, K. N. 1987. Cephalopods of the World. T.F.H. Publications, Inc., Neptune City, NJ. 351 pp.

Potts, W. T. W. 1965. Ammonia excretion in *Octopus dofleini*. Comp. Biochem. Physiol. 16:479–489.

Roper, C. F. E., and Boss, K. J. 1982. The giant squid. Sci. Am. 246:96–105.

Roper, C. E. F., and Brundage, W. L. 1972. Cirrate octopods with associated deep-sea organisms: New biological data based on deep benthic photographs. Smithson. Contrib. Zool. 21:1–46.

Shrock, R. R., and Twenhofel, W. H. 1953. Principles of Invertebrate Paleontology. McGraw-Hill, New York. 816 pp.

Solem, A., and Roper, C. F. E. 1975. Structures of Recent cephalopod radulae. Veliger 18:127–133.

Tompsett, D. H. 1939. *Sepia*. Liverpool Mar. Biol. Comm. Mem. 32:1–184.

Voss, G. L. 1977. Classification of Recent cephalopods. In Nixon, M., and Messenger, J. B. (Eds.): The Biology of Cephalopods. Symposia of the Zoological Society of London, no. 38. Academic Press, London. pp. 75–579.

Ward, P. 1983. The extinction of the ammonites. Sci. Am. 249:136–147.

Ward, P. 1987. The natural history of *Nautilus*. Allen and Unwin, Boston. 267 pp.

Wells, M. J. 1961. What the octopus makes of it; our world from another point of view. Adv. Sci. Lond. 20:461–471.

Young, J. Z. 1976. The nervous system of *Loligo*. II. Suboesophageal centres. Phil. Trans. Roy. Soc. Lond. B 274:101–167.

Internet Sites

http://is.dal.ca/~ceph/TCP/index.html (Cephalopod Page, by J. B. Wood, Dalhousie University.)

www.mnh.si.edu/cephs (Cephalopods at the National Museum of Natural History, by J. Felley, M. Vecchione, C. Roper, M. Sweeney, and T. Christensen, NMNH.)

www.cephbase.utmb.edu (CephBase, by J. B. Wood.)

www.ammonite.ws (Fossils of the Gault Clay and Folkestone Beds of Kent, U.K., by J. Craig.)

www.nrcc.utmb.edu (National Resource Center for Cephalopods, University of Texas.)

BIVALVIA

Andrews, E. B., and Jennings, K. H. 1993. The anatomical and ultrastructural basis of primary urine formation in bivalve molluscs. J. Moll. Stud. 59:223–257.

Atkins, D. 1937. On the ciliary mechanisms and interrelationships of lamellibranchs. II. Sorting devices on the gill. Quart. J. Microsc. Sci. 79:339–370.

Bender, K., and Davis, W. R. 1984. The effect of feeding by *Yoldia limatula* on bioturbation. Ophelia 23:91–100.

Benninger, P. G., Auffret, M., and Le Pennec, M. 1990. Peribuccal organs of *Placopecten magellanicus* and *Chlamys varia:* Structure, ultrastructure and implications for feeding. Mar. Biol. 107:215–223.

Bullock, T. H., and Horridge, G. A. 1965. Structure and Function in the Nervous Systems of Invertebrates. Vol. II. W. H. Freeman, San Francisco. 1391 pp.

Childress, J. J., Fisher, C. R., Favuzzi, J. A., et al. 1991. Sulfide and carbon dioxide uptake by the hydrothermal vent clam, *Calyptogena magnifica,* and its chemoautotrophic symbionts. Physiol. Zool. 64:1444–1470.

Galtsoff, P. S. 1964. The American Oyster. U.S. Fisheries Bull. 64:1–480.

Gilbert, S. F. 1988. Developmental Biology. Second Edition. Sinauer, Sunderland, MA. 726 pp.

Hughes, R. N. 1969. A study of feeding in *Scrobicularia plana*. J. Mar. Biol. Assoc. U.K. 49:805–823.

Jorgensen, C. B. 1966. Biology of Suspension Feeding. Pergamon Press, New York. 357 pp.

Kennedy, W. J., et al. 1969. Environmental and biological controls on bivalve shell mineralogy. Biol. Rev. 44: 499–530.

Lane, C. E. 1961. The teredo. Sci. Am. 204:132–140.

Mackie, G. L. 1984. Bivalves. In Tompa, A. S., Verdonk, N. H., and van den Biggelar, J. A. M. (Eds.): The Mollusca. Vol. 7: Reproduction. Academic Press, Orlando. pp. 351–418.

Morse, M. P., and Zardus, J. D. 1997. Bivalvia. In Harrison, F. W., and Kohn, A. J. (Eds.): Microscopic Anatomy of Invertebrates. Vol. 6A: Mollusca II. Wiley-Liss, New York. pp. 7–118.

Morton, B. 1981. Prey capture in the carnivorous septibranch *Poromya granulata.* Sarsia 66:241–256.

Morton, B. 1996. The evolutionary history of the Bivalvia. In Taylor, J. (Ed): Origins and Evolutionary Radiation of the Mollusca. Malacological Society of London and Oxford University Press, Oxford. pp. 337–359.

Owen, G. 1974. Feeding and digestion in the Bivalvia. Adv. Comp. Physiol. Biochem. 5:1–35.

Owen, G. 1978. Classification and the bivalve gill. Phil. Trans. Roy. Soc. Lond. B 284:377–386.

Parmalee, P. W., and Bogan, A. E. 1998. The Freshwater Mussels of Tennessee. University of Tennessee Press, Knoxville, TN. 328 pp.

Purchon, R. D. 1990. Stomach structure, classification, and evolution of the Bivalvia. In Morton, B. (Ed.): The Bivalvia: Proceedings of a Memorial Symposium in Honour of Sir Charles Maurice Yonge, Edinburgh, 1986. Hong Kong University Press, Hong Kong. pp. 73–82.

Reid, R. G. B., and Reid, A. M. 1974. The carnivorous habit of members of the septibranch genus *Cuspidaria.* Sarsia 56:47–56.

Roberts, L. 1990. Zebra mussel invasion threatens U.S. waters. Science 249:1370–1372.

Silvester, N. R., and Sleigh, M. A. 1984. Hydrodynamic aspects of particle capture by *Mytilus.* J. Mar. Biol. Assoc. U.K. 64:859–879.

Stanley, S. M. 1970. Relation of Shell Form to Life Habits of the Bivalvia (Mollusca). Geological Society of America Memoir, no. 125. Geological Society of America, Boulder, CO. 296 pp.

Stanley, S. M. 1972. Functional morphology and evolution of byssally attached bivalve mollusks. J. Paleontol. 46:165–212.

Tankersley, R. A., and Dimock, R. V. Jr. 1992. Quantitative analysis of the structure and function of the marsupial gills of the freshwater mussel *Anodonta cataracta.* Biol. Bull. 182:145–154.

Trueman, E. R. 1966. Bivalve mollusks: Fluid dynamics of burrowing. Science 152:523–525.

Wilbur, K. H., and Saleuddin, A. S. M. 1983. Shell formation. In Saleuddin, A. S. M., and Wilbur, K. M. (Eds.): The Mollusca. Vol. 4, Pt. 1. Physiology. Academic Press, New York. pp. 235–287.

Yonge, C. M. 1939. The protobranchiate Mollusca, a functional interpretation of their structure and evolution. Phil. Trans. Roy. Soc. Lond. B 230:79–147.

Yonge, C. M. 1953. Form and habit in *Pinna carnea* Gmelin. Phil. Trans. Roy. Soc. Lond. B 237:365–374.

Yonge, C. M. 1975. Giant clams. Sci. Am. 232:96–105.

Internet Sites

www.inhs.uiuc.edu/cbd/collections/mollusk/molluskintro.html (Illinois Natural History Survey Mollusk Collection page.)

http://life.bio.sunysb.edu/marinebio/hotvent.html (Hot Vents, from the State University of New York at Stony Brook.)

http://courses.smsu.edu/mcb095f/gallery (Unio Gallery, maintained by C. Barnhart, Southwest Missouri State University.)

http://sun.science.wayne.edu/~jram/zmussel.htm (The Zebra Mussel Page, by J. L. Ram, Wayne State University.)

SCAPHOPODA

Bilyard, G. R. 1974. The feeding habits and ecology of *Dentalium entale stimpsoni.* Veliger 17:126–138.

Poon, P. 1987. The diet and feeding behavior of *Cadulus tolmiei.* Nautilus 101:88–91.

Reynolds, P. D. 2002. The Scaphopoda. Adv. Mar. Biol. 42:137–236.

Shimek, R. L., and Steiner, G. 1997. Scaphopoda. In Harrison, F. W., and Kohn, A. J. (Eds.): Microscopic Anatomy of Invertebrates. Vol. 6B: Mollusca II. Wiley-Liss, New York. pp. 719–781.

Internet Sites

http://rshimek.com/Scaph1.htm (Scaphopods, by R. Shimek.)

http://academics.hamilton.edu/biology/preynold/Scaphopoda/default.html (The Scaphopod Page, by P. D. Reynolds, Hamilton College.)

13 Annelida^{P}

Ask anyone to imagine a worm and most likely it will be an earthworm, the familiar annelid of garden soils, compost heaps, fishhooks, and sidewalks after rain. But Annelida (= ringed) also includes the leeches as well as myriad marine worms, many of breathtaking beauty and fascinating functional design. Annelids are familiar because they are common and occupy the large-worm niche in marine, freshwater, and terrestrial habitats. The largest annelids, such as the giant earthworms of Australia, reach 3 m in length, whereas the smallest are microscopic animals. With approximately 12,000 species, Annelida is the most diverse taxon of large-bodied worms, and this diversity evolved in conjunction with segmentation, the construction of a body from a series of modular sections. Annelid diversity manifests itself in many ways, but none is more obvious than the manner in which annelids feed. Virtually all feeding modes—suspension feeding, deposit feeding, scavenging, herbivory, carnivory—are represented in annelids. A few annelids are important to humans as an indirect source of food, including earthworms (and marine relatives) used as fish bait, to till soil, and to convert organic matter to compost, thus improving soil fertility. Medicinal leeches are used today to promote the healing of tissue grafts and reattached fingers and other appendages.

FORM AND FUNCTION

SEGMENTATION

From anterior to posterior, the annelid body is composed of three regions: prostomium, trunk, and pygidium. The elongate trunk consists of a longitudinal series of similar body units, the **segments,** which are separated from each other externally by a shallow constriction (Fig. 13-1, 13-2). The annelids are named for the body annulations that result from the ringlike segments. At the anterior end of the trunk is the **prostomium,** which contains the brain and bears the sense organs. The **pygidium** forms the posterior end of the body and includes the anus. The segments lie between the prostomium and pygidium like railroad cars between engine and caboose. The first segment, called the **peristomium,** lies immediately behind the prostomium and ventrally surrounds the mouth.

The prostomium and pygidium, though segmentlike in appearance, are not considered to be segments because they do not develop from the segmental growth zone. The **growth zone** itself is localized to a region immediately in front of the pygidium. The cells in this zone are paired ectodermal and mesodermal teloblast cells that divide and differentiate to form each new segment. Body growth, called **teloblastic growth,** results from the successive addition of segments posteriorly. At any given time, the youngest body segment lies immediately in front of the pygidium, the oldest (peristomium) immediately behind the prostomium.[1]

All internal structures are segmented, but some organs and tissues also span segments, integrating them into an individual worm (Fig. 13-2, 13-6). The chief integrating structures are the nervous system, hemal system, musculature, and digestive system. The remaining organs—appendages (when present), coelomic cavities, nephridia, and gonads—are segmental and thus repeated in each segment.

BODY WALL

Consider the annelid body wall: It consists of a fibrous collagenous cuticle, a glandular monolayered epidermis, a connective-tissue dermis (cutis) of varying thickness, and a

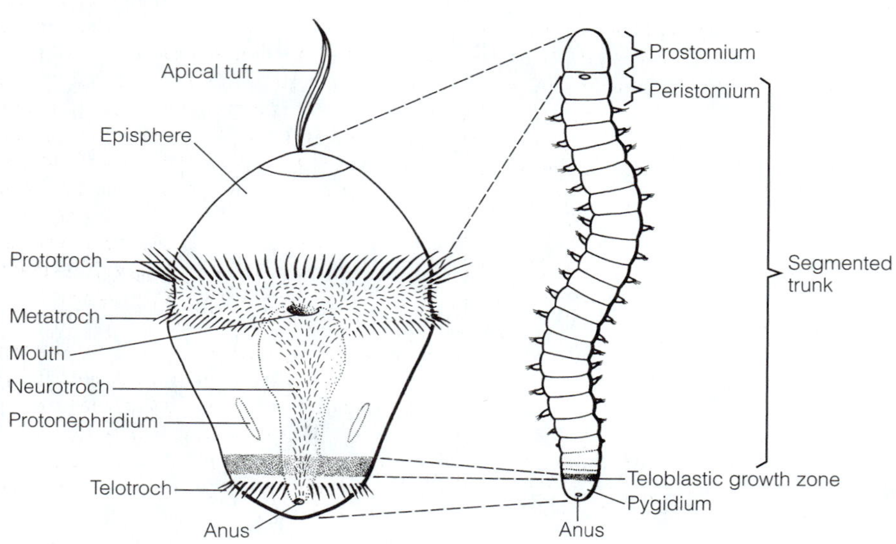

FIGURE 13-1 Corresponding body regions of a trochophore larva and an adult annelid. *(Modified and redrawn from Nielsen, C. 2001. Animal Evolution. Interrelationships of the Living Phyla. 2nd edition. Oxford University Press. Fig. 19.3.)*

[1]Some zoologists consider the peristomium, as well as the prostomium, to be a presegmental region of the body. The peristomium of some polychaetes (certain species of Spionidae and Capitellidae), however, bears chaetae or chaetae and parapodia, which seem to indicate that the peristomium is a segment.

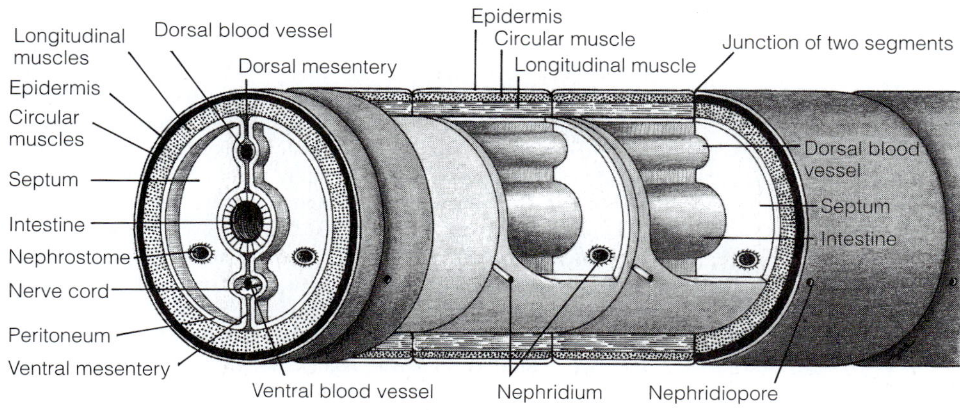

FIGURE 13-2 Annelid segments and anatomy based on Oligochaeta. *(After Kaestner, A. 1967. Invertebrate Zoology, Vol. 1. Interscience Publishers, New York.)*

musculature derived from the coelomic lining (Fig. 13-3A–D). The cuticular fibers, which usually are arranged in a crossed-helical pattern (Fig. 13-3E), toughen the body wall, resist bulges, and often impart an iridescent sheen to the body. A few of the tube-dwelling marine annelids (polychaetes) lack a cuticle, but their secreted tubes resemble the cuticle in structure and composition (Fig. 13-3F). Mucus-secreting gland cells are common in the epidermis. The secreted mucus consolidates the burrow lining and protects the surface of the body.

Chaetae (KEE-tee) are chitinous bristles that project outward from the epidermis to provide traction and perform other tasks (Fig. 13-4). The simplest chaetae are unjointed, tapered bristles (capillary chaetae) in bilaterally paired bundles, one dorsolat-

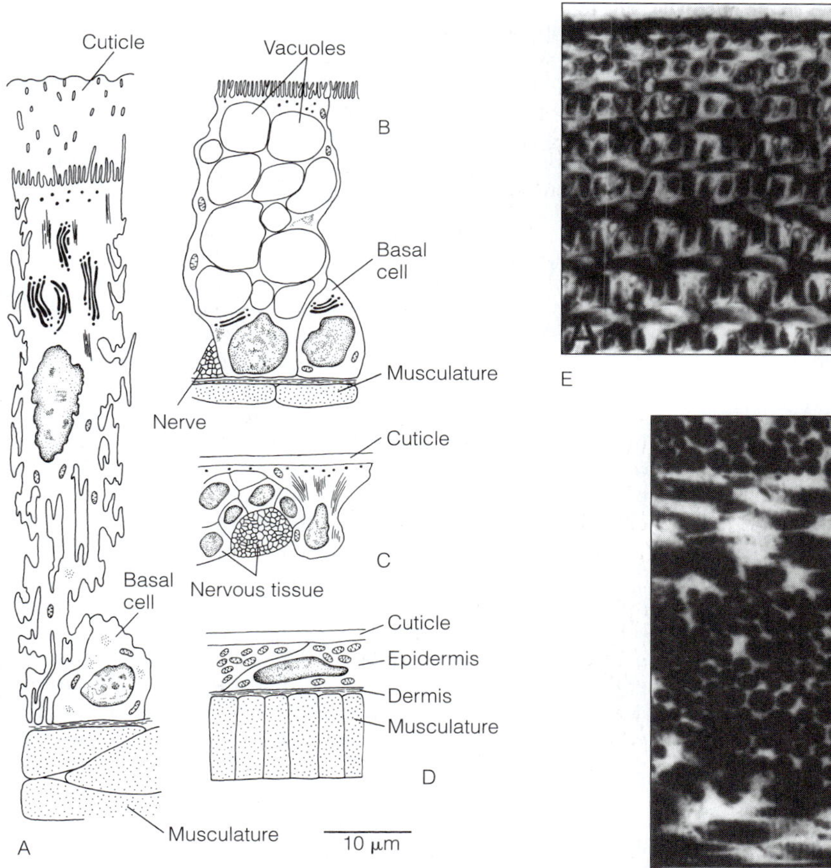

FIGURE 13-3 Annelid body wall **(A–D)**, cuticle **(E)**, and tubes **(F)** based on Polychaeta. **A,** *Lanice conchilega*; **B,** *Spiochaetopterus typicus*; **C,** *Plakosyllis quadrioculata*; **D,** *Streblospio benedicti*. **E,** Crossed helical collagen fibers in the cuticle of *Arabella iricolor*. **F,** Orthogonal collagen fibers in the tube of *Chaetopterus variopedatus*. *(From Storch, V. 1988. Integument. In Westheide, W., and Hermans, C. O. (Eds.): The Ultrastructure of Polychaete. Mikrofauna Marin. 4:13–36. F, From Gaill, F., and Hunt, F. 1988. Ibid. pp. 61–70.)*

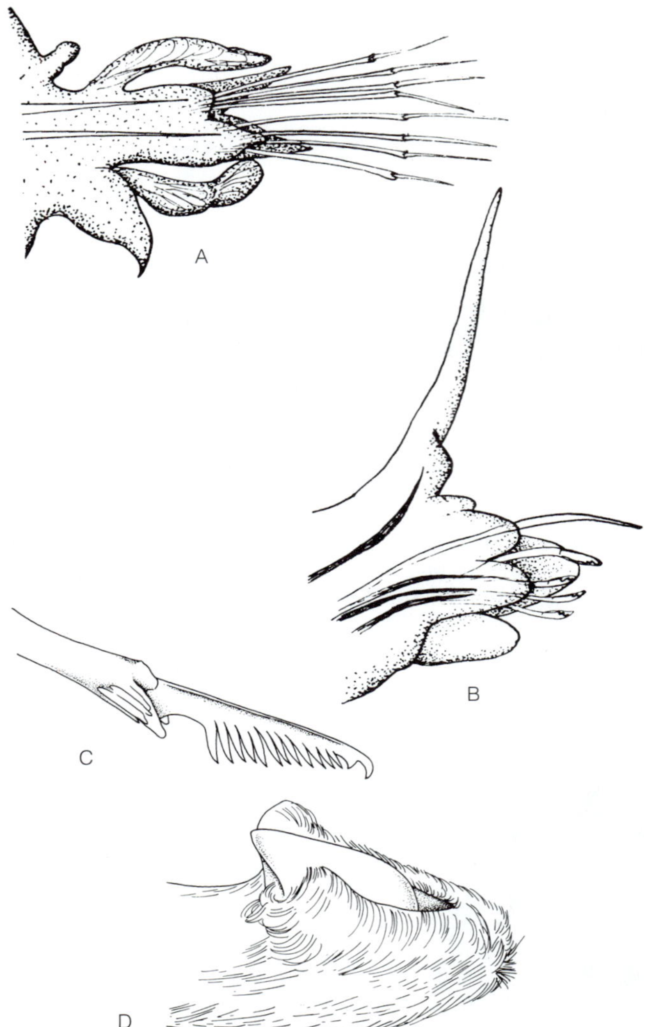

FIGURE 13-4 Annelid chaetae based on Polychaeta. **A,** Parapodium and chaetae of *Glycera dibranchiata,* a burrowing polychaete. **B,** Parapodium and chaetae of *Diopatra cuprea,* an errant tube dweller. **C,** Jointed chaeta from *Typosyllis pulchra,* a crawler in fouling communities. **D,** A hooded, hooked chaeta from the spionid *Tripolydora spinosa,* a sedentary tube-dwelling worm. *(A and B, From Renaud, J. C. 1956. A report on some polychaetous annelids from the Miami-Bimini area. Am. Mus. Novit. 1812:1–40; C, Based on an SEM by A. E. Heacox; D, Drawn from an SEM in Blake, J. A., and Woodwick, K. H. 1981. The morphology of* Tripolydora spinosa: *An application of scanning electron microscopy to polychaete systematics. Proc. Biol. Soc. Wash. 94:352–364.)*

eral and one ventrolateral pair per segment, but many other types and arrangements occur in annelids. A chaeta arises from a pitlike epidermal follicle composed of **follicle cells,** which form the follicle wall, and a single **chaetoblast** cell at the base of the follicle (Fig. 13-5). The chaeta is secreted around long microvilli on the surface of the chaetoblast. The microvilli later withdraw, leaving a bundle of parallel, hollow canals in the completed chaeta. The base of each chaeta remains in the follicle, but the shaft extends beyond it above the surface of the epidermis. The chaetal base is anchored by hemidesmosomes to the follicle cells, which are linked to subepidermal muscles that move the chaeta. Chaetae are composed of β-chitin, a form of chitin in which the polymerized, chainlike molecules are parallel and share the same polarity. β-chitin typically is tough and flexible, but some annelids harden and stiffen their chaetae with sclerotized protein and inorganic materials such as calcium carbonate.

Annelid bristles are called chaetae to distinguish them from the superficially similar **setae** (SEE-tee) of arthropods (crustaceans, insects, and relatives). Arthropod setae do not have a precise definition, but the term typically is applied to hair- or bristlelike sensory projections that pivot on a flexible joint with the exoskeleton. Deflection of the seta stimulates a sensory neuron attached to the setal base. The seta develops from an epidermal follicle composed of one or more **trichogen cells,** which secrete the seta, and one or more **tormogen cells,** which produce the pivot joint. The follicle also includes a sensory cell and its surrounding thecogen. Setae may be hollow or solid, but they lack the internal bundle of parallel canals associated with chaetae. They are composed of α-chitin—a form in which adjacent parallel molecules have opposite polarity. In contrast to the tough flexibility of β-chitin, α-chitin is noted for being hard and stiff, properties that can be further enhanced by sclerotized protein (as in insects) or calcium carbonate (in many crustaceans).

The obliquely striated musculature is arranged in outer circular and inner longitudinal layers. The longitudinal fibers typically are grouped together to form distinct muscle bands. In different annelid taxa, the coelomic lining can be a mesothelium composed solely of epitheliomuscular cells of the body-wall musculature, or the musculature can be separated from the coelom by a peritoneum.

NERVOUS SYSTEM

The central nervous system of annelids consists of an anterior dorsal brain located in the prostomium and a ventral pair of longitudinal nerve cords (Fig. 13-6A). In its simplest form, the brain consists of a pair of dorsal suprapharyngeal ganglia. A pair of circumpharyngeal connectives joins the brain to a pair of ventral subpharyngeal ganglia in the peristomium, which in turn give rise to the nerve cords that extend the length of the body. In each segment, the nerve cords bear a pair of ganglia that are joined together by a transverse nerve (commissure). Because the paired nerve cords and their segmental commissures resemble the sidepieces and rungs of a ladder, the annelids are said to have a ladderlike nervous system (Fig. 13-6A). Each pair of trunk ganglia is essentially a segmental "brain" from which a bilateral pair of **segmental nerves** enters the body wall, arches dorsally, nearly or completely encircles the body, and innervates the body-wall musculature and sensory structures of that segment.

The nervous system of a few annelids is intraepidermal, but in most, it is submerged in the connective tissue below the epidermis.

The innervation pattern of annelid body muscles is similar to that of the arthropods (Chapter 16). In both taxa, one muscle fiber (a cell) is innervated by more than one neuron

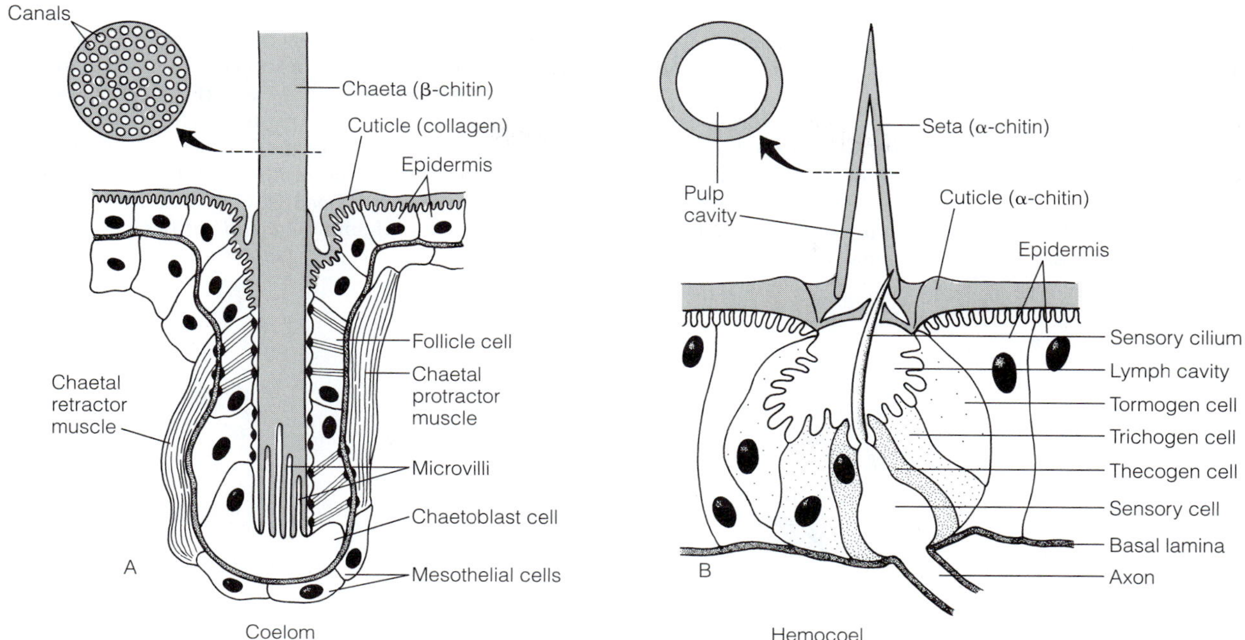

FIGURE 13-5 Contrasting structure of an annelid chaeta and an arthropod seta. **A,** A chaeta in its chaetal follicle. The canals in the chaetal shaft are remnants of withdrawn chaetoblast microvilli. **B,** A seta (bristle sensillum, a mechanoreceptor) from an insect. *(A, Modified and redrawn from Bartolomaeus, T. 1995. Structure and function of the Uncini in* Pectinaria koreni, Pectinaria auricoma *(Terebellida) and* Spirorbis spirorbis *(Sabellida): Implications for annelid phylogeny and the position of Pogonophora. Zoomorphology 115:161–177; B, Modified and redrawn from Keil, T. A. 1998. The structure of integumental microreceptors. In Harrison, F. W, and Locke, M. (Eds): Microscopic Anatomy of Invertebrates. Vol. 11B. Insecta. Wiley-Liss, New York. pp 385–404.)*

(polyneuronal innervation), and the rate (speed) and force of fiber contraction depends on the summed effects of all the neurons, each of which alone elicits a different response from the fiber. The annelid-arthropod innervation pattern contrasts with that of the vertebrates, in which each group of muscle fibers (a motor unit) receives only one neuron and responds to it by contracting all or none of the fibers. Variations in the rate or force of contraction of a vertebrate muscle depend chiefly on recruitment of more or fewer motor units.

Large-diameter giant axons occur in the longitudinal nerve cords of most annelids. As you might expect, the best defense for a soft-bodied worm is a rapid retreat, which is accomplished by a quick shortening of the body. The large diameter (and hence low resistance) of giant axons allows for rapid impulse conduction and thus shorter response times for the worm (Fig. 13-15B,C). For example, the single giant fiber of the polychaete *Myxicola* (1.7 mm in diameter, the largest in the animal kingdom) can be fired at any level along the length of the body and conducts an impulse in both directions. The conduction velocity of the giant axon is 20 m/s, compared with about 0.5 m/s along an axon of average diameter. Branches from the giant axon run to the longitudinal muscles. The giant axons function only in the escape response and not in normal locomotion; if the giant axon is severed, locomotion is unaffected, but the escape response is blocked.

The sensory structures of annelids are primarily unicellular receptors—photoreceptors, chemoreceptors, and mechanoreceptors—and these are distributed on the head, appendages, and elsewhere on the body. The taxon Polychaeta also has a variety of sense organs, such as ocelli and eyes, statocysts, and chemosensory nuchal organs. Some leeches have eyes. In most cases, the sense organs occur on the prostomium, peristomium, and first few segments only, but in some polychaetes (such as *Ophelia, Armandia,* and others with ocelli), they also occur on trunk segments, as well as on the pygidium.

COELOM AND HEMAL SYSTEM

Each annelid segment houses a bilateral pair of coelomic cavities isolated from neighboring segments by transverse septa. The left and right coelomic cavities in each segment are separated from each other by longitudinal mesenteries, one dorsal and one ventral to the gut, in the midsagittal plane of the body (Fig. 13-6B). The placement of the septa coincides with the constrictions on the surface of the body. Each septum is composed of two layers of mesothelium, one from the segment in front and one from the segment behind, and a layer of connective tissue sandwiched between them (Fig. 13-2, 13-6B). A mesentery is structurally similar to a septum, but formed by contact between the left and right coelomic cavities of one segment.

Annelids that move by actively using appendages or whole-body peristalsis typically have well-developed septa that form more or less complete bulkheads between segments. Because the septa isolate the hydrostatic skeleton of each segment, the force of a segmental muscular contraction can be directed and applied solely to that segment, just as "segments" of your bony skeleton restrict the actions of particular muscles to finger, forearm, or forelimb. On the other hand, the septa are incomplete or reduced in semisessile annelids and those that move by other means, such as with cilia (small-bodied annelids), whipping movements, or eversion-retraction of the pharynx (some burrowers).

The paired coelomic cavities of each segment are primitively lined by a ciliated mesothelium. The mesothelium on the body-wall side of the coelom is composed of modified epitheliomuscular cells that form the body-wall muscles. Contractile mesothelial cells may also form a musculature on the septal faces (radial and circular muscles), around the blood vessels (circular muscles), and around the gut (predominantly circular muscles). The mesothelium may be specialized locally as **chlorogogen cells,** which form a yellow or brown layer of tissue around part of the gut wall and certain blood vessels (Fig. 13-6C, 13-61A). This tissue plays a vital role in intermediate metabolism that is similar to that of the liver in vertebrates. Chlorogogen tissue is the chief center of glycogen and fat synthesis and storage. Storage and detoxification of toxins, hemoglobin synthesis, protein catabolism and formation of ammonia, and synthesis of urea also take place in these cells.

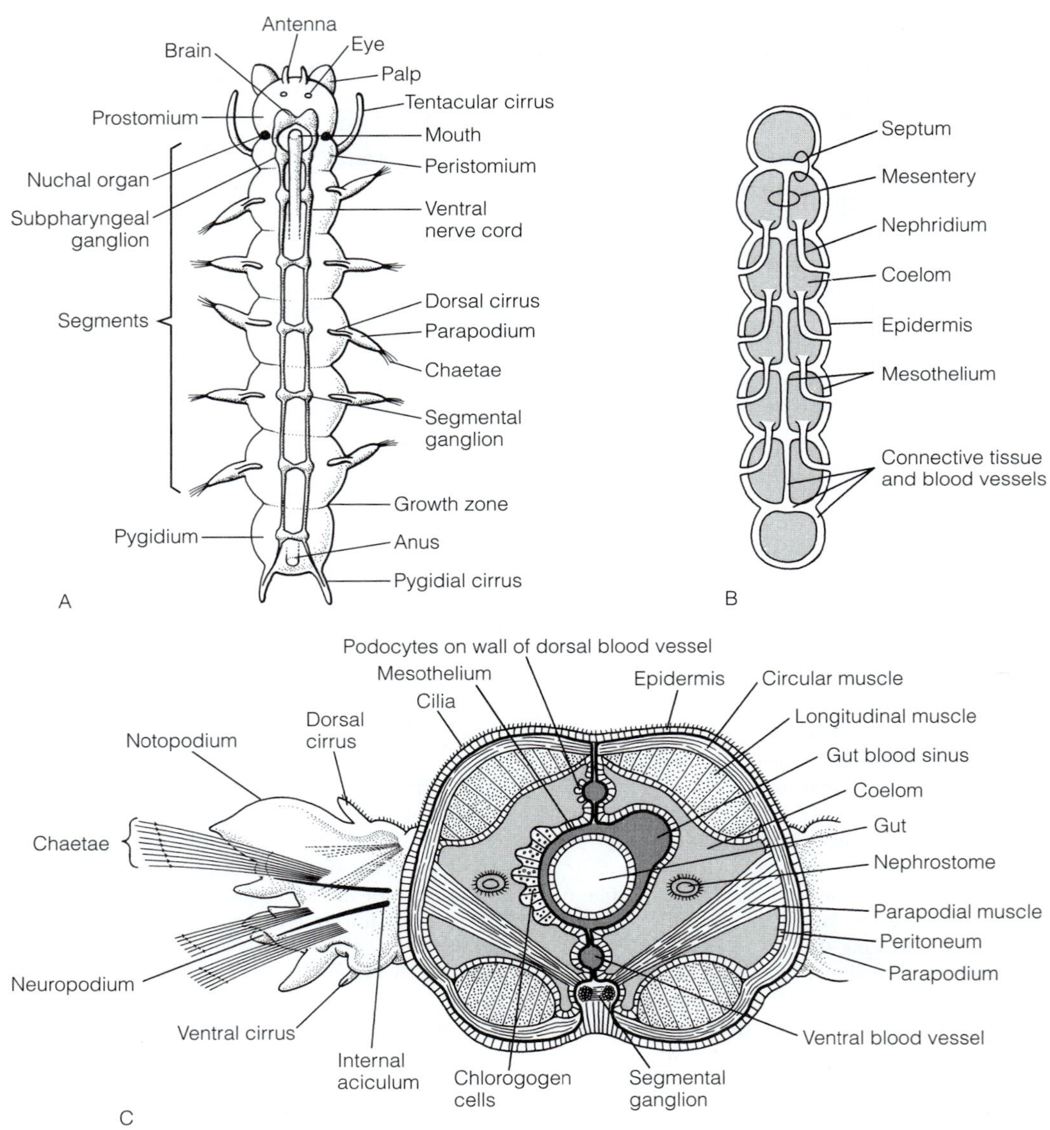

FIGURE 13-6 Annelid anatomy based on Polychaeta. **A** and **B,** Dorsal views. **A,** Body regions, appendages, and nervous system, **B,** Coelomic cavities and nephridia. **C,** Cross section of trunk showing parapodium and internal anatomy. *(Based primarily on Nereis.)*

The coelomic fluid, which is circulated by cilia and the contraction of body-wall muscles, contains coelomocytes that function in internal defense and sometimes gas transport (hemoglobin-containing coelomocytes). The coelom, as already mentioned, is the fluid skeleton (hydrostat) against which the muscles act to change body shape. Contraction of the longitudinal muscles causes the body to widen, whereas contraction of circular muscles results in body elongation (Fig. 9-8). The forces generated enable annelids to penetrate and burrow through soils and sediments.

The coelomic cavities provide for local segmental circulation, but a well-developed hemal system of blood vessels and hearts is responsible for transport throughout the body. Blood vessels tend to be reduced or absent in species with rudimentary or absent septa, however. In these, the coelomic fluid performs the task of whole-body transport. The blood vessels and sinuses are simple fluid channels in the connective-tissue compartment that lack a vertebrate-like endothelium. The principal vessels are the **dorsal blood vessel** in the dorsal mesentery and the **ventral blood vessel** in the ventral mesentery (Fig. 13-6C). Blood flows anteriorly in the dorsal vessel and posteriorly in the ventral vessel. In each segment, blood in the ventral vessel returns to the dorsal vessel via a capillary network (plexus) in the body wall. Blood in the dorsal vessel reaches the ventral vessel by way of a capillary plexus, vessels, or sinuses around the wall of the gut. The major blood vessels, especially the dorsal vessel, are contractile and pump blood peristaltically. In some annelids the enlarged, anterior region of the dorsal vessel forms a muscular heart, whereas in the anterior end of many earthworms, a few of the dorsoventral gut vessels are specialized as hearts.

The common respiratory pigment of annelids, hemoglobin, can occur in coelomic fluid, blood, muscle, and nerve. Hemoglobin is packaged in coelomocytes in the coelom, but dissolved in the plasma of the blood. Annelid gas exchange occurs across the body wall, appendages, and gills.

EXCRETORY SYSTEM

Annelid excretory organs are segmental nephridia, either metanephridial systems or protonephridia. The inner ends of these nephridia (funnels or terminal cells, respectively) are often associated with the anterior face of each septum, from which they project into the coelomic fluid (Fig. 13-2, 13-6B). The nephridial tubule of each nephridium penetrates the septum and passes into the next posterior segment before opening to the exterior. In general, metanephridial systems occur in annelids that have a hemal system, some vessels of which bear podocyte-covered filtration sites, whereas protonephridia are present in annelids that lack a hemal system (see discussion of excretory system in Chapter 9).

DIGESTIVE SYSTEM

The annelid digestive system is a more or less straight tube that extends between the mouth and the pygidial anus (Fig. 13-2). The gut passes through the septa and is supported above and below by the mesenteries (Fig. 13-6C). The gut tube consists of an ectodermal foregut, an endodermal midgut (stomach, intestine), and an ectodermal hindgut or rectum. Frequently, the foregut is specialized into two regions, a muscular, often protrusible or eversible pharynx and a ciliated esophagus that links the pharynx to the midgut.

REPRODUCTION AND DEVELOPMENT

All forms of clonal (asexual) reproduction, including fragmentation, budding, and a form of transverse division (paratomy, stolonization) that resembles scyphozoan strobilation (Fig. 13-29B,C), appear in annelids.

Annelids are primitively gonochoric animals that spawn gametes through their metanephridia. The segmental gonads are submesothelial clusters of germ cells that are released and stored in the coelomic cavities. (The coelomic cavities are packed with gametes in fully ripe individuals.) Once spawned and fertilized externally in seawater, the zygotes undergo spiral cleavage and develop into trochophore larvae (Fig. 12-4). During metamorphosis (Fig. 13-1), the larval episphere (the pretrochal region) becomes the prostomium whereas the body posterior to the telotroch becomes the pygidium. The segments that will form the trunk arise from a growth zone anterior to the telotroch.

DIVERSITY OF ANNELIDA

The 12,000 species of annelids are traditionally distributed among three major taxa—Polychaeta, Oligochaeta, and Hirudinomorpha (or Hirudinea)—considered to be classes in the Linnean system. Polychaeta, the most diverse group with some 8000 species, dominates in the marine environment. Oligochaeta (including earthworms, 3500 spp.) and Hirudinomorpha (leeches, 500 spp.), although present in the ocean, are the annelids that have successfully colonized fresh water and land.

PHYLOGENY OF ANNELIDA

Modern phylogenetic analysis of Annelida has slightly altered the old three-class system of Polychaeta, Oligochaeta, and Hirudinea. At present, Annelida is divided into two monophyletic sister taxa of equal rank, Polychaeta and Clitellata (Fig. 13-7A). Polychaetes have appendages (parapodia) and nuchal organs (described later) whereas clitellates lack appendages and nuchal organs but have highly specialized reproductive structures (clitellum, cocoons) associated with a commitment to hermaphroditism and yolk-rich eggs. Within Clitellata, Hirudinomorpha (leeches) is considered to be monophyletic, but some systematists (including P. Ax) regard Oligochaeta as a paraphyletic taxon, a hodgepodge of worms

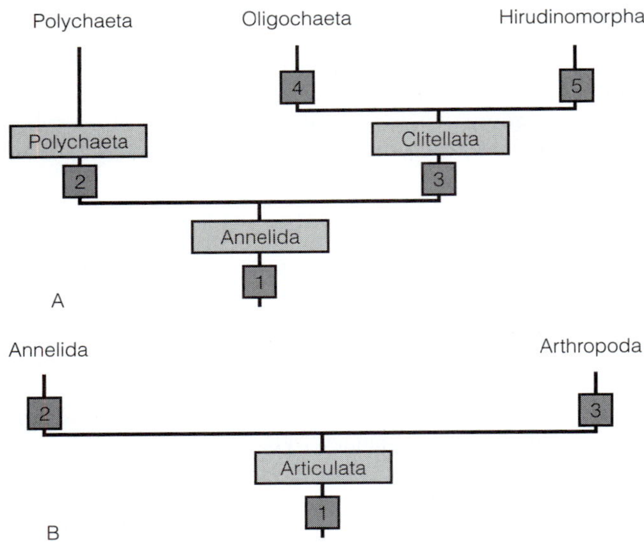

FIGURE 13-7 **A,** Phylogeny of Annelida. **1, Annelida:** segmentation, teloblastic growth (both plesiomorphies), four bundles of simple chaetae per segment. **2, Polychaeta:** parapodia, nuchal organs, pedal ganglia, one pair of pygidial cirri. **3, Clitellata:** clitellum, zygotes deposited in cocoon, hermaphroditism, genital segments, loss of trochophore. **4, Oligochaeta:** muscular pad or bulb in dorsal wall of pharynx. **5, Hirudinomorpha:** posterior sucker, dorsal anus anterior to sucker, superficial annulations obscure segmental divisions, unpaired midventral male gonopore, ectoparasites on animals. **B,** Phylogeny of Articulata. **1, Articulata:** segmentation with presegmental prostomium/acron, postsegmental pygidium/telson; paired coelomic cavities, metanephridia, gonads, nerve cords, and ganglia; ladderlike central nervous system; hemal system; teloblastic growth, *engrailed* expression at posterior margin of developing segments; longitudinal body-wall muscles in distinct bands, not a two-dimensional sheet. **2, Annelida:** simple, undivided chaetae of β-chitin and proteins secreted by ectodermal cells in follicles; each chaeta secreted by one cell on the surface of its long microvilli; chaetae in bundles, one dorsolateral pair and one ventrolateral pair per segment. **3, Arthropoda:** ecdysis (molting) of exoskeleton; exoskeleton of α-chitin and proteins; locomotory cilia absent; head consists of acron (annelid prostomium) and at least three segments; uniramous appendages; body cavity a hemocoel with a horizontal pericardial septum; heart has segmental ostia; saccate nephridia (nephridial sacs are reduced coelomic cavities); superficial cleavage, trochophore absent. *(Modified from Ax, P. 2000. Multicellular Animals. The Phylogenetic System of Metazoa. Vol. II. Springer Verlag, Berlin. 396 pp.)*

PHYLOGENETIC HIERARCHY OF ANNELIDA

Annelida
- Polychaeta
- Clitellata
  - Oligochaeta
  - Hirudinomorpha

that has retained primitive annelid and clitellate characters. Others (such as R. O. Brinkhurst), however, suggest that the muscular pad or bulb uniquely in the *dorsal* wall of the pharynx is an oligochaete autapomorphy, thus establishing the monophyly of Oligochaeta. For the time being, we have adopted the latter viewpoint and consider Oligochaeta to be a monophyletic taxon of Clitellata and sister taxon of Hirudinomorpha.

Identifying the sister taxon of Annelida is a matter of ongoing research and debate. The traditional view, which is grounded in evolutionary and developmental morphology, is that annelids and panarthropods (Onychophora and Arthropoda) are sister taxa of Articulata, the segmented protostomes (Fig. 13-7B). The primary evidence for this hypothesis is the common occurrence in the taxa of segmentation, teloblastic growth, and expression of the gene *engrailed* at the posterior margin in developing segments. Molecular systematists and some morphologists are challenging the Articulata hypothesis, however. Their phylogenetic hypothesis, based on gene sequences and reconsiderations of morphology, places annelids with sipunculans, molluscs, and other protostomes with a trochophore larva in the taxon Trochozoa and with panarthropods and other molting animals in an unrelated taxon, Ecdysozoa (Fig. 9-25, 9-26B). If this hypothesis is correct, then it is likely that segmentation evolved independently in annelids and panarthropods.

EVOLUTION AND SIGNIFICANCE OF SEGMENTATION

Explanations for the evolution of the annelid body attempt to account for the origin and selective advantages of segmentation, a design that is common to but perhaps not homologous in annelids and arthropods. With few exceptions, however, these explanations are little more than hypotheses waiting to be tested. As such, it is fair to say that a satisfactory explanation for the origin of segmentation is one of the Holy Grails of modern evolutionary morphology.

The hypotheses fall into three groups:

1. Gonocoel theory: The articulate ancestor was not segmented, but had at least one set of organs repeated serially along the length of its body. These organs—the gonads, for example, in nemerteans—enlarged and adopted additional functions to become the paired segmental coeloms of each segment.
2. Cyclomerism theory: The ancestor had a few segments (known as oligomery) thought to be in the form of three pairs, with the last pair subdivided to form the segmented trunk. Living oligomeric animals that might provide models for such an ancestor are brachiopods, hemichordates, and echinoderms.
3. Corm theory: The ancestor reproduced clonally to form a chain of zooids, similar to some flatworms undergoing paratomy (Fig. 10-22A). The zooids then lost their "independence" as they became more highly integrated into the body as segments. A related hypothesis regards the annelid trochophore as an axial zooid that produces

buds in bilateral pairs (segments) as it grows. This pattern of growth is reminiscent of modular growth in many colonial cnidarians, especially A-form hydroids and siphonophores (Chapter 7).

Regardless of its evolutionary origin, selection favored segmentation in annelids (as well as in arthropods and vertebrates). What were these selective advantages? One may be that segmentation of the coelomic hydrostat enabled worms to burrow more efficiently than their nonsegmented burrowing relatives, such as nemerteans, echiurans, and sipunculans. This **burrowing hypothesis** can be illustrated by comparing two coelomate worms that are identical except that one has segmental septa that partition its coelom into a series of isolated compartments, whereas the coelom of the other is undivided and continuous throughout the body. Both worms burrow using alternating contractions of longitudinal and circular muscles to generate peristalsis along the body (Fig. 13-8). At any instant, the peristaltic waves of both worms are identical in appearance, but the musculature of the nonsegmented worm is more active than that of the segmented worm and requires more energy to maintain the proper body shape. Along the body of the nonsegmented worm, the pressure of the coelomic fluid is at a maximum where circular or longitudinal muscles are contracted maximally (regions of minimum and maximum body diameter, respectively). Because the coelom is unpartitioned, the elevated fluid pressure is transmitted throughout the coelom and must be antagonized everywhere by the action of body-wall muscles to prevent aneurisms and other deviations from the proper peristaltic wave shape (Fig. 13-8B). Segmented animals, on the other hand, isolate changes in coelomic fluid pressure to individual segments or groups of segments. As a result, body regions between contracted segments do not experience high fluid pressures and need not contract fully, or at all, to maintain the preferred shape of the body (Fig. 13-8C).

Once segmentation evolved, it created the potential for further specialization of the body. This is because segmentation creates a series of compartments, each of which can be regulated more or less independently of the others. As such, segmentation, like multicellularity or coloniality, provided a framework for specialization. In colonies, such specialization is apparent in the polymorphism of zooids, whereas in segmented animals, it results from a regional specialization of segments called **tagmosis.** The simplest examples of tagmosis are found in annelids in which the body is weakly divided into three tagmata: head, thorax, and abdomen (Fig. 13-9). These same three tagmata are especially prominent in insects and many crustaceans.

Tagmosis can result from any one of three processes. The first is the *restriction* of certain segmental structures to only a few segments. For example, gonads are often restricted to

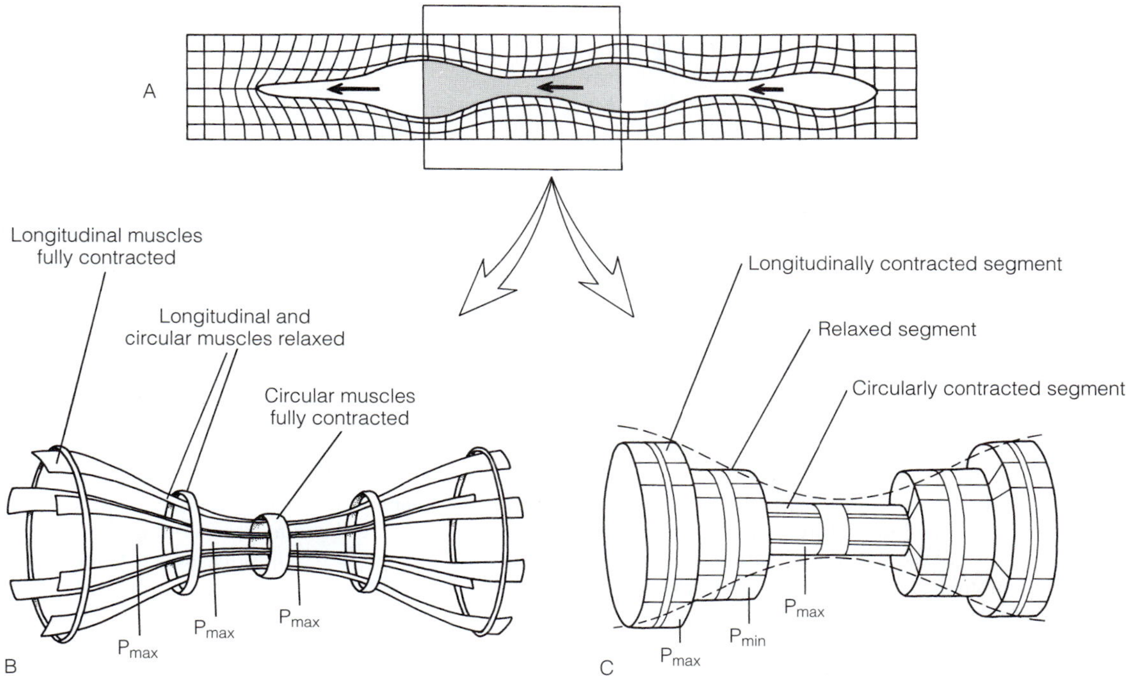

FIGURE 13-8 Peristaltic burrowing **(A)** with **(C)** and without **(B)** segmentation. The peristaltic burrowing wave **(A)** can be sustained in segmented worms **(C)** with less muscular involvement than in nonsegmented worms **(B)** because fluid pressure maxima are localized to a few segments. P_{max}, maximum pressure in hydrostat; P_{min}, minimum pressure in hydrostat. *(Modified and redrawn from Turbeville, J. M. and Ruppert, E. E. 1983. Epidermal muscles and peristaltic burrowing in* Carinoma tremaphoros *(Nemertini): Correlates of effective burrowing without segmentation. Zoomorphology 103:103–120.)*

FIGURE 13-9 Regional specialization (tagmosis) of the annelid body: the polychaete *Americonuphis magna.* The body of this tube-dwelling worm is specialized into a *head* bearing long slender sensory appendages, a *thorax* with anteriorly directed limbs, and an *abdomen* bearing branched gills.

a few specialized **genital segments.** The second is that a segmental structure may be retained in all segments, but may structurally and functionally diverge between the segments. Such *divergence* is common among segmental appendages, which can be regionally specialized for locomotion, grasping, chewing, gas exchange, or reproduction. Third, regional specialization also results from the *fusion* of segments. Although fusion may occur anywhere along the length of a segmented animal, it is commonly expressed as an anterior fusion of one or more segments with the prostomium and peristomium to form a complex head. The head of the common polychaete *Nereis,* for example, consists of a prostomium, peristomium (first segment), and two additional segments, whereas that of the fruit fly *Drosophila* is composed of the acron (equivalent to the annelid prostomium) plus five segments.

Developmentally, segmentation is best understood in insects, especially in *Drosophila.* As a fruit fly develops, a sequence of genes is activated that first defines the anatomical axes—anteroposterior, dorsoventral—of the embryo and then establishes the body segments. Segment morphogenesis proceeds from the general to the specific. First, gap genes define regional differences along the anteroposterior axis. Second, these regional differences control the pair-rule genes, which are expressed periodically as parasegments. (Parasegments are segmentlike stripes of gene expression, not morphological segments, which differentiate later and are offset slightly from the parasegmental pattern. Parasegments establish a periodic pattern on which the differentiation of true segments is controlled.) Third, segment-polarity genes (segmentation genes), such as *engrailed,* establish a series of similar true segments. Fifth, selector genes, including Hox genes, determine the specific fate of each of the segments. This final level of control, the determination of regional identities along the anteroposterior axis by selector genes, has been documented not only in fruit flies but also in other insects, nematodes, and chordates, and thus seems to be a common morphogenetic process in Bilateria. The segmentation gene *engrailed* is expressed similarly in insects and annelids (leeches), suggesting that there is homologous genetic control of segmentation in these two taxa. Segment morphogenesis of leeches, however, differs from that of *Drosophila* in other ways. Such differences are difficult to evaluate because leeches have not been studied as thoroughly as *Drosophila* has, because variability exists within Insecta, and because few data are available from primitive annelid taxa, such as Polychaeta. Additional research is sorely needed to answer the question of whether the genetic control of segmentation is homologous in Annelida and Arthropoda or in all segmented animals, or if it underlies the anteroposterior patterning in all Bilateria.

POLYCHAETA[C]

Polychaetes, or bristleworms, are the diverse, common, and often colorful annelids of the sea. And *diversity* is the key word in describing the form, function, and lifestyles of polychaetes. Of the 8000 species, most are burrowers, but some crawl over the bottom or on the surfaces of attached organisms, others bore into shells or rocks, many can swim when necessary, and a few are permanently pelagic. Still others secrete and occupy tubes, such as the spectacular feather-duster worms whose colorful crowns grace seawater aquaria and a variety of natural habitats. Polychaetes generally are medium-size worms of less than 10 cm in length and 2 to 10 mm in diameter, but many interstitial forms are less than 1 mm, and some species *(Eunice, Nereis, Polyodontes)* are real giants that approach or exceed a length of 1 m and a diameter as thick as your thumb. The largest known polychaete is *Eunice aphroditois,* whose body exceeds 3 m in length.

FORM AND FUNCTION

Polychaetes are annelids with legs. The legs, called **parapodia,** are lateral fleshy outgrowths of the body wall that may be large and complex or little more than low nubbins or ridges, depending on species. Each segment bears one pair of parapodia (Fig. 13-6A). A parapodium is basically biramous, consisting of an upper division, the **notopodium,** and a ventral division, the **neuropodium** (Fig. 13-6C). Each division is supported internally by a chitinous skeletal rod called an **aciculum.** Parapodial muscles that attach to the acicula move the parapodia. Tentacle-like sensory processes, the **cirri,** project from the dorsal base of the notopodium and from the ventral base of the neuropodium. The notopodia and neuropodia assume various shapes in different taxa and may be subdivided into several lobes or even greatly reduced (Fig. 13-9, 13-11, 13-23). A well-developed parapodium is a fleshy, more or less compressed projection of the body wall (Fig. 13-6C).

The parapodial lobes (rami) house the pocketlike **chaetal sacs,** each of which secretes a bundle of chaetae (Fig. 13-4). Primitively, two bundles of chaetae are borne by each parapodium (Fig. 13-4A). New chaetae are produced by the chaetal sac as older chaetae are lost or shed (Fig. 13-5A). Chaetae adopt diverse shapes, and the chaetal bundles of a particular species may be composed of more than one type of chaeta (Fig. 13-4B). Most chaetae are used to improve traction for locomotion through sediment or over surfaces, so their tips are variations on needles, hooks, or serrated blades (Fig. 13-4). Some chaetae, however, are spatulate shovels used for digging (paleae; Fig. 13-49D), others are paddles used for swimming (Fig. 13-30C), still others (uncini) bear numerous microscopic, Velcro-like hooks used to grip the inner walls of tubes and burrows (Fig. 13-35B).

The segments of polychaetes are generally similar, but in some burrowers and tube dwellers there has been a tendency for the trunk to become specialized into distinct regions (thorax and abdomen) as a result of variations in the parapodia or the presence or absence of gills (Fig. 13-9). The number of segments ranges from fewer than 10 to over 200, depending on the species.

The polychaete head can consist of the prostomium alone, prostomium and peristomium, or prostomium, peristomium, and one or more additional segments. In general, however, polychaetes are unique among annelids in having several different types of cephalic sensory appendages, described fully under Nervous System and Sense Organs. The prostomial appendages include anterior or anterodorsal antennae and anteroventral palps (Fig. 13-6A, 13-10A). In many polychaetes, the peristomium bears sensory **tentacular cirri** or two long feeding appendages called **tentacular palps** (as in the spionids), which have migrated from the prostomium, elongated, and adopted a feeding function. The polychaete pygidium primitively bears one pair of **pygidial cirri** (Fig. 13-6A), but some polychaetes have more than one pair.

BODY WALL AND TUBES

The polychaete body wall is similar to that of annelids in general. Polychaetes are unique among annelids, however, because many species secrete and occupy a **tube** (Fig. 13-3F, 13-36B, 13-43). The tube may be open at one or both ends and partially buried in sediment or attached to surfaces. A worm may permanently occupy its tube, enlarging it with growth (as in *Chaetopterus* and feather-duster worms) or it may abandon the tube, crawl to a new location, and secrete another (as *Diopatra* does; Fig. 13-43). The shingle-tube worm, *Owenia fusiformis,* is able to burrow through the sediment in its flexible tube (Fig. 13-49A). So common are these tubes (and the worms that inhabit them) that they often wash ashore in windrows after storms scour them from the sand. For the enthusiast, identifying polychaete tubes on the beach can be as fascinating as associating a bird's nest with its builder.

The tube material is a fibrous protein that has the appearance and texture of cellophane, parchment, or silk. Foreign material such as mud, quartz sand, shell fragments, plant debris, and algae is often incorporated into the wall of the tube for strength, camouflage, and food. A few polychaetes that live on exposed surfaces calcify the tube to form a shell (Serpulidae, Spirorbidae; Fig. 13-51B,C).

The tubes are protective, but they also have other functions. For some polychaetes (*Diopatra,* for example), the tube is a lair from which to emerge and catch passing prey (Fig. 13-43). Like a snorkel projecting above the sediment surface, the tube may provide a source of clean, oxygenated water to its subterranean occupant (Fig. 13-36B). Attachment of the tube enables the worm to inhabit hard, bare surfaces such as rock, shell, or coral (Fig. 13-50B). A few worms brood their eggs and young within the tube (Fig. 13-51C). The parchment-tube worm, *Chaetopterus,* pumps water through its tube for gas exchange and filter feeding (Fig. 13-47).

MUSCULATURE AND LOCOMOTION

Polychaetes have diverse locomotory adaptations related to lifestyle and body design. Peristaltic burrowing, described earlier in this chapter and in Chapter 9, is common among polychaetes having elongate bodies, reduced parapodia and head appendages, and many similar segments. The circular musculature usually is well developed and the septa often are complete, restricting the coelomic fluid to individual segments. Actively burrowing polychaetes typically resemble earthworms (which also burrow using peristalsis) and are well represented in Scolecida (including Arenicolidae, Capitellidae, Orbiniidae; Fig. 13-34).

Other burrowing polychaetes, such as the lugworm, *Arenicola,* and the earthwormlike capitellids, augment or replace whole-body peristalsis with an eversible pharynx. As the pharynx everts, it pushes into the sediment and anchors. The animal then pulls itself forward into the sediment as it retracts the anchored pharynx (Fig. 13-35A, 13-21). The bloodworm *Glycera* has abandoned peristalsis and burrows solely with its prodigious pharynx, which punches into the sediment (Fig. 13-39C). Some members of Nephtyidae, called shimmy worms, and the amphioxus-like *Armandia* and *Ophelia* (Opheliidae) produce rapid lateral undulations of the body and literally swim through loosely consolidated sand (Fig. 13-36A, 13-40). Very small sediment dwellers, such as members of the interstitial Dinophilidae and Diurodrilidae, use cilia for locomotion.

Many polychaetes, such as ragworms (Nereidae), paddleworms (Phyllodocidae), scaleworms (Polynoidae), and pelagic polychaetes (Alciopidae, Tomopteridae), either crawl over surfaces or swim using well-developed parapodia and chaetae. In these, the prostomium is equipped with eyes and other sense organs, the parapodia are well developed, and the segments are generally similar (Fig. 13-10A, 13-11B, 13-18C, 13-38, 13-41B). Movement is the result of the combined action of the

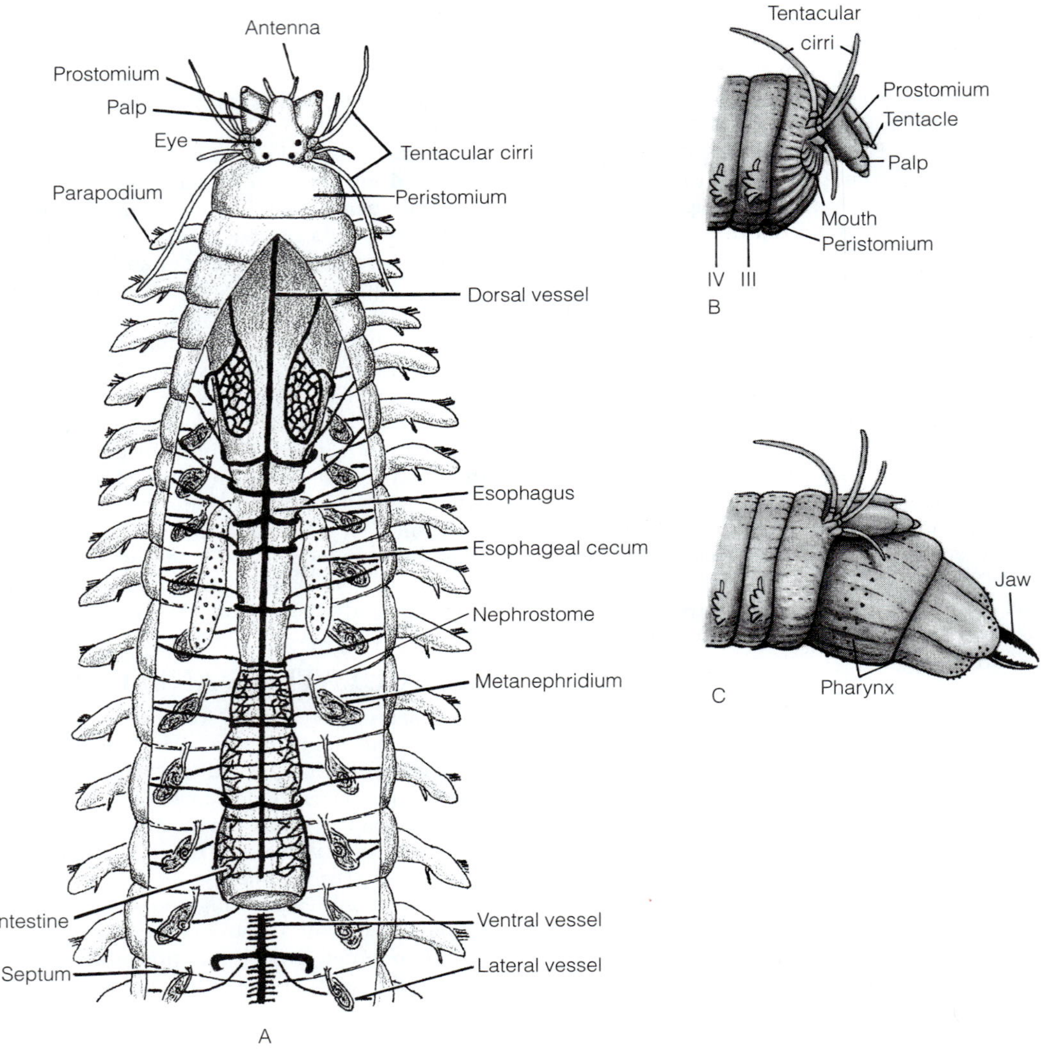

FIGURE 13-10 Polychaete anatomy: *Nereis virens.* **A,** Dorsal view and dissection. **B,** Lateral view of the head. Roman numerals designate segments. **C,** Lateral view of the head with the pharynx everted. *(A, From Brown, F. A. 1950. Selected Invertebrate Types. John Wiley and Sons, New York.)*

parapodia, the body-wall musculature, and the coelomic-fluid skeleton. The longitudinal muscle layer is better developed than the circular layer, the specialized parapodial muscles are well developed, and the septa tend to be incomplete.

During slow crawling (Fig. 13-12A, 13-13A), the parapodia and chaetae move like legs, alternately pushing against the substratum in the power stroke and lifting above the substratum in the recovery stroke. As the power stroke begins, the acicula and chaetae extend, the parapodium makes contact with the substratum, and then the parapodium sweeps rearward. The recovery stroke is initiated as the acicula and chaetae retract, then the parapodium lifts off the substratum and swings forward. The movements of the numerous parapodia are coordinated to avoid interference and move the worm effectively. The parapodia on opposite sides of the same segment are 180° out of phase; if the left-side parapodium is in the power stroke, the right side is in the recovery stroke. The series of parapodia on one side of the body are coordinated in waves, like falling dominoes, that pass from posterior to anterior (a progressive wave). Thus when a parapodium is in the middle of its power stroke, the next anterior parapodium initiates its power stroke as the next posterior parapodium completes its power stroke (Fig. 13-12A, 13-13A). The overall pattern is similar to that of a crawling millipede.

Rapid crawling (as seen in Nereidae or Hesionidae) involves not only the parapodia, but also lateral body undulations produced by waves of longitudinal-muscle contraction (Fig. 13-12B). These waves of contraction progress forward along the body and coincide with the alternating waves of parapodial

FIGURE 13-11 Polychaete heads in dorsal view. **A,** The polynoid scaleworm *Lepidasthenia varia,* which occupies the burrow of its polychaete host, the maitre d' worm, *Notomastus lobatus* (Capitellidae). **B,** The green phyllodocid *Phyllodoce fragilis,* a common crawling inhabitant of oyster reefs and other fouling communities. **C,** The oenonid "opalworm," *Arabella iricolor,* a burrower in fine sands. **D,** The eunicid *Marphysa sanguinea,* an inhabitant of muddy oyster reef communities in the southeastern United States also found in holes in calcareous rock.

activity described above. The parapodial power stroke occurs at the crest of an undulatory wave. With the crest parapodia firmly planted against the substratum, contraction of longitudinal muscles on the opposite, trough side of the body imparts additional thrust to the crest-side parapodia. As the longitudinal muscles shorten and compress the trough side of the body, the crest side expands, thrusting back against the parapodia anchored in the substratum, thus pushing the animal forward (Fig. 13-13B,C).

Many benthic polychaetes, such as *Nereis,* swim by using the rapid-crawling movements just described. The same movements are used for swimming in six exclusively pelagic families of polychaetes that resemble crawling polychaetes, but tend to be transparent like many other planktonic animals. Members of Alciopidae have enormous eyes (Fig. 13-18C), and species of Tomopteridae (Fig. 13-38A), which have lost the chaetae and have membranous parapodial pinnules, are the most highly specialized of the pelagic polychaetes.

NERVOUS SYSTEM AND SENSE ORGANS

The brain and ladderlike nerve cords of polychaetes are like those described for annelids in general (Fig. 13-14, 13-15). The polychaete brain, however, may be large and lobed if the head bears sense organs (Fig. 13-14, 13-15A). Additional ganglia, called pedal ganglia, are commonly associated with the segmental nerve cords at the bases of the parapodia (Fig. 13-14, 13-15). These ganglia, which are unique to polychaetes, are important centers that control the complex parapodial movements. The ventral nerve cord may be paired and ladderlike (Fig. 13-6A), as it is in feather-duster worms, but in most polychaetes, the two cords are fused along the midline to varying degrees (Fig. 13-14, 13-16), and in the tube-dwelling *Owenia,* an unpaired, nonganglionated nerve cord occurs in the epidermis.

FIGURE 13-12 Polychaete locomotion: slow and fast crawling. **A,** The nereid *Ceratonereis irritabilis* showing parapodial movements while crawling slowly. **B,** The phyllodocid *Eulalia macroceros* combining parapodial movements with a progressive undulatory wave while crawling rapidly.

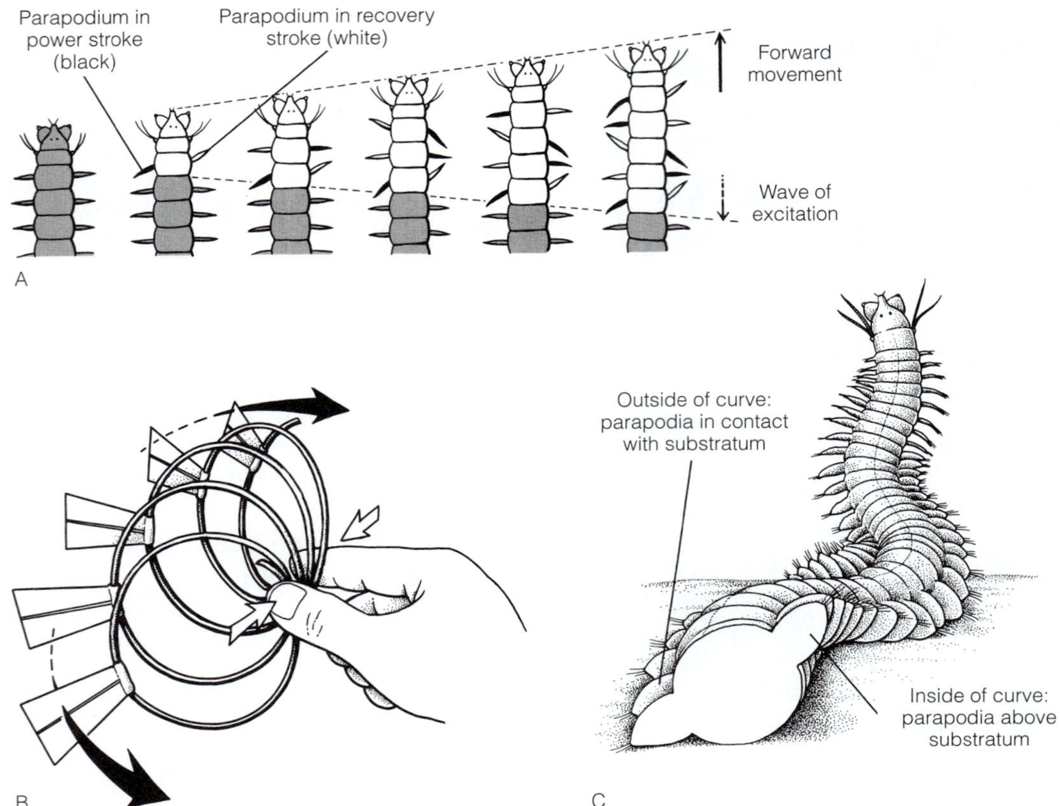

FIGURE 13-13 Polychaete locomotion: slow and fast crawling in nereid polychaetes. **A,** Slow crawling showing alternation of parapodial power and recovery strokes on opposite sides of the same segment, and the retrograde wave of activation along the body. **B,** Fast crawling superimposes a progressive undulatory wave of the body on the stepping movements of parapodia. Parapodia on the wave crests undergo power strokes, which are amplified by the contraction of longitudinal muscles in the troughs. **C,** As the trough side of the body shortens, it creates a backthrust on the crest side in contact with the substratum. *(A, Modified from Clark, R. B. 1964. Dynamics in Metazoan Evolution. Clarendon Press, Oxford. 313 pp.)*

The chief sense organs of polychaetes are nuchal organs, ocelli, and statocysts, but many also have antennae, palps, and cirri. Of these, the nuchal organs are widespread and unique to polychaetes.

Nuchal organs are a pair of ciliated sensory pits or slits situated posterolaterally on the prostomium (Fig. 13-17A,B, 13-37B). These chemoreceptive organs, which often are eversible, are important for detecting food. They attain their greatest development in predatory species, such as some of the fireworms (Amphinomidae), in which they expand to form a convoluted, brainlike crown (caruncle) on the upper surface of the head (Fig. 13-37B).

Eyes (ocelli), which are best developed in errant polychaetes (Aciculata), are found on the surface of the prostomium in two, three, or four pairs (Fig. 13-10A, 13-11B–D, 13-38B, 13-41B). Although prostomial ocelli occur in some species of sedentary polychaetes, they sometimes also have ocelli elsewhere on the body. For example, the feathers (radioles) of some feather-duster worms bear ocelli. These suspension feeders have a shadow reflex and instantly withdraw into their tubes when a shadow is cast over them, as when a fish swims overhead. Some of these same worms also have a pair of ocelli on the pygidium. Under duress, the worms abandon their tubes and crawl away backward as the pygidium becomes the functional head.

In general, the prostomial ocelli are pigment cups, the walls of which are composed of rodlike photoreceptors (modified microvilli), pigment cells, and supporting cells (Fig. 13-18).

The ocelli of most polychaetes probably determine only light intensity and direction, but the huge bulging eyes of the pelagic, carnivorous Alciopidae may be capable of image formation (Fig. 13-18C,D).

Statocysts predominate among sedentary burrowers and tube dwellers. The lugworm *Arenicola* has one pair of cilia-lined statocysts embedded in the body wall of the head, each opening to the exterior via a lateral canal (Fig. 13-16, 13-17C). The statocysts of *Arenicola* contain spicules, diatom shells, and quartz grains, all coated with a chitinous material. *Arenicola* always burrows head downward, and if an aquarium containing

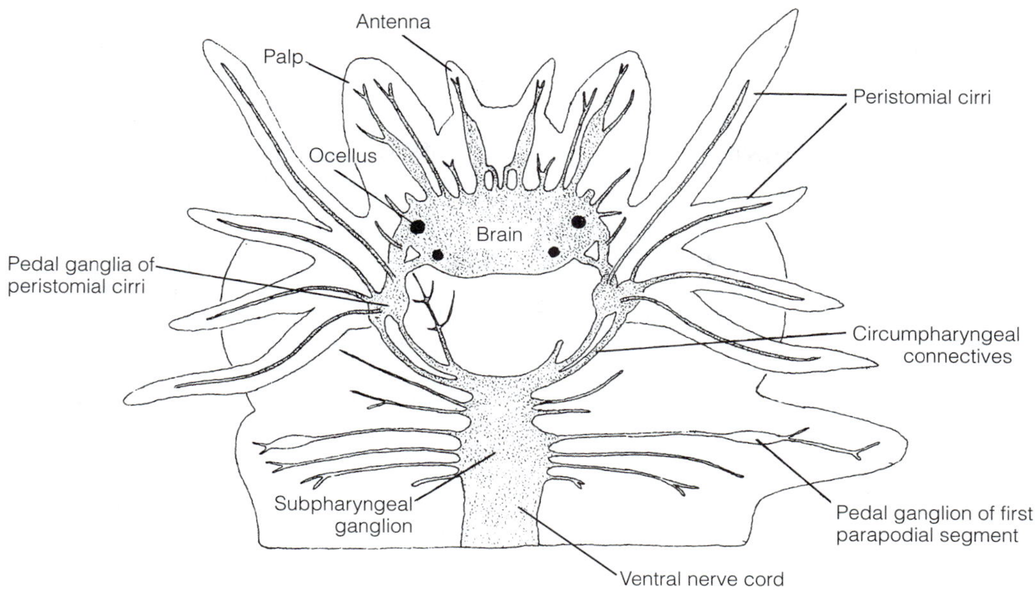

FIGURE 13-14 Brain and anterior nerve cord of *Nereis*. *(After Henry from Kaestner.)*

a worm is tilted 90°, the worm makes a compensating 90° turn in burrowing. If the statocysts are destroyed, this compensating ability is lost.

The sensory appendages of the prostomium (antennae, palps), peristomium (tentacular cirri), parapodia (dorsal and ventral cirri), and pygidium (pygidial cirri), all bear numerous sensory cells (Fig. 13-6A,C). Each appendage is thought to house both mechanoreceptors and chemoreceptors, but distinct sensory roles for the individual appendages have not been determined. Some mechanoreceptors may be tactile, others sensitive to water flow, and still others, whose dendrites are embedded in the cuticle, may be stretch receptors (proprioceptors). The many different types of chemoreceptive cells may be tuned to specific waterborne chemicals or classes of compounds. It has been suggested that some detect pheromones released prior to sexual reproduction.

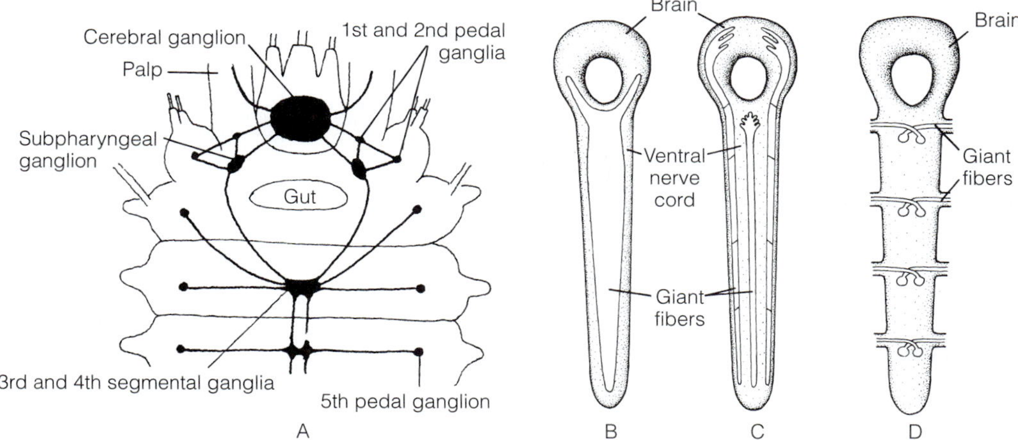

FIGURE 13-15 Polychaete brain, central nervous system, and giant axons. **A,** Anterior nervous system of the scaleworm *Lepidasthenia*. **B–D,** Central nervous systems of three polychaetes, showing arrangement of giant axons in the ventral nerve cord. **B,** *Eunice,* with a single giant axon; **C,** *Nereis,* with medial and lateral giant fibers; **D,** *Thalanessa,* with intrasegmental giant axons. *(A, After Storch from Fauvel; B and C, After Nicol; D, After Rhodes from Nicol.)*

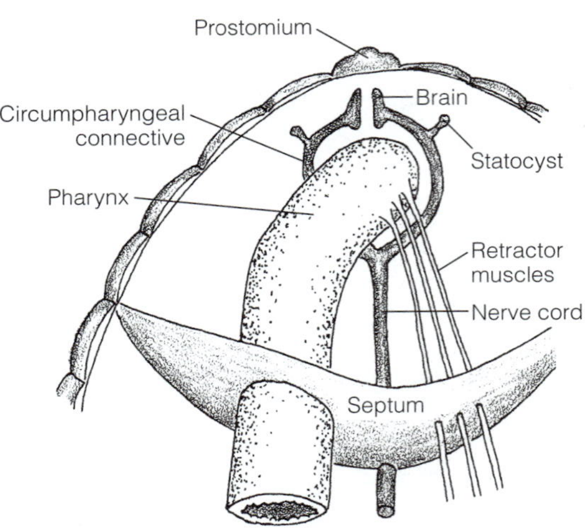

FIGURE 13-16 Brain, statocyst, and foregut in the lugworm *Arenicola* (dorsal dissection). *(After Ashworth from Brown, F. A. 1950. Selected Invertebrate Types. John Wiley and Sons, New York)*

DIGESTIVE SYSTEM

The polychaete gut typically is a straight tube extending from the mouth at the anterior end of the worm to the anus on the pygidium (Fig. 13-2, 13-19). The polychaete digestive system is like that described for annelids in general and is commonly differentiated into a pharynx (or buccal cavity if the pharynx is absent), short esophagus, stomach (in sedentary species), intestine, and rectum (Fig. 13-10A, 13-20). The pharynx can be a protrusible, tonguelike, muscular bulb or an eversible organ (Fig. 13-21). If it is a bulb, it is situated in the midventral wall of the foregut (in contrast to the dorsal position in oligochaetes). Teeth of various forms and functions, sometimes composing grasping jaws, occur in the pharynx of many polychaetes (Fig. 13-10C, 13-22). In *Nereis*, two large, glandular ceca open into the esophagus (Fig. 13-10A). The ceca, along with the anterior end of the intestine, secrete digestive enzymes.

Even the mundane matter of defecation is diverse in Polychaeta. In worms that crawl or burrow, feces are simply released and abandoned. Many tube-dwelling polychaetes, such as *Chaetopterus*, that pump water unidirectionally through their tubes defecate into the downstream exhaust flow (Fig. 13-47B). Such flushing, however, runs afoul when only one end of the tube or burrow is exposed to the surface water. In these cases, some polychaetes avoid fecal contamination by living upside down (anus-end up) in their vertical tubes, as do bamboo worms (Maldanidae; Fig. 13-36B), whereas many others invert themselves temporarily to defecate at the surface. Still other polychaetes that live upright in blind-ended tubes, such as feather-duster worms and mason worms (Sabellariidae), vent their feces without standing on their heads. Mason worms accomplish this feat by permanently bending their body into a U, the posterior half of the body being reduced to little more than a feces-conducting tube (Fig. 13-50A). The dorsal body surface of feather-duster worms has a ciliated groove that transports fecal pellets from the anus anteriorly out of the tube. Many species consolidate their feces into high-density fecal pellets or strings, which tend not to resuspend and reenter their burrows or tubes.

NUTRITION

The feeding methods of polychaetes are closely related to their lifestyles. Many of the burrowers and sedentary burrow- and tube-dwellers are deposit feeders that use the organic material in sediments as food. **Direct deposit feeders** ingest sediment directly with the mouth or a nonmuscular, bulbous, protrusible pharynx; these include the lugworms (Arenicolidae), bamboo worms (Maldanidae), and several families of earthwormlike polychaetes such as Orbiniidae and Capitellidae. **Indirect deposit feeders,** such as spaghetti worms (Terebellidae) or palp worms (Spionidae), also feed on the organic material in sedi-

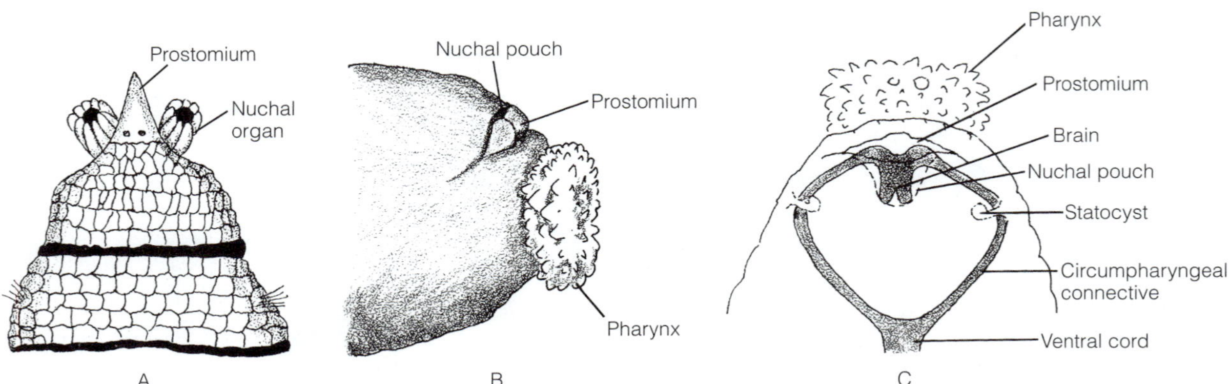

FIGURE 13-17 Polychaete sense organs. **A,** Anterior of *Notomastus latericeus* with everted nuchal organs. **B,** Head of *Arenicola* (lateral view). **C,** Statocysts and anterior nervous system of *Arenicola*. *(A, After Rullier from Fauvel; B and C, After Wells.)*

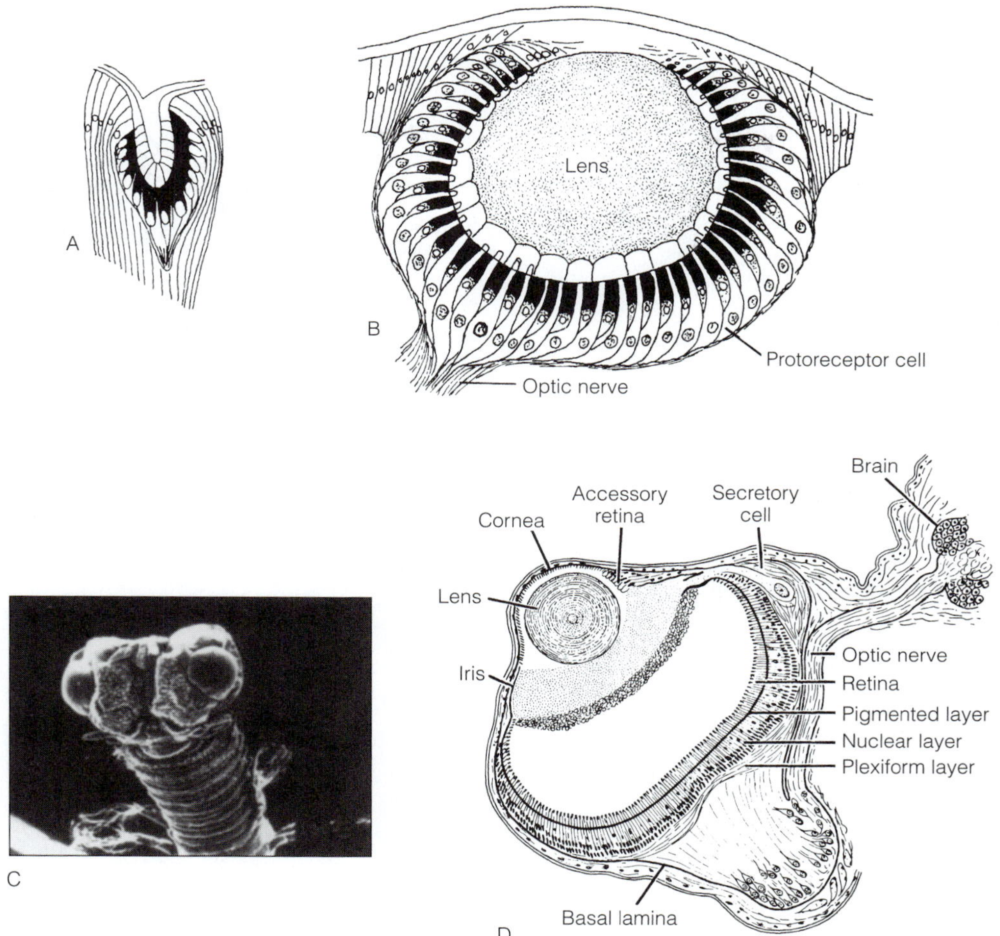

FIGURE 13-18 Polychaete sense organs: ocelli. **A,** Simple pigment-cup ocellus of *Mesochaetopterus.* **B,** Lensed ocellus of *Nereis.* **C,** The head and bulging complex eyes of the pelagic alciopid *Vanadis formosa.* **D,** Section through the complex eye of *Vanadis formosa.* *(A and B, After Hesse from Fauvel; C, From Rice, S. A. 1987. Reproductive biology, systematics, and evolution in the polychaete family Alciopidae. Biol. Soc. Wash. Bull. 7:114–127; D, After Hesse from Hermans, C. O., and Eakin, R. M. 1974. Fine structure of the eyes of an alciopid polychaete* Vanadis tagensis. *Z. Morphol. Tiere. 79:245–267.)*

ments, but collect the material first with a specialized appendage that then conveys the food to the mouth. In spaghetti worms, these appendages are extensible prostomial tentacles, whereas palp worms use a pair of long peristomial palps.

Carnivores, herbivores, and scavengers are typically motile (errant) polychaetes that crawl over surfaces and in crevices or catch plankton in the water (Alciopidae), but some are tube dwellers or active burrowers. Because they seize prey or carrion, these polychaetes typically have a well-developed muscular, eversible pharynx. The everted pharynx may be a muscular grasping tube, as in the paddleworms (Phyllodocidae), that sometimes has a thick, rasping lower lip, as in the fireworms (Amphinomidae), or it may bear jaws with grasping (Nereidae, Eunicidae, Lumbrineridae; Fig. 13-10C, 13-22) or venomous teeth (Glyceridae; Fig. 13-39C). The horny jaws are composed of a cross-linked (tanned) protein. When food is detected, the pharynx rapidly everts, exposing and opening the jaws. The food is seized in the jaws and the pharynx is retracted. Although protractor muscles may be present, increased coelomic pressure resulting from the contraction of body-wall muscles typically everts the pharynx. When these muscles relax and the coelomic fluid pressure drops, the pharynx is withdrawn by retractor muscles (Fig. 13-10B,C).

All suspension feeders are sedentary animals that occupy tubes in sediment or are attached to shells, rocks, and other hard surfaces. They have ciliated appendages with large surface areas for trapping suspended particles, such as the feathery crown of feather-duster worms (Fig. 13-52), or they secrete a mucous net through which water is pumped and

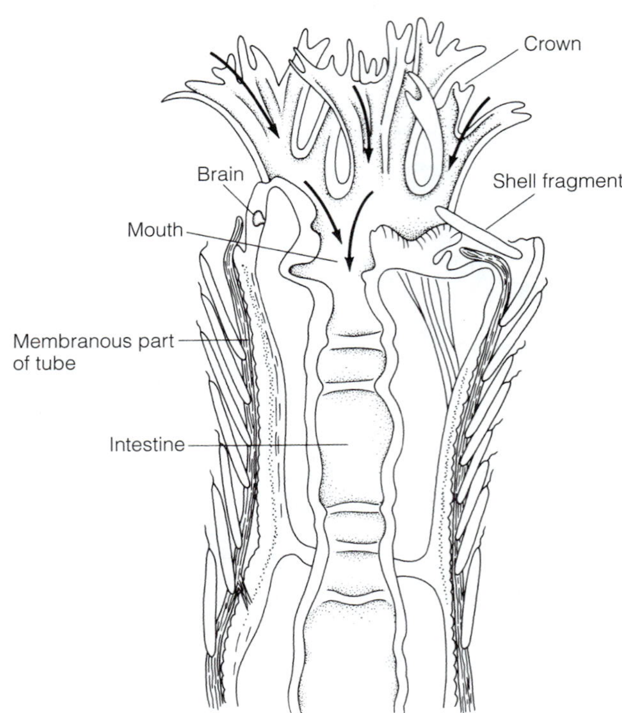

FIGURE 13-19 Polychaete gut. The simple gut of the filter-feeding (route shown by arrows), shingle-tube worm *Owenia fusiformis,* viewed in longitudinal section. *Owenia* is using its peristomial collar to add a shell frament to its tube. *(Modified from Watson.)*

filtered, as in the parchment-tube worm, *Chaetopterus* (Fig. 13-47A,B).

A few species of polychaetes are parasites. *Labrorostratus* (Oenonidae) lives in the coelom of other polychaetes and may be almost as big as its host. Some myzostomids are endoparasites of sea stars. Ectoparasites include species of the bloodsucking Ichthyotomidae, which attach to the fins of marine eels.

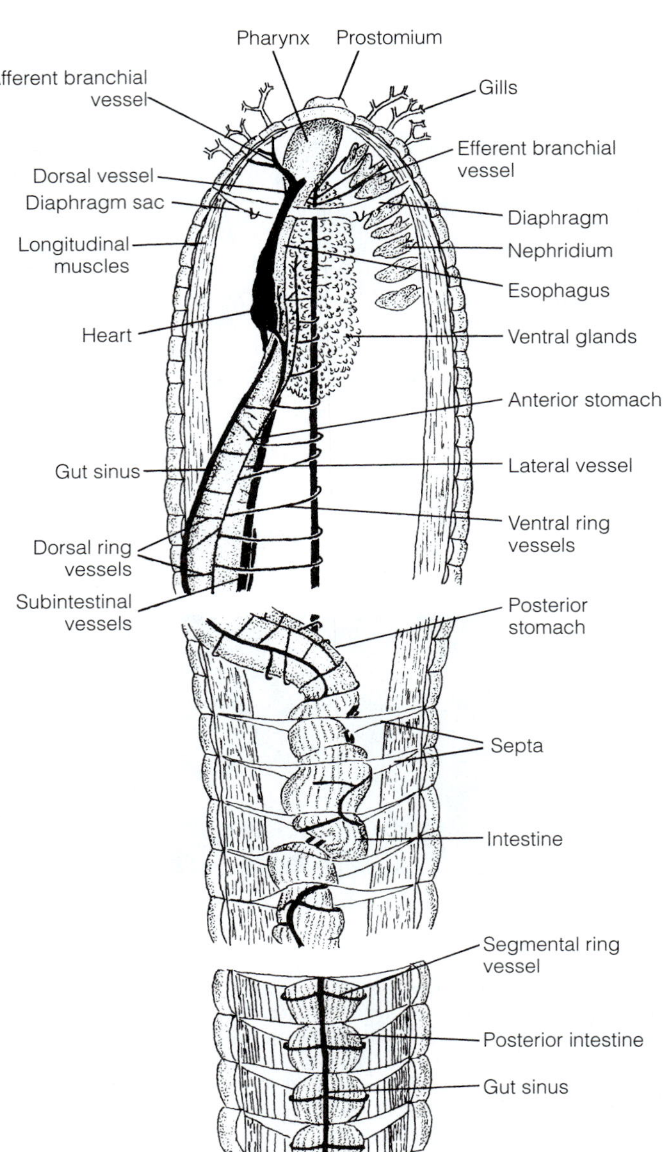

FIGURE 13-20 Polychaete gut and internal anatomy: dorsal dissection of *Amphitrite ornata. (After Brown, F. A. 1950. Selected Invertebrate Types. John Wiley and Sons, New York.)*

GAS EXCHANGE

Gas exchange in polychaetes, as in all annelids, occurs across the general body wall, but often additional, specialized gills are present, especially in the larger-bodied burrowers and tube dwellers. Specialized gills, though thin-walled and delicate, are always unprotected outgrowths of the body surface and never enclosed in gill chambers. Perhaps this is because they already are protected by the worms' tubes and burrows. Specialized gills are absent in tiny polychaetes and those with long, threadlike bodies.

Typically, gills are associated with the parapodia and in many cases are modified parts of the parapodium. The notopodium may be flattened to form a branchial (gill) lobe, as in nereids (Fig. 13-6C). Commonly, the dorsal cirrus of the parapodium is modified as a gill (Fig. 13-23, 13-24C,D), or the gills arise from the base of the dorsal cirrus. Cirratulids have long, contractile, threadlike gills, each one attached to the base of the notopodium (Fig. 13-24A).

Among the several families of polychaetes known as scaleworms, gas exchange is largely restricted to the dorsal body surface, which is roofed over by two rows of overlapping, fishlike scales, the **elytra,** each of which is borne on a short stalk (Fig. 13-41). Cilia on the dorsal body surface create a current of water flowing posteriorly beneath the elytra. The felt-covered sea mouse (*Aphrodite;* Fig. 13-41A,B) lacks cilia, but a similar dorsal water current is produced by rocking its elytra, which are modified dorsal cirri.

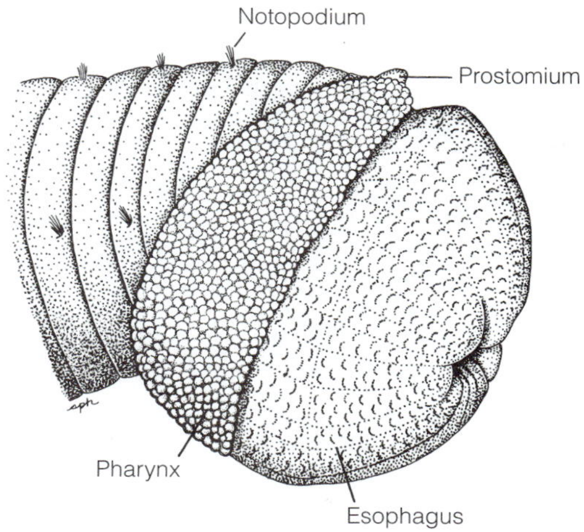

FIGURE 13-21 Polychaete pharynx. Lateral view of the anterior end of the earthwormlike *Notomastus* (Capitellidae), a burrowing direct deposit feeder. The large everted pharynx obscures the tiny prostomium. *(After Michel.)*

FIGURE 13-22 Polychaete pharynx and jaws. Ventral view of the head of the eunicid *Marphysa sanguinea,* showing its complex jaws. Within the expanded beaklike parts (white) is a pair of opposing mandibles (black).

The gills of many sedentary species are branched feather- or treelike structures restricted to the anterior end of the body and independent of parapodia. The spectacular red, arborescent gills of some spaghetti worms (Terebellidae), such as *Amphitrite,* are not only ciliated, but also contractile, and they pulsate rhythmically (Fig. 13-24B). The feather dusters of feather-duster worms function as gills as well as filters for feeding.

Gill ventilation results from ciliary beating, muscular contraction of the gill, or both (Fig. 13-23), but many burrowers and tube-dwellers augment the gill pump by creating a peristaltic water flow through their burrows or tubes. Worms that ventilate using peristalsis typically exhibit a spontaneous ventilating rhythm in which a period of ventilation alternates with a period of rest. The muscular exertion associated with peristaltic ventilation may increase the worm's oxygen consumption by as much as 15-fold, but there is approximately a 20-fold increase in oxygen uptake.

INTERNAL TRANSPORT

Circulation in most polychaetes results from fluid movement in both the hemal system and the coelom. A common variation on this pattern occurs in many polychaetes that have reduced septa, such as the glycerid bloodworms. In these, the coelomic system functionally replaces the hemal system and transports substances throughout the body. Very small species also typically lack a hemal system and sometimes also coelomic cavities (Fig. 13-42B,C).

The basic layout of the polychaete hemal system conforms to that of annelids in general but is complicated by the addition of parapodial and gill circulations (Fig. 13-25A). In each segment, the ventral vessel gives rise to one pair of **parapodial vessels** and several pairs of **intestinal vessels** (Fig. 13-25A). The patterns of blood flow vary from taxon to taxon, but typically the parapodial vessels transport blood to the parapodia, body wall, and nephridia before returning it to the dorsal vessel. The intestinal vessels deliver blood from the gut to the ventral vessel. A capillary plexus in the parapodia and gut wall facilitates exchange in those organs.

The gills usually are provided with afferent and efferent vascular loops permitting a two-way flow, as in the gills of lugworms (Fig. 13-35A) and the branchial, notopodial lobes of nereids (Fig. 13-25A). On the other hand, the "feathers" (radioles) of feather-duster worms each contain only a single vessel, within which blood flows tidally, in and out (Fig. 13-52B). In other polychaetes, such as glycerids, the parapodial gills are irrigated with coelomic fluid instead of blood.

Fewer cells populate the blood than the coelomic fluid. Blood is often colorless in small polychaetes, but in larger species and those that burrow in soft bottoms, large, extracellular molecules of hemoglobin occur in the plasma. Respiratory pigments, like almost everything else in polychaetes, are diverse (Fig. 9-16). Polychaetes have three of the four animal respiratory pigments. Hemoglobin is the most common, but chlorocruorin is characteristic of the blood of serpulid, spirorbid, and sabellid feather-duster worms (the blood of *Serpula* contains both hemoglobin and chlorocruorin) as well as species of Flabelligeridae and Ampharetidae. **Chlorocruorin** is a kind of hemoglobin, but a slight chemical difference makes it green instead of red. The peculiar blood of *Magelona* (Magelonidae) contains enucleated hemocytes that contain a third pigment known as hemerythrin, which has a pink or violet hue. Like hemoglobin, hemerthryin binds an atom of iron at its center, but unlike hemoglobin, it is a nonheme (porphyrin-lacking) protein found elsewhere only in Sipuncula, Priapulida, and inarticulate brachiopods.

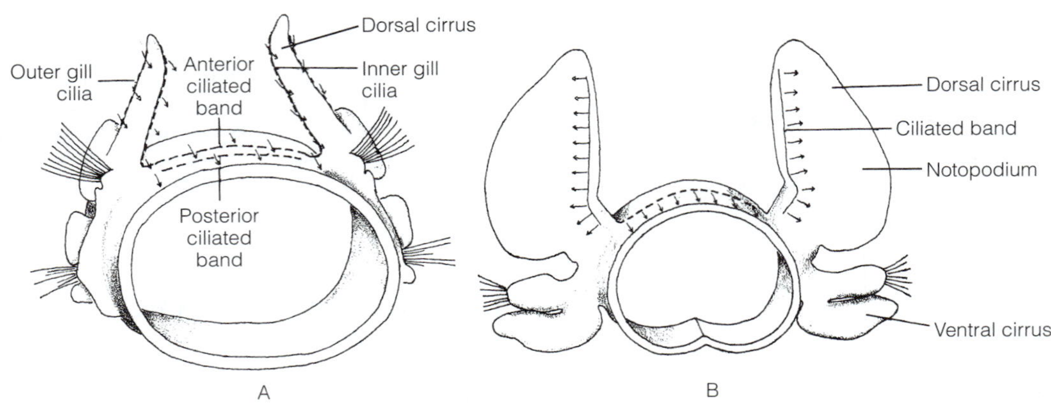

FIGURE 13-23 Polychaete gill surfaces. Arrows indicate direction of water currents. Flows on trunk surface are directed posteriorly (countercurrent to flow in dorsal blood vessel). **A,** *Scolelepis squamata.* **B,** *Phyllodoce laminosa.* In both species the dorsal body wall and dorsal cirri are important gill surfaces. In paddleworm polychaetes (**B;** Phyllodocidae), the notopodium is also flattened to function as a gill. *(After Segrove.)*

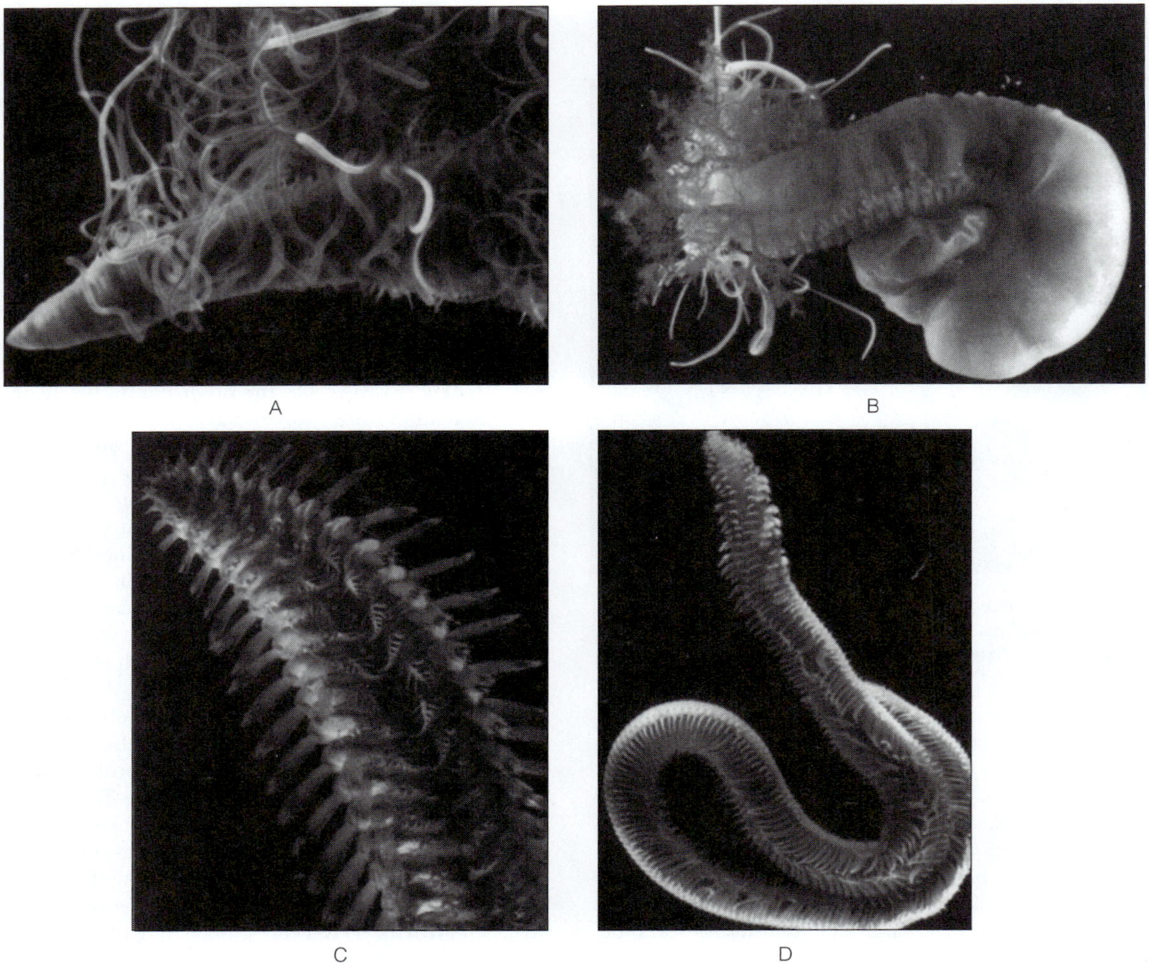

A

B

C

D

FIGURE 13-24 Polychaete gills. **A,** Threadlike gills of *Cirriformia cirriformia* (Cirratulidae). **B,** Arborescent (treelike) gills of the spaghetti worm, *Amphitrite ornata* (Terebellidae). **C,** Featherlike gills of the fireworm *Chloeia viridis* (Amphinomidae). **D,** The comblikelike gills (dorsal cirri) of the burrower *Scoloplos rubra* (Orbiniidae) arch over the animal's dorsum and form a semienclosed ventilatory canal.

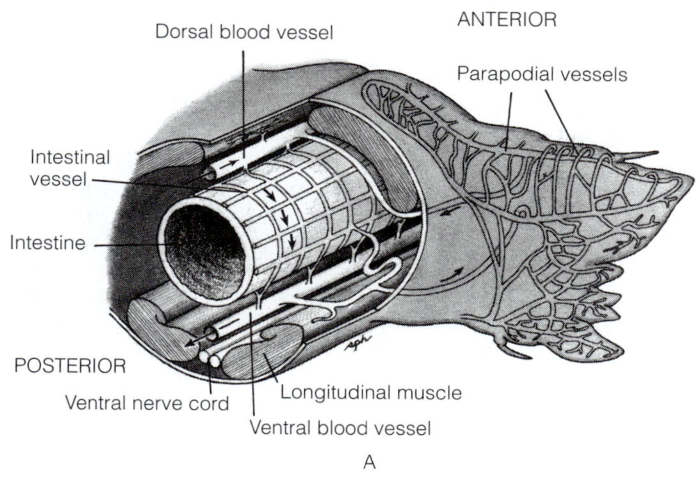

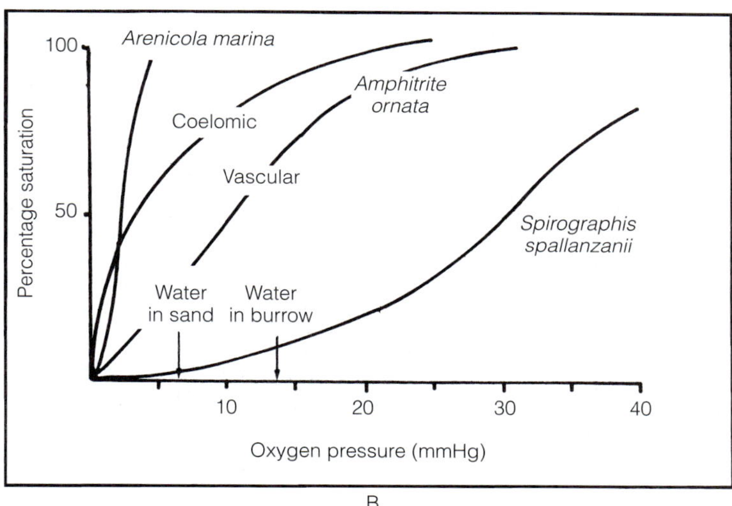

FIGURE 13-25 Polychaete hemal system. **A,** *Nereis virens:* vascular anatomy of one segment. Arrows indicate the direction of blood flow. **B,** Oxygen dissociation curves of the respiratory pigments of three species of polychaetes. *Amphitrite ornata* has two hemoglobins (Hb), one hemal (vascular) and one coelomic. Note that coelomic Hb of *Amphitrite* has a higher O_2 affinity than does the hemal pigment (see text for discussion). The respiratory pigment of the feather-duster worm *Spirographis spallanzanii* is chlorocruorin. The very high affinity of the *Arenicola* Hb means that the O_2 bound to the pigment is not released until the environmental (and tissue) O_2 tension is very low. For the intertidal *Arenicola,* this occurs at low tide, when the worm cannot irrigate its burrow with water. Under such conditions, O_2 is released from the Hb to meet the respiratory demand of the animal until the rising tide returns oxygenated water to the burrow. *(Modified from Dales.)*

The coelomic fluid is circulated by muscular contractions of the body wall or by both muscular contractions and cilia on the coelomic lining. The coelomic fluid may be colorless or contain low-molecular-weight hemoglobin in coelomocytes (erythrocytes). Sometimes the pigment occurs only in the coelom (Capitellidae) and not in the hemal system, but in many others (such as terebellids and opheliids), two different hemoglobins are present, one in the blood and the other in the coelom. In addition, the bloodworm *Glycera,* most scaleworms, and others have hemoglobin (neuroglobin) in or around their nerve cords, which are bright red. Myoglobin occurs in the muscles of many scaleworms.

When hemoglobin is present in both hemal system and coelom as well as in muscle and nerve, the oxygen-binding properties of the pigments differ in each compartment. In general, the hemal-vessel (gill) hemoglobin binds oxygen least tightly, the muscle and nerve hemoglobin binds it most tightly, and the coelomic fluid hemoglobin binds intermediately. This results in a cascade of oxygen transport from gill vessels to coelom to oxygen-demanding cells of the body (Fig. 13-25B). Hemal-system hemoglobin binds oxygen at the gill, only to have it stripped away by the higher-affinity coelomic hemoglobin, which in turn has its oxygen removed by the still-higher-affinity nerve and muscle hemoglobins. The mitochondrial cytochromes, which have the highest affinity of all, are the final destination for the respiratory oxygen.

For many intertidal polychaetes, oxygen availability drops as the tide recedes and water stagnates in their burrows, tubes, or crevices. In such situations, polychaetes may suppress their metabolic rate and become torpid, rely on "stored" oxygen bound to hemoglobin, or both. The intertidal burrower *Euzonus mucronatus,* for example, uses oxyhemoglobin storage to carry it through regular 2- to 4-h periods of sediment anoxia during low tide. If subjected to longer periods of anoxia, the worm switches over to anaerobic respiration, on which it can survive for as long as 20 days.

EXCRETION

Polychaete excretory organs are filtration nephridia, as described for Annelida (Fig. 13-6B,C, 13-10A). Chlorogogen tissue, coelomocytes, and the intestinal wall may play accessory roles in excretion.

Depending on body design, polychaetes have either protonephridia or metanephridial systems. Protonephridia occur in a few adult polychaetes that lack or have a reduced hemal system. Most polychaetes, however, have both hemal and metanephridial systems. The correlation of blood vessels with metanephridia, and their absence with protonephridia, is an indication of how ultrafiltration, the first step in urine formation, occurs in the two functional groups, as described in Chapter 9 (Fig. 9-17).

Polychaetes with metanephridial systems have vascular filtration sites covered with podocytes and septum-associated metanephridia (Fig. 13-26). The preseptal end of the metanephridium in *Nereis,* for example, bears a ciliated, funnel-like nephrostome (Fig. 13-10A). The long postseptal tubule, which extends into the next segment, forms a coiled mass enclosed in a thin mesothelial sheath. Tubule elongation and coiling are probably adaptations that increase the surface area for tubular secretion or reabsorption. The nephridiopore opens at the base of the neuropodium on the ventral side. The entire lining of the tubule is ciliated.

The metanephridia of most other polychaetes differ only slightly from those of *Nereis* (Fig. 13-26B–D), but may be regionally restricted to **excretory segments.** An extreme example of regional restriction occurs in the tube-dwelling feather-duster worms, which have only one pair of anterior nephridia. The ducts of the two nephridia unite in the ventral midline and open through a single nephridiopore on the head (Fig. 13-26C). With its head and nephridiopore extended above the tube mouth, the worm releases urine directly to the external environment and avoids fouling its tube.

Among polychaetes with protonephridia, a protonephridium typically bears a cluster of terminal cells. Each terminal cell, called a **solenocyte,** closely resembles a sponge choanocyte, consisting of a single flagellum in a collar of microvilli (Fig. 13-27).

Many polychaetes—nereids and others—can tolerate low salinities and are adapted to life in brackish sounds and estuaries. Under these conditions, where the relative concentration of salt is low and water high, one might expect the worm to eliminate excess water and take up salt. Such worms typically have a glomerulus-like network of blood vessels around their metanephridia. Presumably, the high density of blood vessels increases the rate of ultrafiltration and urine production, thus "squeezing" excess water out of the body. The gills (notopodial lobes) of *Nereis succinea,* like the gills of freshwater fishes, contain cells specialized for actively absorbing ions. A very few polychaete species, such as the feather-duster *Manyunkia speciosa* (Sabellidae), have managed to colonize fresh water. *Manyunkia* is widely distributed and occurs in enormous numbers in certain regions of the Great Lakes, such as around the mouth of the Detroit River, and in Lake Waccamaw, North Carolina. Even more remarkable are the terrestrial polychaetes of the Indo-Pacific tropics that burrow in soil or live in moist litter.

REPRODUCTION

Regeneration and Clonal Reproduction

Polychaetes readily regenerate missing or damaged parts. Tentacles, palps, "tails," and even heads removed by grazers or predators are soon replaced. Such replacement is a common occurrence in burrowers and tube dwellers. In general, the potential for regeneration is somewhat greater in worms with uniformly differentiated trunks than in those with specialized thoracic and abdominal regions, but there are many exceptions. *Chaetopterus* and *Dodecaceria,* for example, can regenerate the entire body from a single segment. Experimental studies indicate that the nervous system plays an important inductive role in regeneration and that the neuroendocrine system is involved in some way. If the nerve cord alone is severed, a new head forms at the site of the cut. If the nerve cord is cut just behind the subpharyngeal ganglion and then pulled through a hole in the lateral body wall, a second head differentiates at the site.

Clonal reproduction occurs in some polychaetes, including cirratulids *(Dodecaceria),* syllids, feather-duster worms (Sabellidae), and palp worms (Spionidae). Budding and transverse division or fragmentation are the two common modes of asexual reproduction (Fig. 13-29B,C).

Sexual Reproduction

Most polychaetes probably reproduce only sexually, and the majority of species are gonochoric. Polychaete gonads usually are distinct, segmentally paired organs found in the connective tissue associated with such structures as septa, blood vessels, and the lining of the coelom (Fig. 13-28).

In many polychaetes, most segments bear gonads, but in some species they are restricted to genital segments. Gonads are usually limited to the abdomen in polychaetes (Capitellidae) with distinct thoracic and abdominal regions. Among the few hermaphroditic polychaetes, some feather-duster worms have anterior abdominal segments that produce eggs and posterior ones that produce sperm.

Germ cells often are shed into the coelom as gametogonia or primary gametocytes, and growth and differentiation of the gametes is completed in the coelomic fluid (Fig. 13-28A). When the worm is mature, the coelom is packed with ripe eggs or sperm, which often can be seen through the translucent body wall. For example, the abdomen of a ripe male *Pomatoceros* appears white, and that of a female bright pink or orange because of the color of the sperm and eggs, respectively. The blue eggs of the red-white-and-blue worm, *Proceraea fasciata* (Syllidae), show through the red-banded, whitish body.

In general, gametes are either spawned through the metanephridia (Fig. 13-26B) or by rupturing of the body wall. Body-wall rupture is common among species that transform into specialized pelagic swarmers (epitokes; see below), such as members of Nereidae, Syllidae, and Eunicidae. Rupture is also typical of species such as shimmy worms (Nephtyidae) and bloodworms (Glyceridae) that have protonephridia as adults. Because the inner ends of the protonephridia are capped by terminal cells, which do not allow eggs or sperm to pass, the simplest (but fatal) option is to release the gametes like seeds bursting from a milkweed pod.

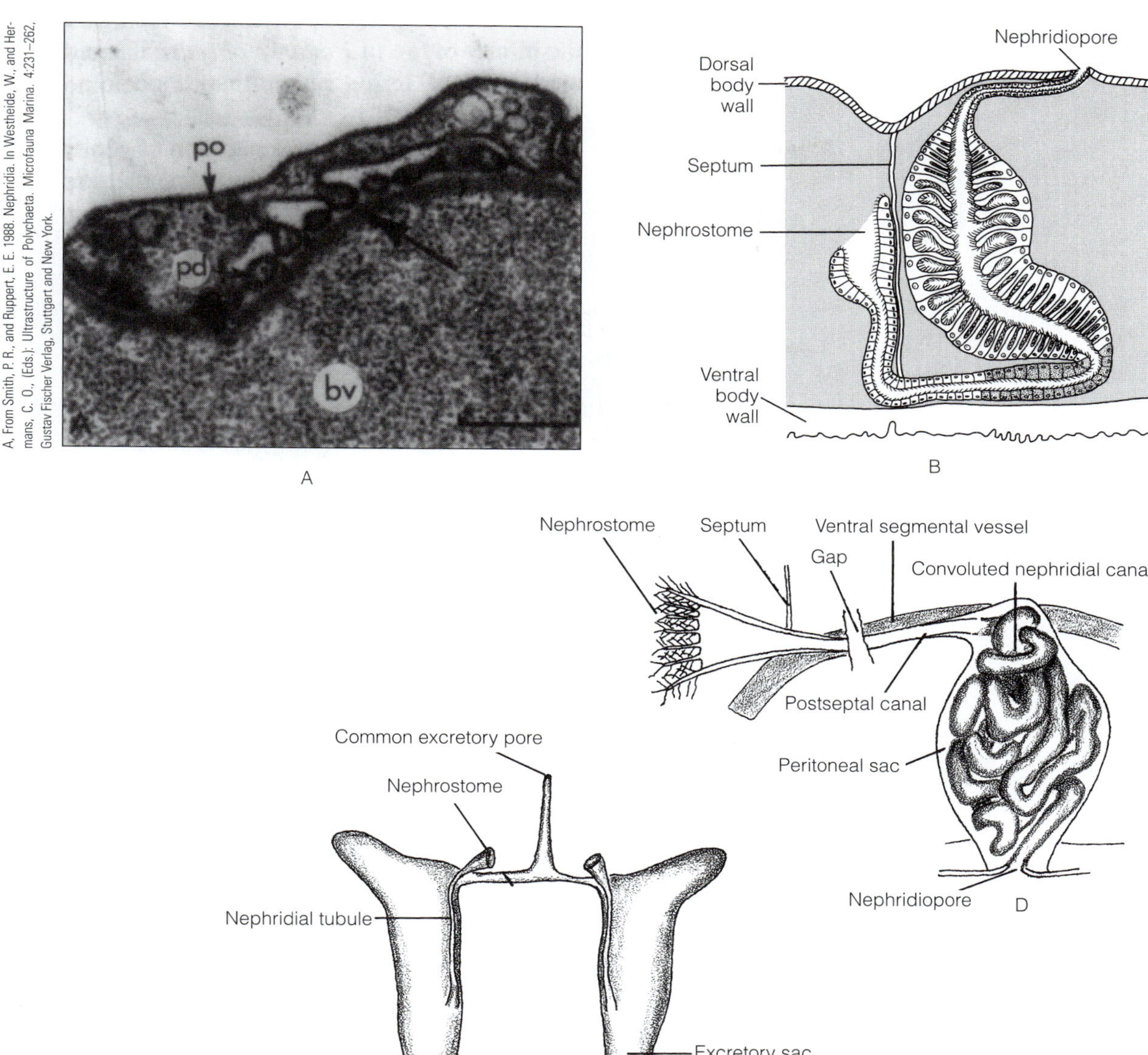

A, From Smith, P. R., and Ruppert, E. E. 1988. Nephridia. In Westheide, W., and Hermans, C. O., (Eds.): Ultrastructure of Polychaeta. Microfauna Marina. 4:231–262, Gustav Fischer Verlag, Stuttgart and New York.

FIGURE 13-26 Polychaete metanephridial systems. **A,** Podocytes (po) on the wall of a segmental blood vessel (bv) of the palp worm *Spio setosa* (Spionidae). **B–D,** Metanephridia of three polychaetes: **B,** The palp worm *Polydora* (Spionidae; this nephridium also fabricates spermatophores); **C,** The feather-duster worm, *Pomatoceros* (Serpulidae); **D,** The ragworm, *Nereis vexillosa* (Nereidae). *(B, From Rice, S. A. 1980. Ultrastructure of the male nephridium and its role in spermatophore formation in spionid polychaetes (Annelida). Zoomorphologie 95:181–194; C, After Thomas; D, After Jones.)*

Epitoky

Epitoky is a reproductive phenomenon characteristic of many polychaetes that is especially well known in nereids, syllids, and eunicids. **Epitoky** is the formation of a *pelagic,* reproductive individual, or **epitoke,** from a *benthic,* nonreproductive individual, the **atoke.** Epitoky coincides with gamete production and sexual maturity, but it also involves changes in nonreproductive structures that adapt the worm for swimming and allow it to detect its mate. Such changes include enlargement of the eyes, modification of the parapodia and chaetae for swimming, enlargement of segments, and enhancement of the segmental musculature.

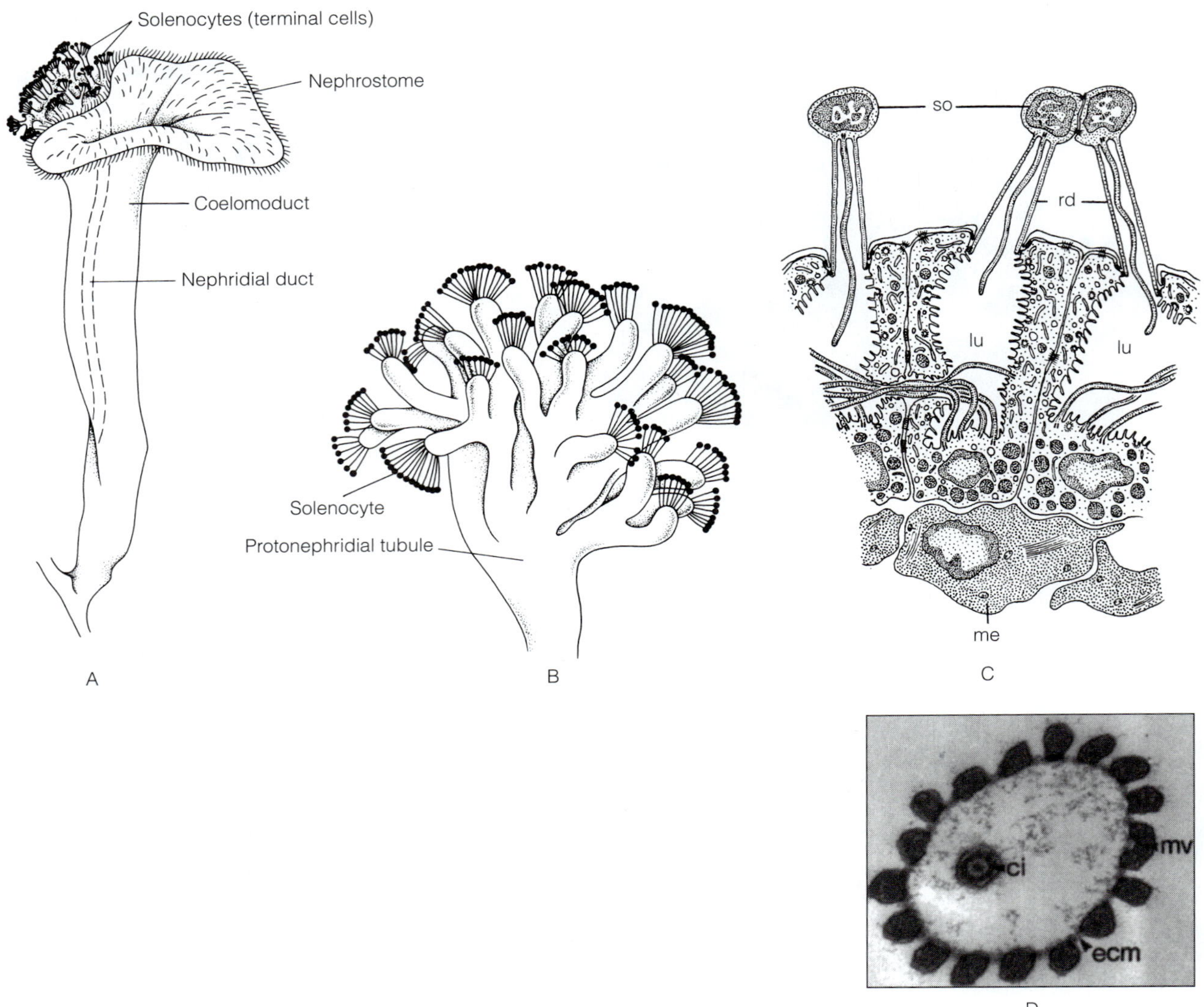

FIGURE 13-27 Polychaete protonephridia. **A,** Protonephridium and coelomoduct (gonoduct) of the paddleworm *Phyllodoce paretti* (Phyllodocidae). **B,** Branched end of protonephridium of *P. paretti.* **C,** Ultrastructure of three terminal cells and the collecting protonephridial tubules (lu) of the bloodworm *Glycera dibranchiata* (Glyceridae). The terminal cells (so = solenocytes) project from the surface of the nephridium into the coelomic fluid and provide a filtration cylinder (cross section in **D**) composed of microvilli (rd, mv). Coelomic fluid is ultrafiltered across the cylinder wall, and reabsorption occurs within the tubule. Reabsorbed metabolites are digested intracellularly and, eventually, stored as glycogen in nearby cells (me). *(A and B, From Goodrich; C, From Smith, P. R. 1992. Polychaeta: Excretory System. In Harrison, F. W., and Gardiner, S. L. (Eds.): Microscopic Anatomy of Invertebrates. Vol. 7: Annelida. pp. 71–108; D, From an unpublished poster by P. R. Smith.)*

One or more epitokes can arise from the atoke either by metamorphosis of the entire individual, as in nereids, or by differentiation and separation (budding) of the epitoke from the posterior half of the atoke's body, as in eunicids and syllids (Fig. 13-29A–C).

Often the gamete-bearing segments of the epitoke are the most strikingly modified, making the body of the worm appear to be divided into two markedly different regions. For example, the epitokes of *Nereis irrorata* and *N. succinea* have large eyes and reduced prostomial palps and tentacles (Fig. 13-30). The anterior 15 to 20 trunk segments are not greatly modified, but the remaining segments, forming the epitokal region and packed with gametes, are much enlarged; their parapodia contain fans of long, spatulate, swimming chaetae. In *Palola (Eunice) viridis,* the Samoan palolo worm, the anterior end of the worm is unmodified and the epitokal region consists of a chain of egg-filled segments (Fig. 13-29A).

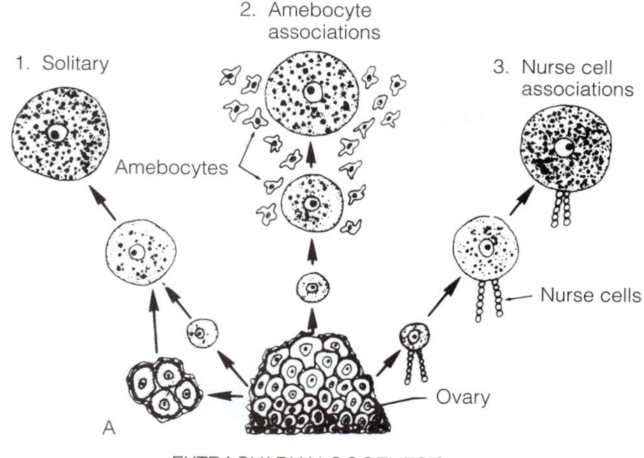

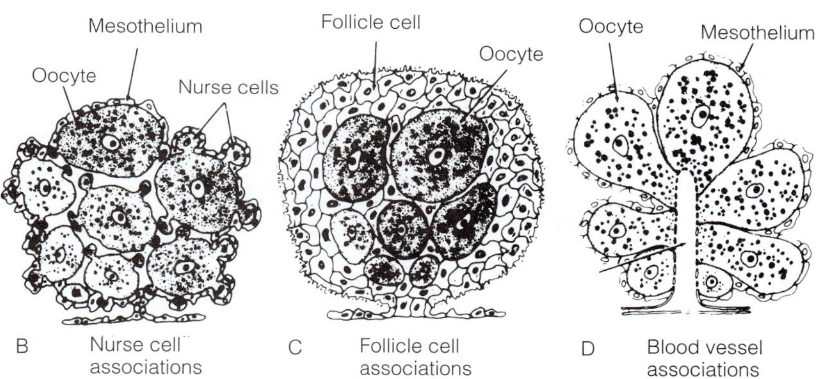

FIGURE 13-28 Polychaete ovaries and testes are compact organs situated beneath the mesothelium. In ovaries, shown here, oogenesis may occur after eggs are released from the ovary into the coelom **(A)**, or it may be completed within the ovary **(B–D)**. *(From Eckelbarger, K. J. 1983. Evolutionary radiation in polychaete ovaries and vitellogenic mechanisms: Their possible role in life history patterns. Can. J. Zool. 61:487–504.)*

Swarming

Epitoky is a polychaete adaptation to synchronize sexual maturation and to ensure the proximity of mates during gamete release. Usually, many epitokes swim to the surface simultaneously and shed eggs and sperm. This synchronous behavior is known as **swarming.** Experimental evidence indicates that the female produces a pheromone that attracts the male and stimulates shedding of sperm. The sperm in turn stimulate the shedding of the eggs. The male syllid *Autolytus,* for example, swims in circles around the female, touching her with his antennae and releasing sperm (Fig. 13-29D).

Light cues are often important in synchronizing swarming. *Autolytus edwardsi,* for example, is induced by changes in the light intensity at dawn and dusk to leave the bottom and swim to the surface. Swarming may also be affected by lunar periods. The Bermuda fireworm, *Odontosyllis enopla* (Fig. 13-37A), of the West Indies and Bermuda, swarms in the summer beginning 3 days after the full moon (for up to 12 days each month) and 56 min after sunset. The females swim in circles at the surface while emitting a steady bioluminescent glow that attracts the males below. The males, called "blinkers," swim rapidly upward toward the female, flashing intermittently until one reaches her. Then both male and female spew gametes in a stunning luminescent finale. The spent males and females return to the bottom and convert back into atokes. Striking examples of swarming lunar periodicity are also displayed by the so-called palolo worms of the South Pacific. The name *palolo* originally referred to the Samoan *Palola viridis* (Eunicidae; Fig. 13-29A), but it is now applied to other species. The Samoan palolo worm occupies rock and coral crevices below the low-tide mark and releases epitokes in October or November at the beginning of the last lunar quarter. Female epitokes can exceed 1 m in length, and local enthusiasts, who consider the epitokes a great delicacy (annelid caviar), eagerly await the predicted night of swarming and scoop great numbers of the gravid worms from the ocean surface.

The West Indian palolo, *Eunice schemacephala,* lives in habitats similar to those of the Samoan palolo. The worm is negatively phototactic and emerges from the burrow to feed only at night. Swarming takes place in July near the last quarter of a lunar cycle. At three or four o'clock in the morning during

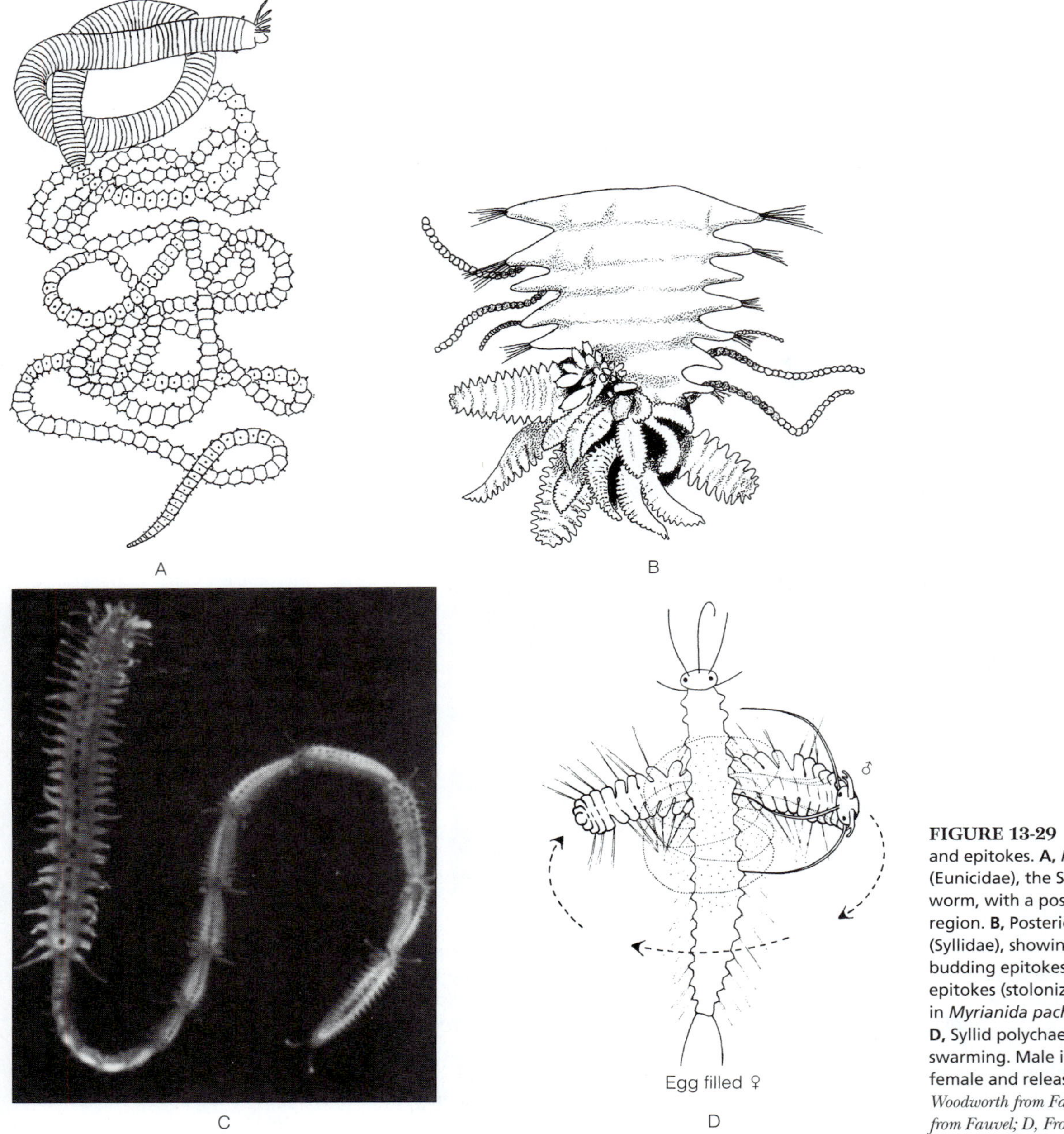

FIGURE 13-29 Polychaete epitoky and epitokes. **A,** *Palola viridis* (Eunicidae), the Samoan palolo worm, with a posterior epitokal region. **B,** Posterior of *Trypanosyllis* (Syllidae), showing a cluster of budding epitokes. **C,** Budding of epitokes (stolonization, paratomy) in *Myrianida pachycera* (Syllidae). **D,** Syllid polychaetes during swarming. Male is swimming around female and releasing sperm. *(A, After Woodworth from Fauvel; B, After Potts from Fauvel; D, From Girdholm.)*

that period, the worm backs out of its burrow, and the caudal, sexual epitokal region breaks free. The epitoke heads for the surface and swims slowly, spiraling. At first light, the ocean is blanketed with ripe sexual bodies, and as the sun rises above the horizon, the epitokes burst and saturate hectares of sea surface with gametes. Fertilization follows quickly and a new generation dawns on the palolos. One day later, the embryos are ciliated larvae, which two days later sink to the bottom and metamorphose into juvenile worms.

Endocrine Control of Reproduction

Polychaete reproductive events are regulated by hormones. The hormones are neurosecretions produced by the brain or, in the case of syllids, by nervous elements of the sucking foregut. In worms such as nereids and syllids that reproduce only once and then die, the hormone regulates the entire reproductive state, both the production of gametes and the development of epitokal features. In worms that breed more

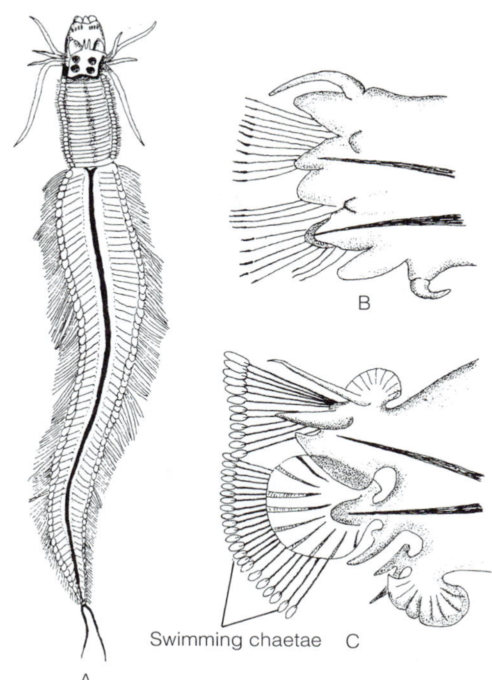

FIGURE 13-30 **A,** Male epitoke of *Nereis irrorata* (Nereidae). The enlarged, posterior reproductive segments bear the sperm. Note also the greater length of the swimming chaetae on the posterior segments. **B** and **C,** Parapodia of atoke **(B)** and epitoke **(C)** of the *N. irrorata* male. *(A, After Rullier from Fauvel; B and C, After Fauvel.)*

than once, a hormone is required for gamete development, especially for eggs, and the hormone effect is largely limited to that process.

The precise mechanisms that control swarming and the relationship between swarming and normal control of reproduction are still poorly understood. The relation of lunar phases to swarming periods differs among species, and swarming occurs even on cloudy nights. This makes any hypothesis based merely on light intensity difficult to support.

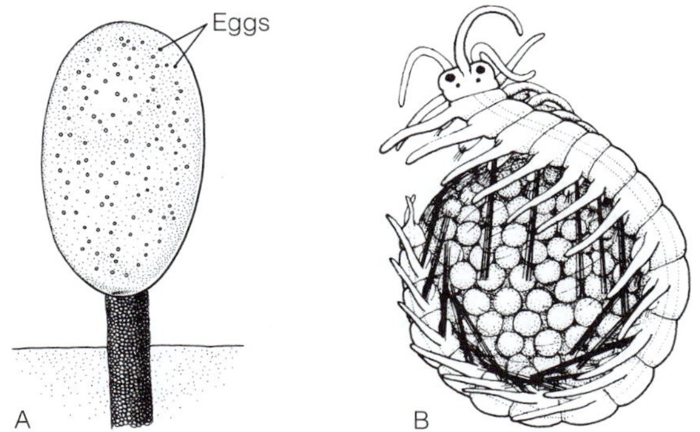

FIGURE 13-31 Polychaete egg capsules and brooding. **A,** Egg jelly mass attached to the end of the tube of the bamboo worm, *Axiothella mucosa* (Maldanidae). **B,** The syllid *Autolytus* carrying the egg mass beneath its body. *(A, From Ruppert and Fox, 1988; B, From Thorson.)*

Oviposition

Many polychaetes shed their eggs freely into the seawater, where after fertilization they become planktonic. Some polychaetes, however, retain the eggs within their tubes or burrows or lay them in jelly masses that are attached to tubes or other objects. For example, *Axiothella* (Fig. 13-31A), a bamboo worm, produces an ovoid egg mass resembling a large transparent grape that is attached to the chimney of its tube.

Many polychaetes brood their eggs. A few tube-dwellers, such as some spionids and serpulids, brood eggs in their tubes. Some species of *Spirorbis* brood in the cavity of the operculum, and *Autolytus* broods its eggs in a secreted sac attached to the ventral surface of the body (Fig. 13-31B). A few species, such as *Nereis limnicola,* gestate eggs in the coelom.

Development and Metamorphosis

The polychaete egg contains a variable amount of yolk, depending on the species, but cleavage is spiral and holoblastic. A displaced blastocoel is usually present, but a stereoblastula develops in *Nereis, Capitella,* and others. Gastrulation takes place by invagination, epiboly, or both.

After gastrulation, the embryo rapidly develops into a top-shaped trochophore larva (Fig. 13-32A,B, 13-33A). The greatest development of larval structures is attained in planktotrophic trochophores (including *Owenia, Polygordius,* Phyllodocidae, and Serpulidae; Fig. 13-33E). The trochophores of many species, however, are lecithotrophic, that is, yolky and nonfeeding (such as Nereidae and Eunicidae), and their short larval existence is spent near the bottom.

Polychaete metamorphosis transforms the trochophore into the juvenile body (Fig. 13-32). The most conspicuous feature of metamorphosis is the gradual lengthening of the body from immediately in front of the growth zone—the region anterior to the telotroch—as trunk segments form and develop (Fig. 13-32). The pretrochal region and apical plate form the prostomium and the brain, while the post-telotrochal region becomes the pygidium.

Metamorphosis may result in the immediate termination of a planktonic existence, but more often the elongate, metamorphosing larvae remain planktonic for varying lengths of time. The metamorphosing stages of spionids, sabellariids, and oweniids even have greatly enlarged, erectile, protective anterior chaetae.

In many polychaetes the trochophore stage is passed in the egg prior to hatching, which occurs at various times during advanced development. In such species metamorphosis is less drastic because larval structures never fully develop. Even when the trochophore stage is bypassed, however, a post-trochophoral stage may hatch and swim for a while in the plankton. For example, in *Autolytus* an elongate larva breaks free from the brood sac of the mother. On the other hand, *Axiothella mucosa* and *Scoloplos armiger* have direct development and the juveniles emerging from the jelly capsule immediately assume the adult mode of existence.

The developmental patterns of the small number of studied polychaetes fall into three categories. Annual species, those that live only one or two years and spawn only once, produce a large number of relatively small eggs. They have well-developed, feeding larvae that are planktonic for a week

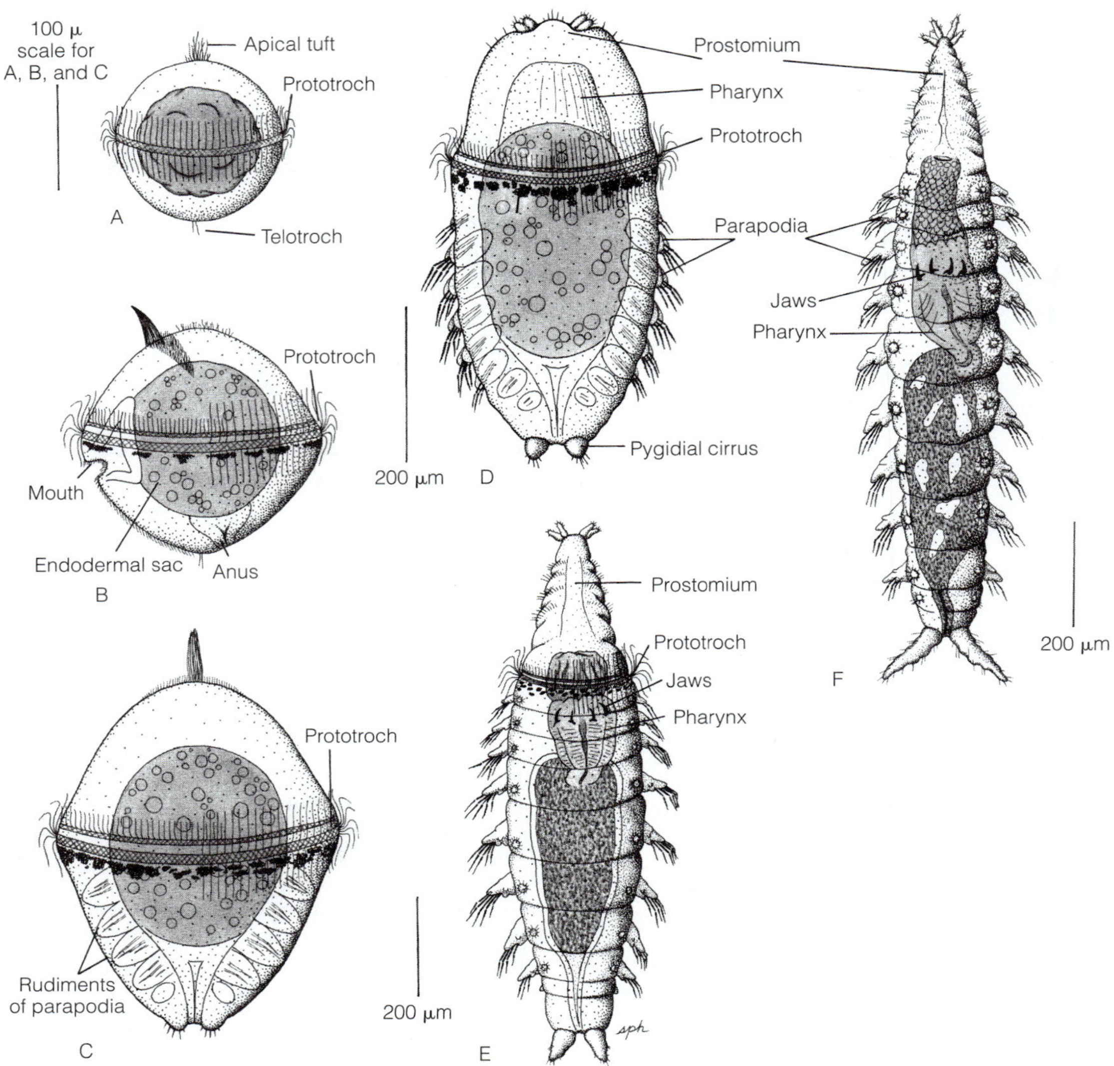

FIGURE 13-32 Polychaete trochophore larvae and metamorphosis. Larval and juvenile stages of the bloodworm *Glycera convoluta* (Glyceridae). **A,** Early trochophore (15 h); **B,** Later trochophore (10 days); **C,** Young metatrochophore (4 weeks); **D,** Metatrochophore at 7 weeks. Although still a swimming stage, it frequently comes to rest on the bottom. Metamorphosis follows this stage. **E,** Juvenile at 8 weeks. **F,** Young worm at 2 months. Compare with Figure 13-39A. *(Redrawn after Cazaux, C. 1967. Development of* Glycera convoluta. *Vie et Milieu 18:559–571)*

or more. Perennial species, which live and breed for more than one year, produce a small number of large, yolky eggs and nonfeeding, benthic larvae. Multiannual species, those with such short life spans that several generations can be produced in one year, produce numerous small batches of large, yolky eggs that become nonfeeding, benthic larvae.

Population Biology

Burrowing and tube-dwelling polychaetes commonly occur in enormous numbers on the ocean floor and compose a major part of the soft-bottom infauna. In Tampa Bay, Florida, for example, the average density of polychaetes is 13,425 individuals per square meter. They belong to some 37 species. On the upper continental slope and the deep ocean floor, polychaetes compose 40 to 80% of the infauna.

Population estimates have been obtained for the little burrowing opheliid *Euzonus (Thoracophelia) mucronatus,* which occurs in dense aggregations on protected intertidal beaches along the Pacific coast of the United States. In such aggregations the number of worms averages 2500 to 3000 per 30 square centimeters. Worms occupying a typical strip of beach 1.5 km long, 3 m wide, and 30 cm thick ingest approximately 13,270 metric tons of sand each year.

In general, such populations do not appear to be limited by food resources, at least not in shallow water. Predation and other pressures usually prevent annelid, mollusc, and other

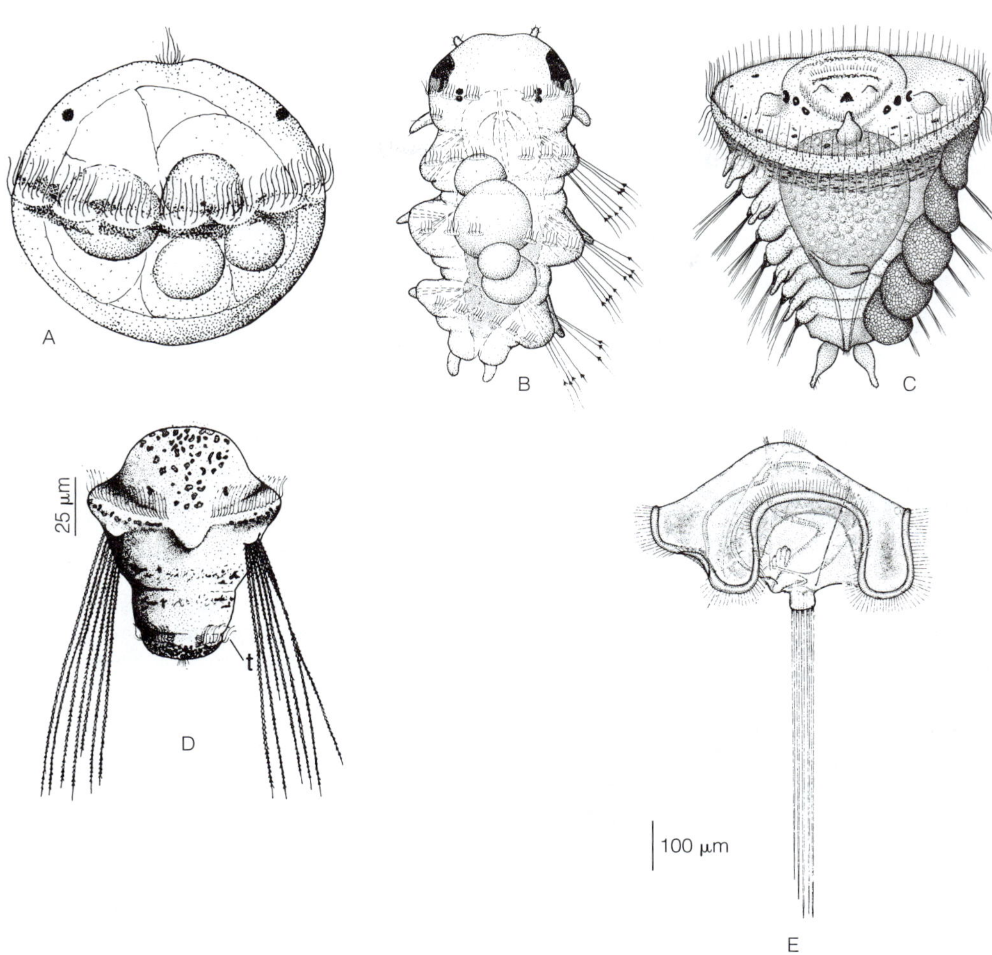

FIGURE 13-33 Polychaete larvae. **A,** Trochophore of the ragworm *Platynereis bicanaliculata* (Nereidae). **B,** Later, three-chaetiger stage (nectochaeta) of *Platynereis.* **C,** Late metatrochophore (nectochaeta) of the scaleworm *Halosydna brevisetosa* (Polynoidae). **D,** Larva of the mason worm, *Phragmatopoma* (Sabellariidae; t = telotroch). **E,** Trochophore, called a mitraria, of the shingle-tube worm *Owenia* (Oweniidae). *(A–C, From Blake, J. D. 1975. The larval development of Polychaeta from the northern California coast. III. Eighteen species of Errantia. Ophelia 14:23–84. D, From Eckelbarger, K. J. 1976. Larval development and population aspects of the reef-building polychaete* Phragmatopoma lapidosa *from the east coast of Florida. Bull. Mar. Sci. 26:117–132.; E, After Wilson from Smith, P. R., Ruppert, E. E., and Gardiner, S. L. 1987. A deuterstome-like nephridium in the mitraria larva of* Owenia fusiformis *(Polychaeta, Annelida). Biol. Bull. 172:315–323.)*

infaunal populations from ever reaching the carrying capacity of the habitat. When areas in the York River estuary of the Chesapeake Bay were protected from fish and crabs by means of wire cages, over half of the species in the polychaete population increased from two to many times their numbers in unprotected conditions.

DIVERSITY OF POLYCHAETA

Scolecida

Prostomial appendages are absent, but two or more pairs of cirri are on the pygidium. Most are burrowers or tube dwellers with a bulbous protrusible pharynx. Sediment is directly ingested with nonmuscular pharynx and mouth (direct deposit feeders; Fig. 13-34, 13-35, 13-36). "Sedentary" polychaetes.

Arenicolidae[F]: Lugworms are common direct deposit feeders. *Arenicola* (and *Abarenicola*) lives in an L-shaped burrow that opens to the surface on the long side (Fig. 13-34, 13-35). The head of the worm is situated at the blind end of the horizontal part of the burrow, where sand is continually ingested by means of a simple pharynx. The ingested sand is rich in surface organic material, which tumbles into the surface depression above the animal's head and slumps downward to the worm. Cilia in the everted pharynx drive loose particles into the gut, and some particle sorting occurs. In a distinct rhythm,

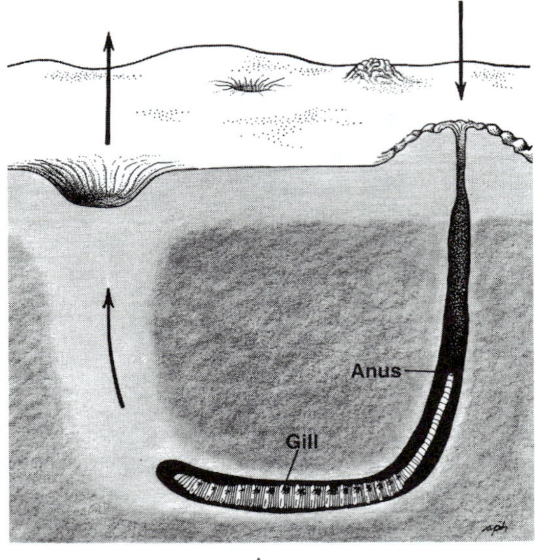

A

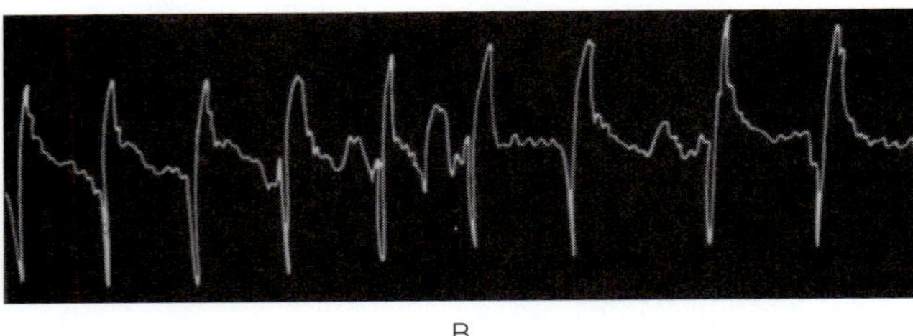

B

FIGURE 13-34 Polychaete diversity (Scolecida). **A,** The lugworm *Arenicola* (Arenicolidae) in its burrow. Arrows indicate the direction of water flow produced by the worm. The worm ingests the column of sand on the left, through which water is filtered. The pile of sand at the burrow opening is defecated castings. **B,** Tracing of activity cycles of *Arenicola* over a period of six hours. The downstroke reflects the worm backing up to the burrow opening to defecate; the sharp upstroke reflects the worm moving back down to the head of the burrow and vigorously resuming ventilation contractions and deposit feeding. Intervals between defecations are about 40 min. *(B, After Wells from Newell, R. C. 1970. Biology of Intertidal Animals. American Elsevier Co., New York.)*

the feeding halts and the worm backs up to the top of the tube to defecate the mineral particles (castings) passed through the gut (Fig. 13-34B). The worm irrigates the burrow by peristaltic contractions that drive water into the burrow opening. Water leaves the burrow by percolating up through the sand.

Maldanidae[F]: Bamboo worms, such as *Clymenella* and *Axiothella,* are deposit-feeding tube dwellers. Their sand-grain tubes, which resemble beverage straws, are common in the intertidal zone. Bamboo worms have wedgelike heads; long, distinct segments and segment joints that have the appearance of bamboo cane and its joints; and parapodia reduced to low ridges (Fig. 13-36B–D). The worms live upside down in their tubes and ingest the substratum from below.

Capitellidae[F]: Earthwormlike with reduced parapodia and no head appendages. Like earthworms, most feed on sediment as they burrow through it. Most capitellids are red because of abundant hemoglobin in the coelomic fluid. Pressurized coelomic fluid everts the thin-walled pharynx (Fig. 13-21), which in living animals resembles a blood blister. Most capitellids are small, but one, the giant (80 cm) *Notomastus lobatus* (maître d'worm), lives in clayey sediment along the southeastern coast of the United States. It occupies a permanent helical burrow that it shares with no fewer than eight species of commensals, including one scaleworm and two other polychaetes, three clams, a crab, and an amphipod. *Capitella.*

Orbiniidae[F]: Earthwormlike in form and habits. They are often red or orange, depending on species. The prostomium lacks appendages. Proceeding from anterior to posterior along the body, the parapodia shift from lateral to dorsal. The eversible pharynx resembles that of capitellids, but orbiniids such as *Orbinia ornata* and *Scoloplos rubra* (Fig. 13-24D) have paired comblike gills, derived from dorsal cirri, that arch over their backs.

Opheliidae[F]: Includes burrowers that ingest sediment with an eversible bulbous pharynx. They often have lateral gills, and sometimes lateral eyes. Some, such as *Travisia,* resemble short, fat, sluggish grubs. *Ophelia* and *Armandia* are fishlike, streamlined, and fast, being adapted to live in current-swept sand (Fig. 13-36A). They move like fish, too, swimming rapidly through either loose sand or water. Among other adaptations for swimming, their pygidium is modified into a tail fin. They so closely resemble amphioxus (Chordata) that collectors often mistake the annelid for the chordate. *Euzonus.*

Palpata

Sister taxon of Scolecida. Have a pair of sensory palps on the prostomium (sometimes translocated to the peristomium).

ACICULATA

Diverse taxon that includes 39 families and many familiar and distinctive polychaetes, such as the bloodworm *Glycera,* the ragworm *Nereis,* and most of the pelagic polychaetes. All have well-developed biramous parapodia with at least one aciculum in the notopodium and neuropodium. Dorsal and ventral cirri are present; prostomium bears four, three (primitive), or two antennae. They have diverse lifestyles and feeding habits, but all are active, mobile worms, formerly called "errant" (Errantia) in contrast to all other polychaetes, which were classified previously as "sedentary" (Sedentaria).

Phyllodocida[O]

Aciculates having a tubular, muscular, eversible pharynx. The pharynx is in line, or coaxial, with the esophagus.

Phyllodocidae[F]: Paddleworms such as *Eteone, Eulalia,* and *Phyllodoce* have greatly flattened, paddlelike notopodia that are used as gills (Fig. 13-11B, 13-12B, 13-23B, 13-37C). Carnivores; actively crawl and seize prey with an eversible, muscular, unarmed pharynx. *Eteone* species find food by tracking the mucous trails laid down by prey on the surface of the mudflat.

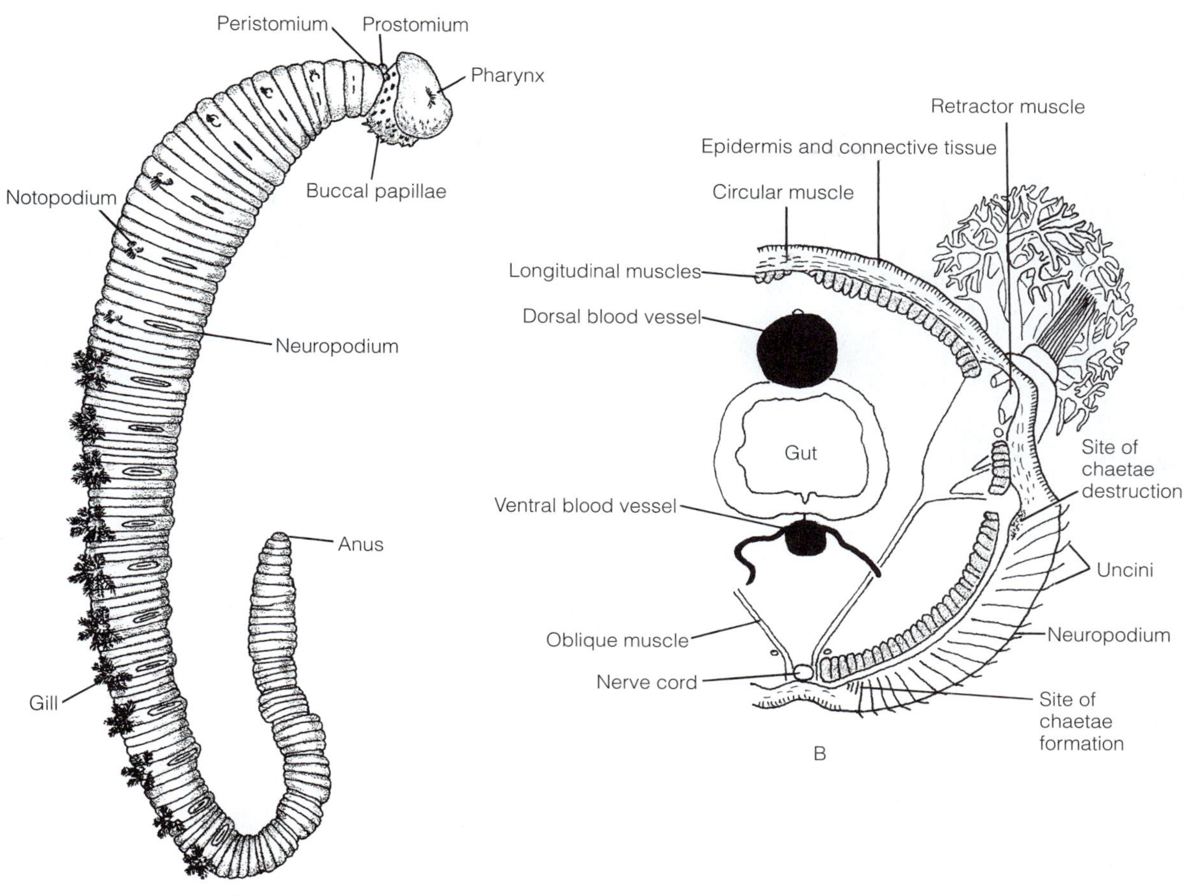

FIGURE 13-35 Polychaete diversity (Scolecida). **A,** Lateral view of the lugworm *Arenicola* (Arenicoloidae). **B,** Transverse section of a chaetigerous segment of *Arenicola marina.* *(A, From Brown, F. A. 1950. Selected Invertebrate Types. John Wiley and Sons, New York; B, Modified from Wells.)*

Tomopteridae[F]**:** Pelagic, transparent carnivores. *Tomopteris* is dorsoventrally flattened, lacks chaetae, and has very long cirri on segment 2 (Fig. 13-38A).

Alciopidae[F]**:** Pelagic, transparent carnivores. *Vanadis* and *Rhynchonerella* resemble paddleworms (Phyllodocidae), but have enormous, bulging eyes with which to locate prey (Fig. 13-18C, 13-38B,C).

Nereidae[F]**:** Ragworms such as *Nereis* (and other genera) actively crawl, burrow, and occasionally swim. Pharynx is muscular, eversible, and has a pair of stout jaws (Fig. 13-10C, 13-12A, 13-30). Head has two antennae, two palps, four eyes, and four pairs of tentacular cirri. Some, such as *Nereis pelagica, Nereis virens,* and *Nereis diversicolor,* are omnivores that feed on algae, other invertebrates, and even detritus. *Nereis succinea* and *Nereis longissima* feed primarily on detritus in the sediments. *Nereis fucata,* a commensal with the hermit crab, is carnivorous. *Nereis brandti,* of the northwest coast of the United States, reaches 1.8 m in length and feeds primarily on green algae. *Ceratonereis, Platynereis.*

Glyceridae[F]**:** *Glycera* (the bloodworm) and other genera occupy a subsurface gallery excavated in muddy bottoms (Fig. 13-39A). The gallery contains numerous loops that open to the surface (Fig. 13-39B). Lying in wait at the bottom of a loop, the worm uses its four tiny, vibration-sensitive antennae to detect surface movements of prey, such as small crustaceans. It then moves to the burrow opening and seizes the prey with its pharynx.

The long, retracted (inverted) pharynx occupies approximately the first 20 body segments. At the rear of the pharynx are four equidistant fangs. The pharynx is attached to an S-shaped esophagus. Septa are absent in the anterior segments (and reduced elsewhere) and the inverted pharynx lies free in the coelom. Just prior to pharynx eversion, longitudinal protractor muscles contract violently, sliding the pharynx forward and straightening out the esophagus. Contraction of body-wall muscles then everts it with explosive force, and the four fanglike jaws are exposed at the tip (Fig. 13-39C). Each hollow jaw

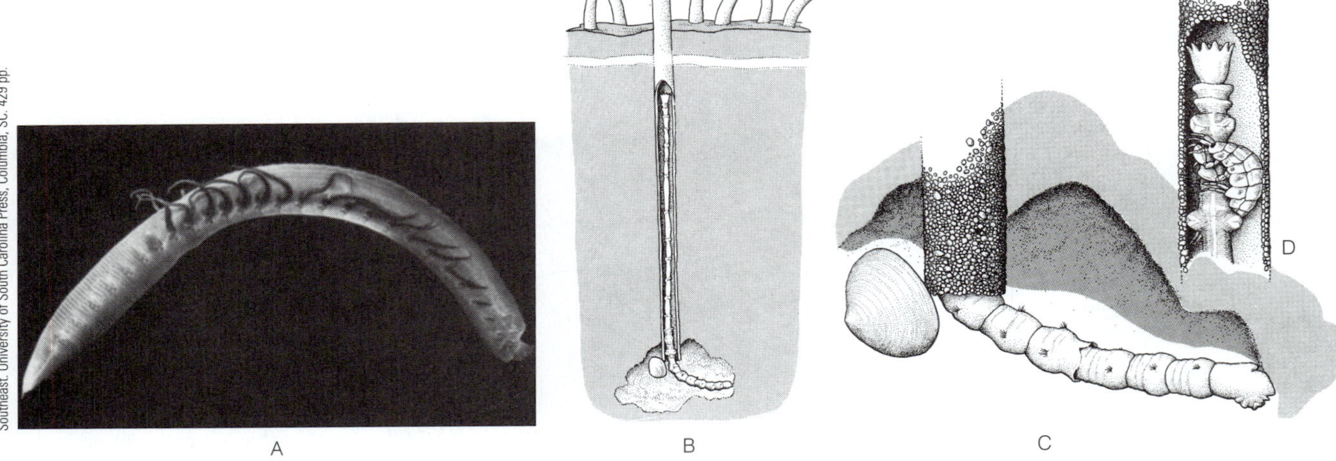

FIGURE 13-36 Polychaete diversity (Scolecida). **A,** The burrowing, amphioxus-like polychaete *Ophelia denticulata* (Opheliidae) in lateral view. Note the small, pointed prostomium at left. The long projections are gills. **B,** The bamboo worm, *Clymenella torquata* (Maldanidae). Bamboo worms live head down in their tubes and excavate subsurface caverns. Two commensals, a tiny clam, *Aligena elevata* **(C),** and an amphipod, *Listriella clymenellae* **(D),** share the protective retreat created by *Clymenella. (B–D, From Ruppert, E. E., and Fox, R. S. 1998. Seashore Animals of the Southeast. University of South Carolina Press, Columbia SC. 429 pp.)*

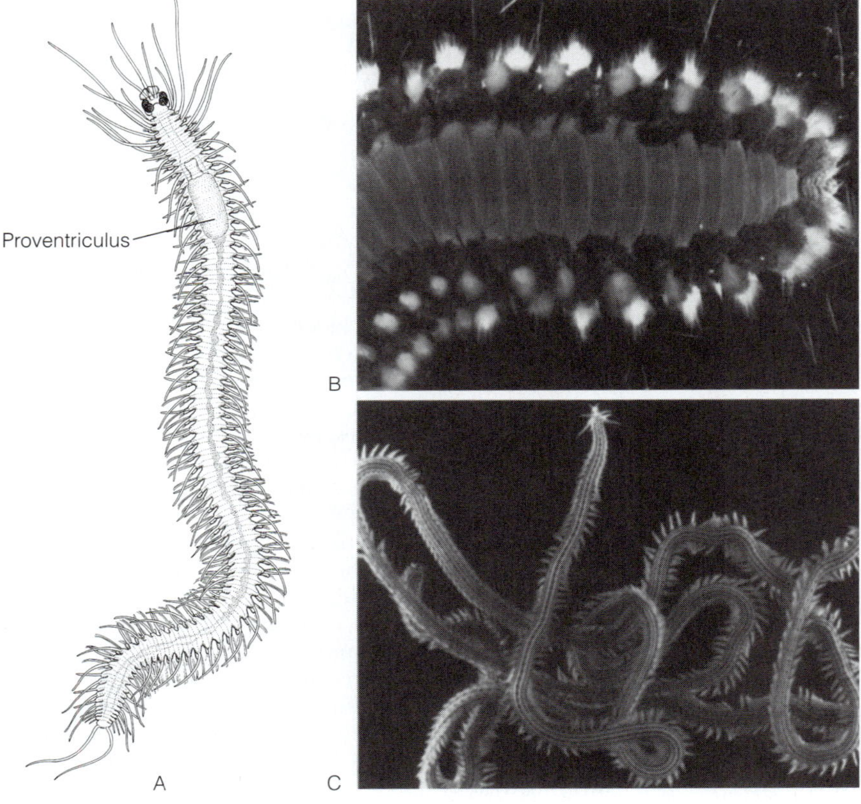

FIGURE 13-37 Errant polychaete diversity (Palpata, Aciculata). **A,** The Bermuda fireworm, *Odontosyllis enopla* (Syllidae; male epitoke). Note large eyes used to spot glowing females. **B,** The fireworm *Hermodice carunculata* (Amphinomidae), from a coral reef in the Florida Keys. When disturbed, these polychaetes shed some of their calcareous chaetae (white tufts), which contain an irritating toxin. The convoluted, brainlike structure (caruncle) on the head is the highly developed nuchal organ. **C,** The paddleworm *Eulalia myriacyclum* (Phyllodocidae) from beneath stones in southern Florida. *(A, From Fischer, A., and Fischer, U. 1995. On the life-style and life-cycle of the luminescent polychaete* Odontosyllis enopla *(Annelida: Polychaeta). Invertebrate Zoology 114:236–247 (Fig. 2). Reprinted with permission.)*

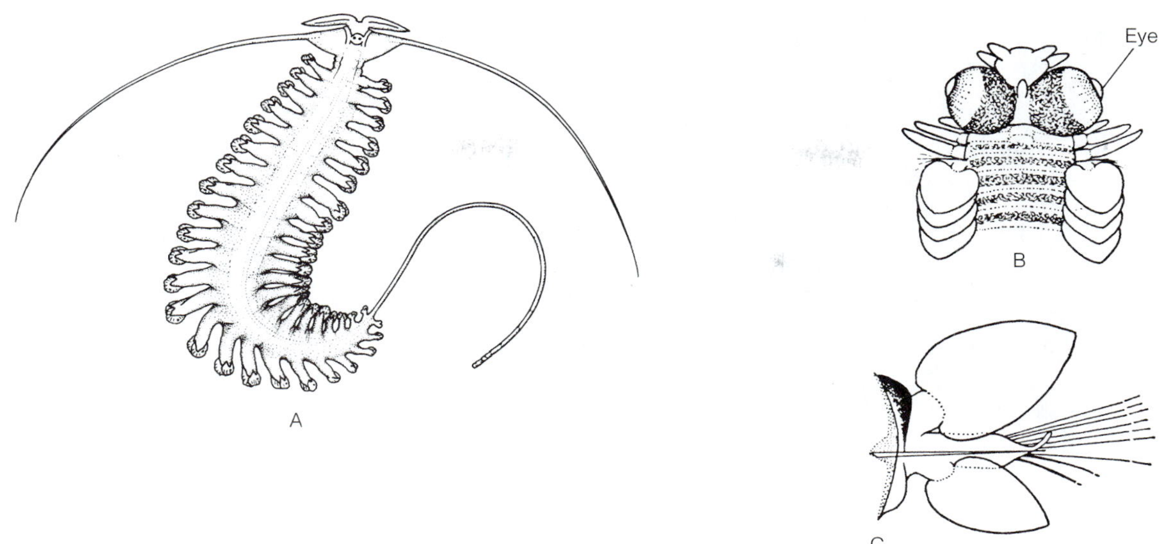

FIGURE 13-38 Errant Polychaete diversity (Palpata, Aciculata). Pelagic polychaetes. **A,** *Tomopteris renata* (Tomopteridae). **B,** Dorsal view of *Rhynchonerella angelina* (Alciopidae). **C,** A paddlelike, swimming parapodium of *Rhynchonerella*. *(From Day, J. 1967. Polychaeta of South Africa. British Museum, London.)*

houses a canal that delivers poison from a gland at the jaw base.

Glycerids are active animals that burrow by using the eversible pharynx to punch into the sediment. They can also swim with complex looping movements of the body. They are called bloodworms because the coelom is richly endowed with hemoglobin-containing cells, which slosh past the vestigial septa from one end of the body to the other. The hemal system is vestigial and the adults have protonephridia.

Nephtyidae[F]: Similar to bloodworms (Glyceridae). Have a long, eversible pharynx, but only two jaws instead of four; carnivores of worms and small crustaceans. *Nephtys* and other nephtyids are called shimmy worms for their rapid, almost vibratory lateral undulations of the body while swimming or burrowing in loose sand (Fig. 13-40); sometimes called cat worms because the anterior prostomial margin bears two triangular palps that resemble a cat's ears.

Aphroditidae[F], Polynoidae[F], Polyodontidae[F], Sigalionidae[F]: Scaleworms, called such because of two rows of overlapping scales on the dorsal surface resembling fish scales (Fig. 13-11A, 13-41C). Each scale is a highly modified dorsal cirrus called an **elytron** (pl., elytra) that is borne on a short stalk (Fig. 13-41B,C). A few scaleworms burrow (sigalionids), others are secretive and occupy tight crevices beneath stones (polynoids), a few live in secreted tubes (polyodontids), but most are commensals (polynoids) in the burrows and tubes of other invertebrates or the shells of hermit crabs. Because most live in tight quarters, the scales may channel the ventilating water current between the scales and body wall. Elytra also bear a variety of sensory structures. In the scaleworm *Aphrodite* (the sea mouse), the entire dorsal surface, including the elytra, is covered with hairlike "felt" composed of long notopodial chaetae that trail back over the dorsal surface of the animal (Fig. 13-41A,B). Scaleworms are considered to be carnivores, but little is known about the feeding habits of commensal species. Aphroditidae: *Aphrodite, Laetmonice;* Polynoidae: *Arctonoe, Halosynda, Harmothoe, Lepidasthenia, Lepidonotus;* Polyodontidae: *Polyodontes;* Sigalionidae: *Sthenelais.* May also include *Thalanessa.*

Hesionidae[F]: Active commensals or free living *(Ophiodromus, Podarke)* and a few tiny species in the interstitial habitat, the labyrinthine water-filled spaces between grains of sand. Of the latter, *Hesionides* (Fig. 13-42B) crawls in typical polychaete fashion using parapodia, but interstitial polychaetes of Saccocirridae[F], Protodrilidae[F], Polygordiidae[F] *(Polygordius),* and Diurodrilidae[F] have lost the parapodia and rely on cilia for locomotion (Fig. 13-42C). Although previously classified as Archiannelida because of their simple body design (and small size), Diurodrilidae is now considered to be a member of Eunicida (discussed below), whereas the remaining families are tentatively placed within Canalipalpata (also discussed below).

Myzostomida[O]

Myzostomidae[F]: Commensals on sea lilies and feather stars or internal parasites of sea stars (Echinodermata). *Myzostoma* and relatives are little worms that rarely exceed 5 mm

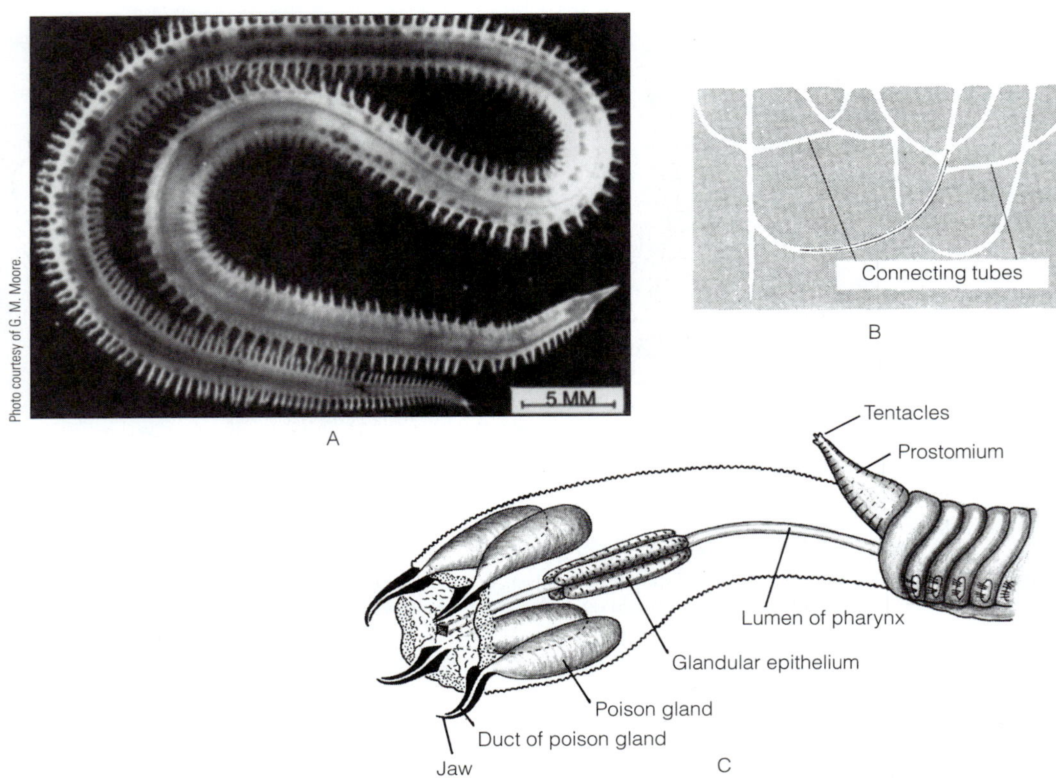

FIGURE 13-39 Errant polychaete diversity (Palpata, Aciculata). **A,** The predatory bloodworm *Glycera americana* (Glyceridae), a common burrowing polychaete found along the east coast of the United States. Note the pointed prostomium. **B,** Burrow system of *G. alba,* showing worm lying in wait for prey. **C,** Everted pharynx showing four hollow teeth (jaws) through which poison is discharged. *(B, After Ockelmann and Vahl; C, After Michel.)*

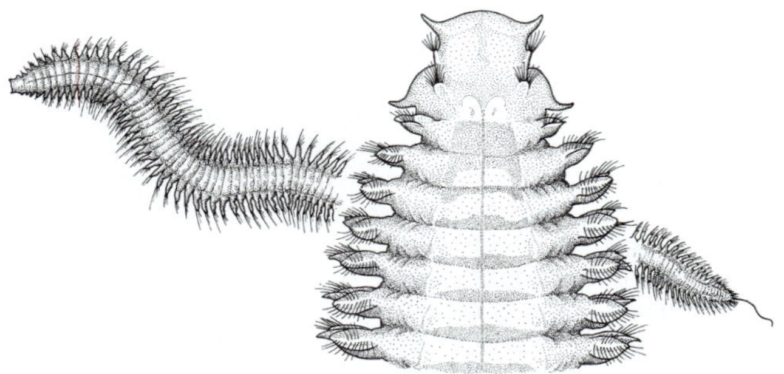

FIGURE 13-40 Errant polychaete diversity (Palpata, Aciculata). The predatory catworm or shimmy worm, *Nephtys bucera* (Nephtyidae). Note the catlike "ears" on the prostomium. These worms undulate rapidly while swimming. *(From Ruppert, E. E., and Fox, R. S. 1988. Seashore Animals of the Southeast. University of South Carolina Press, Columbia. 429 pp.)*

in length and superficially resemble flatworms, but when disturbed they scoot rapidly away on stubby legs, each one terminating in a hooked chaeta (Fig. 13-42A). Oval body is greatly flattened; five pairs of parapodia on the undersurface. Parapodia, chaetae, and segmented nervous system indicate that Myzostomida belongs to Polychaeta, but its sister taxon is unknown; placement in Aciculata is provisional. Some researchers include Myzostomida not with Annelida, but rather with Gnathifera (Chapter 23).

Eunicida[O]

Aciculates with five antennae and usually well-developed parapodial gills. The retracted eversible pharynx is slung ventrally below the level of the esophagus.

Eunicidae[F]: Members have a muscular, eversible pharynx that bears two stout mandibles and other accessory teeth (Fig. 13-22). Includes the longest polychaete, *Eunice aphroditois,* a 3-m-long carnivorore from the tropical Indo-Pacific. *Palola* (Fig. 13-29A), some species of which are known as palolo worms, are famous for epitokal swarming and their savory flavor. *Marphysa* (Fig. 13-11D) burrows in fine muddy sands, and *Lysidice,* like *Palola,* occurs in coralline rock.

Amphinomidae[F]: Mostly large, chiefly tropical, often beautifully colored, caterpillar-like fireworms (Fig. 13-24C, 13-37B) are carnivores living on corals and driftwood, beneath stones, occasionally in sediment. Eat coral polyps, anemones, stalked barnacles, and other sessile invertebrates by prying them loose with a muscular lower lip

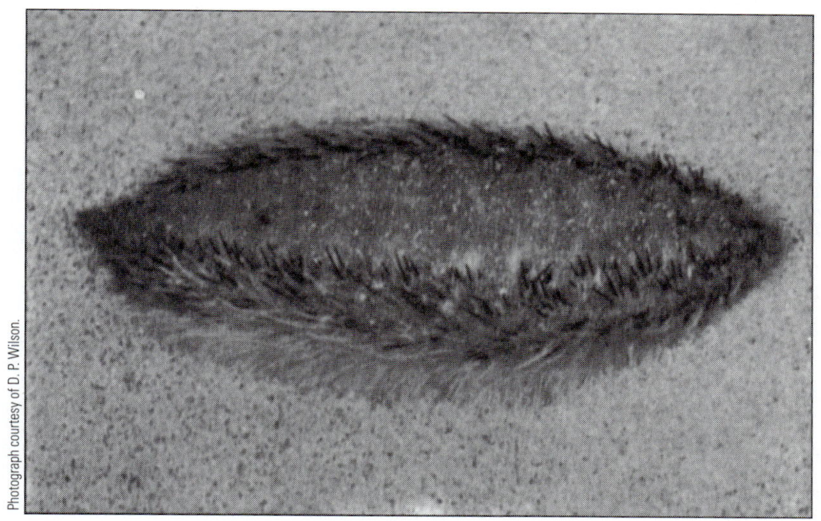

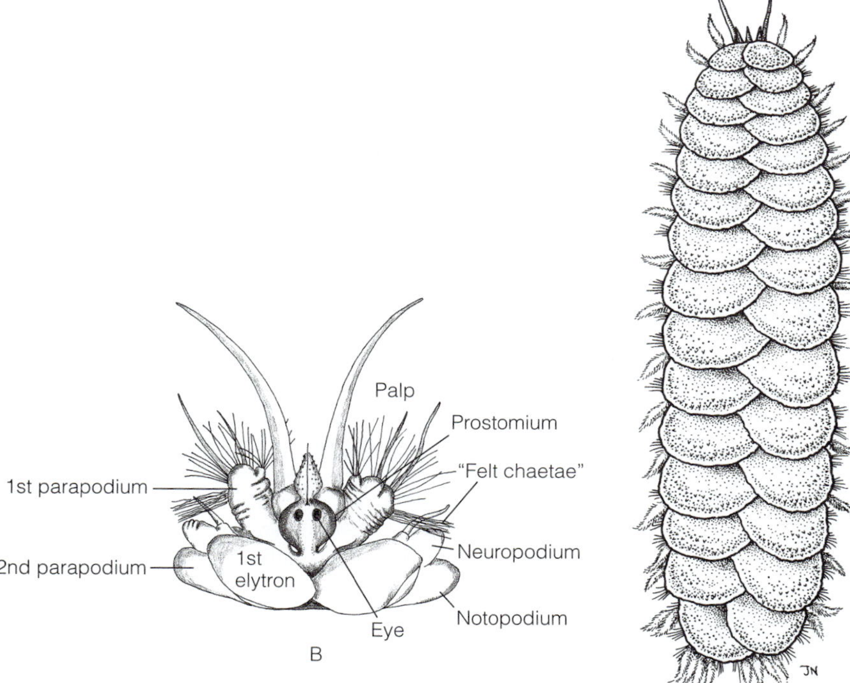

FIGURE 13-41 Errant polychaete diversity (Palpata, Aciculata). Scaleworms. **A** and **B,** The sea mouse, *Aphrodite aculeata* (Aphroditidae). **A,** Dorsal view of the living animal. The dorsal and lateral surfaces are covered by felt chaetae. **B,** Anterior end, including the first pair of dorsal scales, or elytra. Sea mice occur in Europe and on the northwestern and northeastern coasts of the United States. They live offshore, where they burrow in soft-bottom muds, leaving only their posterior end exposed at the surface. The ventral surface is free of felt and forms a flat, muscular, creeping sole. They ventilate their burrow by elevating the ventral surface, which pumps surface water into the burrow and moves it anteriorly along the sole. The sole is then depressed, forcing the water posteriorly in a channel formed by the dorsum and its covering scales. Movement of the scales also helps create the exhaust flow. **C,** *Harmothoe aculeata* (Polynoidae) is a common scaleworm found below rocks and in crevices on the east coast of the United States. The elytra bear many low sensory papillae on their upper surface. *(B, From Fordham; C, From Ruppert, E. E. and Fox, R. S. 1988. Seashore Animals of the Southeast. University of South Carolina Press, Columbia. 429 pp.)*

and sucking them in whole. Goose barnacles *(Lepas)* ingested whole and later defecated as museum-quality barnacle shells. Fireworms are notorious for their defensive chaetae, which are uniquely calcified, brittle, and toxic (Fig. 13-37B). When bare skin brushes against the chaetae, they penetrate the skin, break free of the worm, and create a burning pain. Avoided by fish and experienced scuba divers. *Amphinome, Chloeia, Eurythoe, Hermodice.*

Onuphidae[F]: Tube-dwelling aciculates; tubes are protective retreats and the occupants crawl actively within them by using well-developed parapodia. *Diopatra, Onuphis,* and *Americonuphis* (Fig. 13-9, 13-43A) build tough, conspicuous, membranous tubes that project vertically above the surface of the sand. The tube's chimney is crooked and flared like a ship's ventilator funnel (Fig. 13-43B). The chimneys are covered with bits of shell, seaweed, and other debris that the worm collects and positions with its jaws. Such

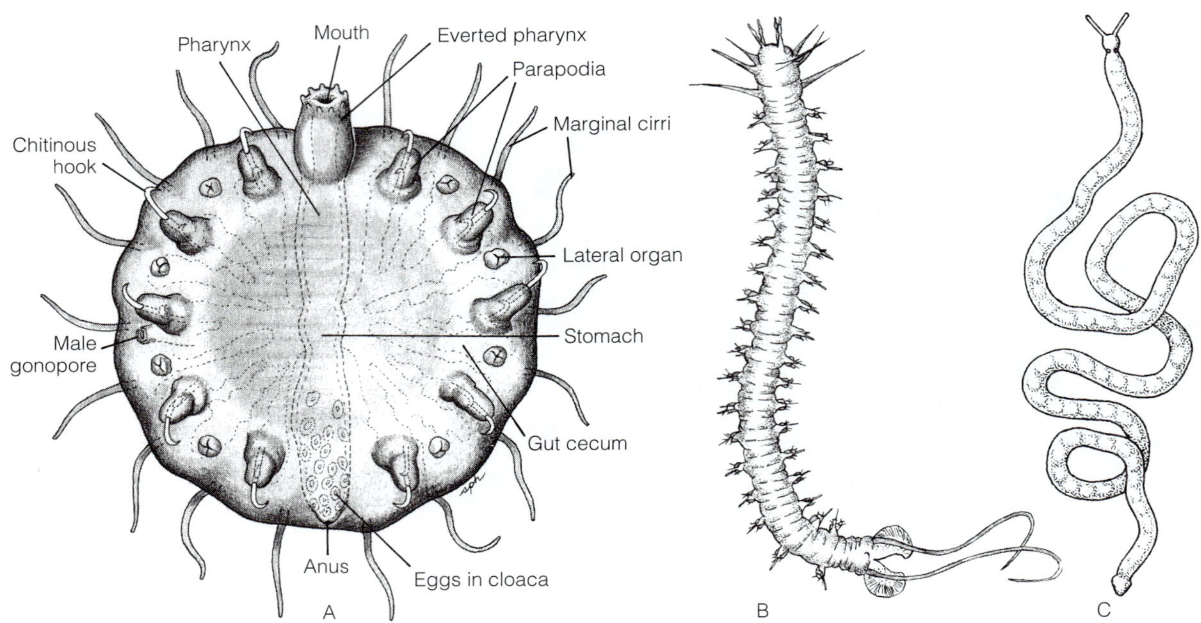

FIGURE 13-42 Errant polychaete diversity (Palpata, Aciculata). Commensal and interstitial polychaetes. **A,** *Myzostoma* (Myzostomida, Myzostomidae), a commensal on crinoids. **B,** *Hesionides arenaria* (Hesionidae), a minute polychaete no more than 2 mm in length that inhabits interstitial spaces of beach sands in northern Europe. The ventrally directed parapodia are adapted for crawling between sand grains. **C,** *Polygordius neopolitanus,* a threadlike interstitial polychaete. Segmentation is poorly indicated externally. The systematic position of *Polygordius* is uncertain, and it may properly belong to the Canalipalpata. *(A, After McIntosh; B, From Ax, P. 1966. Veröffenthlungen des Institut fur Meeresforschung Bremerhaven. Suppl. II. pp. 15–66; C, After Fraipoint.)*

ornamentation provides camouflage, and some worms graze on the algae planted on their own as well as their neighbors' chimneys.

Diopatra uses its tube as a lair (Fig. 13-43A) from which to ambush prey. The worm's chemoreceptors monitor water flowing through the tube and detect the presence of approaching prey. The worm emerges from its tube and seizes the victim with its pharynx and pharyngeal teeth. During feeding the prey may be clasped with the enlarged anterior parapodia. *Diopatra* may also feed on dead animals, organic debris, and small organisms, such as forams, that are near or attached to the tube. *Americonuphis, Kinbergonuphis.*

Syllidae^F: The tiny, abundant, diverse syllids are the polychaete answer to mosquitoes (Fig. 13-37A). They pierce prey with a pharyngeal tooth and then suck out the contents with a specialized, muscular part of the pharynx, the **proventriculus.** Mostly in shallow water and common on coral reefs, in fouling communities. Typical prey are hydroids, sponges, bryozoans, and other sessile invertebrates. *Autolytus, Odontosyllis, Myrianida, Plakosyllis, Proceraea, Syllis, Typosyllis, Trypanosyllis,* and many others.

Oenonidae^F (formerly Arabellidae^F): Includes *Arabella iricolor* (the opalworm; Fig. 13-11C) and *Drilonereis* (the threadworms; Fig. 13-44). Slender, often threadlike burrowers; some are direct deposit feeders, others are carnivores with well-developed pharyngeal jaws. Cuticle usually is glossy and iridescent. Curiously, juveniles of several species are endoparasites in the coelomic cavities of other polychaetes and echiurans. *Labrorostratus.*

Lumbrineridae^F: Slender burrowers with reduced head appendages, well-developed pharynx and jaws. Carnivores. *Lumbrineris.*

Ichthyotomidae^F: Ectoparasitic on eel fins. *Ichthyotomus.*

CANALIPALPATA

Sister taxon of Aciculata. "Sedentary" species in tubes or semipermanent burrows. Longitudinally grooved palps on prostomium, derived from prostomial sensory palps, are used for either deposit or suspension feeding, depending on taxon. In some, the palps have shifted to the peristomium.

Spionida^O

Have palps on peristomium instead of prostomium, which is thought to have occurred by displacement of the original prostomial palps.

Spionidae^F: Tube-dwelling palp worms are either surface deposit feeders or suspension feeders; some may be both. Two long peristomial palps project from the tube opening and extend over the bottom (Fig. 13-45). Particles that adhere to the surface are transported to the mouth in a ciliated channel along the length of each palp. Tubes of *Spio, Scolelepis,* and *Streblospio* generally are soft, straight, and

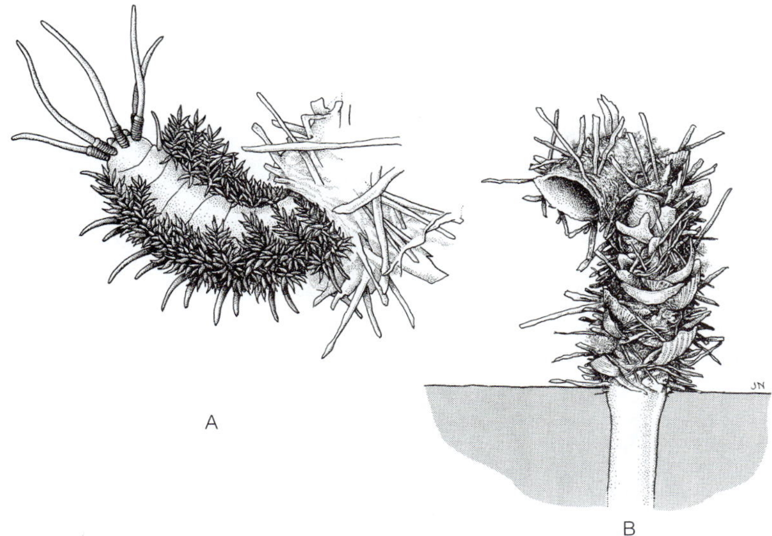

FIGURE 13-43 Errant polychaete diversity (Palpata, Aciculata). **A,** The shaggy tube worm, *Diopatra cuprea* (Onuphidae), a common inhabitant of intertidal mudflats, and its characteristic downturned tube cap **(B)** in which plant fragments and other debris are embedded. Below the surface of the sand, the tube is free of foreign material and smooth. Note the prominent gills on the anterior segments. *(From Ruppert E. E. and Fox, R. S. 1988. Seashore Animals of the Southeast. University of South Carolina Press, Columbia. 429 pp.)*

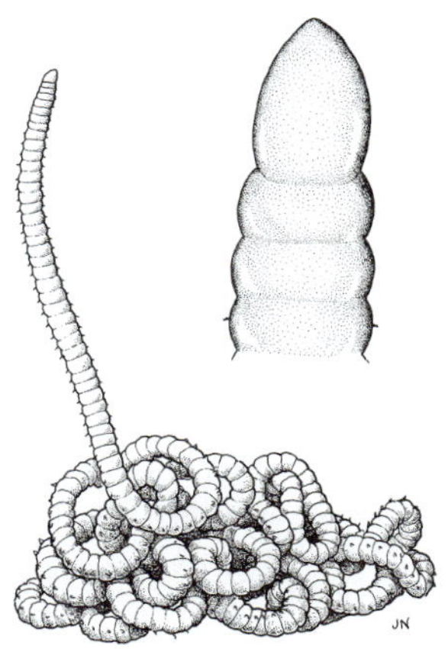

FIGURE 13-44 Errant polychaete diversity (Palpata, Aciculata). The threadworm, *Drilonereis magna* (Oenonidae), a predatory burrower with an eversible pharynx and well-developed jaws. Note the appendageless conical prostomium adapted for burrowing. *(From Ruppert E. E. and Fox, R. S. 1988. Seashore Animals of the Southeast. University of South Carolina Press, Columbia. 429 pp.)*

cylindrical and have incorporated sand or mud, but the suspension-feeding *Polydora* bores into calcareous substrates, including the shells of living oysters, in which they form unsightly shell "blisters."

Magelonidae[F]: Shovel-headed burrowers that have a stiff, flat, spatulate prostomium that overhangs the mouth (Fig. 13-46). The long peristomial palps are covered with short, sticky papillae that collect and pass deposited organic particles to the mouth. The threadlike worms are violet because of hemerythrin-containing hemocytes, unique among polychaetes. Magelonids such as *Magelona* are the only polychaetes with cross-striated muscles, the significance of which has yet to be determined.

Chaetopteridae[F]: Filter feeders that have straight (vertical) or U-shaped tubes partially buried in sand on protected beaches. *Chaetopterus, Mesochaetopterus, Spiochaetopterus* filter feed with one or more mucous nets suspended inside of the tube. As the water passes through, plankton are strained out and periodically the net and food are rolled up into a ball and swallowed.

The bizarre *Chaetopterus variopedatus* (parchment-tube worm) is the largest and best known of all chaetopterids (Fig. 13-47A). It secretes and occupies a tough, opaque, U-shaped tube approximately the diameter and shape of a large banana. The two open, tapered ends project vertically above the surface of the beach and the remainder is buried horizontally below. The soft, pale, and fragile worm is reminiscent of an internal organ, so completely is it adapted to its protected tube environment. The body is strongly regionally specialized and the anterior half is downright weird. The long, paired notopodia on segment 12 are shaped like scythe blades and are called **aliform** (winglike) **notopodia**. Their epidermis is ciliated and richly supplied with mucous glands. Notopodia on segments 14 to 16 are expanded and fused, forming three semicircular **fans** that expand like piston rings against the cylindrical wall of the tube (Fig. 13-47C).

While feeding, the outstretched aliform notopodia arch over the dorsal side of the body and contact each other at their tips to form a hoop, which presses against the inner contour of the tube. Mucus secreted continuously from the notopodial hoop forms a net similar to the net attached to a basketball hoop. The net extends posteriorly until it is grasped and constricted by a **ciliated cup** on the mid-dorsal body wall a short distance behind the aliform notopodia. Rhythmic beating of the notopodial fans generates a water current that enters the inhalant chimney of the U-shaped tube, flows through

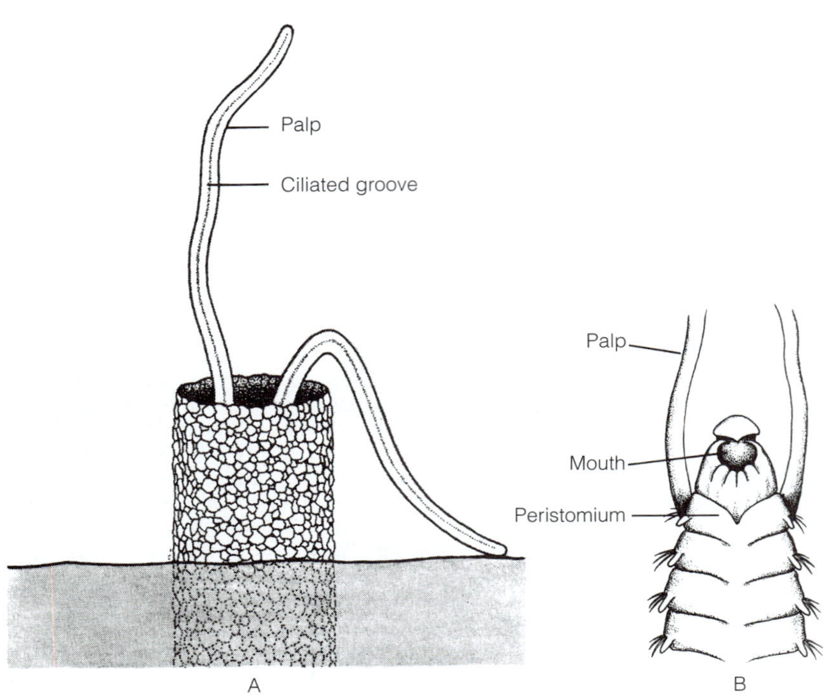

FIGURE 13-45 Polychaete diversity (Palpata, Canalipalpata). Palp worms (Spionidae). **A,** Palps of a spionid polychaete projecting from its sand-grain tube. The ciliated groove of the palp conveys to the mouth detritus picked up from the substratum or suspended in the water. **B,** Ventral view of anterior end of *Spio pettiboneae,* showing the base of palps and mouth.

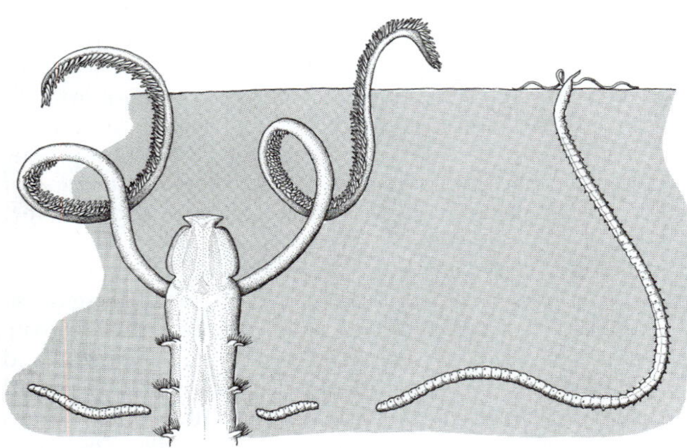

FIGURE 13-46 Polychaete diversity (Palpata, Canalipalpata). The shovel-headed worm *Magelona phyllisae* (Magelonidae). Note the sticky papillae on the palps. Right-side of figure shows worm in feeding position. *(From Ruppert E. E., and R. S. Fox, 1988. Seashore Animals of the Southeast. University of South Carolina Press, Columbia. 429 pp.)*

the mucous net and tube, and then exits through the exhalant chimney. Plankton and detritus are filtered from the water flow by the mucous net. The food-laden mucous net is continuously rolled up into a ball by the ciliated cup. When the ball reaches a certain size, the net is cut loose from the aliform notopodia and incorporated into the ball. The cup then projects forward and deposits the food ball onto a ciliated mid-dorsal groove, which transports it to the mouth at the anterior end of the worm.

A large, potentially fouling object brought into the tube by the water current is detected by peristomial sensory cilia and the aliform notopodia retract to allow the object to bypass the net. An 18- to 24-cm-long specimen of *Chaetopterus* may produce a mucous net at the rate of approximately 1 mm/s, with food balls averaging 3 mm in diameter.

Paradoxically, *Chaetopterus* is strongly bioluminescent although it permanently occupies an opaque tube. The bioluminescence is produced by bacteria and released in the worm's body mucus. Observations of bioluminescence emanating from the submerged tube ends suggests that the worm may discharge its luminous mucus to startle would-be predators nibbling on its tube ends.

Terebellida^O

Terebellidans have numerous peristomial feeding "tentacles" (palps).

Terebellidae^F: Spaghetti worms (Fig. 13-24B, 13-48A) occupy tubes or simple vertical or U-shaped burrows in the sediment. Prostomial sensory appendages are absent, but clusters of highly extensible tentacles are borne on the head (Fig. 13-48A). Movement through the burrow usually is by peristaltic contractions. The parapodia are greatly reduced and are in part represented by transverse ridges provided with uncini, which grip the sides of the burrow. In terebellids such as *Amphitrite,* which occupies a U-shaped burrow, the hooks are oriented oppositely on the anterior and posterior ends of the animal. As a result, attempts to pull the worm from either the head or tail end of its

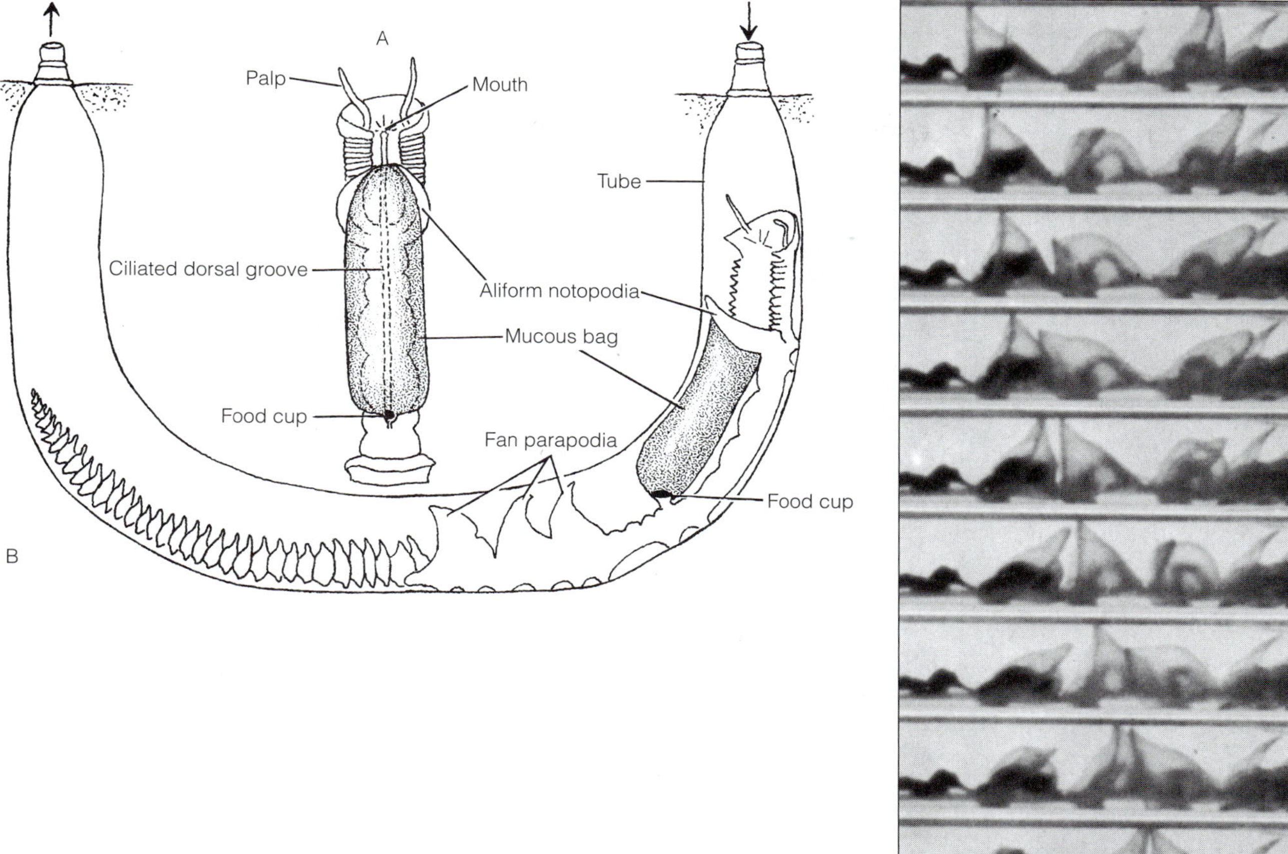

FIGURE 13-47 Polychaete diversity (Palpata, Canalipalpata). The parchment-tube worm, *Chaetopterus variopedatus* (Chaetopteridae), a tube-dwelling, subsurface filter feeder. **A,** Anterior part of the body (dorsal view). **B,** Worm in tube (lateral view). Arrows indicate direction of water current through the tube. **C,** Side views of active fan parapodia of *Chaetopterus* during one pumping cycle. The worm is in a glass tube and anterior is to the left. *(A and B, After MacGinitie; C, From Brown, S. C. 1975. Biomechanics of water-pumping by* Chaetopterus variopedatus *Renier. Skeletomusculature and kinematics. Biol. Bull. 149:136–150.)*

burrow are opposed by the uncini. Other common genera include *Lanice, Lysilla, Terebella, Thelepus.*

When terebellids feed, the tentacles extend outward over the surface of the sediment by ciliary creeping, then each rolls ventrally to form an inverted gutter (Fig. 13-48A–C). Surface detritus adhering to mucus secreted by the tentacle is transported along the ciliated gutter and accumulates at the base of the tentacles, each of which is wiped over the upper lip bordering the mouth. Cilia on the lip then drive the food into the mouth (Fig. 13-48D). The tentacles of some large tropical species, such as *Eupolymnia crassicornis,* reach a meter or more in length and stream over the surface of the sand like active strands of spaghetti.

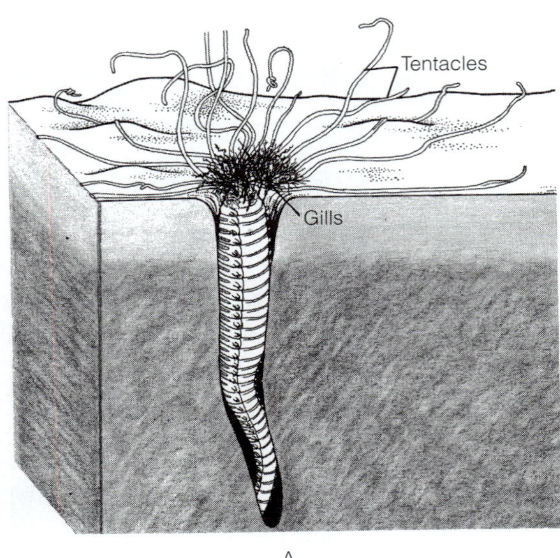

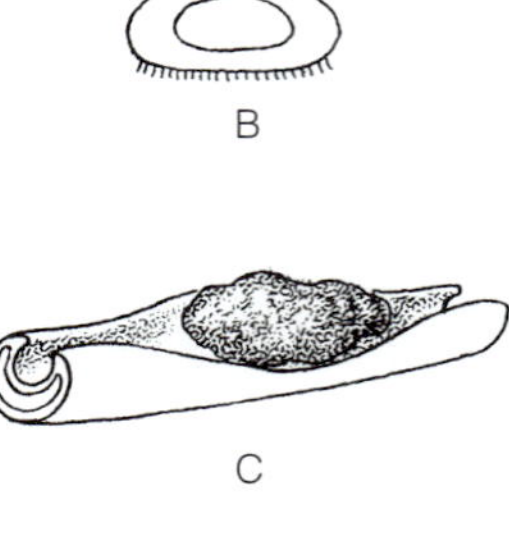

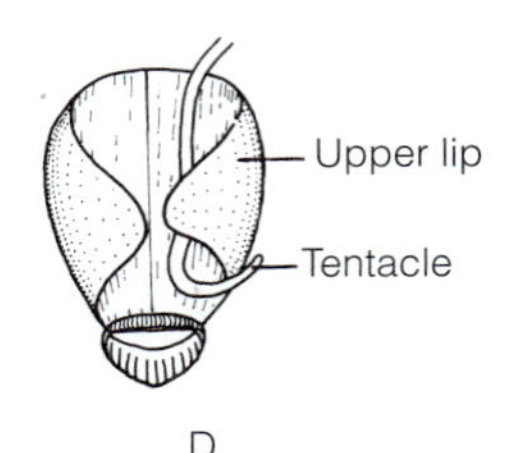

FIGURE 13-48 Polychaete diversity (Palpata, Canalipalpata). Deposit feeding spaghetti worms (Terebellidae). **A,** *Amphitrite ornata* at the aperture of its U-shaped burrow with tentacles outstretched over the substratum. **B,** Cross section through a tentacle of *Terebella lapidaria* creeping over the substratum. **C,** Section of a tentacle of *T. lapidaria* rolled up to form a ciliary gutter, transporting deposited material. **D,** Tentacle being wiped by one of the lips. *(B–D, After Dales.)*

Ampharetidae^{F}: *Ampharete* and *Melinna* (spaghetti-mouth worms) are surface deposit feeders similar to terebellids that use extensible tentacles. Unlike terebellids, however, the tentacles originate from the roof of the mouth and can be withdrawn into it. Ampharetids occupy straight, vertical tubes, from which they emerge headfirst to feed on detritus (Fig. 13-49B).

Cirratulidae^{F}: Fringed worms, such as *Cirratulus* and *Cirriformia,* are burrowers that deposit feed with a single pair, or two anterior clusters, of long feeding palps. A pair of long gill filaments, similar in appearance to the palps, occurs on each of many body segments (Fig. 13-24A). *Dodecaceria.*

Pectinariidae^{F}: *Amphictene, Cistenides,* and *Pectinaria* (scaphopodlike ice cream cone worms) construct a conical, one-layer-thick tube of cemented sand grains (Fig. 13-49C,D). The tube is open at its base and apex and the worm occupies it with its head at the broad basal end. Tube and worm are oriented head down in fine sand; the tube apex projects just above the sediment surface. Organics-rich sand is shoveled into the mouth with short, stout, golden chaetae (paleae), which together function as an operculum to close the tube aperture when the worm withdraws. Feces are released to the surface of the beach through the tube apex, which also provides an inlet for ventilating the gill surfaces.

Flabelligeridae^{F}: Burrowers with cephalic gills that can be retracted with the prostomium into the anterior segments. Extended gills and feeding palps are enclosed in a "cage" of long anterior chaetae. Cuticle has embedded sand grains. Hemocytes have chlorocruorin. *Flabelligera, Piromis.*

Sabellida^{O}

Suspension feeders having a crown of tentacles in the form of a tiara, funnel, spiral, or bilobed fringe. Prostomium and peristomium are fused.

Oweniidae^{F}: *Owenia fusiformis* (shingle-tube worm; Fig. 13-49A) builds a cylindrical tube of flat shell fragments incorporated into the organic tube fabric like overlapping roof tiles or shingles. The tube, head-end up, is oriented vertically in the beach and projects slightly above the sediment surface. The free edges of the shell shingles are oriented upward, making it difficult for a bird or current to dislodge the tube from the sand. The overlapping arrangement of the shell fragments also allows the tube to elongate, shorten, and bend as the worm within it moves. According to an early investigation, *Owenia* is able to burrow through sand while occupying its flexible tube. *Owenia* deposit and suspension feed with a crown of short, branched tentacles (Fig. 13-19). Sediment particles are collected during feeding and those suitable for tube construction are stored in a ventral pouch below the mouth. During construction of the tube, the movable pouch fastens a fragment to the margin of the tube. The membranous tube lining is secreted by paired glands in each of the first seven trunk segments. The secretion is applied by parapodial chaetae as the worm revolves. *Myriochele.*

Sabellariidae^{F}: *Sabellaria* and *Phragmatopoma* (mason worms) build straight or curved cylindrical sand-grain tubes with walls more than one grain thick (Fig. 13-50A,B). The worms typically aggregate, sometimes by the millions, and cement their tubes together to form hard crusts or boulder-size mounds that resemble giant honeycombs. In warm-water areas such as the eastern coast of southern Florida and parts of England, these shallow-water structures form reefs (Fig. 13-50C,D) that often are degraded by storms but quickly rebuilt. Anteriorly, two fused segments have grown forward and dorsally to form an operculum for blocking the tube entrance (Fig. 13-50A,B).

Sabellidae^{F}, Serpulidae^{F}, Spirorbidae^{F}: Among the most beautiful of the sedentary polychaetes are the fan worms,

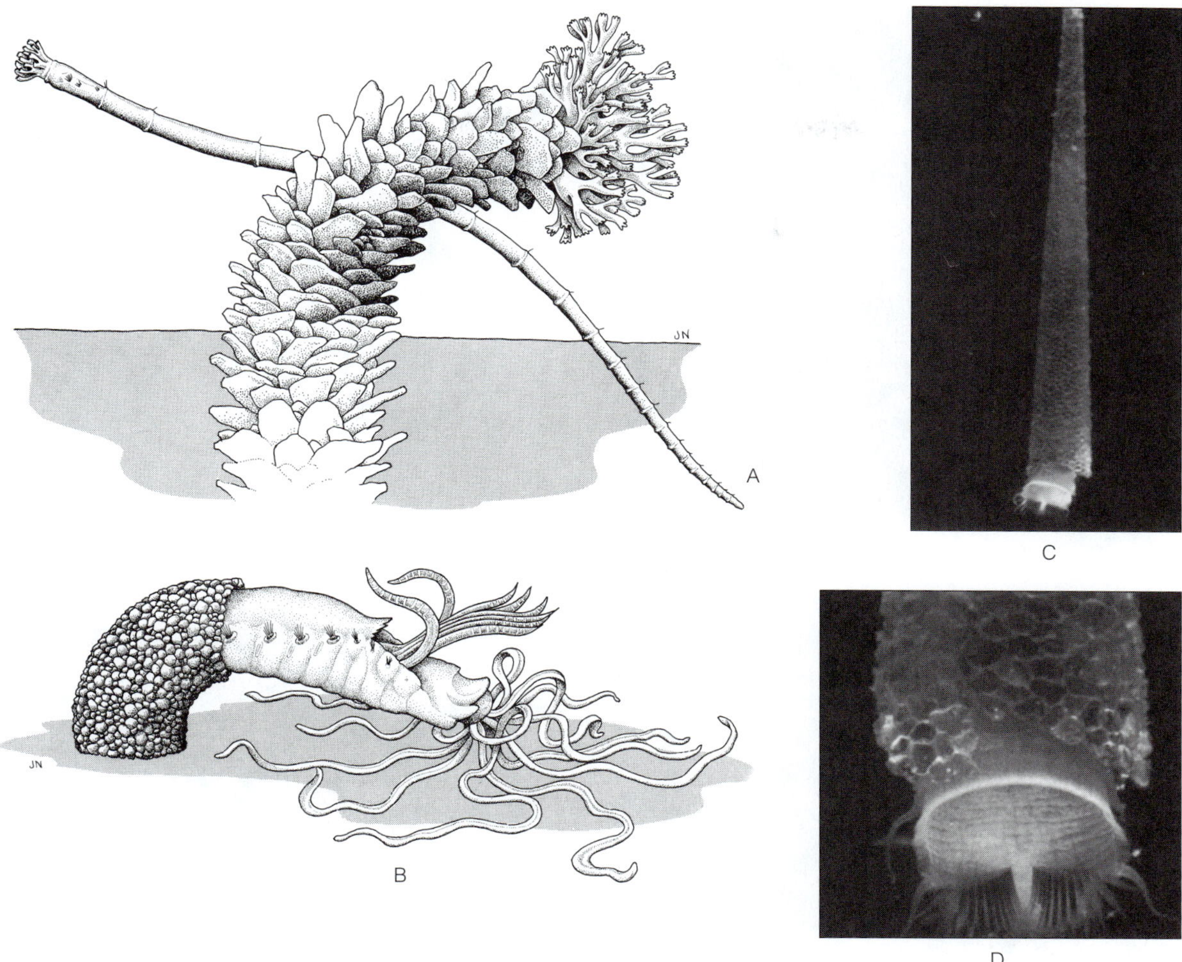

FIGURE 13-49 Polychaete diversity (Palpata, Canalipalpata). Deposit-feeding polychaetes. **A,** The shingle-tube worm, *Owenia fusiformis* (Oweniidae), which also suspension feeds, shown in its tube of flat shell fragments (foreground) and removed from its tube. **B,** The spaghetti-mouth worm, *Melinna maculata* (Ampharetidae), surface deposit feeding using its oral tentacles. **C,** The ice cream cone worm, *Pectinaria gouldii* (*Cistenides gouldi*; Pectinariidae). The hard conical tube is constructed of a single layer of quartz sand grains. **D,** *Pectinaria* digs with its stout chaetae (paleae). *(From Ruppert, E. E., and Fox, R. S. 1988. Seashore Animals of the Southeast. University of South Carolina Press, Columbia. 429 pp.)*

feather-duster worms, or Christmas-tree worms. These families have prostomial palps that form a funnel-shaped or spiral crown of from a few to many bipinnate tentacles called **radioles;** the side branches of each are the **pinnules** (Fig. 13-51A, 13-52A). The radioles are rolled up or folded together as the worm withdraws into its protective tube. Sabellids build membranous or sand-grain tubes. Because serpulids and spirorbids secrete calcareous tubes attached to rocks, shells, or algae, they can live protected on more or less exposed substrata. The most dorsal radiole on one or both sides of a serpulid or spirorbid is modified into a long, stalked knob, the **operculum** (Fig. 13-51B,C), which acts as a protective tube plug when the crown is withdrawn. Sabellids lack an operculum, but when the animal withdraws into its flexible organic tube, the opposite walls of the opening press together like closed lips. Sometimes, the compressed tube's end folds over.

The peristomial region is a distinct collar that folds over the rim of the tube opening and molds secreted tube material onto the growing rim of the tube. When additions are made to the tube, secretions flow out between the collar and the body wall. This space then acts as a mold in which the secretion hardens into a new ring that is grafted onto the end of the tube. The organic tube components are secreted from glandular areas on the ventral surface (shields) of the anterior segments. In serpulids, the collar

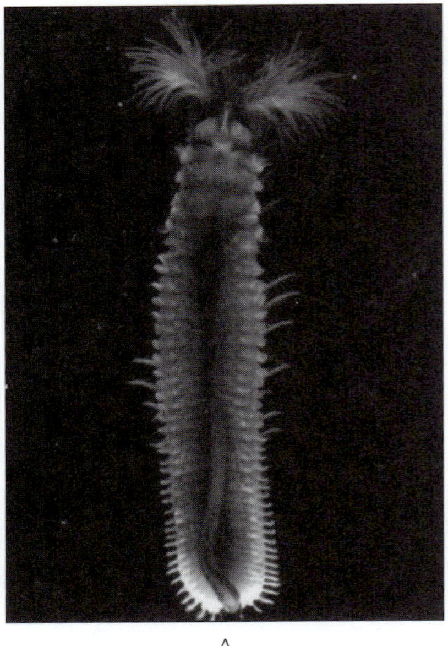

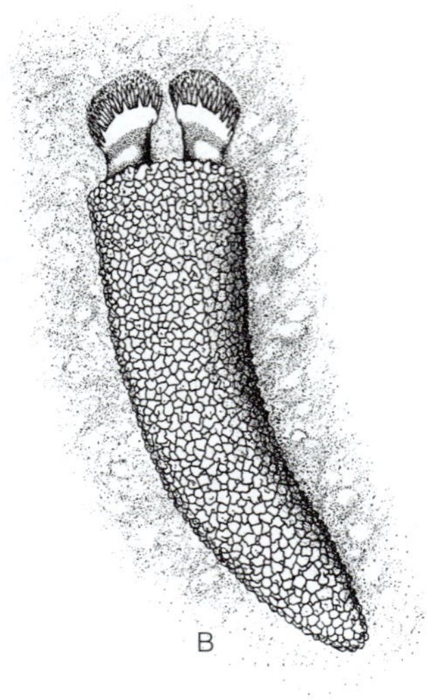

From Wilson, D. P. 1974. *Sabellaria* colonies at Duckpool, North Cornwall, 1971–1972, with a note for May 1973. J. Mar. Biol. Ass. U.K. 54:393–436.

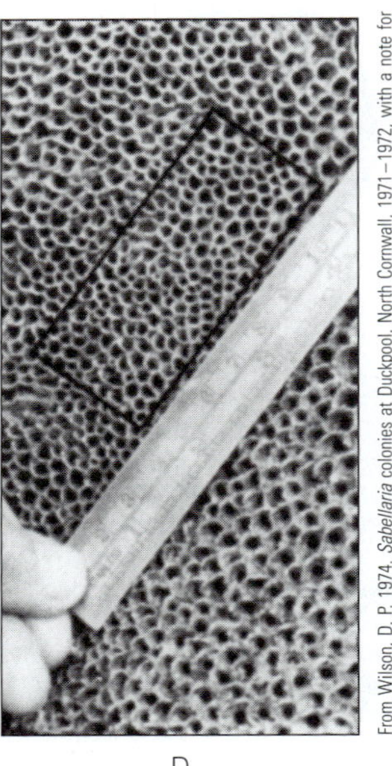

From Wilson, D. P. 1974. *Sabellaria* colonies at Duckpool, North Cornwall, 1971–1972, with a note for May 1973. J. Mar. Biol. Ass. U.K. 54:393–436.

FIGURE 13-50 Polychaete diversity (Palpata, Canalipalpata). Mason worms (Sabellariidae). **A,** Ventral view of *Sabellaria floridana* removed from its sand-grain tube. Its suspension-feeding tentacles are expanded, and behind them are two lobes bearing golden chaetae that form a closing operculum when the animal withdraws into its tube. The soft tubular structure in the animal's midline is the reduced posterior end of the body, which bends anteriorly. It contains only the intestine and is a provision for releasing feces at the mouth of the tube. **B,** Opercular bundles of chaetae projecting from the tube mouth of *S. vulgaris.* Like *S. floridana,* this species builds sand-grain tubes attached to shells, stones, pilings, and other objects along much of the Atlantic coast of North America. Farther south and on parts of the California coast, other species build reefs as shown in **C. C,** An intertidal rock on the Cornish coast of England encrusted with the tubes of *S. alveolata,* a species that forms colonies. Tubes are oriented at right angles to the substratum and are attached to each other like cells in a honeycomb. The scale is 15 cm long. **D,** Enlarged view of a small area of crust surface showing the tube openings.

bears two large calcium-carbonate-secreting glands that release calcareous crystals, which are added to the organic matrix.

Sabella constructs a tube of sand grains embedded in mucus. The worm sorts detritus collected by the ciliated radioles, and sand grains of suitable size for tube construction are stored in a pair of opposing ventral sacs below the mouth. The sac walls produce mucus, which is mixed with the sand grains. To make tube additions (Fig. 13-52A), the ventral sacs deliver a stringlike slurry of mucus and sand grains to a

FIGURE 13-51 Polychaete diversity (Palpata, Canalipalpata). Filter-feeding fan, feather-duster, or peacock worms. **A,** *Sabella pavonina* (Sabellidae), showing the expanded radioles projecting from the apertures of their organic tubes. **B,** *Hydroides* (Serpulidae) with radioles and operculum extended from the end of the calcareous tube attached to a rock. **C,** *Spirorbis* (Spirorbidae), a common animal with a *coiled* calcareous tube found attached to a variety of substrata, including algae and seagrass. Cutaway of shell shows eggs being brooded in the tube.

midventral division of the collar. As the worm rotates slowly in the tube, the collar acts like a pair of hands, molding and attaching the mucus-sand string to the rim of the tube.

Lateral cilia on the pinnules create a water current across the wall of the radiolar funnel from outside to inside and the exhaust is discharged from the upper, wide end of the funnel. Particles are trapped on the downstream side of the pinnules and then transported into a food groove along the length of each radiole. Particles move down the food groove to the base of the radiole, where they are sorted according to size (Fig. 13-52B). The largest particles are rejected and fine material is carried by ciliated tracts into the mouth. Many sabellids sort particles into three grades and store the medium grade for use in tube construction (Fig. 13-52B).

On coral reefs the paired, brightly colored, spiral radioles of Christmas-tree worms (the serpulid *Spirobranchus*) are common on the surface of living corals. The worms bore into and are protected by the coral skeleton. Boring is initiated by a newly settled juvenile, but the boring mechanism is still largely unknown. Sabellidae: *Branchiomma, Manyunkia, Megalomma, Myxicola, Potamilla, Oriopsis, Spirographis;* Serpulidae: *Apotomus, Hydroides, Pomatoceros, Spirobranchus;* Spirorbidae: *Filograna, Spirorbis.*

Pogonophora[O]

The 80-plus species of beard worms are remarkable deepwater polychaetes associated with continental slopes and hydrothermal vents. All are slender (some threadlike), sedentary worms that occupy chitin-protein tubes. The smallest is approximately 5 cm in length and *Riftia pachyptila* of the Galapagos Rift is the giant, measuring up to 1.5 m in length and almost 4 cm in diameter (Fig. 13-53).

Like many sedentary tube dwellers, the pogonophoran body exhibits pronounced regional specialization (Fig. 13-54A,B). The prostomium bears a "beard" of from 1 to many (up to 200,000) long, slender palps ("tentacles") that may be fused together into sheets in some species. The peristomium, called the forepart or vestimentum depending on taxon, secretes the tube. The long trunk is not conspicuously segmented, although it bears girdles of chaetae (uncini) in some species;

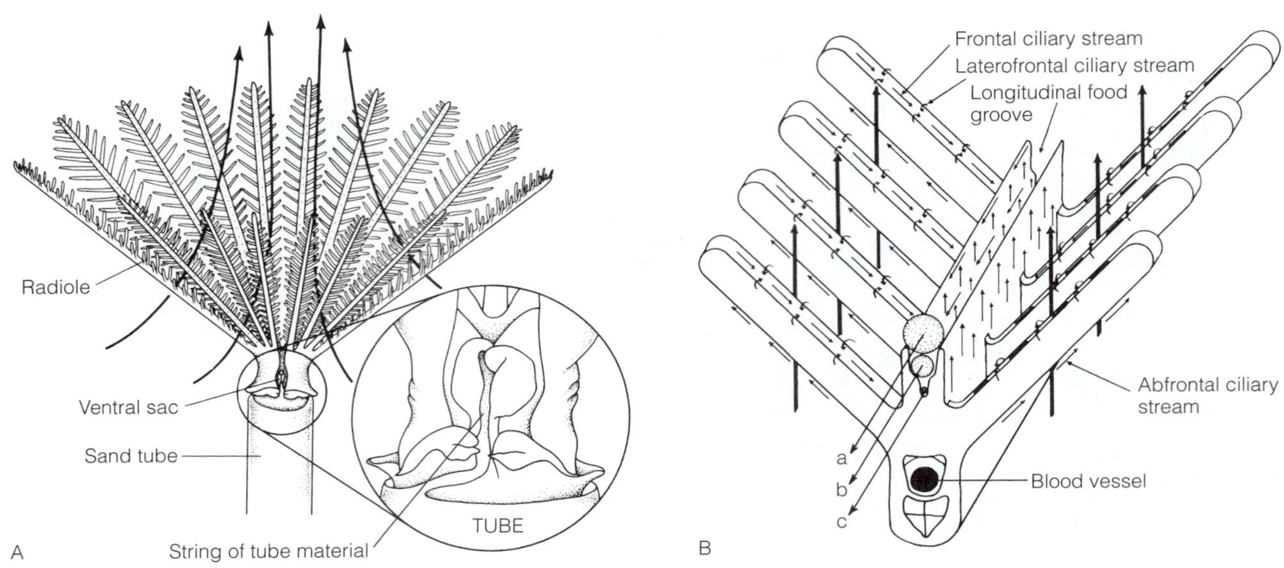

FIGURE 13-52 Polychaete diversity (Palpata, Canalipalpata). Filter-feeding feather-duster worms (Sabellidae). **A,** Anterior end of *Sabella,* showing the filter-feeding currents and tube building. **B,** Filter feeding in *Sabella,* showing water current (large arrows) and ciliary tracts (small arrows) over a section of one radiole. The letters a, b, and c indicate the different sizes of the particles sorted. *(Modified after Nicol from Newell, R. C. 1970. Biology of Intertidal Animals. American Elsevier Co., New York.)*

in others, trunk chaetae are absent. At the posterior end of the body is a short, bulbous **opisthosoma** composed of several short segments (up to 82), each of which has paired bundles of chaetae (also uncini). The opisthosoma and chaetae anchor the worm in its tube (Fig. 13-54C), allowing it to withdraw rapidly.

Internally, coelomic compartments exist in each of the body divisions, and the head coelom extends into the tentacles. In the opisthosoma, septa occur between the segmental coelomic compartments (Fig. 13-55A), but the trunk has no septa and the coelom is continuous in this region. A well-developed hemal system supplies each tentacle with two vessels. Hemoglobin usually occurs in the hemal system and sometimes also in the coelomic fluid. Pogonophorans have a pair of nephridia, probably modified protonephridia, situated in the prostomium or peristomium, but the nature of these organs is not yet clear. The nervous system consists of a nerve plexus at the base of the epidermis and an intraepidermal ventral nerve cord.

Adult pogonophorans lack a mouth, anus, and normal digestive tract. The midgut consists of a mass of tissue called the trophosome that is packed with symbiotic bacteria (Fig. 13-54D, 13-55). The bacteria oxidize sulfur-containing compounds and use the resultant energy to fix carbon. The pogonophoran host obtains its nutrition from the organic compounds synthesized by its bacteria and by lysis and subsequent absorption of the bacteria themselves. The vascular hemoglobin, which accounts for the red color of the tentacular crown of *Riftia* and relatives, binds and delivers the large amounts of oxygen and sulfur-containing compounds required

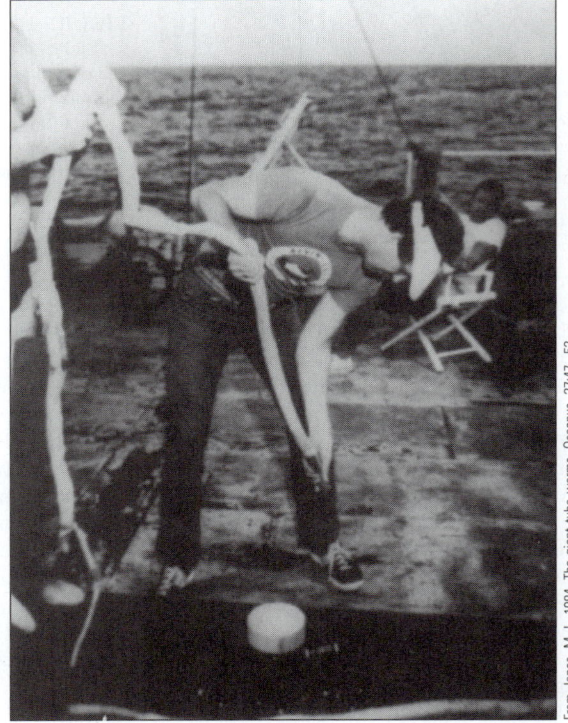

From Jones, M. L. 1984. The giant tube worms. Oceanus. 27:47–52.

FIGURE 13-53 Polychaete diversity (Palpata, Canalipalpata). Pogonophora. The giant vestimentiferan tube worm, *Riftia pachyptila,* from the Galapagos undersea rift.

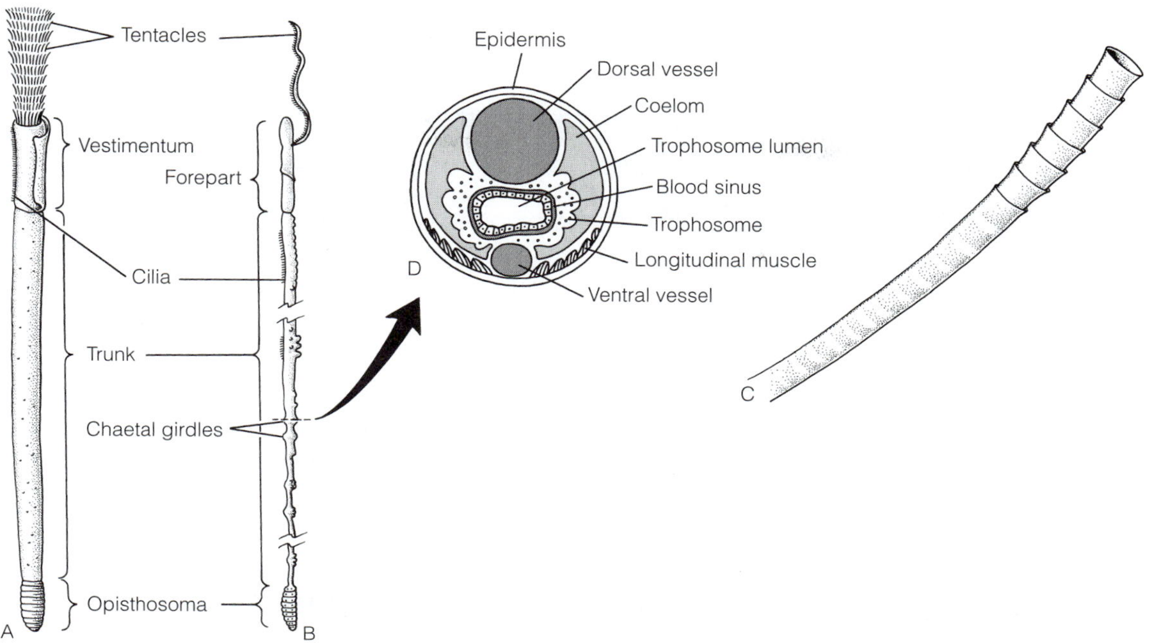

FIGURE 13-54 Polychaete diversity (Palpata, Canalipalpata). Anatomy and tube of Pogonophora. Vestimentiferan **(A)** and perviate pogonophoran organization **(B)**. **C,** Tube of the vestimentiferan *Ridgeia.* **D,** Cross section of the trunk of the perviate *Siboglinum,* showing the trophosome (modified gut). *(A, B, and D, Modified and redrawn from Southward, E. C. 1982. Bacterial symbionts in Pogonophora. J. Mar. Biol. Assoc. U.K. 62:889–906; C, Modified and redrawn from Southward, E. C. 1988. Development of the gut and segmentation of newly settled stages of* Ridgeia *(Vestimentifera): Implications for relationship between Vestimentifera and Pogonophora. J. Mar. Biol. Assoc. U.K. 68:465–487)*

by the chemoautotrophic bacteria. Supplementary nutrition may, however, be supplied by uptake of organic compounds dissolved in the surrounding seawater. Dependence on symbiotic bacteria for nutrition is not unique to pogonophorans. This mode of nutrition is found in certain gutless clams and some other taxa, including other polychaetes and some marine oligochaetes. Pogonophoran juveniles have a complete, annelid-like gut.

Most pogonophorans are gonochoric, having two gonads associated with the trunk coelom. The two male gonopores open dorsally on the anterior end of the trunk or on the peristomium. The female gonopores open onto the peristomium or on the trunk behind the male pores, depending on taxon. Beard worms package their sperm into spermatophores. Sperm transfer, fertilization, and egg deposition have not been observed, but one specimen of *Siboglinum* was seen to move spermatophores to the mouth of the tube with its tentacles. The spermatophores perhaps reach the tubes of neighboring females by floating. The fertilized eggs of many pogonophorans are brooded in their maternal tube.

Cleavage is bilateral, presumably derived from spiral cleavage. The larvae of *Ridgeia* are clearly trochophores (Fig. 13-56). Embryos of many pogonophorans are lecithotrophic and may develop directly in the adult or be released as pelagic lecithotrophs. Whether such larvae are dispersed by currents or quickly sink to the bottom is unknown.

Pogonophora was formerly considered to be the sister taxon of Annelida and was assigned to the Linnean category of phylum. Once thought to be oligomeric coelomates without a gut, the discovery of a segmented opisthosoma with chaetae, a trochophore larva, and a juvenile with a complete gut all established pogonophorans as annelids. The presence of grooved prostomial palps and uncini indicate that they are likely to be canalipalpatan polychaetes. Molecular systematic data also support the inclusion of Pogonophora in Polychaeta.

PHYLOGENY OF POLYCHAETA

The phylogeny shown in Figure 13-57 indicates two major polychaete clades, Scolecida and Palpata. Scolecids are deposit-feeding burrowers without head appendages, much like the presumed ancestor of annelids, while palpate polychaetes have

PHYLOGENETIC HIERARCHY OF POLYCHAETA

Polychaeta
- Scolecida
- Palpata
  - Aciculata
  - Canalipalpata

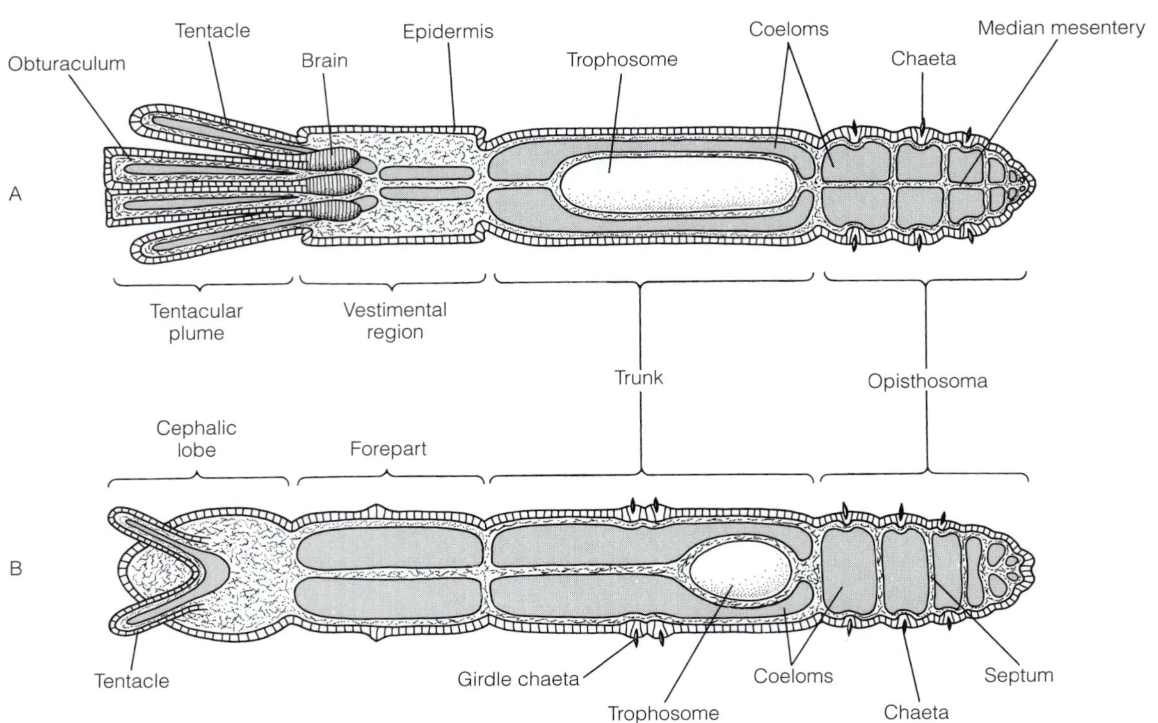

FIGURE 13-55 Polychaete diversity (Palpata, Canalipalpata). Pogonophoran coelomic anatomy. **A,** Vestimentiferan; **B,** perviate. *(Modified and redrawn from Southward, E. C. 1988. Development of the gut and segmentation of newly settled stages of* Ridgeia *(Vestimentifera): Implications for relationship between Vestimentifera and Pogonophora. J. Mar. Biol. Assoc. U.K. 68:465–487.)*

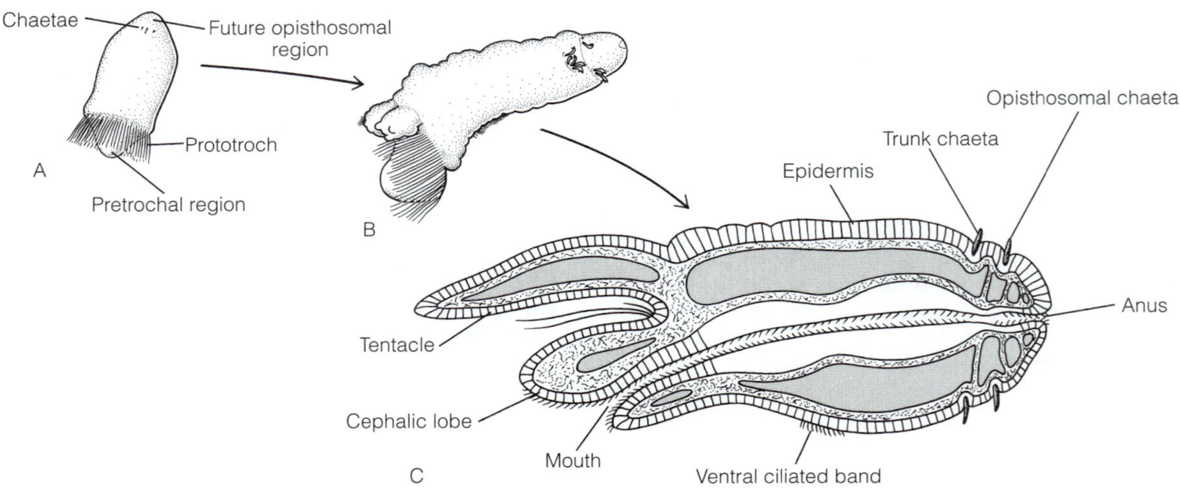

FIGURE 13-56 Polychaete diversity (Palpata, Canalipalpata). Pogonophoran larval development. **A,** Trochophore of the vestimentiferan *Ridgeia* in lateral view. **B,** An early juvenile stage of the *Ridgeia.* **C,** Sagittal section of juvenile *Ridgeia* showing the presence of a typical annelidan gut. Later in development, the gut and associated tissues become modified to form the trophosome. *(A, After a sketch provided by S. L. Gardiner; B and C, Redrawn from Southward, E. C. 1988. Development of the gut and segmentation of newly settled stages of* Ridgeia *(Vestimentifera): Implications for relationship between Vestimentifera and Pogonophora. J. Mar. Biol. Assoc. U.K. 68:465–487.)*

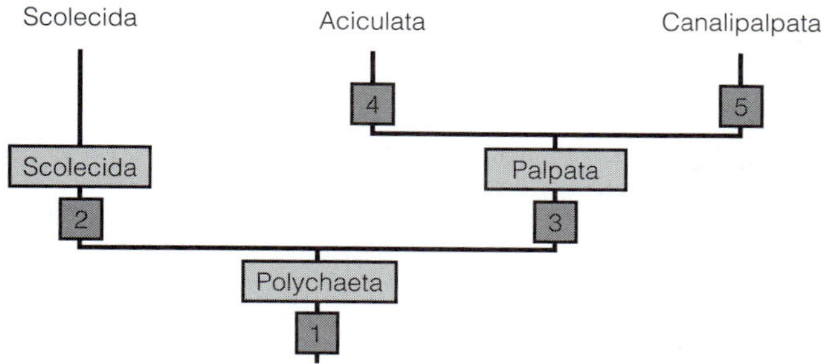

FIGURE 13-57 Phylogeny of Polychaeta. **1, Polychaeta:** parapodia, nuchal organs, pedal ganglia, one pair of pygidial cirri. **2, Scolecida:** two or more pairs of pygidial cirri. **3, Palpata:** one pair of sensory palps on prostomium. **4, Aciculata:** biramous parapodia, each having a notopodium and neuropodium; each parapodial lobe (ramus) has at least one skeletal rod (aciculum) linking parapodial muscle to the lobe and its chaetae; dorsal and ventral cirri on parapodia; antennae (two or three) on prostomium. **5, Canalipalpata:** grooved feeding palps (transformed sensory palps) on prostomium or (secondarily) peristomium. *(Modified after Rouse, G. W., and Fauchald, K. 1997. Cladistics and Polychaetes. Zool. Scripta 26:139–204; and Ax, P. 2000. Multicellular Animals. The Phylogenic System of Metazoa. Vol. II. Springer Verlag, Berlin.)*

head appendages and a wide variety of feeding modes and lifestyles. One can imagine, then, the polychaete progenitor emerging from the mud and radiating into crawlers, swimmers, and tube dwellers. If the first polychaete was a burrower like the parapodia-less annelid ancestor, what was the functional role of its parapodia? Perhaps, along with chaetae, they prevented backslipping as the head or pharynx pushed into the sediment? Or perhaps parapodia originally were little more than low ridges bearing the chaetae? An untested alternative is that the original annelids had parapodia, which were lost in earthworms and leeches (Clitellata) and reduced independently in several taxa of polychaetes.

CLITELLATA

The clitellate annelids, called girdle worms, include the earthworms, many small and inconspicuous worms, as well as leeches and their relatives. All clitellates lack parapodia and head and pygidial appendages. Clitellates have a **clitellum,** a series of anterior segments enclosed in a thick, glandular epidermis that often forms a conspicuous girdle around the body. The clitellum either encompasses the female gonopores or is located behind them. The clitellum secretes mucus for copulation, nutritious albumen for eggs, and a **cocoon** in which eggs and albumen are deposited. In contrast to the gonochoric spawning polychaetes, clitellates are copulating hermaphrodites. Clitellate gonads are always restricted to a few genital segments with the testes anterior to the ovaries. The zygotes develop in the cocoon and emerge as juveniles, so clitellates have direct development. Unlike the ancestral annelid and the polychaetes, the clitellate brain has shifted posteriorly from the prostomium to an anterior trunk segment.

OLIGOCHAETA[C]

The 3500-some species of oligochaetes include the earthworms and many small freshwater and marine species. The small-bodied freshwater oligochaetes burrow in bottom mud and silt whereas others live among submerged vegetation. Of the 200 marine species, most are tiny animals that live between sand grains at depths from shallow intertidal beaches to the deep sea. The largest oligochaetes are the earthworms, including the up-to-3-m-long giant Australian earthworm, *Megascolides australis* (Fig. 13-58).

Form and Function

The oligochaete body is similar to that of the generalized annelid, with well-developed segments, four bundles of chaetae per segment, a small prostomium lacking appendages, and a small pygidium also devoid of appendages (Fig. 13-59A). In a few oligochaetes, however, the prostomium itself tapers into a tentacle-like appendage, as in *Stylaria* (Fig. 13-59C).

Oligochaete chaetae are simple (unjointed) and terminate in a needlelike point or tips that are bifid, pectinate, or otherwise different from the shaft (Fig. 13-59B). Chaetae on the genital segments often are more complex. In general, long chaetae typify aquatic species (Fig. 13-59C), whereas those of earthworms project only a short distance beyond the integument and are commonly sigmoid. On each side of a segment are chaetal sacs (Fig. 13-59B) in which the chaetae are secreted and from which they emerge as groups or bundles. Two chaetal bundles are ventral and two are ventrolateral or dorsolateral (Fig. 13-61A). The number of chaetae per bundle varies from 1 to 25 (Fig. 13-59C). In most earthworms, such as *Lumbricus,* and in some aquatic families, only eight chaetae occur per segment, two chaetae in each of the four bundles (Fig. 13-61A). Whatever the number, there are fewer in these worms than in most polychaetes, hence the origin of the name *Oligochaeta,* meaning "few chaetae." Attached to the base of each chaeta are protractor and retractor muscles that allow the chaeta to be extended or withdrawn (Fig. 13-59B).

Body Wall and Coelom

The oligochaete body wall, especially in terrestrial species, is essentially like that of burrowing polychaetes. A thin cuticle overlies an epidermal layer that contains mucous gland cells. The circular muscles are outermost, the well-developed longitudinal muscles are in four bands, and the septa are relatively complete. Earthworms, which have the best-developed septa, can have sphincters around septal perforations to control the flow of coelomic fluid from one segment to another.

In most earthworms, each coelomic compartment, except at the extremities, is connected to the outside by a mid-dorsal **coelomopore** situated in the intersegmental furrows and provided with a sphincter. These pores exude coelomic fluid, which aids in keeping the integument moist and may deter predators. When disturbed, some earthworms squirt fluid from their pores. The giant Australian earthworm ejects fluid to a height of 10 cm, whereas the "squirter worm" *(Didymogaster sylvaticus)* squirts fluid to a height of 30 cm.

Photograph courtesy of Globe Photos.

FIGURE 13-58 Oligochaeta. An Australian giant earthworm.

Locomotion

Oligochaetes crawl and burrow by peristaltic contractions (Fig. 13-60), as described for burrowing annelids in Evolution and Significance of Segmentation and Figure 13-8. Chaetae are extended as the body shortens (longitudinal muscle contraction) and retracted as the body elongates (circular muscle contraction). Each segment moves forward in steps of 2 to 3 cm at the rate of 7 to 10 steps per minute (Fig. 13-60). The direction of contraction waves can be reversed, enabling the worm to crawl backward. Freshwater species move through bottom debris and algae in the same manner as burrowing earthworms, but the microscopic aeolosomatids swim by means of a ciliated prostomium (Fig. 13-66).

Nervous System

In most oligochaetes, the two nerve cords have fused in the ventral midline and are situated inside the muscle layers of the body wall. As in all clitellates, the oligochaete brain has shifted posteriorly, and in lumbricids it lies in the third segment, above the anterior end of the pharynx (Fig. 13-61B, 13-62).

Within the ventral nerve cord of earthworms are five giant axons, three large-diameter axons in the dorsal part of the cord and two smaller-diameter axons in the ventral half (Fig. 13-61A). The mid-dorsal axon is fired by sensory input from the head, and the two dorsolateral fibers by sensory input from the posterior end of the body. In each segmental ganglion, these giant axons synapse with giant axons in the segmental nerves, which innervate the longitudinal muscles. Appropriate sensory input from either end of the worm elicits an escape response (wriggling and withdrawal into burrow), but it is more rapid and vigorous when the head of the worm is grasped or otherwise disturbed. Earthworms usually emerge from their burrows headfirst, and presumably the head end is most likely to be seized by a bird or other predator. Anyone who has tried to collect nightcrawlers can attest to the speed with which the worms retract into their burrows when touched or exposed to a bright light.

The subpharyngeal ganglion is the principal center of motor control and vital reflexes and dominates the succeeding ganglia in the chain. All movement ceases when the subpharyngeal ganglion is destroyed. Motor control continues normally following removal of the brain (the suprapharyngeal, or cerebral, ganglion), but the worm loses its ability to correlate movement with external environmental conditions. The relation of the subpharyngeal ganglion to the brain is thus somewhat similar to the relation of the medulla to the higher brain centers in vertebrates.

Sense Organs

Oligochaetes lack eyes, except for a few aquatic forms that have simple pigment-cup ocelli. The integument, however, is well supplied with dispersed unicellular photoreceptors situated in the inner part of the epidermis, especially dorsally at the anterior end. The other receptors have a more or less general distribution in the integument. Clusters of sensory cells that form a projecting tubercle with sensory processes extending above the cuticle appear to be chemoreceptors. The **tubercles** form three rings around each segment and are particularly numerous on the more anterior segments, especially the prostomium, where (in *Lumbricus*) there may be as many as $700/mm^2$.

Nutrition and Digestive System

Most species of oligochaetes, both aquatic and terrestrial, are scavengers that feed on dead organic matter, particularly vegetation. Earthworms feed on decomposing matter at the surface and may pull leaves into the burrow. They also deposit feed, ingesting organic material obtained from mud or soil while burrowing. Fine detritus, algae, and other microorganisms are important food sources for many tiny, freshwater species. The common, minute *Aeolosoma* collects detritus with its prostomium (Fig. 13-66). After the ciliated ventral surface of the prostomium is placed against the substratum, the center is elevated by muscular contraction. The partial vacuum dislodges

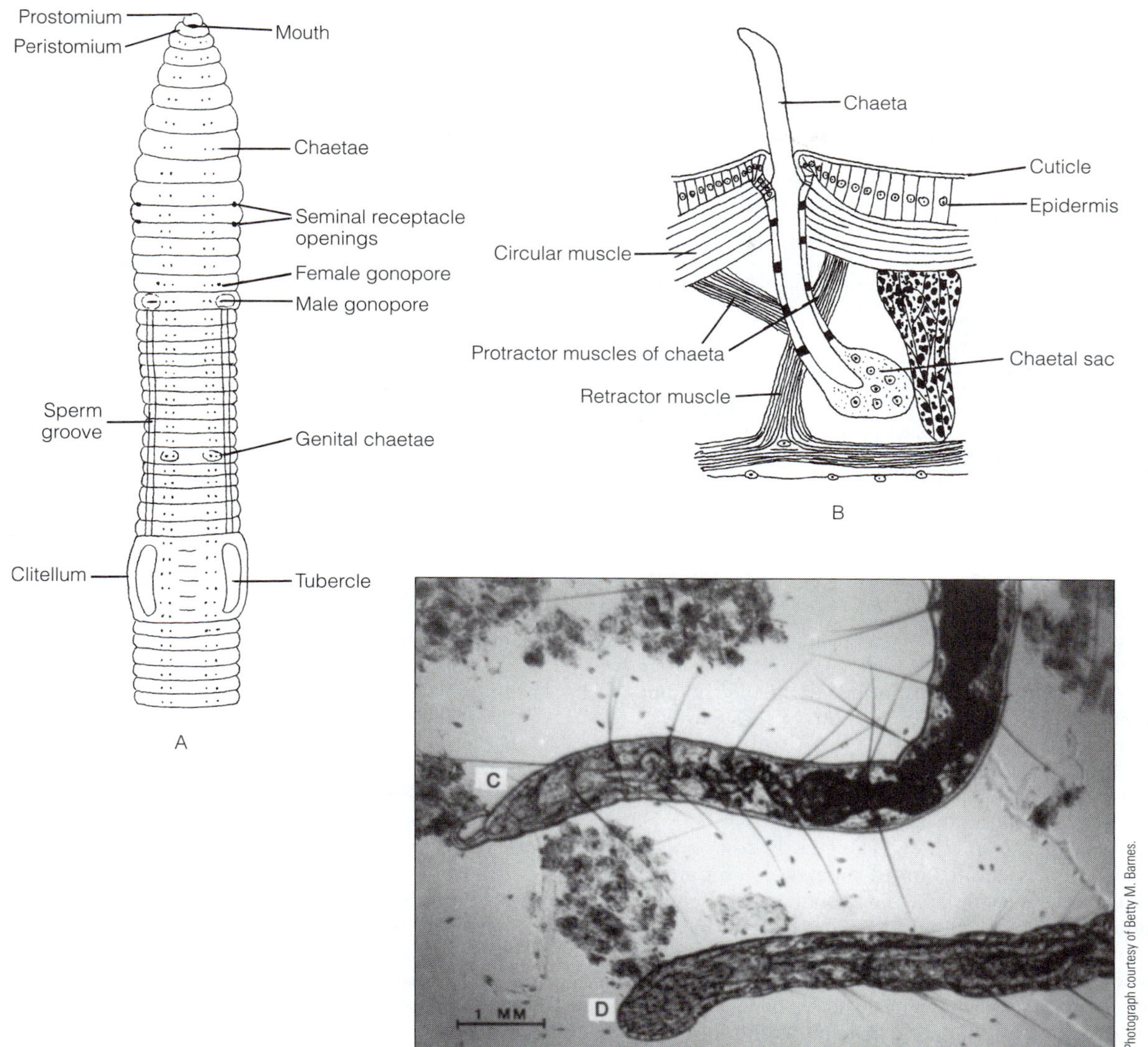

FIGURE 13-59 Oligochaete anatomy and diversity. **A,** Anteroventral surface of the earthworm *Lumbricus terrestris* (Lumbricidae). **B,** Body wall of the earthworm *Pheretima* (transverse section; Megascolecidae). **C** and **D,** Anterior ends of two freshwater oliogochaetes, *Stylaria* **(C)** and *Aeolosoma* **(D).** *Stylaria* has a long, tentacular prostomium. *(A, After Stephenson from Avel; B, After Bahl from Avel.)*

particles, which are then swept into the mouth by cilia. Species of *Chaetogaster,* which feed by using a sucking action of the pharynx, are small, carnivorous oligochaetes that eat amebas, ciliates, rotifers, and trematode larvae, or parasitize freshwater snails.

The oligochaete digestive tract is straight and relatively simple (Fig. 13-61B). The mouth, situated beneath the prostomium, opens into a small buccal cavity, which in turn opens to a more spacious pharynx. The principal ingestive organ is a muscular **mid-dorsal pharyngeal bulb** or pad. In aquatic species, the pharyngeal bulb is everted and collects particles on its adhesive surface (Fig. 13-63). In earthworms, the pharynx acts as a pump. Pharyngeal glands produce a salivary secretion containing mucus and enzymes.

The pharynx opens into a narrow, tubular esophagus, which may be modified at different levels to form a gizzard and, in lumbricid earthworms, a crop (Fig. 13-61B, 13-63A). In some, there is a succession of 2 to 10 gizzards, each occupying a separate segment. The **gizzard,** which is used for grinding food particles, is lined with a chitinous cuticle and is very muscular. The thin-walled **crop** is a storage chamber.

The esophageal wall of many oligochaetes houses **calciferous glands.** When highly developed, the glands bulge from the wall of the esophagus and may appear externally as lateral or

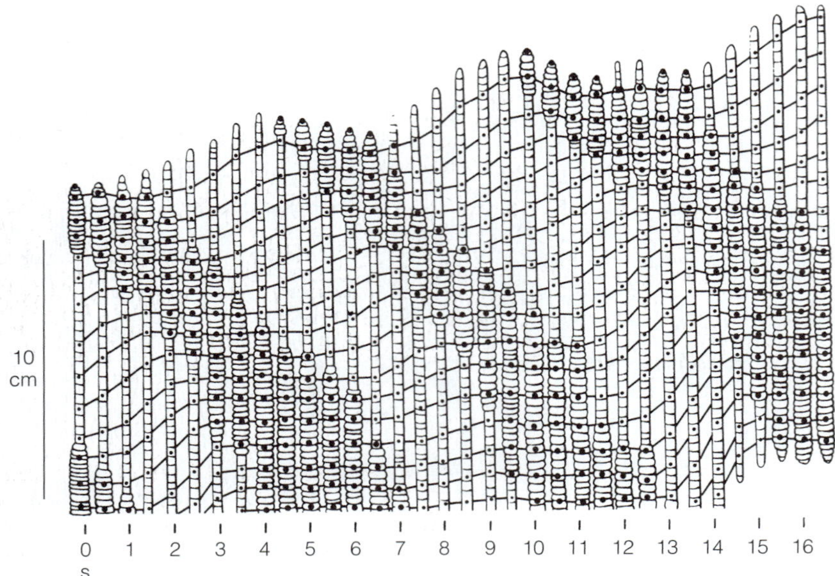

FIGURE 13-60 Oligochaete movement. Diagram of earthworm locomotion. Segments undergoing longitudinal muscle contraction are marked with the larger dot and drawn twice as wide as those undergoing circular muscle contraction. The forward progression of a segment during the course of several waves of circular muscle contraction is indicated by the horizontal lines connecting the same segments. *(From Gray and Lissman, 1938.)*

dorsal swellings (Fig. 13-61B). The calciferous glands secrete calcium carbonate into the esophagus in the form of calcite crystals. The crystals are transported along the gut but not reabsorbed, and eventually they pass out of the body with the feces. The calciferous glands do not play a role in digestion and their function is uncertain, but two hypotheses have been suggested. According to aquatic chemists, soil CO_2 levels can be several hundred times higher than atmospheric levels because of bacterial respiration. When earthworms encounter such surroundings, elimination of their own respiratory CO_2 by diffusion may be hampered by an unfavorable concentration gradient. To overcome this difficulty, CO_2 (as bicarbonate ions) in the blood and other tissues may combine with calcium ions in the calciferous glands to form calcite, thus eliminating CO_2 indirectly via the gut. It has also been suggested that the glands function to eliminate excess calcium taken in with food.

The ciliated intestine forms the remainder of the digestive tract and extends as a straight tube through all but the anterior quarter of the body (Fig. 13-61B). The anterior half of the intestine is the principal site of enzyme secretion and digestion, and the posterior half is primarily absorptive. In addition to the usual classes of digestive enzymes, the intestinal epithelium of earthworms also contains cellulase (to digest plant cell walls) and chitinase (to digest fungal cell walls) produced by symbiotic bacteria. The absorbed food materials are passed to blood sinuses that lie between the gut epithelium and the intestinal muscles. The surface area of the intestine is increased in many earthworms by a ridge or fold called a **typhlosole,** which projects internally from the mid-dorsal wall (Fig. 13-61A).

Surrounding the intestine and investing the dorsal vessel of oligochaetes is a layer of yellowish mesothelial cells called chlorogogen cells (Fig. 13-61A), already discussed with Annelida. Waste-laden chlorogogen cells are released into the coelom, from which they, or their breakdown products, are eliminated through the nephridia or coelomopores. In terrestrial species, silicates obtained from food material and the soil are removed from the body and deposited in the chlorogogen cells as waste concretions.

Hemal System

The hemal system of oligochaetes (Fig. 13-61B) is similar to that described earlier for annelids in general. Earthworms also have integumental capillaries, however, and many oligochaetes have hearts that supplement the contractile dorsal blood vessel (Fig. 13-61B). The hearts are expanded muscular regions of the circumenteric vessels that link the ventral and dorsal longitudinal vessels. The number of such hearts varies. The five pairs of hearts of *Lumbricus* surround the esophagus (in segments 7 through 11; Fig. 13-61B). Only one pair of circumintestinal hearts is present in *Tubifex.* The hearts have valves in the form of folds in the vessel walls, which may also be found in the dorsal vessel at junctions containing segmental vessels.

Gas Exchange

Gas exchange in almost all oligochaetes, both aquatic and terrestrial, takes place by diffusion of gases through the body wall. In larger species, such as the nightcrawler *Lumbricus terrestris,* the integumental capillaries give rise to loops that enter the epidermis. Mucous-gland secretions and fluid discharged through the coelomopores moisten the surface of the epidermis and facilitate gas exchange in the earthworms.

Specialized gills occur in only a few oligochaetes. Species of the aquatic genera *Dero* (Fig. 13-64B) and *Aulophorus* have a circle of fingerlike gills at the posterior end of the body. A tubificid, *Branchiura* (Fig. 13-64A), has gill filaments dorsally and ventrally on the posterior quarter of the body.

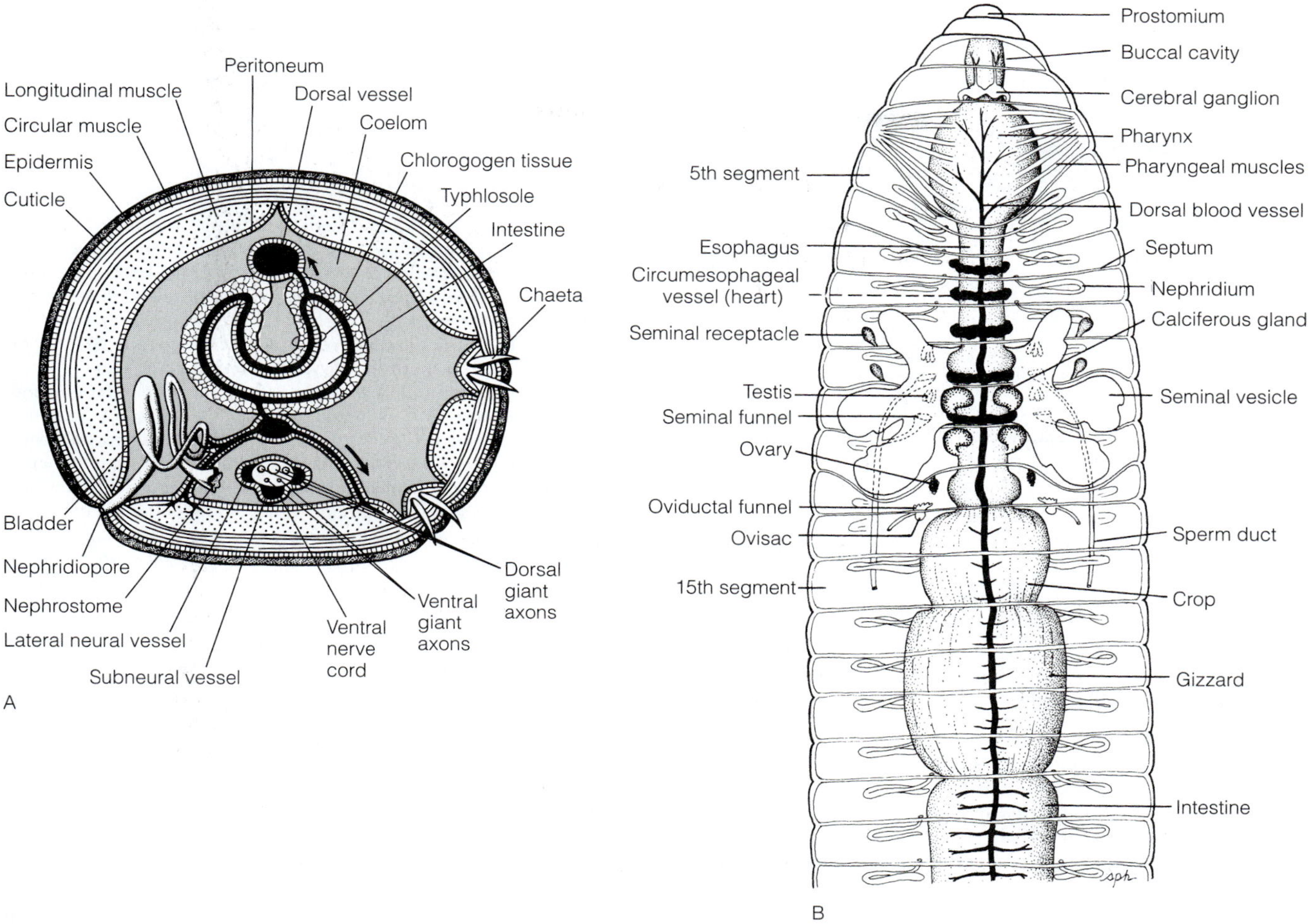

FIGURE 13-61 Oligochaete anatomy. The earthworm *Lumbricus terrestris* (Lumbricidae). **A,** Cross section through trunk. **B,** Dorsal view of the anterior end of the body. *(A Modified and redrawn from F. A. Brown, 1950 Selected Invertebrate Types. ©1950 John Wiley & Sons, Inc. Used with permission of the author.)*

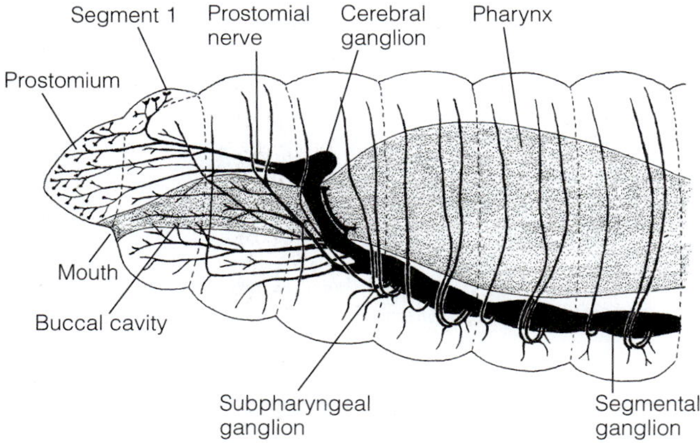

FIGURE 13-62 Oligochaete nervous system. *Lumbricus terrestris* (Lumbricidae) in left lateral view. Note the position of the cerebral ganglion in segment 3. *(After Hess from Avel.)*

The larger oligochaetes usually have hemoglobin dissolved in the blood plasma. The hemoglobin of *Lumbricus* transports 15 to 20% of the oxygen utilized under ordinary burrow conditions, in which the partial pressure of oxygen is about the same as that in the atmosphere aboveground. When the partial pressure drops, the hemoglobin compensates by increasing its carrying capacity. After heavy rains, the oxygen level drops in the saturated soil, perhaps compelling some earthworms to seek atmospheric oxygen at the surface.

Many aquatic oligochaetes tolerate relatively low oxygen levels and, for a short period, even a complete lack of oxygen. Members of Tubificidae, which live in stagnant mud and lake bottoms, are notable examples. Certain members of this family, such as *Tubifex tubifex*, die from long exposure to ordinary oxygen tensions. *Tubifex* ventilates in stagnant water by extending its posterior end into the water and waving it about (Fig. 13-64C).

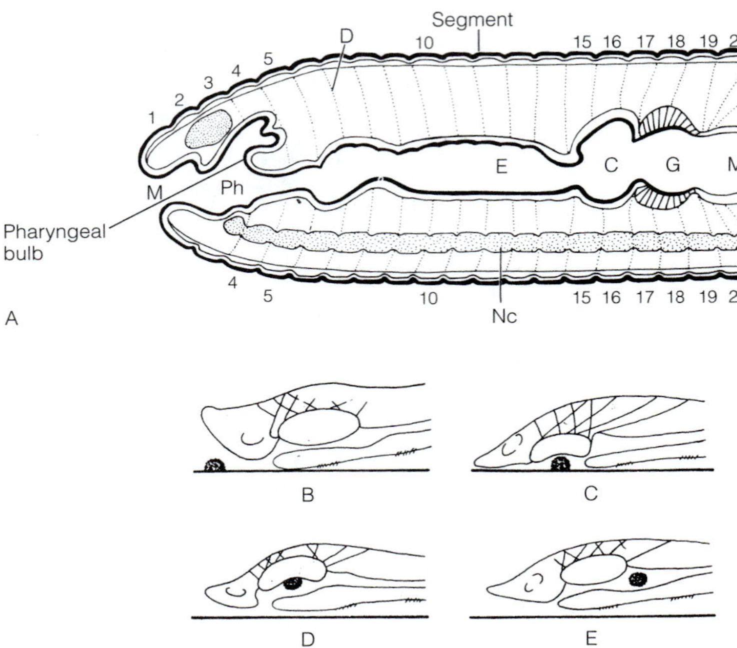

FIGURE 13-63 Oligochaete digestive system. **A,** *Lumbricus terrestris* (Lumbricidae), diagrammatic vertical section of the anterior end of the body. Note the tonguelike dorsal pharyngeal bulb. Note also that the cuticle (thick dark line) not only covers the body, but also lines the digestive tract, ending at the gizzard (G)–midgut (Mg) junction. The mouth (M), pharynx (Ph), and esophagus (E) are foregut and derived from ectoderm and the cuticle is composed of collagen; the crop (C), gizzard (G), and midgut (Mg) are midgut structures derived from endoderm and the cuticle is made of chitin. The chitinous lining of the midgut (not shown) is called a peritrophic membrane. **B,** Ingestion of deposited food using the pharyngeal bulb in *Aulophorus carteri* (Naididae). *(A, From Peters, W., and Walldorf, V. 1986. Endodermal secretion of chitin in the "cuticle" of the earthworm gizzard. Tissue Cell 18:361–374 (Fig. 1); B, C, D, and E, After Marcus and Avel.)*

Excretion and Diapause

Adult oligochaetes have a metanephridial system (Fig. 13-61), and typically, there is one pair of metanephridial tubules per segment except at the extreme anterior and posterior ends. In the segment following the nephrostome, the tubule is greatly coiled, and in some species, such as *Lumbricus,* there are several separate groups of loops or coils (Fig. 13-65B). Before the nephridial tubule opens to the outside, it may be dilated to form a bladder. The nephridiopores are usually situated on the ventrolateral surface of each segment.

In contrast to the majority of oligochaetes, which have in each segment a single typical pair of metanephridia called **holonephridia,** many earthworms of the families Megascolecidae and Glossoscolecidae are peculiar in having additional nephridia that are multiple or branched. Either typical or modified nephridia may open to the outside through nephridiopores, or they may open into various parts of the gut, in which case they are termed enteronephric. A single worm may possess a number of different types of these **enteronephridia,** each being restricted to certain parts of the body.

Earthworms excrete urea, but they are less perfectly ureotelic than other terrestrial animals. Although urea is present in the urine of *Lumbricus* and other earthworms, ammonia remains an important excretory product. The level of urea depends on the condition of the worm and the environmental situation.

Salt and water balance, of particular importance in freshwater and terrestrial environments, is regulated in part by the nephridia (Fig. 13-65C). The urine of both terrestrial and freshwater species is hyposmotic, and considerable reabsorption of

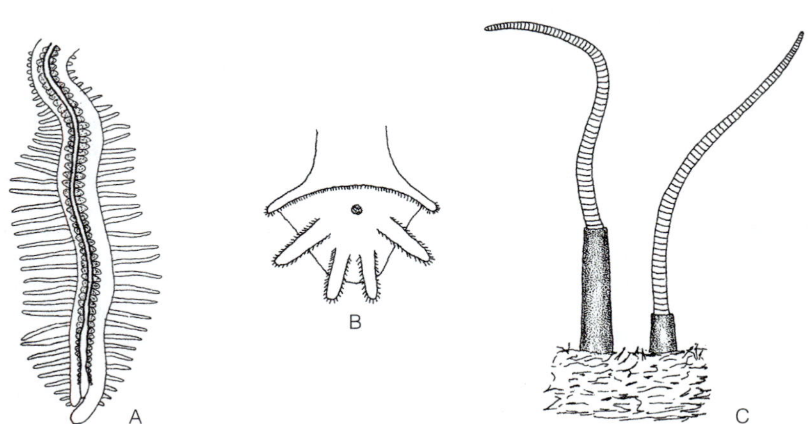

FIGURE 13-64 Oligochaete gill surfaces. **A,** *Branchiura sowerbyi* (Tubificidae), posterior end showing dorsal and ventral gill filaments. **B,** *Dero* (Naididae), posterior end showing gills around the anus. **C,** In *Tubifex tubifex* (Tubificidae), the posterior of the body extends from the tube and waves about in the water, facilitating gas exchange. *(A, After Beddard from Avel; B and C, From Pennak, R. W. 1978. Freshwater Invertebrates of the United States. 2nd Edition. John Wiley and Sons, New York.)*

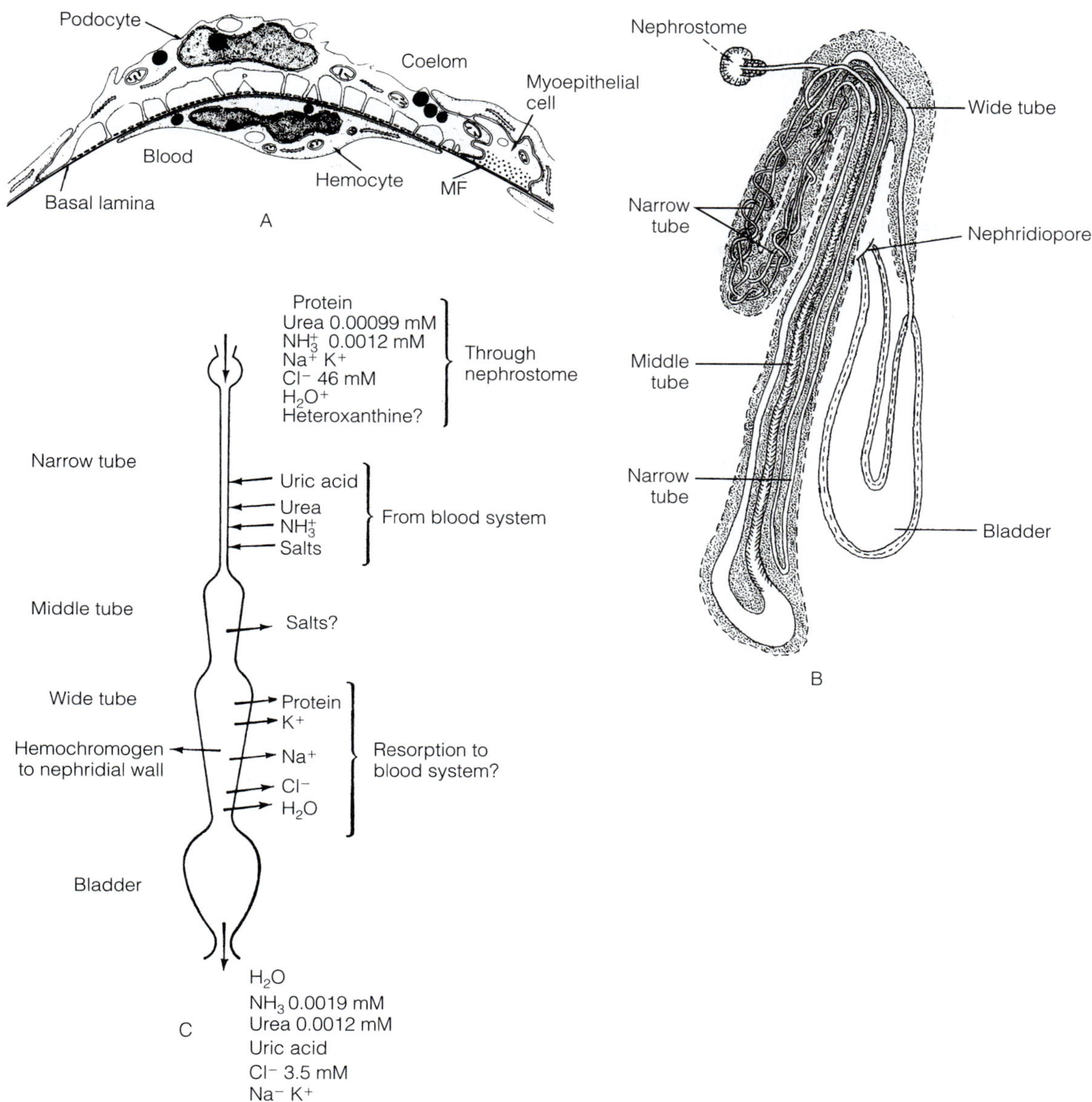

FIGURE 13-65 Oligochaete excretory system and excretion. **A,** Podocyte on the wall of the ventral blood vessel of *Tubifex tubifex* (Tubificidae). **B,** Metanephridium of *Lumbricus terrestris* (Lumbricidae). **C,** Possible functions of various parts of the *Lumbricus* metanephridium. *(A, From Peters, W. 1977. Possible sites of ultrafiltration in* Tubifex tubifex *Mueller (Annelida, Oligochaeta). Cell Tissue Res. 179:367–375; B, After Maziarski from Avel; C, Modified slightly from Laverack, M. S. 1963. The Physiology of Earthworms. Pergamon Press, Oxford, p. 67)*

salts must take place as fluid passes through the nephridial tubule. Some salts are also actively absorbed by the integument.

In the terrestrial earthworms, water absorption and loss occur largely through the integument. Under normal conditions of adequate water supply, the nephridia excrete copious hyposmotic urine. It is not certain whether reabsorption by the holonephridia is important for water conservation, but the enteronephridia do appear to be an adaptation for the retention of water. By first releasing the urine into the digestive tract, much of the remaining water can be reabsorbed in the intestine en route to the anus. Worms with enteronephridia can tolerate much drier soils than can those without them and do not have to burrow so deeply during dry periods.

A few aquatic oligochaetes are capable of encystment during unfavorable environmental conditions. The worm secretes a tough, mucous covering that forms the cyst wall. Some species form summer cysts for protection against desiccation; others form winter cysts when the water temperature becomes low.

During dry seasons and during the winter, earthworms migrate to deeper levels of the soil, down to 3 m in the case of certain Indian species. After moving to deeper levels, an earthworm can enter diapause and lose as much as 70% of its water. Balance is restored and activity resumed as soon as water is again available.

Reproduction and Development

CLONAL REPRODUCTION

Clonal (asexual) reproduction is common among many species of aquatic oligochaetes, particularly the aeolosomatids, naidids, and lumbriculids. In fact, there are many asexually reproducing naidids in which sexual individuals are rare or have never been observed; a clone of *Aulophorus furcatus* was traced through 150 generations for a three-year period with no appearance of sexual individuals and with an undiminished fission rate. Other oligochaetes reproduce asexually in the summer and sexually in the fall. Clonal reproduction is always a transverse division (fission) of the parent worm into two or more new individuals. *Lumbriculus* reproduces by fragmentation, as fission precedes differentiation, but in *Nais* and *Aeolosoma,* which reproduce by paratomy, differentiation precedes the separation of the daughter individuals, resulting in chains of individuals called zooids (Fig. 13-66).

SEXUAL REPRODUCTION

Reproductive System

The hermaphroditic oligochaetes have well-developed gonads, but these are restricted to a few genital segments. The genital segments are situated in the anterior half of the worm, and the female segment or segments are always behind the male segments (Fig. 13-61B). In most aquatic groups, one testicular segment is typically followed by one ovarian segment, whereas in terrestrial taxa, two male segments may be present. The exact location of genital segments along the trunk varies in different taxa.

The paired ovaries and testes occur on the posterior face of a genital septum and bulge posteriorly into the coelom (Fig. 13-67). Gametes are released from the gonads at an early stage of development and enter specialized coelomic pouches called seminal vesicles (for sperm) and **ovisacs** (for eggs). Both arise as septal outpockets in the genital segments, but their number, size, and position vary among taxa.

The genital segments, male and female, each have a pair of gonoducts—sperm ducts for the exit of sperm and oviducts for eggs (Fig. 13-67). The ducts extend backward and pass through one or more segments before opening on the ventral surface of the body. In earthworms, the two pairs of sperm ducts on each side of the body usually fuse before opening to the outside through a single male gonopore, which has a raised border, or lips. In many aquatic species

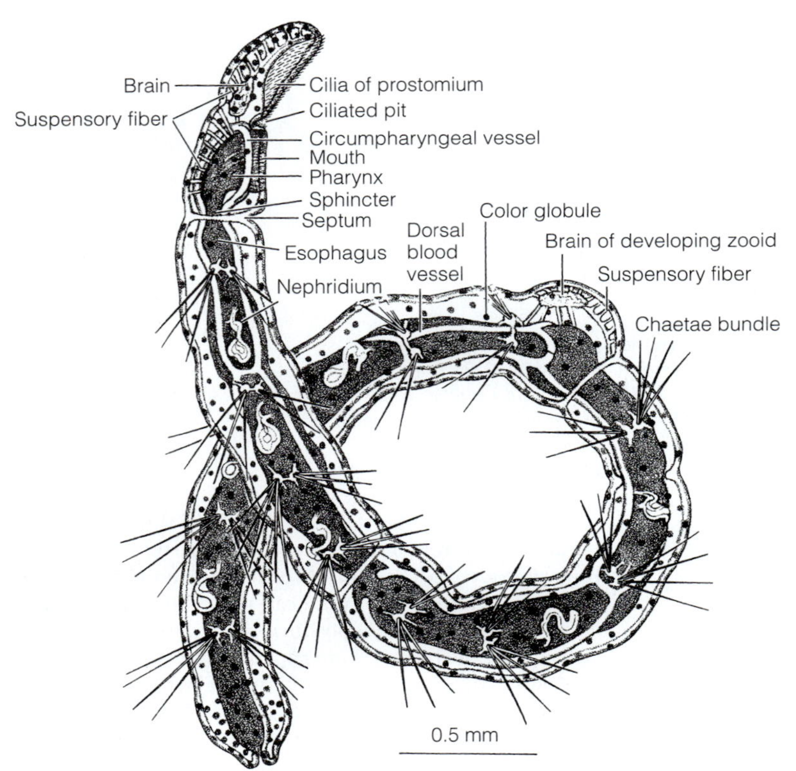

FIGURE 13-66 Oligochaete clonal reproduction. *Aeolosoma* (Aeolosomatidae), paratomy and anatomy. *(Drawing courtesy of R. Singer.)*

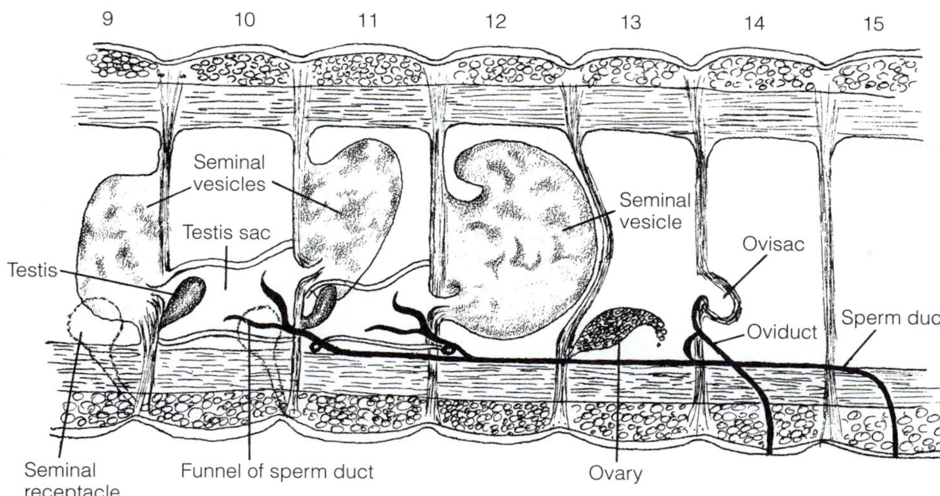

FIGURE 13-67 Oligochaete reproductive system. Lateral view of the reproductive anatomy segments of the earthworm *Lumbricus terrestris* (Lumbricidae). *(After Hess from Avel.)*

a chamber (atrium) preceding the gonopore contains a penis or an eversible area of the body wall (the atrium tip is called the penis; Fig. 13-68A,C). Glandular tissues called **prostate glands** are commonly associated with the male gonoducts. In some megascolecid earthworms, the prostates do not discharge into the vas deferens, but rather open separately onto the ventral surface of segments adjacent to those bearing the male gonopore. Prostates are absent from most lumbricid earthworms.

Forming a part of the female reproductive system but completely separated from the female gonoducts are the seminal receptacles (Fig. 13-61B, 13-67). A seminal receptacle is a sac for storing the partner's sperm prior to fertilization. The pores of the seminal receptacles open ventrally in the adjacent intersegmental groove. The number of seminal receptacles ranges from one to many pairs, each pair commonly in a separate segment. Although they usually are situated in certain segments anterior to the ovarian segment, their position varies with taxon.

The position of reproductive structures in *Lumbricus* is shown in Figure 13-67. The male segment in this species of *Lumbricus* is partitioned so that the testes, the sperm duct funnel, and the opening to the seminal vesicles are enclosed in a special ventral compartment called a **testis sac.** The testis sac is completely separate from the larger, remaining portion of the coelomic cavity. For this reason, the testes are not visible in the usual dorsal dissection.

The general plan of the oligochaete reproductive system is relatively uniform, but the numbers of various structures, the segments in which they are situated, and the segments onto which the genital pores open are variable. This variation is of considerable importance in the classification of oligochaetes.

The only exception to the oligochaete pattern appears in Aeolosomatidae, which, although hermaphroditic, is similar to polychaetes in that they lack distinct gonads and have a large number of genital segments. Also, a gonoduct is absent and the sperm use the nephridia to exit the body. Some zoologists, however, question whether these little worms are really oligochaetes.

Clitellum

The clitellum is a reproductive structure characteristic of clitellates (Fig. 13-59A, 13-68, 13-69). The number of segments composing the oligochaete clitellum varies considerably; there usually are 2 clitellar segments in many aquatic forms, 6 or 7 in *Lumbricus,* and as many as 60 in certain Glossoscolecidae. In aquatic species and megascolecid earthworms, the clitellum is often situated in the same region as the female gonopores; in the lumbricids, the clitellum is considerably posterior to the gonopores. The degree of development of the clitellum varies from taxon to taxon. In aquatic species the clitellum may be only one cell thick, whereas the thick clitellum of many earthworms is composed of three distinct layers (Fig. 13-69). The development of the clitellum also varies from season to season. It generally coincides with sexual maturity, but there are some worms in which the clitellum becomes conspicuous only during the breeding season.

Copulation

Oligochaetes are more or less continuous breeders, in contrast to the intermittent breeding of most polychaetes. Copulation is the rule, and mutual sperm transfer occurs between the hermaphrodites. During copulation, ventral contact is established between the anterior ends of the two oppositely facing worms (Fig. 13-68A,B). In most oligochaetes, except the lumbricids, the male gonopores of one worm directly appose the seminal receptacles of the other. The two worms are held in position by a coat of mucus secreted by the clitella, and by genital chaetae, which hook them together. The genital chaetae are specialized ventral segmental chaetae generally situated in the region of the male gonopore or the seminal receptacles. Transmission of sperm into one pair of seminal receptacles in the earthworm *Pheretima communissima* takes more than 1.5 h and then is repeated for each of the other two pairs of seminal receptacles (Fig. 13-68A).

In the lumbricids, including *Lumbricus,* the male gonopores do not appose the seminal receptacles during copulation, so sperm must swim externally from the anterior male gonopores

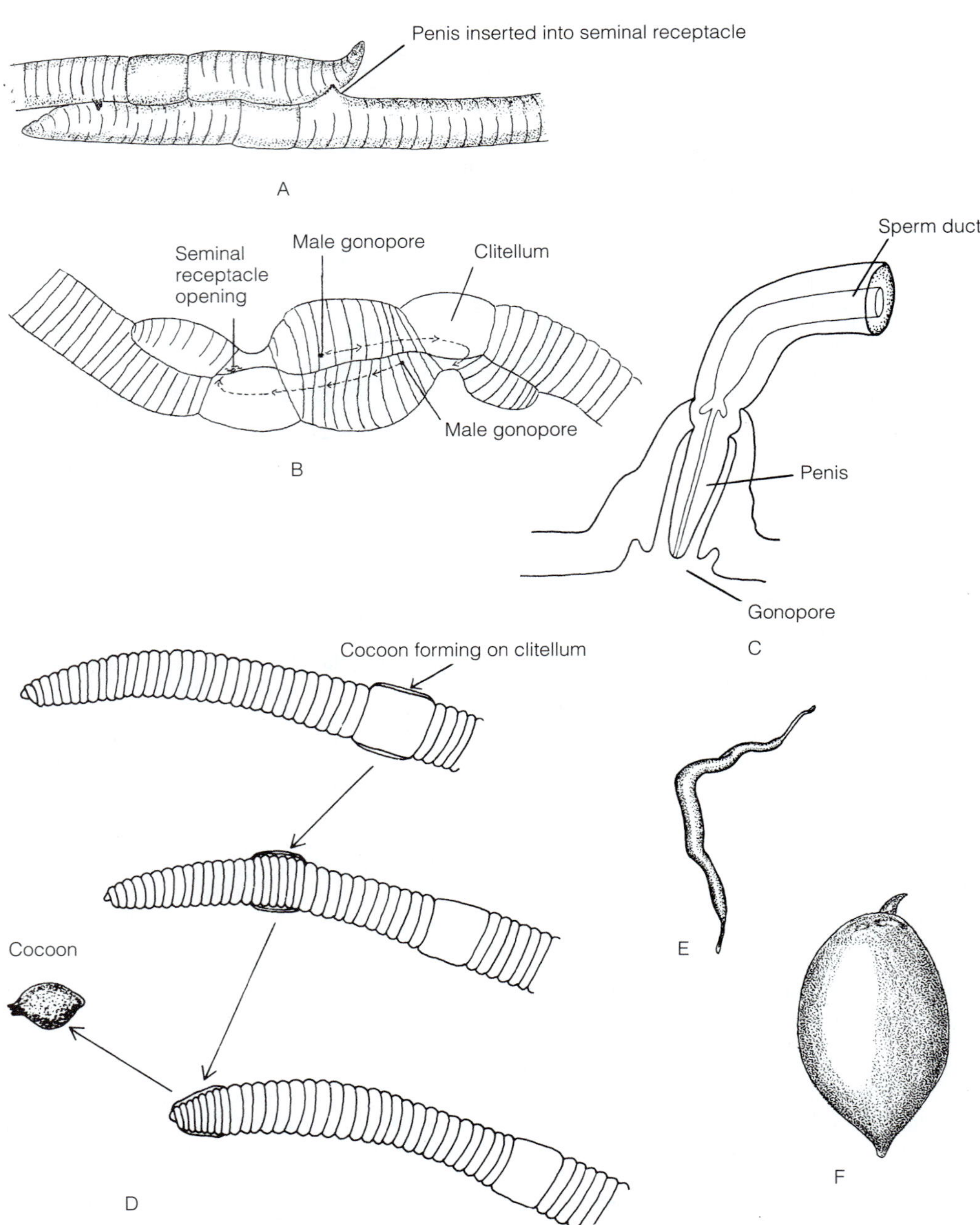

FIGURE 13-68 Oligochaete reproduction. **A,** Copulation by direct sperm transmission in the earthworm *Pheretima communissima* (Megascolecidae). **B,** Copulation by indirect sperm transmission in *Eisenia fetida* (Lumbricidae). Arrows indicate path of sperm from male gonopores to openings of seminal receptacles. **C,** Penis of the lumbriculid *Rhynchelmis* (family membership uncertain). **D,** Cocoon formation in a lumbricid earthworm. **E,** Cocoon of *Alma nilotica* (Glossocolecidae). **F,** Cocoon of *Allobophora terrestris* (Lumbricidae). *(A, After Oishi from Avel; B, After Grove and Cowley from Avel; C and E, From Brinkhurst, R. O., and Jamieson, B. G. M. 1972. Aquatic Oligochaeta of the World. Toronto University Press, Toronto; D, Modified from Tembe and Dubash in Edwards, C. A., and Lofty, J. R. 1972. Biology of Earthworms. Chapman and Hall, London; F, After Avel.)*

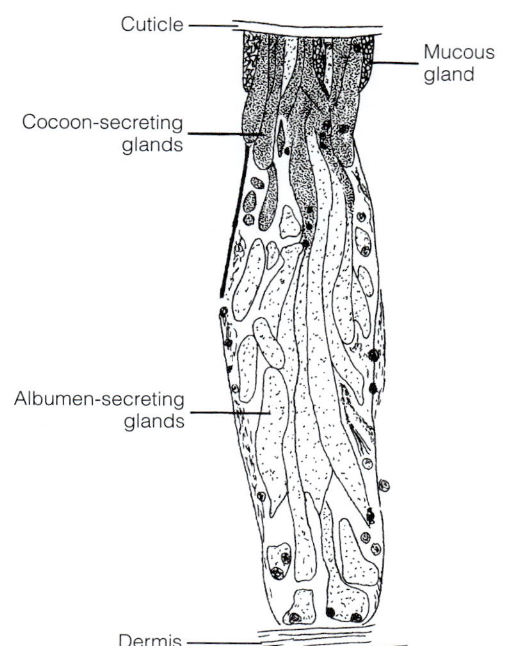

FIGURE 13-69 Oligochaete reproduction. Section through the clitellum of *Lumbricus terrestris* (Lumbricidae). *(After Grove from Avel.)*

of one worm to the more posterior pores of the partner's seminal receptacle (Fig. 13-68B). A favorable environment for the sperm's journey is provided by the clitellum and specialized parts of other segments. During copulation, the clitellum of one worm presses against the seminal receptacle segments of the other and mucus is secreted, enveloping the worms in an aquatic slime tube. Attachment is aided by genital chaetae. As sperm are released from the male pores of each worm, the contraction of specialized muscles along the sperm's route creates two grooves in the body wall between the male gonopores and the clitellum. Sperm move along these slime-filled grooves aided by groove-muscle contractions until they reach their own clitellum. At that point, they cross over to the other worm and enter its seminal receptacle. The emission of sperm may or may not be accomplished simultaneously in both members of the copulating pair. Copulation in *Lumbricus* continues for two to three hours.

Cocoon

A few days after copulation, a **cocoon** is secreted by the clitellum for the deposition of the eggs. First, a mucous tube is secreted around the anterior segments, including the clitellum. Second, the clitellum secretes a tough, encircling, chitin-like material that forms the wall of the cocoon. Third, the deep glandular layer of the clitellum secretes albumen into the space between the wall of the cocoon and the clitellum.

The albumen-filled cocoon slides forward as the worm pulls backward (Fig. 13-68D). As it passes over the female gonopores, eggs are discharged into the cocoon. Continuing on its track, it next moves over the seminal receptacles, from which the partner's sperm are released, cross-fertilizing the eggs externally in the cocoon. Finally the cocoon slips over the head of the worm and is freed from the body, the mucous tube quickly disintegrates, and the ends of the cocoon constrict and seal themselves. The finished cocoon thus contains zygotes and nutritive albumen for their development. The cocoons of terrestrial species are deposited in soil, and those of aquatic species are left in bottom debris or mud or are attached to vegetation.

The ovoid cocoons are yellowish in color (Fig. 13-68D–F). Cocoons of *Tubifex* are 1.60 × 0.85 mm; those of *Lumbricus terrestris* are approximately 7 × 3.5 mm. The largest cocoons, 75 × 22 mm, are produced by the giant Australian earthworm, *Megascolides australis*. A cocoon contains anywhere from 1 to 20 eggs, depending on species, and a succession of cocoons may be produced. Under favorable conditions, lumbricids may mate continually during the spring, with cocoons forming every three or four days. The freshwater tubificids and lumbriculids generally reproduce only once a year. Their reproductive systems are then resorbed and reform the following year.

There is increasing evidence that growth and reproduction in oligochaetes are regulated by neurosecretions of the brain, as is true of polychaetes. In contrast to polychaetes, however, the hormone produced by the brain appears to stimulate rather than inhibit reproduction.

Development

The oligochaetes are direct developers and thus have relatively yolky eggs. In general, the eggs of aquatic groups contain large amounts of yolk. On the other hand, terrestrial species have much smaller eggs with much less yolk because the abundant albumen in the cocoon supplies most of the nutritive needs of the embryo. A larva is absent and development to a juvenile stage occurs in the cocoon.

The cleavage pattern, although retaining traces of the ancestral spiral cleavage, is considerably modified in oligochaetes, especially in earthworms. The juveniles emerge from the end of the cocoon after eight days to several months of development. *Lumbricus* hatches in 12 to 13 weeks. Usually, only some of the eggs deposited in the cocoon hatch; in *Lumbricus terrestris*, only one egg develops.

Many oligochaetes live several years. Earthworms in terraria have been known to live for six years, but their life span is probably much shorter in natural, unprotected conditions. Lumbricid earthworms reach sexual maturity in six months to a year, depending on environmental conditions and species. *Lumbricus terrestris* requires at least 200 days. Some aquatic oligochaetes have shorter generation times. At elevated temperatures, enchytraeids inhabiting sewage trickling filters reach sexual maturity in 13 to 28 days, depending on the species, and the generation time between cocoons ranges from one to two months. On the other hand, some tubificids have a two-year life span, breeding once and then dying.

Diversity of Oligochaeta

At present, oligochaetes are divided into 27 families and several higher taxa based on characteristics of chaetae, clitellum, number and position of genital segments, and reproductive anatomy. Because of the technical nature of these characters and the difficulty of relating most to unique functions, we include only the common, species-rich taxa in relation to their habitat preferences.

COMMON AQUATIC TAXA

Members of the major aquatic families occur throughout the world wherever suitable habitats exist. In freshwater habitats, most burrow in bottom debris, a few construct tubes, and some live on aquatic plants. Oligochaetes are most abundant where the water is shallow, although several families have benthic representatives in deep lakes. The abundance of different species of aquatic oligochaetes can be a good indicator of the degree of water pollution. Aquatic species even inhabit the water reservoirs of bromeliad epiphytes, which attach to the trunks and branches of tropical trees.

Many species are amphibious or transitional between a strictly aquatic and a strictly terrestrial environment. These worms live in marshy or boggy land and on the margins of ponds and streams.

Some 200 species of marine oligochaetes have been described. Most belong to Enchytraeidae and Tubificidae and chiefly inhabit the supratidal and intertidal zones, but subtidal species are known even from abyssal and hadal depths. Marine oligochaetes are members of the interstitial fauna, are shallow burrowers, or live beneath intertidal rocks or in algal drift.

In general, most aquatic oligochaetes are small, delicate animals, especially in comparison with earthworms, and their body walls are often transparent, permitting a clear view of the organs within. Most have a clitellum formed of a monolayer of cells. Such clitella secrete little or no albumen, and their eggs compensate for the lack of nutritional supplementation by being large and yolky.

Aeolosomatidae[F]: *Aeolosoma:* small interstitial worms, usually less than 5 mm in length. Ventrally ciliated prostomium for gliding over and through substratum and for sweeping food particles into mouth (Fig. 13-59D, 13-66). Chaetae on most segments, including anterior segments, but segmental septa are absent. One pair of testes and ovaries. Sexual reproduction is rare; clonal reproduction, by paratomy, is common. Encystment of adults occurs under adverse conditions.

Although included here among oligochaetes, aeolosomatids may actually be polychaetes. The chief argument in favor of a polychaete alliance is the occurrence of nuchal organs on the prostomium (as "ciliated pits"; Fig. 13-66), a polychaete autapomorphy. The brain of *Aeolosoma* is in the prostomium, as it is in polychaetes, not in an anterior segment. *Aeolosoma* also lacks a clitellum, the hallmark of clitellates.

Naididae[F]: Small, aquatic, predominantly freshwater oligohaetes similar to tubificids (below), but differing in living on submerged vegetation and having ocelli. Like the aeolosomatids, naidids have few segments, not more than 40. Prostomium lacks cilia, chaetae are absent from anterior segments, but segmental septa are present. Some have a long proboscis. As in aeolosomatids, there is one pair each of testes and ovaries; sexual reproduction has been infrequently observed. Paratomy is common: under ideal conditions, a new zooid is released every two or three days. *Aulophorus:* a pair of long gills on posterior end. *Branchiura:* a dorsal and a ventral gill filament on each posterior segment (Fig. 13-64A). *Chaetogaster:* a rudimentary prostomium, large mouth and large muscular pharynx for engulfing prey; carnivores of other worms, crustaceans, and insect larvae; parasites (in snails). *Dero:* makes sand-grain tubes and has small posterior gills (Fig. 13-64B). *Stylaria:* prostomium drawn out into a long process (Fig. 13-59C). *Nais, Slavina.*

Tubificidae[F]: Small to large, reddish burrowers in marine and freshwater sediments. The most common oligochaetes in deep freshwater sediments. Body composed of 40 to 200 distinct segments, the number depending on species. Some, such as *Tubifex,* build and occupy vertical tubes from which they extend and wave their tails for gas exchange. *Tubifex tubifex* and *Limnodrilus* spp. thrive on organic pollution and tolerate low oxygen concentrations in the water. In highly polluted lakes and streams, tubificids are the only animals in the benthic sediments, where they form dense, waving beds (Fig. 13-64C). The marine *Phallodrilus albidus* lacks a gut and derives its nutrition in association with bacteria that oxidize reduced inorganic compounds. One pair of testes (segment 10) and one pair of ovaries (segment 11); sexual reproduction is common, clonal reproduction is not.

Lumbriculidae[F]: Freshwater worms, small (not exceeding 14 cm in length), but having many segments, each with four pairs of chaetae. Clonal reproduction by fragmentation dominates. The educationally useful California blackworm, *Lumbriculus variegatus,* is the best-known lumbriculid. In the laboratory, this easily cultured species, with its transparent body wall and rapid rate of asexual reproduction, is used to study functional morphology, regeneration, segment differentiation, locomotion, and other behaviors.

COMMON TERRESTRIAL TAXA (EARTHWORMS)

Of the 10 families of earthworms, 4 contain large numbers of species. **Glossoscolecidae[F], Lumbricidae[F], Megascolecidae[F],** and **Moniligastridae[F]** are all burrowers found everywhere (on land) except in deserts. The first three families have two pairs of testes followed by a sterile segment, followed by one pair of ovaries. The multilayered clitellum secretes albumen, and thus the eggs are relatively nonyolky.

Soils containing considerable organic matter, or at least a layer of humus on the surface, maintain the largest fauna of worms, but other soil factors are important to the distribution of terrestrial species. Acid soils are favorable habitats for most earthworms. Up to 700 lumbricids have been reported from a square meter of farm soil.

A cross section through soil reveals the distinct vertical stratification of worms. The tunnels of larger species, such as the nightcrawler *Lumbricus terrestris,* range from the surface to several meters deep, depending on the nature of the soil. Young worms and small-bodied species are restricted to the few centimeters of upper humus, whereas others have a wider vertical distribution but are still limited to the upper level of the soil that contains some organic matter.

An earthworm constructs its burrow by forcing its anterior end through crevices and by swallowing soil. The egested material and mucus are plastered against the burrow wall, forming a distinct lining. Some egested material is removed from the burrow and deposited at the surface as fecal castings.

The gut turnover time in the lumbricid *Allolobophora rosea* is reported to be 1 to 2.5 h. The burrows may be complex, with two openings and horizontal and vertical branches. *Lumbricus terrestris* always plugs its burrow with debris pulled into the hole. It is not clear why large numbers of *Lumbricus terrestris* leave their burrows after heavy rains, resulting in high mortality, but the low oxygen tension of the water-saturated soil has been cited as a factor.

As first demonstrated by Charles Darwin in his authoritative 1881 book, *The Formation of Vegetable Mould through the Action of Worms, with Observations of Their Habits,* the activities of earthworms have a beneficial effect on the soil. The extensive burrows increase soil drainage and aeration, but more important are the mixing and churning of the soil. Deeper soil is brought to the surface as fecal castings, and organic material is moved to lower levels. Some tropical species produce enormous castings. For example, the towerlike castings of *Hyperoidrilus africanus* may reach 8 cm in height and 2 cm in diameter.

Earthworms' ability to churn soil can be demonstrated in a container half filled with sand and half with potting soil. Five worms will thoroughly mix 500 ml of sand with 500 ml of soil in several months.

Besides cultivating soil and creating compost, earthworms are useful to humans in other ways. Many are raised and sold as fish bait, especially the nightcrawler *Lumbricus terrestris,* the annual sales of which in Canada exceed $50 million. The Maoris of New Zealand eat earthworms and consider them a delicacy, and in parts of Japan, earthworm pies are consumed. The tiger worm, *Eisenia fetida,* occurs naturally in dung and is cultivated in the feces of hogs and other domesticated animals to convert dung into worms, which are then incorporated into animal feed.

Surprisingly, several earthworms occur in tropical trees. These worms live in accumulated humus and detritus in axils of leaves and branches. Although the natural distribution of many species is geographically restricted, humans have spread several worldwide. For example, the earthworm fauna of the larger Chilean cities consists solely of European species, which have displaced the endemic forms. Glossoscolecidae (male gonopores on clitellum; esophageal gizzard): *Glossoscolex, Pontoscolex;* Lumbricidae (male gonopores on segment 15; intestinal gizzard): *Allolobophora, Eisenia, Lumbricus;* Megascolecidae (male gonopores on segment 18): *Driloleirus americanus* (giant Palouse earthworm of Washington), *D. macelfreshi* (giant Oregon earthworm; both *Driloleirus* species reach 1 m in length), *Megascolex, Megascolides, Pheretima;* Moniligastridae (male gonopores on segment behind testicular segment): *Drawidia, Moniligaster.*

HIRUDINOMORPHA[C]

This clitellate taxon includes the tiny branchiobdellids (Branchiobdellida), which cling and creep over the exoskeleton of freshwater crustaceans, especially crayfishes, and the often large, active, and intimidating leeches (Hirudinea). Leeches are common in fresh water, but also occur in the sea and in moist areas on land. All hirudinomorphs have a **terminal** or posterior **sucker** that adheres to the host or the substratum and a dorsal anus located in front of the sucker. The re-positioning of the ancestral anus from terminal to dorsal is related to the origin of the sucker, which is a specialization of the posteriormost segment (and often several others) as well as the pygidium. If the anus had not shifted position, it would open centrally below the sucker. Approximately 500 species of hirudinomorphs have been described.

Hirudinomorph evolution is the story of a worm's transition from a more or less typical oligochaete-like ancestor with coelomic cavities, chaetae, and a detritivorous habit to a carnivorous or bloodsucking leech with a chaetae-less, suckered, and flattened body composed of a fixed number of segments. The branchiobdellids, and especially a leech called *Acanthobdella peledina,* are important chapters in that story, and we will discuss them later in the Diversity of Hirudinomorpha and Phylogeny of Clitellata sections.

Euhirudinea

The 350-some species of Euhirudinea are marine, freshwater, and terrestrial worms commonly known as leeches. Although they are all popularly considered to be bloodsuckers, many are nonbloodsucking carnivores. The leeches share clitellate traits with the Oligochaeta, but also have several general or unique attributes of their own, which are highlighted in the remaining chapter sections.

Leeches are never as small as many polychaetes and oligochaetes. The smallest leeches are 1 cm in length, and most species are 2 to 5 cm long. Some, including the medicinal leech *(Hirudo medicinalis),* may attain a length of 12 cm, but the giant of the class is the Amazonian *Haementeria ghiliani,* which reaches 30 cm. Black, brown, olive green, and red are common colors, and striped and spotted patterns are not unusual.

Although some leeches are marine, most of the aquatic species live in fresh water. Relatively few species tolerate rapid currents; most prefer the shallow, vegetated water bordering ponds, lakes, and sluggish streams. In favorable environments, often those high in organic pollutants, overturned rocks may reveal an amazing number of individuals; more than 10,000 individuals per square meter have been reported in Illinois. Some species estivate during periods of drought by burrowing into the mud at the bottom of a pond or stream and can survive a loss of as much as 90% of their body weight.

Although leeches are found throughout the world, they are most abundant in north temperate lakes and ponds. Much of the North American leech fauna is shared with Europe. The aquatic Hirudinidae, which includes the European medicinal leech, and the terrestrial Haemadipsidae feed primarily on mammals, including humans.

FORM AND FUNCTION

The anatomy of leeches is remarkably uniform. The body typically is dorsoventrally flattened and frequently is tapered at the anterior end (Fig. 13-70). The segments at both extremities have been modified to form suckers. The anterior sucker usually is smaller than the posterior and frequently surrounds the mouth. The posterior sucker is disc-shaped and turned ventrally. Segmentation is very much reduced. Unlike other annelids, leeches have a fixed number of 33 segments, but secondary superficial annulations externally mask the segmentation. Chaetae are absent.

The head consists of a reduced prostomium plus five segments (1 to 5; Fig. 13-70). Dorsally, the head bears several

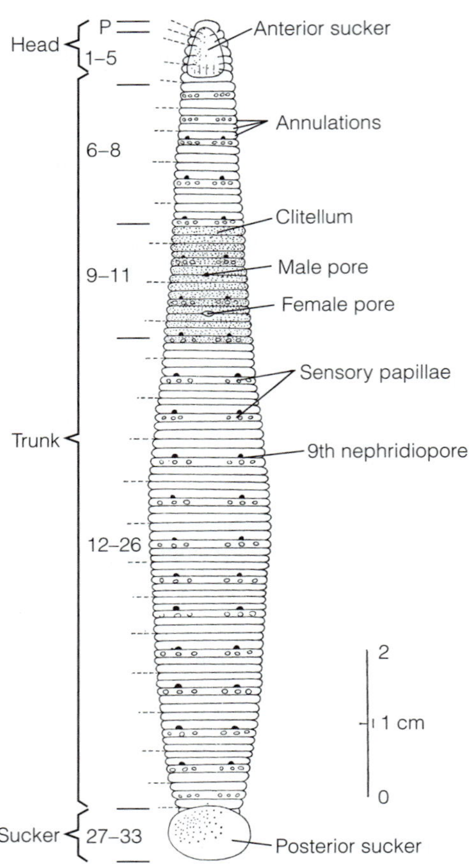

FIGURE 13-70 Leech external anatomy (Euhirudinea). Ventral view of the European medicinal leech, *Hirudo medicinalis* (Arhynchobdellida). *(From Mann, K. H. 1962. Leeches. Pergamon Press, Elmsford, New York.)*

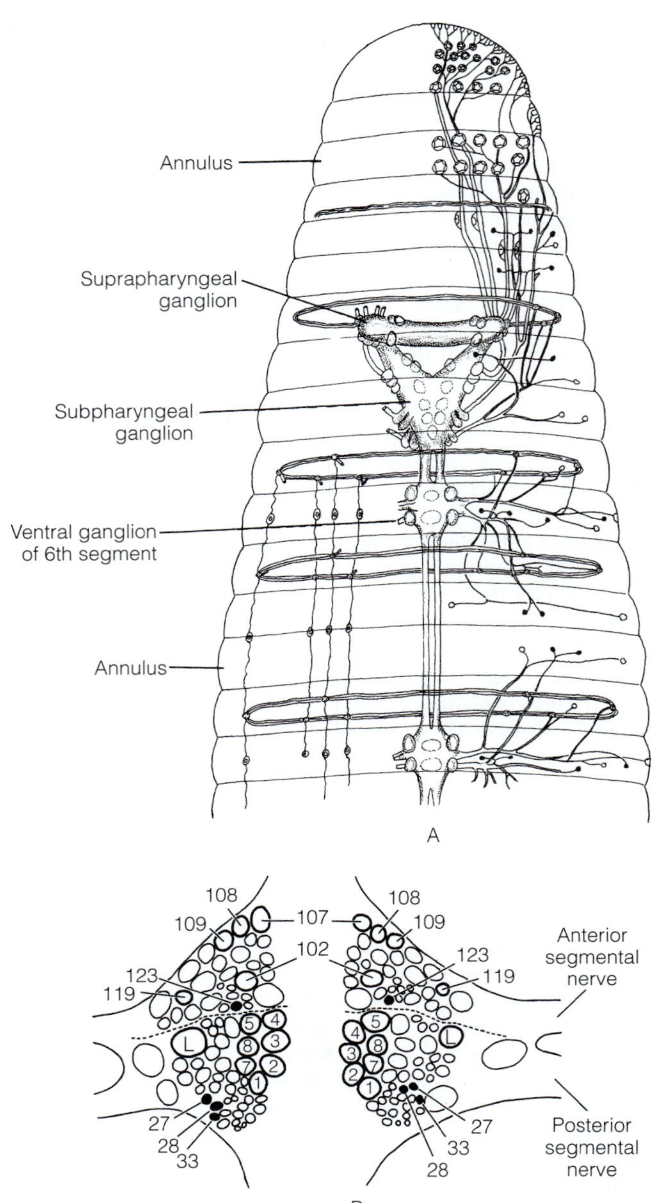

FIGURE 13-71 Leech nervous system (Euhirudinea). **A,** Nervous system of *Erpobdella punctata* (Arhynchobdellida). **B,** Dorsal view of a segmental ganglion of a leech. Circles represent cell bodies; solid black circles are interneurons, and those enclosed by thick lines are motor neurons. *(A, After Bristol from Harant and Grassé; B, From Stent, G. S. 1978. Neuronal generation of the leech swimming movement. Science 200:1348–1356.)*

ocelli, and ventrally it is modified to form the **anterior sucker** surrounding the mouth. The trunk, which encompasses the preclitellar region, clitellum, and postclitellar region, consists of 21 segments (6 to 26). The clitellum, spanning three segments (9 to 11), is conspicuous only during reproductive periods. Behind the trunk is the large, ventral **posterior sucker** derived from 7 segments (27 to 33). The dorsal anus is on or near the last trunk segment in front of the sucker.

The number of annulations per segment varies not only in different regions of the body, but also in different species. The best means of determining the primary segmentation of leeches is by studying the central nervous system and the innervation of the annulations by segmental nerves. The occurrence of a ring of sensory papillae around the first annulation of each segment, the serial repetition of color patterns, and the placement of the ventral nephridiopores, however, also provide segmental landmarks.

Nervous System and Sense Organs

The nervous system of leeches is similar to that of other annelids, but the anterior and posterior ganglia are concentrated into masses because of the segmental modifications forming the suckers. The brain, located in segment 5, consists of paired suprapharyngeal ganglia (the original paired ganglia of the prostomium plus the ganglia of the peristomium). Ventrally, the first four pairs of ganglia (segments 2 to 5) are fused to form the subpharyngeal ganglion (Fig. 13-71A). Behind the subpharyngeal ganglion, the ventral nerve cord consists of 21 "free" pairs of ganglia (segments 6 to 26) that form the nerve

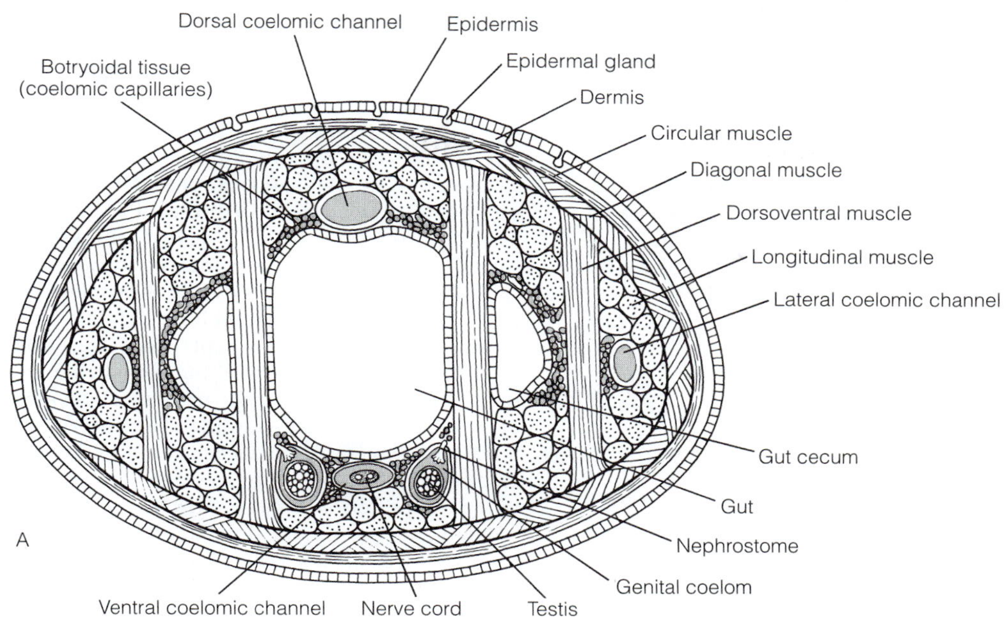

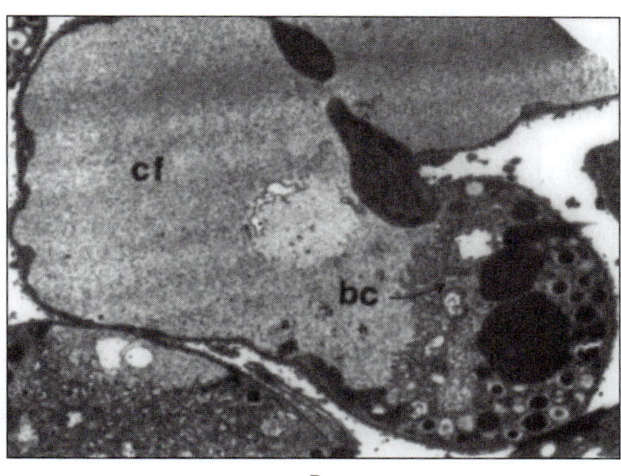

FIGURE 13-72 Leech internal anatomy (Euhirudinea). **A,** Transverse section through the European medicinal leech, *Hirudo medicinalis* (Arhynchobdellida). In arhynchobdellid leeches, the hemal system has been completely replaced by the modified coelomic circulatory system. **B,** Electron micrograph of a section through a botryoidal capillary of the American medicinal leech, *Macrobdella decora* (Arhynchobdellida), showing a botryoidal lining cell (bc); (cf = coelomic fluid). *(A, Modified and redrawn from Bourne, A. G. 1884. Contributions to the anatomy of the Hirudinea. Q. J. Microsc. Sci. 24:419–506 plus plates.)*

cord of the trunk and 7 additional pairs of ganglia (segments 27 to 33) that fuse to form the caudal ganglion associated with the posterior sucker. The entire central nervous system is enclosed in the unpaired ventral coelomic channel (Fig. 13-72A).

The relatively small number and large size of the neurons have made leeches, along with squids, sea hares, and crayfish, favorite invertebrate subjects of neuroanatomists and neurophysiologists. Each of the 21 segmental ganglia in the medicinal leech *Hirudo* contains 175 pairs of neuron cell bodies arranged bilaterally around a central mass of nerve fibers (a neuropil), in which synaptic junctions are made. The cell bodies are large enough to be probed with electrodes and mapped (Fig. 13-71B).

Some rhynchobdellid leeches can change color dramatically as a result of pigment movement in large specialized cells called **chromatophores** that are under neural control. The significance of the color change is not certain, because these leeches do not adapt to background coloration.

The specialized sense organs in leeches consist of 2 to 10 pigment-cup ocelli and sensory papillae. The **sensory papillae** are small, projecting discs arranged in a dorsal row or in a complete ring around one annulation of each segment (Fig. 13-70). Each papilla consists of a cluster of many sensory cells and supporting epithelium.

Despite the lack of highly organized, concentrated sense organs, leeches can detect low levels of many types of stimuli, a sensitivity that is often an adaptation for finding prey or a host. Fish leeches respond to moving shadows and water-pressure vibrations. Both predatory and bloodsucking leeches will attempt to attach to an object smeared with various host or prey substances such as fish scales, tissue juices, oil gland secretions, or sweat. *Hirudo,* a bloodsucker of warm-blooded animals, will swim into waves, which may be generated by a possible host. The same leech is also attracted by body secretions and elevated temperatures and has been reported to swim toward a man standing in water. Supposedly, the terrestrial tropical bloodsuckers (Haemadipsidae), which are attracted by passing warm-blooded mammals, will move over vegetation to converge on a stationary human (as well as other mammals).

Body Wall, Coelom, Hemal System, and Locomotion

The leech body wall consists of a typical annelidan cuticle and epidermis, but unlike polychaetes and most oligochaetes, the fibrous connective tissue beneath the epidermis is very thick and occupies much of the interior of the body. The musculature and other tissues occupy this expanded connective-tissue region (Fig. 13-72). Some of the cell bodies of the enlarged epidermal glands have submerged into the peripheral connective-tissue layer (the dermis). Below the dermis is a layer of circular musculature followed by diagonal muscles and a powerful longitudinal musculature. Dorsoventral muscles are also present (Fig. 13-72).

Expansion of the connective-tissue compartment of leeches is correlated with a reduction of the coelom and septa, a striking difference between leeches and other annelids. Leeches lack the paired segmental coeloms that typify other annelids. (*Acanthobdella peledina,* discussed in Diversity of Hirudinomorpha, is the only exception.) Without septa and mesenteries to partition the coelom into separate compartments, the coelomic cavity is continuous and uninterrupted throughout the leech body (Fig. 13-72, 13-73). Thus the leech coelom became the global circulatory system, overlapping the responsibility of the hemal system and eventually replacing the ancestral hemal system altogether.

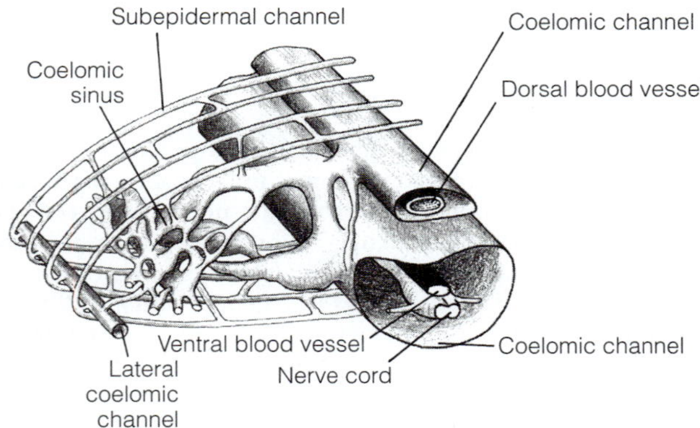

FIGURE 13-73 Leech hemal and coelomic circulatory systems (Euhirudinea). Perspective view of the hemal (black) and coelomic (gray) circulatory systems of *Placobdella costata* (Rhynchobdellida, Glossiphoniidae). *(After Oka from Harant and Grassé.)*

During this takeover, the coelomic system adopted the appearance of the old hemal system—two main longitudinal vessels, arteries, veins, even capillaries—but it betrays its coelomic origin in two ways. The two main longitudinal vessels are lateral rather than dorsal and ventral, and the vessels are lined by mesothelium, unlike hemal vessels, which are lined only by basal lamina. Much of the coelomic mesothelium, especially in the capillaries, is specialized as large nutrient-storage cells called chlorogogen tissue in rhynchobdellids and **botryoidal tissue** in arhynchobdellids (Fig. 13-72B). The coelomic fluid is propelled by the muscular contractions of the lateral longitudinal channels.

The leeches' loss of septa, chaetae, and segmented coelomic cavities is correlated functionally with a change from peristaltic burrowing to new modes of locomotion: Leeches move by inchworm-like crawling (Fig. 13-74) or by swimming, but do not burrow using peristalsis. When crawling, only the anterior and posterior suckers anchor to the substratum. When the posterior sucker is attached, a wave of circular contraction sweeps over the animal, and the body is lengthened and extended forward. The diagonal muscles may be postural, keeping the animal rigid while it is raised and attached posteriorly. The anterior sucker then attaches and the posterior sucker releases. A wave of longitudinal contraction then occurs, shortening the animal and moving the posterior sucker forward (Fig. 13-74). The diagonal muscles may also enable leeches to twist their raised bodies while attached

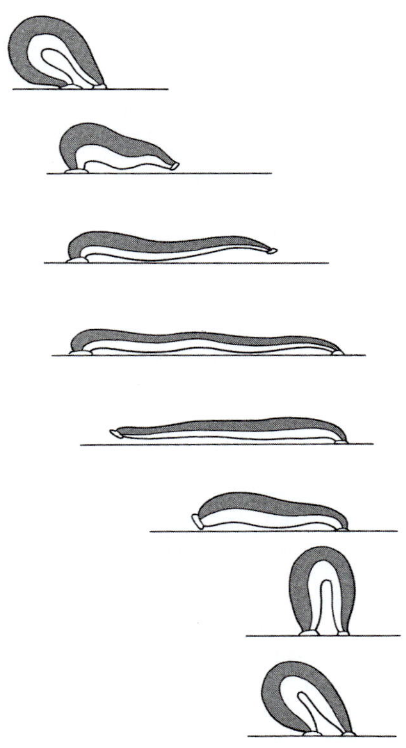

FIGURE 13-74 Leech locomotion. *(Modified and redrawn from Fretter, V., and Graham, A. 1976. A Functional Anatomy of Invertebrates. Academic Press, London and New York.)*

posteriorly. When a leech swims, the body is flattened by contraction of the dorsoventral musculature, and waves of contraction along the longitudinal muscles produce vertical undulations.

Gas Exchange

Gills occur only in the Piscicolidae (fish leeches), with the general body surface providing for gas exchange in other leeches. The piscicolid gills are lateral leaflike or branching outgrowths of the body wall (Fig. 13-82A). The respiratory pigment, an extracellular hemoglobin, is found only in the arhynchobdellid leeches and is responsible for about one-half of the oxygen transport.

Excretory System

The leech excretory system consists of 10 to 17 pairs of metanephridia, 1 pair per segment, situated in the middle third of the body (Fig. 13-72A). As a result of coelom reduction and the loss of septa in the leech body, the nephridial tubules are embedded in connective tissue and the nephrostomes project into the coelomic channels. Each ciliated nephrostome opens downstream into a nonciliated capsule, which joins the nephridial tubule (Fig. 13-75). In most leeches, however, the cavity of the capsule does not open freely into the tubule lumen, and in a few species, the capsule and tubule are entirely separate structures. The nephridial tubule consists of a main duct that receives numerous branched ductules (canaliculi). The main duct typically expands into a **urinary bladder** before opening to the exterior at a ventrolateral nephridiopore.

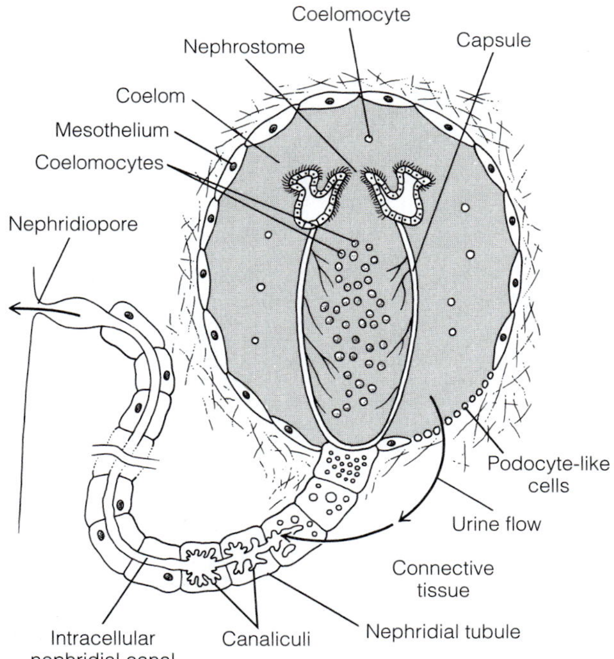

FIGURE 13-75 Leech excretory system (Euhirudinea). Nephridium of *Trocheta* (Arhynchobdellida). The nephrostome and capsule do not communicate with the tubule, but rather collect and break down spent coelomocytes. Urine formation (arrows) apparently results from ultrafiltration of coelomic fluid across the wall of the botryoidal capillary into the connective tissue and tubule. *(Drawn from several sources.)*

The mesothelium of the coelomic channels in the region of the nephrostomes and canaliculi is porous (the lining cells resemble podocytes), and thus the initial phase of urine formation is believed to occur as the hemocoelomic fluid is pressure-filtered across the wall of the channel into the connective tissue around the nephridial tubule or directly into the canaliculi (Fig. 13-75). As the ultrafiltrate enters the main duct, salts are reabsorbed and a watery, hyposmotic urine accumulates in the bladder before being discharged to the exterior. Thus the nephridia are important organs of osmoregulation.

The functions of the structurally isolated nephrostome and nephridial capsule are unrelated to excretion. The capsules, similar to lymph nodes, contain a population of ameboid phagocytes that engulf foreign material and cellular debris. Some phagocytes (coelomocytes) may be released into the coelomic circulation to police the circulating fluid (Fig. 13-75). (Particulate waste is also engulfed by botryoidal and vasofibrous tissue in the hirudinid leeches and by pigmented and coelomic epithelial cells in glossiphoniids and piscicolids.) In the rhynchobdellids, cilia on the nephrostomes beat inward, carrying particulates into the capsule. The nephrostomes of the arhynchobdellids, on the other hand, have cilia that beat outward, away from the capsule, and are believed to promote the circulation of coelomic fluid. The conversion of a nephrostome into a site for phagocytosis and coelomocyte destruction is also common in polychaetes.

Digestive System and Nutrition

Leeches have either a protrusible pharynx (inappropriately called a proboscis) or a nonprotrusible sucking pharynx with or without jaws. The tubular pharynx (Rhynchobdellida) joins the ventral mouth by a short, narrow canal (Fig. 13-76A, 13-77B). The pharynx is highly muscular, has a triangular lumen, and is lined internally and externally with cuticle. Ducts from large, unicellular salivary glands open into the pharynx. When feeding, the animal protrudes the pharynx from the mouth, forcing it into the tissue of the host.

In the gnathobdelliform leeches (Arhynchobdellida), which lack a protrusible pharynx, the mouth is situated in the anterior sucker (Fig. 13-76B,C, 13-77A). Immediately within the mouth cavity of most species are three large, oval, bladelike **jaws,** each bearing many small teeth along the free edge, arranged in a triangle, one dorsally and two laterally. When the animal feeds, the anterior sucker attaches to the prey or host, and the edges of the jaws slice like scalpels through the integument (Fig. 13-76D,E). The jaws swing toward and away from each other, activated by muscles attached to their bases. Salivary glands secrete an anticoagulant called **hirudin.** Immediately behind the teeth, the buccal cavity opens into a muscular, pumping pharynx. The predatory pharyngobdelliforms (Arhynchobdellida) also have a pumping pharynx, but muscular folds replace the jaws.

The remainder of the digestive tract is relatively uniform throughout Euhirudinea. A short esophagus either opens directly into a relatively long stomach or first expands into a crop (Fig. 13-77A). The stomach may be a straight tube, as in the

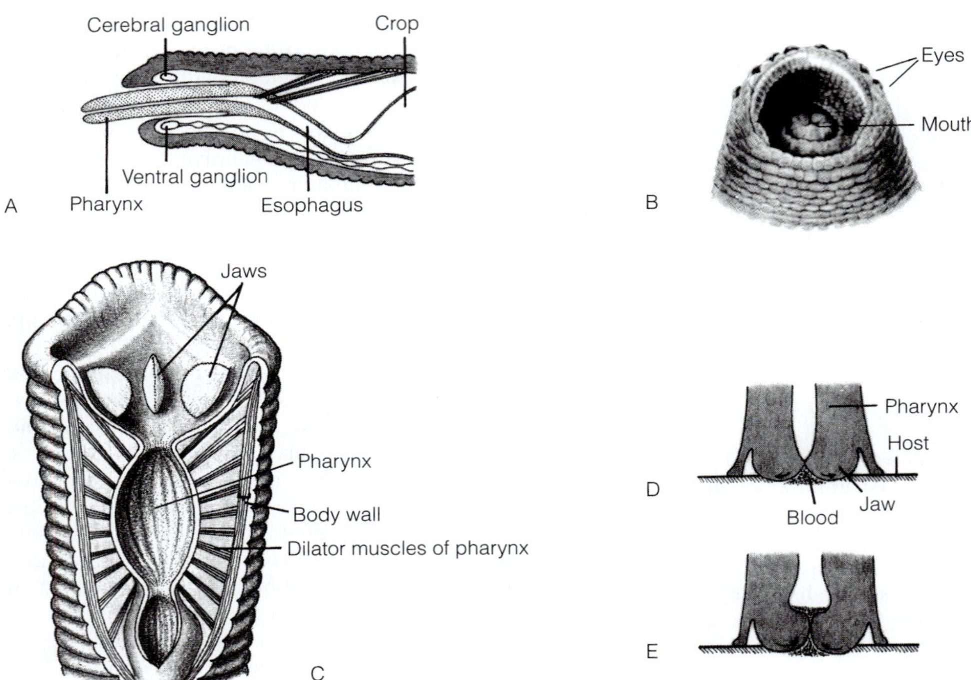

FIGURE 13-76 Leech pharynx and feeding (Euhirudinea). **A,** Sagittal section through anterior end of *Glossiphonia* (Rhynchobdellida), showing an everted tubular pharynx. (Figure 13-77B shows a retracted pharynx.) **B–E,** Arhynchobdellid leeches. **B,** Ventral view of the oral region of a terrestrial, bloodsucking leech (Haemadipsidae). Jaws are not exposed. **C,** Ventral dissection of the anterior end of the European medicinal leech, *Hirudo medicinalis,* showing three slicing jaws. **D** and **E,** Ingestion of blood by *Hirudo.* Outward movement of teeth **(D),** followed by medial movement of teeth and dilation of pharynx **(E).** *(A, Modified after Scribin; B, From Keegan, H. L., et al., 1968. 406th medical laboratory special report. U.S. Army Med. Command, Japan; C, Modified after Pfurtscheller; D and E, After Herter.)*

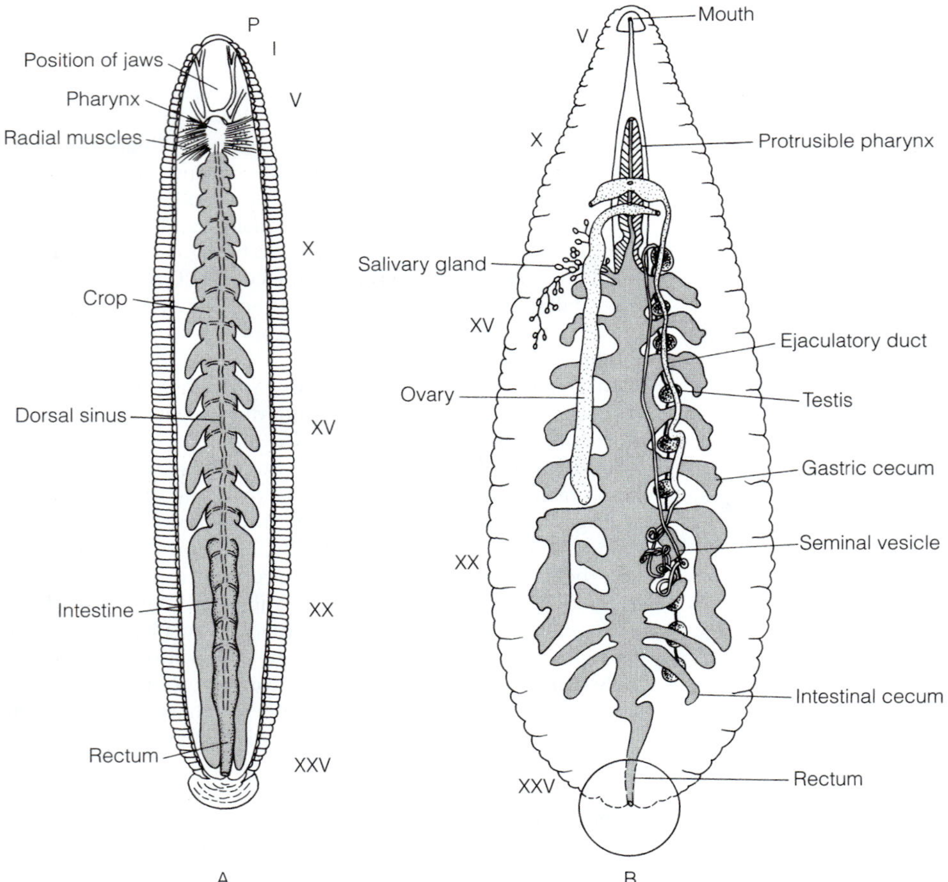

FIGURE 13-77 Leech digestive and reproductive systems (Euhirudinea). **A,** Dorsal view of the digestive system of the European medicinal leech, *Hirudo medicinalis* (Arhynchobdellida). **B,** Ventral view of the digestive and reproductive systems of *Glossiphonia complanata* (Rhynchobdellida). *(A, From Mann, K. H. 1962. Leeches. Pergamon Press, Elmsford, New York; B, After Harding and Moore from Pennak, 1978)*

pharyngobdelliforms, but more commonly it has 1 to 11 pairs of lateral ceca (Fig. 13-72A, 13-77B). Following the stomach is the intestine, which may be a simple tube or, as in the rhynchobdellids, have four pairs of slender lateral ceca. The intestine opens into a short rectum, which empties to the outside through the dorsal anus situated in front of the posterior sucker.

All leeches are carnivores or bloodsucking ectoparasites. The bloodsucking leeches constitute about three-quarters of the known species, and the remainder are predators. The transition from predator to parasite is probably related to the size of the prey. Predatory leeches always feed on small invertebrates, including worms, snails, and insect larvae. Feeding is frequent and the prey usually is swallowed whole. Many predatory Rhynchobdellida, however, suck in the soft parts of their hosts and are thus intermediate between predators and dedicated bloodsuckers. In the laboratory, the predatory arhynchobdellid *Erpobdella punctata* consumed 1.78 tubificid oligochaetes per day and *Helobdella stagnalis* consumed 0.57 per day.

The bloodsucking leeches attack a variety of hosts, including invertebrates and vertebrates. Parasitic leeches are rarely restricted to one host, but usually they are confined to one group of vertebrates. For example, *Placobdella* feeds on almost any species of turtles and even alligators, but they rarely attack amphibians or mammals. On the other hand, mammals are the preferred hosts of *Hirudo*. Furthermore, some species of leeches that are exclusively bloodsuckers as adults are predaceous during juvenile stages.

Bloodsuckers that attack mammals, such as *Hirudo,* attach the anterior sucker tightly to a thin area of the host's skin and then slit it. The jaws of *Hirudo* make about two slices per second. The incision is anesthetized by a substance of unknown origin. The pharynx provides continual suction, and the secretion of hirudin prevents coagulation of the blood. Penetration of the host's tissues is not well understood in the many jawless, pharynx-bearing species that are bloodsuckers. The everted pharynx becomes rigid, and it is possible that penetration is aided by enzymatic action.

Leech digestion is peculiar in a number of respects. The gut secretes no amylases, lipases, or endopeptidases. The presence of only exopeptidases perhaps explains the fact that digestion in bloodsucking leeches is so slow. Also characteristic of the leech gut are mutualistic bacteria that are important in nutrition. In both the bloodsucking European medicinal leech, *Hirudo medicinalis,* and the predaceous *Erpobdella octoculata,* the gut bacteria are responsible for much of the leeches' digestion; they may be significant in the digestion of all leeches. The bacterium *Aeromonas hydrophila* of *H. medicinalis* breaks down high-molecular-weight proteins, fats, and carbohydrates, and the bacterial population increases significantly following the ingestion of blood by the leech. The bacteria may also produce vitamins and other compounds that are used by the leech.

Bloodsuckers feed infrequently, but when the opportunity arises, they gorge themselves on blood. A blood meal for *Haemadipsa* increases its body weight 10-fold, while *Hirudo* doubles or triples its weight in blood. Following ingestion, water is removed from the blood and excreted through the nephridia. A very slow digestion of blood cells ensues, during which the leech can tolerate long periods of fasting. Captive medicinal leeches are known to have fasted for one and a half years, and because they may require 200 days to digest a meal, they need not feed more than twice a year in order to grow.

Reproduction

Unlike many other annelids, leeches do not reproduce asexually, nor can they regenerate lost parts. As clitellates, leeches are hermaphrodites, but they are protandric, not simultaneous hermaphrodites, and the testes mature before the ovaries. The unpaired male gonopore is midventral on segment 10 and the unpaired female gonopore is on segment 11 (Fig. 13-70). There is but a single pair of elongate ovaries and testes that span many segments. Each of the two testes extends for 10 segments and is constricted into lobes, sacs, or follicles (the "testes" in Fig. 13-78A). The accessory reproductive organs are incorporated into the sperm ducts and oviducts (Fig. 13-78), unlike in the oligochaetes, which have separate seminal vesicles and receptacles. Fertilization is internal in all leeches (Fig. 13-79F).

Sperm transfer in the hirudinids, most of which have a penis, is similar to direct sperm transmission in earthworms (Fig. 13-79A). The ventral surfaces of the clitellar regions of a copulating pair come together, with the anterior end of one worm directed toward the posterior end of the other. Thus, the male gonopore of one worm apposes the female gonopore of the other. The penis is everted into the female gonopore and sperm are introduced into the vagina, which probably also acts as a sperm storage area.

Sperm transfer in many rhynchobdellids and pharyngobdelliforms, all of which lack a penis, is by hypodermic impregnation. The two copulating worms commonly intertwine and grasp each other with their anterior suckers. The ventral clitellar regions are in apposition, and by muscular contraction of the atrium (the last part of the sperm duct), a spermatophore is expelled from one worm and penetrates the integument of the other. The site of penetration is usually in the clitellar region, but spermatophores may be inserted some distance away. As soon as the head of the spermatophore has penetrated the integument, perhaps by a combination of expulsion pressure and a cytolytic action by the spermatophore itself, the sperm are discharged into the tissues (Fig. 13-79F). Following liberation from the spermatophore, the sperm migrate to the ovisacs via the coelomic channels or use a tissue pathway, called **target tissue,** that is specialized to receive the spermatophore and conduct the sperm.

Eggs are laid from two days to many months after copulation. At this time, the clitellum becomes conspicuous and usually secretes a cocoon, as in the oligochaetes (Fig. 13-79B,C). The cocoon is filled with nutritive albumen produced by some of the clitellar glands. The cocoon then receives the one to many fertilized eggs as it passes over the female gonopore. As in albumen-secreting oligochaetes, the eggs are small and relatively yolkless. Beginning in May, *Erpobdella punctata* in Michigan, for example, produces some 10 cocoons, each with five eggs that hatch three to four weeks later. The cocoons are affixed to submerged objects or vegetation. Some piscicolids attach their cocoons to their fish hosts. Terrestrial species place them in damp soil beneath stones and other objects, and the hirudinids, such as *Hirudo* and *Haemopsis,* leave the water to deposit their cocoons in damp soil.

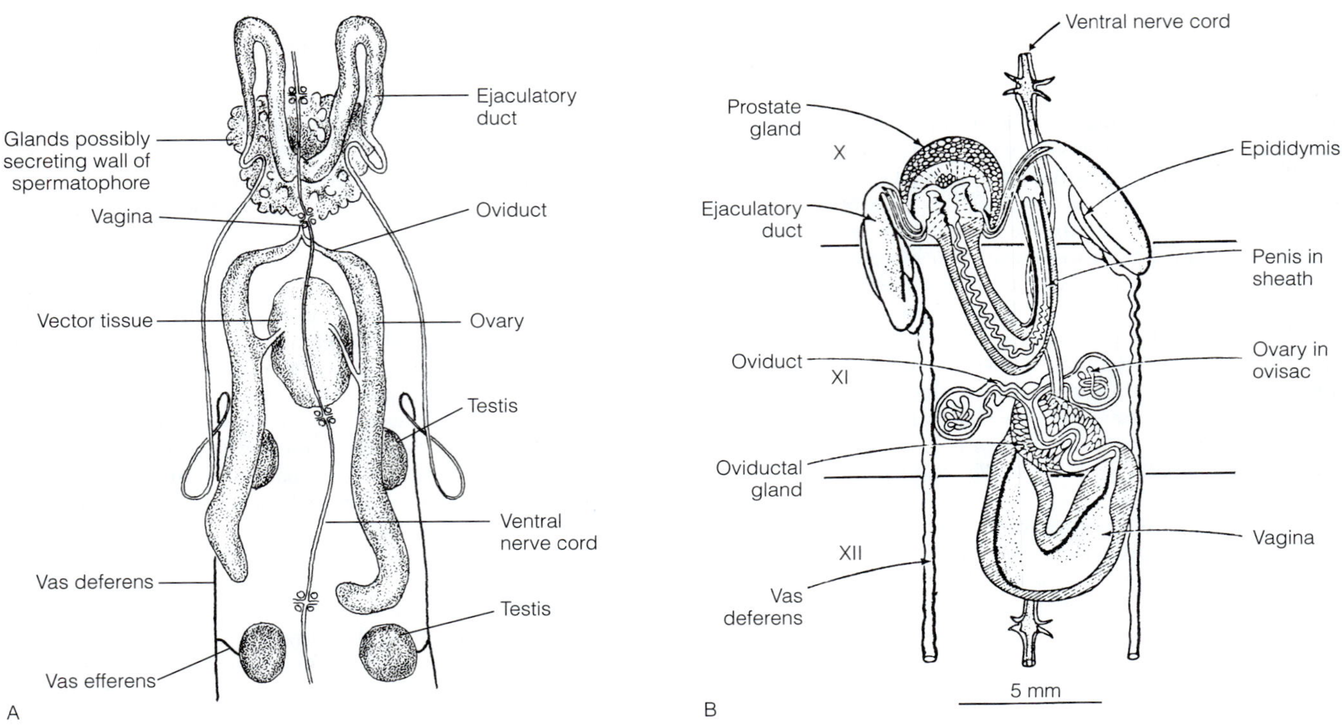

FIGURE 13-78 Leech reproductive system (Euhirudinea). **A,** Reproductive system of the fish leech *Piscicola geometra* (Rhynchobdellida). **B,** Reproductive organs of the European medicinal leech, *Hirudo medicinalis* (Arhynchobdellida). Testis sacs are associated with the vas deferens in more posterior segments. *(A, After Brumpt from Harant and Grassé; B, After Leuckart and Brandes from Mann, 1962)*

The glossiphoniids brood their eggs. In some species the cocoons are attached to the bottom and covered and ventilated by the ventral surface of the worm. In others, the cocoons are membranous, transparent, and attached to the ventral surface of the parent. During the course of development, the embryonic leeches break free of the cocoon and attach themselves directly to the ventral surface of the parent (Fig. 13-79E).

A larva (cryptolarva) develops within the cocoon of arhynchobdellid leeches. It has a pair of protonephridia and a mouth with which it imbibes the nutritive fluid within the cocoon.

Most leeches have an annual or two-year cycle, breeding in the spring or summer and maturing by the following year. Life cycles are correlated in part with feeding habits. Some, such as *Hirudo,* are associated with the host only during actual feeding; others, such as the marine fish leech *Hemibdella,* never leave the host. Most leeches leave the host at least to breed. Less than three months is required to complete the entire life cycle in *Hemibdella.*

Diversity of Hirudinomorpha

BRANCHIOBDELLIDA

The taxon includes 150 species of tiny (1 to 10 mm) ectocommensals or ectoparasites on the surfaces of crayfishes and freshwater crabs. On crayfishes, some are ectoparasites on the gills, whereas others live on the exoskeletal surface and graze on organic debris and microorganisms. Often the crustacean's head, carapace, and other parts are covered with these whitish worms. Chaetae, prostomium, and pygidium usually are absent. Body has 15 segments, the last forming a sucker (Fig. 13-80). Head (peristomium plus three segments) is modified into a sucker with the mouth at its center. Buccal cavity contains two teeth, one dorsal and one ventral. Dorsal anus is on segment 14. Segmented coelom, more or less typical annelidan hemal system, two pairs of metanephridia. Hermaphrodites with internal fertilization. Zygotes deposited in cocoons attached to host. Embryos develop into cryptolarvae, as in arhychobdellid leeches. *Branchiobdella.*

HIRUDINEA

Sister taxon of Branchiobdellida. Includes typical leeches (Euhirudinea) with anterior and posterior suckers; also *Acanthobdella peledina,* a primitive leech with only a posterior sucker and several plesiomorphic traits not inherited by euhirudineans. All hirudineans have crossed-helical (diagonal) body-wall muscles, sperm packaged in spermatophores, an unpaired female gonopore, and sperm storage that does not include seminal receptacles.

Acanthobdella peledina

Sole species in taxon; sister taxon of Euhirudinea. Ectoparasitic on boreal freshwater fishes (Alaska, northern Europe, and Asia). Feeds on host's skin and blood. Although clearly a leech,

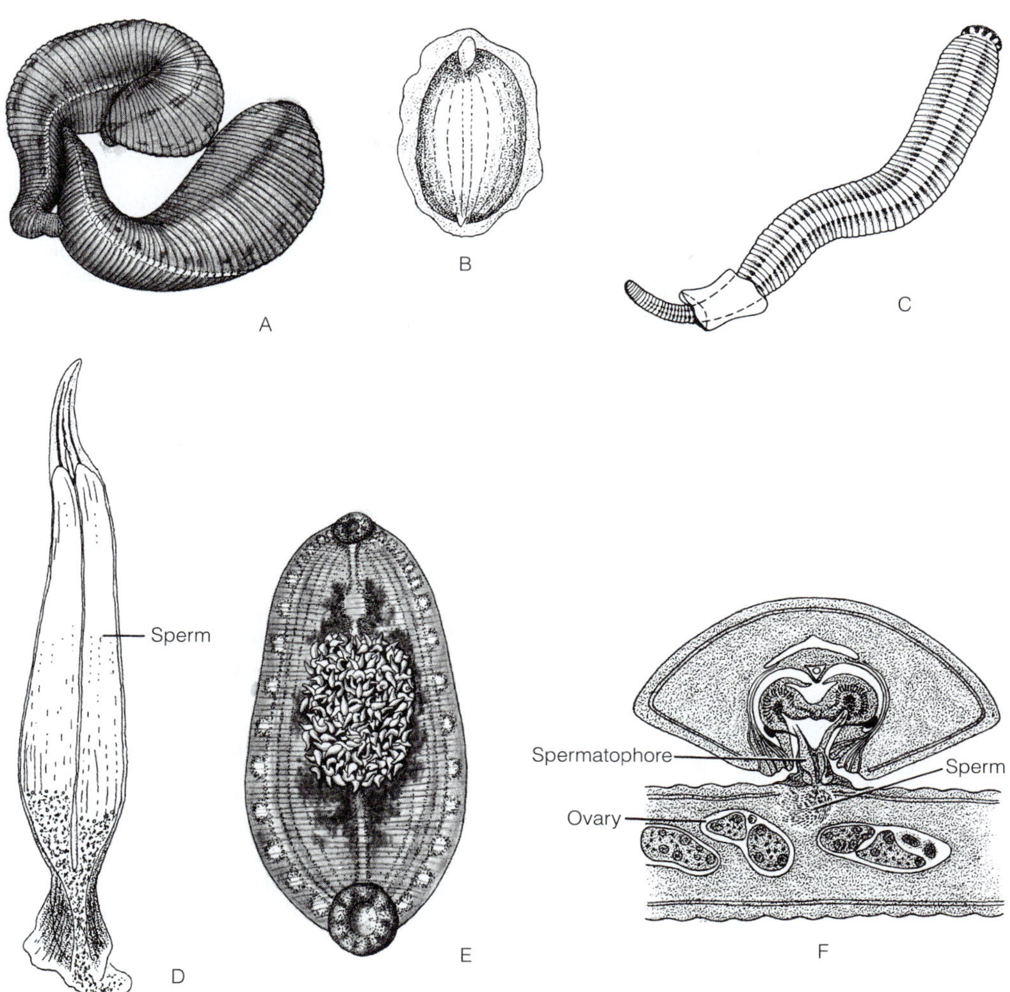

FIGURE 13-79 Leech reproduction (Euhirudinea). **A,** *Hirudinaria* (Arhynchobdellida) copulating. **B,** Cocoon of *Erpobdella octoculata* (Arhynchobdellida). **C,** *Erpobdella* withdrawing from a cocoon. **D,** Spermatophore of *Haementeria* (Rhynchobdellida). **E,** Glossiphoniid brooding young. **F,** Section through two copulating *Erpobdella* individuals. Upper leech is injecting spermatophores into lower individual. *(A, After a photograph by Keegan, et al., 1968; B, After Pavlovsky from Harant and Grassé; C, From Nagao, Z. 1957. J. Fac. Sci. Hokkaido Univ. Ser. VI, Zool. 13:192–196; D, After Pavlovsky from Harant and Grassé; E and F, After Brumpt.)*

Acanthobdella lacks an anterior sucker; instead, it attaches its oral end to the host with hooked chaetae borne on the first five segments (segments 2 to 6; prostomium and peristomium are absent). Unique among leeches in having chaetae, segmental coelomic cavities, and septa (Fig. 13-81). Despite primitive arrangement of coelomic cavities and absence of an anterior sucker, *Acanthobdella* still moves in inchworm fashion and does not use peristalsis. Posterior sucker consists of terminal four segments (27 to 30).

Euhirudinea

Sister taxon of *Acanthobdella peledina.* 350 species of "true" leeches with anterior and posterior suckers. Body composed of prostomium plus 33 segments; posterior sucker of 7 segments (27 to 33). Chaetae absent, segmental coeloms lack septa and mesenteries and have transformed into circulatory channels.

Rhynchobdellida^{O}: Aquatic leeches with a protrusible muscular pharynx (Fig. 13-82A–E), with coelomic and hemal circulatory systems. Feed by protruding pharynx, which seeks a favorable penetration site on the host's surface, and then thrusts deeply into the body-wall tissues. Once inside, radial pharyngeal muscles contract and suck tissue and fluids into the leech's gut. Some, such as *Glossiphonia* and *Helobdella* **(Glossosiphoniidae^{F}),** feed only on invertebrates, (snails, oligochaetes, crustaceans, insects), but vertebrates (amphibians, turtles,

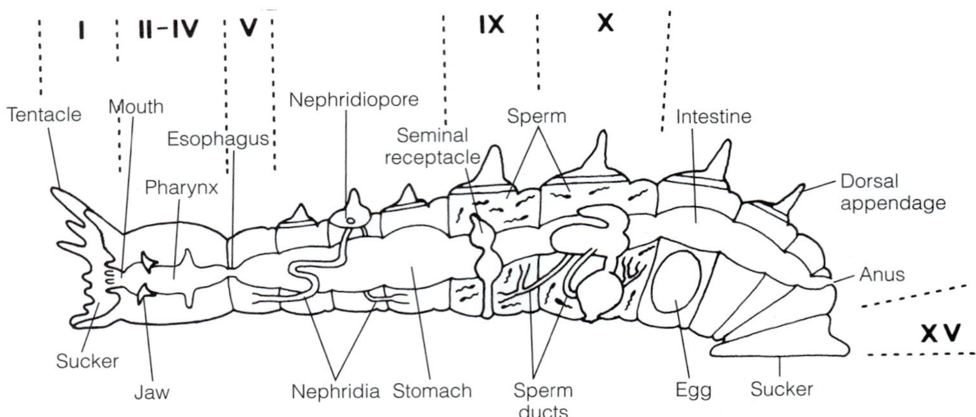

FIGURE 13-80 Leech diversity. Branchiobdellida. Lateral view of branchiobdellidan anatomy. The Roman numerals indicate segments. *(Gelder, S. R., Brinkhurst, R. O. 1990. An assessment of the Branchiobdellida (Annelida, Clitellata), using PAUP. Can. Jour. Zool. 68:1318–1326 (Fig. 1) Reprinted with permission.)*

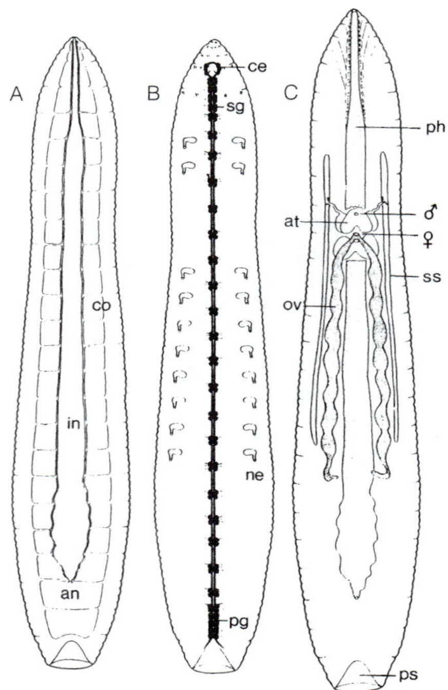

FIGURE 13-81 Leech diversity. Anatomy of *Acanthobdella peledina*. **A,** Gut and coelomic cavities (septa are present, but mesenteries are not). **B,** Central nervous system and nephridia (small dots at anterior end are chaetae). **C,** Gut and reproductive system. Labels: an = anus; at = atrium; ce = cerebrum; co = coelom; in = intestine; ne = nephridia; ov = ovary; pg = posterior ganglionic mass; ph = pharynx; ps = posterior sucker; sg = subpharyngeal ganglionic mass; ss = sperm sac; male and female gonopores also indicated. *(After Purschke et al., 1993, from Ax, P. 2000. Multicellular Animals: The Phylogenetic System of the Metazoa. Vol. II. Springer-Verlag, Berlin. (Fig. 33).)*

snakes, alligators, crocodiles) are hosts for most species. The cosmopolitan *Theromyzon* attaches to the nasal membranes of shore- and waterbirds (Fig. 13-82E). Fish leeches (**Piscicolidae**[F]) parasitize both freshwater and marine fish, including sharks, and rays (Fig. 13-82A–C). Their bodies often bear lateral gills. Glossosiphoniidae: *Haementeria, Hemiclepsis, Placobdella, Theromyzon;* Piscicolidae: *Branchellion, Cystobranchus, Piscicola, Pontobdella, Ozobranchus, Trachelobdella.*

Arhynchobdellida[O]: Hemal system absent, coelom constitutes sole circulatory system. Each segment typically has five annulations. Primarily aquatic, but some are amphibious or even terrestrial. Semiterrestrial *Haemopsis terrestris* occasionally plowed up in fields in the midwestern United States. Fully terrestrial status has been attained in the **Haemadipsidae**[F], whose species inhabit humid rainforests in south Asia and Australia (Fig. 13-82G). *Haemadipsa, Phytobdella.*

Gnathobdelliformes[sO]: Chiefly aquatic or amphibious bloodsuckers. Nonprotrusible pharynx and three pairs of slicing jaws in buccal cavity; salivary glands secrete the anticoagulant hirudin and other compounds. The aquatic *Hirudo medicinalis* (European medicinal leech; **Hirudinidae**[F]; Fig. 13-70, 13-77A) is the best-known leech.

Before the age of modern medicine, medicinal leeches were used for centuries to restore the balance of body humors (blood, bile, phlegm), the imbalance of which was thought to be the cause of disease. Historically, being bled by a leech (bloodletting or leeching) was probably ineffective at best, and at worst may have hastened the patient's departure to the hereafter. Examples of the latter include George Washington, who died after two days of bloodletting for a sore throat, and the Russian novelist Nikolai Gogol, who, in failing health and under physical restraint, had leeches applied to his nostrils. Gogol soon died, and an observer noted that the treatment "probably helped him die much faster."

Today, there is a more sanguine use for leeches in medicine: They are effective aids in the reattachment of skin flaps, severed fingers, and other appendages. After the appendage is surgically reattached, leeches are applied to the injured area. The leech replaces the damaged venous drainage in the injured area, thus allowing arteries to deliver oxygenated blood to the region. The leech saliva not only contains hirudin, which prevents blood clots, but also a vasodilator, which reestablishes blood flow to the newly attached part; a numbing anesthetic; and an antibiotic that reduces infection. Remarkably, the antibiotic is produced by

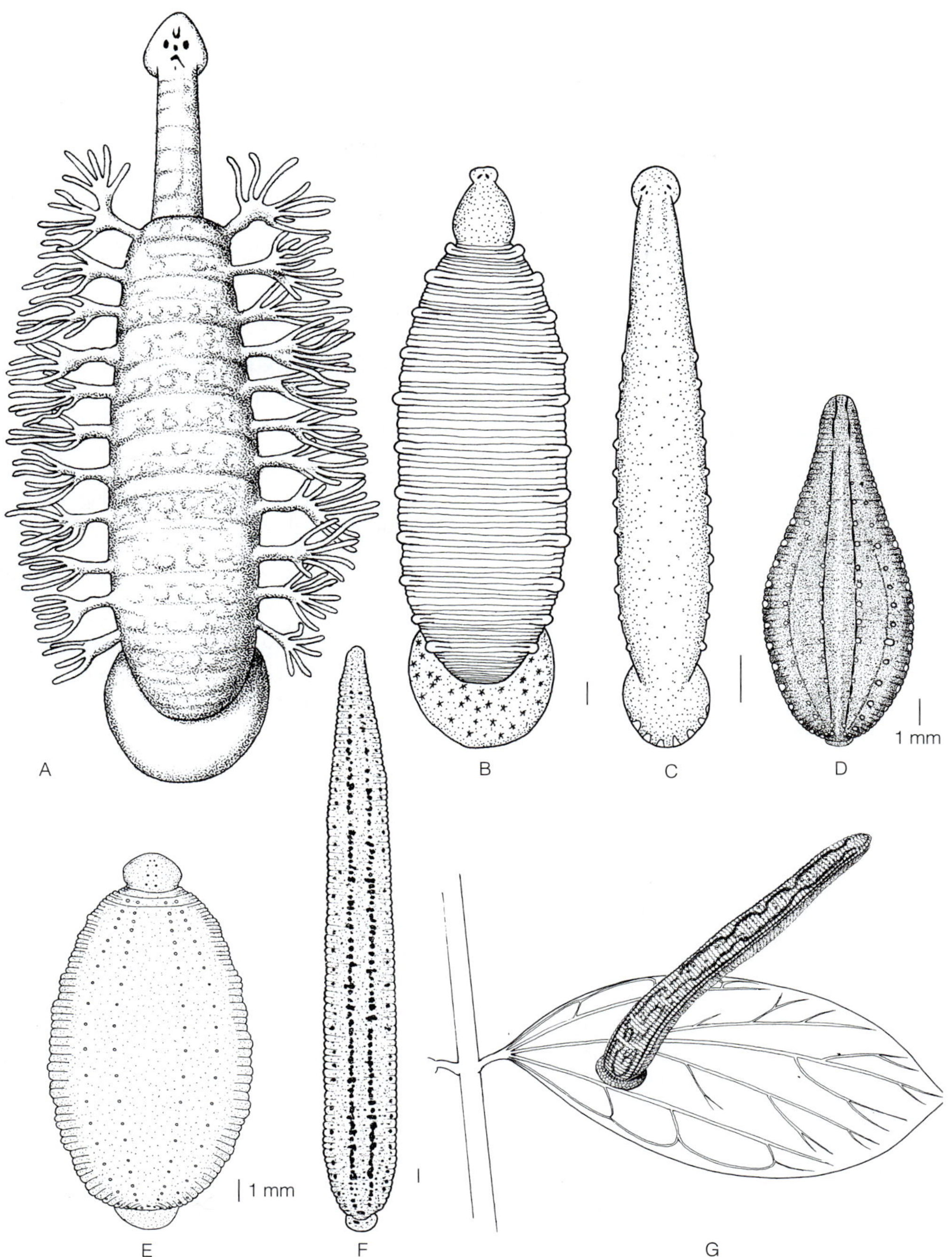

FIGURE 13-82 Leech diversity. Euhirudinea. External dorsal views of different species of leeches. **A–C,** Fish leeches (Rhynchobdellida, Piscicolidae). **A,** *Ozobranchus,* showing lateral gills. **B,** *Cystobranchus.* **C,** *Piscicola.* **D** and **E,** Rhynchobdellid leeches (Glossiphoniidae). **D,** *Glossiphonia complanata,* a common European and North American leech that feeds on snails. **E,** *Theromyzon,* a cosmopolitan genus of leeches that attacks birds. **F** and **G,** Arhynchobdellid leeches. **F,** *Erpobdella punctata* (Erpobdellidae), a common North American scavenger and predatory leech. **G,** *Haemadipsa* (Haemadipsidae), a bloodsucking terrestrial leech of south Asia, poised on a leaf. *(A, After Oka from Mann; B–F, From Sawyer, R. J. 1972. North American Freshwater Leeches. University of Illinois Press, Champaign, IL; G, Adapted from Keegan, et al.)*

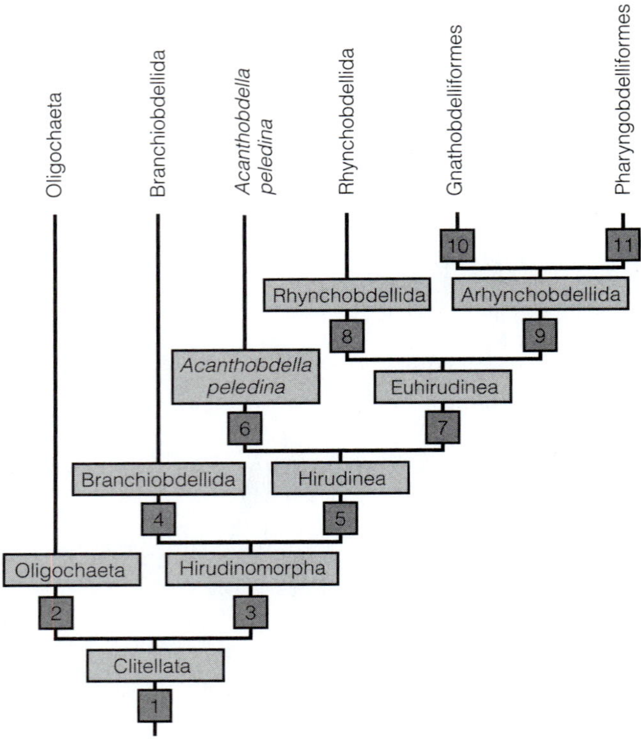

FIGURE 13-83 Phylogeny of Clitellata. **1, Clitellata:** clitellum; cocoon; brain behind prostomium. **2, Oligochaeta:** muscular pharyngeal bulb or pad in mid-dorsal wall of pharynx. **3, Hirudinomorpha:** ectoparasitism; posterior sucker; anus is displaced dorsal and anterior to sucker; secondary superficial annulations overlie segments; unpaired midventral male gonopore. **4, Branchiobdellida:** chaetae are absent, prostomium is absent; 15 segments (head has 4 segments), two pharyngeal jaws, two pairs of metanephridia. **5, Hirudinea:** body composed of prostomium plus 30 segments; posterior sucker is 4 segments (27 to 30); more than two annulations per segment; central nervous system has 31 pairs of ganglia (brain of 2 ganglia, subpharyngeal ganglion of 4 ganglia, ventral nerve cord of 21 unfused ganglia, sucker of 4 fused ganglia); crossed-helical muscles between circular and longitudinal body-wall muscles; spermatophores, unpaired female gonopore. **6, *Acanthobdella peledina:*** lacks a prostomium and peristomium; 29 segments; 40 oral chaetae used for attachment to host; nephridia lack funnels. **7, Euhirudinea:** body of prostomium plus 33 segments, posterior sucker is of 7 segments (27 to 33); central nervous system has 34 pairs of ganglia (brain of 2 ganglia, subpharyngeal ganglion of 4 ganglia, ventral nerve cord of 21 unfused ganglia, sucker of 7 fused ganglia); chaetae are absent; anterior sucker; coelomic septa are absent, cavities converted into circulatory vessels; serial testes. **8, Rhynchobdellida:** protrusible pharynx ("proboscis"). **9, Arhynchobdellida:** hemal system is absent; coelomic circulatory system has vessels and capillaries. **10, Gnathobdelliformes:** slicing jaws like three semicircular scalpel blades in buccal cavity; hirudin (anticoagulant) secreted from jaw-associated glands; direct sperm transfer. **11, Pharyngobdelliformes:** elongate, spirally twisted pharynx adapted to consume large prey. *(Modified from Ax, P. 2000. Multicellular Animals: The Phylogenetic System of the Metazoa. Vol. II. Springer Verlag, Berlin. 396 pp.)*

one of the leech's mutualistic gut bacteria *(Aeromonas hydrophila),* which normally helps the leech digest its blood meal. The antibiotic is used by the bacterium to eliminate competition from other microbes. Hirudinidae: *Haemopsis, Macrobdella, Philobdella.*

Pharyngobdelliformes[sO]: Predatory, best known for the freshwater *Erpobdella,* with its long, nonprotrusible pharynx that occupies the anterior one-third of the body (Fig. 13-82F). Jaws are absent, but the muscular sucking pharynx, with its Y-shaped lumen, enables these leeches to engulf large invertebrate prey. *Dina.*

PHYLOGENY OF CLITELLATA

According to recent phylogenetic research, Clitellata includes the two major sister taxa Oligochaeta and Hirudinomorpha. Within Hirudinomorpha are the sister taxa Branchiobdellida and Hirudinea; within Hirudinea are *Acanthobdella peledina* and the Euhirudinea, or "true" leeches (Fig. 13-83).

Apart from their clitellate characteristics, oligochaetes retain primitive annelid traits, with the possible exception of the muscular dorsal wall of the pharynx, which probably is an autapomorphy of Oligochaeta. According to some authorities, oligochaetes evolved directly from the ancestral marine annelid that radiated in the freshwater environment. Thus, the first oligochaetes probably were burrowers in freshwater sediment. Those burrowers may have given rise in one direction to the strictly freshwater species that invaded loose bottom debris and in another direction to the earthworms that invaded successively drier sediments. The primitive aquatic species had large, yolky eggs; a clitellum that was one cell layer thick; holonephridia; and a well-developed capacity for clonal reproduction. From these, the terrestrial earthworms evolved a yolk (albumen)-secreting clitellum composed of several cell layers; smaller, less yolky eggs; enteronephridia for water conservation; and perhaps a greater reliance on sexual reproduction, although asexual reproduction by parthenogenesis occurs in several earthworms.

The first leeches were probably ectocommensals, as are branchiobdellids, or facultative ectoparasites, as is *Acanthobdella peledina,* on other animals. Perhaps because of the importance of remaining with their hosts, leeches first evolved an adhesive posterior sucker, and later, an anterior sucker around the mouth. Associated with the evolution of suckers came a change in the pattern of locomotion from whole-body peristalsis to inchworm-like creeping, which permitted movement without loss of attachment to the host. Accompanying the new form of locomotion, which relaxed the requirement for a series of paired segmental coelomic cavities (hydrostats), was a redesign of the leech body. This included dorsoventral body flattening and new, specialized muscles that allowed leeches to swim from host to host or to pursue prey, and a complex restructuring of many internal parts, such as the coelom, hemal system, nephridia, and reproductive organs. The loss of septa allowed the coelom to adopt the global circulatory function of the hemal system and eventually to replace it, but the loss of the hemal system, including a vascular filtration site, required a redesign of the nephridia.

PHYLOGENETIC HIERARCHY OF CLITELLATA

Clitellata
- Oligochaeta
- Hirudinomorpha
  - Branchiobdellida
  - Hirudinea
    - *Acanthobdella peledina*
    - Euhirudinea
      - Rhynchobdellida
      - Arhynchobdellida
        - Gnathobdelliformes
        - Pharyngobdelliformes

REFERENCES

SEGMENTATION

Clark, R. B. 1964. Dynamics in Metazoan Evolution: The Origin of the Coelom and Segments. Clarendon University Press, Oxford. 313 pp.

Davis, G. K., and Patel, N. H. 2000. The origin and evolution of segmentation. Trends Genet. 15:M68–M72.

DeRobertis, E. M. 1997. Evolutionary biology: The ancestry of segmentation. Nature 387:25–26.

Kimmel, C. B. 1996. Was Urbilateria segmented? Trends Genet. 12:329–331.

Meier, S. 1984. Somite formation and its relationship to metameric patterning of the mesoderm. Cell Differ. 14:235–243.

Potswald, H. E. 1981. Abdominal segment formation in *Spirorbis moerchi* (Polychaeta). Zoomorphology 97:225–245.

Raff, R. A., and Kaufman, T. C. 1983. Embryos, Genes and Evolution. Pattern Formation. MacMillan, New York. pp. 262–286.

Shankland, M., and Seaver, E. L. 2000. Evolution of the bilaterian body plan: What have we learned from annelids? Proc. Natl. Acad. Sci. USA 97:4434–4437.

Sedgwick, A. On the origin of metameric segmentation and some other morphological questions. Q. J. Microsc. Sci. 24:43–82.

Turbeville, J. M., and Ruppert, E. E. 1983. Epidermal muscles and peristaltic burrowing in *Carinoma tremaphoros* (Nemertini): Correlates of effective burrowing without segmentation. Zoomorphology 103:103–120.

Weisblat, D. A., and Shankland, M. 1985. Cell lineage and segmentation in the leech. Phil. Trans. R. Soc. Lond. B 312:39–56.

Wolpert, L., Beddington, R., Jessell, T., Lawrence, P., Meyerowitz, E., and Smith, J. 2002. Principles of Development. Second Edition. Oxford University Press, Oxford. 542 pp.

ANNELIDA

General

Anderson, D. T. 1973. Embryology and Phylogeny in Annelids and Arthropods. Pergamon Press, Oxford. 495 pp.

Ax, P. 2000. Multicellular Animals: The Phylogenetic System of Metazoa. Vol. II. Springer Verlag, Berlin. 396 pp.

Bartolomaeus, T. 1994. On the ultrastructure of the coelomic lining in the Annelida, Sipuncula, and Echiura. Microfauna Marina 9:171–220.

Bartolomaeus, T. 1999. Structure, function and development of segmental organs in Annelida. Hydrobiologia 402:21–37.

Brinkhurst, R. O. 1982. Evolution in the Annelida. Can. J. Zool. 60:1043–1059.

Brown, F. A. 1950. Selected Invertebrate Types. John Wiley and Sons, New York. 597 p.

Clark, R. B. 1964. Dynamics in Metazoan Evolution: The Origin of the Coelom and Segments. Clarendon Press, Oxford. 313 pp.

Dales, R. P. 1963. Annelids. Hutchinson University Library, London. 200 pp.

Fauvel, P., Avel, M., Harant, H., et al. 1959. Embranchement des Annelides. In Grassé, P. (Ed.): Traité de Zoologie. Vol. 5, Pt. 1. Masson, Paris. pp. 3–686.

Fretter, V., and Graham, A. 1976. A Functional Anatomy of Invertebrates. Academic Press, London. 589 pp.

Harrison, F. W., and Gardiner, S. L. 1992. Microscopic Anatomy of Invertebrates. Vol. 7. Annelida. Wiley-Liss, New York. 418 pp.

Keil, T. A. 1998. The structure of integumental mechanoreceptors. In Harrison, F. W., and Locke, M. (Eds.): Microscopic Anatomy of Invertebrates. Vol. 11B. Insecta. Wiley-Liss, New York. pp. 385–404.

Klemm, D. J. 1985. A Guide to the Freshwater Annelida (Polychaeta, Naidid and Tubificid Oligochaeta, and Hirudinea) of North America. Kendall/Hunt, Dubuque, IA. 226 pp.

Mangum, C. P. 1970. Respiratory physiology in annelids. Am. Sci. 58:641–647.

Mangum, C. P. 1976. Primitive respiratory adaptations. In Newell, R. C. (Ed.): Adaptation to Environment. Butterworth, Boston. pp. 191–278.

Mangum, C. P. 1977. Annelid hemoglobins: A dichotomy in structure and function. In Reish, D. J., and Fauchald, K. (Eds.): Essays in Memory of Dr. Olga Hartman. Allan Hancock Foundation, University of Southern California, Los Angeles. pp. 407–428.

Mangum, C. P. 1983. The function of gills in several groups of invertebrate animals. In Houlihan, D. F., Rankin, J. C., and Shuttleworth, T. J. (Eds.): Gills. Society for Experimental Biology Seminar Series No. 16. Cambridge University Press, Cambridge.

Mangum, C. P. 1985. Oxygen transport in invertebrates. Am. J. Physiol. 248:505–514.

McHugh, D. 2000. Molecular phylogeny of the Annelida. Can. J. Zool. 78:1873–1884.

Mill, P. J. (Ed.): 1978. Physiology of Annelids. Academic Press, London. 684 pp.

Newell, R. C. 1970. Biology of Intertidal Animals. American Elsevier, New York. 555 pp.

Olive, P. J. W., and Clark, R. B. 1978. Physiology of reproduction. In Mill, P. J. (Ed.): Physiology of Annelids. Academic Press, London. pp. 271–368.

Parker, S. P. (Ed.): 1982. Synopsis and Classification of Living Organisms. Vol. 2. McGraw-Hill, New York. 1236 pp.

Pennak, R. W. 1978. Fresh-Water Invertebrates of the United States. 2nd Edition. John Wiley and Sons, New York. 803 pp.

Rieger, R. M. 1985. The phylogenetic status of the acoelomate organization within the Bilateria: A histological perspective. In Conway Morris, S., et al. (Eds.): The Origins and Relationships of the Lower Invertebrates. Systematics Association Spec. Vol. 28. Clarendon Press, Oxford. pp. 101–122.

Seaver, E. C., Paulson, D., Irvine, S. Q., and Martindale, M. Q. 2001. The spatial and temporal expression of *Ch-en,* the *engrailed* gene in the polychaete *Chaetopterus,* does not support a role in body segmentation. Dev. Biol. 236:195–209.

Trueman, E. R. 1975. The Locomotion of Soft-Bodied Animals. Edward Arnold, London. 200 pp.

Internet Sites

http://www.biology.ualberta.ca/facilities/multimedia/index.php?Page=252 (Two animations of annelid locomotion: "Form and Locomotion of a Nereid Worm," and "Form and Locomotion of an Earthworm.")

http://www.kewalo.hawaii.edu/labs/seaver/ (Home page of Elaine Seaver, a developmental biologist studying annelid growth and segmentation.)

POLYCHAETA

General

Anderson, D. T. 1973. Embryology and Phylogeny in Annelids and Arthropods. Pergamon Press, Oxford. 495 pp.

Arp, A. J., and Childress, J. J. 1983. Sulfide binding by the blood of the hydrothermal vent tube worm *Riftia pachyptila.* Science 219:295–297.

Arp, A. J., Childress, J. J., and Fisher, C. R., Jr. 1985. Blood gas transport in *Riftia pachyptila.* In Jones, M. L. (Ed.): Hydrothermal Vents of the Eastern Pacific: An Overview. Bull. Biol. Soc. Wash. 6:289–300.

Ax, P. 1966. Die Bedeutung der interstiellen Sandfauna für allgemeine Probleme der Systematik, Ökologie und Biologie. Veröffentlungen des Institut für Meeresforschung Bremerhaven. Suppl. II. pp. 15–66.

Bakke, T. 1976. The early embryos of *Siboglinum fiordicum* Webb (Pogonophora) reared in the laboratory. Sarsia. 60:1–11.

Bartolomaeus, T. 1995. Structure and formation of the uncini in *Pectinaria koreni, Pectinaria auricoma* (Terebellida) and *Spirorbis spirorbis* (Sabellida): Implications for annelid phylogeny and the position of Pogonophora. Zoomorphology 115:161–177.

Bartolomaeus, T. 1998. Chaetogenesis in polychaetous Annelida: Significance for annelid systematics and the position of Pogonophora. Zool. Anal. Complex Syst. 100:348–364.

Baskin, D. G. 1976. Neurosecretion and the endocrinology of nereid polychaetes. Am. Zool. 16:107–124.

Blake, J. D. 1975. The larval development of Polychaeta from the northern California coast. III. Eighteen species of Errantia. Ophelia 14:23–84.

Blake, J. A., and Woodwick, K. H. 1981. The morphology of *Tripolydora spinosa:* An application of scanning electron microscopy to polychaete systematics. Proc. Biol. Soc. Wash. 94:352–364.

Boilly, B., and Wissocq, J. C. 1977. Occurrence of striated muscle fibers in a contractile vessel of a polychaete: The dorsal heart of *Magelona papillicornis.* Biol. Cell 28:131–136.

Brenchley, G. A. 1976. Predator detection and avoidance: Ornamentation of tube-caps of *Diopatra* spp. Mar. Biol. 38:179–188.

Brown, S. C. 1975. Biomechanics of water-pumping by *Chaetopterus variopedatus* Renier. Skeletomusculature and kinematics. Biol. Bull. 148:136–150.

Caspers, H. 1984. Spawning periodicity and habitat of the palolo worm *Eunice viridis* in the Samoan Islands. Mar. Biol. 79:229–236.

Cavanaugh, C. M., Gardiner, S. L., Jones, M. L., et al. 1981. Prokaryotic cells in the hydrothermal vent tube worm *Riftia pachyptila* Jones: Possible chemoautotropic symbionts. Science 213:340–342.

Cazaux, C. 1967. Development of *Glycera convoluta.* Vie et Milieu A18:559–571.

Chughtai, I., and Knight-Jones, E. W. 1988. Burrowing into limestone by sabellid polychaetes. Zool. Scripta 17:231–238.

Clark, L. B., and Hess, W. N. 1940. Swarming of the Atlantic palolo worm, *Leodice fucato.* Tortugas Lab. Papers 33:21–70.

Dales, R. P. 1957. The feeding mechanism and structure of the gut of *Owenia fusiformis.* J. Mar. Biol. Assoc. U.K. 36:81–89.

Dales, R. P., and Peter, G. 1972. A synopsis of the pelagic Polychaeta. J. Nat. Hist. 6:55–92.

Dauer, D. M. 1985. Functional morphology and feeding behavior of *Paraprionospio pinnata* (Polychaeta: Spionidiae). Mar. Biol. 85:143–151.

Day, J. 1967a. A Monograph on the Polychaeta of Southern Africa. Pt. 1: Errantia. British Museum of Natural History, London. 458 pp.

Day, J. 1967b. A Monograph on the Polychaeta of Southern Africa. Pt. 2: Sedentaria. British Museum of Natural History, London. 420 pp.

Dykens, J. A., and Mangum, C. P. 1984. The regulation of body–fluid volume in the estuarine annelid *Nereis succinea.* J. Comp. Physiol. B 154:607–617.

Eernisse, D. J. 1998. Arthropod and annelid relationships reexamined. In Fortey, R. A., and Thomas, R. H. (Eds.): Arthropod Relationships. Systematics Association Spec. Vol. 55. Chapman and Hall, London. pp. 43–56.

Eernisse, D. J., Alberts, J. S., and Anderson, F. E. 1992. Annelida and Arthropoda are not sister taxa: A phylogenetic analysis of spiralian metazoan phylogeny. Syst. Biol. 41:305–330.

Eckelbarger, K. J. 1976. Larval development and population aspects of the reef-building polychaete *Phragmatopoma lapidosa* from the east coast of Florida. Bull. Mar. Sci. 26:117–132.

Eckelbarger, K. J. 1983. Evolutionary radiation in polychaete ovaries and vitellogenic mechanisms: Their possible role in life history patterns. Can. J. Zool. 61:487–504.

Fauchald, K. 1977. The polychaete worms: Definitions and keys to the orders, families and genera. Nat. Hist. Mus. Los Angeles Co. Sci. Ser. 28:1–190.

Fauchald, K. 1983. Life diagram patterns in benthic polychaetes. Proc. Biol. Soc. Wash. 96:160–177.

Fauchald, K., and Jumars, P. A. 1979. The diet of worms: A study of polychaete feeding guilds. Oceanogr. Mar. Biol. Ann. Rev. 17:193–284.

Fauchald, K., and Rouse, G. W. 1997. Polychaete systematics: Past and present. Zool. Scripta 26:71–138.

Felbeck, H., and Childress, J. J. 1988. *Riftia pachyptila:* A highly integrated symbiosis. In Laubier, L. (Ed.): Actes du Colloque,

"Les Sources Hydrothermales de la Ride du Pacifique Oriental: Biologie et Ecologie." Institut Oceanographique, Paris, 4–7 Novembre 1985. Oceanolog. Acta Spec. Vol. 8. Montrouge: Gauthiers-Villars. pp. 131–138.

Fischer, A., and Fischer, U. 1995. On the life-style and life-cycle of the luminescent polychaete *Odontosyllis enopla* (Annelida: Polychaeta). Invert. Biol. 114:236–247.

Flood, P. R., and Fiala-Medioni, A. 1982. Structure of the mucous feeding filter of *Chaetopterus variopedatus.* Mar. Biol. 72:27–34.

Fransen, M. E. 1980. Ultrastructure of coelomic organization in annelids. I. Archiannelids and other small polychaetes. Zoomorphologie 95:235–249.

Gaill, F., and Hunt, S. 1986. Tubes of deep sea hydrothermal vent worms *Riftia pachyptila* (Vestimentifera) and *Alvinella pompejana* (Annelida). Mar. Ecol. Prog. Ser. 34:267–274.

Gardiner, S. L. 1975. Errant polychaete annelids from North Carolina. J. Elisha Mitchell Sci. Soc. 91:77–220.

Gardiner, S. L., and Jones, M. L. 1993. Vestimentifera. In Harrison, F. W., and Rice, M. E. (Eds.): Microscopic Anatomy of Invertebrates. Vol. 12. Onychophora, Chilopoda, and Lesser Protostomata. Wiley-Liss, New York. pp. 371–460.

George, J. D., and Southward, E. C. 1973. A comparative study of the setae of Pogonophora and polychaetous Annelida. J. Mar. Biol. Assoc. U.K. 53:403–424.

Goerke, H. 1971. Die Ernährungweise der *Nereis*-Arten der deutschen Küsten. Veröff. Inst. Meeresforsch. Bremerh. 13:1–50.

Goodnight, C. J. 1973. The use of aquatic macroinvertebrates as indicators of stream pollution. Trans. Am. Microsc. Soc. 92:1–13.

Gray, J. 1939. Studies in animal locomotion. VIII. The kinetics of locomotion of *Nereis diversicolor.* J. Exp. Biol. 16:9–17.

Grassle, J. F. 1985. Hydrothermal vent animals: Distribution and biology. Science. 229:713–717.

Gupta, B. L., and Little, C. 1975. Ultrastructure, phylogeny and Pogonophora. Z. zool. Syst. Evolut.-forsch. 1975:43–63.

Hartman, O. 1959. Catalog of the Polychaetous Annelids of the World. Allan Hancock Found. Occas. Papers Vol. 23. 628 pp.

Hartman, O. 1965. Supplement and Index: Catalog of the Polychaetous Annelids of the World. Allan Hancock Found. Occas. Papers Vol. 23. 197 pp.

Hermans, C. O. 1969. The systematic position of the Archiannelida. Syst. Zool. 18:85–102.

Hermans, C. O., and Eakin, R. M. 1974. Fine structure of the eyes of an alciopid polychaete *Vanadis tagensis.* Z. Morphol. Tiere. 79:245–267.

Ivanov, A. V. 1963. Pogonophora. Academic Press, New York. 479 pp.

Kay, D. G. 1974. The distribution of the digestive enzymes in the gut of the polychaete *Neanthes virens.* Comp. Biochem. Physiol. 47(A):573–582.

Knight-Jones, P., and Thorp, C. H. 1984. The opercular brood chambers of Spirorbidae. Zool. J. Linn. Soc. 80:121–133.

Kojima, S. 1998. Paraphyletic status of Polychaeta suggested by phylogenetic analysis based on the amino acid sequences of elongation factor-1 alpha. Mol. Phylog. Evol. 9:255–261.

Kudenov, J. D. 1977. The functional morphology of feeding in three species of maldanid polychaetes. Zool. J. Linn. Soc. 60:95–109.

Kuper, M., and Purshke, G. 2001. The excretory organs in *Sphaerodorum flavum* (Phyllodocida, Sphaerodoridae): A rare case of co-occurrence of protonephridia, coelom, and blood vascular system in Annelida. Zoomorphology 120:191–203.

Jones, M. L. 1981. *Riftia pachyptila* Jones: Observations on the vestimentiferan worm from the Galapagos rift. Science 213:333–336.

Jones, M. L. 1984. The giant tube worms. Oceanus 27:47–52.

Jones, M. L. 1985. On the Vestimentifera, a new phylum: Six new species, and other taxa, from hydrothermal vents and elsewhere. Bull. Biol. Soc. Wash. 6:117–158.

Jones, M. L., and Gardiner, S. L. 1988. Evidence for a transient digestive tract in Vestimentifera. Proc. Biol. Soc. Wash. 101:423–433.

Jones, M. L., and Gardiner, S. L. 1989. On the early development of the vestimentiferan tube worm *Ridgeia* sp. and observations on the nervous system and trophosome of *Ridgeia* sp. and *Riftia pachyptila.* Biol. Bull. 177:254–276.

MacGinitie, G. E. 1939. The method of feeding of *Chaetopterus.* Biol. Bull. 77:115–118.

Mangum, C. P. 1976. The oxygenation of hemoglobin in lugworms. Physiol. Zool. 49:85–99.

Mangum, C. P., Woodin, B. R., Bonaventura, C., et al. 1975. The role of coelomic and vascular hemoglobin in the annelid family Terebellidae. Comp. Biochem. Physiol. 51A:281–294.

Martin, N., and Anctil, M. 1984. Luminescence control in the tubeworm *Chaetopterus variopedatus:* Role of nerve cord and photogenic gland. Biol. Bull. 166:583–593.

McConnaughey, B., and Fox, D. L. 1949. The anatomy and biology of the marine polychaete *Thoracophelia mucronata.* Univ. Calif. Publ. Zool. 47:319–339.

McHugh, D. 1997. Molecular evidence that echiurans and pogonophorans are derived annelids. Proc. Natl. Acad. Sci. USA 94:8006–8009.

Morin, J. G. 1983. Coastal bioluminescence: Patterns and functions. Bull. Marine Sci. 33:787–817.

Nakao, T. 1974. An electron microscopic study of the circulatory system in *Nereis japonica.* J. Morphol. 144:217–236.

Newell, R. C. 1970. Biology of Intertidal Animals. American Elsevier, New York. 555 pp.

Nicol, E. A. T. 1931. The feeding mechanism formation of the tube, and physiology of digestion in *Sabella pavonina.* Trans. R. Soc. Edinburgh 56:537–598.

Nørrevang, A. 1970a. The position of Pogonophora in the phylogenetic system. Z. zool. Syst. Evolut.-forsch. 3:161–172.

Nørrevang, A. 1970b. On the embryology of *Siboglinum* and its implications for the systematic position of the Pogonophora. Sarsia 42:7–16.

Nørrevang, A. (Ed.): 1975. The Phylogeny and Systematic Position of Pogonophora. Z. zool. Syst. Evolut.-forsch. Sonderheft. 143 pp.

Nott, J. A., and Parkes, K. R. 1975. Calcium accumulation and secretion in the serpulid polychaete *Spirorbis spirorbis* at settlement. J. Mar. Biol. Assoc. U.K. 55:911–923.

Ockelmann, K. W., and Vahl, O. 1970. On the biology of the polychaete *Glycera alba,* especially its burrowing and feeding. Ophelia 8:275–294.

O'Clair, R. M., and Cloney, R. A. 1974. Patterns of morphogenesis mediated by dynamic microvilli: Chaetogenesis in *Nereis vexillosa.* Cell Tiss. Res. 151:141–157.

Pawlik, J. R. 1983. A sponge-eating worm from Bermuda: *Branchiosyllis oculata* (Polychaeta, Syllidae). Mar. Ecol. 4:65–79.

Pietsch, A., and Westheide, W. 1987. Protonephridial organs in *Myzostoma cirriferum* (Myzostomida). Acta Zool. (Stockh.) 68:195–203.

Purschke, G. 1997. Ultrastructure of nuchal organs in polychaetes (Annelida): New results and review. Acta Zool. (Stockh.) 78:123–143.

Purschke, G. 1999. Terrestrial polychaetes: Models for the evolution of the Clitellata (Annelida)? Hydrobiologia 406:87–99.

Purschke, G., and Tzetlin, A. B. 1996. Dorsolateral ciliary folds in the polychaete foregut: Structure, prevalence and phylogenetic significance. Acta Zool. (Stockh.) 77:33–49.

Renaud, J. C. 1956. A report on some polychaetous annelids from the Miami-Bimini area. Am. Mus. Novit. 1812:1–40.

Rice, S. A. 1980. Ultrastructure of the male nephridium and its role in spermatophore formation in spionid polychaetes. Zoomorphologie 95:181–194.

Rice, S. A. 1987. Reproductive biology, systematics, and evolution in the polychaete family Alciopidae. Biol. Soc. Wash. Bull. 7:114–127.

Roe, P. 1975. Aspects of life history and of territorial behavior in young individuals of *Platynereis bicanaliculata* and *Nereis vexillosa*. Pac. Sci. 29:341–348.

Rose, S. M. 1970. Regeneration: Key to Understanding Normal and Abnormal Growth and Development. Appleton-Century-Crofts, New York. 264 pp.

Rouse, G. W. 1999. Trochophore concepts: Ciliary bands and the evolution of larvae in spiralian Metazoa. Biol. J. Linn. Soc. 66:411–464.

Rouse, G. W., and Fauchald, K. 1995. The articulation of annelids. Zool. Scripta 24:269–301.

Rouse, G. W., and Fauchald, K. 1997. Cladistics and polychaetes. Zool. Scripta 26:139–204.

Rouse, G. W., and Fauchald, K. 1998. Recent views of the status, delineation and classification of the Annelida. Am. Zool. 38:953–964.

Ruby, E. G., and Fox, D. L. 1976. Anerobic respiration in the polychaete *Euzonus (Thoracophelia) mucronata*. Mar. Biol. 35:149–153.

Ruppert, E. E., and Fox, R. S. 1988. Seashore Animals of the Southeast. University of South Carolina Press, Columbia. 429 pp.

Santos, S. L., and Simon, J. L. 1974. Distribution and abundance of the polychaetous annelids in a south Florida estuary. Bull. Mar. Sci. 24:669–689.

Schroeder, P. C. 1984. Chaetae. In Bereiter-Hahn, J., Matoltsy, A. G., and Richards, K. D. (Eds.): Biology of the Integument. I. Invertebrates. Springer-Verlag, Berlin. pp. 297–309.

Schroeder, P. C., and Hermans, C. O. 1975. Annelida: Polychaeta. In Giese, A. C., and Pearse, J. S. (Eds.): Reproduction of Marine Invertebrates. Vol. 3. Academic Press, New York. pp. 1–205.

Smith, P. R. 1992. Polychaeta: Excretory system. In Harrison, F. W., and Gardiner, S. L. (Eds.): Microscopic Anatomy of Invertebrates. Vol. 7: Annelida. Wiley-Liss, New York. pp. 71–108.

Smith, P. R., and Ruppert, E. E. 1988. Nephridia. In Westheide, W., and Hermans, C. O. (Eds.): Ultrastructure of Polychaeta. Microfauna Marina 4:231–262.

Smith, P. R., Ruppert, E. E., and Gardiner, S. L. 1987. A deuterostome-like nephridium in the mitraria larva of *Owenia fusiformis* (Polychaeta, Annelida). Biol. Bull. 172:315–323.

Southward, A. J. 1975. On the evolutionary significance of the mode of feeding of Pogonophora. Z. zool. Syst. Evolut. forsch. 1975:77–85.

Southward, A. J., and Southward, E. C. 1980. The significance of dissolved organic compounds in the nutrition of *Siboglinum ekmani* and other small species of Pogonophora. J. Mar. Biol. Assoc. U.K. 60:1005–1034.

Southward, A. J., Southward, E. C., Dando, P. R., et al. 1981. Bacterial symbionts and low C/C ratios in tissues of Pogonophora indicate unusual nutrition and metabolism. Nature 293:616–620.

Southward, E. C. 1971a. Pogonophora of the Northwest Atlantic: Nova Scotia to Florida. Smithson. Contrib. Zool. No. 88. 29 pp.

Southward, E. C. 1971b. Recent researches on the Pogonophora. Oceangr. Mar. Biol. Ann. Rev. 9:193–220.

Southward, E. C. 1982. Bacterial symbionts in Pogonophora. J. Mar. Biol. Assoc. U.K. 62:889–906.

Southward, E. C. 1988. Development of the gut and segmentation of newly settled stages of *Ridgeia* (Vestimentifera): Implications for relationship between Vestimentifera and Pogonophora. J. Mar. Biol. Assoc. U.K. 68:465–487.

Southward, E. C. 1993. Pogonophora. In Harrison, F. W., and Rice, M. E. (Eds.): Microscopic Anatomy of Invertebrates. Vol. 12. Onychophora, Chilopoda, and Lesser Protostomata. Wiley-Liss, New York. pp. 327–369.

Storch, U., and Alberti, G. 1978. Ultrastructural observations on the gills of polychaetes. Helgol. Wiss. Meeresunters. 31:169–179.

Uebelacker, J. M., and Johnson, P. G. (Eds.): 1984. Taxonomic Guide to the Polychaetes of the Northern Gulf of Mexico. NOAA Tech. Report NMFS Circular 375. U.S. Government Printing Office, Washington, DC. (Seven volumes.)

Warren, L. M. 1976. A population study of the polychaete *Capitella capitata* at Plymouth. Mar. Biol. 38:209–216.

Waxman, L. 1971. The hemoglobin of *Arenicola cristata*. J. Biol. Chem. 246:7318–7327.

Weber, R. E. 1978. Respiration. In Mill, P. J. (Ed.): Physiology of Annelids. Academic Press, London. pp. 369–446.

Wells, G. P. 1950. Spontaneous activity cycles in Polychaeta worms. Symp. Soc. Exp. Biol. 4:127–142.

Wells, G. P. 1959. Worm autobiographies. Sci. Am. 200:132–141.

Westheide, W. 1984. The concept of reproduction in polychaetes with small body size: Adaptations in interstitial species. Fortschr. Zool. 29:265–287.

Westheide, W. 1997. The direction of evolution within the Polychaeta. J. Nat. Hist. 31:1–15.

Westheide, W., and Hermans, C. O. (Eds.): 1988. The Ultrastructure of Polychaeta. Microfauna Marina Vol. 4. Gustav Fischer Verlag, Stuttgart. 494 pp.

Westheide, W., McHugh, D., Purshke, G., and Rouse, G. 1999. Systematization of the Annelida: Different approaches. Hydrobiologia 402:291–307.

Whitlatch, R. B. 1974. Food-resource partitioning in the deposit feeding polychaete *Pectinaria gouldii*. Biol. Bull. 147:227–235.

Wilson, D. P. 1974. *Sabellaria* colonies at Duckpool, North Cornwall, 1971–1972, with a note for May 1973. J. Mar. Biol. Assoc. U.K. 54:393–436.

Woodin, S. A., and Merz, R. A. 1987. Holding on by their hooks: Anchors for worms. Evolution 41:427–432.

Zottoli, R. A., and Carriker, M. R. 1974. Burrow morphology, tube formation, and microarchitecture of shell dissolution by the spionid polychaete *Polydora websteri*. Mar. Biol. 27:307–316.

Zrzavy, J., Hypsa, V., and Tietz, D. F. 2001. Myzostomida are not annelids: Molecular and morphological support for a clade of animals with anterior sperm flagella. Cladistics 17:170–198.

Internet Sites

http://biodiversity.uno.edu/~worms/pc-fam.html (G. B. Read's annotated list of polychaete families wordwide.)

http://biodiversity.uno.edu/~worms/pc-order.html (G. B. Read's annotated table of polychaete higher classification.)

OLIGOCHAETA

General

Bolton, P. J., and Phillipson, J. 1976. Burrowing, feeding, egestion and energy budgets of *Allolobophora rosea*. Oecologia 23:225–245.

Brinkhurst, R. O. 1994. Evolutionary relationships within the Clitellata: An update. Megadrilogica 5:109–112.

Brinkhurst, R. O. 1999. Lumbriculids, branchiobdellidans and leeches: An overview of recent progress in phylogenetic research on clitellates. Hydrobiologia 406:281–290.

Brinkhurst, R. O., and Cook, D. G. 1979. Aquatic Oligochaete Biology. Plenum Press, New York. 529 pp.

Brinkhurst, R. O., and Jamieson, B. G. M. 1972. Aquatic Oligochaeta of the World. Toronto University Press, Toronto. 860 pp.

Brinkhurst, R. O., and Gelder, S. R. 2001. Annelida: Oligochaeta and Branchiobdellida. In Thorp, J. H., and Covich, A. P. (Eds.): Ecology and Classification of North American Freshwater Invertebrates. Second Edition. Academic Press, New York. pp. 431–463.

Darwin, C. R. 1881. The Formation of Vegetable Mould through the Action of Worms with Observations on their Habits. John Murray, London.

Davies, R. W. 1991. Annelida: Leeches, polychaetes, and acanthobdellids. In Thorp, J. H., and Covich, A. P. (Eds.): Ecology and Classification of North American Freshwater Invertebrates. Academic Press, New York. pp. 437–479.

Drewes, C. D. 1996. Those wonderful worms. Carolina Tips 59:17–20.

Edwards, C. A., and Bolen, P. G. 1996. Biology and Ecology of Earthworms. Third Edition. Chapman and Hall, London. 448 pp.

Edwards, C. A., and Lofty, J. R. 1972. Biology of Earthworms. Chapman and Hall, London. 283 pp.

Friedlander, B. 1888. Õber das Kriechen der Regenwürmer. Biol. Zbl. 8:363–366.

Giere, O., and Pfannkuche, O. 1982. Biology and ecology of marine Oligochaeta, a review. Oceanog. Mar. Biol. Ann. Rev. 20:173–308.

Gray, J., and Lissmann, H. W. 1938. Studies in animal locomotion. VII. Locomotory reflexes in the earthworm. J. Exp. Biol. 15:506–517.

Jamieson, B. G. M. 1981. The Ultrastructure of the Oligochaeta. Academic Press, New York. 462 pp.

Jamieson, B. G. M. 1988a. Oligochaete ultrastructure: Some comparisons with the Polychaeta. Microfauna Marina 4:397–428.

Jamieson, B. G. M. 1988b. On the phylogeny and higher classification of the Oligochaeta. Cladistics 4:367–410.

Jamieson, B. G. M. 1992. Oligochaeta. In Harrison, F. W., and Gardiner, S. L. (Eds.): Microscopic Anatomy of Invertebrates. Vol. 7. Annelida. Wiley-Liss, New York. pp. 217–322.

Lasserre, P. 1975. Clitellata. In Giese, A. C., and Pearse, J. S. (Eds.): Reproduction of Marine Invertebrates. Vol. III. Academic Press, New York. pp. 215–275.

Laverack, M. S. 1963. The Physiology of Earthworms. Pergamon Press, New York. 205 pp.

Learner, M. A. 1972. Laboratory studies on the life histories of four enchytraeid worms which inhabit sewage percolating filters. Ann. Appl. Biol. 70:251–266.

Martin, N. A. 1982. The interaction between organic matter in soil and the burrowing activity of three species of earthworms. Pedobiologia 24:185–190.

Meinhardt, U. 1974. Comparative observations on the laboratory biology of endemic earthworm species: II. Biology of bred species. Z. Angew. Zool. 61:137–182.

Peters, W. 1977. Possible sites of ultrafiltration in *Tubifex tubifex* Mueller (Annelida, Oligochaeta). Cell Tiss. Res. 179:367–375.

Peters, W., and Walldorf, V. 1986. Endodermal secretion of chitin in the "cuticle" of the earthworm gizzard. Tiss. Cell 18:361–374.

Pierce, T. G. 1983. Functional morphology of lumbricid earthworms, with special reference to locomotion. J. Nat. Hist. 17:95–111.

Purschke, G., Hessling, R., and Westheide, W. 2000. The phylogenetic position of the Clitellata and the Echiura: On the problematic assessment of absent characters. J. Zool. Syst. Evol. Research 38:165–173.

Rieger, R. M. 1980. A new group of interstitial worms, Lobatocerebridae nov. fam. (Annelida) and its significance for metazoan phylogeny. Zoomorphologie 95:41–84.

Satchell, J. E. (Ed.): 1983. Earthworm Ecology. Chapman and Hall, New York. 512 pp.

Seymour, M. K. 1969. Locomotion and coelomic pressure in *Lumbricus*. J. Exp. Biol. 51:47–58.

Singer, R. 1978. Suction-feeding in *Aeolosoma*. Trans. Am. Microsc. Soc. 97:105–111.

Tynen, M. J. 1970. The geographical distribution of ice worms (Oligochaeta: Enchytraeidae). Can. J. Zool. 48:1363–1367.

Westheide, W., and Müller, M. 1996. Cinematographic documentation of enchytraeid morphology and reproductive biology. Hydrobiologia 334:263–267.

Internet Sites

www.microscopyu.com/moviegallery/pondscum/annelida/chaetogaster/ (Movies of *Chaetogaster*.)

www.microscopyu.com/moviegallery/pondscum/annelida/aeolosomas/ (Movies of *Aeolosoma*.)

HIRUDINOMORPHA

General

Bourne, A. G. 1884. Contributions to the anatomy of the Hirudinea. Q. J. Microsc. Sci. 24:419–506.

Brinkhurst, R. O., and Gelder, S. R. 1991. Branchiobdellida. In Thorpe, G. H., and Covich, A. P. (Eds.): Ecology and Classification of North American Freshwater Invertebrates. Academic Press, New York. pp. 434–441.

Cross, W. H. 1976. A study of predation rates of leeches on tubificid worms under laboratory conditions. Ohio J. Sci. 76:164–166.

Dickinson, M. H., and Lent, C. M. 1984. Feeding behavior of the medicinal leech, *Hirudo medicinalis.* J. Comp. Physiol. Sens. Neural Behav. Physiol. 154:449–456.

Gelder, S. R., and Brinkhurst, R. O. 1990. An assessment of the phylogeny of the Branchiobdellida (Annelida: Clitellata) using PAUP. Can. J. Zool. 68:1318–1326.

Gray, J., Lissmann, H. W., and Pumphrey, R. J. 1938. The mechanism of locomotion in the leech. J. Exp. Biol. 15:408–430.

Haupt, J. 1974. Function and ultrastructure of the nephridium of *Hirudo medicinalis* L. II. Fine structure of the central canal and the urinary bladder. Cell Tiss. Res. 152:385–401.

Hildebrandt, J.-P. 1988. Circulation in the leech, *Hirudo medicinalis* L. J. Exp. Biol. 134:235–246.

Holt, T. C. 1968. The Branchiobdellida: Epizootic annelids. Biologist 50(3–4):79–94.

Jennings, J. B., and Gelder, S. R. 1979. Gut structure, feeding and digestion in the branchiobdellid oligochaete *Cambarinicola macrodonta* Ellis, 1912, an ectosymbiote of the freshwater crayfish *Procambarus clarkii.* Biol. Bull. 156:300–314.

Keegan, H. L., et al. 1968. 406th Medical Laboratory Special Report. U.S. Army Med. Command, Japan.

Klemm, D. J. 1972. Freshwater Leeches of North America. Biota of Freshwater Ecosystems Identification Manual No. 8. Environmental Protection Agency, U.S. Government Printing Office, Washington, DC.

Lent, C. M., Fliegner, K. H., Freedman, E., et al. 1988. Ingestive behavior and physiology of the medicinal leech. J. Exp. Biol. 137:513–527.

Mann, K. H. 1962. Leeches (Hirudinea), Their Structure, Physiology, Ecology, and Embryology. Pergamon Press, New York. 201 pp.

Minkin, B. I. 1990. Leeches in modern medicine. Carolina Tips 53:5–6.

Nicholls, J. G., and Van Essen, D. 1974. The nervous system of the leech. Sci. Am. 230:38–48.

Purschke, G., Westheide, W., Rohde, D., and Brinkhurst, R. O. 1993. Morphological reinvestigation and phylogenetic relationship of *Acanthobdella peledina* (Annelida: Clitellata). Zoomorphology 113:91–101.

Sawyer, R. T. 1972. North American freshwater leeches, exclusive of the Piscicolidae, with a key to all species. Illinois Biol. Monogr. 46:1–154.

Sawyer, R. T. 1984. Leech Biology and Behavior. Oxford University Press, New York. (Three volumes.)

Siddall, M. E., and Burreson, E. M. 1995. Phylogeny of the Euhirudinea: Independent evolution of blood feeding by leeches? Can. J. Zool. 73:1048–1064.

Siddall, M. E., and Burreson, E. M. 1998. Phylogeny of leeches (Hirudinea) based on mitochondrial cytochrome C oxidase subunit I. Mol. Phylog. Evol. 9:156–162.

Stent, G. S., Kristan, W. B., Friesen, W. O., et al. 1978. Neuronal generation of the leech swimming movement. Science 200:1348–1356.

Stent, G. S., and Weisblat, D. A. 1982. The development of a simple nervous system. Sci. Am. 246:136–146.

Wilde, V. 1975. Investigations on the symbiotic relationship between *Hirudo officinalis* and bacteria. Zool. Anz. 195(5/6):289–306.

Young, S. R., Dedwylder, R. D., and Friesen, W. O. 1981. Responses of the medicinal leech *(Hirudo medicinalis)* to water waves. J. Comp. Physiol. Sens. Neural Behav. Physiol. 144:111–116.

Internet Site

http://research.amnh.org/users/siddall/leechindex.html (Mark Siddall's clitellate and leech page concerned chiefly with phylogeny, but including a photo gallery and leech whimsy)

Echiura[P] and Sipuncula[P]

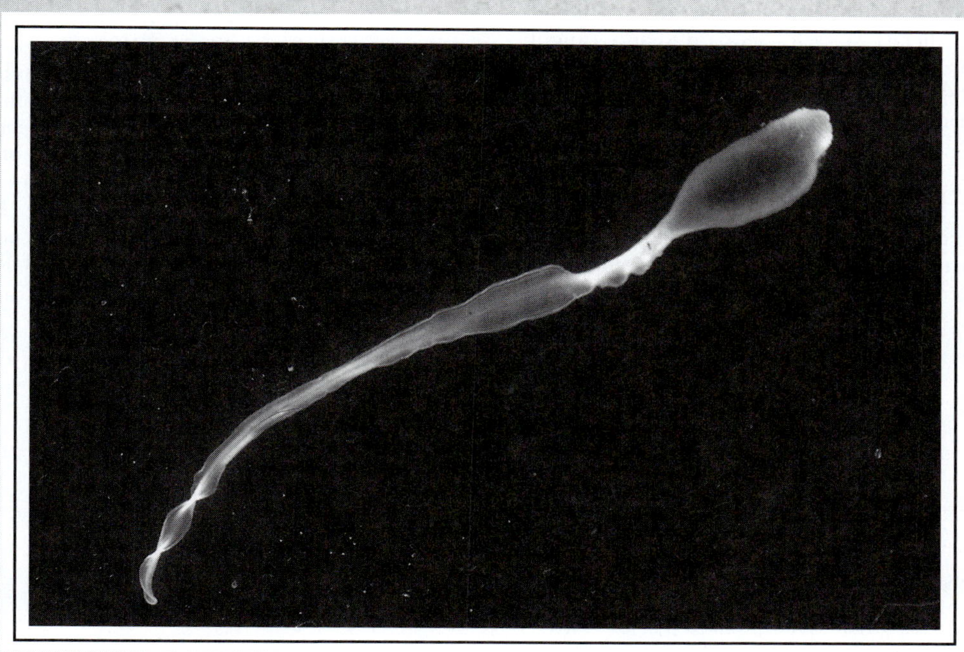

Spoonworms (Echiura) and peanut worms (Sipuncula) are robust, coelomate marine worms of unusual form. They are a thrill to collect, bringing to mind alien beings, and a delight to observe alive, especially because of their peculiar body movements. Echiurans and sipunculans are superficially similar to each other in that both lack segmentation, probably a result of their similar life habits, and long ago they were classified together as Gephyrea. Because they are spiralians that have trochophore larvae, they are members of Trochozoa, but Echiura and Sipuncula probably are not sister taxa despite having affinities with annelids.

ECHIURA^P

Echiurans, or spoonworms, are sausage-shaped marine animals with a tonguelike anterior prostomium, often of prodigious length, that can be rolled into a gutterlike "spoon" (Fig. 14-1). Many species, such as *Thalassema, Urechis,* and *Ikeda,* occupy U-shaped burrows in sand and mud (Fig. 14-2), whereas others live in rock and coral crevices. Echiurans range in size (trunk length) from approximately 1 cm *(Lissomyema)* to over 50 cm *(Urechis).* The majority live in shallow water, but a few occur at great ocean depths. Approximately 150 species have been described. Echiurans may be an important food in the diet of some fishes. In one dietary study of leopard sharks caught along the California coast, large, meaty *Urechis* was their food of choice. Apparently the sharks use suction to slurp the spoonworms from their burrows.

FORM AND FUNCTION

The echiuran body consists of two distinct regions, a nonsegmented cylindrical trunk and a flat, anterior **prostomium** also called a proboscis because it is preoral and prehensile, but it cannot be retracted into the trunk (Fig. 14-1). The whitish prostomium is ciliated ventrally and glandular elsewhere (Fig. 14-1, 14-3, 14-4B). It is highly mobile and capable of extending at least 10 times its retracted length. A specimen of *Ikeda* with a trunk length of 40 cm had a 1.5 m long prostomium, and the prostomium of an 8 cm long female *Bonellia* (Fig. 14-1B) can exceed 2 m. On the other hand, *Urechis* has a stubby prostomium that is always much shorter than its trunk (Fig. 14-1D). Prostomium length is related to feeding mode.

The echiuran trunk may be gray, reddish brown, rose, red, or green. Adult and larval *Bonellia* (and other genera) are green because of **bonellin,** a toxic dermal porphyrin pigment that probably is antibiotic or antipredatory. A pair of short, hooked chaetae of β-chitin occurs ventrally on the anterior part of the trunk (Fig. 14-1A,D). When they are extended from their chaetal sacs, they curve posteriorly and are used for digging as the animal burrows (Fig. 14-2A). In addition to the anterior chaetae, some echiurans, such as *Echiurus* and *Urechis,* have one or two circles of hooked chaetae around the posterior extremity of the trunk (Fig. 14-1D). These are used for burrow maintenance and anchorage.

The body wall of the trunk consists of a thin, collagenous cuticle, a glandular epidermis, a three-layered musculature, and a ciliated peritoneum that lines the coelom (Fig. 14-4B,C). The three muscle layers usually are, from outer to inner, circular, longitudinal, and crossed helical, but in *Ikeda,* the first two layers are reversed. The body wall of the prostomium differs from that of the trunk in being ciliated ventrally, lacking a spacious coelom, and having especially well-developed longitudinal and dorsoventral muscles (Fig. 14-4B). These muscles are responsible for retracting and flattening the prostomium, respectively. The prostomium extends by creeping on its ciliated surface. A ventral transverse muscle is responsible for rolling the proboscis into the shape of a gutter. Echiurans are slow, sluggish animals. Peristaltic waves along the trunk (Fig. 14-3) are used to ventilate the burrow. Extension-retraction movements of the prostomium are employed in feeding and to escape grazing predators.

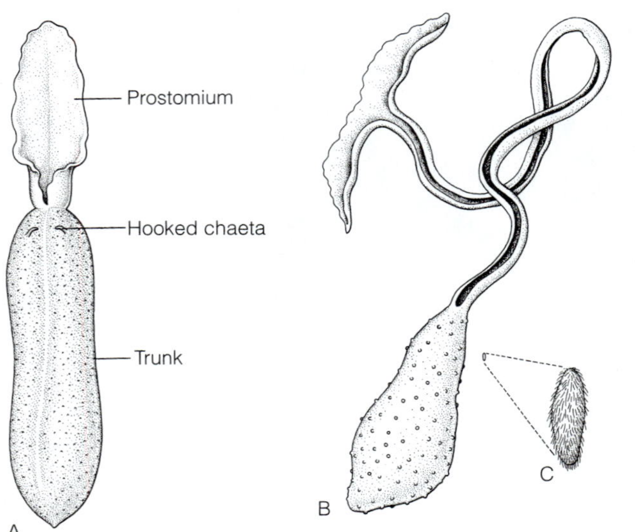

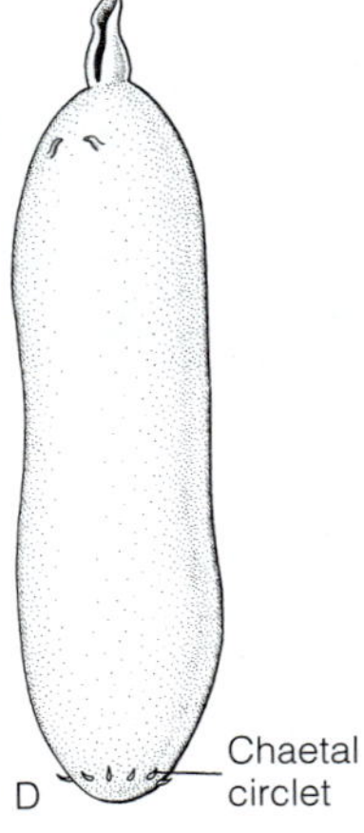

FIGURE 14-1 Echiura: diversity. **A,** *Thalassema hartmani* from the southeastern coast of the United States. **B and C,** The European *Bonellia viridis.* **B,** Female; **C,** dwarf male (removed from female and magnified). **D,** The filter feeder *Urechis caupo,* from the west coast of the United States. *(B and C, Modified and redrawn from MacGinitie, G. E. and MacGinitie, N. 1968. Natural History of Marine Animals, McGraw-Hill Book Co., New York. 523 pp.)*

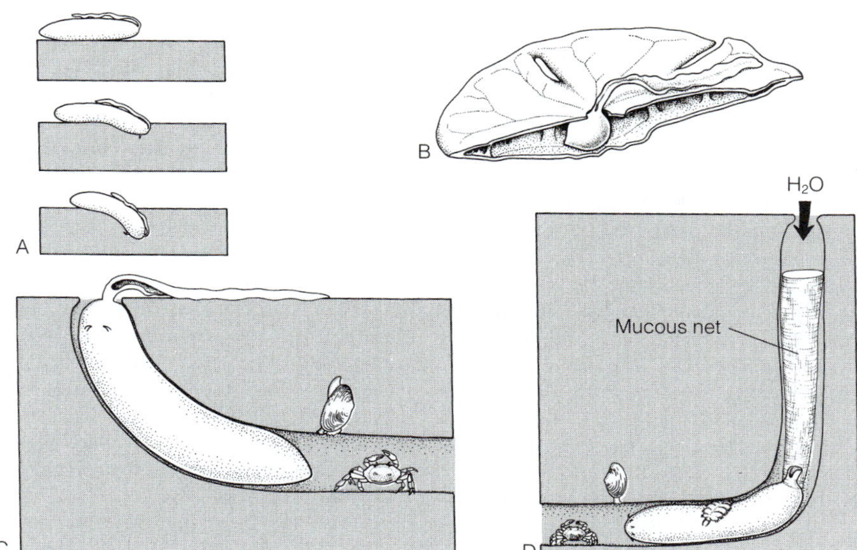

FIGURE 14-2 Echiura: lifestyles and biological associations. **A,** *Thalassema hartmani* uses its anterior chaetae to scoop away sand as the trunk undergoes a slow peristalsis. Sometimes, but not always, the ciliated prostomium is used to gain initial entry into the sand. **B,** The small *Lissomyema mellita* is a crevice dweller and can often be found among mollusc shells and in dead sand dollar tests. **C,** *Thalassema hartmani,* the landlord worm of the east coast of the United States, constructs a U-shaped burrow from which it extends its deposit-feeding prostomium. The security and resources within the burrow attract commensals: a minute clam, *Paramya subovata,* and a small filter-feeding crab, *Pinnixa lunzi.* **D,** On the west coast of the United States, the filter-feeding *Urechis caupo* constructs a U-shaped burrow through which it pumps water and strains out food particles. Its burrow houses at least three commensals, the clam *Cryptomya californica,* the crab *Pinnixa franciscana,* and the scaleworm *Hesperonoe adventor.* *(C, Modified from Ruppert and Fox, 1988; D, Modified and redrawn from MacGinitie, G. E., and MacGinitie, N. 1968. Natural History of Marine Animals, McGraw-Hill Book Co., New York. 523 pp.)*

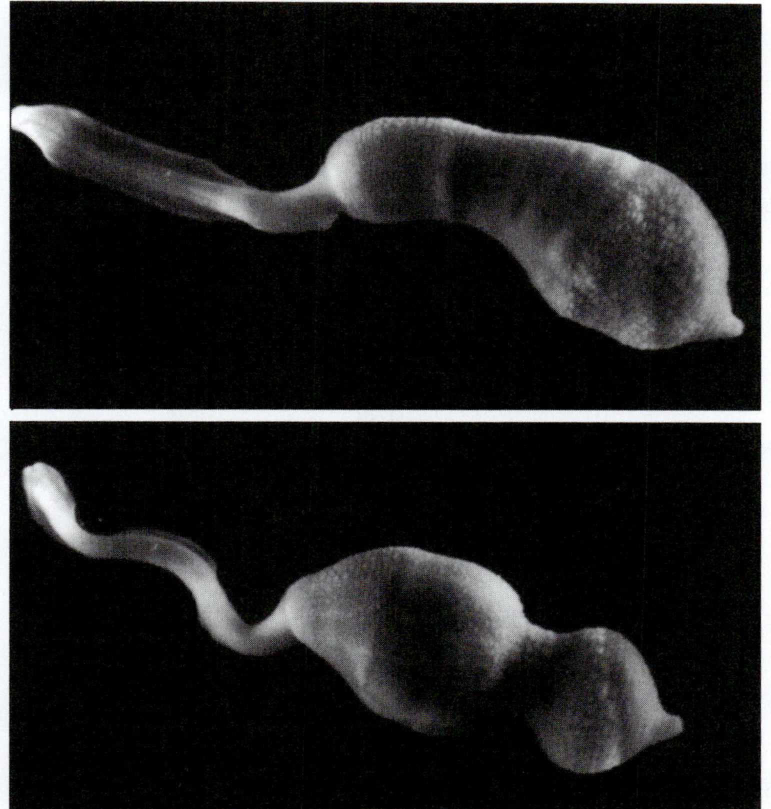

FIGURE 14-3 Echiura: locomotion: *Lissomyema mellita*. **Top,** The extending prostomium is flattened against the substratum as it creeps slowly outward on its ventral cilia. **Bottom,** Once extended, regions of the prostomium curl ventrally to form an inverted gutter through which food is transported to the mouth at the anterior end of the trunk. While in their burrows, echiurans generate slow peristaltic waves, which pass from front to rear along the trunk, as seen in this photograph. Such movements create a ventilating water flow over the body.

The subepidermal nervous system includes a nerve ring around the perimeter of the adult prostomium, but in larval stages the brain consists of supra- and subpharyngeal ganglia joined by connectives, as in annelids. An unpaired ventral nerve cord, which is often pink from the presence of hemoglobin (neuroglobin), extends through the trunk (Fig. 14-4A). The nerve cord gives off segmentally paired lateral branches, each one associated with a ganglion-like cluster of nerve-cell bodies in the cord (Fig. 14-5). The segmental nerves innervate the body-wall musculature. Specialized sense organs, including nuchal organs (see Chapter 13), are absent in adult echiurans.

Most echiurans are deposit feeders, but at least one, *Urechis caupo* (Fig. 14-1D), which lives along the California coast, filter feeds like the polychaete *Chaetopterus* (Chapter 13). *Urechis* has on its anterior trunk a glandular ring, from which it casts a mucous net onto the burrow lining. Peristaltic movements of the body pump water through the burrow and the net at a rate of approximately 1 L/h. Virtually all particles, including typical plankton, are trapped on the net as the water passes through. When loaded with food, the net is detached from the body, seized by the short prostomium, and swallowed. *Urechis* is often called the "innkeeper" because it hosts several commensals, including a polychaete scaleworm, a tiny clam, a filter-feeding crab, and a goby fish (Fig. 14-2D) in its protected, ventilated burrow. Similar commensals are common in the burrows of other species of spoonworms (Fig. 14-2C).

Typical deposit feeders such as *Thalassema* and *Bonellia* extend the prostomium onto the surface of the sediment above their burrows. The prostomium glides outward from the trunk on its ventral ciliated surface (Fig. 14-3). When extended, the prostomium curls ventrally to form a gutter, along which deposited particles are transported to the mouth at its base. Any unusual disturbance to the prostomium causes its rapid retraction.

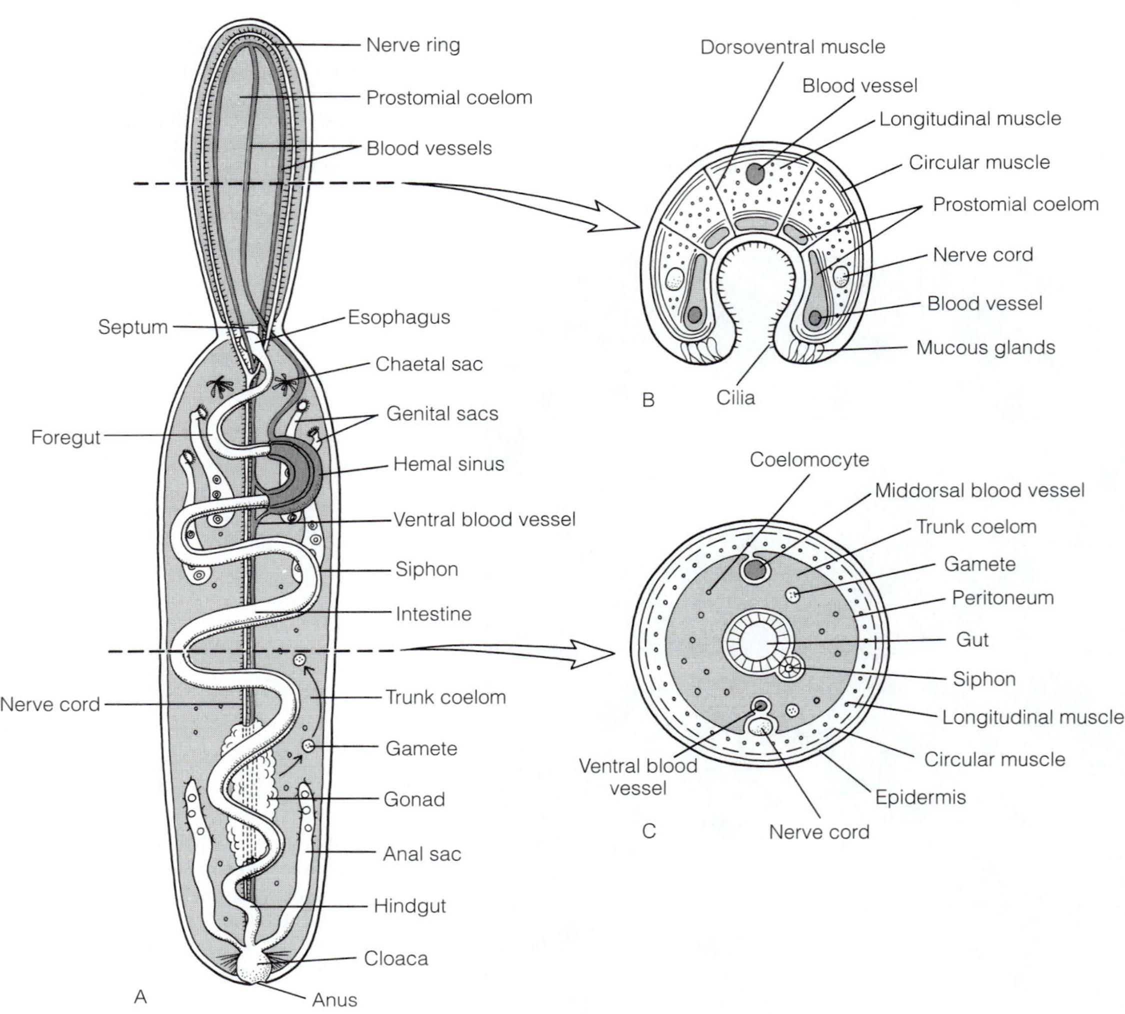

FIGURE 14-4 Echiura: anatomy. **A,** A generalized echiuran. **B,** Cross section of prostomium curled ventrally in feeding position. **C,** Cross section of the trunk.

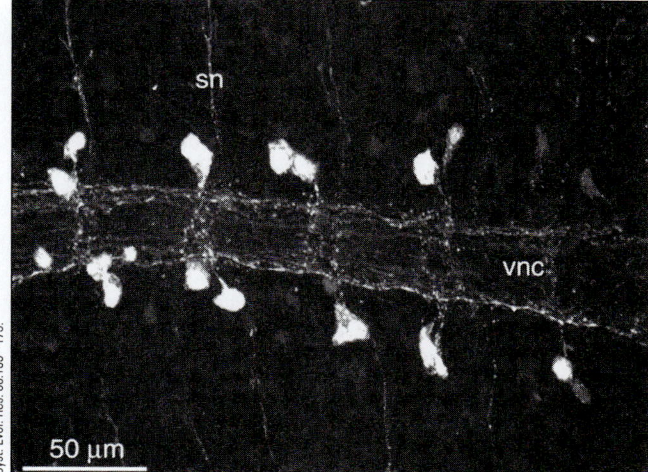

Clitellata and the Echiura — on the problematic assessment of absent characters. J. Zool. Syst. Evol. Res. 38:165–173.

FIGURE 14-5 Echiura: segmentation of the nerve cord. The ventral nerve cord (vnc) of a *Bonellia viridis* premetamorphic larva (female) viewed with a fluorescence microscope. The paired arrangement of fluorescent nerve-cell bodies and segmental nerves (sn) are clearly visible.

The digestive tract is extremely long and coiled and loosely suspended in the coelom (Fig. 14-4A). In one specimen of *Thalassema hartmani,* for example, the uncoiled and outstretched gut was 10 times longer than the trunk. The mouth is situated ventrally at the base of the prostomium, and the anus opens on the posterior end of the trunk. The intestine, which constitutes most of the gut, is the site of digestion (probably extracellular). An accessory gut, or **siphon,** originates from the anterior intestine, runs parallel to the intestine ventrally, and rejoins it posteriorly near the hindgut (14-4A,C). The siphon is an intestinal bypass that probably transports water ingested with food around the digestive region of the gut. It is thus functionally similar to the intestinal ciliary groove of sipunculans, described later, and the intestinal siphon of sea urchins (Chapter 28). Such a bypass circumvents dilution of digestive enzymes by inadvertently ingested water. The hindgut widens into a **cloaca,** which receives the excretory tubules, before opening to the exterior at the anus.

Echiurans, like sipunculans, have two coelomic cavities separated from each other by a septum (Fig. 14-4A). The trunk coelom is voluminous and unpartitioned except by partial mesenteries and radial muscles between the body wall and gut. Fluid is circulated throughout the coelom by muscular contractions of the body and by cilia on the peritoneum. Circulating cells in the trunk coelom include hemoglobin-containing coelomocytes, amebocytes, and germ cells. The prostomial coelom, a small compartment restricted to the ventral part of the prostomium (Fig. 14-4B), contains only fluid; coelomocytes are absent.

A hemal system is present in all echiurans except *Urechis.* The system consists of a hemal sinus around the foregut, from which blood is transported to the prostomium by a middorsal vessel (Fig. 14-4A). Blood returns to the trunk in a pair of lateral prostomial vessels, which unite in the trunk to form a midventral, longitudinal vessel. Branches from the ventral vessel unite with the sinus to complete the circuit. Because there is no respiratory pigment in the colorless blood, it seems unlikely that this hemal system has a gas-transport role. It is more probable that the blood transports nutrients from the gut chiefly to the prostomium, nervous system, and musculature.

Gas exchange in echiurans presumably occurs across the general body wall of both trunk and prostomium. While in the burrow, peristaltic ventilating movements of the body (Fig. 14-2D, 14-3) facilitate exchange across the trunk. Because the prostomium is flat and ribbonlike, oxygen transport to its musculature is probably by simple diffusion. The trunk, however, is thick, and oxygen diffusing across the body wall is transported internally by the hemoglobin-containing coelomocytes. In *Urechis,* body-wall gas exchange is supplemented by exchange across the wall of the cloaca and hindgut, which are enlarged to form a "water lung." Circular and radial muscles on the wall of the cloaca create a tidal flow of water in and out of the "lung." Gas exchange occurs across the cloacal and intestinal walls. Although *Urechis* alone has such a specialized water lung, other echiurans also ventilate the cloaca by inhalation of water through the anus.

Excretion in echiurans is accomplished by two specialized organs known as **anal sacs** (Fig. 14-4A). Each anal sac is a thin-walled, hollow diverticulum from the wall of the cloaca that extends well into the trunk coelom. The anal-sac surface is covered with numerous (sometimes thousands) valved, ciliated funnels, perhaps modified metanephridia, through which coelomic fluid enters the sac. It is not known whether the coelomic fluid contains an ultrafiltrate of the blood, and therefore it is uncertain whether the anal sacs are part of a filtration excretory system. Podocytes have not yet been found. In any case, the anal sacs produce urine and then discharge it into the cloacal outflow.

REPRODUCTION AND DEVELOPMENT

The sexes are separate in echiurans and, in Bonelliidae, sexual dimorphism is pronounced (Fig. 14-1B,C). The retroperitoneal gonad is unpaired and attached to the wall of the ventral blood vessel (Fig. 14-4A). Gametocytes are released from the gonad into the trunk coelom, where they grow and mature into gametes. The gametes are then removed from the coelomic fluid by ciliated funnels on paired storage organs, the **genital sacs** (Fig. 14-4A). The sacs and their funnels, which number up to 20 segmental pairs, are metanephridia specialized for reproduction only. Fully developed gametes are first selected and transported from the coelom into the sac by the funnels and then stored in the sac prior to spawning. When the animals spawn, the genital sacs contract and squeeze the gametes out through ventral pores.

Gametes are shed into seawater, where fertilization occurs. The zygote undergoes spiral cleavage and a typical annelid cross (Fig. 14-15A,B) appears early in development. Development is determinate, mesoderm arises from a mesentoblast, and the coelom forms by schizocoely of paired mesodermal bands. Development that usually is indirect leads to a planktotrophic trochophore larva, often having two eyespots, in which both prototroch and telotroch are well developed (Fig. 14-6A).

Additional trochs (ciliated rings), called metatrochs, circle the larva between the proto- and telotrochs. During metamorphosis, the pretrochal region becomes the prostomium and the growth zone and pygidium become the trunk (Fig. 14-6). The nerve cord originates from paired rudiments that later merge and fuse in the ventral midline. The juvenile nerve cord has a paired segmental (ladderlike) arrangement of nerve-cell bodies and lateral nerves identical to that in many annelids (Fig. 14-5).

Reproduction in *Bonellia* differs from that of other echiurans. A **dwarf male,** whose minute, ciliated body includes only a gonad, a seminal vesicle, and a pair of protonephridia (Fig. 14-1C), permanently inhabits the unpaired genital sac of the female. The male not only fertilizes the eggs, at oviposition it also secretes gelatinous material used to bind the eggs together. The yolky eggs develop into lecithotrophic trochophores. If these short-lived larvae settle on an adult female prostomium, most become dwarf males; if they settle apart from a female, they metamorphose into juvenile females. Larvae induced to become males develop an adhesive organ with which they attach themselves to the prostomium of the female. Later, they enter the genital sacs, where they reside until needed to supply sperm to fertilize eggs. Male bonelliids are necessarily much smaller than females; do not have a prostomium, mouth, anus, or hemal system; and meet their metabolic needs by exchange with the female coelomic fluid in which they are bathed.

DIVERSITY OF ECHIURA

Echiuridae[F]: Six species of *Echiurus*. Chaetae arranged in an anterior pair plus two rings around the posterior end of the trunk. Marine, in shallow temperate and tropical waters.

Urechidae[F]: Four species of *Urechis*. All construct U-shaped burrows through which water is pumped for gas exchange and filter feeding (Fig. 14-2D). Rudimentary prostomium assists in retrieval and ingestion of food-laden mucous net. Hemal system is absent. Anal water inhalation by cloaca and posterior intestine specialized for gas exchange. Two or three pairs of metanephridia in anterior trunk coelom. Trunk has an anterior pair of chaetae and a single posterior chaetal ring.

Thalassematidae[F]: Approximately 70 species of *Thalassema* (Fig. 14-1A, 14-2C), *Lissomyema* (Fig. 14-2B), *Listriolobus,* and *Ochetosoma,* and others; chiefly in warm water. Lack circle of anal chaetae, but have a pair of stout chaetal hooks on the anterior trunk (14-1A). *Ochetosoma* has 4 or 5 segmental pairs of genital sacs and *Ikedosoma pirotansis* has up to 20, but most have 1 to 3 pairs. *Thalassema hartmani:* large spoonworm on southeastern U.S. coast. Its burrow, like that of *Urechis caupo,* provides shelter and food for several commensals (Fig. 14-2C). *Lissomyema mellita:* southeastern U.S. coast, in tests of dead sand dollars and fissures in discarded mollusc shells; it enters the test when small and becomes too large to leave (Fig. 14-2B). *Thalassema dendrorhynchus:* prostomium has gill filaments; in brackish-water lake in India. *Anelassorhynchus abyssalis:* dredged off the California coast from a depth of 2000 m.

Bonelliidae[F]: 70 species of Bonelliidae in warm and cold waters, at all ocean depths, approximately 60% of species below 3000 m, some at abyssal depths. *Bonellia* (Fig. 14-1B,C) and most other species of Bonelliidae: prostomium long, often forked; trunk bears a pair of anterior chaetae, anal chaetae are absent; one to three unpaired genital sacs. Most species green because of bonellin. All exhibit pronounced sexual dimorphism; tiny males parasitic on females.

Ikedidae[F]: Monotypic taxon. *Ikeda taenioides:* large burrower, exceeding 1 m. Prostomium very long, can be autotomized and later regenerated; body-wall muscle layers include outer longitudinal and inner circular layers, opposite that in all other echiurans; up to 400 genital sacs in clusters rather than segmental pairs, suggesting secondary duplication. Sea of Japan.

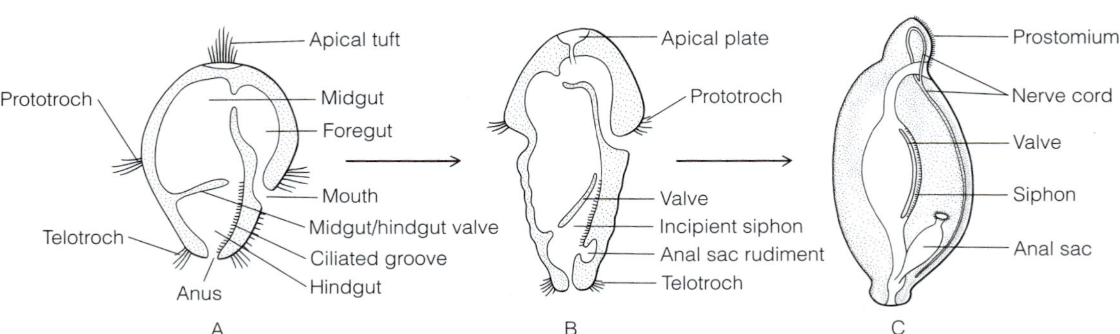

FIGURE 14-6 Echiura: trochophore and larval metamorphosis. **A,** The echiuran trochophore has two well-developed ciliary bands, the prototroch and telotroch. **B** and **C,** During metamorphosis, the preoral region of the trochophore becomes the prostomium and the postoral region becomes the trunk. The larval midgut/hindgut valve, a sheet of tissue, shifts ventrally during metamorphosis and roofs over the ciliated intestinal groove to form the intestinal siphon. Anal sacs develop from hindgut diverticula, and the nervous system, as in most protostomes, develops from the larval apical plate. *(A, After Newby, W. W. 1940. From Pilger, J. F. 1978. Settlement and metamorphosis in the Echiura: A review. In Chia, F.-S., and Rice, M. E. (Eds.): Settlement and Metamorphosis of Marine Invertebrate Larvae. Elsevier North Holland, Inc., New York. pp. 103–122; B and C, Modified and redrawn from Pilger, J. F. 1978. Settlement and metamorphosis in the Echiura: A review. In Chia, F.-S., and Rice, M. E. (Eds.): Settlement and Metamorphosis of Marine Invertebrate Larvae. Elsevier North Holland, Inc., New York. pp. 103–122.)*

PHYLOGENY OF ECHIURA

Molecular analyses of rRNA gene sequences and DNA sequences from the nuclear gene elongation factor 1-α consistently place echiurans within Annelida, interpreting them as derived annelids that have suppressed segmentation. Most morphologists, on the other hand, recognize a sister-group relationship between annelids and echiurans and are reluctant to include Echiura in Annelida because of the supposed absence of segmentation in the former. But morphological evidence for segmentation in echiurans is strong. They have two or more pairs of genital sacs (metanephridia) and a ladderlike series of paired cell bodies and lateral branches from the ventral nerve cord, both indications of segmentation. These traits, along with paired chaetae and a cleavage pattern (annelid cross) identical to those of annelids (Fig. 14-15), suggest that echiurans may well be a taxon of Annelida. If this union is formally adopted in the future, the echiurans may be classified as an annelid family defined by a unique prostomium, anal sacs, and the near loss of trunk segmentation.

On the assumption that spoonworms are annelids that have suppressed segmentation, the phylogeny of echiuran taxa is typically built around the derived loss of segmental characters, especially the loss of chaetae, the outward manifestation of segmentation. Thus Echiuridae, with three "segments" bearing chaetae is regarded as the primitive taxon, whereas thalassematids, bonelliids, and ikedids, which express only the anterior pair of chaetae, are considered to be the evolutionarily derived taxa. Still, the phylogeny of Echiura is incompletely researched and understood, as is evident in Figure 14-7.

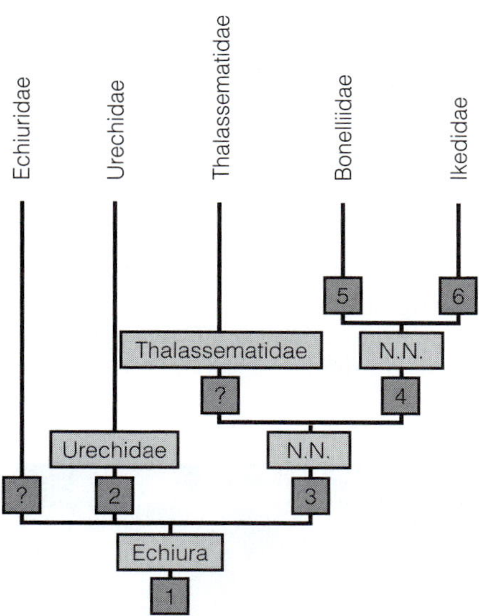

FIGURE 14-7 Echiura: phylogeny. Characters in brackets, although part of the echiuran ground plan, are plesiomorphies. **1, Echiura:** elongate, flattened, deposit-feeding prostomium; prostomium lacks sensory organs; one pair of anterior ventral chaetae, two posterior (anal) rings of chaetae; [outer circular, inner longitudinal muscle layers]; prostomial and trunk coeloms; [hemal system]; segmental metanephridia (reproductive genital sacs) with terminal funnels in anterior trunk; excretory funnels and anal sacs open into cloaca (with ventilatory water flow); intestine has a ventral siphon. **2, Urechidae:** one ring of anal chaetae; reduced prostomium; glandular girdle on anterior trunk (mucous net production for filter feeding); hemal system lost; cloaca is a well-developed "water lung." **3, N.N.:** anal rings of chaetae absent. **4, N.N.:** very long prostomium; unpaired metanephridium (genital sac). **5, Bonelliidae:** forked prostomium; stalked funnels on anal sacs; sexual dimorphism. **6, Ikedidae:** outer longitudinal, inner circular muscle layers; nonsegmental multiplication of metanephridia (genital sacs).

PHYLOGENETIC HIERARCHY OF ECHIURA

Echiura
- Echiuridae/Urechidae
- N.N.
  - Thalassematidae
  - N.N.
    - Bonelliidae
    - Ikedidae

SIPUNCULA[P]

The 150 species of sipunculans are robust, tentaculate marine worms that superficially resemble burrowing sea anemones. Sipunculans are sometimes called "peanut worms," because the plump contracted body of some species resembles a shelled peanut, or "starworms" after the radiating tentacles on the head (Fig. 14-8, 14-9; 14-10C). They range in length from 2 mm to more than 70 cm, although most do not exceed 10 cm. All are cryptic, mostly beige, bottom-dwelling animals, the majority living in shallow water. Some (such as *Sipunculus*) live in sand and mud, whereas others occupy coral crevices, empty mollusc shells *(Phascolion),* or annelid tubes. Several species bore into coralline rock and at least one bores into wood. Borers direct their anterior end toward the opening of the gallery, from which they extend their feeding tentacles. Densities as high as 700 individuals per square meter of coralline rock have been reported in Hawaii. In parts of the tropical Indo-Pacific, large sipunculans are consumed as human food and their mummified bodies are available in Asian food markets.

FORM AND FUNCTION

The nonsegmented sipunculan body is divided into a slender anterior section called the **introvert** and a swollen posterior trunk (Fig. 14-8). The introvert ranges in length from approximately one-half to several times the length of the trunk. Regardless of its length, it can be fully retracted into the trunk (Fig. 14-9). The anterior end of the introvert is a flattened **oral disc** (Fig. 14-10) that bears the mouth, tentacles, and a dorsal, unpaired nuchal organ. The tentacles either surround the mouth or lie above it in an arc that partly encloses the nuchal organ. The tentacles usually are ciliated, and each bears a deep, ciliated groove on its inner, oral side (Fig. 14-10C). Posterior to the mouth and tentacles, the surface of the introvert is sometimes uniformly covered with hard papillae, rings of cat's-claw-shaped hooks, or other cuticular elements.

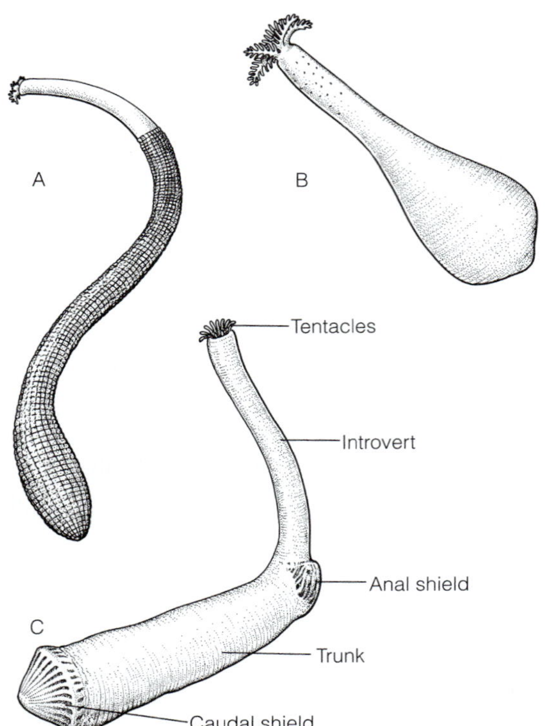

FIGURE 14-8 Sipuncula: diversity. **A,** The sand burrower, *Sipunculus nudus* (10 cm). **B,** The crevice dweller *Themiste lageniformis* (1 cm). **C,** The rock borer *Paraspidosiphon klunzingeri* (2 cm). *(C, Redrawn from Sterrer, W. (Ed.). 1986. Marine Fauna and Flora of Bermuda. John Wiley and Sons, New York. Plate 70, p. 227)*

The anus is located dorsally on the anterior end of the trunk in most species (Fig. 14-9, 14-10). The trunk itself may be smooth, but is often patterned by annulations, papillae, and creases. In some rock-boring sipunculans, the anterior trunk cuticle is thickened and sometimes calcified to form an **anal shield** (Fig. 14-8C). When the introvert is retracted, the shield blocks the entrance of the burrow in the same way that an operculum seals a snail's shell (Fig. 14-11D). Some of these same species also have a thick, rough cuticle known as a **caudal shield** at the broad posterior end of the trunk. The caudal shield may aid in anchorage or in the boring process (Fig. 14-8C, 14-11D). Apparently, rock boring requires both mechanical abrasion and chemical softening.

The sipunculan body wall consists of a cuticle composed of collagen fibers in a crossed-helical array. The hooks, hard papillae, and other structures are locally sclerotized parts of the cuticle. Beneath the cuticle are a glandular epidermis, circular and longitudinal muscle layers, and a noncontractile ciliated peritoneum that encloses the coeloms. One, two, or four **introvert retractor muscles,** depending on the species, originate on the lining of the trunk coelom, span the cavity, and insert on the wall of the esophagus (Fig. 14-10A,B).

The locomotion of adult sipunculans is best developed in the burrowing species of the family Sipunculidae, at least one of which can swim (Fig. 14-11A). In these, burrowing is accomplished by thrusting movements of the introvert and peristalsis of the trunk (Fig. 14-11B). Contraction of circular muscles elevates the fluid pressure of the trunk, causing introvert eversion. Peristalsis results from alternate contraction of circular and longitudinal trunk muscles. The movements of rock, wood, and crevice dwellers, on the other hand, are largely restricted to extension and retraction of the introvert (Fig. 14-9, 14-11C,D). Retraction of the introvert in all species, however, causes the trunk to swell and pressurize substantially. The increase in girth anchors the sipunculan in its burrow.

Sipunculans have a J-shaped digestive tract, and the tubular intestine is long and complexly coiled (Fig. 14-10A). The mouth is at the tip of the introvert and the anus is mid-dorsal on the anterior end of the trunk, except in *Onchnesoma,* in which it opens on the introvert. Sipunculans occupy blind

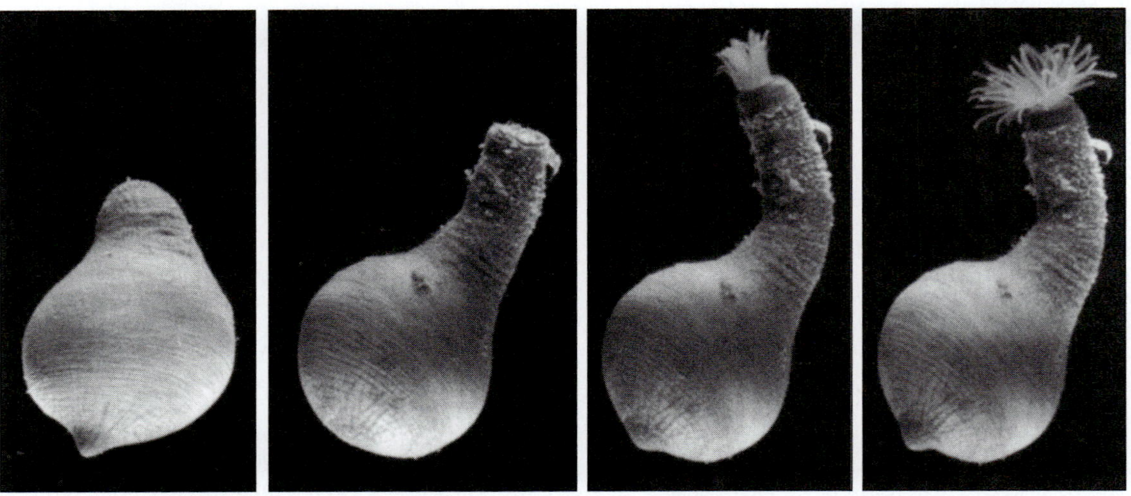

FIGURE 14-9 Sipuncula: introvert eversion in *Themiste lageniformis* (1 or 2 cm). The bump at the base of the introvert is a papilla that bears the anus.

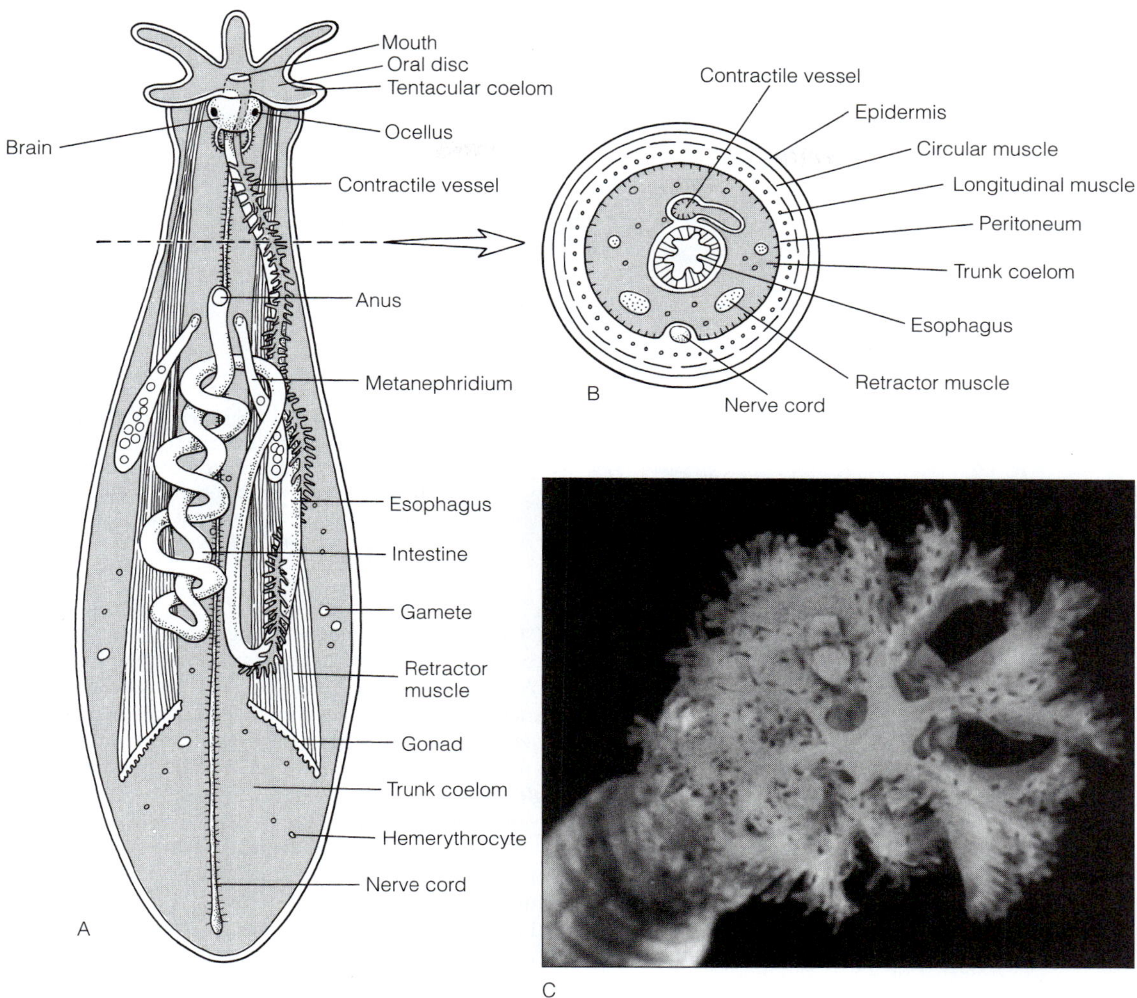

FIGURE 14-10 Sipuncula: anatomy. **A,** A generalized sipunculan with a well-developed contractile vessel. **B,** Cross section of the trunk. **C,** Expanded tentacles of *Themiste alutacea.* The nuchal organ can be seen just above the semicircular notch at the base of the tentacles.

galleries, and the anterior placement of the anus at the gallery opening promotes sanitation. Behind the mouth, the esophagus descends into the trunk and joins the long intestine, which is wound into a double helix consisting of intertwined descending and ascending coils. The ascending coil joins the rectum and anus. A longitudinal **spindle muscle** is in the axis of the coils. As the introvert retracts into the trunk, the contracting spindle muscle presumably compresses the intestinal coils free of kinks. The intestinal coils anchor to the body wall by numerous radial **fixing muscles,** the functions of which are to suspend the gut in the coelom and to help stir the intestinal contents. The intestinal gastrodermis is folded to form a dorsal longitudinal **ciliary groove.** Digestion in sipunculans is extracellular in the intestinal lumen, and the function of the ciliary groove probably is to shunt water, which could dilute the intestinal enzymes, past the digestive region of the gut.

Sipunculans apparently are nonselective suspension or deposit feeders, and particulate material is collected on their ciliated tentacles. In burrowing species such as *Sipunculus,* the worm ingests the sand and silt through which it burrows. Some rock borers, such as *Phascolosoma antillarum,* spread the tentacular crown at the opening of the gallery while suspension feeding. Other hard-bottom species extend the introvert over the rock surface and ingest deposited material. The mechanism of ingestion is not certain, but apparently food collected on the tentacles is ingested as the introvert and tentacles are retracted into the trunk.

Sipunculans have two coelomic cavities, a tentacular coelom and a trunk coelom (Fig. 14-10A). The tentacular coelom forms a **ring canal** around the mouth and at the base of the tentacles. Three branches extend into each tentacle from the ring canal. The large and voluminous trunk coelom is separated from the tentacular coelom by an elaborate septum. The coelomic fluid

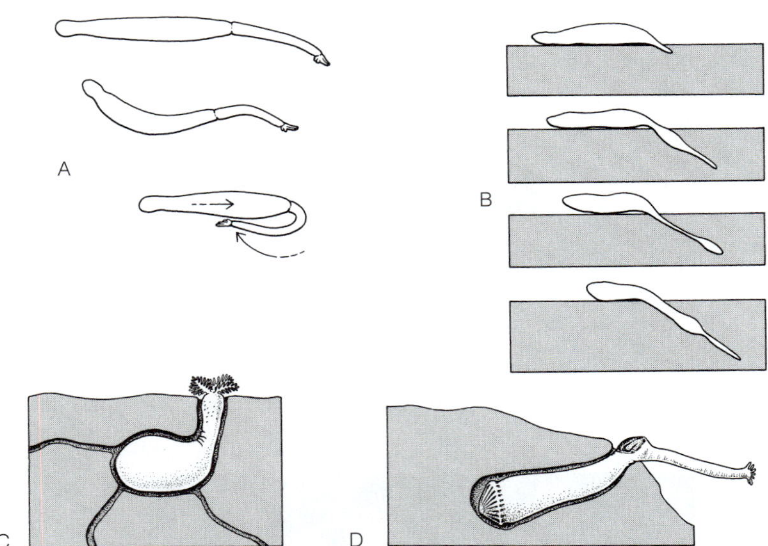

FIGURE 14-11 Sipuncula: locomotion and habitats. Swimming **(A)** and burrowing **(B)** by *Sipunculus nudus.* **C,** *Themiste lageniformis,* a crevice and rock dweller. **D,** *Aspidosiphon,* a rock-boring sipunculan having a calcareous anal shield that caps the opening when the introvert is retracted.

in both cavities is kept in circulation by cilia on some of the peritoneal cells and by contractions of the muscular body. Hemerythrin-containing cells **(hemerthryocytes)** are the most common and conspicuous of the numerous coelomocytes. In addition to these, some sipunculans, such as *Sipunculus,* have micro-organs called **ciliated urns.** These may be attached

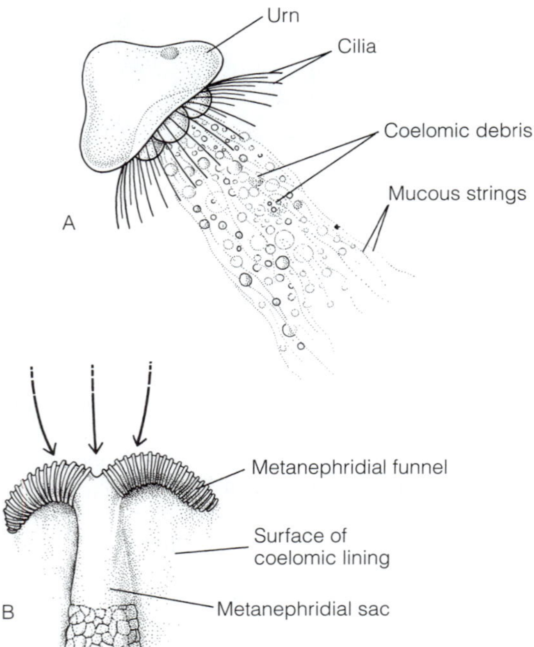

FIGURE 14-12 Sipuncula: ciliated urns and nephridia. **A,** Ciliated urn of *Sipunculus nudus.* **B,** Cilia on the coelomic lining of *Siphonosoma cumanense* direct fluid toward the ciliated funnel of each nephridium. The lower lip of the funnel is attached to the coelomic lining, and the free upper lip (shown) arches upward like a scoop.

("fixed") to the coelomic lining or may circulate ("free") in the coelomic fluid. Free urns swim using a band of cilia and resemble microscopic medusae (Fig. 14-12A). Because of their appearance and mobility, they were first thought to be symbionts rather than specialized organs functioning in internal defense. Swimming urns secrete mucus, which trails behind them and entangles foreign material and exhausted coelomocytes. How the waste-laden urns leave the body, if indeed they do, is unknown.

Sipunculans lack a hemal system, and transport between the two coelomic cavities occurs across the tentacle-trunk septum by diffusion and perhaps by ultrafiltration. To facilitate this exchange, the tentacular coelom has developed one or two diverticula, the **contractile vessels** also called compensation sacs), which project posteriorly along the sides of the esophagus into the trunk coelom (Fig. 14-10A,B). The contractile vessels increase the surface area for exchange between the two coelomic compartments. (Podocytes have been noted on the vessel walls of a species of *Themiste.*) In *Themiste,* these coelomic contractile vessels are long, branched, and superficially resemble the blood vessels of annelids. The contractile vessels may also function as fluid reservoirs for the collapsed tentacles when they are retracted.

In the absence of blood vessels, coelomic fluid is the sole transport medium for gases and nutrients. The tentacles are important respiratory surfaces in sedentary rock and crevice dwellers such as *Themiste* (Fig. 14-9, 14-11C), but the entire body wall is a gill surface in sand-burrowing species. In burrowers such as *Sipunculus* (Fig. 14-8A), branches from the trunk coelom pass through the muscle layers to form a system of **dermal canals** immediately below the thin epidermis. In *Xenosiphon,* outgrowths of these canals form slender, blind-ended gills that project above the body surface, similar to the tube feet of echinoderms. The coelomic fluid and hemerythrocytes circulating through the dermal canals bind oxygen diffusing across the skin and transport it to the trunk coelom and musculature. In *Themiste,* the oxygen affinity of the trunk hemerythrin is higher than that of the tentacle coelom. As a result, environmental oxygen bound at the tentacles and transported to the trunk in the contractile vessels is removed by the trunk hemerythrin and delivered to the tissues of the trunk.

Sipunculans have one (as in *Phascolion*) or two elongate, saclike metanephridia in the anterior part of the trunk (Fig. 14-10A). Each has at its anterior end a ciliated funnel (Fig. 14-12B) and a ventrolateral pore. It is not certain whether sipunculan metanephridia are filtration or secretion kidneys. Podocytes have been noted on the walls of the contractile vessels of *Themiste*, suggesting that filtration occurs there and that sipunculans have a metanephridial system. If so, it would be atypical because the podocytes occur between coelomic compartments and not at the usual interface between the hemal and coelomic spaces, although podocytes also occur between the coelomic cavities of nemerteans. On the other hand, podocytes have also been observed on the outer surface of the nephridium itself, an arrangement functionally similar to a protonephridium. Whatever the arrangement, however, sipunculan metanephridia function in osmoregulation *and* in gamete storage and maintenance prior to spawning.

The nervous system is subepidermal and consists of a dorsal brain above the esophagus and circumesophageal connectives that join the brain to the unpaired ventral nerve cord. The nonganglionated nerve cord gives off a series of lateral branches that innervate the body-wall muscles. In *Sipunculus nudus* and other species, the cord is pink and may contain hemerythrin. Sensory cells are particularly abundant at the end of the introvert, which is used to probe the surrounding environment. They are also present elsewhere on the body. The nuchal organ is probably chemosensory (Fig. 14-10C). A pair of pigmented ocelli typically is embedded dorsally in the brain (Fig. 14–10A).

REPRODUCTION AND DEVELOPMENT

The sexes are separate in most sipunculan species, and inconspicuous retroperitoneal gonads are situated at the base of the introvert retractor muscles (Fig. 14-10A). Gametes are released from the gonads at an early stage and undergo growth and maturation in the trunk coelom. When fully developed, they are removed from the coelomic fluid by the nephridial funnel and stored in the nephridial sacs (Fig. 14-12B).

Sipunculans spawn their gametes into seawater and fertilization is external. The developing zygotes undergo spiral cleavage, and at the 64-cell stage they show a typical **molluscan cross** (Fig. 14-15C,D). Development is determinate, embryos gastrulate by invagination or epiboly, and a mesentoblast gives rise to mesodermal bands, which undergo schizocoely to form the trunk coelom. Embryos can develop directly into small worms, as in *Phascolion cryptus*, indirectly via a lecithotrophic trochophore (Fig. 14-13A), as in *Phascolion strombus*, or indirectly via a secondary larva called a pelagosphera (Fig. 14-13D).

The **pelagosphera** develops from the trochophore (with one exception), is usually planktotrophic and long-lived, and is the agent of long-range dispersal in sipunculans. The pelagosphera swims and feeds using a well-developed metatrochal band of cilia that replaces the earlier prototroch. The larval head and metatroch can be retracted into the trunk (Fig. 14-13B–D). Metamorphosis of the pelagosphera involves the lengthening of the trunk, loss of the terminal organ, lengthening and remodeling of the head to form the introvert, and development of feeding tentacles (Fig. 14-13E).

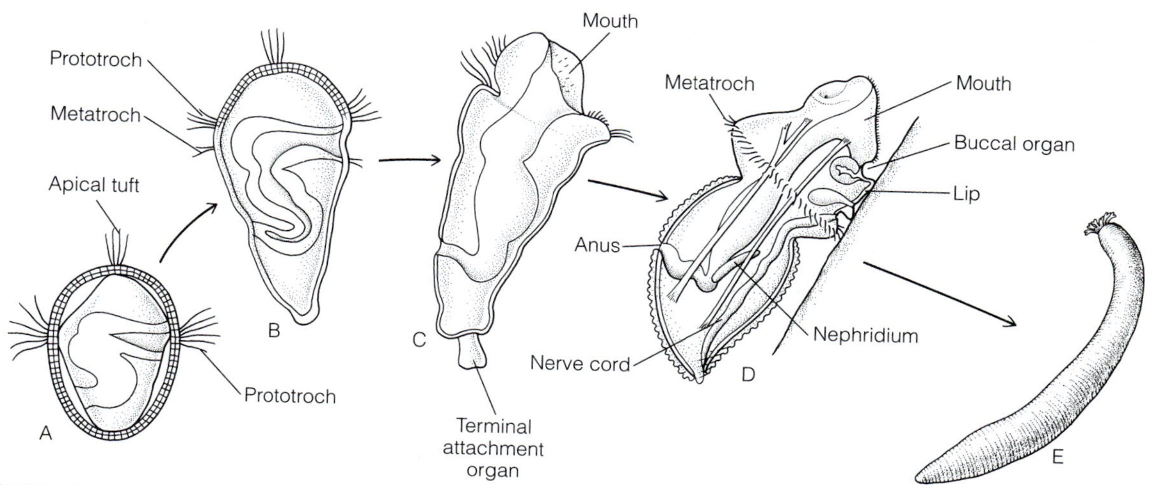

FIGURE 14-13 Sipuncula: larvae and metamorphosis of *Golfingia misakiana*. **A** and **B,** Nonfeeding trochophore larva. **C** and **D,** Feeding pelagosphera larva. The pelagosphera eventually settles and crawls over the bottom on its lip. As it does so, the eversible buccal organ probes the substratum and may dislodge food particles to be ingested by the mouth. The head, lip, and metatrochal region **(D)** of the pelagosphera can be retracted into the trunk for protection. The pelagosphera metamorphoses into a young sipunculan **(E).** Metamorphosis in this species requires approximately two weeks to complete. *(All redrawn from Rice, M. E. 1978. Morphological and behavioral changes at metamorphosis in the Sipuncula. In Chia, F.-S., and Rice, M. E. (Eds.): Settlement and Metamorphosis of Marine Invertebrate Larvae. Elsevier North Holland, Inc., New York. pp. 83–102.)*

DIVERSITY OF SIPUNCULA

The 150 species of sipunculans are divided into two major taxa, Sipunculidea, with approximately 75% of the species, and Phascolosomatidea, including the remaining 25% (Fig. 14-14A). They are found at all depths in polar to equatorial seas and in all benthic habitats—sand, silt, clay, limestone, wood, dead whale skulls, and nestled among fouling organisms.

Sipunculidea[C]: Tentacles encircle the mouth. Most burrow in marine sediments, but *Phascolion strombus* (hermit sipunculan) and similar species occupy discarded snail shells. In some species of *Phascolion,* but not *P. strombus,* the body is permanently coiled to fit the whorls of its shell. *Sipunculus nudus* (Fig. 14-8A, 14-11A,B): best-known sipunculan; widespread, large, robust; burrows actively in sand. *Xenosiphon:* external gills on trunk; *Themiste* (Fig. 14-9, 14-10C): stalked, branched tentacles; shallow water.

Phascolosomatidea[C]: Tentacles dorsal to mouth, in an arc around the nuchal organ. (Associated ring canal still encircles the mouth, but is indented dorsally, and nuchal organ is in that indentation; tentacles borne only on indented part of ring canal and thus are dorsal to the mouth.) Introvert usually bears a series of cuticular rings, each composed of tiny cat's-claw-shaped hooks. Limestone (coral) borers *Aspidosiphon* (Fig.14-11D) and *Lithacrosiphon* both have well-developed anal shields (calcified in *Lithacrosiphon*) and a right-angled introvert (*Paraspidosiphon*; Fig. 14-8C). *Phascolosoma:* often has dark pigment patterns on introvert; includes wood-boring *P. turnerae,* semiterrestrial *P. arcuatum* from Indo-Pacific mangrove swamps, and nearshore *P. agassizii* from U.S. Pacific Northwest.

PHYLOGENY OF SIPUNCULA

From the phylogenetic tree shown in Figure 14-14A, it is apparent that opportunities exist in the study of sipunculan systematics. This is particularly true for the Sipunculidea, which is paraphyletic.

Identifying the sister taxon of Sipuncula has stymied systematists, but most investigators link them to either annelids (Fig. 14-14B) or molluscs (Fig. 14-14C), and a few historically to echinoderms. Although their similarities to annelids include a worm-shaped body, ventral nerve cord, spiral cleavage, and a trochophore larva, these characteristics are widespread among trochozoan protostomes and are not unique to sipunculans and annelids alone. Recent morphological research suggests that a cuticle composed of crossed-helical collagen fibers (absent in molluscs) is a synapomorphy of sipunculans and annelids (Pulvinifera), but this trait also occurs in some flatworms, even nematomorphs, and may thus be a symplesiomorphy. Similarly, a pair of ciliated folds in the larval pharynx of sipunculans and many annelids is another

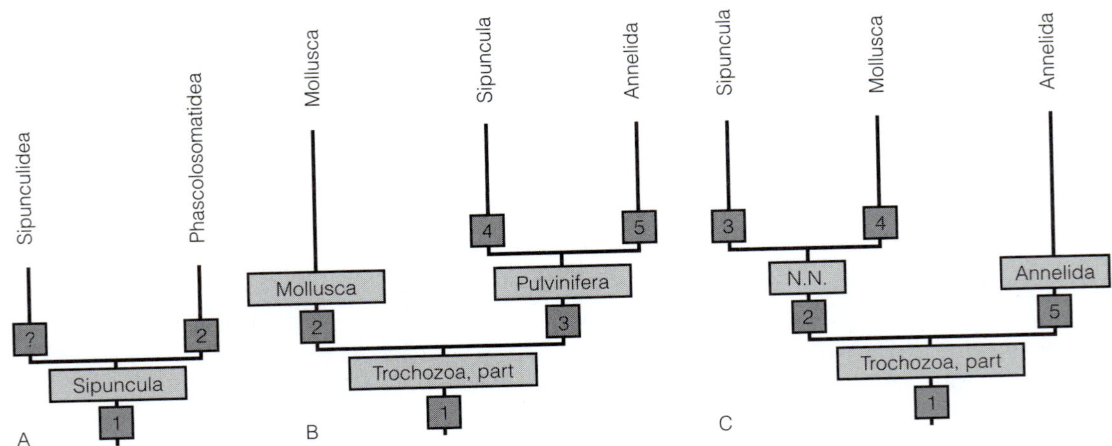

FIGURE 14-14 Sipuncula: phylogeny. **A,** Phylogeny of Sipuncula. **1, Sipuncula:** body composed of a large trunk, part of which is an eversible introvert; introvert bears mouth, tentacles, and unpaired nuchal organ; large, undivided trunk coelom; ring-shaped tentacular coelom with branches into tentacles and trunk (contractile vessels); tentacles surround mouth; two pairs of introvert retractor muscles, two metanephridia in anterior trunk; J-shaped digestive tract, descending and ascending intestine wind together in a helix, anus middorsal at junction of introvert with trunk; dorsal intestinal ciliary groove; collagenous cuticle with fibers in crossed-helical array, cuticle has scattered simple hooks; hemerythrin in cells. **2, Phascolosomatidea:** tentacles in arc above mouth and partly enclosing nuchal organ; cuticle hooks are complex and in rings. **B** and **C,** Competing phylogenies for Sipuncula based on morphology. **B,** Sipuncula and Annelida (Pulvinifera, of Ax, 2000) as sister taxon of Mollusca. **1, Trochozoa, in part:** Trochophore larva; hemal system. **2, Mollusca:** locomotory foot; tetraneury; mantle cavity with gills; calcified exoskeleton; hemocyanin. **3, Pulvinifera:** fibrous collagenous cuticle; dorsolateral ciliary folds in pharynx. **4, Sipuncula:** introvert; tentacular and trunk coeloms; contractile vessels; reflexed, coiled intestine; unpaired nuchal organ; unpaired, nonganglionated ventral nerve cord; hemal system lost. **5, Annelida:** segmentation; teloblastic growth; paired nuchal organs; chaetae of β-chitin. **C,** Sipuncula and Mollusca as sister taxon of Annelida. **1, Trochozoa, in part:** trochophore larva; hemal system. **2, N.N.:** molluscan cross at 64-cell stage. **3, Sipuncula:** introvert; tentacular and trunk coeloms; contractile vessels; reflexed, coiled intestine; unpaired nuchal organ; unpaired nonganglionated ventral nerve cord; hemal system lost. **4, Mollusca:** locomotory foot; tetraneury; mantle cavity with gills; calcified exoskeleton; hemocyanin. **5, Annelida:** segmentation; teloblastic growth; paired nuchal organs; chaetae of β-chitin. *(A, Characters drawn primarily from Cutler, E. B. 1994. The Sipuncula. Their Systematics, Biology, and Evolution. Cornell University Press, Ithaca, NY. 453 pp.; B and C, Based on Ax, P. 2000. Multicellular Animals. Vol. 2. Springer Verlag, Berlin. 396 pp.; Nielsen, C. 2001. Animal Evolution: Interrelationships of the Living Phyla. 2nd Edition. Oxford University Press, Oxford. 563 pp.)*

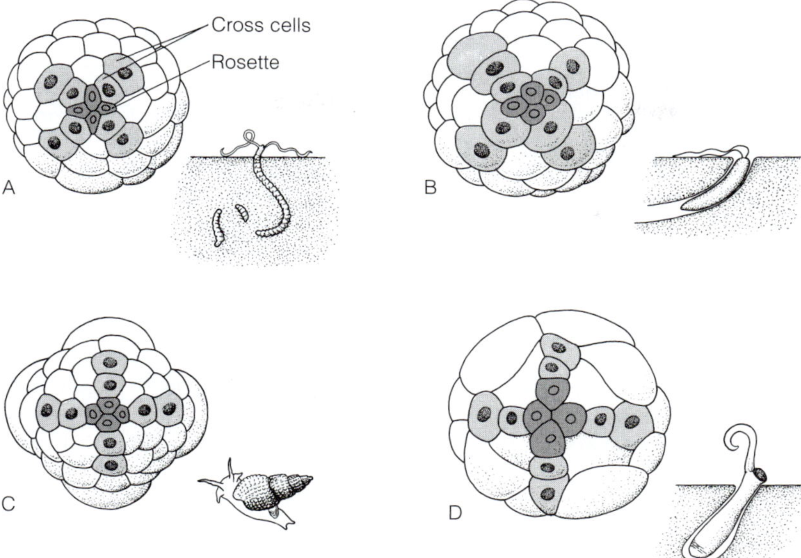

FIGURE 14-15 Echiura and Sipuncula: developmental patterns of possible evolutionary significance. A peculiar but characteristic arrangement of blastomeres appears at the animal pole of some spiralian embryos at the 64-cell stage. The pattern consists of four pairs of cells, each pair radiating outward from the animal pole like a spoke on a bicycle wheel, and a central hub composed of four additional cells. The four radial spokes are equidistant from each other and together form an X called the "cross." The four "hub" cells, the "rosette," are also arranged radially and directly above the animal pole. In annelids, the arms of the cross extend outward along interradial lines, passing *between* the rosette cells, and the resultant pattern is called the **annelidan cross (A).** In molluscs, on the other hand, the four arms of the cross extend outward along radial lines drawn *through* the rosette cells. This figure is known as the **molluscan cross (C).** Other than in annelids and molluscs, rosette and cross patterns have been noted only in embryos of Echiura and Sipuncula. Echiurans have an annelid cross **(B)** and sipunculans, a molluscan cross **(D).** The developmental fates of the rosette and cross cells do not shed further light on the evolutionary relationships of these two taxa: In both annelids and molluscs, the rosette cells become the apical plate, and the cross cells join other cells to form the pretrochal epidermis of the trochophore larva. *(A, Modified and redrawn from Wilson, E. B. 1892. The cell-lineage of Nereis. J. Morphol. 6:361–466.; B, After Newby, W. W. 1932, from Kume, M., and Dan, K. 1968. Invertebrate Embryology. NOLIT Publishing House, Belgrade. 605 pp.; C, Modified and redrawn from McBride, E. W. 1914. Text-book of Embryology. Vol. 1. Invertebrata. Macmillan and Co., Ltd., London. 692 pp.; D, After Gerould, J. H. (1906) from Rice, M. E. 1985. Sipuncula: Developmental evidence for phylogenetic inference. In Conway Morris, S., George, J. D., Gibson, R., et al. The Origins and Relationships of Lower Invertebrates. Oxford University Press, Oxford. pp. 274–296.)*

synapomorphy. Molecular systematic analyses, based on ribosomal and mitochondrial gene sequences, identify Sipuncula alone or Sipuncula plus Nemertea as the sister taxon of Annelida. The hallmark of annelids, however, is a segmented body, an attribute absent in sipunculans except perhaps in the developmental pattern of the ventral nerve cord, which resembles that of echiurans (Fig. 14-5). If sipunculans secondarily suppressed segmentation, perhaps they did so following the pattern of echiurans and, to a lesser extent, leeches. Unlike annelids, sipunculans (and nemerteans) lack a hemal system and rely solely on coelomic cavities for internal transport. Loss of a hemal system by sipunculans may be related to suppressed segmentation. A convergent loss of a hemal system occurs in annelids with reduced segmentation, such as glycerid polychaetes (bloodworms) and higher leeches.

A sister-taxon relationship to Mollusca relies on the presumed homology of mollusclike features expressed during sipunculan development (Fig. 14-14C, 14-15). These include a molluscan cross shared uniquely with molluscs (Fig. 14-15C,D) as well as a protrusible, scraping buccal organ and a muscular, creeping lip in the pelagosphera (Fig. 14-13D), which are reminiscent of the molluscan buccal mass, radula, and foot. If the sipunculans are the nearest living relatives of the molluscs, however, the divergence between them must have been early indeed, well before the adoption of many typical molluscan features.

PHYLOGENETIC HIERARCHY OF SIPUNCULA

Sipuncula
- Sipunculidea
- Phascolosomatidea

REFERENCES

GENERAL

Anderson, D. T. 1973. Embryology and Phylogeny in Annelids and Arthropods. Pergamon Press, New York. 495 pp.

Ax, P. 2000. Multicellular Animals. Vol. 2. Springer Verlag, Berlin. 396 pp.

Bartolomaeus, T. 1994. On the ultrastructure of the coelomic lining in the Annelida, Sipuncula, and Echiura. Microfauna Marina 9:171–220.

Clark, R. B. 1969. Systematics and phylogeny: Annelida, Echiura, Sipuncula. In Florkin, M., and Scheer, B. T. (Eds.): Chemical Zoology. Vol. 4. Academic Press, New York. pp. 1–68.

Giese, A. C., and Pearse, J. S. (Eds.): 1975. Reproduction of Marine Invertebrates. Vol. 3. Annelids and Echiurans. Academic Press, New York. 343 pp.

Giese, A. C., and Pearse, J. S. (Eds.): 1975. Reproduction of Marine Invertebrates. Vol. 2. Entoprocts and Lesser Coelomates. Academic Press, New York. 344 pp.

Kohn, A. J., and Rice, M. E. 1971. Biology of Sipuncula and Echiura. BioSci. 21:583–584.

Kume, M., and Dan, K. 1968. Invertebrate Embryology. NOLIT, Belgrade. 605 pp.

McBride, E. W. 1914. Text-Book of Embryology. Vol. 1. Invertebrata. Macmillan, London. 692 pp.

Nielsen, C. 2001. Animal Evolution: Interrelationships of the Living Phyla. 2nd Edition. Oxford University Press, Oxford. 563 pp.

Rice, M. E., and Todororic, M. (Eds.): 1970. Proceedings of the International Symposium on the Biology of the Sipuncula and Echiura. Vol. 1. Institute for Biological Research, Yugoslavia, and Smithsonian Institution, Washington, DC. 355 pp.

Rice, M. E., and Todororic, M. (Eds.): 1970. Proceedings of the International Symposium on the Biology of the Sipuncula and Echiura. Vol. 2. Institute for Biological Research, Yugoslavia, and Smithsonian Institution, Washington, DC. 254 pp.

Stephen, A. C, and Edmonds, S. J. 1972. The Phyla Echiura and Sipuncula. British Museum of Natural History, London. 528 pp.

Wilson, E. B. 1892. The cell-lineage of *Nereis*. J. Morphol. 6:361–466.

Zrzavy, J., Mihulka, S., Kepka, P., Bezdek, A., and Tietz, D. 1998. Phylogeny of Metazoa based on morphological and 18S ribosomal DNA evidence. Cladistics 14:249–285.

Internet Site

http://biodiversity.uno.edu/~worms/phylum-state.html (Annotated abstracts related to the phylogeny of annelids, echiurans, sipunculans, and pogonophorans.)

ECHIURA

General

Dawydoff, C. 1959. Classes des Echiuriens et Priapuliens. In Grassé, P. (Ed.): Traité de Zoologie. Vol. 5, Pt. 1. Masson, Paris. pp. 855–926.

Harris, R. R., and Jaccarini, V. 1981. Structure and function of the anal sacs of *Bonellia viridis*. J. Mar. Biol. Assoc. U.K. 61:413–430.

Hessling, R., and Westheide, W. 2002. Are Echiura derived from a segmented ancestor? Immunohistochemical analysis of the nervous system in developmental stages of *Bonellia viridis*. J. Morphol. 252:100–113.

Jaccarini, V., Agius, L., Schembri, P. J., et al. 1983. Sex determination and larvae sexual interaction in *Bonellia viridis*. J. Exp. Mar. Biol. Ecol. 66:25–40.

Jaccarini, V., and Schembri, P. J. 1977. Feeding and particle selection in the echiuran worm *Bonellia viridis*. J. Exp. Mar. Biol. Ecol. 28:163–181.

MacGinitie, G. E., and MacGinitie, N. 1968. Natural History of Marine Animals. McGraw-Hill, New York. 523 pp.

McHugh, D. 1997. Molecular evidence that echiurans and pogonophorans are derived annelids. Proc. Natl. Acad. Sci. 94:8006–8009.

McHugh, D. 1999. Phylogeny of Annelida: Siddall et al. (1998) rebutted. Cladistics 15:85–89.

Pilger, J. F. 1978. Settlement and metamorphosis in the Echiura: A review. In Chia, F.-S., and Rice, M. E. (Eds.): Settlement and Metamorphosis of Marine Invertebrate Larvae. Elsevier North Holland, New York. pp. 103–112.

Pilger, J. F. 1993. Echiura. In Harrison, F. W., and Rice, M. E. (Eds.): Microscopic Anatomy of Invertebrates. Vol. 12. Wiley-Liss, New York. pp. 185–236.

Purschke, G., Hessling, R., and Westheide, W. 2000. The phylogenetic position of Clitellata and the Echiura: On the problematic assessment of absent characters. J. Zool. Syst. Evol. Res. 38:165–173.

Schembri, P. J., and Jaccarini, V. 1977. Locomotory and other movements of the trunk of *Bonellia viridis*. J. Zool. 182: 477–494.

Schuchert, P. 1990. The nephridium of the *Bonellia viridis* male (Echiura). Acta Zool. (Stockh.) 71:1–4.

Schuchert, P., and Rieger, R. M. 1990. Ultrastructural observations on the dwarf male of *Bonellia viridis*. Acta Zool. Stockh. 71:5–16.

Internet Site

www.ibss.iuf.net/people/murina/echiura.html (Introduction to Echiura by V. Murina with a list of major taxa and all species.)

SIPUNCULA

Boone, J. L., and Staton, J. L. 2002. The mitochondrial genome of the sipunculid *Phascolopsis gouldii* supports its association with Annelida rather than Mollusca. Mol. Biol. Evol. 19:127–137.

Cutler, E. B. 1994. The Sipuncula. Their Systematics, Biology, and Evolution. Cornell University Press, Ithaca, NY. 453 pp.

Gibbs, P. E. 1977. British Sipunculans. Synopses of the British Fauna, no. 12. Academic Press, London. 35 pp.

Mangum, C. P., and Burnett, L. E. 1987. Response of sipunculid hemerythrins to inorganic ions and CO_2. J. Exp. Zool. 244:59–65.

Moya, J., and Serrano, T. 1984. Podocyte-like cells in the nephridial tube of sipunculans. Cuad. Invest. Biol. (Bilbao) 5:33–37.

Pilger, J. F. 1982. Ultrastructure of the tentacles of *Themiste lageniformis* (Sipuncula). Zoomorphology 100:143–156.

Purschke, G., Wolfrath, F., and Westheide, W. 1997. Ultrastructure of the nuchal organ and cerebral organ in *Onchnesoma squamatum* (Sipuncula, Phascolionidae). Zoomorphology 117:23–31.

Rice, M. E. 1969. Possible boring structures of sipunculids. Am. Zool. 9:803–812.

Rice, M. E. 1970. Asexual reproduction in a sipunculan worm. Science 167:1618–1620.

Rice, M. E. 1978. Morphological and behavioral changes at metamorphosis in the Sipuncula. In Chia, F.-S., and Rice, M. E. (Eds.): Settlement and Metamorphosis of Marine Invertebrate Larvae. Elsevier North Holland, New York. pp. 83–102.

Rice, M. E. 1985. Sipuncula: Developmental evidence for phylogenetic inference. In Conway Morris, S., George, J. D., Gibson, R., et al. (Eds.): The Origins and Relationships of Lower Invertebrates. Oxford University Press, Oxford. pp. 274–296.

Rice, M. E. 1985. Description of a wood dwelling sipunculan, *Phascolosoma turnerae,* new species. Proc. Biol. Soc. Wash. 98:54–60.

Rice, M. E. 1986. Larvae adrift: Patterns and problems in life histories of sipunculans. Am. Zool. 21:605–619.

Rice, M. E. 1989. Comparative observations of gametes, fertilization, and maturation in sipunculans. In Ryland, J. S., and Tyler, P. A. (Eds.): Reproduction, Genetics and Distributions of Marine Organisms. Olsen & Olsen, Fredensborg, Denmark. pp. 167–182.

Rice, M. E. 1993. Sipuncula. In Harrison, F. W., and Rice, M. E. (Eds.): Microscopic Anatomy of Invertebrates. Vol. 12. Wiley-Liss, New York. pp. 237–325.

Ruppert, E. E., and Rice, M. E. 1983. Structure, ultrastructure and function of the terminal organ of a pelagosphera larva (Sipuncula). Zoomorphology 102:143–163.

Ruppert, E. E., and Rice, M. E. Functional organization of dermal coelomic canals in *Sipunculus nudus* (Sipuncula) with a discussion of respiratory designs in sipunculans. Invert. Biol. 114:51–63.

Scheltema, A. H. 1996. Phylogenetic position of Sipuncula, Mollusca and the progenetic Aplacophora. In Taylor, J. (Ed.): Origin and Evolutionary Radiation of the Mollusca. Oxford University Press, London. pp. 53–58.

Sterrer, W. (Ed.): 1986. Marine Fauna and Flora of Bermuda. John Wiley and Sons, New York. 742 pp.

Walter, M. D. 1973. Feeding and studies on the gut content in sipunculids. Helgol. Wiss. Meeresunters. 25:486–494.

Williams, J. A., and Margolis, S. U. 1974. Sipunculid burrows in coral reefs: Evidence for chemical and mechanical excavation. Pac. Sci. 28:357–359.

Onychophora[P] and Tardigrada[P]

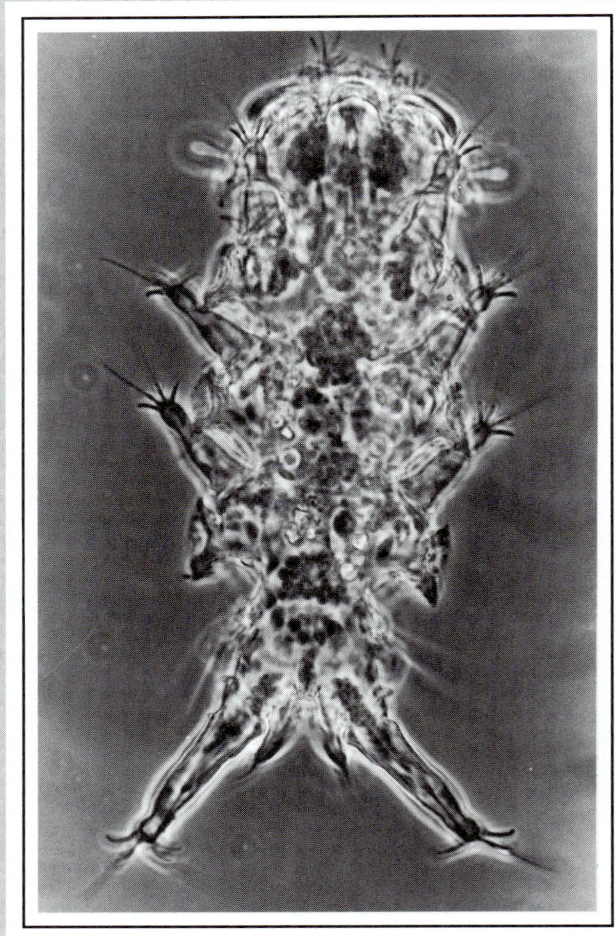

PANARTHROPODA[SP]

Shrimps, spiders, butterflies, centipedes, velvet worms, water bears, and several million other species share similarities in anatomy, development, and nucleotide sequences, suggesting that they should be grouped together in a taxon (superphylum) that has been given the name Panarthropoda. The constituent taxa (phyla) are Arthropoda, Onychophora, and Tardigrada (Fig. 16-15). These myriad species (mostly undescribed) are segmented and have paired segmental appendages and a secreted exoskeleton that must be molted to permit growth. Two small panarthropod taxa, Onychophora and Tardigrada, are the subject of this chapter, whereas the enormous taxon Arthropoda will be covered in several subsequent chapters. General features of panarthropod morphology and evolution, which are mentioned briefly in this chapter, will be discussed more fully in the next chapter.

ONYCHOPHORA[P]

Onychophorans, or velvet worms, are terrestrial, wormlike bilaterians with strong morphological similarities to arthropods and, to a lesser extent, the annelids. Because they resemble worms with legs they are sometimes called "walking worms" (Fig. 15-1A). The 110 described species are restricted to the tropics and the temperate southern hemisphere. Onychophorans are nocturnal, negatively phototactic, and found in humid, usually dark habitats. Superficially they resemble caterpillars, but have also been compared with slugs. They often live in rain forests in leaf litter or under objects on the forest floor. In unfavorable environmental conditions (cold or drought), they move into the soil and become dormant until conditions improve.

Onychophora is of interest for the contributions it makes to our understanding of the evolution of Arthropoda. Onychophorans exhibit a number of traits similar to the arthropods and annelids, but are more closely related to arthropods.

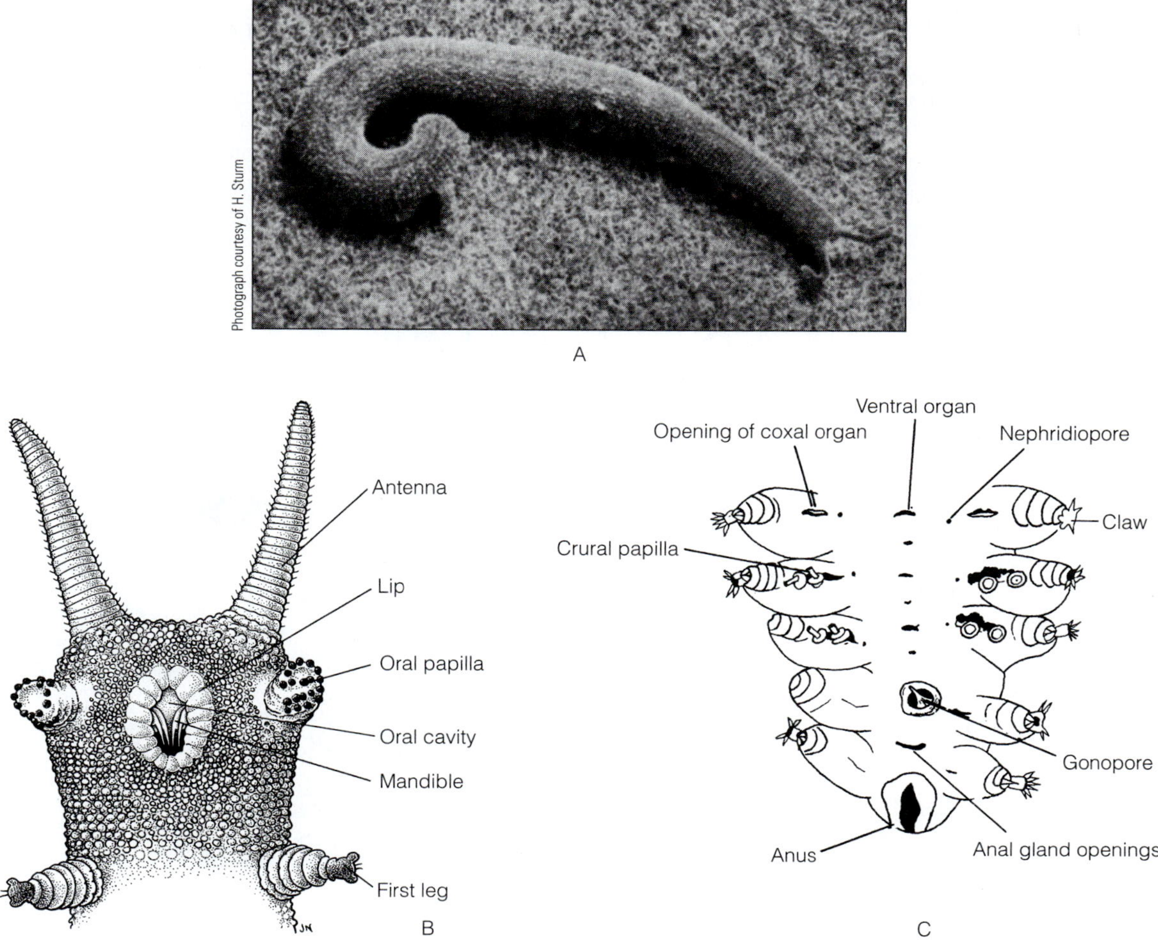

FIGURE 15-1 Onychophora. **A,** *Peripatus.* The papillae of the body surface are conspicuous. The antennae can be seen to the right, and the legs are evident below the animal. **B,** Anterior of *Peripatopsis capensis* (ventral view). **C,** Posterior of a male *Peripatus corradoi* (ventral view). *(B, After Cuénot, L. 1949. 19: Les Onychophores. In Grassé, P.-P. (Ed.): Traité de Zoologie. Vol. 6. Masson et Cie, Paris. pp. 3–75.; C, After Bouvier from Cuénot.)*

FORM

Onychophorans are wormlike with numerous paired, segmental appendages (Fig. 15-1A). They are soft-bodied and the skin is dry and velvety smooth to the touch, resulting in the common name of "velvet worm." Most are 5 mm to 15 cm in length. The body is more or less cylindrical in cross section, although flattened ventrally, and is coated with numerous large and small **papillae** (Fig 15-1A). The papillae are covered with tiny scales, and the larger papillae bear a sensory bristle (Fig. 15-2B).

Onychophorans are segmented animals, as evidenced by the serial repetition of appendages, ostia, nephridia, and ganglia. The body, however, shows no external signs of segmental divisions, other than the regular, segmental spacing of the appendages (Fig. 15-1A). Tagmosis is weak, with the anterior end forming a short head that is scarcely distinct from the long posterior trunk.

The body bears 13 to 43 pairs of segmental trunk appendages plus another 3 pairs of head appendages. All of the trunk appendages are similar, uniramous, stubby, fleshy legs (Fig. 15-1B). These conical, unjointed appendages are **lobopods** with internal musculature. Lobopods, which are saclike and lack articulations, are an unusual type of appendage found in Onychophora and Tardigrada. Males usually have fewer legs than females. They can bend at any position along their length, and each terminates in a pair of sclerotized **claws.** The name Onychophora (onych = claw, phor = to carry) is a reference to the claws. The leg contacts the substratum with a set of spiny pads situated at the base of the claws, and each leg has numerous sensory papillae.

Every leg bears a **coxal organ** (also called a coxal vesicle or coxal sac) on the ventral surface at its base (Fig. 15-1C). The coxal organs are eversible and appear to be used to gather water from the substratum. Similar structures are present in some myriapods. Ectodermally derived **crural glands** occur at the base of some legs in some species. These secretory glands open to the exterior via **crural papillae** (Fig 15-1C) on the ventral surface of the leg.

The head bears three pairs of appendages (Fig. 15-1B). Anteriormost is a pair of annulated, sensory antennae. The sharp, sclerotized, and clawlike **mandibles** that lie deep in the oral cavity lacerate prey. The third pair of appendages is the **oral papillae,** onto which open a pair of very large **slime glands** (Fig. 15-3B). The slime glands are modified crural glands. The milky-white slime ejected from the papillae is used to entangle prey and would-be predators. Homology between onychophoran antennae, mandibles, and oral papillae and their arthropod counterparts is uncertain.

BODY WALL AND LOCOMOTION

The onychophoran body is covered by a thin, chitinous cuticle, or exoskeleton (Fig. 15-2A). The cuticle is flexible and permeable, but resembles that of arthropods in its composition and organization. It is composed of α-chitin and protein

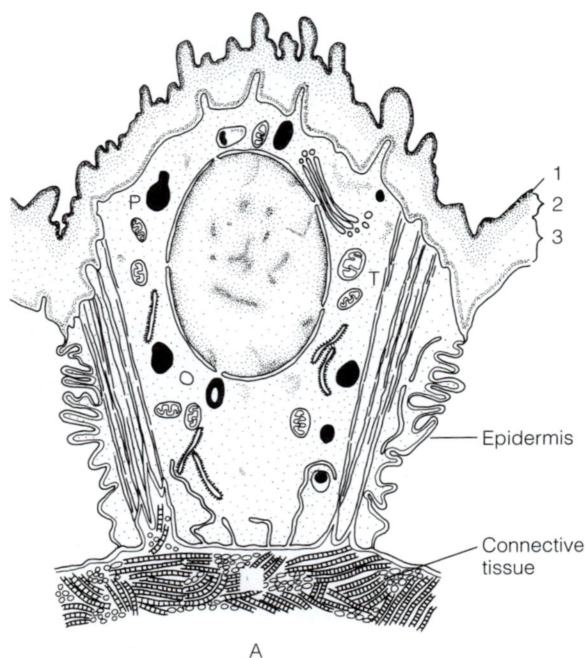

SEM courtesy of Storch, V. and Ruhlberg, H. 1977. Fine structure of the sensilla of *Peripatopsis moseleyi*. Cell Tiss. Res. 177:539–553. Copyright Springer-Verlag.

FIGURE 15-2 Onychophora. **A,** An epidermal cell of *Peripatus acacioi,* showing the three layers of the surface cuticle: epicuticle (1), exocuticle (2), and endocuticle (3). Tonofilaments anchoring the muscle layer to the cuticle are labeled T. P = Pigment granule. **B,** Scanning electron micrograph of the body surface of *Peripatopsis moseleyi,* showing a large papilla with its terminal sensory bristle and several small papillae without bristles. The papillae are covered with small scales. *(A, After Lavallard from Storch, V. 1984. Onychophora. In Bereiter-Hahn, J., Matoltsy, A. G., and Richards, K. S. Biology of the Integument. 1. Invertebrates. Springer-Verlag, Berlin. p. 704.)*

and consists of a thin epicuticle underlaid by a procuticle that consists of an exocuticle and endocuticle. The exocuticle and the outer layer of the epicuticle contain tanned proteins. The onychophorans' highly permeable cuticle requires that they live in habitats or microclimates with high relative humidity. The cuticle is molted frequently, sometimes as often as every two weeks during a lifespan of up to six years. The cuticle is thin and flexible throughout and has no sclerotized plates. Because of its flexibility, onychophorans can squeeze themselves into small crevices. The arthropod cuticle and molting are discussed in more detail in Chapter 16.

The cuticle is secreted by a monolayered epidermis (Fig. 15-2A). Muscles join the cuticle via tonofilaments traversing the epidermis (Fig. 15-2A). There are abundant pigment granules of a variety of colors in the epidermal cells. Onychophorans can be blue, green, orange, or black, and the papillae and scales give the body surface a velvety, iridescent appearance.

A thick layer of connective tissue lies below the epidermis. It consists primarily of layers of collagen fibers in an orthogonal pattern with fibers either parallel or perpendicular to the long axis of the body.

Three uninterrupted layers of smooth, unspecialized body-wall muscles (Fig. 15-3A) lie inside the connective tissue. The outermost layer of circular muscles is adjacent to the connective tissue. The innermost layer consists of longitudinal fibers. Between these two major muscle layers is a thinner sheet of oblique muscles. Dorsoventral muscles divide the hemocoel into compartments—two lateral and one median. The body-wall muscles extend into and are used to move the legs. Most aspects of the body wall, with the exception of the molted, chitinous cuticle, are similar to those of other soft-bodied bilaterians, such as the annelids, and unlike those of arthropods.

The longitudinal muscles border the hemocoel, which is the onychophoran body cavity. The reduced coelom is represented by nephridial sacs and gonads, as in arthropods. The viscera, including the gut, slime glands, salivary glands, heart, saccate nephridia, reproductive system, and nervous system, are immersed in blood in the hemocoel.

Onychophorans hold their body above the substratum and crawl slowly using their legs in combination with extension and contraction of the body. Body length is controlled by the circular and longitudinal body-wall muscles acting in conjunction with the hemocoel.

When a segment is extended by contracting the circular muscles, both of that segment's legs lift off the ground and move forward in a recovery stroke. The effective stroke is accomplished by shortening the segment by contracting the longitudinal muscles. The effective strokes of the two legs of a segment do not alternate as they do in most crawling arthropods.

NUTRITION

Onychophorans prey on small arthropods. Secretions of the two slime glands are ejected from pores on the oral papillae. Jets of slime may extend up to 15 cm from the animal. The proteinaceous slime entangles the prey and is immediately denatured by exposure to air, resulting in a stickiness that is lost after a few minutes. The slime does not stick to the hydrophobic cuticle of the onychophoran itself. Slime glands are large, branched organs located in the middle region of the

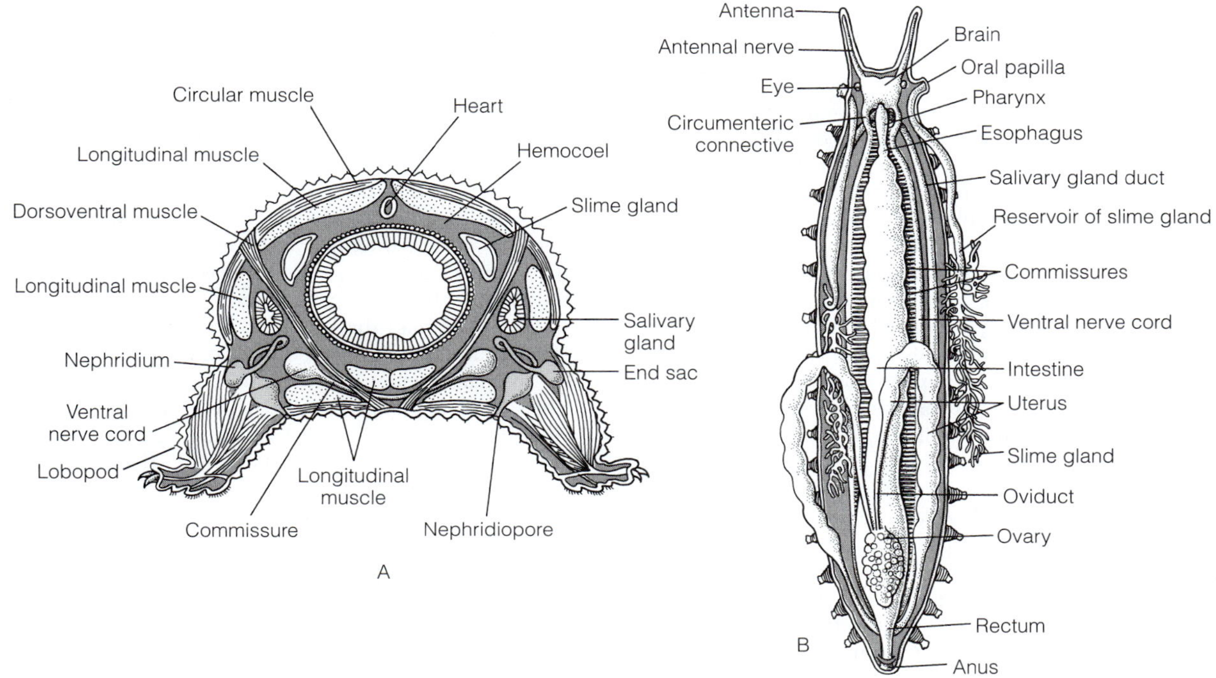

FIGURE 15-3 Onychophora. **A,** Diagrammatic cross section through the body of an onychophoran. **B,** Internal anatomy of a female onychophoran, dorsal view. *(Both after Snodgrass, R. E. 1938. Evolution of the Annelida, Onychophora, and Arthropoda. Smithsonian Misc. Coll. 97(6):50–149.)*

hemocoel (Fig. 15-3B). Each of the two glands connects to pores on the oral papillae by a thick duct that serves as a reservoir for undischarged slime. The slime glands are the modified crural glands of the segment of the oral papillae.

The subterminal mouth is in a shallow prebuccal depression on the ventral surface of the head (Fig. 15-1B). The opening to the depression is surrounded by lips. When feeding, onychophorans lacerate the prey with the mandibles and flood it with saliva. The saliva contains hydrolytic enzymes and mucus secreted by a pair of salivary glands, which open by a single duct into the anterior foregut. The salivary glands are modified saccate nephridia of the head segments and as such consist of an end sac with podocytes and a duct. Digestion begins externally.

The gut consists of a foregut, midgut, and hindgut. The foregut, which includes the pharynx and esophagus, is lined with an epidermal epithelium and its cuticle. The mouth opens into the muscular pumping pharynx (Fig. 15-3B). The pharynx pumps partially predigested liquid food through the esophagus into the midgut. The long midgut (the intestine) extends for most of the length of the animal, and its gastrodermis is absorptive and secretory. As an endodermal derivative, it is not lined with cuticle. A **peritrophic membrane** similar to that of insects and crustaceans is secreted around the food mass as it enters the midgut. This protects the delicate midgut walls from abrasion. The hindgut is the rectum. It is ectodermal and lined with cuticle. The rectum opens to the exterior via the anus on the ventral surface near the posterior end of the animal (Fig. 15-1C).

INTERNAL TRANSPORT, GAS EXCHANGE, AND EXCRETION

The hemal system (Fig. 15-3A), like that of most arthropods, includes a spacious hemocoel and lacks capillaries. The heart is a muscular, dorsal tube penetrated by paired segmental ostia. Its anterior end opens into the hemocoel, but it is not known if the posterior end is open or closed. Blood vessels are absent or, at best, represented by a pair of antennal arteries, so the system may be referred to as "open." The heart wall consists of a monolayered muscular epithelium with a basal lamina and is richly supplied with tracheae.

The hemocoel is a spacious cavity surrounding the internal organs and filled with blood. Like the hemocoel of arthropods, it is divided by a perforated horizontal diaphragm, or pericardial septum, into a dorsal **pericardial sinus** and a ventral **perivisceral sinus** (Fig. 16-7). The pericardial sinus surrounds the heart, whereas the larger perivisceral sinus contains most of the remaining viscera. Blood flow is similar to that of other arthropods. During diastole, blood from the pericardial sinus enters the heart through its ostia. During systole, the heart's circular muscles contract, the ostia close, and blood is propelled from the anterior end of the heart into the perivisceral hemocoel. The blood then flows over the tissues in the hemocoel and makes its way back through the perforated diaphragm into the pericardial sinus and then through the ostia into the heart lumen. The colorless blood lacks respiratory pigment, but contains nephrocytes and amebocytes (hemocytes).

The gas exchange organs are tubular, air-filled **tracheae,** which occur in clumps, or tufts, and deliver oxygen directly to the tissues. Tracheal tufts are scattered over the body surface, but are most abundant dorsally. Each tracheal tuft consists of a short tubular depression, or atrium, in the body surface. The floor of the atrium bears numerous minute openings, the **spiracles,** each of which opens into a single narrow, unbranched, trachea. The tracheae are minute tubes of less than 3 μm in diameter that have an exceedingly thin cuticle. There may be as many as 75 of these tufts per trunk segment.

The segmental excretory organs are paired **saccate nephridia.** Each consists of an end sac with podocytes connected to an exterior nephridiopore via a duct (Fig. 15-3A). The end sac, a coelomic derivative, contains a tiny coelomic space. The duct is derived from the metanephridium, and in Onychophora its nephrostome is ciliated. The annelid nephrostome is also ciliated, but that of arthropods is not. In this respect the onychophoran nephridium is transitional between that of annelids and arthropods. The nephridiopores are situated ventrally at the base of the legs (Fig. 15-1C). Almost all trunk segments have a pair of nephridia.

Onychophoran nephridia function like those of arthropods, with ultrafiltration from the hemocoel into the end sac followed by resorption and secretion in the duct. Some secretory glands, such as the salivary glands and posterior male accessory genital glands (anal glands) are modified nephridia.

NERVOUS SYSTEM AND SENSE ORGANS

The ladderlike nervous system is similar to that of both annelids and arthropods, consisting of a dorsal brain, circumenteric connectives, and a pair of ventral nerve cords connected by commissures (Fig. 15-3A,B). The nervous system arises in the embryo from a series of paired ectodermal placodes.

The nervous system is weakly cephalized, with a concentration of neurons in the anterior brain. The bilobed brain has right and left lobes, or hemispheres, as it does in annelids, and is located dorsally in the head. There is no indication of an arthropodan syncerebrum (three-part brain) in the adult, although such a brain has been reported in embryos. About 15 pairs of nerves arise from the brain and consist of sensory and motor neurons. Its major nerves are the sensory antennal and optic nerves from the antennae and eyes, respectively. Circumenteric connectives exit the brain posteriorly and extend along the ventral body wall as the ganglionated longitudinal ventral nerve cords. Other than the brain, no anterior concentration of segmental ganglia occurs, as it does in most arthropods.

The paired ventral nerve cords are very far apart (Fig. 15-3A). The paired segmental ganglia (neuromeres) are not swollen and consequently are indistinct. Each pair of ganglia is joined by 9 or 10 transverse commissures (Fig 15-3B). Mixed nerves (sensory and motor) supply the appendages and body wall.

Onychophorans have several types of **sensilla** that involve, as in arthropods, a specialized area of the cuticle associated with sensory neurons. The body is covered by sensory papillae (the large papillae mentioned earlier) equipped with a mechanoreceptive sensory bristle at the tip and sensory neurons inside (Fig. 15-2B). Other sensilla that occur on the lips and antennae are chemoreceptive. Some of these sensilla contain modified cilia with irregularly arranged microtubules that do not conform to the typical 9 + 2 pattern.

There is a small eye on the dorsal surface at the base of each antenna. The eyes are direct ocelli with a secreted chitinous lens, a cornea, and a retina composed of photoreceptor and pigment cells. The photoreceptors are modified ciliated cells with a photosensitive pigment on the ciliary membrane. The eyes arise from an epidermal invagination and are innervated by the anterior part of the brain.

REPRODUCTION AND DEVELOPMENT

Onychophorans are always gonochoric and sexually dimorphic, with males being smaller than females. The gonads are paired coelomic derivatives connected by gonoducts (coelomoduct derivatives) to an unpaired gonopore on the posterior ventral surface (Fig. 15-1C). They have internal fertilization and sperm are transferred by spermatophores.

The female reproductive system consists of paired ovaries coalesced along the midline and attached to the horizontal diaphragm (Fig. 15-3B). The paired female gonoducts connect the ovary with the single gonopore located on the posterior ventral midline. Each gonoduct is differentiated into a proximal oviduct and distal uterus (Fig. 15-3B) in viviparous species. The uteri join to form a single vagina that connects with the gonopore. Eggs are gestated in the uteri, which may contain many embryos simultaneously. In females of oviparous species, there is a seminal receptacle for the storage of sperm and the gonopore opens at the end of a large ovipositor.

The male reproductive system is paired proximally and single distally. The testes, seminal vesicles, and vasa efferentia are paired and not fused. The vasa efferentia join to form a single vas deferens, which enlarges to become the ejaculatory duct. The ejaculatory duct opens to the exterior via the male gonopore on the posterior ventral surface (Fig. 15-1C). Sperm are flagellated with axonemal microtubules in the typical 9 + 2 pattern.

Spermatophore transfer varies with species. In some (such as the South American *Peripatopsis,* which lacks seminal receptacles) the spermatophores are attached to the outside of the female's body. Amebocytes from the female's blood secrete enzymes to hydrolyze an opening in the body wall directly under the spermatophore. The sperm enter the hemocoel through this opening and make their way to the ovary, where fertilization occurs. In others (such as *Peripatus*) the spermatophore is placed in the female gonopore. Sperm transfer in onychophorans with seminal receptacles is not understood. Males of some taxa have what appears to be a penis, but it has not been observed in operation. Males also possess two pairs of accessory genital glands (the posterior pair are the anal glands) that open in the vicinity of the gonopore (Fig. 15-1C).

Cleavage may be superficial (intralecithal), as is typical of arthropods, or total (holoblastic). The type of cleavage depends on the amount of yolk in the egg, but there is no spiral cleavage.

Some onychophorans are oviparous, but most are viviparous (matrotrophic or lecithotrophic). Oviparity is thought to be the ancestral reproductive mode and viviparity to have derived from it. In viviparous species the developing embryos are retained in the uterus and the adult gives birth to juveniles. The gestation period may be a year or more. The developing embryos may be of different or the same ages.

In matrotrophic, viviparous taxa, eggs are usually tiny, yolkless, and nourished in the uterus by secretions of the uterine epithelium. Yolk does not interfere with cleavage, which is thus holoblastic rather than superficial. There may be a tissue connection **(placenta),** between the uterine epithelium and the embryo for transport of these secretions. Alternatively, in the absence of a placenta, secretions are released into the uterine lumen and absorbed by the embryo.

Lecithotrophic, viviparous species also retain the embryos in the uterus, but they are nourished by the moderate amounts of yolk present in the egg. These eggs have superficial cleavage.

Oviparous species have large, yolky eggs enclosed in a chitinous eggshell, and their cleavage is superficial. Development in oviparous species occurs outside the mother's body without maternal care.

DIVERSITY OF ONYCHOPHORA

Onychophora includes two allopatric taxa of equal rank.

Peripatopsidae[F]: Relatively primitive, restricted to temperate southern hemisphere (Australia, Tasmania, New Zealand, South Africa, and Chile). 13 to 25 pairs of legs. Salivary gland lacks a reservoir. Coxal organs absent, crural glands in both sexes. Gonopores between the last pair of legs. Ovary has exogenous eggs (ovarian follicles project into hemocoel), no placenta. *Cephalofovea, Euperipatoides, Metaperipatus, Occiperipatoides, Ooperipatellus, Paraperipatus, Peripatopsis, Peripatoides.*

Peripatidae[F]: Relatively derived. Circumtropical distribution (Malaysia, Borneo, the Congo Basin, the West Indies, central Mexico, Central America, and northern South America). Many more legs (22 to 43 pairs). Coxal organs present, crural glands in males only. Salivary glands with reservoir. Gonopore located between penultimate pair of legs; ovary has endogenous eggs. Some with a placenta. *Macroperipatus, Oroperipatus, Peripatus, Typhloperipatus.*

PHYLOGENY OF ONYCHOPHORA

Onychophora is a monophyletic taxon united by common possession of mandibles on the second head segment and oral papillae with slime glands on the third head segment; widely spaced ventral nerve cords connected by numerous segmental commissures; and tufted tracheae. Salivary glands derived from nephridia on the third head segment may be a panarthropod plesiomorphy. The onychophorans probably arose as an exclusively terrestrial line descended from Cambrian panarthropod ancestors.

The onychophorans have morphological characteristics that argue for close relationships with both the annelids and the arthropods, but ties with the arthropods are stronger. Within the taxon there are characters that are annelid-like but unlike the arthropods, characters that are shared with both annelids and arthropods, and characters that are arthropod-like but unlike the annelids.

1. Characters shared with annelids only: body-wall muscles in continuous, uninterrupted, unspecialized sheets; cuticle thin, flexible, and lacking sclerotized plates; weak tagmosis; simple brain lacking segmental neuromeres; and wormlike body.

2. Characters in common with annelids and arthropods: segmented body; paired segmental appendages; paired segmental metanephridia or their derivatives; ladderlike nervous system with dorsal brain and ventral ganglionated double nerve cord.
3. Characters shared with arthropods only: exoskeleton with ecdysis; exoskeleton with α-chitin and noncollagenous protein; coelom reduced to gonadal and nephridial spaces; open hemal system with an ostiate heart; hemocoel divided by horizontal septum into pericardial and perivisceral sinuses; appendages (mandibles) specialized for feeding; respiration via tracheae; and superficial cleavage. Currently, most zoologists believe Onychophora to be closely related to Arthropoda and Tardigrada in the monophyletic taxon, Panarthropoda.

Onychophorans have often been described as intermediate between annelids and arthropods. If the evolutionary relationship between annelids and arthropods is discredited by molecular and other lines of evidence, then the similarities between the onychophorans and the annelids are due to convergence. Alternatively, the common ancestor of protostomes may have been a segmented coelomate, in which case the shared characters are symplesiomorphies.

Because the soft bodies of onychophorans do not fossilize well, and they live in habitats that do not readily create fossils, there are few undisputed onychophoran fossils to aid in unraveling the relationships of this taxon. Some Cambrian fossils, such as *Aysheaia, Hallucigenia,* and *Onychodictyon,* may represent marine ancestors from which the terrestrial Onychophora arose.

PHYLOGENETIC HIERARCHY OF ONYCHOPHORA

Onychophora
- Peripatopsidae
- Peripatidae

TARDIGRADA^P

Tardigrades are tiny metazoans known as water bears or slow walkers (Fig. 15-4). It is difficult to describe living water bears without using words like "cute" or "endearing." They resemble diminutive teddy bears as they clamber over sand grains or moss leaflets on four pairs of stubby legs. The typical body length is 100 to 150 μm, but the grizzlies of the taxon, such as the giant *Macrobiotus,* achieve lengths of 1.5 mm. Tardigrades are common in a variety of aquatic and semiaquatic habitats but are rarely noticed because of their small size and cryptic habits.

The evolutionary relationships of the tardigrades are at present unknown, and they show similarities to cycloneuralians and arthropods. At present, kinship with neither can be demonstrated conclusively, although they seem to be closest to arthropods. The approximately 800 described species are found in marine, freshwater, and intermittently moist terrestrial habitats. Except for one species, all belong to Heterotardigrada and Eutardigrada. A third taxon, Mesotardigrada, contains only one genus, *Thermozodium,* which was described

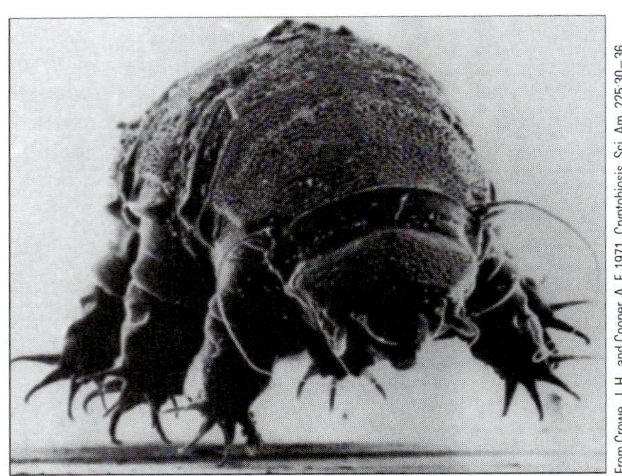

FIGURE 15-4 Tardigrada. Scanning electron micrograph of the heterotardigrade *Echiniscus arctomys* in anterior view. The mouth is located at the end of the snoutlike anterior projection.

and known from a single hot spring in Japan that has since been destroyed by an earthquake.

Tardigrades are found in a variety of wet or moist habitats. Tardigrades are characteristic members of marine sandy interstitial communities. Some tardigrades are found in hot springs and some occur in glaciers. Many inhabit freshwater or marine environments, but a large number are also found in terrestrial mosses, lichens, leaf litter, and soils that experience alternating periods of desiccation and wetting. Like bdelloid rotifers in the same habitats, terrestrial species are active only when a film of water covers the substratum. When this film evaporates, they enter a cryptobiotic state that allows the animal to survive the period of unfavorable conditions. **Cryptobiosis** is characterized by desiccation, reduced metabolic rate, and enhanced resistance to adverse environmental conditions such as drought and temperature extremes. The shriveled, dormant cryptobiotic tardigrade, known as a **tun,** from the German *Tönnchen,* meaning little barrel (Fig. 15-6E), may remain alive for long periods. Tardigrades often live for as long as 10 years and occasionally as long as 100 years, during which time periods of activity alternate with periods of cryptobiosis. Cryptobiosis also facilitates dispersal, as the dried tuns are easily transported by the wind or by other organisms.

FORM

The bilaterally symmetrical body is usually plump and cylindrical (Fig. 15-5). Superficially, it appears to be composed of five segments, including a short head of one segment and a trunk with four segments. Segmentation is indistinct externally, however, and the head may actually consist of three segments and the trunk of four or five. Transverse folds in the cuticle are misleading and do not correspond to actual segments.

Each trunk segment bears a pair of short, stubby legs. The tip of each leg has four to eight retractile claws or adhesive discs (Fig. 15-5). The legs, like those of the onychophorans, are lobopods. Each leg has weakly separated regions that are sometimes designated as coxa and femur. Short, intrinsic muscles are located inside the legs. The legs of many marine species are capable of telescoping.

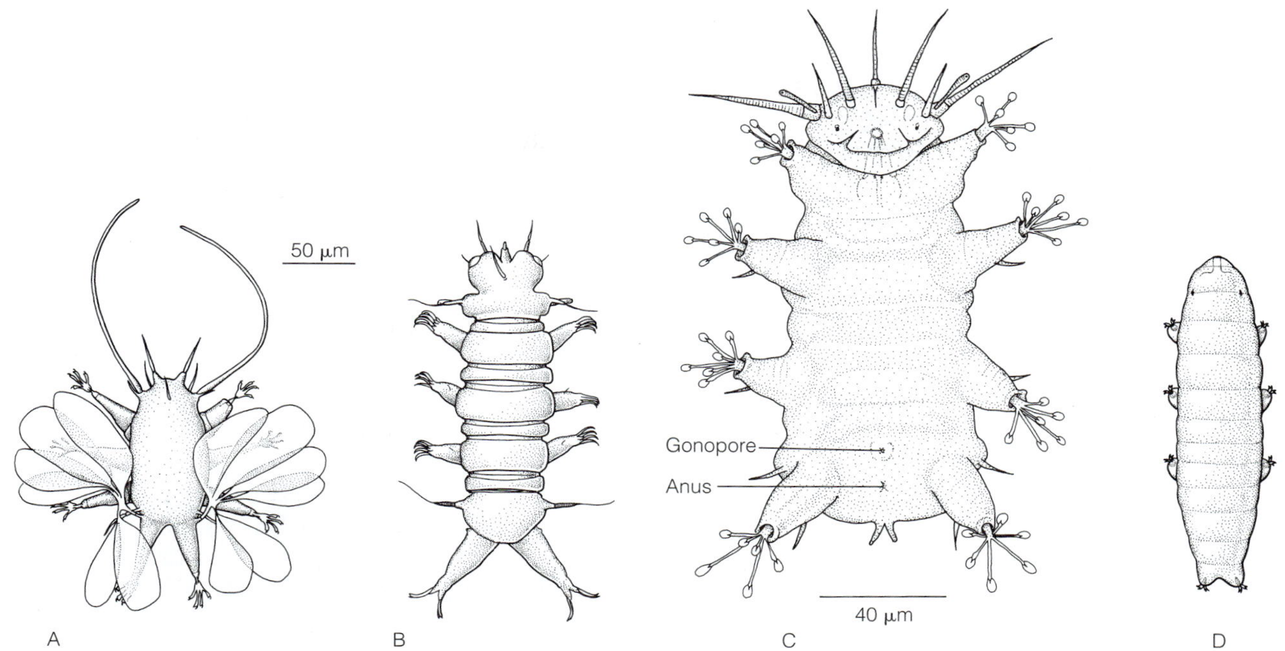

FIGURE 15-5 Tardigrada. **A–C,** Marine heterotardigrades. **A,** *Tanarctus velatus.* **B,** *Megastygarctides orbiculatus.* **C,** *Batillipes noerrevangi.* **D,** The eutardigrade *Macrobiotus hufelandi.* *(A and B, From McKirdy, D., Schmidt, P., and McGinty-Bayly, M. 1976. Interstitielle Fauna von Galapagos. XVI. Tardigrada. Mikrofauna Meeresbodens 58:1–43.; C, From Kristensen, R. M. 1981. Sense organs of two marine arthrotardigrades. Acta Zool. 62:27–41; D, Redrawn after Higgins from Nelson, D. R. 1982a. Developmental biology of the Tardigrada. In Harrison, F. W., and Cowden, R. R. (Eds.): Developmental Biology of Freshwater Invertebrates. Alan R. Liss, New York.)*

BODY WALL

The thin body wall consists of only two layers. The outermost cellular layer is the epidermis (Fig. 15-6B). This monolayered epithelium is **eutelic** (composed of a constant and genetically determined number of cells). Many of the tissues of tardigrades are eutelic, a condition that may be due to their small size. Eutely is also a characteristic feature of some small-bodied gnathiferan and cycloneuralian taxa such as rotifers and nematodes. Tardigrades lack motile cilia, a feature held in common with both nematodes and arthropods.

The epidermis secretes a cuticle, or exoskeleton, composed of four layers designated as, from the outside inward, the epicuticle, exocuticle, mesocuticle, and endocuticle (Fig. 15-6B). The cuticle is complex and composed of an as yet undetermined type of chitin, tanned and untanned glycoproteins, mucopolysaccharides, polysaccharides, lipids, and lipoproteins. In armored species, such as *Echiniscus* and *Megastygarctides,* the dorsal cuticle is thickened, arthropod-like, to form articulated sclerites (Fig. 15-5B). The cuticle is often ornamented with spines, granules, pores, or other sculpture (Fig. 15-4).

The exoskeleton, including the cuticular claws and stylets and the lining of the fore- and hindguts, is molted periodically. During ecdysis, the body contracts and pulls away from the old cuticle, which is then slipped off and left behind as a relatively intact exuvium. The cuticle of the body and gut is secreted by the epidermis, but the stylets and their chitinous supports are secreted by the salivary glands, which are also ectodermal. The stylets are often calcified. New claws are secreted by ectodermal claw glands (Fig. 15-6A).

The epidermis is underlaid by its basal lamina, but there are no muscle layers. The body cavity is a hemocoel lined only with the basal lamina.

MUSCULATURE AND LOCOMOTION

The body wall musculature consists of individual specialized muscles in the trunk and appendages, but, like arthropods and unlike onychophorans, there are no continuous sheets of body-wall muscle. The elongate, discrete muscles are smooth or cross-striated cells that originate and insert on the inner surface of the cuticle (Fig. 15-6A). Muscle attachment to the cuticle is via tonofilaments passing through the epidermal cells, as in arthropods and onychophorans. The muscle cells are surrounded by a basal lamina that lines the hemocoel.

Contraction of the muscles achieves flexion of the body and appendages and is opposed by the hemocoel, which functions as a hydrostatic skeleton. Tardigrades move about slowly by crawling with their legs, using their claws or discs to grasp the substratum. The first three pairs of legs are used in forward locomotion, and the last pair function in retreat or grasping the substratum (Fig. 15-7). The claws of the last pair are reversed to face forward, in contrast with those of the first three pairs.

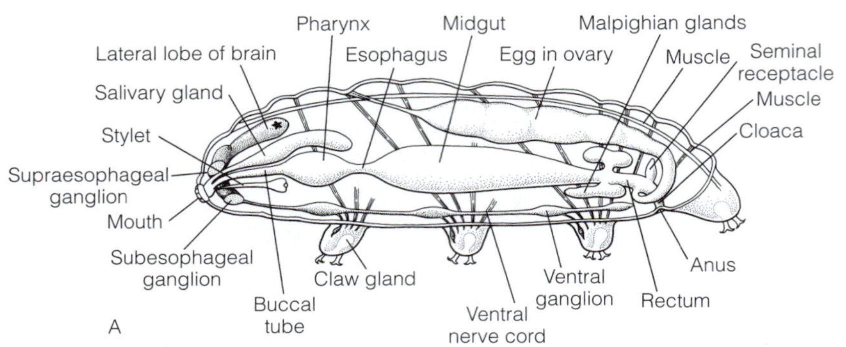

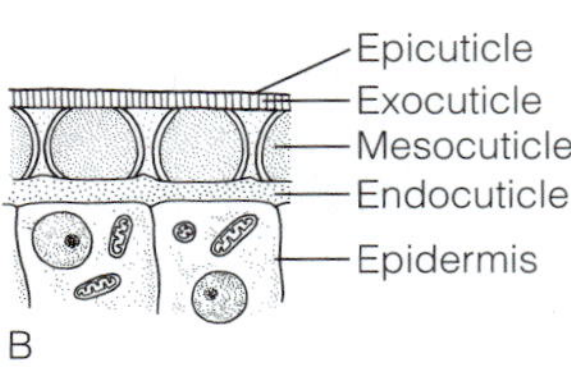

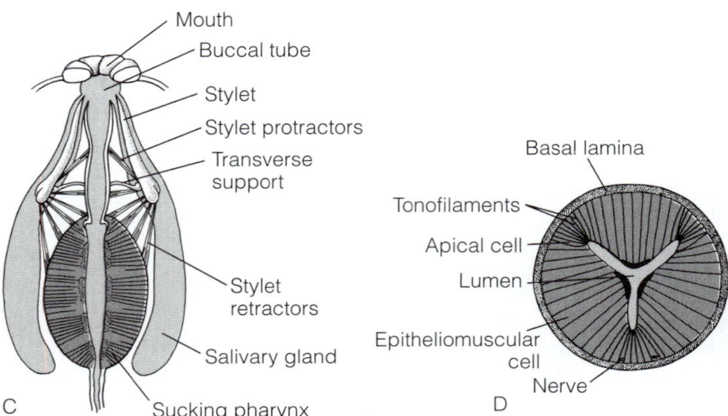

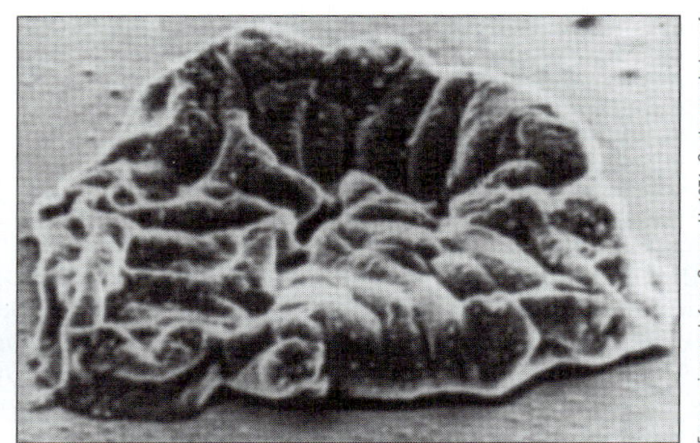

From a micrograph from Greven, H. 1971. On the morphology of tardigrades: A stereoscan study of *Macrobiotus hufelandi* and *Echiniscus testudo*. Forma Functio 4:283–302.

FIGURE 15-6 Tardigrada. **A,** Lateral view of the eutardigrade *Macrobiotus hufelandi*. **B,** Section of the cuticle and epidermis of *Batillipes noerrevangi*. **C,** Dorsal view of the foregut apparatus of a tardigrade. **D,** Cross section of a tardigrade epitheliomuscular pharynx. **E,** Desiccated cryptobiotic tun of *Macrobiotus hufelandi*. *(A, Modified and redrawn from Cuénot, L. 1949. Les Tardigrades. In Grassé, P.-P. (Ed.): Traité de Zoologie. Vol. 6. Masson et Cie, Paris; B, Modified from Kristensen, R. M. 1976. On the fine structure of* Batillipes noerrevangi *Kristensen 1976. Zool. Anz., Jena 197:129–150; C, Combined from Pennak, R. W. 1978. Fresh-Water Invertebrates of North America. 2nd Edition. John Wiley and Sons, New York, and Kristensen, R. M. 1982. The first record of cyclomorphosis in Tardigrada based on a new genus and species from Arctic meiobenthos. Z. zool. Syst. Evolut.-forsch. 20:249–270.)*

Body Cavity

Although the gonad appears to be derived from a coelomic space, the chief body cavity is an expanded connective-tissue compartment, or hemocoel. It is filled with colorless blood containing hemocytes. Much of the hemocoel is occupied by these nutrient storage and phagocytic cells, which may be attached or float freely. There is no heart, and movements of the body and appendages circulate the blood through the hemocoel. Tissues (muscles, epidermis, and neurons, but not hemocytes) in or adjoining the hemocoel are separated from it by their basal laminae.

Nutrition

Most tardigrades feed on the cytoplasmic contents of plant cells. Soil tardigrades feed on algae and probably detritus, and some are predators of nematodes and other minute soil animals, including other tardigrades.

The gut consists of an ectodermal foregut and hindgut connected by an intervening endodermal midgut. The foregut and hindgut are lined with cuticle, which is molted, but the midgut is not. The complicated foregut consists of a heavily cuticularized buccal tube, a pharynx with thick muscular walls, and an esophagus (Fig. 15-6A,C).

The foregut is equipped with a pair of cuticular **stylets** used to puncture the cells of plant or animal prey. The sharp tips of the needlelike stylets protrude into the buccal tube and are extended and withdrawn by a set of protractor and retractor muscles (Fig. 15-6A,C). The stylets are housed in stylet sheaths and braced by transverse **stylet supports,** which are also cuticular and extend from the stylet to the buccal tube. The stylet apparatus probably is homologous to a pair of head appendages and is unique to the tardigrades, although there are similar structures in some herbivorous nematodes and rotifers. Onychophoran mandibles are also modified head appendages.

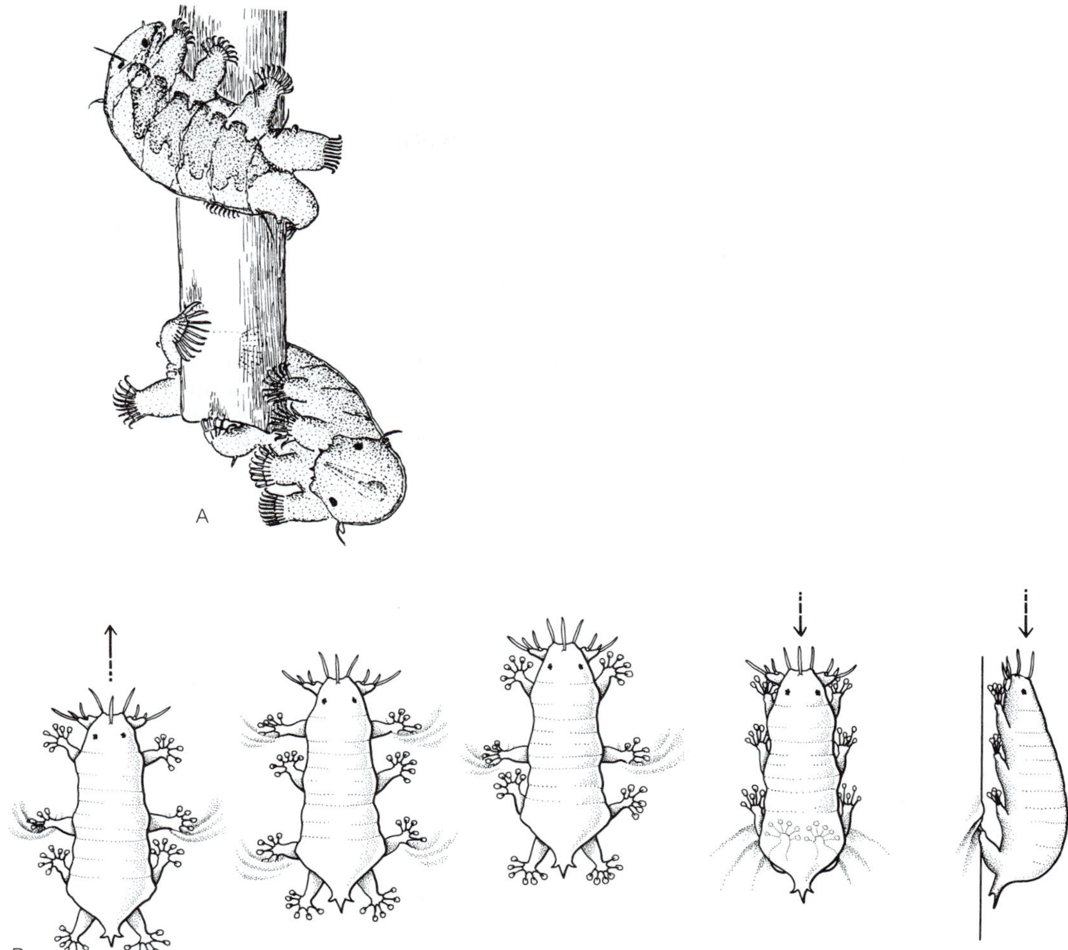

FIGURE 15-7 Tardigrada. **A,** Tardigrades climbing on an algal filament. **B,** Locomotion over a surface (microscope slide) by a marine interstitial tardigrade *(Batillipes)*. *(A, From Marcus, E. 1929. Tardigrada. In Bronn, H. G. (Ed.): Klassen und Ordnungen des Tierreichs. Bd. 5, Abt. IV. Akademische Verlagsgesellschaft, Frankfurt. pp. 1–608.)*

The mouth is an anterior terminal (Fig. 15-6A) or ventral subterminal opening. A unique telescopic mouth cone equipped with the stylets can be extended from the body surface. Numerous sensory structures are associated with the mouth. A pair of salivary glands closely associated with the stylets opens into the buccal tube (Fig. 15-6C). The salivary glands secrete new stylets just prior to ecdysis and also secrete saliva during feeding.

The pharynx is a muscular pump that sucks liquid food into the gut. Its lumen is Y-shaped, or **triradiate,** in cross section, as is that of gastrotrichs and nematodes (Fig. 15-6D). Its walls are composed of radiating cross-striated epitheliomuscular cells. It is surrounded by a basal lamina and its lumen is lined by cuticle. The triradiate pattern is the most efficient configuration for radial muscles surrounding a distensible lumen, and it has perhaps arisen independently in many animal taxa with sucking pharynges (including gastrotrichs, loriciferans, bryozoans, nematodes, and tardigrades).

When feeding, the mouth is placed against the prey and the stylets protruded to puncture the cell or body wall. The contents of the prey are then sucked out by the pharynx. Food passes from the pharynx through the short esophagus to the midgut (Fig. 15-6A).

The endodermal midgut, or intestine, has a monolayered, microvillous epithelium (Fig. 15-6A). It is secretory and absorptive and is the site of hydrolysis and absorption. The food in the midgut may be enclosed in a peritrophic membrane secreted by the midgut epithelium. The intestine leads into the short, cuticle-lined rectum (hindgut), which opens to the outside through the terminal anus on the ventral midline between or a little anterior to the last pair of legs (Fig. 15-5C). Since the rectum receives nitrogenous wastes from the Malpighian glands and in females the oviducts open here, it may be referred to as a cloaca. The hindgut epithelium is equipped for transport and probably plays a role in modifying the rectal contents, including digestive and excretory wastes.

GAS EXCHANGE AND EXCRETION

The favorable surface-area-to-volume ratio associated with organisms as small as tardigrades obviates the need for special gas exchange structures, and so there are none. Gas exchange occurs across the unspecialized body surface.

Discrete excretory organs are absent in many tardigrades (heterotardigrades), but others (eutardigrades) have three large **Malpighian glands** at the midgut-hindgut junction (Fig. 15-6A). These include two lateral and one dorsal gut diverticula. Thought to function in nitrogen excretion, the ultrastructure of these glands is similar to that of insect Malpighian tubules. They are adaptations for a terrestrial existence not present in the primitive marine heterotardigrades and probably arose independently of those in the insects and chelicerates. There are no recognizable saccate nephridia in the tardigrades.

Ventral organs at the bases of the second and third pairs of legs of some terrestrial heterotardigrades may be excretory organs and the general epidermis may play an excretory role, as well.

NERVOUS SYSTEM AND SENSE ORGANS

The central nervous system is ladderlike and similar to that of the annelids, arthropods, and onychophorans (Fig. 15-6A). A ring of ganglia and nerve tracts surrounds the buccal tube and includes the supraesophageal ganglia and a pair of ganglia lateral to the stylets. A pair of large connectives joins the ganglionic ring to the ganglia of the first legs. The supraesophageal ganglia include two pairs of posteriorly oriented lobes. On each side, the large dorsolateral lobe of the brain has an unusual connective with the first ventral segmental ganglion. The entire tardigrade brain may be homologous to the protocerebrum of the arthropod brain. Alternatively, the combination of the supraesophageal ganglion and its lobes has been interpreted by some authors as a syncerebrum with proto-, deuto-, and tritocerebrums.

The trunk is innervated by a double ventral nerve cord with four or five pairs of segmental ganglia. There is one pair of ganglia for each trunk segment and perhaps an additional pair for a legless genital segment. The two ganglia of each pair are joined by a transverse commissure. The unambiguous segmentation of the nervous system supports the hypothesis that tardigrades are composed of a head and four or five segments.

Eutardigrades and some heterotardigrades have a pair of small, simple ocelli. Each ocellus consists of a cup-shaped pigment cell that encloses two photoreceptor cells. The pigment is red or black and the concavity of the cup faces the body surface and the light. This structure is unique.

A wide variety of cuticular sensory structures, or sensilla, are found in some tardigrades. They are presumably chemo- and mechanoreceptive. The head may have a variety of sensory cephalic cirri, clavae, and papillae (Fig. 15-5C). Bristlelike sensilla are found on each segment of some tardigrades.

REPRODUCTION AND DEVELOPMENT

Almost all tardigrades are gonochoric, but there are a few hermaphroditic species. Many are parthenogenetic and males are unknown in some genera (such as *Echiniscus*).

The gonad, either testis or ovary (Fig. 15-6A), is a single, unpaired elongate sac situated above the gut and attached by ligaments to the dorsal body wall. It probably is a coelomic derivative. In the male, two sperm ducts exit the testis and open via a single median gonopore immediately anterior to the anus (Fig. 15-5C). The male reproductive system is independent of the gut. Sperm are flagellated with a typical 9 + 2 arrangement of microtubules.

In females the single oviduct opens on the surface near the anus (in heterotardigrades) or joins the rectum (in eutardigrades), which thus functions as a cloaca (Fig. 15-6A). One (in a few eutardigrades) or two (in many heterotardigrades) seminal receptacles may be present.

Mating and oviposition occur in conjunction with molting by the female (Fig. 15-8). In aquatic species, males may deposit sperm in the recently shed female exuvium, which contains eggs (Fig. 15-8B). In most terrestrial species, however, copulation occurs and the male intromits sperm into the female reproductive tract before her ecdysis is complete (Fig. 15-8A). Fertilization in this case is internal, in the ovary.

One to 30 eggs are laid at a time, depending on the species. The eggs may be deposited in the old exuvium or attached to some substratum. Like rotifers and some gastrotrichs, some aquatic tardigrades produce thin-shelled eggs when environmental conditions are favorable and thick-shelled, resistant eggs when conditions are adverse. The eggs of terrestrial species typically possess a thick, sculptured shell that resists the frequent periods of desiccation to which mosses are subjected (Fig. 15-8C).

Cleavage is holoblastic and equal but neither spiral nor radial. Development is direct, juveniles resemble adults, and there is no real larval stage. The blastocoel is indistinct or absent and the hemocoel develops as a new cavity in the connective-tissue compartment. The gonad develops from a coelomic space. Early instars differ only slightly from older instars. First instar tardigrades have a mouth, but no other openings. Second instars have a mouth and anus. Third instar animals have a mouth, anus, and gonopore. Development is completed within 14 days or less and the little bears hatch by breaking the eggshell with their stylets. Tardigrades are eutelic and further growth is achieved by increasing the size of cells rather than by adding new cells. Mitosis has been reported in adults, so tardigrades may not be strictly eutelic. This mitotic activity may be for the occasional replacement of dead cells. As many as 12 molts may take place over the lifetime of a tardigrade, which has been estimated to be 3 to 30 months. Frequent periods of cryptobiosis may lengthen the life span to decades or more.

DIVERSITY OF TARDIGRADA

Heterotardigrada[C]: Armored tardigrades (Fig. 15-5A–C); consisting of primitive marine arthrotardigrades and the more derived terrestrial and freshwater echiniscoids. Dorsal cuticle sclerotized, divided into sclerites. Conspicuous sensilla on the head. Lobopods terminate in claws or adhesive discs. Gonopore independent of gut. Malpighian glands absent. *Batillipes, Echiniscus, Stygarctus, Tanarctus.*

Mesotardigrada[C]: *Thermozodium* the only genus. Malpighian glands present. Legs with 6 to 10 simple claws.

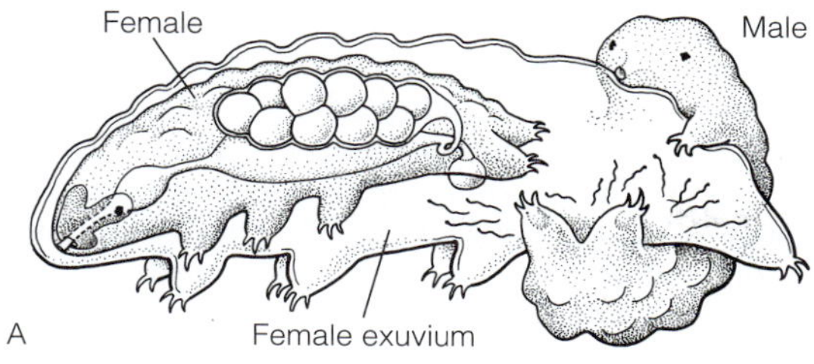

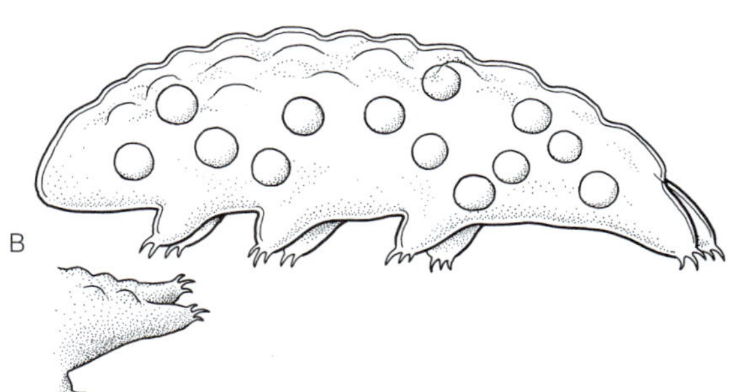

FIGURE 15-8 Tardigrada. **A,** Copulation in the eutardigrade *Hypsibius nodosus.* **B,** Female *Hypsibius* ovipositing in her own exuvium. **C,** Scanning electron micrograph of an attached, dormant egg of *Macrobiotus tonollii. (A and B, Redrawn from Marcus, E. 1929. Tardigrada. In Bronn, H. G. (Ed.): Klassen und Ordnungen Tierreichs. Bd. 5, Abt. IV. Akademische Verlagsgesellschaft, Frankfurt. pp. 1–608.)*

Eutardigrada[C]: Naked tardigrades. Freshwater and terrestrial. Sclerites absent, dorsal cuticle is thin and unsclerotized. Head lacks conspicuous sensilla, legs terminate in claws (Fig. 15-5D, 15-6A). Cloaca formed by junction of intestine and oviduct. Malpighian glands present. *Hypsibius, Macrobiotus, Milnesium.*

PHYLOGENY OF TARDIGRADA

There is consensus that Tardigrada is a monophyletic taxon, but its relationship to other taxa remains a topic of contention. Tardigrades share morphological characteristics with arthropods, onychophorans, and some of the former "aschelminth" taxa. Their strongest morphological ties, however, are with arthropods. Molecular evidence, especially the comparison of 18S ribosomal RNA base sequences, indicates a close relationship with arthropods, but also supports a cycloneuralian relationship.

The tardigrade fossil record is almost nonexistent, consisting of two specimens from Cretaceous amber, one of which is a poorly preserved juvenile. That these specimens resemble modern forms demonstrates that tardigrades have changed little in the last 60 million years. The fossils provide no insight into the evolution of the tardigrades. Some Cambrian fossils, such as *Aysheaia* from the Burgess Shale, have intriguing similarities to tardigrades and also to onychophorans and annelids. These three taxa all have soft unjointed appendages with terminal claws, a terminal mouth, and a posterior body that merges with the last pair of legs. Tardigrades and onychophorans may be related to a stem lineage of arthropods, such as the anomalocarids, by virtue of the shared biradial symmetry of their mouthparts.

PHYLOGENETIC HIERARCHY OF TARDIGRADA

Tardigrada
- Heterotardigrada
- Mesotardigrada
- Eutardigrada

REFERENCES

ONYCHOPHORA

Cuénot, L. 1949. Les onychophores, les tardigrades, et les pentastomides. In Grassé, P. (Ed.): Traité de Zoologie. Vol. 6. Masson, Paris. pp. 3–75.

Pflugfelder, O. 1968. Onychophora. Gustav Fischer Verlag, Stuttgart. 42 pp.

Poinar, G. 2000. Fossil onychophorans from Dominican and Baltic amber: *Tertiapatus dominicanus* n.g, n.sp. (Tertiapatidae n. fam.) and *Succinipatopsis balticus* n.g., n. sp. (Succinipatopsidae n. fam.) with a proposed classification of the subphylum Onychophora. Invert. Biol. 119:104–109.

Ruhberg, H. 1996. Onychophora, Stummelfüßer. In Westheide, W., and Rieger, R. M. (Eds.): Spezielle Zoologie. Gustav Fischer Verlag, Stuttgart. pp. 420–428.

Snodgrass, R. E. 1938. Evolution of the Annelida, Onychophora, and Arthropoda. Smithsonian Misc. Coll. 97(6):50–149.

Storch, V. 1984. Onychophora. In Bereiter-Hahn, J., Matoltsy, A. G., and Richards, K. S. Biology of the Integument. Vol. 1: Invertebrates. Springer-Verlag, Berlin. pp. 703–708.

Storch, V., and Ruhberg, H. 1977. Fine structure of the sensilla of *Peripatopsis moseleyi.* Cell Tiss. Res. 177:539–553.

Storch, V., and Ruhberg, H. 1993. Onychophora. In Harrison, F. W., and Rice, M. E. Microscopic Anatomy of Invertebrates. Vol. 12: Onychophora, Chilopoda, and lesser Protostomata. Wiley, New York. pp. 11–56.

Internet Sites

www.dc.peachnet.edu/~pgore/geology/geo102/burgess/burgess.htm (Includes a color photograph of *Aysheaia* from the Burgess shale.)

http://rbt.ots.ac.cr/onych/news.htm (Organization for Tropical Studies Onychophora Newsletter. Back issues of the Onychophoran Newsletter and a gallery of color photographs.)

www.mnhn.fr/assoc/myriapoda/ONYLIST.HTM (World Checklist of the Onychophora.)

www.mnhn.fr/assoc/myriapoda/GAONYCHO.HTM (Gallery Onychophora.)

www.sciref.org/onychophora (Scientific References Resources Onychophora Homepage.)

www.jcu.edu.au/school/tbiol/zoology/peripat.htm (Tropical Australian Onychophora Research.)

TARDIGRADA

Crowe, J. H., and Cooper, A. F. 1971. Cryptobiosis. Sci. Am. 225:30–36.

Dewel, R. A., and Dewel, W. C. 1998. The place of tardigrades in arthropod evolution. In Fortey, R. A., and Thomas, R. H. (Eds.): Arthropod Relationships. Systematics Association Spec. Vol. 55. Chapman and Hall, London. pp. 109–123.

Dewel, R. A., Nelson, D. R., and Dewel, W. C. 1993. Tardigrada. In Harrison, F. W., and Rice, M. E. Microscopic Anatomy of Invertebrates. Vol. 12: Onychophora, Chilopoda, and lesser Protostomata. Wiley-Liss, New York. pp. 143–183.

Dewel, R. A., Nelson, D. R., and Dewel, W. C. 1996. The brain of *Echiniscus veridissimus* Peterfi, 1956 (Heterotardigrada): A key to understanding the phylogenetic position and the evolution of the arthropod head. Zool. J. Linn. Soc. 116:35–49.

Eibye-Jacobsen, J. 1997. New observations on the embryology of the Tardigrada. Zool. Anzeiger 235:201–216.

Garey, J. R., Nelson, D. R., Mackey, L. Y., and Li, L. 1999. Tardigrade phylogeny: Congruency of morphological and molecular evidence. Zool. Anzeiger 238:205–210.

Greven, H. 1971. On the morphology of Tardigrades: A stereoscan study of *Macrobiotus hufelandi* and *Echiniscus testudo.* Forma Functio 4:283–302.

Kinchin, I. M. 1994. The biology of tardigrades. Portland Press, London. 186 pp.

Kristensen, R. M. 1976. On the fine structure of *Batillipes noerrevangi* Kristensen 1976. Zool. Anzeiger 197:129–150.

Kristensen, R. M. 1981. Sense organs of two marine arthrotardigrades (Heterotardigrada, Tardigrada). Acta Zool. 62:27–41.

Kristensen, R. M. 1982. The first record of cyclomorphosis in Tardigrada based on a new genus and species from Arctic meiobenthos. Z. zool. Syst. Evolut.-forsch. 20:249–270.

Kristensen, R. M., and Neuhaus, B. 1999. The ultrastructure of the tardigrade cuticle with special attention to marine species. Zool. Anzeiger 238:261–281.

Marcus, E. 1929. Tardigrada. In Bronn, H. G. (Ed.): Klassen und Ordnungen des Tierreichs. Bd. 5, Abt. IV. Akademische Verlagsgesellschaft, Frankfurt. pp. 1–608.

McKirdy, D., Schmidt, P., and McGinty-Bayly, M. 1976. Interstitielle Fauna von Galapagos. XVI. Tardigrada. Mikrofauna Meeresbodens 58:1–43.

Nelson, D. R. 1982a. Developmental biology of the Tardigrada. In Harrison, F. W., and Cowden, R. R. (Eds.): Developmental Biology of Freshwater Invertebrates. Alan R. Liss, New York. pp. 363–398.

Nelson, D. R. 2001. Tardigrada. In Thorp, J. H., and Covich, A. P. (Eds.): Ecology and Classification of North American Freshwater Invertebrates. 2nd Edition. Academic Press, New York. pp. 527–550.

Nelson, D. R. (Ed.): 1982b. Proceedings of the Third International Symposium on the Tardigrada. East Tennessee State University Press, Johnson City. 235 pp.

Pollock, L. W. 1976. Tardigrada. Marine Flora and Fauna of the Northeastern United States. NOAA Tech. Report NMFS Circular 394 U.S. Government Printing Office, Washington, DC. 25 pp.

Ramazzotti, G. 1972. Il phylum Tardigrada. 2nd Edition. Mem. Inst. Ital. Idrobiol. 28:1-732.

Renaud-Mornant, J. 1988. Tardigrada. In Higgins, R. P., and Thiel, H. (Eds.): Introduction to the Study of Meiofauna. Smithsonian Institution Press, Washington, DC. pp. 357–364.

Internet Sites

http://member.nifty.ne.jp/angursa/tardigrada (Marine Tardigrada [Heterotardigrada], by Hirokuni Noda. Includes a directory of tardigradologists, a checklist of marine tardigrades with classification, and a PDF of tardigrade classification.)

www.fauna-iberica.mncn.csic.es/htmlfauna/faunibe/zoolist/tardigrada.html (Fauna Iberica: Phylum Tardigrada. An illustrated classification of Spanish/Portugese tardigrades.)

www.geocities.com/RainForest/6135/tardig.html (Line drawings and a photograph of tardigrades.)

www.tardigrades.com (By Martin Mach, Germany, with photo galleries, animations.)

www.microscopy-uk.org.uk/mag/artaug00/ellstard.html (A tardigrade from Scotland by Bill Ells with photographs of an exuvium with eggs.)

16 Introduction to Arthropoda[P]

Familiar and common arthropods include spiders, scorpions, insects, millipedes, crabs, and shrimps in a vast assemblage of segmented animals with exoskeletons and jointed appendages. In fact, most animals are arthropods and although about a million species are known, this is undoubtedly only a small percentage of the total number of living forms, the others as yet undiscovered. Some entomologists believe there are over 30 million species of insects (hexapods) waiting to be found. The number of known species is limited by a shortage of zoologists with the interest, training, and support required to discover and describe them. There are far more arthropods than all other metazoan species combined, and they account for about 80% of all known animals. Their tremendous adaptive diversity has enabled them to survive in virtually every environment, and they are important, and often dominant, in marine, terrestrial, freshwater, and aerial habitats. Arthropods are one of only three extant animal taxa with powered flight (birds, bats, and insects) and one of only two taxa, arthropods and amniotic vertebrates (reptiles, birds, and mammals) with the adaptations necessary for life in dry environments. They are easily the most successful colonizers of terrestrial habitats.

Arthropods are protostome bilaterians and until recently were believed to be closely related to the annelids, with which they share many morphological similarities. New evidence from molecular biology, however, suggests a more distant relationship between these two taxa and, in fact, places the arthropods close to the nematodes and cycloneuralians.

Arthropoda consists of two major extant taxa Chelicerata (horseshoe crabs, sea spiders, and arachnids), and Mandibulata. A third, Trilobitomorpha, is extinct. Mandibulata contains the sister taxa Crustacea (crabs, barnacles, water fleas, and others) and Tracheata (insects, centipedes, millipedes, and others). These taxa share several important characteristics, including a segmented body, a chitinous exoskeleton with ecdysis (molting), jointed and paired segmental appendages, and the absence of locomotory cilia. That these characteristics are shared by all arthropods is evidence of a common ancestor. The evolutionary acquisition of these traits is referred to as **arthropodization.**

Arthropoda, Tardigrada, and Onychophora are generally thought to be closely related and to make up the taxon **Panarthropoda** (Fig. 16-15). Panarthropods are segmented protostomes with a cuticle composed of α-chitin and protein that grow by adding new segments at the posterior end of the body. Onychophora and Tardigrada were discussed in Chapter 15.

FORM

SEGMENTATION

Arthropods, like annelids, are segmented (metameric) animals (Fig. 16-1A, 16-2) composed of a linear series of similar modules, or segments (somites). Segmentation is evident in the embryonic development of arthropods and is a conspicuous feature of most adults. It is especially evident in primitive taxa, but may be obscured in derived arthropods through the loss or fusion of segments. There is a tendency in some arthropods to reduce or eliminate the outward signs of segmentation. This is pronounced in some arachnid taxa and reaches its zenith in the mites, which may show no external indications of segmentation.

In most arthropods, segmentation manifests itself both externally and internally. Typically, the exoskeleton is divided into hardened segmental rings that are clearly visible externally. Successive rings are separated by flexible areas and the pairs of appendages are attached in obvious correspondence with the segments. Internally, the nervous, muscular, hemal, and excretory systems have segmentally repeated components. In derived arthropods there is a tendency for adjacent skeletal rings to fuse together so that individual segments may not be visible externally. Similarly, there is a tendency for internal structures to lose their segmental character.

Within a single individual, structures arising from the equivalent embryonic primordia of different segments and sharing a similar pattern of morphogenesis are said to be **serial homologs.** For example, the antennae and legs of a grasshopper are serial homologs. In the ancestral arthropod, all segments and their appendages were similar or identical in structure, function, and position, a condition known as **homonomy** (Fig. 19-9). Natural selection has resulted in modifications of this initial uniformity and almost all modern arthropods are **heteronomous,** with segments and appendages specialized for many different functions (Fig. 19-19).

The anterior and posterior ends of the arthropod body, the **acron** and **telson,** respectively, are not serially homologous to the segments that lie between them. The first true segment bears the mouth and is immediately posterior to the acron, which is therefore preoral. Similarly, the telson is postanal. As in annelids, growth results from new segments arising from mitotically active teloblast regions immediately anterior to the telson. Segments anterior to the telson are progressively older, and the youngest segment of all is adjacent to the telson. The acron and telson are present in the larva and do not arise from the teloblasts, as true segments do. Many zoologists believe them to be homologous to the annelid prostomium and pygidium, respectively.

TAGMOSIS

Evolution of the arthropod body is characterized by tendencies to tagmosis, heteronomy, cephalization, and fusion of segments. **Tagmosis** (or tagmatization) is the tendency to organize segments into regions having similar structure, function, and appendages. Each group of similar segments is a **tagma** (pl., tagmata). The earliest and simplest tagmosis was the division of the body into an anterior head and a posterior trunk (Fig. 16-1A). This early arthropod condition has been modified in almost all modern taxa, and in most of them, the trunk is divided into a **thorax** and **abdomen,** for a total of three tagmata (Fig. 16-2). In many arthropods, some or all of the thoracic segments unite with the head to form a secondary tagma known as the **cephalothorax.** All hexapods (insects) have a body divided into head, thorax, and abdomen, whereas chelicerates have a cephalothorax and abdomen and crustaceans have various arrangements of head, thorax, and abdomen. These three taxa underwent tagmosis independently and their tagmata are not homologous, even though they bear the same names.

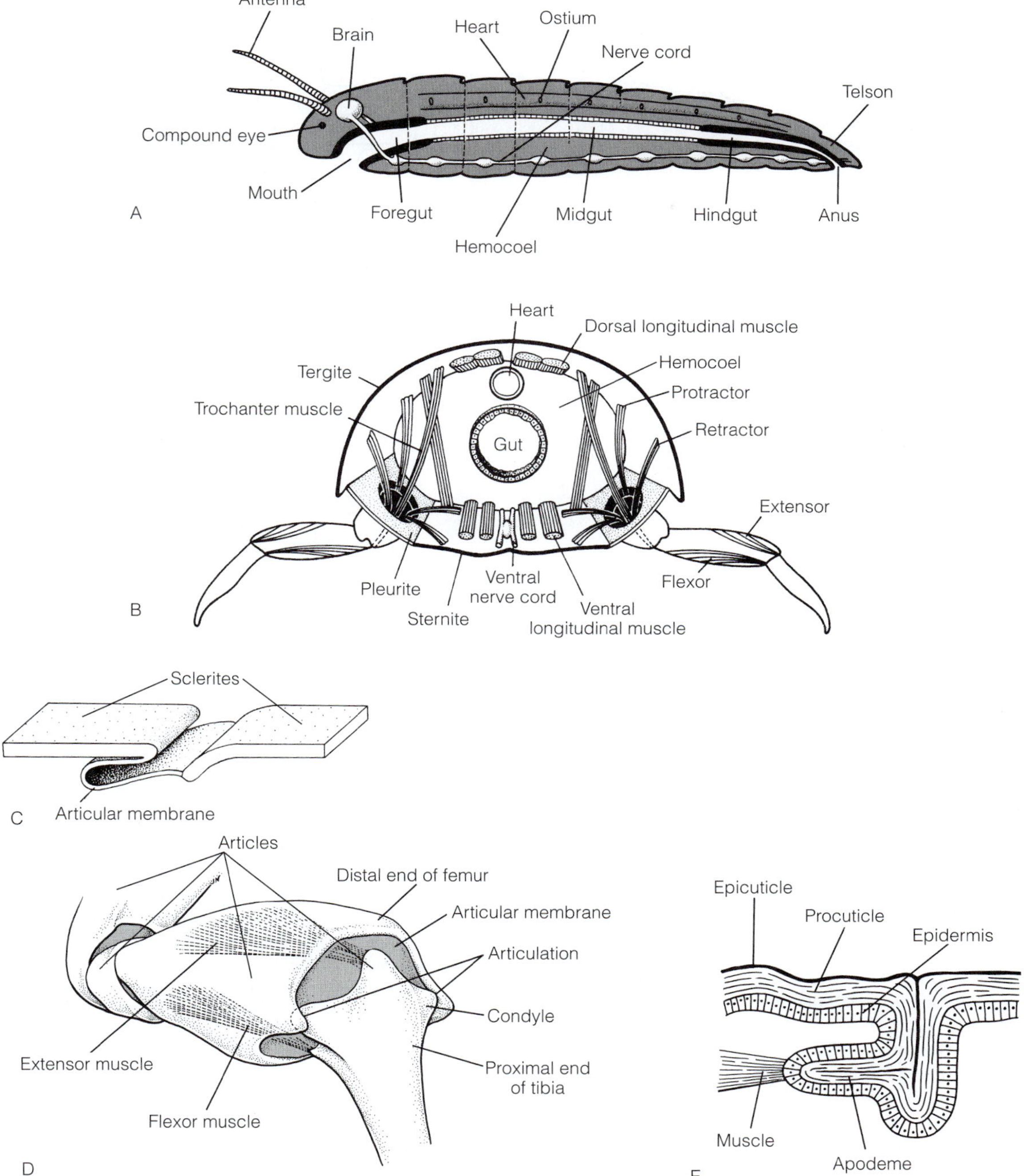

FIGURE 16-1 Structure of a generalized arthropod. **A,** Sagittal section. **B,** Cross section. **C,** Intersegmental articulation. Note the articular membrane folded beneath a sclerite. **D,** Dicondylic leg joint of an insect showing condyles and muscle insertions. **E,** An apodeme. *(C, After Weber from Vandel; B, D, Modified from Snodgrass, 1935; E, After Janet from Vandel)*

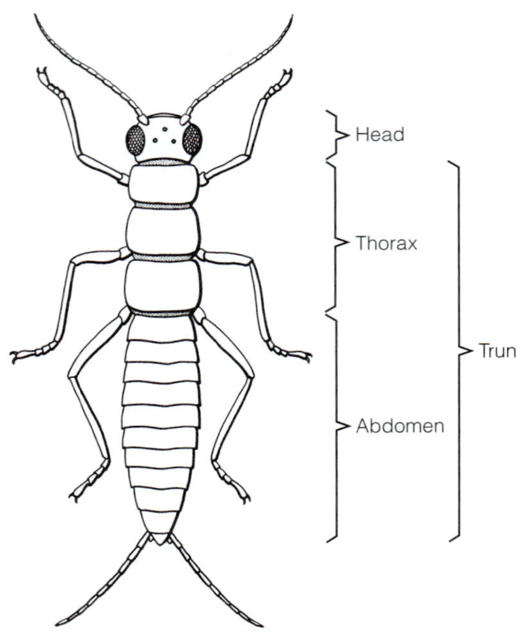

FIGURE 16-2 The basic arthropod tagmata are the head, thorax, and abdomen. The head, as shown here, consists of several fused and externally unrecognizable segments. The thorax here is composed of three appendage-bearing segments, and the abdomen includes eight appendageless segments plus the telson. *(Redrawn from Snodgrass, R. E. 1935. Principles of Insect Morphology. McGraw-Hill, New York. 667 pp.)*

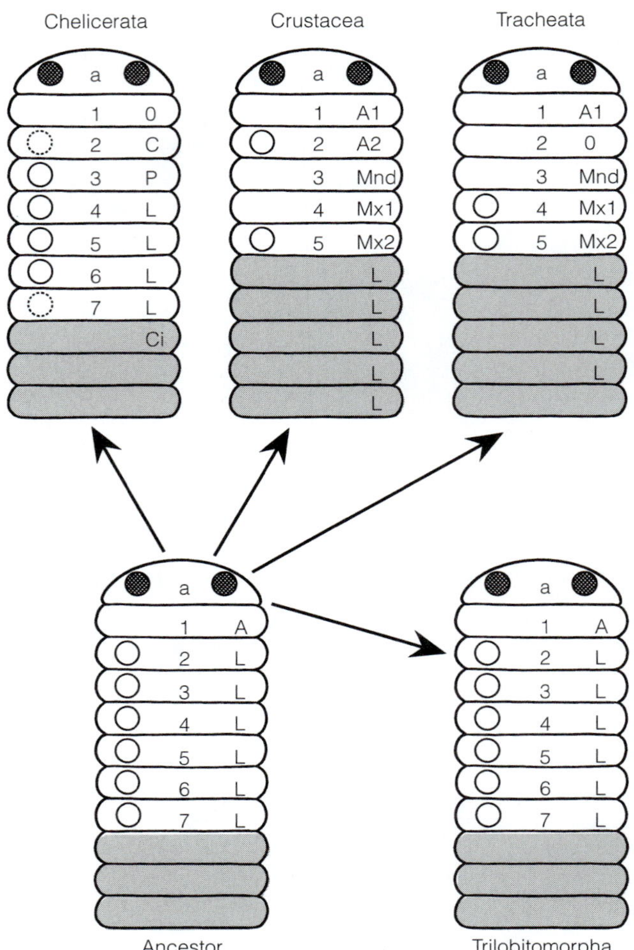

FIGURE 16-3 Evolution of the arthropod head. A = antenna, a = acron, C = chelicera, Ci = chilarium, L = leg, Mnd = mandible, Mx = maxilla, P = pedipalp, and 0 = lost segment. Unshaded segments are those contributing to the head; shaded segments are trunk segments. Solid open circles represent saccate nephridia. Dotted open circles represent embryonic nephridia lost in the adult. Hatched circles are compound eyes. *(Redrawn from Ax, P. 2000. Multicellular Animals: The Phylogenetic System of the Metazoa. Vol. II. Springer, Berlin. 396 pp.)*

CEPHALIZATION

In the early homonomous arthropods the head was scarcely distinguishable from the remainder of the body, and its appendages (other than the antennae) were similar to those of the body segments. **Cephalization,** or the development of an anterior tagma (head, cephalon, or cephalothorax) responsible for sensory reception, neural integration, and feeding, is a strong tendency in arthropod evolution. The head of the ancestral arthropods consisted of the acron and several limb-bearing segments, and the trilobite, chelicerate, and mandibulate heads conform to this pattern (Fig. 16-3). The head or its equivalent (cephalothorax) contains the brain and anterior gut and bears sensory and feeding appendages. The eyes are on the acron.

The anterior appendages and sense organs of the head are served by a concentration of segmental ganglia known as the brain (Fig. 16-10). Because the brain is a neural center responsible for integrating sensory input and motor output, sensory receptors are concentrated on the head to minimize transmission time for sensory input. The mouth and feeding appendages are also located on the head. Each higher taxon of modern arthropods has a characteristic set of head appendages derived from those of the ancestor as indicated in Figure 16-3. The chelicerate cephalothorax includes the acron and six segments that bear the chelicerae, pedipalps, and four pairs of walking legs. The crustacean head comprises five segments with two pairs of antennae, a pair of mandibles, and two pairs of maxillae. The tracheate head consists of four segments that bear the antennae, mandibles, maxilla, and labium. The shared mandibles unite Crustacea and Tracheata in the Mandibulata.

SEGMENTAL APPENDAGES

Paired, jointed, segmental appendages are an arthropod autapomorphy. In the primitive condition, each segment bears a pair of appendages (Fig. 16-1B, 19-9). A similar segmental arrangement is seen in the polychaete annelids, in which the segments bear serially homologous, paired parapodia.

Unlike parapodia, an arthropod appendage is composed of a linear series of **articles** (Fig. 16-1D, 19-3B). Each article is a cylinder of hardened exoskeleton that joins, or articulates, with the body or with another article. Successive articles are connected by flexible cuticle that permits movement of the articles with respect to each other (Fig. 16-1C,D). Flexor and extensor muscles inside the cylinders that extend across the flexible articulations contract to produce movement (Fig. 16-1B).

Although the ancestral arthropod is thought to have been homonomous, most living arthropods are heteronomous and have a variety of specialized appendages (Fig. 19-2B). Specialization is often regional and associated with tagmosis. Head appendages, for example, typically are specialized for sensory reception or feeding. Thoracic and abdominal appendages may be adapted for locomotion, reproduction, respiration, food manipulation, and other functions. Appendages may be lost from some regions or tagmata, and the abdomen in particular tends to have reduced, absent, or highly modified appendages.

One key to the overwhelming success of the arthropods probably is their ability to modify the basic jointed appendage into an almost endless array of specialized structures. A single individual may have a dozen or more different types of appendages, and the taxon as a whole has an enormous variety. Arthropods are like Swiss Army knives, each model equipped with a unique set of specialized tools. A major aspect of arthropod evolution is the functional specialization of their appendages.

The appendages of many arthropods (such as hexapods, myriapods, and arachnids) have a single branch, or **ramus,** and are said to be **uniramous** (Fig. 16-1B, 21-1E). Those of many others (such as crustaceans and trilobites) have two major branches and are said to be **biramous** (Fig. 19-3B). The ancestral arthropod probably had biramous, or perhaps multiramous, appendages. A biramous appendage typically begins with a proximal **protopod,** which often is divided into two articles, a **coxa** and a **basis** (Fig. 19-3B). Forking from the basis are the two rami consisting of a lateral **exopod** and a medial **endopod.** There may be other branches in addition to the endopod and exopod. The uniramous appendages of the modern chelicerates and tracheates probably arose from this ancestral pattern through loss of the exopod.

Arthropod trunk appendages typically are divided into about seven articles, which are given different names in the major taxa. Articles may be lost from some appendages or they may be divided into two or more units in others. The insect tarsus, for example, is often divided. In some cases an article may be divided into numerous short articles to produce a long, flexible, antenniform appendage. The antennae themselves are such appendages, but other anterior or posterior limbs can also be antenniform and serve a sensory role.

The appendages are levers that provide mechanical advantage. In arthropods, appendages with rigid articles are operated by cross-striated muscles capable of contracting much more quickly than the obliquely striated muscles of annelids.

BODY WALL

The outermost layer of the body wall is the nonliving, secreted cuticle (exoskeleton) composed of α-chitin and protein. Collagen, abundant in the cuticle of annelids, is absent from that of arthropods. The living components of the arthropod body wall are reduced and simple. Strength and protection are provided by the exoskeleton, and consequently there is little connective tissue other than the blood. Further, the continuous sheets of unspecialized circular and longitudinal muscles characteristic of annelids have been replaced by specialized muscles that connect adjacent regions of the exoskeleton or extend into appendages.

The cuticle is secreted by the ectodermal epidermis lying immediately below it. The epidermis is a monolayered epithelium that secretes and is underlaid by the basal lamina, or a more elaborate basement membrane, of protein fibers and ground substance (Fig. 16-4) that surrounds and contains the musculature, hemocoel, and viscera. The body cavity of arthropods is a large space in the connective-tissue compartment derived from the blastocoel of the embryo. It is a hemocoel, not a coelom, filled with blood (hemolymph). It is lined not by a mesothelium, but rather by the basal laminae of the gut and epidermis.

CILIA AND FLAGELLA

Because the exoskeleton precludes the use of locomotory cilia on the body surface, such cilia are absent. Flagella are present in the sperm of some taxa and modified cilia are a central feature of most sense organs.

EXOSKELETON

The exoskeleton, or cuticle, is a defining feature of arthropods and its presence has had far-reaching effects on the anatomy, physiology, ecology, and behavior of these animals. The evolution of an exoskeleton compatible with mobility and growth is probably one of the keys to arthropod success. It consists of protein and polymers of α-chitin covalently bonded together. Chitin is a polysaccharide consisting of repeating N-acetylglucosamine monomers. Glucosamine is a glucose molecule with an amine group.

The cuticle protects surfaces from abrasion and from attack by pathogens. It is responsible for structural support and for maintaining the shape of the body. It can be a rigid skeleton that mediates the interactions of antagonistic muscles, but it can also be soft, flexible, and hingelike. The arthropod exoskeleton must be **molted** (shed) for the animal to increase in size or add segments and appendages.

Because cuticle is secreted wherever there is epidermis, it covers the entire outer surface of the body, as well as many surfaces that extend deep into the body. Internal structures derived from invaginations of the embryonic ectoderm are usually lined by epidermis and cuticle. Foregut, hindgut, book lungs, and tracheae, for example, extend deep into the body but are derived from ectoderm and lined with exoskeleton, which must be molted.

The cuticle can be thin and flexible or thick and rigid, and both types may occur in an individual. The cuticle is naturally soft and flexible unless it has been hardened. Hardening can be accomplished by **sclerotization** (*scler-* = hard) or **mineralization.** Sclerotization is accomplished by a chemical reaction known as tanning, which hardens, strengthens, and darkens the cuticle by forming covalent cross-links between protein molecules. Cuticle can also be hardened by the incorporation of calcium salts, a process known as mineralization. Sclerotization occurs in all arthropods and is the only hardening process

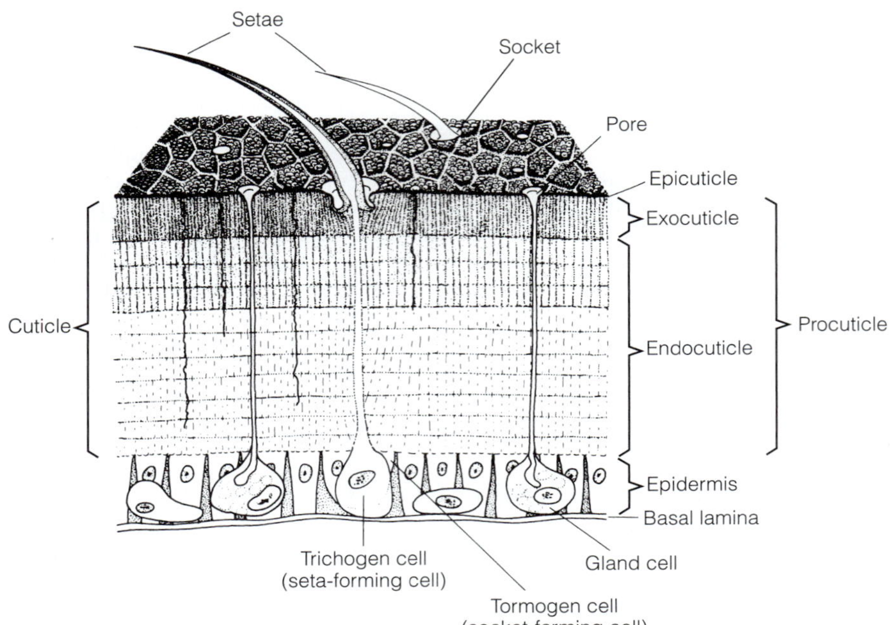

FIGURE 16-4 Diagrammatic section through the arthropod integument. *(From Hackman, R. H. 1971. In Florkin, M., and Scheer, B. T. (Eds.): Chemical Zoology. Vol. VI. Academic Press, New York.)*

in the chelicerates and hexapods. In the crustaceans and diplopods (millipedes), sclerotization may be supplemented by mineralization.

Divisions

Movement is made possible by dividing the hardened cuticle into separate plates known as **sclerites.** Adjoining sclerites and articles are connected by sections of soft, flexible cuticle known as **articular membranes** (Fig. 16-1C,D). Muscles originate on the inner surface of one sclerite, cross an intervening articular membrane, and insert on another sclerite. Contraction of the muscle results in movement of one or both of the sclerites relative to the other.

The cuticle of a typical body segment is divided into four major regions, which may or may not be sclerotized. Articular membranes connect the sclerites to form a complete ring around the segment. The four sclerites of each ring are a dorsal **tergite,** two lateral **pleurites,** and a ventral **sternite** (Fig. 16-1B). Primitively, each body segment has one of these four-part rings, but this pattern is frequently modified by secondary fusion, subdivision, or loss of sclerites. The appendages typically articulate with the lateral pleurites.

In most arthropods the articular membrane between segments is folded beneath the anterior segment to permit expansion when the joint is extended (Fig. 16-1C). Sclerites of adjacent segments may fuse together to form a continuous sheet of rigid, sclerotized exoskeleton. The insect head, for example, is covered by the fused sclerites of the several head segments to form an unbroken, seamless head capsule, or cranium.

The exoskeleton of the appendages is divided into cylindrical articles connected to one another by articular membranes, which create a movable junction (Fig. 16-1B,D). Such a union of skeletal elements is referred to as an **articulation** (or joint) and the two elements are said to be articulated (hence the name "Arthropoda", meaning "jointed feet"). Articulations with flexible articulating membranes permit the articles of the appendages or the segments of the body to move, but articulations can also be rigid. Note that to avoid confusion, the individual units of the body are referred to as segments (or somites), whereas those of the appendages are articles, or podomeres. Segments and articles are quite different and are not homologous.

Articulations usually have specific shapes designed to facilitate a particular type of movement and prevent others. For example, a **monocondylic** joint has one point of articulation (condyle) between its two articles, forming a ball-and-socket joint that permits a wide range of motion similar to that in a human hip or shoulder joint. A **dicondylic** joint has two points of articulation, restricting motion to a single plane (Fig. 16-1D). Dicondylic articulations are hinge joints functionally similar to those of the human elbow or knee.

Although the arthropod skeleton is referred to as an exoskeleton, parts of it form an extensive **endoskeleton.** Sometimes this is due to invagination of the exoskeleton to form an inner projection, or **apodeme** (Fig. 16-1E), continuous with the exoskeleton. Other endoskeletal elements are formed by the sclerotization of connective tissue to form free internal plates that have no connection with the exoskeleton. Both types of endoskeleton provide sites for muscle attachment and transmit forces resulting from muscle contraction.

Composition

The basic composition of the exoskeleton is similar in all arthropods, but there are differences in detail from taxon to taxon. The exoskeleton is secreted by the underlying epidermis in several layers of different chemical composition and function. Its two basic layers are the thin, outer epicuticle and the much thicker inner procuticle (Fig. 16-4).

The **epicuticle** is a thin, complex, water-resistant or waterproof layer of protein, lipoprotein, lipid, and sometimes wax, but no chitin. In terrestrial insects and arachnids, the epicuticle includes an outer layer of wax that renders it impermeable to water and gases. The wax, secreted by gland cells in or near the epidermis and delivered to the surface by ducts, protects the organism from water loss in a dry environment. In freshwater species it prevents or restricts the osmotic influx of water into the organism. In areas such as the tracheae and gill surfaces, where permeability to gases and ammonia is essential, the epicuticle is thin and the wax layer is absent. The waterproof integument is one of the adaptations responsible for the enormous success of the hexapods and chelicerates in terrestrial environments.

The **procuticle** is composed of protein and chitin bound together to form a complex glycoprotein. It is much thicker than the epicuticle and is responsible for most of the thickness and strength of the exoskeleton. It is composed of the outer exocuticle and the inner endocuticle. The proteins of the **exocuticle** may be hardened by sclerotization in places, such as the sclerites, where strength and rigidity are required. In the articular membranes, where flexibility is desired, the exocuticle is not sclerotized. There is much less protein and more chitin in the **endocuticle.** In crustaceans the endocuticle is often mineralized with calcium carbonate.

The body color of arthropods may result from pigments in the cuticle or the subepidermal connective tissue. Cuticular pigments are deposited in the cuticle as it is secreted, and once present their appearance cannot be altered. Subepidermal pigments are contained in cells known as **chromatophores** that are under neurohormonal control (Fig. 19-44). The appearance of the animal can be altered by expanding or contracting the chromatophores. Further, the color of underlying tissues, and even hemoglobin in the blood, may be visible externally if the cuticle is thin and transparent. Finally, the color of the integument is not always due to pigments. Very fine parallel scratches in the epicuticle that refract light of specific wavelengths produce an iridescent color when illuminated by white light, even though no pigments are present.

Setae (sing., seta) are among the most important and varied of the cuticular structures. A seta is an articulated chitinous projection of the exoskeleton (Fig. 13-5B, 16-4, 19-3B) secreted by a trichogen cell in the epidermis. Setae are hairlike, usually flexible, solid or hollow bristles. Their morphological and functional variety is almost limitless. The external sensory receptors of arthropods usually involve modified setae (Fig. 18-8C). Branched, plumose (featherlike) setae are used to increase the surface area of swimming appendages and as filters to extract food particles from water (Fig. 19-5). Dense, feltlike coverings of very fine setae are used by aquatic insects to trap and hold air while the insect is submerged. Heavy, inflexible setae are spines. Arthropod setae differ ultrastructurally and chemically from the β-chitin chaetae of annelids (Fig. 13-5A). Sharp, nonarticulated, immovable projections of the cuticle are not setae and may be referred to as teeth.

Several types of secretory glands or cells are associated with the cuticle. These are located in the epidermis or may be insunk below the basal lamina into the connective-tissue compartment. They connect with the surface via long ducts that traverse the exoskeleton. **Tegumentary glands** may be unicellular or multicellular. They are thought to secrete an enzyme that promotes tanning and are abundant in sclerotized areas of the cuticle. Some tegumentary glands may have other as yet unknown functions. **Trichogen cells** secrete setae, whereas **tormogen cells** create the flexible sockets for the setae (Fig. 13-5B).

Molting

Because of its composition, the arthropod exoskeleton is incapable of stretching to accommodate growth in the animal. Once secreted and hardened it cannot be enlarged. To increase in size, an arthropod must rid itself of its old exoskeleton and replace it with a larger one. This process is referred to as molting, or **ecdysis,** and it is a central feature of arthropod life. Until an animal experiences its final molt, ecdysis is a nearly continuous process, and the animal may spend 90% of its life preparing for a molt, molting, or concluding the preceding molt. Several other animal taxa, notably Nematoda, Nematomorpha, Kinorhyncha, Loricifera, and Priapulida, also have a cuticle that is molted under hormonal control.

Molting has probably been investigated in more detail in decapod crustaceans than in any other arthropods. The crustacean molt cycle has four basic stages designated by letters: intermolt (C), proecdysis (D), ecdysis (E), and postecdysis (A and B).

Intermolt (stage C) is largely free of activities directly involved with molting (Fig. 16-5A) and may be long or short, depending on the species. During intermolt the animal engages in its normal day-to-day activities, including feeding, accumulation of food reserves, and reproduction if it is sexually mature. This is the period in which tissue growth occurs.

Proecdysis (premolt, stages D_1–D_4) is a preparatory phase in which the most extensive changes occur. The animal enters proecdysis from the intermolt condition and exits it with the old cuticle detached but still covering the animal and the new cuticle partially formed. During proecdysis the animal does not feed and subsists on food reserves. Blood calcium levels rise, due largely to the reclamation of calcium from the old exoskeleton. To initiate proecdysis the epidermis secretes a new epicuticle and releases **molting fluid** between the new epicuticle and the old cuticle (Fig. 16-5B). Molting fluid is a mixture of chitinases and proteases. The fluid separates the old cuticle from the incomplete new cuticle by hydrolyzing the old cuticle (Fig. 16-5C,D). The products of hydrolysis are resorbed by the epidermis through the new epicuticle, but the new epicuticle protects the epidermis from the enzymes. A new exocuticle is secreted (Fig. 16-5C) while the old endocuticle is being hydrolyzed throughout this stage (Fig. 16-5D). At this time the animal is encased within old and new exoskeletons, both consisting primarily of epicuticle and exocuticle (Fig. 16-5D), since the new endocuticle has not yet been secreted and the old endocuticle has been hydrolyzed.

Ecdysis (molt, stage E) is the actual shedding of the old exoskeleton. It is usually brief, although difficult and hazardous, and often occurs in a protected retreat or burrow. In certain well-defined areas of the old exoskeleton, the exocuticle is very thin and, once the endocuticle has been removed by hydrolysis, only the thin epicuticle remains. These weak areas are **ecdysial lines** (Fig. 16-5D) and their location is constant for each species.

The uptake of water (in aquatic species) or air (in terrestrial species) raises the blood pressure in the hemocoel and causes the animal to swell, stretching the pliant new cuticle

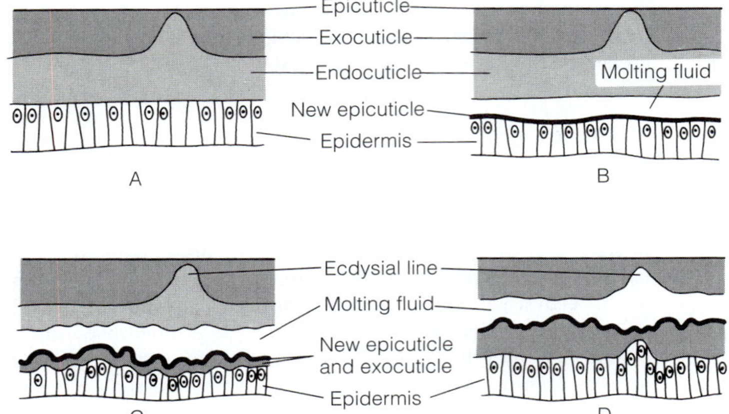

FIGURE 16-5 Molting in an arthropod. **A,** The fully formed exoskeleton and underlying epidermis during an intermolt. **B,** Separation of the epidermis and secretion of molting fluid and the new epicuticle in early proecdysis. **C,** Digestion of the old endocuticle and secretion of the new procuticle. **D,** The animal in late proecdysis, encased within both new and old skeletons. Early postecdysis resembles **D** without the old cuticle.

and splitting the old cuticle along the ecdysial lines. The animal struggles out of the old cuticle and frequently ingests it to reclaim the remaining calcium salts and organic compounds. The new cuticle is soft, wrinkled, and incomplete, and before it hardens it may require additional stretching to accommodate future tissue growth.

Early in **postecdysis** (postmolt, stages A and B) the new cuticle consists of epicuticle and exocuticle and is soft and flexible (stage A), and the animal cannot support itself. The new endocuticle is secreted (stage B) and calcification and sclerotization of the skeleton take place around the water- or air-swollen body). Calcium is transferred from the blood to the cuticle, causing blood calcium levels to drop as the cuticle is calcified. At the same time, proteins are sclerotized. As the new exoskeleton hardens, excess body water is eliminated and the soft tissues shrink away from the now-oversized exoskeleton, leaving room for tissue growth during the upcoming intermolt. The animal remains sequestered in its retreat and does not feed during the first part of this phase. In terrestrial arthropods, epidermal glands secrete wax to waterproof the epicuticle at this time. By the end of postecdysis the new cuticle is almost fully formed, although calcification is not completed until the intermolt.

Molting is a dangerous time for several reasons, and a large percentage of arthropods are unsuccessful in extracting themselves from the exuvium and remain trapped in it until they die or are captured by a predator. Because the muscles have no rigid points of attachment, movement is difficult or impossible immediately during postecdysis, leaving these soft-bodied, feeble animals especially vulnerable to predation, even if they have successfully exited the exuvium. It has been estimated that 80 to 90% of arthropod mortality is in some way associated with molting.

The time between two successive molts is the intermolt, and the organism during this period is an **instar.** The life of an arthropod consists of a series of instars separated by molts. The length of the intermolts becomes longer as the animal becomes older and larger. Some arthropods, such as lobsters and most crabs, continue to molt throughout life, even after achieving sexual maturity. Others, such as insects and spiders, have a more or less fixed number of instars and the animal ceases molting after achieving sexual maturity.

In arthropods, growth is not synonymous with increase in size. True growth is an increase in cells, tissues, and biomass, and it occurs gradually and continuously during the intermolt period, when the animal's size cannot change. In contrast, increase in size is simply an increase in dimensions that occurs abruptly at ecdysis but does not involve the addition of biomass. In most animals the two occur simultaneously, but in arthropods they are sequential. It is during the intermolt speriod that the animal feeds, synthesizes complex polymers, and gradually adds to its store of organic molecules, cells, and tissues. Ecdysis and an increase in size occur abruptly after growth has filled the existing exoskeleton and necessitates its replacement by a larger one.

Molting is under endocrine control**.** The process has been carefully studied and is best understood in insects and crustaceans. The basic pattern seems to be constant throughout the arthropods, but the details differ. The steroid hormone **ecdysone** targets the epidermal cells that secrete the cuticle and controls the molting process in all arthropods. It is secreted by endocrine glands (the prothoracic glands in insects and the Y-organ in crustaceans) and distributed to the epidermis by the blood. The synthesis and release of ecdysone is regulated by neurohormones from sources that vary with taxon.

External sensory receptors always involve a part of the exoskeleton, and these may present special problems during the molting process. Setae associated with the dendrites of sensory neurons form several common types of sensory structures. New setae are secreted beneath the old, but the dendrite remains connected to the old seta until ecdysis.

Muscle connections to the exoskeleton also pose a special problem during ecdysis. The molting animal must maintain a mechanical connection between the muscle and the exoskeleton (old or new) throughout the molting process. Without this connection, it is powerless to move and would be unable to escape from predators or to extract itself from the exuvium, and even with the connection, movement is difficult when the cuticle is flexible. Muscle fibers extend to the epidermis but end there, and do not cross it to insert

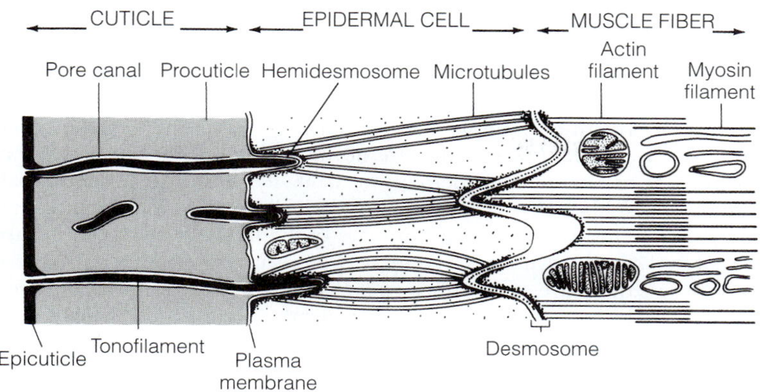

FIGURE 16-6 Diagram of the tendonal attachment of an insect muscle to the skeleton. Not drawn to scale. *(After Caveney, 1969. From Chapman, R. F. 1982. The Insects: Structure and Function. 3rd Edition. Harvard University Press, Cambridge, MA. p. 249.)*

directly on the cuticle. Instead, the muscle is joined to the cuticle by a **tendon** composed of a patch of specialized epidermal cells (tendonal cells) and their secreted fibers (Fig. 16-6). The fibers are two sequential bundles of tension-bearing protein strands spliced together end-to-end. One bundle, composed of intracellular microtubules, spans the tendonal cells and links them to the muscle myofilaments by desmosomes. The second bundle, of extracellular **tonofilaments,** is secreted from the apex of the tendonal cells and attaches them to the cuticle. The tonofilaments are not hydrolyzed during the molting process, and the connection between the muscle and exoskeleton is maintained throughout ecdysis.

MUSCULATURE AND MOVEMENT

The coelomate ancestors of arthropods employed antagonistic sets of circular and longitudinal body-wall muscles working against a hydrostatic skeleton to accomplish motion. For the most part, the arthropods have abandoned the use of a hydrostatic skeleton and instead take mechanical advantage of their leverlike segments and articles. The continuous layers of body-wall muscles have been replaced by a complex array of small, individual, specialized flexor and extensor muscles attached to the exoskeleton (Fig. 16-1B). Arthropod muscles are cross-striated and originate and insert on adjacent sclerites of the body or appendages. Contraction of the muscle results in movement, either flexion or extension, of one sclerite or article with respect to another.

FUNCTIONAL MORPHOLOGY

Each movable articulation of the exoskeleton, whether between body segments or between the articles of an appendage, is equipped with muscles arranged in **antagonistic sets.** Upon contraction, a muscle reverses the action of its antagonist and in the process performs the essential service of restoring the antagonist to its extended condition. Flexors and extensors are the most common antagonistic pairs in arthropods. **Flexors** decrease the angle between two sclerites or articles, whereas **extensors** increase the angle (Fig 16-1D). Flexors usually accomplish work, such as walking or pinching, with respect to the environment, and thus are larger, whereas extensors restore the original condition, and are weaker.

The muscles and skeleton work together in lever systems similar to those of vertebrates. The arthropod musculoskeletal system makes adjustments in the curvature of the body, in the position of the limbs with respect to the body, and in the angles between the articles of a limb. Longitudinal muscles running from segment to segment usually flex the body ventrally and extend it dorsally (Fig. 16-1B). In general, direction of movement of articles or segments is a function not only of the position and orientation of the antagonistic muscles but also of the morphology of the articulation and the relative development of sclerites and articular membranes on opposite sides of the joint. The side with small sclerites and extensive articular membrane, such as the underside of the shrimp abdomen, is flexed by contraction of the powerful ventral longitudinal flexor muscles. The abdomen is extended by the weaker dorsal longitudinal muscles where the extensive dorsal sclerites prevent flexion.

The position of an appendage with respect to the body is adjusted by protractor and retractor muscles that originate on the body sclerites and insert on the proximal articles of the appendage (Fig. 16-1B). The angles between the articles of an appendage are adjusted by intrinsic limb muscles at each articulation. One (or more) of these muscles is an extensor and another is a flexor (Fig. 16-1D). Extension of arthropod appendages may be assisted by the hydrostatic pressure of the hemocoel (the blood pressure). In spiders, for example, extension of the legs is accomplished almost entirely by blood pressure, although flexion is accomplished as usual by flexor muscles. Spiders cannot extend their legs if the hemocoel is ruptured so that the blood pressure cannot be elevated.

Arthropods employ jointed appendages, the body, and wings in locomotion. In aquatic species the appendages may operate as paddles used for swimming, whereas in terrestrial species they are legs for walking. Aquatic species may also use their appendages as legs. Aerial arthropods use wings for flight.

The use of appendages (or wings) for any type of locomotion involves the limb in a cycle of movement consisting of alternating power and recovery strokes. During the **power**

stroke the limb contributes to the motion of the animal, but during the **recovery stroke** the limb is returned to its original position in anticipation of the next power stroke. It is desirable for the limb to have maximal influence on motion during the power stroke but minimal influence during recovery. Accordingly, during the power stroke, contact with the substratum or medium is maximized, whereas it is minimized in the recovery stroke. Otherwise the recovery stroke would undo the work of the power stroke.

Swimming is relatively simple compared to walking or flying. It is well developed in crustaceans, the arthropods best adapted for aquatic life. The appendages of swimming crustaceans typically are ventral and flattened, like oars, to maximize the surface area in contact with the water and therefore the friction. The margins of the appendage are frequently equipped with featherlike setae to further increase the surface area (Fig. 19-15A). On the power stroke, the appendage is moved backward, with its surface perpendicular to the direction of motion and its setae fully extended to maximize surface area and drag. On the recovery stroke, the appendage may be rotated so it is more parallel with the direction of motion and its setae are collapsed to minimize surface area.

Walking also employs segmental appendages and is based on the principles of power and recovery strokes, but in this case contact with the environment is only at the tips of the appendages, where they touch the ground. During the power stroke, the tip is in contact with the ground as the appendage moves backward and the animal forward. During the recovery stroke, the tip is lifted above the ground and the appendage is moved forward, with no effect on the position of the animal.

The power stroke begins with the leg angled forward and the tip extended away from the body and anchored (in contact with the substratum). Early in the power stroke the joints of the leg flex, thus pulling the animal forward toward the anchored tip of the leg. By the conclusion of the power stroke, the leg is angled backward from the body (because the tip of the leg has remained in contact with the ground but the body has moved forward). The leg joints are then extended, pushing the animal forward, away from the still anchored tip, and the leg straightens. During recovery the limb tip is lifted and the limb is moved forward and extended away from the body.

PHYSIOLOGY

Contraction of arthropod skeletal muscles is controlled by several mechanisms and usually differs from that of vertebrates. The many muscle fibers (cells) of a vertebrate skeletal muscle are organized into clusters of fibers known as motor units, each of which is innervated by a single motor neuron. There may be many motor units in a muscle. When the neuron depolarizes, all of the cells of the motor unit contract simultaneously and maximally, following the all-or-none principle. The entire muscle, however, can exhibit a **graded response** in that the strength of its contraction is variable and depends on the number of motor units recruited. Delicate movements are accomplished by using a few motor units whereas actions requiring strength can be accomplished by the same muscle if more motor units are recruited. This system works well in large animals that have large muscles with large numbers of fibers and motor units.

Small animals must also achieve graded responses from their skeletal muscles. Although some large arthropod muscles employ motor units, most do not because they are too small and contain too few fibers. The motor-unit principle requires multiple motor units for recruitment within each muscle. Imagine the size of the muscles of the smallest insect you know of, perhaps a fruit fly from the genetics lab (although many insects are much smaller than fruit flies). Obviously the flexor muscles in the fly's legs, for example, are tiny and must be composed of very, very few fibers. Because there are so few fibers, a graded response cannot be achieved by recruiting motor units. Instead, an arthropod skeletal muscle is innervated by a small number of neurons, each of which may have many junctions with a single fiber or junctions with more than one fiber in the muscle. The muscle fiber *does not* contract completely in response to a single stimulus, as do most vertebrate skeletal motor units. In other words, it does not conform to the all-or-none model. Rather, depolarization of the muscle fiber spreads out from each neuromuscular junction. The distance it spreads depends on the rapidity (frequency) of stimuli received from the neuron. The farther it spreads, the greater the number of sarcomeres recruited and the greater the strength of the contraction. The degree of contraction depends on the extent of muscle depolarization, that is, the number of sarcomeres affected and *not* the number of motor units. A graded response is achieved, but by controlling the number of active sarcomeres rather than the number of active fibers.

Muscle-cell innervation in arthropods is **polyneuronal** in that a given cell typically receives axons from more than one type of motor neuron. Furthermore, there are two quite different types of muscle cells. The two major classes of motor neurons are excitatory and inhibitory, but excitatory neurons can be either phasic or tonic. A given muscle may be innervated by all or a subset of the three neurons. **Excitatory neurons** cause cells to contract, whereas **inhibitory neurons** block contraction. The muscle cell itself may be adapted for either phasic or tonic contractions. **Phasic muscle cells** (fast) are innervated by phasic neurons and **tonic muscle cells** (slow), by tonic neurons. The whole muscle may be made up of one or the other type, but most muscles are mixtures of the two. Phasic and tonic muscle cells differ in ultrastructure and physiology. For example, slow cells have longer sarcomeres and more actin than do fast cells. The terms *fast* and *slow* do not refer to conduction speed, but rather to the rapidity with which the muscle responds to stimulation.

Phasic neurons deliver bursts of closely spaced (high frequency) excitatory impulses that cause phasic muscle cells to respond rapidly with a brief sequence of closely spaced contractions. The muscle contracts quickly and then relaxes quickly. **Tonic neurons** deliver prolonged sequences of low-frequency (widely spaced) impulses having little delay between them. The tonic muscle cell responds with an extended, sustained, powerful contraction. Tonic muscle cells do not respond to isolated nerve impulses. In arthropods, slow contractions are usually much stronger than fast. For example, fast and slow abdominal muscles of the American lobster, *Homarus americanus,* develop tensions of 838 and 4511 g/cm^3, respectively.

Inhibitory neurons hyperpolarize the muscle membrane, thereby reducing the likelihood of contraction even though an excitatory neuron may be sending the opposite command (depolarizing). For example, the movable finger of the crayfish claw is equipped with opener and closer muscles with excitatory and inhibitory neurons. Both muscles are controlled by the same excitatory neuron, but they have different inhibitory neurons. So if, for example, the crayfish wishes to open the claw, a command to contract goes to both muscles, but the closer muscle also receives an inhibitory impulse that prevents it from responding to the excitatory impulse. Only the opener muscle responds, and the finger opens.

COELOM AND MESODERM

The well-developed coelom characteristic of annelids and many other bilaterian taxa has undergone extensive reduction in arthropods in association with the shift from a fluid internal skeleton to a solid external skeleton. The primary functions of the original coelom were locomotion (as a hydrostatic skeleton), excretion (production of an ultrafiltrate of the blood), and reproduction (as a site for the maturation and storage of gametes). In arthropods the exoskeleton replaces the coelomic hydrostatic skeleton in the musculoskeletal locomotory system and thus there is no need to retain a large coelom as a hydrostat. Accordingly, the reduced arthropod coelom is no longer the major body cavity.

Arthropods do not lose the coelom entirely, however, and remnants of it persist in association with the excretory and reproductive systems. During development, segmental, paired, mesodermal masses form from teloblast areas in the posterior embryo. Small coelomic spaces may appear in these mesodermal bands, but most are soon obliterated or coalesce with the blastocoel and contribute to the hemocoel. Some of the embryonic coelomic sacs, however, become excretory organs and gonads. Most of the mesoderm loses its epithelial character and does not form a mesothelium, except as the lining of the gonad and excretory end sac. The mesoderm differentiates into various mesodermal organs and the extensive musculature of the locomotory system.

INTERNAL TRANSPORT

The arthropod hemal system includes a heart, arteries, sinuses, blood, and hemocoel (Fig. 18-2A). The tissues are bathed in the blood that nourishes them. The system functions in transporting nutrients, wastes, hormones, and sometimes gases. In the absence of a coelom, the hemocoel is the functional body cavity of arthropods.

The hemocoel is a space in the connective-tissue compartment derived from the embryonic blastocoel with contributions from the mesoderm. It is filled with blood. The arthropod hemocoel is homologous to the lumina of the hemal systems of other bilaterians, which are also connective-tissue spaces lined by basal lamina and derived from the blastocoel. In the arthropods the hemocoel is not lined by either an endothelium or other cells and usually is not restricted to narrow vessels. A thin, acellular basal lamina typically covers the surface of the tissues of the hemocoel and separates them from the blood. These laminae, when close together, may define channels, or vessels. There are definite patterns of blood flow, and vessels (arteries, veins, and even capillaries) may be present. Since arterial outflow and venous return usually are connected by large hemocoelic spaces, the system is often referred to as being "open," although when capillaries are present the system is "closed" but it is never lined by a cellular endothelium.

The arthropod hemocoel is a large space extending throughout the body and containing the viscera and muscles, which therefore are bathed in blood. It is incompletely partitioned by a perforated horizontal septum known as the **dorsal diaphragm** (Fig. 16-7), but is not divided into segmental compartments. The dorsal diaphragm divides the hemocoel

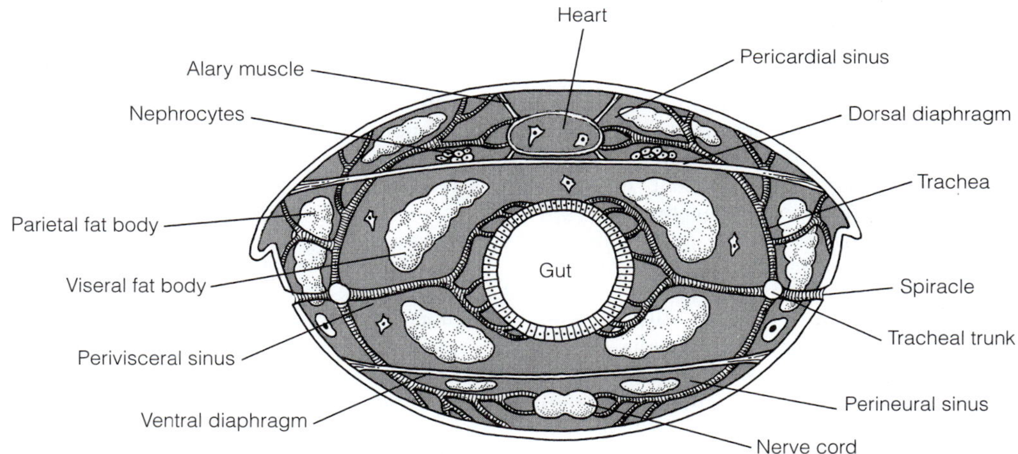

FIGURE 16-7 Cross section of the abdomen of a typical insect. *(Modified from Davies, R. G. 1988. Outlines of Entomology. 7th Edition. Chapman and Hall, London.)*

into two blood-filled sinuses. The dorsal region, known as the **pericardial sinus,** surrounds the heart. It is not a true pericardial cavity, however, since it is not derived from the coelom or lined with mesothelium. The pericardial sinus contains blood, something that is never true of a healthy coelomic pericardial cavity. The ventral region is the **perivisceral sinus,** which surrounds the viscera, including the gut, muscles, fat bodies, gonads, and Malpighian tubules. In some arthropods (such as hexapods), a **ventral diaphragm** separates the larger perivisceral sinus from a small **perineural sinus** lying ventral to it surrounding the ventral nerve cord. The diaphragms are perforated, permitting blood to cross them on its way back to the heart.

The typical arthropod heart is a mid-dorsal, muscular tube in the pericardial sinus (Fig. 18-20). The length of the heart varies, but in the ancestral arthropods it extends the entire length of the body (Fig. 16-1A). The heart wall is composed of circular muscles. Peristaltic contractions of these muscles propagate from posterior to anterior and propel the blood anteriorly in the heart. The heart is perforated by paired **ostia** equipped with one-way valves permitting blood to enter the heart from the pericardial sinus but preventing flow in the reverse direction. In the ancestral condition, each segment has a pair of ostia, but in most arthropods the number is reduced and the segmental arrangement is lost. Elastic **suspensory ligaments** or contractile **alary muscles** extend from the outside of the heart to the nearby exoskeleton or connective tissue (Fig. 16-7). Contraction of these ligaments or muscles dilates the heart.

During early ontogeny, paired coelomic sacs form in each segment and the heart forms in the dorsal mesentery between the two sacs (just as it does in annelids, vertebrates, and other coelomates with a hemal system). Later, the coelomic sacs and mesenteries disappear and are replaced by the hemocoel so that in adult arthropods, the heart is surrounded by hemocoel rather than coelom.

Arteries are tubular extensions from the heart that open into the hemocoel. The largest are sometimes referred to as aortae. The principal arteries are the anterior aorta, paired segmental arteries, and in some taxa, a posterior aorta. In the ancestral condition there is a pair of segmental arteries in each segment, but in derived arthropods the number of pairs is less than the number of segments and the arteries are no longer segmental.

In Crustacea and Chelicerata the blood transports oxygen to the tissues, but in hexapods gas transport employs a tracheal system independent of the hemal system. To assist in gas transport, respiratory pigments are present in many crustaceans, but in few chelicerates and tracheates. The copper-based pigment hemocyanin, by far the most common arthropod respiratory pigment, is widespread in crustaceans and has been reported in some centipedes. Hemoglobin is present in a few crustaceans and insects. As in other invertebrates, but unlike vertebrates, respiratory pigments are in simple solution in the blood plasma, rarely in corpuscles.

The blood transports nutrients from the gut to the tissues, and energy storage compounds may accumulate in the blood. High concentrations of the sugar trehalose are found in insect blood. Amebocytes of various types are present but, in general, there are few cells in arthropod blood. The blood of many arthropods is capable of clotting. The loss of an appendage is a common accidental, or even deliberate, event, and the ability to plug the resulting wound is adaptive. Clotting is initiated by hemocytes when they are subjected to altered conditions. In crustaceans hemocytes release an enzyme that converts the blood protein fibrinogen to fibrin to form the clot.

The pattern of blood flow is relatively constant among the arthropods. **Diastole** is the expansion and filling phase of the heart cycle. Contraction of alary muscles or suspensory ligaments dilates the heart, resulting in negative pressure in the lumen. The ostia open and blood flows from the pericardial sinus into the heart. **Systole,** the contraction and emptying phase of the heart cycle, is caused by contractions of the circular muscles of the heart wall and has several effects. It pressurizes the blood, forcing it out of the heart lumen through the arteries to the hemocoel. The elevated blood pressure in the heart lumen closes the no-return valves of the ostia and prevents backflow into the pericardial sinus. Finally, systole stretches (antagonizes) the alary muscles and suspensory ligaments. The stretched ligaments store energy that will be used to accomplish diastole.

Pressurized blood exits the open ends of the arteries and enters the hemocoel. In the hemocoel it flows posteriorly in the perivisceral (and perineural, if present) sinus, where it bathes the tissues and exchanges materials with them. The blood moves dorsally through perforations in the dorsal (and ventral, if present) diaphragm and eventually enters the pericardial sinus. During diastole it enters the heart through the ostia. Many arachnids and crustaceans have special channels to return oxygenated blood from the respiratory organs directly to the pericardial sinus without allowing it to mix with unoxygenated blood in the perivisceral hemocoel (Fig. 18-20).

EXCRETION

Two quite different types of excretory organs are found in arthropods, and their occurrence is largely correlated with habitat. **Saccate nephridia** are characteristic of aquatic arthropods, especially crustaceans and some chelicerates, but also occur in many terrestrial chelicerates and a few terrestrial tracheates (chilopods, diplopods, and collembolans). In aquatic arthropods, which have no need for water conservation, nitrogen metabolism usually is ammonotelic and nitrogen excretion is accomplished across any permeable surface, typically the gills. The excretory organs function primarily in maintaining ion balance and fluid volume and have little or no role in the elimination of metabolic wastes. They are usually best developed in freshwater species that must counter a heavy osmotic influx of water.

Saccate nephridia are known by a variety of names, including end sac organs, sacculi, coxal glands, green glands, antennal glands, and maxillary glands. They are derived from the coelomic spaces and metanephridial systems of the coelomate ancestors of the arthropods and develop in the embryo from coelomic pouches in paired segmental masses of mesoderm. Saccate nephridia are filtration kidneys consisting of a **tubule,** derived from a metanephridium, draining a small coelomic space, the **end sac.** The nephridium is surrounded by blood, which is pressurized by the heart (Fig. 16-8A, 19-6). The walls of the end sac are derived from the mesothelial basal lamina and podocytes, which form the filter (Fig. 16-8B).

Blood pressure in the hemocoel forces blood across the collagen mesh of the basal lamina and an ultrafiltrate (primary urine) forms in the end sac. The tubule connects the end sac with a nephridiopore on the exterior. The ultrafiltrate is modified as it flows along the tubule to the nephridiopore, and by the time it arrives, its valuable constituents have been reclaimed and additional wastes have been secreted into it. The composition of this final urine is very different from that of the primary urine.

The well-developed saccate nephridia of freshwater crustaceans are adapted primarily for osmoregulation in hyposmotic environments. They conserve ions and eliminate water by producing copious urine. Large volumes of ultrafiltrate form in the end sac and then move through the tubule to the nephridiopore. The tubule epithelium is specialized to reclaim ions, and often a bladder exists for storing the final urine (Fig. 19-6B). The tubule of the crayfish antennal gland, for example, resorbs sufficient ions to reduce the chloride concentration of the urine from about 200 μM/L in the end sac to about 10 μM/L at the nephridiopore.

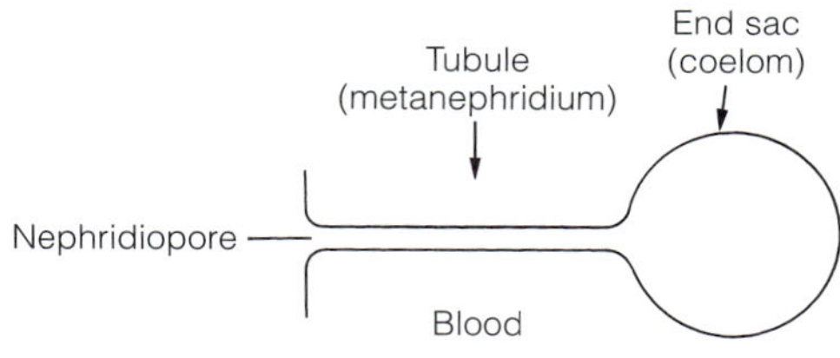

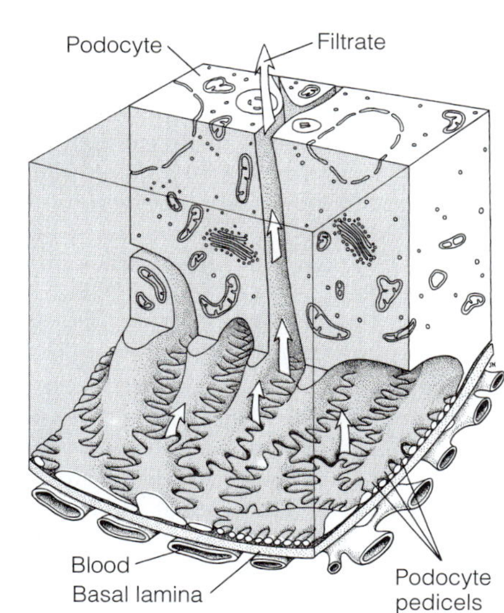

FIGURE 16-8 Diagram of an arthropod saccate nephridium derived from a nephridium and coelom remnant. **A,** A generalized saccule. Podocytes and a basal lamina form part of the end sac lining. **B,** Detail of the end sac wall of a crayfish, showing the flow of ultrafiltrate across the basal lamina and podocytes. *(From Kümmel, G. 1967. Die Podocyten. Zoolog. Beitr. N. F. 13:245–263)*

One of the reasons crustaceans have not been very successful in terrestrial habitats may be their failure to develop a method of eliminating waste nitrogen that does not waste water. Crustaceans, including the terrestrial pill bugs, largely rely on ammonia as the nitrogenous end product. The high toxicity of the ammonia molecule requires rapid dilution with abundant water, which most terrestrial arthropods do not have.

In the majority of terrestrial arthropods, nitrogen metabolism is uricotelic and excretion is accomplished by a second type of excretory organ, the **Malpighian tubules.** These are new structures, invented independently by several taxa of terrestrial arthropods to conserve water.

Malpighian tubules are characteristic of terrestrial arthropods and are the principal or only excretory organs of insects and arachnids, although saccate nephridia are present in some members of both taxa. Myriapods and some arachnids, including some spiders, have both saccate nephridia and Malpighian tubules. An excretory system that does not waste water is a prerequisite to successful colonization of land and such systems are found in terrestrial arthropods and vertebrates. Whereas vertebrates employ nephrons, which are modified metanephridia, terrestrial arthropods invented Malpighian tubules for this purpose. The efficient excretory system of the tracheates, arachnids, and amniotic vertebrates is one reason for their unparalleled success on land.

Malpighian tubules excrete nitrogen in the form of insoluble uric acid or guanine. These nitrogenous bases are synthesized as the end product of nitrogen—especially nucleic acid—metabolism. Both molecules are nearly insoluble in water and are considerably less toxic than ammonia. They do not require dilution to render them nontoxic.

Malpighian tubules are blind-ended tubular diverticula of the gut located at or near the midgut-hindgut junction. They extend into the hemocoel, where they are bathed in blood (Fig. 16-9, 21-8A,C). They are secretion kidneys that require no coelomic space, no source of pressure, and no filter. Metabolizing tissues secrete nitrogenous wastes as soluble uric acid or guanine into the blood. The waste molecules are absorbed from the blood by the tubule epithelium and secreted into the tubule lumen. From there they move to the lumen of the gut, where they precipitate in the low pH environment found there. The precipitate is incorporated into the feces and eliminated through the anus (Fig. 21-9). The muscular Malpighian tubules writhe sinuously in the hemocoel to maintain a high concentration gradient of uric acid across their walls.

The presence of Malpighian tubules in tracheates and arachnids is probably a convergence in these two unrelated terrestrial taxa. These two taxa colonized the land independently, but faced the same challenge of water conservation and evolved similar solutions to the problem.

Nephrocytes are excretory cells that are common in arthropods. They are large pinocytic cells in the hemocoel (Fig. 16-7) that process wastes and toxins that the nephridia or Malpighian

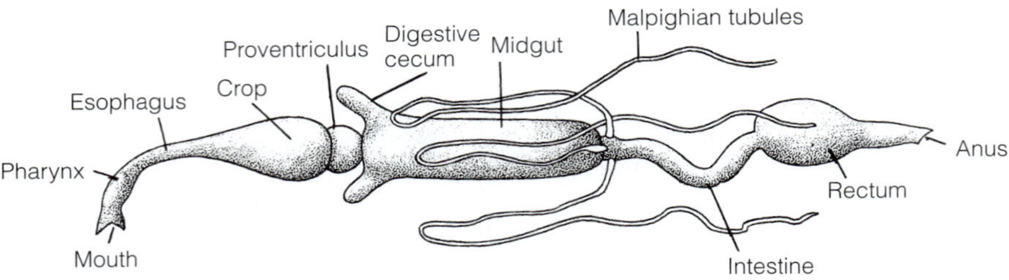

FIGURE 16-9 Diagram of a typical hexapod digestive system. *(After Snodgrass, 1935.)*

tubules are unable to eliminate. Nephrocytes may metabolize the waste and return it to the hemocoel in a form that can be eliminated by the principal kidneys, or they may function as storage kidneys and sequester it indefinitely.

GAS EXCHANGE

In small arthropods (with a high surface area-to-volume ratio) the unspecialized body surface is usually sufficient to supply adequate amounts of oxygen. Larger arthropods, however, require special gas exchange surfaces to provide the area needed to absorb enough oxygen for metabolic needs. Gas exchange structures are extremely varied in arthropods and will be discussed in more detail in subsequent chapters in conjunction with individual taxa. Arthropod respiratory surfaces are epidermal derivatives that must be moist. The cuticle is thin and permeable, and its epicuticle is neither waxed nor waterproof. Aquatic arthropods typically employ gills or book gills, which are evaginations of the body surface, whereas terrestrial species utilize invaginations such as book lungs or tracheae.

Mechanisms for transporting gases between the respiratory surfaces and tissues are also varied. The hemal system may be responsible for this transport, or the respiratory organs themselves may deliver oxygen directly to the tissues.

Gills of many types are found in Crustacea and aquatic insects. In Crustacea, gills are usually modified appendages (Fig 19-36, 19-37). **Book gills,** characteristic of horseshoe crabs, are flat, pagelike lamellae extending from the ventral surface of the abdomen (Fig. 18-2A). Oxygen transport to the tissues is the responsibility of the well-developed hemal system.

Book lungs are present in many arachnids. A book lung is an invaginated pocket of the exoskeleton that contains numerous secondary evaginations. The evaginations are flat, leaflike lamellae whose function is to increase surface area for gas exchange (Fig. 18-20). Oxygen transport from the lungs to the tissues is through the blood.

Tracheae occur in arachnids, onychophorans, and tracheates. These tubular invaginations of the epidermis and cuticle extend inward from an opening, the **spiracle,** on the body surface. Tracheae branch into ever-smaller tubes, ultimately delivering oxygen directly to individual cells (Fig. 21-10) without the mediation of the hemal system in most cases. Tracheae, like Malpighian tubules, are thought to have evolved several times in terrestrial arthropods and their occurrence in the arachnids and tracheates likely resulted from convergence, not common ancestry. Some may have evolved from apodemes, which also are invaginations of the exoskeleton, and others from book lungs.

NUTRITION

Like that of most bilaterians, the arthropod gut consists of three regions (the foregut, midgut, and hindgut), each having a characteristic histology, function, and developmental origin (Fig. 16-1A). The foregut and hindgut are derived from the ectodermal stomodeum and proctodeum, respectively, and only the midgut is endodermal. As ectodermal derivatives, the foregut and hindgut are lined by epidermis and cuticle. The midgut is lined by gastrodermis and has no cuticle.

In general, the foregut is responsible for ingestion, storage, and the initial processing of food prior to chemical digestion. Most mechanical digestion (trituration) occurs in the foregut or prior to ingestion. The midgut functions in enzyme secretion, hydrolysis (chemical digestion), and absorption of the products of hydrolysis. It usually has diverticula, known as digestive ceca (or the digestive glands, hepatopancreas, liver, midgut ceca, or midgut diverticula), to increase the surface area for absorption and intracellular digestion. The ceca may be extremely large and occupy much of the hemocoel. In some arthropods enzymes from the midgut move anteriorly into the foregut or even outside the mouth, initiating hydrolysis in the foregut or prior to ingestion instead of, or in addition to, the midgut. The hindgut is responsible for feces formation, feces storage, and the reclamation of valuable materials, especially water. It sometimes has a respiratory role.

The gut is regionally specialized to accomplish its varied functions (Fig. 16-9, 21-8A,C). The foregut typically includes a pharynx, esophagus, crop, and proventriculus (stomach). Its epithelium often secretes elaborate, sclerotized, cuticular structures such as teeth and setae (Fig. 21-8B). The midgut usually is not subdivided, but has at least one pair of digestive ceca. A valve separates the foregut from the midgut (Fig. 21-8B) and another lies between the midgut and the hindgut. The epithelium in the vicinity of the foregut-midgut valve secretes a thin, permeable **peritrophic membrane** around the food mass. The peritrophic membrane protects the delicate walls of the midgut from abrasion by the food and plays an important role in localizing digestive enzymes. The hindgut may be divided into an intestine and rectum. The hexapod gut is capable of transferring water from its lumen back to the blood, thereby conserving this valuable commodity (Fig. 21-9).

This is another important osmoregulatory adaptation that contributed to the success of the insects in terrestrial habitats. The mammalian colon has the same ability, and for the same reason.

NERVOUS SYSTEM

The arthropod nervous system is similar to that of Annelida, the other major taxon of segmented protostomes. In fact, similarities between arthropod and annelid nervous systems have long been taken as evidence of a close evolutionary relationship between the two taxa. In light of molecular evidence suggesting that annelids and arthropods are not related, it may be that the similarities are convergent and simply reflect the optimal neural design for a segmented protostome.

The arthropod (and annelid) central nervous system consists of a dorsal supraesophageal ganglion, or brain, in the head, a pair of circumenteric connectives that encircle the gut just posterior to the brain, and a paired, ventral, longitudinal nerve cord with paired segmental ganglia, and segmental sensory and motor nerves (Fig. 16-1, 21-7, 16-11).

The arthropod brain (supraesophageal ganglion) is a syncerebrum consisting of the fused, paired ganglia (neuromeres) of two or three anterior head segments (Fig. 16-10). The **protocerebrum,** the paired ganglia of the acron, is the anteriormost region of the brain. It receives sensory nerves (optic) from the various types of arthropod eyes (Fig. 16-11, 21-7) and processes optical input. The commissure connecting the ganglia of the right and left sides of the protocerebrum is preoral, or anterior to the mouth.

The **deutocerebrum,** the second pair of neuromeres, is the integration center for the first head segment (Fig. 16-10A), when present. It connects with the sensory and motor nerves of the antennae of crustaceans and tracheates. Its commissure, like that of the protocerebrum, is preoral. Chelicerates have neither antennae nor deutocerebrum (Fig. 16-10B).

The **tritocerebrum** is the third pair of neuromeres, and it belongs to the second head segment, which is the first postantennal segment. This segment bears the chelicerae in chelicerates and the second antennae in crustaceans, but has no appendages in tracheates. The tritocerebral commissure connecting the right and left neuromeres of the tritocerebrum is postoral, that is, posterior to the foregut. The tritocerebrum is the ganglion of a postoral segment that has shifted anteriorly, as evidenced by its postoral commissure.

Circumenteric connectives extend posteriorly and ventrally from the tritocerebrum, around the gut to the first pair of segmental ganglia. The circumenteric connectives may coalesce with the tritocerebral commissure, which then appears to be a single ring around the gut (Fig. 16-10), or they may be independent of it (Fig. 21-7). The evolution of the head and brain in arthropods is currently an area of active study, and the traditional account is likely to be revised in the next few years.

A ganglion, or neuromere, is a mass of nervous tissue, both cell bodies and axons, as well as glial cells. In general, arthropod ganglia have their cell bodies in a cortical (peripheral) layer whereas the axons are medullary, in the interior of the ganglion. The concentrations of axons—called neuropils—are the locations of most synapses. Most arthropod ganglia occur in pairs connected by a transverse commissure. In an evolutionary trend known as **medial fusion,** the two ganglia of a pair move to the midline and may coalesce to form what appears to be a single, although sometimes bilobed, median ganglion. In reality, the median ganglion is still a pair of ganglia, but one with an exceedingly short commissure. This is true of the major ganglia of the brain, the segmental ganglia of the ventral nerve cord, and the paired connectives of the ventral nerve cord itself.

The arthropod nervous system includes the central nervous system (CNS), consisting of the brain and ventral nerve cord, and the **peripheral nervous system** (PNS), consisting of the segmental nerves connecting muscles and sense organs with the CNS. Sensory and motor neurons in the nerves of the PNS take sensory information to the CNS and carry motor commands

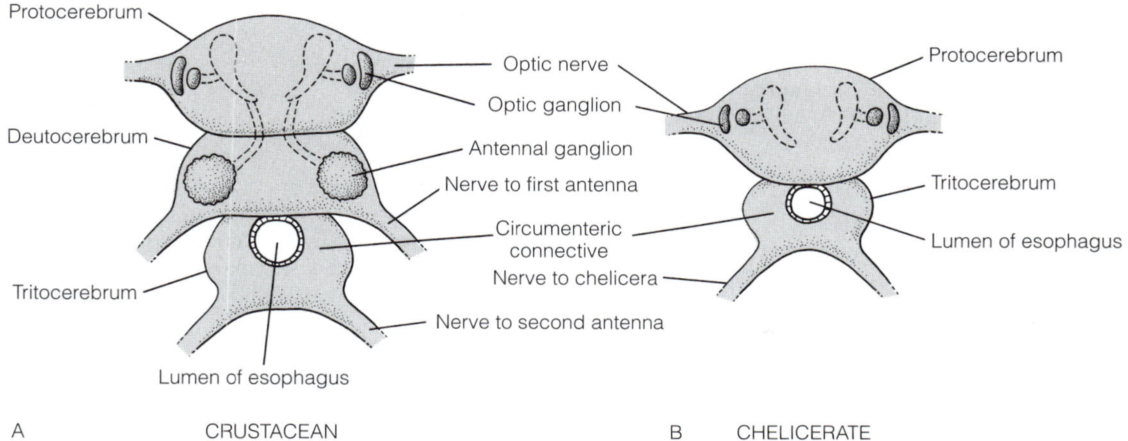

FIGURE 16-10 Arthropod brains. **A,** A crustacean. **B,** A chelicerate. *(Both after Hanstrom from Vandel.)*

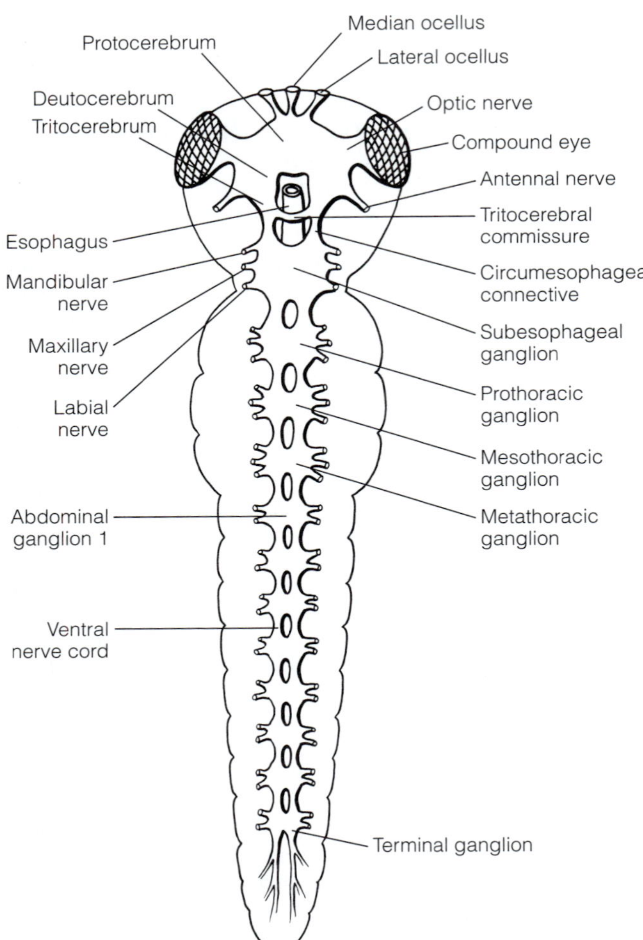

FIGURE 16-11 Ventral view of the central nervous system of a generalized insect. *(Adapted from Klausnitzer, B. 1996. Insects. In Westheide, W., and Reiger, R. M. (Eds.): Spezielle Zoologie I. Einzeller und Wirbellose Tiere. Gustav Fischer Verlag, Stuttgart. pp. 601–681.)*

away from it. Interneurons are restricted to the CNS, where they mediate the interactions between sensory and motor neurons. In addition to the somatic nervous system mediating interactions with the external world, a **somatogastric nervous system,** with its own sensory and motor neurons, monitors the visceral organs and controls internal functions.

The evolution of the arthropod brain was largely a process of cephalization in which the anterior end of the nervous system became increasingly important as a control center. The brain integrates sensory input with motor output and coordinates the activity of the segmental appendages. It is associated with the development of effective, efficient sense organs, such as antennae and eyes, in the head. The brain of the ancestor probably was simply the protocerebrum in the acron, but it enlarged as segmental ganglia of more and more posterior segments became consolidated with the protocerebrum. An early result was the three-part supraesophageal ganglion but the trend has progressed much farther in many arthropod taxa. The ganglia of additional head segments are often associated, both structurally and functionally, with the brain. In most arthropods the ganglia of the posterior head segments (such as those of the mandibles, maxillae, labium, and others) coalesce to form a **subesophageal ganglion** that serves the posterior head appendages (Fig 16-11). The supraesophageal and subesophageal ganglia are joined by the circumenteric connectives that form the nerve ring around the anterior gut. These two large ganglionic masses are closely integrated both structurally and neurologically and function together as a brain.

In most derived arthropods, a variable number of thoracic, and even abdominal, ganglia coalesce with the subesophageal ganglion. The result is an anterior control center consisting of the original three-part supraesophageal ganglion dorsal to the gut and the enhanced subesophageal ganglion ventral to it. When it is this inclusive, the subesophageal ganglion is sometimes referred to as the **thoracic ganglion.** In spiders (and whip spiders) all segmental ganglia are incorporated into the subesophageal ganglion to produce, in conjunction with the supraesophageal ganglion, a massive ganglionic concentration that occupies most of the space in the cephalothorax (Fig. 18-20, 18-7B).

In the ancestral arthropod the trunk segments are homonomous and each is equipped with two jointed appendages. Each appendage is under the control of its own segmental ganglion that lays on the floor of the hemocoel at the base of the appendage. Three pairs of segmental nerves, with both sensory and motor neurons, connect the ganglion with its appendages. Coordination of the right and left appendages of a segment is accomplished by a segmental commissure, a bundle of axons extending transversely across the segment to connect the right and left ganglia of that segment. Coordination of the appendages with those anterior and posterior to the segment is accomplished by longitudinal connectives extending anterior and posterior to connect the ganglia of adjacent segments. The two series of successive connectives are the two parallel, longitudinal ventral nerve cords (Fig. 16-11). Anteriorly, they connect with the subesophageal ganglion or the brain. The result is the characteristic ladderlike nervous system of arthropods (and annelids). The sidepieces of the ladder are the two longitudinal connectives and the rungs are the transverse commissures. The longitudinal connectives in many taxa are coalesced through medial fusion to form a single median ganglionated nerve cord (Fig. 18-7A).

SENSE ORGANS

Arthropods, like other metazoans, depend for survival on the input of sensory information from their external and internal environments. For this they have a sensory system consisting of exoreceptors to detect stimuli from the external environment and endoreceptors to monitor the internal environment.

EXORECEPTORS

Exoreceptors (or exteroreceptors) include receptors for light (photoreception), taste and smell (chemoreception), equilibrium, touch and vibrations (mechanoreception), and pressure. The inert, nonliving arthropod exoskeleton isolates the animal and its nervous system from the external environment. Providing the brain with sensory input from outside the body requires that the exoskeleton be penetrated in some manner. The arthropods have not just penetrated the exoskeleton, they have modified it, in a multitude of ways, so that it has become part of the sensory system. Thus, instead of being a barrier, the cuticle has become an elaborate array of specialized sensory structures for detecting a variety of environmental events.

Arthropod exoreceptors inevitably incorporate a specialized part of the cuticle that helps transduce a stimulus into perturbations of the neuronal membrane potential that are meaningful to the central nervous system. Exoreceptors are sensilla consisting of a modified region of the cuticle associated with a derived cilium, sensory neurons, and support cells. Each is specialized structurally to respond to a specific stimulus such as touch, a pheromone, sound, vibrations, and light, among others. The sensillum is often a hollow seta, but some are cuticular pegs, plates, scales, membranes, or pits. It may be that all setae are sensilla of some type, although this has not been demonstrated. An arthropod's body literally bristles with batteries of specialized sensilla, only a few of which can be mentioned here.

Simple **mechanoreceptors** consist of a hollow seta with one or more sensory neurons inside. Deflection of the seta from its resting position affects the polarity of the neuronal membrane, and the brain (or appropriate ganglion) is informed of the disturbance. The amount of pressure required to deflect the seta and activate the neuron is related to the size of the seta. Fine setae detect gentle pressures such as weak air currents and robust setae respond to heavy pressures. **Trichobothria** are extremely sensitive mechanoreceptors specialized for detection of weak, low-velocity air currents (Fig. 18-8E).

Chemoreceptors detect chemicals from distant (by smell, or olfaction) or contact (by taste, or gustation) sources. Chemoreception requires a thin, permeable or perforated cuticle that permits passage of the chemical to the membrane of a chemosensory neuron. A reaction between the stimulus molecule and specific receptor proteins on the neuronal membrane affects membrane potential and sends a signal to the brain. Chemoreceptive sensilla (Fig. 18-8C, 16-12C), which often resemble mechanoreceptive setae, are modified hollow setae with substrate-specific chemosensory neurons inside (Fig. 18-8C). There may also be more elaborate chemosensory peg organs with numerous sensory neurons (Fig. 16-12A).

Equilibrium receptors for the detection of gravity, balance, and acceleration seem to be absent in most arthropods, with the exception of some malacostracan crustaceans (including crabs, shrimps, lobsters, crayfishes, and mysids). It is useful for a mobile organism to know which way is up, but most arthropods apparently rely on the direction of the light for this. Many malacostracans, however, are equipped with statocysts, analogous to the vestibule of the vertebrate inner ear, for the detection of a gravitational field (Fig. 19-7B).

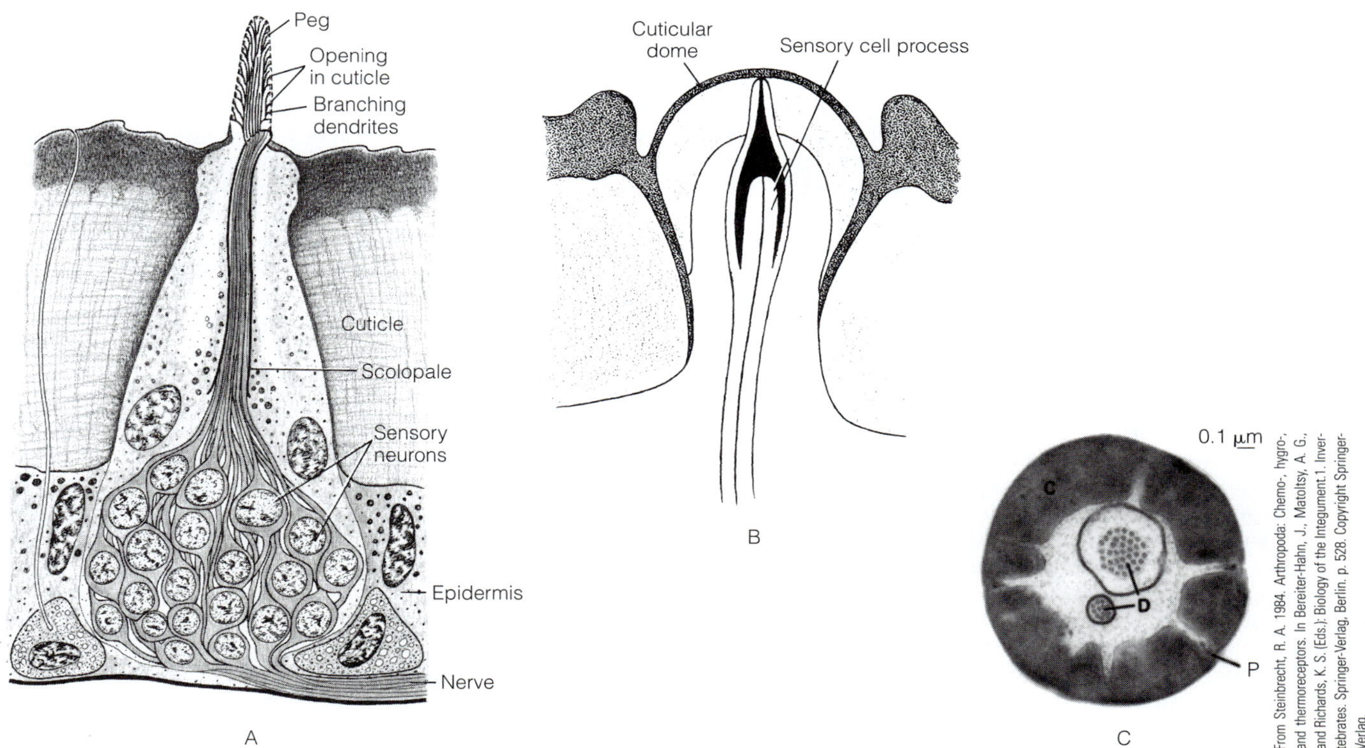

FIGURE 16-12 Sensilla. **A,** A chemosensory peg organ from the antenna of a grasshopper. **B,** A campaniform sense organ. **C,** Cross section of an olfactory seta from the antenna of a male moth. The thick cuticular wall (C) is perforated by pores (P) and contains two dendrites (D). *(A, After Slifer et al.; B, After Snodgrass, 1935)*

Chordotonal organs are sensilla with thin strands of tissue or cuticle stretched between two thicker regions of the cuticle. A dendrite is attached to the strand, and deformation of the strand depolarizes the dendrite membrane. Chordotonal organs detect changes in tension (stretch) due to pressure changes, perhaps in the air in the tracheae, blood in the hemocoel, or water surrounding an aquatic insect.

Some arthropods can detect vibrations in air as sound using modified setae or tympanal organs, which are chordotonal organs adapted for detecting sound waves in air. These sensilla are present in several insect taxa (cicadas, crickets, grasshoppers, and moths) and usually receive acoustic signals from conspecifics in territoriality or courtship interactions. It is no coincidence that most, but not all, insects with tympanal organs are also capable of producing sound.

A tympanal organ is a modified area of the tracheal system that has fused with the exoskeleton. Like the rest of the tracheal system, it is filled with air. One wall of the organ is a **tympanic membrane** (or tympanum) formed of a thin sheet of cuticle that functions like an eardrum in response to vibrations in air. The sensory dendrites attached to the inner surface of the tympanum depolarize in response to vibrations of the membrane. Tympanal organs may be located on the forelegs (as in crickets and long-horned grasshoppers), first abdominal segment (as in short-horned grasshoppers and cicadas), or metathorax (as in moths). Although moths do not generate sound, some have tympanal organs to detect the echolocation hunting signals generated by bats. Upon hearing these signals, the moth takes evasive action to avoid capture. Some arthropods use specialized setae to detect sound. In some flies, the setae are on the antennae.

Campaniform organs are highly modified sensilla in which the seta forms a cuticular dome to which a sensory dendrite is attached (Fig. 16-12B). The dendrite is sensitive to deformation of the dome. In cockroaches campaniform organs on the tarsi are responsible for inhibition and excitation of the "righting response" by which a roach on its back turns over, or rights itself. When the roach is upright, the dome of the campaniform organ is deformed and a continuous signal is sent to the brain to inhibit the righting response. Should the insect be turned on its back, the campaniform organs are no longer deformed and the signal to the brain ceases, causing the insect to begin waving its legs to gain purchase on the substratum to right itself.

Hygroreceptors for the detection of humidity are important in terrestrial arthropods. These sensors typically occur in association with a thermoreceptor cell, and the resulting sense organ is a **thermo-hygroreceptive sensillum.** Present in all hexapod orders that have been studied, they usually are on the antennae, which may have from 2 to 70 of them. The sensillum consists of a hollow cuticular peg containing several support cells and three sensory cells. The sensory cells are a moisture-sensitive hygroreceptor, a dryness-sensitive hygroreceptor, and a thermoreceptor. The moisture-sensitive cell increases its signal frequency with increasing moisture, whereas the dryness-sensitive receptor decreases its impulse frequency with increasing humidity. The thermoreceptor increases its signal frequency with decreasing temperature. The mechanism by which these receptors detect humidity and temperature is not known.

Photoreceptors, or eyes, are widespread and often well developed and sophisticated in arthropods. Two different types of eyes, median pigment-cup ocelli and lateral compound eyes, are found in the arthropods. Both contain rhabdomeric photoreceptor cells connected with the protocerebrum by optic nerves, but they differ in their morphology and function. Both may be present in the same individual.

Pigment-cup ocelli are highly variable among arthropods and range from relatively simple receptors like the naupliar eyes of crustaceans to the complex, image-forming main eyes of spiders. Unlike compound eyes, all the receptor units (retinal cells) of an ocellus share a common lens. Median ocelli are present in arachnids, pycnogonids, crustaceans, and hexapods. Myriapods have lost all median ocelli, and their eyes are derived from lateral compound eyes.

A typical pigment-cup ocellus consists of a cup-shaped layer of pigment-containing cells enclosing a cup-shaped layer of photoreceptor cells known as the retina (Fig. 6-8D,E). The concavity of the cup is oriented toward the environment from which light is received. The pigment layer prevents the entry of light from any other direction so the animal can deduce the direction of the light source. A cuticular lens (cornea) may be associated with the opening of the cup.

Pigment-cup ocelli are physically associated with the exoskeleton and epidermis, from which they are derived. The lens is a thickened area of cuticle (Fig. 18-8A,B) whereas the retina and pigmented layer are derived from the epidermis. Arthropod ocelli usually are small, nonimage forming, and useful only in detecting the source of light, but some, such as the main eyes of spiders, are large, have abundant retinal cells, and form an image.

Ocelli are referred to as median eyes regardless of their location on the animal, and the ancestral arthropod probably had about four pairs. Modern arthropods have some derivative of this number. The naupliar eyes of crustaceans are derived from two pairs of median ocelli, spiders have a single pair (the main eyes; Fig. 18-27C,D), and hexapods usually have three unpaired ocelli (Fig. 21-1A). Ocelli in some taxa are equipped with a layer of reflective pigment known as a **tapetum** (Fig. 18-8A).

Compound eyes are composed of many light-receiving units known as ommatidia, and, in contrast with ocelli, each unit typically (but not always) has its own lens (or cornea). They are referred to as lateral eyes regardless of their position. Compound eyes are present in trilobites, horseshoe crabs, arachnids, crustaceans, insects, and some myriapods, but, of course, not in all members of these taxa. Compound eyes usually are flush with the surface of the head and, if so, are said to be sessile (Fig. 21-3). In some crustaceans (crabs and shrimps, for example), however, the eyes are at the ends of eyestalks, from where they have a much wider field of view (Fig. 19-2B).

The ancestral arthropod may have had as many as five pairs of lateral compound eyes, but in most modern taxa this number has been reduced to a single pair. In the arachnids the number is variable, but their lateral eyes are modified and are not immediately recognizable as being compound. Most spiders, for example, have three pairs of compound eyes (secondary eyes) and one pair of ocelli (main eyes).

The **ommatidium** is the individual, self-contained, and independent light-detecting unit of the compound eye.

A compound eye may contain as few as 15 to as many as several thousand of these units. Ommatidia are elongate and rod-shaped (Fig. 16-13A). They lay side-by-side, perpendicular to the surface of the head, where they are visible as **facets** that usually are hexagonal in shape. Each ommatidium includes its own focusing system, a light-transmitting system, a collection of light-sensitive photoreceptor cells, and screening pigment to isolate the unit optically from adjacent ommatidia. Each has its own field of vision, which overlaps that of adjacent ommatidia.

The exposed outer, or distal, end of the ommatidium is the transparent **cornea** (or lens), which is part of the exoskeleton (Fig. 16-13A,D) and focusing system. The rigid, cuticular cornea is incapable of moving or changing shape and thus has a fixed focal length. It is incapable of accommodation to adjust for changes in the distance to the object. Being cuticular, it is molted with the rest of the exoskeleton during ecdysis. Immediately below the cornea are the epidermal **corneagenous cells,** which secrete it. The transparent **crystalline cone** that lies below the corneagenous cells is composed of the central, transparent regions of four crystalline cone cells. It also is part of the focusing system and functions as a second lens. In some taxa there is a transparent **cone stalk** connecting the cone with the retinula (discussed later). The cone stalk transmits light from the cornea to the light-sensitive rhabdome, functioning like an optical fiber. The myriapod ommatidium has no crystalline cone. The cone and cone stalk are surrounded by **distal pigment cells** that contain dark brown or black screening pigment. Screening pigments are not photosensitive and should not be confused with rhodopsin.

In arthropod (and vertebrate) eyes, light is detected by the light-sensitive pigment **rhodopsin.** This pigment is embedded in the photoreceptor cell membrane. When struck by a photon of light, rhodopsin undergoes a chemical reaction that affects the electrical potential of the cell membrane. This disturbance of the resting polarity is transmitted to the brain, where it is interpreted as light.

The sensory complex of the ommatidium consists of seven or eight light-sensitive **retinular cells** clustered to form a **retinula** (Fig. 16-13A,E). The retinular cells contain rhodopsin arrayed on the membranes of microvilli arising from their axial surfaces (Fig. 16-13E). The microvilli are pointed toward the central axis of the retinula, where they interdigitate with those of other retinular cells to form the **rhabdome** in the core of the retinula (Fig. 16-13A,E). The rhabdome is the mass of microvilli belonging to the cells of the retinula. This axial concentration of microvilli results in a high density of rhodopsin and increases the likelihood that light entering the rhabdome will strike a pigment molecule. Retinular cells also contain proximal screening pigment.

Each retinular cell terminates in an axon at its proximal end. Seven or eight axons thus exit each ommatidium and synapse with the optic ganglion, which usually is in the protocerebrum. The combined axons from all the ommatidia of an eye constitute the **optic nerve,** which extends from the eye to the protocerebrum. Light passes through the cuticular cornea, crystalline cone, and cone stalk (if present) to strike the rhabdome.

The image formed by a compound eye operating in bright light is a **mosaic image,** as is that formed by a vertebrate eye. Each ommatidium responds to a single point of light that originated from an object in the visual field. The brain combines the points from all ommatidia to form a composite image composed of numerous spots of light. Ommatidia are far larger than vertebrate rods or cones, and consequently the image formed is of a coarser grain.

The compound eye is not very good for distance vision compared to the large eyes of birds and mammals. The world within about 20 cm is much more important to most arthropods, however, and compound eyes are effective within this range. Compound eyes are especially good at detecting motion, as you know if you have ever tried to catch a fly with your hand. As an object moves across the visual field, it stimulates a succession of ommatidia, which the brain interprets as motion. The total corneal surface of a compound eye can be strongly convex, resulting in a wide visual field. This is particularly true of stalked compound eyes, in which the cornea may cover an arc of more than 180°.

Compound eyes can be either **light-** or **dark-adapted.** The compound eyes of most arthropods are able to adjust, at least to a limited extent, to both bright and dim light, but in general they tend to be organized to function best in one or the other. Light-adapted eyes are said to be appositional, whereas dark-adapted eyes are superpositional. The two differ in the presence or absence of a cone stalk, deployment of distal and proximal screening pigment, and the distance between the cone and the rhabdome.

Apposition eyes have no cone stalk and the cone cells are close to the rhabdome. Both screening pigments are maximally dispersed in their cells, so the ommatidium is isolated optically from adjacent ommatidia by the resulting curtain of dark pigment (Fig. 16-13B3,C,D). Light entering a given cornea is confined by the screening pigment to one ommatidium, stimulates the rhabdome of that ommatidium, and cannot escape to stimulate the rhabdomes of adjacent ommatidia. In the apposition eyes of honeybees and locusts, for example, less than 1% of the light reaching a rhabdome comes from other ommatidia.

An eye in apposition mode functions well in bright light and permits the greatest resolution of the image. Each point of light stimulates one ommatidium and sends one signal to the brain. That point is registered by the brain as one point in the mosaic image.

Arthropods such as the diurnal terrestrial and shallow-water aquatic species that live in well-lit habitats usually have apposition eyes. The screening pigment is well developed, the length of the crystalline cone is approximately equal to its focal length, and the lower end of the cone and the upper end of the rhabdome are contiguous, or nearly so (Fig. 16-13C,D1). The retinular cells are quite long, extending from the crystalline cone to the basal lamina of the retina. These modifications tend to confine light entering a facet to a single ommatidium and to funnel the light down the axis of the ommatidium to its rhabdome.

The **superposition eye** is modified to collect and concentrate the light from several corneas onto a single ommatidium. The discontiguous cone and rhabdome are connected by a cone stalk. The two screening pigments are concentrated in a small volume of their respective cells, minimizing their effect. They do not form a curtain of pigment separating adjacent ommatidia (Fig. 16-13B1,D2). Light can pass from

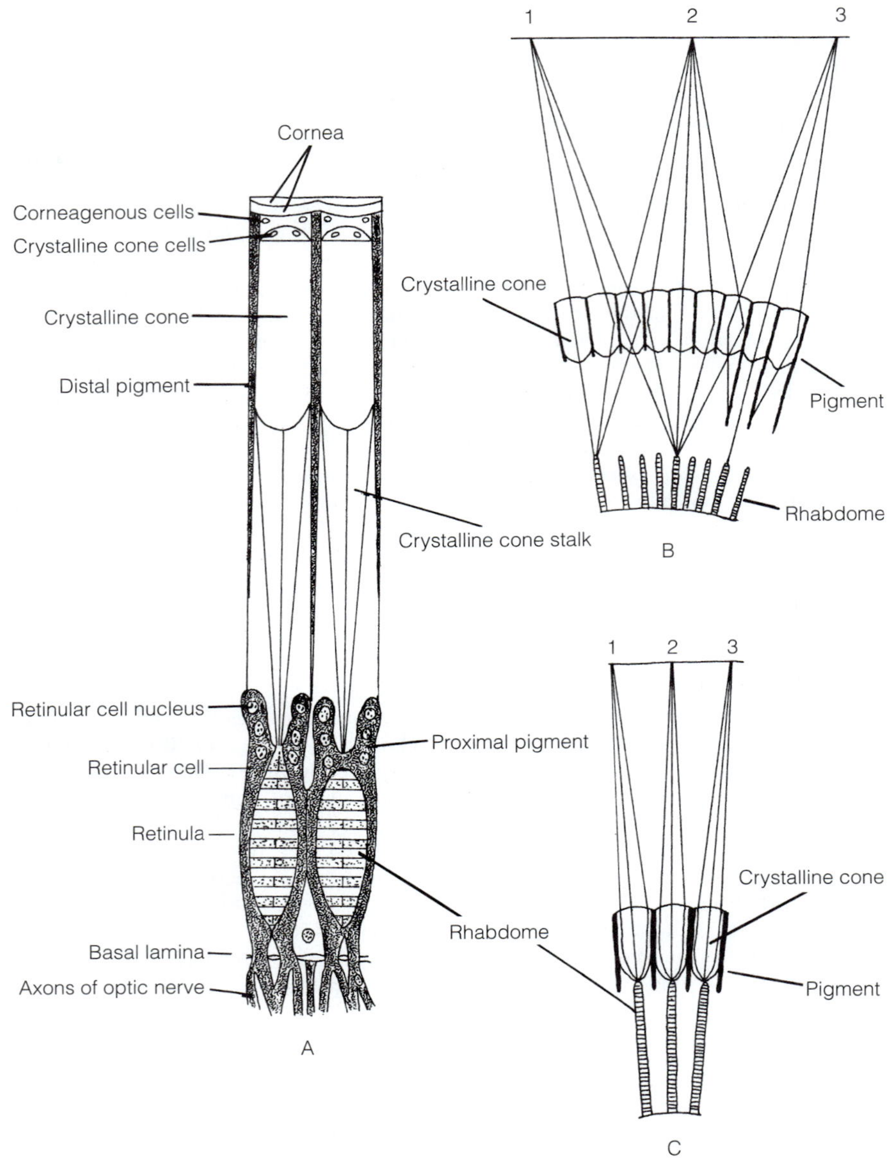

one ommatidium to another, causing a rhabdome to respond to light that entered several different ommatidia. The crystalline cone tends to be twice as long as its focal length, and there is considerable space between the end of the cone and the rhabdome, permitting the bent light rays to cross from the crystalline cone of one ommatidium to the rhabdome of another. The retinular cells are much shorter than those in apposition eyes and are restricted to the base of the ommatidium. These adaptations for low light intensity make it more likely that a rhabdome will be activated, even by a weak light source. In this mode, a single point of light is not associated with a single point in a mosaic image. A blurred image, or no image, is formed, but the brain is aware that there is light in the environment and can tell if it is moving. Each photon has maximal effect on the signal sent to the brain and the animal can be aware of very dim sources of light in its environment. If the few available photons all went to different ommatidia, none might be sufficient to generate an impulse to the brain, but if they enter in a single ommatidium, its threshold may be exceeded and an impulse sent to the brain. In crayfish superposition eyes, about 50% of the light entering a rhabdome originates in other ommatidia. In some insects and arachnids a tapetum of reflecting pigment forms a layer below the retinulae to reflect light back into the ommatidium. Reflections from the tapetum are sometimes visible to human observers as eye-shine.

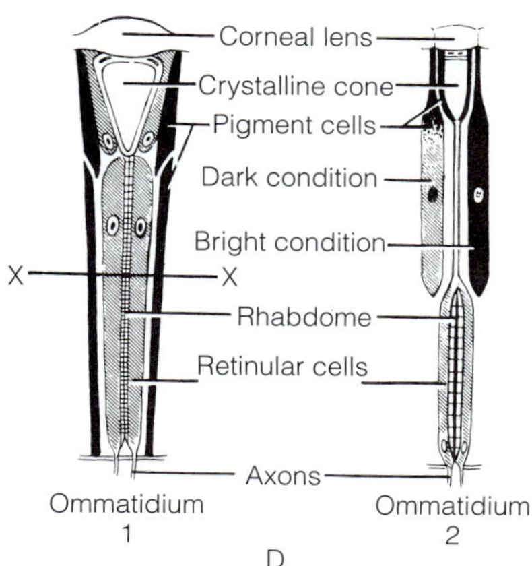

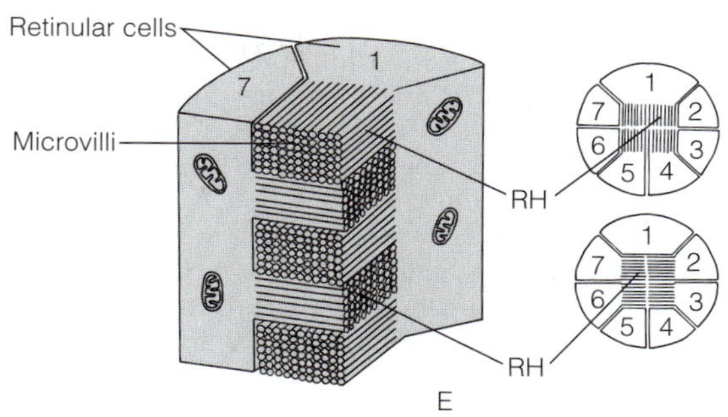

FIGURE 16-13 Compound eyes. **A,** Two generalized ommatidia from the compound eye of the crayfish *Astacus* in longitudinal view. **B,** Compound eye adapted for superposition image formation. Light rays from points 1 and 2 are being received as superposition images. The distal pigment is retracted and light rays, initially received by a number of ommatidia, are concentrated onto a single ommatidium. Point of light 3 is being received as an apposition image by a single isolated ommatidium with its distal pigment screens deployed. The pigment is extended, preventing light rays from crossing from one ommatidium to another. **C,** Compound eye adapted for apposition image formation. Each rhabdome is stimulated only by light received by its own ommatidium. **D,** Insect ommatidia of the diurnal (apposition) type (1) and nocturnal (superposition) type (2). In the nocturnal type, pigment is shown in two positions, adapted for very dark conditions on the left side and for relatively bright conditions on the right. **E,** Stereo diagram of the rhabdome of the compound eye of the crayfish *Procambarus clarkii* showing the perpendicular orientation of the microvilli projecting from two retinular cells. Transverse sections at two levels show the organization of the rhabdome (RH) formed by the microvilli of the seven retinular cells of an ommatidium. *(A, After Bernhards from Waterman; B and C, After Kuhn from Prosser and others; E, After Eguchi from Burr, A. H. 1982. Evolution of eyes and photoreceptor organelles in the lower phyla. In Ali, M. A. (Ed.): Photoreception and Vision in Invertebrates, Plenum Press, New York. pp.131–178.)*

Superposition eyes are found in nocturnal species, deep-water species, and others that live in low light intensities. There are many exceptions, however, and not all superposition eyes are used in dim light. Screening pigment may be reduced or absent in cave-dwelling and bathypelagic species.

The secondary eyes of spiders, although derived from compound eyes, are unlike the typical compound eyes of other arthropod taxa. Spider compound eyes have retinulae and rhabdomes—often many of them—but all the retinulae of an eye share a single lens and there are no facets.

Color vision has been demonstrated in a number of arthropods and has been extensively studied in some insects. The photoreceptive pigments of different retinula cells may be specialized to respond to specific wavelengths of light. For example, in a bee ommatidium two retinula cells respond to green wavelengths and three respond to ultraviolet. Some ommatidia contain four blue-sensitive retinula cells. Variously colored flowers stimulate responses from different combinations of retinula cells.

ENDORECEPTORS

Endoreceptors (or interoreceptors) inform the CNS of conditions inside the body. These have not been well studied in arthropods, and relatively few endoreceptors have been identified. **Proprioceptors,** which are well studied, are an important exception. These provide the CNS with the information needed to coordinate locomotion and maintain posture. They detect and report the degree of contraction (tension) of skeletal muscles and the angle of flexion of body and joints. Vertebrates have receptors with similar functions.

REPRODUCTION

Most arthropods are gonochoric, although there are some important exceptions (such as barnacles). Fertilization may be external (as in horseshoe crabs, sea spiders, and some crustaceans) or internal in aquatic arthropods, but is internal in all terrestrial species. Even when external, the number of eggs usually is relatively small in comparison with many marine invertebrates. The transfer of sperm from male to female is accomplished by a number of quite different mechanisms. There usually is mating or copulation involving some selectivity between the partners (as opposed to the indiscriminate release of gametes). Females usually store sperm in a seminal receptacle and later use it to fertilize the eggs.

Sperm transfer is sometimes direct, in which the male gonopore, usually at the tip of a penis, delivers sperm directly into the female gonopore. The penis may function as an intromittent organ that introduces semen into the gonopore of the female. In many arthropods, however, sperm transfer is indirect and there is no direct contact between male and female gonopores. In indirect transfer some auxiliary structure other than the penis may serve as an intromittent organ to transfer the sperm from the male gonopore to the female gonopore. The nature of this structure varies; in crustaceans it is usually a modified appendage, and in spiders it is the pedipalps. Another common method of indirect sperm transfer employs packets of sperm known as spermatophores.

Indirect sperm transfer via spermatophores was an early response to the challenge of achieving fertilization on land. The sea is a hospitable environment for sperm, eggs, and embryos and it is an easy matter for marine animals to simply broadcast gametes with a fair chance that fertilization will occur and the embryos will survive. In sharp contrast, terrestrial environments are very inhospitable to gametes, and internal fertilization is required for their survival. Indirect sperm transfer, without copulation and with spermatophores, was probably the earliest attempt to meet the challenge of fertilization on dry land, and many arthropods, especially arachnids, still employ it. In the simplest, and presumably ancestral, situation, a spermatophore is deposited by the male somewhere other than in the female gonopore and usually on the ground. The female finds the spermatophore, usually by following chemical cues, and takes the entire packet or, more often, just the sperm, into her gonopore. The transfer is indirect because the sperm passes from the male gonopore to some intermediate substratum and then to the female gonopore. More often, however, the male deposits the spermatophore only after he has located a receptive female, which may require an elaborate courtship ritual.

Many derived terrestrial arthropods have evolved behavior patterns and anatomical equipment for internal fertilization by copulation, in which the male inserts an intromittent organ, which may or may not be the penis, into the reproductive tract of the female and deposits sperm or spermatophores in her seminal receptacle.

Arthropod gonads develop from embryonic coelomic spaces and connect with the exterior via paired gonoducts derived from coelomoducts. The gonads usually are tubular structures lying dorsally in the trunk on each side of the gut. The gonoducts open on characteristic segments in different taxa. The male sperm duct (vas deferens) may have a region specialized for producing spermatophores. The female oviduct often has a seminal receptacle specialized for storing sperm. Sperm transfer and fertilization may occur at widely separated times and sperm are stored in the receptacle until needed for fertilization. Brooding and parental care of eggs, embryos, and juveniles are common in the arthropods.

DEVELOPMENT

Most arthropods have large, yolky (macrolecithal) eggs with the nucleus and a small amount of yolkless cytoplasm enclosed in the center of a mass of yolk. Such eggs are said to be **centrolecithal** (Fig. 16-14A) (even though it is actually the blastula, not the egg, that has yolk in its center). The large amount of yolk and the position of the nuclei interfere with the ancestral spiral cleavage pattern and impose a derived **superficial cleavage** on most arthropod eggs. Superficial cleavage is **meroblastic,** meaning that the cleavage furrows are impeded by the heavy concentrations of yolk and the egg is not completely divided into daughter blastomeres.

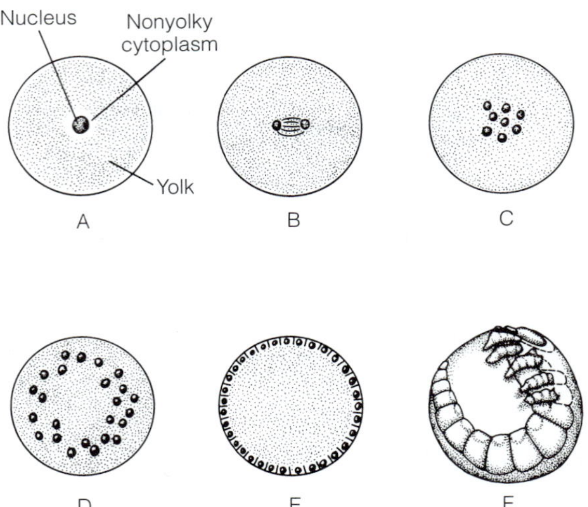

FIGURE 16-14 Superficial cleavage. **A,** Centrolecithal egg. **B** and **C,** Nuclear division. **D,** Migration of nuclei toward periphery of egg. **E,** Blastula. Cell membranes have developed, separating adjacent nuclei. **F,** Embryo of the arachnid *Thelyphonus.* *(After Kaestner, A. 1968. Invertebrate Zoology. Vol. II. John Wiley and Sons, New York.)*

Because we have long believed arthropods to be closely related to annelids, we have assumed spiral, total (holoblastic) cleavage to be the ancestral arthropod developmental pattern. Indeed, some arthropods, such as sea spiders and barnacles, have small microlecithal eggs and total cleavage. Early workers reported spiral cleavage in some of these taxa, but this interpretation has been questioned by some modern zoologists.

Following fertilization the diploid nucleus, located in the center of the zygote, undergoes mitotic divisions to produce many identical daughter nuclei (Fig. 16-14C). These divisions are not accompanied by cytoplasmic divisions and no cell membranes form to isolate the individual nuclei. The uncleaved embryo is still a single cell, but it has multiple central nuclei. These nuclei migrate through the yolk mass to positions in the periphery of the zygote (Fig. 16-14D). This forms a type of stereoblastula with an outer cortex of cytoplasm and nuclei on the periphery of the central yolk mass (Fig. 16-14E). It is this relationship between nuclei and yolk that accounts for the term centrolecithal. Cleavage furrows begin at the surface but do not penetrate into the central yolk. Thus cleavage is superficial and blastomeres form on the surface.

During gastrulation, the superficial cells arrange themselves into a primordial **blastodisc** on the surface of the blastula. The cells of the blastodisc differentiate into endoderm, ectoderm, and mesoderm, and a segmented, wormlike embryo forms on the surface of the yolk mass (Fig. 16-14F). This embryo grows and adds segments like other segmented protostomes.

A pair of mesodermal **teloblast areas** form on the anterior margin of the embryonic telson. These produce pairs of mesodermal masses that become the mesoderm of the segments and their appendages. Mitotic activity of the teloblasts adds new segments and appendages at the anterior margin of the telson.

The blastocoel of the segment becomes the connective-tissue compartment and the hemocoel. Coelomic pouches develop in the mesodermal masses and metanephridial ducts or coelomoducts may appear with them. The sacs and ducts in some segments become the saccate nephridia, gonads, and their ducts. These sacs may also be incorporated into the walls of the heart in some taxa (such as spiders). The mesoderm proliferates to cover the gut and differentiates into the muscles and connective tissue of the segments and their appendages.

The segmental appendages develop from ectodermal evaginations that fill with proliferating mesoderm that differentiates into muscles and connective tissue. Ganglia develop from submerged masses of ectodermal cells that eventually separate from the epidermis. The protocerebrum and deutocerebrum arise from ectodermal masses in two preoral segments, whereas the tritocerebrum arises from the epidermis of the first postoral segment.

Development was probably indirect in the ancestral arthropods and each egg hatched as a larva bearing little or no resemblance to the adult. The original arthropod larva probably hatched with only the anteriormost segments and their appendages, the antennae and the first two pairs of unspecialized biramous head appendages. Pycnogonids and many crustaceans have small eggs that hatch into such larvae with three pairs of legs. The pycnogonid **protonymphon larva** (Fig. 18-46C) hatches with chelicerae, pedipalps, and the first walking legs, all of which are uniramous. The crustacean **nauplius larva** (Fig. 19-8) has the first antennae, second antennae, and mandibles, the latter two of which are biramous. After hatching, growth and the addition of new segments and appendages occur only in conjunction with ecdysis.

At the opposite extreme, many derived arthropods exhibit direct development, in which all larval stages are suppressed or occur in the egg prior to hatching and the young hatch as miniature adults with a full complement of appendages and segments. Between these two extremes are the many arthropods with indirect development that hatch at an advanced larval stage and having a number of segments and appendages intermediate between the nauplius and the adult.

PHYLOGENY OF ARTHROPODA

Our understanding of the evolution of these (and most other) animals is currently in a state of uncertainty due to the conflicting evidence of morphology and molecules. A close relationship of Arthropoda, Onychophora, and Tardigrada is supported by both molecular and morphological data and it is generally agreed that they form the monophyletic taxon Panarthropoda. Relationships within Panarthropoda are less certain, however, and in some analyses Onychophora is thought to be the sister taxon of Tardigrada plus Arthropoda (Fig. 16-15) whereas in others Tardigrada is the sister taxon of Onychophora plus Arthropoda. Some zoologists exclude Tardigrada from Panarthropoda.

Onychophora, Tardigrada, and Arthropoda share a thick cuticle that cannot expand and requires molting to permit growth. The body is segmentally organized and grows by adding new segments from clusters of posterior teloblast cells. The cuticle is unique in containing α-chitin but no collagen. The segments bear paired appendages. Locomotory cilia have been lost, although cilia may persist in specialized applications such as sensilla and sperm. No ciliated larva (trochophore) is present and cleavage is superficial. Protonephridia, and their flagella, are absent and the primary excretory organs are saccate nephridia derived from metanephridia. The body cavity is a hemocoel derived from the blastocoel with contributions from mesenchyme derived from earlier coelomic pouches. The heart is a dorsal tube with paired, segmental ostia.

Onychophora branched early from the arthropod line and is the sister taxon of the remaining panarthropods. These wormlike terrestrial panarthropods possess oral papillae with slime glands. The body-wall muscles are smooth and not organized into discrete, individual muscles. They independently evolved tracheae as an adaptation to terrestrial life. Paired appendages are present, but they are unjointed lobopods lacking intrinsic muscles confined to the appendage.

All other panarthropods have jointed appendages with intrinsic muscles. Individual muscles with cross-striated fibers are present. The brain is a syncerebrum with proto-, deuto-, and tritocerebra.

Tardigrada is the sister taxon of Arthropoda. These small animals share several derived characteristics, such as the absence of a heart and nephridia, presumably as a consequence of their small size. One pair of legs is modified to form a pair of anterior stylets used in suctorial feeding. The ganglia of the first walking legs and the tritocerebrum are connected by a nerve.

Arthropoda (true arthropods, or Euarthropoda) consists of Chelicerata and Mandibulata plus several extinct taxa, of which Trilobitomorpha is the best known. The body is covered by a system of articulating, sclerotized plates that form a flexible, armored exoskeleton. The head primitively bears one pair of preoral and several pairs of postoral appendages. Eyes are associated with the protocerebrum.

The chelicerate head has a pair of prehensile chelicerae on the second segment and there are no antennae. In contrast, the mandibulate head bears antennae on the first segment and mandibles on the third. Mandibulata includes the sister taxa Crustacea and Tracheata.

In the past it has been suggested that Arthropoda might be polyphyletic, meaning that the four great subtaxa are unrelated and have no common ancestor. This hypothesis requires the independent arthropodization of at least four separate protostome lines, culminating in Trilobitomorpha, Chelicerata, Crustacea, and Tracheata. At present, however, there is general agreement that Arthropoda is in fact monophyletic and that its taxa are related and share a common ancestor (Fig. 16-15). In this hypothesis, arthropodization, the evolution of the suite of uniquely arthropod characteristics, thus occurred only once.

Traditionally, the arthropod ancestor is believed to have been a coelomate with a large, well-developed, segmented coelom and a relatively small connective-tissue compartment. The Cambrian *Aysheaia*, which exhibits similarities to annelids, onychophorans, and tardigrades, has been suggested as a possible early panarthropod. During arthropodan evolution the exoskeleton became more important as the coelom was reduced and the connective-tissue compartment expanded. In

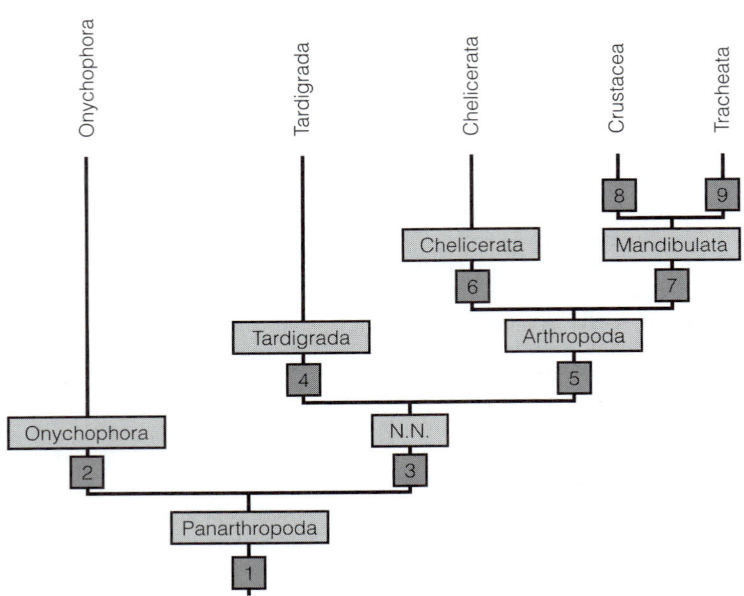

FIGURE 16-15 A phylogeny of the higher arthropodan taxa. **1, Panarthropoda:** cuticle of α-chitin and noncollagenous protein; ecdysis; hemocoel; paired segmental saccate nephridia; locomotory cilia lost; syncerebral brain; open hemal system with dorsal, tubular, ostiate heart and pericardial sinus; coelom reduced to nephridia and gonads. **2, Onychophora:** stubby appendages without articulations; tracheae tufted and unbranched; multiple nerve commissures per segment; oral papillae with slime glands; body-wall muscles in continuous sheets. **3, N.N.:** external segmentation; exoskeleton with articulated plates; articulated appendages with intrinsic muscles; three-part syncerebrum. **4, Tardigrada:** four pairs of short, uniramous, unjointed or weakly jointed legs; protocerebrum with a neural connection to the ganglion of the first legs; sucking pharynx; eutelic; body with head plus four trunk segments; anteriormost appendages vestigial, their claws modified as a stylet apparatus. **5, Arthropoda:** articulated exoskeleton formed of plates; jointed, probably biramous appendages; continuous sheets of body-wall muscles replaced by individual cross-striated muscles; head with acron and probably five segments; one pair of preoral antennae and three additional appendages; one pair of lateral compound eyes, four pairs of median ocelli; six pairs of saccate nephridia (four head, two trunk); larvae have three pairs of appendages. **6, Chelicerata:** body has cephalothorax and abdomen; cephalothorax has acron plus 7 segments and six paired appendages; abdomen probably has 12 segments plus caudal spine; deutocerebrum lost, brain of two parts; first head segment appendages lost, second segment has chelicerae; gonopores on abdominal segment 2. **7, Mandibulata:** head has acron and probably six segments, five pairs of appendages (antennae 1 and 2, mandibles, maxillae 1 and 2); head segment 1 has antenna; head segment 3 has mandibles; three-part brain; ommatidium with crystalline cone, cornea; molting controlled by gland (Y-organ or prothoracic gland). **8, Crustacea:** head segment 2 has antennae, head has five pairs of appendages; naupliar eye derived from four median ocelli; original six pairs of saccate nephridia reduced to two pairs (in segments 2 and 5); appendages biramous. **9, Tracheata:** head segment 2 and appendages lost, head has four pairs of appendages (antennae, mandibles, maxillae, labia); mandible without palp (endopod); tracheae with paired segmental spiracles; Malpighian tubules; nephridia reduced to two pairs (in segments 4 and 5) or lost; appendages uniramous. *(Based on Nielsen, C. 2001. Animal Evolution, Interrelationships of the Living Phyla. 2nd Edition. Oxford Univ. Press. 563 pp.; Ax, P. 2000. Multicellular Animals: The Phylogenetic System of the Metazoa. Vol. II. Springer, Berlin. 396 pp.; Paulus, H. 1996a. Arthropoda. In Westheide, W., and Rieger, R. M. (Eds.): Spezielle Zoologie I. Einzeller und Wirbellose Tiere. Gustav Fischer Verlag, Stuttgart. pp. 411–419.)*

living arthropods the entire interior of the body is the connective-tissue compartment and its spaces are the hemocoel. The embryonic development of onychophorans recapitulates this transition. Adult onychophorans have an extensive connective-tissue compartment and hemocoel similar to those of arthropods. The coelom is represented by small spaces associated with the gonads and excretory organs, as it is in arthropods. The early developmental stages of onychophorans, however, include paired, segmental mesodermal blocks with well-developed coelomic spaces and a small connective-tissue compartment. During development, the coelomic spaces diminish and the connective-tissue compartment (hemocoel) enlarges.

Especially troublesome at the moment is the evolutionary relationship between the arthropods and the remaining protostomes, specifically the annelids and cycloneuralians, and at present no definitive statement regarding this relationship is possible. Morphologically and traditionally, annelids and arthropods have been grouped together in a protostome taxon known as Articulata. Nucleotide sequence analysis does not support the monophyly of Articulata, however, and recent reinterpretation of their presumed morphological similarities suggests they may be convergences and thus not indicative of a genetic relationship. In this hypothesis, older reports of spiral cleavage in some arthropods are seen as misinterpretations, and arthropods thus are not spiralians related to annelids and molluscs.

Most modern molecular studies, primarily the analysis of nucleotide sequences in 18S rDNA, indicate a close relationship between the panarthropods and Nematoda, Nematomorpha, Gastrotricha, Kinorhyncha, Priapulida, and Loricifera. These taxa are combined in a new protostome taxon, Ecdysozoa, most of which undergo ecdysis. Recent simultaneous

PHYLOGENETIC HIERARCHY OF ARTHROPODA

- Panarthropoda
  - Onychophora
  - N.N.
    - Tardigrada
    - Arthropoda
      - Chelicerata
        - Pycnogonida
        - Euchelicerata
          - Xiphosura
          - Arachnida
      - Mandibulata
        - Crustacea
        - Tracheata
          - Hexapoda
          - Myriapoda

consideration of both morphology and 18S rDNA gene sequences support the validity of the Ecdysozoa as a monophyletic taxon of molting animals. Annelida, Mollusca, Sipuncula, and Lophophorata belong to the sister protostome group Lophotrochozoa, the members of which are spiralians. The Ecdysozoa hypothesis is not universally supported by molecular analyses, however. For example, a recent study of nuclear protein alignments revealed arthropods to be closer to vertebrates than to nematodes.

REFERENCES

GENERAL

Adoutte, A., et al. 2000. The new animal phylogeny: Reliability and implications. Proc. Natl. Acad. Sci. USA 97:4453–4456.

Anderson, D. T. 1973. Embryology and Phylogeny in Annelids and Arthropods. Pergamon Press, New York. 495 pp.

Atwood, H. L., and Sandeman, D. C. (Eds.): 1982. Biology of Crustacea. Vol. III. Neurobiology: Structure and Function. Academic Press, New York. 479 pp.

Bater, J. E. 1996. Micro- and macro-arthropods. In Hall, G. S. (Ed.): Methods for the Examination of Organismal Diversity in Soils and Sediments. CAB International, Wallingford, U.K. pp. 163–174.

Blair, J. E., Ikeo, K., Gojobori, T., and Hedges, S. B. 2002. The evolutionary position of nematodes. BMC Evolutionary Biology 2:7–17.

Boardman, R. S., Cheetham, A. H., and Rowell, A. J. (Eds.): 1987. Fossil Invertebrates. Blackwell Scientific, Palo Alto, CA. 713 pp.

Borrer, D. J., Triplehorn, C. A., and Johnson, N. F. 1989. An introduction to the study of insects, 6th ed. Saunders College Pub, Philadelphia. 875 pp.

Briggs, D. E. G. 1991. Extraordinary fossils. Am. Sci. 79: 130–141.

Burr, A. H. 1982. Evolution of eyes and photoreceptor organelles in the lower phyla. In Ali, M. A. (Ed.): Photoreception and Vision in Invertebrates. Plenum Press, New York. pp. 131–178.

Caveney, S. 1969. Muscle attachment related to cuticle architecture in Apterygota. J. Cell Sci. 4:541–559.

Chapman, R. F. 1982. The Insects: Structure and Function. 3rd Edition. Harvard University Press, Cambridge, MA. 919 pp.

Coleman, D. C., and Crossley, D. A. 1996. Fundamentals of Soil Ecology. Academic Press, San Diego. 205 pp.

Conway Morris, S. 1989. Burgess Shale faunas and the Cambrian Explosion. Science 246:339–346.

Damen, W. G. M., Hausdorf, M., Seyfarth, E.-A., and Tautz, D. 1998. A conserved mode of head segmentation in arthropods revealed by the expression pattern of Hox genes in a spider. Proc. Natl. Acad. Sci. USA 95:10665–10670.

Damen, W. G. M., and Tautz, D. 1999. Comparative molecular embryology of arthropods: The expression of Hox genes in the spider *Cupiennius salei*. Inv. Rep. Dev. 36:203–209.

Davies, R. G. 1988. Outlines of Entomology. 7th Edition. Chapman and Hall, London. 408 pp.

Della Cave, L., Insom, E., and Simonetta, A. M. 1998. Advances, diversions, possible relapses and additional problems in understanding the early evolution of the Articulata. Ital. J. Zool. 65:19–38.

Dindal, D. L. 1990. Soil Biology Guide. John Wiley and Sons, New York. 1349 pp.

Eernisse, D. J., Albert, J. S., and Anderson, F. E. 1992. Annelida and Arthropoda are not sister-taxa: A phylogenetic analysis of spiralian metazoan morphology. Syst. Biol. 41:305–330.

Eisenbeis, G., and Wichard, W. 1987. Atlas on the Biology of Soil Arthropods. Springer-Verlag, Berlin. 437 pp.

Fortey, R. A., and Thomas, R. H. (Eds.): 1998. Arthropod Relationships. Systematics Association Spec. Vol. 55. Chapman and Hall, London. 383 pp.

Giribet, G., Edgecombe, G. D., and Wheeler, W. G. 2001. Arthropod phylogeny based on eight molecular loci and morphology. Nature 413:157–160.

Gould, S. J. 1989. Wonderful Life: The Burgess Shale and the Nature of History. W. W. Norton, New York. 347 pp.

Grenier, J. K., Garber, T. L. Warren, R., Whittington, P. M., and Carroll, S. 1997. Evolution of the entire arthropod Hox gene set predated the origin and radiation of the onychophoran/arthropod taxon. Curr. Biol. 7:547–553.

Grosberg, R. K. 1990. Out on a limb: Arthropod appendages. Science 250:632–633.

Hackman, R. H. 1971. In Florkin. M., and Scheer, B. T. (Eds): Chemical Zoology. Vol. VI. Academic Press, New York.

Harrison, F. W., and Foelix, R. F. (Eds.): 1999. Microscopic Anatomy of Invertebrates. Vol. 8A. Chelicerate Arthropoda. Wiley-Liss, New York. 266 pp.

Kaestner, A. 1968. Invertebrate Zoology. Vol. II. John Wiley and Sons, New York. 472 pp.

Klausnitzer, B. 1996. Insecta. In Westheide, W., and Rieger, R. M. (Eds.): Spezielle Zoologie I. Einzeller und Wirbellose Tiere. Gustav Fisher Verlag, Stuttgart. pp. 601–681.

Kümmel, G. 1968. Die Podocyten. Zool. Beitr. N. F. 13:245–263.

Locke, M. 1984. Epidermal cells (Arthropoda). In Bereiter-Hahn, J., Matoltsky, A. G., and Richards, K. S. (Eds.): Biology of the Integument. Vol. 1. Invertebrates. Springer-Verlag, Berlin. pp. 502–522.

Lockwood, A. P. M. 1968. Aspects of the Physiology of Crustacea. Oliver & Boyd, London. 328 pp.

Mangum, C. P. 1985. Oxygen transport in invertebrates. Am. J. Physiol. 248:505–514.

Manton, S. M. 1978. The Arthropoda: Habits, Functional Morphology and Evolution. Oxford University Press, London. 527 pp.

Osorio, D., Averof, M., and Bacon, J. P. 1995. Arthropod evolution: Great brains, beautiful bodies. Trends Ecol. Evol. 10:449–454.

Parry, G. 1960. Excretion. In Waterman, T. H. (Ed.): The Physiology of Crustacea. Vol. I. Metabolism and Growth. Academic Press, New York. pp. 341–366.

Paulus, H. 1996a. Arthropoda. In Westheide, W., and Rieger, R. M. (Eds.): Spezielle Zoologie. I. Einzeller und Wirbellose Tiere. Gustav Fisher Verlag, Stuttgart. pp. 411–419.

Paulus, H. 1996b. Euarthropoda. In Westheide, W., and Rieger, R. M. (Eds.): Spezielle Zoologie. I. Einzeller und Wirbellose Tiere. Gustav Fisher Verlag, Stuttgart. pp. 435–444.

Peterson, K. J., and Eernisse, D. J. 2001. Animal phylogeny and the ancestry of bilaterians: Inferences from morphology and 18S rDNA gene sequences. Evol. Devel. 3:170–205.

Popadic, A., Panganiban, G., Rusch, D., Shear, W. A., and Kaufman, T. C. 1998. Molecular evidence for the gnathobase derivation of arthropod mandibles and for the appendicular origin of the labrum and other structures. Devel. Genes Evol. 208:142–150.

Shear, W. A. 1999. Introduction to the Arthropoda and Cheliceriformes. In Harrison, F. W., and Foelix, R. F. (Eds.): Microscopic Anatomy of Invertebrates. Vol. 8A. Chelicerate Arthropoda. Wiley-Liss, New York. pp. 1–19.

Snodgrass, R. E. 1935. Principles of Insect Morphology. McGraw-Hill, New York. 667 pp.

Snodgrass, R. E. 1952. A Textbook of Arthropod Anatomy. Cornell University Press, Ithaca, NY. 363 pp.

Steinbrecht, R. A. 1984. Arthropoda: Chemo-, hygro-, and thermoreceptors. In Bereiter-Hahn, J., Matoltsky, A. G., and Richards, K. S. (Eds.): Biology of the Integument. Vol. 1. Invertebrates. Springer-Verlag, Berlin. p. 523–553.

Tjønneland, A., Økland, S., and Nylund, A. 1987. Evolutionary aspects of the arthropod heart. Zool. Scr. 16:167–175.

Wills, M. A., Briggs, D. E. G., Fortey, R. A., Wilkinson, M., and Sneath, P. H. A. 1998. An arthropod phylogeny based on fossil and recent taxa. In Edgecombe, G. D. (Ed.): Arthropod Fossils and Phylogeny. Columbia University Press, New York. pp. 33–105.

Zrzavy, J., and Stys, P. 1997. The basic body plan of arthropods: Insights from evolutionary morphology and developmental biology. J. Evol. Biol. 10:352–367.

INTERNET SITES

www.museums.org.za/bio/arthrop.htm (Arthropoda page from the South African Museum.)

http://biosys-serv.biologie.uni-ulm.de/sektion/dieter/dieter.html ("Orsten" Research and Dieter Waloszek's View of Arthropod and Crustacean Phylogeny.)

http://www.geometry.net/science/arthropoda.php Geometry, the Online Learning Center, a searchable index to arthropods.

www-biol.paisley.ac.uk/biomedia/text/txt_arthro.htm (Biomedia Museum, University of Paisley.)

www.geo.ucalgary.ca/~macrae/Burgess_Shale (Photos and short descriptions of Burgess Shale fossils from A. MacRae, University of Calgary.)

www.fao.org/landandwater/agll/soilbiod/soilbtxt.htm (Soil Biodiversity page from the United Nations' Food and Agriculture Organization.)

www.hope.edu/academic/biology/leaflitterarthropods (Hope College Leaf Litter Arthropod Pictorial Key.)

http://web.missouri.edu/~bioscish/ (A Pictographic Key to Leaf Litter Arthropods from the Missouri Ozark Forest Ecosystem Project.)

www.cals.ncsu.edu/course/ent591k/kwikey1.html (North Carolina State University Kwik-Key to Soil-Dwelling Invertebrates.)

http://www.cvm.missouri.edu/cvm/courses/vm556/Arthropods/Arthropo.htm (Parasitic arthropods, by R. M. Corwin and J. Nahm, University of Missouri, Columbia. This is a part of a Veterinary Parasitology Site.)

Trilobitomorpha^{sP}

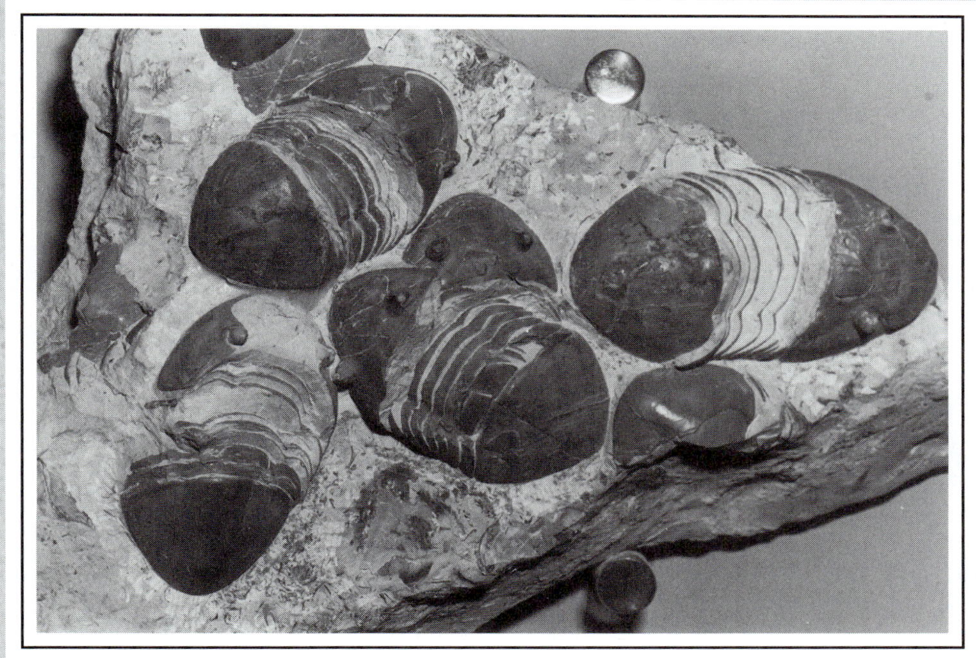

Arthropods make their first appearance in the fossil record during the Cambrian period, along with most other modern invertebrate phyla. The earliest, best known, most common, and most abundant fossil arthropods were the Trilobitomorpha, or trilobites, whose closest living relatives are probably the chelicerates. Trilobitomorpha is monophyletic and has the body plan—including an exoskeleton, segmentation, paired segmental appendages, and compound eyes—expected of early arthropods (Fig. 17-1A). Trilobites were abundant and widely distributed in Paleozoic seas from the Lower Cambrian through the Permian, but are now extinct. They appeared suddenly early in the Cambrian period and reached the height of their distribution and abundance during the Cambrian and Ordovician periods, beginning about 540 mya (million years ago). Trilobites persisted for 300 million years until they disappeared at the end of the Permian, about 260 mya. Their fossils are so common and widespread and their species so characteristic of specific places and times that they are used by geologists as index fossils to identify and associate strata of sedimentary rocks in widely disparate localities and to arrange these strata in their proper geological sequence.

FORM

Most trilobites ranged from 3 to 10 cm in length, although some planktonic species were only 0.5 mm and the largest, *Isotelus,* was 70 cm. The trilobite exoskeleton differs from that of other arthropods in being primarily mineral—a form of calcium carbonate known as calcite—rather than organic, and this probably accounted for their great success.

The exoskeleton consists of a thick, heavily calcified **dorsal shield** (Fig. 17-1A) and a thin, uncalcified **ventral membrane** (Fig. 17-2A) covering the dorsal and ventral surfaces respectively. Around the margins, the thick dorsal shield rolls over the edge to join the thin ventral membrane. The resulting doubly heavy border is the **doublure** (Fig. 17-3, 17-5C). The calcified dorsal shield fossilized readily, but the soft, uncalcified ventral membrane and delicate appendages were rarely preserved. Most of our knowledge of trilobites is based on dorsal shields, but recent studies using new techniques have greatly increased our knowledge of the ventral membrane, appendages, and even internal organs.

The typical trilobite body is dorsoventrally flattened (Fig. 17-2A), especially laterally, and oval in outline (Fig. 17-1A). It is

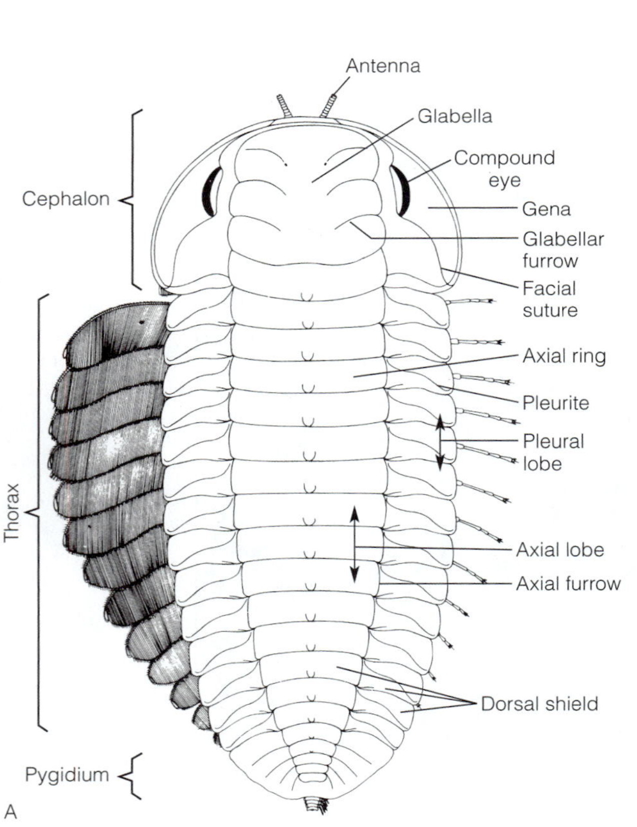

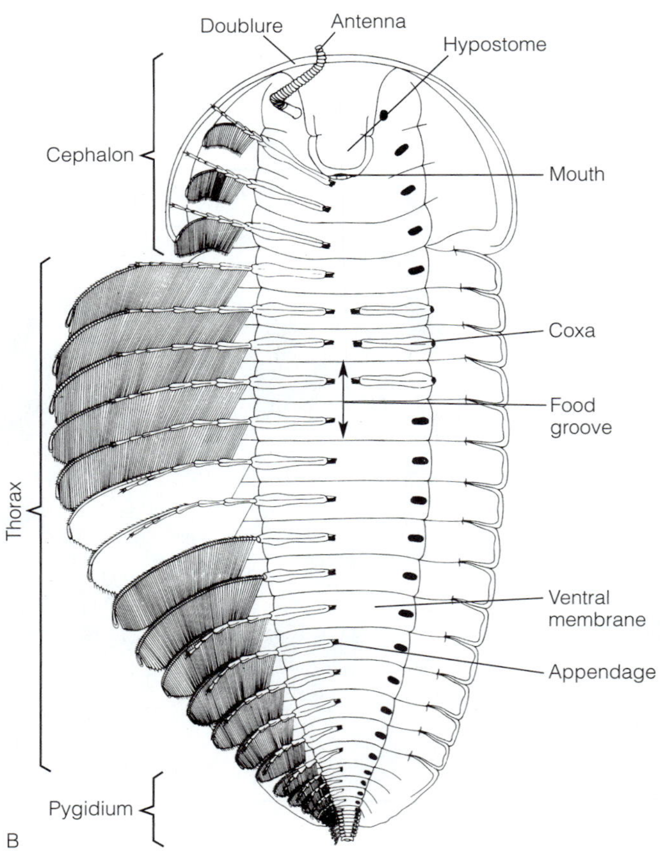

FIGURE 17-1 The ptychopariid trilobite *Triarthrus eatoni.* **A,** The dorsal shield in dorsal view. The facial suture is gonatoparian. **B,** Ventral view. *(From Cisne, J. L. 1975. Anatomy of* Triarthrus *and the relationships of the Trilobita. Fossils and Strata 4:45–63 (Figs. 1 and 2), reproduced from Fossils and Strata www.tandf.no/fossils by permission of Taylor and Francis AS.)*

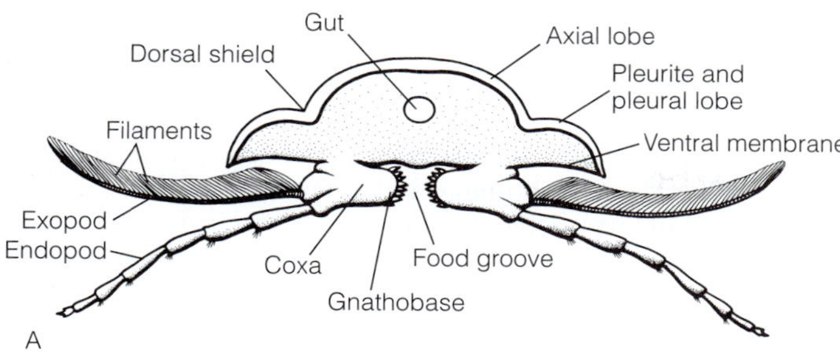

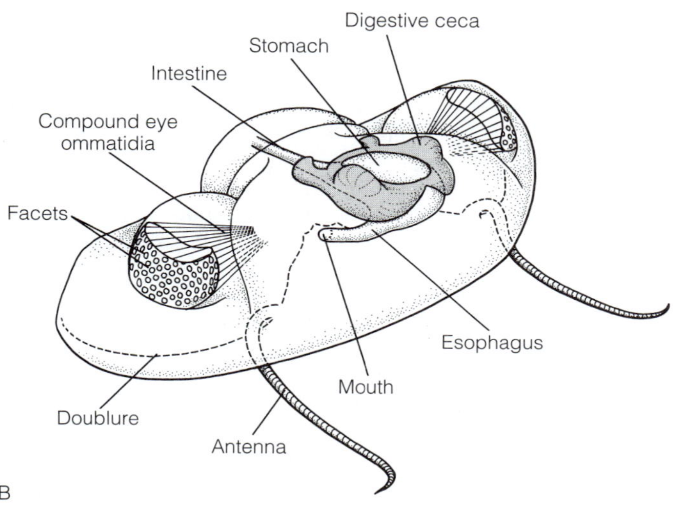

FIGURE 17-2 **A,** Cross section through the thorax of *Triarthrus eatoni.* **B,** Transparent reconstruction of the cephalon of *Phacops.* The schizochroal eyes have widely separated facets. *(A, Redrawn and modified from Müller, K. J., and Walossek, D. 1987. Morphology, ontogeny, and life habit of* Agnosthus pisiformis *from Upper Cambrian of Sweden. Foss. Strata 19:1–124, after Cisne, J. L. 1975. Anatomy of* Triarthus *and the relationships of the Trilobita. Foss. Strata 4:45–63; B, From Stürmer, W., and Bergström, J. 1973. New discoveries on trilobites by X-rays. Paläont. 2. 47:104–141.)*

divided transversely into three tagmata consisting of an anterior cephalon (head), a middle thorax, and a posterior pygidium. Together, the thorax and pygidium are the trunk. In addition, a pair of parallel, longitudinal grooves, the **axial furrows,** divides the body into three side-by-side longitudinal **lobes.** The axial furrows create a raised median **axial lobe,** or axis, flanked on each side by a lateral **pleural lobe** (Fig. 17-1A, 17-2A). The name *trilobite* refers to the three lobes.

The cephalon is composed of four fused segments bearing four pairs of appendages. Dorsally, it is divided into regions consisting of the central, raised **glabella** and two lateral, flatter **genae** (cheeks), one on each side of the glabella (Fig. 17-1A). The glabella is the anterior extension of the axial lobe onto the cephalon. The original segmentation of the cephalon is indistinct, but in some trilobites the glabella is partially divided into lateral lobes by **glabellar furrows** (Fig. 17-1A). In most trilobite fossils the gena bears an **eye lobe** that holds a compound eye, although some were blind. The posterior corners of the gena often have **genal spines** (Fig. 17-7A,E,G) and the glabella may also carry spines. Trilobite spines were hollow with soft tissue inside.

Where adjacent sclerites abut each other, a narrow line, or cephalic suture, can be seen on both surfaces of the cephalon. These served trilobites as ecdysial lines along which the exoskeleton separated during ecdysis to permit the exit of the next instar. They serve taxonomists as important characteristics for classifying and identifying evolutionary lines of trilobites. Among the many cephalic sutures, the **facial sutures** are most important for classification (Fig. 17-1A). Each extends more or less longitudinally along the dorsal cephalon, crossing the eye lobe to divide the gena into lateral and medial regions. Paleontologists recognize three types of facial sutures based on where the posterior end intersects the margin of the gena. Proparian sutures intersect the border laterally (Fig. 17-7F), gonatoparian sutures intersect at the posterior lateral angle of the gena (Fig. 17-1A), and opisthoparian sutures, which are the most common, intersect on the posterior border (Fig. 17-7A,C,D,G).

The segmentation of the cephalon is not obvious dorsally (Fig. 17-1A), but ventrally the segments are clearly evident (Fig. 17-1B). The posteriorly directed mouth is located ventrally in the middle of the cephalon, just behind and beneath a movable, liplike sclerite called the **hypostome** (Fig. 17-1B). The hypostome covers the ventral surface of the central region of the cephalon, the part corresponding to the dorsal glabella. The brain and stomach lie between the glabella and the hypostome and are protected by them (Fig. 17-2B).

Trilobite compound eyes are unique, having mineral lenses of calcium carbonate rather than the organic cuticle characteristic of all other arthropod eyes. The lens of each ommatidium is a single transparent, calcite crystal that selectively transmits light from a preferred angle along its length to strike the photoreceptors at its base. Species inhabiting shallow seas with abundant light had well-developed eyes, whereas those from deeper water had reduced eyes or were blind. The number of ommatidia varies from one to many thousand. Based on what we deduce from the fossilized lenses and their optical properties, trilobite eyes were as sophisticated and effective as those of any modern arthropod. Trilobite eyes can be **holochroal,** with numerous small, hexagonal, closely packed lenses, or **schizochroal,** with a few large, spherical, widely spaced lenses (Fig. 17-2B). The holochroal eye is covered by a common calcitic cornea that is lacking in the schizochroal eye. Most trilobite eyes are holochroal.

The thorax consists of from 2 to 61 (but usually 6 to 15) independent, articulated segments, each consisting of a median **axial ring** from which extend two lateral pleurites (Fig. 17-1A, 17-2A). The combined pleura of successive segments make up the pleural lobes and, similarly, the combined axial rings constitute the axial lobe. In many species the pleura bear **pleural spines** (Fig. 17-7C) that may be quite large. Each segment has a pair of appendages on the ventral surface (Fig. 17-1B).

Successive segments articulate with each other via flexible articular membranes that permit the body varying degrees of ventroflexion. The thorax is the only flexible part of the body, with the cephalon and pygidium being rigid. Many trilobites could **enroll** into a ball (Fig. 17-5A). Two well-known species, *Triarthrus eatoni* (Fig. 17-1) and *Agnostus pisiformis* (Fig. 17-5A), were both capable of complete enrollment. The specific epithet *"pisiformis"* means pea-shaped, in reference to its enrolled shape. Among modern arthropods, enrollment is seen in some terrestrial isopod crustaceans (pill bugs) and millipedes (pill millipedes).

Although the pygidium bears segmental appendages similar to those of the thorax, its segments are fused together to form a single solid unit (Fig. 17-1, 17-5C). The segments regardless remain clearly distinguishable. The size of the pygidium varies according to species and is important taxonomically. Pygidial appendages resemble those of the thorax but rapidly decrease in size posteriorly. The axial and pleural lobes of the thorax extend uninterrupted onto the pygidium and the right and left pleural lobes may join posterior to the axial lobe.

Because of their delicacy, appendages do not fossilize well, and until recently little has been known of their morphology. Worldwide, there are only about 20 sites where trilobites with appendages can be found. The appendages of several species from these sites, including *Agnostus, Olenoides, Phacops,* and *Triarthrus,* are now well known as a result of X-radiography, new acid etching techniques, and the discovery of silicified and pyretized fossils. Occasionally, sulfur bacteria in anoxic environments deposited iron pyrite on a dead trilobite, resulting in preservation of the details of their soft parts and appendages. Pyretized fossils are especially amenable to X-ray studies. Fossils that have been silicified by the replacement of the original calcite exoskeleton with silica can be studied by removing the surrounding calcareous matrix with weak acetic acid.

Trilobite appendages are attached to the ventral membrane and consist of an anterior pair of antennae followed by a series of paired segmental appendages extending the length of the body, from the cephalon to the end of the pygidium. With the exception of the antennae, trilobite appendages are more or less homonomous and show little regional specialization (Fig. 17-1B), although the first two appendages behind the antennae tend to differ somewhat from the others.

The uniramous antennae of the first cephalic segment attach beside the hypostome and are usually long, multiarticulate, and filiform. Although it cannot be confirmed, the antennae are assumed to have been chemo- and mechanosensory in most species. Each postantennal appendage is biramous, consisting of a proximal coxa and two rami. The coxa attaches to the ventral membrane and bears a lateral exopod and a medial endopod. The endopod typically consists of six cylindrical articles and a distal claw (Fig. 17-2A, 17-3) and, since it is leglike, it is presumed to have been used for walking by most species. The shape of the exopod varies considerably with taxon but often bears closely spaced filaments or flattened spines forming a feather- or leaflike ramus (Fig. 17-2A). This ramus usually is short and carried under the pleural lobe, but in *Triarthrus* it is much longer. In *Agnostus* it has large, club-shaped lobes (Fig. 17-3). The endopod may have been used for swimming, gas exchange, generating feeding or respiratory currents, manipulating food, digging, or filtering, or perhaps as a substratum for chemosynthetic bacteria.

The medial borders of the coxae bear teeth or spines and form a **gnathobase** (Fig. 17-2A, 17-3). Gnathobases of the two opposing rows of coxae face each other across the ventral midline to form the **food groove** (Fig. 17-1B, 17-2A). When feeding, movements of the rami moved food particles into the food groove. The accompanying movements of the coxae caused the setae or teeth of the gnathobases to move against each other and simultaneously break up the food and push it anteriorly in the food groove until it eventually arrived at the mouth. The mouth faces posteriorly to open at the anterior end of the food groove (Fig. 17-1B, 17-3, 17-5B, 19-4), where it received food moving anteriorly in the groove.

Two types of postantennal appendages are found, and the difference between them is more pronounced in some trilobites. The second and third cephalic appendages are similar to each other, whereas all the remaining posterior appendages, including the fourth cephalic, resemble each other. This dichotomy is slight in *Triarthrus* (Fig. 17-1B), but more pronounced in *Agnostus* (Fig. 17-3). In the latter, cephalic appendages 2 and 3 have large exopods and vestigial or small endopods, whereas the remaining appendages have large endopods and small exopods. The first three appendages—the uniramous antennae and biramous second and third cephalic limbs—are similar to the crustacean naupliar appendages consisting of uniramous first antennae, biramous second antennae, and biramous mandibles, suggesting homology of these appendages.

Gas exchange in trilobites is poorly understood. Small species probably had no special respiratory surfaces and used any permeable area of the exoskeleton, most likely the thin-walled, uncalcified ventral membrane. Larger species may have augmented this surface, perhaps by using the elaborated filamentous exopod of the appendages (Fig. 17-2A). *Agnostus* had clublike extensions of the endopod, thought to have had a soft, permeable cuticle for gas exchange (Fig. 17-3).

The little we know of internal anatomy is based on X-ray studies of pyretized specimens. The gut is J-shaped, with

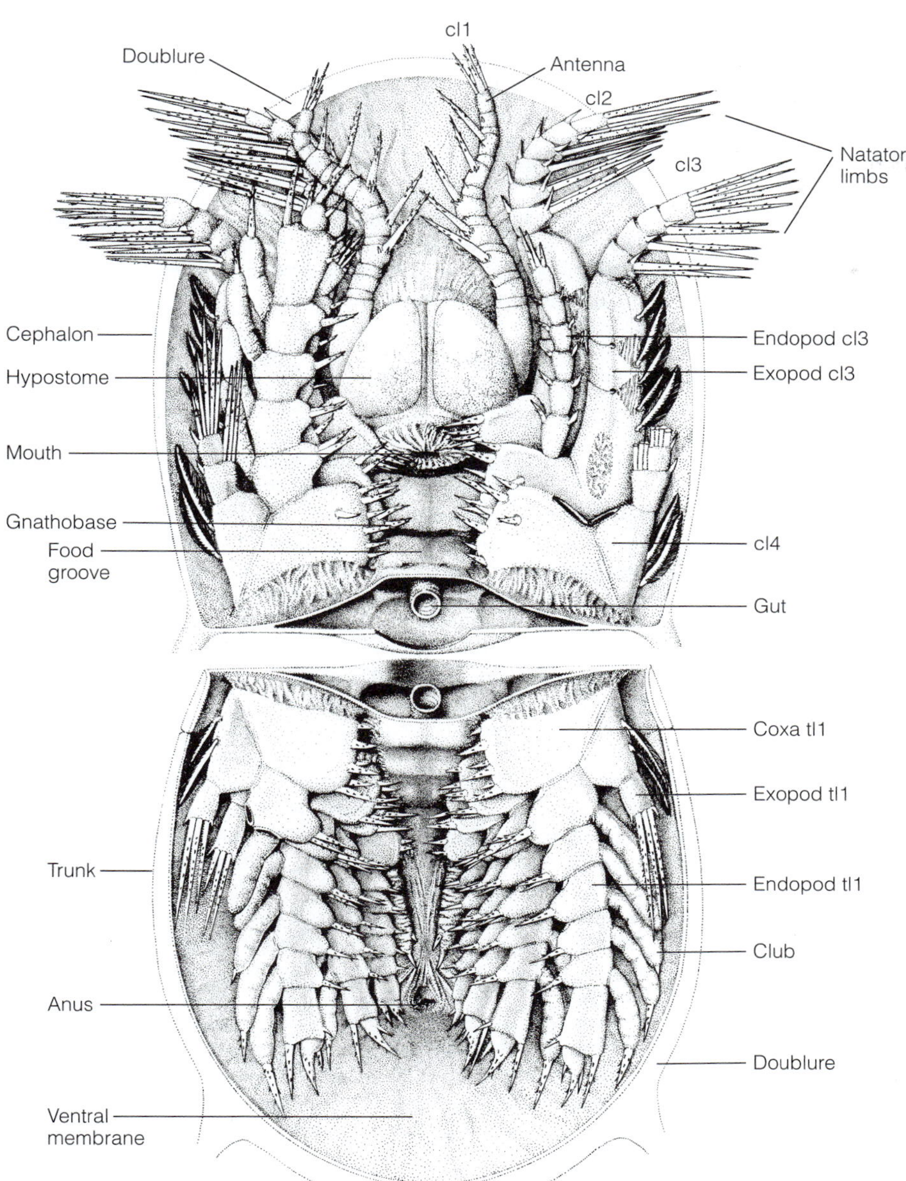

FIGURE 17-3 Ventral view of a late instar meraspis larva of *Agnostus pisiformis.* The cephalon and trunk are drawn disconnected at their articulation. The endopod of the left fourth cephalic appendage and the endopod and a club of the right first trunk appendage have been removed to reveal the underlying structures. The gut is severed in the thoracic region. See Figure 17-5 for explanations of abbreviations. *(From Müller, K. J., and Walossek, D. 1987. Morphology, ontogeny, and life habit of* Agnostus pisiformis *from the Upper Cambrian of Sweden. Foss. Strata 10:1–124 (Fig. 4), reproduced from Fossils and Strata www.tandf.no/fossils by permission of Taylor and Francis AS)*

a stomach or crop in the cephalon between the hypostome and glabella (Fig. 17-2B). The gut extends posteriorly in the axial lobe to the anus on the ventral midline of the pygidium. The cephalon also contains branching structures extending into the genae that probably are digestive ceca. Presumably, there is a hemocoel and nervous system in the axial lobe and a brain in the cephalon.

DEVELOPMENT

Almost nothing is known of reproduction by trilobites, although they are presumed to have been gonochoric. Sexual dimorphism is nonexistent, there are no discernable genitalia, and the sexes are indistinguishable. Fertilization was probably external, with the male fertilizing eggs released by the female, as is done today by the horseshoe crab *Limulus.* Some females may have brooded their eggs. Trilobite development included three larval and one adult stages, with each larval, and perhaps the adult, stage consisting of several instars separated by molts. Additional segments and appendages were added with each molt. The earliest larva was the small (≅ 1 mm) **protaspis larva,** which lacked appendages (Fig. 17-4A). Its dorsal shield consisted of a single plate with no division into tagmata. The protaspis was sometimes the hatching stage, but in some species it was suppressed in the egg and a later stage hatched. It probably was planktonic in most species and the chief dispersal stage in the life cycle. The protaspis was followed by a series of **meraspis larvae,** in which an articulation appeared between the cephalon

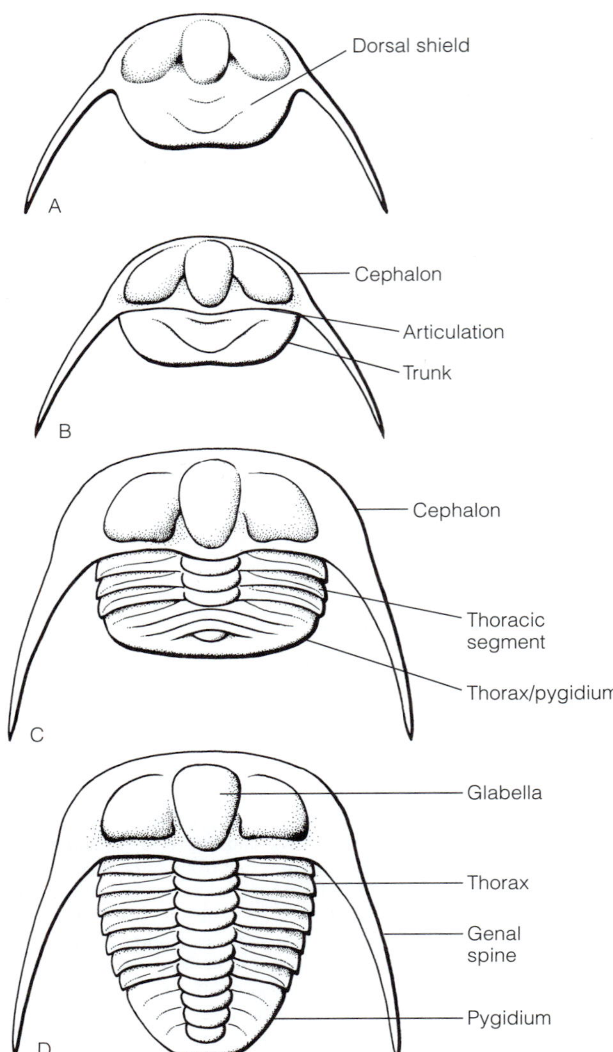

FIGURE 17-4 Four larval instars of a generalized trilobite. **A,** A protaspis. The dorsal shield is undivided. **B,** First instar meraspis with no thoracic segments. The dorsal shield is divided by a single articulation into an anterior cephalon and posterior trunk. **C,** An older meraspis after two thoracic segments have been added. **D,** A holaspis larva with a completed thorax of seven segments and a growing pygidium of four segments. Additional molts in the holaspis stage will complete the pygidium and increase the overall size.

and the trunk (Fig. 17-4B). Successive molts within the meraspis series added appendages to the trunk, and the thoracic segments gradually separated from the pygidium one at a time (Fig. 17-4C). The juvenile became a **holaspis larva** (Fig. 17-4D) when it achieved the adult complement of thoracic segments but lacked a complete pygidium. During the holaspis period pygidial segments and appendages were added and the animal increased in size with each molt. Larval development ended and the adult stage was reached when adult size and the adult complement of pygidial segments were achieved.

Immediately following ecdysis, the new exoskeleton was thin-walled and unmineralized. The exuvium retained most or all of the calcite, which was thus lost to the organism and had to be replaced slowly. Trilobites apparently did not resorb valuable materials from the exoskeleton prior to molting, and until its exoskeleton was remineralized the trilobite was vulnerable.

Adult trilobites are sometimes found with other animals, often brachiopods, attached to their dorsal shields, implying that adults did not molt or molted so infrequently that epizoic organisms had time to settle and grow.

ECOLOGY

Trilobites were exclusively marine, with no freshwater or terrestrial representatives, and most lived in relatively shallow coastal habitats. Differences in size, shape, spination, appendage morphology, cephalon and pygidial morphology, and eye size and position indicate that trilobites exploited a diversity of habitats. Most were presumably epibenthic animals that crawled over the sediment using endopods adapted for walking. Some, however, appear to have been adapted for planktonic, infaunal, and pelagic lives, and some lived in deeper water.

Although most trilobites fed on organic particles, living or dead, collected in the ventral food groove as described earlier, some were predators. Others such as *Trinucleus* may have been filter feeders that used a concavity under their expanded cephalon as a filter chamber. Some trilobites may have cultivated chemoautotrophic bacteria, as do many modern clams and the famous vestimentiferan tube worms of deep-sea hydrothermal vents. Trilobites such as *Triarthrus* and its relatives lived on the interface between hypoxic water and anoxic sediments rich in sulfides. In this depauperate habitat, *Triarthrus* may have maintained populations of sulfur-metabolizing bacteria on the body and exopod surfaces. The chemosynthetic bacteria would have oxidized the sulfides and used the liberated energy to reduce carbon and build organic compounds that then would be shared with the trilobite. It has been suggested that this is why *Triarthrus* exopods were so much larger than usual.

Trilobites may have been the prey of other Paleozoic arthropods such as *Anomalocaris* and of cephalopods and fishes. (*Anomalocaris,* which was probably an arthropod, was a giant Cambrian predator and, at 50 cm, one of the largest Cambrian animals.) Predation pressure probably increased as the Paleozoic advanced and new predators, including the jawed fishes, appeared and became larger and more abundant. The thickness of the calcified exoskeleton increased through the Paleozoic, perhaps in response to increased predation pressure. Spines also provided protection, and some trilobites may have relied on a coating of epizoic organisms for camouflage. Enrolling and burrowing were additional strategies for escaping predation.

Recent advances in our knowledge of trilobite limbs and the ventral membrane of a few species have encouraged speculation about ecology and natural history. Although it was not a typical trilobite, we know more about the anatomy of *Agnostus pisiformis* than any other species, and our knowledge is sufficient to allow meaningful speculation about its biology.

These small animals reached lengths of up to 6 mm (fully extended) and were similar to the agnostid shown in Figure 17-7B. Adults consisted of a large cephalon and pygidium connected by a short thorax of only two segments, with the major articulation between the cephalon and first thoracic segment. They were capable of complete enrollment to form a closed, clamlike ball of up to 3 mm in length. If we pursue the clam analogy, the cephalon would be one valve, the pygidium the other, and the thorax the hinge. The valves would be anterior and posterior, of course, rather than right and left as in a real clam, but the analogy is nevertheless useful. The cephalon and pygidium (the valves) are semicircular and similar to each other in size and shape (Fig. 17-3), so their rims (doublures) fit tightly together when enrolled (Fig. 17-5A). Both are concave ventrally and, when closed, formed a capsule containing all the appendages and the large hypostome. In its presumed **life position** (the typical posture and orientation of the animal in life), *Agnostus* probably was partly enrolled, with the valves gaping to expose the appendages (Fig. 17-5B). It seems unlikely that *Agnostus* could swim, feed, or walk in the fully extended position (Fig. 17-5C) and, of course, it certainly could not do so when fully enrolled.

The appendages consist of the four pairs of cephalic limbs, including the antennae and three pairs of postantennal legs, and five pairs of trunk limbs, two on the thorax and three on

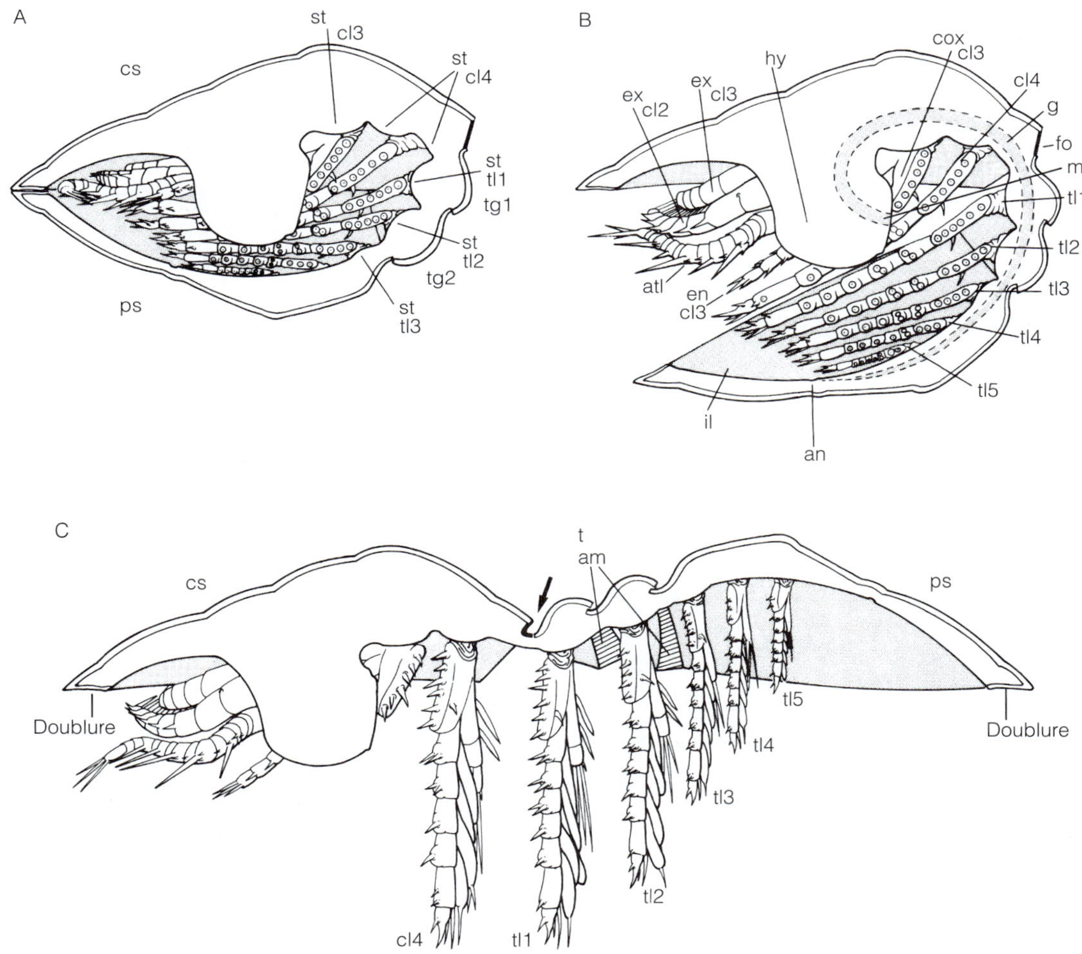

FIGURE 17-5 Enrollment positions of the agnostid trilobite *Agnostus pisiformis.* **A,** Fully enrolled with all appendages protected. **B,** Partly enrolled in the most likely life position. **C,** Fully extended. Anterior is to the left. Abbreviations for 17-3, 17-5, and 17-6: am, articulating membrane; an, anus; atl, antenna; cl, cephalic limb; cox, coxa; cs, cephalon; en, endopod; ex, exopod; fo, articulation between cephalon and thorax; g, gut; hy, hypostome; il, ventral membrane; m, mouth; ps, pygidium; s, seta; st, sternite; t, thorax; tg, tergite; tl, trunk limb. Note the posteriorly directed mouth (unlabeled) on the hypostome. *(All from Müller, K. J., and Walossek, D. 1987. Morphology, ontogeny, and life habit of* Agnostus pisiformis *from the Upper Cambrian of Sweden. Foss. Strata 10:1–124 [Fig. 25], reproduced from Fossils and Strata www.tandf.no/fossils by permission of Taylor and Francis AS.)*

the pygidium. All appendages except the antennae are biramous. The rami bear a variety of setae, spines, and the enigmatic clubs. The antennae are short and thick, lacking the filiform morphology of trilobites such as *Triarthrus,* and probably were not sensory. The endopod of the second cephalic limb is vestigial, but its exopod is strong (Fig. 17-3). The coxae of all appendages except the antennae have heavy medial spines and are gnathobases facing each other across a longitudinal median food groove. In the life position, the appendages all extended anteriorly toward the gape (Fig. 17-5B).

The exopods of the second and third cephalic appendages have long, presumably natatory setae (Fig. 17-3) and resemble the swimming appendages of some crustaceans. *Agnostus* appendages do not appear to be adapted for walking (Fig. 17-5C). These observations imply that, in the partly enrolled configuration, *Agnostus* was capable of swimming. It could have been an epibenthic animal, perhaps occupying the layer of flocculent organic matter on the seafloor and making periodic forays into the water, or it may have been planktonic, spending most or all of its time in the water column. It has been suggested that it was an epibenthic pebble mimic. Other authors have proposed an association with attached or floating seaweeds.

Agnostus probably fed in the partly enrolled position. The appendages are not suitable for filter feeding because there are no filter screens or sucking chambers such as those associated with filter-feeding crustaceans (Fig. 19-12). On the other hand, the antennae do seem to be adapted for grasping food items in front of the gape and transferring them to the endopods of the posterior trunk limbs (Fig. 17-6). From there the food would have moved to the coxae. Rhythmic beating of the limbs caused the coxae to push the food anteriorly in the food groove toward the hypostome. The mouth, which may have been suctorial, is located on the posterior surface of the hypostome and thus faces posteriorly, ready to receive food moving anteriorly from the coxae. The gut has the J shape typical of arthropods known to feed in this manner (Fig. 17-5B) and the anus is located on the ventral membrane of the pygidium (Fig. 17-6).

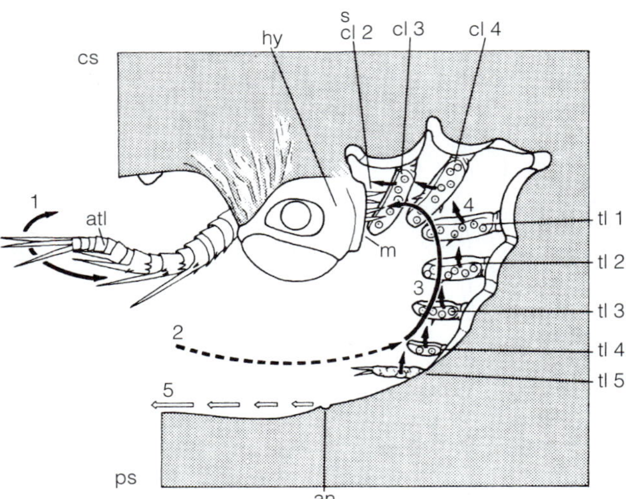

FIGURE 17-6 Diagrammatic representation of the partly enrolled ventral surface of *Agnostus* showing the probable food path. 1 = antennal movement, 2 = food current toward coxae, 3 = food movement toward mouth driven by metachronal movements of limbs, 4 = forward movements of coxae, 5 = exit of feces. Abbreviations as in Figure 17-5. *(From Müller, K. J., and Walossek, D. 1987. Morphology, ontogeny, and life habit of* Agnostus pisiformis *from the Upper Cambrian of Sweden. Foss. Strata 10:1–124 [Fig. 25], reproduced from Fossils and Strata www.tandf.no/fossils by permission of Taylor and Francis AS.)*

If *Agnostus* was planktonic, then feeding was probably combined with the swimming movements of the appendages. If it was epibenthic, it probably rested with its pygidial "valve" on the substratum, with the "valves" gaping, and used rhythmic movements of the appendages to move food particles to the mouth and generate a respiratory current.

No obviously respiratory limbs are present in *Agnostus,* and it may be that, because of its small size, no special respiratory surfaces other than the ventral membrane were necessary. The endopods of the trunk and posterior cephalic appendages, however, bear unique, thin-walled, club-shaped processes (Fig. 17-3) that may have been respiratory. The clubs would have been ventilated by movements of the appendages.

The dorsal shield is penetrated by numerous tiny pores 2 to 3 μm in diameter, presumed to be sensory. Reproductive biology is unknown, but the hatching stage was a meraspis followed by six more meraspis and at least one holaspis instar.

DIVERSITY OF TRILOBITOMORPHA

There is great morphological diversity among the 4000 known species (Fig. 17-7), which are assigned to eight higher taxa (orders). Trilobites inhabited the oceans from the Lower Cambrian through the Ordovician, Silurian, and Devonian periods to the Permian. During the Paleozoic era, 540 to 225 mya, they were dominant or at least important members of most marine communities. The Paleozoic ended with the Permian about 225 mya, accompanied by the mass extinction of 50% of all animal families and 95% of all marine species, including the trilobites, rugose corals, eurypterids, blastoids, and ammonoids along with most brachiopods and crinoids. Trilobites had been declining steadily long before their final extinction and only one order was present in the Carboniferous and Permian. Their heyday was in the Cambrian and Ordovician.

Redlichiida^O (Fig. 17-7A; Cambrian): The most primitive known trilobites; ancestral to the other trilobites. Large, semicircular cephalon, strong genal spines, opisthoparian sutures. Glabella conspicuously segmented by deep glabellar furrows. Thorax well developed, with many segments. Eyes large. Pygidium small or vestigial. Appeared in Lower Cambrian, extinct by Middle Cambrian, having been replaced by their descendants. *Olenellus, Olenoides, Paradoxides, Redlichia.*

Agnostida^O (Fig. 17-7B; Cambrian-Ordovician): Primitive. Very small, < 8 mm long. Cephalon and pygidium large, similar in size and shape. Two or three thoracic segments. Usually blind. No facial sutures. May have been planktonic. Probably diverged early in the Cambrian from primitive redlichiid ancestors. *Agnostus, Eodiscus, Pagetia.*

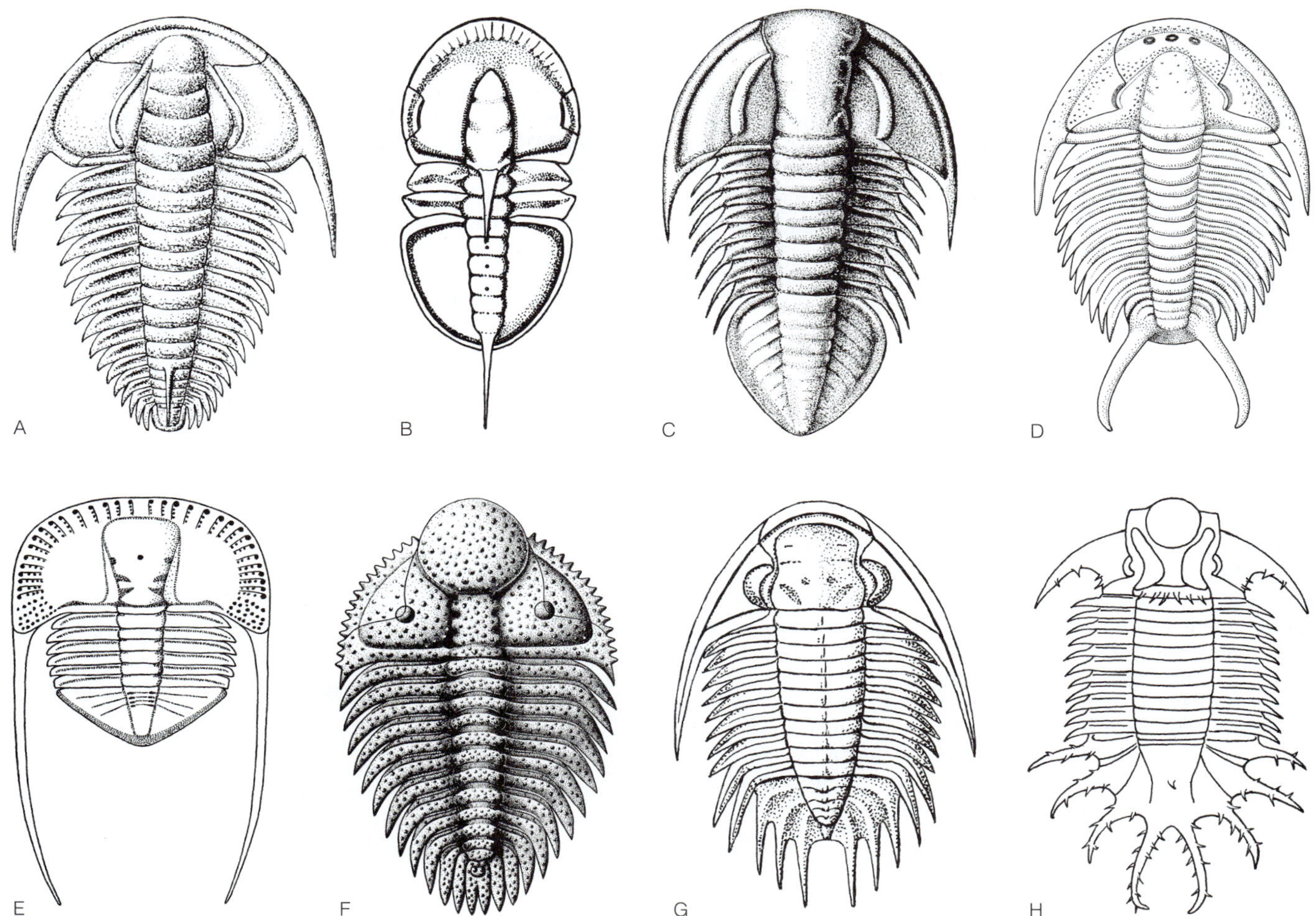

FIGURE 17-7 Representatives of the trilobite taxa (orders). **A,** Redlichiida, *Redlichia.* **B,** Agnostida, *Pagetia.* **C,** Corynexochida, *Polypleuraspis.* **D,** Ptychopariida, *Tricrepicephalus.* **E,** Asaphida, *Trinucleus.* **F,** Phacopida, *Staurocephalus.* **G,** Proetida, *Phaetonellus.* **H,** Lichida, *Terataspis.* *(All from Harrington, H. J., et al., 1959. Treatise on Invertebrate Paleontology. Geological Society of America, New York, and University of Kansas Press, Lawrence, with permission of the Geological Society of America.)*

Corynexochida^O (Fig. 17-7C; Cambrian-Devonian): Long glabella with parallel sides, sometimes flared anteriorly. Hypostome and rostrum (median region of doublure adjacent to hypostome) fused. Opisthoparian facial suture. Thorax of 5 to 11 segments. Pygidium medium to large. Genal spines present in many. Thorax often, pygidium sometimes spiny. Arose from redlichiid ancestors in late Cambrian. *Illaenus, Olenoides, Polypleuraspis.*

Ptychopariida^O (Fig. 17-7D; Cambrian-Devonian): Large taxon with several well-known genera. Facial suture usually opisthoparian. Glabella simple, tapering anteriorly. Number of thoracic segments variable. Pygidium small in early species but generally large in derived species. Arose from redlichiid ancestors in Middle Cambrian. Probably ancestral to several orders that persisted after the Cambrian. *Elrathia, Olenus, Shumardia, Triarthrus, Tricrepicephalus.*

Asaphida^O (Fig. 17-7E; Cambrian-Silurian): Recently established taxon related to Ptychopariida. Most had median ventral suture. Swollen, globular protaspis larva. Arose from ptychopariid ancestors in Middle Cambrian. An important taxon until the end of the Ordovician. A few persisted through Silurian. *Asaphus, Cyclopyge, Isotelus, Ogygiocarella, Trinucleus.*

Phacopida^O (Fig. 17-7F; Ordovician-Devonian): Large taxon. Glabella often widened anteriorly. Facial suture proparian or gonatoparian, but sutures are fused and did not separate during ecdysis. Thorax of 8 to 19 segments. Medium to large semicircular pygidium. Probably evolved from asaphid or ptychopariid ancestors in Lower Ordovician. Phacopina^sO are the only trilobites with schizochroal eyes. *Calymene, Ceraurinella, Ceraurus, Eldredgeops (formerly Phacops), Staurocephalus.*

Proetida^O **(Fig. 17-7G; Ordovician-Permian):** Most have strong genal spines, large glabellae, opisthoparian sutures, 8 to 10 thoracic segments. Pygidium medium-sized. The only taxon surviving the Devonian extinctions to persist into the Carboniferous and Permian. Arose from Upper Cambrian ptychopariid ancestors. *Bathyurus, Ditomopyge, Phaetonellus, Proetus.*

Lichida^O **(Fig. 17-7H; Cambrian-Devonian):** Pygidium larger than cephalon. Usually spiny or tuberculate. Medium to very large. Glabella prominent, with large lobes. Sutures opisthoparian. Thorax has 8 to 11 segments. Pygidium variable. Evolved from Middle Cambrian redlichiid ancestors. *Dicranurus, Lichas, Odontopleura, Terataspis.*

PHYLOGENY OF TRILOBITOMORPHA

Trilobites appeared in the fossil record as fully formed and highly specialized derived arthropods in the Lower Cambrian, about 540 mya. The earliest known trilobites had calcified exoskeletons and presumably were preceded by uncalcified species that left no record. These, in turn, must have had arthropod ancestors in the Precambrian, and before that there must have been the soft-bodied ancestors of the arthropods. These precursors left no record, at least none that has been discovered. Perhaps the protoarthropods were small and had uncalcified exoskeletons that, like those of copepods and cladocerans with similar cuticles, do not fossilize. The major animal taxa are thought to have diverged about 650 to 1000 mya, well back in the Precambrian, but there is little record of it. By the Cambrian the major animal taxa (phyla) were established and there were already many kinds of arthropods, including trilobites. The Middle Cambrian Burgess Shale in British Columbia, the Chinese Chengjiang deposits, and a few other famous fossil beds contain diverse faunae of ancient arthropods that arose during the so-called Cambrian Explosion beginning about 545 mya.

The Burgess Shale, discovered by Charles Walcott in 1909 and made famous outside of paleontological circles by Stephen Jay Gould in 1989, contains a remarkable record of Cambrian invertebrates, including 19 species of arthropods, only 3 of which are trilobites. One of the nontrilobites is a chelicerate and 2 are crustaceans, but the remaining 13 Burgess arthropods belong to no known arthropod taxon, either extant or extinct (Fig. 17-8). Unlike most modern arthropods, these Cambrian species were homonomous and exhibited little regional specialization of their segments or appendages. They had multisegmented trunks, with each trunk segment bearing a pair of appendages like those of all the other trunk segments. Cladistic analysis of these early arthropods reveals the trilobites as being among the most derived, and not primitive at all.

Although undoubtedly related to crustaceans, trilobites are probably closest to the chelicerates. Similarities between the appendages of trilobites and those of xiphosurans and eurypterids support this relationship. Trilobites and chelicerates probably form the taxon Arachnata, which is the sister taxon of Mandibulata (Crustacea plus Tracheata). Among the arachnatans, the trilobites retained the antennae, which were lost in the chelicerates. In ventral view the similarity between *Agnostus* and a horseshoe crab is striking (Fig. 17-3, 18-1), and horseshoe crabs are probably the closest living relatives of trilobites.

The trilobites and the extant arthropod taxa are believed to have had a common ancestor in the Precambrian, but 100 million years of arthropod evolution probably preceded the appearance of the trilobites in the rocks of the Lower Cambrian. During the Cambrian, Arthropoda underwent the adaptive radiation that produced the many different designs of larger, calcified, and fossilizable taxa preserved in the Burgess Shale and Chengjiang faunas. Of this extensive Cambrian arthropod diversity only the chelicerates, crustaceans, and tracheates survive today, but they dominate each of the Earth's four major environments (marine, terrestrial, freshwater, and aerial) and account for most of its biological diversity.

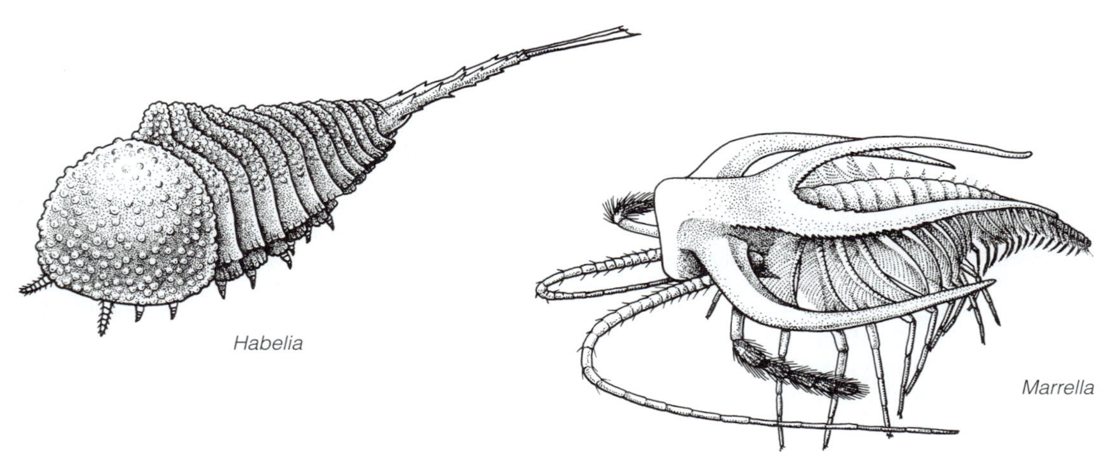

FIGURE 17-8 Reconstructions of two arthropods from the Burgess Shale that belong to no known arthropod taxa. *(Illustrations by Marianne Collins from Wonderful Life, The Burgess Shale and the Nature of History, by Stephen J. Gould, are reproduced by permission of W. W. Norton & Company, Inc. Copyright © 1989 by Stephen Jay Gould.)*

REFERENCES

Bergström, J. B. 1973. Organization, life, and systematics of trilobites. Foss. Strata 2:1–69.

Briggs, D. E. G., Erwin, D. H., and Collier, F. J. 1994. The Fossils of the Burgess Shale. Smithsonian Institution Press, Washington, DC. 238 pp.

Chen, J., and Zhou, G. 1997. Biology of the Chengjiang fauna. Bull. Natl. Mus. Nat. Sci. 10:11–105.

Cisne, J. L. 1974. Trilobites and the origin of arthropods. Science 186:13–18.

Cisne, J. L. 1975. Anatomy of *Triarthrus* and the relationships of the Trilobita. Foss. Strata 4:45–63.

Conway Morris, S. 1989. Burgess Shale faunas and the Cambrian Explosion. Science 246:339–346.

Fortey, R. 2000. Trilobite Eyewitness to Evolution. Vintage, New York. 284 pp.

Fortey, R. 2001. Trilobite systematics, the last 75 years. J. Palaeontol. 75:1141–1151.

Gould, S. J. 1989. Wonderful Life: The Burgess Shale and the Nature of History. W. W. Norton, New York. 347 pp.

Harrington, H. J., et al. 1959. Systematic descriptions. In Moore, R. C. (Ed.): Treatise on Invertebrate Paleontology: Pt. O, Arthropoda 1. Geological Society of America, University of Kansas Press, Lawrence, KS. pp. O170–O539.

Kaesler, R. L. (Ed.): 1997. Treatise on Invertebrate Paleontology. Pt. O, Revised: Arthropoda 1. Geological Society of America, University of Kansas Press, Lawrence, KS. 530 pp.

Levi-Setti, R. 1975. Trilobites: A Photographic Atlas. University of Chicago Press, Chicago. 214 pp.

Moore, R. C. (Ed.): 1959. Treatise on Invertebrate Paleontology. Pt. O: Arthropoda 1. Geological Society of America, University of Kansas Press, Lawrence, KS. 560 pp.

Müller, K. J., and Walossek, D. 1987. Morphology, ontogeny, and life habit of *Agnostus pisiformis* from Upper Cambrian of Sweden. Foss. Strata 19:1–124.

Stürmer, W., and Bergström, J. 1973. New discoveries on trilobites by X-rays. Paläont. Z. 47:104–141.

Whiteley, T. E., Kloc, G. J., and Brett, C. E. 2002. Trilobites of New York: An Illustrated Guide. Cornell University Press, Ithaca. 203 pp.

Whittington, H. B. 1959. Ontogeny of Trilobita. In Moore, R. C. (Ed.): Treatise on Invertebrate Paleontology. Pt. O: Arthropoda 1. Geological Society of America, University of Kansas Press, Lawrence, KS. pp. O127–O144.

Whittington, H. B. 1975. Trilobites with appendages from the middle Cambrian, Burgess Shale, British Columbia. Foss. Strata 4:97–136.

INTERNET SITES

www.aloha.net/~smgon/ordersoftrilobites.htm (A Guide to the Orders of Trilobites, by S. Gon. A large site with photos, links, line drawings, key to orders, and articles.)

http://biosys-serv.biologie.uni-ulm.de/sektion/dieter/dieter.html ("Orsten" Research and Dieter Waloszek's View of Arthropod and Crustacean Phylogeny.)

www.yale.edu/ypmip (Yale University Peabody Museum of Natural History, Invertebrate Paleontology Image Gallery.)

www.biosis.org/zrdocs/zoolinfo/grp_tril.htm (*Zoological Record* Internet Resource Guide for Zoology: Trilobitomorpha. This page links to many other Web pages concerning trilobites.)

18 Chelicerata^sP

Chelicerata and Mandibulata are the two sister taxa of modern arthropods (Fig. 20-14D). Chelicerata, with about 70,000 known living species, includes spiders, scorpions, harvestmen, mites, and horseshoe crabs as well as the sea spiders and extinct sea scorpions. Most modern chelicerates are terrestrial and form one of the two major taxa of land arthropods. The original arthropods were aquatic animals from which descended two distinct evolutionary lines, Chelicerata and Mandibulata, each of which independently produced highly successful terrestrial lineages.

Chelicerata consists of the terrestrial Arachnida (spiders, mites, scorpions, and others), the extinct Eurypterida (sea scorpions), the marine Xiphosura (horseshoe crabs), and many fossil taxa (Fig. 18-47). Together these taxa form Euchelicerata. Pycnogonida (sea spiders) is probably the sister taxon of Euchelicerata in Chelicerata.

FORM

The chelicerate body is divided into two tagmata, the cephalothorax and abdomen. The **cephalothorax,** or prosoma, consists of the acron plus seven segments and bears six pairs of appendages. It forms in the embryo through fusion of the head (with eyes and chelicerae) and the thorax (with pedipalps and walking legs). The appendages of the first cephalothoracic segment have been lost. In other arthropods this segment bears antennae, but they are not present in chelicerates. Associated with the loss of the antennae is the absence of the deutocerebrum from the brain.

The first appendages on the chelicerate cephalothorax, a pair of prehensile **chelicerae,** are on the second, rather than the first, cephalothoracic segment. Each chelicera consists of two or three articles, the distal two of which form a chela, or pincer. The second appendages are **pedipalps** modified to perform various, but often sensory, functions in the different taxa. The remaining cephalothoracic appendages are four pairs of **walking legs.** Primitively the head bears four median pigment-cup ocelli and two lateral compound eyes.

The remaining trunk segments constitute the abdomen, or opisthosoma, consisting of 12 or fewer segments. Abdominal appendages, when present, are primitively platelike and have a respiratory function. The abdomen of derived chelicerates usually has highly modified appendages on some segments or lacks appendages. Primitively the abdomen is divided into an anterior **preabdomen** (mesosoma) of seven segments and a posterior **postabdomen** (metasoma) with five segments and a terminal telson, or spike. This original pattern has been variously modified in the many chelicerate taxa. The gonopores have moved anteriorly to the second abdominal segment.

XIPHOSURA[C]

Xiphosura and Arachnida, sister taxa of Euchelicerata, are represented today by horseshoe crabs, spiders, scorpions, and their relatives. In contrast with that of pycnogonids, the euchelicerate abdomen is well developed and often segmented. The cephalothoracic segments are fused and covered by a continuous sclerite, the carapace. Euchelicerates have retained the ancestral lateral compound eyes, although they may be highly modified. They have lost one of the two pairs of median eyes and their gonopores are on the second abdominal segment.

Horseshoe crabs (Xiphosura) are by far the largest modern chelicerates and, at lengths of up to 75 cm, they dwarf spiders, scorpions, mites, and harvestmen, although they are small in comparison with some extinct chelicerates. Because they retain many characteristics of the stem chelicerate, including the aquatic habitat, they are of special interest in the study of arthropod and chelicerate phylogeny.

Although the fossil record of Xiphosura extends back to the Silurian period, the taxon is almost extinct and only three genera *(Limulus, Tachypleus,* and *Carcinoscorpius),* represented by four species, are alive today. One of these is *Limulus polyphemus,* which is common in shallow water along the North American Atlantic coast. The three other extant xiphosurans occur on Asian coasts from Korea and Japan south through the Philippines and Indonesia.

Horseshoe crabs live in shallow marine water on soft bottoms, where they plow through the upper surface of the sediment. The American species, *Limulus polyphemus,* lives on the continental shelf but migrates into very shallow water at breeding time. Mature females achieve lengths in excess of 60 cm, but males are smaller. Horseshoe crabs are not crabs, of course, and are much closer phylogenetically to spiders than to true crabs.

FORM

Neither the cephalothorax nor the abdomen is obviously segmented, and each is enclosed in a single unbroken, rigid, exoskeletal box (Fig. 18-1). The two tagmata are joined by a movable, hingelike articulation and both bear appendages on the ventral surface. A long, spikelike **tail spine** extends from the posterior end of the abdomen.

The cephalothorax is shaped like a horse's hoof (Fig. 18-1), hence the name *horseshoe crab.* Its dorsal exoskeleton is the **carapace,** whose anterior edge resembles a bulldozer blade and, appropriately, is used to plow through soft sediments. The carapace bears a pair of small ocelli on its anterior midline and a pair of larger lateral compound eyes (Fig. 18-1). Ventrally the cephalothorax is strongly concave and bears six pairs of appendages and the mouth.

The first appendages are the chelate chelicerae composed of three articles (Fig. 18-1). The next four appendages are unspecialized, uniramous walking legs with similar morphology. Each is composed of six articles (coxa, trochanter, femur, patella, tibia, and tarsus), of which the last two form a chela, or pincer. The first walking leg of males, sometimes called a pedipalp, is used to grasp the female prior to mating and is not chelate. The biramous sixth cephalothoracic appendages, also not chelate, are specialized for pushing through soft sediments. Each of these **pusher legs** has four leaflike processes attached to the end of the first tarsal segment. The coxa of this leg bears a lateral spatulate, sensory exopod, the **flabellum** (Fig. 18-1). The leg itself is the endopod. The coxae of legs 2 through 6 each have a spiny medial

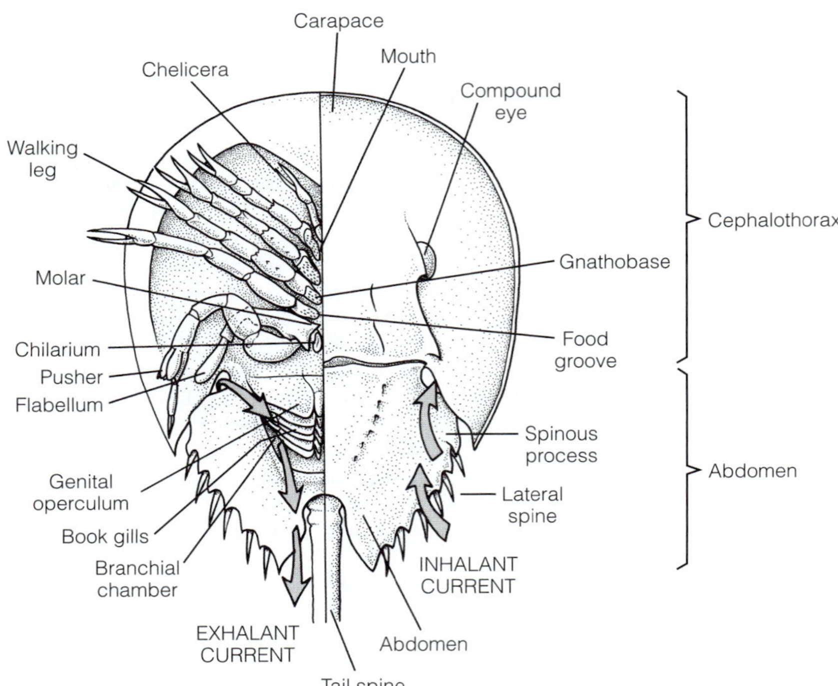

FIGURE 18-1 Xiphosura. A horseshoe crab, *Limulus polyphemus,* in combined ventral (left) and dorsal (right) views. Arrows show the inhalant and exhalant ventilating currents.

gnathobase. Anteriorly the labrum, or hypostome, overhangs the mouth and closes the anterior end of the food groove.

The abdomen also is unsegmented and concave ventrally. It fits into the recess on the posterior edge of the cephalothorax (Fig. 18-1). Its ventral concavity, the **branchial chamber,** houses and protects the abdominal appendages, most of which are book gills. The xiphosuran abdomen is formed of nine fused segments and, while they are not obvious, the external indications of the segments include the appendages, muscle scars, and lateral spines. The ninth segment represents the vestigial postabdomen, whereas segments 1 through 8 are the preabdomen.

The vestigial first abdominal segment is fused with the cephalothorax and no longer appears to be part of the abdomen. The appendages of this segment are the small, one-articled **chilaria** (sing., chilarium).

The second abdominal segment has a pair of large, dorsal spinous processes (Fig. 18-1) and a pair of fused appendages, the broad, flat **genital operculum** (Fig. 18-1, 18-2A, 18-3) with paired male or female gonopores on its posterior surface. These, and the other abdominal appendages, are biramous and have a basal protopod from which arises an endopod and exopod. The protopod and exopod are broad and flat, whereas the endopod is narrow and more leglike.

Segments 3 through 7 each bear a pair of book gills (Fig. 18-1, 18-2A) and a pair of large, movable **lateral spines.** The book gills, which are the appendages of these segments, are swimming legs that have been modified to function as respiratory organs. Young horseshoe crabs still use them for swimming, but adults are less likely to leave the bottom. Gills are biramous appendages similar to the operculum but with abundant pagelike lamellae on the posterior side. The lamellae are elaborations of the exopod, whereas the endopods are sensory.

Segment 8 has lateral spines but no gills and segment 9 has neither spines nor gills. The long tail spine (telson) extends posteriorly from segment 9. The base of the tail spine is equipped with muscles and the spine can be pointed in many directions. It is used for pushing and for righting the body when accidentally turned upside down. The anus is a ventral opening in the soft integument between segment 9 and the tail spine. Horseshoe crabs may appear fearsome with their many pincers and long, sharp tail spine, but they are harmless. The pincers are weak and the spine is not used for defense (or offense).

NUTRITION

Xiphosurans are omnivores that feed on benthic molluscs, worms, and other organisms, including algae. Unlike other living chelicerates, the xiphosuran gut is adapted for processing solid food. The gnathobases of the right and left legs form a food groove along the ventral midline of the cephalothorax (Fig. 18-1, 19-4). The groove begins between the pusher legs and extends anteriorly to the mouth, near the middle of the cephalothorax. The posterior end of the food groove is closed by the chilaria. Because the gut is J-shaped, the mouth faces posteriorly, into the anterior end of the groove (Fig. 18-2A, 18-4). When feeding, the animal uses its many chelae to capture and move small invertebrates to the food groove. Feeding and walking motions of the legs move the coxae, causing the gnathobases to crush and grind food items in the groove. The spines of the moving gnathobases push the food inexorably

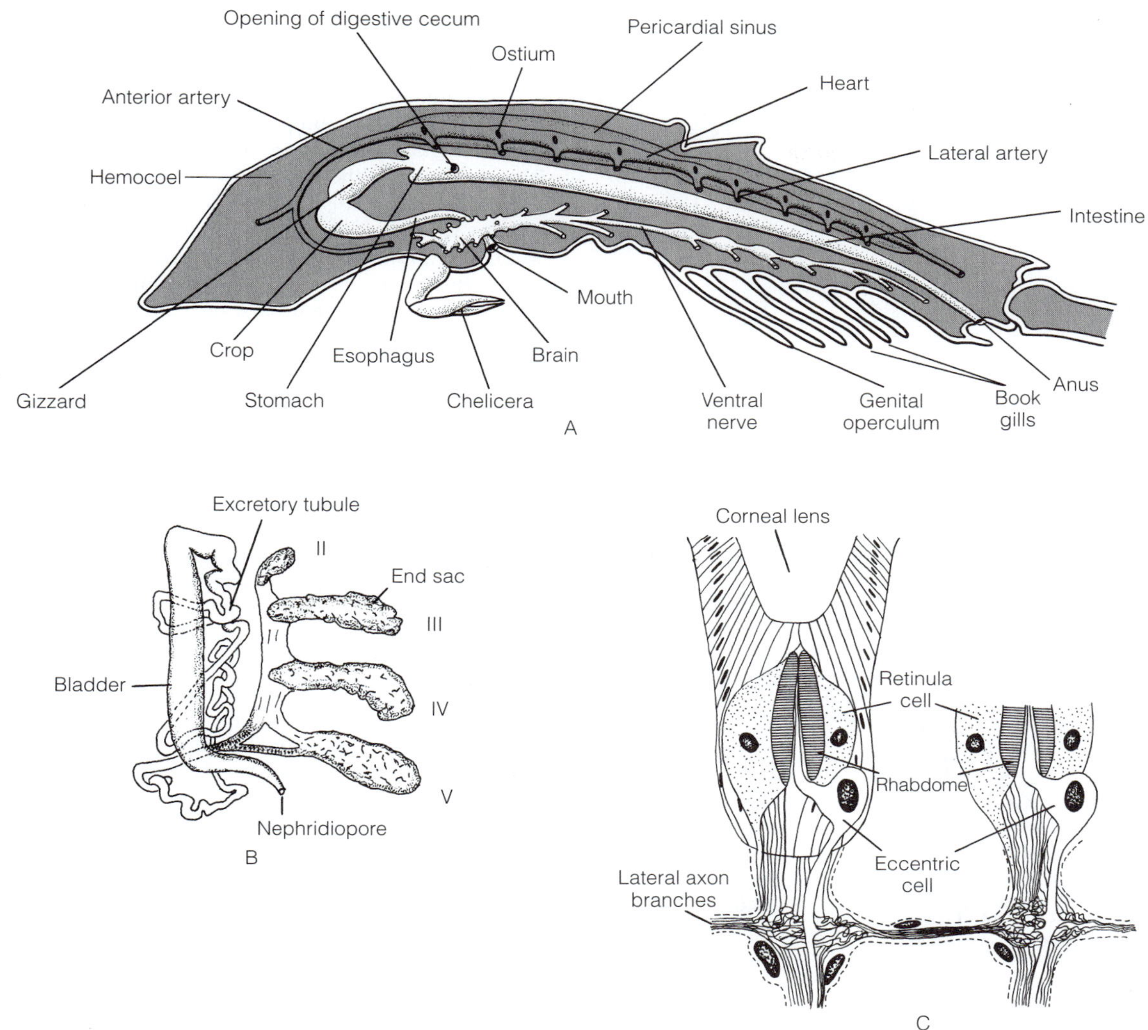

FIGURE 18-2 Xiphosura. **A,** Sagittal section through *Limulus.* The lamellae of the book gills have been omitted for clarity. **B,** Saccate nephridia and excretory duct of the horseshoe crab, *Limulus.* **C,** Longitudinal section through two ommatidia of the eye of *Limulus.* *(A, Greatly modified after Firstman; B, After Patten and Hazen from Fage; C, After MacNichol.)*

toward the mouth. The gnathobase of each pusher leg is equipped with a molar for crushing hard food items such as small clams and crustaceans.

The ectodermal foregut is lined with cuticle and consists of the esophagus and **proventriculus,** the latter consisting of a thin-walled, distensible crop for storage and a muscular, toothed gizzard (Fig. 18-2A, 18-4). Food is triturated in the gizzard and indigestible particles are regurgitated. A pyloric valve separates the gizzard from the midgut.

The endodermal midgut includes a short anterior stomach and a long posterior intestine. Two pairs of large digestive ceca (hepatopancreases) are connected by hepatic ducts to the stomach (Fig. 18-2A, 18-4). Each cecum consists of countless tiny, branched tubules whose lumina are continuous with the midgut lumen. The ceca fill most of the space in the hemocoel of the cephalothorax and abdomen and are the sites of extracellular digestion and absorption. Although typically brown, xiphosuran ceca often have deposits of bright white calcium phosphate. Excess calcium phosphate in the blood is absorbed by the ceca, precipitated in the ceca lumina, and then moved to the lumen of the gut and voided with the feces. The midgut epithelium secretes a peritrophic membrane of chitin and mucoproteins that encloses the food mass and feces. The ectodermal hindgut is a short, cuticularized rectum that opens to the exterior via the anus. Horseshoe crab tissues are potentially lethal to their predators. High concentrations of saxotoxin have been reported from the viscera of the Asian *Carcinoscorpius.*

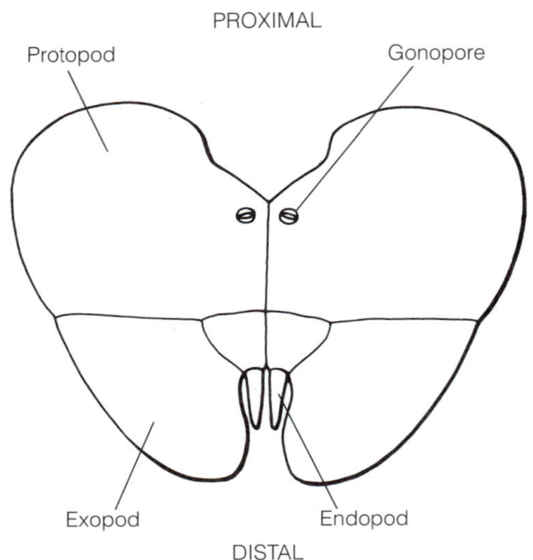

FIGURE 18-3 Xiphosura. Posterior view of the genital operculum of *Limulus.*

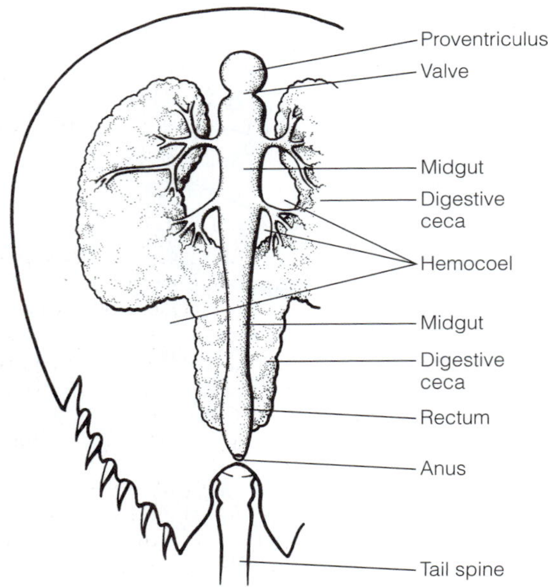

FIGURE 18-4 Xiphosura. Dorsal view of the hemocoel and viscera of *Limulus.*

INTERNAL TRANSPORT

The body cavity is a hemocoel and the coelom is vestigial. The long, tubular heart (Fig. 18-2A) extends from the posterior end of the proventriculus almost to the tail spine. It lies on the midline in the dorsal hemocoel, surrounded by the pericardial sinus.

During diastole blood enters the heart from the pericardial sinus through nine pairs of ostia. During systole, it leaves through a well-developed arterial system consisting of an aorta to the anterior cephalothorax and numerous pairs of lateral arteries. Arteries deliver blood to hemocoelic sinuses surrounding the tissues. Blood then moves ventrally in two large, longitudinal sinuses and from there to the book gills, where it is oxygenated. The same gill movements that create the ventilating current also pump blood through the gill lamellae. From the gills, oxygenated blood returns to the pericardial sinus surrounding the heart.

The blood contains the respiratory pigment hemocyanin in solution as well as two types of hemocytes that function in clotting (granulocytes) and in the production of hemocyanin (cyanocytes). Because of the large amount of blood that can be obtained from a mature horseshoe crab, they are a favorite subject of physiologists and biochemists interested in hemocyanin and other blood components.

GAS EXCHANGE

The respiratory surfaces are the multitudes of flat, pagelike gill lamellae on the posterior surface of the five pairs of book gills located in the branchial chamber under the abdomen (Fig. 18-1, 18-2A). The lamellae are thin, parallel, cuticularized evaginations of the exoskeleton resembling the numerous pages of a book. The 80 to 200 lamellae on the exopod of each gill together create a permeable gas exchange surface of about 2 m^2. The hemocoel extends into the interior of each lamella and gases are exchanged across the thin cuticle between blood and environment.

Rhythmic movements of the gills maintain a flow of water through the branchial chamber to ventilate the lamellae. On each side an inhalant current enters the branchial chamber dorsally through a gap between the cephalothorax and abdomen (Fig. 18-1). The current then flows posteriorly and ventrally over the gill lamellae to exit as an exhalant current on each side of the tail spine. The sensory flabellum of the pusher leg is located in the inhalant current, where it monitors water quality.

EXCRETION

As in most aquatic animals, nitrogen metabolism is ammonotelic and ammonia and other wastes and toxins are eliminated by diffusion across the permeable surfaces of the gill lamellae. The gills are also active in ion regulation, and specialized epithelia on the lamellae pump ions from the environment into the blood. For osmoregulation in brackish water, four pairs of saccate nephridia eliminate some metabolites, toxins, and spent hormones (Fig. 18-2B). The four brick-red saccate nephridia (coxal glands) on each side share a common duct, bladder, and nephridiopore. The nephridiopores open on the coxae of the fifth pair of legs. In embryos there is a pair of saccate nephridia in each cephalothoracic segment, but two pairs are lost in the adult.

NERVOUS SYSTEM AND SENSE ORGANS

The cephalized CNS is dominated by a large brain, or supraesophageal ganglion (Fig. 18-2A), dorsal to the esophagus, two circumesophageal connectives, and a subesophageal ganglion ventral to the esophagus. The brain is a syncerebrum consisting of two pairs of large neuromeres. The eyes are innervated by the protocerebrum and the chelicerae by the second pair of neuromeres, traditionally considered to be homologous to the tritocerebrum. The subesophageal ganglion consists of the fused ganglia of the remaining cephalothoracic and first two abdominal segments. Two longitudinal nerve cords that extend posteriorly from the subesophageal ganglion bear five pairs of segmental ganglia that serve the five pairs of book gills.

The lateral eyes are compound eyes consisting of approximately 1000 ommatidia each. The ommatidia contain retinulae consisting of 8 to 14 retinular cells grouped around a rhabdome (Fig. 18-2C). Each ommatidium has a corneal lens (or cuticular cone) to concentrate light on the rhabdome. Proximal and distal pigment cells surround the ommatidium. Horseshoe crabs probably can detect movement, but there are too few ommatidia to form an image. Because of the simplicity of these eyes, they are often used by neurophysiologists interested in the relationship between stimulus and axonal response in compound eyes.

The median eyes are pigment-cup ocelli with spherical cuticular lenses. There is also a rudimentary ventral eye on the carapace. Additional sensilla function as mechanoreceptors, chemoreceptors, or osmoreceptors.

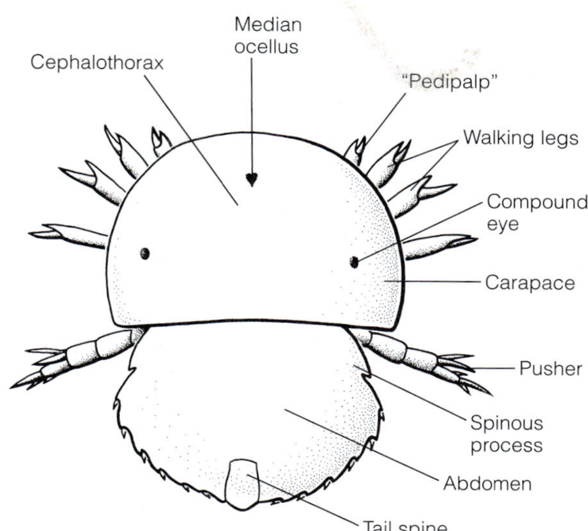

FIGURE 18-5 Xiphosura. Dorsal view of the trilobite larva of *Limulus*.

REPRODUCTION AND DEVELOPMENT

Horseshoe crabs are gonochoric and have true external fertilization, which is unusual in arthropods. The paired gonads are derived from coelomic spaces. Ovaries and testes are composed of branching, arborescent tubes that extend throughout the hemocoel, intertwined with the tubules of the digestive ceca. A gonoduct connects each gonad to a gonopore on the protopod of the genital operculum on the second abdominal segment. Spermatozoa are flagellated with a 9 + 2 or 9 + 0 axoneme, depending on species.

Mating and egg laying in *Limulus* take place during the exceptionally high spring tides of full and new moons in spring and summer. Males and females migrate into shallow water and congregate along the shores of sounds, bays, and estuaries. The smaller male climbs onto the abdominal carapace of the female and maintains its hold with the hook-like first pair of walking legs (pedipalps). Meanwhile, the female excavates a depression in the sand at the high-tide line and deposits 2000 to 30,000 large eggs. As many as 75,000 eggs may be laid in a single season. The male releases sperm onto the eggs after they are deposited. The pair separates and the eggs are covered and left in the sand. The embryos develop into larvae that escape from the depression into the sea when it is flooded by a subsequent spring high tide.

The eggs are macrolecithal, centrolecithal, 2 to 3 mm in diameter, and enclosed in a thick envelope. Meroblastic cleavage forms a stereoblastula in which the yolk is concentrated in the interior cells. A **trilobite larva,** so named because of its similarity to a trilobite, emerges from the egg (Fig. 18-5). This larva, approximately 1 cm long, swims and burrows in the sand. It resembles a miniature adult except that its tail spine is small and only two of the five pairs of book gills are present. All the other appendages are present, as are both types of eyes. With subsequent molts, the remaining book gills appear, the tail spine lengthens, and the young animal achieves the adult form. Juvenile horseshoe crabs, which can be common on intertidal sand flats, attain a carapace width of 4 cm after one year. Sexual maturity is not reached for 9 to 12 years, and the life span may be 19 years.

ARACHNIDA[C]

Almost all recent chelicerates are arachnids, of which there are about 70,000 described species and an estimated million or more awaiting discovery. The taxon includes well-known animals such as spiders, scorpions, mites, and ticks as well as many less familiar taxa. There are about 11 major taxa of arachnids, but over 80% of species are either spiders or mites. Almost all arachnids are terrestrial carnivores.

Arachnida includes Scorpiones (scorpions), Palpigradi (palpigrades, or microwhip scorpions), Uropygi (whip scorpions, or vinegaroons, and schizomids), Araneae (spiders), Amblypygi (amblypygids), Pseudoscorpiones (false scorpions), Solifugae (solifuges, or solpugids), Opiliones (harvestmen, or daddy longlegs), Ricinulei (ricinuleids, or tick spiders), and Acari (mites and ticks), all of which are orders in Linnean classifications.

Arachnida is one of the two extremely successful terrestrial arthropod taxa, with the other being Tracheata (centipedes, millipedes, and insects). The two taxa independently invaded, adapted to, and radiated extensively in dry land habitats. Most of the problems inherent in living in the inhospitable terrestrial habitat were solved separately by arachnids and tracheates, but the two shared a common starting point in the ancestral arthropod body and physiology. The result has been two lineages of similar animals, with some of those similarities inherited from their common ancestor and others due to the evolution of convergent solutions to the problems of living on land.

Arachnida includes all living terrestrial chelicerates, plus a few that have secondarily returned to aquatic habitats. Like all successful conquests of land, their shift from an aquatic to a terrestrial environment required fundamental morphological and physiological modifications of the ancestral aquatic arthropod body, especially for water conservation, that preeminent concern of terrestrial organisms. The epicuticle became waxy to reduce water loss. The book gills were modified for use in air, resulting in the development of the arachnid book lungs and tracheae. Malpighian tubules and uricotely evolved to deal with the end product of nitrogen metabolism. In addition, the appendages became adapted for efficient terrestrial locomotion and an array of sensilla suitable for use in air appeared. Some unique innovations, such as silk in spiders, pseudoscorpions, and some mites as well as poison glands in scorpions, spiders, and pseudoscorpions evolved independently in different taxa.

Fossil representatives of the existing terrestrial arachnid taxa date from the Devonian period, and fossil scorpions have been found in Silurian deposits. The Silurian scorpions were aquatic and contemporaries of the eurypterids, from which they may have evolved. They were large marine animals with compound eyes, and some were more than a meter in length. The first terrestrial arachnid, belonging to a now extinct taxon (Trigonotarbida), appeared in the Upper Silurian and was related to the spiders. It was not a scorpion and it appeared in the fossil record well before the first terrestrial scorpions, which did not appear until the Carboniferous. True spiders, as well as pseudoscorpions, amblypygids, and mites, appeared in the Devonian. Most of the remaining taxa appeared in the Carboniferous.

FORM

Arachnid taxa share the chelicerate ground plan with the body divided into two tagmata, the anterior cephalothorax and the posterior abdomen. The unsegmented cephalothorax is covered dorsally by the carapace, which in most taxa is a single sclerite (Fig. 18-19), although in some it is divided (Fig. 18-31). Although segmentation is often evident in the abdomen, it is not in some important taxa.

Cephalothoracic appendages are similar throughout Arachnida. The anteriormost chelicerae are two- or three-articled, chelate, feeding or defensive structures (Fig. 18-11, 18-19). Sometimes the chelicerae are equipped with poison or silk glands. Chelicerae arise in the embryo in a postoral position, but in adults they are preoral.

The second appendages are the six-articled pedipalps (Fig. 18-11), which may be chelate or leglike and fulfill a variety of functions (such as raptorial, locomotory, sensory, defensive, fossorial, reproductive). In many, the coxa is a gnathobase used to macerate food in the preoral cavity.

The remaining four pairs of appendages are walking legs (Fig. 18-11), each having seven articles, which are, from proximal to distal: coxa, trochanter, femur, patella, tibia, metatarsus, and tarsus. The coxa of the first leg is often a gnathobase (Fig. 18-12). Coxae are frequently displaced to the midline, where they may replace the sternites as the ventral armor of the body. Walking legs are equipped with intrinsic flexor muscles, but in most arachnids extension is accomplished hydrostatically, by elevating hemocoelic blood pressure. Some taxa, however, have extensor muscles (scorpions, pseudoscorpions, harvestmen, and solifuges). Legs may terminate in claws, often bear sensory structures such as trichobothria and slit sense organs, and may be modified for functions other than locomotion.

The cephalothorax and abdomen may be joined by a narrow waist, or **pedicel,** formed of the seventh segment (Fig. 18-18) or they may be broadly connected (Fig. 18-32). In mites they are completely fused and cannot be distinguished (Fig. 18-40A). In most taxa the abdomen is segmented, like that of eurypterids (Fig. 18-9) and scorpions (Fig. 18-10, 18-11), with sclerotized tergites and sternites. It is sometimes divided into an anterior preabdomen and posterior postabdomen (Fig. 18-9, 18-10). In mites and most spiders the abdominal sclerites and all signs of segmentation are lost. Abdominal appendages are usually absent and, when present, do not resemble legs, and are highly modified into, for example, book lungs, spinnerets, genital opercula, and pectines. Although the tail spine, or telson, is usually absent, it is present in the scorpions as a sting and in the palpigrades, uropygids, and schizomids as a sensory flagellum.

NUTRITION

Arachnids, with the exception of some mites and harvestmen, are carnivores. Digestion begins externally, only liquid food enters the mouth, and the gut is adapted for processing fluids. Most of the gut, including the mouth and tubules of the digestive ceca, has a small diameter through which large particles cannot pass. The prey are usually small arthropods that are captured, killed, and masticated, or at least lacerated, by the chelicerae and/or pedipalps. Digestion occurs in the preoral cavity prior to ingestion. The **preoral cavity,** a pocket formed by the surrounding appendages and carapace (Fig. 18-12), is not part of the gut, but rather precedes it. Digestive enzymes from the midgut move anteriorly through the esophagus and out the mouth to flood onto the masticated or lacerated prey in the preoral cavity, where hydrolysis rapidly occurs. The resulting slurry of partially digested food is then sucked into the foregut.

The foregut consists of the cuticularized pharynx, esophagus, and sometimes a stomach. Either the pharynx or the stomach is a muscular pump (Fig. 18-20). A filter of setae in the foregut is often present to exclude particulates that would clog the tubules of the digestive ceca. The thick liquid enters the midgut and its digestive ceca as a mixture of molecules and microscopic particles. Further extracellular digestion, as well as absorption, phagocytosis, and intracellular digestion, occur in the midgut and ceca. The hindgut, which may include an intestine, cloaca, and rectum, connects the midgut with the anus. Water is reclaimed and feces dehydrated and formed in the rectum.

GAS EXCHANGE

Arachnid gas exchange organs are either book lungs or tracheae, both of which are invaginations of the exoskeleton into the hemocoel. Some small arachnids have no special respiratory surfaces, relying instead on cutaneous gas exchange (palpigrades and some mites).

Book lungs are thought to have evolved from book gills when aquatic arachnids colonized land. Some tracheae probably evolved from book lungs (those of spiders, harvestmen, pseudoscorpions) but in others (solifuges, mites) they arose independently, perhaps from apodemes. Book gills are elaborations of the exopods of the abdominal appendages of aquatic chelicerates. Consequently, book lungs, and any tracheae derived from them, are highly modified abdominal appendages.

A book lung is housed in a sclerotized invagination of the integument of the ventral abdominal wall (Fig. 18-20). A slit-shaped spiracle (stigma) connects the invaginated atrium with the exterior (Fig. 18-26). Numerous thin, flat, blood-filled lamellae protrude into the pocketlike atrium from its anterior wall. Short, peglike pedestals between the lamellae prevent them from collapsing onto each other, keeping the interlamellar air spaces open for the flow of air (Fig. 18-6). Blood in intralamellar spaces exchanges gases with air in the interlamellar air spaces across the thin, permeable cuticle of the lamellae. The intralamellar blood spaces are part of the pulmonary sinus that surrounds the lung.

Gases move into and out of the atrium primarily by diffusion through the spiracle. Ventilation may be assisted by a muscle whose contractions expand the atrium and cause air to be inspired through the spiracle. Elastic recoil of the cuticular walls of the atrium compresses the atrium and forces air out of the spiracle. Although book lungs were the original arachnid respiratory organs, they are present today only in scorpions (four pairs), spiders (one or two pairs), amblypygids (two pairs), and uropygids (two pairs). Gases are transported between the book lungs and metabolizing tissues by the blood.

Arachnid tracheal systems arose independently of those of tracheates (myriapods and hexapods), and there are important differences between the two. Like book lungs, tracheae are invaginations of the integument lined with cuticle. They are passageways that arise at spiracles (stigmata) on the body surface and extend into the hemocoel (Fig. 18-26B). In arachnids the tracheae are surrounded by blood and do not make direct contact with metabolizing cells, so the transferring of oxygen to the tissues is performed by the blood. This differs from insect tracheae, which extend to, and even into, their target cells, so that the blood plays no role in oxygen transport. Tracheae conserve water better than book lungs do. Tracheae are best developed in small arachnids because the large surface-to-volume ratio of book lungs results in excessive water loss.

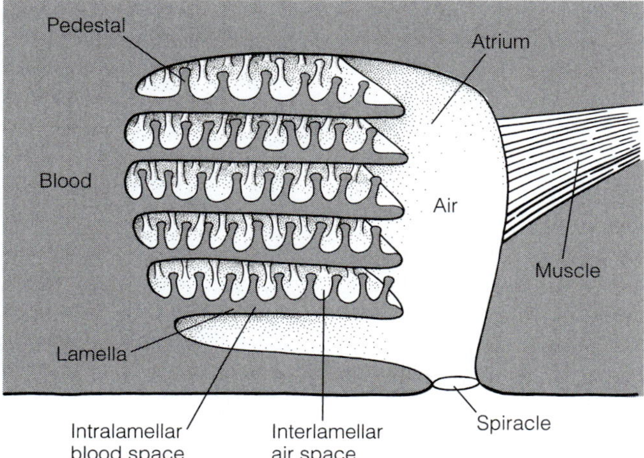

FIGURE 18-6 Arachnida. Diagrammatic sagittal section through a book lung. *(Redrawn from Weygoldt, P. 1996. Chelicerata, Spinnentiere. In Westheide, W., and Rieger, R. (Eds.): Spezielle Zoologie: I. Einzeller und Wirbellose Tiere. Gustav Fischer Verlag, Stuttgart. pp. 449–497.)*

Arachnids have two types of tracheae (Fig. 18-26B). **Sieve tracheae,** derived from book lungs, consist of an atrium and spiracle (as does a book lung), from which arise numerous lamellae (as in a book lung). The lamellae, however, are delicate, branched tubes rather than flat pages. Sieve tracheae occur in a few spiders, pseudoscorpions, and ricinuleids. **Tube tracheae** are branched or unbranched tubes arising individually at surface spiracles. Tube tracheae are found in harvestmen, most spiders, mites, and solifuges. Some evolved from book lungs and others from hollow apodemes. They undoubtedly arose more than once in different arachnid taxa.

INTERNAL TRANSPORT

The heart is a muscular dorsal tube in the anterior abdomen that rarely extends into the cephalothorax. Primitively it is a long tube with seven pairs of segmental ostia, but various degrees of reduction occur in different orders (Fig. 18-20). It is secondarily lost in small arachnids such as the palpigrades and most mites. The heart wall is composed chiefly of circular muscle, but has an outer layer of longitudinal muscle and connective tissue. It is surrounded by the pericardial sinus.

Anterior and posterior aortae exit from opposite ends of the heart and paired segmental arteries leave it from the sides. The anterior aorta delivers blood to the perivisceral sinus of the cephalothorax and supplies the CNS with nutrients and oxygen. The posterior aorta and the segmental arteries supply the perivisceral sinus of the abdomen. Blood is oxygenated by the book lungs and/or tracheae.

Each arachnid book lung is enclosed by a **pulmonary sinus** (Fig. 18-20), which receives blood from within the perivisceral sinus and includes the intralamellar spaces of the book lungs. A **pneumopericardial vein** (or sinus) extends from each pulmonary sinus to the pericardial sinus and carries oxygenated blood from the lungs to the heart. The blood of some arachnids, such as scorpions and many spiders with book lungs, contains the respiratory pigment hemocyanin.

EXCRETION

Excretion of nitrogenous wastes from protein and nucleic acid metabolism is a potential sink for water, and terrestrial animals must use a nitrogenous end product that does not require much water for its disposal. Because ammonia, which is used by most aquatic animals, is very wasteful of water, arachnids rely primarily on guanine, adenine, xanthine, or uric acid. These molecules are relatively nontoxic, highly insoluble, and require very little water for their disposal.

Arachnids employ two types of excretory organs for the elimination of these wastes. Saccate nephridia, or coxal glands (Fig. 18-2B), were inherited from the aquatic ancestors whereas Malpighian tubules were invented at least once and perhaps several times by the arachnids (and also by tracheates). Some arachnid taxa possess both saccate nephridia and Malpighian tubules and some have one or the other. Saccate nephridia derived from coelomic sacs usually are associated with one or two cephalothoracic segments and have ducts opening on the coxae of those segments.

Malpighian tubules are diverticula from the gut in the vicinity of the midgut-hindgut junction (Fig. 18-20). Arachnid Malpighian tubules are endodermal and arise from the midgut, whereas tracheate Malpighian tubules are ectodermal and arise from the hindgut.

Arachnids possess large phagocytic cells called nephrocytes that are localized in clusters in the hemocoel. These cells wander freely in the hemocoel and accumulate metabolic waste products, which they may sequester indefinitely or detoxify and return to the blood.

NERVOUS SYSTEM AND SENSE ORGANS

The arachnid CNS is highly cephalized by the concentration of ganglia in the cephalothorax. The brain consists of only two regions. Its protocerebrum contains the optic centers and is connected with the eyes via the optic nerves (Fig. 16-10B). A pair of cheliceral ganglia, usually considered to be homologous to the tritocerebrum of mandibulates, innervates the chelicerae. The deutocerebrum is absent. Most or all of the segmental ganglia of the cephalothorax and abdomen have moved anteriorly to form a large subesophageal ganglion that fills much of the space in the cephalothoracic hemocoel (Fig. 18-20, 18-7). In all arachnids the subesophageal ganglion sends sensory and motor nerves to the pedipalps and all other cephalothoracic appendages. In most arachnids the abdominal ganglia are also incorporated into the subesophageal ganglion, and only in the relatively primitive scorpions is there a ventral nerve cord on which the abdominal ganglia remain independent of the subesophageal ganglion (Fig. 18-7A). The brain and subesophageal ganglion are connected by circumenteric connectives.

Three types of sense organs are common to most arachnids. These are the many varieties of sensory setae, eyes, and slit sense organs, which are discussed in Chapter 16. Setae, the chief sense organs of most arachnids, include simple mechanoreceptors, chemoreceptors (Fig. 18-8C), and trichobothria.

Trichobothria (trich = hair, bothrium = cup) are mechanoreceptors specialized for detection of weak air movements. A trichobothrium is a long, slender, solid seta (Fig. 18-8E) arising from a cup-shaped **bothrium** in the exoskeleton. The shaft

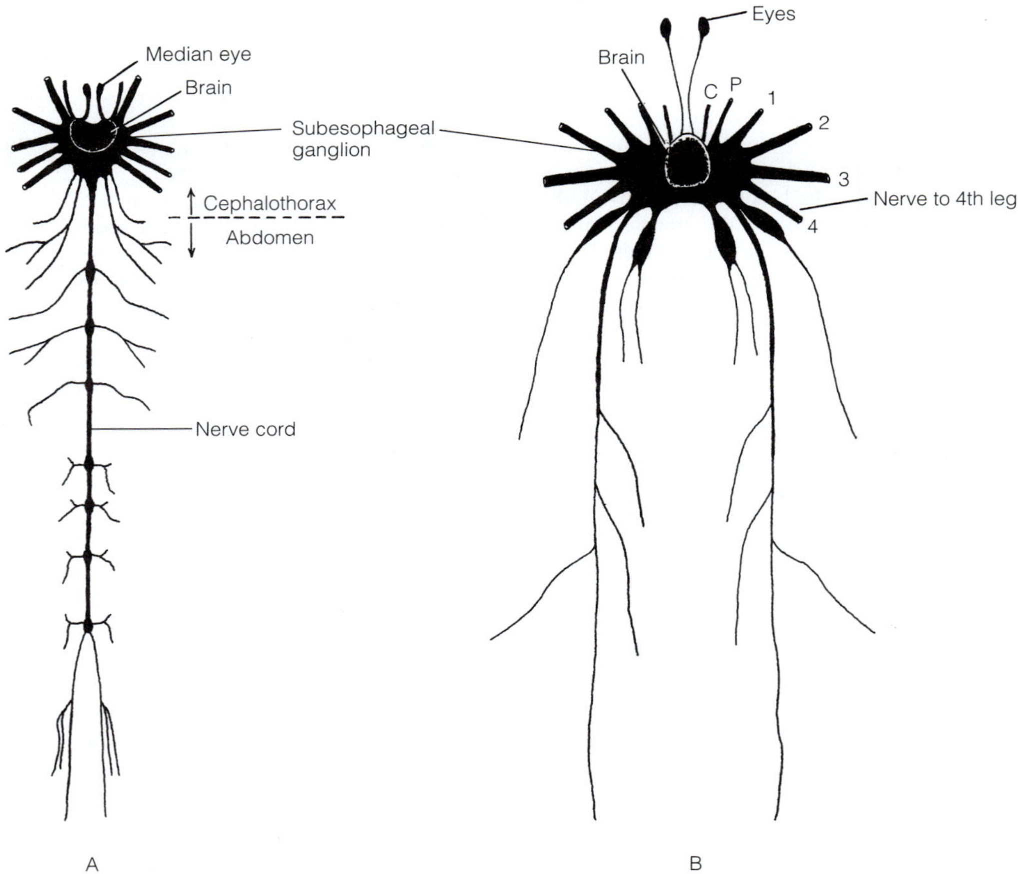

FIGURE 18-7 Arachnid nervous systems. **A,** Scorpion nervous system, in which abdominal ganglia are separate and distinct. **B,** An opilionid nervous system, in which all ventral ganglia have migrated forward and fused to form the subesophageal ganglion. Abbreviations: C, chelicerae; P, pedipalps; numerals refer to walking legs. *(Both after Millot, J. 1949. Traité de Zoologie. Vol. VI. Masson et Cie, Paris.)*

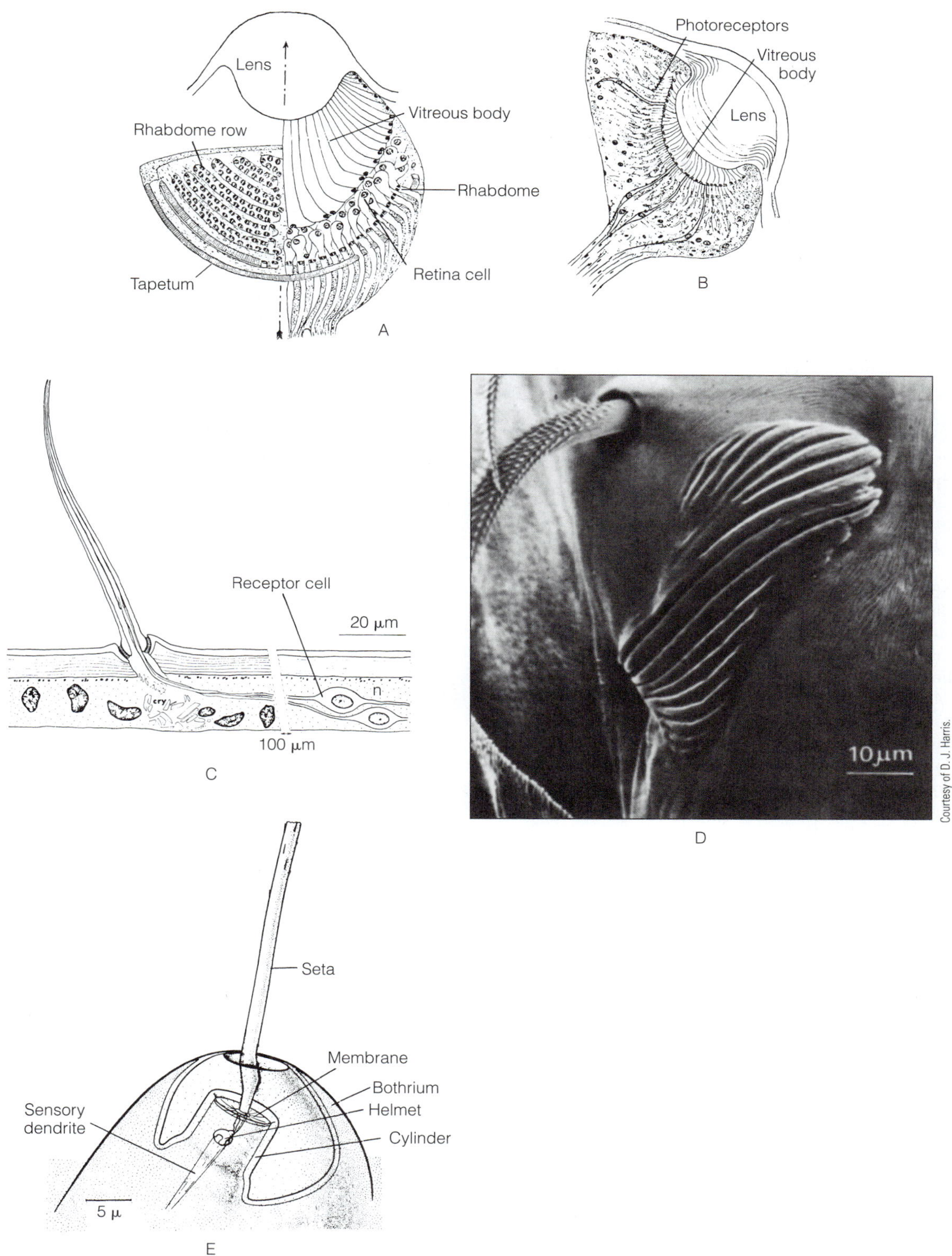

FIGURE 18-8 Arachnid sensory structures. **A,** Diagram of an indirect posterior median eye (secondary eye) of a wolf spider that has a tapetum. Sagittal view to the right of the arrow; three-dimensional view, vitreous body omitted, to the left of arrow. **B,** Sagittal section of a direct eye (main eye) of a spider. **C,** Chemosensory seta of a spider (n = axon). Note that the seta is hollow and open at the tip. **D,** Surface view of a lyriform sense organ. **E,** Detail of a spider trichobothrium. *(A and B, From Homann, H. 1971. Z. Morphol. Tiere. 69(3):201–273; C, From Foelix, R. F. 1970. J. Morphol. 132:313–334; E, From Görner, P. 1965. Cold Spring Harbor Symp. Quant. Biol. 30:69–73.)*

of the seta is attached to the exoskeleton at the bottom of the bothrium by a thin cuticular membrane. This delicate connection permits the seta to move in response to the slightest pressure. The dendrites of at least four sensory neurons are attached to the proximal end of the shaft and confer directional sensitivity to the system. The seta is set into motion by the gentlest of air currents and such motion activates the dendrites. Air currents created by the movement of insect wings can stimulate the trichobothria, allowing a blinded spider to capture flies. Trichobothria are especially common and important in arachnids, but are present in insects also.

The ancestral arachnids had lateral compound eyes and median pigment-cup ocelli. Although some fossil scorpions had large compound eyes similar to those of xiphosurans, modern arachnids have reduced them to three to five small eyes with highly modified ommatidia. The ommatidia of arachnid lateral eyes have fused and share a common lens and single retina, unlike those of other arthropods (Fig. 18-8A). Modern arachnids may have lateral eyes, median eyes, both, or neither.

The photoreceptor cells can be oriented toward the light source (called a direct eye; Fig. 18-8B) or away from the light source (an indirect eye; Fig. 18-8A). In indirect eyes, the pigmented postretinal membrane may function as a reflector, called the tapetum, that reflects the light toward the receptors (Fig. 18-8A). Some arachnids possess only direct or indirect eyes, but many, such as spiders, have both.

Slit sense organs are cuticular strain gauges that detect small changes in the tension on the exoskeleton due to its deformation. They respond to load stress on joints during locomotion and in this respect are proprioceptors. They may also respond to gravity and to airborne vibrations. A single slit sense organ consists of a long, narrow, slit-shaped crevice 8 to 200 μm long and 1 to 2 μm wide in the exoskeleton that is covered by a very thin cuticular membrane. Dendrites of a receptor neuron connect with the inside surface of the membrane. Changes in the tension of the exoskeleton compress the slit so that it narrows, causing the membrane to bow inward. The dendrite depolarizes in response to deformation of the membrane and a signal is sent to the brain. Slit sense organs, present in all arachnids, are most diverse in spiders.

Slit sense organs may be isolated or clustered in groups, in which case they are known collectively as **lyriform organs** (Fig. 18-8D). Lyriform organs are near joints whereas single slit sense organs may be abundant over all the body and appendages of arachnids and are parallel to the long axis of the leg article. The spider *Cupiennius salei*, for example, has about 3000 slit sense organs, some grouped in lyriform organs and some single.

Lacking antennae on the first head segment, many arachnids have compensated by creating the functional equivalent from other appendages, usually a pair of legs (Fig. 18-31). In some taxa the telson, known as a flagellum, is sensory and functions as a posterior antenna (Fig. 18-16A).

REPRODUCTION AND DEVELOPMENT

Arachnids are always gonochoric and fertilization is always internal, but sperm transfer may be direct or indirect. Some arachnids have a copulatory penis for the direct transfer of sperm to the female. Indirect sperm transfer via spermatophores is characteristic of many arachnids (scorpions, uropygids, schizomids, pseudoscorpions, amblypygids). It appears to be the original arachnid mode of sperm transfer and is an adaptation for reproduction on land. Indirect sperm transfer without spermatophores is practiced by spiders. The gonopores of both sexes usually are on the ventral side of the second abdominal segment (Fig. 18-20). The gonads lie in the abdomen and may be either single or paired (Fig. 18-20).

Many species engage in a complex precopulatory **courtship ritual.** The sexes, especially the female, respond to chemical, tactile, or visual cues from the potential partner. Such cues provide for recognition of conspecifics and elicit the complicated behavior, receptivity, and posture that indirect sperm transfer requires. This is especially important in highly predatory species so that the female recognizes the male as a mate rather than a snack.

The macrolecithal eggs are centrolecithal and cleavage is usually superficial. The blastula typically consists of a uniform, segmented blastoderm surrounding a yolk mass. Most arachnids are oviparous, but all scorpions and a few mites are viviparous.

EURYPTERIDA[O]

Eurypterids are extinct euchelicerates that inhabited aquatic environments from the Ordovician to the Permian. These were the predaceous sea scorpions (Fig. 18-9) that lived in marine,

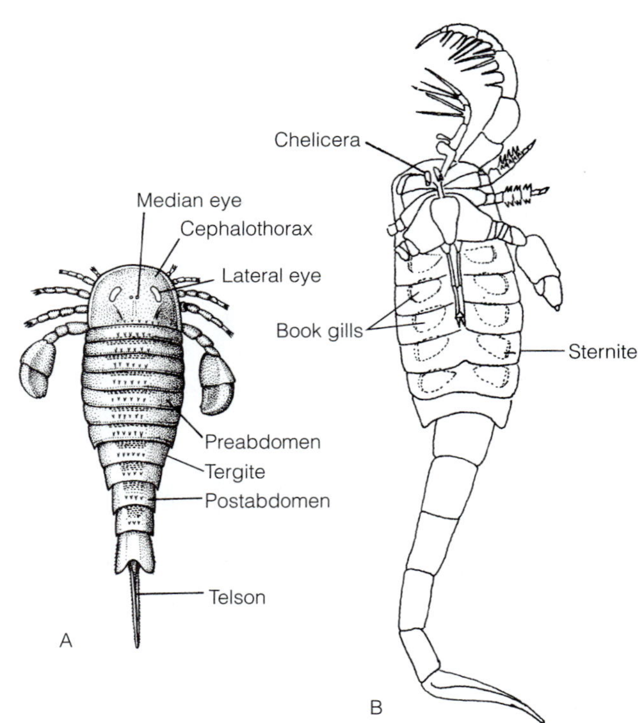

FIGURE 18-9 Eurypterida. **A,** *Eurypterus remipes* (dorsal view). **B,** *Mixopterus kiaeri* showing appendages of one side (ventral view). *(A, After Nieszkowski; B, After Stormer from Fage.)*

brackish, and freshwater habitats. Some probably made short excursions onto land. Eurypterids were the largest arthropods that ever lived, with one species of *Pterygotus* reaching a length of almost 3 m. In some classifications, Eurypterida is ranked as a subclass in the class Merostomata.

Eurypterids are thought to be an early branch of the line leading to the arachnids. Eurypterids share several characteristics with modern scorpions, including a common pattern of tagmosis, and probably are related but not ancestral to them. Eurypterids and scorpions share the apomorphy of a sternum formed of the fused chilaria of the stem chelicerate.

The eurypterid body plan also shares many features with horseshoe crabs, but there are important differences as well. The body is composed of the same two tagmata, the cephalothorax and abdomen (Fig. 18-9). The six cephalothoracic segments are fused and covered by a carapace, but the cephalothorax is smaller relative to the abdomen than it is in xiphosurans. The cephalothorax has six pairs of appendages, the first being the small chelicerae and the last, a pair of large paddles, or walking legs in some species. Eurypterids probably were capable of swimming as well as crawling on the bottom. An important difference between xiphosurans and eurypterids is that in the latter, the abdominal segments are not fused and there is a distinct, seven-segmented preabdomen and a narrower, five-segmented postabdomen. The preabdomen has five pairs of book gills and no genital operculum. The postabdomen has no appendages, but the tail spine extends posteriorly from its last segment. In some species, it may have been a poisonous sting.

SCORPIONES[O]

Scorpions belong to an ancient taxon with a fossil record beginning in the Silurian period. Silurian and Devonian scorpions were aquatic, had gills, and lacked claws on the legs. Terrestrial scorpions appeared in the Devonian. The 1200 described species of Recent scorpions are most common in the tropics and subtropics, but occur on all continents except Antarctica. In the Western Hemisphere, they are found from southern Canada to southern South America and in North America are most abundant in the southeast and southwest.

Scorpions are generally secretive and nocturnal, hiding by day under logs, bark, and stones and in rock crevices or burrows in the ground. Scorpions are found in most terrestrial habitats, including forests, grasslands, and deserts, and some species are associated with vegetation and some live in trees. Some inhabit caves and a few occur in the intertidal zone. Scorpions are often found near or even in human abodes, and the desert-dwellers' practice of shaking out shoes before putting them on is a wise precaution. Many species are characteristic of desert regions, with arid Baja California supporting the greatest diversity of scorpion species in the world. Scorpions are by no means restricted to arid habitats, however; many require a humid environment and some live in tropical forests. Scorpions exhibit a striking green fluorescence under ultraviolet light and can be easily observed at night when illuminated by an ultraviolet lamp.

Scorpions are large arachnids, most ranging from 3 to 9 cm in length. The smallest species is the cave-dwelling *Typhlochactas mitchelli,* which is only 9 mm long, and the largest is the African *Hadogenes troglodytes* at 21 cm. Some Carboniferous scorpions reached 86 cm.

Form

The scorpion body has hardly changed since the Silurian and consists of a cephalothorax and a long, segmented abdomen ending in a telson, or sting (Fig. 18-10). Dorsally the cephalothorax is completely covered by the carapace, in the middle of which is a pair of large, elevated median eyes. In addition, two to five pairs of small lateral eyes are present along the anterior, lateral margin. Some cave-dwelling scorpions are eyeless.

Scorpion chelicerae are small, chelate, and project anteriorly from the front of the body (Fig. 18-11). Along with the labrum, they form the roof of the preoral cavity (Fig. 18-12). A distinctive feature of scorpions is the greatly enlarged pedipalps, which form a pair of large pincers for capturing prey. Their coxae form the sides of the preoral cavity. Four pairs of walking legs and their coxae form the ventral surface of the cephalothorax (Fig. 18-11). These coxae replace the sternites of the anterior cephalothorax, which have been lost. A ventral plate known as the sternum (Fig. 18-11) represents the fused sternites of segments 5 and 6. The coxae of the first two walking legs are gnathobases that form the floor of the preoral cavity (Fig. 18-12).

The cephalothorax and abdomen are broadly connected and there is no pedicel (Fig. 18-10, 18-11). Posterior to the cephalothorax, the long abdomen is composed of a wide, seven-segmented preabdomen and a narrow postabdomen of five segments. That the pre- and postabdomens are well developed and distinct from each other are primitive features held in common with the eurypterids. Dorsally the preabdomen is covered by a series of seven segmental tergites and ventrally, by sternites. The sternites and tergites are connected laterally by flexible, unsclerotized pleural membranes. The anterior sternites form a pair of genital opercula on the first abdominal segment to protect the gonopore. Two unique pectines, the appendages of the second segment, are comblike sensory structures projecting laterally from their attachment near the ventral midline. Abdominal sternites 3 through 7 are unmodified and four of them (3 through 6) each bear a pair of transverse spiracles opening into book lungs (Fig. 18-11). The segments of the postabdomen, sometimes called the tail, are narrow sclerotized rings undivided into individual sclerites. The last segment bears the anus on its posterior ventral side and the characteristic venomous sting.

Nutrition

Scorpions feed on invertebrates, especially insects and other arachnids. Many, perhaps most, sit in an alert position and lie in wait for prey. The pedipalp fingers are open, with the tips of the movable fingers and pectines touching the ground. Burrowing species wait near or at the burrow entrance. Prey is detected by trichobothria on the pedipalps or, in some, by vibrations of the ground sensed by tarsal hairs and slit sense organs. The North American desert scorpion, *Paruroctonus mesaensis,* can locate and dig up a burrowing cockroach 50 cm away in a few seconds. Some scorpions can catch prey in midair, detecting and tracking it with the trichobothria from a distance of 10 cm.

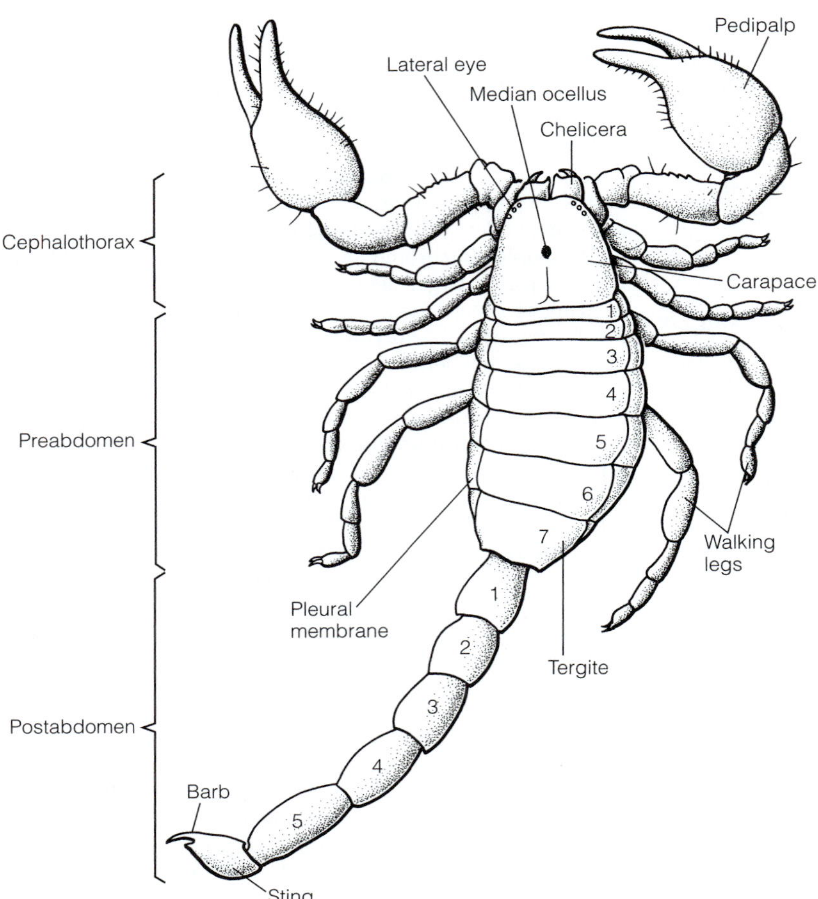

FIGURE 18-10 Scorpion external anatomy. *Vaejovis carolinianus* (dorsal view).

When feeding, the prey is caught and held by the large pedipalps while being killed or paralyzed by the sting (Fig. 18-13), although the sting is not used if the prey can be subdued with the pedipalps alone. Prey is transferred to the preoral cavity, where the chelicerae and gnathobases slowly crush and tear it (Fig. 18-12). Hydrolytic enzymes from the midgut flood the food as it is masticated, beginning digestion outside of the body. The broth of predigested food is sucked into the gut by the pharynx and digestion continues as described for arachnids in general. Metabolic rates for scorpions are among the lowest recorded for animals, and some can survive for a year or more without eating. Many species spend over 90% of their time quiescent in their burrows and their level of activity is very low. Conversion of prey biomass to scorpion tissue is extremely efficient and scorpions can increase their weight by one-third from a single meal.

The **sting** is attached to the posterior end of the last segment and consists of a bulbous base and a sharp, hollow, curved **barb** that injects venom into the prey (Fig. 18-11). The **venom** is produced by a pair of poison glands in the base of the sting. Violent contractions of a muscular envelope surrounding the glands eject venom into a duct to the barb. The scorpion raises the postabdomen over the body so that it is curved forward and stabs the prey (Fig. 18-13).

The venom of most scorpions, although sufficiently toxic to kill many invertebrates, is not dangerous to humans, but it can be painful. The scorpions of the southeastern United States, as well as many of the midwestern and western species, normally are not lethal to humans. Worldwide, however, about 25 species in the family Buthidae possess highly toxic venom that can be fatal to humans. Most notorious are *Androctonus* of North Africa and various species of *Centruroides* in Mexico, Arizona, and New Mexico. Globally, 5000 people a year are estimated to die from scorpion stings. Humans envenomated by *Androctonus australis* may die in 6 to 7 h. In Mexico, species of *Centruroides* have been responsible for deaths, mostly in children. The neurotoxic venom may cause convulsions, paralysis of the respiratory muscles, or cardiac failure. Scorpions are responsible for more human deaths than any other nonparasitic animal other than bees and snakes. Effective antivenins are available for dangerous species. Scorpions are immune to their own venom.

Scorpions are usually most active during the first hours of darkness, but only about 10% of the population appears to be active at any one time. When seeking prey, most burrowing species stay within a meter of the burrow entrance, but *Paruroctonus mesaensis* is known to return to its burrow from distances as great as 8 m.

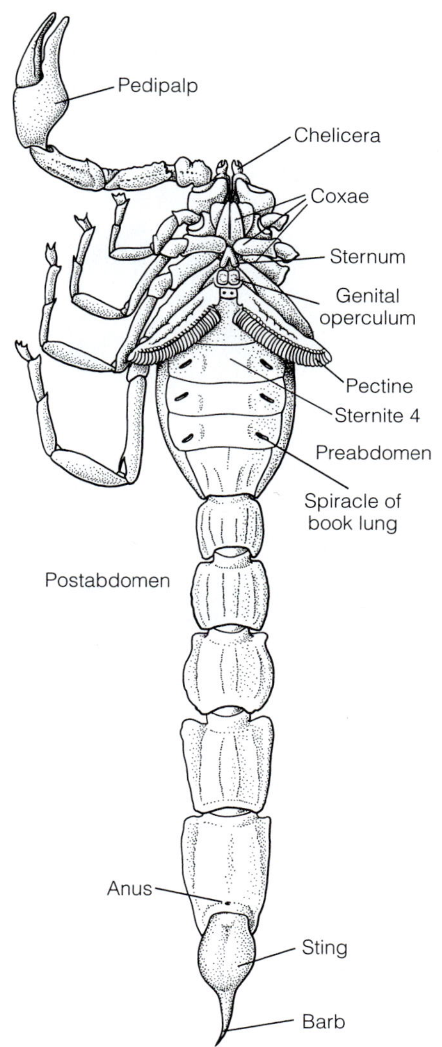

FIGURE 18-11 Scorpion external anatomy. *Androctonus australis* (ventral view). *(After Lankester from Millot and Vachon, 1949.)*

Internal Form and Function

Gas exchange is accomplished solely by the four pairs of book lungs. Water loss from the lungs is reduced by spiracles under muscular control.

The hemal system includes a dorsal tubular heart with seven pairs of segmental ostia in the pericardial sinus of the preabdomen. Transport of oxygen from lungs to tissues is accomplished by the well-developed and efficient hemal system like that described for arachnids in general.

Excretion is accomplished by two pairs of Malpighian tubules and one pair of saccate nephridia, which open on the coxae of the third pair of walking legs. Desert scorpions possess a number of adaptations for life under extreme desiccating conditions. Lethal temperatures are high (45 to 47°C), and evaporative water loss through the almost impervious exoskeleton is extremely low, the lowest of any arthropod (0.01% of body weight per hour at 25°C). Scorpions can tolerate a water loss of up to 40% of their body weight. Like kangaroo rats, some desert species never drink and can live indefinitely on the water from their food, using many of the same mechanisms as kangaroo rats to accomplish this. The integument is impermeable, the feces are dry, excretion produces almost dry nitrogenous wastes, they live in burrows, and they are nocturnal. The burrows of some species, such as *Hadrurus arizonensis* of the Sonoran Desert, may extend 90 cm into the desert floor. Some scorpions at times raise their bodies well off the ground. This behavior, called stilting, permits the circulation of air beneath the animal to maximize convective heat loss to the air and minimize conductive heat gain from the ground.

The scorpion nervous system, unlike that of other arachnids, retains a distinct ventral nerve cord with seven ganglia (Fig. 18-7A). Giant fibers coordinate the rapid pedipalpal grasping of the prey and the stinging thrust that kills it. Trichobothria are probably a scorpion's most important source of sensory input. The role of the pectines still is not clear, but the ventral side of each tooth of the comb is provided with chemo- and mechanoreceptors. When moving, the scorpion holds the pectines out from the sides of its body in a horizontal position so that the teeth touch the ground. Pectines are sensitive to ground vibrations and apparently to some physical or chemical characteristics of the substratum. One function appears to be recognition of an appropriate substratum for placing spermatophores, and amputation of the pectines prevents spermatophore deposition.

Reproduction and Development

Scorpions are gonochoric and use spermatophores for indirect sperm transfer. The gonads, located among the tubules of the midgut diverticula in the preabdomen, have an unusual morphology consisting of four longitudinal tubes with several transverse cross connections. In each sex the gonads empty into a common genital atrium, which opens to the outside via the gonopore on the first abdominal segment. There is a single gonopore in each sex. The female system includes a pair of seminal receptacles for storage of sperm. The male system includes a pair of paraxial organs in which sperm are packaged into complex spermatophores consisting of a lever, a stalk, an ejection apparatus, and a sperm reservoir (Fig. 18-14A). It takes the male three to four days to make a spermatophore.

Details of courtship and mating differ with genus, but in general they conform to the following account. During the breeding season the male wanders about until he encounters a female and initiates an extended courtship (Fig. 18-14B–D). The male grasps the female's pedipalps with his, and together they walk, hand-in-hand, so to speak, backward and forward in a *promenade à deux*. Sometimes the male stings the female. During this dance the male is presumably using the pectines to locate a suitable site for depositing a spermatophore. This behavior may last from a few minutes to hours, depending on how long it takes to locate an acceptable site. Eventually, the male extrudes a spermatophore, which is about 6 mm high, and cements it to the substratum. He then maneuvers the female so that her genital area is over the spermatophore. When her pectines

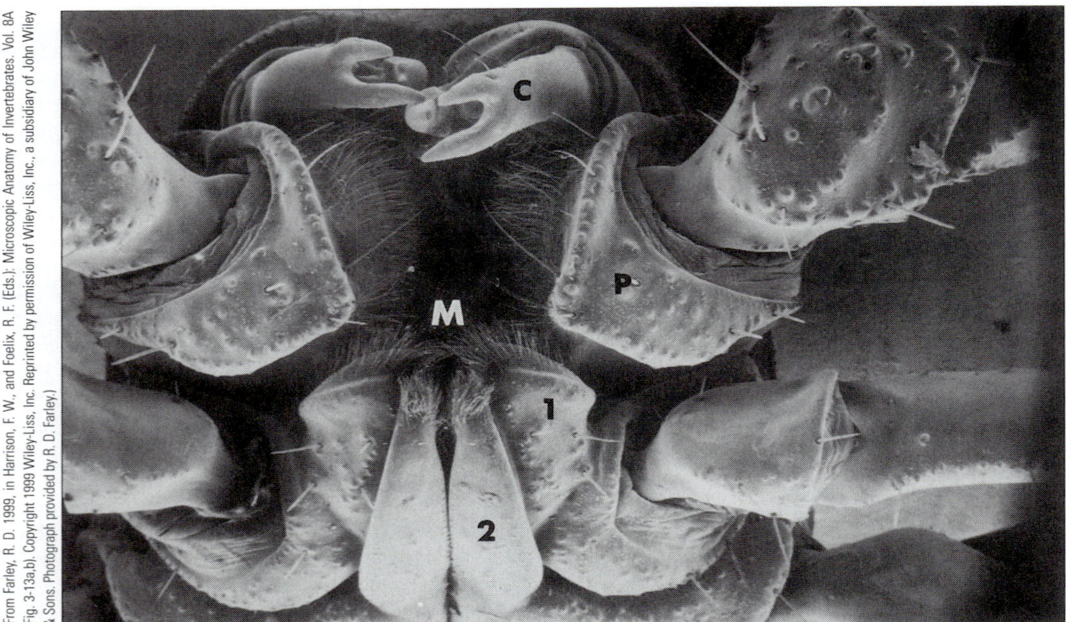

FIGURE 18-12 Anterior view of a scorpion. The preoral cavity is the area around the mouth (M) surrounded by the bases of the appendages. The labrum above the preoral cavity is obscured by the chelicerae (C). Abbreviations: C, chelicerae; M, mouth; P, pedipalp coxa; 1 and 2, gnathobases (coxae) of the first and second legs.

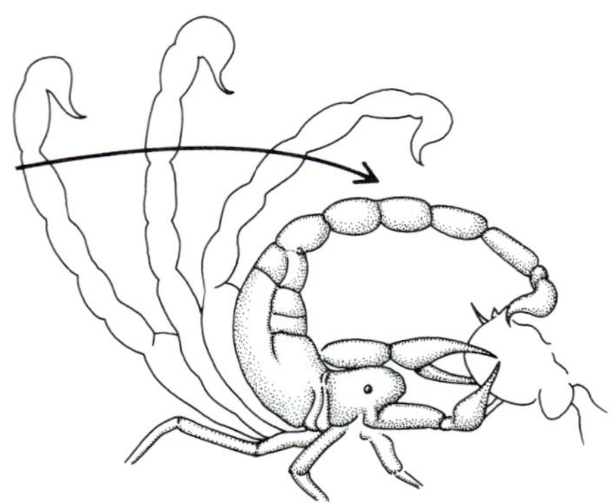

FIGURE 18-13 Scorpiones. Successive positions of the abdomen of the scorpion *Hadrurus* when stinging prey. Total time required for the sequence is about 0.75 s. *(From Bub, K., and Bowerman, R. F. 1979. Prey capture by the scorpion* Hadrurus arizonensis. *J. Arachnol. 7:243–253.)*

touch the spermatophore, her genital opercula move aside and open the gonopore. When the spermatophore touches the gonopore, pressure on the lever ejects the sperm mass into the female's gonopore. The sperm move to the seminal receptacle, where they are stored until used for fertilization at some later time.

Unlike most terrestrial arthropods, scorpions are viviparous and eggs are retained in the female reproductive tract. Scorpions invest much more time and energy in their offspring than is usual among terrestrial arthropods. Development is direct and females give birth to relatively large juveniles that resemble miniature adults. In lecithotrophic species, development of the large, yolky eggs takes place in the lumen of the ovarian tubules. The eggs of matrotrophic species, such as the tropical Asian *Hormurus australasiae*, possess little yolk, and eggs develop in the diverticula of the ovary (Fig. 18-14E). Each diverticulum includes a cluster of absorbing cells at the upper end. These cells rest against the mother's digestive ceca, from which nutritive material is acquired. The food passes through the tubule to nourish the embryo at the base.

Development takes several months, even a year or more, and from 1 to 95 young are produced, depending on the species. At birth the juveniles, a few millimeters long, immediately crawl onto the mother's back (Fig. 18-15), where they remain through the first molt, which usually occurs in one to four weeks. Young scorpions then gradually leave the mother and become independent. They reach sexual maturity in six months to 6 years, molting four to seven times, and some species may live as long as 25 years. These reproductive life history traits sometimes result in an intrinsic rate of natural increase *(r)* lower than that of any other animal, including whales and elephants. Predation by vertebrates such as birds, snakes, and amphibians is a major cause of scorpion mortality.

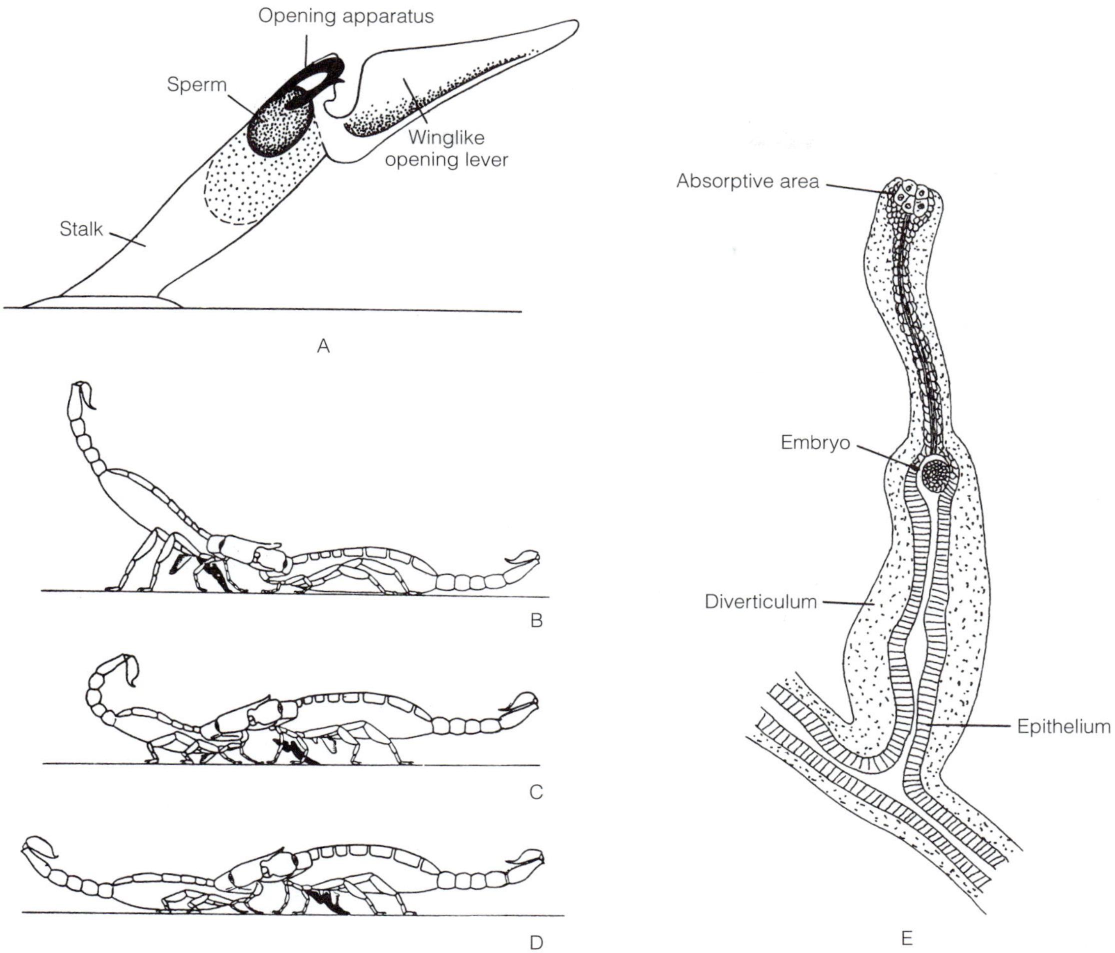

FIGURE 18-14 Scorpion reproduction. **A,** Diagram of a scorpion spermatophore. **B–D,** Sperm transfer in scorpions: **B,** While holding the female's pedipalps in his own, the male (on the left) deposits the spermatophore on the ground. The spermatophore is solid black in the drawing. **C,** The female is pulled over the spermatophore. **D,** The spermatophore is taken up into the female's gonopore. **E,** Diverticulum of the ovary containing a developing embryo in the tropical Asian scorpion, *Hormurus australasiae,* a matrotrophic viviparous species. *(A–D, After Angermann; E, After Pflugfelder from Dawydoff.)*

Photograph courtesy of H. L. Stahnke.

FIGURE 18-15 Female scorpion carrying her young.

UROPYGI[O]

The approximately 100 described species of Uropygi are called whip scorpions. Some are large, and the American vinegaroon *Mastigoproctus giganteus* reaches 80 mm in length (Fig. 18-16A). The cephalothorax is completely covered by a carapace that bears a pair of anterior median eyes and three or five pairs of lateral eyes. Uropygid chelicerae are small and have two articles. The distal article forms a fang that folds against the large basal article. The large, raptorial pedipalps are stout and heavy, and short in comparison with the legs. The last two articles of the pedipalps may be modified to form a pincer used in seizing prey (Fig. 18-16A). During feeding, the prey is seized and torn apart by the pedipalps and then passed to the chelicerae. The animal holds its pedipalps and long, tactile first legs in front as it moves forward. The antenniform, sensory first legs touch the ground repeatedly.

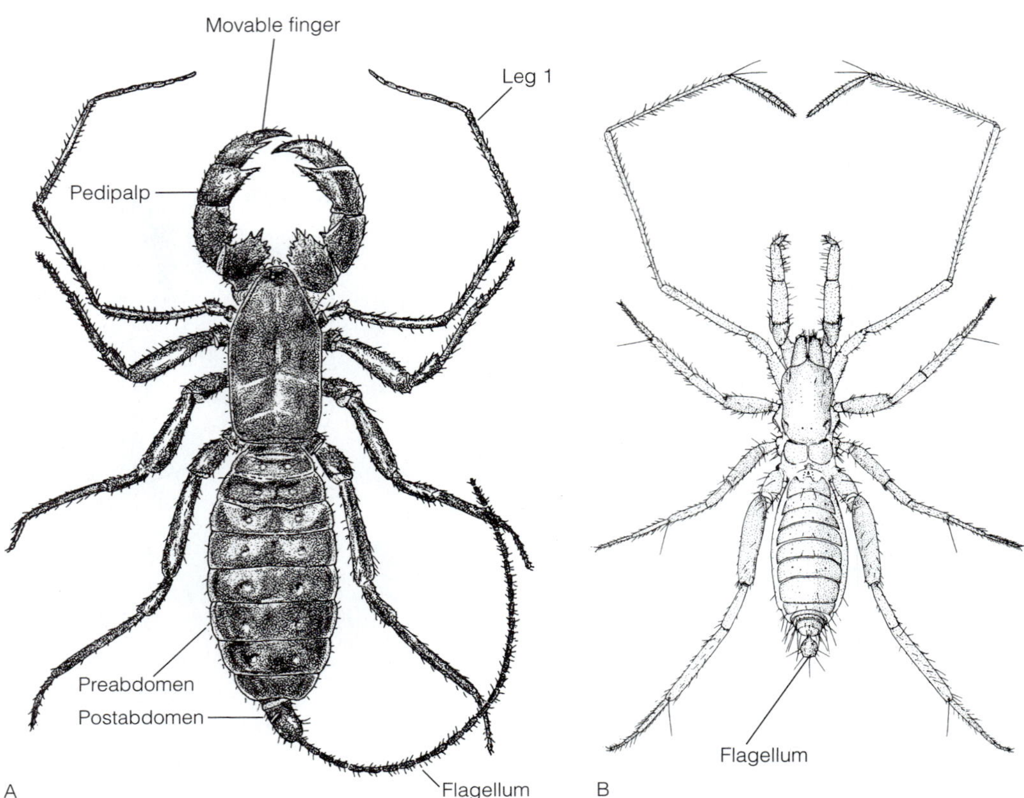

FIGURE 18-16 Uropygi. **A,** American whip scorpion, *Mastigoproctus giganteus.* **B,** A male schizomid, *Schizomus sawadai,* from Asia. The body length is about 5 mm. *(A, After Millot, J. 1949. Traité de Zoologie. Vol. VI. Masson et Cie, Paris; B, From Sekiguchi, K., and Yamasaki, T. 1972. A redescription of "Trithyreus sawadai" from the Bonin Islands. Acta Arachnol. 24(2):73–81.)*

The large, segmented abdomen is divided into two sections, sometimes designated the pre- and postabdomens. The preabdomen is large and consists of nine segments, whereas the postabdomen, with only three segments, is much smaller and narrower. A long, antenniform flagellum extends posteriorly from the postabdomen. Two pairs of book lungs are located in the second and third abdominal segments. A pair of large **repugnatorial glands** open on each side of the anus. When irritated, the animal elevates its postabdomen and sprays the attacker with fluid from these glands. In *Mastigoproctus* the secretion is 84% acetic acid and 5% caprylic acid. The caprylic acid facilitates penetration of the acetic acid through the integument of an arthropod predator. The fluid can burn human skin, and its odor is responsible for the name *vinegaroon.*

Sperm transfer is indirect via a spermatophore. During the complex courtship behavior, the male holds the tips of the long, modified sensory legs of the female with his chelicerae. The female picks up the spermatophores with her genital area and, in *Mastigoproctus* and *Thelyphonellus,* the male uses his pedipalps to push them into her gonopore. The female lays her eggs in a sac attached to her body. She remains in a burrow until they have hatched and undergone several molts.

Schizomida (Fig. 18-16B) is sometimes classified separately although most arachnologists now consider it to be a taxon of specialized uropygids adapted for life as interstitial soil animals. The approximately 180 species are small (8 to 13 mm). The carapace does not cover the entire cephalothorax, and the posterior two segments are covered by independent tergites (Fig. 18-16B). The flagellum is short, with a maximum of three segments. Only one pair of book lungs is present. As in other uropygids, the first leg is antenniform and has a sensory role. Sperm transfer is indirect with spermatophores.

AMBLYPYGI[O]

Amblypygids, or whip spiders, are tropical and semitropical arachnids resembling spiders in many respects. Several of the 100 known species are common, but because they are secretive nocturnal animals, hiding during the day beneath logs, bark, stones, leaves, and similar objects, they typically escape notice. Amblypygids range in length from 4 to 45 mm and have a dark, somewhat flattened, spiderlike body (Fig. 18-17).

The carapace completely covers the cephalothorax and bears one pair of median eyes and three pairs of lateral eyes. The chelate chelicerae are similar to those of spiders but lack poison glands. The distinctive, raptorial pedipalps are heavy and spiny and are used to capture prey, primarily insects. The first legs are especially long and antenniform (Fig. 18-17) but

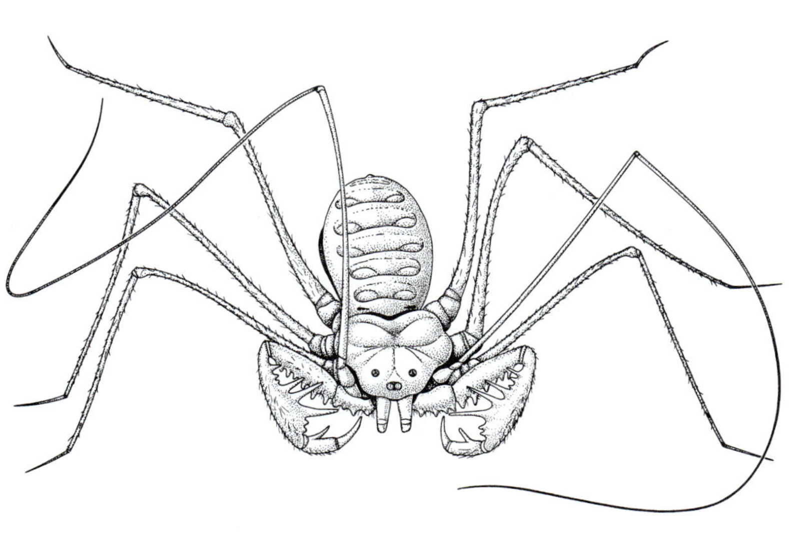

Photograph courtesy of Dr. Robert Mitchell.

A B

FIGURE 18-17 Amblypygi. **A,** The African amblypygid *Charinus milloti.* **B,** Male (top) of the amblypygid *Heterophrynus longicornis,* tapping the body of a female (bottom) with his antenniform legs. *(A, After Millot.)*

all the legs are sensory. One of the long, tactile first legs is always pointed in the direction of movement, whereas the other may explore areas to either side of the animal. The gait of an amblypygid is crablike because of its flattened body and its tendency to move laterally.

The segmented abdomen is connected to the cephalothorax by a narrow pedicel like that of spiders. Two pairs of book lungs are located on the ventral side of the second and third abdominal segments, but there are no other abdominal appendages. Unlike spiders, amblypygids do not produce silk but their internal anatomy resembles that of spiders. The CNS is fully cephalized, with all segmental ganglia moved anteriorly into the subesophageal ganglion. The pharynx and pumping stomach are used to suck liquid food into the gut. Malpighian tubules and one pair of saccate nephridia are responsible for excretion. Spermatophores are used in indirect sperm transmission.

ARANEAE[O]

With the probable exception of mites and ticks, Araneae (spiders) is the largest arachnid taxon. Approximately 40,000 species in about 100 families and 3000 genera have been described, and it has been estimated that there are probably at least 170,000 species alive in the world today. Spider population densities can be very large, and in one study a hectare of undisturbed, grassy meadow in Great Britain was found to support over 5 million spiders. The great success of spiders relative to other arachnids probably results from their innovative use of silk. Many aspects of spider biology depend on silk, and it may be the adaptive equivalent of flight by insects in that both innovations resulted in extensive adaptive radiation culminating in great diversity. Spiders are, with few exceptions, terrestrial and occur on all continents except Antarctica. They are found in most terrestrial habitats but are absent from the sea, and there is only one freshwater species that lives below the water surface. Spiders are easily distinguished from most other arachnid orders by the distinctive pedicel between the cephalothorax and abdomen. Spiders are also the only arachnids with silk-spinning appendages, the spinnerets, on the posterior abdomen. Spiders are strictly carnivorous.

Araneae is divided into three subtaxa, Mesothelae, Mygalomorphae (Orthognatha), and Araneomorphae. Most spiders, including the most derived, are araneomorph, with labidognath fangs that move in the transverse plane so they can oppose each other (Fig. 18-18B). The more primitive mygalomorph spiders (tarantulas, bird spiders, and trapdoor spiders) have orthognath chelae in which the fangs move only in a longitudinal plane and cannot move transversely to oppose each other. Mesothelan spiders are the most primitive of all and have, in addition to the primitive type of chelicerae, a segmented abdomen and other primitive characteristics (Fig. 18-19).

Form

Spiders range from tiny species of less than 0.5 mm in length to the large, tropical spiders with a body length of 9 cm and a much greater leg span. The usual two arachnid tagmata (cephalothorax and abdomen) are present but separated by the pedicel (Fig. 18-18). Because of the flexible

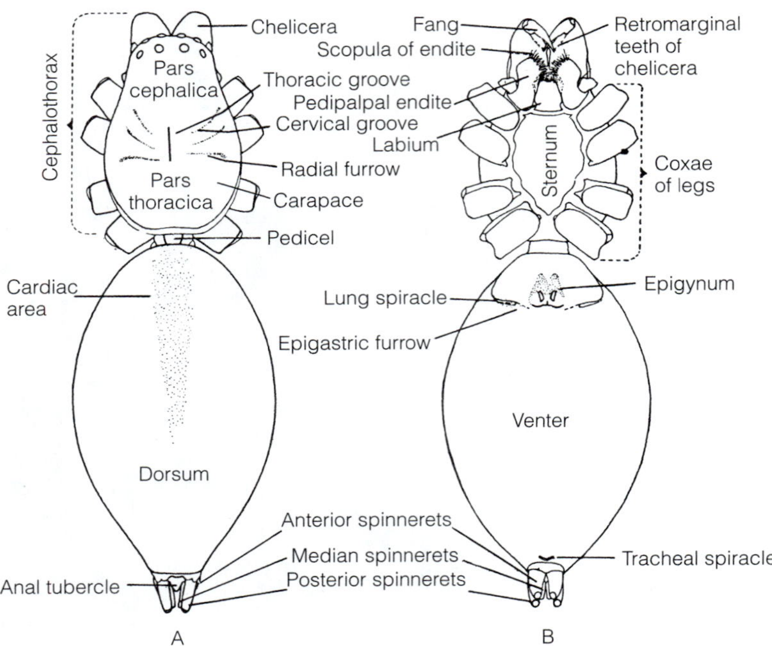

FIGURE 18-18 Araneae. Dorsal **(A)** and ventral **(B)** views of a spider, showing the external structure. *(From Kaston, B. J. 1948. Spiders of Connecticut. State Geol. Nat. Hist. Surv. Bull. 70:13.)*

pedicel, the abdomen can be moved independently of the cephalothorax to distribute silk. The cephalothorax is composed of 6 fused segments, each of which bears a pair of appendages, and the abdomen is composed of 12 segments that, in almost all spiders, are completely fused and without signs of segmentation.

Dorsally the adult cephalothorax is completely covered by a distinctive, convex carapace that usually bears eight anterior eyes (Fig. 18-19). A longitudinal, median **thoracic groove** on the carapace marks the position of the apodeme, on which the dilator muscles of the pumping stomach insert (Fig. 18-18A, 18-20). Ventrally the cephalothorax is covered by two sternites. These are the small anterior labium and the large posterior sternum (Fig. 18-18B). In embryonic development the cephalothorax forms by the fusion of six anterior segments.

Each chelicera consists of a distal fang and a **basal piece** that the fang folds against (Fig. 18-27A). A pore at the tip of the fang is the opening of the duct from the poison gland (Fig. 18-20). The female pedipalps resemble short legs (Fig. 18-20) and are unremarkable, but those of males are modified to serve as sperm transfer organs, with the last segment being greatly enlarged and knoblike (Fig. 18-28A). The coxa of the

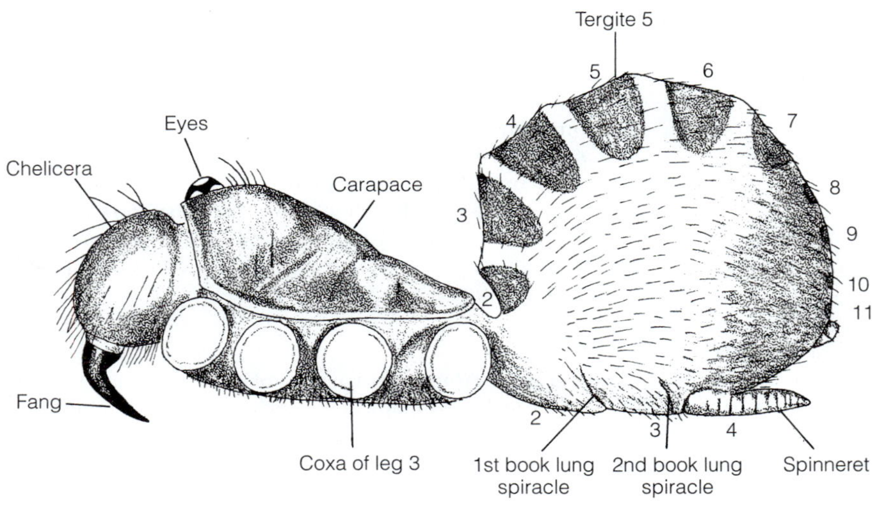

FIGURE 18-19 The primitive Asian mesothelan spider *Liphistius malayanus,* with legs removed for clarity (side view). *(After Millot, 1949. Traité de Zoologie. Vol. VI. Masson et Cie, Paris.)*

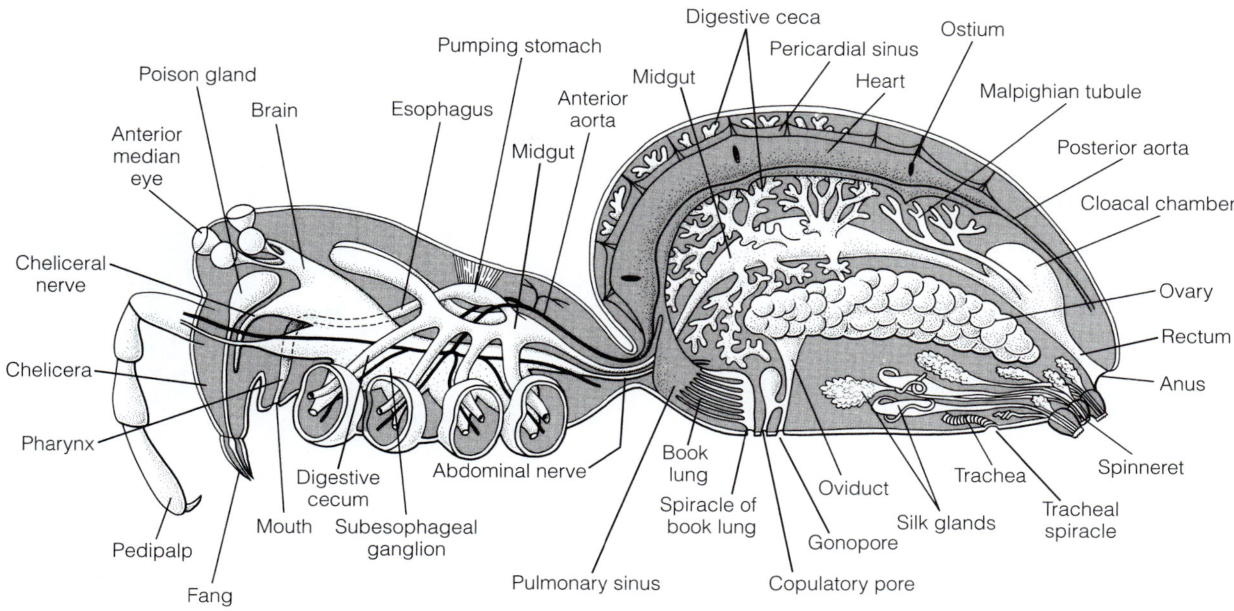

FIGURE 18-20 Internal anatomy of an araneomorph spider. *(After Comstock, 1940.)*

pedipalp has a process known as the **pedipalpal endite** (or gnathobase) that forms the sides of the preoral cavity (Fig. 18-18B). Spiders always have four pairs of legs, which may vary in size, morphology, and function. Each leg is composed of seven articles plus two terminal tarsal claws. Web-spinning spiders have an additional tarsal claw.

The soft, ovoid abdomen is unsegmented, except in the primitive Mesothelae, in which it bears dorsal segmental tergites (Fig. 18-19). The abdomen lacks appendages except for book lungs and spinnerets. Anteriorly, on the ventral side of the abdomen, is a transverse groove known as the **epigastric furrow** (Fig. 18-18B). Associated with this furrow are the epigynum and gonopore. In most spiders, the slit-shaped spiracles of a pair of book lungs lie on each side of the furrow. The posterior end of the abdomen bears a group of modified appendages, the **spinnerets** (Fig. 18-18). A **tracheal spiracle** is situated immediately anterior to the spinnerets and the anus is posterior to them (Fig. 18-20).

Silk

Silk production is a distinctive, but not exclusive, feature of spiders, and many aspects of spider biology and ecology are influenced by it. Fossil spiders with clearly discernable spinnerets are known from as early as the Devonian. Silk is produced by several large abdominal **silk glands** (Fig. 18-20) and extruded from individual **spigots** on the distal tips of the spinnerets (Fig. 18-21). Primitively, spiders probably had four pairs of spinnerets, but in most the number is reduced to

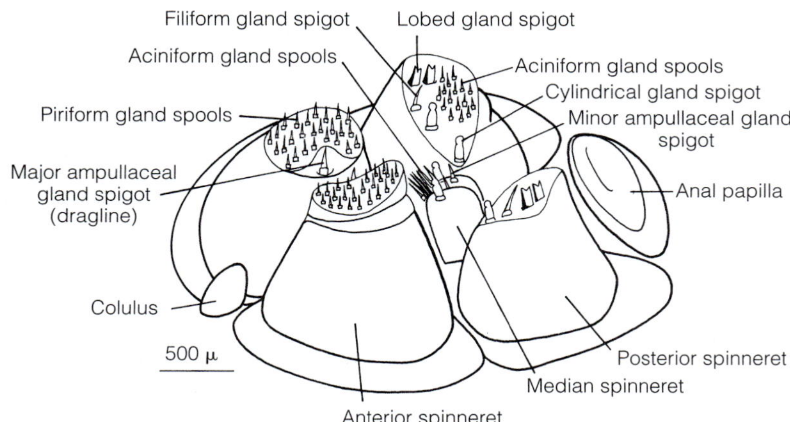

FIGURE 18-21 Spinnerets of the orb-weaving spider *Araneus diadematus,* showing the distribution of spigots for different silk glands. Anterior is to the left. *(From Wilson, R. S. 1969. Am. Zool. 9(1):103–111.)*

three, the anterior, median, and posterior pairs (Fig. 18-21). At least six types of silk glands are known, each producing a different silk. Each spinneret bears many spigots, each of which is connected to a specific silk gland and thus extrudes only one type of silk, but each spinneret may have several types of spigots. Spinnerets are short, conical structures derived from the appendages of abdominal segments four and five. They are very mobile and can move independently. Further, because of the pedicel, the abdomen itself is mobile.

The **cribellum,** found in some araneomorphs, is a modified spinneret with up to 40,000 tiny spigots. Each spigot produces a single very fine fiber. The many fibers are pulled from the cribellum by the **calamistrum,** a comblike row of setae on the fourth metatarsus, and teased into a composite wooly thread, or "hackle-band," especially effective for snagging the setae of insects. The cribellum was present in the earliest araneomorphs and is an autapomorphy of the taxon, although most modern taxa have lost it. It was the first silk capable of snaring insects and it made possible the use of webs for prey capture prior to the development of adhesive silk. Loss of the cribellum and calamistrum seems to have occurred several times in araneomorph spiders.

Spider silk is a protein composed largely of glycine, alanine, and serine and is similar to insect silk. It is emitted as a liquid and hardens not from exposure to air, but from being "drawn out," which changes the protein's conformation. Spider silk is about as strong as nylon but is more elastic. It is similar in tensile strength to other biological structural materials such as cellulose, chitin, and collagen, but is more extensible and can withstand about 10 times the force without breaking. A single thread is composed of several fibers, each drawn out from liquid silk supplied at a separate spigot. Drawing out occurs as the spider moves away from an anchored thread or pulls the threads with its posterior legs.

Silk plays an important role in spider life and is put to a variety of uses, even in the many taxa that do not build webs to catch prey. Originally the function of silk was probably reproductive, and this remains one of its important functions. Spider eggs are always enclosed in a silken egg case (Fig. 18-22) and males spin a silk **sperm web** (Fig. 18-28A) as part of the process of transferring semen to the female. Most spiders continually lay a **dragline** of dry, nonadhesive silk behind them as they move about (Fig. 18-23). At intervals it is fastened to the substratum with adhesive silk. Spiders are often observed suspended in midair at the end of a dragline after being dislodged from their substratum. Many spiders build silken nests beneath bark and stones, which they may use as retreats or in which they may overwinter. The nest can prevent flooding by trapping an air bubble around the spider. After hatching, the young of some species are carried aloft on air currents **(ballooning)** by means of a silken strand. Of course, the most familiar use of silk is the construction of **webs** (Fig. 18-24) for the capture of prey.

FIGURE 18-22 The black widow, *Latrodectus mactans*, with a silk egg case.

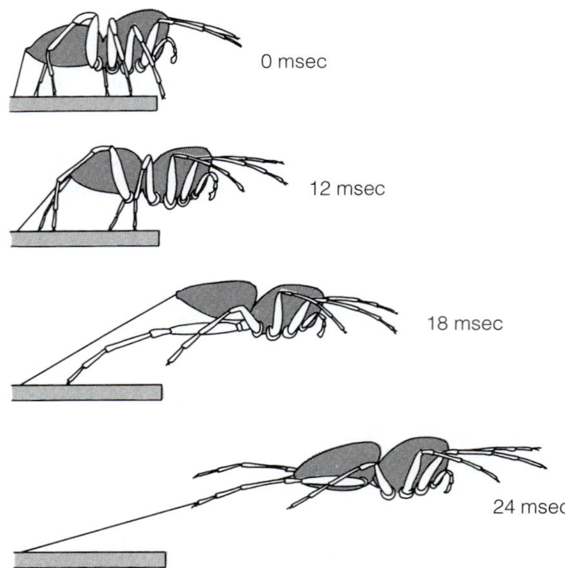

FIGURE 18-23 Tracings from motion picture frames of a jumping spider employing a dragline during a leap. Most of the force is provided by the rapid extension of the fourth pair of legs, which results from sudden elevation of blood pressure. Note that the dragline is anchored before and retained during the jump. *(After photographs by Parry and Brown from Foelix, R. F. 1982. Biology of Spiders. Harvard University Press, Cambridge, MA. p. 155.)*

Prey Capture

Spiders, like most arachnids, are predatory and feed largely on insects, although small vertebrates may be captured by large spiders. The prey is caught in a silken snare (the web) by **web-building** spiders or is pounced on by members of the more active **cursorial** (running) spiders.

The dragline may have been the precursor of the web. In the evolution of spiders, a simple dragline may have been gradually elaborated upon to produce the many types of snare webs used by modern web-building spiders. Most elaborate, and also most familiar, is the circular **orb web** constructed by

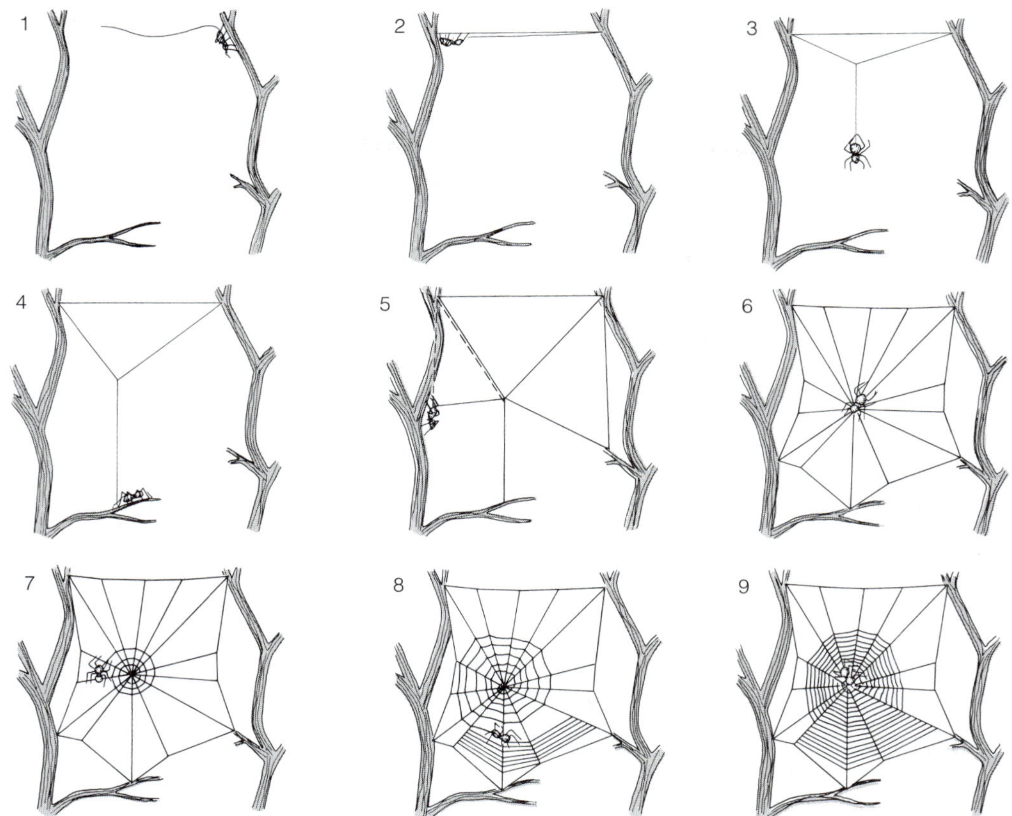

FIGURE 18-24 Stages in the construction of an orb web. The web is usually attached to vegetation by additional frame threads, but these have been omitted for clarity. 1 and 2, the first frame thread; 5, two frame threads and five radial threads; 7, the hub and the start of the catching spiral; 9, catching spiral about halfway completed. The free zone is not apparent in these drawings. See text for further explanation. *(From Levi, H. W. 1978. Orb-weaving spiders and their webs. Am. Sci. 66:734–742.)*

spiders in Araneidae (orb-web spiders), such as the common black-and-yellow garden spider. Most of the several types of threads of an orb web lie in a single plane (Fig. 18-24). A typical orb web consists of five distinct regions (Fig. 18-30C). (1) The central **hub,** or platform, consists of an irregular network of threads where the spider spends most of its time. (2) **Radial** threads extend like a sunburst outward from the hub to the frame. (3) A **frame** of single threads borders the periphery, frames the web, and attaches it to vegetation or other structural features of the environment. (4) The **catching spiral** is the most obvious part of the web. It is an adhesive thread that spirals outward from just outside the hub to the frame so that it occupies most of the space between the hub and frame. It crosses the radial threads repeatedly at close intervals and creates the meshwork of closely spaced threads in which flying insects are caught. Of all the web regions and thread types, only the catching spiral is sticky. The spider carefully avoids touching this thread by using the radial threads to move across the web. (5) Between the hub and the catching spiral there is a **free zone** consisting solely of radial threads. Since there are no sticky threads here, the spider need not worry about catching itself. In this zone the spider can pass safely through the plane of the web from one side to the other.

Eyesight is not well developed in web-building spiders, but their sensitivity to vibrations is exquisite. A web builder can determine from thread vibrations the size and location of the trapped prey, the identity of the prey, and its suitability as food, and many spiders respond to different stimuli with different attack patterns.

The positioning and locations of webs are not random. Some species spin horizontal orbs to catch insects that fly up from underlying vegetation. Others have vertical webs at species-specific elevations to trap insects in horizontal flight. Thus, the same general habitat may support a number of species of orb weavers, which partition the food resource by trapping a different part of the insect community.

Cursorial hunting species include wolf spiders, fishing spiders, crab spiders, jumping spiders, and many mygalomorph spiders. Cursorial spiders are believed to be derived from web-building ancestors through loss of the web-building habit. Species that stalk their prey typically have heavier legs than web builders, and most have a dense tuft of fine hairs behind the tarsal claws that aids in adhering to surfaces and in prey capture. These tufts, known as **scopulae,** are composed of setae whose ends are split into as many as 1000 branches, like a broom. This vastly increases the surface area each seta contacts on the substratum. Spiders with scopulae can walk up smooth vertical walls or glass and even upside down across a ceiling. The mechanism is not fully understood, but it appears to be due to adhesive forces between the seta and an extremely thin layer of water molecules on the substrate surface. Without this thin film, the scopulae cannot adhere.

Some cursorial species roam, actively searching for insects, whereas others utilize a lie-in-wait strategy. Many cryptically colored crab spiders, for example, sit in flowers and ambush visiting insects. Many spiders walk or run over the surface of lakes and ponds to harvest insects trapped in the surface. Wolf spiders and jumping spiders roam and hunt for their prey, detecting it by tactile and visual stimuli, then pounce on it. In these and some other taxa, the eyes are highly developed and of primary importance in prey capture.

Although most spiders are solitary, some degree of social organization has evolved in a few species in nine families. Social spiders share a communal web and cooperate in capturing prey.

Nutrition

Like most arachnids, spiders are carnivores that ingest liquid food after digestion in the preoral cavity. The digestive equipment thus includes the mouthparts and preoral cavity in addition to the gut. Mouthparts include the chelicerae, pedipalpal coxae (pedipalpal endites, palpal gnathobases), and labium (Fig. 18-18B). The foregut includes the mouth, pharynx, and esophagus. The midgut has an extensive system of branching digestive ceca and the hindgut is a short rectum.

Unique to spiders are chelicerae equipped with **poison glands** opening at the tip of the fang (Fig. 18-20). Following capture, be it by web or ambush, spiders bite the prey with the chelicerae and inject it with poison. The glands themselves are located in the basal segment of the chelicerae and usually extend backward into the head. When biting, the fangs are raised and rammed into the prey. Simultaneously, muscles around the poison gland contract and fluid from the gland is discharged from the fang into the body of the prey. Chelicerae may also be used to hold and macerate the tissues during digestion. Being prehensile, they may have additional roles, such as carrying egg cocoons, digging, or stridulation.

Most spider venom is not toxic to humans, but a few species have dangerous bites. Among these are the black widows *(Latrodectus),* which are found in most parts of the world, including the United States and southern Canada (Fig. 18-25A). Black widow venom is neurotoxic and, as in most spiders, composed of a mixture of several proteins. Although the initial bite may go unnoticed, the symptoms become severe and painful. They include pain in the abdomen and legs, high cerebrospinal fluid pressure, nausea, muscular spasms, and respiratory paralysis. Fatal cases are rare, however, and are more apt to result from incorrect diagnosis and treatment (such as removal of the appendix) than from the spider venom. Antivenin and other treatments are available. Black widows are smooth, glossy black with a globose abdomen and a distinctive red hourglass shape on the ventral surface of the abdomen.

The recluse spiders, members of the genus *Loxosceles,* of which there are several species in North America, have a hemolytic venom and produce a persistent local necrosis, or ulceration, that spreads from the bite. The bite of *Loxosceles reclusa,* the brown recluse spider (Fig. 18-25B), which is found in the midwestern and southeastern United States, as well as in other regions in localized populations, can be dangerous. *Loxosceles* has six eyes in three pairs, rather than the eight of most spiders. It is a brownish spider 2 to 4 cm long with a dark, violin-shaped mark on the carapace.

North American mygalomorph tarantulas (Fig. 18-25D), despite their size and reputation, are not toxic, but there are mygalomorphs, such as the Australian *Atrax* and the South American *Trechona,* both funnel-web builders, that are lethal. Many New World tarantulas (Theraphosidae), however, have defensive, **urticating setae** on the abdomen. These setae are heavily barbed, irritating, and readily penetrate the skin of potential predators, such as small mammals, that enter the tarantula's burrow. In humans they may penetrate up to 2 mm into the skin and cause a rash and intense itching.

Although all spiders begin digestion preorally, this may occur in two different ways. Some spiders puncture the otherwise intact prey and pump digestive enzymes from the midgut into it. They then suck the hydrolyzed fluid into the gut, leaving behind a nearly intact husk, which is discarded. Others use the chelicerae to macerate the prey into a shapeless mass and mix it with digestive enzymes from the midgut. Predigestion of this mass occurs in the preoral cavity formed of the chelicerae, pedipalpal coxae, and labium. The pedipalpal coxae have a toothed edge used to lacerate the prey and mix it with digestive fluid. The hydrolyzed fluid is sucked into the gut and the amorphous wad of indigestible, primarily chitinous remains is discarded. In both types of digestion, the fluid is sucked into the midgut, passing through two setal filters to exclude particles. The first is a fairly coarse filter consisting mostly of setae on the pedipalpal endites and the second is a very fine filter with a mesh size of about 1 μm in the pharynx. Suction is provided by the muscular pharynx and pumping stomach (Fig. 18-20), which is a specialized region of the posterior esophagus.

The midgut and its digestive ceca fill almost the entire abdomen, as well as part of the cephalothorax (Fig. 18-20). Extracellular digestion is completed in the ceca and absorption occurs here. The posterior midgut is expanded to form a cloacal chamber, or stercoral pocket, into which the Malpighian tubules empty. Digestive and excretory wastes are collected here and then egested from the anus through a short, sclerotized rectum.

As an adaptation for predation and an uncertain, sporadic food supply, most spiders have relatively low metabolic rates, can tolerate prolonged starvation, and, because of the great capacity of the digestive ceca, can double their body weight with one feeding.

Although spiders are important insect predators, they are preyed on by some insects as well as vertebrates. Two taxa of wasps hunt spiders as food for their larvae. Female wasps paralyze the spider with a sting and lay an egg on its body. Spider wasps (Pompilidae) put the paralyzed spider in an

FIGURE 18-25 Some North American spiders. **A,** Female black widow, *Latrodectus mactans,* hanging in its web. This species of the cosmopolitan taxon *Latrodectus* is found in the southeastern and midwestern United States, where it may be common. **B,** The brown recluse spider, *Loxosceles reclusa,* whose bite produces a necrotic wound that is slow to heal. **C** and **D,** Mygalomorph spiders. **C,** Trap-door spider capturing a beetle. **D,** Tarantulas from the southwestern United States. *(A, From Kaston, B. J. 1970. Trans. San Diego Soc. Nat. Hist. 16(3):33–82.)*

underground chamber. Mud daubers (Sphecidae) put a number of paralyzed spiders in a mud tube attached to elevated objects. The importance of vertebrate predation pressure on spider populations is indicated by the 10-fold increase in spider population densities on vertebrate-free islands.

Gas Exchange

Spiders have book lungs and/or tracheae for gas exchange. The primitive mesothelan and mygalomorph spiders have two pairs of book lungs located ventrally on the second and third abdominal segments (Fig. 18-19, 18-26A) and no tracheae. Most spiders, however, have a pair of anterior book lungs and a posterior set of tubular tracheae (Fig. 18-20), the latter derived from the second pair of book lungs. The spiracles of the tracheae have moved posteriorly and in most spiders open on the ventral midline at a single, posterior tracheal spiracle (Fig. 18-18B). In some spiders the anterior book lungs have also been converted to tubular tracheae and there are no lungs (Fig. 18-26B). A few spiders have sieve tracheae, either exclusively or in combination with tube tracheae (Fig. 18-26B). Spider tracheae are simple, branched or unbranched tubes supported by chitinous taenidia.

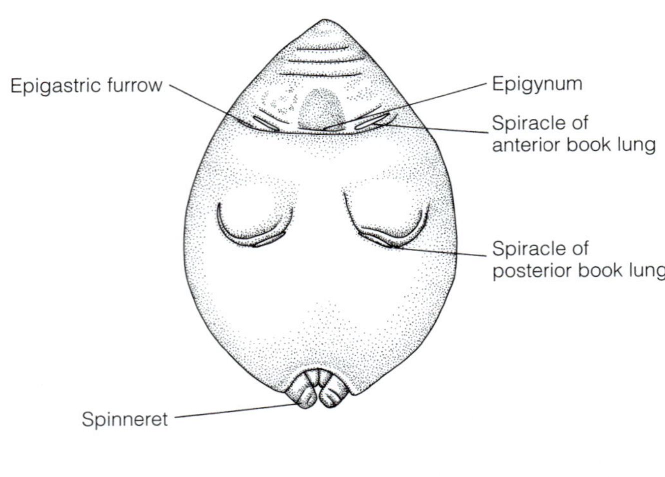

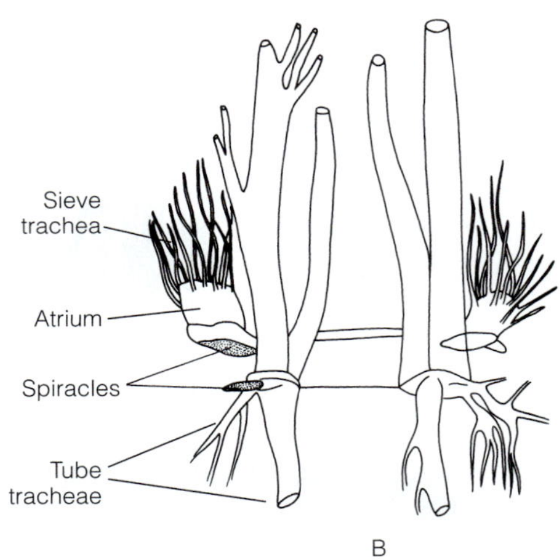

FIGURE 18-26 Spider respiration. **A,** Abdomen of a mygalomorph spider in ventral view, showing the spiracles of the two pairs of book lungs. **B,** Tracheal system of the caponiid spider *Nops coccineus,* in which the ancestral book lungs of segments 2 and 3 have been replaced by both types of tracheae. The two sieve tracheae in the figure are derived from the lungs of segment 2 and have conspicuous atria from which arise fine tubular lamellae. Each of the two tube tracheae consists of four large branched or unbranched tubes arising from a spiracle. The two lateral tubes are derived from the book lungs of segment 3, whereas the two median tubes arise from apodemes. *(A, After Comstock; B, After Bertkau from Kaestner, 1968.)*

Internal Transport

The spider hemal system (Fig. 18-20) is similar to that described for other arachnids such as scorpions. The heart, a dorsal tube in the anterior abdomen, is enclosed in a pericardial sinus and has only two or three pairs of ostia. Pulmonary sinuses surrounding the book lungs connect directly with the pericardial sinus via pneumopericardial veins (Fig. 18-20).

Spiders use blood pressure to extend their legs, which have no extensor muscles. Spiders with a punctured cephalothorax cannot elevate their blood pressure to extend the legs and dead spiders have flexed legs. Jumping spiders (Salticidae) jump by suddenly elevating their blood pressure to forcefully extend the third or fourth pair of legs (Fig. 18-23). Leaps as great as 50 body lengths have been recorded.

Excretion

Nitrogenous end products are chiefly guanine, adenine, and uric acid, which are absorbed from the blood by Malpighian tubules and deposited in the cloacal chamber (Fig. 18-20) for incorporation into the feces. Primitive spiders (Mesothelae and Mygalomorphae), however, have two pairs of well-developed saccate nephridia in the cephalothorax. Their nephridiopores open on the coxae of the first and third walking legs. In derived spiders (Araneomorphae), only the anterior pair is present and usually it is degenerate.

Spiders have other mechanisms for excreting nitrogenous wastes. The epithelium of the intestine and cloacal chamber is similar to that of Malpighian tubules and participates in the transfer of nitrogen from the blood to the gut lumen. Wandering nephrocytes absorb nitrogenous wastes, which are then transferred to the Malpighian tubules or the coxal glands. Guanocytes and interstitial cells located in the epithelium of the digestive ceca absorb and store guanine.

Nervous System and Sense Organs

The CNS is highly cephalized, with all segmental ganglia coalesced in the supraesophageal and subesophageal ganglia (Fig. 18-20, 18-7B) occupying most of the space in the cephalothorax. The branches of one or two **abdominal nerves** that extend from the subesophageal ganglion into the abdomen innervate abdominal structures, but the nerves are not ganglionated.

Most spiders have eight eyes in four pairs (anterior median, anterior lateral, posterior median, and posterior lateral) on the anterior dorsal margin of the carapace (Fig. 18-27A). Each eye has a single lens that serves all its receptor units and there are no facets. Two types of eyes are present in spiders. The anterior median eyes, or **main eyes,** are median ocelli of the direct type (Fig. 18-27A, 18-8B). They do not have a reflective tapetum and are black. The main eyes usually are small, but in the jumping spiders they are very large (Fig. 18-27C). Six-eyed spiders have lost the main eyes.

The anterior laterals, posterior medians, and posterior laterals (Fig. 18-27A) are **secondary eyes.** These indirect eyes are thought to have derived from the lateral compound eyes of the ancestral chelicerate, although they no longer retain the faceted construction of typical compound eyes (Fig. 18-8A). Secondary eyes usually have a tapetum of reflective pigment to improve vision in low light intensities, and they often appear pearly white due to this reflectivity. In some taxa, notably the hunting, cursorial wolf spiders, the reflective tapetum in their

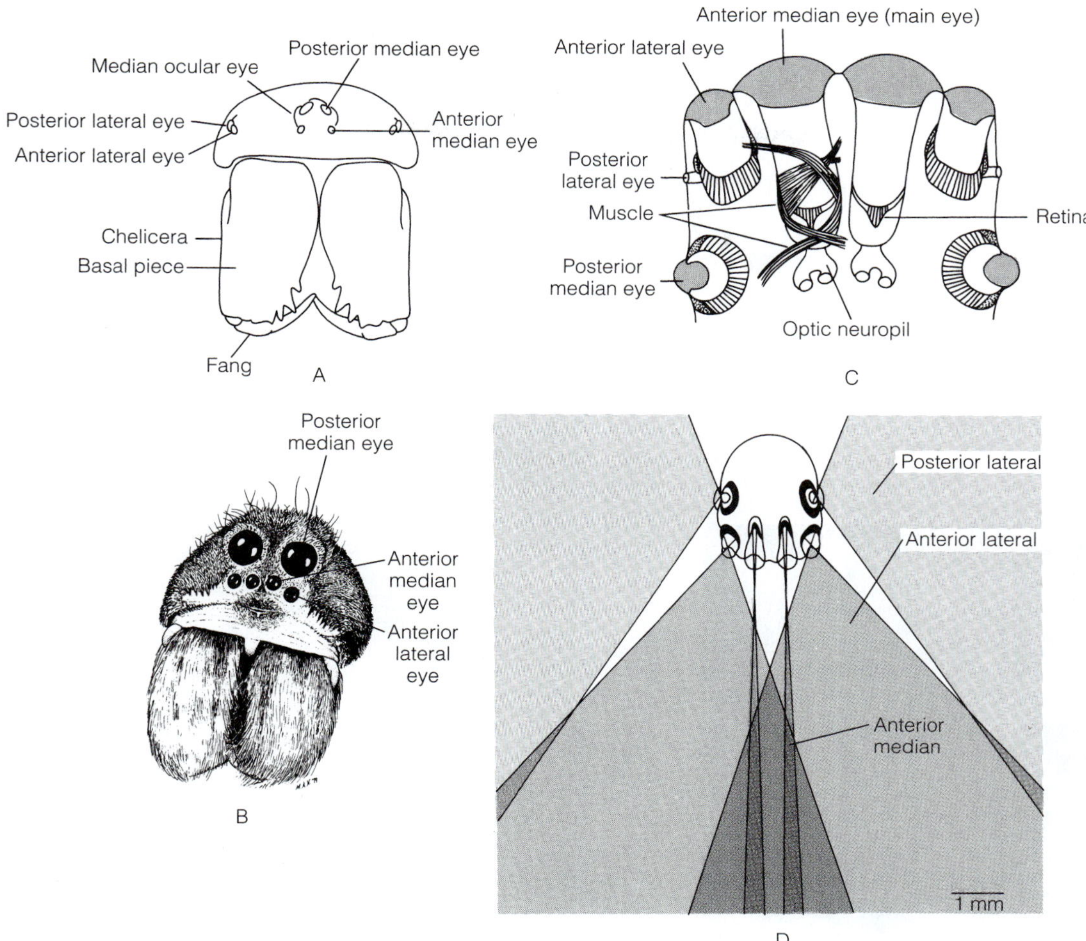

FIGURE 18-27 Eyes. **A,** Anterior view of a spider, showing eight eyes arranged in anterior and posterior rows of four each, a pattern common to many taxa. **B,** Face of a wolf spider. The main eyes are the small anterior median eyes. The posterior median eyes are very large; the posterior lateral eyes are located over the carapace and cannot be seen. **C,** Eyes of a jumping spider. Note the tubular, telephoto structure of the large anterior median eyes, which are the main eyes. The rest of the eyes are secondary eyes. **D,** Visual fields of the eyes of a jumping spider *(Trite planiceps)*. The posterior lateral eyes function in prey detection; the anterior median (main) eyes function in tracking the prey after swiveling the body in the prey's direction. The posterior median eyes are not shown. *(A, After Kaston; C, Modified after Land, M. F. 1969a. Structure of the retinae of the principal eyes of jumping spiders (Salticidae: Dendryphantinae) in relation to visual optics. J. Exp. Biol. 51:443; and after Land, M. F., 1969b. Movements of the retinae of jumping spiders in relation to visual optics. J. Exp. Biol. 51:443; and after Land, M. F. Movement of the retinae of jumping spiders (Salticidae: Dendryphantinae) in response to visual stimuli. J. Exp. Biol. 51:471; D, Modified from Forster, L. M. 1979. Visual mechanisms of hunting behavior in* Trite planiceps, *a jumping spider. N. Z. J. Zool. 6:79–93.)*

large posterior median eyes has developed to such an extent that these spiders can be located at night by using a flashlight to look for the eye shine. Jumping spiders, on the other hand, have no tapeta at all in their indirect eyes.

Vision is far more important in cursorial, hunting spiders, such as jumping and wolf spiders, than in the sedentary web spinners, and the few studies of spider eyes have been mostly restricted to the jumping spiders (Salticidae), which have the most highly developed eyes. The eyes of some jumping spiders surpass those of all other arachnids. In these spiders, the eyes are situated over the surface of the anterior half of the carapace, resulting in a wide visual field (Fig. 18-27D) composed of the overlapping fields of the separate eyes. Secondary eyes detect movement whereas the main eyes of jumping spiders are capable of forming an image. The large anterior median eyes (main eyes) of jumping spiders (Fig. 18-27C,D) are unique in having the photoreceptors arranged in four layers, each of which is believed to respond to different wavelengths of light. Whether this permits color recognition or greater contrast is uncertain.

In cursorial spiders, eyes are important for detecting movement and locating and tracking prey. The main eyes of jumping spiders are capable of perceiving a relatively sharp image. The number of receptors is large, particularly in the main eyes, which have about 1000 photoreceptors. In many species of jumping spiders the depth of the anterior median eyes has increased, resulting in a somewhat tubular, telephoto

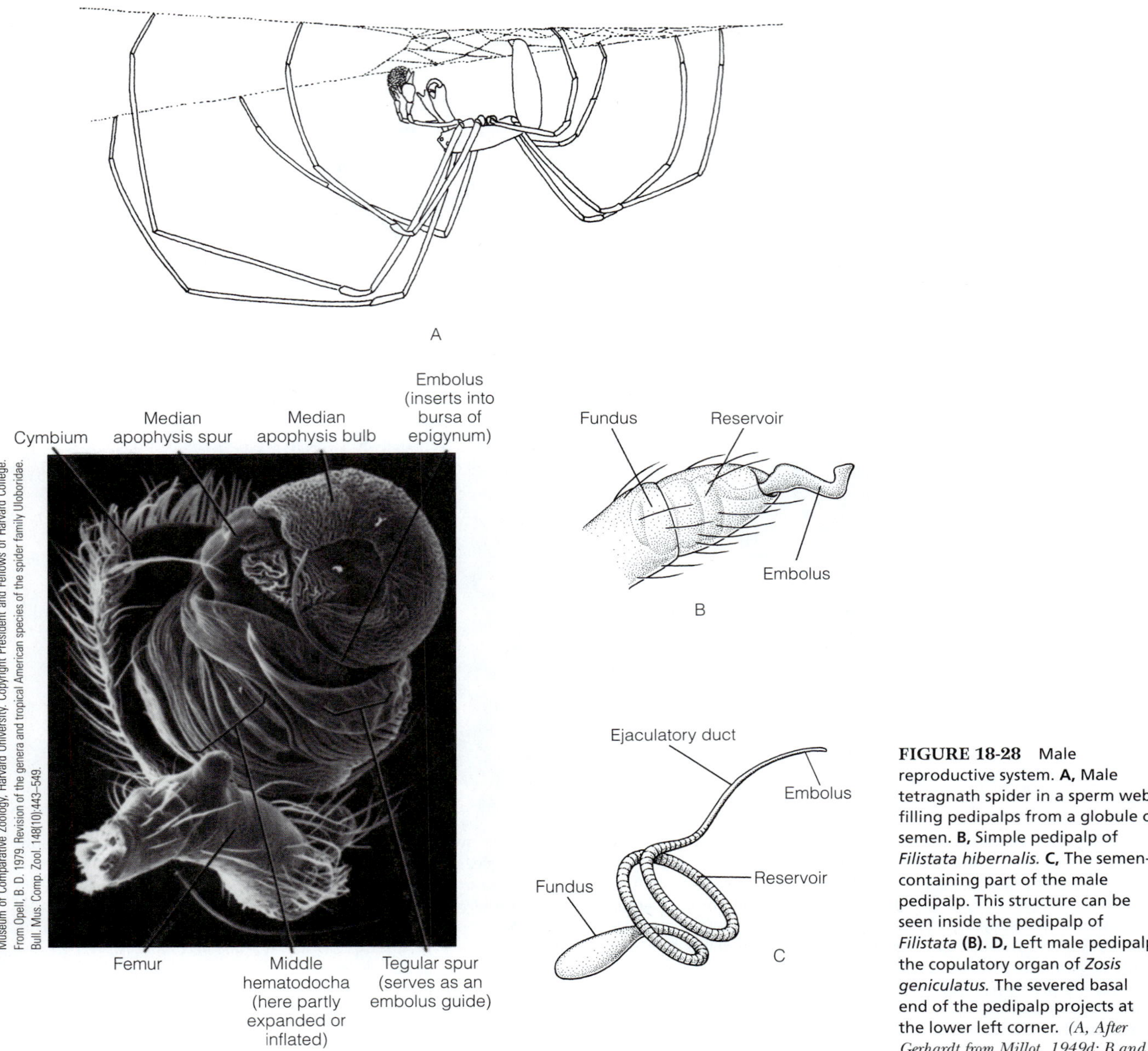

FIGURE 18-28 Male reproductive system. **A,** Male tetragnath spider in a sperm web, filling pedipalps from a globule of semen. **B,** Simple pedipalp of *Filistata hibernalis.* **C,** The semen-containing part of the male pedipalp. This structure can be seen inside the pedipalp of *Filistata* **(B). D,** Left male pedipalp, the copulatory organ of *Zosis geniculatus.* The severed basal end of the pedipalp projects at the lower left corner. *(A, After Gerhardt from Millot, 1949d; B and C, After Comstock.)*

structure (Fig. 18-27C). Muscles attached at the rear can rotate the tubes around the visual axis.

Spiders have a number of cuticular sensilla, including tubular chemoreceptive setae on the tips of appendages, tactile setae, trichobothria, slit sense organs, and tarsal organs. The abundant setae, so conspicuous on many spiders, are probably all equipped with sensory dendrites. These sensilla are important sources of environmental information in all spiders, and in the sedentary, web-building spiders, they are much more important than the eyes. The funnel-web builder, *Agelena,* for example, can determine the position of prey by means of the trichobothria from a distance of 1 cm. Contact chemoreceptive setae (taste receptors) are hollow and have an open distal end and 21 dendrites inside (Fig. 18-8C). Of the 21 dendrites, 19 are chemoreceptors and 2 are mechanoreceptors. Taste receptors are concentrated on the pedipalps and the tarsi of the first legs. A spider touches an object to taste it.

Spiders have the greatest development of slit sense organs of any arachnid. Slit sense organs in the joint between the tarsus and the metatarsus of web-building spiders are especially sensitive and enable the spider to discriminate vibration frequencies transmitted through the silk strands of the web or even through the air. In species in which the spiderlings remain in the parental web after hatching, the parent can discriminate between its offspring and prey. The spider can also discern the size of the entrapped prey and even the kind of insect if it produces buzzing vibrations. The spiders themselves may produce vibration signals to communicate with other spiders. A male may tweak the strands of the female's web, or a mother may produce vibration signals that are detected by her spiderlings. **Tarsal organs,** cuplike structures on the tarsi of the legs, probably are olfactory receptors for pheromones.

Reproduction and Development

Spiders are gonochoric and have internal fertilization with indirect sperm transfer like most arachnids. Unlike most arachnids, however, they do not use spermatophores, and the male uses pedipalps to transfer sperm to the female. Spiders are sexually dimorphic and males, often much smaller than females, have large, highly modified pedipalps (Fig. 18-28). There is variability in the reproductive anatomy of spiders and only the most common situation is described here.

The ovaries consist of two elongate, parallel sacs located in the ventral part of the abdomen (Fig. 18-20). An oviduct exits the anterior end of each ovary and they join together to form a median, tubular, chitinous uterus (vagina) that opens via the single gonopore on the ventral midline, hidden in the epigastric furrow (Fig. 18-18B, 18-29B). Two seminal receptacles are connected with the uterus by fertilization ducts. Fertilization occurs in the uterus and eggs are deposited through the gonopore. In primitive spiders (Mesothelae and Mygalomorphae) the male pedipalp is inserted directly into the gonopore to deposit sperm in the seminal receptacles.

In derived spiders (Araneomorphae) the process is more complicated. Fertilization ducts are present, as in primitive spiders, but there is an additional set of ducts, the **copulatory ducts,** leading to the seminal receptacles from a pair of **copulatory pores** on the epigynum (Fig. 18-29B). The epigynum is a ventral sclerite immediately anterior to the epigastric furrow (Fig. 18-18B, 18-29A). The embolus of the male pedipalp (Fig. 18-28B,C) is inserted into the copulatory pore and extended along the length of the copulatory duct to the seminal receptacles, where sperm is expelled and stored indefinitely. Later, during fertilization, sperm move from the seminal receptacles to the uterus via the fertilization ducts to fertilize eggs in the uterus.

In the male there are two large, tubular testes in the abdomen. A convoluted sperm duct exits each testis and opens at a common gonopore on the ventral midline in the epigastric furrow, as in the female. The intromittent organs, which are the highly modified tips of the pedipalps, are remote from the gonopores. The pedipalpal tarsi are extremely complicated (Fig. 18-28B,C,D), but the essential features are the **sperm reservoir,** a duct, and an **embolus,** the latter being the intromittent organ. Sperm is stored in the reservoir.

Prior to courtship and mating the male spins a small sperm web, into which he ejaculates a droplet of sperm from his gonopore (Fig. 18-28A). He then dips the tip of the pedipalp into the sperm droplet and fills the reservoir. With the pedipalp thus charged, he goes in search of a receptive female.

The highly predatory habits of spiders make identification of the sexual partner especially important so the female does not mistake the male for food. To this end, highly complex precopulatory behavior patterns have evolved in many species. Chemical and tactile cues are of primary importance. On encountering a dragline or web, a male spider may use chemical cues to determine if it was produced by a mature female of his species. The same cues may initiate courtship behavior.

Females respond to a variety of cues produced by the male. In the web builders, the male may pluck the strands of webbing, producing vibrations that are detected and recognized by the female. Orb-weaver males pluck a radial thread held by the

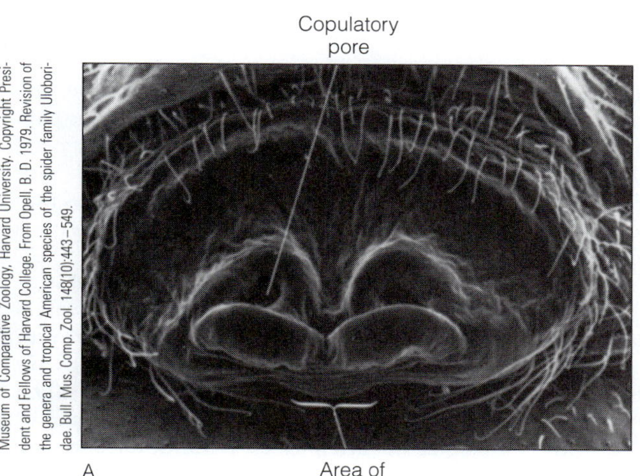

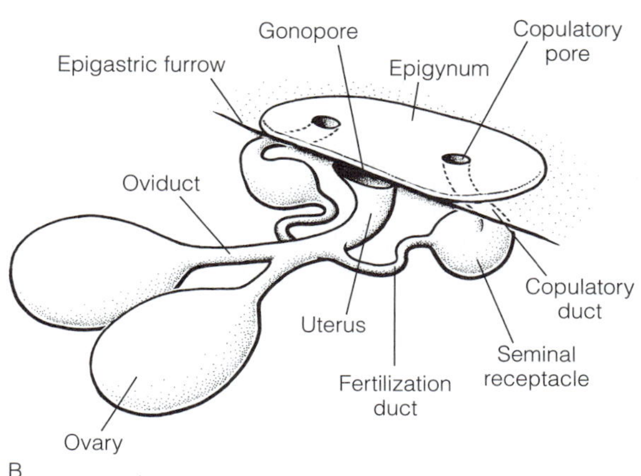

FIGURE 18-29 The reproductive system of a female spider. **A,** The epigynum, a sclerotized plate bearing the reproductive openings of a female *Philoponella republicana*. **B,** The reproductive system of a female spider. During copulation the embolus of the male pedipalp is inserted into a copulatory pore, then through the copulatory duct to the seminal receptacle. From there the fertilization duct carries the sperm to the uterus, where fertilization occurs as the eggs are released through the gonopore. *(B, Redrawn from Kaestner, A. 1968. Invertebrate Zoology. Vol. II: Chelicerata. John Wiley and Sons, New York. 472 pp.)*

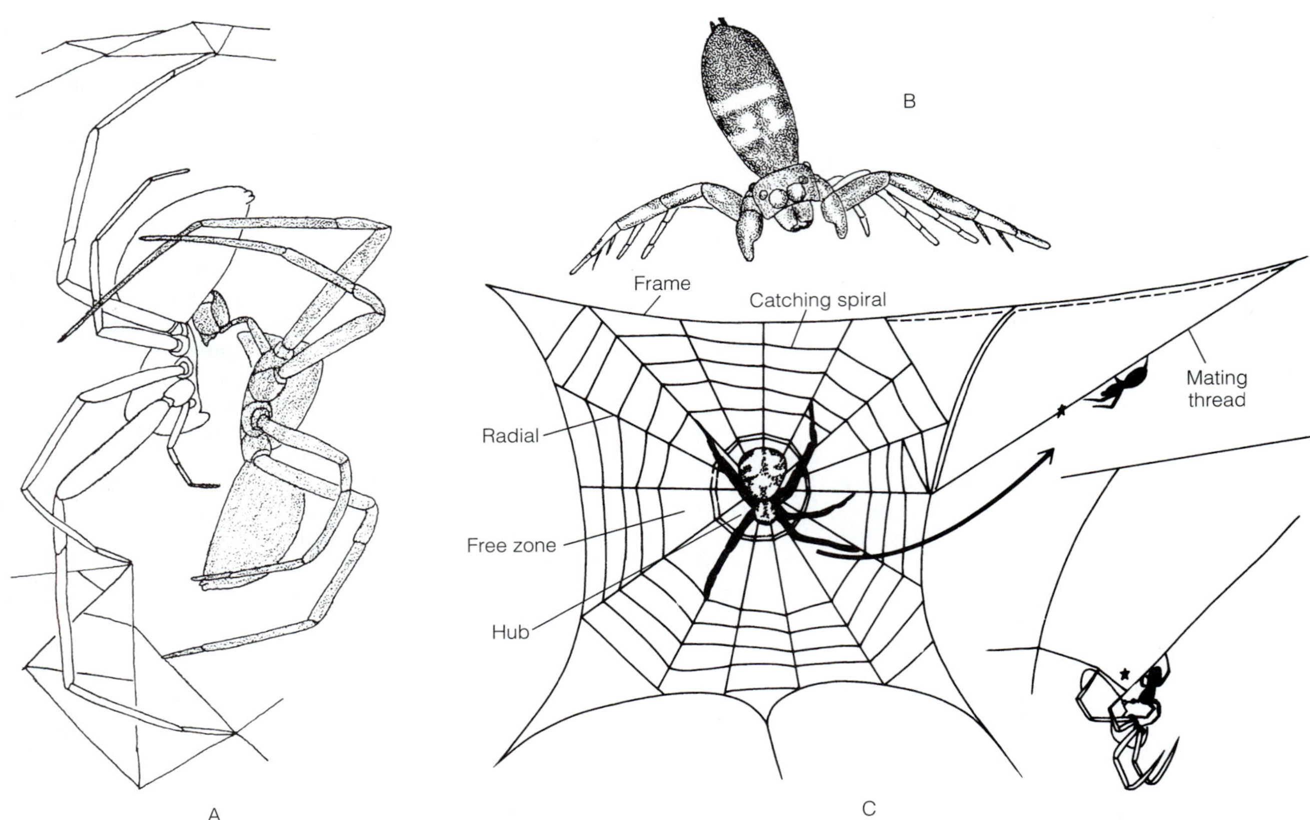

FIGURE 18-30 Reproductive behavior. **A,** Mating position of *Chiracanthium* (male is shaded). **B,** Courting posture of the male jumping spider *Gertschia noxiosa.* **C,** Web morphology and courtship behavior of some orb weavers. The male (silhouette) vibrates the mating thread he has attached to the female's web. The female leaves her position in the hub and moves onto the mating thread, where copulation occurs (star). *(A, After Gerhardt from Kaston; B, After Kaston; C, From Robinson, M. H., and Robinson, B. 1980. Comparative Studies of the Courtship and Mating Behavior of Tropical Araneid Spiders. Pacific Insects Monograph 36. Bishop Museum Press, Honolulu.)*

female or a **mating thread** that he has attached to the female's web (Fig. 18-30C). Species identity is encoded in the number, frequency, and intensity of plucks. In some species the male is so small he is not considered worth eating and the female ignores him as he clambers over her body, going about his business.

Cursorial hunting spiders employ other courtship behaviors. The approach may be direct, with the male pouncing on the female and palpating her body with his pedipalps and legs, causing her to fall into an immobile state. Some male wolf spiders stridulate with a file and scraper located in the tibiotarsal joint of the palp. During oscillations of the joint, a group of heavy spines at the tip of the palp maintain contact with the substratum and the female detects the vibrations through the substratum, rather than as airborne pressure waves.

In taxa with well-developed eyesight, such as the jumping spiders, visual cues are also important, and courtship takes the form of dancing and posturing by the male in front of the female (Fig. 18-30B). This involves various movements and the waving of appendages, which are often brightly colored. Such behavior is highly developed and has been studied most extensively in the colorful jumping spiders.

On attaining the proper position vis-à-vis the female, the male usually scrapes her epigynal surface rapidly with his palp until the appropriate parts on the palp and epigynal plate connect. Then the palp quickly becomes engorged with blood, driving the embolus through the copulatory pores into the copulatory duct to the seminal receptacle, where sperm is deposited (Fig. 18-30A).

Some time after mating, the female lays up to 3000 eggs in one or more silk egg cases. The female spins a small basal sheet and a cuplike well. The eggs are fertilized with sperm from the seminal receptacles. The fertilized eggs are extruded through the gonopore and placed in the silken cup. The female then covers the eggs with another sheet of silk and the entire mass is often given yet another covering of silk to produce a spherical egg sac. The egg sac is attached to the web, hidden in the spider's retreat, carried in the chelicerae, or attached to the spinnerets and dragged about behind the mother. In many spiders the female dies after completing the egg sac, whereas in others she may produce a series of cases and may remain with the young as they hatch.

The **spiderlings** hatch inside the egg sac and remain there until they complete the first molt. Upon exiting the egg case the spiderlings of many species engage in **ballooning,** which disperses the species. A little spider climbs to the top of a twig or blade of grass and spins a strand of silk. When an air current tugs on the strand, the spider releases its hold and is carried away by the wind.

Some spiders care for their young. The female wolf spider carries the young on her back. The mother's dorsal abdominal surface is covered with special knobbed, spiny setae that aid in the attachment of the young. The spiderlings gradually leave the mother and disperse. The mother does not feed while carrying the young.

Although some of the tarantulas (mygalomorph spiders) have lived in captivity for as long as 25 years, most spiders live only 1 to 2 years. The number of molts required to reach sexual maturity varies, depending on the size of the species. Large species undergo up to 15 molts before the final instar, whereas tiny species molt only a few times. Almost half of temperate species overwinter as immatures, usually in soil or leaf litter, and the other half overwinter as eggs or adults. Some overwinter in whatever stage they are in when winter arrives. Temperature and photoperiod are the primary controlling factors in determining the pattern of the life cycle.

Diversity of Araneae

Mesothelae[sO]: Abdomen has segmental tergites and sternites (Fig. 18-19). Seven or eight spinnerets, the first located at the level of the second pair of book lungs. Orthognath chelicerae. About 40 extant species in one taxon, the tropical Asian Liphistiidae. *Liphistius, Heptathela.*

Mygalomorphae[sO] (Orthognatha; Fig. 18-25C,D): Abdominal segmentation not externally apparent. Two pairs of book lungs. Six spinnerets. Fangs are orthognath, moving in a plane parallel to the long body axis. Includes tarantulas, bird spiders (both Theraphosidae; *Aphonpelma, Avicularia, Theraphosa*), purse-web spiders (Atypidae; *Atypus*), funnel-web spiders (Dipluridae; *Atrax, Diplura, Euagrus, Hadronyche, Trechona*), and trap-door spiders (Ctenizidae; *Bothryocyrtum, Ctenizia, Cyclocosmia, Ummidia*). Most stalk or ambush prey, a few are web builders. Many are very large. *Atrax robustus,* with fangs that can penetrate a fingernail, is said to be the world's deadliest spider. About 2200 extant species in 15 families, many North American.

Araneomorphae[sO]: Fangs are not orthognath. Abdomen is not segmented. One pair of book lungs in most species. Cribellum and calamistrum present primitively, but lost in most. 90 families with 32,000 species.

Palaeocribellatae[iO]: Cribellate. Fangs are diagonal, intermediate between orthognath and labidognath. Two pairs of book lungs. Epigynum absent. One family.

Hypochilidae[F]: Oval cribellum on a cone. Calamistrum has two rows of setae. Eight eyes. Nine species. *Ectatosticta, Hypochilus.*

Neocribellatae[iO] (Labidognatha): Fangs are labidognath, moving at right angles to the long axis of the body. Females usually have an epigynum. One or no pairs of book lungs; at least one pair of tracheae.

Dysderidae[F]: Large, primitive spiders. Short legs. Six eyes in two rows of three; anterior median eyes are absent. Book lungs and tracheae; lungs have few leaves; two pairs of spiracles close together. No epigynum. *Dysdera, Sergestria.*

Filistatidae[F]: Small cribellum and reduced calamistrum. Large, branched poison glands, eight eyes. *Kukulcania* (also known as *Filistata*) *hibernalis,* the southern house spider, is sometimes confused with the brown recluse.

Scytodidae[F]: Spitting spiders. Six eyes in three pairs. Three tarsal claws. Carapace is highly arched to accommodate large poison and salivary glands. Cursorial nocturnal hunters. Squirt glue from chelicerae up to 20 mm to entangle prey or predators; harmless to humans. *Scytodes.*

Sicariidae[F]: Includes the recluse spiders (Fig. 18-25B). Six eyes in three pairs, posterior laterals are absent. Cursorial nocturnal hunters that build loose diurnal refuges in dry, dark areas. The bite of the brown recluse, *Loxosceles reclusa,* and others produces serious necrotic lesions. *Sicarius* largely restricted to the Southern Hemisphere. *Loxosceles, Sicarius.*

Pholcidae[F]: Daddy longlegs spiders. Long legs, resemble harvestmen. Small chelicerae. Spin small webs of tangled threads in sheltered recesses. Several species are common in houses and, with other web-building house dwellers, are responsible for cobwebs. *Holocnemus, Pholcus, Physocyclus, Spermophora.*

Caponiidae[F]: Book lungs are absent, replaced by tracheae; two pairs of spiracles. Usually have only two eyes (anterior median). Small chelicerae. Cursorial. *Caponia, Nops, Tarsonops.*

Mimetidae[F]: First legs have a row of long, curved setae. Prey on other spiders. Cursorial, do not make webs. Sometimes ambush and sometimes invade the web of the prey. *Ero, Meta.*

Uloboridae[F]: Cribellate. Lack poison glands and venom. Orb webs made at least partially of cribellate silk, usually positioned horizontally. Some social species have webs in dense colonies. *Philoponella, Uloborus, Zosis.*

Dinopidae[F]: Ogre-faced spiders. Eight eyes in three rows; anterior median eyes are enormous and face anteriorly. *Dinopis, Menneus.*

Theridiidae[F]: Comb-footed spiders. Large taxon of nocturnal, irregular web builders. Includes the venomous black widow, *Latrodectus* (Fig. 18-22, 18-25A), as well as the common house spider *Achaearanea tepidariorum,* which resembles a black widow. *Steatoda, Theridion.*

Linyphiidae[F]: Sheet-web spiders. Webs are horizontal sheets or bowls, sometimes in vegetation. Many tiny species in leaf litter. *Linyphia, Erigone, Halorates.*

Tetragnathidae[F]: Orb weavers similar to Araneidae, but their posterior median eyes lack a tapetum. *Pachygnatha, Tetragnatha* (Fig. 18-28A).

Araneidae[F]: Orb-web spiders. Posterior median eyes have a tapetum. Many are large and colorful, such as the black-and-yellow garden spider, *Argiope. Araneus, Gasteracantha, Gea, Neoscona, Nephila.*

Lycosidae[F]: Wolf spiders. Fast, hairy, dull brown or black, cursorial hunters. Posterior median eyes are the largest of the eight eyes (Fig. 18-27B). Three tarsal claws. *Lycosa* *tarentula* is the Mediterranean tarantula. Common members of the nocturnal ground fauna. *Arctosa, Lycosa, Pardosa, Pirata.*

Pisauridae^F: Fishing and nursery spiders. Large cursorial hunters similar to wolf spiders, but with longer legs. Common around fresh water and move efficiently on, and sometimes under, the water surface to capture prey. *Dolomedes, Pisaura, Thalssius, Tinus.*

Agelenidae^F: Funnel weaver, or grass, spiders. Spin sheet webs with a funnel-shaped retreat. Prey are captured on the sheet. Web is constructed in dense vegetation, in crevices of logs or rocks, and in houses. Easily visible, especially in grass covered with dew. *Agelena consociata* is social. *Agelena, Agelenopsis, Desis, Hololena, Tegenaria.*

Argyronetidae^F: One species. The Eurasian *Argyroneta aquatica* is the world's only aquatic spider. Cursorial, lives and hunts beneath the water surface. Captures aquatic insects, then consumes them in a retreat consisting of an air bubble in a dome-shaped web. Air is replenished from the surface by bubbles trapped in abdominal hydrofuge pubescence.

Oxyopidae^F: Lynx spiders. Diurnal hunters in low vegetation. Capable of jumping onto prey. Good vision, eyes form a hexagon. Long, thin, spiny legs with three tarsal claws. *Oxyopes.*

Clubionidae^F: Sac spiders. Build a cocoonlike silk retreat, sometimes incorporating a folded leaf. Large spinnerets. Nocturnal hunters. *Chiracanthium* (Fig. 18-30A) has a painful but not lethal bite and is often found indoors. Two tarsal claws. *Clubiona, Scotinella, Trachelas.*

Thomisidae^F: Crab spiders. Cursorial ambush predators, often brightly colored and living on flowers. Sideways movement is crablike. Long, raptorial anterior legs are held outstretched like crab chelipeds. Carapace is about as wide as it is long. Two tarsal claws. *Aphantochilus, Synema, Thomisus, Xysticus.*

Salticidae^F: Jumping spiders. The largest spider family. Cursorial, heavy-bodied, furry, diurnal, jumping hunters. The males of many species are brightly colored. Excellent vision. Abundant in temperate and tropical regions, common in houses. Two tarsal claws. *Corythalia, Gertschia* (Fig. 18-30B), *Menemerus, Neon, Phidippus, Salticus, Trite.*

PALPIGRADI^O

The 70 known species of palpigrades are mostly less than 1.5 mm in length and none are larger than 3 mm. Microwhip scorpions (Fig. 18-31) are blind, interstitial soil arachnids occurring primarily in tropical and subtropical habitats but also in European subterranean environments. Many of their unusual characteristics, including their size, are adaptations for life in small spaces.

The cephalothorax is divided into a large anterior region of four fused segments covered by the carapace and two independent posterior segments covered by separate tergites. The chelicerae are three-articulate. A similar division of the cephalothorax is seen in the schizomid uropygids, which live in similar habitats, and the resemblance is thought to be a convergent adaptation to increase flexibility for moving in small spaces.

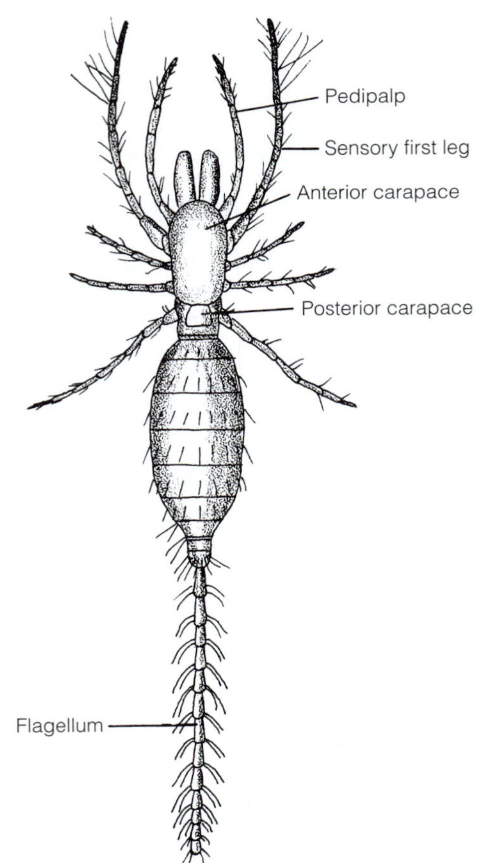

FIGURE 18-31 The palpigrade *Eukoenenia*. *(After Kraepelin and Hansen from Millot, 1949e.)*

The abdomen is composed of an eight-segmented preabdomen and a postabdomen of three segments. The postabdomen terminates with a long, antenniform flagellum. The pedipalps are relatively large and leglike. The first legs are antenniform and have a sensory role. Both pairs of book lungs have been lost, and palpigrades are not known to have respiratory organs. Malpighian tubules are absent, but one pair of saccate nephridia is present. Eyes are absent and the chief sensilla are trichobothria. The hemal system is reduced, but includes a tubular heart. Sperm transfer is indirect and employs a spermatophore.

PSEUDOSCORPIONES^O

This taxon consists of small arachnids (Fig. 18-32) rarely exceeding 7 mm in length and usually less than half that. Pseudoscorpions are widespread in leaf litter (Fig. 18-35), under bark, in soil, and in moss and debris, where they are seldom noticed despite being very common. One species, *Chelifer cancroides*, lives in human houses and is known as the book scorpion. About 2500 species are known worldwide. This distinct and homogeneous taxon has no close ties to other arachnids, except possibly the solifuges. Their resemblance to true scorpions is superficial and due solely to their common (convergent)

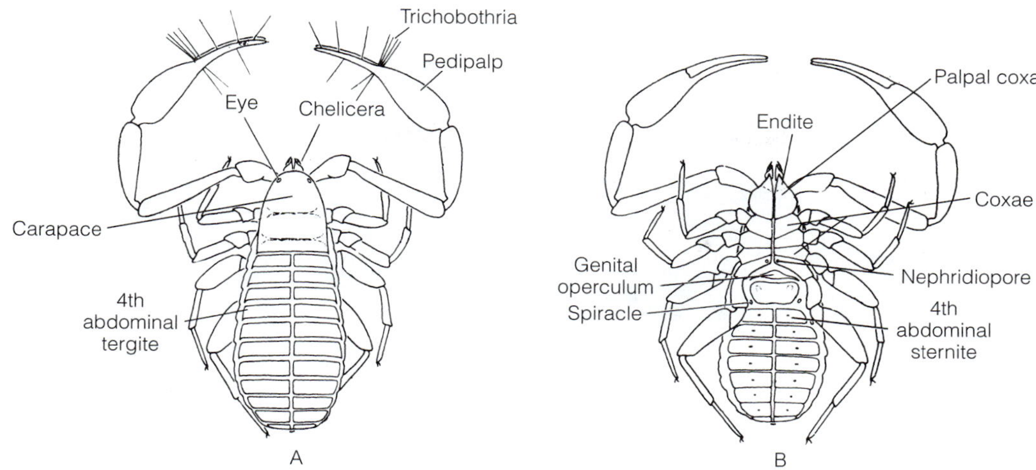

FIGURE 18-32 Pseudoscorpiones. Dorsal **(A)** and ventral **(B)** views of the pseudoscorpion *Chelifer cancroides* (male), showing the external structure. *(After Beier from Weygoldt, P. 1969. Biology of Pseudoscorpions. Harvard University Press, Cambridge, MA.)*

possession of large chelate pedipalps. Pseudoscorpions move rapidly backward, although some also move forward and some, sideways. They exhibit a phenomenon known as **phoresy,** in which they grasp the leg of an insect with their pedipalps and hitch a ride as it flies to a new habitat.

Form

The cephalothorax is covered dorsally by a solid, undivided carapace and laterally by the soft, unsclerotized pleural membrane. Dorsally are usually one or two pairs of white lateral eyes at the anterolateral corners of the carapace (Fig. 18-32A). Anteriorly the cephalothorax has a pair of short, chelate chelicerae (Fig. 18-32), which bear silk-producing spinnerets. The chelicerae consist of only two articles that are used to masticate the prey or to simply tear a hole in it to introduce enzymes. The second pair of appendages is the large and conspicuous pedipalps (Fig. 18-32). Each consists of six articles, the last two of which form a chela. Poison glands in one or both fingers open at pores at the tip of the finger(s). The pedipalps are used to capture and kill food, build nests, court, fight, and receive sensory information. Four pairs of walking legs arise from the cephalothorax. Ventrally the cephalothorax is covered by the five pairs of coxae belonging to the pedipalps and legs. These coxae replace the absent sternites as the ventral covering of the cephalothorax (Fig. 18-32B). The large pedipalpal coxae (endites) form the floor of the preoral cavity.

The junction between the cephalothorax and abdomen is broad—usually as wide as the cephalothorax—although in some species it is slightly narrowed. As in scorpions, the abdomen is conspicuous and lacks locomotory appendages. Dorsally its 12 segments are covered by 11 conspicuous tergites (Fig. 18-32A), which may be whole or divided into right and left halves. Each tergite is surrounded by flexible, unsclerotized exoskeleton. Ventrally the abdomen bears 11 large sternites (Fig. 18-32B). Segment 1 lacks a sternite. The sternites of segments 2 and 3 are fused to form the genital operculum. Paired gonopores open through the genital operculum and the anus is terminal on the tiny 12th segment. Two pairs of tracheal spiracles are located laterally, on the posterior edges of sternites three and four.

Internal Form and Function

Pseudoscorpions feed on small arthropods such as springtails and mites, which are captured, killed or paralyzed by the poisonous pedipalps, and then transferred to the chelicerae. In primitive pseudoscorpions such as *Neobisium* and *Chthonius,* the chelicerae masticate the food into a pulp as digestive fluid from the midgut floods onto it. External digestion takes place in the preoral cavity located between the bases of the chelicerae and the pedipalpal coxae (Fig. 18-33). In derived taxa, the chelicerae tear a hole in the prey and digestive fluid is injected into the hole and then sucked back into the gut after the tissues are liquefied. In these species, the preoral cavity functions like a syringe, injecting digestive fluid into the prey.

The muscular pharynx sucks the fluid into the midgut, leaving a dry, masticated husk behind. The spacious midgut, with its several large ceca, occupies most of the interior of the abdomen and is the site of enzyme secretion, hydrolysis, absorption, and storage. Glycogen and lipids are stored in midgut cells, and pseudoscorpions can subsist on these reserves for months after feeding. A short hindgut extends from the midgut to the anus at the tip of the anal cone.

Gas exchange is via tracheae opening at two pairs of spiracles on abdominal segments 3 and 4. The heart is a dorsal tube in segments 1 through 4 of the abdomen (Fig. 18-33). It has one, two, or four pairs of ostia. The single pair of saccate nephridia have nephridiopores between the coxae of legs 3 and 4. Although there are no Malpighian tubules, guanine granules formed in the digestive ceca are released into the gut lumen and eliminated through the anus. Nephrocytes are present and the silk glands may excrete

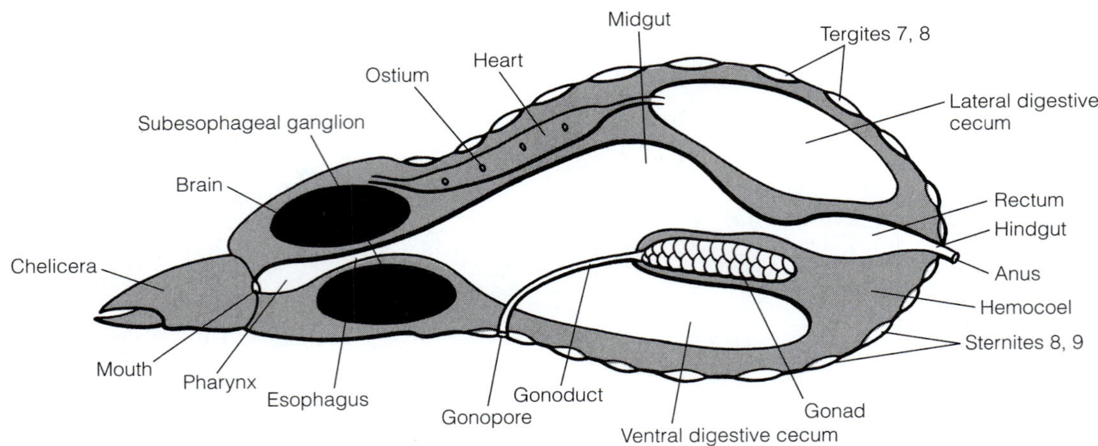

FIGURE 18-33 Sagittal section of the pseudoscorpion *Neobisium.* *(Redrawn from Weygoldt, 1969. Biology of Pseudoscorpions. Harvard University Press, Cambridge, MA.)*

guanine. The CNS is typically arachnid and strongly cephalized, filling most of the cephalothorax (Fig. 18-33). All segmental ganglia are concentrated in the subesophageal ganglion. Pseudoscorpions usually have one or two pairs of lateral eyes, but occasionally have none. Most species probably have well-developed chemoreceptive senses, but the most important sense is undoubtedly the detection of faint air currents by the trichobothria. This is the major sense used in locating and capturing prey, allowing pseudoscorpions to detect prey from as far away as 15 mm.

Pseudoscorpions are gonochoric, with courtship varying from nonexistent to elaborate. The gonad is a single median tube in the abdomen lying between the midgut and its ventral cecum (Fig. 18-33). It connects by a pair of gonoducts to two gonopores located between sternites 2 and 3 on the anterior abdomen. Sperm are transferred in spermatophores, but the process employed is variable. Typically, the male presses his gonopore to the ground and then lifts himself high above the ground and extrudes a spermatophore as he rises (Fig. 18-34A,B). The resulting spermatophore is at the end of a long, slender stalk. An interested female steps over (or is maneuvered over by the male) the spermatophore and removes the sperm with her gonopore. She stores the sperm and fertilization does not occur until oviposition at some later time.

Differences in the process of sperm transmission provide clues regarding the evolution of indirect sperm transfer by spermatophores in arachnids. In the simplest, most primitive, and most wasteful (of sperm) method, the male simply leaves his spermatophores scattered about whether or not a female is present. A female may happen upon one of them, be attracted to it chemotactically, and acquire its sperm. Many of these spermatophores are never found and are wasted. In more derived species, the male produces spermatophores only when a female is present. In the most derived species the two sexes perform a complicated mating dance that assures that the spermatophore will have a willing recipient (Fig. 18-34C).

Females brood and nourish the eggs externally, but there is no placenta. Depending on the species, from 2 to more than 50 young may be brooded. The female constructs a **brood nest** of silk and other materials. She lives in the nest and releases eggs into a membranous **brood sac** attached to her gonopores (Fig. 18-34D). Development takes place within the sac. The eggs are microlecithal, having little yolk, and a nutritive fluid produced by the ovary nourishes the developing embryos in the sac. The female remains in the nest until the first instar juveniles (protonymphs) hatch from the eggs. She then leaves the nest but continues to brood the juveniles in the brood sac. The young emerge from the brood sac during the third instar as tritonymphs.

The life cycle includes three juvenile instars (protonymph, deutonymph, and tritonymph) and the adult. Adults do not molt. Development is direct and juveniles are similar to adults but smaller. Maturity is reached in a year or less, and individuals may live two to five years. Temperate species may produce several generations a year. Pseudoscorpions sometimes overwinter in a silk nest. All pseudoscorpions inhabit small crevices (Fig. 18-35).

SOLIFUGAE[O]

Solifuges form a taxon of about 800 to 1000 tropical and semitropical arachnids, sometimes called sun spiders because of their spiderlike appearance and diurnal habits or wind spiders because of the great speed of the males (Fig. 18-36). In the United States, a few species have been found in Florida and more than 100 have been found in the southwest, some as far north as Colorado. Many solifuges are common in warm, arid regions of the world, but they are also known from grasslands and forests. Solifuges hide under stones and in crevices and many dig burrows (Fig. 18-36). Their length ranges from a few millimeters to as much as 7 cm. The cephalothorax is covered dorsally by a large anterior carapace bearing a pair of closely placed median eyes on the anterior border and a short posterior carapace, or tergum. The hinge

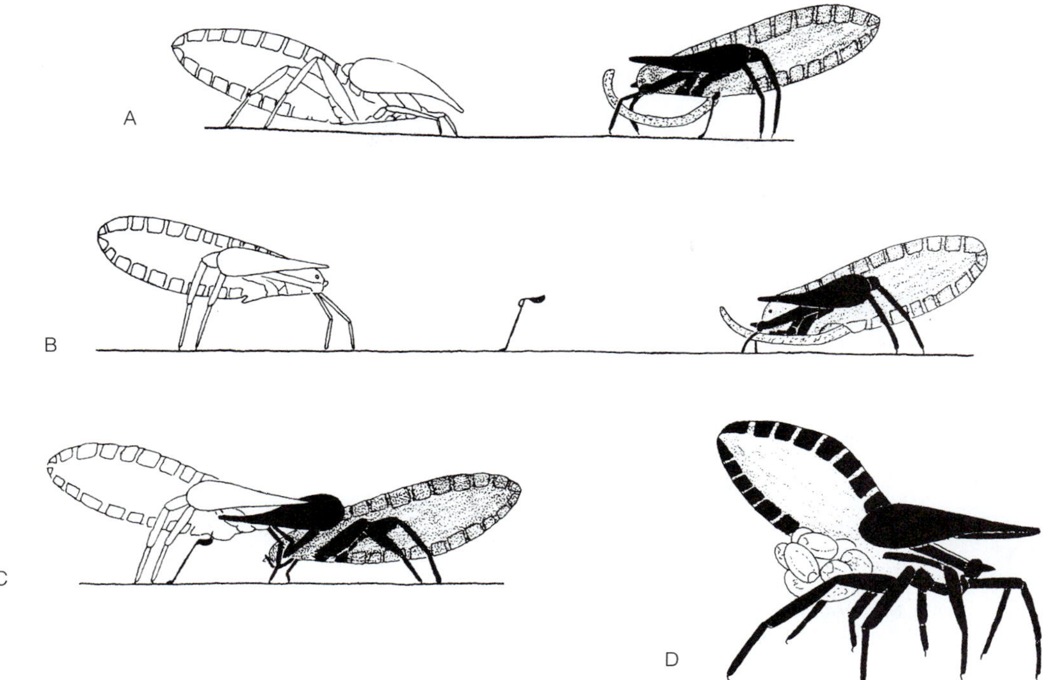

FIGURE 18-34 Pseudoscorpion reproduction. **A–C,** Courtship and sperm transmission in *Chelifer cancroides* (female on the left, male on the right). **A,** Male producing a spermatophore. **B,** Spermatophore attached to the substratum. **C,** Male pressing the female down onto the spermatophore. **D,** Female of *C. cancroides* carrying embryos. *(All after Vachon, M. 1949. Traité de Zoologie. Vol. VI. Masson et Cie, Paris.)*

FIGURE 18-35 Section through a bit of leaf litter showing a few of its many, mostly arthropod, inhabitants.

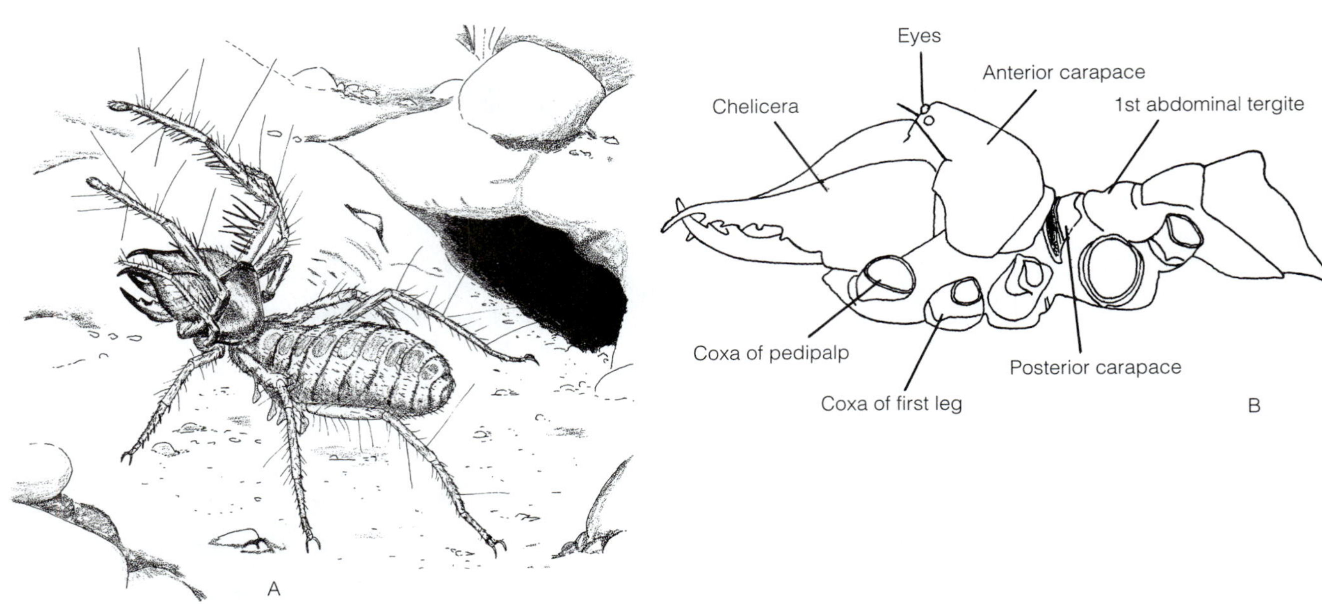

FIGURE 18-36 Solifugae. **A,** North African solifuge *Galeodes arabs.* Four racquet organs are visible on the fourth leg. **B,** Cephalothorax of the solifuge *G. graecus* (lateral view). *(A, After Millot, J., and Vachon, M. 1949b. Traité de Zoologie. Vol. VI. Masson et Cie, Paris; B, After Kaestner, 1968, from Millot and Vachon, 1949b.)*

between the anterior and posterior carapaces is flexible. The segmented abdomen is large and joined to the cephalothorax by a slightly narrowed waist.

Most distinctive are the enormous chelicerae, which extend in front of the cephalothorax and can be directed upward by flexing the hinged cephalothorax (Fig. 18-36). It has been said that, relative to their size, solifuges have the largest jaws in the animal kingdom. Each chelicera is composed of two articles forming a pincer whose finger moves in the vertical plane. Each chelicera has a **stridulating organ** on its inner surface. Long, leglike pedipalps terminate in an **adhesive disc** that is not apparent unless it is everted. Their pedipalpal coxae are gnathobases that macerate the food prior to external digestion. The first legs are used as antennae and the remaining three pairs are used for running. The segmented abdomen is large, soft, ovoid, and distensible. It is broadly connected with the cephalothorax and has obvious tergites and sternites on its segments. No telson is present.

Solifuges are carnivorous or omnivorous, with termites forming an important part of the diet of many American species. Sensory pedipalps locate the prey and the chelicerae kill it and tear apart the tissues. For gas exchange the animal uses highly branched tracheae not derived from book lungs and apparently evolved independently. Solifuge tracheae are more like those of insects than of arachnids. Excretion is accomplished by a pair of saccate nephridia, nephrocytes, and two Malpighian tubules. The heart extends from the abdomen into the cephalothorax and has eight pairs of ostia, two of which are in the cephalothorax. The nervous system consists of the usual brain and subesophageal ganglia, but the latter does not include all the abdominal ganglia. Instead, the ganglia of the posterior abdomen are coalesced to form a ganglion in the anterior abdomen. Two to five sensory **racquet organs** (malleoli) are present on the coxa and trochanter of the fourth legs.

In some solifuges sperm transfer is indirect with spermatophores. The male seizes the female and places her in a passive state by a brief period of stroking and palpation. The male turns the female over and opens her genital orifice with his chelicerae. He emits a spermatophore onto the ground, picks it up with his chelicerae, and inserts it into his partner's genital orifice. The entire act takes only a few minutes and the male then leaps away. In American solifuges, sperm transmission is direct, with sperm passing from male to female gonopores without a spermatophore.

OPILIONES[O]

Opiliones, or Phalangida, contains the familiar long-legged arachnids known as daddy longlegs or **harvestmen** (Fig. 18-37). The 5000 or more described species live in temperate and tropical climates and most prefer humid habitats. Harvestmen are abundant in vegetation, on the forest floor, on tree trunks and fallen logs, in humus, and in caves. They do not produce silk or toxins and they do not bite, although they may (rarely) pinch. The widespread belief that harvestmen have a highly toxic venom is due to their being confused with pholcid spiders, which also go by the name "daddy longlegs." These spiders resemble harvestmen but have a potent toxin, although their chelicerae are too small and weak to envenomate humans. Most harvestmen are predators but, unlike almost all other arachnids, many are herbivores, and many feed on dead organic material. Also unlike other arachnids, they are capable of ingesting particulate food. In fact, since the pharynx

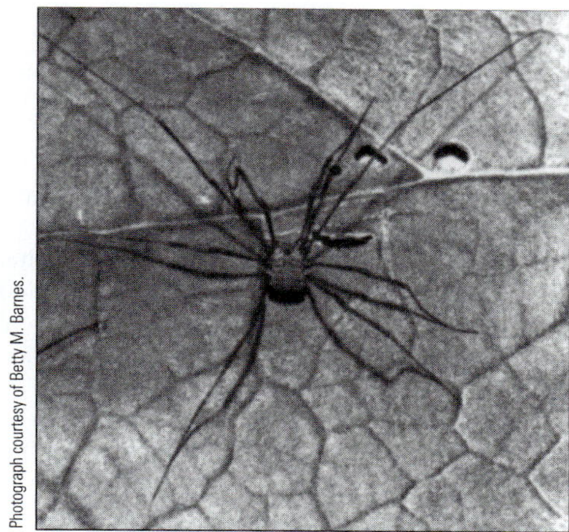

FIGURE 18-37 Opiliones. Dorsal view of a harvestman on a leaf.

lacks the usual setal filter, this cannot be avoided. The average body length is 5 to 10 mm, exclusive of the legs, but some tropical giants reach 22 mm and have a leg length of 160 mm. Some tiny, short-legged, mitelike species are less than 1 mm in length. Most species are shades of brown, but some are red, orange, yellow, green, or spotted. Some have spiny bodies (Fig. 18-38B).

The cephalothorax is broadly joined to the short, segmented abdomen with no constriction between the two (Fig. 18-38). As a result, the body is ovoid. Dorsally the cephalothorax is completely covered by a carapace, in the center of which is an **eye tubercle** with a median eye on each side. Lateral eyes are absent. Along the anterolateral margins of the carapace are the openings to a pair of repugnatorial glands, which produce repellant secretions, often quinones and phenols, with an acrid odor. Some species spray intruders with this substance and others pick up a droplet of secretion mixed with regurgitated gut fluid and thrust it at a would-be predator. Ventrally the cephalothorax is covered by the coxae, labium, and sternites. The small, slender chelicerae have three articles, with the second and third forming a chela (Fig. 18-38). The pedipalps usually are leglike, but in one large taxon they are raptorial and enlarged for capturing prey with a sicklelike claw. The pedipalpal coxae have gnathobases.

The legs of opilionids may be as short as those of spiders, but usually they are extremely long and are the hallmark of the taxon (Fig. 18-37). According to the English arachnologist Theodore Savory, the study of harvestmen is a study of legs. Each leg has the usual seven articles, but contributing to its length is the multisegmented tarsus, which in some species has hundreds of tiny articles that make it flexible and whiplike. Typically the second legs are the longest and are sensory and used as antennae, with the animal moving them about over the substratum, often in front of it. When disturbed, opilionids can move rapidly, and species living in vegetation can climb by wrapping the flexible tarsus around stems or blades of grass. Self-amputation (autotomy) of a leg is an important means of defense against predators, although legs cannot be replaced after being lost. Because of the length of their legs, harvestmen are very effective at maneuvering over uneven terrain.

The abdomen is segmented with some independent tergites and sternites (Fig. 18-38B) and its exoskeleton is sclerotized and hard. The first three to eight abdominal tergites typically are fused to form a single **dorsal shield,** which is usually fused with the carapace. Posterior tergites and sternites are free and unfused (Fig. 18-38B). The anus is terminal.

In general, harvestmen are predatory, but scavenging is more important for them than it is for other arachnids. North American and European opilionids have been observed feeding on small invertebrates, dead animal matter, and pieces of fruits and vegetables. Predatory species feed on other small arthropods and some feed on snails. Food is seized by the

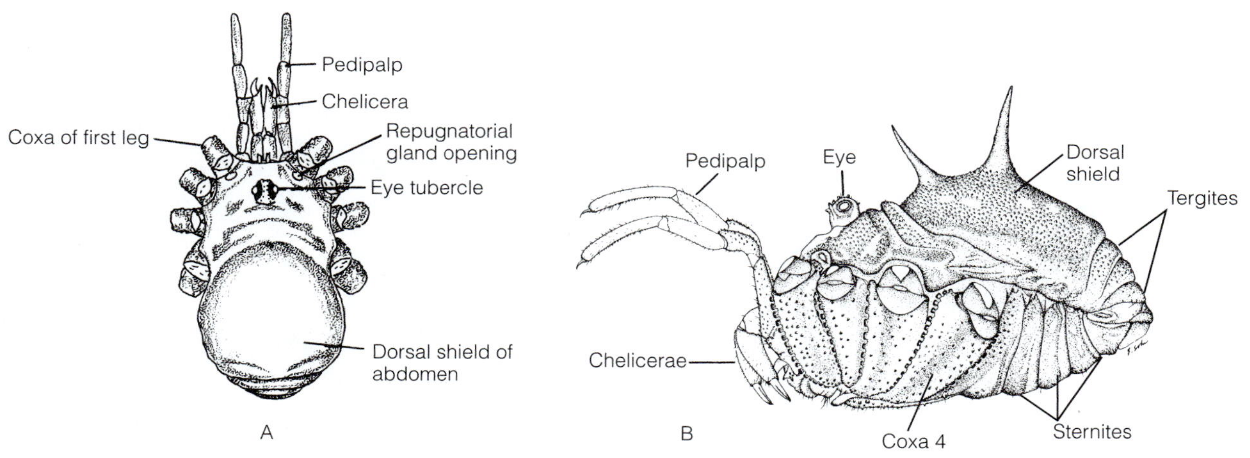

FIGURE 18-38 Opiliones. **A,** The opilionid *Leiobunum flavum,* with only the bases of the legs shown (dorsal view). The abdomen of this species is not as conspicuously segmented as that of many other opilionids. **B,** Lateral view of an opilionid whose abdomen bears two large spines. The legs have been removed to reveal the body. *(A, After Bishop.)*

pedipalps and passed to the chelicerae, which hold and crush it. Unlike other arachnids, the ingested food is not limited to liquid material, but includes small particles. Thus, a greater part of digestion must take place in the midgut. The mouth is preceded by the preoral cavity, which is bounded dorsally by the labrum, ventrally by the labium, and laterally by the gnathobases of the pedipalps and first legs.

Nephrocytes and one pair of saccate nephridia are used for excretion, and Malpighian tubules are absent. Gas exchange takes place through tracheae, which probably are not homologous with those of other arachnids. The spiracles are located on each side of the first abdominal segment. In many active, long-legged harvestmen, secondary spiracles on the tibia of the legs provide oxygen to the distal parts of the leg. The heart has two pairs of ostia. The central nervous system is strongly cephalized, with all segmental ganglia coalesced in the subesophageal ganglion.

Opilionids are gonochoric and have internal fertilization with direct sperm transfer. Gonads are paired but coalesced posteriorly to form a U-shaped organ. The two gonoducts join before reaching the gonopore on the ventral cephalothorax. In both sexes the gonoduct is a long tube with an external extension. In the male the extension is a penis and in the female it is an ovipositor. Mites are the only other arachnids to possess a penis or ovipositor. Mating is not preceded by an elaborate courtship. The male in many species faces the female and projects the tubular penis between the female's chelicerae to the gonopore, where it deposits sperm. Shortly after mating, the female deposits eggs in humus, moss, rotten wood, or snail shells. The number of eggs laid at one time ranges into the hundreds, and several batches are laid during the lifetime of a female. In temperate regions the life span is only one year. An individual may overwinter in the egg or as an immature instar. The name "harvestman" refers to some common temperate species that appear in large numbers in the fall.

RICINULEI[O]

Ricinulei includes about 75 species of small, uncommon arachnids known as tick spiders (Fig. 18-39B) found in Africa *(Ricinoides)* and the Americas *(Cryptocellus)* from Brazil to the southern United States, where they have been collected from leaf litter and in caves. Ricinuleids are heavy-bodied and small, measuring up to 10 mm in length. The cuticle is very thick, heavily sclerotized, and often sculptured. Attached to the anterior margin of the carapace is a unique, hoodlike **cucullus** that can be raised and lowered (Fig. 18-39A). When lowered, it covers the preoral cavity, mouth, and chelicerae. Ricinuleids are blind, but have lateral light-sensitive spots. Unlike the three-part chelicerae of the Acari (mites), tick spider chelicerae have only two articles. The pedipalps are chelate, the second legs are sensory, and the male third legs are used to transfer sperm to the female gonopore. The cephalothorax and 10-segment abdomen are broadly joined. The anterior abdominal segments are wide, but the posterior segments are small and telescoping, with the anus on the 10th segment.

The ostiate heart lies in a partitioned hemocoel. Respiration is via sieve tracheae with a pair of spiracles on the posterior cephalothorax. Excretory organs are one pair each of Malpighian tubules and saccate nephridia (coxal glands). Sperm transfer is similar to that in spiders, but the elaborate third legs of the male are used instead of the pedipalps and there is no sperm web. The life cycle includes one larval, three nymphal, and an adult instar stage. Juveniles have three pairs of walking legs.

ACARI[O]

The extraordinarily diverse taxon Acari contains about 40,000 known species of mites and ticks (Fig. 18-40A), with conservative estimates of another million waiting to be

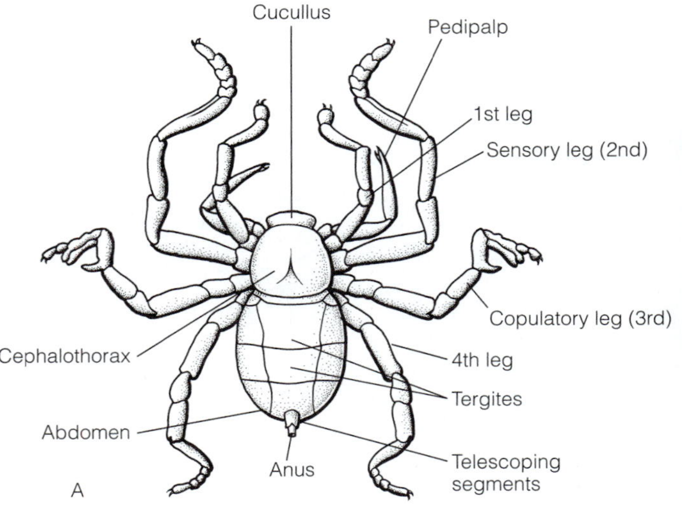

FIGURE 18-39 Ricinulei. **A,** A generalized male ricinuleid in dorsal view. **B,** A ricinuleid from the southwestern United States. *(A, Redrawn from Moritz, 1993, and Savory, 1977, after Hansen and Sörensen.)*

discovered. From the standpoint of health and economics these are the arachnids most important to humans. Although most mites are free-living, numerous species are parasitic on people, domesticated animals, and crops. Many others are destructive to stored food and other products. Mites are among the most ubiquitous animals, and free-living, terrestrial species are extremely abundant, particularly on plants and in mosses, leaf litter, humus, soil, rotten wood, and detritus. Population densities are enormous, certainly surpassing all other arachnid orders. A handful of leaf litter from a forest floor often contains hundreds of individuals belonging to tens of species (Fig. 18-35). Mites are the only arachnids with a significant presence in aquatic habitats and they are common in both marine and fresh waters. Despite their abundance and economic importance, mites are not as well known as the other arachnids.

Some acarologists believe Acari to be polyphyletic with several arachnid ancestors, including opilionids and ricinuleids, and have split the taxon into seven taxa (orders). Most acarologists, however, believe them to be monophyletic and the separation of the body into two unique tagmata, gnathosoma and idiosoma, to be an autapomorphy that distinguishes them from all other arachnids.

Form

Being among the smallest arachnids, mites have achieved their great evolutionary success by exploiting niches for small arthropods. Most adults are 0.25 to 0.75 mm in length, although many are larger and some are as small as 0.1 mm. There are mites small enough to live in the tracheae of honeybees, beneath the wings of beetles, in the quills of feathers, in the lungs of snails, and in human hair follicles and sebaceous glands. Some predatory mites live in specialized leaf pits and pouches of a host plant, from which they emerge to attack their insect prey. Ticks, which are the largest mites, may reach 3 cm when engorged with blood.

An obvious mite characteristic is the apparent lack of tagmata or segmentation (Fig. 18-40A). Abdominal segmentation has disappeared in most species and the abdomen is fused with the cephalothorax. Consequently the positions of the appendages, eyes, and gonopore are the only landmarks that identify the original body regions. In many species the entire body has become covered with a single, sclerotized carapace. Instead of the usual chelicerate tagmata, the body is secondarily divided into two new and unique tagmata, the gnathosoma and idiosoma.

The posterior **idiosoma** is most of the body, including the entire abdomen and most of the cephalothorax. The **gnathosoma,** or capitulum (Fig. 18-40, 18-43A), is the anteriormost part of the cephalothorax, but it contains neither the brain nor eyes and is only the tip of the head, not the entire head. It is a retractable feeding structure consisting of the chelicerae, pedipalpal coxae, preoral cavity, and parts of the anterior exoskeleton. Morphologically, it is almost endlessly varied to serve a remarkable variety of diets. A generalized gnathosoma, however, consists of two parts, chelicerae and infracapitulum, covered dorsally by a tectum (rostrum)

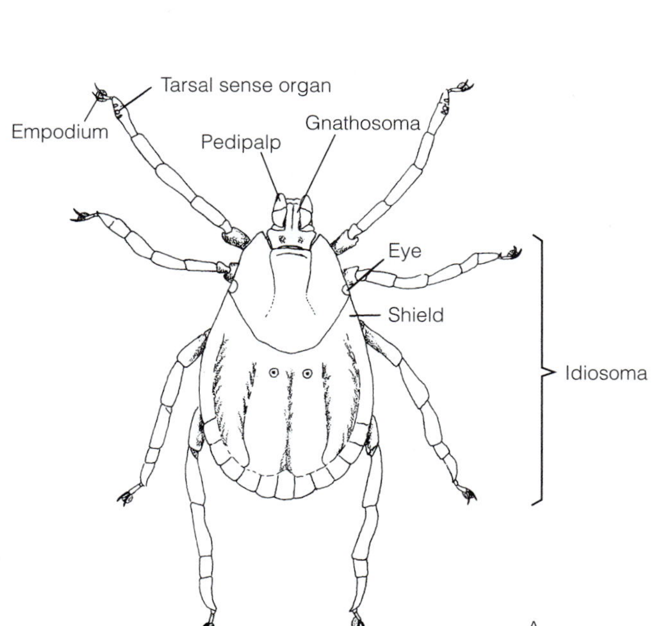

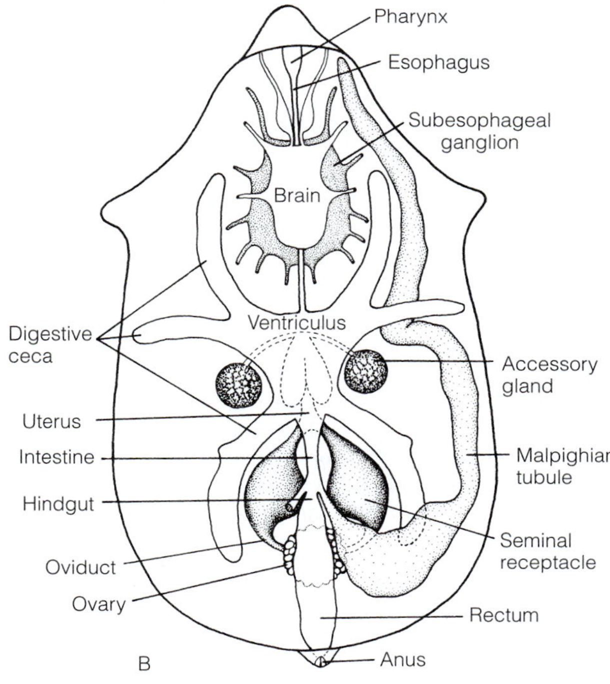

FIGURE 18-40 Acari general anatomy. **A,** External anatomy of the dog tick, *Dermacentor variabilis.* **B,** Internal anatomy of the mesostigmatid mite *Caminella,* in dorsal view. *(A, After Snodgrass, 1952; B, After Ainscough from Krantz, G. W. 1978. Manual of Acarology. 2nd edition. Oregon State University Press, Corvallis.)*

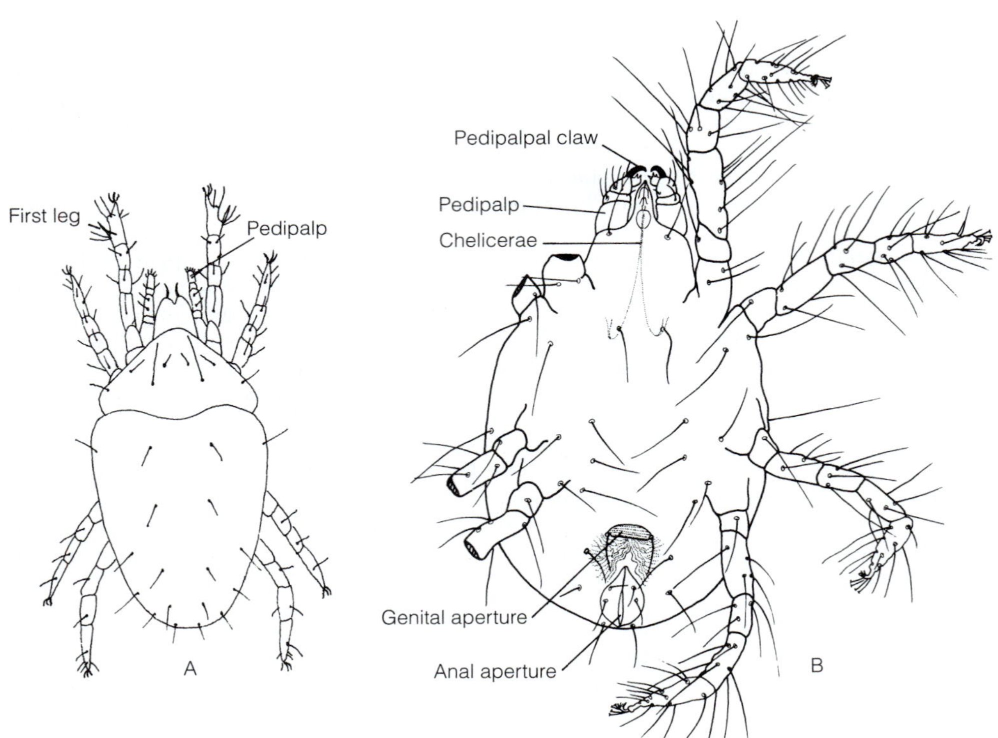

FIGURE 18-41 **A,** A trombidiform mite, *Tydeus starri* (dorsal view). **B,** Ventral view of a spider mite, *Tetranychus.* *(A, After Baker and Wharton; B, After Krantz, G. W. 1978. Manual of Acarology. 2nd edition. Oregon State University Press, Corvallis.)*

extending anteriorly from the carapace (Fig. 18-43A). The retractable chelicerae arise from the anterior body wall between the base of the tectum and the infracapitulum. The **infracapitulum** is formed of the labrum dorsally and the large endites of the pedipalpal coxae ventrally and laterally. Sometimes the palpal endites are fused to form a hypostome, which forms the floor of the infracapitulum. The gnathosoma surrounds the preoral cavity, which contains the chelicerae dorsally and is continuous with the anterior foregut ventrally. Posteriorly the preoral cavity opens into the mouth, which leads to the pharynx.

Chelicerae and pedipalps vary in structure, depending on their function. Both usually are composed of two or three articles and may be chelate (Fig. 18-40A, 18-43). Pedipalps may be relatively unmodified and leglike, or they may be heavy and chelate, resembling chelicerae. In some parasitic forms, they are vestigial. The four pairs of legs usually have six articles each (Fig. 18-41), and in some taxa they are modified for swimming (Fig. 18-42B) or other functions.

The ventral side of the body bears sclerites that vary with taxon in size, morphology, and number. The genital plate, located between the last two pairs of legs, bears the genital aperture (Fig. 18-41B). This location results from a forward migration of abdominal segments, an event that also occurred in harvestmen.

Setae are found on all parts of the mite body. They vary from simple hairs to club-shaped and flattened types, and many are sensory. Most mites are various shades of brown, but many display a wide range of hues, such as black, red, orange, green, or combinations of these colors.

Two taxa of mites, the Oribatida (beetle mites) and several taxa of water mites, deserve special mention. Oribatida, with 145 families and 14,000 species, is the largest and best-studied taxon of free-living mites (Fig. 18-42A). Beetle mites are abundant in leaf litter and moss. These little arachnids usually are globe-shaped, with the dorsal surface covered by a convex, highly sclerotized shield that makes them resemble tiny beetles. The water mites, encompassing some 2800 species, have adapted to an aquatic existence in both fresh and salt water. Marine species have been found from the intertidal zone to abyssal depths, and shallow-water species are common. They do not swim, but rather crawl about over algae, bryozoans, hydroids, and sponges. Most water mites, however, live in fresh water. Some are bright red or other colors, and many are active swimmers with long natatory (swimming) setae on their legs (Fig. 18-42B).

Nutrition

Mites exhibit a tremendous diversity and specialization of diets and feeding habits. Although some utilize particulate food, most ingest fluids and, even when feeding on solid food, rely on initial external digestion and liquefaction. Carnivores living in soil and humus feed on nematodes and small arthropods, including eggs, insect larvae, and other mites. Small crustaceans are the principal prey of water mites. The chelicerae of carnivorous mites are variously modified, depending on the prey. Some mites tear off pieces of prey whereas others suck out the tissues.

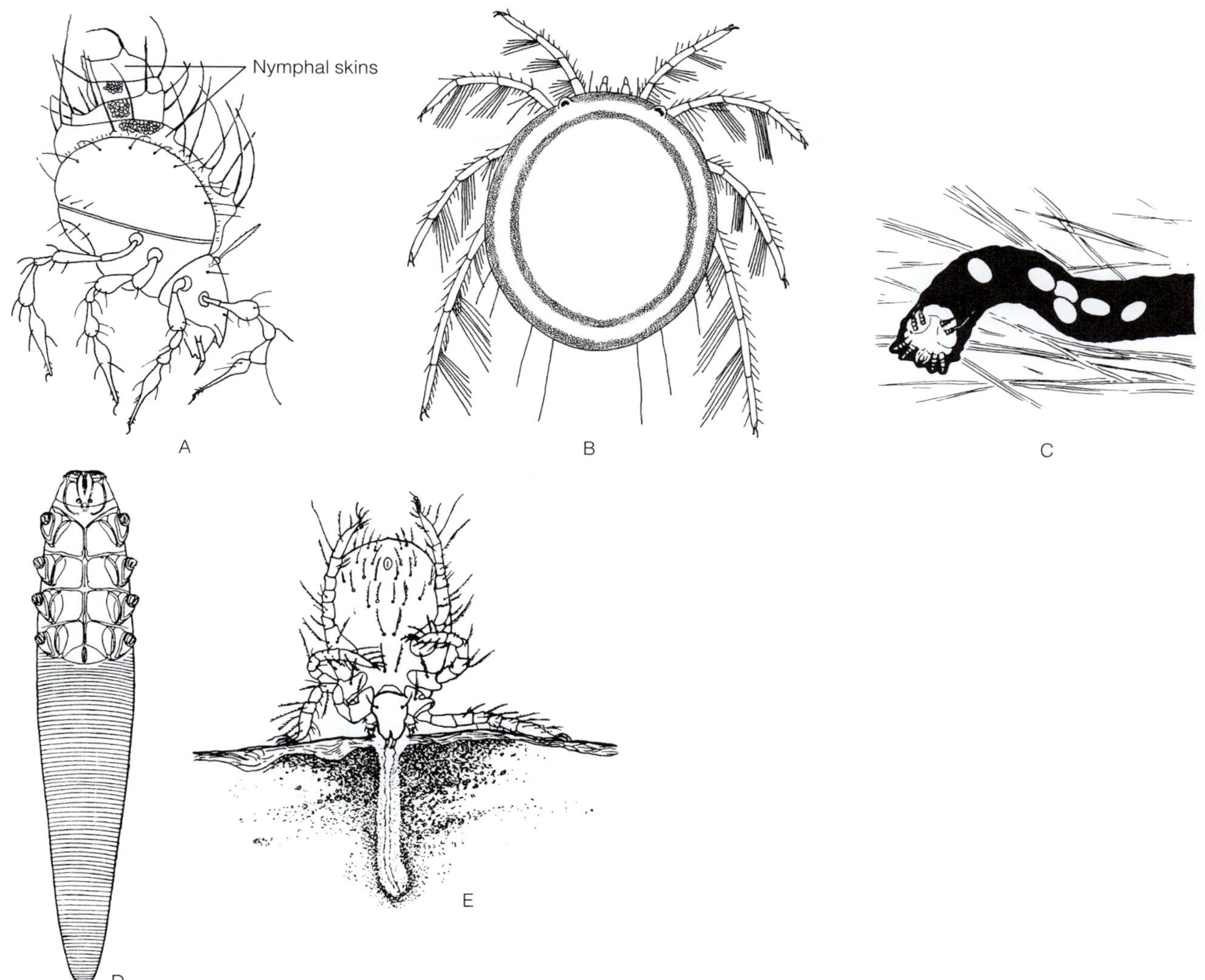

FIGURE 18-42 **A,** The oribatid mite *Belba jacoti,* carrying its nymphal exuvia. **B,** The water mite *Mideopsis orbicularis,* with setose legs adapted for swimming. **C,** The mange mite *Sarcoptes scabiei,* burrowing in the skin and depositing eggs in the tunnel. **D,** Ventral view of a sheep hair-follicle mite, *Demodex ovis* (0.23 mm). **E,** A chigger larva of *Neotrombicula,* feeding on skin (0.25 mm). Note the three pairs of legs characteristic of mite larvae. *(A, After Wilson from Baker and Wharton; B, After Soar and Williamson from Pennak, 1989; C, After Craig and Faust; D, After Hirst from Kaestner, A. 1968. Invertebrate Zoology. Vol. 2. Wiley-Interscience, New York; E, After Vitzthum from Kaestner, 1968.)*

Many herbivorous species, such as spider mites (Tetranychidae), have chelicerae modified as needlelike stylets that pierce plant cells to suck out the contents. A number of spider mites are serious agricultural pests of fruit trees, clover, alfalfa, cotton, and other crops. Spider mites construct protective webs from silk glands that open near the base of the chelicerae. The minute gall mites (Eriophyoidea), which also feed on plant cells and have stylet-shaped chelicerae, include some species that are agricultural pests. Other herbivorous species feed on fungi, algae, and mosses.

Many mites are carrion feeders or scavengers. Most soil-inhabiting oribatid mites feed on fungi, algae, and decomposing plant and animal material. A large number of "scavengers" have highly specialized diets. For example, different species of storage mites (Acaridae) and allied taxa feed on flour, dried fruit, mattress and upholstery stuffing, hay, and cheese. *Dermatophagoides* is commonly associated with house dust, where it feeds on discarded human skin cells. The feather mites and some species that live in the fur of animals are also scavengers, not parasites, subsisting on oil, dead skin, and feather fragments.

The majority of parasitic mites are ectoparasites of animals, both vertebrates and invertebrates, but other forms of parasitism exist. Some mites have become internal parasites through an invasion of the air passages of vertebrates and the tracheal systems of arthropods. Tracheal mites *(Acarapis)* live in the tracheae of honeybees *(Apis mellifera)* and damage tracheal walls to gain access to blood. In the last two decades tracheal mites and *Varroa* mites, which also feed on blood, have made commercial beekeeping much more difficult and expensive and have nearly exterminated feral honeybee populations. Many mites are parasitic only as larvae. For example, the larval stages of freshwater mites are parasitic on aquatic insects and clams.

The juvenile stages of the common harvest mites (Trombiculidae) parasitize the skin of vertebrates. Larvae of species of *Trombicula* are the familiar chiggers, or redbugs. The six-legged larva emerges from an egg that was deposited on the soil. The larva may attack almost any taxon of terrestrial vertebrate, biting the host's skin and feeding on dermal tissue, which is broken down by the external action of proteolytic enzymes (Fig. 18-42E). Feeding takes place for up to 10 days or more; then the larva drops off. After a semidormant stage, the larva molts and becomes a free-living nymph. A later molt transforms the nymph into an adult. Both nymph and adult are predaceous, feeding mostly on insect eggs. The intense itching that results from the bite of a chigger larva is caused by an immune response to its oral secretions and not, as commonly supposed, simply by the presence of the mite. Scratching quickly removes the mite, but the irritation remains for several days or even weeks. Although chiggers can cause severe dermatitis, they are of much greater medical importance as vectors for such pathogens as Asian scrub typhus.

Many mites are parasitic during their entire life cycle but are attached to the host intermittently, only during periods of feeding. The dermanyssid mites of birds and mammals (red chicken and other fowl mites) and the ticks illustrate this type of life cycle. Ticks penetrate the skin of the host by means of the highly specialized, hooked chelicerae (Fig. 18-43B) and feed on blood. The integument, especially that of females, is not highly sclerotized and is capable of great expansion when engorged with blood. With a few exceptions, the tick drops off the host after each feeding and undergoes a molt before attaching to a new host. Many species can live for long periods, well over a year, between feedings. Copulation occurs while the adults are feeding on the host. The fertilized female then drops to the ground and deposits an egg mass. A six-legged "seed" tick (larva) hatches from the egg.

Ticks attack all taxa of terrestrial vertebrates and are sometimes the vectors of pathogens. In humans they are responsible for the transmission of American Rocky Mountain spotted fever, tularemia, Texas cattle fever, relapsing fever, and Lyme disease. Lyme disease is caused by the spirochete *Borrelia burgdorfi,* which is transmitted by the deer tick, *Ixodes scapularis.* Each tick instar feeds on a single host, drops off, and then finds another host. Adult ticks occur predominantly on whitetail deer, whereas the larvae and nymphs attack a number of hosts, including the white-footed mouse, *Peromyscus leucopus,* which is the major reservoir for the spirochete. Any instar may feed on humans, but to be infective an earlier instar must have acquired the spirochete by feeding on an infected white-footed mouse.

Finally, some parasitic mites spend their entire life cycle attached to the host. Included in this group are the wormlike follicle mites (Demodicidae), which live in the hair follicles of mammals (Fig. 18-42D), and the scab- and mange-producing fur mites (Psoroptidae and Sarcoptidae) of mammals. The human itch mite, *Sarcoptes scabiei* (Fig. 18-42C), the cause of scabies, or seven-year itch, tunnels into the epidermis. The female is less than 0.5 mm and the male less than 0.25 mm in length. Irritation is caused by the mite's secretions. The female deposits eggs in the tunnels for a period of two months, after which she dies. Up to 25 eggs are deposited every two or three days. The eggs hatch in several days and the larvae follow the same existence as the adult. The infection can thus be endless. The mite is transmitted to another host by contact with infected areas of the skin.

Internal Form and Function

Most free-living mites have a typical arachnid digestive tract (Fig. 18-40B), although some lack an anus. Mite excretory organs are one pair of saccate nephridia, one or two pairs of Malpighian tubules (Fig. 18-40B), or both. As might be expected in these

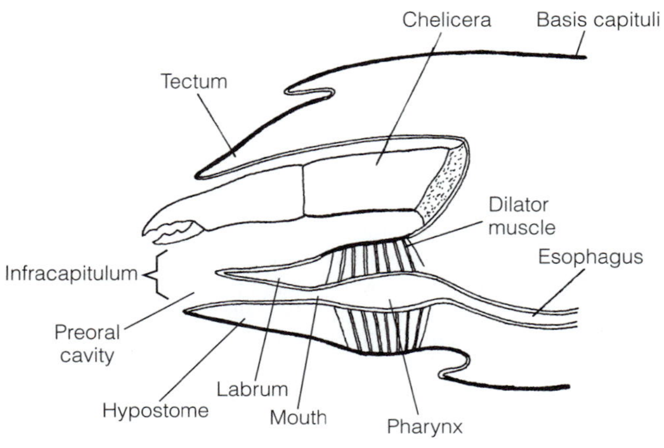

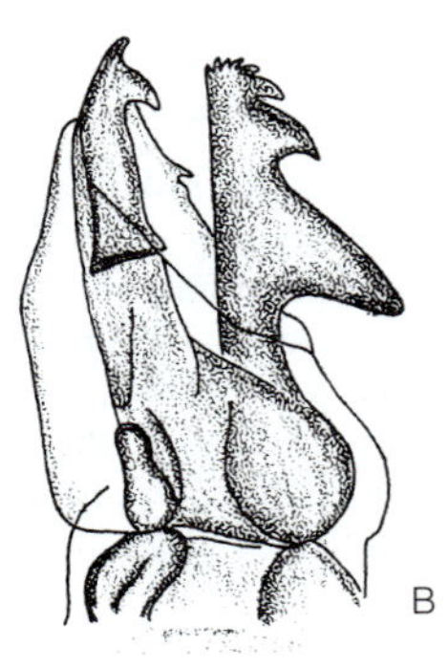

FIGURE 18-43 **A,** Gnathosoma of a mite (sagittal section). **B,** Hooked chelicera of the tick *Ixodes reduvius.* *(A, After Snodgrass; B, After Neumann from Andre.)*

small animals, the hemal system is reduced and, except in a few taxa, consists of a network of sinuses without a heart. Circulation probably results from contraction of body muscles. Although in some mites the gas exchange organs have disappeared, most have tracheae. The spiracles, which vary in number from one to four pairs, are located on the anterior half of the body. Mites also exchange gases across the general body surface.

The CNS is highly cephalized, with all the segmental ganglia concentrated in the subesophageal ganglion (Fig. 18-40B). Sensory setae are probably the most important sense organs. Oribatid mites possess a peculiar form of sensory seta called a pseudostigmatic organ, which is probably a type of trichobothrium. Two such setae located on the cephalothorax are thought to detect air currents, to which the mite responds by moving deeper into the leaf litter, perhaps to avoid desiccation. Median and lateral eyes may be present, but many mites are blind. Innervated pits and slits are common in mites and may be homologous to the slit sense organs of other arachnids.

Reproduction and Development

Acarines are gonochoric. The male reproductive system consists of a pair of lobate testes, located in the midregion of the body. A vas deferens leaves each testis and may join its opposite member ventrally to open through a median gonopore or, in some species, through a chitinous penis that can project through the genital aperture. Females usually have a single ovary connected to the gonopore by an oviduct (Fig. 18-40B). A seminal receptacle and accessory glands are also present.

Sperm are transmitted indirectly in most mites. A spermatophore may be deposited on the substratum and then picked up by the female, or it may be transferred to the gonopore by the male chelicerae or by the third pair of legs. Sperm are transmitted directly into the female by a penis in the Actinotrichida. Eggs are deposited in soil or humus, and some mites enclose their eggs in a case. The oribatid *Belba* attaches its large eggs to the bodies of conspecifics, which carry them about until hatching. Female oribatid mites have an ovipositor. Clutch size varies widely with species.

After an incubation period of two to six weeks, a six-legged larva hatches from the egg. The fourth pair of legs are acquired after a molt transforms the larva into a protonymph. Successive molts produce a deutonymph, then a tritonymph, and finally an adult. During these stages, adult structures are gradually attained. The life span of mites varies greatly depending on the species, but usually it is shorter than that of other arachnids. For example, *Anblyseius brazilli,* a tropical, predatory, mesostigmatid mite, reaches adulthood in 7 days and the female has a life span of 30 days.

Diversity of Acari

Acariformes[sO] (Actinotrichida): The "mitelike" mites. The most diverse mite taxon, with over 30,000 species. Indirect development. Setae have optically active, birefringent actinochitin. Trichobothria are present. Includes Astigmata, Endeostigmata, Oribatida, Sarcoptiformes, and Trombidiformes (Prostigmata). *Belba, Demodex, Dermatophagoides, Mideopsis, Neotrombicula, Sarcoptes, Tetranychus, Trombicula, Trombidium, Tydeus.*

Opilioacarida[sO] (Notostigmata): Large, rare, leathery, primitive mites. Segmented abdomen. Trichobothria are absent. Resemble opilionids. Live in leaf litter and beneath stones. Omnivorous or predatory; ingest solid food. Feed on millipedes, opilionids, mites, pollen. Subtropical and tropical. *Opilioacarus, Panchaertes, Paracarus.*

Parasitiformes[sO] (Anactinotrichida): Ticks and related mites in about 75 families. Setae have no actinochitin. Trichobothria and coxae are absent. Ventral anal aperture covered by sclerotized plates. Cuticularized ring around gnathosoma. Includes Mesostigmata (many parasitic species), Holothyrida (large, armored mites), and Ixodida (ticks: large, hematophagous, have vertebrate hosts). *Caminella, Dermacentor, Ixodes.*

PHYLOGENY OF ARACHNIDA

The ancestral arachnids were aquatic arthropods whose body organization has been inherited, with some modification, by their modern descendants. External predigestion in the preoral cavity and the ingestion of liquid food are important features of the arachnid ground plan that are almost universal among modern arachnids. From the chelicerate ancestor arachnids inherited a pair of prehensile (chelate) chelicerae composed of three articles, a pair of pedipalps, and four pairs of walking legs. The original abdomen included 12 segments and a telson.

Of the ancestral chelicerate's four pairs of saccate nephridia, only two pairs remain in the arachnid ground plan, and they are located on the segments of the first and third walking legs. Malpighian tubules evolved, probably more than once, independently of those of tracheates. The book gills of the aquatic ancestor became four pairs of book lungs adapted for terrestrial respiration. The undivided carapace originally had two pairs of median and perhaps five pairs of lateral eyes, the latter probably derived from the compound eyes of the chelicerate ancestors. Each lateral eye consisted of several ommatidia sharing a single corneal lens and retina. Concentration of the segmental ganglia enlarged the subesophageal ganglion to include all the cephalothoracic ganglia and four pairs of anterior abdominal ganglia. Trichobothria and slit sense organs were present in the stem arachnids. The gonopores on abdominal segment 2 are unpaired. Indirect sperm transfer with spermatophores is ancestral.

The scorpions and Lipoctena are a monophyletic taxon (Arachnida) sharing several derived characteristics (Fig. 18-44). The scorpions have features in common with both eurypterids and Lipoctena, but how these three taxa are related is uncertain. Early scorpions may have been aquatic, and they share with the eurypterids possession of a five-segmented postabdomen. On the other hand, scorpions also have important characteristics, such as slit sensilla and the absence of appendages on the first abdominal segment, in common with Lipoctena. In a widely accepted analysis of chelicerate relationships, scorpions are seen as the most primitive living arachnids and the sister taxon of Lipoctena, with which they form the taxon Arachnida. Eurypterids probably arose from the stem line of arachnids, but are not included in Arachnida. In another study, the scorpions are treated as

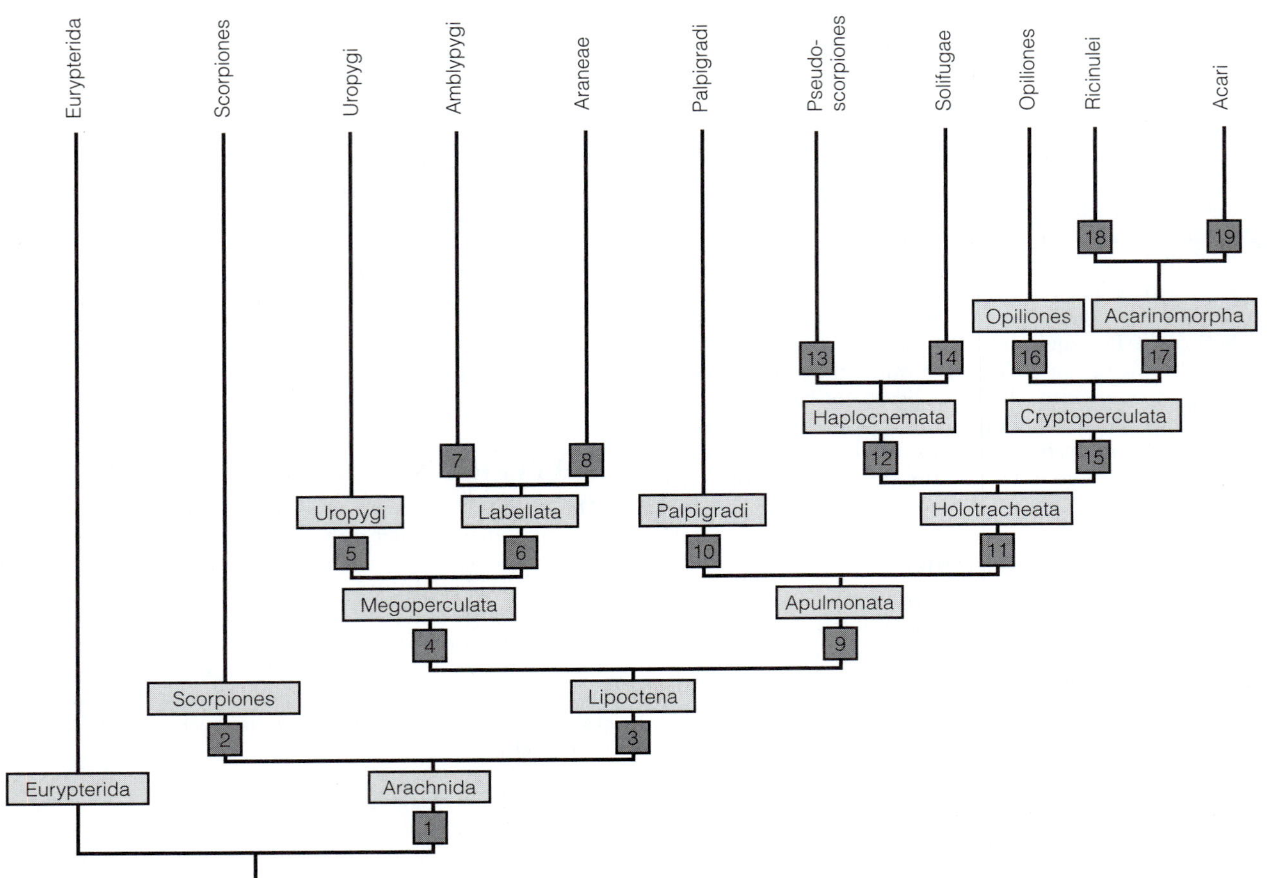

FIGURE 18-44 A phylogeny of arachnids. **1, Arachnida:** Terrestrial, air breathing chelicerates; achelate sensory pedipalps with four pairs of locomotory appendages; carnivorous with liquid diet; digestion extra-intestinal, in preoral cavity; pharynx a precerebral sucking pump; book lungs; two pairs of saccate nephridia; maximum of five pairs of degenerate compound eyes; compound eyes with a single lens serving multiple ommatidia; pigment-cup ocelli; indirect sperm transfer with spermatophores. **2, Scorpiones:** Pectines present; chelate pedipalps; telson has poison glands and sting; one pair of saccate nephridia; four pairs of book lungs; abdomen has 12 segments. **3, Lipoctena:** Lyriform organs; two pairs of book lungs; opisthoma with a flagellum. **4, Megoperculata:** Chelicerae have two articles; sperm flagellum has $9 + 2 \times 3$ microtubular structure. **5, Uropygi:** Massive chelate pedipalps; defensive anal glands; first legs are sensory; five pairs of abdominal ganglia independent of subesophageal ganglion. **6, Labellata:** Cephalothorax and abdomen connected by narrow pedicel formed of the first abdominal segment; gut with postcerebral sucking stomach; sperm flagellum lost; all cephalothoracic and abdominal ganglia coalesced into the subesophageal ganglion. **7, Amblypygi:** First walking leg is antenniform; pedipalps are large and raptorial. **8, Araneae:** Abdominal silk glands, spinnerets, and silk; poison glands in chelicerae; sperm transfer indirect with pedipalp and sperm web; no spermatophores. **9, Apulmonata:** Book lungs lost; lateral eyes reduced to two or three pairs. **10, Palpigradi:** Abdomen has 11 segments; first walking legs are tactile; mouth on a small proboscis; one pair of saccate nephridia; sperm aflagellar. **11, Holotracheata:** Tracheae with one pair of spiracles (evolved independently of those of Araneae); tube tracheae; flagellum reduced; no more than two pairs of lateral eyes. **12, Haplocnemata:** Biarticulate, shearing, chelate chelicerae with ventral movable finger; two pairs of tracheal spiracles. **13, Pseudoscorpiones:** Chelicerae with silk glands; chelate pedipalps have poison glands; median eyes lost; Malpighian tubules lost. **14, Solifugae:** Cephalothorax has movable articulation; rostrum; abdomen of 11 segments; tracheal system with additional spiracles; sensory racquet organs; sperm aflagellar. **15, Cryptoperculata:** Second walking legs are sensory. **16, Opiliones:** Cephalothorax and abdomen broadly fused; abdomen has 10 segments; gonopore anterior, on the ventral cephalothorax; penis or ovipositor present; median eyes only, on a central tubercle; lateral eyes lost; repugnatorial glands; Malpighian tubules lost. **17, Acarinomorpha:** Six-legged larva; three nymphal instars. **18, Ricinulei:** Hood (cucullus) over chelicerae; abdomen has 10 segments; chelicerae and pedipalps each have two articles; second leg is long and sensory; male third leg is copulatory; sieve tracheae; all eyes lost. **19, Acari:** Cephalothorax and abdomen broadly fused; tagmata are gnathosoma and idiosoma; second legs not sensory; aflagellate sperm. *(Redrawn from Weygoldt, P., and Paulus, H. F. 1979. Untersuchungen zür Morphologie, Taxonomie und Phylogenie der Chelicerata. 2. Z. Zool. Syst. Evolut.-forsch. 17:177–200; Weygoldt, P. 1996. Chelicerata, Spinnentiere. In Westheide, W., and Rieger, R. (Eds.): Spezielle Zoologie: I. Einzeller und Wirbellose Tiere. Gustav Fischer Verlag, Stuttgart. pp. 449–497; Ax, P. 2000. Multicellular Animals. Vol. II. Springer, Berlin. 396 pp.)*

derived arachnids closely related to pseudoscorpions and solifuges. Yet another hypothesis holds that scorpions are closer to eurypterids than to arachnids.

In modern phylogenetic systematics, Arachnida is divided into Scorpiones and Lipoctena, with scorpions retaining the four pairs of book lungs of the ground plan. The CNS is the least concentrated of any arachnid and includes a ventral nerve cord with seven ganglia. Scorpions also have chelate pedipalps and a telson with poison glands. In Lipoctena at least two, and sometimes all, of the original four pairs of book lungs are lost. Slit sense organs arranged in clusters to form lyriform organs are an autapomorphy of Lipoctena.

Within Lipoctena are the sister taxa Megoperculata and Apulmonata. In Megoperculata one article is lost from the ancestral triarticulate chelicera. In Apulmonata all four pairs of book lungs in the ground plan have been lost. Triarticulate chelicerae are retained.

Megoperculata includes the sister taxa Uropygi and Labellata. Uropygi has chelate raptorial pedipalps, a unique preoral chamber, antenniform first walking legs, and repugnatorial anal glands. In Labellata the pedicel, consisting of the first abdominal segment, joins the cephalothorax and abdomen.

Labellata consists of Araneae and Amblypygi. Araneae is characterized by silk glands, the use of the male pedipalp to transfer sperm, and cheliceral poison glands. Tracheae arose within the taxon. In Amblypygi the first walking legs are antenniform sensory organs and the pedipalps are large and raptorial.

Apulmonata is made up of the sister taxa Palpigradi and Holotracheata. In Palpigradi the mouth is on a short proboscis, the pedipalps are leglike, and the first walking legs are sensory. In Holotracheata tracheae appeared in conjunction with the loss of book lungs. These arose independently of those of Araneae. No more than two pairs of lateral eyes are present. The flagellum is reduced.

Holotracheata consists of the two sister taxa Haplocnemata and Cryptoperculata. In Haplocnemata the ventral finger of the chelicera moves in the vertical plane and the number of tracheal spiracles increases. Cryptoperculata is not strongly supported by morphological characteristics. Attributed to it are modification of the second pair of walking legs to sensory tactile organs, aflagellar sperm with the acrosome on its concave side, and a broadly joined cephalothorax and abdomen.

Haplocnemata includes the sister taxa Pseudoscorpiones and Solifugae. Pseudoscorpiones is characterized by unique chelate pedipalps with poison glands, chelae with silk glands, median eyes, and the loss of Malpighian tubules. In Solifugae the cephalothorax consists of articulated anterior and posterior tagma with independent tergites. The pedipalp has a distal adhesive organ. Sensory racquet organs are present and sperm are aflagellate.

Cryptoperculata includes Opiliones and Acarinomorpha. In Opiliones the cephalothorax and abdomen are broadly fused and the gonopore has moved anteriorly to the ventral cephalothorax. The median eye is on a tubercle. Males have a penis and females an ovipositor. In Acarinomorpha the life cycle includes a larval and three nymphal stages. Larvae have three pairs of legs.

PHYLOGENETIC HIERARCHY OF ARACHNIDA

- Arachnida
  - Scorpiones
  - Lipoctena
    - Megoperculata
      - Uropygi
      - Labellata
        - Amblypygi
        - Araneae
    - Apulmonata
      - Palpigradi
      - Holotracheata
        - Haplocnemata
          - Pseudoscorpiones
          - Solifugae
        - Cryptoperculata
          - Opiliones
          - Acarinomorpha
            - Ricinulei
            - Acari

Acarinomorpha is Ricinulei and Acari. The small ricinuleids have a movable cucullus covering the chelicerae, an abdomen with 10 segments, biarticulate chelicerae, and chelate pedipalps. The Acari cephalothorax and abdomen are broadly fused and the body is secondarily divided into the gnathosoma and idiosoma.

PYCNOGONIDA[C]

Pycnogonida, or Pantopoda, is a taxon (class or superclass) of about 1000 described species of exclusively marine animals known as sea spiders. Pycnogonids are not spiders, however, and the name is based on superficial similarities, such as eight long legs and a small body (Fig. 18-45). Although largely unnoticed, sea spiders actually are common benthic animals that occur from the intertidal zone to the deep sea. They live in all oceans from the poles to the tropics. Careful examination of bryozoans and hydroids scraped from a wharf piling or rocks usually yields a few specimens. Most are bottom dwellers that crawl about over irregular substrates as if on stilts. Some can swim by flapping their legs alternately up and down. A few are pelagic, some are interstitial, and some are commensal or ectoparasitic on other invertebrates.

FORM

Sea spiders are mostly small animals with a body length ranging from 1 to 10 mm, but polar and deepwater species can attain large sizes. Species of *Colossendeis,* for example, regularly achieve a leg span of 40 cm, and 70 cm has been reported. Although most pycnogonids are drably colored, some are green or purple, and deep-sea forms are often red or orange.

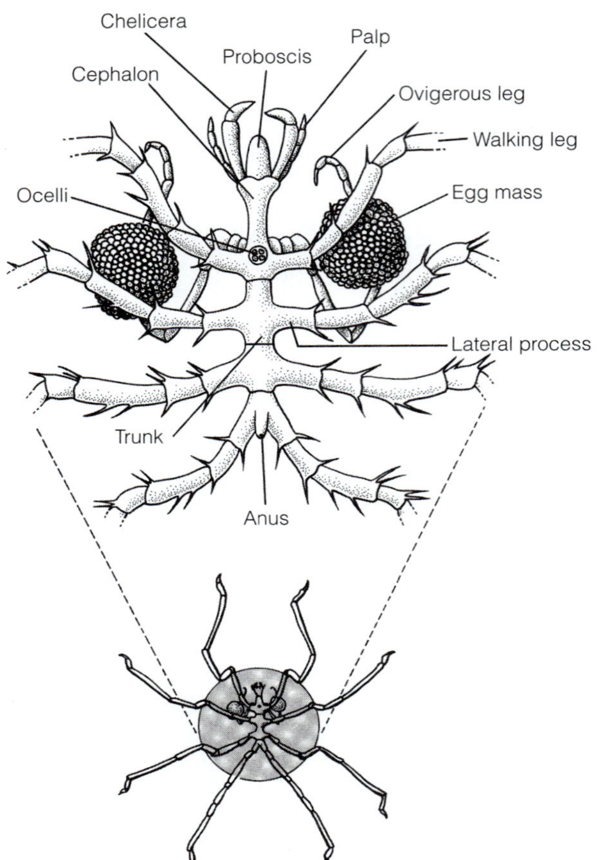

FIGURE 18-45 Pyconogonida. A male sea spider, *Nymphon rubrum*. *(Redrawn after Sars from Fage.)*

Although considerable variation occurs among the taxa, the body usually is narrow, tubular, and composed of seven segments coalesced to form what appear to be four segments (Fig. 18-45). Tagmosis is unlike that of other arthropods, but resembles that of euchelicerates. The anteriormost tagma is the cephalon composed of four segments. The trunk is composed of three, sometimes more, segments. Together, these two tagmata may be homologous to the cephalothorax of euchelicerates. The cephalon bears a dorsal eye tubercle with two pairs of median ocelli. Lateral compound eyes have been lost in the pycnogonids. The vestigial abdomen is a small dorsal process on the last trunk segment.

Protruding anteriorly or ventrally from the cephalon is a large tubular proboscis with the mouth at its distal end (Fig. 18-45, 18-46A). The proboscis is usually at least half the length of the body. The appendages of the cephalon are the chelicerae (chelifores), pedipalps (palps), ovigerous legs (ovigers), and one pair of walking legs (Fig. 18-45). In some taxa either chelicerae or pedipalps are absent. The chelate chelicerae function in food handling, whereas the pedipalps are used for sensory reception, feeding, and cleaning. Ovigerous legs, always present in males and usually in females, are used by the male to carry the egg mass following its deposition by the female. Sometimes they are also used for grooming.

On each side, the cephalon and each trunk segment bear a pair of conspicuous, stumplike lateral processes with which a pair of legs articulates (Fig. 18-45). The posterior end of the cephalon, with its lateral processes and single pair of legs, is regarded as a trunk segment that has fused with the head to form the cephalon. Each leg is composed of nine articles and may be exceedingly long (Fig. 18-46B), but are short in some species. Although most pycnogonids have three trunk segments and thus four pairs of legs, some have four segments and five pairs of legs, and a few have five segments and six pairs of legs. About 10 species of pycnogonids have five or six pairs of legs instead of the usual four. These additional segments do not appear to be phylogenetically important, for there are a number of genera that contain both eight-legged and ten-legged species. The tiny, unsegmented abdomen sits atop the last trunk segment and bears the anus at its distal end. Some fossil pycnogonids from the Devonian have a segmented abdomen.

INTERNAL FORM AND FUNCTION

Most pycnogonids are carnivores that feed on soft-bodied animals such as hydroids, soft corals, anemones, bryozoans, small polychaetes, and sponges. Some feed on algae or microorganisms, or the accumulated detritus on hydroids or bryozoans. Molluscs are the principal hosts for parasitic species.

The foregut includes the mouth at the tip of the proboscis, the large pharynx forming the lumen of the proboscis, and a short esophagus (Fig. 18-46A). Three cuticularized lips that surround the mouth are used to scrape or bite into the prey. Food is pumped into the gut by the muscular pharynx. When feeding, the proboscis is applied directly to the prey to suck up the tissues. The cuticularized pharynx contains a bed of setae that probably masticate and perhaps filter the food. Exiting the posterior pharynx, the esophagus joins the midgut near the middle of the cephalon.

The endodermal midgut (intestine) has long, narrow diverticula, or digestive ceca, extending into each walking leg and sometimes into the chelicerae, pedipalps, and/or ovigers. Hydrolysis and absorption occur in the midgut and its ceca. The midgut connects with the short hindgut (rectum), which is located in the abdomen and opens to the exterior at the anus.

The hemal system is composed of a dorsal heart and a hemocoel. The heart has paired ostia but is secondarily reduced, and circulation is probably assisted by gut peristalsis and movements of the legs. A horizontal diaphragm divides the hemocoel into a dorsal pericardial sinus and a ventral perivisceral sinus. Each leg is similarly divided into dorsal and ventral sinuses that are continuous with those of the trunk. Blood flows anteriorly from the heart into the proboscis, then posteriorly in the perivisceral sinus, and then into the ventral sinus of the legs. At the tips of the legs the blood enters the dorsal sinus and flows back toward the body to enter the pericardial sinus. From there it passes through the ostia back into the heart lumen.

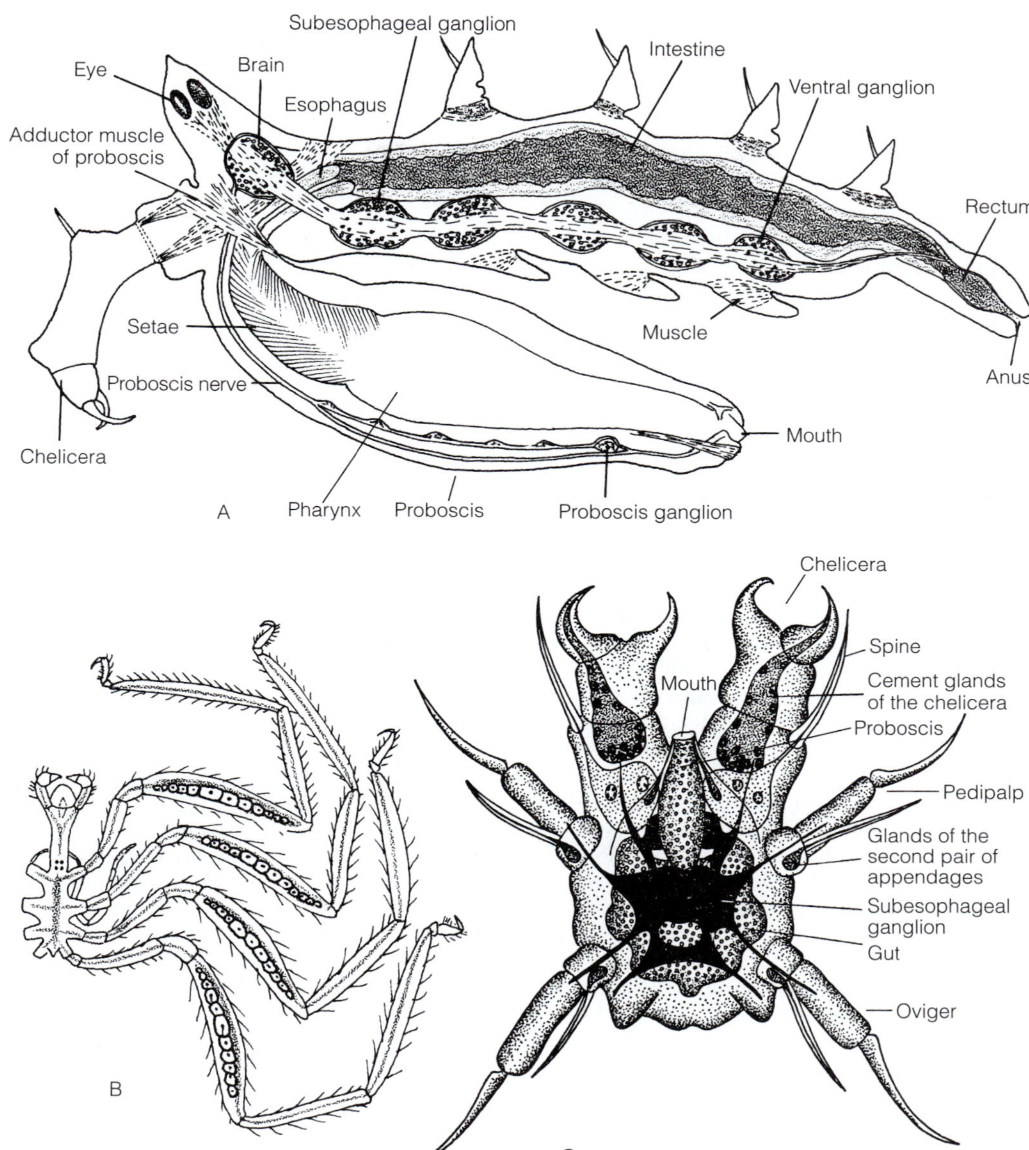

FIGURE 18-46 **A,** The pycnogonid *Ascorhynchus castelli* (sagittal section). **B,** Female of *Pallene brevirostris* with eggs in the femurs. **C,** Protonymphon larva of an *Achelia echinata* in ventral view. *(A, After Dohrn from Fage; B, After Sars from Fage; C, From King, P. E. 1973. Pycnogonids. St. Martin's Press, New York. 137 pp.)*

Excretory organs were unknown in pycnogonids until 2003 when saccate nephridia resembling those of primitive crustaceans were discovered in the chelicerae. These consist of an end sac with podocytes, surrounded by the hemocoel and connected to a nephridiopore by a tubule.

Pycnogonids have no special organs for gas exchange. The general body surface is sufficient for gas exchange, and excretion probably occurs across the gut epithelium and perhaps the body surfaces. The CNS is chelicerate-like, but primitive and largely uncephalized. The dorsal brain (Fig. 18-46A) consists of a protocerebrum and a posterior pair of cheliceral neuromeres, but no deutocerebrum. The protocerebrum receives the optic nerves from the eye tubercle and the posterior neuromeres innervate the chelicerae. Circumenteric connectives connect the brain with a small subesophageal ganglion that innervates the pedipalps and ovigers. The ventral nerve cord has a ganglion for each pair of walking legs.

Pycnogonids are gonochoric and, as expected of aquatic chelicerates, employ external fertilization. Females can usually be distinguished from males by their weak or absent ovigers. The single gonad, either testis or ovary, is located in the trunk above the intestine. Branches of it extend far into the legs.

In both males and females, the reproductive openings are multiple, an unusual condition for arthropods, and are located ventrally on the coxae of the second and fourth pairs of legs in males and of all legs in females. On reaching maturity, the eggs are housed in lobes of the ovary in the femurs of the legs (Fig. 18-46B), and in gravid females the femurs are conspicuously swollen.

In those pycnogonids in which egg laying has been observed, the male suspends himself beneath the female so their ventral surfaces are opposed, or sometimes the male may stand above the female. Eggs are fertilized by the male as they leave the female's gonopores. He then gathers them onto his ovigerous legs, either directly from the female's gonopores or from the female's ovigers (Fig. 18-45). Cement glands in the femurs of the male legs secrete adhesive for holding up to 1000 eggs in a spherical mass. The egg masses are held around the middle joints of the male's ovigerous legs and brooded until they hatch.

The amount of yolk in the eggs varies widely among species, but cleavage is always total. Eggs with little yolk undergo equal cleavage, whereas unequal cleavage is found in macrolecithal eggs. A gastrula forms by the inward growth of cells in small eggs and by epiboly in yolky eggs. In most pycnogonids, a protonymphon larva hatches from the egg (Fig. 18-46C). It has three pairs of appendages (chelicerae, pedipalps, and ovigerous legs), each having only three articles. A short proboscis is present, but the trunk segments are not yet evident. Depending on the species, the larvae either remain on the male's ovigers or leave to take up an independent existence on the substratum, such as hydroids and corals, that will be its food. In any case, the larvae eventually transform into young pycnogonids through a sequence of molts and the addition of new segments and appendages. The eastern Pacific *Propallene longiceps* develops from egg to adult in about five months.

The structure of the brain, the nature of the sense organs, the presence of chelicerae and palps, the absence of antennae, and the hemal system are considered by most zoologists to justify placing the pycnogonids among the chelicerates. The ground plan seems to include seven appendages on the equivalent of a cephalothorax, as is the case with the xiphosurans. Their exact relationship to arachnids and xiphosurans is uncertain because pycnogonids are aberrant in many respects. The presence of multiple gonopores, ovigers, the segmented trunk (the cephalothorax), and the additional pairs of walking legs in many species have no counterparts in the other chelicerate taxa.

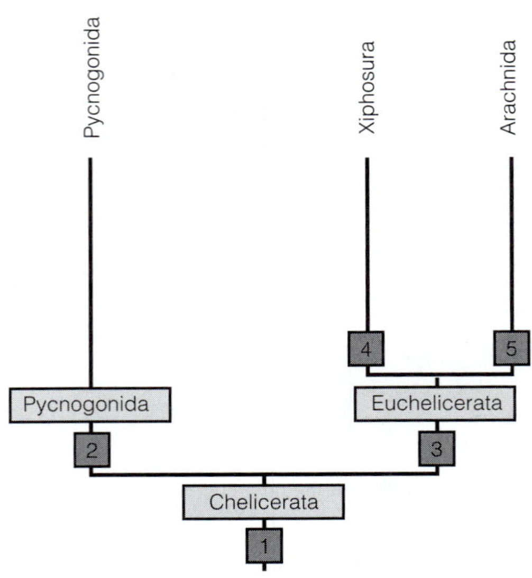

FIGURE 18-47 Phylogeny of Chelicerata. **1, Chelicerata:** Tagmata are cephalothorax and abdomen; cephalothorax is acron plus seven segments; abdomen divided into anterior and posterior regions; appendages absent on first cephalothoracic segment; two-part syncerebrum lacks the deutocerebrum; first appendages are prehensile chelicerae; compound eyes present. **2, Pycnogonida:** Cephalothorax without carapace, segments articulated; abdomen reduced and vestigial; slender, tubular body; anterior proboscis with mouth; gonads extend into legs; four median eyes on an elevated tubercle. **3, Euchelicerata:** Cephalothoracic segments fused, with a dorsal "carapace"; two median eyes. **4, Xiphosura:** Preabdominal segments fused, the first reduced and bearing chilaria; postabdomen reduced; eyes borne on an ophthalmic ridge; abdominal appendages are book gills; abdominal segments fused; body terminates in a spikelike telson; segmental saccate nephridia share a common nephridiopore. **5, Arachnida:** Terrestrial; second appendages are pedipalps; gas exchange via book lungs; compound eyes divided into five pairs of single eyes; internal fertilization; femur of walking legs is biarticulate; abdomen has 12 segments. *(Redrawn from Ax, 2000.)*

PHYLOGENY OF CHELICERATA

Chelicerates evolved in the Ordovician or Cambrian periods from Arachnomorpha, a taxon of ancient arthropods that includes the trilobites and several other extinct taxa. The ancestral chelicerates evolved into two sister taxa (Fig. 18-47), Pycnogonida and the much larger Euchelicerata, which includes Eurypterida, Xiphosura, and Arachnida. In some interpretations, pycnogonids are considered to be the sister taxon of Chelicerata rather than being chelicerates themselves. The ancestral chelicerates were undoubtedly aquatic but, with the exception of the sea spiders, horseshoe crabs, and some mites, modern chelicerates are terrestrial. The great success story of the chelicerates is the adaptive radiation of the terrestrial arachnids, and the dominating theme of chelicerate evolution is their adaptation to life on land.

PHYLOGENETIC HIERARCHY OF CHELICERATA

- Chelicerata
  - Pycnogonida
  - Euchelicerata
    - Xiphosura
    - Arachnida

REFERENCES

GENERAL

Dunlop, J. A., and Selden, P. A. 1998. The early history and phylogeny of the chelicerates. In Fortey, R. A., and Thomas, R. H. (Eds.): Arthropod Relationships. Systematics Association Spec. Vol. 55. pp. 221–235.

Dunlop, J. A., and Webster, M. 1999. Fossil evidence, terrestrialization and arachnid phylogeny. J. Arachnol. 27:86–93.

Kovoor, J. 1977. Silk and the silk glands of Arachnida. Annee Biol. 16:97–172.

Merrett, P. (Ed.): 1978. Arachnology. Symposia of the Zoological Society of London, no. 42. Academic Press, London. 530 pp.

Millot, J. 1949a. Classe des arachnides I: Morphologie général et anatomie interne. In Grassé, P. P. (Ed.): Traité de Zoologie. Vol. VI. Masson, Paris. pp. 263–349.

Moritz, M. 1993. Arachnata. In Gruner, H.-E. (Ed.): Lehrbuch der Speziellen Zoologie. 4 Aufl. Band I, 4 Teil. Gustav Fischer Verlag, Jena. pp. 64–442.

Savory, T. 1977. Arachnida. Academic Press, London. 340 pp.

Schultz, J. W. 1990. Evolutionary morphology and phylogeny of Arachnida. Cladistics 6:1–38.

Shear, W. A. 1999. Introduction to Arthropoda and Cheliceriformes. In Harrison, F. W., and Foelix, R. F. (Eds.): Microscopic Anatomy of Invertebrates. Vol. 8A: Chelicerate Arthropoda. Wiley-Liss, New York. pp. 1–19.

Telford, M. J., and Thomas, R. H. 1998. Expression of homeobox genes shows chelicerate arthropods retain their deutocerebral segment. Proc. Natl. Acad. Sci. USA 95: 10671–10675.

Van der Hammen, L. 1989. An Introduction to Comparative Arachnology. SPB Academic, The Hague. 576 pp.

Wegerhoff, R., and Breidbach, O. 1995. Comparative aspects of the chelicerate nervous system. In Breidbach, O., and Kutsch, W. (Eds.): The Nervous System of Invertebrates: An Evolutionary and Comparative Approach. Birkhäuser, Basel. pp. 159–179.

Weygoldt, P. 1996. Chelicerata, Spinnentiere. In Westheide, W., and Rieger, R. (Eds.): Spezielle Zoologie: I. Einzeller und Wirbellose Tiere. Gustav Fischer Verlag, Stuttgart. pp. 449–497.

Weygoldt, P. 1998. Evolution and systematics of the Chelicerata. Exp. Appl. Acarol. 22:63–79.

Weygoldt, P., and Paulus, H. F. 1979. Untersuchungen zür Morphologie, Taxonomie und Phylogenie der Chelicerata. 1. Morphologische Untersuchungen. Z. Zool. Syst. Evolut.-forsch. 17:85–116.

Weygoldt, P., and Paulus, H. F. 1979. Untersuchungen zür Morphologie, Taxonomie und Phylogenie der Chelicerata. 2. Cladogramme und die Entfaltung der Chelicerata. Z. Zool. Syst. Evolut.-forsch. 17:177–200.

Wheeler, W. 1998. Molecular systematics and arthropods. In Edgecombe, G. D. (Ed.): Arthropod Fossils and Phylogeny. Columbia University Press, New York. pp. 9–32.

Wheeler, W. C., and Hayashi, C. Y. 1998. The phylogeny of the extant chelicerate orders. Cladistics 14:173–192.

Internet Sites

http://medent.usyd.edu.au/ (University of Sydney, Department of Medical Entomology.)

www.arachnology.org/ (Select "Arachnology Pages" from the left-hand menu to go to the Arachnology Home Page.)

www.museums.org.za/bio/arachnids (South African Museum Biodiversity Explorer, Class: Arachnida.)

www.utexas.edu/depts/tnhc/.www/biospeleology/photos.htm (Cave Life Photos.)

www.mov.vic.gov.au/spiders/arach.html (A few photos of arachnid orders.)

XIPHOSURA

Barlow, R. B., Powers, M. K., Howard, H., et al. 1986. Migration of *Limulus* for mating: Relation to lunar phase, tide height, and sunlight. Biol. Bull. 171:310–329.

Bolton, M. L., and Ropes, J. W. 1989. Feeding ecology of horseshoe crabs on the continental shelf, New Jersey to North Carolina. Bull. Mar. Sci. 45:637–647.

Bonaventura, J., Bonaventura, C., and Tesh, S. (Eds.): 1982. Physiology and Biology of Horseshoe Crabs: Studies on Normal and Environmentally Stressed Animals. Alan R. Liss, New York. 334 pp.

Fahrenbach, W. H. 1999. Merostomata. In Harrison, F. W., and Foelix, R. F. (Eds.): Microscopic Anatomy of Invertebrates. 8A: Chelicerate Arthropoda. Wiley-Liss, New York. pp. 21–115.

Owen, R. 1873. On the anatomy of the American king-crab (*Limulus polyphemus,* Latr.). Trans. Linn. Soc. Lond. 28:459–506.

Sekiguchi, K. 1988. Biology of Horseshoe Crabs. Science House, Tokyo. 428 pp.

Tanacredi, J. T. 2001. *Limulus* in the Limelight: A Species 350 Million Years in the Making and in Peril? Kluwer Academic/Plenum, New York. 178 pp.

Internet Sites

www.palaeos.com/Invertebrates/Arthropods/Xiphosura/Xiphosura.htm (Fossil Xiphosurans.)

http://members.tripod.co.uk/Lyall/xipho.html (Fossil Horseshoe Crabs.)

SCORPIONES

Brownell, P. H. 1984. Prey detection by the sand scorpion. Sci. Am. 251:86–97.

Farley, R. D. 1999. Scorpiones. In Harrison, F. W., and Foelix, R. F. (Eds.): Microscopic Anatomy of Invertebrates. Vol. 8A: Chelicerate Arthropoda. Wiley-Liss, New York. pp. 117–222.

Millot, J., and Vachon, M. 1949a. Ordre des Scorpions. In Grassé, P. P. (Ed.): Traité de Zoologie. Vol. VI. Masson, Paris. pp. 386–436.

Polis, G. A. (Ed.): 1990. The Biology of Scorpions. Stanford University Press, Stanford, CA. 587 pp.

Internet Sites

www.science.marshall.edu/fet/euscorpius/INDEX.HTM (Euscorpius: Occasional Publications in Scorpiology, with publications, links, gallery.)

www.ub.ntnu.no/scorpion-files (Scorpion Files, by J. O. Rein.)

http://wrbu.si.edu/www/stockwell/photos/photos.html (Scorpion Photographs, by S. A. Stockwell.)

UROPYGI

Eisner, T., Meinwald, J., Monro, A., and Ghent, R. 1961. Defense mechanisms of arthropods. I. The whipscorpion, *Mastigoproctus giganteus* (Lucas) (Arachnida, Pedipalpida). J. Insect Physiol. 6:272–298.

Millot, J. 1949b. Ordre des Uropyges. In Grassé, P. P. Traité de Zoologie. Vol. VI. Masson, Paris. pp. 533–562.

Sekiguchi, K., and Yamasaki, T. 1972. A redescription of *"Trithyreus sawadai"* from the Bonin Islands. Acta Arachnol. 24:73–81.

Internet Sites

http://entowww.tamu.edu/images/insects/fieldguide/cimg365.html (Vinegaroons page, Texas Cooperative Extension Service.)

www.key-net.net/users/swb/WE.jpg (Photo of female *Mastigoproctus* with egg case.)

AMBLYPYGI

Millot, J. 1949c. Ordre des Amblypyges. In Grassé, P. P. (Ed.): Traité de Zoologie. Vol. VI. Masson, Paris. pp. 563–588.

Weygoldt, P. 1977. Agonistic and mating behaviour, spermatophore morphology and female genitalia in neotropical whip spiders (Amblypygi). Zoomorphologie 86:271–286.

Internet Sites

www.evergreen.edu/ants//alastaxa/amblypygi/genuskey.html (Amblypygi of La Selva Biological Station, Costa Rica.)

www.arachnology.org/Arachnology/Pages/Amblypygi.html (Arachnology: Amblypygi. With gallery and links.)

ARANEAE

Averof, M. 1998. Origin of the spider's head. Nature 395:436–437.

Barth, F. G. (Ed.): 1985. Neurobiology of Arachnids. Springer-Verlag, Berlin. 385 pp.

Buchsbaum, R., and Milne, L. 1960. Lower Animals: Living Invertebrates of the World. Doubleday, New York. 303 pp.

Burgess, J. W. 1976. Social spiders. Sci. Am. 234:101–106.

Coddington, J. A., and Levi, H. W. 1991. Systematics and evolution of spiders (Araneae). Ann. Rev. Ecol. Syst. 22:565–592.

Comstock, J. H. 1940. The Spider Book. Doubleday, New York. 729 pp.

Damen, W. G. M., Hausdorf, M., Seyfarth, E.-A., and Tautz, D. 1998. A conserved mode of head segmentation in arthropods revealed by the expression pattern of Hox genes in a spider. Proc. Natl. Acad. Sci. USA 95:10665–10670.

Damen, W. G. M., and Tautz, D. 1999. Comparative molecular embryology of arthropods: The expression of Hox genes in the spider *Cupiennius salei.* Inv. Rep. Dev. 36:203–209.

Felgenhauer, B. E. 1999. Araneae. In Harrison, F. W., and Foelix, R. F. (Eds.): Microscopic Anatomy of Invertebrates. Vol. 8A: Chelicerate Arthropoda. Wiley-Liss, New York. pp. 223–266.

Foelix, R. F. 1970. Chemosensitive hairs in spiders. J. Morphol. 132:313–334.

Foelix, R. F. 1982. Biology of Spiders. Harvard University Press, Cambridge, MA. 306 pp.

Foelix, R. F. 1996. Biology of Spiders. 2nd Edition. Oxford University Press, Oxford. 330 pp.

Forster, L. M. 1979. Visual mechanisms of hunting behavior in *Trite planiceps,* a jumping spider. N. Z. J. Zool. 6:79–93.

Gertsch, W. J. 1979. American Spiders. 2nd Edition. Van Nostrand Reinhold, New York. 274 pp.

Görner, P. 1965. A proposed transducing mechanism for a multiply-innervated mechanoreceptor (trichobothrium) in spiders. Cold Spring Harbor Symp. Quant. Biol. 30:69–73.

Harwood, R. H. 1974. Predatory behavior of *Argiope aurantia.* Am. Midl. Nat. 91:130–138.

Homann, H. 1971. Die Augen der Araneae. Anatomie, Ontogenie, und Bedeutung für die Systematik (Chelicerata: Arachnida). Z. Morphol. Tiere. 69:201–273.

Kaston, B. J. 1948. Spiders of Connecticut. State Geol. Nat. Hist. Surv. Bull. 70:1–874.

Kaston, B. J. 1970. Comparative biology of American black widow spiders. Trans. San Diego Soc. Nat. Hist. 16:33–82.

Kaston, B. J. 1978. How to Know the Spiders. 3rd Edition. W. C. Brown, Dubuque, IA. 272 pp.

Kaston, B. J. 1981. Spiders of Connecticut. Rev. Edition. State Geol. Nat. Hist. Surv. Bull. 70:1–1020.

Land, M. F. 1969a. Structure of the retinae of the principal eyes of jumping spiders (Salticidae: Dendryphantinae) in relation to visual optics. J. Exp. Biol. 51:443–470.

Land, M. F. 1969b. Movements of the retinae of jumping spiders (Salticidae: Dendryphantinae) in response to visual stimuli. J. Exp. Biol. 51:471–493.

Levi, H. W. 1978. Orb-weaving spiders and their webs. Am. Sci. 66:734–742.

Levi, H. W., and Levi, L. R. 1968. A Guide to Spiders and Their Kin. A Golden Nature Guide. Golden Press, New York. 160 pp.

Millot, J. 1949d. Ordre des Aranéides. In Grassé, P. P. (Ed.): Traité de Zoologie. Vol. VI. Masson, Paris. pp. 589–743.

Opell, B. D. 1979. Revision of the genera and tropical American species of the spider family Uloboridae. Bull. Mus. Comp. Zool. 148:443–549.

Platnick, N. I., and Gertsch, W. J. 1976. The suborders of spiders: A cladistic analysis. Am. Mus. Novit. 2607:1–15.

Robinson, M. H., and Robinson, B. 1980. Comparative Studies of the Courtship and Mating Behavior of Tropical Araneid Spiders. Pacific Insects Monographs, no. 36. Bishop Museum Press, Honolulu, HI. 218 pp.

Roth, V. 1985. Spider Genera of North America. American Arachnological Society, Gainesville, FL.

Savory, T. 1928. The Biology of Spiders. Sidgwick & Jackson, London. 376 pp.

Shear, W. A. (Ed.): 1986. Spiders: Webs, Behavior and Evolution. Stanford University Press, Stanford, CA. 492 pp.

Wilson, R. S. 1969. Control of drag-line spinning in certain spiders. Am. Zool. 9:103–111.

Witt, P. N., and Rovner, J. S. (Eds.): 1982. Spider Communication: Mechanisms and Ecological Significance. Princeton University Press, Princeton, NJ. 440 pp.

Internet Sites

www.amonline.net.au/spiders (Australian Museum OnLine.)

www.angelfire.com/tx4/arachnids (Brent's arachnids, tarantulas, and scorpions.)

www.nhm.org/spiders/article.htm (Common Spiders of Los Angeles, by Blaine Herbert.)

www.xs4all.nl/~ednieuw/Spiders/spidhome.htm (Araneae: Spiders of North-West Europe, by E. Nieuwenhuys.)

www.xs4all.nl/~ednieuw/index.html (Spiders and Immunology, by E. Nieuwenhuys.)

www.xs4all.nl/~ednieuw/australian/Spidaus.html (Spiders of Australia, with links.)

http://research.amnh.org/entomology/spiders/catalog81-87/index.html (World Spider Catalog, by N. I. Platnick, American Museum of Natural History.)

PALPIGRADI

Kovác, L., Mock, A., L'uptácik, P., and Palacios-Vargas, J. G. 2002. Distribution of *Eukoenenia spelaea* (Peyerimhoff, 1902) (Arachnida, Palpigradida) in the Western Carpathians with remarks on its biology and behavior. In *Studies on Soil Fauna in Central Europe.* Proceedings of the 6th Central European Workshop. Institute of Soil Biology, České Budějovice, Czechoslovakia. pp. 93–99.

Millot, J. 1949e. Ordre des Palpigrades. In Grassé, P. P. (Ed.): Traité de Zoologie. Vol. VI. Masson, Paris. pp. 520–532.

Internet Site

www.arachnology.org/Arachnology/Pages/Palpigradi.html (Arachnology Home Page palpigrade page. With links.)

PSEUDOSCORPIONES

Goddard, S. J. 1976. Population dynamics, distribution patterns and life cycles of *Neobisium muscorum* and *Chthonius orthodactylus.* J. Zool. 178:295–304.

Weygoldt, P. 1969. The Biology of Pseudoscorpions. Harvard University Press, Cambridge, MA. 145 pp.

Internet Sites

www.naturserv.org/explorer/speciesIndex/Order_PSEUDOSCORPIONES_102370_1.htm (NaturServe Explorer Species Index for Pseudoscorpions. Database of families, genera, and species with range maps.)

http://insects.tamu.edu/images/insects/fieldguide/cimg375.html (Pseudoscorpion Page of the Texas A & M University Cooperative Extension Service. With a photograph and biological account.)

www.biol.net/Pseudoscorpion.htm (Pictures of a pseudoscorpion. Three photographs.)

www.utexas.edu/depts/tnhc/.www/biospeleology/photos.htm (Cave Life Photos.)

SOLIFUGAE

Millot, J., and Vachon, M. 1949b. Ordre des Solifuges. In Grassé, P. P. (Ed.): Traité de Zoologie. Vol. VI. Masson, Paris. pp. 482–519.

Muma, M. H. 1967. The behavior of North American Solpugida. Florida Entomol. 50:115–123.

Muma, M. H. 1970. A Synoptic Review of North American, Central American, and Western Indian Solipugida. Arthropods of Florida. Vol. 5. Contribution No. 154. Bureau of Entomology, Florida Department of Agriculture and Consumer Services, Gainesville. 62 pp.

Internet Sites

www.museums.org.za/bio/images/scorpions/arachnids/solifugae.htm (The Biodiversity Explorer of Museums Online South Africa. With photographs and biological account.)

http://insects.tamu.edu/images/insects/fieldguide/cimg376.html (Solifugid Page of the Texas A & M University Cooperative Extension Service. With a photograph and short account of biology.)

www.ultimatepethabitat.net/solifugid.htm (Solfugid Gallery, with photographs.)

OPILIONES

Edgar, A. L. 1971. Studies on the biology and ecology of Michigan Phalangida (Opiliones). Misc. Publ. Mus. Zool. Univ. Mich. 144:1–64.

Macías-Ordoñez, R. 2000. Touchy harvestmen. Nat. Hist. 109(8):58–61.

Internet Sites

www.mov.vic.gov.au/spiders/fauna.html (Invertebrates of the *Nothophagus* forest, Victoria, Australia.)

www.david.curtis.care4free/opiliones.htm (Opiliones Notes, by D. Curtis.)

http://members.aol.com/blspecies/opiliones.htm (World Bibliography on Subterranean Arachnida: Opiliones, by B. Lebreton.)

RICINULEI

Cooke, J. A. L. 1967. The biology of Ricinulei. J. Zool. (Lond.) 151:31–42.

Internet Sites

www.arachnology.org/Arachnology/Pages/Ricinulei.html (The Arachnology Home Page ricinuleid page. With links.)

ACARI

Alberti, G., and Coons, L. B. 1999. Acari: Mites. In Harrison, F. W., and Foelix, R. F. (Eds.): Microscopic Anatomy of Invertebrates. Vol. 8B: Chelicerate Arthropoda. Wiley-Liss, New York. pp. 515–1215.

Barr, D. 1973. Methods for the collection, presentation and study of water mites. Royal Ontario Museum of Life Sciences Miscellaneous Publications. Royal Ontario Museum, Toronto. 28 pp.

Binns, E. S. 1983. Phoresy as migration: Some functional aspects of phoresy in mites. Biol. Rev. 57:571–620.

Coons, L. B., and Alberti, G. 1999. Acari: Ticks. In Harrison, F. W., and Foelix, R. F. (Eds.): Microscopic Anatomy of Invertebrates. Vol. 8B: Chelicerate Arthropoda. Wiley-Liss, New York. pp. 267–514.

Evans, G. O. 1992. Principles of Acarology. CAB International, Cambridge. 576 pp.

Krantz, G. W. 1978. A Manual of Acarology. 2nd Edition. Oregon State University Bookstores, Corvallis, OR. 509 pp.

McDaniel, B. 1979. How to Know the Mites and Ticks. W. C. Brown, Dubuque, IA. 335 pp.

Ostfeld, R. S., Jones, C. G., and Wolff, J. O. 1996. Of mice and mast. BioScience 46:323–330.

Sauer, J. R., and Hair, J. A. (Eds.): 1986. Morphology, Physiology, and Behavioral Biology of Ticks. Halsted Press, New York. 510 pp.

Schultz, J. W. 1990. Evolutionary morphology and phylogeny of Arachnida. Cladistics 6:1–38.

Smith, I. M, Smith, B. P., and Cook, D. R. 2001. Water mites (Hydracarina) and other arachnids. In Thorp, J. H., and Covich, A. P. (Eds.): Ecology and Classification of North American Freshwater Invertebrates. Academic Press, San Diego. pp. 551–659.

Walter, D. E., and Procter, H. C. 1999. Mites: Ecology, Evolution, and Behaviour. New South Wales Press, Sydney, Australia. 322 pp.

Woolley, T. A. 1988. Acarology: Mites and Human Welfare. Wiley-Interscience, New York. 484 pp.

Internet Sites

www.lubi.edu.lv/acari.html (Acarology Web Guide. A valuable resource with extensive links.)

www.nhm.ac.uk/hosted_sites/acarology (Acarology Home Page.)

www.fsca-dpi.org/Acari/AcariImages.htm (Images of Acari.)

www.geocities.com/HotSprings/oasis/6455/lyme-links.html (Lots of Links on Lyme Disease.)

www.rothamsted.bbsrc.ac.uk/pie/BrianGrp/VarroaHub.html (*Varroa* Hub.)

http://medent.usyd.edu.au/fact/dustmite.html (University of Sydney, Department of Medical Entomology, Dust Mite fact sheet.)

PYCNOGONIDA

Arnaud, F., and Bamber, R. N. 1987. The biology of Pycnogonida. Adv. Mar. Biol. 24:1–96.

Fahrenbach, W. H. 1994. Microscopic anatomy of Pycnogonida: 1. Cuticle, epidermis, and muscle. J. Morphol. 222:33–48.

Fry, W. G. (Ed.): 1978. Sea spiders (Pycnogonida). Zool. J. Linn. Soc. Lond. 63:1–238.

Hedgpeth, J. W. 1948. The Pycnogonida of the western North Atlantic and the Caribbean. Proc. U.S. Nat. Mus. 97:157–342.

King, P. E. 1973. Pycnogonids. St. Martin's Press, New York. 144 pp.

McCloskey, L. R. 1973. Pycnogonida. Marine Flora and Fauna of the Northeastern United States. NOAA Tech. Report NMFS Circular 386. U.S. Government Printing Office, Washington, DC. 12 pp.

Internet Sites

www-personal.monash.edu.au/~fgodevic//seaspider/ (Sea Spiders [Pantopoda] Web Page, by B. A. Bain and W. H. Fahrenbach. Includes a 2003 paper reporting saccate nephridia in pycnogonids.)

http://scilib.ucsd.edu/sio/nsf/fguide/index.html (Underwater Field Guide to Ross Island and McMurdo Sound, Antarctica: Arthropoda: Chelicerata. A large collection of extraordinary color photographs of Antarctic marine invertebrates, including many pycnogonids.)

Crustacea[sP]

GENERAL BIOLOGY

Crustacea includes about 42,000 described species of crabs, shrimps, crayfishes, lobsters, and wood lice, as well as myriad small animals that mostly go unnoticed. Crustaceans, particularly the small, inconspicuous ones, occupy a central position in global ecology as the major trophic link between primary producers (phytoplankton) and higher-level consumers (fishes) in the sea. Crustacea evolved in the sea and is the only major arthropod taxon that is chiefly aquatic. Several taxa (such as crayfishes, shrimps, crabs, water fleas, copepods, fairy shrimps, and others) have independently invaded fresh water. A few (including wood lice and some crabs) have colonized the land, but are not as well adapted to this life as the arachnids and tracheates.

Crustaceans belong to Mandibulata (Fig. 16-15), one of the two high-level taxa of living arthropods (Chelicerata is the other). Shrimps, crabs, insects, centipedes, and millipedes are all mandibulates, but they fall into two distinct subtaxa, the largely aquatic Crustacea and the mostly terrestrial Tracheata (Hexapoda and Myriapoda).

Mandibulates have antennae on the first head segment, a three-part brain with a deutocerebrum, mandibles on the third head segment, and maxillae on the fourth. The ancestral mandibulates probably were aquatic filter feeders with a ventral food groove to transport food particles to the mouth. The appendages were biramous or multiramous, with gnathobases, and the mouth of the J-shaped gut faced posteriorly into the anterior end of the food groove. The compound eyes of crustaceans and insects (but not myriapods) are similar and differ from those of the chelicerates. The ommatidium has a crystalline cone (lost in myriapods) formed originally of four cone cells and a retinula of eight retinular cells. The cornea is secreted by two corneagenous cells. Median ocelli are also present, except in the myriapods. Molting is controlled by ecdysone and small eggs with total cleavage were probably ancestral.

Crustacea consists of several taxa traditionally considered to be classes, although in cladistic analysis they do not all have the same rank. These are Remipedia, Cephalocarida, Anostraca (fairy and brine shrimps), Phyllopoda (water fleas and clam shrimps), Malacostraca (shrimps, scuds, wood lice, krill, lobsters, and others), Copepoda, Mystacocarida, Tantulocarida, Ascothoracida, Cirripedia (barnacles), Ostracoda (seed shrimps), Branchiura (fish lice), and Pentastomida.

The enormous morphological and ecological heterogeneity of Crustacea rivals that of any other animal taxon. It includes tiny animals less than a millimeter in length and giant spider crabs with a leg span of 3 m. They may resemble shrimps, crabs, annelids, and even molluscs. Most are free-living and can be natant (swimming), reptant (crawling), fossorial (burrowing), or interstitial (living between sand grains). They can be pelagic, epifaunal, infaunal, or terrestrial. A few are sessile and permanently attached to the substratum. Some are so specialized they are unrecognizable as crustaceans without examination of the larva. More than any other arthropods, crustaceans capitalize on the widely varied possibilities offered by specialization of a large number of appendages.

Two pairs of antennae immediately distinguish crustaceans from other arthropods. They are the only arthropods with three or four simple pigment-cup ocelli grouped to form a median naupliar eye. The basic larva is the nauplius. Crustaceans have a maximum of two pairs of saccate nephridia, one in the segment of the second antenna and the other in that of the second maxilla. Beyond these few shared traits, it is difficult to generalize about the crustaceans, and their diversity necessitates reference to an extended classification hierarchy with many intermediate ranks.

FORM

The crustacean body is divided into at least two tagmata, head and trunk (Fig. 19-1). The head consists of the acron plus the anteriormost five segments, which are fused together to form an unsegmented cuticular capsule. The remainder of the body is the trunk. The typical head bears five pairs of appendages (Fig. 19-2A). Anteriormost are the first antennae (antennules), which are considered to be homologous to the antennae of hexapods and other tracheates. Next in line are the second antennae (antennae). These are postoral in embryological origin and probably are homologous to the arachnid chelicerae, both of which are innervated by the tritocerebrum. Flanking and often covering the ventral mouth is the third pair of appendages, the mandibles. These are composed of several articles consisting of a short, heavy, grinding and biting surface and a slender palp. Behind the mandibles are two pairs of feeding appendages, the first maxillae (maxillules) and the second maxillae (maxillae). The head bears lateral compound eyes, median pigment-cup ocelli, or both.

In the ancestral crustacean the trunk was probably homonomous, consisting of a linear series of identical segments with identical appendages, and was not subdivided into tagmata. This idealized ancestral condition is not found in any living crustacean, although some (such as Remipedia) come very close (Fig. 19-9). In all living crustaceans the body consists of a head and trunk and in most the trunk is divided into an anterior thorax and posterior abdomen, each tagma composed of specialized segments with specialized appendages, although in most taxa the abdomen lacks appendages. The number of segments in the thorax and abdomen varies widely with taxon.

In many crustaceans some anterior thoracic segments fused with the head to form a new tagma known as the cephalothorax (Fig. 19-1). The remaining thoracic segments together form the **pereon.** The anterior thoracic appendages, now part of the cephalothorax, are specialized as auxiliary mouthparts known as **maxillipeds,** whereas those of the pereon are **pereopods** (Fig. 19-1). The general name for any thoracic segment is thoracomere and that for any thoracic appendage is thoracopod. Thoracomeres can be part of the cephalothorax or the pereon and, similarly, thoracopods can be either maxillipeds or pereopods.

Crustaceans frequently have a mantlelike fold of the body wall extending posteriorly from the head, usually from the posterior edge of the last head segment, that of the second maxillae. This fold is the carapace (Fig. 19-1, 19-19), and its size varies. It may be fused with some or all of the segments it

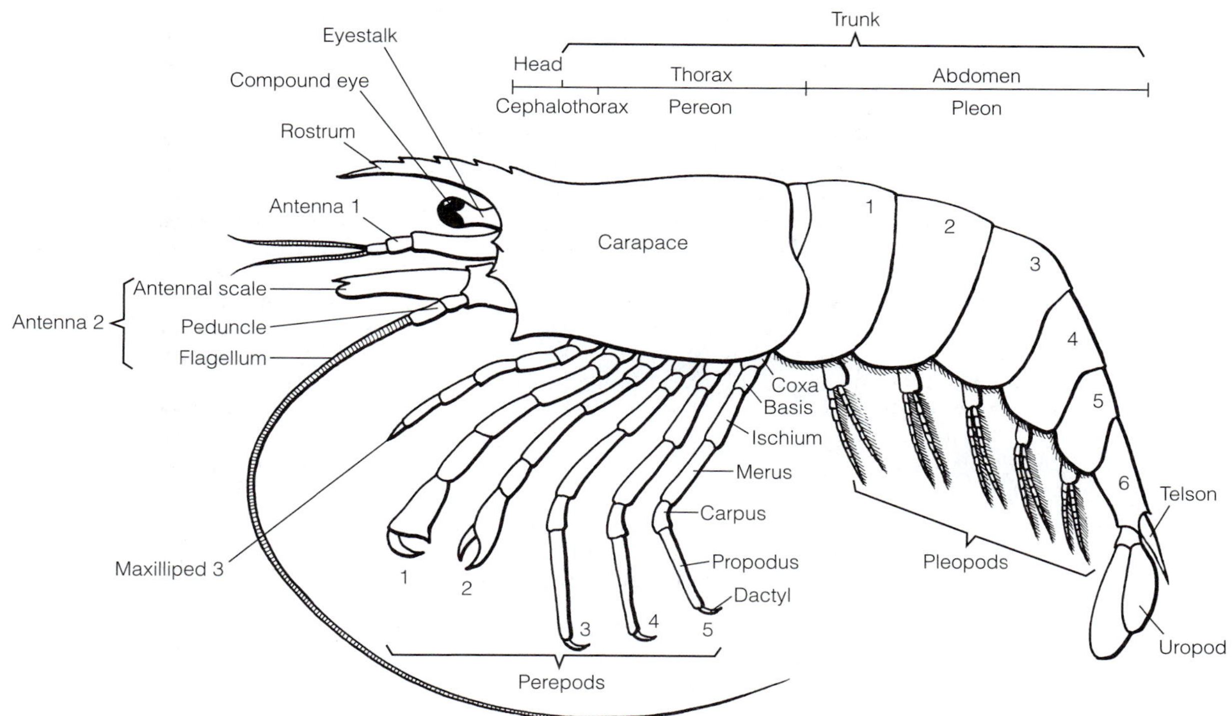

FIGURE 19-1 Crustacea. Tagmata of a shrimp as an example of a generalized crustacean. The head consists of the acron and five fused segments and their appendages. In this example, the thorax has eight segments, the first three of which have become part of the cephalothorax, and their appendages are maxillipeds. The remaining five thoracomeres make up the pereon. The abdomen, or pleon, is composed of six segments plus the telson. The carapace extends posteriorly from the head and, in this example, covers the cephalothorax and pereon. *(Redrawn and modified from Schmitt, W. L. 1965. Crustaceans. Ann Arbor Science Library, Univ. of Michigan Press, Ann Arbor. 204 pp.)*

covers. It may be absent (Fig. 19-11A), enclose the entire animal (Fig. 19-14), or anything in between (Fig. 19-13). Since it is a fold of the body wall, it consists of two layers of integument with an extension of the hemocoel between them.

Early crustaceans had numerous homonomous, multipurpose, biramous appendages specialized for swimming, walking, and feeding. In modern crustaceans the appendages have undergone numerous modifications to alter their original structure and function. A typical biramous appendage consists of a basal **protopod** attached to the ventral surface of the body (Fig. 19-3B, 19-4A) and two branches, or rami. The protopod may be divided into two articles, the proximal **coxa** and distal **basis.** The coxa may be a gnathobase, with spines or teeth on its medial surface for masticating food. The rami, each composed of a linear sequence of articles, attach to the basis. The lateral ramus is the exopod (exopodite) and the medial one is the endopod (endopodite). Either can be modified in many ways. A limb with a slender, leglike ramus resembling a jointed rod is a **stenopod** (Fig. 19-20), whereas a broad, leaflike leg is a **phyllopod** (Fig. 19-11B). A **mixopod** is a combination of both. It is not uncommon for one ramus to be absent, in which case the appendage is uniramous. For example, the uniramous walking legs of malacostracans have lost the exopod and consist solely of a stenopodous endopod (Fig. 19-1). Often there are additional smaller processes, such as exites, endites, and epipods, which are not considered to be rami (Fig. 19-11B). Endites are medial and exites and epipods are lateral.

As in other arthropods, the evolutionary tendency has been toward a reduced number of segments and regionally specialized appendages for specific functions. For example, in a blue crab (Fig. 19-31) two pairs of appendages are sensory, five are mouthparts, one is a pincer, three are for walking, one for swimming, and others for sperm transfer and brooding.

The posteriormost tagma, the abdomen, consists of a variable number of segments that, in most taxa, lack appendages. The abdomen is sometimes called the **pleon,** its appendages **pleopods,** and segments **pleomeres.** The posterior unit of the body is the telson, which bears the anus and lacks the characteristics of a true segment. It often bears a **caudal furca** (tail fork) consisting of two caudal rami (Fig. 19-9B, 19-10A) that are not homologous to segmental appendages.

The crustacean cuticle has four major layers. Outermost is the epicuticle, consisting of lipid and tanned protein. The exocuticle (pigmented layer) consists of an outer layer of tanned protein and other layers of chitin, which may be

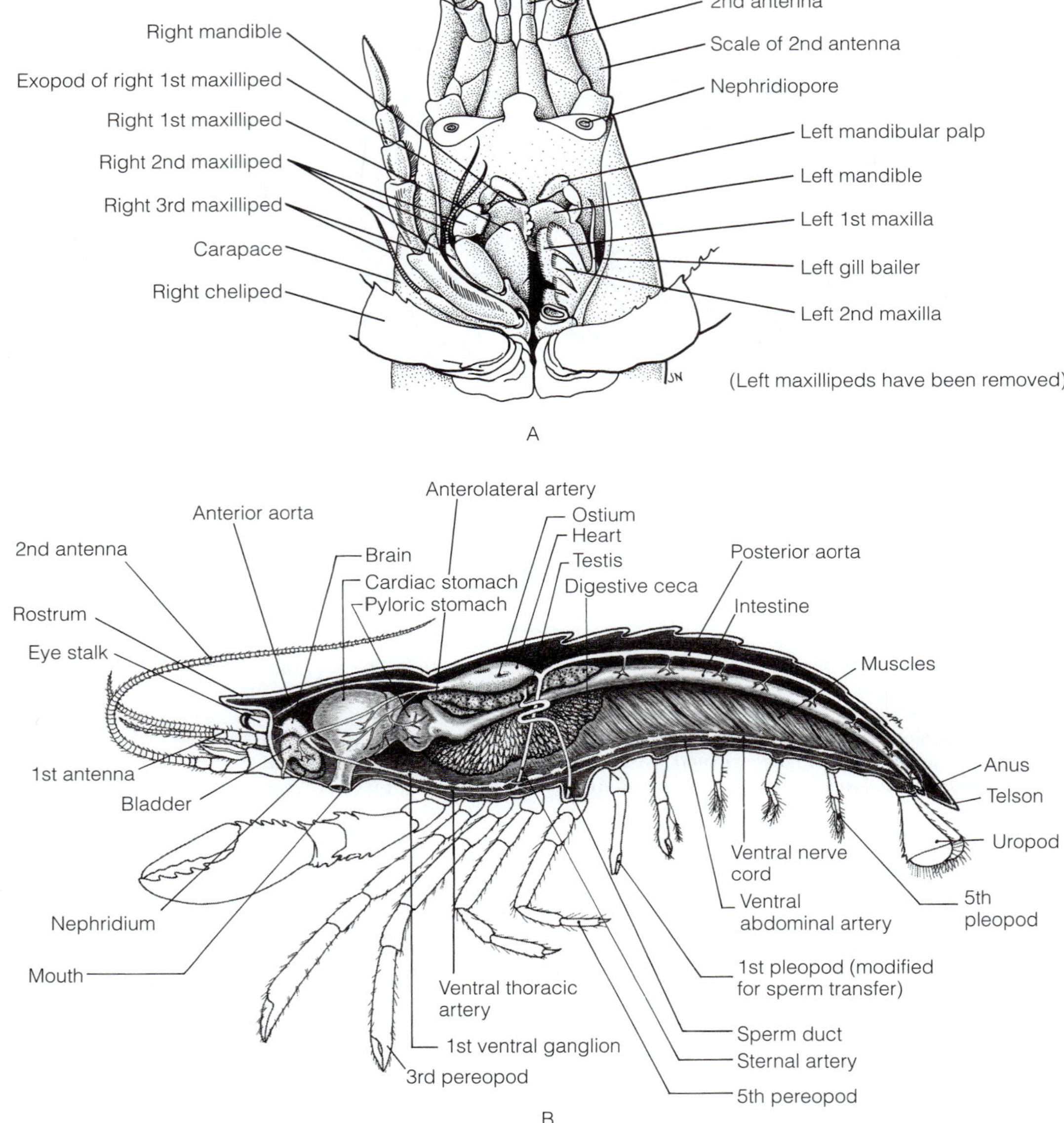

FIGURE 19-2 Crustacea. **A,** Cephalothoracic appendages of a lobster (ventral view). **B,** Internal anatomy of a crayfish (lateral view). *(B, After Howes.)*

calcified. The endocuticle (calcareous layer) consists of untanned protein and large amounts of chitin. It may be heavily calcified. Most of the thickness of the cuticle is endocuticle. Crustaceans have an additional membranous layer adjacent to the epidermis. The membranous layer consists of untanned, uncalcified chitin. The relative importance of sclerotization and calcification vary with taxon. The soft, flexible brine shrimp exoskeleton, for example, is weakly sclerotized and completely uncalcified. That of large malacostracans such as crabs, lobsters, and crayfishes, on the other hand, is heavily calcified and sclerotized. The epidermis secretes all the layers of the cuticle.

NUTRITION

Crustaceans have a great range of diets and feeding mechanisms. The ancestral crustacean probably was a small, epibenthic suspension feeder that used its setose swimming limbs for

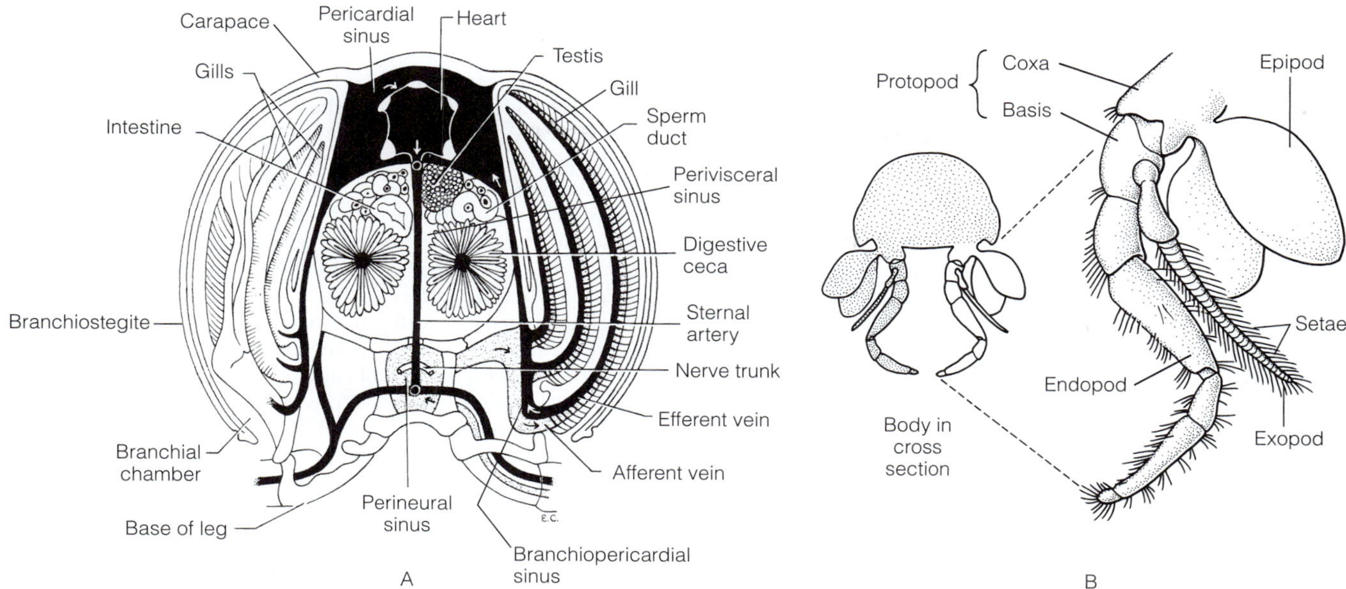

FIGURE 19-3 Crustacea. **A,** Cross section of a crayfish just behind the third pair of legs. **B,** Right fifth pereopod of the syncarid *Anaspides tasmaniae,* showing the basic structure of a crustacean appendage. *(A, After Howes; B, After Waterman and Chace, 1960.)*

filtering. Like the trilobites, this animal probably used mixopods to concentrate particulate food in a ventral food groove for transport to the mouth. The trunk appendages were biramous with a medial gnathobase on the protopod (Fig. 19-4A). The exopod was a leaflike phyllopod used for creating currents and for swimming. The endopod was a leg-like stenopod used for walking. The opposing gnathobases of the trunk appendages formed a longitudinal food groove extending anteriorly for the length of the trunk to end at the posteriorly directed mouth (Fig. 19-4B). Movements of the phyllopods and stenopods moved food to the food groove. Movements of the gnathobases moved the food anteriorly to the mouth and, if the particles were large, masticated it. This mechanism was adaptable for particles of various sizes, from phytoplankton and detritus to invertebrates such as polychaetes, amphipods, and small molluscs. It is found in numerous primitive, living arthropod taxa, including the xiphosurans, cephalocaridans, and phyllopodans.

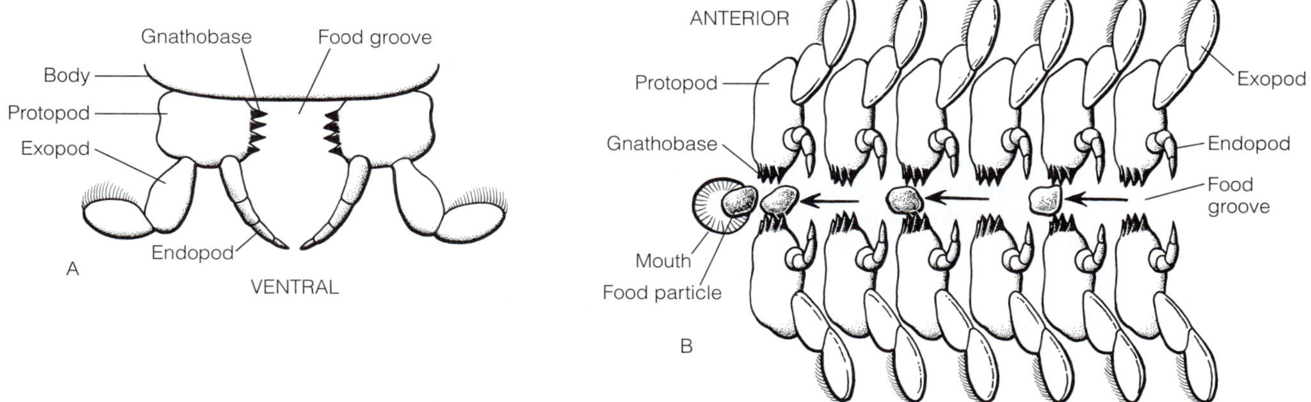

FIGURE 19-4 Feeding mechanism of the ancestral crustacean. **A,** Cross section through a trunk segment showing a typical pair of mixopods. The exopod is a phyllopod, the endopod is a stenopod, and the protopod has a medial gnathobase. **B,** Ventral view of the anterior trunk, showing the food groove and feeding mechanism.

Many modern crustaceans continue to rely on suspension feeding, and mechanisms for this type of feeding evolved independently several times in the crustaceans. Other derived crustaceans adopted feeding mechanisms such as herbivory, carnivory, and parasitism. Malacostraca, for example, consists mostly of large, benthic crustaceans that developed alternatives to suspension feeding. Some of their appendages are heavier and adapted for predation and herbivory, as well as crawling and burrowing. In crustaceans in general, the posterior head and anterior trunk appendages (mandibles, maxillae, and maxillipeds) are adapted for feeding.

The crustacean mouth opens ventrally on the head and the digestive tract is usually straight except for a sharp ventral bend at the anterior end that imparts an overall J- or L-shape (Fig. 19-2B). The foregut usually consists of an esophagus and stomach. In crustaceans that consume large food particles (Malacostraca) the stomach is adapted for grinding, and its walls bear apposing chitinous ridges, teeth, and calcareous ossicles (Fig. 19-34) collectively known as the gastric mill. Setal filters exclude all but the finest particles. Because small crustaceans typically feed on small particles that do not require grinding, a specialized stomach is absent.

The endodermal midgut is responsible for the secretion of digestive enzymes, hydrolysis, and absorption of the products of hydrolysis. These functions are largely accomplished in one or more pairs of digestive ceca (hepatopancreas). The length of the midgut varies; it may be a short region immediately posterior to the stomach or may extend for most of the length of the abdomen. Storage cells for glycogen, lipids, and calcium are also present in the ceca epithelium.

The length of the cuticularized hindgut varies to complement that of the midgut so that if the midgut is short, the hindgut is long and vice versa. It is responsible primarily for water reclamation, feces formation and storage. It opens at the anus, at the base of the telson.

Suspension Feeding

Suspension feeding is the removal of a small volume of suspended particles (food) from a large volume of water by any process that concentrates the food and discards the water. Because crustaceans are primarily aquatic, the taxon includes most of the suspension-feeding arthropods, with the exception of some aquatic insects. Many crustaceans use filtration (the flow of water through a sieve with retention of particles) to separate particles from water. In crustaceans the filter is always composed of closely placed setae or setules (side branches of setae), commonly arranged like the teeth of a comb. Such a **setal comb** is located on one or several pairs of appendages, but the location varies with taxon, indicating that filter feeding evolved numerous times within Crustacea. The rhythmic beating of appendages produces the filtering current.

Filtration is the most obvious and straightforward way to separate particles from water, and until recently it was believed that suspension feeding was usually a process of filtration. The mesh size of the filter (the distance between setae) (Fig. 19-5A)was thought to determine the size of particles that could be collected. Studies on water fleas (Cladocera), however, indicate that this is not always the case.

In the laboratory the water flea *Daphnia magna* was fed a mixture of three sizes of polystyrene spheres (Fig. 19-5A,B), two larger and one smaller than the mesh size of the filter (1 μm between setules). The water flea captured a sizable portion of the smallest spheres, which should have passed between setules acting solely as a filter. The collecting efficiency was about 60% for the smallest spheres, compared to 100% for the two larger sizes. Clearly, the collecting process is not one of simple filtration.

When water flows over or around an object, the adhesive nature of the water molecules causes a boundary layer of water to develop at the object's surface. Water molecules at the surface do not move at all, and those near the surface move more slowly than those farther away. Thus, the current's velocity increases with distance from the surface. This boundary layer of slowly moving water can be significant when the water flows over a very small surface. In *Daphnia,* for example, the boundary layer that develops over one setule of the filter reaches to and extends beyond the next setule. This means that a layer of stationary water extends from setule to setule and very little water flows between the setules (unless it is subjected to considerably higher pressure). The smaller and slower a setose appendage, the less leaky it is. Thus, in *Daphnia,* particles are captured not by sieving, but by the attractive force of opposite charges between the particles and the filter surface. Particles are attracted to the surface by their charge but do not pass through because of the boundary layer. The proportion of polystyrene spheres collected can be experimentally altered by changing the spheres' surface charges.

In copepods, the second maxillae, which are filtering appendages with a setal comb, were long thought to passively filter particles from a forward-moving current created by other appendages. By filming dye-marked water currents around tethered copepods, investigators found that these crustaceans move their second maxillae to actively capture individual particles. When the copepod is not capturing food, the water current generated by other appendages passes over, not through, the sieve of the stationary second maxillae (Fig. 19-5C). When an algal particle in the approaching water stream is detected (probably by sensory setae on the antennae and elsewhere), the other appendages beat asymmetrically so that the copepod turns toward the particle, and the second maxillae are spread apart (Fig. 19-5D). This movement causes the water stream with the particle to move to the midline and the second maxillae clap over it, squeezing out much of the surrounding water (Fig. 19-5E). The captured food is transferred to the mandibles and mouth. In the final analysis, most suspension-feeding organisms and so-called filter feeders probably collect particles by methods other than sieving.

In suspension feeding in general, the required water current is produced by the beating of either the setal comb appendages or, more commonly, other appendages modified for this purpose. Once collected, the particles are removed from the "filter" setae by special combing or brushing setae and transported to the mouthparts by other appendages or, in some species, in a ventral food groove.

Suspension feeding undoubtedly evolved independently a number of times within Crustacea, and virtually every pair of appendages, even the antennae and mandibles, may be modified in one taxon or another for suspension feeding. Suspension feeding probably first arose in connection with

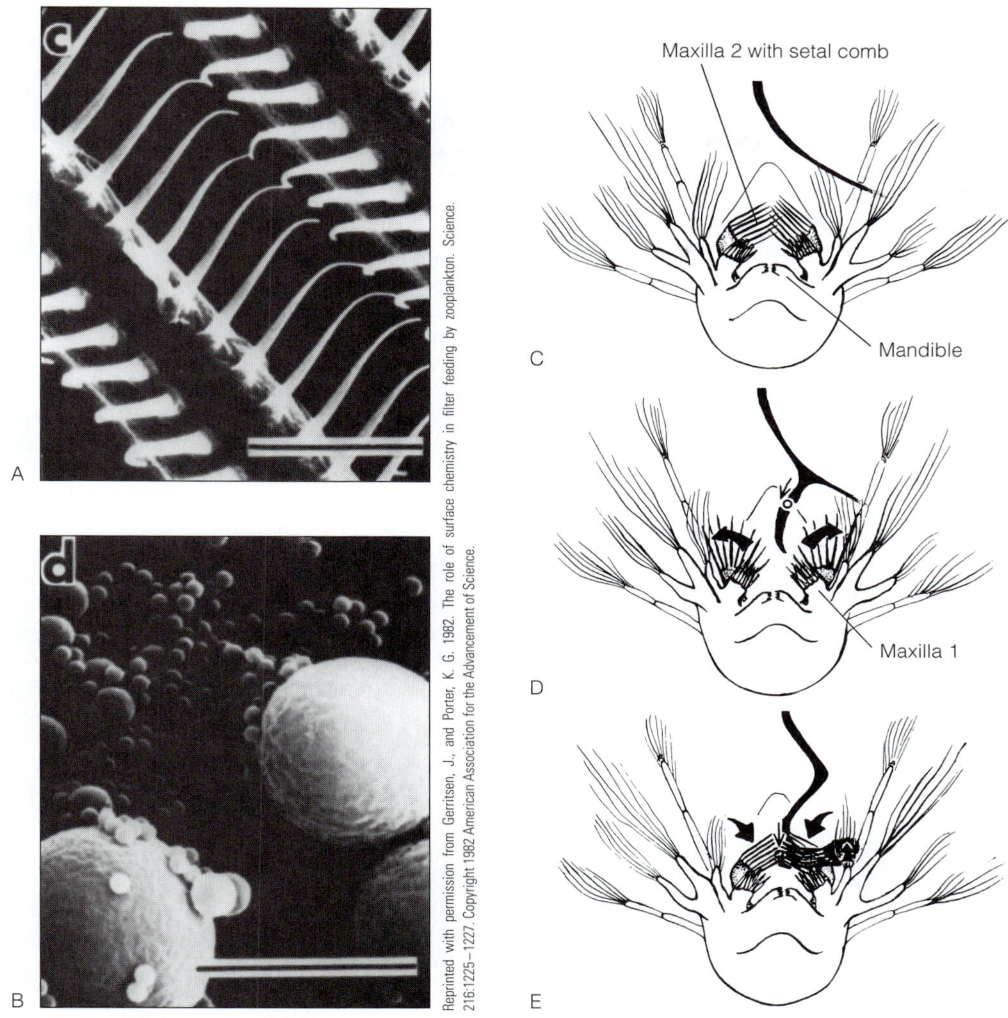

FIGURE 19-5 Crustacea. **A,** The particle-collecting setules of the water flea *Daphnia magna.* **B,** The three sizes of polystyrene beads (0.5, 1.1, and 5.7 μm) used in the experiment. See text for explanation. The scale bar is about 5 μm. **C–E,** Particle collecting in a copepod. The thick black line represents the water current with a food particle. See text for explanation. *(C–E, From Koehl, M. A. R., and Strickler, J. R. 1981. Copepod feeding currents: Food capture at low Reynolds number. Limnol. Oceanogr. 226(6):1062–1073)*

swimming and it is therefore primitively associated with the trunk, the same limbs creating both the swimming and the feeding currents. The tendency in most taxa has been for the feeding apparatus to be shifted anteriorly, nearer the mouth, and to involve only the anterior trunk appendages or head appendages.

INTERNAL TRANSPORT

The hemal system of crustaceans is similar to that of other arthropods. The heart, in the pericardial sinus (Fig. 19-3A), varies from being a long tube extending most of the length of the trunk in the primitive condition to a short, compact vesicle in derived taxa (Fig. 19-2B). The heart is located in the same tagma as the gills and is absent in some small species.

Small crustaceans usually have no blood vessels, although a few may have an anterior aorta to supply the brain with oxygenated blood. In most, the heart pumps blood to the hemocoel to bathe the tissues before flowing back to the heart. Large species may have well-developed arteries and capillaries.

Crustacean blood contains a variety of amebocytes responsible for phagocytosis and clotting. Under certain traumatic conditions, such as self-amputation (autotomy), special amebocytes called **explosive cells** disintegrate and liberate a substance that converts plasma fibrinogen to fibrin. Islands of coagulated plasma then appear and connect with each other to trap blood cells, thus forming a clot.

GAS EXCHANGE

Very small crustaceans often have no special gas-exchange surfaces. In many (such as water fleas and barnacles) the thin, permeable cuticle of the inner surface of the carapace functions in gas exchange. Larger crustaceans, with their unfavorable surface area-to-volume ratios, require specialized expansions of the body surface to provide the necessary oxygen. Such surfaces are gills, of course, and are evaginations of some part of the body wall. Gills are almost always associated with appendages, but they are highly variable regarding location, derivation, and morphology. The gills are typically located in a protected branchial chamber formed of the carapace and/or body wall and ventilated by movements of various appendages.

Gas transport from gills to tissues is by blood in sinuses or veins that direct blood to and from the gills. Gases are exchanged between the medium (water or air) and the blood across the gill surface. Oxygen may be in simple solution in the blood or bound to a respiratory pigment, either hemocyanin or hemoglobin. When present, the pigment is always in solution, never intracellular. Hemocyanin is found in Malacostraca whereas hemoglobin occurs in some other taxa, but both pigments are sporadically distributed.

EXCRETION

Crustacean excretory organs are paired saccate nephridia consisting of a coelomic end sac, with podocytes, that is surrounded by the hemocoel and connected to the exterior by a tubule derived from a metanephridium (Fig. 19-6A). They play only a minor role in nitrogen excretion, however, and their chief function is probably in maintaining the appropriate balance of ions and correct fluid volume in the body.

Paired adult nephridia are located in the segment of either the second antennae or the second maxillae and are accordingly referred to as **antennal** or **maxillary glands,** respectively. Known in some taxa as green glands, shell glands, or coxal glands, they are always located in one (or both) of these two segments. The nephridiopores open on or near the bases of the second antennae or second maxillae. Both antennal and maxillary glands are commonly present in crustacean larvae, but usually only one or the other persists in the adult. Clusters of podocytes have been found at the bases of all the thoracic appendages of cephalocarids. If these are degenerate saccate nephridia, it would be evidence that the stem crustacean had paired segmental nephridia, as the annelids do.

Nitrogen excretion is accomplished primarily by diffusion of ammonia across any permeable body surface, usually the respiratory surface.

Most crustaceans inhabit marine habitats and are isosmotic with their environment. They are osmoconformers that do not face serious osmoregulatory problems. Freshwater crustaceans, on the other hand, are of necessity osmoregulators, and in the crayfishes at least, the saccate nephridia participate in regulation. Crayfish nephridia produce a urine hyposmotic to the blood to compensate for the influx of fresh water across the gill surfaces. In freshwater and brackish-water forms, the gills actively absorb ions from the environment.

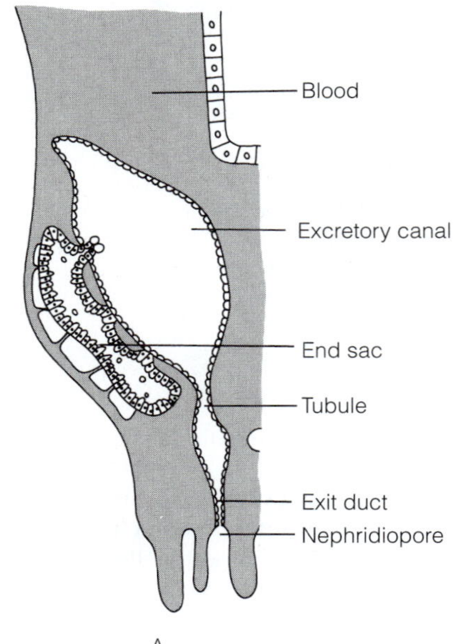

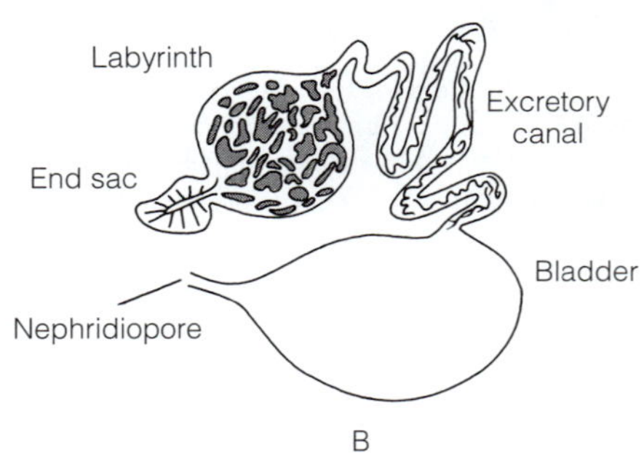

FIGURE 19-6 Crustacea: Saccate nephridia. **A,** Section through the saccate nephridium of a barnacle. **B,** The saccate nephridium of a crayfish. The labyrinth, excretory canal, and bladder are derived from the metanephridial tubule. The nephridiopore opens onto the base of the second antenna. Ultrafiltration of the blood occurs across the wall of the end sac, ion transport and protein resorption occur in the labyrinth, and storage of urine occurs in the bladder. *(A, From White, K. N., and Walker, G. 1981. The barnacle excretory organ. J. Mar. Biol. Assoc. U.K. 61:529–547. Copyrighted and reprinted by permission of Cambridge University Press.)*

Crustaceans, like most other arthropods, have nephrocytes. These phagocytic or pinocytic cells accumulate waste particles. Nephrocytes are usually located in the hemocoelic spaces of the gill axes and in the bases of the legs.

NERVOUS SYSTEM AND SENSE ORGANS

The nervous system is typical of arthropods, and the brain consists of a protocerebrum, deutocerebrum, and tritocerebrum. There is a tendency toward cephalization and fusion of ganglia. All stages in the cephalization of ganglia are represented within Crustacea. In some, all the segmental ganglia are independent and there is no subesophageal ganglion. At the other extreme, all segmental ganglia are coalesced to form a large anterior subesophageal ganglion.

The visceral nervous system is connected to the somatic CNS anteriorly at the tritocerebrum and posteriorly at the posteriormost ganglion of the ventral nerve cord. This system innervates the gut, heart, digestive ceca, and other viscera.

Giant axons, which rapidly conduct impulses, are found in the CNS of many crustaceans. Giant axons are particularly well developed in large species such as shrimps and crayfishes, which dart rapidly backward by suddenly flexing their tail.

Crustacean sense organs include eyes, statocysts, sensory setae, and proprioceptors. Most crustaceans have eyes, which may be median ocelli and/or lateral compound eyes. A cluster of median eyes that is a characteristic feature of the crustacean nauplius larva (Fig. 19-8) is referred to as a **naupliar eye.** It may degenerate or persist in the adult. Each naupliar eye is composed of three or four small pigment-cup ocelli, each containing a few photoreceptors. The cups are located directly over the protocerebrum, to which they connect by an optic nerve. One pigment cup is directed ventrally and two dorsolaterally. They may form a compact mass or be separated somewhat. Naupliar eyes, which do not form images, determine the direction to a light source, orienting the animal with respect to the water surface (source of light and "up") or the sediment surface.

Adults of most species have two lateral compound eyes. The eyes may be at the end of a movable eyestalk (Fig. 19-1) or sessile and flush with the surface of the head (Fig. 19-54). The corneal surface is often greatly convex, resulting in a wide visual field. This is particularly true for stalked eyes, in which the cornea may cover an arc of 180° or more. The number of ommatidia in crustacean compound eyes varies enormously. A single eye of the wood louse, *Armadillidium,* is composed of fewer than 25 ommatidia whereas that of the lobster, *Homarus,* may have 14,000. Crustaceans with well-developed compound eyes, such as some shrimps, lobsters, and crabs, show some ability to discriminate form and size. Most higher crustacean taxa, with the exceptions of cephalocarids, mystacocarids, tantulocarids, and copepods, have compound eyes. Cephalocarids and mystacocarids do not have median eyes, either.

Color discrimination has been demonstrated in a number of crustaceans and may be widespread. For example, the hermit crab, *Pagurus,* can discriminate between snail shells painted yellow and blue and shells colored different shades of gray. The chromatophores of the shrimp *Crangon* adapt to a background of yellow, orange, or red, but not to any shade of gray. Changes in chromatophore color are mediated through the compound eyes.

Various types of sensory setae are located over the body surface, especially the appendages. Among these are the chemoreceptive **aesthetascs,** which are found in the majority of crustaceans, usually on the first antennae (Fig. 19-7A). Aesthetascs are patches of tubular setae with a thin, permeable cuticle. In terrestrial crustaceans, the aesthetascs are tiny plates instead of hairs.

Statocysts are paired equilibrium receptors located in the antennae, abdomen, uropods, or telson of some malacostracans. They are analogous to the labyrinth of the vertebrate inner ear and, like it, are used to orient in a gravitational field. A statocyst is a cavity in which a heavy particle known as a

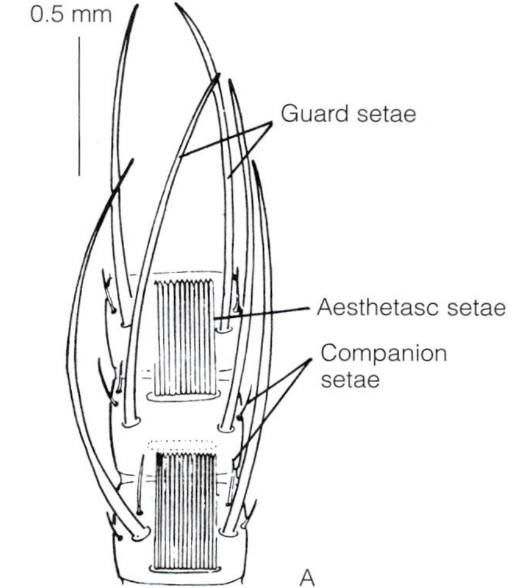

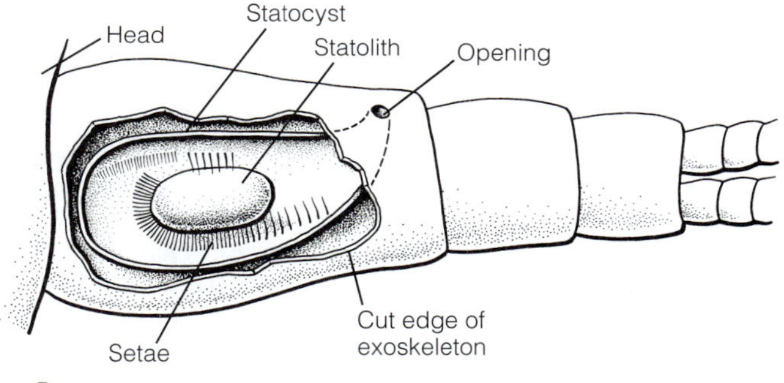

FIGURE 19-7 Crustacea: Sense organs of the first antenna. **A,** Aesthetascs on the first antenna of the spiny lobster *Panulirus.* Each antennal article has two rows of chemoreceptive setae but, for clarity, only one is shown. **B,** The first antenna of the American lobster, *Homarus americanus,* dissected to reveal the statocyst in its basal article. Anterior is to the right. *(A, Laverack, M. S. 1968. Oceanogr. Mar. Biol. Ann. Rev. 6:249–324; B, Redrawn and modified from Lockwood, A. P. M. 1968. Aspects of the physiology of Crustacea. Oliver and Boyd, London. 328 pp., after Cohen, M. J. 1955. The function of receptors in the statocyst of the lobster,* Homarus americanus. *J. Physiol. 130:9–34.)*

statolith rests on a bed of mechanoreceptive setae (Fig. 19-7B) that detect its displacement. The statolith may be secreted, but more often it is a mass of agglutinated sand grains.

REPRODUCTION

Most crustaceans are gonochoric but a few, such as barnacles and remipedes, are hermaphroditic. The gonads typically are paired elongate, tubular organs lying dorsolaterally in the hemocoel of the trunk (Fig. 19-2B, 19-3A). The oviducts and sperm ducts usually are simple, paired tubules that open either at the base of a pair of trunk appendages or on a sternite. The gonopores are located on different segments in different taxa.

Fertilization is usually internal and copulation occurs in most species. Spermatophores are used by many crustaceans and fertilization is often indirect. A penis is frequently present, but is not necessarily the intromittent organ. In the ostracods, fairy and brine shrimps, and barnacles, a penis is used for direct sperm transfer. More commonly, however, a pair of modified appendages, the gonopods, is used to insert the sperm into the female. The male may also have certain appendages, such as the antennae or anterior thoracic appendages, modified for clasping the female during copulation and sometimes for an extended period prior to copulation. In some species the male must wait until the female molts before copulation can occur, and he will often hold her until this happens. In most crustaceans the sperm lacks a flagellum and is nonmotile, but flagellated sperm occur in Phyllopoda, Cirripedia (barnacles), Mystacocarida, and Ostracoda.

A seminal receptacle is sometimes present in the female. It may be located near the base of the oviduct, but frequently it is a pouchlike ectodermal invagination of the genital segment independent of the oviduct.

DEVELOPMENT

The eggs of higher crustaceans are usually centrolecithal and cleavage is superficial. In more primitive taxa, the eggs are small, and holoblastic cleavage is common. In fact, some members of every major taxon exhibit holoblastic cleavage. In some, it is mosaic and may show traces of a spiral pattern (barnacles, copepods, cladocerans) like that of annelids. Most crustaceans brood their eggs and release the young as larvae or juveniles. For brooding, the eggs may be attached to modified appendages, contained within a brood chamber, or retained within a sac secreted when the eggs are extruded.

Crustacean development may be either indirect (anamorphic) or direct (epimorphic). In indirect development a larva hatches from the egg with its embryonic development still incomplete and only a portion of the adult complement of segments and appendages. The larva, which is usually planktonic, undergoes periodic molts, gradually adding additional segments and appendages, until finally the adult condition is achieved. This is the original crustacean condition.

In direct development, there are no larval stages. The eggs hatch as immatures having the adult complement of segments and appendages. The juveniles need only to increase in size and develop gonads to become adults. Direct development is probably a secondary trait in crustaceans.

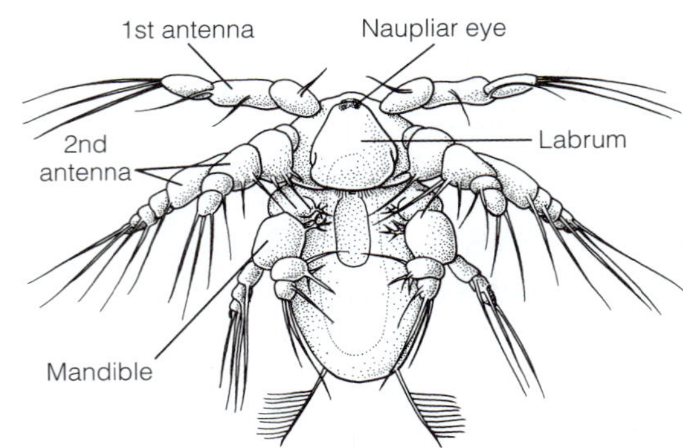

FIGURE 19-8 Crustacea. Ventral view of a nauplius larva.

Associated with large eggs and brooding, it occurs in cladocerans and many peracarids, among others.

In indirect development the individual undergoes development and then leaves the egg at a **hatching stage** typical for the taxon. The hatching stage is characterized by a specific number of segments and appendages and referred to by a specific larval name, such as nauplius or zoea. After hatching, larvae undergo postembryonic development, in which new segments proliferate at the anterior margin of the telson. In the course of successive molts, additional segments and appendages are gradually acquired until the adult condition is achieved. Anterior segments and their appendages are added first.

Crustacean taxa have many larval stages, which tend to have different names in each taxon. The basic crustacean larva is the **nauplius** (Fig. 19-8, 19-48A), which is the earliest developmental stage capable of hatching. This larva consists of the first three head segments and their appendages, the first and second antennae and mandibles, which it uses for swimming. No trunk segments or appendages are present, and an unpaired, median naupliar eye is borne on the head. It is common for the nauplius to be suppressed by the embryo so that the hatching stage is a more advanced larva with additional segments and appendages.

Once the larva hatches, it continues developing, molting, growing, and adding segments and appendages. The process gradually transforms it into other larva types and eventually into an adult. The number of named larval stages and the number of instars in each stage vary. A few of the more important larvae are mentioned below. Each stage may include several instars.

The **metanauplius** stage (Fig. 19-48B) follows the nauplius in many crustacean taxa. Like nauplii, metanauplii have only three pairs of functional (natatory) appendages, but they also have additional segments and nonfunctional appendages. In barnacles, the hatching stage is the nauplius, which is followed by the **cypris** (Fig. 19-79) and then the adult. In copepods, a stage known as the **copepodid** (Fig. 19-72C,D) occurs between the last metanaupliar stage and the adult. Whereas the nauplius does not resemble the adult, the copepodid does.

More complex sequences are characteristic of decapod crustaceans. In primitive decapods (such as penaeid shrimps) the egg hatches as a nauplius (Fig. 19-48A) or metanauplius (Fig. 19-48B). The addition of segments and appendages eventually produces a **protozoea** (Fig. 19-48C) with a complete set of head appendages and the first two maxillipeds. The protozoea swims with its antennae and mandibles, but also uses its new maxillipeds. Upon acquiring its functional thoracic appendages, which it uses to swim, the protozoea becomes a **zoea** (or mysis; Fig. 19-48D). With the appearance of the complete complement of segments and appendages, including the abdomen and pleopods, the developing shrimp becomes a **postlarva** (Fig. 19-48E) that still is small, does not closely resemble the adult, and is not sexually mature, but this will be remedied by additional molts.

In more-derived decapods (some shrimps, most crabs) the nauplius and metanauplius are passed in the egg and the hatching stage is a zoea with a full complement of head appendages plus thoracic segments and natatory maxillipeds (Fig. 19-49A). Subsequent molts add more thoracic segments and appendages until the full set of eight is achieved. After that the animal becomes a postlarva, such as the crab **megalops** larva in Figure 19-49B, with a complete set of head, thoracic, and abdominal appendages and segments. The postlarva becomes larger and more like the adult with subsequent molts. There are many, many other crustacean larvae, such as those in Figure 19-49C,D.

REMIPEDIA[C]

The first known remipede, *Speleonectes lucayensis,* was described in 1981 and the taxon is now represented by 10 species in six genera, all from undersea caves primarily in the Bahamas, West Indies, and Yucatán Peninsula. So far, all representatives have been found in water-filled caves connecting with the sea at one end and with freshwater ponds at the other. The body of these 15- to 45-mm-long marine crustaceans is elongate, resembling that of a polychaete, with numerous biramous appendages (Fig. 19-9). Powerful rowing strokes of the trunk appendages produce rhythmic preferred by MW waves that sweep along the length of the body for swimming. Remipedes are carnivores or scavengers. The body includes a trunk and a cephalothorax, the latter of which is covered by a sclerotized carapace (cephalic shield). Eyes do not seem to be present, although sometimes there are vestigial eyelike structures. Suppression or loss of eyes is a common adaptation in cave-dwelling animals. The large first antennae are biramous (Fig. 19-9B) and bear chemosensory aesthetascs. The much smaller second antennae are paddlelike.

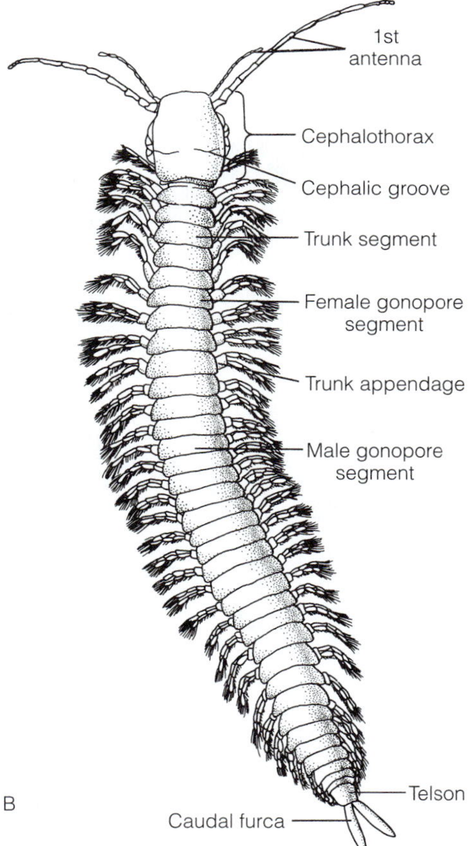

FIGURE 19-9 Remipedia. **A,** *Lasionectes entrichomas* from undersea caves in the Bahamas and West Indies. Note the long, wormlike body with numerous pairs of similar appendages. **B,** Dorsal view of the body of *Speleonectes tulumensis.* *(B, From Felgenhauer, Abele, and Felder, 1992.)*

The trunk consists of up to 38 independent segments and is not divided into thorax and abdomen. The first trunk segment is smaller than the others, bears a pair of uniramous maxillipeds, and is fused with the head to form the cephalothorax. The maxillipeds resemble the second maxillae. The remaining trunk segments are homonomous, and each bears a pair of biramous swimming appendages that are similar along the entire length of the trunk. The two rami of each appendage are nearly identical and consist of three or four flattened articles equipped with a fringe of plumose swimming setae. There is no gnathobase on the protopod. The telson bears a short caudal furca.

The simple gut consists of a cuticularized foregut, a midgut with paired digestive ceca, and a cuticularized hindgut. The midgut extends for most of the length of the trunk. The preoral cavity, or atrium oris, as it is known in crustaceans, is bounded anteriorly by the labrum and laterally by the mouthparts. Excretory organs consist of a pair of maxillary glands. Remipedes are hermaphroditic, with the female gonopores on trunk segment 7 and the male pores on 14. The sperm are flagellated and transferred in spermatophores. It has so far been impossible to keep remipedes alive in the laboratory, and consequently nothing is known of their development or larvae.

The first crustaceans probably were small marine animals that swam or crawled over the bottom. The trunk was composed of numerous unfused and unspecialized segments, each bearing a pair of biramous appendages similar to those on the other segments. Among extant Crustacea, only remipedes and cephalocarids share these primitive characteristics. Although remipedes have plesiomorphies, they also have important apomorphies such as hermaphroditism and carnivory. 18S rDNA base sequence data indicate that remipedes are a derived taxon related to Maxillopoda. At present the phylogenetic position of Remipedia is unresolved, but here they are viewed as an early offshoot of the ancestral crustaceans and a sister taxon of the remaining crustaceans (Fig. 19-90).

CEPHALOCARIDA[C]

Cephalocarids conform closely to the hypothesized ancestral crustacean in that they are epibenthic deposit feeders with undifferentiated trunk appendages. The nine known species have been collected in sand and mud from the high subtidal

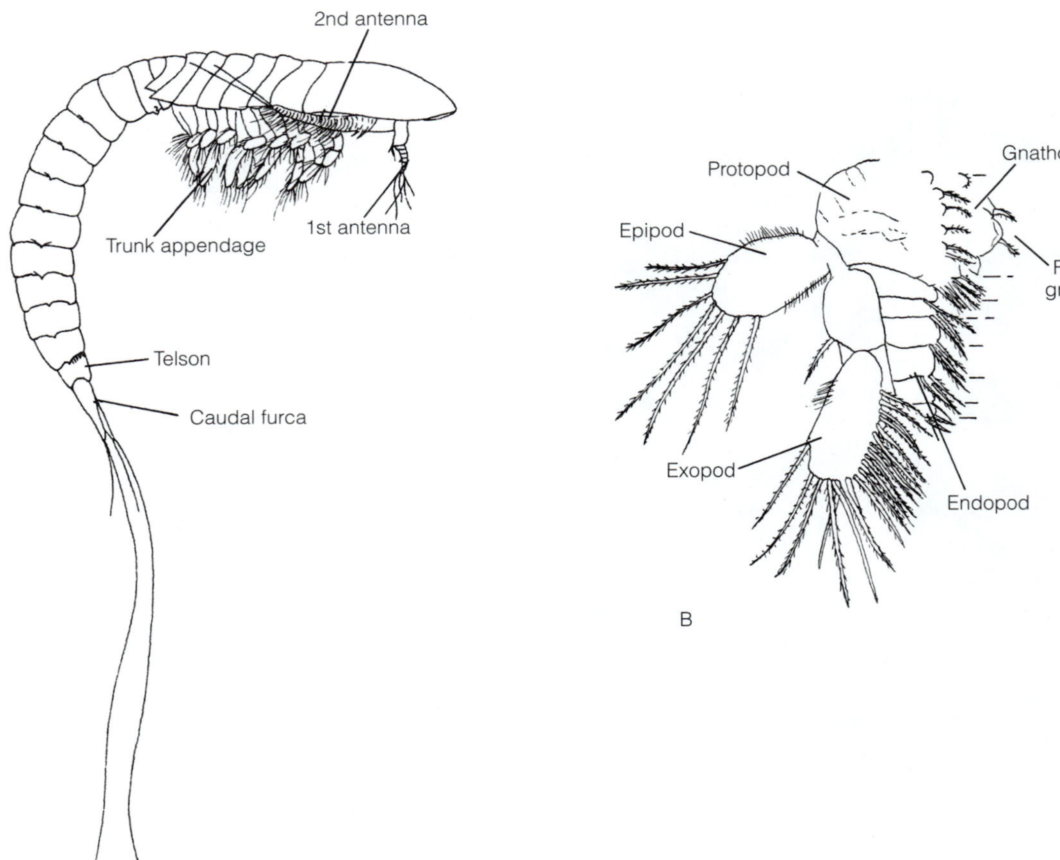

FIGURE 19-10 Cephalocarida. **A,** *Hutchinsoniella macracantha.* **B,** A cephalocarid trunk appendage. The endopod is a stenopod and the exopod is a phyllopod. The protopod bears a gnathobase on its medial margin. *(A, After Waterman and Chace; B, After Sanders.)*

to depths of 1550 m in many parts of the world. The first was found in Long Island Sound in 1953. Cephalocarids are tiny crustaceans of less than 4 mm in length (Fig. 19-10A).

The elongate, wormlike body is divided into a horseshoe-shaped head and an elongate trunk of 19 segments plus the telson and caudal furca. The head is covered by a sclerotized cephalic shield. There are no cephalothorax, maxillipeds, or carapace. The trunk is divided into a thorax of 9 segments and an abdomen of 10. Each thoracic segment bears a pair of appendages and is covered by a sclerotized tergite that extends laterally and ventrally to form lateral plates along the sides of the thorax (Fig. 19-10A). Thoracic appendages are biramous (multiramous) mixopods with a stenopodous endopod and a phyllopodous exopod (Fig. 19-10B). A large epipod on the exopod forms what amounts to a third ramus. The endopod is used for walking and the exopod for creating feeding and swimming currents. The first seven limbs are identical to each other and to the second maxilla. Each protopod bears a median gnathobase, and the combined gnathobases form a midventral food groove. The eighth limb lacks the endopod and is used for manipulation of the egg, whereas the ninth is vestigial and serves as a site for attachment of the egg during brooding. Abdominal segments do not bear appendages. Posteriorly the telson has a caudal furca with long furcal setae (Fig. 19-10A).

A large labrum closes the preoral cavity and food groove anteriorly. The gut has the J shape predicted for the ancestral crustacean, with its mouth facing posteriorly into the anterior end of the food groove. The esophagus forms the terminal curve of the J and connects the mouth with the large midgut, which extends the length of the trunk to connect with a short, cuticularized hindgut. Two small digestive ceca extend from the anterior end of the midgut. The anus opens on the telson between the rami of the caudal furca. Adult excretory organs are maxillary glands, but juveniles have antennal glands.

Other than the large syncerebral brain, the nervous system is not cephalized. Paired segmental ganglia are spaced along the length of the ventral nerve cord, which extends the length of the trunk. Vestigial compound eyes are present but buried in the head, and there are chemoreceptive aesthetascs on the antennae. Cephalocarids are simultaneous hermaphrodites. The ovary and testis share a common duct to the gonopore on the sixth abdominal segment. Sperm are aflagellate and the hatching stage is the metanauplius.

Cephalocarids and remipedes exhibit combinations of primitive and derived characteristics. For example, both are blind and hermaphroditic, characteristics that are probably derived. At the same time, both have weak tagmosis and unspecialized trunk appendages (primitive), but those characteristics differ in degree. Cephalocarids have a differentiated thorax and abdomen (derived) but virtually unspecialized trunk appendages that are identical to the second maxilla (primitive). Remipedes, on the other hand, have an undifferentiated trunk (primitive) but a specialized (derived) first trunk appendage. The triramous mixopods and the feeding mode of cephalocarids are exactly as predicted for the stem crustacean, whereas the remipede appendages and feeding mode are derived. The loss of eyes in the remipedes could easily be an adaptation to a subterranean existence. Vestigial compound eyes have recently been discovered in cephalocarids.

Two families have been erected to contain the four genera. *Chiltoniella*, *Hutchinsoniella*, and *Sandersiella* are in Hutchinsoniellidae, and *Lightiella* is placed in Lightiellidae.

ANOSTRACA[C]

This relatively small taxon inhabits inland, but not necessarily fresh, waters. Anostraca includes the fairy and brine shrimps, both of which are confined to relictual habitats from which fishes are excluded either because the habitat dries up periodically or because the water is too saline. Brine shrimps, whose eggs are creatively marketed with the claim they will hatch into "sea monkeys," are familiar to many people.

The 200 known species occur in salt lakes and temporary wet-weather ponds and ditches. They are relatively large, with most species being 15 to 30 mm in length, although some are much larger (up to 10 cm). All have elongate, shrimplike bodies with up to 27 trunk segments (Fig. 19-11A). The name Anostraca refers to the absence of a carapace. The first antennae are very small and the second antennae are sexually dimorphic, with those of the male being large claspers used to hold the female during copulation. The compound eyes are stalked. The naupliar eye consists of three pigment-cup ocelli. The mandibles are large and triturative and the second maxillae are vestigial.

The trunk is divided into an anterior thorax and a posterior abdomen. Cephalothorax, maxillipeds, and carapace are absent. The thoracopods are homonomous phyllopods and the abdomen lacks appendages. A typical phyllopod consists of a large central protopod. The protopod has exites and an exopod laterally and endites and an endopod medially (Fig. 19-11B). The middle exite, or epipod, is referred to as a gill, but it probably has an osmoregulatory function. The proximal endite bears a setal comb important in suspension feeding. The endopod and exopod are broad, flat, and leaflike; bear plumose setae; and are natatory.

The thoracopods beat rhythmically to produce a continuous propulsive force on the backstroke. Each broad, flat, foliose phyllopod, with its fringe of plumose setae, is an oar with a large surface area (Fig. 19-11B). Fairy shrimp usually swim upside down, but when experimentally lighted from below, they roll over and swim right side up.

Although a few are carnivores, most anostracans are suspension feeders that use the setose phyllopods and a food groove to collect and transport suspended food particles to the mouth (Fig. 19-12). Swimming movements of the phyllopods generate the feeding current, and the shrimp feeds as it swims. The midventral food groove is formed by the protopods. The spaces between successive protopods are called **interlimb spaces.** The medial entrances to the interlimb spaces are guarded by the setal combs of the proximal endites. On the forward stroke, as the limbs move forward, the volume of the interlimb spaces increases and creates suction that pulls water and particles into the spaces from along the midventral line. The water moves laterally, through the setal combs, to enter the interlimb spaces. Particles in the current remain on the medial side of the setal comb. On the backstroke, as the

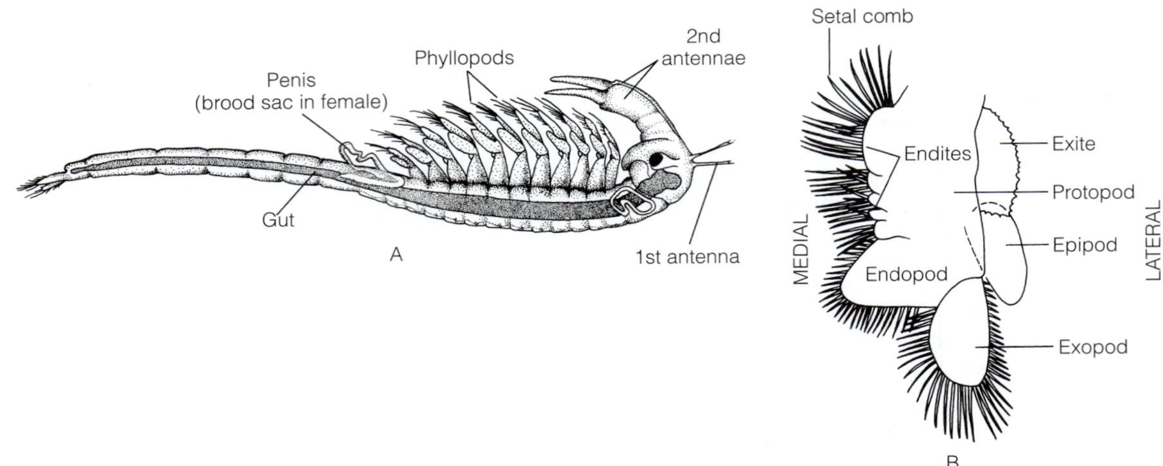

FIGURE 19-11 Anostraca. **A,** The fairy shrimp, *Branchinecta* (lateral view of male). **B,** A phyllopodous trunk appendage of *Bran-chinecta paludosa*. *(A, After Sars from Martin, J. W. 1992. Branchiopoda. In Harrison, F. W., and Humes, A. G. (Eds.): Microscopic Anatomy of Invertebrates. Vol. 9. Crustacea. Wiley-Liss, New York; B, After Sars from Pennak.)*

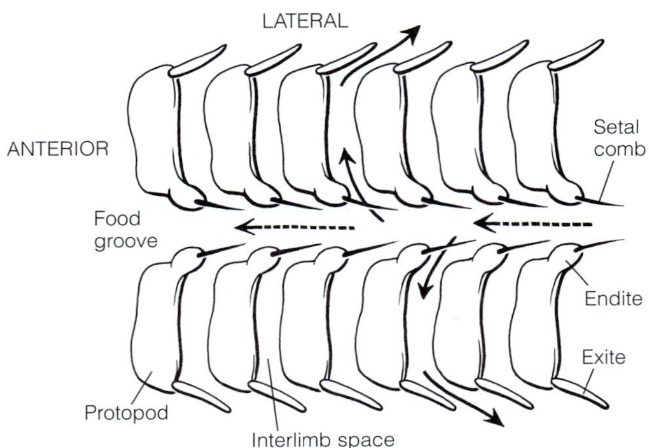

FIGURE 19-12 Anostracan feeding apparatus. Ventral view of the food groove and phyllopods of a typical suspension feeding anostracan or phyllopodan. Water flow is indicated by solid arrows and particle movement by dotted arrows.

limbs move posteriorly, the water is forced laterally out of the now compressed interlimb spaces. Food particles on the setal combs are transferred to the midventral food groove, where they become entangled in mucus and are moved forward, again by the swimming movements of the limbs, to enter the posteriorly facing mouth.

The gut is an unelaborated J-shaped tube with a pair of short digestive ceca in the head. The heart is a dorsal tube with 13 to 18 pairs of ostia that extends the length of the trunk. The nervous system is ladderlike, with minimal cephalization and without medial fusion. There are paired maxillary glands.

The brine shrimp, *Artemia,* has impressive osmoregulatory capabilities and can tolerate salinities ranging from less than 1% up to the saturation point of sodium chloride. The internal osmolarity remains almost constant regardless of the salinity of the environment. Internal osmolarity is maintained by the absorption or excretion of ions through the epipods and the dorsal organ, and perhaps by the excretion of urine hyperosmotic to the blood. For individuals living in brine, urine osmolarity is reportedly four times that of the blood.

Anostracans are gonochoric and the gonopores are located on genital segments at the junction of the thorax and abdomen. The sperm ducts from the testes open on a pair of eversible penes (sing., penis) on the genital segment. Sperm transfer is direct, and during copulation, the male clasps the dorsal side of the female abdomen with his modified second antennae and then, twisting his abdomen around, he inserts the paired penes into the single, median, female gonopore. A special brood sac is secreted as the female extrudes eggs from the glandular uterine chamber at the distal end of the oviduct. The eggs are tolerant of desiccation and hatch as nauplii.

In older classifications Anostraca is included with Conchostraca, Notostraca, and Cladocera in a taxon known as Branchiopoda. In a recent reorganization, the one followed here, Anostraca is removed from this assemblage and the remaining taxa form Phyllopoda (Fig. 19-90). Anostracans are similar to phyllopodans, but differ in having stalked compound eyes and in lacking a carapace. The anostracan naupliar eye has three ocelli whereas that of Phyllopoda has four.

The eight families are Artemiidae *(Artemia),* Branchinectidae *(Branchinecta),* Branchipodidae, Chirocephalidae *(Chirocephalus, Eubranchipus),* Linderiellidae *(Linderiella),* Polyartemiidae *(Polyartemiella),* Streptocephalidae *(Streptocephalus),* and Thamnocephalidae *(Branchinella, Thamnocephalus).*

PHYLLOPODA[C]

Phyllopoda includes about 800 species of mostly freshwater crustaceans. It is convenient to distinguish, for purposes of discussion, between the "large" and "small" phyllopodans. Large phyllopodans are the tadpole shrimps (Notostraca) and clam shrimps (Spinicaudata and Laevicaudata). Like anostracans, large species are restricted to relictual habitats where they are free of fish predation. Such habitats are typically ephemeral (temporary) ponds, ditches, cave pools, and alkaline lakes from which fishes are excluded, either by the temporary nature of the habitat or its chemistry. The life cycle may be adapted to alternating periods of desiccation and flooding, with dry periods being passed in the egg stage. The most successful, however, are the small phyllopodans, the water fleas, or Cladocera. Unlike large phyllopodans, they are not restricted to relictual habitats and are ecologically important members of benthic and planktonic communities of lakes.

The trunk appendages are typically phyllopods, there is a well-developed carapace (Fig. 19-13, 19-14, 19-15), and most have a ventral food groove. A nauplius larva is present in the large phyllopodans, but development is direct in almost all Cladocera. Most, but by no means all, are freshwater suspension feeders.

The body is divided into a head, thorax, and reduced abdomen. The first antennae and second maxillae are vestigial. The number of segments in the thorax and abdomen varies with taxon. Phyllopods are restricted to the thorax, usually bear marginal setae, and are used for suspension feeding, swimming, and gas exchange. Appendages are absent from the abdomen. A caudal furca, or a derivative of it, is found at the posterior end of the abdomen.

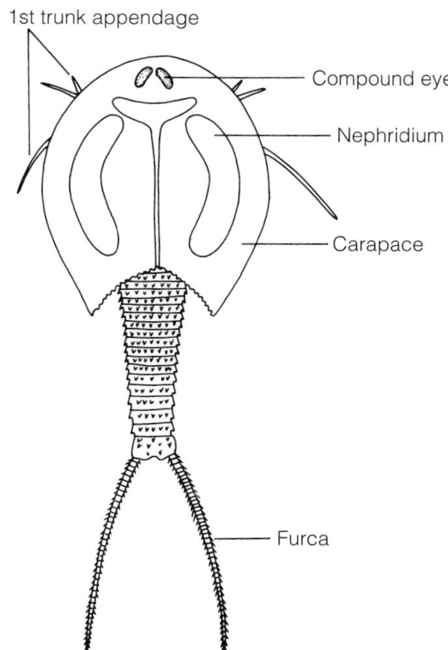

FIGURE 19-13 Notostraca. The tadpole shrimp *Triops* (dorsal view). *(After Pennak, 1989.)*

Phyllopodans have an epidermal **dorsal organ** (salt gland, neck gland, or nuchal organ) on the dorsal midline of the posterior head. Although this structure is not always present in adults, it is present at some stage in the life cycle. In some species it is a salt gland with a demonstrated role in ion regulation and osmoregulation, but in others its function is enigmatic. In some adult ctenopods it is an adhesive organ.

In the current taxonomic organization of the old Branchiopoda, the original four orders, Anostraca, Notostraca, Conchostraca, and Cladocera, are replaced by eight, and Anostraca is removed entirely. The resulting taxon, known as Phyllopoda, more accurately reflects the natural relationships of these animals. Notostraca is unaffected by the revision, but Conchostraca is divided into two new taxa (orders), Laevicaudata and Spinicaudata, probably not sister taxa. Cladocera is replaced by four new related taxa (orders), Anomopoda, Ctenopoda, Onychopoda, and Haplopoda, all belonging to Cladocera (Cladoceromorpha).

LOCOMOTION

Locomotory mechanisms vary with taxon. Cladocerans swim using the powerful second antennae as oars (Fig. 19-15). Because there is only one set of oars, movement is jerky. The animal hangs vertically in the water with its head up so the downstroke of the antennae propels it upward. The animal then sinks slowly, arresting its descent with the deployed antennae functioning as a parachute. The large, plumose, postabdominal setae at the end of the abdomen are stabilizers and if they are removed, the water flea pitches ventral side up.

In large phyllopodans (and anostracans), the trunk appendages are natatory and used for swimming, although conchostracans also use the second antennae. The trunk appendages beat rhythmically to produce a steady propulsive force on the backstroke. The broad, flat, foliose phyllopods, with their fringes of plumose setae, are oars with a large surface area (Fig. 19-14A). Since there are many oars, the motion is smooth.

Many clam shrimp and tadpole shrimp crawl over the bottom, dorsal side up, and some clam shrimp plow through the bottom sediment. Many cladocerans have adopted a benthic or near-bottom existence.

NUTRITION

Feeding is varied among phyllopodans. Notostracans are scavengers and predators. Spinicaudatans are detritivores, scrapers, and suspension feeders. Laevicaudatans are scrapers and detritivores. Among Cladocera, the anomopods and ctenopods are suspension feeders, scrapers, and scavengers. The onychopods and haplopods are predators. In all except the haplopods, the mandible is adapted for trituration.

The anomopods and ctenopods, which have in common an enclosing bivalve carapace, feed on suspended particles of a wide range of sizes, from bacteria to phytoplankters and detritus. *Daphnia,* for example, can grow on a diet consisting solely of bacteria. The feeding mechanism is similar to that

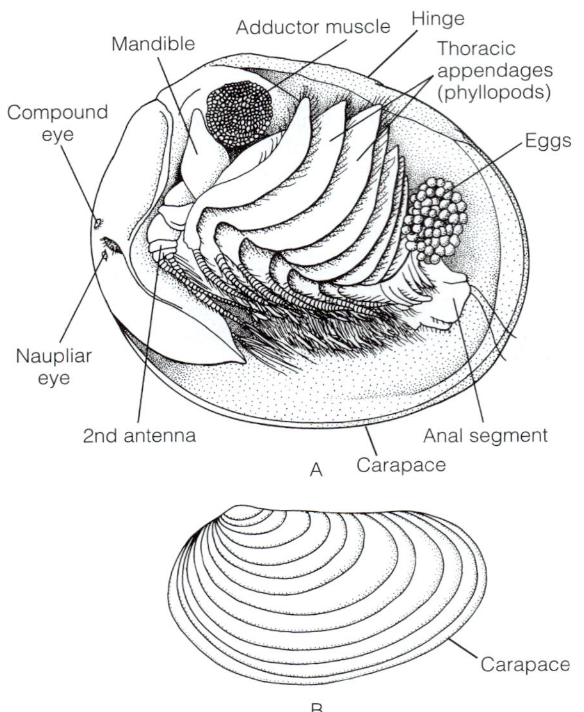

FIGURE 19-14 "Conchostraca." **A,** Lateral view of the laevicaudatan *Lynceus gracilicornis,* with the left valve removed. **B,** Left shell valve of the spinicaudatan *Cyzicus.* *(A, From Martin, J. W., Felgenhauer, B. E., and Abele, L. G. 1986. Redescription of the clam shrimp* Lynceus gracilicornis *from Florida, with notes on its biology. Zool. Scripta. 15(3):221–232; B, After Sars from Calman.)*

used by anostracans except that the number of phyllopods and interlimb spaces is much smaller, the water is channeled in part by the carapace, and the phyllopods are not natatory, so their movements are solely for the purpose of feeding. There is a maximum of six phyllopods, not all dedicated to feeding. In *Daphnia,* for example, there are five trunk appendages, but the first does not participate in feeding. Thus there are only three interlimb spaces.

The phyllopodan gut is J-shaped, with the mouth facing posteriorly into the anterior end of the food groove. The gut is simple and has none of the elaborate modifications of the foregut typical of the malacostracans. The anterior midgut frequently has two small, globose digestive ceca.

GAS EXCHANGE, INTERNAL TRANSPORT, AND EXCRETION

In most phyllopodans the chief gas-exchange organ is probably the inner surface of the carapace, and perhaps the phyllopods as well, but it can be any surface with a thin, permeable cuticle. It was long believed that the epipods of the thoracic appendages (Fig. 19-11B) were the major respiratory organs, a belief reflected in the old name, "Branchiopoda", meaning foot-gill. It is likely, however, that these widely occurring structures are active in ion transport and are osmoregulatory, not respiratory.

The hemal system, as expected, consists of a dorsal heart and a hemocoel. In the large phyllopodans, with their well-developed abdomens, the heart is a long tube with multiple paired ostia. The length of the heart and the number of ostia correlate positively with body length. For example, notostracans have 11 pairs of ostia, spinicaudatans have 4, and laevicaudatans 3. The heart of the short-bodied cladocerans is a small, globular sac with only one pair of ostia (Fig. 19-15A).

Excretion of ammonia occurs across the general body surface and presumably is most intense across the highly permeable respiratory surfaces. The excretory organs of adult phyllopodans are saccate nephridia known as maxillary glands (Fig. 19-15A), which are involved primarily in osmoregulation and fluid balance. Larval nephridia are antennal glands. Nephridia are well-developed in freshwater species and are equipped with a long excretory canal, in which ions are selectively resorbed from the filtrate. In most phyllopodans the duct lies coiled between the folds of the carapace and may be visible externally (Fig. 19-13). The nephridium is often called the shell gland in reference to its position in the carapace, or "shell." The epipods of the thoracopods are probably osmoregulatory also.

The CNS includes a brain (supraesophageal ganglion), circumenteric connectives, and a ventral, ladderlike nerve cord. The system is uncephalized and not fused on the midline so it is truly ladderlike, with a median space between the right and left longitudinal cords. No subesophageal ganglion is present (except in *Leptodora*) and each pair of appendages is served by a pair of ganglia. The ventral nerve cord is unique in that the two ganglia of each pair are connected to each other by *two* commissures, rather than the typical one.

A naupliar eye is present in most phyllopodans (Fig. 19-14A, 19-15A) and consists of four pigment-cup ocelli. Paired lateral compound eyes are found in the large phyllopodans, but in the cladocerans they are fused into a single compound eye located on the midline (Fig. 19-15A). The sessile compound eyes are enclosed in an epidermal vesicle below the surface of the carapace. Cladoceran eyes are unusual in having extrinsic muscles that permit their rotation. The compound eye is used by many phyllopodans for orientation while swimming.

REPRODUCTION AND DEVELOPMENT

Reproduction is varied among phyllopodans, and there are many examples of gonochorism, hermaphroditism, bisexual reproduction, and parthenogenesis. Parthenogenesis is common, and in some species males are scarce or unknown. Phyllopodans brood their eggs for various lengths of time and in various types of brood pouches, depending on the species. Cladocerans (Fig. 19-15, 19-16) and conchostracans (Fig. 19-14A) brood their eggs dorsally beneath the carapace. The brood chamber may be under the dorsal carapace (Cladocera) or associated with the phyllopods (Notostraca). The eggs hatch as nauplii (large phyllopodans) or miniature versions of the adults (most cladocerans).

The gonads usually are paired tubes lying beside the gut, although sometimes they are single. The oviducts open into the brood pouch. Sperm are ameboid and aflagellate and have no acrosome, but do have a nuclear envelope. This combination of characters is unique among crustaceans. In clado-

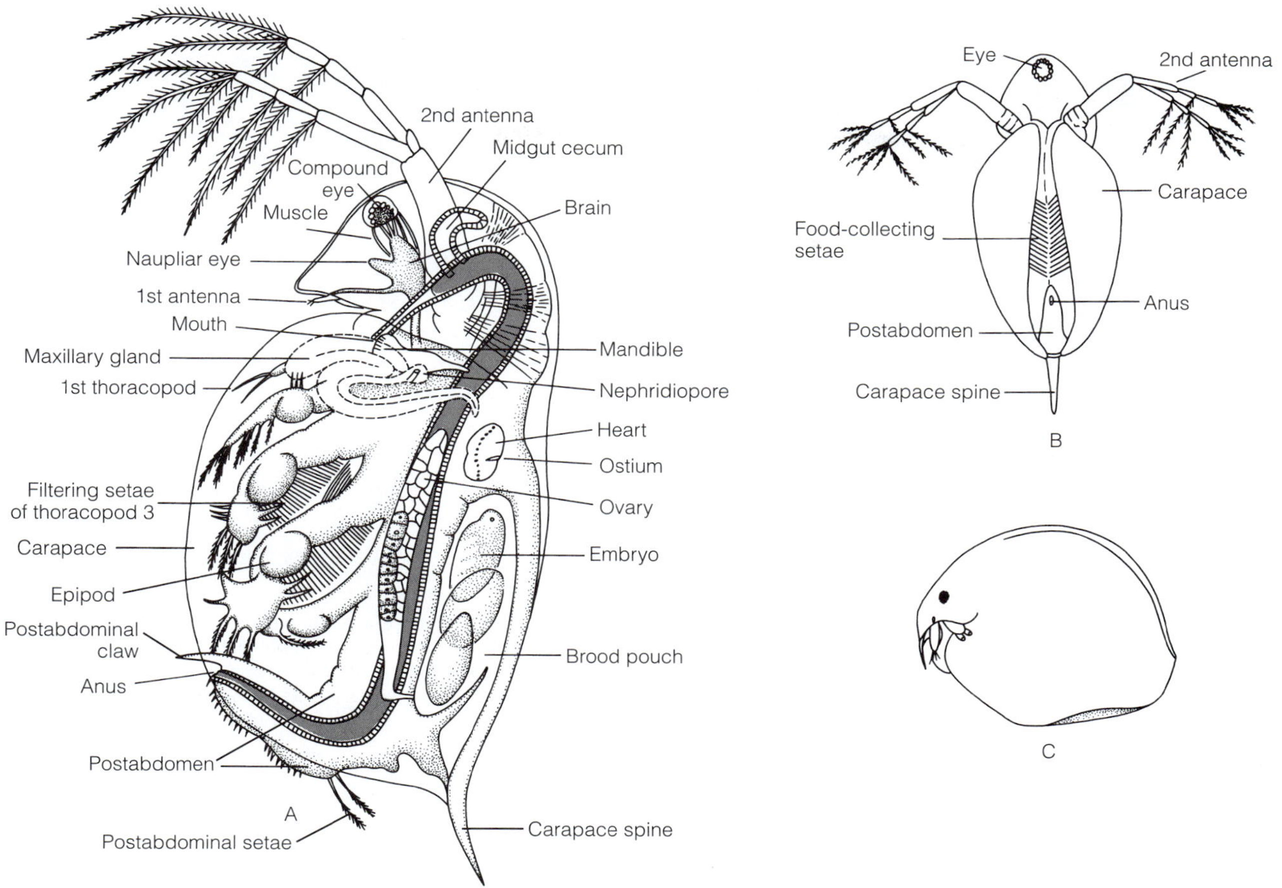

FIGURE 19-15 Cladocera. **A,** Female of the anomopod *Daphnia pulex* (lateral view). **B,** Ventral view of *Daphnia.* **C,** *Chydorus gibbus,* an anomopod with a more rounded body than *Daphnia pulex.* Appendages are not shown. *(A, After Matthes from Kaestner; C, From Pennak, R. W. 1989. Fresh-Water Invertebrates of the United States, Protozoa to Mollusca. 3rd edition. Wiley-Interscience, New York. 628pp.)*

cerans the sperm ducts from the testes open near the anus on the postabdomen, the latter being modified to function as a copulatory organ. In other phyllopodan species, the gonopores open ventrally on various segments.

Phyllopodans produce summer eggs and resting eggs. Both may be either parthenogenetic or fertilized. Some species reproduce solely by resting eggs, whereas others produce such eggs only when environmental conditions are deteriorating, and still others produce no resting eggs.

The thin-shelled **summer eggs,** also known as subitaneous eggs, are produced under benign environmental conditions. Development in summer eggs is rapid, hatching may occur in the brood pouch, and juveniles may briefly remain in the pouch. Summer eggs are produced in clutches of two to several hundred and a single female may produce several clutches. Development in most cladocerans is direct, and young are released from the brood chamber by ventral flexion of the female's postabdomen. In other phyllopodans development is indirect, and the eggs released by the female either fall to the bottom after only a brief brooding period or remain attached to the female and reach the bottom when she dies. These eggs hatch as nauplii.

Resting eggs, or diapausing (dormant) eggs, are an adaptation for life in ephemeral or periodically hostile habitats. The thick shell enables the egg or embryo to survive periods during which normal activity and metabolism are not feasible. Inside the eggshell, water content is less than 1% and the metabolic rate is reduced almost to zero.

Production of resting eggs may be stimulated by a variety of external factors, such as population density, temperature, or photoperiod. A resting egg may undergo an initial period of development and become an embryo prior to entering diapause. Then, embryos are present and ready for rapid hatching when favorable environmental conditions return. Diapause is broken by favorable ecological conditions, which may be signaled by changes in dissolved oxygen tension, salinity, temperature, or illumination. Such signals are species specific and related to the habitat.

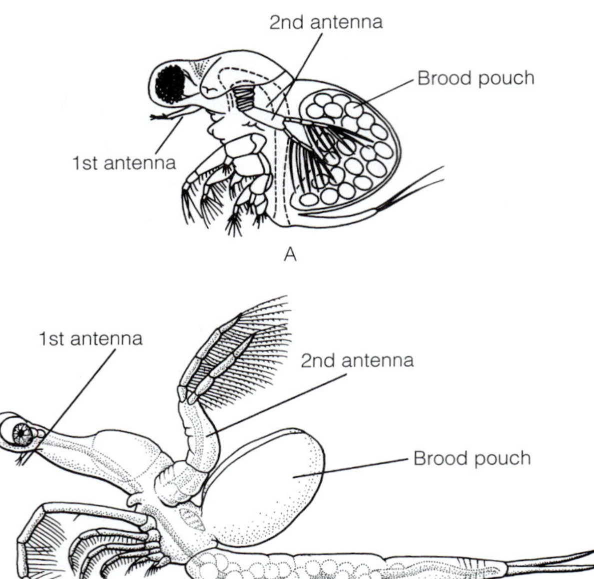

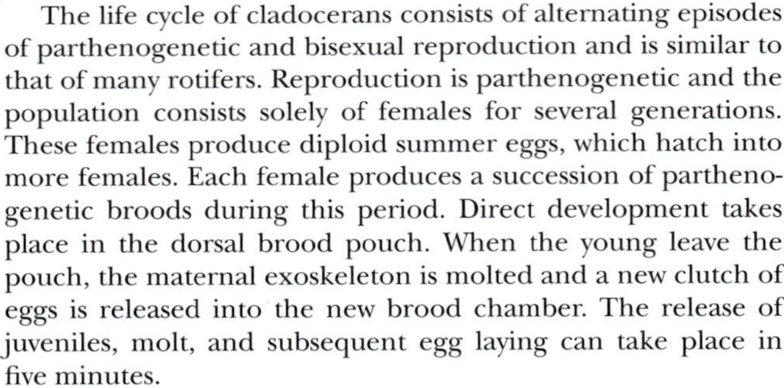

FIGURE 19-16 Cladocera. **A,** The onychopod *Polyphemus pediculus.* **B,** The predatory haplopod *Leptodora kindtii.* *(A, After Lilljeborg from Pennak; B, From Martin, J. W. 1992. Branchiopoda. In Harrison, F. W., and Humes, A. G. (Eds.): Microscopic Anatomy of Invertebrates. Vol. 9. Crustacea. Wiley-Liss, New York. pp. 25–224.)*

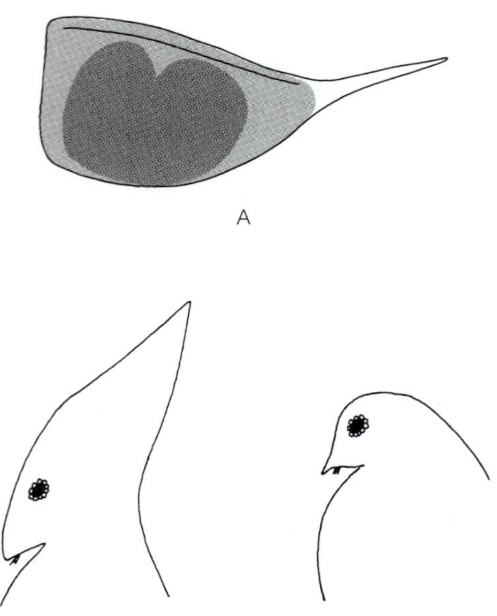

FIGURE 19-17 Cladocera. **A,** Ephippium of the anomopod *Daphnia pulex.* **B** and **C,** Cyclomorphosis in *Daphnia.* **B,** Summer long-headed form. **C,** Spring and fall round-headed form. *(A, From a photograph by Pennak; B and C, After Pennak, R. W. 1989. Fresh-water Invertebrates of the United States, Protozoa to Mollusca. 3rd edition. Wiley-Interscience, New York. 628pp.)*

The life cycle of cladocerans consists of alternating episodes of parthenogenetic and bisexual reproduction and is similar to that of many rotifers. Reproduction is parthenogenetic and the population consists solely of females for several generations. These females produce diploid summer eggs, which hatch into more females. Each female produces a succession of parthenogenetic broods during this period. Direct development takes place in the dorsal brood pouch. When the young leave the pouch, the maternal exoskeleton is molted and a new clutch of eggs is released into the new brood chamber. The release of juveniles, molt, and subsequent egg laying can take place in five minutes.

Eventually, some environmental factor, perhaps a change in photoperiod or water temperature, or a decreasing food supply caused by high population density, induces the females to produce haploid eggs that hatch into males. Once males are present in the population, mating occurs and females produce fertilized, diploid, resting eggs. Only two such eggs, one from each ovary, are produced in a single clutch. They are released into the brood chamber, whose walls are then transformed into a protective capsule known as an **ephippium** (Fig. 19-17A). The ephippium is cast off at the next molt and will either separate from or remain with the rest of the exuvium. The ephippia float, sink to the bottom, or adhere to objects and can withstand drying and freezing and even passage through the guts of planktivorous fishes, fish-eating birds, or mammals. By means of such protected, desiccation-resistant resting eggs, cladocerans may be dispersed over great distances by wind or animals and can overwinter or survive summer droughts.

Some freshwater lake-inhabiting zooplankters undergo seasonal changes in morphology known as **cyclomorphosis.** In some cladocerans, such as *Daphnia dubia* and *Daphnia retrocurva,* the head progressively changes from a rounded, compact shape to a pointed, helmetlike shape between spring and midsummer (Fig. 19-17B,C). Cyclomorphosis may also involve lengthening the tail spine on the carapace and producing lateral teeth on the head. In the spring and early summer, long-headed juveniles can be observed developing in the brood pouch of round-headed females.

Cyclomorphosis is easily observed, and its occurrence is undisputed. Its purpose, on the other hand, is poorly understood. Any response to environmental conditions, such as cyclomorphosis, has causes at two levels. One, the **proximate cause,** is the environmental stimulus that elicits the response. In the case of cyclomorphosis, laboratory and field studies indicate that the long-headed form develops in response to some combination of high temperature, abundant food, turbulence, and the presence of predators. These are conditions that typically occur in the spring in north temperate lakes. The second cause, the **ultimate cause,** is the reason the pattern evolved in the first place and the reason it is advantageous to the animal. In the case of cyclomorphosis, the ultimate cause appears to be predation avoidance, and there is evidence that long-headed morphs suffer less from predation than round-headed ones do.

NOTOSTRACA^O

Notostraca consists of only 10 species of benthic tadpole shrimps in the genera *Triops* and *Lepidurus* that are found in quiet, inland, fish-free waters. Notostracans resemble anostracans in some respects, but the similarities are now thought to be superficial and due to convergence. The body consists of a head, thorax, and abdomen, with no secondary tagmosis. An enormous depressed, shieldlike carapace covers the head, thorax, and most of the abdomen (Fig. 19-13). Sessile compound eyes are present on the anterior carapace. The mandibles are large and triturative.

The trunk consists of up to 40 segments, but tagmosis into thorax and abdomen is ambiguous. Anterior trunk segments bear appendages and form the thorax. Segments anterior to segment 11, which is the genital segment, each bear a single pair of appendages. There is no cephalothorax, but the first trunk appendage is highly modified and unlike those behind it. Posterior to the genital segment, fused thoracic segments form large rings. These rings may bear up to six pairs of appendages each, although most have fewer, and an animal with 40 rings may have as many as 70 pairs of trunk appendages. Most appendages are located on this posterior thoracic region. The cylindrical terminal region of the trunk, which lacks appendages, is the abdomen. It bears the telson and a caudal furca with elongate rami.

Notostracans are distributed worldwide and may reach 10 cm in length, although most are about 20 to 50 mm. They are omnivorous detritivores or predators, rather than suspension feeders. The anterior nine pairs of thoracopods are used to collect sediment particles from the bottom, which is where notostracans, although they are good swimmers, spend most of their time. Adoption of a benthic feeding mode is a derived characteristic.

The gut is an unelaborated J-shaped tube with a pair of short digestive ceca in the head. The heart in the thorax is a long dorsal tube with 11 pairs of ostia. The nervous system is ladderlike, without cephalization or medial fusion. The tubules of the paired maxillary glands are in the carapace (Fig. 19-13). The eggs are tolerant of desiccation and, in fact, a period of desiccation is required in some species. The hatching stage is metanauplius.

LAEVICAUDATA^O AND SPINICAUDATA^O

These two taxa with several superficial similarities were formerly grouped together in Conchostraca, or clam shrimps, but are no longer thought to belong to a monophyletic taxon. Nevertheless, it is still convenient to refer to them unofficially as "conchostraca." The 200 extant described species are characterized by the large, bivalved carapace that encloses the entire body (Fig. 19-14) so that, when viewed from the outside, the animal resembles a small clam. The carapace is laterally compressed, unlike that of notostracans, and equipped with a strong transverse adductor muscle.

The compound eyes are sessile. The first antennae are vestigial and the second antennae are large, biramous, setose, and natatory. The mandibles are large and triturative. Both pairs of maxillae are vestigial. The first and usually the second trunk appendages of males are modified as claspers to hold the female during mating. The remaining trunk appendages are unspecialized phyllopods. All thoracic segments bear appendages and there is a short, limbless abdomen. The gnathobases form a wide ventral food groove. The phyllopods bear an epipodite that is probably osmoregulatory rather than respiratory, although it is known as a gill.

Laevicaudata^O: Carapace lacks concentric growth lines; composed of two independent valves joined by a median dorsal, longitudinal hinge. Head is very large, occupying up to a third of the space inside the valves (Fig. 19-14A), articulated with the thorax, and can be extended through the gape between the valves. Thorax is short, with 10 to 12 segments. Cosmopolitan in ephemeral pools and streams. Up to 8 mm. One family, Lynceidae, with one genus, *Lynceus.*

Spinicaudata^O: Carapace valves formed by folding a single sheet of carapace into right and left halves; thus there is no distinct hinge. Carapace always has concentric growth rings (Fig. 19-14B) representing the persistent outer layer of the carapace, which remains after each molt. Head is small, not articulated with thorax, and cannot be extended outside the carapace. Thorax has 16 to 32 segments, each having a pair of appendages. Found in ephemeral pools and prairie streams; a few in permanent ponds. Up to 18 mm. Five families. *Cyzicus, Leptestheria, Limnadia.*

CLADOCERA^{SO}

Cladocera contains 11 families, with about 600 species, of similar microcrustaceans known collectively as water fleas (Fig. 19-15, 19-16). In the new classification, Cladocera (Cladoceromorpha) has been divided into four new taxa, Anomopoda, Ctenopoda, Onychopoda, and Haplopoda. Cladocerans have a carapace that serves as a dorsal brood pouch (Fig. 19-15A). In most, the carapace is large and bivalved to enclose the trunk, but in some it is reduced to no more than the brood pouch. The compound eyes are fused to form a single movable eye equipped with muscles. There may also be a small median naupliar eye. The second antennae are large and are the major swimming organs. The first antennae are tiny.

The trunk is short, with five or six pairs of appendages. In most taxa the abdomen is reduced and the region posterior to the thorax is known as the postabdomen and is without appendages, although it bears a caudal furca, whose rami are a pair of postabdominal claws, and a pair of large plumose setae (Fig. 19-15A). The anus is at the posterior end of the postabdomen.

Most live in fresh water, but they are not restricted to relictual or temporary habitats as the large phyllopodans are. Most are either planktonic or benthic in inland waters, but some are interstitial, a few live in moist terrestrial environments (tropical leaf litter), and several are marine and planktonic. They are one of the three metazoan taxa that dominate the freshwater zooplankton (rotifers, cladocerans, and copepods). Most are suspension feeders, but a few are predators. Cladocerans are generally small, usually about 4 to 6 mm, although they range from 0.25 to 18 mm. Cladocerans are the most successful of the living phyllopodans.

Diversity of Cladocera

Anomopoda[sO]: Trunk is short and enclosed by a well-developed, hingeless, laterally compressed, bivalve carapace (Fig. 19-15). Head is expanded dorsally and laterally to form a head shield. Trunk is short, its segments fused. Five or six pairs of dissimilar trunk appendages which do not beat in metachronal waves. Postabdomen reflexed anteriorly under trunk. Postabdomen has a pair of long setae and a pair of large terminal claws. Development is direct; hatch as miniature adults. Resting eggs enclosed in ephippium derived from the carapace. Freshwater. Nine families including Bosminidae, Chydoridae, Daphniidae, Macrothricidae, and Moinidae. The majority of cladocerans are anomopods. *Alona, Bosmina, Ceriodaphnia, Chydorus, Daphnia, Macrothrix, Moina, Scapholeberis.*

Ctenopoda[sO]: Similar to anomopods. Short trunk is enclosed by well-developed, laterally compressed, bivalve carapace lacking a true hinge. No head shield. Six pairs of similar trunk appendages beat in rhythmic waves. Postabdomen has a pair of long setae and a pair of large terminal claws. Development is direct; hatch as miniature adults. Chiefly freshwater, but one marine genus. Two families, Holopediidae and Sididae. *Holopedium, Sida.*

Onychopoda[sO]: Do not closely resemble anomopods or ctenopods. Head and trunk are short. Carapace is reduced, represented only by dorsal brood pouch (Fig. 19-16A); does not enclose trunk. Development is direct. Freshwater, marine predators. Trunk appendages raptorial, adapted for grasping rather than filtering. Includes Cercopagidae, Podonidae, Polyphemidae. *Bythotrephes, Evadne, Podon, Polyphemus.*

Haplopoda[sO]: Very different from other cladocerans. Elongate head and body (Fig. 19-16B). Carapace is a dorsal brood pouch in females, absent in males. Freshwater zooplanktivores. Much larger than other cladocerans, up to 18 mm. Anterior trunk appendages are raptorial and adapted for grasping smaller zooplankters. Development indirect, hatching stage is a metanauplius. Leptodoridae the only family, *Leptodora kindtii* the only species.

PHYLOGENY OF PHYLLOPODA

In general, large phyllopodans (notostracans and conchostracans) exhibit primitive features believed to be retained from the ground plan of Phyllopoda, whereas cladocerans are derived (Fig. 19-18). The ground plan includes a pair of sessile compound eyes close to the midline but not contiguous. Each eye is sunk below the surface of the exoskeleton into an epidermal ocular vesicle that remains in communica-

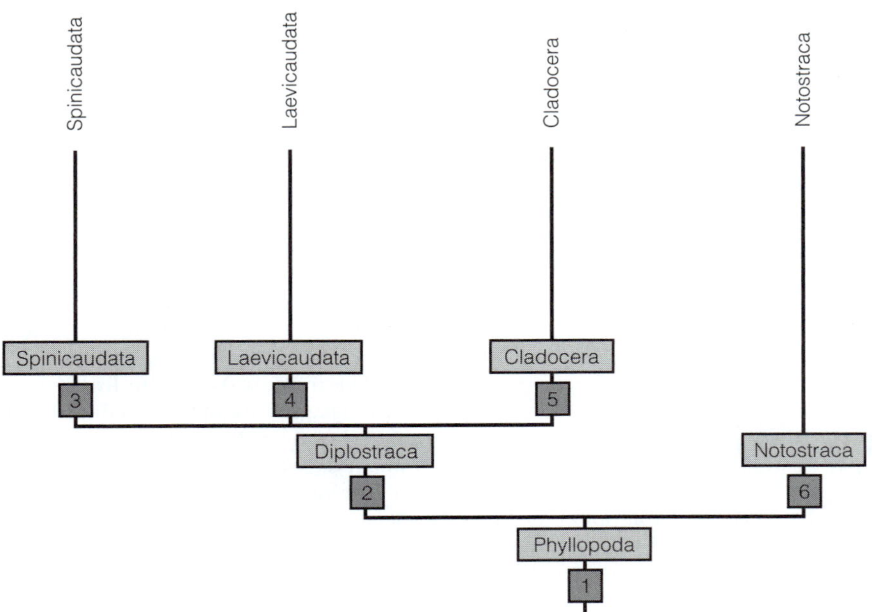

FIGURE 19-18 Phylogeny of Phyllopoda. **1, Phyllopoda:** Compound eye is sessile and enclosed in an epidermal vesicle; about 30 thoracopods in the stem phyllopod. **2, Diplostraca:** Antennae 2 are large, biramous, and the major swimming organ; carapace with growth rings resulting from failure of the outer wall to molt; compound eyes median but not fused; abdomen reduced, short, reflexed ventrally, and limbless; eggs brooded in a dorsal brood chamber under the carapace; furcal rami are stout claws used for grooming. **3, Spinicaudata:** Male has claspers on thoracopods 1 and 2. **4, Laevicaudata:** Carapace has no growth lines, but has a distinct mid-dorsal hinge; head is large and can be extended from the carapace; thorax has fewer than 12 segments. **5, Cladocera:** Compound eyes fused to form a single median eye; life cycle alternates between parthenogenesis and bisexual reproduction; direct development; ephippium encloses dormant eggs. **6, Notostraca:** Benthic feeders; flattened, shieldlike carapace without an adductor muscle; second antennae reduced; first thoracopods have three antenniform, sensory branches; furcal rami is antenniform and very long. *(After Ax, P. 2000. Multicellular Animals, II. Springer Verlag, Berlin.)*

tion with the exterior by a small pore. The naupliar eye includes four pigment-cup ocelli. The trunk is divided into a thorax with about 30 pairs of appendages and a limbless abdomen. A caudal furca with short, stout rami is present. The body is relatively large, about 1 to 2 cm in length. The primitive phyllopodans are planktonic and use the thoracopods for filter feeding in open water. In a departure from the ancestral condition, notostracans have adopted a benthic existence in which they feed on organic particulates on the bottom or are predaceous.

Diverging from the ground plan, conchostraca and Cladocera have developed an enormous compressed carapace with two lateral valves that enclose most of the body. The thorax may be large or small, but the abdomen is reduced to a vestige reflexed below the body and equipped with stout furcal rami used for cleaning debris from under the carapace or from the thoracopods.

In the derived Cladocera, the compound eyes coalesce on the midline to form a single median eye that remains enclosed in an ocular vesicle, but the connection with the exterior is closed. The number of thoracopods is greatly reduced to five or six and body length diminishes to 4 to 6 mm in most.

PHYLOGENETIC HIERARCHY OF PHYLLOPODA

- Phyllopoda
  - Diplostraca
    - Spinicaudata
    - Laevicaudata
    - Cladocera
      - Anomopoda
      - Ctenopoda
      - Onychopoda
      - Haplopoda
  - Notostraca

MALACOSTRACA[C]

Large and successful Malacostraca, with 23,000 living species, contains over half of the known species of crustaceans. It includes most of the large well-known crustaceans, such as crabs, lobsters, crayfishes, and shrimps, in addition to an abundance of small but ecologically important species. Unlike other crustacean taxa, tagmosis, with few exceptions, is standardized.

The head consists of 5 segments, as is true of all crustaceans, and the thorax always has 8 segments and the abdomen (almost) always has 6, plus the telson (Fig. 19-19), for a total of 19 segments. The original malacostracan abdomen included seven segments and, although this condition persists in primitive taxa such as mysids and euphausiaceans, in most species two segments have fused so the abdomen has only six segments. Malacostraca also differs from all other major crustacean taxa, except Remipedia, in having abdominal appendages. There may or may not be a carapace and, if present, the amount of the trunk covered by it varies. The first antennae are often biramous. The exopod of each second antenna is frequently a flattened antennal scale (scaphocerite).

Primitively, the thoracic appendages are similar and the endopod, used for crawling or grasping, is the more highly developed of the two branches (Fig. 19-20). Sometimes the exopod is lost and the resulting pereopod is uniramous. The typical malacostracan endopod has seven articles: from proximal to distal, the coxa, basis, ischium, merus, carpus, propodus, and dactyl (Fig. 19-20). In most malacostracans one, two, or three anterior thoracic segments fuse with the head to form a cephalothorax (Fig. 19-1). Their appendages turn forward and become maxillipeds, which are auxiliary mouthparts used in feeding.

Usually the first five pairs of abdominal appendages are similar, biramous pleopods (Fig. 19-19) used for swimming, burrowing, creating a ventilating or feeding current, brooding eggs, or gas exchange. In the male, the first one or two

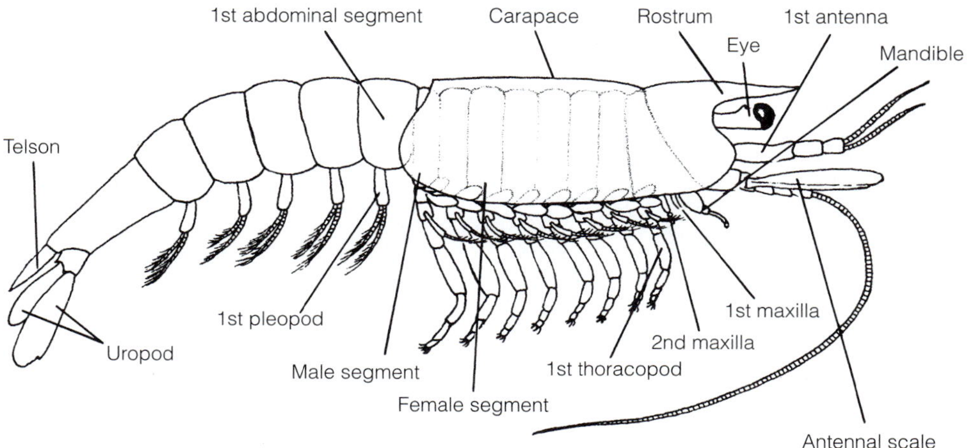

FIGURE 19-19 Malacostraca. Lateral view of body of a generalized malacostracan. In this example, all eight thoracic segments are independent of the head and there are no cephalothorax and no maxillipeds. The carapace covers the entire thorax and thoracic abdominal appendages are biramous. *(After Calman.)*

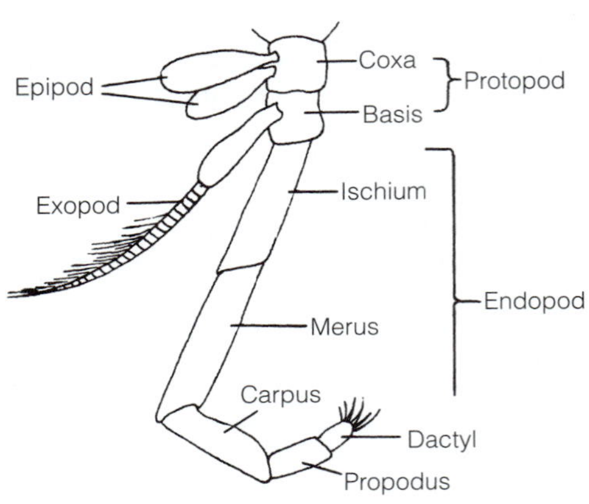

FIGURE 19-20 Malacostraca. A biramous thoracic appendage of a generalized malacostracan. The endopod is an example of a stenopod. *(After Calman.)*

pleopods are often modified, forming copulatory gonopods for the indirect transfer of sperm to the female. In most malacostracans the sixth abdominal appendages are uropods consisting of a protopod and two flat, paddlelike rami (Fig. 19-19). The uropods, in conjunction with a flattened telson, form a **tail fan** which, when flexed, is capable of rapidly generating thrust. Flexion, accomplished by the powerful abdominal muscles, is responsible for the rapid backward escape response of crayfishes, lobsters, and shrimps.

The foregut usually consists of an esophagus and a characteristic two-chambered stomach, or proventriculus, bearing the triturating teeth of the gastric mill and the comblike filtering setae of the filter press. Within the stomach, food is chewed and chemical digestion begins. The soluble or very fine particulate products of this preliminary digestion are sent to the large digestive ceca for absorption.

Female gonopores are always located on the sixth thoracic segment and male gonopores on the eighth. The nauplius larva is usually passed within the egg and the emerging individual is an advanced larva of some type. The naupliar eye has four pigment-cup ocelli.

Malacostraca includes seven major taxa: Leptostraca, Stomatopoda, Decapoda, Syncarida, Euphausiacea, Pancarida, and Peracarida. Of these the ecologically important Decapoda and Peracarida together account for about half of all crustacean species.

LEPTOSTRACA[O]

Leptostracans, or phyllocarids, are small, benthic marine malacostracans. They look like tiny shrimps enclosed in an enormous carapace (Fig. 19-21). About 20 species occur worldwide and range in length from 10 to 40 mm. Except for a single planktonic species, all are benthic and found from the intertidal to the deep sea. Species of *Nebalia* occur in mud and seaweed on the both coasts of North America.

Interest in leptostracans is chiefly from the standpoint of malacostracan evolution, as they are believed, based on morphological and paleontological evidence, to be its most primitive living members. The earliest known malacostracans were leptostracans from the Cambrian period. Leptostracans are probably similar to the first malacostracans and to the stem crustacean.

The body consists of a head, thorax, and abdomen with no secondary reorganization into cephalothorax and pereon. The head has a pair of stalked compound eyes. Thoracic segments are not fused with the head, maxillipeds are absent, and the eight pairs of homonomous thoracopods are biramous phyllopods. The two-part carapace consists of a small, movable, anterior rostrum articulated with the much larger posterior carapace proper (Fig. 19-21). The latter is bivalved and encloses the entire thorax and much of the abdomen. It

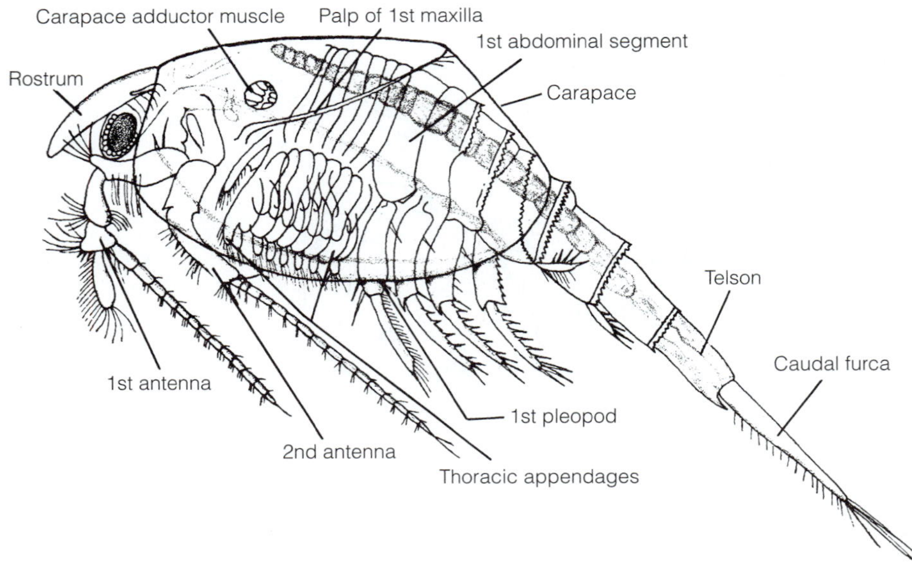

FIGURE 19-21 Leptostraca. Female *Nebalia bipes*. *(After Claus from Calman.)*

arches over the dorsal midline and its right and left valves are held together by an adductor muscle. Contractions of this muscle bring the valves together. There is no specialized hinge, and elastic recoil of the flexed carapace opens the valves when the muscle relaxes. The carapace is not fused with any thoracic segments.

The abdomen is of special interest, as it includes seven rather than the usual six segments (Fig. 19-21). The first six abdominal segments bear pleopods, of which the first four are large, biramous, and natatory. The last two are small and uniramous, and none of them is a uropod. The large telson that extends posteriorly from the seventh segment bears the terminal anus and the large caudal furca.

Feeding methods vary and include deposit feeding and carnivory, although most leptostracans are shallow-water suspension feeders. One planktonic species eats the eggs of other zooplankters. The gut is J-shaped with a posteriorly directed mouth. The foregut consists of the esophagus and stomach, and the long midgut has digestive ceca. The hindgut is a short rectum.

The thoracic appendages and the vascularized carapace are the gas-exchange surfaces. The space under the carapace is a branchial chamber and, in females, a brood pouch. The respiratory current is created by the thoracopods. The four anterior pairs of pleopods are natatory. Paired saccate nephridia are present in the antennal *and* maxillary segments. The heart is a long dorsal tube with seven pairs of ostia. Cephalization of the nervous system is slight, although there is a small subesophageal ganglion. The nerve cord is a chain of 17 segmental ganglia. The male gonopores are at the tips of a pair of penes. The eggs hatch as postlarvae with incomplete pleopods and temporarily remain in the brood pouch under the carapace.

Three families with several genera, including *Nebalia, Nebaliella, Nebaliopsis,* and *Paranebalia,* are known.

STOMATOPODA[O]

The 300 species of marine crustaceans known as mantis shrimps constitute Stomatopoda. Mantis shrimp are highly specialized predators of fishes, crabs, shrimps, and molluscs, and many of their distinctive features are related to their raptorial behavior. The dorsoventrally depressed body is divided into a head, thorax, and abdomen and there is no cephalothorax. The head bears a pair of large, stalked compound eyes and a single median naupliar eye. The first antennae are unusual in being triramous, with three flagella (Fig. 19-22). Each second antenna consists of a peduncle, flagellum, and large antennal scale. Dorsally, a shieldlike carapace (Fig. 19-22) covers the head and first four thoracic segments. A movable median rostrum articulates with the anterior edge of the carapace and covers the bases of the eyestalks.

Each of the eight thoracic segments bears a pair of thoracopods. The first of these are long and slender and are used for grooming. The second are subchelate, prehensile **raptorial claws** for capturing prey (Fig. 19-23). The inner edge of the distal article is armed with long spines or is shaped like the blade of a knife. The claw is extended rapidly for prey capture or defense. Thoracopods 3, 4, and 5 are also subchelate, but are much smaller than the raptorial claws. Thoracopods 6, 7, and 8 are relatively unspecialized walking legs.

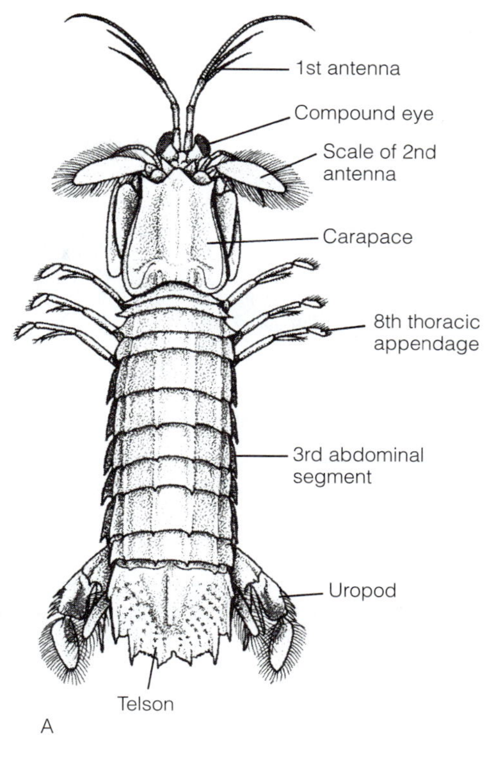

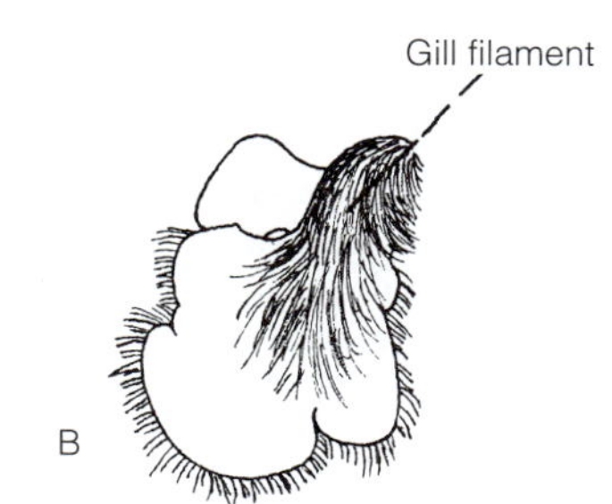

FIGURE 19-22 Stomatopoda. The mantis shrimp *Squilla empusa.* **A,** Dorsal view. **B,** Second pleopod and gill. *(B, After Calman.)*

The abdomen is muscular and well developed. Its five pairs of biramous pleopods bear filamentous gills (Fig. 19-22B). Abdominal gills are unusual in Crustacea, but are also characteristic of isopods. The sixth abdominal segment bears a pair of uropods and a large telson. The uropods and telson together form a tail fan.

The mouth opens directly into a large stomach that fills the anterior cephalothorax. An enormous digestive cecum extends throughout the body, even into the telson. The heart is a long tube extending the length of the body, with 13 pairs of ostia. The hemal system has well-developed arteries and capillaries. Adults have a pair of maxillary glands. The nervous system includes the usual tripartite crustacean brain, a pair of very long circumenteric connectives, and a subesophageal ganglion incorporating the ganglia of the head and first five thoracopods. The remaining appendages are served by the

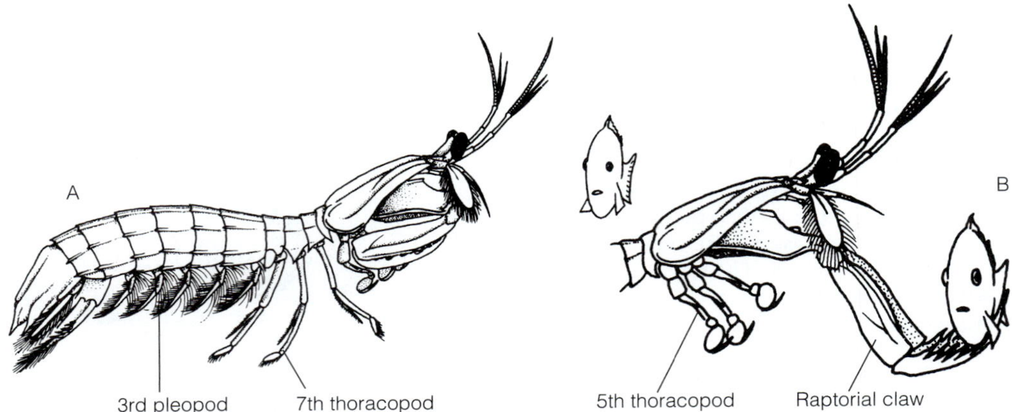

FIGURE 19-23 Stomatopoda. Prey capture by a mantis shrimp that spears its prey. The second thoracic appendage rapidly unfolds and the barbs on the terminal finger are driven into the body of the prey.

segmental ganglia of the ventral nerve cord. Mantis shrimps have the best-developed compound eyes of any crustacean. They not only detect moving objects, they apparently are also capable of depth perception. The antennae are important sites of chemoreception also used in prey detection when the range is short enough.

There is a seminal receptacle on the oviduct and the sperm ducts produce sperm cords rather than spermatophores. Some mantis shrimp pair for life, with the partners sharing the same burrow or retreat, whereas others come together only for mating. The female broods up to 50,000 eggs, which are held together in a globular mass by means of an adhesive secretion. The hatching stage is a zoea and planktonic larval life may last for three months.

Mantis shrimps range in size from small species approximately 5 cm long to giants of 36 cm. Most are tropical, but *Squilla empusa,* which is about 18 cm long, is a common species on the North American Atlantic coast, where it is frequently caught in shrimp trawls. Many stomatopods are brilliantly colored. Green, blue, and red with deep mottling are common, and some species are striped or display other patterns.

Most stomatopods live in rock or coral crevices or in burrows excavated in the bottom. *Squilla empusa* lives in a U-shaped burrow in soft sediments, but for winter quarters it may build a vertical burrow extending 4 m into the sediment. Some stomatopods appropriate burrows excavated by other animals. The short carapace probably provides the flexibility needed to reverse direction and maneuver within the burrow.

Unlike most crustaceans, mantis shrimps are neither scavengers nor omnivores, but rather predators that require living prey. Many leave the burrow to forage for prey whereas others lurk in the opening of the burrow and ambush passing prey. Mantis shrimps spear, slash, or smash their prey with the large, powerful raptorial claws. Many defend their burrows and territories against other stomatopods, and species living in holes and crevices within coralline rock and rubble have complex social behaviors.

The 13 Recent families include Gonodactylidae *(Gonydactylus),* Lysiosquillidae *(Lysiosquilla),* and Squillidae *(Squilla).*

DECAPODA^O

Decapoda includes shrimps, crayfishes, lobsters, and crabs and are certainly the most familiar crustaceans. Most decapods are large and many are edible and support important fisheries. The approximately 10,000 described species represent about one-quarter of the known crustaceans. Because of their large size, abundance, and ecological and economic importance, decapods are the most carefully studied and best-known crustaceans. Most are marine and benthic, but the crayfishes and some shrimps and crabs have invaded fresh water A few crabs are terrestrial and some decapods are planktonic.

The head often bears an anterior, median, immovable rostrum (Fig. 19-1). Both pairs of antennae are biramous. The exopod of each second antenna is a flat antennal scale used to control the direction of water currents. The base of the second antenna contains the statocyst and the nephridiopore. The compound eyes are stalked.

The first three thoracic segments are fused with the head to form a cephalothorax. The appendages of these three segments are maxillipeds, which function as mouthparts. The remaining five pairs of thoracic appendages are the 10 pereopods from which the name Decapoda (= 10 feet) is derived. The first pair of pereopods is frequently enlarged and chelate to form a prehensile **cheliped,** or pincer (Fig. 19-30A) in which the last two articles form opposable fingers, one of which is movable and closes against the other. The remaining pereopods are usually stenopodous walking legs (Fig. 19-31) although some may be chelate. Decapod pereopods usually lack exopods and thus are uniramous, although they typically bear epipods modified as gills.

The carapace is fused dorsally with all eight thoracic segments. The five posterior thoracic segments are independent of the cephalothorax even though the carapace (Fig. 19-1) covers and is attached to them. The sides of the overhanging carapace enclose the gills within well-defined, lateral branchial chambers (Fig. 19-3A, 19-36). The abdomen, or pleon, extends posteriorly from the thorax.

Diversity of Decapoda

The central theme of decapod evolution is the transformation of the ancestral elongate, natant (swimming), shrimplike body to the derived, shortened, depressed, reptant (crawling), crab-like morphology, and decapod classification is based on the extent of this conversion. Although the classification of Decapoda is intricate, for our purposes most of its detail can be avoided. There are three basic decapod body shapes (Fig. 19-24): shrimplike (Penaeidea, Stenopodidea, Caridea; Fig. 19-25*)*, lobsterlike (Astacidea, Thalassinidea, Palinura; Fig. 19-26*)*, and crablike (Anomura, Brachyura; Fig. 19-30).

SHRIMPLIKE DECAPODS

The shrimp body tends to be cylindrical or laterally compressed with a well-developed, muscular abdomen (Fig. 19-25A). The cephalothorax often bears a keel-like, serrated rostrum. The legs are usually slender, and chelipeds may be present or absent. The exoskeleton is usually thin, flexible, and uncalcified. The pleopods, which are large and fringed, are the principal swimming organs, although rapid ventral flexion of the abdomen and its tail fan is used for quick backward darts.

Most shrimps are bottom dwellers that use their pereopods for crawling. Although swimming with the pleopods is possible, it is usually intermittent. They live among algae and sea grasses, beneath stones and shells, and within holes and crevices in coral and rock. Some, including many penaeids and the sand shrimps *(Crangon)*, are shallow burrowers in soft bottoms and use the beating pleopods for digging. Their activity patterns may be coordinated with light or tidal cycles. The edible pink shrimp *Penaeus duorarum* from the east coast of the United States lies quiescent on the sediment surface or buried during the day but actively crawls at night. *Penaeus* and related taxa are the most important commercially harvested shrimps. In the United States, shrimp fisheries are centered primarily along the southeastern Atlantic coast and the Gulf of Mexico. Penaeideans are the most primitive living decapods.

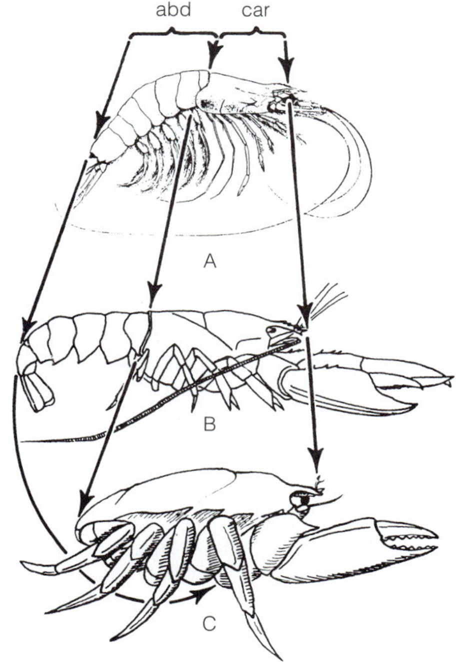

FIGURE 19-24 Decapoda. Comparison of carapace and abdomen in a shrimp, lobster, and crab showing the evolution of the short body form of crabs. *(From Glaessner, M. F. 1969. Decapoda. In Moore, R. C. (Ed.): Treatise on Invertebrate Paleontology. Pt. R. Arthropoda 4. Vol. 2. Geological Society of America and University of Kansas Press, Lawrence. p. 401.)*

The pistol, or snapping, shrimps (Alpheidae; *Alpheus, Synalpheus*) are a common and widely distributed family. Snapping shrimps live in holes and crevices in sponges or beneath shells, rocks, and coral rubble, or construct retreats or burrows. These 3- to 6-cm-long shrimps have one greatly enlarged cheliped (Fig. 19-25B). The base of the movable finger contains a large, tuberculate process that fits into a socket on the immovable finger. The movable finger is locked, or cocked, when contact is made between two specialized discs, one at the base of the elevated finger and one on the immovable finger. The adhesive force between the discs prevents the finger from closing until the contracting muscle has generated a counteracting pull. When the muscle overcomes the adhesive force and the finger closes, it results in cavitation in the vicinity of the discs. The collapse of the cavity is apparently what makes the noise. Although some species have been reported to use the snapping mechanism in predation (cracking small clams and stunning fish), it also functions in agonistic (threat) displays between individuals and probably helps maintain a uniform dispersion pattern among members of a population.

A number of shrimps of several taxa that are collectively known as cleaning shrimps remove ectoparasites and other unwanted materials from the surfaces of reef fishes (Fig. 19-25C). The shrimp may climb over the fish and even enter its mouth and gill chamber in search of parasites, which it removes and eats.

Pelagic shrimps living in the upper 500 m (the epipelagic and upper mesopelagic zones of the ocean) tend to be transparent or semitransparent, whereas those living below about 500 m are red. Many of the latter group also have bioluminescent organs, or photophores, located internally or on the body surface. Among the best-known bioluminescent species are those belonging to *Sergestes* and *Sergia*.

LOBSTERLIKE DECAPODS

The remaining decapods are benthic animals adapted for crawling and are lobster- or crablike, with the body tending to be dorsoventrally flattened. The legs usually are heavier than those of shrimps, and typically the first pair are powerful chelipeds. The pleopods are not adapted for swimming. In many taxa, the more primitive, long-bodied shrimp form (Fig. 19-24A) has been shortened by folding the abdomen ventrally beneath the thorax (Fig. 19-24C). This evolution of the "crab" form occurred independently a number of times in decapods.

Lobsterlike decapods such as lobsters, crayfishes, and burrowing shrimps are intermediate between shrimps and crabs.

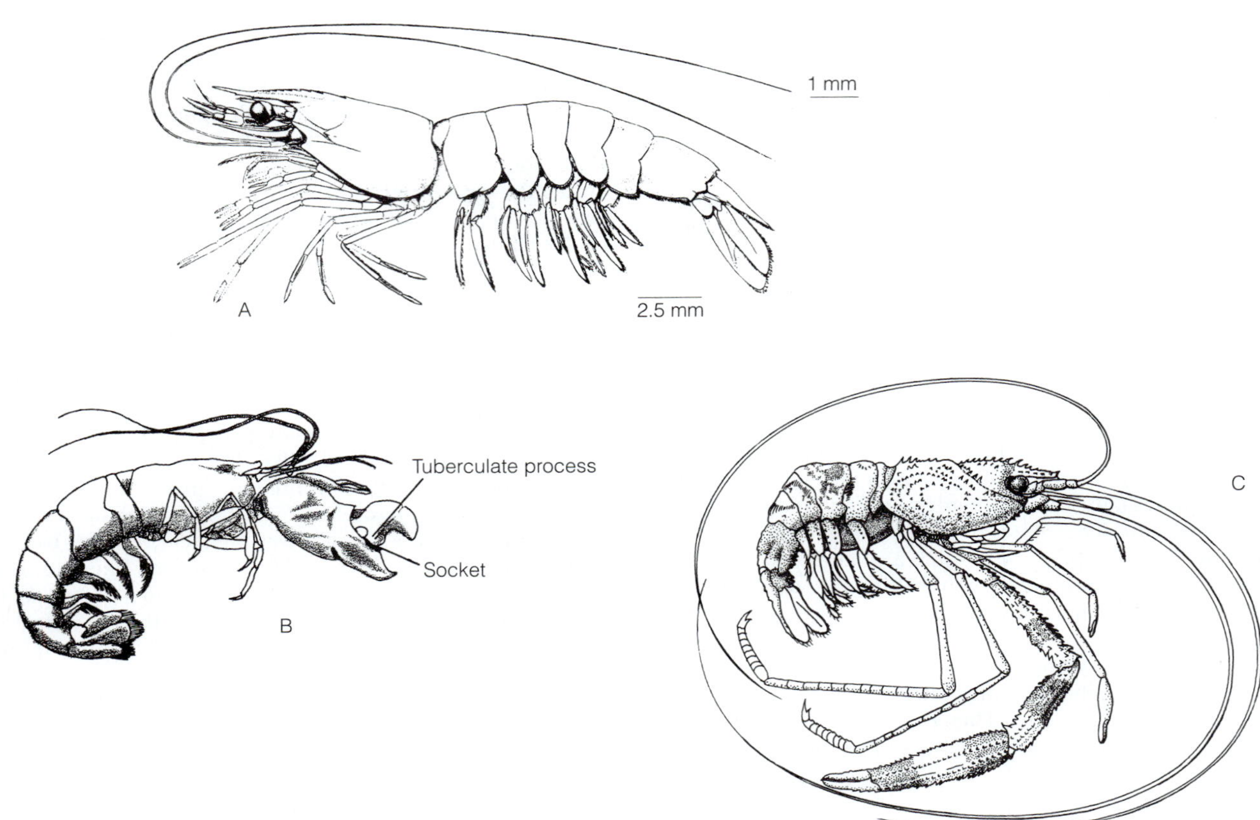

FIGURE 19-25 Decapoda. Representatives of the three higher taxa (infraorders) of shrimps. **A,** *Penaeus setiferus* (Penaeidea), the white shrimp, is a commercially important species along the east coast of the United States. **B,** *Alpheus* (Caridea), a pistol, or snapping, shrimp. **C,** *Stenopus hispidus* (Stenopodidea), the banded coral shrimp, is a cleaner shrimp that occurs primarily near coral reefs. *(A, From Williams, A. B. 1965. Marine decapod crustaceans of the Carolinas. U. S. Fishery Bull. 65(1):1–298; B, After Schmitt from Bruce, A. J. 1976. Shrimps and prawns of coral reefs, with special reference to commensalism. In Jones, O. A., and Endean, R. (Eds.): Biology and Geology of Coral Reefs. Vol. III: Biology 2. Academic Press, New York. p. 49; C, After Limbaugh et al. from Williams, A. B. 1984. Shrimps, Lobsters, and Crabs of the Atlantic Coast of the Eastern United States, Maine to Florida. Smithsonian Institution Press, Washington, DC. 550 pp.)*

They have large, shrimplike abdomens with the full complement of appendages and the carapace is longer than it is broad (Fig. 19-24B, 19-26). Lobsters and crayfishes crawl with the pereopods but can rapidly escape backward by flexing the abdomen and its tail fan ventrally, as do shrimps. The pleopods are used for ventilation.

Lobsters (Fig. 19-27) are heavy-bodied decapods that generally inhabit holes and crevices on rocky and coralline bottoms. Members of Nephropidae have large chelipeds and are similar in morphology to the closely related crayfishes. One of them, the American lobster, *Homarus americanus,* may reach a length of 60 cm and a weight of 22 kg. This species is fished commercially using pots, or traps, which the animal enters seeking shelter or bait. The frozen lobster tails sold in food markets are mostly species of spiny and slipper lobsters shipped from various tropical and subtropical parts of the world. These tropical lobsters do not have chelipeds. Burrowing ghost shrimps or mud shrimps (*Callianassa, Thalassina,* and *Upogebia*) are shallow-water or intertidal decapods that live in long, deep burrows excavated in sand or mud (Fig. 19-26).

CRABLIKE DECAPODS

The remaining decapods are crabs belonging to two taxa, Anomura and Brachyura. Anomura continues the transition from shrimps to true crabs (Brachyura). Some are easily recognized as crabs (Fig. 19-30A), but some are not at all crablike (Fig. 19-29A) and some are intermediate (Fig. 19-28). The abdomen is usually moderately reduced and may or may not be flexed beneath the cephalothorax. Uropods are usually present on the abdomen (Fig. 19-30B). The fifth legs are reduced and located beneath the sides of the carapace or folded above the posterior carapace (Fig. 19-30B).

Most anomurans are hermit crabs (Fig. 19-28). These distinctive decapods appropriate discarded snail shells for use

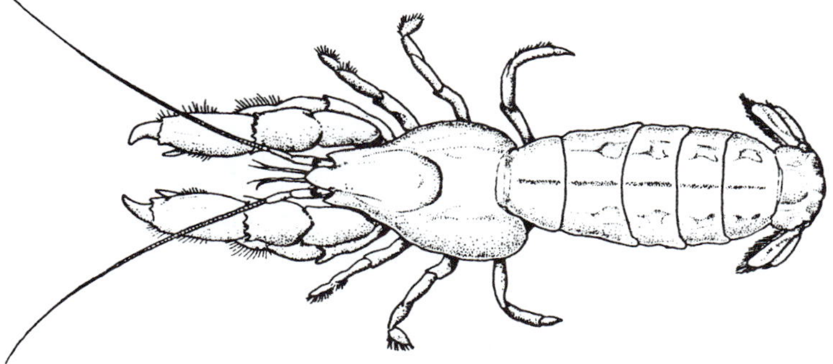

FIGURE 19-26 Decapoda. *Upogebia affinis,* a species of burrowing shrimp from the southeast coast of the United States. *(From Ruppert, E. E., and Fox, R. S. 1988. Seashore Animals of the Southeast. University of South Carolina Press, Columbia.)*

FIGURE 19-27 Decapoda. The American lobster, *Homarus americanus,* the lobster of commercial importance along the northeastern coast of the United States.

as portable domiciles. The hermit crab condition probably originated with ancestors that utilized crevices and holes as protective retreats. In most hermit crabs, the large abdomen is not flexed beneath the cephalothorax. Instead, it is asymmetrical and modified to fit within the spiral chamber of a gastropod shell. The vulnerable abdomen is covered with a thin, soft, nonsegmented cuticle. On the right side of the abdomen the pleopods are reduced or lost, whereas those on the left are retained in the females to carry eggs. The twist of the abdomen is adapted for right-handed (dextral) snails, although left-handed shells also can be used. As hermit crabs grow they must find progressively larger snail shells to accommodate them. A shortage of suitable shells may limit the size of hermit crab populations. Hermit crabs rapidly make the transition from old to new shell. The soft abdomen is defenseless and crabs caught outside their shells are quickly consumed by predators. One or both chelipeds may form an operculum to block the aperture of the shell when

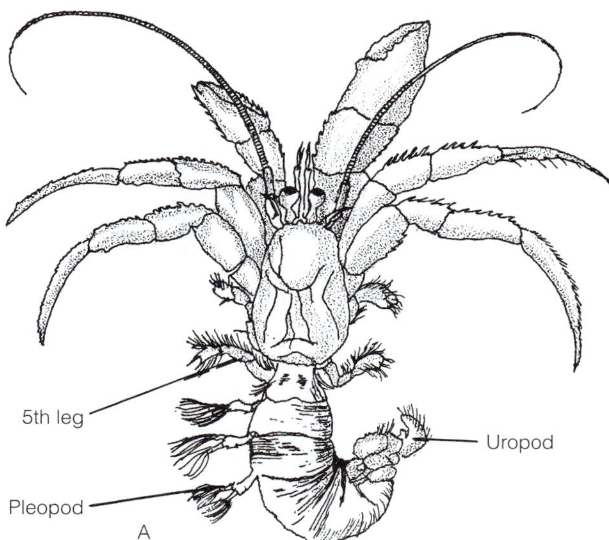

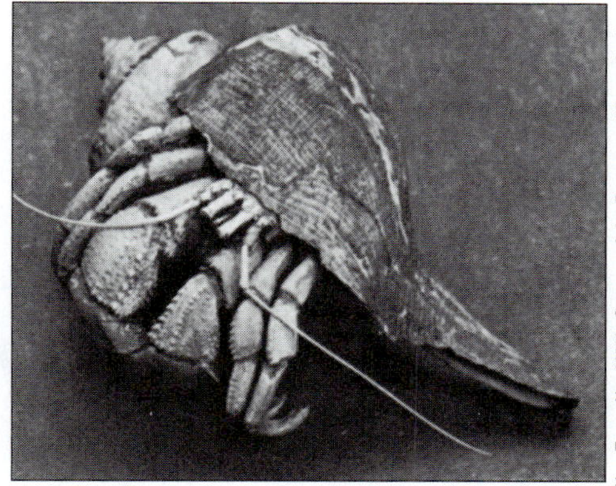

FIGURE 19-28 Decapoda, Anomura: Pagurid hermit crabs. **A,** Dorsal view of *Pagurus* out of its shell, showing the asymmetry of the large abdomen. **B,** A pagurid crab in a gastropod shell.

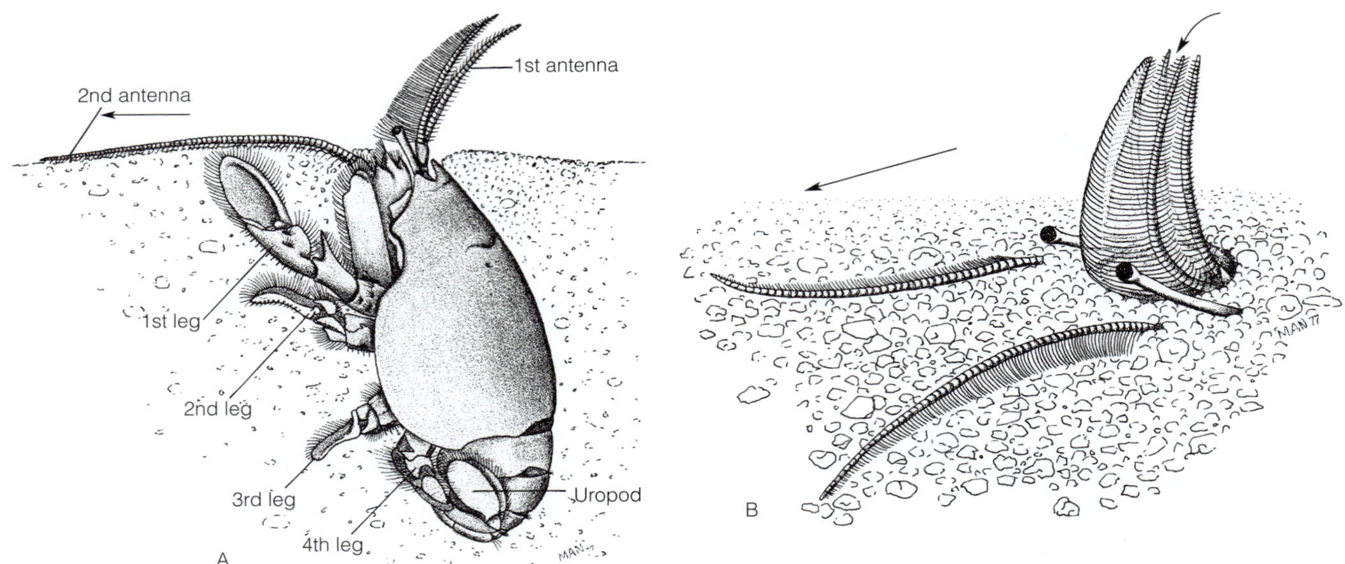

FIGURE 19-29 Decapoda, Anomura: *Emerita talpoida,* the mole crab, in the superfamily Hippoidea. This is a common crab on surf-swept beaches along the east coast of the United States. **A,** Lateral view of the animal buried in the sand. **B,** Surface view of the buried animal. The first antennae form an inhalant siphon to ventilate the gills. The curved arrow indicates the ventilating current. The straight arrow indicates the direction of the receding wave.

the crab is withdrawn. Most hermit crabs inhabit the shells of many different gastropods, depending on availability, and the shell supply can have a high turnover rate.

In sand crabs (mole crabs), the characteristic abdominal flexion and ovoid, streamlined body are adaptations for rapid backward burrowing in sand (Fig. 19-29). Chelipeds are absent in these filter feeders.

The most crablike anomurans are the little porcelain crabs (Fig. 19-30). Their flexed, symmetrical, crablike abdomen is not as reduced as it is in brachyurans. Porcelain crabs are common shallow-water decapods in many parts of the world and often occur in large numbers, sometimes in the hundreds. Although they are much like brachyuran crabs, when disturbed, many species will swim by flapping the abdomen.

The true crabs of Brachyura have the most highly specialized body form and, in terms of species numbers (over 4500), are the most successful decapods. Brachyurans are found in most benthic marine habitats and there are even a few freshwater

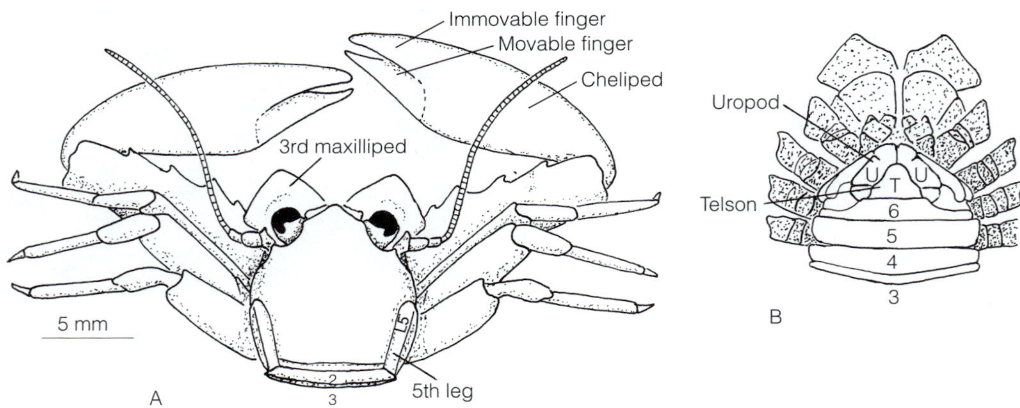

FIGURE 19-30 Decapoda, Anomura: *Petrolisthes,* an anomuran crab belonging to the superfamily Galatheoidea. The members of this genus, like others in the family Porcellanidae (porcelain crabs) are common beneath stones in shallow water on rocky coasts. Note the reduced and folded fifth pair of legs and the retention of the uropods and telson on the folded abdomen. **A,** Dorsal view. **B,** Ventral view. The abdominal segments are numbered.

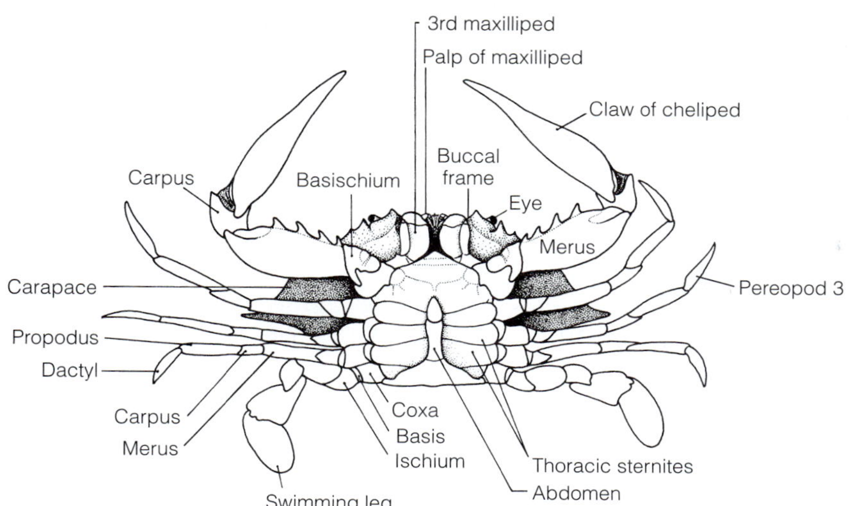

FIGURE 19-31 Decapoda, Brachyura. Ventral view of a male crab in the family Portunidae (swimming crabs). In this family the fifth pereopods are oarlike and adapted for swimming. Note the small abdomen folded inconspicuously under the thorax. The abdomen of the female is much wider but in the same position. *(B, After Schmitt from Rathbun.)*

and terrestrial species. The brachyuran body is strongly depressed and shortened, making it is as wide or wider than it is long. The carapace covers the cephalothorax and pereon dorsally and laterally (Fig. 19-31). Ventrally, the segments of the thorax are easily seen (Fig. 19-31), but they are not visible dorsally. The first pereopods are usually chelipeds and the others are walking, or occasionally swimming, legs. The abdomen is greatly reduced, inconspicuous, and flexed tightly under the pereon (Fig. 19-31). It is not visible unless the crab is held upside down. The uropods have been lost in all but a few primitive species. In the female, all pleopods are retained and used to hold the brooded egg mass (egg sponge) under the flexed abdomen. In males, the pleopods are lost except for two anterior pairs that are gonopods used for intromission. Flexion and reduction of the abdomen shift the center of gravity forward, under the thorax and the locomotory appendages (pereopods). Crabs can crawl forward slowly, but they more typically move sideways, especially when in a hurry. In this gait, the leading legs pull by flexing and the trailing legs push by extending. The chelipeds usually are not used in crawling.

Dissodactylus mellitae, which is commensal on sand dollars, is among the smallest decapods. The carapace of this crab is only a few millimeters wide. At the other extreme, *Macrocheira kaempferi,* a Japanese spider crab, has the greatest leg span of any living arthropod (Fig. 19-32). The carapace may attain a length of 45 cm and the chelipeds may span 3 m. Some lobsters have longer bodies, but none can approach this leg span. In general, decapods tend to be larger, sometimes much larger, than other crustaceans.

Except as larvae, brachyurans are absent from pelagic habitats and most crabs cannot swim. Members of the Portunidae, however, are notable exceptions. This swimming crab family includes the common edible blue crab *(Callinectes sapidus)* of the American Atlantic coast. Swimming crabs are the most powerful and agile swimmers of all crustaceans. The last pair of thoracic legs terminate in broad, flattened paddles (Fig. 19-31) that create thrust by moving in a propeller-like figure-eight pattern. Portunid crabs can swim sideways, backward, and sometimes forward with great rapidity. Nevertheless, like other crabs, they are chiefly benthic animals and swim only occasionally.

Although the chelipeds are important in defense, other protective devices and habits have evolved in many brachyurans. Some species carry sea anemones on their chelipeds. Some spider crabs are covered with hooked setae to which the crab attaches a variety of foreign objects (Fig. 19-33). This "decorating" habit is highly developed in some species and the body may be completely overgrown with living algae, sponges, and other sessile organisms. The camouflaged decorator crab remains relatively immobile during the day, when predators are active.

Members of the pea crab family, Pinnotheridae, are commensal with other invertebrates. Pea crabs are always small, usually less than 1 cm in carapace width. Various species are adapted for living in polychaete tubes and burrows, in the mantle cavities of bivalves and snails, in tunicates, on sand dollars, and with other animals. Often the body is considerably modified for commensal existence. For example, the female oyster pea crab, *Pinnotheres ostreum,* spends her life inside the oyster shell and has a soft, unsclerotized exoskeleton whereas the male, which must leave the safety of the oyster to find a female, has a hard, sclerotized cuticle. These tiny crabs are often encountered when shucking oysters. Pea crabs respond positively to substances produced by the host and can detect and follow a scent trail to find its host.

Many species of crabs support important commercial fisheries throughout the world. The blue crab, *Callinectes sapidus* (Fig. 19-31), is the commercially important edible crab along the eastern and Gulf coasts of the United States, especially in the Chesapeake Bay. Soft-shell crabs, so highly prized by epicures, are simply recently molted blue crabs whose carapace has not yet calcified. On the west and northeast coasts of the United States and in Europe, species of *Cancer* are used as food. *Cancer magister,* the Dungeness or market crab, is the most important edible species on the California coast.

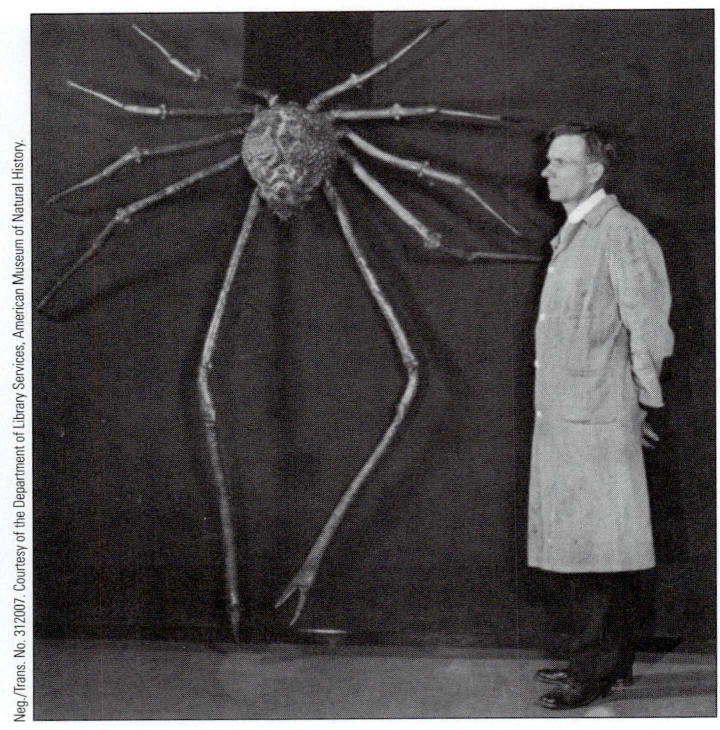

FIGURE 19-32 Decapoda, Brachyura: The Japanese spider crab *Macrocheira kaempferi,* the largest living arthropod. Note the five pairs of pereopods, the first of which are the chelate chelipeds. The abdomen of this brachyuran crab is folded out of sight under the thorax.

Nutrition

FEEDING

The decapod mouth is on the ventral surface of the head surrounded and covered by a multitude of mouthparts. The area around the mouth, including the mouthparts, is the **buccal frame** (Fig. 19-31). Anteriorly and posteriorly it is protected by outgrowths of the body wall, which are not appendages. Of these, the labrum, or upper lip, is anterior to the mouth and the paragnath, or lower lip, is posterior to it. Six pairs of appendages are associated with the decapod mouth. The mandibles flank it, whereas the two pairs of maxillae and three of maxillipeds all attach posterior to it but extend anteriorly to form a stack of five pairs of appendages covering the mandibles and mouth (Fig. 19-2A). The largest, heaviest, and outermost of these are the third maxillipeds, which define the buccal frame and cover and protect the remaining, more delicate appendages.

Decapods exhibit a wide range of feeding habits and diets, but most species combine predation with scavenging. Many decapods are herbivores, omnivores, or detritivores that feed on large food particles. Food is captured or grasped with the chelipeds and passed to the third maxillipeds, which push it between the other mouthparts. While the mandibles hold the food, pieces are torn away by the maxillae and maxillipeds and transferred to the mouth.

The chelipeds are adapted to the feeding habits and food preferences of each species. Those that scrape algae from rocks or feed on detritus from the surface of sand and mud commonly have chelipeds with spoon-shaped fingers. Those that feed on molluscs have dimorphic chelipeds. The heavy crusher claw has blunt, molarlike teeth and is adapted for crushing shells whereas the cutter claw is lighter and adapted for cutting flesh (Fig. 19-27, 19-39).

Detritus feeding and suspension feeding utilize small food particles. Fiddler crabs (Fig. 19-42) are small brachyurans that feed on organic detritus on intertidal sand or mud flats when the tide is out. A fresh supply of detritus is deposited on the mud by each flooding tide. Fiddler crabs have one or two small chelipeds that scoop up mud and detritus. The doorlike third maxillipeds open and the collected material is dumped into the buccal frame. Water from the branchial chamber washes the sediment through setal filters on the second and first maxillipeds to separate organic and mineral particles. The organic particles are ingested and the mineral particles are shaped into small pellets and discarded on the beach (Fig. 19-42). These pellets accumulate and may cover the surface of the mud flat by the time the tide returns.

Many decapods are filter feeders that pass a water current through a sieve and harvest its load of suspended organic particles. The crustacean may generate the feeding current itself or take advantage of an existing current. Burrowing shrimps *(Callianassa, Callichirus, Upogebia)* dig burrows in soft sediments and generate a feeding and ventilating current by rhythmic beating of their pleopods. Some commensal pea crabs take advantage of the respiratory current generated by the host. The maxillipeds are used as filtering appendages by many anomurans and the pea crabs, although some pea crabs living in oysters are known to feed on the mucous food strings collected by the host. Some burrowing shrimps filter with the first two pairs of legs,

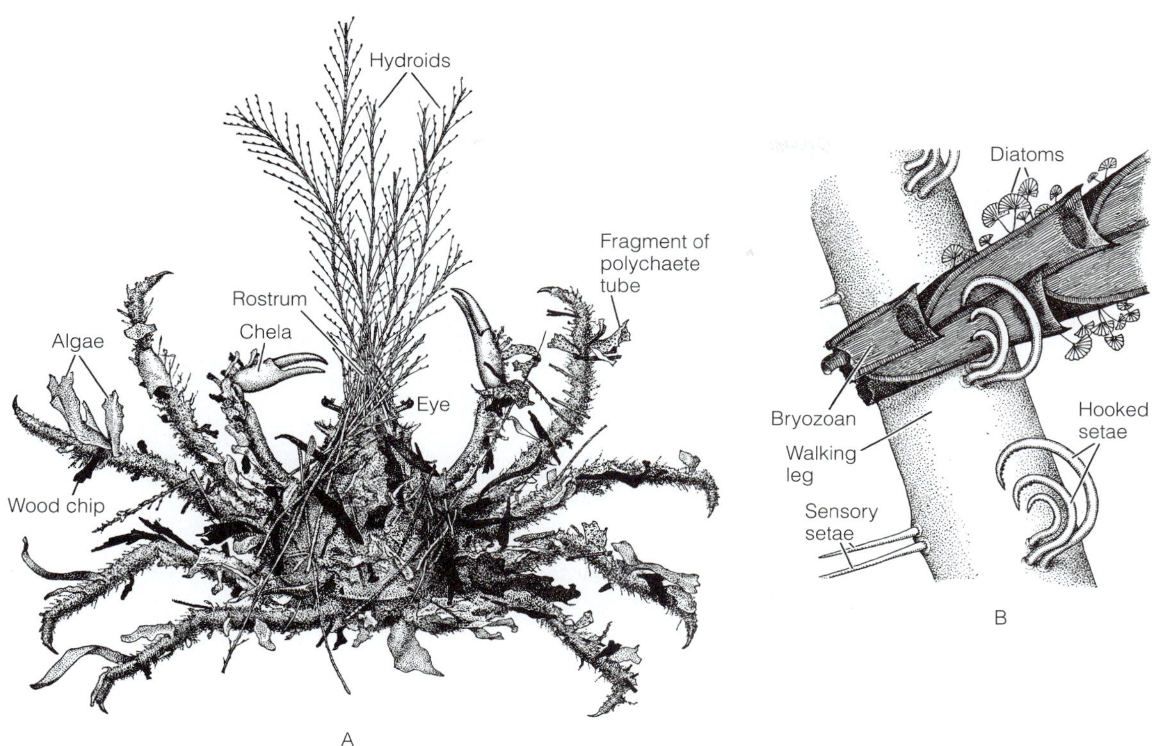

FIGURE 19-33 Decapoda, Brachyura. **A,** A decorator crab (spider crab), *Oregonia gracilis,* camouflaged with attached organisms. **B,** Hooked setae on the leg of the decorator crab *Podochela hemphilli.* *(From Wicksten, M. K. 1980. Decorator crabs. Sci. Amer. 242(2):149.)*

whereas the porcelain crabs use the third pair of maxillipeds. Porcelain crabs also scrape detritus from rocks with their chelipeds.

Among the most interesting filter feeders are the mole crabs, *Emerita* (Fig. 19-29A). These crabs live in the intertidal zone of open, surf-swept beaches and move up and down the beach with the flooding and ebbing tide. As each new wave rushes up the beach, it washes the crab out of the sand and moves it. The crab then uses its uropods and fourth legs to dig rapidly back into the sand. Thus established in the feeding position, with only the antennae and eyestalks extending above the sand (Fig. 19-29B), it feeds in the receding wave. The long, densely fringed, second antennae lie just above the sand surface and make a characteristic V-shaped ripple in the shallow receding water. In species such as *Emerita talpoida,* these antennae filter plankton and detritus from the receding wave. The first antennae together form an inhalant siphon to deliver oxygenated water to the gills.

DIGESTION

The decapod gut consists of an elaborate cuticularized foregut, an endodermal midgut with extensive digestive ceca, and a cuticularized hindgut. The role of the foregut is trituration, hydrolysis, and the separation of small digestible materials from large indigestible particles. Fine particles and solutes are routed to the digestive ceca, whereas coarse materials are either regurgitated through the mouth or sent to the intestine, incorporated into fecal pellets, and voided through the anus.

Typically, the foregut consists of a short esophagus leading into a capacious stomach (proventriculus), which in decapods is divided into a large, anterior cardiac chamber (Fig. 19-2B, 19-34) and a smaller, posterior pyloric chamber. Being ectodermal, both chambers are lined with chitinous exoskeleton having teeth and setae in their walls. The two chambers are separated from each other by a constriction, through which there is bidirectional flow. In general, the dorsal region of each chamber specializes in large, indigestible particles destined for the intestine whereas the ventral portions process small, digestible organic material to be sent to the digestive ceca for absorption

The **cardiac chamber** contains a cuticular gastric mill and a setal screen. The **gastric mill** consists of calcareous ossicles (teeth) that grind the food (Fig. 19-34). The triturating action of the mill and the movement of the stomach walls are accomplished by extrinsic muscles attached to the ossicles. During trituration, extracellular digestion occurs by mixing the food with enzymes from the digestive ceca. The **setal screen** of the cardiac chamber is a filter of finely spaced setae that guards the **ventral channel** passing from the cardiac chamber to the pyloric chamber. After trituration and digestion, fine particles and solutes pass through the setal screen and enter the ventral region of the pyloric chamber. Coarse indigestible particles

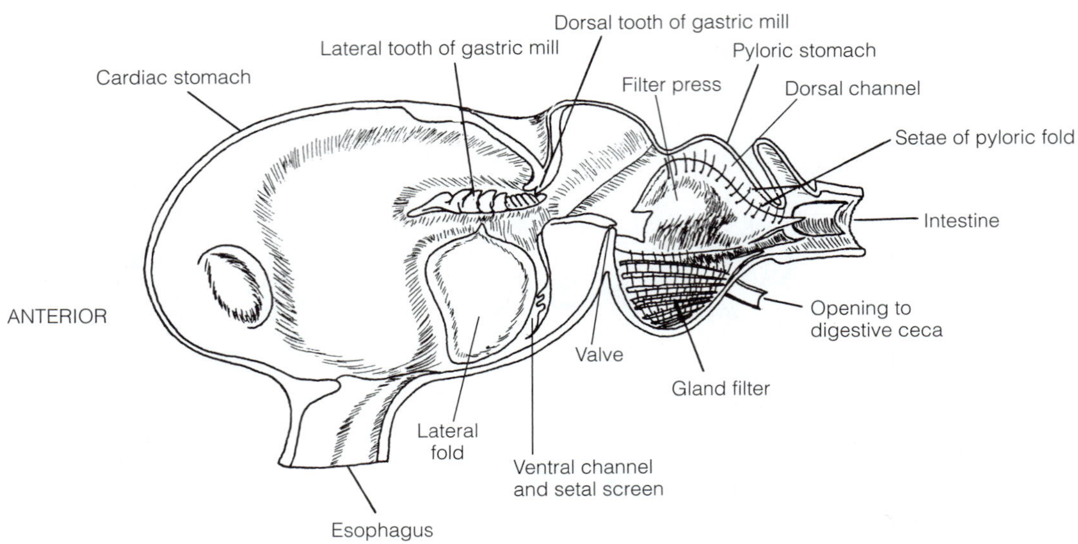

FIGURE 19-34 Decapoda. Stomach of the crayfish, *Astacus*, in sagittal section. *(After Kaestner, A. 1970. Invertebrate Zoology. Vol. 3. Wiley-Interscience, New York.)*

rejected by the setal screen follow the **dorsal channel** out of the cardiac chamber, through the pyloric chamber, and into the intestine to become feces.

The **pyloric chamber** is divided into a dorsal channel that leads directly to the intestine and a complex ventral region through which digestible materials pass on their way to the digestive ceca (Fig. 19-34). The ventral region consists of a filter press and a gland filter. Digested food, which has been prefiltered by the setal screen, passes into a muscular middle region of the pyloric chamber known as the **filter press** (Fig. 19-34). The **gland filter** guards the openings to the digestive ceca and is another setal filter with a mesh size even smaller than that of the setal screen. Contractions of the muscles of the filter press force the fluid across the gland filter and into the digestive ceca, whereas the rejected material is routed to the dorsal channel and into the intestine. The dorsal and ventral portions of the pyloric chamber are separated by the **pyloric fold,** whose setae prevent large particles from entering the gland filter.

The digestive ceca are large organs consisting of numerous blind tubules arising as diverticula of the midgut (Fig. 19-35). The tubule epithelium has many roles in digestion. Its secre-

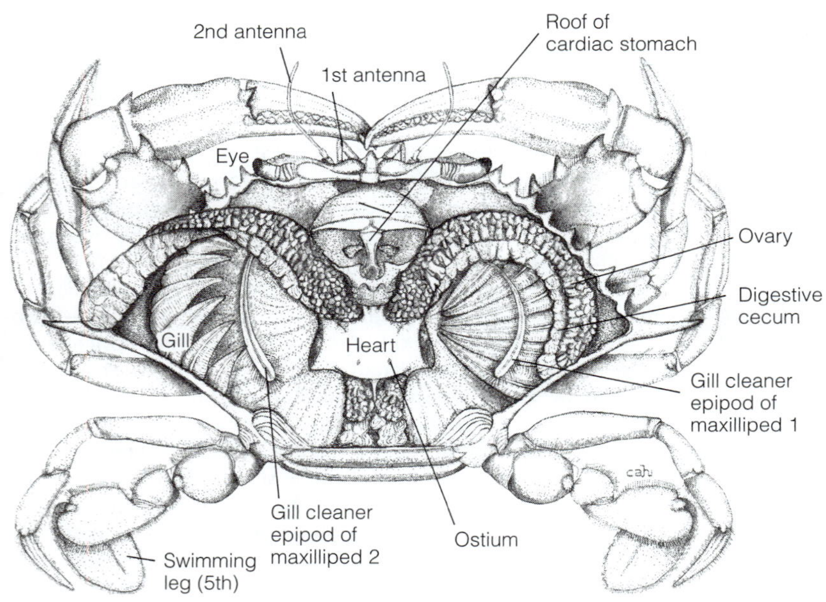

FIGURE 19-35 Decapod, internal anatomy: A female blue crab, *Callinectes sapidus,* in dorsal dissection. *(Drawing by Carolyn Herbert.)*

tory cells produce the digestive enzymes that are sent anteriorly to the cardiac chamber. Its absorptive cells phagocytose organic particles from the setal filter and digest them intracellularly. These cells are also responsible for the pinocytosis of soluble organic molecules. The absorptive cells store organic molecules or release them into the surrounding blood in the hemocoel and package wastes from intracellular digestion in vesicles that undergo exocytosis into the tubule lumen.

Coarse, indigestible material in the intestine is enclosed in a peritrophic membrane secreted by the midgut epithelium. The hindgut (rectum), a simple cuticularized tube connecting the intestine with the anus, is responsible for pumping an antiperistaltic flow of water into the gut from the anus, probably to facilitate the passage of feces through the intestine.

Gas Exchange

Gas exchange is accomplished by up to 24 pairs of gills associated with the thoracic appendages. The gills are epipods that arise in the embryo as tubular evaginations of the coxa, but during development some migrate to the nearby body wall. The gills contain circulating blood and are surrounded by the circulating respiratory medium (water or air). The gill cuticle is very thin and permeable to facilitate the diffusion of gases across it. Because of its permeability, it is also permeable to other materials, such as ammonia and a variety of ions. The gills are enclosed in the branchial chamber, or gill chamber, formed of lateral extensions of the carapace known as **branchiostegites** (Fig. 19-36, 19-3A). The respiratory current is circulated through the chamber.

Each gill consists of a longitudinal axis attached to the body and numerous gill filaments or lamellae (Fig. 19-37) that extend from it. Afferent blood channels (veins) in the axis bring unoxygenated blood to the gills and efferent channels take oxygenated blood away. The thin, permeable cuticle that covers the filaments is molted periodically along with the rest of the exoskeleton. Gill filaments may be branched (dendrobranchiate), unbranched (filamentous or trichobranchiate), or lamellar (phyllobranchiate).

Collectively, the gills form a curtain that divides each branchial chamber into a medial **inhalant** (hypobranchial) **chamber** and a lateral **exhalant** (epibranchial) **chamber** (Fig. 19-36). Water enters the inhalant chamber through the inhalant apertures, flows laterally across the curtain of gill filaments into the exhalant chamber, and then flows anteriorly to exit through the exhalant apertures. The respiratory current is generated by the beating of a paddlelike **gill bailer** (scaphognathite), which is an epipod of the second maxilla in the anterior exhalant chamber. The bailer pulls water into the inhalant chamber through openings above the coxae (Fig. 19-38).

The two exhalant apertures of decapods are always located anteriorly, beside the buccal frame, but the positions of the inhalant apertures vary. In shrimps, the ventral margins of the carapace fit loosely against the sides of the body, and water enters the branchial chamber at any point along the posterior and ventral edges of the carapace (Fig. 19-38A). In other decapods, the carapace fits somewhat more tightly and the entrance of water is limited to the posterior carapace margins and around the bases of the legs (Fig. 19-38B). The inhalant apertures are most restricted in the brachyuran crabs, in which they are confined to a well-defined, anterior opening at the base of each cheliped (Fig. 19-38C). The forward position of the restricted inhalant openings in brachyurans results in water taking a U-shaped course through the gill chambers. Upon entering the inhalant opening, the water passes posteriorly into the medial inhalant side of the chamber and then moves dorsally, passing between the gill lamellae, where gases are exchanged. The exhalant current flows anteriorly in the upper part of the gill chamber and issues from exhalant apertures in the upper lateral corners of the buccal frame.

Because the majority of decapods are bottom dwellers and many are burrowers, various mechanisms have evolved to avoid clogging the gills with sediment. The bases of the chelipeds in crabs and the coxae of the legs of crayfishes and lobsters bear

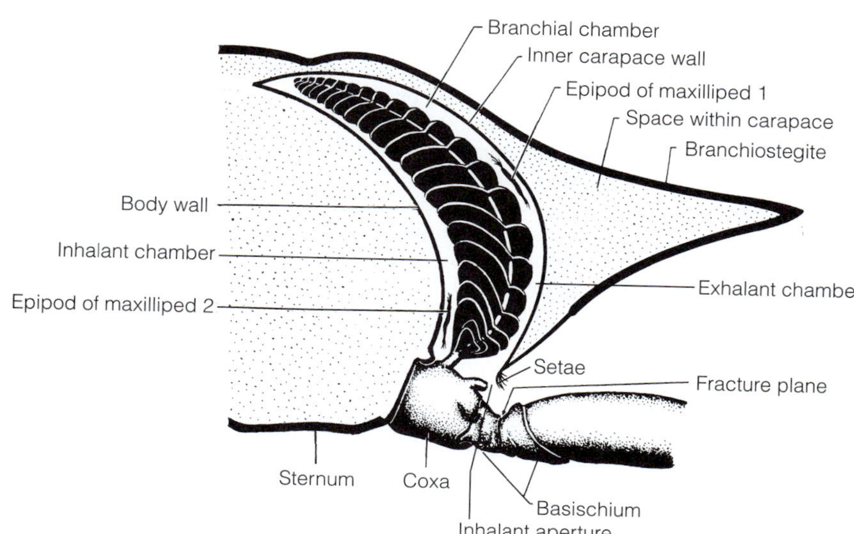

FIGURE 19-36 Decapod gas exchange. Cross section through the gill chamber of a crab showing a gill attached to the coxa. *(From Kaestner, A. 1970. Invertebrate Zoology. Vol. 3. Wiley-Interscience, New York.)*

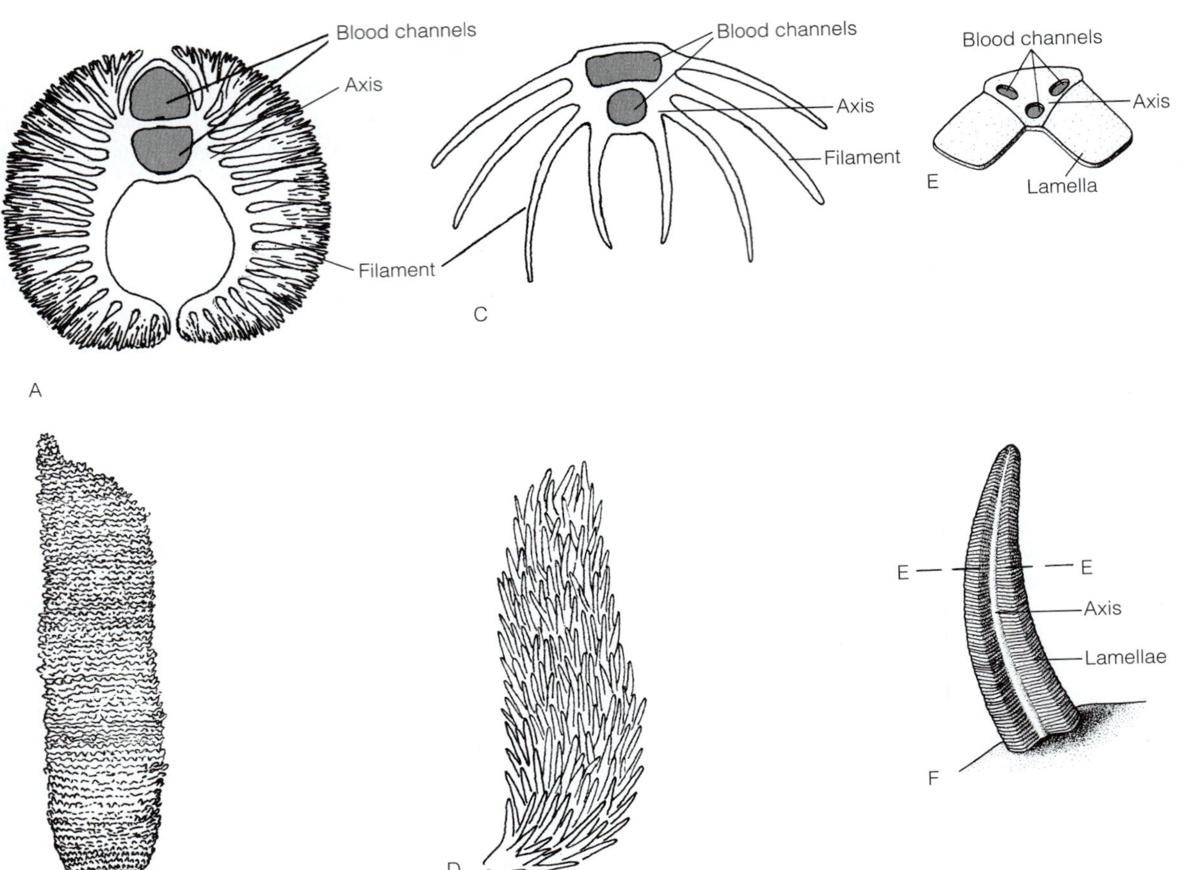

FIGURE 19-37 Decapod gas exchange. **A** and **B,** Dendritic, or dendrobranchiate, gill of penaeid shrimps. The gill filaments are branched. **A,** Transverse section. **B,** Lateral view of the entire gill. **C** and **D,** Filamentous, or trichobranchiate, gill of lobsters, crayfishes, and some other groups. The gill filaments are not branched. **C,** Transverse section. **D,** Lateral view of the entire gill. **E** and **F,** Lamellar, or phyllobranchiate, gill of brachyuran crabs, most anomurans, and shrimps. The gill filaments are platelike. **E,** Transverse section. **F,** Lateral view of the entire gill. *(All after Calman.)*

setae that filter the inhalant stream (Fig. 19-36). In some shrimps the gills are cleaned with the first and second pairs of legs and in some anomurans with the reduced last pair of legs. In crabs the gills are cleaned by the fringed epipods of the three maxillipeds (Fig. 19-35). These elongate epipods, especially those of the first maxillipeds, sweep up and down the surface of the gills, removing detritus. To further clean the gills and branchial chamber, the gill bailer of many decapods periodically reverses its beat and back-flushes the gills. When partially buried in sand, many crabs reverse the direction of the ventilating current so that water enters the more dorsally located "exhalant" apertures beside the buccal frame and exits through the ventral "inhalant" aperture at the base of the cheliped.

Internal Transport

The decapod hemal system consists of a heart, a well-developed arterial system, capillaries, and venous sinuses. The blood, or hemolymph, contains hemocytes and the respiratory pigment hemocyanin in solution in the plasma. In large, active species, such as the swimming crabs, hemocyanin transports about 90% of the oxygen required by the tissues.

The heart has lost its primitive tubular morphology and is a rectangular sac surrounded by the pericardial sinus in the thorax (Fig. 19-35, 19-2B). Three pairs of ostia penetrate its thick muscular walls. Complete circulation has been estimated to take from 40 to 60 s in large decapods.

The hemal system of the blue crab, *Callinectes sapidus,* has been carefully studied using a variety of techniques, including injection with resins and X-ray opaque dyes (Fig. 19-39). These studies have shown that well-defined arteries, arterioles, and abundant capillary-like vessels deliver blood to the tissues, but that most of the venous return is in sinuses. Although the arterial walls lack muscles, they are supported by collagen and elastic fibers and function to some degree as pressure reservoirs. Muscular sphincters in the arteries provide some control over the distribution of blood. Seven major arteries exit the heart and deliver blood to the tissue capillaries.

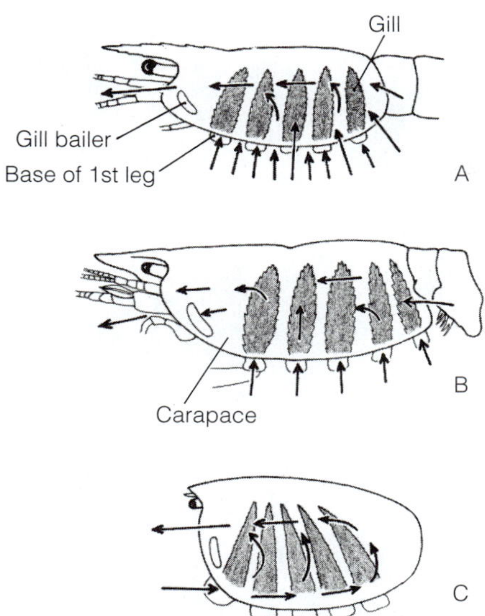

FIGURE 19-38 Decapod gas exchange. Paths of water circulation through the gill chamber of representatives of the three morphological types of decapods, showing progressive restriction of openings into the chamber. **A,** Shrimp: water enters along entire ventral and posterior margin of the carapace. **B,** Crayfish: water enters at the bases of the legs and at the posterior carapace margin. **C,** Crab: water enters only at the base of the cheliped.

Venous blood leaving the capillaries enters tiny lacunae in the tissues and then collects in large sinuses, eventually making its way to a large, median, ventral thoracic sinus. From here it enters a set of well-defined afferent and efferent branchial veins to and from the gill lamellae. Oxygenated blood leaves the gills via the efferent branchial vein to enter the **branchiopericardial vein** (similar to the pneumopericardial sinus of arachnids), which empties into the pericardial sinus. Blood from the pericardial sinus enters the heart through the ostia.

Outflow from the heart is through seven arteries (Fig. 19-2B, 19-3A, 19-39). The anterior aorta extends to a swelling called the **cor frontale,** from which arise arteries to the eyes, X-organ, antennae, and brain. Paired anterior lateral arteries supply the antennal glands, stomach muscles, gonads, epidermis, digestive ceca, and some mandibular muscles. Paired hepatic arteries (lateral arteries) supply the digestive ceca. The unpaired posterior aorta (dorsal abdominal artery) supplies the abdomen and hindgut. This vessel is large in shrimplike decapods with powerful abdominal muscles, but is small in the true crabs. The largest of the seven is the unpaired sternal artery, which runs ventrally to the sternum and then divides into the anterior ventral thoracic artery, which has branches to the pereopods and mouthparts, and the posterior ventral abdominal artery running to the abdomen and hindgut.

Excretion

Most decapods are ammonotelic and excrete nitrogenous wastes across the gill epithelia. The nephridia are osmoregulatory organs responsible for regulating ion concentration and fluid volume. Marine decapods are usually isosmotic osmoconformers that match their blood osmolarity to that of the environment, whereas freshwater decapods are hyperosmotic osmoregulators that maintain internal osmolarities above the ambient level.

The decapod excretory organs are especially well-developed saccate nephridia known as green or antennal glands (Fig. 19-2A,B), with the nephridiopore on the base of the second antenna. The **excretory tubule,** which is more elaborate than in other crustaceans, is regionally specialized into a proximal labyrinth and distal bladder (Fig. 19-6B). In freshwater species an additional long, convoluted excretory canal occurs between the labyrinth and bladder. The greatly folded, glandular, spongy **labyrinth** walls selectively secrete and absorb ions to maintain the desired ion balance in the blood. The **bladder** is a storage sac for the final urine. A short duct connects the bladder with the nephridiopore. In brachyurans, the nephridiopore is covered by a small, movable, cuticular operculum.

The **excretory canal** of freshwater crayfishes resorbs ions from the ultrafiltrate to form a hyposmotic final urine. In the crayfish *Astacus,* the chloride concentration of the urine drops from 200 μM/L to 10 μM/L as it passes through the canal. Resorption of solutes by the canal enables these crustaceans to produce a dilute urine that eliminates excess water while conserving salts. As in freshwater fishes, the gill epithelium absorbs salts from the respiratory current to replace ions lost in the urine and by dialysis across the gill surfaces. The tubule of marine decapods lacks the excretory canal and forms urine that is isosmotic with the sea and blood. Although isosmotic, marine decapods nevertheless must regulate the concentrations of individual ions. For example, the nephridium of most marine decapods retains K^+ and Ca^{2+} while eliminating Mg^{2+} and SO_4^{2-}.

FRESHWATER DECAPODS

Decapods have made several invasions of freshwater and terrestrial habitats but, with the exception of the crayfishes, none has been extensive. Success in fresh water has been considerably greater than on land, largely because of the crayfishes. Freshwater decapods include shrimps, crayfishes, and brachyuran crabs, whereas terrestrial decapods are all crabs.

Freshwater shrimps belong to Atyidae and Palaemonidae. Atyids are scrapers or filter feeders that inhabit streams, pools, lakes, and subterranean waters, mostly in tropical and subtropical regions. Three species are found in the United States, one of them in Mammoth Cave, Kentucky. These are probably relictual populations surviving from more widespread distributions. Palaemonidae includes marine, brackish, and freshwater shrimps. Species of *Palaemonetes* and *Macrobrachium* have representatives in temperate and tropical fresh waters

Crayfishes are the most successful freshwater decapods. Approximately 400 species are found throughout the world in streams, ponds, lakes, and caves and over 300 of these are endemic to North America. Some live beneath stones or among debris. Many species excavate burrows, which they use as retreats and for overwintering. Burrowing crayfishes commonly inhabit river flood plains where the water table is near the ground surface and they may be semiterrestrial. A chimney constructed of excavated mud typically rises above the

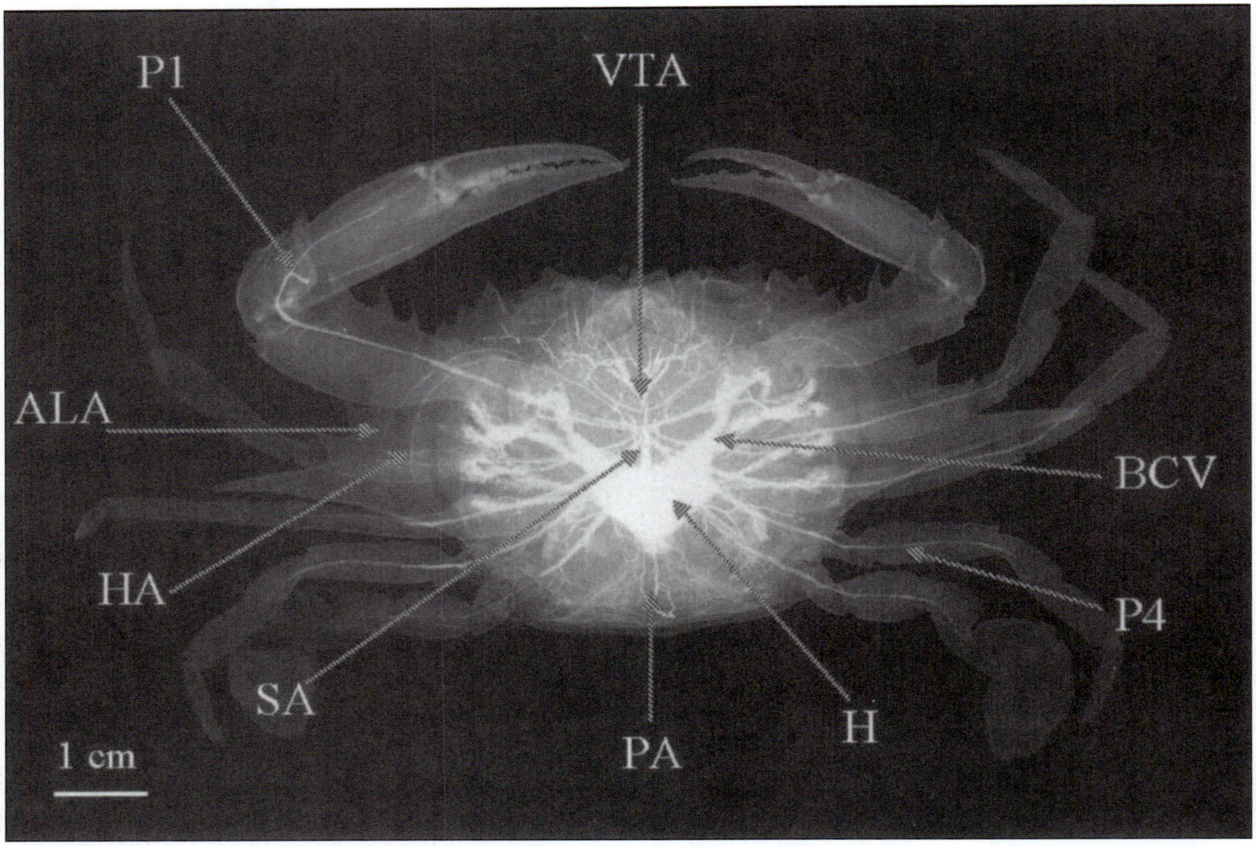

FIGURE 19-39 Decapod hemal system: The arterial system of the blue crab, *Callinectes sapidus.* This is an X-ray photograph taken following injection of barium sulfate into the pericardial sinus. Some arteries are not apparent in this preparation. The arteries of all pereopods (for example, P1, P4) are visible. In this brachyuran crab the posterior aorta (PA) is much smaller than it would be in a decapod with a large, muscular abdomen. Note the extensive branching of the ventral thoracic artery (VTA) to supply the maxillipeds and mouthparts. Note the dimorphism of the chelipeds. The crusher claw is on the left and the cutter claw on the right. The paddlelike morphology of pereopod 5 is apparent. ALA, anterior lateral arteries; BCV, branchiopericardial veins; H, heart; HA, hepatic arteries; P1, cheliped artery; P4, pereopod 4 artery; PA, posterior aorta; SA, sternal artery; VTA, ventral thoracic artery.

burrow opening. Most crayfishes are less than 10 cm long, but some Australian species reach the size of lobsters. Crayfishes are omnivores.

Many brachyuran crabs tolerate brackish or fresh water, but most have marine larvae and must return to salt water to release them. The Chinese mitten crab, *Eriocheir* (Fig. 19-40A), originated in Asian rivers and rice paddies but is now common in Europe, which it invaded around 1912, probably by way of larvae transported in ship ballast water. One of the mud crabs, *Rhithropanopeus,* of the American Atlantic coast, occurs in salt water and in freshwater ditches. It has been introduced into northern Europe and San Francisco Bay, probably via ballast water. Ballast water is a common transport mechanism resulting in the accidental and often ecologically disastrous introduction of exotic species, both marine and freshwater. The blue crab, *Callinectes sapidus,* is broadly euryhaline and ranges far up into brackish estuaries, bays, and rivers, especially those with high calcium concentrations.

Some very successful but poorly known freshwater decapods are the so-called river crabs (Potamidae; Fig. 19-40B). Occurring in all tropical regions, from the mouths of rivers to elevations of 5000 m, these brachyuran crabs have direct development with no larval stage and complete their entire life cycle in fresh water. All brackish and freshwater crabs excrete urine that is isosmotic with the blood and regulate ion balance by ion absorption through the gills.

TERRESTRIAL DECAPODS

Terrestrial decapods are restricted to a few families of anomuran and brachyuran crabs. Anomura is represented by the tropical land hermit crabs, *Coenobita,* which are often sold in pet shops, and the closely related coconut crab, *Birgus.* Both are found in Indo-Pacific regions, and *Coenobita* also occurs in the tropical western Atlantic and Caribbean. The land hermits live close to the shore and use either fresh or seawater to wet

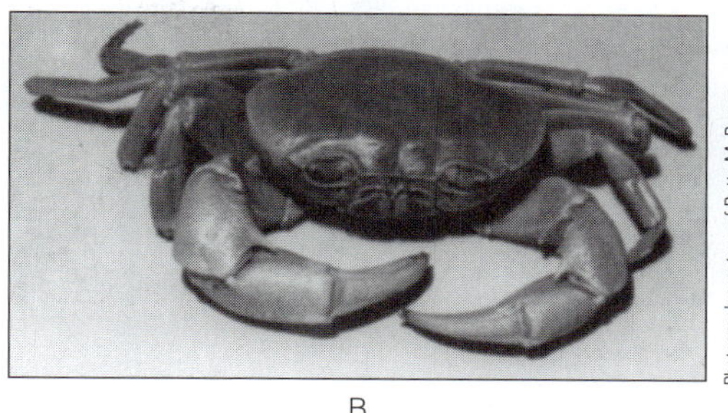

FIGURE 19-40 Decapoda: Freshwater crabs. **A,** The Chinese mitten crab, *Eriocheir sinensis,* in a rice paddy. This amphibious freshwater brachyuran is native to southern Asia but has been introduced into the Rhine and Elbe Rivers of Europe. **B,** A river crab, *Potamon anomalus* (Potamonidae). *(A, After Schmitt)*

their body and respiratory surfaces. Adult coconut crabs have abandoned the hermit crab habit and acquired a crablike form with a flexed abdomen. The adults live in burrows farther back from the sea, but remain close to the coast. Coconut crabs feed on carrion and vegetation, both decaying and fresh, and can dehusk and open coconuts. They obtain water by drinking. Both *Coenobita* and *Birgus* have reduced gills and branchial chambers converted into lungs with highly vascularized areas for gas exchange. At least in *Birgus,* most nitrogenous waste is excreted as water-conserving uric acid. Internal gas-exchange and uricotely are adaptations for water conservation characteristic of terrestrial arthropods.

Most terrestrial decapods are brachyuran crabs probably because, of all the decapods, brachyuran locomotion is best suited for movement on land and their enclosed branchial chamber, with its restricted inhalant and exhalant apertures, is the easiest to protect from desiccation. Although the living terrestrial crabs are derived independently from several different land invasions, most display similar adaptations.

Desiccation is a major problem facing every terrestrial animal and the chief reason for the decapods' lack of success on land is probably their failure to control it. Gills are poorly suited for terrestrial respiration because they are difficult to support in air and are notoriously vulnerable to evaporative water loss. There is a tendency for the branchial chamber to become lunglike, adding new respiratory surface and reducing gill surface to decrease water loss. Fortunately, due to the far higher oxygen concentration of air in comparison with water, less respiratory surface is required. The gill bailer continues to generate the ventilating current, which is now air instead of water. Ammonia is unsuitable as an excretory product for terrestrial animals because its toxicity demands large amounts of water for dilution. A nephridium incapable of conserving water by making a scant, concentrated urine is maladaptive for terrestrial life. No terrestrial crab can excrete a urine hyperosmotic to the blood even though such a capability would aid in water conservation. They can, however, save water by excreting scant urine. Some, such as the coconut crab, excrete uric acid, which is eliminated with the feces and requires no water. A cuticle lacking a waxy waterproof epicuticle subjects its owner to serious desiccation problems. The crustacean cuticle is not waterproof, and evaporation across the cuticle of terrestrial crabs is always much greater than is that of more successful terrestrial arthropods such as arachnids and insects. A terrestrial crab must have a source of water to replace that lost by evaporation and to keep the respiratory surfaces moist. Water is returned to the blood primarily by drinking, and the respiratory surface is often kept moist by taking up water from soil, dew, or rain.

Terrestrial crabs tend to live in microhabitats having a lower temperature and higher humidity than ambient, and consequently evaporation is lower. Most dig into the soil and inhabit burrows or live beneath stones, and they are usually nocturnal. They are primarily vegetarians and scavengers and all can run rapidly. Their eyes are generally well developed.

The name *land crab* usually refers to the members of Gecarcinidae, all of which are terrestrial. They are found in tropical and subtropical America, West Africa, and the Indo-Pacific area. The family includes *Cardisoma* (Fig. 19-41A) and *Gecarcinus,* which live in burrows in coastal fields and woods in Texas, southern Florida, tropical America, and the West Indies. Although terrestrial as adults, reproductive requirements restrict the gecarcinids to habitats within about 10 km of the coast. Females brood the eggs until the embryos reach the protozoea or zoea stage, at which time the females join mass migrations to release the larvae in the sea.

There are also some terrestrial members of the river crab family, Potamidae. These crabs have direct development, without a larval stage, and the young hatch as miniature crabs. The females do not return to the sea to release larvae, as other terrestrial crabs do. Alone among terrestrial crabs, the potamids are independent of the sea and may occur hundreds of miles inland.

Ocypodidae includes some of the most familiar amphibious and terrestrial crabs, including ghost and fiddler crabs. Ghost crabs *(Ocypode)* are common in many parts of the world, including the east coast of the United States (Fig. 19-41B). They are several times larger than fiddler crabs and are

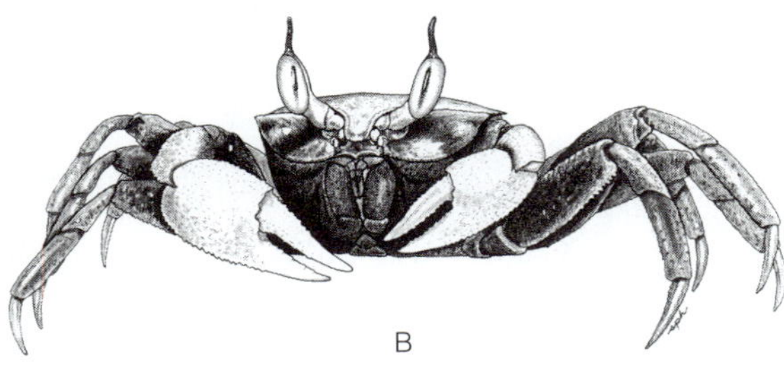

FIGURE 19-41 Brachyura: Terrestrial crabs. **A,** The gecarcinid land crab *Cardisoma guanhumi,* of the West Indies and Florida. **B,** The Indo-Pacific ghost crab, *Ocypode ceratophthalma.* *(B, Based on a photograph by Healy and Yaldwyn.)*

solitary, never occurring in large aggregations. Their widely separated burrows are usually above the high-tide mark on the upper beach or in the dunes. Ghost crabs are largely nocturnal and visit the lower beach to prey on clams, mole crabs, and even sea turtle hatchlings, or to scavenge for dead organisms.

About 65 species of semiterrestrial fiddler crabs *(Uca)* are known. They live on protected, low-energy, sand and mud beaches of bays and estuaries, in brackish marshes, and in mangroves. Although most species are tropical and subtropical, fiddler crabs are found on both the east and west coasts of North America. The crabs dig their burrows in the intertidal or supratidal zones and then close and occupy them during high tide. With the ebbing tide, the crabs come out of their burrows to feed, court, and mate (Fig. 19-42), and large numbers may be present on a beach.

The family Grapsidae contains perhaps the most ecologically diverse assemblage of crabs, including marine, brackish-water, freshwater, amphibious, arboreal, and terrestrial species. Amphibious grapsids include *Sesarma,* which lives beneath drift and stones; the agile, rock-inhabiting *Grapsus;* and numerous mangrove inhabitants such as *Aratus,* which lives in trees. The Jamaican *Metapaulias depressus* spends part of its life in the water-holding leaf axils of arboreal bromeliads.

Nervous System

Decapods display a great deal of variability in the cephalization of the nervous system. In primitive decapods, such as shrimps and crayfishes with large muscular abdomens, the CNS is relatively uncephalized. It consists of a dorsal brain, a subesophageal ganglion serving all the mouthparts except the third maxillipeds, and a ventral nerve cord with segmental ganglia in thoracic segments 3 through 8 and all abdominal segments (Fig. 19-2B). At the other extreme, the brachyurans, with their reduced abdomens, have a highly cephalized nervous system in which all the segmental ganglia have coalesced to form a single very large subesophageal ganglion (thoracic ganglion) and there is no ventral nerve cord. Other decapods with intermediate development of the abdomen, such as the anomurans, exhibit intermediate cephalization.

Sense Organs

As in other crustaceans, the legs and antennae are important sites for the detection of environmental information. Aesthetascs typically are present on the first antennae (Fig. 19-7A)

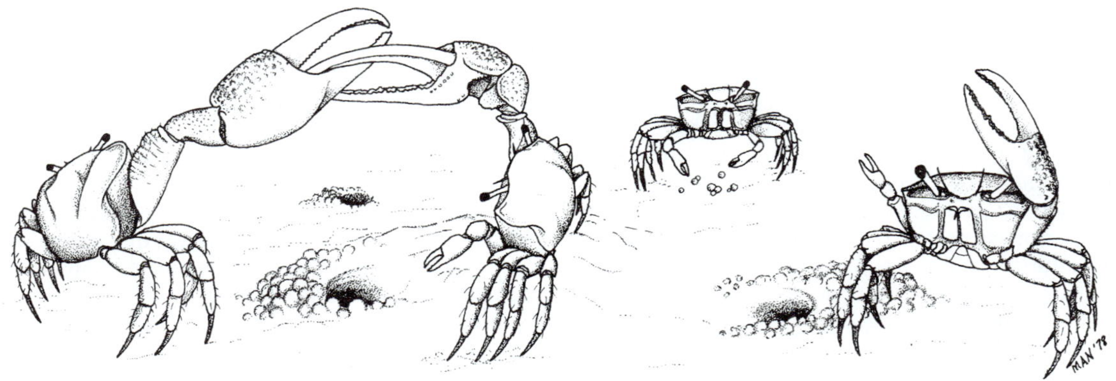

FIGURE 19-42 Brachyura: Fiddler crabs on a beach at low tide. Three burrow openings are shown, each surrounded by large balls of sediment produced during the excavation of the burrow. The crab in the background with two small chelipeds is a feeding female surrounded by discarded small sand pellets. The two males on the left are engaged in ritualized combat. The male on the right is waving the large cheliped in courtship display.

and are important for locating food and recognizing conspecifics and their sexual condition. Many decapods groom the antennae frequently to prevent fouling of the aesthetascs.

Although there are some blind decapods, particularly in the deep sea and among cave-dwelling crayfishes, the stalked compound eyes are usually highly developed and the eyestalk is more mobile than in most other crustaceans. The ommatidia consist of a cuticular cornea, a four-celled cone, a cone stalk (in superpositional eyes), and a retinula of eight retinular cells (seven of which contribute to the rhabdome). Both appositional and superpositional eyes are present in decapods.

Almost all decapods have a statocyst in the basal article of each first antenna (Fig. 19-7B). The decapod statocyst is a water-filled cavity that arises by invagination of the surface ectoderm and retains an opening to the outside. The cavity is lined by tiny mechanoreceptive setae. The dendrites of sensory neurons attached to the setae report any deflection of a seta to the brain. A statolith composed of sand grains cemented together by secretions of the statocyst epithelium rests on the bed of these setae and is unattached and free to move. The heavy statolith, pulled down by gravity, displaces a characteristic set of setae when the animal is in its resting posture, and a steady procession of signals is sent to the brain via the antennular nerve. Any deviation from the resting posture results in movement of the statolith on its bed of setae, stimulating different setae and causing a change in the rate at which signals are sent to the brain. The brain interprets the input from the receptor setae to determine the orientation of the animal in the gravitational field. Some crabs temporarily lose their sense of balance during ecdysis.

Mechanoreceptive setae are **tonic receptors.** This type of receptor sends the brain a constant background of spontaneous impulses, even when no stimulus is present. A stimulus, such as the deflection of a seta by a displaced statolith, causes a change in the frequency of the impulses. The brain interprets this change in frequency, not the ever-present background impulses, as evidence of an event. **Phasic receptors,** in contrast, send the brain an action potential when an event occurs and are quiet at other times.

The statocyst and its setae are exoskeletal structures that must be replaced, along with the statolith, with each molt. The statolith, however, is not epidermal, and after ecdysis it must be reconstructed with new sand grains from the environment. Sand grains for the new statolith enter the statocyst opening when the head is buried in sand or when the animal deliberately inserts sand grains into the cavity. The role of the statocyst was first demonstrated by rearing shrimp in an aquarium containing tiny iron particles but no sand. Upon molting, the shrimp utilized the iron to construct new statoliths and then responded to a magnetic field as if it were a gravitational field. For example, with a magnet held above them, the shrimp behaved as if they were upside down and attempted to right themselves even though they were already in their resting posture with respect to the gravitational field.

Crabs and some other decapods have statocysts capable of detecting acceleration in different directions, as the semicircular canals of the vertebrate inner ear do. These statocysts have additional setae, remote from the statolith, that are deflected by the inertia of the water in the statocyst cavity (as are the neuromasts in vertebrate semicircular canals). For example, when at rest, both the cuticle and the water in the statocyst are motionless and the unstimulated neurons send a steady procession of tonic signals to the brain. If the head moves, the cuticle of the statocyst moves at the same speed as the head, but the water lags behind. This causes the setae to move through the water and be deflected by it. The neurons alter the rate of propagation of the signals being sent to the brain, which interprets this as acceleration.

Endocrine and Neurosecretory System

The endocrine and neurosecretory systems of crustaceans and insects are the best known of any invertebrates, but among the crustaceans only the malacostracans, and especially the decapods, have been well studied. As in vertebrates, many extended or sustained events, such as molting, reproduction, and chromatophore dispersal, are controlled by hormones. Most crustacean hormones are neurosecretory products synthesized in the cell body and released by the axons of secretory neurons, but endocrine organs outside the nervous system also produce some hormones, usually under the control of the nervous system. Hormones are released into the blood for transport to their target organs.

The most important neurosecretory system of decapods is in the eyestalks (Fig. 19-43). (Sessile-eyed crustaceans have equivalent areas near the eyes.) Each eyestalk bears a compound eye at its apex, and within the stalk are the optic nerve and several clusters of secretory neuron cell bodies collectively

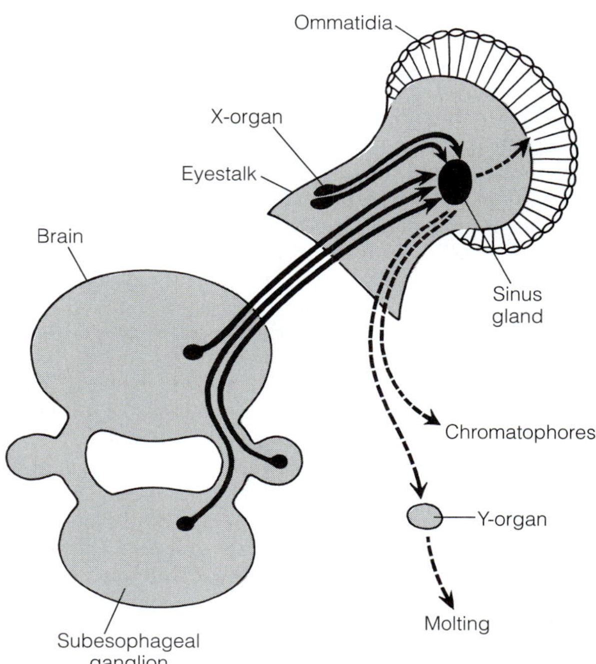

FIGURE 19-43 Crustacea: Eyestalk neurosecretory system of a brachyuran crab. Axonal pathways are indicated by solid lines. Hemal pathways are dashed. *(Redrawn from Welsh, W. H. 1961. Neurohumors and neurosecretion. In Waterman, T. H. (Ed.): The Physiology of Crustacea II. Sense Organs, Integration and Behavior. Academic Press, New York. pp. 281–311.*

known as the **X-organ.** Each cluster of neurons secretes a different hormone, one of which is **molt-inhibiting hormone.** The axons of these neurons extend to and empty into a space known as the **sinus gland,** also in the eyestalk. The sinus gland is a center for the storage and release of hormones, but it is not known to produce any. The sinus gland is separated from the hemocoel by a thin, permeable membrane. Hormones synthesized by the neurosecretory cells of the X-organ are released from their axons into the sinus gland and move from there into the blood in the hemocoel. Normal blood circulation distributes the hormones throughout the body to their target tissues. Additional neurosecretory centers in the brain and subesophageal ganglion also send axons and hormones to the sinus gland.

The only endocrine organs known in Crustacea are the **Y-organ,** ovary, and androgenic gland. The Y-organ in the anterior cephalothorax is controlled, in part, by hormones from the sinus gland delivered by the blood. One of the hormones it secretes is **ecdysone,** which promotes molting. A similar hormone is present in insects.

Chromatophores

Pigment-containing cells known as chromatophores, located in the connective tissue below the epidermis in areas where the overlying exoskeleton is thin or transparent, contribute to the color of crustaceans. The color displayed by each chromatophore is variable, and their role is to change the color of the animal for purposes of thermoregulation or concealment. One mechanism by which chromatophores can alter the color of the animal is by dividing over a period of several days, thereby increasing the number of chromatophores and the total amount of pigment. Of more interest, however, is the ability of chromatophores to change color rapidly and reversibly, without changing the number of cells or the amount of pigment present.

The crustacean chromatophore is a cell with intracellular pigment granules and branched cytoplasmic processes (Fig. 19-44). The cells occur in groups, sometimes as syncytia but usually as independent cells, and form multicellular **chromatophore organs** of cooperating cells. Pigment granules migrate within each cell to alter the visual impact of the entire organ. When the granules are dispersed to occupy the maximum surface, their visual impact is the greatest (Fig. 19-44A). When they are concentrated in a small spot in the center of the organ, their impact is minimized (Fig. 19-44B).

White, red, yellow, blue, brown, and black chromatophore pigments are known. Red, yellow, and blue pigments are carotene derivatives obtained from the diet. The red compound so conspicuous in boiled crabs, lobsters, and shrimps is the carotenoid **astaxanthin.** In the exoskeleton of the living animal, it is conjugated with a protein and appears blue. Boiling denatures the protein and changes the color to the familiar red. Brown and black pigments are melanins. In some shrimps a single chromatophore may have one, two, three, or even four different pigments, which can migrate independently of one another. In general, however, most chromatophores are monochromatic.

The most common type of rapid (physiological) color change is the simple blanching and darkening typical of many crabs, including the fiddler crab, *Uca.* Fiddler crabs darken during the day and lighten at night (Fig. 19-44C). The rhythm is endogenous and persists in the absence of environmental cues. Many crustaceans, however, especially shrimps, are capable of more elaborate color changes and can display a wide spectral range. The little shrimp *Palaemonetes,* for example, has trichromatic chromatophores with red, yellow, and blue pigments. Through the independent movement of these three primary colors, it can adapt to any background color, even black. Other species have similar abilities. The movement of chromatophore pigments is controlled by hormones from the sinus gland.

Physiological Rhythms

Pigment migration in chromatophores and ommatidia and other physiological processes often display a rhythmic activity in crustaceans. Such physiological rhythms, timed by "biological clocks" have been studied extensively in a number of species, especially the green crab, *Carcinus meanas,* and the fiddler crab, *Uca.* Both crabs live on sand flats and in the intertidal zone of protected beaches, but *Carcinus* is active in the water at high tide and *Uca* is active on the exposed sand at low tide. Through dispersion and concentration of chromatophore pigments, both species pale at night and darken during the day (Fig. 19-44C). The endogenous rhythm persists when the crabs are kept in constant light or darkness. Fiddler crabs flown from Woods Hole, Massachusetts, to California continue to blanch and darken on the Woods Hole tide schedule. Removal of the eyestalks, which secrete the pigment-controlling hormone, disrupts the rhythm in *Carcinus* and reduces its amplitude in *Uca.*

Locomotor activity in both crabs follows the tide and is governed by a biological clock entrained to lunar time. After about a week under constant conditions, the rhythm is lost. It can be restored by simply immersing *Uca* in seawater or by cooling *Carcinus.* The clocks of both species can be set to a new tidal rhythm by subjecting the crabs to periods of cooling or high pressure that correspond to the desired tidal rhythm.

Similar rhythms have been recorded for other crustaceans, such as the blue crab, *Callinectes sapidus,* and the semiterrestrial, intertidal isopod *Ligia exotica.* Epigean (surface-dwelling) species of crayfishes display a diurnal rhythm of locomotor activity, but such a rhythm is never present in cave species.

Reproduction

Most decapods are gonochoric, although a few are hermaphroditic. Sperm transfer in decapods is indirect with spermatophores (Fig. 19-45) and usually with copulation. Fertilization may be internal or external.

In general, male decapods have a pair of testes (Fig. 19-2B, 19-3A) connected by a pair of sperm ducts to the male gonopores on thoracic segment 8 (on the coxae of the fifth legs). The testes are located dorsally in the thorax and/or abdomen and their morphology is variable. The degree of development of the sperm duct varies depending on the extent of spermatophore formation, but usually there is a proximal, secretory glandular region and a muscular, distal ejaculatory region. Spermatophores, formed in the secretory region, are thought to provide support and protection for the sperm during transfer and storage. Sperm have neither a midpiece nor flagellum and are tack- or star-shaped (Fig. 19-45D).

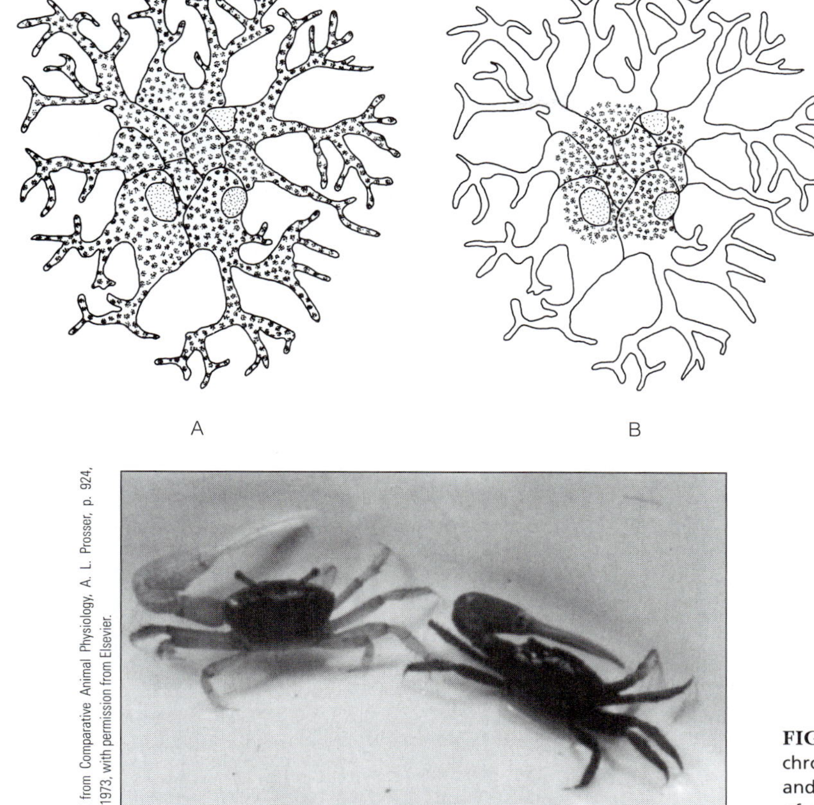

FIGURE 19-44 Crustacea. **A** and **B,** Crustacean chromatophore organs in which pigment is dispersed **(A)** and concentrated **(B).** Each chromatophore organ is a cluster of chromatophore cells. Processes radiate in all directions, not just in the plane of the page. **C,** Fiddler crabs in pale (nighttime) and dark (daytime) phases. *(A and B, Based on a photograph by McNamara, 1981.)*

Penes may be present, but are not used for intromission. Instead, the anterior pleopods, known as **gonopods,** are modified to serve as intromittent organs for the indirect transfer of spermatophores to the female. The penes, with the gonopores at their tips, usually transfer sperm from the testis to the gonopods, or rarely, directly to the female. For example, in many brachyurans the first pleopod is a long, slender tube open at both ends. The second pleopod is a long, slender piston that fits into the tube. The short, flexible penis lies at the base of the cylinder and delivers a spermatophore to the base of the tube. The piston is then rammed through the tube to push the spermatophore to the female.

Females usually have a pair of ovaries (Fig. 19-35) located dorsally in the thorax and abdomen and connected to the exterior by a pair of oviducts. The oviducts open ventrally on the sixth thoracic segment at the base of the third pereopods. Primitively, spermatophores are simply attached to the female sternites, but in most decapods they are injected into an internal or external seminal receptacle, where they are stored, sometimes for several years, until fertilization. Seminal receptacles may be invaginations of the exoskeleton (external) or a dilation of the oviduct (internal). In decapods with internal fertilization (brachyurans), spermatophores are deposited directly into an internal seminal receptacle. In those with external fertilization (most decapods), the spermatophores are placed on the ventral surface of the female or in an external receptacle and fertilization occurs as the eggs exit the gonopores.

Aquatic decapods are attracted to each other by olfactory (pheromonal) and tactile cues, whereas terrestrial species use visual and auditory signals. Precopulatory courtship rituals are common. In many decapods a male's access to the interior of the seminal receptacle is possible only during a brief period immediately following a female molt, when the cuticle is soft. In many brachyurans the male carries the female beneath his sternum while he awaits her upcoming molt. He releases her while she molts and then copulation occurs.

Visual and acoustical signals are of special importance for attraction in terrestrial species. The semiterrestrial fiddler crabs *(Uca)* go through an elaborate courtship behavior that has been studied in considerable detail. Males entice females into their burrows, where mating occurs, and the females remain there, brooding their eggs until they hatch. Male fiddlers use a greatly enlarged cheliped to attract females and to defend their burrow against other males (Fig. 19-42). Male–male encounters are a highly ritualized combat in which the large claw is held like a shield. Combat movements involve variations of pushing and extension.

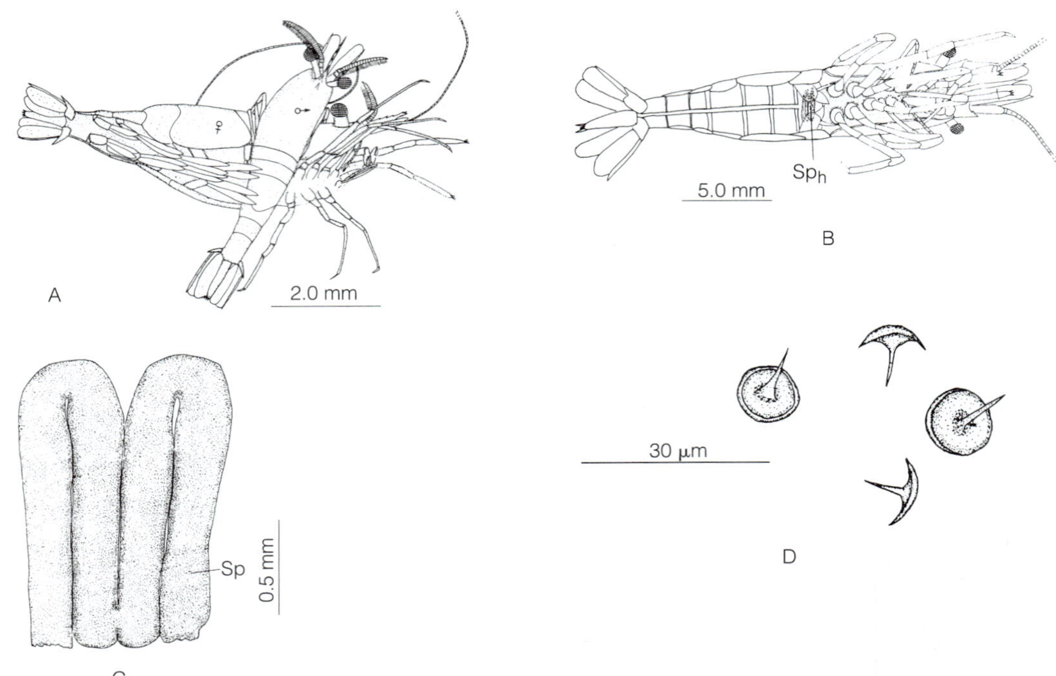

FIGURE 19-45 Decapoda: Reproduction. Sperm and spermatophore transfer in caridean shrimp. **A,** Copulating position. **B,** Ventral view of the female with attached spermatophore (Sph). **C,** Spermatophore (Sp). **D,** Spermatozoa. *(From Bauer, R. T. 1976. Mating behaviour and spermatophore transfer in the shrimp* Heptacarpus pictus. *J. Nat. Hist. 10:415–440.)*

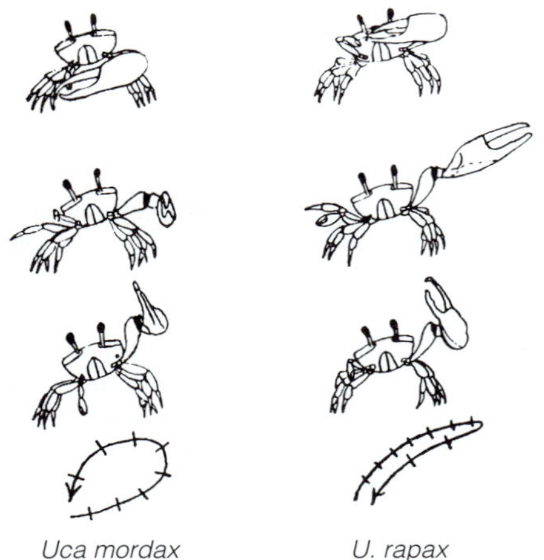

FIGURE 19-46 Brachyura: Reproduction. Cheliped waving by the fiddler crabs *Uca mordax* and *U. rapax.* The starting position is shown at the top of each sequence, and the path of the cheliped is described at the bottom by an arrow. The cheliped is commonly moved in a series of jerks, which are indicated by the crossbar on the arrow path. *(From Salmon, M. 1967. Coastal distribution, display and sound production by Florida fiddler crabs (genus* Uca*). Anim. Behav. 15:449–459.)*

The male fiddler crab attracts a female by waving the large claw in a semaphore fashion with a species-specific pattern of movement (Fig. 19-46). Males may also attract females with acoustical signals produced by rapping the propodus of the large claw against the substratum or by rapidly flexing the ambulatory legs. The number of raps in a series and the interval between series is species-specific. The sounds are transmitted as vibrations through the sediment. Females can detect the signals up to a meter away by means of special myochordotonal organs in the legs.

Copulating shrimps typically orient at right angles with the genital regions of the ventral surfaces opposing one another (Fig. 19-45A). The modified first and second male pleopods are used to transfer a spermatophore to a receptacle between the female's thoracic legs (Fig. 19-45B).

In crayfishes and lobsters, the male turns the female over and restrains her chelipeds with his own (Fig. 19-47A). In crayfishes, the tips of the first pleopods (gonopods) are inserted into the seminal receptacle and spermatophores flow along grooves in the gonopods. If the seminal receptacle is absent, as it is in lobsters, the spermatophores are attached to the body of the female, particularly at the bases of the last two pairs of legs. These are noticeable as "tar spots" on the females of the spiny lobsters *(Panulirus).* Hermit crabs must emerge partially from their shells to mate. Partners place their ventral surfaces together and spermatophores and eggs are released simultaneously.

During copulation in brachyuran crabs, one crab lies beneath the other with their ventral surfaces facing each other (Fig. 19-47B). The male abdomen, which normally is flexed under the cephalothorax, is extended to expose the gonopods.

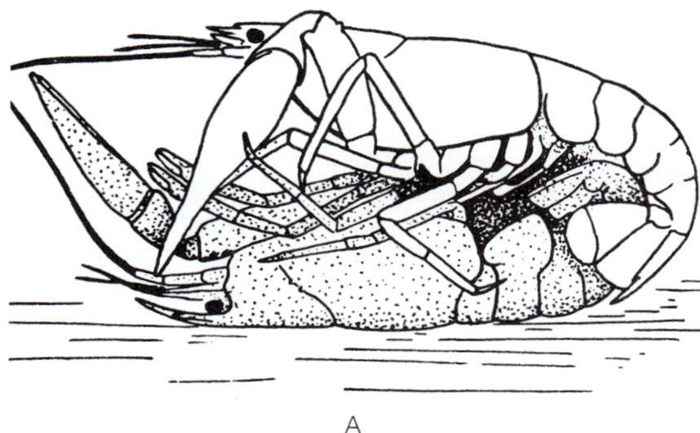

Gecarcinus lateralis (Freminville) in southern Florida. Bull. Am. Mus. Nat. Hist. 160(2):137.

FIGURE 19-47 Decapoda: Reproduction. **A,** Copulating crayfishes. The female is stippled. **B,** Copulating land crabs *(Gecarcinus lateralis).* The female is uppermost. Note the open groove in the ventral sternum of the male indicating that the abdomen is extended to expose the gonopods. *(A, From Andrews in Kaestner, A. 1970. Invertebrate Zoology. Vol. 3. Wiley-Interscience, New York.)*

The first pair of pleopods (gonopods), which conduct the sperm, are inserted into the female gonopore and spermatophores are deposited in her seminal receptacle.

In species without seminal receptacles, eggs are laid soon after copulation. In species with an internal seminal receptacle and internal fertilization (such as brachyuran crabs), the opening to the seminal receptacle is sealed with a plug and fertilization may not occur for a year or more following copulation.

Hormonal Regulation of Reproduction

Although sex is determined genetically in most crustaceans, many aspects of reproduction, including maturation, maintenance, and operation of the gonads and development of secondary sexual characteristics, are under hormonal control. In female malacostracans, the hormonal interrelationship between the ovary and the X-organ/sinus gland complex (Fig. 19-43) is similar to the pituitary/ovary relationship in vertebrates. The sinus gland releases **gonad-inhibiting hormone** (GIH) that inhibits the maturation of eggs in the ovary during the nonbreeding periods of the year. During the breeding season, **gonad-stimulating hormone** (GSH) is secreted, probably by the CNS. As a result of this stimulatory hormone's presence, the blood level of GIH declines and egg development begins. At the same time, and also under the influence of GSH, the ovary elaborates another hormone that initiates secondary anatomical changes preparatory for egg brooding, such as the development of ovigerous setae on the pleopods or the development of oostegites and the marsupium in peracaridans. These characteristics then appear at the next molt.

The development of the testes and male sexual characteristics in malacostracans is controlled by hormones produced in a small mass of secretory endocrine tissue called the **androgenic gland.** This gland is located at the end of the vas deferens, or occasionally in the testis. It is inhibited by neurosecretions from the X-organ (via the sinus gland) but stimulated by neurosecretions of the brain or subesophageal ganglion (also released from the sinus gland). When stimulated, the androgenic gland secretes **androgenic hormone,** which stimulates spermatogenesis and the development of male reproductive equipment. It also stimulates male reproductive behavior and secondary sexual characteristics. Removal of the androgenic gland is followed by a loss of male characteristics and conversion of the testes into ovarian tissue. If an androgenic gland is transplanted into a female, the ovaries become testes and male characteristics develop. If the Y-organ is removed prior to sexual maturity, gonadal development is seriously impaired. If the Y-organ is removed from an adult, the gonads are unaffected. In blue crabs, a female pheromone initiates a courtship response in the male unless his eyestalks have been removed.

Development

Primitive penaeid shrimps and their relatives shed their eggs directly into the sea and abandon them. In all other decapods, the eggs are glued to the female pleopods and brooded. In female crabs, the usually tightly flexed abdomen is extended to provide space for the brooding egg mass, often known as a "sponge."

The hatching stage for decapods varies widely. In primitive shrimps (dendrobranchiates: penaeids and sergestids) that shed eggs directly into the water, the hatching stage is a nauplius or metanauplius larva (Fig. 19-48A,B). In all other decapods, the eggs are brooded on the pleopods of the female and, in marine species, hatching takes place at the protozoea or zoea stage. The zoea and postlarval stages of different decapods have been assigned a number of specialized names (Fig. 19-49C,D). The zoea stages of most crabs are easily recognized by the very long, rostral spine and sometimes by a pair of lateral spines from the posterior margin of the carapace (Fig. 19-49A). At least in some species, the spines appear to reduce predation by small fishes. The crab postlarval stage is a megalops with a full complement of appendages (Fig. 19-49B).

Many decapods with planktonic larvae inhabit shallow coastal waters, estuaries, or salt marshes as adults or during juvenile stages. For example, estuaries and salt marshes are the nursery grounds for many species of penaeid shrimps. There is a tendency for larval life to be shorter in decapods that inhabit cold oceans or abyssal depths. Development is direct and larval stages are usually absent in the strictly freshwater decapods. Brackish-water forms, such as blue crabs (*Callinectes*), and river immigrants, such as *Eriocheir,* return to

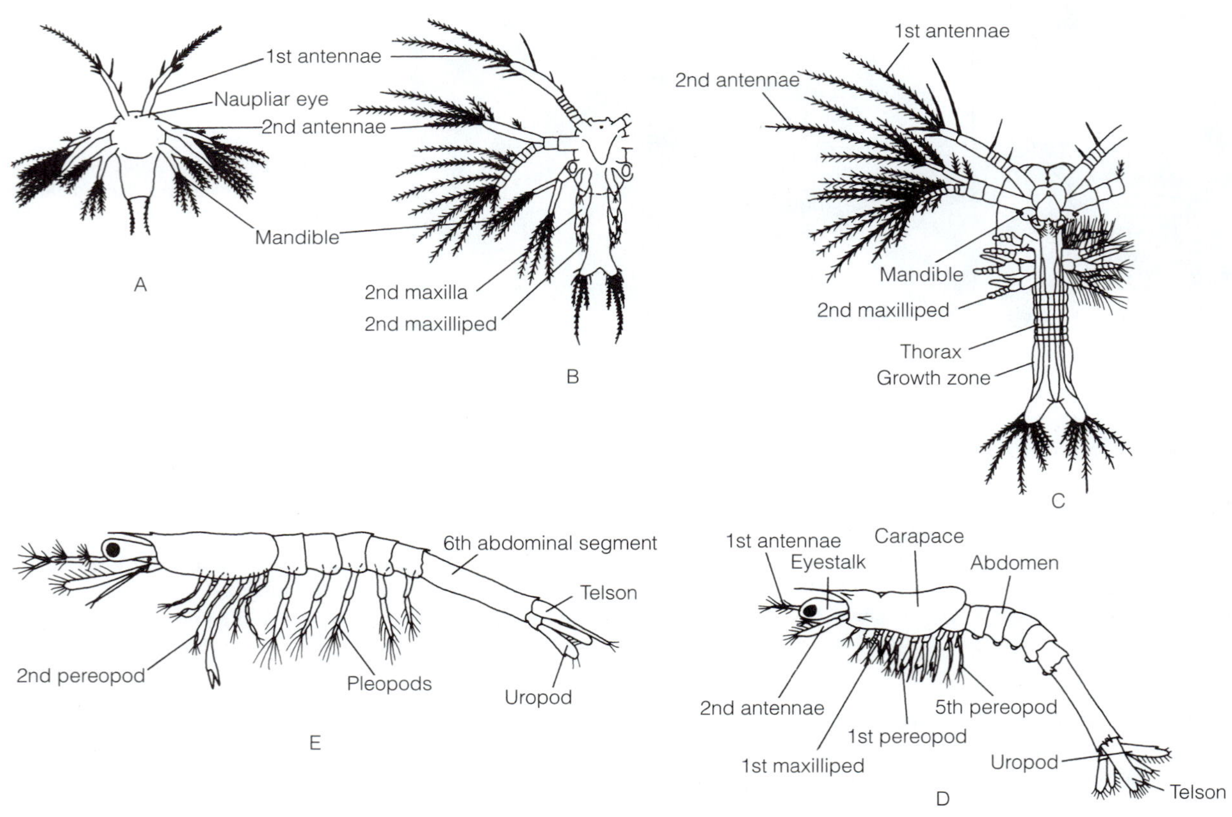

FIGURE 19-48 Decapoda: The major stages in larval development of the shrimp, *Penaeus.* **A,** Early nauplius. **B,** Metanauplius. **C,** Protozoea. **D,** Zoea (mysis). **E,** Postlarva. *(After Dobkin from Anderson, D. T. 1973. Embryology and Phylogeny in Annelids and Arthropods. Pergamon Press, Oxford. p. 495.)*

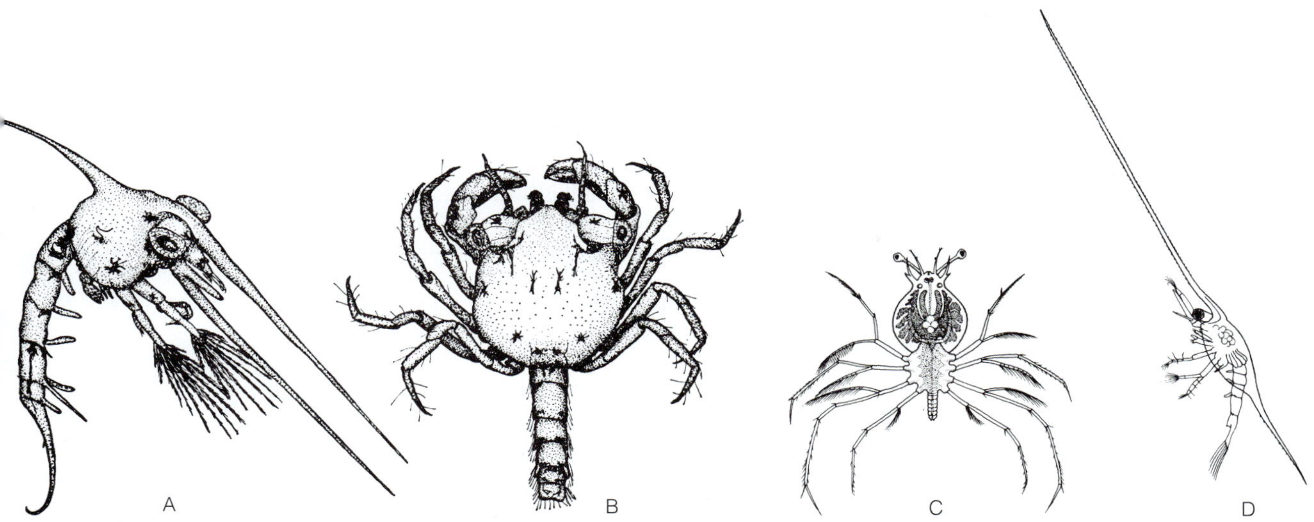

ɜURE 19-49 Decapod larvae. A zoea **(A)** and megalops **(B)** larva of the brachyuran crab, *Rhithropanopeus* *risii.* **C,** Phyllosoma (zoea) larva of the spiny lobster, *Palinurus.* **D,** Metazoea of the anomuran crab, *Porcellana.* *and B, From Costlow, J. D., and Bookhout, C. G. 1971. Fourth European Marine Biology Symposium. Cambridge University* *s, London. p. 214. Copyrighted and reprinted by permission of the publisher; C, Modified after Claus.)*

re-saline water for breeding. Development in almost all restrial anomurans and brachyurans takes place in the sea. e female, carrying her brood of developing eggs beneath abdomen, migrates to the shore and enters the water, at ich time the zoeae hatch and are released into the sea. velopment is direct in the terrestrial potamid crabs, which independent of the sea, although they do return to fresh- er to breed.

olting and Growth

many crustaceans, including, crayfishes, the lobster *Homa-*, and barnacles, molting and growth continue throughout life of the individual, although the intermolts become gressively longer. Such crustaceans may live to be quite old l some may become very large. In others, such as some bs, molting and growth cease after a specific number of tars, when the individual attains either sexual maturity or haracteristic size.

The physiological processes involved in molting are regu- ed by hormonal interactions and are similar to those of ects. The Y-organ (which is analogous to the prothoracic nds of insects) produces ecdysone, which acts on epi- mal cells and the digestive ceca to initiate proecdysis. duction of ecdysone by the Y-organ is inhibited by a hor- ne released from the sinus gland. Thus, experimentally noving the Y-organ prevents molting, whereas removing eyestalks (and sinus gland) initiates premature proecdy- The inhibitory action of a sinus gland hormone on the duction of ecdysone by the Y-organ is an important differ- e from control in insects, where the corpora cardiaca nulate rather than inhibit the prothoracic glands to pro- e ecdysone.

The regulation of molting hormones, and therefore of the actual molt cycle, depends on different stimuli operating on the CNS. In crayfishes, which molt seasonally, photoperiod (day length) is the controlling factor, whereas in the crab *Carcinus,* tissue growth is the controlling factor.

Autotomy

Limb loss is an unavoidable event in the life of many crustaceans, and decapods may even deliberately self-amputate, or **autotomize** (auto = self, tomy = cut) limbs. Severance typically takes place at a preformed **fracture plane** (autotomy plane) bisecting the combined basis and ischium, or **basischium** (Fig. 19-36). Internally at the fracture plane, there is a double, transverse, membranous fold perforated by a nerve and blood vessels. During autotomy, the leg breaks between the two membranes, leaving one membrane attached to the basal stub. The membrane constricts around its perforations, so there is very little bleeding. In some species, autotomy can take place only if the limb is pulled either by the animal itself or by a predator or other outside force. In its most highly developed state, as in most decapods (except some shrimps), autotomy is a reflex. If a leg is caught or damaged by a predator, a reflex is initiated, and an autotomizer muscle contracts violently, breaking the limb along the fracture plane.

Following severance and scab formation, a small **limb bud** grows from the stub. Growth of the limb then halts until the premolt period, when rapid growth and regeneration are completed. The new limb unfolds from a sac at the time of molting, but it is not quite as large as the original. Autotomy of multiple limbs induces molting, but the premolt regeneration phase must first be accomplished. If partially regenerated limbs are removed, the molting cycle is delayed until new limb buds form.

PHYLOGENETIC HIERARCHY OF DECAPODA

- Decapoda
  - Dendrobranchiata
  - Pleocyemata
    - Stenopodidea
    - Caridea
    - Astacidea
    - Thalassinidea
    - Palinura
    - Anomura
      - Paguroidea
      - Galatheoidea
      - Hippoidea
    - Brachyura
      - Dromiacea
      - Archeobrachyura
      - Oxystomata
      - Oxyrhyncha
      - Cancridea
      - Brachyrhyncha

SYNCARIDA[SO]

Syncarida is represented by about 50 species of primitive crustaceans in two taxa, Anaspidacea and Bathynellacea. Some fossil syncaridans are known from marine habitats, but all Recent species are freshwater.

Anaspidacea[O]

Anaspidaceans are shrimplike syncaridans with many primitive characteristics (Fig. 19-50A). Segments and appendages are only moderately specialized. The first thoracic segment bears 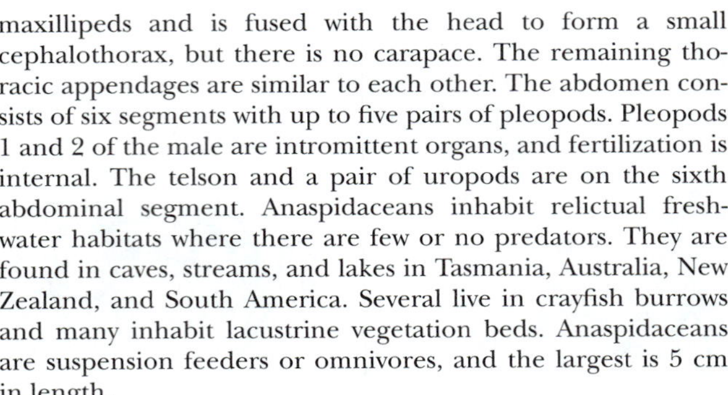maxillipeds and is fused with the head to form a small cephalothorax, but there is no carapace. The remaining thoracic appendages are similar to each other. The abdomen consists of six segments with up to five pairs of pleopods. Pleopods 1 and 2 of the male are intromittent organs, and fertilization is internal. The telson and a pair of uropods are on the sixth abdominal segment. Anaspidaceans inhabit relictual freshwater habitats where there are few or no predators. They are found in caves, streams, and lakes in Tasmania, Australia, New Zealand, and South America. Several live in crayfish burrows and many inhabit lacustrine vegetation beds. Anaspidaceans are suspension feeders or omnivores, and the largest is 5 cm in length.

Bathynellacea[O]

Bathynellaceans are elongate, blind, interstitial or groundwater syncaridans (Fig. 19-50B). Most occur in fresh water or, in a few instances, brackish water. *Thermobathynella* occurs in hot springs at temperatures of up to 55°C. The body is small and vermiform with reduced appendages, as is characteristic of interstitial animals. The tagmata are a head, thorax, and abdomen with no secondary tagmosis. The thorax consists of eight segments, but none is fused with the head. Cephalothorax, maxillipeds, and carapace are absent. Thoracic appendages 1 through 7 are similar and biramous. The eighth thoracopod is reduced or absent. The last of the six abdominal segments is fused with the telson to form a pleotelson. Abdominal appendages, with the exception of a pair of uropods on the pleotelson, are reduced or absent.

Bathynellaceans are among the smallest malacostracans, with the largest, *Bathynella magna,* being only 3.4 mm long and the smallest, *Brasilibathynella fiorianopolis,* being just over 0.5 mm long. Bathynellaceans have a cosmopolitan distribution.

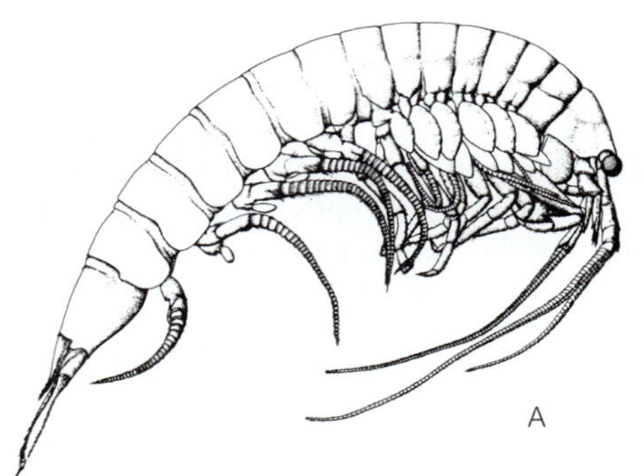

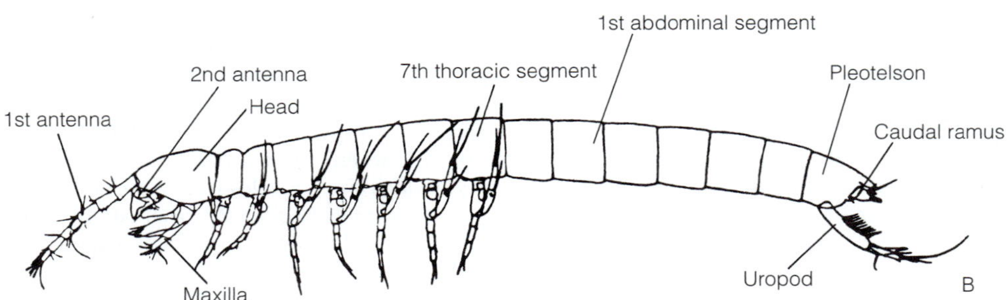

FIGURE 19-50 Syncarida. **A,** The anaspidacean *Anaspides tasmaniae.* This species reaches 5 cm in length. **B,** Lateral view of the bathynellacean *Parabathynella neotropica.* Total length is only 1.2 mm. *(A, From Schminke, H. K. 1978. Die phylogenetische Stellung der Stygocarididae (Crustacea, Syncarida)—unter besonderer Berücksichtigung morphologischer Ähnlichkeiten mit Larvenformen der Eucarida. Z. Zool. Syst. Evolut.-forsch. 16:225–229; B, From Noodt, W. 1965. Natürliches System und Biogeographie der Syncarida. Gewässer und Abwässer. 37–38, 77–186.)*

EUPHAUSIACEA[O]

Euphausiaceans, or krill, are pelagic crustaceans approximately 3 cm in length (Fig. 19-51A). All 85 described species are marine. Like primitive decapods, krill have a shrimplike body with a highly developed carapace fused with all thoracic segments. Unlike the decapods, however, the euphausiacean carapace does not extend ventrally beyond the bases of the legs. It does not form a branchial chamber and the gills are exposed (Fig. 19-51A). The thoracic legs are biramous and the gills are filamentous epipods on the protopod (Fig. 19-51B). The thoracic exopods move rhythmically to ventilate the gills. Although there are eight pairs of thoracic appendages, the eighth, and sometimes the seventh, are reduced. The second pair of thoracopods may be raptorial in carnivorous species, but maxillipeds are absent. Pleopods, uropods, and telson resemble those of shrimps.

The digestive system is similar to that of decapods. Two pairs of ostia perforate the short heart. Maxillary glands are found in larvae, but adults have antennal glands. The compound eyes are stalked and the exoskeleton is thin and flexible.

Krill swim using the large, setose pleopods. Most are suspension feeders specializing in zooplankton, but many consume phytoplankton, some filter bottom sediments, and a few are predaceous. The setose endopods of the thoracic appendages form a funnel-shaped **feeding basket** below the thorax (Fig. 19-51A). Feeding is episodic, with the feeding basket being used like a cast net. The endopods are extended ahead of the animal to enclose a volume of water. As they are retracted, the volume of the basket diminishes as water is forced out through a setal filter. Food remains inside the basket and is transferred by the maxillae to the mandibles and mouth.

Most euphausiaceans are bioluminescent, and the body is speckled with 10 eyelike photophores located on the coxae (Fig. 19-51B), eyestalks, and abdominal sternites. Photophores consist of a lens, pigment cup, reflector, and mass of cells that produces light. The luminescence is probably an adaptation for schooling and courtship.

Sperm are transferred to the female in spermatophores and stored in an external seminal receptacle until needed. Eggs are extruded and fertilization occurs several weeks after copulation. In most euphausiaceans the eggs are released free in the water, but several species attach them to the pleopods and brood them briefly. Cleavage is total and the first three divisions have been interpreted as being spiral. The young hatch as nonfeeding nauplii and pass through a long series of molts before becoming adults. Euphausiaceans are unusual among crustaceans in continuing to molt frequently after reaching sexual adulthood. When food supplies are inadequate, they may diminish in size with successive molts.

Euphausiaceans tend to have cosmopolitan distributions, and about two-thirds of the known species occur in all oceans. Any given oceanic region typically supports many species utilizing similar resources, which they partition spatially. Most species are epipelagic, many are mesopelagic, and a few are bathypelagic. Within each of these zones each species inhabits a preferred depth and many undergo vertical migrations.

Many species are important members of the pelagic fauna. Antarctic species, such as the epipelagic *Euphausia superba,* which is about 5 cm in length, live in enormous schools. They are the chief food of many marine animals, and a blue whale may consume a ton of euphausiaceans at just one of up to four such feedings each day. A euphausiacean swarm may cover an area equivalent to several city blocks that, when seen from the air, looks like a giant ameba slowly moving and changing shape. The several meters of water closest to the surface contain the greatest concentrations and swarms may reach densities of 60,000 individuals/m^3, all of which are adults or near adults. Swarming *Euphausia superba* molt very rapidly (within a second) and, if alarmed, many will literally jump out of their skins. Exuvia remain behind and perhaps function as decoys to foil sight-orienting predators.

Fishing for Antarctic krill is a major industry. Over 395,000 tons were harvested between 1989 and 1990 for human consumption, with Russian and Japanese trawlers accounting for about 95% of the catch.

PANCARIDA[SO]

Thermosbaenacea[O]

The only higher-level pancaridan taxon, Thermosbaenacea, consists of about 35 subterranean species from hot springs, caves, and groundwater and a few from coastal marine interstitial habitats. Thermosbaenaceans are small crustaceans of only a few millimeters in length. Morphology is variable, but the head is fused with one thoracic segment, producing a small cephalothorax and a pereon of seven segments. The small carapace is fused

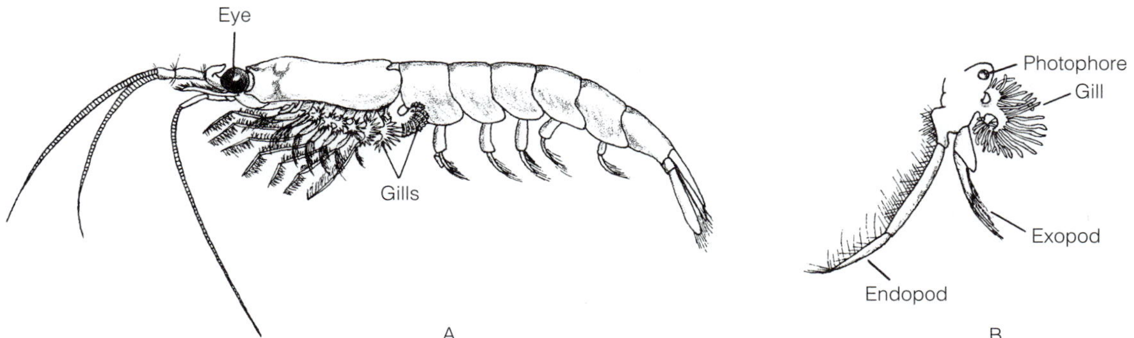

FIGURE 19-51 Euphausiacea. **A,** *Meganyctiphanes.* **B,** Seventh thoracic appendage. *(Both after Calman.)*

with the first thoracic segment and covers a variable number of additional segments. In females the carapace is expanded to form a *dorsal* brood pouch. There are no oostegites. The first thoracopod is a maxilliped, and most pereopods are biramous. Pleopods are present only on the anterior abdominal segments and segment 6 bears the uropods and telson. Thermosbaenaceans are detritivores that swim upside down using the pereopods. The carapace is the chief respiratory surface and the very short heart has only one pair of ostia. Nephridia are unknown. Fertilization is probably internal.

PERACARIDA[SO]

Peracarida, with 12,000 species in seven major taxa (orders), is one of the largest crustacean taxa. Peracaridans usually are smaller than decapods, with most being less than 2 cm in length, but the largest is almost 42 cm. Although they are not as noticeable as decapods, they are abundant and diverse in most marine habitats and are also present in freshwater and even terrestrial environments. Peracarida includes the mysidaceans, amphipods, cumaceans, mictaceans, spelaeogriphaceans, tanaidaceans, and isopods.

The most distinctive peracaridan characteristic is the ventral **marsupium,** or brood pouch, under the thorax of the female (Fig. 19-62B). It is a space formed by large, flexible, platelike oostegites extending medially from some thoracic coxae (Fig. 19-52). The **oostegites** are setose epipods that partition the space below the thorax. The sternum forms the roof of the chamber and the oostegites the floor. Eggs are brooded in the marsupium and development is usually direct, although some hatch as postlarvae. A manca larva, which has all the adult appendages except the eighth thoracopods (seventh pereopods), is characteristic of some peracaridan orders.

The first and sometimes the second and even the third thoracic segments may be fused with the head to form a cephalothorax. The appendages of these segments are maxillipeds and the remaining thoracic segments bear pereopods. The mandible bears a characteristic movable toothed process, the **lacinia mobilis,** between the molar and incisor. A carapace may be present or not. The naupliar eye never persists in the adult and the compound eyes may be sessile or stalked.

Primitively, peracaridans are maxillary suspension feeders, as many modern mysidaceans, cumaceans, and tanaidaceans are, but the tendency in most peracaridans has been to adopt other modes of feeding. The two largest and best-known taxa are Amphipoda and Isopoda.

Mysidacea[O]

Mysidaceans are small, shrimplike peracarids of freshwater, brackish, and marine habitats. Most are 2 to 30 mm in length, but the bathypelagic *Gnathophausia* (Fig. 19-53A) may reach 35 cm. Some are important members of northern lakes, but most are marine; they may be benthic or pelagic.

Mysidaceans have many primitive characteristics, such as the shrimplike shape of primitive malacostracans, and they probably are more like the ancestral peracaridans than other Recent taxa are. Mysidacea comprises two distinct subtaxa, Lophogastrida and Mysida (Fig. 19-53). Lophogastrids are the more primitive of the two, are closest to the origin of the peracarids, and are morphologically similar to the euphausiaceans and the primitive penaeid shrimps. About 1000 described species in several genera, including *Gnathophausia, Heteromysis, Lophogaster, Mysidopsis, Mysis,* and *Neomysis,* are known.

The first one to three segments of the thorax are fused with the head to form the cephalothorax. The carapace is large and, although it encloses the entire thorax, it is physically attached only to the first to third anterior segments, unlike that of shrimps and euphausiaceans. It encloses lateral branchial chambers and may bear a median dorsal, anterior projection, the rostrum. The appendages of the first (in Lophogastrida) or first and second (in Mysida) thoracic segments are maxillipeds. The pereon consists of thoracic segments 3 to 8 or 2 to 8, which are not fused with the head or carapace. The pereopods are biramous and in females bear oostegites. The exopods are filamentous and sometimes natatory.

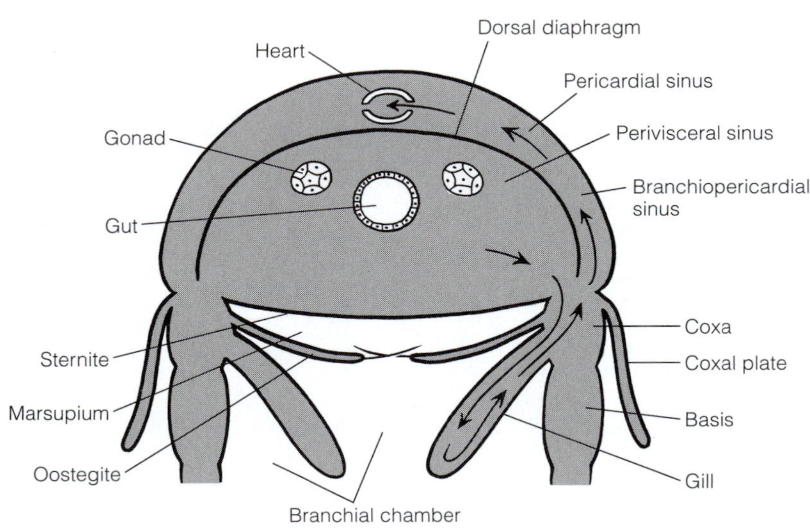

FIGURE 19-52 Peracarida: Diagrammatic cross section through the body of an amphipod showing the relationship between the pereopods, gills, and marsupium. Arrows indicate blood flow. Limb articles distal to the basis are omitted for clarity. *(Redrawn from Kaestner, A. 1970. Invertebrate Zoology. Vol. 3. Wiley-Interscience, New York.)*

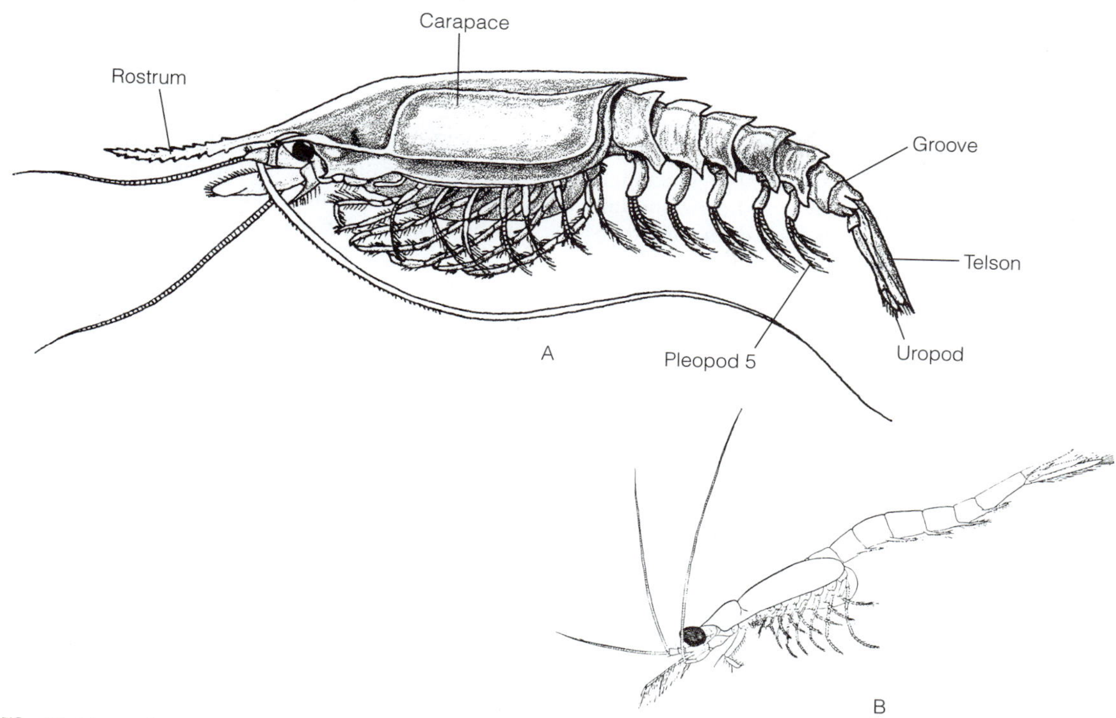

FIGURE 19-53 Mysidacea. **A,** Lateral view of the lophogastrid *Gnathophausia*. **B,** *Mysis relicta*, a freshwater species 3 cm in length belonging to Mysida. *(B, From Pennak, R. W. 1978. Freshwater Invertebrates of the United States. 2nd edition. John Wiley and Sons, New York.)*

The adult abdomen consists of six segments, the sixth of which is about twice as long as the others. Embryos, however, have seven abdominal segments, and the fusion of two of those segments forms the sixth adult segment. In Lophogastrida, the sixth segment is divided by a groove (Fig. 19-53A) representing the line of fusion. The first five abdominal segments are the pleon and may bear biramous pleopods, although pleopods are absent or vestigial in female Mysida. The sixth segment bears the uropods and the telson, which form a tail fan similar to that of shrimps.

Marine mysidaceans occur at all depths from shallow coastal water to the deep sea. Although some species are pelagic, most are associated with the bottom in one way or another. Most swim just above the sediment surface, but some crawl on the bottom, plow through the sediment, or burrow into it. Swimming is the primary means of locomotion, and the pleopods are used for this purpose. Most benthic species leave the bottom during the night to become planktonic. Some marine species are found in shallow algal and grass beds and many others live in large swarms. Mysidaceans can be important in the diet of fish such as shad and flounder.

About 40 species are found in freshwater habitats, including several that inhabit caves and other groundwaters. *Mysis relicta* is confined to cold, deep lakes of the northern United States, Canada, and Europe (Fig. 19-53B), where it is an important food of lake trout.

Most mysidaceans are suspension feeders. A feeding current generated by the pereopod exopods moves food particles into a midventral food groove between the coxae of the pereopods. Water and particles move anteriorly in the groove to a setal filter on the second maxillae, where the particles are separated from the water. Mysidaceans can feed on small particles groomed from the body surface and can also capture relatively large planktonic animals in a basket formed by the thoracic endopods. Many bathypelagic forms are scavengers.

The branchial chamber is enclosed by the carapace, the inner surface of which is the chief gas-exchange surface. The ventilating current is generated by maxillipedal epipodites. In lophogastrids, epipodites on the pereopods are also respiratory.

The heart is a dorsal tube in the thorax with two to three anterior pairs of ostia. A large blood sinus on each side of the carapace drains oxygenated blood from the respiratory surface directly into the pericardial sinus.

Mysidaceans have paired saccate nephridia. In Mysida, they are in the antennal segment, but in Lophogastrida, they are in both the antennal and maxillary segments. Clusters of nephrocytes occur in blood spaces at the bases of the pereopods.

Most mysidaceans have a long subesophageal ganglion consisting of the incompletely fused segmental ganglia of the mouthparts and thoracic appendages. The abdominal ganglia remain independent. Embryos have seven pairs of abdominal ganglia, one for each embryonic segment. Mysidaceans have well-developed, stalked compound eyes. In Mysida there is a statocyst in the base of the endopod of each uropod. With magnification, the statocyst is easily seen through the transparent cuticle of the uropod and is the quickest way to distinguish mysidaceans from other shrimplike crustaceans.

Mysidaceans are gonochoric and fertilization is external, in the marsupium. The gonads are paired tubes in the thorax, with gonopores on the sixth (in females) or eighth (in males) thoracic segment. Males have a pair of long penes on the sternites of the eighth segment and females have oostegites on pleopods 1 through 7. During mating, the male inserts the penes into the marsupium and releases sperm. In less than an hour, eggs are released from the oviducts into the marsupium where fertilization occurs. Eggs are brooded in the marsupium and hatch as miniature adults, there being no postlarva.

Primitive features such as stalked compound eyes, a well-developed carapace, thoracic gills, a long tubular heart, a doubled abdominal segment 6, and the presence of both antennal and maxillary glands suggest the mysidaceans branched off from the ancestral peracaridans early.

Amphipoda[O]

Amphipods are small crustaceans often found in large numbers and high diversity in aquatic habitats. With more than 6000 species, Amphipoda is the largest peracaridan taxon. Most amphipods are marine, but many are found in fresh water and one family has terrestrial and semiterrestrial species. In the sea they are common, diverse, and ecologically important in pelagic and benthic habitats.

The body tends to be laterally compressed, giving the animal a somewhat shrimp-like appearance (Fig. 19-54). Sessile compound eyes replace the stalked eyes of the ancestor. The first and second antennae usually are well developed and the mandibles have palps and a lacinia mobilis.

A carapace is absent, although the first thoracic segment (and sometimes the second) is fused with the head to form a small cephalothorax. The fused appendages of the first thoracomere form a maxilliped. Each of the remaining seven thoracic segments, which make up the pereon, has a pair of uniramous pereopods. The coxae of the pereopods usually have broad, flattened **coxal plates** that are functionally equivalent to the decapod carapace in that they create a protected space ventral to the thorax (Fig. 19-52, 19-54). This space contains the gills and is the branchial chamber and, in females, the marsupium. The first and second pereopods, or gnathopods, usually are enlarged and subchelate for grasping (Fig. 19-54). The first four pereopods are oriented anteriorly, with the dactyls pointing posteriorly, whereas the last three are posteriorly directed with dactyls pointing anteriorly. The gills and oostegites are flat epipods projecting medially into the branchial chamber from the coxae of the pereopods (Fig. 19-52).

The six-segmented abdomen is divided into an anterior **pleosome** and a posterior **urosome** of three segments each (Fig. 19-54). The tergites of the pleosomal segments extend ventrally as **epimera** (sing., epimeron), creating a space ventral to the pleosome continuous with the branchial chamber below the thorax. The three pairs of appendages of the pleosome are pleopods. The biramous pleopods are used for swimming and to create the ventilating current through the branchial chamber. The urosome does not bear epimera and its three pairs of biramous appendages are stiff, heavily cuticularized uropods used for kicking, jumping, swimming, or burrowing (Fig. 19-54).

Most amphipods are between 5 and 15 mm long, but some interstitial species are only 1 mm and a giant, unidentified species 28 cm in length is known from photographs taken at 5300 m in the Pacific. Most amphipods are translucent, brown, or gray, but some species are red, green, or blue-green.

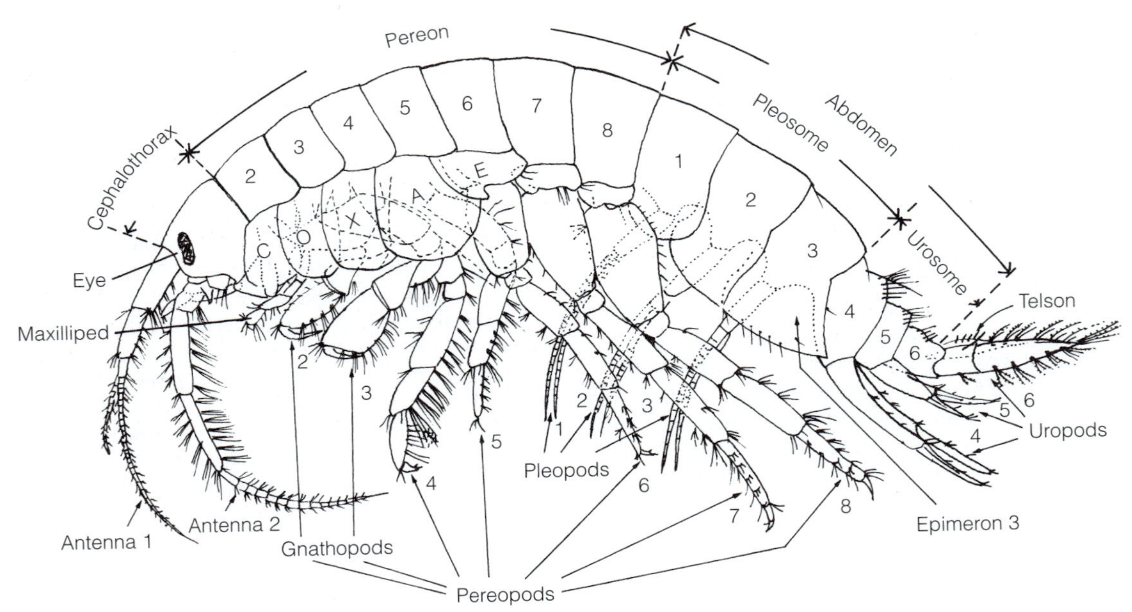

FIGURE 19-54 Amphipoda: External structure of a gammaridean amphipod in lateral view. *(From Bousfield, E. L. 1973. Shallow-Water Gammaridean Amphipoda of New England. Cornell University Press, Ithaca, New York. 344 pp.)*

DIVERSITY OF AMPHIPODA

Amphipoda includes Gammaridea (Fig. 19-55), Hyperiidea (Fig. 19-56), Caprellidea (Fig. 19-57), and Ingofiellidea (Fig. 19-58).

Gammaridea[sO]

The largest amphipod taxon, Gammaridea, includes about 5000 of the 6000 amphipod species (Fig. 19-55). These are the "typical" amphipods most like the stem amphipod. Gammarideans usually are laterally compressed (Fig. 19-54), although some are dorsoventrally depressed (Fig. 19-55E). Most are benthic and usually remain closely associated with the bottom despite being capable of swimming. Typical gammarideans have large coxal plates, a large abdomen with six pairs of appendages, and relatively small compound eyes. The first thoracic segment is fused with the head to form a cephalothorax.

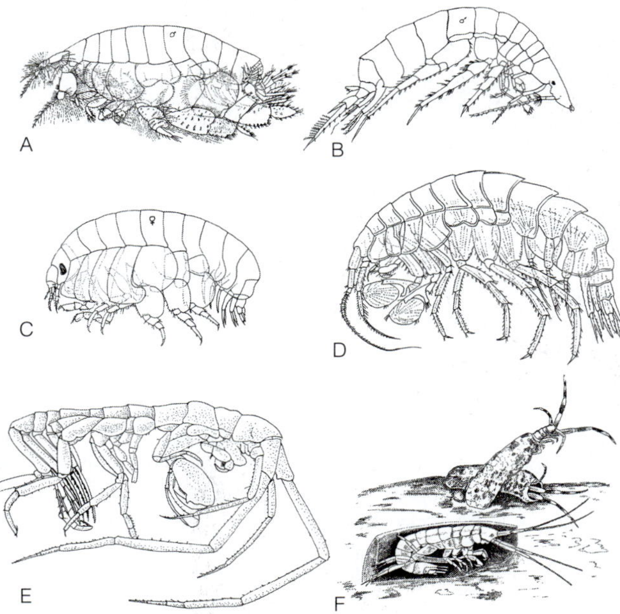

FIGURE 19-55 Amphipoda: Gammaridea. **A,** The burrowing *Haustorius canadensis* (Haustoriidae), with powerful digging legs and filter-feeding mouthparts. **B,** The burrowing *Platyischnopus herdmani* (Platyischnopidae). **C,** *Orchomenella minuta* (Lysianassidae), a flesh scavenger. **D,** *Eusirus cuspidatus* (Eusiridae). **E,** *Dulichia rhabdoplastis* (Podoceridae), a commensal on sea urchins along the Pacific coast of the United States. The amphipod lives on strands of detritus, which it constructs of its own fecal pellets, stretched between sea urchin spine tips. It feeds on diatoms growing on the strands or filters detritus and plankton from the water using its outstretched antennae. **F,** The tubicolous *Cerapus* (Cerapidae). The portable tube is secreted by the animal. *(B, From Barnard, J. L. 1969. The families and genera of marine gammaridean Amphipods. U. S. Nat. Mus. Bull. 27:1–535; D, After Sars, from Bousfield, E. L. 1973. Shallow-Water Gammaridean Amphipoda of New England. Cornell University Press, Ithaca, New York. 344 pp.; E, From McCloskey, L. R. 1970. A new species of Dulichia commensal with a sea urchin. Pac. Sci. 24:90–98. ©1970, By permission of the University of Hawaii Press; F, Modified from Schmitt.)*

Gammarideans are important members of most marine benthic communities, where they often exhibit high population densities and high species diversity. Gammarideans can be infaunal burrowers, epibenthic on complex substrates, or occasionally planktonic. Many species, both infaunal and epifaunal, are tubicolous and construct and inhabit tubes made of various particles, such as clay, sand, and shell and plant fragments cemented together by secretions of glands in the fourth and fifth pereopods. The tube may be stationary or mobile. Many species are commensal with other invertebrates, such as sponges, tunicates, or polychaetes, and a small number are parasitic. Like other malacostracans, gammarideans tend to be scavengers, suspension feeders, or predators.

With the exception of a few interstitial species, gammarideans are the only amphipods in freshwater habitats, where they are also benthic and ecologically important, especially in vegetation. Population densities of $10{,}000/m^2$ have been reported. Lake Baikal in Siberia, the world's oldest and deepest lake, supports a fauna of almost 300 endemic benthic and planktonic species, some over 6 cm in length. Gammarideans are well represented in subterranean habitats and over 600 species have been reported from groundwaters such as those in caves and wells.

Propulsion for swimming typically is provided by the pleopods. Usually, gammarideans swim intermittently between longer periods of crawling and burrowing. Swimming is initiated by an initial backward thrust of the abdomen with the stiff uropods pushing against the substratum. Slow walking over the substratum is accomplished with the pereopods, but rapid locomotion utilizes the pleopods in addition to the pereopods, and the animal often leans far over to one side, resulting in the common name *sideswimmer*. *Scud* is another vernacular name.

Amphipods (unlike isopods) have been unsuccessful in adapting to life on land. The few terrestrial and semiterrestrial amphipods are gammarideans in the beach hopper family, Talitridae. Beach hoppers, which may be abundant on some beaches, live beneath stranded seaweed or driftwood or burrow in the sand above, but near, the high-tide mark. When disturbed, they jump using a sudden backward extension of the abdomen, uropods, and telson. An amphipod 2 cm in length can jump a distance of up to 1 m. A few species of leaf-litter hoppers in the Southern Hemisphere are found in moist humus and soil remote from the shore. Some of these occasionally occur in greenhouses in the Northern Hemisphere. The lack of adaptations to minimize desiccation has severely limited the exploitation of terrestrial habitats by amphipods, and the few terrestrial species are restricted to warm, humid habitats.

The beach hopper, *Talitrus*, uses its eyes to obtain astronomical clues for relocating optimal habitat in the supratidal zone. If displaced either above or below the high-tide mark, the animals migrate accurately back to their preferred zone. The angle of the sun is used as a compass in conjunction with a map sense of the east–west orientation of the particular beach they inhabit. An internal clock mechanism provides interpolation for the changing angle of the sun over the course of the day. Other factors (horizon level, beach slope, sand moisture, and grain size) besides celestial clues are used in orientation, but visual clues are most important. *Alicella, Ampelisca, Aora, Cerapus, Corophium, Cyrtophium, Gammarus, Haploops, Haustorius, Hyalella, Melphidippella, Orchomenella, Talorchestia.*

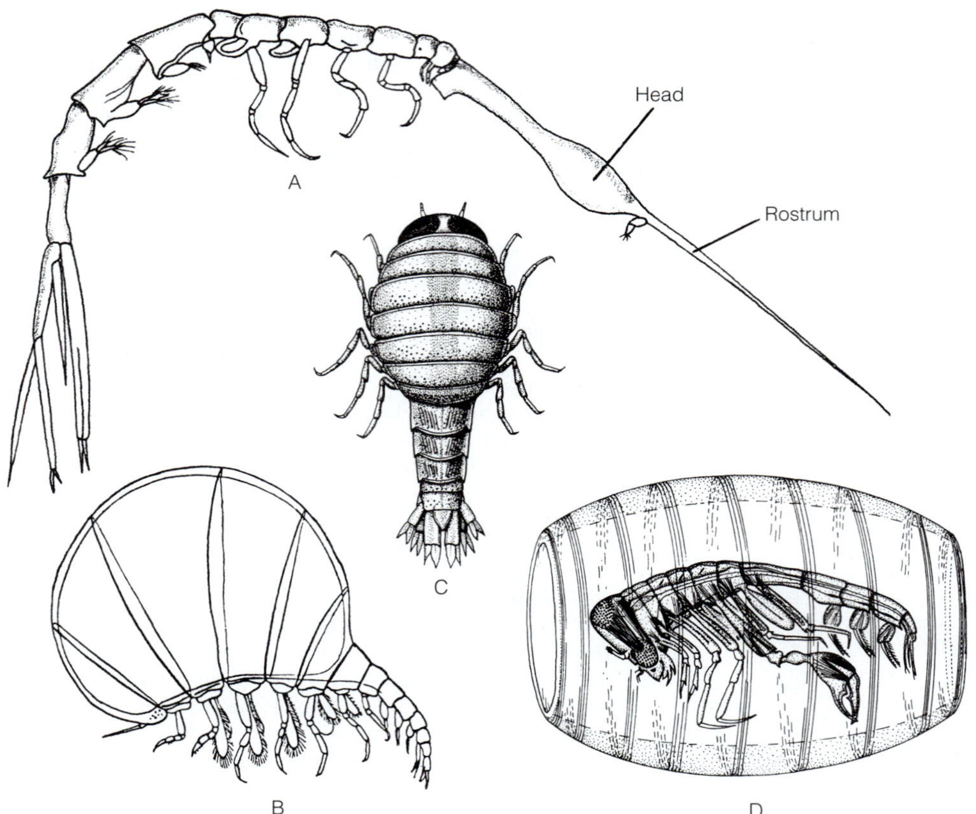

FIGURE 19-56 Amphipoda: Hyperiidea. **A,** *Rhabdosoma,* an amphipod with elongate head and needlelike rostrum. **B,** *Mimonectes,* 2.5 cm long. **C,** *Hyperia,* 2 cm long. Species of this genus are commonly found attached to pelagic coelenterates and ctenophores. **D,** *Phronima sedentaria,* 3 cm long, which lives within salps. *(A, After Stebbing from Calman; B, After Schellenberg from Kaestner, A. 1970. Invertebrate Zoology. Vol. 3. Wiley-Interscience, New York; C, After Sars; D, After Claus.)*

Hyperiidea^sO

Hyperiidea is a taxon of about 500 species of pelagic marine amphipods (Fig. 19-56). What little is known of their life histories suggests that all hyperiideans spend at least part of the life cycle as commensals or parasites of gelatinous zooplankters such as salps, siphonophores, jellyfishes, and ctenophores. For example, members of Phronimidae live in the barrel-shaped tunic of salps (Fig. 19-56D), which they propel through the water.

Hyperiideans resemble gammarideans but have small, weakly developed coxal plates (Fig. 19-56B) and enormous compound eyes that occupy most of the surface of the head (Fig. 19-56C,D). The head is fused with one thoracic segment and its appendages are reduced maxillipeds having only three of the usual seven articles. The Antarctic *Hyperiella dilatata,* which is consumed by fish, avoids predation by carrying a distasteful opisthobranch mollusc, the sea butterfly, *Clione limacina,* on its back. *Hyperia, Mimonectes, Parathemisto, Phronima, Rhabdosoma, Scina.*

Caprellidea^sO

Species of the aptly named skeleton shrimps (Fig. 19-57A) are highly derived, strictly marine benthic amphipods adapted for a life of clinging to complex substrata such as seaweeds, bryozoans, and hydroids. Most are lie-in-wait predators of small invertebrates, but some consume epiphytic diatoms. There are about 350 extant described species. The body, which bears little resemblance to that of other amphipods, is elongate, sticklike, cylindrical, and not laterally compressed. The second thoracic segment is partially fused with the head to create a cephalothorax consisting of the head plus two thoracic segments. The first thoracic appendages are maxillipeds, as usual, but those of the second remain gnathopods. The third thoracic segment bears the second pair of gnathopods. Thoracic segments 4 and 5 bear slender, reduced or vestigial pereopods, gills, and, in females, large oostegites. The marsupium is thus restricted to the middle of the thorax on the fourth and fifth thoracic segments. Thoracic appendages 6 through 8 are well-developed, subchelate prehensile legs used to cling tenaciously to the substratum. Coxae are reduced or vestigial. The abdomen consists of five small segments with vestigial appendages.

Caprellideans move like inchworms, in a looping motion. They grasp the substratum with the gnathopods, arch the pereon to bring the posterior end forward, grasp the substratum with the posterior pereopods, and straighten the pereon to move the anterior end forward.

The genus *Caprogammarus* is intermediate between gammarideans and caprellideans and has an abdomen with functional pleopods and uropods. The whale lice in the family Cyamidae are highly derived epizoic caprellideans specialized for a life on the skin of cetaceans (Fig. 19-57B). The mouthparts are adapted for piercing. Thoracic appendages 4 and

Photograph courtesy of D. P. Wilson.

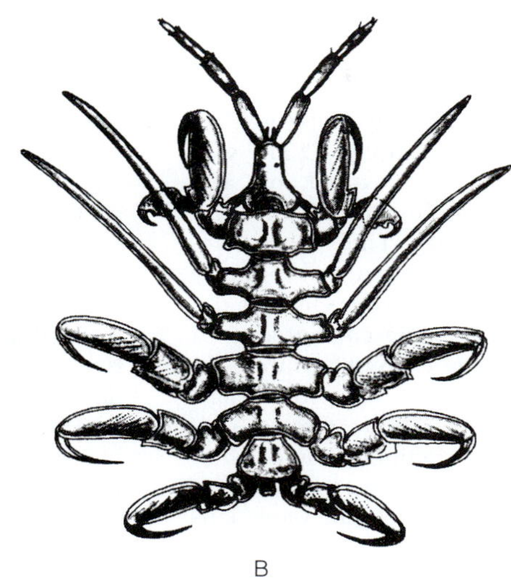

FIGURE 19-57 Amphipoda: Caprellidea. **A,** The skeleton shrimp *Caprella equilibra,* clinging to seaweed. **B,** *Cyamus boopis,* an amphipod commensal on whales. *(B, After Sars.)*

5 are vestigial except for their gills. The remaining thoracic appendages are modified for clinging to the skin of whales, and the abdomen is vestigial. *Cyamus, Caprella, Paracaprella.*

Ingolfiellidea[sO]

Ingolfiellideans, some of which are less than 1 mm long, are adapted for interstitial habitats in both marine and freshwater environments (Fig. 19-58). The body is cylindrical and the thorax and abdomen are well developed. The pleopods are vestigial, but the first two pairs of uropods are strong. The 30 known species occur in hypogean (subterranean), interstitial, and deep-sea habitats. Their small size, loss of eyes and pigmentation, vermiform shape, and reduced appendages are familiar adaptations for life in interstitial habitats. *Ingolfiella.*

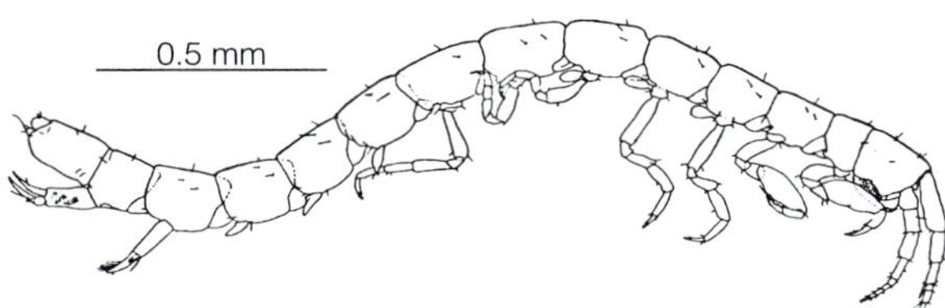

FIGURE 19-58 Amphipoda: Ingofiellidea. *Ingolfiella putealis,* a blind interstitial amphipod with vermiform body. *(From Stock, J. H. 1976. A new member of the crustacean suborder. Ingolfiellidea from Bonaire, with a review of the entire suborder. Vitg. Naturwet. Studiekring Suriname Ned. Antilen. 86:57–75.)*

NUTRITION

Most amphipods are omnivorous detritus feeders or scavengers. Mud and animal or plant remains are picked up with the gnathopods. Some are deposit feeders that use the antennae to gather the surface film of diatoms that covers the sediment. Some burrowing species are "sand lickers" that scrape detritus and diatoms from sand grains. Many amphipods are suspension feeders that filter fine organic particles from the water. Some generate a special feeding current for this purpose, but most utilize the respiratory current. A variety of setose appendages, including the anterior pereopods, maxillae, and antennae, are used as filters in different families. The hyperiideans, many caprellideans, and two families of gammarideans are predaceous and many others supplement their diet by catching small animals.

Parasitism is much less prevalent among amphipods than among their relatives, the isopods. The few ectoparasites of fish have suctorial mouthparts. The cyamids (whale lice) probably are parasites, but they may also feed on diatoms and debris that accumulate on the whale's skin. Some hyperiideans may be parasites that consume parts of their hosts.

Gut morphology varies widely with taxon, but all have the usual foregut, midgut, and hindgut. In gammarideans and caprellideans the gut resembles that of decapods and has an ectodermal stomach with cardiac and pyloric chambers. As in decapods, there is a dorsal channel that leads indigestible material to the intestine and a ventral channel that takes digestible food to the numerous digestive ceca for absorption. The gut of hyperiideans, which are carnivores, is much simpler and lacks the cuticular stomach.

GAS EXCHANGE, INTERNAL TRANSPORT, AND EXCRETION

The chief gas-exchange surfaces are lamellar or saclike gills on the coxae of some or all of pereopods 2 through 7. The gills are on the medial side of the coxa and extend into the branchial chamber formed by the sternites dorsally and the coxae and pereopods laterally (Fig. 19-52). The coxae themselves are also respiratory. The pleopods create the respiratory current, which flows from anterior to posterior through the branchial chamber and is also the swimming current and, in some suspension feeders, the feeding current, as well. In brooding females this current also ventilates the eggs, embryos, or juveniles in the marsupium. The respiratory current is particularly important in those amphipods that dwell in burrows or tubes, where there is no passive water movement. Caprellideans lack pleopods and rely on ambient water currents for ventilation. Oxygen transport is by the blood and is facilitated by extracellular hemocyanin. Although semiterrestrial, talitrids have retained the gills. They are dependent on humid air and thus are restricted to the moist sand beneath drift or damp forest leaf litter. They feed at night, when there is less danger of desiccation.

The amphipod heart is a dorsal tube with one to three pairs of ostia that lies in the thorax above the gills. The arterial system is not highly developed, as it is in decapods. Blood exits the heart via anterior and posterior aortae and flows into the perivisceral sinus, where it bathes the viscera. It collects ventrally and flows into the gills to be oxygenated (Fig. 19-52). It then follows branchiopericardial channels dorsally into the pericardial sinus. The excretory organs are antennal glands, which are well developed in freshwater species and vestigial in terrestrial amphipods.

NERVOUS SYSTEM AND SENSE ORGANS

The subesophageal ganglion of gammarideans and caprellideans includes the ganglia of the mouthparts and maxilliped. In gammarideans the ventral double nerve cord bears independent ganglia for each of the pereonal and pleonal segments. The ganglia of the three urosome segments are fused to form a posterior **urosomal ganglion.**

Like other arthropods, amphipods have a variety of sensory setae. Aesthetascs are found on the first antennae (Fig. 19-7A). **Calceoli,** club-shaped setae found on both pairs of antennae, probably have a chemoreceptive function. Aesthetascs and calceoli are thought to be used by males to locate sexually receptive females. Like all peracarids except mysidaceans, amphipod compound eyes are sessile, apposition-type eyes, although there is no division of the cuticle into individual facets. Gammarideans and hyperiideans have statocyst-like structures (organs of Bellonci) in the head that may function in gravity orientation or detect magnetic fields.

REPRODUCTION AND DEVELOPMENT

Amphipods are gonochoric; have paired, tubular gonads; and are usually sexually dimorphic, with the males larger than the females. The male gnathopods frequently are much larger than those of females and may have an elaborate, derived morphology. Fertilization is external. Sperm ducts from the testes open via gonopores on a pair of soft, flexible penes on the eighth thoracic sternite. Most amphipods have no gonopods.

The oviducts connect the ovaries with a pair of gonopores on the sixth thoracic coxae that open into the marsupium. The oostegites first appear as small projections during juvenile instars and grow larger with successive molts until reaching their final, mature, marginally setose condition with the female's parturial molt, which precedes spawning. Oostegite development is controlled by an ovarian hormone.

There is much variation in mating behavior, but the following is typical for epibenthic gammarideans. Prior to mating, the male locates a preparturial female by using the aesthetascs on his antennae to detect pheromones from the female. The male then rides, or carries, the female, with her dorsum facing his venter, for several days while awaiting her parturial molt. The **parturial molt** is the one that produces complete, functional oostegites and signals the ability of the female to mate, release eggs, and brood young. During the preparturial period the male holds the female by hooking the fingers of his gnathopods under the anterior edge of her first pereonal tergite and the posterior edge of her fifth tergite. Following her parturial molt, the male turns the female so her venter faces his. Sperm strings are released from the penes and are pushed or wafted into the marsupium by the pleopods. The pair separates and a few hours later the female releases eggs from the gonopores and they are fertilized in the marsupium.

Eggs are brooded in the marsupium and ventilated by rhythmic beating of the pleopods. Interlocking marginal setae on the oostegites prevent eggs and juveniles from falling out of the marsupium. The female provides no nutriment, other than yolk, and there is no placenta or nutritive secretion. Development is direct and the juveniles hatch as miniature replicas of the adults, with the adult complement of segments and appendages. They remain in the marsupium for a few days after hatching. Amphipods undergo about 20 molts in the typical, one-year-long life cycle. A female may spawn several times and each spawn is preceded by a parturial molt and attended by a male. The number of eggs in a clutch varies from 2 to 750 depending on species, size, and the physiological condition of the female.

Cumacea^O

Cumaceans are small inhabitants of sand and mud bottoms. The taxon is small, with about 800 species of distinctively tadpole-shaped and easily recognized marine peracaridans (Fig. 19-59). Most are about 1 mm in length, and the largest is

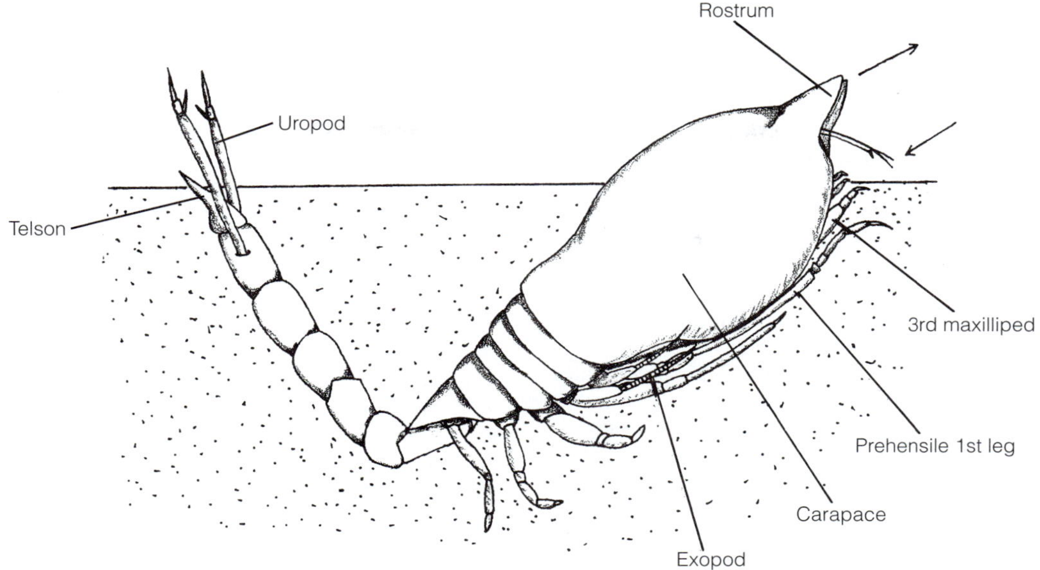

FIGURE 19-59 Cumacea: The cumacean *Diastylis,* buried in sand. Arrows indicate the direction of the feeding-ventilating current. This individual is a female. In this family, the three anterior thoracic segments are fused with the head to form the cephalothorax, which is covered by the carapace. The remaining thoracic appendages form the pereon, whose five free segments and pereopods are visible in the drawing.

about 4 mm. They reach their greatest diversity and ecological significance in the deep sea.

The head and thorax are inflated and markedly wider than the long, slender abdomen. Fusion of the head with three to six anterior thoracic segments forms a cephalothorax, and the first three thoracic appendages are maxillipeds. The remaining thoracic segments, the pereon, are independent of the cephalothorax and carapace. Most thoracic appendages are biramous with a natatory, setose exopod. The first and second antennae of females are small, but the second antennae of males are often longer than the body. The mandibles do not have palps. Females have oostegites on the posterior four pairs of thoracopods.

A large carapace covers the cephalothorax and is fused with its segments. The first maxillipeds bear epipods that are the only gills. The carapace extends laterally and ventrally to form a branchial chamber enclosing the gills. The lateral margins of the carapace approach the ventral midline, leaving only a narrow, longitudinal, midventral cleft between them so that the branchial chamber is almost completely enclosed. The chamber is open anteriorly, in the region of the head. Dorsally, the anterior carapace is extended on the midline to form a short rostrum (or pseudorostrum). Much of the exoskeleton is thick, hard, and opaque.

The narrow, six-segmented abdomen terminates in a pair of slender, elongate uropods and the telson. Each segment is a simple, complete, sclerotized ring. Males usually have biramous pleopods on some of the anterior abdominal segments, but they are absent in females.

Some cumaceans are direct deposit feeders that ingest organic-rich sediments whereas others are carnivores. Some are sand lickers that hold sand grains with the maxillipeds while the maxillae and mandibles scrape away the bacterial or algal biofilm. Many cumaceans are suspension feeders that generate a feeding (and respiratory) current with the first and second maxillae and the maxillipeds. The current enters the branchial chamber anteriorly and ventrally (Fig. 19-59). This posteriorly directed inhalant current reverses direction in the enclosed branchial chamber, becomes the exhalant current, and exits anteriorly and dorsally, just ventral to the rostrum. During its passage through the branchial chamber the water flows over the gills, where gases are exchanged, and through a setal filter on the first maxillae, where food particles are removed. The maxillipeds transfer food from the first maxillae to the mouth. The triturative stomach is similar to that of the decapods. Digestive ceca number one to four pairs.

Gas exchange is accomplished by filamentous epipods of the first maxilliped and by the inner surface of the carapace. The excretory organs are maxillary glands. The short heart has one pair of ostia and is located in the anterior thorax near the gills. Blood flow is similar to that of other small malacostracans. The nervous system is primitive and, other than the brain, uncephalized. The ventral nerve cord has independent segmental ganglia for all trunk segments as well as the first and second maxillae. Sessile compound eyes may be present, but if so, they are small, having less than 10 ommatidia each, and coalesced to form a single dorsal, median eye at the base of the rostrum. Deep-sea species have no eyes.

Cumaceans are chiefly benthic animals that spend most of their time partially buried in soft sediments. Densities as high as 1200/m^2 have been reported. Species distribution patterns and population densities are determined largely by sediment characteristics such as particle size and organic content. When in the sediment, the head is up and slightly exposed, with the rostrum and inhalant and exhalant apertures free so that a feeding and respiratory current can be maintained (Fig. 19-59). Upon leaving the sediment, cumaceans swim using the natatory exopods of the pereopods. The animal burrows backward, using the posterior pereopods, and ends up in its inclined position with its head projecting above the surface. The styliform uropods at the end of the long, mobile abdomen are used for grooming.

Cumaceans are gonochoric and sexually dimorphic. Males tend to have eyes, but females often do not. Males' second antennae are very large, whereas those of females are vestigial. Males tend to be more active than females and are better equipped for swimming. Males usually have natatory exopods on more pereopods than do females, and they have pleopods, which the females lack. The gonads are paired dorsal tubes in the pereon. The female has oostegites on thoracic appendages 3 through 6 and the oviducts open via gonopores on the coxae of the sixth thoracic appendages. The male gonopores are on a pair of penes on the sternite of the eighth thoracic segment.

Little is known of mating and reproduction in cumaceans, and it has never been observed in the field. It is apparently associated with nocturnal swarming behavior in which large numbers of individuals leave the sediment and swim upward in the water column. Females brood the eggs in the marsupium and the young hatch as postlarvae three molts shy of the manca stage. The young remain in the marsupium until they have completed the three molts required to become mancas and then leave.

Mictacea[O]

Mictacea was described in 1985 and is now known from two small, subcylindrical species: *Hirsutia bathyalis* from 1000 m in the North Atlantic and South Pacific and *Mictocaris halope* from Bermudan sea caves. The largest is 3.5 mm in length. The first thoracomere is fused with the head and its appendages are maxillipeds. The carapace covers only the first thoracomere. The seven pereopods are biramous with setose exopods used primarily for swimming. Oostegites are present on most pereopods of females and form a marsupium. The six-segmented abdomen bears reduced pleopods on its first five segments and well-developed uropods and a telson on the sixth. Mictaceans probably are detritivores, but little is known of their biology.

Spelaeogriphacea[O]

This small taxon of subterranean peracaridans includes only three known species: *Spelaeogriphus lepidops* from Bat Cave in Table Mountain, South Africa; the Brazilian *Potiicoara brasiliensis;* and *Mangkurtu mityula* from Western Australia. The body is long and cylindrical and may reach up to 9 mm in length. The first thoracomere is fused with the head and bears maxillipeds. A small carapace extends posteriorly to cover the second thoracomere and part of the third. Most of the seven pereopods are biramous and their exopods are gills. Females have oostegites on the first five pereopods. The abdomen has six segments and the biramous, setose pleopods are mostly well developed and natatory, although the fifth is vestigial. The uropods are biramous. The biology and internal anatomy are poorly known.

Tanaidacea[O]

Tanaidaceans are small, marine, benthic crustaceans (Fig. 19-60). Most of the approximately 700 described species are only 2 to 5 mm in length, although one reaches 12 cm. The body is elongate and cylindrical or depressed in cross section. The head is fused with the first two thoracic segments to form a cephalothorax, which is covered by a carapace. The appendages of the first thoracic segment form a maxilliped whereas those of the second are chelipeds, which may be very large. Chelipeds on the second thoracic segment are a distinctive feature of tanaidaceans. The remaining six thoracic segments form the pereon and their appendages are pereopods. In many tanaidaceans, silk glands in the pereon open at pores at the tips of several anterior pereopods. Silk is used in tube construction

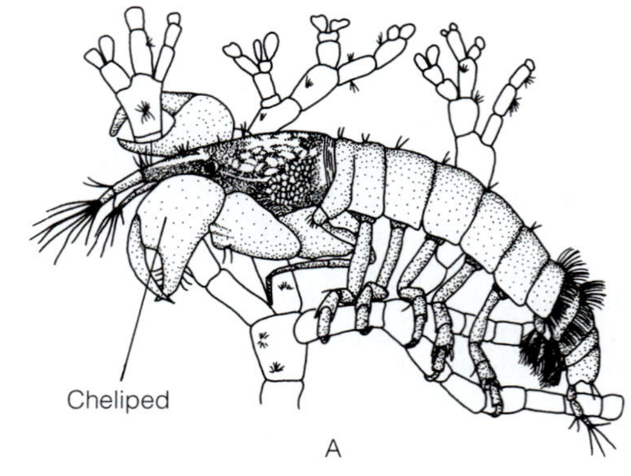

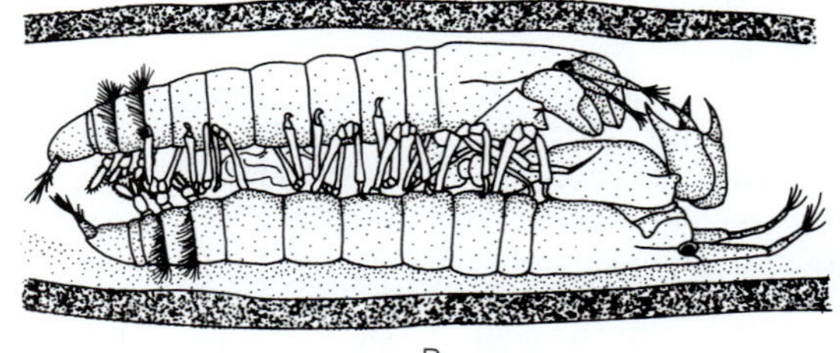

FIGURE 19-60 Tanaidacea: *Tanais cavolinii,* an intertidal tanaidacean from the Norwegian coast. This species lives in tubes constructed on calcareous algae. **A,** This male has left its tube and is crawling over a coralline alga in search of a female. **B,** Copulating pair of *T. cavolinii* within the tube of the female. The female is on top. *(A, From Johnson, S. B., and Attramadal, Y. G. 1982b. A functional morphological model of* Tanais cavolinii *adapted to a tubicolous life-strategy. Sarsia 67:29–42; B, From Johnson, S. B. and Attramdal, Y. L. 1982a. Reproductive behaviour and larval development of* Tanais cavolinii. *Mar. Biol. 71:11–16.)*

and for draglines. Female pereopods 1 to 4 usually bear oostegites that form a marsupium. Some have oostegites only on the fourth pereopods.

The short abdomen consists of six segments, the first five forming the pleon. The sixth segment is fused with the telson to form a pleotelson. The pleonal segments usually bear uniramous or biramous appendages, but these are absent in females of some species. The sixth segment bears uropods, which may be short or long and antenniform.

With few exceptions tanaidaceans inhabit marine benthic habitats from the intertidal zone to the deep sea. Most live in water deeper than 200 m, but they can be common in shallow coastal waters. A few are found in brackish or fresh water and one species has occasionally been found in the plankton. Most tanaidaceans inhabit a burrow, tube, or crevice from which they rarely venture. Reproductive males that must leave their tubes in search of females are the notable exception. Such males are subject to much heavier predation, often by isopods, than are individuals in their tubes. Some tanaidaceans swim using the pleopods. In suitable habitats, population densities may be very high. *Leptochelia dubia,* a widespread tubicolous inhabitant of shallow water, achieves densities of 30,000/m^2 on the soft bottoms of Tomales Bay, California.

The major feeding mode appears to be a combination of detritivory and carnivory. The chelipeds grasp small invertebrate animals or large pieces of detritus. *Leptochelia dubia* and *Tanais cavolinii* use the chelipeds to collect diatoms, algae, and other material from around the mouths of their burrows. Deep-sea tanaidaceans probably feed on detritus. These modes may be supplemented by suspension feeding using a setal filter on the second maxillae. Filtration utilizes a special feeding current generated by movements of the maxillipeds and second maxillae that is supplemented by the respiratory current generated by the maxilliped epipodite. There are one or two pairs of digestive ceca and, in some taxa, a triturative stomach.

The carapace encloses a branchial chamber on each side. The vascularized inner surface of the carapace is the respiratory surface of most species, but it may be supplemented by the large epipod of the maxilliped. In most tanaidaceans the epipod is not itself respiratory, but its rhythmic movements generate the respiratory current through the branchial chambers.

Excretory structures include a pair of maxillary glands and nephrocytes in the hemocoel. The heart usually extends the length of the pereon, but sometimes it is shorter. A network of capillary-like sinuses in the wall of the carapace empties oxygenated blood into the pericardial sinus.

The nervous system includes a subesophageal ganglion with at least the ganglia of the mouthparts and maxillipeds and sometimes those of the chelipeds as well. The remaining ganglia are part of the ventral nerve cord. Compound eyes, if present, are small with few ommatidia. In some taxa the eye is on a short, articulated stalk. Aesthetascs are present on the first antennae.

Most tanaidaceans are gonochoric and usually exhibit sexual dimorphism in the size of the chelipeds, development of the eyes, number of aesthetascs, and development of pleopods in a pattern similar to that of cumaceans. Males often have enhanced sensory systems and reduced or vestigial mouthparts, and mature males of some taxa do not feed, subsisting on energy reserves accumulated as juveniles. The gonads conform to the general malacostracan pattern. Males have a single penis or a pair of penes (cones) on the eighth thoracic sternite.

Fertilization is external and follows the parturial molt of the female. Mating usually occurs in the tube of the female, which the male visits for that purpose. The partners lie facing each other in the tube (Fig. 19-60) and the male releases sperm after inserting his penes into the marsupium. The female immediately extrudes eggs from the gonopores into the marsupium, where fertilization occurs. The eggs are brooded in the marsupium, from which the juveniles exit as manca larvae, which may then remain in the mother's tube for two or three days. Tanaidaceans are protogynous consecutive hermaphrodites, as are some isopods (such as *Cyathura*). In general, newly hatched mancas develop first into reproductive females, produce one or more broods, and then molt to become reproductive males.

Isopoda[O]

Most of the 4000 described species of isopods (Fig. 19-61) live in the sea, where they are widely distributed, diverse, abundant, and ecologically important in most benthic habitats. In addition, many are found in freshwater benthic communities, including interstitial, hypogean, and epigean habitats. The wood lice, or pill bugs, are the most successful terrestrial crustaceans. The taxon includes many parasitic

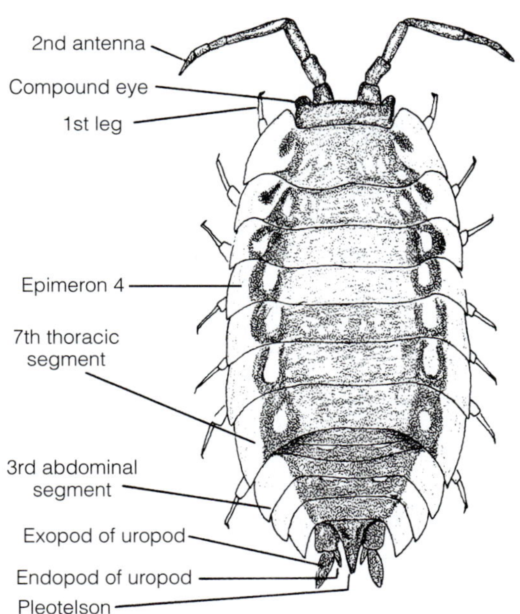

FIGURE 19-61 Isopoda: Oniscoidea. A terrestrial oniscoidean isopod, *Oniscus asellus.* This European species has been introduced into North America and is common in gardens, hothouses, and human habitations. The five free pleonal segments are visible in the drawing. The sixth pleonal segment is fused with the telson to form the small pleotelson. *(After Paulmier from Van Name, 1936.)*

species, including two exclusively parasitic higher taxa. They have, however, no pelagic representatives equivalent to the hyperiidean amphipods. Among crustaceans, Isopoda is exceeded only by the copepods, decapods, and amphipods in number of species. Most isopods are 5 to 15 mm in length, but the giant, deep-sea *Bathynomus giganteus* reaches a length of 42 cm and a width of 15 cm and some interstitial *Microcerberus* are only 1 mm. Isopods resemble amphipods in several respects and are often compared—and sometimes confused—with them, but the similarities are probably convergent. Amphipods tend to be laterally compressed whereas most isopods are dorsoventrally depressed.

FORM

The isopod cephalothorax consists of the head and the first thoracic segment. The first antennae are short and uniramous, and in terrestrial isopods they are vestigial. The compound eyes are sessile. There is one pair of maxillipeds and the mandibles have palps and lacinia mobilis.

The seven free thoracic segments form a pereon with seven pairs of uniramous pereopods. In most isopods there is no specialization of these appendages and the seven are similar (Fig. 19-62B). The name isopod (= equal foot) refers to the similarity, in morphology and orientation, of these appendages. Isopod pereopods are not divided into anteriorly and posteriorly oriented groups, as those of amphipods are. In some isopods, however, the anterior pereopods are specialized as gnathopods (Fig. 19-62A) similar to those of amphipods. The pereopods usually are adapted for walking or crawling. The coxae of the pereopods are fused with the tergites to form epimera (sing., epimeron), or sideplates (Fig. 19-61), that project ventrolaterally and form a protected ventral space below the thorax. The pereopods do not bear gills as they do in amphipods, and so this protected space is not a branchial chamber. Some or all of the pereopods of mature females bear medial oostegites that form a marsupium where eggs are brooded (Fig. 19-62B).

The abdomen consists of six segments of which at least one, and often more, is fused with the telson to form a rigid, sclerotized **pleotelson** (Fig. 19-63D). For example, in Asellota all but the anteriormost one or two abdominal segments are fused with the telson (Fig. 19-62A) whereas in Oniscoidea only the sixth contributes to the pleotelson (Fig. 19-61).

The abdominal appendages are five pairs of biramous pleopods and a pair of biramous uropods similar to those of

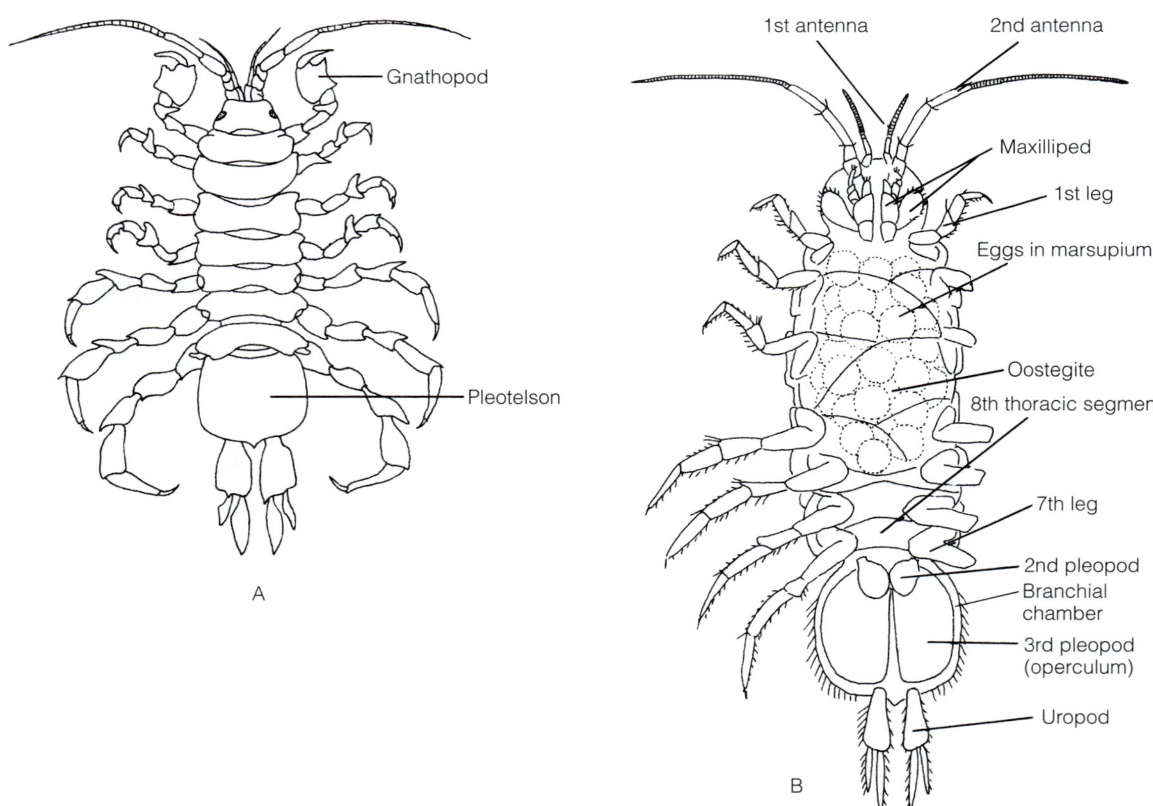

FIGURE 19-62 Isopoda: Asellota. **A,** *Caecidotea (= Asellus)* in dorsal view. **B,** Female *Caecidotea* in ventral view. Appendages omitted from the left side for clarity. Note the overlapping oostegites forming the marsupium around the eggs. *(A, After Pennak, R. W. 1989. Fresh-Water Invertebrates of the United States, Protozoa to Mollusca. 3rd edition. Wiley-Interscience, New York. 628 pp.; B, After Van Name, W. G. 1936. The American land and freshwater isopod Crustacea. Bull. Amer. Mus. Nat. Hist. 71:7.)*

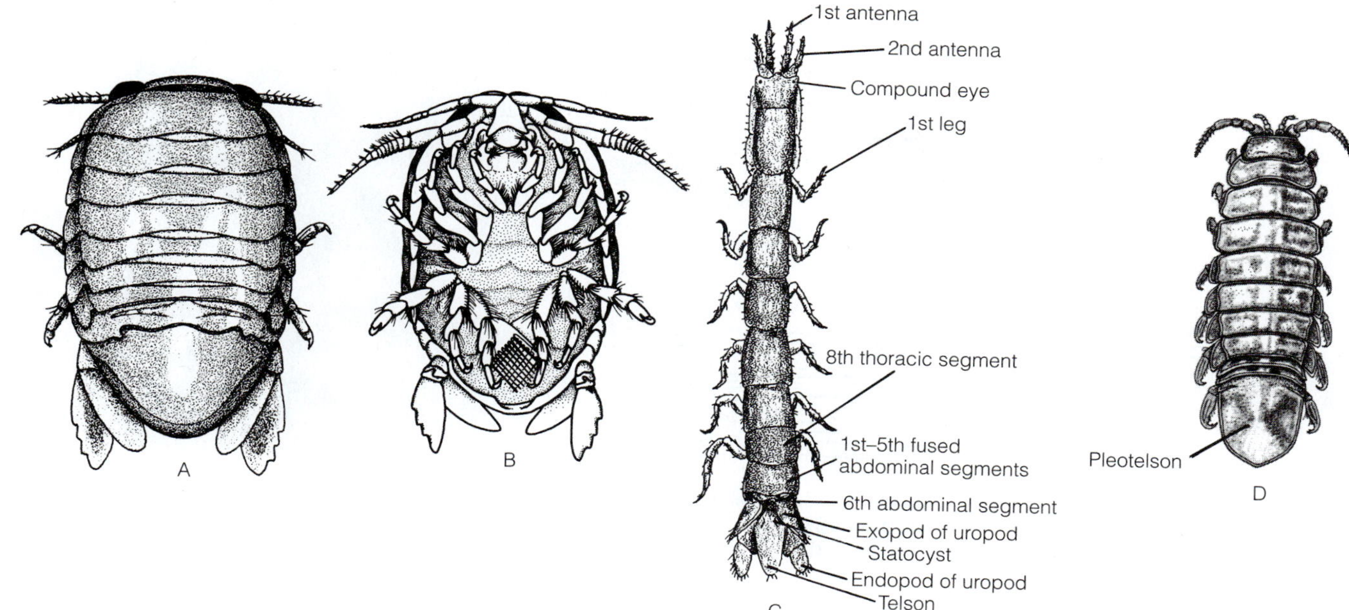

FIGURE 19-63 Isopoda: Representatives of some isopod higher taxa. **A,** Flabellifera: *Sphaeroma quadridentatum,* a common shallow-water marine isopod found on algae and pilings. All but one abdominal tergite are fused with the telson. The members of this genus are capable of rolling into a ball, as are many terrestrial isopods. **B,** Ventral view of *Sphaeroma.* **C,** Anthuridea: *Cyathura,* a genus of marine isopods adapted for living in burrows in mud and sand bottoms. The many other species of this large family have a similar habit. **D,** Valvifera: *Idotea pelagica* (dorsal view). This and other members of the large marine family Idoteidae are common inhabitants of macroalgae in shallow marine waters. *(C, After Gruner from Kaestner, A. 1970. Invertebrate Zoology. Vol. 3. Wiley-Interscience, New York; D, After Sars.)*

decapods. The rami of the pleopods have a thin, permeable cuticle and are the gills (Fig. 19-66A). Abdominal gills are unusual in the Crustacea, but they also occur in stomatopods. The pleopods may also be used for swimming. In most isopods the endopods of the second male pleopods, and occasionally of the first also, are gonopods, or intromittent organs.

Coloration is usually drab, with shades of gray most common, but some are bright colors such as green or orange. Chromatophores adapt the body coloration to the background in many species. The cuticle tends to be more heavily sclerotized than that of amphipods.

DIVERSITY OF ISOPODA

The diversity of Isopoda is reflected in the large number of suborders: nine in comparison with the four of Amphipoda.

Flabellifera[sO]: A large, diverse, important taxon of chiefly marine, benthic isopods (Fig. 19-63A,B, 19-64A, 19-65C). The majority of shallow-water, marine, benthic isopods are flabelliferans. Body usually is depressed, often broadly arched in cross section. Many resemble terrestrial wood lice. A few freshwater species. Many are carnivores or parasites of fishes. *Bathynomus, Cirolana, Cymothoa, Limnoria, Paracerceis, Sphaeroma, Rocinela* (Fig. 12-65C).

Valvifera[sO]: Another important taxon of marine, benthic isopods (Fig. 19-63D). Uropods ventral to the pleotelson form opercula (valves) to enclose and protect branchial chamber and pleopods. Pleotelson often elongate. *Erichsonella, Idotea.*

Asellota[sO]: Chiefly freshwater (Fig. 19-62), but some deep-sea species. Most freshwater isopods are asellotes; both epigean (surface waters) and subterranean. *Caecidotea, Lirceus, Munna.*

Anthuridea[sO]: Chiefly shallow-water, marine, benthic. Mostly Southern Hemisphere. Body elongate, subcylindrical (Fig. 19-63C). Most burrow in soft sediments or occupy crevices in hard substrata. A few in hypogean freshwaters. *Apanthura, Cyathura.*

Phreatoicidea[sO]: Small taxon of amphipod-like freshwater isopods. Only in South Africa, India, New Zealand, and Australia. Feed on plant detritus. *Nichollsia, Phreatoicus.*

Oniscoidea[sO]: Semiterrestrial and terrestrial isopods (Fig. 19-61). Some are so well adapted to terrestrial habitats that they inhabit deserts. Includes supratidal sea slaters and fully terrestrial wood lice, pill bugs, sow bugs. Herbivores. *Armadillidium, Ligia, Oniscus, Porcellio, Tylos.*

Microcerberidea[sO]: Tiny interstitial isopods. Freshwater, shallow marine habitats. Vermiform, cylindrical in cross section, as is typical of interstitial animals. *Microcerberus.*

Gnathiidea[sO]: Small taxon of parasitic marine isopods. Juvenile stages parasitic on fishes; adults do not feed (Fig. 19-65A,B). Adults subsist on fish blood stored in the digestive ceca as

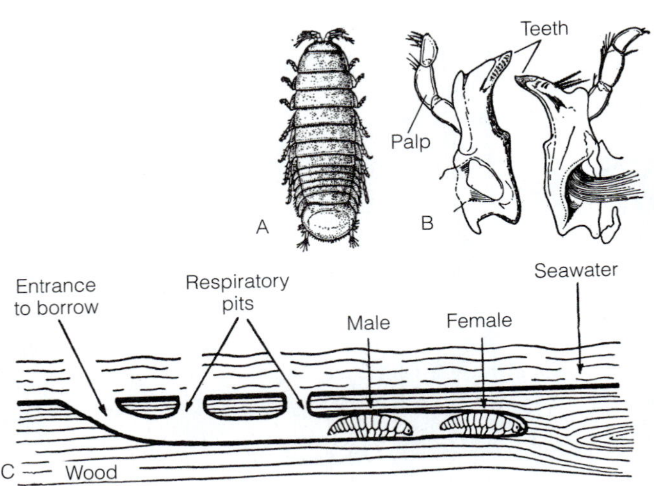

FIGURE 19-64 Isopoda: Flabellifera. **A,** The wood-boring isopod *Limnoria lignorum.* **B,** Mandibles of *L. lignorum.* **C,** Diagrammatic section of a burrow of *L. lignorum.* **D,** Jetty piling nearly eaten through at the base by *L. lignorum.* *(A, After Sars from Yonge; B, After Hoek from Yonge; C, From Yonge, C. M. 1949. The Seashore. Collins, London.)*

juveniles. Adults live in burrows constructed by the male. Some are viviparous, female retains and gestates developing embryos in ovaries. *Gnathia, Paragnathia.*

Epicaridea[sO]: Large taxon of highly modified, parasitic, marine isopods (Fig. 19-65D–I). Sucking mouthparts; feed on blood of crustacean hosts. Asymmetrical female is highly modified for parasitism and barely recognizable as an isopod. Male is a dwarf, usually symmetrical and isopod-like. Some are hermaphroditic. *Asymmetrione, Athelges, Bopyrus Cancricepon, Cyproniscus, Probopyrus.*

LOCOMOTION AND BORING

Most isopods are benthic animals that use their pereopods for crawling. The large, semiterrestrial sea slater *Ligia,* easily the most noticeable isopod along coastlines, runs rapidly over docks, exposed wharf pilings, and rocks. Many aquatic and terrestrial isopods burrow. Some species of the gribble, *Limnoria,* tunnel through wood and can extensively damage docks and pilings (Fig. 19-64), whereas others bore into the holdfasts of kelps. *Sphaeroma tenebrans* bores into the prop roots of mangroves. Although crawling is the primary mode of locomotion, most aquatic isopods can swim using their pleopods. In Sphaeromatidae and Serolidae (Flabellifera) the first three pairs of pleopods are especially adapted for this purpose and gas exchange is restricted to the more posterior pleopods. The ability to enroll into a ball has evolved in many terrestrial Oniscoidea and in the unrelated marine sphaeromatids.

NUTRITION

Most isopods are scavengers and omnivores, although there are also many predators and parasites. Terrestrial and some marine isopods are herbivorous. Deposit feeding is common. Wood lice feed on algae, fungi, moss, bark, and decaying vegetable or animal matter. A few wood lice are carnivorous, as are some marine species, such as the intertidal *Cirolana* and the deep-sea *Bathynomus.*

Wood-boring marine isopods feed on wood by using cellulase secreted by the digestive ceca. Wood-boring species of *Limnoria* are attracted to fungi in the wood that add nitrogen to their otherwise largely nitrogen-free diet of cellulose. In terrestrial wood lice, cellulose digestion is accomplished by symbiotic bacteria, and the hindgut plays a major role in the digestive process.

There are many taxa of parasitic isopods. The larval Gnathiidae (Fig. 19-65A,B) and adult Cymothoidae (Fig. 19-65C) are ectoparasitic on the skin of fishes and have mandibles adapted for piercing. Similar mouthparts are also present in the parasitic taxon Epicaridea, whose members are all bloodsuckers (Fig. 19-65D–I).

The isopod gut consists solely of the foregut and hindgut. The midgut does not contribute to the central tube, but is represented by the all-important digestive ceca arising from the posterior ventral region of the stomach. Beyond these common features, the isopod gut is highly variable, with much of that variability correlated with diet rather than taxonomy.

The ectodermal esophagus is usually muscular and sucks food into the gut. This ability has been enhanced in the bloodsucking parasitic isopods. The posterior end of the foregut is the stomach, where trituration, hydrolysis, and sorting occur. Stomach structure varies but often is similar to that of decapods and some amphipods in having dorsal and ventral channels specialized for separating indigestible material (dorsal) from food (ventral). Most isopods have two or three pairs of digestive ceca, where intracellular digestion and absorption occur. The hindgut begins at the posterior end of the stomach and extends to the anus on the pleotelson.

GAS EXCHANGE

Gas exchange in isopods is abdominal and accomplished across the thin cuticle of some or all of the biramous pleopods. Primitive isopods have five pairs of these. The exopods and endopods are large, flat, overlapping lamellae that function in both gas exchange and swimming. The concave space below the abdomen that houses these pleopods is the branchial

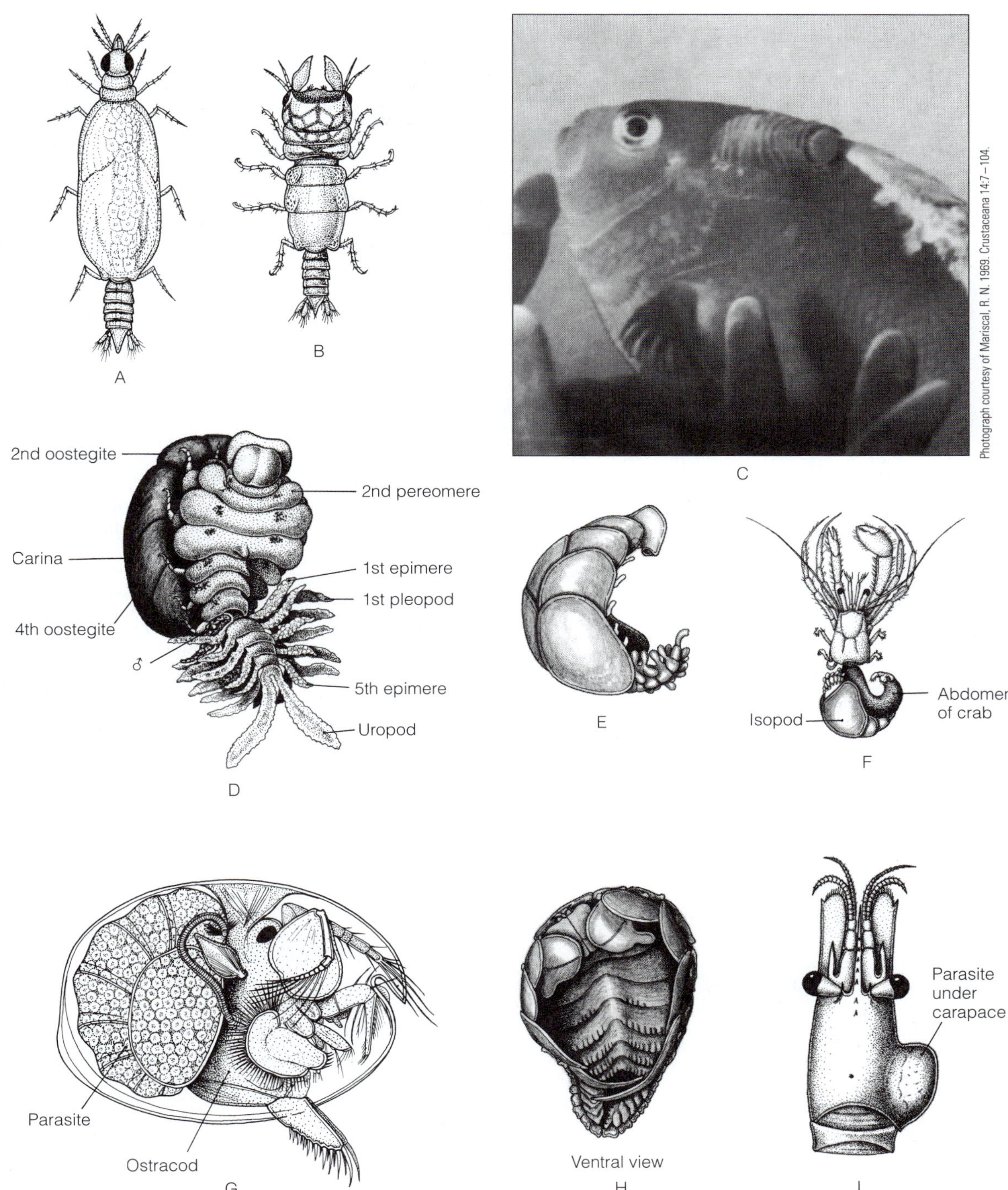

FIGURE 19-65 Parasitic isopods. **A** and **B,** Gnathidea: an aberrant isopod, *Gnathia maxillaris,* that looks like an insect. The larval stage **(B)** is parasitic on fishes and has sucking mouthparts. Both larva and adult are less than 3 mm long. **C,** Flabellifera: an isopod fish louse, *Rocinela,* on a fish living with a sea anemone. **D–I,** Epicaridea: **D,** The bopyrid *Cancricepon elegans,* parasitic in the gill chambers of certain crabs. **E,** *Athlges tenuicaudis* alone. **F,** *A. tenuicaudis* on the abdomen of a hermit crab. **G,** The parasitic isopod *Cyproniscus,* in the ostracod, *Cypridina.* **H,** Ventral view of *Bopyrus squillarum* and **I,** its location in the branchial cavity of a shrimp. *(A and B, Based on living specimens and figures by Sars; D–I, After Sars.)*

chamber. The pleopods typically lie flat against the underside of the abdomen. Most isopods exhibit some variation on this pattern.

The first, second, third, or fourth pleopods or the uropods may be modified as **opercula** to cover the branchial chamber and protect the gills. In the marine Valvifera, the uropods are greatly elongate and meet at the ventral midline to form an operculum covering the branchial chamber, whereas in the freshwater Asellota the third pleopods are the opercula (Fig. 19-62B). In most species only the endopod is respiratory (Fig. 19-66A) and the exopods are natatory. The cuticle of respiratory rami is thinner and more permeable than that of their natatory counterparts.

In some isopods, the anterior pleopods are specialized for swimming and the posterior ones for gas exchange. In some terrestrial isopods, the endopods continue to function as gills but are sequestered in a depression in the accompanying exopod that protects them from desiccation (Fig. 19-66A). In other terrestrial species, a system of epidermal invaginations resembling insect tracheae and known as **pseudotracheae** has developed in the exopod (Fig. 19-66B). Abdominal gas exchange in some isopods is supplemented by the general body surface.

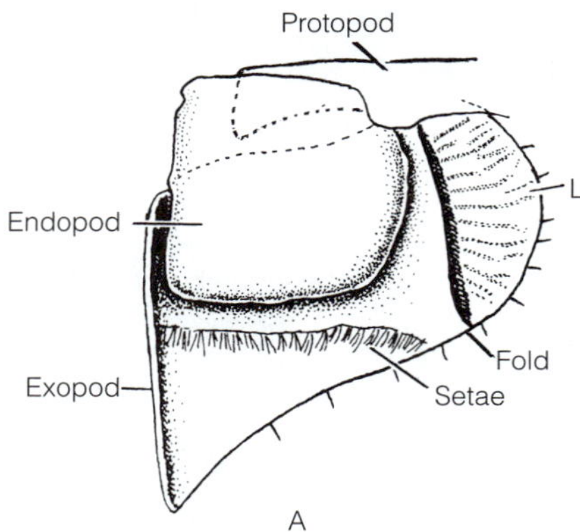

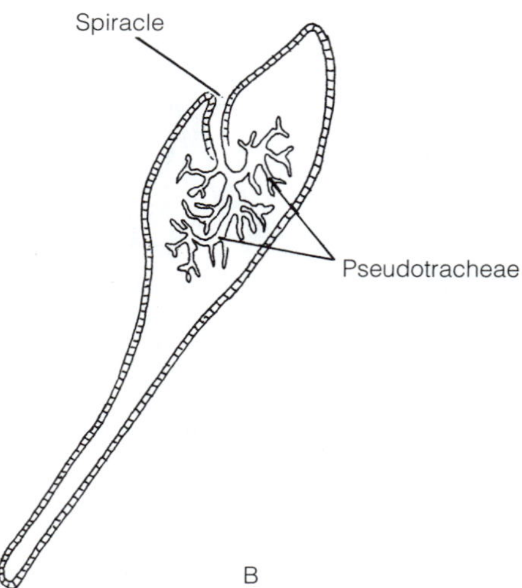

FIGURE 19-66 Isopoda. **A,** Fifth pleopod of the wood louse *Oniscus asellus,* showing the depression in the exopod accommodating the respiratory endopod. The setae exclude particulates. Part of the exopod that is also respiratory is indicated by the L (lateral respiratory surface of the exopod). **B,** Section through the pleopod exopod of the wood louse *Porcellio scaber,* showing spiracle and pseudotracheae. *(Both after Unwin from Kaestner, A. 1970. Invertebrate Zoology. Vol. 3. Wiley-Interscience, New York.)*

INTERNAL TRANSPORT AND EXCRETORY SYSTEM

The isopod hemal system resembles that of other malacostracans except that the heart, like the gills, is abdominal rather than thoracic. The abdominal position of the heart is correlated with the position of the respiratory pleopods. The blood contains hemocyanin.

The excretory organs are maxillary glands. These probably function chiefly in osmoregulation and are larger in freshwater species than in marine or terrestrial isopods. The gills are thought to have an osmoregulatory function and the gut also plays a role in ion and water exchange with the environment. Nitrogen is eliminated across the gills as ammonia. Because their maxillary glands are poorly developed, wood lice release nitrogenous wastes as gaseous ammonia. Terrestrial oniscoideans absorb water across the epithelium of the rectum. Nephrocytes are present.

NERVOUS SYSTEM AND SENSE ORGANS

The nervous system is typically arthropodan, with a subesophageal ganglion consisting of the ganglia of the mouthparts and maxillipeds. Some abdominal ganglia may fuse to form a ganglion in the anterior pleon. The compound eyes are appositional, with the crystalline cones in contact with the rhabdome and the unit shielded by pigment cells. Median ocelli are also present in many isopods. The first antennae bear aesthetascs that are especially numerous in males, which use them to detect females in preparturial condition. In terrestrial isopods (oniscoideans) the first antennae are vestigial and the aesthetascs are on the second antennae, where they probably function as hygroreceptors to detect humidity. Some isopods have statocysts in the pleotelson.

REPRODUCTION AND DEVELOPMENT

Most isopods are gonochoric, although there are some hermaphroditic species. Sexual dimorphism is minimal in most but pronounced in some parasites. Fertilization is internal and sperm transfer is indirect, with modified pleopods (gonopods) being used to transfer sperm to the female. The paired gonads are separate. The paired female gonopores are median, sternal openings on thoracic segment 6. Each of the two oviducts is enlarged distally to form a seminal receptacle, but the receptacle and oviduct are not connected until after the parturial molt.

The male gonopores open on a pair of penes on the sternite of thoracic segment 8. Sometimes the two penes are fused medially. The first two pairs of pleopods are gonopods in most terrestrial isopods. Sperm is transferred from the penes to the second pleopod, in some species with the assistance of the first pleopod. The second pleopod is inserted into a female gonopore and the sperm is injected into a seminal receptacle, where it is stored until fertilization occurs.

Mating usually takes place before the parturial molt, rather than after it as in amphipods, but in some species it occurs during this molt. The parturial molt establishes a connection between the seminal receptacle and the oviduct so that fertilization and spawning can occur. Before this molt the seminal receptacle is isolated from the ovary and eggs and there is no possibility of fertilization even if the female is inseminated. The eggs are brooded in the marsupium (Fig. 19-62B). The marsupium of terrestrial isopods is filled with water, so development of the young is aquatic, despite the terrestrial habitat of the adults. From a few to several hundred eggs are usually brooded, and the hatching stage is a manca postlarva.

ADAPTATIONS FOR TERRESTRIAL HABITATS

Oniscoidea has specialized in the exploitation of terrestrial habitats and its members show various degrees of adaptation for this habitat. Oniscoideans are believed to have invaded land directly from the sea rather than by way of fresh water, and they now occupy a wide variety of terrestrial habits ranging from the moist seashore to the desert. Oniscoidean evolution has been a saga of increasing ability to conserve water and resist desiccation. Although some oniscoideans live at the edge of the sea and others inhabit marshes, most occupy inland environments independent of standing water. Nevertheless, most species require a humid microclimate and take advantage of moist areas beneath stones, in bark, and in leaf mold in both temperate and tropical regions. Desert-inhabiting species are nocturnal burrowers that by day take advantage of the cooler, moister microclimate available in their burrow and venture to the desert surface only at night.

Shore-inhabiting forms include the widespread *Ligia,* which lives on pilings, jetties, and rocks at the water's edge, and *Tylos,* which lives beneath beach drift or sand at the high-tide mark. Among terrestrial isopods, these shoreline species are the least adapted for terrestrial life. Inland taxa, including *Oniscus, Porcellio,* and *Armadillidium,* are better adapted for life on land and are common in suburban yards and gardens.

Most terrestrial isopods have morphological and behavioral adaptations to reduce water loss and, while they are better at it than other crustaceans, they are considerably less well-adapted in this regard than chelicerates and tracheates. Wood lice have not evolved a waxy epicuticle for their thin, ventral exoskeleton to prevent evaporation as insects and arachnids have. Their ability to enroll may be an adaptation to reduce water loss from this surface.

Like aquatic isopods, wood lice use the gills for gas exchange, but in the best-adapted species the gills are enclosed to reduce desiccation (Fig. 19-66A) and pseudotracheae (Fig. 19-66B) are present in the exopod. As protected invaginations of the body surface, pseudotracheae are much better at conserving water than are gills, which are exposed evaginations. Wood lice with pseudotracheae can tolerate much drier air than can those with gills.

In the absence of a waxy epicuticle, terrestrial isopods cannot avoid evaporation of water from the general body surface, although it can be minimized behaviorally and morphologically. To remain permeable to oxygen, the gills must retain a covering film of moisture. That moisture is subject to evaporation, of course, and must be replaced when lost. In most species the water is replaced from food sources and by drinking. In some wood lice, the two uropodal endopods are held together to form a tube that is dipped into a droplet of dew or rain. Capillarity wicks the water up the tube and onto the gills. Other wood lice have a system of surface channels that carry water from the dorsum to the ventral surface and gills.

In general, wood lice are photonegative and strongly positively thigmotactic and can discriminate between relatively slight differences in humidity, resulting in their being active at night and inactive during the day. That the eyes of wood lice are poorly developed is probably related to their nocturnal, secretive behavior and diet of decaying vegetation that does not require vision to locate. Repugnatorial glands are used to defend against such predators as spiders and ants. The ability to enroll probably contributes to this defense.

Phylogeny of Peracarida

Lophogastrid mysidaceans are the most primitive peracaridans and are closest to the peracaridan origins. The remaining taxa arose from mysid-like ancestors.

The more-derived peracaridans form a taxon, Mancoidea, in which the young emerge from the marsupium as manca larvae that resemble miniature adults except that the eighth thoracopod is absent. Mancoidea includes Cumacea, Mictacea, Spelaeogriphacea, Tanaidacea, and Isopoda, with only Mysidacea and Amphipoda being excluded. Amphipoda is the sister taxon of Mancoidea.

The well-developed cumacean carapace almost completely encloses the branchial chamber, but in other mancoideans the carapace has been reduced or lost. Mictacea, consisting of three species from marine caves and abyssal muds, and Spelaeogriphacea, with only three species from freshwater cave ecosystems, both have a short carapace covering the first one or two thoracomeres.

Tanaidaceans and isopods are sister taxa. The cephalothorax of tanaidaceans consists of the head and first two thoracomeres, whereas it is the head and only one thoracomere in the isopods. Tanaidaceans have a carapace, but it has been lost in both isopods and, through convergence, amphipods. Some abdominal segments fused with the telson to form a pleotelson in both tanaidaceans and isopods, but this similarity is probably a convergence, as some fossil tanaidaceans have no pleotelson.

PHYLOGENETIC HIERARCHY OF PERACARIDA

No phylogenetic system has been published for Peracarida. The following incomplete hierarchy is based on comments in Ax (2000).

- Peracarida
  - Mysidacea
  - N.N.
    - Amphipoda
    - Mancoidea
      - Cumacea
      - N.N.
        - N.N.
          - Mictacea
          - Spelaeogriphacea
        - N.N.
          - Tanaidacea
          - Isopoda

PHYLOGENY OF MALACOSTRACA

Leptostraca is the sister taxon of Eumalacostraca, and the latter contains all the remaining malacostracan taxa (Fig. 19-67). In the early eumalacostracans the exopods of the second antennae developed into a scaphocerite, or antennal scale, that is absent in Leptostraca. The phyllopodous thoracopods inherited from the ancestral crustacean were transformed into stenopods by modification of the endopods. The sixth pleopods became uropods, which in conjunction with the enlarged telson formed a tail fan. The caudal furca was lost.

Eumalacostraca is divided into Stomatopoda and Caridoida, the latter including all remaining eumalacostracan taxa. The stomatopods have triramous first antennae, raptorial anterior thoracopods, and respiratory pleopods. The caridoid rostrum is fused with the head and unmovable. There is a statocyst in each of the second antennae, although it is often secondarily lost. The abdominal muscles, acting in conjunction with the tail fan, form a powerful thrusting mechanism permitting rapid rearward escape. The left mandible has a lacinia mobilis, which is a heavy, movable spine.

The incorporation of the three anterior thoracomeres into the cephalothorax and modification of their appendages into three pairs of maxillipeds separates the large and important Decapoda from the Xenommacarida. The decapod pereon is left with five pairs of pereopods. The cephalothorax and pereon are covered dorsally and laterally by a carapace that

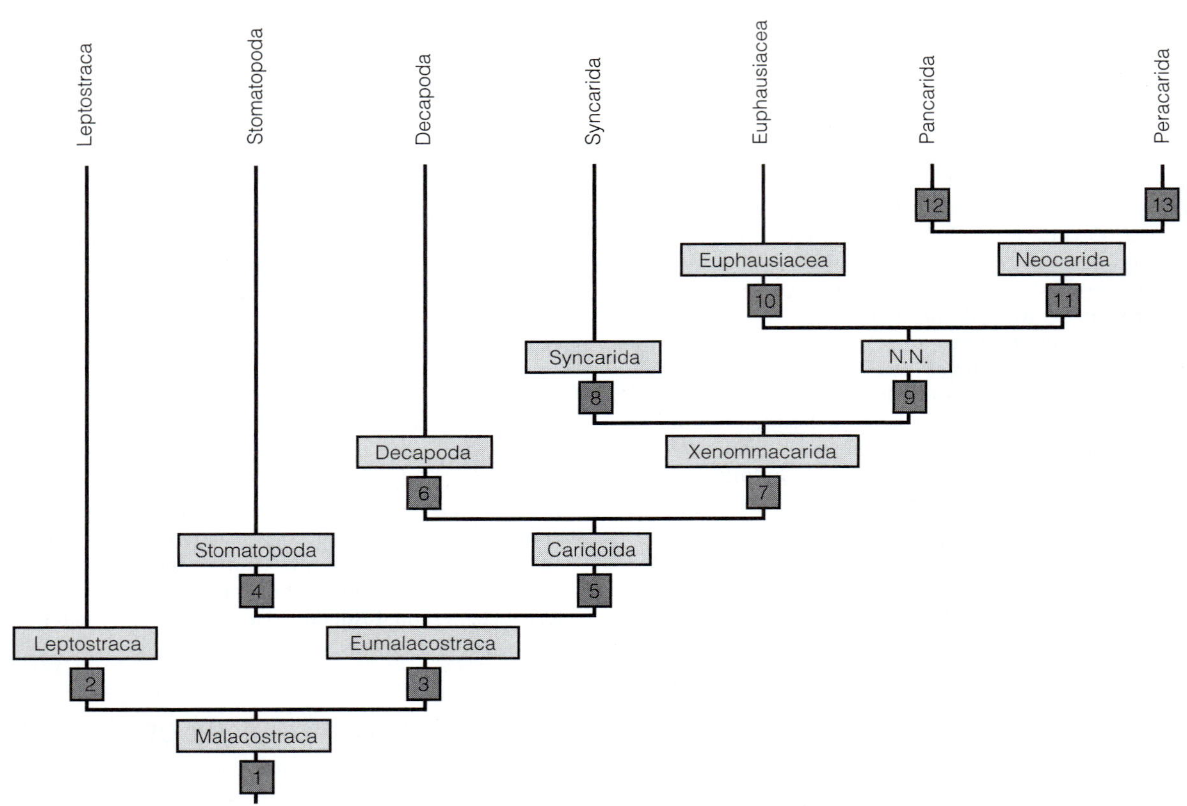

FIGURE 19-67 A phylogeny of Malacostraca. **1, Malacostraca:** Trunk has 15 segments; abdomen and thorax have appendages; thorax has 8 segments; abdomen has 7; female gonopore is on thoracomere 6; male gonopore is on thoracomere 8. **2, Leptostraca:** Antennae 1 have a scalelike appendage; antennae 2 are uniramous; maxillae 2 have a long palp for grooming; brooding with direct development. **3, Eumalacostraca:** Exopod of antennae 2 is a large flat scale; thoracopods are stenopods; uropods and telson form a tail fan. **4, Stomatopoda:** Antennae 2 are triramous; short carapace covers four anterior thoracomeres; thoracopods 1 through 5 are raptorial; pleopods are respiratory; maxillary glands. **5, Caridoida:** Rostrum fused with head, antennal statocyst is present; pleon musculature and tail fan function as an escape mechanism; lacinia mobilis on left mandible only. **6, Decapoda:** Carapace covers and fuses with all eight thoracomeres and encloses the lateral branchial chambers; cephalothorax includes three thoracomeres; thoracopods 1 through 3 are maxillipeds; scaphognathite (epipodite of maxillae 2) generates the respiratory current. **7, Xenommacarida:** Crystalline cone of ommatidium produced by two cone cells; cone stalk is lost; retinula cells extended distally. **8, Syncarida:** Carapace is lost; direct development; fresh water. **9, N. N.:** Thoracic biramous epipodial gills with the medioventral ramus located in a branchial cavity under the thorax. **10, Euphausiacea:** Carapace large and attached to all eight thoracomeres. Telson with a pair of subapical spines. **11, Neocarida:** Both mandibles have lacinia mobilis; cephalothorax is formed of one thoracomere fused with the head; thoracopod 1 is a maxilliped; development is direct. **12, Pancarida:** Dorsal brood pouch; short carapace; saccate nephridia are absent; eyes are absent; pleopods 3–5 absent. **13, Peracarida:** Ventral marsupium composed of oostegites. *(Modified from Ax, P. 2000. Multicellular Animals, II. Springer Verlag, Berlin.)*

PHYLOGENETIC HIERARCHY OF MALACOSTRACA

Malacostraca
- Leptostraca
- Eumalacostraca
  - Stomatopoda
  - Caridoida
    - Decapoda
    - Xenommacarida
      - Syncarida
      - N.N.
        - Euphausiacea
        - Neocarida
          - Pancarida
          - Peracarida

encloses a gill chamber on either side. The exopods of the second maxillae become large gill bailers (scaphognathites) used to generate the respiratory current.

Modification of the ommatidium in the compound eye of the ancestral Xenommacarida reduced the number of cone cells that form the crystalline cone from the ancestral four to two. One group of xenommacarids lost the carapace, abandoned the nauplius larva and adopted direct development, and evolved into the freshwater Syncarida.

The euphausiaceans arose with the development of a carapace covering and fused dorsally with the entire thorax. The carapace does not extend ventrally, however, leaving the eight pairs of unmodified biramous thoracopods exposed. There are no maxillipeds.

The remaining malacostracans belong to Neocarida and are characterized by a lacinia mobilis on each mandible. At least the first thoracomere is fused with the head to form a short cephalothorax, and the first thoracopod is a maxilliped. These crustaceans have lost the nauplius larva and have direct development.

Pancarida is a small neocarid taxon whose members brood the young in a dorsal brood pouch under the carapace. Pancarids are unusual in having no antennal or maxillary glands. Peracarida evolved from neocarid ancestors by developing a ventral marsupium enclosed by setose oostegites extending medially from some thoracopods.

MAXILLOPODA^SC

In most recent classifications, several taxa are grouped together in Maxillopoda. Those included vary with author, but in its broadest and most inclusive sense they are Copepoda, Mystacocarida, Tantulocarida, Ascothoracida, Cirripedia, Ostracoda, Branchiura, and Pentastomida. Maxillopodans are small crustaceans (usually microcrustaceans) with a short trunk.

Several characteristics define Maxillopoda, but the short trunk with 10 or fewer segments is the most important. The naupliar eye has only three ocelli, in contrast to the four of the ancestral crustaceans. Compound eyes are sometimes present, but they are completely absent in Copepoda, Mystacocarida, and Tantulocarida. Sometimes there is a carapace. The thorax includes seven or fewer segments and the abdomen three or less. Some thoracic segments may fuse with the head to form a cephalothorax. The thorax bears biramous (sometimes uniramous) appendages, but the abdomen lacks appendages. Many adult maxillopodan features (short body, naupliar eye, small size, and paucity of thoracopods) are reminiscent of larvae, suggesting that Maxillopoda may have arisen by pedomorphosis when a postlarva attained sexual maturity without developing adult morphology.

COPEPODA^C

Approximately 12,000 described species of extant copepods (Fig. 19-68) make this one of the largest crustacean higher taxa, second only to Malacostraca. Most copepods are marine, but there are many freshwater species and a few terrestrial forms live in moss, soil–water films, and leaf litter. Parasitic copepods attack various marine and freshwater animals, particularly fish. The enormous numbers of marine copepods are almost always the most abundant and conspicuous component of a plankton sample. Copepoda is one of the three taxa that dominate the freshwater zooplankton. It has been suggested that copepods are the most abundant metazoans on Earth, with more individuals (but not species) than insects. Because most planktonic species feed on phytoplankton, copepods are the principal and essential link between the world's most important producers (marine phytoplankton) and higher trophic levels in pelagic marine food chains. Copepods are a major part of the diet of many marine animals.

Most copepods range in length from less than 1 to about 5 mm, although there are larger (17 mm) free-living species. Some parasitic forms are very large, over 32 cm in length with the egg sac and 25 cm without (Fig. 19-73). Although most copepods are pale and transparent, some species may be brilliant red, orange, purple, blue, or black. Many species are bioluminescent.

The body of most free-living copepods is tapered from anterior to posterior and somewhat cylindrical (Fig. 19-69), but there are many exceptions to this generalization. The trunk is composed of a thorax and abdomen. The anterior end is either rounded or pointed. Compound eyes are absent, but the median naupliar eye is a typical and conspicuous feature of most copepods. Also conspicuous are the uniramous first antennae, which generally are long and held outstretched at right angles to the long axis of the body.

The great diversity of Copepoda is reflected in by its 10 higher taxa (orders), 5 of which are free-living and 5 commensal or parasitic. The free-living taxa are Cyclopoida (Fig. 19-68, 19-70C,D), Calanoida (Fig. 19-69), Harpacticoida (Fig. 19-70A,B, 19-72), Gelyelloida, and Platycopioida. The symbiotic taxa are Monstrilloida, Mormonilloida, Misophrioida, Siphonostomatoida (Fig. 19-73, 19-74B–E), and Poecilostomatoida (Fig. 19-74A). Calanoida, Harpacticoida, and Cyclopoida, although not closely related, are overwhelmingly the most ecologically important free-living taxa and will be the focus of this section. Among modern copepods, the calanoids are most like the ancestral copepods.

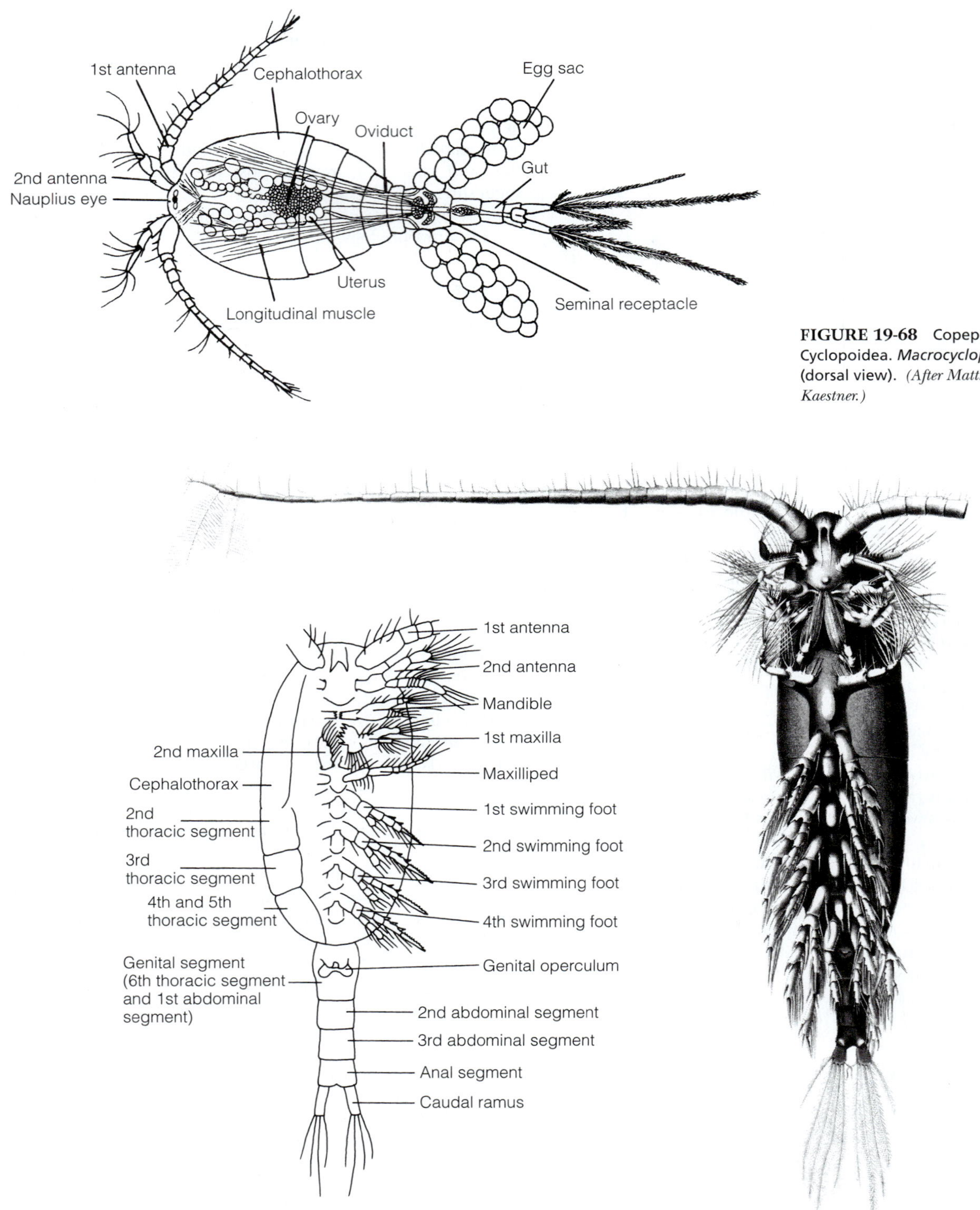

FIGURE 19-68 Copepoda: Cyclopoidea. *Macrocyclops albidus* (dorsal view). *(After Matthes from Kaestner.)*

FIGURE 19-69 Copepoda: Calanoida. **A,** Diagrammatic ventral view of the calanoid *Pseudocalanus,* showing appendages. **B,** Ventral view of *Calanus,* a typical calanoid copepod, with appendages shown. *(A, From Corkett, J., and McLaren, I. A. 1978. The biology of* Pseudocalanus. *Adv. Mar. Biol. 15:2–231; B, After Giesbrecht, W. 1892. Fauna and Flora Golfes Neapel. Monogr. 19:1–831.)*

Of the five free-living taxa, calanoids are largely planktonic; harpacticoids, which account for over 50% of copepod species, are mostly benthic; and cyclopoids include both planktonic and epibenthic species. Variations in body shape are related to the habitat of the species. Planktonic forms tend to have a cylindrical body with a narrow abdomen. Those that live high in the water column tend to be more slender and fusiform than those that swim closer to the bottom. Epibenthic species, which crawl and swim just above the bottom, have somewhat broader bodies. Benthic species that live on algae and sea grasses may be broad and flattened; interstitial species are narrow and vermiform (Fig. 19-70A,B).

Form

The typical copepod body is short and has a head and 10 trunk segments plus a telson (Fig. 19-71). The head bears five pairs of appendages. The first antennae are uniramous and may be long. In males they are **geniculate** (sharply bent) and used to hold the female during copulation. The small second antennae are biramous. The mandibles are biramous and each mandibular coxa bears a large, toothed gnathobase and a biramous palp. The first maxillae are biramous, but often are reduced or lost in parasitic species. The second maxillae are uniramous. The adult retains the naupliar eye with three inverse pigment-cup ocelli, and compound eyes are absent.

The trunk is divided into a thorax of seven segments and a three-segmented abdomen. In the ground plan, the first trunk segment is fused with the head to form a cephalothorax, but in some derived copepods the first two thoracic segments are incorporated into the cephalothorax. No other maxillopodan has such a cephalothorax. The seven thoracic segments bear paired biramous thoracopods whereas the abdominal segments have no appendages. The first thoracic appendages are the maxillipeds of the cephalothorax. Unlike the other thoracopods, the maxillipeds are uniramous and may be reduced or absent. Thoracic segments 2 through 7 are independent of the cephalothorax and form the pereon. Thoracic segments 2 through 6 bear similar biramous pereopods (Fig. 19-71) consisting of a coxa, basis, endopod, and exopod. The right and left coxae of each pair are rigidly joined across the ventral midline by an unpaired median **intercoxal plate.** In both sexes the gonopores are on the seventh thoracic segment. This is the genital segment, and it bears a pair of appendages that form an operculum over the gonopores. In some females the genital segment is fused with the first abdominal segment.

The abdomen includes three segments plus the terminal telson (anal segment). The abdominal segments have no appendages and usually are narrower than the thorax. A caudal furca extends posteriorly from the telson. In some planktonic marine species the furca is elaborately developed. The anus opens dorsally on the telson between the bases of the furcal rami.

In the three major free-living copepod orders (Calanoida, Harpacticoida, and Cyclopoida), between the anterior and posterior portions of the trunk is a major articulation at which the body flexes ventrally. This articulation does not coincide with the junction between the thorax and abdomen, nor is it at the same position in each taxon. In the ancestral copepod and the modern calanoids and platycopoids, the articulation is between thoracomeres 6 and 7. In the other taxa it has shifted from the ancestral position and is between thoracomeres 5 and 6.

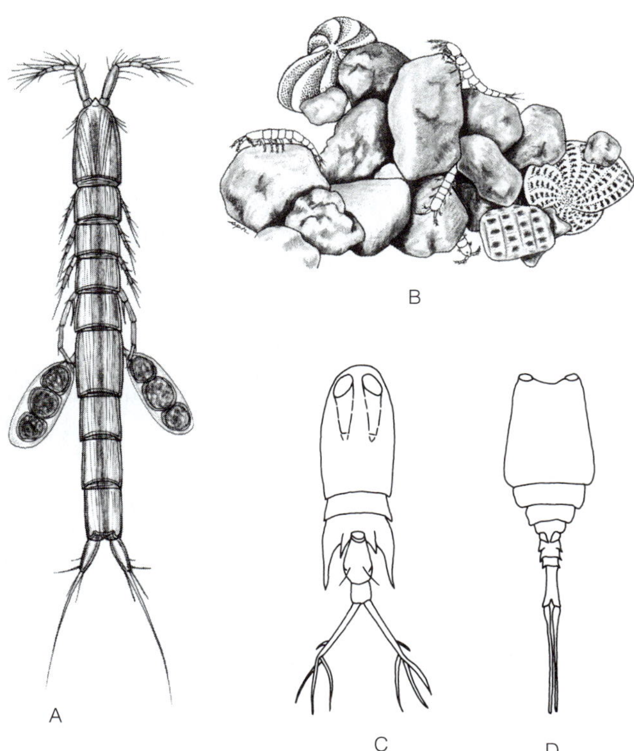

FIGURE 19-70 Copepoda. **A,** A cylindropsyllid harpacticoid copepod (dorsal view). **B,** Interstitial harpacticoid copepods crawling among sand grains. Two foraminiferan shells are among the sand grains. **C** and **D,** Two marine cyclopoid copepods, *Corycaeus* **(C)** and *Copilia* **(D),** in which the naupliar eye is very large and divided. Copepods of these and related genera are often brilliantly colored. *(A, After Sars; C and D, From Smith, D. L. 1977. A Guide to Marine Coastal Plankton and Marine Invertebrate Larvae. Kendall/Hunt Publishing Co., Dubuque, IA.)*

Internal Form and Function

The gut (Fig. 19-71) is typical of crustaceans. Copepods display a range of feeding modes, depending in part on where they live. Planktonic copepods are chiefly suspension feeders, with the second maxillae modified to capture food. Phytoplankton constitute the principal part of the diet of most suspension-feeding species, but some rely heavily on detritus particles as well. Using radioactive diatom cultures, *Calanus finmarchicus,* which is about 5 mm long, was shown to collect and ingest from 11,000 to 373,000 diatoms, depending on their size, every 24 h. When feeding on particles of mixed sizes, larger particles are selected because they can be handled more efficiently, but under natural conditions planktonic copepods probably accept whatever food is available.

Not all planktonic copepods are herbivorous suspension feeders. Some are omnivorous and some are strictly predaceous. Species of *Anomalocera* and *Pareuchaeta* even capture juvenile fishes. Species of the planktonic and epibenthic harpacticoid *Tisbe* swarm over a small fish and eat its fins, immobilizing it. They then devour the body as it falls to the bottom. Some freshwater Cyclopidae are herbivorous suspension feeders utilizing

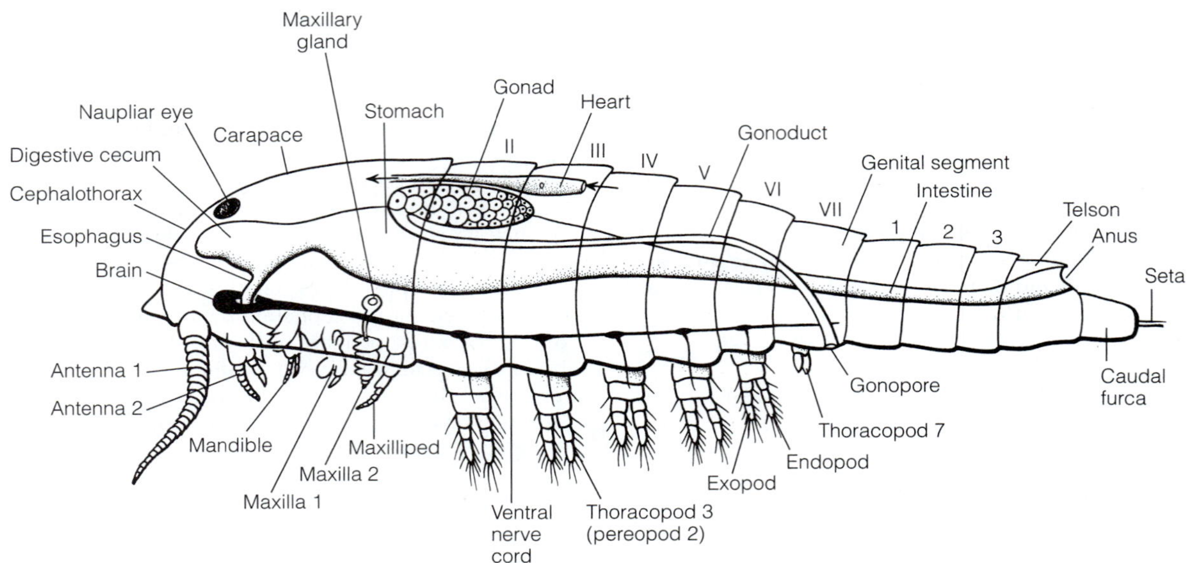

FIGURE 19-71 Copepoda: A generalized copepod in lateral view. The cephalothorax includes the head plus the first thoracic segment. The thorax consists of seven segments, including one that is part of the cephalothorax. Segments 2 through 7 are the pereon. The abdomen begins immediately posterior to the genital segment and consists of three segments plus the telson and caudal furca. Thoracic segments are indicated with Roman numerals, abdominal segments with Arabic numbers. *(Redrawn from Boxshall, G. A., Ferrari, F. D., and Tiemann, H. 1984. The ancestral copepod: Towards a consensus of opinion at the First International Conference on Copepoda 1981. Crustaceana Suppl. 7:68–84.)*

phytoplankton whereas others are zooplanktivores, feeding on other zooplankton. Many cyclopoids, including the common freshwater genus *Cyclops,* are predaceous. Most bottom-dwelling harpacticoids feed on microorganisms and detritus attached to sand grains, algae, and sea grasses.

Planktonic calanoids and some harpacticoids store lipids in a large oil sac that may fill most of the thoracic hemocoel. The lipid is an energy reserve accumulated in the summer and consumed in winter. The oil is often red or blue, and the entire animal may appear to be the color of the oil. Oil makes an important contribution to buoyancy in planktonic species. The oils and waxes of planktonic copepods may have contributed to the formation of petroleum deposits.

Gills are absent in free-living copepods, and gas exchange occurs across the general body surface and the hindgut. Some parasites have gills. A short, dorsal heart is present in the copepod ground plan, but among modern copepods it is present only in calanoids (Fig. 19-71). The heart, when present, lies in a long pericardial sinus separated from the perivisceral sinus by the horizontal diaphragm. In species without a heart, blood is moved by pulsations of the gut. The excretory organs are antennal glands in the larvae and maxillary glands (Fig. 19-71) of the adult. Nephrocytes are also present in the cephalothorax.

The nervous system is variable; in the ground plan it is not cephalized and consists of a ventral nerve cord with a chain of independent segmental ganglia and no subesophageal ganglion. At the other extreme, the nervous system of derived copepods is highly centralized, with all mouthpart and trunk ganglia forming a subesophageal ganglion, which is closely joined by thick connectives to the three-part brain.

Locomotion

The thoracic appendages and second antennae are used in rapid swimming. The second antennae, each of which have two branches that beat in a rotary manner, appear to be more important in calanoid swimming, whereas the thoracic appendages are more important in cyclopoids. The first antennae, which are long and setose in slow-sinking planktonic forms, function as parachutes to decrease settling velocity. When rapid movement is desired, they are held against the body to minimize their surface area.

Carnivorous species cruise continually as they seek potential prey. Herbivorous species, in contrast, alternate periods of cruising with feeding. During a feeding episode, which lasts 10 to 30 s in *Eucalanus crassus,* the first antennae act both as parachutes and as sensors for detecting algae while the other anterior appendages set up a flow field that brings water to the feeding apparatus. A slight tendency to sink helps the copepod maintain the proper orientation for its flow field. After a feeding episode, the copepod cruises or sinks to a new position for another episode. Swimming positions vary greatly among species (upside down, vertical, and so on), and the caudal rami are held in various positions or act as a rudder. For example, *E. crassus* swims backward in a vertical position.

Most planktonic copepods live in the upper 50 m of the sea, but many are found at greater depths. Like other freshwater and marine zooplankters, many planktonic copepods exhibit diel (pronounced DI-al) vertical migration (DVM) up and down in the water column. In the usual pattern, copepods migrate up into the surface water to feed on phytoplankton during the night, when they are relatively safe from sight-orienting predators. During the day they migrate into deeper, darker water, presumably to avoid these predators. Populations of the calanoid *Gaetanus minor* undergo DVM of about 300 m. The calanoid *Calanus tenuciornis* and the cyclopoid *Conaea gracilis* do not undergo DVM and remain at approximately the same depth over a 24-h period. The limnologist G. Evelyn Hutchinson calculated that DVM of zooplankton is the world's greatest animal migration, moving more biomass over greater distances than any other.

The bottom-dwelling harpacticoids and some cyclopoids crawl over or burrow through the substratum, and harpacticoids are a dominant component of the interstitial fauna of sandy bottoms and beaches (Fig. 19-70B). The thoracic limbs are used in crawling, and harpacticoids also use lateral undulations of their wormlike bodies.

Reproduction and Development

The generalized copepod is gonochoric, fertilization is internal, and sperm transfer is indirect and employs spermatophores. The paired gonads connect to the exterior via paired gonoducts, but this pattern is modified in modern copepods and all or part of the system is single. The sixth thoracic appendages of males may be gonopods for the transfer of spermatophores to the female. Copepods, especially parasites, are sexually dimorphic. Males are usually smaller than females, and in some parasitic species the dwarf males have dramatically different morphology than the females (Fig. 19-74D,E). In nonparasitic species, the male's first antennae typically are enlarged to serve as claspers to hold the female during copulation. In females, the distal oviducts form a pair of seminal receptacles for the storage of spermatozoa. During copulation, the male's gonopods glue spermatophores to the female's venter. Spermatozoa, which are aflagellate, move from the spermatophores to the seminal receptacles for storage. Fertilization typically occurs at a later time.

In calanoids, fertilized eggs are released singly into the water, but in most others the eggs are deposited by the female into one or two egg sacs (ovisacs) secreted by the oviduct epithelium (Fig. 19-68, 19-70A, 19-74). Each oviduct forms one egg sac. Eggs are brooded in the egg sacs until the nauplii hatch. Each sac contains a few to 50 or more eggs, and clutches may be produced at frequent intervals. For example, among freshwater copepods, species of diaptomid calanoids alternate between gravid and nongravid conditions every four days, and mating is required for each clutch of eggs.

Development involves six naupliar and five copepodid larval stages preceding the adult instar (Fig. 19-72), with which molting ceases. The first copepodid larva displays the general adult features, but the abdomen is still unsegmented and there may be only three pairs of thoracic limbs. Segmentation and limbs are added with subsequent molts. The entire squence may take as little as one week or almost a year. Six months to a little over a year is the maximum life span of most free-living species. Studies of 20 planktonic species in the Adriatic Sea revealed three to six generations a year.

Many freshwater calanoids and harpacticoids and a few marine calanoids produce both thin-shelled summer eggs and thick-shelled resting, or overwintering, eggs. In freshwater copepods of temporary waters, the copepodid stages (or even adults) secrete an organic, cystlike covering, enter diapause, and become inactive under unfavorable conditions. Such cysts, buried in sediment, are well adapted for enduring desiccation and enable the copepod to survive periods when its pool or pond is dry. They also provide a means of dispersal when carried away on the muddy feet of birds and other animals.

Parasitic Copepods

Copepoda includes a large number of parasites in five orders. Some are ectoparasites on fishes, attaching to the gill filaments, fins, or integument (Fig. 19-73). Others are commensal or endoparasitic with or in polychaete worms, in the intestine

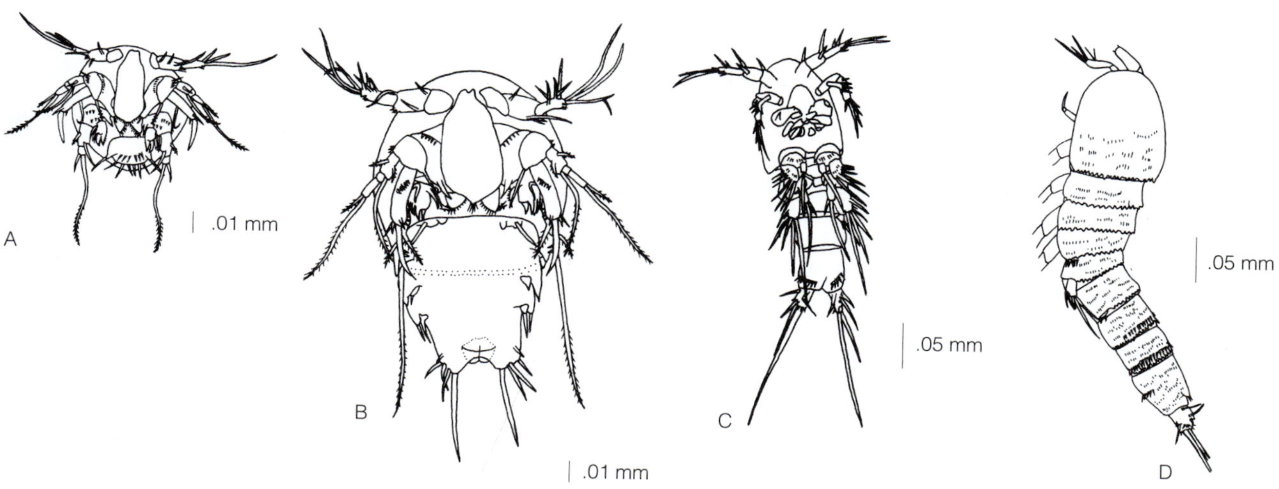

FIGURE 19-72 Copepoda: Representative developmental stages of the harpacticoid copepod *Elaphoidella bidens coronata*. **A–C,** Ventral views. **A,** First nauplius. **B,** Sixth nauplius. **C,** First copepodid. **D,** Fifth copepodid in lateral view. *(Adapted from Carter, M. E., and Bradford, J. M. 1972. Postembryonic development of three species of harpacticoid Copepoda. Smithson. Contrib. Zool. 119:1–26.)*

FIGURE 19-73 Copepoda: Parasitic copepods, *Penella exocoeti* (Siphonostomatoida), on a flying fish. The copepods are in turn host to the barnacle *Conchoderma virgatum.* The copepod body is transversely striped whereas that of the barnacle is indicated by thick longitudinal stripes. Cirri can be seen emerging from the aperture of one of the barnacles. *(Modified after Schmitt.)*

of echinoderms (particularly crinoids), and in tunicates and bivalves. Cnidarians, especially anthozoans, are hosts to many species of copepods. All degrees of modification are exhibited by these parasites. Primitive parasites usually are ectoparasites, such as *Ergasilus,* that resemble free-living species (Fig. 19-74A). Endoparasites are more derived and less like the ancestral copepods. Many highly derived endoparasites, and some ectoparasites as well, bear little or no resemblance to the typical free-living calanoid or cyclopoid (Fig. 19-74). Among the ectoparasitic copepods, some appendages typically are specialized to serve as **holdfast organs** and the mouthparts are adapted for piercing and sucking.

In most parasitic copepods, only the adults are parasitic and the swimming larval stages are usually similar to those of free-living copepods. Contact with the host occurs at various times during the life cycle of the copepod, and modifications appear with each molt. The salmon gill maggot, *Salmincola salmonea* (Fig. 19-74B), which is parasitic on the gills of the Atlantic salmon, *Salmo salar,* has a typical life cycle. When the salmon enters an estuary on its spawning migration into fresh water, the parasite, as a first copepodid larva, attaches to a gill filament by a threadlike process extruded from its head. It then molts and uses its maxillipeds for attachment. At this stage the females are immature but the males are mature. Both are attached by their maxillipeds and can move over the gill. Mating occurs before the female matures and then the male dies. The frontal gland of the female secretes a button-like bulla that adheres to the fish gill filament. She then undergoes her final molt and attaches permanently to the bulla with her second maxillae. The female has achieved the final adult morphology and is no longer mobile. The egg sacs that develop may be up to 11 mm long on an 8 mm female. The salmon spawns and returns to the sea (unlike Pacific salmon). The eggs do not hatch until the salmon returns once again to the estuary, where the resulting larvae infect a new host. A single female produces several clutches with sperm from the original mating.

PHYLOGENETIC HIERARCHY OF COPEPODA

Copepoda
- Platycopioida
- Neocopepoda
  - Calanoida
  - Podoplea
    - N.N.
      - Harpacticoida
      - Mormonilloida
      - Poecilostomatoida
      - Siphonostomatoida
      - Monstrilloida
    - N.N.
      - Cyclopoida
      - Gellyelloida
      - Misophrioida

MYSTACOCARIDA[C]

Mystacocarida is a small taxon of marine interstitial crustaceans believed to be closely related to copepods (Fig. 19-75). The two share their complete lack of compound eyes, the absence of a carapace, and a 10-segmented trunk. Mystaco-

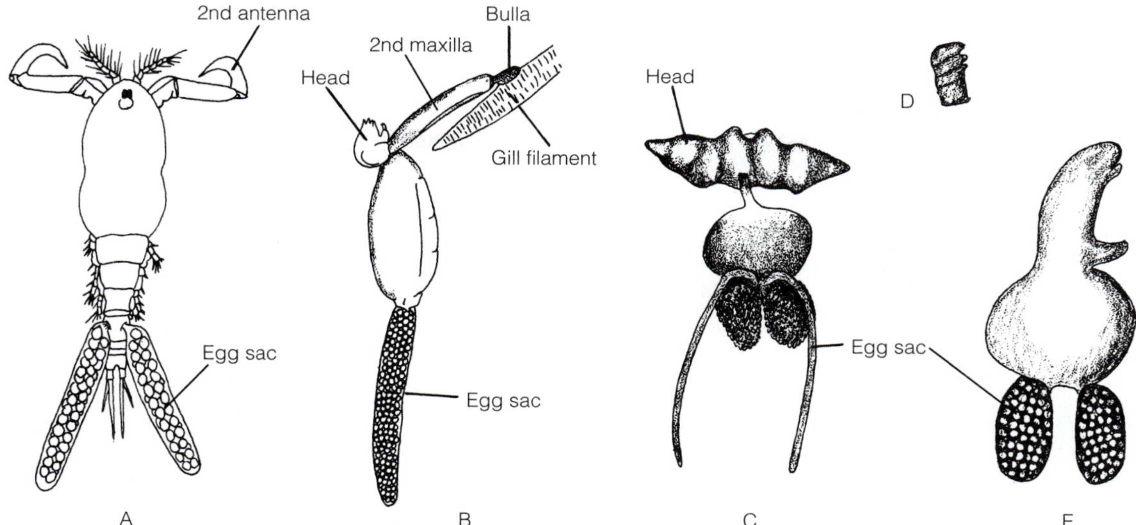

FIGURE 19-74 Copepoda: Parasitic copepods. **A,** Female *Ergasilus versicolor,* a poecilostomatoid parasite that lives on gills of freshwater fish. Only the adult female is parasitic, hooking to fish with the prehensile second antennae. **B,** A *Salmincola salmonea* (Siphonostomatoida) mature female attached to the gill of a European salmon. **C,** *Sphyrion* (Siphonostomatoida): The head is embedded in the skin of a fish and the remainder of the body hangs free. **D** and **E,** Male **(D)** and female **(E)** of *Brachiella obesa* (Siphonostomatoida) live on the gills of red gurnard. *(A, After Wilson from Pennak; B, After Friend; C, From Parker and Haswell; D and E, After Green, J. 1961. A Biology of Crustacea. Quadrangle Books, Chicago. p. 113.)*

carida was first described in 1943 from specimens collected off Massachusetts, and 10 additional species have since been reported from many other coasts, principally around the North and South Atlantic. Most are 0.5 to 1.0 mm in length and adapted for living in the interstices between sand grains in the intertidal zone. The body is long, cylindrical, and vermiform, like that of many other interstitial animals.

The head is divided by a transverse groove into short anterior and long posterior sections. The anterior section bears the first antennae and four independent pigment cups. The posterior region has the second antennae, mandibles, both maxillae, labrum, and first thoracopods. The large first antennae are uniramous and the second antennae and mandibles are biramous. The two pairs of antennae and the mandibles are similar to those of nauplius larvae and are natatory. The distinctive labrum is long and tonguelike, extending along much of the length of the posterior head.

The trunk includes 10 segments, with 7 in the thorax and 3 plus a telson in the abdomen. The first thoracopod functions as a mouthpart, but its segment is not fused with the head. Sometimes it is considered a maxilliped, but sometimes not. There is no cephalothorax or carapace. The next four thoracomeres bear small, vestigial appendages, but the two posterior thoracomeres lack appendages. The abdomen does not have appendages, although the telson has a caudal furca.

The excretory organs and hemal system have not been studied. There are no midgut ceca. The nervous system includes a brain and a ventral chain of large, segmental ganglia. The sexes are separate and the hatching stage is a nauplius.

TANTULOCARIDA[C]

First described in 1983, Tantulocarida includes a growing number (12 species at present) of marine ectoparasites of deep-water copepods, isopods, and ostracods highly specialized for a parasitic life. These are tiny animals, as a copepod parasite would have to be, that do not exceed 1 mm in length, with most being about 0.5 mm. They are gonochoric and sexually dimorphic. The male has a gonopod derived from trunk appendages, and fertilization is thought to be internal. There are sexual and parthenogenetic life cycles connected by a larval stage.

The bodies of sexual individuals (Fig. 19-76A,B) consist of cephalothorax and trunk. The cephalothorax consists of the head and first two thoracomeres whereas the trunk consists of four (in the female) or five (in the male) free thoracomeres plus the telson and caudal furca. Head appendages are lacking except for a pair of rod-shaped first antennae in females. In sexual males (Fig. 19-76B) the first six thoracopods are setose swimming appendages whereas the seventh are fused to form a penis with the gonopore. Sexual females have only two pairs of thoracopods and they do not appear to be natatory (Fig. 19-76A). The female gonopore is on the first thoracomere. Males seem to be adapted for swimming to search for the female, but females seem better suited for a sedentary existence. It is possible that the thoracopods of sexual females hold the male during copulation. Parthenogenetic females (Fig. 19-76E) are sacs of eggs that bear little resemblance to other crustaceans.

Adult sexual males and females are free-swimming but do not feed. All feeding occurs while the larva is attached to the host. The sexual female (Fig. 19-76A) contains numerous

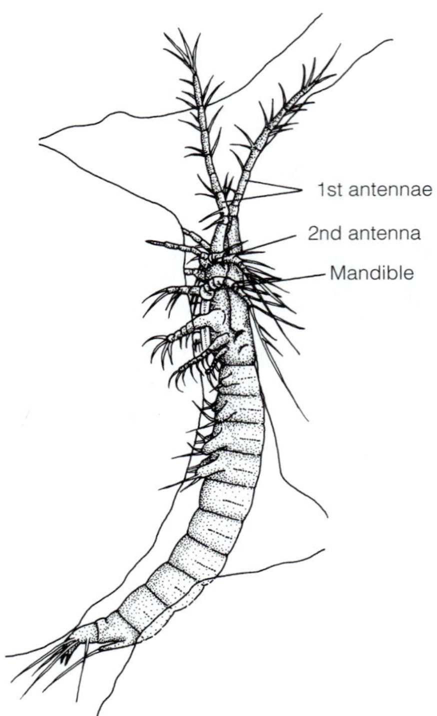

FIGURE 19-75 Mystacocarida: The interstitial mystacocarid, *Derocheilocaris typica*, crawling among sand grains on a beach. The pushing force for movement is produced by the second antennae and mandibles, with the branches of each appendage operating against the substratum both above and below the head of the animal. *(From Lombardi, J., and Ruppert, E. E. 1982. Functional morphology of locomotion in* Derocheilocaris typica. *Zoomorphology 100:1–10.)*

large eggs that, after fertilization, develop into free-swimming **tantulus larvae** (Fig. 19-76C). The unsegmented cephalothorax of the tantulus is covered by a large, dorsal carapace, but it has no segmental appendages. On its ventral surface is the mouth surrounded by an oral attachment disc, a stylet, and a protrusible tube for attaching to the host's tissue. The tantulus attaches to the host (Fig. 19-76D_S) by its oral disc and uses the stylet to make a hole in the host's cuticle. The parasite absorbs nutriment from the host.

In the sexual cycle, the attached tantulus gradually develops into a sexual adult. As it grows, a large **trunk sac** develops posterior to the head and a sexual adult, either male or female, develops within the sac. If the developing adult is a female, the entire segmented trunk of the tantulus is jettisoned and discarded. If it is a male, the trunk appendages remain attached to the large trunk sac. The developing adult is nourished by an umbilicus from the attachment disc. Eventually the sexual adult breaks free of the tantulus exoskeleton, mates, and produces fertilized eggs. Apparently there is no typical arthropod molting process.

In the parthenogenetic cycle, the tantulus attaches to the host (Fig. 19-76C,D_P), jettisons its trunk, and develops into a large, saclike parthenogenetic female (Fig. 19-76E) permanently attached to the host. This female produces eggs that hatch into new tantulus larvae.

ASCOTHORACIDA[C]

Ascothoracida, Acrothoracica, Thoracica, and Rhizocephala constitute the undisputedly monophyletic taxon of Thecostraca. Thoracica, Rhizocephala, and Acrothoracica are highly derived, bear little resemblance to other crustaceans, and are grouped together in Cirripedia, the barnacles. Ascothoracida has crustacean-like males and aberrant females, and is the sister taxon of Cirripedia. The thecostracan larval compound eye has a three-part crystalline cone formed by three cone cells. Compound eyes are absent in adults and the second antennae are vestigial or absent. The first antennae are attachment organs and the carapace tends to be large and bivalved unless it has been secondarily lost. Adults are usually sessile, and commensalism and parasitism have arisen several times. An abdomen is present in only the most primitive taxa. Development includes distinctive nauplius and cypris larvae. Ascothoracidans are the most primitive thecostracans. Rhizocephalans are the most derived and look nothing like crustaceans. A fifth thecostracan taxon, the enigmatic Facetotecta, has been known since 1899 from cosmopolitan, planktonic "Y-larvae," both nauplii and cyprids, for which adults have never been found. The cyprids of all thecostracan higher taxa have five pairs of chemosensory **lattice organs** on the carapace, each consisting of an oval cuticular plate penetrated by a large terminal pore and sometimes by a field of tiny pores. Lattice organs are thought to be highly derived setae used to recognize appropriate substrata prior to settling.

Ascothoracidans are obligate ectoparasites of echinoderms (sea lilies and serpent stars) or endoparasites of alcyonarian corals and in the coelom of sea stars and echinoids. Except for a few free-swimming species *(Synagoga)* that swim from host to host, Ascothoracidans remain permanently attached to their host. There are about 60 species. The largest specimens are up to 4 cm long, but most are much smaller.

Male ascothoracidans with a thorax and abdomen may be shrimplike in appearance (Fig. 19-77), but females may bear little resemblance to other crustaceans. A large, bivalved carapace (mantle) encloses the thorax and sometimes the abdomen as well. The carapace is soft and uncalcified. In some species it forms a sac that encloses the animal, leaving only a small ventral opening. The prehensile first antennae are used to attach to the host. The second antennae are absent in adults. An **oral cone** composed of the labrum, mandibles, and first and second maxillae surrounds the mouth. The labrum is conical and encloses the other mouthparts, which are modified for piercing and sucking. There are no eyes in the adult, but the nauplius larva may have compound or naupliar eyes.

The six or fewer pairs of paddlelike thoracic appendages may be biramous or secondarily uniramous. The thoracopods are not cirriform like those of cirripedes; they resemble the swimming appendages of the cypris larva. The abdomen of four or five segments lacks appendages but has a telson with a caudal furca.

The gut is adapted for sucking. The large midgut has several pairs of branching digestive ceca that extend into the carapace. The mouthparts are inserted into the tissues of the host and fluids are pumped into the gut. A heart is absent. Gas exchange occurs over the body surface, especially the carapace. The

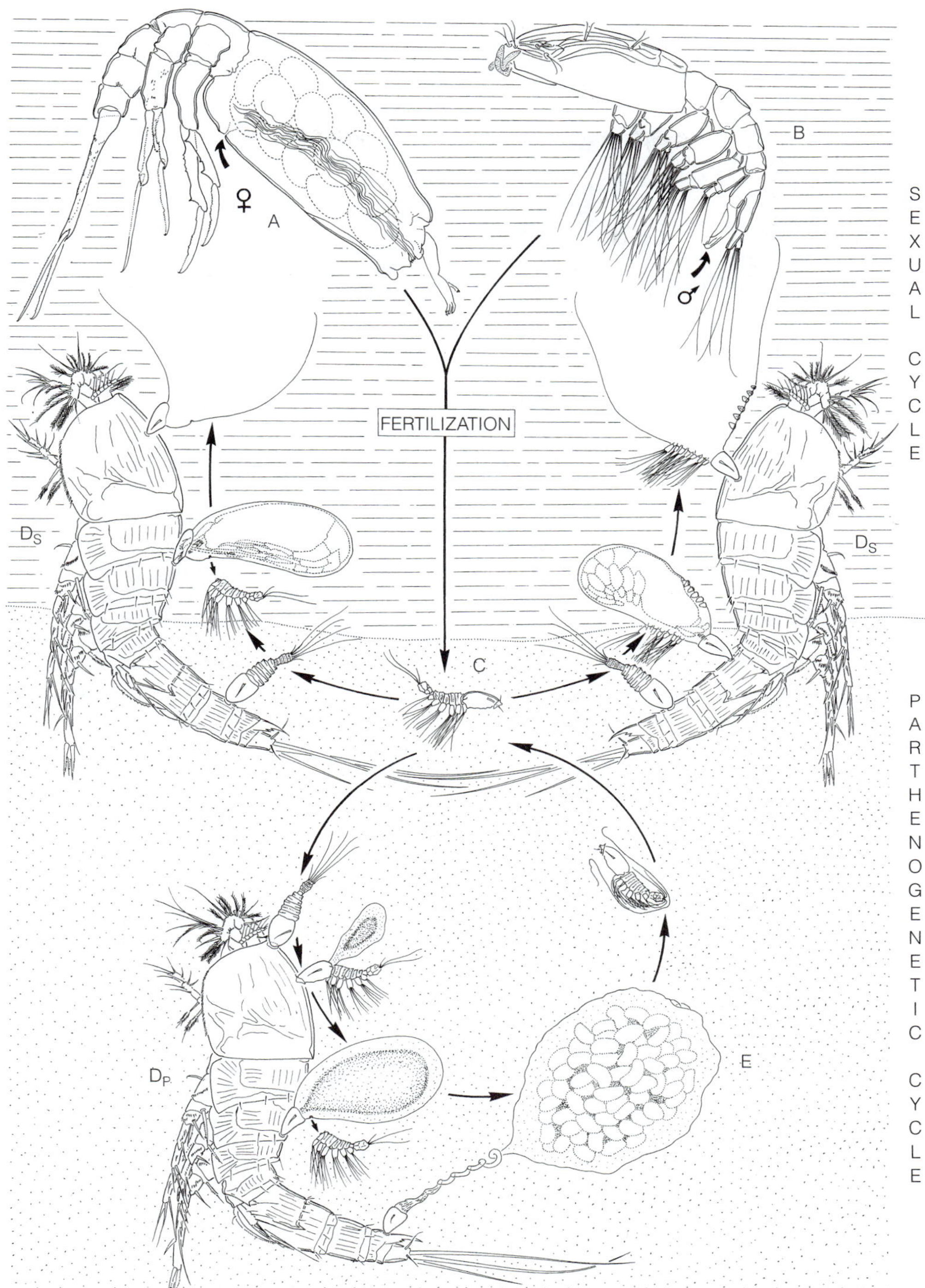

FIGURE 19-76 Tantulocarida: Tantulocarid morphology and life cycle. In the sexual life cycle (top) males **(B)** and females **(A)** copulate and produce fertilized eggs that hatch into tantulus larvae **(C)**. The larva attaches to a host and develops a saclike body. An adult develops within the larval exoskeleton while being nourished by the host. The adult emerges from the larval cuticle ready to mate and begin the cycle anew. In the parthenogenetic cycle (bottom) the tantulus attaches to a host and develops into a female consisting of a sac of eggs **(E)** that bears little resemblance to a crustacean. The eggs develop, without fertilization, into new tantulus larvae. Arrows with female and male symbols indicate position of gonopores. **A,** Sexual female with eggs. **B,** Sexual male. **C,** Tantulus larva. **D_S,** Host of sexual cycle. **D_P,** Host of parthenogenetic cycle. **E,** Parthenogenetic female with eggs. *(From Huys, R., Boxshall, G. A., and Lincoln, R. J. 1993. The tantulocaridan life cycle: The circle closed? J. Crust. Biol. 13:432–442. Reprinted with permission.)*

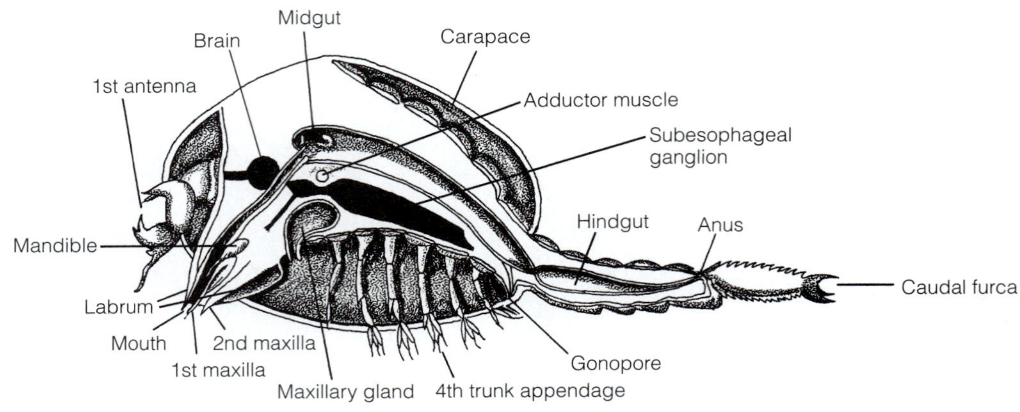

FIGURE 19-77 Ascothoracida: A male *Ascothorax ophioctenis.* This species is parasitic in the genital bursae of brittle stars. *(After Vagin, V. L. 1946.* Ascothorax ophioctenis *and the position of Ascothoracida in the system of the Entomostraca. Acta Zool. 27:155–267.)*

excretory organs are maxillary glands. The CNS consists of the brain and a long subesophageal ganglion from which nerves arise, and there is no ganglionated ventral nerve cord. A transverse carapace adductor muscle pulls the valves of the carapace together. The prehensile first antennae have well-developed muscles.

Most ascothoracidans are gonochoric, although some are hermaphroditic. Sexual dimorphism is pronounced and fertilization is internal. The gonads are usually paired and extend into the carapace. Sperm are flagellated and primitive, and the male gonopore is on a penis on the first abdominal segment. The female gonopore is on the coxa of the first thoracic appendage. Like other thecostracans they have nauplius and cypris (cyprid) larvae. The cypris (Fig. 19-79B) is a bivalved larva that resembles an ostracod (Fig. 19-85A). Ascothoracidan nauplii resemble cirripede nauplii but lack their frontal horns (Fig. 19-79A). Embryos and larvae are brooded under the mother's carapace to the cypris stage.

Ascothoracidans resemble cirripedes in their prehensile first antennae; large, fleshy carapace; female gonopore on thoracic segment 1; absence of second antennae; and possession of a bivalve cypris larva. They differ in having a well-developed abdomen, lacking frontal horns in the nauplius, and having a parasitic, nonfiltering feeding mode.

CIRRIPEDIA[C]

Cirripedia includes the familiar benthic marine animals known as barnacles as well as some unfamiliar and bizarre parasites. These strange and derived creatures have many characteristics that set them apart from other crustaceans. They inhabit a rigid calcareous shell (carapace or mantle) that grows without molting (Fig. 19-78), many are hermaphroditic, and they are sessile. Barnacles have features in common with ostracods and, in fact, resemble an ostracod turned upside down and stuck firmly to the substratum by its head and back. Cirripedes include the only nonparasitic sessile crustaceans, and their anatomy and biology have undergone extensive modifications as a consequence of this unique lifestyle It is not at all obvious on casual examination that they are crustaceans. Although the adult appearance is aberrant, it begins life as an unmistakably crustacean nauplius larva (Fig. 19-79A). The nauplius molts to become a **cypris** larva (Fig. 19-79B) that looks something like the ostracod *Cypris,* for which it is named. The nonfeeding cypris is the settling stage. Like an ostracod, it has a large, bivalved carapace that encloses the entire body. Because of their soft bodies, sedentary habit, and calcareous shells, barnacles were thought to be molluscs until 1830 when Vaughan Thompson made the association between the mollusc-like adult and the unmistakably crustacean nauplius and cypris larvae. Charles Darwin's careful 1851 study of barnacles remains the foundation on which modern barnacle studies are based.

As currently conceived, Cirripedia includes the two nonparasitic thecostracan taxa Thoracica and Acrothoracica and the parasitic Rhizocephala. Acrothoracica is a small taxon of infaunal barnacles that bore into calcareous substrata. Thoracica consists of epibenthic barnacles that live permanently attached to the surface of rocks, shells, coral, timber, whales, crabs, turtles, ships, bottles, and other objects. Most barnacles are thoracicans and this taxon includes the stalked goose barnacles (Fig. 19-80), the stalkless acorn barnacles (Fig. 19-78), and the stalkless wart barnacles. Rhizocephala consists of highly derived sac-shaped parasites (Fig. 19-83).

Thoracica[O]

FORM

The thoracican cypris larva settles out of the plankton to the bottom and attaches to the substratum by means of cement glands located in the base of the first antennae. The larval carapace, which encloses the entire body, as it does in ostracods, persists and becomes an enveloping mantle (carapace) that encloses the adult barnacle. The mantle secretes thick calcareous plates that form a case to enclose and protect the barnacle. Since the cypris attaches by its head, the aperture (ventral opening) of the carapace is directed more or less upward and the trunk of the animal is upside down, with respect to the substratum, with its venter up (Fig. 19-78A).

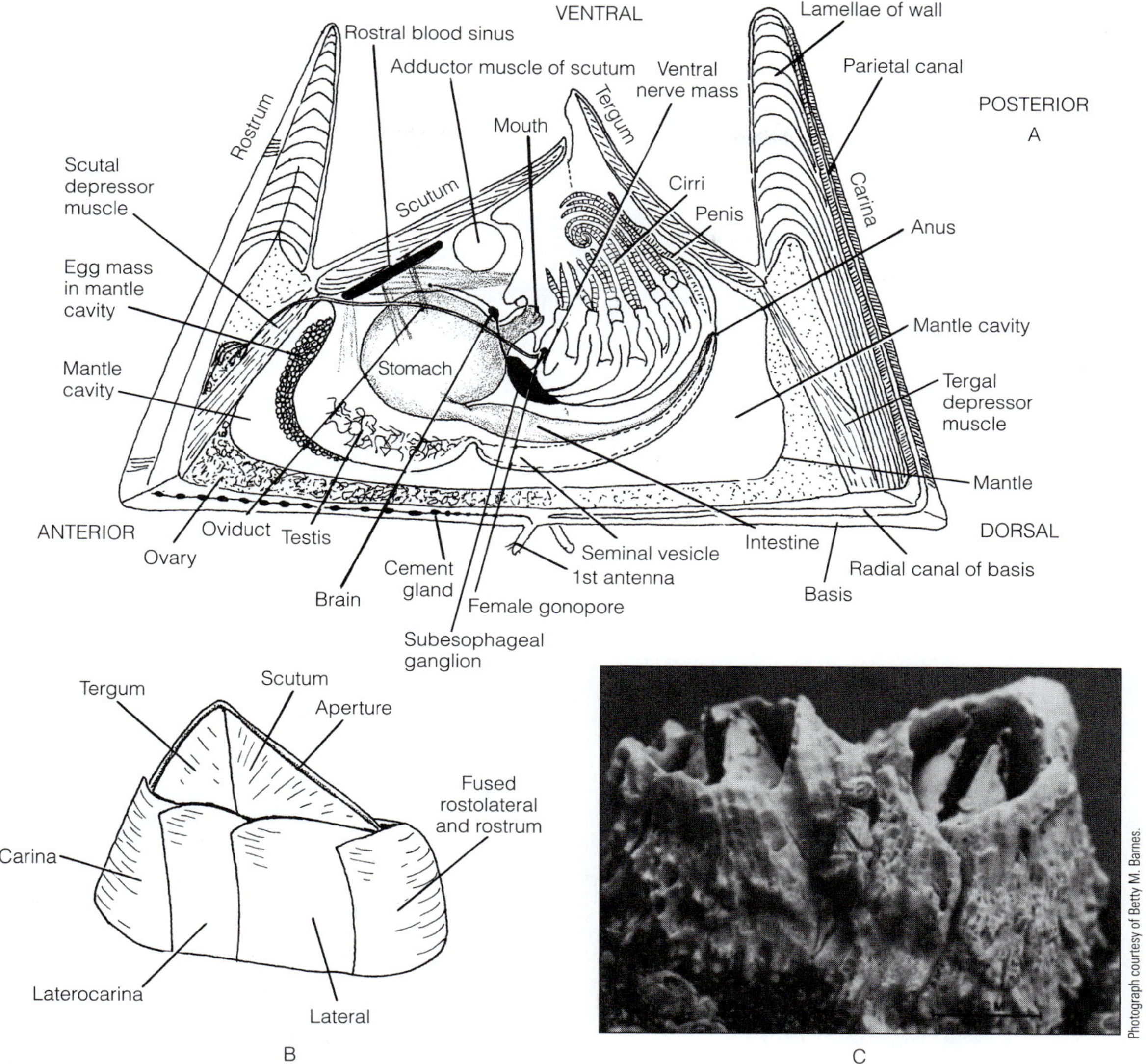

FIGURE 19-78 Cirripedia: Acorn barnacles, Balanomorpha. **A,** *Balanus,* a sessile barnacle in sagittal section viewed from the right. **B,** *Balanus* showing number and position of shell plates in a diagrammatic view from the left side. **C,** *Tetraclita,* a large sessile barnacle, showing the circular calcareous wall and the projecting, movable opercular plates. *(A, After Gruvel from Calman; B, After Broch from Kaestner, A. 1970. Invertebrate Zoology. Vol. 3. Wiley-Interscience, New York.)*

The typical thoracican barnacle has a body consisting of a large head and a thorax, but no abdomen (Fig. 19-78A). The head bears the usual appendages except for the absent second antennae. The preoral region of the head, which is small in most crustaceans, is large and important in thoracican barnacles because it is the region by which the animal attaches to the substratum. It contains the cement glands and bears the first antennae. Since the second antennae are absent, the only preoral appendages are the first antennae, which are vestigial in the adult. The mouthparts include the mandibles and two pairs of maxillae. The thorax bears six pairs of biramous thoracic appendages known as **cirri**. These whiplike appendages have long, setose, multiarticulate rami used for filter feeding (Fig. 19-78A). Posteriorly, the thorax bears a long, highly extensible penis and the anus.

The large fleshy, bilobed carapace extends laterally from either side of the head to arch ventrally around the body and completely enclose it. It is referred to as a mantle because of its resemblance to the similar structure of molluscs. As in the molluscs, the barnacle mantle encloses a water space known as the mantle cavity. The mouth, female gonopore, and anus open into the mantle cavity and the cirri lie in it. The mantle opens to the surrounding sea at the aperture (Fig. 19-78B).

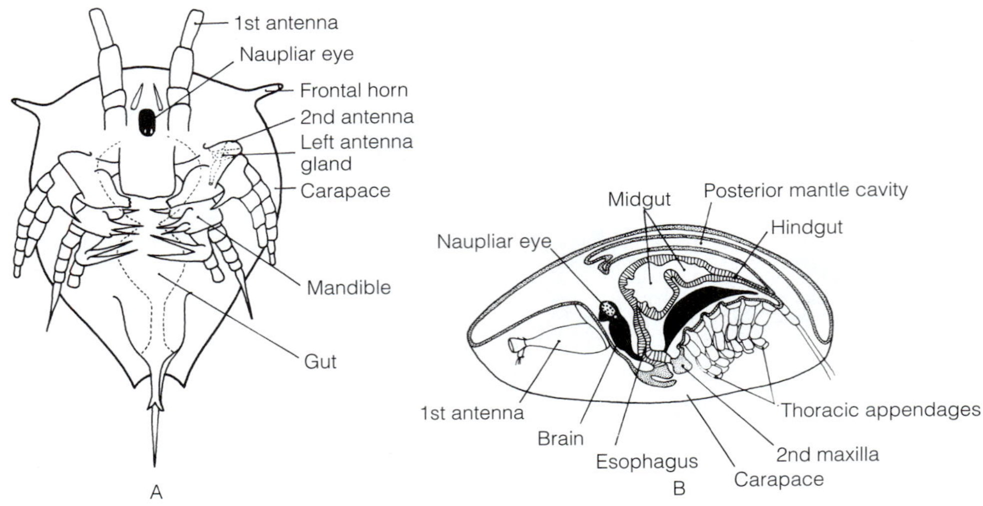

FIGURE 19-79 Cirripedia: Barnacle larvae. **A,** Nauplius of *Balanus,* with setae omitted. **B,** Free-swimming cypris larva of *Balanus.* *(Both from Walley, L. J. 1969. Studies on the larval structure and metamorphosis of the cypris larva of* Balanus balanoides. *Phil. Trans. Roy. Soc. Lond. B 256:237–280.)*

The mantle epidermis secretes calcareous plates in its exoskeleton to enclose the barnacle in a stony box. The plates may vary, but many barnacles have a rostrum at the anterior end and a carina at the posterior end (Fig. 19-78B) of the wall. The aperture is covered by a door, or operculum, made of two pairs of movable plates, the posterior **terga** and anterior **scuta.** A large **scutal adductor muscle** that runs transversely between the two scuta closes the aperture. Several scutal and tergal depressor and abductor muscles move the opercular plates laterally to open the aperture

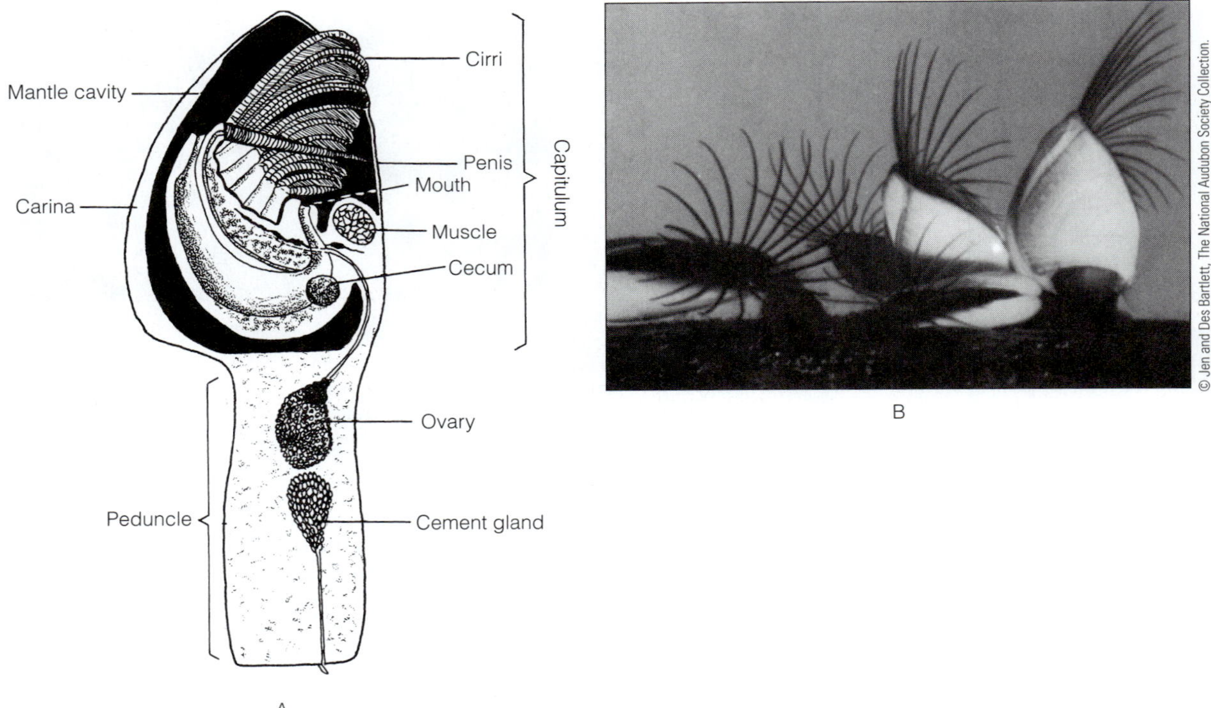

FIGURE 19-80 Cirripedia: Goose barnacles, Lepadomorpha. **A,** *Lepas,* a stalked barnacle, in sagittal section viewed from the left. **B,** A species of *Lepas* with extended cirri. The three major plates of a typical goose barnacle are visible in the individual on the right. The largest plate is the scutum. The tergum is a slender plate beside the apex of the scutum and the carina is another slender plate beside the base of the scutum. *(A, After Broch from Kaestner, A. 1970. Invertebrate Zoology. Vol. 3. Wiley-Interscience, New York.)*

(Fig. 19-78A). When the aperture is open, the cirri can be extended for feeding (Fig. 19-80B).

When not encrusted with other sessile organisms, barnacles are usually white, pink, or purple. Goose barnacles, including the stalk, range from a few millimeters to 75 cm in length. The majority of acorn barnacles are a few centimeters in diameter, but some are considerably larger. *Balanus psittacus* from the west coast of South America reaches a height of 23 cm and a diameter of 8 cm and is a popular local seafood. The calcareous plates of barnacles fossilize readily and an extensive fossil record dates back to the Silurian.

The orientation of barnacles is unusual and can be confusing. The barnacle is attached to the substratum by its head, so anterior is down, toward the substratum, whereas the posterior end extends up into the water. Within the mantle the body is bent dorsally so that the mantle aperture and cirri, which are really ventral, end up at the apex of the body, which is the end opposite the anterior attachment (Fig. 19-78A, 19-80A). External segmentation is indistinct. Except for the calcareous plates, the exoskeleton is soft and flexible.

DIVERSITY OF THORACICA

The ancestral barnacle was probably a cyprislike, bivalved crustacean that attached to the substratum with its first antennae. From this ancestor evolved three distinct lines of thoracicans and one of acrothoracican barnacles. Stalked thoracicans belong to Lepadomorpha and include two main lines, Lepadidae and Scalpellidae. Stalkless thoracicans are Balanomorpha and Verrucomorpha.

Most thoracicans are free-living, but there is a strong tendency toward symbiosis, either commensalism or parasitism, in thecostracans, and commensalism has evolved in all three evolutionary lines of thoracican barnacles. Commensalism tends to be accompanied by loss or reduction of the calcareous plates as the barnacle takes advantage of protection provided by the host.

Lepadomorpha^sO

Of the three types of thoracican barnacles, the stalked barnacles (goose barnacles) are most primitive. The earliest known fossil barnacle is the stalked *Cyprilepis,* which was commensal on the appendages of sea scorpions (Eurypterida). It had a **capitulum** (the major part of the body, exclusive of the stalk; Fig. 19-80A) with a large, bivalved carapace, but there were not yet any calcareous plates in cuticle. To guard the aperture and protect the animal within, its descendants developed chitinous terga and scuta supported by a posterior dorsal plate, the carina. It is from this ancestor that the two principal lines of modern stalked barnacles, lepadids and scalpellids, are believed to have evolved. The stalkless barnacles (acorn and wart) evolved later from a scalpellid ancestor.

The Recent stalked barnacles have a muscular, flexible peduncle (stalk) that attaches to the substratum at one end and bears the capitulum at the other (Fig. 19-80A). The capitulum is enclosed by a series of calcareous plates, including the posterior carina. The movable operculum, which protects the aperture, is composed of the paired scuta and terga. In some, the plates are reduced or lost and in others additional plates, including an anterior rostrum, may be present.

The peduncle—the preoral end of the head—contains the vestiges of the larval first antennae and the cement glands. The capitulum contains all of the body except for the preoral region and thus includes the mouthparts, thoracic appendages, and carapace (mantle).

Lepadidae^F: Peduncle is naked (without calcareous plates); capitulum has only the five original, primary plates (one carina, two terga, and two scuta [Fig. 19-80]). Attach to floating objects, wood, coconuts, bottles, tar balls, and ships or are epizoic commensals. Includes the common, widely distributed *Lepas* species. Commensals include *Conchoderma* (on whales and turtles) and *Alepas* (on jellyfish). A related family of small, stalked barnacles (Poecilasmatidae) includes many common commensals (Fig. 19-81A) such as *Octolasmis muelleri,* which lives on the gills of lobsters and crabs and takes advantage of the safety and flow of water in the branchial chamber of the host.

Scalpellidae^F: Benthic, mostly deep water. Some, such as the common, eastern Pacific *Pollicipes polymerus,* live intertidally in dense aggregations on rocky coasts. Peduncle is covered with calcareous plates or scales, which generally increase in size toward the capitulum. One taxon is adapted for boring into coralline rock (Fig. 19-81C).

Balanomorpha^sO

The stalkless barnacles (balanomorphs and verrucomorphs) are thought to have arisen in the Jurassic from a scalpellid ancestor through the shortening and eventual disappearance of the peduncle.

The acorn barnacles (Balanomorpha; Fig. 19-78) are attached directly to the substratum without a stalk. The attached undersurface of the barnacle, the **basis,** may be either membranous or calcareous, and it is cemented firmly to the substratum. The basis is the preoral region of the barnacle and includes the first antennae and cement glands. It is homologous to the peduncle of the goose barnacles. A vertical, circular wall of stationary calcareous **mural plates** completely rings the animal (Fig. 19-78B,C). The plates composing the wall overlap one another and may be held together by living tissue only or by interlocking teeth, or they may be fused to some extent. They are anchored to the basis by muscle fibers. The aperture is covered by the operculum composed of the paired scuta and terga.

The large attachment surface (the basis) and the low, thick, circular wall adapt these barnacles for life on current-swept and wave-pounded intertidal rocks. The vertical wall of mural plates (Fig. 19-78B) so characteristic of acorn barnacles is believed to have evolved from some of the larger plates covering the base of the capitulum of scalpellid ancestors.

Although there are some deepwater balanomorphs, the taxon is most common in the intertidal or high subtidal, with species typically restricted to particular tidal zones. Intertidal barnacles are among the very few marine animals that can tolerate the stressful conditions (desiccation, temperature extremes, intense wave turbulence, and reduced feeding time) of rocky intertidal habitats. In the intertidal zone they may have little competition from other marine animals and

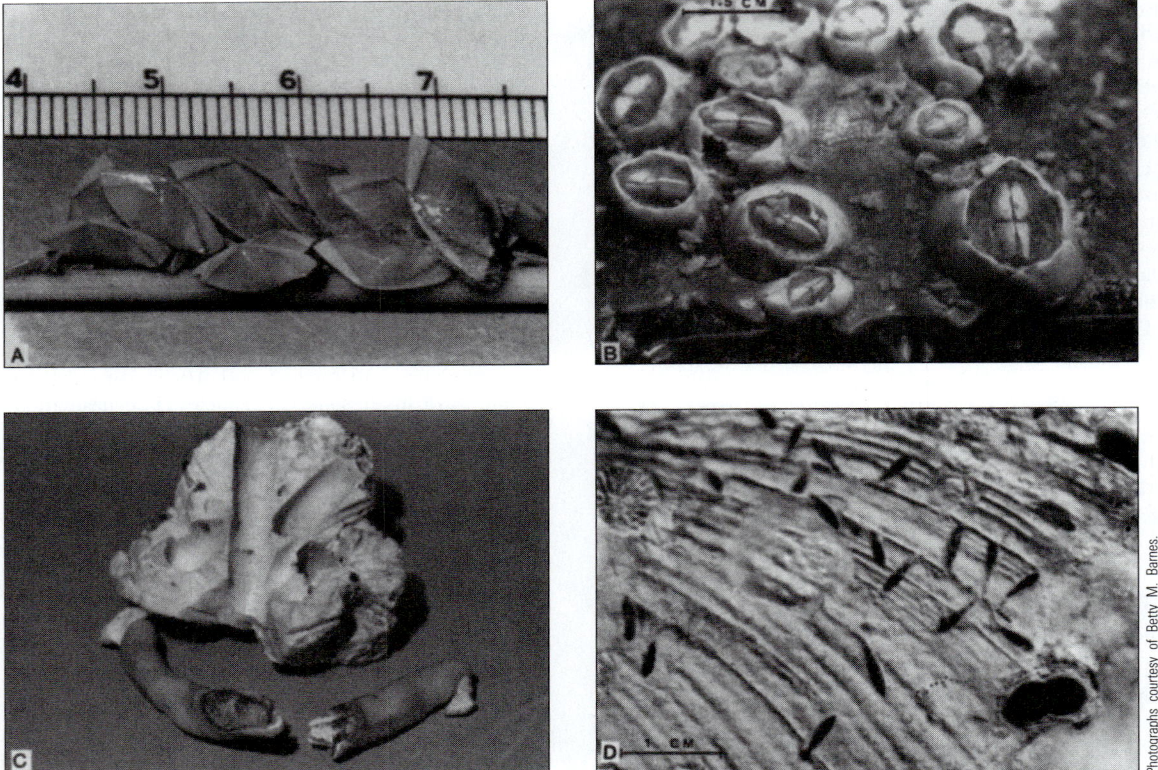

FIGURE 19-81 Cirripedia. **A,** A species of *Poecilasma* on a sea urchin spine. The members of this family of stalked barnacles (Poecilasmatidae) are related to the lepadids but attach to bottom objects, including sea urchins and crustaceans. *Octolasmis,* which is found on crab gills, is also a member of this family. **B,** *Chelonibia* on the carapace of a blue crab. **C,** *Lithotrya,* a scalpellid barnacle that bores into coralline rock. **D,** Slitlike openings of the burrows of the boring acrothoracican barnacle *Kochlorine,* in an old clam shell. The dumbbell-shaped openings are the burrows of a boring clam.

Photographs courtesy of Betty M. Barnes.

often occur in enormous densities. Interspecific competition with other barnacles, however, results in predictable vertical zonation patterns, with each species being favored at a characteristic tide level. A few barnacles are adapted for life in the spray zone at the high-tide mark on wave-splashed rocks. Barnacles are less common on tropical rocky shores, perhaps because of the prolonged high temperatures that must be tolerated at the upper intertidal levels.

Many acorn barnacles have become adapted for life on surfaces other than rock. A number of species have colonized intertidal grasses and mangroves. Many are commensal with a wide range of hosts, including sponges, hydrozoans, alcyonarians, scleractinian corals, crabs (Fig. 19-81B), horseshoe crabs, sea snakes, sea turtles, manatees, porpoises, and whales.

From an economic standpoint, barnacles are among the most detrimental fouling organisms on ship hulls, buoys, and pilings. The speed and fuel efficiency of a barnacle-fouled ship may be reduced by 30%, and much effort and money have been expended to develop special paints and other antifouling measures. Many species have been transported worldwide by global shipping.

Verrucomorpha^{sO}

A third taxon of thoracican barnacles (wart barnacles) contains only about 30 species found chiefly in the deep sea. These stalkless barnacles resemble acorn barnacles but are asymmetrical. The operculum consists of only one tergum and one scutum, with the other tergum and scutum being incorporated into the wall.

INTERNAL FORM AND FUNCTION

Cirripedes are suspension feeders that sweep their cirriform thoracic appendages through the water to capture food. To paraphrase Louis Agassiz, "a barnacle is a shrimp that stands on its head and kicks food into its mouth." When feeding, the paired scuta and terga open and the cirri unroll and extend through the aperture (Fig. 19-80B). When extended, the long posterior cirri on each side form one side of a basket. The two sides of the basket sweep downward and toward each other, with each acting as a cast net. The action is similar to the opening and closing of your two fists simultaneously when the bases of your palms are placed together. On the closing stroke, suspended food particles are trapped by the setae and transferred to the mouthparts by the anteriormost one to three pairs of short cirri. Some barnacles that inhabit currents, such as *Pollicipes polymerus,* extend the cirri and hold them stationary to filter the current as it passes through. The well-developed mandibles and maxillae are used to macerate food. The size of the plankton consumed varies. Some predatory species of *Lepas, Pollicipes,* and *Tetraclita* capture copepods, isopods, amphipods, and other animals. Barnacles such as *Balanus improvisus* and *B. balanoides* feed on both large and small plankton.

Food is masticated by the mandibles and maxillae and passes into a foregut consisting of a pharynx and esophagus (Fig. 19-78A). A cuticular plate in the pharyngeal wall provides a surface against which the mandible grinds food. The midgut (stomach) connects with nine digestive ceca, two of which are glandular and secrete digestive enzymes for use in the midgut whereas the other seven are absorptive. Fecal pellets are formed by the hindgut.

A heart and pericardial sinus are absent, but blood nevertheless circulates along a defined and relatively uniform pathway. Blood collects in a large **rostral sinus** in the head (Fig. 19-78A) and then flows to the anterior/dorsal attachment region (including the peduncle of goose barnacles), then to the mantle, the body, and back to the rostral sinus. The blood probably is propelled by movements of the body and cirri, but there is some evidence that the rostral sinus is contractile and may serve as a heart.

The mantle and cirri are probably the principal sites of gas exchange, but in the acorn barnacles there are also folds of the inner mantle surface that function as gills. When exposed by the ebbing tide, intertidal acorn barnacles fill their mantle cavity with air. A small pneumostome is held open between the opercular plates to allow diffusion of gases into and out of the mantle cavity. If desiccation becomes severe, the pneumostome is closed and the barnacle shifts to anaerobic respiration.

The excretory organs are maxillary glands whose primary role is in ion balance and fluid volume regulation. They may also excrete environmental toxins such as heavy metal ions. Nephrocytes are also present. The brain lies between the esophagus and the adductor muscle and is linked by circumesophageal connectives to the subesophageal ganglion, which in acorn barnacles is coalesced with the thoracic ganglia. In the goose barnacles, there are four or five pairs of free thoracic ganglia on a ventral nerve cord. The typical maxillopodan naupliar eye divides during metamorphosis into its three component ocelli so that the adult has two lateral and one median ocelli. Compound eyes present in late naupliar instars and the cypris are lost in the adult. Small sensory setae associated with the aperture monitor external conditions when the aperture is closed.

REPRODUCTION AND DEVELOPMENT

Most thoracican barnacles are hermaphroditic and are the only higher taxon of crustaceans to be so. Fertilization is external in the mantle cavity. Barnacles usually engage in cross-fertilization. Cross-fertilization is problematic for sessile animals because their immobility makes searching for mates difficult. Many sessile animals solve the problem by broadcasting gametes into the sea, but this is inefficient and wasteful of gametes. The thoracican barnacles live in crowded populations with an abundance of close neighbors, any of which can serve as either male or female. Each barnacle has a highly extensible penis that can wander in search of a mate, even if its owner cannot (Fig. 19-82A).

The ovaries are preoral (Fig. 19-78A, 19-80A) and the paired oviducts open into the anterior mantle cavity. The oviducts include a glandular region known as the **oviductal gland** that secretes a thin, elastic ovisac at the time of egg deposition. As the ovisac receives eggs, it swells and stretches, eventually filling the mantle cavity. The testes are located in the head region and connect via two long sperm ducts with the long, posterior penis (Fig. 19-78A). The extensible penis can protrude from the aperture for distances many times its resting length (Fig. 19-82A). The penis visits the mantle cavity of a neighbor and deposits a mass of sperm near the female gonopore. The spermatozoa penetrate the ovisac and fertilize the eggs within. Functional males can recognize functional females, and any given barnacle may be inseminated by more than one partner. Weak ascorbic acid appears to be the stimulus that elicits penis extension. Sperm are flagellated and cleavage is total and reported to be spiral.

In several lines of otherwise hermaphroditic thoracicans, **complemental males** occasionally appear in the population. It is not known if the appearance of males in hermaphroditic species is environmentally induced or if it has a genetic basis. These small males are attached to a full-size hermaphroditic individual of the same species, and show varying degrees of degeneration. In some species the complemental males are morphologically similar to their hermaphroditic counterparts, but are miniature. At the other extreme, some have lost the mouthparts and cirri, are incapable of feeding, and are parasitic in the mantle cavity of the hermaphroditic conspecific. There is some evidence that the fate of a cypris depends on where it happens to settle. Those that settle on a conspecific become complemental males, whereas those settling on any other substrate develop into normal-size hermaphrodites. Complemental males differ from dwarf males in that the former occur in hermaphroditic populations whereas dwarf males are found in gonochoric species, where females, but not hermaphrodites, are also present.

In all barnacles, the eggs are brooded in the ovisac in the mantle cavity. The ovisac gradually deteriorates during the incubation period, so the larvae are not encumbered by it after they hatch. The nauplius is the hatching stage in most species and can be easily recognized as a cirripede by the triangular, shield-shaped carapace (Fig. 19-79A) with a pair of characteristic, secretory **frontal horns.** The function of the horn secretions is unknown.

The nauplius leaves the mantle cavity and enters the plankton to feed. Six naupliar instars are passed in the plankton and succeeded by a nonfeeding cypris larva (Fig. 19-79B). The entire body of the cypris is enclosed within a bivalve carapace and has a naupliar eye, a pair of sessile compound eyes, and six pairs of thoracic appendages. The cypris is the settling stage. It uses the chemoreceptive first antennae to locate a suitable substratum and attaches, at first temporarily with the secretions of discs located on the first antennae, and then permanently with adhesive secreted by the antennal cement glands.

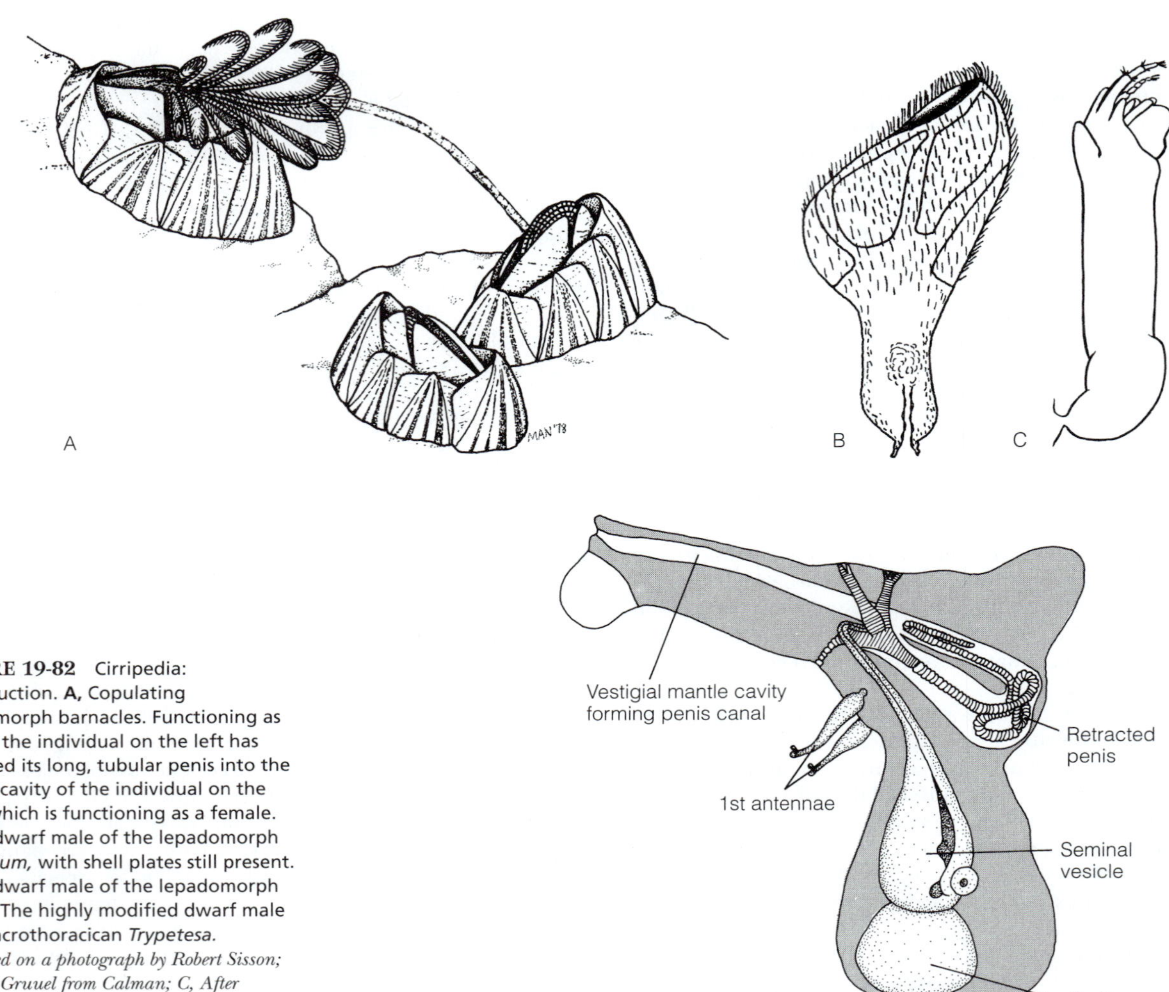

FIGURE 19-82 Cirripedia: Reproduction. **A,** Copulating balanomorph barnacles. Functioning as a male, the individual on the left has extended its long, tubular penis into the mantle cavity of the individual on the right, which is functioning as a female. **B,** The dwarf male of the lepadomorph *Scalpellum,* with shell plates still present. **C,** The dwarf male of the lepadomorph *Ibla.* **D,** The highly modified dwarf male of the acrothoracican *Trypetesa.* *(A, Based on a photograph by Robert Sisson; B, After Gruuel from Calman; C, After Newman; D, After Bernak from Calman.)*

A number of factors appear to increase the likelihood of dense settling, on which future reproduction depends. Simultaneous release of the nauplii by aggregations of individuals has been observed in some species. A protein in the exoskeleton of older attached individuals has been shown to attract settling larvae. Light intensity, substratum texture, position, depth, and bacterial biofilms may also be important. Attachment often does not occur at the first surface contacted by the cypris, and it can delay settling until a suitable substratum is found.

Following attachment, the cypris metamorphoses into the adult shape. The cirri elongate, the body undergoes flexion, larval tissues are destroyed and absorbed, and the first five or six calcareous plates (two scuta, two terga, and a carina, plus a rostrum in some) appear on the new exoskeleton lying beneath the old valves of the cypris.

Newly settled barnacles, referred to as **spat,** molt frequently. Mortality in barnacle spat is very high and often only about 5% survive the first month. Much of the loss is due to intraspecific competition that becomes more intense as the spat grow. Some is also due to interspecific competition with other barnacle species. *Balanus balanoides* and *Chthamalus stellatus,* for example, both inhabit rocky coasts and must partition the habitat. *Chthamalus,* which is more tolerant of desiccation, is favored in the intertidal zone and *Balanus,* which physically interferes with *Chthamalus,* has the advantage in the high subtidal.

Most barnacles that survive the heavy mortality rate immediately following settling probably live from 1 to 10 years. Longevity and growth rates are greatly influenced by environmental conditions, however. Smothering by other individuals and other sessile organisms is a common cause of death.

GROWTH

The calcareous mural and opercular plates of the thoracican shell are secreted by the underlying mantle epidermis. The basis, also secreted by the epidermis, is cemented to the substratum and lies under the rest of the barnacle. The mural plates (side plates) are connected to the basis by fine muscles that hold them and the basis in close, but movable, contact.

Growth and ecdysis are linked in barnacles, as they are in other arthropods, and the sizes of the soft parts of the body increase only with ecdysis. The cuticle, or exoskeleton, that covers the soft parts is molted periodically. Growth of the shell, in contrast, is more or less continuous and independent of body growth and ecdysis. The calcareous plates, which are secreted by the mantle epidermis, are *not* shed at ecdysis, a situation unique among arthropods.

The thickness of the mural plates is increased by the simple secretion of new shell material by the epidermis adjacent to the inner surface of the plate. Increasing the height of the plate is accomplished by simultaneously adding new shell to the periphery of the basis, so it increases in diameter, and to the lower margins of the mural plates, so they increase in length. One consequence of this growth is that the barnacle can insinuate the lower, growing edge of its shell under other sessile objects attached to the same surface. In the high subtidal, *Balanus balanoides* undergrows *Chthamalus stellatus* in this manner and literally pops it off the rock, thereby excluding *Chthamalus* from the zone.

Rhizocephala^O

Rhizocephalans are internal parasites of other crustaceans. Most of the 260 known species are parasitic on decapods (Fig. 19-83), but a few are found on other crustaceans. Rhizocephalans are so highly specialized that almost all traces of arthropod structure have disappeared in the adult, although life begins with typical cirripede nauplius and cypris larvae. The soft, adult body is an amorphous sac and the mantle is not calcified. There is no sign of segmentation or appendages and almost all internal organs are lost. The position of Rhizocephala with respect to Cirripedia has been disputed, but morphological (similar nauplius and cypris larvae) and 18S rDNA nucleotide sequences support a sister relationship with Thoracica. Unlike thoracicans, however, rhizocephalans have no gut, no segmentation, no appendages and no calcareous plates, and are gonochoric.

Rhizocephalans are parasites that transform the host morphologically, behaviorally, and physiologically so that it (the host) becomes wholly dedicated to the service of the parasite. The molting cycle of the host is suppressed, the host is sterilized and no longer reproduces, and in some cases male crabs are feminized and develop, to the benefit of the parasite, the broad abdomen characteristic of brachyuran females. The brood pouch of the parasite occupies the space between the abdomen and the cephalothorax, which is where the crab's own egg mass would normally be. A feminized male crab is better able to carry and protect the parasite than would be a normal male with a narrow abdomen.

Rhizocephala consists of two taxa, Kentrogonida and Akentrogonida, the former having a kentrogon larva. Kentrogonidans are parasites exclusively of decapods, but akentrogonidans are found on stomatopods, cumaceans, isopods, and cirripedes, as well as decapods. In Kentrogonida the life cycle begins with male and female nauplius larvae that molt to become cyprids. The female cypris settles on a host and metamorphoses into a **kentrogon larva.** The kentrogon metamorphoses into the elongate **vermigon stage,** which is injected through the host's cuticle and develops into an extensive, branched, rootlike **interna,** with a mantle, mantle cavity, and ovary, that absorbs nutriment from the host as it grows. Upon maturing, the interna mantle erupts through the cuticle of the host's abdomen and develops into an **externa** outside the abdomen. The externa is the mantle and brood sac of a virgin female parasite. After molting once, the externa is attractive to male cypris larvae. A male cypris molts to become a **dwarf male** (trichogon larva), which develops into a testis parasitic on the female externa. The female ovary and the testis produce gametes, fertilization occurs, and the externa fills with fertilized eggs. The externa grows to become a large, amorphous, egg-filled sac under the crab's abdomen. The host crab, whether female or feminized male, tends this sac as if it were its own egg mass. The eggs are brooded in the externa and released as male and female nauplii to begin the life cycle anew (Fig. 19-93). In Akentrogonida there is no kentrogon and the first antennae of the female cypris penetrate the host's integument and inject infective cells into the hemocoel.

Acrothoracica^O

The 40 known species of acrothoracicans are the smallest cirripedes, usually only a few millimeters in length. They bore into calcareous substrates such as coral, mollusc shells, other barnacles, bryozoans, and limestone using chitinous teeth on the mantle as well as chemical dissolution. The burrow openings are tiny, distinctive slits in the surface of the rock (Fig. 19-81D). The chitinous teeth are renewed with each molt. Unlike thoracicans, but like rhizocephalans, acrothoracicans are gonochoric and sexually dimorphic. Males are aberrant dwarfs that will be described separately. The following description refers to females (Fig. 19-84).

As burrowers, these barnacles are protected by the surrounding rock and have no calcareous plates of their own. The body is enclosed in a large, saclike, uncalcified mantle (carapace) that attaches dorsally to the walls of the burrow. The mantle, which is strengthened by chitinous lateral bars, opens anteriorly through a small aperture that can be closed by an operculum formed of the mantle's thickened lateral lips.

The head bears the mandibles and two pairs of maxillae. The thorax has up to six pairs of cirriform appendages similar to those of other cirripedes. Most of the cirri are long, setose, usually biramous terminal cirri located at the posterior end of the thorax (Fig. 19-84). A smaller pair of mouth cirri (maxillipeds) is situated at the anterior end of the thorax, beside the mouth. A sizable gap separates the mouth cirri and the terminal cirri. There is no abdomen, but in some taxa there is a pair of caudal appendages thought to be its vestiges.

When feeding, the terminal cirri are extended through the slitlike opening of the burrow or used inside the burrow to sweep through the water and capture small food particles. The cirri are held open like a fan and sometimes rotated several

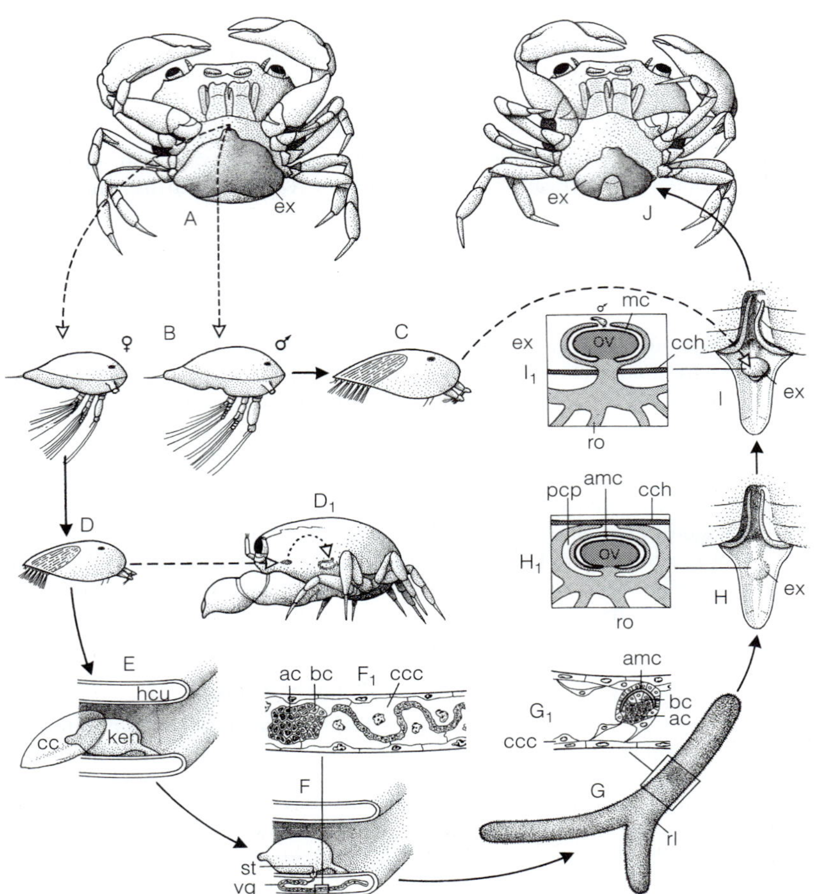

FIGURE 19-83 Cirripedia: Rhizocephala. The life cycle of *Loxothylacus panopaei*, a parasite of the mud crab, *Rhithropanopeus harrisii.* Larval migrations are indicated by dashed arrows and developmental changes by solid arrows. **A,** An adult externa under the abdomen of the host. **B,** Female and male nauplii are released from the externa and, in the plankton, molt to become cyprids. **C,** The male cypris settles at the external aperture of a virgin female externa (**I** and **I_1**) and molts to become a dwarf male. **D** and **D_1,** The female cypris settles on a gill lamella of a new host crab. **E,** A kentrogon develops from the female cypris. **F,** The kentrogon stylet pierces the host cuticle and injects a wormlike vermigon into the hemocoel of the gill lamella. **F_1,** Longitudinal section of the vermigon. The vermigon epidermal cells are not labeled. **G,** Development of an interna from the vermigon about 10 days after injection. **G1,** Longitudinal section of the young interna showing the primordial mantle cavity over the developing ovary. **H,** First appearance of the virgin externa under the cuticle of the host's abdomen about 30 days after injection. **H_1,** Section through the host abdomen immediately prior to eruption of the externa through the host cuticle. **I,** The externa has erupted through the host cuticle. **I_1,** Section of the newly emerged virgin externa showing the now open mantle cavity enclosing the ovary. A male cypris is shown entering the mantle cavity. A dwarf male (trichogon) will develop within the cypris, enter the female mantle cavity, and insert itself into a receptacle of the externa, where it will develop into a functioning testis. **J,** The externa enlarges as it fills with fertilized eggs. ac and bc, cells of primordial ovary; amc, primordial mantle cavity; cc, cyprid carapace; ccc, central core cells; cch, host cuticle; ex, externa; hcu, hemocoel; ken, kentrogon; ov, ovary; rl, rootlet; st, stylet; vg, vermigon. *(From Glenner, H. 2001. Cypris metamorphosis, injection and earliest internal development of the rhizocephalan* Loxothylacus panopaei *(Gissler). Crustacea: Cirripedia: Rhizocephala: Sacculinidae. J. Morphol. 249:43–75. Reprinted with permission.)*

times before being retracted into the mantle cavity. Pumping movements of the body and mantle create a water current into the mantle from which food particles may be removed. The mouth cirri then transfer the particles from the terminal cirri to the mouth. The foregut consists of a pharynx and long esophagus, and sometimes a grinding region, or gizzard (Fig. 19-84). The large midgut may have small digestive ceca. A hindgut and anus may be present. The anus, if present, is at the posterior end of the thorax, lying between the two caudal appendages.

Excretion is accomplished by pair of maxillary glands and nephrocytes. Because there is no heart, blood is circulated through the hemocoel by movements of the body. Specialized respiratory organs are absent and gas exchange occurs across the body and mantle surfaces. The nervous system is reduced and consists of a brain and one or two ventral ganglia (Fig. 19-84). The oviducts of the dorsal ovaries open into the mantle cavity.

Dwarf males (Fig. 19-82D) are even smaller than the cypris larva; have no mantle, gut, mouthparts, or cirri; and do not feed. They are reduced to the essential male features: a large

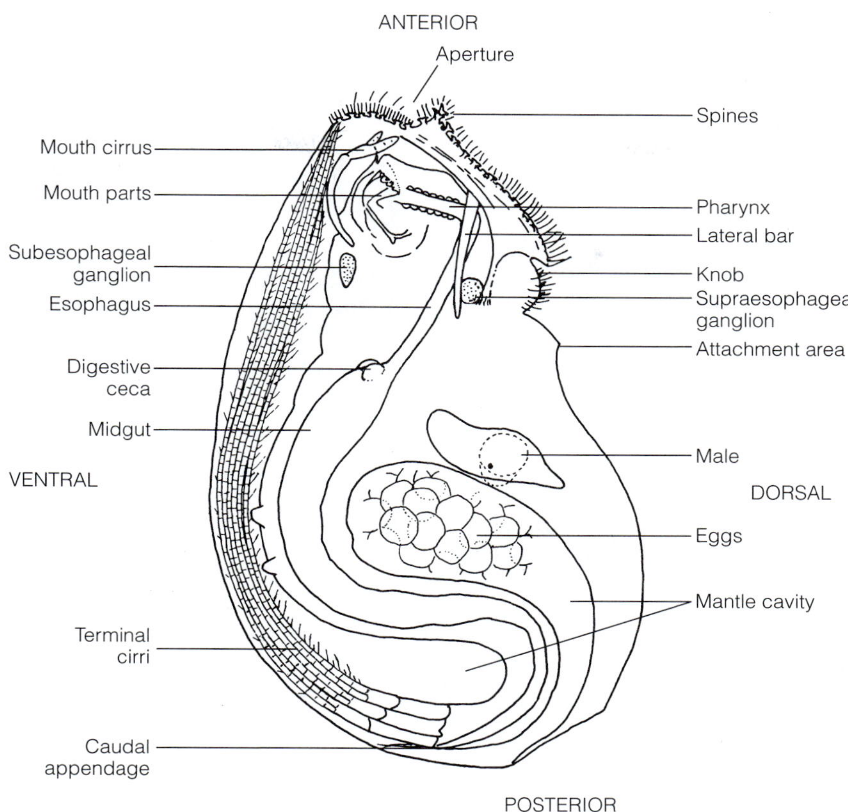

FIGURE 19-84 Cirripedia: Acrothoracica. A female burrowing barnacle, *Kochlorine floridana,* with an attached dwarf male. Viewed from the left side. The female is about 2 mm in length. *(From Wells, H. W., and Tomlinson, J. T. 1966. A new burrowing barnacle from the Western Atlantic. Quarterly Journal of the Florida Academy of Sciences 29:27–37. Used with permission of the Florida Scientist and the Florida Academy of Sciences.)*

testis and a penis. The sperm are flagellated. Males, often many of them, attach by their first antennae to the mantle of the female (Fig. 19-82D, 19-84), but it is not known how or where fertilization occurs. Fertilized eggs are brooded in the mantle cavity and cleavage is thought to be spiral and determinate. The eggs hatch as nauplii that are retained in the atrium until they reach the cypris stage, when they are released. The nauplii have frontal horns and do not feed. The settling stage is the cypris, which finds a suitable substratum and attaches to it with the first antennae. The female cypris metamorphoses into the adult and excavates a burrow, whereas the male attaches to a female, metamorphoses into an adult, and then subsists, without feeding, on its yolk reserves.

Acrothoracica consists of three families: Cryptophialidae *(Cryptophialus),* Lithoglyptidae *(Kochlorine, Lithoglyptes, Weltneria),* and Trypetesidae *(Trypetesa).*

OSTRACODA[C]

Ostracods, or seed shrimps, are tiny microcrustaceans found in most aquatic habitats, both freshwater and marine. Ostracods (Fig. 19-85) are completely enclosed in a bivalve carapace and resemble seeds or minuscule clams. Although some, such as the neutrally buoyant *Gigantocypris,* are planktonic, most live on or near the bottom, where they swim intermittently or crawl over or plow through the upper layer of mud and detritus from the shoreline to great depths. In fresh water they occur in lakes, ponds, and hot springs. They may be suspension feeders, carnivores, detritivores, or herbivores. There are burrowing and interstitial species and species that live on algae, water plants, or other submerged objects. Some are commensal with other animals, living, for example, among the leg setae of crayfishes or with polychaetes or echinoderms. *Sphaeromicola* is commensal with subterranean isopods and *Entocythere* lives on the gills of crayfishes. *Elpidium* lives in the leaf axils of epiphytic bromeliads. Most species are about 1 mm in size, but *Microcythere minuta* is only .25 mm, whereas *Gigantocypris* reaches 25 mm. About 6000 living species have been described, making it one of the largest crustacean taxa. The calcified, shell-like carapace is the source of the name "ostracod," which is derived from the Greek word *ostrakon* (= shell).

The ostracod fossil record, the most extensive of any crustacean taxon, is continuous from the Cambrian period. The heavily calcified carapace and small size undoubtedly have been important factors promoting fossilization. The 10,000 fossil ostracods are known almost exclusively from the valves, so the classification of fossil forms is based entirely on valve morphology.

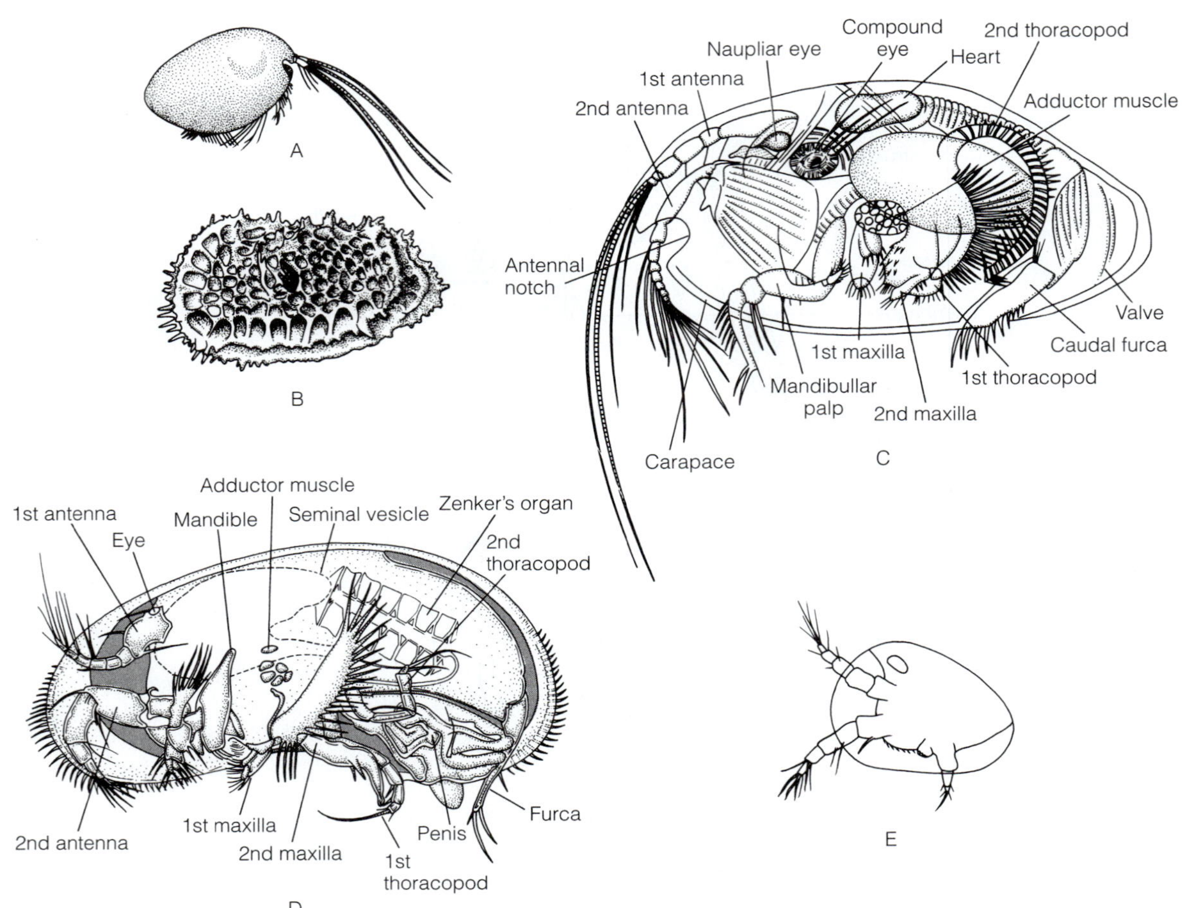

FIGURE 19-85 Ostracoda. **A,** A myodocopid ostracod with antennal notches in the valves (lateral view). **B,** Lateral view of the sculptured valve of *Agrenocythere spinosa,* a benthic marine ostracod. **C,** A female marine myodocopid ostracod, *Skogsbergia,* with the left valve and left appendages removed (lateral view). This marine scavenging carnivore is able to swim over the bottom. **D,** *Candona suburbana,* a very common freshwater podocopid ostracod, with the left valve removed. This nonswimming bottom dweller feeds on algae and decomposing vegetation. Zenker's organ, which is part of the male reproductive system, ejects sperm to the penis. **E,** Lateral view of an ostracod nauplius larva. Only one member of each of the three pairs of appendages is shown. *(C, After Claus from Calman; C and D, From Cohen, A. C. 1982. Ostacoda. In Parker, P. (Ed.): Synopsis and Classification of Living Organisms. Vol. 2. McGraw-Hill Book Co., New York. pp. 191 and 192; E, After Schreiber from Pennak, R. W. 1989. Fresh-water Invertebrates of the United States, Protozoa to Mollusca. 3rd edition. Wiley-Interscience, New York. 628 pp.)*

Form

Most of the ostracod body is head, and the trunk is very much reduced in size (Fig. 19-85C). The body is saclike and without segmental divisions. The most conspicuous feature of ostracod anatomy is the calcified, shell-like, bivalved carapace that completely encloses the animal (Fig. 19-85A). It consists of two lateral valves held together dorsally by a well-developed hinge and a large adductor muscle. The hinge is a longitudinal, mid-dorsal strip of flexible, uncalcified cuticle that in some species may have teeth. Although the valves of benthic species are heavily calcified, this is not the case with planktonic ostracods, in which the extra weight would be unacceptable. The valves of many species are sculptured or ornamented with pits, tubercles, or setae (Fig. 19-85B). The gonads extend into the carapace. Pore canals from the epidermis penetrate the cuticle of the carapace to provide the sensilla with access to the environment. In some species each valve bears an anterior antennal notch that permits extension of the natatory and sensory antennae (Fig. 19-85C).

The head and head appendages are well developed (Fig. 19-85C,D). The uniramous first antennae are chiefly sensory, but can be used for swimming or digging. The biramous second antennae are usually natatory. In some ostracods the second antennae bear the spinnerets of large cephalic silk glands that secrete fine threads used as draglines or to construct a retreat in which ecdysis occurs. The mandibles are equipped with teeth and setae variously adapted for mastication, crawling, and digging. The first maxillae may participate in trituration and in generating a respiratory and feeding current, but the second maxillae are most important in this role. They may also be used for walking.

The trunk is the shortest of any crustacean and consists of a maximum of two segments with two pairs of thoracopods, which may be modified for swimming, walking, feeding, clasping, or grooming (Fig. 19-85C,D). It has lost all external indications of segmentation. The first thoracopods are the chief walking legs of many benthic ostracods. They may also assist in generating the feeding-respiratory current. Sometimes they are absent. The second thoracopods may be walking legs or modified for grooming. A pair of "brush-shaped organs" occur in some male ostracods and are thought to be the vestiges of a third pair of trunk appendages.

The abdomen is vestigial and the anus is located at the posterior end of the trunk. The rami of the caudal furca, which extend posteriorly from the trunk, are considered by some authors to be an additional pair of appendages. The posterior trunk and furca are reflexed anteriorly and held under the body (Fig. 19-85C), much as they are in cladocerans.

Diversity of Ostracoda

Myodocopa^O: Exclusively marine; benthic and planktonic. Valves have anterior antennal notch to accommodate the natatory antennae (Fig. 19-85A,C). Ventral valve margins are convex. Second antennae have well-developed natatory exopods and reduced endopods. Two pairs of trunk appendages. *Cypridina, Gigantocypris, Halocypris, Polycope, Skogsbergia, Vargula.*

Podocopa^O: Marine or fresh water, benthic. No antennal notch. Ventral valve margin is straight or concave (Fig. 19-85D). Two pairs of antennae with well-developed endopods and reduced exopods. Mandibles and first maxillae lack filtering setae. Second maxillae are sometimes leglike. One or two pairs of trunk appendages. All freshwater ostracods are podocopans. Intertidal, commensal, even a few terrestrial species. *Candona, Cypris, Cythereis, Elpidium, Entocythere, Mesocypris, Microcythere, Sphaeromicola.*

Platycopa^O: Closely related to Podocopa. Exclusively marine, benthic. Mandibles and first maxillae have filtering setae. Second antennae have well-developed endopods and exopods. First pair of trunk appendages are not leglike. *Cytherella* the only Recent genus.

Palaeocopa^O: Mostly fossil ostracods, but includes a few Recent species known only from empty valves.

Locomotion

The second antennae are the chief locomotory organs and are used for both swimming and walking. Sometimes the first antennae assist in this role. Benthic species walk on the tips of the antennae and first trunk appendages. Kicking by extending the reflexed trunk thrusts the caudal furca posteriorly and propels the animal forward. Terrestrial species of *Mesocypris* in South Africa and New Zealand use the furca to push their way through humus on the forest floor. Those with silk glands (Cytheridae) use draglines to climb. There is an elaborate and structurally important endoskeleton.

Nutrition

Ostracods exhibit a diversity of feeding modes. Most are suspension feeders, but some are carnivores, herbivores, or scavengers. Some consume leaves that have fallen in the water of forest streams and many feed on carrion. Ostracods may also be either selective or nonselective deposit feeders. The suspension feeders use various appendages to generate a current through a setal filter. Algae are a common food of herbivores, and the prey of carnivorous species includes other crustaceans, small snails, and annelids. The large, predaceous *Gigantocypris* is reported to capture other crustaceans and even small fish with its antennae. Detritus particles, often stirred up from the bottom by the antennae or mandibles, are also a common source of food. In some, the mandibles form a stylet used for piercing the prey and sucking out its contents. The gut is relatively simple and usually does not have a gastric mill, but digestive ceca typically are present.

Internal Form and Function

As they are small animals, most ostracods do not have special respiratory structures and instead exchange gases across the carapace and general body surface. The swimming and feeding currents ventilate the surfaces of the carapace, limbs, and body to facilitate gas exchange.

Small size also reduces the importance of a hemal system for internal transport, and most ostracods do not have a heart. There is, however, a hemocoel partitioned into pericardial and perivisceral sinuses by a muscular septum, the contractions of which circulate the blood. The marine myodocopid ostracods are exceptional and do have a well-developed heart. The excretory organs are antennal glands or maxillary glands. Some ostracods are among the few crustaceans that have both as adults.

The nervous system is compact, as might be expected in a short animal with few segments or appendages. The ventral posterior part of the nerve ring is a subesophageal ganglion consisting of the ganglia of the mouthparts. The two pairs of trunk appendages are innervated from a second ganglion slightly posterior to the subesophageal ganglion. A naupliar eye with three ocelli is present in most ostracods, but compound eyes appear in only some of the Myodocopa (Fig. 19-85C). Deep-sea and subterranean species lack eyes. The naupliar ocelli are inverse pigment cups with a tapetum. The most important sense organs are probably sensory setae on the appendages and valves.

Reproduction and Development

Ostracods are gonochoric and fertilization is internal and direct. Some freshwater species are parthenogenetic. The external genitalia and gonopores of both sexes are located ventrally between the last pair of appendages and the caudal furca, at the posterior end of the trunk. The male gonopores are on a pair of sclerotized penes that extend ventrally just anterior to the caudal furca (Fig. 19-85D). In some, a part of the vas deferens is modified as a peristaltic sperm pump known as Zenker's organ (Fig. 19-85D). The sperm are flagellated and motile, and in some cyprids they are astonishingly large, the largest in the animal kingdom. Some cyprids less than 1 mm in length have sperm 10 mm long, but it is possible they are not functional.

During copulation in some ostracods, the male clasps the female dorsally and posteriorly with his second antennae or first thoracopods and inserts the penes between the valves of the female and into her gonopores. In most ostracods the eggs are shed freely in the water or are attached singly or in clusters to vegetation and other objects on the bottom. In some species the eggs are brooded in the dorsal part of the carapace cavity. Some species, especially those in temporary fresh water, produce desiccation-resistant eggs.

The eggs hatch as nauplii with some of the morphology and behavior patterns of the adult (Fig. 19-85E), including a bivalve carapace. The carapace is an extension of the posterior edge of the head, which, in a nauplius, is the second antennal segment rather than the second maxilla and thus is not homologous to the carapace of other crustaceans. The carapace exoskeleton is lost and replaced at each molt. Molting ceases at adulthood.

Some marine ostracods, such as *Cypridina,* use bioluminescence as a sexual signaling mechanism, much as fireflies do, and ostracods were the first crustaceans in which bioluminescence was observed. The small flashes of bluish light produced externally by the secretions of a gland in the labrum are often patterned as a sequence of distinct flashes lasting 1 or 2 s. Bacteria apparently are not involved in ostracod bioluminescence. On a reef, male *Vargula* may synchronize their flashes, producing a spectacular display.

BRANCHIURA[C]

Branchiurans, or fish lice, are intermittent ectoparasites on freshwater and marine fishes and occasionally tadpoles. The taxon includes a single family with approximately 200 species of flattened crustaceans that feed on fish blood and mucus and range in size between 5 and 30 mm, with most being 3 to 5 mm. *Argulus,* with over 100 species, is the largest genus.

The head has the usual five segments, but their appendages are modified for parasitism (Fig. 19-86). Both pairs of antennae are minute, and the first pair has a claw for attaching to the host. The mouth is surrounded by a tubular proboscis containing the mandibles and openings of the labial ducts that release digestive enzymes. The bladelike mandibles are toothed. In *Argulus* and *Chonopeltis,* the first maxillae form a pair of large suckers that assist in attaching to the host. In the genus *Argulus* there is a hollow, needlelike **preoral spine,** or stylet, between the first maxillae. The spine is connected to a pair of large poison glands and is used to inject a toxic substance of unknown function into the host. *Dolops* lacks both suckers and spine, and *Chonopeltis* has only suckers. The second maxillae are used for grooming. Sessile but movable compound eyes and a naupliar eye are located on the head. A large, bilobed, strongly depressed, shieldlike carapace that extends posteriorly from the head covers most of the thorax. Branchiurans were once thought to be related to copepods, but they are distinguished from them by having compound eyes and a carapace (Fig. 19-86).

Biramous thoracic appendages are present on each of the four segments of the cylindrical thorax (Fig. 19-86). The first thoracic segment is associated with the head, but its appendages are not maxillipeds. An unknown number of segments are fused together to form the short, flat, unsegmented abdomen. This small, bilobed tagma has a deep median notch on its posterior edge and a tiny caudal furca in the cleft, but no appendages.

When not attached to a host, branchiurans swim effectively with the four natatory thoracic appendages. Fish lice readily detach from a host and swim or crawl to another. Branchiuran infestations in confined habitats such as ponds or fish farms can eradicate entire fish populations.

Gas exchange occurs over special, permeable **respiratory areas** of the carapace epithelium and, to a lesser extent, over the general body surface. The gut consists of a foregut, midgut, and hindgut and extends from the mouth on the proboscis to the anus in the notch at the end of the abdomen. The midgut has a pair of branched digestive ceca that extend into the carapace. There is a well-developed heart with a pair of ostia in the posterior thorax. The excretory organs are maxillary glands. The CNS includes a brain and subesophageal ganglion. The ventral nerve cord has five or six ganglia and is confined to the anterior thorax. The compound eyes consist of 20 to 100 ommatidia. The median naupliar eye is composed of three inverse pigment cups.

Branchiurans are gonochoric and in both sexes the gonoducts open on the fourth thoracic segment. Spermatozoa are long, slender, and tapered and do not have an abruptly swollen head. An acrosome is absent and a slender, tapered pseudoacrosome that forms the apical one-third of the sperm contains a pair of long rods. The centriole region connects the pseudoacrosome with the sperm body. The sperm body contains three long filiform mitochondria, a long filiform nucleus, and the flagellar axoneme, which is enclosed in a cylinder formed of dorsal and ventral bands. Pentastomids, which will be discussed next, have nearly identical sperm. Copulation occurs while the parasites are on the host, but the eggs are deposited on a firm substratum other than the host. Eggs hatch as nauplii or, in those with direct development, juveniles.

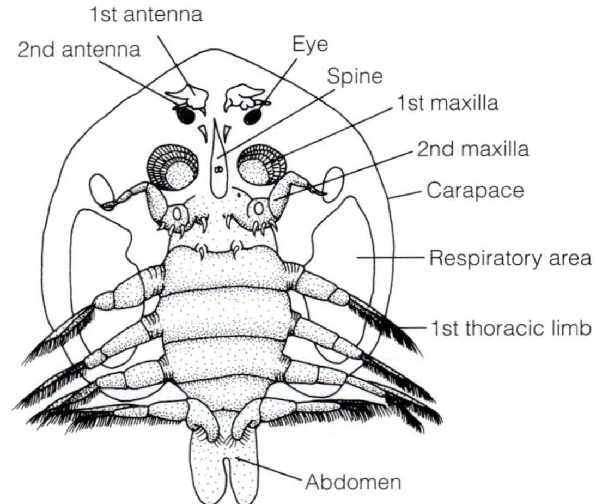

FIGURE 19-86 Branchiura: The fish louse, *Argulus foliaceus.* *(After Wagler from Kaestner, A. 1970. Invertebrate Zoology. Vol. 3. Wiley-Interscience, New York.)*

PENTASTOMIDA[C]

Pentastomids, or tongue worms, are derived, elongate, vermiform parasites of the nasal passages, sinuses, and lungs of vertebrates (Fig. 19-87A). About 100 species of these poorly understood animals have been described. All stages of the life cycle are obligatory parasites and most species require intermediate and definitive hosts. Approximately 90% parasitize reptiles, chiefly snakes, but also crocodiles and lizards. A few infect mammals and birds. Reptile parasites feed on blood, whereas those of mammals consume mucus and cells. There is usually a vertebrate, but sometimes an arthropod, intermediate host. Although largely tropical, pentastomids have been reported from North America, Europe, Australia, and even in Arctic birds.

The wormlike, usually cylindrical body is 1 to 16 cm long and divided into a head and trunk (Fig. 19-87A). Anteriorly, the head bears the mouth and two pairs of chitinous hooks. In some pentastomids the hooks are at the ends of two pairs of leglike extremities, but others lack these extremities and the hooks arise from the surface of the head. The hooks, which are used to attach to the host, are the vestiges of two pairs of segmental appendages. Two pairs of papillae on the head may also be derived from segmental appendages. The elongate trunk is ringed with superficial annulations.

The body is covered by a soft, flexible, nonchitinous cuticle that is molted periodically during larval development. Molting ceases with adulthood. The body wall includes annelid-like layers of circular and longitudinal muscles. The muscles are cross-striated, however, a characteristic of arthropods but not annelids. The gut is complete, consists of a foregut, midgut, and hindgut, and extends from the mouth on the anterior head to the anus at the posterior tip of the trunk (Fig. 19-88). It is a simple tube with a muscular pump (for the host's blood) in the foregut. In some species there is a frontal gland thought to secrete a substance that breaks down host tissues and prevents coagulation of the blood. No special gas-exchange surfaces or excretory organs are present. There is a hemocoel, but no heart.

The CNS is unusual and consists of a subesophageal ganglion as the major neuronal concentration. The two sides of the subesophageal ganglion are joined by a connective that arches *dorsally* over the gut. There is a short, ganglionated ventral nerve cord in the head (Fig. 19-88). The nerve cord does not extend into the trunk, which is instead supplied by nerves, without ganglia, from the subesophageal ganglion. There are no eyes.

Pentastomids are gonochoric and have a well-developed genital system. Fertilization is internal and occurs only once in the lifetime of each female. The gonads and reproductive ducts (Fig. 19-88) occupy most of the space in the trunk. The gonads are usually unpaired with the gonopores located at the junction between the head and trunk. Pentastomid sperm are long, slender, filiform, and essentially identical to those of Branchiura. These sperm are the most derived flagellated crustacean sperm.

Development begins in the uterus and an embryo, the primary larva, develops inside the eggshell. The **primary larva** has two pairs of hooked extremities and resembles a crustacean nauplius (Fig. 19-87B). Embryonated eggs released from the female are moved from the lungs into the digestive system of the host and then outside with the feces. In most pentastomids, the life cycle requires an herbivorous intermediate host. The intermediate host may be any class of vertebrate. For crocodile parasites they are fishes, whereas those for carnivorous mammals such as foxes and wolves are rabbits or rodents. The intermediate host ingests the embryonated egg and the primary larva hatches in the host's gut. An infective larva develops in the intermediate host and, after a number of molts and migrations, arrives in the respiratory tract. Infective larvae have four to six leglike appendages. When the intermediate host is eaten, this larva is transferred to the stomach of the definitive host, from which it migrates to the lungs and nasal passageways via the esophagus and trachea. Pentastomid larvae have been reported in humans but are killed by calcareous encapsulation by the host tissue.

The taxonomic status of the pentastomids has long been disputed and until recently they were placed in a separate phylum of uncertain affinities. There is growing agreement, however, that pentastomids are highly derived crustaceans closely related to branchiurans. Similarities between the two taxa in 18S rDNA base sequences, sperm morphology, and development support this conclusion. The annelid-like body-wall musculature may be a secondary development in response to the loss of a rigid exoskeleton and development of a worm-like body.

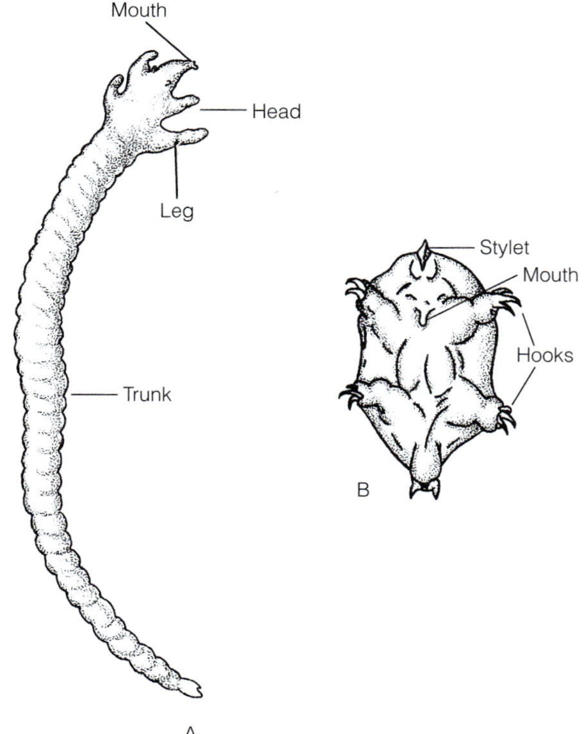

FIGURE 19-87 Pentastomida. **A,** The pentastomid *Cephalobaena tetrapoda,* from the lung of a snake. **B,** Larva of the pentastomid *Porocephalus crotali.* *(A, After Heymons from Cuénot; B, After Penn.)*

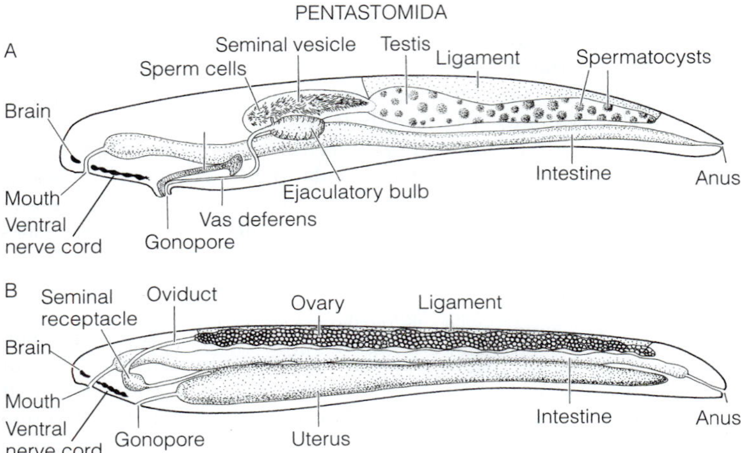

FIGURE 19-88 Pentastomida: Sagittal sections of generalized pentastomids. **A**, Male. **B**, Female. *(From Storch, 1993, after Riley, 1983. Recent advances in our understanding of pentastomid reproductive biology. Parasitology 86:59–83. Copyright © 1993 by John Wiley, reprinted by permission of Wiley-Liss, a subsidiary of John Wiley & Sons, Inc.)*

PHYLOGENY OF MAXILLOPODA

Maxillopoda was proposed as a monophyletic taxon several decades ago, and since then there has been disagreement over its validity, which taxa should be included, and which characters should be used in determining the evolutionary pattern of the taxon. Maxillopoda is derived from an ancestor in which the length of the body was reduced by the loss of all but 10 trunk segments. In this ancestral trunk were seven limb-bearing thoracic segments and three limbless abdominal segments plus a telson and a caudal furca.

Maxillopoda is divided into two sister taxa, Copepodomorpha and Progonomorpha (Fig. 19-89). Copepodomorpha, consisting of Copepoda and Mystacocarida, arose through the complete loss of the compound eyes of the ancestor. Both taxa lack a carapace.

Within Copepodomorpha, Mystacocarida arose through the adult's retention of the larval locomotory functions of the first three head appendages. Maxillae 1 and 2 and the first thoracopods are used for scraping food from sand grains. The three ocelli of the larval naupliar eye dissociate to become the four adult ocelli, presumably after the median cup divides into two.

Copepoda arose from the stem maxillopodan by incorporating the first thoracomere into the cephalothorax and transforming its appendages into maxillipeds. The remaining thoracopods developed a median intercoxal plate rigidly connecting each pair of coxae across the ventral midline. Males produce spermatophores and females secrete an egg sac to enclose the eggs.

In Progonomorpha—at least in those taxa in which the gonopore's position can be determined with some degree of certainty—the gonopore has moved from the seventh thoracomere of the ancestral maxillopodan to the first thoracomere, defining this taxon. Progonomorpha includes the derived, parasitic tantulocaridans, ostracods, thecostracans, branchiurans, and pentastomids.

The highly unusual parasitic larva of tantulocaridans distinguishes them from other maxillopodans. The loss of compound eyes in Tantulocarida is presumed to be independent of the similar loss in Copepodomorpha.

The barnacle-like crustaceans (Thecostraca) include Ascothoracida and Cirripedia. Of them, Ascothoracida is most like other crustaceans, with some taxa retaining a shrimplike morphology and an abdomen. The cirripedes are less like the ancestors and most have developed calcareous plates, which are not molted, in the carapace. Many are hermaphroditic, the abdomen is lost, and the thoracopods have become filter-feeding devices. Within Cirripedia, Acrothoracica and Rhizocephala are gonochoric and lack a calcified carapace.

The ostracods have undergone extreme shortening of the maxillopodan body with additional reduction of the number of locomotory thoracopods to a maximum of two pairs. The nauplius larva and the adult have a bivalved carapace completely enclosing the body.

The evolutionary relationships of Branchiura and Pentastomida with other maxillopodans cannot be determined at present. However, a close relationship between these two morphologically very dissimilar taxa is strongly supported by striking similarities of their unusual sperm, which are long, extremely slender, filiform, and lack a distinct head.

PHYLOGENETIC HIERARCHY OF MAXILLOPODA

- Maxillopoda
  - Copepodomorpha
    - Copepoda
    - Mystacocarida
  - Progonomorpha
    - Tantulocarida
    - Thecostracomorpha
      - Thecostraca
        - Ascothoracida
        - Cirripedia
          - Acrothoracica
          - N. N.
            - Thoracica
            - Rhizocephala
      - Ostracoda

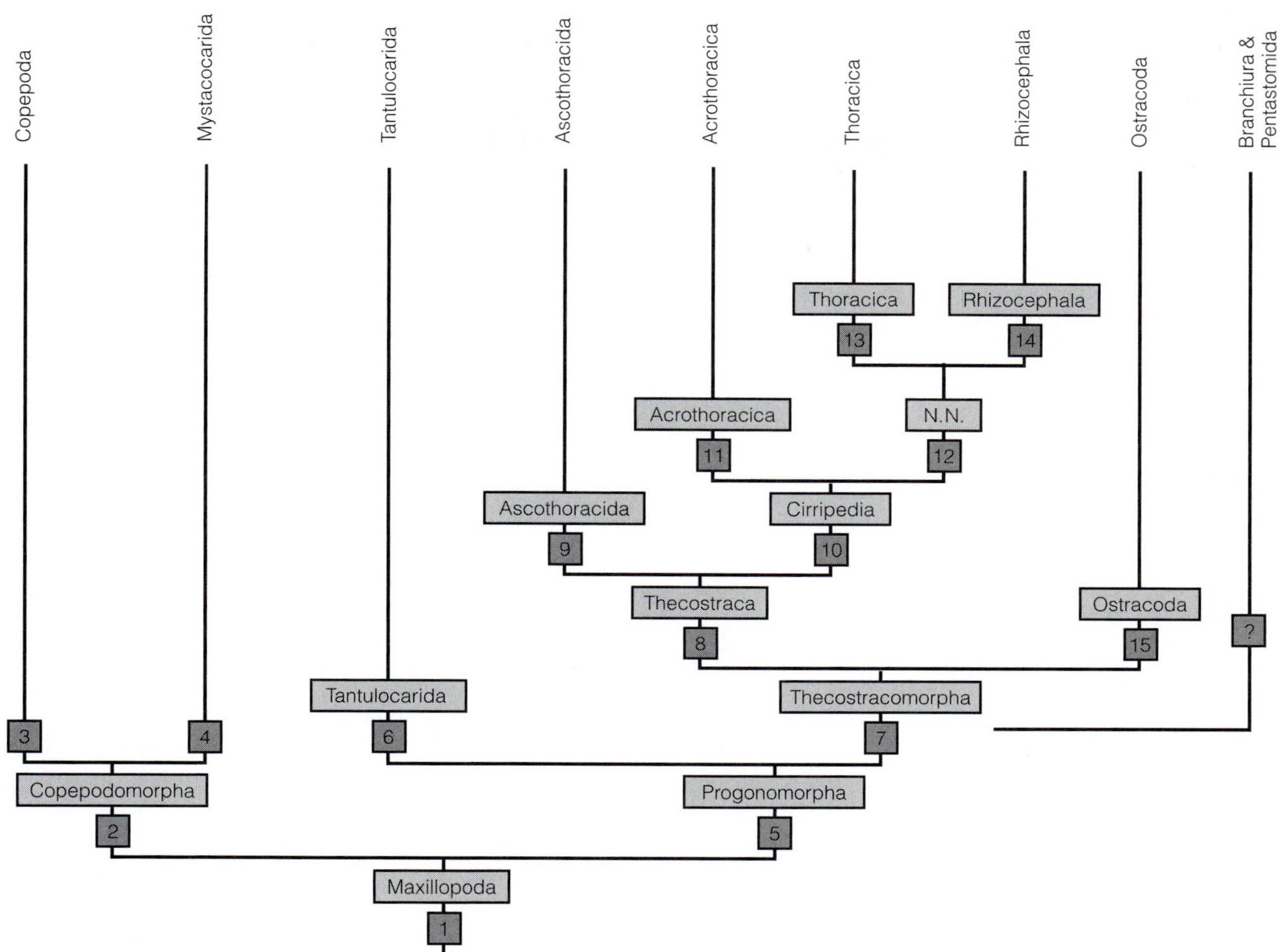

FIGURE 19-89 A phylogeny of Maxillopoda. **1, Maxillopoda:** Trunk has 10 segments; thorax has 7 segments; abdomen has 3 segments, telson, and furca; naupliar eye has three ocelli. **2, Copepodomorpha:** Compound eye is lost in larvae and adults. **3, Copepoda:** Cephalothorax is composed of head and first thoracomere; first thoracopod is a maxilliped; coxae of each pair are joined by a midventral intercoxal plate; male antennae 2 are prehensile; spermatophores are present; female has a seminal receptacle; eggs are enclosed in a secreted egg sac. **4, Mystacocarida:** Head has four pigment-cup ocelli; there is a long linguiform labrum; antennae 2 and mandibles are biramous and locomotory; gonopores are on thoracomere 4. **5, Progonomorpha:** Female gonopores are on thoracomere 1. **6, Tantulocarida:.** Compound eye is lost; larvae are parasitic; reproductive females have reduced head appendages; mouth and gut are reduced. **7, Thecostracomorpha:** Bivalved carapace has an adductor muscle. **8, Thecostraca:** Larval compound eye has a three-part crystalline cone (adult lacks a compound eye); antennae 2 are vestigial or lost; adults are sessile; cyprid carapace has five pairs of lattice organs with posterior terminal pores. **9, Ascothoracida:** Piercing mouthparts; antennae 1 are prehensile; parasitic adults. **10, Cirripedia:** Nauplius larvae have glandular frontal horns and a caudal spine; cypris larva; adult filter feeds; thoracopods are six pairs of biramous, setose cirri; abdomen is lost; second pair of lattice organs have a terminal pore anteriorly; specialized filiform sperm. **11, Acrothoracica:** Bore into calcareous substrates using chitinous epidermal teeth; 180° torsion of the body's long axis during metamorphosis; first pair of cirri are widely separated from the posterior pairs. **12, N.N.:** Lattice organs have pore fields; first and second pairs of lattice organs have anterior terminal pores; cypris lacks an abdominal rudiment. **13, Thoracica:** There are no true morphological apomorphies, although the taxon is strongly supported by molecular evidence. Possible apomorphies include hermaphroditism and mineralized shell plates. **14, Rhizocephala:** Adults are parasitic on other crustaceans; typical crustacean morphology is lost: adult lacks segmentation and consists of an external reproductive body and internal nutrient-absorbing rootlets; female larva infests host via a stylet-bearing kentrogon larva; earliest internal stage is a wormlike vermigon; male implants into female as a trichogon that becomes integrated with and nourished by the female. (These hitherto unpublished apomorphies for the thecostran taxa were provided by Professor Jens Høeg of the University of Copenhagen.) **15, Ostracoda:** Locomotory thoracopods reduced to two pairs; nauplius has uniramous second antennae and mandibles. *(Based on Ax, P. 2000. Multicellular Animals, II. Springer Verlag, Berlin. Perez-Losada, Høeg, Kolbasov, Crandall, 2002; and Høeg, personal communication, 2002.)*

PHYLOGENY OF CRUSTACEA

The earliest known crustacean fossils are Phosphatocopina from the Lower Cambrian. Nearly all higher crustacean taxa are known to have existed since the Cambrian period, but their origin and relationships to the other arthropod taxa are obscure. Several phylogenies of the Crustacea have been proposed in recent years (Fig. 19-90). In the stem Crustacea, the four median eyes of the arthropod ancestor coalesce to form a characteristic naupliar eye that is not present in tracheates. The nauplius larva, characterized by its three pairs of appendages (antennae 1, antennae 2, and mandibles), naupliar eye, and anterior labrum, is another crustacean apomorphy, as is the loss of four of the original pairs of saccate nephridia, leaving only the antennal and maxillary glands.

The ancestral crustacean probably was a small, swimming, epibenthic animal with a head and trunk of numerous similar segments. The head bore two pairs of antennae, a pair of mandibles, two pairs of maxillae, a pair of compound eyes, and a naupliar eye. The mouth was directed backward. The trunk bore appendages on every segment and was not divided into thorax and abdomen. Its appendages, which were numerous and similar (homonomous) not only to each other but probably also to the maxillae, probably served in locomotion, gas exchange, and feeding. They were biramous or multiramous mixopods with coxal gnathobases forming a midventral food

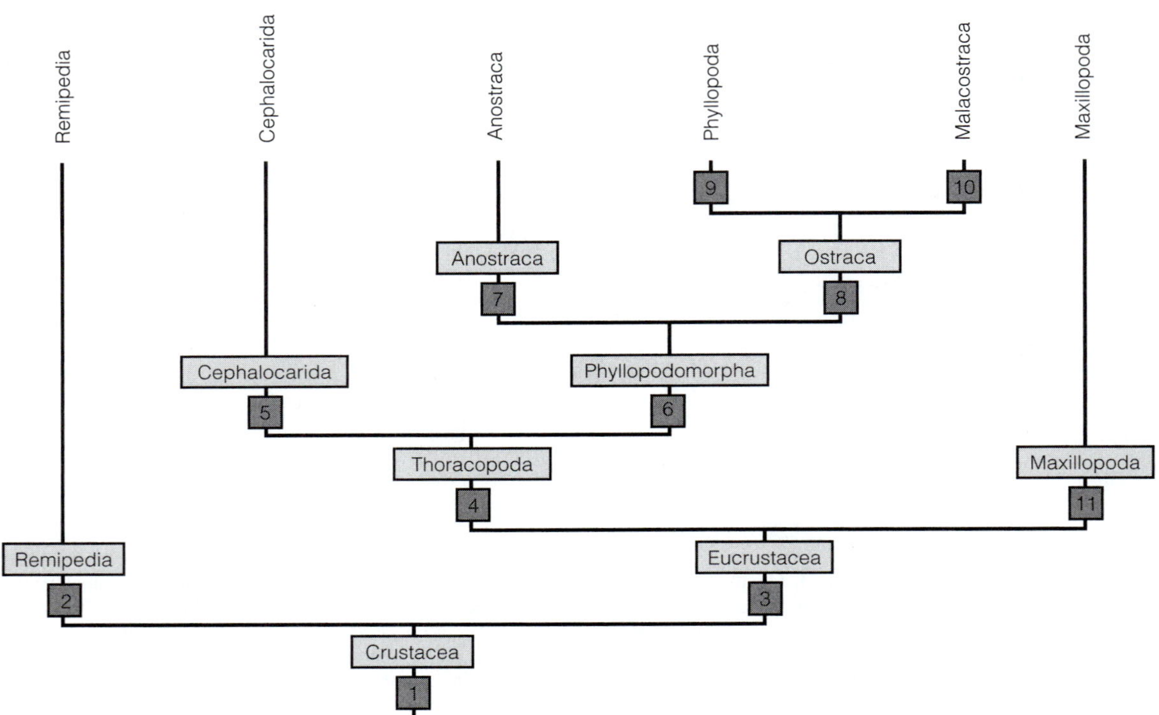

FIGURE 19-90 A phylogeny of Crustacea. **1, Crustacea:** Two pairs of biramous antennae; two pairs of maxillae; hatching stage is a nauplius with antennae 1 and 2 and mandibles; compound eye is present; naupliar eye has four pigment-cup ocelli; antennal and maxillary (and perhaps all) segments have saccate nephridia; tagmata are head and trunk; trunk appendages are biramous, similar to maxillae; all trunk segments have appendages; suspension feeders using mixopods and a ventral food groove. **2, Remipedia:** Cephalothorax has one trunk segment and one maxilliped; maxilliped is similar to the second maxillae; posterior mouthparts are modified for carnivory; hermaphroditic; maxillary nephridia; secondary loss of eyes. **3, Eucrustacea:** Trunk divided into thorax and abdomen, although abdomen may be vestigial; maximum of eight thoracic segments. **4, Thoracopoda:** Thoracopods are turgor appendages with epipodites used for microphagous suspension feeding. **5, Cephalocarida:** Hermaphroditic; nauplius is absent; lacks eyes. **6, Phyllopodomorpha:** Compound eyes are stalked; fine particulates are filtered from open water. **7, Anostraca:** First antennae are reduced; male second antennae are grasping organs; mandible lacks a palp; thoracic segments 12 and 13 are genital segments; naupliar eye has three ocelli. **8, Ostraca:** Carapace forms a bilateral shell that encloses much of the body. **9, Phyllopoda:** Sessile compound eye is sunken below the surface of head and suspended in an ectodermally lined, water-filled vesicle connected to the exterior by a pore; eye has extrinsic muscles and is movable; about 30 thoracopods. **10, Malacostraca:** Antennae 1 are multiramous (usually biramous); abdominal segments have biramous appendages; naupliar eye has four ocelli; primitive malacostracan has 15 trunk segments; thorax has 8 segments; abdomen has 7 (6 in derived taxa) segments; endopods of thoracic legs are stenopodous; last abdominal appendages are uropods; carapace does not enclose limbs; the position of the gonopore is constant, with the female gonopore on thoracomere 6 and the male's on thoracomere 8; stomach has cardiac and pyloric chambers for trituration, hydrolysis, and filtering. **11, Maxillopoda:** Maximum of 10 trunk segments plus a telson and furca; maximum of seven thoracic segments; appendages are restricted to the thoracic segments; abdomen has three appendageless segments; short, ovoid heart; naupliar eye has three pigment-cup ocelli. *(Modified from Ax, P. 2000. Multicellular Animals, II. Springer Verlag, Berlin. 396 pp.)*

groove. Among Recent crustaceans, the remipedes, cephalocaridans, anostracans, and some malacostracans resemble this hypothetical ancestral crustacean to varying degrees, although each has a number of apomorphies in addition to its primitive characteristics. The maxillopodans are least like the ancestor, although even a few of them exhibit some apparently primitive characteristics.

The annelid-like remipedes, which have homonomous biramous appendages on every trunk segment (except the first), may be the most primitive Recent crustaceans and the ones most like the stem crustacean. Remipedia is the sister taxon of all the remaining crustaceans (Eucrustacea). Having a trunk undivided into thorax and abdomen with similar appendages on almost all segments is the condition expected of the stem crustacean and presumably was inherited from that ancestor.

Eucrustacea is divided into the sister taxa Thoracopoda and Maxillopoda. In Thoracopoda, the thoracopods originate as turgor appendages with a filter-feeding role. Thoracopoda includes Cephalocarida, Anostraca, Phyllopoda, and Malacostraca. Like Remipedia, Cephalocarida is a small taxon having many of the characteristics expected of the stem crustacean. The cephalocarid trunk, however, is divided into a limb-bearing thorax and a limbless abdomen and is derived in this regard.

Within Thoracopoda two methods of microphagous suspension feeding evolved to divide Thoracopoda into Cephalocarida and Phyllopodomorpha. The more plesiomorphic Cephalocarida feed by stirring organic particles off the bottom with an endopod of five articulated segments and an exopod of three. In Phyllopodomorpha this original detritus-feeding mode was replaced by filter feeding in open water remote from the bottom. For this, the joints between the articles of the endopods and exopods were lost, resulting in leaflike phyllopods without articulations. In Malacostraca this condition is found only in the primitive Leptostraca.

In many classifications Anostraca and Phyllopoda are combined in Branchiopoda, but the absence of a carapace and other differences may be sufficient reason to view the old Branchiopoda as paraphyletic, necessitating the separation of Anostraca from Phyllopoda. Phyllopoda and Malacostraca, by virtue of their common possession of a carapace, are sister taxa in Ostraca.

Malacostraca has a stabilized tagmosis, almost always with eight thoracic and six abdominal segments. Malacostracans have appendages on the abdominal segments whereas the phyllopodans have lost these plesiomorphic structures. Maxillopoda is characterized by a reduction in the length of the trunk to a maximum of 10 segments.

PHYLOGENETIC HIERARCHY OF CRUSTACEA

- Crustacea
  - Remipedia
  - Eucrustacea
    - Thoracopoda
      - Cephalocarida
      - Phyllopodomorpha
        - Anostraca
        - Ostraca
          - Phyllopoda
          - Malacostraca
    - Maxillopoda

REFERENCES

GENERAL

Anderson, D. T. 1973. Embryology and Phylogeny in Annelids and Arthropods. Pergamon Press, Oxford. 495 pp.

Ax, P. 2000. Multicellular Animals, II. Springer Verlag, Berlin. 396 pp.

Barnes, R. D., and Harrison, F. W. 1992. Introduction. In Harrison, F. W., and Humes, A. G. (Eds.): Microscopic Anatomy of Invertebrates. Vol. 9. Crustacea. Wiley-Liss, New York. pp. 1–8.

Bliss, D. E. (Ed.): 1982–1985. Biology of the Crustacea. Academic Press, New York. (Ten volumes.)

Chang, E. S., and O'Connor, J. D. 1988. Crustacea: Molting. In Laufer, H., and Downer, R. G. H. (Eds.): Endocrinology of Selected Invertebrate Types. Alan R. Liss, New York. pp. 259–278.

Cohen, M. J. 1955. The function of receptors in the statocyst of the lobster, *Homarus americanus*. J. Physiol. 130:9–34.

Fitzpatrick, J. F. 1983. How to Know the Freshwater Crustacea. W. C. Brown, Dubuque, IA. 227 pp.

Fortey, R. A., and Thomas, R. H. (Eds.): 1998. Arthropod Relationships. Systematics Association Spec. Vol. 55. Chapman & Hall, London. 383 pp.

Gerritsen, J., and Porter, K. G. 1982. The role of surface chemistry in filter feeding by zooplankton. Science 216:1225–1227.

Green, J. 1961. A Biology of Crustacea. Quadrangle Books, Chicago. 113 pp.

Harrison, F. W., and Humes, A. G. (Eds.): 1992. Microscopic Anatomy of Invertebrates. Vol. 9. Crustacea. Wiley-Liss, New York.

Hessler, R. R., and Newman, W. A. 1975. A trilobite origin for the Crustacea. Foss. Strata 4:437–459.

Koehl, M. A. R., and Strickler, J. R. 1981. Copepod feeding currents: Food capture at low Reynolds number. Limnol. Oceanogr. 226:1062–1073.

LaBarbera, M. 1984. Feeding currents and particle capture mechanisms in suspension feeding animals. Am. Zool. 24:71–84.

Lockwood, A. P. M. 1968. Aspects of the physiology of Crustacea. Oliver and Boyd, London. 328 pp.

McLaughlin, P. 1980. Comparative Morphology of Recent Crustacea. W. H. Freeman, San Francisco. 159 pp.

McNamara, J. C. 1981. Morphological organization of crustacean pigmentary effectors. Biol. Bull. 161:270–280.

Moore, R. C. (Ed.): 1969. Treatise on Invertebrate Paleontology. Pt. R. Arthropoda 4. Vols. 1 and 2. Geological Society of America, University of Kansas Press, Lawrence, KS. 651 pp.

Morin, J. G. 1983. Coastal bioluminescence: Patterns and functions. Bull. Mar. Sci. 33:787–817.

Schminke, H. K. 1996. Crustacea. In Westheide, W., and Rieger, R. M. (Eds.): Spezielle Zoologie, I. Einzeller und Wirbellose Tiere. Gustav Fisher Verlag, Stuttgart. pp. 501–581.

Schmitt, W. L. 1965. Crustaceans. Ann Arbor Science Library, Univ. Michigan Press, Ann Arbor. 204 pp.

Schram, F. R. 1986. Crustacea. Oxford University Press, New York. 700 pp.

Schram, F. R. (Ed.): 1983. Crustacean Phylogeny. A. A. Balkema, Rotterdam. 372 pp.

Walossek, D. 1999. On the Cambrian diversity of Crustacea. In Schram, F. R., and von Vaupel Klein, J. C. (Eds.): Crustaceans and the Biodiversity Crisis. Proceedings of the 4th International Crustacean Conference, Amsterdam, 1998. Vol. 1. E. J. Brill, Leiden, Netherlands. pp. 3–27.

Waterman, T. H., and Chase, F. A. 1960. General crustacean biology. In Waterman, T. H. (Ed.): The Physiology of Crustacea I. Metabolism and Growth. Academic Press, New York. pp. 1–33.

Waterman, T. H. (Ed.): 1960. The Physiology of Crustacea I. Metabolism and Growth. Academic Press, New York. 670 pp.

Waterman, T. H. (Ed.): 1961. The Physiology of Crustacea II. Sense Organs, Integration and Behavior. Academic Press, New York. 681 pp.

Welsh, J. H. 1961. Neurohumors and neurosecretion. In Waterman, T. H. (Ed.): The Physiology of Crustacea II. Sense Organs, Integration and Behavior. Academic Press, New York. pp. 281–311.

White, K. N., and Walker, G. 1981. The barnacle excretory organ. J. Mar. Biol. Assoc. UK 61:529–547.

Internet Sites

www.bio.uva.nl/onderzoek/cepa (Animal Evolution Pattern Analysis home page.)

http://bromeliadbiota.ifas.ufl.edu/crbrom.htm (Crustacea in Bromeliad Phytotelmata, by W. Janetzky.)

www.zoomschool.com/painting/crustaceans.shtml (Crustaceans to Color Online.)

www.vims.edu/tcs (The Crustacean Society.)

http://biosys-serv.biologie.uni-ulm.de/sektion/dieter/dieter.html ("Orsten" Research and Dieter Waloszek's View of Arthropod and Crustacean Phylogeny.)

www.geocities.com/mediaq/fauna/ (Groundwater Fauna of Italy, by G. L. Pesce.)

www.mov.vic.gov.au/crust/page1a.html (Marine Crustaceans of Southern Australia, Museum Victoria.)

REMIPEDIA

Felgenhauer, B. E., Abele, L. G., and Felder, D. L. 1992. Remipedia. In Harrison, F. W., and Humes, A. G. (Eds.): Microscopic Anatomy of Invertebrates. Vol. 9. Crustacea. Wiley-Liss, New York. pp. 225–247.

Yager, J. 1981. Remipedia, a new class of Crustacea from a marine cave in the Bahamas. J. Crust. Biol. 1:328–333.

CEPHALOCARIDA

Hessler, R. R. 1964. The Cephalocarida: Comparative skeletomusculature. Mem. Conn. Acad. Arts Sci. 16:1–97.

Hessler, R. R., and Elofsson, R. 1992. Cephalocarida. In Harrison, F. W., and Humes, A. G. (Eds.): Microscopic Anatomy of Invertebrates. Vol. 9. Crustacea. Wiley-Liss, New York. pp. 9–24.

Sanders, H. L. 1955. The Cephalocarida, a new subclass of Crustacea from Long Island Sound. Proc. Nat. Acad. Sci. USA 41:61–66.

Sanders, H. L. 1963. The Cephalocarida. Mem. Conn. Acad. Arts Sci. 15:1–180.

ANOSTRACA AND PHYLLOPODA

Dodson, S. I., and Frey, D. G. 2001. Cladocera and other Brachiopoda. In Thorp, J. H., and Covich, A. P. (Eds.): Ecology and Classification of North American Freshwater Invertebrates. Academic Press, San Diego. pp. 849–913.

Fryer, G. 1987. A new classification of the branchiopod Crustacea. Zool. J. Linn. Soc. Lond. 91:27–124.

Martin, J. 1992. Branchiopoda. In Harrison, F. W., and Humes, A. G. (Eds.): Microscopic Anatomy of Invertebrates. Vol. 9. Crustacea. Wiley-Liss, New York. pp. 25–224.

Martin, J. W., Felgenhauer, B. E., and Abele, L. G. 1986. Redescription of the clam shrimp *Lynceus gracilicornis* from Florida, with notes on its biology. Zool. Scripta 15:221–232.

Internet Sites

www.cladocera.fsnet.co.uk (Cladocera: A Short Course, by C. Duigan, University College, London.)

www.cladocera.uoguelph.ca (Cladocera, by C. L. Rowe and P. D. N. Hebert, University of Guelph.)

www.natureserve.org/getData/dataSets/animalDataSets/shrimpus.htm (Fairy, Clam, and Tadpole Shrimp of the United States, by NatureServe.)

http://mailbox.univie.ac.at/erich.eder/UZK/index2.html (Large Branchiopods, by E. Eder.)

http://perso.wanadoo.fr/nicolas.rabet (Les Grands Branchiopodes, by N. Rabet.)

www.microscopy-uk.org.uk/mag/artjun99/wflea.html (Water Fleas, by J. Parmentier and W. van Egmond.)

http://royal.okanagan.bc.ca/newsletr/v2n1/ggindex.html (Water Fleas, by G. Green, Royal British Columbia Museum.)

www.homepages.ucl.ac.uk/~ucfagls/thewaterflea/ (The Water Flea, by G. Simpson.)

MALACOSTRACA: STOMATOPODA

Caldwell, R. L., and Dingle, H. 1976. Stomatopods. Sci. Am. 234:80–89.

Manning, R. B. 1974. Crustacea: Stomatopoda. Marine Flora and Fauna of the Northeastern United States. NOAA Tech. Report NMFS Circular 386. U.S. Government Printing Office, Washington, DC. 6 pp.

Internet Sites

www.blueboard.com/mantis (The Lurker's Guide to Stomatopods, by A. San Juan.)

www.mov.vic.gov.au/crust/stombiol.html (Marine Crustaceans of Southern Australia: Mantis Shrimps, Museum Victoria.)

http://dmoz.org/Science/Biology/Flora_and_Fauna/Animalia/Arthrpoda/Crustacea/Stomatopoda/ (Stomatopoda: DMOZ Open Directory Project.)

MALACOSTRACA: DECAPODA

Barnes, R. D., and Harrison, F. W. 1992. Introduction to the Decapoda. In Harrison, F. W., and Humes, A. G. (Eds.): Microscopic Anatomy of Invertebrates. Vol. 10. Decapod Crustacea. Wiley-Liss, New York. pp. 1–6.

Bauer, R. T. 1976. Mating behaviour and spermatophore transfer in the shrimp *Heptacarpus pictus.* J. Nat. Hist. 10:415–440.

Bliss, D. E., van Montfrans, J., van Montfrans, M., et al. 1978. Behavior and growth of the land crab *Cecarcinus lateralis* (Freminville) in southern Florida. Bull. Am. Mus. Nat. Hist. 160:113–151.

Bruce, A. J. 1976. Shrimps and prawns of coral reefs, with special reference to commensalism. In Jones, O. A., and Endean, R. (Eds.): Biology and Geology of Coral Reefs. Vol. III: Biology 2. Academic Press, New York. pp. 37–94.

Cameron, J. N. 1985. Molting in the blue crab. Sci. Am. 252:105–109.

Crane, J. 1975. Fiddler Crabs of the World. Princeton University Press, Princeton, NJ. 660 pp.

Felgenhauer, B. E. 1992a. External anatomy and integumentary structures. In Harrison, F. W., and Humes, A. G. (Eds.): Microscopic Anatomy of Invertebrates. Vol. 10. Decapod Crustacea. Wiley-Liss, New York. pp. 7–43.

Felgenhauer, B. E. 1992b. Internal anatomy of the Decapoda: An overview. In Harrison, F. W., and Humes, A. G. (Eds.): Microscopic Anatomy of Invertebrates. Vol. 10. Decapod Crustacea. Wiley-Liss, New York. pp. 45–75.

Fingerman, M. 1992. Glands and secretion. In Harrison, F. W., and Humes, A. G. (Eds.): Microscopic Anatomy of Invertebrates. Vol. 10. Decapod Crustacea. Wiley-Liss, New York. pp. 345–394.

Glaessner, M. F. 1969. Decapoda. In Moore, R. C. (Ed.): Treatise on Invertebrate Paleontology. Pt. R. Arthropoda 4. Vol. 2. Geological Society of America, University of Kansas Press, Lawrence. pp. R400–R532.

Gurney, R. 1942. Larvae of Decapod Crustacea. Ray Society Vol. 129. Ray Society, London. 306 pp.

Harrison, F. W., and Humes, A. G. (Eds.): 1992. Microscopic Anatomy of Invertebrates. Vol. 10. Decapod Crustacea. Wiley-Liss, New York. 459 pp.

Hobbs, H. H. 2001. Decapods. In Thorp, J. H., and Covich, A. P. (Eds.): Ecology and Classification of North American Freshwater Invertebrates. Academic Press, San Diego. pp. 955–1001.

Icely, J. D., and Nott, J. A. 1992. Digestion and absorption: Digestive system and associated organs. In Harrison, F. W., and Humes, A. G. (Eds.): Microscopic Anatomy of Invertebrates. Vol. 10. Decapod Crustacea. Wiley-Liss, New York. pp. 147–201.

Johnson, P. T. 1980. The Histology of the Blue Crab, *Callinectes sapidus.* Praeger, New York. 440 pp.

Martin, G. G., and Hose, J. E. 1992. Vascular elements and blood (hemolymph). In Harrison F. W., and Humes, A. G. (Eds.). Microscopic anatomy of Invertebrates, 10. Decapod Crustacea. Wiley-Liss, New York. pp. 117–146.

McGaw, I. J., and Reiber, C. L. 2002. Cardiovascular system of the blue crab, *Callinectes sapidus.* J. Morphol. 251:1–21.

Omori, M. 1974. The biology of pelagic shrimps in the ocean. Adv. Mar. Biol. 12:233–324.

Peterson, D. R., and Loizzi, R. F. 1974. Ultrastructure of the crayfish kidney-coelomosac, labyrinth, nephridial canal. J. Morphol. 142:241–263.

Powell, P. R. 1974. The functional morphology of the foreguts of the thalassinid crustaceans, *Callianassa californiensis* and *Upogebia pugettensis.* Univ. Calif. Publ. Zool. 102:1–41.

Prosser, C. L. 1973. Comparative Animal Physiology. W. B. Saunders, Philadelphia. 924 pp.

Salmon. M. 1967. Coastal distribution, display and sound production by Florida fiddler crabs (genus *Uca*). Anim. Behav. 15:449–459.

Taylor, H. H., and Taylor, E. W. 1992. Gills and lungs: The exchange of gasses and ions. In Harrison, F. W., and Humes, A. G. (Eds.): Microscopic Anatomy of Invertebrates. Vol. 10. Decapod Crustacea. Wiley-Liss, New York. pp. 203–293.

Warner, G. F. 1977. The Biology of Crabs. Van Nostrand Reinhold, New York. 202 pp.

Wicksten, M. K. 1980. Decorator crabs. Sci. Am. 242:146–154.

Williams, A. B. 1965. Marine decapod crustaceans of the Carolinas. U.S. Fish. Bull. 65:1–298.

Williams, A. B. 1984. Shrimps, Lobsters, and Crabs of the Atlantic Coast of the Eastern United States, Maine to Florida. Smithsonian Institution Press, Washington, DC. 550 pp.

Wolcott, T. G. 1984. Uptake of interstitial water from soil: mechanisms and ecological significance in the ghost crab *Ocypode quadrata* and two gecarcinid land crabs. Physiol. Zool. 57:161–184.

Internet Sites

http://crayfish.byu.edu/crayhome.htm (Crayfish Home Page, by K. A. Crandall and J. W. Fetzner Jr.)

www.denison.edu/~stocker/crayfish.html (Crayfish Page, by G. W. Stocker.)

www.kheper.auz.com/gaia/biosphere/arthropods/malacostraca/Decapoda.html (Order Decapoda: Crabs, Lobsters, Shrimps, etc.)

www.tlu.edu/academics/departments/biology/aquatic/Decapoda/decapod.html (Decapods.)

www.biology.ualberta.ca/facilities/multimedia/index.php?Page=252 (Decapod Claw Form and Function, animation by H. Kroening, University of Alberta.)

www.fiddlercrab.info/ (Fiddler Crabs, by M. S. Rosenberg.)

MALACOSTRACA: EUPHAUSIACEA

Ivanov, B. G. 1970. On the biology of the Antarctic krill *Euphausia superba.* Mar. Biol. 7:340.

Mauchline, J. 1980. The biology of mysids and euphausids. Adv. Mar. Biol. 18:3–677.

MALACOSTRACA: MYSIDACEA

Grossnickle, N. E. 1982. Feeding habits of *Mysis relicta:* An overview. Hydrobiologia 93:101–108.

Storch, V. 1989. Scanning and transmission electron microscopic observations on the stomachs of three mysid species (Crustacea). J. Morph. 200:17–27.

Internet Sites

www.geocities.com/CapeCanaveral/Hangar/1167/fauna/mysids.html (Groundwater Mysids.)

http://allserv.rug.ac.be/~tdeprez/Mysidacea/index.html (Mysida Information Site, by T. Deprez.)

MALACOSTRACA: AMPHIPODA

Barnard, J. L. 1969. The families and genera of marine gammaridean Amphipoda. U.S. Nat. Mus. Bull. 271:1–535.

Barnard, J. L., and Barnard, C. M. 1983. Freshwater Amphipoda of the World: I. Evolutionary Patterns. II. Handbook and Bibliography. Hayfield, Alexandria, VA. 830 pp.

Bousfield, E. L. 1973. Shallow-water Gammaridean Amphipoda of New England. Cornell University Press, Ithaca, NY. 344 pp.

Bousfield, E. L. 1978. A revised classification and phylogeny of amphipod crustaceans. Trans. Roy. Soc. Can. (Ser. IV) 16:343–390.

Caine, E. A. 1974. Comparative functional morphology of feeding in three species of caprellids from the northwestern Florida Gulf Coast. J. Exp. Mar. Biol. Ecol. 15:81–96.

Caine, E. A. 1978. Habitat adaptations of North American caprellid Amphipoda. Biol. Bull. 155:288–296.

Fox, R. S., and Bynum, K. H. 1975. The amphipod crustaceans of North Carolina estuarine waters. Chesapeake Sci. 16:223–237.

Holsinger, J. R. 1972. The Freshwater Amphipod Crustaceans (Gammaridae) of North America. Biota of Freshwater Ecosystems, Identification Manual 5. US Environmental Protection Agency, Washington, DC. 89 pp.

Laubitz, D. R. 1970. Studies on the Caprellidea of the American North Pacific. Nat. Mus. Can. Publ. Biol. Oceanogr. 1:1–89.

Laval, P. 1980. Hyperiid amphipods as crustacean parasitoids associated with gelatinous zooplankton. Oceanogr. Mar. Biol. Ann. Rev. 18:11–56.

McCain, J. C. 1968. The Caprellidae (Crustacea: Amphipoda) of the western North Atlantic. Bull. U.S. Nat. Mus. 278:1–147.

McCloskey, L. R. 1970. A new species of *Dulichia* commensal with a sea urchin. Pacific Sci. 24:90–98.

Schmitz, E. H. 1992. Amphipoda. In Harrison, F. W., and Humes, A. G. (Eds.): Microscopic Anatomy of Invertebrates. Vol. 9. Crustacea. Wiley-Liss, New York. pp. 443–528.

Stock, J. H. 1976. A new member of the crustacean suborder: Ingolfiellidea from Bonaire, with a review of the entire suborder. Stud. Fauna Curacao 50:56–75.

Watling, L. 1988. Amphipoda. In Higgins, R. P., and Theil, H. (Eds.): Introduction to the Study of Meiofauna. Smithsonian Institution Press, Washington, DC. pp. 409–412.

Internet Sites

www.odu.edu/sci/biology/amphome/ (The Amphipod Homepage, with links.)

www.kbinirsnb.be/general/sections/amphi/carciliens.htm (Amphipod links.)

www.imv.uit.no/ommuseet/enheter/zoo/wim/amp_main.html (Database on Amphipod Literature.)

http://web.odu.edu/sci/biology/amphome/anmain.htm (Amphipod Newsletter.)

www.imv.uit.no/ommuseet/enheter/zoo/wim/amp_main.html (Amphipod Page, Tromso Museum, Wim Vader. Includes a Database on Amphipod Literature and the Amphipod Newsletter.)

www.krapp.org/flohkrapp/amphi.html (Amphipod Resources, by T. Krapp.)

www.bio.uva.nl/onderzoek/cepa//IngolfiellideorumCatalogus.html (World Catalogue and Bibliography of the Ingolfiellidea, by F. R. Schram and R. Vonk.)

www.odu.edu/sci/biology/amphipod/index.html (Subterranean Amphipod Database, by J. R. Holsinger, Old Dominion University.)

MALACOSTRACA: CUMACEA

Alfonso, M. I., Bandera, M. E., López-González, P. J., and García-Gómez, J. C. 1998. The cumacean community associated with a seaweed as a bioindicator of environmental conditions in the Algeciras Bay (Strait of Gibraltar). Cahiers Biol. Mar. 39:197–205.

Valentin, C., and Anger, K. 1977. *In situ* studies on the life cycle of *Diastylis rathkei.* Mar. Biol. 39:71–76.

Watling, L. 1979. Crustacea: Cumacea. Marine Flora and Fauna of the Northeastern United States. NOAA Tech. Report NMFS Circular 423. U.S. Government Printing Office, Washington, DC. 22 pp.

Internet Sites

www.ims.usm.edu/cumacean/index.html (Cumacean Newsletter.)

http://nature.umesci.maine.edu/Cumacea/key.html (Key to Families of Cumacea, by P. Haye.)

http://nature.umesci.maine.edu/cumacea.html (Cumacean Page, by I. Kornfield and L. Watling.)

www.rrz.uni-hamburg.de/biologie/zim/niedere2/Uteengl.htm (Taxonomy, Phylogeny, and Zoogeography of Cumacea and Mysidacea, by U. Mühlenhardt-Siegel.)

MALACOSTRACA: TANAIDACEA

Bird, G. 1999. A new species of *Pseudotanais* (Crustacea, Tanaidacea) from cold seeps in the deep Caribbean, collected by the French submersible *Nautile.* Zoosystema 21:445–451.

Johnson, S. B., and Attramadal, Y. G. 1982a. Reproductive behaviour and larval development of *Tanais cavolinii.* Mar. Biol. 71:11–16.

Johnson, S. B., and Attramadal, Y. G. 1982b. A functional-morphological model of *Tanais cavolinii* adapted to a tubicolous life-strategy. Sarsia 67:29–42.

Larsen, K. 2001. Morphological and molecular investigations of polymorphism and cryptic species in tanaid crustaceans: Implications for tanaid systematics and biodiversity estimates. Zool. J. Linn. Soc. 131:353–379.

Mendoza, J. A. 1982. Some aspects of the autecology of *Leptochelia dubia.* Crustaceana 43:225–240.

Internet Site

http://tidepool.st.usm.edu/tanaids (Tanaidacea Home Page, by R. W. Heard, G. Anderson, and K. Larsen, University of Southern Mississippi.)

MALACOSTRACA: ISOPODA

Brusca, R. C., and Wilson, G. D. F. 1991. A phylogenetic analysis of the Isopoda with some classificatory recommendations. Mem. Queensland Mus. 31:143–204.

Schultz, G. A. 1969. How to Know the Marine Isopod Crustaceans. W. C. Brown, Dubuque, IA. 359 pp.

Sutton, S. L., and Holdich, D. M. (Eds.): 1984. The Biology of Terrestrial Isopods. Oxford University Press, Oxford. 518 pp.

Van Name, W. G. 1936. The American land and freshwater isopod crustaceans. Am. Mus. Bull. 71:1–535.

Wägele, J.-W. 1992. Isopoda. In Harrison, F. W., and Humes, A. G. (Eds.): Microscopic Anatomy of Invertebrates. Vol. 9. Crustacea. Wiley-Liss, New York. pp. 529–617.

Williams, W. D. 1972. Freshwater Isopods (Asellidae) of North America. Biota of Freshwater Ecosystems, Identification Manual 7. US Environmental Protection Agency, Washington, DC. 45 pp.

Yonge, C. M. 1949. The Seashore. Collins, London. 311 pp.

Internet Sites

www.mov.vic.gov.au/crust/isopogal.html (Isopods of Southern Australia, Museum Victoria.)

http://mavicka.ru/directory/rus/17802.html (MavicaNet Isopod Page with links.)

www.vims.edu/~jeff/isopod.htm (Epicaridea: The Parasitic Isopods of Crustacea.)

www.nmnh.si.edu/iz/isopod (World List of Marine, Freshwater and Terrestrial Isopod Crustaceans, Smithsonian National Museum of Natural History.)

OTHER MALACOSTRACA

Boutin, C. 1998. Thermosbaenacea. In Juberthie, C., and Decu, V. (Eds.): Encyclopedia Biospeleologica II. Société de Biospéologie, Moulis. pp. 887–888.

Bowman, T. E., Garner, S. P., Hessler, R. R., Iliffe, T. M., and Sanders, H. L. 1985. Mictacea, a new order of Crustacea Peracarida. J. Crust. Biol. 5:74–78.

Brooks, H. K. 1962. On the fossil Anaspidacea, with a revision of the classification of the Syncarida. Crustaceana 4:229–242.

Fryer, G. 1964. Studies on the functional morphology and feeding mechanism of *Monodella argentarii.* Trans. Roy. Soc. Edinb. 64:49–90.

Martin, J. W., and Christiansen, J. C. 1995. A morphological comparison of the phyllopodous thoracic limbs of a leptostracan *(Nebalia)* and a spinicaudate chonchostracan *(Leptestheria),* with comments on the use of the term Phyllopoda as a taxonomic category. Can. J. Zool. 73:2283–2291.

Martin, J. W., Vetter, E. W., and Cash-Clark, C. E. 1996. Description, external morphology, and natural history observations of *Nebalia hessleri,* new species (Phyllocarida, Leptostraca), from southern California, with a key to the extant families and genera of the Leptostraca. J. Crust. Biol. 16:347–372.

Noodt, W. 1965. Natürliches System und Biogeographie der Syncarida: Gewässer und Abwässer. 37–38:77–186.

Schminke, H. K. 1981. Adaptation of Bathynellacea to life in the interstitial ("Zoea Theory"). Int. Rev. ges. Hydrobiol. 66:575–637.

Wagner, H. P. 1994. A monographic review of the Thermosbaenacea (Crustacea: Peracarida). Zool. verhand. Leiden 291:1–388.

Internet Sites

http://crustacea.nhm.org/peet/leptostraca/index.html (The Leptostraca: NSF PEET Program, by T. Haney.)

www.tamug.tamu.edu/cavebiology/fauna/thermosbaenaceans/T_unidens.html (Caribbean Thermosbaenaceans: *Tulumella unidens* Bowman & Iliffe, 1988, by G. L. Pesce, University of L'Aquila.)

www.univaq.it/~sc_amb/syncarid.html (Groundwater Fauna of Italy: Syncarids, by G. L. Pesce.)

www.geocities.com/CapeCanaveral/Hangar/1167/thermo.html (Groundwater Biology Home Page: Thermosbaenaceans.)

www.geocities.com/~mediaq/fauna/therm.html (Thermosbaenaceans, by G. L. Pesce.)

MAXILLOPODA: COPEPODA

Boxshall, G. A. 1977. The depth distributions and community organization of the planktonic cyclopods of the Cape Verde Islands region. J. Mar. Biol. Assoc. UK 57: 543–562.

Boxshall, G. A. 1983. A comparative functional analysis of the major maxillopodan groups. In Schram, F. R. (Ed.): Crustacean Issues 1: Crustacean Phylogeny. A. A. Balkema, Rotterdam. pp. 121–143.

Boxshall, G. A. 1992. Copepoda. In Harrison, F. W., and Humes, A. G. (Eds.): Microscopic Anatomy of Invertebrates. Vol. 9. Crustacea. Wiley-Liss, New York. pp. 347–384.

Boxshall, G. A., Ferrari, F. D., and Tiemann, H. 1984. The ancestral copepod: Towards a consensus of opinion at the First International Conference on Copepoda 1981. Crustaceana Suppl. 7:68–84.

Boxshall, G. A., and Jaume, D. 2000. Making waves: The repeated colonization of fresh water by copepod crustaceans. Adv. Ecolog. Res. 31:61–79.

Carter, M. E., and Bradford, J. M. 1972. Postembryonic development of three species of harpacticoid Copepoda. Smithson. Contrib. Zool. 119:1–26.

Corkett, C. J., and McLaren, I. A. 1978. The biology of *Pseudocalanus.* Adv. Mar. Biol. 15:1–231.

Coull, B. C. 1977. Copepoda: Harpacticoida. Marine Flora and Fauna of the Northeastern United States. NOAA Tech. Report NMFS Circular 399. U.S. Government Printing Office, Washington, DC. 48 pp.

Gruner, H.-E. 1993. Klasse Crustacea, Krebse. In Gruner, H.-E. (Ed.): Lehrbuch der Speziellen Zoologie. 4. Aufl. Band I. 4. Tiel: Arthropoda (ohne Insecta). Gustav Fischer Verlag, Jena. pp. 448–1030.

Hicks, G. F., and Coull, B. C. 1983. The ecology of marine meiobenthic harpacticoid copepods. Oceanogr. Mar. Biol. Ann. Rev. 21:67–175.

Huys, R., and Boxshall, G. A. 1991. Copepod Evolution. Ray Society Vol. 159. Ray Society, London. 468 pp.

Roe, H. S. J. 1972. The vertical distributions and diurnal migrations of calanoid copepods collected on the Sond Cruise, 1965, pt. II. J. Mar. Biol. Assoc. UK 52:315–343.

Williamson, C. E., and Reid, J. W. 2001. Copepoda. In Thorp, J. H., and Covich, A. P. (Eds.): Ecology and Classification of North American Freshwater Invertebrates. Academic Press, San Diego. pp. 915–954.

Internet Sites

http://copepods.interfree.it/index.html (Copepods Web Portal, by G. L. Pesce, University of L'Aquila.)

http://copepods.interfree.it/ground.htm (Groundwater Copepods, by G. L. Pesce.)

www.uni-oldenburg.de/zoomorphology/Biology.html (Zoosystematics and Morphology Section: Copepod Biology, Carl von Ossietzky University of Oldenberg.)

www.uni-oldenburg.de/monoculus/ (Monoculus Copepod Newsletter.)

www.meiofauna.org/ (International Association of Meiobenthologists.)

http://jaffeweb.ucsd.edu/pages/celeste/copepods.html (Virtual Copepod Page, by C. Fowler, J. Yen, and J. Jaffe, Scripps Institution of Oceanography.)

http://members.aol.com/blspecies/copepoda.htm (World Bibliography on Subterranean Crustacea, Copepoda, by B. Lebreton and D. Defaye.)

www.nmnh.si.edu/iz/copepod (The World of Copepods, Smithsonian National Museum of Natural History.)

MAXILLOPODA: MYSTACOCARIDA

Lombardi, J., and Ruppert, E. E. 1982. Functional morphology of locomotion in *Derocheilocaris typica.* Zoomorphology 100:1–10.

Pennak, R. W., and Zinn, D. J. 1943. Mystacocarida, a new order of Crustacea from intertidal beaches in Massachusetts and Connecticut. Smithson. Misc. Collect. 103:1–11.

MAXILLOPODA: TANTULOCARIDA

Boxshall, G. A. 1991. A review on the biology and phylogenetic relationships of the Tantulocarida, a subclass of Crustacea recognized in 1983. Verh. Dtsch. Zool. Ges. 84:271–279.

Boxshall, G. A., and Lincoln, R. J. 1983. Tantulocarida, a new class of Crustacea ectoparasitic on other crustaceans. J. Crust. Biol. 3:1–16.

Boxshall, G. A., and Lincoln, R. J. 1987. The life cycle of the Tantulocarida (Crustacea). Phil. Trans. Roy. Soc. Lond. B. Biol. Sci. 315:267–303.

Huys, R. 1991. Tantulocarida (Crustacea: Maxillopoda): A new taxon from the temporary meiobenthos. Mar. Ecol. 12:1–34.

Huys, R., Boxshall, G. A., and Lincoln, R. J. 1993. The tantulocaridan life cycle: The circle closed? J. Crust. Biol. 13:432–442.

MAXILLOPODA: ASCOTHORACIDA

Vagin, V. L. 1946. *Ascothorax ophioctenis* and the position of Ascothoracida in the system of the Entomostraca. Acta Zool. 27:155–267.

MAXILLOPODA: CIRRIPEDIA

Anderson, D. T. 1994. Barnacles: Structure, function, development, and evolution. Chapman and Hall, London. 357 pp.

Crisp, D. J., and Bourget. E. 1985. Growth in Barnacles. Adv. Mar. Biol. 22:199–244.

Darwin, C. 1851–1854. A Monograph on the Subclass Cirripedia. Ray Society, London. (Two volumes.)

Glenner, H. 2001. Cypris metamorphosis, injection and earliest internal development of the rhizocephalan *Loxothylacus panopaei* (Gissler). Crustacea: Cirripedia: Rhizocephala: Sacculinidae. J. Morphol. 249:43–75.

Høeg, J. T. 1992. Rhizocephala. In Harrison, F. W., and Humes, A. G. (Eds.): Microscopic Anatomy of Invertebrates. Vol. 9. Crustacea. Wiley-Liss, New York. pp. 313–345.

Høeg, J. T., and Lutzen, J. 1995. Life cycle and reproduction in the Cirripedia Rhizocephala. Oceanogr. Mar. Biol. Ann. Rev. 33:427–485.

Newman, W. A. 1989. Juvenile ontogeny and metamorphosis in the most primitive living sessile barnacle, *Neoverruca,* from abyssal hydrothermal springs. Bull. Mar. Sci. 45:467–477.

Newman, W. A., and Ross, A. 1976. Revision of the balanomorph barnacles; including a catalog of the species. San Diego Soc. Nat. Hist. Mem. 9:1–108.

Perez-Losada, M., Høeg, J. T., Kolbasov, G. A., and Crandall, K. A. 2002. Reanalysis of the relationships among the Cirripedia and the Ascothoracida and the phylogenetic position of the Facetotecta (Maxillopoda: Thecostraca) using 18S rDNA sequences. J. Crust. Biol. 22:661–669.

Ritchie, L. E., and Høeg, J. T. 1981. The life history of *Lernaeodiscus porcellanae* and coevolution with its porcellanid host. J. Crust. Biol. 1:334–347.

Stubbings, H. G. 1975. *Balanus balanoides.* LMBC Memoirs on Typical British Marine Plants and Animals, no. 37. Liverpool University Press, Liverpool. 174 pp.

Tomlinson, J. T. 1969. The burrowing barnacles (Cirripedia: Order Acrothoracica). Bull. U.S. Nat. Mus. 269:1–162.

Walker, G. 1992. Cirripedia. In Harrison, F. W., and Humes, A. G. (Eds.): Microscopic Anatomy of Invertebrates. Vol. 9. Crustacea. Wiley-Liss, New York. pp. 249–311.

Walley, L. J. 1969. Studies on the larval structure and metamorphosis of the cypris larva of *Balanus balanoides.* Phil. Trans. Roy. Soc. Lond. B 256:237–280.

Wells, H. W., and Tomlinson, J. T. 1966. A new burrowing barnacle from the Western Atlantic. Q. J. Florida Acad. Sci. 29:27–37.

Internet Sites

www.vattenkikaren.gu.se/fakta/arter/crustace/cirriped/cirrnaup/cirrnae.html (Aquascope: Cirripedia Nauplius Larvae, Tjärnö Marine Biological Laboratory.)

www.museum.vic.gov.au/crust/barngall.html (Barnacles of Southern Australia. With a photo gallery.)

www.barnacle.com (Barnacle Dive Center: About Barnacles: Moacir Apolinário: Brief History.)

www2.adm.ku.dk/kutest/puf_www3.forf_pub_personid?p_person_id=4767 (Bibliography of J. T. Hoeg, University of Copenhagen.)

www2.adm.ku.dk/kutest/puf_www3.forf_pub_personid?p_person_id=8035 (Bibliography of J. G. Lytzen, University of Copenhagen.)

MAXILLOPODA: OSTRACODA

Cohen, A. C. 1982. Ostracoda. In Parker, P. (Ed.): Synopsis and Classification of Living Organisms. Vol. 2. McGraw-Hill, New York. pp. 191–192.

Delorme, L. D. 2001. Ostracoda. In Thorp, J. H., and Covich, A. P. (Eds.): Ecology and Classification of North American Freshwater Invertebrates. Academic Press, San Diego. pp. 811–848.

Maddocks, R. F. 1992. Ostracoda. In Harrison, F. W., and Humes, A. G. (Eds.): Microscopic Anatomy of Invertebrates. Vol. 9. Crustacea. Wiley-Liss, New York. pp. 415–441.

Internet Sites

www.uh.edu/~rmaddock/IRGO/irgohome.html (International Research Group on Ostracoda, by R. F. Maddocks, University of Houston.)

www.gre.ac.uk/schools/nri/earth/ostracod/introduction.html (Introduction to Ostracoda, by the Ostracod Research Group.)

www.ru.ac.za/academic/departments/zooento/Martin/aostracoda.html (The Non-Marine Ostracods (Ostracoda) of Southern Africa, by K. Martens, Royal Belgian Institute of Natural Sciences.)

MAXILLOPODA: BRANCHIURA

Cressey, R. F. 1972. The genus *Argulus* (Crustacea: Branchiura) of the United States. Biota of Freshwater Ecosystems, Identification Manual 2. U.S. Environmental Protection Agency, Washington, DC. 14 pp.

Cressey, R. F. 1978. Crustacea: Branchiura. Marine Flora and Fauna of the Northeastern United States. NOAA Tech. Report NMFS Circular 414. U.S. Government Printing Office, Washington, DC. 10 pp.

Mikheev, V. N., Mikheev, A. V., Pasternak, A. F., and Valtonen, E. T. 2000. Light-mediated host searching strategies in a fish ectoparasite, *Argulus foliaceus* L. (Crustacea: Branchiura). Parasitology 120:409–416.

Overstreet, R. M., Dyková, I., and Hawkins, W. E. 1992. Branchiura. In Harrison, F. W., and Humes, A. G. (Eds.): Microscopic Anatomy of Invertebrates. Vol. 9. Crustacea. Wiley-Liss, New York. pp. 385–413.

Internet Sites

www.tamut.edu/~allard/Biology/photos/argulus.htm (*Argulus* page, Texas A & M University.)

www.fishdoc.co.uk/disease/argulus.htm (FishDoc: *Argulus*—the Fish Louse.)

www.science.siu.edu/zoology/gradstudents/poly/argulus.html (*Argulus*, by W. J. Poly, Southern Illinois University at Carbondale.)

www.nmnh.si.edu/iz/copepod (The World of Copepods: Smithsonian National Museum of Natural History.)

http://www-biol.paisley.ac.uk/home.htm (Biomedia Zoology Museum: Crustacea: Class Branchiura.)

MAXILLOPODA: PENTASTOMIDA

Able, L. G., Kim, W., and Felgenhauer, B. E. 1989. Molecular evidence for inclusion of the phylum Pentastomida in the Crustacea. Mol. Biol. Evol. 6:685–691.

Barnes, R. D., and Harrison, F. W. 1993. Introduction to the Onychophora, Chilopoda, and lesser Protostomata. In Harrison, F. W., and Rice, M. E. (Eds.): Microscopic Anatomy of Invertebrates. Vol. 12. Onychophora, Chilopoda, and Lesser Protostomata. Wiley-Liss, New York. pp. 1–9.

Riley, J. 1983. Recent advances in our understanding of pentastomid reproductive biology. Parasitology 86:59–83.

Riley, J. 1986. The biology of pentastomids. Adv. Parasitol. 25:45–128.

Storch, V. 1993. Pentastomida. In Harrison, F. W., and Rice, M. E. (Eds.): Microscopic Anatomy of Invertebrates. Vol. 12. Onychophora, Chilopoda, and Lesser Protostomata. Wiley-Liss, New York. pp. 115–142.

Internet Site

http://www.geocities.com/parasitepics/ (Pentastomida: ParasitePics.)

Myriapoda^{SC}

TRACHEATA^{iP} AND MYRIAPODA^{SC}
Chilopoda^{C}

PHYLOGENETIC HIERARCHY OF CHILOPODA

Symphyla^{C}
Diplopoda^{C}

PHYLOGENETIC HIERARCHY OF DIPLOPODA

Pauropoda^{C}

PHYLOGENY OF TRACHEATA

PHYLOGENETIC HIERARCHY OF TRACHEATA

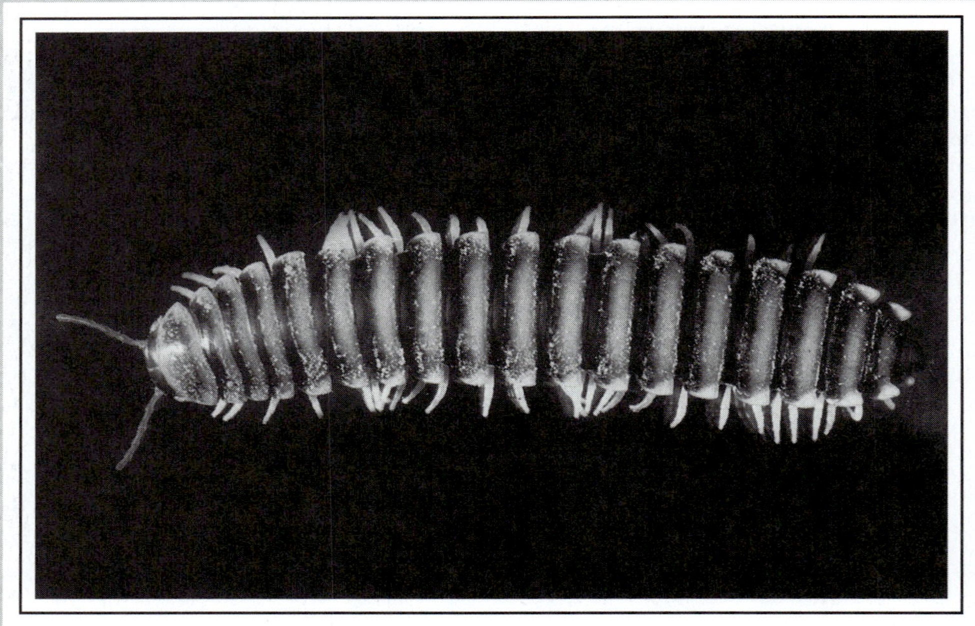

TRACHEATA[iP] AND MYRIAPODA[SC]

Hexapods (insects), centipedes, and millipedes, along with their less well-known relatives the pauropods and symphylans, form a taxon known as Tracheata, which is one of the two great taxa of terrestrial arthropods (the other being Arachnida). The ancestral tracheate was a terrestrial mandibulate with tracheae and a body consisting of a series of similar, unspecialized segments like those of some Recent centipedes.

In the tracheate ground plan, the head bears a single pair of antennae on its first segment (Fig. 20-1). Uniquely, the second head segment (intercalary segment) is vestigial and in the adult bears no appendages, although they appear briefly during development. Thus, the second pair of antennae, characteristic of crustaceans, is absent (Fig. 16-3). Gas exchange is via tracheae, and a pair of spiracles serves each segment. Excretion is accomplished by Malpighian tubules in the trunk and paired saccate nephridia (coxal glands) in head segments 4 and 5, which are the segments of the maxillae/labium. **Tömösváry organs** (TIM-ish-vary), or their homologs, on the head are probably hygroreceptive (humidity) or chemoreceptive. Sexes are separate, fertilization is internal, and sperm transfer is indirect with spermatophores. Malpighian tubules, tracheae, and internal fertilization with indirect sperm transfer and spermatophores evolved independently in tracheates and arachnids as adaptations for life on land.

Four tracheate taxa (traditionally considered to be classes), the centipedes, millipedes, pauropods, and symphylans, form a taxon of about 13,000 species of related animals known as Myriapoda. These tracheates have a body composed of a head and an elongate trunk with many leg-bearing segments, but the trunk is not divided into a thorax and abdomen. On the head is one pair of antennae. Median ocelli are absent in myriapods, but simple lateral eyes, perhaps derived from compound eyes, are present in centipedes and millipedes. These consist of a few loosely scattered ommatidia, but there is no crystalline cone. Symphylans and pauropods have no eyes.

Anteriorly on the head is the labrum (upper lip), forming the roof of the preoral cavity (Fig. 20-2A,C, 20-3). The mouthparts lie on the ventral side of the head. The lower lip, which closes the preoral cavity posteriorly, is formed by the first or second maxillae, depending on the taxon. Two mandibles flank the mouth and a fleshy hypopharynx lies posterior to the mouth in the preoral cavity (Fig. 20-3). The hypopharynx is a median, unpaired fold of the body wall, not an appendage.

Gas exchange is accomplished by a tracheal system in which the spiracles usually cannot be closed, thus providing an ever-present avenue for the unavoidable loss of water. Myriapods are not as well adapted to terrestrial environments as insects are. Excretion takes place through Malpighian tubules, but saccate nephridia (coxal glands) of uncertain function are also present. The heart is a dorsal tube extending the length of the trunk and having a pair of ostia in each segment, but branching arteries are rarely present. The typically arthropodan nervous system is not strongly cephalized. It includes a tripartite syncerebrum, circumenteric connectives, and a small subesophageal ganglion. The ventral nerve cord contains a pair of ganglia in each segment. Indirect sperm transfer by spermatophores is highly developed, and the myriapods parallel the arachnids in many aspects of this process.

Most myriapods require a humid environment because their relatively permeable epicuticle is not waterproof and usually lacks the lipid and wax characteristic of spiders and insects. What lipids are present may be more important in repelling water from the outside (functioning as a hydrofuge) than in reducing water loss from the inside. The widely distributed myriapods live beneath stones and wood and in soil and humus in both temperate and tropical regions.

CHILOPODA[C]

Although the name promises 100 legs, chilopods, or centipedes (centi = 100, ped = foot) have between 15 and 191 pairs of legs but, since the number of pairs is always odd, none has exactly 100. Centipedes are, along with millipedes, the most familiar myriapods. They are distributed throughout the world in both temperate and tropical regions from sea level to high elevations in soil and humus; beneath stones, bark, and logs; and in caves and mosses. Most are nocturnal, some are intertidal. Centipedes are poison-fanged raptors, and some large species will bite humans. The 2800 described species are assigned to five taxa.

Diversity of Chilopoda

Scutigeromorpha[O]: Long legs, long antennae, well-developed pseudofaceted eyes (Fig. 20-1C). Some live in and around buildings. *Scutigera coleoptrata,* the cosmopolitan house centipede, is sometimes found trapped in an (empty) bathtub or washbasin or seen scurrying across the floor. Have 15 pairs of legs; extremely fast, fast enough to catch flies. Spiracles are dorsal, on the tergites (unique). Open-ground hunters rely on speed to capture prey. Earliest known centipede fossils are scutigeromorphs from the late Silurian.

Lithobiomorpha[O] (stone centipedes; Fig. 20-1D): Robust, flat. Inhabit enclosed spaces among stones, in soil, under bark. *Lithobius.*

Geophilomorpha[O] (earth centipedes): Long, threadlike, adapted for living in soil (Fig. 20-1B). Have 27 to 191 pairs of legs, the most of any centipede. Blind. Unlike other centipedes, all tergites are the same length rather than alternately long and short. *Dicellophilus, Stigmatogaster.*

Scolopendromorpha[O] (Fig. 20-1A): Large, heavy-bodied, flattened. In crevices, beneath stones, under bark and logs, in soil. *Otocryptops, Scolopendra.*

Craterostigmomorpha[O]: One described species, *Craterostigmus tasmanianus,* from Tasmania, and an undescribed species from New Zealand. Have 21 tergites and 15 pairs of legs.

Form

Scolopendra gigantea, a tropical American scolopendromorph that reaches 30 cm in length, is the largest known centipede. Many other tropical species, usually scolopendromorphs, exceed 20 cm, but most North American and European centipedes are 3 to 6 cm long and the smallest are about 1 cm. Temperate-zone centipedes typically are reddish brown, but many tropical species, especially the scolopendromorphs, are red, green, yellow, blue, or a combination of colors.

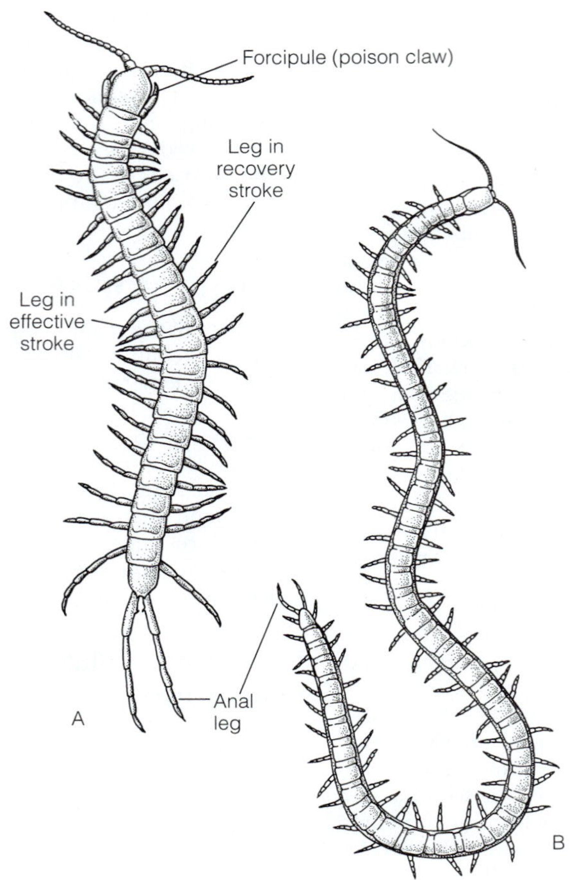

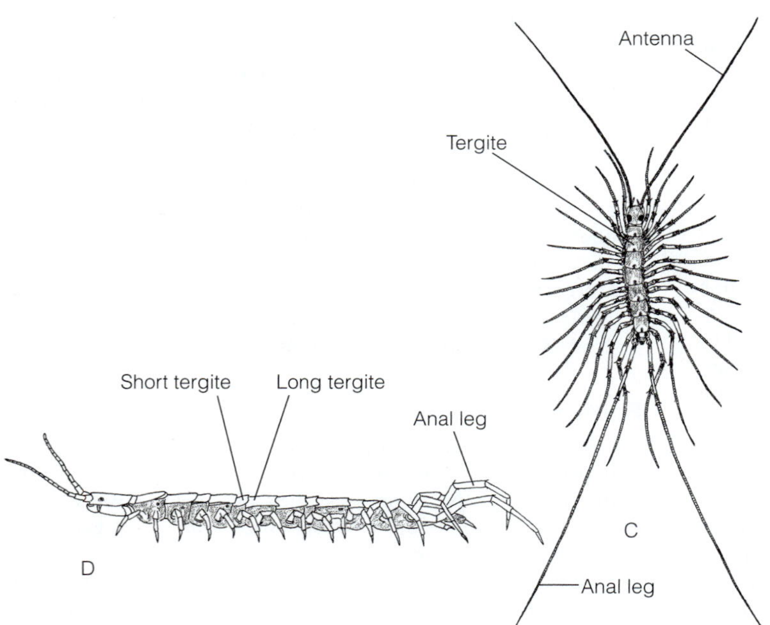

FIGURE 20-1 Chilopoda. Representatives of the four common centipede taxa. **A,** *Otocryptops sexspinnosa,* a scolopendromorph. **B,** An unidentified geophilomorph. **C,** *Scutigera coleoptrata,* the common house centipede, a scutigeromorph. The small black spot at the posterior margin of each tergite is a spiracle. **D,** *Lithobius,* a lithobiomorph. *(All after Snodgrass, R. E. 1952. A Textbook of Arthropod Anatomy. Cornell University Press, Ithaca, New York)*

The head is covered by a rigid, sclerotized **cephalic shield** (Fig. 20-2A) and tends to be flattened in most centipedes, although the scutigeromorph head is nearly spherical. Antennae may be long or short. Centipedes are **trignathic** (= three "jaws," or mouthparts), having mandibles, first maxillae, and second maxillae (Fig. 20-3). The second maxillae are larger than the first and bear clawlike **telopodites** (the movable portion of an appendage). The mouth opens from a preoral cavity bounded by the anterior labrum, the lateral mandibles, the posterior hypopharynx, and first maxillae (Fig. 20-3). Centipedes usually have lateral clusters of loosely arranged ommatidia, but many (such as geophilomorphs) are blind, and scutigeromorphs have ommatidia closely packed to form pseudofaceted eyes.

The appendages of the first trunk segment are enlarged **forcipules** (maxillipeds, poison fangs; Fig. 20-2A,C), unique to centipedes. Each is curved toward the midventral line and bears a terminal, pointed fang that connects via a duct with a poison gland usually located in the telopodite. The combined coxa and sternite of the forcipule form a large, platelike **coxosternite** that covers the underside of the head (Fig. 20-2C, 20-3).

The **anal legs,** on the last trunk segment, are variously modified and, unlike other legs, are not locomotory. Anal legs may be sensory, defensive, or aggressive, with antenniform (Fig. 20-1C) or pincerlike (Fig. 20-1A) morphology. The last two trunk segments are the pregenital and genital segments, the former of which may either bear a pair of gonopods or be limbless. In female lithobiomorphs and scutigeromorphs, the gonopods are used to manipulate the eggs after laying them. The genital segment bears the telson.

Between the anterior forcipules and the posterior anal legs are a variable number of similar uniramous legs, usually one, but sometimes none, per segment (Fig. 20-1). Except for the forcipules, anal legs, and gonopods, all of the trunk appendages are walking legs (Fig. 20-2D) consisting of a coxa, trochanter, prefemur, femur, tibia, two tarsal articles, and a claw. The coxa is incorporated into the pleural membrane on the side of the segment (Fig. 20-2E).

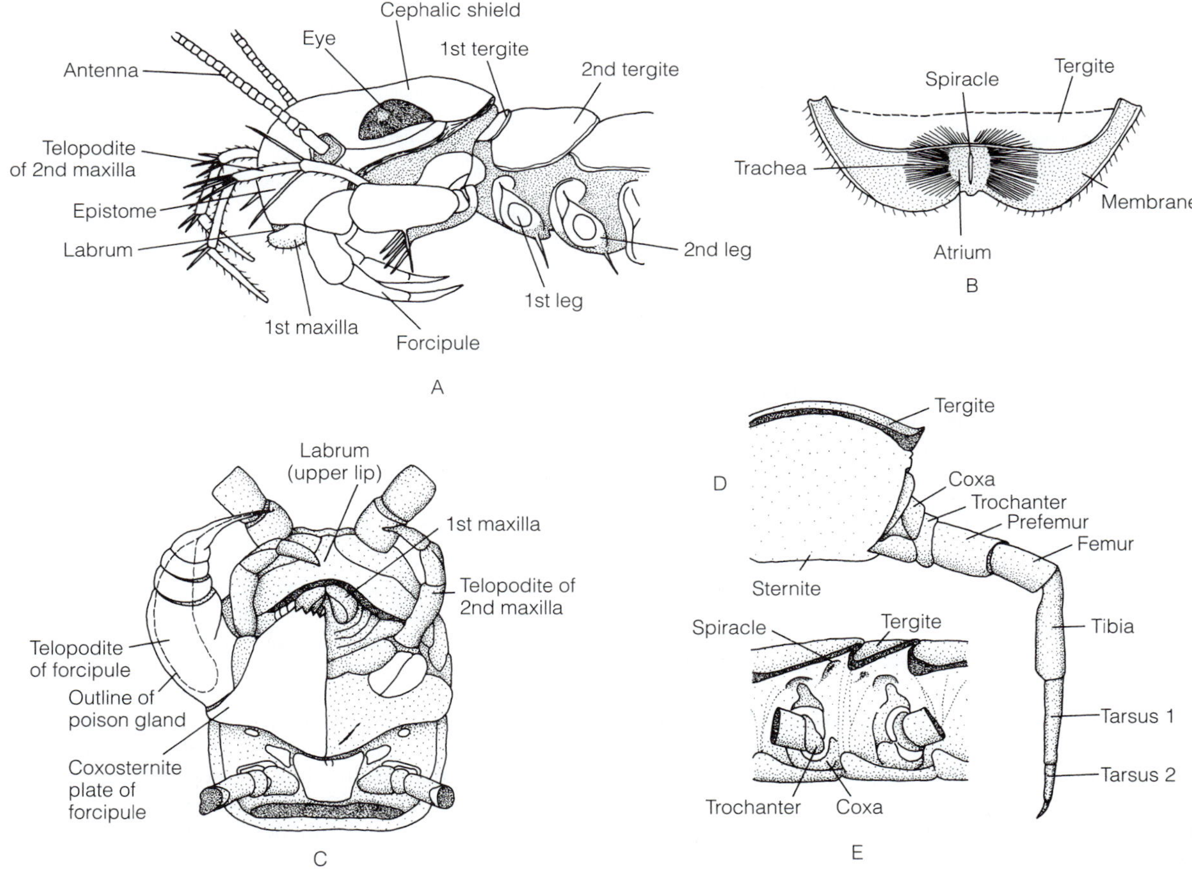

FIGURE 20-2 Chilopoda. **A,** The head of *Scutigera coleoptrata* (lateral view). **B,** A pair of *Scutigera* tracheal lungs seen from below between an inflected tergal membrane and the tergite itself. **C,** Ventral view of head of *Lithobius forficatus.* Only one forcipule (poison claw) is shown. **D,** Cross section of *L. forficatus.* **E,** Lateral view of two segments of *L. forficatus.* Only the basal leg segments are shown. *(A and B, After Snodgrass, 1952; C, After Rilling, 1968; D and E After Manton, 1965; E, From Lewis, J. G. E. 1981. The Biology of Centipedes. Cambridge University Press, pp. 9, 11. Copyrighted and reprinted by permission of the publisher.)*

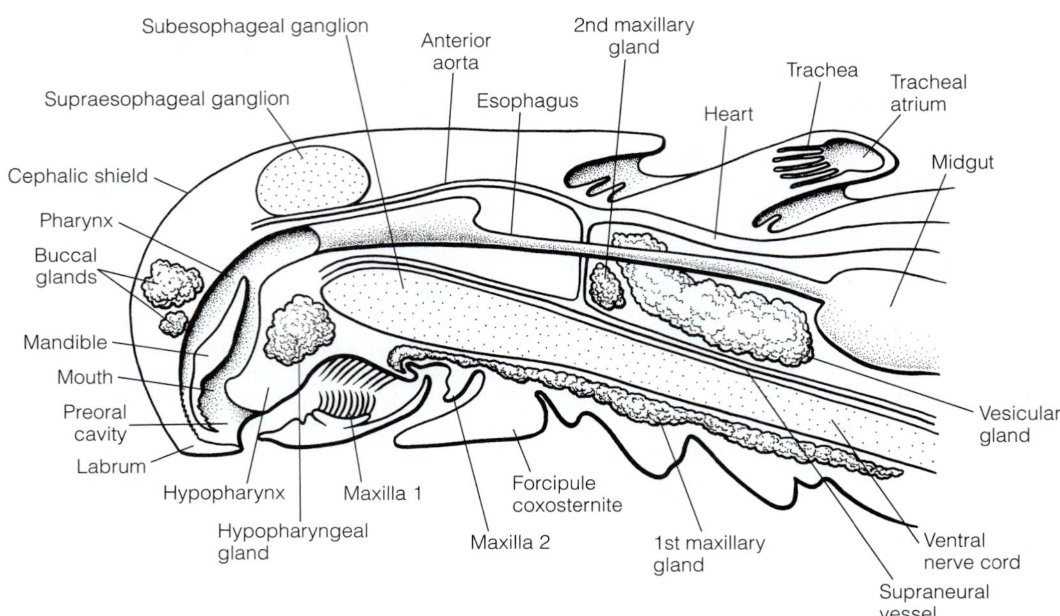

FIGURE 20-3 Chilopoda. Sagittal section through the head and anterior trunk of the centipede *Scutigera*. *(Adapted from Manton, S. M. 1965. The evolution of arthropod locomotory mechanisms. Pt. 8. Functional requirements and body design in Chilopoda, together with a comparative account of their skeletomuscular systems and an appendix on the comparison between burrowing forces of annelids and chilopods and its bearing upon the evolution of the arthropodan haemocoel. J. Linn. Soc. (Zool.) 46:251–483.)*

Each trunk segment is enclosed by a dorsal tergite, ventral sternite, and two lateral pleurites (Fig. 20-2D,E). The considerable variation in the size and number of the tergites is correlated with locomotor habits. Geophilomorphs resemble annelids in the homonomy of their trunk segments and appendages (Fig. 20-1B), and there is little variation, except for the forcipules and anal legs, from segment to segment. Unlike those of other centipedes, successive tergites are similar in size. In contrast, the tergites of scutigeromorphs, lithobiomorphs, and craterostigmomorphs are heteronomous. Alternating tergites vary markedly in size so that long tergites alternate with short (Fig. 20-1D). In scutigeromorphs, the short tergites are fused indistinguishably with the long tergites. (Fig. 20-1C). Scolopendromorph tergites are only weakly heteronomous.

Defense

Few predators specialize in centipedes, but they are the occasional prey of some birds, moles, rodents, snakes, ants, slugs, and other centipedes. Centipede hiding places provide protection from predators and desiccation. Most centipedes, other than scutigeromorphs, are nocturnal and emerge at night to hunt for food or new living quarters. Scolopendromorphs construct a burrow system in soil or beneath stones and logs, with a chamber into which the animal retreats.

In addition to the poison forcipules, centipedes have other adaptations for defense. Some scolopendromorphs pinch with their anal legs. Many species autotomize legs to escape predators. Anal legs are especially likely to be cast off. Autotomized appendages often continue to twitch, presumably to distract a predator while the centipede quietly escapes. Lost legs are replaced by the next molt. Many scolopendromorphs and geophilomorphs have repugnatorial glands on the ventral side of each segment. Most centipedes have coxal organs (not to be confused with coxal glands, which are saccate nephridia) in the posterior coxae; their function is unknown, but may be defensive. Some lithobiomorphs produce sticky secretions from glands in the anal legs. Centipedes usually are drab colors that blend with the environment.

Locomotion

Two taxa are adapted for running. The scolopendromorphs (Fig. 20-1A) have long legs, all of approximately the same length, and correspondingly long strides. The scutigeromorphs (Fig. 20-1C) can run three times faster, however, and many of their structural peculiarities are associated with the evolution of a rapid gait. *Scutigera* is capable of speeds of at least 40 cm/s. In scutigeromorphs, the leg's effective stroke is faster than the recovery stroke, unlike in the scolopendromorphs, whose strokes are equal in duration. Moreover, scutigeromorphs have a markedly progressive increase in leg length from anterior to posterior, which enables the posterior legs to step to the outside of the anterior legs, reducing interference. For example, in *Scutigera* the posterior legs are twice as long as the first pair.

To overcome the tendency to undulate laterally, the trunk is strengthened by alternately long and short tergites in the lithobiomorphs and by a reduced number of large, overlapping, tergites in the scutigeromorphs. Finally, the annulated, flagella-like distal leg segments of the scutigeromorphs enable

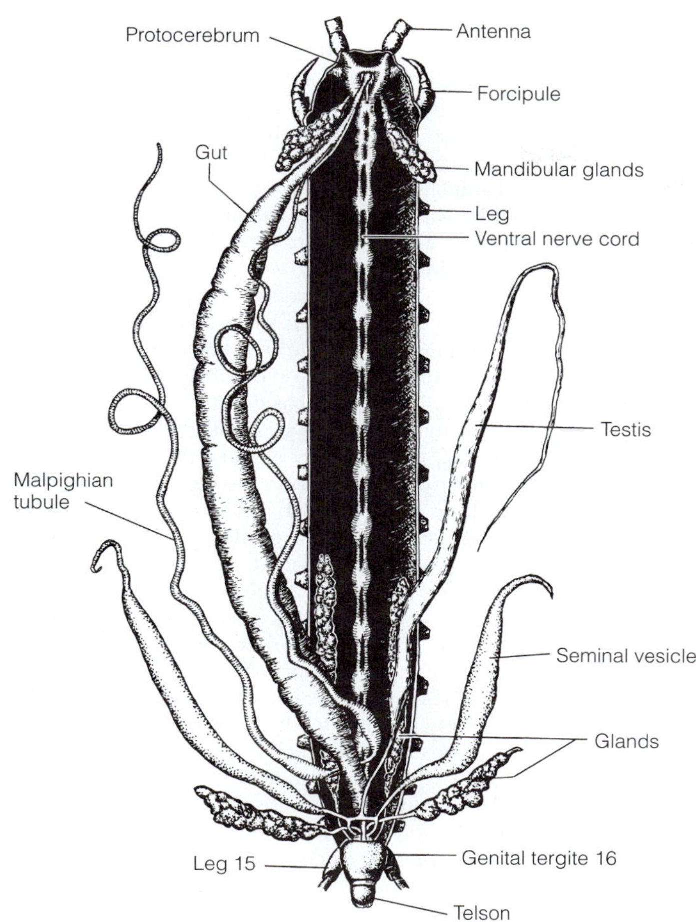

FIGURE 20-4 Chilopoda. Internal structure of the centipede *Lithobius.* *(From Kaestner, A. 1968. Invertebrate Zoology. Vol. II. Wiley-Interscience, New York.)*

the animal to place a long section of the end of the leg against the substratum, very much like a foot, to increase friction and reduce slippage.

In contrast to other centipedes, the wormlike geophilomorphs are adapted for burrowing through loose soil or humus (Fig. 20-1B). The pushing force required is provided not by the legs, as in millipedes, but by extension and contraction of the trunk, as in earthworms. A British species of *Stigmatogaster,* for example, can increase its body length by as much as 68%. Powerfully developed longitudinal muscles of the body wall, an elastic pleural wall, an increased number of segments, and small sclerites between the large sclerites all facilitate great extension and contraction of the trunk in burrowing. The short legs anchor the centipede when it is in its burrow, as the chaetae of earthworms do. Geophilomorphs walk with their legs, but there is little overlapping leg movement.

Internal Form and Function

Almost without exception, centipedes are predaceous. Small arthropods form the major part of their diet, but there are numerous reports of large scolopendromorphs feeding on vertebrates such as frogs, toads, snakes, lizards, birds, and mice. Some centipedes, especially geophilomorphs, feed on earthworms, snails, and nematodes. Prey is detected and located with the antennae, or with the legs and eyes in *Scutigera,* and then captured and killed or stunned with the forcipules. *Scolopendra* attacks glass beads smeared with food extract immediately upon making contact with its antennae, and *Lithobius* will not feed if deprived of its antennae. *Scutigera* uses its long, flexible, multiarticulated tarsi as lassos to capture prey. An individual may capture several flies at once and hold them with the tarsi while it dispatches each, in turn, with its forcipules.

Large tropical centipedes are often feared, but the venom of most species, although painful, is not lethal to healthy adult humans, and usually not to children. In fact, the forcipules of most centipedes are incapable of penetrating human skin. The bite of large species is rarely serious and is generally similar to a severe sting from a yellowjacket or hornet. Reports of fatalities from older literature are difficult to authenticate, but a large species of *Scolopendra* in the Solomon Islands apparently causes human fatalities every year. Little is known of the composition of the venom other than that it contains serotonin and histamine in species of *Scolopendra.*

Following capture, the prey is held by the second maxillae and the forcipules while the mandibles and first maxillae manipulate it prior to ingestion. Some centipedes chew the prey whereas others suck it into the gut using the muscular pumping pharynx. Some geophilomorphs, which have small teeth on their less mobile mandibles, may partially digest their prey in the preoral cavity prior to ingestion.

The gut is a straight tube, with the foregut (pharynx and esophagus) occupying up to two-thirds of the body length, depending on the species (Fig. 20-4). Most of the remainder is midgut and the hindgut is short. Digestive secretions are provided by several buccal, mandibular, maxillary (not the same as maxillary nephridia), and vesicular glands associated with the preoral cavity and pharynx (Fig. 20-3). The midgut secretes a peritrophic membrane.

Gas exchange in centipedes is accomplished by tracheae that, in all except the scutigeromorphs, deliver oxygen directly to the tissues without mediation of the blood. Spiracles, with scutigeromorphs excepted, are segmental, paired, and situated laterally in the membranous pleural region near the coxae (Fig. 20-2E). Typically each segment has one pair of spiracles, but some segments lack them and the pattern of distribution varies in different taxa. Spiracles open into an atrium lined with cuticular hairs (trichomes), which may reduce water loss or prevent the entrance of dust particles.(Fig 20-5) Tracheal tubes arise from the base of the atrium. Depending on the taxon, the tracheal system may be a series of longitudinal trunks, a network of interconnected tubes, or unconnected tubes. The tracheae terminate in fine, blind-ended tracheoles in the tissues. Centipede spiracles are not equipped with muscles and cannot be closed to prevent water loss.

Perhaps because of their more active habits and corresponding higher metabolic rate, the tracheal system of scutigeromorphs differs in several respects from those of other centipedes and may have evolved independently. Similar tracheae are found in some spiders. In contrast to those of other centipedes, scutigeromorph tracheae apparently lack chitinous taenidia (supporting rings), and they supply the blood, not the tissues, with oxygen. Further, the spiracles are unpaired and located mid-dorsally, near the posterior margin of the tergites of the leg-bearing segments (Fig. 20-1C, 20-2B). Each spiracle opens into an atrium, from which extend two large tufts of short tubular tracheae bathed in the blood of the pericardial sinus. Blood is oxygenated as it passes over the tracheae immediately prior to entering the heart.

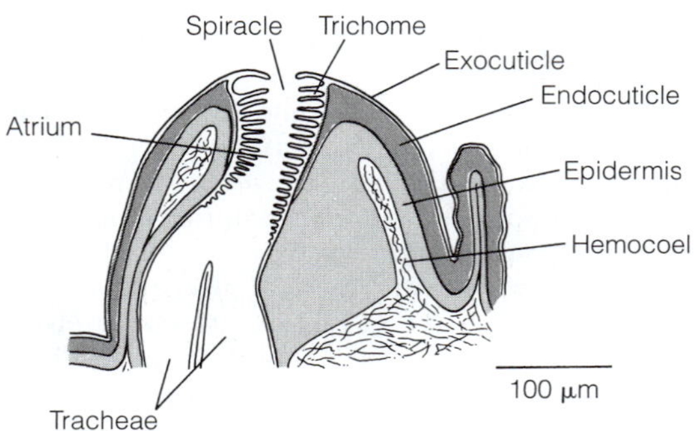

FIGURE 20-5 Chilopoda. Longitudinal section through the spiracle of *Lithobius forficatus*. *(From Curry, A. 1974. The spiracle structure and resistance to dessication of centipedes. In Blower, J. G. (Ed.): Myriapoda. Academic Press, London. p. 368.)*

As terrestrial animals with a permeable integument and spiracles that cannot be closed, water conservation is a primary concern for centipedes. Malpighian tubules (Fig. 20-4), the chief excretory organs, produce uric acid as waste. A pair of maxillary nephridia (not the same as the secretory, and presumably digestive, maxillary glands) is homologous to the saccate nephridia of crustaceans. While they presumably are excretory, this has not been demonstrated, and they are found only in lithobiomorphs and scutigeromorphs. Each gland has two ducts to the exterior, indicating, perhaps, that the glands are what remain of two pairs of saccate nephridia, one pair for each of the two maxillary segments. That the two ducts open on the segments of the first and second maxillae is consistent with this hypothesis.

The body cavity is a hemocoel divided by two horizontal diaphragms into pericardial, perivisceral, and perineural sinuses, as in insects (Fig. 16-7), and filled with blood (hemolymph). The heart is a dorsal ostiate tube extending most of the length of the trunk (Fig. 20-3). From the anterior end of the heart the anterior aorta extends into the head and a pair of vessels extend ventrally to a large, longitudinal supraneural vessel (Fig. 20-3). Paired segmental vessels exit laterally from the heart. In each segment the heart is supported and dilated by a pair of alary muscles. The blood of *Scutigera* contains a respiratory pigment with an absorbance spectrum consistent with hemocyanin. Although common in crustaceans, this is the only known occurrence of this pigment in tracheates.

Nervous System and Sense Organs

The nervous system is similar to that of other myriapods (Fig. 20-3, 20-4). Cephalization is weak and the subesophageal ganglion includes only the neuromeres of the three mouthparts (the mandibles and maxillae 1 and 2). All the ganglia of the leg-bearing trunk segments are spaced along the ventral nerve cord (Fig. 20-4).

All geophilomorphs, some scolopendromorphs, and some lithobiomorphs lack eyes of any sort. Other centipedes have from few to many lateral ommatidia. In most, scutigeromorphs excepted, the ommatidia are few in number and loosely clustered, but never touching, on the side of the head. Median eyes are absent in centipedes. Most centipedes are nocturnal and negatively phototactic.

In scutigeromorphs, which are diurnal hunters, the ommatidia are tightly clustered and organized to form **pseudofaceted eyes** resembling, but probably not homologous with, the compound eyes of insects. The contiguous ommatidia, of which there may be as many as 600, form a compact group on each side of the head. In *Scutigera,* the combined corneal surface is strongly convex (Fig. 20-2A), as it is in the compound eyes of insects and crustaceans. There is no evidence, however, that the compound eyes of *Scutigera* or the eyes of any other centipedes do more than simply detect light, dark, and movement. *Scutigera* makes effective use of its eyes to detect motion and capture fast-moving insects.

A pair of Tömösváry organs is on the head at the base of the antennae in lithobiomorphs, scutigeromorphs, and craterostigmomorphs. Each consists of a sunken cuticular disc surrounded by a groove and a central pore in which sensory dendrites converge. Although long assumed to be olfactory, they have recently been shown to be **hygroreceptors** for detecting humidity. Hygroreception is a valuable ability indeed for a terrestrial animal dependent on moisture to avoid death by dehydration.

Centipedes have various types of sensory setae and other cuticular sensory structures on the antennae, mouthparts, body surface, and other parts of the body. These sensilla have been shown to have chemoreceptive, mechanoreceptive, and thermoreceptive capabilities. The long, antenniform anal legs of many centipedes have a sensory function, especially in lithobiomorphs and scutigeromorphs (Fig. 20-1C).

Reproduction and Development

Females have a single, tubular ovary located above the gut. Its oviduct opens through a median genital atrium and gonopore on the ventral surface of the genital segment. A pair of seminal receptacles also opens into the genital atrium. In the male, from one to many testes are located above the midgut (Fig. 20-4). Testes are connected to a pair of sperm ducts, which open through a median gonopore on the ventral side of the genital segment. In scutigeromorphs, lithobiomorphs, and geophilomorphs, the genital segment of both sexes carries modified appendages called gonopods that function in reproduction.

Sperm transmission is indirect in centipedes, as it is in other myriapods. Except in scutigeromorphs, the male constructs a little web of silk strands (Fig. 20-6A) secreted by a spinneret located on the genital segment. A spermatophore, which may be several millimeters in diameter, is placed on the web. The female picks up the spermatophore with her gonopods and puts it in her gonopore. In both sexes the gonopods aid in handling the spermatophore.

Species-specific courtship behavior occurs in many centipedes. The posterior coxal organs of lithobiomorphs produce a phenolic pheromone attractive to conspecifics. Males and females produce different pheromones, suggesting that attraction is also gender specific. Males usually do not produce a spermatophore until a female is encountered. Sexual partners may palpate each other's posterior ends with their antennae while moving in a circle. This behavior may continue for as long as an hour before the male spins his web and deposits a spermatophore. Following spermatophore deposition, the male "signals" the female in various ways. For example, in species of *Lithobius,* the male straddles the web and spermatophore with his anal legs while stroking the female's antennae with his own (Fig. 20-6B). She responds by crawling across the web and picking up the spermatophore.

Scolopendromorphs and geophilomorphs brood their eggs in clusters of 15 or more. A female excavates in decayed wood or soil a cavity in which she lays her eggs. Inside the cavity she wraps herself around the egg mass and guards the eggs during their development through hatching and dispersal of the young (Fig. 20-6C). Female lithobiomorphs and scutigeromorphs carry the eggs between the gonopods for a short time and then deposit them singly in the soil.

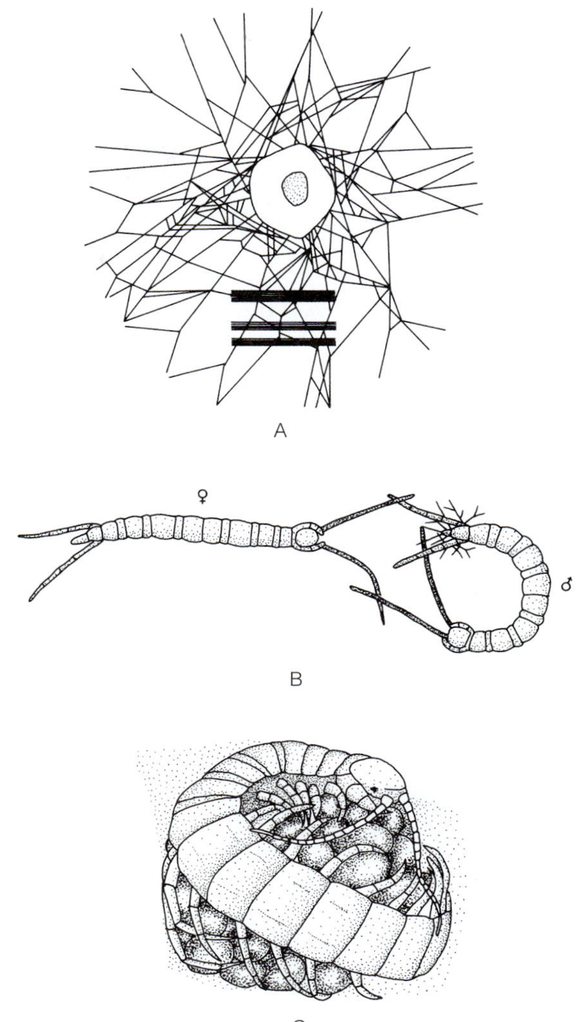

FIGURE 20-6 Chilopoda. **A,** Web and spermatophore of *Lithobius forficatus.* **B,** Male *L. forficatus* with its posterior end over the web, signaling the female to pick up the spermatophore. **C,** A female of *Scolopendra* brooding her eggs. *(A and B, After Klingel, 1960, from Lewis, J. G. E. 1981. The Biology of Centipedes. Cambridge University Press, London. p. 281; C, After Brehm, from Lewis, J. G. E. 1981. The Biology of Centipedes. Cambridge University Press, London. p. 272.)*

In the brooding taxa Scolopendromorpha and Geophilomorpha, development is direct (epimorphic) and the young have the full complement of segments when they hatch. Development in Lithobiomorpha and Scutigeromorpha is indirect (anamorphic), but the hatchlings have only some of the adult complement of segments. For example, newly hatched *Scutigera* have 4 pairs of legs and in the subsequent six molts pass through stages with 5, 7, 9, 11, and 13 pairs until they become adults with 15 pairs. The craterostigmomorph hatching stage has all legs except the last pair. The life span of many centipedes is from four to six years or more.

PHYLOGENETIC HIERARCHY OF CHILOPODA

Chilopoda
- Scutigeromorpha
- Pleurostigmomorpha
  - Epimorpha *sensu lato*
    - Craterostigomorpha
    - Epimorpha *sensu stricto*
      - Scolopendromorpha
      - Geophilomorpha
  - Lithobiomorpha

SYMPHYLA[C]

Symphyla is a small taxon of approximately 160 described species of moist soil and leaf litter found in most parts of the world. Their permeable cuticle restricts them to humid habitats.

Symphylans are small, 1 to 8 mm in length, white, weakly sclerotized, and superficially similar to centipedes in being elongate and multisegmented (Fig. 20-7A). Symphylans also resemble hexapods in several respects, but the similarities may be due to convergence. The antennae are situated laterally on the head (Fig. 20-7B). Like chilopods, symphylans are trignathic, having three pairs of mouthparts. Ventrally the

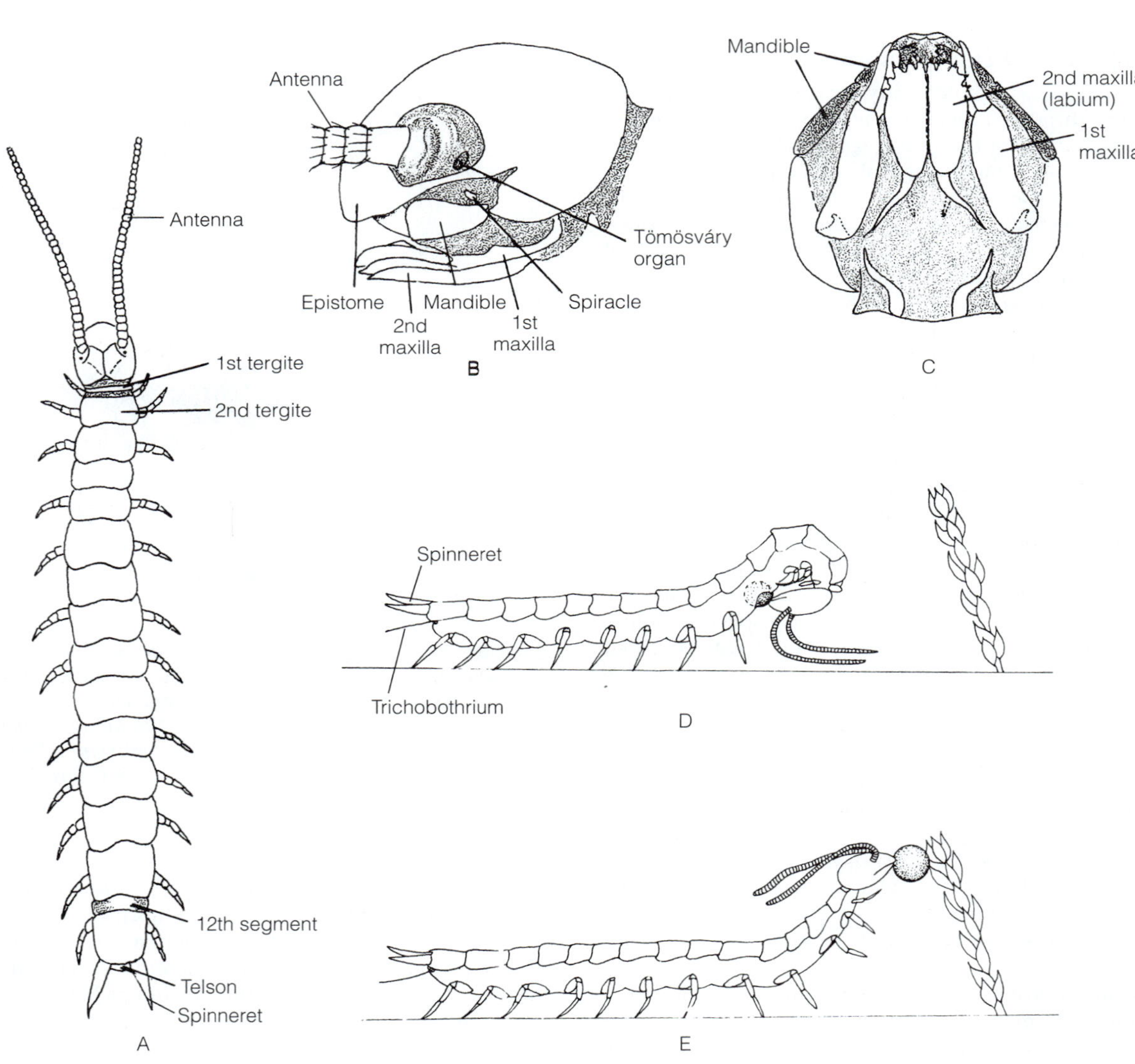

FIGURE 20-7 Symphyla. **A,** *Scutigerella immaculata,* dorsal view. **B,** The head of *Hanseniella,* lateral view. **C,** Ventral view of the head of *S. immaculata.* **D–E,** Female of *Scutigerella* removing an egg from the gonopore with her mouthparts and attaching it to moss. When carried by the mouthparts, the egg is smeared with semen stored in buccal pouches. *(A–C, After Snodgrass, 1952; D and E, After Juberthie-Jupeau.)*

weak, biarticulate mandibles are covered by a pair of long first maxillae. The second maxillae are fused to form a labium (Fig. 20-7C) similar to that of the hexapods, although this may be a convergence. The mouthparts enclose a preoral cavity. Eyes are absent, but a well-developed Tömösváry organ is present at the base of each antenna (Fig. 20-7B).

All symphylans have a trunk with 14 segments covered by 15 to 24 tergites. The additional tergites are thought to increase the flexibility of the trunk and facilitate movement through small spaces. Trunk segments 1 through 12 bear walking legs, segment 13 bears a pair of spinnerets, and 14 bears a pair of trichobothria (Fig. 20-7D). Silk glands open through pores on the spinnerets, which are thought by some to be homologous to insect cerci. The trunk terminates in a tiny, oval telson (Fig. 20-7A). Attached to the body wall at the base of most legs are an eversible **coxal sac** and a small process, the **stylus,** both of which are present in primitive hexapods. The coxal sacs, which take up moisture, may also have a respiratory function. The function of the stylus is unknown, although it may be sensory. Symphylans are vulnerable to low humidity and desiccation.

The trunk structure, especially the additional tergites, which increase dorsoventral flexibility, is undoubtedly correlated with the locomotor habits of these animals. Many symphylans run rapidly and can twist, turn, and loop their bodies when crawling through crevices within humus. This ability is probably an adaptation for escaping predators in the network of living and decayed vegetation in which they live and feed. *Scutigerella* feeds on plant roots and can be a serious pest to vegetable and flower crops, especially in greenhouses. It has been widely, albeit inadvertently, distributed by the worldwide shipping of plants.

Two Malpighian tubules are the major excretory organs. One or two pairs of saccate nephridia located in the segments of the first and second maxillae are known as maxillary nephridia. The heart is a dorsal tube extending from segments 6 through 12, but it has only one pair of ostia. The two spiracles of the only pair of tracheae are on the sides of the head (Fig. 20-7B), an unusual position for them. The tracheae supply only the head and first three trunk segments. Posterior tracheae, presumably present in the tracheate ground plan, have been lost. There being no eyes, the optic nuclei of the protocerebrum are vestigial, but the protocerebrum also serves the Tömösváry organs. Cephalization is weak, and the subesophageal ganglion includes only the ganglia of the three pairs of mouthparts. Twelve pairs of segmental ganglia, one for each pair of legs, are arrayed along the ventral nerve cord.

Symphylans are gonochoric and sperm transmission is indirect with spermatophores. Fertilization is external. The unpaired, median gonopore is located ventrally on the fourth trunk segment. Copulatory behavior in *Scutigerella* is most unusual. The male deposits 150 to 450 spermatophores each at the end of a stalk. Upon encountering a spermatophore, the female takes it into her preoral cavity and bites off the stalk. Sperm are stored in a pair of seminal receptacles associated with the preoral cavity. Later the female extrudes an egg from the gonopore on segment 4 and glues it to a moss gametophyte, a lichen, or in a crevice and then fertilizes it with sperm from the seminal receptacle (Fig. 20-7D,E). Eggs are usually laid in clusters of 8 to 12. Parthenogenesis is common in symphylans. The role of the silk glands and spinnerets in reproduction is unknown. Development is indirect and, on hatching, the young have only six or seven pairs of legs. A new pair of appendages is added with each molt. *Scutigerella immaculata* lives for as long as four years and molts throughout life.

DIPLOPODA[C]

Diplopods are commonly known as millipedes, a name that implies possession of 1000 legs, although no known species has more than 710 and most have far fewer. Even so, millipedes have the most legs of any terrestrial animal. Diplopods are secretive, nocturnal detritivores that subsist on decaying vegetation. Negatively phototactic, they avoid light and live beneath leaves, stones, bark, and logs. Millipedes are common in soil and leaf litter and many are cave dwellers. They vary from 2 mm to almost 30 cm in length. Estimated to contain about 10,000 extant described species, Diplopoda is the largest myriapod taxon, and specialists estimate there could be as many as 80,000 species alive today, 70,000 of them as yet unknown. Their distribution is cosmopolitan, but they are especially abundant in the tropics.

Diversity of Diplopoda

Penicillata[SO] (bristly millipedes, Polyxenida): Very small (< 4 mm), distinctly bristly due to unique lateral and posterior tufts of serrate setae (Fig. 20-8B). The most primitive millipedes and those most like the diplopod ground plan. Cuticle is uncalcified, soft, flexible, unlike that of all other millipedes. Trichobothria are present on the head (but absent in all other diplopods, although present in symphylans and pauropods). Repugnatorial glands, which are present in at least some members of all other higher diplopod taxa, are absent. Sperm transfer is indirect, with spermatophores left on a silk web. *Polyxenus.*

Pentazonia[SO] (pill millipedes): Unlike other millipedes; superficially look more like crustacean pill bugs, or wood lice (Fig. 20-8D,E). Legs are inconspicuous. Body is short, strongly convex dorsally. Capable of rolling the body into a sphere (like pill bugs), with vulnerable ventral parts tucked inside and protected by the heavily sclerotized tergites on the outside. The resulting ball is invulnerable to most would-be predators. Last pair of male legs with enlarged telopods is used to hold the female's vulva during sperm transfer. *Glomeris.*

Colobognatha[SO] (sucking millipedes): Slow-moving colobognaths are even more wormlike than the juliforms (see below), although at 3 to 50 mm long they are usually much smaller. Body is arched dorsally. Legs are so small that they can be mistaken for worms. Pointed head is narrower than body. Mouthparts are reduced, mandibles are stylets used for piercing and sucking. Food is assumed to be decayed and partially liquefied vegetation. Females and occasionally males protect the eggs by curling around them. *Polyzonium.*

Nematophora[SO]: Body is cylindrical, may appear flattened if paranota are well developed. Three setae on the dorsal posterior margin of the telson ring are spinnerets. Have a mandibular molar with a strong posterior cusp, unknown in other millipedes. *Chordeuma.*

Merochaeta[SO] (flat-backed millipedes, Polydesmida): Common, widespread. Large lateral paranota create a depressed

Photograph courtesy of Betty M. Barnes.

A

Photograph courtesy of K. H. Schomann.

B

C

From Schaller, F. 1968. Soil Animals. University of Michigan Press, Ann Arbor. Copyright University of Michigan Press.

D

From Schaller, F. 1968. Soil Animals. University of Michigan Press, Ann Arbor. Copyright University of Michigan Press.

E

FIGURE 20-8 Diplopoda. **A,** A polydesmid millipede (Merochaeta), *Sigmoria.* **B,** Bristly millipedes (Penicillata) of the genus *Polyxenus.* These tiny millipedes are only 4 mm in length. The legs are obscured by the large, scalelike spines. **C,** A flat-backed millipede (Merochaeta). **D,** The pill millipede (Pentazonia) *Glomeris.* **E,** *Glomeris* enrolled.

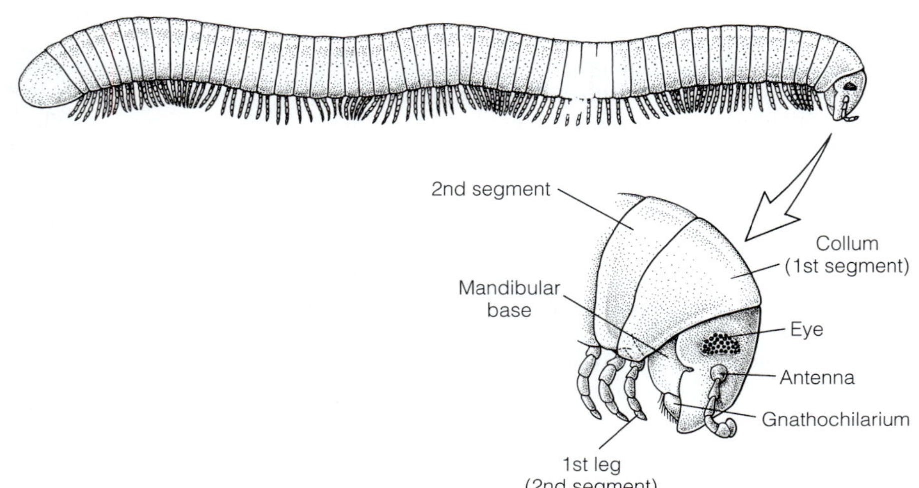

FIGURE 20-9 Diplopoda. A juliform millipede (Juliformia).

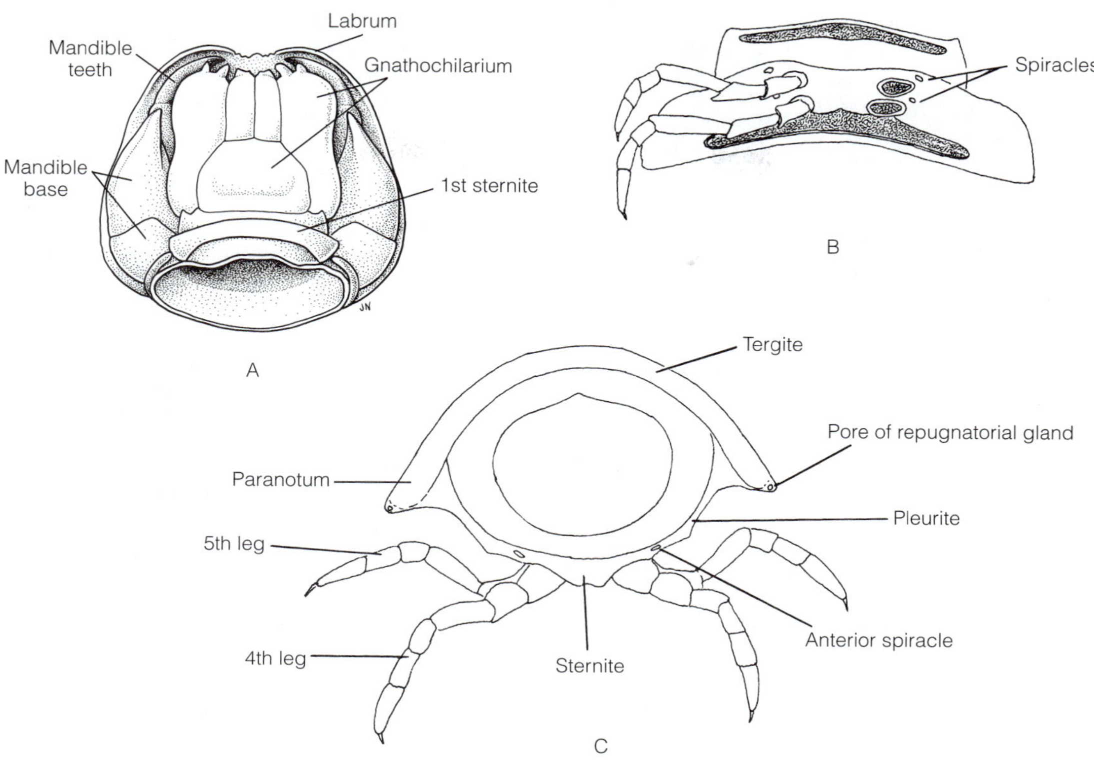

FIGURE 20-10 Diplopoda. **A,** Ventral view of the head of *Habrostrepus,* a juliform millipede. **B,** Ventral view of a diplosegment of *Apheloria,* a flat-backed millipede. **C,** A diplosegment of *Apheloria* (transverse section). *(All after Snodgrass, 1952.)*

appearance (Fig. 20-8A; 20-10C). Three to 130 mm long. Most have 20 diplosegments. *Apheloria, Polydesmus, Sigmoria.*

Juliformia^SO (worm millipedes): The most common, familiar millipedes. Cylindrical, smooth, shiny, often large, with elongate, wormlike bodies (Fig. 20-9). The largest millipedes, with some reaching 30 cm in length. To facilitate movement through the substratum, the collum is enlarged and covers part of the head capsule and anterior trunk. Repugnatorial secretions contain benzoquinones that are unique in millipedes but common in other arthropods. *Cylindroiulus, Narceus.*

Form

The exoskeleton of most millipedes is heavily calcified. The body typically is long and cylindrical (Fig. 20-9), but in many it may appear to be depressed because of shelflike **paranota** (sing., paranotum) projecting laterally from the tergites (Fig. 20-10C). Like that of other myriapods, the body is divided into a head and multisegmented trunk. The head tends to be convex dorsally and flattened ventrally (Fig. 20-9). It bears Tömösváry organs and, in some species, lateral clusters of from 4 to 90 loosely arranged, poorly developed lateral ommatidia. The first head appendages are the short antennae, which always have eight articles with four large, chemoreceptive cones at the tip of the eighth.

Dignatha, a taxon consisting of millipedes and pauropods, has two pairs of mouthparts, rather than the three of the stem tracheates. These are the mandibles and the **gnathochilarium** (in millipedes; Fig. 20-10A) or lower lip (in pauropods). Specialists disagree about how the number of mouthparts became reduced in Dignatha. Some believe the gnathochilarium to be the fused first maxillae and the second maxillae to have been lost, whereas others think it is formed of the combined first and second maxillae. That only one pair of nerves innervates it supports the first hypothesis. Whatever its origin, the characteristic diplopod gnathochilarium is a single piece formed of fused right and left appendages (Fig. 20-9, 20-10A). It is a broad, flattened plate attached to the posterior, ventral surface of the head. It bears distal sensory palps and forms the floor and posterior wall of the preoral cavity. It is sensory and bears six short apical pegs covered with chemoreceptive sensilla. The heavily sclerotized, biarticulate mandibles are responsible for trituration (Fig. 20-10A). Their medial edges bear teeth and a rasping surface whereas the heavy convex bases cover the sides of the head (Fig. 20-11).

Externally, the trunk consists of a series of segmentlike cuticular **rings.** Most rings are **diplosegments,** derived during development from the fusion of two segments. Most diplosegments bear two pairs of legs, from which the name Diplopoda is derived, and two pairs of spiracles (Fig. 20-10B,C). This doubled condition is also evident internally, for there are two

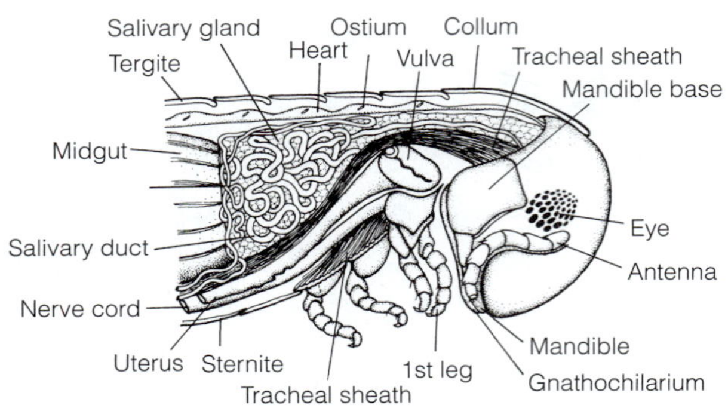

FIGURE 20-11 Diplopoda. Lateral view of the head and anterior trunk segments of the juliform millipede *Narceus.* *(After Buck and Keister, 1950)*

pairs of ganglia and two pairs of ostia within each ring. Ring 1 is the legless **collum,** which is covered by an enlarged tergite forming a large collar behind the head (Fig. 20-9). It is not a diplosegment. Rings 2, 3 and 4 each bear a single pair of appendages, ganglia, ostia, and spiracles and probably are not diplosegments. The legs consist of the typical myriapod articles. In most taxa, one or both legs of the seventh ring are gonopods modified to transfer sperm from the male to the female. The body terminates in the telson, on which the anus opens ventrally. Some posterior rings are limbless.

The exoskeleton of each ring is formed of the usual sclerites (Fig. 20-10C), and in primitive diplopods the tergite, sternite, and two pleurites of each ring may be separate and distinct. In the flat-backed and juliform millipedes the four sclerites are fused, and in the juliforms, they form a nearly cylindrical ring.

In most millipedes the integument, especially the tergites, is hard and, like that of many crustaceans, impregnated with calcium salts. The surface is often smooth, but in some taxa, the tergites have ridges, tubercles, spines, or isolated bristles and in the little, soft-bodied polyxenids (Penicillata) the sides bear conspicuous tufts of setae (Fig. 20-8B).

Diplopods vary greatly in size. Penicillata contains minute forms, with some species of *Polyxenus* being only 2 mm long. The largest millipedes are tropical species of the family Spirostreptidae (Juliformia), which may be 30 cm long.

Locomotion

Most millipedes are like bulldozers crawling slowly about over the ground or, more frequently, pushing their way through loose particulate substrata. Unlike the alternating pattern of stepping that characterizes centipedes and most other arthropods, the effective stroke of the legs on one side of the body of millipedes coincides with those on the opposite side. Diplopod gaits, although slow, exert a powerful force, enabling the animal to push through humus, leaves, and loose soil. The force is exerted entirely by the legs, and the diplosegmented structure is probably an adaptation for maximizing the force exerted. The backward, pushing stroke is activated in waves along the length of the body and is of a longer duration than the forward stroke. Thus, at any moment, more legs are in contact with the substratum than are raised. The number of legs involved in a single wave is proportional to the amount of force required for pushing. When the animal is running, for example, 12 or fewer legs may compose a wave, but when pushing, a single wave may involve as many as 52 legs in some juliform millipedes.

Locomotion and pushing mechanisms are correlated with the shape of the animal, and five ecomorphological types of millipedes, roughly corresponding to the higher millipede taxa, are recognized.

1. **Bulldozers:** The head-on pushing habit is most highly developed in the juliform millipedes (Fig. 20-9), which burrow headfirst into relatively compact leaf litter and soil. This habit is reflected in their smooth, fused, rigid cylindrical segments that have no projections to snag leaves or debris; the rounded head; and the placement of the legs close to the midline of the body.
2. **Flat wedgers** (flat-backed millipedes, Merocheata: Polydesmida): These powerful millipedes push with the flat dorsal surface of the back to widen cracks that open along a single plane, such as between leaves or under bark. The lateral, shelflike paranota increase the dorsal surface available for pushing (Fig. 20-8A,C, 20-10C). The paranota provide a protected working space for the more laterally placed legs. The tapered anterior end is pushed into a small crevice and then the legs push the rest of the animal into the crack, wedging it wider so that the animal can advance into the space.
3. **Borers** (round wedgers, Colobognatha): These are wedgers of a different type, adapted for pushing into substrata composed of small, irregular mineral or organic particles that do not split along planes but can be pushed aside. The body is tapered anteriorly and the legs force the small head and anterior rings into a circular crevice. Larger posterior rings are pulled forward to telescope over the anterior rings. Then the legs are used to anchor the millipede and push the small anterior rings farther into the crevice.
4. **Rollers** (pill millipedes, Pentazonia): Essentially bulldozers, these animals have the additional ability to enroll into a ball (Fig. 20-8E) in a manner reminiscent of terrestrial pill bugs (Isopoda). They do this for protection or to decrease desiccation.
5. **Bark dwellers** (bristly millipedes, Penicillata: Polyxenida): Less than 4 mm in length, the body of bark dwellers have 11 to 13 rings (Fig. 20-8B). They are softer than other millipedes and, unlike almost all other millipedes, they are weak burrowers not adapted for pushing. They are small and depressed and the cuticle is soft and uncalcified. They inhabit preexisting crevices under bark.

The ability to climb is striking in some nematophorans in rocky habitats. These millipedes can climb up smooth surfaces by gripping with opposite legs. Some of these rock dwellers are the swiftest millipedes. Their speed is correlated with their predatory and scavenging feeding habits and the need to cover great distances to find food.

With the exception of the bristly millipedes, all are able to coil the body into a tight, flat planispiral or even a ball. This is possible because the sternites are shorter than the tergites.

Defense

Several invertebrates, including some ants, spiders, scorpions, assassin bugs, parasitic mites, and glowworms, prey on millipedes. In addition, some vertebrates, including toads, lizards, tortoises, rodents, mongooses, and birds, occasionally eat them. To compensate for the lack of speed, natural selection has provided millipedes with other protective mechanisms. The drab, cryptic coloration of many diplopods reduces the likelihood of discovery by predators. The hard, smooth, calcified cuticle provides protection for the upper and lateral aspects of the body and makes them difficult to grasp and manipulate. The long, many-segmented millipedes, such as the polydesmid and the juliform taxa, protect the more vulnerable ventral surface by rolling the trunk into a planispiral coil when at rest or disturbed. Pill millipedes, as well as some others, can roll into a ball (Fig. 20-8D,E). When enrolled, some tropical giant pill millipedes are larger than golf balls. Mongooses have learned to throw large pill millipedes against rocks to break the exoskeleton. Unlike centipedes, millipedes do not attempt to bite in defense.

Repugnatorial glands are present in all the major taxa except the bristly millipedes. Usually there is one pair of glands per diplosegment and the openings are located on the sides of the tergites or, in the flat-backed millipedes, on the margins of the paranota (Fig. 20-10C). Each gland consists of a large secretory sac, which empties into a duct leading to the external pore. The composition of the secretion varies by species, and aldehydes, quinones, phenols, iodine, chlorine, and hydrogen cyanide have been identified. Hydrogen cyanide is generated on demand by mixing a nontoxic precursor with an enzyme. The cyanide released by the reaction has the odor of bitter almonds and is toxic to invertebrates and repellent to small vertebrates.

The caustic secretions of some large tropical species can blister human skin. *Narceus americanus*, a common North American juliform, also is capable of producing blisters on human skin. Defensive secretions usually are exuded slowly, but large, tropical juliforms can discharge it as a spray or jet for distances of up to 30 cm. Malaysian natives use these secretions as arrow poison.

Most interesting of all is the sedative produced by a European pill millipede, *Glomeris marginata*. This species secretes a cocktail consisting of a substance that inhibits feeding by the predator, usually a wolf spider, and another compound similar to the sedative Quaalude. Once sprayed with this secretion, the spider looses interest in feeding and falls asleep, often for several days.

Most diplopods are black or brown, but many are striking, bright, glossy colors, including orange, red, yellow, blue, and green. In general, toxic, distasteful, or otherwise dangerous animals are brightly and distinctively colored so that potential predators may more quickly learn to recognize and avoid them. Such warning, or **aposematic, coloration** is common throughout the living world in bees, butterflies, nudibranchs, coral snakes, berries, tropical frogs, and millipedes, to name a few. One southern California millipede is bioluminescent, and it has been suggested that this is a way for a nocturnal animal to be aposematic.

Internal Form and Function

Most millipedes are detritivores, feeding primarily or exclusively on decomposing vegetation. A few consume healthy vegetation. Food is usually moistened by secretions of the salivary glands and chewed or scraped by the mandibles. In the tropical Siphonophoridae (Colobognatha), however, the labrum and gnathochilarium are modified to form a long, piercing beak for feeding on plant juices. A carnivorous or omnivorous diet has been adopted by some rock-inhabiting Nematophora. As predators, these millipedes are relatively agile and fast. Their prey includes harvestmen, hexapods, centipedes, and earthworms, which are held with the first legs when being consumed.

Although earthworms are the principal detritivores in most woodland ecosystems, millipedes share this niche and contribute importantly to the recycling of plant debris on the forest floor. Up to 90% of detritus that enters the millipede gut is egested back into the habitat as feces. Passage of this material through the gut facilitates later microbial assimilation of its organic compounds and nutrients. In typical woodland communities, millipedes may process up to 10% of the annual leaf fall, but up to 25% if earthworms are absent. Coprophagy (ingestion of feces) is common in millipedes and may be a mechanism for recovering additional nutriment from food by passing it through the gut twice. Some species (such as *Apheloria montana*) die if prevented from ingesting their own feces. Like earthworms, some millipedes ingest soil, from which they digest the organic matter.

The gut typically is a straight tube with a long midgut. Salivary glands open into the preoral cavity. Digestion and absorption occur in the midgut, which produces a peritrophic membrane to surround the food, as hexapods and crustaceans do. The midgut is covered by a layer of tissue known as the "liver" that appears to be similar to the chlorogogen tissue of annelids. Its cells store glycogen and may also function as accumulation kidneys where toxins are sequestered. A fat body in the hemocoel stores glycogen, lipids, protein, and uric acid.

Gas exchange is entirely via the segmentally organized tracheae. Each diplosegment has two pairs of spiracles located on the sterna (Fig. 20-10B,C), an unusual position for them. Each spiracle opens into an internal atrium, from which arises a bundle of unbranched tracheae that extend to the tissues. The atria are hollow, rigid apodemes that also function as sites for muscle attachment. The spiracles of most millipedes cannot be closed, but those of *Glomeris* are an exception.

The heart (Fig. 20-11) is a dorsal tube extending the length of the trunk, and the hemocoel is divided into pericardial, perivisceral, and perineural sinuses. The heart is perforated by two pairs of ostia per diplosegment, but the anterior segments, which are not diplosegments, have only one pair of ostia each. The heart ends blindly at the posterior end of the trunk, but anteriorly a short cephalic aorta extends into the head.

The two Malpighian tubules that arise from the midgut-hindgut junction often are long and looped. Like centipedes, millipedes excrete both ammonia and uric acid. Also like centipedes, millipedes have saccate nephridia in the maxillary segments, but their function is unknown.

Although most millipedes cannot tolerate desiccating conditions, some live in arid habitats. These species have eversible coxal sacs, which supposedly take up water from sources such as dewdrops. The ability of many millipedes to enroll may reduce evaporation from the ventral spiracles during periods of inactivity. The sudden appearance of large numbers of ground millipedes on tree trunks, rocks, or walls probably is related to humidity, as the animals tend to move upward, out of the litter and into the air, when the humidity is high. Nocturnal and burrowing habits also contribute to water conservation.

Nervous System and Sense Organs

The CNS consists of the brain, subesophageal ganglion, and ganglionated ventral nerve cord. The subesophageal ganglion consists of the neuromeres of the mandibles and gnathochilarium only. The long, uncephalized ventral nerve cord has a pair of ganglia for each of the first four trunk segments and two pairs for each diplosegment.

Eyes may be totally lacking, as in the flat-backed millipedes, or there may be from 2 to 80 lateral eyes derived from ommatidia that are loosely clustered beside the base of each antenna (Fig. 20-9). In some cases each eye represents a single ommatidium, whereas others are the result of a few ommatidia fusing. Most millipedes are negatively phototactic, and even those without eyes have integumental photoreceptors and respond negatively to light. The antennae contain tactile setae and unique sensory cones richly supplied with what are probably chemoreceptors. The animal continually taps the substratum with the antennae as it walks. Many millipedes have Tömösváry organs that probably monitor humidity but may be olfactory. Some have trichobothria on the head.

Reproduction and Development

Millipedes are gonochoric and sperm transfer is indirect with spermatophores. In both sexes the tubular gonads extend most of the length of the trunk posterior to the third, or genital, segment. In females, a pair of long, fused ovaries lies between the midgut and the ventral nerve cord. Two oviducts (uteri) extend anteriorly to the genital segment, where each opens into a protractible, pouchlike vulva behind the coxa of the second pair of legs (which are on the third ring since the collum has no legs; Fig. 20-11, 20-12C,D). When retracted, the vulva is covered and protected by the sclerotized, hoodlike operculum (Fig. 20-12C). Inside the vulva, a groove leads into a seminal receptacle.

The testes occupy positions corresponding to those of the ovaries but are joined by transverse connections. Anteriorly, near the region of the genital segment, each testis tapers into a sperm duct, which opens through a penis (or two penes) on the genital segment. Note that in contrast to the posteriorly located gonopores of centipedes, those of millipedes, like those of symphylans and pauropods, are located at the anterior end of the trunk (hence the name Progoneata). Millipede sperm are unlike those of other myriapods in that they are aflagellate.

The copulatory organs are the gonopods (modified legs) of the seventh trunk ring (Fig. 20-12A,B). As in many other arthropods, the gonopods have species-specific morphology and are valuable for distinguishing and identifying species.

The flat-backed millipede *Apheloria* exemplifies millipede reproductive behavior. The male charges the gonopods with sperm by bending the anterior part of his body ventrally to transfer sperm from the two penes (on the third ring) into a sperm reservoir at the base of the gonopods (on the seventh ring). Some millipedes have two pairs of gonopods on the seventh ring, but usually there is only one.

Males communicate their identity and intent to females in a variety of ways. The signal is tactile in most juliform millipedes, with the male climbing onto the back of the female and clinging there using special leg pads. Antennal tapping, head drumming, and stridulating are utilized by various other millipedes. Many millipedes produce pheromones, which either initiate mating behavior or prolong the courtship initiated by other signals.

During copulation the male's body is twisted or stretched out against that of the female so that the gonopods on the male's seventh ring are opposite the vulvae on the female's third ring. The male holds the female with his legs. The gonopods are extended, their telopodites are inserted into the vulva, and sperm from the reservoir are transferred to the vulva and then to the seminal receptacle for storage. The female has been inseminated, but fertilization of the eggs does not occur until later.

Various millipedes exhibit other behaviors. The tiny *Polyxenus* male deposits sperm in a silken sperm web and then spins a silk trail for the female to find (Fig. 20-12E). She follows the trail, which is about seven times her length of 2 mm, and picks up the sperm with her vulvae. Pentazonian millipedes have no

PHYLOGENETIC HIERARCHY OF DIPLOPODA

- Diplopoda
  - Penicillata
    - Polyxenida
  - Chilognatha
    - Pentazonia
      - Glomeridesmida
      - Sphaerotheriida
      - Glomerida
    - Helminthomorpha
      - Colobognatha
        - Platydesmida
        - Siphonophorida
        - Polyzoniida
      - Eugnatha
        - Nematophora
          - Stemmiulida
          - Callipodida
          - Chordeumatida
        - Merochaeta
          - Polydesmida
        - Juliformia
          - Spirobolida
          - Spirostreptida
          - Julida

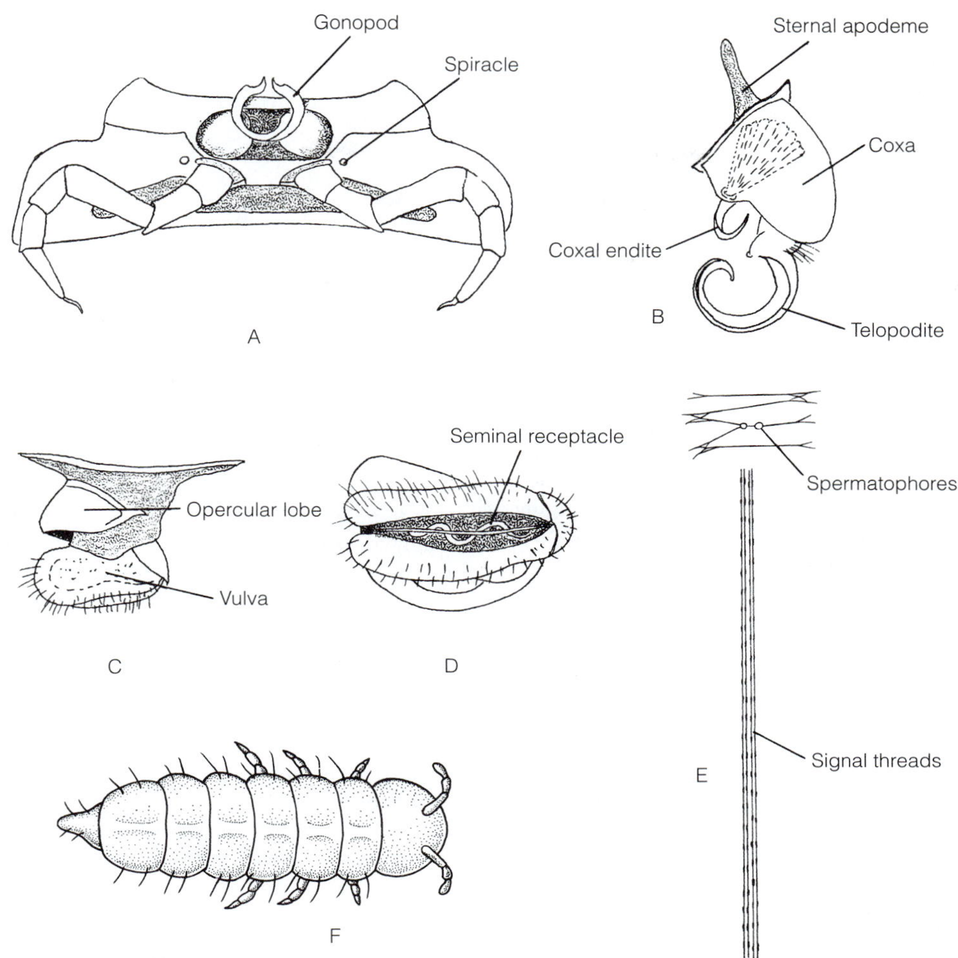

FIGURE 20-12 Diplopoda: millipede reproduction. **A,** Seventh diplosegment of a male *Apheloria,* showing gonopods and legs (ventral view). **B,** Left gonopod of *Apheloria.* **C,** Right vulva of the third segment of *Apheloria* in a lateral view. **D,** Vulva (ventral view). **E,** Signal threads leading to a spermatophore web of *Polyxenus.* **F,** A newly hatched millipede. *(A–D, After Snodgrass, 1952; E, After Schomann; F, After Cloudsley-Thompson, J. L. 1958. Spiders, Scorpions, Centipedes and Mites. Pergamon Press, New York.)*

gonopods, so the male uses his mouthparts to transfer sperm. Parthenogenesis is common in Penicillata (bristly millipedes) and males are rare.

At some time following copulation, the female deposits eggs. Diplopod eggs are fertilized as they are laid and anywhere from 10 to 300 are produced at one time, depending on species. Some deposit their eggs in clusters in soil or humus. Many millipedes construct a nest for the eggs. *Narceus* fashions a cup of regurgitated material into which a single egg is laid. The cup is sealed, deposited in humus or a crevice, and then consumed by the young millipede upon hatching. The European pill millipede, *Glomeris,* has similar habits but forms the capsule with excreta. Some flat-backed species and colobognaths also construct the nest from excrement, building a thin-walled, domed chamber topped by a chimney. The vulvae are applied to the chimney opening and the eggs fall into the chamber as they are laid. The opening is then sealed and the chamber is covered with grass and other debris. The female, and in some species the male, may remain coiled around the nest for several weeks. Nematophorans enclose their eggs in silk cocoons.

In most species the eggs hatch in several weeks. Development is indirect and the newly hatched young usually have only the first three pairs of legs and not more than seven trunk rings (Fig. 20-12F). With each molt, additional rings and legs are added to the trunk. Many millipedes undergo ecdysis in specially constructed molting chambers that are similar to the egg nests. Many tropical species estivate in a molting chamber during the dry season. Following ecdysis the exuvium usually is ingested, probably to reclaim its calcium. Millipedes live from 1 to 10 or more years, depending on the species.

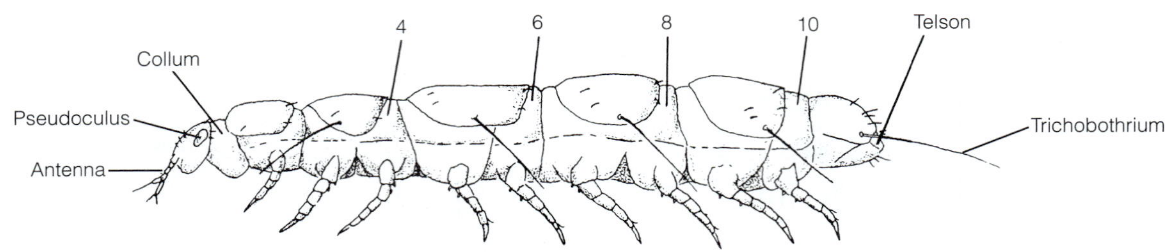

FIGURE 20-13 Pauropoda. The pauropod *Pauropus silvaticus* in lateral view. *(After Tiegs from Snodgrass, 1952.)*

PAUROPODA[C]

Pauropoda is a small taxon of soft-bodied, nocturnal animals (Fig. 20-13). These poorly known arthropods are widely distributed in both temperate and tropical regions in leaf litter, soil, damp wood, and under stones. All are minute, ranging from 0.5 to 1.9 mm in length. Although once considered rare, pauropods have now been found to be abundant in forest litter and approximately 500 species are known. The small number of known species is due to the small number of taxonomists who study them.

Pauropods are dignathans similar to millipedes in a number of ways. The trunk usually contains 11 segments, 9 of which bear a pair of legs. Segments 1 (the collum) and 11 and the telson are legless. The number of tergites is less than the number of segments. Some of the dorsal tergites are very large and overlap adjacent segments. Five tergites carry a pair of long, laterally placed trichobothria. In contrast with the diplopod collum, that of pauropods is inconspicuous dorsally and expanded ventrally.

On each side of the head is a peculiar, disclike **pseudoculus** that is perhaps homologous to the Tömösváry organ of other myriapods. Eyes are absent. The six-articled antennae branch into three flagella, a feature that quickly distinguishes pauropods from all other tracheates. One branch of the antenna bears a club-shaped sense organ of unknown function. The weak, uniarticulate mandibles are adapted for grinding or piercing. The lower lip, also weak, is probably formed by the fusion of the two first maxillae and is homologous to the diplopod gnathochilarium. The second maxillae are absent, as they are in all dignathans. The cuticle usually is weakly sclerotized, soft, and permeable. Pauropods are intolerant of desiccation and depend absolutely on a humid environment.

Most pauropods feed on fungi or decomposing plant tissue, but some are predaceous. The foregut is equipped with pumping muscles, and some species have been observed biting fungal hyphae to suck their contents.

There is neither heart nor (except in some primitive species) tracheae, and their absence probably is associated with the small size of these animals. A pair of Malpighian tubules is present. Two pairs of glands in the head are derived from coelomic pouches and at least one, and perhaps both, is probably homologous to the saccate maxillary nephridia of other tracheates. Their function is unknown. The brain lies in the head and first trunk segment. The subesophageal ganglion innervates only the mandibles and the lower lip.

Pauropods are gonochoric. As in diplopods, the third trunk segment is the genital segment and bears the gonopore (two in males, one in females) and, in males, the penis. Sperm are transferred indirectly via spermatophores, which, along with two signal threads, are deposited by the male in the female's absence. The eggs are laid in humus, either singly or in clusters. Development is indirect and, as in diplopods, the young hatch with only three pairs of legs. In *Pauropus sylvaticus,* development to sexual maturity takes about 14 weeks.

PHYLOGENY OF TRACHEATA

Phylogenetic relationships within Tracheata (Atelocerata, Monoantennata) and even its validity as a monophyletic taxon are currently under scrutiny. Tracheata is divided into three high-level taxa, Hexapoda, Chilopoda, and Progoneata (Diplopoda, Symphyla, and Pauropoda), each of which is widely agreed to be monophyletic. Controversy, however, arises over the pattern of relationship among these three taxa, and three phylogenies, known as the Myriapoda, Labiophora, and Opisthogoneata hypotheses, have been proposed. The Myriapoda hypothesis holds that Chilopoda and Progoneata are sister taxa in Myriapoda and that Myriapoda and Hexapoda are sister taxa (Fig. 20-14A). This traditional viewpoint is adopted here. In the Labiophora hypothesis, Hexapoda and Progoneata are thought to be sister taxa in a taxon known as Labiophora that has Chilopoda as its sister taxon (Fig. 20-14B). Finally, according to the Opisthogoneata hypothesis, Chilopoda and Hexapoda are sister taxa in the Opisthogoneata, which has Progoneata as its sister taxon (Fig. 20-14C).

In the Myriapoda hypothesis, Myriapoda is monophyletic but in both the Labiophora and Opisthogoneata hypotheses, it is paraphyletic. Although there is agreement that Hexapoda, Progoneata, and Chilopoda are each monophyletic, the same cannot be said about Myriapoda. The evidence for the monophyly of Myriapoda is weaker than that supporting the monophyly of each of its constituent taxa.

Furthermore, that Tracheata itself is a monophyletic taxon (Fig. 20-14D) has been questioned due to molecular and morphological evidence based on nervous system, brain, and eye morphology. This evidence suggests that Hexapoda and Crustacea are sister taxa more closely related to each

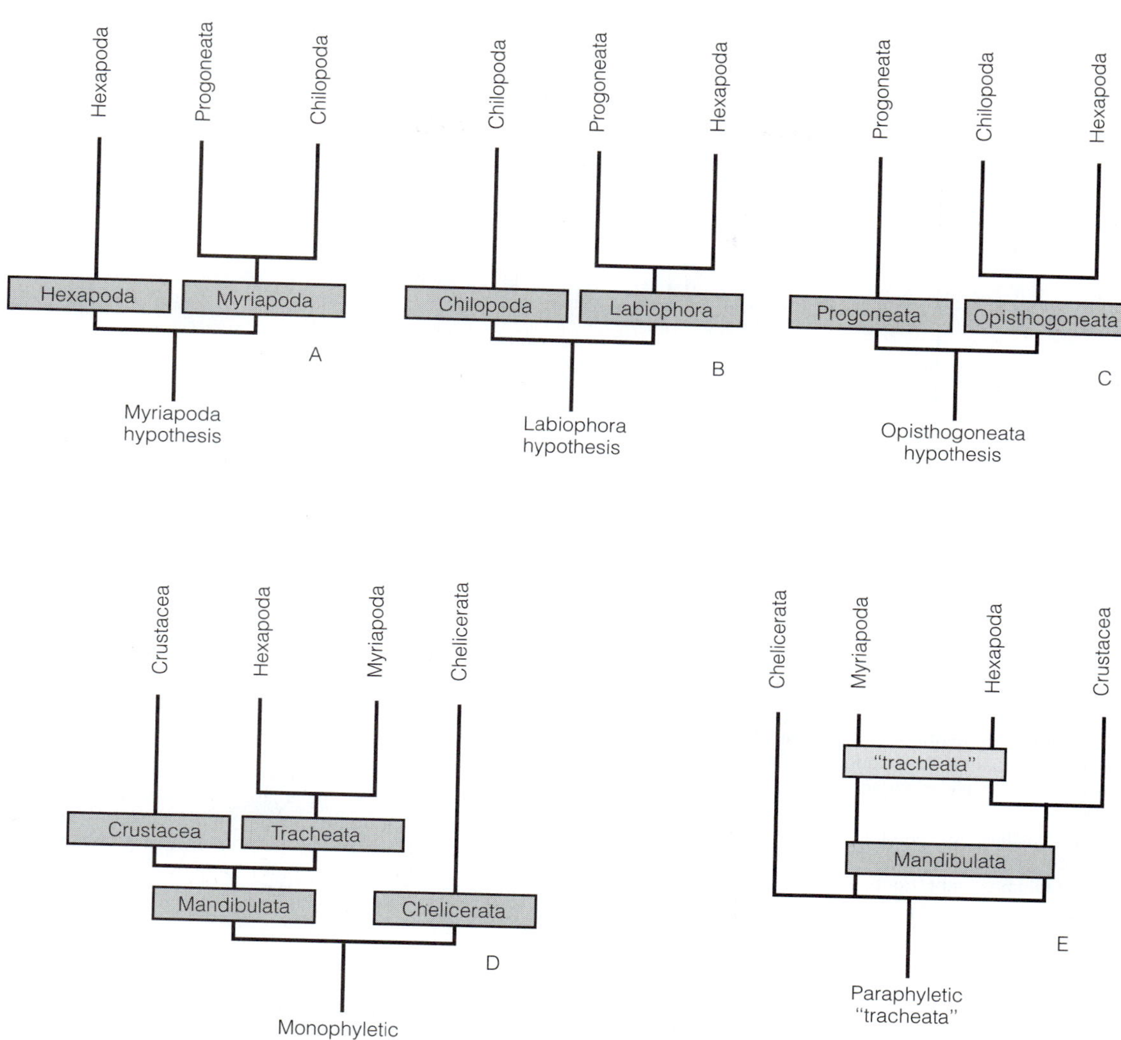

FIGURE 20-14 Myriapod and tracheate phylogeny. **A–C,** Alternative hypotheses of the evolution of Tracheata. **D–E,** Alternative hypotheses of the evolution of Mandibulata. *(A–C, From from Ax, P. 2000. Multicellular Animals, II. Springer Verlag, Berlin; D, From Wheeler et al., 1993; E, From Friedrich, M., and Tautz, D. 1995. Ribosomal DNA phylogeny of the major extant arthropod classes and the evolution of myriapods. Nature 376:165–167.)*

other than either is to Myriapoda (Fig. 20-14E), thus rendering Tracheata paraphyletic. In this text we adopt the more traditional Myriapoda hypothesis and consider both Myriapoda and Tracheata to be monophyletic, but the next few years may bring important changes in the higher classification of the terrestrial mandibulates.

Tracheata is united by a number of autapomorphies, many of which are adaptations for terrestrial life convergent with similar adaptations in arachnids. These include tracheae and Malpighian tubules, both of which are presumed to have arisen in the ancestral tracheate. Spermatophores and internal fertilization are also adaptations for terrestrial life developed independently by tracheates and arachnids. The tracheates are further united by the possession of one pair of antennae and the absence of appendages on the vestigial second head segment.

Myriapoda, the sister taxon of Hexapoda (Fig. 20-14A,D), is united by division of the body into two tagmata (the head and trunk), the absence of median eyes, and ommatidia without a crystalline cone. The derived ommatidium of

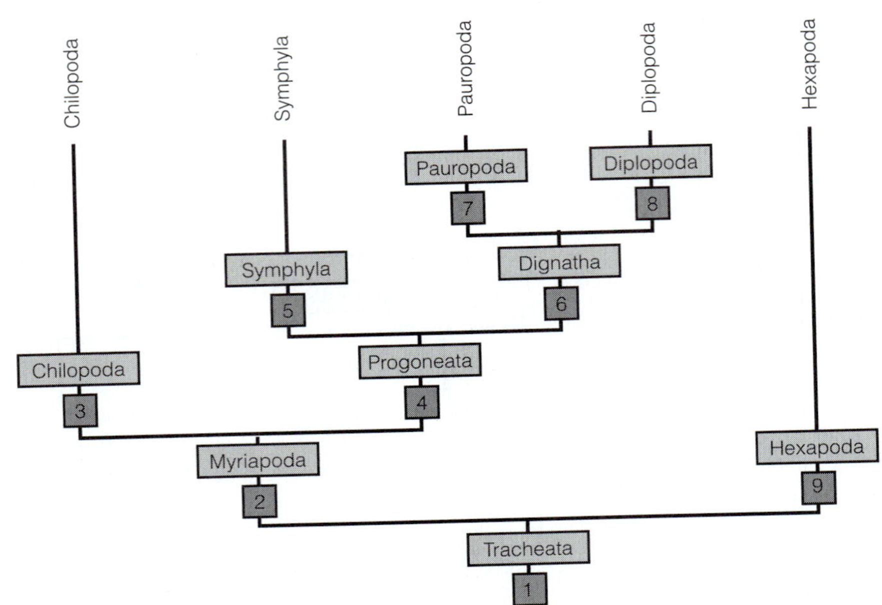

FIGURE 20-15 A phylogeny of the tracheates. **1, Tracheata:** Paired, segmental tracheae; Tömösváry organs; one pair of antennae; mandibles lack palps; head segment 2 has no trachea; head segments 4 and 5 have saccate nephridia; Malpighian tubules; terrestrial. **2, Myriapoda:** Median ocelli lost; compound eyes lack crystalline cones; ommatidia are few and loosely arranged. **3, Chilopoda:** Gonopores are posterior; first trunk appendages are maxillipeds (forcipules); raptorial feeders; second maxillae of embryo have a hatching spine. **4, Progoneata:** Anterior gonopores; maxilla 1 has no palp. **5, Symphyla:** First maxillae and labium present; gonopores are unpaired; one pair of head spiracles; compound eyes are absent; have a pair of preanal silk glands. **6, Dignatha:** Only one pair of head appendages posterior to mandibles; gonopores are paired in some species; spiracles open into the tracheal pouch. **7, Pauropoda:** Twelve trunk segments; antennae are branched, with six articles; compound eyes are absent; have one pair of tracheae and one pair of spiracles; unpaired female gonopore. **8, Diplopoda:** Last head appendage is a gnathochilarium; most trunk segments fused in pairs to form diplosegments; antennae are unbranched, with eight articles, and four apical cones; paired female gonopores. **9, Hexapoda:** Trunk is divided into thorax and abdomen; thorax has three pairs of limbs. *(From Ax, P. 2000. Multicellular Animals, II. Springer Verlag, Berlin.)*

myriapods arose from the more typical compound eyes of the ancestral mandibulate, and the median ocelli were lost secondarily. Because the oldest fossil tracheates are terrestrial Devonian species similar to centipedes and millipedes, the few marine millipedes and insects and the numerous freshwater insects are probably secondary invaders of the aquatic environment.

Myriapoda includes Chilopoda (centipedes), Diplopoda (millipedes), Symphyla, and Pauropoda. In most phylogenies, Chilopoda is seen as being distinct from the other myriapod taxa, and here we consider it to be the sister taxon of Progoneata (Fig. 20-14A, 20-15). Chilopoda arose from the stem myriapod with the development of poisonous forcipules derived from the first trunk appendages. The remaining taxa together form Progoneata, named in reference to the migration of the gonopores anteriorly from their ancestral posterior position. In Symphyla, two pairs of mouthparts, the first and second maxillae, are present posterior to the mandibles. Dignatha, the sister taxon consisting of Diplopoda and Pauropoda, has only one pair of mouthparts in that location. It is not known if the reduction in the number of mouthparts is due to the complete absence of the second maxillae or to the fusion of the first and second maxillae. Diplopoda arose by fusing consecutive trunk segments into diplosegments having two pairs of legs, two pairs of spiracles, two pairs of ostia, and two paired segmental ganglia. Diplopod sperm are aflagellate, and each antenna has four unique sensory apical cones. Pauropoda retains the ancestral solitary segments in a trunk of 12 segments and 11 pairs of legs.

PHYLOGENETIC HIERARCHY OF TRACHEATA

- Tracheata
  - Myriapoda
    - Chilopoda
    - Progoneata
      - Symphyla
      - Dignatha
        - Pauropoda
        - Diplopoda
  - Hexapoda

REFERENCES

GENERAL

Blower, J. G. (Ed.): 1974. Myriapoda. Academic Press, London. 712 pp.

Camatini, M. (Ed.): 1979. Myriapod Biology. Academic Press, London. 456 pp.

Dohle, W. 1996. Antennata (Tracheata, Monoantennata, Atelocerata). In Westheide, W., and Rieger, R. M. (Eds.): Spezielle. Zoologie. I. Einzeller und Wirbellose Tiere. Gustav Fischer Verlag, Stuttgart. pp. 582–584.

Dohle, W. 1997. Myriapod-insect relationships as opposed to an insect-crustacean sister group relationship. In Fortey, R. A., and Thomas, R. H. (Eds.): Arthropod Relationships. Systematics Association Spec. Vol. 55. pp. 305–315.

Friedrich, M., and Tautz, D. 1995. Ribosomal DNA phylogeny of the major extant arthropod classes and the evolution of myriapods. Nature 376:165–167.

Kraus, O., and Kraus, M. 1996. On myriapod/insect interrelationships. Mém. Mus. Natn. Hist. Nat. 169:283–290.

Lawrence, R. F. 1984. The centipedes and millipedes of Southern Africa. A. A. Balkema, Cape Town. 147 pp.

Manton, S. M. 1965. The evolution of arthropod locomotory mechanisms. Pt. 8. Functional requirements and body design in Chilopoda, together with a comparative account of their skeletomuscular systems and an appendix on the comparison between burrowing forces of annelids and chilopods and its bearing upon the evolution of the arthropodan haemocoel. J. Linn. Soc. (Zool.) 46:251–483.

Moore, R. C. (Ed.): 1969. Treatise on Invertebrate Paleontology. Pt. R. Arthropoda 4. Vol. 2. Geological Society of America, University of Kansas Press, Lawrence, KS. 651 pp.

Schaller, F. 1968. Soil Animals. University of Michigan Press, Ann Arbor, MI. 144 pp.

Snodgrass, R. E. 1952. A Textbook of Arthropod Anatomy. Cornell University Press, Ithaca, NY. 363 pp.

Štys, P., and Zrzavý, J. 1994. Phylogeny and classification of extant Arthropoda: Review of hypotheses and nomenclature. Eur. J. Entomol. 91:257–275.

Turbeville, J. M., Pfeifer, D. M., Field, K. G., and Raff, R. A. 1991. The phylogenetic status of arthropods, as inferred from 18S rRNA sequences. Mol. Biol. Evol. 8:669–686.

Wheeler, W. C., Cartwright, P., and Hayashi, C. 1993. Arthropod phylogeny: A combined approach. Cladistics 9:1–39.

Internet Sites

www.mnhn.fr/assoc/myriapoda/WELCOME.HTM (Centre International de Myriapodologie.)

www.biosis.org/zrdocs/zoolinfo/grp_arac.htm#Myriapoda (Myriapoda: Zoological Record Resource Guide for Zoology.)

CHILOPODA

Albert, A. M. 1983. Life cycle of Lithobiidae, with a discussion of the r- and K-selection theory. Oecologia 56:272–279.

Borucki, H. 1996. Evolution und Phylogenetisches System der Chilopoda (Mandibulata, Tracheata). Verh. naturwiss. Ver. Hamburg 35:95–226.

Curry, A. 1974. The spiracle structure and resistance to desiccation of centipedes. In Blower, J. G. (Ed.): Myriapoda. Academic Press, London.

Edgecombe, G. D. 2002. Centipedes: The great Australian bite. Nature Australia 26: 42–51.

Edgecombe, G. D., Giribet, G., and Wheeler, W. 1999. Phylogeny of Chilopoda: Combining 18S and 28S rRNA sequences and morphology. Bol. Soc. Entomol. Aragonesa 26: 293–331.

Giribet, G., Carranza, S., Riutort, M., Baguñà, J., and Ribera, C. 1999. Internal phylogeny of the Chilopoda (Myriapoda, Arthropoda) using complete 18S rDNA and partial 28S rDNA sequences. Phil. Trans. Roy. Soc. Lond. B Biol. Sci. 354:215–222.

Klingel, H. 1960. Die Paarung des *Lithobius forficatus*. Verh. Dtsch. Zool. Ges. 23:326–332.

Lewis, J. G. E. 1981. The Biology of Centipedes. Cambridge University Press, London. 476 pp.

Littlewood, H. 1991. The water relations of *Lithobius forficatus* and the role of the coxal organs. J. Zool. 223:653–665.

Littlewood, H., and Blower, J. G. 1987. The chemosensory behavior of *Lithobius forficatus*. 1. Evidence for a pheromone released by the coxal organs. J. Zool. 211:65–82.

Minelli, A. 1993. Chilopoda. In Harrison, F. W., and Rice, M. E. (Eds.): Microscopic Anatomy of Invertebrates. Vol. 12. Onychophora, Chilopoda, and Lesser Protostomata. Wiley-Liss, New York. pp. 57–114.

Rilling, G. 1968. *Lithobius forficatus*. Grosses Zoologisches Praktikum, Pt. 13b. Gustav Fischer Verlag, Stuttgart. 136 pp.

Shear, W. A., and Bonamo, P. M. 1988. Devonobiomorpha, a new order of centipedes from the Middle Devonian of Gilboa, New York State, USA, and the phylogeny of centipede orders. Am. Mus. Novitates 2927:1–30.

Shelley, R. M. 1999. Centipedes and millipedes with emphasis on North American fauna. Kansas School Naturalist 45:1–16.

Internet Sites

www.bioimages.org.uk/HTML/T2810.HTM (Bioimages, the Virtual Field Guide [UK]: Chilopoda.)

www.palaeos.com/Invertebrates/Arthropods/Chilopoda/Chilopoda.htm (Class Chilopoda: M. Alan Kazlev.)

www.ento.csiro.au/Ecowatch/Insects_Invertebrates/chilopoda.htm (Chilopoda: CSIRO Ecowatch.)

http://insects.tamu.edu/images/insects/fieldguide/cimg379.html (Chilopoda Images: Texas A & M University.)

http://earthlife.net/insects/chilopod.html (Chilopoda (Centipedes), by G. Ramel.)

www.ipm.iastate.edu/ipm/iiin/housece.html (House Centipede: Iowa Insect Information Notes, Iowa State University.)

SYMPHYLA

Manton, S. M. 1966. Body design in Symphyla and Pauropoda. J. Linn. Soc. Lond. 46:103–141.

Michelbacher, A. E. 1938. The biology of the garden centipede, *Scutigerella immaculata*. Hilgardia 11:55–148.

Scheller, U. 1982. Symphyla. In Parker, S. (Ed.): Synopsis and Classification of Living Organisms. Vol. 2. McGraw-Hill, New York. pp. 688–689.

Internet Site

www.ent3.orst.edu/kgphoto/imagedisplay.asp?Order=Symphyla (Symphyla Images, by K. Gray.)

DIPLOPODA

Buck, J. B., and Keister, M. L. 1950. *Spirobolus marginatus.* In Brown, F. A. (Ed.): Selected Invertebrate Types. John Wiley and Sons, New York. pp. 462–475.

Cloudsley-Thompson, J. L. 1958. Spiders, Scorpions, Centipedes and Mites. Pergamon Press, New York. 228 pp.

Enghoff, H. 1984. Phylogeny of millipedes: A cladistic approach. Z. zool. Syst. Evolut.-forsch 22:8–26.

Enghoff, H., Dohle, W., and Blower, J. G. 1993. Anamorphosis in millipedes (Diplopoda): The present state of knowledge with some developmental and phylogenetic considerations. Zool. J. Linn. Soc. 109:103–234.

Hopkin, S. P., and Read, H. J. 1992. The Biology of Millipedes. Oxford Science, Oxford. 233 pp.

Shear, W. A. 1999. Millipedes. Am. Sci. 87:232–239.

Internet Site

www.ento.csiro.au/Ecowatch/Insects_Invertebrates/diplopoda.htm (Diplopoda: Millipedes: CSIRO Ecowatch.)

PAUROPODA

Starling, J. H. 1943. Pauropoda from Duke Forest. Proc. Entomol. Soc. Wash. 45:183–200.

Internet Site

www.kheper.auz.com/gaia/biosphere/arthropods/myriapoda/Pauropoda.htm (Class Pauropoda, by K. Heper, University of Sydney.)

21 Hexapoda[SC]

GENERAL BIOLOGY

It has been said, with perhaps a little exaggeration, that on first approximation all animals are insects. Hexapoda contains at least 870,000 and maybe as many as 1,200,000 described species, almost all of which are insects, and is by far the largest animal taxon (of those of equivalent rank). No other taxon comes close and it is, in fact, over twice as large as all other animal taxa combined. Only a cursory treatment of this gigantic taxon is possible here, and we must forego the more extensive discussions accorded to other taxa in this text. This chapter is intended to meet the requirements of invertebrate zoology courses that wish to include a brief overview of hexapods.

The enormous diversity of hexapods results largely from their exquisite adaptation to life on land, flight, and coevolution with flowering plants. Although they are essentially terrestrial animals and have occupied virtually every possible ecological niche on land, hexapods have also invaded freshwater habitats and are absent only from the subtidal waters of the sea (which, admittedly, is most of the planet). Their success can be attributed to a number of factors, but surely the arthropod body plan, resistance to desiccation, flight, and holometabolous (indirect) development are the most important. In holometabolous insects, immatures and adults are morphologically and ecologically very different, permitting resource partitioning, avoidance of competition, and specialization for different but complementary roles in the life cycle. It is no coincidence that the four most speciose insect taxa (beetles; butterflies and moths; wasps, bees, and ants; and flies, for a total of 740,000 species) are holometabolous.

Several features distinguish hexapods from other tracheates. The trunk is divided into a thorax and abdomen. Adults, and usually immatures (juveniles), have three pairs of legs, hence the name "hexapod." Adults of most taxa have two pairs of wings (Fig. 21-1C), although wings are not present in the most primitive hexapods, extinct or living. The legs and wings are always on the thorax. Only one pair of antennae is present (Fig. 21-1A,B). A tracheal system provides for gas exchange, the gonoducts open at the posterior end of the abdomen, and Malpighian tubules are the excretory organs.

The armored cuticle of the arthropod body plan provides protection from predators, pathogens, and desiccation as well as attachment sites for skeletal muscles. Adaptations for water conservation such as tracheae, Malpighian tubules, and a waterproof cuticle and eggshell made hexapods one of the few taxa of well-adapted terrestrial organisms. Colonization of the land exposed hexapods to a variety of new habitats that favored speciation and adaptive radiation to an extent not experienced by the crustaceans in their more uniform marine environment. The evolution of flight endowed them with a distinct advantage over other terrestrial invertebrates. Dispersal, escape from predators, and rapid access to food and optimum habitat are facilitated by flight. Powered flight also evolved in reptiles, birds, and mammals, but the first flying animals were hexapods. Primitive hexapods (Entognatha) lack both flight and efficient mechanisms for avoiding desiccation and have not shared in the success of their winged descendants (Insecta).

Hexapods are of great ecological, economic, and medical importance. Two-thirds of all flowering plants depend on them for pollination. Mosquitoes, lice, fleas, bedbugs, and a host of biting flies are responsible for much human misery, and many are vectors of important diseases of humans and domesticated animals. A partial list includes mosquitoes (malaria, elephantiasis, encephalitis, West Nile virus, and yellow fever), the tsetse fly (sleeping sickness), lice (typhus and relapsing fever), fleas (bubonic plague), and the housefly (typhoid fever and dysentery). Crops are damaged by the abundance of herbivorous species. Vast sums are expended to control insect pests, which can greatly reduce the agricultural yields necessary to support an exponentially expanding human population. The overzealous use of pesticides, however, can be counterproductive and hazardous to the environment and human health.

Entomologists today divide the six-legged tracheates (the hexapods) into two sister taxa, Entognatha and Insecta (Fig. 21-23). The primitive Entognatha consists of 4 taxa (orders) whereas the derived and much larger Insecta includes 30.

FORM

The hexapod body consists of three tagmata, a head, thorax, and abdomen (Fig. 21-1C). The head consists of an unknown number of segments, believed to be between three and seven, fused to form a rigid unit. In a recent interpretation based on studies of the fruit fly *Drosophila,* the head consists of an archecerebrum and seven segments, which are, in order from anterior to posterior, the labral, ocular, antennal, intercalary, mandibular, maxillary, and labial segments. The last three are postoral, bear mouthparts, and are known collectively as the gnathal segments. The segments bear the appendages or structures indicated by their names. For example, the antennal segment bears a pair of antennae. The intercalary segment appears in embryos and then regresses without developing appendages. In the ancestor, however, this segment bore a second pair of antennae homologous to those of crustaceans, but these have been lost in Hexapoda.

The head skeleton is a complete, rigid, external **head capsule** enclosing and protecting the soft inner tissues. An elaborate system of apodemes, the **tentorium,** braces the interior of the head capsule and provides attachment sites for muscles.

The head typically and plesiomorphically bears a pair of compound eyes and up to three unpaired ocelli (Fig. 21-1A). Hexapods have a single pair of antennae (Fig. 21-1A,B) located on the anterior head. Antenna morphology is varied, but the simplest form is a filament (Fig. 21-1A).

The mouth opens ventrally from a preoral cavity formed by the surrounding mouthparts (Fig. 21-7). Anteriorly, the preoral cavity is bordered by the labrum, which is a sclerotized, platelike extension of the head sometimes referred to as the upper lip (Fig. 21-1A,B) that is attached by a flexible articulation to the front of the head. It may represent the fused appendages of the labral segment. Its muscles can raise and lower it. Posterior to the mouth is an unpaired hypopharynx (Fig. 21-7, 21-1A). This tonguelike structure is a soft median outgrowth of the body wall that may aid in pushing food into the mouth, but it is not a segmental appendage. A pair of heavily sclerotized mandibles lies lateral to the mouth and forms the sides of the preoral cavity (Fig. 21-1A,B). The first maxillae, which are the least

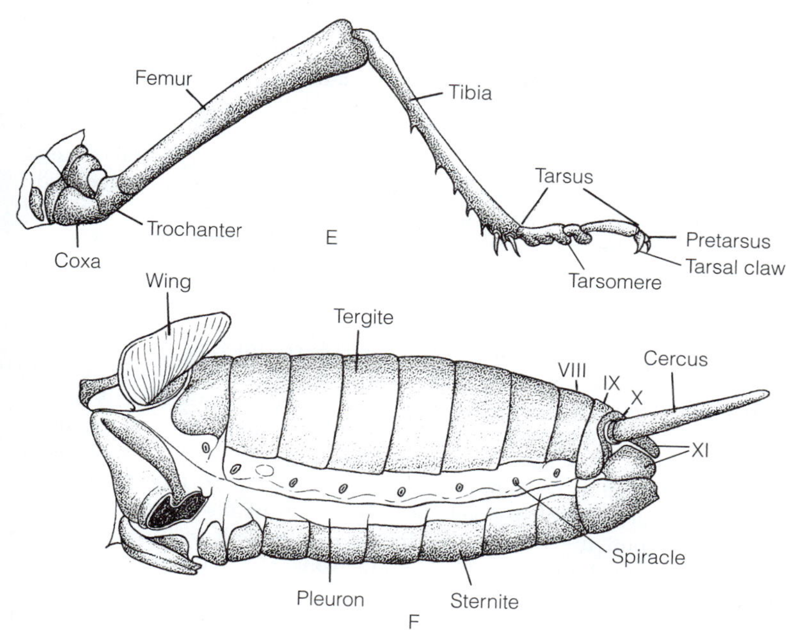

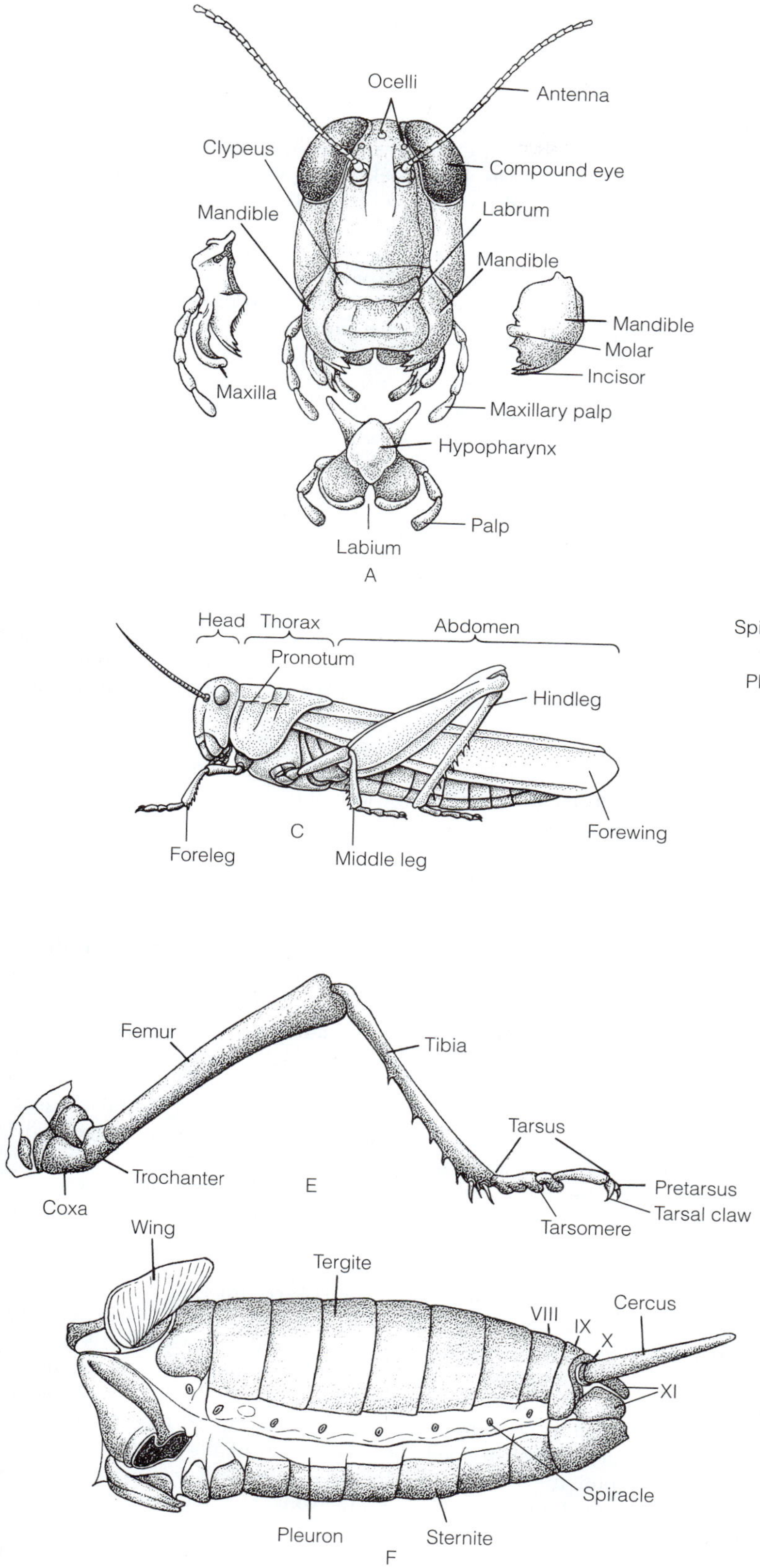

FIGURE 21-1 External morphology. **A,** Anterior surface of the head of a grasshopper showing the typical mandibulate mouthparts. **B,** Lateral view of the head of a grasshopper. **C,** Lateral view of the body. **D,** Lateral view of a wingless thoracic segment. **E,** Leg of a grasshopper. **F,** Lateral view of the abdomen of a male cricket. *(A and B, After Snodgrass from Ross. D and E, From Snodgrass, 1935.)*

modified of the mouthparts, are toothed in many species and may assist in macerating the food. The labium, formed by the fusion of the two second maxillae, is the posterior wall of the preoral cavity. The maxillae and labium bear sensory, antenna-like palps (Fig. 21-1A,B).

The thorax always consists of three segments: the **prothorax, mesothorax,** and **metathorax** (Fig. 21-14). Tergites (known as nota when on the thorax) and sternites are usually sclerotized, but the pleurites are usually soft except for a few small sclerotized areas (Fig. 21-1D). A pair of legs (**forelegs, middle legs,** and **hindlegs,** respectively) articulates with each of the three segments (Fig. 21-1C). A hexapod leg usually consists of six articles, which, in order from proximal to distal, are the **coxa, trochanter, femur, tibia, tarsus,** and **pretarsus** (Fig. 21-1E). The tarsus may be divided into as many as five smaller units and the pretarsus usually consists of a pair of **tarsal claws** with a padlike **arolium** between them. Articulations between leg articles may be mono- or dicondylic.

The legs of hexapods generally are adapted for walking or running, and during their effective stroke, the forelegs pull while the middle and hindlegs push. The middle legs step outside of the other two pairs, reducing interference. One or more pairs may be modified for such functions as grasping prey or a mate, jumping, swimming, or digging.

Most adult insects have two pairs of wings, **forewings** on the mesothorax and **hindwings** on the metathorax (Fig. 21-19B). In many taxa one or both pairs of wings is absent, and frequently one pair is modified for a function other than flight.

The abdomen usually is composed of 9 to 11 segments (Fig. 21-1F) plus a telson, but the telson is complete only in embryos and some primitive taxa. A typical abdomen consists of the pregenital (segments 1 through 7), genital (8 and 9), and postgenital (10 and 11) segments. The female gonopore is on segment 8, the male on 9. The postgenital segments are reduced and represented by a few small sclerites (Fig. 21-1F, 21-11B, 21-12B).

Most abdominal segments lack appendages. The pregenital segments of higher insects do not have appendages, although some Entognatha have small sensory styli that may be derived from appendages. Adult abdominal appendages include a terminal pair of sensory **cerci** (sing., cercus) on the 11th segment (Fig. 21-1F) and the external genitalia of some taxa. Cerci in some species are long, antenniform, and sensory. A variety of abdominal appendages serving various functions are present in many larvae.

WINGS AND FLIGHT

Even though wings are a characteristic feature of insects, many hexapods are wingless. In some, the lack of wings is secondary. For example, ants and termites have wings only at certain periods of the life cycle, and workers, although adult, always lack wings. Some parasitic insect orders, such as lice and fleas, have lost the wings completely. On the other hand, winglessness is primary in primitive hexapods (Entognatha, Zygentoma, and Archaeognatha; Fig. 21-18). These taxa arose from ancestors that never had wings and are known collectively as **apterygotes.** Most insects, however, either have wings or, if wingless, had winged ancestors and are **pterygotes.**

Wings are not segmental appendages or homologous to legs and mouthparts, nor are they constructed of articles. Instead, they are very thin folds of the body wall. As such, each wing consists chiefly of a double sheet of epidermis and cuticle strengthened by sclerotized areas known as **veins** (Fig. 21-19B). Veins are chiefly structural, are hollow, and contain tracheae, nerves, hemocoel, and blood. Wings articulate with the tergal and pleural sclerites of the meso- and metathorax. Fully developed functional wings are present only in adults, although thick, nonfunctional precursors called **wing pads** are present in the juveniles of some insects.

The evolution of wings and flight in insects is a matter of conjecture. The earliest known fossil hexapod is a bristletail, an apterygote from the early Devonian period. Winged insects appear later, but no intermediates have been discovered. The most widely accepted hypothesis is that wings were originally flat, lateral flanges of the tergum that helped the insect to alight upright after jumping. Subsequent enlargement of the flanges made gliding possible. Finally, the development of specialized muscles and hinges permitted the wings to move.

Primitively, wings are held outstretched, as in dragonflies (Fig. 21-19B), and cannot be folded. The evolution of sclerites in the wing base, which permits the insect to fold the wings over the abdomen and thus move them out of the way when at rest, was an important evolutionary event. This innovation enabled radiation into interstitial microhabitats beneath bark and stones and in soil, dung, and wood, where outstretched wings would be a serious handicap. Wing folding was probably accompanied by reduction in the body size of many taxa.

Each wing articulates with the edge of a thoracic tergite and rests on a vertical, sclerotized pleural process that acts as a fulcrum (Fig. 21-2A), making the wing resemble an off-center seesaw. Wing movement is accomplished by the large **flight muscles,** which may fill most of the space in the adult thorax. Flight muscles may insert on the wings, in which case they are **direct muscles.** Alternatively, **indirect muscles** have no direct connection with the wings. A set of **vertical indirect muscles** extends from the sternum to the tergum and an antagonistic set of **longitudinal indirect muscles** extends between apodemes at the anterior and posterior ends of the thorax (Fig. 21-2A,B). **Posterior direct muscles** insert on the posterior edge of the base of the wing, distal to the fulcrum (pleural process), and **anterior direct muscles** insert on the anterior edge of the base of the wing, also distal to the fulcrum.

Because early flying insects lacked indirect muscles, power was provided by direct muscles. The most derived modern insects, in contrast, provide power to the wings by indirect muscles and the direct muscles serve only to control wing orientation. In Recent insects, contraction of the vertical indirect muscles lowers the tergum and elevates the wing, which is attached to it (Fig. 21-2A). Downward movement is accomplished in different ways in different taxa, depending on their stage in the evolution of flight. In primitive insects such as dragonflies and roaches downward movement is still accomplished solely by contraction of direct muscles, which pull the wing down. In some intermediate insects, such as grasshoppers and beetles, the wing is moved down by a combination of indirect longitudinal muscles and direct muscles. In derived insects such as wasps, bees, and flies, contraction of

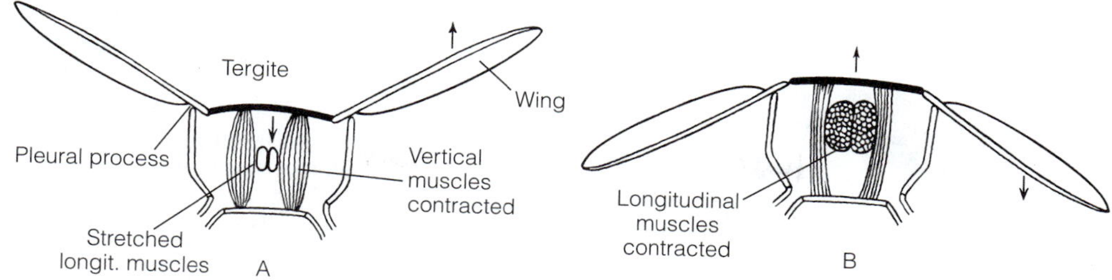

FIGURE 21-2 Flight: Diagrams showing relationship of wings to tergite and pleura and the mechanism of the basic wing strokes in an insect. **A,** Upstroke resulting from the depression of the tergite by contracting the vertical muscles. **B,** Downstroke resulting from the arching of the tergite by contracting the longitudinal muscles. *(After Ross, H. H. 1965. A Textbook of Entomology. 3rd Edition. John Wiley and Sons, New York.)*

the longitudinal indirect muscles shortens the thorax and antagonizes the vertical muscles. This pushes the tergum up and moves the wing down (Fig. 21-2B) so that the propulsive force is supplied entirely by indirect muscles.

Simple up and down movement alone is not sufficient for flight. The wings must at the same time be moved forward and backward and their angle changed. These movements are accomplished by the anterior and posterior direct muscles. The anterior muscle pronates the wing (by depressing the leading edge and elevating the trailing edge) and the posterior muscle supinates the wing (by depressing the trailing edge to elevate the leading edge). In addition to altering the angle at which the wing is held, insects can also achieve lift by raising and lowering selected wing veins, thereby changing the wing shape, or contour.

Flying ability varies greatly among insects. Many butterflies and damselflies have a relatively slow wing beat and limited maneuverability. At the other extreme, some flies, bees, and moths are very agile and can hover in one place or dart rapidly. From the standpoint of maneuverability, a housefly can outperform any bird. Not only can a housefly fly a rapid straight course and hover, it can fly upside down and turn in the distance of one body length. The fastest flying insects are hummingbird moths and botflies, which have been timed at 40 kph (25 mph). Honeybees can cruise at 24 kph (15 mph). Gliding, which is important in birds, occurs in only a few large insects.

Horizontal stability is maintained in part by a **dorsal light reaction** in which the insect keeps the dorsal ommatidia of the compound eyes under maximum illumination from above. Deviation because of rolling is corrected by slightly changing the wing position to bring the dorsal ommatidia back to maximum illumination. Many insects exhibit a general positive phototaxis in which they orient toward the light by adjusting the direction of flight so that illumination is equal in both eyes. A consequence is that they fly, with great persistence, into lamps.

Members of Diptera (flies, gnats, and mosquitoes) have hind wings reduced to stalked knobs called **halteres** (Fig. 21-22D). These beat with the same frequency as the forewings and function as gyroscopes to offset flight instability. Receptors on the haltere base detect tendencies to deviate from course due to pitch, roll, and yaw, and compensating corrections in wing position can be made. The flight muscles are striated and the fibers are large and powerful, their mitochondria are huge, and the respiratory rate is high. The high oxygen demand is met by a profusion of tracheae.

NUTRITION

There is great diversity in hexapod diets and in modification of mouthparts to obtain and process food. Herbivory is common, but many are carnivores, parasites, or parasitoids. The original hexapod mouthparts were adapted for cutting and chewing and are said to be mandibulate, but in taxa with liquid diets they are modified.

Mandibulate mouthparts (Fig. 21-1A,B) occur in many taxa, including apterygotes, dragonflies, crickets, grasshoppers, beetles, wasps, ants, termites, and many others. The mandibles, or jaws, are hard and heavily sclerotized, with the distal end equipped with a molar for chewing and an incisor for cutting. The molar predominates in herbivores but carnivores have a well-developed incisor. Insect mandibles are dicondylic whereas those of entognathans are monocondylic. The dicondylic articulation restricts the mandibles to motion in the transverse plane (in the plane of the page as you look at Figure 21-1A and perpendicular to the page when you look at 21-1B). The cutting and grinding surfaces of the two mandibles meet across the midline ventral to the mouth.

Many insects specialize in a liquid diet consisting of either plant or animal juices, for which the original mandibulate mouthparts are unsuited. These insects have mouthparts specialized in many ways for sucking fluids in conjunction with a muscular pump. Sucking evolved several times in insects, and consequently there are many morphological variations sometimes adapted for more than one function, such as chewing and sucking, cutting and sucking, piercing and sucking, or just plain sucking.

The **sucking mouthparts** of moths and butterflies (Lepidoptera) are adapted for pumping nutritious liquids, without piercing, from various sources. A long tube, or proboscis, formed from the highly modified maxillae (Fig. 21-3). At rest, the proboscis is coiled under the head and then unrolled to feed. Except for the large labial palps, the other mouthparts are absent or vestigial. The morphology and function of hexapod mouthparts and the digestive tract may

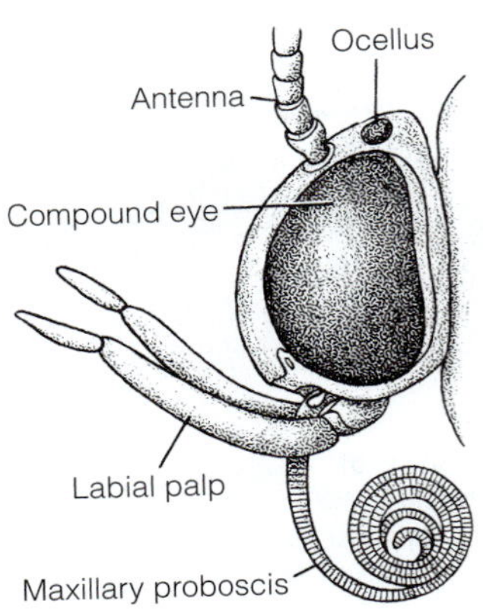

FIGURE 21-3 Sucking mouthparts: Lepidoptera. Lateral view of the head of a moth *(After Snodgrass, 1935.)*

change with life-history stage. In Lepidoptera, for example, the larvae are herbivorous and have chewing mouthparts and a gut adapted for processing solid plant material. Adults, on the other hand, have sucking mouthparts and a gut specialized for digesting liquid food. Although lepidopterans are well known for utilizing the nectar from flowers, such a diet, while rich in carbohydrates and energy, is deficient in nitrogen and salts. The nectar diet is supplemented by drinking the liquid from tears, urine, slug slime, feces, animal carcasses, and mud puddles.

Piercing and sucking mouthparts are characteristic of herbivorous insects such as aphids, leafhoppers, cicadas, and stinkbugs (Hemiptera), which feed on plant juices, but they are also found in many predaceous insects, such as assassin bugs (Hemiptera) and mosquitoes (Diptera), which feed on blood. These mouthparts are organized in different ways in different taxa to form a **beak,** which supports a **stylet** that is inserted into the prey. For example, the beak of bugs (Hemiptera) consists of a piercing stylet supported in a groove on the heavier labium (Fig. 21-4). The stylet is derived from the highly modified maxillae and mandibles and contains a **salivary channel** for the outward passage of saliva and a **food channel** for the inbound flow of liquid food. Only the stylet penetrates the prey. The labial portions support the stylet but remain outside.

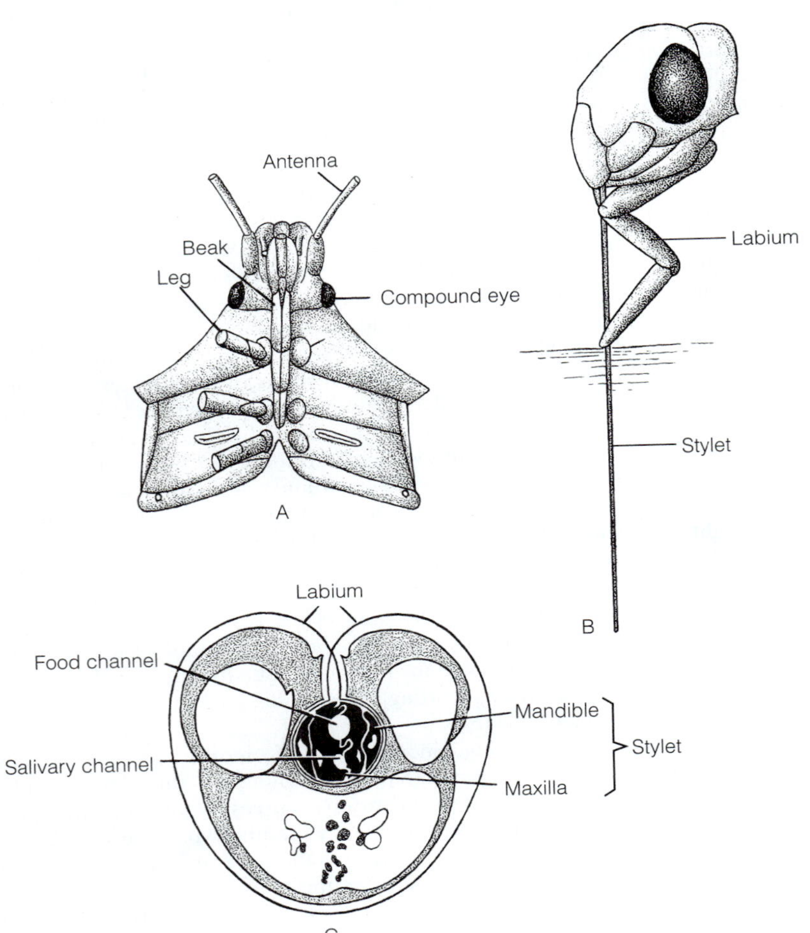

FIGURE 21-4 Piercing and sucking mouthparts: Hemiptera. **A,** Ventral view of anterior half of a hemipteran showing the beak. **B,** A hemipteran penetrating plant tissue with its stylet. The labium supports the delicate stylet but remains outside the prey. **C,** Cross section through a hemipteran beak, showing the food and salivary channels in the stylet formed by the maxillae and mandibles, which are enclosed by the labium. *(A, After Hickmann. B, After Kullenberg. C, After Poisson.)*

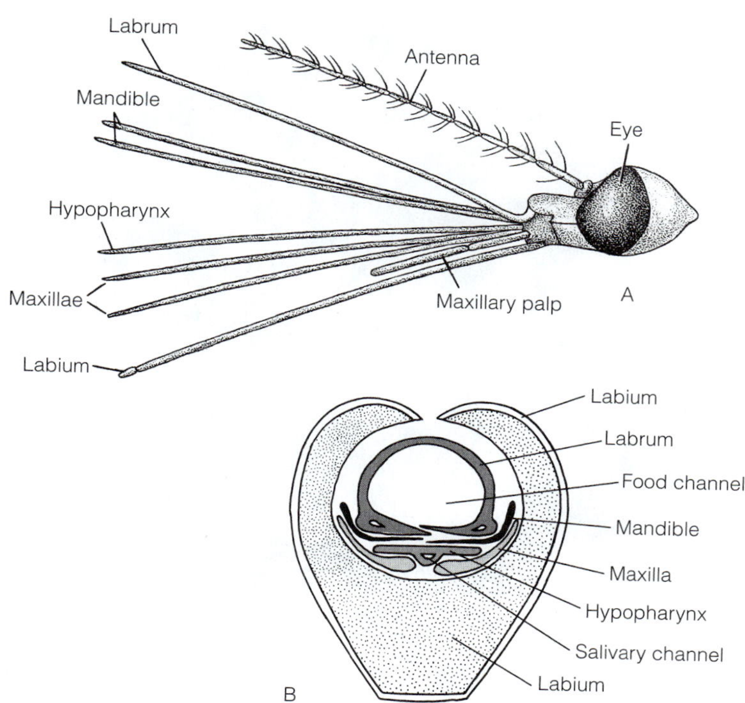

FIGURE 21-5 Piercing and sucking mouthparts: Diptera. **A,** Lateral view of the head of a mosquito, showing the mouthparts widely separated for clarity. **B,** Cross section of mouthparts of a mosquito in their normal functional position with the parts combined to form a stylet. *(After Waldbauer from Ross.)*

There are many families of biting flies (Diptera) that feed, at least some of the time, on the blood of vertebrates. In most, but not all instances it is only the female that takes a blood meal, which she requires to produce eggs. Males (and females, part of the time) feed on plant juices—primarily nectar—but both sexes utilize liquid diets. The biting flies include mosquitoes, horseflies, deerflies, blackflies, no-see-ums (punkies), and others having a variety of mechanisms for feeding on blood.

Mosquitoes have piercing and sucking mouthparts in which the elongate labrum forms a food channel for incoming blood and is supported by the labium, mandibles, and maxillae (Fig. 21-5). The hypopharynx forms a small outbound channel for saliva, which contains an anticoagulant to facilitate the flow of blood through the narrow food channel. The itching associated with a mosquito bite is a reaction to proteins in the saliva.

Blackflies, horseflies, and deerflies have **cutting and sponging mouthparts,** and their bites can be more painful than those of mosquitoes. The mandibles are tiny scalpels (Fig. 21-6A) used to slice the skin of the prey. The spongelike labium (Fig. 21-6B) soaks up the blood as it oozes from the wound. From the sponge, blood moves into the mouth via a short tube formed from the hypopharynx and labrum. Houseflies, by the way, do not bite, but many closely related and similar flies, such as face flies and stable flies, do.

Sponging alone, without cutting or piercing, is a feeding strategy employed by several taxa of nonbiting flies, including the houseflies (Fig. 21-6C) and blowflies. These dipterans have a spongelike labium, but the reduced mandibles and maxillae are not formed into blades. The labium is used to absorb liquid from various sources, such as juicy meat or fruit, without first lacerating the source. These insects, which, like all dipterans, are restricted to liquid foods, can liquify solid food with the enzymatic activity of their saliva. Saliva is exuded through the labium onto the solid food and the resulting fluid then can be sucked into the preoral cavity and mouth.

Bee and wasp mouthparts are adapted for both **chewing and sucking.** A bee, for example, sucks nectar through the elongate maxillae and the labium. Pollen and wax are processed by the labrum and mandibles, which have retained the chewing morphology.

The hexapod gut is preceded by the preoral cavity, which is outside the gut tube (Fig. 21-7). The preoral cavity is divided into the space between the labrum and hypopharynx (the **cibarium**) and the space between the hypopharynx and labium (the **salivarium**). The mouth opens from the dorsal end of the cibarium and the salivary duct opens into the salivarium. The muscles of the cibarium form a sucking **cibarial pump** in many hexapods with liquid diets.

Food in the preoral cavity is mixed with saliva and sometimes with regurgitated enzymes from the midgut. Digestion may begin here prior to ingestion. The mouth opens into the foregut, or stomodeum, which usually is divided into an anterior pharynx, esophagus, crop, and narrower proventriculus (Fig. 16-9, 21-8A). Food is moved through the gut with peristaltic contractions of circular muscles, and in sucking insects the pharynx may be a muscular pump. The narrow tube of the esophagus connects the pharynx with the proventriculus. In many species the posterior region of the esophagus is expanded to form a crop for the storage of undigested food, although some digestion may occur here. In hexapods that eat solid food, the proventriculus usually is a muscular gizzard with elaborate cuticular teeth for grinding food (Fig. 21-8B).

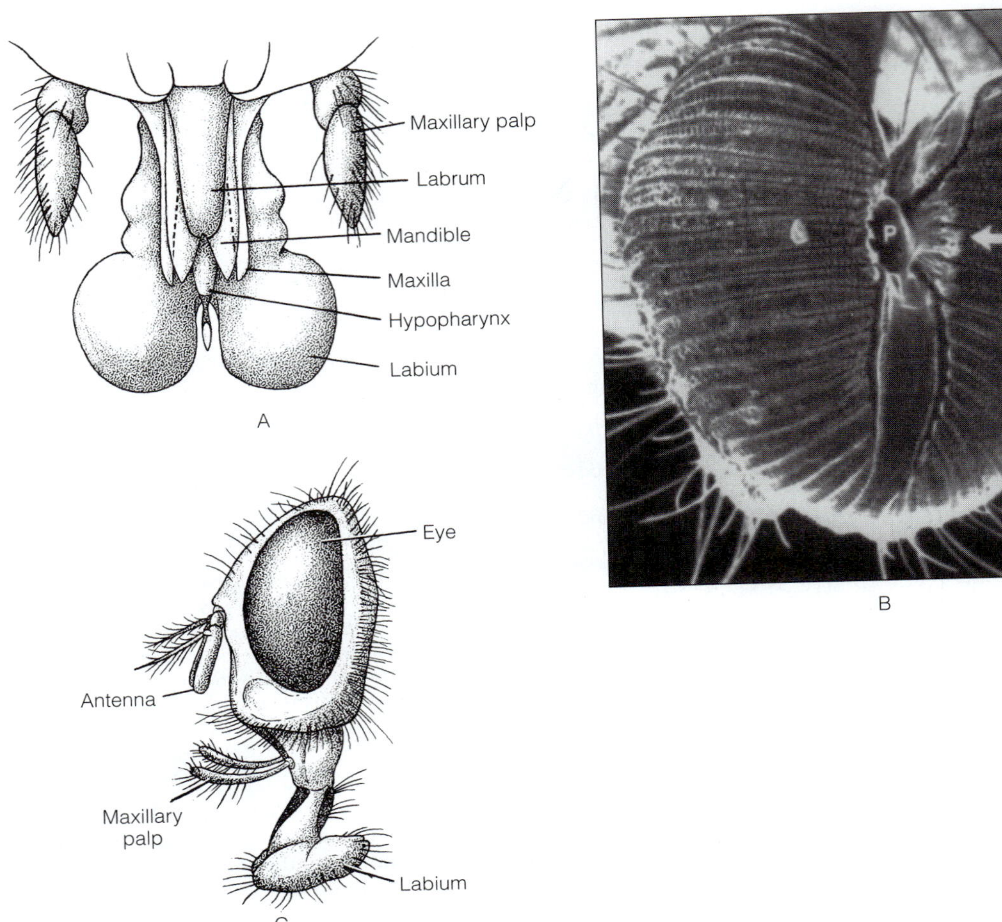

FIGURE 21-6 Sponging mouthparts. **A,** Mouthparts of a blackfly, adapted for cutting and sponging. **B,** Scanning electron micrograph of the ventral surface of the spongelike labium of a face fly. *P* marks the opening to the preoral cavity. The mouth is at the opposite end of the preoral cavity. **C,** Lateral view of the head and sponging mouthparts of a housefly. *(A, After Ross, H. H. 1965.* A Textbook of Entomology. *3rd Edition. John Wiley and Sons, New York. C, After Snodgrass, 1935.)*

In sucking insects with a liquid diet, the proventriculus is only a simple valve opening into the midgut (Fig. 21-8C). Sometimes it is a filter. In fleas, which feed on blood, spines that project posteriorly from the foregut into the midgut are used to rupture blood cells.

Most hexapods have a pair of salivary glands (labial glands) lying below the midgut and opening by a common duct into the salivarium (Fig. 21-7, 21-8A). Mandibular glands, in addition to salivary glands, are functional in apterygotes and a few pterygotes. (Salivary glands are not homologous to the labial nephridia of Entognatha.) The function of salivary glands varies, but they usually secrete saliva, which moistens the mouthparts and lubricates the food. Salivary glands may also produce digestive enzymes, especially an amylase that is mixed with the food mass before it is swallowed. Other secretions include silk, mucus, hyaluronidase to hydrolyze connective tissue, a pectinase (in aphids) to hydrolyze the pectin of cell walls, venom (in some predators), agglutinins and anticoagulants, and the antigen that produces the mosquito-bite reaction in humans.

In lepidopterans (moths), trichopterans (caddisflies), psocids, and hymenopterans (bees, ants, and wasps), the salivary glands are silk glands whose secretion, silk, is used to make pupal cells, cocoons, filters, draglines, and domiciles. The proteinaceous silk is secreted through a single median pore at the apex of the labium. In addition to salivary glands, hexapods have other types of silk-secreting glands, such as modified accessory sex glands, tarsal glands, and Malpighian tubules.

A valve (Fig. 21-8B) separates the foregut from the midgut. The hexapod midgut (the ventriculus, or stomach) usually is tubular and, as in other arthropods, it is the principal site of enzyme synthesis, secretion, hydrolysis, and absorption. There is some evidence that, although the entognathan midgut is endodermal as expected, that of the insects may be ectodermal.

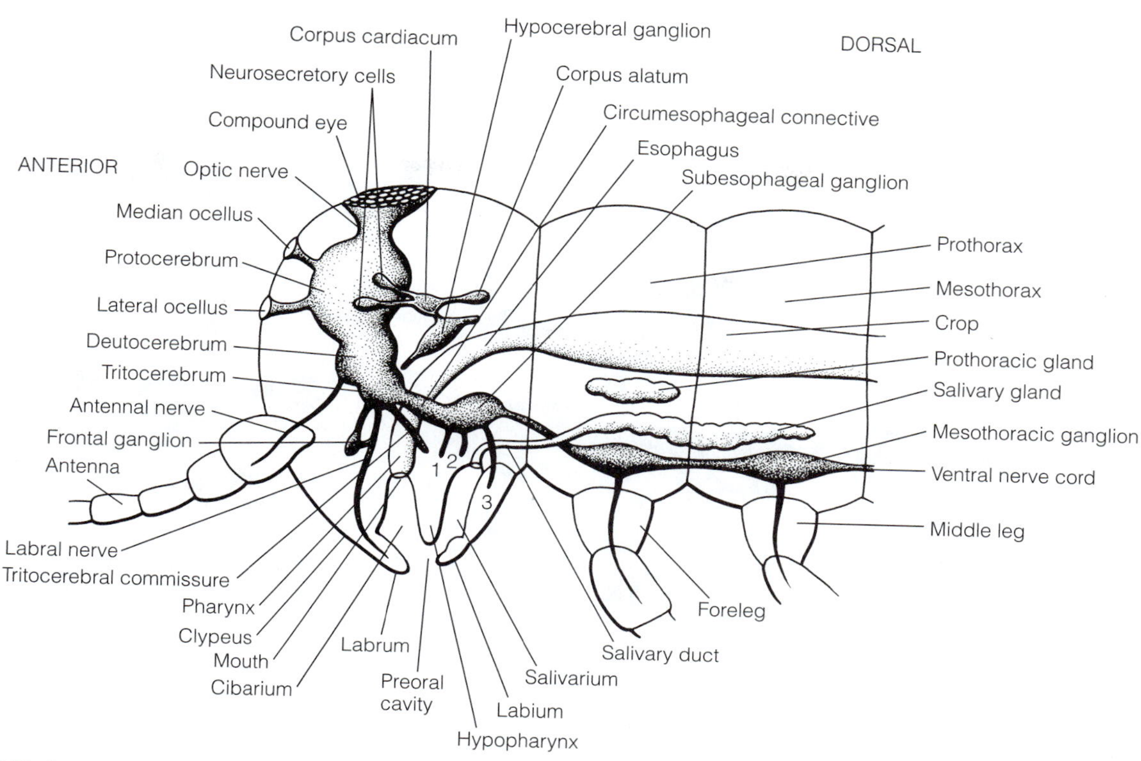

FIGURE 21-7 Anatomy. Lateral view of the left side of the head and anterior thorax of a typical hexapod. The mandible and maxilla have been removed to reveal the hypopharynx and preoral cavity. The compound eye is displaced dorsally. 1, mandibular nerve; 2, maxillary nerve; 3, labial nerve. *(Modified from several authors, including Klausnitzer, B. 1996. Insekta (Hexopoda), Insekten. In Westheide, W., and Rieger, R. M. (Eds.): Spezielle Zoologie, Erster Tiel: Einzeller und Wirbellose Tiere. Gustav Fischer Verlag, Stuttgart. pp. 601–681; and Snodgrass, R. E. 1935. Principles of Insect Morphology. McGraw-Hill, New York. 667 pp.)*

The peritrophic membrane is a characteristic feature of the midgut of most hexapods, as it is of many other arthropods (Fig. 21-9). This thin, porous, acellular, secreted membrane forms a tube inside the gut tube that encloses the food mass as it moves through the mid- and hindguts. The membrane protects the delicate midgut walls (which are not lined with cuticle, as the rest of the gut is) from abrasion by the food mass. It tends to be poorly developed or absent in insects with liquid diets.

The peritrophic membrane consists of a matrix of chitin microfibrils supporting a ground substance of proteins, glycoproteins, and proteoglycans. It is usually a mesh rather than a solid sheet and is perforated by regularly spaced pores of about 0.2 μm in diameter. Microvilli from the midgut epithelium sometimes protrude through these pores into the lumen of the membrane. The cells lining the midgut usually secrete the peritrophic membrane, but in some hexapods it originates in the proventriculus.

The peritrophic membrane partitions the gut lumen into two compartments. The lumen of the peritrophic tube is the endoperitrophic space whereas the area between the membrane and gut wall is the ectoperitrophic space (Fig 21-9). The membrane probably plays an important role in digestion by partitioning enzymes into compartments. Some enzymes secreted by the midgut epithelium pass through the membrane into the endoperitrophic space, where initial hydrolysis occurs. The products of this hydrolysis then pass into the ectoperitrophic space, where intermediate digestion occurs using different enzymes. The products are then absorbed by epithelial cells, where final, intracellular digestion occurs.

The midgut usually has two to six digestive ceca (gastric ceca) at its anterior end (Fig. 21-8A,B), and these can be important sites of hydrolysis and absorption. Water is absorbed by the cecal epithelium and transferred to the hemocoel, although this also occurs in the rectum (Fig. 21-9). In some hexapods, populations of symbiotic bacteria are maintained in the ceca.

The hindgut, or proctodeum, consists of an anterior intestine and a posterior rectum, both of which are lined with cuticle (Fig. 21-8A, 21-9). The anus opens at the posterior end of the abdomen (Fig. 21-11B). A muscular pyloric sphincter may separate the intestine from the midgut. The hindgut functions in the egestion of digestive and excretory waste and in water and salt balance. In most hexapods, **rectal pads** or similar organs that occur in the epithelium are the principal sites of water reabsorption from the feces. In terrestrial hexapods of arid habitats, the hindgut reabsorbs all water and produces completely dry feces. Digestion of cellulose by termites utilizes cellulases produced by mutualistic flagellates that inhabit the hindgut.

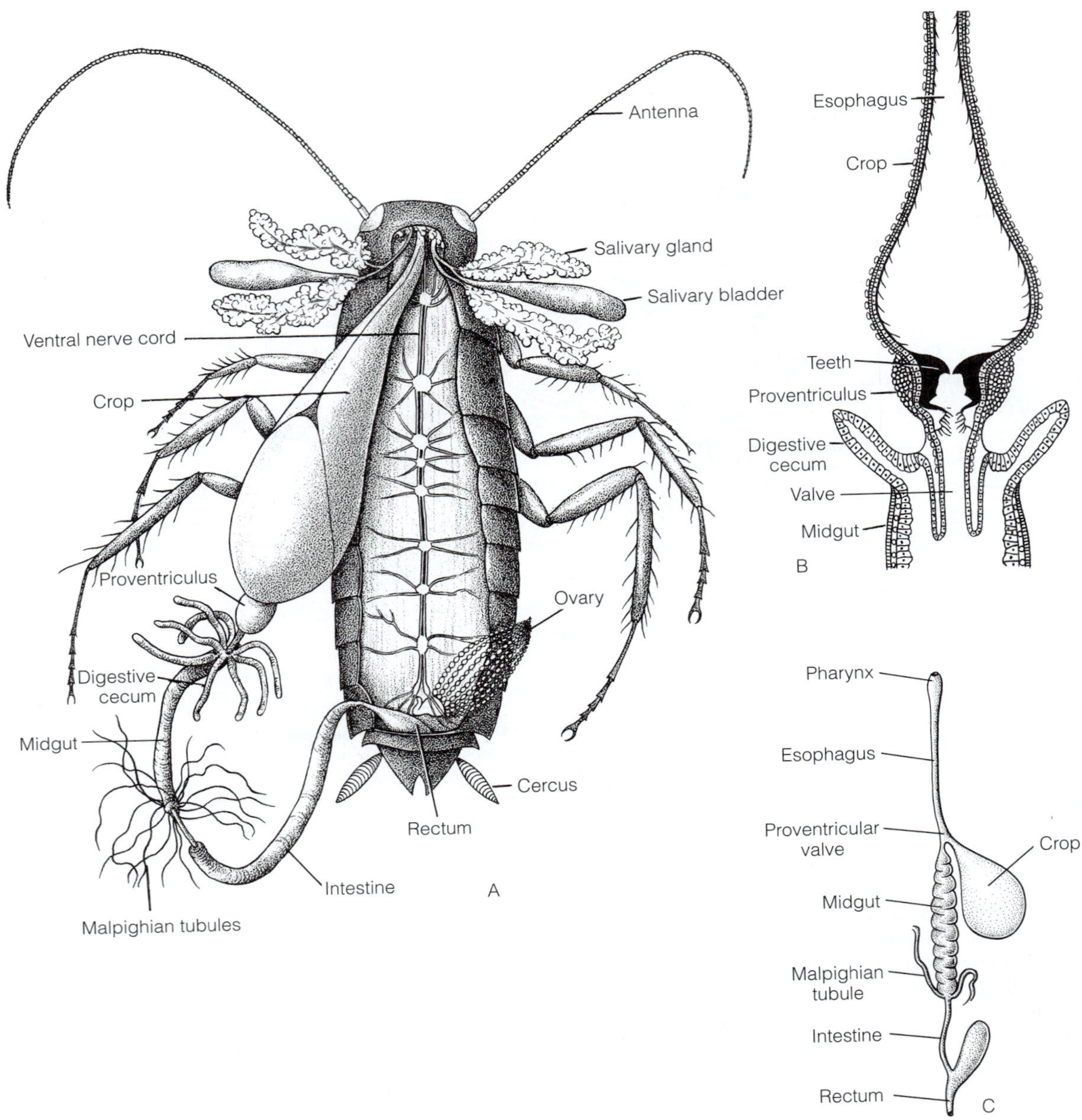

FIGURE 21-8 Digestion. **A,** Internal anatomy of a cockroach. **B,** Longitudinal section through the foregut and anterior part of the midgut of a cockroach. **C,** Digestive tract of a moth, which uses sucking mouthparts to feed on fluids. *(A, After Rolleston. B, After Snodgrass, 1935. C, From Elzinga, R. J. 1987. Fundamentals of Entomology. 3rd Edition. Prentice-Hall, Englewood Cliffs, NJ.)*

A **fat body,** analogous in many respects to annelid chlorogogen tissue and the vertebrate liver, are present in the hemocoel (Fig. 16-7). Synthesis and long- and short-term storage of lipids, carbohydrates, and blood and other proteins occur in the fat body. Glycogen reserves in the fat body can be rapidly mobilized to release glucose into the blood in response to hormones. The many insects that do not feed as adults rely on fats, proteins, and glycogen stored in the fat bodies during their immature stages. The fat body arises from the mesoderm of the embryonic coelomic pouches.

INTERNAL TRANSPORT

The hexapod hemal system consists of the heart, accessory hearts, hemocoel, and blood. With the exception of the aorta, blood vessels are absent. The heart is a contractile middorsal

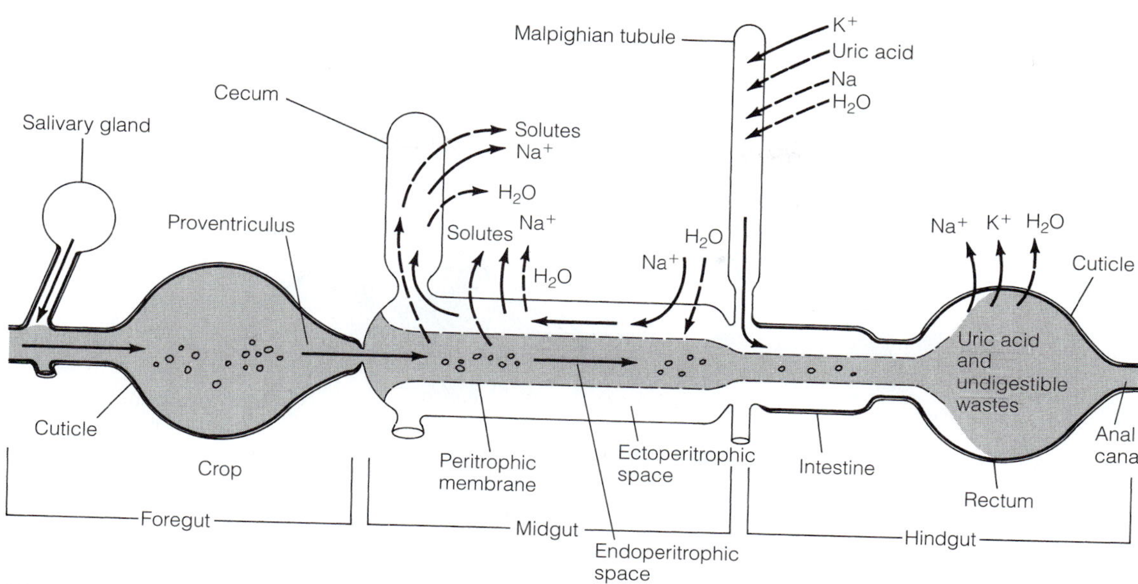

FIGURE 21-9 Digestion: Diagram of the digestive tract of a hexapod, showing the passage of food (small circles) through the gut, the absorption of food products in the cecum, and the secretion of wastes by a Malpighian tubule. Active transport of salts out of the gut (solid arrows) leads to passive diffusion of water and other substances (dashed arrows). *(Modified from Berridge, M. J. 1970. A structural analysis of intestinal absorption. In Neville, A. C. (Ed.): Insect Ultrastructure. Symp. Roy. Ent. Soc. 5:135–151; and Evans, H. E. 1984. Insect Biology: A Textbook of Entomology. Addison-Wesley Publishing Co., Reading, MA. p. 85.)*

tube in a pericardial sinus similar to that of other arthropods (Fig. 16-7). It extends through the first nine abdominal segments and is perforated by 1 to 12 pairs of segmental ostia. Anteriorly it narrows and reaches into the head as the aorta, which lacks ostia and may or may not be pulsatile. Paired, segmental alary muscles that extend laterally from the heart to the body wall (Fig. 16-7) expand the heart lumen during diastole. Systole is a wave of contractions of the myoepithelial cells of the heart.

Periodic heartbeat reversal is common among hexapods. In Lepidoptera, Diptera, and Coleoptera, blood is pumped into the head hemocoel during forward beating and into the abdominal hemocoel during backward beating of the heart. Special paired excurrent ostia in the walls of the posterior heart accommodate the exit of blood during backward flow.

In the absence of arteries to deliver blood to extremities such as the antennae, mouthparts, legs, wings, and cerci, hexapods rely on a muscular **accessory heart** located at the base of the extremity it serves. These saclike hearts pump blood from the hemocoel into the appendage.

Many hexapods in temperate regions survive subfreezing temperatures by accumulating in the blood such compounds as glycerol, sorbitol, and trehalose, which depress the freezing point and function as antifreeze. Some species can supercool the blood and cellular fluids to −30°C without freezing, whereas others exhibit controlled freezing, permitting ice crystals to form only in extracellular spaces. Special proteins may be produced to act as nuclei for the formation of ice crystals in the desired locations.

GAS EXCHANGE

Gas exchange in hexapods employs a system of air-filled tracheae opening to the atmosphere by spiracles and delivering oxygen directly to the tissues independently of the hemal system. The typical (holopneustic) system includes 10 pairs of segmental spiracles—2 in the thorax, 8 in the abdomen (Fig. 21-1D,F, 21-10B), and none in the head. Tracheae arise at the spiracles, branch repeatedly, and become progressively smaller as they approach their target cells. The spiracles are connected by two major **longitudinal tracheal trunks** that extend the length of the body and branch to the head tissues, heart, nerve cord, muscles, and gut. The longitudinal trunks are connected by transverse tracheae. The extent of the tracheal supply to each organ is related to its metabolic activity and oxygen requirements.

The simplest spiracles, as found in some entognaths, are merely holes in the integument. In most insects, however, the spiracles open into a pit, or atrium, from which the trachea arises (Fig. 21-10A). The spiracle usually has a closing mechanism and filtering devices. Closing reduces water loss and filtering prevents the entry of dust and parasites and also inhibits water loss.

Helical cuticular rings, the **taenidia** (Fig. 21-10A), allow the trachea to expand but prevent its collapse. The epicuticle of the tracheal exoskeleton lacks the waxy waterproofing layer typical of external parts of the exoskeleton. Tracheae are sometimes expanded in various places to form internal air sacs.

The smallest extremities of the tracheal system are **tracheoles** about 1 µm in diameter (Fig. 21-10C). These fine intracellular,

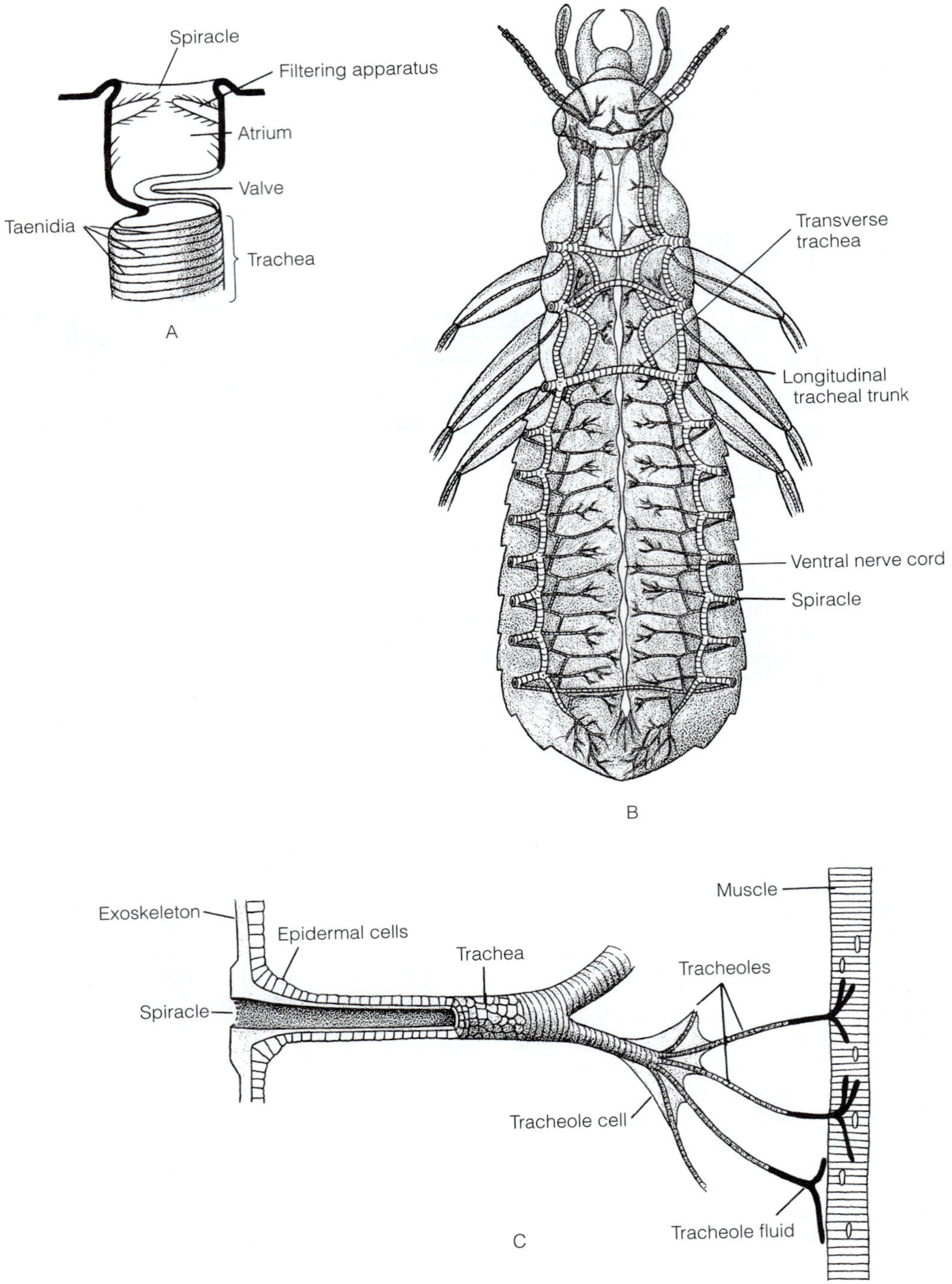

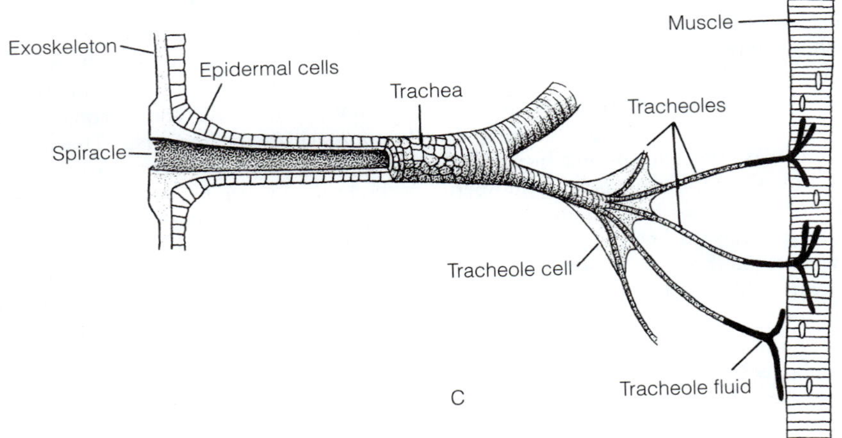

FIGURE 21-10 Gas exchange. **A,** A spiracle with the atrium, filtering apparatus, and valve. **B,** A holopneustic tracheal system of a hexapod. **C,** Diagram showing the relationships of a spiracle and tracheoles to a trachea. *(A, After Snodgrass, 1935. B and C, After Ross, H. H. 1965. A Textbook of Entomology. 3rd Edition. John Wiley and Sons, New York.)*

cuticular tubes are located in the cytoplasm of tracheole cells. Tracheoles extend to and physically contact their target cells and in some cases, such as flight muscles, even push into the cytoplasm and deliver oxygen directly to the target mitochondria. The tracheal cuticle is shed during molting after hydrolysis of the taenidia. Because the tracheoles are intracellular, they do not molt and instead retain their cuticle.

Gas movement in the tracheae is accomplished by simple diffusion and muscular ventilation. Ventilation usually moves air through the large tracheae near the spiracles whereas diffusion is responsible for transport in the small tracheoles near the tissues. Because hexapods must balance oxygen need and water loss, spiracles are kept closed as much as possible. Some very small hexapods of moist surroundings, such as proturans and some collembolans, lack tracheae, and gas exchange occurs over the general body surface. Early instars of some aquatic nymphs lack tracheae, but they are present in most pterygotes.

Aquatic juveniles usually have tracheae and aquatic adults always do. Many aquatic nymphs, such as stoneflies and mayflies, respire underwater and have a **closed tracheal system** without spiracles (apneustic). Blind-ended but air-filled tracheae arise in tracheal gills, which usually are lamellae or filaments extending from the body. Oxygen diffuses from the water, across the gill surface, and into the tracheae for distribution. Some aquatic larvae (such as mosquitoes), however, have an open tracheal system with one or two spiracles at the end of a breathing tube. The larva rises to the surface periodically and obtains air from the atmosphere through the spiracles. Aquatic adults visit the surface to breathe directly from the atmosphere via open tracheae or to replenish an air bubble held against the body surface by special, unwettable hydrofuge hairs. Spiracles open into the air bubble and the insect breathes from it when it is submersed. In some cases, as the oxygen in the bubble is depleted, more oxygen diffuses in from the surrounding water, allowing some aquatic insects with open tracheae to achieve independence, or near independence, of the atmosphere.

EXCRETION

Protein and nucleotide catabolism produces waste ammonia, but in terrestrial species the fat body converts it into uric acid or sometimes insoluble ammonium salts for excretion. Uric acid has the advantage of being insoluble, relatively nontoxic, and requiring little water for its elimination. The major excretory organs are Malpighian tubules and the rectum. Malpighian tubules are absent or vestigial in entognaths but present in most insects with the exception of aphids. Between 2 and 250 tubules, depending on taxon, lie free in the hemocoel, where they are bathed in blood rich in uric acid (Fig. 21-8A, 21-9).

Malpighian tubules are secretion kidneys that indiscriminately absorb ions, uric acid, water, and toxins from the hemolymph and secrete them into the tubule lumen to form the primary urine, which then moves into the hindgut. Desirable ions, sugars, and water are selectively reclaimed by the rectal epithelium and returned to the hemolymph (Fig. 21-9). Uric acid and a minimum of water remain in the rectum and are voided through the anus with the feces.

Storage excretion by specialized cells is important in some apterygotes with poorly developed or absent Malpighian tubules, or it may supplement excretion by Malpighian tubules. Urate cells in the fat body concentrate uric acid intracellularly. Nephrocytes, typically located on or near the heart (Fig. 16-7), absorb waste molecules from the blood by pinocytosis. Wastes may be stored or detoxified and returned to the blood.

A pair of saccate nephridia, the **labial kidneys,** is found in the head of some apterygotes such as Collembola and Zygentoma. Each consists of an end sac with podocytes and a tubule. A nephridiopore opens near the base of the labium. Since Malpighian tubules are poorly developed or absent in apterygotes, the saccate nephridia may be important excretory organs.

Of the terrestrial arthropods, insects are the best-adapted for water conservation. The epicuticle is impregnated with waxy compounds that reduce surface evaporation, and spiracle closure reduces evaporation from the tracheal system. The excretion of uric acid reduces the loss of water that usually accompanies protein metabolism and deamination. The reabsorption of water by the rectum further conserves water that would be lost through excretion and egestion. Hexapods are one of the few taxa of invertebrates that can produce urine that is hyperosmotic to the blood. As might be predicted, aquatic insects excrete ammonia rather than uric acid and salts are reclaimed by the hindgut.

NERVOUS SYSTEM

The nervous system is like that of other arthropods, although there tends to be less fusion of segmental ganglia (Fig. 21-7, 16-11). The protocerebrum receives the optic nerves from the eyes, the deutocerebrum serves the antennae, and the tritocerebrum innervates the labrum. The subesophageal ganglion is always composed of at least three pairs of fused ganglia controlling the mandibles, maxillae, and labium, the salivary glands, and some cervical muscles. The brain and subesophageal ganglia are connected by circumesophageal connectives. In some apterygotes and many larval pterygotes, the ventral nerve cord is a chain of three thoracic and eight abdominal ganglia, with the last one being the **terminal ganglion** consisting of three or four fused ganglia (Fig. 16-11, 21-8A). In derived insects, however, the nervous system is cephalized and the number of independent segmental ganglia is reduced. In the most derived insects all thoracic and abdominal ganglia are fused to form a large **thoracoabdominal ganglion.**

The **visceral nervous system** consists of the somatogastric system, unpaired ventral nerves, and a caudal sympathetic system. The somatogastric system includes the hypocerebral ganglion (Fig. 21-7), frontal ganglion, and nerves to the fore- and midguts. The ventral nerves are unpaired median nerves that arise from each trunk ganglion and send branches to the spiracles. The caudal sympathetic system consists of nerves running from the terminal ganglion of the nerve cord to the hindgut and sex organs.

The principal components of the **endocrine system** are neurosecretory regions in the protocerebrum, corpora cardiaca, corpora alata, and prothoracic glands (Fig. 21-7). Molting, reproduction, water resorption, heartbeat, and metabolic processes are under endocrine control.

SENSE ORGANS

Sensilla of many types are scattered over the hexapod body, but they are especially numerous on the appendages, most notably the antennae, palps, and legs. Most are bristles, pegs, scales, domes, and plates derived from simple setae. They include mechanoreceptors of many types, campaniform organs, chemoreceptors, tympanic organs, proprioceptors, chordotonal organs, hygroreceptors, and thermoreceptors, which were discussed in Chapter 16, Introduction to Arthropoda.

The antennae are major centers for sensory reception in hexapods and have olfactory, gustatory, mechanosensory, and thermo/hygroreceptors in most species. The tarsi also bear numerous sensilla, including gustatory and tactile receptors. Flies taste their food by walking on it.

The photoreceptors are ocelli and compound eyes. Ocelli are absent in many adult hexapods, but when present, there usually are two or three on the anterior dorsal surface of the head (Fig. 21-1A). The photoreceptor cells of each ocellus are organized more or less like those of a single ommatidium of a compound eye. Ocelli can detect changes in light intensity and may be very sensitive to low intensities. They function in orientation and appear to have a general stimulatory effect on sensory function by enhancing the reception of other receptors. The number of facets in the compound eyes is greatest in flying raptors that depend on vision for prey capture, and the facets are larger in nocturnal than in diurnal insects.

The visible spectrum for hexapods tends to be shifted toward shorter wavelengths in comparison with human vision. Most are sensitive to ultraviolet wavelengths of 300 to 400 nm. Phototaxis, both positive and negative, is often cued by UV rather than visible wavelengths. Color vision has been demonstrated in a few insects. Some can navigate using polarized light.

REPRODUCTION

Hexapods are gonochoric, fertilization is internal, and sperm transfer is usually direct with copulation, although it is indirect in some apterygotes. The use of spermatophores is plesiomorphic in hexapods and occurs in all apterygotes and most pterygotes. The reproductive system is in the abdomen.

A typical female reproductive system consists of two dorsolateral ovaries with oviducts (Fig. 21-11A). Each ovary is a bundle of tubular **ovarioles,** but the bundle usually is not enclosed by a common sheath. Each ovariole, however, is surrounded by a cellular tunic with muscles and tracheoles. An ovariole consists of a small-diameter upstream germarium, where mitotic divisions of primordial germ cells produce new oogonia, and a much larger downstream vitellarium, where yolk is transferred to the oocytes as they pass through to the oviduct. Each developing oocyte is surrounded by a cellular follicle that secretes a waterproof eggshell (chorion). In most holometabolous insects, yolk is transferred from follicular nurse cells to the oocyte, but primitive hexapods typically lack such cells. Meiosis is not completed until after the oocyte leaves the ovariole, and sometimes not until after fertilization.

The lateral oviducts unite to form a common oviduct, which leads into a vagina. The vagina in turn opens via a ventral gonopore on the eighth segment. Diverticula of the oviduct or vagina include a seminal receptacle for the storage

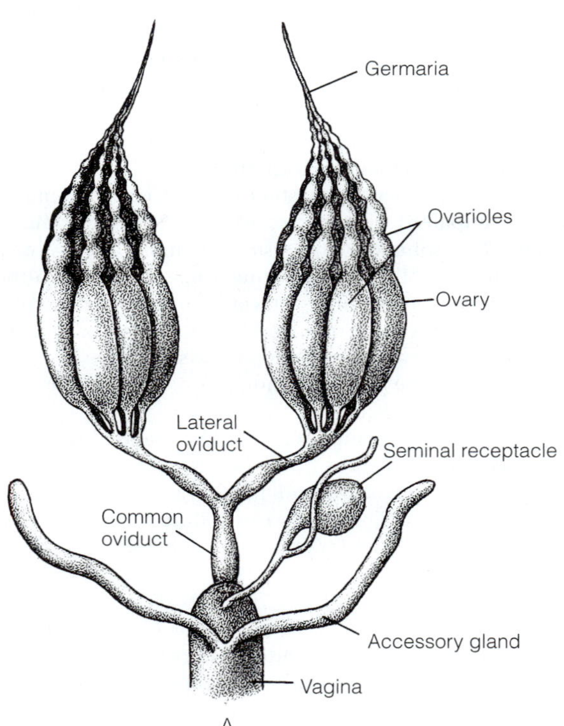

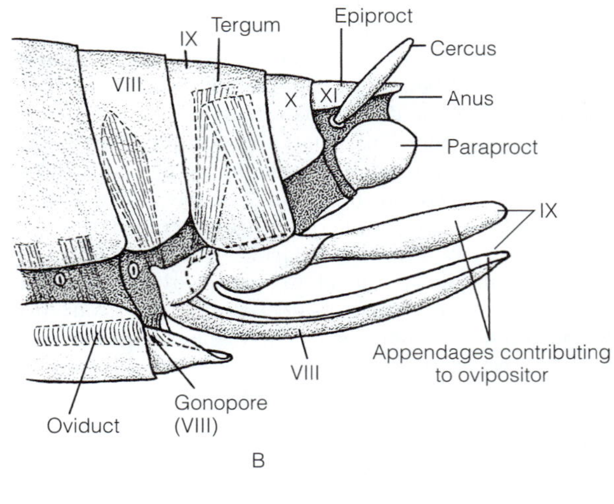

FIGURE 21-11 Reproductive system of a female hexapod. **A,** Internal anatomy. **B,** Lateral view of the posterior end of the abdomen, showing the reproductive opening and appendages that form the ovipositor. Abdominal segment numbers are indicated by Roman numerals. *(Both after Snodgrass, 1935.)*

of sperm or spermatophores and paired accessory glands that have a variety of functions, including secretion of an adhesive for attaching eggs to the substratum. Fertilization occurs, using sperm stored in the seminal receptacle, as the egg passes through the vagina. There is much variation in this plan.

Female external genitalia, when present, are on segments 8 and 9 (Fig. 21-11B), and in many taxa (Odonata, Orthoptera, Hemiptera, Hymenoptera) the modified appendages of those segments form a more or less tubular ovipositor used to deposit eggs on or in an appropriate substratum. During oviposition, fertilized eggs extruded from the gonopore pass through the lumen of the ovipositor. The ovipositor may be modified to insert eggs into the soil, glue them to a leaf surface, inject them into another insect, or bore into wood to deposit them deep in the trunk of a tree. In some Hymenoptera (bees, wasps, ants) the ovipositor is a sting equipped with poison glands and is not used for egg laying. Only females have a sting. In these insects, eggs leaving the gonopore do not pass through the ovipositor. In some orders there is no true ovipositor, but the posterior abdominal segments, usually 6 through 9, telescope into and out of each other to form an extensible device used for oviposition. When not in use, it is withdrawn by telescoping it into segment 5.

The male reproductive system includes a pair of testes, a pair of lateral sperm ducts (vasa deferentia), and a median ejaculatory duct that opens through a ventral penis (aedeagus) on the ninth abdominal segment (Fig. 21-12). The gonopore is on the penis. Each testis consists of a bundle of tubular follicles in a connective-tissue capsule. The follicles contain germ cells in various stages of spermatogenesis and maturation. The follicles are zoned longitudinally, with spermatogonia in the uppermost end of the follicle, followed by primary spermatocytes, secondary spermatocytes, spermatids, and finally sperm in the lower, distal end of the follicle.

The follicles deliver sperm to a sperm duct, which joins its counterpart from the other side to form a single, muscular **ejaculatory duct.** The distal portion of each sperm duct is usually enlarged into a seminal vesicle, where sperm are stored. Accessory glands, which secrete seminal fluid for packaging spermatophores, are usually present as diverticula from the sperm ducts or the upper end of the ejaculatory duct.

The male external genitalia (Fig. 21-12B) are located chiefly on the ninth abdominal segment. Male dragonflies, true flies, butterflies, moths, and many others have claspers for holding the female during copulation (Fig. 21-12C). Claspers are thought to be derived from the appendages of the ninth or perhaps the tenth segment and are extremely variable at all taxonomic levels. These variations are valuable in the recognition of species. The penis is an intromittent organ used to deposit sperm or spermatophores in the female reproductive tract. Although these structures are referred to as "external" genitalia, they may be retracted when not in use and may not be visible without dissection.

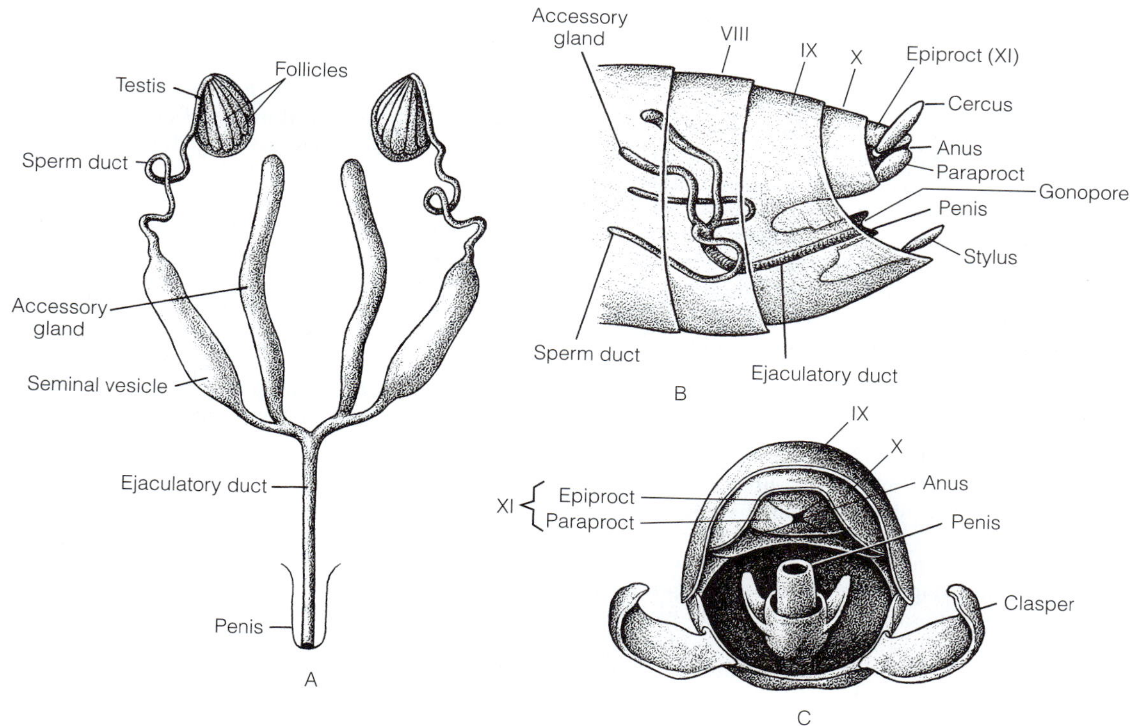

FIGURE 21-12 Reproductive system of a male hexapod. **A,** Internal anatomy. **B,** Lateral view of the posterior end of the abdomen, showing the reproductive opening and other structures. **C,** Posterior view of the abdomen, showing the penis and claspers. Abdominal segment numbers are indicated by Roman numerals. *(All after Snodgrass, 1935.)*

During copulation, claspers may be used to hold the female while the penis is inserted into the vagina. In most hexapods sperm are transferred in spermatophores. Among some apterygotes, such as zygentomans and collembolans, the spermatophore is deposited on the ground and then taken up by the female without copulation. In most, however, spermatophores are injected directly into the female vagina during copulation. Following insemination, sperm are released from the spermatophores and stored in the seminal receptacle. Fertilization occurs during oviposition as the eggs pass through the oviduct and vagina. At each mating, a large number of sperm sufficient for fertilization of more than one clutch of eggs are stored in the seminal receptacle. Many hexapods mate only once in their lifetime and none mate more than a few times.

By the time the eggs reach the oviduct, they have been surrounded by a shell-like membranous chorion secreted by ovarian follicle cells. The shell is perforated by one or more **micropyles** through which sperm enter. Another adaptation that has contributed to the success of hexapods in terrestrial environments is the waterproof shell that protects the egg from desiccation during development. The site of oviposition varies tremendously and depends chiefly on the food and habitat requirements of the immatures.

DEVELOPMENT

Superficial cleavage is characteristic of most hexapods. Eggs usually are macro- and centrolecithal, although those of derived insects have less yolk and others (such as some parasitoid wasps) are yolkless and nourished entirely by their surroundings. Immatures vary in the degree of development they have achieved at hatching.

The basic life-history stages are egg, juvenile, and adult, although other stages may be present. The stages are divided into substages, or **instars,** separated by molts. Usually there is only one adult instar, the **imago.** Only the imago is sexually mature and has functional wings. Upon hatching from the egg, the first instar juvenile begins a series of alternating instars and molts that eventually culminate in adulthood.

In many taxa the responsibility for feeding and growth lies with juvenile instars, whereas adults are responsible for reproduction and dispersal. Some hexapods spend the greater part of the life span, sometimes several years, as a juvenile and the adult stage may be brief, sometimes only a few hours, although these are extremes. In such cases the juvenile accumulates the food reserves that support the adult that follows. In some instances, adults have neither mouthparts nor gut and do not feed.

There are four basic developmental patterns by which this transformation can occur. Simple **ametabolous development** is characteristic of apterygote hexapods. The young are identical to the adults except in size and sexual maturity. No instar has wings, so the only externally visible difference is size. The adult form is reached gradually through successive molts. Each instar, of which there may be over 50, is slightly larger that its predecessor. Adult instars are larger and have functional reproductive systems. Unlike other hexapods, ametabolous species continue molting and growing after reaching sexual maturity through several adult instars. Adults and juveniles have similar niche characteristics and compete for resources.

Hemimetabolous development is characteristic of mayflies, dragonflies, damselflies, and stoneflies. Juveniles lack wings, are sexually immature, and do not closely resemble adults. The juveniles of hemimetabolous and paurometabolous insects, referred to as **nymphs,** have wing pads that gradually increase in size and become functional wings only with the final molt. Adults have wings, are sexually mature, and do not molt or grow (mayflies are the only exception and have two winged instars, the sexually immature subimago and mature imago). Hemimetabolous juveniles are aquatic whereas the adults are aerial. Nymphs and adults occupy different ecological niches and do not compete for resources.

Paurometabolous development (gradual or incomplete metamorphosis), characteristic of grasshoppers, cockroaches, earwigs, bugs, and others, is similar to hemimetabolous development except that the adults and nymphs are similar morphologically. Adults are larger and have wings and functional reproductive equipment (Fig. 21-13). The adult form is reached gradually through a series of molts. Usually there is a characteristic number of juvenile molts, often four or five, and molting ceases with the attainment of sexual maturity and wings. Adults do not molt or grow. External wing pads appear in late instars and increase in size with each molt (Fig. 21-13). Adults and nymphs may be in competition with each other for food and other resources.

Holometabolous development (complete metamorphosis, indirect development), a dramatic departure from the development patterns of the other hexapods, is characteristic of Holometabola, including bees, wasps, ants, flies, moths, butterflies, caddisflies, beetles, and others. The life cycle of holometabolous insects includes an additional stage, the **pupa,** between the last juvenile instar and the imago (Fig. 21-14). It may be thought of as a highly modified final juvenile instar. Imagoes are winged, sexually mature, and do not molt or grow. The wormlike juveniles known as larvae (maggots, grubs, and caterpillars, for example) do not resemble the adults (Fig. 21-14, 21-21D). Larvae do not have wing pads at any stage. They pass through a series of instars in which they feed and increase in size but do not become more adultlike. Larval food may be quite different from that of the adult, and the two often have different kinds of mouthparts so that competition for food is avoided. For example, butterfly larvae have chewing mouthparts, whereas the adults have sucking mouthparts. The number of molts required to reach the adult instar varies with taxon. Flies, for example, have four larval instars, although one may be suppressed.

At the end of the sequence of larval instars, the juvenile becomes a nonfeeding, usually quiescent pupa in a protected location, such as a cocoon or cavity in the ground. Inside the pupa the larva undergoes metamorphosis, during which larval organs are destroyed (hydrolyzed) and adult structures develop anew, phoenixlike, from clusters of embryonic reserve cells known as **imaginal discs.** Wing rudiments, for example, develop internally from imaginal discs and the wings, which have no larval precursors, appear *de novo* with the emergence of the imago from the pupa.

The larva of holometabolous insects probably is a specialized (rather than primitive) condition that evolved by suppressing the appearance of adult features. The development of

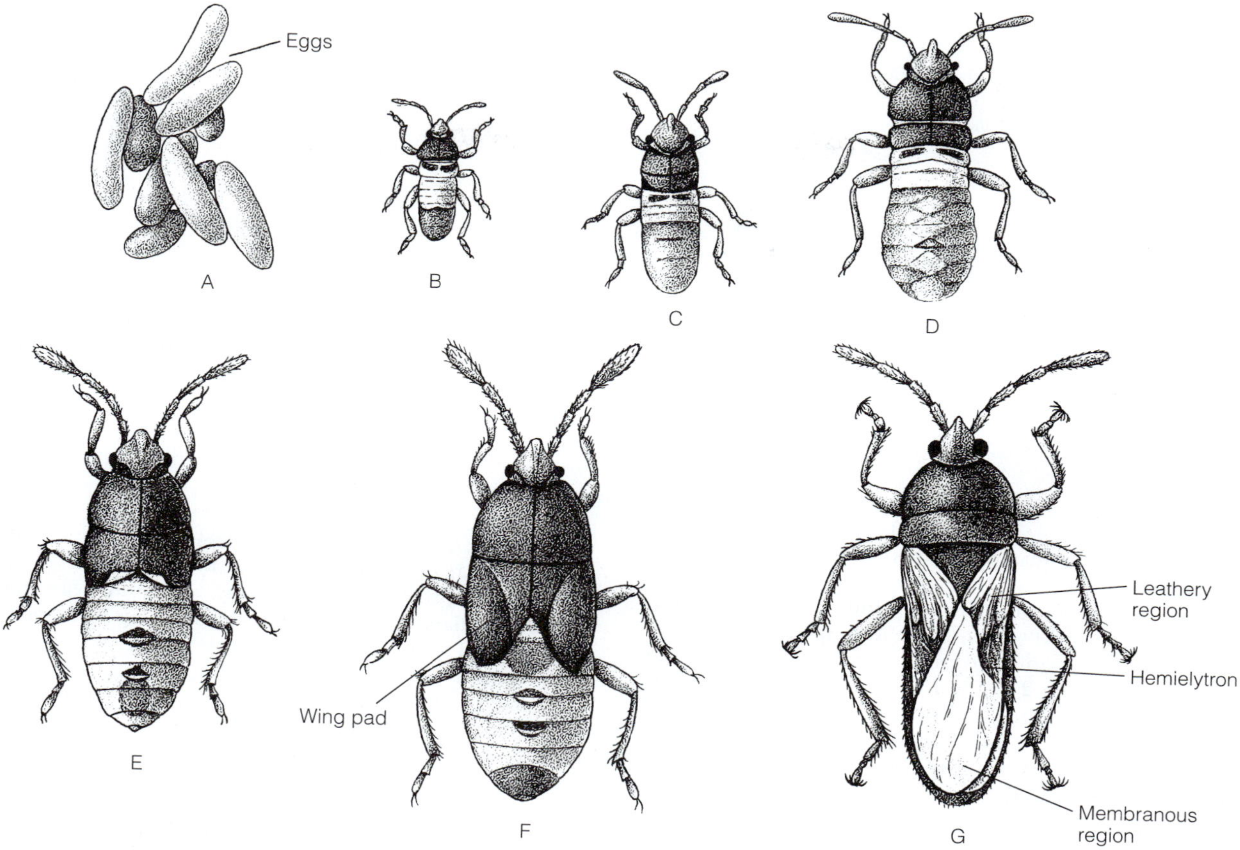

FIGURE 21-13 Development: Stages in the paurometabolous development of the chinch bug, *Blissus leucopterus* (Hemiptera). *(After Ross, H. H. 1965. A Textbook of Entomology. 3rd Edition. John Wiley and Sons, New York.)*

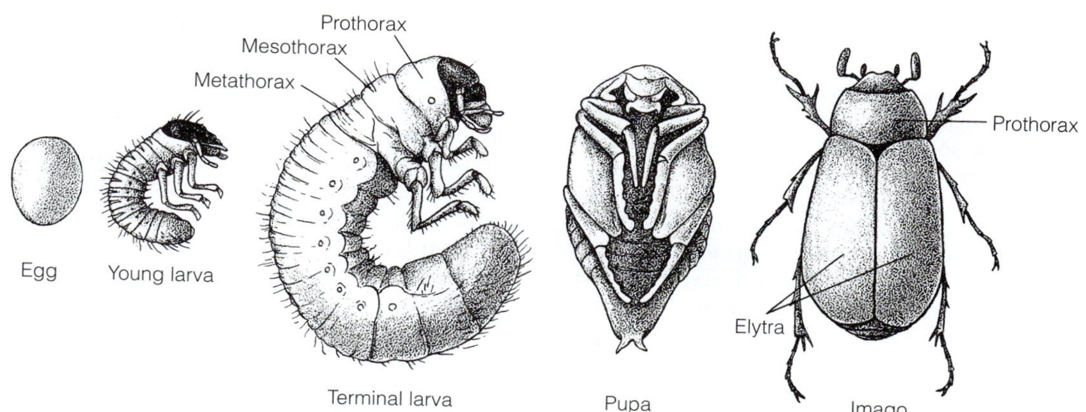

FIGURE 21-14 Development: Stages in the holometabolous development of a beetle (Coleoptera). *(After Ross, H. H. 1965. A Textbook of Entomology. 3rd Edition. John Wiley and Sons, New York.)*

holometabolism was of great adaptive significance in the evolution of insects partly because it allows larvae to utilize different food sources, habitats, and life styles than those of adults, thus reducing or eliminating competition between life-history stages. Of the 33 taxa of Recent hexapods, only 11 (Coleoptera, Neuroptera, Megaloptera, Raphidioptera, Hymenoptera, Trichoptera, Lepidoptera, Strepsiptera, Diptera, Siphonaptera, and Mecoptera) have holometabolous development, but these are by

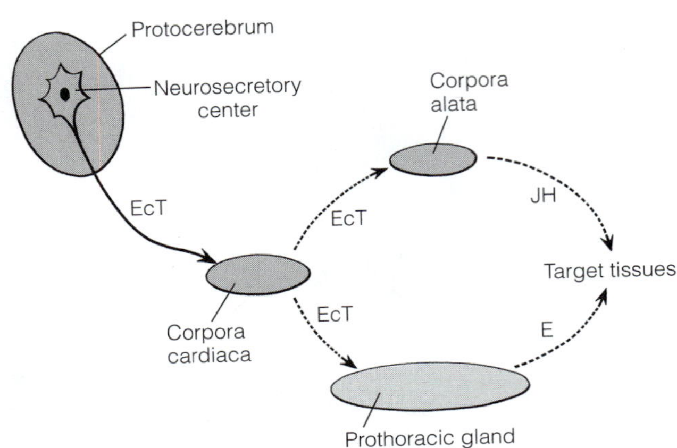

FIGURE 21-15 Endocrine system: Diagram of a generalized hexapod endocrine system. Axonal transport is indicated by the solid arrow and blood transport by dotted arrows. EcT = ecdysiotropin, E = ecdysone, JH = juvenile hormone. Explanation in text.

far the most successful hexapods and account for over 80% percent of all species.

Molting and metamorphosis are under endocrine control. Axons from neurosecretory centers in the protocerebrum deliver the tropic hormone **ecdysiotropin** to the corpora cardiaca (Fig. 21-7, 21-15), which are neurohemal organs associated with the aorta that store ecdysiotropin and then release it into the blood. Ecdysiotropin in the blood has two effects. It directs the **corpora alata** to synthesize and release **juvenile hormone** and the **prothoracic glands** to release **ecdysone,** both into the blood. Ecdysone initiates a new molt, during which juvenile hormone perpetuates the juvenile condition.

Under the influence of ecdysone, the epidermis initiates the molting cycle. The corpora alata secrete juvenile hormone into the blood during the molt. Juvenile hormone can exert its effect only after the molting process has been initiated and thus acts only in conjunction with ecdysone. Once the molt cycle begins, its outcome is determined by the level of juvenile hormone in the blood. Juvenile hormone suppresses the expression of genes controlling the development of adult characteristics. If the blood juvenile-hormone level is high, then the molt will be larva → larva; if it is intermediate, larva → pupa; and if low, pupa → adult. In the absence of juvenile hormone, adult characteristics are free to develop and metamorphosis can occur. In addition, without juvenile hormone, the prothoracic glands degenerate and, without this source of ecdysone, molting ceases. Once molting ceases the effects of juvenile hormone are not expressed because it acts only during a molt.

Environmental or physiological events may initiate the onset of a molt cycle by stimulating or inhibiting the release of ecdysiotropin from the brain. Stretching of the gut wall due to feeding or changing temperatures or photoperiod (day length), for example, may stimulate the release of ecdysiotropin and initiate a molt.

The life cycle of many hexapods includes diapause, a period of physical and metabolic inactivity that enables the organism to survive such adverse environmental conditions as long dry or cold periods. Diapause is under hormonal control and may be passed in any stage of development: egg, juvenile, pupa, or imago, depending on the species or environmental conditions. It typically is passed in the ground, a cocoon, or some other protected place. Many hexapods cease molting and enter diapause in response to declining photoperiod in the fall and resume molting in the spring, with increasing photoperiod.

Many variations on the basic reproductive pattern can be found. Aphids (Hemiptera) undergo parthenogenetic reproduction during most of the growing season, then reproduce sexually to produce diapausing eggs that overwinter and hatch in the spring. Many hexapods of several orders are viviparous, and the female retains the developing embryos in a uterus. In the simplest form of viviparity, lecithotrophy, the female retains unmodified, yolky eggs without providing additional nutrients. In the derived type of viviparity, the matrotrophic mother provides all nutrients via a placenta. Female fleshflies retain their embryos and deposit first instar larvae, rather than eggs, on dead animal tissue. The larvae begin feeding immediately to take advantage of an ephemeral food source. **Polyembryony,** in which the initial mass of embryonic cells from an egg divides to give rise to more than one embryo, is highly developed in some parasitoid wasps. From two to hundreds of larvae are formed from a single egg deposited in the body of an insect host. Usually the eggs are yolkless and the developing embryos are nourished by the host. The tiny chalcid wasp *Litomastix* deposits a few eggs into the body of a large moth caterpillar and over a thousand larvae develop from these eggs.

ECOLOGY

COEVOLUTION

Plants and insects interact in ways that are vital to both, the most important of which are herbivory and pollination. In these relationships both plant and insect evolve in response to adaptations by the other in a stepwise process known as **coevolution,** and each is a major selective force in the other's evolution. In the case of herbivory, the plant evolves defenses and the insect responds with counteroffensives to overcome the defenses. Then the plant develops refinements in the defenses that neutralize the counteroffensives and so on, like an arms race. In the case of pollination, which is mutually beneficial, the goals of both participants are similar, but each evolves to maximize benefits and minimize costs to themselves.

Many hexapods are herbivores that consume plants as their sole or chief food resource, and insects are the chief plant predators. The principal plant defenses against herbivory are chemical and mechanical. Plants have evolved a chemical arsenal of toxic or repellant compounds to discourage predation. These include nicotine, caffeine, cardiac glucosides, atropine, solanine, tannins, cyanide, mustard oils, various alkaloids, and more. Sometimes plant tissues contain high concentrations of indigestible compounds such as tannins, resins, and silica or they may be protected mechanically by spines, bristles, or fine hairs. These plant defenses have led to various reciprocal evolutionary strategies among insect herbivores. Some utilize less nutritious parts to avoid toxins in the more desirable tissues. Many insects are restricted to the one or a few plant species for which they have evolved enzyme systems that detoxify nutritious but protected tissues. Some insects that feed on milkweeds cut the vascular bundle delivering the toxic alkaloid and then feed downstream of the severed vein. The life cycles of many insects may be adapted to coincide with the availability of the plant food source.

Ironically, in many plants the substance developed to repel most herbivores may become an attractant for a species that has developed defenses against it. Such a species is likely to specialize in this plant because it will have few or no competitors for it. The plant family Apiaceae includes a number of species, such as carrot, caraway, parsley, fennel, and dill, that are protected from most insects by linear furanocoumarins. The female black swallowtail butterfly, *Papilio polyxenes,* is attracted by these chemicals, however, and oviposits selectively on plants of this family. The larvae, known as "carrot worms," have enzymes that rapidly detoxify the furanocoumarins that kill other insects. In the latest coevolutionary gambit in this conflict, some Apiaceae have started synthesizing angular furanocoumarins, which, unlike the linear form, are toxic to carrot worms. The family Solanaceae includes a number of plants, such as tobacco, tomato, potato, deadly nightshade, jimsonweed, and Jerusalem cherry, that contain powerful toxins, such as nicotine, to prevent insect herbivory. Some insects, such as tobacco and tomato hornworms and potato beetles, detoxify the poisons and, instead of being repelled, are attracted to and feed exclusively on these plants.

Many herbivorous insects induce their host plants to form **galls,** and in North America, about 2000 species of gall-inducing insects are known. The females of gall wasps (Hymenoptera) and gall gnats (midges; Diptera) deposit their eggs in plant tissues. The plant tissue adjacent to the egg grows abnormally to form a swollen, easily visible gall that has a shape characteristic of the specific insect and plant. The gall forms a protective chamber for the developing eggs, larvae, and pupae, and the larvae feed on the gall tissue. Gall tissue forms in response to substances secreted by the ovipositing female and the larva.

Coevolution is particularly striking between flowering plants and the insects that pollinate them, a mutually beneficial arrangement. Insects pollinate almost 70% of flowering plants, and the great diversity of floral structure largely reflects adaptations for facilitating pollination. Bees, wasps, butterflies, moths, and flies are the principal insect pollinators.

Colors, odors, and nectaries that secrete sugar solutions are developed by the plant to attract and reward or bribe insects. The flower provides nectar and pollen as foods to attract an insect that will inadvertently transfer pollen to another flower as it visits a succession of flowers. The morphology of the flower is adapted to maximize the probability of pollination, often by a specific insect. Concealment of the nectaries deep in the flower forces the pollinator to brush past the anthers, from which they pick up pollen, or touch the stigma, where pollen must be deposited for fertilization. Flowers are pigmented to maximize their visibility to their pollinator, and many reflect UV radiation to take advantage of the insect's ability to detect these wavelengths. Moth-pollinated flowers often have large, pale flowers to maximize their visibility on moonlit nights. Many flowers have guide marks in contrasting colors to direct the insect to the nectaries. The life-history characteristics of the pollinator and the blooming season coincide. Similarly, the activity period of the pollinator coincides with the time of day at which the plant blooms.

PARASITISM

Parasitism is common among insects and has evolved many times. Usually, only one life-cycle stage, juvenile or adult, is parasitic so that different stages exploit different food resources and habitats and thus minimize competition. In many species, the adults are parasitic and juveniles feed by more conventional methods. For example, adult fleas (Fig. 21-22A) are bloodsucking ectoparasites on the skin of birds and mammals. The larvae, however, develop off the host in its nest or den. Adults of several families of flies, including louse flies and tsetse flies, are also bloodsucking parasites of birds and mammals. Females are viviparous and give birth to final instar larvae that immediately fall off the host and pupate. The larvae do not feed directly on the host, but rather are nourished by the female, which does.

In contrast, sometimes the larva is parasitic and adults exploit conventional resources. Many species of wasps and flies are parasites as larvae. Infestation of livestock and wild mammals by several families of fly larvae (maggots) is a common veterinary condition known as myiasis, which occurs in humans, as well. The screwworm fly *Callitroga hominivorax,* a species of blowfly and a serious parasite of domestic animals (and humans), lays its eggs in wounds and the maggots feed on living tissue. The horse botfly *Gasterophilus* causes gastrointestinal myiasis by invading the digestive tract of horses and other mammals. The adult fly lays its eggs on the hairs of the host, and when the host licks the hairs, the eggs hatch and the maggots enter the gut. They develop while attached to the wall of the stomach or intestine and eventually exit with feces and then pupate. Warble fly maggots (Cuterebridae) live in swellings known as warbles just beneath the skin of mammals, where they feed on tissues and fluids. The larvae breathe through a small opening in the skin. Females of one species, *Dermatobia hominis,* the human botfly, employ an unusual strategy for dispersing eggs to suitable hosts (humans, cattle, dogs). The female botfly captures a stable fly or mosquito, both of which are biting flies that feed on mammals. The botfly oviposits on the captured fly, and then releases it unharmed but now carrying a clutch of botfly eggs. The first instar botfly larva develops within the egg. When the carrier fly lands on and bites a host, the larva drops off and penetrates the host's skin through the small opening made by the bite of the carrier. The larva passes through three instars in a warble, then falls to the ground and pupates.

A small number of insects spend the entire life cycle on the host. Lice, most of which are bloodsucking parasites of birds and mammals (Fig. 21-16), are always in contact with their host. Bedbugs are bloodsuckers throughout their life cycle, but they do not live continually on the host.

PARASITOIDISM

The insects just discussed are conventional parasites, in which a small parasite feeds on a larger host over a long period of contact, usually without killing it. Other insects are conventional predators, in which a larger species captures, kills, and consumes smaller prey in a brief period of contact. Three very large families of insects, Braconidae (Hymenoptera), Ichneumonidae (Hymenoptera), and Tachinidae (Diptera), as well as some others, exploit other insects in an interaction referred to as **parasitoidism** that combines features of parasitism and predation. Parasitoids are smaller than their hosts and remain in contact with them over long periods, but they eventually kill and consume the host. It is likely that most, and maybe all, hexapod species have at least one parasitoid, and this has fueled speculation that there may be as many species of parasitoids as of all other insects combined. Parasitoidism is so widespread it is thought to be the most important mechanism controlling the size of insect populations. **Hyperparasitoids** are parasitoids that attack parasitoids.

Female parasitoids oviposit on or in their host (prey) and the larvae consume the host gradually, avoiding vital organs, so that the host remains alive, and even feeding, until the very end, when new adult parasitoids emerge and the host finally dies. Parasitoids can be ecto- or endoparasitoids and may specialize in feeding on any life-history stage, including the egg. There are many tiny egg parasitoids smaller than the insect eggs they inhabit and consume. Adult parasitoids are usually free-living and feed on nectar or plant juices and only the larvae are parasitoidic. Female parasitoid wasps use the slender needlelike ovipositor to oviposit in or near the host. Some ichneumonid wasps (Fig. 21-21B) specialize in wood-boring beetle larvae and use the long ovipositor to drill through wood to oviposit on their hosts. Tachinid flies lack the ovipositor and lay their eggs outside the host, so the first instar larvae must find the host and penetrate it.

Most vegetable gardeners are familiar with the conspicuous white cocoons of the braconid wasp *Apanteles* that appear on the surface of the tomato hornworm, *Manduca quinquemaculata* (a sphinx moth larva). The female wasp oviposits in the host caterpillar and the tiny wasp larvae, sometimes as many as 500 of them, feed and grow inside. Eventually they bore through the host's cuticle and spin a cocoon attached to the outside of the caterpillar. Although the white ovals may look like eggs,

From Elzinga, R. J. 1987. Fundamentals of Entomology. 3rd Edition. Prentice-Hall, Englewood Cliffs, NJ.

A

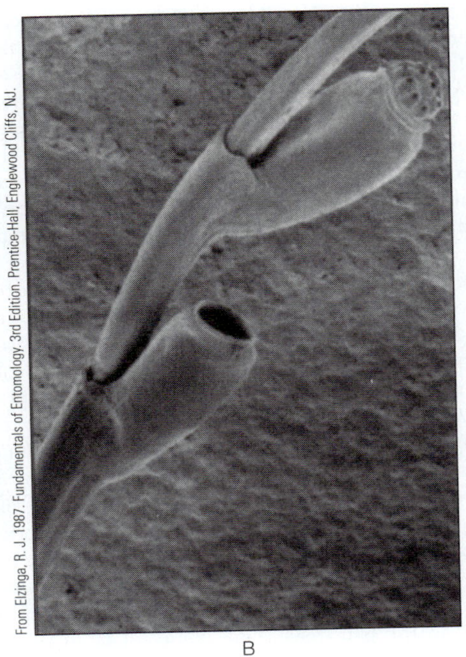

From Elzinga, R. J. 1987. Fundamentals of Entomology. 3rd Edition. Prentice-Hall, Englewood Cliffs, NJ.

B

FIGURE 21-16 Phthiraptera: The human crab louse (pubic louse), *Pthirus pubis.* **A,** An adult gripping two pubic hairs. **B,** Two egg cases attached to a pubic hair. The bottom case is empty; the upper one contains an unhatched nymph.

they are cocoons, and a tiny adult wasp, about 3 mm long, will emerge from each one. Other braconids parasitoidize aphids, clothes and gypsy moth caterpillars, and cabbageworms, and are important agents for biological control of pests.

COMMUNICATION

Both social and nonsocial hexapods use chemical, tactile, visual, and auditory signals to communicate with other animals. **Pheromones** are chemicals used to signal conspecifics. Many insects use pheromones to attract one sex to the other, and the much-studied gypsy moth, *Lymantria dispar,* is a classic example. Females release a pheromone in minuscule concentrations that is carried downwind in a plume of decreasing concentration. Male moths can detect the pheromone in concentrations as low as 1 molecule in 10^{17} of atmospheric gases. Males fly up the plume, using wind direction and concentration gradient as aids, to find the female.

Pheromones also mark trails or territories in some species. For example, chemicals deposited on the ground by ants returning from a successful foraging trip serve as trail markers that other ants follow to the newly discovered food source. Substances produced by the death and decomposition of the body of an ant within the colony stimulate other workers to remove the body. If a live ant is painted with an extract from a decomposing body, the painted ant is carried alive and struggling from the nest by other ants. Some pheromones produce alarm responses in other individuals of an aggregation or colony.

In some cases the release of chemicals is detrimental to the source organism, but beneficial to the target. Predators and parasites employ chemoreception to find prey, and similarly, prey may escape capture if they are warned by odors released by the predator. Some parasitoids locate their caterpillar prey by detecting compounds from the prey's mandibular glands. The females of species with herbivorous juvenile stages also use chemicals to locate their preferred plants for oviposition. Female squash borers find zucchini, corn earworms find young ears of corn, black swallowtails find parsley, and bean beetles find beans by using compounds the plants inadvertently released into the atmosphere.

An **allomone** is a compound released by a source organism that has an adverse effect on the target organism. Some lacewing larvae (Neuroptera) release a gas that subdues the prey, and most Hymenoptera produce venom to kill or immobilize the prey.

Among the more unusual visual signals are the bioluminescent flashings of fireflies, which function in sexual attraction. In species of *Photinus,* for example, flying males flash at species-specific intervals and females located on vegetation flash in response. The male then redirects his flight toward her and further flashing and then mating occur.

Sound can be produced either by stridulation or the vibration of a membrane. In stridulation, a rough, rasplike scraper is rubbed over a file to produce sound and is characteristic of grasshoppers, katydids, and crickets. Each species of cricket produces songs that differ from those of other species and function in sexual attraction and aggression. The sound is received by membranous tympanal organs. The static-like sounds of cicadas, which serve to aggregate individuals, are produced by vibrations of special chitinous, abdominal membranes.

SOCIAL INSECTS

Colonial organization, characterized by integration and interaction of joined or independent specialized, conspecific zooids or individuals, has evolved in many animals. Only among some insects, vertebrates, and a few spiders, however, are individuals functionally interdependent although physically unconnected with each other. Social organization, or eusociality, the highest development of colonial organization, has evolved in only two taxa of insects, the Isoptera (termites) and the Hymenoptera (bees, ants, and wasps), but apparently has been responsible for the great success of both taxa. All termites and all ants are eusocial, as are many bees and wasps. Eusociality evolved independently in termites and hymenopterans, but it has culminated in similar states of development in both.

Eusocial societies are recognizable by three chief characteristics. Members of the society cooperate in caring for the juveniles. There is a reproductive division of labor, with sterile individuals performing the physical labor of the society while reproduction is left to fertile individuals. There is an overlap of at least two generations so parents can be assisted by their progeny. In addition, no individual can exist independent of the colony nor can it be a member of any colony but the one in which it developed. All eusocial insects exhibit some degree of polymorphism, and the different types of individuals in a colony are termed **castes.** The principal castes are reproductive male, reproductive female (or queen), sterile worker, and sterile soldier. Males function in the insemination of the queen, whose role is to produce new individuals for the colony. The workers are responsible for the support and maintenance of the colony and soldiers for its defense. Caste is determined by the ploidy of the egg and on the quality and quantity of the diet.

Termites live in a nest usually constructed in soil, wood, or above the ground. The nest may be huge, rising two meters or more above the ground surface, and structurally complex. A colony consists of reproductive, worker, and soldier castes (Fig. 21-17). Termites differ from social hymenopterans in that workers and soldiers are sterile individuals of both sexes and, since termites are paurometabolous (juveniles are physically similar to adults), workers and soldiers may be either juveniles or adults. The reproductive male (king) is fertile, of course, and is a permanent member of the colony, mating periodically with the reproductive female (queen). The colony's single reproductive female has enormously swollen ovaries and an enlarged abdomen to accommodate them. She may produce up to 3000 eggs daily. The nest is built and maintained by workers. Soldiers have large, heavily sclerotized heads and either large mandibles or a frontal gland for defense of the colony. The frontal gland releases or ejects a sticky, toxic, or repugnatorial secretion from a pore or nozzle on the front of the head capsule. Workers and soldiers are wingless, but the reproductive castes have wings, which are used only for a brief nuptial flight during which pairing and dispersal occur.

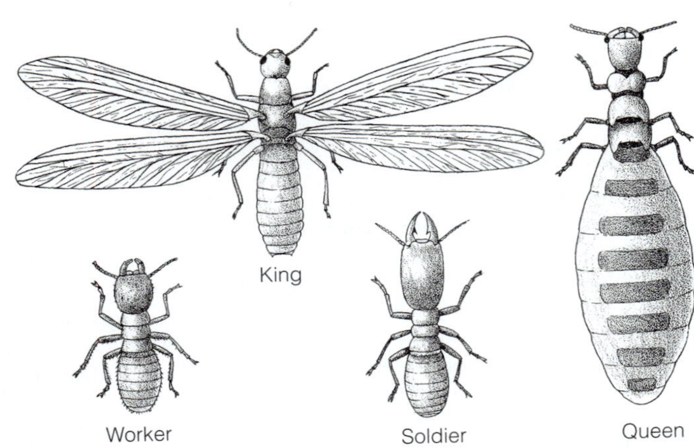

FIGURE 21-17 Isoptera: Castes of the common North American termite *Reticulitermes flavipes*. Note the simple moniliform antennae, the fracture planes at the base of the male wings, and the sclerotized abdominal tergites of the otherwise soft-bodied queen. *(After Lutz.)*

Primitive termites construct simple tunnels and galleries in wood buried in or in contact with the soil. Wood remote from the soil may also be utilized but, if so, an enclosed passageway from the soil to the wood must be maintained to provide access to moisture. Many tropical termites build subterranean nests with conical extensions several meters above the ground surface. The aboveground portions are constructed of carton, a mixture of soil and secretions of the frontal gland of workers. Dry-wood termites, in contrast, build galleries in wood above the ground surface, have no subterranean nest, and need not maintain covered contact with the ground. This makes them especially difficult to control when they infest buildings.

A large percentage of a plant's photosynthetic output is invested in the polysaccharides cellulose and lignin, which are used for the construction of cell walls, including the wood of woody plants. Although potentially a vast source of energy for animals, these two molecules are notoriously difficult to digest, and very few metazoans can synthesize the enzymes required to hydrolyze either compound. Termites, however, have specialized in the utilization of the cell walls of wood, grass, leaves, humus, or herbivore dung as food sources. To accomplish this, they apparently are capable of some cellulase synthesis themselves, but they also enlist the aid of flagellates, bacteria, and fungi. Most termites rely on endosymbiotic flagellates in the hindgut as the primary source of the cellulase needed for the hydrolysis of wood. Anaerobic respiration of cellulose by the flagellates produces acetic acid, which is then absorbed by the hindgut epithelium and transferred to the blood for distribution to cells, where it will be respired aerobically. Termites are generally thought to be closely related to cockroaches, some of which also harbor cellulose-digesting symbionts, and the Harvard entomologist E. O. Wilson has referred to termites as "social cockroaches."

Some termites have a gut flora of anaerobic bacteria instead of flagellates. Fungus-growing termites culture a basidiomycete fungus on "combs" of decaying plant material and termite feces. The fungi break down lignin prior to ingestion of the comb by the termites. Following ingestion, the products of lignin metabolism are metabolized further by the gut bacteria into compounds that can be absorbed and utilized by the termite. Young termites must acquire their bacterial or protozoan symbionts via **proctodeal feeding** from the anus of another termite, a process that requires social interaction. The termite-symbiont association thus may have been important in the evolution of social behavior in termites.

Hymenopteran societies, like those of termites, are based on a caste system with a division of labor, but there are differences. Soldiers and workers are always sterile, diploid females. They may be winged or not. Reproductive females are fertile and diploid. Fertile, reproductive males develop from unfertilized eggs and consequently are haploid. Following copulation with the queen, who mates only once, the male, his contribution made, dies and never becomes a functional part of the new colony. In many societies reproduction is parthenogenetic.

Ant colonies resemble those of termites and are usually housed within a gallery of tunnels in soil or wood or beneath stones. There may be a soldier caste in addition to workers. Workers and soldiers are wingless. Reproductives have wings prior to completing their nuptial flight, but then they lose them. Some species of ants raid the nests of other species and carry away their larvae and pupae to be raised as "slaves." Since hymenopterans are holometabolous and the larvae are grubs, only the adults are contributing members of the society and the helpless larvae and pupae must be tended and fed until they metamorphose. Ants exhibit a wide range of diets. Many are scavengers, but some, like the harvester ants, feed on seeds, which they store. Leafcutting ants carry bits of fresh leaves to their subterranean gardens and use them to grow the fungus on which they subsist. The cultivation of fungal gardens evolved independently in ants and termites. In contrast to most ants, tropical army ants do not live in permanent nests and instead "bivouac" together in the open. They are predatory, and during their nomadic mass movements they consume virtually all invertebrates they encounter. Ants account for 10 to 15% of animal biomass in most terrestrial communities.

In comparison with ants, polymorphism is less highly developed in wasps and bees. There is no soldier caste and workers are winged, but many of these insects exhibit remarkable adaptations for a social organization. The honeybee, *Apis mellifera,* is the best-studied social insect. This species is believed to have originated in Africa and been relatively recently introduced to temperate regions worldwide. Unlike other temperate-zone social bees and wasps, honeybee colonies survive the winter. Multiplication occurs by the division of the colony, a process called **swarming.** Stimulated at least in part by the crowding of workers (20,000 to 80,000 in a single colony), the mother queen leaves the hive along with many of the workers (a swarm) to found a new colony. The old colony is left with developing queens. On emerging, a new queen takes several nuptial flights during which copulation with males (drones) occurs, and she stores enough sperm to last her lifetime. The male dies following copulation, for his reproductive organs literally explode into the female.

Honeybee colonies are large. The workers' life spans are not long, but a queen may lay 1000 eggs per day to replace workers as they die. The eggs hatch into larvae that reside in hexagonal wax cells constructed by the workers. The nursing workers provide these larvae with a diet that results in their developing into sterile females, all of which are workers. The nursing behavior of the workers is elicited by a pheromone (queen substance) produced by the queen's mandibular glands. At the advent of swarming, or when the vitality of the queen diminishes, the production of this pheromone declines. In the absence of the inhibiting effect of the pheromone, the nursing workers construct royal cells, into which eggs, royal jelly, and a greater amount of food are placed. Royal jelly is a secretion from the hypopharyngeal glands of the worker nurses, and larvae that feed on it develop into queens in about 16 days. At the same time, unfertilized eggs are deposited into cells similar to those for workers and develop into haploid drones.

DIVERSITY OF HEXAPODA

Entognatha[C]: Apterygote. Mouthparts sunk into a depression on the surface of the head (entognathous). Poorly equipped for dry habitats; most are obligatory inhabitants of moist environments. Ametabolous.

Collembola[O] (springtails; Fig. 21-18A): Small, wingless, with abdominal furcula (jumping organ), collophore for water uptake. Abundant (but rarely noticed) in leaf litter, soil, rotten wood. Some are tolerant of dry conditions. Most feed on decaying plant material and fungi. Of all entognaths, collembolans are the most unlike the insects. *Anurida, Hypogastrura, Isotoma, Neelus, Onychiurus, Podura, Sminthurus, Tomocerus.*

Protura[O] (Fig. 21-18B): Small taxon. <2 mm long. Antennae and eyes lost. Sensory forelegs are held forward, antenna-like. Head is conical. Restricted to moist habitats, such as leaf litter and soil, by intolerance of desiccation. Probably feed on fungi. *Acerentulus, Eosentomon.*

Campodeina[O]: Sometimes combined with Japygina in Diplura. Both have long, multiarticulate antennae, styli on most abdominal segments, coxal organs on some. Usually only a few millimeters long. In damp humus, decomposing wood, beneath stones, bark. *Campodea.*

Japygina[O] (Fig. 21-18C): Carnivores; use pincerlike abdominal cerci to capture prey. *Anajapyx, Japyx.*

Insecta[C]: Includes almost all hexapods (Fig. 21-23); overwhelmingly successful in terrestrial and freshwater habitats. Mouthparts are ectognathous, not recessed into a pocket. Compound eyes and ocelli usually are present. Typically are well adapted for terrestrial life.

Archaeognatha[O] (jumping bristletails; Fig. 21-18D): Cylindrical, scale-covered body. Mandibles are monocondylic. Abdomen has three terminal processes. Although insects, they are apterygote. Require moist environments beneath bark, stones, in leaf litter. Jump when disturbed. Feed on algae, lichens, and moss. Ametabolous. *Machilis, Petrobius.*

Zygentoma[O] (silverfishes, firebrats; Fig. 21-18E): Apterygote. Styli on most abdominal segments. Mandibular articulation is dicondylic, like the remaining insects. Have three abdominal processes. Most inhabit moist habitats among dead leaves, in wood, around stones. Some absorb atmospheric moisture and are tolerant of dry environments. Some occur in houses, where they eat books and clothing. *Atelura, Lepisma, Nicoletia, Thermobia, Tricholepidion.*

Pterygota[sC]: Derived from a winged common ancestor; the adults of most retain these wings, although they have been modified or lost in many taxa. Sperm transfer is direct, with copulation.

Ephemeroptera[O] (mayflies; Fig. 21-19A): Adult is a delicate, graceful flying insect with net-veined wings. Most have three terminal abdominal filaments. Nymphs have abdominal tracheal gills. Adults do not feed, short-lived. Hemimetabolous; aquatic nymphs, aerial adults. Have two imaginal instars, the only pterygotes to have more than one imago. *Baetis, Ephemerella, Hexagenia.*

Odonata[O] (dragonflies, damselflies; Fig. 21-19B): Voracious predators as adults and nymphs. Hemimetabolous; aquatic nymphs, aerial adults. Larvae have extensible raptorial labial mask.

Anisoptera[sO] (dragonflies): Heavy-bodied, strong fliers that hold wings horizontally at rest. *Anax, Cordulegaster, Gomphus, Macromia.*

Zygoptera[sO] (damselflies): Slender, delicate, weak fliers. Hold wings over the abdomen at rest. *Calopteryx, Coenagrion, Lestes.*

Neoptera[iC]: The remaining insects. Fold wings flat over the abdomen; have lost the styli and coxal organs.

"Polyneoptera" (orthopteroids, or lower Neoptera): Artificial group with cerci, biting mouthparts, and abundant Malpighian tubules. Most are paurometabolous.

Orthoptera[O] (grasshoppers, katydids, crickets, locusts; Fig. 21-1, 21-19C): Relatively large insects. Large pronotum, large compound eyes. Usually have enlarged hind femora for jumping. Female ovipositor is large; male genitalia are not visible externally. Most have stridulating and auditory organs. About 20,000 species; diverse, widespread, and common especially in the tropics. Most are herbivores. Some, such as locusts and Mormon crickets, can severely damage crops.

Ensifera[sO]: Long-horned grasshoppers, katydids, mole crickets, camel crickets (Fig. 21-19C), true crickets. *Anabrus, Ceuthophilus, Conocephalus, Gryllus, Scudderia.*

Caelifera[sO]: Short-horned grasshoppers (Fig. 21-1), monkey grasshoppers, locusts, pygmy mole crickets. *Melanoplus, Romalea.*

Phasmida[O] (walking sticks, leaf insects): Slow-moving, nonjumping, cryptic. Mimic sticks or vegetation. Repugnatorial glands. Body usually is elongate, cylindrical, and sticklike. Wings are reduced or absent except in leaf insects, in which they are large and leaflike. Chiefly herbivores. Most abundant and diverse in the tropics. The longest Recent insects are phasmids, and some are over 30 cm in length. *Anisomorpha, Aplopus, Diapheromera, Timema.*

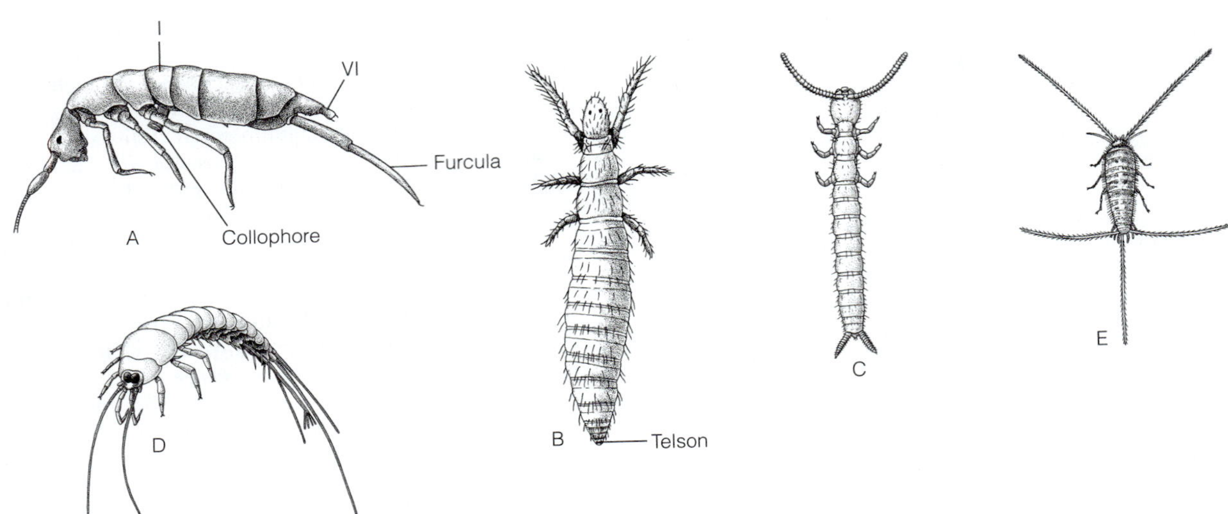

FIGURE 21-18 Apterygote Hexapoda. **A–C,** Entognatha. **A,** Collembola: a springtail. **B,** The proturan *Acerentulus barberi.* **C,** Japygina: *Anajapyx vesiculosus.* **D** and **E,** Apterous insects. **D,** Archeognatha, a jumping bristletail. **E,** Zygentoma, a silverfish. *(A, After Willem. B, After Ewing, 1940, from Ross. C, After Borror. D, After Borror. E, After Lutz.)*

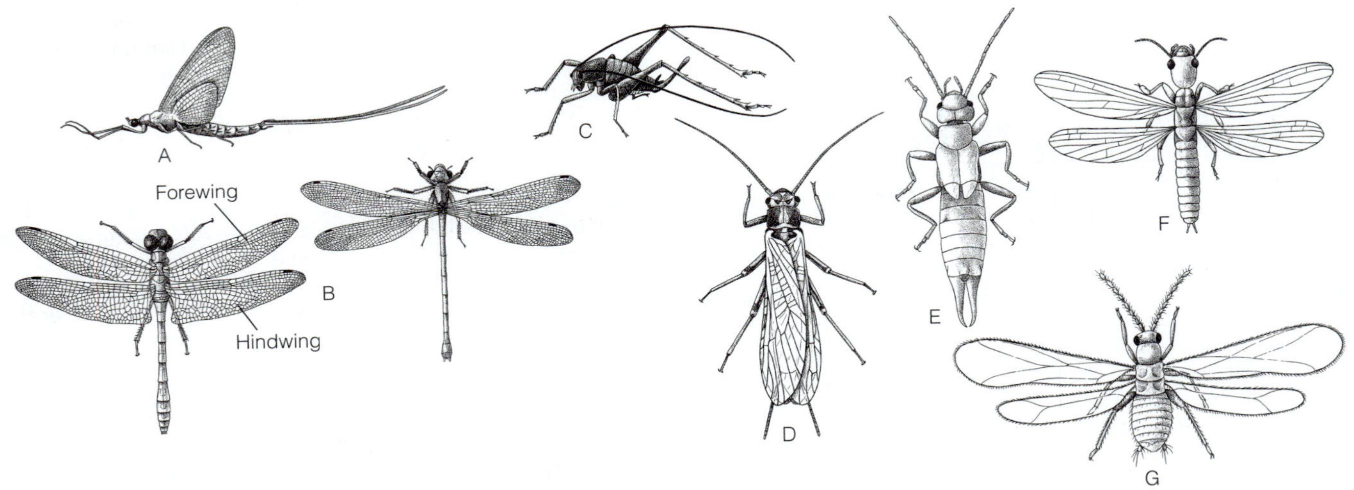

FIGURE 21-19 **A** and **B,** Primitive Pterygota. **A,** Ephemeroptera: a mayfly. **B,** Odonata: a dragonfly (left) and a damselfly (right). **C–G,** "Polyneoptera." **C,** Orthoptera (sO Ensifera): a camel cricket. **D,** Plecoptera: a stonefly. **E,** Dermaptera: the earwig, *Forficula auricularia.* **F,** Embioptera: a male web spinner, *Embia sabulosa.* **G,** Zoraptera: *Zorotypus hubbardi.* *(A, After Ross. B, After Kennedy from Ross. C, After Lutz. D, After Ross from Illinois Nat. Hist. Survey. E, After Fulton from Borror and De Long, 1971. F, After Enderlein from Comstock, 1930. G, After Caudell, 1920, Proc. Ent. Soc. Wash. Vol. 22 from Comstock, 1930.)*

Grylloblattaria^O (rock crawlers, ice crawlers): Elongate. Secondarily wingless. All are carnivores in cold, wet habitats, often at high elevations. Found in leaf litter, under stones or logs, in ground. Many inhabit interface between snow cover and ground surface. Twenty species. *Grylloblattaria.*

Mantophasmatodea^O (gladiators): Described in 2002; the first new insect order discovered since 1914. Adults are secondarily wingless. Elongate, slender trunk. Raptorial forelegs, carnivorous. Probably allied with Mantodea and Phasmida. Four species (two extinct) in two genera (one extinct). *Mantophasma, Raptophasma* (extinct).

Dermaptera^O (earwigs; Fig. 21-19E): Elongate, depressed, cosmopolitan. Superficially resemble some beetles. Large, sometimes pincerlike cerci are used for defense and prey capture. Nocturnal, omnivorous. *Doru, Forficula, Labia.*

Plecoptera^O (stoneflies; Fig. 21-19D): Aerial adults, aquatic nymphs. Two multiarticulate, caudal cerci; never have a median caudal filament. Nymphal tracheal gills, if present, tend to be thoracic. Nymphs require clean, cool, well-oxygenated water and, along with mayflies and caddisflies, are important water-quality indicators. Hemimetabolous. *Allocapnia, Peltoperla, Perla, Pteronarcys.*

Embioptera^O (Fig. 21-19F): Feed on plants. Live in silken tunnels, which they weave ahead of themselves to create routes. Silk glands in enlarged foretarsi. Gregarious; many individuals live together in connected tunnels, although there is no social organization. Tropical, only about 10 North American species. *Anisembia, Embia, Oligembia, Oligotoma.*

Isoptera^O (termites; Fig. 21-17): Soft, unsclerotized, pale or dark body. Abdomen is broadly joined to the thorax, without narrowing of the waist. Antennae composed of multiple beadlike annuli. Forewings and hindwings of equal size. Easily distinguished from ants, which are darker and have elbowed antennae, narrow waists, and hind wings that are smaller than the forewings. *Cryptotermes, Nasutitermes, Reticulitermes, Zootermopsis.*

Mantodea^O (mantids): Large, elongate, cryptically colored, relatively slow moving. Ambush predators with raptorial forelegs, elongate prothorax, freely movable head, well-developed eyes and ocelli. *Brunneria, Mantis, Stagmomantis, Tenodera.*

Blattaria^O (cockroaches): Fast, usually nocturnal, omnivorous runners. A few feed on wood and harbor endosymbiotic zooflagellates similar to those of termites. Termites, cockroaches, and mantids are closely related. *Blaberus, Blatta, Blatella, Periplaneta, Parcoblatta.*

Zoraptera^O (angel insects; Fig. 21-19G): Rare; about 30 species in one genus, *Zorotypus.* Resemble tiny termites. Live in colonies under dead wood, often in sawdust piles in warm climates. Thought to feed on fungal spores and small, already dead arthropods.

Eumetabola (higher Neoptera): Have a greatly reduced number of Malpighian tubules. Juvenile ocelli lost in adult.

Paraneoptera^SO (hemipteroid insects related to true bugs): Most have suctorial mouthparts. Probably arose from a psocidlike ancestor. Paurometabolous.

Psocoptera^O (psocids, book lice, bark lice; Fig. 21-20E): Small, fragile, pale. Probably primitive and similar to stem hemipteroids. Have chewing mouthparts, not suctorial. Hypopharynx can extract water from atmosphere. Chiefly herbivorous. Live under bark, in foliage, under stones. The book louse, *Liposcelis,* is found in old, musty books, where it feeds on fungi. *Archipsocus.*

Phthiraptera^O (lice; Fig. 21-20A,B): Bird and mammal ectoparasites. Body is depressed. Secondarily wingless. Legs are prehensile, adapted for clinging (Fig. 21-16A). Chewing or sucking mouthparts. Unlike in fleas, all stages of the life cycle are spent on the host. Transfer from host to host requires physical contact of the hosts; may occur during sexual contact, in the nest, or while nursing or grooming. Probably arose from psocidlike ancestors.

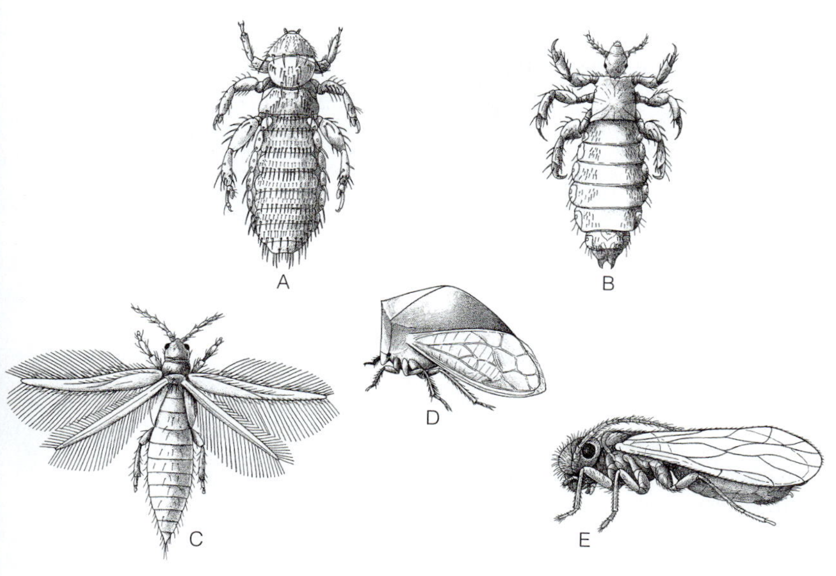

FIGURE 21-20 Paraneoptera (hemipteroid) insects. **A,** Phthiraptera ("mallophaga"): a guinea pig louse. **B,** Phthiraptera (Anoplura): the human body louse, *Pediculus humanus.* **C,** Thysanoptera: a thrips. **D,** Hemiptera ("homoptera"): a buffalo treehopper. **E,** Psocoptera: a psocid. *(A and B, From Grassé, P. (Ed.): Traité de Zoologie. C, After Moulton from Grassé, P. (Ed).: Traité de Zoologie. D, After Irving from Curran, H. 1954. Golden Playbook of Insect Stamps. Simon and Schuster, New York. E, After Sommerman from Ross.)*

Anoplura^SO (sucking lice; Fig. 21-16A, 21-20B): Mammal ectoparasites. Suck blood using sucking mouthparts and a pharyngeal pump. Some are disease vectors. Human crab louse (pubic louse, *Phthirus pubis*), human body louse *(Pediculus humanus)*, human head louse *(Pediculus capitis)*.

"Mallophaga"^SO (chewing lice; Fig 21-20A): Parasitize mostly birds. Chewing mouthparts, well-developed mandibles. Most feed on skin, hair, feather particles. *Cuclotogaster, Felicola, Menacanthus, Menopon. Trichodectes canis* is an intermediate host for the dog tapeworm.

Hemiptera^O (true bugs): The largest order (80,000 species) of nonholometabolous insects. The only animals correctly referred to as "bugs." Suctorial mouthparts (Fig. 21-4). Diet, either animal or plant, is liquid. Economically significant as herbivores and vectors of plant and animal diseases.

Heteroptera^SO (stinkbugs, squash bugs, water striders, chinch bugs [Fig. 21-13], many more): Forewings (hemielytra) have distinctive proximal sclerotized and distal membranous regions. Hind wings are entirely membranous. Most have repugnatorial (stink) glands. Most are terrestrial, but there are several important aquatic and semiaquatic families. *Corixa, Dysdercus, Gerris, Lygaeus, Murgantia, Notonecta, Reduvius, Triatoma.*

Homoptera^SO (aphids, adelgids, cicadas, treehoppers [Fig. 21-20D], leafhoppers): Strictly herbivorous; feed on plant juices. Both pairs of wings are membranous, no hemielytra. *Adelges, Aetalion, Aphis, Magicicada, Philaenus, Stictocephala, Tibicen.*

Thysanoptera^O (thrips; Fig. 21-20C): Small (0.5 to 15 mm). Suctorial mouthparts, cibarial pump. May be economically important as herbivores, plant-disease vectors, or pollinators. Wings are distinctive, narrow, have few veins, and are fringed with setae. *Aeolothrips, Heterothrips, Idolothrips, Merothrips, Thrips.*

Holometabola^SO: These 11 taxa, which have holometabolous development and complete metamorphosis, are by far the most successful hexapods, including, for example, 200,000 species of moths and butterflies; 9000 caddisflies; 100,000 flies; 130,000 wasps, bees, and ants; and 300,000 beetles. All have a pupal stage during which the larva undergoes a radical transformation into the imago.

Coleoptera^O (beetles; Fig. 21-14): The largest order of insects, with impressive ecological and morphological diversity. Beetles carry sclerotization to an extreme not seen in other insects, with most body surfaces hard and armor-clad. Forewings usually are heavily sclerotized elytra, or wing covers; hind wings are soft and membranous. Most are herbivores as adults and larvae, but many, such as the beloved ladybird beetles, are carnivores. Almost all are terrestrial, but a few families are important in fresh water. *Copris, Dermestes, Dytiscus, Gyrinus, Photinus, Popillia, Rodolia, Stenelmis.*

Neuroptera^O (ant lions, lacewings, spongillaflies, mantidflies [Fig. 21-21A]): Adults are delicate, graceful, predaceous insects with chewing mouthparts, large compound eyes, and long antennae. Larvae of the only aquatic family (Sisyridae) feed on freshwater sponges. The familiar doodlebugs, or ant lions, as larvae construct conical pits in dry sand to trap small arthropods. *Chrysopa, Climacia, Climaciella, Mantispa, Myrmeleon, Oliarces, Sisyra.*

Megaloptera^O (alderflies, dobsonflies, fishflies): Tend to be large. Chewing mouthparts, long antennae. Two pairs of similar membranous wings with primitive venation are held

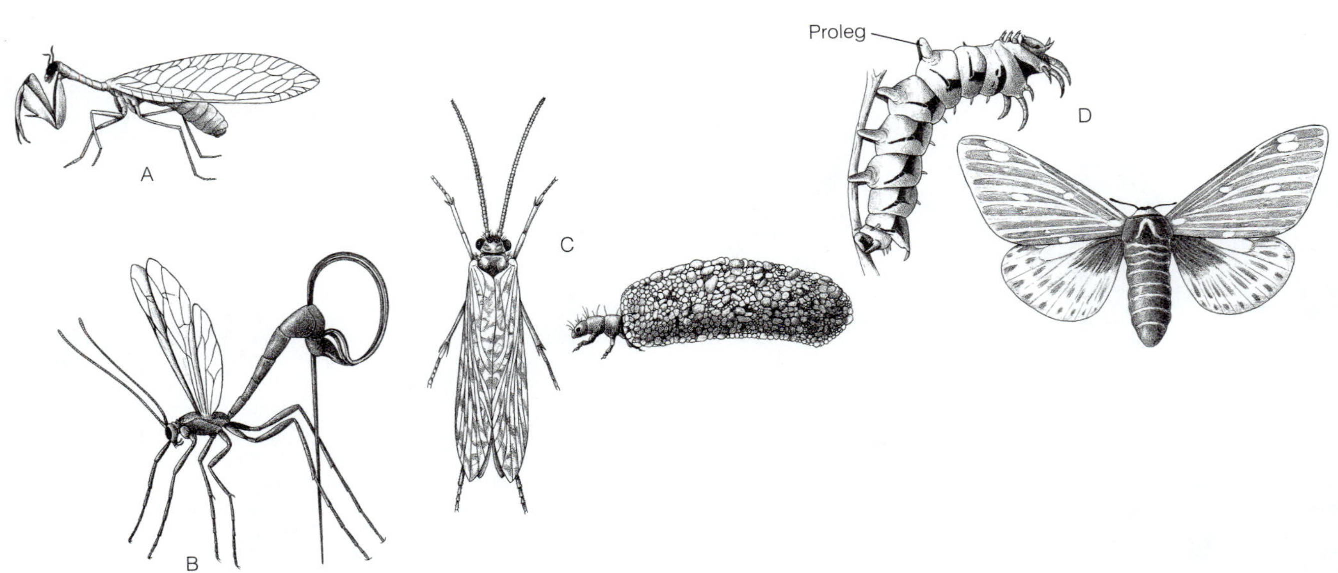

FIGURE 21-21 Holometabola. **A,** Neuroptera: the mantispid *Mantispa cincticornis.* **B,** Hymenoptera: a parasitoid female ichneumon wasp. Note the large ovipositor. **C,** Trichoptera: an adult caddis fly and larva in case. **D,** Lepidoptera: the royal walnut moth and its caterpillar, the hickory-horned devil. *(A, After Banks from Borror and De Long, 1971. B, After Lutz. C, After Mohr;* Illinois Nat. Hist. Survey, *from Ross. D, After Lutz.)*

tentlike over abdomen. Compound eyes are well developed and the mouthparts are mandibulate. Larvae are predaceous, aquatic, and have tracheate gills. *Corydalus, Sialis.*

Raphidioptera^O **(snakeflies):** Adults are similar to megalopterans except for having an elongate head and prothorax protruding far anterior to the legs and wings. Adults are predaceous; live in vegetation. Larvae are predaceous; most are terrestrial, living under the bark of conifers. *Raphidia.*

Hymenoptera^O **(sawflies, ants, bees, and wasps; Fig. 21-21B):** Large and varied taxon. All have chewing mouthparts, but they are modified for sucking in many taxa. Two pairs of membranous, transparent wings are present in most, but absent in some important exceptions, such as worker ants. First abdominal segment usually is broadly joined to the thorax and separated from the remaining abdomen by a narrow petiole, or waist, creating a secondary tagmosis. In females the ovipositor may be a sting. Larvae are maggotlike, with chewing mouthparts. Most adults feed on fluids such as nectar and honeydew and are found near flowers, where they may be important pollinators. *Apanteles, Aphidius, Apis, Bombus, Eciton, Solenopsis, Vespa, Vespula, Xylocopa.*

Trichoptera^O **(caddisflies; Fig. 21-21C):** Adult is mothlike. Have two pairs of hairy wings. Poorly developed mouthparts. Adults either do not feed or subsist on a liquid diet of water and nectar and are relatively short-lived. Like those of mayflies and stoneflies, caddisfly larvae are aquatic microhabitat specialists often present in great diversity in suitable habits. Species diversity is greatest in cool, swift mountain streams. Larval life revolves around the use of silk for construction of cases and portable retreats of stones and vegetation, as anchor lines, and as filters for suspension feeding. *Beraea, Helicopsyche, Hydropsyche, Polycentropus, Sericostoma.*

Lepidoptera^O **(butterflies, moths; Fig. 21-21D):** Adult is soft-bodied. Wings, body, and appendages are covered with deciduous pigmented scales or hairlike setae. Mandibles are absent, maxillae form a coiled extensible proboscis (Fig. 21-3) for sucking nectar and fruit juices; some species do not feed as adults. Larvae are herbivorous caterpillars. Salivary glands produce silk, the uses of which include construction of a cocoon in which the last larval instar pupates. *Actias, Danaus, Heliconius, Heliothis, Lymantria, Manduca, Melittia, Papilio, Sibine.*

Mecoptera^O **(scorpionflies; Fig. 21-22B):** Mouthparts are extended into a prominent ventral beak. Most have large, membranous wings. Adults are omnivorous; grublike larvae feed on organic matter. *Bittacus, Boreus, Panorpa.*

Siphonaptera^O **(fleas; Fig. 21-22A):** Small, secondarily wingless pterygotes with heavily sclerified, strongly compressed bodies. Bird and mammal ectoparasites. Legs adapted for jumping. Hematophagous, with piercing and sucking mouthparts. Some are vectors of bubonic plague *(Xenopsylla)* and typhus *(Nosopsyllus).* Mating takes place on the host or in its nest and eggs fall into the nest, where they hatch. Larvae feed on organic matter in the nest and are not parasitic. Eventually the larvae pupate and emerge as adults, which then infest the host. *Ctenocephalides, Orchopeas, Pulex.*

Strepsiptera^O **(twisted-wing parasites; Fig. 21-22C):** These tiny, bizarre insects are, as females and larvae, parasites on a variety of other insects. Females usually are larvalike, wingless, and never leave the host. They live embedded in the host's integument, with only their anterior end, with spiracles, protruding as a low bump above the host's exoskeleton. Nutriment from the host's blood is absorbed directly across the permeable cuticle of the parasite's embedded abdomen. In contrast, males have large, unusual, fanlike hind wings. The forewings are reduced to club-shaped, haltere-like appendages that may have a role in balance, and the hind wings are large. Adult males are short-lived, nonparasitic, and do not feed. Male's compound eyes are large, protruding. Antennae typically have processes extending from the flagellum. *Eoxenos, Halictophagus, Stylops, Triozocera.*

Diptera^O **(true flies; Fig. 21-22D):** Large taxon including mosquitoes, horseflies, houseflies, blowflies, botflies, midges, gnats, craneflies, blackflies, and many others. Compound eyes are large, well developed. Functional,

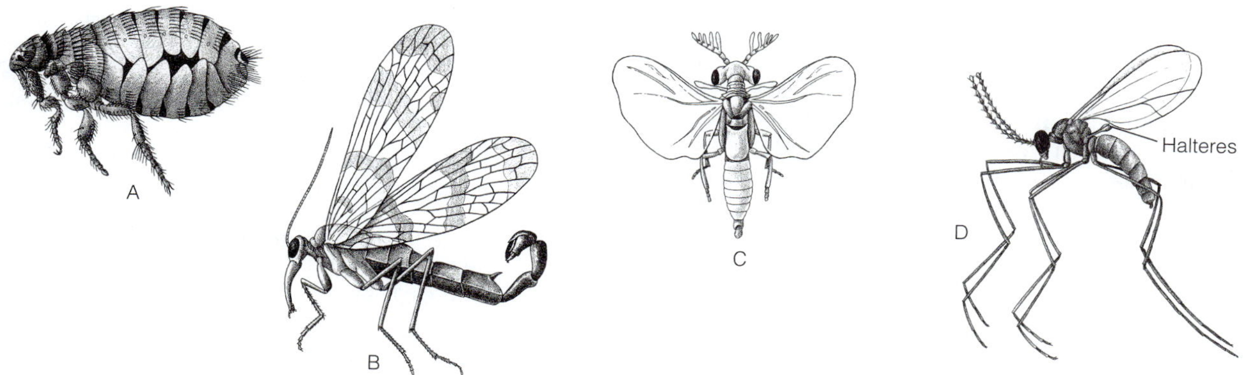

FIGURE 21-22 Holometabola. **A,** Siphonaptera: a flea. **B,** Mecoptera: a male scorpion fly, *Panorpa helena.* **C,** Strepsiptera: a male *Halictophagus serratus.* **D,** Diptera: the gall gnat, *Aphidolestes meridionalis.* *(A, After Bouche from Borror and De Long, 1971. B, After Taft from Borror and De Long, 1971. C, After Bohart; Entomological Society of America from Borror and De Long, 1971. D, After USDA from Borror and De Long, 1971.)*

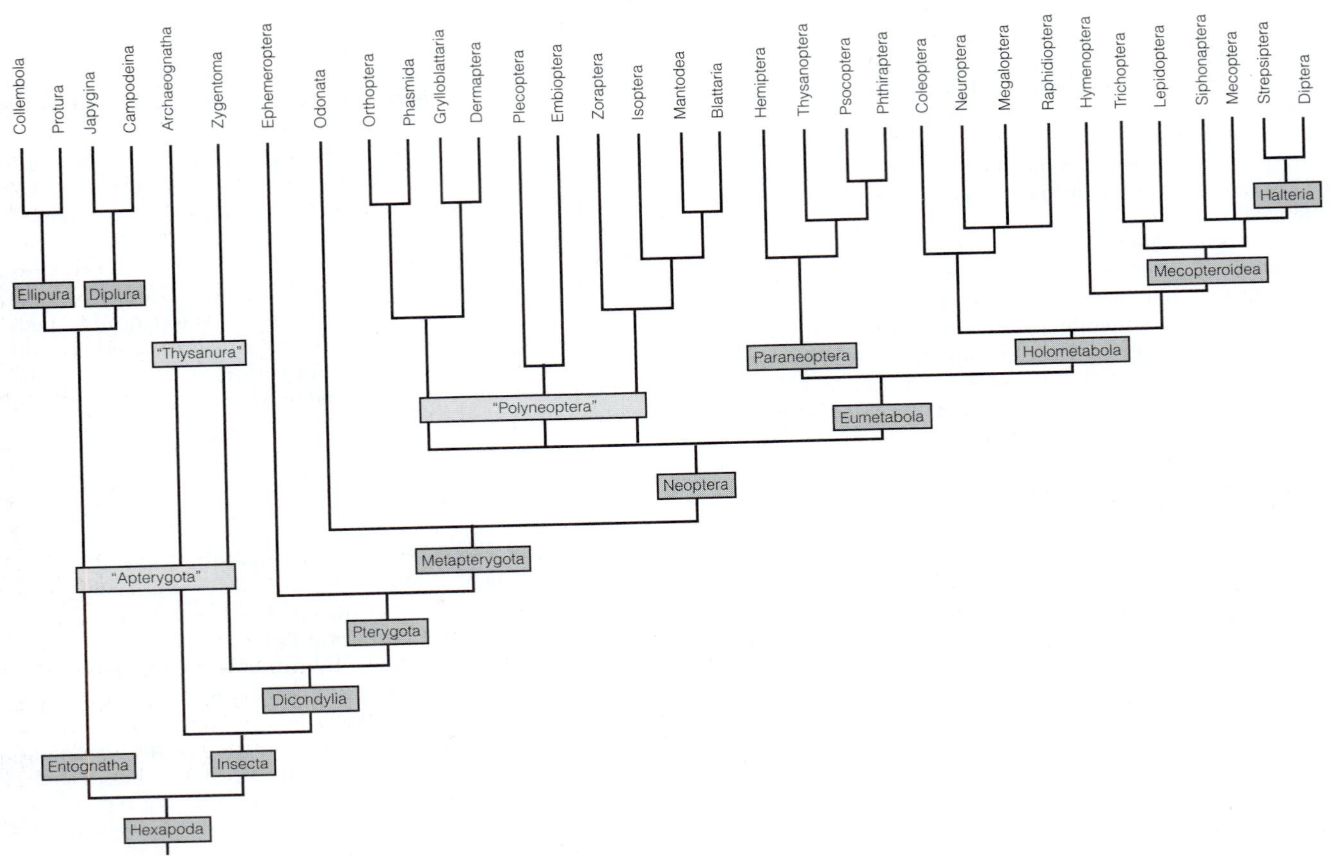

FIGURE 21-23 A phylogeny of hexapods based on 350 morphological characters, 1000 18S rRNA bases, and 350 28S rRNA bases. Mantophasmatodea, the most recently described order, is not included. The 33 low-level taxa of this analysis are the traditional hexapod orders, although in cladistic analysis they do not have equivalent ranks. *(Redrawn from Wheeler, W. C., Whiting, M., Wheeler, Q. D., and Carpenter, J. M. 2001. The phylogeny of the extant hexapod orders. Cladistics 17:113–1691, with modifications based on Wheeler, W. C., personal communication, 2002.)*

membranous forewings; reduced, knoblike hind wings (halteres). Mouthparts are variable, but basically are suctorial or sponging; used in conjunction with a cibarial pump. Almost all adult flies feed on plant or animal fluids. In many taxa known collectively as biting flies, females suck blood from vertebrate hosts, usually birds or mammals. Adults so frequently are vectors of human and animal diseases that this is the most important hexapod order in terms of human health. Larvae are maggots and are always found in wet substrata such as flesh, fruit, water, or mud. Females typically oviposit in a suitable moist food source for the larvae, which then live in their food as they consume it. *Aedes, Aphidolestes, Callitroga, Chaoborus, Dermatobia, Eristalis, Gasterophilus, Musca, Psychoda, Tabanus.*

This brief survey chapter does not allow space for a discussion of phylogeny such as appears in other chapters. As a substitute, a cladogram, without apomorphies, showing the probable relationships of the above taxa is provided (Fig. 21-23).

REFERENCES

Adis, J., Zompro, O., Moombolah-Goagoses, E., and Marais, E. 2002. Gladiators: A new order of insects. Sci. Am. Nov:60–64.

Arnett, R. H. 2000. American Insects: A Handbook of the Insects of North America. 2nd Edition. CRC Press, Boca Raton, FL. 1003 pp.

Arnett, R. H., Downie, N. M., and Jaques, H. E. 1980. How to Know the Beetles. 2nd Edition. W. C. Brown, Dubuque, IA. 416 pp.

Askew, R. R. 1971. Parasitic Insects. Elsevier-North Holland, New York. 316 pp.

Bennington, K. C., Lehane, M. J., and Beaton, C. D. 1998. The peritrophic membrane. In Harrison, F. W., and Locke, M. (Eds.): Microscopic Anatomy of Invertebrates. Vol. 11B: Insecta. Wiley-Liss, New York. pp. 747–758.

Berenbaum, M. R. 1995. Bugs in the System: Insects and Their Impact on Human Affairs. Addison-Wesley, New York. 377 pp.

Berridge, M. J. 1970. A structural analysis of intestinal absorption. In Neville, A. C. (Ed.): Insect Ultrastructure. Symposia of the

Royal Entomological Society Vol. 5. Blackwell Scientific, Oxford. pp. 135–151.

Bland, R. G., and Jaques, H. E. 1978. How to Know the Insects. 3rd Edition. W. C. Brown, Dubuque, IA. 409 pp.

Borror, D. J., and De Long, D. M. 1971. An Introduction to the Study of Insects. 3rd Edition. Holt, Rinehart, and Winston. 812 pp.

Borror, D. J., Triplehorn, C. A., and Johnson, N. F. 1989. An Introduction to the Study of Insects. 6th Edition. W. B. Saunders, Philadelphia. 875 pp.

Chapman, R. F. 1998. The Insects: Structure and Function. 4th Edition. Cambridge University Press, Cambridge. 770 pp.

Chu, H. F. 1949. How to Know the Immature Insects. W. C. Brown, Dubuque, IA. 234 pp.

Comstock, J. H. 1930. An Introduction to Entomology. Comstock, Ithaca, NY. 1044 pp.

Covell, C. V. 1984. A Field Guide to the Moths of Eastern North America. Peterson Field Guide Series. Houghton Mifflin, Boston. 496 pp.

Curran, H. 1954. Golden Playbook of Insect Stamps. Simon and Schuster, New York.

Daly, H. V., Doyen, J. T., Purcell, A. H., and Daly, B. 1998. Introduction to Insect Biology and Diversity. 2nd Edition. Oxford University Press, Oxford. 680 pp.

Elzinga, R. J. 1987. Fundamentals of Entomology. 3rd Edition. Prentice-Hall, Englewood Cliffs, NJ. 456 pp.

Evans, D. L., and Schmidt, J. O. (Eds.): 1990. Insect Defenses. State University of New York Press, Albany, NY. 482 pp.

Evans, H. E. 1984. Insect Biology: A Textbook of Entomology. Addison-Wesley, Reading, MA. 436 pp.

Ewing, H. E. 1940. The Protura of North America. Ann. Entomol. Soc. Am. 33:495–551.

Gillott, C. 1995. Entomology. 2nd Edition. Plenum, New York. 816 pp.

Goldsworthy, G. J., and Wheeler, C. H. 1989. Insect Flight. CRC Press, Boca Raton, FL. 352 pp.

Harrison, F. W., and Locke, M. (Eds.): 1998a. Microscopic Anatomy of Invertebrates. Vol. 11A: Insecta. Wiley-Liss, New York. 384 pp.

Harrison, F. W., and Locke, M. (Eds.): 1998b. Microscopic Anatomy of Invertebrates. Vol. 11B: Insecta. Wiley-Liss, New York. 455 pp.

Harrison, F. W., and Locke, M. (Eds.): 1998c. Microscopic Anatomy of Invertebrates. Vol. 11C: Insecta. Wiley-Liss, New York. 462 pp.

Hölldobler, B., and Wilson, E. O. 1990. The Ants. Harvard University Press, Cambridge, MA. 732 pp.

Hopkin, S. P. 1997. Biology of the Springtails (Insecta: Collembola). Oxford University Press, Oxford. 330 pp.

Klass, K.-D., Zompro, O., Kristensen, N. P., and Adis, J. 2002. Mantophasmatodea: A new insect order with extant members in the Afrotropics. Science 296:1456–1459.

Klausnitzer, B. 1996. Insekta (Hexapoda), Insekten. In Westheide, W., and Rieger, R. M. (Eds.): Spezielle Zoologie, Erster Tiel: Einzeller und wirbellose Tiere. Gustav Fischer Verlag, Stuttgart. pp. 601–681.

Kristensen, N. P. 1991a. Phylogeny of extant hexapods. In Nauman, I. D. (Ed.): The Insects of Australia. Cornell University Press, Ithaca, NY. pp. 125–140.

Kristensen, N. P. 1991b. The groundplan and basal diversification of the hexapods. In Fortey, R. A., and Thomas, R. H. (Eds.): Arthropod Relationships. Chapman and Hall, London. pp. 281–293.

Paclt, J. 1957. Diplura. Gen. Insect. 212:1–123.

Romoser, W. S., and Stoffolano, J. G. 1997. The Science of Entomology. 4th Edition. McGraw-Hill, New York. 624 pp.

Ross, H. H. 1965. A Textbook of Entomology. 3rd Edition. John Wiley and Sons, New York. 539 pp.

Ross, H. H., Ross, J. R. P., and Ross, C. A. 1982. A Textbook of Entomology. 4th Edition. John Wiley and Sons, New York. 666 pp.

Schmidt-Ott, U., González-Gaitán, M., Jäckle, H., and Technau, G. M. 1994. Number, identity, and sequence of the *Drosophila* head segments as revealed by neural elements and their deletion patterns in mutants. Proc. Natl. Acad. Sci. USA 91:8363–8367.

Snodgrass, R. E. 1935. Principles of Insect Morphology. McGraw-Hill, New York. 667 pp.

Stehr, F. W. 1987. Immature Insects. Vol. 1. Kendall/Hunt, Dubuque. IA. 754 pp.

Stehr, F. W. 1991. Immature Insects. Vol. 2. Kendall/Hunt, Dubuque, IA. 974 pp.

Sullivan, D. J. 1987. Insect hyperparasitism. Ann. Rev. Entomol. 32:49–70.

Terra, W. R. 1990. The evolution of digestive systems of insects. Ann. Rev. Entomol. 35:181–200.

Wheeler, W. C., Whiting, M., Wheeler, Q. D., and Carpenter, J. M. 2001. The phylogeny of the extant hexapod orders. Cladistics 17:113–169.

Wilson, E. O. 1971. The Insect Societies. Harvard University Press, Cambridge, MA. 548 pp.

Wooton, R. J. 1990. The mechanical design of insect wings. Sci. Am. 263:114–120.

Internet Sites

http://web.missouri.edu/~bioscish/coll.html (The Collembola, by S. Heyman and J. Weaver, University of Missouri.)

www.ent.iastate.edu/imagegallery (Iowa State University Entomology Image Gallery.)

http://entweb.clemson.edu/entsites (Entomology and Related Web Sites, Clemson University.)

http://insects.ummz.lsa.umich.edu (Insects: University of Michigan Museum of Zoology.)

www.sungaya.de/oz/gladiator (Mantophasmatodea page, by O. Zompro and J. Ardis.)

www.museums.org.za/bio/insects/mantophasmatodea/index.htm (Biodiversity Explorer of Museums OnLine South Africa, with many photographs and extensive biological account of Mantophasmatodea.)

http://medent.usyd.edu.au (Department of Medical Entomology Web site, by R. Russell and S. Doggett, University of Sydney.)

http://entowww.tamu.edu/research/collection/hallan/all.txt (Synopsis of the Described Taxa of the World, Texas A & M University.)

Cycloneuralia[SP]

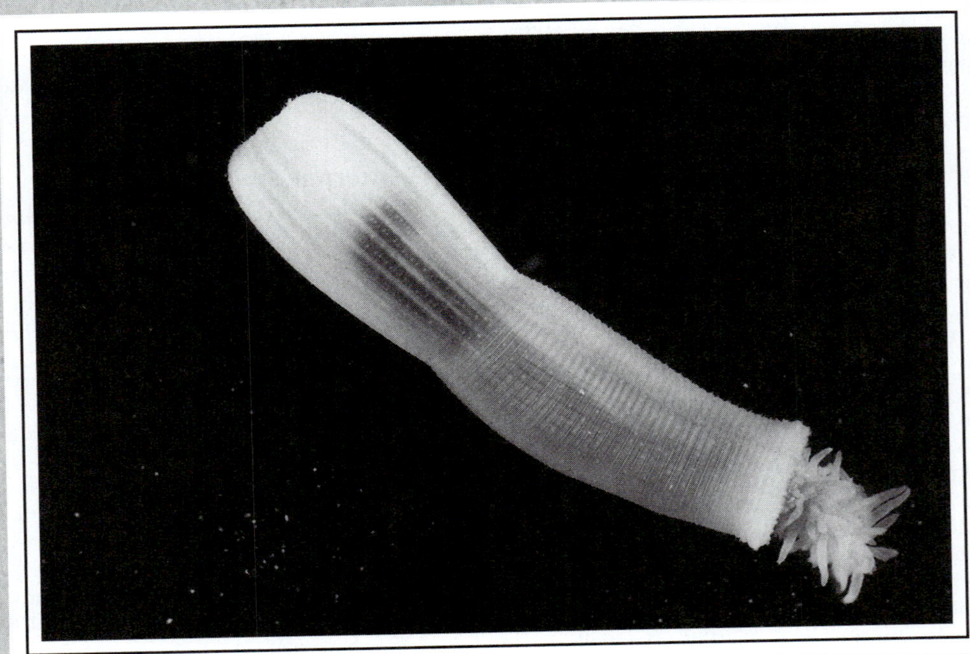

Cycloneuralia consists of six taxa of strange and fascinating animals that, with the important exception of nematodes, have not received the attention they deserve. The taxon includes marine, freshwater, and terrestrial metazoans in Gastrotricha (hairy bellies), Nematoda (nematodes or roundworms), Nematomorpha (horsehair worms), Priapulida, Kinorhyncha (mud dragons), and Loricifera. Some are important parasites of special relevance to the well being of humankind, but most are small, obscure inhabitants of out of the way habitats where they are rarely encountered. The majority are tiny worms, ranging from microscopic up to a centimeter in length, although some may be much larger.

Cycloneuralians have a secreted cuticle that, in all except the gastrotrichs, is molted, and sometimes contains chitin. Except for gastrotrichs, locomotory cilia are lacking and movement is accomplished with muscles. Derived, nonlocomotory cilia are present, however, in sensory structures, protonephridia, and rarely, in the gastrodermis. The eponymous brain, or nerve ring, a circular band surrounding the anterior gut, is composed of three consecutive and contiguous rings known as the forebrain, midbrain, and hindbrain (Fig. 22-2A). The forebrain and hindbrain are ganglionic and composed primarily of cell bodies (perikarya), whereas the midbrain is a neuropil consisting almost exclusively of nerve axons and synapses. The mouth has shifted from the median ventral location characteristic of most protostomes to a terminal position at the anterior tip of the body. A radially symmetrical pharynx has evolved to replace the ancestral bilateral pharynx. It is cylindrical and triradiate, with three bundles of epitheliomuscular cells or mesodermal myocytes radiating outward from the center (Fig. 22-15). Contractions of the radial muscles dilate the lumen of the pharynx, which functions as a sucking pump. A mesodermal musculature is probably plesiomorphic for the taxon and an epitheliomuscular pharynx probably arose independently in Gastrotricha, Nematoda, and Loricifera.

Most cycloneuralians are compact (acoelomate) animals without a body cavity, and the interior tends to be filled with connective tissue consisting of cells and extracellular matrix. Any fluid-filled space present arises through persistence of the embryonic blastocoel, is unlined by a coelomic epithelium, and is thus a hemocoel. A few large cycloneuralians, such as some parasitic nematodes and most priapulids, have an expanded hemocoel filled with blood (hemolymph). The blood is a hydrostatic skeleton and a medium for internal transport, but a heart is absent and body movements circulate the fluid irregularly. A coelom is absent in most.

Species with a thin, flexible cuticle have antagonistic circular and longitudinal muscles acting on a hydrostatic skeleton. The hemocoel, whether fluid- or tissue-filled, is the hydrostat. Circular musculature tends to be absent in taxa with a thick cuticle, which itself antagonizes the longitudinal muscles. - Adhesive glands are characteristic of many species and often open to the outside of the body through projecting cuticular tubes. Although the anterior tip of the body bears the mouth and sense organs, there is no well-formed head. The gut is usually a straight, complete tube. Protonephridia are the typical excretory organs. The body is composed of relatively few cells, often around 1000. Many species, including the nematodes, have an invariant, species-specific, and genetically fixed number of cells, a phenomenon known as eutely, or cell constancy. The male of the roundworm *Caenorhabditis elegans,* for example, has 1031 cells. In eutelic animals, mitosis ceases following embryonic development and thereafter growth results from increases in cell size. Most cycloneuralians are gonochoric and development is strongly determinate.

GASTROTRICHA[P]

Gastrotricha is a small but diverse taxon of some 500 species in 50 genera. These tiny metazoans inhabit interstitial spaces of marine and freshwater sediments, superficial detritus, the surfaces of submerged plants and animals, and the water film covering soil particles in terrestrial habitats. Gastrotrichs are common animals of ponds, streams, and lakes. The intertidal zone of a marine beach may harbor 40 or more species living in the spaces between sand grains, and as many as 1000 individuals have been reported from 20 ml of sand. Gastrotricha is divided into the marine Macrodasyida and the marine and freshwater Chaetonotida (Fig. 22-1).

Most gastrotrichs are microscopic, ranging in length from 50 to 1000 μm, although some species approach 4 mm. The bilaterally symmetrical, bowling-pin- or strap-shaped body is flattened ventrally and arched dorsally (Fig. 22-2). The anterior head bears the sense organs, brain, and pharynx, whereas the elongate trunk contains the midgut and reproductive organs. There is no introvert or extensible mouth cone.

Locomotion is by ciliary gliding, and the locomotory cilia are always restricted to the flattened ventral surface of the trunk and head, although the pattern varies widely with taxon.

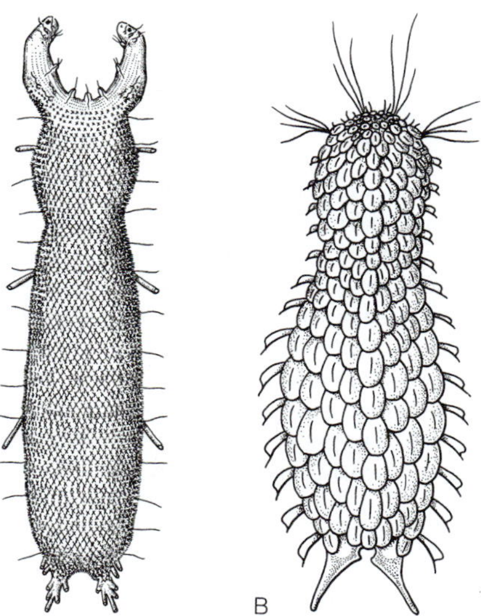

FIGURE 22-1 Gastrotricha. **A,** A marine macrodasyan, *Pseudostomella.* **B,** A marine chaetonotan, *Halichaetonotus.* *(A, From Ruppert, E. E. 1970. On* Pseudostomelia *Swedmark 1956 with descriptions of* P. plumosa *nov. spec.,* P. cataphracta *nov. spec. and a form of* P. roscovita *Swedmark 1956 from the West Atlantic coast. Cah. Biol. Mar. 11:121–143.)*

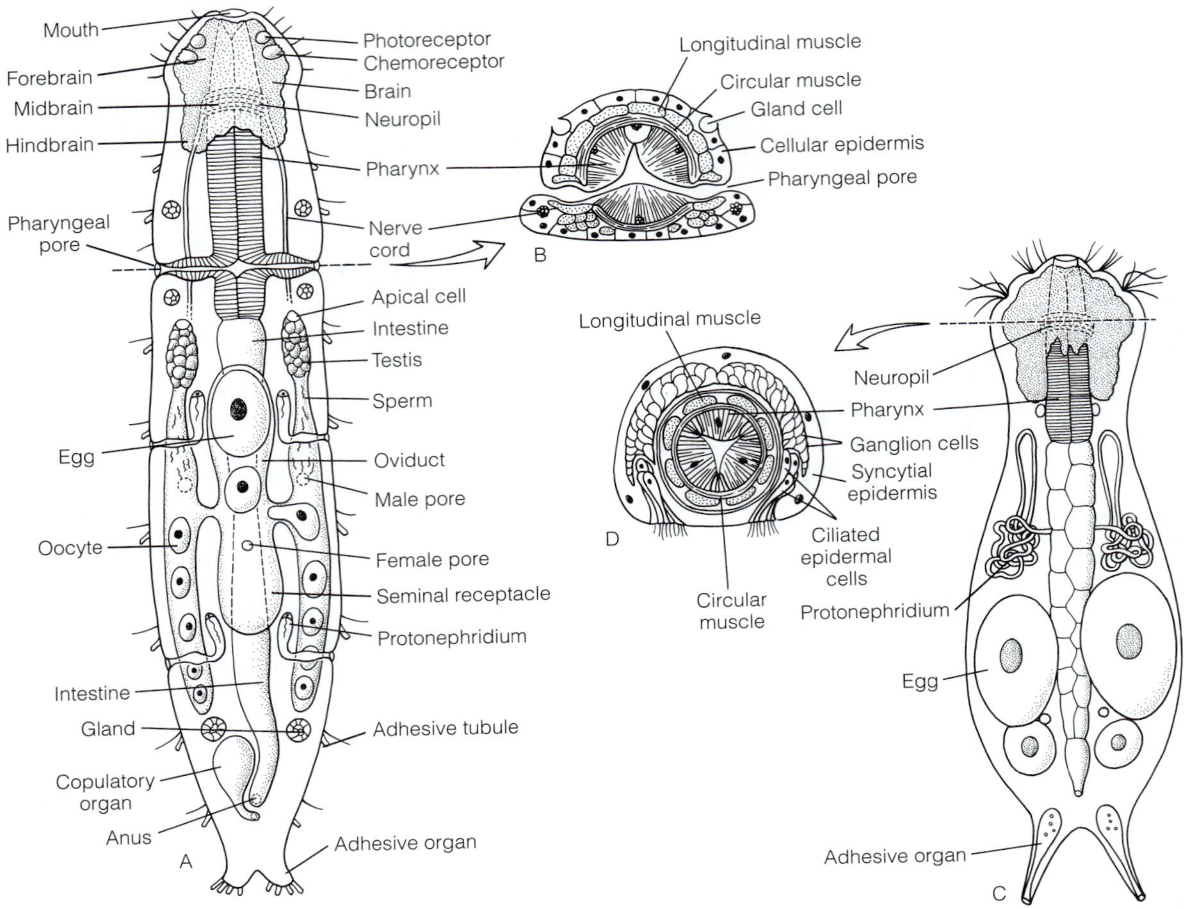

FIGURE 22-2 Gastrotricha: anatomy. **A** and **B,** A generalized macrodasyan in frontal section **(A)** and a cross section through the pharyngeal region **(B). C** and **D,** A generalized chaetonotan in frontal section **(C)** and a cross section through the pharyngeal region **(D).** *(Redrawn and modified from Ruppert, E. E. 1991. Gastrotricha. In Harrison, F. W., and Ruppert, E. E. (Eds.): Microscopic Anatomy of Invertebrates. Vol. 4. Aschelminthes. Wiley-Liss, New York. pp. 41–109.)*

The entire ventral surface may be ciliated or the cilia may be arranged in longitudinal bands, transverse rows, or in patches, or grouped into cirri, like those of hypotrichous ciliates (Fig. 22-3). The primitive taxa of gastrotrichs have monociliated epidermal cells, a feature shared only with the gnathostomulans among the acoelomate protostomes. The name Gastrotricha refers to the ventral ciliation (*gastro* = stomach, *trich* = hair).

Adhesive tubes provide reversible adhesion to the substratum and contain duo-gland systems consisting of a viscid gland coupled with a releasing gland, as in the turbellarians. Adhesive tubes may be numerous and located beneath the head, along the sides of the body, or on a pair of posterior terminal adhesive organs (Fig. 22-2A,C, 22-4).

The body wall is composed of an external cuticle, epidermis, and circular and longitudinal muscle fibers. The epidermal basal lamina is weakly developed. There is no body cavity and the connective-tissue compartment is filled with cells, so the organization of the body is compact. The connective tissue itself is poorly developed. In the macrodasyans, large, vacuo-

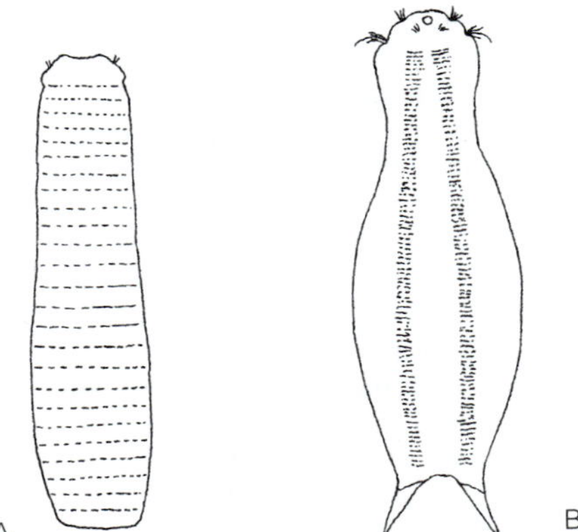

FIGURE 22-3 Gastrotricha. **A,** Ventral ciliation pattern of the macrodasyan *Thaumastoderma.* **B,** Ventral ciliation pattern of the chaetonotan *Chaetonotus.* *(After Remane, A. 1936. Gasrotricha. In Bronn, H. G. (Ed.): Klassen Ordn. Tierreichs. 4:1–242.)*

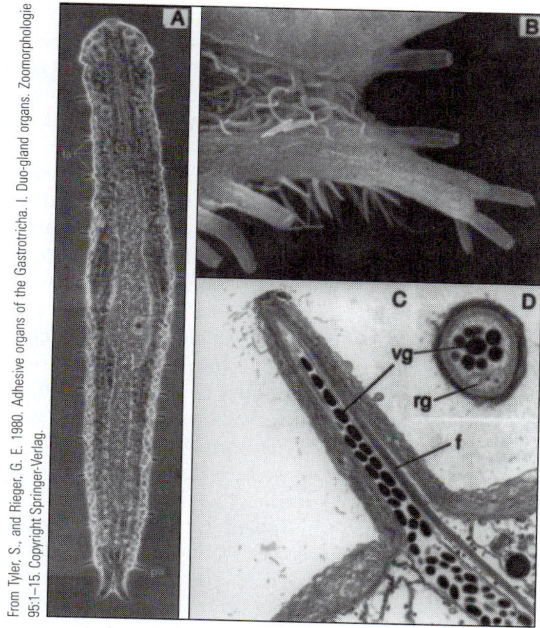

FIGURE 22-4 Gastrotricha: adhesive tubes of the macrodasyan gastrotrich, *Turbanella*. **A,** Photograph of a living animal, showing lateral and posterior adhesive tubes. Each lateral adhesive tube is accompanied by a sensory cilium. **B,** Enlarged view of posterior adhesive tubes. **C** and **D,** Longitudinal and cross sections of an adhesive tube. pa, posterior adhesive tube; la, lateral adhesive tube; f, fiber; vg, viscid gland; rg, releasing gland.

lated Y-cells surrounding the gut may function as a hydrostatic skeleton. There is no hemal system or specialized respiratory structures in these small animals.

The cuticle consists of a fibrous inner endocuticle and an outer epicuticle (exocuticle) composed of numerous bilayers of alternating electron-dense and electron-lucent layers (Fig. 22-5A). These multiple bilayers surround the entire body, including each of the cilia, and probably function as a physiological barrier (Fig. 22-5A,B). In some gastrotrichs the endocuticle is locally thickened and specialized to form scales, spines, and hooks (Fig. 22-1B, 22-5C). The cuticle of the general body surface lacks chitin, but traces have been found in that of the pharynx. Unlike all other cycloneuralians, gastrotrichs do not molt.

The epidermis can be cellular, as in macrodasyans, or syncytial, as in most chaetonotans. It is ciliated ventrally and, in some chaetonotans, the ciliated cells are arranged in two longitudinal rows that bulge inward to form thickened ridges, or cords (Fig. 22-2D, 22-3B).

The musculature may be cross- or obliquely striated or smooth depending on the species. The body-wall musculature is usually arranged as rings of outer circular and bands of inner longitudinal muscle (Fig. 22-2B,D). Two ventrolateral bands of longitudinal muscle are strongly developed and extend the length of the body, enabling it to shorten and curl ventrally. The circular musculature allows for hydrostatic re-extension of the body in most species of Macrodasyida, which have a soft, flexible cuticle. In many Chaetonotida, however, a circular body-wall musculature is absent and a thick, elastic cuticle antagonizes the longitudinal muscles. The innervation of gastrotrich muscle fibers, like that of nematodes (and cephalochordates), is accomplished by axonlike **innervation processes**

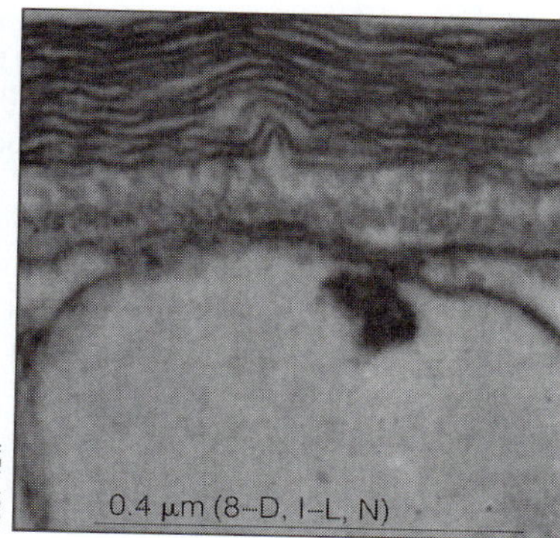

A

B

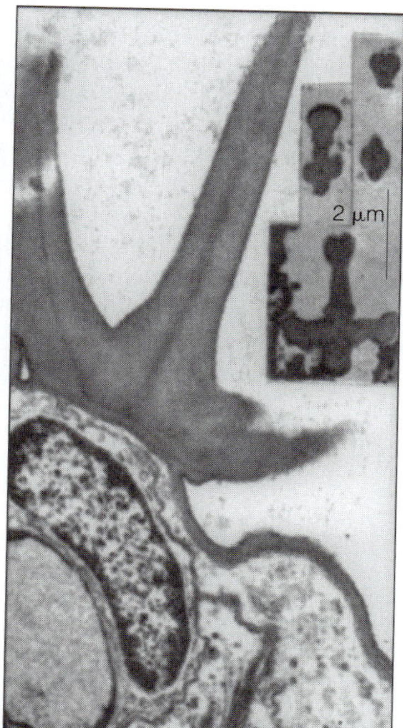

C

FIGURE 22-5 Gastrotricha. **A,** Cuticle and part of the epidermis of *Turbanella ocellata,* showing the thick, pale endocuticle overlying and adjacent to a large epidermal vesicle. The outer epicuticle composed of numerous parallel bilayers is adjacent to the endocuticle and at the top of the image. **B,** Cross section of a locomotory cilium of *Urodasys nodostylis,* showing the many layers of the epicuticle (epc) surrounding the cilium. **C,** Cuticular spines of *Tetranchyroderma* sp. Inset shows cross sections of spines at various levels.

from muscle cells to motor neurons in the nerve cord rather than by axons from motor neurons to muscle cells.

The intraepidermal nervous system consists of a brain and a pair of ventrolateral longitudinal nerve cords (Fig. 22-2A,D). The brain is a typical cycloneuralian triannular collar. The nerve cords are near the large ventrolateral longitudinal muscle bundles and, when present, the thickened epidermal cords. Sense organs include cerebral organs as well as sensory bristles located over the general body surface. The cerebral sensory organs consist of ciliary bristles and tufts (mechanoreceptors), ciliated pits, fleshy appendages (chemoreceptors), and simple ciliary ocelli (photoreceptors). All sensory receptors are modified monociliated cells.

Forward movement in gastrotrichs is the result of smooth ciliary gliding. Muscular action is important, however, in specialized movements, such as escape responses, turning, and copulation (Fig. 22-6). An escape response can be a rapid rearward withdrawal of the head and trunk to an attachment made by the posterior adhesive organ or a series of inchworm-like retreating movements. During copulation, the two animals use muscles to twist their posterior ends together.

The gut includes the mouth, pharynx, intestine, and anus. The terminal mouth usually opens directly into the pharynx, which is an elongate, glandular, muscular tube (Fig. 22-2). The pharyngeal wall is composed largely of a thick layer of epitheliomuscular cells. In the pharynx circular and longitudinal muscle layers surround a thick layer of radial muscles. The lumen is triradiate (triangular or Y-shaped, as in Fig. 22-2D) and lined with cuticle. In macrodasyans, the pharynx opens to the exterior through a pair of **pharyngeal pores** (Fig. 22-2A,B) before joining the intestine (midgut) via a valve that prevents backflow. The intestine is a cellular, cylindrical tube that tapers to join a posterior ventral anus. The monolayered midgut epithelium generally lacks cilia, but is lined with a brush border of microvilli. Primitively the midgut wall includes circular and longitudinal muscles, but in chaetonotans the circulars are absent.

Gastrotrichs feed on small, dead or living organic particles, such as bacteria, diatoms, and small protozoa, all of which are sucked into the mouth by the pumping action of the pharynx. In macrodasyans the paired pharyngeal pores vent excess water ingested with food. Digestion, which is probably both extracellular and intracellular, and absorption take place in the intestine.

A single pair of protonephridia occurs in the anterior trunk of marine and freshwater chaetonotans, but several pairs are arranged serially along the length of the body in the marine macrodasyans (Fig. 22-2A,C). They probably are chiefly osmoregulatory in function. The nephridiopores open on the ventrolateral surface of the body.

In contrast with other cycloneuralians, gastrotrichs are hermaphroditic and sperm transfer is indirect. Reproductive anatomy and behavior vary widely in Gastrotricha and are not yet well understood. In *Macrodasys*, which probably approximates the ancestral plan, there is a pair of hermaphroditic gonads, each with an anterior testis and a posterior ovary (Fig. 22-2A). Sperm ducts carry the sperm, which may be packaged in spermatophores, to a pair of ventral male gonopores located near the middle of the body. Gonopores, both male and female, are often temporary. A copulatory organ is located at the posterior end of the body. Immediately prior to copulation, this organ is brought into contact with the male gonopores and loaded with sperm. The copulatory organ then transfers the sperm into the partner's seminal receptacle through the female gonopore. Fertilization occurs internally in the female part of the system and fertilized eggs are released by rupturing through the body wall. All species have complex accessory reproductive organs. There are many variations on this pattern and at least one species is viviparous.

The male system of the freshwater chaetonotans is degenerate and nonfunctional, so all individuals are believed to be female and to reproduce parthenogenetically (Fig. 22-2C). Recently, however, sperm have been discovered in the common and widespread *Lepidodermella squamata* and in many other species, but it is not known if they are functional.

In the freshwater parthenogenetic chaetonotans, two types of eggs are produced and attached to the substratum. One type is the dormant, or resting, egg, which, like those of rotifers, cladocerans, and freshwater flatworms, can withstand desiccation and low temperatures. The other type hatches in one to four days. Although tiny in absolute terms ($< 50\ \mu m$), the eggs, which occupy most of the trunk and are produced and laid one at a time, are enormous compared to the size of the animal.

Cleavage is bilateral and development is determinate, direct, and benthic. Young gastrotrichs have most of the adult structures upon hatching and reach sexual maturity in about three days. Growth results primarily in elongation of the trunk

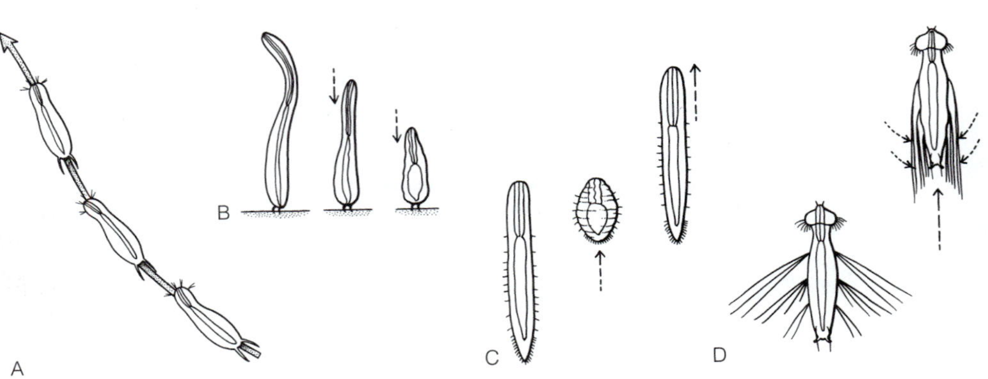

FIGURE 22-6 Gastrotricha. **A,** Ciliary gliding over a surface in *Chaetonotus*. **B,** Retraction toward anchored posterior adhesive organs in *Neodasys*. **C,** Forward creeping in *Macrodasys*. **D,** Skipping locomotion using muscle-actuated spines in the pelagic *Stylochaeta*. *(A, C, and D, Redrawn from Remane, A. 1936. Gasrotricha. In Bonn, H. G. (Ed.): Klassen Ordn. Tierreichs. 4:1–242.)*

and morphogenesis of the reproductive system. Laboratory studies of *Lepidodermella* indicate a maximum life of about 40 days, during the first 10 of which four or five eggs are produced.

DIVERSITY OF GASTROTRICHA

Macrodasyida[C]: Body usually is wormlike, elongate, dorsoventrally flattened. Adhesive tubes located at anterior and posterior ends and along the sides of body. Pharyngeal lumen is inverted-Y shape; pharyngeal pores present. Cellular epidermis. Hermaphroditic. In marine and estuarine sediments. *Dactylopodola, Macrodasys, Pseudostomella, Tetranchyroderma, Turbanella, Urodasys.*

Chaetonotida[C]: Body is bowling-pin-shaped. Adhesive tubes are restricted to posterior end. Pharyngeal lumen is Y-shaped; pharyngeal pores are absent. Syncytial epidermis. Freshwater and marine. Freshwater species are parthenogenetic. The primitive marine *Neodasys,* with a wormlike body, lateral and posterior adhesive tubules, and a cellular epidermis, is exceptional. *Chaetonotus, Halichaetonotus, Lepidodermella, Neodasys, Xenotrichula.*

NEMATODA[P]

Nematodes, or roundworms, are the most diverse cycloneuralian taxon and possibly the largest animal phylum. At present only about 20,000 species have been described, but there is speculation that millions more await discovery. Nematodes are among the most widespread and abundant of all metazoans. These worms inhabit moist interstitial environments in all habitats, including the bodies of plants and animals. They are abundant in marine and freshwater benthic habitats, in the soil, and as parasites in a great variety of plant and animal hosts. They occur from the poles to the tropics in all types of environments, including deserts, high mountains, and the deep sea. Nonparasitic nematodes are meiofaunal animals that live in the interstitial spaces of algal mats, aquatic sediments, and terrestrial soils, where they may be present in staggering numbers. One square meter of mud off the Dutch coast contains 4,420,000 nematodes, a hectare of good farm soil can contain billions of terrestrial nematodes, and a single decomposing apple lying on the ground in an orchard yielded 90,000 roundworms in several species.

Unusual aquatic habitats include hot springs in which the water temperature reaches 53°C and water in the leaf axils of epiphytic bromeliads high in the canopy of a tropical rain forest. In large lakes, there is often a distinct zonation of benthic nematode species from the shoreline into deeper water. Terrestrial species live in the film of water surrounding each soil particle, making them, like other nematodes, aquatic. Some species, in fact, are found in both soil and fresh water. Although terrestrial nematodes exist in enormous numbers in the upper soil, population density decreases rapidly with increasing depth. Moreover, densities are highest near plant roots. In addition to the more typical terrestrial habitats, nematodes have also been reported from accumulations of detritus in leaf axils and in the angles of tree branches. Mosses and lichens maintain a characteristic nematode fauna with the ability to withstand periodic desiccation. During such times the worm passes into a state of suspended animation called cryptobiosis.

In addition to free-living species, there are many parasitic nematodes displaying all degrees of parasitism and attacking virtually all plants and animals, often in species-specific relationships. The numerous species that infest humans, food and horticultural crops, and domesticated animals make Nematoda one of the most important parasitic animal taxa. Nematoda also contains one of the most intensely studied laboratory animals, *Caenorhabditis elegans,* whose every cell has been traced throughout the course of development and whose genome is among the best known of all organisms.

That nematodes are ubiquitous, abundant, and habitat- or host-specific was famously noted long ago by the nematologist N. A. Cobb, who said, "If all the matter in the universe except the nematodes were swept away, our world would still be dimly recognizable, and if, as disembodied spirits, we could investigate it, we should find its mountains, hills, vales, rivers, lakes and oceans represented by a thin film of nematodes." (Cobb, N. A. 1915. Nematoda and their relationships. Year Book Dept. Agric., 1914. USDA, Washington, DC. pp. 457–490.)

FORM

In spite of the large number of species, nematode morphology is unusually homogeneous and most share a common body plan. Nematodes are the quintessence of vermiformity, being elongate, cylindrical, and tapered at both ends (Fig. 22-7A). This shape, an important adaptation for moving in small spaces, is typical of interstitial animals. The majority of free-living nematodes are less than 2.5 mm in length and most are about 1 mm. Many are microscopic, but some soil nematodes are as long as 7 mm, some marine species attain lengths of 5 cm, and some parasites reach 50 cm or more.

The body is cylindrical, hence the common name *roundworm.* It is not divided into regions, although one genus, *Kinonchulus,* has an introvert with six double, longitudinal rows of cuticular spines. The anterior end is radially symmetrical, but most of the body and organ systems have bilateral symmetry. The mouth is terminal, located at the anterior end, and surrounded by various lips and sensilla (Fig. 22-8A). In many marine nematodes, including the most primitive species, six lips border the mouth, with three on each side; as a result of fusion, however, often there are only three in the more derived terrestrial and parasitic species. Primitively, the lips and the adjacent surface bear sensilla and a variety of cuticular projections.

A caudal gland (spinneret) is typical of many free-living nematodes, including most marine species. The gland opens at the posterior tip of the body, which is sometimes drawn out to form a cylindrical tail. In some nematodes it is known to be a duo-gland adhesive organ.

Nematode tissues may be cellular or syncytial and typically are characterized by a species-specific number of cells or nuclei. There are no locomotory cilia in nematodes, but some primitive species have ciliated gastrodermal cells.

BODY WALL

The body wall consists of a cuticle, epidermis, and longitudinal muscles. Members of the marine interstitial Stilbonematidae have body surfaces clothed with symbiotic filamentous blue-green bacteria and consequently appear to be hairy.

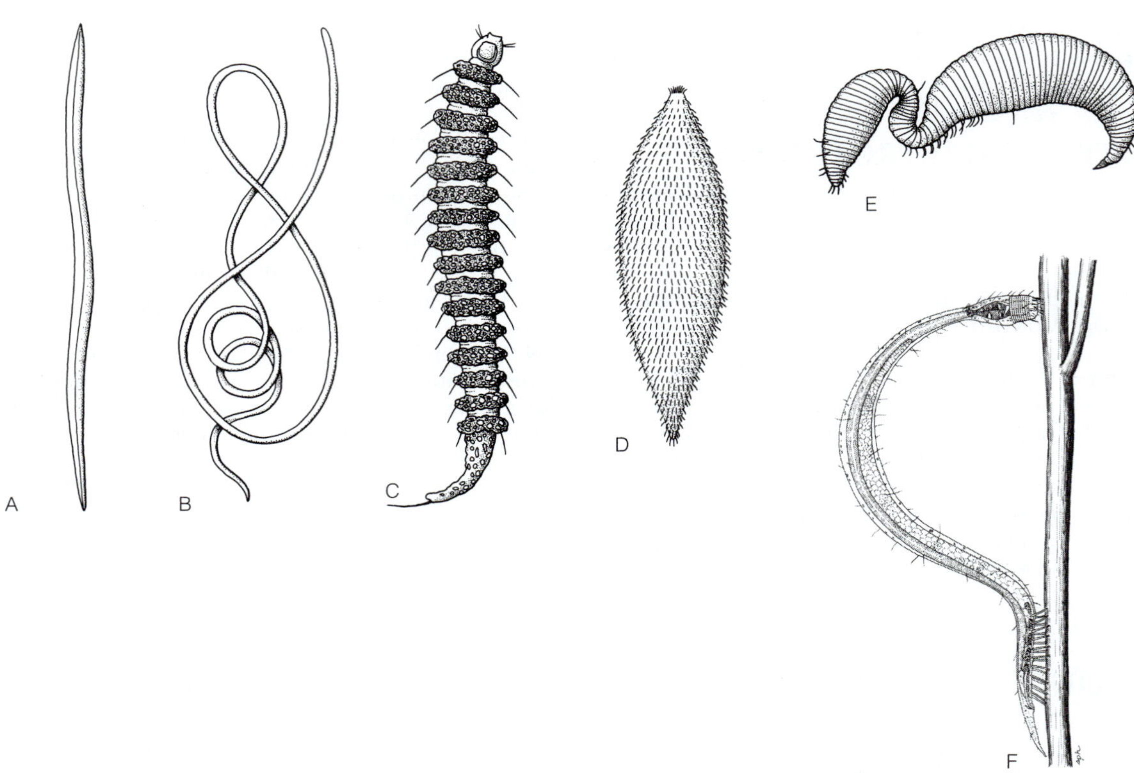

FIGURE 22-7 Nematode diversity. **A,** The marine *Echinotheristus.* **B,** A parasitic mermithid nematode. **C,** The marine interstitial *Desmoscolex.* **D,** The marine *Greeffiella.* **E,** The marine interstitial *Epsilonema.* **F,** The marine *Draconema.* *(A, C, and D, Redrawn from Riemann, F. 1988. Nematoda. In Higgins, R. P., and Theil, H. (Eds.): Introduction to the study of meiofauna. Smithsonian Institution Press. Washington, DC. pp. 293–301; B, Drawn from a photograph in Pearse, V., Pearse, J., Buchsbaum, M., et al. 1987. Living Invertebrates. Boxwood Press. Pacific Grove, CA. p. 284.)*

The cuticle covers the body and lines the pharynx and hindgut. As in most cycloneuralians, it is molted and is often sculptured or ornamented in various ways (Fig. 22-7). In nematodes it is more complex than in other cycloneuralians and includes several layers (Fig. 22-9). Outermost is a thin epicuticle of largely unknown composition that probably includes lipids and carbohydrates. It resembles the gastrotrich epicuticle but has only one electron-lucent layer enclosed on each side by an electron-dense layer. Inside the epicuticle are three additional, largely collagenous layers that account for most of the thickness of the cuticle. The **cortex,** which is outermost, is sometimes annulated (ringed) and may contain elastin. The **median layer,**

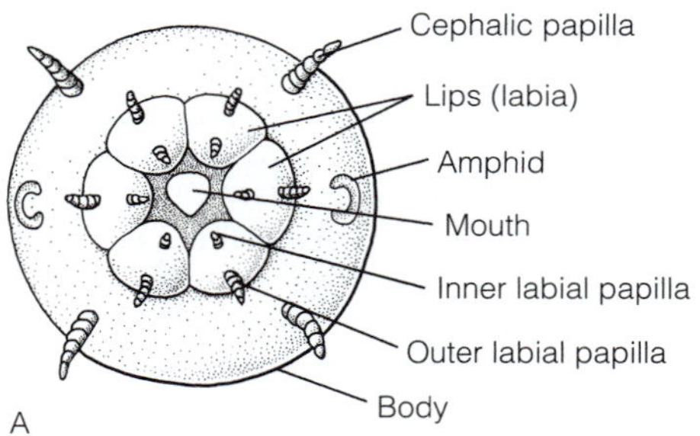

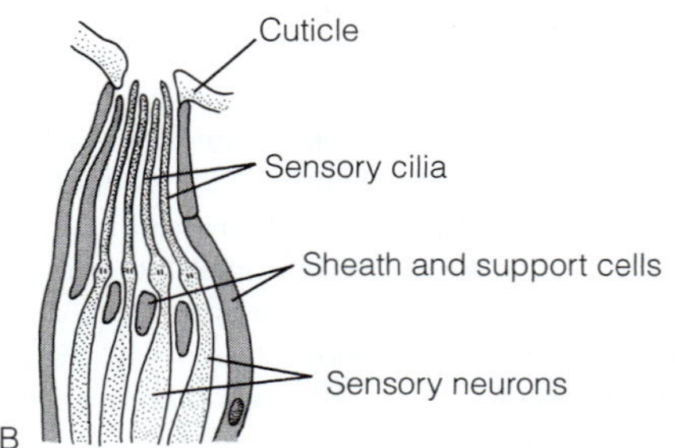

FIGURE 22-8 Nematoda. **A,** Oral view of a generalized nematode, showing typical sensory structures. **B,** Longitudinal section of an amphid of *Caenorhabditis elegans.* *(A, Modified and redrawn after Jones from Lee, D. L., and Atkinson, H. J. 1977. Physiology of Nematodes. Columbia University Press, New York. p. 161; B, Modified from Ward, S., Thomson, J. G., and Brenner, S. 1975. Electron microscopical reconstruction of the anterior sensory anatomy of the nematode* Caenorhabditis elegans. *J. Comp. Neurol. 160:313–338.)*

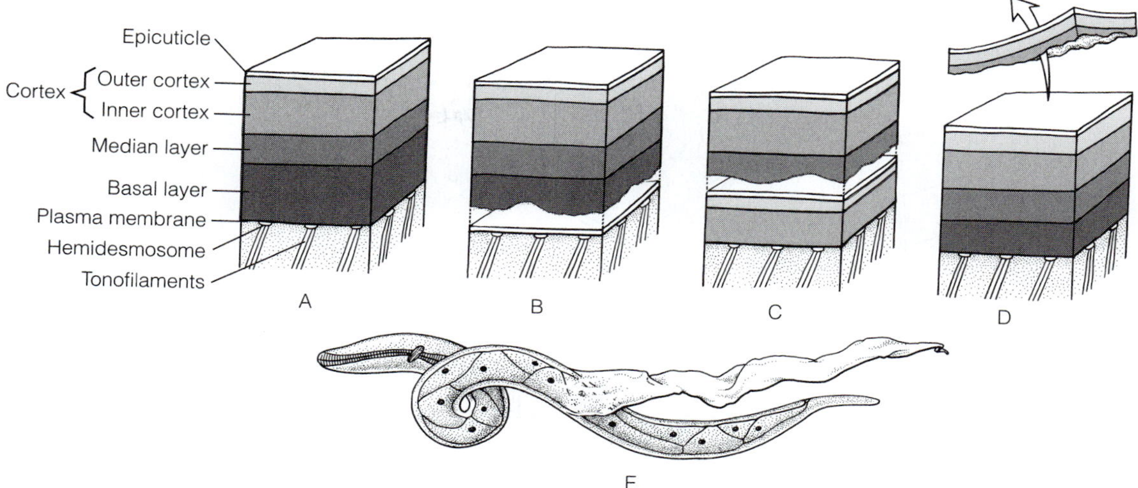

FIGURE 22-9 Nematoda ecdysis: Although details vary with species, molting often occurs in the sequence shown. **A,** Intact cuticle. **B,** Separation of the old cuticle from the epidermis, digestion of the old basal layer, and secretion of a new epicuticle. **C,** Secretion of new outer and inner cortex and digestion of old median layer. **D,** Secretion of new basal layer to complete the new cuticle. The remaining old cuticle is shed (**D** and **E**).

which is variable and may be fluid or structured, accounts for most of the differences between cuticles of different species. The innermost **basal layer** may be striated or laminated or contain fibers in a crossed-helical arrangement. There is no chitin in the body cuticle, although it is present in that of the pharyngeal lining and eggshell of some species.

Nematodes molt four times as they grow, although some growth also occurs without molting. The old cuticle separates from the underlying epidermis, a new cuticle is secreted, and the old cuticle is shed, sometimes in fragments (Fig. 22-9E). Molting ceases when adulthood is reached, but the animal continues to grow and the existing cuticle expands. Molting occurs in three stages. First, the old cuticle separates from the epidermis and its hydrolysis begins with its innermost (basal) layer (Fig. 22-9B). Second, secretion of the new cuticle (Fig. 22-9B–D) is accompanied by the continued hydrolysis of the old cuticle from the inside out. The new cuticle, conversely, is secreted beginning with the outermost layer (epicuticle) first (Fig. 22-9B). Finally, during ecdysis the remains of the old exocuticle are shed (Fig. 22-9D). Ecdysone, the hormone that controls molting in arthropods, is present in nematodes and presumably has the same role.

The epidermis may be cellular or syncytial, depending on the species. It secretes the cuticle, stores nutrients, bears fibers (tonofilaments) that attach the musculature to the cuticle (Fig. 22-9), and, in some endoparasitic species, is an important surface for the absorption of nutrients from the host. The nematode epidermis expands internally to form ridges along the mid-dorsal, midventral, and midlateral lines of the body (Fig. 22-10C). These four **epidermal cords** extend the length of the body and contain the epidermal nuclei, excretory organs, and longitudinal nerve cords.

The muscle layer of the body wall is composed entirely of longitudinal fibers and there are no circular muscles. The longitudinal muscles are in four bands separated by the epidermal cords (Fig. 22-10B,C, 22-11). Each nematode muscle fiber, like those of gastrotrichs, is divided into a basal region containing contractile proteins and an axonlike **innervation process** that lacks contractile fibers. This cytoplasmic process extends to the closest dorsal or ventral nerve cord, where it synapses with motor neurons. Motor signals from the CNS are transmitted to the muscle's contractile region by these processes, rather than by axons. In addition to the body-wall musculature, nematodes have muscles associated with the reproductive organs and sometimes a few muscle cells with the intestine. The body-wall muscles attach to the inside of the cuticle by tonofilaments extending from the muscles through the epidermal cells to hemidesmosomes on the cuticle (Fig. 22-9A, 22-11B).

The nematode body cavity is a hemocoel (Fig. 22-10C, 22-11A), but it is small or nonexistent in most small, free-living nematodes, which are thus compact. It may be spacious in large species, such as the parasitic *Ascaris.* When present, it occupies the space between the body-wall musculature and the gut and surrounds the reproductive organs. Fluid in the cavity is pressurized and functions as a hydrostat. The fluid contains a variety of organic metabolites, including hemoglobin in some species, but no circulating cells. A few phagocytic cells, however, are permanently attached to the walls of the cavity and are important in internal defense. There is no specialized gas-exchange surface.

NERVOUS SYSTEM AND SENSE ORGANS

The entire nematode nervous system is intraepithelial in the epidermis, pharynx, and hindgut. The brain is a collarlike, tri-annular, circumpharyngeal nerve ring (Fig. 22-10, 22-11, 22-14) like that of other cycloneuralians. Sensory nerves extend anteriorly from the brain to innervate the many cephalic sensilla.

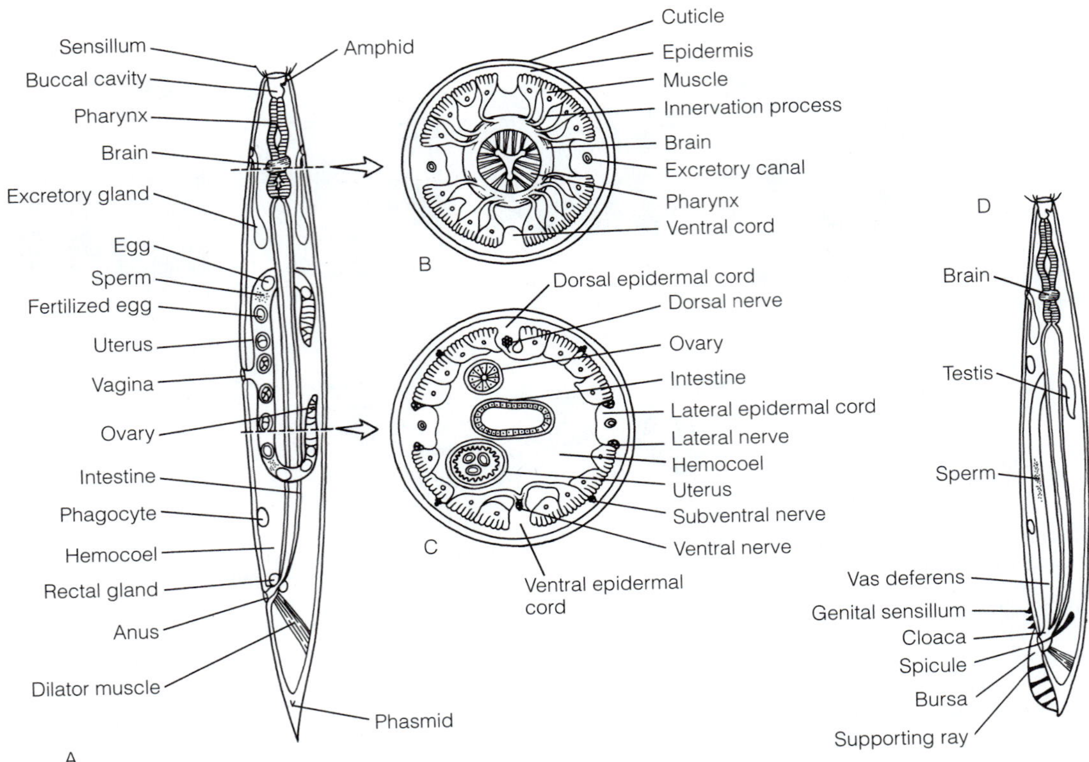

FIGURE 22-10 Nematode general anatomy. **A–C,** Generalized female nematode in lateral view **(A)** and cross sections (**B** and **C**). **D,** Generalized male nematode in lateral view. The volume of the hemocoel has been exaggerated for clarity. *(Redrawn from Lee, D. L., and Atkinson, H. J. 1977. Physiology of Nematodes. Columbia University Press, New York. p. 161.)*

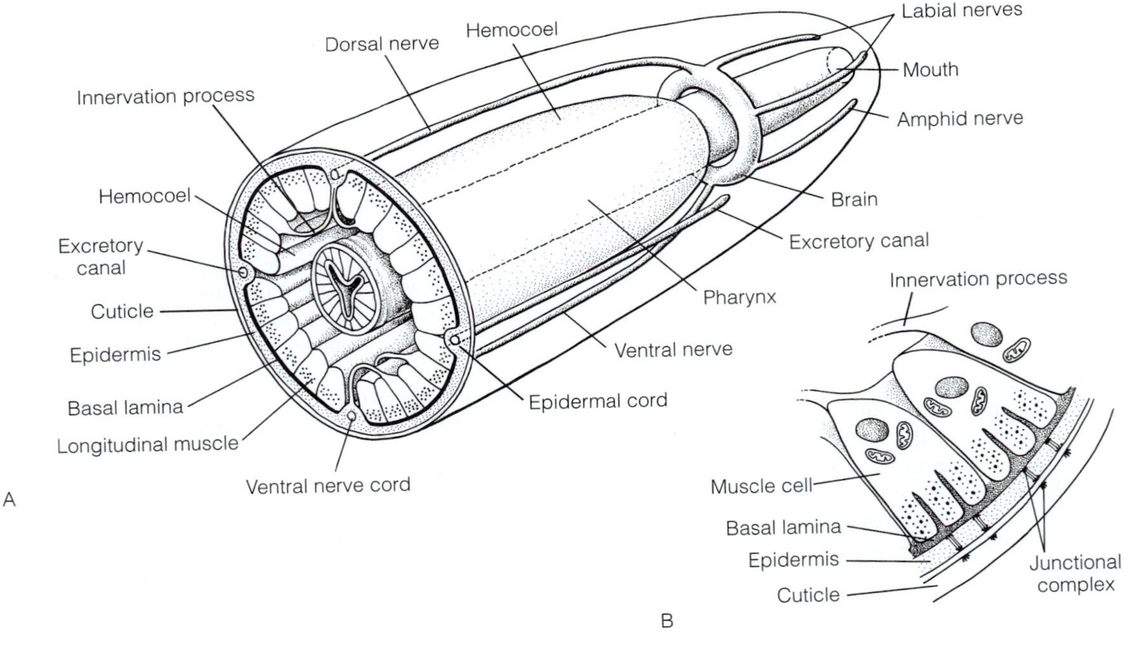

FIGURE 22-11 Nematoda neuromuscular anatomy. **A,** Stereogram of the anterior end of a generalized nematode. The nervous system is situated entirely within the epidermis and the musculature (detail in **B**) resembles an epithelium. The size of the hemocoel has been exaggerated for clarity.

Dorsal, lateral, and ventral nerve cords extend posteriorly in the epidermal cords, but motor output to the body-wall muscles is from the dorsal and ventral cords only. Two short cords exit the brain ventrally and quickly join to form the ventral nerve cord. Some motoneurons from the ventral nerve cord extend dorsally via commissures to join the dorsal nerve cord, but most run posteriorly in the ventral cord. The dorsal nerve cord is motor, but the ventral cord is sensory and motor. The lateral cords are chiefly sensory and serve the excretory canals.

The principal external sensilla are papillae, setae, amphids, and phasmids, all of which bear a ciliated dendrite enclosed in a specialized, cuticularized part in the body wall (Fig. 22-8). The papillae are arranged in three rings around the mouth. The labial (two rings of six each) and cephalic (one ring of four) papillae are low projections of the cuticle on the lips and head, respectively. The outer labial papillae and cephalic papillae are believed to be mechanoreceptors. The inner labial papillae, amphids, and phasmids all open to the exterior via a small cuticular pore, thus exposing the sensory cilia directly to the external environment. These sensilla are believed to be chemosensory, but each probably responds to a different set of chemical cues. Males also have sensilla surrounding the anus, which is also the male gonopore. Setae are elongate cuticular bristles on the head and body. They are touch receptors that, when stimulated, cause the animal to withdraw from the stimulus.

Amphids are best developed in free-living aquatic, and especially marine, nematodes. They are pouch- or tubelike invaginations of the cuticle open to the exterior and containing ciliated receptor cells. Amphids are paired, with one on each side of the head (Fig. 22-8), and are mechano- and chemosensory. In the tail region of some nematodes (such as Secernentea) is a pair of enigmatic unicellular glands called **phasmids** (Fig. 22-18A) that may be chemosensory, secretory, or excretory. They are best developed in parasitic nematodes.

Two simple pigmented ocelli are located laterally, one on each side of the pharynx, on some marine and freshwater nematodes, but their function is uncertain. Some are equipped with cuticular lenses and are probably photoreceptive. Internal stretch receptors have been found in the epidermal cords and probably regulate locomotor movements.

LOCOMOTION

Most nematodes move forward and rearward using sinuous, eel-like undulations of the body. The undulations are in the dorsoventral plane, rather than side to side as in eels, and are produced by the alternate contraction of dorsal and ventral longitudinal muscles (Fig. 22-12A). In the absence of circular muscles, the longitudinal muscles are antagonized by the elastic cuticle. Annulations in the cuticle of many nematodes may

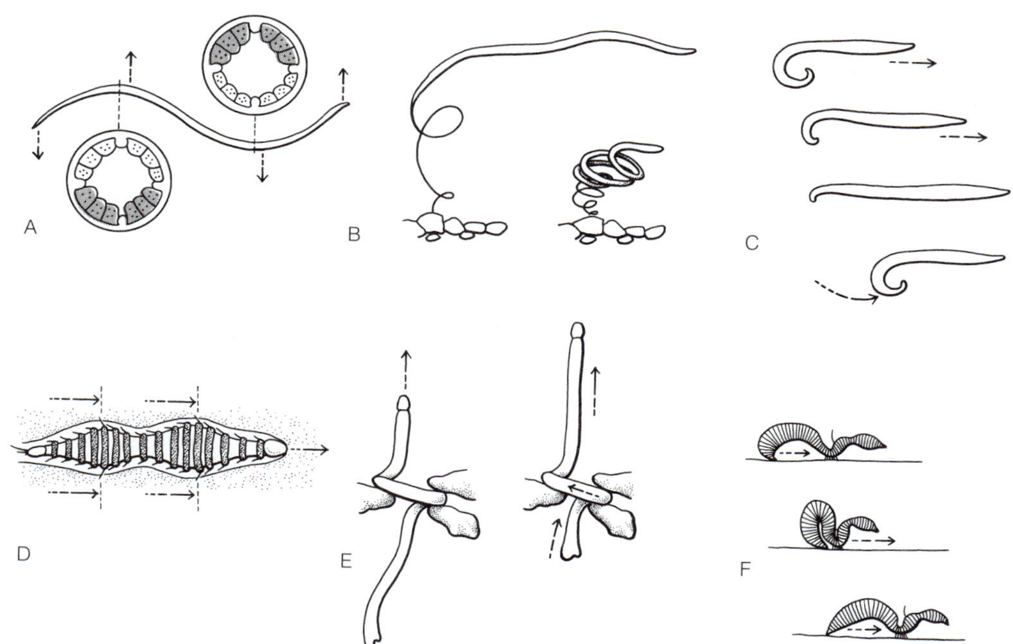

FIGURE 22-12 Nematode locomotion. **A,** Typical dorsoventral undulatory locomotion in a generalized nematode. Cross sections show contracted muscles (shaded) used to bend the body in the indicated regions. **B,** Retraction movement similar to that found in the ciliate *Vorticella,* shown here in the marine semisessile nematode *Trefusia.* **C,** Springing or thrusting locomotion in the marine *Theristus caudasaliens.* This nematode forcibly flicks its posterior end and springs forward. **D,** Earthwormlike progression in the marine genus *Desmoscolex.* **E,** Single spiral wave used by the gut parasite *Nippostrongylus* to move among intestinal villi. **F,** Inchwormlike locomotion in the marine interstitial *Epsilonema.* This nematode alternates attachment of its adhesive toe and bristles on the trunk. *(B, Redrawn from Riemann, F. 1974. On hemisessile nematodes with flagelliform tails living in marine soft bottoms and on micro-tubules found in deep sea sediments. Mikrofauna Meeresbodens 40:1–15; C, Redrawn from Adams, P. J. M., and Tyler, S. 1980. Hopping locomotion in a nematode: Functional anatomy of the caudal gland apparatus of* Theristus caudasaliens *sp. n. J. Morphol. 164:265–285; E, Lee, D. L., and Atkinson, H. J. 1977. Physiology of Nematodes. Columbia University Press, New York. p. 161.)*

improve flexibility, whereas crossed-helical fibers prevent kinking as the body flexes and herniation as the hydrostatic pressure rises.

If nematodes are removed from their natural environment, as, for example, during laboratory observations, movement is degraded to ineffective whipping and thrashing motions and the animal cannot make forward progress. Efficient undulatory locomotion requires a substratum, such as a sand-grain or surface film, to push against and is ineffective in a pelagic habitat. Most free-living nematodes are interstitial and can move rapidly and efficiently by pushing against the confines of the small spaces they inhabit (Fig. 22-12E). The pore size that allows for optimum undulatory locomotion is about 1.5 times the worm's diameter and for most soil nematodes, soil-pore sizes of 15 to 45 μm are ideal. In a variation of undulatory motion, some nematodes contract the posterior end of the body and thrust against the substratum to spring forward (Fig. 22-12C).

Many nematodes may swim intermittently for short distances. This is true, for example, of moss-inhabiting species when the moss is flooded following a rain. A few species can crawl, taking advantage of a cuticle sculptured to grip the surface. The crawling of one species, *Desmoscolex,* which has a ringed cuticle, is similar to that of earthworms (Fig. 22-12D). Some crawl like caterpillars, and others move like inchworms (Fig. 22-12F). The caudal gland, which is present in most marine nematodes, is used for temporarily attaching to the substratum. It may also be used for tail-anchored escape movements similar to those employed by the ciliate protozoan *Vorticella* (Fig. 22-12B). Rapid contractions of the caudal organ and posterior body pull the animal close to the substratum and perhaps remove it from harm's way.

NUTRITION

Many free-living nematodes are carnivorous and feed on small metazoan animals, including other nematodes, but many others are herbivores. Many marine and freshwater species feed on diatoms, algae, fungi, and bacteria. Algae and fungi frequently are important food for terrestrial species. A large number of terrestrial nematodes pierce the cells of plant roots and suck the contents. In the United States, such nematodes are annually responsible for billions of dollars of damage to agricultural crops. There are also many deposit-feeding marine, freshwater, and terrestrial species that ingest particles of substratum. Deposit feeders and the many nematodes that ingest dead organic matter such as dung or the decomposing bodies of plants and animals are actually utilizing the associated bacteria and do not digest the dead organic matter itself. This is true, for example, of the common vinegar eel, *Turbatrix aceti,* which lives in the sediment of nonpasteurized vinegar. Nematodes are the largest and most ubiquitous group of organisms feeding on fungi and bacteria and as such are a major link between decomposers and higher trophic levels. Although many nematodes feed on fungi, the tables are sometimes turned, for there are a few fungi that prey on nematodes. The worms are snared when they unwittingly pass through a noose made of a loop of a threadlike fungal hypha. The noose contracts and traps the worm, which is then digested.

The gut is typically cycloneuralian, with a foregut, midgut, and hindgut and an anterior terminal mouth. The ectodermal foregut consists of the buccal cavity and pharynx, the endodermal midgut is the intestine, and the ectodermal hindgut is the rectum. The mouth opens into a tubular buccal cavity (Fig. 22-10A, 22-13). The cuticle is often strengthened with ridges, rods, or plates, or it may bear teeth. The structural details of the buccal cavity are correlated with feeding habits and are important for identifying species. Teeth are common in carnivorous nematodes and may be small and numerous or limited to a few large, jawlike processes. The toothed terrestrial nematode *Mononchus papillatus* (Fig. 22-13C), which has a large dorsal tooth opposed by a buccal ridge, consumes as many as 1000 other nematodes during its life span of approximately 18 weeks. In feeding, this nematode attaches its lips to the prey and makes an incision with the large tooth. The contents of the prey are then pumped out by the pharynx.

In some carnivores, as well as in many herbivorous species that feed on the contents of plant cells, the buccal cavity is equipped with a long, hollow or solid cuticular stylet (Fig. 22-13B), which can protrude from the mouth to puncture the cell. A hollow stylet functions like a hypodermic needle through which the contents of the prey are sucked out by the pharynx. In contrast, a solid stylet is thrust rapidly forward and backward through the cell wall to release the contents.

The buccal cavity leads into a tubular pharynx (Fig. 22-10A,D, 22-14). The pharyngeal lumen is triradiate in cross section and lined with cuticle (Fig. 22-15). The wall is composed of three

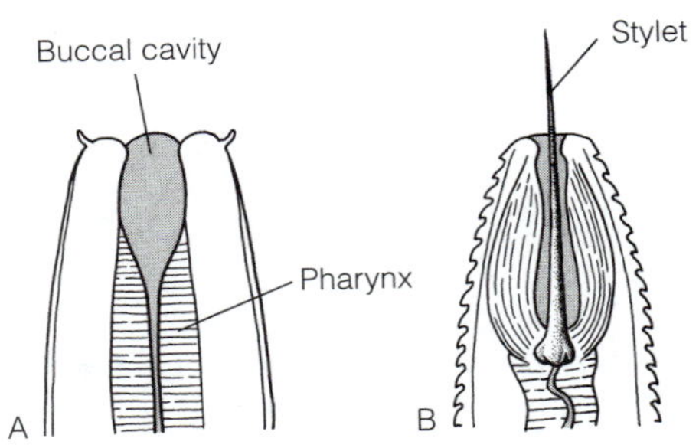

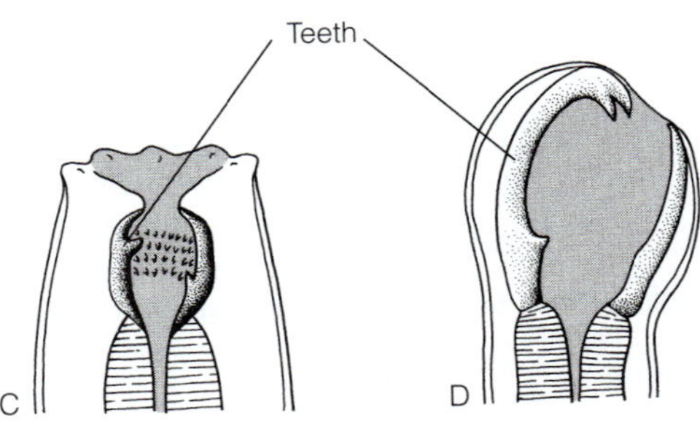

FIGURE 22-13 Nematode buccal cavities. **A,** The bacteriovore *Rhabditis.* **B,** The plant root cell predator *Criconemoides.* **C,** The protozoan and micrometazoan carnivore *Mononchus.* **D,** The intestinal parasite *Ancylostoma.* *(A, C, and D, Redrawn from Lee, D. L., and Atkinson, H. J. 1977. Physiology of Nematodes. Columbia University Press, New York. p. 161; B, Redrawn from Nicholas, W. L. 1984. The Biology of Free-Living Nematodes. Clarendon Press, Oxford. p. 28.)*

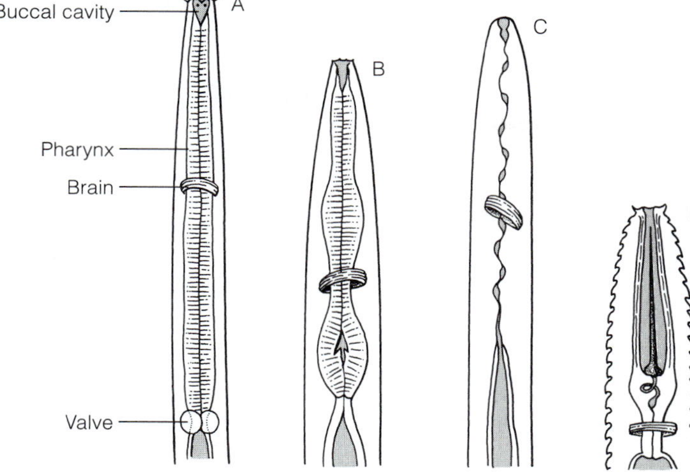

FIGURE 22-14 Nematode pharynges. **A,** *Monhystera,* marine and freshwater species that feed on bacterial aggregates. **B,** *Rhabditis,* marine and terrestrial species that feed on bacteria. **C,** Juvenile *Gastromermis* species parasitize the hemocoel of insects, absorbing nutrients directly through the body wall rather than with the vestigial pharynx. **D,** The stylet-bearing *Criconemoides,* which punctures plant roots and sucks out the contents. *(A, B, and D, Modified and redrawn from Nicholas, W. L. 1984. The Biology of Free-Living Nematodes. Clarendon Press, Oxford. p. 28.)*

bands of radial epitheliomuscular cells and associated gland cells, as in gastrotrichs. The pharynx is a pump that sucks food from the mouth into the intestine. The triradiate arrangement is an adaptation for sucking. Three bands of muscle are required in order for contraction to expand the lumen. Valves are frequently present to control the direction of flow.

A long, tubular intestine consisting of a single layer of epithelial cells without muscles or connective tissue extends posteriorly from the pharynx for the length of the body (Fig. 22-10A). In primitive species such as *Eudorylaimus* (Fig. 22-16A) the epithelium is ciliated. In most nematodes, however, cilia are absent and the epithelium has a brush border of microvilli (Fig. 22-16C). Absorption occurs in the intestine and is facilitated by the microvilli. A valve located at each end prevents food from being forced out of the intestine by the hydrostatic pressure of the hemocoel. In mermithid nematodes, which are parasites of invertebrates, the midgut is syncytial, lacks a lumen, and functions in food storage but not digestion (Fig. 22-16D,E). Nutrients absorbed through the body wall are stored in the syncytium, which thus functions as a liver.

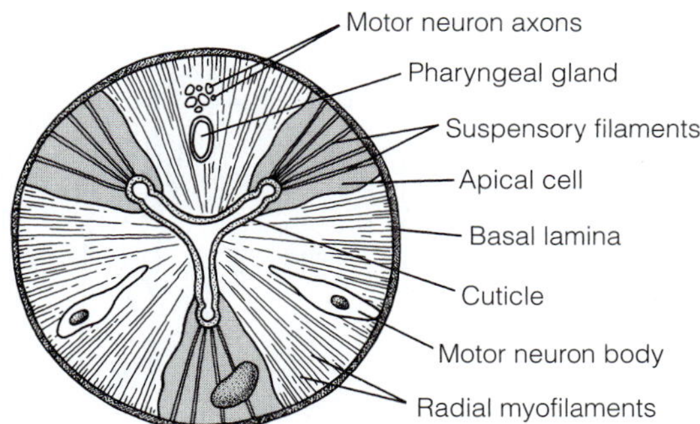

FIGURE 22-15 Nematoda. Transverse section of the epitheliomuscular pharynx of *Caenorhabditis elegans.* *(Drawn from an electron micrograph from Albertson, D. G., and Thomson, J. N. 1976. The pharynx of* Caenorhabditis elegans. *Philos. Trans. R. Soc. Lond. B. Biol. Sci. 275:299–325.)*

Digestive enzymes are produced by the pharyngeal glands (Fig. 22-15) and the intestinal epithelium. Digestion begins extracellularly in the intestinal lumen but is completed intracellularly. The intestine is also an important organ of nutrient storage and yolk synthesis for developing oocytes. In *Caenorhabditis elegans,* the intestine synthesizes yolk proteins and exports them, via the hemocoel, to the ovary. A short, cuticle-lined rectum connects the intestine with the anus, which is on the midventral line just anterior to the posterior tip of the body (Fig. 22-10A).

EXCRETION

Nematodes excrete ammonia, which is lost by diffusion across the body wall. Osmoregulation, ionic regulation, and perhaps the excretion of other waste metabolites seem to be associated with two types of specialized structures unique to nematodes. These are one or more excretory glands and the excretory canal system, or both (Fig. 22-17), but they are poorly understood and may not be excretory after all. When gland and canal cells occur in the same worm, they share a common pore. They do not resemble protonephridia or metanephridia, do not operate by filtration, and may be secretion kidneys. A few nematodes lack any recognizable excretory organs.

The **excretory gland** (ventral cell or renette cell) occurs by itself in Adenophorea and in conjunction with an excretory canal system in some members of Secernentea. This large cell, or cells, protrudes into the hemocoel and has a necklike duct that opens at a midventral pore (Fig. 22-17). The cell is secretory, but the function of its secretions may not be excretory and several alternatives have been suggested or demonstrated for it. In the citrus nematode *Tylenchulus semipenetrans,* it may secrete a gelatinous matrix around the eggs. It may secrete the outer glycoprotein coat of the cuticle of the root-knot nematode *Meloidogyne javanica,* and it may produce exoenzymes to initiate digestion of host tissues in some animal parasitics. In some it may secrete a molting fluid that initiates the separation of the old cuticle from the epidermis.

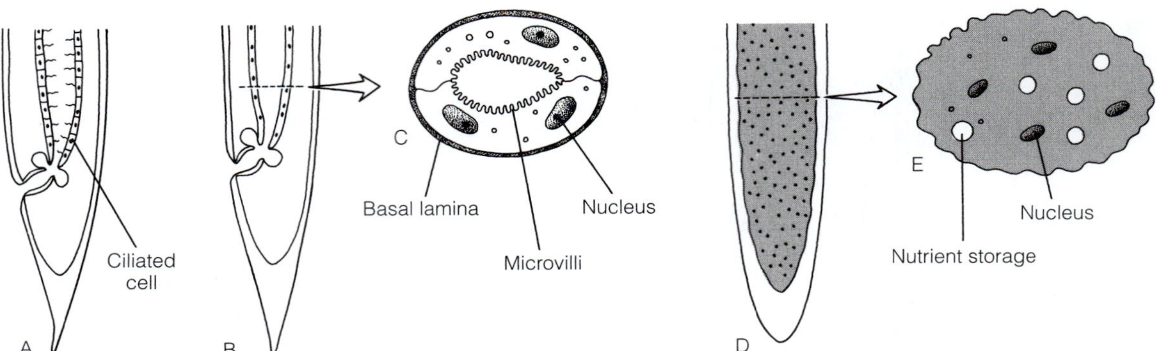

FIGURE 22-16 Nematode midgut specializations. **A,** Although unusual in nematodes, the midgut is ciliated in *Eudorylaimus.* **B** and **C,** The midgut of most nematodes lacks cilia and is lined only by a brush border of microvilli. **D** and **E,** In the parasitic mermithid nematodes, the midgut is syncytial and lacks a lumen. Mermithids absorb host nutrients directly across the body wall, the gut playing no role in ingestion or digestion. Instead, it functions as a nutrient storage organ, called a trophosome. *(A, After Zmoray, I., and Guttekova, A. 1969. Ecological conditions for the occurrence of cilia in intestines of nematodes. Biol. Bratisl. 24:97–112; B, Redrawn after Lee, D. L., and Atkinson, H. J. 1977. Physiology of Nematodes. Columbia University Press, New York, p. 37; D and E, Modified and redrawn after Batson, B. S. 1979. Ultrastructure of the trophosome, a food-storage organ in* Gastromeris boophthoorae *(Nematode: Mermithidae). Int. J. Parasitol. 9:505–514.)*

All members of Secernentea, which includes many terrestrial species, have an **excretory canal system,** often in addition to an excretory gland. The entire canal system lies within a single elaborate cell, the largest in the animal's body (Fig. 22-17A,B). Generally the cell is laid out in the form of the letter H, with two long canals (the uprights of the H) embedded in the lateral epidermal cords and joined together by a short transverse canal (the crossbar of the H). A short excretory duct leads anteriorly from the transverse canal to a midventral pore in the pharyngeal region of the body. In some nematodes, the duct is enlarged to form an ampulla (bladder), which fills and empties rhythmically at a rate correlated with the osmolarity of the environment. The canal system has been shown to have an osmoregulatory function in *Caenorhabditis elegans,* but the mechanism by which water enters the canals from the body is unknown; osmosis may play a role. Cilia and muscles are absent from the canals.

REPRODUCTION AND DEVELOPMENT

Most nematodes are gonochoric, but hermaphroditic species such as *Caenorhabditis elegans* are not uncommon, and some species are parthenogenetic. Fertilization is internal, with copulation. Many are sexually dimorphic, with males smaller than

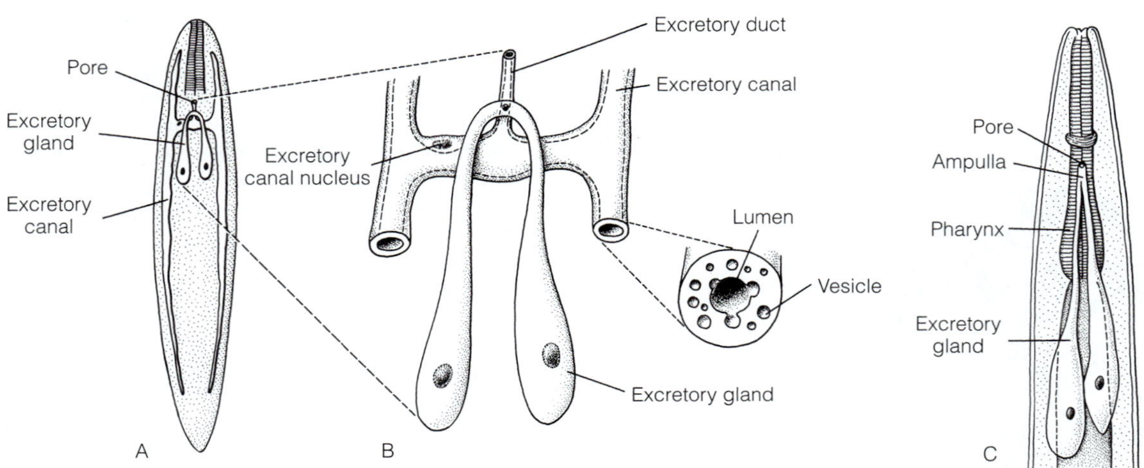

FIGURE 22-17 Nematode excretion. **A,** Generalized anatomy of the binucleate excretory gland and excretory canal of secernentean nematodes in ventral view. **B,** Enlargement of the system in A. **C,** The excretory gland of the adenophorean nematode *Rhabdias.* *(A and B, Based chiefly on* Caenorhabditis elegans. *Modified from Nelson, F. K., Albert, P. S., and Riddle, D. L. 1983. Fine structure of the* Caenorhabditis elegans *secretory–excretory system. J. Ultrastruct. Res. 82:156–171; C, Redrawn after Chitwood, 1931, from Hyman, L. H. 1951. The Invertebrates. Vol. 3. McGraw-Hill Book Co., New York. p. 241.)*

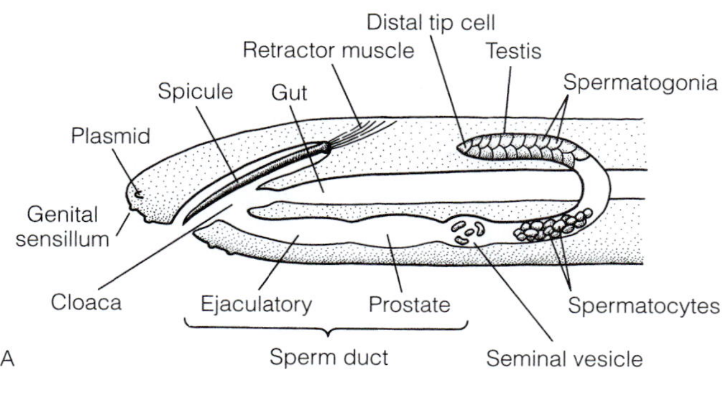

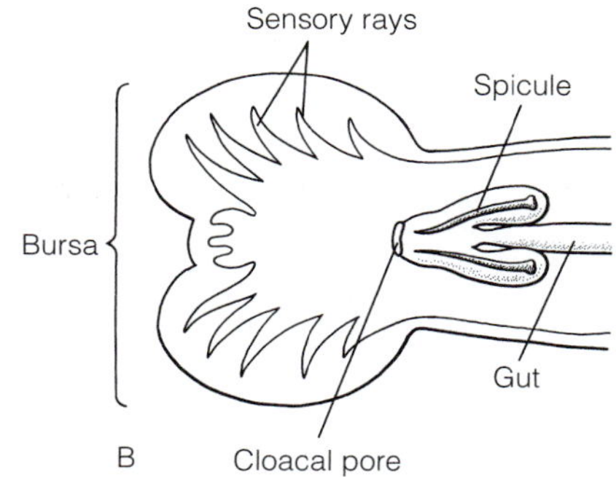

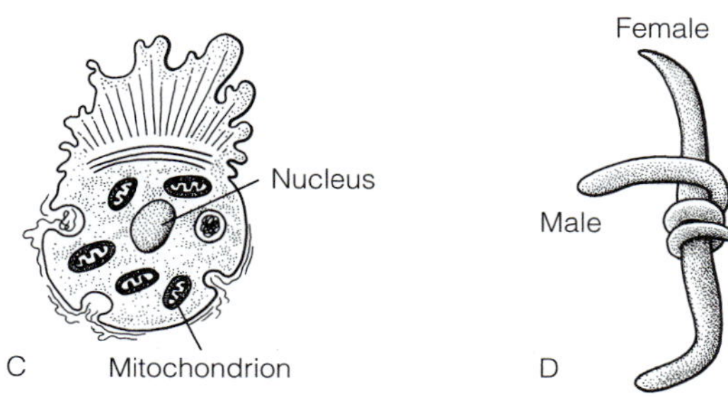

FIGURE 22-18 Nematode reproduction, male. **A,** Lateral view of the male reproductive system of a generalized nematode with one testis. **B,** Copulatory bursa of a parasitic trichostrongyle nematode. During copulation, the bursa wraps around the body of the female like a glove and secures the male firmly to the surface of her body. **C,** Ameboid sperm of *Caenorhabditis elegans.* **D,** Nematode copulation. *(A and B, Redrawn from Lee, D. L., and Atkinson, H. J. 1977. Physiology of Nematodes. Columbia University Press, New York. pp. 116, 118; C, Redrawn from Ward, S., Argon, Y., and Nelson, G. A. 1981. Sperm morphogenesis in wild-type and fertilization-defective mutants of* Caenorhabditis elegans. *J. Cell Biol. 91:26–44.)*

females. The posterior end of the male is curled like a hook or broadened into a fan-shaped copulatory aid called a **bursa** (Fig. 22-18B), and the male reproductive system joins the rectum to form a common cloaca, which opens to the exterior via the anus.

The gonads are tubular and may be paired (Fig. 22-19A) or single (Fig. 22-18A). When paired, one is anterior and the other posterior (Fig. 22-10A) rather than to the right and left. In free-living species that produce small numbers of gametes, each gonad typically bends back on itself to form a simple C shape (Fig. 22-19A). In parasitic species that produce vast numbers of gametes, the gonads often are extremely long and coil on themselves. In general, the upper region of a gonad is a **germinal zone** where gonial cells are produced by mitosis (Fig. 22-19A). The remainder of the tube is regionally specialized for various functions, including the meiotic divisions (gametogenesis) that convert the gonia into gametes. In most nematodes gametogenesis occurs as the gonial cells move down the tube toward the gonopore (Fig. 22-18A, 22-19A).

Males may have one or two tubular testes (the germinal zone), each of which widens gradually into a long sperm duct (Fig. 22-18A). Moving downstream, each sperm duct expands to become a seminal vesicle where spermatozoa from the testis are stored. The seminal vesicle joins a prostatic region with glandular, secretory walls. Prostatic secretions are adhesive and supposedly aid in copulation. The final region of the sperm duct is a muscular ejaculatory duct that empties into the cloaca. The male rectum is a multipurpose cloaca that receives digestive wastes from the intestine and sperm from the sperm duct. *Cloaca* is Latin for sewer. The wall of the cloaca is evaginated to form two pouches, which join before they open into the cloaca (Fig. 22-18A,B). Each pouch contains a **copulatory spicule,** which usually is short and shaped like a pointed, curved blade. Special protractor muscles extend the spicules through the cloaca and out the anus. In many nematodes, the walls of the pouch bear special cuticular pieces, the gubernaculum, that guide the spicules through the cloaca.

Females may have one or, more typically, two gonads, which usually are oriented in opposite directions (Fig. 22-19A). The ovary (germinal zone) at the upper end of the gonad grades into a tubular oviduct and then into a much wider, elongate uterus. The two uteri join to form a short common vagina, which opens through the female gonopore located on the ventral midline near the middle of the body. The female reproductive system is independent of the gut and there is no cloaca. A raised cuticular vulva surrounds the gonopore. A part of the female duct usually functions as a seminal receptacle for the storage of sperm whereas other regions are secretory or for storage of fertilized eggs.

The females of some nematodes are known to produce a pheromone that attracts males, and further study will probably prove this to be widespread. During copulation, the curved

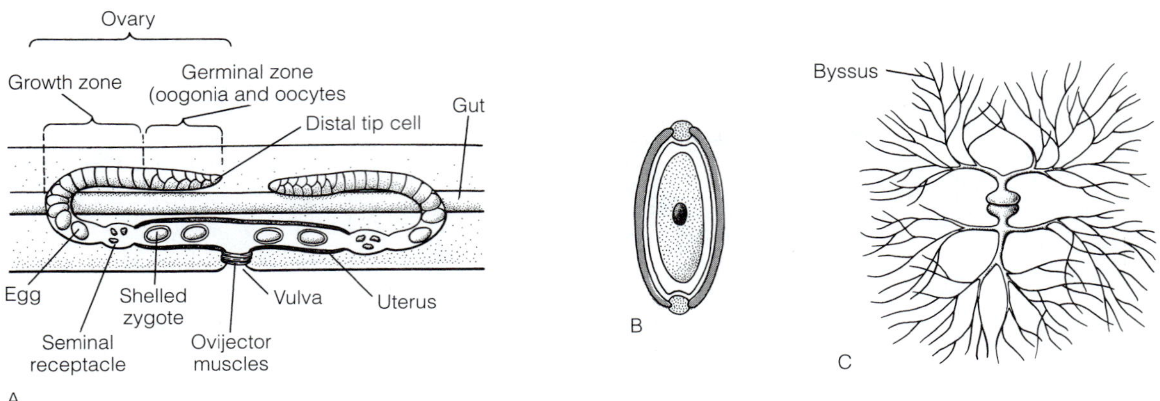

FIGURE 22-19 Nematode reproduction, female. **A,** Lateral view of the reproductive anatomy of a generalized female nematode with two ovaries. **B,** Shelled egg of *Trichuris.* **C,** Egg of the mermithid nematode *Mermis,* whose juveniles parasitize grasshoppers. Adults live in soil and, when ripe, the female attaches her eggs to grass blades and other vegetation using the proteinaceous byssus. *(A and B, Redrawn and modified from Lee, D. L., and Atkinson, H. J. 1977. Physiology of Nematodes. Columbia University Press, New York. p. 121, 123; C, Redrawn after Christie, 1937, from Hyman, L. H. 1951. The Invertebrates. Vol. 3. Mc-Graw Hill Book Co., New York. p. 253. Reprinted with permission.)*

posterior end of the male is usually coiled around the female so that the male anus is in contact with the female gonopore (Fig. 22-18D). The male copulatory spicules are extended from the anus into the vulva, where they hold the female gonopore open during the transmission of the sperm into the vagina. Nematode sperm are aflagellate and some move in a more or less ameboid manner (Fig. 22-18C).

After copulation the sperm migrate to the seminal receptacle, where fertilization takes place. The fertilized egg (zygote) secretes a thick fertilization membrane that hardens to form the inner part of the shell. Onto this inner shell the uterine epithelium secretes an outer layer. The outer surface of nematode eggs is sculptured in species-specific ways. A medical or veterinary technician can identify the eggs of parasitic nematodes by examination of a stool sample from the host (Fig. 22-19B,C). Nematode eggs may be stored in the uterus until deposition and, not infrequently, embryonic development begins while the eggs are still in the female. Many parasitic nematodes and some free-living species, such as the vinegar eel, are viviparous.

Some terrestrial nematodes, such as *Caenorhabditis elegans,* are hermaphroditic. Sperm develop before the eggs within the same gonad (an ovotestis) and are stored until eggs are produced. Self-fertilization takes place after the formation of the eggs. Although hermaphroditic individuals do not cross-fertilize, small numbers of males arise periodically in hermaphroditic populations and cross-fertilize the hermaphrodites so there is some gene recombination. Parthenogenesis also occurs in some nematodes, especially terrestrial species, and there are some species for which males are unknown.

Egg deposition by free-living nematodes still is not well studied. Marine species rarely produce more than 50 eggs, which are often deposited in clusters. Terrestrial species may produce up to several hundred eggs, which are deposited in the soil. A single female of the parasitic genus *Ascaris* may produce millions of eggs at a rate of 200,000 per day.

Early cleavage is neither spiral nor radial, but follows a fixed asymmetrical pattern that soon becomes bilateral. Development is determinate with early commitment of future germ cells and somatic cells. Nematode embryos undergo **chromosome diminution** (Fig. 22-20), in which the blastomeres destined to become somatic cells (designated EMS, A, and B) lose the genetic material their descendants will not need. Only the cells of the germ line (P_2) retain the complete genome.

Nematodes exhibit eutely, in which cell divisions (except in the gonad) cease near the end of embryonic development so the number of cells of the adult is constant and characteristic of the species. The various organs and tissues also contain a fixed number of cells, most of which are present by the time hatching takes place. For example, in certain *Rhabditis* adults there are 200 nerve cells, 120 epidermal cells, and 172 gut cells. In the female *Caenorhabditis elegans,* there are always exactly 959 somatic nuclei. Only the cells of the germ line continue to divide after adulthood and have a variable number of cells. Although there is a limited increase in the number of cells during juvenile stages, most growth in nematodes results from increases in cell size, not cell number.

The life cycle includes the egg, three juvenile instars, and the adult. Development is direct and the juveniles have almost all of the adult structures when they hatch, except for gametes and parts of the reproductive system. Growth through the three juvenile and one adult instars is accompanied by four molts, the first two of which may occur within the eggshell before hatching. The adult is produced by the fourth molt, after which molting ceases, although growth may continue.

PARASITISM

A large number of parasitic nematodes attack virtually all taxa of animals and plants and display all degrees and types of parasitic relationships with their hosts. Some major taxa of nematodes contain both free-living and parasitic species. Parasitism probably evolved many times within the taxon, and episodes of nematode radiation into parasitic niches

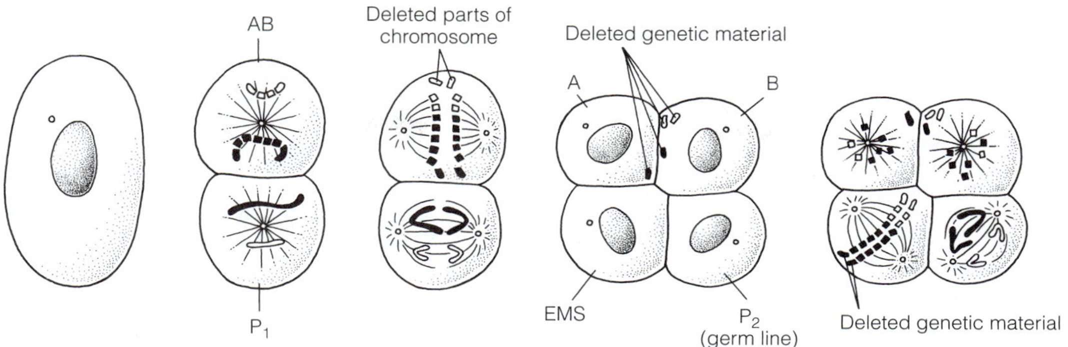

FIGURE 22-20 Nematoda cleavage. Chromosome diminution in *Parascaris aequorum* during early development. During the first cleavages of some nematodes, the chromosomes in all cells, except the stem cell of the germ line (P_2), fragment and jettison some pieces, which degenerate. The deleted fragments presumably contain genetic information needed only by the germ line and not the somatic cells. Thus, in nematodes such as *Parascaris,* there is genetic differentiation as early as the two-cell stage. Blastomeres A and B proliferate primarily ectodermal cells whereas blastomere EMS produces endodermal, mesodermal and stomodeal cells. P_2 produces the gametes. *(Adapted from Boveri, T. 1904. Ergebnisse über die Konstitution der chromatischen Substanz des Zellkerns. Verlag Gustav Fischer, Jena, Germany. p. 27.)*

accompanied the evolution of the flowering plants, the insects, and the amniotic vertebrates, all of which are subject to extensive nematode parasitism.

Almost every conceivable relationship between host and parasite exists. At one extreme are the completely free-living nematodes and at the other are nematodes with complex life cycles involving multiple hosts. Some nematodes are ectoparasites and some are endoparasites. Some parasitize plants, some animals, and some both. Sometimes the juveniles are parasitic, sometimes the adults, sometimes all stages of the life cycle are parasitic. In some, only the female is parasitic. In the most highly adapted parasites, the entire life cycle of both sexes is spent in the host as a parasite. There may a single host or, in the most specialized parasites, two hosts, one the intermediate host and the other the definitive host. Nematodes of medical and veterinary importance are highly specialized species in which adults of both sexes parasitize a vertebrate host or hosts.

Medically important parasites with only **one host** include *Ascaris,* hookworms, pinworms, and trichinella. The **ascaroid nematodes,** which feed on the intestinal contents of humans, dogs, cats, pigs, cattle, horses, chickens, and other vertebrates, include the largest nematodes. The entire life cycle, except for the egg, is spent within a single host. Transmission to a new host is achieved by the ingestion of eggs or juveniles passed in the feces of the original host. After ingestion, the eggs hatch and the juvenile stages usually penetrate the host's intestinal wall to enter the hemal system, where they are carried to the lungs. Here they break into the alveoli and migrate back to the small intestine by way of the trachea, pharynx, esophagus, and stomach.

The human ascaroid, *Ascaris lumbricoides,* reaches a length of 50 cm, although most are smaller, and is one of the best-known parasitic nematodes (Fig. 22-21). The species is widely distributed throughout the world, including the southeastern United States. The eggs can be as ubiquitous as bacteria and occur in soil, on paper money, and on other frequently handled objects. Children are especially likely to be infected due to their habit of putting things into their mouths. The developing eggs are notoriously resistant to adverse environmental conditions and may remain viable in soil for 20 years. The worms feed on fluid gut contents, not blood, but heavy infections can cause malnutrition in children and a massive infection can result in intestinal blockage and death. Physiological studies suggest that *Ascaris* secretes enzyme inhibitors that protect the worm from the host's digestive enzymes. If this is true, such a mechanism is probably utilized by most other gut-inhabiting nematodes.

A closely related ascarid of pigs, *Ascaris suum,* probably had a common ancestor with the species in humans. The common ancestor spread either from pigs to humans or vice versa (the direction is unknown) during the long period of about 10,000 years when humans and pigs lived in close association, often in the same hut or shelter, on primitive farms. With improved living conditions (for humans), and a more effective separation

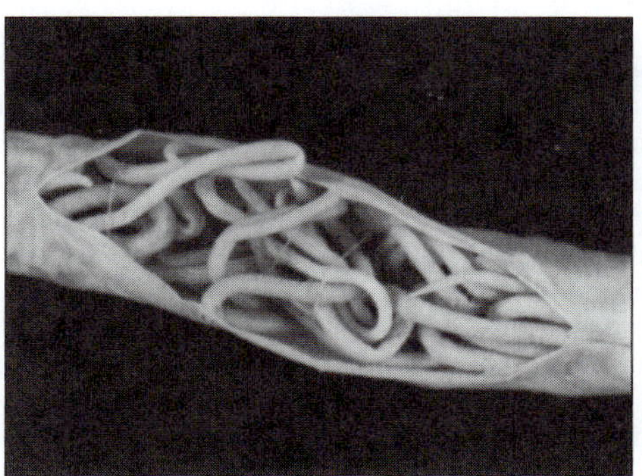

FIGURE 22-21 Parasitic Nematoda. Adult specimens of *Ascaris suum* within the small intestine of a pig. *Ascaris lumbricoides,* which parasitizes humans, is similar.

of human and pig populations, gene flow between the two *Ascaris* populations ceased, and the ancestral *Ascaris* diverged into the two very similar species we know today.

Toxocara canis and *Toxocara cati* are two small ascaroids common in dogs and cats, respectively. It is for these species that puppies and kittens are routinely dewormed. These worms may also infect humans, but they become confused during their migration in this inappropriate host and can end up in any tissue. This medical condition is known as **visceral larval migrans** and its seriousness depends on which tissue is invaded.

The **hookworms**, *Ancylostoma* and *Necator*, also inhabit the digestive tract of vertebrates, but unlike the ascarids, most hookworms feed on the host's blood. The buccal cavity usually has cutting plates, hooks, teeth, or combinations of these structures for attaching to and lacerating the gut wall (Fig. 22-13D). An infection of more than about 25 worms produces symptoms of hookworm disease, and a heavy infection can be serious because of blood loss and tissue damage. An adult worm may live as long as two years in the intestine. Hookworm infection is widespread in humans and it is estimated that over 380 million people are infected with Necator americanus, the most important species throughout the tropics (despite the specific name). The hookworm life cycle involves an indirect migratory pathway by the juveniles, as in ascarids. The fertilized eggs leave the host in its feces and hatch outside the host's body on the ground. The juvenile gains entry to a new host by penetrating the host's skin (on the feet in humans) and is then carried in the blood to the lungs. From the lungs the juvenile migrates up the trachea to the pharynx, where it is swallowed and passes to the intestine. In the intestine it attaches to the gut wall and feeds on blood.

Oxyuridan nematodes, known as **pinworms,** have a simpler life cycle. These small nematodes are parasitic in the gut of vertebrates and invertebrates. Infection usually occurs with ingestion of eggs passed in the feces of a previous host. Following ingestion, the eggs hatch and juveniles develop within the gut of the new host. There is no indirect migratory pathway in the host. The human pinworm, *Enterobius vermicularis,* affects children throughout the world. The female worm crawls out of the host's anus at night and deposits eggs in the perianal region. The movement of the female as she deposits eggs tickles, and scratching by the child contaminates the fingernails and hands with eggs. The eggs thus easily spread to the mouth of other children or reinfect the same child.

Trichinelloids are also parasites of the digestive tracts of vertebrates, especially birds and mammals. The whipworms, *Trichuris,* which infect humans, dogs, cats, cattle, and other mammals, are relatively small (the human whipworm, *Trichuris trichura,* is about 4 cm) and have a life cycle similar to that of pinworms. The most familiar trichinelloid is *Trichinella spiralis,* which infects mammals and is the cause of the disease **trichinosis.** The minute worm, which lives in the intestinal wall, is viviparous, and its juveniles are carried in the blood to the striated (skeletal) muscles. There the juveniles form calcified cysts and, if infection is severe, can produce pain and stiffness (Fig. 22-22). Transmission to another host occurs only if flesh containing encysted juveniles is ingested. Thus, in some animals, such as rats, this can be a one-host parasite as rat eats rat; in others, such as man and the pig, it would normally require two hosts. For humans, pigs are typically the intermediate host. For this reason pork should always be well cooked.

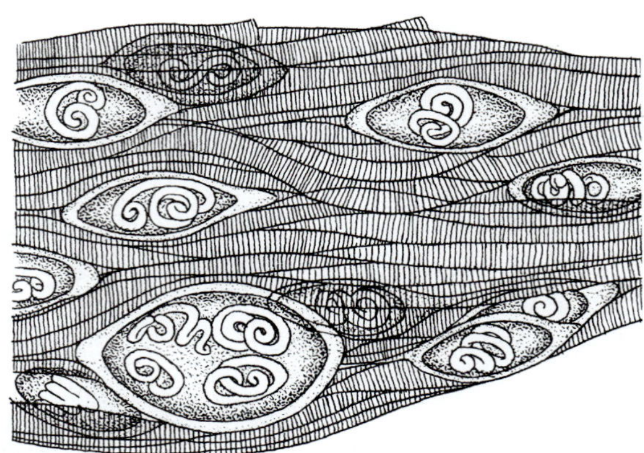

FIGURE 22-22 Parasitic Nematoda. Juveniles of *Trichinella spiralis* within calcareous cysts in the skeletal muscle tissue of the host. *(After Chandler, A. C., and Read, C. P. 1961. Introduction to Parasitology. John Wiley and Sons, New York.)*

The other common life cycle for human nematode parasites requires **two hosts,** an intermediate host and the definitive host, which can be a human. This type of life cycle is characteristic of many nematode parasites, including the familiar filarioids and dracunculoids.

Filarioids are threadlike worms that inhabit lymph glands and some other tissues of the definitive (vertebrate) host, especially birds and mammals. The female is viviparous, and the juveniles are called microfilariae. Bloodsucking insects such as fleas, some flies, and especially mosquitoes are the intermediate hosts. A number of species parasitize humans, resulting in **filariasis.**

The chiefly African and Asian *Wuchereria bancrofti* has a typical two-host life cycle. The intermediate host is a mosquito and the definitive host is a human. Adults are long, but very slender—about 40 to 90 $\times$ 0.1 to 0.2 mm—and live in lymph ducts adjacent to the lymph glands of the definitive host. Here the female produces eggs, which hatch into microfilariae that live in the blood. The microfilariae migrate to surface blood vessels at night, when mosquitoes are most likely to be biting. The mosquito is infected by these microfilariae when it takes a blood meal from an infected human. Development within the mosquito involves a migration of the microfilariae through the gut to the thoracic muscles and eventually to the proboscis. The proboscis injects microfilariae into the next definitive host when the mosquito feeds. In severe filariasis, blockage of the lymph vessels by large numbers of worms results in serious short-term lymphatic inflammation marked by pain and fever. Over a long period, the increase in the mass of connective tissue affected may result in the terrible enlargement of the legs, arms, breasts, and scrotum known as **elephantiasis** (Fig. 22-23). Extreme cases are no longer common, although they do still occur.

Dirofilaria immitis, the **heartworm,** which lives as an adult in the heart or pulmonary artery of dogs, wolves, and foxes, is also transmitted by mosquitoes and has a life cycle similar to that of *Wuchereria. Loa loa,* the African **eye worm,** lives in the

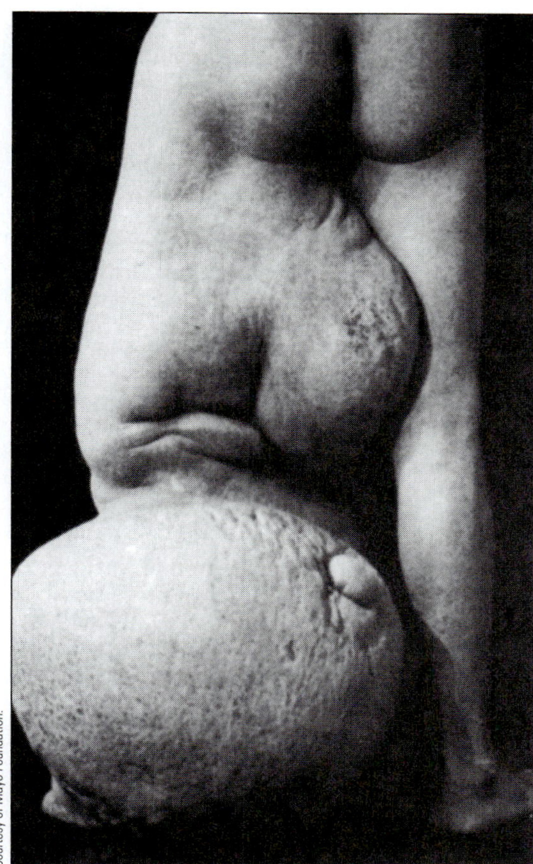

Courtesy of Mayo Foundation.

FIGURE 22-23 Parasitic Nematoda. A patient with elephantiasis, a severe form of filariasis caused by *Wuchereria bancrofti.*

subcutaneous tissues of humans and baboons. The worm migrates unnoticed through the connective tissue beneath the skin but occasionally passes across the eyeball, where it is visible to an observer (Fig. 22-24).

The **dracunculoids** are also slender, threadlike worms found in connective tissue and body cavities of vertebrate definitive hosts. A notable example is the **guinea worm,** *Dracunculus medinensis,* which parasitizes humans and many other mammals, especially in Asia and Africa. The definitive host is a human and the intermediate host is a freshwater copepod crustacean. The female is about 1 mm in diameter and up to 120 cm in length. After a period of development in the body cavity and connective tissue of the definitive host, the gravid female migrates to the subcutaneous tissue and produces an ulcerated lesion in the skin (Fig. 22-25). When the lesion comes in contact with water, juveniles are released into the water. After a short free-living stage, the juveniles are ingested by a copepod and continue their development in its hemocoel. When a new definitive host (a human) swallows infected copepods in drinking water, the nematode juveniles are released and penetrate the intestinal wall to reach the coelom or subcutaneous tissue, where they mature. Today a worm can be removed surgically, but the traditional method, still practiced, is to *slowly and carefully* wind it out of the lesion on a small stick. Breaking the worm causes severe inflammation and a skilled practitioner is required for success. Removal of these worms by ancient healers is represented today by the caduceus, the symbol of the medical profession. The caduceus, however, has been misinterpreted as a staff with snakes, instead of a nematode on a stick. Furthermore, *Dracunculus* is thought to be the "fiery serpent" with which the biblical Israelites were afflicted on their exodus from Egypt.

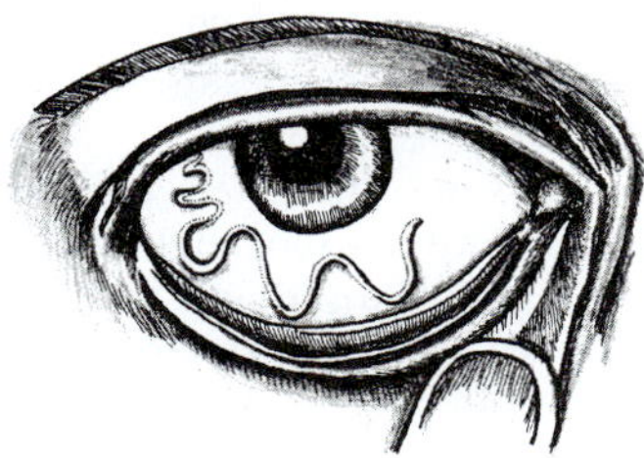

FIGURE 22-24 Parasitic Nematoda. The eye worm *Loa loa* crossing the cornea. *(From Chandler, A. C., and Read, C. P. 1961. Introduction to Parasitology. John Wiley and Sons, New York.)*

DIVERSITY OF NEMATODA

Adenophorea^{C} (= Aphasmidia): Polyphyletic. Most are free-living; some are parasitic. Free-living species include some terrestrial and almost all freshwater and marine nematodes. Variously shaped amphids located behind lips. Lack phasmids.

Photograph courtesy of the Institute of Public Health Research, Tehran University School of Public Health.

FIGURE 22-25 Parasitic Nematoda. A guinea worm, *Dracunculus medinensis,* being removed from an ulcerated opening of the arm by slowly winding it on a matchstick.

Epidermal cells are uninucleate. Excretory glands are present. No excretory canals. Some taxa, with some representative genera, are Chromadorida^{O} *(Chromadora, Stilbonema, Desmodora, Epsilonema),* Desmodorida^{O} *(Draconema),* Desmoscolecida^{O} *(Desmoscolex, Greeffiella),* Dorylaimida^{O} *(Dorylaimus),* Enoplida^{O} *(Oncholaimus),* Mermithida^{O} *(Mermis),* Mononchida^{O} *(Mononchus),* Monhysterida^{O} *(Monhystera, Gammarinema),* Trefusiida^{O} *(Onchulus, Kinonchulus, Trefusia),* Trichocephalida^{O} *(Trichinella, Trichuris).*

Secernentea^{C} (= Phasmidia): Almost all are terrestrial. Many are important parasites. Most free-living species live in soil. Amphids are porelike, located on lateral lips. Phasmids are present. Epidermal cells are uni- or multinucleated. Excretory glands and/or excretory canals are present. Includes Ascaridida^{O} *(Ascaris, Enterobius, Toxocara, Parascaris),* Camallanida^{O} *(Dracunculus),* Diplogasterida^{O} *(Odontopharynx),* Rhabditida^{O} *(Caenorhabditis, Rhabditis, Strongyloides, Turbatrix),* Spirurida^{O} *(Wuchereria, Spirura, Loa, Dirofilaria),* Strongylida^{O} *(Strongylus, Ancylostoma, Necator, Nippostrongylus),* Tylenchida^{O} *(Tylenchulus, Meloidogyne, Heterodera, Globodera).*

NEMATOMORPHA^{P}

Superficially resembling Nematoda is Nematomorpha, a small taxon of astonishingly long, but very slender cycloneuralians known as horsehair worms or hairworms. The adults are free-living and short-lived, whereas the parasitic larvae are the dominant stage of the life cycle. About 320 of the 325 nematomorph species belong to the taxon Gordioida and live as adults in fresh water and damp soil (Fig. 22-26). Adult gordioids live in all types of freshwater habitats in temperate and tropical regions of the world. Their larvae parasitize terrestrial arthropods, especially insects. Another four species, all in the genus *Nectonema,* constitute the taxon Nectonematoida, which, as adults, are pelagic in the coastal zones of most of the world's oceans (Fig. 22-27E). *Nectonema* larvae parasitize marine crustaceans.

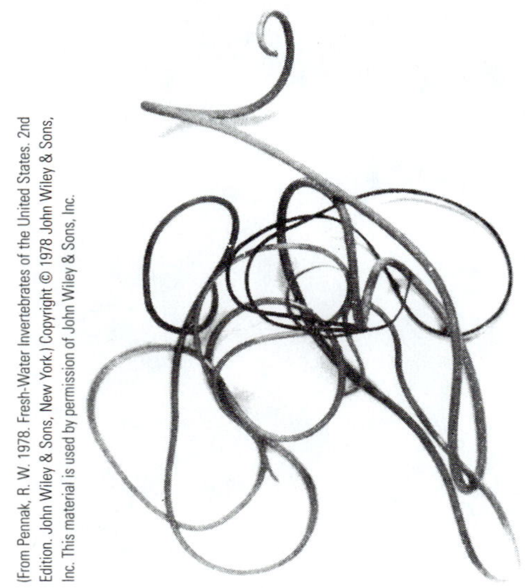

FIGURE 22-26 Nematomorpha. Three female gordioid nematomorphs, or horsehair worms.

The nematomorph body (Fig. 22-27A,E) is cylindrical, threadlike, slender (1 to 3 mm), and exceedingly long (up to 1 m), but lengths of 5 to 10 cm are typical. Externally it is nearly featureless and lacks a distinct head. The posterior end may be rounded or forked with two caudal lobes (Fig. 22-27A). The hairlike nature of these worms is so striking that they were once thought to arise spontaneously when a hair from a horse's tail fell into water. This belief was reinforced by the frequent occurrence of these worms in livestock watering tanks. The pelagic *Nectonema* has dorsal and ventral rows of swimming bristles (Fig. 22-27E).

The nematomorph body wall consists of a cuticle, epidermis, and muscle layer. The thick, iridescent, multilayered cuticle contains collagen fibers in crossed-helical layers (Fig. 22-27A). The cuticle is molted. Its surface typically is sculptured with crisscrossed grooves and covered with a coat of detritus, diatoms, or bacteria. Chitin is present in the larval, but not the adult, cuticle.

The epidermis is a monolayered, nonciliated, cellular epithelium. In gordioids it forms a thickened ventral epidermal cord containing the nerve cord and dividing the muscle layer (Fig. 22-27C). In *Nectonema* there is also a dorsal epidermal cord. The basal lamina is continuous with the collagenous matrix surrounding the muscle layer and extending through the connective-tissue compartment. Nematomorphs are not eutelic.

The musculature is a sheath of obliquely striated longitudinal muscle enclosed in a collagenous matrix. Like nematodes, nematomorphs have no circular muscles. The connection between muscle cells and their motor neurons is uncertain, but it does not appear to be via innervation processes, as it is in nematodes and gastrotrichs. The muscles attach to the cuticle with tonofilaments and hemidesmosomes, as they do in nematodes and other cycloneuralians. Adult nematomorphs, especially males, swim with whiplike undulations of the body. The cuticle and longitudinal muscles antagonize each other by using the mesenchymal hemocoel as a hydrostatic skeleton.

Unusual muscle cells include peripheral paramyosin filaments, which, like those in the closing muscles of clams, are especially thick and long and may be important in sustained muscular contractions. Adults coil tightly around streamside vegetation, and each other, for extended periods of time. During copulation, a male forms and sustains tight coils around the female's posterior end (Fig. 22-27A). If crowded together, nematomorphs coil together in masses that resemble the legendary Gordian knot.

The intraepidermal nervous system consists of an anterior, circumenteric brain and a nonganglionated ventral nerve cord. The brain is highly derived and lacks the obvious triannular structure characteristic of other cycloneuralians. The nerve cord, although deeply submerged into the hemocoel, remains enclosed by the epidermal basal lamina (Fig. 22-27C). Excretory organs are absent, but the vestigial midgut may function as a secretion kidney, like an insect Malpighian tubule. No specialized gas-exchange surfaces are present.

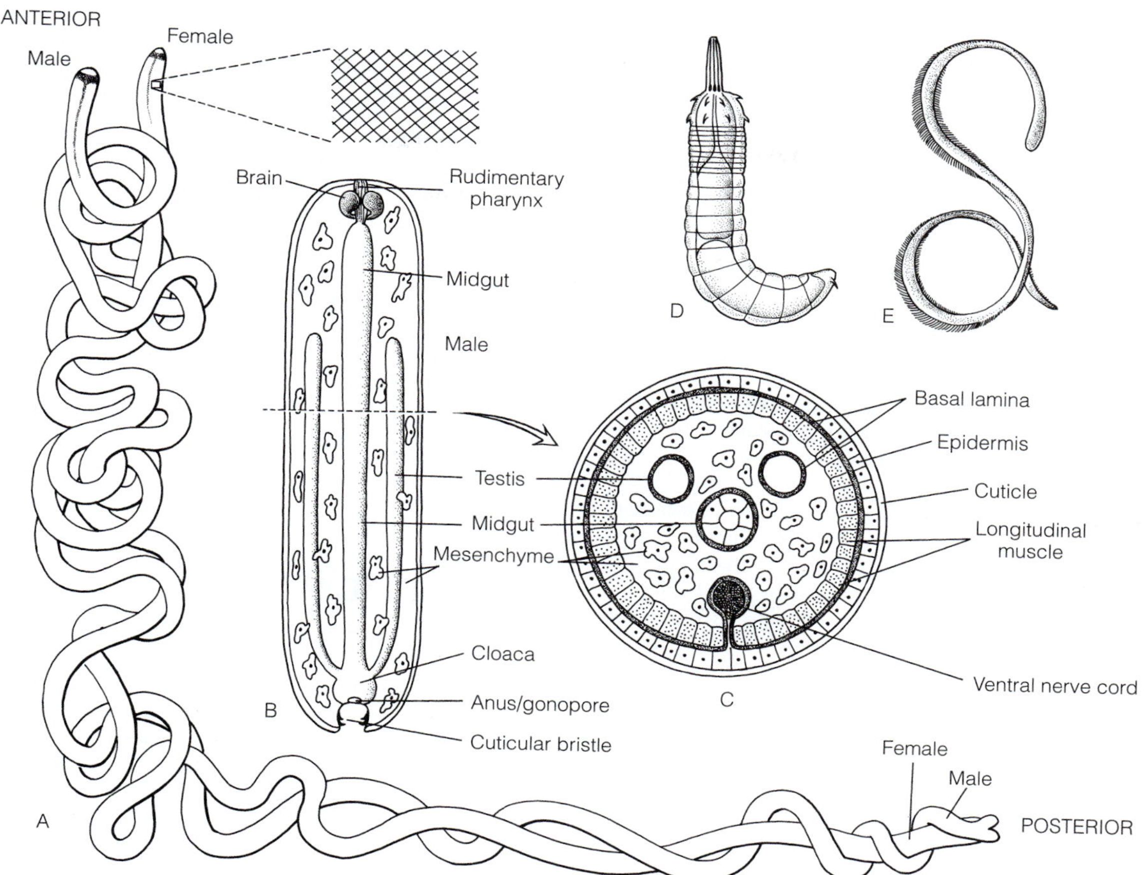

FIGURE 22-27 Nematomorpha anatomy. **A,** Male and female of the nematomorph *Gordius robustus.* Posterior ends show typical copulatory position. Inset shows crossed-helical fibers in the cuticle. The fiber angle with respect to the longitudinal body axis is approximately 55°. **B,** Generalized anatomy of a gordioid nematomorph. The worm is greatly shortened for clarity. **C,** Cross section of B. **D,** Nematomorph larva of the gordioid *Paragordius.* **E,** The marine nectonematoid worm, *Nectonema agile,* showing its dorsal and ventral swimming setae. *(A, Modified from May, H. G. 1919. Contributions to the life histories of* Gordius robustus *Leidy and* Paragordius varius *(Leidy). Ill. Biol. Monogr. 5:1–118. C, Redrawn after Bresciani, J. 1991. Nematomorpha. In Harrison, F. W., and Ruppert, E. E. (Eds.): Microscopic Anatomy of Invertebrates. Vol. 4. Aschelminthes. Wiley-Liss, New York. pp. 197–218. D, After Mühldorf, 1914, and Montgomery, 1904, from Hyman, L. H. 1951. The Invertebrates. Vol. 3. McGraw-Hill Book Co., New York. p. 467. Reprinted with permission.)*

The body cavity is a hemocoel filled to varying degrees with mesenchyme and a fibrous, collagenous matrix similar to the parenchyma of large flatworms (Fig. 22-27C). In *Nectonema* it is more spacious and a transverse septum divides it into a small anterior and large posterior compartment.

The gut consists of a mouth, buccal cavity, pharynx, midgut, and cloaca, but it is reduced and the mouth and anus are usually absent. The midgut (intestine) consists of a simple microvillous gastrodermis with no muscles. The gut plays no direct role in the digestion of food, but it appears to be indirectly involved in the digestive process as a storage site for food molecules absorbed across the cuticle, as in mermithid nematodes (which are also invertebrate parasites as juveniles).

Nematomorphs are gonochoric and the gonads are paired, cylindrical sacs extending the length of the body. The unique sperm are aflagellate and rod-shaped. In both sexes the gonoducts empty into a cuticular cloaca (Fig. 22-27B) at the posterior end of the gut. During copulation, the male wraps his posterior end tightly around that of the female (Fig. 22-27A) and transfers a spermatophore to the female's cloacal region or inserts sperm directly into the cloaca. Either way, the sperm migrate to the seminal receptacle, where they are stored. Fertilization is internal in the uterus. Females deposit an egg string in the water or among plant roots on the shore. Cleavage is holoblastic, equal or unequal, apparently indeterminate, and results in a coeloblastula. Gastrulation is by invagination and creates the gut.

The life cycle includes the egg, larva (often encysted), parasitic larva, and free-living adult. The hatching stage is a tiny larva (Fig. 22-27D) that resembles an adult cephalorhynchan. Its anterior end is a eversible introvert armed with a central, spiny proboscis used to penetrate the host. The introvert is everted by hydrostatic pressure and withdrawn by retractor muscles, but the proboscis is protruded, not everted.

Hosts include a wide variety (at least 20 families) of aquatic and terrestrial insects, including beetles, crickets, grasshoppers, dragonflies, caddis flies, bugs, moths, and cockroaches, as well as centipedes, millipedes, amblypygids, scorpions, snails, and leeches. In the simplest life cycle, only one host (the definitive host) is required. The nematomorph larva either penetrates the host's body wall or is ingested as a free-swimming larva or a cyst, penetrates the host's gut wall, and becomes an active larva. In many cases, the larva encysts in a drying pool and the desiccation-resistant cyst is later ingested by a terrestrial arthropod after the water is gone. The larva enters the host's hemocoel, where it passes through an uncertain number of molts to become wormlike and achieve adulthood. Feeding apparently is accomplished by direct absorption of food materials from the blood through the body wall. After several weeks to several months of development in the hemocoel, the worms attain the morphology of adults, resulting in a very long worm coiled in a much shorter insect.

Worms inhabiting terrestrial arthropods induce the host to migrate to water, at which time they emerge and enter the water. It is not known how the worm induces this migratory behavior in the host, perhaps hormonally or by desiccating the host and inducing thirst. Worms living in aquatic invertebrates simply emerge and no migration is necessary. Upon emergence, free-living adult worms quickly achieve sexual maturity and mate, producing new larvae, during the brief, free-living, aquatic, adult phase of the life cycle. In most nematomorphs, development is more complicated than the simple life cycle described above and involves two hosts, one intermediate and one definitive. The definitive host is always an invertebrate but the intermediate host is sometimes a tadpole or fish.

There have been rare instances of human infections by nematomorphs, but these are probably accidental. Worms have been recovered from human digestive and urogenital tracts and larval hairworms have infected facial tissue and produced orbital tumors. The larvae of *Nectonema*, the only marine hairworms, parasitize true crabs, hermit crabs, and shrimps.

DIVERSITY OF NEMATOMORPHA

Nectonematoida^C: Pelagic, marine. One genus, *Nectonema*, with four species. Body is 20 cm long. Have dorsal and ventral swimming setae, internal dorsal and ventral epidermal cords, fluid-filled hemocoel. Larvae and juveniles parasitize decapod crustaceans.

Gordioida^C: Includes almost all nematomorphs. Adults are freshwater or semiterrestrial in damp soil. Juveniles mostly parasitize insects. Hemocoel is mostly filled with connective tissue. Only the ventral epidermal cord is present. *Chordodes, Gordius, Parachordodes, Paragordius.*

PRIAPULIDA^P

Priapulida consists of only 18 living species of marine worms (Fig. 22-28). Priapulids are benthic and live in sand or mud bottoms in shallow and deep water. Most of the large-bodied species live in cold water, often at high latitudes, including the Arctic Ocean and the Southern Ocean surrounding Antarctica. Some families show a distinctly bipolar distribution, being absent from low latitudes. On North American coasts they occur northward from Massachusetts and central California. Some small priapulids, such as *Tubiluchus*, are more widely distributed, with ranges including the tropics. Priapulids were important members of the marine benthos in the middle

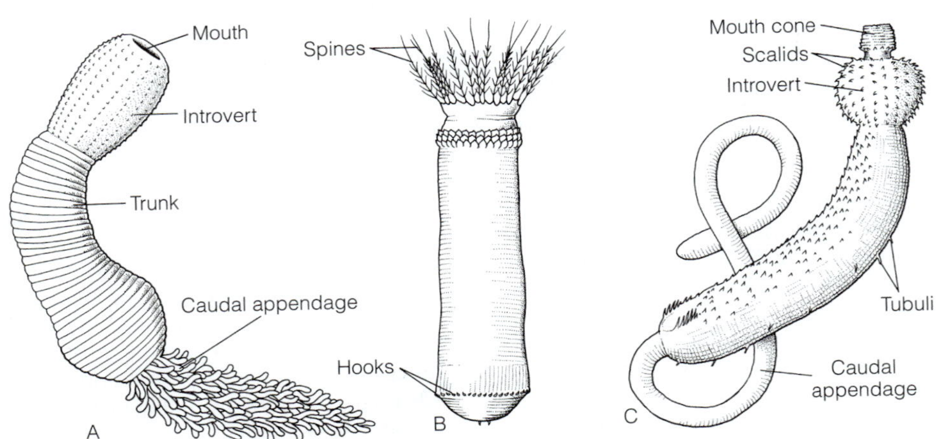

FIGURE 22-28 Priapulida. **A,** The large (8 cm) cold-water *Priapulus caudatus*. **B,** The small (2 to 3 mm) tube-dwelling *Maccabeus tentaculatus* from the Mediterranean. This animal apparently uses its spiny tentacles as a spring trap to snare small prey. **C,** The tiny (less than 2 mm) *Tubiluchus corallicola* lives in tropical coralline sands. *(All redrawn and modified. A, After Theel from Hyman, L. H. 1951. The Invertebrates. Vol. 3. McGraw-Hill Book Co., New York. p. 467; B, Combined from Por, F. D. 1972. Priapulida. In Parker, S. P. (Ed.): Synopsis and Classification of Living Organisms. McGraw-Hill Book Co., New York. pp. 941–944, and Calloway, C. B. 1982. Priapulida. In Parker.; C, From a drawing by Brian Marcotte.)*

Cambrian period and, although only 11 fossil species have been described, they were among the dominant invertebrates of Cambrian seas. The ecological importance of the taxon has declined precipitously, but modern species have changed little from their Cambrian ancestors. Priapulida is named for Priapus, the Greek god of male fertility.

Priapulids are only now beginning to receive the attention required to understand their evolution. They probably are cycloneuralians, a conclusion based on their collar-shaped, triannulate, circumenteric brain and anterior terminal mouth. Their closest relatives are kinorhynchs and loriciferans, with which they share several apomorphies and constitute Cephalorhyncha (Fig. 22-35).

The cylindrical body ranges in length from 0.5 mm to 40 cm and is divided into a large trunk with an anterior introvert and, sometimes, one or two posterior **caudal appendages** (Fig. 22-28A,C). Priapulids have both bilateral and radial symmetry. The introvert and body wall are radial, but some organ systems, such as the nerve cord and urogenital system, are bilateral.

In the interstitial species *Tubiluchus corallicola,* the trunk has a long, terminal caudal appendage (tail) that is probably used to anchor the body in the surrounding sediment (Fig. 22-28C). In the much larger *Priapulus,* the posterior end of the trunk bears one or two caudal appendages, each consisting of a hollow stalk from which extend many short, fingerlike diverticula (Fig. 22-28A). The cuticle of the appendages is thin, and their lumina are continuous with the body cavity of the trunk. Gas exchange, osmoregulation, and chemoreception have been suggested as functions of these structures, but none have been demonstrated.

The outermost layer of the body wall is the extracellular, chitinous cuticle (Fig. 22-29C, 22-30). Consisting of three layers, epicuticle, exocuticle, and endocuticle, it is secreted by the underlying epidermis. The epicuticle is collagenous, the exocuticle is chiefly proteinaceous, and the endocuticle is chitinous. The cuticle is molted periodically as the animal grows.

The cuticle exhibits a variety of specializations, the most common of which are the many types of **scalids** (Fig. 22-28C) similar to those of kinorhynchs and loriciferans. Scalids are chitinous setae, scales, or spines variously adapted for sensory reception, locomotion, feeding, and other functions. Sensory scalids are hollow, contain monociliated sensory neurons, and have an apical pore opening to the environment. Scalids are most abundant on the introvert, where they typically occur in longitudinal rows. In one species, *Maccabeus tentaculatus,* the

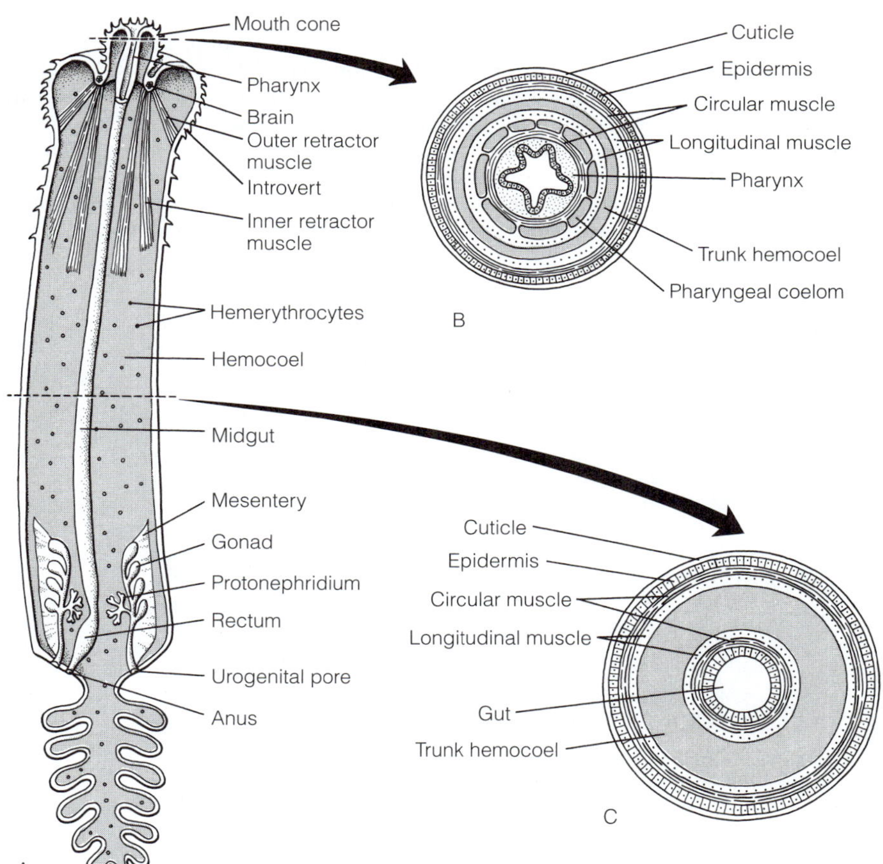

FIGURE 22-29 Priapulida anatomy. **A,** Priapulid body organization, based primarily on *Meiopriapulus.* **B** and **C,** Cross sections of the pharynx and trunk, respectively. In *Meiopriapulus,* a specialized peritoneum lines the pharyngeal coelom. *(A and B, Modified and redrawn from Storch, V., Higgins, R. P., and Morse, M. P. 1989. Internal anatomy of* Meioriapulus fijiensis *(Priapulida). Trans. Am. Micros. Soc. 108:245–261.)*

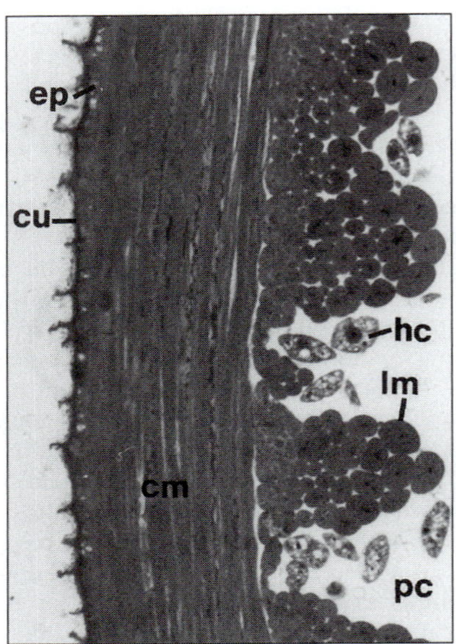

FIGURE 22-30 Priapulida. Cross section of a portion of the body wall of *Priapulus caudatus.* cu, cuticle; ep, epidermis; cm, circular muscle; hc, hemerythrocyte, lm, longitudinal muscle; pc, hemocoel.

mouth is surrounded by a crown of setose tentacles and spines in addition to the customary scalids (Fig. 22-28B). The trunk also bears cuticular projections of many types. Setaelike hooks on the posterior end of the body of *M. tentaculatus* and *Meiopriapulus fijiensis* help grip the sediment as the animals burrow tail first (Fig. 22-28B).

The monolayered, nonciliated epidermis rests on a basal lamina. Inside the lamina are two well-developed layers of obliquely striated muscles consisting of an outer circular muscle and an inner longitudinal muscle. The longitudinal muscles border the body cavity (Fig. 22-30).

In the anterior end of the body, special longitudinal muscles form two sets of introvert retractor muscles arranged in a characteristic pattern. These muscles extend from the body wall to insert on the circumenteric brain at the base of the mouth cone (Fig 22-29A). The short outer retractor muscles originate at the base of the introvert whereas the longer inner muscles originate on the trunk wall.

The large priapulid body cavity is a hemocoel (Fig. 22-30). There are additional minor spaces in priapulids, such as the pharyngeal coelomic chambers of *Meiopriapulus fijiensis* (Fig. 22-29B). The lumen of the gonad may also be a coelomic space.

The hemocoel contains phagocytic amebocytes and hemerythrocytes, the latter containing the respiratory pigment hemerythrin. The blood is circulated solely by the activity of the body wall muscles. The hemerythrocytes give the blood a pinkish color.

The hemocoel is a hydrostatic skeleton with a role in both introvert extension and burrowing through soft sediments. Contraction of circular muscles pressurizes the blood, causes the introvert and pharynx to extend, and stretches the retractor muscles. Contraction of the retractor muscles withdraws the introvert into the anterior end of the trunk and stretches the circular muscles. The larger priapulids burrow through sediments using the trunk as a penetration anchor and the introvert as a terminal anchor (Fig. 22-31).

Large priapulids, such as *Priapulus* and *Halicryptus,* are thought to be carnivores, whereas some small species, such as *Tubiluchus* and *Meiopriapulus,* probably are deposit-feeding bacteriovores. A few, such as *Maccabeus,* may be suspension-feeding carnivores. The carnivores feed on soft-bodied, slow-moving invertebrates, particularly polychaete worms, and

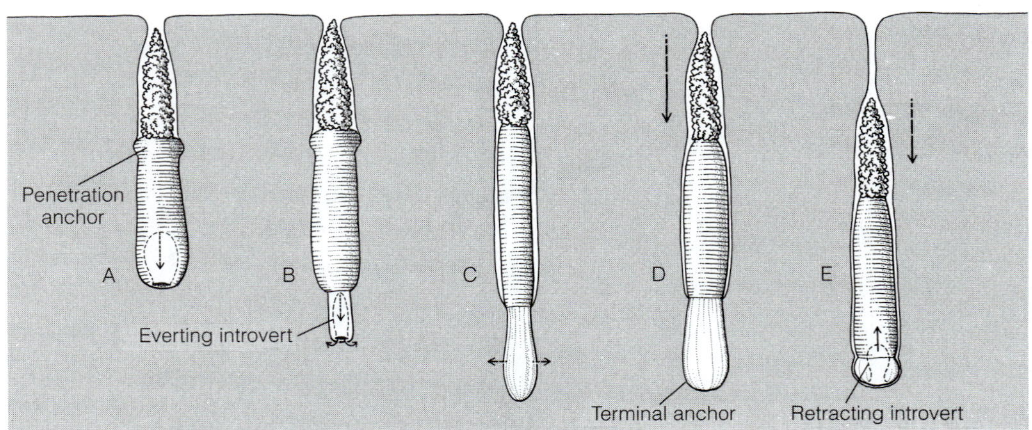

FIGURE 22-31 Priapulida locomotion. Priapulids, such as *Priapulus caudatus* (shown), burrow primarily by using eversion/retraction movements of the introvert. In **A** and **B,** the introvert pushes downward into the sediment as the expanded body forms a penetration anchor. The introvert then swells to form a terminal anchor **(C)** as the trunk is drawn forward **(D).** The trunk continues to advance downward as the introvert is retracted **(E)** in preparation for the next burrowing cycle. *(Modified and redrawn from Elder, H. Y., and Hunter, R. D., 1980. Burrowing of* Priapulus caudatus *(Vermes) and the significance of the direct peristaltic wave. J. Zool. Lond. 191:333–351.)*

sometimes each other. During feeding the pharynx can be everted through the mouth (Fig. 22-28C) and heavy pharyngeal teeth may be used to seize prey.

The gut is divided into foregut (mouth, pharynx, esophagus, and polythyridium), midgut (intestine), and hindgut (rectum). The foregut and hindgut are ectodermal and have cuticular linings. The midgut is endodermal. In cross section, the gut is composed of a gastrodermis, circular muscles, longitudinal muscles, and external lamina, which lines the hemocoel. The epidermal lining of the foregut and hindgut secretes a cuticle, which in the pharynx usually includes teeth of various designs. Pharyngeal teeth are acellular and should not be confused with scalids.

The terminal mouth is located at the anterior end of the resting (everted) introvert (Fig. 22-28A) and opens into the muscular, cuticularized pharynx. The pharynx lumen is often lined with strong cuticular teeth. During feeding, the pharynx can be everted (turned inside out) through the mouth, forming the **mouth cone** (Fig. 22-29A, 22-28C). This places the pharyngeal teeth on the outside, where they can scrape against the substratum or otherwise capture food. Retraction of the pharynx brings the teeth, along with any captured bacteria or small metazoans, back into the pharynx lumen. In some priapulids a second muscular and cuticularized region of the foregut, the polythyridium, may be a gizzard.

The straight, tubular midgut (intestine) lacks cuticle or cilia but is lined with a brush border of microvilli for absorbing food molecules. Both extracellular and intracellular digestion occur in the midgut. The midgut joins a short terminal rectum (hindgut) that, like the pharynx, is ectodermal and lined with cuticle. The anus is located at the posterior end of the trunk (Fig. 22-29A).

The nervous system is intraepidermal and consists of a triannular circumenteric brain around the mouth at the anteriormost end of the introvert (Fig. 22-29A). A single midventral, nonganglionated nerve cord extends posteriorly from the ring and ends in a caudal ganglion. Nerves extend from the brain to the body wall and gut. A variety of cuticular sensory structures, including flosculi (sensory pits), tubuli, and scalids, all with monociliated sensory receptors, occur on the trunk and introvert (Fig. 22-28C, 22-32C).

The urogenital system includes of a pair of urogenital ducts shared by the excretory and reproductive systems. Each duct is suspended by a mesentery and lies beside the intestine in the posterior trunk hemocoel (Fig. 22-29A). The urogenital duct is formed by the union of the protonephridial tubule and the gonoduct. Each protonephridium consists of many monociliated terminal cells (solenocytes) separated from the hemocoel by a basal lamina. Each urogenital duct opens to the exterior at a pore on the posterior end of the trunk.

Priapulids are gonochoric and fertilization is external in large species, but internal fertilization probably occurs in most small meiofaunal species. Some are sexually dimorphic. Development is poorly known, but the small egg is yolky and cleavage is holoblastic, equal, and radial (Fig. 22-32A,B). Gastrulation occurs by polar ingression, and the hatching stage is a nonciliated stereogastrula. Priapulid larvae (Fig. 22-32C,D) are benthic and most are dorsoventrally flattened. Morphologically the larvae are more uniform than the adults and consist of an

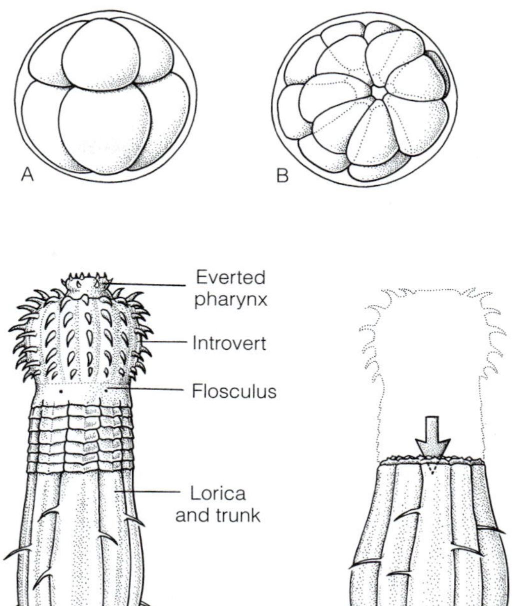

FIGURE 22-32 Priapulida. The few observations of priapulid early development (**A** and **B**) indicate that cleavage is radial, holoblastic, and nearly equal. The larva (**C** and **D**) is dorsoventrally flattened and has a thick trunk cuticle, the lorica, into which the introvert can be retracted. *(A and B, Redrawn after Ginkine from Grassé.)*

introvert and trunk. The trunk is enclosed in a thick cuticular lorica into which the introvert can be retracted. The lorica disappears at metamorphosis. Direct development is reported in some species.

DIVERSITY OF PRIAPULIDA

Priapulomorpha[O]: Macrobenthic or meiobenthic. With caudal appendages. External or internal fertilization.

Priapulidae[F]: Large, macrobenthic. One or two caudal appendages. No polythyridium. No sexual dimorphism. Fertilization is external. Sperm are primitive. *Acanthopriapulus, Priapulopsis, Priapulus.*

Tubiluchidae[F]: Small, meiobenthic, interstitial. One elongate caudal appendage. Trunk has tubuli. Distinct neck connects trunk and introvert. Has one set of retractor muscles. Pharyngeal teeth are pectinate. Foregut has a polythyridium. Fertilization is internal, sexes are dimorphic. *Tubiluchus.*

Halicryptomorpha[O]: Large, macrobenthic carnivores. Lack caudal appendages. Fertilization is external. Larvae are meiobenthic.

Halicryptidae[F]: *Halicryptus higginsi,* at 39 cm long, is the largest priapulid.

Meiopriapulomorpha[O]: Meiobenthic detritivores, < 2 mm. Pharynx lacks teeth, is noneversible. Polythyridium is present. Viviparous, with direct development.

Meiopriapulidae^F: *Meiopriapulus.*

Seticoronaria^O: Small, suspension-feeding, < 3 mm. Anterior scalids form a crown of tentacles around the mouth. Pharynx has pectinate teeth and several sets of retractor muscles. Preanal trunk has a circle of hooks. A caudal appendage is absent.

Chaetostephanidae^F: *Maccabeus.*

LORICIFERA^P

Loricifera is a small taxon of minuscule (< 0.5 mm) interstitial, marine animals related to priapulids and kinorhynchs and sharing many apomorphies with them. They are about the size of large ciliate protozoans but are multicellular, being composed of up to 10,000 tiny cells. Loriciferans adhere tenaciously to the substratum and consequently eluded discovery until 1983, when the first species, *Nanaloricus mysticus,* was described by the Danish zoologist R. M. Kristensen. Since then almost 100 additional species have been discovered, although most have not yet been described. They are widely distributed on a variety of sediment types over a wide range of depths and from the polar regions to the tropics.

Because loriciferans are tiny metazoans living in low population densities in large volumes of sand or other substrata, they are difficult to find and collect. They adhere so tightly to the substratum that they can be removed only by brief immersion in fresh water. The osmotic shock kills the animals and causes them to release their grip on the substratum. They can then be separated from the water by filtering. Unfortunately, because of this technique, loriciferans are always dead when studied and almost nothing is known of their behavior, locomotion, feeding, physiology, or development. Only one specimen, a larva, has been observed alive.

The body is divided into an anterior introvert and large posterior trunk separated by a thorax (Fig. 22-33A). There is a protrusible, but not eversible, mouth cone at the anterior end of the introvert. The introvert bears up to 300 large, elaborate scalids assumed to have sensory and locomotory functions. Some scalids are equipped with intrinsic muscles. The introvert can be retracted into the anterior end of the lorica. About 30 small retractor muscles in two sets are responsible for retracting the introvert. It is extended by hydrostatic pressure.

The body wall consists of an epidermis, which secretes a chitinous cuticle, and individual muscle cells. The cuticle is best developed in the plates of the lorica, where it consists of a sclerotized epicuticle, intracuticle, and procuticle. The cuticle joining the plates is flexible and unsclerotized. The trunk is enclosed in a cuticular lorica similar to that of kinorhynchs. It is superficially similar to the rotifer lorica but is extracellular, whereas that of rotifers is intracellular. The name Loricifera means "lorica bearer." The lorica is composed of either 6 (Fig. 22-33A) or 22 long, side-by-side cuticular plates that may be heavily or lightly sclerotized. The epidermis is a monolayered epithelium separated from the underlying connective-tissue compartment by a basal lamina. The body-wall muscles are under the basal lamina and include longitudinal fibers in combination with diagonal, dorsoventral, or circular fibers. The muscles occur as individual fibers rather than continuous layers. Body-wall and pharyngeal muscles are cross-striated, whereas some others, such as the retractors, are obliquely striated.

The gut includes an anterior cuticularized foregut, a middle noncuticularized midgut, and a posterior cuticularized hindgut. The terminal mouth is at the anterior end of the mouth cone on the introvert (Fig. 22-33C). It opens into a buccal tube that leads posteriorly into a bulbous pharynx composed of epitheliomuscular cells. The pharyngeal lumen is triradiate in cross section (Fig. 22-33C inset). A large midgut, with a microvillous and presumably absorptive gastrodermis, forms most of the gut tube. A short, cuticularized esophagus connects the pharynx with the midgut. The hindgut (rectum) opens to the outside through the anus near the posterior end of the body. The muscular, sucking pharynx and the morphology of the mouth cone suggest that loriciferans may be carnivores that suck fluids from their prey, but no observations support this supposition.

The nervous system is intraepidermal but submerged into the connective-tissue compartment. A very large triannular brain fills much of the space in the introvert (Fig. 22-33C). The ganglionated forebrain innervates the scalids and introvert. The midbrain is a fibrous neuropil without ganglia. The hindbrain is a ring of 10 ganglia from which extend 10 longitudinal nerve cords. The two cords of the midventral pair are larger and form a double, ganglionated ventral nerve cord.

Loriciferans are gonochoric and some exhibit sexual dimorphism in scalid morphology and arrangement. The gonads and excretory organs are combined in a urogenital system, as in the priapulids. The gonads are paired sacs containing not only the germinal epithelium but the protonephridia as well (Fig. 22-33C). The terminal cells of the protonephridia are monociliated solenocytes. Gametes and urine are released into a common urogenital duct, which opens with or beside the anus. Seminal receptacles that have been found in one species imply internal fertilization.

Embryonic development is unknown, but it is followed by a larval stage called a **Higgins larva** that is similar to the adult (Fig. 22-33B). The posterior end of the larval trunk carries a pair of toes. In some species (nanaloricids) the toes are wide paddles equipped with muscles and probably used for swimming. In others (pliciloricids), large adhesive glands open at the tips of the slender toes, which are thought to be used for adhesion to the substratum.

DIVERSITY OF LORICIFERA

Pliciloricidae^F: Lorica is lightly sclerotized, with 22 or more plates and no anterior marginal spines. Foregut is simple. No sexual dimorphism. Anus and gonopore usually are ventroterminal. Toes are adhesive. *Pliciloricus, Rugiloricus.*

Nanaloricidae^F: Lorica is heavily sclerotized, with 6 plates and 15 hollow anterior marginal spines. Foregut is complex. Sexes are dimorphic. Anus and gonopore are dorsoterminal. Toes are paddles. *Nanaloricus.*

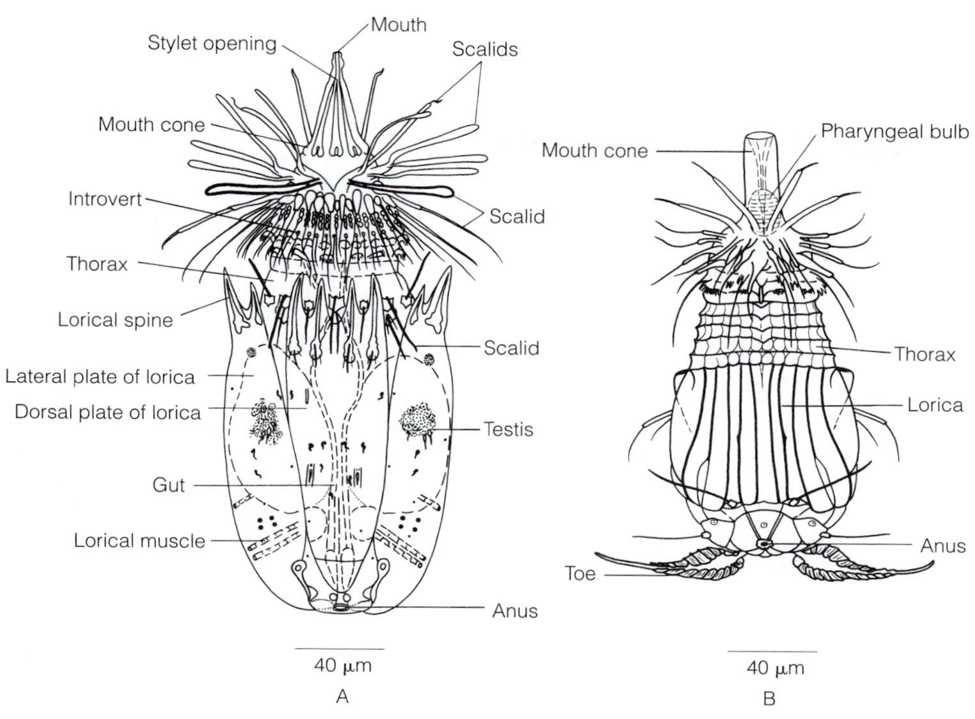

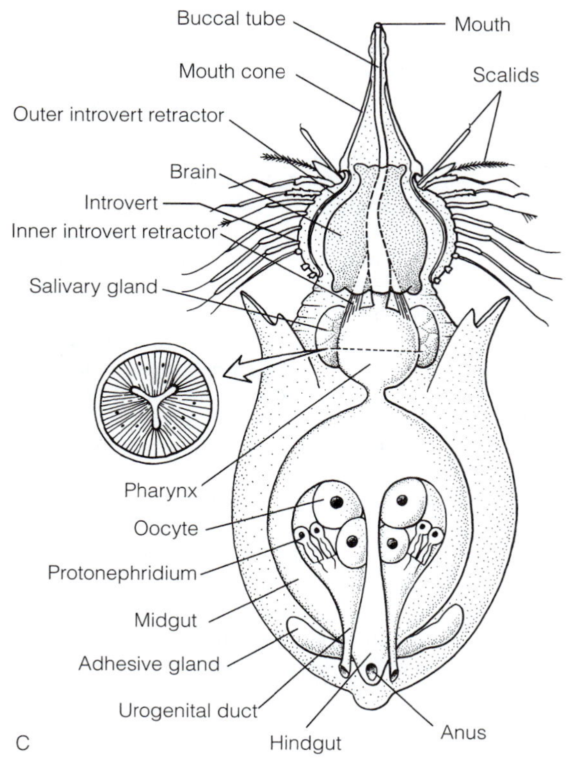

FIGURE 22-33 Loricifera: anatomy of Loricifera exemplified by the nanaloricid *Nanaloricus mysticus*. **A,** Dorsal view of an adult male. **B,** Dorsal view of the Higgins larva. **C,** Internal anatomy of the adult in dorsal view (inset shows cross section of epitheliomuscular pharynx). *(A and B, From Kristensen, R. M. 1983. Loricifera, a new phylum with Aschelminthes characters from the meiobenthos. Z. Zool. Syst. Evolut.-forsch. 21:163–180; C, Modified and redrawn from Kristensen, R. M. 1991. Loricifera In Harrison, F. W., and Ruppert, E. E. (Eds.): Microscopic Anatomy of Invertebrates. Wiley-Liss, New York. pp. 334, 335.)*

KINORHYNCHA[P]

Kinorhyncha (mud dragons) consists of about 150 described species of small marine metazoans that burrow in the surface layer of mud or live in the interstitial spaces of sand. They have been found from the intertidal zone to depths of several thousand meters and usually are less than 1 mm in length. The short body is flattened ventrally, like that of gastrotrichs, but kinorhynchs lack locomotory cilia and, except for the absence of paired appendages, superficially resemble interstitial harpacticoid copepods, with which they are sometimes confused.

The segmentation of the cuticle, body-wall musculature, epidermal glands, and nervous system is a distinguishing feature. The body is divided into 13 segments, of which the first is the introvert (head) and the second is the neck (Fig. 22-34A). The remaining 11 segments make up the trunk, which often is triangular in cross section (Fig. 22-34B).

The mouth is anterior and terminal, as in other cycloneuralians, and is situated at the end of a protrusible mouth cone (Fig. 22-34A). The mouth cone can be retracted and protruded (Fig. 22-34C) but, unlike that of priapulids, cannot be everted (turned inside out). The mouth is surrounded by a circle of nine cuticular **oral styles** (Fig. 22-34A). The introvert itself bears up to 90 protruding spinelike cuticular scalids arranged in seven concentric rings around the introvert. The entire introvert can be withdrawn into the neck or first trunk segment, hence the name Kinorhyncha, meaning "movable snout." A set of cuticular plates, or **placids** (pronounced PLAK-ids), on the second or third segment close over the retracted introvert.

A thin, monolayered cellular epidermis underlies and secretes the cuticle. The epidermis contains segmental mucous glands, which open through pores in the cuticle and secrete a mucous layer over the surface of the cuticle. The chitinous cuticle is usually thick and sclerotized and is molted. The cuticle of each segment is divided into two or three sclerotized plates consisting of a dorsal tergite and one or two ventral sternites (Fig. 22-34B). The cuticle includes a thin outer epicuticle, a middle sclerotized intracuticle, and a fibrillar procuticle. Adhesive tubules may be present, often on segment 4, but are sometimes restricted to males.

The body musculature is located beneath the epidermal basal lamina. It consists of segmental bundles of longitudinal, diagonal, and dorsoventral muscles, all of which are cross-striated, like those of arthropods. Circular muscles are absent in most of the body wall but are present in the mouth cone. Body-wall muscles attach to the inner surface of the cuticular plates by an attachment complex like that of other cycloneuralians. Tonofilaments from the muscle cells cross the intervening epidermal cells and attach to the cuticle by hemidesmosomes. The anterior margins of the cuticular plates bear apodemes similar to those of arthropods for the attachment of longitudinal muscles.

A reduced fluid-filled hemocoel is located between the body wall and gut tube (Fig. 22-34B). It contains amoebocytes of unknown function, which may occupy most of its volume. The two sets of introvert retractor muscles are similar to those of priapulids in function and location (Fig. 22-29A). They originate on the cuticular plates of the trunk segments and insert on the brain. Contraction of these muscles retracts the introvert. Extension is accomplished by the contraction of body-wall muscles with the hemocoel acting as a hydrostatic skeleton.

A kinorhynch burrows by alternately everting and retracting its spiny introvert. The body moves forward during eversion of the introvert and remains stationary during retraction. As the introvert everts, the scalids unfurl and sweep rearward to pull the animal forward (Fig. 22-34C). When the introvert is fully everted, the mouth cone, bearing a terminal mouth surrounded by the oral styles, protrudes into the sediment.

The intraepidermal nervous system consists of a triannular brain and a ventral nerve cord. The brain is a wide, collarlike band around the anterior of the pharynx similar to that of the other cycloneuralians. The forebrain innervates the mouth cone and introvert scalids. Eight longitudinal nerves exit the hindbrain to innervate the neck and trunk. The double ventral nerve cord extends posteriorly from the brain (Fig. 22-34A,B) and bears paired segmental ganglia connected by commissures. The sense organs include a variety of cuticular structures, including the scalids and flosculi, which contain monociliated receptor cells. A few species have anterior ocelli of unusual structure, and all appear to be sensitive to light.

Kinorhynchs feed on diatoms, fine organic detritus, or both. The gut consists of a foregut, midgut, and hindgut. The foregut is lined with cuticle and consists of a buccal cavity that probably has a filtering function, a sucking pharynx with walls composed of mesodermal radial and circular muscles, and a short esophagus that joins the midgut. The midgut is lined with a presumably absorptive, microvillous gastrodermis and enclosed by circular and longitudinal muscles. The midgut opens into a short, cuticle-lined hindgut that itself opens to the exterior via the terminal anus on segment 13. The physiology of digestion has not been studied.

Two protonephridia, each consisting of three biflagellated terminal cells, lie in the hemocoel and open by ducts to nephridiopores on the lateral surface of the 11th segment. The reproductive and excretory systems are independent of each other, unlike those of priapulids and loriciferans.

Kinorhynchs are gonochoric and have paired sac-shaped gonads (Fig. 22-34A,B). Each gonad is connected to the exterior by a gonoduct and a gonopore between segments 12 and 13. Seminal receptacles are present in the female gonoducts and fertilization is assumed to be internal. Copulation has never been observed, but species in two genera have been found to extrude spermatophores, which are transferred to the female by the penal spines. Very little is known of development in kinorhynchs, but it is direct and the young hatch with 11 segments and look very much like adults. The juveniles undergo periodic molts to attain adulthood, after which molting ceases.

Kinorhynchs have many features in common with arthropods, but until recently these have been dismissed as convergences. Studies of nucleotide sequences, however, suggest a sister-group relationship in Ecdysozoa between arthropods and cycloneuralians. Arthropods and kinorhynchs are segmented and have a chitinous exoskeleton with ecdysis. The body cavity is a hemocoel. The body-wall muscles are cross-striated in both taxa. The longitudinal muscles originate and insert on the inner surfaces of the cuticular plates in arthropods and kinorhynchs, although there are differences in the attachment mechanism. Both have apodemes to improve the mechanical advantage of the muscle attachments. Both have an anterior

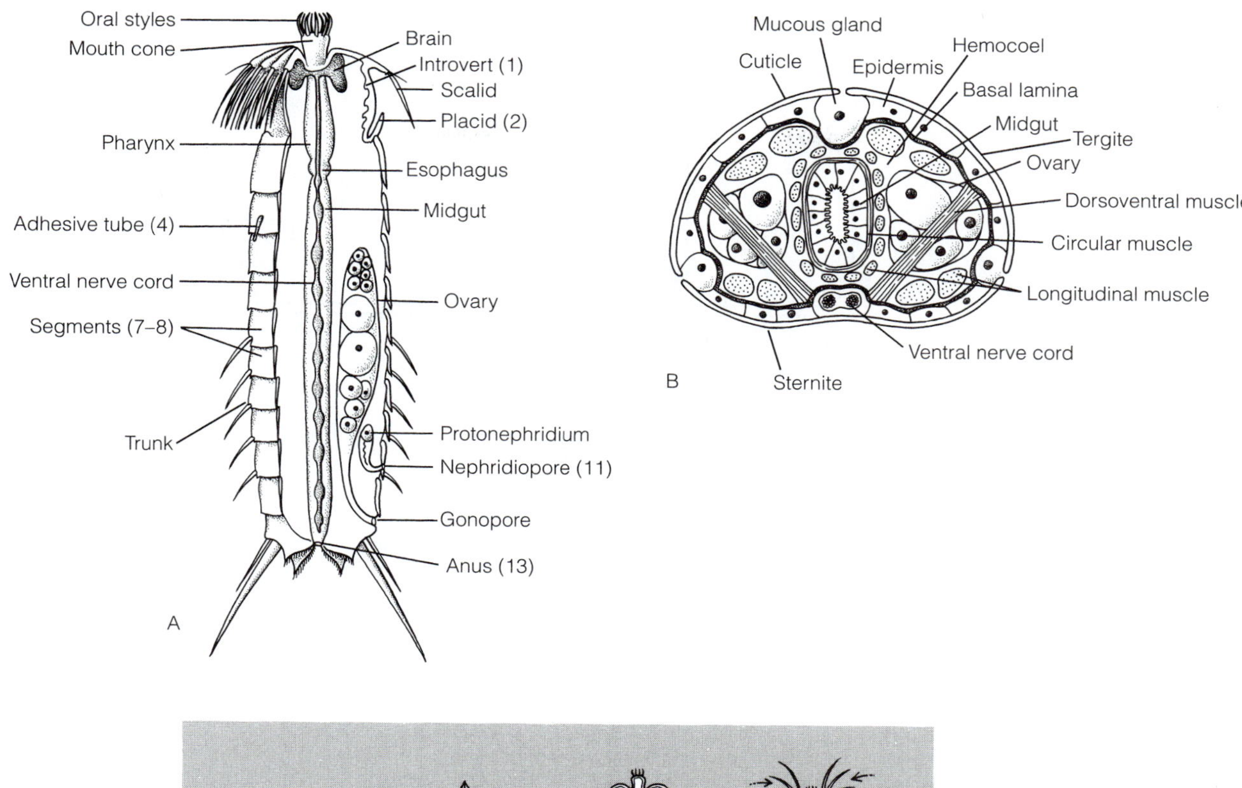

FIGURE 22-34 Kinorhyncha. **A,** Anatomy of a generalized kinorhynch in ventral view. The numbers in parentheses are segment numbers. **B,** Cross section of the trunk of a female kinorhynch. **C,** Introvert eversion and locomotion. A kinorhynch moves forward through sediment as the introvert is everted and the scalids stroke rearward. The body does not move when the introvert retracts. Note that the mouth cone, although protruded and retracted by this process, is not inverted. *(A and B, Modified and redrawn from Kristensen, R. M., and Higgins, R. P. 1991. Kinorhyncha. In Harrison, F. W., and Ruppert, E. E. (Eds.): Microscopic Anatomy of Invertebrates. Wiley-Liss, New York. pp. 378, 379.)*

circumenteric brain and a paired ventral nerve cord with paired segmental ganglia connected by commissures.

DIVERSITY OF KINORHYNCHA

Cyclorhagida[C]: The most widely distributed and diverse kinorhynch taxon. Most have 14 to 16 placids in the neck region. Trunk is spiny and circular to triangular in cross section. Common in marine muds and subtidal sands. *Campyloderes, Cateria, Centroderes, Condyloderes, Echinoderes, Semnoderes, Sphenoderes.*

Homalorhagida[C]: Relatively large (up to 1 mm). Have six to eight placids in the neck region. Trunk has few spines. Distinctly triangular in cross section. Common in marine muds. *Kinorhynchus, Neocentrophyes, Paracentrophyes, Pycnophyes.*

PHYLOGENY OF CYCLONEURALIA

Cycloneuralia is provisionally considered to be a monophyletic taxon (Fig. 22-35) and is treated here as being independent of the panarthropods, although the evolutionary relationships among its component taxa and its relationship with Panarthropoda in Ecdysozoa remain unresolved.

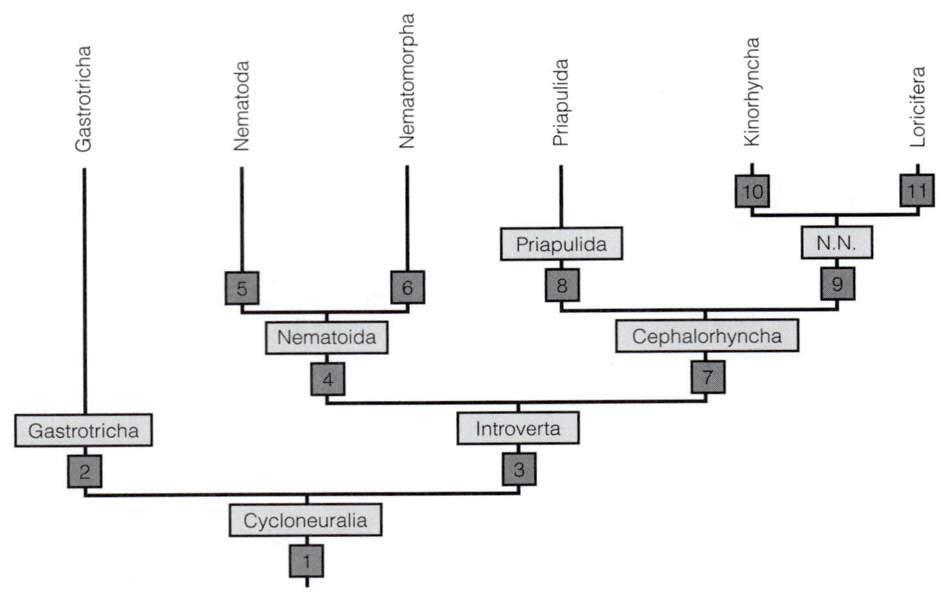

FIGURE 22-35 A phylogeny of Cycloneuralia based on morphological characters. **1, Cycloneuralia:** With anterior terminal mouth; a triradiate, sucking pharynx; and a triannular brain with longitudinal cords. **2, Gastrotricha:** Multilayered cuticle, an epitheliomuscular pharynx; are hermaphroditic. **3, Introverta:** With introvert and ecdysis, motile cilia are lost. **4, Nematoida:** Cuticle is reinforced with collagen and lacks microvilli; has longitudinal epidermal cords; circular muscles are lost. **5, Nematoda:** Three circumoral rings with six, six, and four sensilla in successive rings; most have a pair of amphids; an epitheliomuscular pharynx is present. **6, Nematomorpha:** Adults have a vestigial gut. **7, Cephalorhyncha:** Cuticle is reinforced with chitin; have two rings of pharynx retractor muscles that penetrate the brain; chitinous scalids on the introvert; sensory flosculi. **8, Priapulida:** Spacious hemocoel. **9, N.N.:** Noninversible mouth cone with cuticular ridges and spines. **10, Kinorhyncha:** Body segmented. **11, Loricifera:** Scalids with muscles, an epitheliomuscular pharynx. *(Modified from Nielsen, C. 2001. Animal Evolution. Interrelationships of the Living Phyla. 2nd ed. Oxford University Press, Oxford. 563 pp.)*

Zoologists agree that Kinorhyncha, Loricifera, and Priapulida form a natural taxon, Cephalorhyncha, that is well supported by morphological and molecular evidence. A sister-taxon relationship of Nematoda and Nematomorpha and the inclusion of Gastrotricha in Cycloneuralia, however, are questioned by many zoologists. Unlike all other cycloneuralians, gastrotrichs have locomotory cilia, lack ecdysis, are hermaphroditic, and lack an introvert, all of which suggest that they may not be part of this assemblage. The remaining five taxa make up Introverta, which, in the cycloneuralian hypothesis, is the sister taxon of Gastrotricha and characterized by the absence of locomotory cilia and the presence of ecdysis.

If Ecdysozoa is accepted as a taxon consisting of Panarthropoda and Cycloneuralia as sister taxa, then the inclusion of Gastrotricha in Cycloneuralia becomes less likely and would require that gastrotrichs lost ecdysis and redeveloped cilia as locomotory devices. Zoologists that accept Ecdysozoa usually exclude Gastrotricha from Cycloneuralia and believe the gastrotrichs arose early from the ecdysozoan stem before the development of ecdysis and the reduction and loss of cilia.

An anterior retractable region, or introvert, is the defining apomorphy of Introverta. Such a structure is present in at least some members of each major Introverta taxon, although only one genus of nematode has it. The cycloneuralian hypothesis maintains that an introvert was present in the first nematodes but has been in lost in almost all their descendants, a proposition some zoologists find unlikely.

Two monophyletic taxa evolved from the ancestral introvertans. Nematoida (nematodes and nematomorphs) have a collagenous cuticle without microvilli, longitudinal epidermal cords, a body wall without circular muscles, and an elongate vermiform shape. Their sister taxon, Cephalorhyncha (priapulids, kinorhynchs, and loriciferans), has a chitinous cuticle, no epidermal cords, and both circular and longitudinal body-wall muscles. Whereas the monophyly of Cephalorhyncha is generally accepted, that of Nematoida is on shakier ground and is not supported by some molecular and morphological evidence. Nematomorphs differ from nematodes in their unusual muscle structure (paramyosin), degeneration of the gut, the derived nature of the brain, and their unique sperm, which is unlike that of any other animal. Nematodes have an epitheliomuscular pharynx and a characteristic radially symmetric array of sensilla surrounding the mouth, both of which are absent in nematomorphs. Almost all nematodes lack the introvert and instead have specialized in using a sucking pharynx for feeding. Nematomorphs have retained the ancestral mesodermal pharynx and the larva has an introvert.

The three cephalorhynchan taxa are united by numerous synapomorphies, including a chitinous cuticle and rings of sensory and locomotory chitinous setae (scalids) on the introvert. They have unique sensory flosculi consisting of microvillous

PHYLOGENETIC HIERARCHY OF CYCLONEURALIA

Cycloneuralia
- Gastrotricha
- Introverta
  - Nematoida
    - Nematoda
    - Nematomorpha
  - Cephalorhyncha
    - Priapulida
    - N.N.
      - Kinorhyncha
      - Loricifera

monociliated cells. Two sets of introvert retractor muscles extend from the body wall to the oral epidermis and insert through the mediation of ectodermal tanycytes that pass through the brain. Some zoologists include Nematomorpha in Cephalorhyncha.

Relationships within Cephalorhyncha, however, are less certain. Sometimes Kinorhyncha and Loricifera are considered to be sister taxa on the basis of their common possession of a retractable but noninversible mouth cone. The alternative case for a sister-taxon relationship between kinorhynchs and priapulids is based on their common possession of mesodermal rather than epitheliomuscular pharynges. The former position is adopted here.

Aschelminthes was for decades a large taxon (phylum or superphylum) into which it was convenient to relegate protostome taxa with uncertain relationships. Membership in Aschelminthes has varied over the years, but at one time or another it has included rotifers, nematodes, nematomorphs, priapulids, acanthocephalans, tardigrades, kinorhynchs, loriciferans, and even gnathostomulids. At present it is believed to be polyphyletic and its taxa have been distributed elsewhere. Tardigrada has gone to Panarthropoda; Gnathostomulida, Rotifera, and Acanthocephala form the new taxon Gnathifera; and Gastrotricha, Nematoda, Nematomorpha, Priapulida, Kinorhyncha, and Loricifera constitute Cycloneuralia. Nevertheless, the persistent problem of Aschelminthes still has not been solved to everyone's satisfaction.

REFERENCES

GASTROTRICHA

d'Hondt, J. L. 1971. Gastrotricha. Oceanogr. Mar. Biol. Ann. Rev. 9:141–192.

Remane, A. 1936. Gastrotricha. In Bronn, H. G. (Ed.): Klassen Ordn. Tierreichs. 4:1–242.

Rieger, G. E., and Rieger, R. M. 1977. Comparative fine structure study of the gastrotrich cuticle and aspects of cuticle evolution within the Aschelminthes. Z. Zool. Syst. Evolut.-forsch. 15:81–124.

Ruppert, E. E. 1970. On *Pseudostomella* Swedmark 1956 with descriptions of *P. plumosa* nov. spec., *P. cataphracta* nov. spec. and a form of *P. roscovita* Swedmark 1956 from the West Atlantic coast. Cah. Biol. Mar. 11:121–143.

Ruppert, E. E. 1988. Gastrotricha. In Higgins, R. P., and Thiel, H. (Eds.): Introduction to the Study of Meiofauna. Smithsonian Institution Press, Washington, DC. pp. 302–311.

Ruppert, E. E. 1991. Gastrotricha. In Harrison, F. W., and Ruppert, E. E. (Eds.): Microscopic Anatomy of Invertebrates. Vol. 4. Aschelminthes. Wiley-Liss, New York. pp. 41–109.

Strayer, D. L., and Hummon, W. D. 2001. Gastrotricha. In Thorp, J. H., and Covich, A. P. (Eds.): Ecology and Classification of North American Freshwater Invertebrates. 2nd Edition. Academic Press, San Diego. pp. 181–194.

Tyler, S., and Rieger, G. E. 1980. Adhesive organs of the Gastrotricha. I. Duo-gland organs. Zoomorphologie 95:1–15.

Internet Site

www.microscopyu.com/moviegallery/pondscum/gastrotrich/chaetonotus (Nikon MicroscopyU. With a still photo and two movies of *Chaetonotus*.)

NEMATODA

Adams, P. J. M., and Tyler, S. 1980. Hopping locomotion in a nematode: Functional anatomy of the caudal gland apparatus of *Theristus caudasaliens* sp. n. J. Morphol. 164:265–285.

Albertson, D. G., and Thomson, J. N. 1976. The pharynx of *Caenorhabditis elegans*. Philos. Trans. Roy. Soc. Lond. B. Biol. Sci. 275:299–325.

Batson, B. S. 1979. Ultrastructure of the trophosome, a food-storage organ in *Gastromeris boophthoorae* (Nematoda: Mermithidae). Int. J. Parasitol. 9:505–514.

Boveri, T. 1904. Ergebnisse über die Konstitution der chromatischen Substanz des Zellkerns. Gustav Fischer Verlag, Jena, Germany. p. 27

Chandler, A. C., and Read, C. P. 1961. Introduction to Parasitology. John Wiley and Sons, New York. 822 pp.

Chitwood, B. G. 1931. A comparative histological study of certain nematodes. Ztschr. Morphol. Ökol. Tiere 23:237–284.

Christie, J. R. 1937. *Mermis subnigrescens,* a nematode parasite of grasshoppers. J. Agricult. Res. 55:353–364.

Lee, D. L., and Atkinson, H. J. 1977. Physiology of Nematodes. 2nd Edition. Columbia University Press, New York. 215 pp.

Maggenti, A. 1981. General Nematology. Springer Verlag, New York. 372 pp.

Nelson, F. K., Albert, P. S., and Riddle, D. L. 1983. Fine structure of the *Caenorhabditis elegans* secretory–excretory system. J. Ultrastruct. Res. 82:156–171.

Nicholas, W. L. 1984. The Biology of Free-Living Nematodes. 2nd Edition. Oxford University Press, London. 251 pp.

Poinar, G. O. 2001. Nematoda and Nematomorpha. In Thorp, J. H., and Covich, A. P. (Eds.): Ecology and Classification of North American Freshwater Invertebrates. 2nd Edition. Academic Press, New York. pp. 255–295.

Riemann, F. 1974. On hemisessile nematodes with flagelliform tails living in marine soft bottoms and on micro-tubes found in deep sea sediments. Mikrofauna Meeresbodens 40:1–15.

Riemann, F. 1988. Nematoda. In Higgins, R. P., and Thiel, H. (Eds.): Introduction to the Study of Meiofauna. Smithsonian Institution Press, Washington, DC. pp. 293–301.

Schmidt, G. D., and Roberts, L. S. 1985. Foundations of Parasitology. 3rd Edition. Times Mirror/Mosby College, St. Louis, MO. 775 pp.

Ward, S., Argon, Y., and Nelson, G. A. 1981. Sperm morphogenesis in wild-type and fertilization-defective mutants of *Caenorhabditis elegans*. J. Cell Biol. 91:26–44.

Ward, S., Thomson, J. G., and Brenner, S. 1975. Electron microscopical reconstruction of the anterior sensory anatomy of the nematode *Caenorhabditis elegans*. J. Comp. Neurol. 160:313–338.

Wood, W. B. (Ed.): 1988. The nematode, *Caenorhabditis elegans*. Cold Spring Harbor, New York. 667 pp.

Wright, K. A. 1991. Nematoda. In Harrison, F. W., and Ruppert, E. E. (Eds.): Microscopic Anatomy of Invertebrates. Vol. 4. Aschelminthes. Wiley-Liss, New York. pp. 11–195.

Zmoray, I., and Guttekova, A. 1969. Ecological conditions for the occurrence of cilia in intestines of nematodes. Biol. Bratisl. 24:97–112.

Internet Sites

http://elegans.swmed.edu (*Caenorhabditis elegans* WWW Server, by L. Avery. A resource for *Caenorhabditis* researchers. Includes links.)

www.nyu.edu/projects/fitch/index.html (Fitch Nematode Lab page, by D. H. A. Fitch, New York University. Molecular systematics. Includes links.)

http://flnem.ifas.ufl.edu/nemadoc.htm (Florida Nematology, by K. B. Nguyen, University of Florida.)

www.ucmp.berkeley.edu/phyla/ecdysozoa/nematoda.html (Introduction to the Nematoda: The Roundworms. Museum of Paleontology, University of California at Berkeley. A short, general account with an SEM photo.)

http://nematode.unl.edu/wormgen.htm (What Are Nematodes? University of Nebraska. Includes identification aids, literature, and links.)

*www.life.sci.qut.edu.au/*LIFESCI*/darben/paramast.htm* (Parasitology Images List, by P. Darben, Queensland University of Technology. A collection of clinically oriented parasite photographs.)

http://nematode.unl.edu/ (Plant and Insect Parasitic Nematodes Home Page. University of Nebraska. Includes links, biological control, literature, projects, and more.)

www.barc.usda.gov/psi/nem/home-pg.html (USDA Nematode Lab. Includes links, collections, plant parasite list, and laboratory techniques.)

NEMATOMORPHA

Bresciani, J. 1991. Nematomorpha. In Harrison, F. W., and Ruppert, E. E. (Eds.): Microscopic Anatomy of Invertebrates. Vol. 4. Aschelminthes. Wiley-Liss, New York. pp. 197–218.

May, H. G. 1919. Contributions to the life histories of *Gordius robustus* Leidy and *Paragordius varius* (Leidyi). Ill. Biol. Monogr. 5:1–119.

Montgomery, T. 1904. Development and structure of the larva of *Paragordius*. Proc. Acad. Nat. Sci. Phila. 56.

Mühldorf, A. 1914. Beiträge zur Entwicklungsgeschichte der *Gordius* larve. Ztschr. Wiss. Zool. 111.

Poinar, G. O. 2001. Nematoda and Nematomorpha. In Thorp, J. H., and Covich, A. P. (Eds.): Ecology and Classification of North American Freshwater Invertebrates. 2nd Edition. Academic Press, New York. pp. 255–295.

Schmidt-Rhaesa, A. 1998. Phylogenetic relationships of the Nematomorpha: A discussion of current hypotheses. Zool. Anz. 236:203–216.

Internet Site

www.inhs.uiuc.edu/chf/pub/surveyreports/mar-apr95/page4.html (Horsehair worms in Illinois, by M. J. Wetzel and D. J. Watermolen, Illinois Natural History Survey. General account with a photograph.)

PRIAPULIDA

Aarnio, K., Bonsdorff, E., and Norkko, A. 1998. Role of *Halicryptus spinulosus* (Priapulida) in structuring meiofauna and settling macrofauna. Mar. Ecol. Prog. Ser. 163:145–153.

Adrianov, A. V., and Malakhov, V. V. 1996. Priapilida (Priapulida): Structure, development, phylogeny, and classification. KMR Scientific Press, Moscow. 266 pp. (In Russian with English summary.)

Calloway, C. B. 1982. Priapulida. In Parker, S. P. (Ed.): Synopsis and Classification of Living Organisms. McGraw-Hill, New York. pp. 941–944.

Elder, H. Y., and Hunter, R. D. 1980. Burrowing of *Priapulus caudatus* (Vermes) and the significance of the direct peristatic wave. J. Zool. Lond. 191:333–351.

Higgins, R. P., and Storch, V. 1991. Evidence for direct development in *Meiopriapulus fijiensis* (Priapulida). Trans. Am. Microsc. Soc. 110:37–46.

Por, F. D. 1972. Priapulida from deep bottoms near Cyprus. Israel J. Zool. 21:525–528.

Shirley, T. C., and Storch, V. 1999. *Halicryptus higginsi* n. sp. (Priapulida): A giant new species from Barrow, Alaska. Invert. Biol. 118:404–413.

Storch, V. 1991. Priapulida. In Harrison, F. W., and Ruppert, E. E. (Eds.): Microscopic Anatomy of Invertebrates. Wiley-Liss, New York. pp. 333–350.

Storch, V., Higgins, R. P., Anderson, P., and Svavarsson, J. 1995. Scanning and transmission electron microscopic analysis of the introvert of *Priapulopsis australis* and *Priapulopsis bicaudatus* (Priapulida). Invert. Biol. 114:64–72.

Storch, V., Higgins, R. P., and Morse, M. P. 1989. Internal anatomy of *Meiopriapulus fijiensis* (Priapulida). Trans. Am. Microsc. Soc. 108:245–261.

Internet Site

www.gwdg.de/~clembur/prialifr.htm (Literature List for Priapulida, by C. Lemburg. An extensive bibliography of priapulids. Backtrack to home page for more priapulid information.)

LORICIFERA

Higgins, R. P., and Kristensen, R. M. 1986. New Loricifera from southeastern United States coastal waters. Smithson. Cont. Zool. 438:1–70.

Higgins, R. P., and Kristensen, R. M. 1988. Loricifera. In Higgins, R. P., and Thiel, H. (Eds): Introduction to the Study of Meiofauna. Smithsonian Institution Press, Washington, DC. pp. 319–321.

Kristensen, R. M. 1983. Loricifera, a new phylum with Aschelminthes characters from the meiobenthos. Z. zool. Syst. Evolut.-forsch. 21:163–180.

Kristensen, R. M. 1991. Loricifera. In Harrison, F. W., and Ruppert, E. E. (Eds.): Microscopic Anatomy of Invertebrates. Vol. 4. Aschelminthes. Wiley-Liss, New York. pp. 351–375.

Internet Site

www.gwdg.de/~clembur/lorlitfr.htm (Literature List for Loricifera, by C. Lemburg. An extensive bibliography of loriciferans.)

KINORHYNCHA

Higgins, R. P. 1971. A historical overview of kinorhynch research. In Hulings, N. C. (Ed.): Proceedings of the First International Conference on Meiofauna. Smithson. Contrib. Zool. 76:25–31.

Kristensen, R. M., and Higgins, R. P. 1991. Kinorhyncha. In Harrison, F. W., and Ruppert, E. E. (Eds.): Microscopic Anatomy of Invertebrates. Vol. 4. Aschelminthes. Wiley-Liss, New York. pp. 377–404.

Müller, C. M. and Schmidt-Rhaesa, A. 2003. Reconstruction of the muscle system in *Antygomonas* sp. (Kinorhyncha, Cyclorhagida) by means of phalloidin labeling and cLSM. J. Morphol. 256:103–110.

Nebelsick, M. 1993. Nervous system, introvert, and mouth cone of *Echinoderes capitatus* and phylogenetic relationships of the Kinorhyncha. Zoomorphology 113:211–232.

Neuhaus, B. 1994. Ultrastructure of the alimentary canal and body cavity, ground pattern, and phylogenetic relationships of the Kinorhyncha. Microfauna Marina 9:61–156.

Neuhaus, B. and Higgins, R. P. 2002. Ultrastructure, biology, and phylogenetic relationships of Kinorhyncha. Integ. Comp. Biol. 42:619–632.

Internet Sites

www.gwdg.de/~clembur/kinlitfr.htm (Literature List for Kinorhyncha, by C. Lemburg. An extensive bibliography of kinorhynchs.)

www.nat.ku.dk/as/ASdk-RMK26.htm (*Pycnophyes groenlandicus* photomicrograph by R. M. Kristensen.)

www.inhs.uiuc.edu/chf/pub/surveyreports/mar-apr95/page4.html (Horsehair worms in Illinois, by M. J. Wetzel and D. J. Watermolen, Illinois Natural History Survey. General account with a photograph.)

Gnathifera[SP]

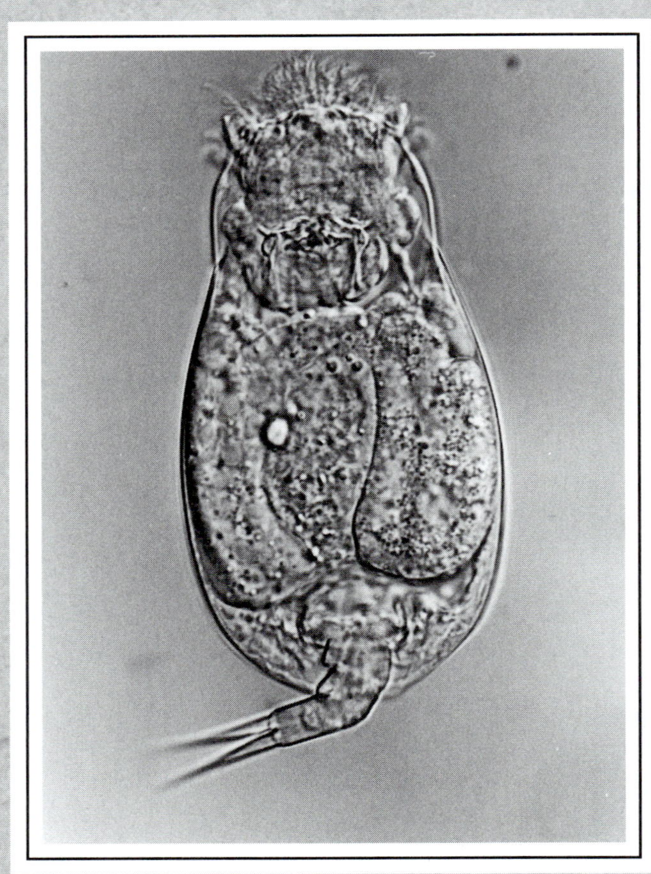

Gnathostomulida, Rotifera, Seisonida, Acanthocephala, and the newly discovered Micrognathozoa are thought to have evolved from ancestors with a complex pharynx featuring a unique cuticular jaw apparatus. These taxa have heretofore been classified with other groups, such as platyhelminths, annelids, and aschelminths, but are now believed to form a monophyletic taxon, which has been named Gnathifera. In these taxa the pharynx, when present, consists of a ventral muscular pharyngeal bulb on the floor of the pharyngeal lumen. The jaws, constructed of cuticular rods with electron-dense cores and electron-lucent sheaths, are secreted by the pharyngeal epithelium.

Most gnathiferans are less than 1 mm in length and exhibit the features you would predict for small animals. Internal transport is usually by diffusion and there is no heart. Locomotion is ciliary and specialized gills are absent. Fertilization is internal and development is direct. Cross-striated muscles rapidly operate the jaw apparatus.

GNATHOSTOMULIDA^P

Gnathostomulida is a small taxon of minute worms that live in the interstitial spaces of fine marine sands, especially those with little or no oxygen and a high concentration of hydrogen sulfide. Gnathostomulida was discovered in 1956, and since then more than 80 species in 18 genera have been found worldwide, especially along the east coast of North America.

Gnathostomulans can reach almost 4 mm, but most are between 0.5 and 1.0 mm in length and 50 μm in diameter. All are elongate and some are threadlike (Fig. 23-1A). The cylindrical body consists of a head separated from the trunk by a slightly constricted neck. In some the head is extended anteriorly as a mobile, sensory rostrum. The trunk tapers posteriorly into a tail.

The epidermis lacks a cuticle and its cells are monociliated, although other gnathiferans have multiciliated epithelia. A thin basal lamina underlies the epidermis, and some epidermal cells secrete mucus. Synchronous beating of epidermal cilia accomplishes a ciliary gliding locomotion. The body-wall musculature of cross-striated myocytes lies below the epidermal basal lamina (Fig. 23-2A). It consists of a sheath of weak outer circular fibers and usually three pairs of longitudinal muscles and is used not for locomotion, but to shorten the body and make turning movements. Gnathostomulans have no connective tissue and as a result, muscles, protonephridia, and reproductive organs are simply sandwiched between the epithelia of the epidermis and gut (23-2A). The organization of the connective-tissue compartment is compact (acoelomate) and there is no fluid-filled body cavity.

Gnathostomulans have a blind gut composed of a single layer of microvillous, nonciliated epithelial cells. It consists of a mouth, buccal cavity, pharynx, esophagus, and intestine. The mouth is a midventral, subterminal pore near the anterior end (Fig. 23-3). The laterally compressed pharynx includes a complex ventral, muscular **pharynx bulb.** The bulb includes striated muscles, a secretory epithelium, a jaw apparatus consisting of hard cuticular **trophi** (sclerites), and a ganglion. The trophi consist of a comblike **basal plate** on the ventral lip posterior to the mouth and a complex set of toothed jaws embedded in the lateral pharyngeal walls (Fig. 23-2B,C). The basal plate can be protruded through the mouth, and the tips of the jaws extend into the buccal cavity (Fig. 23-3). The jaws, which are secreted by pharyngeal epithelial cells, are made of a core of electron-dense material surrounded by a sheath of an electron-lucent material. This tubular arrangement is similar to the trophi of rotifers, seisonidans, and micrognathozoans. Some of the pharyngeal muscles may be epitheliomuscular cells, but most are myocytes. Epitheliomuscular cells and myocytes are on opposite sides of the epidermal basal lamina. One species, *Agnathiella beckeri,* has no jaws.

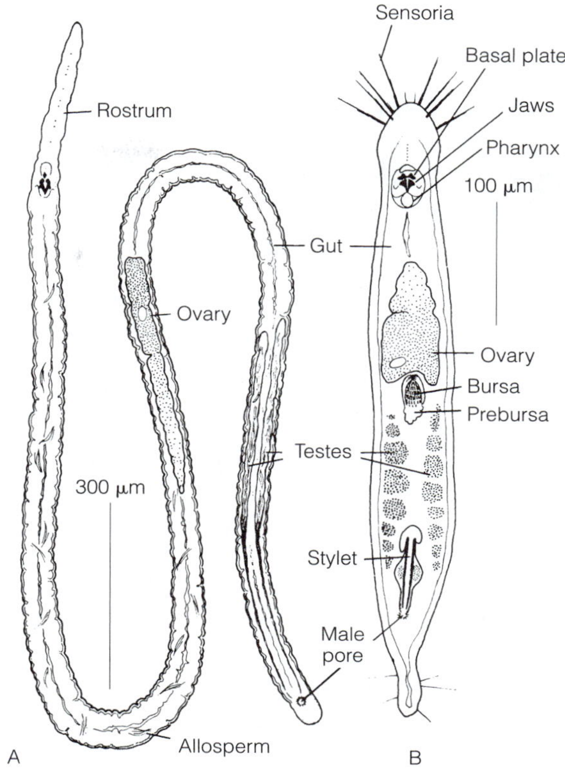

FIGURE 23-1 Gnathostomulida. **A,** The filospermoidan *Haplognathia simplex.* **B,** The bursovaginoidan *Gnathostomula jenneri.* *(From Sterrer, W. 1972. Systematics and evolution within the Gnathostomulida. Syst. Zool. 21:151–173.)*

The remainder of the gut consists of an anterior esophagus and a large posterior intestine. There is no anus, but in some species a specialized structure consisting of interdigitating epidermal and gastrodermal cells with no intervening basal laminae is located at the posterior end of the intestine. The function of this area is not known, but it may be a temporary, periodically functional anus or a vestigial, no longer functional anus. The food appears to be bacteria and fungi that are scraped from the substratum by the basal plate and passed into the gut with snapping movements of the jaws (Fig. 23-2C).

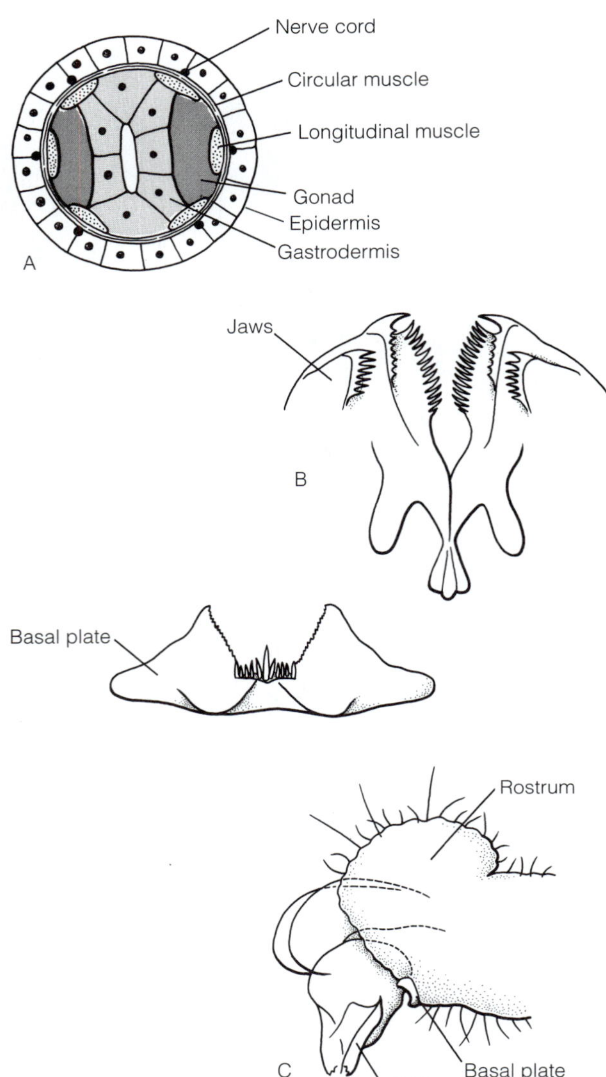

FIGURE 23-2 Gnathostomulida. **A,** Simplified cross section of a gnathostomulan trunk. **B,** Jaws and basal plate of *Gnathostomula mediterrane,* in dorsal view. **C,** Jaws and basal plate of *Problognathia minima* in feeding position, viewed from the side. *(B, From Sterrer, W. 1972. Systematics and evolution within the Gnathostomulida. Syst. Zool. 21:151–173; C, From Sterrer, W., and Farris, R. A. 1975.* Problognathia minima *n. g., n. sp. A representative of a new family of Gnathostomulida, Problognathiidae n. fam. from Bermuda. Trans. Am. Micros. Soc. 94:357–367.)*

Two to five pairs of protonephridia are arranged serially along the sides of the body. Each protonephridium consists of a monociliated, microvillar terminal cell, a canal cell, and a pore cell at the surface. Each has its own short duct to the exterior.

The intraepidermal nervous system consists of an anterior cerebral ganglion dorsal to the gut (Fig. 23-3), a buccal ganglion innervating the pharyngeal bulb ventral to the gut, and one to three pairs of longitudinal cords. The cerebral and buccal ganglia are joined by a pair of connectives passing around the gut. A ganglion serves the penis and the caudal ganglion innervates the posterior end of the body. The sensory organs are ciliary pits and sensoria, which are especially well developed on the head. **Sensoria** are stiff, bristlelike clusters of cilia.

Gnathostomulans are hermaphroditic. The female reproductive system usually consists of a single dorsal ovary and an associated seminal receptacle for storage of allosperm. A vagina is present in some species. The male system consists of one or two posterior testes and usually a penis, which in some species bears a hardened stylet composed of several intracellular rods. The male gonopore is at the posterior end. Copulation has not been observed, but transfer of the uniflagellate or aflagellate sperm probably occurs by hypodermic impregnation of the body wall, by attachment of the male gonopore to the integument of the partner, or by injection into the vagina. A single 50 μm egg, which is enormous in comparison with the gnathostomulan body but small compared with the eggs of other animals, is laid at each oviposition. The egg ruptures through the body wall (which quickly heals) and adheres to the substratum. Cleavage is spiral and development is direct.

DIVERSITY OF GNATHOSTOMULIDA

Filospermoidea^{C}: Long, slender, up to 3.5 mm in length (Fig. 23-1A) with relatively simple, compact jaws and filiform sperm. Anterior end is a long, tactile rostrum with abundant sensory receptors. Two families. *Haplognathia, Pterognathia.*

Bursovaginoidea^{C}: Most gnathostomulans belong to this taxon (Fig. 23-1B). Body is relatively short, up to 1.5 mm long, 50 μm wide. Anterior end is not elongated to form a rostrum, but has sensory receptors. Elaborate, basket-type jaws. Sperm are spherical or pyriform rather than filiform, lack flagella and centrioles. Female system includes a seminal receptacle; penis has a stylet. Seven families in two orders. *Austrognathia, Gnathostomaria, Gnathostomula, Labidognathia, Mesognatharia, Paragnathiella, Problognathia, Rastrognathia.*

MICROGNATHOZOA

In 2000, R. M. Kristensen and P. Funch described Micrognathozoa as a new class from moss pillows in a cold, freshwater spring on Disko Island, Greenland, at 70° N latitude. The authors did not assign the new taxon to a phylum, although they did place it in Gnathifera (Fig. 23-26). Micrognathozoa is of particular interest because the structure of its pharyngeal jaw apparatus is similar to that of both gnathostomulans and monogonont rotifers and supports the hypothesis that these taxa are related. At present there is only one known species, *Limnognathia maerski,* which is epiphytic on mosses.

Micrognathozoans are microscopic compact bilaterians reaching lengths of up to 150 μm. The body is divided into a short head, pleated thorax, and abdomen (Fig. 23-4). The epidermis is cellular, like that of gnathostomulans, and unlike the syncytium of other gnathiferans. In fact, no

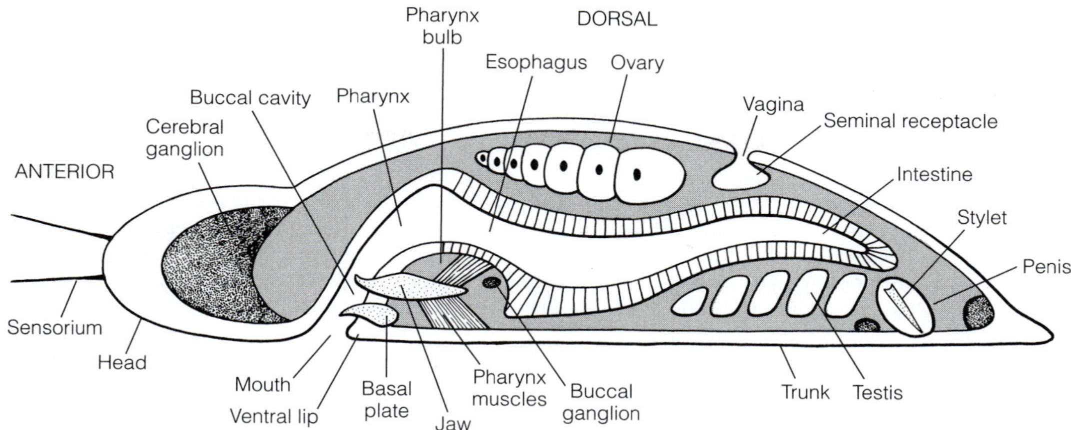

FIGURE 23-3 Gnathostomulida. Sagittal section of the bursovaginoidean *Gnathostomula paradoxa*. *(Redrawn from Lammert, V. 1991. Gnathostomulida. In Harrison, F. W., and Ruppert, E. E. (Eds.): Microscopic Anatomy of Invertebrates. Vol. 4. Aschelminthes. Wiley-Liss, New York. pp. 19–39.)*

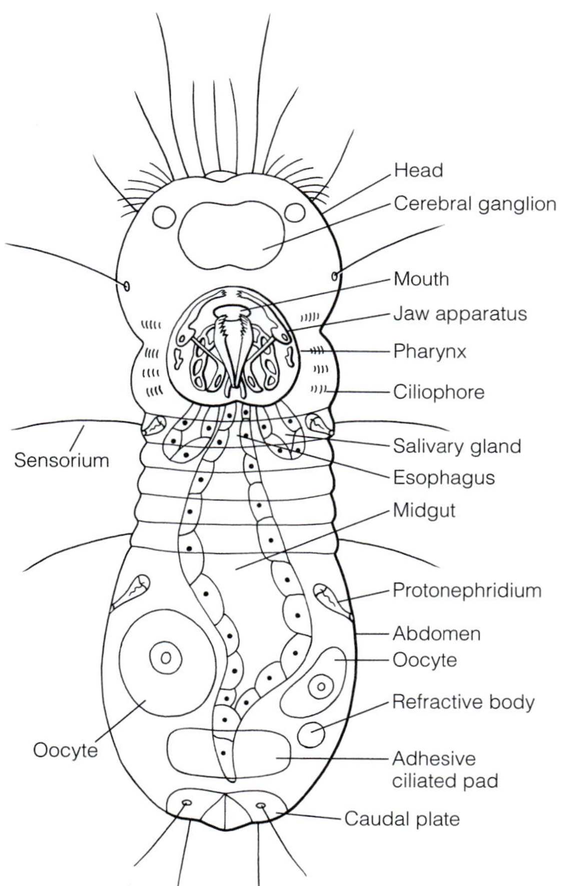

FIGURE 23-4 Micrognathozoa. Ventral view of *Limnognathia maerski*. The drawing shows both internal and external features. *(Redrawn from Kristensen, R. M., and Funch, P. 2000. Micrognathozoa: A new class with complicated jaws like those of Rotifera and Gnathostomulida. J. Morphol. 246:1–49, and Harrison, F. W. 2000. A note from the editor. J. Morphol. 246:50–52.)*

tissues are syncytial, although there are indications that the epidermis is eutelic. Both multi- and monociliated cells are present. The dorsal and lateral epidermis is nonciliated and bears an **intracellular lamina** of cytoskeletal plates. The plates are composed of a conspicuous double layer of intracellular matrix. Overlapping lateral plates create the accordion-like pleats of the thorax.

The ventral epidermis bears a well-developed and complex ciliation, much of which arises from multiciliated cells known as **ciliophores** that bear stiff compound cilia. Four pairs of ciliophores lie beside the mouth (Fig. 23-4) and two long rows of 18 pairs of ciliophores extend along the trunk. These are the chief locomotory organs and their synchronous ciliary beating is used for crawling and swimming. A posterior, ventral, ciliated **adhesive pad** consists of 10 ciliophores. Gland cells associated with the pad produce a secretion that, based on observations of living animals, appears to be adhesive.

The blind-ended gut consists of ventral mouth, pharynx, esophagus, and midgut. A cuticular oral plate surrounds the mouth, which opens into a muscular, cuticularized pharynx (Fig. 23-4). The pharynx is similar to that of gnathostomulans and consists of a muscular bulb with cuticular jaws, the epidermal cells that secrete the jaws and the cuticular lining, cross-striated muscles, a nervous system, and sensory cells. The complicated cuticular jaw apparatus consists of nine paired and one unpaired extracellular trophi (sclerites). The trophi are composed of an inner electron-dense core surrounded by an electron-lucent sheath like those of gnathostomulans and rotifers. The jaws can be protruded from the mouth and are used to grasp food particles. *Limnognathia* has been observed using its jaws to pick up individual diatoms, which are then crushed. There is no evidence that the cuticular parts are molted, in fact, adult epidermal cells do not appear to be capable of secreting new cuticle. The jaws apparently are secreted during development and the process is not repeated. The jaw apparatus is much more complicated than that of any rotifer.

A short esophagus connects the pharynx with the large endodermal midgut (Fig. 23-4). Two salivary glands that open into the midgut probably secrete digestive enzymes. There is no cuticularized hindgut. The midgut tapers posteriorly and ends blindly at a dorsal anal plate. The posteriormost midgut cells interdigitate with overlying epidermal cells and are associated with a pair of small muscles. They probably serve as a periodically functioning anus. A similar system is found in gnathostomulans. Defecation has not been observed.

The musculature consists of the pharyngeal muscles, which operate the jaws, and the somatic muscles. The somatic musculature consists of a few longitudinal cells and some dorsoventral cells. Tiny longitudinal muscle fibers extend between adjacent dorsolateral epidermal plates to contract the thorax. There are no circular muscles. The pharyngeal muscles are cross-striated whereas somatic muscles are obliquely striated, but all seem to have a mesodermal origin. Pharyngeal muscles attach to the cuticular jaw trophi indirectly by means of filaments passing through pharyngeal epidermal cells. Somatic muscles attach to the intracellular lamina of the dorsolateral epithelium, but the connection is indirect through the epidermal cell.

The two pairs of protonephridia, located laterally in the thorax and abdomen, probably are osmoregulatory. Each protonephridium consists of four terminal cells, two duct cells, and a nephridiopore cell. All seven cells are monociliated.

The nervous system includes a large cerebral ganglion in the head and a pair of ventrolateral nerve cords. Buccal nerves are present in the pharyngeal bulb, but buccal ganglia have not been demonstrated. A pair of ganglia is found in the thorax and there is a single caudal ganglion at the posterior end of the body. It is not known if the cerebral ganglion is intra- or subepidermal, but the nerve cords appear to be subepidermal. Sensoria similar to those found on the gnathostomulan head occur in pairs on all parts of the body, but are most abundant on the head.

No males have been discovered, and *Limnognathia* probably reproduces parthenogenetically, although hermaphroditism or the existence of undiscovered dwarf males is possible. Females have paired ovaries consisting solely of oocytes. No oviduct or gonopore has been found. There is no vitelline gland and yolk is synthesized in the oocyte. The oocytes mature one at a time and are very large in relation to the animal. Two refractive bodies of unknown function are located near the ovary. Cleavage has not been observed. Development is direct.

SYNDERMATA

Three taxa formerly included in the large, probably polyphyletic Aschelminthes are now placed in Gnathifera. Rotifera, Seisonida, and Acanthocephala, although superficially dissimilar, have significant morphological and molecular features in common and are believed to constitute a monophyletic taxon usually referred to as Syndermata (Fig. 23-26). Syndermata is characterized by a syncytial epidermis that contains within its cytoplasm a skeleton known as the **intrasyncytial lamina.** This skeleton is not molted. This is in sharp contrast to the extracellular, secreted cuticle of Arthropoda and Cycloneuralia. Further, the anterior end of the body can be retracted into the trunk, an action that requires the presence of a fluid-filled hemocoel to receive the head.

Syncytial tissues arise from normal cellular tissues during development when the lateral membranes between adjacent cells break down and the cytoplasm becomes continuous. Such lateral fusion of cells produces a single multinucleated "cell," or syncytium. Syncytial epithelia present an uninterrupted plasma membrane to the environment and thus form an effective barrier capable of regulating the passage of materials. There may also be structural advantages to a syncytial epithelium, for example in Syndermata. In this taxon the skeleton is a continuous sheet *inside* the epithelial cytoplasm, an arrangement that would not be possible if the cytoplasm were broken up into individual cells. Syncytial epithelia also occur in some unrelated taxa such as flatworms and sponges.

Syndermata includes four taxa, Monogononta, Bdelloidea, Seisonida, and Acanthocephala, but an understanding of their evolutionary relationships has not yet been achieved. In some modern classifications Acanthocephalans are considered to be derived rotifers. In this text Monogononta and Bdelloidea constitute Rotifera, Acanthocephala and Rotifera are sister taxa, and Seisonida is the sister taxon of Acanthocephala + Rotifera. Based on the evidence currently available, other arrangements are equally likely.

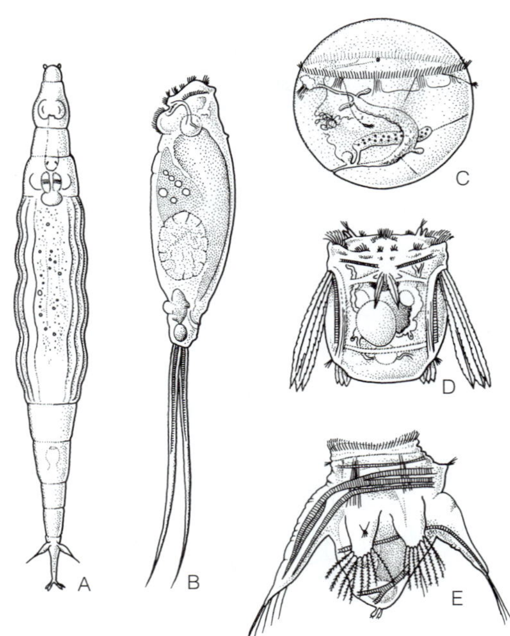

FIGURE 23-5 Representative rotifers. **A,** The bdelloid rotifer *Dissotrocha.* **B–E,** Monogonont rotifers: **B,** *Monommata.* **C–E,** Planktonic rotifers: **C,** *Trochosphaera.* **D,** *Polyarthra.* **E,** *Hexarthra.* *(A, Redrawn from Donner, J. 1966. Rotifers. Frederick Warne & Co., Ltd., London. p. 61; B–E, Redrawn from Ruttner-Kolisko, A. 1974. Plankton Rotifers. In Elster, H.-J. and Ohle, W. (Eds.): Die Binnengewässer 26 (Suppl.):1–146.)*

ROTIFERA[P]

Rotifera contains the microscopic metazoans known as rotifers, or wheel animalcules. Rotifers are common in freshwater habitats, such as lakes and streams, and in terrestrial habitats, such as the film of water covering mosses and soil particles. Some are found in the sea. Although most rotifers are benthic, the relatively few planktonic species are so abundant that they are one of the three most important metazoan taxa in freshwater zooplankton communities. Although only about 50 rotifer species occur exclusively in marine waters, some, such as *Synchaeta,* are disproportionately important in nutrient cycling in marine ecosystems. In freshwater systems, rotifers typically reach densities of between 200 and 1000 individuals per liter, sometimes much more. Soil rotifers may achieve densities as high as 2 million/m^2 and play an important role in nutrient cycling. Almost 2000 species have been described, and most have widespread distributions.

Most rotifers are 0.1 to 1 mm in length, only a little larger than most ciliate protozoans and smaller than some of them. Rotifers are strongly bilaterally symmetrical and the body cavity is a hemocoel. The body is composed of about 1000 cells, and both species and organ systems are eutelic. Rotifers hatch with the species-specific number of cells or nuclei and do not add to that number as adults. The rotifer *Epiphanes senta,* for example, has 958 nuclei. Most rotifer tissues are syncytial and all ciliated epithelial cells are multiciliated. Most are solitary, free-swimming or crawling animals (Fig. 23-5), but there are also sessile (Fig. 23-6) and colonial species (Fig. 23-7E). Colonies are aggregations of solitary individuals that arise parthenogenetically. The body usually is transparent, although some rotifers appear green, orange, red, or brown due to coloration of the digestive tract. Reproduction usually is parthenogenetic and males are rare or absent. Unless specified otherwise, discussions of rotifers apply to females.

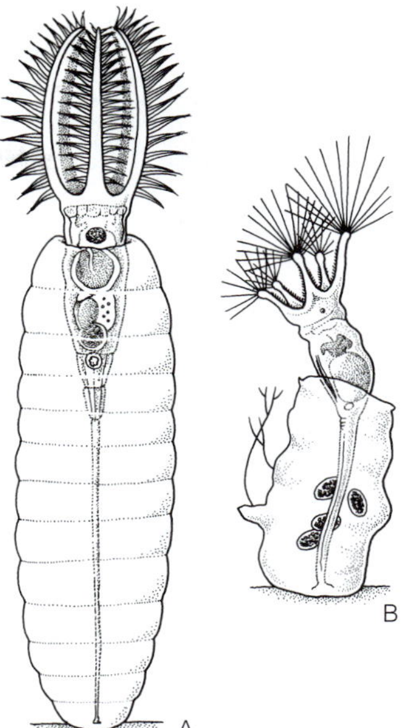

FIGURE 23-6 Sessile rotifers. **A,** *Stephanoceros.* **B,** *Collotheca.* *(Redrawn from Ruttner-Kolisko, A. 1974. Plankton Rotifers. In Elster, H.-J. and Ohle, W. (Eds.): Die Binnengewässer 26 (Suppl.):1–146.)*

Rotifera includes two taxa, Monogononta and Bdelloidea, or three if Seisonida is included.

Form

The elongate, cylindrical or saccate body is, in most species, divided into a short anterior head region, a narrow connecting neck, a large trunk that is the major part of the body, and a terminal foot (Fig. 23-8A,B). The surface may be sculptured

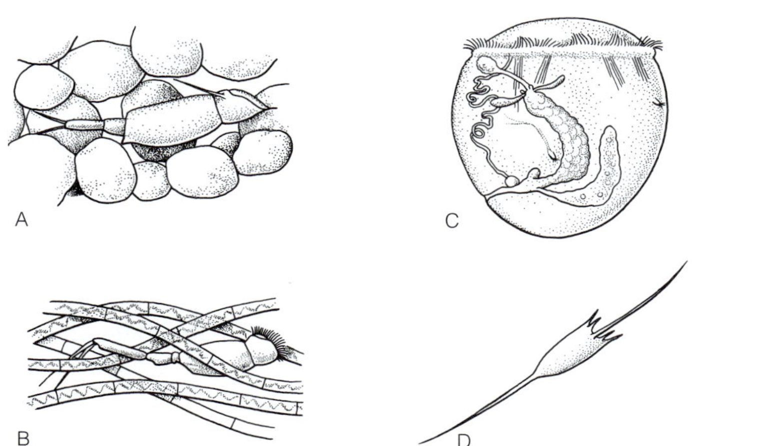

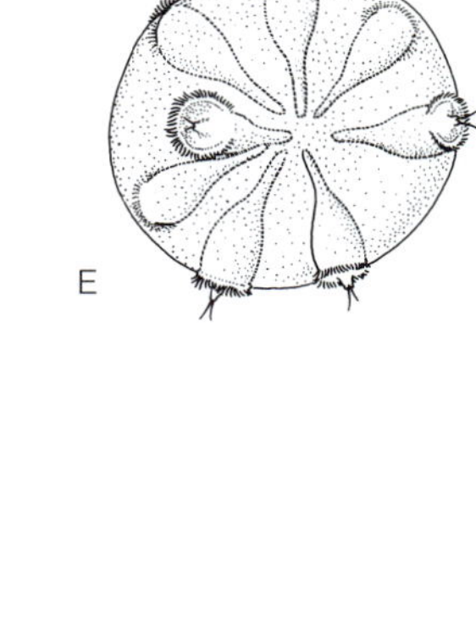

FIGURE 23-7 Rotifera. **A** and **B,** Meiofaunal rotifers: **A,** *Bryceella tenella* in sand. **B,** *Scaridium longicaudum* in algae. Both have long, slender bodies and well-developed adhesive toes. **C–E,** Planktonic rotifers: The voluminous pseudocoel of *Trochosphaera solstitialis* **(C)** and the gelatinous cuticle of the colonial *Conochilus unicornis* **(E)** may be adaptations to lower the density and improve buoyancy. In *Kellicottia longispina* **(D),** on the other hand, long spines retard the rate of sinking. *(A–E, Redrawn from Ruttner-Kolisko, A. 1974. Plankton Rotifers. In Elster, H.-J. and Ohle, W. (Eds.): Die Binnengewässer 26 (Suppl.):1–146.)*

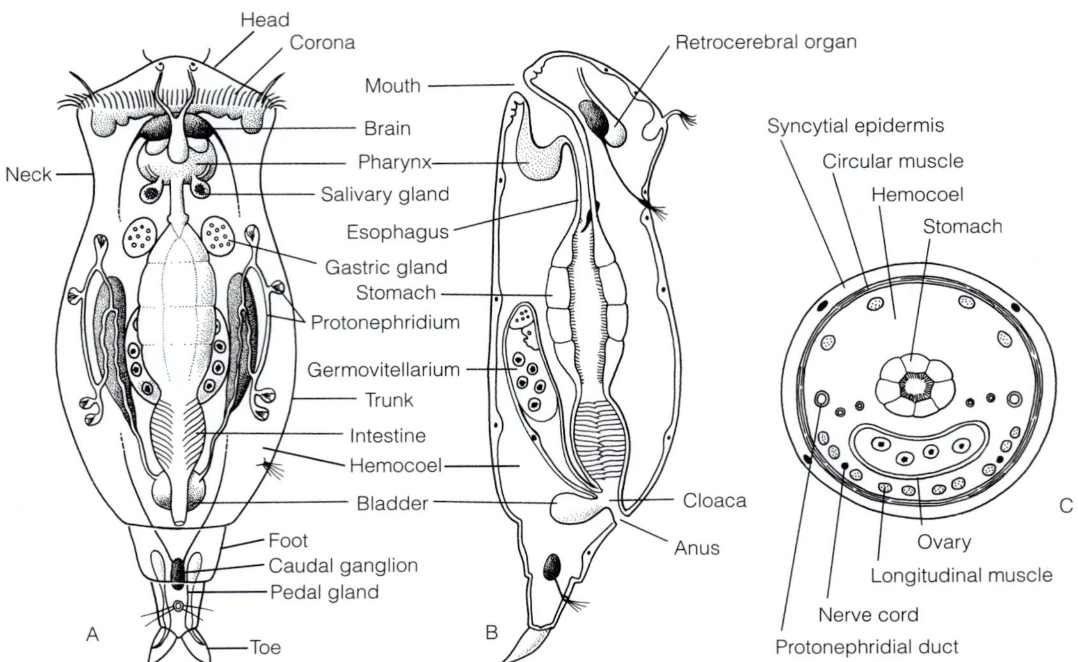

FIGURE 23-8 Rotifer anatomy. **A,** Dorsal view. **B,** Sagittal section. **C,** Cross section. *(A and B, Redrawn from Remane, A. 1929–1933. Rotatoria. In Bronn, H. G. (Ed.): Klassen Ordn Tierreichs 2:1–576; C, Redrawn from Beauchamp, P. de. 1965. Classe des Rotifères. In Grassé, P.-P. (Ed.): Traité de Zoologie. Vol. 4. Masson et Cie., Paris. p. 1235.)*

or ornamented in various ways. The body is frequently divided into a series of transverse rings that decrease in diameter anteriorly and posteriorly from the largest ring in the center of the trunk (Fig. 23-17A). Longitudinal muscles cause the rings of the foot to telescope into each other and into the trunk to shorten the body. A pair of retractor muscles pulls the head into the trunk.

HEAD AND NECK

A defining characteristic of rotifers is the ciliary **corona** (rotatory organ, wheel organ) on the head at the anterior end of the body (Fig. 23-8A). It is present in various forms in almost all rotifers. The corona is involved in locomotion and feeding, for which it is modified in many ways. It is a ring of ciliated cellular epidermis that encircles the head. In a generalized corona (Fig. 23-9), the ventral part of the ring is the **buccal field,** which surrounds the mouth. A dorsal **circumapical band** arches from the buccal field over the top of the head. The nonciliated anterior end of the head, encircled by the corona, is the **apical field.** Eyes, antennae, and the openings of the retrocerebral glands are situated in the apical field (Fig. 23-9). The coronal epithelial cells are multiciliated.

There are at least seven major variations on the basic coronal pattern, and any of the parts may be reduced, enhanced, modified, or absent in various rotifers. The entire corona is lost in a few, such as *Acyclus* and *Cupelopagis.* A row of large compound cilia (cirri) known as the **trochus** (pseudotrochus, preoral ciliary band; Fig. 23-9, 23-17A) may

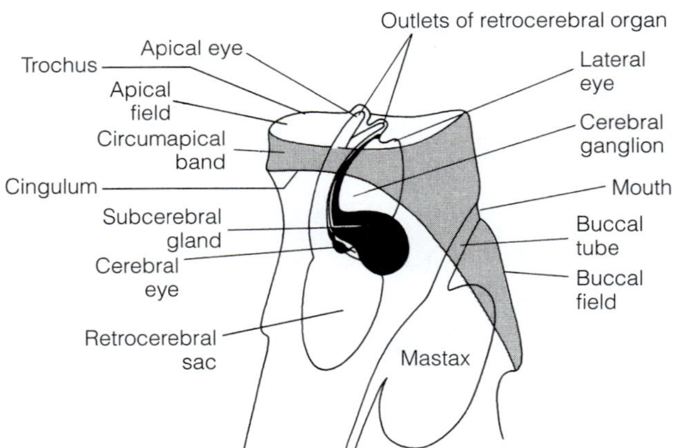

FIGURE 23-9 Primitive rotifer corona (shaded) in lateral view. The retrocerebral organ consists of the retrocerebral sac and subcerebral gland. *(After Beauchamp, 1907, from Hyman, 1951.)*

be present along the anterior border of the circumapical band and the posterior border may be delimited by another row of cirri known as the **cingulum** (postoral ciliary band). The trochus functions in suspension feeding by moving food particles to the mouth whereas the cingulum is the chief swimming organ. Cilia may be absent from the region between the trochus and cingulum. In males the mouth and trochus are absent.

In *Polyarthra* and related species (Fig. 23-5D,E) the corona consists entirely of the trochus and cingulum. The corona of bdelloid rotifers, which include many common terrestrial species, also have these two ciliated bands, but the trochus is raised on a pedestal and divided into two lateral rings called **trochal discs** (Fig. 23-17A). The cingulum passes around the base of the pedestal and runs beneath the mouth. The beating cirri of the trochal discs resemble two rotating wheels, from which the names *rotifer* and *wheel animalcule* are derived. The pedestals can be retracted when the discs are not in use.

The buccal field in *Collotheca* and related species is modified into a funnel with the mouth at the bottom, and ciliation is reduced (Fig. 23-6B). The edges of the buccal field expanded to form tentacle-like lobes bearing long retractile bristles that form a basket in which food is trapped.

The benthic rotifer *Dicranophorus* uses the cilia of the buccal field for creeping over the substratum. The raptorial, planktonic *Asplanchna* has a well-developed cingulum forming a nearly complete ring, but the trochus is reduced to two small ciliary tufts (Fig. 23-10B).

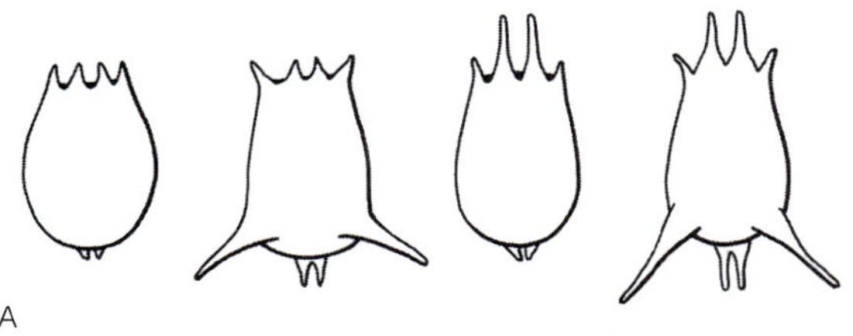

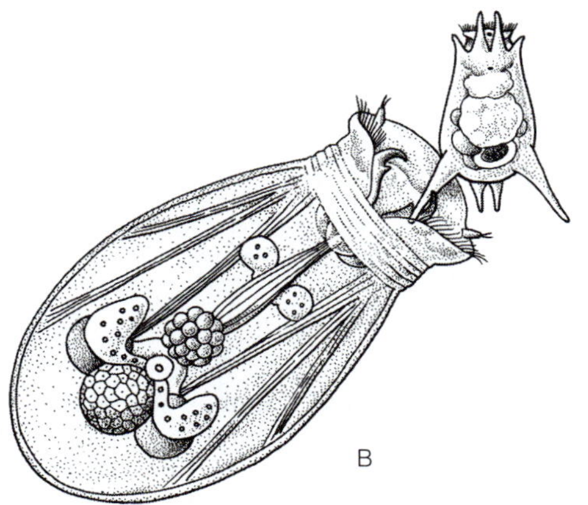

FIGURE 23-10 Cyclomorphosis in rotifers. **A,** Seasonal variability in spines of the lorica of the planktonic rotifer *Brachionus,* a favorite prey of *Asplanchna* **(B).** An unidentified kairomone released by *Asplanchna* induces the offspring of *Brachionus* to grow long defensive spines. The spines increase the body size of *Brachionus* and make it difficult for *Asplanchna* to swallow them. The *Asplanchna* in **B** is engaged in a futile attempt to ingest a long-spined *Brachionus.* Note the forcepslike incudate trophi typical of *Asplanchna.* The corona of the *Brachionus* is extended from the lorica but its foot is retracted. *(Modified and redrawn from Koste, W. 1978. Rotatoria. Gebrüder Bornträger, Berlin.)*

The head in bdelloid and some monogonont rotifers carries a mid-dorsal projection called the rostrum (Fig. 23-13). This little process bears cilia and sensory bristles at its tip and is also adhesive.

The **retrocerebral apparatus** of the head and neck consists of a retrocerebral sac and subcerebral gland adjacent to the brain (Fig. 23-9). It opens by pores in the apical field. Its function is uncertain, but it probably secretes mucus to lubricate the coronal cilia.

TRUNK AND FOOT

The trunk—the largest part of the body—contains most of the body cavity and viscera (Fig. 23-8). Its diameter usually is greater that that of the head or foot and its epidermis may be ringed.

The terminal portion of the body, the **foot,** is considerably narrower than the trunk (Fig. 23-5A, 23-8A,B). Its epidermis may be ringed (Fig. 23-17A). The foot is reduced or absent in many planktonic rotifers. The posterior end of the foot usually bears one to four projections called **toes** (Fig. 23-8A). Most rotifers, but especially the free-living benthic and terrestrial species, use the foot as an attachment organ. In these groups, the foot contains secretory **pedal glands** that open by ducts on the tips of the toes or elsewhere on the foot. The glands produce an adhesive substance for temporary attachment. Duo-gland organs like those of flatworms and gastrotrichs have not yet been found in rotifers.

Cyclomorphosis

Many pelagic rotifers undergo seasonal changes in body shape or proportions, a phenomenon known as cyclomorphosis (Fig. 23-10A). For example, during one season of the year individuals of some species have longer or shorter spines than those developed by their genetically identical (cloned) descendants during another season. In *Brachionus calyciflorus,* spines in subsequent generations can be induced by starvation, low temperature, and kairomones from the predaceous rotifer *Asplanchna.* The enhanced spines protect *Brachionus* from predation (Fig. 23-10B). Similar morphological changes are observed in cladocerans and some dinoflagellates.

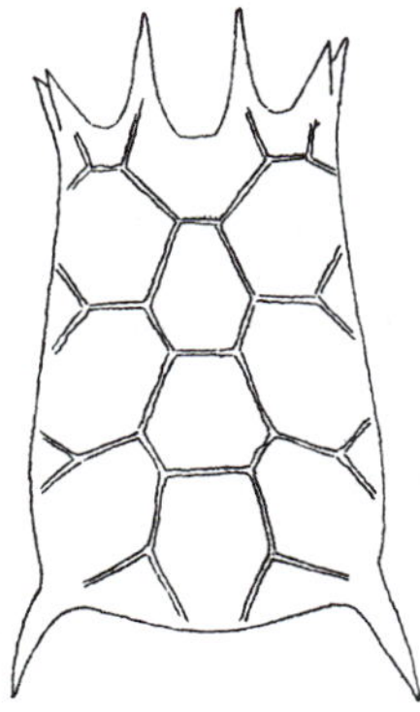

FIGURE 23-11 Lorica of the rotifer *Keratella quadrata. Keratella* has no foot. The head and corona, which are not covered by the lorica, are not included in this drawing. *(From Koste, W. 1978. Rotatoria. Gebrüder Bornträger, Berlin.)*

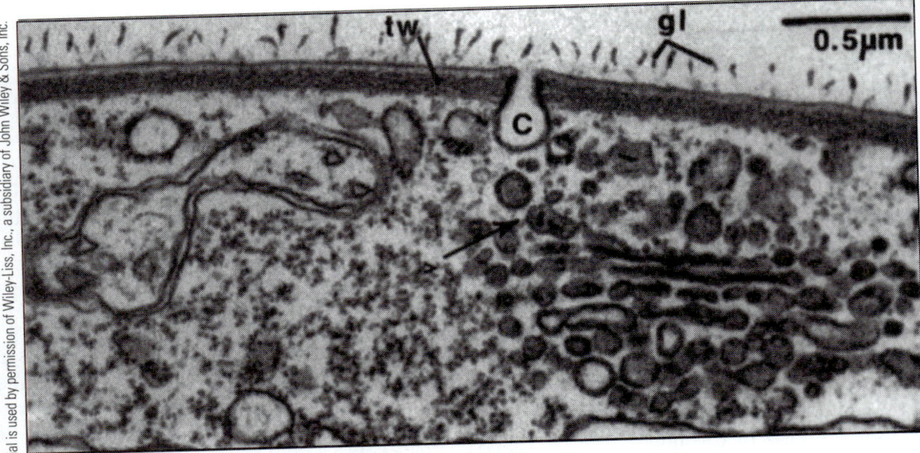

FIGURE 23-12 Rotifera. Electron micrograph of the syncytial epidermis of *Asplanchna sieboldi,* showing the intrasyncytial lamina (tw) and one surface pore (C). The lamina forms the skeleton of the body wall and is much thicker in loricate species such as *Keratella. Asplanchna* is illoricate and has a thin lamina. The true cuticle (gl) is visible outside the plasma membrane.

After Koehler from Ruppert, E. E. 1991. Introduction to the aschelminth phyla: A consideration of mesoderm, body cavities and cuticle. In Harrison, F. W., and Ruppert, E. E. (Eds.): Microscopic Anatomy of Invertebrates. Vol. 4. Aschelminthes. Wiley-Liss, New York, p. 14. Copyright © 1991 John Wiley & Sons, Inc. This material is used by permission of Wiley-Liss, Inc., a subsidiary of John Wiley & Sons, Inc.

Body Wall

Most of the rotifer epidermis is syncytial although that of the trochus and cingulum is cellular. The epidermis is thin and always eutelic (Fig. 23-8C). The epidermis is secretory and pores on its apical surface release its products to the outside (Fig. 23-12).

A true cuticle is an extracellular secretion of the epidermis that often functions as a skeleton for protection and muscle attachment. The rotifer cuticle, however, contains no collagen or chitin and is a thin, glycoproteinaceous layer of no skeletal significance (Fig. 23-12). As the functional equivalent of a cuticle, the rotifer epidermis produces a dense intrasyncytial lamina (terminal web) of keratin-like protein fibers immediately *inside* the plasma membrane of the epidermis. The lamina functions like a skeleton, providing protection and a site for muscle attachment, although it is on the inside, not the outside, of the plasma membrane.

In many rotifers, especially planktonic genera such as *Brachionus, Keratella,* and *Kellicottia* (Fig. 23-10A, 23-11, 23-7D) the lamina of the trunk is especially thick and forms a rigid skeleton known as the **lorica.** Such rotifers are said to be loricate and those without a lorica are illoricate. The head and foot of loricate species usually have a thin lamina and flexible epidermis and can be retracted into the trunk lorica for protection. Most of the so-called cuticular structures, such as spines, stiff body sections, and the lorica, are stiffened solely by the intrasyncytial lamina and not by the true cuticle. The lorica may be divided into distinct plates or ringlike sections and is usually ornamented with ridges, spines, or articulated appendages. The spines may be long, and in some rotifers they are articulated, equipped with muscles, and movable. In *Polyarthra* the appendages are 12 long, flat skipping paddles in four clusters of three each (Fig. 23-5D). The pharyngeal lining, extracellular jelly coats, and tubes of rotifers such as *Stephanoceros* and *Collotheca,* however, are true cuticles (Fig. 23-6A,B).

Beneath the epidermis is a well-developed body musculature (Fig. 23-8C) consisting of cross-striated, obliquely striated, and smooth fibers (Fig. 23-13). Although both circular and longitudinal fibers are present, they are individual muscles and are not in continuous sheets. Fibers extend from one unit of the epidermal skeleton to another and attach by tonofilaments that pass through the epidermis to hemidesmosomes on the intrasyncytial lamina. Contractions change the position of the skeletal unit with respect to adjacent units.

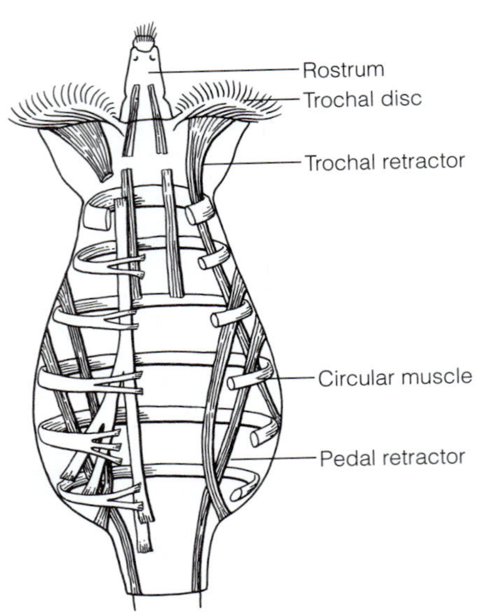

FIGURE 23-13 Rotifera: Musculature of *Rotaria* in ventral view. *(Redrawn after Brakenhoff, 1937, from Hyman, L. H. 1951. The Invertebrates. Vol. 3. McGraw-Hill Book Co., New York, p. 83. Reprinted with permission.)*

Hemocoel

A spacious, unlined hemocoel lies between the epithelia of the body wall and gut (Fig. 23-8C). It lacks structural connective tissue and consequently is filled with fluid (blood, hemolymph). It is a regulated compartment containing the gut, musculature, nervous system, protonephridia, and reproductive system. The hemolymph is a fluid-transport system mediating the distribution of food molecules from the gut to the tissues. The appropriate osmolarity is maintained by protonephridia. In *Asplanchna,* cilia associated with the protonephridia project into the hemocoel and circulate its fluid.

The hemocoel also is a hydrostatic skeleton. Contraction of longitudinal muscles shortens the body. A pair of longitudinal muscles inserts on the head and retracts the corona into the trunk hemocoel. Contraction of circular muscles pressurizes the hemolymph and lengthens the body, extending the corona and antagonizing the longitudinal muscles.

Locomotion

Rotifers move (Fig. 23-14) by creeping, ciliary swimming using the corona, or skipping through water using specialized appendages. Most aquatic rotifers inhabit the bottom sediments or live on submerged vegetation and other firm substrata. Some of these benthic species never swim, but many both creep and swim. In crawling, as exemplified by the common bdelloids, the corona is retracted when the animal creeps and the foot adheres to the substratum using the adhesive secretion of the pedal glands. The animal then extends the body, attaches the rostrum, detaches the foot to move it forward, and again grips the substratum with the foot. The result is an inchworm- or leechlike locomotion (Fig. 23-14C).

Most rotifers are benthic and intersperse periods of temporary attachment with swimming. During swimming, which for bdelloids is only for short distances, the corona is extended and the foot is retracted. Swimming speeds are typically 1 mm/s, but a few species, such as the planktonic monogononts *Polyarthra* and *Hexarthra,* skip rapidly forward at velocities of up to 35 mm/s using sudden flicks of their swimming appendages.

Planktonic rotifers swim continually and are adapted for maintaining a favorable position in the water column. Usually, the body is globose, the body wall thin and flexible, and the hemocoelic volume large (Fig. 23-5C,D,E). Oil droplets may be present in the hemocoel to decrease density. Long spines may be present to retard the rate of sinking, and the foot and toes may be absent. Among the many strictly pelagic rotifers are a few colonial species, such as *Conochilus,* whose members resemble trumpets radiating from the center of a communal gelatinous ball (Fig. 23-7E). The ciliary action of the combined coronae propels the colony through the water.

Many sessile rotifers attach to vegetation and display a marked substratum specificity, not only for a species of alga or plant, but also for the site of attachment on the plant. Many species of Flosculariacea are immobile and live in vaselike tubes typically composed of foreign particles embedded within a secreted material (Fig. 23-6A,B). Sessile species swim only as larvae, but as adults they attach permanently to the substratum.

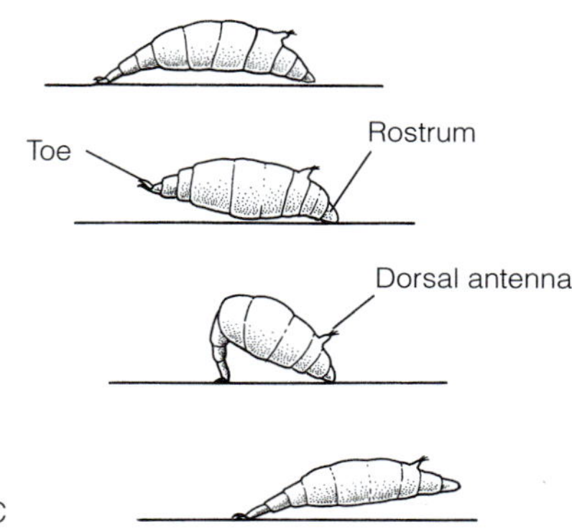

FIGURE 23-14 Rotifera. **A,** Ciliary swimming in *Brachionus pala* at approximately 1 mm/s. **B,** Skipping in *Polyarthra* using muscular movements of appendages. Each skip moves the animal 12 body lengths at a velocity of 35 mm/s. **C,** Inchwormlike locomotion in a bdelloid rotifer. Alternate attachments (anchors) are made with the toe and rostrum. The body-wall muscles act against the hydrostatic hemocoel to extend and contract the body. *(A, After Viaud, 1940, from Hyman, L. H. 1951. The Invertebrates. Vol. 3. McGraw-Hill Book Co., New York, p. 143; B, Greatly modified from Gilbert, J. J. 1985. Escape response of the rotifer* Polyarthra: *A high-speed cinematolographic analysis. Oecologia. 66:322–331. Reprinted with permission.)*

Nutrition

The rotifer mouth is anterior and ventral (Fig. 23-8B) and usually surrounded by part of the corona. It opens into the ciliated buccal tube or, in suspension feeders, directly into the pharynx (Fig. 23-15A). The pharynx contains a set of jaws known as the **mastax** (Fig. 23-15), a characteristic feature of rotifers. The pharynx is ventral, usually oval or elongate, and muscular. Its epithelium secretes a cuticular mastax similar to the jaw apparatuses of the Gnathostomulida and Micrognathozoa. The mastax is composed of a set of chitinous **trophi** connected by ligaments and operated by the mastax muscles. The trophi apparently are not a part of the intrasyncytial laminar skeletal system and instead appear to be derivatives of the true

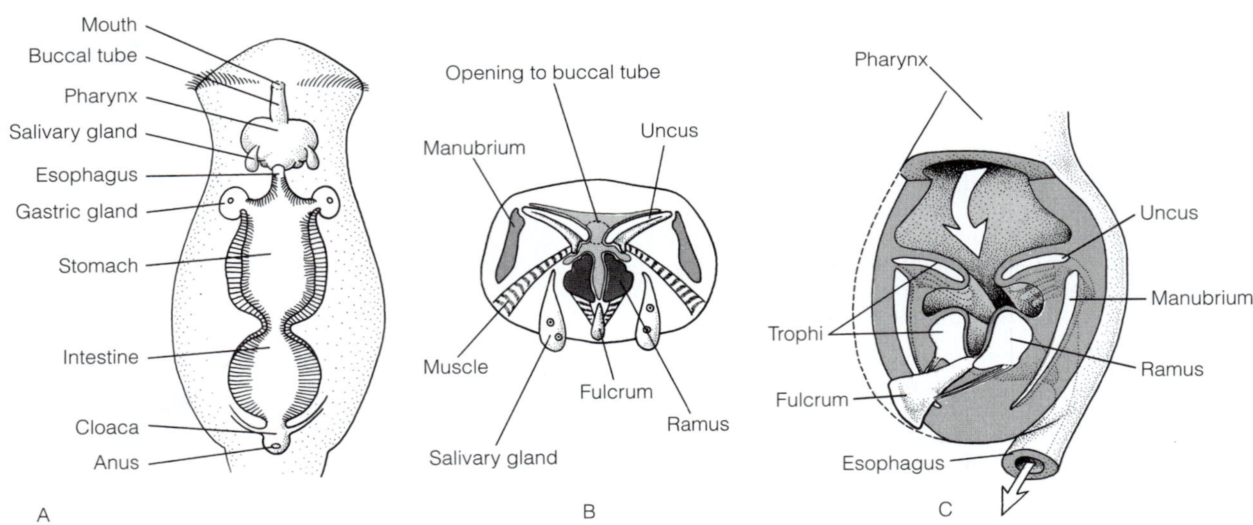

FIGURE 23-15 Rotifera. **A,** Anatomy of the digestive system in dorsal view. **B,** Enlargement of the pharynx and mastax in A. **C,** Three-dimensional view of mastax and trophi. *(A, Redrawn after Remane from Ruttner-Kolisko, A. 1974. Plankton Rotifers. In Elster, H.-J., and Ohle, W. (Eds.): Die Binnengewässer 26 (Suppl.):8.)*

cuticle. They are extracellular secretions of the ectodermal epithelium of the pharynx (Fig. 23-15B,C). The terms mastax and trophi are used inconsistently in the rotifer literature. In this text the mastax is the cuticular jaw apparatus and the trophi are the separate cuticular pieces of which the mastax is composed. The pharynx is the muscular region of the gut in which the mastax is secreted and housed.

The mastax is used for capturing and mechanically processing food and, as is typical of mouthparts, its structure varies considerably with taxon, depending on the type of food and feeding behavior. There are nine different types of mastax, two of which are illustrated in Figure 23-16. A typical mastax is

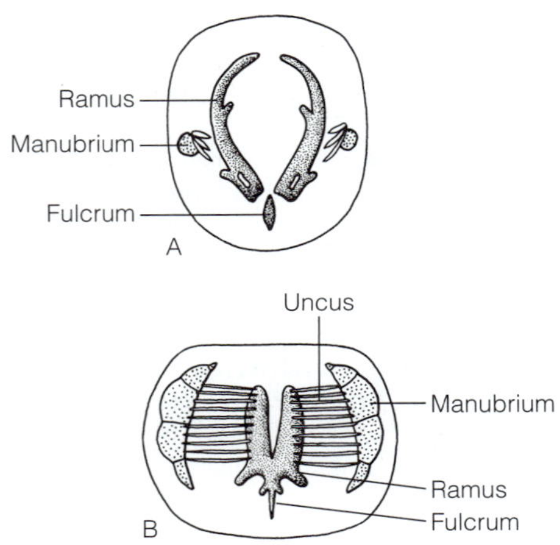

FIGURE 23-16 Rotifera. **A,** Incudate mastax of the carnivore *Asplanchna* used for seizing prey. **B,** Malleoramate mastax of *Filinia* used for grinding. *(Redrawn after Beauchamp, P. de. 1965. Classe des Rotifères. In Grassé, P.-P. (Ed.): Traité de Zoologie. Vol. 4. Masson et Cie, Paris. p. 1235. Reprinted with permission.)*

composed of seven trophi of four types. In the center are two **rami** (sing., ramus; Fig. 23-15B) held together by flexible ligaments and operated by muscles. The **fulcrum** is an unpaired, median trophus posterior to the rami. The **unci** (sing., uncus, or hook) are paired and usually bear teeth or grinding plates. The unci and rami are responsible for contacting and processing food items. The two **manubria** (sing., manubrium, or handle) are each attached to an uncus by a ligament. The trophi have dense cores and lucent sheaths, as do those of Gnathostomulida and Micrognathozoa.

Most rotifers are suspension feeders, raptors, or omnivores. The suspension feeders, of which the bdelloids are good examples, feed on minute organic particles brought to the mouth by the water current produced by the coronal cilia. In the bdelloids, which have trochal discs, the large cilia of the trochus produce the principal feeding current, which is directed backward. Food particles brought in by the water current are swept by the trochus and cingulum into a food groove that lies between them. Food-groove cilia then carry the particles to the mouth. The mastax of suspension feeders (Fig. 23-16B) is adapted for grinding. The unci are large, platelike, and ridged. The two plates oppose each other and the ridges form a grinding surface. The pharynx of suspension feeders probably also acts as a pump, sucking in particles that have collected at the mouth. Food intake can be regulated in various ways. In *Brachionus,* for example, the ciliated buccal field can be screened or uncovered by large coronal cirri, the buccal field's ciliary beat can be reversed, or the pharynx can reject particles.

Carnivorous species feed on larger food items such as protozoa, rotifers, and other small metazoans, and capture their prey either by trapping or pharyngeal suction. The forcepslike trophi of suction feeders are used to hold or manipulate prey in the pharynx cavity (Fig. 23-16A, 23-10B). *Asplanchna* uses its forcepslike rami to remove indigestible materials from the stomach and then egest them from the mouth.

Some rotifers, such as *Collotheca,* trap prey in a basket formed of long, retractable bristles that surround the mouth (Fig. 23-6B). The mechanism is similar to that used by the well-known insectivorous plant, the Venus flytrap, to capture insects. When small protozoa accidentally swim into the basket, the bristle-bearing lobes of the basket fold inward, preventing escape. The captured organisms are then sucked into the foregut by the pumping pharynx, which is called the proventriculus in this genus. The mastax of trapping rotifers is often very much reduced and apparently plays little or no role in food handling. Some rotifers, especially the notommatids, feed on the cell contents of filamentous algae. In the bdelloid *Henoceros* the trophi are extended from the mouth to puncture algal cells.

A pair of salivary glands located in the wall of the pharynx (Fig. 23-15A) secretes digestive enzymes into the pharyngeal lumen via ducts. A short, tubular esophagus connects the pharynx with the stomach. At the junction of the esophagus and stomach is a pair of syncytial, enzyme-secreting gastric glands (Fig. 23-15A), each of which opens by a pore into the gut. The digestive and absorptive stomach is a large sac or tube. It is cellular and ciliated in monogonont rotifers but syncytial in bdelloids. Although the lumen often is not visible with light microscopy, its existence has been demonstrated with the electron microscope. The stomach epithelium contains large vacuoles of uncertain function. A short, ciliated, microvillar syncytial intestine follows the stomach. The posterior end of the intestine is muscular, receives ducts from the protonephridia and gonad, and functions as a cloaca. The cloaca opens through the anus, which is situated dorsally near the junction between the trunk and foot (Fig. 23-8B, 23-15A). The intestine and anus are absent in the large predatory species of *Asplanchna,* in which the stomach is a blind sac. In some sessile tube-dwelling species, such as *Stephanoceros* and *Collotheca* (Fig. 23-6A,B), the anus has shifted anteriorly to allow egestion of wastes at the tube mouth to prevent fouling of the tube.

Epizoic rotifers live in obligate relationships with other metazoans and sometimes may be parasitic. The marine bdelloid *Zelinkiella* attaches to the gills of the polychaete worm *Amphitrite* and the tube feet of some brittle stars and sea cucumbers. The freshwater bdelloid *Embata parasitica* lives on the gills and pereopods of amphipods and the gills of crayfish. Several species of the monogonont genus *Proales* are commensals and parasites. *Proales werneckii* is parasitic *within* the filaments of the freshwater alga *Vaucheria,* which responds with gall-like swellings that enclose the invader. The rotifer feeds on algal cytoplasm and lays eggs in the gall. *Proales gigantea* bores into a snail egg, where it lives as it eats the embryo and lays its own eggs. *Ascomorphella* lives inside colonies of the green alga *Volvox,* where it consumes the cells of the host. The wormlike *Drilophaga* has a reduced corona and foot and lives attached to the surface of freshwater annelids. *Albertia* is parasitic in the intestine and coelom of earthworms.

Excretion

Rotifers are ammonotelic and nitrogen excretion occurs across the general body surface. Typically, two protonephridia are present in the hemocoel, one on each side of the body (Fig. 23-8A). Each has few to many multiciliated terminal cells (flame bulbs) that discharge into a collecting tubule. In some taxa the protonephridia are syncytial. The collecting tubules usually empty into a bladder, which opens into the ventral side of the cloaca. In the bdelloids, there is a constriction between the intestine and the cloaca, and the cloaca itself receives the ducts and functions as a bladder (Fig. 23-17A). The contents of the bladder or cloaca are emptied through the anus by contraction of their muscular walls. The protonephridia of freshwater rotifers function in osmoregulation and ionic balance. The excreted fluid is hyposmotic to the blood and the rate of bladder discharge is determined by the osmolarity of the environment. Its pulsation rate is about one to four times per minute.

Most terrestrial rotifers are associated with soil, leaf litter, mosses, and lichens and are active only during the brief time just after a rain when these surfaces are covered with a film of water. During this time the rotifers swim about in the film. They are capable of surviving desiccation, usually

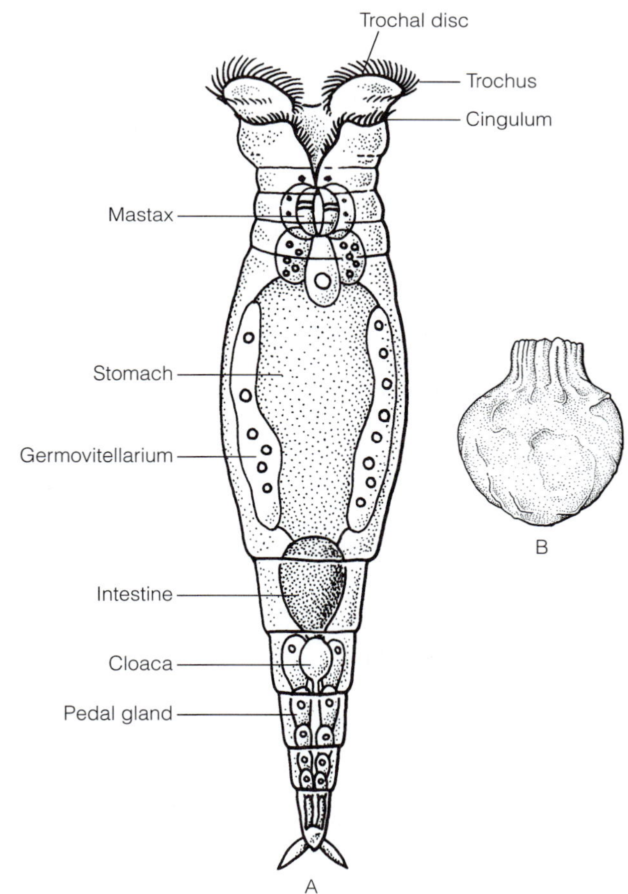

FIGURE 23-17 Cryptobiosis in rotifers. **A,** Active hydrated form of the bdelloid rotifer *Philodina roseola.* **B,** Cryptobiotic form of the rotifer *Habrotrocha rosa.* *(A, Redrawn after Hickernell from Pennak, R. W. 1953. Fresh-Water Invertebrates of the United States. Ronald Press, New York; B, Redrawn from Wallace, R. L., and Snell, T. W. 1991. Rotifera. In Thorp, J. H., and Covich, A. P. 1991. (Eds.): Ecology and Classification of North American Freshwater Invertebrates. Academic Press, New York. pp. 187–248.)*

without the formation of cysts, and can remain in a dormant, highly resistant state of cryptobiosis for as long as three to four years (Fig. 23-17A,B).

Cryptobiosis

Metazoans inhabiting mosses, lichens, temporary ponds, and damp soil regularly face the threat of dehydration and other physical extremes. Such animals often evolve reproductive adaptations such as resistant eggs, which allow their progeny to survive extreme conditions even though the adults perish. Although unusual, the ability to survive adverse conditions occurs in the adults of some species of rotifers, nematodes, and tardigrades.

As the environment deteriorates, the animals enter a death-like state of suspended animation called cryptobiosis ("hidden life"), or anhydrobiosis in reference to the loss of water, and are resurrected when conditions again become favorable. Cryptobiotic animals can remain dormant for months or years, or even a century, before reviving. Recovery may require 10 minutes to several hours. While dormant, they survive desiccation and temperature extremes that seem incompatible with life. There are records of cryptobiotic animals surviving temperatures ranging from 150°C (for a few minutes) to −200°C (for days). In one experiment, a rotifer and a tardigrade recovered after a brief exposure to near absolute zero (0.008° K).

To successfully enter cryptobiosis, the animal must desiccate slowly, usually over several days, as would be typical in natural habitats. As the animals dry, the body shape changes to decrease volume and minimize surface area, presumably to retard the rate of desiccation (Fig. 23-17B). Rotifers shrink to about 25% of their normal size. Many nematodes tightly coil their bodies, rotifers retract head and foot and adopt a spherical shape, and tardigrades contract into the shape of a barrel (Fig. 23.17B, 15-6E).

Cryptobiosis is truly a near-death state. Metabolism is almost immeasurably low, 0.01% of normal, or may be absent entirely, and the water content of the body decreases to less than 1%. In the absence of metabolism and in near total dehydration, cryptobiotic animals are essentially dead, but revivable. If metabolic processes actually cease, then somehow the structural order essential to life must be preserved in a sort of stable crystallized condition until water returns and life resumes. Two molecules, glycerol and the disaccharide trehalose, appear to be important in the formation and maintenance of the crystalline state. Both are synthesized as the animals slowly enter cryptobiosis. Glycerol is believed to protect the tissues from oxidation and to replace the water bound to biological macromolecules. Trehalose replaces water in membranes. In substituting for water, glycerol and trehalose act as spacers to preserve macromolecular organization during dehydration.

Nervous System and Sense Organs

The cellular, subepithelial CNS includes a cerebral ganglion (brain) dorsal to the pharynx (Fig. 23-18), a **mastax ganglion** ventral to the mastax, and a **caudal ganglion** (pedal ganglion) at the posterior end of the body. A pair of ventral nerve cords extends from the cerebral ganglion to the caudal ganglion (Fig. 23-8C). Nerves from the cerebral ganglion extend to the anterior sense organs and to other parts of the body. The cerebral ganglion is eutelic and consists of about 200 cells, depending on the species. The mastax ganglion innervates the muscles of the mastax and the caudal ganglion innervates the foot and cloaca. Most muscle cells are supplied by axons but a few muscles, such as the coronal retractors, send innervation processes to the brain.

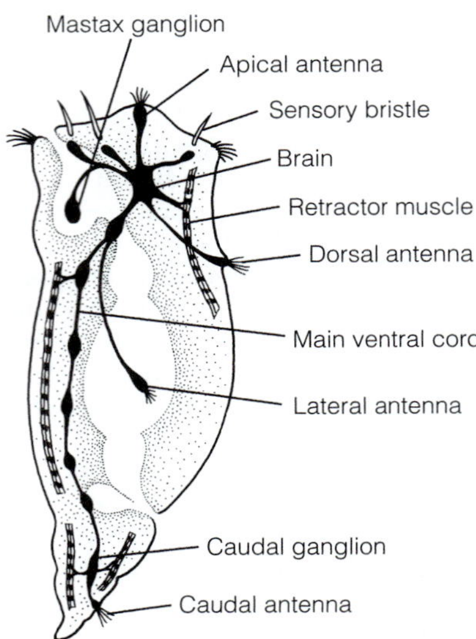

FIGURE 23-18 Rotifera. Nervous system and sensory structures (in lateral view) of a typical rotifer. *(Redrawn after Remane from Ruttner-Kolisko, A. 1974. Plankton Rotifers: Biology and Taxonomy. Binnengewässer 26 (Suppl.):10.)*

The sense organs, most of which are in the apical field, consist of mechanoreceptive bristles (Fig. 23-18) and antennae, chemoreceptive pores, and photoreceptive eyes (Fig. 23-9). The antennae consist of a small number of sensory cells and a few cilia. They may be apical, dorsal, lateral, or caudal. The eyes are simple pigment-cup ocelli composed of one or two photoreceptor cells plus an accessory cell containing red screening pigment. Eyes are close to or embedded in the cerebral ganglion. The eyes provide input for phototaxis and for photoperiodic control of life-cycle events.

Reproduction

ANATOMY

Rotifers are either gonochoric and bisexual or parthenogenetic females, mostly the latter. Among bisexual species, males are always dwarfs and those organs not required for reproduction are degenerate.

Parthenogenesis is characteristic of most rotifers (in contrast to the bisexual reproduction in Seisonida). Bdelloids have no males and reproduction is exclusively parthenogenetic. In monogononts males are briefly present in the population at restricted times and most reproduction is parthenogenetic. Such populations are exclusively female most of the year and even when males are present, they are rare. Even among the monogononts males are unknown for many species, although they are assumed to exist.

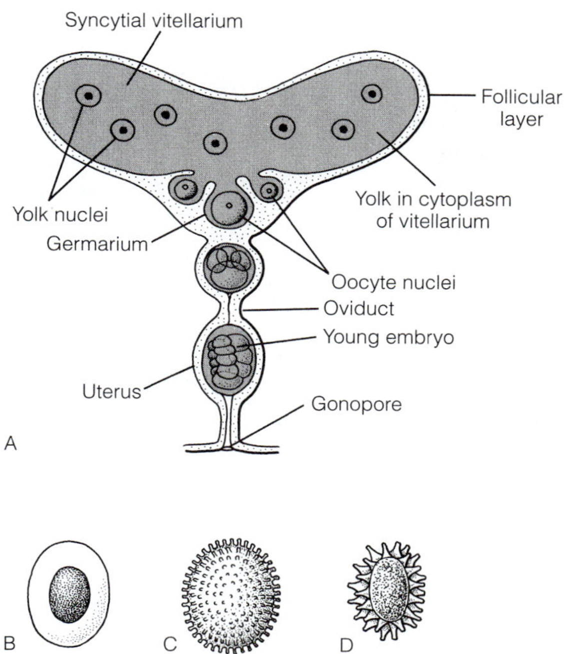

FIGURE 23-19 Rotifera. **A,** Germovitellarium of *Asplanchna brightwelli.* **B,** Rapidly developing amictic egg of *Ploeosoma hudsoni.* **C,** Resting egg of *Rhinoglena frontalis.* **D,** Resting egg of *Ploeosoma truncatum.* *(A, Modified and redrawn from Bentfield, 1971. B–D, Redrawn from Ruttner-Kolisko, A. 1974. Plankton Rotifers: Biology and Taxonomy. Binnengewässer 26 (Suppl.):59, 107.)*

The female reproductive system usually consists of one (as in monogononts) or two (as in bdelloids) syncytial germovitellaria in the hemocoel. Each germovitellarium consists of a thin, syncytial, follicular layer enclosing a small, inconspicuous germarium (ovary), where the oocyte nuclei originate, and a large, syncytial, yolk-producing vitellarium (Fig. 23-19A) with large polyploid nuclei. The follicular layer sometimes forms an oviduct leading to the cloaca. During egg formation, yolk from the vitellarium accumulates around an oocyte nucleus and the combination pinches off from the syncytium to form a mature egg, which passes through the oviduct into the cloaca (or to a genital pore if there is no intestine).

Female rotifers hatch with 8 to 20 oocyte nuclei which, since they are eutelic, is all they will ever have. Each oocyte nucleus will be incorporated into one egg. Thus each female will produce 8 to 20 eggs in her lifetime. Each egg is enclosed in a chitinous shell and a number of egg membranes, all of which are secreted by the egg itself. Benthic rotifers usually attach their eggs to the substratum. Planktonic rotifers, in a habitat with no obvious substratum, exhibit several adaptations to maintain their eggs in the water column rather than letting them sink to the bottom. Many brood the eggs by attaching them to their own bodies. A few rotifers, such as *Asplanchna* and *Rotaria,* are viviparous, with their eggs developing internally. Some attach their eggs to other rotifers.

Males, if present at all, are dwarf, aberrant, short-lived, and little more than a swimming testis and oversized penis (Fig. 23-20A). It has only what it needs to find and inseminate a female. The gut is vestigial or absent, but the testis and penis are large, accounting for about half of the body. The sensory system and locomotory corona are well developed (to find females). The single saclike testis connects with the exterior by a ciliated sperm duct. Because a gut and cloaca are absent in the male, the sperm duct runs directly to a gonopore homologous to the anus in the female and in the same dorsal position above the foot. Two or more glandular masses called accessory (prostate) glands are associated with the sperm duct. The end of the sperm duct is usually modified to form a penis. Rotifer sperm are unusual in having an anterior flagellum that pulls, rather than pushes, the sperm forward (Fig. 23-20C).

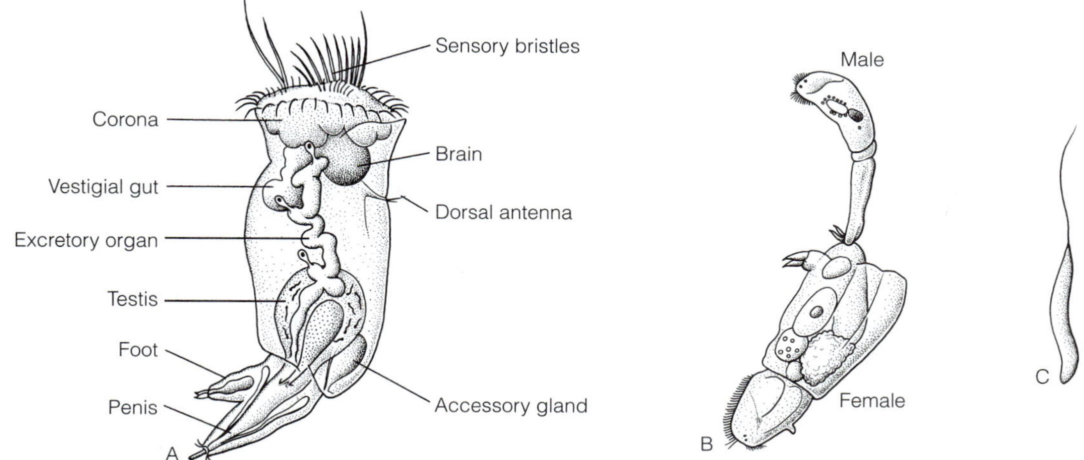

FIGURE 23-20 Rotifera. Dwarf male of *Brachionus calyciflorus* in lateral view. **B,** Cloacal copulation in *Cephalodella catellina.* **C,** Typical rotifer sperm with leading flagellum (at top). *(A, Redrawn after Beauchamp, from Ruttner-Kolisko, A. 1974. Plankton Rotifers: Biology and Taxonomy. Binnengewässer 26 (Suppl.):59, 107. B, Redrawn after Wulfert from Beauchamp, P. de. 1965. Classe des Rotifères. In Grassé, P.-P. (Ed.): Traité de Zoologie. Vol. 4. Masson et Cie, Paris. p. 1266.)*

Copulation is usually by hypodermic impregnation or, rarely, by insertion of the penis into the female's cloaca. In **hypodermic impregnation,** the male stabs any part of the female's body with his penis. Penetration is typically in a region where the epidermis is thin and flexible, such as the corona, and sperm are injected into the hemocoel. In some species copulation occurs within a few hours of hatching, when the female's epidermis is still relatively pliable. Sperm make their way to the germovitellarium to fertilize an oocyte. Male rotifers hatch with a species-specific number of sperm, about 30 in *Brachionus plicatus,* and use 2 or 3 with each mating.

LIFE CYCLES

Among rotifers, bdelloids have the simplest life cycle. Populations are composed exclusively of diploid females that produce only one type of egg. Males are absent and reproduction is entirely clonal (asexual) by parthenogenesis. Females produce diploid eggs by mitosis and the eggs hatch into more diploid females.

In monogonont rotifers, the life cycles are more complex, reproduction can be either bisexual or parthenogenetic, and several types of eggs can be produced (Fig. 23-19B–D). In the parthenogenetic, or amictic life cycle, an amictic diploid female (Fig. 23-21) produces an amictic egg (subitaneous egg or "summer" egg). It is thin-shelled, cannot be fertilized, and develops into an amictic female genetically identical to the mother. Typical meiosis does not take place in maturation and these eggs are diploid. **Amictic**, which means "without mixing," refers to the absence of genetic recombination (no meiosis or fertilization). Amictic eggs develop rapidly and hatch in 12 to 24 hours. This process is similar to the parthenogenetic reproduction of bdelloids.

A second type of monogonont egg, called a **mictic egg,** is also thin-shelled, but it is the product of meiosis and is haploid. If these eggs are not fertilized, they develop into haploid males with haploid sperm.

Mictic (haploid) eggs fertilized by sperm from one of the males secrete heavy, resistant, often sculptured shells (Fig. 23-19C,D). These fertilized eggs are **dormant eggs** (resting eggs, "winter" eggs). Dormant eggs are diploid and will eventually hatch into diploid females. In contrast to the thin-shelled, unfertilized amictic and mictic eggs, which hatch in about a day, dormant eggs contain a diapausing embryo capable of withstanding desiccation, extreme temperatures, and other adverse conditions and may not hatch for several months or even years. Dormant eggs over 20 years old have been hatched in the laboratory. At any one time, a female may produce amictic or mictic eggs, but not both, and the type of egg produced appears to be determined at the time of oocyte development.

The reproductive pattern of monogonont rotifers is cyclic and seasonal, but varies with species. In a typical cycle, the warmer temperatures of spring induce dormant winter eggs to break diapause, complete their development, and hatch into amictic females (Fig. 23-21). During spring and early summer these amictic females produce several generations of parthenogenetic females each having a life span of one to two weeks. Monogonont populations spend most of their time in this amictic phase of the life cycle. In the late spring or early summer, when this population reaches a maximum, some stimulus causes females to become mictic and produce haploid mictic eggs that, if they are not fertilized, hatch into haploid males. These males fertilize the mictic females and diploid dormant eggs are produced. These eggs carry the species through the winter until the next season and, if the pond or stream dries up, can be dispersed by birds or wind.

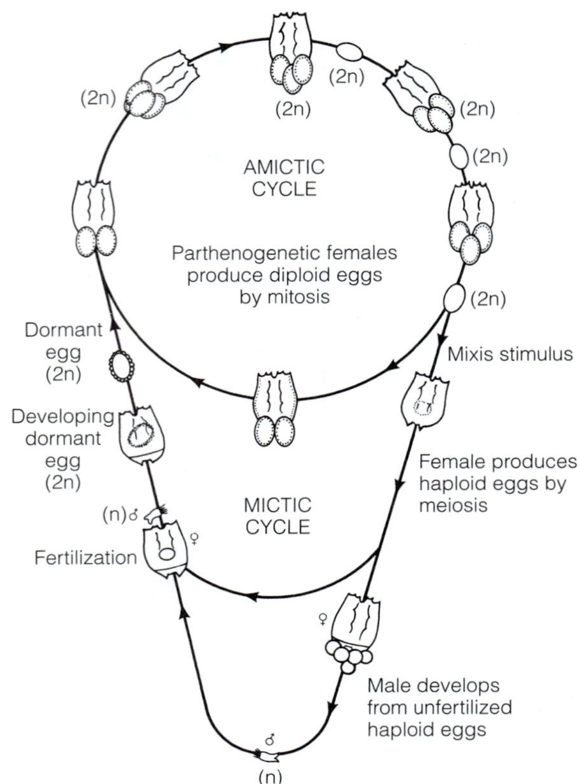

FIGURE 23-21 Life cycle of a monogonont rotifer. This particular species, *Brachionus leydigi,* lives in temporary ponds. Dormant eggs hatch with melting snows and spring rains to begin the first amictic cycle. In the summer, stagnating water stimulates the production of mictic eggs. Dormant eggs carry the species through the summer, when the pond dries. With autumn rains, there is a second amictic cycle. Frost stimulates the production of mictic eggs, and dormant eggs carry the species over the winter. *(Modified from Koste, W. 1978. Rotatoria. Gebrüder Bornträger, Berlin. p. 34.)*

Rotifers inhabiting large, permanent bodies of fresh water may display a number of cycles or population peaks during the warmer months or may be present during the entire year. The onset of the mictic cycle is induced by specific environmental factors, such as high population density or changes in photoperiod, temperature, or amount or type of food. The importance of these factors varies with species, as does the relationship between the two cycles and the seasons. Amictic eggs, although often called "summer" eggs, can be produced at any time of the year, as can resting eggs, depending on the particular adaptations of each species' life cycle. An alternative to the typical life cycle is described in the caption of Figure 23-21.

This process of alternating clonal with sexual reproduction is known as **heterogony** and in rotifers it is thought to be an adaptation for life in freshwater habitats with alternating periods of suitable and unsuitable environmental conditions. The amictic phase makes possible very rapid recolonization of a habitat such as a lake or pond following a period of adverse environmental conditions, such as winter or drought. Some species can double their population every two days. The sexual phase confers the advantages of sexual reproduction on the population and also provides a mechanism for surviving adverse conditions and for dispersal as resistant dormant eggs. The widespread distribution of rotifer species is due, in part, to the ease of overland dispersal of resting eggs. Cladocera, another important taxon in the freshwater zooplankton, are also heterogonous and parthenogenetic and have resistant eggs and widespread distributions.

Development

Rotifers exhibit modified spiral cleavage. Nuclear division is completed early in development and individuals hatch with the full complement of cells or nuclei characteristic of their species. Postembryonic growth is the result of increasing cell size or spacing, not the addition of new cells. Development is determinate and direct, with no true larval stage. Males are sexually mature when they leave the egg and do not undergo a growth period. Females of free-living species hatch with all adult features and the species-specific number of nuclei. They attain sexual maturity after a growth period of a few days. For example, the common bdelloid rotifer *Philodina roseola* has an average life span of 48 days, and the adult, despite having the same number of cells, is 28 times heavier than a newly hatched individual. A female lays 45 eggs and the generation time—the time between hatching of an individual and that of its first progeny—is four days.

Sessile rotifers hatch in a free-swimming stage sometimes referred to as a larva that is morphologically very similar to free-swimming species. After a short period it settles, attaches permanently to the substratum using adhesive secretions of the pedal glands, and develops the characteristics of the sessile adults.

Diversity of Rotifera

Monogononta[C]: Taxon of about 1600 species in 95 genera includes most rotifers. Reproduction is chiefly by parthenogenesis but sometimes is bisexual with dwarf, nonfeeding males. Have one germovitellarium, heterogonous life cycles. Gametes are produced by mitosis or meiosis. Stomach is cellular. Most are benthic, either free-living or attached, but approximately 100 freshwater species are planktonic.

Ploima[O]: Large taxon of 17 families contains the majority of rotifer species. Aquatic; benthic and planktonic (many). Lorica is present or absent, sometimes short or saclike. Usually have two toes. Corona lacks buccal field; circumapical band is not delimited by a trochus and cingulum. Important families include Asplanchnidae, Brachionidae, Synchaetidae, Notommatidae, and Trichoceridae, and genera include *Albertia, Asplanchna, Brachionus, Chromogaster, Dicranophorus, Drilophaga, Euchlanis, Gastropus, Keratella, Notommata, Polyarthra, Proales, Synchaeta.*

Flosculariacea[O]: Some are sessile with secreted tubes; others are planktonic and free-swimming. Many important planktonic rotifers belong to this order. Usually are toeless. Corona has a trochus and cingulum; mastax is malleoramate. The six families include Flosculariidae, Hexarthridae, Conochilidae, Testudinellidae. *Conochilus, Floscularia, Hexarthra, Testudinella, Trochosphaera.*

Collothecacea[O]: Most are sessile. Mouth is at the bottom of a shallow, flaring, anterior concavity. Anterior end often surrounded by arms with bundles of extensible bristles. Some taxa have a corona. Foot is enclosed in gelatinous envelope. Mastax is reduced, uncinate. Pedal glands are present, but toes are lacking. Collothecidae. *Acyclus, Collotheca, Cupelopagis, Stephanoceros.*

Bdelloidea[C]: Paired germovitellaria; males are absent; reproduction is always parthenogenetic. Benthic or terrestrial; about 350 species in 18 genera. Some inhabit the film of capillary water covering soil particles and mosses. Anterior end is retractile; usually have two trochal discs and a cingulum. Mastax is adapted for grinding with flattened uncini and a reduced fulcrum. Stomach is syncytial. Have two germovitellaria. Telescoping cylindrical body usually has 16 rings. Includes swimming and creeping species. Males are absent. Gametes (amictic eggs) are diploid, produced mitotically. Contains four families, including Philodinidae and Habrotrochidae. *Adineta, Embata, Henoceros, Philodina, Rotaria, Zelinkiella.*

SEISONIDA[C]

Seisonida contains only two species, which are sometimes considered to be rotifers. Both species are ectoparasites on the carapace of the primitive malacostracan crustacean *Nebalia.* They feed on debris and the eggs of the host. Both belong to the genus *Seison* (Fig. 23-22) and, although they

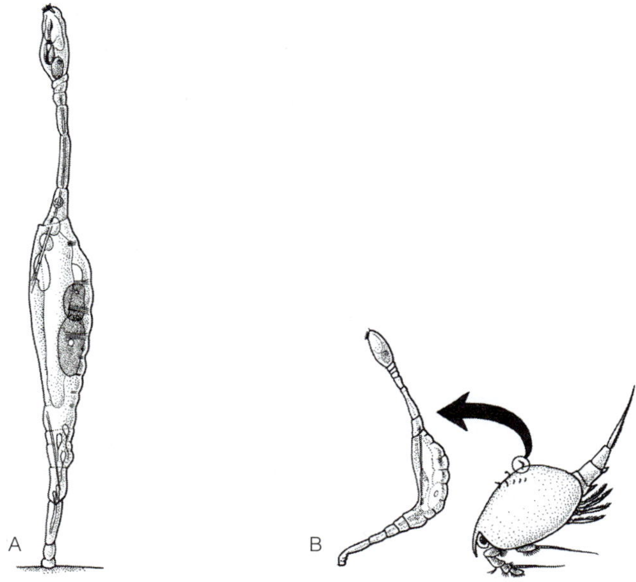

FIGURE 23-22 Seisonida. **A,** The seisonid *Seison.* **B,** The marine ectoparasite *Seison annulatus* attaches securely to the carapace of the crustacean *Nebalia* using a well-developed toe. *(A, Redrawn from Hyman, L. H. 1951. The Invertebrates. Vol. 3. McGraw-Hill Book Co., New York, p. 106. Reprinted with permission.)*

share many characteristics with rotifers, there are important differences. Seisonids, at about 3 mm, are very large in comparison with rotifers, which are usually less then 0.5 mm. The corona is reduced to a few clusters of bristles and, unlike that of rotifers, does not function in feeding or locomotion. The mastax is aberrant and difficult to compare with those of rotifers. It is fulcrate, a type unique to seisonids, and its trophi are difficult to homologize with those of rotifers. It apparently consists of a large fulcrum, two manubria, and several small trophi.

Unlike any rotifer, the gonochoric *Seison* has fully developed separate sexes and reproduction is exclusively bisexual. Males are a little smaller and less numerous than females. The ovaries lack vitellaria and there is no penis in the male. Only one type of egg is formed and it is produced sexually by fertilization. The female attaches the egg by a stalk to the gills of the host. There have been no studies of the development of *Seison*. The gonads are paired in both sexes and sperm are packaged in spermatophores. *Seison* moves with a leechlike looping motion using the anterior corona and posterior toe to attach to the substratum. Gonochorism, the paired gonads, absence of vitellarium and penis, near absence of sexual dimorphism, and the possession of only one type of egg are primitive characteristics that separate Seisonida from Rotifera.

ACANTHOCEPHALA[P]

Acanthocephala (spinyheaded worms; Fig. 23-23A) is a taxon of some 1150 species of derived, relatively large, parasitic, wormlike gnathiferans. The body cavity is a spacious hemocoel, there is no gut, and many tissues are syncytial and eutelic. Food is absorbed across the epidermis and distributed by a unique intraepidermal lacunar system. All are endoparasites and require two hosts to complete the life cycle. The larvae are parasitic in arthropods and the adults live in the digestive tract of vertebrates, especially fishes. The cylindrical adult body is elongate and composed of a trunk and a short, anterior, spiny, retractable proboscis (introvert) joined to the trunk by a neck.

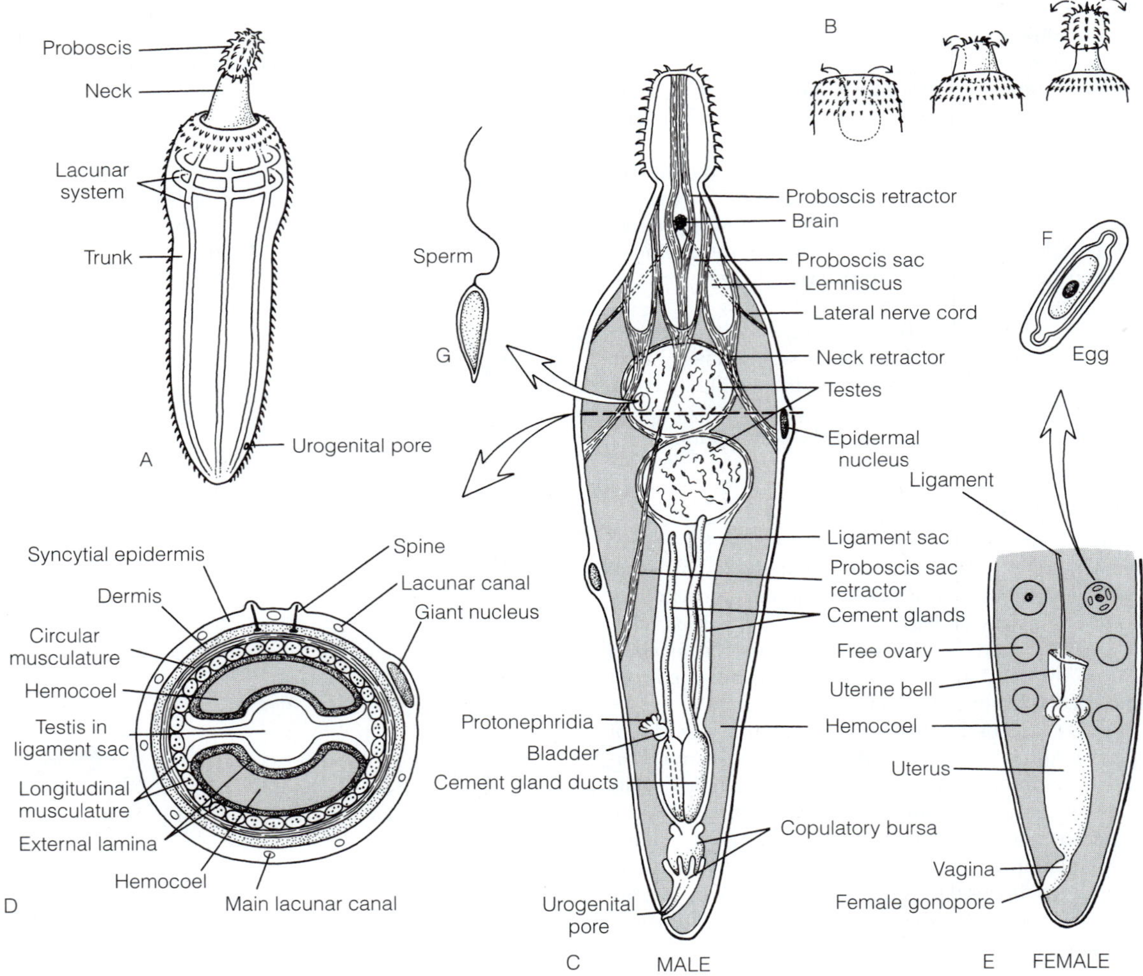

FIGURE 23-23 Acanthocephala. **A,** External view of a generalized acanthocephalan showing lacunar canals in the epidermis. **B,** Eversion of the proboscis. **C,** Internal anatomy of a male acanthocephalan in lateral view. **D,** Cross section through the trunk of a male. **E,** Posterior internal anatomy of a female in lateral view. **F,** Egg. **G,** Sperm with anterior flagellum.

The trunk is often covered with spines and in some is superficially divided into segments. Most are white, but some are red to brown. The length is usually 1 to 2 cm, although one species, *Macracanthorhynchus hirudinaceus,* the giant spiny-headed worm of pigs, may attain 80 cm. The proboscis is covered with distinctive recurved spines, hence the name "Acanthocephala" (= spiny head). Anatomical and molecular similarities indicate a close evolutionary relationship between acanthocephalans, rotifers, and seisonids (Fig. 23-26).

The body wall includes the epidermis, the connective tissue dermis, a circular muscle layer, and a longitudinal muscle layer. The epidermis is a thick, syncytial epithelium consisting of several layers. It is eutelic with a more or less constant and species-specific number of nuclei (6 to 20). Although few in number, the nuclei are very large (up to 5 mm in diameter!) and may produce conspicuous bulges in the body surface (Fig. 23-23C,D).

The syncytial epidermis secretes a thin external glycocalyx to the exterior but, as in rotifers, there is no true, structurally significant cuticle. Instead, the strength of the epidermis and body wall is provided by a thin, intrasyncytial lamina (terminal web) of protein filaments immediately inside the plasma membrane (Fig. 23-24). The plasma membrane is invaginated by numerous blind **crypts** whose necks penetrate the skeletal lamina and extend a short distance into the syncytium (Fig. 23-24). The crypts open to the surface at pores, are thought to provide additional surface area for food absorption, and may be sites for the formation of pinocytotic vesicles. Both roles are consistent with the uptake of food molecules across the integument. The abundance of crypts and their canals gives this outermost layer of cytoplasm a striated appearance. A thicker layer beneath the striated layer, the feltwork layer, consists primarily of smooth endoplasmic reticulum and fibers running in different directions. The innermost layer of the cytoplasm is the radial layer consisting of radial fibers that anchor the epidermis to its underlying basal lamina. The epidermis secretes a thick basal lamina that is continuous with the connective-tissue dermis from which arise the hooked spines that cover the proboscis (Fig. 23-23D).

The **lacunar system** is a unique fluid-transport system arising in the radial layer of the epidermis and extending into the body-wall musculature and lemnisci (Fig. 23-23A,D). Details vary with taxon, but the system typically consists of dorsal, ventral, and sometimes lateral longitudinal channels in the epidermis, ring canals also in the epidermis that connect the dorsal and ventral canals, extensive anastomosing canals around and in the hollow muscles, and canals in the lemnisci (which are described later). In electron micrographs, each canal is

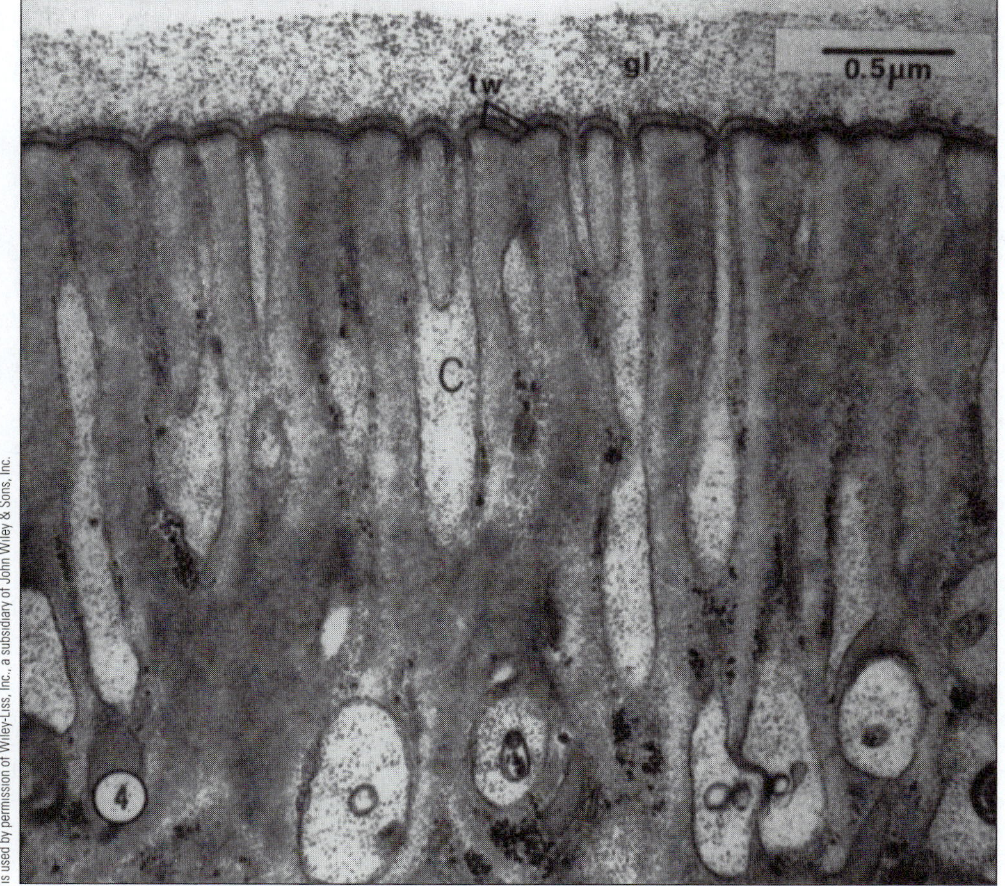

FIGURE 23-24
Acanthocephala. Electron micrograph of the outermost (striated) layer of the syncytial epidermis of the acanthocephalan *Moniliformis dubius.* This image shows only a small part of the total thickness of the epidermis. As in rotifers, the epidermis is supported apically by an intrasyncytial skeletal lamina (tw, terminal web) of proteinaceous filaments and bears a number of invaginations (crypts, C), which open via pores onto the surface of the body. The felt layer, radial layer, and lacunar canals are not visible in this photograph. Also labeled is the extracellular glycocalyx (gl), or cuticle, outside the epidermis.

revealed as simply an unlined, organelle-free region of the epidermal cytoplasm. Respiratory pigments are absent from this fluid of the canals. Contractions of the body-wall muscles probably provide the propulsive force for moving fluid in the system. In fact, this may be the chief function of the body-wall muscles. In these gutless animals nutrients are thought to be absorbed by diffusion or endocytosis from the crypts and transferred to the lacunar system for distribution to the muscles and hemolymph. The giant epidermal nuclei are situated deep in the cytoplasm beside the dorsal and ventral longitudinal canals, by which they are nourished. The lacunar system can function as a through-transport system because the syncytial organization of acanthocephalan tissues lacks intervening cell membranes.

Beneath the epidermis is a thick, combined basal lamina and connective-tissue dermis (Fig. 23-23D). The body-wall musculature, located just inside the dermis, consists of consecutive sheaths of circular and longitudinal smooth muscles (Fig. 23-23D). Both muscle layers are embedded in a thick, fibrous extracellular matrix that lines the body cavity. The muscle cells are hollow, containing large lumina that are continuous with the canals of the lacunar system. Locomotion is accomplished primarily by the proboscis, which is equipped with proboscis, proboscis sac, and neck retractor muscles.

The proboscis and neck can be retracted into a muscular **proboscis sac** in the anterior trunk (Fig. 23-23B,C). The retractable proboscis and anchoring hooks provide for locomotion and attachment in the gut of the definitive host and also enable the larva to migrate within the intermediate host. The proboscis is used chiefly to maintain position in the host's intestine in the face of peristaltic contractions and, when everted, the recurved hooks grasp the intestinal wall. As the proboscis is retracted, the spines withdraw from the tissue and release the worm. The proboscis and neck each have sets of longitudinal retractor muscles whose action is to invert the proboscis and neck into the proboscis sac. Contraction of the circular proboscis sac and body-wall muscles pressurizes the proboscis sac and hemocoel and everts the proboscis and neck.

Acanthocephalans have no functional gut, mouth, or anus. Food is absorbed across the integument and distributed by the lacunar system. A strand of endodermal cells known as the **ligament** extends the length of the hemocoel from neck to gonopore and is thought to be a vestigial gut. In most acanthocephalans the ligament is elaborated into one or two **ligament sacs** (Fig. 23-23D). These hollow, elongate sacs are suspended in the hemocoel from an anterior attachment on or near the proboscis sac to the posterior accessory reproductive organs. The wall of the ligament sac is acellular and composed of collagenous connective tissue. The sacs house the gonads and open posteriorly into the cavities of the accessory reproductive organs (Fig. 23-23C,D). The large, fluid-filled body cavity between the body wall and the endodermal ligament sacs has no epithelial lining and is a hemocoel (pseudocoel).

The **lemnisci** are two large invaginations lying on each side of the neck beside the proboscis sac (Fig. 23-23C). They extend into the trunk hemocoel and are filled with fluid with high concentrations of glycogen and lipids. Their function is unknown, but several roles have been suggested, including that of participating in the proboscis hydraulic system, nutrient (especially lipid) uptake, excretion, and fluid transport. Lemnisci are enclosed in the interior of the neck retractor muscles, which also compress them. They contain an extension of the lacunar system.

Excretory organs are unknown in most acanthocephalans, but two syncytial protonephridia, which open by a common excretory canal into the reproductive ducts, are present in one family (Fig. 23-23C). The 30 or more terminal cells are multiciliated flame bulbs located in the hemocoel.

The nervous system is composed of an anterior cerebral ganglion, paired genital ganglia, two lateral longitudinal nerves, and many small, specialized nerves to body-wall muscles and apical organs. The eutelic cerebral ganglion has 50 to 110 cells, depending on species. Its structure is typical of invertebrate cerebral ganglia, with an outer layer of cell bodies surrounding an inner neuropil of axons and synapses. It innervates the sense organs and muscles of the proboscis. In the absence of a gut and other reliable landmarks it is difficult to know if the brain is dorsal or ventral. Surprisingly, many specialists consider it to be ventral, which is in contrast with the dorsal brain of almost all bilaterally symmetrical animals, including the closely related rotifers. Two lateral longitudinal nerves exit the cerebral ganglion and extend to the body wall, where they divide into anterior and posterior longitudinal nerves. The posterior lateral nerves innervate the posterior body-wall muscles and extend to the genital ganglia. Little is known of the sensory system except that it is weakly developed, as it often is in parasites. An apical organ at the tip of the proboscis probably has secretory and sensory functions. A pair of lateral sense organs is located at the base of the proboscis. Other sensory structures are associated with the penis and bursa of males.

Acanthocephalans are gonochoric and the gonopore is located near the posterior end of the trunk. Fertilization is internal with copulation. The male system includes two testes, a copulatory bursa, a protrusible penis, cement glands, and other accessory structures (Fig. 23-23C). The testes are contained in and supported by the ligament sac and are connected with the copulatory bursa and penis by a sperm duct formed from the ligament sac. This arrangement is similar to that in male rotifers, in which the testis is supported by the vestigial gut. Sperm, like those of rotifers, have an anterior flagellum (Fig. 23-23G).

The penis is a digitiform evagination of the base of the copulatory bursa inside the bursa. During copulation, the bursa and then the penis are protruded through the male urogenital pore. The bursa turns inside out and engulfs the posterior end of the female, including her gonopore. Eversion of the bursa places the penis on the outside, where it can enter the female gonopore and discharge sperm into the uterus. The sperm then move to the female hemocoel and then to the ovary, where fertilization occurs. Following copulation, the male applies a temporary mucilaginous fertilization cap over the female gonopore. The cap is secreted by the cement glands.

In females, the ovary breaks up into fragments called **free ovaries** (ovarian balls), which are unattached in the hemocoel (Fig. 23-23E). The oocyte is fertilized while in the ovary, meiosis is completed, and a shelled zygote is released into the hemocoel. Development of the zygote takes place in the hemocoel and results in the formation of **acanthor larvae,** which are encapsulated in a shell and referred to as eggs, that are expelled. Intervening cell membranes regress and most embryonic tissues become syncytial early in development. Each acanthor has an anterior crown or rostellum with hooks. The eggs enter a funnel-shaped muscular structure called the **uterine bell,** which opens into the muscular uterus (Fig. 23-23E). A muscular sphincter known as the vagina regulates the release of eggs from the uterus. The female gonopore is subterminal, but because of the absence of landmarks, it is not known if it is dorsal, like that of rotifers, or ventral. The gonopore opens on the same side as the brain so, if the brain is dorsal, the gonopore is also.

"Eggs" (encapsulated larvae) are released from the gonopore into the intestine of the definitive host and are passed to the exterior with the host's feces. If the capsules are eaten by certain insects, such as roaches or beetle larvae, or by aquatic crustaceans, such as amphipods (Fig. 23-25), isopods, or ostracods, the acanthor larva emerges from the capsule, bores through the gut wall of the host, and becomes lodged in the intermediate host's hemocoel. The acanthor uses its hooked rostellum to penetrate the host's tissues. In the intermediate host, the acanthor passes through an **acanthella** stage and then becomes an encysted **cystacanth.** When the intermediate host is eaten by a fish, bird, or mammal, the cystacanth excysts and the juvenile worm attaches to the intestinal wall of the vertebrate using the spiny proboscis. The worm can release its hold on the intestine and change its position by retracting the proboscis.

Compared with tapeworms and roundworms, acanthocephalans are rare, but sometimes vertebrate hosts have heavy parasite loads that can do considerable damage to the intestinal wall. As many as 1000 acanthocephalans have been reported in the intestine of a duck and 1154 in the intestine of a seal.

Although all modern acanthocephalans are parasites, one extinct species, *Ottoia prolifica,* from the mid-Cambrian Burgess shale may have been a free-living predator or scavenger. The assignment of *Ottoia* to this phylum, however, is controversial and many paleontologists regard it as a fossil priapulid.

Acanthocephalans and rotifers share the syncytial epidermis with an intrasyncytial skeletal lamina. Both taxa have sperm with an anterior flagellum. Both have invaginations of the epidermal plasma membrane and a retractable anterior end. The testis supported by the vestigial gut is another synapomorphy. There is molecular evidence (18S rRNA nucleotide sequences) that the acanthocephalans are rotifers related to the bdelloids. Acanthocephala also has many characteristics in common with priapulids, but the intrasyncytial skeletal lamina is radically different from the true extracellular priapulid cuticle and the two taxa probably are not related. The loss of the gut is a derived character (autapomorphy) in the Acanthocephala.

Diversity of Acanthocephala

Archiacanthocephala[C]: Bird and mammal parasites; intermediate hosts are centipedes, millipedes, insects. Main lacunar canals are dorsal and ventral. Two ligament sacs, which break down in reproductive females. Protonephridia present in one family. *Macracanthorhynchus, Mediorhynchus, Moniliformis, Oligacanthorhynchus, Prosthenorchis.*

Eoacanthocephala[C]: Parasites of fishes, amphibians, reptiles; crustaceans are intermediate hosts. Main lacunar canals are dorsal and ventral. Two ligament sacs, which persist in reproductive females. *Neoechinorhynchus.*

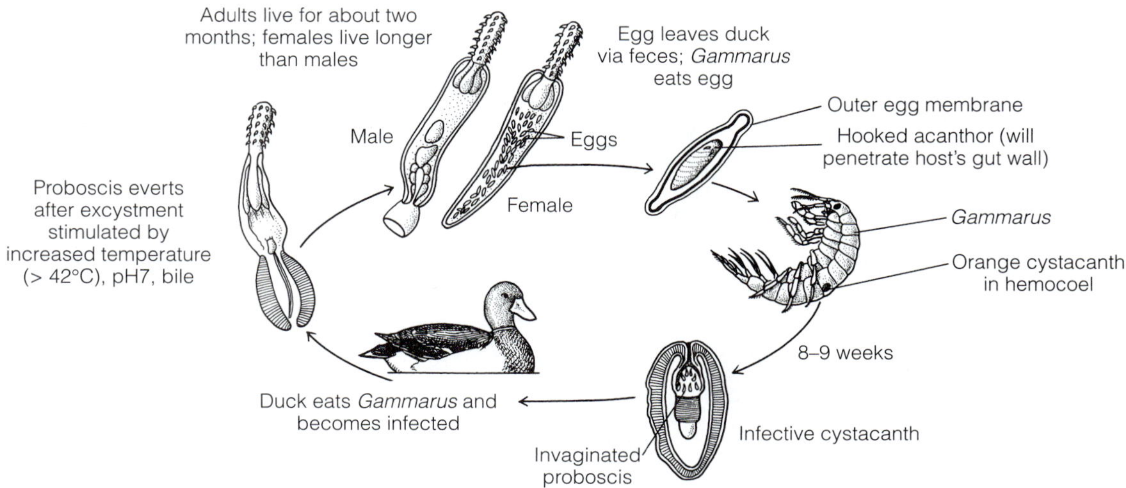

FIGURE 23-25 Acanthocephala. Life cycle of *Polymorphus minutus,* an acanthocephalan that parasitizes the gut of wildfowl and domestic ducks. The embryonated eggs of *Polymorphus* pass out of the duck with feces and enter the water. The eggs are then eaten by the amphipod *Gammarus.* An acanthor larva hatches from the egg in the shrimp's gut and bores through the gut wall into the hemocoel. After passing through an acanthella stage, the parasite becomes an encysted cystacanth, which is bright orange and can be seen through the transparent body wall of the amphipod. Ducks are reinfected when they eat the amphipods. *(From Lyons, K. M. 1978. The Biology of Helminth Parasites. Edward Arnold Publishers, Ltd., London. p. 7.)*

Palaeacanthocephala[C]: The largest, most diverse taxon. Parasites of all vertebrate classes; intermediate hosts usually are crustaceans. Main lacunar canals are lateral. One ligament sac, which breaks down in reproductive females. *Acanthocephalus, Echinorhynchus, Leptorhynchoides, Polymorphus.*

PHYLOGENY OF GNATHIFERA

Gnathifera includes Gnathostomulida, Micrognathozoa, Rotifera, Seisonida, and Acanthocephala (Fig. 23-26). The jaw apparatus, located in a muscular pharyngeal bulb ventral to the pharyngeal lumen and composed of tubular cuticular rods, is a chief synapomorphy uniting these taxa. Its trophi are secreted by pharyngeal epidermal cells, but there is no indication that they are molted in any of these taxa. Because acanthocephalans have no gut, it is impossible to compare their jaws (which they also don't have) with those of other gnathiferans.

The strongest evidence for the monophyly of Gnathifera is the skeleton. In Cycloneuralia and Panarthropoda the skeleton is a secreted, structurally important extracellular cuticle, but the thin cuticle of Gnathifera makes no structural contribution. Instead, Rotifera, Seisonida, and Acanthocephala have a syncytial epidermis with an intracytoplasmic (intracellular or intrasyncytial) skeletal lamina. This feature is a synapomorphy of these three taxa and they are accordingly grouped together in Syndermata within Gnathifera.

In contrast, Gnathostomulida has a cellular epidermis *without* a cytoplasmic skeleton and Micrognathozoa has a cellular epidermis *with* a cytoplasmic skeleton. Both taxa are thus excluded from Syndermata, but their relationship with each other is unclear. Micrognathozoa has characteristics of both Syndermata and Gnathostomulida, appears to be related to both, and supports a relationship between Gnathostomulida and Syndermata and the monophyly of Gnathifera.

PHYLOGENETIC HIERARCHY OF GNATHIFERA

Gnathifera
- Gnathostomulida
- N.N.
  - Micrognathozoa
  - Syndermata
    - Seisonida
    - N.N.
      - Rotifera
      - Acanthocephala

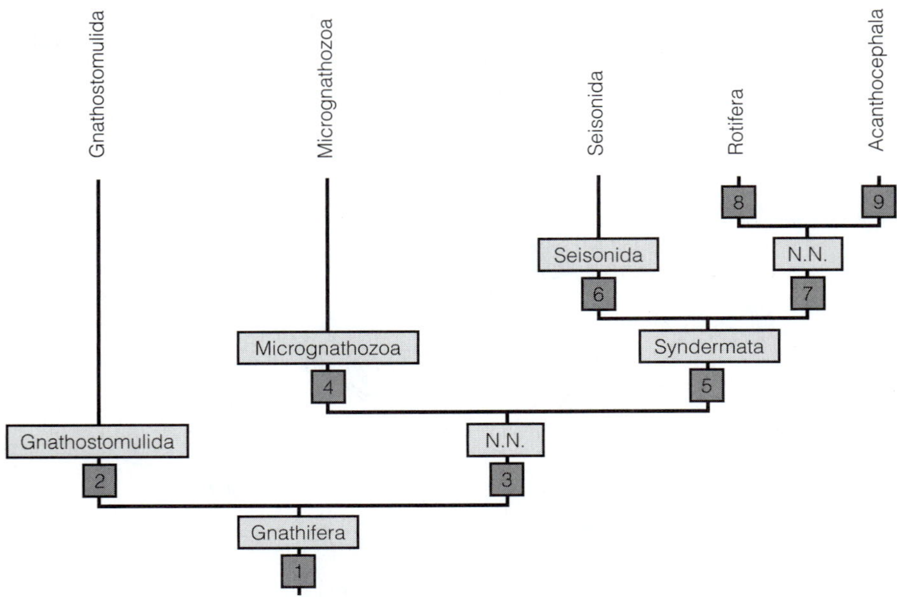

FIGURE 23-26 A phylogeny of Gnathifera. The position of Micrognathozoa is uncertain, as are relationships within Syndermata. Reasonable alternatives include Micrognathozoa as the sister taxon of Gnathostomulida and Acanthocephala as the sister taxon of Seisonida. **1, Gnathifera:** Cuticular pharyngeal jaws with an electron-lucent sheath and dense core. **2, Gnathostomulida:** Hermaphroditic, with a complex reproductive system. **3, N.N.:** Epidermis has an intracellular skeletal lamina. Locomotory cilia are absent on the dorsal and lateral trunk. **4, Micrognathozoa:** With ventral ciliophores, a cuticular oral plate. Epidermis is cellular, skeletal lamina is intracellular. Cuticular jaw apparatus is very complex, with three pairs of jaws and supports. **5, Syndermata:** Epidermis is syncytial, with intrasyncytial skeletal lamina. Epidermis is eutelic. Sperm has anteriorly inserted flagellum. **6, Seisonida:** Epizoic. Head ciliation consists of a few tufts of bristles. Jaw apparatus forms an anterior cuticular sheath. **7, N.N.:** Sperm have no acrosome. **8, Rotifera:** Head has a ciliary corona, unpaired retrocerebral glands. Female has a vitellarium. Parthenogenetic reproduction. **9, Acanthocephala:** Endoparasites with a two-host life cycle. Gut is lost. *(From Kristensen, R. M., and Funch, P. 2000. Micrognathozoa: A new class with complicated jaws like those of Rotifera and Gnathostomulida. J. Morphol. 246:1–49.)*

REFERENCES

GNATHOSTOMULIDA

Herlyn, H., and Ehlers, U. 1997. Ultrastructure and function of the pharynx of *Gnathostomula paradoxa* (Gnathostomulida). Zoomorphology 117:135–145.

Lammert, V. 1991. Gnathostomulida. In Harrison, F. W., and Ruppert, E. E. (Eds.): Microscopic Anatomy of Invertebrates. Vol. 4. Aschelminthes. Wiley-Liss, New York. pp. 19–39.

Rieger, R. M., and Tyler, S. 1995. Sister-group relationship of Gnathostomulida and Rotifer-Acanthocephala. Invert. Biol. 114:186–188.

Sorenson, M. V., Funch, P., Willerslev, E., Hansen, A. J., and Olesen, J. 2000. On the phylogeny of the Metazoa in light of Cycliophora and Micrognathozoa. Zool. Anz. 239: 297–318.

Sterrer, W. 1972. Systematics and evolution within the Gnathostomulida. Syst. Zool. 21:151–173.

Sterrer, W., and Farris, R. A. 1975. *Problognathia minima* n.g., n.sp., a representative of a new family of Gnathostomulida, Problognathiidae n. fam. from Bermuda. Trans. Am. Micros. Soc. 94:357–367.

MICROGNATHOZOA

Harrison, F. W. 2000. A note from the editor. J. Morphol. 246:50–52.

Kristensen, R. M., and Funch, P. 2000. Micrognathozoa: A new class with complicated jaws like those of Rotifera and Gnathostomulida. J. Morphol. 246:1–49.

Internet Site

www.zmuc.dk/InverWeb/Dyr/Limnognathia/Limno_intro_UK.htm (Micrognathozoa: A new microscopic animal from Greenland, by R. M. Kristensen, Zoological Museum, University of Copenhagen. Includes literature, information on ecology, morphology, and phylogeny, and a color reconstruction of the jaw apparatus of *Limnognathia*.)

SEISONIDA

Aldrich, W. H. 1993. On the protonephridia of *Seison annulatus* (Rotifera). Zoomorphology 113:245–251.

Aldrich, W. H. 1997. Epidermal ultrastructural of *Seison nebaliae* and *Seison annulatus*, and a comparison of epidermal structures within the Gnathifera. Zoomorphology 117:41–48.

Aldrich, W. H. 1998. Spermatogenesis and ultrastructure of the spermatozoa of *Seison nebaliae* (Syndermata). Zoomorphology 118:255–261.

Ricci, C., Melone, G., and Sotiga, C. 1993. Old and new data on Seisonidea (Rotifera). Hydrobiologia 255/256:495–511.

ROTIFERA

Beauchamp, P. de. 1965. Classe des Rotifères. In Grassé, P.-P. (Ed.): Traité de Zoologie. Vol. 4. Masson, Paris. p. 1235.

Bentfield, M. E. 1971. Studies of oögenesis in the rotifer *Asplanchna*. I. Fine structure of the female reproductive system. Z. Zellforsch. 115:165–183.

Brakenhof, H. 1937. Zur Morphologie der Bdelloidea. Zool. Jahb. Abt. Anat. 63.

Clément, P. 1993. The phylogeny of rotifers: Molecular, ultrastructural, and behavioural data. Hydrobiologia 255/256: 527–544.

Clément, P., and Wurdak, E. 1991. Rotifera. In Harrison, F. W., and Ruppert, E. E. (Eds.): Microscopic Anatomy of Invertebrates. Vol. 4. Aschelminthes. Wiley-Liss, New York. pp. 219–297.

Gilbert, J. J. 1985. Escape response of the rotifer *Polyarthra:* A high-speed cinematographic analysis. Oecologia 66: 322–331.

Koste, W. 1978. Rotatoria. Die Rädertiere Mitteleuropas. Gebrüder Bornträger, Berlin. (Two volumes.)

Pourriot, R. 1979. Soil rotifers. Rev. Ecol. Biol. Sol. 16:279–312.

Remane, A. 1929–33. Rotatoria. Klassen Ordn. Tierreichs 2: 1–576.

Rieger, R. M., and Tyler, S. 1995. Sister-group relationship of Gnathostomulida and Rotifer-Acanthocephala. Invert. Biol. 114:186–188.

Ruppert, E. E. 1991. Introduction to the aschelminth phyla: A consideration of mesoderm, body cavities and cuticle. In Harrison, F. W., and Ruppert, E. E. (Eds.): Microscopic Anatomy of Invertebrates. Vol. 4. Aschelminthes. Wiley-Liss, New York. pp. 1–17.

Ruttner-Kolisko, A. 1974. Plankton rotifers: Biology and taxonomy. Binnengewässer 26 (Suppl.):1–146.

Wallace, R. L. 1980. Ecology of sessile rotifers. Hydrobiologia 73:181–193.

Wallace, R. L., and Snell, T. W. 1991. Rotifera. In Thorp, J. H., and Covich, A. P. (Eds.): Ecology and Classification of North American Freshwater Invertebrates. Academic Press, San Diego. pp. 187–248.

Welch, D. B. M. 2000. Evidence from a protein coding gene that acanthocephalans are rotifers. Invert. Biol. 119:17–26.

Zrsavý, J., Mihulka, S., Kepka, P., Bezdek, A., and Tietz, D. 1998. Phylogeny of the Metazoa based on morphological and 18S ribosomal RNA evidence. Cladistics 14:249–285.

Internet Sites

www.microscopy-uk.org.uk/mag/wimsmall/extra/rotif.html (Gallery of Rotifers, by W. van Egmond. Excellent color images.)

www.gate.net/~hltcompr (Rotifer Study Methods, by H. L. Taylor. Offers access to print rotifer study procedures.)

http://dmc.utep.edu/rotifer/main.html (Rotifer Systematic Database, by E. Walsh, University of Texas, El Paso. A well-organized, comprehensive site with, among other features, an extensive database of images, classifications, and references for rotifer species.)

www.microscopy-uk.org.uk/mag/artnov99/rotih.html (Wonderfully Weird World of Rotifers, by R. L. Howey and W. van Egmond. Text and images.)

ACANTHOCEPHALA

Conway Morris, S., and Crompton, D. W. T. 1982. The origins and evolution of the Acanthocephala. Biol. Rev. 57:85–115.

Crompton, D. W. T., and Nichol, B. B. (Eds.): 1985. Biology of Acanthocephala. Cambridge University Press, New York. 500 pp.

Dunagan, T. T., and Miller, D. M. 1991. Acanthocephala. In Harrison, F. W., and Ruppert, E. E. (Eds.): Microscopic Anatomy of Invertebrates. Vol. 4. Aschelminthes. Wiley-Liss, New York. pp. 299–332.

Garey, J. R., Near, T. J., Nonnemacher, M. R., and Nadler, S. A. 1996. Molecular evidence for Acanthocephala as a sub-taxon of Rotifera. J. Mol. Evol. 43:287–292.

Gee, R. J. 1988. A morphological study of the nervous system of the praesoma of *Octospinifer malicentus* (Acanthocephala: Neochinorhynchidae). J. Morphol. 196:23–31.

Lyons, K. M. 1978. The Biology of Helminth Parasites. Edward Arnold Publishers, London. p. 7.

Nicholas, W. L. 1973. The biology of the Acanthocephala. Adv. Parasitol. 11:671–706.

Welch, D. B. M. 2000. Evidence from a protein-coding gene that acanthocephalans are rotifers. Invert. Biol. 119:17–26.

Internet Sites

www.biosci.ohio-state.edu/~parasite/acanthocephala.html (Graphic Images of Parasites, Ohio State University. Three light photomicrographs.)

www.inhs.uiuc.edu/chf/pub/surveyreports/nov-dec96/acanth.html (Acanthocephalans and Rotifers Provide Clues for the Study of Evolution of Animal Parasites, by T. J. Near, Center for Biodiversity, Illinois Natural History Survey. Text and a photograph of *Echinorhynchus* from rainbow trout.)

Kamptozoa[P] and Cycliophora[P]

Understanding the evolutionary relationships of Kamptozoa has always been a problem for zoologists and, although they are protostomes (trochozoans), kamptozoans have no unambiguous affinities with other taxa and their sister taxon has not been identified. Because they share some characteristics with the newly discovered taxon Cycliophora, which also has uncertain affinities, we have chosen to consider the two in the same chapter until more is known about both. They are small animals less than 1 mm long and exhibit most of the features predicted for animals with small bodies as discussed in Chapter 9, Introduction to Bilateria.

KAMPTOZOA^P

Kamptozoa (= Entoprocta) includes approximately 150 known, living species of small, sessile, mostly marine, mostly colonial animals. Kamptozoans are sometimes known as "nodders" because of their amusing habit of bobbing on the end of their stalk. The name comes from the Greek *kampte,* meaning bending, hence "nodders." Kamptozoa is a monophyletic trochozoan taxon with no obvious affinities with any other taxon, save possibly Cycliophora.

Except for a single freshwater genus, *Urnatella* (Fig. 24-1A), all kamptozoans are marine and live attached to firm substrata such as rocks, shells, pilings, or other animals. Many are epizoic or commensal on sponges, polychaetes, bryozoans, ascidians, and other marine invertebrates. The cosmopolitan,

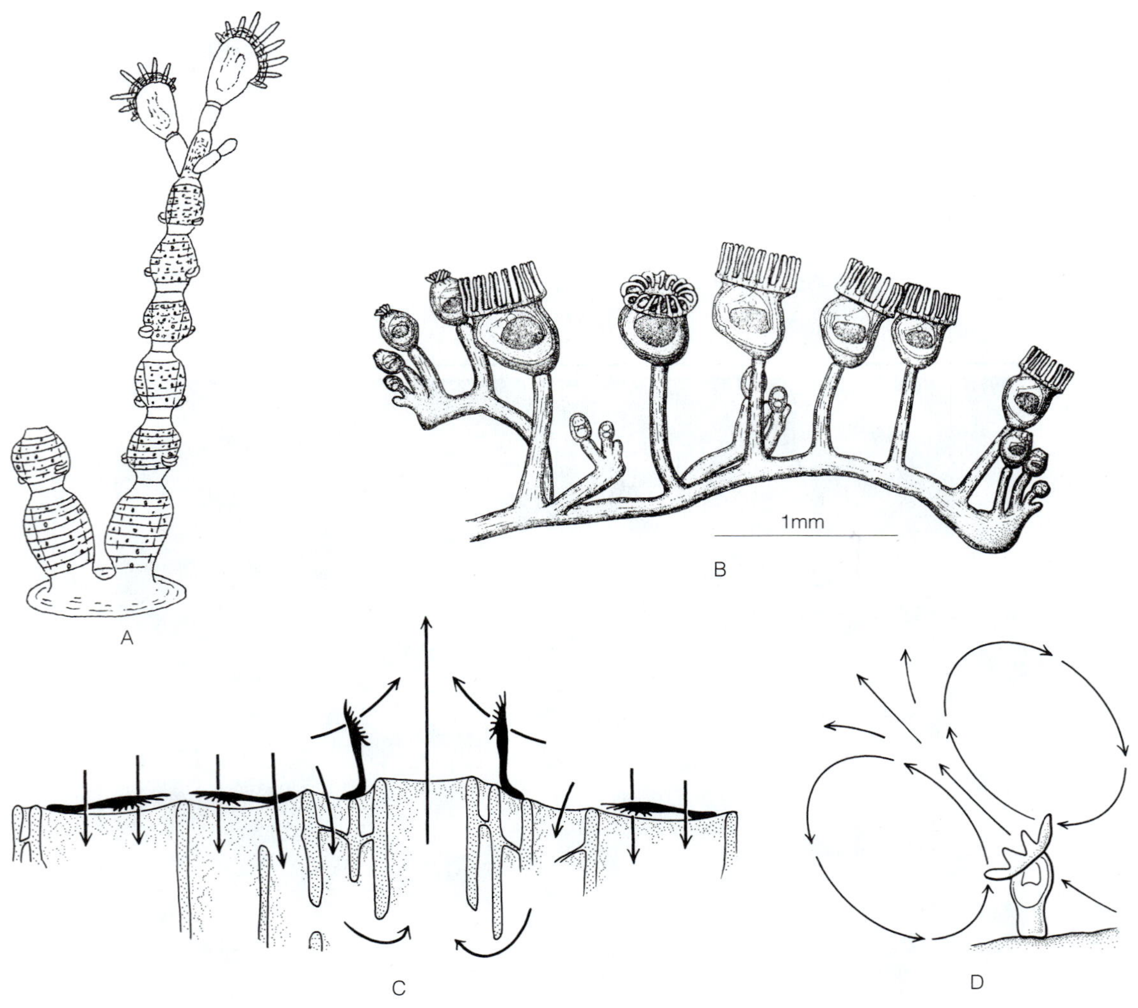

FIGURE 24-1 Kamptozoa. **A,** *Urnatella gracilis,* a freshwater species. **B,** Part of a colony of the marine *Pedicellina.* **C,** Diagram of five individuals of the commensal *Loxosomella vivipara* on a marine sponge. Although capable of generating their own feeding currents, these commensals profit from the water flow produced by the sponge (arrows). **D,** Water currents around a feeding zooid. *(A, Modified after Leidy, 1884, from Pennak, 1953. B and D, From Nielsen, C. 1964. Studies on Danish Entoprocta. Ophelia 1:1–76; C, From Nielsen, C. 1966. On the life-cycle of some Loxosomatidae (Entoprocta). Ophelia 3:221–247.)*

freshwater *Urnatella gracilis* can be found on almost any hard substratum, including sticks, docks, shells, plants, and rocks in streams or the littoral zone of lakes and reservoirs. Members of the commensal family Loxosomatidae are solitary whereas the other three families are colonial.

FORM

Kamptozoan colonies consist of interconnected zooids, which individually are about 1 mm long but range from 0.1 to 7 mm. The body (Fig. 24-2C) consists of a somewhat ovoid or boat-shaped **calyx,** which contains the viscera, and a **stalk** by which the calyx is attached to the substratum (Fig. 24-1A,B). The underside of the calyx is considered, on embryological grounds, to be the dorsal surface (proximal). The upper, ventral (distal) margin of the calyx bears an encircling **tentacular crown** of 8 to 30 solid tentacles, which are extensions of the body wall.

The space, or atrium, enclosed by the tentacles contains the mouth at the anterior end and the anus posteriorly (Fig. 24-3). Note that, in contrast with lophophorates, both mouth and anus are located within the ring of tentacles, hence the alternative and widely used name, Entoprocta, meaning "inside anus." The bases of the tentacles are united by a circumferential membrane that pulls partly over the crown when the tentacles contract and folds inward over the atrium.

There may be a single stalk, as in the solitary *Loxosomella* (Fig. 24-1C,D), or several stalks may originate from a common attachment disc (Fig. 24-1A) or stolon. In *Pedicellina,* numerous stalks arise from a horizontal, creeping stolon (Fig. 24-1B) or from upright, branching stems. In some, the stalk is separated from the calyx by a septumlike fold of the body wall and is commonly partitioned into short cylinders, or segments (Fig. 24-1A). The stalk is equipped with longitudinal muscles. In many species (such as *Barentsia*), certain segments are swollen with longitudinal muscle fibers that, on sudden contraction, produce the nodding motion of the calyx (Fig. 24-2C). Many solitary species of *Loxosoma* and *Loxosomella* can move about on their stalks, which have a sucker at the basal end, whereas others, such as *Loxosoma agile,* move by somersaulting (Fig. 24-2D).

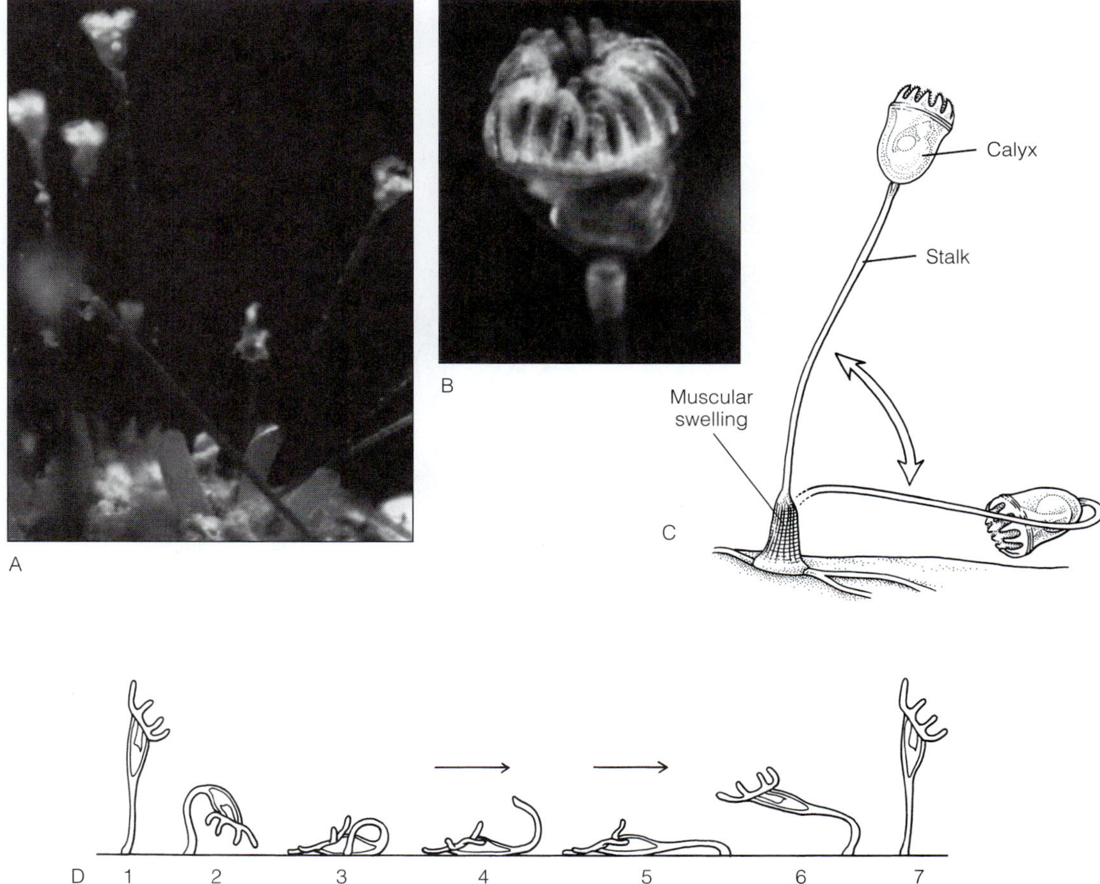

FIGURE 24-2 Kamptozoa. **A** and **B,** Living colony of *Barentsia laxa.* **A,** Complete zooids, showing the swollen bases. **B,** View of the calyx and tentacular crown. **C,** The muscular stalks and bases enable zooids to nod and twist. **D,** Somersaulting in *Loxosoma agile.* *(D, From Nielsen, C. 1964. Studies on Danish Entoprocta. Ophelia 1:1–7.)*

INTERNAL FORM AND FUNCTION

Body Wall and Hemocoel

The body wall consists of a thick or thin, secreted, extracellular, proteinaceous, fibrous cuticle and its underlying monolayered, microvillous epidermis. Epidermal microvilli penetrate the cuticle, which is thin over most of the body. Some areas may be thickened and some contain chitin. The ciliated epithelial cells of adult kamptozoans are multiciliated, as are those of bryozoans, although some larval cells are monociliated. The muscle layer is limited to longitudinal fibers along the inner wall of the tentacles, the tentacular membrane, some areas of the calyx, and the stalk. The epidermis is underlaid by a basement membrane, and the well-developed connective-tissue compartment fills the interior of the body. The body organization is compact (acoelomate) and the body cavity consists of small, interconnected, fluid-filled hemocoelic (pseudocoel, blastocoel) spaces filled with mesenchyme, extracellular matrix, and blood (hemolymph). There is no coelom, mesothelium, or peritoneum.

Nutrition

Kamptozoans are filter feeders that use the tentacular crown to capture suspended organic particles and small phytoplankton. Lateral cilia on the sides of the tentacles generate a feeding current that enters the atrium between the bases of the tentacles (Fig. 24-3, 24-1C,D). The current then curves apically and exits at the open end of the crown. Suspended food particles passing between the tentacles are trapped by the lateral cilia and transferred to frontal cilia on the inside face of the tentacles. The downward-beating frontal cilia transport the particles to the circumferential atrial groove at the base of the tentacles. Cilia in the groove move the particles to the mouth, where they are ingested. Note that because the frontal cilia are downstream of the lateral cilia and tentacles, the system is said to be a downstream collecting mechanism. This is the reverse of the upstream collecting systems of lophophorates (Fig. 25-25), sipunculans, crinoids, and pterobranchs. Unique **lime-twig glands** in the tentacular crown secrete sticky, entangling threads to capture large particles that might otherwise escape.

The gut is U-shaped (Fig. 24-3), with a mouth, esophagus, stomach, intestine, rectum, and anus. The large, bulbous stomach is the largest region and probably is the site of enzyme secretion and digestion. The gastrodermis is monolayered and multiciliated. Phagocytosis has not been demonstrated in any part of the gut. The stomach and intestine have abundant microvilli and are presumably the sites of absorption. The mouth and anus are equipped with circular sphincter muscles.

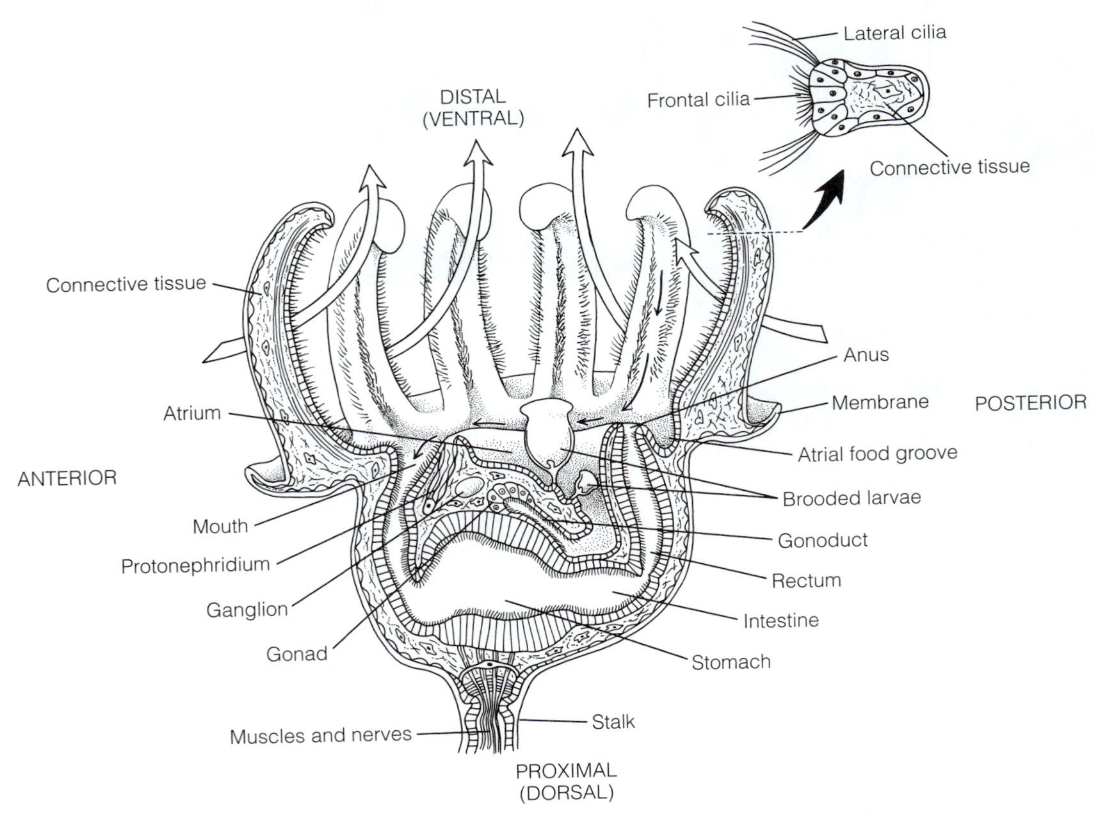

FIGURE 24-3 Kamptozoan anatomy. Open arrows show the water current generated by the lateral cilia. Solid arrows show path of food moving down the downstream frontal face of the tentacles and then along the atrial groove to the mouth.

Transport, Excretion, and Nervous System

The ciliated gut transports nutritive materials to within effective diffusion distance of all parts of the calyx of these small animals. Contractile cells in the **star-shaped organ** at the junction of the stalk and calyx in some species cause the blood to flow through the hemocoelic sinuses, perhaps to distribute nutrients from the calyx to the stalk. Gases exchanged across the general body surface are transported by diffusion.

In most kamptozoans two ectodermal protonephridia (Fig. 24-3), each consisting of two terminal cells, open into the atrium by nephridiopores beside the mouth. The terminal cells are multiciliated flame cells. The freshwater species *Urnatella gracilis,* which inhabits a hyposmotic environment, has multiple protonephridia in both the calyx and stalk.

The nervous system is subepidermal in the connective tissue. It consists of a median pair of large ganglia connected by a transverse commissure. The ganglia are between the stomach and the atrium, which is distal, or ventral, to the gut; this is an unusual location that could be the result of the loss of a supraesophageal ganglion and retention of the subesophageal. From the ganglia, nerves extend to the tentacles, stalk, sense organs, and calyx (Fig. 24-3). Sensory receptors are present on the tentacles, stalk, and calyx.

Reproduction and Development

Clonal reproduction by budding, common in all kamptozoans, is capable of forming large colonies. In most species, the buds arise from the stolon (Fig. 24-1B) or from the upright branches (Fig. 24-1A). In solitary species, the buds develop from the parental calyx, become separated, and then attach to the substratum as new individuals.

Kamptozoans may be gonochoric or, more often, hermaphroditic, and there are some protandric species, especially among the solitary loxosomatids. The one or two pairs of saclike gonads are located between the atrium and the stomach (Fig. 24-3), and the common gonoduct opens into the atrium. In some species the gonoduct is equipped with a **shell gland** that secretes an egg membrane. Fertilization is thought to occur internally, in the ovaries. Most species are oviparous and have planktotrophic larvae. Some, however, brood embryos in the atrium on a stalk secreted by the shell gland (Fig. 24-3). Brooders with small eggs may nourish the embryos via a placenta, whereas those with large eggs are lecithotrophic.

Cleavage is spiral and a ciliated, free-swimming larva hatches from the egg. Major developmental features are typically spiralian and mesoderm arises from the 4d mesentoblast cell. No coelom appears in any stage.

The larva is a trochozoan trochophore more or less like that of annelids and molluscs. It has a ciliated apical tuft at the anterior end, a ciliated frontal organ, a ciliated prototroch around the ventral margin of the body, a ciliated foot (Fig. 24-4A), a pair of protonephridia, a downstream collecting system, and a pair of eyes with a pigment cup and lens. Most are planktotrophic. After a short free-swimming existence (usually only a few hours), the larva settles to the bottom, where it creeps over the surface with the ciliated foot and eventually attaches with the frontal organ. Many larvae undergo a complex metamorphosis, including in extreme cases the 180° rotation of the future calyx to attain the inverted condition of the adult (Fig. 24-4B–E). In some species, the larva does not develop directly into an adult, but rather produces buds from which the adults are derived.

PHYLOGENY OF KAMPTOZOA

Early biologists believed kamptozoans to be bryozoans but, when it was discovered that they lacked a coelom, they were reclassified with the now defunct pseudocoelomate aschelminths. At present they are thought to be trochozoans, but some zoologists continue to support a bryozoan relationship and consider Kamptozoa and Bryozoa to be sister taxa. This relationship is supported by their common possession of multiciliated epithelia, in contrast to the monociliated epithelia of other lophophorates, and their similar larval eyes, which are unique to the two taxa. A comparison of 18S rRNA sequences indicates protostome affinities for both Kamptozoa and Bryozoa but does not support a close relationship between the two. Most zoologists believe that the similarities between Kamptozoa and Bryozoa are convergent adaptations for sessile filter-feeding by small animals. The differences between upstream and downstream collecting systems argue strongly that suspension feeding evolved independently in the two taxa. This conclusion is supported by the absence of a coelom, the location of the anus *within* the crown of tentacles, the presence of protonephridia, and spiral cleavage in Kamptozoa, none of which are true of bryozoans. Nodders share with other spiralians such synapomorphies as spiral cleavage, a trochophore larva, mesoderm formation from the 4d mesentoblast, and protonephridia, but their sister-taxon relationship remains, for the time being, uncertain.

DIVERSITY OF KAMPTOZOA

The 150 species of kamptozoans are currently placed in four families that align themselves into two higher taxa. Loxosomatidae and Loxokalypodidae are solitary and lack the star cell complex. The calyx and stalk are not divided by a septum and longitudinal muscles pass uninterrupted between the two. Pedicellinidae and Barentsiidae are united by possession of a star cell complex. The longitudinal muscles of the stalk and calyx are separated by a septum. These kamptozoans are colonial.

Loxosomatidae[F]**:** Most are commensal with marine invertebrates. Anus is elevated above the floor of the atrium by a tall anal cone. Protonephridia with separate nephridiopores are present. *Loxomespilon, Loxosoma, Loxosomella.*

Loxokalypodidae[F]**:** Monotypic family. Solitary. Musculature extends uninterrupted between the calyx and stalk. Star cell complex and septum are absent. *Loxokalypus.*

Pedicellinidae[F]**:** No anal cone; anus and mouth are on the floor of the atrium. Blood is circulated by contractile cells. Protonephridia share a common nephridiopore. Unsegmented stalk and its longitudinal muscles are continuous. Larvae undergo a complex metamorphosis after settling. *Myosoma, Pedicellina, Pedicellinopsis, Sangavella.*

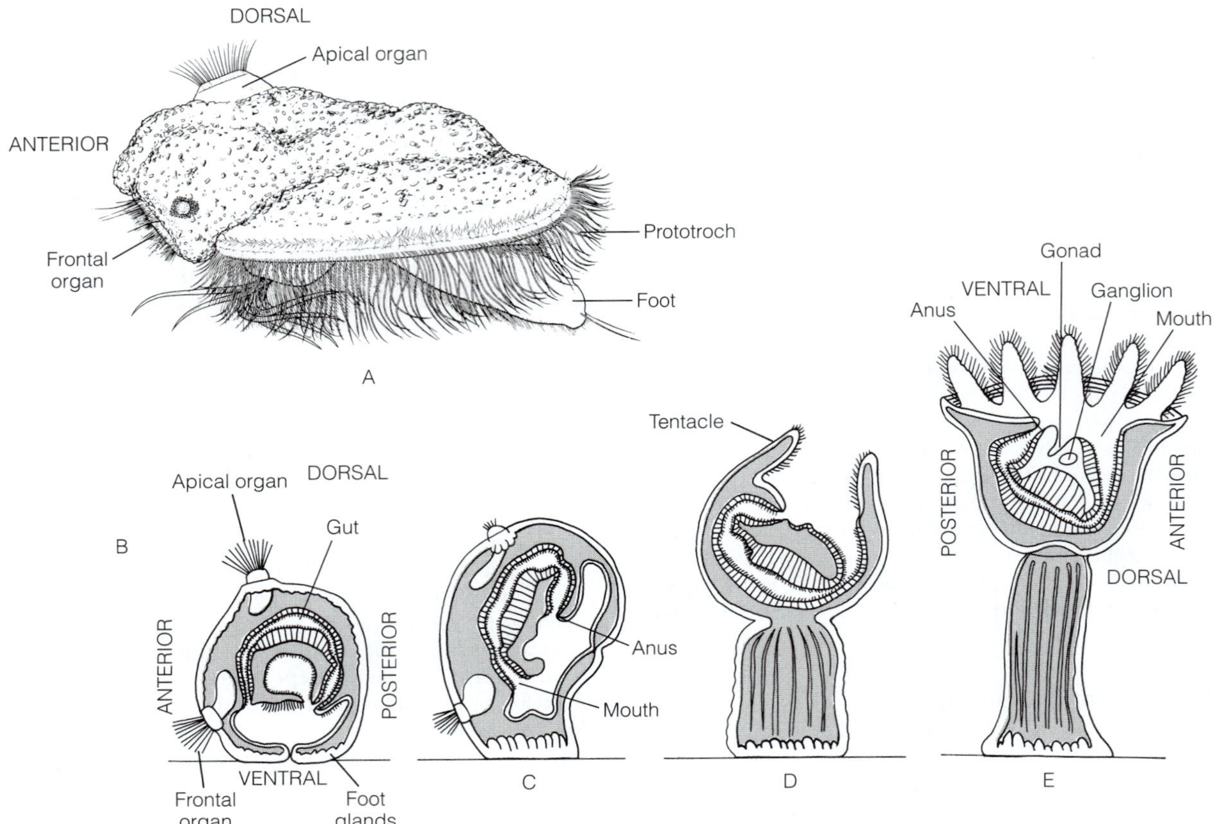

FIGURE 24-4 Kamptozoa. **A,** Larva of *Loxosomella harmeri.* **B–E,** Metamorphosis of *Pedicellina.* During metamorphosis the body rotates 180° and the dorsoventral axis is inverted, as are the mouth and anus. The mouth and anus in B face downward and the U of the gut arches upward. In C, the mouth and anus have rotated to the right (in this image), and in D and E they point upward, with the U of the gut arched downward. *(A, From Nielsen, C. 1971. Entoproct life-cycles and the entoproct-ectoproct relationship. Ophelia 9:209–341; B–E, Highly modified after Cori from Hyman, L. H. 1951. The Invertebrates. Vol. 3. McGraw-Hill Book Co., New York. 572 pp.)*

Barentsiidae[F]: Anal cone is lacking. Blood is circulated by contractile cells. Protonephridia have a common nephridiopore. Stalk is divided into segments. Proximal and distal segments are muscular, but intervening segments lack muscles. *Ascopodaria, Barentsia, Urnatella.*

CYCLIOPHORA[P]

Cycliophora, one of the most recently discovered major animal taxa, currently consists of a single described species, *Symbion pandora,* that is commensal with lobsters. Cycliophorans were first found in 1995 by P. Funch and R. M. Kristensen on the mouthparts of the lobster *Nephrops norvegicus,* collected in Denmark. Other species have been collected but have not yet been described. All are epizoic with crustaceans, including the American lobster, *Homarus,* and are tiny, compact (acoelomate), bilaterally symmetrical, microscopic protostomes with a superficial resemblance to rotifers.

The life history is complex, but essential to understanding cycliophoran biology. It includes sexual and asexual cycles, sessile and motile stages, feeding and nonfeeding stages, and three kinds of larvae. The asexual and sexual cycles are linked by the sessile, asexual **feeding stage** (Fig. 24-5), which is common to both. This is the dominant and central stage of the life cycle and is the subject of most of the following discussion. Of the many life-history stages, only this one has a gut and is capable of feeding. At almost 0.5 mm long, the feeding stage is the largest. It lives attached to setae on the mouthparts of its host. The other stages are motile and short-lived.

FORM

The body of the feeding-stage individual consists of a buccal funnel, a trunk, a short stalk, and an adhesive disc (Fig. 24-5). The **buccal funnel** resembles an inverted bell with the mouth at the large, open, anterior end, surrounded by a mouth ring of multiciliated epithelium. The cilia create a downstream suspension-feeding system. The buccal funnel is connected to

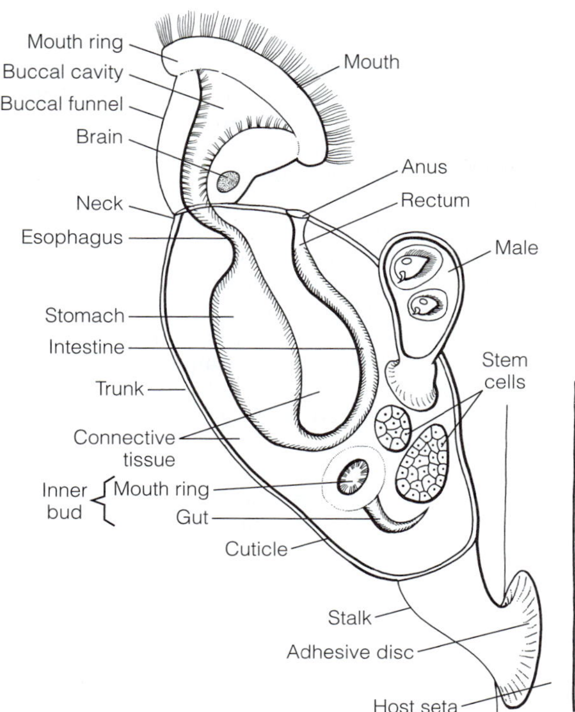

FIGURE 24-5 Cycliophora. Sessile feeding stage of *Symbion pandora* with an attached adult male, an inner bud, and clusters of undifferentiated stem cells. The developing but still incomplete mouth ring and gut of an inner bud are visible in the posterior trunk. The inner bud will replace the current digestive system of this individual. The undifferentiated stem cell masses will develop into a motile stage that could be a Pandora larva, Prometheus larva, or mature female, but that, in the case illustrated, will become a female. The attached male will mate with the female when she matures and emerges. The fertilized female will produce a chordoid larva, which will develop into a new feeding stage individual. Cells have been omitted for clarity. *(Redrawn from Funch, P., and Kristensen, R. M. 1997. Cycliophora. In Harrison, F. W., and Woollacott, R. M. (Eds.): Microscopic Anatomy of Invertebrates. Vol. 13. Lophophorates, Entoprocta, and Cycliophora. Wiley-Liss, New York. pp. 409–474.)*

the trunk by a narrow, flexible neck. The large ovoid trunk is joined to the basal **adhesive disc** by a narrow stalk. The disc attaches the feeding-stage individual to the host. The trunk contains the gut and various developing stages, both sexual and asexual.

The body wall of the feeding-stage individual consists of a cuticle, epidermis, and basal lamina. The epidermis is a monolayered, microvillous epithelium underlaid by a basal lamina. It may be nonciliated or multiciliated. An extracellular cuticle secreted by the epidermis covers the body. The cuticle includes a trilaminar epicuticle similar to that of cycloneuralians and a fibrillar procuticle. There is no chitin in the cuticle. The nonliving stalk and adhesive disc are extracellular secretions consisting entirely of cuticle. No continuous sheaths of longitudinal or circular muscles are present, although there are individual muscles.

The U-shaped gut begins at the anterior, flaring end of the buccal funnel with a large mouth surrounded by a mouth ring of nonciliated epitheliomuscular cells and multiciliated epidermal cells. The mouth opens into a conical buccal cavity that tapers into a narrow esophagus, which empties into the large stomach in the trunk. A long, tubular, ciliated intestine extends anteriorly from the stomach to the short, cuticle-lined rectum, which opens anteriorly at the anus on the midline at the base of the buccal funnel. With the exception of the rectum, which is epitheliomuscular, the gastrodermis is ciliated, microvillar, and without a cuticle. There is no gut or endoderm in any life-history stage other than the feeding stage. The short-lived motile stages subsist on food stored by the feeding stage.

The buccal funnel and digestive system of each feeding stage individual are continuously degraded and renewed by **inner buds** that develop into a succession of replacements for the gut. This **asexual inner budding** is not reproduction, for it produces only a new buccal funnel and gut for an existing feeding-stage individual and does not produce a new organism. It resembles the polypide regeneration characteristic of bryozoans.

The connective-tissue compartment between the epidermis and gastrodermis is filled with turgid connective-tissue cells and there is no fluid-filled body cavity. The body organization is compact. The mesenchymal cells contain large, gelatinous vacuoles and probably serve as a cellular hydrostatic skeleton and for storage of food molecules.

Most of the muscles are mesodermal myocytes that lie between the basal laminae of the epidermis and gastrodermis. A single myocyte encircles the esophagus and forms a sphincter at the junction of the buccal cavity and esophagus. Four obliquely striated muscles that extend from the anal area across the trunk to insert near the stalk move the buccal funnel. Some muscles of the male are cross-striated. Protonephridia with multiciliated terminal cells are present in the chordoid larva but have not been observed in other life-history stages.

The nervous system is not well understood, and the position of the brain with respect to the gut is uncertain. Indications of nervous tissue have been found in the base of the buccal funnel and also beside the esophagus. The strongest evidence suggests the former site as the location of the brain (Fig. 24-5), and this is consistent with its position—between mouth and anus—in kamptozoans. A pair of longitudinal nerve cords from this ganglion extends posteriorly through the trunk. All the swimming stages have a well-developed bilobed brain but, since they have no digestive system, they provide no clues regarding the position of the brain vis-à-vis the gut. No sense organs have been identified in the feeding stage, but all motile stages have numerous external sensory receptors.

REPRODUCTION AND LIFE CYCLES

The life history includes both sexual and asexual cycles linked by the sessile, nonswimming feeding stage. In addition to the inner bud, which only produces new digestive equipment, each feeding-stage individual contains masses of undifferentiated **stem cells** (Fig. 24-5) that, in different individuals, can differentiate into three different types of **motile stages.** These are the Pandora larva, the Prometheus larva, and the mature female, but each feeding-stage individual produces only one of them. In a feeding-stage individual, each motile stage develops within a cuticular brood pouch, where it is nourished via a pseudoplacenta. When its development is complete, it leaves the feeding stage.

Asexual Life Cycle

In the asexual life cycle, the stem cells of a feeding-stage individual produce an asexual **Pandora larva** with a buccal funnel and gut. Once mature, the Pandora larva, which is about 100 μm long, exits the feeding-stage individual and creeps about using a ventral ciliary field until it attaches and develops into a new feeding-stage individual, still on the original lobster. This larva was given its name by Funch and Kristensen, to whom its investigation was like opening Pandora's box. At first they thought there was only one motile stage, the chordoid larva, but soon they discovered the Pandora, then the male and female, and, most recently, a new male larva, the Prometheus, which has not yet been described.

Sexual Life Cycle

Cycliophorans are gonochoric and undergo sexual reproduction involving males and females produced asexually by different feeding-stage individuals. Stem cells in one feeding-stage individual produce a male larva (Prometheus larva) and stem cells in another individual produce a mature female.

The **Prometheus larva** is an immature male without reproductive organs. It emerges from its feeding-stage individual, quickly settles nearby, and degenerates. Some cells from the degenerating Prometheus larva develop into a functional mature male, which swims to a different feeding-stage individual, one that will produce a female, and attaches to its surface with a cuticular adhesive disc (Fig. 24-5). Here he awaits the eventual release of a mature female from the feeding-stage individual. The waiting male contains sperm and has a cuticular penis, but does not yet have a mate. He is about one-quarter of the length of the feeding-stage individual.

Meanwhile, a female with a single oocyte develops from stem cells inside the feeding-stage individual to which the male is attached. At maturity, the female resembles a Pandora larva but has an oocyte. Upon emerging from this feeding-stage individual, she is fertilized by the waiting male. The male uses his penis to achieve internal fertilization. The fertilized female, now with a zygote, swims away from the feeding-stage individual that produced her and to which the male is still attached. She settles and attaches nearby, still on the same lobster, and begins to degenerate.

A **chordoid larva** (Fig. 24-6) develops from the zygote and swims away from the remains of the degenerated female. The chordoid resembles a kamptozoan larva and is a primary larva considered to be a trochophore. It is the dispersive stage and will settle on a new lobster and metamorphose into a new feeding-stage individual. New cycliophorans in the feeding stage thus can be produced asexually by Pandora larvae or sexually by chordoid larvae.

A key feature of the chordoid larva is its longitudinal, ventral **chordoid organ.** This axial rod, so named for its notochord-like morphology and ultrastructure, is a column of about 50 disc-shaped muscle cells. The center of each cell is occupied by a large vacuole around which contractile filaments are arrayed in a peripheral circle. The rod is surrounded by a basal lamina. Longitudinal muscles insert on the organ and flex it by contracting. They are antagonized by the elasticity of the rod. Similar structures are present in urochordates, cephalochordates, vertebrates, and gastrotrichs. The chordoid larva has a pair of protonephridia and a bilobed brain. The secretions of its anterior ventral adhesive glands are used to attach the chordoid to the exoskeleton of the new host.

ECOLOGY

Symbion apparently is common on the lobster *Nephrops,* with 75% of 42 lobsters examined being infected. It was found attached only to the setae of the maxillae and maxillipeds, but thousands of feeding stages may be present on a single host. *Symbion* is a suspension feeder, and diatoms, bacteria, protozoa, and mineral particles have been found in the gut.

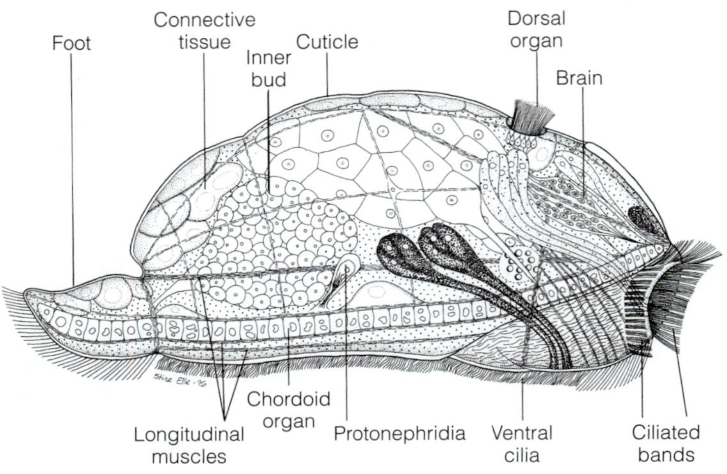

FIGURE 24-6 Cycliophora. Chordoid larva of Symbion pandora in lateral view, with anterior to the right. *(From Funch, P. 1996. The chordoid larva of* Symbion pandora *(Cycliophora) is a modified trochophore. J. Morphol. 230:231–263. Used with permission from Wiley-Liss, Inc., a subsidiary of John Wiley & Sons, Inc.)*

PHYLOGENY OF CYCLIOPHORA

Cycliophora was established as a distinct taxon on the basis of the sessile, asexual feeding stage with internal buds and gut regeneration: The three types of free-swimming larvae include the unique chordoid and motile males and females. Cycliophorans share morphological and developmental similarities with kamptozoans, bryozoans, cycloneuralians, rotifers, and even chordates, whereas studies of 18S ribosomal RNA nucleotide sequences suggest a relationship with rotifers and acanthocephalans. Bryozoans and cycliophorans share a U-shaped gut, suspension feeding, small size, polypide regeneration, and multiciliated epithelial cells, of which polypide regression is a possible synapomorphy. Like kamptozoans, cycliophorans have compact body organization, a downstream collecting system, and no indications of trimery; are epizoic; and have a U-shaped gut and multiciliated epithelial cells. Their strongest ties appear to be with Kamptozoa, with which they share the possible synapomorphies of epizoic commensalism and similarities in their larvae, such as the apical and frontal organs. It is possible that cycliophorans support the hypothesized link between Bryozoa and Kamptozoa discussed in Chapter 25, Lophophorata.

REFERENCES

KAMPTOZOA

Leidy, J. 1884. *Urnatella gracilis,* a freshwater polyzoan. J. Acad. Nat. Sci. Phila. 9:5–16.

Nielsen, C. 1964. Studies on Danish Entoprocta. Ophelia 1:1–76.

Nielsen, C. 1966. On the life-cycle of some Loxosomatidae (Entoprocta). Ophelia 3:221–247.

Nielsen, C. 1971. Entoproct life cycles and the entoproct/ectoproct relationship. Ophelia 9:209–341.

Nielsen, C. 1989. Entoprocts. Syn. Brit. Fauna N.S. 41:1–131.

Nielsen, C, and Jespersen, Å. 1997. Entoprocta. In Harrison, F. W., and Woollacott, R. M. (Eds.): Microscopic Anatomy of Invertebrates. Vol. 13. Lophophorates, Entoprocta, and Cycliophora. Wiley-Liss, New York. pp. 13–43.

Riisgård, H., Nielsen, C., and Larsen, P. S. 2000. Downstream collecting in ciliary suspension feeders: The catch-up principle. Mar. Ecol. Prog. Ser. 207:33–51.

Wasson, K. 1997. Sexual mode in the colonial kamptozoan genus *Barentsia.* Biol. Bull. 193:163–170.

Wood, T. 2001. Bryozoans. In Thorp, J. H., and Covich, A. P. (Eds.): Ecology and Classification of North American Freshwater Invertebrates. 2nd Edition. Academic Press, San Diego. pp. 505–525.

CYCLIOPHORA

Funch, P. 1996. The chordoid larva of *Symbion pandora* (Cycliophora) is a modified trochophore. J. Morphol. 230:231–263.

Funch, P., and Kristensen, R. M. 1995. Cycliophora is a new phylum with affinities to Entoprocta and Ectoprocta. Nature 378:711–714.

Funch, P., and Kristensen, R. M. 1997. Cycliophora. In Harrison, F. W., and Woollacott, R. M. (Eds.): Microscopic Anatomy of Invertebrates. Vol. 13. Lophophorates, Entoprocta, and Cycliophora. Wiley-Liss, New York. pp. 409–474.

Morris, S. C. 1995. A new phylum from the lobster's lips. Nature 378:661–662.

Sørensen, M. V., Funch, P., Willerslev, E., Hansen, A. J., and Olesen, J. 2000. On the phylogeny of the Metazoa in light of Cycliophora and Micrognathozoa. Zool. Anz. 239:297–318.

Winnepenninckx, B. M. H., Backejiau, T., and Kristensen, R. M. 1995. Relations of the new phylum Cycliophora. Nature 393:636–638.

Internet Site

www.microscopy-uk.org.uk/mag/articles/pandora.html (A Lobster's Microscopic Friend: *Symbion pandora*—a new life form and a new phylum, by D. Walker, Micscape Magazine, Microscopy UK. A general account of the biology of *Symbion.*)

25 Lophophorata[SP]

Three taxa of aquatic coelomate bilaterians, Phoronida, Brachiopoda, and Bryozoa, are traditionally classified together in Lophophorata (Fig. 25-34, 25-35). The elongate and wormlike phoronids resemble slender sea anemones or featherduster worms, the bivalved brachiopods look like clams, and the colonial bryozoans can resemble seaweeds, colonial sea squirts, or miniature corals.

Lophophorates are sessile suspension feeders enclosed in a secreted exoskeleton, shell, or tube with a single aperture. Because all communication with the external environment (feeding, defecation, excretion, spawning) is restricted to this aperture, the mouth, anus, nephridiopores, and gonopores are at the apertural (anterior) end of the body, close to the opening of the enclosing tube or exoskeleton. The gut is bent into a U, bringing the anus into proximity with the mouth at the apertural end (Fig. 25-25A). The anus is dorsal to the mouth. One or more pairs of metanephridia or coelomoducts may be present and, if so, their pores are also anterior.

Lophophorates feed using a **lophophore,** or crown of hollow, ciliated tentacles encircling the mouth (Fig. 25-2). The dorsal anus lies *outside* the ring of tentacles, but nearby. The lophophore is an **upstream collecting system** for suspension feeding. Its ciliated tentacles are arranged as a funnel with the small end surrounding the mouth and its large end open to the water (Fig. 25-17, 25-25). Lateral cilia on the sides of the tentacles create a flow of water into the large, open end of the funnel, then outward between the tentacles to exit the funnel. Suspended particles striking the tentacles become entangled in mucus and are moved to the mouth by frontal cilia located on the upstream (inside) surface of the tentacles.

The body is divided into two (perhaps three) regions, each containing a coelomic cavity. The first body region is the mesosome containing the mesocoel, or lophophoral coelom. The lophophore is part of the mesosome. Most of the body is the posterior metasome, or trunk, with the metacoel, or trunk coelom. A tiny anterodorsal lobe, the **epistome,** may precede the first region and overhang the mouth (Fig. 25-2). The epistome has been considered to be homologous to the protosome, and its cavity was once thought to be a coelomic space, the protocoel.

Lophophorate zygotes undergo radial cleavage, develop into pelagic larvae, and metamorphose into sessile benthic adults. Kamptozoans are sometimes included in Lophophorata, but they are compact (acoelomate) spiralians, and, although they do have a crown of ciliated tentacles around the mouth, its tentacles are solid and the anus is within it. Furthermore, the tentacles operate as a downstream collecting system, in sharp contrast to those of lophophorates.

PHORONIDAP

Phoronida consists of only two genera, *Phoronis* and *Phoronopsis,* and 14 species of wormlike, sessile, benthic animals found only in the sea. Externally they are bilaterally symmetrical, but internally they are asymmetrical with left-side dominance. Phoronids live in secreted chitinous tubes buried in sand or attached to rocks, shells, and other objects in shallow water (Fig. 25-1, 25-2). Some have bright green lophophores. A few species, such as the primitive *Phoronis ovalis,* bore into mollusc shells or calcareous rock. *Phoronis vancouverensis* along the Pacific coast of North America often forms masses of intertwined individuals. Phoronids are 2 to 20 cm in length but only 1 to 2 mm in diameter, making them exceedingly slender. Most are found in the relatively shallow water of the continental shelves.

The cylindrical body lacks appendages or regional differentiation except for the conspicuous anterior lophophore and a swollen posterior **ampulla** (Fig. 25-2). The anterior epistome is a small dorsal process overhanging the mouth (Fig. 25-2B, 25-3B) and containing a space often referred to as the protocoel, although it is probably a blastocoel derivative, not a coelom. The mesosome, immediately posterior to the epistome, bears the lophophore and mouth. The **lophophoral coelom** (mesocoel) forms a ring at the base of the lophophore and sends a branch into each tentacle. Most of the body is the trunk (metasome) and contains the **trunk coelom** (metacoel).

Primitively the lophophore is a circular tentacle-bearing ridge around the mouth and epistome. Frequently, however, the circular ring is collapsed inward dorsally to form a crescent, or horseshoe, of two parallel rows of tentacles (Fig. 25-2B, 25-3B). The crescent arches ventrally (Fig. 25-3B), with one ridge passing above and the other below the mouth. The horns of the crescent are directed dorsally and may be rolled up as a spiral on each side. The lophophore is used in gas exchange and feeding. Folding and coiling of the lophophore increases the number of tentacles and the area for food collection and gas exchange. The amount of coiling is positively correlated with body size (Fig. 25-4). The anus is dorsal, close to but outside and immediately posterior to the lophophore (Fig. 25-3B).

FIGURE 25-1 *Phoronis hippocrepia* from the coasts of Europe, South Africa, and Brazil. Aggregations of tubes encrust rocks, shells, and coral. Note the obviously horseshoe-shaped lophophore in the upper right of the lower photo. *(From Emig, C. C. 1974. The systematics and evolution of the phylum Phoronida. Z. Zool. Syst. Evolut.-forsch. 12:128-151.)*

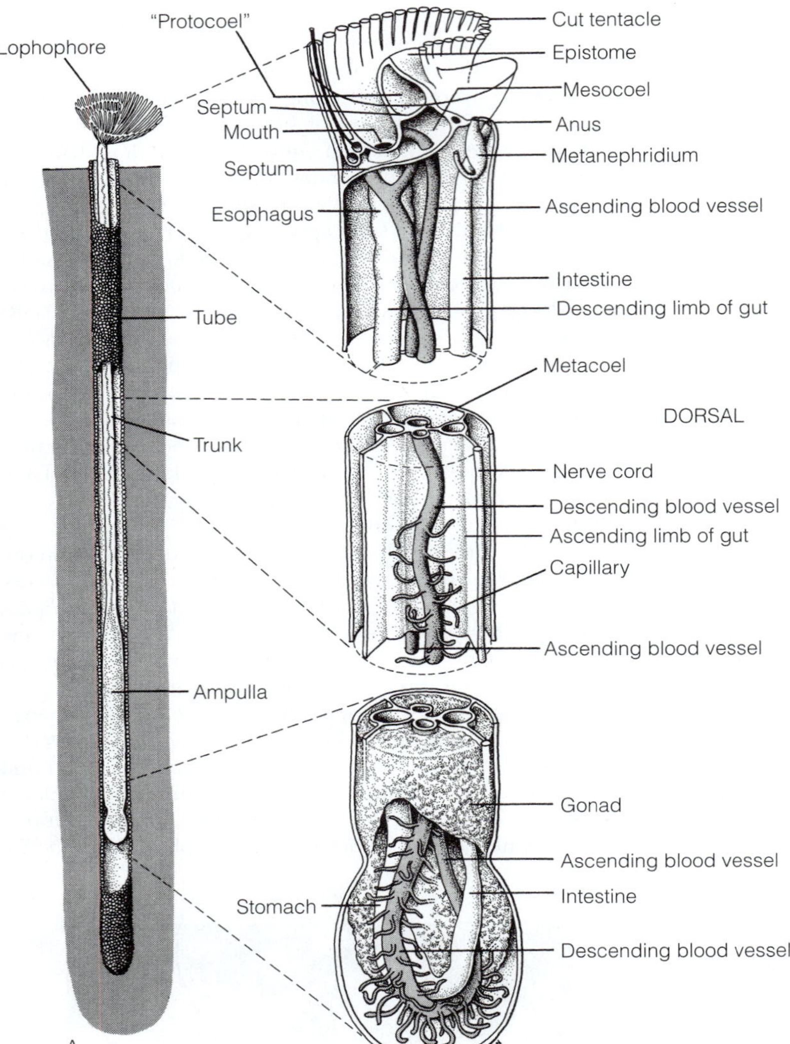

FIGURE 25-2 Phoronid anatomy. **A,** In life position. **B,** Internal anatomy. *(A and B, Modified from Vandergon, T. L., and Colacino, J. M., 1991. Hemoglobin function in the lophophorate* Phoronis architecta *(Phoronida). Physiol. Zool. 64:1561–1577. University of Chicago Press. Reprinted with permission.)*

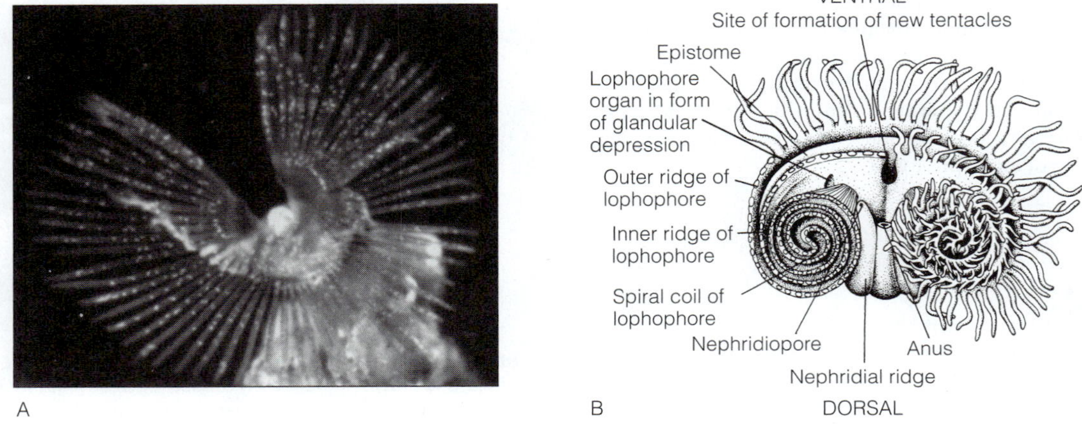

FIGURE 25-3 Phoronida. The lophophores of **A,** *Phoronis architecta* and **B,** *Phoronis australis* (anterior view). The mouth, not visible in this view, is on the midline, under the overhang of the epistome and between the two rows of tentacles. *(B, After Shipley.)*

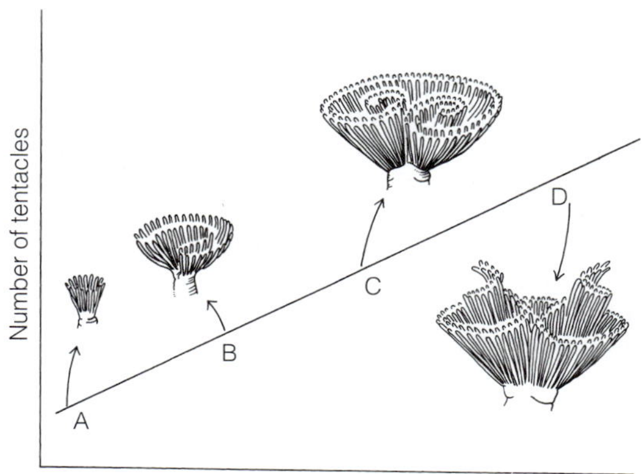

FIGURE 25-4 Phoronida: relationship between body size and lophophore shape. **A,** *Phoronis ovalis,* with a circular lophophore. **B,** *Phoronis architecta,* with a horseshoe-shaped lophophore. **C,** *Phoronis australis,* with a simple spiral lophophore. **D,** *Phoronopsis californica,* with a more elaborate spiral lophophore. *(Modified and redrawn from Abele, L. G., Gilmour, T., and Gilchrist, S. 1983. Size and shape in the phylum Phoronida. J. Zool. Lond. 200:317–323.)*

The glandular epidermis (Fig. 25-5) is underlaid by a basement membrane. The lophophore epidermis, in contrast to that of bryozoans but like that of brachiopods, is monociliated. There is no cuticle, but the epidermis secretes the chitinous tube the animal occupies. Sediment particles are incorporated into the tube during its construction to stiffen the enclosure. Phoronids move within the tube, but it is closed at the posterior end and they never leave it.

The coelom is lined with a mesothelium, which, in the body wall, is differentiated into a thin layer of circular muscles and a thicker, well-developed longitudinal muscle layer (Fig 25-5). These epitheliomuscular cells work against the coelomic hydrostatic skeleton to move the body in the tube and to extend and retract the lophophore from the tube opening. The longitudinal muscles make possible the rapid retraction of the lophophore and withdrawal of the trunk. Extension of both, using the weaker circular muscles, is slower. The posterior end of the trunk is swollen to form an ampulla, which anchors the worm in the tube. The mesothelium covering the gut, mesenteries, and blood vessels consists of noncontractile cells, epitheliomuscular cells, and nutrient-storage cells. The metacoel is divided by four longitudinal mesenteries into four chambers (Fig. 25-5), but the mesocoel is undivided. The mesenteries support the gut and longitudinal blood vessels.

Except for retraction into the tube, phoronid movements are slow and limited to partial emergence from the tube, bending movements of the extended trunk, and flicking movements of the tentacles. When disturbed, phoronids may autotomize the lophophore and later regenerate it. Freshly collected specimens often have lophophores in various stages of regeneration.

Like all lophophorates, phoronids are suspension feeders that use the upstream collecting system described in the introduction. The digestive tract is a long, U-shaped loop extending the length of the metacoel (Fig. 25-2B). The **descending limb** of the gut begins at the mouth and consists of an esophagus and stomach in which extracellular digestion occurs. The **ascending limb** is the intestine ending at the anus (Fig. 25-2B, 25-5).

The hemal system consists of blood confined to definite vessels. A **ring vessel** at the base of the lophophore gives rise

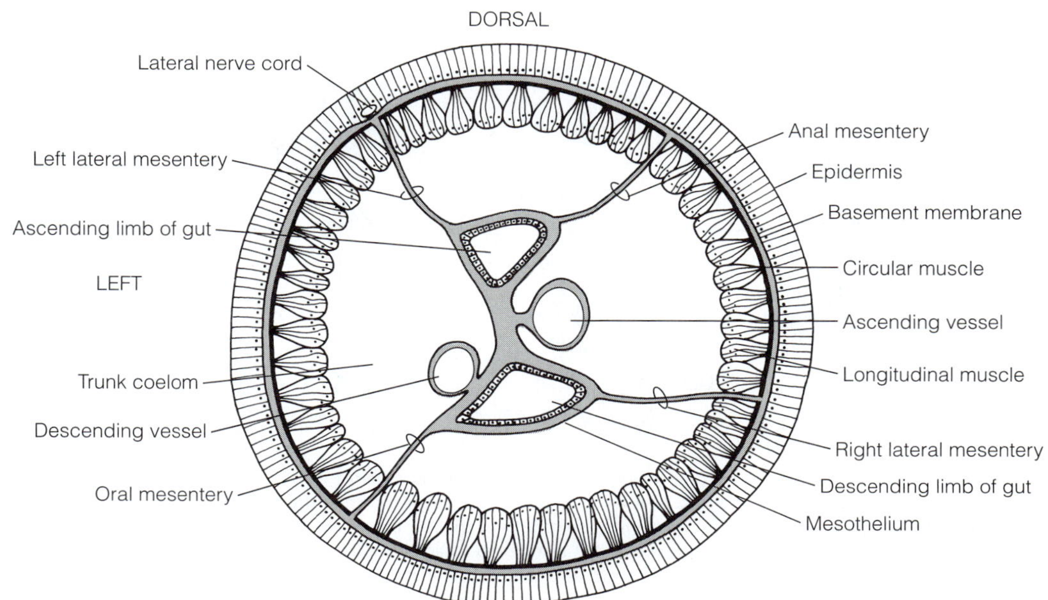

FIGURE 25-5 Phoronid anatomy: cross section. *(Redrawn from Hyman, L. H. 1959. The Invertebrates. Vol. V. The lesser coelomates. McGraw-Hill Book Co., New York, after Pixell, H. 1912. Two new species of Phoronidea from Vancouver Island. Quart. J. Microsc. Sci. 58, and Herrmann, K. 1997. Phoronida. In Harrison, F. W., and Woollacott, R. M. (Eds.): Microscopic Anatomy of Invertebrates. Vol. 13. Lophophorates, Entoprocta, and Cycliophora. Wiley-Liss, New York. pp. 207–235.)*

to blind-ended **tentacular vessels,** one extending into each tentacle. A large **descending vessel** (lateral vessel; Fig. 25-2B, 25-5) extends from the ring vessel to the posterior end of the body, where it reverses direction and returns as the **ascending vessel** (median vessel). Blood flows anteriorly in the ascending vessel and posteriorly in the descending vessel, propelled by contractions of epitheliomuscular cells in the vessel walls. Blind-ended capillaries that extend from the vessels into the ampulla supply the stomach and gonad. The blood vessels are retromesothelial.

The blood has hemocytes that contain hemoglobin. The presence of hemoglobin in such small animals apparently is an adaptation to life in anoxic or hypoxic habitats, such as fine marine sediments. The trunk extends into such sediments whereas the lophophore, as the gas-exchange surface, is exposed to oxygenated water above the sediment. Unlike many tubicolous animals, phoronids do not ventilate their tubes with oxygenated water. Oxygen to support aerobic metabolism in trunk tissues must diffuse into the lophophore and be transported throughout the body bound to the hemoglobin in the hemal system. The blood of *Phoronis architecta* has an oxygen-carrying capacity equal to that of most vertebrates and its volume in relation to the body volume is twice that of humans.

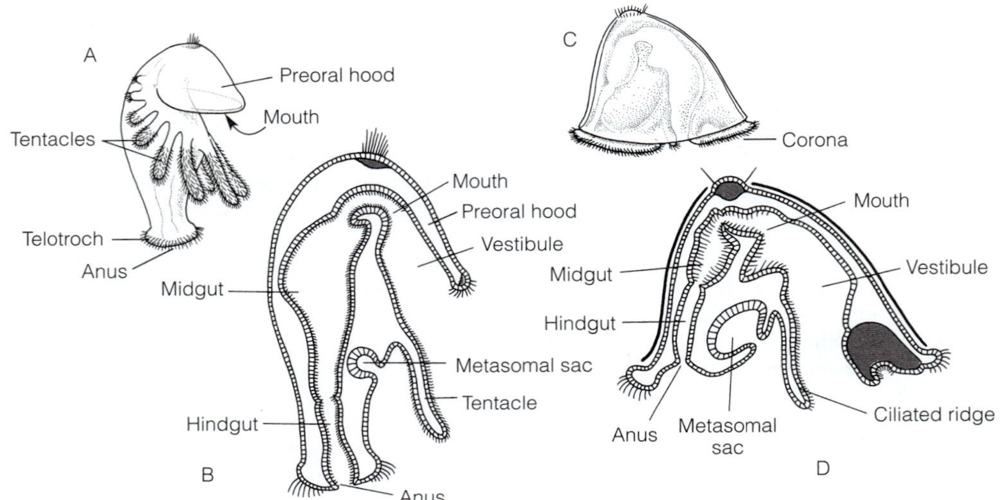

FIGURE 25-6 Lophophorate larvae viewed from the right. **A** and **B,** Phoronid actinotrocha larva. **A,** External view. **B,** Sagittal section. Note the small metasomal sac near the middle of the straight gut. **C** and **D,** Bryozoan cyphonautes larva. **C,** External view. **D,** Sagittal section. *(D, Simplified and redrawn from Stricker, S. A., Reed, C. G., and Zimmer, R. L. 1988. The cyphonautes larva of the marine bryozoan* Membranipora membranacea. *I. General morphology, body wall, and gut. Can. J. Zool. 66:368–383.)*

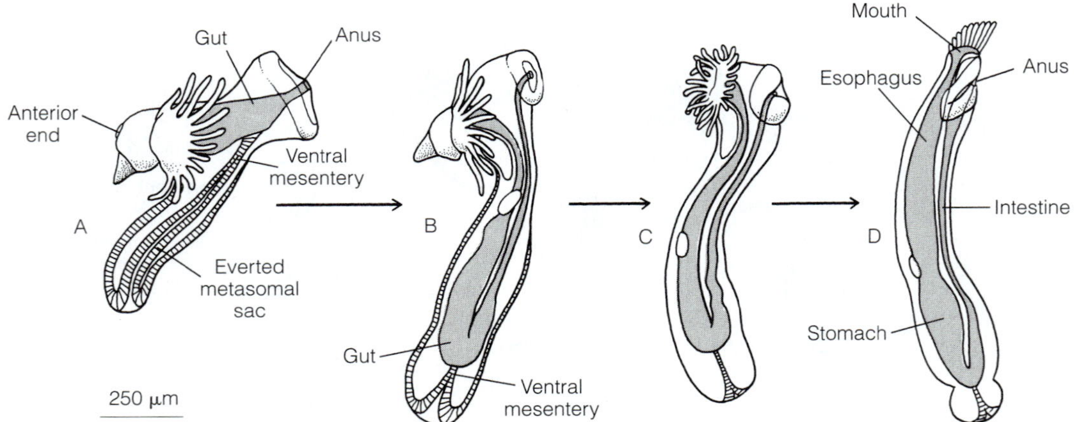

FIGURE 25-7 Phoronid metamorphosis from planktonic actinotrocha larva to benthic juvenile is accomplished in 15 to 30 min. Larvae viewed from the left. **A,** A larva, slightly more advanced than the one in Figure 25-6A,B, beginning metamorphosis. The gut is straight, with the mouth at the left of the image, obscured by the ring of larval tentacles. The metasomal sac evaginates ventrally, remaining attached to the gut by the ventral mesentery. **B,** Evagination of the metasomal sac pulls the gut into the characteristic U shape of the adult and establishes the worm shape of the adult trunk. The ectoderm of the metasomal sac becomes the adult epidermis whereas its mesoderm becomes the circular and longitudinal body-wall muscles. **C,** Larval tentacles regress as those of the adult begin to form. **D,** Larval tentacles are replaced by adult lophophoral tentacles. *(Modified and redrawn from Hermann, K. 1975. Einfluss von Bacterien auf die Metamorphose-Auslösung branchiata* (Phoronis muelleri). *Verh. Dtsch. Zool. Ges. 67:112–115.)*

Among invertebrates, hemoglobin-containing cells are widespread in the coelom but almost unheard of in blood. Hemoglobin in invertebrate blood is almost always extracellular, dissolved in the plasma. Extracellular hemoglobins are very large molecules whereas cellular hemoglobins are small. At equal concentrations, small molecules bind more oxygen than large ones because they have greater surface area on which to expose binding sites. On the other hand, many small molecules have a greater osmotic pressure than do a few large ones. By packaging small hemoglobins in cells, phoronids may be enhancing oxygen transport while minimizing the effect on blood osmolarity.

Two metanephridia open from the trunk coelom by nephrostomes and to the exterior by nephridiopores beside the anus (Fig. 25-2B, 25-3B). The left nephridium often is larger than the right. Ultrafiltration of blood into the trunk coelom apparently occurs across the walls of the contractile blood vessels, on which podocytes have been found.

The intraepidermal nervous system consists of a nerve ring at the base of the lophophore from which nerves arise to supply the tentacles and body-wall muscles. *Phoronis ovalis* has two lateral, nonsegmented nerve cords, but all other phoronids have only the left cord (Fig. 25-2B, 25-5). Giant axons are present in the cord.

Phoronids reproduce clonally and sexually. A few species form colonies by clonal budding and transverse fission. Phoronids have well-developed powers of regeneration. In response to unfavorable conditions, phoronids routinely autotomize the mesosome and lophophore, including the nerve ring, metanephridia, and mouth. The metasome replaces the missing parts in two to three days. Some species fragment into numerous pieces, each of which becomes a new, complete phoronid.

Most phoronids are hermaphroditic. The gonad occupies the coelomic space around the stomach in the ampulla and receives numerous capillaries from the descending blood vessel (Fig. 25-2B). Gametes are shed into the trunk coelom and escape to the outside by way of the metanephridia, which function as gonoducts. The sperm are packaged into spermatophores produced by a pair of **lophophore organs** (Fig. 25-3B) and released into the water. The spermatophores are captured by the lophophore of other individuals, whereupon the sperm become ameboid and penetrate the body wall to enter the metacoel. The eggs are fertilized internally and zygotes exit through the metanephridia to be brooded in the lophophore or released into the plankton.

Development is a mix of deuterostome and protostome characters. Cleavage is holoblastic and radial. The blastopore becomes the mouth. Development is regulative, and experimentally isolated blastomeres develop into complete larvae. Cleavage leads to a coeloblastula that gastrulates by invagination, but coelom formation is schizocoelous. No 4d mesentoblast is present, and mesoderm originates as mesenchyme from the archenteron, which becomes epithelial and encloses the larval trunk coelom. Recent work has demonstrated that the cavity of the epistome, long believed to be a protocoel, is in fact not lined by an epithelium, is filled with ECM, and is a blastocoel derivative, not a coelom. The lophophore coelom develops later from mesodermal cells in the adult's tentacle buds.

Development is indirect in almost all phoronids and the gastrula becomes a characteristic, elongate, tentaculate, ciliated larva called an **actinotrocha** (Fig. 25-6A,B). The gut is straight with the mouth anterior and anus posterior. A large **preoral hood** overhangs the mouth and a ring of ciliated tentacles partially encircles the mouth. The tentacles function chiefly in feeding and the ciliated telotroch around the anus is probably the principal locomotory organ. The larva has a pair of solenocytic protonephridia. The **metasomal sac,** which is an invagination of ectoderm associated with the mesoderm, lies near the middle of the ventral surface (Fig. 25-6B) and connects to the gut by the **ventral mesentery.** During metamorphosis the metasomal sac everts and develops into the body wall of the adult trunk (Fig. 25-7). After a long or short free-swimming, planktotrophic period, the actinotrocha undergoes a rapid metamorphosis in which the gut becomes U-shaped and the larval tentacles are replaced by adult tentacles. It sinks to the bottom, secretes a tube, and becomes a benthic adult (Fig. 25-7).

Among the 14 phoronid species, *Phoronis ovalis* is believed to embody many primitive traits. It has a simple oval lophophore and its hemal, nervous, and excretory systems are bilaterally symmetrical. Other phoronids show a greater or lesser reduction of these organs on the right side of the body. Such left-side dominance is reminiscent of several taxa of deuterostomes, including enteropneust hemichordates, echinoderms, and cephalochordates, all of which have larval and adult asymmetries favoring the left side of the body.

BRACHIOPODA^P

Brachiopods are commonly known as *lamp shells* because their shells sometimes resemble ancient Roman oil lamps (Fig. 25-8). More importantly, these animals also resemble bivalve molluscs in being suspension feeders with a mantle, mantle cavity, and bivalved calcareous shell. In fact, until the middle of the 19th century they were thought to be molluscs. The resemblance to clams, though striking, is superficial, and brachiopods are now placed in their own taxon far removed from Mollusca. In both taxa, two mineralized valves enclose a soft body, but in brachiopods the valves are dorsal and ventral (Fig. 25-8C) whereas in bivalve molluscs they are right and left.

All brachiopods are marine, benthic suspension feeders found from the intertidal zone to the deep sea, but most live on the continental shelf. They are solitary and most species live attached to the surface of rocks or other hard substrata (Fig. 25-12) but some, such as *Lingula,* live in vertical burrows in sand or mud bottoms (Fig. 25-9A). Although fossil species were widely distributed in ancient seas, modern species are largely inhabitants of cold waters.

Brachiopods are relatively large and thick-bodied, with an average diameter of about 5 cm and an unfavorable surface area-to-volume ratio. Their exceptionally large lophophore is presumably a mechanism for increasing surface area to compensate for the increased volume.

The approximately 350 species of extant brachiopods are but a fraction of the 12,000 known extinct species that flourished in Paleozoic and Mesozoic seas. Because of their mineral shells, they are well represented in the fossil record.

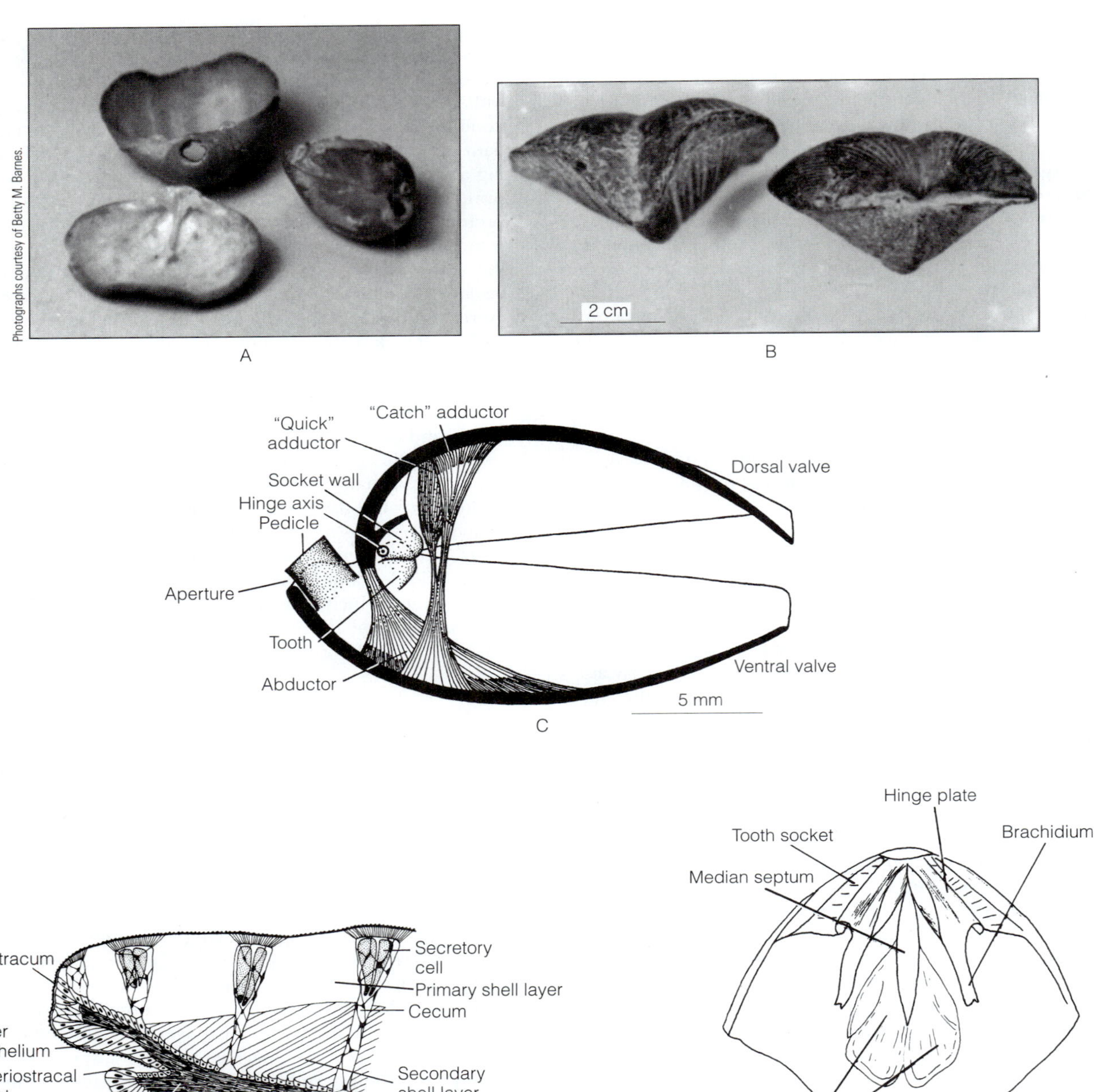

FIGURE 25-8 Brachiopoda. **A,** The articulate brachiopod *Terebratella.* On the left are dorsal (bottom of image) and ventral (top of image) valves. Note the conspicuous foramen for the pedicle in the ventral valve. On the right are two valves in the closed position. The foramen is evident in this image, also. **B,** *Paraspirifer,* a fossil articulate brachiopod. **C,** Sagittal section through the terebratulid *Waltonia,* showing the relationship of valves, muscles, and pedicle. **D,** Section through the edge of the shell and mantle of an articulate brachiopod. **E,** Internal surface of the dorsal valve of an articulate brachiopod. *(C, From Rudwick, M. J. S. 1970. Living and Fossil Brachiopods. Hutchinson and Co., London; D, From Williams and Rowell, 1965; E, After Davidson, 1887, from Hyman, L. H. 1959. The Invertebrates. Vol. V. The lesser coelomates. McGraw-Hill Book Co., New York.)*

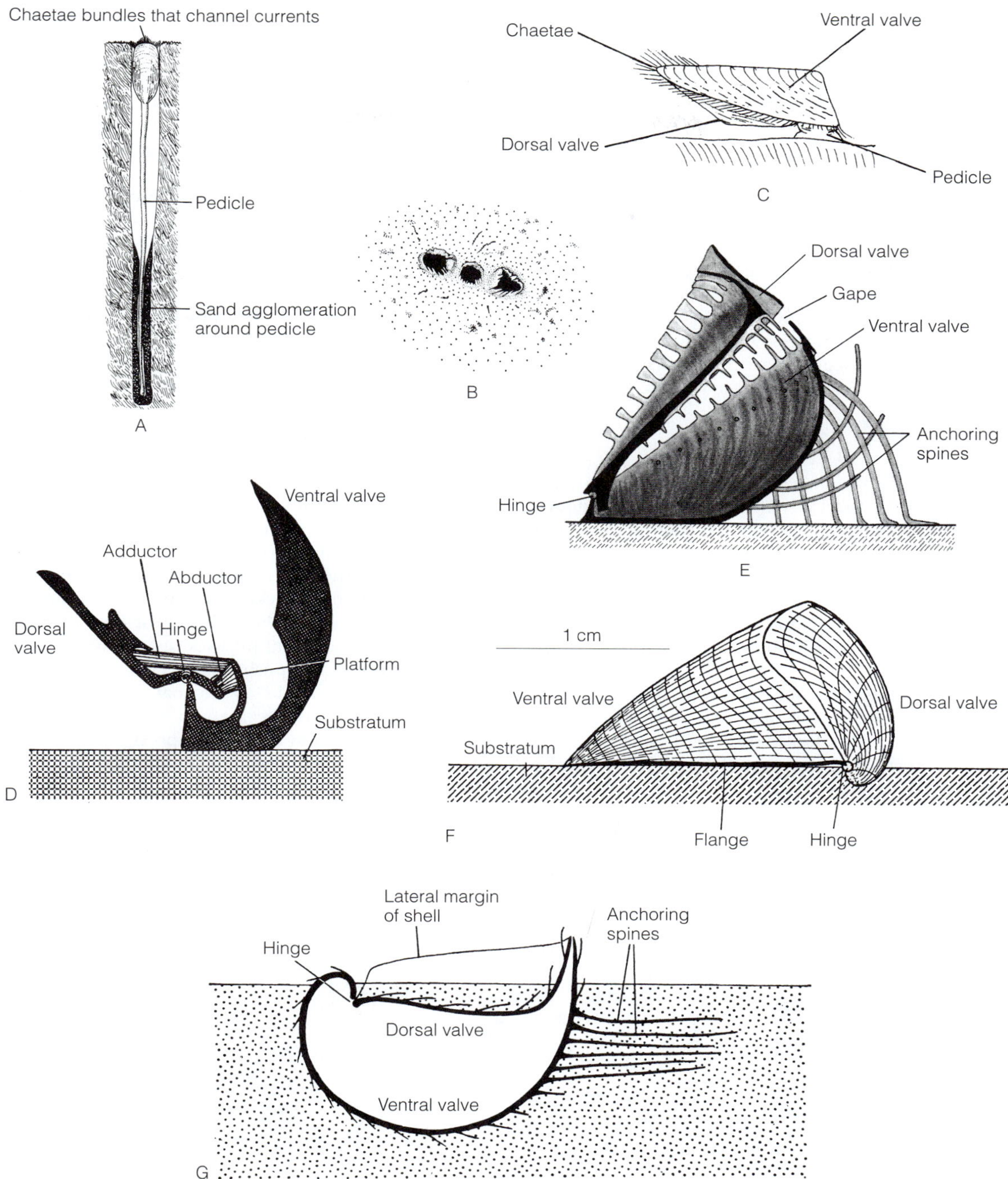

FIGURE 25-9 Brachiopoda. **A,** The inarticulate *Lingula* in the feeding position within its burrow. **B,** Burrow openings of *Lingula* when feeding. Chaetae show around the middle exhalant and lateral inhalant apertures. **C,** A living epibenthic inarticulate *Discinisca,* attached by the pedicle. **D,** Diagrammatic lateral view of a living articulate *Lacazella,* which lives attached directly to the substratum by its ventral valve. **E,** Sagittal section through the Permian articulate *Chonosteges,* which has cemented its ventral valve directly to the substratum. **F,** The Devonian articulate *Spyringospira,* which is believed to have lived unattached on soft bottoms. **G,** The Permian articulate *Waagenoconchia,* which is believed to have lived partially buried in soft bottoms. *(A, Modified from François, 1891; B, Modified from Rudwick, 1970; C, After Morse, 1902, from Hyman, L. H. 1959. The Invertebrates. Vol. V. The lesser coelomates. McGraw-Hill Book Co., New York; D, F, and G, From Rudwick, M. J. S. 1970. Living and Fossil Brachiopods. Hutchinson and Co., London; E, Modified from Rudwick, 1970.)*

Brachiopoda made its appearance in the lower Cambrian period and reached its peak of generic diversity during the Devonian. Since then they have apparently suffered relative to bivalve molluscs, with which they compete for sedentary, benthic, bivalved suspension-feeder niches. Brachiopods declined substantially in the great Permian extinction and the limited numbers and restricted distribution of Recent species are consequences of that episode. Following this setback, the number of genera increased modestly during the Cenozoic. The living genus *Lingula* dates back to the Ordovician.

Brachiopoda includes two taxa, Inarticulata and Articulata, that differ in many morphological features, especially the articulating mechanism of the hinge and its muscles. Most modern brachiopods are articulates. Although morphological features suggest that inarticulates are more primitive, both taxa are present from the beginning of the fossil record.

FORM

A typical brachiopod has a body with a large anterior lophophore (Fig. 25-10). Two folds of the body wall, the **dorsal** and **ventral mantle lobes,** extend anteriorly from the body and partly enclose a water space, the mantle cavity, in which the lophophore is located. Each lobe consists of an **outer mantle epithelium,** which secretes the shell, and an **inner mantle epithelium,** which lines the mantle cavity (Fig. 25-8D). The entire body, mantle lobes, and lophophore are enclosed in a bivalved shell.

Each of the two valves is bilaterally symmetrical and has a convex outer margin. The ventral valve typically is larger than the dorsal (Fig. 25-9E). The **dorsal** (brachial) **valve** supports the lophophore and fits over the **ventral** (pedicle) **valve,** the apex of which sometimes contains a posterior aperture (Fig. 25-8A,C) for the exit of an attachment stalk, the **pedicle** (Fig. 25-9C). In the burrowing lingulids, the valves are flattened and similar in size and shape. The valves may be ornamented with concentric growth lines and a fluted, ridged, or even spiny surface. The shells of most living brachiopods are dull yellow or gray, but some species are orange or red.

The two valves articulate with one another along a posterior hinge (Fig. 25-9D,E,F). The characteristics of this hinge are the basis for the division of Brachiopoda into Inarticulata and Articulata. In inarticulate brachiopods, such as *Lingula* and *Glottidia,* the valves are held together only by muscles, and the hinge region of the valves is simple and unelaborated. In articulate brachiopods, the posterior edge of the ventral valve bears a pair of hinge teeth that fit into complementary sockets on the hinge of the dorsal valve (Fig. 25-8C,E). This articulating mechanism locks the valves securely together and allows a slight anterior gape of only about 10° (Fig. 25-9E).

Because muscles exert force only when shortening, bivalve animals such as clams and brachiopods are faced with the difficult task of opening the valves, while keeping all soft parts safely within the confines of the shell. It is an easy matter to close (adduct) the valves with a muscle extending from one valve to the other but opening (abducting) is more difficult and brachiopods use two different mechanisms, neither the same as that used by clams. In inarticulate brachiopods, contraction of longitudinal muscles pressurizes the coelom, causing it to expand and push the valves outward. Remember that a gape of only about 10° is the goal. Although the hinge itself is simple, lingulids have a complex musculature not only to adduct the valves but also to compress the coelom and abduct the valves. Some of these muscles can also cause the valves to

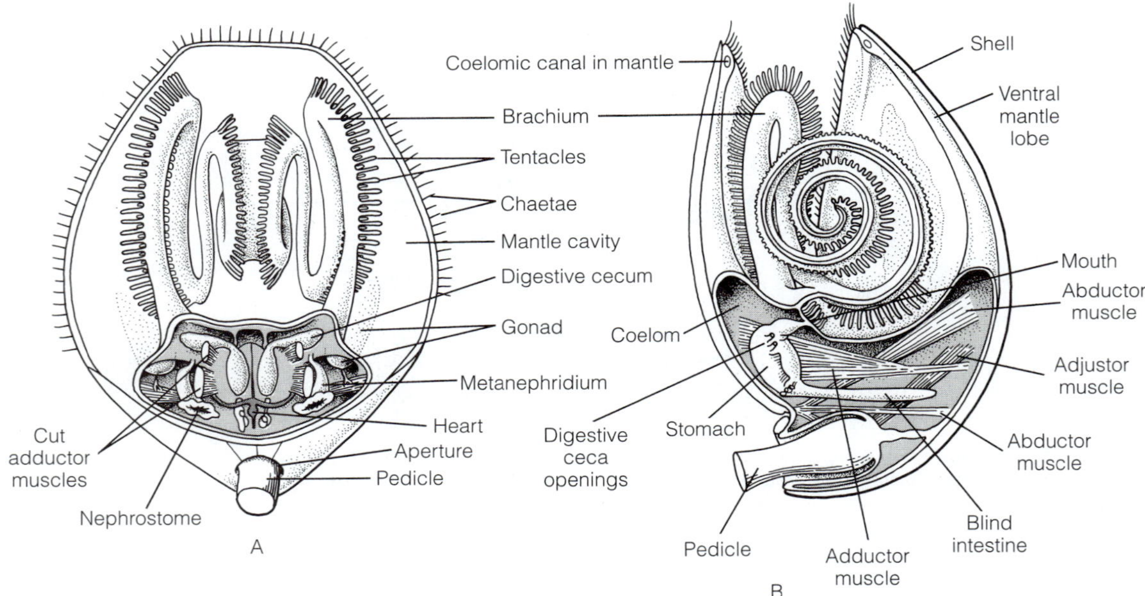

FIGURE 25-10 Brachiopod anatomy. **A,** Dorsal view. **B,** Sagittal section. *(Modified and redrawn from Delage and Hérouard modified from Beauchamp, P. de. 1960. Classe des Brachiopodes. In Grassé P.-P. (Ed.): Traité de Zoologie. Vol. 5. Part 2. Masson et Cie, Paris. Reprinted with permission.)*

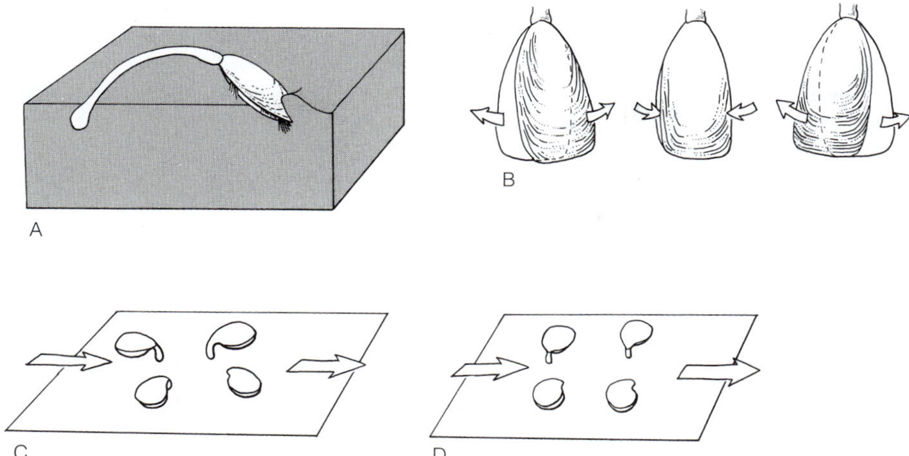

FIGURE 25-11 Brachiopod movements. Burrowing *Lingula* and *Glottidia* **(A)** arch the body using the pedicle and slice into the sediment using scissorslike movements of the dorsal and ventral valves **(B).** Individuals of the articulate *Terebratalia transversa* placed randomly in a unidirectional current **(C)** rotate on their pedicles, positioning themselves with the longitudinal body axis perpendicular to the water flow **(D).** In this orientation, the water currents aid, rather than oppose, the ciliary current produced by the lophophore. *(A, Modified from Trueman, E. R., and Wong, T. M. 1987. The role of the coelom as a hydrostatic skeleton in lingulid brachiopods. J. Zool. Lond. 213:221–232, and personal observations of E. E. R.; B, Based on data and discussion by LaBarbera, M. 1977. Brachiopod orientation to water movement. I. Theory, laboratory behavior, and field orientations. Paleobiology 3:270–287.)*

shear past each other in a scissorslike motion (Fig. 25-11). Lingulids burrow using these shearing movements to slice through sediment.

In inarticulates and articulates, adductor muscles extend from valve to valve with both insertion and origin *anterior to* the hinge. These muscles pull the valves together and close the shell (Fig. 25-8C). In articulates, a pair of **abductor** (diductor) **muscles** originates on the ventral valve and inserts on the dorsal valve *posterior to* the hinge. Contraction of these muscles opens the valves (Fig. 25-8C, 25-9D). Although effective, this arrangement requires an opening in the shell and a muscle or tendon partially exposed and vulnerable to predation. Clams, you may recall, have no abductor muscle and rely instead on the elastic recoil of the hinge ligament to open the valves.

Like those of bivalve molluscs, brachiopod adductor muscles are divided into quick and catch portions (Fig. 25-8C). The quick muscles are striated fibers capable of rapid contraction to close the valves quickly. The catch muscles, composed of smooth fibers, are slower, but can remain contracted for long periods without fatigue.

Each valve (Fig. 25-8D) consists of an inner biomineralized layer covered by an outer organic (proteinaceous) periostracum. The periostracum is secreted by epidermal cells of the mantle margin whereas the mineral portion is secreted by the outer mantle epidermis. The mineral layer is divided into outer primary and inner secondary layers. The layers of the shell are secreted in sequence, beginning with the periostracum and ending with the secondary mineral layer. In articulates the primary layer consists of calcite and the secondary layer is organocalcitic, including protein as well as calcite. The chitinophosphatic shell of inarticulates seems to be the more primitive and appears first in the fossil record, in the Cambrian. It is more complex and not as well understood as the articulate shell. It has an organic matrix consisting chiefly of glycosaminoglycans (GAGs). The flexible primary layer consists of apatite (another form of calcium carbonate) crystals embedded in GAGs whereas the secondary layer is a complex of collagen and other proteins, GAGs, chitinophosphate, and apatite. The inarticulate periostracum contains traces of chitin.

In most brachiopods, the outer mantle epidermis extends diverticula, or ceca, through the calcareous shell to the periostracum (Fig. 25-8D). The ceca may be storage sites, secrete repellants to discourage fouling organisms, or function in gas exchange.

In addition to secreting the shell, the mantle edge in most species bears long, chitinous chaetae that may have a protective and perhaps a sensory function (Fig. 25-9C, 25-10). All are equipped with muscles and are movable, some more so than others. The chaetae have a core of parallel tubular lumina and are similar morphologically to those of annelids but unlike the setae of arthropods (Fig. 13-5). They channel the flow of water into and out of the mantle cavity in some brachiopods (Fig. 25-9A,B).

The body occupies only a third of the space between the two valves (Fig. 25-10). Anteriorly, the space between the mantle lobes is the mantle cavity, which is filled with seawater and occupied by the lophophore (Fig. 25-10A).

Most brachiopods are attached to the substratum by a cylindrical extension of the body wall called the pedicle. It is covered by a thick, chitinous cuticle secreted by the epidermis and extends posteriorly from the aperture in the hinge region of the shell (Fig. 25-9A,C, 25-10). The pedicles of articulates and inarticulates develop from different larval primordia and thus are not homologous. Although they have similar functions, they differ morphologically.

The pedicle of the inarticulate lingulids (*Lingula* and *Glottidia*) is long and muscular (Fig. 25-9A). Its core is an extension of the metacoel that contains hemerythrocytes and is lined by a mesothelium. A layer of longitudinal muscle lies between the coelom and the basement membrane of the pedicle epidermis.

The articulate pedicle may be short, rigid, and lack muscles (Fig. 25-9C) or it may be a flexible, muscular tether (Fig. 25-10). Its core is mostly connective tissue and lacks a coelomic cavity. Furthermore, the pedicle emerges either from a notch in the hinge line of the ventral valve or through a hole at the upturned apex (Fig. 25-8A, 25-10A). This means, of course, that the pedicle emerges from the dorsal side of the ventral valve, which extends posteriorly well beyond the dorsal valve (Fig. 25-8C). An articulate brachiopod may be attached to the substratum upside down with its valves in a horizontal position (Fig. 25-9C) or with its hinge end down and the gape directed upward. Muscles within the valves, which insert on the pedicle base, permit erection, flexion, and even rotation of the animal on the pedicle (Fig. 25-11C,D). The distal end of the pedicle adheres to the substratum by means of rootlike extensions or short papillae.

The pedicle has been completely lost in a few brachiopods of both classes, such as *Crania* (Inarticulata) and *Lacazella* (Articulata). Such species cement their ventral valve directly to the substratum and thus are oriented with the dorsal side more or less up (Fig. 25-9D,E,F). The posterior part of the ventral valve forms the point of attachment and the anterior margin is directed somewhat upward and clear of the substratum. Some species of fossil brachiopods that attached by cementation were important components of Paleozoic reefs. A few deep-sea species, such as *Chlidonophora,* anchor into sediment with a branched, rootlike pedicle.

The shell shape of a number of fossil taxa suggests that they were adapted for living free on the surface of soft bottoms. Spines, long "wings," flanges, and flattened ventral surfaces appear to have been devices used to prevent sinking into the sediment (Fig. 25-9F,G). Among living species, the New Zealand *Neothyris lenticularis* lives free on gravel or coarse sand bottoms and *Terebratella sanguinea* lives either attached to rock or free on sand and mud. Most modern species attach to hard substrata (Fig. 25-12).

FIGURE 25-12 Brachiopoda. A rock wall on the coast of New Zealand supporting three species of brachiopods. The white, branching organism is a stylasterine hydrocoral. Photograph covers approximately 1 m^2.

LOPHOPHORE AND FEEDING

As in other lophophorates, the brachiopod lophophore is a crown of hollow tentacles encircling the mouth. In order to increase the surface area, however, the lophophore is extended anteriorly as two arms, or **brachia,** from which the name *brachiopod* is derived. In its simplest form the lophophore is horseshoe-shaped and each arm, or brachium, projects anteriorly into the mantle cavity. In larger species, the arms may be looped or spiraled in complex arrays, greatly increasing the collecting surface area of the lophophore (Fig. 25-10, 25-13) to provide an adequate respiratory and feeding surface for the relatively massive body. Each brachium bears a row of lophophoral tentacles with a ciliated **brachial groove** (Fig. 25-13B) parallel to its base. Each brachium contains a cartilage-like axial support and two coelomic canals. Branches of the canals extend into the tentacles. The coelomic spaces in the brachia and tentacles function together as a hydrostatic skeleton that maintains the turgid, erect conformation of the tentacles. In many articulate brachiopods, the dorsal valve bears a pair of complex, delicate, calcareous processes known as **brachidia** that support the brachia. The inner valve surface may be grooved and ridged to make space for the lophophore (Fig. 25-8E).

The upstream particle collecting system of the tentacles functions as described for lophophorates in general and is used for suspension feeding. During feeding, water enters and leaves the anterior valve gape through distinct inhalant and exhalant apertures and channels (Fig. 25-13A). Particles, especially fine phytoplankton, are stopped by the lateral cilia of adjacent tentacles and then transported down the inside (upstream) edge of the tentacles to the brachial groove. The brachial groove conducts food from the base of the tentacles to the mouth (Fig. 25-13B). Rejected particles are carried away by the exhalant current.

Lingulids live in burrows in soft sediments. The anterior ends of the valves are directed upward, toward the burrow opening, and the posterior pedicle extends downward toward the bottom of the burrow and is encased in sand (Fig. 25-9A). When the animal is in the feeding position the gaping valves are near the opening of the burrow. The long mantle chaetae are bunched together in three clusters to form two lateral inhalant siphons and a median exhalant siphon that project slightly above the sediment surface and prevent sediment particles from entering the mantle cavity (Fig. 25-9B). When the animal is disturbed, the pedicle contracts and pulls the animal downward into the burrow.

INTERNAL FORM AND FUNCTION

The adult brachiopod body consists of a mesosome and metasome. These contain an anterior lophophore coelom with extensions into the tentacles and a large, spacious trunk coelom in which the gut, metanephridia, gonads, shell muscles, and

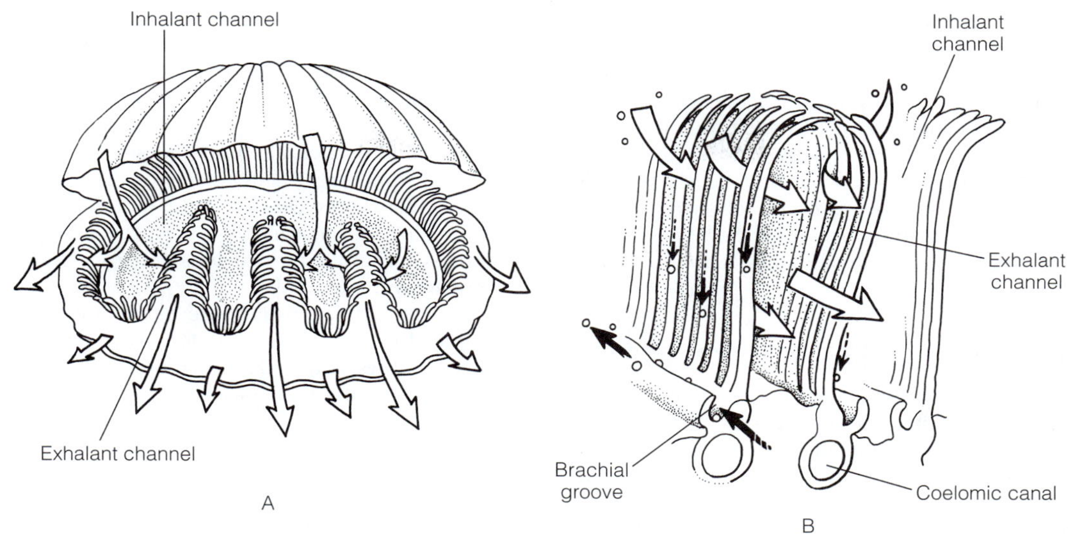

FIGURE 25-13 Brachiopoda. **A,** View into the anterior gape of the brachiopod *Megathyris,* showing the lophophore and water currents. Inhalant currents are indicated by the two forked arrows and exhalant flow by the simple arrows. **B,** Enlargement of an exhalant channel of *Megathyris.* Open arrows indicate water flow; small black arrows represent particle transport on the tentacles; large black arrows indicate particle transport to the mouth by the brachial groove. *(A, Modified and redrawn Rudwick, M. J. S. 1970. Living and Fossil Brachiopods. Hutchinson and Co., London. B, After Atkins, 1959.)*

hemal system are located. The mesothelial lining is ciliated. The presence of a protosome is doubtful, although a small, dorsal, preoral epistome is present anterior to the mouth. In articulates it is solid but in inarticulates it contains a space that is continuous with the lophophore coelom; there is no evidence that it is a protocoel.

The body wall consists of an outer epidermis, a layer of connective tissue, and the coelomic mesothelium. All brachiopod epithelia are monolayered and all ciliated cells are monociliated. Those body-wall muscles lying beneath the shell are poorly developed or specialized for opening and closing the valves. Epitheliomuscular cells in the mesocoel mesothelium move the brachidia and tentacles. Brachiopod muscles are either smooth or cross-striated.

Diverticula from the metacoel extend into the mantle lobes as **mantle channels** in the connective tissue between the inner and outer mantle epithelia. The coelomic fluid contains coelomocytes of several sorts and in the inarticulates, some are hemerythrocytes containing hemerythrin. There are no specialized gas-exchange surfaces other than the lophophore and mantle lobes. Oxygen transport is probably provided by the coelomic fluid, for there is a definite circulation of this fluid through the mantle channels and oxygen is carried, at least in part, by hemerythrin in the coelomocytes of some species. Circulation of the coelomic fluid is accomplished by the mesothelium, either by contractions of its epitheliomuscular cells or beating of its cilia.

In addition to coelomic spaces and channels, all brachiopods have a hemal system consisting of a heart, colorless blood, and vessels. The muscular, contractile heart is in the dorsal mesentery over the stomach and from it extend anterior and posterior vessels (Fig. 25-10A). These vessels branch to supply various parts of the body, including the gut and lophophore. The function of the hemal system is uncertain, but the chief role may be delivering nutrients to tissues.

The gut consists of a mouth, pharynx, esophagus, stomach, digestive ceca, intestine, and sometimes an anus. The gut wall has layers of circular and longitudinal muscles and the gastrodermis is ciliated throughout. The mouth opens into a muscular pharynx that connects with the esophagus. The esophagus extends dorsally, then turns posteriorly to join the stomach (Fig. 25-10B). Large, highly branched digestive ceca surround the stomach and connect with it by one to three ducts on each side. Digestion is chiefly intracellular in the digestive ceca. In some inarticulates, the gut is complete and U-shaped and the intestine extends anteriorly from the stomach to the rectum, which opens to the outside via an anterior anus located between the valves on the right side. In others the anus is on the posterior midline. In articulates, however, the intestine extends ventrally from the stomach but ends blindly, and there is no anus (Fig. 25-10B).

Brachiopods are ammonotelic and most nitrogen excretion takes place by diffusion across the body surface, especially the lophophore and mantle. The metanephridia are gonoducts with no obvious role in excretion. Careful searching has failed to demonstrate podocytes on any blood vessel or the heart.

A circumesophageal nerve ring with a small, dorsal, supraesophageal ganglion and a larger, ventral, subesophageal ganglion form the CNS. The supraesophageal ganglion is not present in adult inarticulates, resulting in a "brain" ventral to the gut, as in kamptozoans. From the ganglia and their commissures, nerves extend anteriorly and posteriorly to innervate the lophophore, the mantle lobes, and the valve muscles.

Like that of clams, the mantle margin is probably the most important site of sensory reception. The mantle chaetae, although not directly associated with sensory neurons, probably transmit tactile stimuli to receptors in the adjacent mantle epidermis. In some brachiopods, the chaetae are long and form a "sensory grille" over the gaping valves. Many brachiopods exhibit a "shadow response" in which they respond to a passing shadow by closing the valves, but obvious photoreceptors have not been found and it is not known what cells are responsible or where they are located. Statocysts are present in some brachiopods.

REPRODUCTION AND DEVELOPMENT

With a few exceptions, brachiopods are gonochoric, but sexual dimorphism is known in only one species. The gonads, usually four in number, with a pair in each valve, are masses of developing gametes behind the mesothelium of the coelomic mantle channels in articulates (Fig. 25-10B) or in the gut mesenteries of inarticulates. When ripe, gametes are released into the metacoel and discharged to the exterior through the metanephridia. Most brachiopods have a pair of large metanephridia (Fig. 25-10A), but some articulates have two pairs. Their funnel-like nephrostomes open from the metacoel on each side of the posterior end of the stomach and the tubules then extend anteriorly to empty into the mantle cavity through nephridiopores situated posteriorly and to each side of the mouth. The metanephridia are gonoducts for the transport of gametes from the metacoel to the exterior and in some species are closed until sexual maturity is achieved.

Except for brooding and viviparous species, the eggs are shed into the seawater and fertilized at the time of spawning. Many articulates brood developing embryos in the mantle cavity or sometimes gestate them in the metanephridia. Females of the only dimorphic species, *Lacazella mediterranea,* have a hypertrophied ventral valve that serves as a brood pouch.

Cleavage is radial, holoblastic, nearly equal, and produces a coeloblastula that usually undergoes gastrulation by invagination, but sometimes by delamination. Development is regulative. The blastopore closes and the mouth and anus arise anew. Coelom formation, which is interpreted as schizocoelous in inarticulates and enterocoelous in articulates, produces the mesocoel and metacoel.

The embryo eventually develops into a free-swimming juvenile. In inarticulates, the long-lived planktotrophic juvenile resembles a minute brachiopod (Fig. 25-14E). Paired mantle lobes and valves enclose the body and the lophophore. The ciliated lophophore can be extended outward from the mantle cavity and is the juvenile swimming organ. The pedicle, which in this taxon is derived from the mantle, is coiled in the back of the mantle cavity. As additional shell is secreted, the juvenile becomes heavier and eventually sinks to the bottom. There is no metamorphosis in *Lingula* and development is essentially direct. The pedicle attaches to the substratum and the young brachiopod takes up an adult existence.

Development is indirect in articulates, and the lecithotrophic larva experiences a brief planktonic life prior to metamorphosis. The larva, which does not resemble the adult, has a ciliated anterior lobe that will become the body and lophophore, a posterior lobe that forms the pedicle, and a middle, skirtlike mantle, the hem of which is directed backward (Fig. 25-14A). Four bundles of chaetae, two dorsal and two ventral, are present. In *Terebratulina,* the larva settles after a short free-swimming existence of approximately

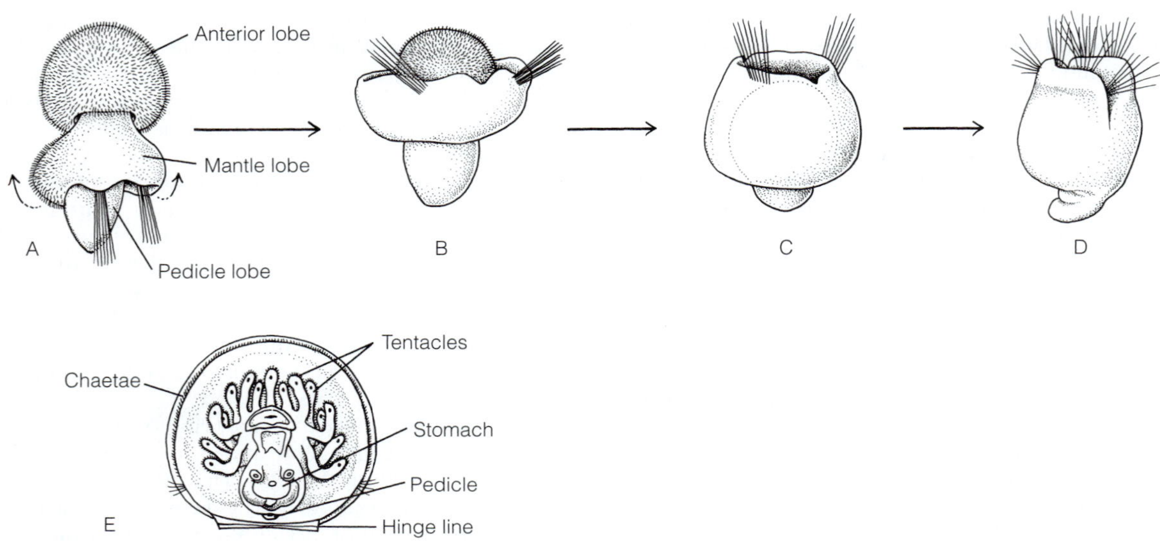

FIGURE 25-14 Brachiopod larva, juvenile, and larval metamorphosis. **A–D,** Metamorphosis of an articulate larva in lateral view **(A),** showing reversal of the mantle. **E,** Pelagic juvenile of *Lingula* and *Glottidia.* The ciliated tentacles project from the shell as the animal swims. *(E, After Yatsu.)*

24 to 30 h and then undergoes metamorphosis. The mantle lobe reverses position to begin secreting the valves and the adult structures develop from their larval precursors (Fig. 25-14B-D).

DIVERSITY OF BRACHIOPODA

Inarticulata[C]: Possibly not monophyletic. Hinge is simple; valves are held together only by muscles and connective tissue; no hinge teeth or sockets. Valves are closed by direct muscular action, opened by hydrostatic coelomic pressure. Valve musculature is complex. Gut is complete, U-shaped, includes an anus. No brachidia from the dorsal valve support the lophophore. Development is direct; larva resembles a tiny adult, with a bivalve mantle and body lobe.

Lingulida[O]: Burrow in soft sediments. Have hemerythrin as respiratory pigment. Most have a chitinophosphate shell; the periostracum has chitin. Anus is on the right. *Lingula, Glottidia.*

Discinida[O]: Attach to hard surfaces with a short pedicle. *Discina, Discinisca.*

Craniida[O]: Pedicle is absent; ventral valve attaches directly to the hard substratum. Shell is calcareous; periostracum contains chitin. Anus is on posterior midline. *Crania, Neocrania.*

Articulata[C]: Complex hinge with interlocking teeth and sockets. Calcite (calcium carbonate) shell. Periostracum lacks chitin. Dorsal valve usually has calcareous brachidia to support the lophophore. Shell is opened and closed by direct action of the adductor and abductor muscles. The pedicle, when present, is supported by connective tissue. Gut ends blindly; anus is absent. Larva has three body regions: body, mantle, and pedicle lobes. Most attach to hard substrates but some (such as *Chlidonophora*) anchor in soft sediments with a branching rootlike pedicle.

Terebratulida[O]: The most abundant Recent brachiopods. Brachidium is a complex loop. *Argyrotheca, Calloria, Chlidonophora, Gryphus, Gwynia, Lacazella, Liothyrella, Megathyris, Neothyris, Pajaudina, Terebratalia, Terebratella, Terebratula, Terebratulina, Thecidellina.*

Rhynchonellida[O]: Brachidia are simple, a strongly convex shell, and two pairs of metanephridia. *Hemithyris, Notosaria.*

BRYOZOA[P]

Bryozoa (= Ectoprocta, Polyzoa, "moss animals"), with approximately 5000 living species, is the largest, best known, and most widespread of the lophophorate taxa (Fig. 25-15, 25-16). Bryozoans are benthic and colonial, with most species living attached to firm substrata, but one genus is solitary and a few form motile colonies. The colonies, which do not look much like animals, may be large, but each is composed of numerous tiny zooids that are unmistakably animal-like. Bryozoa

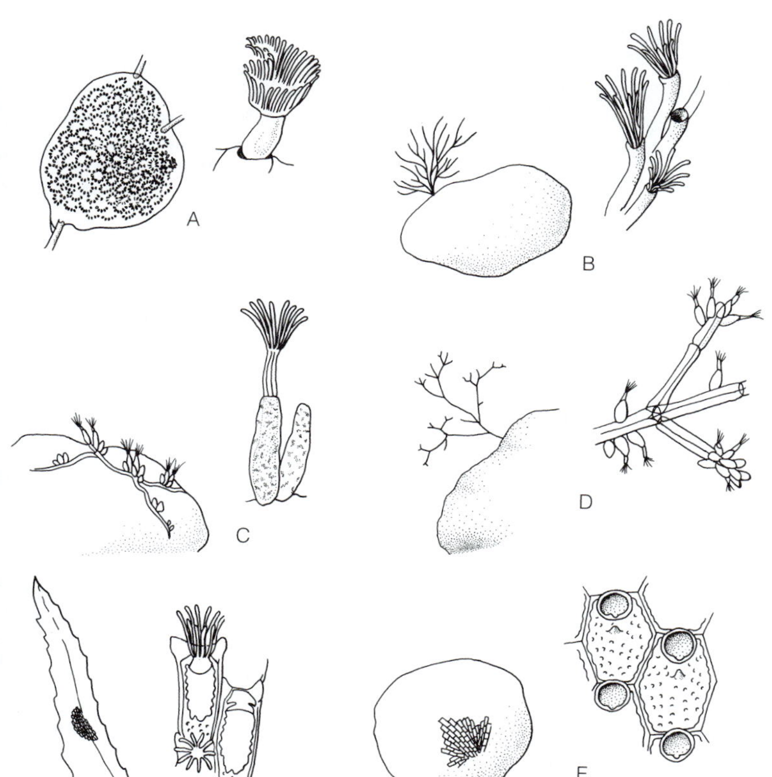

FIGURE 25-15 Bryozoan diversity. **A,** The freshwater bryozoan *Pectinatella magnifica* (Phylactolaemata). **B–F,** marine bryozoans (Gymnolaemata). **B,** The marine bryozoan *Crisia eburnea* (Cyclostomata). **C,** The stolonate ctenostome *Bowerbankia maxima.* **D,** The erect ctenostome *Zoobotryon verticillatum,* showing the jointed stolon composed of tubular kenozooids. **E,** Frontal surface of the anascan cheilostome *Membranipora tuberculata* encrusting a leaflet of *Sargassum.* **F,** Frontal surface of the ascophoran cheilostome *Schizoporella unicornis* encrusting the surface of a rock.

FIGURE 25-16 Living bryozoan colonies. **A,** Biserially arranged autozooids of the erect cheilostome species *Bugula neritina*. **B,** Autozooids on the surface of a branch (kenozooid) of the stolonate ctenostome *Zoobotryon verticillatum*. **C,** Erect calcified colony of *Scrupocellaria regularis*, an ascophoran cheilostome. **D,** Surface view of the encrusting ascophoran *Watersipora subovoidea*, showing one fully expanded and one partly expanded polypide.

is a major animal taxon, but due to the colonies' inconspicuous, often plantlike appearance, they go unnoticed by most people, who assume them to be seaweeds or mosses. Most are marine, but about 50 species live in fresh water. Because many bryozoans have a calcareous exoskeleton, there is an extensive fossil record.

Being less than 1 mm long, zooids might be predicted to have neither coelom nor internal transport system, yet they have both. The coelom is the space into which the introvert is retracted and is also the hydrostatic skeleton necessary for its protrusion. The fluid transport system connects otherwise isolated zooids and is an interzooidal system permitting communication and transport among zooids. The zooids, being repeating modules, might be compared with the segments of an annelid, and the bryozoan colony, not the zooids, should be thought of as the individual.

Traditionally Bryozoa has been divided into three taxa, Phylactolaemata, Gymnolaemata, and Stenolaemata, but a recent reorganization has placed the cyclostome stenolaemates in Gymnolaemata, leaving only two extant higher taxa, Phylactolaemata and Gymnolaemata.

FORM

Bryozoan colonies are composed of zooids, each about 0.5 mm in length. Zooids are often polymorphic, but the body of a typical, feeding zooid consists of a trunk and an eversible introvert, which bears the lophophore (Fig. 25-17). Zooid shape varies with taxon and may be boxlike, oval, or tubular (Fig. 25-15). The epistome is a small dorsal lobe overhanging the mouth of phylactolaemates, but it is absent in gymnolaemates. The mesosome forms an anterior ring around the mouth, contains the lophophore coelom (mesocoel), and bears the lophophore. Diverticula of the lophophore coelom extend

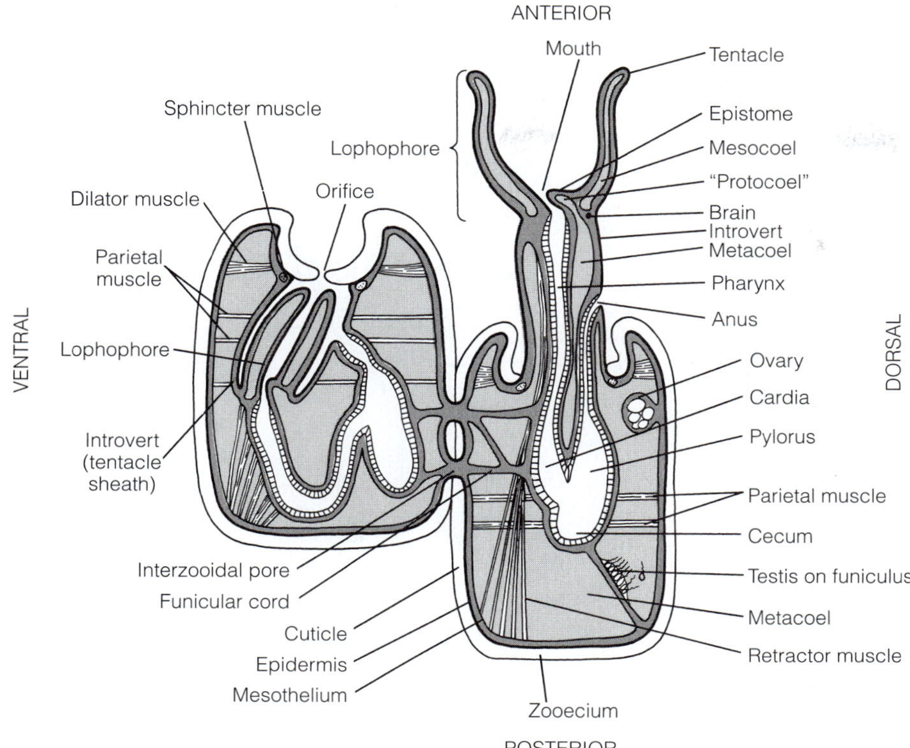

FIGURE 25-17 Bryozoa. Organization of two generalized bryozoan zooids. The zooid on the left has retracted the introvert and lophophore whereas on the right they are extended and in the feeding position. *(Modified and redrawn after Marcus, 1926, from Hyman, L. H. 1959. The Invertebrates. Vol. V. The lesser coelomates. McGraw-Hill Book Co., New York. Reprinted with permission.)*

into the tentacles of the lophophore. The much larger trunk (metasome) constitutes most of the body and contains the spacious metacoel, or trunk coelom (perivisceral coelom). Pores penetrate the septum separating the two coeloms. The introvert, consisting of the lophophore and part of the trunk, can be retracted into the metacoel. The metacoel contains the gut and is crossed by muscle fibers and a strand of mesothelial tissue, the **funiculus,** which functions in nutrient transport. Heart, excretory organs, and specialized gas-exchange surfaces are absent in these tiny animals.

In most regions the body wall is very thin and consists of epidermis, basal lamina, connective tissue, muscles, and the coelomic mesothelium. The epidermis of the metasome secretes a protective exoskeletal **zooecium,** which may be organic (chitin, polysaccharide, or protein) or mineral (calcium carbonate). The zooecium (= animal house), also known as the cuticle, exoskeleton, or ectocyst, encloses and protects the zooid (Fig. 25-17) and has an orifice through which the introvert and lophophore can be extended. If it is calcareous, the zooecium is rigid, but an organic zooecium may be flexible or rigid. Areas of the body with a flexible zooecium may have body-wall muscles arranged in outer circular and inner longitudinal layers. Muscle layers are absent where the body wall underlies a rigid zooecium. The basement membrane, which may be thin or thick, consists of irregularly arranged collagen fibrils. The living portion of the body wall, including the epidermis, basement membrane, muscles, and mesothelium, is sometimes called the endocyst. Ectocyst and endocyst together form the **cystid.**

The remainder of the zooid consists of the lophophore, introvert, gut, gonad, funiculus, specialized muscles, and splanchnic mesothelium, known collectively as the **polypide.** The lophophore encircles the mouth and bears hollow, ciliated tentacles (Fig. 25-17). In phylactolaemates it is horseshoe-shaped (Fig. 25-18A), but in the gymnolaemates it is a simple circle (Fig. 25-24) of 8 to 30 tentacles. When they are retracted, the tentacles are bunched together, but when protruded they fan out, forming a bell-shaped funnel with the mouth at its narrow end (Fig. 25-17).

The multiciliated epidermal cells of the lophophoral tentacles bear frontal and lateral cilia that form an upstream suspension-feeding collecting system, as in other lophophorates. Below the epidermis of each tentacle are a longitudinal nerve, a basal lamina, and a mesothelium enclosing the mesocoel. Some of the mesothelial cells form a longitudinal muscle tract on the frontal edge of the tentacle.

The lophophore is carried on a necklike introvert (tentacle sheath) that can be extended or retracted through the orifice of the zooecium. When extended (Fig. 25-17, right), the introvert is held erect above the body and the conical lophophore is deployed in the feeding position. Retractor muscles pull the introvert and lophophore back into the body (Fig. 25–17, left). In many marine bryozoans, the orifice has a hinged lid, or operculum, that is closed by an **occlusor muscle** after the lophophore is withdrawn (Fig. 25-24D–G).

The mouth is at the center of the lophophore and the anus is on the dorsal surface of the introvert outside the lophophore.

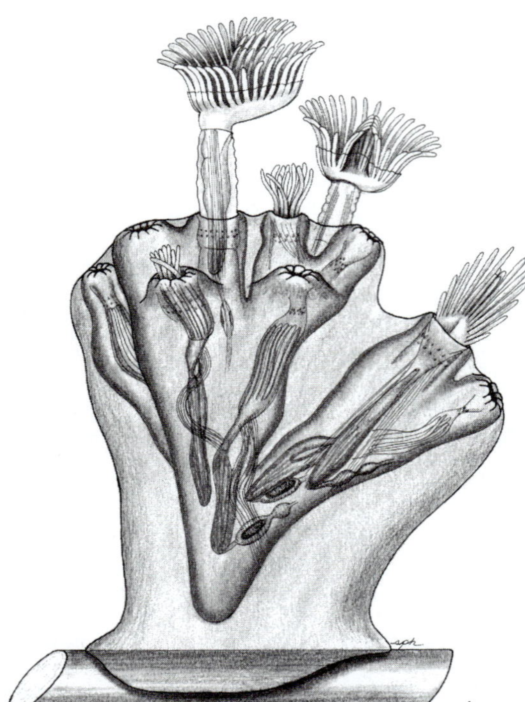

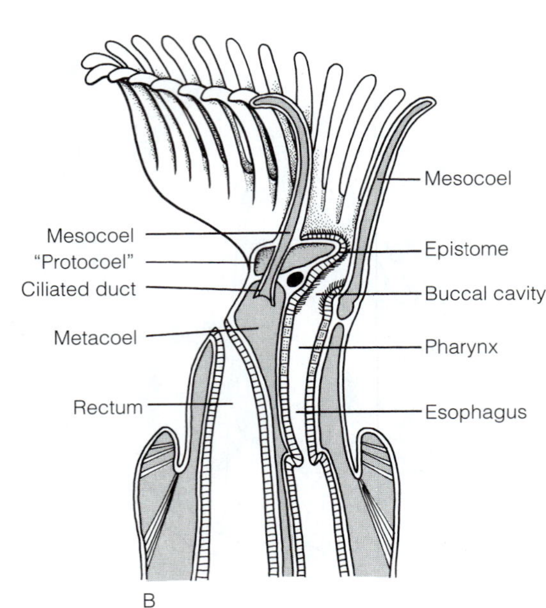

FIGURE 25-18 Phylactolaemate (freshwater) bryozoans. **A,** Small colony of *Lophopus crystallinus* attached to a water plant. **B,** Organization of *Plumatella fungosa,* shown in parasagittal section. One of the two ducts linking the metacoel and mesocoel is shown. *(A, After Allman; B, Modified and redrawn from Brien, P. 1960. Sous-Classe des Phylactolèmes ou Phylactolémages. In Grassé, P.-P. (Ed.): Traité de Zoologie. Vol. 5. No. 2. Masson et Cie, Paris. Reprinted with permission.)*

The alternative name Ectoprocta (meaning "outside anus") refers to that position.

Zooids of the freshwater Phylactolaemata differ from gymnolaemates in many ways. The phylactolaemate body wall contains circular and longitudinal muscle layers and the epidermis is covered by a flexible cuticle that, in *Pectinatella,* for example, is a thick, gelatinous layer, although in some it is a thin protein and chitin tube. The lophophore of freshwater phylactolaemates (with the exception of *Fredericella,* which has a circular lophophore) is horseshoe-shaped, bears 16 to 106 tentacles, and arches dorsally around the mouth (Fig. 25-18). The horseshoe shape provides a greater food-collecting surface for these zooids, which tend to be larger than the marine gymnolaemates.

The phylactolaemate body consists of an epistome, mesosome, and metasome. The epistome, a dorsal hollow lobe overhanging the mouth, is absent in gymnolaemates. The epistome contains a cavity that was once thought to be a protocoel but, since it is connected with the lophophore coelom, probably is not (Fig. 25-17, 25-18B). The epistome is responsible for the name "Phylactolaemata," which means "covered throat." "Gymnolaemata" means "naked throat."

COLONY FORMS

Almost without exception, bryozoans are colonial, or modular, with the colony consisting of physically and physiologically integrated zooids. The form of a bryozoan colony, or **zoarium,** depends on the pattern of asexual budding of the zooids, the degree of polymorphism, the arrangement of the polymorphs within the colony, and the type and amount of secreted skeletal material. In addition to the support provided by the substratum on which the colony grows, there are three sources of skeletal support in bryozoan colonies. These are the turgor pressure of the coelomic fluid (approximately two atmospheres in *Zoobotryon*), calcification (in most bryozoans), and the production of a chitinous, gelatinous (as in *Pectinatella*) or rubbery (as in *Alcyonidium*) extracellular material.

Gymnolaemates exhibit a wide range of colonial forms, including stolonate, fruticose, encrusting, and foliaceous. Members of such genera as *Bowerbankia, Amathia,* and *Zoobotryon* form **stolonate** colonies that resemble hydroids (Fig. 25-15C, D, 25-16B). The feeding zooids arise from a stemlike creeping or erect stolon. The stolons, which have a jointed appearance, are composed of a linear series of specialized tubular zooids. Unspecialized feeding zooids are often well separated from one another but attached by their posterior ends to the common stolon. The exoskeleton of stolonate bryozoans usually lacks calcium carbonate, but may be chitinous.

The vast majority of marine bryozoans are not stolonate, with the colony consisting instead of adjacent, fused feeding zooids. The growth patterns of nonstolonate bryozoan colonies vary greatly. The orientation of the body to the substratum is different from that of stolonate species: The dorsal surface is attached to the substratum, or to other zooids, and the ventral surface, now called the **frontal surface,** is exposed to the overlying water (Fig. 25-15E,F, 25-24D–G).

Many slightly calcified species, such as the common Atlantic *Bugula neritina,* form erect, **fruticose** (bushy) colonies that resemble seaweeds. In *Bugula,* the fruticose form is characteristically attained through a biserial attachment of zooids (Figs. 25-16A, 25-21A), in which the staggered zooecia of side-by-side zooids form erect and branching stems.

The most common type of colony is the **encrusting** form, in which side-by-side and end-to-end zooids are united in a two-dimensional crust, or sheet, one zooid thick, that is attached to rocks, seaweeds, and shells. The exoskeleton is usually calcareous and, because the lateral and end walls of the zooids are fused to those of other zooids, the orifice is located on the exposed upper frontal surface. *Membranipora, Microporella,* and *Schizoporella* are common encrusting genera (Fig. 25-15E,F, 25-16D, 25-19A).

Erect **foliaceous** colonies are composed of a single sheet of zooids or two sheets attached back-to-back that rise leaflike above the substratum. The cheilostome *Thalamoporella* resembles an open head of lettuce, each "leaf" of which is composed of a double sheet of calcareous zooids and is as rigid and fragile as a potato chip. Other colonies are tuftlike and in some, the zooids are radially arranged.

Photo by Judith Winston, Courtesy the Library, American Museum of Natural History.

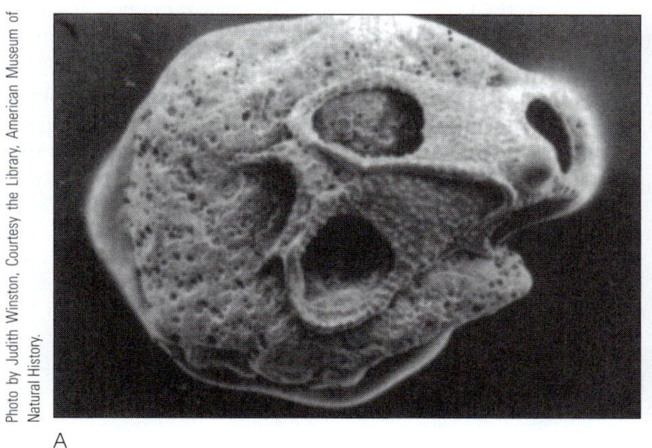

A

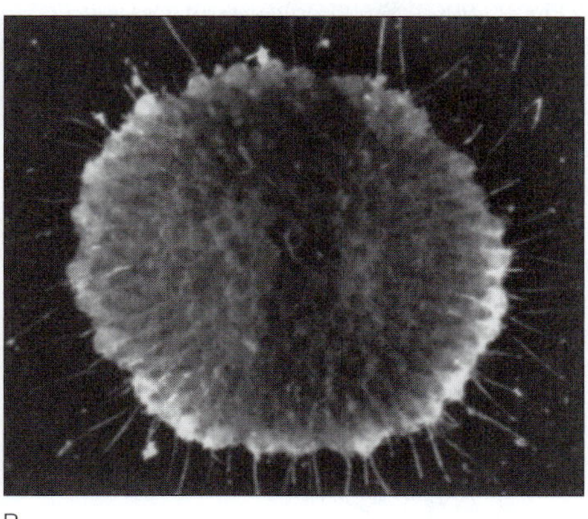

B

FIGURE 25-19 Bryozoans living on or in sand. **A,** A small colony of *Membranipora triangularis* encrusts a single grain of sand. Such colonies can spread from grain to grain by asexual division. Although not a common habitat for bryozoans, the surface of sand grains is exploited by at least 35 species. **B,** A domed colony of *Cupuladria doma* from an offshore sand bottom near Ft. Pierce, Florida. The slender vibracula enable the species to remove sand from its surface and to move over and through the sand.

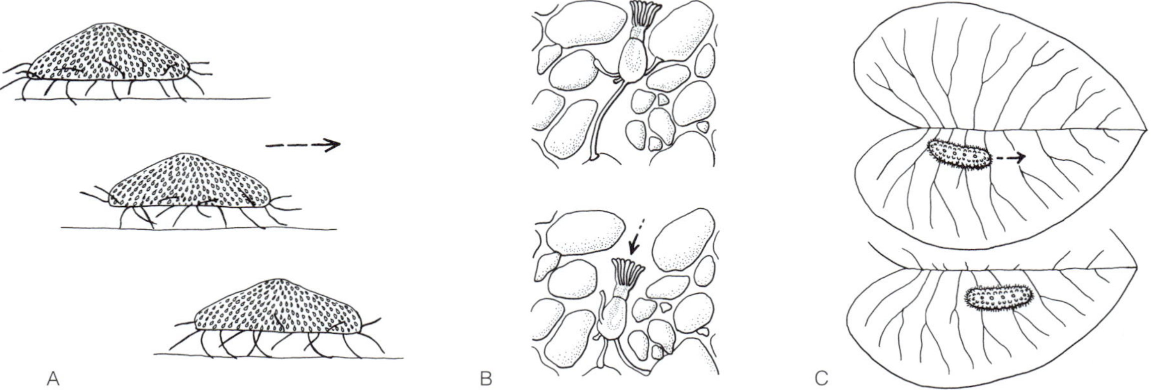

FIGURE 25-20 Bryozoan locomotion. Although of rare occurrence, a few species of bryozoans can move over or through the substratum. *Selenaria maculata* is a small, discoid colony that moves over sand using its marginal vibracula as stepping appendages. As the leglike vibracula move, the colony lurches ahead in 3 mm long increments. **B,** *Monobryozoon ambulans* is a tiny interstitial zooid that uses contractile processes bearing sticky tips to anchor and pull itself through the spaces between sand grains. New buds typically form on the processes, which are sometimes called *pseudostolons.* **C,** A colony of the phylactolaemate *Cristatella mucedo* creeps on its flattened lower surface. The creeping mechanism is unknown, but muscles are thought to be involved. *(A, Drawn from photographs and description of Cook, P. L., and Chimonides, P. J. 1978. Observations on living colonies of* Selenaria *(Bryozoa, Cheilostomata). I. Cah. Biol. Mar. 19:147–158; B, Modified from Swedmark, B. 1964. The interstitial fauna of marine sand. Biol. Rev. 39:1–42. C, After Wesenburg-Lund, 1896, from Hyman, L. H. 1959. The Invertebrates. Vol. V. The lesser coelomates. McGraw-Hill Book Co., New York. Reprinted with permission.)*

The colonies of freshwater phylactolaemates are of two types. In forms such as *Lophopus, Cristatella,* and *Pectinatella,* the zooids project from one side of a soft, globular, saclike zooecium and resemble the fingers of a glove (Fig. 25-18A). Colonies of *Pectinatella* secrete a thick, gelatinous matrix up to a meter in diameter, to which the zooids adhere. The other type of colony, exemplified by *Plumatella, Fredericella,* and *Stolella,* has a more or less stolonate, plantlike growth form, with either erect or creeping branches composed of a succession of zooids. Freshwater bryozoan colonies are attached to vegetation, submerged wood, rocks, and other objects. *Cristatella,* in which the colony is a flattened gelatinous ribbon, is not fixed and creeps over the substratum at a rate of up to 10 cm a day (Fig. 25-20C). Small colonies of *Pectinatella magnifica* and *Lophopus crystallinus* can also move, but only about 2 cm per day. The locomotory mechanism is not known.

Although the zooids are microscopic, the colonies themselves are typically one to several centimeters in diameter or height and may be much larger. Some stolonate species reach a meter or more in size, and some encrusting colonies may attain a diameter of more than 50 cm and contain as many as 2,000,000 zooids. Most colonies are white or pale tints, but darker colors, especially orange, also occur. The taxonomy of marine bryozoans is based almost exclusively on the structure of the exoskeleton and colonial organization.

ZOOID POLYMORPHISM

Although phylactolaemate colonies are strictly monomorphic and those of cyclostomes are almost so, eurystomes (especially cheilostomes) are polymorphic. In polymorphic colonies, the typical feeding zooids, or **autozooids,** make up the bulk of the colony. Modified, nonfeeding zooids specialized to serve a variety of other functions are known collectively as **heterozooids.** Heterozooids typically have a reduced or absent polypide and consist chiefly of the cystid.

Kenozooids are heterozooids modified to serve as stolons, attachment discs, rootlike holdfasts, and defensive spines. In *Membranipora,* for example, spines protecting the vulnerable frontal membrane are produced in response to kairomones

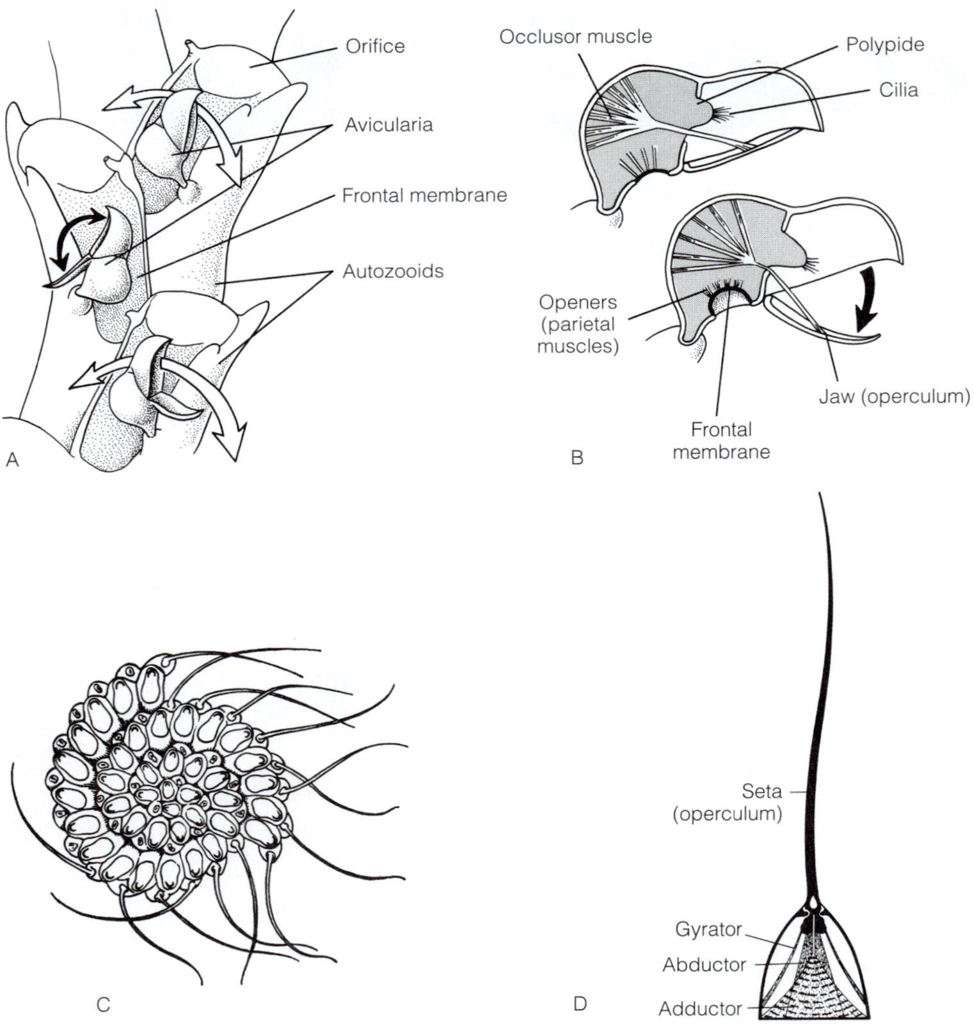

FIGURE 25-21 Bryozoan heterozooids. **A,** Bird's head avicularia of the anascan cheilostome *Bugula fulva.* Open arrows show avicularium movement; solid arrow, movement of the jaw (operculum). The lophophores of the autozooids are retracted. **B,** Simplified anatomy of a bird's head avicularium. Compare with Figure 19-12D,E. **C,** Part of a colony of *Heliodoma,* showing marginal vibracula. **D,** Anatomy of a vibraculum. *(A, Modified and redrawn from Maturo, F. J. S., Jr. 1966. Bryozoa of the southeast coast of the United States: Bugulidae and Beaniidae (Cheilostomata:Anasca). Bull. Mar. Sci. 16:556-583; B, Modified and redrawn after Calvet from Brien, P. 1960. Classe des Bryozoaires. In Grassé, P.-P. (Ed.): Traité de Zoologie. Vol. 5. Part 2. Masson et Cie, Paris; C, From Moore, R. C. 1953. Treatise on Invertebrate Paleontology. Geological Society of America and University of Kansas Press, Lawrence; D, Based on Ryland, 1970.)*

(chemicals) released by predatory nudibranchs. In the absence of the predator, these heterozooids do not develop. The stolons of stolonate bryozoans, such as *Bowerbankia* and *Zoobotryon,* are a succession of cylindrical kenozooids (Fig. 25-15C,D). Kenozooids consist of little more than the body wall and strands of funicular tissue passing through the interior.

Two types of defensive heterozooids, avicularia and vibracula, are found in many cheilostomes. An **avicularium** is usually smaller than an autozooid and its polypide is greatly reduced (Fig. 25-21A,B). The operculum is highly modified to serve as a movable lower jaw, and its occlusor muscle has become a mandibular adductor for closing the jaw, which it can snap shut like a mousetrap. It is abducted (opened) indirectly by contraction of opener muscles on the frontal membrane. These cause the reduced polypide to bulge outward and push the jaw open.

Avicularia may be sessile or stalked. When stalked, they make repetitive nodding motions. Stalked avicularia are found in most *Bugula* species (but not in the common *Bugula neritina*) and resemble diminutive bird heads attached to the colony. Avicularia typically defend the colony against small organisms, including the settling larvae of other animals. The avicularia of *Bugula,* however, appear to be more important in defending against larger crawling animals (0.5 to 4 mm), such as tube-building amphipods and polychaetes, whose appendages are seized by the jaws of the avicularium. The sessile avicularia of *Reptadeonella costulata* can immobilize and strangle syllid polychaetes.

The operculum of a **vibraculum** is modified to form a long movable bristle with muscles that can move it in more than one plane (Fig. 25-21D). A combination of muscles inserting on the base of the seta is used to sweep it in an arc over the surface of the colony. The vibraculum is used by the sand-dwelling *Discoporella, Cupuladria,* and *Diplosolen* to sweep particles away from the surface of the colony (Fig. 25-22). Vibracula are also used for locomotion in these motile colonies. Their combined flicking movements propel the colony vertically through the sand and even horizontally over the surface (Fig. 25-21C, 25-19B).

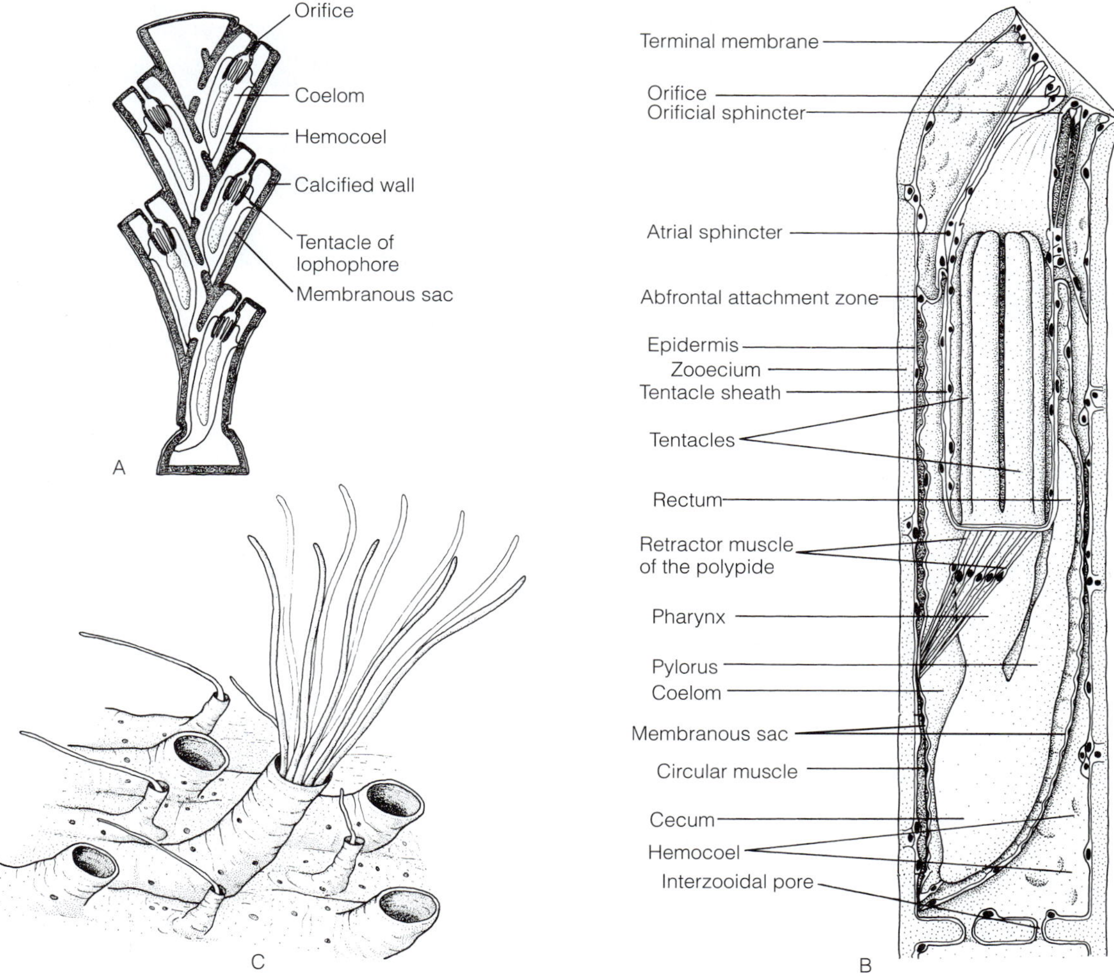

FIGURE 25-22 Cyclostome bryozoans. **A,** Diagram of a colony of *Crisia.* **B,** Diagram of one zooid of *Crisia.* **C,** Five autozooids and four heterozooids of *Diplosolen.* The sweeping motion of the single tentacle of the heterozooid is believed to keep the surface of the colony clean. *(A and B, From Nielsen, C., and Pedersen, K. J. 1979. Cystid structure and profusion of the polypide in* Crisia *(Bryozoa, Cyclostomata). Acta Zool. 60:65–88; C, From Silen, L., and Harmelin, J.-G. 1974. Observations on living Diastoporida, with special regard to polymorphism. Acta Zool. 55:81–96.)*

Species of the discoid *Selenaria* move efficiently over sand by coordinating the movements of its marginal vibracula (Fig. 25-20A). Heterozooids modified to function as reproductive **ovicells** are common and will be discussed later.

INTERZOOIDAL PORES

Adjacent zooids of a colony are connected by **interzooidal pores** passing though the intervening body walls (Fig. 25-17, 25-22B). Interzooidal communication and transport are accomplished through these pores, but by different mechanisms in different taxa. In phylactolaemates and cyclostome gymnolaemates the pores are open and fluid flows freely through them, but in eurystomes the pores are plugged with cells and the flow is regulated.

In phylactolaemates, the epidermis and mesothelium of adjacent zooids are continuous through the open pores (Fig. 25-23C). Consequently, coelomic fluid passes freely through the pores of adjacent zooids.

In cyclostomes, too, the pores are open, but the mesothelium has pulled away from the epidermis and does not pass through the pores, which thus are lined only with epidermis. Consequently, adjacent coeloms are not in communication (Fig. 25-23D) and coelomic fluid cannot flow through the pores. The coelom is reduced in size and enclosed by the mesothelium, which forms a **membranous sac** (Fig. 25-22B). Because the mesothelium is not adjacent to the epidermis, there is a large, fluid-filled connective-tissue compartment, or hemocoel, between them (Fig. 25-22B). As a result, the interzooidal pores connect the hemocoels, not the coeloms, of

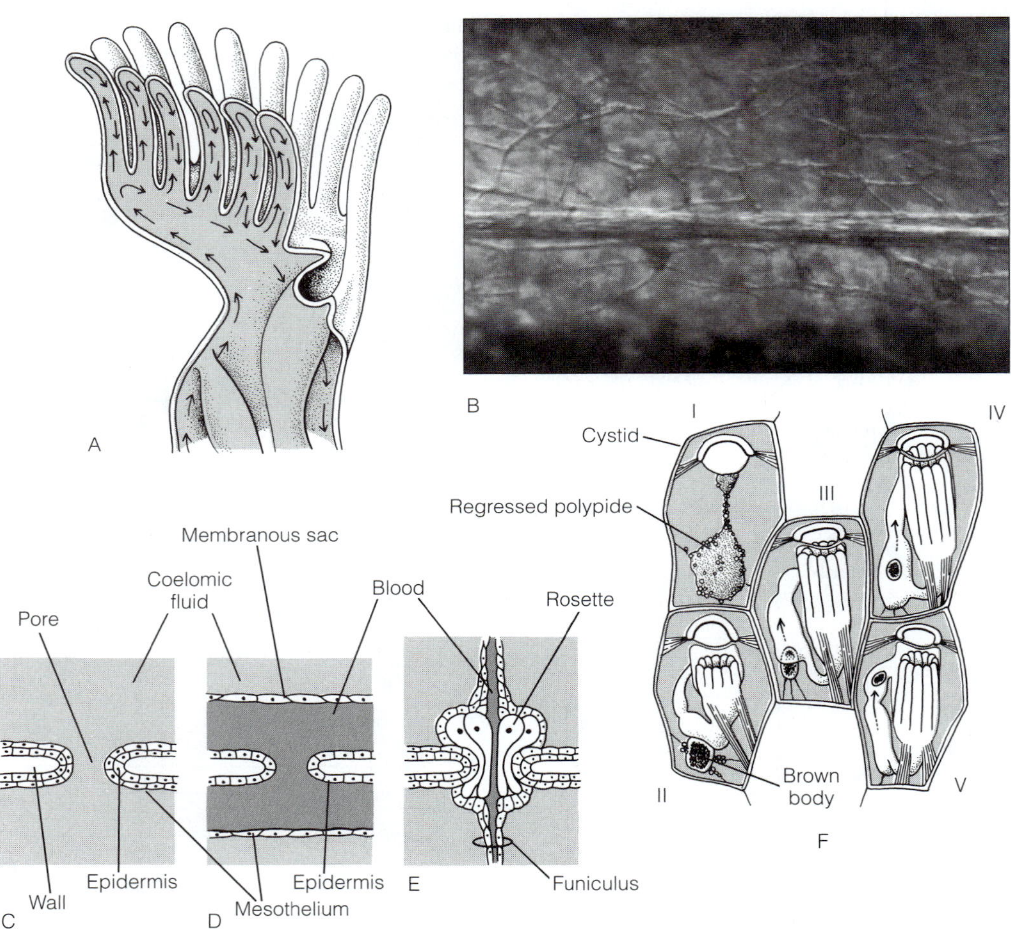

FIGURE 25-23 Bryozoa: internal transport and polypide regression. **A,** Circulation of coelomic fluid using cilia in the phylactolaemate *Pectinatella.* **B,** Photograph of the main funiculus and its branches in a kenozooid (stolon zooid) of the ctenostome *Zoobotryon verticillatum.* The bases of a few attached autozooids can be seen at the top of the photograph. **C–E,** Communication pores between two zooids of a: phylactolaemate **(C),** cyclostome **(D),** and eurystome **(E). F,** Polypide regression (I) and brown body formation and expulsion (II–V) in the anascan cheilostome *Electra pilosa.* *(A, After Oka, 1891, from Hyman, L. H. 1959. The Invertebrates. Vol. V. The lesser coelomates. McGraw-Hill Book Co., New York; C–E, Modified and redrawn from Mackie, G. O. 1986. From aggregates to integrates: Physiological aspects of modularity in colonial animals. Phil. Trans. R. Soc. Lond. B 313:175–196; F, Redrawn and rearranged after Marcus from Brien, P. 1960. Classe des Bryozaires. In Grassé, P.-P. (Ed.): Traité de Zoologie. Vol. 5. Part 2. Masson et Cie, Paris. Reprinted with permission.)*

adjacent zooids and it is blood (hemolymph), not coelomic fluid, that flows through them.

In **eurystomes** the interzooidal pores are closed by a plug of **rosette cells** that regulate the flow of materials, and neither coelomic fluid nor blood passes freely through the pores. Instead, branches of the funiculus extend to the interzooidal pores (Fig. 25-23E). Organic materials are transported from the funiculus of one zooid through the rosette cell cytoplasm to the funiculus of the zooid on the opposite side. All zooids in a colony are connected by the funicular system, making it possible for nonfeeding zooids to be nourished by those with direct access to food.

FUNICULAR SYSTEM

A fluid-filled, tubular **funicular system** is found in the coelom of all bryozoans (Fig. 25-17, 25-23B). Its mesothelial wall surrounds a connective-tissue space filled with fluid, presumably blood. It always connects the mesothelium of the gut with that of the body wall, and in some taxa other structures are also included. The testis forms on the funiculus, and if a uterine chamber or placenta is present, it too is always served by a funiculus. In phylactolaemates the overwintering statoblasts form on the funiculus. Because it connects the source of nutritive materials (the gut) with tissues having a demand for those materials, it is believed to be a transport system for food, but the mechanism of blood flow has not been determined.

In eurystomes (ctenostomes and cheilostomes) the funiculus is associated with internal organs, as described above, but it also extends to the interzooidal pores and rosette cells and transports materials to and from contiguous zooids. It thus functions in both inter- and intrazooidal transport. In phylactolaemates and cyclostomes the funiculus is not associated with these pores and probably has a purely intrazooidal transport role.

MUSCULATURE

Bryozoan musculature consists chiefly of a pair of lophophore retractor muscles that are present in all bryozoans, circular and longitudinal layers found in the body wall of phylactolaemates and cyclostomes, parietal muscles in eurystomes, and opercular occlusor muscles in cheilostomes. These muscles may all be involved, directly or indirectly, in extension and retraction of the lophophore and operation of the operculum or its derivatives. Both smooth and striated muscles are present.

Lophophore Protraction

Preparatory to feeding, the introvert and lophophore are extended (protruded) outward through the orifice (Fig. 25-17). The tentacles then expand, forming a bell-shaped funnel, and feeding commences. Lophophore protrusion is accomplished in all bryozoans by muscular elevation of coelomic hydrostatic pressure, although the mechanism by which this is accomplished varies.

In phylactolaemates, which have a flexible cystid, coelomic fluid pressure is elevated by contracting the circular body-wall musculature, which compresses the coelom and extends the introvert. The mechanism of introvert protrusion in cyclostomes, such as *Crisia,* depends on the contraction of circular body-wall muscles, as in the phylactolaemates. Unlike the freshwater bryozoans, however, the cyclostome cuticle is calcified and rigid and the entire body cannot be compressed to accomplish protrusion. Instead, the body-wall muscles are located in the membranous sac that, along with the coelom it encloses, is widely separated from the rigid cystid by the hemocoel. Contraction of circular muscles of the membranous sac pressurizes the coelom and extends the lophophore (Fig. 25–22A,B).

In eurystomes with a flexible, chitinous zooecium *(Bowerbankia, Amathia),* contraction of circular **parietal muscles** compresses the body and elevates the coelomic pressure to protract the lophophore (Fig. 25-17, 25-24A–C).

In some calcareous eurystomes with a rigid zooecium (anascan cheilostomes, such as *Membranipora*) and some species with a rigid chitinous exoskeleton, the otherwise rigid frontal wall includes a thin, flexible, chitinous **frontal membrane** (Fig. 25-24D,E, 25-15E). When parietal muscles that insert on the inner side of the membrane are contracted, the frontal membrane bows inward, increasing the coelomic pressure and ejecting the lophophore.

An uncalcified frontal membrane, however, is vulnerable to attack by predators and many calcareous eurystomes (Ascophora) protect the frontal surface by calcifying it. In these species, the flexible pressure-regulating membrane is internalized in the form of a sac called an **ascus,** which opens to the exterior via one or two tiny pores. The parietal muscles attach to the ascus and upon contraction increase its volume, causing water to enter through the pore. As the volume of the ascus increases, the coelomic fluid pressure is raised and the lophophore protrudes (Fig. 25-24F,G).

Lophophore Retraction

Retraction of the lophophore and introvert is more straightforward and in all bryozoans is accomplished directly by the contraction of the two **lophophoral retractor muscles** (Fig. 25-17, 25-24A,D). These bundles of individual muscle fibers extend from the cystid to the lophophore or anterior gut. Retraction is rapid, often requiring less than 60 msec to complete. Re-extension is slower, and may be hesitant. Zooids may partly extend the tightly bundled tentacles and test the water with sensory cilia at their tips before fully deploying the lophophore. In cheilostomes the operculum closes following retraction of the lophophore. It is pushed open again by the extending lophophore.

NUTRITION

Capture

Most bryozoans are thought to be suspension feeders chiefly utilizing small phytoplankton. When the lophophore is protruded, the lateral cilia on its tentacles create a current that sweeps downward into the open end of the funnel and then passes outward between the tentacles (Fig. 25-25A). Small particles are drawn into the funnel with the water current, trapped on the tentacles, and delivered to the mouth by frontal cilia on the inner surface of each tentacle.

The mechanism by which small food particles are separated from the water is not fully understood, but two hypotheses have been advanced. The **ciliary reversal hypothesis** suggests

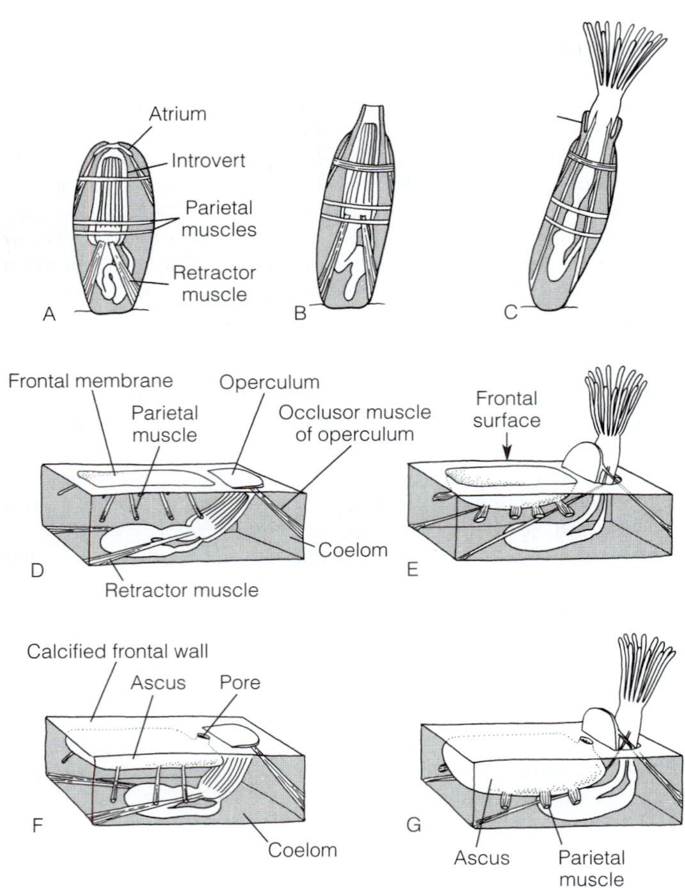

FIGURE 25-24 Bryozoa: polypide protrusion in gymnolaemate bryozoans. **A–C,** Contraction of parietal muscles in ctenostomes compresses the body to elevate the coelomic pressure, protruding the polypide. The collar in many ctenostomes **(C)** is pleated longitudinally, and each pleat resembles a tooth on a comb. The name *ctenostome,* meaning *comb mouth,* is a reference to this feature. **D** and **E,** Although the cystid wall is largely calcified in the anascan cheilostomes such as *Membranipora,* one surface, the frontal membrane, remains uncalcified and flexible. Contraction of the parietal muscles bows the frontal membrane inward **(E),** which elevates the coelomic pressure and protrudes the polypide. **F** and **G,** In the ascophoran cheilostomes, the frontal wall is calcified and the flexible surface is internalized as a sac, or ascus. The action of the parietal muscles is to bow one wall of the ascus inward **(G),** which elevates the coelomic pressure and extends the polypide. Water enters the ascus via the surface pore during polypide protrusion. In the cribrimorph cheilostomes (not illustrated) the frontal wall is membranous but protected by an overarching canopy of partially fused spines. Parietal muscles operate the frontal membrane, as they do in anascans, but the otherwise vulnerable membrane is protected by a calcified plate, as it is in ascophorans.

that when particles touch the lateral cilia they cause a local reversal of beat, which kicks the particle back onto the upstream, frontal side of the tentacle for transport to the mouth by the frontal cilia. The **impingement hypothesis,** on the other hand, proposes that the momentum of particles entering the funnel causes them to continue straight into the mouth when the water current bends sharply to exit between the tentacles. Regardless of the process that separates particles from the water, the feeding mechanism is an upstream ciliary collecting system. Particle rejection may be accomplished by closing the mouth, tentacle flicking, funnel closure, or simply being passed between the tentacles (Fig. 25-25E).

Many species scan for particles by rotating or bending the lophophore (Fig. 25-26). Multiple, adjacent, similarly oriented, expanded funnels may form an extensive filtering array. In encrusting species, the large volume of water passing through and then below the filters exits the colony through "chimneys," which are open, lophophoreless areas. Chimneys form either where lophophores tilt away from each other or in the spaces created by nonfeeding heterozooids.

Bryozoans also employ accessory feeding mechanisms. They absorb dissolved organic molecules across the microvillous tentacle epidermis. In **tentacular flicking,** which is exhibited by all bryozoans, an individual particle is batted toward the mouth with a rapid, inward flick of a tentacle (Fig. 25-25B, 25-26). Some bryozoans are part-time predators exploiting zooplankters that stray into the lophophore. *Bugula neritina* captures zooplankton by closing the tips of its tentacles to form a cage around the prey.

Digestion

The mouth at the apex of the lophophore opens into a muscular, sucking pharynx, which leads to the tubular esophagus (Fig. 25-17). Most of the U-shaped gut is the large, three-part stomach consisting of the cardia, cecum, and pylorus. The esophagus joins the anterior **cardia** at a sphincter valve. A second sphincter separates the posterior **pylorus** from the intestine and rectum. The large pouch-like **cecum** projects backward from the central part of the stomach. In some bryozoans, such as *Bowerbankia* and *Amathia,* the cardia is a muscular **gizzard** that crushes diatoms, exposing their contents to digestive enzymes (Fig. 25-25D). The gizzard has a well-developed circular muscle layer and its gastrodermis bears chitinous chaetae similar to those of annelids and brachiopods.

Captured food particles accumulate in a bolus beneath the epistome in phylactolaemates and within the expanded mouth of gymnolaemates. When the bolus reaches a critical

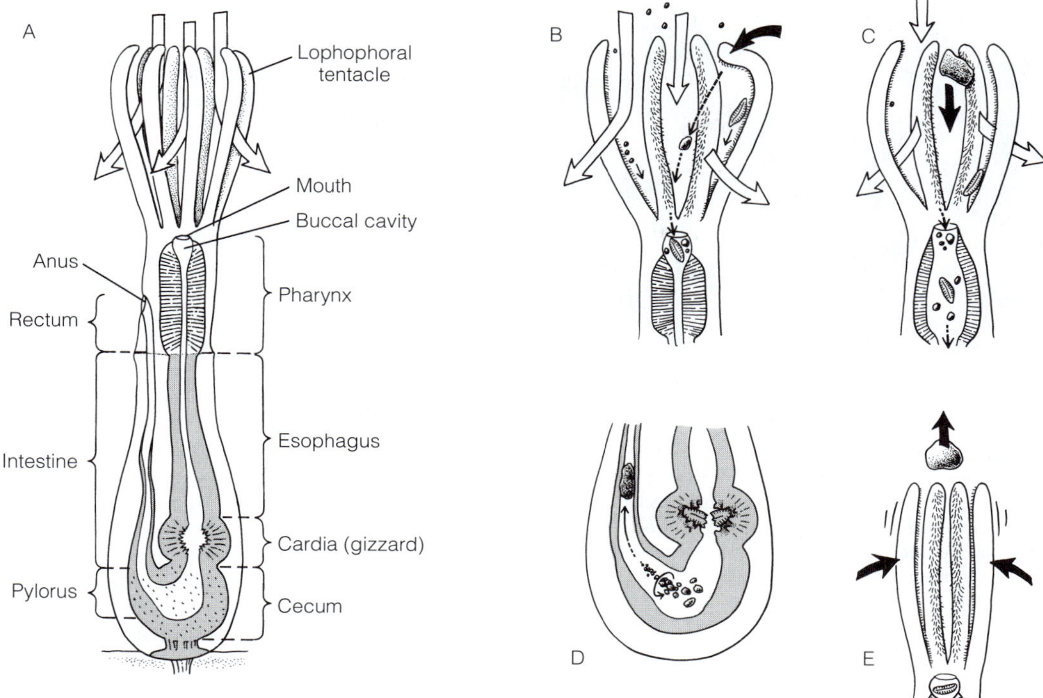

FIGURE 25-25 Bryozoa. Digestive system and feeding in gymnolaemates (based on *Zoobotryon*). **A,** Water currents (open arrows) and digestive system. **B–D,** Water currents (open arrows) bring in food particles **(B)**, which are either batted toward the mouth by a tentacular flick (large solid arrow, dashed arrow) or caught on the tentacle surface and moved toward the mouth by frontal cilia (small arrows). The collected particles accumulate in the ciliated buccal cavity **(B)** and are engulfed by a rapid dilation of the muscular pharynx **(C)**. The food particles move rapidly down the esophagus and pass through the muscular gizzard, which crushes diatoms, before entering the cecum **(D)**. Once in the cecum, the food is rotated by cilia and digested; indigestible material is compacted into fecal pellets in the intestine **(D)**. Rejected particles may pass between the tentacles and be carried away in the exhaust, but sometimes the tentacles bunch together rapidly and particles are rejected as shown in **E.**

size, the pharynx dilates rapidly and sucks it in. Subsequent contraction of the pharynx pushes the bolus into the stomach (Fig. 25-25B,C).

Digestion is both extracellular and intracellular within the stomach, chiefly in the cecum. Peristaltic contractions move food through the stomach, but ciliary activity in the pylorus rotates and compacts waste materials, which then pass into the intestine (Fig. 25-25D) and ultimately out the anus.

GAS EXCHANGE, INTERNAL TRANSPORT, NERVOUS SYSTEM

Gas exchange occurs across the exposed body surface, especially the lophophore. Intrazooidal transport of gases, some food, and wastes is provided by the coelomic fluid. Ciliary circulation of coelomic fluid is known only in the relatively large (2 to 4 mm) zooids of some phylactolaemates *(Paludicella, Pectinatella;* Fig. 25-23A). The coelomic fluid contains coelomocytes (but no respiratory pigments), which engulf and store waste materials. In cyclostomes, it is likely that both coelomic and hemocoelic fluids are involved in transport (Fig. 25-23D). The funicular system provides for at least some nutrient transport and is the main system for the colonywide dispersal of metabolites in ctenostomes and cheilostomes (Fig. 25-23B,E). There is no heart or typical blood vessels.

The nervous system consists of a nerve ring around the pharynx with a ganglionic mass (the brain) on its dorsal side (Fig. 25-17). The ganglion and ring give rise to nerves that extend into each of the tentacles and to other parts of the body. There are no specialized sense organs in bryozoans, but individual sensory cilia occur on the tentacles. The growth of the branches of *Bugula* is positively phototropic, so there must be a photoreceptive capability.

Electrophysiological studies confirm that conduction of nerve impulses occurs in colonies of some nonstolonate bryozoans such as *Electra* and *Membranipora.* Nerves that encircle the zooid wall enter the interzooidal pores and make connections with similar nerves in adjacent zooids. This colonial nervous system coordinates lophophore feeding activity, retraction, and orientation within the colony.

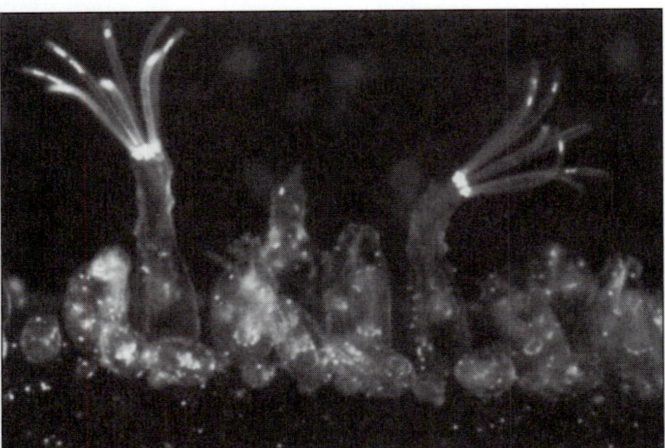

FIGURE 25-26 Bryozoa: living zooids of the ctenostome *Zoobotryon verticillatum.* The two fully extended zooids are actively orienting their lophophores while feeding. The right zoid is using a tentacle flick to bat a food particle (not visible) toward the mouth. Several retracted autozooids as well as new autozooid buds are visible in the photograph. All of the autozooids have arisen from and are attached to a large horizontal kenozooid (stolon segment) at the bottom of the photograph.

EXCRETION

Nephridia are not present in bryozoans and ammonia presumably diffuses across the body and lophophore surface. Other wastes, such as uric acid, may be stored temporarily or indefinately in body tissues.

Polypide regression and brown-body formation is probably a type of storage excretion (Fig. 25-23F) and is characteristic of bryozoans. Regardless of the age of a colony, the polypide (lophophore and gut) of each zooid regresses (degenerates) after a few weeks, leaving the cystid (zooecium and body wall) unaffected. Some polypide components are phagocytosed and their molecules reused, but a large residual mass of necrotic cells containing accumulated waste products remains lodged in the coelom as a conspicuous dark sphere called a **brown body.** Regression is followed by the regeneration of a new polypide from the tissues of the cystid. In some species, the brown body remains permanently in the coelom, whereas in others it is incorporated in the stomach of the new polypide and expelled at the first defecation (Fig. 25-23F).

REPRODUCTION AND DEVELOPMENT

All freshwater and most marine bryozoans are hermaphroditic. The one or two ovaries and the one to many testes are bulging masses of developing gametes covered by mesothelium (retromesothelial). The ovaries are located in the distal end of the zooid whereas the testes occur on the funiculus in the basal end (Fig. 25-17). Gametes rupture from the gonads into the metacoel, but conventional gonoducts for their exit are absent. Simultaneous production of eggs and sperm may take place in some, but a tendency toward protandry is more common and self-fertilization is probably rare. A colony may thus be composed of functionally male and female zooids in different stages of protandry. Fertilization between zooids of the same colony is probably common, but sufficient cross-fertilization between colonies occurs to ensure outbreeding.

Fertilization

Sperm must exit the coelom and travel to another zooid to achieve fertilization. Eggs, on the other hand, may either remain in the coelom, where they are fertilized internally (in which case sperm must enter the coelom), or be released from the coelom to be fertilized externally.

To exit the coelom, eggs move along a ciliated groove in the mesothelium leading to an opening, the **coelomopore,** between the bases of the two dorsalmost tentacles (Fig. 25-27A,B). The coelomopore may be a simple opening or it may be at the apex of a projection called the **intertentacular organ.** Note that the position of the coelomopore is similar to that of the nephridiopores of phoronids, suggesting that it may be the vestige of a metanephridium.

In all species that have been studied (such as *Electra, Membranipora, Schizoporella, Bugula,* and others), sperm are shed from the coelom though **terminal pores** at the tips of two or more lophophoral tentacles (Fig. 25-27A). The liberated sperm, when caught in the feeding currents of other individuals, adhere to the tentacles (for external fertilization) or enter the intertentacular organ (for internal fertilization). Eggs may be fertilized externally as they leave the coelomopore or internally in the intertentacular organ (as occurs in *Electra* and *Membranipora;* Fig. 25-27B) or coelom.

Development

Cleavage is radial or biradial (Fig. 25-28), holoblastic, equal or subequal, regulative, and results in a coeloblastula. **Biradial cleavage** is a form of radial cleavage in which the early embryo is bilaterally, rather than radially, symmetrical, although the blastomeres are aligned more or less on top of each other as they are in radial cleavage. Early cleavages first produce two layers of cells (Fig. 25-28B), one on top of the other, with the upper being the animal plate and the lower the vegetal. A small blastocoel lies between the two layers. Gastrulation occurs when four central, presumptive mesendodermal blastomeres in the vegetal plate divide and contribute four daughter blastomeres to the blastocoel. These four daughter cells divide to fill the blastocoel and eventually give rise to endoderm and mesoderm. The method of coelom formation is unclear. The blastopore closes and the mouth develops from a new opening.

Brooding and Gestation

The vast majority of bryozoans retain their eggs, which are almost always large, yolky, and few in number, and development may take place internally or externally. In the few viviparous species gestation occurs in the coelom, where the gut and lophophore may degenerate to provide the necessary space. Most, however, brood in a variety of external brood chambers.

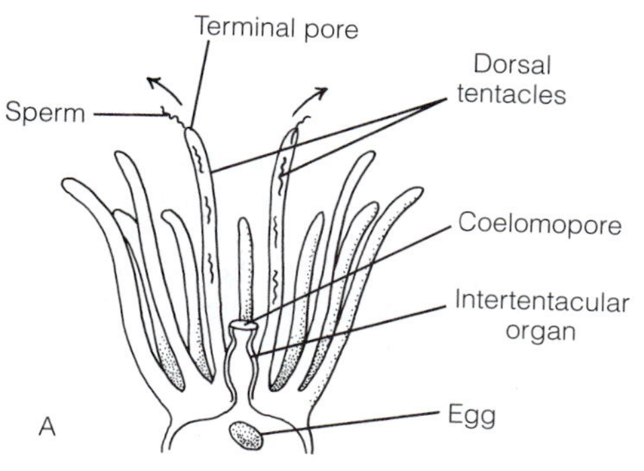

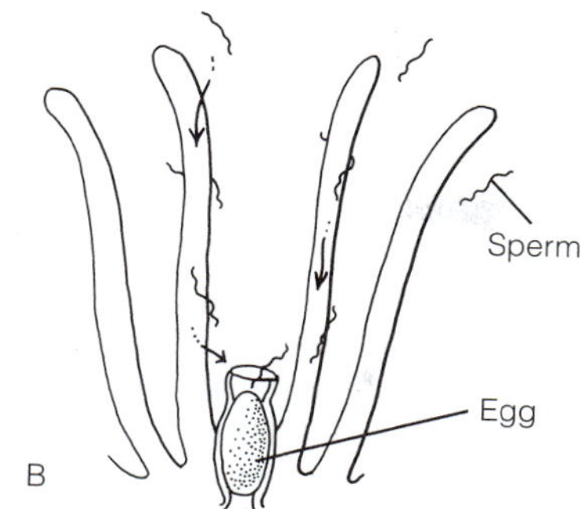

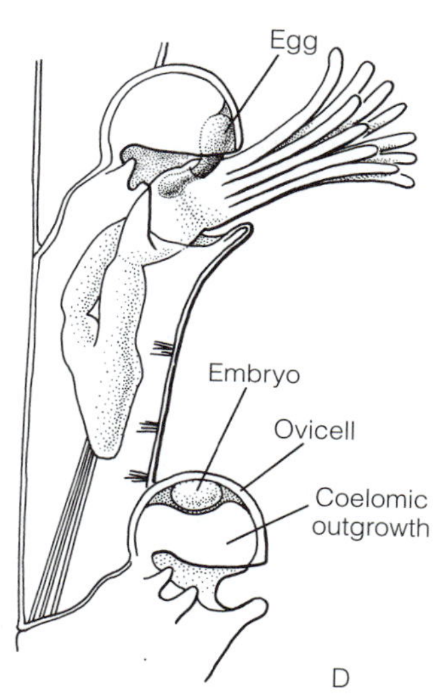

FIGURE 25-27 Bryozoan reproduction: sperm release and egg brooding in bryozoans. **A** and **B,** Dorsal view of the lophophore of *Electra posidoniae.* **A,** Sperm release and entry of egg into the intertentacular organ. **B,** Sperm entry and fertilization. **C,** Autozooid and nearby ovicell containing a developing embryo in *Bugula neritina.* **D,** *Bugula avicularia,* showing extrusion of an egg from the coelomopore of an autozooid into the ovicell (above) and an egg positioned within an ovicell (below). *(A and B, Modified and redrawn from Silen, L. 1966. On the fertilization problem in the gymnolaematous bryozoa. Ophelia 3:113–140; D, Modified and redrawn from Brien, P. 1960. Classe des Bryozaires. In Grassé, P.-P. (Ed.): Traité de Zoologie. Vol. 5. Part 2. Masson et Cie, Paris. Reprinted with permission.)*

The cavity of the introvert, invaginations of the distal wall, or special heterozooids known as ovicells are typical sites for brooding. Zygotes or embryos exit the coelom via the coelomopore and enter the brood chamber. In species with ovicells, the malleable egg deforms and oozes through the coelomopore before resuming its characteristic spherical shape within the ovicell (Fig. 25-27D).

Ovicells are characteristic of many cheilostomes, such as *Bugula* (Fig. 25-27C,D). The cystid at the distal end of the maternal zooid grows outward to form a large hood, which is the ovicell. A second, smaller evagination that arises from the maternal coelom bulges into the space formed by the ovicell. A single egg is released from the coelomopore and brooded in the ovicell. The developing embryo may derive its nutrition entirely from yolk, but in many species, including *Bugula,* placental connections from the funiculus of the maternal zooid to the ovicell provide food to the embryo.

Cyclostomes also brood their larvae and are unusual in exhibiting polyembryony, in which the original embryo asexually produces numerous secondary embryos, which may in turn produce tertiary embryos, all of which are genetically identical (clones) and capable of founding a new colony.

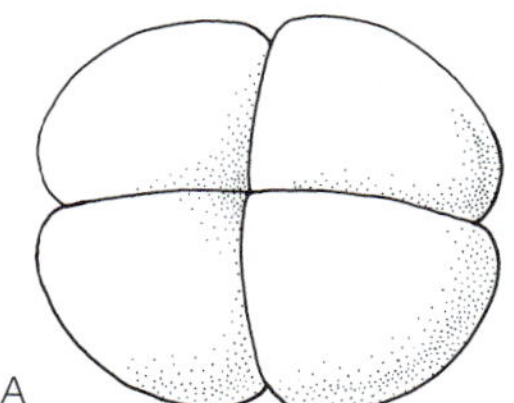

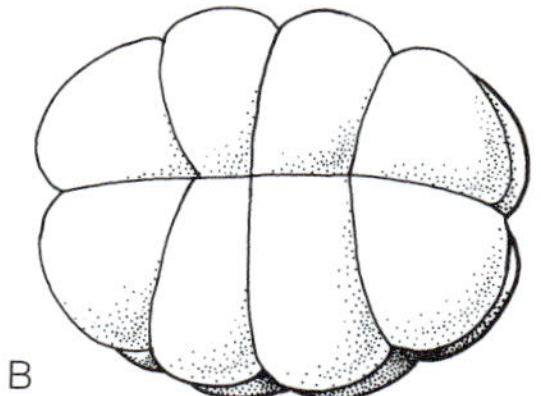

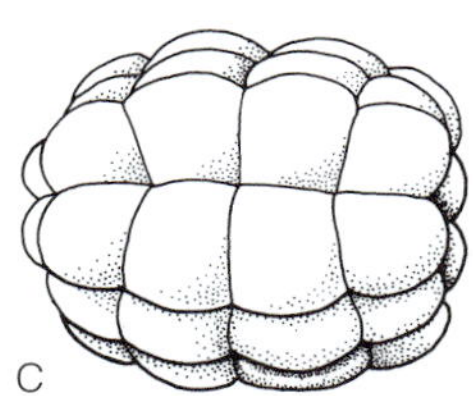

FIGURE 25-28 Bryozoan development: biradial cleavage in the gymnolaemate *Schizoporella erata.* Animal pole views. **A,** Eight-cell embryo. The four blastomeres of the vegetal hemisphere are not visible, but lie directly under the four visible cells. **B,** Sixteen-cell embryo. **C,** Thirty-two-cell stage. *(Redrawn after Zimmer from Reed, C. G. 1991. Bryozoa. In Giese, A. C., Pearse, J. S., and Pearse, V. B. (Eds.): Reproduction of Marine Invertebrates. Vol. VI. Echinoderms and Lophophorates. The Boxwood Press, Pacific Grove, CA. pp 85–245.)*

Up to 100 embryos may be produced from a single zygote by this process.

Phylactolaemates brood their embryos in **embryo sacs** in the cystid. The embryo sac originates as an external invagination of the cystid wall that bulges into the parental coelom.

Larvae

Bryozoan larvae vary considerably in form (Fig. 25-29), but all have a locomotory ciliated **corona** that girdles the body, an anterior tuft of apical cilia, and a ventral, adhesive metasomal sac (Fig. 25-6D). During settling, the metasomal sac everts and fastens to the substratum with adhesive secretions. The larval structures undergo histolysis and degeneration, which is followed by development into an autozooid.

A few gymnolaemates, such as *Electra* and *Membranipora,* produce small, microlecithal eggs that develop into long-lived planktotrophic **cyphonautes larvae** that are triangular, laterally compressed, and enclosed in a chitinous bivalved shell (Fig. 25-29A, 25-6C,D) having right and left valves. The cyphonautes closely resembles the phoronid actinotroch in many respects, including the metasomal sac from which the adult epidermis, or part of it, is derived (Fig. 25-6). Both have a preoral hood overhanging a large vestibule that leads into the mouth and both have ciliated tentacles (ciliated ridge) near the mouth. In addition, the cyphonautes has a pair of protonephridia (referred to as "a network of ciliated tubules" by S. A. Stricker, C. G. Reed, and R. L. Zimmer, 1988a) between the membranous sac and the posterior gut, as does the actinotroch.

Among Bryozoa, only planktotrophic larvae have a functional gut and feed during their larval existence. Such larvae may take several months to settle. The lecithotrophic larvae of brooding species, which is most bryozoans, do not feed during their brief larval existence prior to settling (Fig. 25-29B,C).

In phylactolaemates a placenta develops between the maternal zooid and the embryo in the embryo sac. Development leads to the formation of a larva whose ciliated surface (Fig. 25-29D) may correspond to the coronal cilia of gymnolaemate larvae. One end of the larva invaginates to form a **vestibule** in which budding produces one to several precocious polypides. The larva thus contains a prefabricated young colony. The larva swims about for a short time and then settles and attaches, after which the once-ciliated outer epithelium of the larva degenerates, leaving the young colony, which continues to bud more zooids until an adult colony is formed. A parent zooid dies after producing a number of daughter zooids. Thus, in branching colonies, only the tips of branches contain living zooids, whereas in flattened, gelatinous colonies, living zooids are restricted to the periphery.

Metamorphosis

At the end of its planktonic life a larva settles on an appropriate substratum and attaches with its everted metasomal sac (Fig. 25-30). The sac secretes a **cocoon** that encloses the metamorphosing larva. Within the cocoon the larval epidermis migrates inward to become the lining of the adult coelom. The polypide and cystid develop from larval **rudiments** and the cystid epidermis secretes a zooecium. The **residual larval cells** form a food reserve that nourishes the developing ancestrula until it begins feeding.

Colony Development

The first zooid to develop from the settled larva is the **ancestrula** (Fig. 25-30E, 25-31B,C). Although a colony may eventually consist of thousands of zooids, all are the asexual descendants of a single original ancestrula. By means of clonal budding, the ancestrula gives rise to a series of new zooids that, although genetically identical, may be polymorphic, exhibiting different shapes and sizes. Continuous asexual reproduction increases the size of the colony (Fig. 25-31).

Clonal budding involves the growth and subsequent segregation of a part of a parent zooid by the formation of a body-wall partition. The new daughter zooid begins as a cystid with only its body wall. A new polypide develops from the ectoderm and mesothelium of the cystid. The exact pattern of budding (the number and location of buds on their parents) determines the growth pattern and morphology of the colony. In

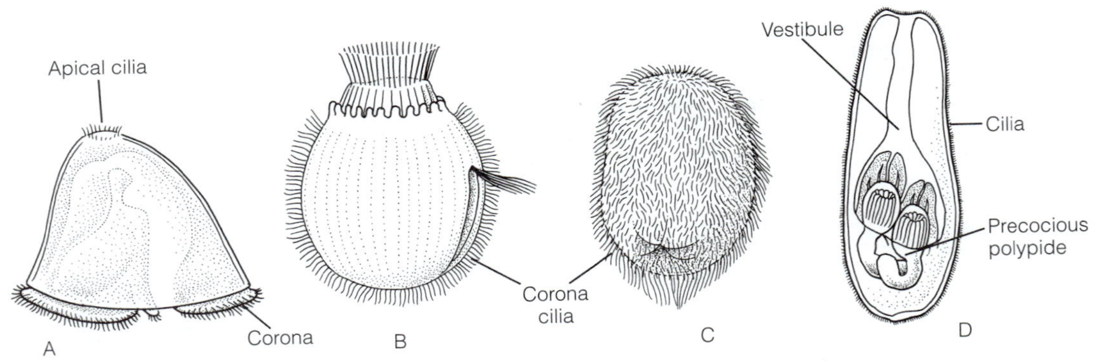

FIGURE 25-29 Bryozoan larvae. **A,** A planktotrophic cyphonautes larva, as found in *Membranipora* and *Electra.* The larva is laterally compressed and bears a shell (see also Figure 25-6D). **B–D,** Lecithotrophic larvae of the gymnolaemate *Bugula neritina* **(B),** the cyclostome *Crisia eburnea* **(C),** and a phylactolaemate **(D).** In phylactolaemates, autozooids differentiate precociously within the larval body. *(B, Redrawn from Nielsen, C. 1971. Entoproct life-cycles and the entoproct/ectoproct relationship. Ophelia 9:209–341; C, Redrawn from Nielsen, C. 1970. On metamorphosis and ancestrula formation in cyclostomatous bryozoans. Ophelia 7:217–256; D, Redrawn from Brien, P. 1960. Classe des Bryozaires. In Grassé, P.-P. (Ed.): Traité de Zoologie. Vol. 5. Part 2. Masson et Cie, Paris. Reprinted with permission.)*

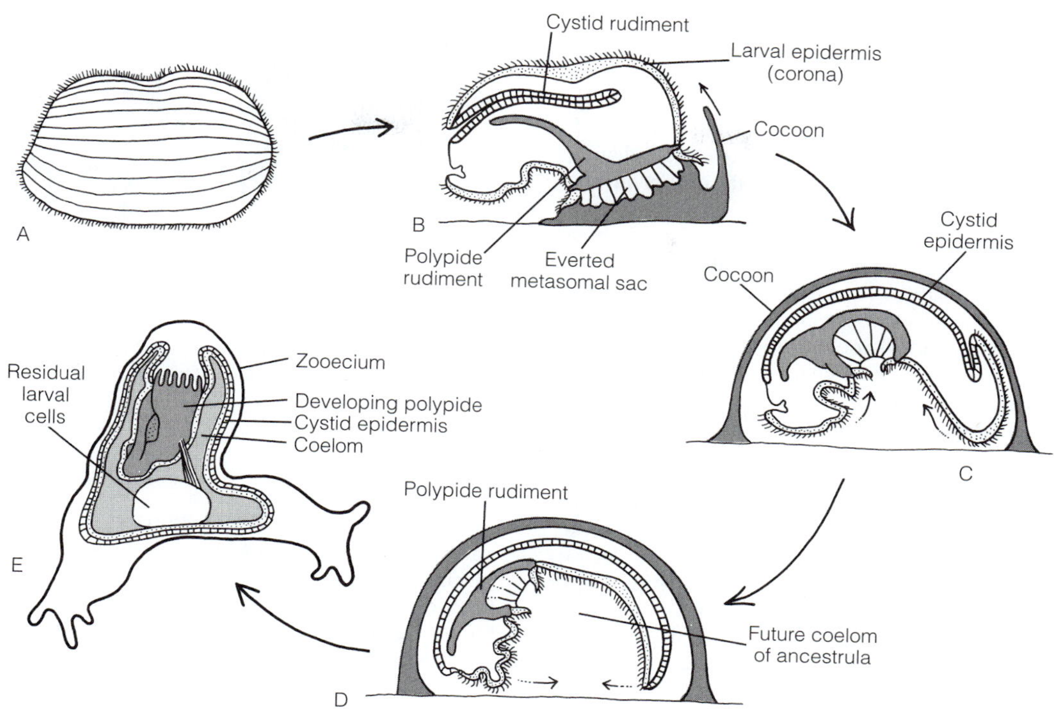

FIGURE 25-30 Bryozoa: larval metamorphosis of the ctenostome gymnolaemate *Bowerbankia.* The settling larva **(A)** attaches to a substratum using its everted metasomal sac **(B)**, which also secretes a substance that forms a cocoon around the settled larva **(C)**. Once enclosed in the cocoon, the larva detaches the metasomal sac from the substratum and rolls its ciliated epidermis inward **(C** and **D)**. The inrolled larval epidermis eventually encloses a cavity that gives rise to the ancestrula's coelom **(D** and **E)**. Residual larval cells provide nutrition for the developing ancestrula. Metamorphosis from B to D requires approximately 2 min to complete. The ancestrula begins to feed five days after the onset of metamorphosis. *(A–D, Simplified from Reed, C. G. 1991. Bryozoa. In Giese, A. C., Pearse, J. S., and Pearse, V. B. (Eds.): Reproduction of Marine Invertebrates. Vol. VI. Echinoderms and Lophophorates. The Boxwood Press, Pacific Grove, CA. pp 85–245; E, Modified according to the description of Reed from d'Hondt, J.-L. 1977. Structure larvaire et histogènese post-larvaire chez* Bowerbankia imbricata *(Adams, 1798) bryozoaire cténostome (vesicularines). Arch. Zool. Exp. Gen. 118:211–243.)*

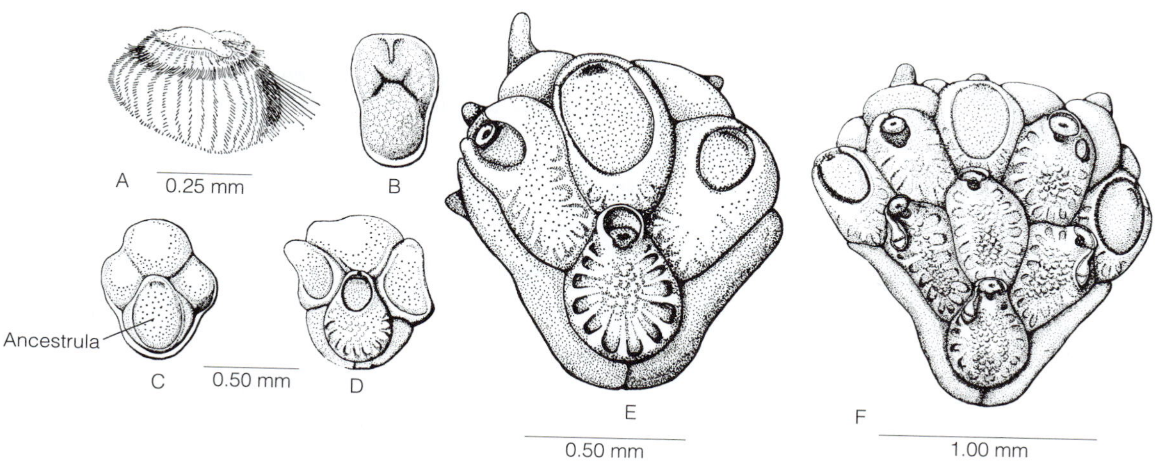

FIGURE 25-31 Bryozoa: ancestrula and early colony formation in *Metarabdotos unquiculatum,* an encrusting, shallow-water bryozoan found on stones on the west coast of Africa. **A,** Larva. **B** and **C,** Formation of the ancestrula, which is the oldest of an initial tetrad of zooids. **B** is 2 h after settlement; **C** is 28 h after settlement. **F,** Formation of additional zooids 140 h after settlement. *(From Cook, P. 1973. Settlement and early colony development in some Cheilostomata. In Larwood, G. P. (Ed.): Living and Fossil Bryozoa. Academic Press, London. pp. 65–71.)*

the erect, dendritic *Bugula,* growth occurs at the tips of each branch. In stolonate bryozoans, new buds arise from the surface and tips of the stolons.

Encrusting species have a peripheral growing edge (Fig. 25-32). In *Membranipora,* whose growth has been carefully studied, the zooids occur in subparallel, side-by-side rows radiating outward from the central ancestrula. Growth occurs at the peripheral end of each row. Here, **giant buds** consisting of a tip and a succession of differentiating zooids grow peripherally, leaving behind a trail of developing zooids. The lateral epidermis and mesothelium invaginate to form transverse partitions that divide the giant bud into successive compartments that will become the cystids of individual zooids. Polypides then develop from the cystids. The new exoskeleton is uncalcified and flexible, with calcification of the zooecium to take place later well proximal to the distal tip and after the maturing zooid has reached its adult size.

The life span of bryozoan colonies varies greatly. Some live only a single year and in temperate-zone species, growth takes place when the water temperature rises. Liberation of larvae typically occurs in the summer at the end of the life of the colony. Annual life spans are especially characteristic of epiphytic species living on algae. Some of these may pass through several generations in one season. Many species live for 2 or more years (up to 12 years in the European *Flustra foliacea*), with growth slowing down or halting during the winter. The colonies of some species, such as *Bugula,* may die back to the stolons or holdfast and re-form the following season. Sexual reproduction may occur during a restricted period or, as is common in perennial species, it may occur throughout the growing period.

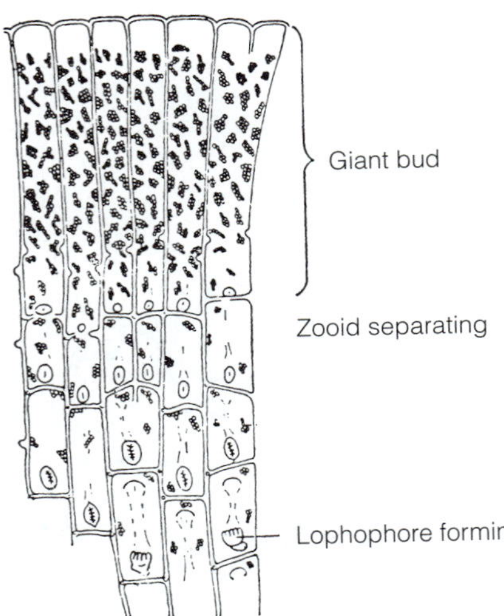

FIGURE 25-32 Bryozoa. Overhead view of the growing edge of the encrusting *Membranipora,* showing six rows of zooids. The zooids in each row are in various stages of development, with the youngest, undifferentiated giant buds at the periphery (top of the image) of each row. The zooids farthest from the giant buds (bottom of the image) are the most developed. New zooids are cut off at the rear of the giant buds. *(From Lutaud, G. 1961. Contribution à l'étude du bourgeonnement et de la croissance des colonies chez* Membranipora membranacea *(Linné), Bryozoaire Cheilostome. Ann. Soc. Zool. Belg. 91:157–300.)*

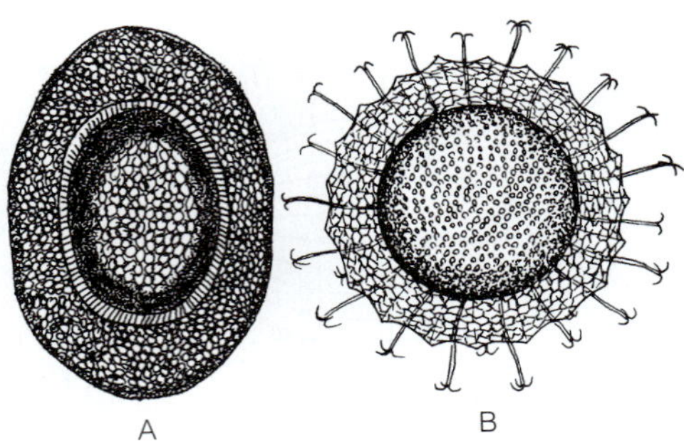

FIGURE 25-33 Bryozoa: statoblasts of freshwater bryozoans. **A,** A floating statoblast of *Hyalinella punctata.* **B,** A statoblast with hooks from *Cristatella mucedo.* *(A, After Rogick from Pennak, 1978; B, After Allman.)*

Statoblasts

In addition to sexual reproduction and asexual budding, freshwater phylactolaemates also reproduce clonally by means of special resting bodies called **statoblasts** (Fig. 25-33), which are functionally similar to the gemmules of freshwater sponges. One to several statoblasts develop on the funiculus, which nourishes it, and bulge into the coelom to form masses of mesothelial cells containing stored food. The developing statoblast is eventually covered by migrating epidermal cells. The young statoblast secretes upper and lower chitinous valves that form a protective covering for the internal cells. Because the rims of the valves often project peripherally to a considerable extent, the statoblasts are usually flattened and somewhat disc-shaped with a bulge in the center of the disc. Statoblasts are continuously formed during the summer and fall. Some types of statoblasts adhere to the parent colony or fall to the bottom **(sessoblasts)** whereas others contain air spaces and float **(floatoblasts).** Floatoblasts may have attachment hooks around their margin (Fig 25-33B).

Statoblasts are resistant to unfavorable environmental conditions such as drought and cold and can remain dormant for variable lengths of time. While dormant, they may attach to and be spread across considerable distances by animals, floating vegetation, currents, or other agents and are able to withstand desiccation and freezing. When environmental conditions become favorable, such as in the spring, the two valves separate and a zooid develops from the internal mass of cells. The number of statoblasts produced by a freshwater bryozoan colony is enormous. Windrows of statoblast valves over 1 m wide have been reported along the shore of Douglas Lake, Michigan. *Plumatella repens* colonies in a 1 m^2 patch of littoral lake vegetation were estimated to have produced 800,000 statoblasts.

DIVERSITY OF BRYOZOA

Phylactolaemata[C]: Exclusively in fresh water. Widely distributed in lakes, reservoirs, and streams without excessive mud or silt. Many species (such as *Fredericella sultana* and *Plumatella repens*) are cosmopolitan. Zooids are cylindrical, the lophophore is horseshoe-shaped (except *Fredericella*). An epistome is present. Body-wall musculature consists of circular and longitudinal fibers. Zooecium is compressible, uncalcified, thin or thick, sometimes gelatinous. Trunk coelom is continuous between zooids. Colonies are monomorphic; all zooids are autozooids, no heterozooids. *Fredericella, Plumatella, Pectinatella, Lophopus, Cristatella.*

Gymnolaemata^C: Colonies are polymorphic. Zooids are cylindrical or flattened. Body wall may be calcified or not. Lophophore is circular. Epistome is absent. Intrinsic body-wall musculature is absent. Almost entirely marine.

Most live in coastal waters attached to rocks, pilings, shells, wood, algae, other animals, and the surface of individual sand grains (Fig. 25-19A). A few solitary species of the genus *Monobryozoon* live and move about in the interstitial spaces of marine sand (Fig. 25-20B). Some bore in calcareous substrata. Stolons of boring species are in tunnels in the substratum, with the autozooids emerging at the surface. A few bryozoans live on soft bottoms. Shield-shaped colony of *Cupuladria doma* reaches the size of a small coin. It rests free on the bottom with the frontal walls of the zooids directed upward (Fig. 25-19B).

From an economic standpoint, marine bryozoans are one of the most important taxa of fouling organisms on ship bottoms. About 120 species, of which species of *Bugula* are among the most abundant, have been taken from the hulls of ships. On U.S. northeastern and northwestern coasts the introduced lace bryozoan *Membranipora membranacea* weakens kelp fronds, causing them to break and thus destroying habitat for fishes such as juvenile cod and various invertebrates, including the green sea urchin *Strongylocentrotus drobachiensis.*

Along with hydroids, bryozoans are among the most abundant marine epiphytic animals. Large brown algae are colonized by many species, which display distinct preferences for certain algal species. At the time of settling, the larvae of epiphytic species are attracted to the algal substratum, presumably by substances released by the algae. The widely distributed, encrusting *Membranipora,* one species of which is abundant on floating *Sargassum* (gulfweed), displays a number of adaptations for an epiphytic life. The frontal wall is uncalcified and the lateral walls of adjacent zooids are articulated, permitting some flexibility when the algal thallus bends. Although colonies encrust over large portions of kelp, growth of the colonies is predominantly in the direction of the stalk of the kelp thallus and is controlled by the movement of water over the algal surface.

Cyclostomata^SO: Entirely marine. Mesothelium separated from body wall to form a membranous coelomic sac. Large, spacious hemocoel is continuous with that of adjacent zooids. These are tubular zooids with calcified walls that are fused with those of adjacent zooids. Orifices are circular and terminal. Lophophore protrusion is dependent on muscular deformation of the membranous sac, not the body wall. *Crisia, Lichenopora, Stomatopora, Tubulipora.*

Eurystomata^SO: Body wall lacks muscles. Lophophore protrusion depends on body-wall deformation. Interzooidal pores are plugged with cellular rosettes and have an associated funicular system.

Ctenostomata^O: Stolonate or compact colonies. Zooecium is noncalcified membranous, chitinous, or gelatinous. Orifice usually is terminal. No operculum is present.

Stolonifera^sO: Colonies consist of branching stolons with solitary or clustered vase-shaped zooids. Ancestrula develops stolons from which autozooids arise. Zooids bud from the stolon, not from each other. *Amathia, Aeverrillia, Bowerbankia, Zoobotryon, Walkeria.*

Carnosa^sO: Ancestrula produces zooids, which then produce more zooids; zooids bud from other zooids, not from stolons. Zooids are flat, not circular, in cross section. Colonies may be compact or diffuse. *Alcyonidium, Flustrellidra, Monobryozoon, Paludicella* (freshwater).

Cheilostomata^O: Most Recent bryozoans are cheilostomes. Colonies are composed of contiguous boxlike zooids with separate calcareous walls. Orifice has an operculum (lost in *Bugula*). Avicularia, vibracula, or both may be present. Eggs typically are brooded in ovicells.

Anasca^sO: Frontal (ventral) wall is membranous, flexible. Lophophore is protruded by parietal muscles acting on the frontal membrane. *Aetea, Callopora, Electra, Flustra, Membranipora, Tendra, Bugula, Scrupocellaria, Cupuladria, Discoporella, Thalamoporella, Cellaria.*

Cribrimorpha^sO: Almost extinct. Frontal wall is a vault of fused overarching spines, which form a porous calcified plate over the flexible frontal membrane. Lophophore is protruded by parietal muscles acting on the frontal membrane. Sometimes included in Anasca or Ascophora. *Cribrilina.*

Ascophora^sO: Frontal wall is calcified, underlaid by a flexible invaginated sac, the ascus. Lophophore protrusion is achieved through dilation of the ascus by parietal muscles. *Microporella, Schizoporella, Smittina, Watersipora.*

PHYLOGENY OF BRYOZOA

Phylactolaemata and Gymnolaemata are sister taxa in Bryozoa. Phylactolaemate autapomorphies include statoblasts, a reduced hemal-funicular system, and open coelomic pores between zooids. Gymnolaemate autapomorphies include the loss of body-wall muscles and loss of the epistome.

The freshwater phylactolaemates are usually considered to be the more primitive, although they have many derived characteristics. The cylindrical zooids, anterior orifice, horseshoe-shaped lophophore, presence of an epistome, and monomorphic colonies may all be primitive features. In many other ways, however, phylactolaemates are specialized, especially in their reproduction.

Unfortunately, there are no fossil phylactolaemates and thus paleontology contributes no information regarding their relationship with the marine taxa. The earliest known marine bryozoan is a questionable fossil species from the late Cambrian. Beginning in the Ordovician, however, there is a rich fossil record, and thousands of fossil species have been described. Cyclostomes dominate the Paleozoic fauna, although ctenostomes were also present. Cheilostomes, the dominant marine forms today, made their appearance in the late Jurassic.

PHYLOGENY OF LOPHOPHORATA

In most phylogenies Phoronida, Brachiopoda, and Bryozoa are united in Lophophorata, but the relationships between the three taxa and between Lophophorata and the protostomes and deuterostomes are not well understood and are the subject of an ongoing debate among zoologists (Fig. 9-25). Morphological and molecular evidence often are not in agreement. In general, morphological data tend to support a deuterostome alliance, embryological evidence is ambivalent, and nucleotide sequences indicate protostome affinities. Here, we conservatively retain the three phyla in the monophyletic taxon Lophophorata and provide alternative phylogenies in which it is allied in turn with Lophodeuterostomia (Fig. 25-34) on one hand and the protostome Lophotrochozoa (Fig. 25-35) on the other.

Phoronida and Bryozoa share similar actinotroch and cyphonautes larvae, both of which have a metasomal sac and a pair of protonephridia, and here we consider them to be sister taxa. In some classifications Phoronida and Brachiopoda, with shared metasomal metanephridia and a monociliated epidermis, are grouped as sister taxa, with Bryozoa excluded.

Lophophorates have long been considered to be trimeric, with bodies and coeloms composed of three regions; the anterior protosome (epistome) with a protocoel, the middle mesosome with its mesocoel, and the posterior metasome (trunk) with its metacoel. Trimery would be a synapomorphy with the deuterostomes, and the two have traditionally been viewed as sister taxa. New studies and reexamination of older reports indicate, however, that the cavity of the epistome, when present, is not a protocoel, thus weakening the argument for trimery and casting doubt on the affinity with deuterostomes (although not refuting it). Furthermore, the sister-taxon relationship between Lophophorata and Deuterostomia (= Lophodeuterostomia, Radialia) is not supported by molecular data. Thus new morphological and molecular data indicate a protostome affinity for the lophophorates in Lophotrochozoa (Fig. 25-35, 9-26B) whereas older data

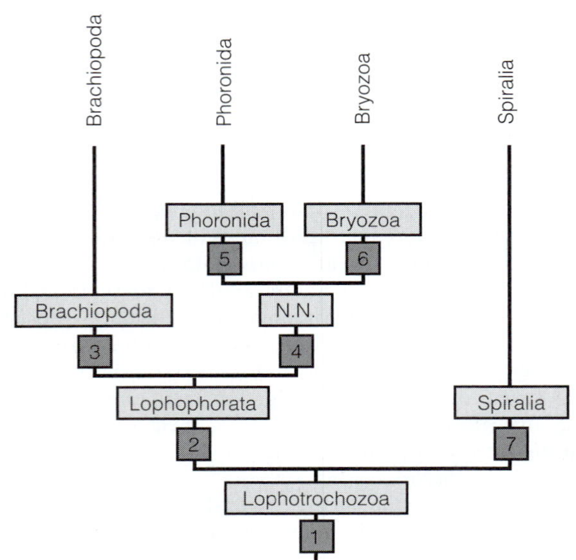

FIGURE 25-35 Lophophorate phylogeny assuming protostome relationships. **1, Lophotrochozoa:** β-chitin. Chaetae, larval protonephridia, possibly monociliated cells. **2, Lophophorata:** Lophophore, U-shaped gut. Anus is outside the lophophore. Upstream collecting system **3, Brachiopoda:** Bivalve shell and mantle. Brachia. Heart. Digestive ceca. **4, N.N.:** Metasomal sac. Adult epidermis forms from that of metasomal sac. **5, Phoronida:** Corpuscular hemoglobin in the hemal system. Actinotroch larva with unique metamorphosis. **6, Bryozoa:** Introvert. Funicular system (modified hemal system). Colonial. Multiciliated cells. Coelomopore (a metanephridial homolog?). **7, Spiralia:** Spiral cleavage. Mesoderm arises from 4d (mesentoblast).

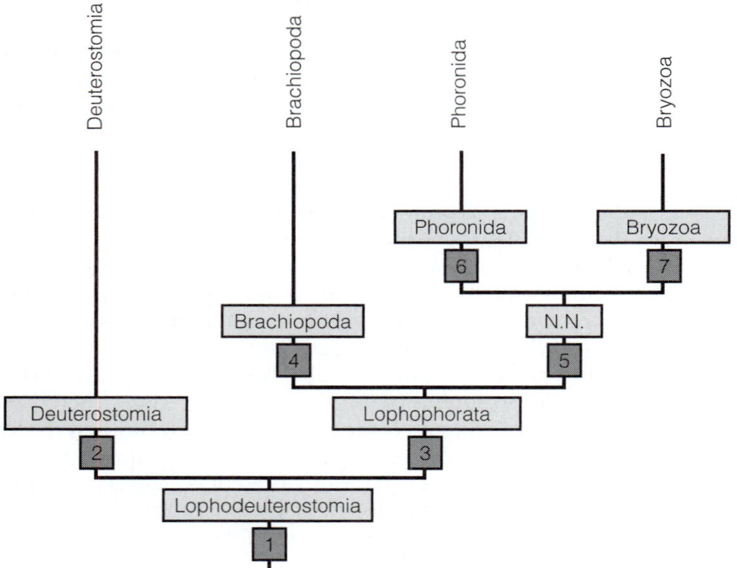

FIGURE 25-34 Lophophorate phylogeny assuming deuterostome relationships. **1, Lophodeuterostomia:** Dimeric, without protosome or protocoel. With lophophore, U-shaped gut, dorsal anus, upstream collecting system. **2, Deuterostomia:** Trimeric. Protocoelomic heart-kidney complex. Blastopore becomes the anus. **3, Lophophorata:** There are no unequivocal lophophorate apomorphies. Possibilities are a pair of metasomal metanephridia and the extracellular skeletal covering secreted by the metasomal epidermis. **4, Brachiopoda:** Bivalve shell and mantle. Brachia. Heart. Digestive ceca. **5, N.N.:** Metasomal sac. Adult epidermis forms from that of metasomal sac. Possible schizocoely and larval protonephridia. **6, Phoronida:** Corpuscular hemoglobin in the hemal system. Have an actinotroch larva with unique metamorphosis. **7, Bryozoa:** Introvert. Funicular system (modified hemal system). Colonial. Multiciliated cells, a coelomopore (a metanephridial homolog?).

PHYLOGENETIC HIERARCHY OF LOPHOPHORATA

Lophophorata
- N.N.
  - Phoronida
  - Bryozoa
- Brachiopoda

and interpretations argue for deuterostome relationships (Fig. 25-34, 9-26A). At present there is not sufficient evidence to place lophophorates with certainty in either of these two major metazoan taxa.

Absence of a protosome and protocoel suggests that the lophophorate ground plan is dimeric rather than trimeric. If lophophorates are lophodeuterostomes, this implies that dimery is ancestral for Lophodeuterostomia and trimery is a derived autapomorphy of Deuterostomia. The recognition of Lophophorata as a monophyletic taxon distinct from other lophodeuterostomes is then problematical. Lophophorates have no certain autapomorphies establishing them as distinct from other lophodeuterostomes, all of which share the lophophore, upstream collecting system, and other characters that have traditionally been used to define Lophophorata. A pair of metasomal metanephridia and an extracellular skeletal covering (shell, tube, or zooecium) secreted by the metasomal epidermis might be lophophorate autapomorphies.

As lophotrochozoans, however, these characters would be lophophorate autapomorphies (as they have always been considered to be) and Lophophorata could be a valid monophyletic taxon. The suite of typical lophophorate characters (dimeric body, lophophore, U-shaped gut, upstream collecting system, and dorsal anus) would be convergent with the similar suite found in deuterostomes.

Morphological evidence for lophophorate relationships is contradictory. That the classic lophophorate character suite is common to Lophodeuterostomia suggests kinship with deuterostomes. The chitinous chaetae of brachiopods and bryozoans, on the other hand, are ultrastructurally like those of the trochozoan annelids, molluscs, and echiurans. Chitin is very unusual in deuterostomes and its presence in all lophophorates suggests protostome ties. In fact, if lophophorates are protostomes, then the absence of chitin would be a deuterostome autapomorphy.

Developmental evidence is similarly equivocal. All three taxa exhibit radial cleavage (a deuterostome trait). The blastopore becomes the mouth in phoronids (as in protostomes) whereas it closes and its fate is uncertain in brachiopods and bryozoans. Coelom formation is probably enterocoelous (as it is in deuterostomes) in brachiopods and schizocoelous (as in protostomes) in phoronids and uncertain in bryozoans. The phoronid actinotroch and bryozoan cyphonautes larvae both have a pair of protonephridia reminiscent of those found in some annelid larvae (protostomes).

Finally, 18S rDNA nucleotide sequences place the three lophophorate taxa solidly in the protostome taxon, Lophotrochozoa, in the company of molluscs and annelids (Fig. 25-35, 9-25)

REFERENCES

GENERAL

Bartolomaeus, T. 2001. Ultrastructure and formation of the body cavity lining in *Phoronis muelleri* (Phoronida: Lophophorata). Zoomorphology 120:135–148.

Conway Morris, S., Cohen, B. L., Gawthrop, A. B., Cavalier-Smith, T., and Winnepenninckx, B. 1996. Lophophorate phylogeny. Science 272:283.

Halanych, K. M., Bacheller, J. D., Aguinaldo, A. M. A., Liva, S. M., Hillis, D. M., and Lake, J. A. 1995. Evidence from 18S ribosomal DNA that the lophophorates are protostome animals. Science 267:1641–1643.

Halanych, K. M., Bacheller, J. D., Aguinaldo, A. M. A., Liva, S. M., Hillis, D. M., and Lake, J. A. 1996. Lophophorate phylogeny. Science 272:282.

Harrison, F. W., and Woollacott, R. M. (Eds.): 1997. Microscopic Anatomy of Invertebrates. Vol. 13. Lophophorates, Entoprocta, and Cycliophora. Wiley-Liss, New York. 500 pp.

Nielsen, C. 1977. The relationship of Entoprocta, Ectoprocta, and Phoronida. Am. Zool. 17:149–150.

Nielsen, C. 1985. Animal phylogeny in the light of the trochaea theory. Biol. J. Linn. Soc. 25:243–299.

Strathmann, R. 1973. Function of lateral cilia in suspension feeding of lophophorates. Mar. Biol. 23:129–136.

Zimmer, R. L. 1973. Morphological and developmental affinities of the lophophorates. In Larwood, G. P. (Ed.): Living and Fossil Bryozoa: Recent Advances in Research. Academic Press, New York. pp. 593–599.

Woollacott, R. M., and Harrison, F. A. 1997. Introduction. In Harrison, F. W., and Woollacott, R. M. (Eds.): Microscopic Anatomy of Invertebrates. Vol. 13. Lophophorates, Entoprocta, and Cycliophora. Wiley-Liss, New York. pp. 1–11.

PHORONIDA

Abele, L. G., Gilmour, T., and Gilchrist, S. 1983. Size and shape in the phylum Phoronida. J. Zool. Lond. 200:317–323.

Bartolomaeus, T. 1989. Ultrastructure and relationship between protonephridia and metanephridia in *Phoronis muelleri* (Phoronida). Zoomorphology 109:113–122.

Emig, C. C. 1974. The systematics and evolution of the phylum Phoronida. Z. zool. Syst. Evolut.-forsch. 12:128–151.

Emig, C. C. 1977. Embryology of Phoronida. Am. Zool. 17:21–37.

Emig, C. C. 1982. The biology of Phoronida. Adv. Mar. Biol. 19:1–89.

Herrmann, K. 1997. Phoronida. In Harrison, F. W., and Woollacott, R. M. (Eds.): Microscopic Anatomy of Invertebrates. Vol. 13. Lophophorates, Entoprocta, and Cycliophora. Wiley-Liss, New York. pp. 207–235.

Pixell, H. 1912. Two new species of Phoronidea from Vancouver Island. Quart. J. Microsc. Sci. 58.

Vandergon, T. L., and Colacino, J. M. 1991. Hemoglobin function in the lophophorate *Phoronis architecta* (Phoronida). Physiol. Zool. 64:1561–1577.

Zimmer, R. L. 1991. Phoronida. In Giese, A. C., Pearse, J. S., and Pearse, V. B. (Eds.): Reproduction of Marine Invertebrates. Vol. VI. Echinoderms and Lophophorates. Boxwood Press, Pacific Grove, CA. pp. 1–45.

Internet Site

www.com.univ-mrs.fr/DIMAR/Phoro/PhoEng.htm (Phoronid@2003, by C. C. Emig. Bibliographies, systematics and evolution, biology.)

BRACHIOPODA

Atkins, D. 1959. The growth stages of the lophophore and loop of the brachiopod *Terebratalia transversa* (Sowerby). J. Morphol. 105:401–426.

Beauchamp, P. de. 1960. Classe des Brachiopodes. In Grassé P. P. (Ed.): Traité de Zoologie. Vol. 5, Pt. 2. Masson et Cie, Paris.

Davidson, T. 1887. A monograph of recent Brachiopoda, II. Trans. Linn. Soc. Lond. Zool. Ser. 2, Vol. 4.

Emig, C. C. 1981. Observations on the ecology of *Lingula reevei*. J. Exp. Mar. Biol. Ecol. 52:47–62.

François, P. 1891. Observations biologiques sur les lingues. Arch. Zool. Exp. Gén. Ser. 2, Vol. 9.

Gould, S. J., and Calloway, C. B. 1980. Clams and brachiopods—ships that pass in the night. Paleobiology 6:383–396.

Gustus, R. M., and Cloney, R. A. 1972. Ultrastructural similarities between the setae of brachiopods and polychaetes. Acta Zool. 53:229–233.

James, M. A. 1997. Brachiopoda: Internal anatomy, embryology, and development. In Harrison, F. W., and Woollacott, R. M. (Eds.): Microscopic Anatomy of Invertebrates. Vol. 13. Lophophorates, Entoprocta, and Cycliophora. Wiley-Liss, New York. pp. 297–407.

James, M. A., Ansell, A. D., Curry, G. B., Collins, M. J., Peck, L. S., and Rhodes, M. C. 1992. The biology of living brachiopods. Adv. Mar. Biol. 28:175–387.

LaBarbera, M. 1977. Brachiopod orientation to water movement. 1. Theory, laboratory behavior, and field observations. Paleobiology 3:270–287.

Long, J. A., and Stricker, S. A. 1991. Brachiopoda. In Giese, A. C., Pearse, J. S., and Pearse, V. B. (Eds.): Reproduction of Marine Invertebrates. VI. Echinoderms and Lophophorates. Boxwood Press, Pacific Grove, CA. pp. 47–84.

Morse, E. 1902. Observations on living Brachiopoda. Mem. Boston Nat. Hist. Soc. 5.

Nielsen, C. 1991. The development of the brachiopod *Crania (Neocrania) anomala* (O. F. Mueller) and its phylogenetic significance. Acta Zool. 72:7–28.

Paine, R. T. 1963. Ecology of the brachiopod *Glottidia pyramidata*. Ecol. Monogr. 33:187–213.

Rudwick, M. J. S. 1970. Living and Fossil Brachiopods. Hutchinson, London. 199 pp.

Trueman, E. R., and Wong, T. M. 1987. The role of the coelom as a hydrostatic skeleton in lingulid brachiopods. J. Zool. Lond. 213:221–232.

Williams, A. 1997. Brachiopoda: Introduction and integumentary system. In Harrison, F. W., and Woollacott, R. M. (Eds.): Microscopic Anatomy of Invertebrates. Vol. 13. Lophophorates, Entoprocta, and Cycliophora. Wiley-Liss, New York. pp. 237–296.

Williams, A., and Rowell, A. J. 1965. Brachiopod anatomy. In Williams, A., and Brunton, C. H. C. (Eds.): Treatise on Invertebrate Paleontology. Vol. H. Brachiopoda. Geological Society of America, University of Kansas Press, Lawrence, KS. pp H6–H57.

Internet Site

http://paleopolis.rediris.es/BrachNet/ (BrachNet, by C. C. Emig and M. V. Pardo. With annual bibliographies, classification, images, and links.)

BRYOZOA

Bobin, G. 1977. Interzooecial communications and the funicular system. In Woollacott, R. M., and Zimmer, R. L. (Eds.): Biology of Bryozoans. Academic Press, New York. pp. 307–333.

Brien, P. 1960. Sous-Classe des Phylactolèmes ou Phylactolémates. In Grassé, P.-P. (Ed.): Traité de Zoologie. Vol 5, No. 2. Masson, Paris.

Carle, K. J., and Ruppert, E. E. 1983. Comparative ultrastructure of the bryozoan funiculus: A blood vessel homologue. Z. zool. Syst. Evolut.-forsch. 21:181–193.

Cook, P. 1973. Settlement and early colony development in some Cheilostomata. In Larwood, G. P. (Ed.): Living and Fossil Bryozoa. Academic Press, London. pp. 65–67.

Cook, P. L., and Chimonides, P. J. 1978. Observations on living colonies of *Selenaria* (Bryozoa, Cheilostomata). I. Cah. Biol. Mar. 19:147–158.

Edwards, D. D. 1987. Home on the grain. Sci. News 131:156–157.

Gordon, D. P. 1975. Ultrastructure and function of the gut of a marine bryozoan. Cah. Biol. Mar. 16:367–382.

d'Hondt, J.-L. 1977. Structure larvaire et histogenèse post-larvaire chez *Bowerbankia imbricata* (Adams, 1798) bryozoaire cténostome (vesicularines). Arch. Zool. Exp. Gen. 118:211–243.

Larwood, G. P., and Abbott, M. B. (Eds.): 1979. Advances in Bryozoology. Systematics Association Spec. Vol. 13. Academic Press, London. 638 pp.

Lidgard, S., and Jackson, J. B. C. 1989. Growth in encrusting cheilostome bryozoans: I. Evolutionary trends. Paleobiology 15:255–282.

Lutaud, G. 1961. Contribution à l'étude du bourgeonnement et de la croissance des colonies chez *Membranipora membranacea* (Linné), Bryozoaire Cheilostome. Ann. Soc. R. Zool. Belg. 91:157–300.

Mackie, G. O. 1986. From aggregates to integrates: Physiological aspects of modularity in colonial animals. Phil. Trans. Roy. Soc. Lond. B 313:175–196.

Marcus, E. 1926. Bryozoa. In Grimpe, E., and Wagler, E. (Eds.): Die Tierweld der Nord- und Ostsee. Teil VII. Cl.

Moore, R. C. (Ed.): 1953. Treatise on Invertebrate Paleontology: Bryozoa, Pt. G. Geological Society of America, University of Kansas Press, Lawrence, KS. 253 pp.

Moore, R. C. (Ed.): 1965. Brachiopoda: Pt. H. Geological Society of America, University of Kansas Press, Lawrence, KS. (Two volumes.)

Mukai, H., Terakado, K., and Reed, C. R. 1997. Bryozoa. In Harrison, F. W., and Woollacott, R. M. (Eds.): Microscopic Anatomy of Invertebrates. Vol. 13. Lophophorates, Entoprocta, and Cycliophora. Wiley-Liss, New York. pp. 45–206.

Nielsen, C. 1971. Entoproct life-cycles and the entoproct/ectoproct relationship. Ophelia 9:209–341.

Nielsen, C., and Pedersen, K. J. 1979. Cystid structure and protrusion of the polypide in *Crisia*. Acta Zool. 60:65–88.

Oka, A. 1891. Observations on the fresh-water Polyzoa. J. Coll. Sci. Univ. Tokyo. 4.

Reed, C. G. 1991. Bryozoa. In Giese, A. C., Pearse, J. S., and Pearse, V. B. (Eds.): Reproduction of Marine Invertebrates. Vol. VI. Echinoderms and Lophophorates. Boxwood Press, Pacific Grove, CA. pp. 85–245.

Reed, C. G., and Cloney, R. A. 1982. The settlement and metamorphosis of the marine bryozoan *Bowerbankia gracilis* (Ctenostomata: Vesicularioidea). Zoomorphology 101:103–132.

Ryland, J. S. 1970. Bryozoans. Hutchinson, London. 175 pp.

Ryland, J. S. 1976. Physiology and ecology of marine bryozoans. Adv. Mar. Biol. 14:285–443.

Silen, L. 1966. On the fertilization problem in the gymnolaematous bryozoa. Ophelia 3:113–140.

Silen, L. 1972. Fertilization in the Bryozoa. Ophelia 10:27–34.

Strathmann, R. R. 1982. Cinefilms of particle capture by an induced local change of beat of lateral cilia of a bryozoan. J. Mar. Biol. Ecol. 62:225–236.

Stricker, S. A. 1988. Metamorphosis of the marine bryozoan *Membranipora membranacea:* An ultrastructural study of rapid morphogenetic movements. J. Morphol. 196:53–72.

Stricker, S. A. 1989. Settlement and metamorphosis of the marine bryozoan *Membranipora membranacea.* Bull. Mar. Sci. 45:387–405.

Stricker, S. A., Reed, C. G., and Zimmer, R. L. 1988a. The cyphonautes larva of the marine bryozoan *Membranipora membranacea.* I. General morphology, body wall, and gut. Can. J. Zool. 66:368–383.

Stricker, S. A., Reed, C. G., and Zimmer, R. L. 1988b. The cyphonautes larva of the marine bryozoan *Membranipora membranacea.* II. Internal sac, musculature, and pyriform organ. Can J. Zool. 66:384–398.

Swedmark, B. 1964. The interstitial fauna of marine sand. Biol. Rev. 39:1–42.

Todd, J. A. 2000. The central role of ctenostomes in bryozoan phylogeny. In Herrera Cubilla, A., and Jackson, J. B. C. (Eds.): Proceedings of the 11th International Bryozoology Association Conference. Smithsonian Tropical Research Institute, Balboa, P.R. pp. 104–135.

Wesenberg-Lund, C. 1896. Biologiske Studier over Ferskvandsbryozoer. Vidensk. Meddel. Dansk. Naturhist. Foren. Ser. 5, Vol. 8.

Winston, J. E. 1978. Polypide morphology and feeding behavior in marine ectoprocts. Bull. Mar. Sci. 28:1–31.

Winston, J. E. 1984. Why bryozoans have avicularia—a review of the evidence. Am. Mus. Novitates 2789:1–26.

Wood, T. 2001. Bryozoans. In Thorp, J. H., and Covich, A. P. (Eds.): Ecology and Classification of North American Freshwater Invertebrates. 2nd Edition. Academic Press, San Diego. pp. 505–525.

Woollacott, R. M., and Zimmer, R. L. (Eds.): 1977. Biology of Bryozoans. Academic Press, New York. 566 pp.

Zimmer, R. L. 1997. Phoronids, brachiopods, and bryozoans, the lophophorates. In Gilbert, S. F., and Raunio, A. M. (Eds.): Embryology: Constructing the organism. Sinauer, Sunderland, MA. pp. 279–305.

Zimmer, R. L., and Woollacott, R. M. 1989. Larval morphology of the bryozoan *Watersipora arcuata* (Cheilostomata: Ascophora). J. Morphol. 199:125–150.

Internet Site

www.civgeo.rmit.edu.au/bryozoa/default.html (Bryozoa Home Page: Recent and Fossil Bryozoa, by P. Bock. Comprehensive bryozoan site for students and professionals. Includes links, images, a glossary, lists of specialists and meetings, conference abstracts.)

26 Chaetognatha[P]

FORM AND FUNCTION

REPRODUCTION AND DEVELOPMENT

PHYLOGENY OF CHAETOGNATHA

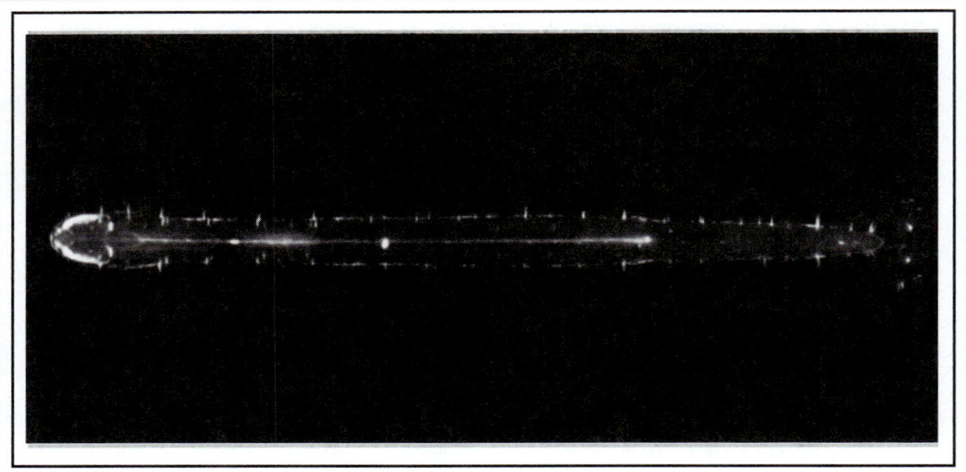

Most chaetognaths are small, transparent, ambush predators of marine zooplankton. Called arrowworms, their slender, shaftlike bodies bear horizontal fins reminiscent of those on a torpedo or the feathered vanes of an arrow. The body is characteristically straight and stiff, except during sudden swimming bursts, which thrust the animal to a new location in the blink of an eye. Like an arrow shot from a bow, they reach the zenith of their flight, arc downward headfirst, and gradually descend. It is during the slow descent that they have the opportunity to detect the swimming vibrations of their prey, mostly small crustaceans, which they grasp and devour. Although most chaetognaths are adapted to life in the plankton, some are benthic animals that attach to macroalgae and seize passing prey.

All of the approximately 150 species of Chaetognatha are marine. They reach their highest diversity in the tropics, but are common in all oceans worldwide. Most arrowworms are 1 to 2 cm long, but species range in size from a few millimeters to 12 cm *(Pseudosagitta gazellae)*. The sister taxon of chaetognaths has not yet been determined, but recent evidence indicates that they are prostostomes, perhaps allied with molting animals (Ecdysozoa; Fig. 9-26B).

FORM AND FUNCTION

Arrowworms are striking for their perfect bilateral symmetry. The slender body is divided into a head, cylindrical trunk, and dorsoventrally flattened, postanal tail. Externally, the head-trunk division is marked by a slender neck, but internally the body regions are separated by septa (Fig. 26-1). The septa occur at the divisions of the head-trunk and trunk-tail.

The ovoid head bears on its underside a depression, the **vestibule,** which leads into the ventral mouth (Fig. 26-2). On each side of the head and flanking the vestibule is a row of 4 to 14 large, curved, prey-seizing **grasping spines** composed of α-chitin (Fig. 26-1, 26-2). The spines are hollow and filled with epidermal (pulp) cells. New spines are formed at the anterior end of each row and presumably shed from the posterior end. Anterior to the spines and also flanking the mouth are usually either one or two rows of short teeth. The spines and teeth are used to capture prey, introduce venom into the prey, and shove the prey into the mouth. A pair of eyes is located posteriorly on the dorsal side of the head (Fig. 26-1B). In one species, each of the 70 to 600 retinal cells has its own lens. Together the retinal cells surround and radiate from a large, central pigment cell. The more or less radial symmetry of the eye enables it to determine the orientation of light arriving from any direction, a clear advantage for a transparent animal. The entire head, including the grasping spines, is normally ensheathed by a foreskinlike fold of body wall called the **hood** (Fig. 26-1B, 26-2A). The hood is withdrawn rapidly to expose the spines and teeth during feeding, then reextended over the head to streamline it for swimming.

The elongate trunk and tail bear the **lateral fins** (Fig. 26-1, 26-3), each of which has conspicuous fin rays. In many arrowworms there are two pairs of lateral fins (Fig. 26-1B), but most species have but a single pair (Fig. 26-1A). Posteriorly, a large, fanlike **caudal fin** encompasses the end of the tail. Each fin is a fold of the epidermis that consists of upper and lower epidermal layers with an extracellular matrix between them (Fig. 26-3). The **fin rays** are concentrations of intracellular filaments in specialized epidermal cells.

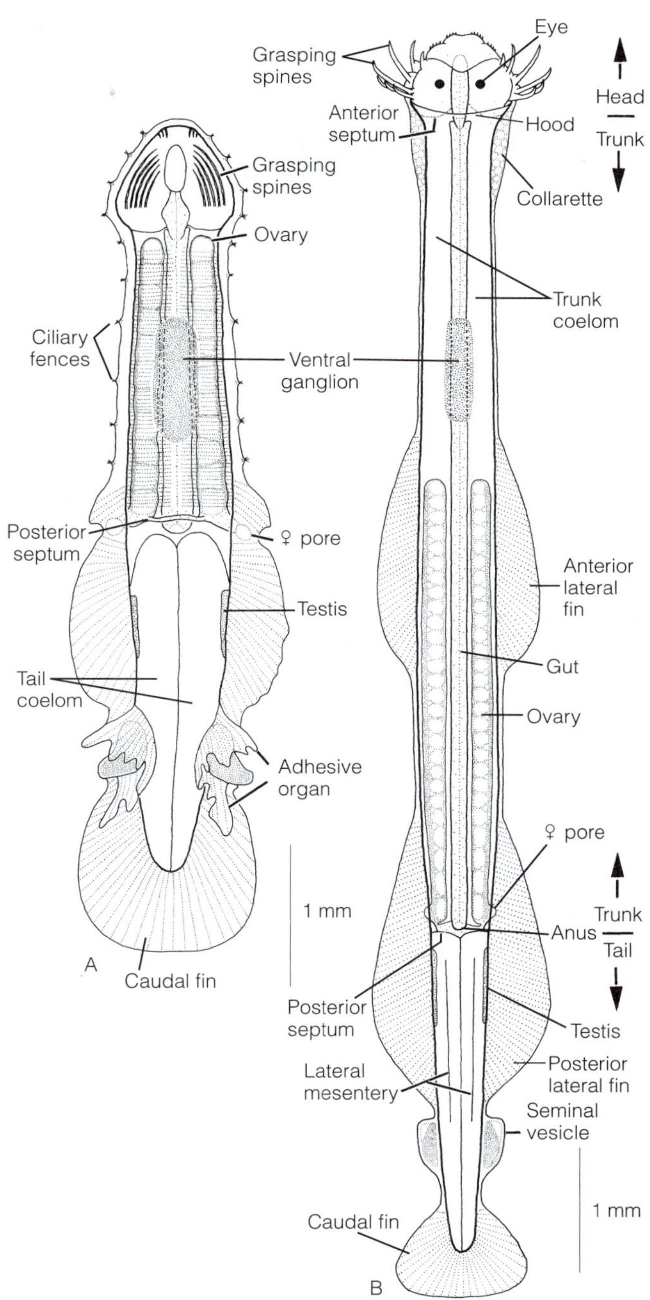

FIGURE 26-1 Chaetognatha: anatomy. **A,** *Paraspadella gotoi* (ventral view). **B,** *Adhesisagitta hispida* (dorsal view). *(Modified from Shinn, G. L. 1997. Chaetognatha. In Harrison, F. W., and Ruppert, E. E. (Eds.): Microscopic Anatomy of Invertebrates. Vol. 15. Hemichordata, Chaetognatha, and the Invertebrate Chordates. Wiley-Liss, New York. pp. 103–220. Reprinted with permission.)*

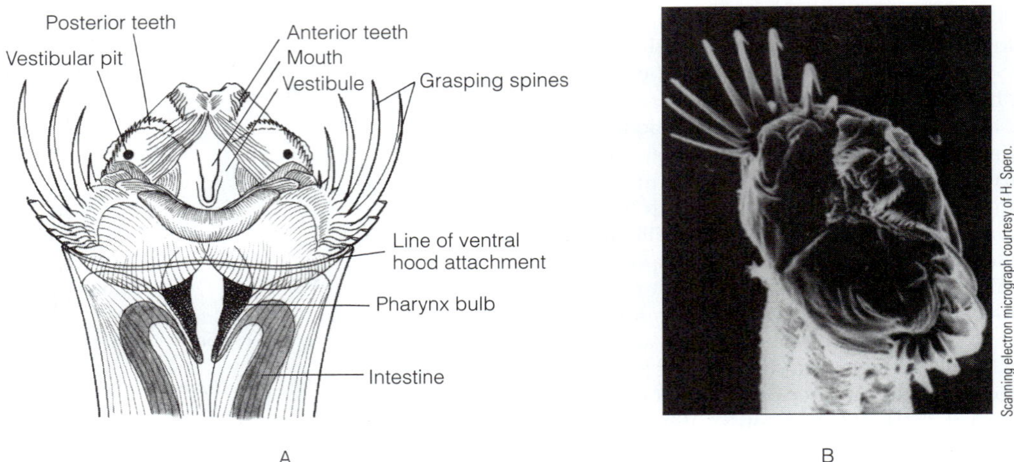

FIGURE 26-2 Chaetognatha: head anatomy. **A,** Ventral view of *Parasagitta (Sagitta) elegans.* **B,** Anterior and ventral view of head with hood retracted, spines and teeth exposed (scanning electron micrograph). *(A, After Ritter-Zahony from Hyman, 1959. Reprinted with permission.)*

The body wall is covered with a multilayered, or **stratified, epidermis** three to five cells thick, except on the ventral surface of the head and on the inside surface of the hood (Fig. 26-3). Cuticle occurs only on the head, where it forms a protective mask on the ventral, lateral, and dorsal surfaces. The spines and teeth are localized specializations of the head cuticle. Although the trunk and tail epidermis lack cuticle, the subsurface epidermal cells are chock full of microscopic skeletal fibers (tonofilaments) that crisscross around the body. The collagen fibers in the prominent, epidermal basement membrane are arranged in crossed helices. These and the epidermal filaments probably toughen the body wall, resist aneurisms, prevent kinking, and play other skeletal roles.

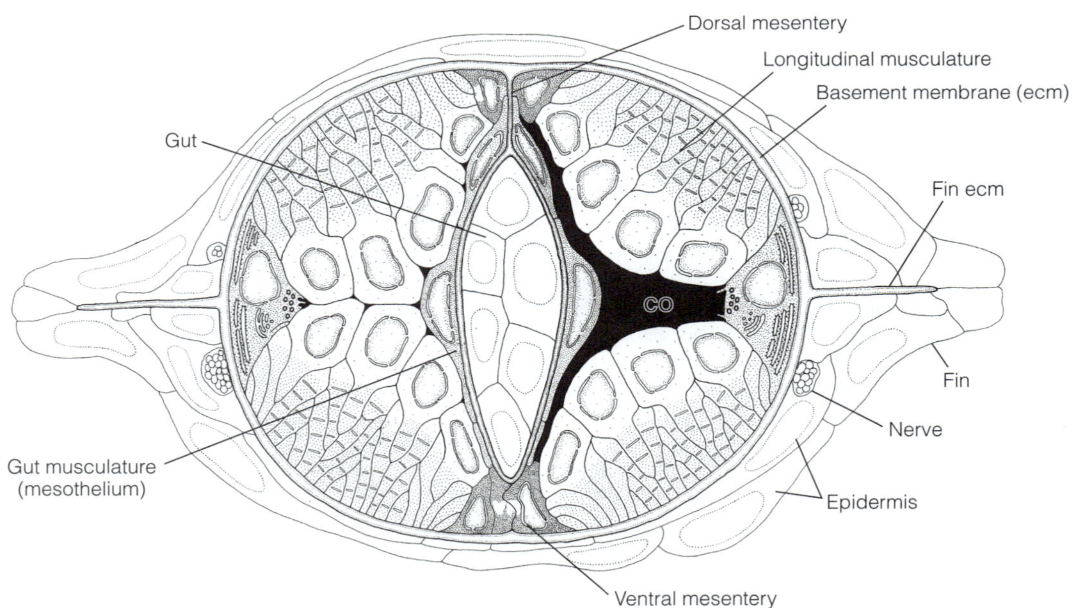

FIGURE 26-3 Chaetognatha: anatomy and histology. Diagrammatic cross section of *Adhesisagitta (Sagitta) hispida* hatchling (juvenile). Left side shows coelom closed and right side shows coelom (co) as it begins to open. The hemal system and peritoneum covering the longitudinal musculature are not yet formed. *(Adapted from Shinn, G. L. 1997. Chaetognatha. In Harrison, F. W., and Ruppert, E. E. (Eds.): Microscopic Anatomy of Invertebrates. Vol. 15. Hemichordata, Chaetognatha, and the Invertebrate Chordates. Wiley-Liss, New York. pp. 103–220. Reprinted with permission.)*

Beneath the epidermal basement membrane of the trunk and tail is a well-developed longitudinal musculature arranged in two dorsolateral and two ventrolateral bundles (Fig. 26-3). These muscles generate the dorsoventral swimming undulations of the body. Most species have small bundles of longitudinal muscles in the lateral body wall and some species have oblique transverse muscles in the trunk. The musculature of the head is complex and specialized for the movement of the hood, teeth, grasping spines, and other structures (Fig. 26-2A). All of these muscles are cross-striated and fast contracting.

Chaetognaths are tricoelomate: The head houses an unpaired coelom, and the trunk and tail have a pair of coeloms each (Fig. 26-1). Septa separate the three coelomic regions and dorsal and ventral mesenteries divide the trunk and tail coeloms into left and right halves. Each left and right tail coelom, moreover, is partitioned into medial and lateral halves by an additional lateral mesentery. That mesentery, however, is incomplete anteriorly and posteriorly, allowing for fluid flow between medial and lateral chambers.

The coelom is lined by a mesothelium that is regionally specialized into muscle cells (longitudinal musculature), ciliated epithelial cells, and circular epitheliomuscular cells around the gut (Fig. 26-3). The coelomic face of the longitudinal musculature is covered by a thin, noncontractile peritoneum that elsewhere is either absent or incomplete. The ciliated cells create circulation in the coelomic cavities. The peritoneum separates the longitudinal muscles from the coelomic fluid, but its functional role is unknown.

Chaetognaths have a hemal system, but it is confined to the trunk. Its chief components are a gut sinus between the gastrodermis and mesothelial muscles (gut musculature) and a bilateral pair of blood-filled sinuses in the ovaries (Fig. 26-4). The gut musculature probably pumps the blood, which is colorless without a respiratory pigment. From its close association with the gut wall and ovaries (which synthesize yolk), it is likely that its functional role is nutrient transport.

Chaetognaths are hermaphrodites and the testes also require nutrients for the production of sperm, but curiously they occur in the tail, which lacks both gut and hemal system. How nutrients reach the sperm, which develop in the tail coelom, across the trunk-tail septum is unknown, but there are clues that suggest a mechanism. The first is that blood and tail (but not trunk) coelomic fluid stain identically after treatment with certain histological stains. The second is that the trunk-tail septum has a blood sinus on its anterior face and podocytes on its posterior face (Fig. 26-4). The presence of podocytes suggests that a blood ultrafiltrate containing nutrients may cross the septum and enter the coelomic fluid of the tail for uptake by the developing male gametes (and tail tissues).

The presence of coelomic cavities, podocytes, and a hemal system predicts the existence of a metanephridial system in chaetognaths, but none has been identified, nor has any other form of excretory structure. For this reason, it is assumed that arrowworm excretion occurs entirely by simple diffusion of waste products from its cells and tissues. If this is correct, then chaetognaths are the only adult bilaterians to lack specialized excretory organs or tissues. On the other hand, the blood sinus, podocytes, and tail coelom may be part of a metanephridial system now integrated with the male reproductive system. If so, each ciliated sperm duct, which conveys sperm from the coelom to the seminal vesicles and exterior, corresponds to a metanephridial duct. In support of this hypothesis is the observed fact that reabsorption of metabolites, a key function of metanephridia, occurs in the chaetognath sperm duct. Similarly, the paired oviducts of the trunk may also be transformed metanephridia.

The digestive tract is a straight but regionally specialized tube (Fig. 26-1). The ventral mouth leads into a muscular, glandular, nonciliated pharynx that penetrates the head-trunk septum to join a straight, ciliated intestine. The posterior end of the pharynx may form a swollen bulb in some species. The laterally compressed intestine extends the length of the trunk to a short ciliated rectum at its posterior extremity. The intestinal lumen is collapsed except when occupied by food. The ventral anus occurs at the trunk-tail septum (Fig. 26-1B).

Planktonic chaetognaths such as *Tenuisagitta setosa* and *Adhesisagitta hispida* alternately swim and sink, with the fins acting as stabilizers during swimming movements. When the body sinks, the longitudinal muscles contract rapidly, producing a dorsoventral undulation, and the animal darts swiftly forward. This forward motion is followed by an interval of passive sinking. Some planktonic species have specializations that enhance neutral buoyancy. In *Parasagitta elegans,* for example, the enlarged gastrodermal cells obliterate the coelom and are filled with lightweight NH_4^+. The benthic *Spadella* adheres to bottom objects by means of special posterior adhesive papillae, but it can swim short distances.

All arrowworms are carnivorous and feed on other planktonic animals, particularly copepods, which they detect from vibrations. They are thus of great importance as a trophic link between the primary and higher consumers in the sea. Some planktonic chaetognaths consume young fish and other arrowworms that may be as large as themselves and of their own species. Arrowworms are voracious feeders. *Zonosagitta nagae,* for example, consumes 37% of its own weight each day. As a prey passes beside a chaetognath, a rapid flex of the body quickly repositions the arrowworm, enabling it to seize the prey with its grasping spines and pierce it with its teeth. Initially, prey immobilization may result from sticky oral secretions, but these are aided by a bacteria-derived tetrodotoxin, a potent sodium-ion channel blocker, whose chief function may be to quiet or kill the prey before ingestion. Immediately after capture, the prey is swallowed and then surrounded by a peritrophic membrane before passing into the intestine for extracellular digestion.

The chaetognath nervous system is well developed, ganglionated, and located primarily within the epidermis. A prominent cerebral ganglion is located on the dorsal side of the head, and from it a nerve ring arises to encircle the foregut just behind the mouth. The ring bears a pair of large vestibular ganglia laterally and a pair of esophageal ganglia ventrally. The vestibular ganglia control the grasping spines whereas the esophageal ganglia innervate the gut musculature. From each side of the cerebral ganglion, a nerve extends posteriorly to the very large **ventral ganglion** in the trunk (Fig. 26-1). Numerous pairs of lateral nerves issue from the ventral ganglion and join an extensive intraepidermal nerve net that innervates the body-wall musculature and sensory cilia. The sensory cilia are

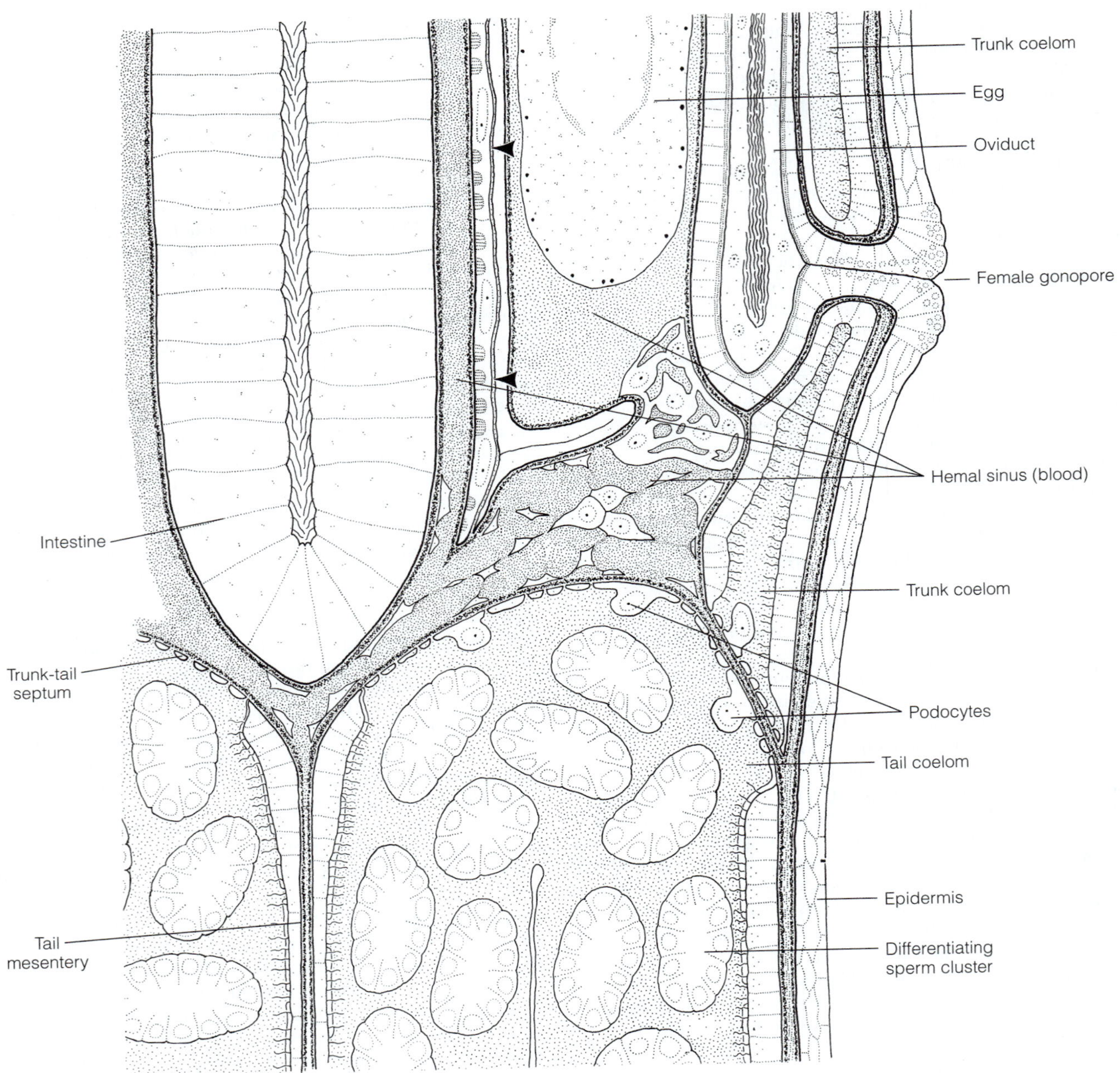

FIGURE 26-4 Chaetognatha: anatomy and histology. Horizontal (frontal) section of *Adhesisagitta (Sagitta) hispida* at the level of the trunk-tail septum. The follicle-cell layer on the egg is not shown. *(Adapted from Shinn, G. L. 1997. Chaetognatha. In Harrison, F. W., and Ruppert, E. E. (Eds.): Microscopic Anatomy of Invertebrates. Vol. 15. Hemichordata, Chaetognatha, and the Invertebrate Chordates. Wiley-Liss, New York. pp. 103–220. Reprinted with permission.)*

distributed over the surface of the body and arranged in short, fanlike rows called **ciliary fences** that detect waterborne vibrations (Fig. 26-1A). Some ciliary fences are oriented parallel and others perpendicular to the long axis of the body.

Arrowworms are relatively stiff shafts that bend or undulate dorsoventrally, but do not shorten along the longitudinal axis. Most likely, the body stiffness results from coelomic turgor pressure contained within the fiber sheaths of the epidermis and basement membrane. The fibers may be arranged in a manner that opposes shortening of the body. Vacuolated epidermal cells in the collar region of some species may also have a skeletal function.

REPRODUCTION AND DEVELOPMENT

All chaetognaths are hermaphrodites with internal fertilization. A pair of elongate ovaries occurs in the trunk anterior to the trunk-tail septum and a pair of testes is situated in the tail behind the septum (Fig. 26-1). Sperm leave the testis as spermatogonia and spermatogenesis is completed in the tail coelom, from which a sperm duct extends posteriorly to terminate in a seminal vesicle embedded in the lateral body wall. When mature, the sperm enter the ciliated funnel of the sperm duct and then pass into the seminal vesicle, in which the sperm are formed into a single spermatophore. The spermatophore is released by rupture of the seminal vesicle and body walls.

Within each ovary, the eggs are in a blood sinus (Fig. 26-4) and are separated from the blood solely by a thin, fenestrated (porous) follicle-cell layer. An oviduct runs along the lateral side of each ovary and opens to the exterior through two gonopores, one on each side of the body just in front of the trunk-tail septum (Fig. 26-1, 26-4). The eggs do not begin to mature until after spermatogenesis has begun in the tail coelom.

Mating occurs after a preliminary visual signaling behavior used for species recognition and perhaps to reduce the risk of being eaten by the partner. When contact is made, a spermatophore is released from the seminal vesicle and attached to the surface of the partner's body. Sperm then leave the spermatophore, migrate into the female gonopores and oviducts, and fertilize the eggs internally. Under optimal conditions, chaetognath pairs will mate each day and release fertilized eggs from the female's gonopores into the sea or attach them to the sea bottom, except for species of *Eukrohnia,* which brood their eggs in pouches formed by the lateral fins. Each egg is surrounded by a thin shell and jelly coat.

Development is direct and rapid, often as short as one day, and a miniature chaetognath, called a hatchling, escapes from the eggshell. Because cleavage is holoblastic and equal and leads to a spherical coeloblastula, it has been regarded as radial. After the second cleavage, however, two opposite blastomeres are displaced toward the animal pole and the other two are displaced toward the vegetal pole. Contact between the polar blastomeres at each pole creates a furrow called a cross furrow (Fig. 26-5B) that is typical of spirally cleaving embryos at the four-cell stage. Furthermore, as the embryo cleaves from two into four cells (second cleavage), the two blastomeres displaced to the animal pole move counterclockwise with respect to the vegetal-pole blastomeres (Fig. 26-5A,B). This suggests that the mitotic spindles involved in this division are tilted away from the animal-vegetal axis, which is another hallmark of spiral cleavage. Gastrulation is by invagination, and the blastocoel is obliterated. Bilateral folds of the anterior archenteron wall grow rearward, cutting off two lateral coelomic sacs from the medial archenteron (Fig. 26-5C,D). Thus the coelom originates by an unusual form of enterocoely. The mouth arises anteriorly at the opposite end of the embryo from the blastopore.

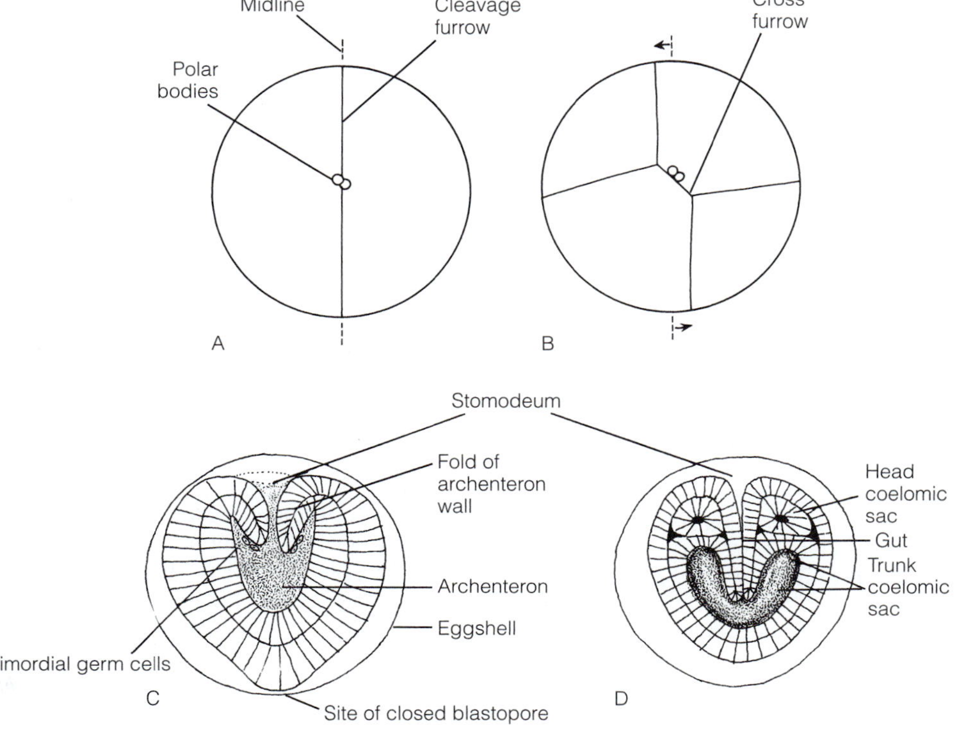

FIGURE 26-5 Chaetognatha: development. **A,** Two-cell stage in animal-pole view. **B,** Four-cell stage in animal-pole view. Note cross furrow typical of spiralian (protostome) embryos. Slight counterclockwise (leiotropic) shift (see arrows) of animal-pole cross-furrow cells is also typical of spiralian embryos at the four-cell stage. **C** and **D,** Dorsal views of coelom formation in a *Sagitta* embryo (anterior above, posterior below). **C,** Initial folding of anterior archenteron wall. **D,** Posterior growth of folds, establishment and separation of head and trunk coelomic sacs, and early morphogenesis of gut. *(A and B, Modified after Shimotori and Goto, 2001; C and D, After Burfield from Hyman, 1959.)*

PHYLOGENY OF CHAETOGNATHA

Chaetognatha is a distinctive taxon of Bilateria because of its many unique traits (autapomorphies), such as its grasping spines, hood, ciliary fences, doubled-walled lateral fins, and stratified epidermis, among others. But the absence of a definite synapomorphy between Chaetognatha and any other metazoan taxon except Bilateria itself has cast the arrowworms into an enduring isolation. Systematists have nevertheless drawn cautious comparisons based on characters of uncertain homology. Some of these characters suggest an alliance between chaetognaths and protostomes. Included among these are a head cuticle of α-chitin, shedding and replacement (perhaps molting?) of spines, compoundlike eyes, a peritrophic membrane, spiral-like cleavage (all characters of panarthropods), and the ventral nervous center (as in protostomes in general). A tricoelomate body, enterocoely, and a secondary morphogenesis of the mouth (deuterostomy), however, point to a relationship with deuterostomes. Finally, a coelomate body design with a hemal system and metanephridia (assuming that the gonoducts are modified metanephridia) may all have originated in the common ancestor of the Bilateria. Biomolecular data based on 18S rDNA sequences indicated a sister-group relationship between chaetognaths and nematodes, but the results are plagued by a possible artifact called "long-branch attraction" and thus are by no means definite.

Molecular phylogenies are built by computer programs that join taxa based on similarities in the sequence of molecules, such as nucleotides in a gene. Because such sequences contain only four nucleotides (A, T, C, and G) and substitutions can occur randomly, two rapidly evolving lineages in which the substitution rate is high (thus they are on long branches) tend to have convergently similar sequences and are erroneously joined. This false claim of close relationship, a form of homoplasy, is called long-branch attraction.

REFERENCES

GENERAL

Ahnelt, P. 1984. Chaetognatha. In Bereiter-Hahn, J., Matoltsky, A. G., and Richards, K. S. (Eds.): Biology of the Integument. Vol. 1. Invertebrates. Springer-Verlag, Berlin. pp. 746–755.

Alvariño, A. 1965. Chaetognaths. Oceanogr. Mar. Biol. Ann. Rev. 3:115–194.

Bergey, M. A., Crowder, R. J., and Shinn, G. L. 1994. Morphology of the male system and spermatophores of the arrowworm *Ferosagitta hispida* (Chaetognatha). J. Morphol. 221:321–341.

Bone, Q., Kapp, H., and Pierrot-Bults, A. C. 1991. The Biology of Chaetognaths. Oxford University Press, Oxford. 173 pp.

Bieri, R., and Thuesen, E. V. 1990. The strange worm *Bathybelos*. Am. Sci. 78:542–549.

Feigenbaum, D. L. 1978. Hair-fan patterns in the Chaetognatha. Can. J. Zool. 56:536–546.

Ghirardelli, E. 1968. Some aspects of the biology of the chaetognaths. Adv. Mar. Biol. 6:271–375.

Giribet, G., Distel, D. L., Polz, M., Sterrer, W., and Wheeler, W. C. 2000. Triploblastic relationships with emphasis on the acoelomates and the position of Gnathostomulida, Cycliophora, Plathelminthes, and Chaetognatha: A combined approach of 18S rDNA sequences and morphology. Syst. Biol. 49:539–562.

Goto, T., and Yoshida, M. 1984. Photoreception in Chaetognatha. In Ali, M. A. (Ed.): Photoreception and Vision in Invertebrates. Plenum Press, New York. pp. 727–742.

Goto, T., and Yoshida, M. 1985. The mating sequence of the benthic arrowworm *Spadella schizoptera*. Biol. Bull. 169:328–333.

Halanych, K. M. 1996. Testing hypotheses of chaetognath origins: Long branches revealed by 18S ribosomal DNA. Syst. Biol. 45:223–246.

Hyman, L. H. 1959. The Invertebrates. Vol. 5. Smaller Coelomate Groups. McGraw-Hill, New York. pp. 1–71.

Jordan, C. E. 1992. A model of rapid-start swimming at intermediate Reynold's number: Undulatory locomotion in the chaetognath *Sagitta elegans*. J. Exp. Biol. 163:119–137.

Nagasawa, S., and Marumo, R. 1972. Feeding of a pelagic chaetognath, *Sagitta nagae,* in Suruga Bay Central Japan. J. Oceanogr. Soc. Japan 32:209–218.

Parry, D. A. 1944. Structure and function of the gut in *Spadella cephaloptera* and *Sagitta setosa*. J. Mar. Biol. Assoc. U.K. 26:16–36.

Reeve, M. R., and Cosper, T. C. 1975. Chaetognatha. In Giese, A. C., and Pearse, J. S. (Eds.): Reproduction of Marine Invertebrates. Vol. 2. Academic Press, New York. pp. 157–184.

Shimotori, T., and Goto, T. 1999. Establishment of axial properties in the arrow worm embryo, *Paraspadella gotoi* (Chaetognatha): Developmental fate of the first two blastomeres. Zool. Sci. 16:459–469.

Shimotori, T., and Goto, T. 2001. Developmental fates of the first four blastomeres of the chaetognath *Paraspadella gotoi:* Relationship to protostomes. Develop. Growth Differ. 43:371–382.

Shinn, G. L. 1992. Ultrastructure of somatic tissues in the ovaries of a chaetognath *(Ferosagitta hispida)*. J. Morphol. 211:221–241.

Shinn, G. L. 1997. Chaetognatha. In Harrison, F. W., and Ruppert, E. E. (Eds.): Microscopic Anatomy of Invertebrates. Vol. 15. Hemichordata, Chaetognatha, and the Invertebrate Chordates. Wiley-Liss, New York. pp. 103–220.

Sullivan, B. K. 1980. *In situ* feeding behavior of *Sagitta elegans* and *Eukrohnia hamata* (Chaetognatha) in relation to the vertical distribution and abundance of prey at Ocean Station "P." Limnol. Oceanogr. 25:317–326.

Thuesen, E. V., and Bieri, R. 1987. Tooth structure and buccal pores in the chaetognath *Flaccisagitta hexaptera* and their relation to the capture of fish larvae and copepods. Can. J. Zool. 65:181–187.

Thuesen, E. V., Kogure, K., Hashimoto, K., et al. 1988. Poison arrowworms: A tetrodotoxin venom in the marine phylum Chaetognatha. J. Exp. Mar. Biol. Ecol. 116:249–256.

Internet Site

http://academic.evergreen.edu/t/thuesene/chaetognaths/chaetognaths.htm (E. V. Thuesen's chaetognath Web page.)

27 Introduction to Deuterostomia and Hemichordata[P]

Deuterostomia, the sister taxon of Protostomia, includes the acorn worms and sea angels (Hemichordata; Fig. 27-1), sea stars and their relatives (Echinodermata), as well as the sea squirts, lancelets, and vertebrates (Chordata). Lancelets and vertebrates are segmented animals but, in the deuterostome ground plan, the body is trimeric being divided into three distinct regions. These are an anterior preoral lobe, the **protosome,** a mouth-bearing middle region, or **mesosome,** and a trunk, the **metasome,** which bears the gut and gonads. Each of these regions houses a pair of coeloms, the **protocoels, mesocoels,** and **metacoels,** respectively (Fig. 27-3A, 27-5C). The coelomic mesothelium differentiates into the musculature, which primitively is composed of epitheliomuscular cells. Tubular outgrowths of the coelom, or **coelomic diverticula,** extend from one body region into other regions to create specialized muscles (see below), a specialized circulatory system (such as the echinoderm water-vascular system), or other functionally unique structures. Primitive deuterostomes, including hemichordates, have a **heart-kidney,** or axial complex, in the protosome. The heart-kidney is a metanephridial system consisting of the heart, pericardial cavity, glomerulus, protocoel, and a pair of metanephridia (Fig. 27-3A). Deuterostome zygotes undergo radial cleavage, the embryonic blastopore becomes the posterior anus, and the mouth forms secondarily at the anterior end (**deuterostome** = second mouth). Coelomic cavities arise by enterocoely, as outpockets from the archenteron (Fig. 9-23G), but other modes of coelom differentiation, such as schizocoely, occur in hemichordates.

Hemichordates express all the attributes (autapomorphies) of Deuterostomia and at least one unique structure, a median, unpaired diverticulum of the foregut that projects into the protocoel (unpaired in hemichordates) and supports the heart-kidney. This diverticulum, called a **stomochord** (Fig. 27-5), was previously regarded as a rudimentary notochord, a trait associated with Chordata, among which are the vertebrates. In addition to this notochord-like structure, hemichordates also express two other traits typical of chordates. The first is one or more pairs of gill slits, the general name for perforations of the pharyngeal wall, regardless of shape, which extend through the body wall and open to the exterior. The second is the so-called **collar cord** (neurocord; Fig. 27-5A), a short section of sometimes hollow dorsal nerve cord reminiscent of the chordate dorsal hollow nerve cord (spinal cord). The hemichordate gill slits are probably homologous to those of chordates and may even be part of the deuterostome ground plan, but the collar cord and stomochord are likely unique to hemichordates. We will return to these evolutionary issues later in the chapter.

Two distinctly different body forms typify Hemichordata. Representing one design are the large, solitary acorn worms (Enteropneusta) that inhabit soft sea bottoms at all water depths. Sea angels (Pterobranchia), on the other hand, are bryozoan-like colonies of tiny zooids that grow on hard marine surfaces.

ENTEROPNEUSTA[C]

The 70 species of Enteropneusta, or acorn worms, are chiefly benthic inhabitants of shallow water, but a few occur in the deep sea and in association with hydrothermal vents. Some, such as species of *Ptychodera,* live under stones and shells, and many common species, including those of *Saccoglossus* and *Balanoglossus,* construct mucus-lined burrows in mud and sand. Exposed tidal flats are frequently dotted with the coiled, ropelike fecal castings of these animals (Fig. 27-4F). One unidentified species from the Sea of Japan is reported to swim.

FORM

Acorn worms are large, soft, and fragile animals, the majority being between 10 and 45 cm in length. *Saccoglossus kowalevskii* typically is 15 cm long, and the common *Balanoglossus aurantiacus* of the southeastern coast of the United States reaches approximately 1 m in length (Fig. 27-1A). The largest described acorn worm, *B. gigas,* which ranges

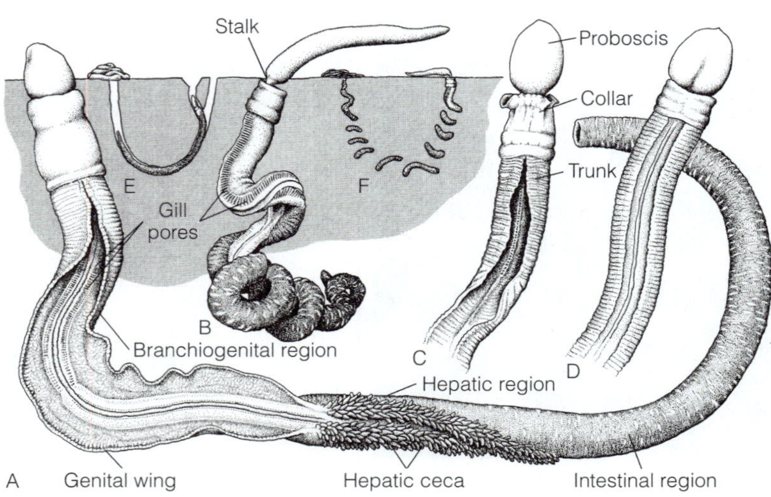

FIGURE 27-1 Enteropneust diversity and burrow structure. **A,** *Balanoglossus aurantiacus.* **B,** *Saccoglossus (Dolichoglossus) kowalevskii.* **C,** *Ptychodera jamaicensis.* **D,** *Schizocardium brasiliense.* The intestinal region of *B. aurantiacus* has been shortened. **E,** *B. aurantiacus:* burrow arrangement and feeding position showing, from left to right, fecal cast, feeding funnel, and ventilation shaft. **F,** *Saccoglossus kowalevskii:* burrow arrangement and feeding position with proboscis extended on sediment surface. The burrow is helical, throwing the trunk into a permanent twist. *(Adapted from Ruppert, E. E., and Fox, R. S. 1988. Seashore Animals of the Southeast. University of South Carolina Press, Columbia. 429 pp.)*

from Brazil to Cape Hatteras, North Carolina, exceeds 2.5 m in length and constructs 3 m long burrows.

In practice, it is difficult to establish the length of acorn worms because the fragile body breaks easily when handled. Unbroken, the three regions of the cylindrical body are the anterior **proboscis** (protosome), a short middle **collar** (mesosome), and a long posterior **trunk** (metasome; Fig. 27-1C), but the trunk is rarely collected intact.

The muscular proboscis is usually short and more or less conical (Fig. 27-1A,C,D), but in *Saccoglossus,* it is long and slender (Fig. 27-1B). The proboscis attaches to the collar dorsally by a slender but sturdy **stalk** (Fig. 27-1B). The collar is a short cylinder that opens anteriorly to form the large mouth (Fig. 27-4E). The collar surrounds and overlaps the proboscis stalk and the posterior end of the proboscis. In some species, the proboscis and collar together resemble an acorn and its cap, and the common name, acorn worm, was coined for this similarity.

The regionated trunk constitutes the major part of the body. From anterior to posterior, its three sections are called the branchiogenital, hepatic, and intestinal regions. The **branchiogenital region** bears a longitudinal row of **gill pores** on each side of the dorsal midline (Fig. 27-1A). Gill pores are the external body-wall openings that correspond to the internal pharyngeal gill slits. Lateral to the two rows of gill pores are either two low, longitudinal **genital ridges** (as in *Saccoglossus, Schizocardium,* and others) or two (as in *Balanoglossus, Ptychodera*) or four (as in *Stereobalanus*) large, flaplike **genital wings,** which arch dorsally over the branchiogenital region (Fig. 27-1A, 27-5B). A specialized **hepatic region** follows the branchiogenital region posteriorly. In many large-bodied species, it bears numerous fingerlike, hollow evaginations of gut and body wall called **hepatic ceca** that project dorsally above the surface of the body (Fig. 27-1A). The hepatic ceca are lined by intestinal gastrodermis. The long, cylindrical **intestinal region** completes the trunk and extends rearward to the terminal anus.

The enteropneust epidermis lacks a cuticle but is densely multiciliated, richly glandular, and bears numerous short microvilli (Fig. 27-2). The gland cells, which are especially abundant at the tip of the proboscis and on the collar, are responsible for secreting the mucus coating of the body. Some of the gland cells produce bromine-containing compounds that impart a strong medicinal odor to acorn worms and may protect them from bacterial infection or animal predation. A basal lamina and a connective-tissue dermis lie beneath the epidermis, but the dermal fibers are not well developed or well organized, perhaps accounting for the fragility of the body.

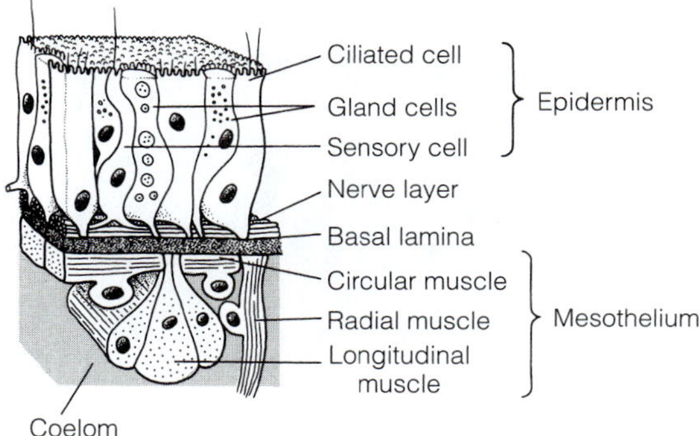

FIGURE 27-2 Hemichordate body wall. The body wall lacks a cuticle and is either monociliated (in pterobranchs) or multiciliated (in enteropneusts). The coelom is lined by a mesothelium that forms the musculature; a peritoneum is absent.

COELOMS, MUSCULATURE, AND LOCOMOTION

Acorn worms, like echinoderms and chaetognaths, are tricoelomate. A single unpaired protocoel occupies the proboscis, a pair of mesocoels is found in the collar, and a pair of metacoels occurs in the trunk (Fig. 27-3A). Septa separate the three coelomic regions and dorsal and ventral mesenteries divide the coeloms of the collar and trunk into right and left halves. The protocoel opens to the exterior on the left side of the proboscis stalk via one small opening, the **proboscis pore** (a nephridiopore; Fig. 27-3A). The mesocoels house a pair of metanephridia, the **mesocoel ducts,** that pass through the collar-trunk septum and discharge into the first gill pores (Fig. 27-3A). The metacoel lacks coelomic ducts, except for the gonoducts discussed below.

Specialized coelomic diverticula extend from one body region into another. Two **perihemal coeloms,** each a metacoelic diverticulum, originate on the collar-trunk septum and extend anteriorly through the collar region, ending on the proboscis stalk (Fig. 27-3A). They are called perihemal coeloms because they flank the dorsal blood vessel in the collar region, but they probably play little or no role in circulation. Instead, their mesothelium is a well-developed longitudinal musculature whose chief function is to retract the proboscis into the collar, thus closing the mouth.

The general coelomic lining is a monolayered mesothelium composed predominantly of smooth epitheliomuscular cells. Each of these cells bears a single cilium that helps to circulate the coelomic fluid. The muscle cells are pseudostratified into a poorly developed outer circular musculature and a well-developed inner longitudinal musculature (Fig. 27-2). In addition, **radial muscles,** each a single fiber, span the coelom in all body regions and may functionally supplement the weak circular muscles (Fig. 27-2, 27-3A, 27-5A). The musculature is best developed in the proboscis and trunk. The longitudinal musculature of the trunk usually is divided into a pair of dorsal muscles and a well-developed pair of ventrolateral muscles, which are the principal retractor muscles that shorten the trunk. Generally, acorn worms are sluggish animals that crawl or burrow slowly using retrograde peristaltic contractions of the proboscis. After the proboscis advances and anchors, the trunk and collar are pulled forward and the cycle is repeated (Fig. 27-4A–D).

SKELETON

Muscles act against the hydrostatic coelomic cavities during locomotion and most other movements. In addition to the coelomic hydrostat, however, enteropneusts have two specialized, rigid, collagenous skeletons, the **branchial skeleton** for

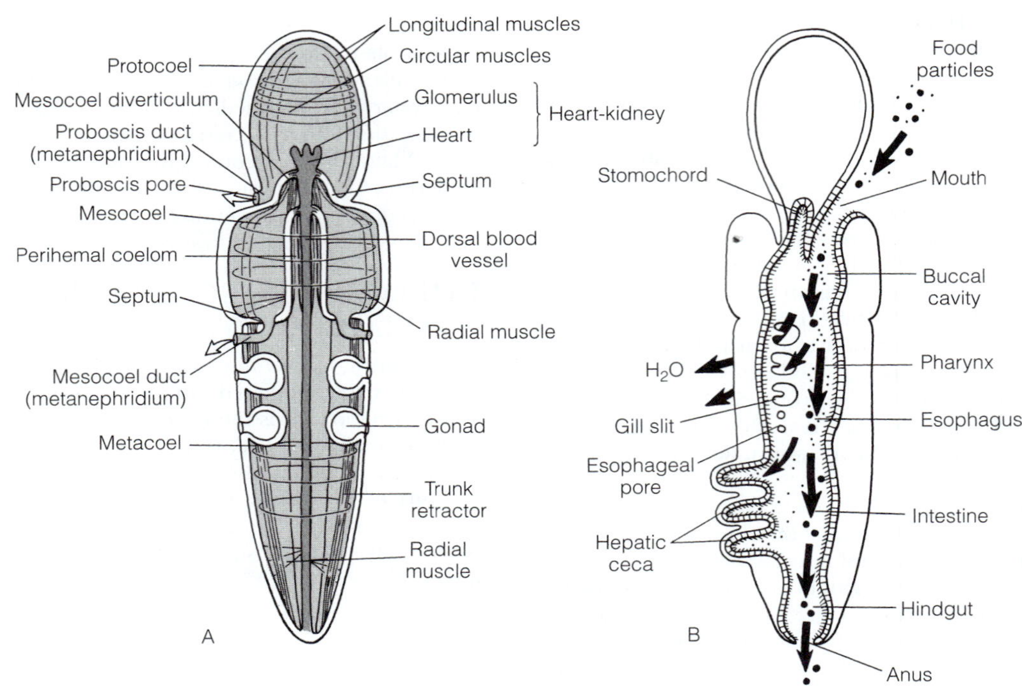

FIGURE 27-3 Enteropneust internal anatomy. **A,** Musculature, hemal, and coelomic organization, dorsal view. **B,** Digestive system showing the path of food and removal of water via the pharyngeal gill slits.

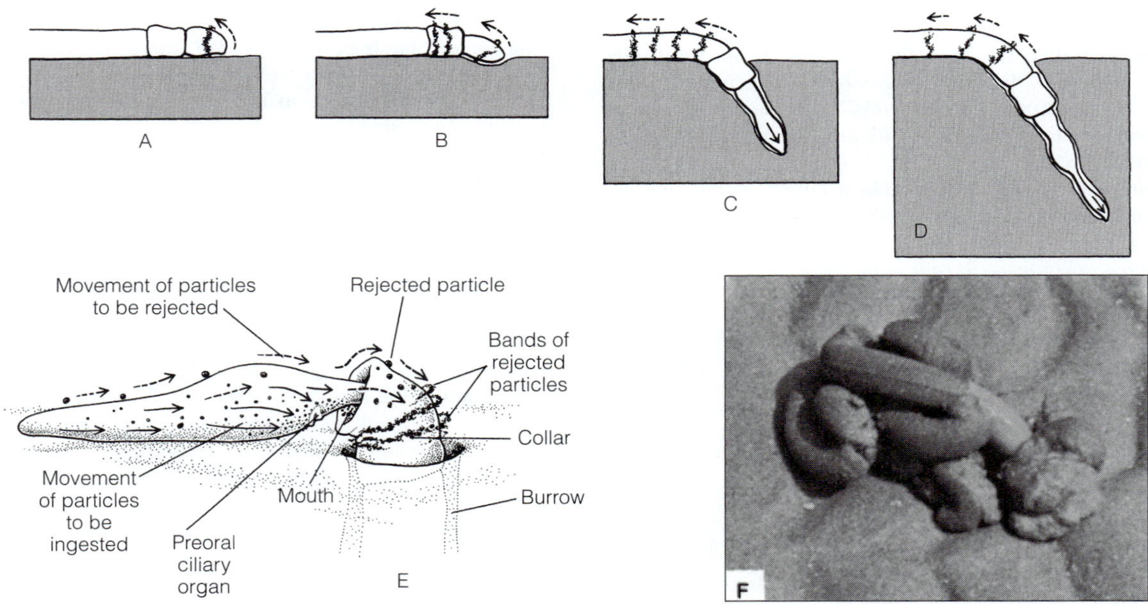

FIGURE 27-4 Enteropneust burrowing, feeding, and fecal casting. **A–D,** Acorn worms burrow using retrograde peristaltic waves along the muscular proboscis assisted by surface cilia, which transport sand rearward along the body. **E,** While feeding, particles are trapped in surface mucus and transported by cilia to the mouth. Rejected particles are carried over the rim of the collar and moved posteriorly in rings. **F,** The posterior end and ropy fecal cast of the giant acorn worm *Balanoglossus gigas* on the surface of an exposed sand flat at low tide. The cast is approximately 1 cm in diameter. *(E, Modified and redrawn after Burdon-Jones, C. 1956. Observations on the enteropneust,* Protoglossus kohleri. *Proc. Zool. Soc. Lond. 127:35.)*

support of the pharynx and gill slits (Fig. 27-5B, 27-7B) and the **proboscis skeleton** (Fig. 27-5A, 27-6). The Y-shaped proboscis skeleton anchors the proboscis to the collar and strengthens the slender proboscis stalk. The base of the Y inserts in the stalk and the two arms of the Y anchor in the dorsal wall of the collar.

DIGESTIVE SYSTEM AND NUTRITION

The digestive tract is a straight tube differentiated into several regions (Fig. 27-3B). The mouth—the large anterior opening of the cylindrical collar—can be closed by withdrawing the proboscis into the collar, like inserting a plug in a drain. The mouth leads into a **buccal cavity** within the collar (Fig. 27-3B). The stomochord projects anteriorly from the mid-dorsal wall of the buccal cavity, passes through the proboscis stalk, and enters the proboscis, where it lies directly below the heart and kidney (described below; Fig. 27-3B, 27-5A, 27-6). The buccal cavity joins the pharynx, which occupies the branchial region of the trunk and is laterally perforated by the paired gill slits. In many acorn worms, such as species of *Balanoglossus* and *Ptychodera,* the pharynx is divided into an upper **branchial channel** with gill slits and a respiratory water flow and a ventral **food channel** that transports food posteriorly to the esophagus and intestine (Fig. 27-5B).

The pharynx joins a short esophagus, in which food particles and mucus are molded into a cord. In species of *Schizocardium* and related genera, simple, oval **esophageal pores** that open from the esophagus directly to the exterior may help to remove excess water from the food cord (Fig. 27-3B).

The intestine constitutes the remainder of the gut. Its anterior hepatic region is the site of extracellular digestion, followed by intracellular digestion and nutrient storage (in hepatic ceca, if present). The posterior intestine compacts the fecal material and conveys it to the terminal anus, which releases the feces in a long, often ropy casting (Fig. 27-4F).

Acorn worms may be deposit feeders, suspension feeders, or both. Burrowing species are predominantly deposit feeders, consuming sand and mud and digesting the organic matter, but a few species protrude the proboscis from the burrow opening and collect suspended material from the water. The recently described burrower *Harrimania planktophilus* of the Pacific Northwest is a subsurface suspension feeder.

Among deposit feeders, detritus and sediment are trapped in mucus on the proboscis and transported by cilia to the mouth. Some of the material enters the mouth ventrally after passing over a shallow, ciliated groove, the **preoral ciliary organ** (POCO), located on the posterior face of the proboscis (Fig. 27-4E). The POCO traps the finest particles and then directs them into the mouth. Particles that are not ingested are transported rearward over the collar and trunk and then discarded (Fig. 27-4E). Suspension feeders trap and transport suspended material in mucus on the proboscis, like deposit feeders, but the inhalant water current generated by the gill-slit cilia also transports suspended particles directly into the pharynx. Once the particles enter the pharynx, they are trapped in mucus secreted by the pharynx lining. The trapped particles are transported ventrally to the pharyngeal food channel, if present (Fig. 27-5B), or to the ventral midline of the pharynx (hypobranchial ridge). Once in the food channel or on the hypobranchial ridge, cilia move the mucus and particles posteriorly into the esophagus.

Two common burrowing enteropneusts, *Saccoglossus kowalevskii* (Fig. 27-1B) and *Balanoglossus aurantiacus* (Fig. 27-1A), are both surface deposit feeders that occupy U-shaped burrows, but they have markedly different feeding styles. *Saccoglossus* extends its long proboscis from the burrow opening and directs it radially over the sediment surface (Fig. 27-1F). While it is extended, cilia on the proboscis convey the surface detritus back to the mouth. After feeding for a time, the proboscis is retracted and then reextended in a new direction. After several cycles of proboscis movements, *Saccoglossus* has created a spokelike series of radiating furrows, a **feeding rosette,** around the mouth of its burrow. The burrow of *Balanoglossus* typically has two surface openings and a third, craterlike depression called a **feeding funnel,** which lies above the head end of its burrow (Fig. 27-1E). While feeding, *Balanoglossus* positions its proboscis at the bottom of the feeding funnel. As it removes sand from the base of the funnel, surface detritus, rich in organic matter, tumbles into the funnel and is ingested by the worm. Thus, without leaving the safety of its burrow, *Balanoglossus* is able to feed on the organic surface detritus. Periodically, *Saccoglossus* and *Balanoglossus* back up in their burrows, extend the anus at the surface, and extrude a coiled fecal cast (Fig. 27-4F).

GAS EXCHANGE

The pharyngeal gill slits are narrow U-shaped perforations in the wall of the pharynx that function in gas exchange, removal of excess water from ingested food, and, in some species, suspension feeding (Fig. 27-5A, 27-7). The number of slits ranges from a few to as many as 200 or more pairs in different species. The serial arrangement of bilaterally paired gill slits and associated structures is called **branchiomery** (= gill segments).

Gill slits are supported internally by a rigid, collagenous branchial skeleton. The skeleton consists of a series of gill bars alternating with tongue bars. Each hairpin-shaped **gill bar** supports a gill slit laterally and arches around it dorsally. The **tongue bar** is a downgrowth from the dorsal arch of the gill bar that divides the originally oval gill slit into a U (Fig. 27-7A,B). In many species, but not in *Saccoglossus,* cross braces called **synapticles** bridge the gill slit and join the gill and tongue bars. The synapticles help to support the gill slits and are vascularized, like the gill arches and tongue bars, to supply blood to the gills.

Each gill slit opens into an internal saclike **atrium,** a pouch of developmentally invaginated body wall that, in some species, forms a water jacket around the pharynx (Fig. 27-5B). In most species, each atrial sac opens to the exterior at a dorsolateral gill pore. The gill pores occur in a longitudinal row on each side of the body. The walls around the gill slits are ciliated and contain a plexus of blood vessels that are involved in gas exchange. The coelomic surface of some of the vessels bears podocytes (see Internal Transport and Excretion). The beating gill-slit cilia produce a stream of water that enters the mouth and pharynx, and then exits through the gill slits, atria, and gill pores.

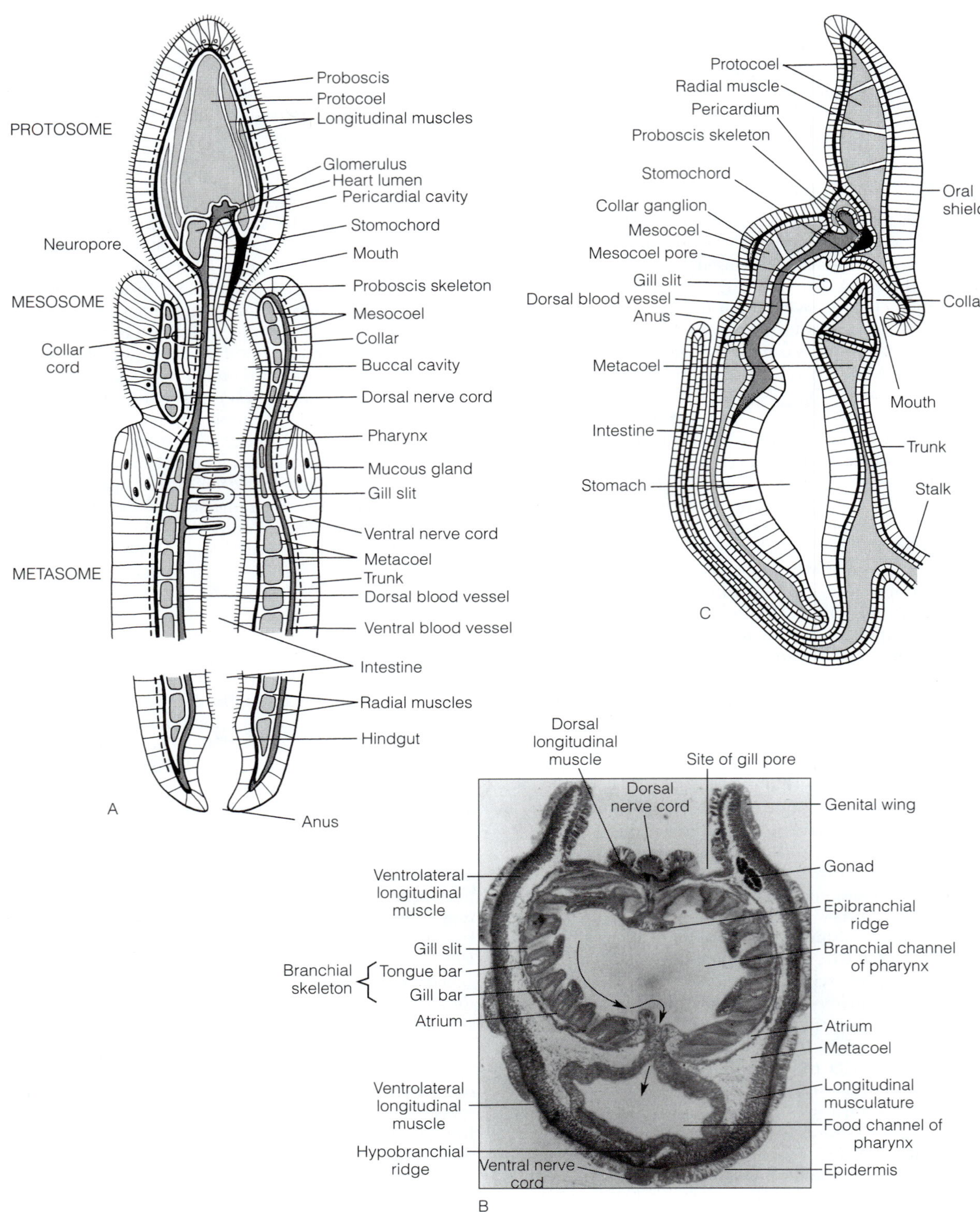

FIGURE 27-5 Diagrammatic internal anatomy of Enteropneusta and Pterobranchia. **A,** Sagittal section of an enteropneust. **B,** Cross section through the branchiogenital region (pharynx) of the enteropneust *Balanoglossus aurantiacus.* Gill pores are not seen in this section, but their location is indicated. Arrows indicate the direction of particle movement. The opposite walls of the pharynx are held together but not fused at the transition between the branchial and food channels. **C,** Sagittal section of a pterobranch.

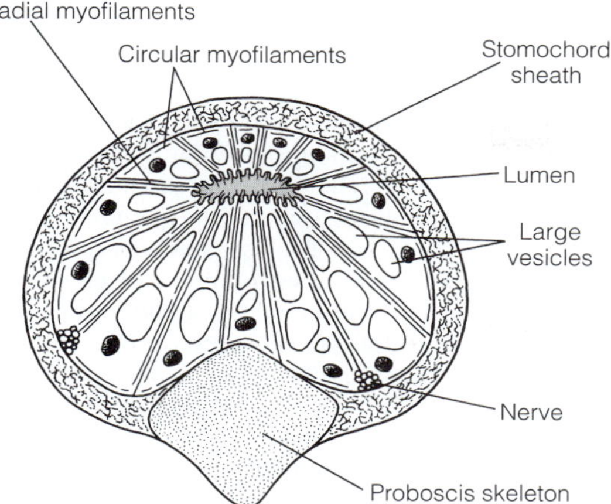

FIGURE 27-6 Enteropneust stomochord. Cross section of the stomochord. The stomochord owes some of its stiffness to turgid intracellular vesicles and muscle filaments, but a highly thickened and cross-linked part of the extracellular sheath, the proboscis skeleton, is the principal skeletal element. It secures the proboscis to the collar and, with the stomochord, forms a rigid platform to support the heart.

INTERNAL TRANSPORT AND EXCRETION

The hemal system of enteropneusts (and hemichordates in general) consists of an anterior heart and two main contractile vessels, the dorsal blood vessel carrying the colorless blood forward and the ventral blood vessel transporting it posteriorly. Smaller vessels supply the body wall, gill slits, and major organs.

The heart-kidney is located posteriorly in the protocoel (Fig. 27-3A). It is supported from beneath by the stomochord and pressurized from above by a small, contractile coelomic cavity, the pericardial cavity with its muscular wall (Fig. 27-5A). The stomochord is a semirigid platform against which the heart is compressed by the cross-striated, muscular pericardium. Blood enters the heart lumen posteriorly between the stomochord and pericardium. As the blood is pressurized by the slow contractions of the pericardium, at approximately 6 beats/min, it flows into efferent vessels for transport elsewhere in the body and into numerous blind-ended folds of the heart wall. These blood-filled folds form the **glomerulus** (Fig. 27-3A, 27-5A) and greatly increase the area of heart wall in contact with the protocoel. The glomerular wall is composed of podocytes. As the heart contracts, blood is ultrafiltered across the wall of the heart into the protocoel to form primary urine, which may be modified by the muscle cells of the protocoel mesothelium. Urine then passes from the protocoel to the exterior through the proboscis duct (metanephridium) and pore on the left side of the proboscis stalk (Fig. 27-3A). Some enteropneusts have a bilaterally symmetric pair of ducts and pores on the stalk.

A pair of ciliated mesocoel ducts (metanephridia) open into the first gill pores and transport mesocoelic fluid to the exterior, but an associated hemal filtration site has not yet been identified (Fig. 27-3A). The paired metacoels of the trunk, on the other hand, lack metanephridia, but mesothelial podocytes occur on gill-bar blood vessels. The podocyte layer separates the blood from the trunk coelomic fluid. Lacking metanephridial outlets, these **branchiomeric podocytes** have no obvious functions except perhaps to deliver blood ultrafiltrate (nutrients) to the trunk musculature via the coelomic fluid or to play a role in ion regulation. The physiology of excretion in hemichordates needs further study.

NERVOUS SYSTEM

Adult enteropneusts are noteworthy for their lack of sensory organs and more or less decentralized nervous system. All of the described sensory structures are individual sensory cilia that are widespread on the epidermis. A definite, concentrated,

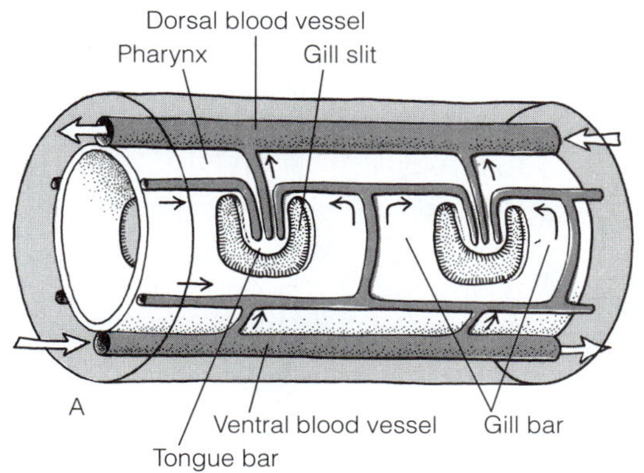

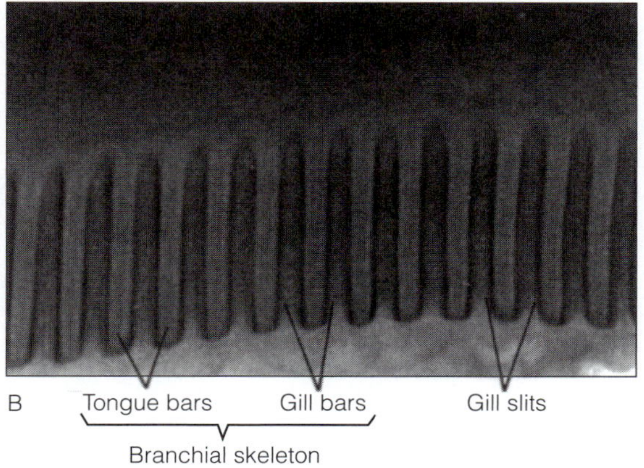

FIGURE 27-7 Enteropneust gill circulation and gill slits. **A,** Diagrammatic section of pharyngeal circulation; anterior is to the left. Open arrows show blood flow in dorsal and ventral blood vessels. Solid arrows indicate presumed direction of blood flow in and around the gills. The distance between the U-shaped gill slits is greatly exaggerated. **B,** Photograph of gill slits from the interior of the pharynx of a living acorn worm.

ganglionic brain is absent. The CNS consists of intraepidermal dorsal and ventral nerve cords (Fig. 27-5A,B) linked at intervals along the length of the body by circumferential nerve rings. These components join peripheral nerve nets in the epidermis and gut. In the trunk, the prime function of the dorsal cord is to innervate the two small dorsal longitudinal muscles, whereas the larger ventral cord innervates the better developed ventrolateral pair of longitudinal muscles. In the short collar region, the dorsal cord joins a more or less hollow, subepidermal collar cord (neurocord; Fig. 27-5A). The cavity of the collar cord, when present, opens to the exterior at an anterior **neuropore** located immediately inside the dorsal margin of the collar (Fig. 27-5A). The neurons of the collar cord are primarily responsible for conducting impulses between the proboscis and trunk, but some innervate the musculature of the perihemal coeloms (see above). This specialized musculature, which is in direct contact with the collar cord, retracts the proboscis into the collar. Ciliated sensory cells of unknown function occur in the cavity of the collar cord.

REPRODUCTION AND DEVELOPMENT

Most acorn worms are fragile animals that frequently tear or break when handled, and damage also can occur under natural conditions. For example, the Atlantic auger snail, *Terebra dislocata,* lingers on the sandy fecal mounds of *Balanoglossus aurantiacus.* When the worm exposes its tender posterior end to defecate, the snail bites off a piece and eats it. Acorn worms, however, readily regenerate damaged or lost parts and clonal reproduction by fragmentation has been reported for several taxa, including species of *Glossobalanus* and *Balanoglossus.*

Acorn worms are gonochoric animals that free-spawn eggs and sperm into the surrounding seawater (or their burrows), where fertilization occurs. The numerous saccate gonads are situated in the connective-tissue space of the genital ridges or genital wings (Fig. 27-3A, 27-5B). Each sac has its own surface duct and pore. Male gonads are often orange and female gonads, lavender or gray. Early development of the zygotes and the fate of the blastopore follow the deuterostome pattern (Fig. 27-8E,F). The mesoderm and coelomic cavities originate from the archenteron, but all morphogenetic modes—enterocoely, schizocoely, and other patterns—are represented in Enteropneusta in association with particular species.

A few taxa, such as species of *Saccoglossus,* have large, yolky eggs that develop into short-lived, planktonic, lecithotrophic larvae before settling and undergoing metamorphosis (Fig. 27-8A–D). The recently settled juvenile of *Saccoglossus* has a long **postanal tail,** reminiscent of the pterobranch stalk (Fig. 27-8D; see below), that helps to anchor it in the sand. Most enteropneusts, however, have small, nonyolky eggs that develop into transparent, long-lived, planktotrophic larvae called **tornaria** larvae (Fig. 27-8G). In addition to the meandering circumoral band of suspension-feeding cilia, a posterior ring of multiciliated cells, the telotroch, is used for locomotion

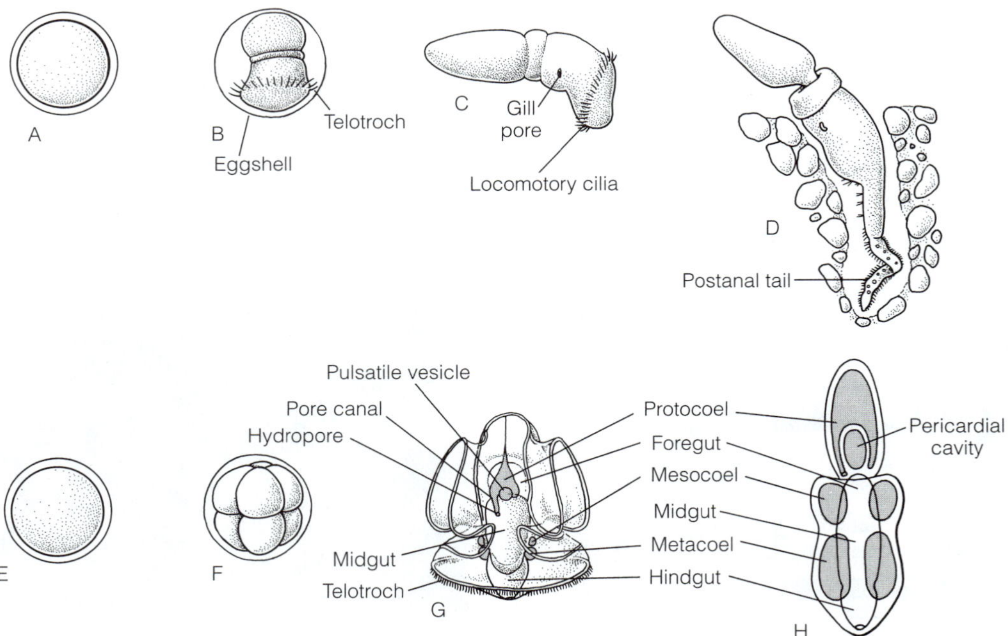

FIGURE 27-8 Enteropneust development. **A–D,** Indirect development in *Saccoglossus kowalevskii.* The large, yolky eggs of this species develop into short-lived, lecithotrophic, swimming larvae **(C).** After settlement **(D),** the juvenile develops a postanal tail. **E** and **F,** Indirect development in *Schizocardium brasiliense.* The small, nonyolky eggs develop into long-lived planktotrophic larvae called tornaria **(G)** before undergoing a rapid metamorphosis to form a juvenile worm **(H).** *(D, Redrawn after Burdon-Jones, C. 1952. Development and biology of the larva of Saccoglossus horsti. Phil. Trans. Roy. Soc. Lond. B. 236:553–590.)*

(Fig. 27-8G). The anatomy of the tornaria is strikingly similar to some echinoderm (dipleurula) larvae and provides the chief structural evidence for assuming a close evolutionary relationship between hemichordates and echinoderms. Like many echinoderm larvae, the tornaria has an anterior protocoel from which a short **pore canal** extends to a **hydropore** on the left side of the dorsal midline (Fig. 27-8G). These structures together with a small pulsatile vesicle (the presumptive pericardium) form the larval nephridium and are the developmental anlagen of the adult heart-kidney.

PTEROBRANCHIA[C]

The pterobranchs, or sea angels, are colonial, tube-dwelling hemichordates that attach to hard substrata and superficially resemble bryozoans and hydroids. As in these other two taxa, pterobranch zooids are small, not exceeding 5 mm in body length *(Cephalodiscus),* and many species are less than 1 mm long. Despite their small zooid size, pterobranch tissues and organs are similar to those of their larger relatives, the enteropneusts. Because of these similarities, only the distinctive features of pterobranchs will be described in the following sections.

Pterobranchs are uncommon marine animals that occur infrequently in shallow water on the surfaces of shells and rocks. More often, they are dredged from deep water and are best known from the seas around Antarctica. There is growing evidence, however, that pterobranchs are widespread, but easily overlooked, in shallow tropical and temperate seas, including along the coasts of Florida and Bermuda. The 22 species of pterobranchs belong to three genera: *Cephalodiscus, Atubaria,* and *Rhabdopleura.* The tubeless *Atubaria,* however, closely resembles a *Cephalodiscus* zooid that has abandoned its tube, which *Cephalodiscus* zooids often do, and is known only from preserved collections dredged from a depth of 200 to 300 m off the coast of Japan.

ZOOID FORM AND FUNCTION

Each tripartite zooid consists of an anterior oral shield (protosome), a short collar (mesosome), and a saccate trunk (metasome) (Fig. 27-5C, 27-9C). The broad **oral shield** is a disclike glandular creepsole on which the zooid glides over the inner surface of its tube. The short collar region dorsally bears two or more **arms,** each with pinnately arranged tentacles. In *Cephalodiscus,* 2 to 18 arms, varying with body size and species, radiate from the collar to form a spherical basket above the zooid (Fig. 27-10C), whereas in *Rhabdopleura* (Fig. 27-9C), two long arms are held in an upright V above the tube opening. These two arms with their tentacles, arising from the shoulderlike collar, resemble angel wings and are

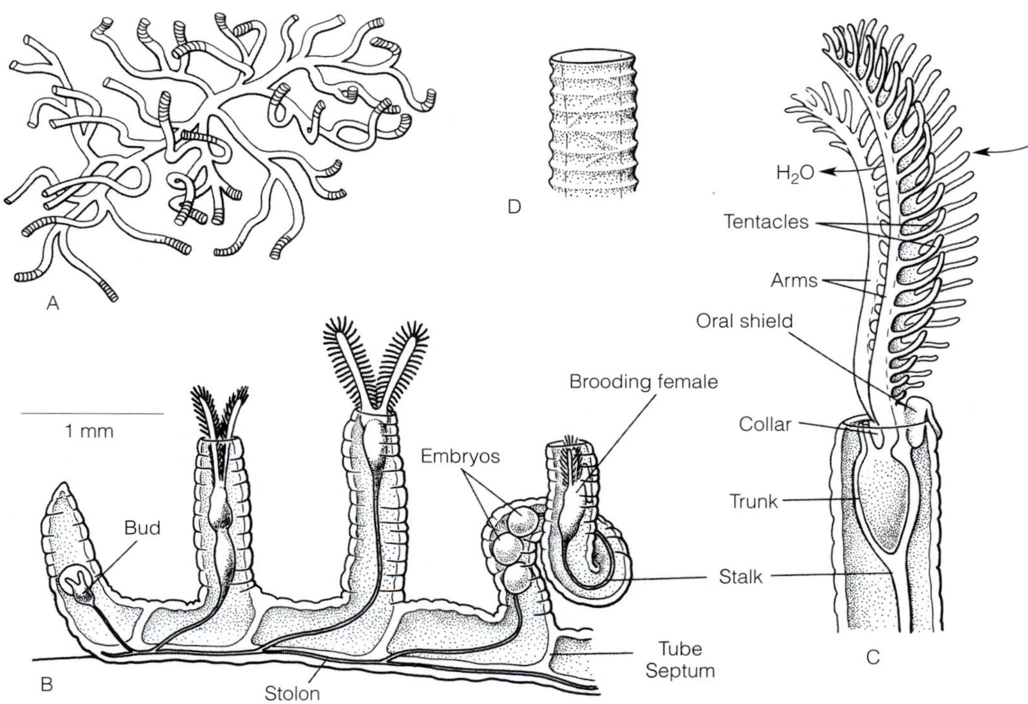

FIGURE 27-9 The pterobranch *Rhabdopleura.* **A** and **B,** Colonial organization. **C,** Feeding position of a zooid with the oral shield folded over the tube rim. While in this position, the oral shield secretes a new tube section, several of which are shown in **D.** *(A, Redrawn after van der Horst, C. J. 1939. Hemichordata. In Bronn, H. C. (Ed.): Kl. Ord. Tierreichs. 4:1–735.)*

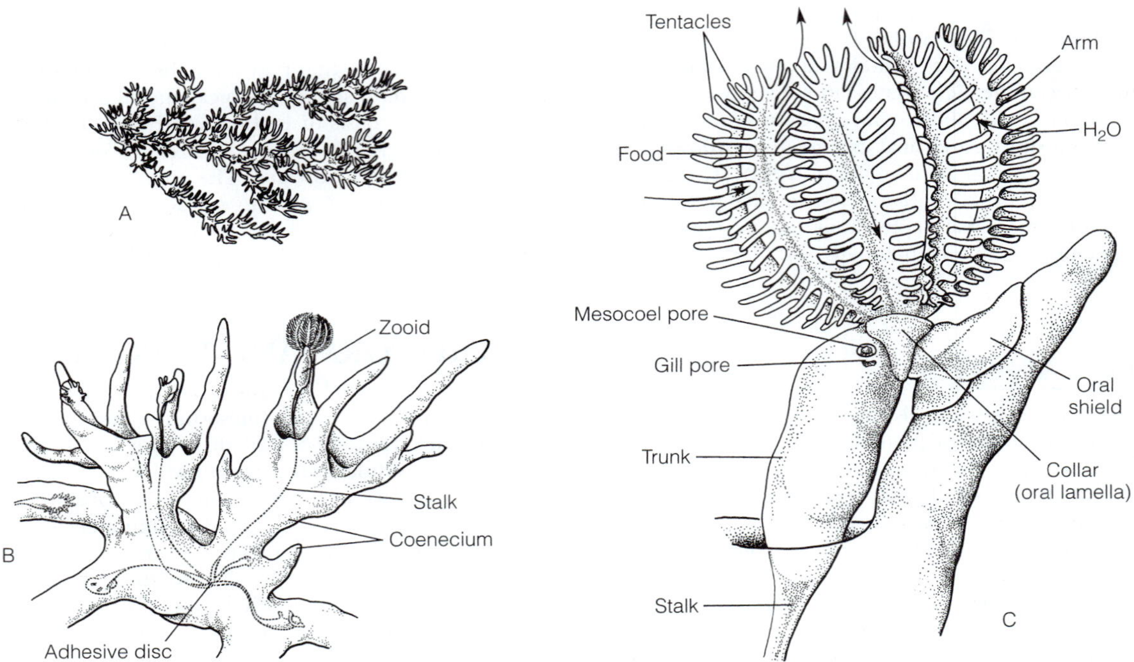

FIGURE 27-10 The pterobranch *Cephalodiscus*. **A** and **B**, Colonial organization. Zooids emerge from openings in the coenecium and perch on branch tips while feeding **(B)**. The leftmost zooid in **B** occupies the common coenecium but belongs to a colony separate from that of the other zooids. **C**, Feeding posture. *(A, Modified and redrawn from van der Horst, C. J. 1939. Hemichordata. In Bronn, H. C. (Ed.): Kl. Ord. Tierreichs. 4:1–735. B and C, Modified and redrawn after Lester, S. M. 1985. Cephalodiscus sp. (Hemichordata: Pterobranchia): Observations of functional morphology, behavior and occurrence in shallow water around Bermuda. Mar. Biol. 85:263–268.)*

responsible for the common name "sea angel." A pair of ciliated collar folds, the **oral lamellae,** occurs at the base of the arms, one lamella on each side of the mouth (Fig. 27-10C). A pair of large, valved, ciliated pores, the **mesocoel pores,** one from each of the paired collar coeloms, opens laterally at the collar-trunk boundary (Fig. 27-10C). In addition, a single pair of gill pores opens immediately beside the mesocoel pores in *Cephalodiscus.* The surface gill pores correspond to a pair of circular gill slits in the pharyngeal wall, but pterobranchs lack an atrium between the slits and pores. Gill slits and pores are absent in *Rhabdopleura.* Ventrally, the trunk bears a conspicuous fleshy, muscular **stalk,** which quickly retracts zooids into their tubes and unites zooids into a colony (Fig. 27-5C, 27-9B, 27-10B,C).

COLONY FORM AND LOCOMOTION

Pterobranch colonies are composed of few to many zooids. The colonies inhabit a network of cuticular tubes called a **coenecium** that is secreted by gland cells on the oral shields of the zooids. In *Rhabdopleura,* the coenecium is a network of prostrate tubes on the substratum that gives rise to unbranched vertical tubes, each of which has a simple circular opening and is occupied by a single zooid (Fig. 27-9A,B). The zooids are separated from each other by coenecial bulkheads called **tube septa,** which are bridged solely by a threadlike stolon of living tissue that runs through the prostrate tubes and connects the zooids (Fig. 27-9B). New zooids bud from the stolon. Most colonies of *Rhabdopleura* are between 1 and 25 cm in diameter and 5 to 10 mm in height.

The coenecium of some *Cephalodiscus* species resembles that of *Rhabdopleura,* but typically is larger (Fig. 27-10A). In others, however, the coenecium is a bulky, cakelike mass that reaches 30 cm in diameter and 10 cm in height. The tube openings may have smooth, circular rims, as in *Rhabdopleura* (Fig. 27-9D), but in many species the rim bears one or more long spines on which the zooids perch while feeding (Fig. 27-10B,C). A *Cephalodiscus* coenecium is occupied by many non-interjoined clusters (colonies) of zooids. Thus, a single coenecium of *Cephalodiscus* contains an aggregation of colonies. It is not known if these are genetically related. The zooids in each cluster are connected by their stalks to a common **adhesive disc,** which anchors them to the interior of the tube (Fig. 27-10B).

Individual zooids move to and from their feeding apertures and within the coenecium by gliding on the ciliated ventral surface of the oral shield. As they move about, they remain tethered to the stolon (as in *Rhabdopleura*) or adhesive disc (as in *Cephalodiscus*) by their contractile stalks, which can be stretched to 10 times their contracted lengths. The adhesive disc of each *Cephalodiscus* colony is motile, like a zooid's oral shield, and glides over surfaces with its zooids in tow. Under certain conditions, the adhesive disc and its attached zooids abandon the coenecium, creep over the substratum, and secrete a new coenecium.

BODY WALL AND INTERNAL FORM

The pterobranch body wall is similar to that of enteropneusts, but in pterobranchs, the cells are always monociliated (Fig. 27-2). The musculature consists of longitudinal and radial fibers only; circular body-wall muscles are absent (Fig. 27-5C). The radial muscles are common in the oral shield, collar, and arms. The chief longitudinal muscles are a pair of smooth stalk retractors that draw the zooid into the coenecium, cross-striated tentacular muscles that flick the tentacles toward the food groove on the arm, and cross-striated dilator muscles of the mesocoel ducts that open duct valves.

All of the cells lining the coelom, including the epitheliomuscular cells, are monociliated and help to circulate the coelomic fluid. Some mesothelial cells on the gut wall store nutrients. A heart-kidney, supported by a stomochord, occurs in the unpaired protocoel, which opens to the exterior via two tiny protocoel pores, one on each side of the dorsal midline. Cilia in the large mesocoel ducts and pores transport coelomic fluid to the exterior and may have an excretory function.

The anatomy of the pterobranch hemal system is imperfectly known, but the major vessels include a heart, a dorsal blood vessel, an enlargement of the dorsal vessel in the trunk to form a genital blood sinus around the gonad, and a ventral vessel (Fig. 27-5C). Smaller blood vessels are known to occur (for example, one extends into each tentacle), but details of the circuit and flow direction of the colorless blood are unknown. All blood vessels and sinuses are simple, unlined channels in the connective tissue between epithelia. Gas exchange most likely occurs across the arms, tentacles, and general body surface, all of which are ciliated and generate water flow over the body surface.

The nervous system is entirely intraepidermal and a hollow, submerged collar cord is absent, although a **collar ganglion** occurs dorsally in the mesosome (Fig. 27-5C). A dorsal nerve extends anteriorly and posteriorly from the collar ganglion. Each arm receives a branch from the anterior nerve, which in turn provides a nerve to each tentacle. The tentacular nerves are probably both motor and sensory. From the collar ganglion, a nerve ring (or rings) surrounds the pharynx and joins a nerve net in the ventral oral shield and a ventral nerve cord that extends posteriorly into the stalk.

Innervation of the muscles apparently occurs by diffusion of neurotransmitters across the epidermal basal lamina to the coelomic epitheliomuscular cells. Individual ciliated sensory cells are scattered on the surface of the tentacles and perhaps elsewhere on the body.

DIGESTIVE SYSTEM AND NUTRITION

Pterobranchs suspension feed using their hollow, ciliated arms and tentacles, which constitute an upstream collecting system. Lateral cilia on the tentacles create a flow of water and particles over the frontal surface of the tentacles and arms (Fig. 27-10C). Small particles are intercepted on the frontal upstream surfaces or batted onto them by the flicking tentacles. Once captured, frontal cilia transport the particles down the length of the arms to a ciliated groove that passes under the oral lamellae before joining the mouth. The undulating oral lamellae probably direct food into the mouth and remove excess water, thus concentrating the food before it is swallowed. The downstream (abfrontal) surface of the arms transports large inedible particles away from the mouth to the arm tips, where they are released into the exhaust flow.

Feeding zooids of *Rhabdopleura* grip the rim of the tube opening with the oral shield and extend the arms and tentacles, which resemble two feathers, into the surrounding water (Fig. 27-9C). As feeding continues, each zooid rotates around the tube rim, thus exposing the frontal surface of its feeding apparatus to all compass directions. Like *Rhabdopleura, Cephalodiscus* also collects suspended particles from 360° around the zooid body, but it does not rotate around its tube aperture. In *Cephalodiscus,* the multiple arms are arranged, like meridians, to form a radially symmetrical feeding sphere that gathers particles from all directions.

Pterobranchs have a U-shaped gut (Fig. 27-5C). The mouth is located ventrally under the posterior margin of the oral shield and the anus is mid-dorsal on the trunk, immediately behind the collar. Cilia transport food through the gut. Food entering the mouth enters a short pharynx, which in *Cephalodiscus* has one pair of gill slits, presumably to eliminate excess water from the food. Food passes from the pharynx into the saccate stomach situated posteriorly and ventrally in the trunk. From the posterior end of the stomach, the slender intestine bends dorsally and then extends anteriorly to a short hindgut that joins the anus. Digestion probably occurs in the stomach, and fecal pellets are formed in the intestine. Absorption may occur in the stomach and part of the intestine.

REPRODUCTION AND DEVELOPMENT

Pterobranch zooids arise asexually by budding from the stolon (*Rhabdopleura;* Fig. 27-9B) or adhesive disc (*Cephalodiscus;* Fig. 27-10B) during colony growth. Moreover, an individual adhesive disc of *Cephalodiscus* is capable of fragmentation or fission to form new colonies.

Pterobranch colonies are hermaphroditic, but their component zooids are usually male, female, or neuter. The fertile zooids, which have one (as in *Rhabdopleura*) or two (as in *Cephalodiscus*) gonads, are often sexually dimorphic. In *Cephalodiscus sibogae,* for example, male zooids have only rudimentary arms and tentacles and much of the trunk is occupied by the large testes. Females function as feeding zooids and support the males. Male and neuter zooids of *Rhabdopleura normani,* however, retain fully developed arms and tentacles, but those of female zooids are reduced (Fig. 27-9B) and the males are the feeding zooids. The details of fertilization are not fully known, but the ripe males probably release sperm or spermatophores into the water and females oviposit in the coenecium. Fertilization is probably external in the coenecium, but there are unconfirmed reports of internal fertilization. In any case, developmental stages are brooded in

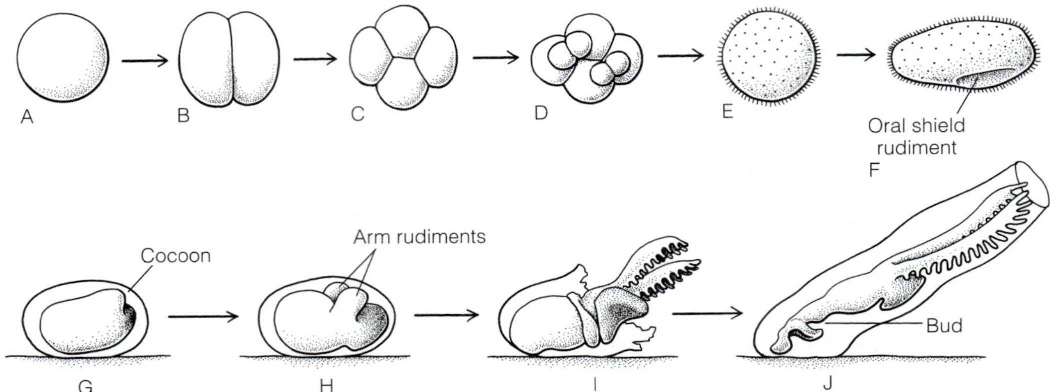

FIGURE 27-11 Pterobranchia: development, larvae, and larval metamorphosis. **A–D,** The yolky eggs of *Rhabdopleura* show a form of biradial cleavage **(D)** leading to a ciliated, lecithotrophic larva (**E** and **F**). The ventral furrow of the larva is not the blastopore, but rather the invaginated oral shield rudiment. **G–J,** After a brief swimming existence, the larva settles, secretes a cocoon (a coenecium rudiment, **G**), and gradually differentiates into a feeding zooid that produces other colony members by budding. *(Modified and redrawn after Lester, S. M. 1988. Settlement and metamorphosis of* Rhabdopleura normani *(Hemichordata: Pterobranchia). Acta Zool. 69:111–120.)*

a modified part of the coenecium. In *Rhabdopleura normani,* embryos are brooded in the lower part of a coiled vertical tube occupied by the nonfeeding female zooid (Fig. 27-9B). Cleavage is initially biradial but soon becomes bilateral and results in a coeloblastula (Fig. 27-11A–E). Gastrulation is by ingression or delamination, and the coelom forms by schizocoely before the gut differentiates.

The embryos develop into short-lived, uniformly ciliated, lecithotrophic larvae that are released from the coenecium into the sea (Fig. 27-11F). After a brief swimming existence of a day or two, each larva settles and encapsulates itself in a cocoon (a coenecium rudiment) that is attached to the substratum. A gradual metamorphosis results in a single zooid, which breaks through the upper wall of the cocoon, feeds, secretes a new coenecium, and founds a new colony (Fig. 27-11G–J).

PHYLOGENY OF HEMICHORDATA AND DEUTEROSTOMIA

Perhaps because we humans are especially curious about our own deuterostome ancestry, researchers have sustained an intense interest in the phylogeny of Deuterostomia for 150 years. With only three major taxa to consider—Hemichordata, Echinodermata, Chordata—you might think that deuterostome phylogeny would have been settled long ago, but nothing could be further from the truth. Even today, deuterostome phylogeny remains unresolved and key questions center on the pattern of relationships among the three major taxa, the nature of the deuterostome ancestor, the evolutionary origin of chordates, and the relationships of Chaetognatha and Lophophorata to deuterostomes. As discussed in Chapter 26, chaetognaths are probably protostomes despite their tricoelomic anatomy and a deuterostome-like secondary morphogenesis of the mouth. The uncertain alliance of lophophorates to deuterostomes or protostomes or, for that matter, to each other, is discussed in Chapters 9 and 25. A discussion of chordate evolution will be considered in Chapter 29, but here we provide two (of several) alternative phylogenies of Deuterostomia based on classical and contemporary research.

Phylogeny One (Fig. 27-12A) positions echinoderms and pterobranchs at the base of the tree. Both of these taxa primitively suspension feed using outgrowths of the mesosome (arms and tentacles in pterobranchs, arms and tube feet in echinoderms). The assignment of a planktotrophic dipleurula larva to the deuterostome ancestor suggests that the adult was large and perhaps more like an echinoderm than a pterobranch, because the latter produces only a few eggs that undergo lecithotrophic development. The remainder of the tree depicts a gradual, stepwise evolution of gill slits, beginning with pterobranchs and culminating in the filter-feeding pharynx of chordates. The tree, therefore, implies an evolutionary transition from suspension-feeding tentacles to a filter-feeding pharynx. Phylogeny One also denies the monophyly of Hemichordata, recognizing it instead as a paraphyletic taxon.

Phylogeny Two, although based chiefly on morphology, incorporates recent data from molecular systematics and other molecule-based disciplines (Fig. 27-12B). It divides Deuterostomia into two sister taxa, Hemichordata + Echinodermata (N.N.) and Chordata. In this phylogeny, a planktotrophic, dipleurula-like larva (the tornaria of enteropneusts) is regarded as a synapomorphy of hemichordates and echinoderms. A lecithotrophic larva, on the other hand, is hypothesized to have been in the deuterostome ground plan. Like Phylogeny One, Phylogeny Two accepts recent evidence that

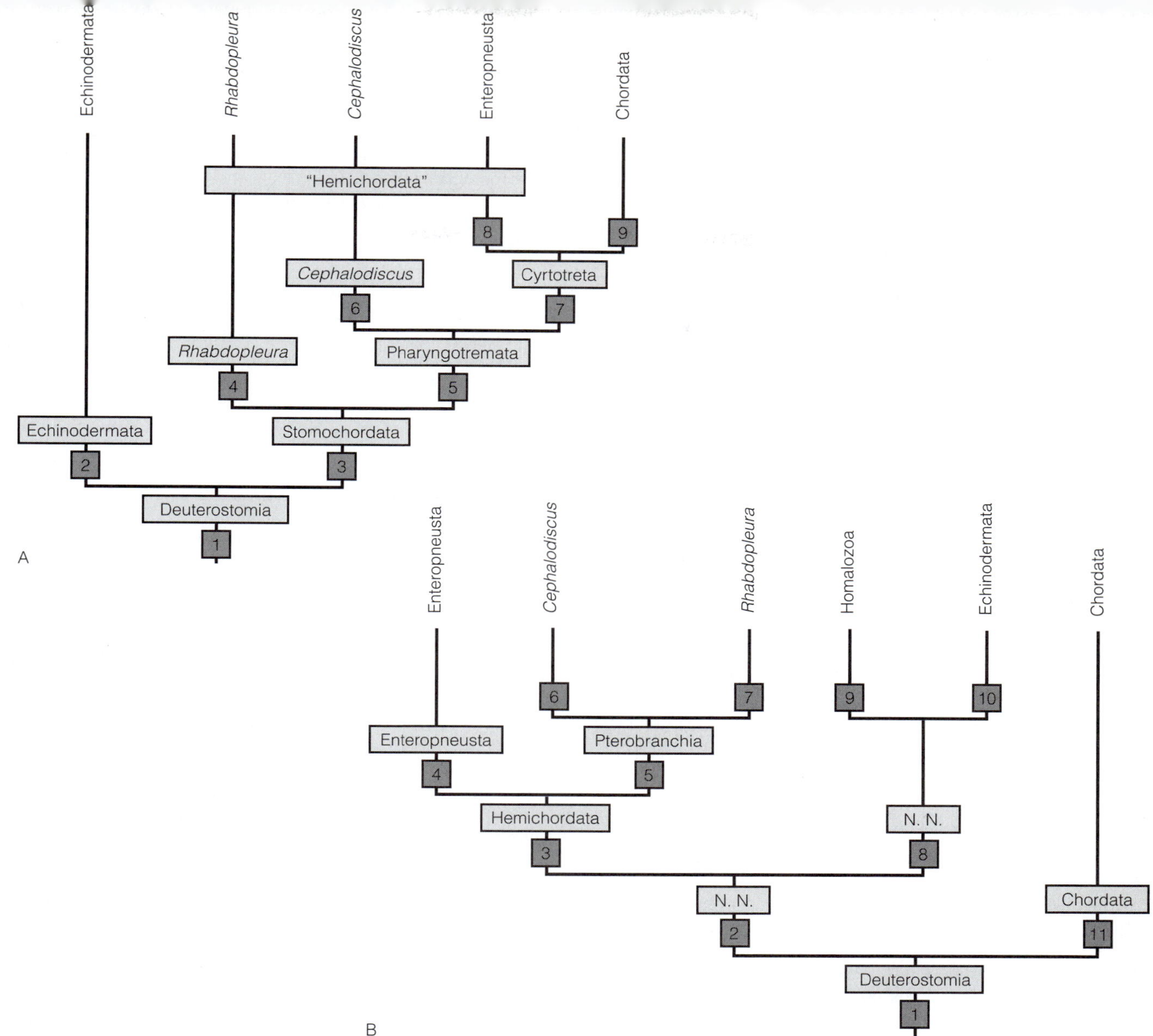

FIGURE 27-12 Competing phylogenies of Hemichordata and Deuterostomia. **A, Phylogeny One. 1, Deuterostomia:** Blastopore becomes anus, heart-kidney is in protocoel, trimeric body, dipleurula. **2, Echinodermata:** Has pentamerous symmetry, stereom ossicles, water-vascular system, mutable connective tissue. **3, Stomochordata:** Has a stomochord. **4, *Rhabdopleura:*** Colonial zooids arise singly from a stationary creeping stolon. **5, Pharyngotremata:** Has gill slits. **6, *Cephalodiscus:*** Colonial zooids arise from a common, motile adhesive disc. **7, Cyrtotreta:** Branchiomery (multiple paired gill slits). **8, Enteropneusta:** Has a proboscis skeleton, brominated mucus. **9, Chordata:** Has a notochord, dorsal hollow nerve cord, endostyle. **B, Phylogeny Two. 1, Deuterostomia:** Has a trimeric body, endodermal suspension-feeding pharynx with multiple paired gill slits (branchiomery) discharging into an ectodermally derived atrium, branchiomeric metanephridia, a protocoelic metanephridial system with a pericardial cavity (heart-kidney), mesocoel ducts open in common with first gill pores, dorsal gonads and gonoducts, enterocoely, blastopore becomes anus, lecithotrophic larva. **2, N. N.:** Has planktotrophic larva with circumoral ciliary band (dipleurula), upstream removal of food particles. **3, Hemichordata:** Has dorsal and ventral longitudinal nerve cords, stomochord antagonizes contractile pericardium, valved collar (mesocoel) ducts, muscular locomotory-secretory protosome. **4, Enteropneusta:** Has a proboscis skeleton, collar cord (neurocord), brominated mucus, a locomotory telotroch on a dipleurula-like larva (tornaria), loss of branchiomeric nephridioducts (or appropriation by gonads?). **5, Pterobranchia:** Colonial growth with small zooids, coenecium, suspension-feeding tentacles replace suspension-feeding pharynx, one pair of gill slits. **6, *Cephalodiscus:*** Zooid tentacles form suspension-feeding sphere, colonies motile on common adhesive disc and can vacate coenecium. **7, *Rhabdopleura:*** Gill slits absent, unpaired gonad, one pair of zooid tentacles, oral shield folds over tube rim and rotates body while feeding, colonies fixed in coenecium, separated by tube septa. **8, N. N.:** Bilateral asymmetry, calcareous endoskeleton that antagonizes pericardium, calcitic stereom ossicles. **9, Homalozoa:** Gill slits on only one side of body. **10, Echinodermata:** Pentamerous symmetry, water-vascular system, mutable connective tissue, loss of gill slits and atrium. **11, Chordata:** Has a dorsal hollow nerve cord with an open anterior neuropore, notochord, pharynx with endostyle, protocoelic heart-kidney is absent, has a ventral postoral heart, fins, cross-striated longitudinal muscles produce lateral undulations for swimming. *(A, Modified and redrawn from Ax, P. 2001. Das System der Metazoa III. Ein Lehrbuch der phylogenetischen Systematik. Spektrum Akademischer Verlag, Heidelberg. 283 pp.)*

PHYLOGENETIC HIERARCHY OF HEMICHORDATA AND DEUTEROSTOMIA

Phylogeny One	Phylogeny Two
Deuterostomia	Deuterostomia
Echinodermata	N. N.
Stomochordata	Hemichordata
Rhabdopleura	Enteropneusta
Pharyngotremata	Pterobranchia
Cephalodiscus	N. N.
Cyrtotreta	Homalozoa (extinct)
Enteropneusta	Echinodermata
Chordata	Chordata

deuterostome gill slits are homologous structures. However, it proposes the additional hypothesis that an extinct echinoderm-like taxon (Homalozoa; Fig. 28-61), species of which lived 600 to 400 million years ago, may have had gill slits that were subsequently lost in their living echinoderm descendants. If homologous gill slits occur in hemichordates, extinct echinoderms, and chordates, then the likelihood is high that gill slits occurred in the deuterostome ancestor. Thus, Phylogeny Two suggests an enteropneust-like ancestor for the deuterostomes. It further implies that the pterobranchs may have departed from pharyngeal feeding in favor of tentacular feeding in response to a reduction in body size.

REFERENCES

GENERAL

Armstrong, W. G., Dilly, P. N., and Urbanek, A. 1984. Collagen in the pterobranch coenecium and the problem of graptolite affinities. Lethaia 17:145–152.

Ax, P. 2001. Das System der Metazoa III. Ein Lehrbuch der phylogenetischen Systematik. Spektrum Akademischer Verlag, Heidelberg. 283 pp.

Balser, E. J., and Ruppert, E. E. 1990. Structure, ultrastructure and function of the preoral heart–kidney in *Saccoglossus kowalevskii* (Hemichordata, Enteropneusta) including new data on the stomochord. Acta Zool. 71:235–249.

Barnes, R. D. 1977. New record of a pterobranch hemichordate from the Western Hemisphere. Bull. Mar. Sci. 27:340–343.

Barrington, E. 1940. Observations of feeding and digestion in *Glossobalanus minutus*. Q. J. Microsc. Sci. 82:227–260.

Barrington, E. 1965. The Biology of Hemichordata and Protochordata. W. H. Freeman, San Francisco. 176 pp.

Benito, J., and Pardos, F. 1997. Hemichordata. In Harrison, F. W., and Ruppert, E. E. (Eds.): Microscopic Anatomy of Invertebrates. Wiley-Liss, New York. pp. 15–101.

Burdon-Jones, C. 1952. Development and biology of the larvae of *Saccoglossus horsti* (Enteropneusta). Proc. Roy. Soc. Lond. B Biol. Sci. 236:553–589.

Burdon-Jones, C. 1956. Observations on the enteropneust, *Protoglossus kohleri* (Caullery and Mesnil). Proc. Zool. Soc. Lond. 127:35–59.

Burdon-Jones, C. 1962. The feeding mechanism of *Balanoglossus gigas*. Bol. Fac. Filos. Cienc. Letr. Univ. Sao Paulo, No. 261. Zoologia 24:255–280.

Cameron, C. B. 2002. Particle retention and flow in the pharynx of the enteropneust worm *Harrimania planktophilus:* The filter-feeding pharynx may have evolved before the chordates. Biol. Bull. 202:192–200.

Cameron, C. B. 2002. The anatomy, life habits, and later development of a new species of enteropneust, *Harrimania planktophilus* (Hemichordata: Harrimaniidae) from Barkley Sound. Biol. Bull. 202:182–191.

Cameron, C. B., Garey, J. R., and Swalla, B. J. 2000. Evolution of the chordate body plan: New insights from phylogenetic analyses of deuterostome phyla. Proc. Natl. Acad. Sci. USA 97:4469–4474.

Cameron, C. B., and Mackie, G. O. 1996. Conduction pathways in the nervous system of *Saccoglossus* sp. (Enteropneusta). Can. J. Zool. 74:15–19.

Colwin, A. C., and Colwin, L. H. 1953. The normal embryology of *Saccoglossus kowalevskii* (Enteropneusta). J. Morphol. 92:401–453.

Corgiat, J., Dobbs, F. C., Burger, M., and Scheuer, P. 1993. Organohalogen constituents of the acorn worm *Ptychodera bahamensis*. Comp. Biochem. Physiol. 106B:83–86.

Dawydoff, C. 1948. Embranchement des Stomocordes. In Grassé, P. (Ed.): Traité de Zoologie. Vol. 11. Echinodermes, Stomocordes, Procordes. Masson, Paris. pp. 367–551.

Dilly, P. N., Welsch, U., and Rehkämper, G. 1986a. Fine structure of heart, pericardium and glomerular vessel in *Cephalodiscus gracilis* M'Intosh, 1882 (Pterobranchia, Hemichordata). Acta Zool. 67:173–179.

Dilly, P. N., Welsch, U., and Rehkämper, G. 1986b. Fine structure of the tentacles, arms and associated coelomic structures of *Cephalodiscus gracilis* (Pterobranchia, Hemichordata). Acta Zool. 67:181–191.

Dilly, P. N., Welsch, U., and Rehkämper, G. 1986c. On the fine structure of the alimentary tract of *Cephalodiscus gracilis* (Pterobranchia, Hemichordata). Acta Zool. 67:87–95.

Dobbs, F. C., and Guckert, J. B. 1988. Microbial food resources of the macrofaunal deposit feeder *Ptychodera bahamensis* (Hemichordata: Enteropneusta). Mar. Ecol. Prog. Ser. 45:127–136.

Duncan, P. B. 1987. Burrow structure and burrowing activity of the funnel-feeding enteropneust *Balanoglossus aurantiacus* in Bogue Sound, North Carolina, USA. PSZNI Mar. Ecol. 8:75–95.

Gilmour, T. H. J. 1979. Feeding in pterobranch hemichordates and the evolution of gill slits. Can. J. Zool. 57:1136–1142.

Hadfield, M. G. 1975. Hemichordata. In Giese, A. C., and Pearse, J. S. (Eds.): Reproduction of Marine Invertebrates. Vol. II. Academic Press, New York. pp. 185–240.

King, G. M., Giray, C., and Kornfield, I. 1995. Biogeographical, biochemical, and genetic differentiation among North American saccoglossids (Hemichordata; Enteropneusta; Harimaniidae). Mar. Biol. 123:369–377.

Knight-Jones, E. W. 1952. On the nervous system of *Saccoglossus cambrensis*. Philos. Trans. Roy. Soc. Lond. B 236:315–354.

Lester, S. M. 1985. *Cephalodiscus* sp. [*gracilis*] (Hemichordata: Pterobranchia): Observations of functional morphology, behavior and occurrence in shallow water around Bermuda. Mar. Biol. 85:263–268.

Lester, S. M. 1988a. Ultrastructure of adult gonads and development and structure of the larva of *Rhabdopleura normani* (Hemichordata: Pterobranchia). Acta Zool. 69:95–109.

Lester, S. M. 1988b. Settlement and metamorphosis of *Rhabdopleura normani* (Hemichordata: Pterobranchia). Acta Zool. 69:111–120.

Morgan, T. H. 1891. The growth and metamorphosis of tornaria. J. Morphol. 5:407–458.

Ogasawara, M., Wada, H., Peters, H., and Satoh, N. 1999. Developmental expression of Pax 1/9 genes in urochordate and hemichordate gills: Insight into the function and evolution of the pharyngeal epithelium. Development 126: 2539–2550.

Okai, N., Tagawa, K., Humphreys, T., Satoh, N., and Ogasawara, M. 2000. Characterization of gill-specific genes of the acorn worm *Ptychodera flava*. Dev. Dyn. 217:309–319.

Pardos, F. 1988. Fine structure and function of pharynx cilia in *Glossobalanus minutus* Kowalewsky (Enteropneusta). Acta Zool. 69:1–12.

Pardos, F., and Benito, J. 1988. Blood vessels and related structures in the gill bars of *Glossobalanus minutus* (Enteropneusta). Acta Zool. 69:87–94.

Petersen, J. A., and Ditadi, A. S. F. 1971. Asexual reproduction in *Glossobalanus crozieri* (Ptychoderidae, Enteropneusta, Hemichordata). Mar. Biol. 9:78–85.

Romero-Wetzel, M. B. 1989. Branched burrow-systems of the enteropneust *Stereobalanus canadensis* (Spengel) in deep-sea sediments of the Vöring-Plateau, Norwegian Sea. Sarsia 74:85–89.

Ruppert, E. E., and Fox, R. S. 1988. Seashore Animals of the Southeast. University of South Carolina Press, Columbia. 429 pp.

Spengel, J. W. 1909. Pelagisches Vorkommen von Enteropneusten. Zool. Anz. 34:54–59.

van der Horst, C. J. 1939. Hemichordata. Kl. Ord. Tierreichs. 4:1–735.

Welsch, U., Dilly, P. N., and Rehkämper, G. 1987. Fine structure of the stomochord in *Cephalodiscus gracilis* M'Intosh 1882 (Hemichordata, Pterobranchia). Zool. Anz. 218:209–218.

Internet Sites

http://faculty.washington.edu/bjswalla/Hemichordata/Enteropneusta/class%20enteropneusta.html (B. J. Swalla's color images of living hemichordates.)

http://cluster3.biosci.utexas.edu/faculty/cameronc/Images.htm (C. B. Cameron's taxonomic key, classification, and color images of living hemichordates.)

Echinodermata[P]

For a curious beachcomber, the discovery of a stranded sea star or sea urchin, two symbols of the sea, places the mind at the threshold of wonder. Echinoderms could be variations on celestial stars that fell to Earth as extraterrestrials, so extraordinary is their form and function. But echinoderms are descendants of a bilaterally symmetrical deuterostome that over the course of eons adopted radial symmetry and became very different from all other bilaterians. A glimpse of that evolutionary journey excites the intellect as much as the discovery of a sand dollar delights the senses. And between first discovery and evolutionary perspective are opportunities to solve the myriad mysteries of echinoderm biology and ecology.

GENERAL BIOLOGY

The approximately 6000 living species of sea stars (Fig. 28-1), brittle stars, sea urchins, sand dollars, sea cucumbers, and sea lilies are all marine and mostly bottom dwellers, as were the approximately 13,000 extinct species that flourished as long ago as the early Cambrian, 545 mya. Echinoderms fossilize well because their endoskeleton is composed of calcareous ossicles. The ossicles may articulate with one another, as in the flexible sea stars and brittle stars, or may be fused together to form a rigid skeletal test, as in sea urchins and sand dollars. Commonly, skeletal elaborations project outward as spines or warts that are responsible for the name echinoderm, meaning "spiny skin." The ossicles are located in a well-developed connective-tissue dermis, which itself provides skeletal support.

FIGURE 28-1 Echinodermata: pentamerous symmetry and the water-vascular system. A sea star attached by its five rows (ambulacra) of tube feet to the glass wall of an aquarium.

Unlike any other animals, echinoderms can reversibly vary the rigidity of their dermis and general connective tissue. Echinoderms are thus said to have catch, or **mutable, connective tissue** (Fig. 28-2). The extremes of rigidity are almost as great as water and ice. When a sea star arches over a clam to feed, it stiffens its connective tissue and the body then becomes a rigid scaffold against which the attached tube feet pull on the clam's valves. After feeding, the sea star softens its connective tissue, becomes flexible, flattens against the substratum, and moves away. Sea urchins, by stiffening ligaments at the base of their movable spines, can fix the spines like pikes to ward off attackers or to wedge themselves tightly in rock crevices. Under conditions of stress, brittle stars and sea cucumbers voluntarily cast off (autotomize) arms or cast out viscera (eviscerate) by a local softening of their connective tissue that causes the affected region of the body to loosen and detach. In the extreme, as when some sea cucumbers are removed from water and exposed to air, the entire body softens and disintegrates and the animal dies (Fig. 28-2).

The physiological control of mutable connective tissue and the molecular mechanisms of hardening and softening are active areas of modern research with clinical implications for certain types of human connective-tissue disorders, such as arthritis. Although the echinoderm dermis contains muscle, nerve, and other types of cells, it is the extracellular matrix that changes in stiffness, and nerves have been found to terminate in the extracellular matrix itself. Apparently, two types of nerves are present: one whose action may harden the matrix, and another whose action softens it. The stiffness of the matrix is affected by changes in Ca^{2+} concentrations and by experimental manipulations of other cations. In general, an increase in the Ca^{2+} concentration stiffens the matrix; a decrease softens it, suggesting that Ca^{2+} may form bridges between macromolecules in the matrix.

The radial symmetry of echinoderms is called **pentamerous symmetry** because the body can be divided into five similar parts arranged around a central axis (Fig. 28-1). This radial symmetry, however, evolved within the echinoderms (as an autapomorphy) and does not indicate a close evolutionary relationship with, for example, the cnidarians.

Like the members of their sister taxon, Enteropneusta, echinoderms are tricoelomate, and some of the coelomic cavities develop diverticula that adopt unique functions. The most conspicuous of these is the **water-vascular system,** which includes tubular outgrowths of the body wall (the tube feet; Fig. 28-1) that are used for locomotion, feeding, gas exchange, and other tasks. The hemal system has little or no role in gas transport and the blood ebbs and flows tidally in most vessels rather than flowing in the unidirectional circuit of most other bilaterians. Some echinoderms have an organ (the functionally integrated heart, axial canal, and axial hemal vessel) that is homologous with the hemichordate heart-kidney, but its role in excretion is dubious. Other echinoderms lack this organ altogether, and all echinoderms seem to have tissue-level excretion.

Echinoderms are generally large animals that develop many types of specialized gills to supplement the gas-exchange surface of the body. Tube feet are gill surfaces found in all echinoderms

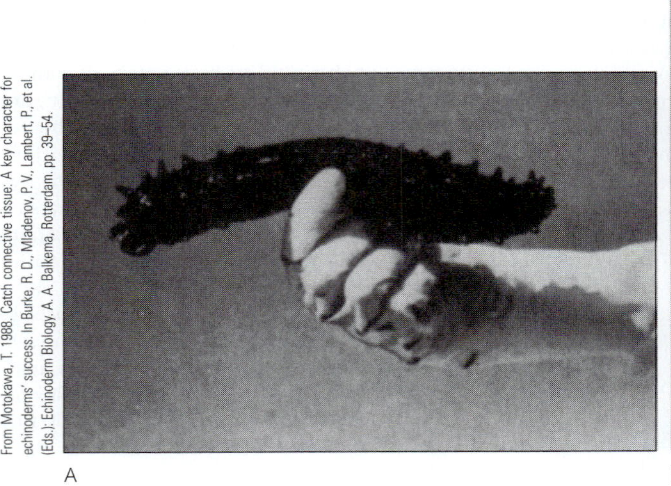

From Motokawa, T. 1988. Catch connective tissue: A key character for echinoderms' success. In Burke, R. D., Mladenov, P. V., Lambert, P., et al. (Eds.): Echinoderm Biology. A. A. Balkema, Rotterdam. pp. 39–54.

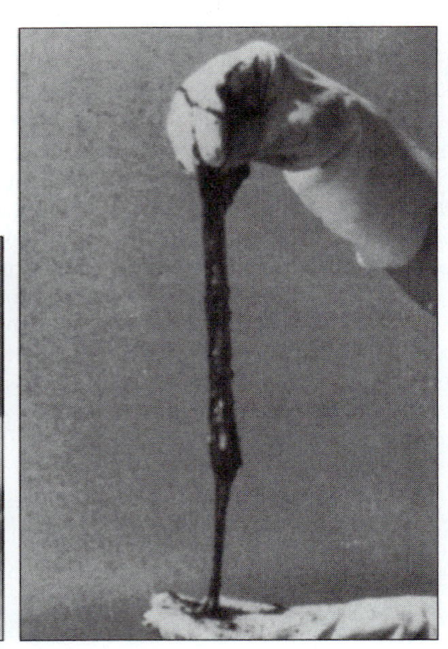

From Motokawa, T. 1988. Catch connective tissue: A key character for echinoderms' success. In Burke, R. D., Mladenov, P. V., Lambert, P., et al. (Eds.): Echinoderm Biology. A. A. Balkema, Rotterdam. pp. 39–54.

FIGURE 28-2 Echinoderm mutable connective tissue: the sea cucumber *Stichopus chloronotus.* **A,** When first touched or handled, the animal stiffens the connective tissue and hardens its body wall, but after being rubbed vigorously, the body wall becomes so soft that it flows between the fingers of the experimenter **(B).**

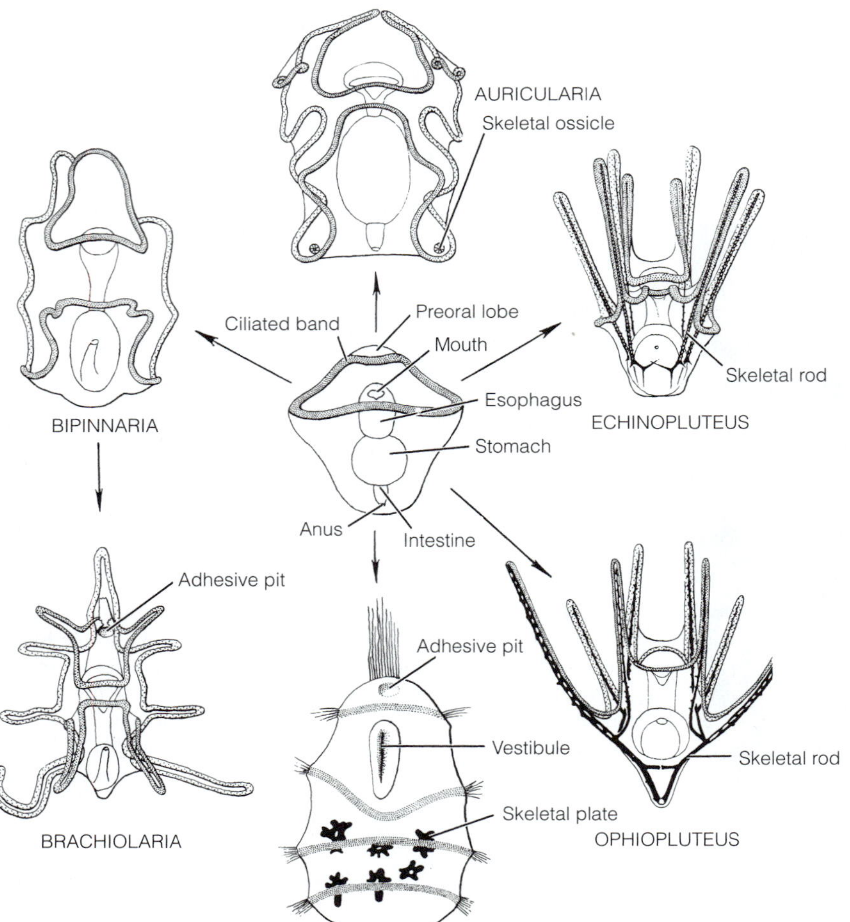

FIGURE 28-3 Hypothetical generalized echinoderm dipleurula larva (center) and actual echinoderm larvae, all in ventral view. Counterclockwise from top: Auricularia of a sea cucumber (Holothuroidea); early larva (bipinnaria) of a sea star (Asteroidea); settling-larva (brachiolaria) of a sea star; doliolaria larva of a feather star (Crinoidea); ophiopluteus larva of a brittle or serpent star (Ophiuroidea); echinopluteus larva of a sea urchin (Echinoidea). All larvae are planktotrophs, except the lecithotrophic crinoid doliolaria. All larvae swim toward the top of page. *(Modified from Ubaghs, G. 1967. General characteristics of Echinodermata. In Moore, R. C. (Eds.): Treatise on Invertebrate Paleontology. Pt. S. Vol. 1. Courtesy of the Geological Society of America and the University of Kansas, Lawrence. pp. S3–S60.)*

and each major taxon usually has its own additional specialized gill. Most echinoderms are gonochoric with external fertilization. Their planktotrophic larva, known generally as a dipleurula, closely resembles the enteropneust tornaria, except that the dipleurula lacks a telotroch and ocelli (Fig. 28-3).

DEVELOPMENTAL ORIGIN OF PENTAMEROUS SYMMETRY

Echinoderm development, as in deuterostomes in general, includes radial cleavage, regulative development, and enterocoelous origin of the coelomic cavities. The blastopore becomes the anus and the mouth forms as a secondary invagination of ectoderm that fuses with the advancing tip of the archenteron to complete the gut. The planktotrophic dipleurula adopts a unique form (and name) in each of the echinoderm taxa, but it is always bilaterally symmetric externally (Fig. 28-3), in contrast to the pentamerous symmetry of the metamorphosed juvenile and adult (Fig. 28-1). The unique adult symmetry can best be understood by tracking the changes that occur during larval metamorphosis.

It is difficult to generalize about echinoderm metamorphosis because the patterns differ in each major taxon and sometimes even between species. The following generalized description, although based on the metamorphosis of a larval sea star (Asteroidea), is partly hypothetical and should be regarded chiefly as a learning aid. The names of many larval and adult structures in this description will be unfamiliar at this point in the chapter, but the general pattern of metamorphosis should be evident from the text and illustration (Fig. 28-4). Later, after studying the appropriate chapter sections, you can reread this description and better understand the link between larval and adult anatomy.

Echinoderm metamorphosis typically includes two major events, a **body-axis shift** and a twisting, or **torsion,** of the body around the new axis. For example, within a sea star larva, the developing juvenile, called a **juvenile rudiment,** differentiates from two tissue sources: a mass of embryonic reserve cells and transformed larval tissue. The body axis of the differentiating rudiment is shifted 90% from the larval anterior-posterior axis (Fig. 28-4A,C). As development proceeds, most of the bilaterally symmetric larval coeloms (paired protocoels, mesocoels, and metacoels) are modified and incorporated into the rudiment. The left mesocoel and both metacoels undergo torsion, winding into rings around the new body axis as the right mesocoel degenerates (Fig. 28-4B). The ringlike left mesocoel and its outgrowths become the water-vascular system (WVS) and the fused doughnutlike metacoels enclose the viscera and occupy the hollow arms as the perivisceral coelom (Fig. 28-4C). A diverticulum from the left metacoel twists around the body axis to form the genital ring canal associated with the gonads (Fig. 28-4B,C). The left larval protocoel, pore canal, and hydropore transform into the juvenile stone canal and madreporite, which link the water-vascular system to the exterior; they also produce the axial canal (axial sinus), axial hemal vessel (axial gland), and hyponeural coelomic cavities (Fig. 28-4). The right protocoel largely degenerates, but a small vestige, called the pulsatile vesicle (dorsal sac), forms the heart. In echinoderm terminology, the protocoel is called the axocoel, the mesocoel is the hydrocoel, and the metacoel is the somatocoel.

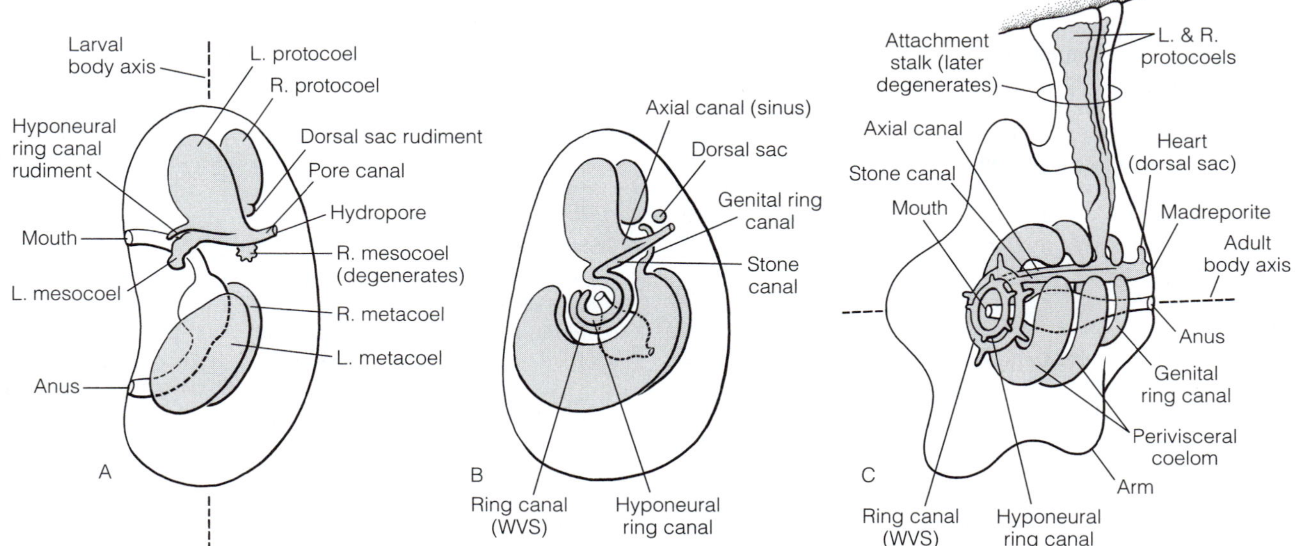

FIGURE 28-4 Generalized echinoderm larval metamorphosis: developmental origin of pentamerous symmetry. **A,** Left-side view of the larval gut and coeloms. **B,** Beginning of torsion (left mesocoel and left and right metacoels twist). **C,** Metamorphosing larva attaches to the substratum as the torsional twists of the coeloms are completed and the juvenile rudiment adopts the new adult body axis.

The transformation of the bilateral larva to a pentamerous adult is shown diagrammatically in Figure 28-4. Although the details of metamorphosis differ among echinoderm taxa, there is little doubt that the developmental change in body symmetry reflects a similar evolutionary shift in the remote past. The question then is, "Under what conditions is radial symmetry superior to bilateral symmetry?" If your answer is, "When a taxon adopts a sessile, suspension-feeding lifestyle," you are in the good company of generations of biologists. Indeed, the standard explanation for echinoderm pentamerous symmetry is that the echinoderm ancestor, although ultimately descended from a motile bilaterian, was a sessile suspension feeder, and paleontology and biology both support this interpretation. Sessility and suspension feeding are primitive echinoderm traits retained by the living crinoid sea lilies.

The final step in the evolutionary scenario is the secondary readoption of motility by the once-sessile echinoderms. Today, all major living taxa of echinoderms are at least partly motile, a lifestyle that favors bilateral symmetry. Few motile echinoderms, however, have readopted bilateral symmetry, although examples are found among the sea cucumbers, sand dollars, and heart urchins.

ELEUTHEROZOA

Eleutherozoans are the sea stars, brittle stars, sea urchins, and sea cucumbers. These echinoderms are all motile and primitively oriented with their oral surface against the substratum. The sister taxon of Eleutherozoa is Crinoidea, the sessile sea lilies, which will be discussed later in this chapter.

ASTEROIDEA[C]

The 1500 living species of asteroids are the sea stars, or starfishes. Their star-shaped, free-moving body consists of hollow arms, or rays, projecting from a central **disc** (Fig. 28-5). Asteroids are common animals that crawl about over rocks and shells or live on sandy or muddy bottoms. They occur worldwide, largely in coastal waters, but the northeast Pacific Ocean, particularly from Puget Sound to the Aleutian Islands, has the greatest diversity of asteroids with over 70 species. Sea stars commonly are red, orange, blue, purple, or green or exhibit combinations of colors.

Form

Sea stars typically are pentamerous animals and most species have 5 arms, although sun stars have 7 to 40 or more (Fig. 28-5B). Most asteroids range from 12 to 24 cm in diameter, but some are less than 2 cm, and the many-rayed star *Pycnopodia helianthoides* of the northwest coast of the United States and the west coast of Canada may measure almost 1 m across.

Asteroid arms generally widen at their base and grade smoothly into the central disc. In the cushion stars, *Plinthaster* and *Goniaster*, each arm has the shape of an isosceles triangle, in *Culcita* (Fig. 28-5C), the arms are so short that the disc is a pentagon, and among the tiny deepwater sea daisies (*Xyloplax*), the armless body is a circular disc (Fig. 28-8A).

The mouth is located centrally on the underside of the disc, which is called the **oral surface,** and is in contact with the substratum. A wide furrow, or **ambulacrum** (ambulacral groove), extends radially from the mouth along each arm (Fig. 28-1). Each ambulacrum contains two or four rows of small, tubular projections called **tube feet,** or podia. The margins of the ambulacra are guarded by movable spines that can close over it (*ambulacrum* = covered path). The tip of each arm terminates in one or more small, tentacle-like **sensory tube feet** and a red **eyespot** (Fig. 28-6A).

The upper, **aboral surface** bears the inconspicuous anus (when present) in the center of the disc and a large, button-like **madreporite** on one side of the disc between two of the arms. The general body surface may be smooth or covered with spines, tubercles, or ridges. In some species the arms and disc are bordered by large **marginal plates** (Fig. 28-5A).

Body Wall and Skeleton

The body wall consists of a thin cuticle, a monolayered epidermis, a thick connective-tissue dermis, and the coelomic epithelium of myoepithelial cells, which form the musculature, and a peritoneum (Fig. 28-7). The ciliated cells of the epidermis and peritoneum typically are monociliated collar cells. The epidermis also includes nonciliated cells, secretory mucous cells, and sensory cells. Detritus that falls on the body is trapped in mucus and swept away by the epidermal cilia. The sensory cells and other neurons create an intraepidermal net, which is part of the **ectoneural system** (Fig. 28-7). The dermis houses the skeletal ossicles, described below, and various connective-tissue cells, including the sclerocytes that produce the skeletal ossicles (Fig. 28-7, 28-14). Within the coelomic lining is a network of chiefly motor neurons that constitute part of the **hyponeural system** (Fig. 28-7), which innervates the muscles and mutable connective tissue (discussed later).

The echinoderm endoskeleton is located in the dermis, as are the scales of fishes, and covered by the epidermis. It has two components: the collagenous connective tissue and the calcareous ossicles. The presence of ossicles in the dermis increases the rigidity of the dermis and provides attachment sites for muscles, but the connective tissue also plays an important skeletal role.

OSSICLES

Asteroid **ossicles** adopt a variety of shapes—spines, rods, plates, crosses (Fig. 28-8A)—and microscopically consist of a three-dimensional lattice called a **stereom** (Fig. 28-8B). The honeycombed structure of the ossicles may reduce weight, increase strength, and prevent running fractures. Although each ossicle is a composite of calcite microcrystals, it behaves optically (under polarized light) as a single crystal because all the microcrystal axes are parallel. The ossicle forms intracellularly in a syncytium of fused dermal (mesodermal) sclerocytes. The labyrinthine spaces in the ossicles, collectively called **stroma,** allow the entry of sclerocytes for growth and modification of the stereom and for attachment of collagenous ligaments. The ligaments suture ossicles together to create the skeletal framework (Fig. 28-9A, 28-10). Specialized phagocytes are capable of reabsorbing calcite from ossicles. All ossicles, including those that

FIGURE 28-5 Echinoderm symmetry and asteroid diversity. **A,** *Astropecten duplicatus,* a burrowing sea star with marginal plates and aboral paxillae (aboral view). It feeds intraorally on clams and snails. **B,** The sun star *Crossaster papposus* (aboral view). *Crossaster* feeds extraorally on sea pens, molluscs, and echinoderms. **C,** Oral view of *Culcita,* a sea star with very short arms that feeds on corals and other attached invertebrates. **D,** The Pacific coral-eating sea star, *Acanthaster planci.* It also feeds on sea anemones.

project above the body surface, such as tubercles and spines (including the spines of sea urchins), are endoskeletal and thus are covered by epidermis.

Some sea stars have specialized, compound, articulated ossicles called paxillae and pedicellariae. **Paxillae** cover the aboral surface of many asteroids such as *Luidia, Astropecten,* and *Goniaster* that burrow in sediment (Fig. 28-5A, 28-9C,D). Each paxilla resembles a miniature parasol with its handle attached to the surface of the animal and its umbrella-like crown unfurled above (Fig. 28-9C). The perimeter of the crown bears a fringe of tiny, movable spines. The crowns of adjacent paxillae abut one another and together create a "second skin" above the body wall proper. The water-filled space between this second skin and the body proper is protected from intrusion by sediment and contains the delicate gills and madreporite. The branchial space is ventilated by epidermal cilia (Fig. 28-9C).

Pedicellariae are movable compound ossicles that function as forceps. They occur on the surface of asteroids such as

FIGURE 28-6 Asteroidea: sense organs and ambulacra of *Asterias forbesi.* **A,** Upturned arm tip showing the dark eyespot and slender sensory tube feet, which bear white suckers. **B,** Ambulacral (oral) surface of the disc and one arm.

Asterias, Pycnopodia, and *Pisaster* and are used primarily to defend the aboral surface against settling larvae or other small animals. These pedicellariae usually consist of a short, fleshy stalk surmounted by usually two, but sometimes three, small, movable ossicles arranged to form grasping jaws (Fig. 28-10, 28-11C). The pedicellariae may be scattered over the aboral surface or localized to the surface of spines, commonly forming a wreath around the base of each spine (Fig. 28-11A,B, 28-16B). Stalkless, sessile pedicellariae, each with two jaws, occur in a few species (Valvatida).

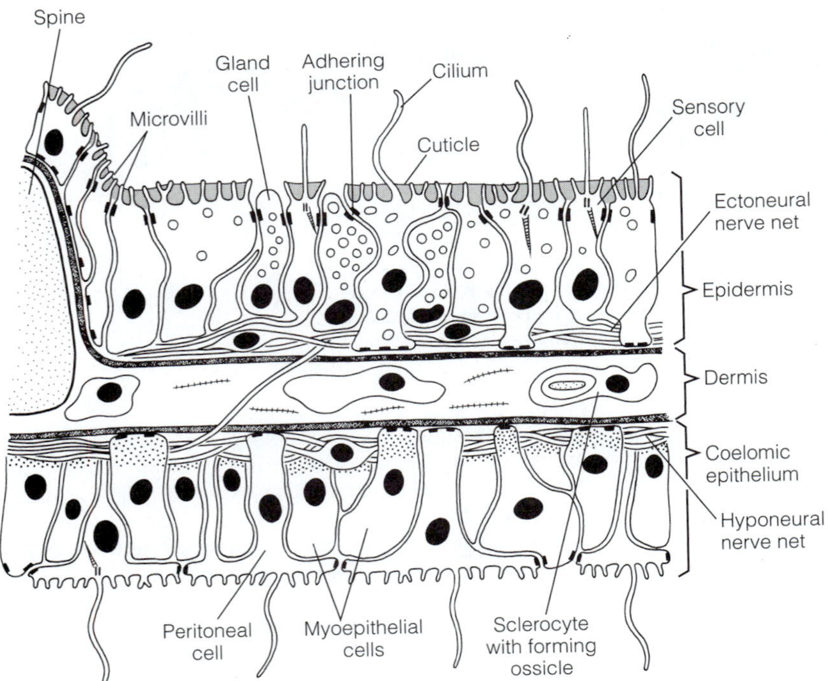

FIGURE 28-7 Diagrammatic section of the echinoderm body wall.

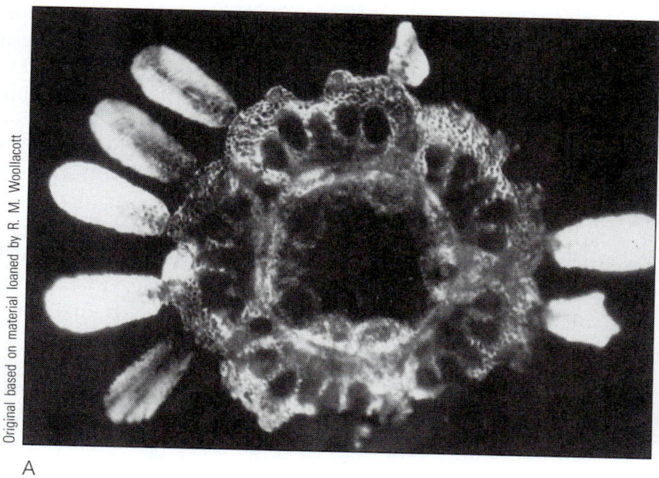

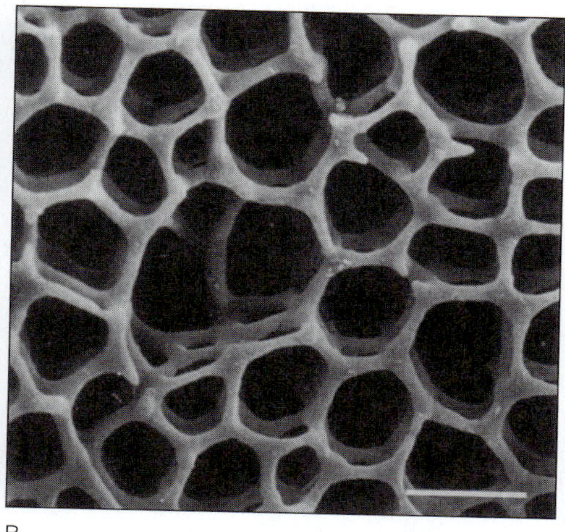

FIGURE 28-8 Echinoderm skeleton. **A,** Skeletal framework of a juvenile sea daisy, the asteroid *Xyloplax turnerae,* under polarized light. **B,** Honeycombed stereom structure of a crinoid ossicle (electron micrograph). The spaces within the ossicle constitute the stroma.

Water-Vascular System (WVS)

The asteroid water-vascular system, like that of all echinoderms, consists of the hydraulic locomotory tube feet and a pentamerous arrangement of internal coelomic canals (Fig. 28-12A). The internal canals include a circumoral **ring canal,** from which a **radial canal** extends into each arm. Perpendicular to the radial canal is a bilateral sequence of **lateral canals,** each of which terminates in a bulbous **ampulla,** tube foot, and sometimes, a sucker. A **stone canal** ascends aborally from the ring canal and opens into a small chamber, the **madreporic**

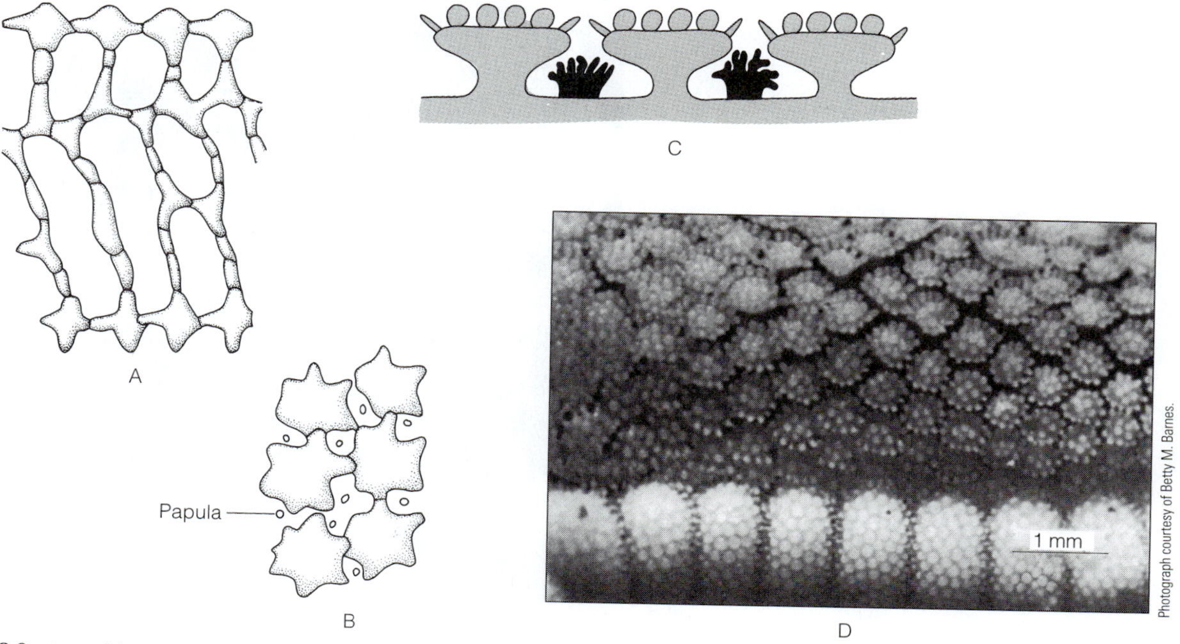

FIGURE 28-9 Asteroidea endoskeleton. **A,** Lattice of skeletal ossicles in the arm of a sea star (Asteriidae). **B,** Small section of sea star endoskeleton (Paxillosida). **C,** Diagrammatic vertical section through the aboral body wall of *Luidia* showing three paxillae. The raised, table-shaped ossicles bear small, rounded spines on the surface and flat, movable spines around the edge. Branched papulae (black) are located in the spaces below the paxillae. **D,** Surface view of the paxillae of *Astropecten* (compare with Fig. 28-5A, 28-9C). *(A, After Fisher from Hyman, L. H. 1955; B, After Hyman, 1955. The Invertebrates. Vol. IV. McGraw-Hill Book Co., New York. Reprinted with permission.)*

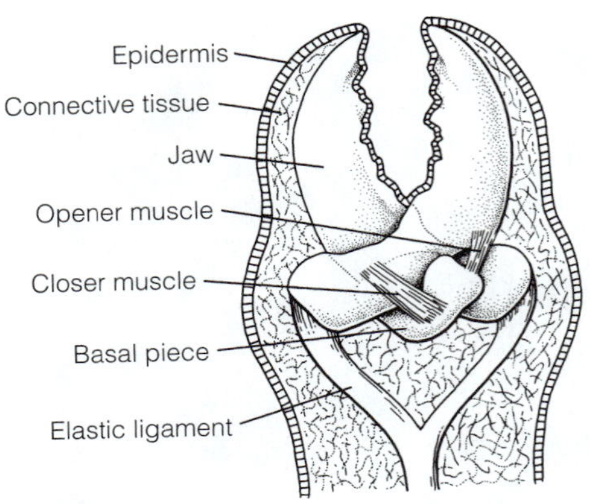

FIGURE 28-10 Asteroidea: pedicellaria. Functional anatomy of an *Asterias* pedicellaria. *(Modified and redrawn from Hyman, L. H. 1955. The Invertebrates. Vol. IV. McGraw-Hill Book Co., New York. Reprinted with permission.)*

ampulla, which joins the **madreporite,** a specialized, porous ossicle on the surface of the disc. The pores of the madreporite open to the surrounding seawater. The often colorful and conspicuous madreporite is irregularly furrowed and resembles a solitary stony (madreporarian) coral. The entire WVS is lined by a myoepithelium and thus is ciliated and muscular. The muscles are best developed in the ampullae and tube feet, and the ciliary flows in the internal canals function in transport. The WVS is well developed in asteroids and used for locomotion, adhesion, prey manipulation, and gas exchange.

Each lateral canal has a valve and terminates in an ampulla and a tube foot (Fig. 28-12B,C). The ampulla is a small, muscular sac that bulges into the perivisceral coelom. The ampulla opens directly into a canal that passes downward between the ambulacral ossicles into the tube foot. The **ambulacral ossicles** form the floor of the ambulacral groove and bulge inward to form the ambulacral ridge (Fig. 28-16A,C); a radial canal lies on the oral side of these ossicles (Fig. 28-12B,C).

The tube feet project into the ambulacral groove (Fig. 28-6B) and are arranged in two rows when the lateral canals are all of equal length or in four rows when they are alternately long and short. A well-developed longitudinal musculature lines

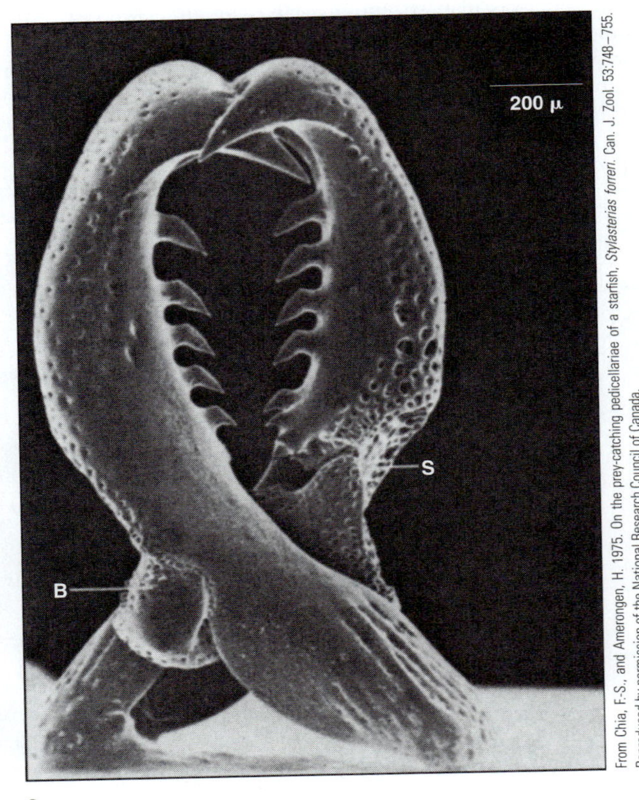

FIGURE 28-11 Asteroidea: pedicellariae. Fish-catching pedicellariae of *Stylasterias forreri*. **A,** A wreath of pedicellariae at rest around the spine. **B,** The wreath is raised when stimulated by potential prey. **C,** Scanning electron micrograph of jaws. B, basal piece; S, muscle attachment scar. *(A and B, From Chia, F.-S., and Amerongen, H. 1975. On the prey-catching pedicellariae of a starfish,* Stylasterias forreri. *Can. J. Zool. 53:748–755. Reproduced by permission of the National Research Council of Canada.)*

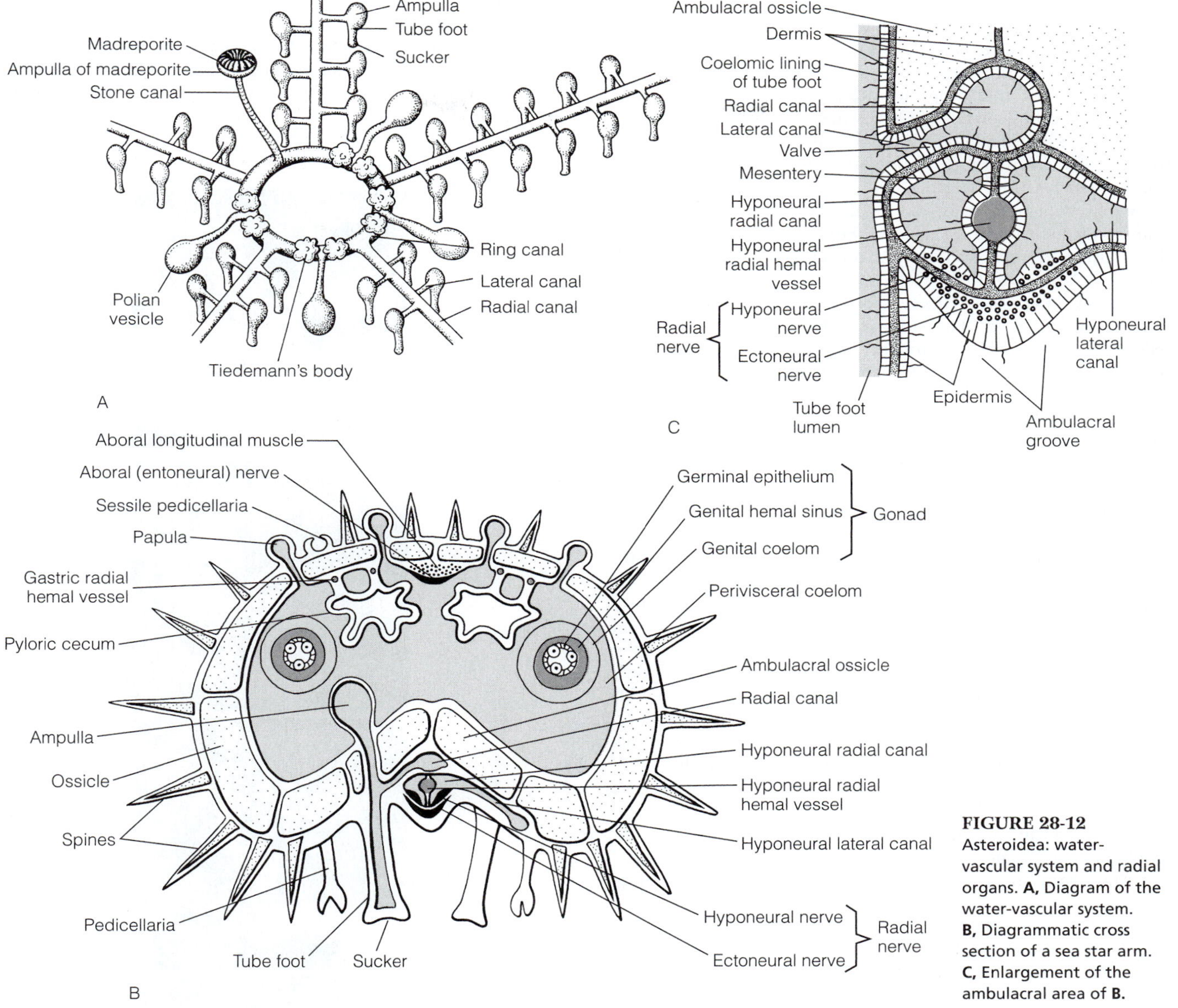

FIGURE 28-12 Asteroidea: water-vascular system and radial organs. **A,** Diagram of the water-vascular system. **B,** Diagrammatic cross section of a sea star arm. **C,** Enlargement of the ambulacral area of **B.**

the ampulla and tube foot. Commonly, the tube-foot tip is expanded to form a **sucker** (Fig. 28-6B, 28-12A,B).

The cyclic movement of each tube foot consists of extension-attachment, force generation (pulling), and detachment-retraction. Extension occurs as the longitudinal musculature of the ampulla contracts, the valve in the lateral canal closes, and water is forced into the tube foot, which then elongates. When the tube foot contacts the substratum, the sucker adheres. Adhesion is largely chemical, with the tube foot secreting a substance that bonds with the surface. Another secretion later breaks the bond and releases the tube foot. (Recall the duo-gland adhesive systems of flatworms and gastrotrichs.) When attachment is prolonged or a large force is generated, adhesion by suction may also occur. Once attached by its sucker, longitudinal muscles contract and exert a pulling force for locomotion or other purposes. The sucker then detaches and the contraction of longitudinal muscles retracts the tube foot as fluid is displaced into the ampulla. Contraction of longitudinal muscles on only one side of the tube foot bends the appendage.

The internal canals of the WVS have two primary functions: circulation and fluid-volume maintenance for hydraulic function. All ciliary flows in the WVS, including the tube feet, are bidirectional, so each canal or tube foot is simultaneously analogous to both artery and vein. This circulation probably transports oxygen from the tube feet to the internal tissues of the disc, and perhaps nutrients from the gut to the musculature of the tube feet (see Internal Transport).

The ring canal is the hub of the system, where all the radial vessels intersect. Two organs that play a role in system maintenance are also situated here. On the inner circumference of the ring canal are five pairs of spongy pouches called **Tiedemann's bodies** (Fig. 28-12A, 28-15). They are located interradially, between the arms. Tiedemann's bodies function as lymph nodes, removing and destroying bacteria and other unwanted particulates in the WVS fluid. Also attached interradially to the inner side of the ring canal of many asteroids, but not *Asterias,* are one to five elongate, muscular sacs known as **polian vesicles** (Fig. 28-12A). Their function is uncertain, but they may be a fluid reservoir for the WVS as a whole or perhaps only for the oralmost tube feet, which often are much larger than the feet on the arms but have normal-size ampullae.

The stone canal and madreporite help maintain the fluid volume of the WVS. The stone canal originates on the ring canal, ascends vertically to the aboral body surface, and joins the madreporic ampulla (shared by the axial canal; see Internal Transport), which opens to the exterior via the madreporite (Fig. 28-12A, 28-15). The stone canal is named for the calcareous ossicles in its wall. They support the canal wall as do the cartilagenous rings in your trachea. The madreporite is perforated by pores that lie in ciliated surface furrows. The ciliary beat in the stone canal is bidirectional, but the inward- (oralward-) beating cilia are strongly developed and overwhelm the opposing cilia. This results in an inflow of seawater across the madreporite, down the stone canal, and into the remainder of the WVS. The stone canal is thus a ciliary pump that maintains the fluid volume of the WVS. Under normal conditions only a trickle of water enters the WVS via the madreporite and stone canal because the system ends blindly at the tube feet and back-pressure opposes inflow. But the pressurized, thin-walled tube feet are leaky and some fluid recovery is required. The cilia on the surface of the madreporite knock away particles that might clog the pores or enter the WVS. Those that slip through the madreporite filter are removed by the Tiedemann's bodies or phagocytic coelomocytes.

The WVS fluid is similar to seawater except that it contains coelomocytes, some protein, and a higher concentration of K^+. The dissolved proteins and elevated K^+ concentration creates an osmotic influx of water across the tube feet. Thus osmotic recovery of water by the tube feet and physical pumping by the stone canal combine to offset water losses from muscular pressurization of the tube feet.

Locomotion

During movement each tube foot performs a stepping motion. The tube foot swings forward, grips the substratum, and then moves backward. In a particular section of an arm, most of the tube feet are performing the same step and the animal moves forward. The action of the tube feet is highly coordinated. During progression one or two arms lead and the tube feet in all the arms move in the leading direction, but not all in unison. The combined action of the tube-foot suckers exerts a powerful adhesive force and enables the sea star to climb vertically over rocks or up the side of an aquarium. In general, sea stars move slowly and tend to remain within a more or less restricted area.

If a sea star is turned over, it undergoes a **righting reflex** during which it regains its normal oral-side-down position (Fig. 28-13). The tips of one or two arms twist, bringing their tube feet in contact with the substratum. Once the substratum is gripped, these arms slowly crawl below the body, rolling it over onto its oral surface. A sea star may also right itself by arching its body, rising on the tips of its arms, and rolling onto its oral surface.

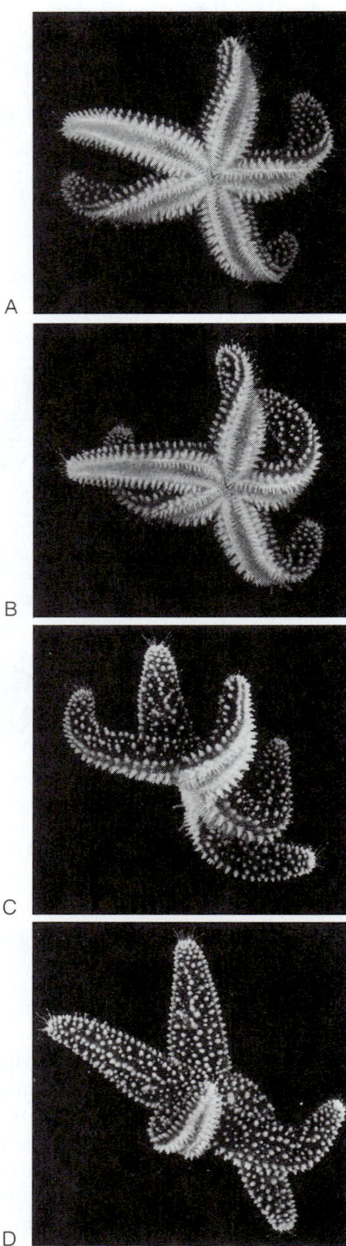

FIGURE 28-13 Asteroidea: righting reflex. **A,** *Asterias forbesi* experimentally inverted to place the oral side up twists the tips of two arms to bring its tube feet into contact with the substratum. **B,** The two arms in contact with the substratum crawl under the inverted disc and the other three arms to flip the body onto its oral surface (**C** and **D**).

The tube feet of some soft-bottom inhabitants, such as *Astropecten* and *Luidia,* lack suckers because they would be useless on soft sediment. Instead the tip is pointed, adapting it to be thrust into the sand. These animals pole themselves along with their tube feet and some move relatively quickly for sea stars. Associated with this adaptation is the presence of bilobed ampullae, which provide additional force for driving the tube feet into the substratum.

Gas Exchange

Echinoderms in general are large animals that have little anaerobic capacity and thus are sensitive to oxygen availability. To enhance gas exchange, all have gills. Gills are hollow, thin, body-wall evaginations (or invaginations) that are ventilated with seawater externally and coelomic fluid internally. Coelomic fluid, not blood, is the transport medium for gases. Typically, each major coelomic cavity has its own separate, specialized gills. Tube feet are the gills for the WVS of all echinoderms, including asteroids.

In asteroids, the specialized gills associated with the perivisceral coelom are called **papulae,** which are on the aboral surface of the thick arms (Fig. 28-9B,C, 28-12B, 28-14, 28-15). A papula is similar to a tube foot, but it lacks a sucker and is a direct outgrowth of the perivisceral coelom, not of the WVS (Fig. 28-14). Sometimes papulae are branched, and in species with paxillae, they are in the water-filled branchial space below the "second skin" (Fig. 28-9C). Papulae supply oxygen to the perivisceral coelom, ventilating the gut, gonads, and muscles of the disc and arms.

Nervous System

The pentamerous echinoderm CNS consists of a circumoral nerve ring and five **radial nerves** that originate on the nerve ring and extend outward along the ambulacra (Fig. 28-15). The peripheral nervous system includes two intraepithelial

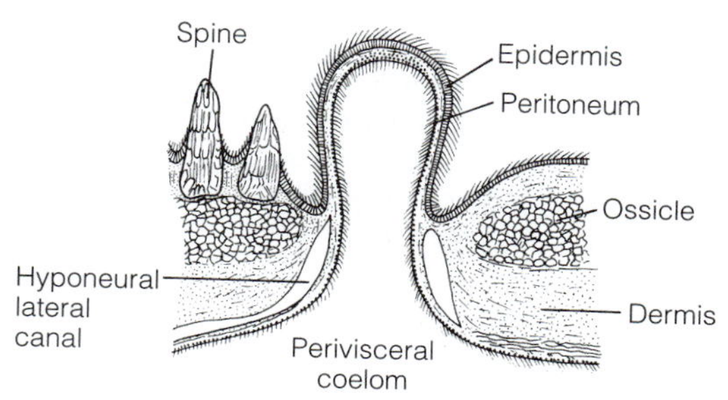

FIGURE 28-14 Asteroidea: gills and body wall. A papula, which functions as a gill, and the aboral body wall in vertical section. *(Modified from Cuénot, L. 1948. Échinodermes. In Grassé, P. (Ed.): Traité de Zoologie, Vol. XI. Masson et Cie, Paris. Reprinted with permission.)*

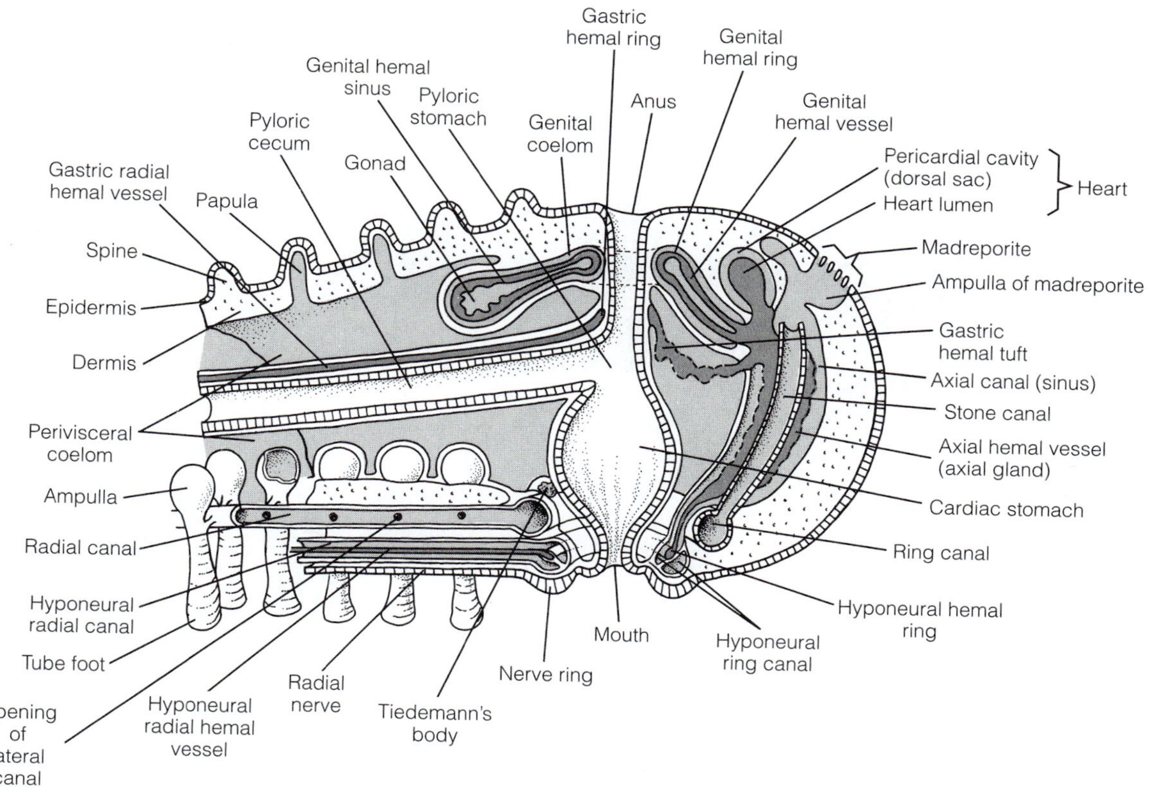

FIGURE 28-15 Asteroidea: diagrammatic internal anatomy of gut, coeloms, and hemal system.

nets, the sensory ectoneural system in the epidermis and the motor hyponeural system in the coelomic lining (Fig. 28-7, 28-12B,C). The two nets are connected by neurons that cross the dermis (Fig. 28-7). The nervous system lacks ganglia and glia.

The nerve ring and radial nerves also have ectoneural (sensory) and hyponeural (motor) components. The ectoneural ring and radial nerves are in the epidermis whereas the hyponeural ring and radial nerves are in the coelomic lining of the hyponeural ring and radial canals (Figs. 28-7, 28-12B,C). The motor component of the radial nerve innervates the ampullae, tube feet, and body-wall muscles, while the sensory part receives input from the sense cells and organs.

Experimental studies indicate that coordinated movement of the tube feet requires intact radial nerves and nerve ring. These structures initiate tube-foot stepping and control the direction of stepping. Each arm has a motor center, probably at the junction of the radial nerve and nerve ring. A leading arm exerts a temporary dominance over the nerve centers of the other arms. In the majority of sea stars, including *Asterias,* any arm can act as a dominant arm, and such dominance is determined by reaction to external stimuli. In a few species one arm is permanently dominant. Of all the reactions to external stimuli, contact of the tube feet with the substratum appears to be the most important and probably accounts for the righting reaction.

The asteroid sense organs are the eyespots and tentacle-like sensory tube feet, both located at the arm tips (Fig. 28-6A). The eyespot is composed of a mass of 80 to 200 pigment-cup ocelli. Most asteroids are positively phototactic, but reactions vary among species. Individual sensory cells are widespread throughout the epidermis and probably function for the reception of light, tactile, and chemical stimuli. These cells are prevalent on the suckers of the tube feet and along the margins of the ambulacral groove, where 70,000 sensory cells/mm^2 have been reported.

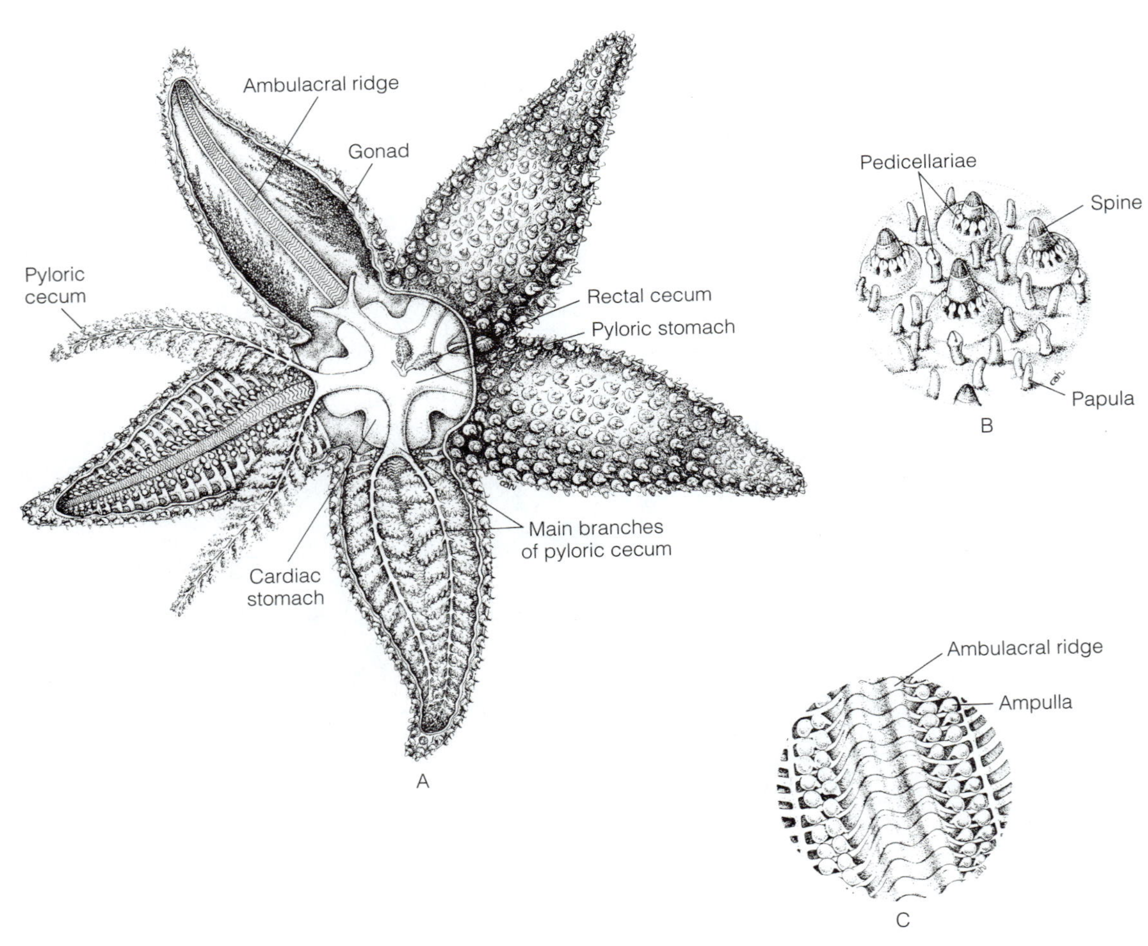

FIGURE 28-16 Asteroidea: anatomy of *Asterias.* **A,** Aboral view with the arms in various stages of dissection. **B,** A small area of the aboral surface enlarged. **C,** The interior of the arm, showing the ambulacral ridge and ampullae on each side.

Digestive System

The pentamerous asteroid digestive system occupies most of the space in the disc and arms (Fig. 28-15, 28-16A). The mouth is located in the center of a tough, circular **peristomial membrane** that is muscular and has a sphincter. The mouth opens into a short esophagus that leads to a large stomach, which occupies most of the disc. The stomach is divided by a horizontal constriction into a large, oral **cardiac stomach** and a smaller, flattened, aboral **pyloric stomach.** The walls of the glandular cardiac stomach are pouched and connected to the ambulacral ossicles of each arm by a pair of triangular mesenteries called gastric ligaments. The smaller, aboral pyloric stomach, which is often star-shaped, receives ducts from a pair of pyloric ceca in each arm (Fig. 28-15, 28-16A). The **pyloric ceca** are elongate, hollow, branched extensions of the pyloric stomach (Fig. 28-12B). Each cecum is suspended by a pair of dorsal mesenteries in the perivisceral coelom of the arm. From the pyloric stomach, a short, tubular intestine joins the rectum, which opens at a minute anus in the center of the aboral disc. The rectum commonly bears several small or large outpockets called **rectal ceca** (Fig. 28-16A). The entire digestive tract is lined with a ciliated gastrodermis, and the cilia in the ducts of the pyloric ceca create both incoming and outgoing flows. Gland cells occur throughout, but are particularly abundant in the cardiac stomach and pyloric ceca.

Many asteroids digest food extraorally—outside of the body—by everting the cardiac stomach. Contraction of body-wall muscles elevates coelomic pressure and causes the cardiac stomach, with its digestive surface outermost, to evert through the mouth. The everted stomach, which is anchored by the gastric ligaments, releases digestive enzymes and engulfs the prey. The partially digested prey can be brought into the stomach by retraction, or the thick prey-broth can enter the body along ciliated gutters of the stomach. When feeding ends, the stomach muscles contract, retracting the stomach into the interior of the disc. In primitive asteroids, including *Astropecten* and *Luidia,* which cannot evert their stomachs, the prey (often small clams) is swallowed whole and digested intraorally in the stomach. Shells and other indigestible material are then cast out of the mouth.

Digestion in asteroids is chiefly extracellular, and enzymes are produced by the stomach and pyloric ceca. The products of partial extracellular digestion in the stomach pass into the pyloric ceca, where digestion is completed and absorption occurs. Absorption also occurs in the rectal ceca. Nutrients may be stored in the cells of the pyloric ceca or released into the coelom or hemal system for distribution. Indigestible wastes from the stomach or pyloric ceca enter the rectum, where they are formed into feces and released through the anus.

Nutrition

Most asteroids are scavengers and carnivores that feed on a variety of animals, especially snails, bivalves, crustaceans, polychaetes, other echinoderms, and even fish. Some have restricted diets, whereas others capture a wide range of prey but may exhibit preferences, depending on availability. For example, the Chilean *Meyenaster* feeds on 40 species of echinoderm and molluscan prey. Most asteroids detect and locate prey with substances the prey releases into the water, and many prey species have evolved escape responses to the slow-moving asteroids. Some soft-bottom sea stars, including species of *Luidia* and *Astropecten,* can locate buried prey and then dig into the substratum to reach it. *Stylasterias forreri* and *Astrometis sertulifera* on the west coast and *Leptasterias tenera* of the east coast of the United States catch small fish, amphipods, and crabs with the pedicellariae when the prey comes to rest against the aboral surface of the sea star.

Asterias and relatives (Asteriidae) feed chiefly on protected prey, such as clams, oysters, and barnacles. When feeding, the sea star arches over the clam, holding the gape upward against its mouth while applying its arms and tube feet to the sides of the valves. The pull of the attached tube feet may part the valves slightly, or the sea star may find a natural gap between the valves. In either case, once an entry is found the sea star everts its stomach through the opening and secretes digestive enzymes onto the soft parts of the clam, digesting it in its own shell. The everted cardiac stomach of some sea stars can squeeze through a space as slight as 0.1 mm. The shell gape increases over time as the clam's adductor muscles are digested. Japanese species of *Asterias* require 2.5 to 8 h to consume a bivalve, depending on the species of bivalve.

Asteroids are of considerable economic importance as predators of oysters and sometimes are removed from commercial oyster beds with a large, moplike apparatus that is dragged over the bottom. The sea stars grasp or become entangled in the mop threads with their pedicellariae and are brought to the surface and destroyed.

Some asteroids feed on sponges, sea anemones, and the polyps of hydroids and corals. The tropical Pacific *Acanthaster planci* (crown-of-thorns sea star) is notorious for its consumption of coral polyps. High densities of this sea star, as great as 15 adults per square meter, have temporarily devastated large numbers of reef corals in some areas. Branching and plate corals are preferred over massive and encrusting types, and with its everted stomach one star can consume an area as great as its own disc in one day. An outbreak of this sea star may be part of a natural cycle or the result of environmental modification by humans. Current evidence suggests that the outbreaks coincide with plankton blooms. The blooms are related to nutrient runoff, both natural and agricultural, following heavy rains.

A few sea stars suspension feed on plankton and detritus *(Echinaster, Henricia, Porania)* whereas others feed on deposited material *(Ctenodiscus, Goniaster)* that contacts the body surface. This material is trapped in mucus and then swept toward the oral surface by the epidermal cilia. On reaching the ambulacral grooves, the food-laden mucous strands are carried by ciliary currents to the mouth. Some asteroids, such as *Astropecten* and *Luidia,* which are primarily carnivores, use ciliary feeding as a supplement.

The everted stomach is an effective feeding organ for many omnivores and nonpredaceous sea stars. The American west coast bat star, *Patiria miniata,* spreads its stomach over the bottom, digesting the organic matter it encounters. In the same manner the tropical cushion star, *Culcita,* and *Oreaster,* which inhabit reef flats, feed on sponges, algal felt, and organic films.

Internal Transport

Asteroids, like other echinoderms, rely chiefly on coelomic circulation for internal transport of gases and nutrients. The echinoderm hemal system is rudimentary in asteroids, although it may play a role in nutrient transport. Both systems, however, share the fundamental radial symmetry of the echinoderm body. This means that radial branches to the arms extend from a ring, or common cavity, in the disc. Pumps for both the hemal system (heart) and the WVS (stone canal) are axially located in the disc. The radial branches of both systems extend into the arms but end blindly at the arm tips, which raises the question of how the coelomic fluid and blood return to the disc. In coelomic systems, cilia create a bidirectional flow in each of the canals, so each is simultaneously "artery" and "vein." Blood flow in the hemal system results from muscle contraction, but the flow pattern is unknown in asteroids. Research on blood flow in sea cucumbers, however, indicates that blood ebbs and flows in each of the vessels.

COELOMIC CAVITIES

Asteroids have four coelomic circulatory systems (Fig. 28-15): the WVS, already described, which supplies the locomotory muscles of the tube feet; the perivisceral coelom, which occupies the disc and arms and supplies the viscera; the hyponeural coelom, which circulates fluid to the hyponeural nervous system; and the genital coelom, which supplies the gonads. Cilia on the coelomic lining circulate the fluid bidirectionally in all four compartments.

The voluminous **perivisceral coelom** of the disc and arms surrounds and bathes the gut, the outer walls of the gonads, and the tube-feet ampullae. The papulae are aboral evaginations of this coelom. The perivisceral coelom is undoubtedly important for the transport of gases and perhaps also of nutrients.

The canals (also called sinuses) of the **hyponeural coelom** are anatomically similar to those of the WVS, but the equivalents of ampullae and tube feet are absent (Fig. 28-15) and most of the canals of the hyponeural coelom are paired, with the left being separated from the right by a mesentery (Fig. 28-12B,C, 28-15). The components of the hyponeural coelom are the **hyponeural ring canal** (hyponeural ring sinus) that accompanies the hyponeural nerve ring, and a **hyponeural radial canal** (hyponeural radial sinus) along the hyponeural component of the radial nerve in each arm. The hyponeural (motor) neurons of the nerve ring and radial nerves lie in the coelomic lining of the hyponeural ring and radial canals (Fig. 28-12B,C). Branches of the hyponeural radial canal called **hyponeural lateral canals** (marginal canals) pass laterally into the dermis (Fig. 28-12B,C, 28-14). A vertical canal called the **axial canal** (axial sinus) is situated beside the stone canal and joins the hyponeural ring canal to the madreporic ampulla and madreporite (Fig. 28-15). Because the adjacent stone canal and axial canal both open into the ampulla of the madreporite, the two systems, the WVS and hyponeural coelom, are united at this point with each other and with the exterior (Fig. 28-15). The two systems exchange coelomic fluid at this junction, but flow is primarily from the axial canal into the stone canal and WVS.

The hyponeural coelom is sometimes called the perihemal coelom because a hemal (blood) vessel is suspended in the mesentery between the two halves of each canal (Fig. 28-12B,C). However, hemal vessels also occur in mesenteries of the perivisceral coelom, especially in association with the pyloric ceca, and in the genital coelom (Fig. 28-15).

The **genital coelom** (Fig. 28-15) consists of a small, aboral **genital ring canal** from which **genital radial canals** extend into the gonads. The genital coelom originates developmentally from the left metacoel, but it is independent in adults, with no open connection with any other coelom.

HEMAL SYSTEM

The asteroid hemal system includes three interjoined radial sets of vessels, each set consisting of a circumenteric (around the gut) hemal ring in the disc and five sometimes branched radial vessels. From oral to aboral, the three rings are the hyponeural hemal ring, the gastric hemal ring, and the genital hemal ring (Fig. 28-15). The three rings are united in the disc by a common vertical axial hemal vessel that aborally joins the heart (Fig. 28-15).

The **hyponeural hemal ring** is suspended in the mesentery between the two halves of the hyponeural ring canal (Fig. 28-15). In the disc, it joins the oral end of the axial hemal vessel (Fig. 28-15). A **hyponeural radial hemal vessel** branches from the hemal ring into each of the five arms. Each radial hemal vessel is in the mesentery between the halves of the hyponeural radial canal (Fig. 28-12B,C). Because the hyponeural (motor) component of the nervous system is in the wall of the hyponeural coelom (Fig. 28-12C), it is assumed that the hyponeural hemal ring and radial vessels supply these nerves.

The **gastric hemal ring** encircles the gut on the aboral side of the pyloric stomach (Fig. 28-15). From the ring, a pair of radial hemal vessels extends into each arm, then branches into two pairs, one pair for each branch of the pyloric cecum (Fig. 28-12B). Two specialized vessels called **gastric hemal tufts** cross the perivisceral coelom and unite the gastric hemal ring with the axial hemal vessel and heart (Fig. 28-15). The surface of these vessels bears a tuft of blind-ended diverticula covered by podocytes. The functional significance of the gastric hemal tufts is unknown, but if they are sites of ultrafiltration, the ultrafiltrate will enter the perivisceral coelom, which lacks a metanephridial outlet. Perhaps such an ultrafiltrate, if it exists, would carry nutrients from the pyloric ceca to the coelomic fluid for distribution to its lining tissues.

The aboral **genital hemal ring** (Fig. 28-15) and its five pairs of **genital radial hemal vessels** supply the ten gonads. On reaching the gonad, each radial vessel forms a **genital hemal sinus** around the gonad in the space between the gonad's germinal epithelium and the genital coelom (Fig. 28-12B).

The three hemal rings are united by the vertical **axial hemal vessel** (axial gland) that lies in the wall of the axial canal (Fig. 28-15). The axial hemal vessel, axial canal, and stone canal are integrated together in the same interradius and thus do not interfere with *radial* branches of any of the hemal or coelomic systems. The axial hemal vessel, like the gastric hemal tufts, is covered with podocytes and is probably a site of ultrafiltration. If an ultrafiltrate of blood is produced, it will enter the axial canal and be transported by ciliary action throughout the hyponeural coelom, perhaps supplying nutrients to the hyponeural nerves.

The heart is situated atop the axial hemal vessel, above all three hemal rings and immediately below, but to one side of, the madreporite. It originates developmentally as a hollow sphere (the larval pulsatile vesicle) that invaginates in a manner similar to a gastrula. The invaginated cavity is the blood-filled heart lumen, its wall is the myocardium, and the outermost wall is the pericardium. Between the pericardium and myocardium is the fluid-filled pericardial cavity (Fig. 28-15). In echinoderm terminology, the heart lumen is called the central blood sinus and the remainder of the heart is the dorsal sac.

The heart beats rhythmically (at approximately 6 beats/min at 25°C in *Asterias forbesi*), but the pattern of circulation of the colorless blood is unknown. Limited evidence indicates that the blood has a role in nutrient transport (and perhaps in nutrient storage).

The body fluids of all asteroids, as well as those of all echinoderms, are isosmotic with seawater. Their inability to osmoregulate prevents most species from inhabiting estuaries and fresh water. The coelomic fluid contains phagocytic coelomocytes produced by the peritoneum that clot in response to tissue damage.

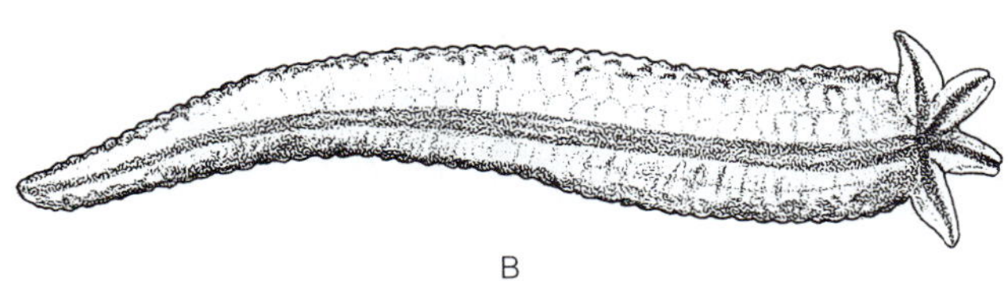

FIGURE 28-17 Asteroidea: clonal reproduction and regeneration. **A,** Regenerating arms in a specimen of *Coscinasterias,* which reproduces by fission of the disc. **B,** Autotomized arm of *Linckia* regenerating the disc and missing arms. At this stage, the regenerands are called "comets." *(B, After Richters from Hyman, L. H. 1955. The Invertebrates. Vol. IV. McGraw-Hill Book Co., New York.)*

Excretion

The asteroid heart, axial canal, and axial hemal vessel correspond, respectively, to the hemichordate heart, proboscis coelom, and glomerulus (the heart-kidney). The echinoderm heart-kidney, however, plays an uncertain role, if any, in excretion. Although the only possible outlet for urine is the madreporite, there are no confirmed reports of fluid outflow from its pores, but rather the opposite, as discussed under Water-Vascular System. Thus it seems that the heart-kidney has a nonexcretory function. Perhaps instead it has a role in nutrient transport, as mentioned above. This is one of many functional topics in need of creative research.

Asteroids and other echinoderms excrete nitrogen primarily in the form of ammonia, which diffuses across thin areas of the body wall, such as the tube feet and the papulae. Other nitrogen-containing compounds (urates), as well as particulates, may be taken up and stored by coelomocytes. Waste-laden cells migrate to the tips of the papulae, and perhaps to the tube feet. The tip of the papula then constricts and pinches off, discharging a package of coelomocytes into the sea.

Reproduction

REGENERATION AND CLONAL REPRODUCTION

Damaged asteroids regenerate readily, rebuilding lost arms and damaged parts of the disc. Species of *Asterias* will autotomize an arm if they are appropriately disturbed. Studies on *Asterias vulgaris* indicate that regeneration of a complete star will occur when starting with as little as one arm and one-fifth of the disc. If the disc fragment includes the madreporite, regeneration is successful with even less than a fifth of the disc. Once the disc and gut have healed, the animal can resume feeding, even before the gut and arms are fully regenerated. Complete regeneration can be slow and may require as long as a year to complete.

Several asteroids normally reproduce by fission by softening the mutable connective tissue occurring at the fission plane. The most common form of fission is disc division, which separates the animal into two halves. Each half then regenerates missing parts of the disc and arms, although extra, fully functional arms are often produced in the process (Fig. 28-17A). Species of *Linckia,* a genus of common sea stars in the Pacific and other parts of the world, are unique in being able to cast off entire arms. Each severed arm can regenerate a complete new body (Fig. 28-17B). Several species of asteroids reproduce clonally in the larval stage. These larvae develop buds on their larval arms (see Development), which differentiate into new larvae.

SEXUAL REPRODUCTION

With few exceptions, asteroids are gonochoric. The ten gonads, two per arm, resemble tufts or clusters of grapes (Fig. 28-16A). In nonreproductive individuals the shriveled gonads are restricted to the arm bases, but the expanded gonads of ripe specimens fill the arms. Each gonad discharges gametes through its own gonopore, usually located between the bases of the arms, although the gonopores of some asteroids open serially along the arms or on the oral surface. There are a few hermaphroditic species, such as the common European sea star *Asterina gibbosa,* which is protandric.

Most asteroids spawn eggs and sperm into the seawater, where fertilization takes place. There usually is only one breeding season per year, and a single female may shed as many as 2,500,000 eggs.

In most asteroids the liberated eggs and later developmental stages are planktonic. Some sea stars, especially cold-water species, brood large, yolky eggs beneath the arched body, in aboral pockets of the disc, in brooding baskets formed by spines between the arm bases, under paxillae, or even in the cardiac stomach. In all brooding species, development is direct. Although it is not a brooding species, *Asterina gibbosa* is unusual in that it attaches its eggs to stones and other objects.

Development

The early stages of development conform to the pattern described in the introduction of this chapter. Most asteroid embryos hatch and begin to swim at the blastula stage. The coelom arises from the tip of the advancing archenteron as two lateral pouches that extend rearward toward the blastopore (anus). A minute tubular outgrowth of the left coelom (protocoel + mesocoel = axohydrocoel) opens on the dorsal surface as the **hydropore,** which is the larval nephridiopore. As the larval coelomic cavities and gut are completed, the surface cilia become restricted to a circumoral ciliary band. This convoluted

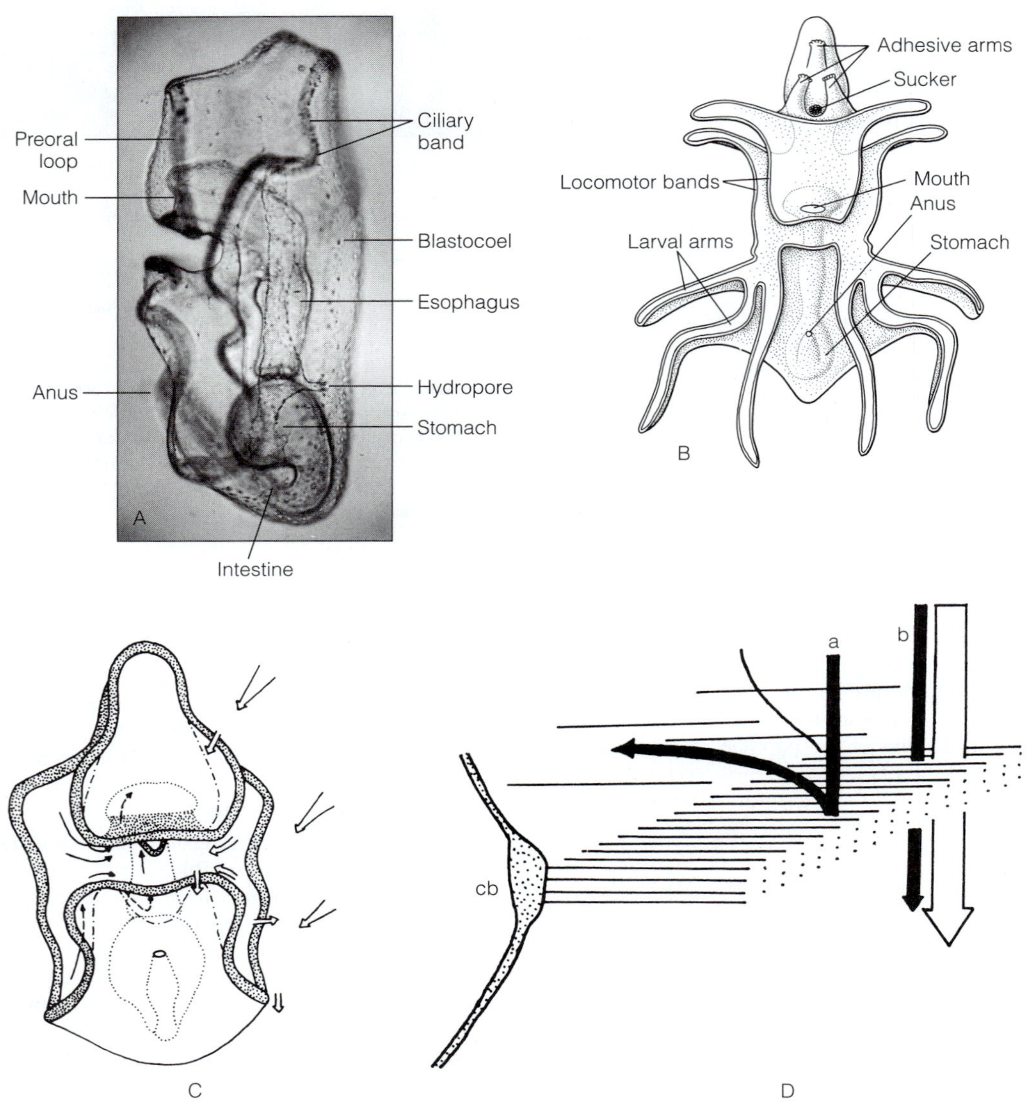

FIGURE 28-18 Asteroidea: larvae and larval feeding. **A,** Bipinnaria larva of *Asterias forbesi,* lateral view. **B,** Brachiolaria larva of *Asterias,* ventral view. **C,** Water currents (open arrows) and paths of food particles (closed arrows) produced by the ciliated band of a bipinnaria larva. **D,** Diagrammatic view of a section of the ciliated band (cb) of an echinoderm larva, showing the water current (open arrow), an uncollected particle (solid arrow b), and a particle collected by a local reversal of the ciliary beat (solid arrow a). *(B, Modified and redrawn with permission after Agassiz from Cuénot, L. 1948. Échinodermes. In Grassé, P. (Ed.): Traité de Zoologie, Vol. XI. Masson et Cie, Paris; C and D, From Strathmann, R. R., Jahn, T. L., and Fonesca, J. L. C. 1972. Suspension feeding by marine invertebrate larvae: Clearance of particles by ciliated bands of a rotifer, pluteus, and trochophore. Biol. Bull. 142:505–519, and Strathmann, R. R. 1975. Larval feeding in echinoderms. Am. Zool. 15:717–730.)*

band winds over the surface of the larva and later onto outgrowths called arms. Eventually, the anterior ventral part of the ciliary band separates from the remainder of the band and forms a distinct **preoral loop**. The bilaterally symmetric, suspension-feeding larva is then known as a **bipinnaria** (Fig. 28-18A,C).

The ciliated bands function in both locomotion and feeding, and the larval arms increase the area of the bands. Phytoplankton and other tiny food particles are collected on the upstream side of the ciliary bands and transported to the mouth (Fig. 28-18C,D).

The bipinnaria becomes a **brachiolaria** with the appearance of three additional arms at the anterior end (Fig. 28-18B). These arms are short, ventral in position, and covered with adhesive cells at the tip. Between the bases of the three arms is a glandular adhesive disc or sucker. The three arms and the adhesive disc are used for attachment during settlement. Generally, the brachiolaria is the larval stage that settles and metamorphoses, but in some asteroids, such as *Luidia* and *Astropecten,* the bipinnaria is the settling stage.

Metamorphosis

Asteroid larvae typically switch from positive to negative phototaxis as the larva prepares to settle and metamorphose. When the brachiolaria settles, it attaches its anterior end to the bottom with its brachiolar arms and sucker, which form an attachment stalk (Fig. 28-4C).

Metamorphosis converts the bilaterally symmetric larva into a pentamerous juvenile. The transition is radical, involving the loss and remodeling of larval tissues and a morphogenesis of new structures to form a juvenile rudiment on the posterior left side of the body (Fig. 28-19). As the rudiment differentiates and grows into a juvenile sea star, the left side of the larval body becomes the oral surface and the right side becomes the aboral surface. Part of the larval gut is retained by the juvenile, but the larval mouth and anus are lost and formed again later in their new positions. The larval right mesocoel degenerates, but the left protomesocoel (axohydrocoel), including the pore canal and hydropore, is retained and modified into the WVS (Fig. 28-4). The fates of other coeloms have been described earlier in the chapter. The arms of the juvenile star are new outgrowths of the body wall unrelated to the larval arms. Eventually, a young sea star, less than 1 mm in diameter, detaches from its larval stalk and crawls away on short, stubby arms.

Growth rates and life spans are variable, as illustrated by two intertidal species from the Pacific coast of the United States. *Leptasterias hexactis* broods a small number of yolky eggs during the winter and the young mature sexually in two years, when they weigh about 2 g. The average life span is 10 years. *Pisaster ochraceus* releases a large number of eggs each spring, and development is planktonic. Sexual maturity is reached in five years, at which time the animal weighs 70 to 90 g. Individuals may live 34 years, reproducing annually.

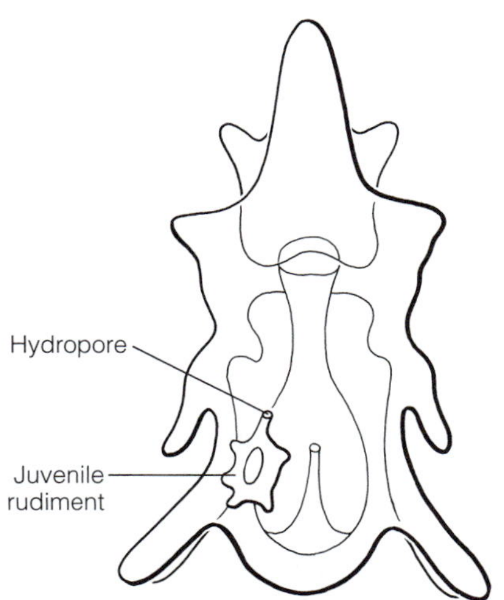

FIGURE 28-19 Asteroidea: larval metamorphosis. Dorsal view of a larva showing the juvenile rudiment on the left side of the body.

Diversity of Asteroidea

The 1500 species of asteroids are divided among six taxa (the deep-sea Notomyotida[O] is omitted).

Paxillosida[O]: Primitive asteroids that lack tube-foot suckers and anus, but have marginal plates and paxillae; have papulae on aboral surface. Intraoral feeders. On soft bottoms. *Astropecten, Luidia, Platyasterias.*

Valvatida[O]: Have suckered tube feet, few large marginal plates; some spp. have paxillae and sessile valvate (clamlike) pedicellariae. Often have rigid pentagonal bodies. Extraoral feeders. *Acanthaster, Ctenodiscus, Culcita, Goniaster, Linckia, Oreaster, Porania, Xyloplax.*

Velatida[O]: Papulae are widely distributed on upper surface; have suckered tube feet and spiniform or sessile pedicellariae. Extraoral feeders. *Crossaster, Pteraster, Solaster.*

Spinulosida[O]: Have suckered tube feet; aboral surface bears short spines; marginal plates and pedicellariae are absent. Extraoral feeders. *Asterina, Dermasterias, Echinaster, Henricia, Patiria.*

Forcipulata[O]: Have suckered tube feet, usually four rows per arm; stalked pedicellariae with two jaws. Extraoral feeders. *Asterias, Heliaster, Leptasterias, Pisaster, Pycnopodia, Zoroaster.*

Brisingida[O]: Have more than five long, slender spiny arms; abundant pedicellariae with crossed jaws are used to catch suspended prey. Deepwater animals. *Brisinga, Freyella, Midgardia.*

CRYPTOSYRINGIDA

The remaining eleutherozoans—brittle stars, sea urchins, and sea cucumbers—are members of this taxon. They differ from asteroids in having internalized the ectoneural (sensory) components of the nerve ring and radial nerves. Instead of being located in the oral epidermis of the disc and arms, as in asteroids (Fig. 28-12B,C), these nerves now occur in the epithelial lining of an internal, hollow **epineural canal** ("Cryptosyringida" = hidden tube; Fig. 28-20). Developmentally, the epithelium and neurons arise from ectoderm and move inward in a process similar to chordate neurulation (Fig. 29-2). The epithelium that lines the epineural canal is derived from and homologous to the epidermis. Cilia line the epineural canals and presumably circulate the fluid within them. The development of an epineural canal creates a permanently closed

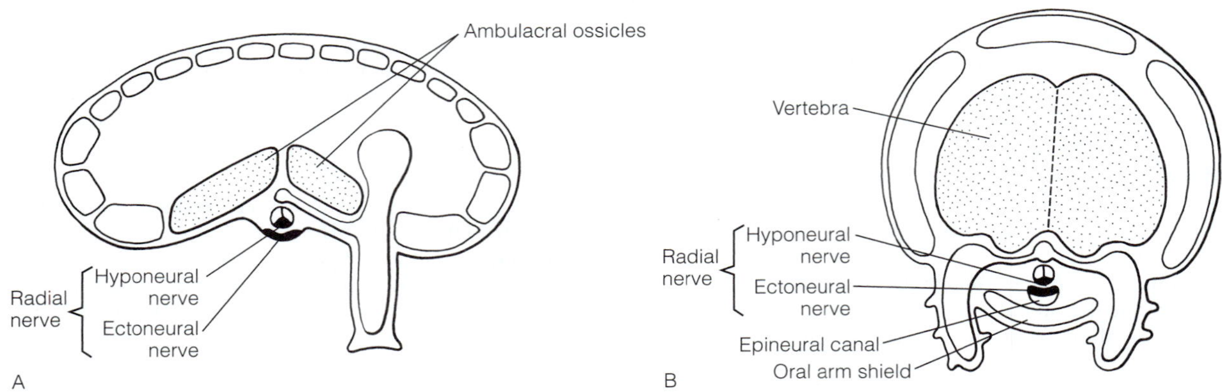

FIGURE 28-20 Cryptosyringida: origin of the epineural canal from an ancestral asteroid-like echinoderm. **A,** Cross section of an asteroid arm showing the ectoneural part of the radial nerve in the epidermis and ambulacral ossicles. **B,** Cross section of an ophiuroid arm showing the internalization of the ectoneural radial nerve in the epineural canal and the internalized ambulacral ossicles modified to form vertebrae. *(Greatly modified and redrawn from Nichols, D. 1969. Echinoderms. 4th Edition. Hutchinson University Library, London. 200 pp.)*

ambulacrum, in contrast to the open ambulacrum of asteroids (Fig. 28-6B, 28-20A), which can only be closed and protected by the active movement of covering ossicles. Such behavioral protection may explain why the otherwise exposed ectoneural system persists in the asteroid epidermis.

Ophiuroidea[C]

The 2000 species of snaky-armed ophiuroids are known as serpent stars or brittle stars, and those with branched arms are called basket stars (Fig. 28-21). As the most diverse group of echinoderms, brittle stars are adapted to a wide variety of mostly cryptic lifestyles and for that reason usually are not conspicuous. Nevertheless, brittle stars are common in many benthic marine habitats and a search for them is rarely unfulfilled. Along with sea lilies and feather stars (Crinoidea), they are the most graceful, delicate, and striking echinoderms. Some species are covered with colorful, faceted crystalline spines and seem to be made of cut glass.

The ability of ophiuroids to live on, under, and between rocks, shells, and living organisms, as well as in sediments, undoubtedly contributes to their high species diversity. Under favorable conditions, ophiuroids such as the European *Ophiothrix fragilis* can reach densities of up to 2000/m^2. The only symbiotic echinoderms are ophiuroids, some of which are commensals in sponges or on corals, feather stars, and sand dollars.

FORM

Ophiuroids, like asteroids, are mostly five-armed stars, but their slender jointed arms are distinctly set off from the central disc, which may be circular or pentagonal (Fig. 28-21A,B). Sometimes, especially in burrowers, the arms are exceptionally long and thin. Although many ophuiroids are small, the armspan of other species equals that of common asteroids. Their small disc and slender arms make the less-massive brittle stars seem small despite their long arms. The disc of most species ranges only from 1 to 3 cm in diameter, but a 12 cm diameter disc occurs in some cold-water brittle and basket stars with armspreads of 1 m (Fig. 28-21C). In basket stars, the arms are not only long, but also highly branched (Fig. 28-21C). Ophiuroid arms lack an ambulacral groove and the suckerless tube feet are rarely used for locomotion.

BODY WALL AND SKELETON

The ophiuroid epidermis differs from that of asteroids in its lack of cilia, except in localized areas. A few species, such as *Ophiocoma wendti* of the West Indies, have epidermal chromatophores that enable them to change the color and pattern of the body. As in other echinoderms, the dermis produces and contains the skeletal ossicles, but pedicellariae are absent. The ossicles may be in the form of plates, called **shields,** when large and associated with the arms and mouth rim. Other ossicles are spines, vertebrae, tubercles, and small scales (Fig. 28-22, 28-23).

Arms

Ophiuroid arms are composed of a series of segmentlike sections, or articles. Each article is formed by a peripheral ring of four shields: a **lateral shield** on each side, an **aboral shield** above, and an **oral shield** below (Fig. 28-23). Internally, each article is supported by a large ossicle called a **vertebra** (Fig. 28-20B, 28-23, 28-24). Not infrequently, the oral and aboral shields are reduced by encroachment of the enlarged lateral shields, which may even meet on the oral and aboral surfaces (Fig. 28-22A). Each lateral shield usually bears 2 to 15 large **arm spines** arranged in a vertical row (Fig. 28-23). These spines vary considerably in size and shape in different species.

Ophiuroid arms lack an ambulacral groove because the original groove, as well as the radial nerve, radial canal, and ambulacral ossicles forming the roof of the groove, have all been internalized (Fig. 28-20, 28-23). The now internal ambulacral ossicles are modified and enlarged to form the vertebrae. The enclosed remnant of the ambulacral groove has

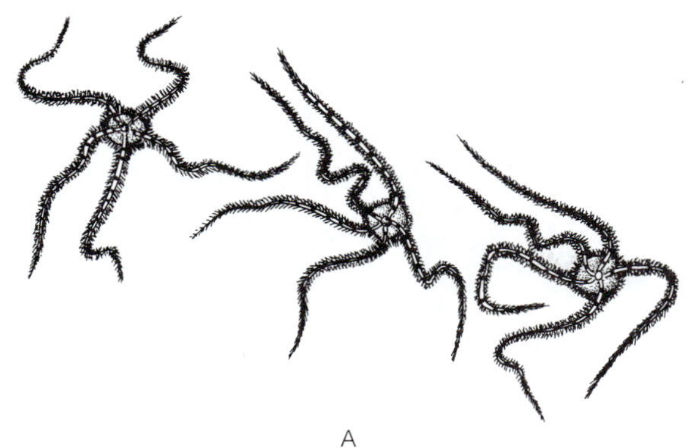

A

B

C

FIGURE 28-21 Ophiuroidea: form and locomotion. **A,** A Caribbean brittle star, shown in a drawing based on repetitive flash images, rowing itself from left to right with its two lateral arms, led by one arm, and trailing two. Ophiuroids are far more agile and flexible than sea stars. **B,** A giant Antarctic brittle star, *Astrotoma agassizii.* **C,** A living basket star, *Astrophyton muricatum,* from Beaufort, North Carolina. *(A, Based on photographs by Fitz Goro.)*

become the epineural canal, which lies, as before, oral to the radial nerve. The radial canal of the WVS is between the radial nerve and the vertebra. The tentacle-like tube feet, usually one pair per segment, emerge between the oral and lateral shields. In some species a small protective covering, the **tentacle scale,** is associated with each tube foot (Fig. 28-23). Movable compound spines such as pedicellariae are absent in ophiuroids.

The vertebrae of each arm are arranged single file in a column from one end of the arm to the other and occupy nearly all of the arm's internal space (Fig. 28-24). The contact surfaces of each vertebra bear nodes and sockets that articulate with corresponding surfaces on adjacent vertebrae, as well as pits for the insertion of large **intervertebral muscles** that move the arm (Fig. 28-23). When these muscles are positioned laterally, contraction bends the arm from side to side; muscles that link adjacent vertebrae aborally and orally bend the arm up and down. In many brittle stars lateral arm mobility predominates, but in basket stars and related brittle stars, the arms can bend and coil in any direction.

Disc

The oral surface of the disc is occupied by a complex arrangement of oral shields that frame the mouth and form five interradial **jaws,** each of which bears teeth, or papillae (Fig. 28-22A). In most ophiuroids, one oral shield is modified to form the madreporite. The oral position of the ophiuroid madreporite contrasts with the aboral madreporite of asteroids. Aborally, the disc is often covered with fishlike scales, small spines, or both. Flanking the aboral base of each arm like doubled epaulets is a pair of **radial shields** (Fig. 28-22B). Muscles attached to the radial shields and other skeletal parts enable the flexible disc to make pumping ventilatory movements.

NERVOUS SYSTEM

The nervous system is composed of a circumoral nerve ring and radial nerves, as it is in asteroids, but the sensory components of the ring and the ectoneural cord of each radial nerve are internal in the epithelial lining of an epineural canal, as described above (Fig. 28-20B, 28-23). Specialized sense organs are absent, and individual sensory cells compose the sensory system. Most ophiuroids are negatively phototaxic, and are also able to detect food without contact.

LOCOMOTION

The ophiuroids are the most agile echinoderms. During movement the disc is held above the substratum while one or two arms extend forward and one or two trail behind. Rowing movements of the remaining two lateral arms propel the body forward in leaps or jerks. The arm spines provide traction (Fig. 28-21A). Brittle stars have no arm preference and can move in any direction. In clambering over rocks or among attached organisms, the supple arms often coil around objects (Fig. 28-25C). Although most ophiuroids use the arms to move, a few, such as *Ophionereis annulata* of the American west coast, creep about on their tube feet.

Burrowing occurs primarily in members of the family Amphiuridae (including *Amphiura, Amphiodia,* and *Microphiopholis*).

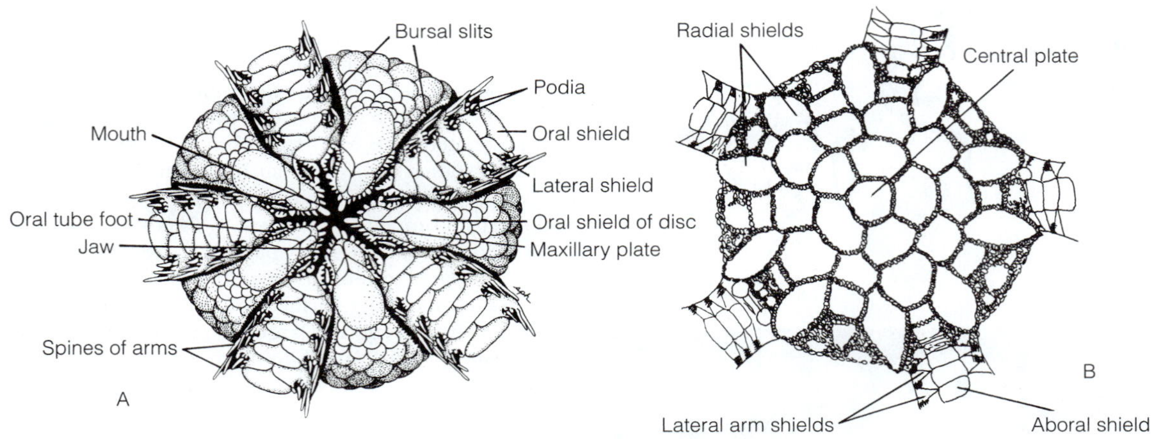

FIGURE 28-22 Ophiuroidea: external anatomy of the disc and arm bases. **A,** The disc of *Ophiura sarsi* (oral view). **B,** The disc of *Ophiolepis* (aboral view). *(A, After Strelkov. B; After Hyman, L. H. 1955. The Invertebrates. Vol. IV. McGraw-Hill Book Co., New York. Reprinted with permission.)*

Using its tube feet and arm undulations, the animal digs a mucus-lined burrow with tubular channels to the surface of the mud or sand. The animal never leaves the burrow unless dislodged, but two or three arms extend into the water above the burrow for feeding and gas exchange (Fig. 28-25D). Arm undulations in the burrow ventilate it and the subterranean parts of the body.

WVS AND OTHER COELOMS

The ophiuroid WVS is basically similar to that of asteroids except the madreporite is positioned orally rather than aborally. The oral shield that forms the madreporite usually has but one pore and canal. The stone canal ascends from the madreporite to the ring canal, which is located in a groove on the aboral surface of the jaws. The ring canal bears four interradial polian vesicles (none in the madreporite interradius) and also gives rise to the radial canals, which extend along the lower side of the vertebrae (Fig. 28-23). The polian vesicles may function as ampullae for the buccal tube feet, which surround the mouth. A radial canal gives rise to a series of pinnately arranged lateral canals. Each arm segment houses a pair of lateral canals and two tube feet. The suckerless tube feet bear numerous adhesive **papillae** (Fig. 28-23, 28-26A). Ampullae are absent, perhaps due to insufficient space in the

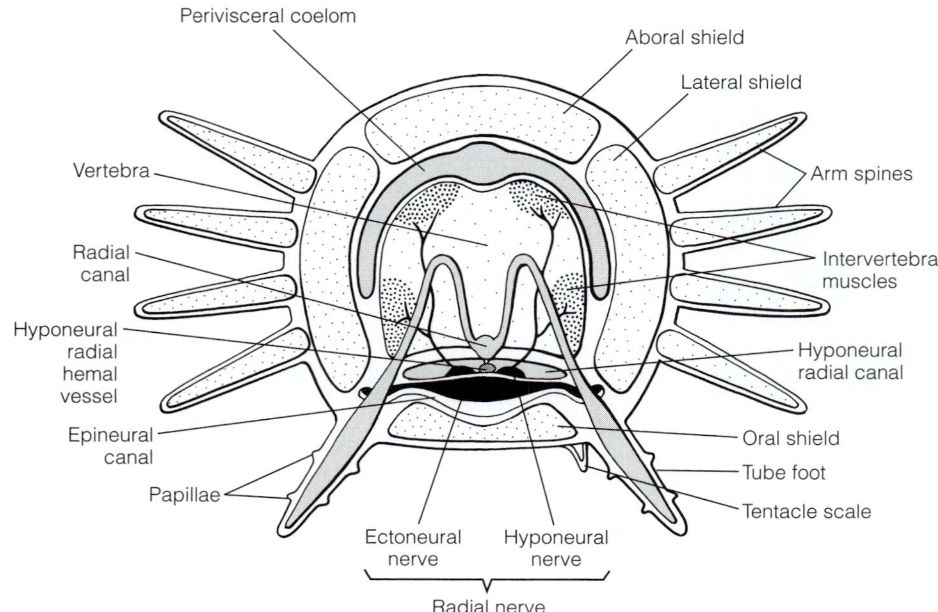

FIGURE 28-23 Ophiuroidea: radial organs. Diagrammatic cross section through the arm of a brittle star.

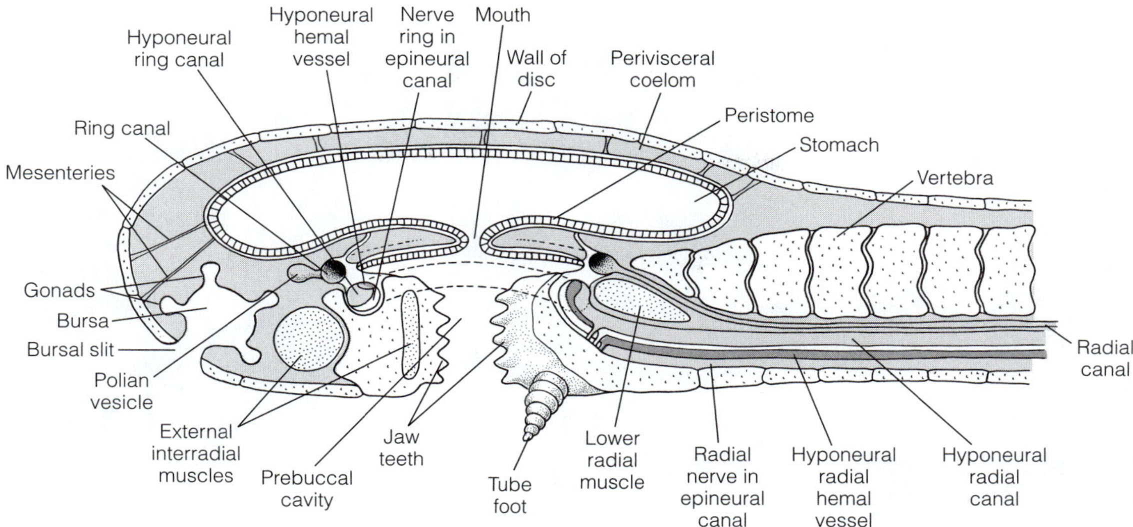

FIGURE 28-24 Ophiuroidea: internal anatomy. Vertical section of the disc and arm base.

arm, but a valve occurs between the tube foot and the lateral canal. Fluid pressure for protraction is generated by a dilated, ampulla-like section of the tube-foot canal and in some forms by localized contraction of the radial canal. Ophiuroid tube feet commonly make flicking movements that enable them to kick food particles toward the mouth or to displace sediment particles while burrowing.

The perivisceral coelom is much reduced compared with that of other echinoderms. The vertebral ossicles restrict the coelom to the aboral part of the arms (Fig. 28-23); the stomach, bursae, and gonads leave only enough room for small coelomic spaces in the disc. The hyponeural coelom is similar to that of asteroids. The epineural canal in the disc and arms superficially resembles a coelomic cavity, but it is lined by an ectodermally derived epithelium, not a peritoneum, and is not a coelom.

DIGESTIVE SYSTEM AND NUTRITION

The ophiuroid gut is a blind sac restricted to the disc. The mouth lies at the center of a peristomial membrane and is preceded by five jaws surrounding a prebuccal cavity (Fig. 28-22A, 28-24). A short, thick esophagus joins the mouth with a large, saclike stomach (Fig. 28-24). The stomach, which occupies most of the interior of the disc, typically is folded at the margins to form 10 pouches. Intestine and anus are absent. Extracellular and intracellular digestion as well as absorption occur in the stomach. Ophiuroids are carnivores, scavengers, deposit feeders, or suspension feeders. Most use several feeding modes, but one is generally predominant.

When suspension feeding, the Atlantic brittle star *Ophiocomina nigra* lifts its arms off the bottom and waves them in the water. Plankton and detritus adhere to mucous strings cast between adjacent arm spines. The trapped particles may be swept downward toward the tentacle scale by ciliary currents or collected from the spines by the tube feet, which extend upward for this purpose (Fig. 28-26A). The tube feet are then scraped across the tentacle scales, depositing collected particles in front of the scale (Fig. 28-26B). This is also the destination for particles transported by specialized ciliary tracts. Tube-foot pairs along the arm cooperate to gather particles, compact them into a pellet, and transport it toward the disc on the aboral surface of the arm. Once the food pellet reaches the disc, cilia transport it to the mouth. *Ophiothrix fragilis* feeds similarly, but twists its spiny arms to expose their oral surface to the water current. Its long arm spines and even longer outstretched tube feet form a comb on which suspended particles are trapped (Fig. 28-27B).

The suspension-feeding basket stars, such as *Astrophyton* (Fig. 28-21C) and *Gorgonocephalus,* capture zooplankton of relatively large size (10 to 30 mm: crustaceans, polychaetes, and others). Perched above the bottom, often on a gorgonian coral, the basket star extends its branched arms in a parabolic fan with the concave aboral surface directed toward the water current (Fig. 28-27A). Prey is seized with the ends of the many arm branches, which coil around the catch, and minute surface hooks prevent escape. Periodically, the basket star removes the collected plankton from the arms by passing them through comblike oral papillae. Unlike brittle stars, basket-star tube feet do not play a role in food transport.

Deposit feeding in *Ophiocomina* is performed by the tube feet alone. They collect the particles from the substratum, compact them into food balls, and move them toward the mouth.

Carnivores, such as species of *Ophioderma,* lasso prey with their highly mobile and flexible arms. Seizure of a small, waterborne crustacean can be as rapid as a snake's initial coil around a mouse. Once embraced, the arm flexes and transfers the helpless prey to the underside of the disc, jaws, and mouth (Fig. 28-27C–E). The mobile disc "hugs" the prey and quickly swallows it. *Ophioderma* is a voracious carnivore and, if offered sufficient and appropriate food, it will eat until the disc ruptures.

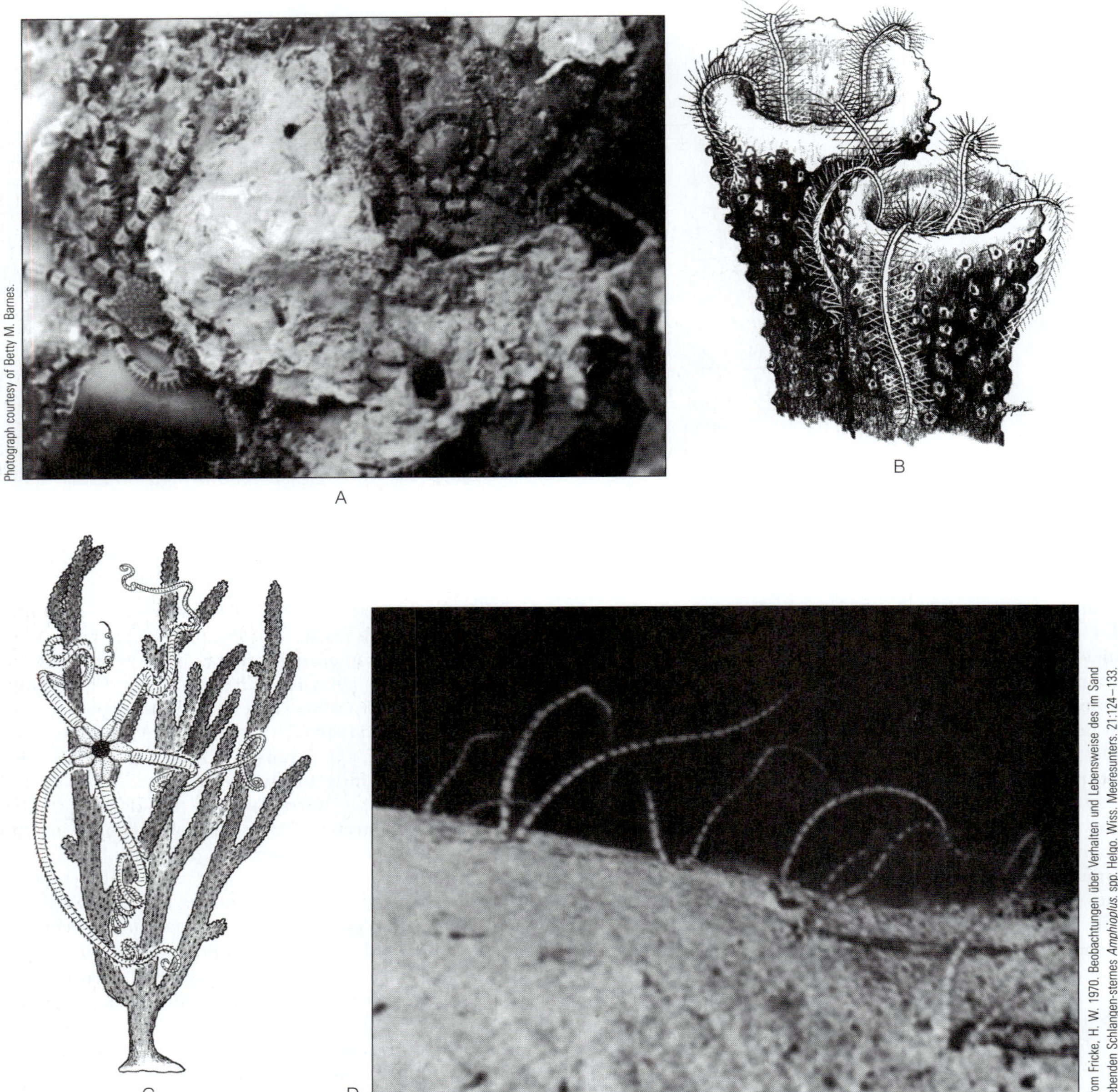

FIGURE 28-25 Ophiuroidea: habitats and lifestyles. **A,** Two West Indian brittle stars *(Ophionereis)* lodged in crevices below a coral head. **B,** Two brittle stars in a sponge. **C,** A brittle star climbing on a gorgonian coral. These brittle stars are related to the basket stars and can coil their arms vertically. **D,** Members of *Amphioplus* projecting two arms from their tubelike burrows and trapping suspended particles in the passing water current. *(C, Modified and redrawn from Hyman, L. H. 1955. The Invertebrates. Vol. IV. McGraw-Hill Book Co., New York. Reprinted with permission.)*

GAS EXCHANGE, EXCRETION, AND INTERNAL TRANSPORT

Ophiuroid gills are tube feet and specialized invaginations, called **bursae** or genital bursae, of the disc's oral surface. The bursae are blind sacs, 10 in number, that open orally at slits, 1 on each side of an arm base (Fig. 28-22A). Cilia on the bursal lining create a water current that enters the end of the slit at the disc margin, passes through the bursa, and exits the slit at its opposite end nearer the mouth (Fig. 28-22A, 28-24). Many species also pump water through the bursae by raising and lowering the oral or aboral disc wall or by contracting disc muscles associated with the bursae.

A few species of ophiuroids, such as the burrowing *Hemipholis elongata,* have hemoglobin in coelomocytes of the WVS. Instead of ventilating its burrow with arm movements, as

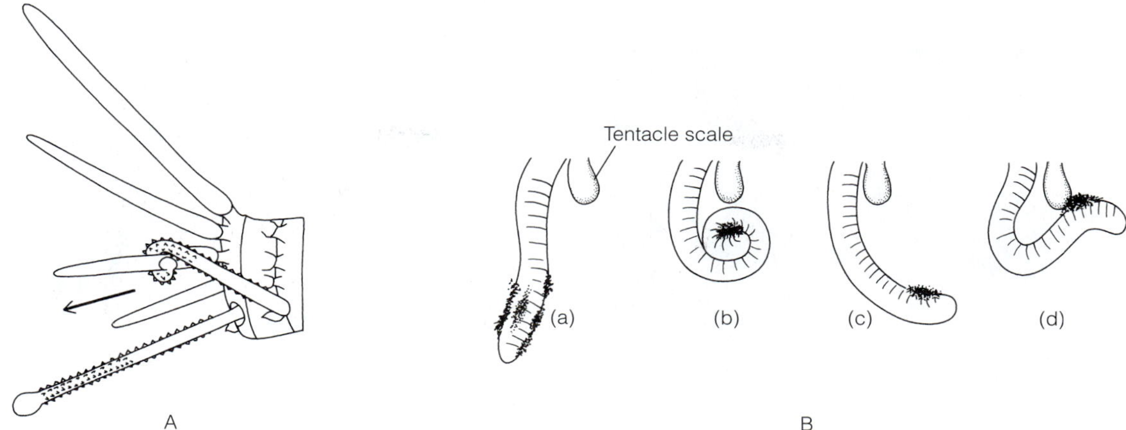

FIGURE 28-26 Ophiuroidea: feeding action of the tube feet. **A,** Spine wiping in *Ophiocoma wendti,* a West Indian brittle star. Note that the tube foot of one side wipes food from the spines on the opposite side of the arm. **B,** Particle consolidation and transfer in the suspension-feeding brittle star *Ophionereis fasciata.* **(a),** Particles collected by the tube foot from the spines. **(b),** Particles consolidated by the tube foot into one mass. **(c** and **d),** The mass is transferred from the tube foot to the tentacle scale. *(A, From Sides, E. M., and Woodley, J. D. 1985. Niche separation in three species of* Ophiocoma *in Jamaica, West Indies. Bull. Mar. Sci. 36:701–715; B, From Pentreath, R. J. 1970. Feeding mechanisms and the functional morphology of podia and spines in some New Zealand ophiuroids (Echinodermata). J. Zool. 161:395–429.)*

most other burrowers do, *Hemipholis* holds a few of its arms in the water above the burrow and transports oxygen internally by binding it in its red cells. The cells flow bidirectionally in the WVS, to and from the disc. *Hemipholis* lacks bursae, but has branches off its ring canal that extend into the tissues of the disc like a capillary plexus. The thin-walled respiratory bursae may be sites for release of waste-laden coelomocytes. No other excretory provision has yet been identified in ophiuroids.

The hemal system is essentially like that of asteroids, but even less is known of its function. In the absence of functional information on the hemal system, the coeloms are assumed to account for circulation in the ophiuroid body. The ectodermally derived ciliated epineural canal (Fig. 28-20B, 28-23), however, surely plays a circulatory role.

REPRODUCTION

Many ophiuroids can voluntarily cast off one or more arms, a part of an arm, or the aboral surface of the disc if disturbed or seized by a predator or collector. The lost part is then regenerated. Their readily breakable arms account for the common name, brittle stars. The arms break, however, not because they are brittle, but as a result of a local softening of mutable connective tissue, specifically the collagenous intervertebral ligaments. Some ophiuroids, notably six-armed species of *Ophiactis,* reproduce clonally by fission of the disc into two pieces, each with three arms. Each fragment then regenerates the missing half, but can feed long before the missing arms are fully regenerated. Remarkably, at least two species of ophiuroids have larvae that can reproduce clonally. The larva casts off a larval arm (see below), which then regenerates the remainder of the larval body.

The majority of ophiuroids are gonochoric. The gonads are small sacs attached to the coelomic side of the bursae near the bursal slit (Fig. 28-24). There may be one, two, or numerous gonads per bursa at various positions of attachment. Hermaphroditic species are not uncommon. Some simultaneously bear separate testes and ovaries whereas others are protandric.

When the gonads are ripe, they discharge into the bursae, probably by rupturing the bursal wall, and the sex cells are carried out of the body in the ventilating water current. Fertilization and development take place in the sea in many species, but brooding is common. The bursae are commonly used as brood chambers, as in the common *Axiognathus squamata (Amphipholis squamata),* but the female of some viviparous species gestates her eggs in the ovary or coelom. Development takes place within the mother until the juvenile stage is reached, when the young crawl out of the bursal slits. In most brooding species only a few young occur in each bursa.

DEVELOPMENT AND METAMORPHOSIS

In nonbrooding, oviparous ophiuroids, early development is similar to that in asteroids. The planktotrophic larva of many species, called an **ophiopluteus,** displays four pairs of elongate arms bearing ciliated bands and supported internally by calcareous rods. The distinctive V-shaped larva swims with its arm tips forward (Fig. 28-28). Metamorphosis is similar to that in asteroids, but occurs while the larva is free-swimming, and there is no attachment stage. The tiny brittle star sinks to the bottom and takes up an adult existence. Development of oviparous species takes 14 to 40 days, while a brooder may require 3 to 7 months.

DIVERSITY OF OPHIUROIDEA

Oegophiurida[O]: Mostly extinct ophiuroids, with only one extant Indonesian species, *Ophiocanops fugiens.* Lacks genital bursae, oral and aboral arm plates; madreporite on disc margin, not oral.

Phrynophiuroida[O]: Have a fleshy integument, glandular mucus-secreting epidermis; often lack aboral arm shields;

Photograph courtesy of Meyer, D. L., and Lane, N. G. 1976. The feeding behavior of some Paleozoic chrinoids and recent basket stars. J. Paleontol. 50(3)472–480. Courtesy of The Paleontological Society.

A

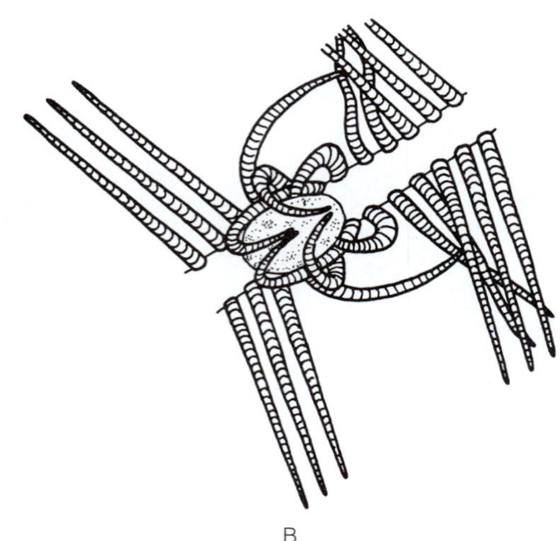

B

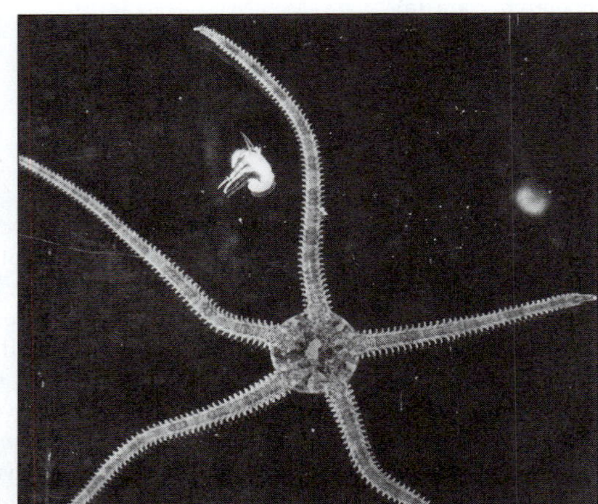

C

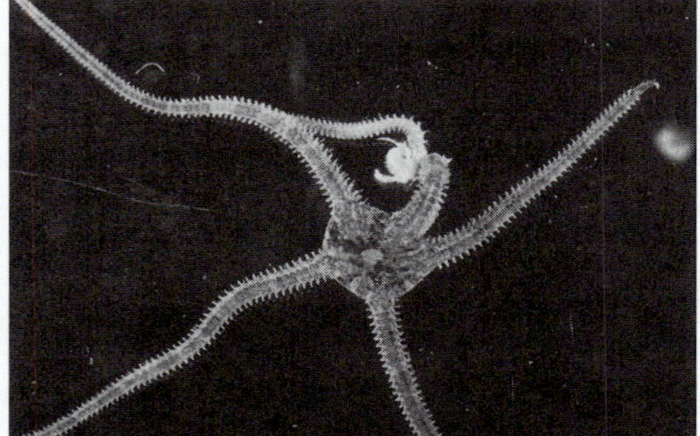

D

E

FIGURE 28-27 Ophiuroidea: feeding and food transport. **A,** Suspension feeding position of the basket star *Astrophyton.* The arms form a parabolic fan with the tips directed toward the current, which in this photograph is moving away from the viewer. **B,** Oral view of an arm of the suspension feeder *Ophiothrix fragilis,* showing transport of a food mass by the tube feet. Particles are added to the mass, which is about 1 mm in diameter upon reaching the mouth. **C–E,** Prey capture by *Ophioderma brevispinum,* which uses its arm to seize and then shove the prey under its disc, where it will be swallowed. This species can also use an arm to lasso swimming prey. *(B, from Warner, G. F., and Woodley, J. D. 1975. Suspension-feeding in the brittle star* Ophiothrix fragilis. *J. Mar. Biol. Assoc. 55:199–210. Copyrighted and reprinted by permission of Cambridge University Press.)*

coil arms vertically. Basket stars (Gorgonocephalidae) and Ophiomyxidae: *Asteronyx, Astrophyton, Gorgonocephalus, Ophiomyxa.*

Ophiurida[O]: Have a thin epidermis, well-developed dermal disc ossicles and arm shields; arms move horizontally. Includes most species of ophiuroids. *Amphipholis, Amphiura, Hemipholis, Ophiactis, Ophiocoma, Ophioderma, Ophiolepis, Ophionereis, Ophiothrix, Ophiura.*

Echinozoa

Echinozoa includes Echinoidea (urchins) and Holothuroidea (sea cucumbers). Both taxa lack arms, but the ambulacra and oral surface have expanded aborally to cover most of the body, except for an aboral anus and **periproct,** a small region around the anus (Fig. 28-30B). Primitively, the oral-aboral axis is elongate and the body is bulbous or cylindrical. The pharynx is surrounded by a **calcareous ring** of fused ossicles, and ossicles also occur in the tube feet. Unlike other echinoderms, echinozoans have a well-developed hemal system.

ECHINOIDEA[C]

The 950 species of echinoids are mobile echinoderms known as sea urchins, heart urchins, sea biscuits, and sand dollars. The conspicuous sea urchins graze on hard substrates and have been known since antiquity. Incessant urchin grazing

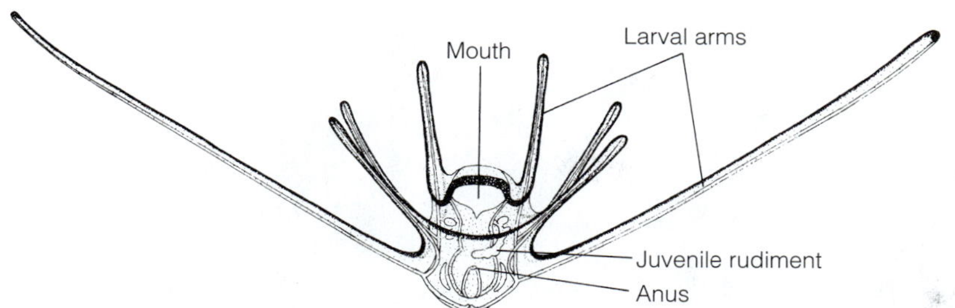

FIGURE 28-28 Ophiuroidea: larva. Ophiopluteus larva of *Ophiomaza* (oral view). *(Modified and redrawn after Mortensen from Hyman, L. H. 1955. The Invertebrates. Vol. IV. McGraw-Hill Book Co., New York.)*

controls the growth of algae, but can decimate large areas if populations become large. In Caribbean coral-reef areas, urchin grazing accounts for 90% of bioerosion. Some urchins are valued as food by, among other animals, sea otters and humans. People especially value their roe for use in sushi (uni) and other culinary specialties, but collections of urchins for this purpose, in quantities reaching 50,000 metric tons per year worldwide, have devastated many populations. The skeletons of dead and, alas, living sand dollars are collected with enthusiasm for their unique form and curious markings, which have mystical significance for some Christians.

The name "Echinoidea," meaning hedgehog (porcupine), refers to the movable spines that cover the body. Body-wall ossicles are fused to form a test, which resembles a shell but is internal. Arms are absent and the body is more or less spherical or flattened to a dome or disk. Pentamerous symmetry is evident in the pattern of ambulacra and tube feet, which occupy much of the body surface. The entire tube-foot-bearing surface is oral and the aboral surface technically is limited to a small area at the aboral pole of the body. Flattened species such as sand dollars have distinct lower and upper surfaces referred to in practice as oral and aboral, respectively, but both are actually oral because they bear tube feet.

The more or less spherical sea urchins are known as "regular urchins" whereas the flattened echinoids are called "irregular urchins." The general description of Echinoidea is based on sea urchins. The irregular echinoids are discussed later in the chapter.

Form

Sea urchins have a more or less spherical body bearing long spines (Fig. 28-29). Sea urchins are brown, black, purple, green, white, or red, and some are multicolored. Most are 6 to 12 cm in diameter, but some Pacific species may attain a diameter of 36 cm. For descriptive purposes, the sea urchin body is divided into aboral and oral hemispheres with the parts arranged radially around the polar axis. The oral pole bears the mouth and is directed against the substratum. The anus is situated at the opposite, aboral pole of the body (Fig. 28-30).

The globose body surface can be divided into 10 radial sections that converge at the oral and aboral poles (Fig. 28-30). Five sections that bear tube feet are called **ambulacral areas.** The ambulacral areas alternate with sections devoid of tube feet, the **interambulacral areas.** The skeletal plates are in rows from the oral pole to the aboral pole. Each ambulacral area is composed of two rows of ambulacral plates, and each interambulacral area is composed of two rows of interambulacral plates. Thus there are 20 rows of plates: 10 ambulacral and 10 interambulacral (Fig. 28-30B). The ambulacral plates are pierced by paired holes for the passage of canals, each pair connecting the internal ampulla with its external tube foot (Fig. 28-30B). These **pore pairs** are unique to echinoids (Fig. 28-31B, 28-37A), for in sea stars and brittle stars, the tube-feet canals pass between ossicles rather than through them and there is only one canal per tube foot.

The mouth is surrounded by a flexible **peristomial membrane** that covers a large test aperture in the center of the oral pole. The membrane bears several radially arranged structures, including five pairs of short, stocky tube feet called **buccal podia** and five pairs of bushy gills (Fig. 28-30A). In addition, the peristomial area includes small spines and pedicellariae.

The anus, periproct, and apical system are situated at the aboral pole (Fig. 28-30B, 28-31A). The **periproct** is a small, circular membrane that bears the anus at its center and a variable number of embedded plates, depending on species (Fig. 28-31A). The **apical system** is a ring of specialized plates around the periproct. It consists of five large **genital plates,** one of which is the porous madreporite, and five smaller **ocular plates** (Fig. 28-31A). The genital plates, each of which bears a gonopore, align with the interambulacral areas and alternate with the ocular plates, which coincide with the ambulacral areas (Fig. 28-30B).

The spines are arranged more or less symmetrically in the ambulacral and interambulacral areas. The spines are longest around the equator and shortest at the poles. Most sea urchins have long **primary spines** and short **secondary spines,** the two types being essentially equally distributed over the body surface (Fig. 28-29C, 28-34). *Arbacia punctulata,* a common sea urchin along the Atlantic coast of North America, however, has only primary spines (Fig. 28-29A).

A ball-and-socket joint at the base of each spine enables it to move. The socket is at the spine base, and the ball, or tubercle, is on the surface of the test (Fig. 28-31B, 28-32A, 28-34). Between the ball and socket are two sheaths of fibers. Contraction of the outer muscular sheath inclines the spine to one direction or another. The inner sheath of collagen fibers ("catch" fibers), a mutable connective tissue, can reversibly shift from soft to rigid, thus locking the spine in place.

The spines usually are cylindrical and taper to a point, but many species depart from this generalization. Species of *Diadema,* which are common on tropical reefs, have very long, needlelike spines, which can be tilted rapidly and waved in the direction of intruders that cast a shadow on the urchin (Fig.

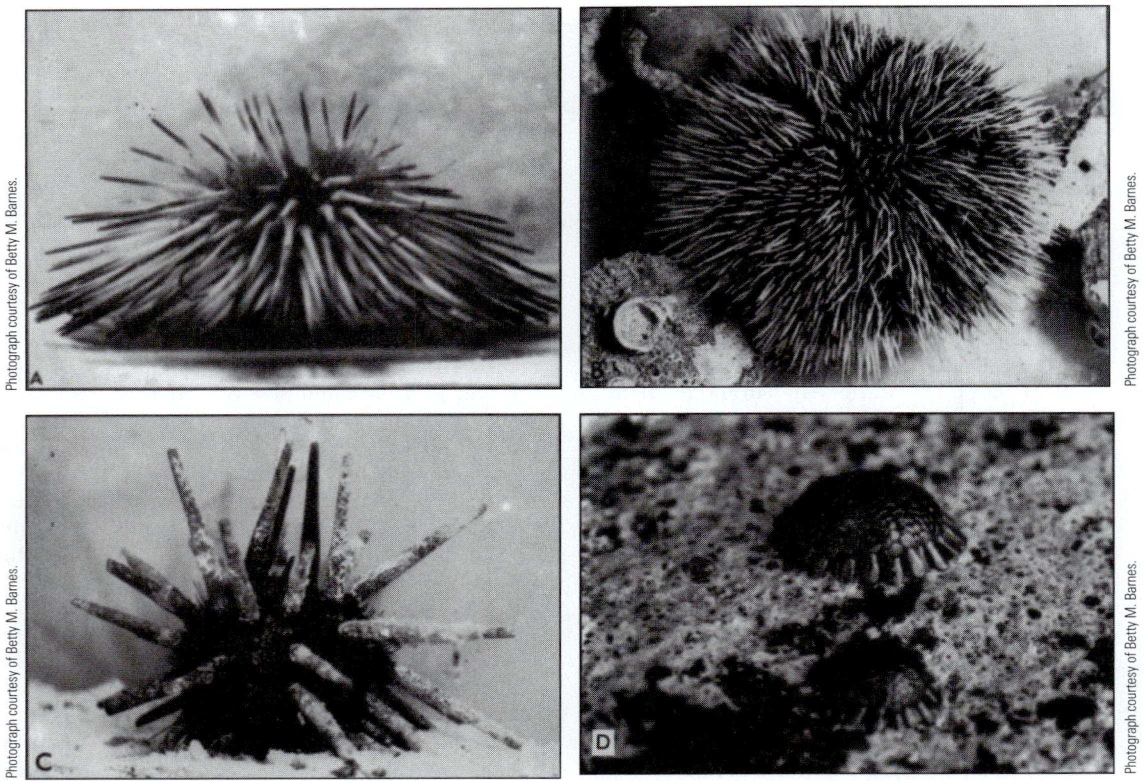

FIGURE 28-29 Echinoidea: regular urchin diversity. **A,** Side view of the common Atlantic sea urchin *Arbacia punctulata* showing the long spines and tube feet. **B,** A West Indian species of *Tripneustes* viewed from above. **C,** A pencil urchin, *Eucidaris tribuloides,* with very small secondary spines around the base of the heavy primary spines. **D,** *Colobocentrotus,* a Pacific sea urchin with blunt aboral spines that fit together to form a smooth surface. Such spines are perhaps an adaptation for living on intertidal rocks.

28-33A). The spines, which can be regenerated, are hollow, brittle, and loaded with an irritant, and the outer surface is covered with circlets of small barbs directed toward the spine tip (Fig. 28-33B). This urchin can inflict serious, painful wounds if stepped on. The thick, blunt, primary spines of the slate-pencil urchins, species of *Heterocentrotus,* are collected and used in decorative mobiles. The spines of pencil urchins (Cidaroida) may also be heavy and blunt (Fig. 28-29C). The aboral spines of the intertidal Indo-Pacific genus *Colobocentrotus* are short, heavy, and polygonal in cross section (Fig. 28-29D). These spines fit together like tiles, providing an effective wave-resistant surface and protection against desiccation. The deepwater leather urchins of Echinothuridae, which have a flexible test, bear special poison spines on the aboral surface (Fig. 28-32B).

Pedicellariae, which are characteristic of all echinoids, are located over the general body surface as well as on the peristome. An echinoid pedicellaria is composed of a long, movable stalk surmounted by jaws. The stalk may contain a supporting skeletal rod, and usually there are three opposing jaws (Fig. 28-34). Muscles at the base of the stalk permit movement.

Any one species usually possesses several types of pedicellariae, and at least one type may be poisonous. **Poisonous pedicellariae** are best developed in members of the widespread warm-water Toxopneustidae, which includes common species of *Lytechinus* and *Tripneustes* (Fig. 28-34). The outer side of each jaw is surrounded by one or two large poison sacs whose ducts open below the terminal tooth of the jaw. The poison has a rapid paralyzing effect on small animals and discourages larger enemies. The body spines frequently incline away from the poison pedicellariae to better expose the latter to intruders. The poison pedicellariae of some tropical species of *Tripneustes* and *Toxopneustes* resemble tiny parasols that, when expanded, form a defensive "second skin." They produce a painful reaction in humans.

Nonpoisonous pedicellariae are used for defense or for cleaning the body surface, biting, and breaking up small particles of debris, which are then removed by the surface cilia. When the pedicellariae are touched on the outside, they snap open; when touched on the inside, they snap shut. Pedicellariae also respond to chemical stimuli.

Body Wall

The body wall of echinoids is composed of the same layers as in asteroids. A ciliated epidermis permanently covers the outer surface, including the spines, except for in pencil urchins (Cidaroida), in which the epidermis wears off the

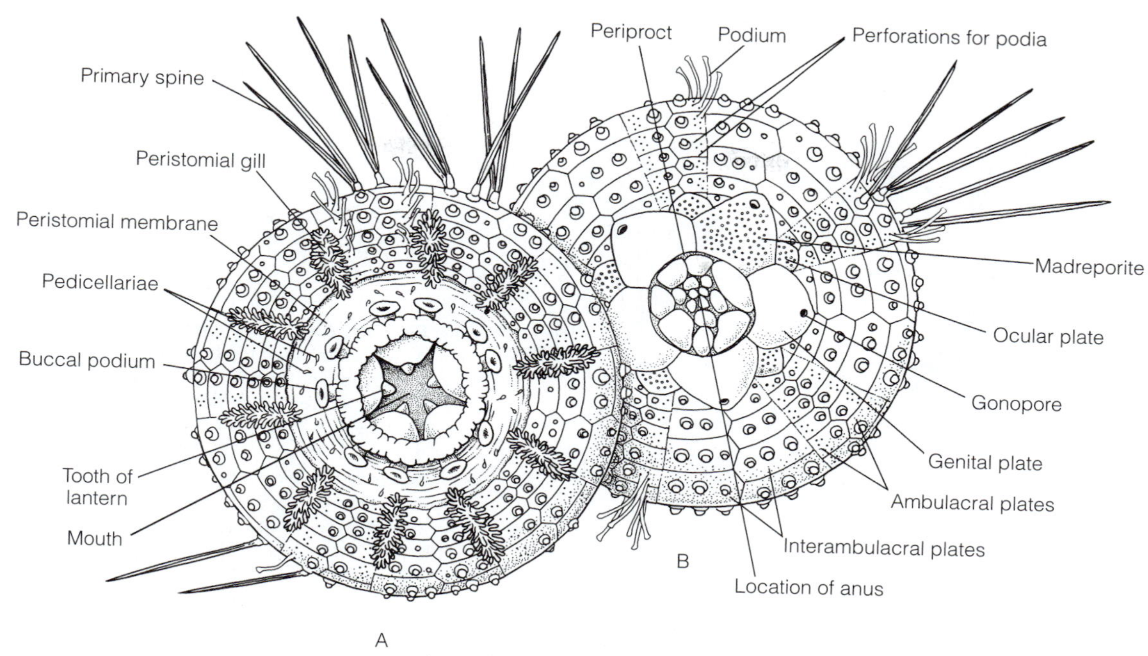

FIGURE 28-30 Echinoidea: external anatomy of the regular urchin, *Arbacia punctulata*. **A,** Oral view. **B,** Aboral view. *(Modified and redrawn after Reid, W. M. 1950.* Arbacia punctulata. *In Brown, F. A. (Ed.): Selected Invertebrate Types. John Wiley and Sons, New York. pp. 528–538.)*

spines. Basally in the epidermis is a nerve layer and below a connective-tissue dermis that contains the flattened and fused skeletal plates. A muscle layer is absent because the ossicles are immovable, and the inner surface of the test is covered by the ciliated epithelial lining of the perivisceral coelom.

Locomotion

Sea urchins are adapted for life on both hard and soft bottoms, and spines and tube feet are used in movement. The tube feet function in the same manner as those of the sea stars, and

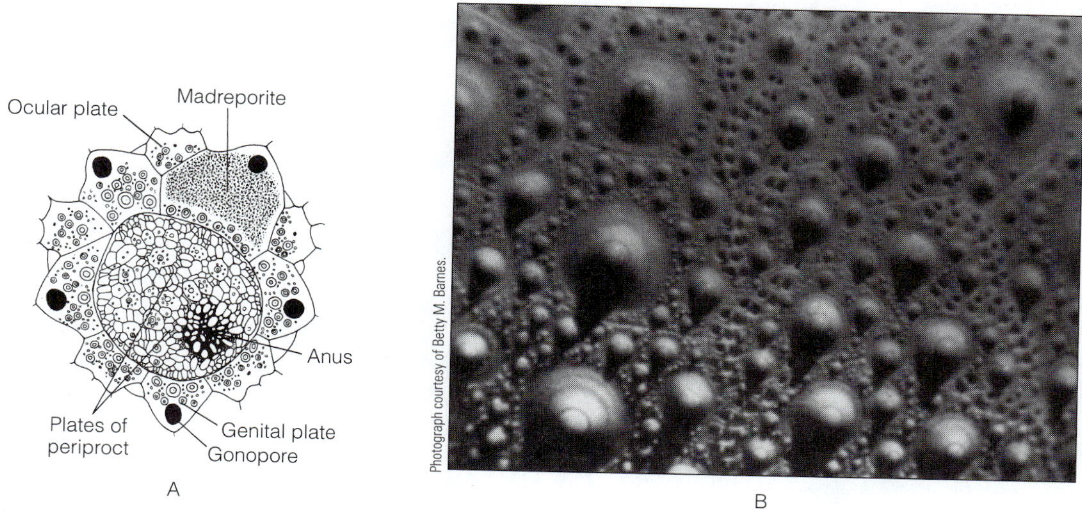

FIGURE 28-31 Echinoidea: the test of regular urchins. **A,** The periproct and surrounding plates (apical system) of the regular urchin *Strongylocentrotus droebachiensis*. **B,** Surface view of part of a sea urchin test, showing the tubercles (balls) to which the spine sockets were attached, the pore pairs for tube feet, and the junction lines (sutures) of adjacent plates. Compare with Figure 28-37B. *(A, Modified and redrawn after Loven from Ludwig.)*

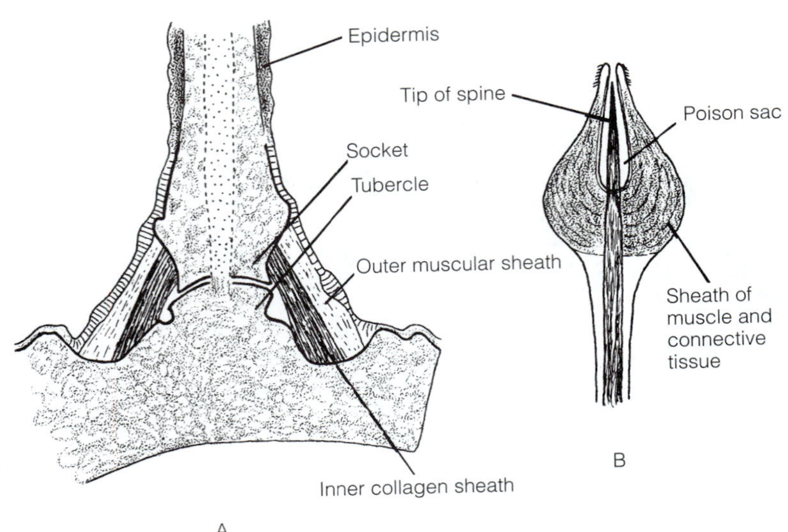

FIGURE 28-32 Echinoidea: spines of regular urchins. **A,** Section through the base of a pencil urchin *(Cidaris)* spine, showing muscle and collagen sheaths. **B,** The poison spine of *Asthenosoma varium,* an Indo-Pacific species. *(A, From Cuénot, L. 1948. Échinodermes. In Grassé, P. (Ed.): Traité de Zoologie, Vol. XI. Masson et Cie, Paris. Reprinted by permission; B, After Sarasins from Hyman, L. H. 1955. The Invertebrates. Vol. IV. McGraw-Hill Book Co., New York. Reprinted by permission.)*

spines may be used for pushing and raising the oral surface off the substratum. Sea urchins can move in any direction without turning, and any one of the ambulacral areas can act as the leading section. If overturned, urchins right themselves by successively attaching, from aboral to oral, the tube feet of one ambulacral area, thus rolling the body upright. Righting may also involve specialized movements of the spines.

Movement of sea urchins is closely related to feeding activity. For example, *Strongylocentrotus franciscanus* in kelp beds off the California coast exhibits movement of about 7.5 cm per day, but where food supplies are low, movement may be as great as 50 cm per day.

Some sea urchins tend to seek rocky depressions (Fig. 28-35A), and some species are capable of increasing the depth of such depressions or even boring into rock and other hard surfaces (Fig. 28-35B). Boring is performed largely by the scraping action of the jaw apparatus.

Boring behavior appears to be an adaptation to counteract excessive wave action, and these species are largely found in habitats exposed to rough water. One of the most notable boring sea urchins is *Paracentrotus lividus,* which lives along the coast of Europe. This sea urchin riddles rock walls with burrows. When the burrows are shallow, the animal leaves to feed, but it remains permanently within deeper burrows,

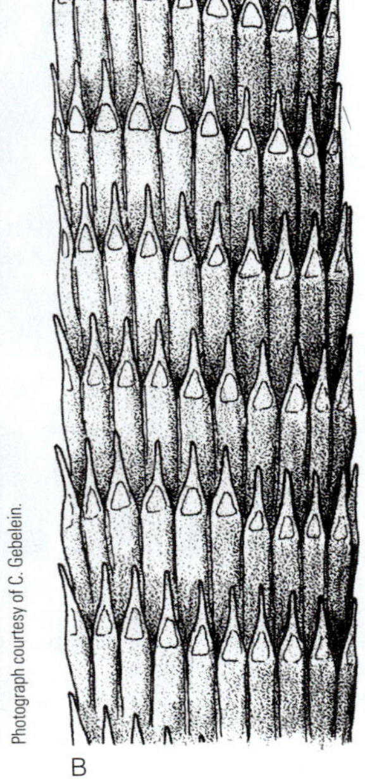

FIGURE 28-33 Echinoidea: spines of the regular urchin *Diadema.* **A,** Species of *Diadema* of the Caribbean and Indo-Pacific have long, hollow, needlelike spines that can inflict painful punctures when the urchin is handled or stepped on. This West Indian species is common on reefs, where it lives in sheltered or protected recesses on exposed or sand bottoms. **B,** Section of a spine of *Diadema.* Circlets of barbs are directed toward the spine tip.

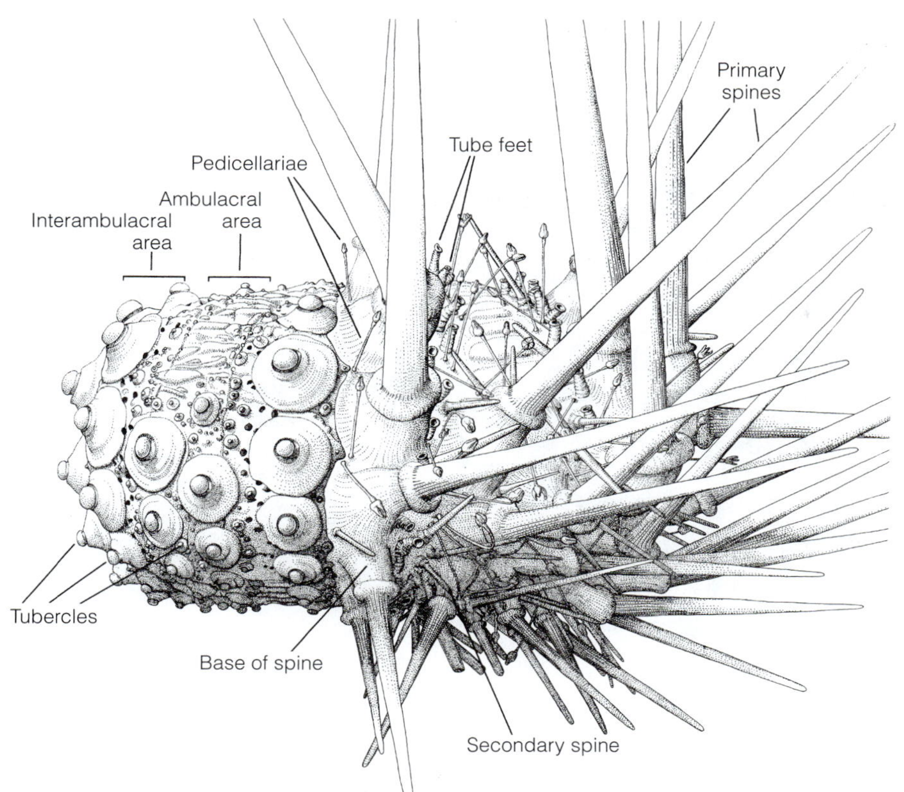

FIGURE 28-34 Echinoidea: external anatomy of a regular urchin in lateral view. Left side: test only; right side: spines, tube feet, and epidermis intact. *(After Messing from Hendler, G., Miller, J. E., Pawson, D. L, and Kier, P. M. 1995. Sea Stars, Sea Urchins, and Allies. Echinoderms of Florida and the Caribbean. Smithsonian Institution Press, Washington and London. p. 199.)*

which often have entrances too small to permit exit. Echinometrids are common boring species on tropical reefs. The urchins can usually be seen within their shallow, irregular grottoes, but are difficult to remove without breaking the surrounding rock. The West Indian *Echinometra* honeycombs coralline rock in surge areas, but nonburrowing populations of this species are sometimes encountered. *Strongylocentrotus purpuratus* is a surge-loving sea urchin found along the Pacific coast of North America that commonly burrows in soft rock.

WVS and Other Coeloms

The WVS of echinoids is essentially like that of the sea stars (Fig. 28-36). One of the genital plates around the periproct is modified as the madreporite (Fig. 28-31A). A slender stone canal, typically less calcified than that of asteroids, descends orally to join the ring canal. The ring canal encircles the esophagus and lies in the aboral membrane of the jaw apparatus. (In the jawless heart urchins it is on the inner surface of the peristomial membrane.) Radial canals extend from the

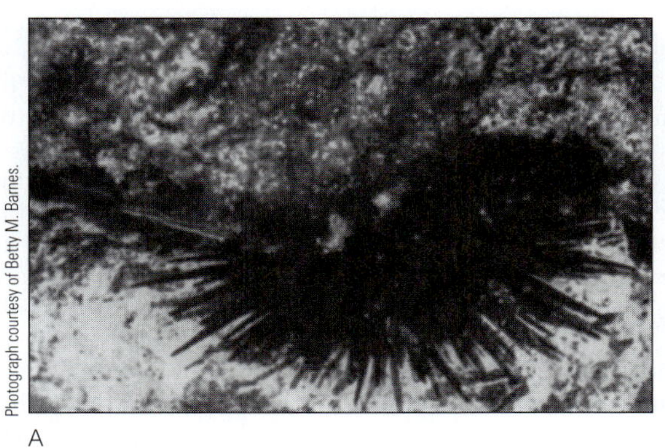

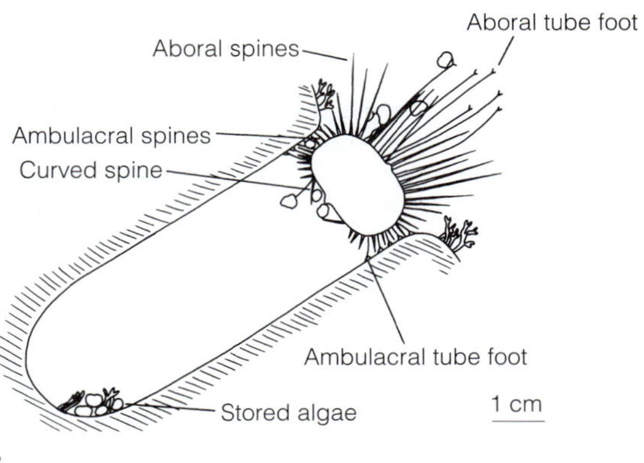

FIGURE 28-35 Echinoidea: rock-boring regular urchins. **A,** The Caribbean *Echinometra lacunter,* which bores into limestone with its teeth and spines, in its burrow. **B,** The Indo-Pacific *Echinostrephus molaris* feeding on pieces of algae collected at the entrance to its burrow. *(B, From De Ridder, C., and Lawrence, J. M. 1982. Food and feeding mechanisms: Echinoidea. In Jangoux, M., and Lawrence, J. M. (Eds.): Echinoderm Nutrition. A. A. Balkema, Rotterdam. p. 90.)*

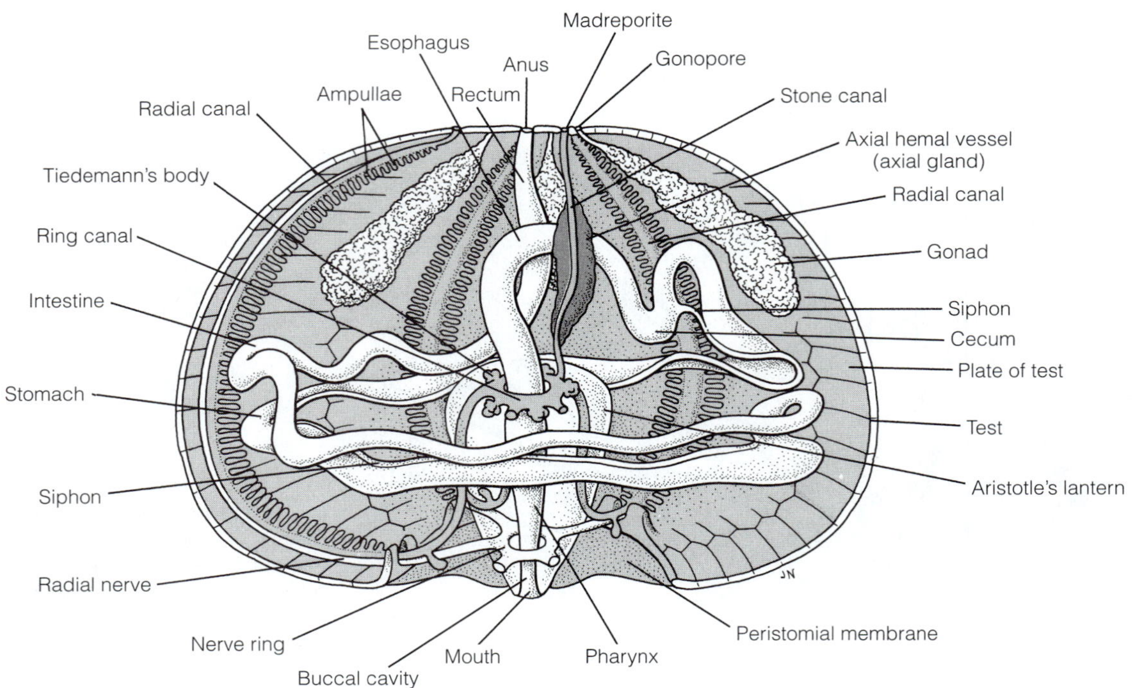

FIGURE 28-36 Echinoidea: internal anatomy of the regular urchin *Arbacia punctulata*. *(Modified and redrawn after Petrunkevitch from Reid in Brown, F. A. (Ed.): Selected Invertebrate Types. John Wiley and Sons, New York. pp. 528–538.)*

ring canal and run along the inside of the ambulacral areas of the test. Each radial canal terminates aborally in a small terminal tentacle, which penetrates the apicalmost ambulacral plate. The lateral canals of one side of the radial canal alternate with those of the other side and terminate in ampullae and tube feet. These canals are also doubled: From each ampulla two canals pass through a pore pair, piercing the ambulacral plate to join the tube foot (Fig. 28-37A). Sea urchins have long tube feet that must be able to reach beyond the spine tips to be effective. Internally, the tube feet are often partitioned lengthwise by a **septum,** which separates the bidirectional flow of coelomic fluid into outgoing and incoming streams (Fig. 28-37A). The two ampullar canals accommodate these two streams. The muscular suckers of sea urchin tube feet have an internal frame of specialized **frame ossicles** that oppose the tendency of the sucker to buckle under load (Fig. 28-37A).

The perivisceral coelom is spacious, occupying up to 75% of the body volume. The hyponeural coelom and the genital coelom are similar to those of asteroids.

Nutrition

Sea urchins feed using a highly developed protrusible jaw apparatus called **Aristotle's lantern,** composed of a complex set of circumpharyngeal ossicles and muscles (Fig. 28-38). The jaw apparatus is absent in heart urchins, but present in other irregular echinoids (to be discussed later). The pentamerous apparatus consists of five large, calcareous plates, the **pyramids,** each of which is shaped like an arrowhead with the point directed toward the mouth. Each pyramid is joined to adjacent pyramids by transverse muscles. Along the vertical midline on the inner side of each pyramid is a long, calcareous **tooth band.** The oral end of the band projects beyond the tip of the pyramid as an extremely hard, pointed **tooth.** Because there is one tooth band for each pyramid, five teeth project from the oral end of the lantern (Fig. 28-30A). The curled, upper end of the tooth band is enclosed in a dental sac, the region of tooth growth. In *Paracentrotus lividus,* new tooth material is produced at a rate of about 1 to 1.5 mm per week. In addition to the teeth and pyramids, Aristotle's lantern includes several smaller, rodlike pieces at the aboral end.

Specialized muscles protrude and retract the lantern and teeth through the mouth, and rock the lantern from side to side. Other muscles control the teeth, which can be opened and closed. These lantern and teeth movements allow urchins to scrape, grasp, pull, and tear.

The majority of sea urchins graze the substratum with their teeth. Although algae are the key food, most sea urchins are generalists and include a wide range of plant and animal material in their diets. Moreover, the diet of a particular species varies from area to area, depending on availability. *Lytechinus variegatus,* an inhabitant of turtle-grass beds, consumes about 1 g of grass per day. The ecological significance of urchin grazing was dramatically revealed following the 1983 crash in Caribbean populations of the long-spined *Diadema antillarum.* In the absence of this urchin, the algal mat on dead coral increased in thickness from 1 to 2 mm to 20 to 30 mm.

Boring sea urchins feed on algae growing on the walls of their shallow burrows, as well as on suspended algal fragments

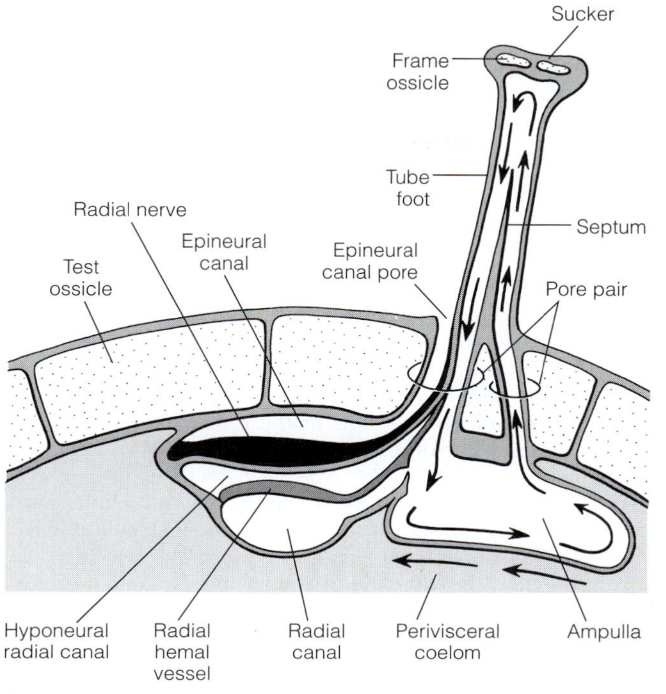

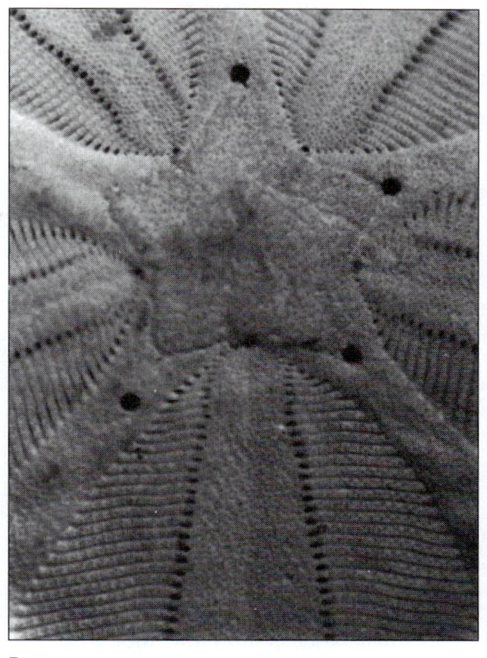

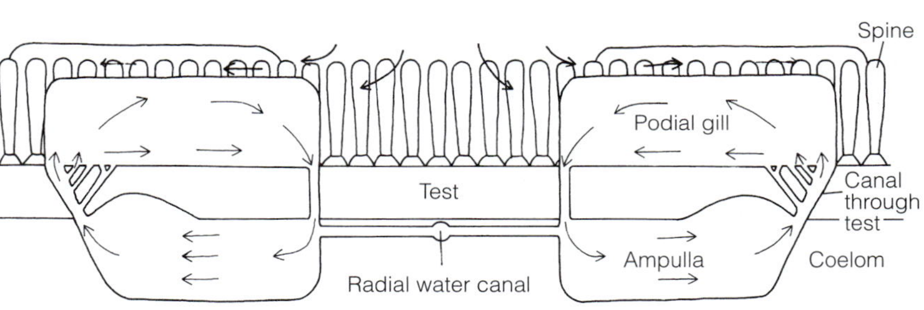

FIGURE 28-37 Echinoidea: tube feet as gills in regular and irregular echinoids. **A,** Section of the body wall, ampulla, and tube foot of *Strongylocentrotus.* All tube feet are suckered and partially septate, separating the oxygenated and deoxygenated ciliary flows. **B,** Aboral center of the test of the sand dollar *Mellita.* The four large holes at the edges of the central star (madreporite) are gonopores. The fifth gonopore (and gonad) is absent in the interambulacrum that bears the rectum and anus. Below the star is the base of a petaloid with a double row of transverse grooves. Each groove corresponds to a compressed petaloid tube foot used for gas exchange. Two small holes, one at each end of a groove, are the pore pair for the petaloid tube foot. **C,** Diagram of two petaloid tube feet across one petaloid of a sand dollar. Arrows show countercurrent flow of external seawater and internal coelomic fluid. *(A, Modified and redrawn from Lang, 1894; C, Adapted from Fenner, D. H. 1973. The respiratory adaptations of the podia and ampullae of echinoids. Biol. Bull. 145:323–339.)*

and other organic debris captured by the tube feet. *Echinostrephus molaris* of the Indo-Pacific sits at the mouth of its burrow and, when debris touches the spines or long tube feet, the spine tips converge and grasp it (Fig. 28-35B).

Aristotle's lantern encloses the buccal cavity and pharynx (Fig. 28-36). The pharynx joins the esophagus. The esophagus descends along the outer side of the lantern and then enters the widened stomach. At the junction of the esophagus and stomach, a blind pouch, or cecum, is usually present. The stomach, which is suspended by a mesentery, circles (counterclockwise in aboral view) the body cavity in the equatorial plane. It then turns 180° and joins the intestine, which circles the body cavity clockwise in aboral view. The intestine is arranged in drapelike festoons suspended by mesenteries. It joins a vertical rectum, which empties through the anus.

In most echinoids a slender, tubular bypass, the **siphon,** parallels the inner edge of the stomach for some or all of its length (Fig. 28-36). The siphon originates from the esophagus

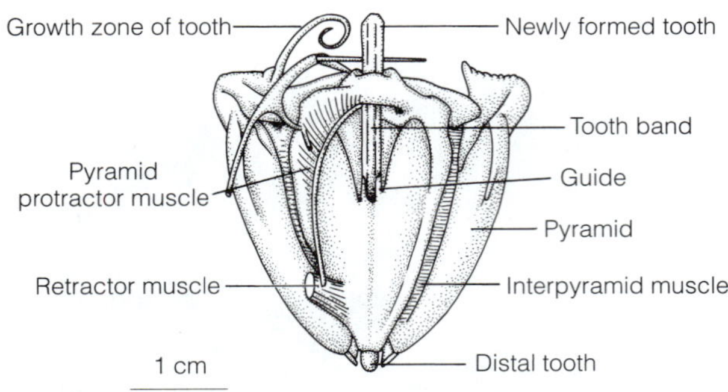

FIGURE 28-38 Echinoidea: Aristotle's lantern of regular echinoids. Lateral view of the lantern of *Sphaerechinus granularis.* *(From De Ridder, C., and Lawrence, J. M. 1982. Food and feeding mechanisms: Echinoidea. In Jangoux, M., and Lawrence, J. M. (Eds.): Echinoderm Nutrition. A. A. Balkema, Rotterdam. p. 81.)*

near its junction with the stomach and rejoins the gut at the union of the stomach and intestine.

Chewing movements of the teeth remove bits of food and compact them in the buccal cavity, which secretes mucus that binds the particles into a pellet. Radial muscles cause the pharynx to dilate and swallow the food pellets. Additional mucus supplied by the pharynx lubricates the pellets, facilitating transport through the remainder of the gut. The pluglike pellets are moved through the esophagus by peristalsis and water is removed by the siphon before the pellets enter the stomach. The stomach secretes enzymes and is responsible for extracellular digestion, endocytosis, intracellular digestion, and nutrient storage. By removal of excess water from the food pellets, the siphon probably prevents dilution of the stomach's digestive enzymes. Although some extracellular digestion, carried over from the stomach, may occur in the intestine, it functions chiefly in endocytosis, intracellular digestion, and nutrient storage. The rectum and anus form and release feces, respectively.

Internal Transport, Gas Exchange, and Excretion

Coelomic fluid is the principal circulatory medium, and coelomocytes are abundant. The echinoid hemal system is well developed and there is a complex network of vessels in the wall of the gut mesenteries, but next to nothing is known about the function of this system or its colorless blood. It has the same basic design as that of asteroids, but gastric hemal tufts are absent.

Most sea urchins have five pairs of **peristomial gills** on the peristomial membrane (Fig. 28-30A). Each is a highly branched outpocket of the peripharyngeal (lantern) coelom and lined with ciliated peritoneum. Coelomic fluid is pumped to and from the gills by a system of muscles and ossicles associated with Aristotle's lantern. The peristomial gills are probably the gas-exchange surface for the extensive lantern musculature.

As in other echinoderms, echinoid tube feet are gills that supply the WVS and perivisceral cavities. The physical separation of opposing fluid streams by a septum in the foot as well as separate canals between foot and ampulla reduces the diffusive loss of oxygen from the incoming O_2-rich stream to the outgoing O_2-depleted flow (Fig. 28-37). In many sea urchins the more aboral tube feet are specially modified to enhance gas exchange. The aboral tube feet may lack suckers, and in *Arbacia* they are flattened.

Coelomocytes remove particulate wastes and carry these accumulations to the gills, tube feet, and axial organ for disposal or storage. Some coelomocytes form clots and are important in wound healing. The heart-kidney (axial organ complex) is similar to that of asteroids.

Nervous System

The nervous system is similar to that of ophiuroids (Fig. 28-36). A circumoral nerve ring encircles the pharynx inside the lantern, and the radial nerves pass between the pyramids of the lantern and run along the underside of the test. The ectoneural components of the radial nerves lie in epineural canals just below the radial canals of the WVS and hyponeural coelom (Fig. 28-37A).

The numerous sensory cells in the epithelium, particularly on the spines, pedicellariae, and tube feet, compose the major part of the echinoid sensory system. The buccal podia of sea urchins, the tube feet around the circumference of heart urchins, and the tube feet of the oral surface of sand dollars are all important in sensory reception.

Except for pencil urchins, echinoids have statocysts, which are located in spherical bodies called **spheridia.** In sea urchins the few to many spheridia are stalked and located at various places along the ambulacra. In sand dollars spheridia are situated near the peristome and buried in the test. The spheridia function in gravitational orientation.

Echinoids are, in general, negatively phototactic and many tend to seek the shade of crevices in rocks and or under shells. Some species of sea urchins, such as *Tripneustes, Lytechinus, Strongylocentrotus,* and the sea biscuit, *Clypeaster,* cover themselves with shell fragments and other objects using the tube feet (Fig. 28-39). The adaptive significance of this behavior is uncertain, but clearly it is a light response. *Tripneustes,* for example, covers more in the summer than in the winter and

FIGURE 28-39 Echinoidea: sea urchin sunscreen. *Lytechinus variegatus* covers its aboral surface with shells, stones, and algae in response to light.

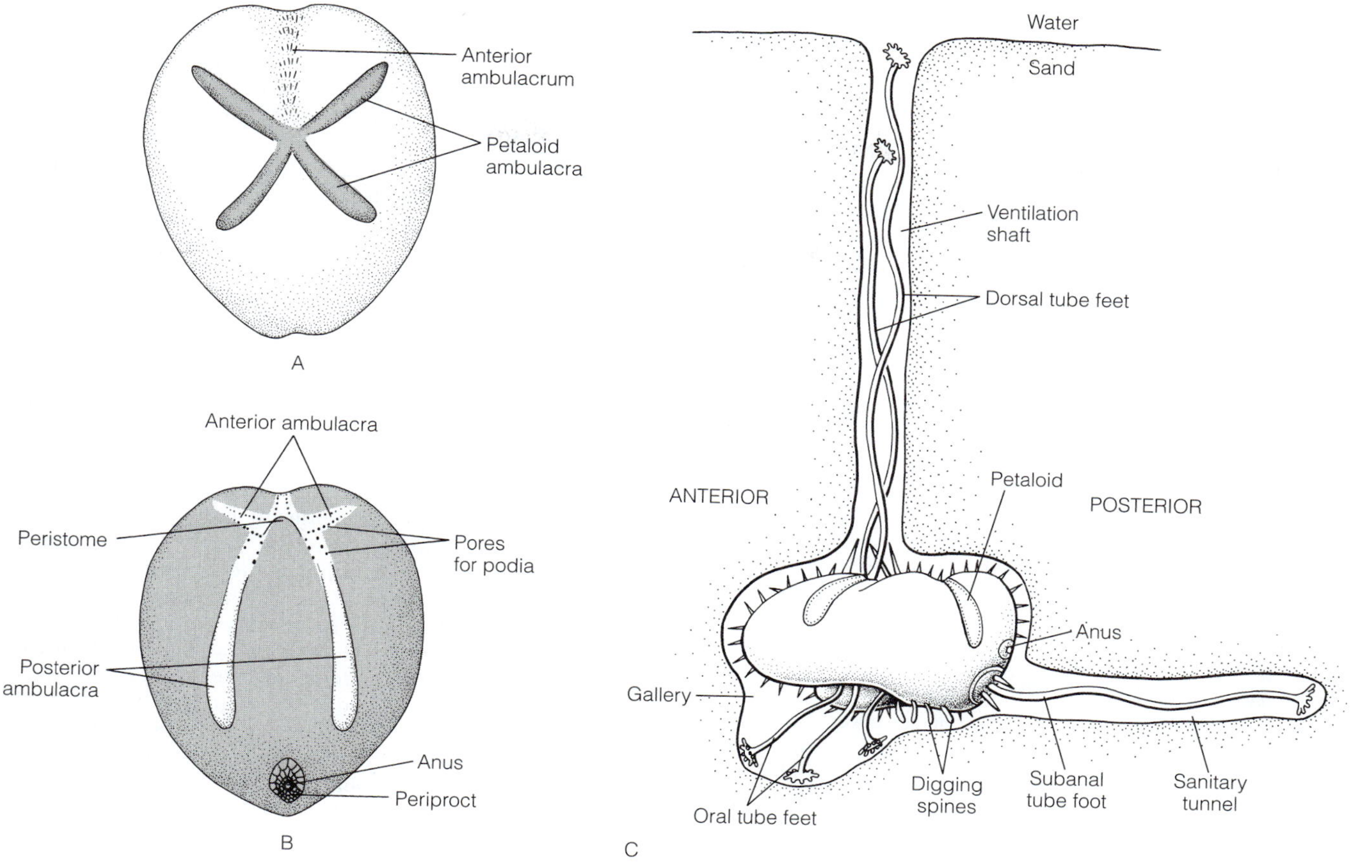

FIGURE 28-40 Echinoidea: irregular urchins. Heart urchins (Spatangoida). **A,** *Meoma ventricosa* from the West Indies (aboral view of the test). **B,** *Meoma* (oral view of the test). **C,** *Moira atropos* in its burrow. *(A and B, Modified and redrawn from Hyman, L. H. 1955. The Invertebrates. Vol. IV. McGraw-Hill Book Co., New York. Reprinted with permission.)*

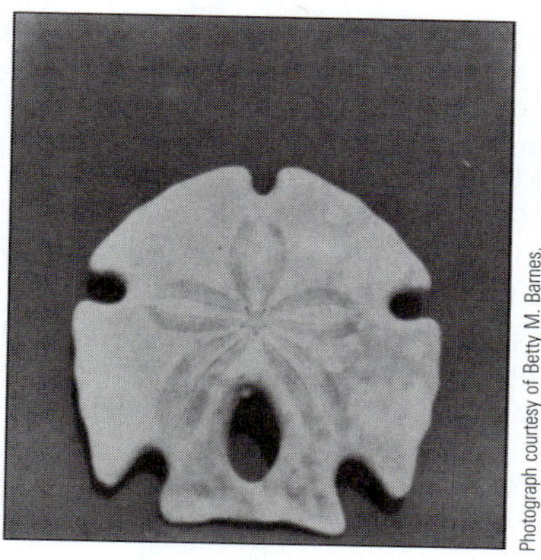

FIGURE 28-41 Echinoidea: irregular urchins. Sea biscuit and sand dollar (Clypeasteroida). **A,** Side view of the test of the sea biscuit *Clypeaster.* **B,** The aboral surface of the test of the arrowhead sand dollar, a species of *Encope.*

the related *Lytechinus* drops its cover at night. When a *Lytechinus* in darkness is partly illuminated experimentally, it covers only the lighted part of the body.

Irregular Echinoids

Irregular echinoids are depressed along the oral-aboral axis so that the body resembles an egg on its side, a low dome, or a flat disc, as represented by heart urchins, sea biscuits, and sand dollars, respectively (Fig. 28-40C, 28-41, 28-42C). Depression of the body creates an equatorial rim, like the edge of the pre-Columbian flat Earth, and the anus is relocated to that rim or midway between the pole and rim (Fig. 28-40B, 28-42B,D). Depression of the body, and especially the repositioning of the anus, imposes a bilateral symmetry on the pentamerous body. This new symmetry is not only formal, but also functional: Irregular urchins move forward along a fixed anterior-posterior axis, with the anterior end opposite of the posterior, anus-bearing end. The lower surface of the body is functionally ventral and modified differently from the upper dorsal surface.

Irregular echinoids are adapted to move over or through sand or mud. In contrast to sea urchins, the test is carpeted with short spines, which are used instead of tube feet for locomotion (Fig. 28-40C, 28-41A, 28-42C,F). They also are employed to remove sediment from the body surface and for other tasks.

As in sea urchins, tube feet in ambulacra cover the surface of the body, but on the ventral (oral) and dorsal surfaces they are specialized for different functions. The ventral ambulacral areas, called **phyllodes,** bear tube feet modified for food handling or adhesion (Fig. 28-42B, 28-43). The dorsal ambulacral areas resemble the petals of a flower and are known, appropriately, as **petaloids** (Fig. 28-37B, 28-40A, 28-42A). The petaloid tube feet are broad, flat gills (Fig. 28-37C). (Irregular echinoids lack peristomial gills.) The ciliary water current over the petaloid tube feet flows in the opposite direction to the flow of WVS coelomic fluid within the gill. This countercurrent exchange ensures a favorable gradient for the uptake of oxygen. A similar countercurrent exchange occurs between the fluid in the ampulla and the surrounding fluid of the perivisceral coelom.

Irregular urchins are primarily selective deposit feeders, ingesting organic material as they burrow through the sediment. The alimentary canal of irregular echinoids is more or less like that described for sea urchins, although the rectum extends posteriorly to join the repositioned anus.

CLYPEASTEROIDA The clypeasteroids are the sea biscuits (cake urchins) and sand dollars, which live on or immediately below the surface of sand. The domed sea biscuits, such as *Clypeaster* (Fig. 28-41A), have an oval outline and a very dense, sturdy skeleton, whereas the sand dollars are circular, highly flattened, fragile discs (Fig. 28-42). The mouth, five jaws, and peristome are at the center of the undersurface, but the anus is repositioned posteriorly along an interambulacrum. The madreporite is a single pentagonal plate at the north pole, from which radiate the five petaloids. In most clypeasteroids, the four interambulacral corners of the madreporic plate bear a genital pore, but the fifth genital pore and gonad are absent, having been displaced by the repositioned rectum and anus (Fig. 28-37B).

The body margin of some common sand dollars, such as species of *Mellita,* has conspicuous elliptical slots known as **lunules** (Fig. 28-42A,B). Lunules vary in number from two to many and are symmetrically arranged. In most cases the lunules develop from notches on the perimeter of the body that, with growth, become incorporated into the test (Fig. 28-41B). In lunulate sand dollars, the anus occurs in the anal lunule, whereas in dollars lacking lunules, the anus is on the posterior rim. Sand dollars live immediately below the sand surface, and the five-lunuled *Mellita quinquiesperforata* of the southeastern United States occupies current-swept beaches. The winglike profile of their lightweight bodies subjects them to lift and dislodgment by water currents. By creating holes in the "wing" and equalizing pressure above and below, the lunules may help to reduce lift and keep the dollars in place. *Mellita* feeds on particles the tube feet pick up from the substratum beneath the oral surface of the animal. These tube feet are irregularly scattered across the oral surface (the radial water canal has lateral lobes). Particles are passed from tube foot to tube foot, in bucket-brigade fashion, to the food grooves and then down the grooves to the mouth (Fig. 28-43). The jaws crush diatoms and other particles, even sand grains.

The slotless sand dollars of the U.S. west coast, *Dendraster excentricus* (Fig. 28-42D), are suspension feeders that live in aggregations (beds) of up to 2000/m^2. Although *Dendraster* feeds like other dollars, under moderate to high water-flow conditions, it will stand on edge and align itself with the current. Food includes not only particles that pass between the spines but also diatoms and algal fragments collected by the tube feet and small crustaceans caught by the pedicellariae.

Some sand dollars can right themselves if turned over. In righting itself, the animal burrows its anterior end into the sand, gradually elevates its posterior end, and eventually flips its body over. *Mellita quinquiesperforata* partially elevates its body and then apparently depends on water currents to flip it onto its oral surface.

The massive sea biscuit *Clypeaster rosaceus* of Florida and the West Indies does not burrow, but rather rests on the surface of sandy bottoms. Its weight holds it in place while its sturdy skeleton, about as impregnable as rock, deters predation.

SPATANGOIDA Heart urchins are fragile, egg-shaped animals that resemble small porcupines (Fig. 28-40C). The long dimension of the body is the anteroposterior axis (Fig. 28-40A,B). The oral surface is flat and the aboral surface is arched. On the dorsal side of the body, the five ambulacra are modified into four petaloids and one anterior ambulacrum that is submerged into a heartlike cleft (Fig. 28-40A). The anterior ambulacrum bears specialized tube feet, described below. Ventrally, the mouth and peristome are shifted anteriorly and the anus and periproct are repositioned along an interambulacrum to the posterior margin of the body. The anterior three ambulacra are short and the posterior two are long (Fig. 28-40B). The few tube feet are localized, long, and specialized. Spatangoids lack jaws. They occupy soft or muddy bottoms and deposit feed on fine particles.

A heart urchin burrows into sediment by inclining its anterior end downward and digging with specialized paddle-shaped spines on the ventral side of the body (Fig. 28-40C). Eventually, most heart urchins, such as *Moira atropos,* excavate

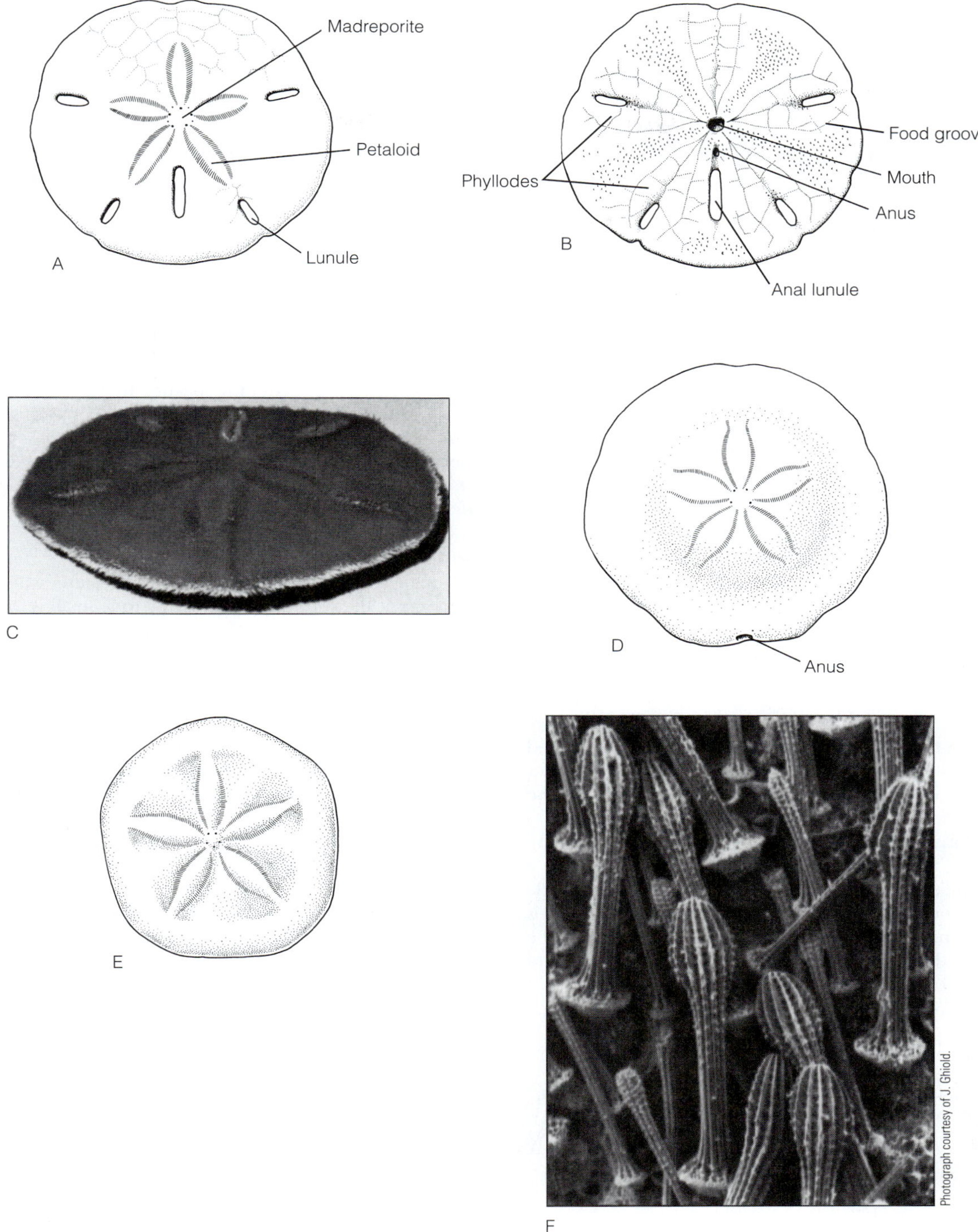

FIGURE 28-42 Echinoidea: irregular urchins. Sand dollars (Clypeasteroida). **A–C,** *Mellita quinquiesperforata,* the five-slotted sand dollar of the Atlantic coast of the United States. **A,** Aboral (dorsal) view of the test. **B,** Oral (ventral) view of the test. **C,** Anterior view of a specimen with the spines intact. **D,** Aboral (dorsal) view of the test of *Dendraster excentricus* from the Pacific coast of the United States. **E,** Aboral view of the test of a species of *Laganum* from the Indo-Pacific. **F,** Scanning electron micrograph (SEM) of the aboral spines of *Mellita.*

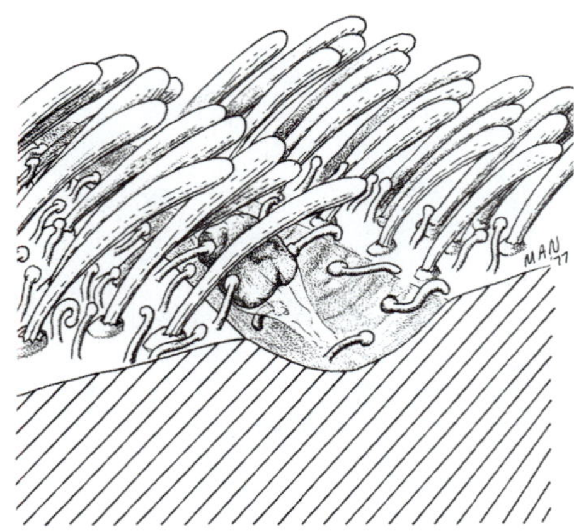

FIGURE 28-43 Echinoidea: irregular urchins. Section through a food groove on the oral surface of a sand dollar (Clypeasteroida), showing a food mass being moved by the tube feet. Many movable, club-shaped spines rise above the tube feet.

a subterranean gallery and feed as they slowly tunnel forward. (Some heart urchins, however, remain buried in one spot below the surface of the sand.) To maintain contact with oxygenated surface water, they dig a vertical ventilation shaft with incredibly long and large tube feet that extend upward from the dorsal anterior ambulacrum (Fig. 28-40C). These dorsal tube feet, which have expanded and elaborate tips, also are used to plaster the shaft and gallery walls with stabilizing mucus. Specialized dorsal spines assist the tube feet and protect them when the feet are withdrawn. Tracts of small, ciliated spines **(fascioles)** pump surface water into the gallery, over the body, and out through the posterior tunnel into the porous sand.

Heart urchins obtain food using modified tube feet on the ventral surface near the mouth (Fig. 28-40C). These large tube feet grope about on the sand surface of the gallery, picking up fine food particles. A short, blind sanitary tunnel is excavated behind the body by a group of large subanal tube feet. Feces are deposited in the tunnel, which is downstream of the ventilatory flow. Some intertidal species, such as the Indo-Pacific *Lovenia,* come to the surface at low tide and rebury themselves when the tide returns. *Lovenia* can bury itself in 1 min, but other species may take as long as 50 min.

Reproduction

All echinoids are gonochoric. A regular echinoid has five gonads suspended along the interambulacra on the inside of the test (Fig. 28-36), but most sand dollars and sea biscuits have only four and heart urchins have three or two. A short gonoduct extends aborally from each gonad and opens at a gonopore located on one of the five genital plates (Fig. 28-30B, 28-31A, 28-37B). Some burrowing sand dollars have long genital papillae that permit the release of eggs or sperm above the sand surface.

Sperm and eggs are shed into the seawater, where fertilization takes place. Brooding occurs in some cold-water sea urchins and heart urchins, and there is a brooding species of sand dollar. Brooding sea urchins retain their eggs on the peristome or around the periproct and use the spines to hold the eggs in position. The irregular forms brood their eggs in deep concavities on the petaloids.

Development and Metamorphosis

Cleavage is equal up to the eight-cell stage, after which the blastomeres at the vegetal pole generate four small micromeres. A typical blastula ensues and becomes ciliated and free-swimming within 12 h after fertilization.

Gastrulation is typical but preceded by an interior proliferation of mesenchymal cells originating from the micromeres. The mesenchyme forms some larval muscles and the larval skeleton. The coelom arises by separation of the free end of the archenteron. It then divides into right and left pouches, or lateral divisions may appear before separation is complete. The gastrula becomes somewhat cone-shaped (prism stage) and gradually develops into a planktotrophic larva, the **echinopluteus.** Like the ophiuroid pluteus, the echinopluteus bears four pairs of long larval arms, each supported by an internal skeletal rod (Fig. 28-44A). The spatangoid echinopluteus is distinctive for its two additional pairs of lateral arms and an especially distinctive unpaired posterior arm. The echinopluteus swims

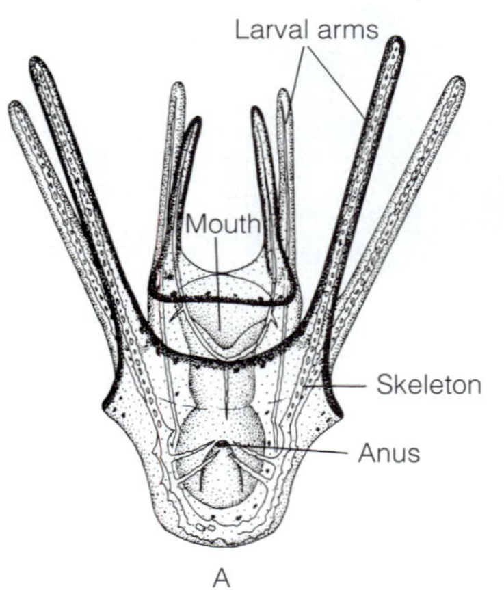

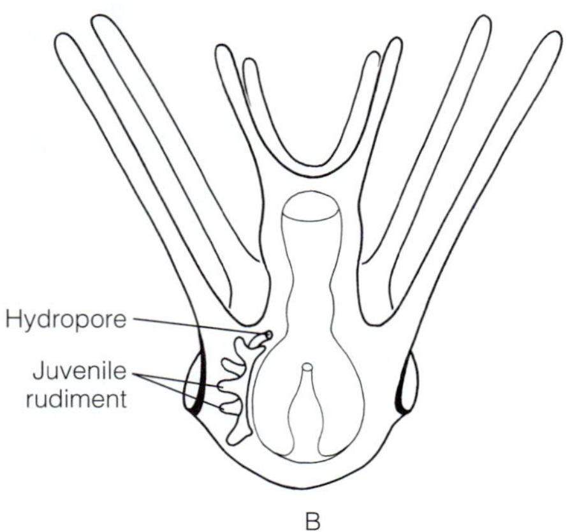

FIGURE 28-44 Echinoidea: larva and larval metamorphosis. **A,** Echinopluteus larva of the sand dollar *Fibularia craniola.* **B,** An echinopluteus showing the juvenile rudiment. *(A, After Mortensen from Hyman, L. H. 1955. The Invertebrates. Vol. IV. McGraw-Hill Book Co., New York.)*

and feeds for as long as several months. During later larval life the juvenile rudiment forms on the left side of the larval body, its oral-aboral axis perpendicular to the larval body axis (Fig. 28-44B). The echinopluteus gradually sinks to the bottom, but unlike the asteroid brachiolaria, which attaches with specialized larval arms, it adheres with tube feet extending from the juvenile rudiment. Metamorphosis is rapid, taking place in about an hour. Young urchins are no larger than 1 mm.

The larvae of *Dendraster excentricus,* the common sand dollar along the U.S. Pacific coast, settles and metamorphoses in response to a substance released by adults. Such preferential settlement explains why this species, and many other sand dollars, occurs in sand beds with high population densities. The life span of *Dendraster* is about eight years and sand-dollar beds persist for decades.

Growth rates are known for only a few echinoids. Two sand dollars from the Gulf of California, *Encope grandis* and *Mellita grantii,* which reach diameters of 74 and 38 mm, respectively, require five years to attain 95% of their maximum size. The annual mortality rate is 18% for *Encope* and 58% for *Mellita. Strongylocentrotus purpuratus* from off the California coast, one of the best-known sea urchins, reaches sexual maturity during its second year when only 25 mm in diameter, but may live 30 years or more.

Diversity of Echinoidea

Cidaroida[sC]: Have simple (noncompound) ambulacral plates, each with one pore pair; widely separated primary spines, small secondary spines; jaw muscles attached to interambulacral plate processes; gills and spheridia absent. 150 spp. Pencil urchins *Cidaris, Eucidaris, Notocidaris.*

Euechinoida[sC]: Ambulacral plates are compound, each formed of three or more primary plates with two or more pore pairs; jaw muscles attach to ambulacral plate processes. 800 spp.

Echinothuroida[O]: Mostly deep-water leather urchins with a flexible test and poison sacs near the spine tips. *Asthenosoma.*

Diadematoida[O]: Regular urchins with a rigid or flexible test with hollow spines; have 10 oral plates on peristomial membrane. *Diadema, Plesiodiadema.*

Echinacea[O]: Regular urchins with a rigid test and solid spines. Gills are present. *Arbacia, Echinometra, Echinostrephus, Echinus, Lytechinus, Paracentrotus, Psammechinus, Toxopneustes, Tripneustes, Strongylocentrotus.* Slate pencil urchins *Colobocentrotus, Heterocentrotus.*

The remaining two taxa constitute the "irregular" (bilaterally symmetric) urchins.

Clypeasteroida[O]: Sea biscuits and sand dollars specialized to live on or just below sediment surface. Test is slightly depressed to flat; jaws and petaloid tube feet are present; anus is displaced posteriorly. *Clypeaster, Dendraster, Encope, Fibularia, Mellita, Rotula.*

Spatangoida[O]: Oval, elongate heart urchins. Mouth is shifted anteriorly, anus posteriorly. Jaws are absent. *Echinocardium, Lovenia, Meoma, Moira, Spatangus.*

HOLOTHUROIDEA[C]

The holothuroids, or holothurians, are the sea cucumbers, a taxon of only 1200 species that has radiated into a greater number of different habitats than any other group of echinoderms. Many cucumbers burrow, some crawl on sand, others occupy crevices, still others attach to hard surfaces, a few climb on algae, and a handful are pelagic. Burrowers and crevice dwellers may resemble a cucumber, banana, or worm; crawlers imitate a giant caterpillar, those on hard surfaces can mimic a chiton, and swimming species resemble a parasail or umbrella. Ranging in size from a few millimeters for the interstitial *Leptosynapta minuta* to 2 m for the tropical *Holothuria thomasi,* most cucumbers are greater than 10 cm in length and often have robust, meaty bodies. Approximately one-third of the species live in the deep sea (more than 60% of the Earth's surface), where they can account for up to 90% of the benthic biomass.

Form

Holothuroids lack arms and the oral surface and ambulacra are expanded aborally along the elongate polar axis. The great length of the polar axis requires the animal to lie with the side of the body, rather than the oral pole, against the substratum. The body side on the substratum, the functional ventral surface, includes three ambulacra (trivium), and is

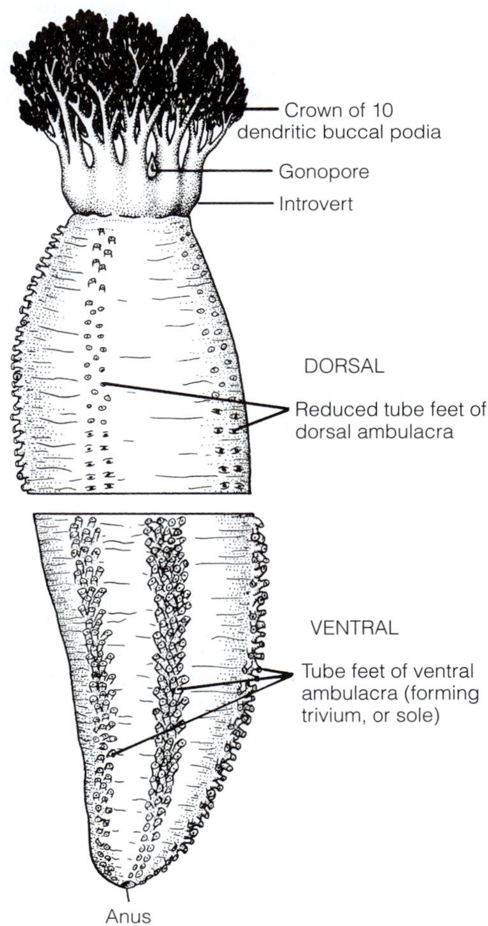

FIGURE 28-45 Holothuroidea: The North Atlantic sea cucumber *Cucumaria frondosa.*

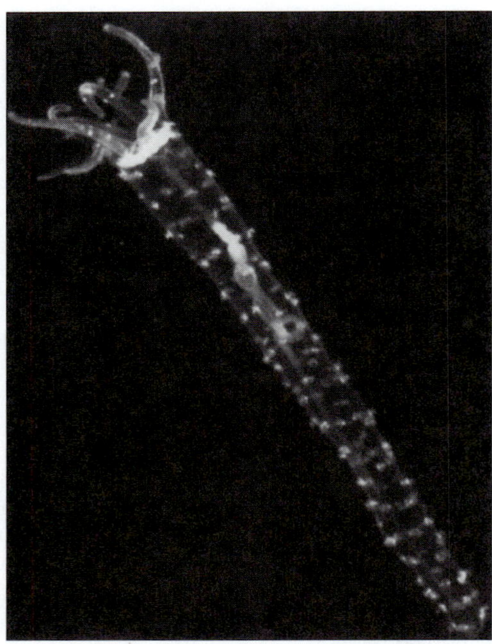

FIGURE 28-46 Holothuroidea: form of the tiny tropical *Synaptula hydriformis* (Apodida), a wormlike holothuroid that lacks tube feet but clings to algae using anchor-shaped ossicles (see Figure 28-47), seen as white patches in this photograph.

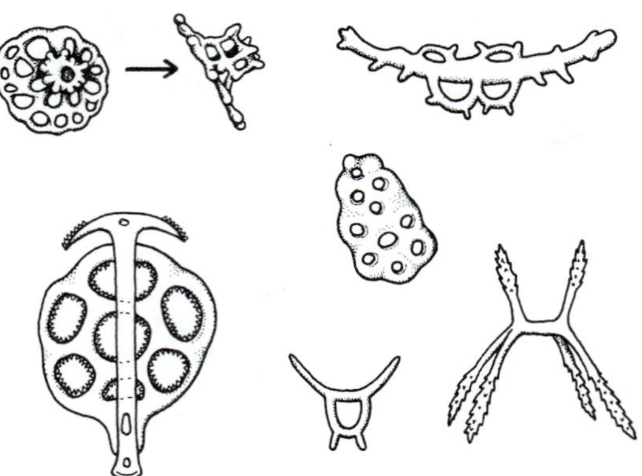

FIGURE 28-47 Holothuroidea: ossicles. An anchor and its anchor plate are shown at lower left. *(After Bell.)*

commonly called the **sole,** whereas the dorsal side includes two ambulacra (bivium; Fig. 28-45). This arrangement superimposes bilateral symmetry on the body of many sea cucumbers, but it is unlike that of echinoids. Burrowing species generally lack a sole and exhibit pentamerous symmetry.

Depending on the species of cucumber, tube feet may be concentrated in ambulacral rows (*Thyonella, Cucumaria;* Fig. 28-45), scattered more or less uniformly over the surface *(Thyone, Sclerodactyla),* or absent (*Leptosynapta, Synaptula;* Fig. 28-46). When present, those on the ventral surface usually have suckers. The dorsal tube feet, on the other hand, may bear suckers, but often they are reduced to warts or slender papillae, or are absent altogether.

The mouth and anus are at opposite ends of the long body and the mouth is surrounded by large, sometimes branched **buccal podia** (tentacles), which are specialized oral tube feet (Fig. 28-45, 28-50). There are 10 to 30 buccal podia, depending on taxon. Unless the animal is feeding, the buccal podia usually are retracted and a sphincter muscle constricts the oral end of the body. In some species, a short, collarlike section of the trunk, the **introvert,** is retracted with the tentacles (Fig. 28-45). A sphincter muscle also closes the anus. When both openings are closed, it can be difficult to distinguish between the oral and anal ends of the body.

Body Wall

The typically soft and leathery body wall includes a thin cuticle, a nonciliated epidermis, and a thick dermis. The dermis, sometimes also the connective-tissue sheath of the tube feet, contains microscopic ossicles, which occur in a variety of shapes, often in a single animal (Fig. 28-47). Large, aboral protective plates, like those of chitons, occur in some species *(Psolus, Ceto)* that attach to exposed rock surfaces. Ossicles are used by sea cucumbers to toughen and protect their body walls and by taxonomists to classify holothuroids. A pair of distinctive ossicles, called an **anchor** and **anchor plate,** occurs in the body wall of *Leptosynapta* and relatives (Apodida), all of which lack tube feet (Fig. 28-47). The sharp flukes of the anchor protrude from the body wall, catch on sediment or algae, and help secure the cucumber. When handled by humans, *Leptosynapta* sticks to skin like a bur, or sticktight, to clothing. The collagenous dermis is a mutable connective tissue that under duress can be cross-linked to nearly rock hardness in some species or softened to such a degree that the body literally disintegrates and drips between the fingers, killing the animal, as in the Indo-Pacific *Stichopus chloronotus* (Fig. 28-2) or *Isostichopus badionotus* of the West Indies.

Internal to the dermis is a thin sheet of circular musculature followed by five longitudinal muscle bands. Each well-developed longitudinal band lies along an ambulacrum, on the coelomic side of a radial canal. All muscles are smooth. The dermis itself encloses several canals, all of which are located below the longitudinal muscle bands. From outside to inside, the canals are the epineural canal and its ectoneural radial nerve, the hyponeural radial canal and its hyponeural radial nerve, and the radial canal of the WVS.

Large sea cucumbers are an important food in Asian cuisine. Japanese and Korean cucumber fisheries harvest over 13,000 metric tons per year of *Stichopus japonicus,* which is also aquacultured for human consumption. Another 625 metric tons per year are provided by species of *Parastichopus* from the United States and Canada. The five muscle bands are removed, combined with other ingredients, cooked, and consumed. Gutted and dried species of *Holothuria* and *Actinopyga,* which resemble giant insect pupae, are rehydrated, cooked, and eaten, partly as an aphrodisiac, in China and Malaysia. These dried cucumbers are known locally as *trepang* or *bêche-de-mer.*

Locomotion and Lifestyle

A cylindrical body, flexible body wall, antagonistic muscle layers, and a large, fluid-filled hydrostat (perivisceral coelom) are all reminiscent of other soft-bodied animals, such as sea anemones and worms. Like worms, some holothuroids burrow using peristalsis, and like anemones, some expand and contract the body, change the body volume, or rhythmically flex it and swim. Cucumbers with a sole, such as *Stichopus,* creep on their tube feet in the manner of a sea star and, if dislodged, right themselves by twisting the oral end until the tube feet touch the substratum. Some tiny species, such as *Synaptula* or *Leptosynapta,* drag themselves along with their buccal podia. All of these movements, however, are slow, and sea cucumbers in general are sluggish animals. Many, in fact, are nearly sessile, particularly those that live in rock crevices *(Cucumaria, Holothuria),* on hard surfaces *(Psolus),* or in semipermanent burrows (*Sclerodactyla, Leptosynapta,* molpadiids; Fig. 28-48).

Swimmers occur among approximately one-half of the species of the deep-sea Elasipodida (Fig. 28-49B,C). The remaining species are benthic and have greatly enlarged tube feet on which they walk over the seabed. The flat sole and downturned buccal podia and mouth convey a bilateral symmetry on the body (Fig. 28-49A). Pelagic species, most of which are facultative swimmers, have papillae webbed together in various ways to

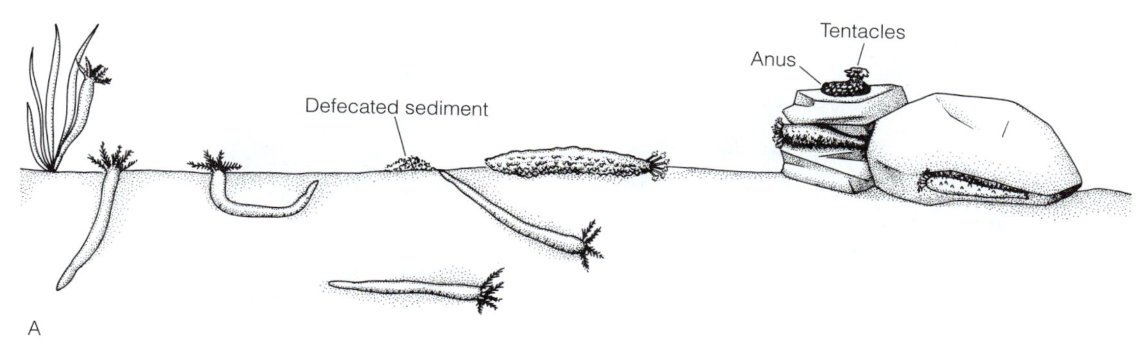

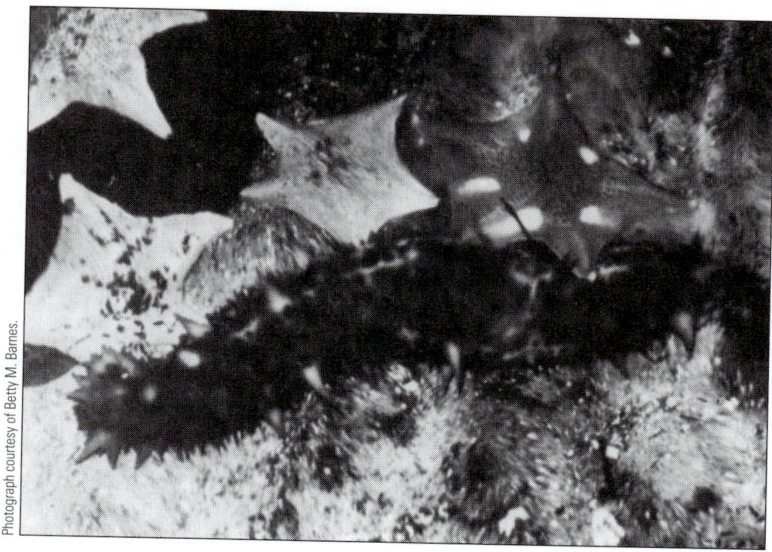

FIGURE 28-48 Holothuroidea: habitats and lifestyles. Animals are not drawn to scale. The genera named below have common species that illustrate a particular lifestyle. **A,** Inhabitants of rock crevices *(Cucumaria, Holothuria);* inhabitants of the undersurfaces of large stones *(Holothuria, Euapta);* rock and sediment-surface dwellers (*Stichopus, Holothuria,* some apodidans); inhabitants of marine-plant surfaces *(Synaptula);* burrowers that dig two openings to the sediment surface *(Thyone, Sclerodactyla, Thyonella, Cucumaria, Echinocucumis);* burrowers with tentacles that project to *(Synapta)* or just below the surface *(Leptosynapta);* burrowers with the anus projecting to the surface *(Leptopentacta, Caudina).* **B,** The north Pacific *Stichopus californicus* in a tank with bat stars, *Patiria miniata,* both from the west coast of the United States. Species of the widely distributed genus *Stichopus* are large sea cucumbers (over 36 cm) with tough body walls.

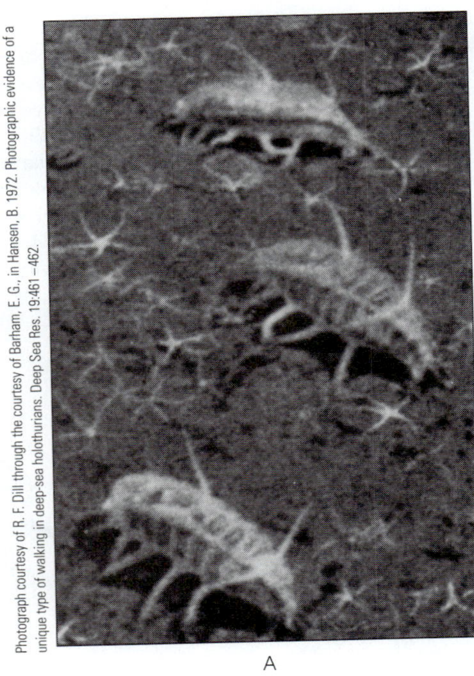

Photograph courtesy of R. F. Dill through the courtesy of Barham, E. G., in Hansen, B. 1972. Photographic evidence of a unique type of walking in deep-sea holothurians. Deep Sea Res. 19:461–462.

A

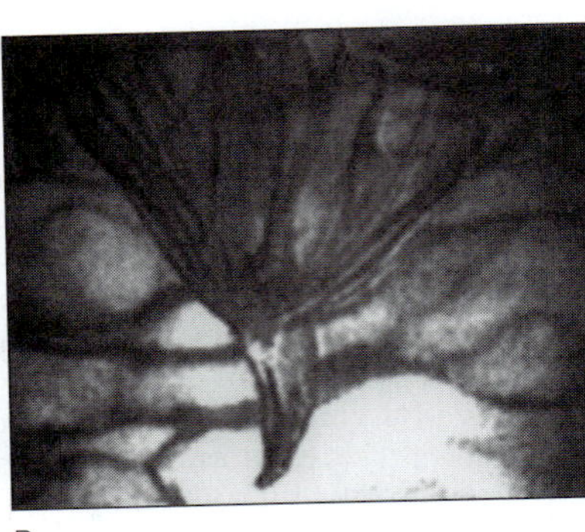

From Miller, J. E., and Pawson, D. L. 1990. Swimming sea cucumbers (Echinodermata: Holothuroidea): A survey, with analysis of swimming behavior in four bathyal species. Smithson. Contr. Mar. Sci. 35:1–18.

B

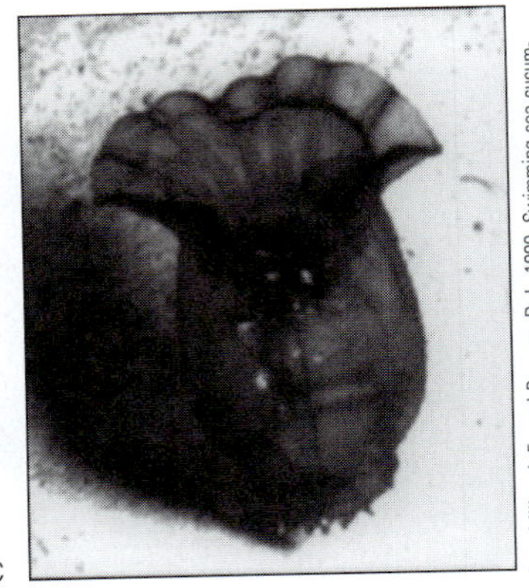

From Miller, J. E., and Pawson, D. L. 1990. Swimming sea cucumbers (Echinodermata: Holothuroidea): A survey, with analysis of swimming behavior in four bathyal species. Smithson. Contr. Mar. Sci. 35:1–18.

C

FIGURE 28-49 Holothuroidea: diversity (Elasipoda). **A,** Three specimens of *Scotoplanes* crawling over the bottom in the San Diego trough (1060 m deep). The leglike structures are tube feet. **B,** *Pelagothuria natatrix,* a pelagic species having a circlet of long, webbed papillae behind the mouth and buccal podia. This species alternately hangs suspended in the water and swims using a rearward thrust of the web. **C,** The benthic species *Enypniastes eximia* leaving the bottom on a brief swimming excursion. This species and *Pelagothuria* are medusa-like in behavior and in the gelatinous consistency of their bodies.

form fins, sails, or medusa-like bells. The transparent bathypelagic *Peniagone diaphana* lives up to 70 m off the bottom, holding its body in a vertical position, buccal podia upward.

Digestive System and Nutrition

The digestive system is basically a long tubular gut, often coiled, with mouth and anus at opposite ends of the body. The mouth is in the center of a buccal membrane at the base of the buccal podia. The mouth opens into a usually muscular pharynx, which is surrounded anteriorly by a **calcareous ring** of 10 ossicles (Fig. 28-50). The calcareous ring provides support for the pharynx and the ring canal, an attachment site for the longitudinal muscles of the body wall and, when present, the retractor muscles of the buccal podia. In most species, the buccal podia and mouth can be retracted completely into the trunk when the animal is disturbed. Protrusion results from circular-muscle contraction and the resulting elevation of the coelomic fluid pressure.

The pharynx opens into a short, inconspicuous esophagus and then into a stomach (Fig. 28-50, 28-51). The stomach is absent in many holothuroids, but when present, it is muscular and lined with a thick cuticle. The remainder of the gut is the long endodermal intestine, which is thrown into a single major S-shaped bend before joining the anus. From the stomach, the **anterior intestine** extends posteriorly as the descending anterior intestine, turns 180° anteriorly and continues as the ascending anterior intestine, which then turns 180° posteriorly to become the **posterior intestine**, which joins the cloaca and anus. The **cloaca** (= sewer) is an expanded rectumlike chamber preceding the anus. It receives not only feces from the posterior intestine, but also the ventilatory flow from the respiratory trees, described below. Cucumbers that lack respiratory trees also lack a cloaca and have a conventional rectum instead. The cloaca is suspended in the perivisceral coelom by radiating **cloacal dilator muscles,** which originate on the body wall and insert on the outer wall of the cloaca (Fig. 28-50). The role of these muscles is described below under Gas Exchange, Internal Transport, and Excretion. The wall of the cloaca is perforated by minute, valved **cloacal ducts** that connect the cavity of the cloaca with the perivisceral coelom (Fig. 28-50). Their significance is discussed under WVS and Other Coeloms. Like the pharynx, esophagus, and stomach, the cloaca is derived from ectoderm and is lined with a cuticle.

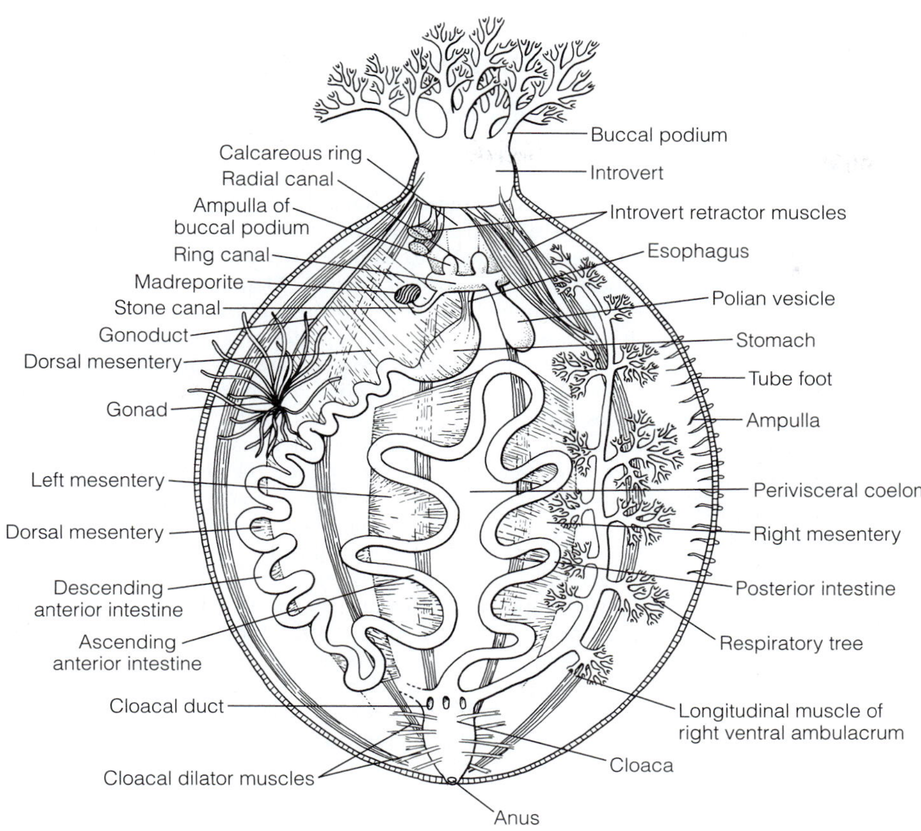

FIGURE 28-50 Holothuroidea: internal anatomy of *Sclerodactyla briareus* (= *Thyone briareus*), a common sea cucumber that inhabits North Atlantic coastal waters. *(Modified and redrawn after Coe from Hyman, L. H. 1955. The Invertebrates. Vol. IV. McGraw-Hill Book Co., New York.)*

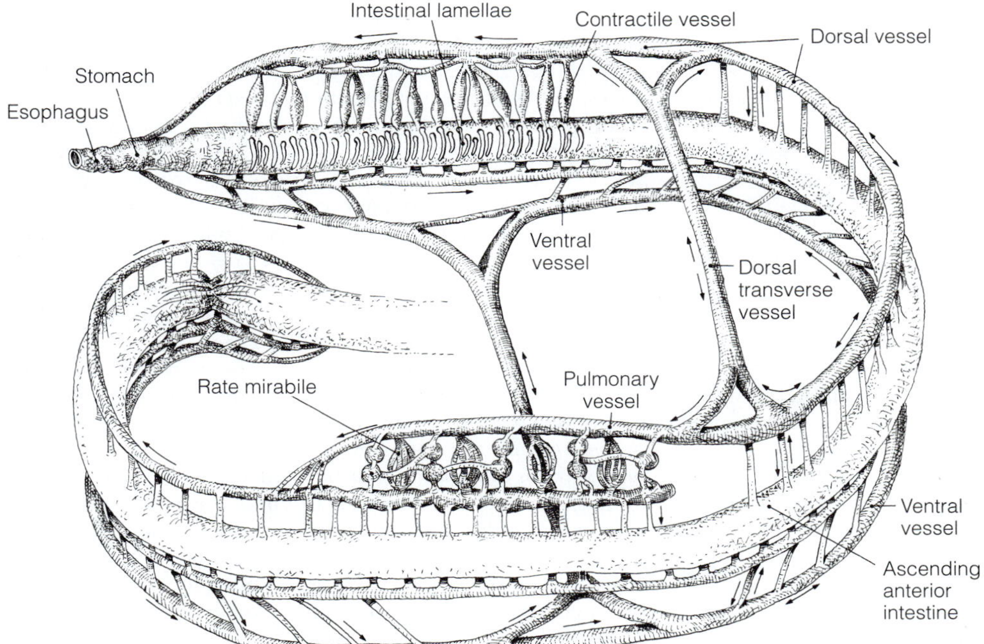

FIGURE 28-51 Holothuroidea: hemal system. The hemal system of *Isostichopus badionotus.* Arrows indicate direction of blood flow. *(Modified from Herreid, C. F., La Russa, V. F., and DeFesi, C. R. 1976. Blood vascular system of the sea cucumber,* Stichopus moebii. *J. Morphol. 150:423–451.)*

The pharynx and esophagus secrete mucus and function chiefly in transporting food to the stomach. If present, the gizzardlike stomach triturates the food; if absent, the pharynx may perform this function. Endocytosis and intracellular digestion occur in the intestine, and extracellular digestion is suspected but unconfirmed. Feces are formed in the cloaca or rectum.

In general, sea cucumbers deposit or suspension feed using their extended buccal podia. Once the podia have gathered food, they are stuffed into the pharynx, one at a time, and the adhering food particles are wiped off as the podia are withdrawn from the mouth. Suspension feeders have branched (dendritic) buccal podia and are represented by many sedentary burrowers and crevice dwellers, including species of *Sclerodactyla, Thyone, Eupentacta,* and *Cucumaria.* Mobile deposit feeders are generally epibenthic animals. Most have soles and stout buccal podia with broad moplike ends (peltate podia), such as *Stichopus, Holothuria, Actinopyga,* and the deep-sea elasipods. Sedentary infaunal deposit feeders lack tube feet and are represented by two groups. One taxon (Apodida) has pinnately branched buccal podia. Species of *Leptosynapta,* for example, dig and occupy ventilated U-shaped burrows, ingesting sand as they slump into the head end of the burrow. The other taxon (Molpadiida), which includes *Molpadia* and *Caudina,* burrow in mud head down, with the anus at the surface. They shovel mud into their mouths with the buccal podia. Deposit-feeding sea cucumbers significantly churn benthic sediments. The sand-dwelling *Holothuria arenicola,* for example, dredges 47 kg/m^2/year.

WVS and Other Coeloms

Although the pentamerous WVS of holothuroids is similar to that of other echinoderms, in most species the madreporite is unique in having lost its connection with the body surface. Instead, it is suspended in the perivisceral coelom at the end of a short stone canal (Fig. 28-50). This location means that perivisceral coelomic fluid, rather than seawater, enters the WVS across the madreporite. In some species, several stone canals and madreporites are attached to the ring canal.

If water loss from the WVS is replenished by perivisceral coelomic fluid that enters an internal madreporite, then perivisceral coelomic fluid apparently is replaced by seawater that enters the anus and cloaca and then flows through the ciliated cloacal ducts into the perivisceral coelom. Under certain conditions, as when the body contracts strongly, coelomic fluid also is expelled from the cloacal ducts and anus, to be replenished later when the body relaxes and extends.

The ring canal surrounds the pharynx immediately behind the calcareous ring and bears one or more polian vesicles (Fig. 28-50). The five radial canals give off canals to the buccal podia before turning posteriorly to supply the ampullae and tube feet. Each radial canal lies in the dermis below a longitudinal muscle band. Each buccal podium and tube foot typically has its own ampulla (Fig. 28-50), although when the tube feet are reduced or absent, there is a corresponding reduction or loss of their ampullae. In the Apodida, which lack tube feet, the WVS is limited to the ring canal, polian vesicles, and buccal podia (tentacles). Molpadiids also lack tube feet, but have radial canals as well as other components of the WVS. Tiedemann's bodies are absent, but they are functionally replaced by specialized blind funnels attached to the wall of the perivisceral coelom in *Leptosynapta* and relatives.

Gas Exchange, Internal Transport, and Excretion

Buccal podia and tube feet function as gills, but the chief gas-exchange organs of holothuroids are a bilateral pair of internal **respiratory trees,** or water lungs (Fig. 28-50). Each tree arises as a diverticulum from the wall of the cloaca and branches repeatedly to form a system of hollow, blind-ended tubes in the perivisceral coelom. Mammalian lungs are also highly branched diverticula of the gut, but they are situated at the gut's anterior, rather than posterior, end.

The respiratory trees are ventilated by muscular pumping of the cloaca and by contraction of the respiratory trees themselves. Contraction of the cloacal dilator muscles dilates the cloaca, causing an inrush of water through the anus (Fig. 28-50). The anal sphincter then closes, the cloaca contracts, and water is forced into the respiratory trees. Contraction of the trees expels water from the system. Typically, several short cloacal inhalations are required to fill the trees and one prolonged exhalation empties them.

The respiratory trees are the principal gas-exchange surfaces of large-bodied species with thick body walls. As long as the anus is exposed to oxygenated seawater, these animals can occupy habitats deficient in oxygen, such as burrows or crevices. The trees also allow cucumbers to achieve a large body size without resorting to extensive, vulnerable gills on the body surface. The slender tropical pearlfish (*Carapus* spp.), which is about 15 cm long, makes its home in the respiratory tree trunk of certain sea cucumbers, such as species of *Holothuria, Actinopyga,* and *Isostichopus.* The fish leaves the host at night to search for food and then returns to the security of the respiratory tree as the cucumber dilates its anus to inhale. Up to 15 pearlfish have been found in a single large cucumber. Negative effects on the host have not yet been detected.

Pelagic and benthic deepwater Elasipodida and burrowing, thin-walled Apodida lack respiratory trees. They obtain oxygen through the general body surface. The tube feet of elasipods are especially important as sites of gas exchange.

In holothuroids, both the coelomic cavities and hemal system are well developed and used for internal transport. The coelomic cavities include the spacious perivisceral coelom, the WVS, and the hyponeural coelomic canals associated with the hyponeural nerves. Coelomocytes abound in the coelomic cavities, especially in the perivisceral coelom and WVS. The hemal system is better developed than it is in any other group of echinoderms, but a heart is absent and the vessels themselves are contractile. Hemoglobin-containing (red) cells can occur in the WVS, perivisceral coelom, hemal system, and even the gut wall of some species, especially in large-bodied, sedentary burrowers or crevice dwellers. Species of *Sclerodactyla, Cucumaria, Eupentacta,* and the molpadiids, for example, have red cells in one or more internal compartments. Fluid flow is bidirectional in the coeloms and also to some extent in the hemal vessels.

Holothuroids, especially the Aspidochirotida *(Holothuria, Cucumaria, Stichopus),* have the best-developed hemal system of any echinoderm (Fig. 28-50). As in other echinoderms, the hyponeural hemal ring and radial hemal vessels parallel the ring canal and radial canals of the WVS. The conspicuous parts of the hemal system, however, are a dorsal and a ventral vessel that accompany the intestine. Branches from the dorsal vessel supply the intestinal wall with numerous smaller vessels. In the loop formed by descending and ascending sections of the small intestine, the dorsal vessel produces an extensive capillary bed, called the **rete mirabile** (REE-tee mir-ABA-lee; = wondrous blood network; Fig. 28-51). The rete is a portal system that links the dorsal vessel of the descending intestine with the dorsal vessel of the ascending intestine with capillaries in the tree. The left respiratory tree is tightly intermingled with the wondrous blood network. The function of the rete is unknown, but all portal systems are specialized for exchange and the anatomy of the rete suggests physiological exchange occurs between rete, gut, and respiratory tree. Perhaps oxygen is supplied to the gut wall via the tree and rete, or perhaps metabolic wastes are released from the rete to the tree, or perhaps both. Adult sea cucumbers lack a heart, axial hemal vessel, and axial canal. Excretion occurs at the level of tissues and cells, but details of the processes are incomplete. Ammonia probably diffuses across the body wall and respiratory trees. Particulate waste, as well as nitrogenous material in crystalline form, is carried by coelomocytes from various parts of the body to the gonadal tubules, the respiratory trees, and the intestine. Wastes then leave the body through these organs.

Nervous System and Sense Organs

The anatomy of the holothuroid nervous system is similar to that of ophiuroids and echinoids. The circumoral nerve ring, which lies in the buccal membrane near the base of the tentacles, supplies nerves to the tentacles and also to the pharynx. The five radial nerves each consist of an ectoneural nerve in an epineural radial canal and a hyponeural nerve in a hyponeural radial canal.

The burrowing members of Apodida, which tend to keep the oral end directed downward, have one statocyst adjacent to each radial nerve near the point where the nerve leaves the calcareous ring. Some apodidans have an eyespot at the base of each tentacle.

Cuvierian Tubules, Evisceration, and Regeneration

As large, fleshy, and often exposed animals, sea cucumbers would seem to be ideal prey for foraging fishes and crustaceans. Although unarmored, some cucumbers defend themselves with chemical toxins, with the functional equivalent of flypaper (Cuvierian tubules), or perhaps by the forceful ejection of all or part of their viscera (evisceration).

A few cucumbers, species of *Holothuria* and *Actinopyga*, eject sticky tubules called **Cuvierian tubules** (koo-VARE-ian) from the anus and aim them at a would-be predator. After discharge the tubules are regenerated. The undischarged tubules are attached to the base of one (frequently the left) or both respiratory trees or to the common trunk of the two trees (Fig. 28-52). When a tubule-equipped cucumber is irritated or attacked by a predator, the anus is directed toward the intruder, the body wall contracts, and the detached tubules shoot out of the anus. Once ejected, the liberated tubules, which can extend to 20 times their original length, become as sticky as spider's silk and quickly foul their target. Small crabs and lobsters are entangled and immobilized, dying slowly as the indifferent cucumber abandons them to their fate. The tubules are not only sticky, but also toxic because of a substance called **holothurin** (a saponin) that also occurs in the

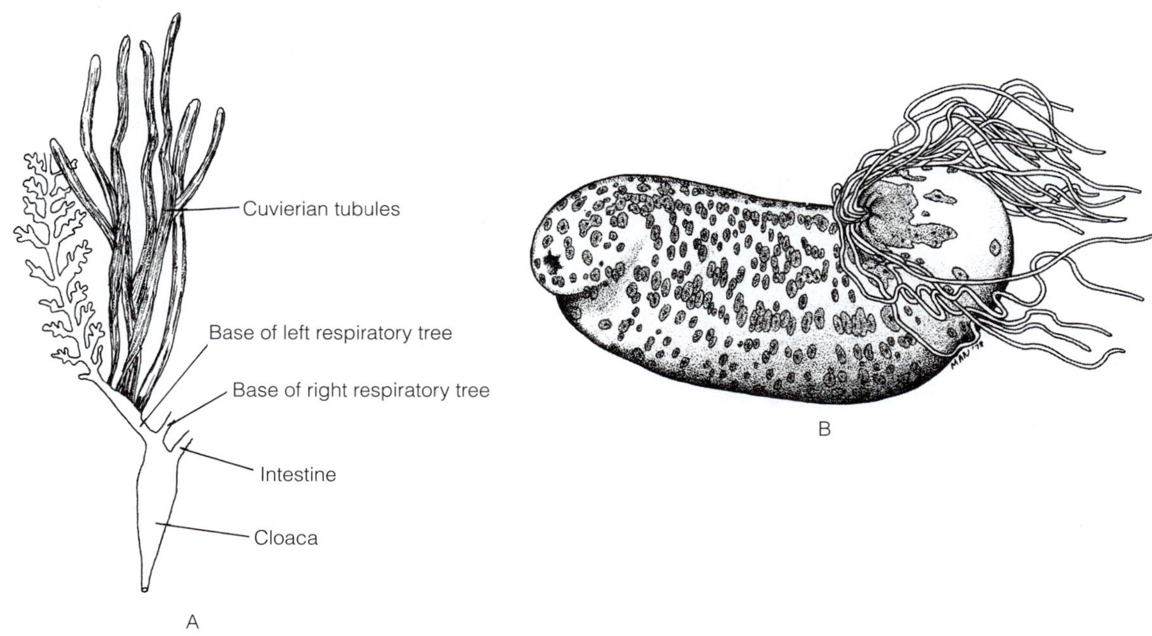

FIGURE 28-52 Holothuroidea: Cuvierian tubules. **A,** The base of a respiratory tree of *Holothuria impatiens,* showing the Cuvierian tubules. **B,** A specimen of *Holothuria* releasing tubules. *(A, After Russo from Hyman, L. H. 1955. The Invertebrates. Vol. IV. McGraw-Hill Book Co., New York; B, Drawn from a photograph by Isobel Bennett.)*

body wall of some species. Being aware of their toxicity, South Pacific islanders have long used the macerated bodies of certain sea cucumbers to catch tide-pool fish.

Sometimes confused with the discharge of the Cuvierian tubules is a more common phenomenon called **evisceration** that occurs as a stress response in many holothuroids. Depending on taxon, the anterior or posterior end ruptures and parts of the gut and associated organs are forcibly expelled. Eviscerated individuals in the process of regeneration have been reported from natural habitats during certain times of the year, and it is a normal seasonal phenomenon in some species, perhaps initiating a period of inactivity when food supply is low or serving to eliminate wastes stored in internal tissues. Evisceration is later followed by regeneration of the lost parts.

Reproduction

Holothuroids differ from all other Recent echinoderms in having a single gonad, but that gonad typically branches into many tubules that, when ripe, may occupy much of the perivisceral coelom. Most cucumbers are gonochoric, and the gonad is attached anteriorly in a mesentery beneath the mid-dorsal interambulacrum (Fig. 28-50). The mid-dorsal gonopore opens between the bases of two buccal podia or just behind the introvert (Fig. 28-45).

Most sea cucumbers spawn their gametes and fertilization occurs in seawater, but some 30 primarily cold-water species brood their young. During spawning, brooders catch the eggs with their buccal podia and transfer them to the sole or to the dorsal body surface for incubation. Even more remarkable is viviparity, which takes place in the Californian *Thyone rubra,* in *Leptosynapta* from the North Sea, in *Synaptula hydriformis* from Florida and the Caribbean, and in a few species from other parts of the world. The eggs pass from the gonads into the coelom and are fertilized in an unknown manner. Development takes place within the coelom and the young leave the body of the mother through ruptures in the body wall.

Development and Metamorphosis

Except in brooding and viviparous species, development of the planktonic embryo occurs externally in the seawater. Development through gastrulation is like that of asteroids. The anterior half of the archenteron separates as a coelomic cavity, leaving a shorter, posterior part to become the gut. The right protomesocoel (axohydrocoel) never forms.

Later in development the embryo differentiates into a planktotrophic larva called an **auricularia** (Fig. 28-53A). The auricularia is similar to the asteroid bipinnaria. Both lack the arm ossicles of pluteus larvae and have convoluted ciliary bands. The preoral ciliary band of the auricularia, however, is continuous with bands elsewhere on the body, whereas the ventral preoral loop of the bipinnaria is usually separate. The coelomic cavities of the auricularia are well developed on the left side only, whereas those of asteroid larvae are nearly bilaterally symmetric. The auricularia larva is eventually succeeded by a barrel-shaped stage called a **doliolaria,** in which the original ciliated band breaks up into three to five ciliated hoops that encircle the body (Fig. 28-53B). Metamorphosis begins in the doliolaria stage. Indirect development with a planktotrophic auricularia stage probably is primitive in holothuroids, but most species have large, yolky eggs that develop directly into lecithotrophic doliolaria-like larvae.

Holothuroid larvae do not form a juvenile rudiment and larval metamorphosis gradually transforms the doliolaria into a juvenile sea cucumber. During the transformation, the larval body axis is retained and five buccal podia differentiate quickly, before the tube feet. At this stage, the juvenile cucumber is called a **pentactula.** Eventually, the pentactula settles to the bottom and assumes the adult mode of existence. The life span of many sea cucumbers is between 5 and 10 years.

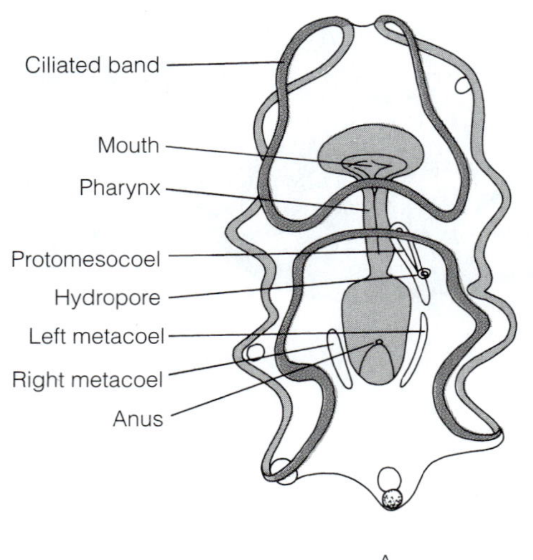

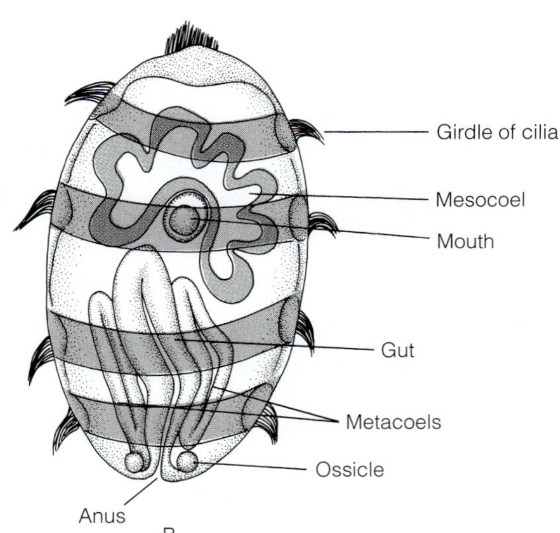

FIGURE 28-53 Holothuroidea: larvae. **A,** An auricularia larva (ventral view). **B,** The doliolaria larva of *Leptosynapta inhaerens,* a common North Atlantic holothuroid (ventral view). *(A, After Mortensen from Hyman, L. H. 1955. The Invertebrates. Vol. IV. McGraw-Hill Book Co., New York; B, After Runnström from Cuénot, L. 1948. Échinodermes. In Grassé, P. (Ed.): Traité de Zoologie, Vol. XI. Masson et Cie, Paris. Reprinted by permission.)*

Diversity of Holothuroidea

Dactylochirotida^O: Have a U-shaped body enclosed in flexible test and simple, unbranched buccal podia. *Echinocucumis, Sphaerothuria.*

Dendrochirotida^O: Have a cucumber, banana, or U-shaped body; branched (dendritic) buccal podia lack ampullae; usually have a retractile introvert. *Cucumaria, Psolus, Sclerodactyla, Thyone, Thyonella.*

Aspidochirotida^O: Have bilateral symmetry, a creeping sole, and peltate (shield-shaped) buccal podia; deposit-feeders. *Actinopyga, Holothuria, Isostichopus, Stichopus.*

Elasipodida^O: Deepwater or pelagic with gelatinous bodies adapted for swimming. Have peltate buccal podia; lack respiratory trees; ossicles are reduced or absent. *Enypniastes, Pelagothuria, Peniagone, Scotoplanes.*

Apodida^O: Wormlike, small-bodied (except *Euapta*) animals with a thin body wall; buccal podia are pinnate or fingerlike; lack respiratory trees, radial canals, tube feet; anchor and anchor-plate ossicles function like Velcro hooks. *Euapta, Labidoplax, Leptosynapta, Synapta, Synaptula.*

Molpadiida^O: Includes head-down mud burrowers with digitate buccal podia, respiratory trees, and radial canals, but tube feet are absent. *Molpadia, Caudina.*

CRINOIDEA^C

The 700 species of crinoids are represented by 100 species of sea lilies and 600 species of colorful feather stars. There are many more extinct species known only from fossils. All are radially symmetric, sessile or semisessile suspension feeders that, unlike other echinoderms, live with the mouth oriented upward. Crinoids collect food with the tube feet on their long, often branched arms. The arms constitute most of the crinoid body, with the disc being small and often inconspicuous. A long stalk attaches sea lilies to the substratum, but the stalk tissues are primarily skeletal and may, like those of the disc, demand only a modest amount of metabolic energy. The sedentary habits or slow pace of crinoids, their heavily ossified, jointed appendages, and their floral appearance confer on them a primordial elegance that, by good fortune (theirs and ours), has been retained in the present, especially by the deepwater sea lilies. Feather stars occur in shallow water and are common and spectacular members of coral-reef communities.

Of the living taxa of echinoderms, crinoids have the longest fossil record, extending back to the early Cambrian, some 570 mya. This is not to say that living crinoids embody primitive traits only, although pentamerous symmetry and a sessile suspension-feeding habit are good candidates for plesiomorphic echinoderm characters.

FORM

The crinoid body consists of an attachment **stalk** that supports a pentamerous **crown** (Fig. 28-54A), which includes the disc and arms. The well-developed stalk of sea lilies is lost during the post-larval development of the free-moving feather stars (Fig. 28-54B). The conspicuously jointed stalk contains a column of well-developed skeletal ossicles. The sea lily stalk may reach almost 1 m in length, but it is usually much shorter. (Some fossil species have 20 m stalks.) The stalk attaches via a flattened attachment plate or by jointed rootlike projections called cirri that are arranged in whorls (Fig. 28-55A). Although the stalk is lost in feather stars (comatulids), the most proximal cirri remain to form one or more circles around the base of the crown (Fig. 28-54B). Comatulid cirri resemble a bird's foot and are used for grasping tightly to a perch (Fig. 28-54B, 28-55D). They are long and slender in crinoids that rest on soft bottoms and stout and curved in species that cling to rocks, seaweed, and other objects.

The pentamerous crown is equivalent to the body of other echinoderms and, like that of asteroids and ophiuroids, it is drawn out into arms. The stalk attaches to the aboral surface of the crown and thus the oral surface is directed upward, in contrast to other living echinoderms. The crinoid disc is composed of two parts, the aboral **calyx,** which is a heavily calcified cup that houses the viscera, and the membranous oral wall of the disc called the **tegmen,** which is weakly calcified and covers the calyx cup like a drumhead (Fig. 28-54, 28-55B). The mouth is at or near the center or near the margin of the tegmen, according to species. Five ambulacral grooves radiate from the mouth, across the tegmen, and onto the arms. Like the mouth, the anus opens on the oral surface, but in an interambulacrum. Often, the anus is at the summit of a prominence called the **anal cone,** which elevates it above the feeding current and helps to prevent the fouling of food with feces (Fig. 28-55B).

The arms radiate from the perimeter of the crown and have a jointed appearance similar to that of the stalk. Although some primitive species have only 5 arms (Fig. 28-54A), the arm bases of most crinoids bifurcate to form a total of 10 arms (Fig. 28-55D). Subsequent branching may produce additional arms, such as some species of comatulids that have 80 to 200 arms. The arms are usually less than 10 cm in length but may reach almost 35 cm in certain species.

Each arm bears a bilateral series of pinnately arranged, jointed appendages called **pinnules** (Fig. 28-54, 28-55C). An arm and its pinnules resemble a feather and are responsible for the name *feather star,* although both feather stars and sea lilies have featherlike arms (as do pterobranch hemichordates). The five ambulacral grooves on the tegmen bifurcate and extend onto the oral surface of the arms and pinnules. The groove margins bear movable flaps (lappets) that can expose or cover the groove (Fig. 28-55C). On the inner side of each lappet are three tube feet united at their bases into a triplet (Fig. 28-56B).

BODY WALL

The crinoid body is enclosed by a cuticle and epidermis, the latter of which is nonciliated except in the ambulacral grooves. Below the epidermis is a connective-tissue layer that does not form a distinct dermis; rather, it is continuous with the deeper connective tissues of the body. Skeletal ossicles occur throughout the connective tissue. The stalk, cirri, arms, and pinnules are nearly solid, with their interiors being occupied almost entirely by a series of thick, vertebra-like ossicles, which accounts for the jointed appearance of these appendages (Fig. 28-54). The articulating surfaces of the arm ossicles (brachials) permit limited movement, similar to the ossicles of ophiuroid arms. The ossicles of the stalk

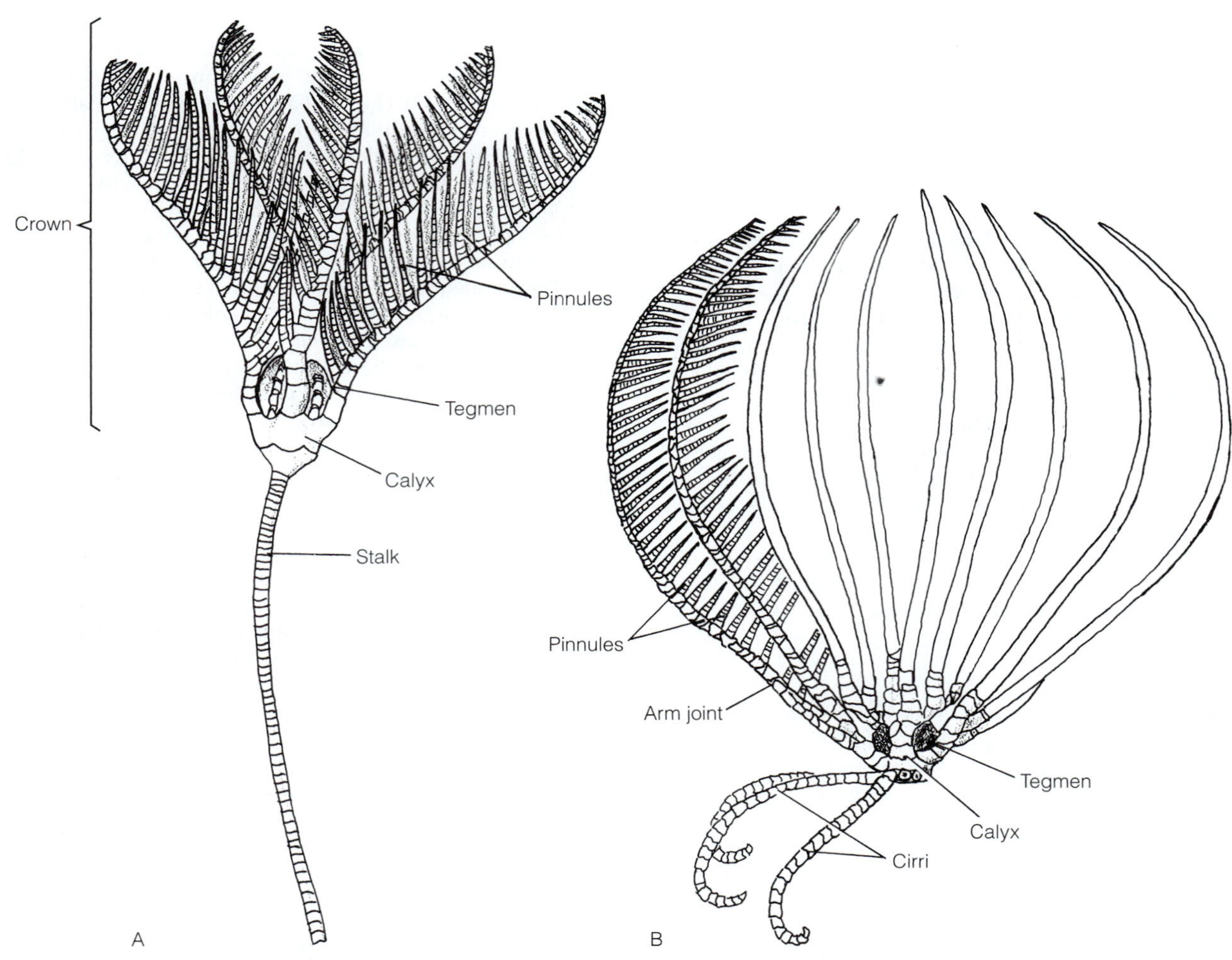

FIGURE 28-54 Crinoidea: external form. **A,** *Ptilocrinus pinnatus,* a stalked crinoid (sea lily) with five arms. **B,** A Philippine 30-armed comatulid (feather star), *Neometra acanthaster.* Not all of the arms are shown. *(Both after Clark from Hyman, L. H. 1955. The Invertebrates. Vol. IV. McGraw-Hill Book Co., New York.)*

(columnals) are more securely interlocked than are those of the arms, but even here some bending is possible. The ossicles of the stalk, cirri, and arms are bound together by collagen fibers called **ligaments** that penetrate into the porous skeletal material. As in other echinoderms, these ligaments can rapidly change from a soft to a rigid state, permitting the animal to roll up the arms using the flexor muscles or lock them in an extended feeding position.

The interior of the crinoid disc houses the gut as well as coelomic and hemal channels. Between these compartments are calcified mesenteries, a unique attribute of Crinoidea.

MUSCULATURE AND LOCOMOTION

Distinct muscles are confined to the arms and pinnules, although epitheliomuscular cells line parts of the various coelomic spaces and individual muscle fibers extend into the connective tissue of the body. Obliquely striated **flexor muscles** that extend between the oral edges of the arm and pinnule ossicles (brachials; Fig. 28-59B) flex (curl) the arms and pinnules. Extension of these appendages apparently results from elastic recoil of interbrachial ligaments.

The sea lilies are mostly limited to bending movements of the stalk and flexion and extension of the arms and pinnules, but a few can crawl slowly over the seabed. The stalkless comatulids, however, move freely and are capable of both swimming and crawling. The oral surface is always directed upward during locomotion unless the animal is accidentally overturned. The righting reflex apparently is controlled by the thigmotactic cirri.

Feather stars swim by alternately raising and lowering their arms. In 10-armed species, every other arm sweeps downward while the remaining arms move upward. In species with more than 10 arms, the arms still move in sets of 5, but sequentially. To crawl, the animal lifts its body from the substratum and moves around on its downward-flexed arms. The animal often uses the arms and pinnules, which have minute terminal hooks, to grasp and pull itself over irregular and vertical surfaces.

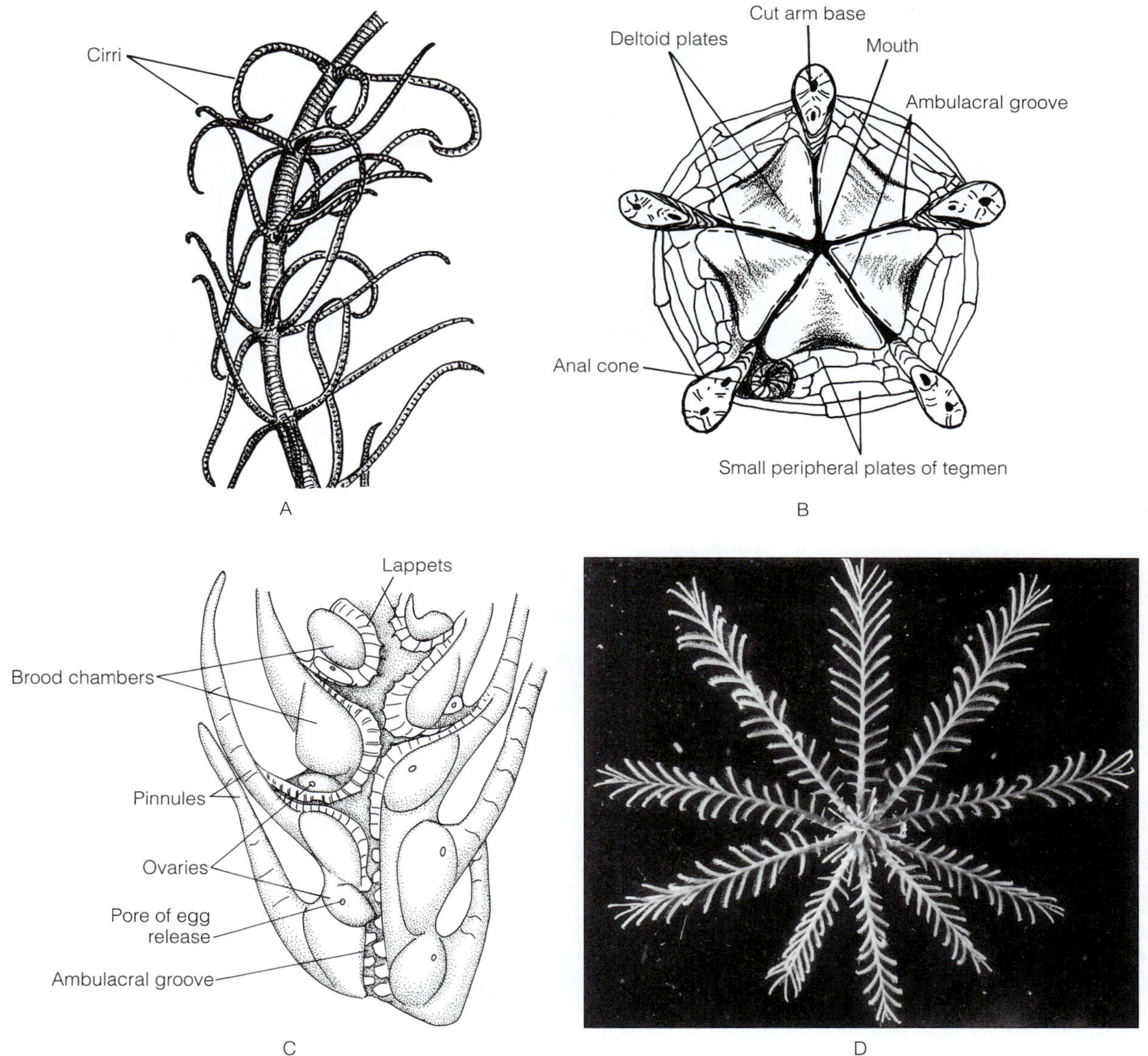

FIGURE 28-55 Crinoidea: external form. **A,** Part of the stalk of the West Indian crinoid *Cenocrinus asteria,* showing whorls of cirri. **B,** The tegmen of *Hyocrinus,* a stalked crinoid (oral view). **C,** Part of an arm of *Notocrinus virile,* a comatulid feather star (oral view). Tube feet are not shown. **D,** Photo of a living *Comactinia echinoptera,* a feather star with 10 arms. *(A and B, After Carpenter from Hyman, L. H. 1955. The Invertebrates. Vol. IV. McGraw-Hill Book Co., New York. Reprinted with permission; C, Modified from Hyman, 1955. The Invertebrates. Vol. IV. McGraw-Hill Book Co., New York. Reprinted with permission.)*

Feather stars swim and crawl for short distances only, and swimming is largely an escape response. Normally, they cling to the bottom for long periods by means of the grasping cirri. Many shallow-water species are nocturnal. Three coral-reef species of *Lamprometra, Capillaster,* and *Comissia* in the Red Sea hide during the day in coralline crevices, keeping their arms tightly rolled (Fig. 28-57A). Stimulated by the lower light intensities at sunset, they crawl upward out of their hiding places to exposed positions, where the arms are extended to feed. Species inhabiting deeper water may be stationary. Many specimens of two species of *Decametra* and *Oligometra* were reported to be clinging in the same position to a gorgonian coral at about 30 m for several months. They roll their arms in response to daytime illumination.

DIGESTIVE SYSTEM AND NUTRITION

Crinoids are passive suspension feeders that rely on ambient currents to deliver plankton and detritus. During feeding the arms and pinnules are held outstretched and the tube feet are erect. The suckerless tube feet resemble tiny tentacles and bear mucus-secreting papillae along their length (Fig. 28-56).

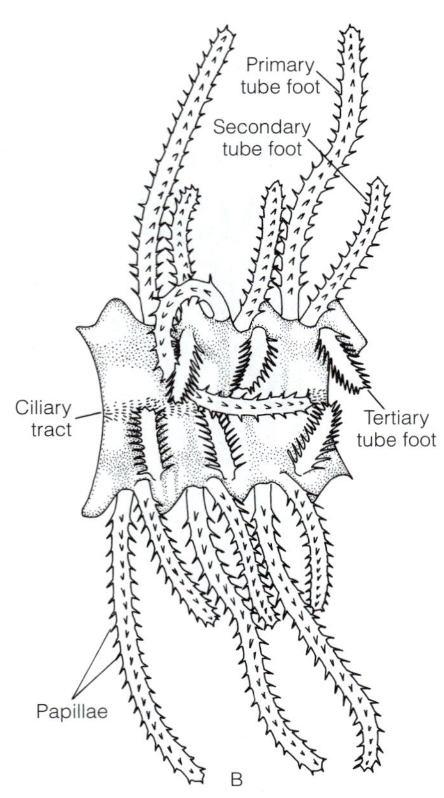

FIGURE 28-56 Crinoidea: functional morphology of tube feet in suspension feeding. **A,** The extended position of the primary tube feet along the pinnules (papillae not visible) of the feather star *Florometra serratissima.* **B,** View of the ambulacral groove along part of a pinnule of *Antedon bifida.* One tube foot is wiping against the ciliary tract of the ambulacral groove and against a tertiary tube foot. *(B, From Lahaye, M. C., and Jangoux, M. 1985. Functional morphology of the podia and ambulacral grooves of the comatulid crinoid* Antedon bifida. *Mar. Biol. 86:307–318.)*

Just as they do elsewhere along the ambulacral grooves, the tube feet on the pinnules occur in a series of triplets. The primary tube foot of each triplet is long, conspicuous, and outstretched, and the combined primary tube feet of a pinnule form a food-catching mesh (Fig. 28-56A). Food particles that adhere to a tube foot are flicked into the mucus-lined ambulacral groove (Fig. 28-56B). Depending on the species, the particle may be transferred directly to the groove, wiped free against the tertiary tube foot, or scraped loose between adjacent lappets (Fig. 28-56B). Cilia transport particles down the groove to the mouth. The lappets function largely as sidewalls for the deep ambulacral groove. The total length of the food-trapping ambulacral surface may be enormous. The Japanese stalked crinoid, *Metacrinus rotundus,* which has 56 arms (plus pinnules), each one 24 cm in length, has a total ambulacral groove length of 80 m.

Feeding crinoids hold their arms in a plane or parabola and orient it perpendicular to the water current (Fig. 28-57C). In these positions, the adjacent pinnules and tube feet form a relatively tight mesh for trapping suspended food. Shallow-water comatulids may also remain in protective crevices and extend the arms outward in several directions (Fig. 28-57B).

The digestive system is confined to the disc. The mouth leads into the short esophagus, which then opens into the intestine (Fig. 28-58). The intestine descends and makes a complete turn around the inner side of the calyx wall. The terminal part then passes upward (orally) into the short rectum, which opens through the anus at the tip of the anal cone.

Little is known about digestive physiology, although there are claims of both extracellular and intracellular digestion in the intestine. Wastes are egested as large, compact, mucus-cemented pellets that fall from the anal cone onto the surface of the disc and then drop off the body.

WVS AND INTERNAL TRANSPORT

The crinoid WVS is similar to that of a generalized echinoderm with two significant differences. One is that crinoids lack a madreporite. Instead, the oral surface bears numerous (often hundreds) of separate **tegmenal pores** that lead, via short integumental canals, directly into the perivisceral coelom (Fig. 28-59A). Second, the ring canal gives rise to numerous stone canals, several in each interradius, which also open into the perivisceral coelom. Presumably, seawater enters the tegmenal pores and perivisceral coelom and then is pumped into the WVS by the stone canals. This arrangement is reminiscent of the cloacal ducts, perivisceral coelom, and internal stone canals of Holothuroidea. From the ring canal, a radial canal bifurcates to enter each of the 10 arms and sends a lateral canal into each pinnule (Fig. 28-59). A small branch from the lateral canal supplies each trio of tube feet. Ampullae are absent and hydraulic pressure is generated to extend the tube feet by contracting the radial canal, which has muscle fibers spanning it.

The perivisceral coelom occupies the disc and extends into the arms and pinnules (Fig. 28-59). It is complexly subdivided

From Fishelson, L. 1974. Ecology of the northern Red Sea crinoids and their epi- and endozoic fauna. Mar. Biol. 26:183–192.

From Fishelson, L. 1974. Ecology of the northern Red Sea crinoids and their epi- and endozoic fauna. Mar. Biol. 26:183–192.

Photograph courtesy of C. Neumann, from Macurda, D. B., and Meyer, D. L. 1976. The identification and interpretation of stalked crinoids from deep-water photographs. Bull. Mar. Sci. 26(2):205–215. Copyright Springer-Verlag.

C

FIGURE 28-57 Crinoidea: arm positions in living crinoids. **A,** Diurnal, nonfeeding, rolled arm position of the feather star *Heterometra.* **B,** The Red Sea feather star, *Commissia,* which inhabits crevices in coral reefs, feeding with its arms extended. **C,** Feeding crinoids in the Strait of Florida at 600 to 700 m. A feather star atop a sponge is holding its arms in the form of a vertical fan. Three inclined sea lilies *(Diplocrinus)* form parabolic fans with the arm tips directed toward the current.

by mesenteries into a network of communicating spaces, most of them called by a special name based on anatomical position or association with a specific organ.

Within the disc, the perivisceral coelom surrounds the gut and a unique structure, the axial organ (Fig. 28-59A). The **axial organ** is a large blood vessel that houses a bundle of blind-ended epithelial tubes of unknown function. In stalked species, the axial organ passes through the center of the stalk and is surrounded by five equal-sized compartments of the perivisceral coelom. Together these are called the **chambered organ.** Associated with the axial organ is a podocyte-covered ultrafiltration site, the **spongy body,** which may be homologous to the eleutherozoan axial hemal vessel (axial gland).

Branches of the perivisceral coelom enter the arms and pinnules. Each arm houses a bilateral pair of oral coelomic cavities and an unpaired aboral cavity (Fig. 29-59B). Where a pinnule joins the arm, it receives a branch from one oral cavity and also from the aboral cavity. The two cavities are united by canals along the length of the pinnule. Ciliary flow

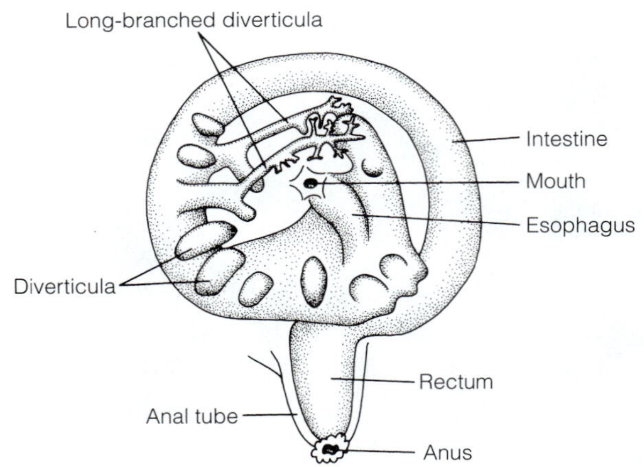

FIGURE 28-58 Crinoidea: digestive tract of the feather star *Antedon bifida*, in oral view. *(After Chadwick from Hyman, L. H. 1955. The Invertebrates. Vol. IV. McGraw-Hill Book Co., New York.)*

is outbound from the disc in the aboral cavity and inbound in the oral cavities. Flow rates of 1 mm/s have been measured. Thus the perivisceral coelom is the principal circulatory system in the disc, arms, and pinnules of crinoids.

The hyponeural coelom is represented by the hyponeural ring and radial canals, both closely associated with the ring and radial canals of the WVS, but they are not well developed and their function is unknown. A genital coelom, which is incompletely separated from the oral and aboral branches of the perivisceral coelom, accompanies the gonad in each of the arms (see Reproduction, below; Fig. 28-59B).

The center of the hemal system is the oral hemal ring, which gives rise to the axial organ, a network of hemal vessels in the coelomic mesenteries, and one or two radial hemal vessels for each arm. A heart is absent, although the axial organ is said to pulsate, and the blood is colorless.

GAS EXCHANGE AND EXCRETION

The tube feet are the principal sites of gas exchange, and the great surface area presented by the branching arms makes unnecessary any other special respiratory surfaces, such as those found in most other echinoderms. No specialized excretory organs of crinoids are known. They probably are ammonotelic.

NERVOUS SYSTEM

The complex crinoid nervous system is composed of three interconnected divisions. The oral **ectoneural system,** which is sensory, is homologous to the system of the same name found in other echinoderms. It consists of an intraepidermal nerve ring around the mouth, and five radial nerves, one for each arm (Fig. 28-59A). The ectoneural system innervates sensory cells of the epidermis and tube feet. The remaining two divisions of the nervous system, the hyponeural and entoneural systems, are motor in function and both may be homologous to the hyponeural system of other echinoderms. Neither, however, is located in the hyponeural coelomic lining, but rather in the connective tissue of the body. The center of the **hyponeural system** is an oral ring that produces radial nerves to the arms and pinnules (Fig. 28-59B). These nerves supply the musculature of the tube feet and other structures. The chief motor system is an **entoneural system,** the center of which is a large, cup-shaped mass in the apex of the calyx (Fig. 28-59A). From this mass, five **brachial nerves** extend to the flexor muscles of the arms and pinnules (Fig. 28-59A,B).

REPRODUCTION

Crinoids regenerate well and, in this respect, are similar to the asteroids and ophiuroids. Part or all of an arm can be cast off if it is seized or subjected to unfavorable environmental conditions. The lost arm is then regenerated. The visceral mass within the calyx can be regenerated in several weeks; such regeneration may be important in surviving fish predation. Clonal reproduction is absent in crinoids.

Crinoids are gonochoric, with gonads in their pinnules (common) or arms (uncommon) (Fig. 28-55C, 28-59B). Not all pinnules are reproductive, only those along the proximal half of the arm. Each gonad consists of a genital tubule (rachis) enclosed in a hemal (blood) sinus, which in turn is surrounded by the genital coelom. When the eggs or sperm are mature, spawning takes place by rupturing the pinnule walls, and the eggs and sperm are shed into the seawater. In *Antedon* and others, the eggs are cemented to the outer surface of the pinnules by the secretion of epidermal gland cells. Hatching takes place at the larval stage. Cold-water crinoids (many Antarctic forms) brood, as other echinoderms do. The brood chambers are saclike invaginations of the arm or the pinnule walls adjacent to the genital canals, and the eggs probably enter the brood chamber by rupturing the wall (Fig. 28-55C).

DEVELOPMENT

Development through the early gastrula stage is similar to that in asteroids and holothuroids. During the formation of the coelomic sacs the embryo elongates and then proceeds to a free-swimming larval stage. The crinoid larva, a nonfeeding doliolaria that resembles the doliolaria of holothuroids, is barrel-shaped and has an anterior apical tuft and four or five hooplike ciliated bands (Fig. 28-60A).

After a free-swimming existence the doliolaria settles to the bottom and attaches by employing a glandular midventral depression (the adhesive pit) located near the apical tuft. There ensues an extended metamorphosis resulting in the formation of a minute, stalked, sessile crinoid. In the comatulids, metamorphosis also results in a stalked sessile stage (the **pentacrinoid**) that resembles a minute sea lily (Fig. 28-60B). The pentacrinoid of *Antedon* is a little over 3 mm long when the arms appear, and it requires about six weeks from the time of attachment as a doliolaria to attain this stage. After up to several months as a pentacrinoid, during which time the cirri are formed, the crown breaks free of the stalk and the young animal assumes the adult, free-living existence.

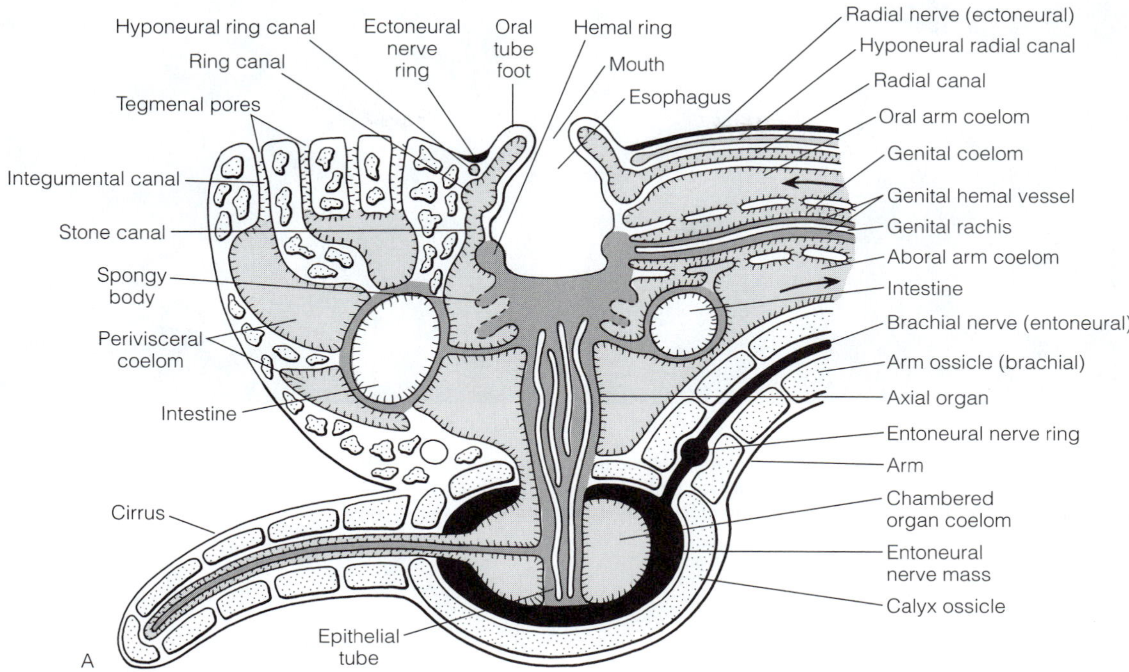

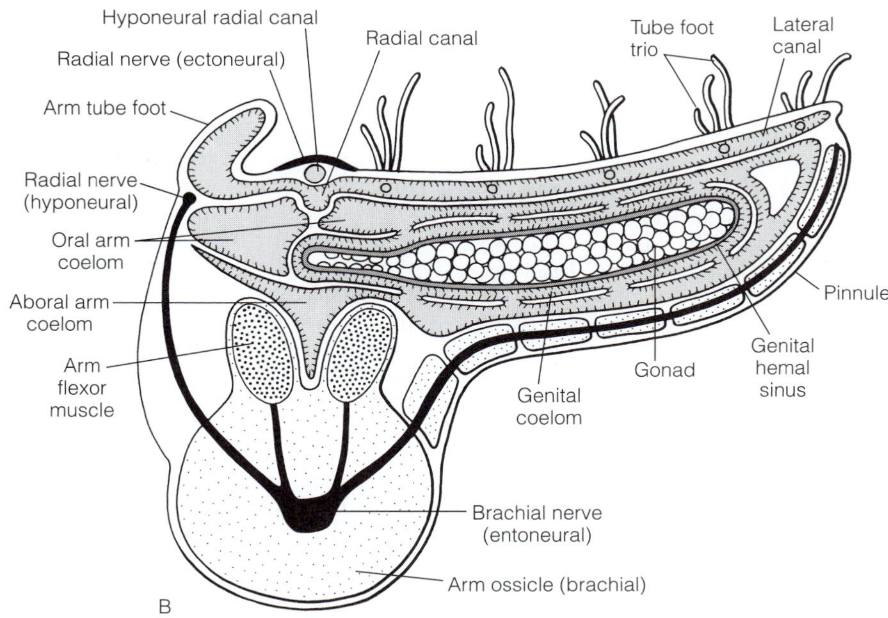

FIGURE 28-59 Crinoidea (feather star): diagrammatic internal anatomy. **A,** Vertical section of the disc, showing the base of one arm (right) and one cirrus (left). **B,** Cross section of an arm and longitudinal section of one pinnule (shortened). *(Based on Lang, A. 1894. Lehrbuch der Vergleichenden. Anatomie der Wirbellosen Thiere. Gustav Fischer Verlag, Jena. 1197 pp.)*

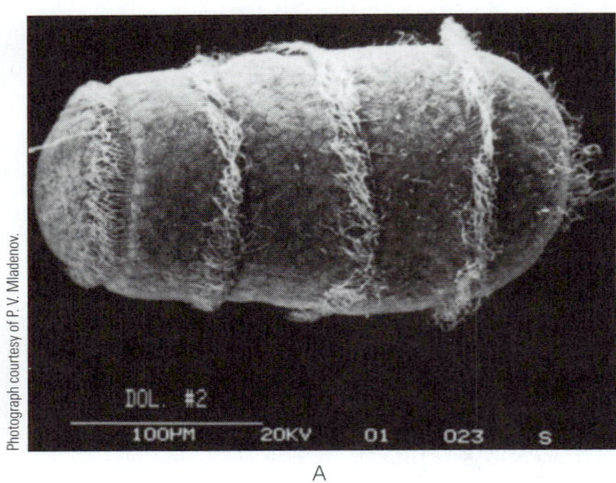

A B

FIGURE 28-60 Crinoidea: larva and juveniles. **A,** Scanning electron micrograph of the doliolaria larva of the feather star *Florometra serratissima*. Four ciliated bands encircle the body and an apical tuft of sensory cilia projects to the right. **B,** A cluster of pentacrinoids attached to a cirrus of an adult *Comactinia echinoptera*.

DIVERSITY OF CRINOIDEA

Isocrinida^O: Deepwater, stalked crinoids with evenly spaced whorls of five cirri each along the stalk; large crown has some 50 arms; can crawl over the bottom using cirri and crown arms. *Diplocrinus, Metacrinus.*

Millericrinida^O: Deepwater animals with a stalk lacking cirri; have a long, conical crown with five arms. *Hyocrinus.*

Cyrtocrinida^O: Deepwater, small-bodied crinoids with a short, irregular stalk that lacks cirri; have 10 arms. Three species of *Holopus.*

Bourgueticrinida^O: Have a slender stalk that ends in an attachment base and lacks cirri. *Conocrinus.*

Comatulida^O: Ranges from shallow water to great depths. Stalk is present only in pentacrinoid stage, absent in adult, which has grasping cirri and can swim with the arms. Six hundred species of *Antedon, Comactinia,* and *Florometra,* among others.

PALEONTOLOGY AND PHYLOGENY OF ECHINODERMATA

At the outset of this chapter, the bilateral symmetry of nonechinoderm deuterostomes and echinoderm larvae was contrasted with the pentamerous symmetry of adult echinoderms. The explanation offered for the evolution of radial symmetry by echinoderm adults was their historical adoption of a sessile suspension-feeding habit. Presumably, once that symmetry change was made and the body was reorganized around a radial design, a shift back to bilateral symmetry was nearly impossible despite the return to a motile lifestyle by most living echinoderms. So far, we have presented little actual evidence in support of this scenario, but we can now examine the fossil record of echinoderms, remaining mindful of two questions: Is there fossil evidence for pre-pentamerous (bilateral), motile echinoderms that might establish a link between Echinodermata and the bilateral deuterostomes? Was there a time in the remote past when most echinoderm taxa were sessile and pentamerous? The fossil record answers "yes" to both questions.

Echinoderms rank with molluscs, brachiopods, and arthropods in having one of the richest and oldest fossil records of any group of animals. Echinoderms first appeared in the early Cambrian period, some 545 mya, and were abundant during the later periods of the Paleozoic era, when several extinct taxa reached their zenith. The echinoderms of Paleozoic seas included not only the 5 Recent taxa, but 15 other taxa of equal rank (classes). Of these 20 taxa, 15 were either suspension feeders or both suspension and deposit feeders, and of these, 11 were pentamerous. Thus suspension feeding and pentamerous symmetry predominated in echinoderms of the phytoplankton-rich Paleozoic ocean.

The earliest echinoderm fossils, from the early to middle Cambrian, were all suspension feeders, or are presumed to have been so because they are chiefly sessile, attached forms with upwardly directed ambulacra and tube feet. These early fossils belong to Crinoidea (the only ancient taxon with living members), Eocrinoidea, Helicoplacoidea, Edrioasteroidea, and four additional taxa included in Homalozoa.

Homalozoans are unique in lacking any trace of pentamerous symmetry, an attribute that sets them apart from all other adult echinoderms. Various species of homalozoans were either bilaterally symmetric or asymmetric and presumably motile, crawling on or even swimming over the seabed. Also unlike other echinoderms, homalozoans may have lacked ambulacra and fed instead by direct uptake through the mouth. If this is correct, then the only echinoderm attribute of homalozoans was their stereom ossicles. An intriguing trait of one homalozoan taxon (Cornuta), which includes *Cothurnocystis* (Fig. 28-61), was a row of pores, perhaps gill slits, to one side of the

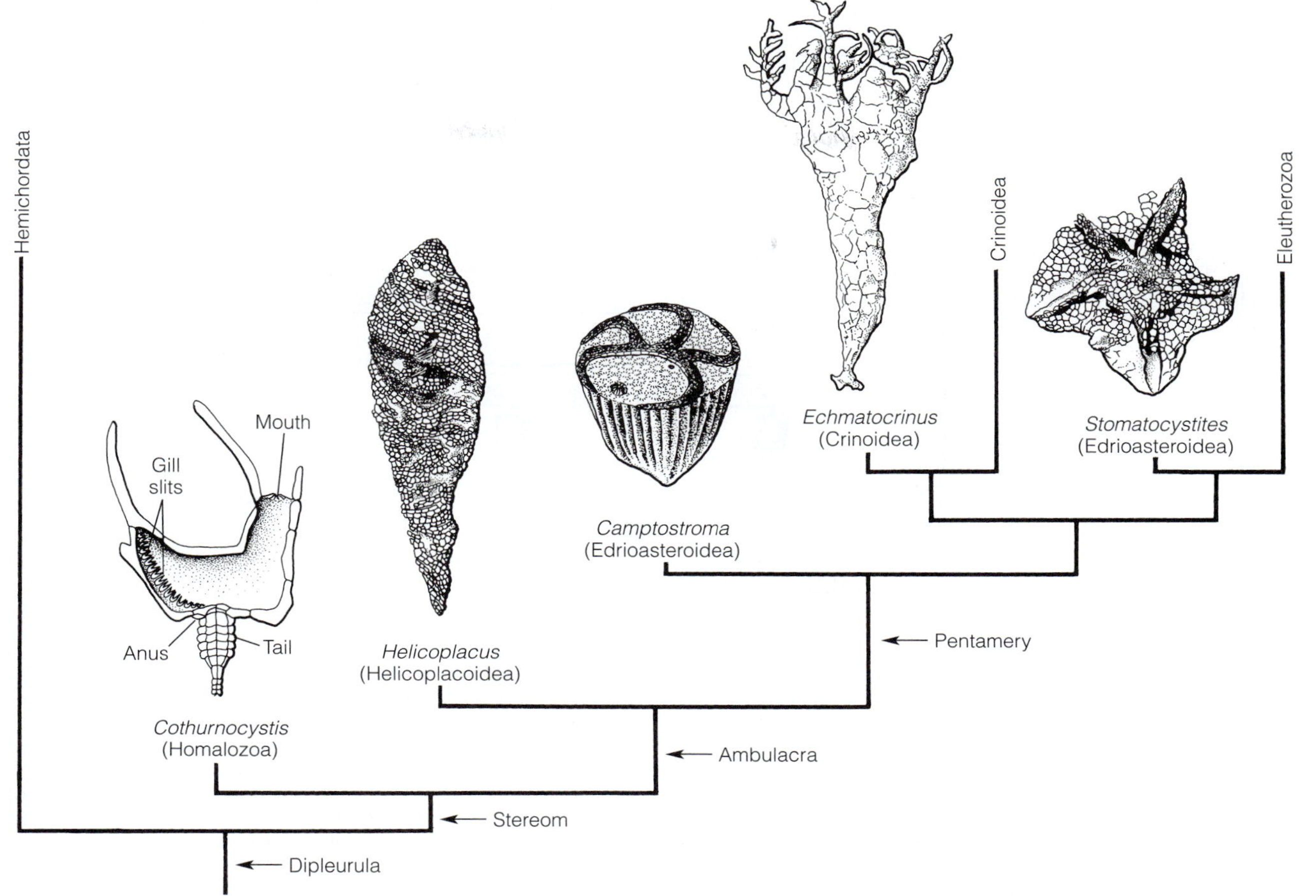

FIGURE 28-61 An echinoderm phylogeny with emphasis on extinct taxa. The homalozoan, *Cothurnocystis elizae* (1 to 2 cm), was a middle Cambrian species, which together with its relatives (all members of Cornuta) was asymmetrical and had a series of openings in the body wall that may have been gill slits. The skeleton is composed of calcitic, stereom ossicles. The single arm, only part of which is shown here, may have been a locomotory tail. Cornutes are thought to have lived on soft bottoms over which they moved by wriggling or flexing the arm-tail. If *Cothurnocystis* actually had gill slits, then cornutes may have been pre-pentamerous echinoderms with a chordatelike pharynx. This would indicate that a pharynx with gill slits is a common feature (synapomorphy) of hemichordates, echinoderms, and chordates—all deuterostomes. The interpretation of these openings as gill slits in *Cothurnocystis,* however, is controversial, as are the functional interpretations of other anatomical features. The lower Cambrian Helicoplacoidea were spindle-shaped burrowers with a skeleton composed of small, platelike ossicles that spiraled around the body and allowed for its expansion and contraction. This flexibility perhaps permitted the animal to retract its body, which may have been situated vertically in the sand. The mouth was located at one end and a single branched ambulacrum was present. The test of early Cambrian to middle Pennsylvanian edrioasteroids (including *Camptostroma*) was coin-sized, globular, discoid, or clublike with pentamerous symmetry. The mouth was on the upper surface at the center of five curved or straight ambulacra; the anus was off-center on the same surface. Edrioasteroids attached to hard surfaces and probably were suspension feeders. The earliest-known crinoid, *Echmatocrinus* from the Middle Cambrian Burgess Shale, is significant for two, apparently primitive, crinoid traits: The arms are unbranched and the stalk, a tapering extension of the calyx, is composed of numerous plates instead of a series of disclike columnals. Presumably, arm branching and a stalk built on a series of columnals evolved later, in the late Cambrian. *(Modified from Ax, P. 2001. Das System der Metazoa. III. Spektrum Akademischer Verlag, Berlin. 283 pp.)*

midline on the upper surface of the body. If these were indeed gill slits, then cornutes may be stem echinoderms that retained an ancestral deuterostome trait (gill slits) and evolved a new one (stereom skeleton). The asymmetry of some homalozoans can also be viewed as intermediate between an ancestral bilateral symmetry and the derived pentamerous symmetry of most echinoderms. This viewpoint, although by no means the only one, is incorporated into Figure 28-61.

Living echinoderms can be separated into two major taxa, the attached, oral-end-up Crinoidea and the motile, oral-end-down Eleutherozoa (Fig. 28-62). Of the two sister taxa of Eleutherozoa, Asteroidea retains the plesiomorphic ectoneural nerve ring and radial nerve in the epidermis of the mouth frame and roof of each ambulacrum. All other eleutherozoans—ophiuroids, echinoids, holothuroids—have internalized these nerves in a closed epineural canal, a synapomorphy that unites

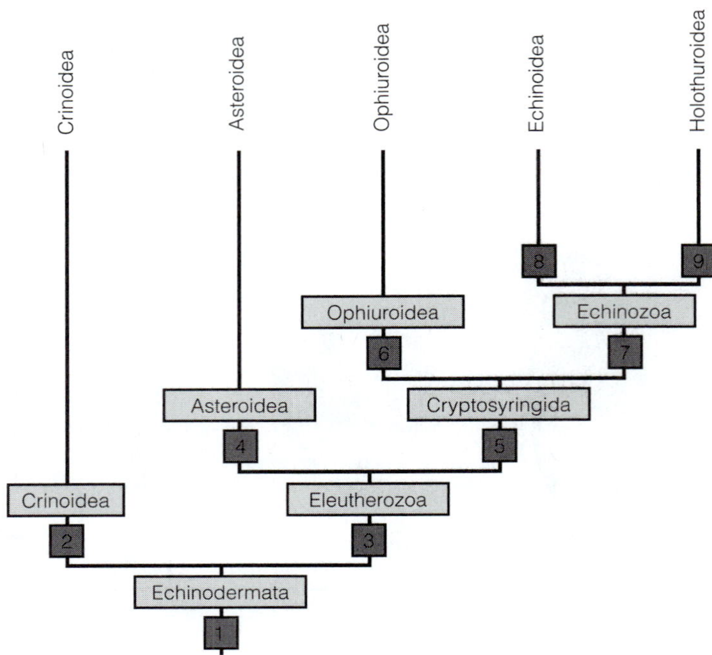

FIGURE 28-62 Echinodermata: phylogeny of living taxa. **1, Echinodermata:** Have pentamerous symmetry, calcareous stereom ossicles, mutable connective tissue, water-vascular system, functionally transformed heart-kidney, dipleurula. **2, Crinoidea:** Body is divided into crown and stem; arms have pinnules, ring canal bears many stone canals, heart-kidney is absent. **3, Eleutherozoa:** Motile lifestyle, oral surface is on the substratum; has locomotory tube feet, a madreporite, polian vesicles on the ring canal, may have Tiedemann's bodies on the ring canal, has movable spines. **4, Asteroidea:** Has an eyespot at each arm tip, a paired pyloric cecum in each arm, gastric hemal tufts, gonads in arms; larva is a bipinnaria. **5, Cryptosyringida:** Radial nerves are internalized in the epineural canal. **6, Ophiuroidea:** Arms are sharply set off from the disc; has an oral madreporite, vertebral ossicles in arms, arm-plate ossicles; larva is an ophiopluteus. **7, Echinozoa:** The oral surface and ambulacra have expanded aborally to cover most of the body, except for a small aboral anus and periproct; ossicles form a ring around the pharynx, has ossicles in tube feet, suckered tube feet, hemal system is well developed and has rete mirabile. **8, Echinoidea:** Has a test of fused ossicles in five paired ambulacral and five paired interambulacral rows, two skeletal pores (a pore pair) for each tube foot, periproct has 10 ossicles (apical system: five ocular plates at the ends of the ambulacral areas, five genital plates at the ends of the interambulacral areas), Aristotle's lantern; polian vesicles are lost; larva is an echinopluteus. **9, Holothuroidea:** Has minute ossicles in a thick dermis, body-wall musculature is in five longitudinal bands, pharyngeal skeleton has five radial and five interradial ossicles, has microphagous buccal podia, longitudinal muscle bands attach to the pharyngeal skeleton and retract the buccal podia, has respiratory trees, heart-kidney absent. *(Based chiefly on Ax, 2001.)*

these three taxa in the monophyletic Cryptosyringida. Within Cryptosyringida, the echinoids and holothuroids share a calcified ring around the pharynx, an expansion of the oral field and amubulacra aborally, a general elongation of the body along the oral-aboral axis, and ossicles in their tube feet. These synapomorphies justify the establishment of the taxon Echinozoa for the echinoids and holothuroids.

PHYLOGENETIC HIERARCHY OF EXTANT ECHINODERMATA

- Echinodermata
  - Crinoidea
  - Eleutherozoa
    - Asteroidea
    - Cryptosyringida
      - Ophiuroidea
      - Echinozoa
        - Echinoidea
        - Holothuroidea

REFERENCES

ECHINODERMATA

General

Binyon, J. 1972. Physiology of Echinoderms. Pergamon Press, Oxford. 264 pp.

Cuénot, L. 1948. Échinodermes. In Grassé, P. (Ed.): Traité de Zoologie. Vol. XI. Masson et Cie, Paris. pp. 1–363.

Emson, R. H., and Wilkie, I. C. 1980. Fission and autotomy in echinoderms. Oceanogr. Mar. Biol. Ann. Rev. 18:155–250.

Harrison, F. W., and Chia, F.-S. (Eds.): 1994. Microscopic Anatomy of Invertebrates. Vol. 14. Echinodermata. Wiley-Liss, New York. 510 pp.

Hendler, G., Miller, J. E., Pawson, D. L., and Kier, P. M. 1995. Sea Stars, Sea Urchins and Allies. Echinoderms of Florida and the Caribbean. Smithsonian Institution Press, Washington and London. 390 pp.

Jangoux, M. (Ed.): 1980. Echinoderms: Present and Past. Proceedings of the European Colloquium on Echinoderms,

Brussels, 3–8 September 1979. A. A. Balkema, Rotterdam. 480 pp.
Jangoux, M., and Lawrence, J. M. (Eds.): 1982. Echinoderm Nutrition. A. A. Balkema, Rotterdam. 654 pp.
Jangoux, M., and Lawrence, J. M. (Eds.): 1983. Echinoderm Studies. Vol. I. A. A. Balkema, Rotterdam. 203 pp.
Lawrence, J. M. 1987. A Functional Biology of Echinoderms. Johns Hopkins University Press, Baltimore, MD. 340 pp.
Lawrence, J. M. 2001. Function of eponymous structures in echinoderms: A review. Can. J. Zool. 79:1251–1264.
Motokawa, T. 1988. Catch connective tissue: A key character for echinoderms' success. In Burke, R. D., Mladenov, P. V., Lambert, P., et al. (Eds.): Echinoderm Biology. A. A. Balkema, Rotterdam. pp. 39–54.
Nichols, D. 1969. Echinoderms. 4th Edition. Hutchinson University Library, London. 200 pp.
Ruppert, E. E., and Balser, E. J. 1986. Nephridia in the larvae of hemichordates and echinoderms. Biol. Bull. 171:188–196.
Strathmann, R. R., Jahn, T. L., and Fonesca, J. L. C. 1972. Suspension feeding by marine invertebrate larvae: clearance of particles by ciliated bands of a rotifer, pluteus, and trochophore. Biol. Bull. 142:505–519.
Strathmann, R. R. 1975. Larval feeding in echinoderms. Am. Zool. 15:717–730.
Ubaghs, G. 1967. General characteristics of Echinodermata. In Moore, R. C. (Ed.): Treatise on Invertebrate Paleontology. Pt. S. Vol. 1. Geological Society of America, University of Kansas, Lawrence. pp. S3–S60.
Welsch, U., and Rehkämper, G. 1987. Podocytes in the axial organ of echinoderms. J. Zool. 213:45–50.
Wilkie, I. C. 1984. Variable tensility in echinoderm collagenous tissues: A review. Mar. Behav. Physiol. 11:1–34.
Yoshida, M., Takasu, N., and Tamotsu, S. 1984. Photoreception in echinoderms. In Ali, M. A. (Ed.): Photoreception and Vision in Invertebrates. Plenum Press, New York. pp. 743–772.

Internet Sites

http://tolweb.org/tree?group=Echinodermata&contgroup=Animals (R. Mooi and G. A. Wray's general introduction to echinoderm phylogeny.)

http://scilib.ucsd.edu/sio/nsf/fguide/echinodermata-1.html and *http://scilib.ucsd.edu/sio/nsf/fguide/echinodermata-2.html* (Color images of living Antarctic echinoderms.)

http://depts.washington.edu/fhl/zoo432/plankton/plechinodermata/plEchinoderms.html (Color images of living echinoderm larvae.)

http://jrscience.wcp.muohio.edu/FieldCourses00/PapersMarineEcologyArticles/MassMortalityinDiademaant.html (Discussion, with references, of mass mortality of the coral-reef urchin *Diadema.*)

www.starfish.ch/Korallenriff/Stachelhauter.html (Text is in German, with excellent images of tropical echinoderms.)

ASTEROIDEA

Anderson, J. M. 1978. Studies on functional morphology in the digestive system of *Oreaster reticulatus.* Biol. Bull. 154:1–14.
Baker, A. N., Rowe, W. E., and Clark, H. E. S. 1986. A new class of Echinodermata from New Zealand. Nature 321:862–864.
Binyon, J. 1964. On the mode of functioning of the water vascular system of *Asterias rubens.* J. Mar. Biol. Assoc. U.K. 44:577–588.
Burnett, A. L. 1960. The mechanism employed by the starfish *Asterias forbesi* to gain access to the interior of the bivalve *Venus mercenaria.* Ecology 41:583–584.
Chia, F.-S., and Amerongen, H. 1975. On the prey-catching pedicellariae of a starfish, *Stylasterias forreri.* Can. J. Zool. 53:748–755.
Christensen, A. M. 1970. Feeding biology of the sea star *Astropecten irregularis.* Ophelia 8:1–134.
Eylers, J. P. 1976. Aspects of skeletal mechanics of the starfish *Asterias forbesi.* J. Morphol. 149:353–368.
Feder, H. M. 1955. On the methods used by the starfish *Pisaster ochraceus* in opening three types of bivalved mollusks. Ecology 36:764–767.
Ferguson, J. C. 1989. Rate of water admission through the madreporite of a starfish. J. Exp. Biol. 145:147–156.
Ferguson, J. C. 1990. Seawater inflow through the madreporite and internal body regions of a starfish *(Leptasterias hexactis)* as demonstrated with fluorescent microbeads. J. Exp. Zool. 255:262–271.
Jaeckle, W. B. 1994. Multiple modes of asexual reproduction by tropical and subtropical sea star larvae: An unusual adaptation for gene dispersal and survival. Biol. Bull. 186:62–71.
Laxton, J. H. 1974. Aspects of the ecology of the coral-eating starfish, *Acanthaster planci.* Biol. J. Linn. Soc. 6:19–45.
Mauzey, K. P., Birkeland, C., and Dayton, P. K. 1968. Feeding behavior of asteroids and escape responses of their prey in the Puget Sound region. Ecology 49:603–619.
Menge, B. 1975. Brood or broadcast? The adaptive significance of different reproductive strategies in the two intertidal sea stars *Leptasterias hexactis* and *Pisaster ochraceus.* Mar. Biol. 31:87–100.
Moran, P. J. 1990. *Acanthaster planci* (L.): Biographical data. Coral Reefs 9:95–96.
Motokawa, T. 1984b. Connective tissue catch in echinoderms. Biol. Rev. 59:255–270.
Scheibling, R. E. 1980. The microphagous feeding behavior of *Oreaster reticulatus.* Mar. Behav. Physiol. 7:225–232.
Shick, J. M., Edwards, K. C., and Dearborn, J. H. 1981. Physiological ecology of the deposit-feeding sea star *Ctenodiscus crispatus:* Ciliated surfaces and animal-sediment interactions. Mar. Ecol. Prog. Ser. 5:165–184.
Sloan, N. A. 1980. Aspects of the feeding biology of asteroids. Oceanogr. Mar. Biol. Ann. Rev. 18:57–124.
Sloan, N. A., and Northway, S. M. 1982. Chemoreception by the asteroid *Crossaster papposus.* J. Exp. Mar. Biol. Ecol. 61:85–98.
Thomas, L. A., and Hermans, C. O. 1985. Adhesive interactions between the tube feet of a starfish, *Leptasterias hexactis,* and substrata. Biol. Bull. 169:675–688.
Thomassin, B. A. 1976. Feeding behavior of the felt-, sponge-, and coral-feeding sea stars, mainly *Culcita schmideliana.* Helgol. Wiss. Meeresunters. 28:51–65.
Wilkinson, C. R., and Macintyre, I. G. (Eds.): 1992. The *Acanthaster* debate. Coral Reefs 11:51–124.

OPHIUROIDEA

Balser, E. J. 1998. Cloning by ophiuroid echinoderm larvae. Biol. Bull. 194:187–193.

Brehm, P., and Morin, J. G. 1977. Localization and characterization of luminescent cells in *Ophiopsila californica* and *Amphipholis squamata*. Biol. Bull. 152:12–25.

Broom, D. M. 1975. Aggregation behavior of the brittle-star *Ophiothrix fragilis*. J. Mar. Biol. Assoc. U.K. 55:191–197.

Emson, R. H., and Woodley, J. D. 1987. Submersible and laboratory observations on *Asteroschema tenue*, a long-armed euryaline brittle star epizoic on gorgonians. Mar. Biol. 96:31–45.

Fontaine, A. R. 1965. Feeding mechanisms of the ophiuroid *Ophiocomina nigra*. J. Mar. Biol. Assoc. U.K. 45:373–385.

Fricke, H. W. 1970. Beobachtungen über Verhalten und Lebensweise des im Sand Lebenden Schlangen-sternes *Amphioplus* sp. Helgol. Wiss. Meeresunters. 21:124–133.

Gage, J. D. 1990. Skeletal growth bands in brittle stars: Microstructure and significance as age markers. J. Mar. Biol. Assoc. U.K. 70:209–224.

Hendler, G. 1975. Adaptational significance of the patterns of ophiuroid development. Am. Zool. 15:691–715.

Hendler, G. 1982. Slow flicks show star tricks: Elapsed-time analysis of basketstar *(Astrophyton muricatum)* feeding behavior. Bull. Mar. Sci. 32:909–918.

Meyer, D. L., and Lane, N. G. 1976. The feeding behavior of some Paleozoic crinoids and recent basket stars. J. Paleontol. 50:472–480.

Pentreath, R. J. 1970. Feeding mechanisms and the functional morphology of podia and spines in some New Zealand ophiuroids (Echinodermata). J. Zool. 161:395–429.

Sides, E. M., and Woodley, J. D. 1985. Niche separation in three species of *Ophiocoma* in Jamaica, West Indies. Bull. Mar. Sci. 36:701–715.

Tyler, P. A. 1980. Deep-sea ophiuroids. Oceanogr. Mar. Biol. Ann. Rev. 18:125–153.

Warner, G. F., and Woodley, J. D. 1975. Suspension-feeding in the brittle star *Ophiothrix fragilis*. J. Mar. Biol. Assoc. U.K. 55:199–210.

Wilkie, I. C., Emson, R. H., and Mladenov, P. V. 1984. Morphological and mechanical aspects of fission in *Ophiocomella ophiactoides* (Echinodermata, Ophiuroidea). Zoomorphology 104:310–322.

Woodley, J. D. 1975. The behavior of some amphiurid brittle stars. J. Exp. Mar. Biol. Ecol. 18:29–46.

ECHINOIDEA

Chia, F.-S. 1969. Some observations of the locomotion and feeding of the sand dollar, *Dendraster excentricus*. J. Exp. Mar. Biol. Ecol. 3:162–170.

De Ridder, C., and Lawrence, J. M. 1982. Food and feeding mechanisms: Echinoidea. In Jangoux, M., and Lawrence, J. M. (Eds.): Echinoderm Nutrition. A. A. Balkema, Rotterdam. pp. 57–115.

Ebert, T. A., and Dexter, D. M. 1975. A natural history study of *Encope grandis* and *Mellita grantii*, two sand dollars in the northern Gulf of California. Mar. Biol. 32:397–407.

Ellers, O., and Telford, M. 1984. Collection of food by oral surface tube feet in the sand dollar, *Echinarachnius parma*. Biol. Bull. 166:574–582.

Fenner, D. H. 1973. The respiratory adaptations of the podia and ampullae of echinoids. Biol. Bull. 145:323–339.

Ferber, I., and Lawrence, J. M. 1976. Distribution, substratum preference, and burrowing behavior of *Lovenia elongata* in the Gulf of Elat and Red Sea. J. Exp. Mar. Biol. Ecol. 22:207–225.

Ghiold, J. 1979. Spine morphology and its significance in feeding and burrowing in the sand dollar, *Mellita quinquiesperforata*. Bull. Mar. Sci. 29:481–490.

Greenway, M. 1976. The grazing of *Thalassia testudinum* in Kingston Harbor, Jamaica. Aquat. Bot. 2:117–126.

Hidaka, M., and Takahashi, K. 1983. Fine structure and mechanical properties of the catch apparatus of the sea-urchin spine, a collagenous connective tissue with muscle-like holding capacity. J. Exp. Biol. 103:1–14.

Lawrence, J. M. 1975. On the relationship between marine plants and sea urchins. Oceanogr. Mar. Biol. Ann. Rev. 13:213–286.

Lewin, R. 1988. Sea urchin massacre is a natural experiment. Science 239:867.

Lewis, J. B. 1968. The function of the sphaeridia of sea urchins. Can. J. Zool. 46:1135–1138.

Millott, N. 1975. The photosensitivity of echinoids. Adv. Mar. Biol. 13:1–52.

Ogden, J. C., Brown, R. A., and Salesky, N. 1973. Grazing by the echinoid *Diadema antillarum* Philippi: Formation of halos around West Indian patch reefs. Science 182:715–717.

Reid, W. M. 1950. *Arbacia punctulata*. In Brown, F. A. (Ed.): Selected Invertebrate Types. John Wiley and Sons, New York. pp. 528–538.

Smith, D. S., Wainwright, S. A., Baker, J., et al. 1981. Structural features associated with movement and "catch" of sea-urchin spines. Tiss. Cell 13:299–320.

Telford, M. 1983. An experimental analysis of lunule function in the sand dollar *Mellita quinquiesperforata*. Mar. Biol. 76:125–134.

Telford, M., Mooi, R., and Ellers, O. 1985. A new model of podial deposit feeding in the sand dollar, *Mellita quinquiesperforata:* The sieve hypothesis challenged. Biol. Bull. 69:431–448.

Timko, P. 1976. Sand dollars as suspension feeders: A new description of feeding in *Dendraster excentricus*. Biol. Bull. 151:247–259.

HOLOTHUROIDEA

Baker, S. M., and Terwilliger, N. B. 1993. Hemoglobin structure and function in the rat-tailed sea cucumber, *Paracaudina chilensis*. Biol. Bull. 185:115–122.

Bakus, G. J. 1973. The biology and ecology of tropical holothurians. In Jones, O. A., and Endean, R. (Eds.): Biology and Geology of Coral Reefs. Vol. II. Biology 1. Academic Press, New York. pp. 326–368.

Byrne, M. 1985. Evisceration behaviour and the seasonal incidence of evisceration in the holothurian *Eupentacta quinquesemita*. Ophelia 24:75–90.

Eylers, J. P. 1982. Ion-dependent viscosity of holothurian body wall and its implications for the functional morphology of echinoderms. J. Exp. Biol. 99:1–8.

Fankboner, P. V. 1981. A re-examination of mucus feeding by the sea cucumber *Leptopentacta (Cucumaria) elongata*. J. Mar. Biol. Assoc. U.K. 61:679–683.

Fankboner, P. V., and Cameron, J. L. 1985: Seasonal atrophy of the visceral organs in a sea cucumber. Can. J. Zool. 63:2888–2892.

Fish, J. D. 1967. Biology of *Cucumaria elongata*. J. Mar. Biol. Assoc. U.K. 47:129–143.

Feral, J.-P., and Massin, C. 1982. Digestive systems: Holothuroidea. In Jangoux, M., and Lawrence, J. M. (Eds.): 1982. Echinoderm Nutrition. A. A. Balkema, Rotterdam. pp. 191–212.

Hansen, B. 1972. Photographic evidence of a unique type of walking in deep-sea holothurians. Deep Sea Res. 19:461–462.

Herreid, C. F., La Russa, V. F., and DeFesi, C. R. 1976. Blood vascular system of the sea cucumber, *Stichopus moebii*. J. Morphol. 150:423–451.

Miller, J. E., and Pawson, D. L. 1990. Swimming sea cucumbers (Echinodermata: Holothuroidea): A survey, with analysis of swimming behavior in four bathyal species. Smithson. Contr. Mar. Sci. 35:1–18.

Shinn, G. L., Stricker, S. A., and Cavey, M. J. 1990. Ultrastructure of transrectal coelomoducts in the sea cucumber *Parastichopus californicus*. Zoomorphology 109:189–199.

Smith, G. N., Jr., and Greenberg, M. J. 1973. Chemical control of the evisceration process in *Thyone briareus*. Biol. Bull. 144:421–436.

Stricker, S. A. 1985. The ultrastructure and formation of the calcareous ossicles in the body wall of the sea cucumber *Leptosynapta clarki* (Echinodermata, Holothuroida). Zoomorphology 105:209–222.

VandenSpiegel, D., and Jangoux, M. 1987. Cuvierian tubules of the holothuroid *Holothuria forskali* (Echinodermata): A morphofunctional study. Mar. Biol. 96:263–275.

CRINOIDEA

Balser, E. J. 2002. Phylum Echinodermata: Crinoidea. In Young, C. M., Sewell, M., and Rice, M. E. (Eds.): Atlas of Marine Invertebrate Larvae. Academic Press, New York. pp. 463–482.

Balser, E. J., and Ruppert, E. E. 1993. Ultrastructure of axial vascular and coelomic organs in comasterid featherstars (Echinodermata: Crinoidea). Acta Zool. (Stockh.) 74:87–101.

Byrne, M., and Fontaine, A. R. 1981. The feeding behaviour of *Florometra serratissima*. Can. J. Zool. 59:11–18.

Fishelson, L. 1974. Ecology of the northern Red Sea crinoids and their epi- and endozoic fauna. Mar. Biol. 26:183–192.

LaHaye, M. C., and Jangoux, M. 1985. Functional morphology of the podia and ambulacral grooves of the comatulid crinoid *Antedon bifida*. Mar. Biol. 86:307–318.

La Touche, R. W. 1978. The feeding behavior of the featherstar *Antedon bifida*. J. Mar. Biol. Assoc. U.K. 58:877–890.

La Touche, R. W., and West, A. B. 1980. Observations on the food of *Antedon bifida*. Mar. Biol. 61:39–46.

Macurda, D. B., and Meyer, D. L. 1974. Feeding posture of modern stalked crinoids. Nature 247:394–396.

Macurda, D. B., and Meyer, D. L. 1976. The identification and interpretation of stalked crinoids from deep-water photographs. Bull. Mar. Sci. 26:205–215.

Macurda, D. B., and Meyer, D. L. 1983. Sea lilies and feather stars. Am. Sci. 71:354–364.

Meyer, D. L., LaHaye, C. A., Holland, N. D., et al. 1984. Time-lapse cinematography of feather stars on the Great Barrier Reef, Australia: Demonstrations of posture changes, locomotion, spawning and possible predation by fish. Mar. Biol. 78:179–184.

Meyer, D. L., and Lane, N. G. 1976. The feeding behavior of some Paleozoic crinoids and recent basket stars. J. Paleontol. 50:472–480.

Mladenov, P. V., and Chia, F.-S. 1983. Development, settling behaviour, metamorphosis and pentacrinoid feeding and growth of the feather star *Florometra serratissima*. Mar. Biol. 73:309–323.

Rutman, J., and Fishelson, L. 1969. Food composition and feeding behavior of shallow water crinoids at Eilat (Red Sea). Mar. Biol. 3:46–57.

PALEONTOLOGY AND EVOLUTION

Durham, J. W., and Caster, K. E. 1963. Helicoplacoidea, a new class of echinoderms. Science 140:820–822.

Fell, H. B. 1965. The early evolution of the Echinozoa. Breviora 219:1–19.

Fell, H. B. 1966. Ancient echinoderms in modern seas. Oceanogr. Mar. Biol. Ann. Rev. 4:233–245.

Janies, D. 2001. Phylogenetic relationships of extant echinoderm classes. Can. J. Zool. 79:1232–1250.

Jefferies, R. P. S. 1986. The Ancestry of the Vertebrates. British Museum (Natural History), London. 376 pp.

Littlewood, D. T. J., Smith, A. B., Clough, K. A., and Emson, R. H. 1997. The interrelationships of the echinoderm classes: Morphological and molecular evidence. Biol. J. Linn. Soc. 61: 409–438.

Nichols, D. 1967. The origin of echinoderms. In Millott, N. (Ed.): Echinoderm Biology. Academic Press, New York. pp. 209–229.

Paul, C. R. C., and Smith, A. B. (Eds.): 1988. Echinoderm Phylogeny and Evolutionary Biology. Clarendon Press, Oxford. 373 pp.

Smith, A. B., Paterson, G. L. J., and Lafay, B. 1995. Ophiuroid phylogeny and higher taxonomy: Morphological, molecular, and paleontological perspectives. Zool. J. Linn. Soc. 114:213–243.

Wada, H., and Satoh, N. 1994. Phylogenetic relationships among extant classes of echinoderms as inferred from sequences of 18S rDNA coincide with relationships deduced from the fossil record. J. Mol. Evol. 38:41–49.

Chordata^P

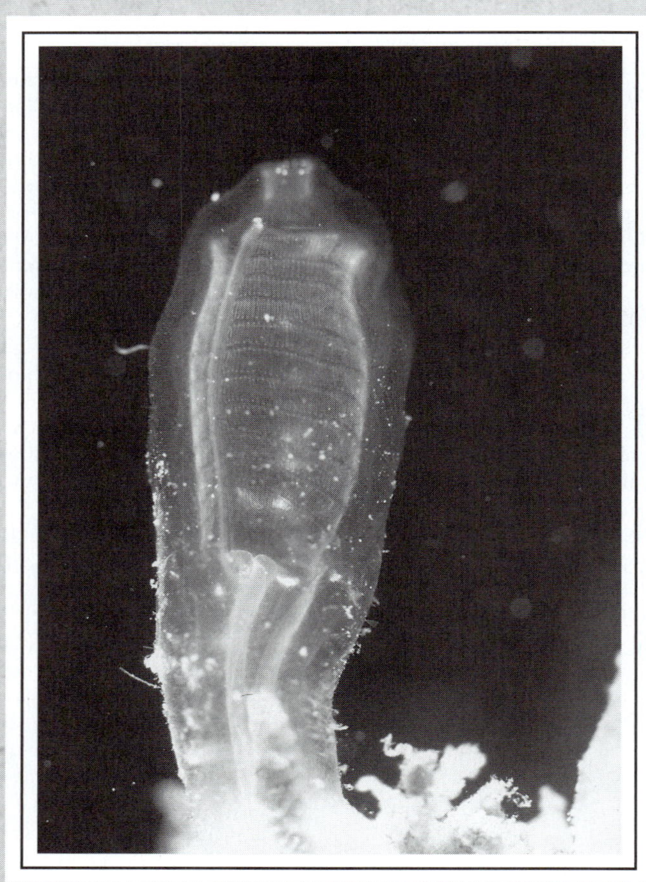

Chordata, with more than 50,000 species in all of Earth's environments, is the largest and most successful taxon of deuterostomes. Most of this diversity and adaptive radiation resides with Vertebrata (fishes, amphibians, reptiles, birds, mammals), a major chordate taxon beyond the scope of this book. The subjects of this chapter are the two other chordate taxa, Cephalochordata (lancelets) and Tunicata (= Urochordata; sea squirts and relatives), both of which are invertebrates and entirely marine. These two taxa of invertebrate chordates account for approximately 2180 species. Although the vertebrates are omitted as a practical necessity, this chapter will highlight the chordate ground plan and its evolution in the invertebrate chordates.

GROUND PLAN OF CHORDATA

The ancestral chordate was a tadpolelike or fishlike animal that swam with lateral undulations of its muscular trunk and tail and filter fed with its pharynx. Of the several attributes associated with this functional design, the four most prominent are a notochord, a dorsal hollow nerve cord, gill slits (pharyngeal clefts), and a continuation of the body beyond the anus called the **postanal tail** (Fig. 29-1). The **notochord** is a flexible, but incompressible longitudinal axial rod that causes the body to bend rather than shorten during contraction of the longitudinal musculature. Alternating right- and left-side contractions of a cross-striated, longitudinal musculature produce the fishlike swimming movements

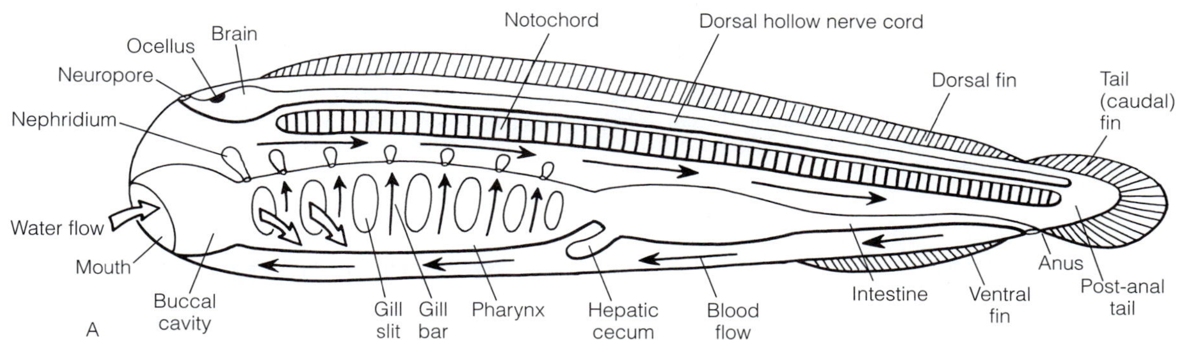

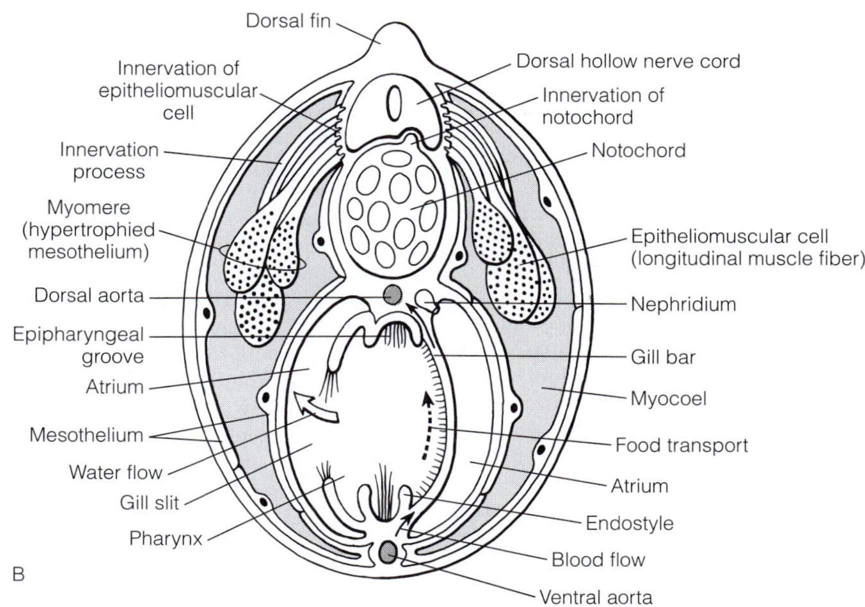

FIGURE 29-1 Chordata: generalized chordate anatomy. **A,** Lateral view; direction of blood flow indicated by arrows. **B,** Transverse section at the level of the pharynx. Left side shows a gill slit, right side shows a gill bar. Reproductive system is omitted.

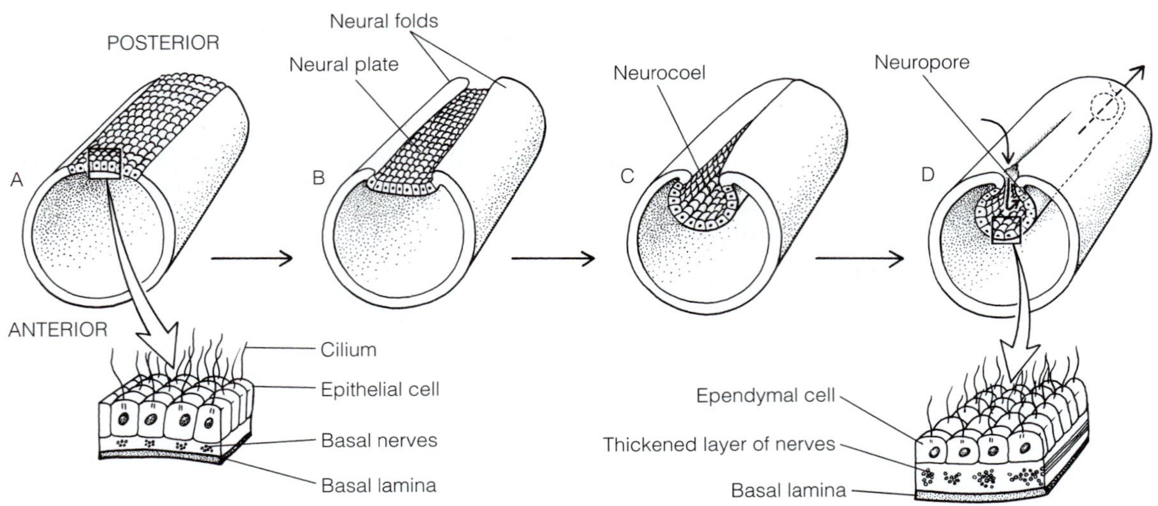

FIGURE 29-2 Chordata: neurulation. **A–D**, Neural plate cells submerge below the surface of the dorsal midline, roll into a tube, and are covered over by the neural folds. The original ciliated surface epidermis in **A** of the protochordate ancestor persists as the ciliated lining (ependymal cells) of the neural tube, whereas the intraepidermal nerves persist as the nervous layer of the nerve cord (enlargement, **D**). In the vertebrates, the neuropore **(D)** is a transient structure present only during development, but among the invertebrate chordates it persists throughout the life of the animal.

of the body. These movements are aided by the additional length and musculature provided by the postanal tail and by rudderlike outgrowths of the body wall called median **fins.** Immediately above and in contact with the notochord is the **dorsal hollow nerve cord,** which extends from the anterior brain into the tail.

The nerve cord is hollow because it develops as a longitudinal mid-dorsal epidermal furrow that rolls up into a tube as its sinks below the surface of the embryonic skin (Fig. 29-2). The furrow remains open at the anterior end of the neural tube to form a **neuropore,** which links the neural canal with the external environment (Fig. 29-1A, 29-2). The neuropore persists throughout the life of invertebrate chordates, but closes during development in the vertebrates.

A filter-feeding pharynx with gill slits, perhaps inherited from the deuterostome ancestor, is the key chordate organ for feeding and gas exchange. The ancestral chordate pharynx pumped water using gill-slit cilia and trapped the suspended food particles in a mucous net cast on the pharyngeal lining. This net is produced by the **endostyle** (Fig. 29-1B), a secretory furrow in the ventral midline of the pharynx.

In addition to these hallmarks, the chordates also inherited deuterostome traits as part of their ground plan. Among these are paired coeloms lined by a monolayered mesothelium that forms the body musculature, a metanephridial system, radial cleavage (probably a eumetazoan apomorphy), and coelom morphogenesis by enterocoely. The chordates also inherited a hemal system, but the direction of blood flow in the circuit is opposite that of hemichordates and protostomes. In chordates, blood moves anteriorly through the ventral vessel, then dorsally through the gills, and posteriorly in the dorsal vessel (Fig. 29-1A).

CEPHALOCHORDATA[sP]

The 30 described species of cephalochordates are glossy slivers that swim rapidly and slip through sand or fingers like tiny eels. Also known as lancelets, or amphioxus, these small, fish-like chordates occur in tropical and temperate oceans worldwide. Unlike most fish, however, lancelets burrow in porous intertidal or shallow subtidal sand and rarely emerge to swim in the water. While in their burrows, they expose the mouth to water at the sediment surface and filter out suspended food particles. Although widespread globally, their patchy local distributions are influenced by sediment suitability and the abundance of phytoplankton food. When these conditions are met, lancelet populations can be impressive. For example, *Branchiostoma caribaeum* occurs in densities of up to 5000 individuals per m^2 of sand in Discovery Bay, Jamaica.

Cephalochordates are important as human food in parts of southeast Asia. Despite their modest size, lancelets are highly nutritious animals that, unlike fish, lack bones. Fresh animals may be cooked for immediate consumption or dried for later use. A now defunct amphioxus fishery near Xiamen, China, once recorded an annual catch of 35 tons, or approximately 1 billion lancelets *(Branchiostoma belcheri)* from a fishing ground 1 mile wide and 6 miles long. In parts of Brazil, chickens are herded onto beaches at low tide to feed on amphioxus.

FORM AND LOCOMOTION

The small lancelet body, 4 to 8 cm in adult length, is laterally compressed and tapered to a point at both head and tail ends (Fig. 29-3A,B). It is from this characteristic that the common name, *amphioxus,* meaning "opposite points," was coined

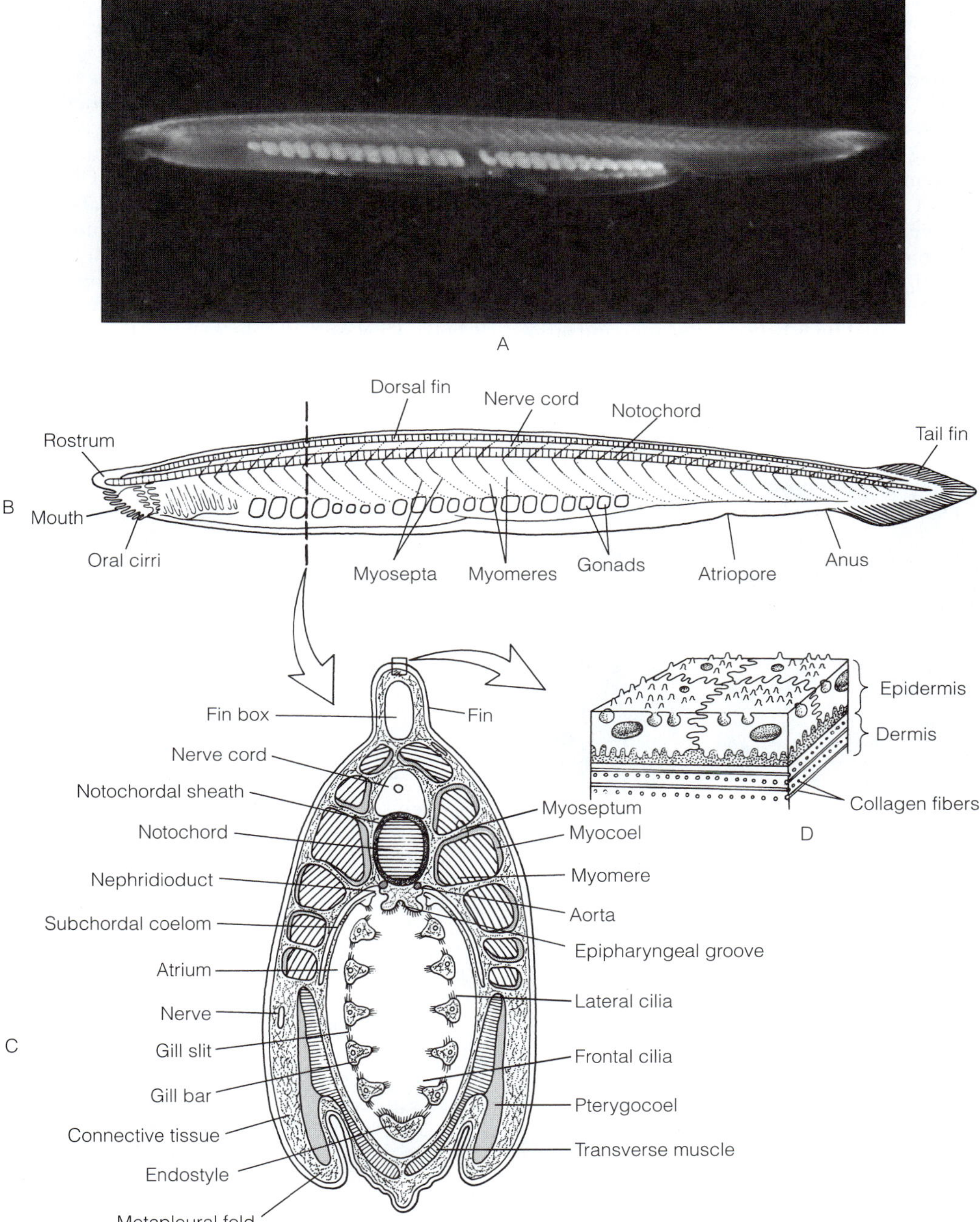

FIGURE 29-3 Cephalochordata: anatomy. **A,** Photograph of a living cephalochordate, *Branchiostoma virginiae,* in left lateral view. **B,** Left lateral view of a generalized lancelet. **C,** Cross section through the pharynx of a generalized lancelet; gonads omitted. **D,** Enlargement of body wall, showing the absence of cuticle but presence of a well-developed, fibrous dermis.

(Fig. 29-3A,B). In life, many of the internal organs and tissues can be seen through the slightly pink, translucent, and iridescent body wall.

The body is divided into a poorly developed head, a long trunk, and a short tail (Fig. 29-3B). The head terminates anteriorly in a short, blunt snout, or **rostrum,** which helps the animal push aside sand while burrowing. The neuropore is a tiny pit at the base of the left side of the rostrum (Fig. 29-5B). Immediately posterior to the rostrum, the head bears a ventral mouth that is surrounded by a veil of fingerlike projections, the **oral cirri** (Fig. 29-3B). The generic name, *Branchiostoma,* meaning "gill mouth," refers to the oral cirri, which were once thought to be gills. An **oral hood** forms the "cheeks" on both sides of the head, frames the mouth, and encloses the buccal cavity.

The head grades smoothly into the trunk, which houses the gut and gonads (Fig. 29-3, 29-6). Internally, the trunk is divided into a branchiogenital region occupied by the pharynx and gonads and an intestinal region that contains the stomach and intestine. A large midventral **atriopore,** the outlet for the filter-feeding water stream, opens on the ventral midline approximately three-quarters of the body length from the head (Fig. 29-3B). The ventral anus is located just anterior to the **tail fin.** A short **ventral fin** is between the anus and atriopore and a continuous **dorsal fin** (Fig. 29-3B) extends from the head to the tail. The trunk is triangular in cross section, and two longitudinal, finlike ridges, the **metapleural folds,** are located ventrolaterally at the lower angles of the triangle (Fig. 29-3C).

Cephalochordates lack a cuticle or tunic, and the epidermis, never stratified as in the vertebrates, is a monolayered, nonciliated glandular epithelium (Fig. 29-3D). A basal lamina lies below the epidermis, and beneath the basal lamina is a thick **dermis** of extracellular matrix. The dermis is laced with cross-helically wound collagen fibers that support and toughen the body and are responsible for its iridescence.

Although they are primarily sedentary burrowers (Fig. 29-4B), cephalochordates can vacate their burrows and swim in short bursts of rapid, eel-like, lateral undulations of the body. They swim equally well in either the headfirst or tail-first direction (Fig. 29-4C,D). The same swimming movements are used for burrowing in sand. Living specimens confined to an

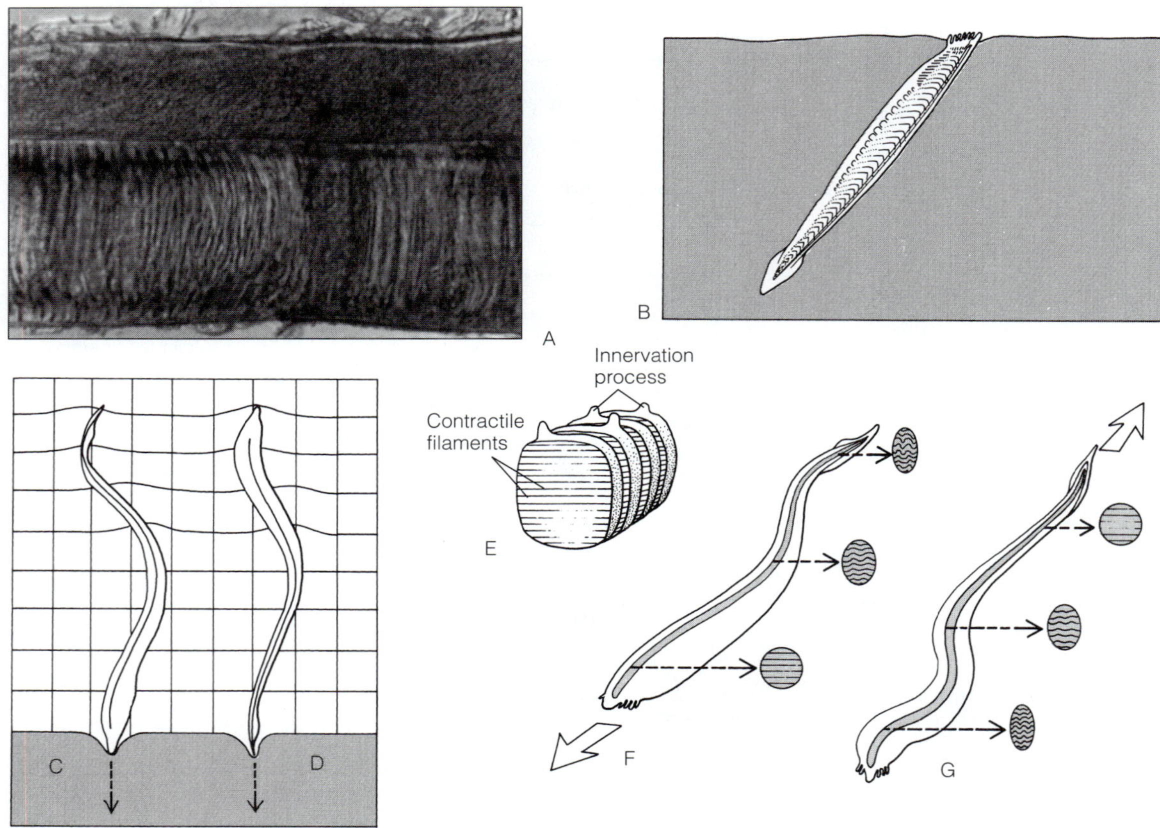

FIGURE 29-4 Cephalochordata: nerve cord, notochord, and burrowing. **A,** Photograph of the nerve cord (above) and notochord (below) dissected from a living lancelet; lateral view. Several of the thin, platelike notochordal muscle cells, in edge-on view, can be seen. **B,** Head-up, inverted feeding position of amphioxus. Headfirst **(C)** and tail-first **(D)** burrowing in amphioxus. **E,** Four muscle plates from the notochord. The myofilaments in each plate run transversely from left to right. The dorsal projection from each plate is an innervation process that makes contact with the dorsal nerve cord. **F** and **G,** Hypothetical changes in the cross-sectional profile of the notochord during forward **(F)** and rearward **(G)** swimming. While swimming forward, the posterior notochordal muscle may be contracted as shown to lower the bending resistance of the tail and increase its amplitude, thus creating greater forward thrust. Conversely, the anterior notochord may be contracted while the animal swims rearward **(G).**

aquarium are nearly impossible to catch with fingers or forceps because of their streamlined shape, speed, and slippery skin, and must instead be netted like small fish.

NERVOUS SYSTEM AND SENSE ORGANS

The CNS of amphioxus consists of a rudimentary brain, a dorsal hollow nerve cord, and segmental sensory (and a few motor) nerves, which leave the cord dorsally and innervate a variety of structures, especially the rostrum, oral cirri, velum, and tail (Fig. 29-5B). The cord extends from the base of the rostrum to nearly the tip of the tail. Anterior to the brain, the neural canal opens to the exterior via the permanent, ciliated neuropore (Kölliker's pit; Fig. 29-5B). Cilia in the neuropore create a flow of surface water that dips shallowly into the pore and then reemerges. Sensory cilia, originating from cells lining the brain cavity, extend into the base of the neuropore and may be exposed to the water flow.

A short distance behind the neuropore, the cord enlarges slightly to form a small, hollow brain. An unpaired pigment-cup ocellus is located in the anterior wall of the brain. Additional ocelli, up to 1500 or more in *Branchiostoma lanceolatum,* are located in the cord posterior to the brain. Curiously, the ocelli on the left side of the cord face dorsally, and those on the right and ventral sides face ventrally. Amphioxus is negatively phototactic, and continuous illumination seems to arrest locomotion, causing the animals to remain in their burrows. A sudden pulse of bright light in dark-adapted animals, however, will cause them to leave the sand and swim. The nerve cord ends in a slight expansion at the tip of the tail. Many cells lining the neural tube bear motile flagella that are directed posteriorly through the canal.

MUSCULATURE AND NOTOCHORD

The notochord is located directly below the nerve cord and extends from the tip of the tail forward into the rostrum (Fig. 29-3B, 29-4A, 29-5). The name *cephalochordate,* meaning "head notochord," refers to the unique continuation of the notochord anterior to the brain in amphioxus. The notochord consists of a longitudinal series of disc-shaped cells arranged like a stack of coins (Fig. 29-4A,E). The entire structure is enclosed in a fibrous extracellular matrix, the **notochordal sheath,** that resembles the epidermal basal lamina and dermis (Fig. 29-3D) and probably helps to stiffen the notochord and control its shape (Fig. 29-3C). Each notochordal cell is a specialized epitheliomuscular cell in which the contractile filaments run transversely, from left to right. Thus, the amphioxus notochord is a muscle designed for a skeletal function. A regional contraction of the discoid muscle cells might locally adjust the stiffness of the notochord to achieve optimal performance while swimming forward, rearward, through water, or through sand (Fig. 29-4F,G). Each notochordal cell is innervated via a short, slender outgrowth of the muscle cell itself the innervation process, that makes contact with the overlying nerve cord (Fig. 29-4E).

Cephalochordates are segmented, coelomate animals, like annelids (and vertebrates). Lancelet segmentation is obvious in the arrangement of the swimming muscles into a longitudinal series of 50 to 75 V-shaped **myomeres** (= muscle segments). The myomeres, which are out of register on opposite sides of the body, are separated from each other by **myosepta** (Fig. 29-3B,C). The myomeres form the longitudinal cross-striated musculature of the body wall and are responsible for the undulatory swimming and burrowing movements of the body. A functional explanation for the V-shape of the myomeres has not yet been advanced.

Circular muscles are absent, but a well-developed **transverse muscle** (pterygial muscle) spans across the belly between the metapleural folds (Fig. 29-3C). Contraction of this muscle pulls together the two stiff metapleural folds (similar to closing the two valves of a clam). This action compresses the atrium and pharynx and forces the water within to squirt from the mouth in a so-called cough reflex. Compression of the pharynx also closes the gill slits (and presumably arrests ciliary water pumping) by rocking rearward the free ventral side of the pharynx.

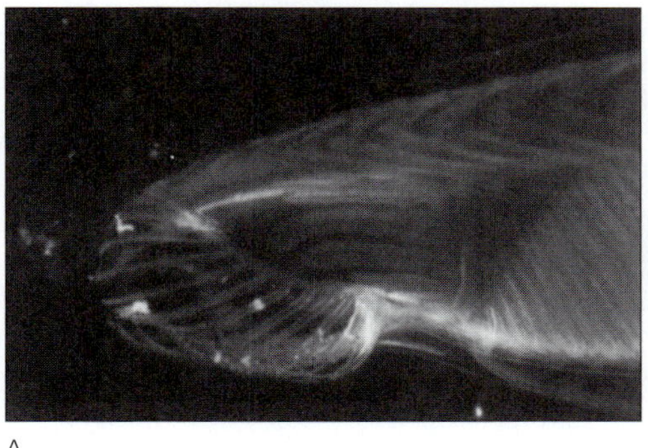

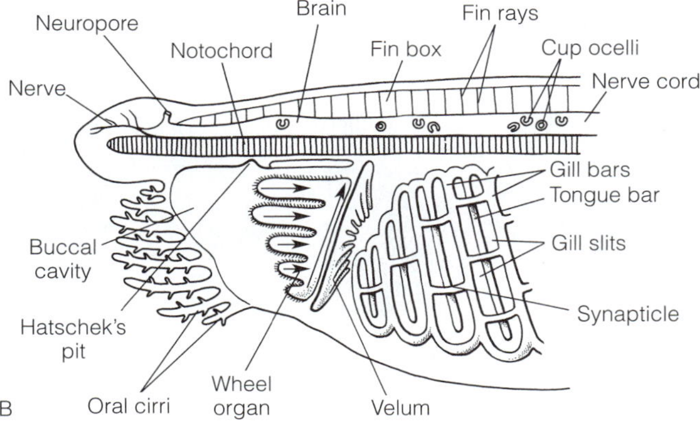

FIGURE 29-5 Cephalochordata: head and oral region. **A,** Left lateral view of the head of a living *Branchiostoma caribaeum,* showing the oral cirri, wheel organ, velum, and anterior pharynx with gill slits. **B,** Left lateral view of the lancelet head.

COELOM

Segmentation in lancelets is more like that of annelids than is obvious upon superficial examination. Each amphioxus myomere is actually a segmental coelomic cavity lined by a monolayered mesothelium. The mesothelial cells in contact with the nerve cord are greatly enlarged epitheliomuscular cells that bulge into the coelomic cavity and together form the swimming muscles (Fig. 9-11F,G, 29-1B). Because lancelet myomeral muscle cells rest directly on the outer wall of the nerve cord, they are innervated by contact with the cord, in contrast to the vertebrates, in which motor neurons leave the cord to innervate distant muscles. The enlarged myocytes nearly fill the coelomic space, but a small cavity called the **myocoel** remains (Fig. 9-11F,G, 29-1B, 29-3C). That part of the myocoel not in contact with the nerve cord is lined by a mesothelium of noncontractile, ciliated cells that may help to circulate the myocoelomic fluid. The myosepta occur at areas of transverse contact between the myomeral coelomic cavities. Like the annelid septa, they consist of a sandwich of extracellular matrix between two layers of mesothelium.

The dorsal and ventral fins, but not the tail fin, are supported by a longitudinal series of small, unpaired coeloms called **fin boxes** (Fig. 29-5B). The number of fin boxes in the dorsal fin, for example, often exceeds 200, and thus the number of fin "segments" does not correspond to the number of myomeres. The septa between the fin boxes form the **fin rays.**

A paired, nonsegmented **perivisceral coelom** surrounds the entire gut from the pharynx to the anus, but nowhere is it voluminous, and most of the extracellular fluid volume around the pharynx and gut is provided by the atrium (discussed later). In the pharyngeal region, the perivisceral coelom forms a set of small, complexly interconnected spaces. One of these, the **endostylar coelom,** surrounds the contractile ventral aorta (the endostylar artery, or "heart") and is essentially a pericardial cavity. Another branch of the perivisceral coelom, the **subchordal coelom,** covers the outer wall of the pharynx above the gill slits and is associated with the excretory organs, discussed below (Fig. 29-3C).

The **pterygocoel** occupies the interior of the metapleural folds (Fig. 29-3C). The coelomic fluid in the pterygocoels is gelatinous and contributes to the stiffness of the folds.

DIGESTIVE SYSTEM AND NUTRITION

The digestive system of amphioxus is strikingly similar to that of filter-feeding enteropneusts, tunicates, and primitive vertebrates. As in these other groups, lancelets ingest large quantities of particles from which they extract organic material while eliminating water through the gill slits and mineral particles through the anus (Fig. 29-6). Water entering the pharynx passes through the gill slits into a circumpharyngeal water jacket, the **atrium** (Fig. 29-3C), which opens to the exterior posteriorly at a midventral atriopore and then continues posteriorly to the anus as the blind-ended **post-atrioporal atrium** (Fig. 29-6). Water flow through the pharynx and atrium not only delivers food particles, but also ventilates the internal organs for gas exchange. Lateral cilia on the gill slits are chiefly responsible for generating the water flow, but cilia also occur on the lining of the atrium.

The large mouth, functionally an inhalant siphon, is surrounded by oral cirri that form a coarse screen and prevent the entry of large particles. Sensory cells on each cirrus probably provide the animal with information on water quality. The buccal cavity (vestibule), which is enclosed by the oral hood, is a short section of the foregut that receives particles and water from the mouth (Fig. 29-6), and its inner ciliated surface bears two specialized structures. The **wheel organ** is a set of ciliated grooves in the lining of the buccal cavity. The grooves trap very fine particles in mucus and then convey the mucus-bound particles into the pharynx (Fig. 29-5B). **Hatschek's pit** is a shallow pocket in the mid-dorsal wall of the buccal cavity. Recent research using immunocytochemical techniques indicates that Hatschek's pit contains endocrine cells, which are believed to secrete pituitary-like hormones into the blood.

Separating the buccal cavity from the pharynx is a muscular iris diaphragm called the **velum** (Fig. 29-5B, 29-6). The aperture of the velum bears a ring of sensory velar tentacles. One function of the velum is to mediate the so-called **cough reflex,** a sudden and rapid contraction of the atrium that expels water from the pharynx through the mouth.

The pharynx, a long spacious tube perforated by 180 or more pairs of narrow gill slits, is surrounded by the atrium. Early in development, the gill slits are segmental, corresponding in number to the myomeres, but later the number of gill slits (branchiomeres) exceeds that of the myomeres. Each gill

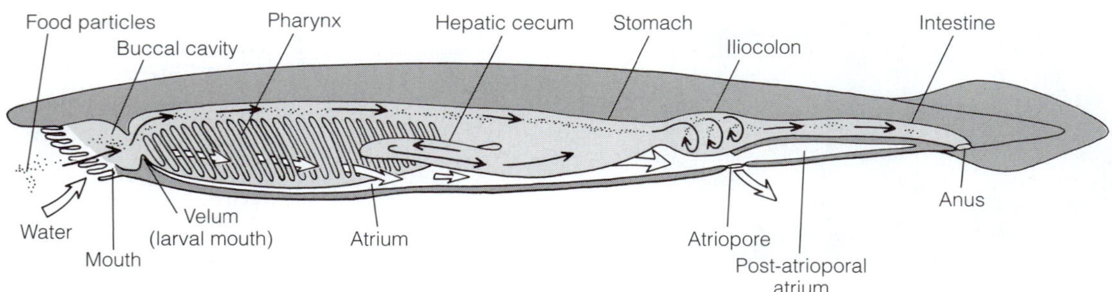

FIGURE 29-6 Cephalochordata: functional organization of the digestive system. Open arrows show the passage of water, solid arrows show the path of food particles and digestive enzymes. The hepatic cecum, a right-side structure, is shown on the left side of the body for clarity.

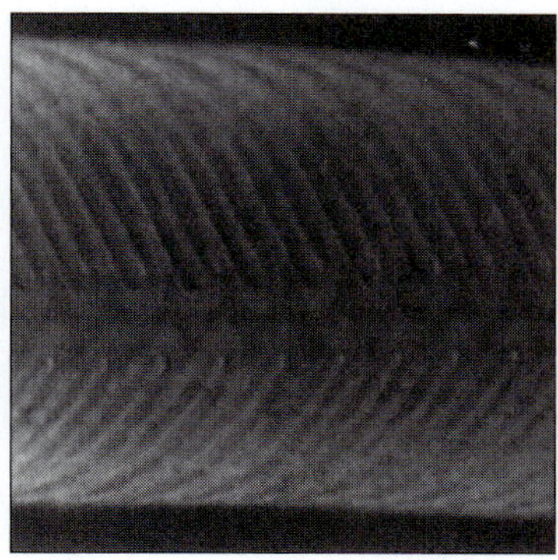

A B

FIGURE 29-7 Cephalochordata: gill slits and gill skeleton. **A,** Ventral view of the outer wall of the pharynx from a living lancelet showing the elongate, U-shaped gill slits and slender tongue bars. The dark zigzag band along the ventral midline is probably a blood vessel. **B,** High-magnification view of part of four gill slits showing the transverse synapticles that support the tongue bars.

cleft develops as a circular or oval opening in the pharyngeal wall, but it soon elongates vertically and is divided into two **gill slits** by the downgrowth of a **tongue bar** (Fig. 29-5, 29-7). **Gill bars** surround and support each pair of gill slits. The endostyle is a ciliated midventral groove, U-shaped in cross section, in the floor of the pharynx. Opposite the endostyle, in the mid-dorsal line of the pharynx, is the ciliated **epipharyngeal groove,** which extends rearward into the short, narrow esophagus (Fig. 29-3C). At the anterior end of the pharynx, near the velum, the two ciliated arms of the U-shaped endostyle diverge and encircle the pharyngeal lining as **peripharyngeal bands.** The ciliated peripharyngeal bands join the epipharyngeal groove dorsally well behind the velum.

The esophagus joins a slightly inflated stomach. A single unbranched diverticulum, the **hepatic cecum,** arises from the stomach and projects anteriorly into the atrium along the right side of the pharynx (Fig. 29-6). The stomach is followed posteriorly by an expanded tubular section of the gut called the **iliocolon.** The iliocolon joins the intestine posteriorly. The intestine opens to the exterior at a ventral anus between the ventral and tail fins.

All lancelets are filter feeders. Lateral cilia on the gill-slit wall pump water from the pharynx, through the gill slits, into the atrium, and out the atriopore. The outflow from the pharynx draws water and suspended food particles into the mouth, buccal cavity, and pharynx. Fine suspended particles are trapped on the wheel organ, as already mentioned, but most particles are separated from the water in the pharynx. The pharynx traps particles on a mucous net secreted by the endostyle.

The endostyle, which is the evolutionary forerunner of the vertebrate thyroid gland, is the principal center for the elaboration of a specialized mucus—a complex mucoprotein containing iodine bound to the amino acid tyrosine. Iodination of the protein occurs in two localized regions of the endostyle, near the upper ends of its U-shaped profile (Fig. 29-3C). Endostylar flagella cast mucus onto the lining of the pharynx in two sheets, one on each side of the endostyle. Together, the two sheets, which microscopically resemble mosquito netting, constitute the **mucous net.** The mucous net is transported slowly from the endostyle (ventral) to the opposite side of the pharynx (dorsal) by frontal cilia on the pharyngeal lining (Fig. 29-15D). As the net moves over the slits, water passes through it and food particles are trapped. When the food-laden net reaches the dorsal midline of the pharyngeal lining, it is rolled into a single threadlike cord by the epipharyngeal groove (Fig. 29-6).

Cilia in the epipharyngeal groove convey the food cord into the esophagus, then the food enters the stomach and iliocolon. Well-developed cilia in the iliocolon rotate the food cord while digestive enzymes from the hepatic cecum are released onto the food. Extracellular digestion occurs in the stomach and iliocolon and absorption occurs primarily in the hepatic cecum. Indigestible material passes into the intestine, is compacted into a fecal cord, and then egested through the anus. Lancelets filter particles in the size range of 0.06 to 100 μm, which includes phytoplankton, bacteria, and even colloidal particles (proteins).

Like the proboscis and branchial skeletons of acorn worms, the branchial skeleton of cephalochordates is composed of cross-linked collagen fibers.

HEMAL SYSTEM AND INTERNAL TRANSPORT

The hemal system is well developed, consisting of arteries, veins, and smaller vessels, but no typical, compact heart (Fig. 29-8). Instead, the ventral blood vessel, consisting of the sinus venosus and **ventral aorta** (endostylar artery), is strongly contractile and heartlike in function. Contraction of the ventral aorta pushes blood through capillaries in the gills to the paired dorsal aortae, which unite in the post-pharyngeal region of the body to form an unpaired **dorsal aorta.** The aortae transport blood posteriorly. The dorsal aortae give rise to three primary branches: segmental arteries leading to capillaries in the myosepta, segmental arteries leading to capillaries in the atrium and gonads, and intestinal arteries extending to capillaries in the gut wall (Fig. 29-8). Blood returning from these capillary beds in the myosepta, atrium, and gonads passes into the cardinal veins, then into the common cardinal veins, which unite anteriorly to form the sinus venosus which then becomes the ventral aorta. Blood from capillaries in the gut wall drains into the hepatic portal vein and enters a capillary bed in the hepatic cecum before being collected by the hepatic vein, which joins the sinus venosus and ventral aorta.

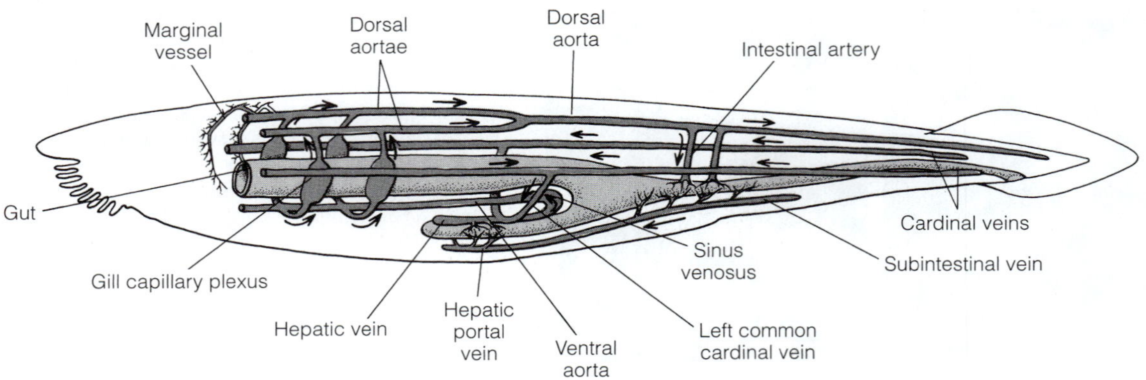

FIGURE 29-8 Cephalochordata: simplified anatomy of the circulatory system. *(Modified and redrawn from Rähr, H. 1979. The circulatory system of amphioxus* (Branchiostoma lanceolatum *Pallas). A light-microscopic investigation based on intravascular injection technique. Acta Zool. 60:1–18.)*

The blood is colorless, contains few cells, and is unlikely to play a significant role in gas transport. Instead, water flow through the atrium, including the post-atrioporal atrium, ventilates the tissues of the body. The notochord and muscles contain hemoglobin, which may facilitate the diffusion of oxygen or, by storing it, may extend the performance time of these tissues. Despite these provisions, lancelets are able to swim actively for only short periods of time.

The blood vessels of lancelets lack the endothelium typical of vertebrates and are lined only with the basal lamina of overlying cell layers or are simply open channels in the connective tissue. Thus although the gross anatomy of the lancelet circulatory system is like that of vertebrates, it is histologically similar to the hemal systems of invertebrates.

EXCRETION

Although cephalochordates are probably ammonotelic animals, they have paired filtration nephridia associated with the pharynx and an unpaired nephridium called **Hatschek's nephridium** in the head. Each of the paired nephridia has a short, ciliated duct that discharges into the atrium. The duct from Hatschek's nephridium empties into the pharynx immediately behind the velum. From the atrium, urine leaves the body through the atriopore, but uric acid, an end product of purine metabolism, is reportedly stored in the tissues associated with the gonads.

The paired nephridia are in the subchordal coelom near the dorsal side of each gill slit (Fig. 29-9). Each nephridium consists of a cluster of podocytes on a branch of the dorsal aorta. However, the apical end of each podocyte bears a flagellum encircled by a collar of long microvilli and thus resembles a protonephridium. Similar to a protonephridium, the tube-like collar inserts into the excretory duct so that urine must pass through the collar before entering the duct (Fig. 29-9). Cephalochordates are thought to have inherited a metanephridial system, but its structure seems to indicate combined metanephridial and protonephridial function. One hypothesis for the presence of these dualistic filtration cells is that they are downstream of the gill capillaries, in which the blood pressure drops, and rest on a noncontractile blood vessel (a branch of the dorsal aorta). Perhaps the podocyte flagella supplement the insufficient filtration pressure of the dorsal aorta.

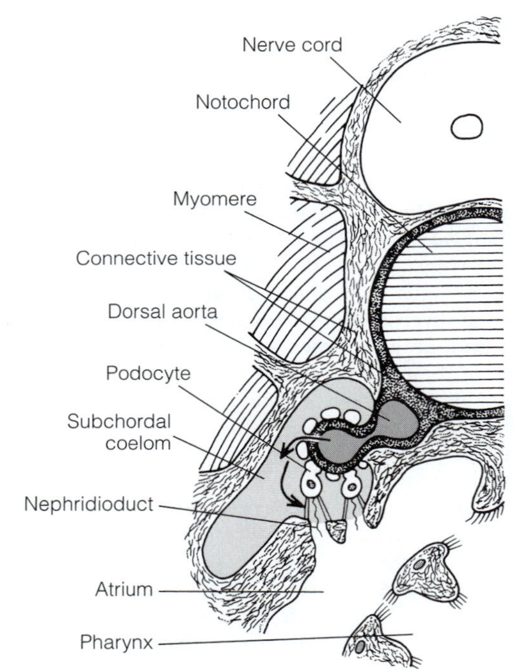

FIGURE 29-9 Cephalochordata: nephridia. Amphioxus has paired branchiomeric nephridia, and one such nephridium is shown here in cross-sectional view. Presumably, primary urine is formed by ultrafiltration of the blood into the subchordal coelom. Apparently, the urine is modified in the subchordal coelom before being swept by cilia into the nephridioduct and atrium, from which it eventually leaves the body via the atriopore. The cells at the filtration site are intermediate in form between podocytes and protonephridial terminal cells.

REPRODUCTION AND DEVELOPMENT

Cephalochordates are gonochorists with external fertilization and a complex life cycle that includes a benthic adult and a planktotrophic larva. *Branchiostoma* usually has 26 pairs of gonads (Fig. 29-3A,B), often corresponding to myomeres 10 through 35. Spawning occurs at sunset in the European *Branchiostoma lanceolatum,* when ripe gametes rupture from the gonads into the atrium and are carried to the exterior in the exhaust from the atriopore. The zygotes undergo equal, holoblastic, radial cleavage to form a coeloblastula. Gastrulation is by invagination, and the anterior coelomic cavities, myomeres, and notochord all form by enterocoely (Fig. 29-10).

The planktotrophic larva is a tiny, ciliated, tadpolelike creature (Fig. 29-11). Oral cirri, an oral hood, an atrium, and dorsal and ventral fins are absent, but there is a well-developed tail fin. Anteriorly, the body exhibits a striking asymmetry: a large mouth is located on the left side of the head, and the first gill slits are found on the right side. The endostyle is vertical in the dorsoventral axis at the anterior end of the pharynx. The preoral pit, which is the precursor of the wheel organ, and Hatschek's pit (neuropore) open on the left side of the head. The **club-shaped gland** is a glandular ciliated duct that opens externally on the left side of the head below the mouth and internally into the anterior pharynx near its dorsal midline. Its function is uncertain, but most evidence suggests a role in larval feeding.

Using cilia for locomotion, the larva adopts the same head-up and slightly inverted orientation in the plankton as the adult does in its burrow (Fig. 29-4B). The gill-slit cilia pump water transversely, from left to right, through the mouth, pharynx, and gill slits (Fig. 29-11B). The large mouth and gill slits, as well as the short path through the pharynx, provide a low-resistance channel through which a substantial water volume may be pumped. Suspended particles are trapped on an endostyle-secreted mucous net cast across the path of water flow (Fig. 29-11B). The preoral pit, like the adult wheel organ, contributes to particle capture. During the gradual metamorphosis, the larval mouth becomes the velum and the oral hood develops to enclose the buccal cavity and form the adult mouth and oral cirri (making the adult mouth and buccal cavity, strictly speaking, preoral structures). The metapleural folds grow ventrally and medially to enclose the pharynx and form the atrium. The developing gill slits become bilaterally distributed along the pharynx (Fig. 29-11C).

DIVERSITY OF CEPHALOCHORDATA

Most species (23) of cephalochordates are members of the genus *Branchiostoma,* including *B. lanceolatum* in Europe, *B. virginiae* of the southeastern coast of the United States, *B. floridae* in the Gulf of Mexico, and *B. californiense* on the southwestern coast of the United States. These species are all similar in appearance and habits and have been widely studied as "amphioxus." In contrast to these "amphioxi" are the few species (seven) of *Epigonichthys,* formerly known as *Asymmetron.* These species are asymmetric in the placement of their gonads, which occur only on the right side of the body.

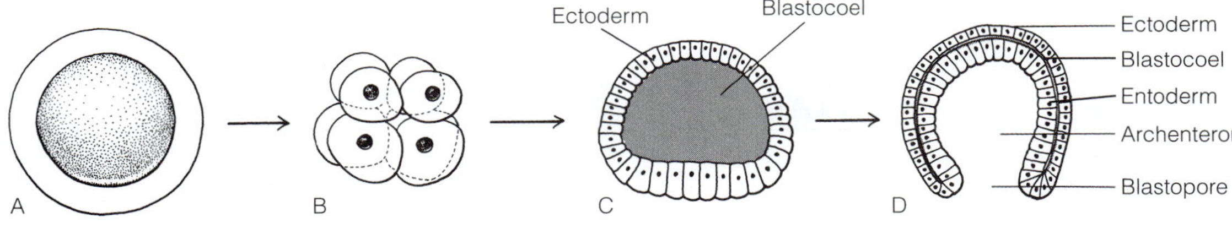

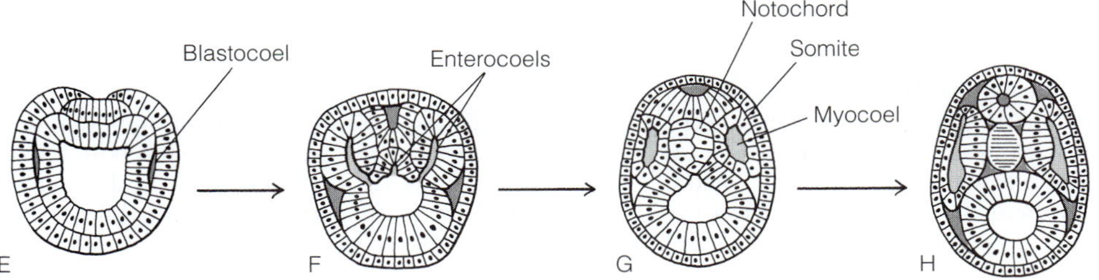

FIGURE 29-10 Cephalochordata: development. **A,** Zygote. **B,** Eight-cell stage, showing radial cleavage pattern. **C,** Blastula. **D,** Gastrula. **E,** Onset of neurulation and somitogenesis (segment formation). **F,** Onset of notochordogenesis and later stages of neurulation and somitogenesis. **G** and **H,** Completion of organogenesis. *(Modified and redrawn from Willey, A. 1894. Amphioxus and the Ancestry of the Vertebrates. MacMillan and Co., New York. 316 pp., and Conklin, E. G. 1932. Embryology of amphioxus. J. Morphol. 54:69–151.)*

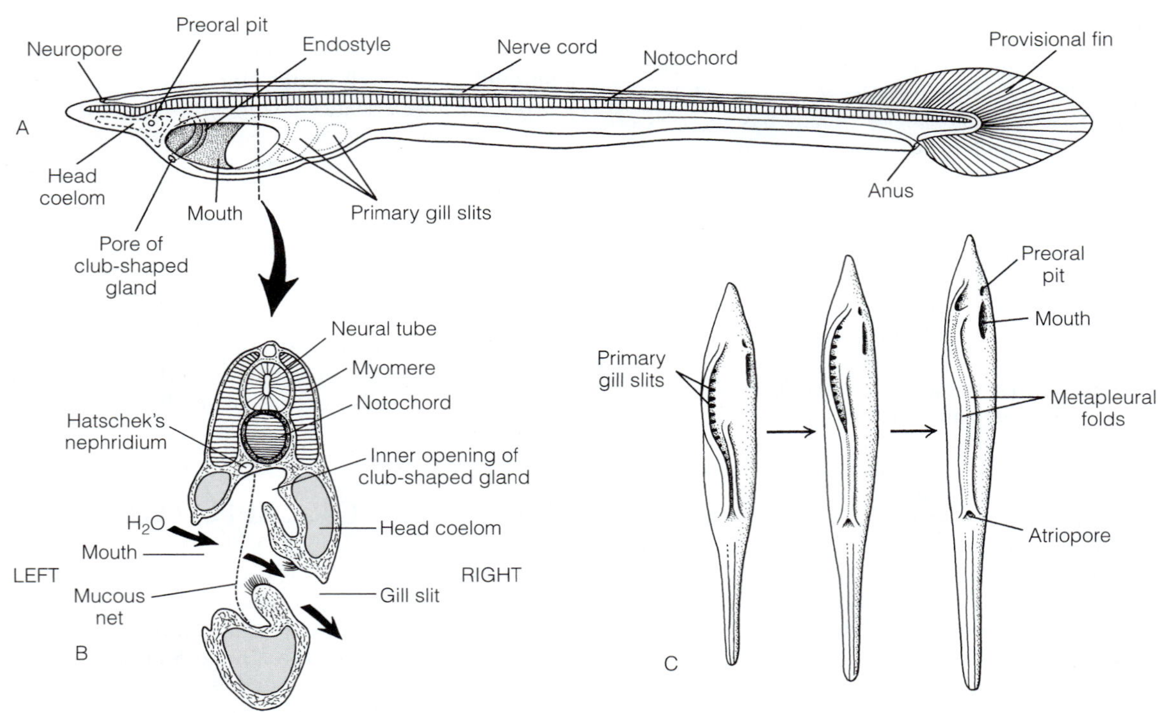

FIGURE 29-11 Cephalochordata: larval anatomy and metamorphosis. **A,** Left lateral view of the larva. **B,** Cross section through the mouth of the larva in A, showing asymmetry, the path of water flow, and the filter-feeding mucous net. **C,** Ventral views of stages in metamorphosis showing the enclosing of the larval gill slits by the metapleural folds to form the atrium. Later, the mouth and preoral pit migrate ventrally from their positions on the left side of the head and become enclosed in the buccal cavity, which has not yet formed. The larval mouth becomes the adult velum, and the preoral pit differentiates into Hatschek's pit and the wheel organ. *(C, Modified and redrawn from Willey, A. 1894. Amphioxus and the Ancestry of the Vertebrates. MacMillan and Co., New York. 316 pp.)*

TUNICATA[sP] (UROCHORDATA)

The sea squirts and their planktonic relatives, although clearly chordates, have diverged farthest from the chordate ground plan. Tunicates are common marine animals that little resemble other chordates, or other animals for that matter. Most are sessile, attached organisms with the body covered and attached to the substratum by a complex tunic. Many species are colonial, the only examples of coloniality among the chordates. The pharynx is well developed, but the notochord, dorsal hollow nerve cord, and postanal tail are absent in adults, except in one taxon (Appendicularia). With the exception of Appendicularia, the four main chordate characteristics occur only in the transitory larval stage, which both resembles and is called a **tadpole.** Both the tadpole and the adult lack a coelom and a metanephridial system and the gut is absent from the tail. Instead, the U-shaped gut is located solely in the head and foreshortened trunk. Excretory structures, when present, are of a type unique to tunicates. The tunicate musculature is composed of myocytes, not the epitheliomuscular cells of the chordate ground plan. Tunicate connective tissues, including the blood and tunic, are well developed and often contain functionally diverse cells.

The tunicates are divided into three taxa (classes): Ascidiacea, Thaliacea, and Appendicularia (Larvacea). The benthic ascidians contain the majority of species and are the most common tunicates. The other two classes are adapted for life in the plankton. Approximately 2150 species of tunicates have been described.

ASCIDIACEA[C]

Ascidians, often called sea squirts, are common sessile, marine invertebrates throughout the world. Species are either solitary or colonial, with few to thousands of tiny zooids per colony (Fig. 29-12). Most ascidians occur in shallow water, where they attach to rocks, shells, pilings, and ship bottoms or sometimes affix themselves in mud and sand by filaments or a stalk. A great diversity of species inhabit shallow tropical seas, and many colonial species occupy crevices in old coral heads and the undersides of coralline rock. Still others form in large, conspicuous clusters on gorgonian corals and mangrove roots or in thick, rubbery mats on rocks, pilings, and floating docks (Fig. 29-13). Some ascidians are drab lumps, but many are brightly colored. Solitary species can be the color of cherries, limes, or peaches, and colonies may be blood red, kiwi green,

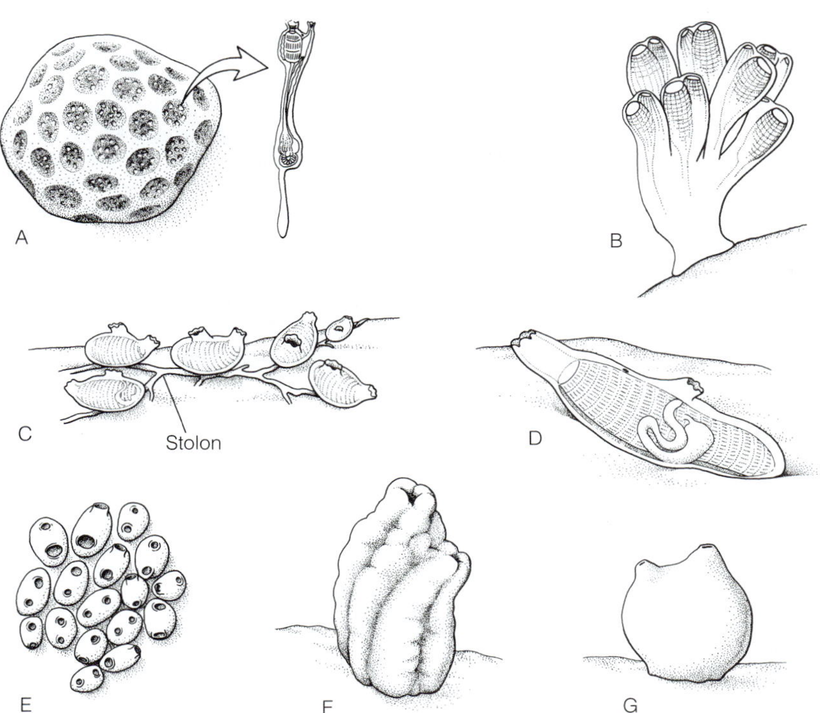

FIGURE 29-12 Tunicata: ascidian diversity. **A,** Colonial aplousobranchs: the compound ascidian *Aplidium stellatum* (sea pork). **B,** The light bulb tunicate *Clavelina oblonga.* **C,** The colonial stolonate phlebobranch *Perophora viridis* (honeysuckle tunicate). **D,** The solitary phlebobranch *Ascidia curvata.* **E,** The colonial stolidobranch *Symplegma rubra.* **F,** Solitary stolidobranchs *Styela plicata* and **G,** the sea grape, *Molgula manhattensis.*

FIGURE 29-13 Tunicata: ascidian diversity. The sea liver, *Eudistoma hepaticum* (Aplousobranchia), a compound ascidian, grows into tough, massive colonies on floats, jetties, and pilings along the southeastern coast of the United States.

or patterned like Persian rugs. Their striking colors and forms are rivaled only by corals and sponges, the two other major groups of epibenthic, suspension-feeding macroinvertebrates. Of the approximately 2000 species, only 100 have been collected from depths greater than 200 m (Fig. 29-14), and a few from the interstices of sandy beaches.

Solitary Body Form

Solitary ascidians such as *Styela, Ascidia,* and *Molgula* are called **simple,** or solitary, **ascidians** (Fig. 29-12D,F,G) in contrast to the many species of colonial ascidians (Fig. 29-12A–C,E). The size and shape of solitary species can resemble a seed, grape, peach, or potato, and some are irregular. *Halocynthia pyriformis* of the northeast U.S. coast and *H. aurantium* on the Pacific coast are known as sea peaches because of their size, shape, and color. The widespread sea grape, *Molgula manhattensis,* not only resembles a grape, but often grows in grapelike clusters. One of the largest solitary ascidians is *Pyura pachydermatina,* a 1 m tall, stalked species of the southern hemisphere. But the largest ascidians are colonial species, discussed below, whose bodies may easily exceed one meter.

One surface of the body is attached to the substratum, and on the opposite are two openings, each borne on a tubular projection. These are the **buccal** and **atrial siphons,** the inlet and outlet, respectively, for the filter-feeding water stream (Fig. 29-15A).

In contrast to shallow-water ascidians, 95% of which live attached to firm surfaces, most deep-sea species anchor in soft bottoms and project upward into the water. Most are tiny, spherical zooids anchored by fibrils, but some are large and stalked. A few transparent, irregular species float above their tethers like rags on a line (Fig. 29-14).

Colonial Body Form

A colonial body form has evolved independently several times in the Ascidiacea. Colonies occur in a range of designs, from those in which the zooids are nearly independent to colonies with a high degree of zooidal integration. In general, the zooids

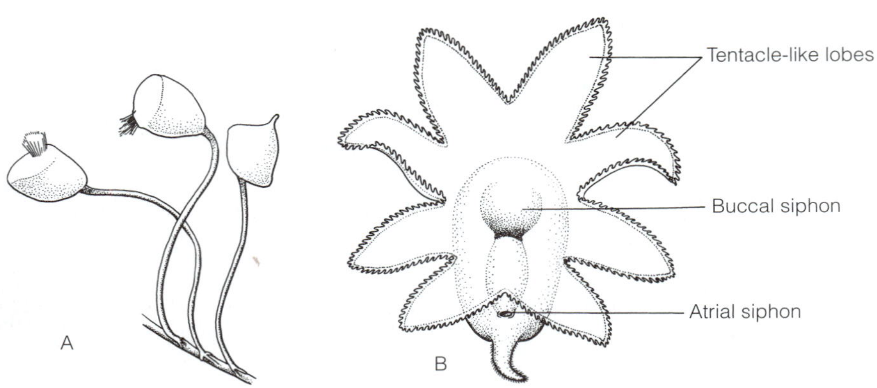

FIGURE 29-14 Tunicata: deep-sea ascidians. Most live on soft bottoms, in contrast to predominately hard-bottom species of shallow water. Note the long, slender stalk or rootlike process by which fixation in soft bottoms is made possible. **A,** *Coleolus suhmi.* **B,** *Octonemus ingolfi.* This last species, unlike most other tunicates, is believed to be predaceous. Prey are trapped by means of the large, tentacle-like siphon lobes. *(Drawn from photographs by Monniot and Monniot, 1975.)*

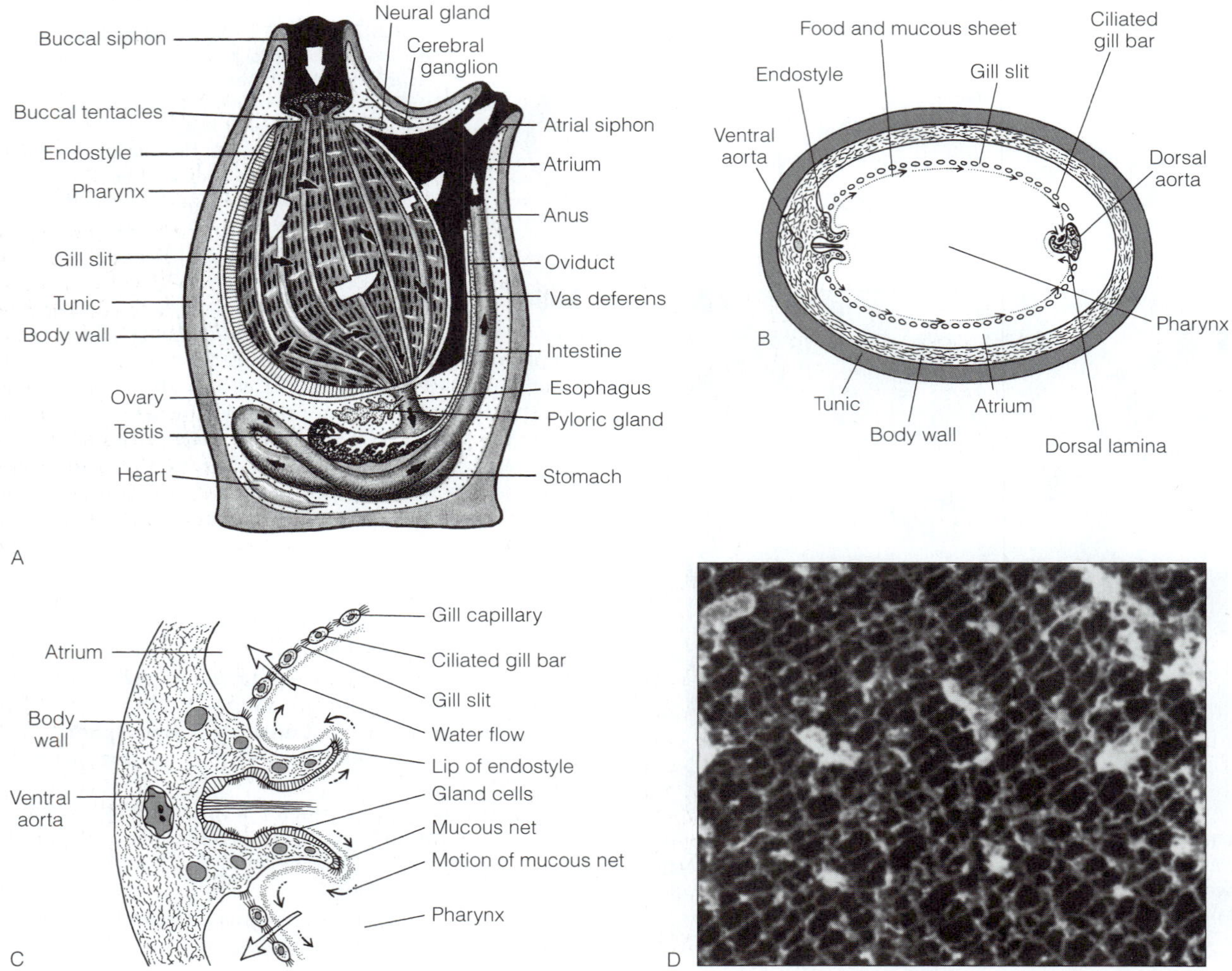

FIGURE 29-15 Tunicata: functional anatomy of the pharynx and gut. **A,** Diagrammatic cutaway view of a solitary ascidian, showing internal organs. Large arrows show the path of water flow, small arrows that of the mucous net. **B,** Cross section of the pharynx in A. **C,** Enlargement of the endostylar region of **B** to show transport of the mucous net and the water flow through the gill slits. **D,** Scanning electron micrograph of the mucous net of *Ciona intestinalis.* *(D, From Flood, P. R., and Fiala-Medioni, A. 1979. Filter characteristics of ascidian food trapping mucous films. Acta Zool. 60:271.)*

that compose a colony are small and often microscopic, but sometimes they are so numerous that the colonial body may be enormous. Such colonies often form thick, irregular mats on rocks, floating docks, boat hulls, and pilings (Fig. 29-13).

In the simplest colonies, the zooids are spaced apart but united by stolons. For example, in *Perophora,* the colony is like a vine with attached globular zooids (Fig. 29-12C). In others, such as *Ecteinascidia turbinata,* the stolons are short and the zooids are clustered in tufts (Fig. 10-18). A more intimate association occurs in *Clavelina.* Here the zooids are united not only by stolons, but also by fusion of the lower half of the zooidal tunics (Fig. 29-12B). In the most specialized colonies, the **compound ascidians,** all zooids are completely embedded, except for their openings, in a common tunic.

Compound ascidians may be small or large, encrusting and thin, or massive and thick (Fig. 29-12A, 29-13). Within the tunic of compound species, the siphons of each zooid may open independently of those of other zooids or the atrial siphons of neighboring zooids may share a common exhaust chamber, the cloaca. Such highly integrated zooids are usually arranged in orderly **systems.** One common system is a circle of buccal siphons (and their zooids) around a central cloaca. A large colony may be dotted with hundreds or thousands of these systems, each appearing like a star against the tunic firmament. This starlike character is reflected in the names of a few species, such as *Aplidium stellatum* and *A. constellatum* (Fig. 29-12A). Another common and colorful species with this arrangement is *Botryllus schlosseri* (Fig. 29-16A). A vertical section through the center of a *Botryllus* system reveals how the atrial siphon of each zooid opens into the cloaca, which then discharges at the surface (Fig. 29-16B). The zooids of *Botryllus* are only a few millimeters in diameter, but because a single colony may contain many star-shaped systems, the colonial body may reach 12 to 15 cm.

Body Regions

Within the tunic, the ascidian zooid can be divided into three body regions: an anterior (or distal) **thoracic region** containing the pharynx, an **abdominal region** containing the digestive tract and associated structures, and the **postabdominal region** (Fig. 29-17). The postabdomen, which may be long and threadlike, is the most basal (posterior) part of the body and contains the heart and reproductive organs. These three regions are conspicuous only in the microscopic zooids of some colonial species (aplousobranchs). Most ascidians lack a postabdomen, and many species with large solitary zooids *(Ascidia, Styela, Molgula, Microcosmos)* even lack an abdomen. With the loss of the postabdomen, abdomen, or both, the viscera shift anteriorly. For example, in species with a thorax only, the heart, gonads, and gut share the thorax with the pharynx (Fig. 29-17).

Tunic, Body Wall, and Musculature

The body of an ascidian is covered by a monolayered epidermis overlaid with a complex **tunic** from which the name tunicate is derived (Fig. 29-18). The tunic usually is thick but varies from a soft, delicate consistency to one that is tough and similar to cartilage. The tunic of *Aplidium stellatum* (called sea pork) has both the appearance and texture of stiff fat. On the other hand, the transparent, gelatinous tunic of *Diplosoma* species encloses numerous colonial zooids that resemble frog eggs in jelly.

Like most cuticles and exoskeletons, the tunic is composed of various proteins and carbohydrates, but it also has several features unique to tunicates (Fig. 29-18). One is the presence of structural fibers composed of a kind of cellulose called **tunicin.** These parallel fibers typically occur in successive thin sheets, each of which has its fibers oriented at a different angle from that preceding it and is stacked within the tunic like successive plies of plywood. This fiber arrangement, which is similar to that of some arthropod cuticles, gives the tunic toughness and strength. Other striking features are the presence of blood cells (hemocytes) in the tunic and, in some species, **tunic vessels,** which develop as tubular outgrowths of the hemal system and body wall. The tunic vessels are uniquely lined by epidermis and epidermal basal lamina. The presence of hemocytes in the tunic means that it is a living tissue akin to a connective tissue in the body. A third unique characteristic is

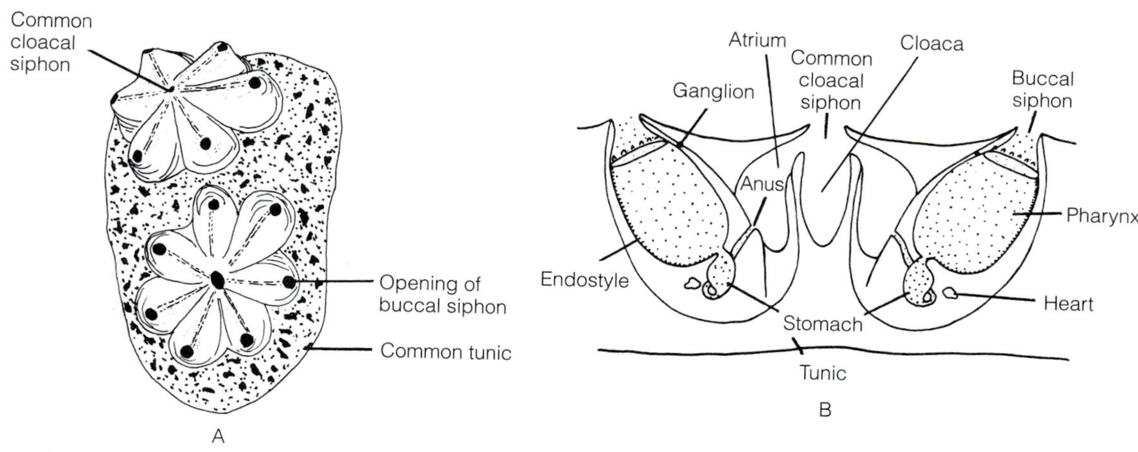

FIGURE 29-16 Tunicata: integration of zooids in colonial ascidians. **A,** The compound ascidian *Botryllus schlosseri,* showing two starlike systems of zooids. **B,** Vertical section through the center of one system. *(A, After Milne-Edwards from Yonge; B, After Delage and Hérouard.)*

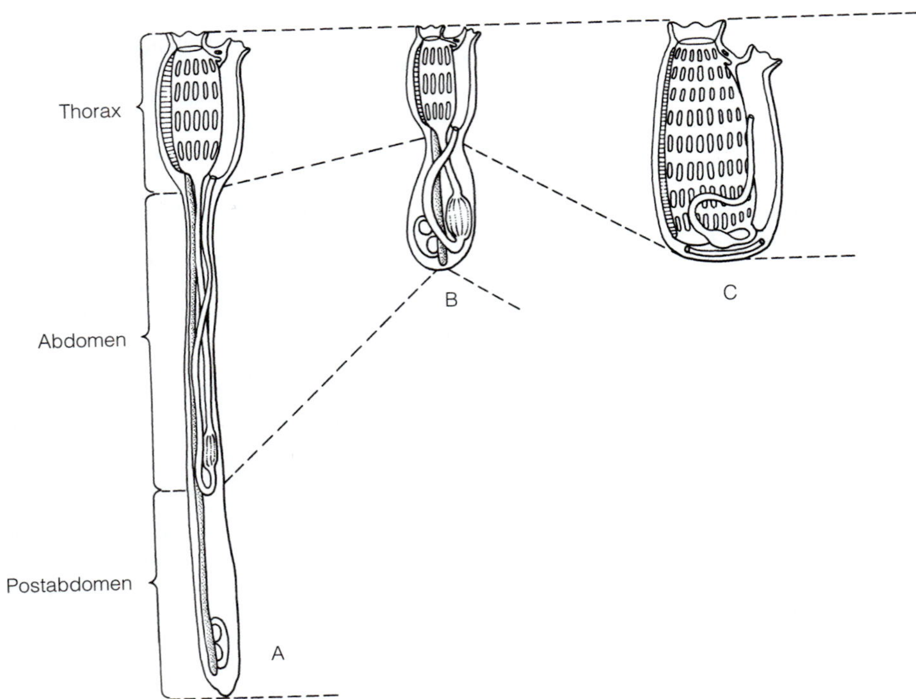

FIGURE 29-17 Tunicata: ascidian body regions. **A,** Many aplousobranch colonies are composed of zooids that each have three body regions: the thorax (pharynx and atrium), abdomen (gut loop), and postabdomen (heart and gonads). **B,** Some phlebobranchs have a thorax and abdomen, but no postabdomen. **C,** Most phlebobranchs and all stolidobranchs have only one body region, the thorax. The body of all ascidian juveniles (oozooids) is composed of thorax only. Posterior elongation to form the abdomen or abdomen and postabdomen results from postmetamorphic growth and is correlated with tunic thickness. Aplousobranch zooids having all three body regions form compound colonies with a very thick common tunic, as shown in Figure 29-13. *(Modified and redrawn from Berrill, N. J. 1935. Studies in tunicate development. Part IV—Asexual reproduction. Phil. Trans. Roy. Soc. Lond. B 225:327–379.)*

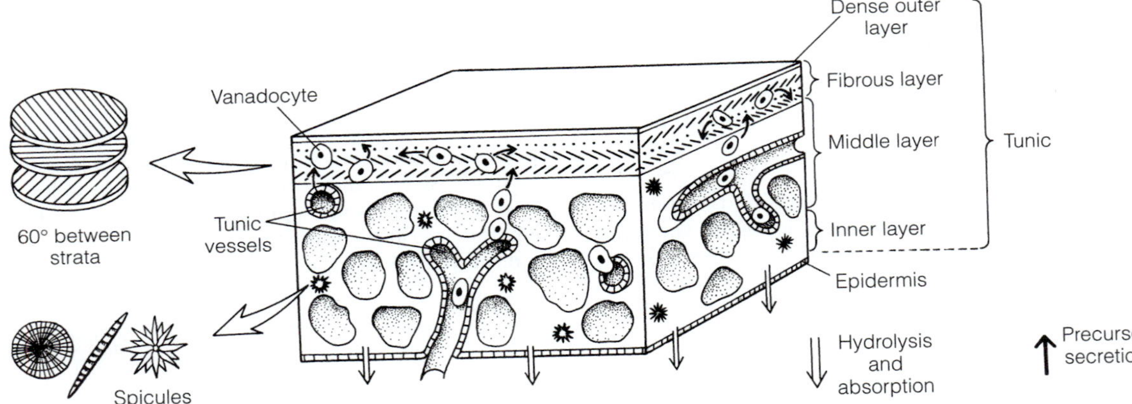

FIGURE 29-18 Tunicata: generalized ascidian tunic. The extracellular tunic of ascidians is a dynamic exoskeleton that enlarges, without being molted, as the body grows. Although the process of tunic growth is not fully understood, tunic precursors are transported to the tunic by the blood in species having a vascularized tunic and released into the outer surface of the tunic. Secretion and removal of tunic by the epidermis may also occur.

that ascidians are the only animals that do *not* molt as they grow *inside* a thick exoskeleton. This means that the tunic must grow to keep pace with the growth of the zooid. Tunic growth apparently is the responsibility of both the epidermis and the tunic hemocytes.

The tunic clearly has supportive and protective functions that are enhanced in some species by the addition of hard structures. Species of didemnids and some pyurids secrete calcareous spicules or plates embedded in the tunic (Fig. 29-18). Some species, such as *Eudistoma carolinense,* incorporate sand grains in the tunic to such an extent that their bodies are rock hard. Others, for example *Didemnum psammathodes,* whose colonies spread over rocks like melted chocolate, incorporate their cocoa-brown feces into the tunic, perhaps for camouflage or to increase its stiffness. The tunic attaches ascidians to the substratum and often is thick and expanded at the point of attachment. Rootlike stolons may ramify from the attached base to provide better anchorage, and these, too, are covered by tunic.

A basal lamina and a thick, gelatinous connective tissue underlie the epidermis. The muscles, nerves, blood vessels, and ameboid cells occur in the connective-tissue layer. The musculature consists of distinct outer circular and inner longitudinal bands of smooth muscle. The longitudinal fibers extend from the body to the siphons and on contraction withdraw the siphons (Fig. 29-19). The circular muscles predominate on the siphons as sphincters that regulate the size of the openings. Contraction of the body-wall musculature occurs periodically, compressing the body, forcing jets of water from the siphons, and eventually closing the siphons. The **periodic squirting** of water from the siphons, characteristic of all ascidians, and often in response to unwanted matter in the water or other disturbances, is the reason for the common name *sea squirt.*

Pharynx and Atrium

The anterior buccal siphon opens internally into a large pharynx. At the junction of siphon and pharynx, the mouth is encircled by a ring of **buccal tentacles.** The tentacles project into the water-flow path to prevent the entry of large, unwanted particles (Fig. 29-15A).

The walls of the pharynx are perforated by small, ciliated gill slits (stigmata; sing., stigma) that permit water to pass from the pharynx into the surrounding atrium. Lateral cilia project into the gill slits and frontal cilia cover the interior surface of the pharynx. On the ventral side of the pharynx—the side opposite the atrial siphon—is the endostyle, which extends the length of the pharynx (Fig. 29-15A–C). Anteriorly, at the junction of the pharynx and buccal siphon, the endostyle divides into two ciliated peripharyngeal bands that encircle the opening into the pharynx. These converge in the dorsal midline to form a ciliated ridge, the **dorsal lamina** (Fig. 29-15B) in some taxa or a row of tonguelike **languets** in others, that extends posteriorly into the esophagus.

The pharynx is surrounded by the water-filled atrium except where the pharynx is attached to the body wall on its ventral endostyle side. Elsewhere, cordlike mesenteries cross the atrium between the atrial and pharyngeal walls, suspending the unattached part of the pharynx, which lacks a rigid skeleton, and preventing its collapse. Dorsally, the atrium opens to the exterior through the atrial siphon. The cavity just inside the atrial siphon is sometimes called the cloaca because it receives both wastes from the anus and gametes from the gonoducts. The epithelium lining the atrium and atrial siphon is derived from ectoderm. Traditionally, the pharynx is believed to be derived from embryonic endoderm, but a recent developmental study of the edible Japanese ascidian *Halocynthia roretzi* indicates that the pharynx, like the atrium, originates from ectoderm.

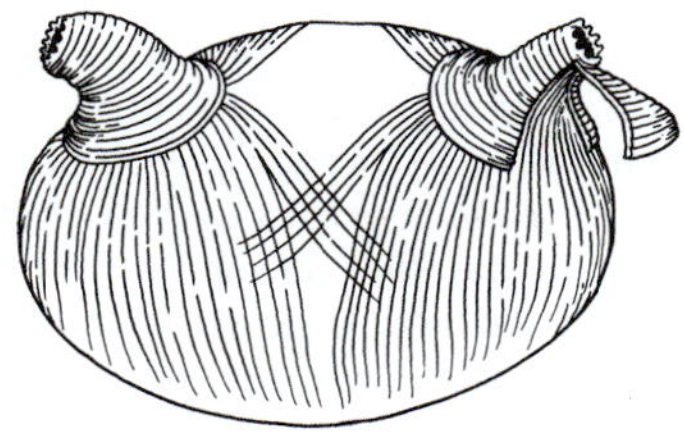

FIGURE 29-19 Tunicata: ascidian musculature. Musculature of the stolidobranch *Pyura ostreophila. (Modified and redrawn from Kott, P. 1989. Form and function in the Ascidiacea. Bull. Mar. Sci. 45:253–276.)*

Digestive System and Nutrition

Ascidians are filter feeders that remove plankton from the water current pumped through the pharynx. The water current is produced by the lateral cilia on the margins of the gill slits (Fig. 29-15C). As in most filter feeders, an enormous quantity of water must be filtered to provide adequate food. On average, this amounts to 1 body volume of water per second. A specimen of *Ascidia nigra,* itself only a few centimeters in length, pumps 173 liters of water through its body in 24 h (Fig. 29-20A).

As water is pumped through the pharynx, suspended food particles are trapped on the lining of the pharynx in a mucous net secreted by the endostyle. Frontal cilia on the pharyngeal lining slowly transport the continuously secreted net from the ventral endostyle dorsally (Fig. 29-15D). When the food-laden net reaches the dorsal midline of the pharynx, it is rolled into a threadlike cord by the dorsal lamina or dorsal languets and then conveyed into the esophagus.

As in cephalochordates, the endostylar mucous net of ascidians (and other tunicates) contains a complex iodoprotein. When examined with an electron microscope, the net is a regular meshwork with openings of approximately 0.5 μm (Fig. 29-15D). Such a filter readily removes the smallest plankton, including bacteria, from the water. Ascidians can halt feeding by closing the buccal siphon, by arresting ciliary beat, or by stopping the flow of mucus from the endostyle.

Downstream of the pharynx, the digestive tract is a U-shaped loop (Fig. 29-15A). The esophagus, the descending arm of the U, joins the stomach, an enlargement at the turn of the U. The stomach is lined with secretory cells and is the site of extracellular digestion. The ascending arm of the digestive tract is the intestine, the terminal end of which is a rectum and anus. The intestine forms feces and is probably the site of absorption (Fig. 29-15A). The anus discharges fecal strings or pellets into the atrium for transport out of the body through the atrial siphon.

In all tunicates, a network of tubules, the **pyloric gland,** covers the outer wall of the anterior intestine (Fig. 29-15A). By way of one or many collecting canals, the pyloric gland opens into the intestine near its attachment to the stomach. The secretory products of the pyloric glands may be digestive enzymes and substances involved in pH regulation. The gland also stores glycogen and plays a role in the removal of blood-borne toxins.

A few soft-bottom ascidians feed on deposited material from the surface of the surrounding sediment. Remarkably, some deep-water species are carnivores that feed on small animals such as nematodes and small, epibenthic crustaceans (Fig. 29-14B). These are caught with muscular lobes around the buccal siphon. The pharynx is small and glandular and has only a few gill slits.

Some tropical members of the colonial family Didemnidae have symbiotic algae in their tunics and atrial linings. One of these symbionts, *Prochloron,* is found only in ascidians. Structurally, *Prochloron* is a prokaryote, but its photosynthetic pigments are typical of eukaryotic green algae. These same ascidians also may have cyanobacterial symbionts embedded in the tunic. Excess photosynthate produced by the symbionts is assumed to

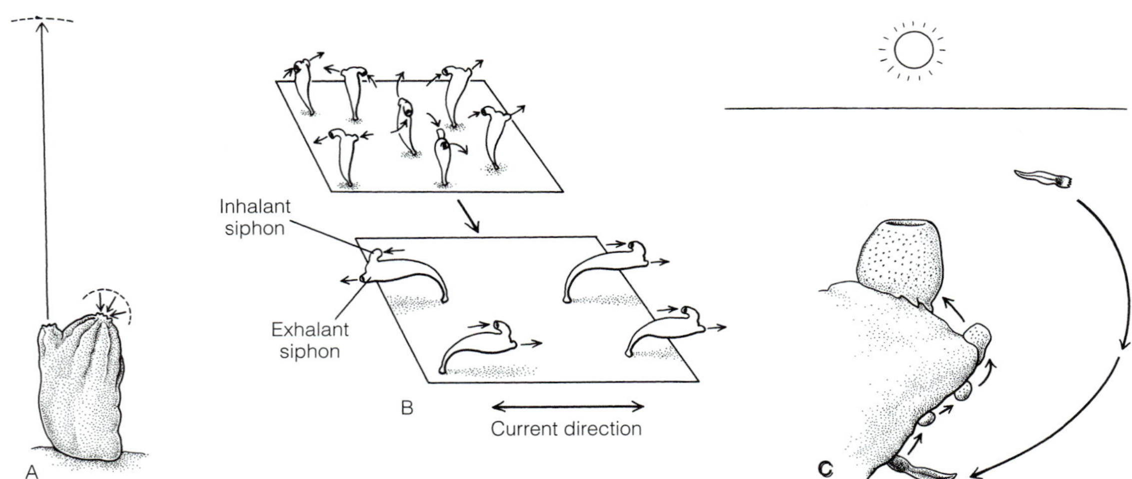

FIGURE 29-20 Tunicata: adaptive water pumping and body orientation in ascidians. **A,** Water pumping by the tropical *Ascidia nigra.* A powerful flow from the atrial siphon carries wastewater beyond the zone of water intake by the buccal siphon. **B,** Recently metamorphosed juveniles of *Styela montereyensis* may settle with their siphons oriented randomly in a bidirectional water current (arrows), but through differential mortality, those remaining have siphons oriented in the plane of the current, which enhances water flow through the animal. **C,** Movement and light adaptation in the tropical compound ascidian *Didemnum molle.* This species harbors the photosynthetic symbiont *Prochloron.* Larvae settle and transparent juvenile colonies develop on the shaded undersides of coral rubble, but as the colony grows it migrates slowly over the substratum into the fully lighted waters on the surface of the rock. This migration, which may take a month or more to complete, coincides with the development in the ascidian of masking pigments and calcareous spicules. These presumably filter the sunlight and control the intensity of light falling on the symbiotic algae. *(A, Redrawn from Berrill, N. J. 1950. The Tunicata. Quaritch, Ltd. London. 354 pp.; B, Modified and redrawn from Young, C. M., and Braithwaite, L. F. 1980. Orientation and current-induced flow in the stalked ascidian* Styela montereyensis. *Biol. Bull. 159:428–440; C, Redrawn after Olson, R. R. 1983. Ascidian–*Prochloron *symbiosis: The role of larval photoadaptations in midday larval release and settlement. Biol. Bull. 165:221–240.)*

be used by the tunicate as an auxiliary food source. At least one *Didemnum* species migrates over the substratum as it grows, perhaps to optimize the light intensity for photosynthesis by its *Prochloron* symbionts (Fig. 29-20C).

Internal Transport and Gas Exchange

Ascidians have a well-developed hemal system that includes a heart, vessels, and small sinuses through which blood flows in a definite circuit. The blood has a rich diversity of hemocytes, but lacks a respiratory pigment.

The heart is a short, curved, or U-shaped cylinder at the base of the digestive loop (Fig. 29-15A, 29-21A). On close examination, it consists of a cylinder with its outer wall folded inward along its length to form a tube within a tube (Fig. 29-21B). The outer tube is the fluid-filled pericardial cavity and the inner tube encloses the blood-filled heart lumen. The wall of the heart lumen (the inner tube), formed by the muscular invaginated wall of the pericardium, is the myocardium; the wall of the outer tube is the pericardium. The contractile filaments of the myocardium are arranged in two crossed helixes with an angle of 60° between them. Their action causes the heart to contract in a twisting, wringing motion that passes from one end of the heart to the other, pushing the blood forward and preventing backflow.

The blood vessels lack endothelia and are simple channels in connective tissue. Although the blood is frequently pigmented—often pale green—none of the pigments is respiratory in nature, and gases are transported in the plasma. The large surface area of the pharynx, the large volume of water pumped, and low metabolic activity all probably account for the absence of a respiratory protein. The hemocytes have a variety of other functions, however, including tunic synthesis, internal defense, and even defense against predators and competitors.

The blood circuit is not clearly established for any ascidian, but the placement of major vessels and the flow pattern through the pharynx are well understood. From the anterior end of the heart, blood enters a large ventral aorta (endostylar vessel) that lies below the endostyle and pharynx (Fig. 29-21A, 29-15B,C). The ventral aorta supplies blood to the gill capillaries of the pharynx (Fig. 29-15C). The capillaries join a mid-dorsal vessel, the dorsal aorta, that carries blood posteriorly to the gut and other viscera (Fig. 29-15B). Vessels from these viscera eventually return blood to the heart. In ascidians with a tunic circulation, such as *Ciona* and *Ascidia,* the tunic vessels originate at both ends of the heart (Fig. 29-21A).

There is growing evidence in tunicates (based largely on thaliacean anatomy) that the organs and tissues—gut, gonads, pharynx, and parts of the nervous system—are arranged in series, rather than in parallel, along the blood circuit (Fig. 29-21D,E). This unusual arrangement may explain the periodic reversal of blood flow, known elsewhere only in some arthropods. In general, **heartbeat reversal** occurs every few minutes when the heartbeat briefly stops and then resumes in the opposite direction. Because the body tissues supplied by blood are arranged in a series, one after another, those first in

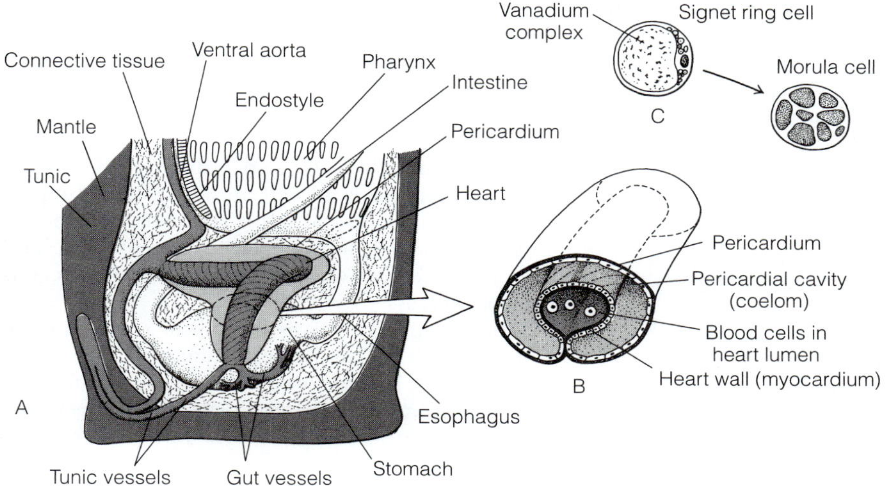

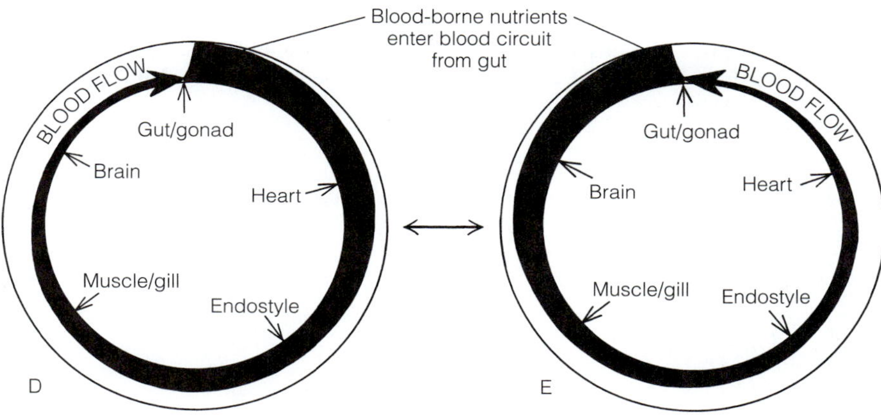

FIGURE 29-21 Tunicata: ascidian hemal system. **A,** The heart and major vessels of *Ciona*. **B,** Section of the heart wall from **A. C,** Immature and mature vanadocytes (morula cells) from *Ascidia*. **D** and **E,** Diagram of presumed series arrangement of organs in the blood circuit in forward **(D)** and reverse **(E)** blood flow. Shaded arrows show the depletion of blood-borne substances as blood flows through the circuit. Reversal of blood flow may be a means of averaging the distribution of substances such as oxygen and nutrients to the organs and tissues. *(A, Modified and redrawn from Berrill, N. J. 1936. Studies in tunicate development. Part VI—The evolution and classification of ascidians. Phil. Trans. Roy. Soc. Lond. B 226:43–70; B, Modified and redrawn from Oliphant, L. W., and Cloney, R. A. 1972. The ascidian myocardium: Sarcoplasmic reticulum and excitation–contraction coupling. Z. Zellforsch. 129:395–412; C, Drawn from an electron micrograph in Pirie, B. J. S., and Bell, M. V. 1984. The localization of inorganic elements, particularly vanadium and sulfur, in haemolymph from the ascidians* Ascidia mentula *(Muller) and* Ascidiella aspersa *(Muller). J. Exp. Mar. Biol. Ecol. 74:187–194; D and E, Based on a discussion of thaliacean circulatory system in Heron, A. C. 1975. Advantages of heart reversal in pelagic tunicates. J. Mar. Biol. Assoc. U.K. 55:959–963, and unpublished data of L. Corley.)*

line access a high concentration of blood-borne nutrients and oxygen, whereas those last in line receive a lower concentration of these substances. But when the heartbeat and blood flow reverse, tissues formerly at the end of the line are now at the beginning, where they can receive substances in high concentration. Thus, periodic heartbeat reversal may be an adaptation for averaging the supply of metabolites to tissues (Fig. 9-15B, 29-21D,E).

Heartbeat is myogenic and pacemakers at each end of the heart alternate in dominance and initiate the opposite directions of beat. Pacemaker fatigue or rising back pressure may generate the stimulus for beat reversal.

The blood of tunicates contains several kinds of hemocytes that can be grouped into three categories: totipotent **lymphocytes** (hemoblasts), which form in the connective tissue and give rise to all other types of cells; **amebocytes,** including phagocytic macrophages; and **vacuolated cells,** including morula cells and nephrocytes (discussed under Excretion).

Morula cells concentrate heavy metals in intracellular clusters of vesicles (Fig. 29-21C). Members of Pyuridae concentrate iron, others concentrate niobium or tantalum, and species of Ascidiidae and Perophoridae accumulate and store high concentrations of vanadium in their yellowish green morula cells, then called **vanadocytes.** The concentration of vanadium in seawater is approximately 5×10^{-8} M, whereas in vanadocytes it reaches 1 M (a 100 millionfold increase in vanadium concentration). The only other organism known to concentrate vanadium above the very low levels found in animals in general is the death cap mushroom, *Amanita muscaria.*

Vanadium itself is a potent inhibitor of many enzymes, including those responsible for the sodium-potassium pump, ciliary motility, and muscular contraction. It is also a powerful reducing agent. In ascidians, vanadium occurs in the +3 oxidation state, which is stable only at a pH of less than 2, and concentrated sulfuric acid is found in the cellular vesicles containing vanadium. Typically in ascidians, the vanadocytes enter the tunic from the blood and may align themselves immediately below the outer tunic surface, where they are sometimes called bladder cells (Fig. 29-18). The vanadocytes eventually disintegrate in the tunic, releasing their contents, or if the tunic is damaged by some external agency, vanadium and sulfuric acid are discharged to the exterior.

Vanadium in tunicates appears to have at least two functions. The reducing power of vanadium (and iron) is used in the polymerization of tunicin filaments, and hence in tunic synthesis. Second, it has been suggested that the toxicity of vanadium and the unpalatability of sulfuric acid may be used to discourage predation on tunicates and to prevent fouling of the tunic by the attachment and growth of other organisms.

Excretion

Unique among deuterostomes, the tunicates, including ascidians, lack conventional nephridia. Most lack excretory organs entirely and probably rely instead on the diffusion of ammonia across the pharynx and other tissues to dispose of nitrogenous wastes. Other metabolic byproducts, however, such as uric acid and calcium oxalate (urates) are stored internally and released only at the death of the zooid. This phenomenon is known as **storage excretion.** These urates may be stored in **nephrocytes,** which then accumulate in various tissues of the body, including the body wall, digestive loop, or gonads. Urates may also be secreted by epithelia into extracellular compartments such as the pericardial cavity (as in *Ascidia;* Fig. 29-22A) or two specialized structures.

One of these specialized excretory structures is the **epicardium,** which occurs in many aplousobranchs and a few phlebobranchs *(Ciona).* It consists of one (sometimes two) long, blind-ended pharyngeal diverticulum that extends posteriorly and lies beside the heart. Urate crystals, presumably secreted by the epicardial walls, accumulate in the epicardium, which remains open to the pharyngeal cavity (Fig. 29-22B).

In the stolidobranch *Molgula,* a specialization of the epicardium called a **renal sac** is located on the anterior (upper) surface of the heart (Fig. 29-22C). It is a closed, bean-shaped sac that lacks any opening to the exterior or interior. Urates are secreted into the sac, where they form a conspicuous whitish or yellowish mass.

A unicellular fungus called **nephromyces** occurs in the renal sacs of molgulids and in the pericardium of *Ascidia.* The nephromyces, in turn, harbors bacterial symbionts. The functional role of nephromyces is uncertain, but it has been suggested that they, perhaps in collaboration with their bacteria, may break down the stored urates.

Nervous System

Adult ascidians lack a dorsal hollow nerve cord and the CNS thus departs from the chordate ground plan. Modified remnants of the neural tube are the hollow cerebral ganglion (brain) and a hollow organ called the neural gland.

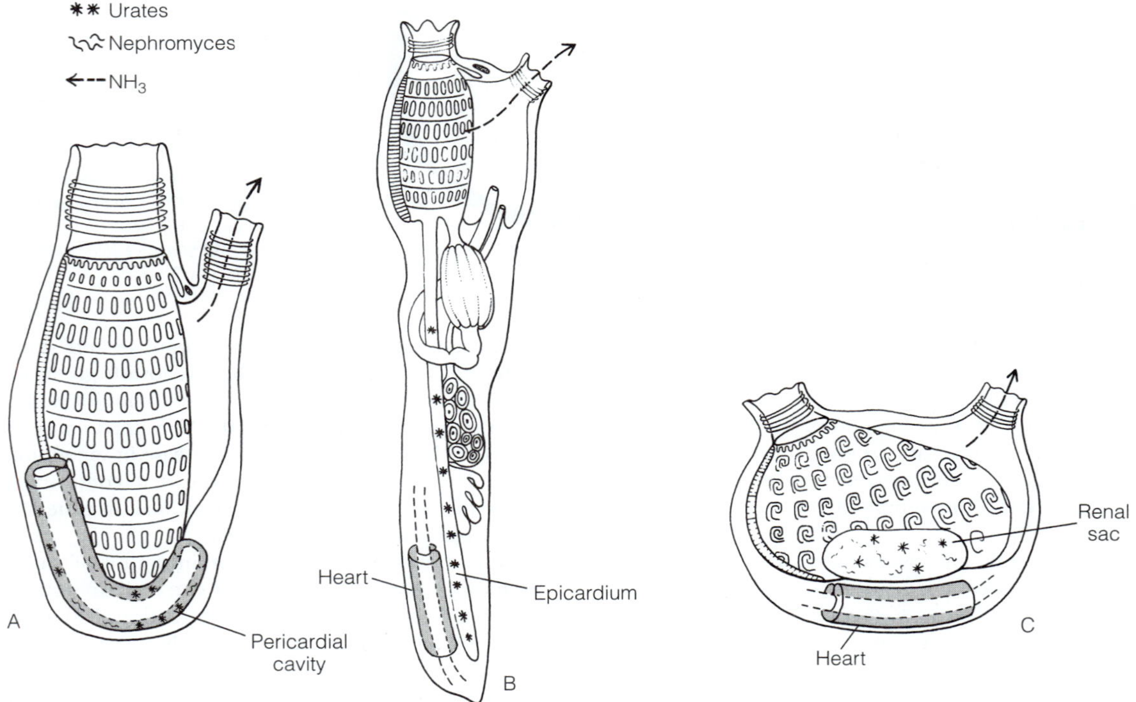

FIGURE 29-22 Tunicata: storage excretion in ascidians. **A,** In *Ascidia,* crystallized excretory wastes (urates) are stored in the pericardial cavity in association with a symbiotic fungus called *Nephromyces.* **B,** In some aplousobranchs, this storage function is adopted by one or two diverticula of the pharynx called epicardia, the lining of which secretes the urates into the epicardial cavity. The epicardial cavity remains open to the pharynx in aplousobranchs, but in the stolidobranch *Molgula* **(C)** and its relatives, the epicardium pinches free of the pharynx during development and becomes a closed renal sac. The renal sac contains urates and *Nephromyces,* which may help to break down the stored urate crystals.

The **cerebral ganglion** is a hollow cylindrical or spherical "brain" located in the connective tissue between the two siphons (Fig. 29-15A, 29-23). The nerves arising from the anterior end of the ganglion supply the buccal siphon and musculature; those from the posterior end innervate the greater part of the body—the atrial siphon, body-wall musculature, pharynx (in which they control ciliary arrest), and visceral organs.

In some, perhaps all, colonial ascidians, the tunic vessels propagate nerve impulses and are responsible for initiating vessel contraction and coordinated retraction of the zooids. These blood vessels are uniquely lined with epidermis (Fig. 29-18), which lacks neurons but nevertheless conducts impulses along the epithelium. It has been suggested that tunicate epithelial tissue in general, like the heart in particular, may have rhythmic electrical activity and conductive ability.

Beneath the cerebral ganglion lies a hollow, blind sac called the **neural gland** (Fig. 29-15A, 29-23), which, like the cerebral ganglion, originates from the embryonic neural tube. Contrary to its name, nerves are absent from the neural gland, and the presence of gland cells is questionable. A **ciliated duct,** also a neural tube remnant, extends from the anterior end of the neural gland and opens into the pharynx. This opening, a modified neuropore, is a large, often complexly coiled, ciliated funnel called the **dorsal tubercle.** Cilia in the duct create an incurrent flow of water that enters the gland, crosses its wall, and enters the branchial blood vessels. Cilia on the dorsal tubercle prevent large particles, such as sand grains, from entering the duct, while phagocytes lining the gland remove bacteria and other small particulates from the water. The dorsal tubercle, ciliated duct, and neural gland together function to restore and maintain the fluid volume of the blood, and are thus functionally similar to the madreporite, stone canal, and Tiedemann's bodies of some echinoderms. It has long been speculated that the neural gland, like Hatschek's pit in lancelets, is an evolutionary precursor of the vertebrate anterior pituitary gland, and there are reports of pituitary-like hormones in neural gland tissue.

Sensory organs are absent in adult ascidians, but sensory cells are abundant on the internal and external surfaces of the siphons, on the buccal tentacles, and in the atrium. These most likely play a role in controlling the current of water passing through the pharynx.

Reproduction and Development

With few exceptions, tunicates are hermaphroditic and cross-fertilization is typical, but a few species self-fertilize. Either a single testis and ovary (sometimes combined into an ovotestis) are located in the gut loop (Fig. 29-15A) or one to many gonads are located in the connective tissue of the body wall. The separate oviduct and sperm duct run parallel to the intestine and open into the atrium near the anus.

Solitary ascidians are oviparous and typically have small eggs with relatively little yolk. Eggs are shed from the atrial siphon and fertilization takes place in the sea. Often, the eggs are surrounded by special membranes that aid in buoyancy. Most colonial species are viviparous with yolky eggs and, like solitary species, lecithotrophic development. In a few viviparous species, however, the zooid supplies nutrition to the embryo and thus development is matrotrophic. The eggs are gestated in the oviduct or in specialized pockets in the atrium. Viviparous species release tadpole larvae.

Cleavage is holoblastic and bilateral and, unlike primitive deuterostomes, development is highly determinate (Fig. 9-22C). The coeloblastula gastrulates by epiboly as well as invagination and a large archenteron obliterates the blastocoel. The blastopore marks the posterior end of the embryo and closes as the embryo elongates along the anteroposterior axis. Mid-dorsally, the archenteron gives rise to the notochord (Fig. 29-24B, 29-25A). Laterally, the archenteron proliferates a solid band of mesodermal cells along each side of the body. The mesodermal bands neither arise by enterocoely nor later develop coelomic cavities. In this respect, ascidian (and tunicate) development departs from that of other deuterostomes. There is also no hint of mesodermal segmentation. The ectoderm along the mid-dorsal line differentiates as a neural plate, sinks inward, and rolls upward to form the neural tube. Later development leads to a lecithotrophic tadpole larva, described below (Fig. 29-24, 29-25). Because it is lecithotrophic, its larval life is short, usually

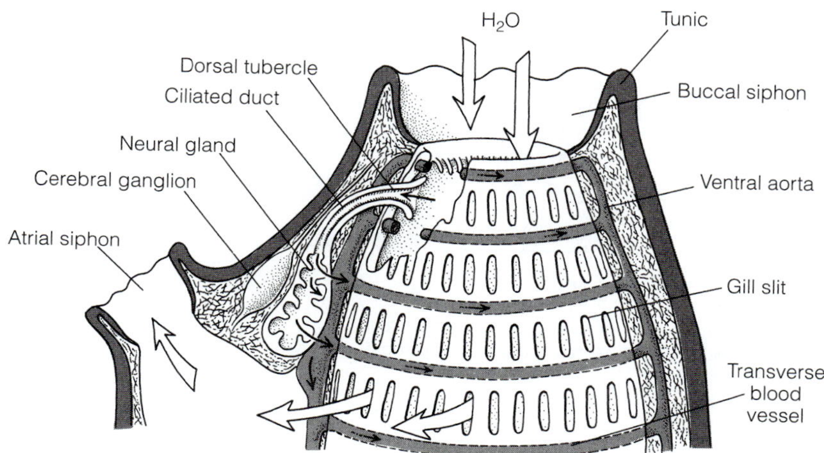

FIGURE 29-23 Tunicata: ascidian neural gland and volume regulation. The neural gland and its associated dorsal tubercle (the ciliated funnel) and ciliated duct pump seawater into the blood and thus help to regulate the fluid volume of the blood. The dorsal tubercle excludes large particles from the incurrent and the gland removes small particles by phagocytosis. *(Modified and redrawn from Ruppert, E. E. 1990. Structure, ultrastructure, and function of the neural gland complex of* Ascidia interrupta *(Chordata, Ascidiacea): Clarification of hypotheses regarding the evolution of the vertebrate anterior pituitary. Acta Zool. 71:135–149.)*

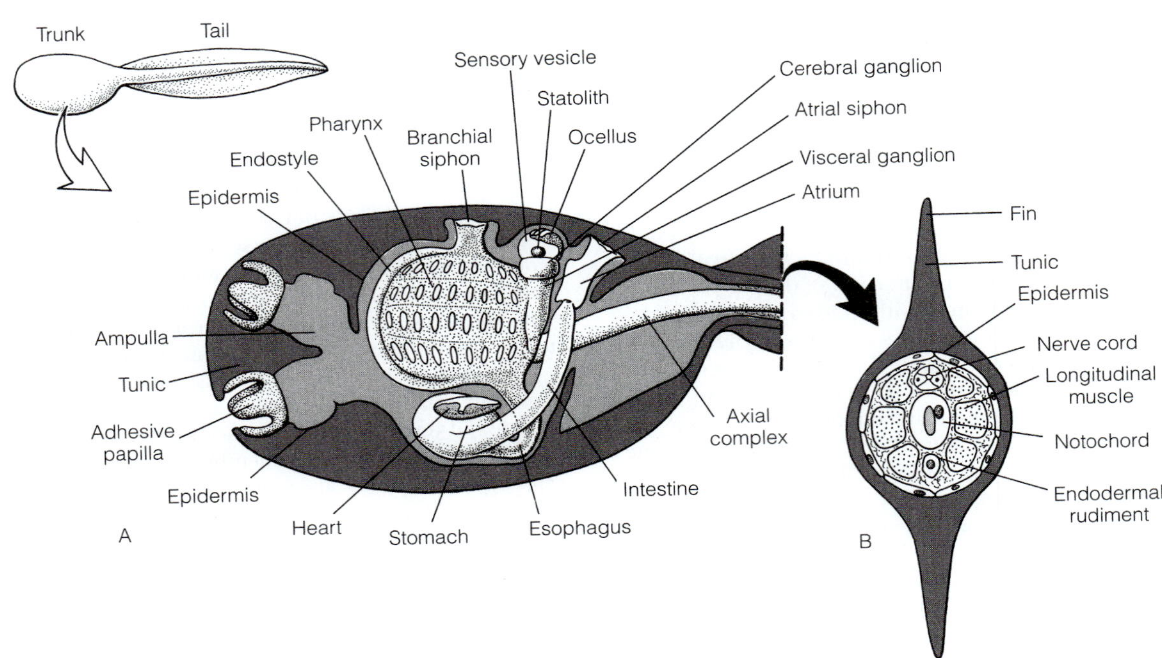

FIGURE 29-24 Tunicata: generalized ascidian tadpole. **A,** Left lateral view of the body emphasizing the anatomy of the prefabricated juvenile rudiment. The "axial complex" is the nerve cord and notochord. **B,** Cross section of the tail. The endodermal rudiment is a solid strand of endodermal cells, perhaps the evolutionary vestige of a once-functional gut. *(A, Modified and redrawn from a laboratory handout of Richard A. Cloney.)*

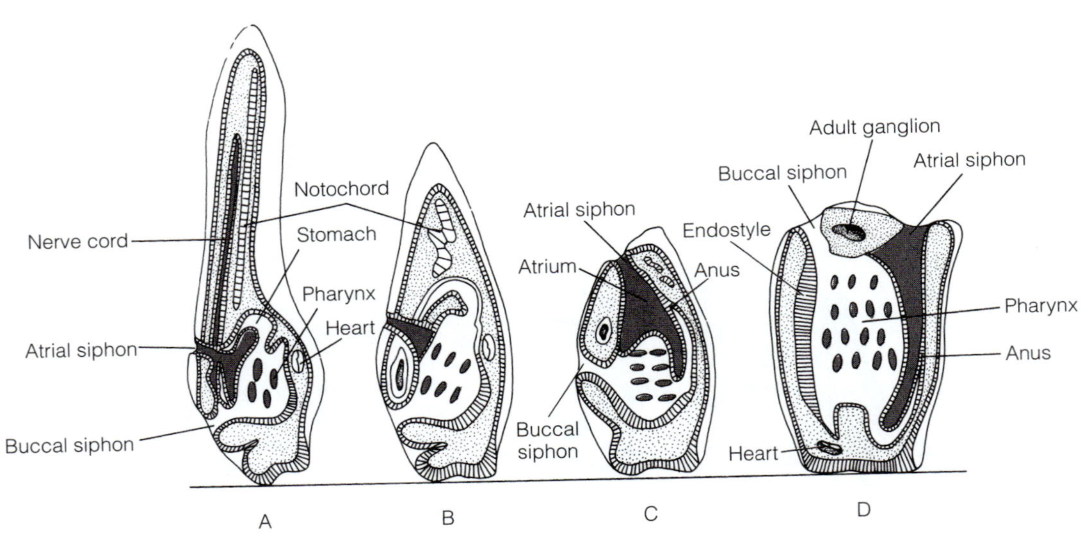

FIGURE 29-25 Tunicata: metamorphosis of the ascidian tadpole. **A,** Diagrammatic lateral view of an ascidian tadpole larva that has just attached to the substratum by its anterior end. **B** and **C,** Metamorphosis. **D,** Oozooid (juvenile) immediately after metamorphosis. *(Modified and redrawn from Seeliger.)*

not more that 36 h and sometimes as brief as a few minutes. During that period of time, the larva must find a suitable substratum for attachment and metamorphosis.

Tadpole Larva and Metamorphosis

Unlike the adult, the tadpole larva expresses all the key chordate traits, but it differs from the ancestral chordate in that it does not feed. In the absence of feeding, the larval gut and associated structures are nonfunctional and undifferentiated, or differentiated precociously to form structures that will function only in the juvenile and adult after metamorphosis. Thus the ascidian tadpole anatomy has a dual nature: a locomotory larval body that reflects a chordate ancestry and a prefabricated juvenile rudiment unique to ascidians.

The tadpole body is divided into a visceral trunk and locomotory tail (Fig. 29-24). These regions, however, may not correspond to the regions of the same name in cephalochordates and vertebrates. The trunk bears the cerebral vesicle and viscera and is chiefly a prefabricated juvenile zooid. The propulsive tail, which disappears during metamorphosis, contains the swimming musculature, notochord, dorsal neural tube, and an endodermal rudiment (Fig. 29-24). The notochord is a hollow tube, and its cellular sheath encloses a fluid-filled extracellular cavity. The tail has continuous dorsal and ventral fins that arise as folds of the larval tunic. Curiously, the tail of some colonial ascidians rotates 90° to the left during development, and thus the fins are horizontal rather than vertical in position (as in appendicularians, discussed later). The larval cerebral vesicle, which later gives rise to the adult cerebral ganglion, contains an ocellus and a statocyst, both within the dilated anterior end of the neural tube. The siphons, pharynx, and digestive system resemble those of the postmetamorphic juvenile, but are nonfunctional. Neither the buccal nor the atrial siphon opens to the exterior because they are covered by the larval cuticle (Fig. 29-24A). The endodermal rudiment in the tail is a solid band of endodermal cells located where one would expect to find a functional larval gut, if the tadpole had one (Fig. 29-24).

At the close of the free-swimming stage, the larva settles to the bottom and attaches with three anterior **adhesive papillae** (Fig. 29-24A). A radical metamorphosis ensues, during which the notochord loses its turgor and the tail is retracted and absorbed (Fig. 29-25). Tail retraction results in the loss of the notochord, dorsal hollow nerve cord, swimming musculature, and endodermal rudiment. As a result of rapid growth of the region between the adhesive papillae and buccal siphon, the body rotates 90°. This rotation positions the siphons upwards, away from the substratum. The atrium expands to enclose the anus and the pharynx. The number of gill slits rapidly increases. The siphons open to the exterior as the larval cuticle is molted, and the juvenile begins to feed.

Budding

Modular growth in colonial ascidians occurs by budding, but the site of bud formation differs among taxa. In general, the zooid that develops from a metamorphosed tadpole is called an **oozooid** and subsequent buds are known as **blastozooids.** The oozooid rarely, if ever, becomes sexually mature; sexual reproduction occurs in the blastozooids.

Stolonate species, such as *Perophora,* bud blastozooids from the stolon. In other taxa, the buds can arise from the abdomen, postabdomen, or even precociously in the larval stage, as in species of *Diplosoma* (= two bodies), in which the larval trunk houses two prefabricated zooids.

Diversity of Ascidiacea

Ascidian diversity manifests itself most obviously in solitary and colonial body forms and in patterns of modular growth. But this diversity has little value in systematics because coloniality apparently has evolved independently on several occasions. The taxonomically useful traits are all internal. Most important are the anatomical position of the gonads and the structure of the pharynx, as well as larval characters. Variations in pharynx structure include the number and form of gill slits and the degree of folding and vascularization of the pharyngeal wall (Fig. 29-26). Since these are characteristics that increase the pharyngeal area for food capture and gas exchange, they are correlated with body size.

Enterogona[O]: Gonads located in or beside gut loop; neural gland is below cerebral ganglion.

Aplousobranchia[sO] (= simple gills): Colonial, primarily compound species. Epicardia and sometimes a postabdomen are present. *Clavelina* (light bulb tunicates; Fig. 29-12B) zooids are not compound, but interjoined at bases. Zooids are tiny (except *Clavelina*). Pharynx lining is more or less flat, but transverse blood vessels protrude and help support the mucous feeding net (Fig. 29-26A). The few gill slits usually are oval, arranged in transverse rows. Tadpole tail is horizontal, rotated 90°. Juvenile rudiment is well differentiated in the tadpole. *Aplidium* (Fig. 29-12A; *A. stellatum* [sea pork]), *Didemnum* (paint-splash tunicate), *Diplosoma* (frog-egg tunicate), *Distaplia* (*D. bermudensis* [color wheel tunicate]), *Eudistoma* (Fig. 29-13; *E. hepaticum* [sea liver]), among others.

Phlebobranchia[sO] (= veined gills): Colonial or solitary, with or without an epicardium, lack a postabdomen. Lining of phlebobranch pharynx is either flat or undulating. Numerous raised papillae that protrude from the transverse pharyngeal vessels help to support the mucous feeding net and may improve water flow through it. In some taxa, papillae fuse in longitudinal rows to form longitudinal blood vessels that bulge above the pharyngeal lining like varicose veins (Fig. 29-26B). Number of gill slits ranges from few to hundreds. Tadpole tail is vertical, except in Perophoridae, which have a horizontal tail rotated 90°. Solitary forms include *Ascidia* (Fig. 29-12D), the well-known *Ciona intestinalis* of Europe and cold North American waters, the tropical *Phallusia.* Colonial forms include *Diazona, Ecteinascidia, Perophora* (Fig. 29-12C; *P. viridis* [honeysuckle tunicate]).

Pleurogona[O]: Gonad, or gonads, situated on inner surface of body wall. Epicardium is absent, except as a specialized renal sac in Molgulidae (Fig. 29-22C).

Stolidobranchia[sO] (= robed gills): Solitary or colonial. Pharyngeal lining is highly pleated with both transverse and longitudinal vessels (Fig. 29-26C,D). Gill-slit number varies with body size. Molgulidae has renal sacs and numerous spiral gill slits (Fig. 29-26D). Tadpole has a vertical tail and

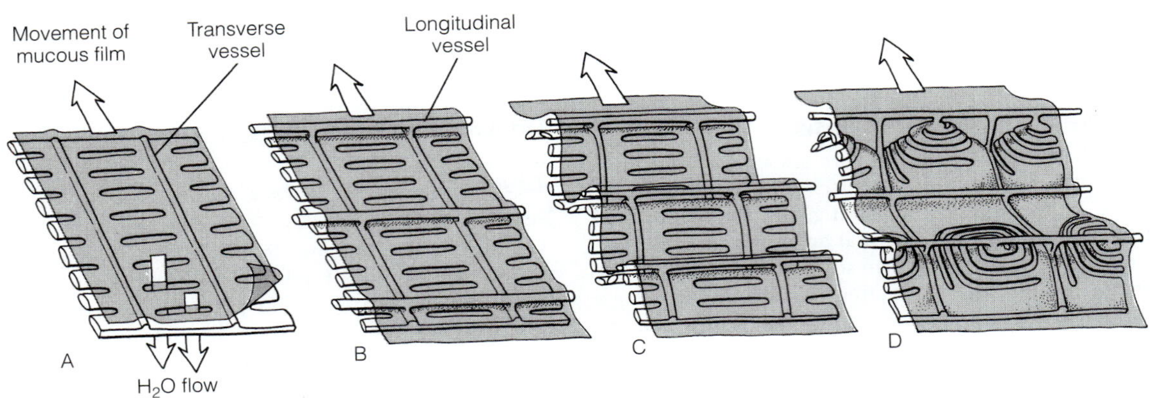

FIGURE 29-26 Tunicata: pharyngeal lining in relation to body size and systematics. The ascidian pharynx exhibits a size- and taxon-related trend to increase the area of its lining, gill slits, and blood vessels. **A,** The planar linings of aplousobranchs, such as *Aplidium* and *Eudistoma,* have only transverse vessels. **B,** Those of some phlebobranchs, for example *Ciona,* have both transverse and longitudinal vessels. The new longitudinal vessels not only increase the blood supply to the pharynx but also support and transport the mucous net across the lining. **C** and **D,** In stolidobranchs, both transverse and longitudinal vessels are present and the pharyngeal lining is pleated to increase its area. **D,** In molgulids, the gill slits are coiled. *(Modified and redrawn from Kott, P. 1989. Form and function in the Ascidiacea. Bull. Mar. Sci. 45:253–276.)*

an undifferentiated juvenile rudiment. Common solitary forms are *Halocynthia roretzi* (a Japanese edible ascidian), *Molgula* (sea grapes; Fig. 29-12G), *Pyura, Styela* (Fig. 29-12F). *Gasterascidia, Hexacrobylus, and Octonemus* (Fig. 29-14B) are solitary deep-sea carnivores. Compound forms include colorful *Botryllus* (Fig. 29-16A), *Botrylloides,* and *Symplegma* (Fig. 29-12E).

THALIACEA[C]

The other two tunicate taxa, Thaliacea and Appendicularia, are both specialized for a free-swimming, planktonic existence. Thaliacean zooids, which include the luminescent pyrosomes, barrel-shaped doliolids, and chainlike salps, differ from most ascidians in having the buccal and atrial siphons at opposite ends of the body. The water current is thus used not only for feeding and gas exchange, but in most also for jet propulsion. Like many other planktonic animals, thaliaceans are transparent. The tunic, which may be thin or thick, and the connective tissue are gelatinous and buoyant. Like ascidians, thaliaceans are mucous-net filter feeders that remove large quantities of minute suspended material from seawater. They may occur in enormous numbers over large areas of the sea, but most species occur in warm water. The approximately 75 species of thaliaceans are divided into three taxa, Pyrosomida, Doliolida, and Salpida.

Pyrosomida[O]

The 10 species of tropical pyrosomes are brilliantly luminescent colonies in the form of a hollow tube closed at one end (Fig. 29-27A,B). The colony length ranges from a few centimeters (as in *Pyrosoma atlanticum*) to over 20 m (as in *Pyrostremma spinosum*). The zooids are embedded in the wall (common tunic) of the tube, similar to the zooids of a compound ascidian. The buccal siphon of each zooid opens at the outer surface of the colony and the atrial siphon discharges into the hollow interior, or common cloaca (Fig. 29-27C). The exhaust of water from the cloaca provides thrust and moves the colony through the water (Fig. 29-27B). The diameter of the cloaca and cloacal aperture in *P. spinosum* can reach 2 m, large enough for a diver to swim inside the colony. Pharyngeal cilia produce both the feeding and locomotory currents. The pharynx has several gill slits and both transverse and longitudinal blood vessels. Two **luminescent organs,** one on each side of the pharynx, contain bioluminescent bacteria and are responsible for producing an intense light when the colony is disturbed (Fig. 29-27C).

Pyrosome colonies grow by adding new zooids, which originate as buds from an epicardial sac near the posterior end of the endostyle of the parent zooids (Fig. 29-27C). Sexual reproduction begins when the single egg in each zooid is fertilized internally. Development is internal, direct, and condensed. A tadpole larva is absent. The rudimentary oozooid (cyathozooid) undergoes precocious budding to form four blastozooids—the first four colonial zooids (Fig. 29-27E). The oozooid soon degenerates, except for its atrial siphon, which is retained by the growing colony as its exhaust aperture (Fig. 29-27F).

Doliolida[O]

The 23 species of doliolids have barrel-shaped zooids ranging in size from 1 to 50 mm (Fig. 29-28). The body-wall muscles are eight or nine hooplike bands around the circumference of the body. Contraction of these cross-striated muscles results in a jet thrust from the atrial siphon. Backflow through the buccal siphon, at the opposite end of the body, is prevented by a one-way valve. Although this muscular pumping is responsible for locomotion, the filter-feeding current is generated by gill-slit cilia, as it is in ascidians and pyrosomes.

The doliolid life cycle is the most complex of any free-living animal. It includes clonal (asexual) and sexual phases, solitary

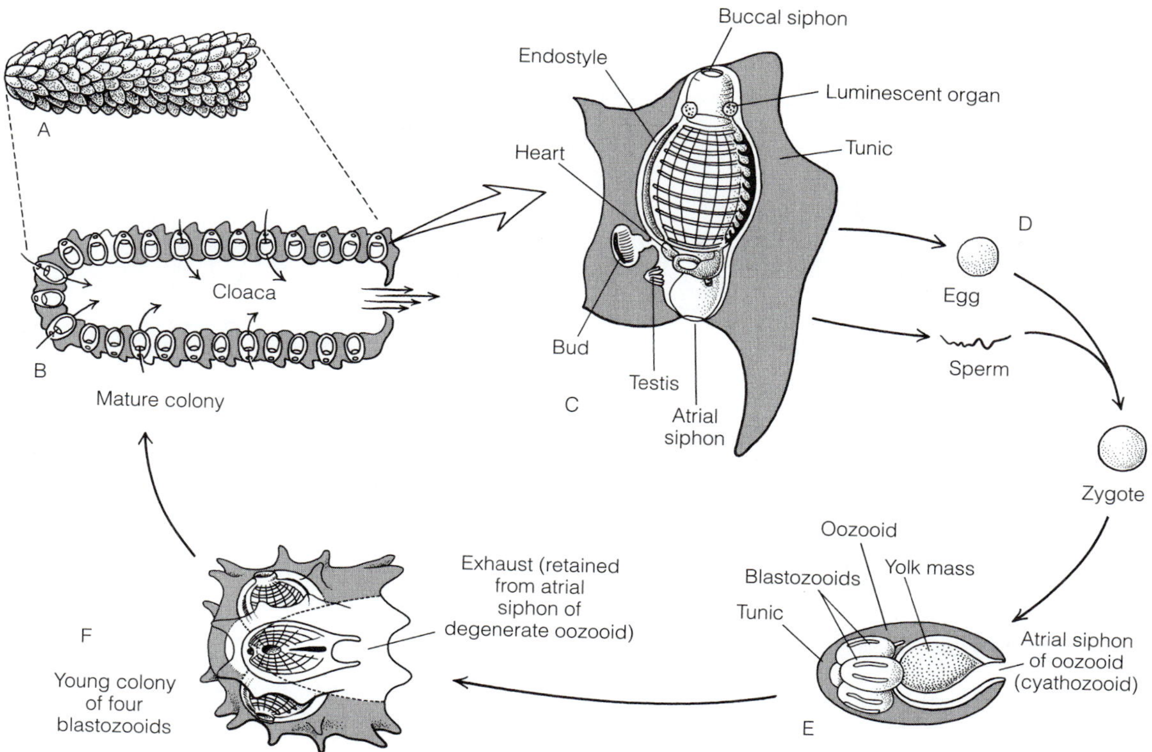

FIGURE 29-27 Tunicata: Pyrosomida. **A,** Adult colony of *Pyrosoma atlanticum.* **B,** Longitudinal section of **A** showing zooids, the common cloaca, and the exhaust aperture. Arrows indicate the path of water flow through the colony. **C,** Enlargement of one zooid from **B. D,** Spawned gametes. **E,** Lecithotrophic oozooid (cyathozooid) and its four precocious buds (blastozooids). **F,** Young colony composed of four zooids.

and colonial forms, and even a tadpole larva. The asexual phase enables an individual to increase its numbers rapidly in response to temporarily favorable conditions in the water.

The doliolid oozooid has nine muscle bands and a trailing tail-like appendage called a **spur.** A stolon grows out from an epicardial sac near the posterior end of the oozooid's endostyle and produces buds (blastozooids). Before the buds differentiate, however, they break free of the stolon, migrate over the surface of the oozooid, and insert on the spur (Fig. 29-28A). Once the buds have attached to the oozooid's spur, the oozooid is called a **nurse.** The attached blastozooid buds differentiate into one of three different kinds of zooids. **Trophozooids** have an enormous mouth and pharynx and are responsible for feeding the colony (the digestive system of the nurse degenerates; Fig. 29-28E). **Phorozooids** are locomotory zooids with eight muscle bands. They have their own short spur with buds that differentiate into **gonozooids,** the sexually reproducing members of the colony (Fig. 29-28B,C). When fully developed, the phorozooids and their attached gonozooids break free of the nurse's spur and adopt an independent existence (Fig. 29-28B). Fertilization is internal and development in the viviparous gonozooid is lecithotrophic. A single tadpole larva is released into the plankton from each gonozooid (Fig. 29-28C,D). The swimming tadpole has a muscular tail with a notochord, but a neural tube is absent and muscle contractions are myogenic. Metamorphosis of the tadpole gives rise to the oozooid and completes the cycle.

Salpida[O]

The 40-some species of salps have prism-shaped zooids with hooplike muscles, like doliolids, but some adjacent muscle bands are partly fused (Fig. 29-29). The zooids range in size from a few millimeters to 4 cm (as in *Salpa fusiformis*). The muscle bands of salps pump water for jet propulsion and, uniquely, also for filter feeding. The salp pharynx is peculiar in having only a single mid-dorsal gill bar and no side walls (Fig. 29-29A). The gill bar runs obliquely from dorsal to ventral and separates the pharynx from the atrium. They feed using an endostylar mucous net, but the net is rolled up and passed into the esophagus by esophageal cilia and cilia on the gill bar, which, functionally, is a dorsal lamina. They are found in the upper levels of all oceans but are more common in warmer seas.

Salps, like doliolids, alternate clonal and sexual generations. The salp oozooid produces a long stolon (up to several meters) that trails ventrally from the body (Fig. 29-29A). Along the stolon, blastozooid buds differentiate in groups separated from each other by division planes called **deploying**

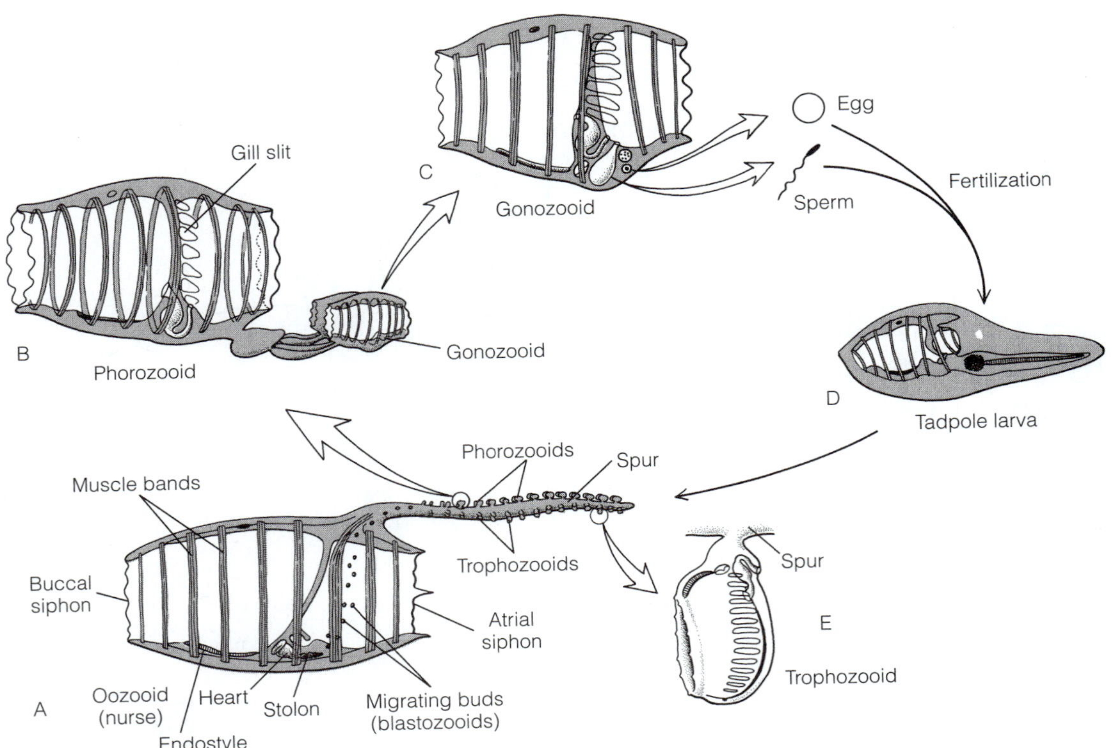

FIGURE 29-28 Tunicata: Doliolida. **A,** The doliolid nurse (oozooid) is a colony of polymorphic zooids. The nurse herself developed from a fertilized egg and subsequent tadpole larva **(D),** but the other members of the colony arose by budding from the nurse's stolon. The undifferentiated buds migrate from the stolon of the nurse and then lodge in her trailing spur, which may reach 50 cm or more in length. Once attached to the spur, the buds differentiate into trophozooids **(E),** which are specialized for feeding the colony (the nurse's digestive system degenerates), or phorozooids **(B),** which eventually break free of the spur and jet away under their own power. Buds attached to the phorozooids differentiate into the sexually reproductive gonozooids **(C).** Fertilization is probably internal in the gonozooids, but a free-swimming tadpole **(D)** is released to metamorphose into a young nurse in the plankton, thus completing the life cycle.

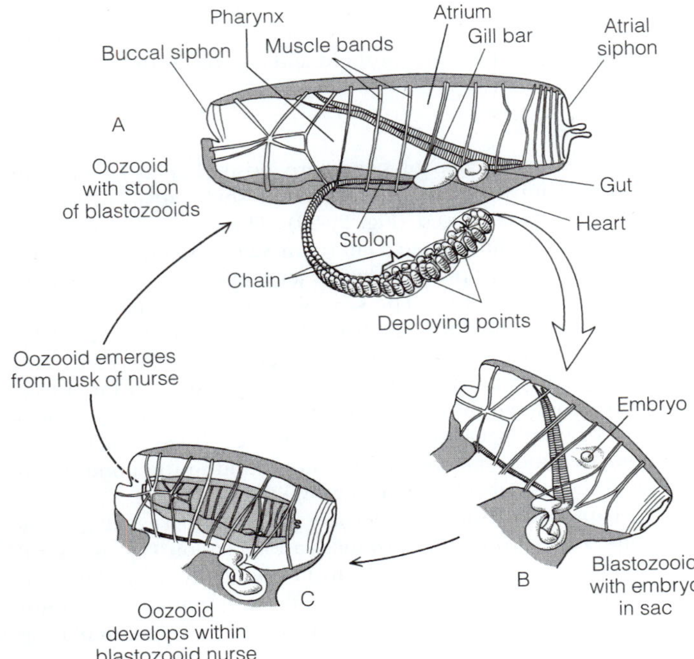

FIGURE 29-29 Tunicata: Salpida. Salp organization and life cycle. **A,** An oozooid of *Cyclosalpa* trailing its stolon of differentiating buds and blastozooids. The stolon breaks at predetermined points and deploys clusters, or chains, of blastozooids, which swim away from the oozooid and other members of the parent colony. Each blastozooid **(B)** bears a single egg that is fertilized internally and develops in a special brood sac, complete with a placental connection to the circulatory system of the blastozooid. The growing embryo eventually occupies the entire volume of the blastozooid-nurse's body **(C)** and then breaks free as a young oozooid, thus completing the life cycle.

points (Fig. 29-29A). Deployment liberates a group of blastozooids, now called a **chain.** Sexual reproduction occurs in the blastozooid. The blastozooids are viviparous and embryonic nutrition is matrotrophic. A single egg is fertilized internally and develops directly within a special pouch provided with a placental attachment to the parent's blood (Fig. 29-29B). The juvenile oozooid grows within the parent (nurse) and eventually occupies its entire body before casting off the nurse's hull and becoming an independent oozooid (Fig. 29-29C).

APPENDICULARIA^C (LARVACEA)

Most of the 70 species of Appendicularia are tiny filter-feeding tadpoles that abound in the surface and midwater plankton of the world's oceans. Under ideal conditions, their numbers may be astronomical: 25,620 individuals of *Oikopleura dioica* per cubic meter of water were reported from British Columbia. The name Appendicularia refers to their tail, which persists throughout the life of the animal. The alternative name, Larvacea, indicates their resemblance to the tadpole larva of ascidians (Fig. 29-30). Appendicularians are the only tunicates to have all the key chordate characteristics as adults.

Appendicularians resemble ascidian larvae in size as well as form (Fig. 29-30). Most adults are only a few millimeters in length, but the giant, cosmopolitan, midwater species *Bathochordaeus charon* and *B. stygius* reach 80 to 90 mm. The tail has dorsal and ventral fins, but because the entire structure is rotated 90° to the left, the fins are horizontal and the tail undulates in the vertical plane, as it does in some ascidian larvae. The root of the appendicularian tail, however, is shifted ventrally below the trunk. Often, as in *Oikopleura,* the tail is flexed anteriorly and projects forward in front of the mouth. Appendicularian fins are folds of epidermis, not outgrowths of the tunic, as in ascidian tadpoles. In fact, appendicularians lack a cellulose tunic. The mouth is located at the anterior tip of the body, and the ventral anus opens directly to the outside. The pharynx bears only *two* circular gill slits (spiracles), one on each side, and each opens directly to the exterior. Siphons and atrium are absent.

All appendicularians secrete a prefabricated mucous **"house,"** which they inflate by pumping it full of water with their tail (Fig. 29-31). Once inflated, the animal either attaches to the house with its mouth or occupies it with the entire body. *Fritillaria* attaches to the outside of the house, but species of *Oikopleura,* whose spherical or oval houses are the size of a pea or walnut, live within it. The giant *Bathochordaeus* builds houses that can reach 2 meters in diameter.

The house is the primary filter in the filter-feeding process. By pumping water through the house with its tail, the animal can filter more water and trap more food than it could with its tiny pharynx alone (Fig. 29-31). The house is necessary to collect sufficient food to meet the animal's energy demand. Feeding only with its pharynx may be analogous to drinking water through a capillary tube to quench your thirst. The concentration of food in the house can be 900 times greater than that in the surrounding seawater. An appendicularian actually ingests a highly concentrated broth.

The structure and function of the appendicularian house is best known for species of *Oikopleura.* The house contains a number of interconnecting passageways, two water inlets, and an unpaired outlet (Fig. 29-31B). Attached inside the house by its mouth, the undulating tail of the animal creates a water flow through the house. The two inlets are covered by a mucous screen that excludes all but the finest plankton ($\leqslant$ 0.1 to 8.0 μm, depending on the species). During its passage through the house, the water is strained a second time through two very fine feeding filters that unite in a food-collecting tube to which the animal's mouth is attached. The concentrated plankton is then transported to the mouth in a continuous stream resulting primarily from action of the ciliated gill slits. Food entering the pharynx is trapped on an endostyle-secreted mucous net. The net and food are gathered into a food cord and transported into the esophagus by a pair of peripharyngeal bands. Appendicularians feed primarily on fine phytoplankton and even colloidal particles that are unavailable to many other planktonic filter feeders.

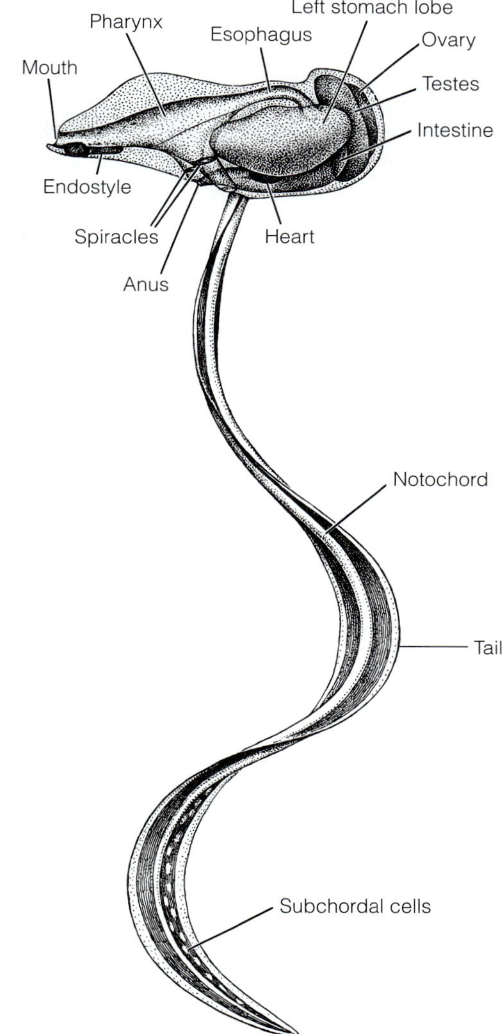

FIGURE 29-30 Tunicata: appendicularian anatomy. Lateral view of *Oikopleura albicans.* *(From Alldredge, A. 1976. Appendicularians. Sci. Am. 235:95–102.)*

The house is shed periodically—no more than every 4 h in *Oikopleura dioica*—and replaced. House building occurs in

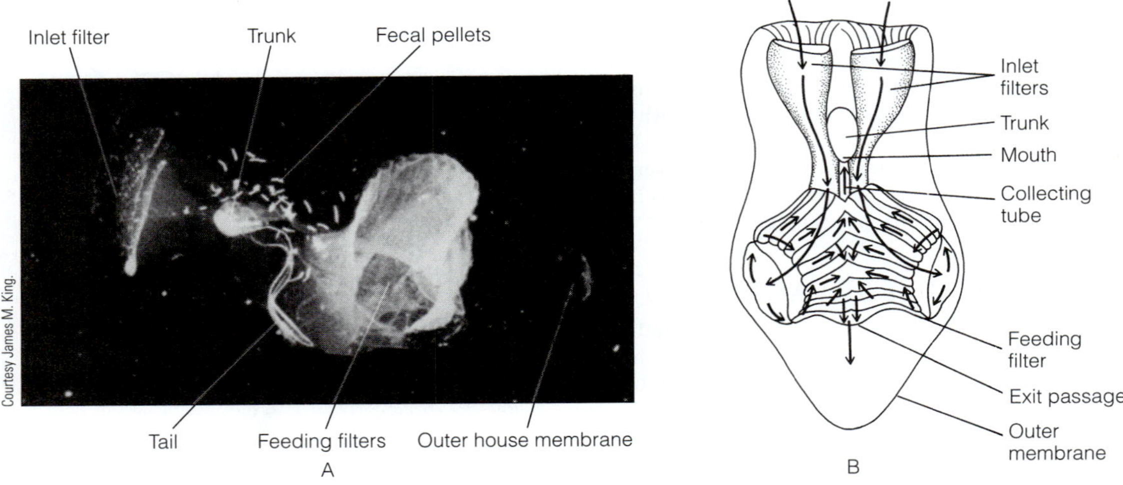

FIGURE 29-31 Tunicata: appendicularian house. **A,** *Megalocercus* in its house. **B,** Diagram of the house of *Megalocercus.* Long arrows indicate the path of the water flow; short arrows show the path of filtered particles. The oval represents the trunk of the animal. *(B, From Alldredge, A. Appendicularians. 1976. Sci. Am. 235:95–102.)*

several stages. First, the house is secreted by a specialized epidermis, the **oikoplast epithelium,** which is composed of a variety of secretory cells. Remarkably, the characteristic mosaic pattern of these cells in the epithelium is related to the pattern of structures in the house, as if the epithelium was a template on which the house was cast. As soon as the old house is abandoned, the new house rudiment swells and covers the oral end of the trunk like a muzzle. Then the animal enlarges its house slightly by nodding its trunk to and fro. This slight enlargement of the house allows the beating tail to slip in and complete the expansion. House inflation requires about 1 minute to complete. One animal can produce 4 to 16 houses a day, depending on temperature and food availability, and up to 46 houses can be secreted during the life span of one individual. The discarded houses are a significant contribution to marine "snow" (suspended organic particulates that resemble snow) and is an important substrate for microbial decomposition and nutrient recycling in the sea. House production ceases at the time of spawning.

Only sexual reproduction occurs in the Larvacea. All species are protandrous hermaphrodites, except *O. dioica,* which is gonochoric. Development is direct and rapid. At 22°C, it takes only 7 h for *O. dioica* to develop from zygote to a feeding juvenile. Hatchlings (after 3 h) are enclosed in a simple cuticle and have a posterior tail. Later, the tail shifts to its ventral adult position. The hatchling body is composed of a small fixed number of cells (eutely). Growth occurs by cell enlargement rather than by cell division.

PHYLOGENY OF CHORDATA

One must marvel at the evolutionary plasticity of the chordate ground plan. Vertebrates aside, the invertebrate chordates alone are responsible for the spongelike ascidians, the pelagic thaliaceans, some of which resemble jet engines, the sleek burrowing lancelets, and surely the most delicate, elegant, and elaborate of all filter feeders, the house-building appendicularians. In the process of evolving this incredible range of functional designs, the invertebrate chordates have modified many of their ancestral chordate traits, or dispensed with them altogether. At the extreme, as represented by many ascidians and thaliaceans, the body has become little more than a giant pharynx. The impressive range of variation in chordate design, size, and life history presents a challenge to the evolutionary biologist who attempts to draw relationships among such dissimilar creatures. Fortunately, headway is being made thanks to a combination of new morphological and biomolecular discoveries, but much new research is needed before a clear picture can emerge.

At a general level, the phylogeny of Chordata is clear-cut and noncontroversial. Cephalochordata and Vertebrata are sister taxa within Metameria, the segmented chordates (Fig. 29-32). The sister taxon of Metameria is Tunicata (Urochordata). Relationships within the Tunicata and the nature of the ancestral chordate, however, are a matter of spirited debate. One issue is the origin of the fishlike tadpole. From which nonchordate ancestor did it arise? And was its original form similar to that of an amphioxus, a vertebrate fish, an ascidian larva, or an appendicularian adult? Various evolutionary scenarios claim that echinoderms, hemichordates, and even some protostome taxa are ancestral to the chordates. The different models specify that the chordate ancestor resembled amphioxus, an ascidian tadpole, or an appendicularian.

Outside the chordates themselves, no deuterostome larva or adult resembles a tadpole or fish, and only the hemichordates have gill slits and hints of other chordate characters. Recent morphological and biomolecular evidence indicates an alliance between hemichordates and chordates. For example, it has been shown that some enteropneusts suspension feed with their pharynx, and Pax 1/9 genes are expressed similarly during gill-slit

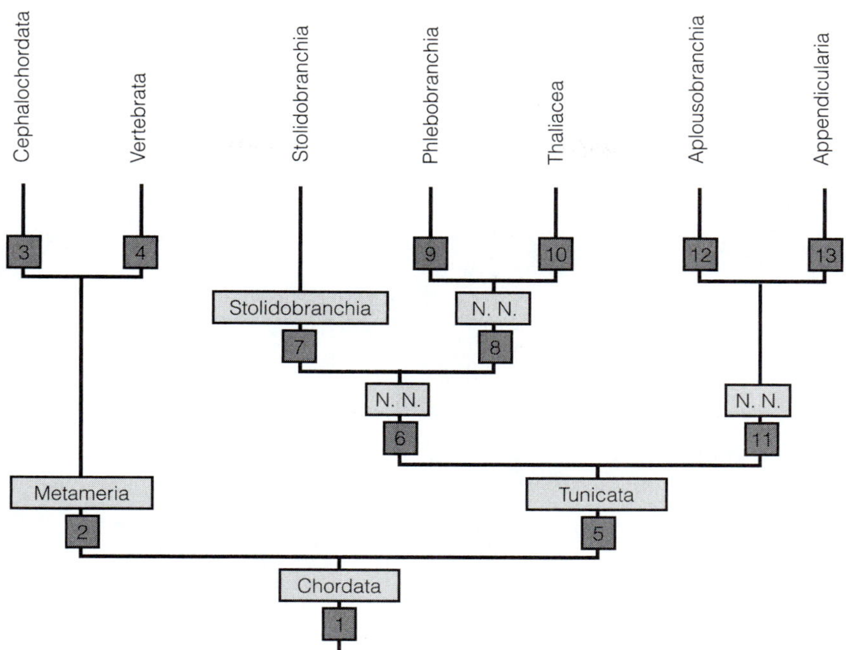

FIGURE 29-32 A chordate phylogeny. **1, Chordata:** Notochord has intracellular vacuoles, dorsal hollow nerve cord has an anterior neuropore. Pharynx has gill slits (a plesiomorphy) and an endostyle. Have a ventral heart and pericardial cavity, transverse (dorsoventral) branchial blood vessels; locomotory trunk and tail with cross-striated longitudinal musculature; the longitudinal gut extends posteriorly far into the trunk. The anus is ventral, there is a short postanal tail, fins are vertical. Plesiomorphies include monolayered epithelia, an atrium, paired coeloms, a mesothelium of epitheliomuscular cells, muscle cells innervated by diffusion of transmitters across the epidermal basal lamina (no motor fibers or end plates), a hemal system, blood vessels without endothelia, a metanephridial system, radial cleavage, enterocoely, and direct development. **2, Metameria:** Segmentation (metamerism). **3, Cephalochordata:** Muscular notochord extends into rostrum (anterior extremity), has myoglobin; staggered myomeres; cyrtopodocytes; and asymmetric planktotrophic larva. **4, Vertebrata:** Have multilayered (stratified) epithelia; blood vessels have endothelia, hemoglobin-containing hemocytes. Many complex structures are derived from embryonic mesenchyme (neural crest, dermatome, sclerotome). Have a glomerular metanephridial system (kidneys). **5, Tunicata:** Have a tunic, buccal tentacles, heartbeat reversal, a neural gland ventral to the cerebral ganglion, a pyloric gland. Nephridia and a coelom are absent. Body consists of a thorax only. Hermaphroditic; triradial larval adhesive papillae are noneversible; have a tubular notochord with an extracellular lumen; paired atrial invaginations later fuse. Have a vestigial ancestral gut (an endodermal rudiment), a novel juvenile rudiment with a U-shaped gut that differentiates during metamorphosis, gonads in the loop of the gut, bilateral cleavage, determinate development. **6, N. N.:** Have longitudinal branchial blood vessels. **7, Stolidobranchia:** Pharynx lining is pleated, have a neural gland dorsal to the cerebral ganglion, gonads in the body wall lateral to the pharynx, single atrial invagination during development. **8, N. N.:** Have papillae associated with the longitudinal branchial vessels. **9, Phlebobranchia:** No known autapomorphies. **10, Thaliacea:** Larvae are reduced, budding arises from epicardial sac at the posterior end of the endostyle. **11, N. N.:** Tail is horizontal, rotated 90° to left. Juvenile rudiment differentiates precociously in the tadpole, three eversible larval papillae are aligned in the midsagittal plane (these three characters also apply to the phlebobranch Perophoridae). **12, Aplousobranchia:** Zooid body is divided into two (includes also the phlebobranch Diazonidae) or three regions. Gonad is in the abdomen or postabdomen. **13, Appendicularia:** Constructs a mucous filter-feeding house. Exhibits the loss of the atrium, a sexually mature tadpole (pedomorphosis), and the loss of the ancestral adult. *(Based on Stach and Turbeville, 2002, and other sources.)*

morphogenesis in both taxa. For these reasons, the enteropneust-chordate link currently is attracting research attention.

Perhaps the chordate ancestor resembled an adult enteropneust. Like amphioxus, it was a burrowing suspension feeder, trapping food with its pharynx and gill slits supplemented by preoral structures (the proboscis and POCO in enteropneusts and the buccal cavity and wheel organ in amphioxi). But mucus-trapped food particles are transported dorsally in the chordate pharynx and ventrally in enteropneusts. The blood-flow direction also is opposite in the two groups: anterior in the dorsal vessel of enteropneusts and posterior in the dorsal vessel (aorta) of amphioxi. These and other diametrically opposite features of enteropneusts and chordates, however, can be resolved by inverting the body of the chordate in relation to the enteropneust.

A **dorsoventral axis inversion** in the evolution of chordates not only brings discrepant morphology into agreement, but also the localized expression of various genes during development (Fig. 29-33). At first glance, body inversion in chordate evolution may seem to be a fantastic concept, but recall that the normal feeding posture of both larval and adult cephalochordates is inverted—with the dorsal side down—so lancelets are, in fact, inverted (Fig. 29-4B). A careful examination of Figure 29-33A,B will also suggest possible evolutionary precursors of the chordate notochord, nerve

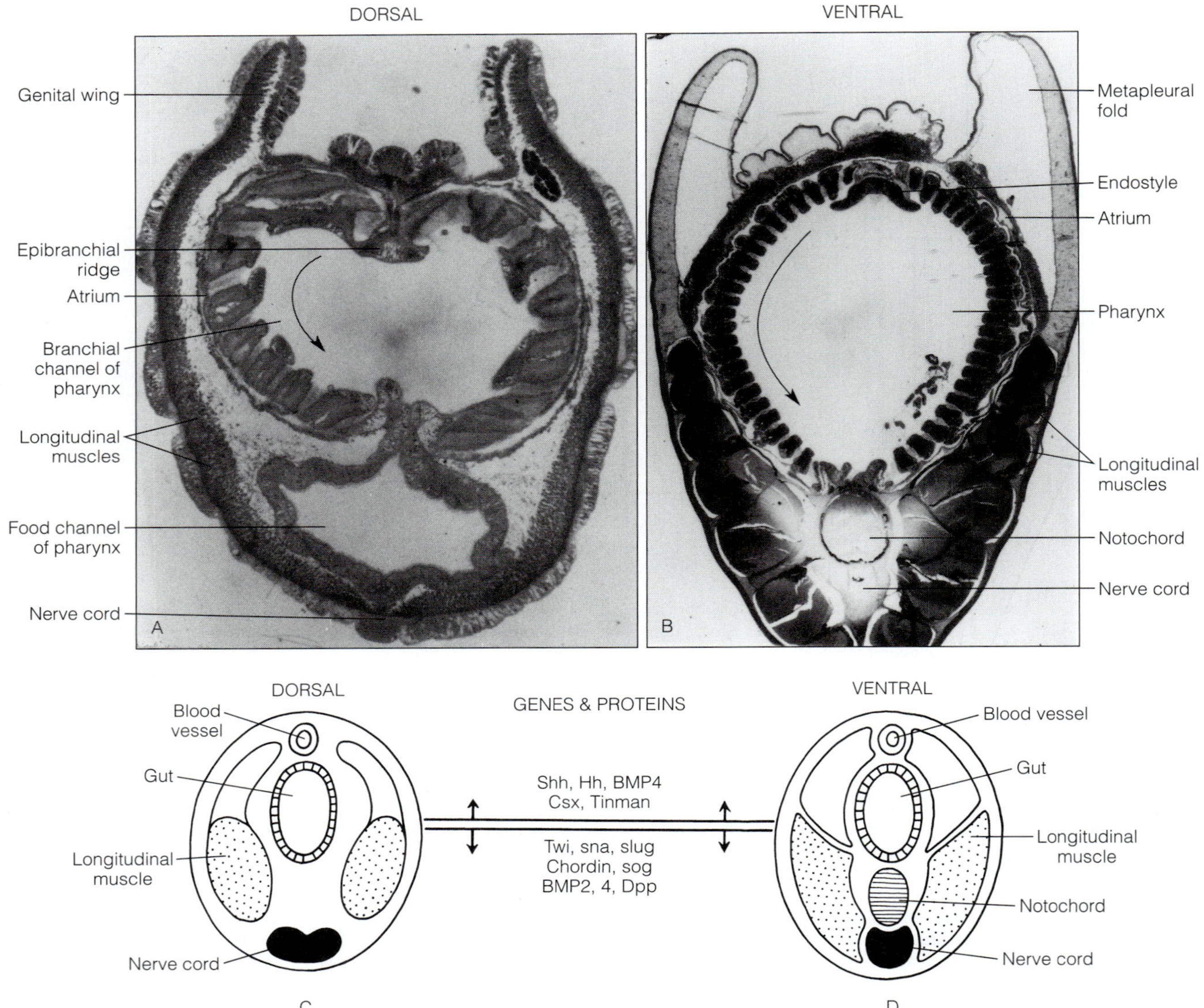

FIGURE 29-33 Dorsoventral axis inversion in chordate evolution. Compare the corresponding parts in the cross sections of the pharyngeal region of the enteropneust *Balanoglossus aurantiacus* **(A)** and the cephalochordate *Branchiostoma virginiae* **(B).** The enteropneust section is in a dorsal-side-up orientation, but the cephalochordate is inverted, with the dorsal side down (its *natural* orientation; see Fig. 29-4B). Arrows indicate the transport direction of food particles trapped in mucus on the pharyngeal lining. **C** and **D,** Generalized cross sections of a protostome arthropod (germ-band embryo); **(C)** and an *inverted* embryonic chordate **(D)** showing correspondences in the expression of various genes and proteins in relation to the dorsal and ventral sides of the two bodies. The biomolecules expressed dorsally in the protostome **(C)** are expressed ventrally in the deuterostome **(D),** whereas those expressed ventrally in the protostome **(C)** appear dorsally in the deuterostome **(D).** These data suggest that chordates are inverted with respect to nonchordates, including their nearest relatives, the hemichordates. *(C and D, Modified and redrawn from Gerhart, J., and Kirschner, M. 1997. Cells, Embryos, and Evolution. Blackwell Science, Malden, MA. 642 pp.)*

cord, and endostyle. The dorsoventral axis inversion hypothesis, however, rejects a homology between the enteropneust stomochord-collar cord and the chordate notochord-nerve cord. The stomochord and collar cord probably are unique to enteropneusts.

Most likely, the notochord evolved from a strip of gut wall, as it does developmentally in amphioxus. Internalization of the dorsal nerve cord, as occurs during chordate neurulation (Fig. 29-2), may have evolved to innervate a muscular notochord, as in amphioxus (Fig. 29-4E), or to provide a large surface on which to innervate the epitheliomuscular cells of the myomeres, also as indicated by amphioxus (Fig. 29-1B). The result of these changes to preexisting structures was the establishment of the tadpole or fishlike body form.

Once the tadpole body was established, the cephalochordates and vertebrates evolved segmentation, perhaps by expanding the genetic control of branchiomeric patterning to include the coelomic musculature. In contrast to the progressive evolution of lancelets and vertebrates, the tunicates generally are typified by the loss of traits associated with the chordate ground plan (Fig. 29-34). These include the loss of filtration nephridia, the absence (except for remnants) of the gut and anus in the tail, the absence of a coelom, and the loss of the notochord and nerve cord in adult ascidians and thaliaceans. How can these absences be explained in the evolutionary line to the tunicates?

One hypothesis is that tunicates evolved, pedomorphically, from an early developmental stage, perhaps a lecithotrophic larva, of the chordate ancestor. Because lecithotrophic larvae feed only on yolk stored in their endodermal and mesodermal cells, the gut typically does not differentiate until late in larval life or during metamorphosis. If tunicates evolved from a larva with such an undifferentiated tissue rudiment, they may have been "free" to restructure that rudiment in a new form and a new location, namely outside of the tail. This localized tissue became the juvenile (oozooid) rudiment in the trunk of the ascidian tadpole (Fig. 29-34B). Differentiation of that rudiment into a novel adult anatomy coincides with the loss of larval traits (the tail, notochord, and so on) in the same manner as differentiation of a starfish from a juvenile rudiment coincides with the loss of its larval body. The same argument applies to nephridia and coeloms: most lecithotrophic larvae lack both. Thus a pedomorphic origin of tunicates may explain their glaring departures from typical chordate and deuterostome design.

Whereas ascidians and thaliaceans evolved a new adult body form, the appendicularians capitalized on the larval body. With their few body cells, rapid rate of development, and absence of metamorphosis, appendicularians seem to have evolved from an ancestral ascidian larva that dispensed with the adult stage and adopted a specialized form of filter feeding (the mucous house). Here again is another example of pedomorphosis (Fig. 29-34C). The common 90° left rotation of the tail in appendicularians and aplousobranch (except *Clavelina*) and perophorid ascidians suggests a sister-taxon relationship between aplousobranch ascidians and Appendicularia. The juvenile rudiment in larvae of these ascidians undergoes precocious differentiation and thus a pedomorphic descendant of such a larva would have a functional gut. Although tunicate phylogeny is imperfectly understood, as reference to Figure 29-32 will indicate, it seems likely that pedomorphosis has played a role in the origin of both the tunicates and one of its most distinctive taxa, Appendicularia.

Ascidiacea is probably a paraphyletic taxon. Recent molecular and morphological analyses indicate that Tunicata consists of three monophyletic taxa: Stolidobranchia, Phlebobranchia plus Thaliacea, and Aplousobranchia plus Appendicularia (Fig. 29-32). If this hypothesis is supported by additional research, then the old Ascidiacea will lose its monophyletic status and be relegated to a primitive grade of organization that included a sessile benthic adult and a pelagic tadpole larva.

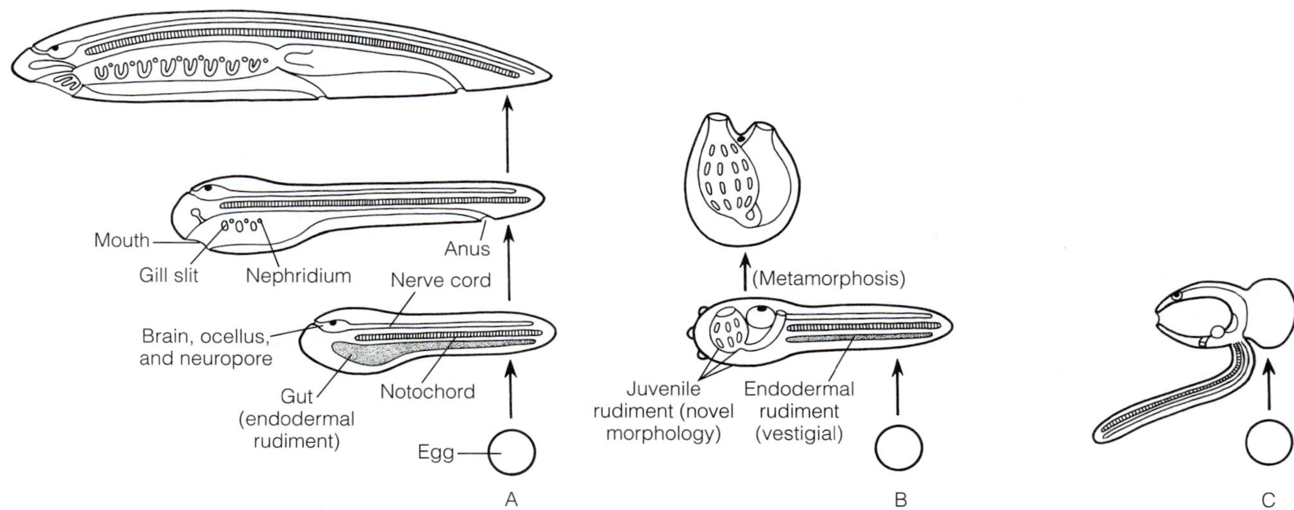

FIGURE 29-34 Chordate development and evolution. **A,** Developmental stages of a hypothetical ancestral chordate resembling a lancelet (Cephalochordata). The first stage following the egg is a lecithotroph with undifferentiated endoderm. Later stages show the progressive differentiation of typical chordate traits, including a long, straight gut and filtration nephridia. **B,** Developmental stages of Ascidiacea. The first stage following the egg is a lecithotrophic tadpole larva with a precociously differentiated juvenile rudiment, as found in aplousobranch (and perophorid) ascidians. Because the larva is lecithotrophic and does not feed with a differentiated mouth and gut, its undifferentiated endoderm is "free" to adopt a new form, the unique U-shaped gut of the juvenile ascidian zooid. The endodermal rudiment may be a vestige of the ancestral straight gut. It does not differentiate and is lost later in development. **C,** Developmental stages of Appendicularia. The form and short developmental period of appendicularians suggests that they may have evolved pedomorphically from an aplousobranch-like tadpole with a horizontal fin, but before the atrial invaginations occurred.

PHYLOGENETIC HIERARCHY OF CHORDATA

Chordata
- Metameria
  - Cephalochordata
  - Vertebrata
- Tunicata (Urochordata)
  - N. N.
    - Stolidobranchia
    - N. N.
      - Phlebobranchia
      - Thaliacea
  - N. N.
    - Aplousobranchia
    - Appendicularia

REFERENCES

CEPHALOCHORDATA

Conklin, E. G. 1932. The embryology of amphioxus. J. Morphol. 54:69–151.

Flood, P. 1970. The connection between spinal cord and notochord in amphioxus *(Branchiostoma lanceolatum).* Z. Zellforsch. 103:115–128.

Flood, P. 1975. Fine structure of the notochord of amphioxus. Symp. Zool. Soc. Lond. 36:81–104.

Frick, J. E., and Ruppert, E. E. 1997. Primordial germ cells and oocytes of *Branchiostoma virginiae* (Cephalochordata, Acrania) are flagellated epithelial cells: Relationship between epithelial and primary egg polarity. Zygote 5: 139–151.

Frick, J. E., and Ruppert, E. E. 2001. Preliminary nutritional analysis of lancelets, a promising seafood with aquacultural potential. J. Aquat. Food Prod. Technol. 10:63–75.

Gans, C., Kemp, N., and Poss, S. (Eds.): 1996. The lancelets (Cephalochordata): A new look at some old beasts. Israel J. Zool. 42 (Suppl.):S1–S446.

Gilmour, T. 1996. Feeding methods of cephalochordate larvae. Israel J. Zool. 42 (Suppl.):S87–S95.

Holland, N. D., and Holland, L. Z. 1991. The histochemistry and fine structure of the nutritional reserves in the fin rays of a lancelet, *Branchiostoma lanceolatum* (Cephalochordata = Acrania). Acta Zool. 72:203–207.

Lacalli, T. C. 2001. New perspectives on the evolution of protochordate sensory and locomotory systems, and the origin of brains and heads. Phil. Trans. Roy. Soc. Lond. B 356: 1565–1572.

Ogasawara, M. 2000. Overlapping expression of amphioxus homologs of the thyroid transcription factor-1 gene and thyroid peroxidase gene in the endostyle: Insight into evolution of the thyroid gland. Dev. Genes Evol. 210: 231–242.

Rähr, H. 1979. The circulatory system of amphioxus (*Branchiostoma lanceolatum* Pallas). Acta Zool. 60:1–18.

Ruppert, E. E. 1997. Cephalochordata (Acrania). In Harrison, F. W., and Ruppert, E. E. (Eds.): Microscopic Anatomy of Invertebrates. Vol. 15. Wiley-Liss, New York. pp. 349–504.

Ruppert, E. E., Nash, T. R., and Smith, A. J. 2000. The size range of suspended particles trapped and ingested by the filter-feeding lancelet *Branchiostoma floridae* (Cephalochordata: Acrania). J. Mar. Biol. Assoc. U.K. 80:329–332.

Stach, T., and Eisler, K. 1998. The ontogeny of the nephridial system of the larval amphioxus *(Branchiostoma lanceolatum).* Acta Zool. 79:113–118.

Stach, T. 1999. The ontogeny of the notochord of *Branchiostoma lanceolatum.* Acta Zool. 80:25–33.

Stokes, M. D. 1997. Larval locomotion of the lancelet *Branchiostoma floridae.* J. Exp. Biol. 200:1661–1680.

Stokes, M. D., and Holland, N. D. 1995. Ciliary hovering in larval lancelets. Biol. Bull. 188:231–233.

Stokes, M. D., and Holland, N. D. 1995. Embryos and larvae of a lancelet, *Branchiostoma floridae,* from hatching through metamorphosis: Growth in the laboratory and external morphology. Acta Zool. (Stockh.) 76:105–120.

Webb, J. E. 1973. The role of the notochord in forward and reverse swimming in the amphioxus, *Branchiostoma lanceolatum.* J. Zool. Lond. 170:325–338.

Whittaker, J. R. 1997. Cephalochordates, the lancelets. In Gilbert, S. F., and Raunio, A. M. (Eds.): Embryology: Constructing the Organism. Sinauer, Sunderland, MA. pp. 365–381.

Willey, A. 1894. Amphioxus and the Ancestry of the Vertebrates. MacMillan, New York. 316 pp.

TUNICATA

General

Alldredge, A. L. 1976. Appendicularians. Sci. Am. 235:94–102.

Alldredge, A. L. 1977. House morphology and mechanisms of feeding in the Oikopleuridae (Tunicata, Appendicularia). J. Zool. Lond. 181:175–188.

Alldredge, A. L., and Madin, L. P. 1982. Pelagic tunicates: Unique herbivores in the marine plankton. BioScience 32:655–663.

Arkett, S. A., Mackie, G. O., and Singla, C. L. 1989. Neuronal organization of the ascidian (Urochordata) branchial basket revealed by cholinesterase activity. Cell Tiss. Res. 27:285–294.

Barham, E. G. 1979. Giant larvacean houses: Observations from deep submersibles. Science 205:1129–1131.

Berrill, N. J. 1935. Studies in tunicate development. Part IV—Asexual reproduction. Phil. Trans. Roy. Soc. Lond. B 225:327–379.

Berrill, N. J. 1936. Studies in tunicate development. Part V. The evolution and classification of ascidians. Phil. Trans. Roy. Soc. Lond. B 226:43–70.

Berrill, N. J. 1950. The Tunicata. Quaritch, London. 354 pp.

Berrill, N. J. 1975. Chordata: Tunicata. In Giese, A. C., and Pearse, J. S. (Eds.): Reproduction of Marine Invertebrates. Vol. II. Academic Press, New York. pp. 241–282.

Bone, Q., Braconnot, J. C. and Carre, C. 1997. On the heart and circulation in doliolids (Tunicata: Thaliacea). Sci. Mar. 61:189–194.

Bone, Q., Braconnot, J. C., Carre, C., and Ryan, K. P. 1997. On the filter-feeding of *Doliolum* (Tunicata: Thaliacea). J. Exp. Mar. Biol. Ecol. 214:179–194.

Bone, Q. (Ed.): 1998. The Biology of Pelagic Tunicates. Oxford University Press, Oxford. 340 pp.

Brien, P. 1948. Embranchement des Tuniciers. In Grassé, P. (Ed.): Traité de Zoologie. Vol. 11. Echinodermes, Stomocordes, Procordes. Masson, Paris. pp. 553–930.

Bullough, W. S. 1958. Practical Invertebrate Anatomy. 2nd Edition. Macmillan, New York. 483 pp.

Burighel, P., and Cloney, R. A. 1997. Urochordata: Ascidiacea. In Harrison, F. W., and Ruppert, E. E. (Eds.): Microscopic Anatomy of Invertebrates. Vol. 15. Hemichordata, Chaetognatha, and the Invertebrate Chordates. Wiley-Liss, New York. pp. 221–347.

Burighel, P., Brena, C., Martinucci, G. B., and Cima, F. 2001. Gut ultrastructure of the appendicularian *Oikopleura dioica* (Tunicata). Invert. Biol. 120: 278–293.

Burighel, P., Sorrentino, M., Zaniolo, G., Thorndyke, M. C., and Manni, L. 2001. The peripheral nervous system of an ascidian, *Botryllus schlosseri,* as revealed by cholinesterase activity. Invert. Biol. 120:185–198.

Cloney, R. A. 1982. Ascidian larvae and the events of metamorphosis. Am. Zool. 22:817–826.

Deibel, D. 1986. Feeding mechanism and house of the appendicularian *Oikopleura vanhoeffeni.* Mar. Biol. 93:429–436.

Fenaux, R. 1985. Rhythm of secretion of oikopleurid's houses. Bull. Mar. Sci. 37:498–503.

Fenaux, R. 1986. The house of *Oikopleura dioica* (Tunicata, Appendicularia): Structure and function. Zoomorphology 106:224–231.

Flood, P. R. 1982. Transport speed of the mucous feeding filter in *Clavelina lepadiformis* (Aplousobranchia, Tunicata). Acta Zool. 63:17–23.

Flood, P. R. 1990. Visualization of the transparent, gelatinous house of the pelagic tunicate *Oikopleura vanhoeffeni* using *Sepia* ink. Biol. Bull. 178:118–125.

Flood, P. R. 1991. A simple technique for preservation and staining of the delicate houses of oikopleurid tunicates. Mar. Biol. 108:105–110.

Flood, P. R. 1991. Architecture of, and water circulation and flow rate in, the house of the planktonic tunicate *Oikopleura labradoriensis.* Mar. Biol. 111:95–111.

Flood, P. R., and Fiala-Medioni, A. 1979. Filter characteristics of ascidian food trapping mucous films. Acta Zool. 60:271.

Gibson, D. M., and Paffenhofer, G. A. 2000. Feeding and growth rates of the doliolid, *Dolioletta gegenbauri* Uljanin (Tunicata, Thaliacea). 22:1485–1500.

Godeaux, J. E. A. 1989. Functions of the endostyle in the tunicates. Bull. Mar. Sci. 45:228–242.

Goodbody, I. 1974. The physiology of ascidians. Adv. Mar. Biol. 12:1–149.

Harbison, G. R., and McAlister, V. L. 1979. The filter-feeding rates and particle retention efficiencies of three species of *Cyclosalpa.* Limnol. Oceanogr. 24:875–892.

Heron, A. C. 1975. Advantages of heart reversal in pelagic tunicates. J. Mar. Biol. Assoc. U.K. 55:959–963.

Heron, A. C. 1976. A new type of excretory mechanism in the tunicates. Mar. Biol. 36:191–197.

Holland, L. Z. 1989. Fine structure of spermatids and sperm of *Dolioletta gegenbauri* and *Doliolum nationalis* (Tunicata: Thaliacea): Implications for tunicate phylogeny. Mar. Biol. 101:83–95.

Holland, L. Z., and Holland, N. D. 2001. Evolution of neural crest and placodes: Amphioxus as a model for the ancestral vertebrate? J. Anat. 199:85–98.

Jeffery, W. R. 1997. Evolution of ascidian development. BioScience 47:417–425.

Jeffery, W. R., and Swalla, B. J. 1997. Tunicates. In Gilbert, S., and Raunio, A. M. (Eds.): Embryology: Constructing the Organism. Sinauer, Sunderland, MA. pp. 331–364.

Jones, J. C. 1971. On the heart of the orange tunicate, *Ecteinascidia turbinata.* Biol. Bull. 141:130–145.

Katz, M. J. 1983. Comparative anatomy of the tunicate tadpole, *Ciona intestinalis.* Biol. Bull. 164:1–27.

Kott, P. 1984. Prokaryotic symbionts with a range of ascidian hosts. Bull. Mar. Sci. 34:308–312.

Kott, P. 1989. Form and function in the Ascidiacea. Bull. Mar. Sci. 45:253–276.

Lambert, C. C., Lambert, G., Crundwell, G., and Kantardjieff, K. 1998. Uric acid accumulation in the solitary ascidian *Corella inflata.* J. Exp. Zool. 282:323–331.

Lambert, G. 1998. Spicule formation in the solitary ascidian *Bathypera feminalba* (Ascidiacea, Pyuridae). Invert. Biol. 117:341–349.

Lambert, G., Lambert, C. C., and Waaland, J. R. 1996. Algal symbionts in the tunics of six New Zealand ascidians (Chordata, Ascidiacea). Invert. Biol. 115:67–78.

Lane, D. J. W., and Wilkes, S. L. 1988. Localisation of vanadium, sulfur and bromine within the vandocytes of *Ascidia mentula* Muller: A quantitative electron probe X-ray microanalytical study. Acta Zool. 69:135–145.

Macara, I. G. 1980. Vanadium—An element in search of a role. Trends Biochem. Sci. 5:92–94.

Mackie, G. O. 1978. Luminescence and associated effector activity in *Pyrosoma* (Tunicata: Pyrosomida). Proc. Roy. Soc. Lond. B 202:483–495.

Mackie, G. O. 1983. Coordination of compound ascidians by epithelial conduction in the colonial blood vessels. Biol. Bull. 165:209–220.

Mackie, G. O. 1986. From aggregates to integrates: Physiological aspects of modularity in colonial animals. Phil. Trans. Roy. Soc. Lond. B 313:175–196.

Markus, J. A., and Lambert, C. C. 1983. Urea and ammonia excretion by solitary ascidians. J. Exp. Mar. Biol. Ecol. 66:1–10.

McHenry, M. J. 2001. Mechanisms of helical swimming: Asymmetries in the morphology, movement and mechanics of larvae of the ascidian *Distaplia occidentalis.* J. Exp. Biol. 204:2959–2973.

Meinertzhagen, I. A., and Okamura, Y. 2001. The larval ascidian nervous system: The chordate brain from its small beginnings. Trends Neurosci. 24:401–410.

Millar, R. H. 1970. British Ascidians. Synopses of the British Fauna, no. 1. Academic Press, London. 92 pp.

Millar, R. H. 1971. The biology of ascidians. Adv. Mar. Biol. 9:1–100.

Monniot, C., and Monniot, F. 1975. Abyssal tunicates: An ecological paradox. Ann. Inst. Oceanogr. 51:99–129.

Monniot, C., and Monniot, F. 1978. Recent work on the deep-sea tunicates. Oceanogr. Mar. Biol. Ann. Rev. 16: 181–228.

Nishida, H. 1987. Cell lineage in ascidian embryos by intracellular injection of a tracer enzyme. III. Up to the tissue restricted stage. Dev. Biol. 121:526–541.

Nishino, A., Satou, U., Morisawa, M., and Satoh, N. 2001. *Brachyury* (T) gene expression and notochord development in *Oikopleura longicauda* (Appendicularia, Urochordata). Dev. Genes Evol. 211:219–231.

Oliphant, L. W., and Cloney, R. A. 1972. The ascidian myocardium: Sarcoplasmic reticulum and excitation–contraction coupling. Z. Zellforsch. 129:395–412.

Olsson, R. 1965. Comparative morphology and physiology of the *Oikopleura* notochord. Israel J. Zool. 14:213–220.

Olson, R. R. 1983. Ascidian–*Prochloron* symbiosis: The role of larval photoadaptations in midday larval release and settlement. Biol. Bull. 165:221–240.

Parry, D. L., and Kott, P. 1988. Co-symbiosis in the Ascidiacea. Bull. Mar. Sci. 42:149–153.

Pennachetti, C. A. 1984. Functional morphology of the branchial basket of *Ascidia paratropa* (Tunicata, Ascidiacea). Zoomorphology 104:216–222.

Petersen, J. K., and Svane, I. 2002. Filtration rate in seven Scandinavian ascidians: Implications of the morphology of the gill sac. Mar. Biol. 140:397–402.

Pirie, B. J. S., and Bell, M. V. 1984. The localization of inorganic elements, particularly vanadium and sulfur, in haemolymph from the ascidians *Ascidia mentula* (Muller) and *Ascidiella aspersa* (Muller). J. Exp. Mar. Biol. Ecol. 74:187–194.

Rinehart, K. L. 2000. Antitumor compounds from tunicates. Med. Res. Rev. 20:1–27.

Rinkevich, B., Shlemberg, Z., and Fishelson, L. 1995. Whole-body protochordate regeneration from totipotent blood cells. Proc. Natl. Acad. Sci. USA 92:7695–7699.

Ruppert, E. E. 1990. Structure, ultrastructure and function of the neural gland complex of *Ascidia interrupta* (Chordata, Ascidiacea): Clarification of hypotheses regarding the evolution of the vertebrate anterior pituitary. Acta Zool. 71:135–149.

Saffo, M. B. 1981. The enigmatic protist *Nephromyces*. BioSystems 14:487–490.

Saffo, M. B., and Nelson, R. 1983. The cells of *Nephromyces:* Developmental stages of a single life cycle. Can. J. Bot. 61:3230–3239.

Satoh, N. 1994. Developmental Biology of Ascidians. Cambridge University Press, New York. 234 pp.

Sawada, H., Yokosawa, H., and Lambert, C. C. 2001. The Biology of Ascidians. Springer Verlag, Tokyo. 465 pp.

Stoecker, D. 1978. Resistance of a tunicate to fouling. Biol. Bull. 155:615–626.

Stoecker, D. 1980. Chemical defenses of ascidians against predators. Ecology 61:1327–1334.

Takamura, K., Fujimura, M., and Yamaguchi, Y. 2002. Primordial germ cells originate from the endodermal strand cells in the ascidian *Ciona intestinalis.* Dev. Genes Evol. 212:11–18.

Thompson, E. M., Kallesoe, T., and Spada, F. 2001. Diverse genes expressed in distinct regions of the trunk epithelium define a monolayer cellular template for construction of the oikopleurid house. Dev. Biol. 238:260–273.

Young, C. M., and Bingham, B. L. 1987. Chemical defense and aposematic coloration in larvae of the ascidian *Ecteinascidia turbinata.* Mar. Biol. 96:539–544.

Young, C. M., and Braithwaite, L. F. 1980. Orientation and current-induced flow in the stalked ascidian *Styela montereyensis.* Biol. Bull. 159:428–440.

Internet Sites

http://nsm.fullerton.edu/~lamberts/ascidian/ (Home page of the *Ascidian News,* a valuable source of current publications and research on tunicates by G. and C. Lambert.)

www.biology.ualberta.ca/facilities/multimedia/index.php?Page=252 (Click on "Sea Squirt Anatomy and Feeding" for an excellent animation by H. Kroening.)

CHORDATE PHYLOGENY

Arendt, D., and Nübler-Jung, K. 1997. Dorsal or ventral: Similarities in fate maps and gastrulation patterns in annelids, arthropods, and chordates. Mech. Dev. 61:7–21.

Bateson, W. 1886. The ancestry of the Chordata. Q. J. Microsc. Sci. 26:535–571.

Berrill, N. J. 1955. The Origin of Vertebrates. Clarendon Press, Oxford. 257 pp.

Berrill, N. J. 1987. Early chordate evolution. Part 1. Amphioxus, the riddle of the sands. Int. J. Invert. Reprod. Dev. 11:1–14.

Bone, Q. 1960. The origin of the chordates. J. Linn. Soc. Lond. 44:252–269.

Bone, Q. 1972. The Origin of Chordates. Oxford Biology Readers, no. 18. Oxford University Press, London. 16 pp.

Bone, Q. 1981. The neotenic origin of chordates. Atti Conv. Lincei 49:465–486.

Gans, C., and Northcutt, R. G. 1983. Neural crest and the origin of vertebrates: A new head. Science 220:268–274.

Garstang, W. 1928. The morphology of the Tunicata and its bearing on the phylogeny of the Chordata. Q. J. Microsc. Sci. 72:51–187.

Gee, H. 1996. Before the Backbone: Views on the Origin of Vertebrates. Chapman and Hall, London. 346 pp.

Gerhart, J., and Kirschner, M. 1997. Cells, Embryos, and Evolution. Blackwell Science, Malden, MA. 642 pp.

Gislen, T. 1930. Affinities between the Echinodermata, Enteropneusta, and Chordonia. Zool. Bidrag. Uppsala 12:199–304.

Holland, L. Z., Gorsky, G., and Fenaux, R. 1988. Fertilization in *Oikopleura dioica* (Tunicata, Appendicularia): Acrosome reaction, cortical reaction and sperm-egg fusion. Zoomorphology 108:229–243.

Holland, L. Z., and Holland, N. D. 2001. Evolution of neural crest and placodes: Amphioxus as a model for the ancestral vertebrate? J. Anat. 99:85–98.

Jefferies, R. P. S. 1986. The Ancestry of the Vertebrates. Cambridge University Press, Melbourne. 376 pp.

Jollie, M. 1982. What are the "Calcichordata"? And the larger question of the origin of chordates. Zool. J. Linn. Soc. 75:167–188.

Lacalli, T. C. 1994. Apical organs, epithelial domains, and the origin of the chordate central nervous system. Am. Zool. 34:533–541.

Lacalli, T. C. 1996. Mesodermal pattern and pattern repeats in the starfish bipinnaria larva, and related patterns in other deuterostome larvae and chordates. Phil. Trans. Roy. Soc. Lond. B 351:1737–1758.

Lacalli, T. C. 1997. The nature and origin of deuterostomes: Some unresolved issues. Invert. Biol. 116:363–370.

Lacalli, T. C. 1999. Tunicate tails, stolons, and the origin of the vertebrate trunk. Biol. Rev. 74:177–198.

Mackie, G. O., and Singla, C. L. 2002. The capsular organ of *Chelyosoma productum* (Ascidiacea: Corellidae): A new tunicate hydrodynamic sense organ. Brain Behav. Evol. 61:45–58.

Nielsen, C. 2001. Animal Evolution: Interrelationships of the Living Phyla. Oxford University Press, New York. pp. 450–489.

Nishino, A., and Satoh, N. 2001. The simple tail of chordates: Phylogenetic significance of appendicularians. Genesis 29:36–45.

Nübler-Jung, K., and Arendt, D. 1996. Enteropneusts and chordate evolution. Curr. Biol. 6:352–353.

Ogasawara, M., Wada, H., Peters, H., and Satoh, N. 1999. Developmental expression of Pax 1/9 genes in urochordate and hemichordate gills: Insight into function and evolution of the pharyngeal epithelium. Development 126:2539–2550.

Ruppert, E. E. 1997. Introduction: Microscopic anatomy of the notochord, heterochrony, and chordate evolution. In Harrison, F. W., and Ruppert, E. E. (Eds.): Microscopic Anatomy of Invertebrates. Vol. 15. Wiley-Liss, New York. pp. 1–13.

Ruppert, E. E., Cameron, C. B., and Frick, J. E. 1999. Endostyle-like features of the dorsal epibranchial ridge of an enteropneust and the hypothesis of dorsal-ventral axis inversion in chordates. Invert. Biol. 118:202–212.

Schaeffer, B. 1987. Deuterostome monophyly and phylogeny. Evol. Biol. 21:179–235.

Sedgwick, A. 1886. The original function of the canal of the central nervous system of Vertebrata. Stud. Morphol. Lab. Univ. Camb. 2:160–164.

Shu, D.-G., Chen, L., Han, J., and Zhang, X.-L. 2001. An early Cambrian tunicate from China. Nature 411:472–473.

Stach, T., and Turbeville, J. M. 2002. Phylogeny of Tunicata inferred from molecular and morphological characters. Mol. Phylog. Evol. 25:408–428.

Swalla, B. J. 2001. Phylogeny of the urochordates: Implications for chordate evolution. In Sawada, H., Yokosawa, H., and Lambert, C. (Eds.): The Biology of Ascidians. Springer Verlag, Tokyo. pp. 219–224.

Swalla, B. J., Cameron, C. B., Corley, L. S., and Garey, J. R. 2000. Urochordates are monophyletic within the deuterostomes. Syst. Biol. 49:52–64.

Wada, H. 1998. Evolutionary history of free-swimming and sessile lifestyles in urochordates as deduced from 18S rDNA molecular phylogeny. Mol. Biol. Evol. 15:1189–1194.

Willey, A. 1894. Amphioxus and the Ancestry of the Vertebrates. Columbia University Biological Series II. MacMillan, New York. 314 pp.

Winchell, C. J., Sullivan, J., Cameron, C. B., Swalla, B. J., and Mallatt, J. 2002. Evaluating hypotheses of deuterostome phylogeny and chordate evolution with new LSU and SSU ribosomal DNA data. Mol. Biol. Evol. 19:762–776.

INDEX

Italicized page numbers refer to illustrations; page numbers followed by "t" refer to tables. Page numbers in boldface refer to definitions of terms boldfaced in text.

D

H

T